THE STEPHEN BECHTEL FUND

IMPRINT IN ECOLOGY AND THE ENVIRONMENT

The Stephen Bechtel Fund has established this imprint to promote understanding and conservation of our natural environment.

The publisher gratefully acknowledges the generous contribution to this book provided by the Stephen Bechtel Fund.

ENCYCLOPEDIA OF THEORETICAL ECOLOGY

ENCYCLOPEDIA OF THEORETICAL ECOLOGY

EDITED BY

ALAN HASTINGS
University of California, Davis

LOUIS J. GROSS
University of Tennessee, Knoxville

UNIVERSITY OF CALIFORNIA PRESS
Berkeley Los Angeles London

University of California Press, one of the most distinguished university presses in the United States, enriches lives around the world by advancing scholarship in the humanities, social sciences, and natural sciences. Its activities are supported by the UC Press Foundation and by philanthropic contributions from individuals and institutions. For more information, visit www.ucpress.edu.

Encyclopedias of the Natural World, No. 4

University of California Press
Berkeley and Los Angeles, California

University of California Press, Ltd.
London, England

Library of Congress Cataloging-in-Publication Data

Encyclopedia of theoretical ecology / edited by Alan Hastings, Louis J. Gross.
p. cm.
Includes bibliographical references.
ISBN 978-0-520-26965-1 (cloth : alk. paper)
1. Ecology--Encyclopedias. I. Hastings, A. (Alan), 1953- II. Gross, Louis J.
QH540.4.E526 2012
577.03--dc23
2011027208

19 18 17 16 15 14 13 12
10 9 8 7 6 5 4 3 2 1

The paper used in this publication meets the minimum requirements of ANSI/NISO Z39.48-1992 (R 1997) (*Permanence of Paper*). ♾

Cover photographs: Coral-dominated reef, © Bent Christensen/Azote.se. Insets, from left: Glanville fritillary butterfly (*Melitaea cinxia*), from Robinson R. (2006): *Genes Affect Population Growth, but the Environment Determines How*. PLoS Biol 4/5/2006: e150 http://dx.doi.org/10.1371/journal.pbio.0040150; aspen grove, Fishlake National Forest, Utah, photo by Mark Muir, courtesy U.S. Forest Service; crab among coral (*Trapezia rufopunctata*), courtesy Adrian Stier; A. *coelestinus*, courtesy Duncan J. Irschick.

Title page photo: Images from nature have inspired many studies in ecology and evolution, yielding important theoretical insights. Here tropical army ants (*Eciton burchellii*) cooperate to form a bridge over one another on Barro Colorado Island, Panama. Images like this are a fertile source of hypotheses and testing, suggesting possible theoretical approaches to the understanding of such topics as colony formation and demography, movement ecology, and spatial dynamics. Photograph by Christian Ziegler/STRI.

CONTENTS

Contents by Subject Area / xi

Guide to the Encyclopedia / xiii

Preface / xv

Adaptive Behavior and Vigilance / 1
Peter A. Bednekoff

Adaptive Dynamics / 7
J. A. J. Metz

Adaptive Landscapes / 17
Max Shpak

Age Structure / 26
Tim Benton

Allee Effects / 32
Caz M. Taylor

Allometry and Growth / 38
Andrew J. Kerkhoff

Apparent Competition / 45
Robert D. Holt

Applied Ecology / 52
Cleo Bertelsmeier; Elsa Bonnaud; Stephen Gregory; Franck Courchamp

Assembly Processes / 60
James A. Drake; Paul Staelens; Daniel Wieczynski

Bayesian Statistics / 64
Kiona Ogle; Jarrett J. Barber

Behavioral Ecology / 74
B. D. Roitberg; R. G. Lalonde

Belowground Processes / 80
James Umbanhowar

Beverton–Holt Model / 86
Louis W. Botsford

Bifurcations / 88
Fabio Dercole; Sergio Rinaldi

Biogeochemistry and Nutrient Cycles / 95
Benjamin Z. Houlton

Birth–Death Models / 101
Christopher J. Dugaw

Bottom-Up Control / 106
John C. Moore; Peter C. De Ruiter

Branching Processes / 112
Linda J. S. Allen

Cannibalism / 120
Alan Hastings

Cellular Automata / 123
David E. Hiebeler

Chaos / 126
Robert F. Costantino; Robert A. Desharnais

Coevolution / 131
Brian D. Inouye

Compartment Models / 136
Donald L. DeAngelis

Computational Ecology / 141
Stuart H. Gage

Conservation Biology / 145
H. Reşit Akçakaya

Continental Scale Patterns / 152
Brian A. Maurer

Cooperation, Evolution of / 155
Matthew R. Zimmerman; Richard McElreath; Peter J. Richerson

Delay Differential Equations / 163
Yang Kuang

Demography / 166
Charlotte Lee

Difference Equations / 170
Jim M. Cushing

Discounting in Bioeconomics / 176
Ram Ranjan; Jason F. Shogren

Disease Dynamics / 179
Giulio De Leo; Chelsea L. Wood

Dispersal, Animal / 188
Gabriela Yates; Mark S. Boyce

Dispersal, Evolution of / 192
Marissa L. Baskett

Dispersal, Plant / 198
Helene C. Muller-Landau

Diversity Measures / 203
Anne Chao; Lou Jost

Dynamic Programming / 207
Michael Bode; Hedley Grantham

Ecological Economics / 213
Sunny Jardine; James N. Sanchirico

Ecosystem Ecology / 219
Yiqi Luo; Ensheng Weng; Yuanhe Yang

Ecosystem Engineers / 230
Kim Cuddington

Ecosystem Services / 235
Fiorenza Micheli; Anne Guerry

Ecosystem Valuation / 241
Stephen Polasky

Ecotoxicology / 247
Valery Forbes; Peter Calow

Energy Budgets / 249
S. A. L. M. Kooijman

Environmental Heterogeneity and Plants / 258
Gordon A. Fox; Bruce E. Kendall; Susan Schwinning

Epidemiology and Epidemic Modeling / 263
Lisa Sattenspiel

Evolutionarily Stable Strategies / 270
Richard McElreath

Evolutionary Computation / 272
James W. Haefner

Facilitation / 276
Michael W. McCoy; Christine Holdredge; Brian R. Silliman; Andrew H. Altieri; Mads S. Thomsen

Fisheries Ecology / 280
Elliott Lee Hazen; Larry B. Crowder

Food Chains and Food Web Modules / 288
Kevin McCann; Gabriel Gellner

Food Webs / 294
Axel G. Rossberg

Foraging Behavior / 302
Thomas Caraco

Forest Simulators / 307
Michael C. Dietze; Andrew M. Latimer

Frequentist Statistics / 316
N. Thompson Hobbs

Functional Traits of Species and Individuals / 324
Duncan J. Irschick; Chi-Yun Kuo

Game Theory / 330
Karl Sigmund; Christian Hilbe

Gap Analysis and Presence/ Absence Models / 334
Jocelyn L. Aycrigg; J. Michael Scott

Gas and Energy Fluxes Across Landscapes / 337
Dennis Baldocchi

Geographic Information Systems / 341
Michael F. Goodchild

Harvesting Theory / 346
Wayne Marcus Getz

Hydrodynamics / 357
John L. Largier

Individual-Based Ecology / 365
Steven F. Railsback; Volker Grimm

Information Criteria in Ecology / 371
Subhash R. Lele; Mark L. Taper

Integrated Whole Organism Physiology / 376
Arnold J. Bloom

Integrodifference Equations / 381
Mark Kot; Mark A. Lewis; Michael G. Neubert

Invasion Biology / 384
Mark A. Lewis; Christopher L. Jerde

Landscape Ecology / 392
Jianguo Wu

Marine Reserves and Ecosystem-Based Management / 397
Leah R. Gerber; Tara Gancos Crawford; Benjamin Halpern

Markov Chains / 404
Louis J. Gross

Mating Behavior / 408
Patricia Adair Gowaty

Matrix Models / 415
Eelke Jongejans; Hans De Kroon

Meta-Analysis / 423
Michael D. Jennions; Kerrie Mengerson

Metabolic Theory of Ecology / 426
James F. Gillooly; April Hayward; Melanie E. Moses

Metacommunities / 434
Marcel Holyoak; Jamie M. Kneitel

Metapopulations / 438
Ilkka Hanski

Microbial Communities / 445
Thomas G. Platt; Peter C. Zee; Keenan M. L. Mack; James D. Bever

Model Fitting / 450
Perry De Valpine

Movement: From Individuals to Populations / 456
Paul R. Moorcroft

Mutation, Selection, and Genetic Drift / 463
Brian Charlesworth

Networks, Ecological / 470
Anna Eklöf; Stefano Allesina

Neutral Community Ecology / 478
Stephen P. Hubbell

Niche Construction / 485
John Odling-Smee

Niche Overlap / 489
Howard V. Cornell

Nicholson-Bailey Host Parasitoid Model / 489
Cheryl J. Briggs

Nondimensionalization / 501
Roger M. Nisbet

NPZ Models / 505
Peter J. S. Franks

Ocean Circulation, Dynamics of / 510
Christopher A. Edwards

Optimal Control Theory / 519
Hien T. Tran

Ordinary Differential Equations / 523
Sebastian J. Schreiber

Pair Approximations / 531
Joshua L. Payne

Partial Differential Equations / 534
Nicholas F. Britton

Phase Plane Analysis / 538
Shandelle M. Henson

Phenotypic Plasticity / 545
Mario Pineda-Krch

Phylogenetic Reconstruction / 550
Brian C. O'Meara

Phylogeography / 557
Scott V. Edwards; Susan E. Cameron Devitt; Matthew K. Fujita

Plant Competition and Canopy Interactions / 565
E. David Ford

Population Ecology / 571
Michael B. Bonsall; Claire Dooley

Population Viability Analysis / 582
William F. Morris

Predator–Prey Models / 587
Peter A. Abrams

Quantitative Genetics / 595
Paul David Williams

Reaction–Diffusion Models / 603
Chris Cosner

Regime Shifts / 609
Reinette Biggs; Thorsten Blenckner; Carl Folke; Line Gordon; Albert Norström; Magnus Nyström; Garry Peterson

Reserve Selection and Conservation Prioritization / 617
Atte Moilanen

Resilience and Stability / 624
Michio Kondoh

Restoration Ecology / 629
Richard J. Hall

Ricker Model / 632
Eric P. Bjorkstedt

Sex, Evolution of / 637
Jan Engelstädter; Francisco Úbeda

Single-Species Population Models / 641
Karen C. Abbott; Anthony R. Ives

SIR Models / 648
Lewi Stone; Guy Katriel; Frank M. Hilker

Spatial Ecology / 659
Alan Hastings

Spatial Models, Stochastic / 666
Stephen M. Krone

Spatial Spread / 670
Alan Hastings

Species Ranges / 674
Kevin J. Gaston; Hannah S. Smith

Stability Analysis / 680
Chad E. Brassil

Stage Structure / 686
Roger M. Nisbet; Cheryl J. Briggs

Statistics in Ecology / 691
Kevin Gross

Stochasticity (Overview) / 698
Matt J. Keeling

Stochasticity, Demographic / 706
Brett A. Melbourne

Stochasticity, Environmental / 712
Jörgen Ripa

Stoichiometry, Ecological / 718
James J. Elser; Yang Kuang

Storage Effect / 722
Robin Snyder

Stress and Species Interactions / 726
Ragan M. Callaway

Succession / 728
Herman H. Shugart

Synchrony, Spatial / 734
Andrew M. Liebhold

Top-Down Control / 739
Peter C. De Ruiter; John C. Moore

Transport in Individuals / 744
Vincent P. Gutschick

Two-Species Competition / 752
Priyanga Amarasekare

Urban Ecology / 765
Mary L. Cadenasso; Steward T. A. Pickett

Glossary / 771

Index / 803

CONTENTS BY SUBJECT AREA

MAJOR THEMES

Applied Ecology
Behavioral Ecology
Computational Ecology
Ecosystem Ecology
Epidemiology and Epidemic Modeling
Population Ecology
Spatial Ecology
Statistics in Ecology

PHYSIOLOGICAL AND BIOPHYSICAL ECOLOGY

Allometry and Growth
Belowground Processes
Energy Budgets
Functional Traits of Species and Individuals
Hydrodynamics
Integrated Whole Organism Physiology
Plant Competition and Canopy Interactions
Transport in Individuals

POPULATION DYNAMICS

Age Structure
Allee Effects
Apparent Competition
Beverton–Holt Model
Cannibalism
Chaos
Demography
Disease Dynamics
Dispersal, Animal
Dispersal, Plant
Ecosystem Engineers
Facilitation
Food Chains and Food Web Modules
Foraging Behavior
Invasion Biology
Mating Behavior
Metapopulations
Movement: From Individuals to Populations
Nicholson–Bailey Host Parasitoid Model
NPZ Models
Predator–Prey Models
Ricker Model
Single-Species Population Models
SIR Models
Spatial Spread
Species Ranges
Stage Structure
Stochasticity, Demographic
Stochasticity, Environmental
Stress and Species Interactions
Synchrony, Spatial
Two-Species Competition

EVOLUTIONARY ECOLOGY

Adaptive Behavior and Vigilance
Adaptive Dynamics
Adaptive Landscapes
Coevolution
Cooperation, Evolution of
Dispersal, Evolution of
Evolutionarily Stable Strategies
Evolutionary Computation
Mutation, Selection, and Genetic Drift
Niche Construction
Phenotypic Plasticity
Phylogenetic Reconstruction
Phylogeography
Quantitative Genetics
Sex, Evolution of

COMMUNITY ECOLOGY

Assembly Processes
Bottom-Up Control
Diversity Measures
Food Webs
Metacommunities
Microbial Communities
Neutral Community Ecology
Niche Overlap
Regime Shifts
Resilience and Stability
Storage Effect
Succession
Top-Down Control

ECOSYSTEM ECOLOGY

Biogeochemistry and Nutrient Cycles
Compartment Models
Continental Scale Patterns
Environmental Heterogeneity and Plants
Forest Simulators
Gas and Energy Fluxes across Landscapes
Geographic Information Systems
Landscape Ecology
Metabolic Theory of Ecology
Ocean Circulation, Dynamics of
Stoichiometry, Ecological

MATHEMATICAL APPROACHES

Bayesian Statistics
Bifurcations
Birth–Death Models
Branching Processes
Cellular Automata
Delay Differential Equations
Difference Equations
Dynamic Programming
Frequentist Statistics
Game Theory
Individual-Based Ecology
Information Criteria in Ecology
Integrodifference Equations
Markov Chains
Matrix Models
Meta-Analysis
Model Fitting
Networks, Ecological
Nondimensionalization
Optimal Control Theory
Ordinary Differential Equations
Pair Approximations
Partial Differential Equations
Phase Plane Analysis
Reaction–Diffusion Models
Spatial Models, Stochastic
Stability Analysis
Stochasticity (Overview)

APPLICATIONS

Conservation Biology
Discounting in Bioeconomics
Ecological Economics
Ecosystem Services
Ecosystem Valuation
Ecotoxicology
Fisheries Ecology
Gap Analysis and Presence/Absence Models
Harvesting Theory
Marine Reserves and Ecosystem-Based Management
Population Viability Analysis
Reserve Selection and Conservation Prioritization
Restoration Ecology
Urban Ecology

GUIDE TO THE ENCYCLOPEDIA

The *Encyclopedia of Theoretical Ecology* is a comprehensive, complete, and authoritative reference dealing with the many ways in which theory is used in ecology. Articles written by researchers and scientific experts provide a broad overview of the current state of knowledge with respect to ways in which theoretical ecology has been used by biologists, ecologists, environmental scientists, geographers, botanists, and zoologists. Almost every aspect of ecological and environmental research has been affected by theory, and this volume reflects this diversity of impact. The contributed reviews are intended for students as well as the interested general public but are uncompromising regarding the breadth and depth of scholarship.

In order for the reader to easily use this reference, the following summary describes the features, reviews the organization and format of the articles, and is a guide to the many ways to maximize the utility of this *Encyclopedia.*

SUBJECT AREAS

The *Encyclopedia of Theoretical Ecology* includes 129 topics that review how theory has been employed to reveal patterns and understand processes that operate across and within landscapes, habitats, ecosystems, communities, and populations. The *Encyclopedia* comprises the following subject areas:

- Major Themes
- Physiological and Biophysical Ecology
- Population Dynamics
- Evolutionary Ecology
- Community Ecology
- Ecosystem Ecology
- Mathematical Approaches
- Applications

ORGANIZATION

Articles are arranged alphabetically by title. An alphabetical table of contents begins on page vii, and another table of contents with articles arranged by subject area begins on page xi.

Article titles have been selected to make it easy to locate information about a particular topic. Each title begins with a key word or phrase, sometimes followed by a descriptive term. For example, "Stoichiometry, Ecological" is the title assigned rather than "Ecological Stoichiometry," because *stoichiometry* is the key term and, therefore, more likely to be sought by readers. Articles that might reasonably appear in different places in the *Encyclopedia* are listed under alternative titles—one title appears as the full entry; the alternative title directs the reader to the full entry. For example, the alternative title "Environmental Toxicology" refers readers to the entry entitled "Ecotoxicology."

ARTICLE FORMAT

The articles in the *Encyclopedia* are all intended for the interested general public. Therefore, each article begins with an introduction that gives the reader a short definition of the topic and its significance. Here is an example of an introduction from the article "Fisheries Ecology":

> Fisheries ecology is the integration of applied and fundamental ecological principles relative to fished species or affected nontarget species (e.g., bycatch). Fish ecology focuses on understanding how fish interact with their environment, but fisheries ecology extends this understanding to interactions with fishers, fishery communities, and the institutions that influence or manage fisher behaviors. Traditional fisheries science has focused on single-species stock assessments and management with the goal of understanding population dynamics and variability. But in the past few decades, scientists and

managers have analyzed the effects of fisheries on target and nontarget species and on supporting habitats and food webs and have attempted to quantify ecological linkages (e.g., predator–prey, competition, disturbance) on food web dynamics. Fisheries ecology requires understanding how population variability is influenced by species interactions, environmental fluctuations, and anthropogenic factors.

Within most articles and especially the longer articles, major headings help the reader identify important subtopics within each article. The article "Species Ranges" includes the following headings: "Establishment," "Range Limits," "Extinction," and "Practical Implications."

CROSS-REFERENCES

Many of the articles in this *Encyclopedia* concern topics for which articles on related topics are also included. In order to alert readers to these articles of potential interest, cross-references are provided at the conclusion of each article. At the end of "Ecosystem Ecology," the following text directs readers to other articles that may be of special interest:

SEE ALSO THE FOLLOWING ARTICLES

Biogeochemistry and Nutrient Cycles / Environmental Heterogeneity and Plants / Gas and Energy Fluxes across Landscapes / Regime Shifts

Readers will find additional information relating to Ecosystem Ecology in the articles listed.

BIBLIOGRAPHY

Every article ends with a short list of suggestions for "Further Reading." The sources offer reliable in-depth information and are recommended by the author as the best available publications for more lengthy, detailed, or comprehensive coverage of a topic than can be feasibly presented within the *Encyclopedia*. The citations do not represent all of the sources employed by the contributor in preparing the article. Most of the listed citations are to review articles, recent books, or specialized textbooks, except in rare cases of a classic ground-breaking scientific article or an article dealing with subject matter that is especially new and newsworthy. Thus, the reader interested in delving more deeply into any particular topic may elect to consult these secondary sources. The *Encyclopedia* functions as ingress into a body of research only summarized herein.

GLOSSARY

Almost every topic in the *Encyclopedia* deals with a subject that has specialized scientific vocabulary. An effort was made to avoid the use of scientific jargon, but introducing a topic can be very difficult without using some unfamiliar terminology. Therefore, each contributor was asked to define a selection of terms used commonly in discussion of their topic. All of these terms have been collated into a glossary at the back of the volume after the last article. The glossary in this work includes over 600 terms.

INDEX

The last section of the *Encyclopedia of Theoretical Ecology* is a subject index consisting of more than 3,500 entries. This index includes subjects dealt with in each article, scientific names, topics mentioned within individual articles, and subjects that might not have warranted a separate stand-alone article.

ENCYCLOPEDIA WEB SITE

To access a Web site for the *Encyclopedia of Theoretical Ecology,* please visit:

http://www.ucpress.edu/book.php?isbn=9780520269651

This site provides a list of the articles, the contributors, several sample articles, published reviews, and links to a secure website for ordering copies of the *Encyclopedia*. The content of this site will evolve with the addition of new information.

PREFACE

Ecologists face almost daily ecological and environmental crises caused by human activities and by natural disasters. Changes across our planet add to a sense of urgency and a need to predict how these catastrophes and abuses, as well as everyday imperceptible events, affect natural systems. Our ability to predict depends on our understanding of today's ecological systems, how they functioned in the past, and how they will respond in the future. Prediction depends on data, and many different kinds of data—from microcosms to continents, from days to millions of years, and from individuals to entire lineages. Ecologists gather data, they synthesize, and they make predictions—this is theoretical ecology. The function of theory in any discipline, including ecology, is to guide data gathering, find the best ways to synthesize diverse sources of information, reveal the impact of alternative assumptions, and ascertain possible impacts upon the future. Ecology is scientifically diverse. Thus, it is difficult for an active researcher to keep up, let alone others who are interested but not actively involved in ecological or environmental research. The need for this *Encyclopedia of Theoretical Ecology* is clear.

Well-known giants—Lotka, Volterra, Gause, Lindeman—were the pathfinders for theoretical ecology, and their findings still play an essential role in all fundamental aspects of ecology. The exciting progress of recent years augurs well for continued growth in the depth and breadth of theoretical ecology. This volume presents an introduction to all of theoretical ecology while also providing background on both new developments and classical results. We hope it will appeal to students seeking understanding of current research, to professional research ecologists needing succinct summaries of developments across different areas of ecology, and to scientists in other fields who need to understand the fundamentals of ecology.

The coverage is of necessity broad, matching theoretical ecology. For this we thank dedicated and generous contributors. No individual could deal with all of the theoretical developments in ecology. Therefore, our goal has been to assemble the perfect team of individuals to create a comprehensive, scientifically uncompromising, and yet still succinct review of theoretical ecology. In only one volume, not all areas can be treated deeply, but we have tried to be as complete as possible. We are pleased with the results and hope the reader will be equally pleased.

Theoretical ecology is not simply mathematical ecology; nevertheless, mathematics plays a central role in theoretical ecology. It is a part of the conceptual and computational foundation created to explain ecological systems and assess the efficacy of theory. Thus, many contributions focus on ecological questions and others must cover topics in mathematics, statistics, and computation, because these topics are useful in addressing diverse issues of theoretical ecology.

These pages include contributions dealing with a range of organizational scales—small/large, complex/simple, rapid/slow—encompassed by ecology. Some focus on the within-organism level, and others on organisms through to populations, communities, ecosystems, and landscapes. There is great emphasis on population ecology and related areas because substantial theoretical developments in this area have the longest history and have attracted the greatest attention.

Computational and statistical issues also play a central role in this volume. Such methods are important in theoretical developments because assessing efficacy depends on mathematical descriptions. Furthermore, recent advances in mathematical and statistical matters, greatly enhanced computational power, and new computational approaches are changing the way ecology is done. Methodological entries tend to be more varied in the level of presentation. Some are appropriate to readers with very limited formal mathematical training, whereas others are more challenging and require more background. All contributors tried to draft entries helpful to both novices and experts. Recently, the predictions of theoretical ecologists are matching empirical observations of ecological systems, and this success is mirrored in this volume.

We are grateful for the participation and assistance of many who have been kind enough to offer their insights, time, and support. Gail Rice has been tireless and patient with us. She has encouraged contributors, reminded us to complete tasks, and has kept the quality of the entire volume high. Chuck Crumly at UC Press has helped immensely in this project from conception to completion. AH would like very much to thank Elaine Fingerett for helping with humor about the subject. LG thanks Marilyn Kallet for helping him savor the poetry all around us.

Alan Hastings
University of California, Davis

Louis J. Gross
University of Tennessee, Knoxville

A

ADAPTIVE BEHAVIOR AND VIGILANCE

PETER A. BEDNEKOFF

Eastern Michigan University, Ypsilanti

Adaptive behavior is behavior that raises an animal's fitness in its biotic and abiotic environment. The study of adaptive behavior is mainly the study of how behavior changes with environmental changes. Fitness is the relative contribution of an organism to genes in future generations. Because understanding fitness across the entire life of an organism is a daunting task (and tracking it across several generations even more so), researchers have commonly assumed a particular relationship between behavior in the short term and fitness. This has involved examining many behaviors using many fitness proxies and modeling techniques. This entry explores the simplifying assumptions about the relationship between behavior and fitness while concentrating on a single type of behavior, watching for predators. Though the focus here is on such vigilance behavior, the approach employed may be applied to many other types of behavior.

BASELINE MODELS OF VIGILANCE BEHAVIOR

Anti-predator vigilance involves pauses in other behavior, such as feeding, in order to scan the environment for predators. Researchers commonly assume that that foraging animals lift their heads independently of one another and share information about detected attacks. Shared information about attacks is known as collective detection. The original model of anti-predator vigilance showed that in bigger groups each individual could scan less while maintaining the same probability of an undetected attack. If each individual is vigilant v proportion of the time, then an attack goes undetected when no individual is vigilant, which occurs a fraction $(1 - v)^n$ of the time, where n is the number of individuals in the group. This model did not relate this directly to fitness but determined that animals could scan less in larger groups at the same risk of an undetected attack. Subsequent studies have built upon the assumptions of independent scanning and collective detection of attacks but added dilution of risk. Here, the predator can only effectively attack one individual (or perhaps a small portion of the group), even when all individuals are unaware of the impending attack.

Other researchers have supposed that animals maximize survival while maintaining a required level of food intake. This second criterion can make sense if food increases do not impact future reproduction very much, for example, during the nongrowing season for animals that have small growth during this period. These two criteria for adaptive behavior—maximizing food intake while maintaining a set probability of surviving and minimizing risk of attack and maintaining a set level of food intake—obviously cannot both hold true simultaneously. They can, however, both be approximations of a larger truth. In general, fitness will depend on the value of food and the probability of surviving. In simple models, vigilance increases survival but decreases food intake, and thereby future reproductive output. This entry discusses models of the optimal level of vigilance that build upon the work of Parker and Hammerstein but differ from their models by allowing multiple attacks and assuming future reproduction is a linearly increasing function of foraging intake.

To develop heuristic models of vigilance in groups, assume that foragers feed for some extended time, T, before reproducing. Animals can forage at any rate between 0 and 1, and there is a direct tradeoff with vigilance—food intake is proportional to $1 - v$. An animal's total foraging intake across the period is thus proportional to $(1 - v)T$, and reproductive success at the end of the period is proportional to foraging intake—$V = k(1 - v)T$. In order to reproduce, however, animals must survive this period. If they are attacked at rate α according to a Poisson process, then the probability of survival decreases exponentially with the attack rate, period length, and the mortality per attack—$S = \text{Exp}[-\alpha TM]$. The probability of dying in an attack increases with foraging rate. For a single individual the mortality rate is $M = (1 - v)$. In a group, the mortality rate may depend on the actions of others. This is the topic of the next section.

PERFECT COLLECTIVE DETECTION

When a predator attacks a group, the focal individual dies if it is not vigilant and is not warned by other members of the group. For the effects of others, the classic assumption is perfect collective detection: the focal individual is warned of the attack if any group member is vigilant at the time. Thus, the attack succeeds only if no member of the group is vigilant at the time. If the focal individual is joined by $n - 1$ other individuals, each vigilant $\hat{v}$ of the time, this equals $(1 - v)(1 - \hat{v})^{n-1}$. If the predator must choose among these n unaware prey and does so equally, the risk is $1/n$ for each group member. Thus, the probability that the focal individual dies in an attack is

$$M = \frac{(1 - v)(1 - \hat{v})^{n-1}}{n}.$$

As we would expect, animals are safer when they are more vigilant and when they are in larger groups. Because the fitness of the focal individual depends on the vigilance of others, game theory is needed to find the solution. The evolutionarily stable strategy (ESS) can be found by differentiating the fitness function with respect to v, setting it to zero, and solving for v. This gives v as a function of $\hat{v}$—in this case:

$$v^* = 1 - \frac{n}{\alpha T(1 - \hat{v})^{n-1}}.$$

Because an individual can be warned by others, it can rely on their vigilance to some extent. In a group where everyone else is very vigilant, the best response is to be less vigilant. In a group where others are not vigilant, the best response is to be more vigilant. In between, there is a level of vigilance that is the best response to itself. To find this evolutionarily stable strategy, set $\hat{v} = v$ and again find v. In this case the evolutionarily stable level of vigilance is

$$v^* = 1 - \left(\frac{n}{\alpha T}\right)^{\frac{1}{n}}.$$

The evolutionarily stable strategy here is equivalent to the Nash equilibrium from game theory. For comparison, the optimal cooperative strategy, or Pareto equilibrium, can be found by substituting $\hat{v} = v$ in the fitness function, differentiating, setting this equal to zero, and then solving for v. With this first model, the optimal cooperative strategy is

$$v_c^* = 1 - \left(\frac{1}{\alpha T}\right)^{\frac{1}{n}}.$$

These expressions differ only in the numerator of the ratio in the second right-hand term. Since this second right-hand term equals the feeding rate, individuals feed at $n^{1/n}$ times the cooperative rate under the ESS. These results replicate findings that ESS levels of vigilance are lower than cooperative when collective detection is perfect. Analyzing $n^{1/n}$ as n varies shows that the difference between the ESS and cooperative solutions is greatest for $n = 3$ and declines with larger group sizes.

NO COLLECTIVE DETECTION

A null model indicates how anti-predator behavior would be different without collective detection. In this model, individuals that are vigilant at the start of the attack escape, and the predator targets one of the nondetectors. As the number of detectors, i, goes up, the effective group size for dilution decreases by the same number. As before, our focal individual is in danger only when it is not vigilant $(1 - v)$ but now the effects of others must be summed across all possible numbers of detectors:

$$M = (1 - v)\sum_{i=0}^{n-1}\left(\frac{(n - 1)!}{(n - 1 - i)!i!}\right)\frac{\hat{v}^i(1 - \hat{v})^{n-1-i}}{n - i}.$$

The possible number of detectors ranges from zero to $n - 1$—all group members other than our focal individual. Inside the summation, the factorial gives the number of ways of having i detectors, the numerator of the right-hand term gives the probability of any one such combination, and the denominator gives the dilution of risk among the $n-1$ nondetectors (including the focal individual). This sum simplifies such that the overall mortality is

$$M = \frac{(1 - v)(1 - \hat{v}^n)}{n(1 - \hat{v})}.$$

Once again the optimal response depends on the vigilance of others,

$$v^* = 1 - \frac{n(1 - \hat{v})}{\alpha T(1 - \hat{v}^n)},$$

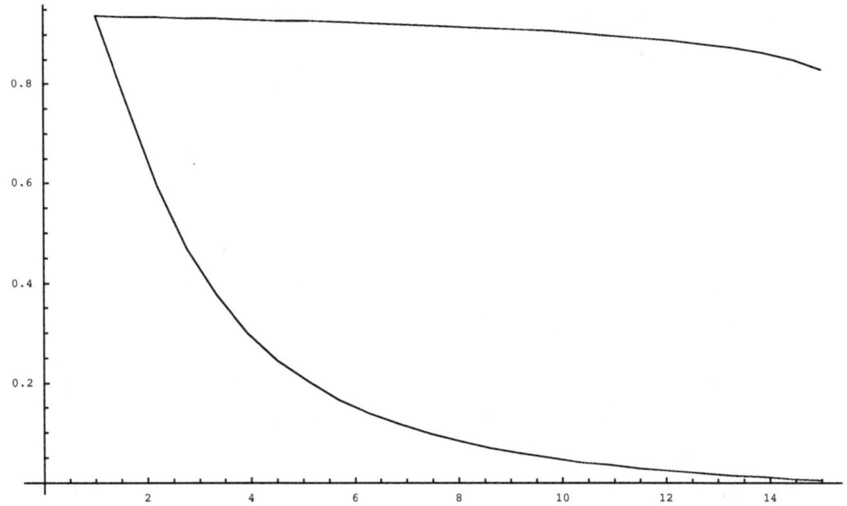

FIGURE 1 The evolutionarily stable level of vigilance declines steeply with group size when collective detection is perfect (lower curve) but very weakly when collective detection is absent (upper curve). Here the expected number of attacks (αT) is 16.

and the ESS occurs in the case when the optimal response is to match the vigilance of others, which yields

$$v^* = \left(1 - \frac{n}{\alpha T}\right)^{\frac{1}{n}}.$$

This expression is negative or undefined when the expected number of attacks is less than the group size, $\alpha T < n$. In this situation, zero vigilance is the best solution.

In this null model, a group size effect is present, but generally weak (Fig. 1). Since a strong effect of group size on vigilance has been observed many times in nature, this result suggests that the observed group size effect on vigilance probably depends upon collective detection to some extent.

The cooperative solution for the null model is difficult to state in general, but for a pair of animals it is

$$v_c^* = \frac{1}{2} + \sqrt{\frac{\alpha T - 4}{2\alpha T}}.$$

When fewer than four attacks occur on average ($\alpha T < 4$), this expression is undefined and zero vigilance ($v = 0$) is the best available vigilance level. Compared to the ESS solution for a pair (Fig. 2), without collective detection the cooperative vigilance level is lower than the ESS vigilance level.

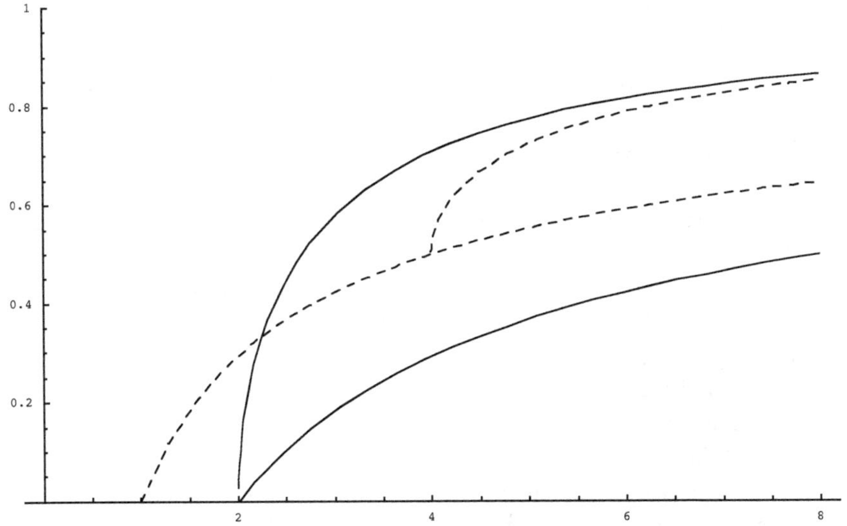

FIGURE 2 Vigilance levels increase with the expected number of attacks whether vigilance is cooperative or not, and whether collective detection is perfect or absent. The upper two curves are when collective detection is absent. Here ESS levels (solid lines) are higher than cooperative levels (dashed lines). The lower two curves assume collective detection is perfect. Here cooperative vigilance levels (dashed lines) are higher than ESS levels (solid lines). All results are for pairs of animals.

The models in this entry differ from those of Parker and Hammerstein first in allowing for multiple attacks rather than a single attack. Their equivalent model produced an ESS of zero vigilance for all group sizes greater than 1. In the models described here, which include perfect collective detection, the ESS level of vigilance falls to zero when the number of expected attacks is no bigger than the group size. Thus, the two models produce similar results under the correct simplifying assumptions, but quite different results are possible when multiple attacks are considered.

While earlier models suggested ESS vigilance should be zero for pairs or larger groups, observations showed nonzero levels of vigilance in groups. This mismatch led to consideration of cooperative models of vigilance. The conclusions of these early models apply only in some cases. With perfect collective detection, cooperative vigilance is generally higher than the ESS level. Without collective detection, however, cooperative vigilance is generally lower. In the model with no collective detection, the cooperative response is zero vigilance, whereas the ESS solution is moderate levels of vigilance when between two and four attacks are expected (Fig. 2) and the ESS level of vigilance is generally higher than the cooperative level anytime two or more attacks are expected.

PARTIAL COLLECTIVE DETECTION

The models in this entry have thus far only examined the extremes of perfect collective detection and no collective detection. Empirical studies indicate that collective detection is real yet slow and potentially faulty. For large groups, a model of imperfect collective detection would need to keep track of the overall contagion of information. This has not been done and is beyond the scope of this entry. For a pair of animals, however, a model only needs to track the flow of information from a primary detector to the other member of the pair. Here, the danger for the focal animal is

$$M = (1 - v)\left(\frac{(1 - \hat{v})}{2} + \hat{v}(1 - i)\right).$$

The focal animal is in danger when it fails to detect an attack $(1 - v)$. If the other animal fails to detect the attack $(1 - \hat{v})$, the two are equally likely to be targeted by the predator. If the other animal detects the attack (with probability $\hat{v}$), the focal animal escapes if it gets information, but is the sole target for attack if it fails to get information $(1 - i)$ from the other individual, where i is the probability that the focal animal receives a warning from the other animal.

Study of vigilance generally includes the separate advantages for collective detection and risk dilution, where risk dilution occurs when the predator chooses among of a group of uninformed prey. This distinction blurs, however, when collective detection is imperfect and takes time. If different members of the group have different amounts of information during the course of the attack, dilution depends strongly on when and how predators target prey for attack. If predators target prey early in an attack, risk is spread equally among the n group members. If predators wait until later in the attack, however, they can choose among the subset of the group that has not yet detected them. Thus, dilution depends on how many other individuals are similarly uninformed of the attack. The formulation given here assumes that the predator chooses a target after primary detection but before secondary detection is complete. Thus, the predator ignores any individual with their head up at the time of the attack, and chooses with equal probability among the other individuals. This gives an advantage to individuals who are vigilant at the start of the attack. Other formulations would be appropriate if predators could target individuals that were the very last to become informed. Such formulations would place even greater advantage on personal detection of attacks.

Working with the mortality expression for imperfect collective, we solve and find that the ESS level of vigilance for a member of a pair is

$$v^* = \left(\frac{1}{2i - 1}\right)\left(\frac{i - \sqrt{4i - 2 + \alpha T(1 - i)^2}}{\sqrt{\alpha T}}\right),$$

while the cooperative level is

$$v_c{}^* = \left(\frac{1}{4i - 2}\right)\left(\frac{3i - 1 - \sqrt{8i - 4 + \alpha T(1 - i)^2}}{\sqrt{\alpha T}}\right).$$

For small numbers of attacks, these expressions may be negative—so that the internal optimum is zero vigilance. With more attacks, vigilance tends to increase under either the ESS or the cooperative criterion. The effects of information flow are more complicated. With increasing information passing from detector to nondetector, the ESS level of vigilance generally decreases. Cooperative vigilance, however, increases with greater information flow at moderately low attack rates but then decreases at higher attack rates. In all cases, cooperative vigilance is higher than noncooperative vigilance only with $i > 1/2$ (Fig. 3). Other work has shown that $i > 1/2$ is also the criterion for coordinated vigilance to be advantageous. As pointed out by Steve Lima, the widespread observation that vigilance is not coordinated is evidence against

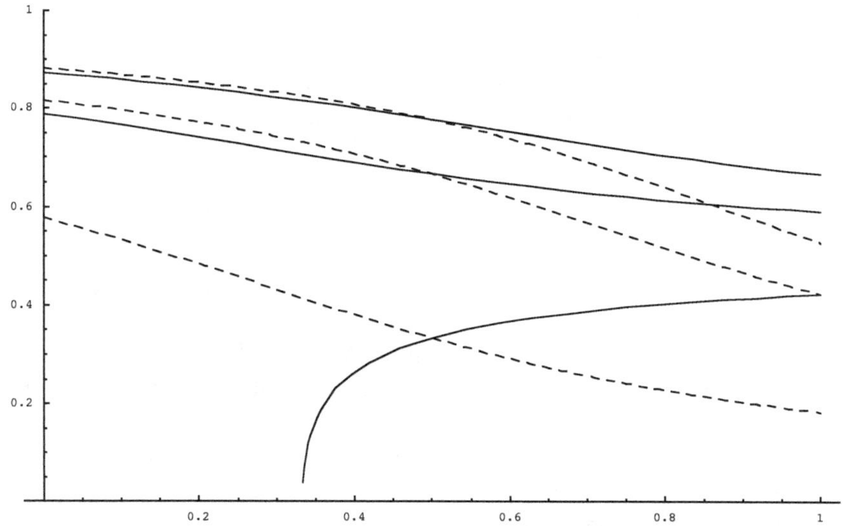

FIGURE 3 ESS and cooperative levels of vigilance as a function of information (i) passing from an animal detecting an attack to one that has not. The three pairs of curves are for 3, 6, and 9 expected attacks. Vigilance levels increase when more attacks are expected. Within each pair of curves, cooperative levels of vigilance (dashed lines) are lower than ESS levels (solid lines) for $i < 1/2$ and higher for $i > 1/2$.

cooperation. The potential fruit of cooperation, the public goods in this case, depend critically on the information that nondetectors get from detectors of attacks. In nature this information may be incomplete or unreliable, so each individual needs to gather information for itself. Individual vigilance is both more valuable and more costly than relying on social information. Thus with multiple attacks and imperfect collective detection, the norm is for each individual to invest in vigilance.

The ESS level of vigilance decreases with greater information passing from detector to nondetector. This is seen by examining the best response to the vigilance of the other group member (Fig. 4). With little or no information, the best reaction approaches the level for a single bird (7/8). When information only has a probability one-half of reaching a nondetector, the reaction does not depend on the vigilance of the other group member. Here any increase in detection with the vigilance of the other animal is exactly offset by decrease in dilution when information fails to get through. With $i > 1/2$, the best response is considerably lower than for a lone animal, and lower when the other animal is more vigilant. Observations show that

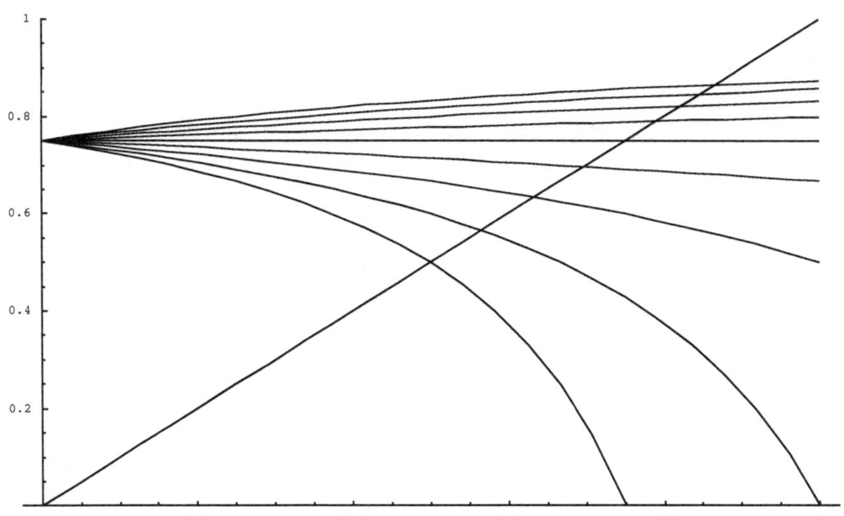

FIGURE 4 Optimal vigilance level in response to vigilance of the other animal in a pair. The curves differ in the probability of information about an attack passing from a detecting individual to a nondetector. The top curve is for no information ($i = 0$) and the bottom one for perfect information ($i = 1$) with i changing by 1/8 for each curve in between. ESS levels of vigilance lie along the diagonal line where the best response equals the vigilance of the other animal. The expected number of attacks (αT) equals 8.

animals are often much less vigilant in pairs than alone, suggesting considerable information flow and collective detection. On the other hand, there is much less information that animals react directly to the vigilance rates of others. Perhaps the danger to one member of a group is little influenced by the vigilance rates of others, either because information flow is moderate ($i \approx 1/2$) or due to some other factor not included in this model.

RISK ALLOCATION

Classic models of vigilance behavior generally considered perfect collective detection and a very limited number of attacks. Some outcomes of the models differ with imperfect collective detection and multiple attacks. Multiple attacks might occur over a substantial period of time, and conditions might change during this period. Consider the situation when group size changes over time. For simplicity, assume that the focal animal spends some portion of its time, p, alone and the rest of the time, q, paired with some other animal. Other assumptions are the same as above. If individuals face various situations with different levels of risk, they may allocate vigilance and feeding behavior across the situations, with v_1 being the vigilance in situation 1 and v_2 the vigilance in situation 2.

In contrast to previous models of risk allocation, the models presented here use a linear relationship between feeding rate and detection behavior. This assumption simplifies the models, and also leads to the solution of zero vigilance in the safer situation. The results are striking in that the safer situation in a pair switches at $i = 1/2$. This model of risk allocation does not yield a simple ratio of feeding rates, because one feeding rate is always the maximum, 1. The model, nonetheless, illustrates that vigilance depends upon the schedule of other situations an animal faces. With $i < 1/2$, the ESS level of vigilance is zero when alone, and $1 - (1/\alpha Tp)$ when in a pair. Thus, allocation has reversed the modest group size effect seen above with little collective detection. With moderate to high collective detection ($i > 1/2$), the vigilance in the pair goes to zero with the vigilance of an animal feeding alone, depending on how often it is in a pair:

$$v_2{}^* = 0,\ v_1 = \frac{2\alpha T - 1}{2\alpha Tp}.$$

The result is remarkable and counterintuitive in that collective detection leads to an evolutionary stable solution where neither individual is vigilant, detection does not occur, and so collective detection never happens. Thus in any attack on a pair, one individual or the other will die. Because this model assumes that the proportion of time in a pair is not affected by vigilance levels, it implicitly assumes that group members that die are replaced before the next attack. The next section explores whether this combination of assumptions leads to the surprising results.

PARTNER LOSS AND MUTUAL DEPENDENCE

Most models of vigilance have assumed a stable group size. Models of risk allocation assume changes in risk factors, such as group size, but that the schedule of risk factor changes is not influenced by their feeding or vigilance rates. Now consider a model where group size changes as a result of mortality. Assume all individuals start in pairs and that collective detection is perfect ($i = 1$). The previous results for risk allocation provide an interesting foundation in that they do not predict vigilance in pairs. Therefore vigilance in pairs is due to protecting an irreplaceable partner—which is one kind of mutual dependence in fitness.

The assumptions for what happens when a predator attacks a pair are exactly as before. If the other animal dies in an attack, however, our focal animal faces all future attacks alone. The solution involves two vigilance rates, one in a pair and one alone. Survival is summed across all routes in which the focal individual survives, each weighted by its probability. To find foraging success, note that if when the other animal dies in the ith of j attacks, the focal animal forages in a pair for on average $i/(j + 1)$ of the total time period and alone the rest of the period:

$$S = e^{-\alpha} + \sum_{i=1}^{\infty} \frac{\alpha^i e^{-\alpha}}{i!} \sum_{j=1}^{i} (v_2 + \hat{v}_2 + v_2\hat{v}_2)^{j-1} \times \left(\frac{(1 - v_2)(1 - \hat{v}_2)}{2}\right)(1 - v_1)^{i-j}\, k\left(v_2\frac{j}{i + 1} + v_1\frac{1 - j}{i + 1}\right).$$

The most important result of this model is that the ESS result is to be vigilant in a pair. With simple risk allocation, the ESS solution is zero vigilance. Thus each individual invests in the survival of its irreplaceable partner because having a partner increases its own survival. Potentially, this is a widespread phenomenon for animals in the nonbreeding season because animals lost to predation cannot be replaced until the next breeding season. It will only have a large effect, however, when loss of another individual strongly affects the fitness of a focal individual. This could happen if animals spend the nonbreeding season in small, closed groups. Given that nonbreeding groups are often fluid, mutual dependence might apply more widely to mated pairs that stay together year-round. If mates are difficult to replace or mating increases with experience, then each member of the pair would do well to ensure that its partner survives.

Thus individuals might often be "mate guarding" but in the broad sense of protecting their partner's survival and ability to reproduce. Perhaps mutual dependence might contribute to the year-round territorial behavior of tropical birds.

CONCLUSIONS

In this entry the initial models of vigilance were very simple and yet considered multiple attacks. This exercise was very successful because it was able to re-create results from previous models while showing that these results are not general to multiple attacks. Extending these models to partial information worked well, though general and tractable models were only found for pairs of individuals and not for larger groups. For risk allocation and mutual dependence, the rationale for the first models is limited. The models for risk allocation developed here provide a baseline for analyzing mutual dependence. Here again, the models of mutual dependence were neither as simple nor as general as the models of vigilance they build upon, and other models might more readily demonstrate mutual dependence. Thus simplifying assumptions for one purpose may complicate extending the model to another purpose. To explain the logic of potential contributing factors, it can be very helpful to examine a series of very simple models, each resting on its own foundations. When examining the relative importance of and interactions between contributing factors, more complicated models may be necessary. Simple and complex models inform each other, and a range of tools can be useful on any modeling job.

SEE ALSO THE FOLLOWING ARTICLES

Behavioral Ecology / Cooperation, Evolution of / Evolutionarily Stable Strategies / Foraging Behavior / Game Theory / Predator–Prey Models

FURTHER READING

Bednekoff, P. A. 2001. Coordination of safe, selfish sentinels based on mutual benefits. *Annales Zoologici Fennici* 38: 5–14.

Bednekoff, P. A. 2007. Foraging in the face of danger. In D. W. Stephens, J. S. Brown, and R. C. Ydenberg, eds. *Foraging*. Chicago: University of Chicago Press.

Bednekoff, P. A., and S. L. Lima. 1998. Randomness, chaos, and confusion in the study of anti-predator vigilance. *Trends in Ecology & Evolution* 13: 284–287.

Bednekoff, P. A., and S. L. Lima. 1998. Re-examining safety in numbers: Interactions between risk dilution and collective detection depend upon predator targeting behavior. *Proceedings of the Royal Society B: Biological Sciences* 265: 2021–2026.

Bednekoff, P. A., and S. L. Lima. 2004. Risk allocation and competition in foraging groups: reversed effects of competition if group size varies under risk of predation. *Proceedings of the Royal Society of London Series B: Biological Sciences* 271: 1491–1496.

Lima, S. L. 1990. The influence of models on the interpretation of vigilance. In M. Bekoff and D. Jamieson, eds. *Interpretation and explanation in the study of animal behavior: vol. 2. explanation, evolution, and adaptation.* Boulder, CO: Westview Press.

Lima, S. L., and P. A. Bednekoff. 1999. Temporal variation in danger drives antipredator behavior: the predation risk allocation hypothesis. *American Naturalist* 153: 649–659.

Parker, G. A., and P. Hammerstein. 1985. Game theory and animal behaviour. In P. J. Greenwood and P. H. Harvey, eds. *Evolution: Essays in honour of John Maynard Smith*. Cambridge, UK: Cambridge University Press.

Pulliam, H. R. 1973. On the advantages of flocking. *Journal of Theoretical Biology* 38: 419–422.

Treisman, M. 1975. Predation and the evolution of gregariousness. I. Models for concealment and evasion. *Animal Behaviour* 23: 779-800.

ADAPTIVE DYNAMICS

J. A. J. METZ

International Institute for Applied Systems Analysis

Laxenburg, Austria

Adaptive dynamics (AD) is a mathematical framework for dealing with ecoevolutionary problems that is primarily based on the following simplifying assumptions: clonal reproduction, rare mutations, small mutational effects, smoothness of the demographic parameters in the traits, and well-behaved community attractors. However, often the results from AD models turn out to apply also under far less restrictive conditions. The main AD tools are its so-called canonical equation (CE), which captures how the trait value(s) currently present in the population should develop over evolutionary time, and graphical techniques for analyzing evolutionary progress for one-dimensional trait spaces like pairwise invasibility plots (PIPs) and trait evolution plots (TEPs). The equilibria of the CE, customarily referred to as evolutionarily singular strategies (ess's), comprise, in addition to the evolutionary equilibria (or ESSs), points in trait space where the population comes under a selective pressure to diversify. Such points mathematically capture the ecological conditions conducive to adaptive (Darwinian) speciation.

CONTEXT

Micro-, Meso-, and Macroevolution

Adaptive dynamics was initiated as a simplified theoretical approach to mesoevolution, defined here as evolutionary changes in the values of traits of representative individuals and concomitant patterns of taxonomic diversification. This is in contrast to microevolution, i.e., changes in gene frequencies on a population dynamical time scale, and macroevolution, a term that then can be reserved for changes like anatomical innovations, where one cannot even speak in terms of a fixed set of traits.

Mesoevolution is more than microevolution writ large, and a similar statement applies to macro- versus mesoevolution. Each of these levels has its own emergent phenomena and its own explanatory frameworks, which in the end should be based on idealized summaries of the outcomes of lower-level mechanisms. Trait changes result from the microevolutionary process of mutant substitutions taking place against the backdrop of a genetic architecture and developmental system as deliverers of mutational variation, internal selection keeping the machinery of a body in concert, and ecological selection due to the interactions of individuals with their conspecifics, resources, predators, parasites, and diseases. AD focuses on these encompassing mechanisms.

To get a clean story, AD assumes a time-scale separation between ecology and evolution. In reality, this assumption holds only rather rarely. The idea is that arguments based on it may often lead to outcomes that are fair approximations, provided one applies them selectively and takes a sufficiently gross look at reality. The time-scale argument considerably eases the transition from population genetics to the perspective of ecologists, morphologists, and taxonomists. AD aims at addressing that larger picture at the cost of being wrong in the details.

The Fitness Landscapes of Mesoevolution

Mesoevolution proceeds by the selective filtering by the ecology of a continual stream of mutations. AD concentrates on the ecological side of this process, as at that end there are clearer a priori mathematical structures. The basic theory assumes clonal reproduction, and only a subset of the results extends to the Mendelian case—for monomorphic populations directly and for polymorphic populations after appropriate modification.

When approaching the evolutionary process from ecology, one sees immediately that fitnesses are not given quantities. They depend not only on the traits of individuals but also on their environment. The ecological feedback loop means that in the monomorphic and clonal cases necessarily the fitnesses of types permanently present on the ecological time scale are zero. Only the fitnesses of potential mutants can be positive or negative. The signs and sizes of these mutant fitnesses determine the direction and speed of evolution. Evolution corresponds to uphill movement in fitness landscapes that keep changing so as to keep the resident types at zero (see Fig. 1).

The main insight from the mathematical analysis of this picture has been the discovery of a potential mechanism for adaptive speciation that appears with a certain ubiquity in ecological models. Apart from that, the theory has produced effective tools for analyzing special families of ecoevolutionary models.

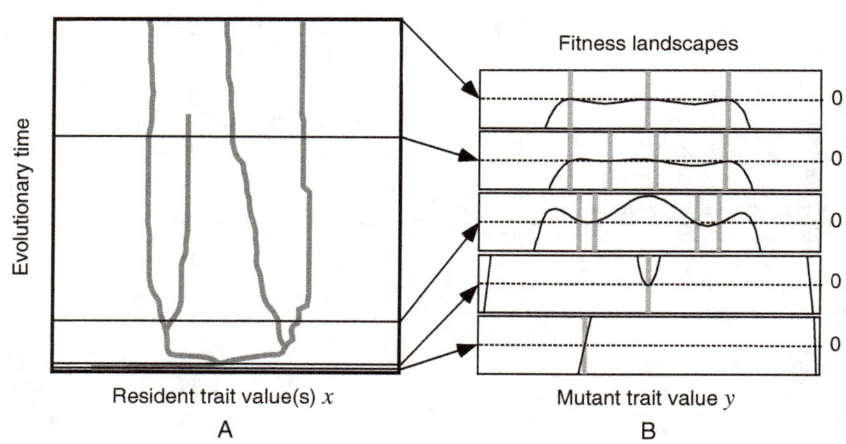

FIGURE 1 (A) Evolutionary path simulated on the basis of a population dynamical model, assuming clonal reproduction. Only the traits that are dominantly present in the population are shown. The second ascending branch finishes since its subpopulation went extinct. (B) The fitness landscapes for five population compositions as these occurred at the indicated times. The vertical bars indicate the types that at that moment were present in the population. At the second selected time, the population resided at a branching point. At the final time, the remaining three subpopulations reside at an evolutionarily steady coalition.

In the section that follows, first clonal reproduction is assumed, where relevant followed with a discussion of the Mendelian case. In line with the landscape analogy, zero fitness is referred to as sea level, etc.

PRELIMINARIES

Fitness

HISTORY

The concept of fitness as a quantitative measure of competitive prowess is a recent invention. Darwin never used the term in this way, and neither did the founding fathers of population genetics. (Except in one of Fisher's early papers; elsewhere Fisher, Haldane, and Wright use terms like selective advantage.) In modern population genetics, fitness is generally seen as the probability to survive to reproduction. However, this only works for relatively simple ecological scenarios with the different life phases neatly separated and synchronized. In ecology, one has to account for a less simple world where populations have age, size, spatial, and other structures and where demographic properties vary with the weather and local conditions.

POPULATION STRUCTURE AND EVOLUTIONARY ENVIRONMENTS

In AD, anything outside an individual that influences its population dynamical behavior (which by definition consists of impinging on the environment, giving birth, and dying) is called environment. It is then always possible in principle to find a representation of that behavior in terms of a state space, transition probabilities that depend on the course of the environment, and outputs that depend on the state of the individual and the condition of the environment at the time. Given the course of the environment, individuals independently move through their state spaces, the population state is a frequency distribution over this space, and the mathematical expectation of this frequency distribution, which is again a frequency distribution, moves according to what mathematicians call a positive linear dynamics (linear due to the independence, positive since we cannot have negative numbers of individuals). Mathematics then says that generally the expected size of a population in an ergodic environment will in the long run on average grow or decline exponentially. (The technical term ergodic means roughly that the environment may fluctuate but that these fluctuations have no persistent trend.) This growth rate, ρ, is what ecologists call fitness. It is necessarily a function of two variables—the type of the individuals parameterized by their traits, Y, and the environment, E, written as $\rho(Y \mid E)$, pronounced as "the fitness of Y when the environment is E" (the vertical bar as separator of the arguments is a notation borrowed from probability theory). The mathematical theory of branching processes (a mathematical discipline that deals with independently reproducing objects), moreover, says that a population starting from a single individual will, barring some technical conditions, either eventually go extinct or grow exponentially, with the probability of the latter being positive if and only if its fitness is positive. For constant environments, ρ is usually written as r and referred to as intrinsic rate of natural increase or Malthusian parameter.

In the theory of longer-term adaptive evolution, one is only interested in populations in which the number of individuals exposed to similar environments is sufficiently large that the internal workings of these populations can be modeled in a deterministic manner. In nature, populations are necessarily bounded. A mathematical consequence is that the community dynamics will converge to an attractor, be it an equilibrium, a limit cycle, or something more complicated. The corresponding environments are not necessarily ergodic, but the exceptions tend to be contrived. So here it will be assumed that community attractors generate ergodic environments.

Let the environment generated by a coalition of clones $C = \{X_1, \ldots, X_n\}$ be written as $E_{\text{attr}}(C)$. Then $\rho(Y \mid E_{\text{attr}}(C))$ is the invasion fitness of a new type Y in a C-community.

It follows immediately that all residents, i.e., types that are present in a community dynamical attractor, have zero fitness, since resident populations by definition do not in the long run grow or decline (see Fig. 1). This fact is basic to the following considerations. For ease of exposition, it is assumed throughout that $E_{\text{attr}}(C)$ is unique; the main results extend to the general case with small modifications.

MENDELIAN DIPLOIDS

The extension of the previous framework to Mendelian populations is easier than perhaps expected (although implementing it can be difficult). For the community dynamics, one has to distinguish individuals according to their genotypes and incorporate their mating opportunities with different genotypes into the description of the environment (in the case of casual matings, with pair formation it becomes necessary to extend the state space of individuals to keep track of their marriage status). Alleles reproduce clonally and as such have invasion fitnesses. It is also possible to define a (mock) fitness of phenotypes by introducing a parallel clonal model with individuals passing through their lives like their Mendelian counterparts and having a reproduction equal to the average of the contributions through the micro- and macrogametic routes (for humans, semen and ova). With such a definition, some essential, but not all, fitness-based deductions for the clonal case extend to the case of Mendelian inheritance. In particular, for genetically homogeneous populations a resident also has fitness zero (since genetically homogeneous populations breed true.) Moreover, the invasion of a new mutant into a homogeneous population is correctly predicted, as that mutant initially only occurs in heterozygotes that breed true by backcrossing with the homogeneous resident.

Mesoevolutionarily Statics: ESSs

Evolution stays put whenever the community produces an environment such that all mutants that differ from any of the residents have negative fitness. In the special case of a single resident type, we speak of an evolutionarily steady strategy (ESS). (The old name evolutionarily stable strategy is a bit of a misnomer since ESSs need not be evolutionarily attractive.) The general case when there may be more than one resident type is called an evolutionarily steady coalition (ESC). ESCs are the equilibria of evolution.

One way of calculating ESSs is depicted in Figure 2. For each environment as generated by a possible resident, the maximum of the invasion fitness landscape $\rho(Y \mid E_{\text{attr}}(X))$ is calculated. Next, one intersects the resulting curve (or surface) $Y = Y_{\text{opt}}(X)$ with the line (or surface) $Y = X$ to get the ESS $X^* = Y^*$. As monomorphic residents have fitness zero, all potential mutants $Y \neq Y^*$ have negative fitness.

The situation for ESCs is more complicated, as there may be so-called genetic constraints, with heterozygote superiority as generic example. However, in the so-called ideal free (IF) case all phenotypes comprising an ESC have fitness zero (at least when there is a single birth state and the ESC engenders a community dynamical equilibrium). This IF case is defined by the requirement that there are no genetic constraints whatsoever, that is, mutants can occur that produce any feasible type as heterozygotes in the genetic background of the resident population.

Fitness Proxies

The existence of a well-defined fitness forms the basis for the calculation of ESSs and AD. However, given its existence it is often possible to replace ρ by some more easily determined quantity that leads to the same outcome for the calculations of interest. For example, in

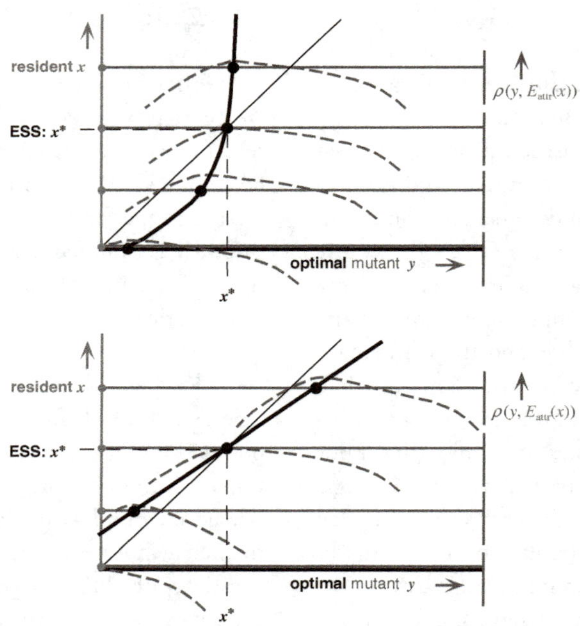

FIGURE 2 Scheme for calculating ESSes: For each of the possible resident populations, here characterized by a scalar trait, the invasion fitness of all potential mutants is calculated (interupted curves). The mutant axis is drawn on the same scale as the resident axis. From these fitness curves, the optimal strategy for the corresponding resident environment is calculated (fat curve). The ESS is the optimal reply to itself, to be calculated by intersecting the fat curve with the 45° line. In (A), the ESS attracts evolutionarily as can be seen from the fact that for other trait values the fitness landscape increases in the direction of the ESS. In (B), the fitness landscapes decrease in the direction of the ESS. Hence it repels.

optimization calculations ρ can be replaced with any quantity monotonically related to it, and for the graphical methods of AD one may replace ρ with any sign-equivalent quantity. Such quantities are referred to as fitness proxies. An example of a fitness proxy of the first type is the average rate of energy intake. Being a fitness proxy is always predicated on additional assumptions. For instance, it may help a forager little to increase its energy intake in an environment where this drastically increases its exposure to predation. A fitness proxy of the second type, restricted to nonfluctuating environments, is the logarithm of the average lifetime offspring number $\ln(R_0)$. If individuals may be born in different birth states, as is the case for spatial models, where birth position is a component of the birth state, R_0 is defined as the dominant eigenvalue of the next generation matrix (or operator in the case of a continuum of birth states). This matrix is constructed by calculating from a model for the behavior of individuals how many offspring are born on average in different birth states dependent on the birth state of the parent.

Optimization Principles

Often biologists try to predict evolutionary outcomes by pure optimization. Of course, if one measures only the environment one may predict the evolutionarily steady trait values that go with that environment by maximizing fitness in that environment. This is why predictions from optimization work. In general, optimization procedures do not predict the outcome of evolution, for that entails also predicting the environment that goes with the ESS, but they often satisfactorily predict the strategies that go with that environment. However, such limited predictions are of little use when considering the consequences of environmental change like increasing fishing mortality or global warming.

Many papers on ESS theory also derive optimization principles by which ESSs may be calculated. Hence, it is of interest to know when there exist properties of phenotypes that are maximized at an ESS. It turns out that this is the case if and only if one of two equivalent conditions holds good: the effect of the trait (environment) can be summarized in a single variable such that for each environment (trait) there exists a single threshold above which fitness is positive. These statements can be paraphrased as follows: the trait (environment) should act in an effectively one-dimensional monotone manner. (Think, for example, of the efficiency of exploiting a sole limiting resource.) Given such a one-dimensional summary $\psi(E)$ of the environment (which is often more easily found), it is possible to construct a matching summary of the traits $\phi(X)$, and vice versa, through

$$\phi(X) = -\psi(E_{\text{attr}}(X)). \tag{1}$$

The previous statements hinge on the interpretation of the term optimization principle. The latter should be interpreted as a function that attaches to each trait value a real number such that for any constraint on the traits the ESSs can be calculated by maximizing this function. This proviso mirrors the practice of combining an optimization principle derived from the ecology with a discussion of the dependence of the evolutionary outcome on the possible constraints.

If an optimization principle ϕ exists, each successful mutant increases $\phi(X)$ and hence any ESS attracts. Moreover, $\psi(E_{\text{attr}}(X))$ decreases with each increase in $\phi(X)$. Since fitness increases with ψ where it counts, i.e., around zero, ψ may be dubbed a pessimization principle. When a pessimization principle exists, in the end the worst attainable world remains, together with the type(s) that can just cope with it.

Optimization principles come closest to the textbook meaning of fitness, which generally fails to account for the fact that the fitnesses of all possible types are bound to change with any change in the character of the residents. However, optimization principles, although frequently encountered in the literature, are exceptions rather than the rule.

In evolutionary ecology textbooks, the maximization of r or R_0 takes pride of place without mention of the reference environment in which these quantities should be determined. Hence, all one can hope for is that the same outcome results for a sufficiently large collection of reference environments. It turns out that this is the case if and only if r can be written as $r(X \mid E) = f(r(X \mid E_0), E)$ respectively. $\ln(R_0)$ can be written as $\ln(R_0(X \mid E)) = f(\ln(R_0(X \mid E_0)), E)$ for some function f that increases in its first argument, where E_0 is some fixed but otherwise arbitrary environment. These two criteria are relatively easy to check. A fair fraction of textbook statements, if taken literally, apply only when these special conditions are met. This happens, for instance, when the only influence of the environment is through an additional state- and type-independent death probability, or rate, respectively, when the life history can be decomposed into stages that are entered through single states (so that no information carries over), of which one comprises all states after the onset of reproduction, and no stage is affected by both X and E.

ADAPTIVE DYNAMICS (AD)

Traits, PIPs, MIPs, and TEPs

Paleontologists and taxonomists are interested in the change of traits on an evolutionary time scale. What are traits to taxonomists are parameters to ecologists. So in AD one is after a dynamics in the parameter space of a community dynamics. The trick to arriving at such a simple picture is to assume that favorable mutants come along singly after a community has relaxed to an attractor. Another trick is to assume clonal reproduction, on the assumption that this way one can find out where the ecology would drive evolution if the latter were not hampered by the constraints of Mendelian genetics.

To get at a purely trait-oriented picture, any reference to the environment should be removed from the expression for invasion fitness:

$$s(Y\,|\,C) = \rho(Y\,|\,E_{\text{attr}}(C)) \qquad (2)$$

(often this is written as $s_C(Y)$ to emphasize the interpretation as a family of fitness landscapes).

This subsection focuses on scalar traits, starting with the case where there is only a single clonally reproducing resident, $C = x$. The first step in the analysis is plotting a contour plot of $s(y|x)$. Usually this is simplified to plotting only the zero contours, as those matter by far the most. The result is called a pairwise invasibility plot (PIP) (see Fig. 3).

Note that the diagonal is always a zero contour, as residents have fitness zero. A point where some other contour crosses the diagonal is referred to as an evolutionarily singular strategy (ess). The ESSes are a subset of the ess-es.

Now assume that mutational steps are small and that in the beginning there is only one resident trait value $x(0)$. Plot this value on the abscissa of the PIP, say, in the left panel of Figure 3. After some random waiting time, mutation creates a new trait value y. This trait value can invade only when it has positive fitness, i.e., is in a plus area of the PIP. It can be proved that an invading type replaces its progenitor if the latter is not too close to an ess or a bifurcation point of the community dynamics and the mutational step is not too large. If such a replacement has occurred, we call the new trait value $x(1)$. In the PIP under consideration, if $x(0)$ lies to the left of the ESS, then $x(1)$ lies to the right of $x(0)$, and vice versa. Hence, repeating the process pushes the evolutionary path to a neighborhood of the ESS. Upon reaching that neighborhood, it may become possible that ecologically the mutant and its progenitor persist together.

To see how such coexisting pairs of strategies fare evolutionarily, it is necessary to consider the set of protected dimorphisms, i.e., pairs of strategies that can mutually invade, denoted as (x_1, x_2). Its construction is shown in Figure 4.

The evolutionary movement of the pair (x_1, x_2) is governed by $s(y \mid x_1, x_2)$. Under the assumption of small mutational steps, a good deal of information can be extracted from the adaptive isoclines, calculated by setting the selection gradient

$$g_i(x_1, x_2) = \mathrm{d}s(y \mid x_1, x_2)/\mathrm{d}y|_{y=x_i} \qquad (3)$$

equal to zero. As depicted in Figure 5, x_1 will move to the right when g_1 is positive and to the left when it is negative, and x_2 will move up when g_2 is positive and down when it is negative.

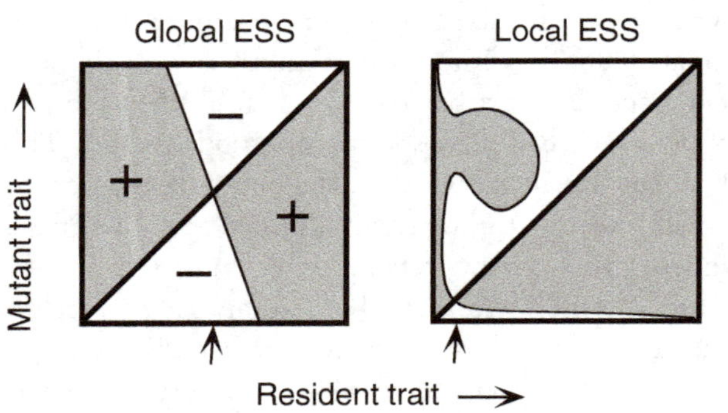

FIGURE 3 Pairwise invasibility plots: sign of the fitness of potential mutants as a function of the mutant and the resident traits.

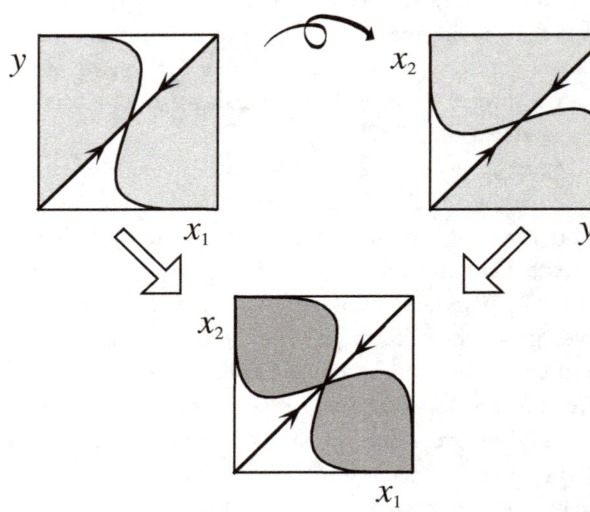

FIGURE 4 The construction of a *mutual invasibility plot* (MIP), depicting the set in (trait space)^2 harboring protected dimorphisms. Not all polymorphisms occurring in AD are protected, but unprotected polymorphisms have the habit of never lying close to a diagonal.

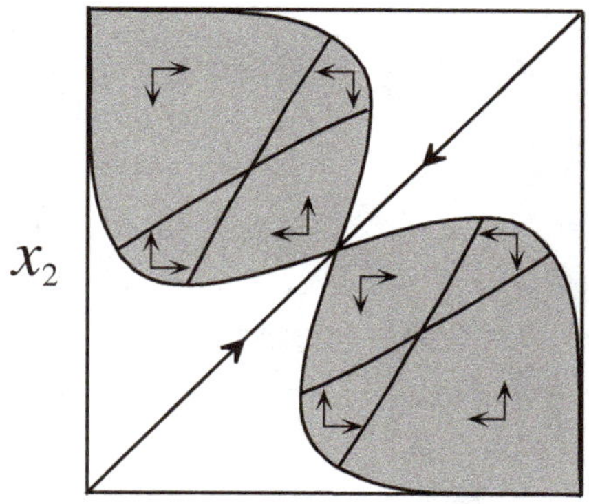

FIGURE 5 *Trait evolution plot* (TEP), i.e., MIP together with arrows that indicate the direction of the small evolutionary steps that result from the invasion by mutants that differ but little from their progenitor, and adaptive isoclines.

From the classification of the possible dynamics near an ess in Figure 7, it can be seen that the ess in the PIPs from Figure 3 also attracts in the dimorphic regime.

The Canonical Equation

BASICS

The dependence of the dynamical outcomes on no more than the sign of the invasion fitness hinges on the ordering properties of the real line. For vectorial traits, one must proceed differently. The workhorse is the so-called canonical equation (CE) of AD, a differential equation that captures how the trait vector changes over evolutionary time, on the assumption of small mutational steps, accounting for the fact that favorable mutants do not always invade due to demographic fluctuations. For unbiased mutational steps, the evolutionary speed and direction are given as the product of three terms: from left to right, (1) the effective population size N_e (as in the diffusion equations of population genetics), (2) the probability of a mutation per birth event ε times a matrix C consisting of the variances and covariances of the resulting mutational step, and (3) the selection gradient (the "curly d" notation stands for differentiating for that variable while treating the other variables as parameters):

$$G(X) = \left. \begin{pmatrix} \partial s(Y \mid X)/\partial y_1 \\ \vdots \\ \partial s(Y \mid X)/\partial y_n \end{pmatrix} \right|_{Y=X}, \qquad (4)$$

which is a vector pointing from the position of the resident in the steepest uphill direction. The uphill pull of selection is thus modified by the differential directional availability of mutants expressed in the mutational covariance matrix. For diploid Mendelian populations, an additional factor 2 appears, since the substitution of a mutant allele leads to a mutant homozygote twice as far removed (at the considered order of approximation). The equilibrium points of the CE are the ess-es mentioned previously.

In reality, many mutants attempt to substitute simultaneously. Luckily, for small mutational steps this appears to affect the environment only in the higher-order terms that in the derivation of the CE disappear. However, in the clonal case the effects of the mutants do not add up, since a mutant may be supplanted while invading by a better mutant from the same parent type. In the Mendelian case, the CE will do a better job as substitutions occur in parallel on different loci, which to the required order of approximation act additively.

THE LINK WITH EVO-DEVO

From an AD perspective, the link with evolutionary developmental biology (Evo-Devo) is through the mutational covariance matrices. Unfortunately, Evo-Devo has yet little to offer in this area. Therefore, at present the most that AD researchers can do is work out how the outcomes of a specific ecoevolutionary model depend on the possible forms of the mutational covariance matrix. The answers from AD thus become Evo-Devo questions: is the mutational covariance matrix for these traits expected to fall within this or that class?

To show the importance of the missing Evo-Devo input in AD: mutational covariance matrices have an (often dominating) influence on the time scales of evolution (Fig. 6A) and the basins of attraction of ess-es (Fig. 6B), even to the extent that they often determine whether an ess attracts or not.

On a more philosophical level, it bears noting that the selection gradient points only in a single direction, while the components of the trait vector orthogonal to that gradient hitchhike with the selectively determined motion thanks to a developmental coupling as expressed in the mutational covariance matrix. The higher the dimension of the trait space, the larger the contribution of development as a determinant of evolutionary motion. The dimensions of the trait spaces that are routinely considered thus make for the contrast in attitudes of behavioral ecologists and morphologists, with the former stressing selection and the latter developmental options for change.

Evolutionarily Singular Strategies

Evolutionarily singular strategies X^* can be calculated by setting the fitness gradient equal to zero. Figure 7 shows their classification according to dynamical type for scalar

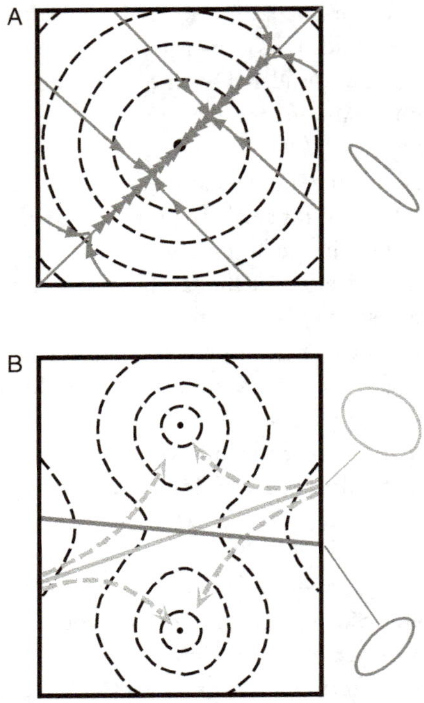

FIGURE 6 Two fitness landscapes that are supposed to keep their shape and sink only when the adaptive trajectory moves uphill (as is the case if and only if the population regulation is through an additional state-independent death rate). Distributions of mutational steps are symbolized by ovals. (A) The shape of the mutation distribution induces a time scale separation between the movement along the diagonal and anti-diagonal direction. (B) The difference in mutation distributions causes a difference in the domains of attraction of the two ESSes.

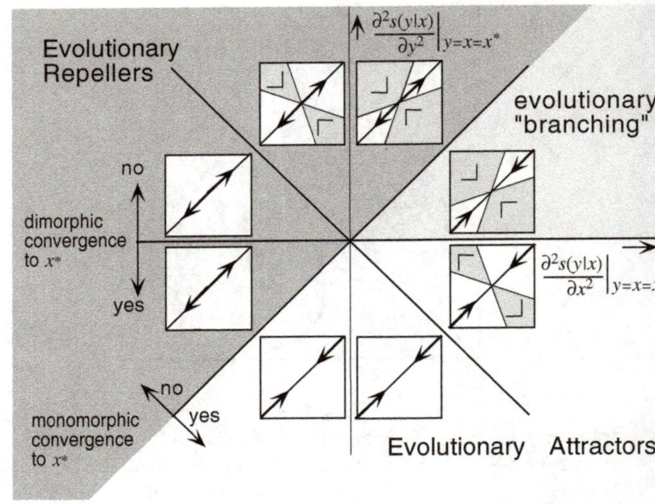

FIGURE 7 A classification of the ESSes for scalar traits. The cases in the lower half are all ESSes. The leftmost of these repels, the others attract. The latter ESSes are thus genuine evolutionary attractors. The branching points in the rightmost upper sector attract monomorphically but repel dimorphically.

traits. (Note that this classification was derived deductively from no more than some mathematical consistency properties shared by well-posed ecoevolutionary models.) Devising a good classification for higher-dimensional ess-es is an open problem. One reason is that in higher dimensions the attractivity of a singular point crucially depends on the mutational covariance matrix except in very special cases.

Adaptive Speciation

Perhaps the most interesting ess-es are branching points, where the ecoevolutionary process starts generating diversity. When approaching such points, the evolutionary trajectory, although continually moving uphill, gets itself into a fitness minimum. More precisely, it is overtaken by a fitness minimum. This is due to the population dynamics; think of the following analogy: Somewhere gold has been found, attracting people to that spot. However, after the arrival of too many diggers it becomes more profitable to try one's luck at some distance.

The buildup of diversity can take very different forms. In the clonal case, the population just splits into two as depicted in Figure 1. In the Mendelian case, the diversification starts with a broadening of the variation in the population. The fitness landscape locally has the shape of a parabola that increases away from x^*. This means that types more on the side have a higher fitness than those in the center. It therefore pays not to beget kids near the center. The Mendelian mixer has the contrary tendency to produce intermediate children from dissimilar parents. Fortunately, there are all sorts of mechanisms that may thwart this counterproductive mixing. The most interesting of these is the buildup of some mechanism that allows like extremes to mate only among themselves, thus ensuring that the branches become separate genetic units. A simple mechanism occurs in insects that diversify in their choice of host plants, with mating taking place on those hosts.

The author's conviction is that in cases where there is no automatic mating barrier, a buildup of other mechanisms engendering assortative mating is not unexpected. Present-day organisms are the product of 3.5 billion years of evolution. During that time, their sensory and signaling apparatus has been evolutionarily honed for finding the most advantageous mates. Hence, there will be an abundance of template mechanisms. These mechanisms, once recruited to the task, will probably have a tendency to enhance each other in their effect. One may thus expect that the available generalized machinery can often

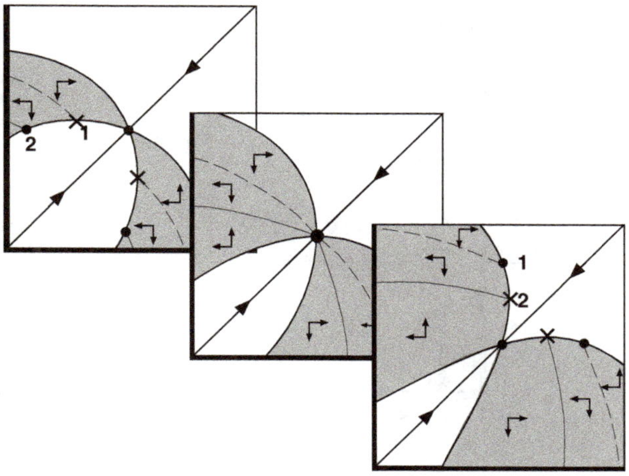

FIGURE 8 Three TEPs corresponding to a bifurcation of an ESS to a branching point. In this case, the adaptive tracking of the ESS in the wake of slow global environmental change stops due to a change in character of the ESS. In the fossil record, this scenario would correspond to a punctuation event that starts with speciation.

easily be adapted so as to genetically separate the branches whenever evolution brings the population to a branching point. However, most scientists working on the genetics of speciation do not appear to agree with this view.

Bifurcations

The bifurcations of AD encompass all the classical bifurcations found in ecological models. In addition, there is a plethora of additional bifurcations. An example is the transition from an ESS to a branching point depicted in Figure 8.

JUSTIFYING THE AD APPROXIMATION

Traits and Genotype-to-Phenotype Maps

The real state space of the mesoevolutionary process is genotype space, while the phenotypic trait spaces of AD are but convenient abstractions. Phenotypic mutational covariances reflect both the topology of genotype space, as generated by mutational distances, and the genotype-to-phenotype map generated by the developmental mechanics. This reflection can only be expected to be adequate locally in genotype space, and therefore locally in evolutionary time.

AD focuses on small mutational steps. A partial mechanistic justification comes from the expectation that the evolutionary changes under consideration are mostly regulatory. Coding regions of genes are in general preceded by a large number of short regions where regulatory material can dock. Changes in these regulatory regions lead to changes in the production rate of the gene product. The influence of a single regulatory region among many tends to be rather minor. Note, moreover, that phenotypes should in principle be seen as reaction norms, i.e., maps from microenvironmental conditions to the characteristics of individuals (another term is conditional strategies). The phenotypes of AD are inherited parameters of these reaction norms. Only in the simplest cases is a reaction norm degenerate, taking only a single value.

Internal Selection

Functional morphologists usually talk in terms of whether certain mechanisms work properly or not and discuss evolution as a sequence of mechanisms all of which should work properly, with only slight changes at every transformational step. Translated into the language of fitness landscapes, this means that only properly working mechanisms give fitnesses in the ecologically relevant range, while the improperly working ones always give very low fitnesses. This leads to a picture of narrow, slightly sloping ridges in a very high-dimensional fitness landscape. The slopes on top of the ridges are the domain of ecology; their overall location is largely ecology independent.

As a simple example, one may think of leg length. The left and right leg are kept equal by a strong selection pressure, which keeps in place a developmental system that produces legs of equal length, notwithstanding the fact that during development there is no direct coupling between the two leg primordia. Hence in a trait space spanned by the two leg lengths, ecologists concentrate on only the diagonal.

The trait spaces dealt with in morphology are very high dimensional so that the top of a ridge may be higher dimensional, while away from the ridge the fitness decreases steeply in a far larger number of orthogonal directions. A picture similar to that of functional morphologists emerges from Evo-Devo. The long-term conservation of developmental units (think of the phylotypic stage or homology) can only be due to strong stabilizing selection, since mutations causing large pattern changes have many side effects with dire consequences. As a result, ecological selection generally acts only on quantitative changes in the shapes and sizes of homologous body parts.

Since the fitness differences considered in functional morphology and Evo-Devo largely result from the requirement of an internal coherence of the body and of the developmental process, people speak of internal selection. Here, the term is used to label features of the fitness landscape that are roughly the same for all the environments that figure in an argument.

The Assumptions of AD

THEORY

Figure 9 illustrates an argument by Fisher showing that the higher the dimensionality of the trait space, the more difficult becomes the final convergence to an adaptive top. This argument extends to the movement in a ridgey fitness landscape: the higher the number of orthogonal off-ridge directions, the more rare it is for a mutational step to end up above sea level, and small mutational steps have a far higher propensity to do so than large ones. Together, these two arguments seem to underpin the requirements of AD that mutations in ecologically relevant directions are scarce and the corresponding mutational steps are small.

Unfortunately the above arguments contain a biological flaw: the assumed rotational symmetry of the distribution of mutational steps. Real mutation distributions may be expected to show strong correlations between traits. Correlation structures can be represented in terms of principal components. The general experience with biological data is that almost always patterns are found like the ones shown in Figure 10B. Figure 10C shows that the existence of mutational correlations will in general enhance the rareness and smallness of the mutational steps that end up above sea level.

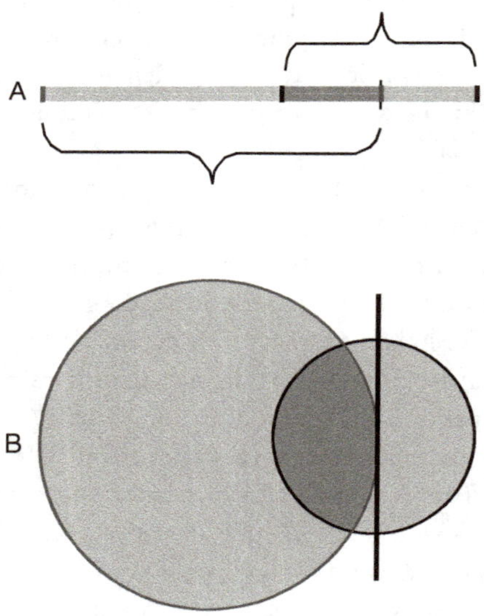

FIGURE 9 (A) Two balls in $\mathbb{R}^1$, with the center of the smaller ball on the boundary of the larger ball. The ratio of the volume of their intersection to the volume of the smaller ball is $\frac{1}{2}$. (B) A similar configuration in $\mathbb{R}^2$. The volume of the intersection is now a smaller fraction of the volume of the smaller ball. For similar configurations in $\mathbb{R}^n$, this fraction quickly decreases to zero for larger n. Now think of the larger ball as the part above sea level of a fitness hill and of the smaller ball as a mutation distribution. Clearly the fraction of favorable mutants will go to zero with n.

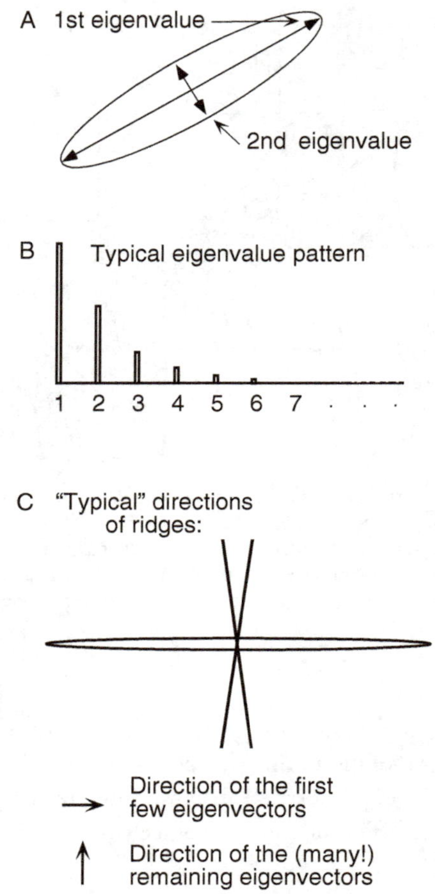

FIGURE 10 (A) Contour line of a bivariate distribution, supposedly of mutational steps. The lengths of the two axes of the ellipse, called principal components, are proportional to the square root of the eigenvalues of the mutational covariance matrix. (B) Typical eigenvalue pattern found for large empirical covariance matrices. (C) The mutation distribution will rarely be fully aligned with the fitness ridges. If in a high dimensional fitness landscape one takes one's perspective from the mutation distribution and looks at the orientation of the ridges relative to the first few principal axes of this the mutation distribution, then, when the number of the dimensions of the trait space is very large and the ridge has a relatively low dimensional top, the ridge will typically extend in a direction of relatively small mutational variation.

DATA

The conclusions from the previous subsection seem to underpin nicely the assumptions of AD. Unfortunately, various empirical observations appear to contradict these conclusions. Populations brought into the lab always seem to harbor sufficient standing genetic variation to allow quick responses to selection, and a few loci with larger effect (quantitative trait loci, or QTL) are often found to underlie the variation in a trait. However, these empirical observations may have less bearing on the issue than one might think.

First, given the speed of evolution relative to the changes in the overall conditions of life, populations in the wild are probably most often close to some ESS. Moreover, in noisy environments, fitness maxima tend to be flat. This means that near-neutral genetic variation will accumulate, which is exploited first when a population gets artificially selected upon. At ESSes, the mutation limitation question is largely moot. Beyond its statics, AD's main interest is in the larger-scale features of evolutionary trajectories after the colonization of new territory or, even grander, a mass extinction. The scale of these features may be expected to require a further mutational supply of variation. Second, directional selection on an ecological trait may be hampered by stabilizing internal selection on pleiotropically coupled traits. In the lab, this stabilizing selection is relaxed. This means that far more variation becomes available for directional selection than is available in the wild.

Finally, AD-style theory has shown that in the absence of assortative mating the initial increase of variability after the reaching of a branching point tends to get redistributed over a smaller number of loci with increasing relative effect The end effect will be QTL, but it will be produced through the cumulative effect of small genetic modifications.

SEE ALSO THE FOLLOWING ARTICLES

Adaptive Landscapes / Branching Processes / Coevolution / Evolutionarily Stable Strategies / Mutation, Selection, and Genetic Drift

FURTHER READING

Dercole, F., and S. Rinaldi. 2008. *Analysis of evolutionary processes: the adaptive dynamics approach and its applications.* Princeton: Princeton University Press.

Dieckmann, U., M. Doebeli, J. A. J. Metz, and D. Tautz, eds. 2004. *Adaptive Speciation.* Cambridge Studies in Adaptive Dynamics vol. 3. Cambridge, UK: Cambridge University Press.

Dieckmann, U., and R. Law. 1996. The dynamical theory of coevolution: a derivation from stochastic ecological processes. *Journal of Mathematical Biology* 34: 579–612.

Diekmann, O. 2004. A beginner's guide to adaptive dynamics. *Mathematical Modelling of Population Dynamics* 63(4): 47–86.

Durinx, M., J. A. J. Metz, and G. Meszéna. 2008. Adaptive dynamics for physiologically structured models. *Journal of Mathematical Biology* 56: 673–742.

Geritz, S. A. H., É. Kisdi, G. Meszéna, and J. A. J. Metz. 1998. Evolutionarily singular strategies and the adaptive growth and branching of the evolutionary tree. *Evolutionary Ecology* 12: 35–57.

Gyllenberg, M., J. A. J. Metz, and R. Service. 2011. When do optimisation arguments make evolutionary sense? In F. A. C. C. Chalub and J. F. Rodrigues, eds. *The mathematics of Darwin's legacy.* Basel, Switzerland: Birkhauser.

Leimar, O. 2009. Multidimensional convergence stability. *Evolutionary Ecology Research* 11: 191–208.

Metz, J. A. J. 2008. Fitness. In S. E. Jørgensen and B. D. Fath, eds. *Evolutionary ecology.* Vol. 2 of Encyclopedia of Ecology. Oxford: Elsevier.

Metz, J. A. J. 2011. Thoughts on the geometry of meso-evolution: collecting mathematical elements for a post-modern synthesis. In F. A. C. C. Chalub and J. F. Rodrigues, eds. *The mathematics of Darwin's legacy.* Basel, Switzerland: Birkhauser.

Metz, J. A. J., S. A. H. Geritz, G. Meszéna, F. J. A. Jacobs, and J. S. van Heerwaarden. 1996. Adaptive dynamics, a geometrical study of the consequences of nearly faithful reproduction. In S. J. van Strien and S. M. Verduyn Lunel, eds. *Stochastic and spatial structures of dynamical systems.* Amsterdam: North-Holland.

ADAPTIVE LANDSCAPES

MAX SHPAK

University of Texas, El Paso

The concept of an adaptive (or fitness) landscape was introduced by Sewall Wright as a metaphor for the genetic "space" on which adaptive evolution occurs. The metaphor also drew on analogies from the physical sciences, in that the adaptive landscape was conceptualized as a sort of inverted potential energy surface. Essentially, the view that evolution takes place on an adaptive landscape entails representing natural selection as deterministic "hill climbing"—with the peaks representing local fitness optima and valleys representing fitness minima. The metaphor proved to be an enduring one, principally because of its intuitive visual appeal. However, it has also caused a great deal of confusion, in part because of its use to describe two entirely different entities, one of which is dynamically sufficient for a description of natural selection, the other only under certain limiting assumptions (the first being the space of possible genotypes; the second, mean population fitness as a function of allele frequencies). Furthermore, it will be argued that excessive reliance on the three-dimensional metaphor of a rugged landscape results in flawed analyses that do not hold up to more formal treatments of biologically realistic multilocus model systems.

THE SELECTION EQUATIONS

In order to make sense of adaptive landscapes, we first consider the selection equations. For the sake of notational simplicity and without loss of generality, we begin with a population of haploid asexual organisms, starting with the standard scenario of a single locus with two alleles in an effectively infinite population. Denoting the frequency of one allele as p and the other as $1 - p$, we assign respective fitness values (defined as viability, fecundity, or the product of the two) to be W_1, W_2. The mean population fitness is $\overline{W} = pW_1 + (1 - p)W_2$, while the expected change in the frequency of the first allele in discrete time is

$$E[\Delta p] = \frac{p(1 - p)}{\overline{W}}(W_1 - W_2) \quad (1)$$

where $\Delta p = p(t + 1) - p(t)$.

For instance, when $W_1 > W_2$, Equation 1 states that the frequency of the first allele increases in proportion to the fitness difference between it and the competing allele. This equation is a special case of a more general expression for the change in the frequency of any true-breeding trait value x, i.e.,

$$E[\Delta x] = \frac{var(x)}{\overline{W}} \frac{\partial \overline{W}}{\partial x} \quad (2)$$

where var(x) is the variance of the trait value and the derivative measures the change in mean fitness with respect to the heritable trait. Equation 2 provides the basis for one concept of the adaptive landscape, i.e., the mean population fitness $\overline{W}(x)$ as a function of allele or trait frequency in the population. In part, this interpretation was motivated by Fisher's fundamental theorem, which states that in the absence of frequency dependence or the confounding effects of genetic linkage, the mean population fitness increases at a rate proportional to its variance; i.e.,

$$\Delta \overline{W} = \frac{var(\overline{W})}{\overline{W}}. \quad (3)$$

Furthermore, in a finite population subject to genetic drift as well as natural selection, we can approximate the probability density of the allele frequency at equilibrium as

$$F(p) \approx \overline{W}(p)^N. \quad (4)$$

These results underscore the importance of the mean fitness in evolution and lead Wright to conceptualize the fitness landscape in terms of mean population fitness rather than as the fitness values of individual alleles or genotypes. This distinction is illustrated for the simple diallelic diploid example in Figures 1A and 1B, where the first figure plots representative fitness values for the two alleles and the second shows mean population fitness as a function of the frequency of the first allele.

The principal difficulty with the mean fitness representation is that in more realistic evolutionary models that incorporate frequency dependence or mutation, $\overline{W}$ is not necessarily maximized by natural selection. Furthermore, if we include the complications of linkage and recombination, we no longer generally have a dynamically sufficient description of $\overline{W}(p_1 \ldots p_n)$ as a function of allele frequencies at each locus $1 \ldots n$, because the marginal fitness of an allele depends on which particular allelic states it is associated with at other loci.

In contrast, one can always derive a dynamically sufficient description of selection (analogous to Eq. 1) from the fitness values of individual genotypes, either for the individual fitness values or for mean population fitness as a function of genotype frequency. Much of the confusion in the interpretation of adaptive landscapes arose from confounding the two representations. Unless otherwise indicated, this entry will use the term adaptive landscape to refer to the genotype space representation. It will be argued that this form is often more fruitful in addressing a number of questions relevant to our understanding of adaptation, speciation, and in bridging the conceptual divide between macro and microevolution.

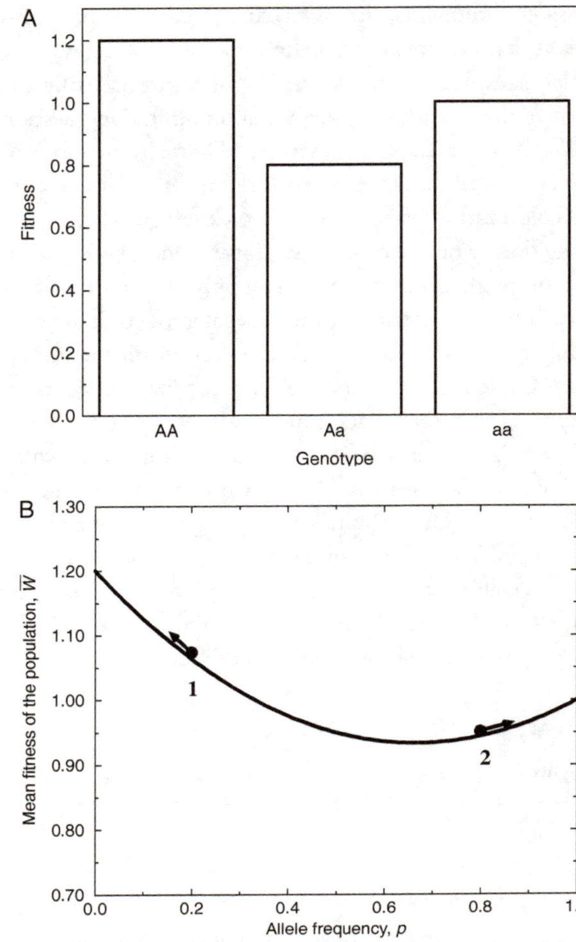

FIGURE 1 (A) The adaptive landscape as a function of genotypes, in this case for a single diploid locus. (B) The adaptive landscape as a function of genotype frequency, based on the fitness values of individual diploid genotypes in part (A).

MUTATION AND MULTIPLE LOCI

Understanding fitness functions and the structure of adaptive landscapes for multilocus genotypes requires an increase in the dimensionality of the model system. This is particularly the case if sexual reproduction, and consequently recombination, introduce additional dynamical complexity into the models.

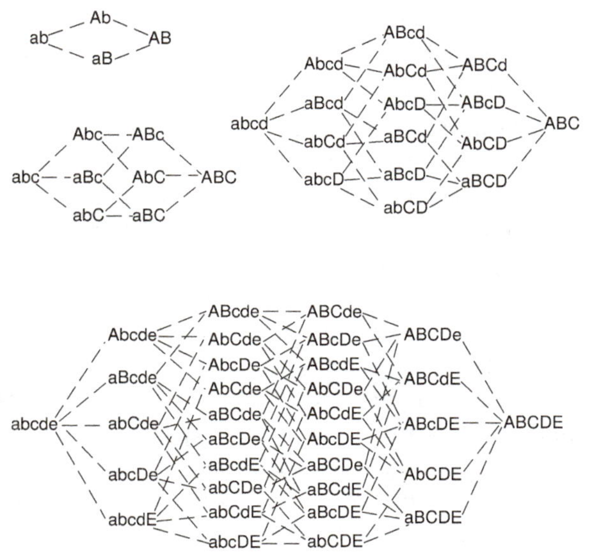

FIGURE 2 Graphs in haploid genotype space for n = 2, 3, 4, and 5 loci and 2 alleles, illustrating the increase in the number of paths between any two genotypes as a power function of the number of loci.

Begin by considering the simplest scenario—a haploid asexual organism with n loci and 2 alleles (results are generalizable to a 4-letter nucleotide or 20-letter amino acid alphabet, so a binary model is chosen for the sake of simplicity). In this model, there are 2^n possible genotypes, and if we assign a fitness value W_i to the ith genotype, we can describe the changes in their frequency with a trivial modification of Equation 1. However, a complete model of evolution requires both a selection and a transmission operator, where the latter involves mutation and/or recombination. The transmission operator imposes a topology onto the space of 2^n genotypes by defining a neighborhood of each genotype x_i. For example, if mutations are rare, a first-order approximation assumes that only single-point mutations are possible. Representing every genotype as a point on a lattice with n neighbors, the genotype space defined by point mutation is a hypercube, shown in two through five dimensions, corresponding to 2–5 loci, in Figure 2.

If we allow multiple mutations, the graph's connectivity increases, up to the case where every vertex has an edge with all $2^n - 1$ others. In a model life cycle where selection occurs before reproduction and mutation, the dynamics on a general fitness landscape are defined by the change in the frequency x_i of every genotype:

$$x_i(t+1) = \frac{1}{\overline{W}}\left(w_i x_i(t) + \sum_{j \neq i} (\mu_{ji} w_j x_j(t) - \mu_{ij} w_i x_i(t)) \right) \quad (5a)$$

where μ_{ij} is the probability of mutating from state i to state j in a single time interval. If one interprets x as absolute rather than relative frequencies, we can write Equation 5a without the mean fitness term. In this case, Equation 5 becomes a system of linear equations involving the mutation and selection operators, i.e.,

$$\vec{x}(t+1) = A\vec{x}(t) \quad (5b)$$

for **x** a vector of genotype absolute frequencies, and **A** = **MW**, where **M** is a matrix of mutation probabilities (i.e., for point mutations, $M_{ii} = (1 - n\mu)$, $M_{ij} = \mu$ if the Hamming distance between i and j is equal to unity, otherwise $M_{ij} = 0$) and **W** is a diagonal matrix with genotype fitness $W_{ii} = W_i$.

In an infinite population, one can calculate the mutation–selection equilibrium distribution of genotype frequencies from Equation 5b as the dominant eigenvector of **A**. This distribution is often referred to as the quasispecies, following the terminology of Manfred Eigen and Peter Schuster, who derived the results for model systems of self-replicating macromolecules.

For a complete dynamical characterization, the system of equations requires as many variables as there are genotypes. Because the number of genotypes increases as a power of the number of loci, finding analytical solutions becomes intractable because of the high dimensionality. However, if one assumes certain symmetry properties in the fitness function W_i, the description of mutation–selection dynamics can be greatly simplified. We begin by considering some canonical examples of "simple" fitness landscapes—the additive fitness function and the flat (neutral) landscape, before proceeding to more general models.

Additive Landscapes

The additive landscape is so called because it is assumed that the fitness contribution of an allele at every locus is independent of the allelic state at other loci, so that the fitness of any genotype is just the sum (or product) of the value of each allele, i.e., $W_i = s_1 + s_2 \cdots + s_n$, where s_k is the fitness contribution of allelic state s at locus k. This generates a landscape with a single optimum or fitness peak, as shown in Figure 3.

Biologically, one can interpret the single-peak landscape as a scenario where there is one most fit genotype and that every point mutation away from it decreases the fitness incrementally. If we make the further assumption that selection is relatively weak, higher-order interaction terms across loci can be ignored, so that the contributions from each site will be approximately additive, which

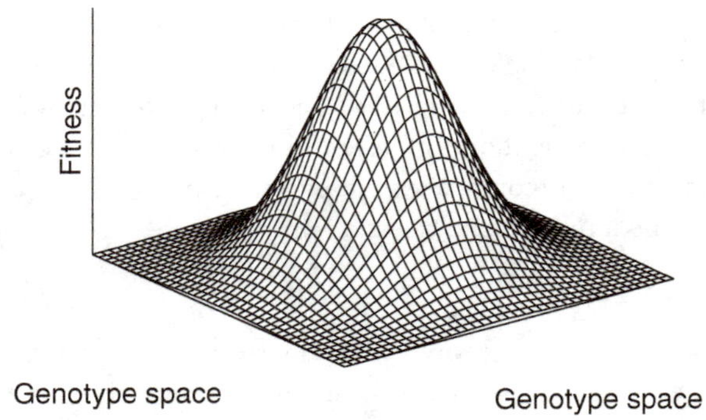

FIGURE 3 An additive fitness landscape in genotype space with a single optimum, under the assumption that each point mutation away from the local optimum proportionately decreases genotype fitness.

simplifies the representation of selection dynamics still further.

Quantitative Traits

A special case of an additive fitness landscape involves stabilizing selection on an additive quantitative trait. If we assume that the number of loci n is large and the phenotypic/fitness contributions of alleles are additive and identically distributed random variables at every locus, then we can approximate the phenotype x by a Gaussian distribution with a mean at the optimum (rescaled to $x_{opt} = 0$) and a genetic variance σ^2 determined by the sum of the variance contributions from individual loci. Standard models of stabilizing selection postulate a quadratic or a Gaussian selection function about the mean value, i.e.,

$$W(x) = Exp\left[-\frac{x^2}{2V}\right]$$

where V is a variance parameter that measures the intensity of stabilizing selection (i.e., small V indicating strong selection against nonoptimal phenotypes). Combining selection and mutation under the assumption of weak mutational phenotypic effects at each locus with an approximately normal distribution, the selection equations on the mean phenotype $\bar{x}$, rescaled with an optimum at 0, are

$$E[\Delta\bar{x}] = -\frac{\sigma_A{}^2}{\sigma_A{}^2 + \sigma_B{}^2 + V}\bar{x} \quad (6)$$

where $\sigma_A{}^2$ is the additive variance due to mutation and $\sigma_B{}^2$ is the variance due to environmental noise. For larger mutational variances, the normal approximations break down and results based on mixtures of distribution, such as the "house of cards" model, need to be introduced.

Neutral Landscapes

A flat or neutral landscape is simpler still, in that every genotype, regardless of dimensionality, is assumed to have the same fitness. This model, which stemmed from the assumption (as argued by Motoo Kimura) that the majority of mutations are either extremely deleterious, and therefore rapidly removed from any population, or were effectively neutral (e.g., mutations in noncoding and nonregulatory sequences, synonymous nucleotide substitutions, substitutions of functionally similar amino acids, and the like).

The deterministic dynamics of allele and genotype frequency on a neutral landscape depend only on the mutation rates. In the absence of any directional bias to mutation, the evolutionary process can be modeled as a random walk on an unweighted graph, converging to a uniform equilibrium distribution of genotype frequencies in the limit of infinite time. If finite population effects such as Fisher–Wright genetic drift (i.e., approximation of finite samples from an effectively infinite gamete pool) are introduced, the expected genotype frequencies follow a multinomial distribution on the K genotypes in a population of size N,

$$F(v_1 \ldots v_K) = \binom{N}{v_1 \ldots v_K} f_1^{v_1} \ldots f_K^{v_K}, \quad (7)$$

where f_i are the initial frequencies of each genotype while v_i are the number of each genotype in the current sample (the sum of all v_i values is N, by definition), implying no expected (average) directional change in the frequency of any particular genotype. This mode of sampling implies that the first and second moments of change in the frequency of any genotype f are given by

$$E[\Delta f] = 0; \quad \text{Var}[\Delta f] = \frac{f(1-f)}{N}. \quad (8)$$

If mutation is nondirectional, then there is no contribution to the first moment term, while the second moment term includes the sample variance in allele frequency and mutation probabilities.

We next consider extensions of these simple models to more realistic scenarios incorporating epistatic interactions.

EPISTASIS AND RUGGED LANDSCAPES

Equation 5a provides an entirely general and dynamically sufficient description of natural selection and point mutation when there are n loci with k alleles. The problem of dimensionality arises from the fact that there are as many fitness coefficients and frequency values (state variables) as there are genotypes. This requires parameterizations that on the one hand can capture some of the complexity of nonadditivity and epistasis while remaining tractable.

Epistasis involves interaction of allelic effects at multiple loci on phenotype and fitness. From this definition, it follows that one can approximate a fitness function up to arbitrary order using Taylor or Fourier series expansions. Denoting a genotype **z** as an n-vector of allelic states $z_1 \ldots z_n$, we can write the fitness in series form:

$$W(\vec{z}) = a_0 + \sum_i a_i S_i + \sum_{i,j} a_{i,j} S_i S_j + \cdots + a_{1,n} S_1 S_2 \cdots S_n. \quad (9)$$

The first term in the series represents the zeroth-order baseline fitness value, it would be the only term in a neutral landscape. The coefficients a_i, $a_{i,j}$, and so on, represent, respectively, the first-order additive contribution of each allele to fitness, the second-order term contributions, and so forth. The polynomial terms S_i associated with alleles z_i in the expansion can have a range of interpretations. In a Taylor expansion, they represent the deviations from mean allelic effects, whereas in a Fourier approximation we have $S_i = \pm 1$. This is a generalization of the model of epistasis familiar to quantitative geneticists, who divide phenotypic variance into additive, additive by additive, and higher-order epistatic component terms.

In heuristic terms, the zeroth-order term generates a flat landscape and the first-order terms give a single-peak landscape, so it follows that higher-order epistatic interactions can generate increasingly more rugged landscapes with multiple local optima, as illustrated in Figure 4.

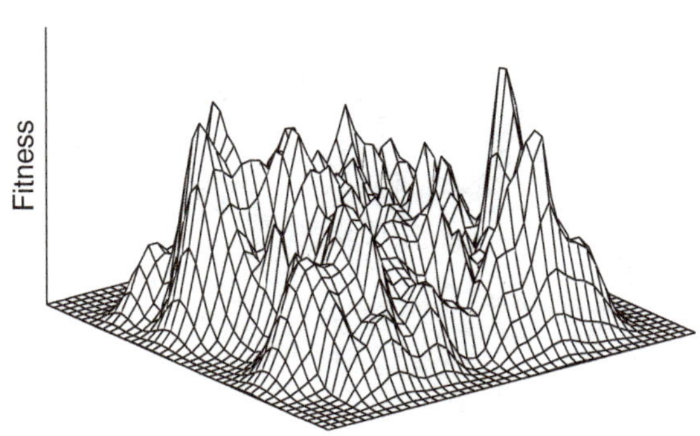

FIGURE 4 A typical rugged adaptive landscape with multiple local optima separated by local minima. Such a landscape has the canonical topography for shifting balance theory and other models, because it requires populations to initially decrease in fitness in order to eventually attain higher fitness.

This iconic image of the rugged fitness landscape has been the source of an extensive research program in evolutionary theory. The specific problem raised by the multipeaked adaptive landscape is the following: the strictly deterministic process of natural selection always favors genotypes with higher fitness. As a result, if the mutation rate is comparatively low (i.e., anything other than single point mutations are negligibly rare), selection will take the population distribution to the nearest peak, behaving as a greedy search algorithm. As a consequence, any population located in the neighborhood of a local optimum would have no way of reaching the global optimum, or for that matter other local optima that have higher fitness. The implication, of course, is that with natural selection and low mutation rates alone, adaptive evolution would rapidly stagnate. One of the principal goals of research in this area is to determine why, in fact, such stagnation at or near local optima does not occur in natural and artificial populations.

Kauffman's *n–k* Model

A numerical approach to understanding the dynamics on rugged landscapes was proposed by Stuart Kauffman. He introduced the so-called *n–k* model for epistasis, which can be thought of as a special, random-variable case of Equation 9. In the *n–k* model, there are n loci contributing fitness effects and k is the order of epistatic interactions. When $k = 1$, a random number (sampled from a uniform or normal distribution) is chosen at each of the n loci, and the genotype fitness is the sum of fitness contributions across loci. With $k = 2$, the fitness contribution of each locus is determined by the sum or mean value of two random variables: its own, and the contribution of one of the randomly chosen $n - 1$ remaining loci. This can be generalized to arbitrary k, where the fitness contribution of each locus is the mean of its own contribution and $k - 1$ chosen at random from the other $n - 1$ sites. It is clear that $k = 1$ corresponds to an additive, single-peak landscape, while higher values of k give a random-variable, kth-order representation of epistasis.

In a series of simulation studies, Kauffman contrasted the outcomes of different search strategies with increasing orders of epistasis. The simplest search strategy involved single point mutations and a random walk that continued for as long as more fit genotypes were accessed in each step. On a $k = 1$ single-peaked landscape, this greedy algorithm always found the global optimum. In the other limiting case, $k = n - 1$, one has a completely uncorrelated landscape such that the mutational neighbors of a high-fitness genotype are no more likely to have a similar fitness than any random genotype on the lattice. Since random walks terminate once no higher fitness genotypes are accessed, it can be shown that the probability of a random walk of length R is

$$P_R = \prod_{k=0}^{R}\left(1 - \left(\frac{2^k - 1}{2^k}\right)^{n-1}\right), \tag{10}$$

which gives an an expected walk length of

$$E[R] = \mathrm{Log}_2(n - 1). \tag{11}$$

If larger steps are allowed, corresponding to mutations at multiple loci (i.e., $\sim n$ mutations per time step), it can similarly be shown that on an uncorrelated landscape the probability of finding a genotype with higher fitness decreases as a factor of 2 with every step, or, equivalently, that the waiting time to find a more fit genotype doubles with each transition.

Numerical studies have shown that for rugged correlated landscapes ($1 < k << n - 1$), the performance of large transitions is qualitatively similar, entailing a high probability of accessing the neighborhood of a new local optimum initially, but decreasing as a power function with each subsequent step. James Crutchfield and his colleagues have argued that an optimal search strategy for such landscapes is the "Royal Road" algorithm. In this model, the initial search involves mutations at multiple sites, followed by point mutations, with natural selection retaining genotypes along sample paths with higher fitness than the initial state. Essentially, the initial large steps permit the population to access the neighborhoods of local optima with higher fitness, whereas more conservative point mutations allow the population to attain the optimum itself. The model is analogous to a simulated annealing process, in which high-energy particles are allowed to move freely (i.e., passing through regions with lower free energy or high potential energy) initially, followed by a cooling phase in which they settle in the nearest minimum free energy state.

These results raise the question: What biological processes can drive a search strategy that produces large transitions initially before settling into a more conservative (point mutation and selection driven) approach? Interpreted at face value, the large transition steps would involve macromutations, as opposed to the single point mutations during the local hill-climbing phase. Others considering the problem of transitions between local optima have emphasized the importance of genetic drift.

Counteradaptive Evolution and Dimensionality

Genetic drift in finite populations has been proposed as a mechanism that could allow evolving populations to access distant adaptive peaks via low-fitness valleys. Wright's shifting-balance model for evolution was an important elaboration on this theme, as it influenced speculations about the speciation process (see below).

The idea behind these claims is that barring macromutation, the only way to make a transition between peaks along a low-fitness valley is for unfit intermediate genotypes to temporarily increase in frequency through genetic drift. It was shown by Kimura that in sufficiently small populations, genotypes with lower fitness have a nonnegligible probability of increasing in frequency relative to more fit genotypes and even become fixed as a consequence of sampling error in a finite gamete pool.

Although the potential significance of genetic drift cannot be dismissed outright, much of the emphasis on its importance in major evolutionary transitions has been an artifact of our spatial intuition about the shape of adaptive landscapes. In a two- or three-dimensional space, local peaks separated by low-fitness valleys are the de facto norm. However, if the genetic architecture of a trait or fitness value involves higher dimensionality, there is increasing evidence that isolated peaks are a pathological case because in multidimensional genotype or phenotype space there will usually be adaptive "ridges" connecting optima. These results suggest that the topological space that evolution occurs on probably resembles Figure 5 more than the iconographic Figure 4.

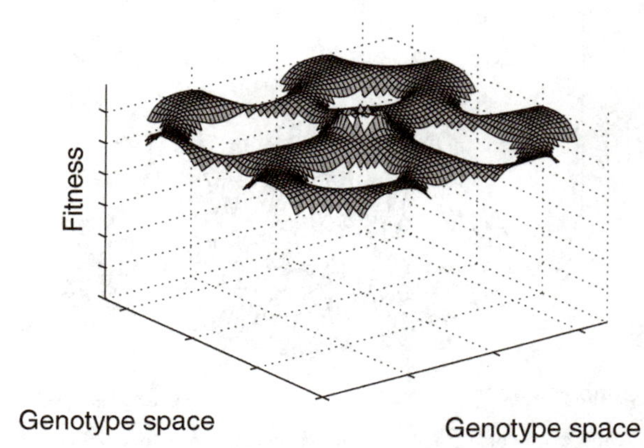

FIGURE 5 An example of a holey landscape, in which all fitness values are either 1.0 or 0. The connected fitness 1.0 genotypes generate a neutral network that does not require counteradaptive evolution in order to access one genotype from another along a mutational path.

One approach that has provided much insight into the topology of multidimensional adaptive landscapes was adopted by Christian Reidys and by Sergey Gavrilets, who made the simplifying assumption of a threshold model of fitness where a genotype in an n-locus, two-allele scenario is either viable or nonviable. If only point mutation is permitted, the sole means of reaching a genotype with a fitness of 1 is via a mutational path along other genotypes with fitness 1. This "Russian Roulette" landscape is constructed by assigning either a fitness of 1 or 0 to each genotype with respective Bernoulli probabilities P, $1 - P$.

As the number of loci (and therefore genotypes) tends to infinity, several patterns in these "holey" landscapes are revealed. First, if $P > 1/2$, the landscape consists of a fully connected neutral network, so that every fitness 1 genotype can be accessed from any other in the landscape through single point mutation. When $P < 1/2$, there are almost always several non-connected neutral networks, so by definition there are several clusters of fit genotypes that are not mutually accessible. However, it was shown by Reidys that when $P > 1/n$, the system is characterized as reaching a percolation threshold, in which the set of fitness 1 genotypes is dominated by a single large component network of viable genotypes, surrounded by comparatively small (including single genotype) viable clusters (see Fig. 6).

In other words, provided that the probability of viability is not very small (i.e., less than the reciprocal of the number of loci), there will always be a dominant "giant" connected component containing $\sim 2^n/n$ genotypes followed by a next largest component of order $\sim n$. The giant component has a radius of size n, meaning that there are viable paths involving mutations at all n loci. These results can be generalized to a landscape where the fitness values can assume values between 0 and 1.

Altogether, these investigations have demonstrated that the problem of accessing optima and moving between local maxima largely disappears with high dimensionality, and that there is no need to invoke genetic drift, macromutation, or other nonadaptive mechanisms to account for the major transitions in evolution.

THE COMPLICATION OF RECOMBINATION

So far, we have only considered mutation as a transmission operator in the evolutionary process. Sexually reproducing organisms produce offspring whose genotypes differ from their own through recombination, which, depending on the number of crossover points and the genetic distance between homologous chromosomes from each parent, can potentially span the entire genotype space.

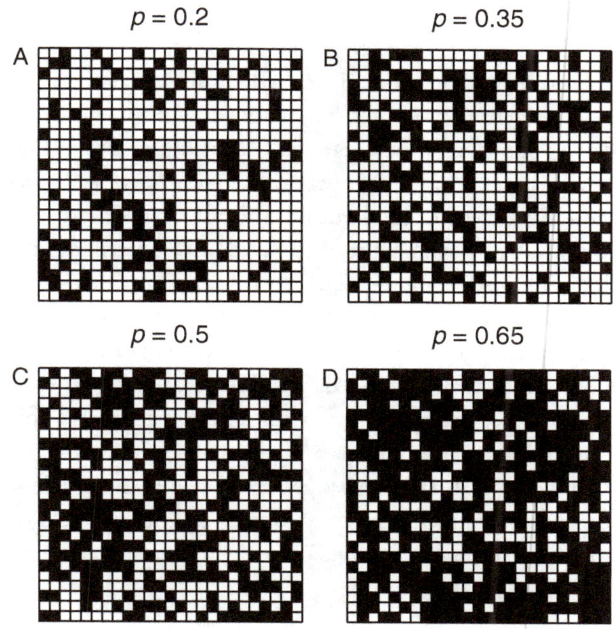

FIGURE 6 Holey landscapes generated by a model in which genotypes have probability P of being viable and $1 - P$ of being inviable. When $P > 0.5$ (i.e., $p = 0.6$ in part (D)), the percolation threshold is attained, so that every viable genotype is a member of a single connected subgraph. For $P < 0.5$, there are multiple subgraph neutral networks, with a giant component dominating when $P > 1/n$ (e.g. $P = 0.45$ in part (C)). As I decreases, the giant component disappears, and the neutral networks become increasingly small and isolated (i.e., $P = 0.15$ and 0.3 in parts (A) and (B), respectively).

Apart from the fact that the genetic (Hamming) distance between recombinants is often greater than that derived through point mutation, there is the additional complication that the transmission-selection dynamics under recombination are inherently quadratic. Assuming that selection occurs before reproduction and recombination in the life cycle, we have

$$X_i(t + 1) = \frac{1}{\bar{W}} \left(\sum_{j,k} R_{j,k \rightarrow i} w_j w_k X_j X_k \right), \qquad (12)$$

where R is the recombination operator describing the probability that parents with genotypes j, k produce an offspring of type i. Unlike Equation 5, there is no global linearization of the recombination–selection equations.

This nonlinear dynamical system was first analyzed by Richard Lewontin and Ken-Ichi Kojima. The state variables are multilocus haplotypes, with frequencies denoted by $X_{ij\ldots}$ where i, j and other indices denote the allelic state at each of the $1 \ldots n$ loci. Consider the simplest case of two loci and two alleles in a population of haploid, sexually reproducing organisms, such that X_{11}, X_{12}, X_{21}, X_{22} denote the frequencies of genotypes **AB**, **Ab**, **aB**, and **ab**, respectively. If the probability of recombination between the two loci is given by r, then (for instance) a cross between **AB** and **ab** produces each parental type with frequency $(1 - r)/2$ and each recombinant type (**Ab** and **aB**) with frequency $r/2$.

The selection–recombination equations for the two-locus, two-allele model can be obtained by substituting $(1 - r)$ and r for R, and genotype fitness for w in Equation 10. Except in the limit of nonepistatic selection (i.e., weak selection with approximately additive effects), significant linkage disequilibria across loci are generated, such that in general the frequency $X_{IJ} \neq p_I p_J$ at loci I, J. Consequently, neither the haplotype frequencies nor the mean fitness of the population can be predicted from allele frequencies at each locus. Thus, genetic systems with recombination and epistatic selection offer a clear example of an evolutionary system where allele frequencies are not dynamically sufficient predictors of fitness landscape topology, either in terms of the distribution of genotypes or of mean fitness.

SPECIATION AND SHIFTING BALANCE RECONSIDERED

New insights into the topology of adaptive landscapes have put a new perspective on long-standing notions about speciation and macroevolution. Consider Wright's "Shifting Balance" model for major adaptive transitions. Based on the intuition of a rugged fitness landscape, Wright proposed that the first phase of the process requires genetic drift in a population subdivided into small demes in order to shift genotype frequencies away from a local optimum. This allows genotype frequencies to pass through a low-fitness valley into the neighborhood of a higher adaptive peak. As the genotype frequencies in the population approach a new local optimum, the second phase of shifting balance is largely deterministic, dominated by natural selection taking the subpopulation to the local optimum (i.e., fixation or near fixation of the locally optimal genotype within each subpopulation). The final phase involves a combination of interdemic selection and migration, with populations at higher fitness peaks displacing those occupying lower optima.

Shifting Balance theory was introduced by Wright in large part because it offered a solution to the problem of peak shifts that synthesized selection, mutation, genetic drift, reproductive isolation, and migration into a single process. In spite of its intellectual and aesthetic appeal, a number of recent papers have shown that although it is not impossible for natural populations to follow the shifting balance scenario, they can only do so under rather restrictive parameters. Especially problematic is the first phase. Following Equation 4, the probability of genetic drift traversing a fitness valley is negligible unless the fitness differences are small and the population size is very small. The former condition is problematic because if the fitness differences are minor, then there is no substantial peak shift to speak of, whereas in the latter case there is often little segregating genetic variation for selection (or drift) to act on to begin with.

Perhaps an even stronger critique of the shifting balance theory comes from the fact that in multidimensional genotype space, the peak and valley intuition is misleading as ridges and networks become the dominant topological features (see the section "Epistasis and Rugged Landscapes," above). In the absence of adaptive valleys, there is no need to postulate genetic drift in small populations as a crucial driving force in evolution.

The same issues also arise in population genetic models for speciation. Many contributors to the "New Synthesis," particularly Ernst Mayr, emphasized that speciation necessarily involved a special case of the type of peak shifts associated with phase one of shifting balance. Specifically, Mayr maintained that genetic drift was necessary to traverse fitness valleys and establish reproductive isolation. Consequently, concepts such as "founder flush" speciation and genetic revolutions (sensu Hampton Carson) were introduced as models for allopatric speciation, where drift in very small, peripheral populations lead to drastic changes in the genetic composition of the population, thereby facilitating the peak shift.

Founder flush models, while plausible in theory, suffer from the same problems as phase one of shifting balance—low probability for realistic combinations of parameters, and the fact that they are not necessary if we allow for viable paths between optimal genotypes. As a counterexample, consider the two-locus, two-allele haploid genotypes **AB** and **ab**, both of which have a high fitness. If we assume that recombinants **aB** or **Ab** have low fitness, then there will be strong selection against hybrids and in favor of reproductive isolation. However, this raises the question of how, if mutation rates are low, one attains the **ab** genotype to begin with from an **AB** ancestor, because a single mutation always results in a low fitness intermediate genotype. The simplest solution to this problem was independently proposed by William Bateson, Theodosius Dobzhansky, and Hermann Muller, namely, that one of the intermediate genotypes, for instance **Ab**, was as fit as either **AB** or **ab**, so that there is no need to cross a fitness valley. **Ab** can then mutate to **ab**, which in turn will lead to selection for reproductive isolation with **AB** on account of the unfit **aB** hybrid. An analogous example of an adaptive landscape for two-diploid loci leading to Bateson–Dobzhansky–Muller

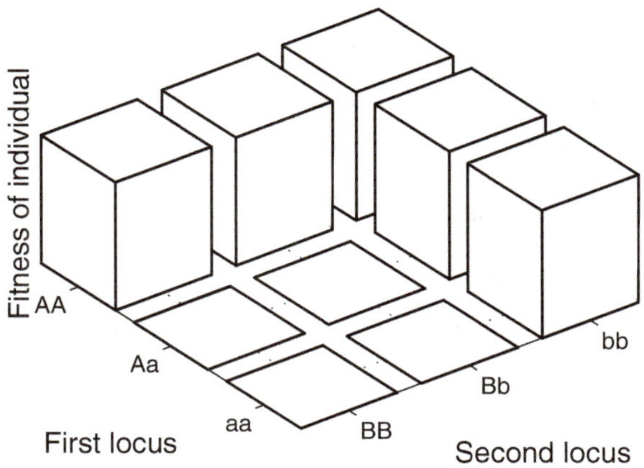

FIGURE 7 An example of a diploid two-locus, two-allele Bateson-Dobzhansky-Muller landscape. Note that the genotypes *AABB* and *aabb* have high fitness, as do *AAbb* and *Aabb*, while the recombinants *aaBB* and *aaBb* are inviable.

(BDM) incompatibilities is shown in Figure 7. As the number of loci increases, the number of possible BDM incompatibilities increase exponentially, as do the number of possible mutational paths connecting viable genotypes. Consequently, contra-adaptive mechanisms such as genetic drift and macromutation need not be invoked as the principal driving forces behind speciation, either.

DISCUSSION

Small changes to parameters determining the order and extent of epistatic interactions can lead to marked changes to the topology of an adaptive landscape. The simplest illustration of this can be seen with an increase in the frequency of viable versus inviable genotypes in the percolation models. However, in order to assess what parameterizations and topologies are biologically relevant, empirical data is necessary.

Measuring the fitness of different genotypes is an extremely difficult task, both in laboratory settings and even more so in nature. It is a nontrivial problem for a combination of reasons—i.e., the statistical difficulties of inferring fitness components from competition experiments, the problem of genotyping (becoming less of an issue today on account of high-throughput sequencing technology), and above all the issue of controlling for extraneous environmental variables that confound survival or reproductive advantage in competition experiments.

Consequently, some of the most robust studies of fitness landscape structure don't involve whole organisms, and instead focus on in vitro studies of the catalytic performance or the thermodynamic and/or structural stability of macromolecules such as RNA and proteins. In the case of RNA, the biophysics of folding are sufficiently well understood to predict fitness (i.e., mean square distance to an optimal structure, or to a free energy minimum). Walter Fontana and colleagues analyzed the topology of RNA landscapes using a random walk autocorrelation measure ρ, which provides a measure of the landscape ruggedness (i.e., a high autocorrelation score suggests a relatively smooth fitness surface with few peaks, while a low score implies a rugged landscape where a single nucleotide substitution can lead to radical changes in secondary structure). The authors compared the autocorrelation scores to those obtained on landscapes using Kauffman's n–k model and found that the RNA landscape was consistent with a model where $k \approx 7$, for $n \sim 100$.

Another approach to the problem of inferring the shape of adaptive landscapes is indirect, involving comparative sequence analysis across taxa. For example, comparing human protein sequences to those of other mammals, Alexey Kondrashov and his collaborators found that amino acid substitutions at sites associated with genetic diseases in humans are viable in other mammals such as the mouse. This suggests that epistatic interactions across sites are crucial in maintaining (or restoring) function. Since the mammalian sequences in the sample shared common ancestors with humans ~100 million years (and many speciation events) ago, this suggests a path of viable point mutations connecting the orthologs and provides evidences for BDM-type incompatibilities discussed in the previous section.

These examples are cited as two case studies that are representative of a growing literature among researchers using similar strategies—i.e., attempts to predict fitness from first principles based on macromolecular biophysics, and comparative approaches of homologous sequences across different species. In the future, as our knowledge of gene regulatory networks and quantitative trait loci in model organisms improves, it should be possible to make predictions about the structure of adaptive landscapes for more complicated epistatic model systems.

In addition to these empirical studies, there remains the important task of characterizing the evolutionary dynamics on multidimensional landscapes as our understanding of their structure improves. As was the case at the field's inception, there remains the daunting problem of high dimensionality in mutation–selection on landscapes in multilocus epistatic systems. The problem is even more pronounced for recombination–selection systems because of their nonlinearity. Most recent studies have either made use of approximations involving the first few moments of quantitative trait distributions or assumptions about the symmetry properties of fitness functions to reduce the numbers of variables and parameters necessary to characterize evolution. These methods will be increasingly valuable as more elaborate, data-driven models of epistatic interactions and pleiotropic effects are generated.

SEE ALSO THE FOLLOWING ARTICLES

Adaptive Dynamics / Evolutionary Computation / Mutation, Selection, and Genetic Drift / Quantitative Genetics

FURTHER READING

Coyne, J. A., N. H. Barton, and M. Turelli. 1997. A critique of Wright's shifting balance theory of evolution. *Evolution* 51: 643–671.

Eigen, M., J. MacCaskill and P. Schuster. 1989. The molecular quasispecies. *Advances in Chemical Physics* 75: 149–63.

Fontana, W., P. F. Stadler, P. Tarazona, E. O. Weinberger, and P. Schuster. 1993. RNA folding and combinatory landscapes. *Physical Review E*: 47: 2083–2099.

Gavrilets, S. 2004. *Fitness landscapes and the origin of species*. Princeton, NJ: Princeton University Press.

Kauffman, S.A. 1993. *The origins of order*. Oxford: Oxford University Press.

Kimura, M. 1980. *The neutral theory of molecular evolution*. Cambridge, UK: Cambridge University Press.

Kondrashov, A. S., S. Sunyaev, and F. A. Kondrashov. 2002. Dobzhansky–Muller incompatibilities in protein evolution. *Proceedings of the National Academy of Sciences* 99: 14878–14883.

Reidys, C. M., and P. F. Stadler. 2001. Neutrality in fitness landscapes. *Applied Mathematics and Computation* 117: 321–350.

van Nimwegen, E., J. P. Crutchfield, and M. Mitchell. 1999. Statistical dynamics of the Royal Road genetic algorithm. *Theoretical Computer Science* 221: 41–102.

Wright, S. 1932. The roles of mutation, inbreeding, crossbreeding and selectin in evolution. In D. F. Jones, *Proceedings of the 6th International Congress on Genetics,* Vol. 1. pp. 356–366. Austin, TX.

AGE STRUCTURE

TIM BENTON

University of Leeds, United Kingdom

An individual's age is often a strong determinant of its life history. Age can be used to stratify individuals into different age classes, with the rationale that individuals within the same age class will have similar life histories and therefore similar demographic rates. Such a population is termed age structured. Modeling age structure in discrete time is commonly undertaken with matrix models; in continuous time, with partial or delay differential equations. Age- (and stage-) structured models have been influential in many areas of evolutionary, population, and conservation ecology.

WHAT DOES AGE DO?

Predicting population dynamics with any precision requires some sort of population model. The simplest conceptualization is an "unstructured" model, where the future population size is some function of the current (or past) total population size; i.e., $N_{t+1} = f(N_t)$. This formulation implies that variation between individuals in age does not matter. This may be because there is an assumption that on the scale of the model individual differences are averaged out or age may not be relevant. For example, modeling populations of annual plants based on annual census data does not need to take into account the age of the plants. However, in other circumstances, age differences between individuals are crucial. All individuals go through a life cycle: they are born (sprout or hatch) small, grow as they develop, mature and reproduce as prime-aged adults, and then decline during senescence in reproductive rates and survival. At some spatial or temporal scale, a population's demographic rates will always depend on its age profile or structure, but it is sometimes possible to ignore this—for the example of annual plants, age structure can be ignored at the annual temporal scale, but it would become important when looking at their dynamics at a monthly scale.

To illustrate the importance of age structure, two human populations are contrasted in Figure 1: the population of India has a greater proportion of young people than the UK (India has 43% of its population between 15 and 39, whereas the UK has 33%), and far fewer older people (8% over 60 versus 22%). These differences in age structure are reflected in the populations' demographic rates (i.e., per capita reproduction and survival) and ultimately cause differences in the population growth rates: India's population is growing almost twice that of the UK.

Age is not the only factor that differentiates between the N individuals in a population, and therefore it is not the only factor that influences demographic rates and creates population structure. Sex ratio may be important, and for some systems, there may also be differential performance between individuals depending on their body size, quality, or condition; so population structure can arise not only due to age, but it can also be size structured, or even size-and-age structured (see below in the discussion of Integral Project Models below). In many invertebrates, adults and larvae may have very different demographic rates, and so populations may be structured according to the developmental stage they are in. In many invertebrate species, developmental duration can be very plastic depending on food availability and temperature, so knowing an individual is a certain age will tell you less about their vital rates than knowing they are an adult or a pupa.

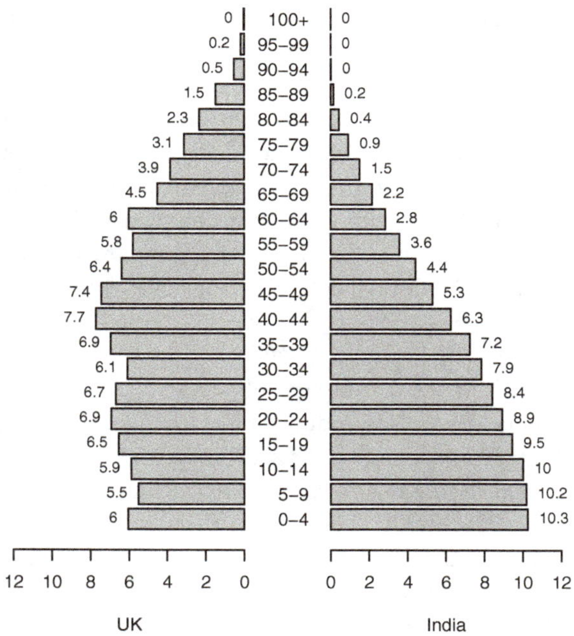

FIGURE 1 Population pyramids for India and the UK, 2009. Each bar shows the proportion of the total population in an age group. The Indian population has a much greater proportion of young people and people of reproductive age. Data from the U.S. Census Bureau, International Database (http://www.census.gov/ipc/www/idb/index.php). The population growth rate for the UK is 0.7 (though some of this is driven by immigration); for India it is 1.34 (data from http://data.worldbank.org/).

The importance of incorporating the life cycle into population models (whether as stage or age) is that it explicitly incorporates history: individuals born at time t cannot reproduce until they have developed and passed into the correct stage, age, or size class. This "history in the life history" is particularly important when modeling the way populations respond to changes in the environment (whether the biotic environment, such as density dependent processes, or the abiotic environment, such as changes in the climate), as the population's response to a change may lag behind the change. The characteristics of the lagged response will be determined by the population's structure and may therefore change with time. For example, if food suddenly becomes available in a population with many adults, there could be a quick increase in reproduction leading to a fast response. Conversely, a population with many juveniles would respond more slowly, as the juveniles would have to grow to adulthood in order to reproduce and therefore boost population growth rate. Lags in dynamical systems also have the property of introducing instability, and so lags due to generation time are one of the common causes behind population cycles. In general, structured models may have quite different dynamical properties from unstructured models with the same average demographic rates.

Age-structured models incorporate age structure into the model and describe the dynamics of the whole population as the sum of the subpopulations in the different age classes. Age-structured models can naturally be formulated in discrete or continuous time. Discrete-time models are often appropriate for species that reproduce annually and may be censused each year (which includes populations of many vertebrates). Such discrete-time models encapsulate an equation that gives transition rates (i.e., survival) from one age class to the next, and also reproductive rates for each age class. The reproductive rates are strictly called fertilities, as they incorporate both reproduction and the survival of the offspring until the next time step. The classical, discrete-time, age-structured model is called a Leslie matrix model (see the section below) and is an example of the broader family of matrix models. In continuous time, the McKendrick–von Foerster equation is an analogue of the Leslie matrix, but because age structure introduces lags into the dynamics, another natural way to deal with age structure is via delay-differential equations (DDEs).

THE CLASSIC AGE-STRUCTURED LESLIE MATRIX MODEL

The classic age-structured, discrete-time, matrix model is often called a Leslie matrix model, after P. H. Leslie (a colleague of Charles Elton at the Bureau of Animal Populations at Oxford), who was an early pioneer of their use in ecology. The Leslie matrix contrasts with the Lefkovitch matrix, which is a stage-structured matrix model.

At time t, the population is represented by a vector n_t with each element n_i corresponding to the number of individuals in age class, i $(1 \geq i \geq \max(age))$; so the total population size $N = \sum_1^{\max(age)} n_i$. The population at time $t + 1$ is projected using a recurrence equation,

$$\mathbf{n}_{t+1} = \mathbf{A}\mathbf{n}_t, \quad (1)$$

where $\mathbf{A}$ is a square matrix. For a simple three-age class model, with individuals growing to age 3 (then dying), reproducing at ages 2 and 3 (a life cycle similar to many small passerine birds perhaps), the life-cycle diagram and corresponding matrix would be as follows:

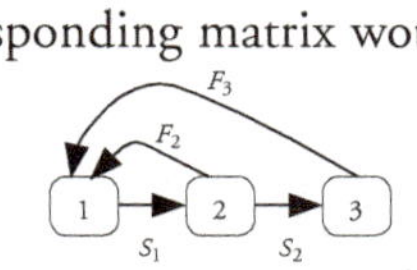

$$\mathbf{A} = \begin{pmatrix} 0 & F_2 & F_3 \\ S_1 & 0 & 0 \\ 0 & S_2 & 0 \end{pmatrix}. \quad (2)$$

F_i is the fertility, the number of offspring that are born to an adult in age class i and which survive to the next census, S_i is the proportion of individuals in age class i that survive to age class $i + 1$ at the next census. An age-structured matrix is always one that has nonzero elements only in the top row and in the subdiagonal. In the classic model, the values of F_i and S_i are constant. If Equation 1 is iterated forward sufficiently, the influence of any arbitrary starting vector **n** disappears and the population grows at a constant rate, λ, and the age structure (i.e., the proportion of individuals in each age class) becomes constant: this is the stable age distribution (see Fig. 2). This property that initial conditions make no impact on the long-run (asymptotic) behavior is known as ergodicity. Rather than iterate Equation 1 to find the asymptotic properties of the matrix, they are available as the solution of the characteristic equation of Equation 1 (i.e., $\det(\mathbf{A} - \lambda\mathbf{I}) = 0$, where "det" is the determinate, and $\mathbf{I}$ is the identity matrix). The solution gives a vector of eigenvalues (of which there are as many as age classes), but the largest, or principal, eigenvalue is λ, the population growth rate. When the eigenvalues are known, the equation $\mathbf{A}\mathbf{w} = \lambda\mathbf{w}$ can then be solved. The vector **w** is the right eigenvector (as it sits on the right of the matrix **A** in the equation). The left eigenvector, **v**, is found from $\mathbf{v}'\mathbf{A} = \lambda\mathbf{v}'$. The right eigenvector corresponds to the stable age distribution, and the left eigenvector corresponds to the reproductive value of each age class (i.e., the relative contribution to population growth from each age class and which typically increases up to maturity and then declines). Clearly from Equation 1 if the matrix **A** changes, the characteristic equation changes and so do the eigenvalues and eigenvectors as well as, therefore, their biological interpretations: population growth, stable age distribution, and age-related reproductive values. In fact, some of the appeal of matrix modeling arises from the fact you can easily analyze the way λ will change with changes in the matrix. As each element of **A** (the a_{ij}) is varied, λ will change and the rate of change in λ with respect to a change in a_{ij} is given by

$$\frac{\partial\lambda}{\partial a_{ij}} = \frac{v_i w_j}{\langle \mathbf{w}, \mathbf{v} \rangle}. \tag{3}$$

Here, v_i is the ith element of vector **v**, and $\langle \mathbf{w}, \mathbf{v} \rangle$ represents the scalar product of the two vectors. The quantity $\partial\lambda \backslash \partial a_{ij}$ is called the sensitivity of λ and is the absolute rate of change in λ with a change in each matrix element. The elasticity is the sensitivity rescaled to be the proportional change in λ with a change in the matrix element. The mathematical formulation is the same whether the matrix model is age, sex, size, or stage structured. Figure 2 gives an example of an arbitrary Leslie matrix iterated over a number of time steps, plus the simple analytical output.

The classical Leslie matrix model (and its related stage-structured formulations) has been hugely influential in both population management and evolutionary ecology. Much of the popularity comes from the work of Hal Caswell (see Further Reading for Caswell's iconic book). Parameterizing a matrix model from

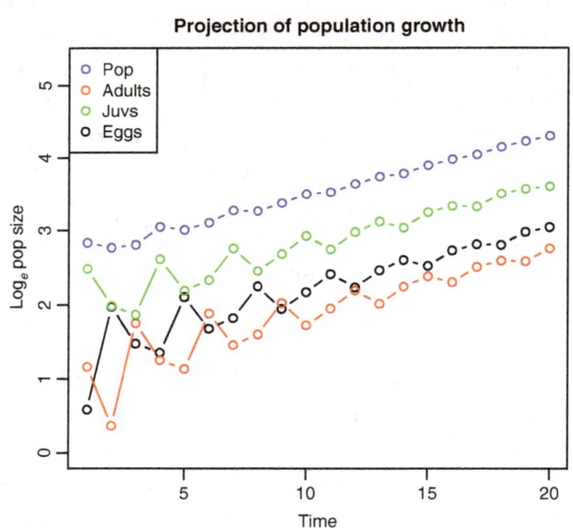

$$A = \begin{pmatrix} 0 & 0.5 & 2 \\ 0.6 & 0 & 0 \\ 0 & 0.8 & 0 \end{pmatrix}$$

$$v = \begin{pmatrix} 1.00000 \\ 0.55171 \\ 0.40584 \end{pmatrix}$$

$$w = \begin{pmatrix} 0.36114 \\ 0.45518 \\ 0.33483 \end{pmatrix}$$

$$s = \begin{pmatrix} 0.364 & 0.201 & 0.148 \\ 0.660 & 0.364 & 0.268 \\ 0.669 & 0.369 & 0.271 \end{pmatrix}$$

$$\lambda = 1.087534$$

FIGURE 2 Example of an age-structured model. Matrix A is multiplied by an arbitrary initial vector (n = (2, 4, 7)′), and this is iterated for 20 time steps. The population growth settles down to a constant rate (linear gradient on a log scale), λ = 1.088 (on a log scale that is a gradient of 0.084); with the age distribution being given by w, the right eigenvector (i.e., 36% of the population in age class 1, 46% in 2, and 33% in 3); the reproductive value being given by v; and the matrix of sensitivities by s. Thus, λ is most sensitive to changes in the survival of individuals from age 1 to 2 (s_{21} = 0.66). This suggests, all being equal, that a management intervention that increases S_1 would have the largest impact on λ compared to changing any other vital rate. Note, the largest sensitivity a_{31} = 0.669 is for an impossible transition (moving from age 1 to age 3 without passing through age 2).

census data may often be quite striaghtforward, and in common software (like MATLAB, Mathematica, or R) the eigenanalysis and calculation of sensitivities may only take a couple of lines of code. Hence, matrix models in their simplest form are characteristically uncomplicated; one can estimate population growth rate and how, by altering different vital rates, it may respond most productively to interventions. Typically, in Leslie models λ is most sensitive to changes in survival rates rather than fertilities. Sensitivities have been used, for example, in turtle conservation to switch some effort from protecting eggs on beaches to preventing mortality of adults via fishing. Furthermore, fitness can be identified as the population growth rate of a gene's lineage, and so a matrix with a larger λ will invade a population whose λ is smaller. This means that the sensitivity of λ to changes of the a_{ij} measures the selection pressure on the matrix element. If a small change in the element causes a large change in fitness, it will be strongly selected; if, conversely, a small change in the element has no effect on λ, it will not be selected. Therefore, this simple modeling framework has been influential in studies of evolutionary ecology as well as population and conservation ecology.

EXTENSIONS OF THE LESLIE MODEL

The classic Leslie matrix model is time invariant, in that the vital rates (the a_{ij}) are assumed to be constant. Such a formulation is useful for population projection (i.e., asking "What would happen *if* . . . ?" questions) but are less good for population prediction (asking "What *will* happen?"). This is because in any real-world situation demographic rates vary over time. Time-varying rates occur because the environment varies (and varies over multiple time scales: daily, seasonally, annually, decadally, and also may change permanently due to a range of factors, such as habitat destruction, climate change, and so on). Also, any realistic system will be regulated by density dependence around a dynamical attractor; any small perturbation from the attractor will be followed by a return that is governed by density-dependent processes. Time-varying vital rates at a population level correspond with changes in individuals' life histories. Many of these changes will reflect phenotypically plastic changes in the life history that results from changes in resources to different traits (for example, a severe winter may reduce population density, which, in turn, reduces resource competition in the spring, giving greater access to resources and allowing an increase in survival, growth, or reproduction).

The simplest way to build in variability in the vital rates is conceptually to assume that, for each time step, the vital rates are drawn from some probability distribution, with the mean being the observed parameter. A number of studies have conducted simulations that have resampled matrices at each time step and iterated the population dynamics over a long time period to give population trajectories in stochastically varying environments. From the observed dynamics, one can estimate both the stochastic population growth rate (λ_s) and the stochastic sensitivities by numerical differentiation. Conceptually, simulating population trajectories with time-varying vital rates underpins population viability analysis (PVA), in that PVA estimates the distribution of population sizes for a stochastic model at some point in the future; so PVA is an important application of stochastic structured modeling. In addition to simulation studies, considerable efforts have been made in recent years to develop stochastic matrix theory, and there is now a range of formulae (backed up by freely available computer code) that allows approximate analytical estimation of the standard quantities. For example, Shripad Tuljapukar published a formula in 1990 for estimating the stochastic population growth rate. The simplified version is given here:

$$\log \lambda_s \approx \log \lambda - \frac{1}{2}\left(\frac{\tau^2}{\lambda^2}\right). \qquad (4)$$

Log λ_s is therefore approximated by the log of the population growth rate of the mean matrix (λ) minus a quantity that is a function of τ^2, which is itself a summation across every possible pair of matrix elements of the covariances between those elements, multiplied by the sensitivity of λ for each of those elements. τ^2 is therefore the approximate variance of the population growth rate that is caused by variation in the a_{ij}. From this formula a general point emerges: temporal variation in the vital rates will typically lead to a reduction in population growth rate (or fitness). Those rates with a high sensitivity will disproportionately decrease growth rate if they vary, so selection should tend to minimize variation in those rates. However, this general conclusion depends on the way that the vital rates covary. If there is strong negative covariation between rates (e.g., a good year for fecundity is a bad year for survival), the net effect can be that variation in some traits may be selected for, as variation in one trait can counteract the variation in another.

The stochastic demography above assumes that, although they may covary, traits vary independently of each other: changes in one vital rate variance can occur independently of changes in any other. This is clearly not necessarily the case. Life history theory is built around resource allocation tradeoffs (where to increase one vital rate necessarily reduces another); these can readily be incorporated into age-structured models. Likewise, density dependence will create covariation in vital rates: if the density goes down, per capita access to resources goes up, and so individuals may be able to increase more than one vital rate simultaneously. In theory, it is quite straightforward to introduce density dependence into vital rates, but in practice this may be fraught with methodological difficulties. For example, there is rarely sufficient high-quality information about local density changes coupled to changes in vital rates, so knowledge is incomplete. In addition, the estimation of the temporal or spatial scale of density measurements is complex. The performance of a juvenile may be related to the density in its current neighborhood, but it may also be related to the density in its mother's neighborhood when she was allocating her resources to him. So identification of the "correct" density relationship with one or more vital rates may be problematic. Nonetheless, density-dependent versions of structured models have been investigated, and one interesting result emerges in the time-varying case: the sensitivities of λ_s predict the effects of changes on fitness but not on population size. An increase in a vital rate can cause a decrease in population size (for example, an increase in fecundity, all things being equal, will increase fitness, but if it increases the population density of the juveniles so much it reduces their survival, it may decrease the total population size). Density-dependent models have been explored numerically, but recently Caswell has developed considerable theory for the estimation of elasticities of nonlinear models and transient dynamics (see Further Reading).

THE INTEGRAL PROJECTION MODEL

The Leslie matrix model structures the population by age and ignores all other information. In reality, within each discrete age class (or stage class in the stage-structured case) there will inevitably be variation between individuals in other important characteristics. For example, adults of the same age may differ continuously in size or quality, both of which may cause predictable change in their survival and reproductive rates. Such variation in characters and traits can be incorporated into an age-and-stage–based model (e.g., adults of age i can be subcategorized as large and small), but anything more than a few categories quickly becomes unmanageable as the matrix increases dimension. Integral projection models (IPMs) are a way to combine trait, or character, variation into a structured model. IPMs are constructed around functions that relate variation in characters or traits to demographic rates. These functions can vary with age (or stage) and may vary over time due to environmental variation or changes in population density. In general, there are only four main relationships needed to fully describe demographic rates in a population: the associations between the structuring variables (e.g., age and size) and (a) survival, (b) development among those that survive, (c) fertility, and (d) the variation in offspring size that results from parents with the observed values of age and size.

An important concept in IPMs is that the population structure comprises a set of components, which are a combination of discrete classes (such as age classes) and continuous domains (such as size functions within each discrete class). The vector of population sizes, **n**, is replaced by a set of distribution functions $n_i(x, t)$, where $n_i(x, t)dx$ is the number of individuals in age class i at time t with their state variable (such as size) within the interval $[x, x + dx]$. The projection matrix **A** is replaced with a set of kernels, $K_i(y, x)$, that represent, within age i, the survival and growth of state x individuals to state y, or the production of state y offspring by state x parents. The dynamics of the $n_i(x, t)$ are given by a set of coupled integral equations:

$$n_i(y, t + 1) = \sum_{j=1}^{C} \int_{\Omega_j} K_{ij}(y, x) n_j(x, t) dx, \qquad (5)$$

where C is the total number of components (e.g., age classes or age x stage classes), Ω is a closed interval characterizing the size domain, and $1 \geq i, j \geq N$. This model is density independent, but it can easily be made density dependent by making the kernels responsive to a measure of population density (i.e., $K(y, x, d)$), or by making the kernels time variant, or even both (i.e., $K(y, x, d, t)$). For a set of assumptions, there will be stable population growth, and equivalent calculations can be made to calculate population growth rate, stable age distributions, reproductive values, and sensitivities.

So, it is possible to build models based on trait-demography associations that describe age-specific functions linking the trait to survival, fertility, development of the trait in individuals that survive, and reproductive allocation to offspring. These models can be used to describe a range of metrics, including population growth, selection differentials, descriptors of the life history (such as generation time), and even estimates of heritabilities of traits. This approach has recently been illustrated by Tim Coulson and colleagues (see Further Reading). This indicates that the coupled life history, phenotypic, population, and evolutionary dynamics can be extremely complex. Perturbing a vital rate, or a trait, at one age can have effects later on in the life history and impact upon later generations, with consequences both ecological (i.e., changes in population structure and growth rate) and evolutionary (changes in selection and the response to it). In Soay sheep, for example, an increase in reproduction by young adults leads to a decrease in mean body size and also a decrease in the selection for larger body size. Understanding how populations respond to environmental change (perturbations) in ecological and evolutionary terms therefore requires modeling the associations between traits, ages, and population structure, and, without a comprehensive model, accurate prediction of how a system will respond will not be possible. The complexity of causation of dynamics is increasingly being recognized by both life historians and dynamicists, and it is backed up from our in-depth understanding of a range of well-studied model systems. The IPM framework therefore provides a detailed way of modeling this complexity and therefore fully understanding, and predicting, the system responses to environmental change. The simplicity of the classic Leslie matrix is a double-edged sword in that it is very easy to implement but also only really useful for projection, not prediction.

This entry has focused on discrete time models, partly because we naturally discretize age (incrementing it by 1 unit annually on our birthday), and partly because our understanding of the population dynamics of many systems is based on annual censuses, with the biological systems having set, discrete reproductive seasons (rather than continuous reproduction throughout the year). It is, of course, perfectly possible to model age-structured populations using a continuous time framework. McKendrick introduced a continuous time model in 1926 that describes the dynamics of the age distribution using a model now referred to as the McKendrick–von Foerster equation. If $n(t, a)$ represents the age distribution at time t, the McKendrick–von Foerster equation is

$$\frac{\partial n(t,a)}{\partial t} + \frac{\partial n(t,a)}{\partial a} = -\mu(t,a)n(t,a), \qquad (6)$$

where $\mu(t, a)$ is the specific death rate for age a and time t. The McKendrick–von Foerster equation expresses the dynamics of the population with mortality and aging, but has no reproduction. To complete the formulation, one needs a boundary condition for the birthrate:

$$n(t,0) = \int m(a,t)n(a,t)da,$$

where $m(a, t)$ is the birthrate of individuals at time t from age a mothers.

This model has been extended such that it can be used to describe dynamics in fluctuating and density-dependent environments. It can also be applied, much like the IPM, when more than age (e.g., age and size) has a strong impact on an individual's vital rates. Unlike matrix models, these general physiologically structured models are complex to analyze because they are based on partial differential equations. There is an intermediate type of model, called the escalator box car, that can span the continuum from a classic matrix model to a continuous time version of the McKendrick–von Foerster equation; this has the benefit of being able to combine the computational ease of matrix models with the flexibility of continuous time models.

It is also possible to model in continuous time an age-structured population using coupled delay differential equations, but logically these are perhaps best thought of as stage-structured models and are well described in that entry.

INTRODUCTION TO THE LITERATURE

Age-structured theory is advancing apace. The definitive text for matrix models remains Caswell (2001), which also includes an introduction to the problem of parameter estimation for the models and some of the basics of stochastic and density-dependent formulations. A general review about the importance of individual variation, and hence the need for structured models, is given by Benton et al. (2006). The classic reference for stochastic matrix analysis is found in Tuljapurkar (1990). A good review of stochastic matrix models and their use in applied biology can be found in Fieberg and Ellner (2001). Caswell has developed a considerable body of theory recently on sensitivity analysis of nonlinear models (Caswell 2008) and models with transient dynamics following a perturbation (Caswell 2007). A worked example, using numerical approaches, of a density dependent matrix model applied to the well studied flour beetle, *Tribolium,* is given in Grant and Benton (2003). Integral projection models are reviewed in Ellner and Rees (2006), which both discusses the general theory and gives a thoroughly explored example for a thistle. Well-studied ungulate systems have been particularly instrumental in highlighting applications of the theory discussed above. For example, Coulson et al. (2003), on red deer, apply Van Tienderan's extension of sensitivity analysis and link selection gradients between a phenotypic trait and multiple fitness components with the effects of these fitness components on the population growth rate (i.e., mean absolute fitness). One of the most synthetic papers of recent years is Coulson et al. (2010), which uses the Soay sheep data from St. Kilda, Scotland, and investigates the joint eco-evolutionary dynamics by using a trait (size)-based, age-structured model to estimate heritabilities, selection differentials, and life-history descriptors in addition to population growth parameters.

SEE ALSO THE FOLLOWING ENTRIES

Delay Differential Equations / Matrix Models / Population Ecology / Partial Differential Equations / Population Viability Analysis / Stage Structure / Stochasticity

FURTHER READING

Benton, T. G., S. J. Plaistow, and T. N. Coulson. 2006. Complex population dynamics and complex causation: devils, details and demography. *Proceedings of the Royal Society B: Biological Sciences* 273: 1173–1181.

Caswell, H., 2001. Matrix Population Models—construction, analysis, and interpretation. Sinauer Associates, Inc. Publishers, Sunderland, Massachusetts.

Caswell, H. 2007. Sensitivity analysis of transient population dynamics. *Ecology Letters* 10: 1–15.

Caswell, H. 2008. Perturbation analysis of nonlinear matrix population models. *Demographic Research* 18: 59–113.

Coulson, T., L. E. B. Kruuk, G. Tavecchia, J. M. Pemberton, and T. H. Clutton-Brock. 2003. Estimating selection on neonatal traits in red deer using elasticity path analysis. *Evolution* 57: 2879–2892.

Coulson, T., S. Tuljapurkar, and D. Z. Childs. 2010. Using evolutionary demography to link life history theory, quantitative genetics and population ecology. *Journal of Animal Ecology* 79: 1226–1240.

Ellner, S. P., and M. Rees. 2006. Integral projection models for species with complex demography. *American Naturalist* 167: 410–428.

Fieberg, J., and S. P. Ellner. 2001. Stochastic matrix models for conservation and management: a comparative review of methods. *Ecology Letters* 4244–266.

Grant, A., and T. G. Benton. 2003. Density-dependent populations require density-dependent elasticity analysis: An illustration using the LPA model of *Tribolium*. *Journal of Animal Ecology* 72: 94–105.

Tuljapurkar, S. 1990. *Population dynamics in variable environments*. New York: Springer-Verlag.

ALLEE EFFECTS

CAZ M. TAYLOR

Tulane University, New Orleans, Louisiana

In the 1930s, Warder C. Allee demonstrated that cooperation between individuals can be as important as competition. Individuals can suffer decreased fitness when they are in groups in which there are too few conspecifics or in which the individuals are too diffuse. This simple observation, now called the Allee effect, has far-reaching consequences, affecting population dynamics, extinction rates, and invasion rates, as well as evolution. The most dramatic dynamical consequence of an Allee effect happens in cases when density levels lower than a threshold drive the per capita growth rate below replacement rate and the population begins to decline, eventually to extinction.

CONCEPT AND DEFINITIONS

The formal definition of a component Allee effect is the existence of a positive relationship between a component of fitness and population size or population density. This means that survival or reproduction is decreased when density of conspecifics is low. A component Allee effect can lead to a demographic Allee effect when the overall per capita growth rate is also decreased at low densities and shows a positive relationship with population density or size. A demographic Allee effect can be either strong or weak. When a population experiences a weak demographic Allee effect, the per capita growth rate is lowered at low densities but never drops below replacement rate so that the population does not actually decline but increases more slowly. When a population experiences a strong demographic Allee effect, the per capita growth rate drops below replacement rate when the population density is lower than a threshold value (called the Allee threshold). Below the Allee threshold, a population subject to a strong Allee effect declines to extinction.

Allee effects are also known as positive density dependence, in contrast to (negative) density dependence, the negative relationship between density and fitness due to competition between conspecifics. At the demographic level, competition is a negative feedback (or compensatory) because the growth rate drops as the population increases, usually stabilizing the population at a carrying capacity. Allee effects, by contrast, are a positive feedback and are called depensation or depensatory dynamics, especially in the fisheries literature, as opposed to compensation or compensatory dynamics.

MECHANISMS

There are multiple mechanisms by which an individual can suffer from insufficient numbers or density of conspecifics. Broadly speaking, for an Allee effect to occur, there must be a positive relationship between a component of fitness and population density at low densities. Fitness is the contribution an individual makes to future populations and has many components that can be classified roughly into those that affect survival and those that affect reproduction. Thus any mechanism that lowers any component of survival or reproductive success at low densities is a mechanism of a component Allee effect. The two most commonly cited general causes of Allee effects are the difficulty of finding a mate (or viable gametes) at low densities in sexually reproducing species and the higher likelihood of being depredated at low densities. Other cooperative behaviors can also cause Allee effects when there are too few conspecifics to convey the benefit of the cooperative behavior. Some Allee effects are caused by the behavior of other, interacting species like pollinators or predators, and sometimes Allee effects are attributable to genetic causes.

Mate Finding

Perhaps the most commonly cited mechanisms of Allee effects are based on the idea that, at low population density, sexually reproducing species may have difficulty finding a (or enough) suitable, receptive mates (or gametes) and reproductive output is decreased. This occurs in both mobile organisms that actively seek mates and in sessile organisms that rely on transport of gametes through surrounding air or water. When the transport is largely passive, a cloud of sperm or pollen diffuses and becomes more dilute with increasing distance from the source organism. Consequently, in sparse populations, widely separated individuals suffer decreased fertilization.

Predation

Another very common mechanism of an Allee effect is the predation dilution effect caused by aggregation in order to reduce the per capita probability of being depredated. Further reduction in per capita mortality from predation, and therefore more benefit from being in large groups, can be achieved in species that form such large aggregations that predators are swamped or satiated. Aggregation can occur both in space and in time and potentially includes phenomena such as masting, where seed production of all individual plants in a population is synchronized. Additionally, terrestrial predators may attack the edge of an animal group (e.g., a herd of ungulates). There will tend to be proportionally fewer individuals on the edge of a large group than a small group, so per capita mortality is reduced in a larger group. Furthermore, animals in large groups may be able to devote less time to vigilance than in a small group but achieve the same or better protection by relying on their conspecifics to raise an alarm if a predator approaches. Time saved in vigilance can be spent foraging or on other activities that improve fitness. Animals may also engage in other cooperative behaviors that are designed to confuse predators or otherwise reduce predation.

Cooperative Behaviors

Allee effects resulting from decreased predation risk are caused by aggregative or more active cooperative behaviors on the part of the species. Other types of cooperative behaviors can also result in Allee effects. Examples include cooperative foraging, in which animals hunt in packs and large groups and are therefore more efficient at catching or finding prey; reproductive facilitation, in which individuals are stimulated to reproduce by the presence of reproductive conspecifics; cooperative breeding, in which large groups improve juvenile survival; female choice, where females may choose not to mate at all if the selection of males is too small; and environmental conditioning, in which species are able, when in large groups, to modify or condition their environment via several different mechanisms (for example, thermoregulation) in such as way as to improve survival or reproduction.

Behaviors of Interacting Species

Sometimes an Allee effect is caused by the behavior of an interacting species rather than by the species experiencing the Allee effect. A species that benefits or provides a service to another species will cause an Allee effect if it prefers to service individuals in large groups. The most common example, and one that could be classified under the mate-finding section, is pollinator limitation. Some pollinators are less likely to visit small groups of flowering plants than to visit larger groups. When this is the case, animal-pollinated plants in smaller groups may experience lower pollination and seed production than those in larger groups.

In the reverse mechanism, a harmful species like a predator that preferentially preys on individuals in small groups will cause an Allee effect in the prey. The clearest example of this type of Allee effect occurs when humans act as the predator either hunting or collecting species. It has been shown that humans put a high value on rarity for its own sake and species that are rare or in danger of extinction often experience increased hunting or collection pressure or simply disturbance from people wanting to own a specimen or get a glimpse of a disappearing species.

Genetic Mechanisms

In small populations, and particularly in suddenly reduced populations, genetic diversity tends to be lower than in large populations. Low genetic diversity often leads to a decrease in individual fitness. These two facts put together lead to genetic Allee effects. Mechanisms that lead to low genetic diversity in small populations are genetic drift, the sampling effect, and increased inbreeding. The sampling effect is a loss of allelic richness due to a relatively abrupt reduction in the number of individuals in the population, and genetic drift is the continuing loss of alleles in small populations because chance mating events mean that some alleles drop out in each generation. Increased inbreeding leads to increased homozygosity. A drop in allelic richness allows the fixation of mildly deleterious alleles and a drop in beneficial alleles and a resulting decline in individual fitness. Increased homozygosity reduces fitness because it allows the increased expression of deleterious alleles and also when homozygotes are less fit than heterozygotes.

POPULATION MODELS AND DYNAMICS

To include an Allee effect in a mathematical population model, there are two basic approaches. One is to model the demographic Allee effect directly by creating a phenomenological model, which means creating a population model in which some part of the per capita growth rate has a positive slope when plotted against population density (Fig. 1). The other approach is to separate the different components of fitness and include each separately in the model and then to include the Allee effect only in the component of fitness in which it is know to be present. Component Allee effects do not always lead to demographic Allee effects, and this can be shown by this second approach. On the other hand, multiple component Allee effects are possible in a single population and can interact with one another, giving interesting population dynamics.

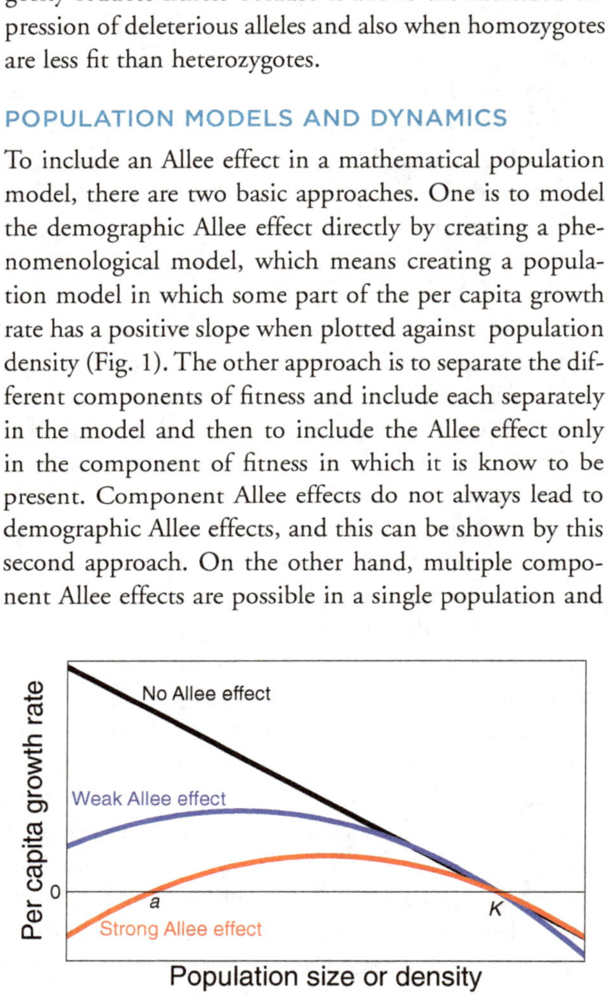

FIGURE 1 *Per capita* growth rate of a population. With no Allee effect (black line), the *per capita* growth rate declines with population size or density. In populations with both any kind of Allee effect (red and blue lines), there is a positive relationship with density at low densities. A strong Allee effect (red line) creates a threshold (a) below which the *per capita* growth rate is negative and the population will become extinct. A weak Allee effect (blue line) has no threshold only a slower growth rate at low densities than at higher densities.

Models of Demographic Allee Effects

A very general population model in continuous time can be written

$$\frac{dn}{dt} = nf(n),$$

where n is population density and $f(n)$ is the density-dependent per capita growth rate. Any functional form of $f(n)$ that produces the hump-shaped curve of the red or blue lines in Figure 1 can be a model of an Allee effect. This means that there are an almost unlimited number of potential Allee effect models, and several have been used in the literature. Only one model is described here to provide an example. In most cases, it is easiest to start with a model that has only negative density dependence and modify it to incorporate positive density dependence. For instance, one of the simplest mathematical models of population growth in continuously reproducing population is the logistic growth model, where the per capita growth rate is

$$f(n) = r\left(1 - \frac{n}{K}\right).$$

In logistic growth, the population growth rate dn/dt is a quadratic function so that the population grows most slowly when the population is very small and when the population is high and near carrying capacity. The *per capita* growth rate $f(n)$ declines linearly with population density (Fig. 1), and this negative relationship between per capita growth rate and population density shows that there is no Allee effect in this model. The fixed points of this model are at $n = 0$ and $n = K$, but $n = 0$ is an unstable equilibrium, which means that the population from any positive starting density will never go extinct but will increase and always reach the carrying capacity K. In logistic growth, $n = K$ is the only stable equilibrium so that from any starting density above or below K, the population will shrink or increase toward K (Fig. 2A).

We incorporate a strong Allee effect into this model by adding a parameter, a, representing the Allee threshold, $0 < a < K$, and the *per capita* growth rate now becomes

$$f(n) = r\left(1 - \frac{n}{K}\right)\left(\frac{n - a}{K}\right).$$

There are now three fixed points, $n = 0$, $n = a$, and $n = K$. The population growth rate is a cubic function of n, and the population grows slowly near the fixed points. Between 0 and a, the population growth rate is negative; above a, the population growth rate is positive. The *per capita* growth rate is quadratic (Fig. 1). The left-hand

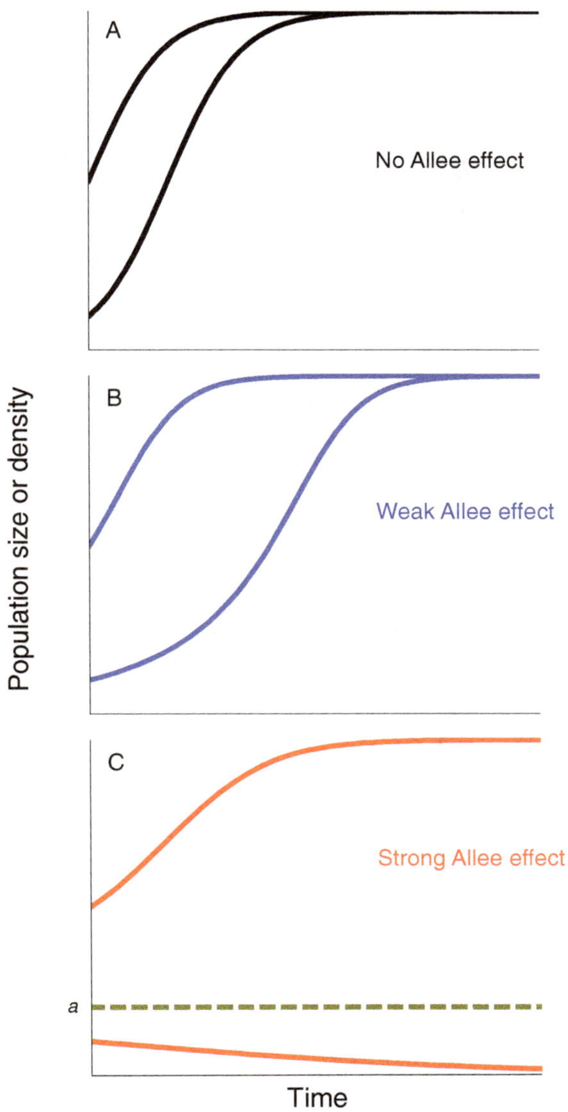

FIGURE 2 Trajectories of populations shown in Figure 1. (A) With no Allee effect, the population always grows toward carrying capacity, *K*. (B) With a weak Allee effect, the population also grows toward *K* but more slowly than with no Allee effect. (C) With a strong Allee effect, the population grows toward *K* if it is initially above the Allee threshold *a* but declines to extinction if it is initially above *a*.

side of this curve is the positive relationship between per capita growth rate and n that defines the Allee effect. Mathematically, the addition of the strong Allee effect stabilizes the $n = 0$ point (both $n = 0$ and $n = K$ are stable equilibria, and $n = a$ is unstable). Biologically, this means that population extinction is now possible. The population will decline to extinction if the starting density is below the threshold, a, and will grow to carrying capacity, K, if the population density starts above the threshold (Fig. 2C).

The same model can be used to incorporate a weak Allee effect by simply making $a < 0$. When a is negative, the population growth rate is still a cubic function of population density but never becomes negative (for positive, biologically reasonable values of n), so the population does not decline at low densities but always grows to carrying capacity (Fig. 2B). There is still a positive relationship between per capita growth rate and density, and dynamically this means that the population grows more slowly at low densities but will eventually reach carrying capacity as it does in the logistic growth model without an Allee effect.

Models of Component Allee Effects

MODELS OF MATE FINDING

One of the most commonly modeled component Allee effect mechanisms is mate finding. Usually to incorporate a mate-finding Allee effect, we need a sex-classified population model in which we have two variables, the number of males (M) and the number of females (F). The birth rate then depends upon the female mating rate $P(M, F)$, which is affected both by the number of females (F) in the population as well as the number of males (M). $P(M, F)$ can be any function that is zero if there are no males ($P(0, F) = 0$) and increases as the number of males increases, approaching 1 as the number of males gets very large relative to the number of females.

MODELS OF PREDATION

We can use a single-species framework to model predation if we can assume that the predator is not affected by the prey numbers, as would be the case for a generalist predator that has multiple other prey species. In this case, a component Allee effect is created in any system in which mortality is affected by density such that per capita mortality is higher when density is lower (in other words, a component of fitness, survival, is positively affected by density). A 2004 paper by Joanna Gascoigne and Romuald Lipcius shows that predation is a general mechanism that creates an Allee effect in prey species (Fig. 3). When predators consume prey, the consumption rate varies as the density of the prey changes. This is called a functional response and there are three types of functional response (Figs. 3A–C). A type I function response, used often for suspension feeders or predators that catch prey in traps, is a linear response in which consumption rate increases linearly with density. A type II functional response is a saturating function so that the consumption rate increases with prey density up to some density but then levels off. A type III

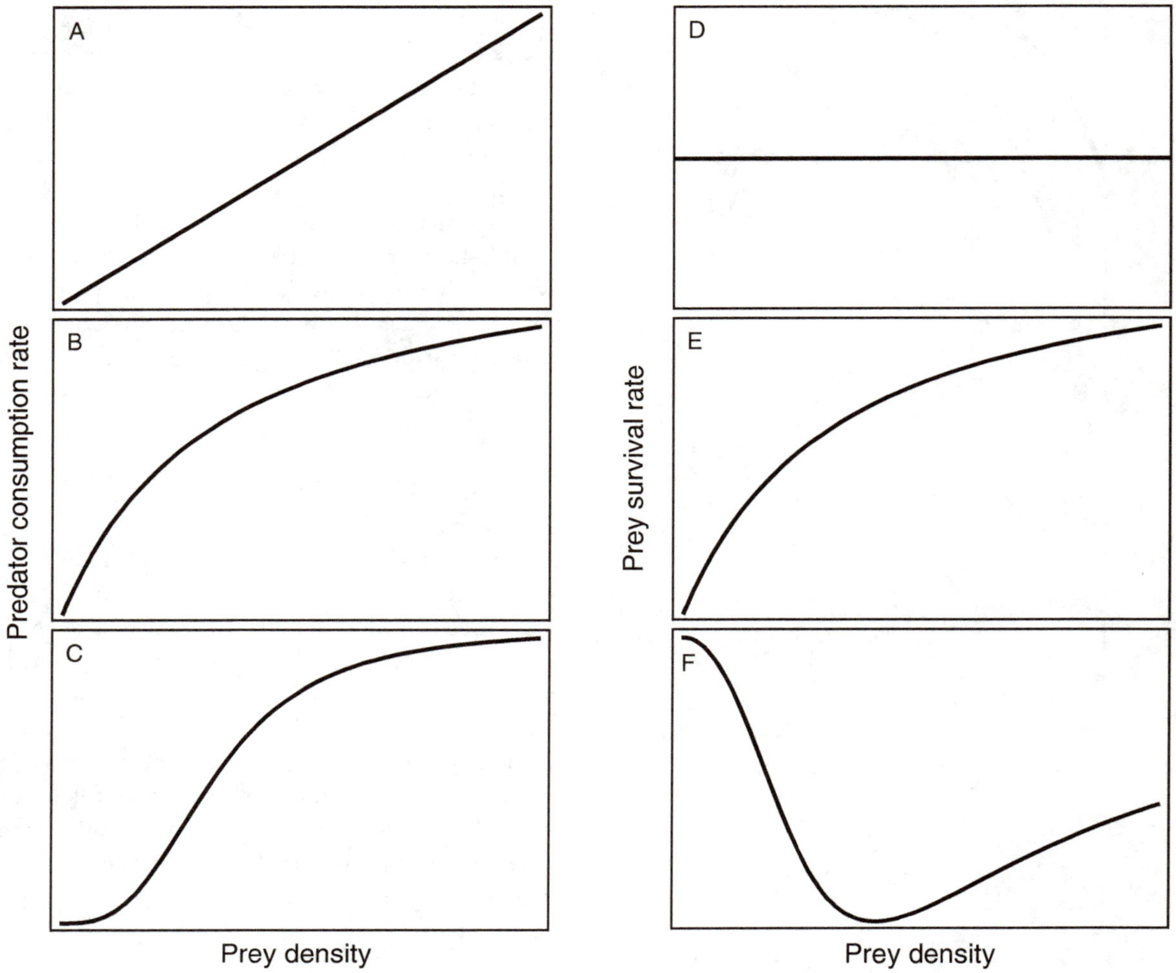

FIGURE 3 (A–C) The prey consumption rate of a predator following a type I, type II, and type III functional response, respectively. (D–F) The resulting prey survival rate. An Allee effect is created by a type II functional response but not by a type I or type III functional response. Adapted from Gascoigne and Lipcius (2004).

functional response has a sigmoid shape and is seen in predators that switch to different prey species when one species reaches low densities or in cases where small numbers of prey have refuges and can hide from predators at low densities. The resulting per capita survival of the prey as a function of density for the three types of functional response is shown in Figures 3D–F. The survival of the prey of a type I predator is does not vary with density (Fig. 3D). A type II response creates a positive relationship between prey survival and prey density (Fig. 3E), and a type III functional response produces a negative relationship between prey survival and density at low density (Fig. 3F). Hence, a type II functional response creates an Allee effect in the prey species, whereas a type I and a type III do not. The positive relationship between survival and density seen in the survival of prey with type III predators (Fig. 3F) is not considered an Allee effect because it occurs at higher densities. Predator density or abundance sometimes changes in response to prey density. This is called an aggregative response and can follow the same three types described above for functional response. An Allee effect caused by a type II functional response can be eliminated by including a type III aggregative response.

MODELING OTHER COMPONENT ALLEE EFFECTS

Predation and mate finding are the most commonly studied mechanisms for Allee effects, but there are many others and there have been many types of population models created to investigate them, including population genetics models, simulation models, competition models, predator–prey models, and metapopulation models. An extended discussion of all models of component and demographic Allee effects can be found in a 2008 book about Allee effects by Frank Courchamp, Ludek Berec, and Joanna Gascoigne.

Metapopulation-Level Allee Effects

A classic metapopulation simply models the dynamics of patch occupancy. Each patch is occupied or not, and patches can become occupied at a colonization rate and become unoccupied at an extinction rate. In the classic formulation (the Levins model), colonization rate is proportional to the product of the fraction of occupied patches (p) and the fraction of unoccupied patches ($1 - p$). The per-patch colonization rate decreases linearly with p, whereas the per-patch extinction rate is constant and the metapopulation grows to a stable equilibrium fraction of patches occupied. It is possible, however, for situations to occur in which the per-patch colonization rate also drops with the number of occupied patches, creating what is termed an Allee-like effect, essentially an Allee effect that operates at the level of the metapopulation rather than at the level of local population. An Allee-like effect is usually analogous to a strong demographic Allee effect, creating a critical threshold fraction of patches occupied that needs to be exceeded for the metapopulation to persist. This occurs, for example, in parasite dynamics when infection of new hosts depends on there being more than a critical number of other infected hosts. Another example is in animals that form packs. When the number of packs drops below a critical threshold, not enough dispersers are produced and the rate of colonization or formation of new packs is reduced. Allee-like dynamics at the metapopulation level do not necessarily depend on Allee dynamics at the local population level, but Allee dynamics at the local level can lead to Allee-like dynamics at the metapopulation level.

DYNAMICAL IMPLICATIONS

Including Allee effects in population or metapopulation models has large consequences on how the population behaves. The most obvious consequence of a strong Allee effect is the introduction of a critical threshold for population level below which the population will go extinct. In stochastic models there is always some probability of extinction, but in stochastic models that include a strong Allee effect the probability of extinction declines sharply at the Allee threshold. In spatial models, this threshold is also spatial; the population must occupy an area larger than a critical threshold to remain viable (or a minimum proportion of patches in a discrete landscape). Also in spatial populations the rate at which a population spreads is reduced by both strong and weak demographic Allee effects. If habitat is patchy (discrete space), a strong Allee effect can produce pulsed invasions as local populations overcome Allee effects and colonize patches. In some cases, when local populations fail to overcome Allee thresholds this can prevent the population from expanding into available habitat patches (range pinning). In a continuous habitat, expansion of a population with a strong Allee effect can result in patchy distribution.

EVOLUTIONARY IMPLICATIONS

An Allee effect imparts a strong selection pressure and is responsible for many evolutionary adaptations. These include adaptations that cause the species to remain above the Allee threshold, like gregariousness, and adaptations that lower or remove the Allee threshold, for instance, adaptations that improve mate finding, like songs, pheromones, dispersal, mating synchronicity, and many others. Additionally, adaptations that decrease the frequency at which mating has to occur can be driven by Allee effects. These include traits as diverse as sperm storage and lifetime pair bonding, as well as adaptations that improve fertilization efficiency. We particularly expect species that are naturally rare to have evolved so as to minimize Allee effects. Practically, this means that we are most likely to be able to actually detect Allee effects in species that have become rare due to anthropogenic activities or are forced into low densities in manipulative experiments or lab situations as in many of Allee's original experiments.

CONSERVATION IMPLICATIONS

With the wide range of mechanisms described, it is likely that Allee effects or positive density dependence are very common and potentially have large effects on population dynamics and on the viability of populations. An Allee threshold, if one exists, can increase the likelihood of extinction of a rare species, and conservation biologists are likely to be most concerned about Allee effects in rare or endangered species or populations, especially those that are made rare by humans. Conservation of such species requires keeping the population above the Allee threshold. Invasive species are also affected by Allee effects. Sometimes the Allee threshold can be exploited for management if an invasive species needs only to be reduced to below its Allee threshold to be eradicated. For harvested species such as fisheries, a constant fishing effort does not create an Allee effect but does strengthen an existing Allee effect. However, using a constant yield management plan generates a component Allee effect, since per capita mortality decreases as population size increases.

SEE ALSO THE FOLLOWING ARTICLES

Conservation Biology / Invasion Biology / Metapopulations / Population Ecology / Predator–Prey Models / Single-Species Population Models

FURTHER READING

Allee, W. C. 1931. *Animal aggregations: a study in general sociology.* Chicago: University of Chicago Press.

Allee, W. C. 1941. *The social life of animals,* 3rd ed. London: William Heineman.

Courchamp, F., L. Berec, and J. Gascoigne. 2008. *Allee effects in ecology and conservation.* Oxford: Oxford University Press.

Gascoigne, J. C., and R. N. Lipcius. 2004. Allee effects driven by predation. *Journal of Animal Ecology* 41: 801–810.

Stephens, P. A., W. J. Sutherland, and R. Freckleton. 1999. What is the Allee effect? *Oikos* 87: 343–61.

Stephens, P. A., and W. J. Sutherland. 1999. Consequences of the Allee effect for behaviour, ecology and conservation. *Trends in Ecology & Evolution* 14: 401–405.

Taylor, C. M., and A. Hastings. 2005. Allee effects in biological invasions. *Ecology Letters* 8: 895–908.

ALLOMETRY AND GROWTH

ANDREW J. KERKHOFF

Kenyon College, Gambier, Ohio

Allometry is the study of how organism body size affects various aspects of form and function. The allometric scaling of metabolic rate and other physiological processes explicitly links an individual organism's energy budget to changes in size. Several models of ontogenetic growth assume that organism growth trajectories arise from differential allometric changes in different components of metabolism.

A BRIEF OVERVIEW AND HISTORY OF ALLOMETRY

Individual organisms span an amazing size range. The ratio of the mass of a blue whale to that of a bacterium is approximately 10^{21}; that is, a blue whale is ten-thousand-million-million-million times heavier than the smallest bacteria. To put this ratio into perspective, it is similar to the mass ratio of the Moon to a typical human, that of a human to a single molecule of cytochrome-c oxidase (a protein that facilitates cellular respiration), and that of the known universe to our Sun! The scales of ecological interactions between species are yet broader, spanning over 30 orders of magnitude (powers of 10) in mass, from the smallest interacting microbes to the entire biosphere (which is estimated to weigh in at $\sim 1.8 \times 10^{19}$ g). Even among the more familiar land mammals, an elephant is almost six orders of magnitude (i.e., 10^6, or one million times) heavier than the smallest mouse. Thus, understanding biodiversity and the ecological complexity of life on Earth is, at least in part, a matter of understanding how life processes change across this range of scale.

While this observation seems obvious, it has important ramifications for almost every aspect of biology and ecology, because the size of an organism fundamentally affects its morphology and physiology, as well as its interactions with the physical environment and other species. For example, because they are "small," mice can easily climb a vertical surface and survive a fall from great height without injury. The same is clearly not true for humans, and a surprisingly modest tumble can fatally disable an elephant. The huge differences we see between small and large organisms are matched by equally intriguing constancies. For example, a mouse's heart beats 500 times per minute, while an elephant's only beats 28 times, but over their lifetimes, both will experience (on average) approximately the same number of heartbeats—about 1.5 billion, as will humans. Why 1.5 billion? Understanding how life must change with size is one of the keys to explaining these sorts of mysteries and to developing a more complete understanding of how nature works.

The term allometry (derived from the Greek *allos* = other, *metros* = measure) was coined by the biologists Julian Huxley and George Tessier in 1936. However, the changes in form and function that accompany changes in size have long intrigued scientists, dating back at least to Galileo's examination of the changing dimensions of bones and da Vinci's quantification of area-preserving branching in trees. One of the breakthroughs made by Huxley and Tessier was the standardized quantification of allometric relationships in the form of a power law,

$$Y = y_0 X^z,$$

where Y is a measure of some aspect of form or function, X is a measure of organism size (generally either length or mass), and y_0 (the allometric coefficient) and z (the scaling exponent) are constants that describe the proportionality of the relationship between size and the response. Box 1 presents more information on the mathematical aspects of allometric power laws.

BOX 1. EXTRACTING EIGENVALUES AND EIGENVECTORS FROM TRANSITION MATRICES

The allometric power law,

$$Y = y_0 X^z,$$

provides a simple mathematical model for the relationship between two measurable biological variables. If we take the logarithm of both sides of the allometric equation, we find an equation for a straight line relating $\log Y$ to $\log X$:

$$\log Y = \log y_0 + z \log X.$$

Thus, a power function relationship between two variables X and Y means that their logarithms are linearly related, with a slope equal to the exponent of the power function. Recall that a slope is defined as the "rise over run" or, in this case (since we are working with base-10 logarithms), the change in the magnitude of Y for every 10-fold change in X.

Looking at some graphs of power functions on both standard arithmetic and logarithmic scales can help to understand them better. The upper graph shows power functions with five different exponents. In all cases, the coefficient (y_0) is set to 1, for simplicity. Note that when the exponent is less than 1, the line climbs at an ever-decreasing rate, while when the exponent is greater than 1, it climbs at an ever-increasing rate. In all cases, the curves are not straight lines, that is, they are nonlinear, except when the exponent is equal to 1, for which $Y = X$. Finally, note that when viewed on an arithmetic scale, the five lines all appear to converge toward zero.

Now examine the same five power functions on logarithmic scales (lower graph), on which each equal increment is a power of ten. As we would predict from the derivation above, each function is linear, and since the coefficient a is always one, the lines only differ in their slopes, and they all cross at the point [1, 1]. The constant slopes indicate that for every 10-fold increment in X, Y changes 10^z-fold. Viewed on logarithmic scales, we can also see that the curves diverge on the small end as well as on the large end. These differences are not apparent on the upper graph, because they are compressed into the space between 0 and 1.

In a way, these two graphs represent different ways of looking at the world, and some of the differences are subtle. For example, the way in which the upper graph "minimizes" the differences between the power functions near the origin shows that arithmetic scales are sensitive to the units of measure: 1 mg is very different from 1 kg, but the space between 0 and 1 is always the same on the graph. Logarithmic scales, on the other hand, present proportional changes that are insensitive to the units of measure, because ten times larger is ten times larger, whether in mg or kg. This difference does not make one quantitative view superior to another, but it does mean that one or the other may be more appropriate, depending on the situation. Because allometric analyses are generally concerned with proportional changes, logarithmic scales are generally the appropriate choice.

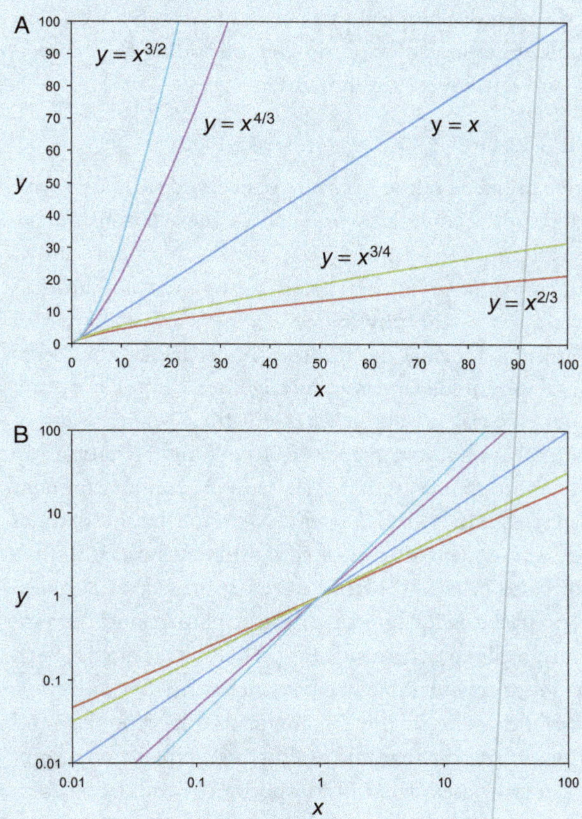

BOX FIGURE 1 Plot of power functions of the form $Y = X^z$ on arithmetic (upper) and logarithmic (lower) axes.

The importance of organism size for biological form and function and the long history of its study has led to applications of allometry in many different areas of biology. The portability of the mathematics of allometric power laws (Box 1) means that very different biological applications can take advantage of the same mathematical tools, but it is very important to recognize differences in biological context, especially when trying to develop generalized explanations for the relationships between size and biological form and function.

At least two fundamentally different approaches to allometry can be distinguished, based principally on the units of data used to develop the allometric relationship.

Comparative (also called evolutionary or interspecific) allometry studies how form or function changes across species that vary in size. Species are the unit of observation, and data generally consist of species mean values of adult size and the biological variable of interest. The physiological allometry relating basal metabolic rate to body mass in mammals (see "The Allometry of Metabolism," below) is a classic example. In contrast, intraspecific (or ontogenetic) allometry examines variation across individuals of a single species that vary in size, either developmentally or ecologically. The patterns that arise in different sorts of allometric studies do not always agree, in part because the observed differences in both size and biological form and function reflect different combinations of evolutionary, physiological, ecological, and developmental processes.

THE ALLOMETRY OF METABOLISM

Early studies of allometry were concerned primarily with patterns of relative growth, such as the relationship between brain size and body size among mammals. However, from the beginning, biologists applied similar ideas to patterns in the physiology, ecology, and behavior of organisms. During the 1930s, Max Kleiber and other animal physiologists brought together careful measurements of basal metabolic rates (B) for a wide variety of mammals. The data displayed a very tight relationship with adult animal mass (M), with a scaling exponent of approximately 3/4, *i.e.*, $B = b_0M^{3/4}$. In the decades since, subsequent studies of very different taxa, from fish to birds to insects to plants, have confirmed similar allometric patterns for almost all multicellular (and perhaps even unicellular) organisms (Fig. 1). While there has been decades of considerable debate concerning the "universality" of the value of the metabolic scaling exponent and a number of controversial attempts to explain its origin, for our purposes it is only important to note two facts of metabolic allometry, neither of which depend on the exact value of the exponent or its basis.

First, most (but perhaps not all) analyses, whether interspecific, intraspecific, or ontogenetic, find relationships with exponents less than 1. This implies that as organisms get larger, their mass specific rate of metabolism decreases, since, for example, if $B = b_0M^{3/4}$, $\frac{B}{M} = b_0M^{3/4}M^{-1} = b_0M^{-1/4}$. Thus, even though organisms large and small share a common biochemistry and cellular structure, the cells of larger organisms run their biochemistry at slower rates than the cells of smaller organisms.

Second, in addition to direct measurements of respiration, a variety of other biological rates, including rates

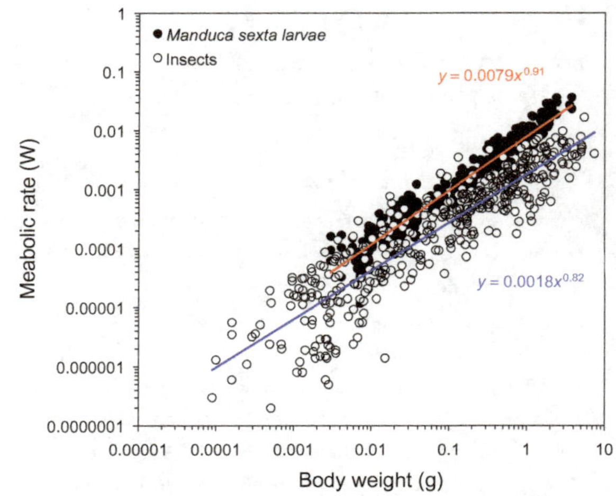

FIGURE 1 Allometric scaling of metabolic rate as a function of body mass for insects showing evolutionary (interspecific) allometry across 392 species of adult insects (open circles with blue line, compiled from the literature by Chown et al., 2007, *Functional Ecology* 21: 282–290) as well as ontogenetic (intraspecific) allometry measured daily during development for *N* individual larvae of the tobacco hornworm, *Manduca sexta* (solid circles and red line, unpublished data from A. Boylan, H. Itagaki, and A. Kerkhoff). Note that the scaling exponents differ between the two relationships. Differences in the height of the relationship (the scaling coefficient) likely derive from differences in temperature. The adult insect data were all corrected to 27 °C, while the *Manduca* larvae were reared at 27 °C.

of ingestion, excretion, reproduction, somatic growth, and even mortality have been shown to share similar patterns of allometric scaling. This implies a principle of similitude (or similarity) linking these different physiological and demographic processes. That is, if different processes scale similarly with body size (i.e., with the same exponent), their ratio is approximately constant, or at least does not vary systematically with the size of the organism. It is exactly this kind of similitude that produced the example above of 1.5 billion heartbeats per lifetime in mammals. Heart rate, like mass-specific metabolic rate, scales with an exponent of about $-1/4$. Average lifespans, which are the inverse of mortality rates, scale with an exponent of 1/4. The number of beats per lifetime is simply the product of these two relationships, and since $X^{-1/4}X^{1/4} = X^0$, we can predict, on average, that all mammals experience about the same number of heartbeats over the course of their life, regardless of their size.

The generality of metabolic scaling and the similitude that it shares with so many other biological rates and times has broad implications for the ecology and life history of organisms, the dynamics of populations, and the functioning of ecosystems. Some of these implications are explored in greater depth elsewhere in this volume. Here, we explore their implications for modeling the growth of individual organisms.

ALLOMETRIC MODELS OF ANIMAL GROWTH AND PRODUCTIVITY

Most models of organismal growth begin with the balanced growth assumption, which applies the first law of thermodynamics (conservation of mass and energy) to biological systems. That is, any change in the size of the animal must result from the balance of material and energetic inputs and outputs, and

$$\text{Growth} = \text{Ingestion} - \text{Egestion} - \text{Excretion} - \text{Respiration} - \text{Reproduction}.$$

This basic balanced growth model is cast entirely in terms of rates. Since most biological rates exhibit allometric scaling, growth trajectories represent the balance of allometric changes of the various input and output rates.

Typically, growth trajectories have a roughly sigmoid pattern, with accelerating growth early in development (i.e., near their initial mass, m_0) followed by a progressive slowing of growth as the organism approaches its asymptotic mass, M (see Fig. 2). In some organisms (e.g., insect larvae), the curve appears nearly exponential and then abruptly ceases without any gradual slowing. The sigmoidal pattern is similar to the pattern of logistic growth observed in natural populations growing under some sort of resource limitation. A generalized mathematical function that captures the pattern of sigmoid growth is

$$\frac{dm}{dt} = am^{\alpha} - bm^{\beta},$$

where m is organism mass (and dm/dt is thus the growth rate). To conform to a sigmoid pattern of growth, the two allometric relationships (with coefficients a and b, and exponents α and β, respectively) must have the additional constraints that $a > b$ (which assures positive initial growth at $m = m_0$) and $\alpha < \beta$ (which leads to a cessation of growth at $m = M$). The asymptotic adult mass of the animal, M, is the point at which $0 = \frac{dm}{dt} = am^{\alpha} - bm^{\beta}$; thus, $M = \left(\frac{a}{b}\right)^{\frac{1}{\beta-\alpha}}$. Likewise, the inflection point, i.e., the size at which the growth rate is maximized, is a constant fraction of the asymptotic mass, $\left(\frac{\alpha}{\beta}\right)^{\frac{1}{\beta-\alpha}}M$. Note this may, however, be less than the initial size m_0, in which case maximum growth rate occurs at the initial size.

Depending on the parameter values assigned (or derived), this general equation (which is sometimes called the Pütter equation) corresponds to several prominent models of ontogenetic growth. For example, $\alpha = 1$ and $\beta = 2$ gives the classical logistic growth model.

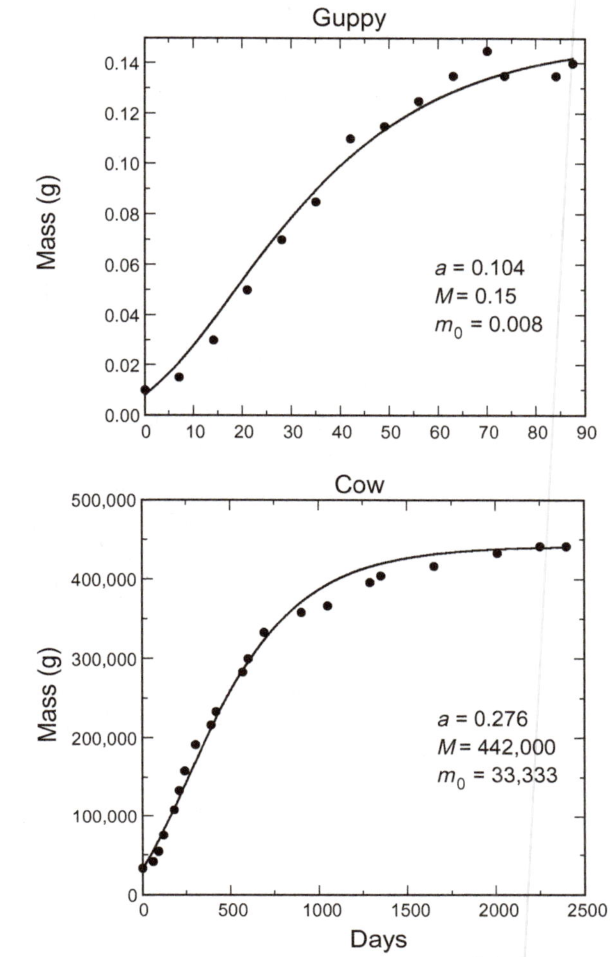

FIGURE 2 Ontogenetic growth trajectories and fitted model parameters for two vertebrates, the guppy and the cow (redrawn from West et al., 2001, *Science* 413: 628–631). The model was the form of the Pütter equation assumed by West et al., i.e., with $\alpha = 3/4$ and $\beta = 1$. Despite the vast evolutionary divergence and differences in adult size (M) and size at birth (m_0), both taxa show a common sigmoidal pattern of growth. According to the model, differences in the parameter a principally reflect differences in the scaling coefficient of the metabolic allometry (b_0) between fish and mammals.

While the logistic model sometimes provides a reasonable fit to growth trajectory data, it is difficult to assign biological meaning to the model parameters based on the balanced growth assumption, because most of the biological rates and times involved tend to exhibit scaling exponents less than 1. This entry will review three approaches that differ not just in parameter values but also in the method by which those parameter values are derived and thus assigned biological meaning. For an example of another approach based on many of the same principles, see the discussion of dynamic energy budgets elsewhere in this volume.

In 1947, Ludwig von Bertalanffy published a model of ontogenetic growth based on the balance of anabolic processes (represented by am^{α}) and catabolic processes (bm^{β}). Further, he assumed that anabolic processes were limited by the surface area over which organisms assimilate resources, and that the costs of catabolism were directly proportional to the mass of the animal, i.e., $\beta = 1$. Because mass is proportional to volume (length cubed), while areas are length squared, the surface area assumption leads to $\alpha = 2/3$. Bertalanffy based this assumption on an interpretation of the metabolic allometry that pre-dates Kleiber's work and still persists today. Surface areas are thought to be a critical constraint on the allometry of metabolism because they limit the rate at which organisms can assimilate resources (e.g., across the gut surface) or eliminate wastes (e.g., heat dissipation in homeotherms). The resulting model $dm/dt = am^{2/3} - bm$ has been applied widely, especially to the growth of marine fishes. Subsequently, Bertalanffy recognized different growth types corresponding distinguished by whether their metabolic rates scaled with surface area ($\alpha = 2/3$), mass ($\alpha = 1$), or in some intermediate way (e.g., $\alpha = 3/4$). In all cases, he assumed that the exponent α reflected the scaling of metabolic rate.

In 1989, Michael Reiss developed an alternative interpretation, still based on the structure of the Pütter equation. He assumed that, instead of the balance of anabolism and catabolism, the first term represents the scaling of resource input (which is Ingestion − Egestion in the balanced growth model), while the second term represents the metabolic cost of living, which couples respiration and excretion. While Reiss recommends parameterizing the resulting model based on empirical allometric relationships for particular taxa, he utilizes $\alpha = 2/3$ and $\beta = 3/4$ as a general solution, based on a broad survey of empirical allometries for ingestion rates and metabolic rates, respectively. Note that although the forms of their models are quite similar, Bertalanffy assumes that metabolic allometry is reflected in the first term, while Reiss assumes it is reflected in the second term. Thus, even if the two models provide reliable fits to empirical growth data, their biological interpretation is very different. More generally, distinguishing between the two models cannot be accomplished simply by comparing their fit-to-growth trajectories alone. Instead, their underlying assumptions must be addressed.

In 2001, West, Brown, and Enquist presented a model that while conforming to the same overall form, had yet another derivation. They began with the assumption that the total metabolic rate of an organism (B) is simply the sum of the energy devoted to growth (i.e., the synthesis of new biomass) plus the energy devoted to maintaining existing biomass,

$$B = E_m \frac{dm}{dt} + B_m m$$

where E_m is the energy required to synthesize a unit of biomass (e.g., in J g^{-1}) and B_m is the metabolic rate required to maintain a unit of biomass (e.g., in W g^{-1}). Rearranging to solve for the growth rate, and taking into account the allometry of metabolic rate (i.e., $B = b_0 m^{\alpha}$), they arrive at a Pütter-style model,

$$\frac{dm}{dt} = am^{\alpha} - bm,$$

where

$$a = \frac{b_0}{E_m} \text{ and } b = \frac{B_m}{E_m}.$$

Thus, a is the ratio of the metabolic scaling coefficient (b_0) to the cost of biomass synthesis, and b provides a measure of tissue turnover rate. In the original derivation, they made the additional assumption that $\alpha = 3/4$, as in the interspecific metabolic allometry documented by Kleiber, but later versions were generalized to any metabolic scaling exponent. While their model is thus almost identical in form to Bertalanffy's, the important advance is that they provide an unambiguous (and thus testable) biological interpretation not just of the exponent (α) of the growth model (i.e., it should match the observed exponent for metabolism) but also of the two coefficients (a and b), which are related to the mass-specific metabolic parameters (b_0 and B_m) and the energetic costs of biomass synthesis (E_m). Although estimates of these parameters from the original analysis of their growth model appear to be biologically reasonable, further formal tests of the model are required.

These three examples serve to illustrate three points concerning ontogenetic growth models and their relation to allometry. First, while all of the models are based on the balanced growth assumption, they make very different assumptions about what the different terms in the model represent, leading to different applications of allometric principles. For example, whereas Reiss expects metabolic allometry to be reflected in the second term, the other two models associate it with the first term. Second, because models based on very different assumptions can take on very similar or even identical mathematical forms, distinguishing between the models as alternative hypotheses cannot be accomplished by comparing their ability to fit growth trajectory data alone. Instead, comparative studies must address the underlying assumptions of the models and their ancillary predictions to assess whether the models capture growth trajectories for the right reasons. Finally, despite a rather long history of employing allometric principles to the study of ontogenetic growth, open questions remain that require careful mathematical and biological reasoning and the confrontation of abstract models with carefully collected data. Further developments in ontogenetic growth modeling have included the effects of temperature and the elemental stoichiometry of organismal growth, and ongoing experimental approaches have begun to address model assumptions by taking measurements of growth, assimilation, and metabolism on the same organisms.

ALLOMETRIC INSIGHTS ON ANIMAL GROWTH AND LIFE HISTORY

The discussion above ignored one component of the balanced growth model that is critical both for the persistence of organisms and for understanding ontogenetic growth models in an evolutionary context: reproduction. Including this fundamental biological process in the model requires mathematical modifications, the details of which depend on whether the organism in question exhibits determinate or indeterminate growth. For determinate growth (as observed in most mammals and birds), the size at maturity (first reproduction) corresponds to the asymptotic size, *M*. All growth takes place before reproduction begins and reproductive investments are either assumed to be supported by metabolic scope (represented by temporal variation in *a*) or to tradeoff against the costs of maintenance, e.g., tissue turnover, *b*. For indeterminate growers, like many fish, perennial plants, and some invertebrates, the issue is more complicated. From birth to the age (and size) at maturity, the models outlined above apply directly. However, once the organism begins reproducing, the parameters of the equations need to be adjusted to reflect the allocation to reproduction vs. growth, because any materials and energy allocated to reproduction are by definition unavailable for growth. Thus, indeterminate growers may slow or even cease growth well below their asymptotic size.

Although the dynamics of life history can complicate the modeling of growth, many aspects of life history display allometric scaling, themselves. Based on principles of similitude like those described above, Eric Charnov and others have focused on particular dimensionless numbers in life history that are independent of size. For example, the ratio of turnover rate (*b* in the West et al. and Bertalanffy growth models, above) to adult mortality rate (*Z*) have both been shown to scale across species within a taxonomic group such that their ratio (*b*/*Z*) is invariant with respect to size. This invariance is essentially the same as the invariant number of heartbeats described above; both represent an invariance in the average total lifetime energy flux observed of an organism. Although this invariance is itself remarkable, it is perhaps even more interesting that different taxonomic groups exhibit different average values. For example, the estimate of *b*/*Z* for fish (0.13) is an order of magnitude lower than that of mammals (14), and it is likely even higher for birds, which at a given size have both a faster metabolic rate (higher *b*) and a longer lifespan (lower *Z*) than mammals.

Life history invariants like *b*/*Z* provide a means of extending models of ontogenetic growth to higher-order, population-level processes. By combining a model of ontogenetic growth with life history invariants and assuming a stable age distribution, it is possible to predict how the production to biomass ratio (*P*:*B*) of whole populations changes with average adult body size (Fig. 3). Moreover, measured differences in the values of the life history invariants correspond to observed quantitative differences in the scaling of *P*:*B* between taxonomic groups, e.g., mammals and fish.

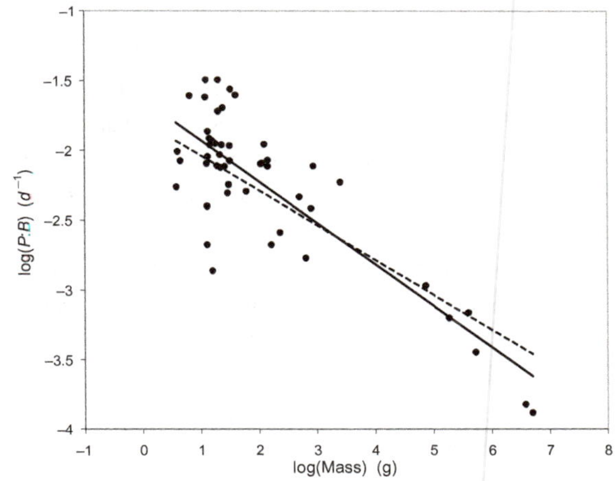

FIGURE 3 Population-level scaling of the ratio of production to biomass (*P:B*) as a function of average adult size for mammals (redrawn from Economo et al., 2005, *Ecology Letters* 8: 353–360). The dashed line is the regression fit to the data, and the solid line is the theoretical prediction obtained from combining an ontogenetic growth model with principles drawn from the study of life history invariants, principally the invariance of maintenance costs of tissue turnover (*b*) relative to mortality rate (*Z*).

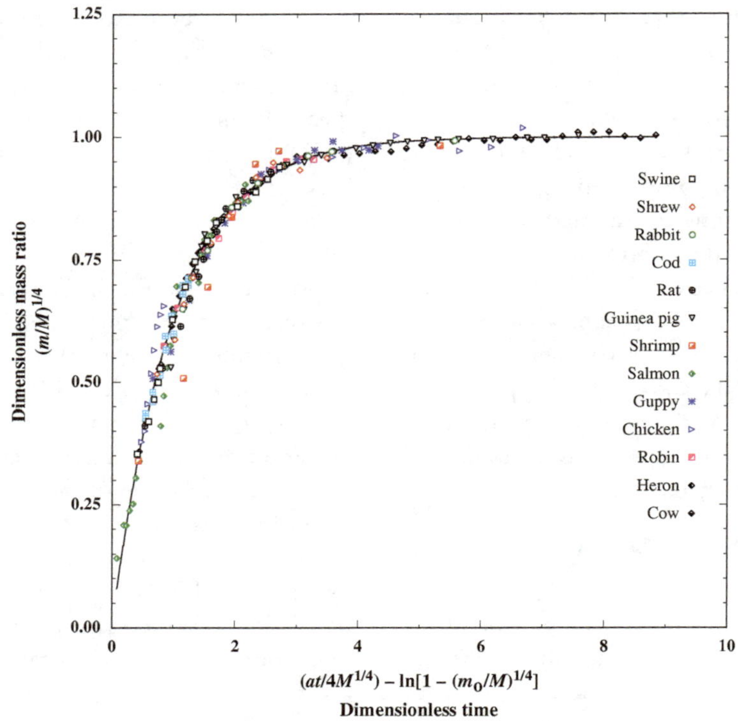

FIGURE 4 Plot of the "universal growth curve" derived by West et al. The dimensionless mass and time variables calculated for a wide variety of taxa exhibiting both determinate and indeterminate growth all fall quite near the same curve, $r = 1 - e^{-\tau}$, which is shown in the solid line. According to the West et al. model, the dimensionless mass ratio $r = (\frac{m}{M})^{\frac{1}{4}}$ is the proportion of metabolic power used for maintenance and other activities not contributing directly to growth.

The quantitative regularity of allometric scaling relationships makes them a powerful predictive tool for organizing and understanding variation in biological form and function. Part of the power of this point of view is that any particular animal (or species of animal) can be seen as a scale model of any other; Kleiber's famous relationship $B = b_0M^{3/4}$ tells us literally how to use a mouse as a scale model for an elephant, energetically speaking. Likewise, the application of allometric principles to patterns of ontogenetic growth implies that individual growth trajectories are simply examples of a more general process that has been allometrically rescaled by the particular evolved physiology and ecology of the taxa being examined. In their 2001 paper, West, Brown, and Enquist make this rescaling explicit when they renormalize the parameters of the growth trajectories for a wide variety of species to show that animals as different as cod, shrimp, chicken, and dog all fall along a single, rescaled growth curve (Fig. 4). The scaling properties of allometries and ontogenetic growth models may represent deep biological symmetries—pervasive "rules" that generate, guide, or constrain biological form and function. In addition to providing powerful predictive tools, they suggest an elegant unity underlying the fascinating diversity of life.

SEE ALSO THE FOLLOWING ARTICLES

Energy Budgets / Metabolic Theory of Ecology / Stoichiometry, Ecological

FURTHER READING

Bonner, J. T. 2006. *Why size matters.* Princeton: Princeton University Press.

Calder, W. A. 1984. *Size, function, and life history.* Cambridge, MA: Harvard University Press.

Charnov, E. L. 1993. *Life history invariants: some explorations of symmetry in evolutionary ecology.* Oxford: Oxford University Press.

Kleiber, M. 1961. *The fire of life.* New York: Wiley.

McMahon, T. A., and J. T. Bonner. 1983. *On size and life.* New York: W. H. Freeman.

Niklas, K. J. 1994. *Plant allometry: the scaling of form and process.* Chicago: University of Chicago Press.

Peters, R. H. 1986. *The ecological implications of body size.* Cambridge, UK: Cambridge University Press.

Reiss, M. J. 1989. *The allometry of growth and reproduction.* Cambridge, UK: Cambridge University Press.

Schmidt-Nielsen, K. 1984. *Scaling: why animal size is so important.* Cambridge, UK: Cambridge University Press.

West, G. B., J. H. Brown, and B. J. Enquist. 2001. A general allometric model of ontogenetic growth. *Nature* 413: 628–631.

Whitfield, J. 2006. *In the beat of a heart: life, energy, and the unity of nature.* Washington, DC: Joseph Henry Press.

ANIMAL DISPERSAL

SEE DISPERSAL, ANIMAL

APPARENT COMPETITION

ROBERT D. HOLT
University of Florida, Gainesville

Apparent competition is an indirect negative interaction between species mediated through the action of a shared natural enemy. The concept of apparent competition illuminates how natural enemies at times constrain the species richness of communities but at other times help maintain diversity. It is an integral part of any community theory focused on predators and their prey, parasites and their hosts, herbivores and the plants they consume, or entire food webs. In applied ecology, apparent competition can influence conservation risks, the success of invasive species, epidemiological patterns, and the efficacy and dangers of pest control. Finally, apparent competition exemplifies the general theme that ecological communities are not a haphazard assemblage of species, each interacting separately with the external environment, but instead exhibit complex and at times surprising chains of interactions.

ECOLOGICAL THEORY CAN HELP UNIFY ECOLOGICAL UNDERSTANDING OF DISPARATE SYSTEMS

Natural historians glory in life's rich diversity. Ecological theory can help tease out commonalities among disparate systems and thus help unify understanding. What do the following real-world stories have in common, and how can theory help explain the patterns they reveal?

Case Studies: Ecological Puzzles from Studies of Interacting Species

DO RABBITS "EAT" VOLES?

In the Grampian Mountains of Scotland, the native water vole, *Arvicola amphibius,* resides in lush vegetation along streams and other water bodies. During the eighteenth century, the European rabbit, *Oryctolagus cuniculus,* colonized the area, and its warrens of deep burrows are mainly found in dry fields of short grass. The two species live in different habitats and have different diets, and the rabbit invasion had no obvious impact on the water vole. Sadly, the water vole has plummeted in abundance over the last 50 years from millions to a few hundred thousand across the entire UK. In the Grampians it has disappeared from sites near upland fields containing rabbits. If the two species do not compete, how can the presence of one affect the fate of the other?

ANOTHER PUZZLE IN AN INVASION

In the United Kingdom, the native red squirrel, *Sciurus vulgaris*, once widespread, has been largely supplanted by the introduced grey squirrel, *Sciurus carolinensis*. The two squirrel species overlap in diet and habitat and so probably compete. But there are puzzling aspects of the decline: sometimes the red squirrel disappears from a locale even before the grey squirrel has built up its numbers. Is this just a straightforward story of competitive exclusion?

COUNTERINTUITIVE EFFECTS OF PROTECTING WILDLIFE

In the dry savanna of northern Kenya, humans, livestock, and wildlife have coexisted for millennia. To boost ecotourism, there have been localized shifts from cattle ranching to wildlife conservation. Some species, in particular the plains zebra, *Equus quagga burchellii*, increased greatly post-protection, but other ungulates such as hartebeest (*Alcelaphus buselaphus*) have severely declined. Are these declines driven mainly by competitive interactions among these large mammalian herbivores, or something else?

AN ENDANGERED ISLAND SPECIES

Feral pigs (*Sus scrofa*) were introduced into the California Channel Islands, which later saw abrupt crashes toward near-extinction of an endemic predator, the island fox, *Urocyon littoralis*. How did the pig introduction endanger the fox—could it for instance be habitat degradation driven by the destructive rootings of the pigs?

Simple Models of Apparent Competition

NATURAL ENEMIES ARE SIGNIFICANT FACTORS IN MOST SPECIES' LIVES

At first glance, several of these case studies are consistent with the hypothesis that species are directly competing—and this may be true. But to understand what forces drive these systems, it turns out one must consider species or populations beyond those that at first glance seem to be the main players—and all these empirical patterns reflect apparent competition.

Communities are complex webs of interacting species, affecting each other not just one on one but via interlocking chains of indirect interactions mediated through other species. Most species suffer from natural enemies—a term broadly encompassing predators, herbivores, parasites, pathogens, and indeed any species that to make a living inflicts harm on other species, taking resources (energy and nutrients) from living organisms so as to survive or reproduce itself. Even the fiercest top predator—lion, tiger, or bear—can be laid low by an infectious disease agent, and the smallest quasi-organisms of all—viruses—themselves can be attacked by other viruses or consumed when their infected hosts fall prey to natural enemies.

Many natural enemies are generalists, exploiting more than just one victim species (prey or host). The level of damage imposed on any particular prey or host species (as assessed by reduced growth rates, fitness, or abundance) may depend upon the availability, traits, and productivity of alternative prey or hosts. In other words, there is an indirect interaction between these species. Formally, in a mathematical model, an indirect interaction exists between species I and III, mediated through II, if the dynamical equation describing growth rates in I contains a variable referring to II (e.g., abundance), and the equation for II has a variable referring to III. Changes in III lead to changes in II, which in turn affects I. When an indirect interaction between two victims mediated by a natural enemy is negative, it is called apparent competition. This term was coined by the author in 1977 because ecological patterns that appear to be due to competition for resources, such as nonoverlapping spatial distributions, can also emerge from impacts of a shared natural enemy.

A COMPARISON OF EXPLOITATIVE AND APPARENT COMPETITION

Figure 1 depicts community modules corresponding to exploitative competition and apparent competition. Each node represents a species, and arrows represent directions of effects. On the left, two natural enemies are consumers of the same resource, and any resource gathered by species A is thus unavailable to species B, which therefore suffers a reduction in its growth rate. Apparent competition is a mirror image of this familiar indirect interaction. Victim species A has a positive effect on the abundance or activity of a natural enemy, and because the natural enemy has a negative effect upon species B, species A indirectly has a negative effect upon species B.

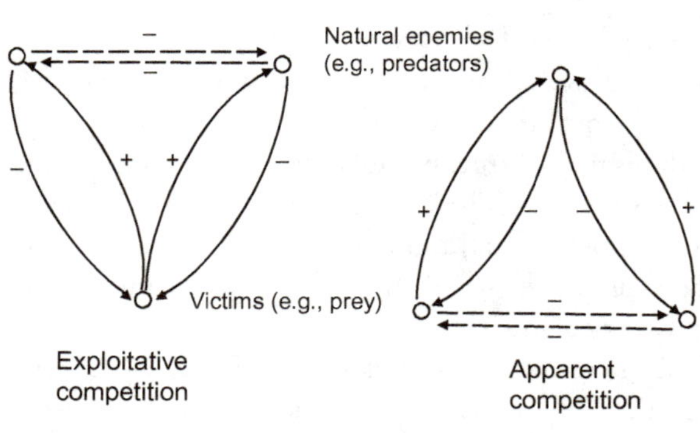

FIGURE 1 Exploitative and apparent competition. Each node is a species. Solid arrows are signed direct interactions; dashed arrows, emergent indirect interactions. Higher species in the figure are higher in the food web. On the left, two consumers (e.g., predators) share a resource. Multiplying the signs, each consumer has a negative indirect effect upon the other—competition for resources. On the right, two victim species (e.g., prey) share a natural enemy (predator). Again, multiplying the signs shows that each victim indirectly harms the other, via their shared natural enemy.

A GRAPHICAL MODEL OF APPARENT COMPETITION

A simple graphical model exploring this process provides one step towards a formal theory of apparent competition. Assume a predator of abundance P has an instantaneous population growth rate dependent only on availability of two prey species, of abundance R_1 and R_2:

$$\frac{1}{P}\frac{dP}{dt} = F_P(R_1, R_2). \tag{1}$$

Assume the system reaches equilibrium, so $F_P(R_1, R_2) = 0$. We plot this (the predator zero-growth isocline) as a curve on a graph with axes of prey abundance (Fig. 2A). Outside the isocline, the predator grows; inside, it declines.

Assume each prey species when alone can sustain the predator. The predator reaches equilibrium when prey abundances match the open circles in the figure. Now assume an equilibrium exists with both prey species present. This equilibrium must lie on the predator isocline, for instance at the closed circle. If we compare the equilibrial abundance of each prey when alone to its abundance when the two coexist, note that at the joint equilibrium each prey species is depressed in abundance. The little arrows in the figure show that adding a small amount of either prey to the system permits predator growth. Mathematically, this is equivalent to saying the predator isocline has negative slope. Thus, each prey species indirectly depresses the equilibrial density of the other because each benefits the predator, increasing its numbers and thus predation upon the alternative prey.

A SIMPLE RULE FOR DOMINANCE IN APPARENT COMPETITION

But maybe no equilibrium exists with both prey species. Imagine the predator depresses prey species i to a level where it experiences only density-independent per capita growth at rate r_i and predation at rate a_iP (a_i is the attack rate, and P predator abundance). The net per capita growth of species i is $r_i - a_iP$. For equilibrium, the predator must settle to $P_i^* = r_i/a_i$ (the asterisk indicates equilibrium, and the index on P indicates just prey i is present). Suppose prey 1 is at equilibrium with the predator, and prey 2 tries to invade. Invasion succeeds only if $r_2 - a_2P_1^* > 0$, which implies $P_2^* > P_1^*$. But if this holds, then when prey species 2 in turn is present at equilibrium, and prey species 1 tries to invade, it cannot—predation upon it exceeds its intrinsic $>k$ growth rate.

If attack rates are constant but intrinsic growth rates and predator densities vary over time, one can replace the r's and P's with time averages, and this conclusion still holds. In other words, if the predator is the only regulatory factor limiting each prey species, one expects to see exclusion of one prey species by the other, and that prey which sustains higher average predator abundance wins in apparent competition. Shared predation can constrain prey species diversity.

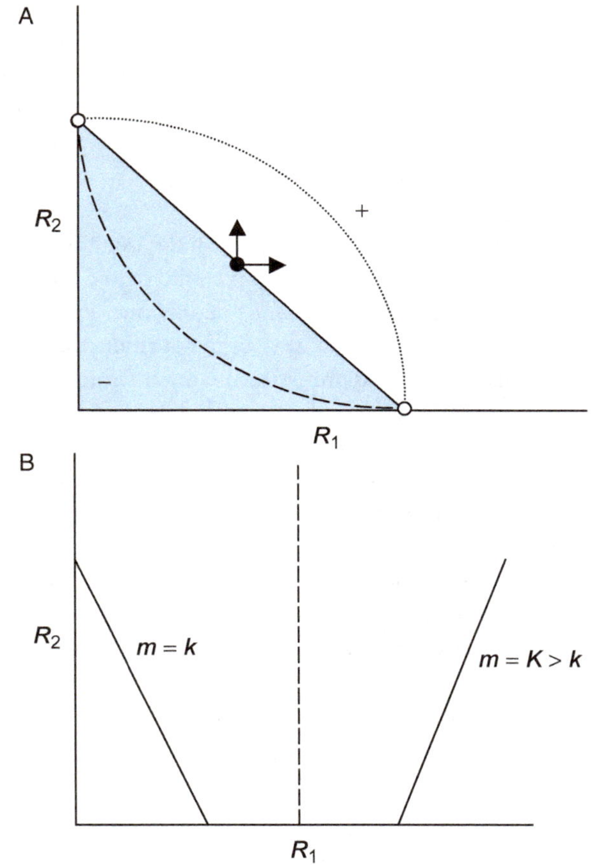

FIGURE 2 A graphical model of shared predation. (A) The solid straight line is the zero growth isocline of a predator. Between this line and the origin, the predator declines in abundance because of a shortage of food; outside the line, it thrives and grows in numbers. The small arrows show that increases in either prey leads to predator growth, and this leads to a negative isocline slope, and apparent competition (compare open circles to closed circles; see text). Apparent competition is even stronger if the predator thrives better on a mixed diet (dashed line), and weaker if it does not benefit from the mix (dotted line). (B) The model is for a food-limited predator with a saturating response to each of two prey:

$$\frac{1}{P}\frac{dP}{dt} = F_P(R_1, R_2) = \sum_{i=1}^{2} \frac{a_i b_i R_i}{1 + a_1 b_1 R_1 + a_2 b_2 R_2} - m,$$

where P and R_i are respectively the abundances of the predator and prey species i; a_i, b_i, h_i are attack rates on prey i, the benefit the predator gains per prey caught, and handling times; and m, the predator death rate. We assume one prey is better than the other (as measured in b/h, benefit gained per unit handling time). If the predator is a generalist for all prey densities, at low abundances it benefits from each prey, so the isocline has a negative slope, but at high abundances, it can waste time on the poor prey (see text). At low predator mortality (the line indicated with $m = k$), the two prey experience apparent competition, but at higher mortality (marked by $m = k$), the isocline has a positive slope, so the indirect interaction between prey is (−, +).

All the Above Ecological Puzzles Involve Apparent Competition.

Apparent competition arises in many different ecological systems. The solution to each ecological puzzle above turns out to involve apparent competition. Each illuminates different features and realistic complications of apparent competition and points out directions for theory development.

RABBITS DO NOT EAT VOLES—BUT MINK DO

In the first example above, there is another invasive species that also moved into Scotland, but well after the rabbits—the American mink, *Neovison vison*. This species is a smart, voracious predator, and it can follow the water vole into all its normal refuges. Minks do consume rabbits, but they cannot readily eliminate them from their protective warrens. In short, mink numbers get boosted by a species that they cannot control—the rabbit—which then permits the mink to impose very heavy mortality on a more vulnerable species—the water vole. Rabbits in no way directly compete with water voles, but nonetheless by sustaining a predator—the mink—they have a strong, albeit indirect, deleterious impact upon the persistence of water voles. This appears to be an excellent (if sad) example of apparent competition in action.

The water vole–rabbit–mink interaction matches expectations of our simple model. Mink populations respond strongly to increased food availability, so are food-limited. Rabbits have notoriously high fecundity and can reach high abundance; they have a high intrinsic growth rate. Many rabbits are protected from mink predation in their warrens and so have an average lower (but nonzero) attack rate than the water voles. Rabbits, we can reasonably infer, have a higher r/a than do water voles and so should tend to dominate in apparent competition. This appears to be what happens—the water vole is excluded from sites near rabbit warrens.

Spatial Processes and Apparent Competition This example can be used to glean other insights that have been explored in theoretical models. The theory of apparent competition has been extended from closed to open communities, comprised of habitat patches linked by dispersal. If different prey species occupy different habitats (and so do not compete), their dynamics may be nonetheless coupled by predator movement. It turns out that alternative prey species can coexist, if prey species segregate among habitats and predator mobility is constrained. Water voles and rabbits do live in quite different habitats. The reason their fates are joined at all in the Scottish landscape is that minks are highly mobile. The farther a water vole population is located from rabbit populations, the less likely it is that mink (sustained by feeding on rabbits) will wander through. Remnant populations of water voles in the Grampians tend to be those at some distance from rabbit habitat, presumably because they suffer less "spillover" predation.

A critical dimension of apparent competition is thus the nature of spatial processes connecting different habitats. Natural enemies are often mobile, permitting apparent competition to act at a distance. This effect has been studied in laboratory microcosms consisting of patch arrays containing two moth species. Arrays were set up so that moths lived in different patches and did not directly compete, but a mobile wasp parasitoid could move freely among them. The wasp coexisted just fine with either host when alone, but the three-species combination collapsed, and one host species went extinct from elevated parasitism—apparent competition. In like manner, movement of prey across space can subsidize predators in a community, permitting them to more effectively limit resident prey.

WITH A LITTLE HELP FROM MY FRIENDS

Apparent competition is implicated in the conservation risks experienced by many endangered species, including the replacement of red by grey squirrels. The squirrels do seem to compete for resources, and the grey squirrel seems more effective at using acorn resources. Hence, grey squirrels can potentially have a higher carrying capacity and so may dominate in competition for shared resources. But on top of this, the grey squirrel harbors a poxvirus (SQPV)—an infectious disease agent to which they are seemingly immune but to which the red squirrel is highly vulnerable. Mathematical models suggest this shared pathogen greatly speeds up the demise of the red squirrel, even in advance of an increase in grey squirrel numbers.

Apparent Competition Mediated by Parasitism This example illustrates the potential importance of pathogens as conduits of apparent competition. This is a very important issue for conservation and public health. In California, the invasive pathogen *Phytophthora ramorum* is a highly generalized natural enemy, inflicting serious damage on many native tree species. The California bay laurel (*Umbellularia californica*) is a heavy producer of pathogen spores but is not itself greatly harmed. The spillover of spores from this species onto more vulnerable species such as tanoak (*Lithocarpus densiflorus*) leads to high mortality. Because trees compete, as vulnerable trees decline the bay laurel increases and hammers its competitors via the pathogen even harder.

Asymmetric Apparent Competition These examples reveal strong asymmetry in the effect of the shared natural enemy—the indirect interaction is mainly one-way. This is a common (not universal) feature of apparent competition. Explaining why asymmetries are often strong is still an open question and probably reflects both ecological factors and evolutionary history. In general, whichever species has the highest productivity (high r) and is not limited strongly by factors other than predation tends to dominate. If grey squirrels more effectively utilize a resource such as acorns, this could accentuate their dominance over reds in apparent competition. A full explanation of species extinctions from local communities often involves both competition (in the usual sense of the term) and apparent competition. These are not alternative, incompatible explanations but processes that can occur simultaneously and interact in various ways.

PROTECTING PREDATORS CAN PERMIT APPARENT COMPETITION TO OCCUR AMONG PREY

In the Kenyan example, humans historically suppressed zebras (which compete with livestock) and predators such as lions. Limiting livestock and reducing hunting in the interest of wildlife conservation allowed zebras to surge to high numbers, and predators such as lions reappeared. Predators do not substantially limit zebra numbers; zebras form large herds that provide protection from predation. But these herds do provide a steady supply of young, sick, or injured individuals that are easy pickings and can sustain predator populations. Other ungulates do not necessarily enjoy this kind of protection, and intensified predation appears to account for their declines.

Apparent Competition Does Not Arise Only in Disturbed Ecosystems In contrast to the invasion case studies, this African system involves species that have lived together for a very long time. Apparent competition is not just a process that shows up as a brief transient in unnatural situations created by human disturbance and transplantation of species around the globe but can be important in natural ecosystems. The reason it was detected was that there was in effect a large-scale inadvertent experiment driven by a shift in land use patterns by humans. Stronger evidence for apparent competition in natural assemblages comes from deliberate experimental manipulations. In the rain forests of Belize, a rich community of leaf-mining insects (flies and beetles) sustains a high diversity of parasitoids. Experimental removal of some hosts led to lower parasitism in the remaining hosts, showing strong apparent competition in this natural community of herbivorous insects. Apparent competition could play a significant role in the dynamics of biodiversity over evolutionary time scales as well. Shared predation provides novel niche axes for specialization and diversification, and localized coadaptation of natural enemies and victims can lead to a sorting out of species among habitats or along gradients. Understanding the evolutionary dimensions of apparent competition is a largely unexplored area of theory.

Availability of Refuges Is a Key Element in Apparent Competition Another general message can be gleaned from the Kenyan example: the zebra has a partial refuge from predation by virtue of grouping behavior. Refuges come in many forms, ranging from permanent physical locations providing escape (e.g., rabbit warrens), to transient refuges in space (as in metapopulation dynamics), to escapes in time, to plastic adaptations that lower predation or parasitism rates. Stage structure (e.g., an invulnerable adult class) provides a particularly important form of refuge in some systems, and leads to rich complexities in theoretical models, because of the multiplicity of feedbacks that are possible (e.g., alternative stable states, and complex dynamics). If refuges protect some but not all individuals in a species, natural enemies can be sustained by this species without endangering it. Such species can dominate in apparent competition over species lacking refuges. Hosts can evolve to tolerate parasites or herbivores, without eliminating them. Such hosts could then exert strong apparent competition on alternative hosts that are not so well adapted. Understanding how population structure and evolutionary processes affect the strength of interspecific interactions, including apparent competition, is an active and growing area of ecological theory.

Multiple prey or host species can coexist, despite strong apparent competition, if each has its own refuge—a kind of niche partitioning in enemy-free space. An important subtlety is that this works if species are more likely to be in such a refuge when rare than when common. Theoretical models suggest that such refuge-mediated coexistence is greatly amplified if natural enemies behaviorally aggregate, spending more time where their victims are common.

PREDATOR BEHAVIOR GENERATES OTHER MECHANISMS OF APPARENT COMPETITION

Behavior—foraging tactics of predators and escape maneuvers of prey—is an important element in apparent competition. In the Channel Islands story, the presence of an abundant prey species, feral pigs, led to colonization by a few pairs of Golden Eagles, *Aquila chrysaetos.* Foxes on their own are far too scarce to warrant eagles setting up house on the island, but they are easy to casually catch as tasty morsels by eagles lured to the islands by an abundant alternative prey. This is called incidental predation in the literature.

In this case study, predator behavior is a key driver of apparent competition. The handful of Golden Eagles present could have nested on the mainland but instead chose to reside on the islands. In Figure 2, rather than interpreting the isocline as describing the dynamics of an entire predator population (births and deaths), we can view it as a model of predator use of a small habitat patch (i.e., numbers change via immigration and emigration). When prey are scarce, predators leave; when prey are flush, predators arrive and stay. This aggregative numerical response is expected from optimal foraging theory; models predict that it leads to apparent competition between prey species within a patch, even if predator populations as a whole do not respond to shifts in prey numbers.

Flexible behavior (including patch use) can lead predators to ignore rare prey species, at least if those prey require special foraging tactics or specific recognition cues (search images), or if they are found in sites lacking other, more common prey. Many mechanistic processes lead to such switching by generalist predators. Theoretical models suggest that switching often promotes the persistence of multiprey assemblages. But counterexamples are also suggested by theory, dependent on the temporal scale and accuracy of switching relative to changes in prey abundance.

Changes in predator behavior that alter the indirect interaction between alternative prey exemplify what are called trait-mediated indirect interactions. There can be a range of effects alternative species have on each other, and the net effect is often context specific. One species can indirectly negatively impact another not by feeding a natural enemy but instead by in some other way facilitating its presence or activity levels. Invasive shrubs can foster their own invasion by sheltering native herbivores from predators, so the herbivores more effectively reduce their native food plants, freeing resources for the invader. This is still apparent competition, albeit via a kind of ecological engineering. One direction for future theory will be to develop mechanistic models incorporating the multiplicity of potential channels of interactions among species.

A KIND OF APPARENT COMPETITION CAN OCCUR WITHIN INFECTED HOSTS

A major defense that vertebrates have against parasites is acquired immunity. To boil a complex story down to some basic elements, the immune system works by the stimulation of the proliferation of particular cell lines—a population of predatory cells—tailored in their attacks to parasites with certain attributes. If parasite A invades a host body and that host mounts an immune response, this may help fend off invasion by a relatively similar parasite B if it is recognized as being hostile by the host immune system. In this case, the natural enemy is the host immune system, and its numerical response (comparable to Eq. 1) is the growth of a particular population of defensive cells. The victims are different species or categories of parasites attempting to invade the host. Such apparent competition among similar strains of parasites is a powerful force selecting for parasite diversity.

This example illustrates the general point that abstract theoretical models can help illuminate comparable processes found in seemingly radically different systems. There are important differences between theoretical immunology and the theory of, say, vertebrate predator–prey interactions, but recognizing commonalities can help point to a unification of perspectives and approaches across levels of biological organization guided by powerful theoretical insights.

APPARENT COMPETITION IS ONLY PART OF THE STORY OF SHARED PREDATION

Does shared predation *always* lead to apparent competition? When it does, does it always tend to reduce prey species diversity? The short answer to both questions is "No!"

Let us go back to the basics. The simple theoretical model sketched above, part of which is shown in Figure 2A, makes many assumptions. Relaxing these assumptions can lead to a shift in theoretical predictions.

First Assumption We assumed an increase in prey numbers, for either of two prey species, benefits the predator (as assessed by a boost in its per capita growth rate). This might not hold, and for two reasons, one having to do with the prey themselves, and one having to do with the predator, largely independent of its prey.

Equation 1 assumes predator growth depends on prey availability. More generally, predator dynamics can depend on its own density. For instance, for successful reproduction weasels might require specialized nest sites, which could be in short supply and lead to direct competition. This could constrain the numerical response of weasels to mice and shrews, weakening apparent competition between these prey. Even if this is not the case, higher-order predators such as owls might limit weasel abundance. A pulse in mouse numbers might lead to a temporary increase in weasels, suppressing shrews, but in the long-term owl predation might bring the weasels back to their original levels. So food-web interactions can at times temper apparent competition or make it a transient response in system dynamics.

Equation 1 and Figure 2A assume consumption of each prey benefits the predator. This is not always true. Some prey contain toxins, harming consumers. Fish that die in the mass fish kills of red tides presumably are not very good at discriminating poisoned food from safe food. More subtly, even if a certain prey species is not absolutely bad, it might be bad in a relative sense, compared to other prey types. One generality about predators is that the rate at which they feed always saturates, due to limitations in handling time, gut capacity, or attention span. When food is abundant, time spent handling a low-quality prey type is in a sense wasted; there is an opportunity cost, because higher-quality prey are being ignored. Figure 2B shows how this alters the slope of the predator isocline. At sufficiently high abundance of the good prey, boosting numbers of the poor prey actually depresses predator growth; the indirect interaction shifts from (−, −) to (−, +).

This may seem unlikely, but something like this is believed to be of great importance in vector-borne infectious diseases involving multiple potential host species for the vector. Isoclines with partial positive slopes arise quite naturally in theoretical models of such systems. Some hosts are poor for pathogen reproduction, and if they can draw off attacks by vectors from more competent hosts, the net effect is a reduction in pathogen load across the entire system. Biodiversity can thereby moderate the risk of infectious disease.

Second Assumption We assumed interacting species reach equilibrium. Saturating responses, demographic stochasticity, and environmental fluctuations can lead to sustained oscillations. Theoretical studies show that alternative prey can sometimes boost each other's average numbers, when one averages over the nonequilibrial dynamics of these highly nonlinear systems. Understanding how environmental variability and complex dynamics modify indirect interactions is a largely open area of theoretical ecology.

Third Assumption In our examples, apparent competition drove particular prey species to lower abundance or even extinction. Apparent competition via shared natural enemies can sometimes facilitate coexistence—if prey species also compete by interference or exploitatively for limiting resources. For instance, two prey species can persist on a single resource if a predator is present and one prey is better at resource competition and the other is better at apparent competition.

For a single generalist predator to enhance prey diversity beyond this effect requires some mechanism in effect permitting prey species to each have its own refuge from shared predation. As noted above, switching behavior by the predator is one such mechanism, as is habitat segregation among prey (given limited predator mobility). Another is to imagine that there is not just one predator species but instead a number, each to a degree specialized in its attacks to different prey. There can then be a kind of codependence of diversity across different trophic levels, and coexistence of numerous species in each trophic level becomes possible. Recent theoretical studies have explored this idea in some detail, highlighting the symmetry between shared predation and exploitative competition illustrated in Figure 1.

The symmetry is, however, incomplete. All species need resources, so there is always a potential for competition, within or among species. Whether or not shared predation leads to apparent competition is contingent (as we have seen) on many details. Moreover, there can be a difference in time scales over which dynamics play out. Resource levels can change very rapidly in response to changes in consumption (e.g., plants competing for light), whereas there can be significant time lags in responses of predators to their prey. Pathogen loads, however, can change rapidly, relative to host generation lengths, and for some purposes the impacts of shared pathogens and parasites might almost be viewed as a form of interference competition.

APPARENT COMPETITION PLAYS A ROLE IN HUMAN HISTORY AND OUR IMPACTS ON THE BIOSPHERE

Ecological theories such as apparent competition have implications for understanding ourselves and our impacts upon the rest of the biosphere. All the above case studies exemplify important problems in applied ecology. A clear appreciation of apparent competition theory can inform issues and management strategies in many practical disciplines, from conservation, to natural resource management, to pest control, to epidemiology, to invasion biology.

Apparent competition among peoples arguably has carved major channels in our own history. Parallel to the interaction between red and grey squirrels of the United Kingdom, plagues (in combination with other forms of more direct competition) have tragically influenced the waxing and waning of different peoples around the globe. For instance, as western Europeans conquered much of the world they had a hidden weapon—shared infectious diseases such as smallpox, to which indigenous populations were much more vulnerable. Further back in time, one hypothesis to explain the mass extinction of large mammals in North America is that it was overkill by a wave of colonizing people sweeping across the continent. Theoretical models suggest this hypothesis may work, if smaller species with a higher reproductive rate sustained the population of hunters, who could continue to preferentially attack large mammals such as mastodons even when the latter became vanishingly rare. In other words, apparent competition, mediated through humans, may be implicated in sculpting major features of the current fauna of entire continents.

Nearer the present, the ability of humans to hunt to extinction the Passenger Pigeon and Dodo reflects the fact that we do not depend on these species alone for sustenance. Humans are the ultimate generalist consumer, able to crack almost any victim's defenses, and our burgeoning population is sustained in terms of calories and nutrients by a quite small number of species, permitting us to wreak havoc upon the rest. The biodiversity crisis across the globe, seen through the lenses of ecological theory, may be an example of apparent competition, writ large.

SEE ALSO THE FOLLOWING ARTICLES

Conservation Biology / Ecosystem Engineers / Food Webs / Invasion Biology / Movement: From Individuals to Populations

FURTHER READING

Abrams, P. E. 2010. Implications of flexible foraging for interspecific interactions: lessons from simple models. *Functional Ecology* 24: 7–17.

Bonsall, M. B., and M. P. Hassell. 1999. Parasitoid-mediated effects: apparent competition and the persistence of host-parasitoid assemblages. *Researches on Population Ecology* 41: 59–68.

Chesson, P., and J. J. Kuang. 2008. The interaction between predation and competition. *Nature* 456: 235–238.

DeCesare, N. J., M. Hebblewhite, H. S. Robinson, and M. Musiani. 2010. Endangered, apparently: the role of apparent competition in endangered species conservation. *Animal Conservation* 13: 353–362.

Harmon, J. P., and D. A. Andow. 2004. Indirect effects between shared prey: predictions for biological control. *BioControl* 49: 605–626.

Holt, R. D., and J. H. Lawton. 1994. The ecological consequences of shared natural enemies. *Annual Review of Ecology and Systematics* 25: 495–520.

Mideo, N. 2009. Parasite adaptations to within-host competition. *Trends in Parasitology* 25: 261–269.

Oliver, M., J. J. Luque-Larena, and X. Lambin. 2009. Do rabbits eat voles? apparent competition, habitat heterogeneity and large-scale coexistence under mink predation. *Ecology Letters* 12: 1201–1209.

APPLIED ECOLOGY

CLEO BERTELSMEIER, ELSA BONNAUD, STEPHEN GREGORY, AND FRANCK COURCHAMP

University of Paris XI, Orsay, France

Applied ecology aims to relate ecological concepts, theories, principles, models, and methods to the solving of environmental problems, including the management of natural resources, such as land, energy, food or biodiversity.

DEFINITION AND SCOPE

What Is Applied Ecology?

Despite its somewhat restrictive name, applied ecology is more than simply the application of fundamental ecology. In a nutshell, ecological management requires prediction, and prediction requires theory. Applied ecology is a scientific field that studies how concepts, theories, models, or methods of fundamental ecology can be applied to solve environmental problems. It strives to find practical solutions to these problems by comparing plausible options and determining, in the widest sense, the best management options.

One particular feature of applied ecology is that it uses an ecological approach to help solve questions concerned with specific parts of the environment, i.e., it considers a whole system and aims to account for all its inputs, outputs, and connections. Of course, accounting for everything is no more possible in applied ecology than it is in fundamental ecology, but the ecosystem approach of applied ecology is both one of its characteristics and one of its strengths.

Indeed, one could view the overall objective of applied ecology as to maintain the focal system while altering either some of the elements we take from the system (i.e., ecosystem services or exploitable resources) or some of those we add to the system (i.e., exploitation regimes or conservation measures) through an educated management strategy. Since those two types of elements are not mutually independent, long-term management strategies are best aimed at optimizing rather than maximizing exploited items. This is more efficiently achieved through an adequate understanding of theoretical ecology, which generally considers all parts of the system rather than a limited set of its components.

What Are the Fields Covered?

The word "applied" implies, directly or indirectly, human use or management of the environment and of its resources, either to preserve or restore them or to exploit them. Humans influence the Earth at all levels: the atmosphere, the hydrosphere (oceans and fresh water), the lithosphere (soil, land, and habitat), and the biosphere. Understandably, questions related to human populations (notably its demography) fall within the scope of applied ecology, as most impacts on ecosystems are directly or indirectly anthropogenic.

Aspects of applied ecology can be separated into two broad study categories: the outputs and the inputs (Fig. 1). The first contains all fields dealing with the use and management of the environment for its ecosystem services and exploitable resources. These can be very diverse and include energy (fossil fuel or renewable energies), water, or soil. They can also be biological resources—for their exploitation—from fish to forests, to pastures and farmland. They might also, on the contrary, be species we wish to control: agricultural pests and weeds, alien invasive species, pollutants, parasites, and diseases. Finally, they can be species and spaces we wish to protect or to restore.

The fields devoted to studying the outputs of applied ecology include agro-ecosystem management, rangeland management, wildlife management (including game), landscape use (including development planning of rural, woodland, urban, and peri-urban regions), disturbance management (including fires and floods), environmental engineering, environmental design, aquatic resources management (including fisheries), forest management, and so on. This category also includes the use of ecological knowledge to control unwanted species: biological invasions, management of pests and weeds (including biological control), and epidemiology.

The inputs to an applied ecology problem consist of any management strategies or human influences on the target ecosystem or its biodiversity. These include conservation biology, ecosystem restoration, protected area design and management, global change, ecotoxicology

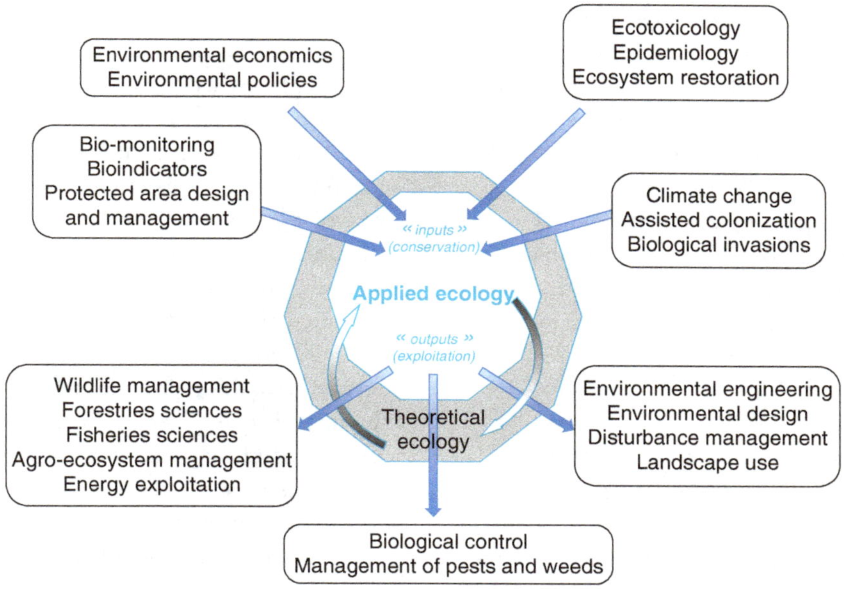

FIGURE 1 The relationships between the different fields of applied and theoretical ecology.

and environmental pollution, bio-monitoring and bioindicators of environmental quality and biodiversity, environmental policies, and economics. Of course, these outputs and inputs are intimately connected. For example, the management of alien invasive species is relevant to both natural resource management (e.g., agriculture) and the protection of biodiversity (conservation/restoration biology).

In addition to using fundamental ecology to help solve practical environmental problems, applied ecology also aspires to facilitate resolutions by nonecologists, through a privileged dialogue with specialists of agriculture, engineering, education, law, policy, public health, rural and urban planning, natural resources management, and other disciplines for which the environment is a central axiom. Indeed, some of these disciplines are so influential on environmental management that they are viewed as inextricably interlinked. For example, conservation biology should really be named "conservation sciences" because it encompasses fields that are not very biological, such as environmental law, economics, administration and policy, philosophy and ethics, resources management, psychology, sociology, biotechnologies, and more generally, applied mathematics, physics, and chemistry.

And obviously, as we are dealing with environment and ecosystems, everything is connected, all questions are interrelated, all disciplines are linked, and all answers are interwoven. Understanding a process through one field of applied ecology will allow advancing knowledge in other fields.

What Are the Links between Applied and Theoretical Ecology?

Because theoretical ecology may be defined as the use of models (in the widest sense) to explain patterns, suggest experiments, or make predictions in ecology, it is easy to see that relations between theoretical and applied ecology are bidirectional. Simply put, theory feeds application, but application also allows for the testing of theory. Indeed, applied ecological problems are used to assess and develop ecological theory. In this regard, these two aspects of ecology complement and stimulate each other.

Yet the links between theoretical and applied ecology can fray. Theoretical ecology operates within the bounds of plausibility. What is theoretically plausible, however, is not always ecologically realistic. And if it is not ecologically realistic, then theoretical ecology cannot be applied to interrogate real ecological situations. In other words, theoretical ecology cannot always be used in applied ecology. The links between theoretical and applied ecology range from spurious to robust and have been used with varying success in different fields. Fields that have benefited from theory include fisheries and forestry management and veterinary sciences and epidemiology (both human and nonhuman). However, some other fields of applied ecology have not (or not yet) benefited fully from ecological theories, concepts, and principles, and that aspect is the focus of this entry.

Three such examples are presented here. The first is an illustration of a major field of ecology that has a strong applied branch but has not, until very recently, fully taken advantage of theoretical ecology: invasion biology. The second describes a theoretical process important to many aspects of applied ecology (including epidemiology, fisheries, biological control, conservation biology, and biological invasions) but which has nonetheless been underused so far in applied ecology: the Allee effect. The third depicts an emerging field that is now, by obligation, mostly theoretical but which has an applied future and for which it is hoped that ecological applications will emerge rapidly: climate change.

APPLIED ECOLOGY WITHOUT THEORETICAL ECOLOGY: BIOLOGICAL INVASIONS

The study of biological invasions is one striking case where applied ecology has historically been at the core of the discipline's development without fully benefiting from theoretical ecology. Alien invasive species are those species introduced beyond their ordinary geographical range by human agency and that are infamous for causing ecological and economical damage. Globally, they are considered as the second biggest cause of species extinctions after habitat degradation, but they reign as the single most devastating impact in island ecosystems.

The seventheenth century marked the start of the vigorous and uninterrupted human colonization of the world's islands, initially by sailors, pioneers, settlers, hunters and fishermen, and more recently by militaries and tourists. Whether by a poor understanding of ecosystem functioning or a scarce consideration for the environment, this colonization led to many ecological disturbances, in particular the introduction—deliberate or accidental—of many exotic plant and animal species. Although not confined to islands (exotic organisms have been introduced to continents, lakes, and even oceans), the introduction of exotic organisms to islands has often led to dramatic changes in their communities. This is in part because insular ecosystems are fragile and characterized by simple trophic webs, vacant ecological niches, and low species diversity. Perhaps more important, however, is the fact that native species often evolve in the absence of strong competition, herbivory, parasitism, or predation and are less adapted to these newcomers. Moreover, islands harbor high levels of endemism, several times higher than comparable continents ecosystems, which makes every population extirpation a probable species extinction. As a result, alien invasive species on islands have affected countless island plant and animal communities and have been responsible for numerous native species extirpations and extinctions (nearly 60% of modern species extinctions have occurred on islands).

The first attempts at controlling alien invasive species included mechanical (trapping, hunting) and chemical (poisoning) methods, but the majority were biological: deliberate introductions of the alien invasive species' natural enemies to efficiently find and kill them. For example, numerous attempts at controlling invasive rodents have involved releasing cats, dogs, or mongooses. Prior to many of these attempts, it had not been realized that even the most efficient predators very seldom remove an entire prey population but instead control it at reduced numbers. This is because invasive rodents have evolved with cat predation and have developed antipredator behaviors and high reproductive rates. Unfortunately for the native species that evolved in the absence of these predators and whose populations were impacted by them, this problem could have been anticipated since it is a fundamental prediction of predator–prey models and evolutionary ecology theory.

A famous example of such misguided attempts is the case of the repeated endeavors at controlling rodents in sugarcane fields in Jamaica during the nineteenth century. There, cane growers introduced ants (*Formica omnivora*), which did not reduce rat numbers but soon became a problem themselves. To remove rats and ants together, it was then decided to introduce toads (*Bufo marinus*). But toads still did not control rats, and they too became a pest themselves. Finally, small Indian mongooses were introduced to control rats and toads. Mongooses failed to control either and began preying on native birds, posing new threats to wildlife.

Another illustration of using biological methods in the absence of a careful theoretical framework is the use of pathogens to control invasive rabbits. The introduction of the myxoma virus into both Europe and Australia to reduce rabbit numbers, and the more recent introduction of rabbit haemorrhagic disease (RHD) into New Zealand for the same goal, were both designed without a proper ecological framework and performed neither by ecologists nor by conservation managers. These two cases point out how difficult it may be to fully control the intentional introduction of microorganisms by persons who are not fully aware of (and/or interested in) the potential ecological effects of such actions. Theoretical epidemiology was, however, sufficiently advanced to predict these unfortunate outcomes. Sadly, history now shows that the absence of a rigorous theoretical framework often renders these empirical attempts to apply simple ecological principles (e.g., predator–prey or host–parasite interactions) to real ecological problems prone to failure.

Unfortunately, early conservation managers, too, used trial and error to design island restoration projects, to the point that the whole subdiscipline of alien invasive species control has been long dominated by empirical development of techniques and methods at the expense of the use of ecological theory. For example, chemical advances have led to powerful, specific poisons, and ingenious trapping devices have outwitted even the most cunning invasive mammals. But these are technological developments that do not rely much on theory. With the notable exception of Virus-Vectored immuno-Contraception, in which empirical and theoretical studies were developed in parallel, most of the progress made in this field has neglected the potential benefits of theoretical ecology. This understandably might have been because there was an urgent need for a specific application. Yet this general trend has also led to many delays (and even some failures) in eradication programs.

Recent advances to address failing eradication programs have spawned adaptive control programs. This second generation of control programs capitalizes on theory by using ecological concepts from population dynamics and behavioral ecology. Such theory has been used in a number of ways. For example, it has advised that the timing or order of species eradications can be used maximize their success, that techniques such as the "Judas goat" technique can exploit herding behavior to locate the last individuals, or that helicopters and GIS can be used to deliver baits more accurately. More recently, it has allowed us to develop a better understanding of spatial ecology in the first stages of rodent invasions on islands, in order to better detect and control them at this crucial step.

Despite the improvements in the second-generation adaptive control programs, they still often lack an ecosystem perspective, leading to a general underappreciation of the importance of chain reactions following sudden alien invasive species eradications. As, understandably, nongovernmental organizations, conservation managers, and wildlife bodies have to react faster than the rate at which fundamental research can operate, some eradication plans have suffered from a lack of pre-eradications studies aimed at understanding the direct and indirect biotic interactions linking native and alien invasive species. As a result, some exotic species that were held in low densities by alien invasive species exploded once the pressure from the alien invasive species was removed following their eradication, leading to further damage to the native ecosystem. For example, invasive goats were recently removed from the threatened native forest on Sarigan Island, in the tropical western Pacific Ocean. Unfortunately, goats were selectively suppressing an exotic vine (*Operculina ventricosa*), which was highly competitive compared to other native plants and eventually exploded following the release from goat browsing.

Although a few empirical examples exist, the study of these "surprise" secondary effects of alien invasive species eradications has been largely theoretical. Take, for example, the mesopredator release effect. In theory, when a native species is a shared prey of introduced predators (e.g., cats and rats), the removal of the introduced top predator (cat) might result in a numerical increase of the introduced mesopredator (rat) population. This increase may be highly detrimental for the native species. In this case, these biotic interactions were studied theoretically through a mathematical model, and there have been few empirical studies of this "surprise" effect. This situation seems to occur only under certain ecological circumstances (e.g., absence of alternative food resources) and depends on which species are affected. Similarly, the competitor release effect suggests that when controlling for a higher competitor (e.g., a rat), the population of the lower competitor (e.g., a mouse) may erupt, even if these are controlled at the same time. This is simply because under certain conditions the direct effect of the removal is lower than the indirect, positive effect of the competitor's removal. In that case, the more severe the control, the more sudden the competitor release. This may explain why in several cases the eradication of rats was followed by a dramatic explosion of hitherto overlooked introduced mice.

Ideally, eradication programs should be based on a thorough knowledge of the interactions between the species involved, in particular regarding other introduced species. This principle has been applied to a long-term study that included removal invasive black rats from Surprise Island in the Entrecasteaux Reef, New Caledonia. Over a 4-year period, the island flora and fauna was characterized qualitatively and quantitatively to study the impact of rats on native species but also to reveal the presence of other introduced species: mice, ants, and plants. A thorough study of the trophic relationships within the invaded ecosystem, both empirical and theoretical (including mathematical modeling), led to the design of a case-specific eradication strategy that enabled removal of the rats without triggering a release of other invaders (e.g., simultaneous mouse removal was planned into the case-specific protocol). Then, this ecosystem has been the subject of a long-term post-eradication survey to follow both the recovery of the local communities and the potential emergence of surprise effects following the removal of rodents. This shows how theoretical ecology (here trophic ecology) may be helpful in imporving the efficiency of applied ecology programs.

Because it is not possible to conduct long pre-eradication studies before each eradication control, as was done for Surprise Island, the establishment of robust control programs rooted deeply in theoretical ecology will be necessary to achieve the full potential of this central area of applied ecology.

THEORETICAL ECOLOGY WITHOUT APPLIED ECOLOGY: THE ALLEE EFFECT

In contrast to the case of biological invasions, population dynamics, and in particular the domains developing around the Allee effect, is a discipline deeply rooted in theoretical ecology, but it still lacks the full transfer to applied ecology.

A demographic Allee effect describes a reduced per capita population growth rate in a population reduced in size (or density). Theory suggests that many, or even most, species could be susceptible to a demographic Allee effect, either directly, or indirectly through interaction with another species. As a consequence of this perceived ubiquity, it has become a cornerstone concept of ecology, with many potential applications to population management. Since its origins in the 1940s, the Allee effect has been the subject of roughly two types of studies: those aiming at a better understanding of the mechanisms underpinning it and its theoretical implications, and those trying to assess the presence of this process in given taxonomic groups. Probably because the concept is intellectually exciting, but also because empirical and experimental approaches are generally more difficult, the theoretical line of study has been the most active.

Another reason for the dominance of theory-based studies in this field is that empirical demonstrations of Allee effects are scarce and have mostly been restricted to mechanisms affecting individual fitness (component Allee effects) rather than manifestations in population dynamics (demographic Allee effects). There are sound biological reasons why demographic Allee effects might be the exception rather than the rule, although these are irrelevant to the point made here. What matters is that the majority of Allee effect studies, particularly over the last decade, have been theoretical studies. Consequently, there is now a thorough understanding of the Allee effect based on its conditions, mechanisms, and implications in a variety of contexts, including single populations and metapopulations, interspecific interactions, and spatially explicit frameworks. The importance of this process for different fields of applied ecology, including conservation biology, species management, and population exploitation, is also well understood. Yet studies on Allee effects conducted in the realm of applied ecology remain unfortunately scarce.

The fact is that managers and conservationists should be concerned about Allee effects, even in populations that are not apparently exhibiting them, because (i) they can create critical thresholds in size or density below which a population will crash to extinction, and (ii) even if there is no threshold, they increase extinction probability (due to stochasticity) and reduce per capita population growth rate in low-density or small populations. Whether extinction is a desirable or dreaded outcome, such knowledge may be critical for effective population management. For example, field studies of the spread of the wind-pollinated invasive cordgrass *Spartina alterniflora* in Willapa Bay, USA, reveal that it has been slowed by the existence of an Allee effect affecting plants in low-density areas, such as those that characterize the leading edge of the invasion, and this information has been central to control strategies.

Clearly, there is an important need for a whole line of work aimed at applying our theoretical understanding of Allee effects to solve real practical problems, particularly in those areas for which theory suggests it will have its greatest influence. One pressing situation that would deeply benefit from this transfer is the study of the conditions needed to maintain the population above the extinction threshold. This task will not be easy, especially since the extinction threshold is difficult to estimate and will vary according to both species and circumstance—habitat quality, mortality rates from predator or exploitation, and so on. But the benefits would likely be worth the effort. For example, it might be applied to invasion biology to enhance the efficiency of alien invasive species control programs. Theory states that targeting every individual of an introduced population might be unnecessary if the species exhibits a demographic Allee effect, with a substantial financial saving to be gained from leaving the last individuals to die out by themselves. Similarly, Allee effect theory can be used for the control of insect pests in agroscience. Several studies have investigated minimum propagule sizes for successful establishment of agents used in biological control programs.

Apart from management of introduced populations, Allee effect theory can also be used, somewhat obviously, in the management of threatened populations, affording managers an early warning system to avert their imminent extinction. For example, Allee effects interact synergistically with mortality from exploitation, which not only reduces the population size but can also create or increase an extinction threshold. Exploitation of populations with possible Allee effects needs to be considered and managed very carefully. Examples of fishery collapses (and lack of recovery) show how the neglect of Allee effect theory in the implementation of maximum sustainable yields and quotas resulted in disaster. In 2006, Courchamp and colleagues showed that exploitation itself can also act as a mechanism to create an Allee effect, particularly if the economic value of a species is found to be increasing with rarity, and the implications of this new process remain to be studied. In addition to trying to preserve small and decreasing populations, it might be better to augment their size or distribution via translocation or population reinforcement. Just as Allee effects affect alien invasive species eradication campaigns, they will also alter reintroduction strategies. For example, reintroductions of quokkas and black-footed wallabies to Western Australia were met with failure because the sizes of the populations released were too small to overcome the sustained predation pressure from introduced cats and foxes.

Even studies of the second type, those aiming at finding an Allee effect in specific taxonomic groups, have suffered from a failure to use theory. It is now common knowledge that Allee effects may occur in a wide number of species and that the presence of this process may dramatically alter the conditions and chances of population viability. Yet entire taxonomic groups of conservation interest, including, for example, bats, primates, and felids, remain to be investigated in this context. Whether for conservation purposes or in relation to an economic interest, eusocial insects should also be a focus. For example, bees are of commercial importance and ants are of major relevance as biological invaders, and both are major providers of crucial ecosystem services. Yet these and the other tens of thousands of eusocial species of ecological importance have so far not been the subject of Allee effect studies (Fig. 2). Perhaps the only example of such a subject for an Allee effect study is a colonial spider (*Anelosimus eximius*), whose lifetime reproductive success (a composite measure of offspring production and survival) increases with colony size.

In many contexts, such as the study of alien invasive species, applied ecology has advanced thorough empirical studies despite the existence of a wealth of relevant theory. In other contexts, such as the study of Allee effects, theoretical ecology has been at the forefront of our understanding. Now, the time is right to use our theoretical understanding of processes such as the Allee effect to benefit applied ecology.

FIGURE 2 Argentine ants (*Linepithema humile*), as an illustration of a species fitting the three examples detailed in this entry. They are one of the worst biological invaders in several parts of the world. In southern Europe, this species forms a supercolony from Spain to Italy and is likely to present an Allee effect. There, it seems unable to expand its range northward of its current area of invasion because of unsuitable climatic conditions at higher latitudes, but models predict that this is likely to change with global climate change. Photograph by Alex Wild.

THEORETICAL ECOLOGY FOR APPLIED ECOLOGY: CLIMATE CHANGE

A final example concerns an emerging area that has, up to now, mostly benefited from theoretical ecology, but which promises to be of major importance for future developments in applied ecology: climate change.

Over the last two centuries, anthropogenic activities have caused rapid climate change, with the mean global annual temperature expected to rise by 1–3.5 °C by 2050. Although climate change encompasses a wide range of phenomena like changes in precipitation regimes, increases in extreme event frequency, ocean acidification, and sea level rise, the vast majority of studies concern rises in temperature, which currently figures among the biggest threats to biodiversity. Species' responses to climate change have already been observed in many taxa. They include physiological changes, phenological changes (i.e., the timing of life cycle events such as plant flowering and fruiting or animal migration and reproduction), and geographic range shifts where species migrate to an area with a more favorable climate. If a species fails to adapt or to shift their ranges in response to climate change, then they may go extinct. This is the major prediction of many studies that have projected global biodiversity loss due to climate change. In one of the most widely cited studies, Chris Thomas and colleagues estimate that 15–37% of species are "committed" to extinction. It is clear that there is an urgent need to revise conservation strategies in the face of global climate change, and because these predictions are based on events yet to pass, we must turn to the theory of climate change science to serve as the basis of these revisions. Because it is based on projections, the field of climate change biology has taken off by mainly developing a foundation of mathematical models that aims to predict species' potential ranges and fates under different scenarios.

There are two main categories of predictive models commonly used in climate change science. The first category represents the most widely used bioclimatic envelope models. They relate a current species' range to multiple environmental variables (such as temperature, precipitation, and so on) and thereby define the climatic niche of that species (its potential spatial distribution based on these variables), making the assumption that it reflects its ideal climatic conditions. Under different climate scenarios, the species' predicted future geographical distribution can be identified. Species whose predicted distribution will be no longer be within their climatic envelope are predicted to go extinct.

The second category of models is based on a species' physiology. For example, dynamic global vegetation models estimate plant biodiversity loss following climate-induced changes in geochemical cycles and CO_2 concentrations because of their physiological constraints. Physiology-based models that can be applied to animals include degree-day models, which estimate the number of days per year that the temperature is above a critical minimum temperature—the temperature below which the species cannot survive. These models are particularly useful for ectotherms.

These models can, of course, be complemented to some extent by experimental studies (based on increased temperature) or comparison with past data (some effects of global warming have been observable for the last several decades), but for now the major source of ecological knowledge comes from modeling. Yet the major interest for ecology in this domain will probably be in applied ecology. For example, bioclimatic models can help to reevaluate the current set of protected areas. They can help to identify climatic refuges and heterogeneous microclimatic conditions that may be important to save a species from extinction. Furthermore, models can help to prioritize protected regions. For example, islands that are less susceptible to sea-level rise should be favored both for protection and as refuges for threatened species. A widespread view is that an important strategy is to enhance landscape connectivity to enable species to move through a matrix of interconnected habitats in order to escape from unsuitable climatic conditions. However, as haunts the previous debates on corridors, enhancing landscape connectivity can also be argued as counterproductive because populations in the lagging edge of their advancing range would compete with arriving populations and communities. On the other hand, species range shifts might be the unavoidable consequence of climate change because a species' existing range might not overlap with its predicted future range. As this situation could lead to extinction, climate change has been a major argument for the proponents of human-assisted colonization. In this increasingly hot debate, it is likely that the solution might come from a solid blend of robust theoretical ecology and educated case-by-case decisions based on the known ecological characteristics of the concerned species and the socioeconomic context in which conservation is taking place. Surely theory has a major role to play here, too.

Bioclimatic models can also be used to explore the impact of different drivers of global change on biodiversity. For instance, it is very important to get a better understanding of how alien invasive species will react to climate change and whether they might benefit from shifting climatic envelopes to invade new areas previously outside their distribution. Models can help to identify species posing the greatest risk for invasion of a particular region under a scenario of climate change (Fig. 3).

Moreover, climatic models can also serve to evaluate the risks of disease spread following a rise in temperature, for example. Many disease vectors are very sensitive to climatic conditions, and there is a general fear that tropical diseases might extend to higher latitudes, leading to a dramatic increase of infectious diseases, especially when local hosts are naïve to the new parasites. This concerns plant, wild animal, domestic animal, and human populations.

These theoretical studies can help to estimate risks to biodiversity (invasions, diseases, species losses) and to prioritize protected areas. However, as all these projections of future biodiversity are based on models, they include the uncertainties inherent in their input variables, including projected climate change and the type of species' responses. They can thus be validated only indirectly, by extrapolating from observed changes over the last decades and through independent data sets of current species' occurrences. Even then, these models do not account for possible behavioral, phenological, or physiological adaptations and do not adopt an ecosystem perspective (communities will react differently than the sum of species

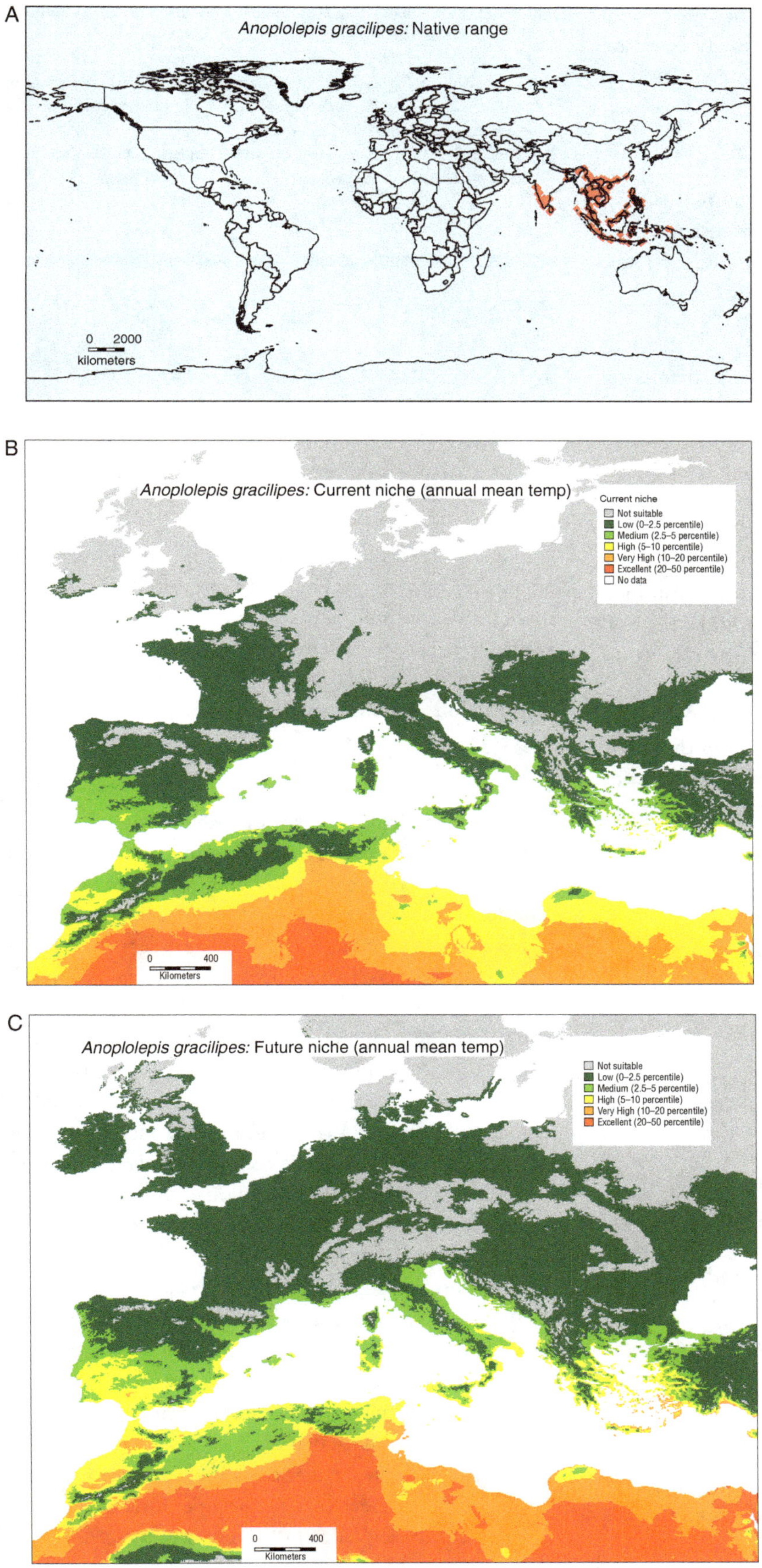

FIGURE 3 Bioclimatic envelope models for the yellow crazy ant (*Anoplolepis gracilipes*) that relate the species' native distribution (A) to climatic variables (e.g., temperature) in order to determine the species' current climatic range that can then be projected on a climate map to determine its potential range based on climatic conditions in 2010 (B) and under a scenario of climate change, for example in 2050 (C).

reactions). Paradoxically, one may claim that the current theoretical approaches should be more closely based on theoretical ecology. Indeed, there is an urgent need to improve model predictions because many future conservation decisions will have to be made rapidly with the help of climate change model predictions.

Climate change science is, by its very nature, a predictive science, using today's concepts and models in developing a robust theoretical understanding of future climate change to advance the applied ecology of tomorrow.

CONCLUSION: WHAT NOW?

This entry deals with the antithesis of theoretical ecology—applied ecology. A vast number of disciplines fall within its realm, and applied ecology provides theoretical ecology with a *raison d'être.* Applied ecology can (and should) benefit more from theoretical ecology, either by investigating new fields of applied ecology or by working to apply new concepts of theoretical ecology. This means not only that theoretical ecologists can look forward to a whole new set of questions, approaches, and tools to study, but also that they will have new subjects to explore and new colleagues to interact with.

The relationships between theoretical and applied ecology are better defined by a feedback loop than by a simple, unidirectional supply link. Enhancing the connections between these two complementary aspects of ecology will not only help solve pressing environmental issues but will also further stimulate theoretical ecology.

SEE ALSO THE FOLLOWING ARTICLES

Allee Effects / Conservation Biology / Demography / Ecosystem Ecology / Fisheries Ecology / Invasion Biology / Restoration Ecology

FURTHER READING

Bellard, C., C. Bertelsmeier, P. Leadley, W. Thuiller, and F. Courchamp. 2012. Impacts of climate change on the future of biodiversity. *Ecology Letters* 15(33). doi: 10.1111/j.1461-0248.2011.01736.x.

Cadotte, M. W., S. M. McMahon, and T. Fukami, eds. 2006. *Conceptual ecology and invasion biology.* Berlin: Springer.

Caut, S., E. Angulo, and F. Courchamp. 2009. Avoiding surprise effects on Surprise Island: alien species control in a multi-trophic level perspective. *Biological Invasions* 11(7): 1689–1703.

Courchamp, F., E. Angulo, P. Rivalan, R. Hall, L. Signoret, L. Bull, and Y. Meinard. 2006. Value of rarity and species extinction: the anthropogenic Allee effect. *PLoS Biology* 4(12): e415.

Courchamp, F., J. Gascogne, and L. Berek. 2008. *Allee effects in ecology and conservation.* Oxford: Oxford University Press.

Lovejoy, T., and L. Hannah. 2005. *Climate change and biodiversity.* New Haven: Yale University Press.

Newman, E. I. 2001. *Applied ecology & environmental management.* London: Blackwell Science.

Shigesada, N., and K. Kawasaki. 1997. *Biological invasions: theory and practice.* Oxford Series in Ecology and Evolution. Oxford: Oxford University Press.

Thomas, C. D., A. Cameron, R. E. Green, M. Bakkenes, L. J. Beaumont, Y. C. Collingham, B. F. N. Erasmus, M. Ferreira de Siqueira, A. Grainger, L. Hannah, L. Hughes, B. Huntley, A. S. Van Jaarsveld, G. E. Midgely, L. Miles, M. A. Ortega-Huerta, A. T. Peterson, O. L. Phillips, and S. E. Williams. 2004. Extinction risk from climate change. *Nature* 427: 145–148.

Williamson M. 1997. *Biological invasions.* Population and Community Biology Series. Berlin: Springer.

ASSEMBLY PROCESSES

JAMES A. DRAKE AND PAUL STAELENS

University of Tennessee, Knoxville

DANIEL WIECZYNSKI

Yale University, New Haven, Connecticut

Ecological systems are dynamically nonlinear, self-organizing, spatially extended, and multistable, driven by species colonization and extinction events, all operating against a backdrop of environmental and spatial variation, disturbance, noise, and chance. System assembly or construction is fundamentally important to understanding ecological systems.

ECOLOGICAL REALITIES

The nature of ecological systems creates a number of epistemological difficulties for the observer. When historical events have shaped the current system state, it becomes difficult for the observer to assign cause without knowledge of such events. For example, variation in the order of arrival and timing of species colonization has been shown to generate alternative community states. In fact, Drake, in 1990, found that assembling communities not only result in irrevocable differences in structure, but if perchance the developing trajectories cross (e.g., identical food webs), at some time they respond differently to a common invader. Such behavior is a consequence of both where each system came from and where they are going. This historical disconnect creates a very real danger when fundamental principles or cause are assigned based on our necessarily brief analyses of presently operating mechanisms, processes, and cycles. It is also unclear whether any given system is riding along a transient heading elsewhere or has reached some solution. This limits the power of experimentation to temporal slices of assembly time. How can the observer know whether the mechanisms and processes experimentally documented in a system are causal or are simply those permitted at a given time along some trajectory of development? Understanding these permissions is key to understanding biological nature. But how to proceed?

Manipulating system development from its inception to some end state is an approach to understanding ecological systems termed assembly. Assembly experimentation and theory seek to describe the creation of novel attractors that accompany colonization and extinction. This approach builds dynamics and explores the structural consequences of traversing alternative dynamical realms. For example, simply by comparing systems created by permuting the order of species arrival, the multistable character of ecological nature has been exposed. By analogy, consider a simple jigsaw puzzle. Here, the arbitrary placement of any given piece at any time does not alter the final outcome. There is but a single solution. However, should the puzzle pieces exhibit dynamics of their own, as is the case with species populations, multiple solutions arise and a finite set of parts produces multiple images.

Assembly processes are the events and subsequent dynamical response that arise during the construction, succession, and self-organization of a system. These events initiate and drive the trajectory of a system by redefining the systems attractor, its basin, and the behavior of the system along alternative transients. Therefore, assembly processes generate a map of species colonization, interactions, extinction, and the development of a topology or network (e.g., food web). It is important to note that this transient stage of development can be very long lived. An ecological attractor might be of a simple sort, the system arriving in uneventful fashion to some stationary state (Fig. 1). Just as likely, however, an attractor may itself contain complex

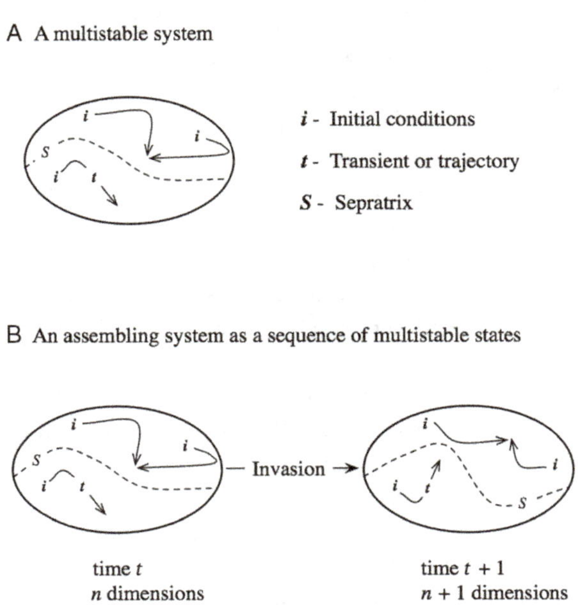

FIGURE 1 System trajectories in a multistable landscape (A), and a higher-order trajectory (B) created by increasing system dimensionality.

dynamics being composed of a series of dynamically varies states. Here, for example, the system proceeds through a period of regular dynamics, followed by period doubling and chaos, and residency on a strange attractor.

The assembly approach exploits the fact that complex systems exhibit sensitive dependence to initial conditions. Here, arbitrarily small changes in initial population sizes can create novel trajectories to a systems solution. Sensitive dependence creates a situation where a given system, variously initialized, experiences regions of state space unique to particular sets of initial conditions. Consequently, system properties and the action of mechanisms may be fundamentally different as a function of history. Further, ecological systems are open in the sense that they can be colonized. System dimensionality and degrees of freedom are not fixed but vary with time and over space. For example, a species might successfully colonize a system, but fail to invade that system if the system developed from a slightly different set of initial conditions. Because species composition varies over the course of community assembly, ecological systems experience a series of initial conditions and subsequent restructuring of dynamical character as a function of dimensionality changes.

PATTERN AND ANTECEDENT CAUSE

The idea that historical influences play a strong if not defining role in the determining species composition of a community is hardly new. Henry Cowles, for example, wrote in 1901:

> Antecedent and subsequent vegetation work together toward the common end. Where there is no antecedent vegetation, *Ammophila* and other herbs appear first, and then a dense shrub growth. . . .

Cowles had surely observed sequence and priority effects, both hallmarks of community assembly. Even earlier, Dureau de la Malle (1825) visualized alternative themes in forest regeneration and suggested that such themes could reflect fundamental natural laws. Despite these early conceptual advances, the temporal scale of community assembly and succession has hindered direct experimentation in the real world.

Inferential approaches to revealing historical effects are based on the assumption that recurrent patterns are a consequence of dynamical commonalities among developing or assembling communities. Both Elton (in 1946) and Hutchinson (in 1959) sought explanations for observed patterns in the action of mechanisms like competition occurring in the past. Allied to these ideas, as community development proceeds, rules are invoked perhaps tuning invasion success around a particular species-to-genus ratio or patterning gap sizes, are exceedingly difficult to document in nature. This is as it should be because system assembly is occurring in an environment subject to disturbances of various magnitudes and at multiple scales. Should disturbance ameliorate competitive interactions, system development is reset or comes under the control of a new attractor, and vulnerability to invasion is modified. Of course, it is also possible that system dynamics and the species that generate those dynamics have variously incorporated disturbance as a resource.

In 1975, Jared Diamond took this approach further by suggesting that presently observed patterns among a set of communities could be used to infer dynamical rules that govern system development. Such rules are ostensibly a higher-level summary of the mathematical intricacies that underly species interactions and topological development in space and time. For example, Diamond observed that particular species pairs never coexist, a phenomenon seductively titled forbidden combinations. When species in some forbidden combination exhibit priority effects, where the first species to colonize wins, an assembly rule emerges. While Diamond's arguments were met with fervent resistance on statistical grounds, simple null models were capable of generating similar patterns; his discourse stimulated a vigorous and ongoing experimental and theoretical effort.

A more direct approach was taken by Piechnik, Lawler, and Martinez in 2008. They reexamined Simberloff and Wilson's arthropod recolonization data and found that trophic breadth was reflected in species colonization order during community assembly. Early invaders functioned as generalists, while successful colonists late during community development were specialists. This result could fit nicely with Petermann and colleagues' 2010 study implicating pathogens as potential drivers of community assembly, should pathogen load increase with time among these arthropods. At an evolutionary scale, Fukami and colleagues were able to show in 2007 that alternative assembly routes were capable of enhancing and retarding the rate of evolutionary change among community members. Not only does assembly generate local and regional species diversity by creating alternative states, but it also offers nature a variety of scenarios against which evolution proceeds in varied fashion.

Interestingly, recent numerical studies of networks of identical coupled oscillators have shown that pattern in the form of subsets of rogue oscillators (chimera states) are capable of breaking synchrony during system evolution. In systems such as these, additional pattern can emerge if coupled and uncoupled sets respond differently to events like species invasion attempts. It would seem that in light of the extraordinary behaviors recently observed in nonlinear dynamical systems, our null model approach must be recast.

ASSEMBLY OF COMPLEX DYNAMICAL SPACES

Until recently, modeling in ecology focused on the simplest of systems where asymptotic behavior and equilibrium conditions can be precisely evaluated, the assumption being that little of consequence occurs during the transient phase, the period of time between the initiation of dynamics and the systems arrival at equilibrium. This might be true if nature contained one or two species and met the assumptions needed for analytical modeling. Nevertheless, analytical solutions simply do not exist for systems of containing many species, and there is no reason to believe that the behaviors observed here are extensible to systems of higher dimension. In fact, the dynamics seen in many species systems suggest little to no role for a purely analytical approach. At present, numerical simulation or approaches based on graphs and network theory would appear to be our best tools.

The basic process that drives the assembly of any system is coupling or invasion and decoupling or extinction. While it is easy enough to add a row and column to an array, representing an invader colonizing a community, spatial variation in coupling as the invader colonizes is exceedingly difficult to model, yet it is likely an essential feature of assembly.

Clearly, coupling and decoupling have significant consequences for subsequent system development, and herein reside the keys to understanding the genesis of emergent properties and structure. Unfortunately, very little is known about the earliest events in system coupling and decoupling, despite the fact that much of nature spends its existence in this far-from-equilibrium, readily disturbed situation. Immediately following a coupling or decoupling event, a new dynamical system is created that enters into a transient phase, as it begins moving toward its new solution or attractor. However, the dynamical control that previously operated had placed the system somewhere in a particular state space creating unique initial conditions, possibly outside the trajectory field of the new basin. Hence, some period of time exists before the new dynamics fully assume control and pattern exists that reflect those dynamics. During this period there is incomplete coordination or asynchrony among the elements that comprise the new system. Additional coupling to this system during this stage could very well be successful, but impossible as synchronization proceeds, or vice versa. Colonization success and failure depend not only on species composition and community organization of the community but also on where that community has come from and where it is going.

Consider, for example, a system where a predator population is mediating competitive interactions among its prey, thereby thwarting competitive exclusion and maintaining higher species diversity. Should coupling with another system occur, perhaps through some novel metapopulation or metacommunity interaction or direct invasion, the role of the predator may change. However, the initial distribution and abundance of prey species still reflect previous controls, but now under the control of new dynamical rules. Because nature is spatially extended, expression of this control varies due to pattern previously created, further altering pattern.

Recent numerical work exploring the long-term dynamics of nonlinear systems has produced a tantalizing array of behaviors that appear to have biological analogues. Such studies offer potential explanations for the intricacies of community assembly. For example, attractor ghosts have been discovered that function ostensibly as a switch by delaying system collapse, thus enhancing the possibility of a species invasion and rescuing a system from collapse. Should this occur, a novel state space with new assembly rules emerges, a state potentially derivable only through this route. Similarly, the collective behavior exhibited by coupled oscillators (e.g., metapopulations, metacommunities, population and community patches along some gradient), moving from an unsynchronized, uncorrelated state to a synchronized state, creates a sequence of radically different phenotypes. Here, the reference to phenotypes is based on variation in the response of potential colonists to various stages of synchrony in the system being invaded.

Despite considerable progress, many questions remain unanswered, and needed concepts have yet to be developed. Where is nature going and how might it get there? Can assembly be thought of and cast in terms of self-organization? How do the relative roles of determinism and indeterminism in dynamics drive system assembly? Do dynamical phenomena like basin riddling and ghosting play a role in the developmental trajectories of developing ecological systems? Do assembly processes play a fundamental role in influencing global patterns in species diversity? Finally, could many of the observed regularities seen in nature be a direct function of assembly? For example, the SLOSS (single large or several small) debate in ecology, based on species–area relationships and cast as $S = cA^z$, may have less to do with area than the effect multiple initializations and attendant initial conditions in generating species diversity.

SEE ALSO THE FOLLOWING ARTICLES

Food Webs / Metacommunities / Networks, Ecological / Stability Analysis / Succession

FURTHER READING

Diamond, J. M. 1975. Assembly of species communities. In M. L. Cody and J. M. Diamond, eds. *Ecology and evolution of communities.* Cambridge, MA: Belknap Press.

Dureau de la Malle, A. 1825. Mémoire sur l'alternance ou sur ce problème: la succession alternative dans la reproduction des espèces végétales vivant en société, est-elle une loi générale de la nature. *Annales des sciences naturelles* 15: 353–381.

Elton, C. 1946. Competition and the structure of ecological communities. *Journal of Animal Ecolology* 15: 54–68.

Fukami, T., H. J. E. Beaumont, Xue-Xian Zhang, and P. B. Rainey. 2007. Immigration history controls diversification in experimental adaptive radiation. *Nature* 466: 436–439.

Howles, H. C. 1901. The physiographic ecology of Chicago and vicinity; a study of the origin, development, and classification of plant societies (concluded). *Botanical Gazette* 31: 145–182.

Hutchinson, G. E. 1959. Homage to Santa Rosalia, or why are there so many kinds of animals? *American Naturalist* 93: 145–159.

Petermann, J. S., A. J. F. Fergus, C. Roscher, L. A. Turnbull, A. Weigelt, and B. Schmid. 2010. Biology, chance, or history? The predictable reassembly of temperate grassland communities. *Ecology* 91: 408–421.

Piechnik, D. A., S. P. Lawler, and N. D. Martinez. 2008. Food-web assembly during a classic biogeographic study: species' "trophic breadth" corresponds to colonization order. *Oikos* 117: 665–674.

Simberloff, D. S., and E. O. Wilson. 1969. Experimental zoogeography of islands: the colonization of empty islands. *Ecology* 50: 278–296.

Simberloff, D. S., and E. O. Wilson. 1970. Experimental zoogeography of islands: a two-year record of colonization. *Ecology* 51: 934–937.

B

BAYESIAN STATISTICS

KIONA OGLE AND JARRETT J. BARBER
Arizona State University, Tempe

Bayesian statistics involves the specification of a joint probability model to describe the dependence among observable quantities (e.g., observed, or unobserved but predictable, data) and unobservable quantities (e.g., ecological model parameters, treatment effects, variance components). Inference about unobserved quantities is based on the posterior distribution (or posterior), which is the conditional probability distribution of the unobserved quantities given, or posterior to observing, the observed quantities. In principle, once a joint probability model is specified, the posterior follows via Bayes theorem, a well-known result from probability theory. In practice, except for relatively simple models, numerical methods must be used to approximate the posterior. Methods consist largely of algorithms to sample the unknown quantities from the posterior, hence effectively obtaining a histogram, whose approximation to the posterior improves as this sample size increases; inference is often reduced to summarizing this histogram with means, medians, and credible intervals. The majority of these numerical methods are broadly referred to as Markov chain Monte Carlo (MCMC) methods. MCMC has become so popular in the context of Bayesian statistics that it is often mistaken as synonymous with Bayesian, but many Bayesian statistical problems may be seen simply as an area of application of MCMC methodology with an objective to sample from a distribution of interest.

THE FUNDAMENTALS

Subjective Degrees of Belief, Probabilities, and Coherency

Aside from the well-established mathematical definition of probability, how do we interpret probability? We adopt the interpretation of probability as subjective or as a personal degree of belief about the occurrence of events. Among several other interpretations, the relative frequency interpretation is undoubtedly the most influential in statistics, wherein the probability of an event is viewed as the relative frequency of occurrence of the event in a large number of (hypothetical) repeated experiments. This view forms the basis of much of statistics as practiced over the past 80 years, including the hypothesis testing framework(s) of R. A. Fisher, J. Neyman, and E. S. Pearson. The subjective view allows us to specify probabilities for any events whatsoever, not just for those events for which the relative frequency interpretation is applicable (e.g., the event that it will rain tomorrow somewhere in the Great Plains of North America); yet it allows us to call upon frequentist, or other, notions of events to help us specify *our* probabilities.

Why adopt a Bayesian approach? Aside from more pragmatic considerations, a Bayesian approach (or a procedure consistent with a Bayesian approach) is essentially the only statistical approach to statistical inference that is coherent. Coherency refers to a calculus of (i.e., method for computing) degrees of belief that is self-consistent and noncontradictory. Coherency is often discussed in terms of gambling odds and illustrated by showing that a lack of coherency can result in a wager that ensures certain loss (or gain, depending on perspective; a so-called Dutch book). Alternatively, foundational developments of degrees of belief show that lack of coherency violates a set of principles that many people find to be self-evident. And, if our degrees of belief and the manipulation thereof obey the coherency principle, then they coincide essentially with the modern mathematical definition of probability. Thus, in practice, we may simply use the term *probability*, and all of the properties that follow from the modern definition. In this entry, the term *quantification of uncertainty* is used frequently in place of *subjective probability* or *degree of belief*. For those who do not find coherency compelling, a Bayesian approach remains a way to implement the likelihood principle, which many people do find compelling. Finally, Bayes theorem provides a coherent mechanism for updating our prior probabilities with information in observed data.

Basic Probability Rules and Bayes Theorem

Bayesian statistics is based on a simple, uncontested probability result: Bayes theorem. To derive Bayes theorem, consider two random variables X and Y (i.e., quantities that are uncertain and whose potential values are described by a probability distribution), which, in the Bayesian context, may refer to data or to parameters. We are interested in the joint probability that X assumes value x and Y assumes value y, which we express as $P(X = x, Y = y)$, or, for convenience, $P(x, y)$. (Throughout, we use P to denote probability or probability density.) Given $P(x, y)$, then, in principle, we have its marginal distributions, $P(x)$ and $P(y)$, and its conditional distributions, $P(x \mid y)$ and $P(y \mid x)$. The marginal distribution of X, for example, describes the distribution of potential values after having summed or integrated over all possible values of Y; we might use the marginal distribution of X when we do not know about the other variable, Y. We would use the conditional distributions when we know the values of the conditioning variables, as in the distribution of x given y, $P(x \mid y)$, with y being the conditioning variable such that the conditional distribution of X depends on the particular value of $Y = y$. Using basic results, the joint probability distribution of two quantities can be written as the product of the conditional probability of one quantity given the other and the marginal probability of the other: $P(x, y) = P(x \mid y) P(y) = P(y \mid x)P(y)$.

The operational appeal of this result is that the statistical modeler may not know how to specify, or model, the joint distribution directly, but may know how to model a conditional distribution and a marginal. For example, we may be satisfied with a regression model of the distribution of tree height, Y, given tree diameter, X, and we may have a model for tree diameter. In this case, the above result ensures that our inferences are based on a valid joint probability distribution given by $P(y \mid x)P(x)$. The order of conditioning is left to the modeler, and it is important to realize that the joint model built from a specification of $P(y \mid x)$ and $P(x)$ generally is not the same as the joint model built from a specification of $P(x \mid y)$ and $P(y)$; this should not be confused with having a joint model, $P(x, y)$, for which both conditionals and both marginals are determined, in principle.

With a joint distribution specified, Bayes theorem offers a mechanism by which to implement the principle of conditional inference, i.e., inferring unknowns given knowns. Assuming we wish to infer about events involving X, given what we know, i.e., conditional on, $Y = y$, the object of our inference is the conditional distribution, $P(x \mid y)$. We simply rearrange $P(x \mid y)P(y) = P(y \mid x)P(x)$ to get Bayes theorem,

$$P(x \mid y) = \frac{P(y \mid x)P(x)}{P(y)}. \tag{1}$$

Typically, $P(y \mid x)$ is a model for observed data y given unobserved parameters x, and $P(x)$ is a (prior) distribution, together, by the previous result, giving a full (joint) probability model for all quantities in the numerator of Equation 1. A simple application of Bayes theorem is illustrated in Box 1.

BOX 1. USING BAYES THEOREM WITH DISCRETE RANDOM VARIABLES

Consider the following example that illustrates the application of Bayes theorem to compute a posterior probability in a discrete setting. Assume you are hiking through a forest that has been affected by bark beetle infestations. For simplicity, we will define two discrete random variables: let B denote the beetle status of a tree in the forest ($B = 0$ if no evidence of beetle attack, $B = 1$ if evidence of beetle attack) and D the state of the tree ($D = 0$ if tree is living, $D = 1$ if tree is dead). You notice a dead ($D = 1$) tree in the distance, and you ask: "what is the probability that the tree was attacked by beetles?" That is, you wish to compute, for this tree, the following conditional probability: $P(B = 1 \mid D = 1)$. There is a deep ravine that prevents you from hiking to the tree to directly evaluate its beetle status. But, coincidently, you have with you a publication by a local researcher that reports the results of a census whereby he inventoried the beetle status and state of

BOX 1 *(continued)*.

thousands of trees in several nearby locations, giving the following probabilities:

$P(B = 1) = 0.3$ (i.e., 30% of the trees were attacked by beetles)

$P(D = 0 | B = 0) = 0.8$ (i.e., of the trees that were not attacked, 80% were living)

$P(D = 0 | B = 1) = 0.1$ (i.e., of the trees that were attacked, 10% were living)

You use this information to compute the probability of interest, $P(B = 1 | D = 1)$. Using Bayes theorem, you have

$$P(B = 1 | D = 1) = \frac{P(D = 1 | B = 1) \cdot P(B = 1)}{P(D = 1)} \quad (1.1)$$

From the inventory publication, you know $P(B = 1)$ and you compute $P(D = 1 | B = 1) = 0.9$ since $P(D = 1 | B = 1) + P(D = 0 | B = 1) = 1$, and you are given $P(D = 0 | B = 1)$. Thus, the numerator in Equation 1.1 is equal to $0.9 \times 0.3 = 027$. You are not given $P (D = 1)$, but using basic probability rules, you compute it as

$$\begin{aligned} P(D=1) &= \sum_{b=0}^{1} P(D = 1, B = b) = \sum_{b=0}^{1} P(D = 1 | B = b) \cdot P(B = b) \\ &= P(D = 1 | B = 0) \cdot P(B = 0) \\ &\quad + P(D = 1 | B = 1) \cdot P(B = 1) \quad (1.2) \\ &= (1 - 0.8) \times (1 - 0.3) + 0.9 \times 0.3 \\ &= 0.2 \times 0.7 + 0.9 \times 0.3 = 0.41 \end{aligned}$$

Thus, the probability of interest is $P(B = 1 | D = 1) = 0.27/0.41 \cong 0.659$. And the probability that the tree was not attacked by beetles is $P(B = 0 | D = 1) \cong 1 - 0.659 = 0.341$.

Here, we can think of $P(B = 1)$ as the prior probability that a tree was attacked by beetles, which is simply based on the inventory data and the overall proportion of trees that were attacked; $P(D = 1 | B = 1)$ as the likelihood of a tree being dead if it was attacked by beetles; and $P(B = 1 | D = 1)$ as the updated or posterior probability that the tree was attacked by beetles given that it is dead.

In the Bayesian statistical context, we denote observed quantities by *Data*, unobserved quantities by θ. In Equation 1, θ plays the role of x, *Data* the role of y. In principle, θ includes all unobserved quantities about which we wish to make inference. This may include unobservable parameters as in, say, a regression model, or unobserved but potentially observable data values that we may wish to infer (e.g., missing or future observations). Thus, Bayes theorem is used to obtain the conditional probability distribution of θ given observations *Data*:

$$P(\theta | Data) = \frac{P(Data | \theta)\, P(\theta)}{P(Data)}. \quad (2)$$

Here, $P(\theta | Data)$ is the posterior (probability) distribution of θ given *Data*, which may be a density function, a mass function, or a combination of both, depending on the nature of the components of θ. (Often, as in a regression setting, covariates are assumed known and are notationally suppressed.) Because inference proceeds conditional on *Data*, and because the likelihood function associated with $P(Data | \theta)$ also is viewed as a function of θ given *Data*, $P(Data | \theta)$ is often referred to as the likelihood. This may cause confusion among those who are familiar with the notion of a likelihood function: if we call $P(Data | \theta)$ the likelihood, this suggests conditioning on *Data*. This perspective, however, is inconsistent with the interpretation of the numerator of Equation 2 as a joint distribution of ($Data$, θ) being built from the conditional distribution of *Data* given θ. For this reason, we prefer to call $P(Data | \theta)$ the data model. Here, $P(\theta)$ is the prior (probability) distribution for θ, which quantifies our prior understanding or uncertainty about θ before observing *Data*, and $P(Data)$ is the prior predictive distribution of *Data* because this is the distribution we would use to predict *Data* before it is observed; we cannot use $P(Data | \theta)$ directly, because θ is unknown. Uncertainty about θ is incorporated into the prior predictive by marginalizing over θ with respect to $P(\theta)$, giving $P(Data) = \int P(Data | \theta) P(\theta) d\theta$ if θ is a continuous random variable.

If we want to predict unobserved data, denoted *Data′*, we use Bayes theorem in Equation 2 to obtain the posterior of all unobserved quantities, now ($Data'$, θ), conditional on observed *Data*. That is, $P(Data', \theta | Data) \propto P(Data', Data | \theta) P(\theta)$. Note that we follow the common practice of denoting unobserved data separately from other unknowns in θ, though this is done merely for convenience of interpretation, and all unknowns may be denoted by θ. With this latter note, we see the *posterior predictive* distribution simply as the marginal posterior distribution, $P(Data' | Data) = \int P(Data', \theta | Data) d\theta$. If our data model is constructed hierarchically as $P(Data', Data | \theta) = P(Data' | Data, \theta) P(Data | \theta)$, then we may write $P(Data' | Data) = \int P(Data' | Data, \theta) P(\theta | Data) d\theta$, which simplifies to $\int P(Data' | \theta) P(\theta | Data) d\theta$ if *Data′* and *Data* are independent.

Sequential processing of data is, in principle, straightforward with Bayes theorem and is one of the strengths of this approach. Assume we observe *Data* followed by *Data′* at a later time. We may obtain the posterior $P(\theta | Data)$, then, when *Data′* is available, simply use

$P(\theta \mid Data)$ as the prior for θ in $P(Data' \mid Data, \theta)$, updating the posterior to $P(\theta \mid Data, Data')$. That is,

$$P(\theta \mid Data', Data) = \frac{P(Data' \mid Data, \theta) P(\theta \mid Data)}{P(Data' \mid Data, \theta)}. \quad (3)$$

We recognize the denominator as the posterior predictive distribution for *Data'*, given the first data set, *Data*. The above procedure can be repeated indefinitely as data arrive. Or, we may choose to obtain the posterior at once when both data sets are available; the result is the same posterior, $P(\theta \mid Data, Data')$, which can be shown by applying Bayes theorem again with a bit of algebra.

Except for relatively simple applications, the integral in the denominators in Equation 2 or 3 typically is not available analytically, and, hence, an analytical solution for the posterior is not available. In practice, this generally is not an issue, because once we have observed *Data* and defined $P(Data \mid \theta)$ and $P(\theta)$, $P(Data)$ is a normalizing constant (or constant of proportionality) with respect to the unknown quantity, θ, and many algorithms are available to sample from distributions known up to a constant, thereby allowing us to approximate the actual distribution (see the section on numerical methods below). Thus, the joint distribution defined by the numerator of Equation 2 is often the main focus of modeling efforts. For this reason, the posterior is often expressed as

$$P(\theta \mid Data) \propto P(Data \mid \theta)\, P(\theta). \quad (4)$$

We also refer to the right-hand side of Equation 4 as the unnormalized posterior since division by $P(Data)$ is required for $P(\theta \mid Data)$ to integrate (θ continuous) or sum (θ discrete) to 1. Equation 4 is the full probability model for observable and unobservable quantities and is the essence of a Bayesian statistical model.

GRAPHICAL MODELS

Graphical models, especially directed acyclic graphs (DAGs), are useful for depicting and aiding the construction of probability models. In particular, DAGs are useful for building complicated, full probability models from relatively simple, conditionally independent components (Box 2). Indeed, graphical models play an important role in the development of the popular software programs WinBUGS and OpenBUGS. We introduce only enough material to write an expression for a full probability model in terms of conditionally independent model components and to aid in the construction of full conditional distributions, to which we return later.

BOX 2. GRAPHICAL MODELS TO FACILITATE CONSTRUCTION OF BAYESIAN MODELS

We can use a directed acyclic graph (DAG) to describe the conditional dependencies between model quantities. We may find it easier or more appropriate to express *B* as conditional on *A* (or depends upon *A*) as in DAG (i). Conversely, to express *A* as conditional on *B*, we would draw DAG (ii). In (i), we may refer to *A* as parameters (θ) and *B* as data (*Data*) in our model, and thus DAG (i) becomes (iii) in our typical notation. The DAG defines the joint distribution for the quantities of interest (e.g., *A* and *B*, or θ and *Data*) as the product of conditional and marginal probability distributions. In a Bayesian data analysis, we are interested in the posterior distribution of some quantities (e.g., θ) given other quantities (e.g., *Data*), and this conditional distribution, $P(\theta|Data)$, is proportional to the joint distribution described by the corresponding DAG in (iii).

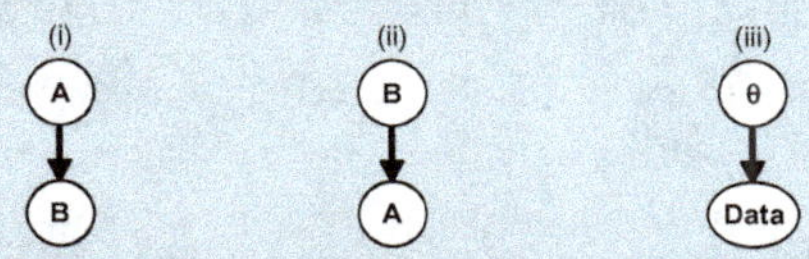

(i)	(ii)	(iii)
$P(A,B) = P(B \mid A) \cdot P(A)$	$P(A,B) = P(A \mid B) \cdot P(B)$	$P(\theta, Data) = P(Data \mid \theta) \cdot P(\theta)$
$P(A \mid B) \propto P(A,B)$	$P(B \mid A) \propto P(A,B)$	$P(\theta \mid Data) \propto P(Data \mid \theta) \cdot P(\theta)$
$P(A \mid B) \propto P(B \mid A) \cdot P(A)$	$P(B \mid A) \propto P(A \mid B) \cdot P(B)$	

The DAG consists of circular or elliptical nodes that represent different stochastic quantities in the model that are described by probability distributions (see "Graphical Models" section). Square or rectangular nodes represent fixed quantities (e.g., covariate data that we may assume to be measured without error). The nodes may be referred to as child or parent nodes (see the "Graphical Models" section) and as root, internal, or terminal nodes. A root node is a node that does not have parents, e.g., *A* in (i) and θ in (iii); a terminal node is a node that does not have any children, e.g., *A* in (ii) and *Data* in (iii); and an internal node is a node that gives rise to child nodes and that has its own parent nodes. The nodes are connected by unidirectional edges (or arrows) that indicate the conditional dependency between two nodes *A*.

A graph consists of a set of nodes, typically depicted by circles and representing stochastic quantities, including data (before they are observed) and parameters; relationships between nodes are specified with a set of directed edges or arrows. A DAG is directed because edges have single tips and is acyclic because following the arrows never allows a node to be revisited. Sometimes, squares are used to denote fixed quantities that do not receive a stochastic specification (see Box 3), like covariates in a regression, but such fixed quantities are typically not relevant to the qualitative probabilistic properties represented by the DAG and are often notationally suppressed. Also, some authors use squares to denote observed values of stochastic quantities, but circles are used here.

Let v denote a node in the graph, and let V denote the set of all nodes. Referring to our previous notation, V includes θ and *Data* (before it is observed; Box 2). Define a parent of v as any node having an arrow emanating from it and pointing to v. Denote the set of parents of v as $parents[v]$, and let V_{-v} denote all nodes except v. Now, an otherwise complicated full (joint) probability model of all stochastic quantities can be written as a product of relatively simple, conditionally independent components,

$$P(V) = \prod_{v \in V} P(v \mid parents[v]). \tag{5}$$

BOX 3. DAGS FOR THE TREE MORTALITY EXAMPLE

The Bayesian model for the beta-binomial tree mortality example can be expressed as a DAG in (i). In this example, the DAG is associated with two levels: level 1 is the stochastic data level for the X's, and level 2 is the parameter level (representing the prior for θ). In (i), each observation is explicitly indicated such that separate nodes are shown for the number of trees that survived (X_j) and the total number of trees (N_j) in each plot j, $j = 1, 2, \ldots, M$. As the model becomes more complicated, with more levels and more quantities, the DAG can become complicated. Thus, we may express the DAG in (i) in a more compact form whereby **X** and **N** denote the vectors of all data on the number of trees that survived and the total number of trees, respectively: DAG (ii).

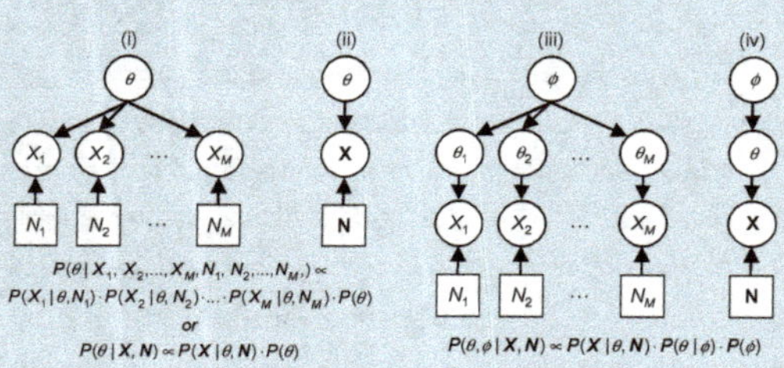

The above model and DAGs (i) and (ii) assume the probability of mortality (θ) is the same for all plots. However, θ may vary by plot due to plot-level differences in, for example, stand density, soil properties, and exposure. Thus, we may want to include plot specific θ's such that $\theta = (\theta_1, \theta_2, \ldots, \theta_M)$. We may specify a hierarchical prior for the θ's and treat each θ_j as coming from a parent population defined by hyperparameters (e.g., ϕ), as in (iii) and (iv); see the "Hierarchical Bayesian" section. This model has three levels: level 1 is the stochastic data level, level 2 is the first parameter level (here, for θ), and level 3 is the second parameter level (representing the hyperprior for ϕ).

In the tree mortality example (see Box 3), $parents[X_i] = \{\theta\}$, and the factors corresponding to the right-hand side of $P(V)$ above are $P(X_i \mid \theta, N_i)$. Note that the N_i have no bearing on the conditional independence of the X_i, since the N_i are fixed covariates, and we may also write $P(X_i \mid \theta)$. Incidentally, if the N_i were stochastic, still, the X_i would be conditionally independent given $parents[X_i] = \{\theta, N_i\}$ in this case.

We should realize that $P(V)$ is the numerator in Bayes theorem, before any conditioning on observations is done, and we will likely not recognize its form, except in relatively simple models. In this case, we may attempt to approximate the posterior via numerical methods (discussed below), for which we need the conditional distributions of subvectors of V or of individual nodes v, which are called full conditional distributions or just full conditionals. For this, let the children of node v, denoted $children[v]$, be those nodes that are pointed to directly by the arrows emanating from v. Then the form of the full conditional for v is given by

$$\begin{aligned} P(v \mid V_{-v}) &\propto P(v, V_{-v}) \\ &\propto (\text{factors in } P(V) \text{ containing } v) \\ &= P(v \mid parents[v]) \prod_{w \in children[v]} P(w \mid parents[w]). \end{aligned} \tag{6}$$

In other words, to obtain (an expression proportional to) the full conditional of v, simply look at the right-hand side of the conditional representation of $P(V)$ in Equation 5, which, again, occurs as the numerator of Bayes theorem, and use only those factors depending on v. Note that this works for node v, which may be a vector, or for collections of nodes, but application of numerical methods (discussed below) to sample larger vectors is generally more challenging in practice.

Estimation and Testing

In principle, the posterior, or its approximating histogram (discussed below), contains all of the information we need for inference. We may compute various summaries such as means, medians, modes, quantiles, variances, and intervals or regions containing θ with specified probability (credible intervals). Still, how "good" are these summaries? To answer this question, we introduce the decision theoretic notions of a loss function and minimum expected loss as a goodness criterion.

Let θ be the quantity we wish to estimate, and let $\delta(y)$ denote the procedure, which depends on data, y, that we use to estimate θ. We call δ an estimator (of θ), and, for a particular value of y, $\delta(y)$ is an estimate of θ. Note that δ may be referred to as a (decision) procedure, an estimator, or a rule. The objective is to find an estimator that is somehow optimal. For this, we introduce a loss function, $L(\theta, \delta(y))$, defined on the Cartesian product of the ranges of θ and $\delta(y)$, which are typically the same. The loss function assumes nonnegative values, with greater values indicating a larger discrepancy, or loss, between target, θ, and estimator, $\delta(y)$. The loss function is almost never sufficient to allow us to choose a best estimator since, in general, the loss depends on θ and y, and we don't know θ. For some θ and y, δ_1 may minimize loss, and, for other values, δ_2 may minimize loss. To address this ambiguity, we may compute an average or expected loss.

In particular, we may average $L(\theta, \delta(y))$ with respect to $P(y|\theta)$ to get frequentist risk, the objective being to find δ that minimizes frequentist risk. Still, frequentist risk depends on the unknown θ, and additional criteria, such as admissibility or minimaxity, which we do not define here, may be used to obtain an optimal estimator. Or, unbiasedness may be introduced to help find an optimal estimator, though this criterion is relatively uncommon from a decision theoretic point of view.

For a familiar example, consider squared error loss, $L = (\theta - \delta(y))^2$, the most ubiquitous loss function. (Assume θ and δ are scalar valued for simplicity of presentation.) Then, frequentist risk is commonly known as mean square error (MSE), and, if θ is the expected value of y, frequentist risk reduces to the variance of y. In some cases, $\delta(y) = y$ minimizes frequentist risk and is the best unbiased estimator of its mean, the sample average being best unbiased, in many cases, if we have a sample of y values from $P(y|\theta)$.

Alternatively, we may average $L(\theta, \delta(y))$ with respect to $P(\theta|y)$ to get posterior expected loss. Again, the optimality criterion is to find the procedure, δ, to minimize expected loss. Posterior expected loss still depends on y, but this is less of a problem than dependence on θ since we observe y. This criterion says, for each y, choose the procedure $\delta(y)$ that minimizes posterior expected loss. Such a procedure is called a Bayes rule, procedure, or estimator. (To avoid confusion, note that Bayes theorem is sometimes called Bayes rule.) It turns out that Bayes rules are usually admissible estimators, and admissible rules are Bayes rules or limits of Bayes rules. Thus, in principle, a frequentist (using frequentist risk) looking for admissible rules may adopt a Bayesian approach.

As yet another alternative, we may choose to average $L(\theta, \delta(y))$ over both y and θ (assuming we introduce a prior) to obtain integrated (frequentist) risk. It can be shown that the procedure minimizing integrated risk is the same as the Bayes rule. Thus, again, in principle, a frequentist may adopt a prior and use the criterion of posterior expected loss to find an optimal estimator in terms of integrated risk, called Bayes risk when evaluated at the Bayes rule.

The upshot of the current discussion is that common summaries of the posterior are often Bayes rules with respect to some loss function and using the posterior expected loss criterion. For example, the posterior mean is the Bayes rule under squared error loss, and the posterior median is the Bayes rule under absolute error loss, $|\theta - \delta(y)|$. Similar results hold for the posterior mode, quantiles, and other summaries of the posterior for other losses.

For testing hypotheses about θ, Bayes rules (decisions) to reject or accept a hypothesis can be framed as an estimation problem, leading again to minimizing posterior expected losses. But, roughly speaking, hypothesis testing in a Bayesian framework is a relatively delicate matter, especially with simple point null hypotheses like H_0: $\theta = \theta_0$. In some sense, the Bayes factor, which is the ratio of posterior to prior odds, may be considered a Bayesian response to the ubiquitous *p*-value in frequentist significance testing.

PRIORS

The prior distribution has been criticized for not being objective, but the data model and, perhaps, loss are subjective components of a Bayesian analysis as well. Likewise, the choice of the likelihood in a classical analysis is subjective. When we do have prior information about θ, then a Bayesian analysis is difficult to ignore. When we have no prior information, we may choose a noninformative or reference prior if we wish to exploit the conditional nature of Bayesian inference or if we prefer to think (and behave coherently) in terms of probabilities. In this case, or when prior information is vague, then a sensitivity analysis is appropriate, wherein we adjust the prior within a reasonable range to determine the effects on the posterior. If the posterior exhibits little sensitivity to the prior, perhaps being dominated by a large data set, then we may feel satisfied with our analysis. In cases where the posterior is sensitive to a vague prior, then we may want to work harder to obtain prior information, more or better data, or both. The literature on prior distributions is vast, and here we highlight notions of (conditionally) conjugate priors, noninformative priors, and prior and posterior propriety.

A prior is said to be conjugate with respect to some class of distributions if the form of the posterior is the same as the prior. Hence, the form of the posterior is known, and the computation of the posterior is usually a matter of computing a few parameters with simple formulae. For a large class of commonly used data models (exponential families), we can, in principle, get arbitrarily close to any prior using a mixture of conjugate priors. So, in principle, conjugate priors can be useful approximations to our true prior, and they permit useful analytical simplifications. In the case where we do not have prior information, use of conjugate priors is often for analytical simplicity or convenience. Aside from analytical conveniences, conjugate priors often aid interpretation via the "device of imaginary observations." That is, they can be interpreted as contributing prior information in the form of an imaginary sample size; this is discussed in greater detail in the simple example given below.

Only in the simplest cases, as in the example below, can we hope to use conjugacy to obtain the posterior. In more complicated models, we may not have overall conjugacy, but may still retain conditional conjugacy. In this case, a prior can be chosen for a subvector of θ so that the resulting full conditional distribution is of the same form as the prior, and, hence, the full conditional may be sampled from directly (see the numerical methods section, below). In the context of the DAG notation, this is a matter of looking at the factors of $P(V)$ that include v, ignoring the prior, $P(v)$, for the moment, then choosing $P(v)$ so that the form of the full conditional $P(v|V_{-v})$ is the same as that of $P(v)$ (generally with different parameter values, of course). For example, we may have a logistic regression model with linear predictor $\alpha + \beta x + \varepsilon$, for some unknown parameters α and β, fixed covariate x, and error $\varepsilon \sim N(0, \sigma^2)$. In this case, an inverse-gamma prior for σ^2, independent of the prior for α and β, is conditionally conjugate to the factor (containing σ^2) arising from the normal distribution in the linear predictor, yielding an inverse-gamma full conditional for σ^2. Typically, in practice, we simply consult a catalog of conjugate data model prior pairs rather than actually implement this procedure for finding a conjugate prior; the work has already been done for us in many cases.

In many cases, we do not have prior information, and we may appeal to noninformative priors. Intuitively, noninformative priors are meant to embody a lack of information about a parameter. There are various senses in which this may be true, and only Jeffreys prior is mentioned here. Jeffreys prior is obtained from the Fisher information matrix, $I^F(\theta)$, or number as $P(\theta) \propto |I^F(\theta)|^{1/2}$, where $|A|$ denotes the determinant of matrix A. Under some technical conditions—often referred to as regularity conditions—which hold for many familiar distributions, including exponential families, $I^F(\theta) = -E\left[\frac{\partial^2}{\partial\theta\partial\theta^T}\log(P(y|\theta))\right]$. For example, Jeffreys (joint) prior for the mean and variance, $\theta = (\mu, \sigma^2)$, of a normal distribution is $P(\theta) \propto (\sigma^2)^{-3/2}$, which, incidentally, is not the same as the product of Jeffreys priors obtained for the mean and variance, separately, $P(\theta) \propto 1/\sigma^2$, which we may use if we are thinking that the mean and variance are *a priori* independent. Jeffreys prior is invariant to transformation. Thus, specifying Jeffreys prior for θ and then obtaining the prior for the transformation $\eta = h(\theta)$ is the same as specifying Jeffreys prior for η. Jeffreys prior represents an ad hoc method for obtaining a prior and, technically, is outside the realm of Bayesian statistics since it involves expectation over unobserved data values, y, violating the likelihood principle by not conditioning on observed data, hence, strictly speaking, violating the Bayesian paradigm.

In what sense is Jeffreys prior noninformative? Fisher's information is a commonly used measure of information about the parameters contained in the data model. A larger information number discriminates θ from $\theta + \Delta\theta$ more than a smaller information number, and similarly for $|I^F(\theta)|^{1/2}$. Thus, choosing a prior proportional to $|I^F(\theta)|^{1/2}$ is noninformative in the sense of not changing the discriminating action of this measure of information. Jeffreys prior can also be thought of as the prior that is equally noninformative for all transformations of θ.

The Jeffreys priors given in the normal examples above are improper distributions, which means that their integrals are infinite, as is often the case with Jeffreys priors. If the prior is improper, then the interpretation of a full probability model, conditional distribution, and marginal distribution no longer holds technically. However, the posterior may still be proper, which is the most important thing to check when specifying improper priors. Note also that improper priors can create problems with Bayes factors, and we recommend avoiding improper priors in this case unless much care is taken to avoid their pitfalls when computing Bayes factors.

A SIMPLE EXAMPLE

Here is a simple example to illustrate some of the concepts and equations presented above. Refer to the DAGs in Box 3 (i and ii) as we develop the example. Consider the problem of estimating the probability, θ, of a tree dying during a severe drought. A researcher conducts a study and counts the number of trees that died, X_i, and the total number of trees, N_i, in $i = 1, 2, 3, \ldots, M$ randomly located plots within a forested region. We define the different components or levels of the model, which in this simple example means we define the data model for the data, $P(\mathbf{X}|\theta, \mathbf{N})$, and the prior for θ, $P(\theta)$. We treat $\mathbf{N}$ as a known, fixed covariate.

If we assume θ is the same for all plots and the number of trees that died in plot i (X_i) is independent of the number that died in other plots, then the logical choice for a data model is a binomial distribution for the X_i, and we write

$$X_i|\theta \sim binomial(\theta, N_i). \tag{7}$$

In defining the model for X_i, we can ignore the factors that do not depend on θ, expressing the data model as proportional to the kernel of the binomial pmf (i.e., the factors that contain θ):

$$P(X_i = x_i|\theta, N_i) = \binom{N_i}{x_i}\theta^{x_i}(1-\theta)^{N_i-x_i} \propto \theta^{x_i}(1-\theta)^{N_i-x_i}. \tag{8}$$

Next, we may assume that the X_i are conditionally independent given θ, and the complete data model or the likelihood of all data is

$$P(\mathbf{X}|\theta, \mathbf{N}) \propto \prod_{i=1}^{P}\theta^{x_i}(1-\theta)^{N_i-x_i} = \theta^{\sum_{i=1}^{P}x_i}(1-\theta)^{\sum_{i=1}^{P}(N_i-x_i)}. \tag{9}$$

Of course, this does not mean that the X_i are (unconditionally) independent, because counts of dead trees depend on θ.

Next, we define $P(\theta)$, and, in doing so, we should think about constraints on θ. In this example, θ is a probability parameter, and we should consider picking a prior that is defined for $0 \le \theta \le 1$. We consider a conjugate prior, and, without previous experience, we seek to identify a pdf (θ is a continuous random variable) with a kernel that matches the form of the likelihood in Equation 9, which will give a posterior of the same form as the prior (see the previous discussion about obtaining conjugate priors). In this case, the conjugate prior is the beta distribution, and we write $\theta \sim beta(\alpha, \beta)$, where α and β are the (hyper)parameters for which we assign specific values shortly. The beta pdf is

$$P(\theta) = \frac{\Gamma(\alpha+\beta)}{\Gamma(\alpha)\Gamma(\beta)}\theta^{\alpha-1}(1-\theta)^{\beta-1} \propto \theta^{\alpha-1}(1-\theta)^{\beta-1}. \tag{10}$$

Again, since we may ignore the factors that do not contain θ, we can express the prior as proportional to the kernel of the beta pdf.

Finally, we combine the data model and prior via Bayes theorem to arrive at the posterior for θ:

$$\begin{aligned} P(\theta|\mathbf{X}, \mathbf{N}) &\propto P(\mathbf{X}|\theta, \mathbf{N})P(\theta) \\ &\propto \left(\theta^{\sum_{i=1}^{P}x_i}(1-\theta)^{\sum_{i=1}^{P}(N_i-x_i)}\right)\left(\theta^{\alpha-1}(1-\theta)^{\beta-1}\right) \\ &= \theta^{\sum_{i=1}^{P}x_i+\alpha-1}(1-\theta)^{\sum_{i=1}^{P}(N_i-x_i)+\beta-1}. \end{aligned} \tag{11}$$

Via kernel matching, we recognize the unnormalized posterior as the kernel of a beta such that

$$\theta|\mathbf{X}, \mathbf{N} \sim beta\left(\sum_{i=1}^{P}x_i + \alpha, \sum_{i=1}^{P}(N_i - x_i) + \beta\right). \tag{12}$$

The role of α and β should now be clear, but their interpretation may vary slightly. Returning to the "device of imagining observations," if we examine Equation 12, α is equivalent to the imaginary or prior number of trees that died, and β is equivalent to the prior number of trees that survived; $\alpha + \beta$ is interpreted as the prior sample size. If we examine Equation 11 from the perspective of the likelihood, then $\alpha - 1$, $\beta - 1$, and $\alpha + \beta - 2$ may be interpreted as the prior number of trees that died, the prior number that survived, and the prior sample size, respectively.

We can also compare properties of the prior and posterior to further evaluate the role of the prior. For example, the prior and posterior means for θ, $E(\theta)$ and $E(\theta|\mathbf{X}, \mathbf{N})$, respectively, and for comparison, the maximum likelihood estimate (MLE) for θ ($\hat{\theta}_{MLE}$) are

$$\begin{aligned} E(\theta) = \frac{\alpha}{\alpha+\beta}, \quad E(\theta|\mathbf{X}, \mathbf{N}) &= \frac{\sum_{i=1}^{P}x_i + \alpha}{\sum_{i=1}^{P}N_i + \alpha + \beta}, \\ \hat{\theta}_{MLE} &= \frac{\sum_{i=1}^{P}x_i}{\sum_{i=1}^{P}(N_i - x_i)}. \end{aligned} \tag{13}$$

$E(\theta|\mathbf{X}, \mathbf{N})$ may be viewed as a compromise between $E(\theta)$ and $\hat{\theta}_{MLE}$, and $E(\theta|\mathbf{X}, \mathbf{N}) = \hat{\theta}_{MLE}$ if $\alpha = \beta = 0$. The beta pdf is undefined for $\alpha = \beta = 0$, resulting in an improper (Haldane) prior, but the posterior is proper as long as $\sum x_i \neq 0$ and $\sum x_i \neq \sum N_i$. If we want a proper, noninformative prior for θ, we may chose a uniform prior on the interval (0, 1), equivalent to $beta(1, 1)$. This would contribute an effective prior sample size of 2 ($\alpha = 1$ dead tree, $\beta = 1$ living), with $E(\theta) = 0.5$. The Jeffreys prior, $beta(½, ½)$, would contribute a prior sample size of 1. Both priors are conjugate and contribute some amount of information, but their influence will depend on the data, $\sum N_i$ and $\sum x_i$. If $\sum x_i$ and $\sum N_i$ are "large," then the data overwhelm the prior and $E(\theta \mid \boldsymbol{X}, \boldsymbol{N})$ will be very close to $\hat{\theta}_{MLE}$.

HIERARCHICAL BAYESIAN

The simple model in Equation 4 can be extended to many levels by, for example, incorporating hierarchical parameter models. For example, in the tree mortality example presented previously, we may wish to model the probability of mortality (θ) as varying by plot i to account for plot-to-plot variability associated with differences in environmental conditions or to include information about spatial correlation in the θ_i. One approach is to model the θ_i as coming from a population of θ's, with the population distribution described by hyperparameter ϕ, introducing a third level to the DAG (see Box 3, iii and iv). Such hierarchical parameter models are often relevant for modeling data obtained from nested sampling designs or representing different levels of aggregation. Here, we specify a joint prior for $P(\theta, \phi)$, and, based on the DAG and probability rules, we can write this as $P(\theta \mid \phi)P(\phi)$.

The simple Bayesian framework can also be extended to explicitly partition measurement or observation error from ecological process uncertainty, adding another level to the model. This extension is often referred to as a hierarchical Bayesian model (although, the previous example also results in a hierarchical model) or the process sandwich. Here, we define *Process* as the underlying latent, unobservable process that generates the data, and we partition θ into θ_D ("data-related" parameters) and θ_P ("process-related" parameters) and write

$$P(\theta_D, \theta_P, \mathit{Process} \mid \mathit{Data}) \propto P(\mathit{Data} \mid \mathit{Process}, \theta_D) \times P(\mathit{Process} \mid \theta_P)P(\theta_D, \theta_P). \quad (14)$$

The data model is now given by $P(\mathit{Data} \mid \mathit{Process}, \theta_D)$, and for simplicity, we can think of the data as varying around the latent process plus observation error such that the expected value of the data, $E(\mathit{Data})$, is equal to (some function of) *Process*, and θ_D typically contains parameters describing measurement error variances, covariances, and/or biases (e.g., instrument drift). Here, *Process* is a stochastic quantity, and $P(\mathit{Process} \mid \theta_P)$ is referred to as the stochastic process model; we can think of the true, latent process as varying around some expected process, $E(\mathit{Process})$, plus process error. Thus, θ_P may contain parameters in the $E(\mathit{Process})$ model as well as process error (co)variance terms. Note that $E(\mathit{Process})$ is typically deterministic, conditional on θ_P, and may be described by different types of models ranging from, for example, a simple linear regression to more complex models such as a difference equation or matrix model of population dynamics, a differential equation describing the spread of a pathogen or invasive species, or a nonlinear, biochemical-based model of plant photosynthesis. Thus, it is via $E(\mathit{Process})$ that we have the opportunity to explicitly incorporate ecological theory/models into the probabilistic Bayesian framework.

In the vast majority of applications, once we have conditioned the data on the latent process, it is fair to assume that the measurement or observation errors are independent. Likewise, one approach to modeling process errors is to assume that they are independent. However, the assumption of independence for both error components often leads to identifiability problems such that process and observation variance terms cannot be separated. One solution is to specify a tight prior for the variance terms associated with observation error, which is reasonable when existing information is available. If such information is lacking, identifiability of the different variances terms may be facilitated by incorporation of different error structures for each component. Since we cannot measure and account for all factors affecting *Process*, we cannot model it perfectly via $E(\mathit{Process})$, and we are led to consider modeling the process error structure. These unobserved factors are almost invariably temporal, spatial, or biological in nature, thus leading us to incorporate temporal, spatial, or biological structure into the process errors.

NUMERICAL METHODS AND MARKOV CHAIN MONTE CARLO (MCMC)

In practice, an analytical solution for the posterior typically is unavailable because most realistic models are nonlinear or otherwise too complex to permit easy calculation of the joint posterior. But this typically is not a problem, because effective computational strategies are available for sampling from the joint posterior despite such complexities. For models that involve conjugate priors, we may obtain analytical summaries or straightforwardly simulate values of θ from the posterior. If we cannot simulate directly from the joint posterior, we can simulate directly from known full conditionals, particularly if we are able to specify conditionally conjugate priors; this is known as Gibbs sampling. However, when Gibbs does not perform well, or when we do not recognize the form of the full conditionals, we may be able to use other methods for simulating from the posterior, such as adaption–rejection, Metropolis–Hastings, or slice sampling. Nearly all algorithms work with the normalized and/or unnormalized full conditionals for the unknowns. Consider the DAG in Box 3iii, which indicates the full joint posterior as

$$P(\theta_1,\theta_2,\ldots,\theta_M,\phi \mid X_1,X_2,\ldots,X_M,N_1,N_2,\ldots,N_M) \\ \propto P(X_1 \mid \theta_1,N_1)\cdot P(X_2 \mid \theta_2,N_2)\cdot \ldots \cdot P(X_M \mid \theta_M,N_M) \\ \cdot P(\theta_1 \mid \phi)\cdot P(\theta_2 \mid \phi)\cdot \ldots \cdot P(\theta_M \mid \phi)\cdot P(\phi). \tag{15}$$

If we are interested in a particular θ_j, we can work directly with its full conditional, and, since it only depends on X_j, N_j, and ϕ, we can simply write the full conditional for θ_j as

$$P(\theta_j \mid \phi, X_j, N_j) \propto P(X_j \mid \theta_j, N_j)\cdot P(\theta_j \mid \phi). \tag{16}$$

For the example in Box 3iv, we choose the binomial pmf for $P(X_j \mid \theta_j, N_j)$ and a conjugate beta prior for $P(\theta_j \mid \phi)$ with hyperparameter $\phi = (\alpha, \beta)$. Thus, the full conditional for θ_j is recognizable: it is a beta distribution, and we can use Gibbs to sample directly from it. However, we cannot identify conjugate priors for α and β based on knowledge of common distributions. And we do not recognize the full conditional for $\phi = (\alpha, \beta)$, so we cannot use Gibbs to simulate values of ϕ. In this case, we may use Metropolis–Hastings (M–H) or some other algorithm that will allow us to sample from the unnormalized full conditional.

Both the Gibbs and M–H algorithms are part of a more general class of Markov chain Monte Carlo (MCMC) algorithms. The general idea behind MCMC methods is to simulate a sequence of values, e.g., $\theta^1, \theta^2, \ldots, \theta^T$, from the joint posterior. The draws are from a Markov chain because the probability of drawing a particular value of θ at iteration t depends on the value of θ at iteration $t-1$. The general procedure is to sample θ sequentially from a proposal distribution that may depend on the last value of θ in the sequence, on the data, or both, and the MCMC draws will eventually approximate or converge to the target (posterior) distribution in that the histogram formed from these values can be made arbitrarily close to the actual posterior with increasing T.

To illustrate an MCMC approach, a simple M–H algorithm is outlined in Box 4. The M–H algorithm is based on an accept/reject rule, and it requires us to specify a proposal distribution, Q. If we know the normalized full conditional for θ, $P(\theta \mid \phi, X, N)$, then we may use this for Q, though this may not be the optimal choice in terms of MCMC behavior. In this case, if we let $Q_t\ (\theta_j^* \mid \theta_j^{t-1}) = P(\theta_j^* \mid \phi^{t-1}, N_j, X_j)$, then r in Equation 2.1 (Box 4) reduces to $r = 1$, and we accept every proposed value, yielding the Gibbs algorithm, a special case of the M–H algorithm.

Key issues common to both the M–H and Gibbs samplers, and other MCMC methods, include choosing starting values (e.g., θ^0; Box 4), evaluating convergence and burn-in, and determining the length of the simulation, T. One must understand and consider these issues when implementing MCMC algorithms, but it is beyond the scope of this chapter to define these terms and discuss these issues; we refer the reader to relevant texts

BOX 4. GENERAL METROPOLIS-HASTINGS (M-H) ALGORITHM

The general M-H sampling algorithm is outlined as follows. Let θ be a vector of P elements, and denote θ_j as jth element and θ_j^0 as the starting value for θ_j in the MCMC sequence. The M-H algorithm proceeds as follows:

- For $t = 1, 2, 3, \ldots T$ iterations,
  - For $j = 1, 2, \ldots, P$ parameters,
    - Sample a proposal value θ_j^* from a proposal distribution $Q_t(\theta_j^* \mid \theta_j t-1)$.
    - Calculate the acceptance ratio as
    $$r = \frac{P(\theta_j^* \mid \phi, X, N)\cdot Q_t(\theta_j t-1 \mid \theta_j^*)}{P(\theta_j^{t-1} \mid \phi, X, N)\cdot Q_t(\theta_j^* \mid \theta_j^{t-1})} \tag{2.1}$$
    - Set $\theta_j^t = \begin{cases} \theta_j^* & \text{With probability min } (1, r) \\ \theta_j^{t-1} & \text{otherwise} \end{cases}$ (2.2)

There are four main points to note. First, the accept/reject rule results in the proposed value being accepted if it increases the posterior density relative to the previous value (i.e., when $r > 1$). If the proposed value decreases the posterior density ($r < 1$), then we keep the proposed value with probability r. Second, multiplication by the ratio of the proposal densities evaluated at θ_j^{t-1} and θ_j^* satisfies the "detailed balance condition" such that the algorithm is guaranteed to converge to the posterior distribution. Third, there are many methods for choosing the proposal. For example, Q may be chosen to be independent of the previous θ_j value, and a common choice for Q is the prior for θ_j. The random walk proposal is very common, and it is defines $\theta_j^* = \theta_j^{t-1} + \varepsilon^t$. A common choice for ε^t is $\varepsilon^t \sim N(0, \nu)$, yielding a symmetric proposal for Q, and thus the acceptance ratio reduces to the ratio of the posterior densities evaluated at θ_j^* and θ_j^{t-1}. Fourth, the posterior for θ and the proposal for θ should be expressed on the same scale. For example, if it is more convenient to specify a prior for θ, but propose on a different scale, e.g., for $f(\theta) = \log(\theta)$, then a transformation of variables must be applied to either the prior (and thus the posterior) or to the proposal so that both are defined for the same quantity. See "Further Readings" for more details about different M-H algorithms.

in the "Further Reading" section. High-level software packages are available for implementing most Bayesian models that require MCMC methods. Popular software includes OpenBUGS, its predecessor WinBUGS, and JAGS. One must still be familiar with evaluating MCMC output (i.e., convergence, burn-in, mixing, and so on), and these software packages have built-in tools to help facilitate the process.

OTHER CONSIDERATIONS

This entry has only skimmed the surface of Bayesian statistics. In practice, ecologists often deal with challenging problems that merge ecological models with one or more datasets which may vary at different scales or represent different levels of completeness, intensity, accuracy, or resolution. Implementation of such models may require that we consider "tricks" for improving MCMC behavior such as, but not limited to, reparameterization of a model or model components, which may also facilitate the interpretation of model parameters, parameter expansion in hierarchical parameter models (applied in certain situations when small hierarchical variance terms cause poor MCMC behavior), or reversible jump MCMC to accommodate models characterized by a variable dimension parameter space. Finally, we see model diagnostics as an underdeveloped field in Bayesian statistics, and classical diagnostic tools are often used. Regardless of whether one chooses a Bayesian versus classical approach, one should always consider conducting some sort of diagnostics, a common one being evaluating model goodness-of-fit by comparing, for example, posterior results for replicated data to observed data, but there are many other avenues to explore.

CONCLUSIONS

The Bayesian approach represents a different way of thinking compared to the classical or frequentist approach. For example, unknown quantities such as model parameters are treated as random variables in a similar fashion to the treatment of data. The flexibility of the Bayesian framework allows the scientific problem to drive the modeling and data analysis, whereas classical methods may often force the analysis to fit within a relatively restrictive framework.

A major challenge that ecologists face is how to integrate diverse datasets representing complex ecological phenomena with process models designed to learn about these complexities. Bayesian statistics enables such integration, which has otherwise been difficult to achieve via traditional approaches. An important aspect of the Bayesian framework is the ease with which multiple data sources can be integrated. And it also allows us to explicitly deal with common issues experienced by ecologists, such as missing data, unbalanced sampling designs, temporally or spatially misaligned data, temporal or spatial correlation, and multiple sources of uncertainty. Thus, there is tremendous potential for Bayesian methods in ecology, and the probabilistic, hierarchical modeling framework presents an excellent opportunity to integrate ecological theory and models with empirical information.

SEE ALSO THE FOLLOWING ARTICLES

Computational Ecology / Frequentist Statistics / Information Criteria in Ecology / Markov Chains / Model Fitting / Statistics in Ecology

FURTHER READING

Berger, J. O. 1985. *Statistical decision theory and Bayesian analysis,* 2nd ed. New York: Springer-Verlag.

Bernardo, J. M., and A. F. M. Smith. 1994. *Bayesian theory.* Chichester: John Wiley & Sons.

Casella, G., and R. L. Berger. 2002. *Statistical inference.* Pacific Grove, CA: Duxbury.

Clark, J. S., and A. E. Gelfand. 2006. *Hierarchical modelling for the environmental sciences: statistical methods and applications.* Oxford: Oxford University Press.

Gamerman, D., and H. F. Lopes. 2006. *Markov chain Monte Carlo.* Boca Raton: Chapman and Hall/CRC Press.

Gelman, A., J. B. Carlin, H. S. Stern, and D. B. Rubin. 2004. *Bayesian data analysis.* Boca Raton: Chapman and Hall/CRC Press.

Gelman, A., and J. Hill. 2006. *Data analysis using regression and multilevel/hierarchical models.* New York: Cambridge University Press.

Gilks, W. R., S. Richardson, and D. J. Spiegelhalter. 1995. *Markov chain Monte Carlo in practice.* Boca Raton, FL: Chapman and Hall/CRC.

Ogle, K., and J. J. Barber. 2008. Bayesian data-model integration in plant physiological and ecosystem ecology. *Progress in Botany* 69: 281–311.

Robert, C. P. 2001. *The Bayesian choice,* 2nd ed. New York: Springer-Verlag.

BEHAVIORAL ECOLOGY

B. D. ROITBERG

Simon Fraser University, Burnaby, British Columbia, Canada

R. G. LALONDE

University of British Columbia, Okanagan, Kelowna, Canada

Behavioral ecologists study a very wide variety of behaviors, as can be seen in almost every issue of the journals *Behavioral Ecology* and *Behavioral Ecology and Sociobiology,* but these behaviors can easily be categorized in four main behavioral classes, each of which has a high association with fitness variation. These are energy acquisition (feeding), aggression (fighting), reproduction (copulation), and predator avoidance (fleeing). Regardless of the category, behavioral ecologists evaluate the adaptive nature of these behaviors in an ecological context.

THE CONCEPT

Behavioral ecology uses the concept of evolutionary fitness: the relative contribution of a heritable strategy set to the population gene pool. In contrast to most other evolutionary disciplines, behavioral ecology recognizes that each organism expresses a range of phenotypic values associated with a particular trait that are responses to perceptual changes in the environment. This perspective is important because a number of phenotypic traits (for example, level of aggression) are, in nature, highly plastic in their expression. Thus, in studying such plastic traits, behavioral ecologists seek to quantify the fitness consequences of particular patterns of behavioral variation. This can take the form of a surrogate of fitness, such as rate of energy acquisition in a forager or more ideally some more direct measure such as reproductive value or intrinsic rate of increase (see below).

Behavioral ecologists typically do not focus on the underlying genetic basis of behavioral variation and instead assume "The Phenotypic Gambit," where selection on behavioral phenotypes happens because of some nonzero level of heritability. Behavioral ecology, when done well, focuses both on the behavioral phenomenon itself (the "how" or proximate aspect, i.e., the mechanism) as well as the functional aspects of that behavior (the "why" or ultimate aspect, i.e., the function).

THE BEHAVIORAL ECOLOGIST'S TOOLBOX

Measurements of Fitness

The ideal measure of fitness is direct computation of the effect of a strategy set on the intrinsic rate of increase, r, which can be approximated by

$$r = \frac{\ln(R_0)}{T}. \quad (1)$$

Here, R_0 is the cohort replacement rate, which is the summation of age-specific proportionate survivorship (l_x) and fecundity (m_x), and T is the generation time of a strategy set.

When more than one age class reproduces, this parameter can solved more exactly using Euler's formula:

$$1 = \int_{x=0}^{\infty} lx \cdot mx \cdot e^{-rx} dx. \quad (2)$$

However, in order to calculate r using either method, it is necessary to obtain the relevant demographic parameters (i.e., age-specific survivorship and fecundity of a behavioral strategy set and, ideally, the same for other strategy sets if relative fitness is to be calculated). If incomplete demographic data are available, it is sometimes possible to calculate some portion of the reproductive value of a strategy set:

$$V_x = m_x + \sum_{y=x+1}^{\infty} \frac{l_y}{l_x} m_y. \quad (3)$$

Here, v_x is the reproductive value of a given age class, x. This parameter has the advantage of being calculable for any given age class as long as parameter values are available for the later age classes. Thus, it is possible to calculate v_x for a strategy set at its time of expression. As a measure of fitness, it falls short because it does not account for differences in generation time or life history tradeoffs if such a partial calculation is made (see below).

Within the context of experimental studies of behavior, obtaining appropriate demographic data is often not practical or even possible. Consequently, behavioral ecologists employ a number of fitness surrogates when comparing behavioral strategies. Fitness surrogates generally measure the proximate consequences of the behavior of interest. For example, optimal foraging studies typically use rate of resource acquisition as a surrogate of strategy fitness. Other fitness surrogates include survivorship, frequency of successful agonistic encounters, mating success, and success in raising offspring to independence.

Use of fitness surrogates can overlook life-history tradeoffs with unmeasured parameters. For example, bumblebees compensate for wing wear by increasing foraging effort, but at the cost of longevity. Vigilance behavior often trades off with foraging success and mating strategists that sneak copulations often show early maturation and suffer consequent reduced longevity. A further problem with fitness surrogates is the default assumption that a direct linear relationship exists between surrogate value and fitness. In fact, the utility of a fitness surrogate is often a curvilinear function of an individual's state or recent past. For example, utility of millet seeds for yellow-eyed juncos changes with the satiation level and state of individual birds in a curvilinear manner. This nonlinearity in the relationship between a fitness surrogate and its utility can also affect a forager's choice of resource type, if associated variances differ. Because of all of these considerations, to avoid problems in interpretation great care must be taken to establish the relationship between a fitness surrogate and fitness itself.

Theoretical Aspects

Many theoretical tools that are employed by behavioral ecologists have their origins in economic theory (e.g., marginal analysis, risk analysis, game theory, portfolio theory, dynamic programming models). This makes sense in that it is a relatively small conceptual leap to replace dollars with fitness as the currency of choice. From there, standard economic tools can easily be applied. Below are some of the main tools that are used.

OPTIMIZATION

Here, the behavioral ecologist seeks the optimal solution to some problem that frequently requires maximizing some utility (e.g., fitness from energy accrual) given some set of constraints. For example, in optimal diet theory food items are accepted or rejected based upon maximizing rate of energy intake. For example, suppose a snail-eating fish forages at a site where two types of snail prey are available. Snail 1 contains 8 calories of energy (E_1) and requires 20 minutes to be subdued, its shell broken, and the body eaten; we call this handling time, η_1. In contrast, snail 2 contains $E_2 = 10$ calories of energy but harbors a complex, very hard shell and thus a handling time, η_2, of 30 minutes. According to theory, food items should be ranked by $\frac{E_i}{\eta_i}$, where i refers to prey type. In our example, $\frac{E_1}{\eta_1} > \frac{E_2}{\eta_2}$, and thus snail 1 ranks first even though it harbors less total energy. Clearly snail 1 should always be eaten when encountered, but what about snail 2? Since the optimal diet breadth should maximize rate of energy intake, the decision whether to include snail 2 in the diet depends upon encounter rates with two snail types, defined as λ_1 and λ_2, respectively. The solution to this diet breadth problem is to employ an inequality to determine whether the lower-ranked item should be included in the diet. This model, which has been derived independently several different times, has the important feature that foraging time T comprises two components, time spent searching, T_s, and time spent handling prey, T_h. In general, the jth item should only be added to the diet i if

$$\frac{E_i}{T_i} < \frac{E_i + \lambda_j E_j}{T_i + \lambda_j h_j}. \quad (4)$$

The optimal diet model has been modified many times to make quantitative predictions for explicit diet problems with unique criteria. For example, in host-attack decisions by parasitic wasps on aphid hosts, body size of the forager determines both handling time and probability of a successful attack across different hosts. Consistent with this finding, a review of the diet breadth literature showed that prey mobility can explain deviation from classic diet theory mostly through its impact on attack success.

Aside from diet breadth, optimality models have been applied to a range of problems, including optimal nest defense, optimal patch exploitation, foraging paths, optimal web size for spiders, optimal thermal regulatory behavior, and parental care decisions. In a large majority of these studies, the focal organism faces some tradeoff that it must solve (e.g., risk or starvation vs. risk of predation).

Within the optimization approach there are two basic approaches, rate maximization, wherein the average rate of some utility is maximized (e.g. energy accrual), and dynamic state variable models (DSVs), where the state of the organism is an explicit constraint that is considered while maximizing that utility. Returning to the diet breadth example above, suppose that our focal organism has largely depleted its energy reserves. Here, we might expect that individual to accept low-quality food resources simply to maintain somatic function (i.e., avoid starving). By contrast, an individual with full energy reserves would likely be discriminating and would only accept those items that would maximize long-term rate of energy accrual with little risk of starving.

GAME THEORY

When optimal decisions (e.g., diet selection) are independent of the behavior of other individuals, behavioral ecologists call such decisions tactical; however, when payoffs from decisions are contingent upon the response of other individuals, then the appropriate decisions are strategic and a different approach is required, that of game theory (see entry on Game Theory). Here, rather than solving for the optimal, one seeks the Evolutionarily Stable Strategy (ESS): an equilibrium condition, once adopted by the population, is not invasive to individuals with alternate behaviors. For example, in the earlier diet selection model there was an assumption that prey encounter rates were constant. Imagine, however, that our focal forager searches for prey in a depleting patch and that competing fish are also present in that patch. The presence of even a single competitor can cause the focal forager to adopt a broader diet than it would otherwise, largely because it has lost total control of resources and failure to consume can lead to loss of such resources to others. Similarly, when optimal habitat choices are made independent of others, then some habitat-specific set of criteria (e.g., food, shelter) are used to determine the absolute value of the habitat. By contrast, when the relative value of the habitat depends upon the presence and actions of others, then individuals will distribute themselves across habitats according to an Ideal Free Distribution wherein all individuals achieve equal performance and cannot improve their performance by moving to another habitat.

Behavioral games may be state dependent or independent, static (simultaneous decisions) or dynamic (repeated decisions such that the game evolves). When payoffs are state dependent, the problem is best solved through the application of dynamic games; for example, in a predator–prey game, wherein the optimal level of antipredator behavior depends upon prey energy levels and predator threat and predator attack rates depend upon predator energy state and costs associated with attack.

Game theory has been applied to a wide range of problems in behavioral ecology, including habitat selection, mating games, predator-inspection games, resource defense games, fighting games, cooperation, producer-scrounger games, and aggregation just to name a few. This is not surprising in that payoffs from expressing a wide range of behaviors are likely to depend upon actions of others.

BEHAVIORAL RULES

When behavioral ecologists discuss how and why their study organisms make decisions and choose particular options, they do not imply that high levels of cognition are employed. Rather, individuals are assumed to use simple rules or rules of thumb that approximate optimal or ESS behaviors. Thus, the study of such rules is an important area of behavioral ecology; i.e., how do they do it? For example, considerable discussion has focused on rules that foragers use for exploiting patches (Fixed Time Rule, Fixed Number Rule, Fixed Giving Up Time, Adjustable Giving Up Time, Count Down Rule, etc.). The success of such rules will often depend on the distribution of resources, but none of them require sophisticated calculations by the forager to achieve optimal or near optimal results.

Phylogenetics

The advent of molecular methodologies has allowed inclusion of more accurate phylogenetic evidence in the study of behavioral ecology. Among other benefits, this has helped elucidate the thorny question of sexually selected ornaments in males. Experimental work has shown that selection for such features is often facilitated by preexisting sensory biases in females for particular colors and for particular exaggerated anatomical features (Fig. 1). Prior to this insight, explanations for sexually selected ornaments had to account for the independent fixation of a male feature and female preferences for that feature.

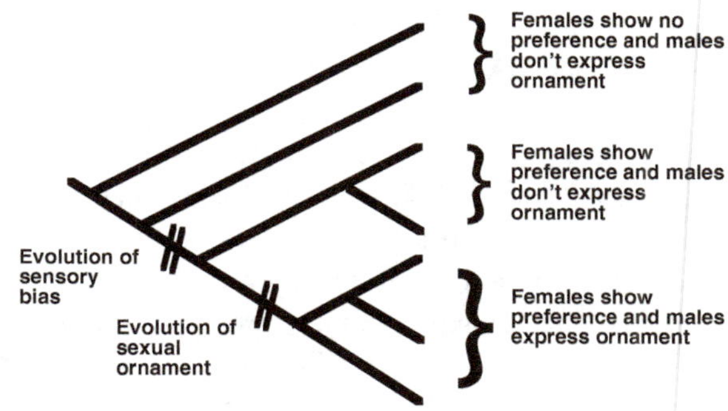

FIGURE 1 A generalized scheme showing inferred evolution of a preexisting sensory bias followed by the evolution of a sexual ornament.

More broadly, a behavioral strategy may be present in a species either because of phylogenetic baggage or because of *de novo* natural selection. In the latter case, measures of functionality may have more relevance than the former, since a behavioral strategy that is expressed but of neutral functionality would not be especially amenable to experimental manipulation. In general, phylogenetic analysis has demonstrated that behavioral characteristics are more malleable (i.e., show relatively little phylogenetic autocorrelation) in comparison to other traits. Finally, phylogenetic analyses can elucidate underlying behavioral tendencies that can, in turn, drive morphological evolution. Work on the arms race between male and female waterstriders has strongly supported a hypothesis that intersexual conflict over the control of copulation initiation and duration has driven the evolution of secondary sexual armaments on both sexes within that clade.

APPLICATIONS

Below, we consider three different applications of behavioral ecology. In each case, behavioral decisions are scaled up to some higher-level process. For example, attack decisions by predators are scaled up to recruitment at the population level via birth and death rates. Because behavioral decisions are often nonlinear (e.g., step functions), it is often not obvious what the impact of behaviors will be on higher-order processes. For example, population dynamics can range from stable limit cycles to chaos with strange attractors as a result of the feedback between flexible exploitation behaviors and resultant changes in resource availability.

Conservation Biology

Conservation biology is a field of biology that is employed to reduce the risk of extinction for one or more species. There are several means by which this can be accomplished, including (i) retaining, restoring, or expanding critical habitats, (ii) providing protection through reserves, (iii) reducing the impact of natural enemies on focal species (cf. biological control), and (iv) mitigating anthropogenic disturbance. Behavioral ecology comes into play when individuals of focal species modify their behavior in response to such interventions and, therefore, impact population processes. Take the effect of natural enemies. In a simple Lotka–Volterra model of predator–prey interactions, the dynamics can be described by

$$\begin{aligned} \frac{dV}{dt} &= (b_v - aP)V, \\ \frac{dP}{dt} &= (acV - d_p)P, \end{aligned} \tag{5}$$

where V is the prey population, P is the predator population, bv is the birth rate of the prey, a is the attack rate on prey by predators, c is the conversion rate of prey into predator offspring, and dp is the death rate of the predator. This mass action model assumes that both prey and predator behaviors are invariant, i.e., death rates of prey and birth rates of predators are products of constant interaction terms and population densities. On the other hand, behavioral ecologists posit and frequently observe such organisms to alter their behavior either to avoid natural enemies or more readily exploit resources (prey). As noted above (see theory section), however, such modifications often come with costs (e.g., hiding from predators can lead to reduced feeding rates and ultimately lower fecundity), suggesting that there is an optimal level of investment in antipredation. This means that predators can impact prey populations in two different ways, either through removal of prey (direct consumptive effects) or by effecting changes in prey behavior that impacts their reproduction (nonconsumptive effects). These two effects are often termed density-mediated and trait-mediated effects, respectively.

One can now rewrite the traditional Lotka–Volterra model taking nonconsumptive effects into account:

$$\begin{aligned} \frac{dV}{dt} &= (b_v f(P) - af(P)P - ef(P))V, \\ \frac{dP}{dt} &= (af(P)cV - d_P)P. \end{aligned} \tag{6}$$

Now, birth rates for prey will be reduced (the trait-mediated effect) due to (optimal) investment by prey in antipredator behaviors (e.g., hiding) as will capture rates (the density mediated effect), which will in turn impact predator birth rates. In addition, prey may emigrate from their local population as a function of predator threat. Notice that birth rates and capture rates are now functions of predator density. This implies that prey will adjust their level of antipredation investment according to the current risk of predation. Together, these modified equations can generate different dynamics from the behavior-invariant models, and in fact such NCE models have recently been described as a possible missing link for describing population dynamics. Determining both consumptive and nonconsumptive effects may suggest different interventions than those based upon consumptive effects alone. Finally, since predators and their prey rarely live in isolation from the rest of the community, inclusion of adaptive behavior at different trophic levels is also an essential component of conservation-based behavioral ecology.

It is also important to remember that many predators also have their own predators and would be expected to invest in antipredation behaviors as well. This can lead to reduced foraging by the focal predators and, by extension, reduced predation threat on the focal at-risk prey. With the new, lower predation risk, those prey now need not invest so much to offset predation risk and can invest more in reproduction. An example of this is the scarecrow effect, wherein the presence of eco-tourists can cause mongooses to forage less in areas where sea turtle eggs are vulnerable.

Finally, behavioral ecology may be a particularly important tool for identifying ecological traps. Ecological traps exist when environmental change (e.g., through anthropogenic disturbance) causes organisms to settle in poor quality habitats because they misinterpret habitat quality cues. For example, some species of mosquitoes and beetles preferentially choose to colonize habitats that are treated with particular pesticides whereas other species are either indifferent to or discriminated against such sites. This differential colonization and survival may have important effects on community structure. Where behavioral ecology may play an important role is in predicting the kinds of cues that organisms use to assess sites (see "Behavioral Rules," above) and how they might be altered to mitigate ecological traps.

Darwinian Medicine

Medical practitioners, for the most part, employ a proximate perspective with their craft. They determine the causal effect of some disease and then seek cures to mitigate disease symptoms. Therefore, they ask, does the cure work? Rarely do medics concern themselves with the ultimate or functional reason for presence or severity of any given disease. By contrast, behavioral ecologists study both the proximate and ultimate explanation for disease patterns in nature. In general, this latter approach falls under the rubric of Darwinian medicine.

Darwinian medicine is a relatively young field. It was first defined by George Williams and Randolph Nesse in 1991 as the application of evolutionary theory to health and disease. It has been applied to a wide range of diseases, both infectious and self-generated.

One good example of Darwinian medicine is the hygiene hypothesis that explains immune system disorders. It is now established that chronic inflammatory disorders (IDs; e.g., allergies, arthritis, atherosclerosis) have become more prevalent of late in developed countries. At the same time, rates of infectious diseases in those countries have plummeted. Is there a connection? Is it coincidental, for example, that immigrants to Israel from developing nations are far less likely to show IDs than immigrants from Europe?

The hygiene hypothesis states that some of this increased prevalence is the result of defective regulation of the immune system resulting from diminished exposure to some classes of microorganism. This is a proximate explanation, but it doesn't explain why such an effect should exist. Underlying mechanisms have been sought, but none of them provide consistent explanations for global patterns of ID.

A functional explanation goes by the name of the "Old Friends" Hypothesis. The idea here is that in less hygienic times, most humans were exposed to and carried many parasites, including parasitic helminth worms, many of which are (i) relatively innocuous and (ii) difficult to eliminate without causing serious collateral damage (e.g., lymphatic damage). As a result, the ultimate hypothesis predicts that humans will evolve to tolerate such parasites. Moreover, according to the theory, a parasite–host relationship has evolved wherein such parasites suppress the host immune system from producing an inflammatory response to those parasites as well gut flora, allergens, and host cells. In the absence of helminthes, due to hygienic practices, inflammatory responses are no longer suppressed and expected to greatly increase. Thus, this behavioral ecology approach to ID provides both an ultimate explanation and a mechanism. At the moment, there is no formal theory for the Old Friends Hypothesis.

Nausea and vomiting during the first trimester of pregnancy (NVP), often referred to as morning sickness or *nausea gravidarum*, is a disease that appears to be unique to humans. Many proximate mechanisms have been proposed, including increased sensitivity to odors, but only of late have behavioral ecologists attempted to explain the presence of this unique and often debilitating disease.

Darwinian medicine proposes an adaptive explanation for NVP: prophylaxis. Morning sickness causes its victims to generally avoid and/or reject meat and strong-tasting vegetables, two foods that historically may have harbored parasites and toxins. There are two adaptive functions here: (i) to limit exposure of the developing fetus when it is most susceptible to substance that cause abnormal growth (teratogens), and (ii) to limit mother's exposure to parasites when they are immunosuppressed to limit rejection of the only partially related fetus. Data on eating habits for first-trimester women are consistent with this hypothesis wherein foods such as cereals, pulses, and spices are preferred over meat, eggs, and fish. Another evolutionary but nonadaptive explanation is that NVP arises as a side effect of a coevolutionary arms race between mothers and offspring.

Biological Control

Classical biological control is a technique employed to suppress pest species (including both animals and plants) below some defined economic threshold. Usually, the target pest species are exotics and candidate agents are typically selected from the pest's region of origin. This last is predicated on the assumption that a long association increases the likelihood that an agent will be more likely to efficiently exploit the target pest species. Biological control efforts have become less phenomenological over the years, and there has been an increasing emphasis on understanding the population dynamics of the agent–host relationship. This has led to increasing interest in the behavioral mechanisms that underlie population-level phenomena. This is important because the goal of classical biological control is the establishment of stable populations of the agent and the pest (kept below its economic threshold) in order to maintain long-term, low-cost control. Because of thorough theoretical and empirical work on exploiter–victim systems, the influence of host narrowness or host breadth, propensity to over- or under-exploit host patches, greater or lesser searching efficiency, and longer or shorter handling time all have fairly predictable effects on system persistence and on an agent's tendency to exploit nontarget hosts. Behavioral assays are now a necessary first step in the screening process used to select agents, inasmuch as they can provide insight into the likelihood that a system will be persistent and on an agent's tendency to exploit nontarget hosts.

SEE ALSO THE FOLLOWING ARTICLES

Adaptive Behavior and Vigilance / Adaptive Dynamics / Cooperation, Evolution of / Evolutionarily Stable Strategies / Foraging Behavior / Game Theory / Mating Behavior

FURTHER READING

Arnqvist, G., and L. Rowe. 2005. *Sexual conflict*. Princeton: Princeton University Press.

Clark, C.W., and M. Mangel. 2000. *Dynamic state variable models in ecology. Methods and applications*. New York: Oxford University Press.

Clutton-Brock, T. H. 1991. *The evolution of parental care*. Princeton: Princeton University Press.

Fisher, R. 1930. *The genetical theory of natural selection*. Oxford: Oxford University Press.

Grafen, A. 1984. Natural selection, kin selection and group selection. In J. R. Krebs and N. B. Davies, eds. *Behavioural ecology: an evolutionary approach*, 2nd ed. Oxford: Blackwell Scientific.

Gosling, L. M., and W. J. Sutherland, eds. 2000. *Behaviour and conservation*. Cambridge, UK: Cambridge Univeristy Press.

Krebs, J. R., and N. B. Davies, eds. 1996. *Behavioural ecology: an evolutionary approach*. Cambridge, UK: Blackwell Science.

Nesse, R., and G. C. Williams. 1995. *Why we get sick*. New York: Time Life Education.

Stephens, D. W., and J. Krebs. 1987. *Foraging theory*. Princeton: Princeton University Press.

Wajnberg, E., C. Bernstein, and J. Van Alphen. 2007. *Behavioural ecology of insect parasitoids: from theoretical approaches to field applications*. Chichester, UK: Wiley-Blackwell.

BELOWGROUND PROCESSES

JAMES UMBANHOWAR

University of North Carolina, Chapel Hill

A variety of ecological processes occur belowground. Many are analogous to aboveground processes, but some only occur belowground. These belowground processes have been less emphasized in the general and theoretical ecological literature than the activity aboveground, despite the recognition of the importance of these processes in ecological systems. Several factors have hindered a more full development and testing of ecological theory for belowground processes. Prominent are the technical difficulties of direct measurement and observation of soil organisms in their natural habitat. It is the nature of soil that sampling in it greatly disturbs the structure to which organisms living in it are adapted. Further, the bulk of belowground biomass is in microbes that have historically only been identified by culturing and quantified by means of bulk chemical analysis. Finally, a rich development of theory covering belowground processes has been hindered by the historical disciplinary split between ecosystem ecology and community and population ecology. Despite these hindrances, much theory has been developed covering important belowground processes.

NUTRIENT DYNAMICS AND PLANT COMPETITION

General Theory

Plant growth and primary productivity in all habitats is limited by mineral nutrients such as nitrogen (N) and phosphorus (P), though water, light, temperature, and CO_2 are all important variables that alter rates of growth and productivity. In terrestrial ecosystems, nearly all of the mineral nutrients and water are taken up from the soil either directly by plant roots or indirectly by fungal symbionts. Ecological theory has elucidated the impacts of plant nutrient uptake and nutrient dynamics in the soil on various aspects of plant competition. Specifically, resource competition theory relates the ability of plants to grow at different nutrient concentrations in the soil, R, to their competitive ability. As plants grow in abundance and biomass, they consume resources, thereby lowering the nutrient concentration in the soil. The theory posits that each species has a critical nutrient concentration, termed the R^*, where its growth rate is exactly zero. This R^* value is also the equilibrium resource concentration of a system with only that plant. The final conclusion of this theory is that the plant species having the lowest R^* will eventually outcompete other species as it reduces resource concentrations in the soil below those able to sustain less competitive species (those with higher R^*'s). Resource Ratio Theory extends this theory to detail how the presence of multiple limiting resources constrain the number of species that can persist to equal the total number of limiting nutrient. The eventual outcome of competition depends on the relative supply of these resources and the relative ability of plants to take up and use the resource, hence the name Resource Ratio Theory. These general theories of resource competition apply most specifically to well-mixed systems, such as aquatic systems where turbulence tends to mix nutrients widely across spatial scales greater than the size of the volume of plants depleting them.

Spatial Effects

LOCAL INTERACTIONS

In contrast to the general assumptions of resource competition theory, terrestrial plants are sessile and nutrients are relatively immobile in the soil. This combination creates the potential for spatial heterogeneity in resource concentrations, supplies, and uptake. This heterogeneity provides opportunities for plant species that have higher R^* values than those of the best competitor to maintain themselves in the community. Various modeling strategies have been developed to account for differences in spatial and temporal scales of movement of nutrients (and therefore competitive effects) and plants (mostly seed dispersal) and their effects on plant competition. Most commonly, models of spatial processes of plant resource competition have assumed strong local competition, with a regional mixing of either nutrient or plants. Coexistence on one limiting nutrient has been demonstrated with at least two different regional mixing mechanisms. The first framework assumes that nutrient movement is minimal and plant competition only occurs at the scale of one plant. Plants disperse via seeds at a much larger spatial scale. Coexistence among plants occurs when tradeoffs occur between plants' R^* and their ability to colonize bare earth, which is an amalgam of seed production, seed movement, and growth rates. The second framework assumes a small amount of regional mixing of nutrients such that plants can compete with distant plants. This approach generally does not examine the effect of variation in dispersal by plants and suggests that significant heterogeneity in soil nutrient levels can prevent, at least in the short term, the competitive exclusion of plants that are not good competitors for resources.

Finally, this local-regional framework provides an understanding that the impacts of spatial heterogeneity in resource supply rate can facilitate coexistence to a greater extent than Resource Ratio Theory would predict. For example, while simple, nonspatial Resource Ratio Theory predicts that only two plant species can persist in an environment with two limiting resources, spatial heterogeneity allows for many (possibly an infinite number of) species to persist. Key to this theory is that there is little movement of either resources or plants across space. Further, resource supply rates cannot spatially covary in a positive fashion, and plant species need to have tradeoffs in their resource use capacity.

NEIGHBORHOOD INTERACTIONS

Relaxing the assumption that competition for resources only occurs at local scales has been less studied, as models of these types of systems are resistant to analysis. One method of accommodating some implicit movement of resources is through the use of neighborhood models of plant competition. These models assume that plants compete more strongly with a set of nearby plants. While these models do not explicitly track soil nutrient flow, one justification for this approach is based on the fact that soil nutrients are not mobile and plant roots can only forage in a small area near the main stem of the plant and the scale of nutrient movement can be encompassed in the neighborhood size. It needs to be noted that this type of competition could also be interpreted as competition for aboveground resources such as light. Given the various scales of competition involved in these issues, analysis frequently relies on computational simulation of individual-based models that have been parameterized to represent particular plant communities. This type of approach has been used to explore the varying strategies in terms of allocation to growth and dispersal that plants can use in competition and the role that these strategies play in determining the diversity of plant communities, demonstrating that there exists more than one dispersal and competition strategy tradeoff that allows for plant coexistence.

DECOMPOSITION

Much theory about nutrient dynamics and plant competition assumes that nutrients instantly transform from plant biomass into mineral nutrients suitable for uptake by other plants. Obviously this assumption is not real, as decomposition is a slow biological process. There is a large body of theory that tries to predict the variation in decomposition rate of dead material. Given that vast majority of biomass in terrestrial systems is in plants, most of this theory has focused on the decomposition of dead plant material (litter). When plants and animals die, these nutrients are eventually released back into the soil via decomposition, mostly by microbes such as bacteria and fungi. The simplest phenomenological models of litter decomposition assume a first-order decay of decomposed material that release CO_2 and mineral nitrogen. These simple models do not adequately capture the dynamics of dead matter for a variety of reasons, described below. Despite this fact, this simple model is frequently used in models of community and population dynamics that incorporate decomposition as a process in them.

One important aspect of plant litter that explains why simple exponential decay is not an adequate model is that it is chemically diverse. This diversity manifests itself within and between individual plants and across plant species, and theoretical studies have taken advantage of both types of diversity. The major determinants of quality differences in plant tissue that explain variation in decomposition rates are nutrient content expressed as C:N ratios and the size of molecules. Fast-decomposing material contains high concentrations of N and smaller C molecules, while slower-decomposing material has low concentrations of N and contains complex C molecules such as lignin. A variety of modeling approaches have been used to describe this variation and its impacts. The phenomenological approach to incorporating this chemical diversity divides litter into multiple fractions, each with a different decay rate. These models provide a better fit to experimental data about the decay of individual litter cohorts than the simple decay models, and they have proved to fit data about long-term dynamics of soil organic matter better than simple one-quality models.

Long-Term Dynamics of Soil Organic Matter

More mechanistic approaches to modeling decomposition follow not just decay of plant litter but the dynamics of soil organic matter (SOM) split into at least three categories of quality. Discrete compartment models of SOM also explicitly assign carbon-to-nutrient ratios to different pools. These frameworks allow for not just evaluation of the decay of organic matter but the mineralization and immobilization of nutrients such as N and P that are then used by soil microbes and plants for growth (see Fig. 1). These complex models were designed to incorporate most ecosystem processes and provide quantitative fits to systems to which they have been parametrized. These efforts have been successful, most notably for the Rothamsted and CENTURY models. Their success in producing accurate dynamics of soil organic matter and mineral nutrients has led to their inclusion as modules in full ecosystem models and models of climate change to determine feedbacks between climate and ecosystems. However, the design underlying these models makes them relatively resistant to analysis and hence unsuitable for use in theoretical exercises. More theoretical examinations of soil and ecosystem dynamics use reduced versions of these models containing just two or three SOM compartments of varying decomposability to facilitate analysis. This simplifying assumption has been demonstrated to be reasonable by analyzing linearized versions of the more complex models for many estimated parameter values. These analyses show that soil organic matter, as a good approximation, essentially follows a one-directional process from litter through microbes to the atmosphere in the form of CO, while releasing any nitrogen into the soil in mineral form.

A qualitatively different approach to modeling soil carbon dynamics is to view quality as a continuously varying quantity. SOM is then described as a distribution of qualities, and the dynamics of this distribution are caused by soil microbes ingesting the material and releasing it as they die. Matter is typically transformed to organic matter of lower quality while some is released as CO_2 due to respiration. These models are represented mathematically as integro-differential equations, which, without various types of simplifying assumptions, are difficult to analyze and use in full ecosystem models. A significant theoretical advantage of these models, however, is that they include nearly all other models of decomposition as subsets of various functional and parameter combinations and may be useful for comparing various models, such as those compartment models discussed above.

Indirect Interactions Between Plants and Decomposers

COMPETITION AND MUTUALISM BETWEEN PLANTS AND DECOMPOSERS

Plants and decomposing microbes engage in an interaction that is fairly unusual among ecological interactions. For the most part, it is an indirect interaction that can vary from mutualistic to competitive. Mutualism arises because plants produce energy-rich organic carbon that microbes use for

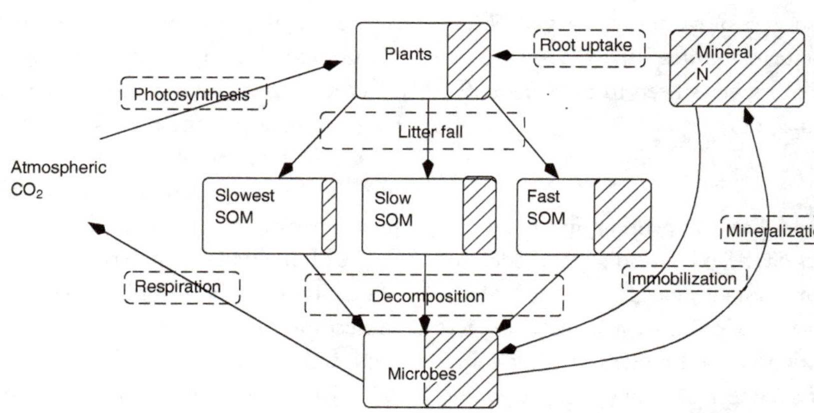

FIGURE 1 Diagram of the general structure of major material fluxes belowground. Relative nutrient content is demonstrated by size of the patterned section of pools. Major processes are identified by dashed boxes.

growth and respiration and the action of microbes mineralizes nutrients from dead plant matter. The potential for competition arises because both organisms require mineral nutrients for growth. Plants get most of their nutrients from inorganic sources in the soil, while microbes access nutrients from both the inorganic sources in the soil and from organic detritus. When detritus does not contain sufficient nutrients, microbes immobilize mineral nutrients in the soil. Theory suggests that the relative strength of mutualism and competition in ecosystems differs among open and closed systems. In particular, in closed systems where there is no source of mineral nutrient other than the release from dead organic matter by decomposers, the system needs to be a mutualistic system at equilibrium or both species will go extinct. A necessary condition for mutualism in closed and open systems is that decomposers need to be limited by carbon from organic matter, and this condition will be met when decomposers are better competitors for the mineral nutrients for which they compete. Open systems do not require a net mutualism between plants and microbes, as inputs of carbon or nutrients could support decomposers or plants, respectively, if they were to exclude the other. Still, in this case, the absence of recycling/producing of these nutrients will greatly reduce abundances of the species that wins in competition.

ABIOTIC FORCING AND THE RESPONSE OF TERRESTRIAL SYSTEMS TO ABIOTIC FORCING

Given that there is long-term persistence of plants and decomposers, there is a large body of theory that seeks to understand how this interaction responds to forcing by changes in the abiotic environment. Key abiotic forcing variables are changes in CO_2 caused by human activities, the concomitant increase in temperature caused by CO_2 increase, and nutrient addition, such as increases in N deposition caused by burning of fossil fuels. These forcings are interesting in that they provide tests of different theories about how ecosystems work, but they are the source of key feedbacks between the biotic systems and climate. Soil organic matter is a large pool of carbon, while production and decomposition are large fluxes of carbon; depending on the net effects of these forcings on different aspects of ecosystems, soil organic pools can either be increased and become a net sink, or decreased to become a net source of carbon. Theory suggests key factors that need to be better known to accurately predict ecosystem responses to abiotic forcings. First, all models of varying SOM quality make assumptions about whether a class of SOM is wholly resistant to degradation or whether it still degrades slowly. If a wholly resistant class does not exist, increases in temperature will increase the decomposition rate of all soil organic matter, potentially leading to a significant source of new CO_2 in the atmosphere and creating a positive feedback between climate and biotic processes in warming. Second, a key factor is the ability of ecosystems to respond to the limitation caused by nutrients after CO_2 increases the productivity of plants. Physiological and compositional changes in plants and microbes could allow productivity to continue to increase despite no increase in nutrient supplies in the ecosystem. If this ecosystem adaptation is strong enough, terrestrial ecosystems could be sinks for CO_2 despite increases in decomposition rate caused by increased temperature.

ROLE OF DETRITUS IN STABILITY OF COMMUNITIES

Consumer–resource interactions have the potential to generate cyclic dynamics. Theory suggests that the presence of a detrital compartment stabilizes community dynamics when plant growth is limited by nutrients held in detritus. This is despite the fact that most models that this theory is based on include only a single detritus compartment where decomposition occurs according to a first-order decay process. In these cases, detritus alters the properties of stability as it provides a buffer against perturbations to the system. An interesting corollary to this is that these communities are particularly prone to perturbations of detritus, which can temporarily destabilize their dynamics.

SPECIES INTERACTIONS IN SOILS

While most general theory of species interactions should hold for interactions in the soil, theory about interactions in the soil has diverged from other interaction theory. First, soil ecologists have embraced the importance of interactions such as parasites and mutualists more than aboveground ecologists. Plants are afflicted by many belowground pathogens. Plants also interact with many microorganisms in the soil in a (mostly) mutualistic fashion. Most commonly studied is the nearly ubiquitous mycorrhizal symbioses. Mycorrhizal symbioses are interactions where fungi grow both in the roots of plant and in the soil. The fungi take soil nutrients (both N and P) and give it to the plant. In turn, they receive photosynthate to be used for growth and respiration. While generally viewed as a mutualism, under some common circumstances it can be viewed as a parasitic interaction where fungi receive benefits from plants but the resources that plants receive are insufficient to overcome the cost of photosynthates negatively impacted by the presence of fungi. Several types of mycorrhizas are recognized; the two most common are ectomycorrhizal fungi that associate with woody plants in a small set of families and provide phosphorus and different forms of nitrogen, and arbuscular mycorrhizal fungi that associate more generally and provide primarily phosphorus compounds.

Feedback Theory

The difficulty of quantifying soil organisms has led to the development of a theory that more generally addresses the interactions between soil organisms and plants. This theory, under the rubric of feedback dynamics, has been developed to explore the impact of plant–microbe interactions on plant community dynamics. The feedback approach envisions that plant species culture a particular soil community that can have a variety of effects, either directly positive or negative, on the culturing plant or other plant species. Key to determining the impact of these plant–soil microbe community feedbacks is the relative impact of each soil community on plants and the relative ability of plants to culture these soils. Three qualitative outcomes for long-term community dynamics can be characterized from this theory, determined by the relative benefit that each plant of a class receives from its soil community. In the first, competitive dominance can occur if one plant receives higher relative benefits from both of the soil communities. For example, if both soil communities increase the abundance of plant A, plant A will tend to out-compete plant B. The second outcome is competitive coexistence, which occurs when benefits are distributed in a fashion that creates a negative feedback among species in their relative abundance (Fig. 2a). This feedback occurs when the soil microbe community cultured by plant species A provides a significant relative benefit to from plant species B and, conversely, the soil microbe community cultured by plant species B leads to a high relative benefit accruing to plant species A. The soil microbe community of this two-plant species community will be a combination of those generated under both communities. In contrast, the third qualitative outcome that arises from these dynamics occurs if the soil communities cultured by plants give relatively larger benefits to the culturing plants than to the nonculturing plants. This leads to a situation where one plant species will dominate, and the identity of that plant species will depend on its initial abundance in the community (Fig. 2b). Extensions of this theory to a spatial context suggest that allowing for explicit spatial structure with strong local interactions in relation to movement rates allows for coexistence of species when globally interacting species do not persist.

While the mathematical model of these feedbacks is relatively simple, its utility has been its relative tractability to experimental testing and generalization. The theory can be tested by conditioning soil by growing plants in that soil and testing the effect of that conditioned soil on the growth of each plant species. If the theory is correct, the outcome of competition should be determined by the relative growth response of plants grown in soil conditions by themselves versus soil conditioned by heterospecifics. This theory has been tested and demonstrated to have predictive power in many communities. An intriguing part of the general theory is that it is relatively catholic in regards to the type of interactions it can be applied to, but it is generally thought to incorporate important interactions such as soil pathogens and mutualists such as mycorrhizal fungi. Its explanatory value relies once again on the relatively small movement rates of soil microbes in relation to their growth rate.

Microbes and Nutrient Dynamics

Soil feedback theory does not explicitly examine interactions between soil nutrient dynamics and soil microbe communities, despite strong evidence that many soil microbes do alter these dynamics. In particular, mycorrhizal fungi increase uptake of soil nutrients. Several intuitive results have been advanced by arguing that mycorrhizal fungi act

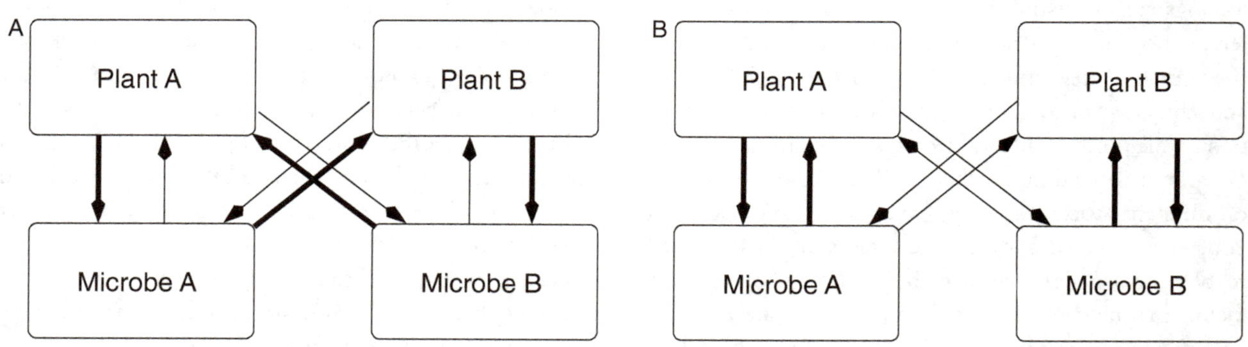

FIGURE 2 Schematic diagram of feedbacks between plants and soil microbe communities that they culture. Arrow thickness represents the relative positive impact of a plant/microbe on the community. (A) Representation of negative feedback where soil microbe communities tend to reduce the relative fitness of the plant under which they are hosted. (B) Representation of positive feedback where soil microbe communities tend to reduce the relative fitness of the plant under which they are hosted.

to alter plant characteristics within the confines of simpler models. These applications assume that mycorrhizal fungal populations can, when present, be assumed to be static. For example, if mycorrhizal fungi, by altering the resource uptake of a plant, increase that plant's competitive ability on that resource, these mycorrhizal fungi can be responsible for the competitive dominance of that species or may facilitate species coexistence by allowing for niche differentiation of plant species on different nutrients. Fewer studies have examined how feedbacks between plant density and mycorrhizal fungal density may alter the interaction between plants and nutrient dynamics. Initial theory has been useful in demonstrating that a single mycorrhizal fungi can facilitate the coexistence of two plant species competing for one nutrient in the soil. This result is topologically similar to predator-mediated competition. Current theory has yet to differentiate among different types of mycorrhizal fungi, though their function and effect on plant communities can vary. The two most widely distributed classes have affinities for nutrients and plant hosts: ectomycorrhizal fungi that associate with woody plants in a small set of families and provide phosphorus and different forms of nitrogen, and arbuscular mycorrhizal fungi that associate more generally and provide primarily phosphorus compounds.

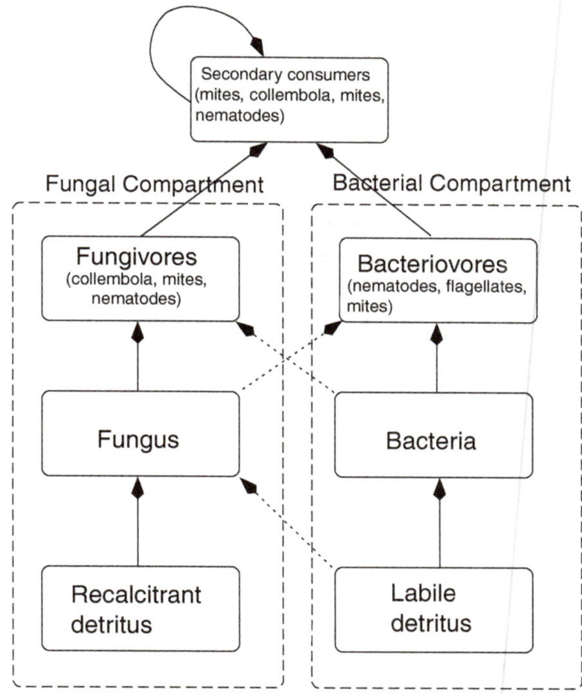

FIGURE 3 Schematic diagram of the general structure of a soil detritus food web. Solid arrows represent relatively strong links, and dotted arrows represent weaker links. Labile detritus and recalcitrant detritus form roughly independent chains that are unified by tertiary consumers. Tertiary consumers undergo large amounts of intraguild predation.

SOIL FOOD WEBS

In contrast to most aspects of soil biology, soil food webs have been an important model system for the development of broad ecological theories. Soil ecologists have been better able than aboveground or aquatic ecologists to generalize a structure of food webs across different systems (Fig. 3). The basis of this structure arises from the aforementioned variability in quality of different sources of energy in litter and the consumers who specialize on those different sources. In particular, the variation in quality of detritus produce two sources of energy that in turn create compartments within food webs within which interactions are strong compared to the interactions between those compartments. These compartments arise from relatively decomposable detritus, which is primarily fed upon by bacteria, and less decomposable detritus, which is a source of food for fungus. Primary consumers of bacteria and fungi tend to only feed on one or the other and in turn are fed on by secondary consumers that specialize within these compartments. However, at higher trophic levels, soil fauna tend to generalize so that tertiary consumers are feeding on secondary consumers from both compartments. This structure has been demonstrated to have a high potential to generate stable long-term dynamics in speciose food webs, in contrast to the general property that randomly structured food webs become more unstable as they increase in diversity. The mechanisms generating this stability are that any fluctuations in compartments that are generated are likely to out of sync with each other and the higher-level consumers that link these compartments tend to average out the asynchronous fluctuations of the different compartments.

An important characteristic of soil food webs that also tends to generate stability is that the bases of these food webs are donor controlled. Detritivores do not (directly) alter the production of new detritus, and this fact lends stability to this interaction as the influx of new detritus tends to be uncorrelated with any internally generated fluctuations in abundance. A standing crop of detritus also tends to buffer systems from external nutrient shocks. Comparisons of food chains based on detritus compared to food chains based on plants that are not donor controlled show that the presence of donor control in the basal consumer–resource interaction stabilizes detrital food webs in comparison to nondonor-controlled webs.

SUMMARY

A wide variety of ecological theory has been developed to describe belowground processes. This theory spans disciplinary boundaries within ecology, providing a common cause for ecosystem ecologists and community and population ecologists. Future developments will further integrate processes of material flow and changes in community and population structure. The other major factor that underlies successful ecological theory regarding belowground processes is theory that provides measurable quantities in the difficult empirical conditions that the soil provides. Finally, theoretical studies will provide guidance in the future for a fuller integration of ecological processes occurring both above- and belowground.

SEE ALSO THE FOLLOWING ARTICLES

Biogeochemistry and Nutrient Cycles / Dispersal, Plant / Environmental Heterogeneity and Plants / Food Webs / Microbial Communities

FURTHER READING

Ågren, G. I., and E. Bosatta. 1998. *Theoretical ecosystem ecology: understanding element cycles.* Cambridge UK: Cambridge University Press.

Bolker, B. M., S. W. Pacala, and W. J. Parton, Jr. 1998. Linear analysis of soil decomposition: insights from the CENTURY model. *Ecological Applications* 8: 425–439.

Coleman, D. C., D. A. Crossley, and P. F. Hendrix. 2004. *Fundamentals of soil ecology.* New York: Elsevier.

De Angelis, D. L. *Dynamics of nutrient cycling and food webs.* 1992. New York: Chapman and Hall.

Moore, J. C., E. L. Berlow, D. C. Coleman, P. C. de Ruiter, Q. Dong, A. Hastings, N. C. Johnson, K. S. McCann, K. Melville, P. J. Morin, K. Nadelhoffer, A. D. Rosemond, D. M. Post, J. L. Sabo, K. M. Scow, M. J. Vanni, and D. H. Wall. 2004. Detritus, trophic dynamics and biodiversity. *Ecology Letters* 7: 584–600.

Reynolds, H. L., A. Packer, J. D. Bever, and K. Clay. 2003. Grassroots ecology: plant-microbe-soil interactions as drivers of plant community structure and dynamics. *Ecology* 84: 2281–2291.

Wardle, D. A. 2000. Communities and ecosystems: linking the aboveground and belowground components. Princeton: Princeton University Press.

BEVERTON–HOLT MODEL

LOUIS W. BOTSFORD

University of California, Davis

The Beverton–Holt model is a mathematical representation of the dependence of the annual recruitment to a population on the number of eggs produced by the population. It is most frequently used in age-structured population models, where it plays a fundamental role in determining population dynamic behavior. It had its origins in, and is most commonly associated with, marine fisheries, though it does appear in other contexts (e.g., waterfowl).

THE ORIGINS

The Beverton–Holt model was first presented in Beverton and Holt (1957) in a comprehensive description of the mathematical analysis of fisheries motivated by the declining fish stocks in the North Sea as fishing increased following World War II. That publication is so comprehensive that, in spite of its age, it is still worth reading. In fact, there is a saying among fishery analysts that if you ever think you have thought of something new in the mathematics of fisheries, you should check Beverton and Holt because it is quite likely that they said something about it (and made some useful calculations, even with the mechanical calculators of the day).

Beverton and Holt (1957) made the observation that populations usually fluctuated about a constant value. Mathematically, they noted that if the number of eggs produced in the lifetime of each recruit and the number of recruits that resulted from each egg were constant, then a population would rarely remain constant, but rather would increase or decrease geometrically. From this they concluded that there must be some kind of density dependence in one of these, and they assumed that it was in the survival from eggs to recruits. They proposed that as individuals developed from eggs to recruits, abundance, N, in a series of sequential stages declined according to

$$\frac{dN}{dt} = -(\mu_{1,s} + \mu_{2,s}N)N, \tag{1}$$

where the values of $\mu_{1,s}$ and $\mu_{2,s}$ could vary from stage to stage (denoted by s). Solving Equation 1 for an arbitrary number of sequential stages, they showed that the relationship between the initial value, total egg production, E, and the final value, recruitment, R, could be written as

$$R(E) = \frac{E}{\frac{1}{\alpha} + \frac{E}{\beta}}, \tag{2}$$

where α was an expression containing only values of $\mu_{1,s}$ for the various stages, and β contained values of both $\mu_{1,s}$ and $\mu_{2,s}$. According to this relationship, recruitment increases with egg production at a diminishing rate and asymptotically approaches a constant (Fig. 1). In the form that Equation 2 is written here, α is the slope of the egg-recruit function at the origin, and β is the asymptotic value of R at high egg production.

POSSIBLE MECHANISMS

The distinctive aspect of the Beverton–Holt model is its shape and how that is related to the possible mechanisms

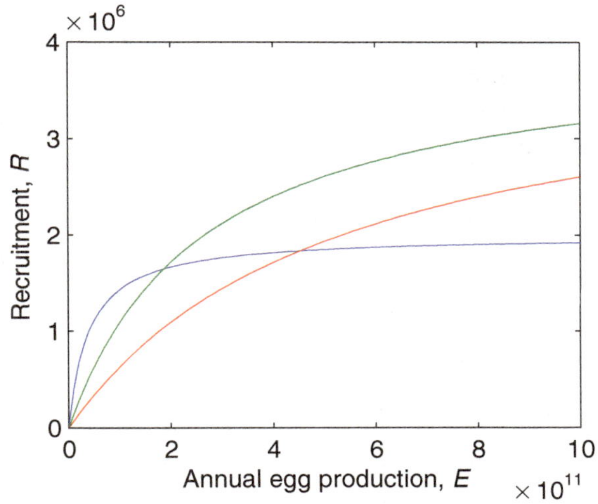

FIGURE 1 Examples of the Beverton–Holt model representing the dependence of recruitment to a fish population on annual egg production. The two parameters of this equation represent the slope of the function at the origin and the asymptotic value of recruitment at high egg production. The curves shown are for $\alpha = 5 \times 10^{-5}$ and $\beta = 2 \times 10^6$ (blue line), $\alpha = 1.5 \times 10^{-5}$ and $\beta = 4 \times 10^6$ (green line), and $\alpha = 7.5 \times 10^{-6}$ and $\beta = 4 \times 10^6$ (red line).

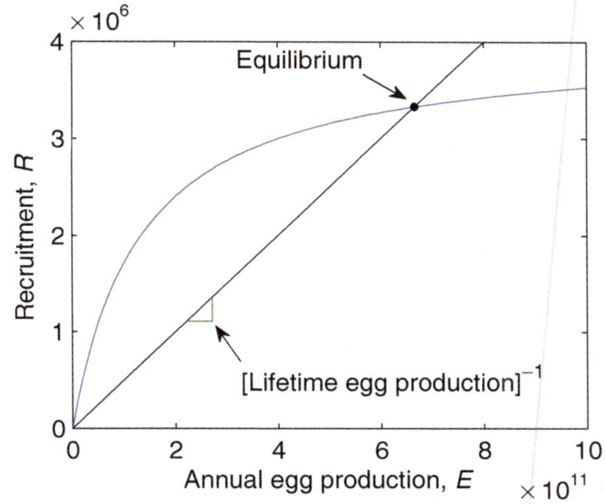

FIGURE 2 The graphical representation of the equilibrium condition of an age-structured model with density-dependent recruitment, using the Beverton-Holt model to describe the density dependence (blue line). The equilibrium is where a straight line through the origin with slope equal to the inverse of the lifetime egg production (black line) intersects the egg-recruit curve. If a population is exposed to increased adult mortality from fishing or pollution, as examples, lifetime egg production will decrease causing the equilibrium point to move to the left and downward. If the lifetime egg production is reduced to the point that it equals the inverse of the slope of the egg-recruit function at the origin, the population will collapse.

underlying density-dependent recruitment. The Beverton–Holt model would apply to mechanisms involving a limited amount of space or some other resource in which increased limitation of recruitment with increased egg production did not ever cause a decline in recruitment. An example would be the limited number of nesting sites for a hole-nesting bird. If the nesting behavior of a species involved a competitive mechanism that reduced recruitment with greater egg production, the Beverton–Holt mechanism would not apply. An example would be some sort of nesting behavior in which a likely outcome of the competition was lack of reproduction by both competitors. Competitive mechanisms involving declining recruitment at high egg production would be best modeled by the Ricker model or some other stock-recruitment model.

POPULATION DYNAMIC IMPLICATIONS

The theoretical importance of the Beverton–Holt model is tied to the kind of population behavior to which it leads. Different egg-recruit relationships lead to different kinds of population behavior. For example, an age-structured model with the egg-recruit relationship determined by the Ricker model would be less stable than one with the Beverton–Holt model.

Equilibrium

The first population characteristic of interest would be the equilibrium level of an age-structured population model with a Beverton–Holt egg-recruit relationship. Fortunately, the equilibrium condition can be represented graphically (Fig. 2): the equilibrium levels of egg production and recruits lie at the intersection of the egg-recruit function and a straight line through the origin with a slope of (lifetime egg production)$^{-1}$. This relationship has the intuitive interpretation that a population will be constant at the point where the number of recruits per egg in reproduction exactly equals the inverse of the number of eggs produced in the lifetime of each recruit.

This equilibrium condition also provides a useful interpretation of the effects of increased adult mortality upon a population, such as might be caused by fishing or pollution. As adult survival declines, lifetime egg production will decline, and the slope of the straight line in Figure 2 will become steeper, causing the equilibrium to move to the left. One can see that initially an increase in adult mortality might not cause much of a change in recruitment, but as the adult mortality declined further, recruitment could decline substantially.

The equilibrium condition also provides a convenient interpretation of persistence in terms of the concept of replacement. The condition for persistence, i.e., that lifetime egg production be greater than 1/(slope of $R(E)$ at the origin), is consistent with the notion that a population will persist if individuals reproduce enough in their lifetime to replace themselves. In this case the inverse of the slope of $R(E)$ at the origin is the required threshold for persistence. This value is $1/\alpha$ in the Beverton–Holt model.

A completely collapsed population, i.e., $R(0) = 0$, is, of course, also an equilibrium. Note that it is the other point where the straight line in Figure 2 intersects the egg-recruit function.

Stability

A second informative characteristic of population dynamics is stability. Here, we consider only local stability, i.e., whether a population when perturbed from one of the equilibria will return to that equilibrium. The condition for local stability of the nonzero equilibrium indicates the equilibrium will be stable if the slope of the egg-recruit function is greater than zero. Since the slope of the Beverton–Holt model is greater than zero everywhere, an age-structured model with Beverton–Holt recruitment will always be stable. The common unstable mode for this equilibrium involves cyclic behavior, so we can conclude that the Beverton–Holt egg-recruit relationship will not cause such cyclic behavior.

Consideration of local stability about the zero equilibrium is a way of viewing population persistence. For a population to be persistent, it would have to be unstable about that equilibrium, i.e., it would have to tend to grow away from zero if perturbed away from zero. This local instability occurs when lifetime egg production is greater than the inverse of the slope of the egg-recruit function at the origin, which is $1/\alpha$ in the Beverton–Holt relationship.

PRACTICAL APPLICATIONS

The dominant practical use of the Beverton–Holt model is in the management of populations to avoid their collapse to low abundance, and the greatest such use is in the management of fisheries. Fisheries scientists and managers attempt to determine the value of the slope of the egg-recruit function at low abundance so that they can manage the fishery to maintain lifetime egg production at a value greater than the inverse of that slope. Because populations are seldom allowed to decline to very low abundance, the value of that slope is highly uncertain. This represents a fundamental uncertainty in fishery management: knowing the fishery removal rate that will cause the population to collapse (i.e., when lifetime egg production is greater than $1/\alpha$ in the Beverton–Holt model of the egg-recruit relationship).

Because the implementation of marine reserves is a spatially specific change in fishing mortality rate, it is not surprising that the Beverton–Holt model is used in the analysis of marine reserves also. In that context, the model most often represents the dependence of the number of recruits on the number of settling larvae, which can vary over space. In consideration of spatial heterogeneity of fishing, the calculation of replacement at each point becomes more complex, having to account for replacement through all possible larval dispersal paths.

SEE ALSO THE FOLLOWING ARTICLES

Age Structure / Fisheries Ecology / Harvesting Theory / Marine Reserves and Ecosystem-Based Management / Resilience and Stability / Ricker Model / Stability Analysis

FURTHER READING

Beverton, R. J. H., and S. J. Holt. 1957. On the dynamics of exploited fish populations. *Fisheries Investigations* 2(19), London: UK Ministry of Agriculture and Fisheries. (Reprinted by Chapman and Hall, 1993).

Botsford, L. W., A. Hastings, and S. D. Gaines. 2001. Dependence of sustainability on the configuration of marine reserves and larval dispersal distance. *Ecology Letters* 4: 144–150.

Gurney, W. S. C., and R. M. Nisbet. 1988. *Ecological dynamics* (Chapter 5). Oxford: Oxford University Press.

Hilborn, R., and C. Walters. 2001. *Quantitative fisheries stock assessment: choice, dynamics, and uncertainty.* Boston: Kluwer Academic.

Mace, P. M., and M. P. Sissenwine. 1993. How much spawning per recruit is enough? *Canadian Special Publications in Fisheries and Aquatic Sciences* 120: 101–118.

Quinn II, T. J., and R. Deriso. 1999. *Quantitative fish dynamics.* Oxford: Oxford University Press.

Sissenwine, M. P., and J. G. Shepherd. 1987. An alternative perspective on recruitment overfishing and biological reference points. *Canadian Journal of Fisheries and Aquatic Sciences* 44: 913–918.

BIFURCATIONS

FABIO DERCOLE

Polytechnic University of Milan, Italy

SERGIO RINALDI

International Institute for Applied Systems Analysis, Laxenburg, Austria

Ecological models are often described by difference or differential equations in which variables are population densities and parameters are constant environmental or demographic characteristics, such as water temperature, birth and death rates, carrying capacities, and so forth. Bifurcations are parameter values at which the long-term behavior of the model has a significant change. They correspond to collisions of equilibria, cycles, or more complex attractors, and they can be obtained through numerical analysis. Bifurcations are of paramount importance in ecology, because they identify parameter combinations at which ecosystems collapse or regenerate or change their behavior—for example, from stationary to cyclic. For brevity, the discussion is limited to continuous-time systems.

CONTINUOUS-TIME SYSTEMS AND STATE PORTRAITS

Continuous-time systems are described by n ordinary differential equations (ODEs) called *state equations,* i.e.,

$$\dot{x}_i(t) = f_i(x_1(t), x_2(t), \ldots, x_n(t)), \quad i = 1, \ldots, n, \quad (1)$$

where $x_i(t) \in \mathbf{R}$ is the ith state variable at time $t \in \mathbf{R}$, $\dot{x}_i(t)$ is its time derivative, and the functions $f_1, \ldots, f_n$ are assumed to be smooth. In ecology, x_i is typically the density of the ith population.

In vector form, the state equations are

$$\dot{x}(t) = f(x(t)), \quad (2)$$

where x, $\dot{x}$, and f are n-dimensional vectors.

Given the initial state $x(0)$, the state equations uniquely define a trajectory of the system, i.e., the state vector $x(t)$ for all $t \geq 0$. A trajectory is represented in state space by a curve starting from point $x(0)$, and the vector $\dot{x}(t)$ is tangent to the curve at point $x(t)$. Trajectories can be easily obtained numerically through simulation (numerical integration) and the set of all trajectories (one for any $x(0)$) is called the state portrait. If $n = 2$ (second-order or planar systems) the state portrait is often represented by drawing a sort of qualitative skeleton, i.e., strategic trajectories (or finite segments of them), from which all other trajectories can be intuitively inferred. For example, in Figure 1A the skeleton is composed of 13 trajectories: three of them (A, B, C) are just points (corresponding to constant solutions of Eq. 2) and are called equilibria, while one (γ) is a closed trajectory (corresponding to a periodic solution of Eq. 2) called limit cycle. The other trajectories allow one to conclude that A is a repellor (no trajectory starting close to A tends or remains close to A), B is a saddle (almost all trajectories starting close to B go away from B, but two trajectories tend to B and compose the so-called stable manifold; the two trajectories emanating from B compose the unstable manifold, and both manifolds are also called saddle separatrices), while C and γ are attractors (all trajectories starting close to C [γ] tend to C [γ]).

The skeleton of Figure 1A also identifies the basin of attraction of each attractor: in fact, all trajectories starting above (below) the stable manifold of the saddle tend toward the limit cycle γ (the equilibrium C). Notice that the basins of attraction are open sets since their boundaries are the saddle and its stable manifold. Often, the full state portrait can be more easily imagined when the skeleton is reduced, as in Figure 1B, to its basic elements, namely attractors, repellors, and saddles with their stable and unstable manifolds. From now on, the reduced skeleton is called the characteristic frame.

The characteristic frame is often very useful in ecology, even for devising promising management actions. For example, if x_1 and x_2 in Figure 1 are algae and zooplankton densities in a lake, then the characteristic frame tells us that there are two possible long-term regimes in that lake. One is a turbid-water regime, because at point C algae are very dense, while the other is a clear-water regime, because algae are at low density, at least periodically, along the cycle γ. Thus, a shock cure for switching from a turbid- to a clear-water regime requires a substantial increase of the zooplankton population, so that the state of the lake, which is initially at point C, or close to it, can be pushed above the stable manifold of the saddle B. In principle, this shock cure can be performed by stocking carnivores into the lake, because this reduces the population of small planktivorous fish and hence stimulates the growth of zooplankton.

The asymptotic behaviors of continuous-time second-order systems are quite simple, because in the case $n = 2$ attractors can be equilibria (stationary regimes) or limit cycles (cyclic or periodic regimes). But in higher-dimensional systems, i.e., for $n \geq 3$, more complex behaviors are possible since attractors can also be tori (quasi-periodic regimes) or strange attractors (chaotic regimes). For simplicity, in the following the discussion is mainly focused on second-order systems.

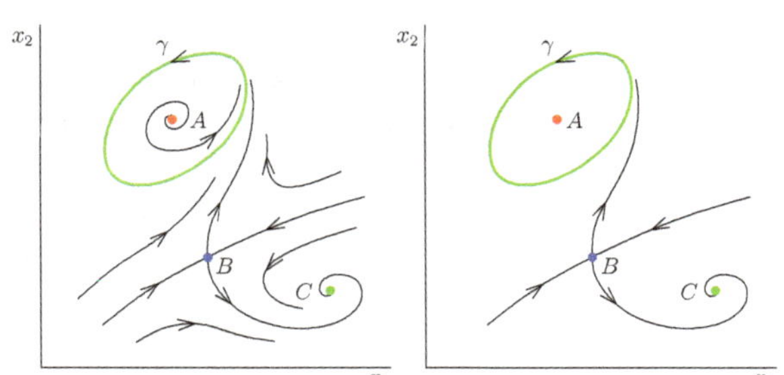

FIGURE 1 Skeleton of the state portrait of a second-order system: (A) skeleton with 13 trajectories; (B) reduced skeleton (characteristic frame) with 8 trajectories (attractors in green; repellors in red; and saddles in blue, with stable and unstable manifolds).

STRUCTURAL STABILITY

Structural stability is a key notion in the theory of dynamical systems, since it is needed to understand interesting phenomena like catastrophic transitions (e.g., the collapse of a forest due to a small decrease of rainfall pH), bistability (e.g., the exsistence of alternative long term regimes as in the lake described in Figure 1), hysteresis (e.g., discontinuous jumps from a desired to an undesired regime, and viceversa, obtained by varying back and forth a strategic control parameter), as well as many others. The final target of structural stability is the study of the asymptotic behavior of parameterized families of dynamical systems of the form

$$\dot{x}(t) = f(x(t), p), \tag{3}$$

where p is a vector of constant parameters. In ecology, p is typically a set of environmental and demographic parameters, like carrying capacity, growth and mortality rates, harvesting effort and so on.

Structural stability allows one to rigorously explain why a small change in a parameter value can give rise to a radical change in the system behavior. More precisely, the aim is to find regions $\mathcal{P}_i$ in parameter space characterized by the same qualitative behavior of system 3, in the sense that all state portraits corresponding to values $p \in \mathcal{P}_i$ are topologically equivalent (i.e., they can be obtained one from the other through a simple deformation of the trajectories). Thus, varying $p \in \mathcal{P}_i$, the system conserves all the characteristic elements of the state portrait, namely, its attractors, repellors, and saddles. In other words, when p is varied in $\mathcal{P}_i$, the characteristic frame varies but conserves its structure.

If p is an interior point of a region $\mathcal{P}_i$, system 3 is said to be structurally stable at p since its state portrait is qualitatively the same as those of the systems obtained by slightly perturbing the parameters in all possible ways. By contrast, if p is on the boundary of a region $\mathcal{P}_i$ the system is not structurally stable, since small perturbations can give rise to qualitatively different state portraits. The points of the boundaries of the regions $\mathcal{P}_i$ are called bifurcation points, and, in the case of two parameters, the boundaries are called bifurcation curves. Bifurcation points are therefore points of degeneracy. If they lie on a curve separating two distinct regions $\mathcal{P}_i$ and $\mathcal{P}_j$, $i \neq j$, they are called codimension-1 bifurcation points, while if they lie on the boundaries of three distinct regions they are called codimension-2 bifurcation points, and so on.

BIFURCATIONS AS COLLISIONS

A generic element of the parameterized family of dynamical systems 3 must be imagined to be structurally stable because if p is selected randomly it will be an interior point of a region $\mathcal{P}_i$ with probability 1. In generic conditions, attractors, repellors, saddles, and their stable and unstable manifolds are separated from each other. By continuity, small parametric variations will induce small variations of all attractors, repellors, saddles, and their stable and unstable manifolds, but these will remain separated if the parametric variations are sufficiently small. Thus, in conclusion, starting from a generic condition, it is necessary to vary the parameters of a finite amount to obtain a bifurcation, which is generated by the collision of two or more elements of the characteristic frame, which then changes its structure at the bifurcation, thus involving a change of the state portrait of the system.

When there is only one parameter p and there are various bifurcations at different values of the parameter, it is often advantageous to represent the dependence of the system behavior upon the parameter by drawing in the three-dimensional space (p, x_1, x_2), often called control space, the characteristic frame for all values of p. This is done, for example, in Figure 9 for a tritrophic food chain model.

MOST COMMON BIFURCATIONS

In this section, we discuss seven important bifurcations. The first three, called transcritical, saddle-node, and pitchfork, can be viewed as collisions of equilibria, while the others involve limit cycles. Three of them can occur in second-order systems, namely, the Hopf bifurcation (i.e., the collision of an equilibrium with a vanishing cycle), the tangent of limit cycles (collision of two cycles), and the homoclinic bifurcation (collision between the stable and unstable manifolds of the same saddle). The last bifurcation, the flip (or period-doubling), is more complex because it can occur only in three- (or higher-) dimensional continuous-time systems.

Transcritical, Saddle-Node, and Pitchfork Bifurcations

Figure 2 shows three different types of collisions of equilibria in first-order systems of the form of Equation 3. The state x and the parameter p have been normalized in such a way that the bifurcation occurs at $p^* = 0$ and that the corresponding equilibrium is zero. Green lines in the figure represent stable equilibria, while red lines indicate unstable

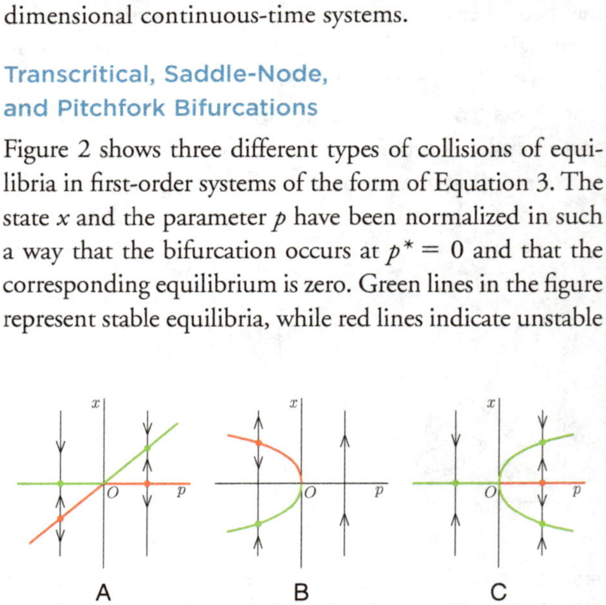

FIGURE 2 Three local bifurcations viewed as collisions of equilibria: (A) transcritical; (B) saddle-node; (C) pitchfork.

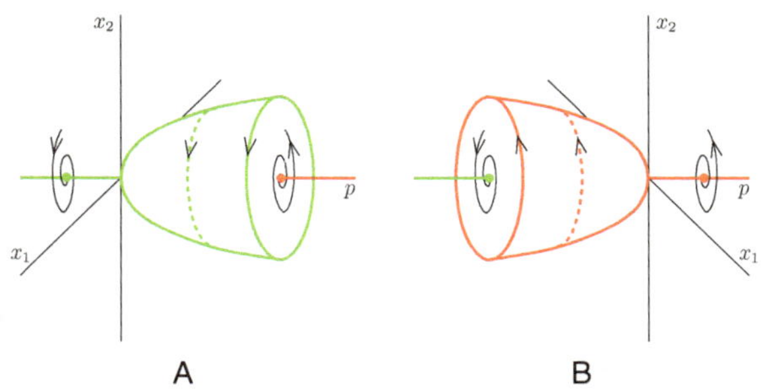

FIGURE 3 Hopf bifurcation: (A) supercritical; (B) subcritical.

equilibria. In Figure 2A, the collision is visible in both directions, while in Figures 2B and 2C the collision is visible only from the left or from the right. The three bifurcations are called, respectively, transcritical, saddle-node, and pitchfork, and the three most simple state equations (called normal forms) giving rise to Figure 2 are

$$\dot{x}(t) = px(t) - x^2(t), \quad \text{transcritical,} \tag{4a}$$

$$\dot{x}(t) = p + x^2(t), \quad \text{saddle-node,} \tag{4b}$$

$$\dot{x}(t) = px(t) - x^3(t), \quad \text{pitchfork.} \tag{4c}$$

The first of these bifurcations is also called exchange of stability since the two equilibria exchange their stability at the bifurcation. The second is called saddle-node bifurcation because in second-order systems it corresponds to the collision of a saddle with a node, but it is also known as fold bifurcation, in view of the form of the graph of its equilibria. Due to the symmetry of the normal form, the pitchfork has three colliding equilibria, two stable and one unstable in the middle.

Hopf Bifurcation

The Hopf bifurcation (actually discovered by A. A. Andronov for second-order systems) explains how a stationary regime can become cyclic as a consequence of a small variation of a parameter, a rather common phenomenon not only in physics but also in biology, economics, and life sciences. In terms of collisions, this bifurcation involves an equilibrium and a cycle that shrinks to a point when the collision occurs. Figure 3 shows the two possible cases, known as supercritical and subcritical Hopf bifurcations, respectively. In the supercritical case, a stable cycle has in its interior an unstable focus. When the parameter is varied, the cycle shrinks until it collides with the equilibrium and after the collision only a stable equilibrium remains. By contrast, in the subcritical case the cycle is unstable and is the boundary of the basin of attraction of the stable equilibrium inside the cycle. Thus, after the collision there is only a repellor.

Determining if a Hopf bifurcation is supercritical or subcritical is not easy. The mathematical procedure is rather involved and is therefore not reported here (see, e.g., pp. 99–102 of Kuznetsov 2004, for an application to a standard prey–predator model used to interpret the "paradox of enrichment").

Homoclinic Bifurcation

The most common homoclinic bifurcation is the collision between the stable and unstable manifolds of the same saddle, as depicted in Figure 4. The figure also shows that the bifurcation can be viewed as the collision of the cycle $\gamma(p)$ with the saddle $S(p)$. When p approaches p^*, the cycle $\gamma(p)$ gets closer and closer to the saddle $S(p)$, so that the period $T(p)$ of the cycle becomes longer and longer, since the state of the system moves very slowly when it is very close to the saddle. Thus, $T(p) \to \infty$ as $p \to p^*$, and this property is often used to detect homoclinic bifurcations through experiments and simulation. Another property used to detect homoclinic bifurcations is related to the form of the limit cycle that becomes "pinched" close to the bifurcation, the angle of the pinch being the angle between the stable and unstable manifolds of the saddle.

Looking at Figure 4 from the right to the left, we can recognize that the homoclinic bifurcation explains the

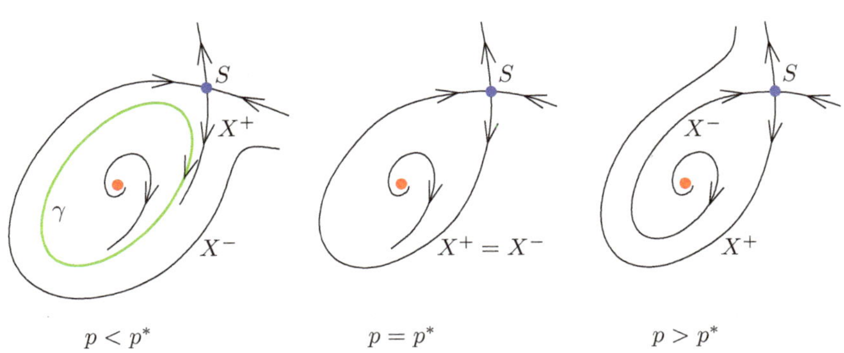

FIGURE 4 Homoclinic bifurcation to standard saddle: for $p = p^*$ the stable manifold X^- of the saddle S collides with the unstable manifold X^+ of the same saddle. The bifurcation can also be viewed as the collision of the cycle γ with the saddle S.

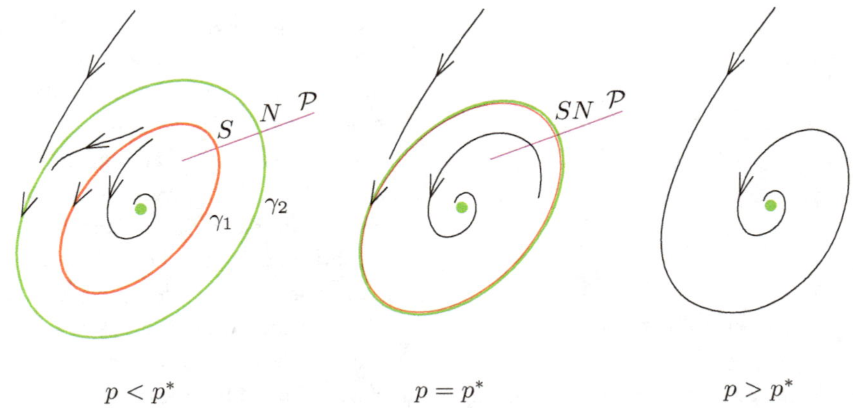

FIGURE 5 Tangent bifurcation of limit cycles: two cycles γ_1 and γ_2 collide for $p = p^*$ and then disappear.

birth of a limit cycle. As in the case of Hopf bifurcations, the emerging limit cycle is degenerate, but this time the degeneracy is not in the amplitude of the cycle but in its period, which is infinitely long. The emerging cycle is stable in Figure 4 (the gray region is its basin of attraction), but reversing the arrows of all trajectories the same figure illustrates the case of an unstable cycle.

As proved by Andronov and Leontovich, this result holds under a series of assumptions that essentially rule out a number of critical cases. A very important and absolutely not simple extension of Andronov and Leontovich theory is Shil'nikov's theorem concerning homoclinic bifurcations in three-dimensional systems.

Tangent Bifurcation of Limit Cycles

Other bifurcations in second-order systems involve only limit cycles and are somewhat similar to transcritical, saddle-node, and pitchfork bifurcations of equilibria. The most common of them is the saddle-node bifurcation of limit cycles, more often called fold or tangent bifurcation of limit cycles, where two cycles collide for $p = p^*$ and then disappear, as shown in Figure 5.

Varying the parameter in the opposite direction, this bifurcation explains the sudden birth of a pair of cycles, one of which is stable. While in the cases of Hopf and homoclinic bifurcations the emerging cycles are degenerate (zero amplitude and infinite period), in this case the emerging cycle is not degenerate.

Flip (Period-Doubling) Bifurcation and the Feigenbaum Cascade

The flip bifurcation is the collision of two particular limit cycles, one tracing twice the other and therefore having double period, in a three-dimensional (or higher) state space. In the supercritical (subcritical) case, it corresponds to a bifurcation of a stable (unstable) limit cycle of period T into a stable (unstable) limit cycle of period $2T$ and an unstable (stable) limit cycle of period T, as sketched in Figure 6 for the supercritical case.

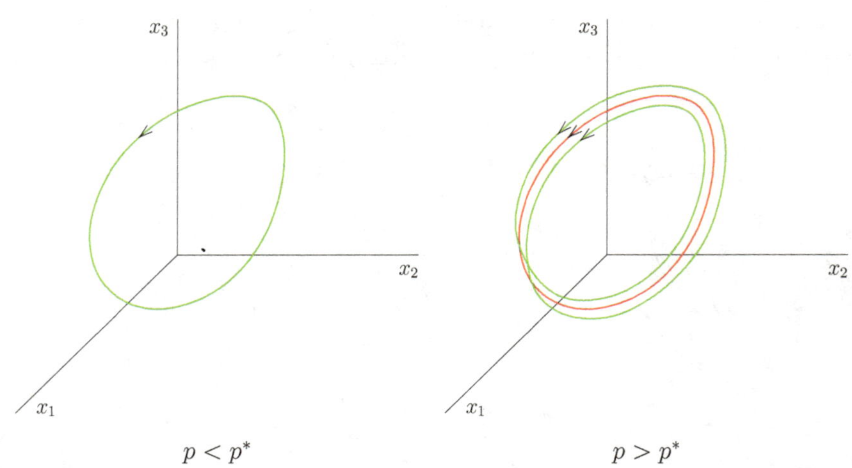

FIGURE 6 Flip bifurcation: ($p < p^*$) stable limit cycle of period T; ($p < p^*$) unstable limit cycle of period T and stable limit cycle of period $2T$.

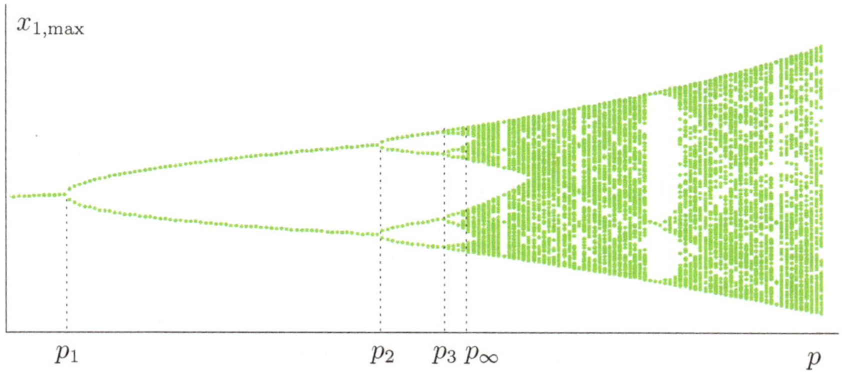

FIGURE 7 The Feigenbaum route to chaos: for $p \leq p_1$ the attractor is a cycle; for $p_1 < p < p_\infty$ the attractor is a longer and longer cycle; for $p \geq p_\infty$ the attractor is a strange attractor. The variable on the vertical axis measures a heritable phenotypic trait (e.g., body size) characterizing the top species in a tritrophic coevolution model (genetic mutations and natural selection, see F. Dercole et al., 2010, *Proceedings of the Royal Society of London B: Biological Sciences* 277: 2321–2330).

Physically speaking, the key feature is that the period of the limit cycle doubles through the bifurcation. In other words, if before the bifurcation the graph of one of the state variables, say, x_1, has a single peak in each period T, just after the bifurcation the graph has two slightly different peaks in each period $2T$.

Very often, after a first flip bifurcation at $p = p_1$, an infinite sequence $\{p_i\}$ of flip bifurcations occurs. In this sequence, known as the Feigenbaum cascade, the p_i's accumulate at a critical value p_∞ after which the attractor is a genuine chaotic attractor. Very often, this route to chaos is depicted by plotting the local peaks of a state variable, say, x_1, as a function of a parameter p, as shown in Figure 7 for a coevolution model. Physically speaking, the attractor remains a cycle until $p = p_\infty$, but the period of the cycle doubles at each bifurcation p_i, while unstable cycles of longer and longer periods accumulate in state space. This route to chaos points out a general property of strange attractors, namely, the fact that they are basically composed of an aperiodic trajectory visiting a bounded region of the state space densely filled of unstable cycles. Such cycles are actually of saddle type, i.e., they repel along some directions (stretching) and attract along others (folding). The Feigenbaum cascade is present in almost all models pointing out chaos in ecology, in particular, in many studies concerning the dynamics of tritrophic food chains.

CATASTROPHES AND HYSTERESIS

We can now specify what catastrophic transitions in dynamical systems are. Reduced to its minimal terms, the problem of catastrophic transitions (related with resilience of ecosystems) is the following: assuming that a system is functioning in one of its asymptotic regimes, is it possible that a microscopic variation of a parameter triggers a transient toward a macroscopically different asymptotic regime? When this happens, we say that a catastrophic transition occurs.

To be more specific, assume that an instantaneous small perturbation from p to $p + \Delta p$ occurs at time $t = 0$ when the system is on one of its attractors, say, $\mathcal{A}(p)$, or at a point $x(0)$ very close to $\mathcal{A}(p)$ in the basin of attraction of $\mathcal{A}(p)$. The most obvious possibility is that p and $p + \Delta p$ are not separated by any bifurcation. This implies that the state portrait of the perturbed system $\dot{x} = f(x, p + \Delta p)$ can be obtained by slightly deforming the state portrait of the original system $\dot{x} = f(x, p)$. In particular, if Δp is small, by continuity, the attractors $\mathcal{A}(p)$ and $\mathcal{A}(p + \Delta p)$, as well as their basins of attraction, are almost coincident. This means that when the perturbation has ceased, a transition occurs from $\mathcal{A}(p)$ (or $x(0)$ close to $\mathcal{A}(p)$) to $\mathcal{A}(p + \Delta p)$. In conclusion, a microscopic variation of a parameter has generated a microscopic variation in system behavior.

The opposite possibility is that p and $p + \Delta p$ are separated by a bifurcation. In such a case, it can happen that the small parameter variation triggers a transient, bringing the system toward a macroscopically different attractor. When this happens for all initial states $x(0)$ close to $A(p)$, the bifurcation is called catastrophic. By contrast, if the catastrophic transition is not possible, the bifurcation is called noncatastrophic, while in all other cases the bifurcation is said to be undetermined.

We can now revisit all bifurcations we have discussed in the previous section. Let us start with Figure 2 and assume that p is small and negative*, i.e., $p = -\epsilon$, that $x(0) \neq 0$ is very close to the stable equilibrium, and that $\Delta p = 2\epsilon$ so that, after the perturbation, $p = \epsilon$. In case of Figure 2A (transcritical bifurcation), $x(t) \to \epsilon$ if $x(0) > 0$ and $x(t) \to -\infty$ if $x(0) < 0$. Thus, this bifurcation is undetermined because it can, but does not always, give rise to a catastrophic transition. In a case like this, the noise acting on the system has a fundamental role. We must notice, however, that in many cases the sign of $x(0)$ is a priori fixed. For example, if x represents the density of a population, then for physical reasons $x(0) > 0$ and the bifurcation is therefore noncatastrophic. Similarly, we can conclude that the saddle-node bifurcation (Fig. 2B) is catastrophic and that the pitchfork bifurcation (Fig. 2C) is noncatastrophic.

From Figure 3, we can immediately conclude that the supercritical Hopf bifurcation is noncatastrophic, while the subcritical one is catastrophic. This is why the two Hopf bifurcations are sometimes called catastrophic and noncatastrophic. Finally, Figures 4 and 5 show that homoclinic and tangent bifurcations are catastrophic.

When a small parametric variation triggers a catastrophic transition from an attractor $\mathcal{A}'$ to an attractor $\mathcal{A}''$, it is interesting to determine if it is possible to drive the system back to the attractor $\mathcal{A}'$ by suitably varying the parameter. When this is possible, the catastrophe is called reversible. The most simple case of reversible catastrophe is the hysteresis, two examples of which (concerning first-order systems) are shown in Figure 8. In Figure 8A, the system has two saddle-node bifurcations, while Figure 8B, there is a transcritical bifurcation at $p_1{}^*$ and a saddle-node bifurcation at $p_2{}^*$. All bifurcations are catastrophic (because the transitions $A \to B$ and $C \to D$ are macroscopic), and if p is varied back and forth between $p_{\min} < p_1{}^*$ and $p_{\max} > p_2{}^*$ through a sequence of small steps with long time intervals between successive steps, the state of the system follows closely the cycle $A \to B \to C \to D$ indicated in the figure and called a hysteretic cycle (or, briefly, hysteresis). The catastrophes are therefore reversible, but after a transition from $\mathcal{A}'$ to $\mathcal{A}''$ it is necessary to pass through a second catastrophe to come back to the attractor $\mathcal{A}'$. This simple type of hysteresis explains many phenomena not only in physics, chemistry, and electroinechanics but also in biology and social sciences. For example, the hysteresis of Figure 8B was used by Noy-Meir (1975, *Journal of Ecology* 63: 459–483) to explain the possible collapse (saddle-node bifurcation) of an exploited resource x (e.g., density of grass) where p is the number of exploiters (e.g., cows). If p is increased step by step (e.g., by adding one extra cow every year), the resource declines smoothly until it collapses to zero when a threshold $p_2{}^*$ is passed. To regenerate the resource, one is obliged to radically reduce the number of exploiters to $p < p_1{}^*$.

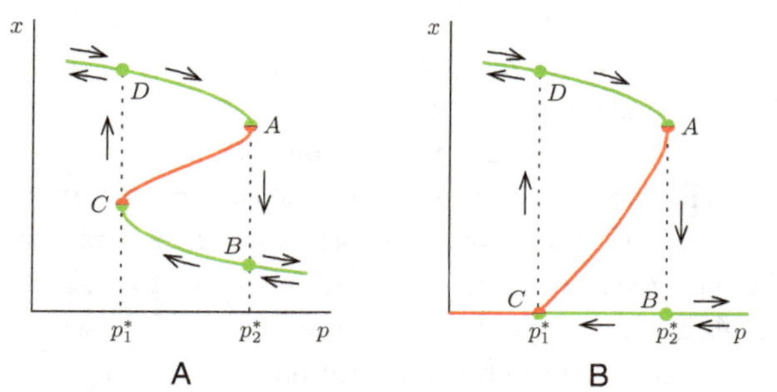

FIGURE 8 Two systems with hysteresis generated by two saddle-node bifurcations (A), and a saddle-node and a transcritical bifurcation (B).

Hysteresis can be more complex than in Figure 8, not only because the attractors involved in the hysteretic cycle can be more than two, but also because some of them can be cycles. To show the latter possibility, consider a standard three-species food chain, where x_1 is the prey, x_2 is the predator, and p is a constant superpredator. Without entering into the details of the analysis of this second-order model (see Yu. A. Kuznetsov et al., 1995, *Nonlinear Analysis, Theory, Methods & Applications* 25: 747–762), Figure 9 shows the equilibria and the cycles of the system in the control space (p, x_1, x_2) for a specified value of all other parameters. The figure points out five bifurcations: a transcritical (TR), two homoclinic (h_1 and h_2), a supercritical Hopf (H), and a saddle-node (SN). Two of these bifurcations, namely, the homoclinic h_2 and the saddle-node SN, are catastrophic and irreversible. In fact, catastrophic transitions from h_2 and SN bring the system toward the trivial equilibrium (K, 0) (extinction of the predator population), and from this state it is not possible to return to h_2 or SN by varying p step by step. By contrast, the two other catastrophic bifurcations, namely, the homoclinic h_1 and the transcritical TR, are reversible and identify a hysteretic cycle obtained by varying back and forth the superpredator p in an interval slightly larger than $[p_{TR}, p_{h_1}]$. On one extreme of the hysteresis, we have a catastrophic transition from the equilibrium (K, 0) to a prey–predator limit cycle. Then, increasing p, the period of the limit cycle increases (and tends to infinity as $p \to p_{h_1}$), and on the other extreme of the hysteresis we have a catastrophic transition from a homoclinic cycle (in practice a cycle of very long period) to the equilibrium (K, 0). Thus, if p is varied smoothly, slowly, and periodically from just below p_{TR} to just above p_{h_1}, one can expect that the predator population varies

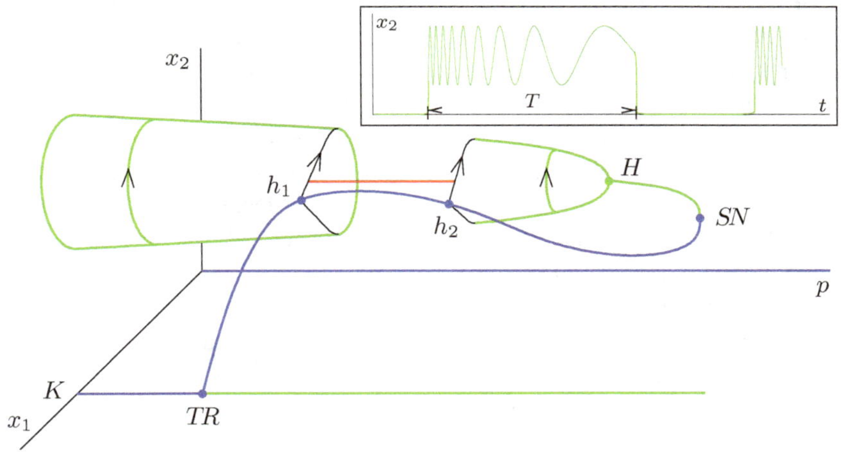

FIGURE 9 Equilibria and limit cycles (characteristic frame) of a standard tritrophic food chain model with constant superpredator population p. Inset: Periodic variation of a predator population induced by a periodic variation of the superpredator.

periodically in time, as shown in the inset of Figure 9. In conclusion, the predator population remains very scarce for a long time and then suddenly regenerates, giving rise to high-frequency prey–predator oscillations that slow down before the crash of the predator population occurs. Of course, tritrophic food chains do not always have such wild dynamics. In fact, many food chains are characterized by a unique attractor and therefore cannot experience catastrophic transitions and hysteresis.

NUMERICAL METHODS AND SOFTWARE PACKAGES

All effective software packages for numerical bifurcation analysis are based on *continuation* (see Allgower and Georg, 1990).

The two most widespread software packages for bifurcation analysis are AUTO and MATCONT. AUTO (http://indy.cs.concordia.ca/auto) is a very efficient Fortran code but limited to codimension-1 bifurcations, while MATCONT (http://www.matcont.ugent.be) also supports codimension-2 bifurcations and is developed in the MATLAB environment.

SEE ALSO THE FOLLOWING ARTICLES

Chaos / Difference Equations / Ordinary Differential Equations / Phase Plane Analysis / Stability Analysis

FURTHER READING

Allgower, E. L. and K. Georg. 1990. *Numerical continuation methods: an introduction.* Berlin: Springer-Verlag.

Alligood, K. T., T. D. Sauer, and J. A. Yorke. 1996. *Chaos: an introduction to dynamical systems.* New York: Springer-Verlag.

Guckenheimer, J., and P. Holmes. 1997. *Nonlinear oscillations, dynamical systems and bifurcations of Vector Fields,* 5th ed. New York: Springer-Verlag.

Kuznetsov, Yu. A. 2004. *Elements of applied bifurcation theory,* 3rd ed. Berlin: Springer-Verlag.

Strogatz, S. H. 1994. *Nonlinear dynamics and chaos.* Reading, MA: Addison-Wesley.

BIOGEOCHEMISTRY AND NUTRIENT CYCLES

BENJAMIN Z. HOULTON

University of California, Davis

Nutrient biogeochemical cycles bear the imprint of life. Three basic principles—thermodynamics, feedback, and stoichiometry—affect both biogeochemical and nutrient cycles of ecosystems. When combined, these principles provide a theoretical framework for examining the role of nutrient cycling and limitation in global environmental change.

INTRODUCTION TO THREE PRINCIPLES

This entry focuses on biogeochemical cycles and nutrient dynamics of ecosystems. Biogeochemistry is a discipline that includes many different aspects of ecology; it examines interactions between Earth's chemical cycles—particularly their biocycles—along the interface of living and nonliving systems—drawing from key concepts in biology, geology, hydrology, chemistry, and physics. Biogeochemistry is scale independent in that it spans elements to molecules, seconds to eons, microbes to the biosphere. Here the emphasis is on scales ranging from ecosystems to the globe—with concepts (largely) centered on those of plant–microbe–soil interactions in terrestrial environments.

Nutrient cycling is concerned with interactions among chemical elements, plants, microbes, and soils on land; plants, microbes, and water in water; and how resources cycle between these different subsystems (both aquatic and terrestrial) in the Earth system. Nutrients are chemical elements that serve important biological functions: they include macronutrients that are vital to life in relatively large quantities (e.g., nitrogen and phosphorus) as well as essential micronutrients, which are needed in much smaller amounts, typically trace metals found in enzyme systems (e.g., zinc, molybdenum, iron, and the like). Nutrient cycles are affected by such things as species competition (inter- and intra-), biotic/abiotic interactions, evolution, kinetics, and energetics. This entry focuses on terrestrial nutrient cycles—largely at plant–soil–microbe scales—particularly the influence of nutrient limitations on the biology of populations.

Not unlike ecology, there is at present no single guiding theory (or premise) unique to biogeochemistry or nutrient cycling. Rather, nutrient biogeochemical cycles have many theoretical underpinnings. Therefore, this entry does not seek to provide an exhaustive set of theories; neither does it set out to derive a single cohesive one. Instead, the focus will be on three basic principles that have a strong history in shaping the biogeosciences: (1) thermodynamics, (2) feedback, (3) and stoichiometry. Through a basic understanding of these principles, one can begin addressing an array of problems in biogeochemistry and nutrient cycling, both conceptually and quantitatively.

PRINCIPLE I: THERMODYNAMICS

Thermodynamics is based on analysis of physical systems—it covers two basic laws, both concerned with energy- and mass-flux transfer. The first law evokes conservation of energy (and mass): energy can be neither created nor destroyed, but it can be converted from one form to the next. Thus, the first law of thermodynamics tells us that everything goes somewhere, nothing disappears—though the forms that energy and mass take can and do change during a given transformation. The three basic kinds of systems about which transformations occur are isolated systems, closed systems, and open systems. Open systems will be emphasized here as the vast majority of biogeochemical and nutrient systems are open and dynamic. By definition, open systems exchange both mass and energy with other systems, in contrast to isolated (which exchange neither mass nor energy) and closed ones (which exchange energy but not appreciable mass).

The second law of thermodynamics deals with entropy, or movement toward greater disorder. Entropy (S) opposes enthalpy (i.e., internal energy, H) in energetics (G) as described by the following equation:

$$\Delta G = \Delta H - \Delta S^{*}T,$$

in which G is often referred to as free energy, in that it posits the likelihood for a given reaction to occur spontaneously. (Note that free energy is descriptive and should not be taken as truly "free." The sun is the ultimate source of energy on Earth, and photosynthesis requires substantial energy input for carbon fixation.) The Δs refer to the difference between products and reactants in a given chemical transformation, and units are energy over mass (e.g., kilojoules (kJ)/mole). Hence, if ΔS is high (at a given temperature, T) relative to ΔH, the reaction has a higher degree of free energy and is therefore predicted to occur spontaneously (energy yield is high). In terms of sign, the propensity toward disorder in the forward reaction is high when ΔG is negative, and so the reaction is more likely to occur in that direction; positive ΔGs predict the opposite, that the reaction will not occur, all else being equal. Organisms can and do capitalize on such (negative) free energy (see below), thereby reducing energy expenditures relative to energy yield in a given biogeochemical transformation.

The two laws of thermodynamics (and associated energetic principles) have long shaped the fields of ecosystem biogeochemistry and nutrient cycling. The first clear application of thermodynamics as it relates to ecosystem science can be traced back to Tansley (1935). In his landmark paper on succession, Tansley demarcates natural ecological systems as follows:

> [T]he more fundamental conception is, as it seems to me, the whole system (in the sense of physics), including not only the organism-complex, but also the whole complex of physical factors forming what we call the environment of the biome—the habitat factors in the widest sense. The organisms may claim our primary interest when we are trying to think fundamentally we cannot separate them from their special environment, which they form one physical system.

Note the use of "in the sense of physics" and "one physical system." With these phrases Tansley defines an ecosystem as having thermodynamic properties, with physical

TABLE 1

Free energy (ΔG) calculations for a given set of microbial biogeochemical transformations

Process	Reaction	Free energy (kJ)
(A) Decreasing redox potential		
(1) Aerobic respiration	$CH_2O + O_2 = CO_2 + H_2O$	−501
(2) Denitrification	$CH_2O + (4/5)NO_3^- + (4/5)H^+ = CO_2 + (2/5)N_2 + (7/5)H_2O$	−476
(3) Sulfate reduction	$CH_2O + (1/2)SO_4^{2-} + (1/2)H^+ = (1/2)HS^- + H_2O + CO_2$	−102
(4) Methanogenesis	a) $CH_2O = (1/2)CH_4 + (1/2)CO_2$	−93
	b) $(1/2)CO_2 + 2H_2 = (1/2)CH_4 + H_2O$	−66
(B) Increasing redox potential		
	$O_2 + (1/2)CH_4 = (1/2)CO_2 + H_2O$	−408
	$O_2 + (1/2)HS^- = (1/2)SO_4^{2-} + (1/2)H^+$	−399
	$O_2 + (1/2)NH_4^+ = (1/2)NO_3^- + H^+ + (1/2)H_2O$	−181

boundary conditions and rules that constrain flux and energy transfers among organisms and between the broader environment, a system with living and nonliving properties. This lays the foundation for the ecosystem concept of today.

Since Tansley's early work, the first law—mass and energy conservation—has substantially shaped our understanding of both biogeochemistry and nutrient cycling. A classic illustration involves work at the Hubbard Brook Experimental Forest (HBEF), a series of well-studied watersheds that have been subject to myriad treatments/experiments aimed at understanding temperate forest responses to various stressors, nutrient biogeochemical responses in particular. The watersheds are underlain by relatively watertight bedrock, meaning that material fluxes into the watersheds are transported to streams or the atmosphere as opposed to leaching to groundwater in deep bedrock fractures. Thus, by measuring stream water chemistry one can begin applying input–output mass balance principles. Based on the conservation of mass (first law of thermodynamics), the following equation has been used to examine watershed element gains and losses at the HBEF:

$$\Delta\text{accumulation} = \Delta\text{input} - \Delta\text{output},$$

which at steady state reduces to

$$\Delta\text{input} - \Delta\text{output} = 0.$$

This elegant set of equations has led to profound insights in ecosystem biogeochemical cycles at the HBEF and beyond—especially in relation to understanding differences between elements with substantial rock vs. atmospheric sources. Elements with active rock input pathways—e.g., phosphorus, base cations—show an imbalance between atmospheric inputs and stream losses; this reflects rock inputs via chemical weathering as well as changes in gains and losses in soil, litter, and plant pools. As applied to calcium, for example, input–output budgets have revealed net losses of calcium from the HBEF that have been accelerated by acid–base interactions with other chemicals, particularly sulfate in acid rain. Elements that lack substantial rock inputs at HBEF—especially nitrogen—often show higher atmospheric inputs than mass fluxes observed in streamwaters. This imbalance points to a significant loss term (via gaseous nitrogen removal) or accumulation in growing biomass, or both. Thus, as illustrated by this elegant application of thermodynamics, one can easily develop testable hypotheses for which to explore biogeochemical and nutrient cycles from an ecosystem perspective.

Another example includes application of ΔG in specific biogeochemical transfers. As electron acceptors are used by heterotrophic microorganisms, systematic transformations in the abundance of ions and gases follow predictions based on ΔG. Because aerobic respiration has the lowest ΔG (most spontaneous), utilization of O_2 during the breakdown of organic carbon yields the greatest amount of energy to heterotrophs. The electron acceptor nitrate is next, followed by manganese, iron, sulfate, and ultimately carbon dioxide and organic carbon (fermentation; Table 1).

This thermodynamic sequence, known as the electron acceptor sequence, suggests predictable patterns of concentrations of electron acceptors (electron poor molecules) and donors (electron rich molecules), moving from well-aerated to poorly aerated soil. Visually, this is clearly seen following soil flooding events, whereby the concentrations of electron acceptors follow thermodynamic predictions (Fig. 1). Flooding slows ventilation of O_2 from the atmosphere such that the order of disappearance of electron acceptors (O_2, NO_3^-) and increased concentrations of reduced metals (exchangeable Mn^{2+}, Fe^{2+})

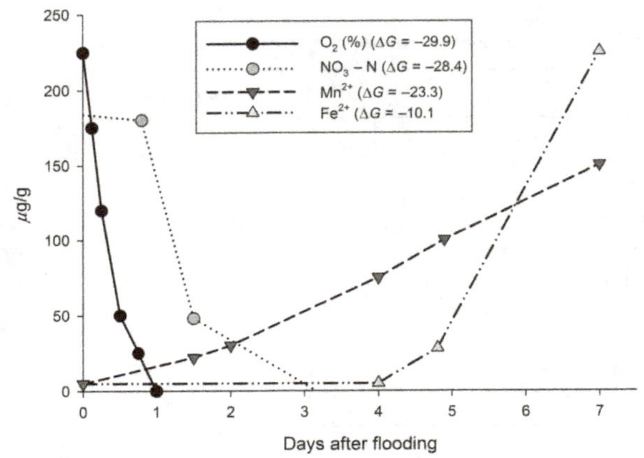

FIGURE 1 Changes in concentrations of chemical constituents found in soil waters as a function of days after soil flooding. Theoretical ΔG (kcal/mole) calculations are based on reduction with hydrogen at standard conditions of pH = 7 and 25 °C. The loss of O_2 and NO_3^- and gain in concentrations of reduced metals (Mn^{2+} and Fe^{2+}) follow predictions of thermodynamics (i.e., ΔG). Redrawn from Turner and Patrick (1968). See also Schlesinger in Further Readings.

follows predictions based on ΔG (Table 1). Moreover, similar patterns can be observed spatially along transects from upland soils (well-aerated) to water-saturated (poorly aerated) riparian zones.

PRINCIPLE II: FEEDBACK

In addition to input–output (i.e., system-level) controls, biogeochemical/nutrient transfers are influenced by within system dynamics as well. Within system processes concern transfers of chemical elements within and among organisms (plants, animals, microbes), often resulting in feedback that either amplifies (positive feedback) or down-regulates (negative feedback) the initial state or given process of interest. Feedback can be seen at many different scales with implications that can differ depending on scale; the definition of positive or negative feedback is arbitrary to the starting point.

A classic example in biogeochemistry and biophysics is Lovelock's so-called daisy-world model, which envisaged a heuristic negative feedback whereby the albedo (amount of sunlight reflected back to space) of a given organism affects both the climate and overall fitness of the organism. The model, consisting of black vs. white daisies, reveals that the physical features of organisms can and do affect the entire climate system. Fictitious black daisies have an advantage under cooler conditions when the sun's luminosity is relatively low; they retain heat within the Earth system and thus keep the planet from freezing. As more of the sun's heat energy is trapped, however, competitive advantage is conferred to white daisies, which reflect more of the sun's light back to space than black daisies, thus keeping the climate from moving toward inhospitably high temperatures. In this way, stabilization and equilibrium is reached through *negative* feedback.

A positive feedback is also apparent in albedo/biogeochemical interactions. Currently, at high latitudes, warming of the climate has been substantial, partly owing to the role of darker-colored forest vegetation (low albedo) in absorbing more incoming solar radiation than the bare ground, which is for much of the season covered by snow (high albedo). As trees expand into higher latitudes with climate warming, they heat high-latitude environments even more, thus resulting in positive feedback and amplified warming. Darker-colored trees absorb more incoming solar radiation than snow, which warms the climate, promoting more tree growth, which in turn leads to more trees and a lower albedo . . . and so on and so forth. Moreover, this positive feedback can lead to accelerated releases of greenhouse gases such as methane, as permafrost thaws and thermo-karst erosion results in the formation of carbon rich and O_2 poor melt ponds (following thermodynamic predictions, Table 1). As this case illustrates, *positive* feedback results in destabilization and rapid environmental change.

With specific focus on nutrient cycles, particularly nitrogen cycles, positive feedback is often seen within the plant–microbe–soil subsystem. Many ecosystems on Earth are limited by nitrogen availability; this macronutrient is a significant component of protein and DNA, and it is often in low supply relative to biological demands. Plants in nitrogen-poor sites have thus evolved toward N conservation such that productivity is maximized per unit of nutrient investment where nitrogen is most limiting. This widens the carbon/nitrogen ratio of leaves and litter fall, meaning that soil microbes living on plant litter become more nitrogen than carbon limited, retaining more of the nitrogen for themselves as opposed to mineralizing it to plants. Consequently, the rate of release of nitrogen back to plants (nitrogen mineralization) is slowed in nitrogen-limited sites, further promoting higher carbon/nitrogen ratios and plants and more substantial nitrogen conservation in ecosystems. Consequently, this positive feedback results in more profound nitrogen limitation to plants—and the converse can also happen in nitrogen-rich sites, in which positive feedback and lower plant carbon/nitrogen leads to accelerated nitrogen cycles and enhanced nitrogen richness (see Summary and Synthesis, below).

PRINCIPLE III: STOICHIOMETRY

The final principle discussed in this entry is known as stoichiometry, or ecological stoichiometry—as it relates to interactions between chemical element cycles and organisms. Stoichiometry builds on the foundation of mass balance (thermodynamics) and influences feedback strength (described above), pointing out that the quantity of chemical elements in reactants and products must sum equally. Consider, for instance, photosynthesis (aerobic respiration when reversed):

$$6CO_2 + 6H_2O = C_6H_{12}O_6 + 6O_2.$$

In the forward direction, this equation shows that for every six CO_2 molecules assimilated by plants from the atmosphere, six O_2 molecules are produced, or a 1:1 stoichiometry between CO_2 and O_2. This is a cellular-level equation, derived from steady-state mass-balance principles, but it also holds at large space and time scales. For instance, in the global ocean over the scale of millions of years, the photosynthesis–respiration equation indicates that the amount of O_2 in the atmosphere is immutably coupled to C burial in marine sediments; for every mole of C buried, an equivalent amount of O_2 remains in the atmosphere, thus affecting the oxidizing potential of the entire Earth system.

Beyond individual chemical reactions, stoichiometric concepts can be applied to complex biomolecules to gain understanding of constraints on organisms and nutrient cycles. In their book, *Ecological Stoichiometry,* Sterner and Elser assembled information on the stoichiometry of molecules that serve important biological functions, from DNA to proteins, from organelles to cells, and so on. Based on this assessment, one can see that the carbon, nitrogen, and phosphorus stoichiometric demand of different biomolecules varies substantially. DNA, for instance, has a stoichiometry that is rich in phosphorus relative to many other biomolecules; proteins (enzymes) by comparison are rather rich in nitrogen. mRNA, involved in growth and protein synthesis, is rich in phosphorus, yet the amino acids and proteins that are synthesized have a much higher nitrogen/phosphorus ratio than does mRNA. Thus, a comparatively phosphorus-rich compound is requisite to the synthesis of nitrogen-rich biomolecules, representing a coupling that underpins widespread nutrient limitations by nitrogen and phosphorus globally. These types of nutrient couplings and their molecular differences set the stage for nutrient cycling and ecological competitions that play out in all ecosystems.

At the larger scales, canonical relations between ecological stoichiometry and nutrient biogeochemical cycles have led to major insights into the functioning of the Earth system. The seminal example is the Redfield ratio, named after its discoverer Alfred Redfield. The molar "Redfield ratio" of nitrogen/phosphorus in phytoplankton biomass is 16/1, observed across many different environments in the global ocean. By itself, the constancy of nitrogen/phosphorus among phytoplankton is remarkable; but perhaps even more interesting is that the nitrogen/phosphorus ratio of phytoplankton matches that of dissolved nitrate and phosphate in the ocean. Thus, plankton exhibit the same average nitrogen/phosphorus ratio as that found in the dissolved nutrients in which they assimilate. This coupling points to large-scale homeostatic regulation of nutrient cycles in the ocean, mediated by inputs and outputs of nitrogen and phosphorus over the long term.

In terrestrial ecosystems, such large-scale stoichiometric analyses are rather nascent when compared to marine and freshwater systems. Nevertheless, patterns of nutrient cycling, stroichiometry, and biogeochemistry have enhanced understanding of terrestrial ecosystem functioning for decades. Vitousek noted that nitrogen/phosphorus ratios of vegetation leaves are on average higher in nitrogen-rich tropical sites than nitrogen-poor temperate ones. This pattern is also seen in the nitrogen/phosphorus ratio of falling leaves, though even more substantially pronounced, pointing to nutrient conservation strategies and enhanced resorption of limiting nutrients by plants that follows patterns of ecosystem nutrition. This pattern has been validated with much larger datasets since Vitousek's earlier work, showing again that leaf nitrogen/phosphorus ratios track available nutrient supplies across global ecosystems (Fig. 2).

SUMMARY AND SYNTHESIS

The three principles described—thermodynamics, feedback, and stoichiometry—provide a means toward investigating the complexity of both biogeochemical cycles and nutrient cycles. These principles are not exhaustive in scope, nor are they necessarily mutually exclusive (see above), nor is this entry particularly deep on any of these subjects. Nevertheless, the case studies and applications described provide a snapshot of what is known and what can be known, when viewing

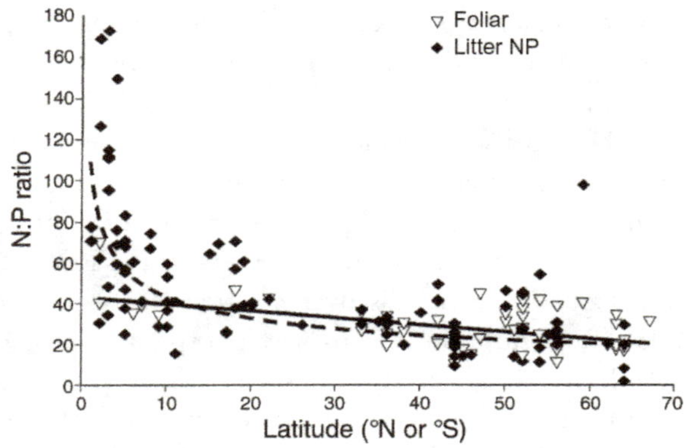

FIGURE 2 Global-scale patterns of nutrient stoichiometry of live forest leaves and litter fall. The significant rise in N/P of forest vegetation with decreasing latitude points to P limitation and N richness of tropical compared to extra-tropical forests on average. The more substantial rise in N/P of litter vs. live foliage results from plant resorption of P as this nutrient becomes scarce in old tropical forest soils. From McGroddy et al., 2004.

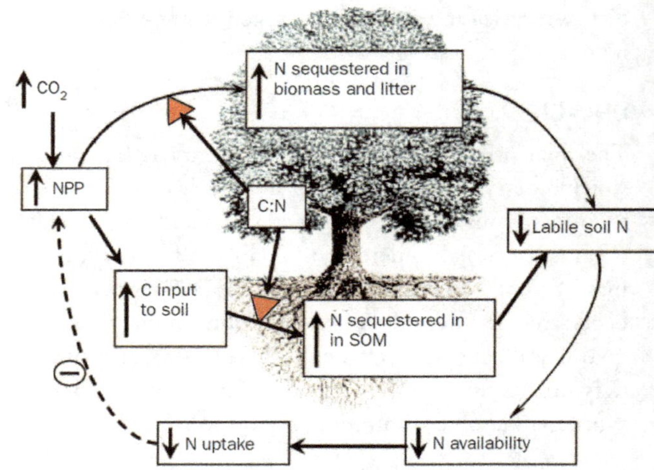

FIGURE 3 Proposed feedbacks between carbon and nitrogen cycles associated with the "Progressive Nitrogen Limitation" hypothesis. Initially, elevated carbon dioxide concentrations stimulate plant productivity leading to carbon storage in plants and soils. As new carbon stocks increase, nitrogen becomes locked up in pools with slow turnover times, particularly plant woody materials. This leads to less nitrogen available to soil microbes (and potentially higher carbon/nitrogen ratios of plant materials), such that net nitrogen mineralization rates decrease to the point that nitrogen limitation of plant productivity is enhanced. This feedback to less carbon dioxide uptake continues unless nitrogen inputs (via nitrogen fixation or atmospheric deposition) increase or nitrogen losses decrease, or both. Ultimately, the extent of progressive nitrogen limitation depends on the flexibility in plant and soil carbon/nitrogen ratios and nitrogen balances at the ecosystem scale. From Luo et al., 2004.

nutrient and biogeochemical cycles through a theoretical lens.

A final case study illustrates how integration of thermodynamic, feedback, and stoichiometric principles has led to new understanding of climate–nutrient–ecosystem interactions and ecosystem responses to global environmental change. Many models have been assembled to understand how ecosystems will both respond to and feedback on contemporary climate change; but only recently have models begun to consider vital interactions between ecology, nutrient biogeochemical cycles, and the climate system.

One of the emerging paradigms for how nutrient cycles will affect climate change is known as progressive nitrogen limitation (Fig. 3). This hypothesis states that elevated CO_2 stimulates photosynthesis, resulting in initial increases in ecosystem productivity, carbon storage, and negative feedback to CO_2-induced climate warming. However, as productivity increases so do nutrient demands of plants and soil microbes, with plants storing more carbon and nitrogen in biomass, in some cases even increasing the carbon/nitrogen ratio of new plant matter. This in turn reduces the size of available soil nitrogen pools and reduced rates of nitrogen mineralization (see also nitrogen limitation feedback, above). This positive feedback and stoichiometric constraint ultimately leads to more severe nitrogen limitation, thereby reducing CO_2 uptake and storage by ecosystems, eliminating altogether the negative feedback on climate warming associated with the initial CO_2 fertilization effect. This feedback breaks down, however, if nitrogen inputs increase or nitrogen losses decrease—thus pointing ultimately to the importance of thermodynamic principles of mass conservation in determining the fate of carbon–nitrogen interactions in affecting global climate change. Moreover, evolved strategies of plant nutrition and increased root/shoot ratios under conditions of nitrogen limitation may allow for mining of nitrogen from pools that are deeper in the soil, which might also reduce the severity of progressive nitrogen limitation.

This final synthetic case, like all others, shows the power and utility of linking multiple theoretical principles to gain knowledge on nutrient biogeochemical cycles. It also speaks to the complexity of ecosystem processes and the difficulties in trying to identify one single theory (or set of theories) to encapsulate biogeochemistry.

ALSO SEE THE FOLLOWING ARTICLES

Belowground Processes / Gas and Energy Fluxes across Landscapes / Microbial Communities / Ocean Circulation, Dynamics of / Stoichiometry, Ecological

FURTHER READING

Hedin, L. O. 2004. Global organization of terrestrial plant-nutrient interactions. *PNAS* 101(30): 10849–10850.

Likens, G. E., F. H. Bormann, R. S. Pierce, J. S. Eaton, and N. M. Johnson. 1977. *Biogeochemistry in a forested ecosystem.* New York: Springer-Verlag.

Schlesinger, W. H. 1997. *Biogeochemistry: an analysis of global change.* 2nd ed. San Diego: Academic Press.

Sterner, R. W., and J. J. Elser. 2002. *Ecological stoichiometery: the biology of elements from molecules to the biosphere.* Princeton: Princeton University Press.

Tansley, A. G., The use and abuse of vegetational concepts and terms. *Ecology*, vol. 16, no. 3, pp 284–307.

Vitousek, P. 2004. *Nutrient cycling and limitation: Hawai'i as a model system.* Princeton: Princeton University Press.

Walker, T. W., and J. K. Syers. 1976. Fate of phosphorus during pedogenesis. *Geoderma* 15(1): 119.

Wang, Y. P., and B. Z. Houlton. 2009. Nitrogen constraints on terrestrial carbon uptake: implications for the global carbon-climate feedback. *Geophysical Research Letters* 36, L24403, doi:10.1029/2009GL041009.

BIRTH-DEATH MODELS

CHRISTOPHER J. DUGAW

Humboldt State University, Arcata, California

Birth–death models represent a population using a whole number. Change occurs in discrete jumps of positive or negative 1 (i.e., births and deaths). Birth and death events occur stochastically in continuous time, and the probability of these events occurring is a function of the current population size. This model formulation is particularly relevant for small populations and is used in conservation biology and invasion biology. Birth–death models have also been applied to metapopulations, epidemics, and population genetics.

MATHEMATICAL MODEL

Classic differential equation–based population models are deterministic and assume that populations can be represented by fractional numbers that change continuously in time. However, populations are made up of discrete individuals and are more naturally represented by whole numbers. Furthermore, populations change in increments of 1 at discrete events (e.g., births, deaths, and migration). Birth–death models track a single integer-valued population size in continuous time (discrete-time birth–death models exist but have limited application in ecology because multiple births and deaths can occur in a period of time). Births and deaths necessarily occur at random instances with probabilities determined by the population size. Randomness occurs only in the form of demographic stochasticity and there are no external random forces (i.e., environmental stochasticity). Because of their stochastic nature, birth–death models do not uniquely determine the population size over time. For a given set of parameters and a fixed starting population, multiple outcomes (realizations) are possible. The stochastic model determines the probability distribution of the population through time, $p_i(t)$ (i.e., the probability that the population size at time t is i). Thus, the fraction of a large number of realizations with population size i at time t is approximately $p_i(t)$.

It is often impossible to find an explicit formula for this probability distribution, but other quantities give valuable information about the population. The mean of $p(t)$ gives the expected population size as a function time t, and the variance of $p(t)$ gives a measure of variability across realizations. In many applications, $p_0(t)$, the probability of population extinction by time t, and probability of eventual extinction ($\lim_{t\to\infty} p_0(t)$) are particularly important. Birth–death models may have nonzero extinction probabilities even when the the expected population approaches infinity. Furthermore, when environmental variability is included, paradoxically extinction may be certain even when the mean population approaches infinity.

In a birth–death model, the probability that a birth occurs in a small time interval of length Δt is approximately $\lambda_i \Delta t$, and the probability of a death is $\mu_i \Delta t$. More precisely:

$$\lambda_i = \lim_{\Delta t \to 0} \frac{\textbf{Pr}\,[\text{population increases by 1 in } (t, t+\Delta t]]}{\Delta t}$$

and

$$\mu_i = \lim_{\Delta t \to 0} \frac{\textbf{Pr}\,[\text{population decreases by 1 in } (t, t+\Delta t]]}{\Delta t}.$$

In these definitions, the term death may include emigration of an individual out of the population, and birth may include the immigration of an individual into the population.

Every birth–death model corresponds to a deterministic differential equation. On average the change of a population in a Δt interval of time is $\lambda_i \Delta t - \mu_i \Delta t$. If N represents the population in the deterministic model, then

$$\frac{dN}{dt} = \lambda_N - \mu_N, \tag{1}$$

where λ_N and μ_N are extensions of the birth and death rates to functions of continuous population values. Note that multiple birth–death models can lead to the same deterministic model and that these stochastic models can exhibit different behaviors because of cancellation in the subtraction of the birth and death terms.

If the birth rate at some population size N is zero ($\lambda_N = 0$), then we can restrict the possible population sizes to $i = 0, 1, 2, \ldots, N$ and the system will have a finite number of states. Otherwise the population may take on an infinite number of values ($i = 0, 1, 2, \ldots$). In the absence of immigration, $\lambda_0 = 0$, and once the population reaches zero it will never leave that state (i.e., 0 is an absorbing state). To maintain nonnegative population size, $\mu_0 = 0$ in all cases.

Expressions for the birth and death rates define the particular birth–death model. Two main examples will be used to illustrate various aspects of birth–death models.

Example 1: Simple Birth–Death Model

A simple birth–death model ignores density dependence and assumes that the birth rates and death rates are proportional to the population size, i.e., $\lambda_i = \lambda i$ and $\mu_i = \mu i$. The simple birth–death model is linear and has an infinite number of possible states. This model corresponds to the deterministic exponential growth model $\frac{dN}{dt} = (\lambda - \mu)N$, which has the solution $N(t) = N_0 e^{(\lambda - \mu)t}$ for a starting population N_0.

Example 2: A Logistic Birth–Death Model

The birth–death model with rates

$$\lambda_i = ri - \frac{ri^2}{2K} \text{ and } \mu_i = \frac{ri^2}{2K}, \; i = 0, 1, \ldots, 2K$$

is a logistic birth–death model. It is a nonlinear, density-dependent growth model with a finite number of possible states. Substituting the expressions for λ_i and μ_i into Equation 1 reveals that this model corresponds to the deterministic logistic model $\frac{dN}{dt} = rN\left(1 - \frac{N}{K}\right)$. There are several other birth–death models that also correspond to this deterministic model (see "Further Reading" section).

MATHEMATICAL ANALYSES

Monte Carlo Simulations

Individual realizations of a birth–death process can be easily simulated on a computer. The probability that an event (a birth or death) occurs in any small interval of length Δt is approximately $(\lambda_i + \mu_i)\Delta t$. Between events this rate is constant, and therefore the time between events follows an exponential distribution with mean $\frac{1}{\lambda_i + \mu_i}$. To determine which event occurs, contemplate births and deaths separately. The time until a birth occurs is exponentially distributed with mean $\frac{1}{\lambda_i}$, and the time until a death occurs is exponentially distributed with mean $\frac{1}{\mu_i}$. The probability that the second time exceeds the first is $\frac{\lambda_i}{\lambda_i + \mu_i}$. Thus, given that an event occurs, the probability that the event is a birth is $\frac{\lambda_i}{\lambda_i + \mu_i}$ and the probability that the event is a death is $\frac{\mu_i}{\lambda_i + \mu_i}$. Then the basic algorithm for simulation is as follows:

1. Initialize the population level to some initial value and initialize time to zero.
2. Calculate μ_i and λ_i based on the current population level.
3. Simulate an exponential random variable with mean $\frac{1}{\lambda_i + \mu_i}$ and increase time by this amount.
4. With probability $\frac{\lambda_i}{\lambda_i + \mu_i}$, increase the population by 1, otherwise decrease it by 1.
5. Repeat steps 2–4 until the desired time has elapsed or until the population has reached an absorbing state.

Monte Carlo simulations of the simple birth–death model and a logistic birth–death model are shown in Figure 1.

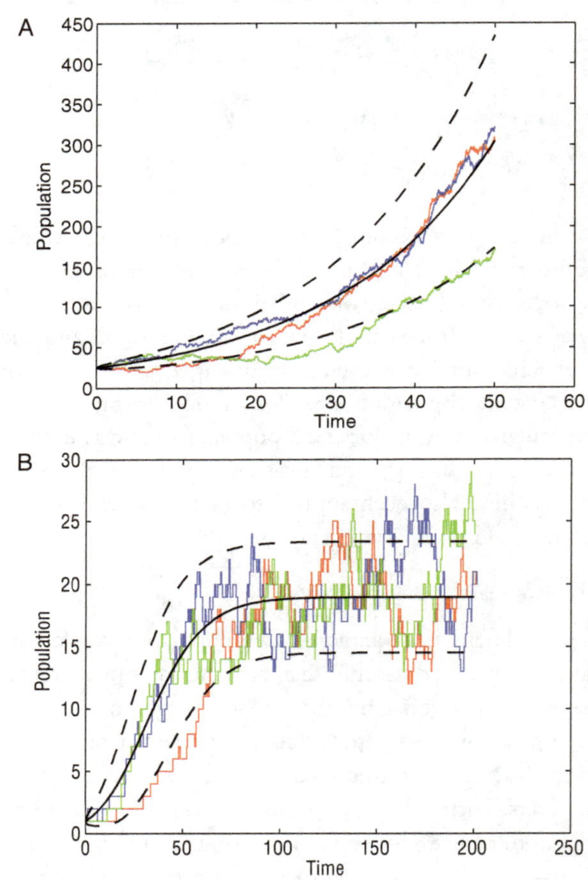

FIGURE 1 Results of birth–death models. Solid black line shows mean population size through time, dashed lines show mean ±1 standard deviation, and colored lines show three separate realizations. Panel (A) shows results for simple birth–death process with $\lambda = 0.15$ and $\mu = 0.1$. Panel (B) shows results for logistic birth–death model with $r = 0.1$ and $K = 20$.

Differential Equations for Probability Distribution

The probability distribution for the state of the population changes over time and is derived using differential equations that describe the rate of change of probability for each state. The probability of having i individuals at time $t + \Delta t$ is $p_i(t + \Delta t)$ and can be computed using conditional probability. Then

Pr[pop. is i at $t + \Delta t$]
$\approx$ **Pr**[pop. is $i - 1$ at t and birth occurs in $(t, t + \Delta t)$]
$+$ **Pr**[pop. is i at t and no births or deaths in $(t, t + \Delta t)$]
$+$ **Pr**[pop. is $i + 1$ at t and death occurs in $(t, t + \Delta t)$],

or using more notation,

$$p_i(t + \Delta t) \approx \lambda_{i-1}\Delta t p_{i-1}(t) + (1-(\mu_i + \lambda_i)\Delta t)p_i(t) + \mu_{i+1}\Delta t p_{i+1}(t), \quad i = 1, 2, \ldots$$

and

$$p_0(t + \Delta t) \approx (1 - \lambda_0\Delta t)p_0(t) + \mu_1\Delta t p_1(t).$$

Rearranging these equations and taking the limit as Δt goes to 0 yields

$$\frac{dp_i(t)}{dt} = \lambda_{i-1}p_{i-1}(t) - (\mu_i + \lambda_i)p_i(t) + \mu_{i+1}p_{i+1}(t), \quad i = 1, 2, \ldots \tag{2}$$

and

$$\frac{dp_0(t)}{dt} = -\lambda_0 p_0(t) + \mu_1 p_1(t). \tag{3}$$

For most cases, it is intractable to obtain an explicit solution for this potentially infinite set of coupled differential equations. However, when there are a finite number of possible states, this system of differential equations can be solved numerically, and some cases allow analytic solutions using a computer algebra system. If there are an infinite number of possible states, close approximations of the probability distribution may be obtained by assuming $p_i(t) = 0$ for all i greater than some value. Such an approximation will be appropriate when there is a very small probability that the population becomes exceedingly large, which is the case in most density-dependent population growth models. The numerical solution of Equations 2 and 3 the probability distribution for the logistic birth–death model is shown in Figure 2.

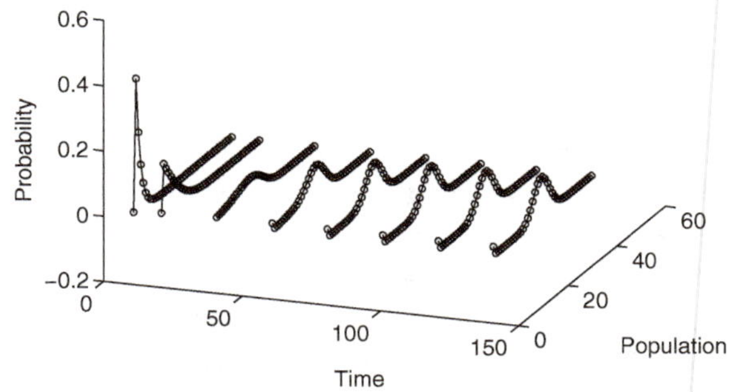

FIGURE 2 Probability distribution for a logistic birth-death model with $r = 0.1$ and $K = 20$ at times separated by 20 time units.

Mean and Variance of Population

The mean $m(t)$ and variance $v(t)$ of a birth–death model represent the mean and variance in population size of a large collection of realizations at time t. Then $m(t)$ and $v(t)$ can be computed from the probability distribution as

$$m(t) = \sum_i ip_i(t) \quad \text{and} \quad v(t) = \sum_i (i - m(t))^2 p_i(t).$$

Using these definitions can be difficult because the formula for $p_i(t)$ may not be available. However, other techniques exist for computing the mean and variance. For a simple birth–death model starting with n_0 individuals, the mean is $m(t) = n_0 e^{(\lambda-\mu)t}$ and variance is $v(t) = \frac{\lambda+\mu}{\lambda-\mu} e^{(\lambda-\mu)t}(1 - e^{(\lambda-\mu)t})$. Notice that in this case the mean is the same as the solution to the corresponding deterministic model. In contrast, the solution of the deterministic logistic model is greater than the logistic birth–death model as shown in Figure 3. The difference is a result of the nonlinearity of the logistic birth–death model and is common in nonlinear models. However, the error is typically small. The mean and variance for

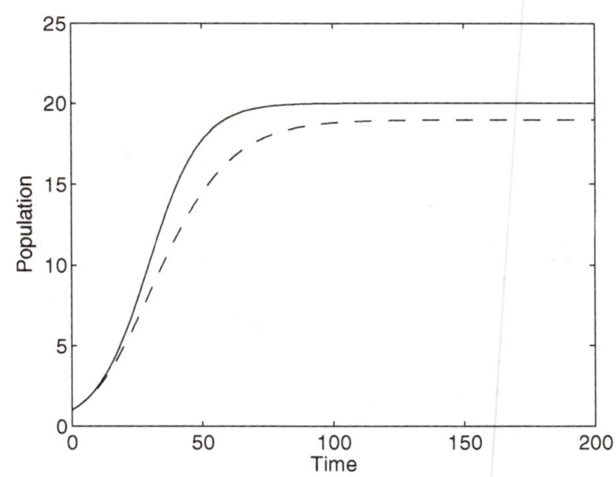

FIGURE 3 A comparison of a logistic birth-death model and the corresponding deterministic model. The solid line is the solution of the deterministic model, and the dashed line is the mean of the birth-death model. Parameters are the same for both models: $r = 0.1$ and $K = 20$.

the logistic and simple birth–death models are shown in Figure 1.

Long-Term Behavior

An approach to analyzing many mathematical models is to examine their long-term behavior, but this can be problematic in birth–death models. For example, in a birth–death model with a finite number of states and no immigration, eventual population extinction is guaranteed. Even with an infinite number of states, extinction may also be certain. With immigration, extinction does not occur and it is possible to compute the long-term probability distribution for the population, the stationary distribution. Without immigration, it is necessary to compute the quasi-stationary distribution of the population, which is the conditional distribution of the population given that it has not yet gone extinct. Thus, examinations of the long-term behavior of birth–death models focus on (1) the probability of extinction, (2) average time to extinction in the case that it is certain, (3) the stationary distribution, and/or (4) the quasi-stationary distribution.

EXTINCTION PROBABILITY

In the absence of immigration ($\lambda_0 = 0$), population extinction is possible. Furthermore, if the death rate is nonzero for all positive populations, then 0 is the only absorbing state. Thus, in the long run ($t \to \infty$) the population either goes extinct or approaches infinity. The probability of eventual extinction in birth–death models with a finite number of states and no other absorbing states is 1. The probability of eventual extinction for models with an infinite number of states and a starting population value n_0 is

$$\textit{Prob. of eventual extinction} = \begin{cases} 1 & \displaystyle\sum_{i=1}^{\infty} \frac{\mu_1\mu_2\cdots\mu_i}{\lambda_1\lambda_2\cdots\lambda_i} = \infty \\ \dfrac{\sum_{i=n_0}^{\infty} \frac{\mu_1\mu_2\ldots\mu_i}{\lambda_1\lambda_2\ldots\lambda_i}}{1+\sum_{i=1}^{\infty} \frac{\mu_1\mu_2\ldots\mu_i}{\lambda_1\lambda_2\ldots\lambda_i}} & \displaystyle\sum_{i=1}^{\infty} \frac{\mu_1\mu_2\cdots\mu_i}{\lambda_1\lambda_2\cdots\lambda_i} < \infty \end{cases}.$$

For a simple birth–death process ($\lambda_i = \lambda i$ and $\mu_i = \mu i$), this formula reduces to

$$\textit{Prob. of eventual extinction} = \begin{cases} 1 & \mu \geq \lambda \\ \left(\dfrac{\mu}{\lambda}\right)^{n_0} & \mu < \lambda \end{cases}.$$

Note that the simple birth–death model is the stochastic analogue of the deterministic differential equation $\frac{dN}{dt} = (\lambda - \mu)N$, which has the solution $N(t) = N_0 e^{(\lambda-\mu)t}$. The deterministic model predicts exponential growth when $\lambda > \mu$ and decay when $\lambda < \mu$. Thus, there is a nonzero probability of extinction even when the corresponding deterministic model predicts exponential growth.

Mean Time to Extinction

If eventual population extinction is certain, it is natural to investigate how soon population extinction expected. In this case, the mean time to extinction starting with a n_0 individuals τ_{n_0} is:

$$\tau_{n_0} = \begin{cases} \dfrac{1}{\mu_1} + \sum_{i=2}^{N} \dfrac{\lambda_1\lambda_2\cdots\lambda_{i-1}}{\mu_1\mu_2\cdots\mu_i} & n_0 = 1 \\ \tau_1 + \sum_{s=1}^{n_0-1} \left[\dfrac{\mu_1\mu_2\cdots\mu_s}{\lambda_1\lambda_2\cdots\lambda_s} \times \sum_{i=s+1}^{N} \left(\dfrac{\lambda_1\lambda_2\cdots\lambda_{i-1}}{\mu_1\mu_2\cdots\mu_i} \right) \right] & n_0 = 2, 3, \ldots, N \end{cases}.$$

This formula holds when the maximum population size N is finite or infinite, but in the latter case the formula can be difficult or impossible to compute. However, the mean time to extinction starting with one individual in a simple birth–death process reduces to $\tau_1 = -\frac{1}{\lambda}\ln\left(1 - \frac{\lambda}{\mu}\right)$. This formula gives an underestimate for the extinction time of larger initial populations. For birth–death models that have a finite number of states, this formula can be calculated using a computer. The mean time to extinction for a logistic birth–death process is shown in Figure 4. Notice that extinction times are on the order of 10^{11} time units, which is likely longer than any relevant ecological time scale. Thus, a population in reality may be safe from extinction even if eventual extinction is predicted by the model.

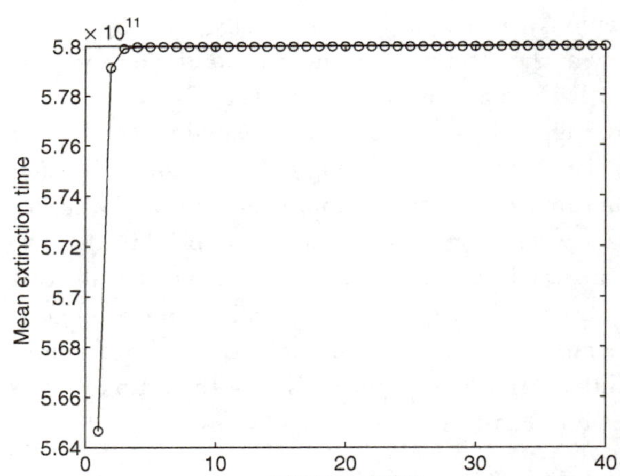

FIGURE 4 Mean time to extinction as a function of initial population size for a logistic birth-death model with $r = 0.1$ and $K = 20$.

The Stationary Distribution

A stationary distribution is a probability distribution of the population that does not change through time. It can be found by setting the right-hand sides of Equations 2 and 3 to equal zero and solving. However, this large or infinite system of algebraic equations can be difficult or impossible to solve exactly. Some birth–death models will approach their stationary distribution as time increases, but others like the simple birth–death model can grow without bound. When zero is an absorbing state, $p_0(t) = 1$ and $p_i(t) = 0$ for all other i is the stationary distribution. As with the logistic birth–death model, extinction may be certain, but the average time to reach this state may be very long.

The Quasi-stationary Distribution

Because zero is an absorbing state in many birth–death models, it is natural to consider the probability distribution of the population conditional on the event that the population has not yet gone extinct. The conditional distribution is

$$q_i(t) = \frac{p_i(t)}{1 - p_0(t)}.$$

The quasi-stationary distribution is a set of $q_i(t)$ that does not change in time. It is often difficult to find a quasi-stationary distribution, but many approximate solution methods exist. The quasi-stationary distribution for the logistic birth–death model is shown in Figure 5. Notice that the distribution is centered about a value just below K, the equilibrium of the deterministic logistic model. This is because the mean of the stochastic model is less than the value of the deterministic model.

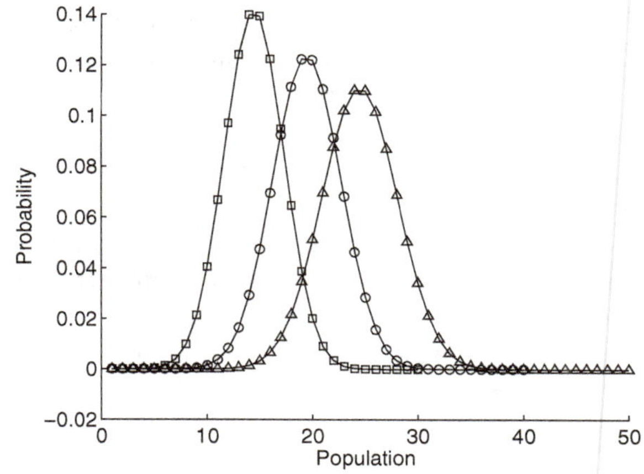

FIGURE 5 Quasi-stationary distribution for a logistic birth-death model with $r = 0.1$ for all lines; $K = 15$ for squares, $K = 20$ for circles, and $K = 25$ for triangles.

PRACTICAL USE OF THEORY

Any time a population is small, its discreteness is apparent and birth–death models are potentially relevant. Birth–death models are used for predicting and understanding population extinction. MacArthur and Wilson used birth–death models in their theory of island biogeography. The ability of species to colonize islands was studied extensively using birth–death models. They discussed the properties of a successful colonizer and the effect of island size on colonization success using these models.

Subpopulations in a metapopulation are by definition small and prone to extinction. The rate of extinction of subpopulations is an important parameter in metapopulation models. Birth–death models have been used to estimate extinction rates caused by demographic stochasticity in metapopuation models.

At the start of an epidemic, a few infectious individuals enter a population. These individuals may come in contact with susceptible individuals and spread the disease to them. Classic theory based on deterministic models predicts that an epidemic will occur if R_0 exceeds 1, where R_0 is the average number of secondary infections when an infected individual is introduced into a susceptible population. However, at the start of an epidemic the population of infected individuals is small and stochasticity can play a significant role. Thus, birth–death models that track the population of infected individuals are used to model this scenario.

Birth–death models play a central role in population viability analysis, which seeks to find the minimum population size needed to make the risk of extinction insignificant. Birth–death models are a natural starting point for such analyses because they incorporate demographic stochasticity. However, all modern population viability analyses include environmental stochasticity in some form. Birth–death–catastrophe models build on classic birth–death models by adding random catastrophic events where multiple individuals die simultaneously and are used in population viability analysis. Extinction probability and time to extinction are obviously important questions in population viability analysis.

Birth–death models have many other applications where the state of the system measures some quantity other than population size. For example, birth–death models have been used to model metapopulations where the state represents the number of extant subpopulations. In this case, births correspond to colonization of habitat patches to form new subpopulations, and deaths represent the extinction of a subpopulation. The quasi-stationary distribution represents the probability distribution for the number of subpopulations, and absorption to the zero state represents the extinction of the entire metapopulation.

Birth–death models are also used in population genetics. Moran modeled the frequency of an allele at a single locus in a population of haploid individuals. He assumed that there were only two alleles and that the population has a fixed size; when an individual dies, they are replaced by an individual with an allele chosen at random with probability equal to the fraction of each allele in the population. This process occurs in continuous time, and the frequency of an allele can only change in increments of 1. The model can incorporate selection and mutation of one allele to another. Natural questions that can be answered with this model are the probability that an allele is removed from the system (i.e., fixation of the allele), the mean time to fixation, and probability distribution for the frequency of the allele.

SEE ALSO THE FOLLOWING ARTICLES

Branching Processes / Markov Chains / Mutation, Selection, and Genetic Drift / Ordinary Differential Equations / Population Viability Analysis / Stochasticity, Demographic

FURTHER READING

Allen, L. J. S. 2003. *An introduction to stochastic processes with applications to biology*. Upper Saddle River, NJ: Pearson Education.

Goel, N. S., and N. Richter-Dyn. 1974. *Stochastic models in biology*. New York: Academic Press.

Mangel, M., and C. Tier. 1994. Four facts every conservation biologist should know about persistence. *Ecology* 75(3): 607–614.

MacArthur, R. H., and E. O. Wilson. 1967. *The theory of island biogeography*. Princeton: Princeton University Press.

Moran, P. A. P. 1958. Random processes in genetics. *Mathematical Proceedings of the Cambridge Philosophical Society* 54: 60–71.

Nåssell, I. 2001. Extinction and quasi-stationarity in the Verhulst logistic model. *Theoretical Population Biology* 211: 11–27.

Nisbet, R. M., and W. S. C. Gurney. 1982. *Modelling fluctuating populations*. New York: John Wiley.

Novozhilov, A. S., G. P. Karev, and E. V. Koonin. 2006. Biological applications of the theory of birth-and-death processes. *Briefings in Bioinformatics* 7(1): 70–85.

Quinn, J. F., and A. Hastings. 1987. Extinction in subdivided habitats. *Conservation Biology* 1(3): 198–208.

BOTTOM-UP CONTROL

JOHN C. MOORE

Colorado State University, Fort Collins

PETER C. DE RUITER

Wageningen University Research Centre, The Netherlands

Bottom-up control is a concept that refers to the influences that organisms and resources at the first trophic level of a food web have on the structure and dynamics of other organisms within the food web. The concept is often discussed in conjunction with its companion concept of top-down control wherein consumers at upper trophic positions regulate via consumption the population densities and dynamics of organisms at lower trophic positions. Food webs exhibit some degree of both bottom-up and top-down control, the relative amounts of which may shift through time and over space as the availability and quality of resources and environmental conditions change.

THE CONCEPT

Bottom-up control, if not by name, is an old concept. For historical and heuristic purposes, its modern interpretations can be traced back to an address given by G. Evelyn Hutchinson to the American Society of Naturalists in 1957 and subsequently published in 1959 in the *American Naturalist* entitled "Homage to Santa Rosalia, or why are there so many kinds of animals?" Plants and other primary producers form the energetic base of communities providing sustenance to higher trophic positions in a bottom-up manner, while organisms at higher trophic positions (particularly the top predators) regulate the densities of organisms at lower trophic positions in a top-down manner. This concept of bottom-up control was grounded in thermodynamics by focusing on the efficiencies of trophic interactions and the diminishing availability of energy with each transfer.

In 1960, Hairston, Smith, and Slobodkin published in the *American Naturalist* a rejoinder of sorts to the paper by Hutchinson that furthered the argument by asking *why the world was green?* and *why detritus did not accumulate now as in the past eras?* In doing so, they invoked the notions of trophic levels, community organization, and multiple factors regulating populations and communities. What emerged was the HSS hypothesis, which starts from the premise that herbivores have not devoured green plants, detritus is not accumulating, and that the planet is no longer producing fossil fuels. These observations led to following three suppositions:

1. Populations of producers, predators, and decomposers are all limited by the availability of resources in a density-dependent manner.
2. Interspecific competition regulates the distributions and abundances of producers, predators, and decomposers.
3. Herbivores are not limited by resources, but rather are limited by predators, and therefore are not likely to compete for resources.

Bottom-up control emerged from the first of the three suppositions, while its top-down counterpart was embodied in the third supposition. The HSS hypothesis posits that plants are limited by nutrients and water (bottom-up), decomposers are limited by the availability of detritus (bottom-up control), and predators are limited by the availability of their herbivore prey (bottom-up control), whereas herbivores are limited by consumption from their predators (top-down control). These limitations depend on the assumption that interspecific competition among producers, decomposers, and predators affected their respective distributions, but that herbivores were largely immune from competition. These ideas were products of their time, as competition for limited resources and predation were viewed as the primary regulators of community structure. Since the publishing of HSS, many authors have pointed out that control mechanisms are numerous, concluding that the regulation of populations and communities was the interplay among resource availability, predation, and a host of other mechanisms. Populations self-regulate, seasonality impacts species and population densities, resource amount may not equate to availability, and temporal and spatial heterogeneity of species and populations distributions and activities abound.

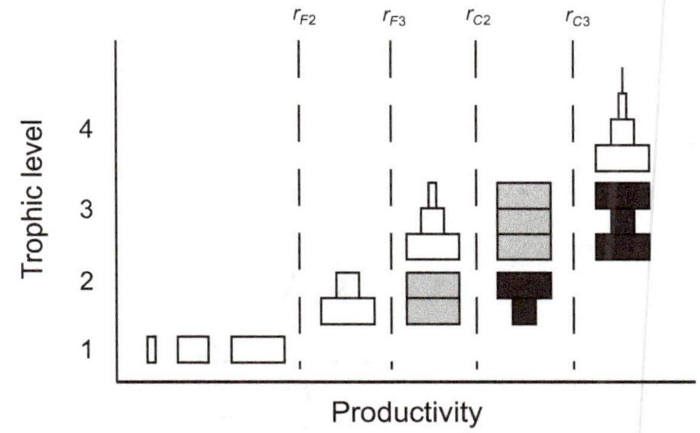

FIGURE 1 Caricature of the changes in trophic structure that occur in response to increased productivity achieved by increasing the intrinsic rate of increase, r, of a primary producer or the input rate of detritus, R_d (see Eqs. 2–3). The vertical dashed lines indicate levels of input where either an additional trophic level can be supported (r_{Fn}) or, in the case where a higher-order consumer is not present, the biomass of the top trophic level equals and then exceeds the level below it as a cascade of biomass (r_{Cn}) (see Eqs. 5–8). A third threshold, not depicted here but discussed in the text, represents the level of productivity wherein the dynamics of a system with a given number of trophic levels in response to a minor disturbance changes ($r_{\lambda n}$).

Types of Bottom-Up Control

Bottom-up control can arise in several ways, but it can be sorted into two broad categories—one that deals with enrichment or the influences of the rates of inputs of basal resources and one that addresses the influences that the types and quality of basal resources have on species at higher trophic positions. The rate of primary production or input of detritus from allochthonous sources is the primary means of initiating bottom-up control. The rates of input can potentially affect both the structure and dynamics of the food web.

From a structural standpoint, the rates of input govern the likelihood of there being sufficient energy to support successive trophic levels, given the energetic efficiencies of organisms. This feature generates a stepwise increase in biomass at successive trophic positions. For a fixed number of species or trophic levels, increasing the rates of inputs (i.e., enrichment) of primary production or detritus moves biomass to higher trophic levels with the distribution of biomass at each trophic level depended on the number of trophic levels present (Fig. 1). Enrichment affects both the likelihood of supporting additional trophic positions and the distribution of biomass within trophic levels. An increase in the rate of primary production or inputs of detritus results in an increase and movement of biomass up the trophic pyramid to higher trophic positions, until which time a species at the next trophic level enters the community. Once the new species enters, the distribution of biomass shifts to back to the lower trophic positions. The details of the model are presented in the section below.

From a dynamic standpoint, bottom-up control through enrichment can affect the responses of species to disturbance and can lead to transitions in the underlying dynamics of species (Fig. 2). The rate that the densities of populations recover from minor disturbances, *aka* resilience, is tied to productivity. The "paradox of enrichment" presented by Michael Rosenzweig and the biological route to chaos presented by Robert May, discussed in greater detail below, demonstrated how enrichment affected the densities of populations at steady-state in a manner described above, and their dynamic properties. In these cases, enrichment-initiated transitions from a stable equilibrium to an alternative dynamic state (e.g., instability, stable limit cycle, chaos).

Bottom-up control can also be expressed through the influence that the types of primary producers or sources of detritus have on the environment chemically or structurally, and how these attributes affect higher trophic positions. For example, in terrestrial and aquatic ecosystems, a shift in the dominant plant and algal communities, respectively, can result in changes in the assemblage of herbivore species and higher-order consumers. Allelopathy within the plant kingdom is common either through the

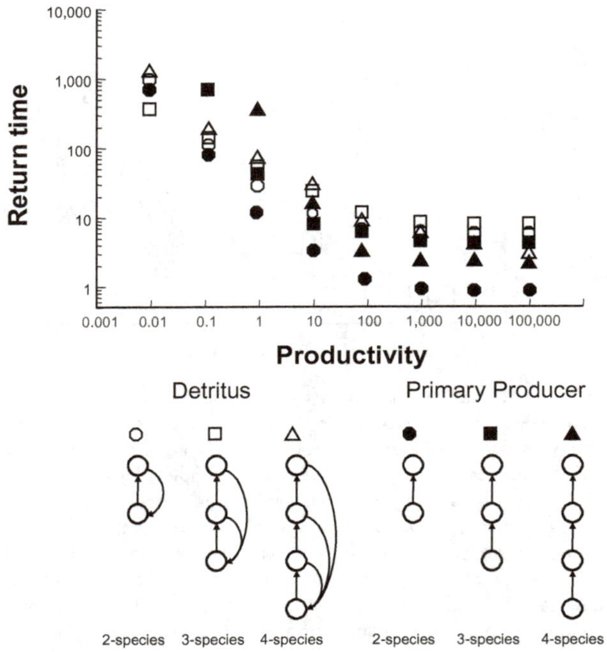

FIGURE 2 The return-times of models of simple detritus-based and primary producer-based food chains described using Equations 2–4 with increased productivity. Productivity was varied by adjusting the intrinsic rate of increase, r, of the primary producer, or the rate of detritus input, R_d, of detritus. Return-times are defined for models that are stable, where in all eigenvalues of the Jacobian matrix A are negative. Return time is calculated as $-1/\lambda_{(max)}$, where $\lambda_{(max)}$ is the largest negative eigenvalue, i.e., the negative eigenvalue closest to zero.

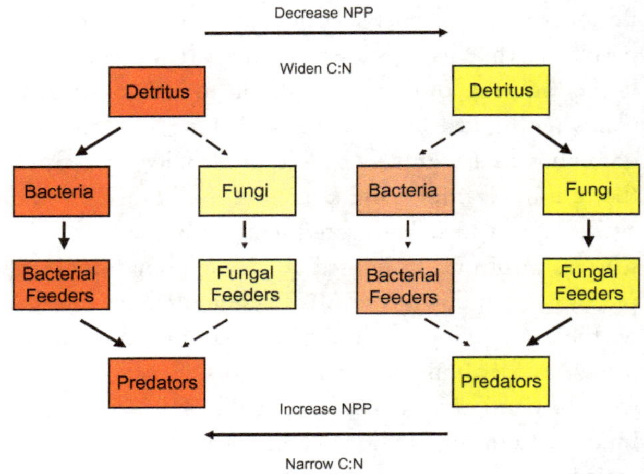

FIGURE 3 Simple representations of the bacterial and fungal energy channels found in soil system. Bottom-up control in soil systems occurs through changes in the input rate of detritus (NPP) and the quality of detritus (C:N). Shifts towards the bacterial energy channel increase C and N mineralization rates, while shifts toward the fungal pathway decrease the rates of these functional attributes.

direct release of toxins by the living plant or through the chemical composition of its litter. These chemicals affect the growth and establishment of additional plants, the structure of the decomposer communities, and the rates of decomposition and nutrient cycling in the vicinity of the host plant. The chemical composition of detritus directly impacts the structure and activity of the decomposer community (Fig. 3). Detritus with high C:N ratios (>30:1) tends to promote fungi and their consumers (fungal energy channel), while substrates with low C:N ratios (<30:1) tend to promote bacteria and their consumers (bacterial energy channel). An increase in the activity of the bacterial energy channel is generally accompanied by increased rates of N-mineralization and C-respiration, and increased rates of rates of the decomposition of detritus, while the pattern is observed if the fungal energy channel is enhanced. Hence, changes in the chemical composition or source of detritus have the potential to induce shifts in the distribution of biomass and material flow through systems and functional attributes of the system. Lastly, canopy structure provided by living plants and the complex structural milieu offered by the accumulation of detritus also serve as habitats and refuges for consumers.

Many vascular plants engage in different forms of mutualistic interactions with fungi and bacteria that involve the exchange of limiting nutrients between the partners. The interactions occur within plant roots or at the interface of the plant root and the soil. Examples include the obligate associations of nitrogen-fixing bacteria and leguminous plants and the obligate and facultative mycorrhizal associations between fungi and vascular plants. Vascular plants generally provide carbon and shelter to the microbes, while the microbes provide nitrogen and phosphorus to the plants. The presence or absence of these symbiotic interactions can affect the structure and productivity of the plant community and nutrient dynamics within the ecosystem.

MATHEMATICAL MODELS

Simple models can illustrate how bottom-up control driven by energy acquisition and trophic interactions can influence the structure and dynamics of food webs. The foundations of the structural and dynamic aspects of bottom-up control are evident in Robert May's study of population dynamics using the following one-dimensional discrete logistic equation:

$$x_{t+1} = rx_t(1 - x_t). \qquad (1)$$

Increasing the intrinsic rate of increase (r) of the model organism (x) and following the changes in its population density at each time step (t) generated complex changes

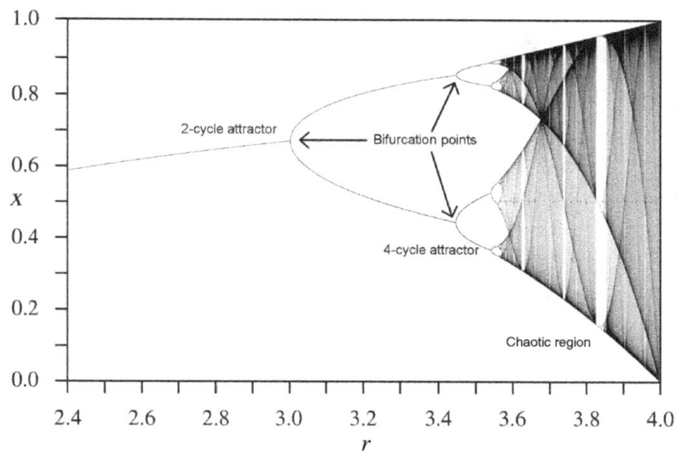

FIGURE 4 The logistic map, also known as the bifurcation diagram, presented by Robert May for the simple logistic difference equation for a single population (Eq. 1). The graph depicts the behavior of the population density, X_t, at different values of the intrinsic rate of increase, r, as time, $t \to \infty$. For low values of r the population densities converge to a single level. At $r = 3$, a shift in the dynamic state occurs as the population density does not converge on a single value, but undergoes a bifurcation after which it oscillates between two values. Between $r = 3.0$ and $r = 3.57$, additional bifurcations occur. At $r = 5.57$ and beyond, chaos ensues followed by oscillations of period 3, 6, 12, 24, and then back to chaos.

in its abundance and dynamics that have the potential to influence their consumers. At low levels of r the system possessed a single point attractor, or stable equilibrium, but as r increased, the population density (x) increased to a point after which a bifurcation occurred giving rise to 2-cycle attractor consisting of a maximum and alternative minimum (Fig. 4). At still large values of r the system transitioned to a 4-cycle attractor, then an 8-cycle attractor, to an attractor that possessed multiple (possibly infinite) points.

Simple mathematical models that follow the familiar Lotka–Volterra form capture many of aspects of bottom-up control described above and are simple heuristic tools to illustrate these points. The dynamics of a primary producer in a primary producer-based model with n species is modeled as follows:

$$\frac{dX_i}{dt} = r_iX_i - \sum_{i=1}^{n} f(X_i)X_j, \qquad (2)$$

where X_i and X_j represent the population densities of the primary producer and consumers, respectively, r_i is the specific growth rate of the primary producer, and $f(X_i)$ represents the functional response of the interaction between (species $i \neq j$) or within the populations (species $i = j$).

The dynamics of detritus in a detritus-based model of similar form are as follows:

$$\frac{dX_d}{dt} = R_d - \sum_{i=1}^{n} \sum_{j=1}^{n} (1 - a_j)f(X_i)X_j + \sum_{i=1}^{n} d_iX_i - \sum_{j=1}^{n} f(X_d)X_j. \qquad (3)$$

The subscript d highlights and keeps track of where detritus arises. Here, X_i and X_j represent the population densities of living species, and X_d is the density of detritus. The model includes a single allochthonous source, R_d, and possesses two autochthonous sources, one from the unassimilated fractions of prey that are killed by consumers, $\sum_{i=1}^{n}\sum_{j=1}^{n}(1-a_j)f(X_i)X_j$, which includes feces and leavings, and a second source from the corpses that die from causes other than predation, $\sum_{i=1}^{n} d_iX_i$. Consumption of detritus, $-\sum_{j=1}^{n} f(X_d)X_i$, resembles the consumption within the primary producer-based equation (Eq. 1).

The dynamics of consumers within primary producer-based systems and detritus-based systems are treated as follows:

$$\frac{dX_i}{dt} = -dX_i + \sum_{j=1}^{n} a_jp_jf(X_i)X_j - \sum_{j=1}^{n} f(X_j)X_i. \qquad (4)$$

Consumer growth is offset by natural deaths represented by a specific death rate, d_i, and as a function of the being consumed by other consumers, $-\sum_{j=1}^{n} f(X_j)X_i$, or is offset as a result of intraspecific competition (when $i = j$). Consumer populations grow in a density-dependent manner as a function of the living or nonliving prey that is consumed. This involves the summation of the consumption of individual prey types, $\sum_{j=1}^{n} a_jp_jf(X_i)X_j$, that includes the functional response, $f(X_i)$, and the assimilation efficiency, a_j, and consumption efficiency, p_j, of the consumer for each prey type.

A common approach to demonstrate the impacts of bottom-up control in a food web is to increase r or R_d, the consumption rates within in the functional responses, and energetic efficiencies of consumers, or decrease the d and follow changes in the system's structure and dynamics. Three thresholds are as follows:

1. r_F, the point at which an additional trophic level is energetically feasible,

2. r_C, the point at which the pyramid of biomass disappears giving rise to a cascade of biomass,

3. r_λ, the point at which oscillations ensue.

The first two thresholds, r_F and r_C, indicate that trophic structure is a function of productivity and that food chains, and by extension food webs, will undergo transitions in their trophic structure along a gradient of productivity. The first threshold, r_F, represents the level of primary production that there is sufficient for all species maintain positive steady-state densities, i.e., all $X_n{}^* > 0$. For a two-species primary producer-based food chain modeled with a Holling Type I functional response for the consumer on the prey and to describe intraspecific competition among the prey, i.e., $f(X_2) = c_{12}X_1$ and $f(X_1) = c_{11}X_1$, respectively, this condition is met if $r_1 = r_{F2}$:

$$r_{F2} > \frac{d_2}{a_2 p_2}\frac{c_{11}}{c_{12}}, \tag{5}$$

where c_{11} is the coefficient of intraspecific competition and c_{12} is the consumption coefficient of the consumer on the prey. The second threshold, r_C, represents the level of primary production where the equilibrium density of the species at the upper trophic (herbivore) equals the lower trophic position (plant), i.e., $X_2{}^* = X_1{}^*$. For the two-species primary producer food chain, this occurs when $r_1 = r_{C2}$:

$$r_{C2} = \frac{d_2}{a_2 p_2}\left(1 + \frac{c_{11}}{c_{12}}\right). \tag{6}$$

The level of productivity where the three-species food chain is feasible ($r_1 = r_{F3}$):

$$r_{F3} = r_{C2} - \frac{d_2}{a_2 p_2} - \frac{c_{11}^2 d_3}{c_{12}c_{23}a_3 p_3}. \tag{7}$$

The level of productivity where the three-species food chain exhibits a cascade of biomass (r_{C3} when $X_3{}^* = X_2{}^*$) is

$$r_{C3} = r_{F3} + \frac{d_3}{a_3 p_3 a_2 p_2}. \tag{8}$$

From Equations 5–8, the level of productivity where the three-species food chain first exhibits a cascade is greater than the level of productivity where the chain becomes feasible. Furthermore, the point at which the three-species food chain becomes feasible is near the point that that the two-species food chain cascades, i.e., $r_{F2} < r_{F3} < r_{C2} < r_{C3}$. If we extend this reasoning to four levels and beyond, the differences between the productivity that insures feasibility of four levels and the rate that initiates cascades for each trophic position precludes the widespread existence of trophic cascades of biomass (see Fig. 1).

The third threshold, $r_{\lambda 2}$, represents the level of productivity where we see a shift in dynamics. We estimated $r_{\lambda 2}$ by determining the level of productivity that marks the point at which point the dominant eigenvalue, λ_{max}, becomes complex. The dominant eigenvalue for the two-species primary producer–based food chain is

$$\lambda_{max} = \frac{\alpha_{11} + \sqrt{\alpha_{11}^2 + 4\alpha_{12}\alpha_{21}}}{2}, \tag{9}$$

where the α_{ij} represent the interaction strengths or elements of the Jacobian matrix (A) defined as the partial derivatives near equilibrium: $\alpha_{ij} = \left(\partial \frac{dX_i}{dt} / \partial X_j\right)$. To obtain the interaction strengths in this form, the differential equations (Eqs. 2–4 in this example) are redefined in terms of small deviations from equilibrium, where $x_i = X_i - X_i^*$. The redefined equations are rewritten using the Taylor expansion and neglecting terms with an order larger than 1. When the partials are taken from the two-species primary producer–based equations, the Jacobian matrix A is as follows:

$$A = \begin{bmatrix} \alpha_{11} = -c_{11}X_1^* & \alpha_{12} = -c_{12}X_1{}^* \\ \alpha_{21} = a_2 p_2 c_{12} X_2{}^* & \alpha_{22} = 0 \end{bmatrix}. \tag{10}$$

Here, α_{21} represents the bottom-up effect of the primary producer on the consumer, while α_{12} depicts the top-down effect of the consumer on the primary producer.

The real part of λ_{max} (Eq. 9) obtained from the Jacobian matrix A (Eq. 10) describes the direction and rate of decay of the perturbation; as $r_1 \rightarrow \infty$, RT $\rightarrow 2/\alpha_{11}$. The imaginary part of λ_{max} describes the oscillations that the system undergoes during the return to steady state. The term that is responsible for the introduction and increases in the oscillations with increased productivity is α_{21} within the discriminant, u, where $u = \alpha_{11}{}^2 + 4\alpha_{12}\alpha_{21}$. All the α_{ij} terms are functions of the steady-state biomass ($X_1{}^*$ or $X_2{}^*$) of the species involved. The α_{21} term, which describes the bottom-up effect of the prey on the consumer, is a function of the equilibrium density of the consumer, $X_2{}^*$, which in turn is dependent on the specific rate of increase (r_1) of the prey. As $r \rightarrow \infty$, u transitions from positive values, to zero, to negative values, with the following dynamic properties:

$$\begin{aligned} &\alpha_{ii} > 2(\alpha_{ij}\alpha_{ji})^{1/2} && \text{over-damped (monotonic damping),} \\ &\alpha_{ii} > 2(\alpha_{ij}\alpha_{ji})^{1/2} && \text{critically damped,} \\ &\alpha_{ii} > 2(\alpha_{ij}\alpha_{ji})^{1/2} && \text{under-damped oscillations.} \end{aligned} \tag{11}$$

The overdamped recovery occurs when $u > 0$. In this situation, the decay of the disturbance is governed by the real part of λ_{max} and the disturbance decays without oscillation to system's original steady-state condition. At some level of productivity, defined here as r_λ, $u = 0$, the over-damped recovery gives way to the critically damped recovery where the disturbance decays in a monotonic manner but over-shoots the steady state once and then returns to the steady-state without oscillations. At still greater levels of productivity, when $u < 0$ the system develops damped oscillations with a quasi-period of $T_d \approx 2\pi/(u)^{1/2}$.

For the two-species primary producer–based food chain, $r_{\lambda 2}$ can be estimated as the level of productivity where the discriminant (u) becomes negative. We obtained the following relationship:

$$r_{\lambda 2} = r_{C2} + \frac{c_{11} r_{F2}}{4 c_{12} a_2 p_2} - \frac{d_2}{a_2 p_2}. \qquad (12)$$

The placement of this inflection point ($r_{\lambda 2}$) relative to the points of feasibility (r_{F2}) and cascade (r_{C2}) is not readily apparent. It is clear from Equation 12 that whether the onset of oscillations occurs before or after the cascade depends on the death rate and ecological efficiency of the predator and the self-regulation of the prey. We find the following three possibilities:

$$\begin{aligned} &1.\ \text{If}\quad c_{11}/c_{12}(a_2p_2)^{1/2} = 2 \quad \text{then} \quad r_{\lambda 2} = r_{C2}.\\ &2.\ \text{If}\quad c_{11}/c_{12}(a_2p_2)^{1/2} > 2 \quad \text{then} \quad r_{\lambda 2} < r_{C2}. \qquad (13)\\ &3.\ \text{If}\quad c_{11}/c_{12}(a_2p_2)^{1/2} < 2 \quad \text{then} \quad r_{\lambda 2} > r_{C2}. \end{aligned}$$

Since ecological efficiencies range from 0.03 for large mammals to a theoretical maxima for 0.80 for bacteria, and the self-limiting terms are often less than 0.01, from the relationships presented in Equations 11 and 13, the following could be deduced to occur with increased productivity: (1) the oscillatory behavior in the dynamics that would follow a minor disturbance initiates within the window of productivity between feasibility (r_{F2}) and cascade (r_{C2}), i.e., $r_{F2} < r_{\lambda 2} < r_{C2}$ (see Fig. 1); (2) the frequency of the oscillations increases and the period decreases; and (3) the cascade of biomass for the two-species system would more likely experience wider fluctuations in dynamics in response to minor disturbances when compared to the pyramid of biomass.

The above examples illustrate a tight connection between population densities, the distribution of biomass within the system, and the dynamic state of the system. Analyses of the patterning of biomass, the fluxes of materials through trophic interactions, and the interaction strengths within the Jacobian matrices with trophic position using models patterned after Equations 2–4 parameterized with field data reveal the importance of these features to dynamics and stability. The systems under study possessed a pyramid of biomass structure and material flows with biomass and flows being greatest at the base of the food web and diminishing at increased trophic positions. The systems also possessed asymmetry in the distribution of the pairwise interaction strengths associated with each trophic interaction with increased trophic position that was important to their stability. Bottom-up enrichment moves biomass to higher trophic positions and alters the stable asymmetric configuration of interaction strengths within the Jacobian matrix, leading to instability.

PRACTICAL USE OF THE CONCEPT

The concept of bottom-up control has important implications to ecological assessment, monitoring, restoration, and preservation efforts. The examples presented above illustrate the following: (1) enrichment leads to a change in dynamic states; (2) enrichment alters the distribution of biomass within the food chain; (3) enrichment moves biomass to higher trophic levels and this movement of biomass can lead to shifts in dynamic states, some of which are unstable; (4) given the conclusions 1–3, it may be that resource limitations have a stabilizing affect on systems; and (5) if systems were to exist in a resources-limited state, there are clear advantages to the species within that system having the capacity to survive in a resources-limited environment or at a low level of productivity and yet have the capacity to respond to enrichment.

These five points, of which there are more, coupled with the conceptual and quantitative modeling discussed above offer guidance into the variables that one would collect and how one might gauge the structural and dynamic status of a community over time. Changes in the rates of input of primary production or detritus or shifts in the structure of the community of primary producers form the bases of bottom-up control that can lead to changes in the dynamics of the species within the community and the processes that they mediate and influence (e.g., nutrient cycles, respiration, and decomposition).

SEE ALSO THE FOLLOWING ARTICLES

Bifurcations / Biogeochemistry and Nutrient Cycles / Food Chains and Food Web Modules / Microbial Communities / Top-Down Control

FURTHER READING

Carpenter, S. R., J. F. Kitchell, and J. R. Hodgson. 1985. Cascading trophic interactions and lake productivity. *BioScience* 35: 634–639.

Collins-Johnson, N., J. D. Hoeksema, L. Abbott, J. Bever, V. B. Chaudhary, C. Gehring, J. Klironomos, R. Koide, R.M Miller, J.C. Moore, P. Moutoglis, M. Schwartz, S. Simard, W. Swenson, J. Umbanhowar, G. Wilson, and C. Zabinski. 2006. From Lilliput to Brobdingnag: extending models of mycorrhizal function across scales. *Bioscience* 56: 889–900.

DeAngelis, D.L. 1980. Energy flow, nutrient cycling, and food webs. *Ecology* 61: 764–771.
de Ruiter, P. C., A.-M. Neutel, and J.C. Moore. 1995. Energetics, patterns of interaction strengths, and the stability of real ecosystems. *Science* 269: 1257–1260.
Hairston, N.G., F. E. Smith, and L.B. Slobodkin. 1960. Community structure, population control and competition. *American Naturalist* 94: 421–425.
May, R.M. 1976. Simple mathematical models with very complicated dynamics. *Nature* 261: 459–467.
Moore, J.C., E.L. Berlow, D.C. Coleman, P. C. de Ruiter, Q. Dong, A. Hastings, N. Collins-Johnson, K.S. McCann, K. Melville, P. J. Morin, P. J., K. Nadelhoffer, A.D. Rosemond, D.M. Post, J.L. Sabo, K.M. Scow, M. J. Vanni, and D. Wall. 2004. Detritus, trophic dynamics, and biodiversity. *Ecology Letters* 7: 584–600.
Oksanen, L., S.D. Fretwell, J. Arruda, and P. Niemel. 1981. Exploitative ecosystems in gradients of primary productivity. *American Naturalist* 118: 240–261.
Rosenzweig, M.L. 1971. The paradox of enrichment: destabilization of exploitation ecosystems in ecological time. *Science* 171: 385–387.

BRANCHING PROCESSES

LINDA J. S. ALLEN

Texas Tech University, Lubbock

The study of branching processes began in the 1840s with Irénée-Jules Bienaymé, a probabilist and statistician, and was advanced in the 1870s with the work of Reverend Henry William Watson, a clergyman and mathematician, and Francis Galton, a biometrician. In 1873, Galton sent a problem to the *Educational Times* regarding the survival of family names. When he did not receive a satisfactory answer, he consulted Watson, who rephrased the problem in terms of generating functions. The simplest and most frequently applied branching process is named after Galton and Watson, a type of discrete-time Markov chain. Branching processes fit under the general heading of stochastic processes. The methods employed in branching processes allow questions about extinction and survival in ecology and evolutionary biology to be addressed. For example, suppose we are interested in family names, as in Galton's original problem, or in the spread of a potentially lethal mutant gene that arises in a population, or in the success of an invasive species. Given information about the number of offspring produced by an individual per generation, branching process theory can address questions about the survival of a family name or a mutant gene or an invasive species. Other questions that can be addressed with branching processes relate to the rate of population growth in a stable versus a highly variable environment.

BACKGROUND

The fundamental tools required for studying branching processes are generating functions. As the name implies, a generating function is a function that "generates" information about the process. For example, a probability generating function is used to calculate the probabilities associated with the process, whereas a moment generating function is used to calculate the moments, such as the mean and variance. This section introduces some notation and defines some terms in probability theory.

Let X be a discrete random variable taking values in the set $\{0, 1, 2, \ldots\}$ with associated probabilities,

$$p_j = \text{Prob}\{X = j\}, \quad j = 0, 1, 2, \ldots,$$

that sum to 1,

$$p_0 + p_1 + p_2 + \cdots = \sum_{j=0}^{\infty} p_j = 1.$$

Let the expectation of $u(X)$ be defined as

$$\mathbb{E}(u(X)) = p_0u(0) + p_1u(1) + p_2u(2) + \cdots = \sum_{j=0}^{\infty} p_j u(j).$$

If $u(X) = X$, then $\mathbb{E}(X)$ is the expectation of X or the mean of X,

$$\mathbb{E}(X) = p_1 + 2p_2 + 3p_3 + \cdots = \sum_{j=0}^{\infty} jp_j = \mu.$$

The summation reduces to a finite number of terms if, for example, $p_j = 0$ for $j > n$.

With this notation, three generating functions important in the theory of branching processes are defined. The probability generating function (pgf) of X is

$$\mathcal{P}(s) = \mathbb{E}(s^X) = p_0 + p_1s + p_2s^2 + \cdots = \sum_{j=0}^{\infty} p_js^j,$$

where s is a real number. Evaluating $\mathcal{P}$ at $s = 1$ yields $\mathcal{P}(1) = 1$. The moment generating function (mgf) of X is

$$M(s) = \mathbb{E}(e^{sX}) = p_0 + p_1e^s + p_2e^{2s} + \cdots = \sum_{j=0}^{\infty} p_je^{js}.$$

Evaluating M at $s = 0$ yields $M(0) = 1$. The cumulant generating function (cgf) of X is the natural logarithm of the moment generating function,

$$K(s) = \ln[M(s)].$$

Evaluating K at $s = 0$ yields $K(0) = \ln[M(0)] = \ln 1 = 0$.

As noted previously, the pgf generates the probabilities associated with the random variable X. For example, if X is the random variable for the size of a population, then the probability of population extinction p_0 can be found by evaluating the pgf at $s = 0$:

$$\mathcal{P}(0) = \text{Prob}\{X = 0\} = p_0.$$

Differentiation of $\mathcal{P}$ and evaluation at $s = 0$ equals p_1, the probability that the population size is 1.

The mgf generates the moments of the random variable X. Differentiation of M and evaluation at $s = 0$ yields the moments of X about the origin. For example, the mean of X is the first derivative of M evaluated at $s = 0$. That is,

$$\left.\frac{dM(s)}{ds}\right|_{s=0} = M'(0) = \sum_{j=0}^{\infty} jp_j = \mathbb{E}(X).$$

The second moment is

$$\mathbb{E}(X^2) = M''(0).$$

The variance of X is $\sigma^2 = \mathbb{E}[(X - \mu)^2]$. Using properties of the expectation, the variance can be expressed as $\sigma^2 = \mathbb{E}(X^2) - [\mathbb{E}(X)]^2$. Written in terms of the mgf,

$$\sigma^2 = M''(0) - [M'(0)]^2 = \sum_{j=0}^{\infty} j^2 p_j - \left[\sum_{j=0}^{\infty} jp_j\right]^2.$$

Formulas for the mean and variance of X can be computed from any one of the generating functions by appropriate differentiation and evaluation at either $s = 0$ or $s = 1$. They are defined as follows:

$$\mu = \mathcal{P}'(1) = M'(0) = K'(0)$$

and

$$\sigma^2 = \begin{cases} \mathcal{P}''(1) + P'(1) - [P'(1)]^2, \\ M''(0) - [M'(0)]^2, \\ K''(0). \end{cases}$$

The preceding expressions for the mean and variance will be applied in the following sections.

GALTON–WATSON BRANCHING PROCESS

Galton–Watson branching processes are discrete-time Markov chains, that is, collections of discrete random variables, $\{X_n\}_{n=0}^{\infty}$, where the time $n = 0, 1, 2 \ldots$ is also discrete. The random variable X_n may represent the population size of animals, plants, cells, or genes at time n or generation n. The term chain implies each of the random variables are discrete-valued; their values come from the set of nonnegative integers $\{0, 1, 2, \ldots\}$. The name Markov acknowledges the contributions of probabilist Andrei Markov to the theory of stochastic processes. The Markov property means that the population size at time $n + 1$ is only dependent on the population size at time n and is not dependent on the size at earlier times. In this way, the population size X_n at time n predicts the population size in the next generation, X_{n+1}.

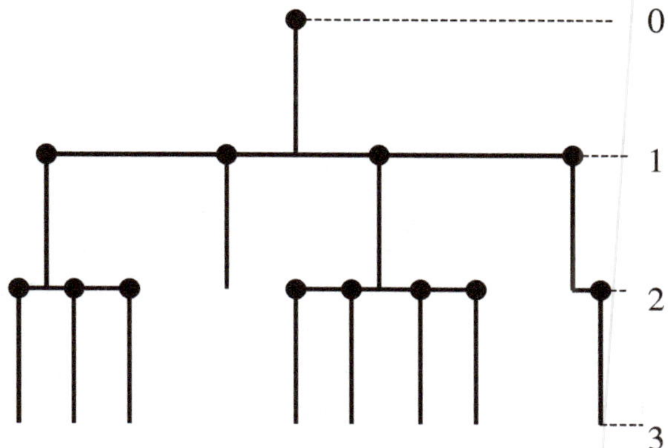

FIGURE 1 One sample path or stochastic realization of a branching process, beginning with $X_0 = 1$. The parent population produces four offspring in generation 1, $X_1 = 4$. These four parents produce three, zero, four, and one progeny in generation 2, respectively, making the total population size eight in generation 2, $X_2 = 8$. The sample path is 1, 4, 8, ...

The graph in Figure 1 illustrates what is referred to as a sample path or stochastic realization of a branching process. Figure 1 also shows why the name "branching" is appropriate for this type of process. Beginning with an initial size of $X_0 = 1$ (such as one mutant gene), in the next two generations the number of mutant genes are $X_1 = 4$ and $X_2 = 8$, respectively. The parent genes are replaced by progeny genes in subsequent generations. The graph illustrates just one of many possible sample paths for the number of mutant genes in generations 1 and 2. Associated with each parent population is an offspring distribution that specifies the probability of the number of offspring produced in the next generation. The following three assumptions define the Galton–Watson branching process more precisely.

Three important assumptions about the Markov process $\{X_n\}_{n=0}^{\infty}$ define a single-type Galton–Watson branching process (GWbp).

1. Each individual in the population in generation n gives birth to Y offspring of the same type in the next generation, where Y is a discrete random variable that takes values in $\{0, 1, 2 \ldots\}$. The offspring probabilities of Y are

$$p_j = \text{Prob}\{Y = j\}, \quad j = 0, 1, 2, \ldots.$$

2. Each individual in the population gives birth independently of all other individuals.
3. The same offspring distribution applies to all generations.

The expression "single-type" refers to the fact that all individuals are of one type such as the same gender, same cell type, or same genotype or phenotype. The pgf of X_n will be denoted as $\mathcal{P}_n$ and the pgf of Y as g.

If, in any generation n, the population size reaches zero, $X_n = 0$, then the process stops and remains at zero, and population extinction has occurred. The probability that the population size is zero in generation n is found by setting $s = 0$ in the pgf,

$$\text{Prob}\{X_n = 0\} = \mathcal{P}_n(0).$$

To obtain information about population extinction in generation n, it is necessary to know the pgf in generation n, $\mathcal{P}_n(s)$. It can be shown that if the assumptions 1 through 3 are satisfied, then the pgf in generation n is just an n-fold composition of the pgf of the offspring distribution g:

$$\mathcal{P}_n(s) = g(g(\cdot \cdot \cdot(g(s))\cdot \cdot \cdot)) = g^n(s).$$

To indicate why this is true, note that if the initial population size is 1, $X_0 = 1$, then the pgf of X_0 is $\mathcal{P}_0(s) = s$. Then the pgf of X_1 is just the offspring of the one individual from the first generation, that is, $X_1 = Y$ so that $\mathcal{P}_1(s) = g(s)$. In the case of two generations, $\mathcal{P}_2(s) = g(g(s)) = g^2(s)$ so that the probability of population extinction after two generations is simply $\mathcal{P}_2(0) = g^2(0)$. In general, the pgf in generation n is found by taking n compositions of the offspring distribution g, as given above.

The preceding calculation assumed $X_0 = 1$. The probability of extinction can be calculated when the initial population size is greater than 1, $X_0 = N > 1$. In this case, the pgf of X_n is just the n-fold composition of g raised to the power N,

$$\mathcal{P}_n(s) = [g^n(s)]^N.$$

Then the probability of population extinction in generation n can be found by evaluating the pgf at $s = 0$:

$$\mathcal{P}_n(0) = [g^n(0)]^N.$$

In most cases, it is difficult to obtain an explicit simple expression for $\mathcal{P}_n(s)$ in generation n. This is due to the fact that each time a composition is taken of another function, a more complicated expression is obtained. Fortunately, it is still possible to obtain information about the probability of population extinction after a long period of time (as $n \to \infty$). Ultimate extinction depends on the mean number of offspring produced by the parents. Recall that the mean number of offspring can be computed from the pgf $g(s)$ by taking its derivative and evaluating at $s = 1$. We will denote the mean number of offspring as m,

$$m = g'(1) = \sum_{j=1}^{\infty} jp_j.$$

The following theorem is one of the most important results in GWbp: if the mean number of offspring is less than or equal to 1, $m \le 1$, then eventually the population dies out, but if $m > 1$, the population has a chance of surviving. The probability of extinction depends on the initial population size N and the offspring distribution g. The theorem also shows how to calculate the probability of extinction if $m > 1$. A fixed point of g is calculated: a point q such that $g(q) = q$ and $0 < q < 1$. Then the probability of extinction is q^N.

Fundamental Theorem I

Let the initial population size of a GWbp be $X_0 = N \ge 1$ and let the mean number of offspring be m. Assume there is a positive probability of zero offspring and a positive probability of more than two offspring.

If $m \le 1$, then the probability of ultimate extinction is 1,

$$\lim_{n\to\infty} \text{Prob}\{X_n = 0\} = \lim_{n\to\infty} \mathcal{P}_n(0) = 1.$$

If $m > 1$, then the probability of ultimate extinction is less than 1,

$$\lim_{n\to\infty} \text{Prob}\{X_n = 0\} = \lim_{n\to\infty} \mathcal{P}_n(0) = q^N < 1,$$

where the value of q is the unique fixed point of g: $g(q) = q$ and $0 < q < 1$.

The GWbp is referred to as supercritical if $m > 1$, critical if $m = 1$, and subcritical if $m < 1$. If the process is subcritical or critical, then the probability of extinction is certain. But if the process is supercritical, then there is a positive probability, $1 - q^N$, that the population will survive. As the initial population size increases, the probability of survival also increases. We apply the Fundamental Theorem I to the question of the survival of family names.

EXAMPLE 1: SURVIVAL OF FAMILY NAMES

In 1931, Alfred Lotka assumed a zero-modified geometric distribution to fit the offspring distribution of the 1920s American male population. The theory of branching processes was used to address questions about survival of family names. In the zero-modified geometric distribution, the probability that a father has j sons is

$$p_j = bp^{j-1}, \quad j = 1, 2, 3, \ldots,$$

and the probability that he has no sons is

$$p_0 = 1 - (p_1 + p_2 + \cdots) = 1 - \sum_{j=1}^{\infty} p_j.$$

Lotka assumed $b = 1/5$, $p = 3/5$, and $p_0 = 1/2$. Therefore, the probability of having no sons is $p_0 = 1/2$, the probability of having one son is $p_1 = 1/5$, and so on. Then the offspring pgf has the following form:

$$g(s) = p_0 + p_1 s + p_2 s^2 + \cdots = p_0 + \sum_{j=1}^{\infty} bp^{j-1}s^j$$

$$= p_0 + \frac{bs}{1 - ps} = \frac{1}{2} + \frac{s}{5 - 3s}.$$

The mean number of offspring is $m = g'(1) = 5/4 > 1$. According to the Fundamental Theorem I, in the supercritical case $m > 1$ there is a positive probability of survival, $1 - q^N$, where N is the initial number of males. Applying the Fundamental Theorem I, the number q is a fixed point of g, that is, a number q such that $g(q) = q$ and $0 < q < 1$. The fixed points of g are found by solving the following equation:

$$\frac{1}{2} + \frac{q}{5 - 3q} = q.$$

There are two solutions, $q = 1$ and $q = 5/6$, but only one solution satisfies $0 < q < 1$, namely, $q = 5/6$. It should be noted that $q = 1$ will always be one solution to $g(q) = q$ due to one of the properties of a pgf, $g(1) = 1$.

To address the question about the probability of survival of family names, it follows that one male has a probability of 5/6 that his line of descent becomes extinct and a probability of 1/6 that his descendants will continue forever.

The zero-modified geometric distribution is one of the few offspring distributions where an explicit formula can be computed for the pgf of X_n, that is, the function $\mathcal{P}_n(s) = [g^n(s)]^N$. This particular case is referred to as the linear fractional case because of the form of the generating function. For $X_0 = 1$, the pgf of X_n for the linear fractional case is

$$\mathcal{P}_n(s) = \begin{cases} \dfrac{(m^n q - 1)s + q(1 - m^n)}{(m^n - 1)s + q - m^n}, & \text{if } m \neq 1, \\[2ex] \dfrac{[(n+1)p - 1]s - np}{nps - (n-1)p - 1}, & \text{if } m = 1, \end{cases}$$

where the value of q is a fixed point of g, $g(q) = q$. If $m > 1$, then q is chosen so that $0 < q < 1$ and if $m < 1$, q is chosen so that $q > 1$.

EXAMPLE 2: PROBABILITY OF EXTINCTION

Applying the preceding formula to Lotka's model in Example 1, for $X_0 = 1$, the probability of population extinction at the nth generation or the probability that no sons are produced in generation n is found by evaluating the pgf at $s = 0$:

$$\mathcal{P}_n(0) = \frac{q(m^n - 1)}{m^n - q}, \quad m > 1.$$

Substituting $m = 5/4$ and $q = 5/6$ into $\mathcal{P}_n(0)$, the probabilities of population extinction in generation n can be easily calculated. They are graphed in Figure 2. The probability of extinction approaches $q = 5/6 \approx 0.83$ for a single line of descent, $X_0 = 1$. For three lines of descent $X_0 = 3$, $q^3 \approx 0.58$.

Formulas for the mean and variance of X_n depend on the mean and variance of the offspring distribution, m and σ^2. The expectation of X_n is

$$\mathbb{E}(X_n) = m\mathbb{E}(X_{n-1}) = m^n \mathbb{E}(X_0),$$

and the variance is

$$\mathbb{E}\left[(X_n - \mathbb{E}(X_n))^2\right] = \begin{cases} \dfrac{m^{n-1}(m^n - 1)}{m - 1}\sigma^2, & m \neq 1, \\[2ex] n\sigma^2, & m = 1. \end{cases}$$

In the subcritical case, $m < 1$, the mean population size of X_n decreases geometrically. In the critical case, $m = 1$, the mean size is constant but the variance of the population size X_n increases linearly, and in the supercritical case, $m > 1$, the mean and variance increase geometrically.

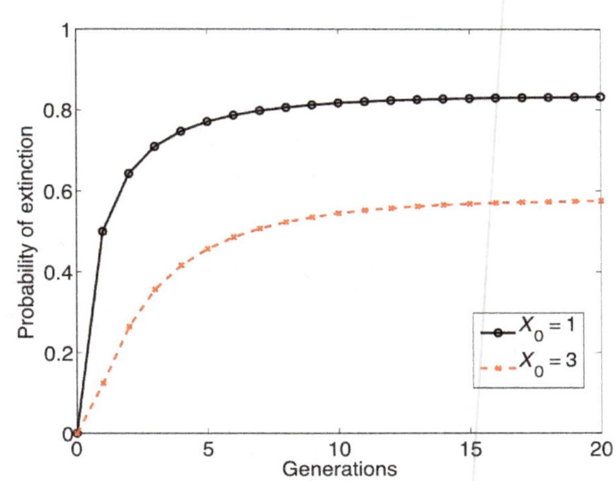

FIGURE 2 Probability of extinction of family names with one line of descent, $X_0 = 1$, and three lines of descent. $X_0 = 3$, based on Lotka's model.

This is an interesting result when one considers that in the supercritical case, there is a probability of extinction of q^N. In the next example, branching processes are used to study the survival of a mutant gene.

EXAMPLE 3: SURVIVAL OF A MUTANT GENE

Suppose the population size is very large. A new mutant gene appears in N individuals of the population; the remaining individuals in the population do not carry the mutant gene. Individuals reproduce according to a branching process, and those individuals with a mutant gene have offspring that carry the mutant gene. Suppose the mean number of offspring from parents with a mutant gene is m. If $m \leq 1$, then the line of descendants from the individuals with a mutant gene will eventually become extinct with probability 1. But suppose the mutant gene has a mean that is slightly greater than 1,

$$m = 1 + \epsilon, \epsilon > 0.$$

There then is a probability $1 - q^N$ that the subpopulation with the mutant gene will survive. The value of q can be approximated from the cgf $K(s) = \ln M(s)$ using the fact that $K(0) = 0$, $K'(0) = m$, and $K''(0) = \sigma^2$. Let $q = e^\theta$, where θ is small and negative, so that q will be less than but close to 1. Then $e^\theta = g(e^\theta) = M(\theta)$, or equivalently, $\theta = \ln M(\theta) = K(\theta)$. Expanding $K(\theta)$ in a Maclaurin series about zero leads to

$$\theta = K(\theta) = 0 + m\theta + \sigma^2\frac{\theta^2}{2!} + \cdots.$$

Truncating the preceding series gives an approximation to θ:

$$\theta \approx m\theta + \sigma^2\frac{\theta^2}{2}.$$

Solving for θ yields $\theta \approx -2\epsilon/\sigma^2$, and an approximation to $q = e^\theta$ is

$$q \approx e^{-\frac{2}{\sigma^2}\epsilon}.$$

The probability that the mutant gene survives in the population is $1 - q^N \approx 1 - e^{-\frac{2}{\sigma^2}\epsilon}$. Suppose the offspring distribution has the form of a Poisson distribution, where the mean and variance are equal, $m = \sigma^2 = 1 + \epsilon$, and $\epsilon = 0.01$. Then if $N = 1$, the probability of survival of the mutant gene is $1 - q \approx 0.02$, but if $N = 100$, the probability of survival is much greater, $1 - q^{100} \approx 0.86$.

RANDOM ENVIRONMENT

Plant and animal populations are subject to their surrounding environment, which is constantly changing. Their survival depends on environmental conditions such as food and water availability and temperature. Under favorable environmental conditions, the number of offspring increases, but under unfavorable conditions the number of offspring declines. With environmental variation, assumption 3 in the GWbp no longer holds. Suppose the other two assumptions still hold for the branching process:

1. Each individual in the population in generation k gives birth to Y_k offspring of the same type in the next generation, where Y_k is a discrete random variable.
2. Each individual in the population gives birth independently of all other individuals.

In some cases, the random environment can be treated like a GWbp. For example, suppose the environment varies periodically, a good season followed by a bad season, so that the offspring random variables follow sequentially as Y_1, Y_2, Y_1, Y_2, and so on. As another example, suppose the environment varies randomly between good and bad seasons so that a good season occurs with probability p and a bad one with probability $1 - p$.

In general, if m_n is the mean number of offspring produced in generation $n - 1$, the expectation of X_n depends on the previous generation in the following way:

$$\mathbb{E}(X_n) = m_n\mathbb{E}(X_{n-1}).$$

Repeated application of this identity leads to an expression for the mean population size in generation n,

$$\mathbb{E}(X_n) = m_n \cdots m_1\mathbb{E}(X_0).$$

If the mean number of offspring $m_i = m$ is the same from generation to generation, then this expression is the same as in a constant environment, $\mathbb{E}(X_n) = m^n\mathbb{E}(X_0)$.

Suppose the environment varies periodically over a time period of length T. The offspring distribution is Y_1, $Y_2, \ldots, Y_T$, and then repeats. The preceding formulas can be used to calculate the mean size of the population after nT generations. Using the fact that the expectation $\mathbb{E}(Y_i) = m_i$, and $\mathbb{E}(X_T) = m_T \cdots m_2m_1\mathbb{E}(X_0)$, it follows that after nT generations,

$$\begin{aligned}\mathbb{E}(X_{nT}) &= (m_T \cdots m_2m_1)^n\mathbb{E}(X_0)\\ &= \left[(m_T \cdots m_2m_1)^{1/T}\right]^{nT}\mathbb{E}(X_0).\end{aligned}$$

The mean population growth rate each generation in an environment that varies periodically is $\mu_r = (m_T \cdots m_2m_1)^{1/T}$, which is the geometric mean of $m_1, m_2, \ldots, m_T$. It is well known that the geometric mean is less than the arithmetic mean (average),

$$\mu_r = (m_1 m_2 \cdots m_T)^{1/T} \le \frac{1}{T}(m_1 + m_2 + \cdots + m_T) \le \mu.$$

The mean population growth rate in a periodic environment is less than the average of the growth rates, $\mu_r \le \mu$. If $\mu_r < 1$, then the population will not survive.

Suppose the environment varies randomly and the random variable for the offspring distribution is Y_k. Let $m_k = \mathbb{E}(Y_k)$ be the mean number of offspring in generation $k - 1$. For example, suppose there are two random variables for the offspring distribution that occur with probability p and $1 - p$ and their respective means are equal to μ_1 and μ_2. The order in which these offspring distributions occur each generation is random, that is, m_k could be either μ_1 or μ_2 with probability p or $1 - p$, respectively. Thus, m_k is a random variable with expectation $\mathbb{E}(m_k) = p\mu_1 + (1 - p)\mu_2 = \mu$. The set $\{m_k\}_{k=1}^{\infty}$ is a sequence of independent and identically distributed (iid) random variables, that is, random variables that are independent and have the same probability distribution. To compute the mean population growth rate, the same technique is used as in the previous example, but the limit is taken as $n \to \infty$. Rewriting the geometric mean using the properties of an exponential function leads to

$$(m_n \cdots m_2 m_1)^{1/n} = e^{(1/n)\ln[m_n \cdots m_2 m_1]} = e^{(1/n)[\ln m_n + \cdots + \ln m_2 + \ln m_1]}.$$

It follows from probability theory that the average of a sequence of iid random variables $(1/n)[\ln m_n + \cdots + \ln m_2 + \ln m_1]$ approaches its mean value $\mathbb{E}(\ln mk) = \ln \mu_r$. Thus, in the limit, the mean population growth rate in a random environment is

$$\lim_{n\to\infty} (m_n m_{n-2} \cdots m_1)^{1/n} = e^{\mathbb{E}(\ln mk)} = e^{\ln \mu_r} = \mu_r.$$

As in the periodic case, the mean population growth rate in the random environment is less than the average of the growth rates: $\mu_r \le \mu$:

$$\mu_r = e^{\ln \mu_r} = e^{\mathbb{E}(\ln m_k)} \le e^{\ln \mathbb{E}(m_k)} = e^{\ln \mu} = \mu.$$

Interestingly, a population subject to a random environment may not survive, $\mu_r < 1$, even though the mean number of offspring for any generation is greater than 1, $\mu > 1$, that is, $\mu_r < 1 < \mu$. A random environment is a more hostile environment for a population than a constant environment.

MULTITYPE GALTON-WATSON BRANCHING PROCESS

In a single-type GWbp, the offspring are of the same type as the parent. In a multitype GWbp, each parent may have offspring of different types. For example, a population may be divided according to age, size, or developmental stage, representing different types, and in each generation, individuals may age or grow to another type, another age, size, or stage. In genetics, genes can be classified as wild or mutant types, and mutations change a wild type into a mutant type.

A multitype GWbp $\{\vec{X}(n)\}_{n=0}^{\infty}$ is a collection of vector random variables $\vec{X}(n)$, where each vector consists of k different types, $\vec{X}(n) = (X_1(n), X_2(n), \ldots, X_k(n))$. Each random variable $X_i(n)$ has k associated offspring random variables for the number of offspring of type $j = 1, 2, \ldots, k$ from a parent of type i.

As in a single-type GWbp, it is necessary to define a pgf for each of the k random variables $X_i(n)$, $i = 1, 2, \ldots, k$. One of the differences between single-type and multitype GWbp is that the pgf for X_i depends on the k different types. The pgf for X_i is defined assuming initially $X_i(0) = 1$ and all other types are zero, $X_l(0) = 0$. The offspring pgf corresponding to X_i is denoted as $g_i(s_1, s_2, \ldots, s_k)$. The multitype pgfs have properties similar to a single-type pgfs. For example, $g_i(1, 1, \ldots, 1) = 1$, $g_i(0, 0, \ldots, 0)$ is the probability of extinction for X_i given $X_i(0) = 1$ and $X_l(0) = 0$ for all other types, and differentiation and evaluation when the s variables are set equal to one leads to an expression for the mean number of offspring. But there are different types of offspring, so there are different types of means. The mean number of j-type offspring by an i-type parent is denoted as m_{ji}. A value for m_{ji} can be calculated from the pgf g_i by differentiation with respect to s_j and evaluating all of the s variables at 1:

$$m_{ji} = \left.\frac{\partial g_i(s_1, s_2, \ldots, s_k)}{\partial s_j}\right|_{s_1=1,\, s_2=1,\, \ldots,\, s_k=1}.$$

There are k^2 different means m_{ji} because of the k different offspring and k different parent types. When the means are put in an ordered array, the $k \times k$ matrix is called the expectation matrix:

$$\mathcal{M} = \begin{pmatrix} m_{11} & m_{12} & \cdots & m_{1k} \\ m_{21} & m_{22} & \cdots & m_{2k} \\ \vdots & \vdots & \cdots & \vdots \\ m_{k1} & m_{k2} & \cdots & m_{kk} \end{pmatrix}.$$

Population extinction for a multitype GWbp depends on the properties of the expectation matrix $\mathcal{M}$. Therefore, some matrix theory is reviewed. An eigenvalue of a square matrix $\mathcal{M}$ is a number λ satisfying $\mathcal{M}\vec{V} = \lambda\vec{V}$, for some nonzero vector $\vec{V}$, known as an eigenvector. If all of the entries m_{ij} in matrix $\mathcal{M}$ are positive or if all of the entries are nonnegative and $\mathcal{M}^2$ or

$\mathcal{M}^3$ or some power of $\mathcal{M}$, $\mathcal{M}^n$, has all positive entries, then $\mathcal{M}$ is referred to as a regular matrix. It follows from matrix theory that a regular matrix $\mathcal{M}$ has a positive eigenvalue λ that is larger than any other eigenvalue. This eigenvalue λ is referred to as the dominant eigenvalue. It is this eigenvalue that determines whether the population grows or declines. The dominant eigenvalue λ plays the same role as the mean number of offspring m in the single-type GWbp.

The second fundamental theorem for GWbp extends the Fundamental Theorem I to multitype GWbp. If $\lambda \leq 1$, the probability of extinction is one as $n \to \infty$, but if $\lambda > 1$, then there is a positive probability that the population survives. This latter probability can be computed by finding fixed points of the k generating functions.

Fundamental Theorem II

Let the initial sizes for each type be $X_i(0) = N_i$, $i = 1, 2, \ldots, k$. Suppose the generating functions g_i for each of the k types are nonlinear functions of s_j with some $g_i(0, 0, \ldots, 0) > 0$, the expectation matrix $\mathcal{M}$ is regular, and λ is the dominant eigenvalue of matrix $\mathcal{M}$.

If $\lambda \leq 1$, then the probability of ultimate extinction is 1,

$$\lim_{n\to\infty} \text{Prob}\{\vec{X}(n) = \vec{0}\} = 1.$$

If $\lambda > 1$, then the probability of ultimate extinction is less than 1,

$$\lim_{n\to\infty} \text{Prob}\{\vec{X}(n) = \vec{0}\} = q_1^{N_1} q_2^{N_2} \cdots q_k^{N_k},$$

where $(q_1, q_2, \ldots, q_k)$ is the unique fixed point of the k generating functions $g_i(q_1, \ldots, q_k) = q_i$ and $0 < q_i < 1$, $i = 1, 2 \ldots, k$. In addition, the expectation of $\vec{X}(n)$ is

$$\mathbb{E}(\vec{X}(n)) = \mathcal{M}\mathbb{E}(\vec{X}(n-1)) = \mathcal{M}^n\mathbb{E}(\vec{X}(0)).$$

The mean population growth rate is λ.

The following example is an application of a multitype GWbp and the Fundamental Theorem II to an age-structured population.

EXAMPLE 4: AGE-STRUCTURED POPULATION

Suppose there are k different age classes. The number of females of age k are followed from generation to generation. The first age, type 1, represents newborn females. A female of age i gives birth to r females of type 1 with probability $b_{i,r}$ and then survives, with probability p_{i+1}, i to the next age $i + 1$. There may be no births with probability $b_{i,0}$ or one birth with probability $b_{i,1}$, and so on. The mean number of female offspring by a female of age i equals $b_i = b_{i,1} + 2b_{i,2} + 3b_{i,3} + \cdots$. Age k is the oldest age; females do not survive past age k. The probability generating functions are

$$g_i(s_1, s_2, \ldots, s_k) = [p_{i+1,i}s_{i+1} + (1 - p_{i+1,i})] \sum_{r=0}^{\infty} b_{i,r}s_1^r, \quad i = 1, \ldots, k-1,$$

$$g_k(s_1, s_2, \ldots, s_k) = b_{k,0} + b_{k,1}s_1 + b_{k,2}s_1^{\,2} + \cdots$$
$$= \sum_{r=0}^{\infty} b_{i,r}s_1^{\,r}.$$

Note that $g_i(1, 1, \ldots, 1) = 1$ and $g_i(0, 0, \ldots, 0) = (1 - p_{i+1,i})b_{i,0}$ except for the oldest age, where $g_k(0,0,\ldots,0) = b_{k,0}$. The expression $(1 - p_{i+1,i})b_{i,0}$ is the probability of extinction given $X_i(0) = 1$ and $X_l(0) = 0$. The term $1 - p_{i+1,i}$ is the probability an individual of age i does not survive to the next age, and $b_{i,0}$ is the probability an individual of age i has no offspring.

Taking the derivative of the generating functions and evaluating the s variables at 1 gives the the following expectation matrix $\mathcal{M}$:

$$\mathcal{M} = \begin{pmatrix} b_1 & b_2 & \cdots & b_{k-1} & b_k \\ p_{21} & 0 & \cdots & 0 & 0 \\ 0 & p_{32} & \cdots & 0 & 0 \\ \vdots & \vdots & \ddots & \vdots & \vdots \\ 0 & 0 & \cdots & p_{k,k-1} & 0 \end{pmatrix},$$

where the mean birth probabilities are on the first row and the survival probabilities are on the subdiagonal. In demography, matrix $\mathcal{M}$ is known as a Leslie matrix in honor of the contributions of Patrick Holt Leslie. The mean population growth rate is the dominant eigenvalue of $\mathcal{M}$.

As a specific example, consider two age classes with generating functions

$$g_1(s_1, s_2) = [(1/2)s_2 + 1/2][1/2 + (1/6)s_1 + (1/6)s_1^2 + (1/6)s_1^3],$$

$$g_2(s_1, s_2) = 1/4 + (1/4)s_1 + (1/4)s_1^2 + (1/4)s_1^3.$$

The mean number of offspring for ages 1 and 2 is, respectively,

$$b_1 = b_{1,1} + 2b_{1,2} + 3b_{1,3} = (1/6) + 2(1/6) + 3(1/6) = 1$$

and

$$b_2 = b_{2,1} + 2b_{2,2} + 3b_{2,3} = (1/4) + 2(1/4) + 3(1/4) = 3/2.$$

The expectation matrix is regular,

$$\mathcal{M} = \begin{pmatrix} 1 & 3/2 \\ 1/2 & 0 \end{pmatrix},$$

with a dominant eigenvalue equal to $\lambda = 3/2$, which is the mean population growth rate. According to the Fundamental Theorem II, there is a unique fixed point (q_1, q_2), one q_i for each type satisfying $g_1(q_1, q_2) = q_1$ and $g_2(q_1, q_2) = q_2$. This fixed point can be calculated and shown to be equal to $(q_1, q_2) \approx (0.446, 0.443)$. For example, if there are initially two females of age 1 and three females of age 2, $X_1(0) = 2$ and $X_2(0) = 3$, then according to the Fundamental Theorem II the probability of ultimate extinction of the total population is approximately

$$(0.446)^2(0.433)^3 \approx 0.016.$$

On the other hand, if $X_1(0) = 1$ and $X_2(0) = 0$, then the probability of population extinction is 0.446.

Suppose a plant or animal population that has several developmental stages and is subject to a random environment is modeled by a branching process. Environmental variation causes the expectation matrix $\mathcal{M}$ to change from generation to generation, $\mathcal{M}_1$, $\mathcal{M}_2$, and so on. Matrix $\mathcal{M}$ may change in a periodic fashion such as seasonally or randomly according to some distribution. As shown in the case of a single-type GWbp, if the environment varies periodically, a good season followed by a bad season, where $\mathcal{M}_1$ is the expectation matrix in a good season and $\mathcal{M}_2$ in a bad season, then after $2n$ seasons the mean population size for each stage is

$$\mathbb{E}(\vec{X}(2n)) = (\mathcal{M}_2\mathcal{M}_1)^n\mathbb{E}(\vec{X}(0)).$$

The mean population growth rate is λ_r, where $(\lambda_r)^2$ is the dominant eigenvalue of the product of the two matrices $\mathcal{M}_2\mathcal{M}_1$. Note that $(\lambda_r)^2$ does not necessarily equal $\lambda_2\lambda_1$, where λ_1 and λ_2 are the dominant eigenvalues of $\mathcal{M}_1$ and $\mathcal{M}_2$. In general,

$$\lambda_r \leq \frac{\lambda_1 + \lambda_2}{2} = \lambda.$$

The mean population growth rate in a random environment is less than the average of the growth rates. If the environment varies randomly each generation, and if the expectation matrices $\{\mathcal{M}_i\}_{i=1}^{\infty}$ are drawn from a set of iid regular matrices, then a similar relationship holds. In this case, if the expectation in each generation $\mathbb{E}(\mathcal{M}_k)$ has a dominant eigenvalue of λ, then there exists a mean population growth rate λ_r, where $\lambda_r \leq \lambda$. The population growth rate in the random environment is generally less than the growth rate in a constant environment.

The previous discussion was confined to discrete-time processes, where the lifetime is a fixed length $n \rightarrow n + 1$, which for convenience was denoted as one unit of time. At the end of that time interval, the parent is replaced by the progeny. In continuous-time processes, an individual's lifetime is not fixed but may have an arbitrary probability distribution. In the case of an exponentially distributed lifetime, the branching process is said to have the Markov property because the exponential distribution has the memoryless property; the process at time t only depends on the present and not the past. If the lifetime distribution is not exponential, then it is referred to as an age-dependent branching process, also known as a Bellman–Harris branching process, where the name is in honor of the contributions of Richard Bellman and Theodore Harris. Similar results as the Fundamental Theorem I and II can be derived for single-type and multitype branching processes of the continuous type. Additional biological applications of branching processes to discrete-time processes and applications to continuous-time processes can be found in the "Further Reading" section.

SEE ALSO THE FOLLOWING ARTICLES

Age Structure / Birth-Death Models / Markov Chains / Matrix Models / Stochasticity, Environmental

FURTHER READING

Allen, L. J. S. 2003. *An introduction to stochastic processes with applications to biology*. Upper Saddle River, NJ: Prentice Hall.

Athreya, K. B., and P. E. Ney. 1972. *Branching processes*. Berlin: Springer-Verlag.

Caswell, H. 2001. *Matrix population models: construction, analysis and interpretation*, 2nd ed. Sunderland, MA: Sinauer Associates.

Haccou, P., P. Jagers, and V. A. Vatutin. 2005. *Branching processes variation, growth, and extinction of populations*. Cambridge Studies in Adaptive Dynamics. Cambridge, UK: Cambridge University Press.

Harris, T. E. 1963. *The theory of branching processes*. Berlin: Springer-Verlag.

Jagers, P. 1975. *Branching processes with biological applications*. London: John Wiley & Sons.

Kimmel, M., and D. Axelrod. 2002. *Branching processes in biology*. New York: Springer-Verlag.

Mode, C. J. 1971. *Multitype branching processes theory and applications*. Oxford: Elsevier.

Tuljapurkar, S. 1990. *Population dynamics in variable environments*. Berlin: Springer-Verlag.

C

CANNIBALISM

ALAN HASTINGS

University of California, Davis

Cannibalism is the consumption of one life stage of a species by another life stage. Although this may seem somewhat specialized, cannibalistic interactions are very common in natural systems. This is a subject that has played a very important role in the development of theoretical ecology because of the early experiments by Thomas Park with *Tribolium* and the development of corresponding models. More recent models of cannibalism have specifically included age or stage structure and have been used to understand population cycles as well as more complex dynamics including chaos.

CANNIBALISM AND *TRIBOLIUM*

Some of the earliest laboratory experiments investigating population dynamics were carried out with species of the flour beetle, *Tribolium castaneum* and *Tribolium confusum*, by Thomas Park and others. In these species, adults and larvae cannibalize eggs, and adults cannibalize pupae. Additionally, the respective stages of the other species are eaten as well. In early experiments with single species, it was noted that cannibalism was a regulator of population density. Additionally, investigators noted the prevalence of cycles of population numbers through time, especially in the egg, larval, and pupal stages. However, detailed models explaining this behavior were developed later.

Another intriguing set of experiments were two species experiments with *confusum* and *castaneum*, where it was clear that cannibalism and consumption of each species by the other played a significant role in the lack of coexistence.

THE SIMPLEST MODEL OF CANNIBALISM: THE RICKER MODEL

The role of cannibalism in population dynamics has been explored in a variety of models. One of the simplest models including cannibalism is the Ricker model, which implicitly includes the effects of cannibalism in the recruitment process. Although in the fisheries literature this is often considered as a description of recruitment into a population with age structure, this section will present the model as one for a population with nonoverlapping generations. Letting $N(t)$ be the number of adults in generation t, the model assumes that there would be RN(t) individuals in generation $t + 1$ without cannibalism. Cannibalism is included by assuming there is a fixed time over which cannibalism occurs at a rate proportional to the cannibalistic individuals, the adults of the previous generation. Thus, the probability of a potential recruit escaping cannibalism is given by $e^{-aN(t)}$, where a is a parameter taking into account the cannibalism rate times the vulnerable period. Thus, the model takes the form

$$N(t+1) = RN(t)\, e^{-aN(t)}, \qquad (1)$$

This model can have extremely complex dynamical behavior, and it is important to note that one possibility is cycles of period 2—twice the generation time. This feature of cannibalism-producing cycles in population abundance through time is a key outcome of cannibalism models, with the relationship to the generation time one aspect that may help to relate data to the underlying models and potentially elucidate details of the mechanism producing the cycles. Another feature of the Ricker model is that the stage doing the cannibalizing receives no benefit from the consumption, which is a typical assumption in cannibalism models, though there have been some exceptions. Note as well that increasing R in the Ricker model would produce dynamics more complex than simple cycles including chaos.

CONTINUOUS-TIME AGE-STRUCTURED MODELS OF CANNIBALISM

Another approach to descriptions of cannibalism is to take into account more detailed descriptions of different life stages. A basic model of this form would be built using the classic McKendrick–von Forster description of age-structured population dynamics, where $n(t, a)$ is a density function on age a at time t for the population density. One typical version of a model then takes the form

$$\frac{\partial n(t,a)}{\partial t} + \frac{\partial n(t,a)}{\partial a} = -\mu(C(t), a),$$

$$C(t) = \int_0^\infty c(a)n(t,a)da, \qquad (2)$$

$$\mu(C(t), a) = \begin{cases} \mu_0(a) + C(t) & \text{for } a < a_0 \\ \mu_0(a) & \text{for } a \geq a_0 \end{cases},$$

where the first equation describes aging and survival, the second equation is a measure of the overall cannibalism pressure C, and the third equation describes how the mortality rate is affected by cannibalism. Note that to this set of equations one would need to add an equation describing the overall birth rate at time t, $B(t)$, which would provide the boundary condition $n(t, 0)$. Letting $b(a)$ be the age dependent birth rate, we would use

$$n(t,0) = B(t) = \int_0^\infty b(a)n(t,a)da, \qquad (3)$$

where this equation reflects the assumption that the birth rate depends on adult age (sex is ignored) and is density independent. Also note that typically the function describing the dependence of cannibalism on the age of the individual doing the cannibalizing, $c(a)$, would be zero for ages a smaller than some value (typically greater than a_0, the maximum age of a cannibalized individual). The model as written here assumes that all individuals subject to cannibalism have the same risk; straightforward modifications could be used to make this risk age dependent.

Hastings and Costantino used a model essentially of the form of Equation 2 to show that in *Tribolium*, cycles essentially equal to the total time spent in the cannibalizing stage were produced—essentially generation 1 cycles. The analysis was a combination of showing the existence of a Hopf bifurcation plus numerical work. Note that the dynamics are similar to that produced by competition where an older stage preempts resources that could be used by a younger stage, since in this cannibalism model there is no benefit to the stage being cannibalized.

The general model of Equation 2 is of a form that is too complex (especially when the life cycle is completed by adding an equation for reproduction) to permit general analysis. Additionally, numerical solutions of this model require care since the straightforward approach of taking small time steps can lead to numerical problems due to the instability (in a numerical sense) of this scheme. As discussed by de Roos, there is a better procedure that essentially uses the mathematical approach of integrating along characteristics of the partial differential equation that he called the *escalator box car train.* This method essentially follows the fate of each cohort through time from birth.

Simplifications of models of the form of Equation 2 have been used to look at different aspects of cannibalism. One appealing idea is to think of the length of the stage being cannibalized as short and to simplify the model by taking an appropriate limit as a_0 approaches zero. In this case, cannibalism is essentially a modification in the birth rate that depends on the population in a nonlinear fashion. As noted in the literature by Diekmann and colleagues, this has to be done very carefully so that the total overall cannibalism rate remains constant.

CANNIBALISM AND CHAOS

As noted above, the Ricker model can produce both cycles and complex dynamics. However, the number of examples in natural systems or laboratory systems of observed chaotic dynamics in population numbers that have been clearly matched to an underlying mechanistic model is quite limited. One notable exception has been a series of studies of dynamics of *Tribolium* developed by Costantino, Cushing, Dennis, and Desharnais (and others more recently). They combine laboratory studies of population dynamics with a simple model for population dynamics defined in terms of three classes in discrete time: a larval class, a pupal class, and an adult class. In this LPA model, larvae and pupa invariably move to the next class at the next time step if they survive. This model, too, produces cycles and complex dynamics. Modern statistical approaches were used to connect the model and data. The simplest version of the model is

$$L_{t+1} = bA_t \exp(-c_{el}L_t - c_{ea}A_t),$$
$$P_{t+1} = (1 - m_l)L_t,$$
$$A_{t+1} = P_t \exp(-c_{pa}A_t) + (1 - m_a)A_t,$$

where the dynamics have been approximated by a model that uses three classes: larvae, pupae, and adults whose numbers at time t are denoted by L_t, P_t, and A_t, respectively. Note that here units of t are not generation times, but are chosen both to match experimental laboratory censuses and the biology of the system. In this LPA model, the larvae at time $t + 1$ arise from eggs laid during the previous time interval that have not been cannibalized (cannibalism by larvae occurs at a rate $c_{el}L_t$ and by adults at a rate $c_{ea}A_t$). Pupae are simply those larvae from the precious census that survive since the time scale is chosen so that all larvae at time t become pupae at time $t + 1$—both very large relatively immobile larvae and very newly emerged not yet reproductive adults are counted as pupae. Finally, adults at time $t + 1$ arise either from adults at time t that survive or pupae at time t that are not cannibalized by adults. In this, as in the previous model of cannibalism, there is no benefit to the class that is doing the cannibalism.

This model has been carefully analyzed and parameterized with laboratory data in an extensive series of investigations. The first important point is that in the absence of cannibalism, this is essentially a slight variation of a Leslie matrix model. The difference is that the oldest class, adults, feeds into the adult class. Thus, the only solutions are extinction or exponential growth. Any bounded dynamics that persist can only be the result of cannibalism. With cannibalism, from computer simulations one sees that solutions of period 2 are possible, as well as more complex dynamics including chaos.

To match the model to data, a stochastic version of the model is needed. The beauty of working with this laboratory system is that all counts of population levels are essentially exact, so the assumption was made that the only sources of stochasticity were demographic stochasticity (the randomness of timings of transitions) and environmental stochasticity. Very good fits to data and estimates of parameters were possible by fitting this stochastic LPA model using maximum likelihood approaches.

One of the most dramatic results was then the demonstration of chaotic dynamics by artificially altering the cannibalism rates of adults on pupae (by experimentally adding or removing individuals). The existence of chaos was predicted in the model and confirmed in the time series of experimental results.

INTRAGUILD PREDATION

Even in *Tribolium* populations in the laboratory, in cultures with more than one species, not only will adults and larvae of one species cannibalize their own eggs and (the adults will cannibalize pupae), but additionally individuals of the other species will be consumed. Thus, the two species are not only competitors for resources, but they also eat each other. This interaction was found by Park in his classic experiments to be an important part of understanding the dynamics of two interacting species of *Tribolium*.

More generally, the interaction between species in nature often includes aspects of both consumption and competition. This class of interactions is now known as intraguild predation, which actually refers to a more general kind of interaction. There are many predator species that interact with a species both through exploitative competition (consuming shared resources) and as a predator. This kind of interaction, which is often age dependent (consuming the young of another species and competing with the adults) shares many dynamic similarities with cannibalism, and the models of the two situations have much in common. One striking result is that the dynamics of intraguild predation can be very unstable, so explaining coexistence of species interacting this way remains a challenge.

CANNIBALISM IN FOOD WEBS

For understanding larger-scale patterns of coexistence of species, the approach of food webs has often been used. Here, species are connected in ways that display the feeding relationships among species, and investigations focus on issues of what interactions are present as well as relationships to stability. Obviously, cannibalism is a different kind of feeding relationship than is typically portrayed in food web dynamics, and the inclusion of cannibalism can change the interpretation of food web dynamics. The dynamics of intraguild predation can be thought of as a simple food web module that exhibits some of the issues that arise when including cannibalism in food webs. As noted by Polis, among others, the ubiquity of cannibalism in natural systems provides a challenge for understanding the dynamics of food webs solely in terms of competitive or exploitative interactions. Cannibalism introduces cycles into the feeding relationships in food webs in the simplest fashion.

EVOLUTION OF CANNIBALISM

One important question is why cannibalism is so ubiquitous in so many taxa. To approach this question one cannot proceed with the kinds of models explored most here, which do not include a benefit to the cannibalistic stage, since without a benefit the costs of cannibalism in reduced survival would lead to its elimination through selection. Thus, the models that have been used to explore the evolution of cannibalism, such as the approach using evolutionarily stable strategies (ESS) by Stenseth, have explicitly included a benefit to the cannibalistic stage. Cannibalism is expected to be an ESS if the benefit of increasing cannibalism from zero outweighs the cost. Thus, cannibalism is expected to evolve if the reproductive value (in the sense of Fisher) of the stage being cannibalized is low relative to the resulting increase in the reproductive value of the stage doing the cannibalizing. For example, very low expected survivorship of the cannibalized stage to a reproductive age would be more likely to lead to cannibalism. Note, however, that for *Tribolium*, at least, reducing cannibalism of eggs would require active avoidance. Thus, more complete models might have to include other constraints and factors.

CONCLUSIONS AND OTHER DIRECTIONS

The implications of cannibalism for single-species dynamics are relatively well understood, with this interaction producing cycles that scale with the generation time of the organism. A variety of models, both discrete and continuous in time, have been carefully investigated. However, despite this and the relative importance of cannibalism in natural systems, much less is known about the influence of cannibalism in the dynamics of interacting species.

SEE ALSO THE FOLLOWING ARTICLES

Age Structure / Chaos / Evolutionarily Stable Strategies / Food Webs / Partial Differential Equations / Ricker Model / Stage Structure

FURTHER READING

Cushing, J., R. Costantino, B. Dennis, R. A. Desharnais, and S. M. Henson. 2003. *Chaos in ecology: experimental nonlinear dynamics.* Theoretical Ecology Series. San Diego: Academic Press/Elsevier.

Diekmann, O. 1986. Simple mathematical models for cannibalism: a critique and a new approach. *Mathematical Biosciences* 78: 21–46.

Dong, Q., and G. A. Polis. 1992. The dynamics of cannibalistic populations: a foraging perspective. In M. A. Elgar and B. J. Crespi, eds. *Cannibalism: ecology and evolution among diverse taxa.* New York: Oxford University Press.

Hastings, A., and R. Costantino. 1987. Cannibalistic egg-larva interactions in Tribolium: an explanation for the oscillations in population numbers. *The American Naturalist* 130: 36–52.

Holt, R. D., and G. A. Polis. 1997. A theoretical framework for intraguild predation. *The American Naturalist* 149: 745–764.

Polis, G. A. 1981. The evolution and dynamics of intraspecific predation. *Annual Review of Ecology and Systematics* 12: 225–251.

Polis, G. A. 1991. Complex trophic interactions in deserts: an empirical critique of food-web theory. *The American Naturalist* 138: 123–155.

CELLULAR AUTOMATA

DAVID E. HIEBELER

University of Maine, Orono

Cellular automata are spatially explicit lattice models. They were originally applied primarily to systems from physics and computer science, but they have also found applications to topics in biology such as population dynamics, epidemiology, pattern formation, and genetics. One of the main advantages of cellular automata are that complex spatial features can be readily incorporated into the models.

DEFINITIONS

Traditional cellular automata models are spatial models that are deterministic and discrete in space, time, and state. Formally, a cellular automaton consists of a lattice or grid, a neighborhood, and an update function or rule. In the (typically one-, two-, or three-dimensional) lattice, each site (or cell) is in one of a finite number of states; the number of possible states k per site is typically small. For example, cellular automata can be used to model a spatial predator–prey system. In such a model, each site could represent a small patch of land; states could be 0 = empty, 1 = occupied by only prey, 2 = occupied by only the predator, and 3 = occupied by both species. A more complex model could include age structure for the populations as part of the states. On every time step, all sites simultaneously update their states according to a rule. The rule specifies the new state of a site as a function of the current state of that site along with states of other nearby sites in a local neighborhood. Rectangular/square lattices are often used, but triangular or hexagonal lattices are also sometimes used, particularly for applications in hydrodynamics. In a two-dimensional rectangular lattice, the neighborhood can be specified via a set of coordinate offsets $(\Delta x, \Delta y)$ of other sites relative to the site being updated. The neighborhood for a site often consists of the site along with its four adjacent sites, called the von Neumann neighborhood, or the site with its eight adjacent sites (including diagonally adjacent sites), called the Moore neighborhood (Fig. 1). For the predator–prey example, the rule would incorporate events such as birth onto neighboring sites and death by each species, as well as consumption of prey by predators.

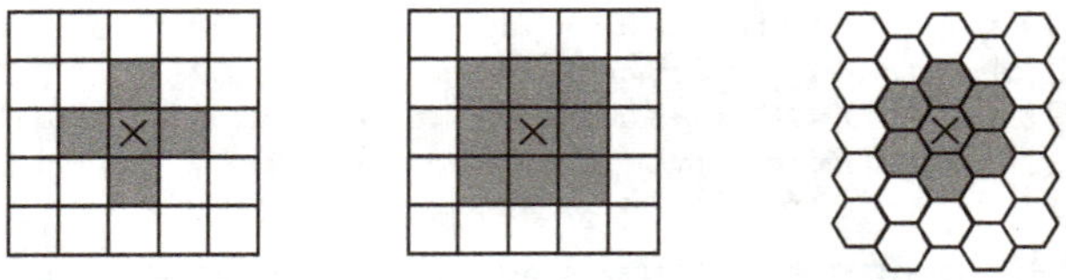

FIGURE 1 Commonly used neighborhoods (indicated by the shaded cells) are shown for a given cell (marked with an "X"). Left: the von Neumann neighborhood; center: the Moore neighborhood; right: nearest neighbors in a hexagonal lattice.

When studied via simulations, a choice must be made for how to treat the boundaries or edges of the lattice. "Wraparound" boundaries are often used, where a site on the right edge of the lattice looking toward its neighbor to the right "sees" the site on the corresponding opposite (left) edge. In two dimensions, this connects the opposite edges of the lattice in a way equivalent to drawing the lattice on the surface of a torus (doughnut). In many models, this reduces the edge effects caused by running simulations on a finite lattice. Another way to interpret the use of wraparound boundaries is that it's equivalent to assuming that the simulated lattice is just one of many identical lattices tiled in space—information that moves off the right edge of one lattice moves onto the left edge of the next (identical) lattice. Other possible boundary conditions used include fixed or truncated boundaries (cells beyond the edges of the lattice are assumed to be in fixed states) and reflecting boundaries (a cell looking beyond the edge of the lattice sees a copy of itself). The latter two boundary conditions are analogous to Dirichlet and Neumann boundary conditions for partial differential equations.

EXAMPLES

A simple example of a cellular automata rule is the "majority" rule. In the simplest deterministic version of the rule, each site in the lattice can be in one of k states labeled $0, 1, \ldots, k-1$. On each time step, each site changes its state to match the majority state in the neighborhood. If there is a tie for the majority in the neighborhood, a tie-breaker rule must be used; for example, the site may not change its state, or it may randomly choose the largest state from among those tied for the majority (Fig. 2). Variations of this rule have been used as very simple models of genetic drift in spatially distributed populations, where, for example, the different states represent different alleles.

Perhaps the most famous cellular automaton is the mathematician John Conway's "Game of Life." Each site in the lattice is in one of two states, dead or alive. On each time step, each site counts how many of its eight neighbors (four orthogonal and four diagonal) are alive. If a site is currently alive, it will die from isolation if it has fewer than two live neighbors or from overcrowding if it has more than three live neighbors; otherwise, it remains alive. If a site is currently dead, it becomes alive if it has exactly three live neighbors; otherwise, it remains dead. Conway originally conceived of the rule as a simple model of biological reproduction, but it was later found to have very rich dynamics and surprisingly complex computational properties, which has made it probably the most well-studied cellular automaton rule.

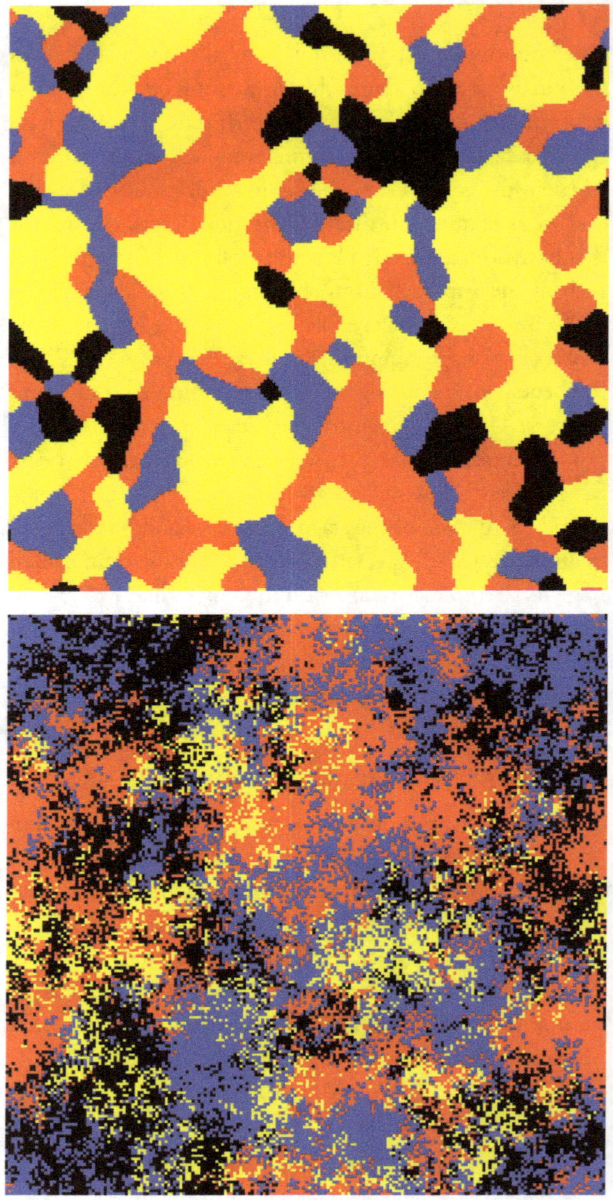

FIGURE 2 Configurations of the lattice running the "majority" cellular automaton. (A) A four-state model where a site's neighborhood consists of all sites within distance 2 of the site being updated (i.e., a 5 x 5 block centered at the site). (B) The stochastic four-state model using the Moore neighborhood.

GENERALIZATIONS

Basic cellular automata are often generalized; some of the common generalizations are listed below.

- Interactions may occur over long distances: the rule may also include as inputs the global densities of sites in various states, or the states of particular sites that are far away from the site being updated.
- The state of each site may be a continuous value. For example, a simple rule where each site updates its state to be a weighted average of its own state and its neighbors' states corresponds to an explicit finite difference method for approximating the partial differential equation describing heat flow. Continuous-state cellular automata are also applied to reaction-diffusion systems.
- The rule may be stochastic rather than deterministic. A stochastic variation of the majority rule described above is one where each site chooses a random neighbor and copies its state. The most common state in the neighborhood is most likely to be chosen, but the strict majority no longer always "wins" due to the randomness.
- Time may be continuous, rather than discrete. In this variation, sites in the lattice are updated asynchronously, rather than all at once. For example, in a discrete-time population model, each occupied site would potentially reproduce and/or die simultaneously on each time step. In a continuous-time version, individual events corresponding to a single site reproducing or dying would occur; these events are typically simulated in a way that allows a corresponding set of differential equations to be derived which approximate the dynamics of the spatial cellular automata model.
- Cellular automata models may be coupled with individual-based or agent-based models, where agents move around on and interact with the lattice. For example, the agents may represent organisms, while the lattice represents the environment.

Many spatially explicit models in epidemiology and population ecology can be thought of as continuous-time stochastic cellular automata. For example, in a spatial Susceptible–Infectious–Recovered (SIR) epidemiological model, a susceptible site may transition to an infectious state at a rate proportional to the number of infectious sites in the neighborhood; each infectious site then transitions to the recovered state at a fixed rate.

The classic Levins metapopulation model is equivalent to a two-state cellular automaton where each site is empty or occupied (a "patch-occupancy model"); each empty site transitions to the occupied state at a rate proportional to the global density of occupied sites, and each occupied site becomes empty at a fixed rate. Many other patch-occupancy lattice models, where each site is either empty or occupied by some species of interest, can also be represented using cellular automata. The population size is given by the total number of occupied sites in the lattice.

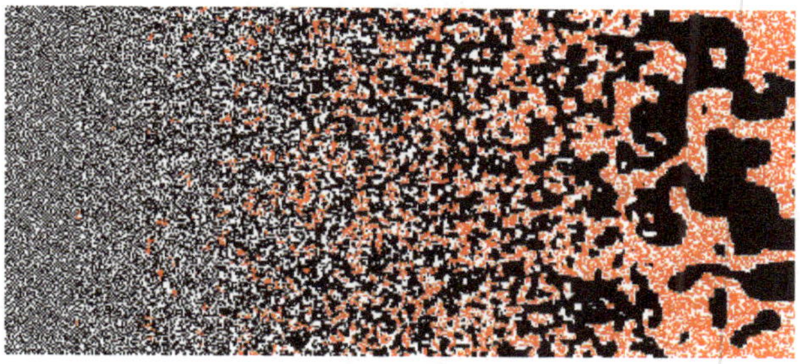

FIGURE 3 A stochastic patch-occupancy population model with structured heterogeneous landscape. Colors: white = empty suitable sites; black = unsuitable sites; red = occupied suitable sites. Occupied sites reproduce onto nearest neighbors and die at fixed per capita rates. The proportions of suitable and unsuitable sites are constant across the landscape, but clustering of suitable sites increases from left to right. The model shows that locally dispersing populations persist at higher densities on landscapes with spatially clustered habitat.

One of the primary advantages of cellular automata over mathematical approaches is that complex boundary conditions or spatial features may easily be included in the models. For example, for a model of a population with local dispersal, environmental heterogeneity with complex spatial structure may be present in the lattice (Fig. 3). The difficulty of simulating the model is not affected by the spatial complexity of the environment, such as the arrangement of different habitat types.

ANALYSIS

Cellular automata are most often studied via spatially explicit computer simulations. Because the rule is identical at each site and generally fairly simple, the simulations are well suited to run in software platforms supporting vectorized computations, such as R or MATLAB. They are also well suited for parallel-processing computing environments.

They are also sometimes studied via mathematical approximations, typically to try and predict macroscopic statistical properties of the configuration of the lattice, such as the densities of various states in the lattice or the frequencies of combinations of states among pairs of adjacent sites. For discrete-time cellular automata, these approximations take the form of difference equations; for continuous-time models, differential equations are used. Mean-field approximations (which neglect all spatial correlations in the lattice), pair approximations, and other moment-closure approximations are often used to perform these mathematical analyses.

SEE ALSO THE FOLLOWING ARTICLES

Epidemiology and Epidemic Modeling / Metapopulations / Pair Approximations / Reaction–Diffusion Models / SIR Models

FURTHER READING

Chopard, B., and M. Droz. 1998. *Cellular automata modeling of physical systems*. Cambridge, UK: Cambridge University Press.

Deutsch, A., and S. Dormann. 2005. *Cellular automaton modeling of biological pattern formation*. Boston: Burkhäuser.

Ermentrout, G. B., and L. Edelstein-Keshet. 1993. Cellular automata approaches to biological modeling. *Journal of Theoretical Biology* 160: 97–133.

Molofsky, J., and J. D. Bever. 2004. A new kind of ecology? *BioScience* 54: 440–446.

Poundstone, W. 1985. *The recursive universe*. New York: William Morrow and Company.

Schiff, J. L. 2008. *Cellular automata*. Hoboken, NJ: John Wiley & Sons, Inc.

Toffbli, T., and N. Margolus. 1987. *Cellular automata machines*. Cambridge, MA: MIT Press.

CHAOS

ROBERT F. COSTANTINO

University of Arizona, Tucson

ROBERT A. DESHARNAIS

California State University, Los Angeles

Chaos is a type of dynamical behavior with population trajectories that are deterministic (the computational rules are fixed and have no random elements), aperiodic (never repeat the same population size), bounded (do not increase to infinity), and sensitive to initial conditions (two trajectories that are initially close will separate with time). Chaos is a relative newcomer to ecology, having been hypothesized as an explanation for the observed fluctuations in population abundance since the mid-1970s.

BACKGROUND

Mathematical research in chaos started with the French mathematician Henri Poincaré at the end of the nineteenth century. Some 80 years later, the beginning of ecological research in chaos occurred with the landmark papers of Robert May. What is particularly fascinating about the introduction of chaos in ecology is that it was accomplished solely with elementary mathematics; there was no supporting biological data.

The Malthusian formulation for population growth states that the size of the new population x' (for example, an annual plant) is given by the old population size x times the rate of population growth r, namely, $x' = rx$. The scheme May used to introduce chaos was the nonlinear discrete-time logistic model $x' = rx(1 - x)$, where the modification $(1 - x)$ to Malthusian growth introduces density dependence due to crowding. (A parameter representing the population carrying capacity is often included, but it has no effect on the qualitative dynamics.) While the logistic equation is a biologically naïve description of population growth, it provides a vivid illustration of the potential of nonlinear systems to exhibit a variety of long-term population patterns including extinction, equilibrium, periodic cycles, and chaos.

Ecologists were intrigued by the concept of chaos, but cautious. Existing data were suggestive but the time series were short; there were hints of chaotic dynamics rather than strong, convincing evidence. Almost immediately after its introduction, controversy arose about the existence of chaos in biological populations. Thirty-five years after its introduction, May's hypothesis remains the subject of lively debate.

BIFURCATIONS

The dynamic behavior of a nonlinear system depends—subtly—on the magnitude of the parameters. In the case of the nonlinear logistic equation, behavior depends on the magnitude of the growth rate parameter r. What is surprising is that small changes in the value of a parameter can abruptly alter the population's behavior. For example, a slight increase in the magnitude of a parameter can cause a shift from equilibrium behavior to a cycle. The form of the equation remains intact; it is a change in the value of a parameter that gives rise to the sudden emergence of a new dynamic.

The change from one kind of dynamic behavior to another as a parameter shifts is called a bifurcation. A graph summarizing the final states of a system as a parameter varies is a bifurcation diagram. It is obtained by plotting the set of long-term values of a dynamic variable (after discarding transients) versus the value of the bifurcation parameter. Figure 1 shows the bifurcation diagram for the logistic model as the value of the growth rate parameter is varied. When r is small, the final state is equilibrium (the single line on the left side of the diagram). As the parameter increases, the equilibrium bifurcates to a period 2-cycle (two branches) with the population size alternating between two different values. As r increases further, the bifurcations continue with the 2-cycle bifurcating to a 4-cycle and then to an 8-cycle, 16-cycle, 32-cycle, and so on. However, the parameter shifts needed for the next bifurcation become geometrically smaller until a critical point is reached where the dynamics become aperiodic. This represents the onset of chaotic oscillations, which are indicated by the solid vertical bars on the right side of the graph. The banding pattern is caused

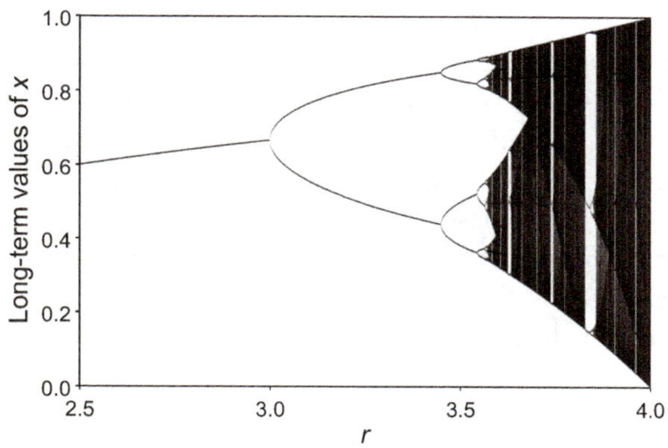

FIGURE 1 A bifurcation diagram for the nonlinear logistic model using the growth rate r as the bifurcation parameter.

by period locking, where chaotic behavior gives rise again to cycles and more period-doubling cascades.

The bifurcation diagram is a powerful tool for understanding nonlinear dynamics. One idea that resonates from the diagram is the possibility of manipulating a bifurcation parameter experimentally to demonstrate transitions in dynamic behavior, for example, from an equilibrium to a cycle. The diagram also opens the door to the possibility of experimentally documenting transitions to chaotic oscillations.

SENSITIVITY TO INITIAL CONDITIONS

A characteristic feature of chaotic systems is sensitive dependence on initial conditions. This means that two populations with similar initial sizes will diverge from each other over time. An example of this is provided by the logistic model (Fig. 2). Although the two trajectories are initially close, they quickly diverge. The exponential rate of divergence (or convergence if the dynamics are not chaotic) is called the dominant Lyapunov exponent. A positive Lyapunov exponent is a signature of chaos.

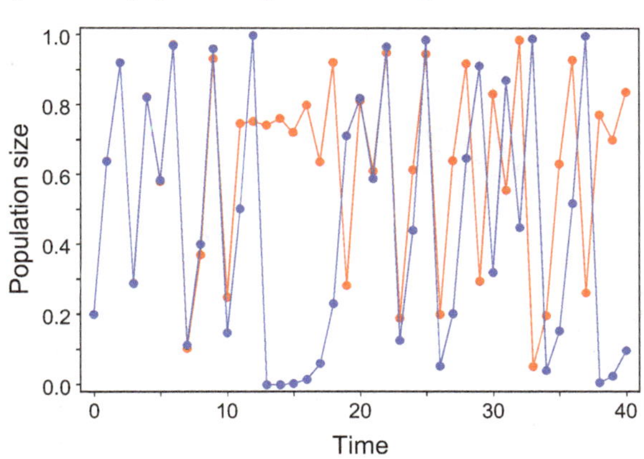

FIGURE 2 In chaotic systems, the time series of two populations starting with nearly identical numbers diverge, demonstrating sensitivity to initial conditions.

Sensitivity to initial conditions has important scientific implications. If an ecological system is chaotic, then our ability to predict the future state of the system depends critically on our current knowledge of the system. Any errors in our knowledge will lead to increasingly erroneous predicted values. For example, the two trajectories in Figure 2 could represent the true trajectory of the population and our predicted trajectory based on a slightly flawed assessment of the initial state. Since most ecological estimates are subject to errors, long-term forecasting of chaotic systems seems impossible.

On the other hand, it has also been suggested that sensitive dependence on initial conditions might have some practical advantages for the control of chaotic systems. One idea is to "nudge" the parameters or state variables at points in the trajectory where the system is sensitive to changes, producing a large desired effect from small perturbations. This approach has been demonstrated in a variety of physical and chemical systems and in vitro cardiac and brain tissues.

PHASE SPACE AND STRANGE ATTRACTORS

Another way to examine chaotic systems is to visualize the long-term values of the state variables. Although the exact sequence of these values is sensitive to initial conditions, the state variables will converge to the same system attractor for different starting values. These attractors can be visualized in phase space by using the values of the system variables as the dimensions of a phase space plot. Chaotic systems will often display attractors with an enormous amount of structure and detail, often referred to as strange attractors.

An example of a strange attractor in phase space can be obtained from a simple extension of the discrete logistic model to allow for a predator and a prey:

$$x' = rx(1 - x - \alpha y),$$
$$y' = y(\beta x - d).$$

Here, the state variables x and y represent the densities of the predator and prey, r is the population growth rate of the prey, α is the rate at which predators consume prey, β is the rate at which predators convert consumed prey into new predators, and d is the death of predators. A phase space representation of the attractor is obtained by plotting the long-term values of x and y for a given set of parameters. Figure 3 shows an example of a strange attractor for parameter values that lead to chaotic dynamics. The detail and fine structure of the strange attractor distinguishes it from a stochastic system.

Phase-space plots have a practical use. Time-series plots of chaotic systems often look random. It is often easier to

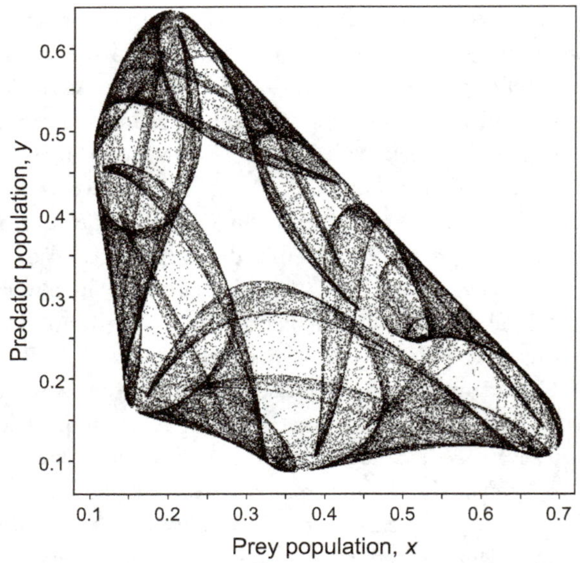

FIGURE 3 Strange attractor for the chaotic predator–prey system with $r = 3.45$, $\alpha = 1$, $\beta = 3.5$, and $d = 0.02$.

distinguish among different types of dynamics by examining the data in phase space. This is especially true if there is noise or measurement error associated with the system.

ORDER IN CHAOS

There is order to chaotic dynamics. A chaotic strange attractor can be characterized as a dense set of unstable cycles. Some of these unstable cycles may be saddle cycles, which are stable when approached from some directions but unstable in other directions (much like a ball on an equestrian saddle, which will roll toward the saddle center from the front or back directions but away from the center along the sides). Chaotic systems display evidence of these cycles in their trajectories as they wander from one unstable cycle to the next. Cycles that are closer to being stable will have a greater influence and some of these cycles may be observed in population data.

METHODS OF DETECTING CHAOS

Following Robert May's discovery of chaos in the nonlinear logistic equation, many other deterministic population models were identified that displayed chaotic dynamics. Clearly chaos was not restricted to the particular details of the logistic scheme. The widespread recognition of chaos in ecological models led to the search for chaos in existing population time-series data.

There are three methods that have been used to detect chaos. Model-free approaches take advantage of the sensitivity to initial conditions that is characteristic of chaotic systems. One looks for pairs of data points near one another in phase space and then follows the trajectories in time to measure the Lyapunov exponent. A positive exponent implies chaos. This method requires long time series, which are rare in ecology. Approaches using nonmechanistic models take advantage of Takens' theorem, which allows the properties of a dynamical system to be reconstructed from time-lagged observations of one or more system variables. One can use various statistical techniques to fit a curve or surface to the time-lagged data and see if the fitted function predicts chaos. Approaches that use mechanistic models start with the derivation of a mathematical model based on biological knowledge. The model parameters are estimated from data, and the fitted model is checked for chaotic dynamics. Each of these methods has been used to examine historical data sets; none of them have provided incontrovertible evidence for population chaos. Connecting nonlinear theory to observed population fluctuations proves to be a formidable challenge.

CHAOS AND POPULATION EXTINCTION

There has been much interest in the possible role of chaos as cause of population extinction. Since chaotic dynamics often involve large fluctuations in population numbers, it has been suggested that chaotic populations may be doomed to extinction. In fact, extinction has been proposed as an explanation for why chaotic dynamics are unlikely to occur in natural populations. One hypothesis suggests that many populations live on the "edge of chaos." The idea is that adaptations due to natural selection lead to higher rates of reproduction, which moves populations across the bifurcation diagram toward chaos. However, once a population crosses into the chaotic realm, it goes extinct. Therefore, most populations exhibiting nonlinear dynamics should be nonchaotic, but near the edge of chaos. On the other hand, theoretical studies have also suggested that chaos may prevent extinction by allowing metapopulations composed of geographically distinct populations linked by dispersal to fluctuate asynchronously, reducing the likelihood that the entire metapopulation goes extinct. Given the paucity of data linking chaos and extinction, these conjectures remain unresolved.

ECOLOGICAL EVIDENCE FOR CHAOS

Claims for chaos have been made for a variety of species, including measles, bacteria, protozoa, marine plankton, fruit flies, blowflies, grouse, muskrats, hares, and voles. Most of the claims are based on the analysis and interpretation of historical data and have been met with some skepticism.

Laboratory studies of population chaos have also been conducted. The first and most successful of these involved the use of flour beetles of the genus *Tribolium*. A mechanistic model of the dynamics of these insects was derived and validated using historical data. The LPA model involved a set of three difference equations to describe the abundances of the larval, pupa, and adult life stages:

$$L' = bA_t \exp(-c_{ea} A - c_{el} L),$$
$$P' = (1 - \mu_l)L,$$
$$A' = P \exp(-c_{pa} A) + (1 - \mu_a) A.$$

Here, L is the number of feeding larvae, P is the number of nonfeeding larvae, pupae, and callow adults, and A is the number of sexually mature adults. The unit of time is 2 weeks, which is the approximate amount of time spent in each of the L and P stages under experimental conditions. The average number of larvae recruited per adult per unit time in the absence of cannibalism is $b > 0$, and the fractions μ_a and μ_l are the adult and larval probabilities of dying from causes other than cannibalism in one time unit. In the species *T. castaneum*, larvae and adults eat eggs and adults eat pupae. The exponential expressions represent the fractions of individuals surviving cannibalism in one unit of time, with cannibalism coefficients c_{ea}, c_{el}, c_{pa}. Parameters were estimated from historical data, and a bifurcation diagram was generated using c_{pa} as the bifurcation parameter with $\mu_a = 0.96$ (Fig. 4). This bifurcation diagram formed the basis of the experimental design. Adult recruitment and mortality were manipulated to establish experimental treatments, which correspond to the arrows in Figure 4. Qualitatively different dynamics were forecast; equilibrium (controls), period-2 cycles ($c_{pa} = 0$), period-3 cycles ($c_{pa} = 0.5$), and chaos ($c_{pa} = 0.35$) were among the predicted attractors. Phase space plots of the data (Fig. 5) resembled the predicted attractors and supported the case for chaos.

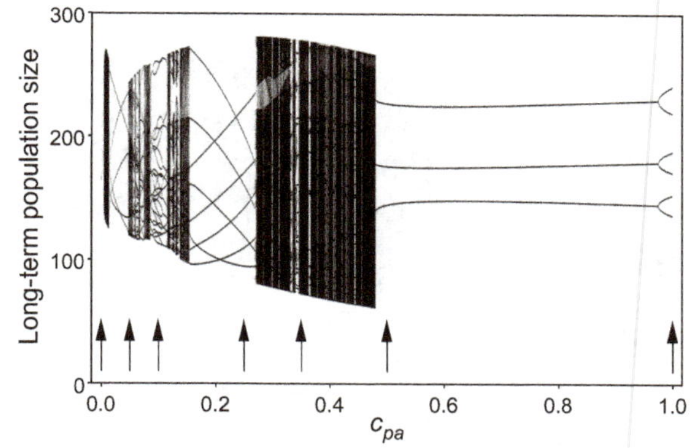

FIGURE 4 Bifurcation diagram for the LPA model using c_{pa} as the bifurcation parameter. Arrows correspond to the treatments for a *Tribolium* experiment to study nonlinearity in population dynamics.

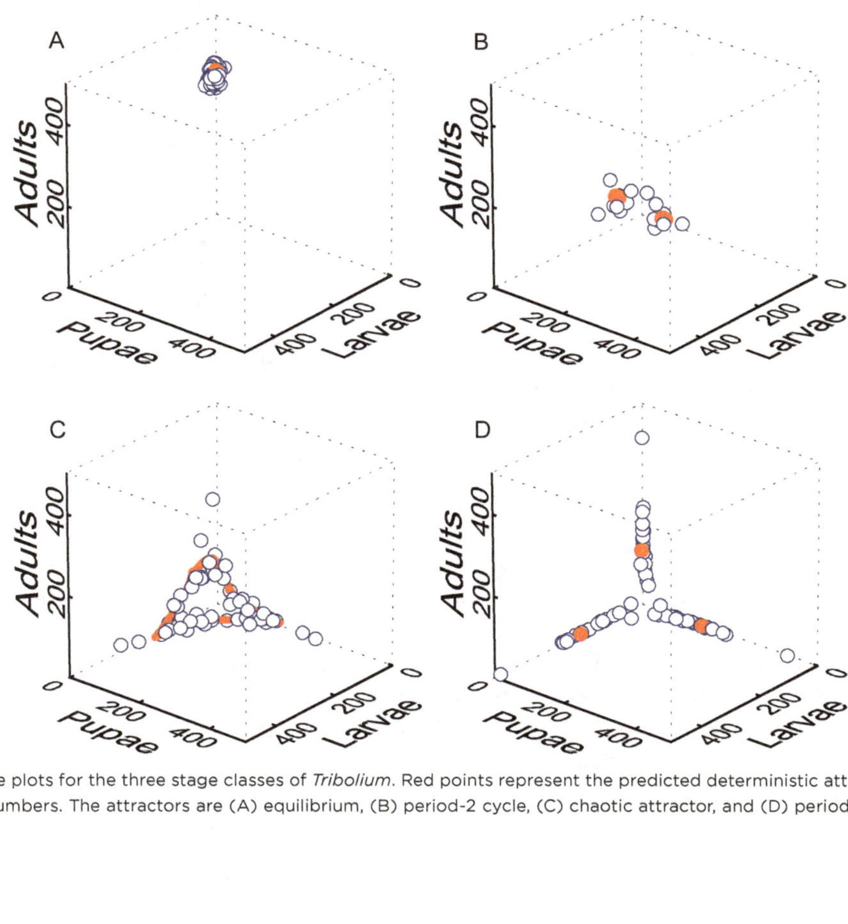

FIGURE 5 Phase space plots for the three stage classes of *Tribolium*. Red points represent the predicted deterministic attractors. Open circles are the observed insect numbers. The attractors are (A) equilibrium, (B) period-2 cycle, (C) chaotic attractor, and (D) period-3 cycle.

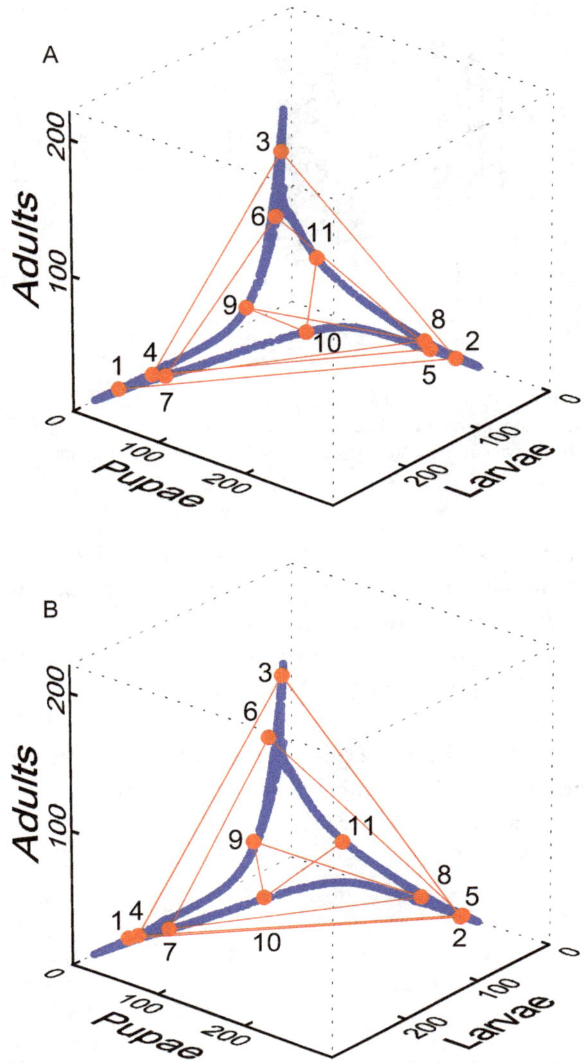

FIGURE 6 The LPA model with chaotic dynamics predicts (A) an influential unstable saddle cycle of period 11 that can be detected in (B) population data.

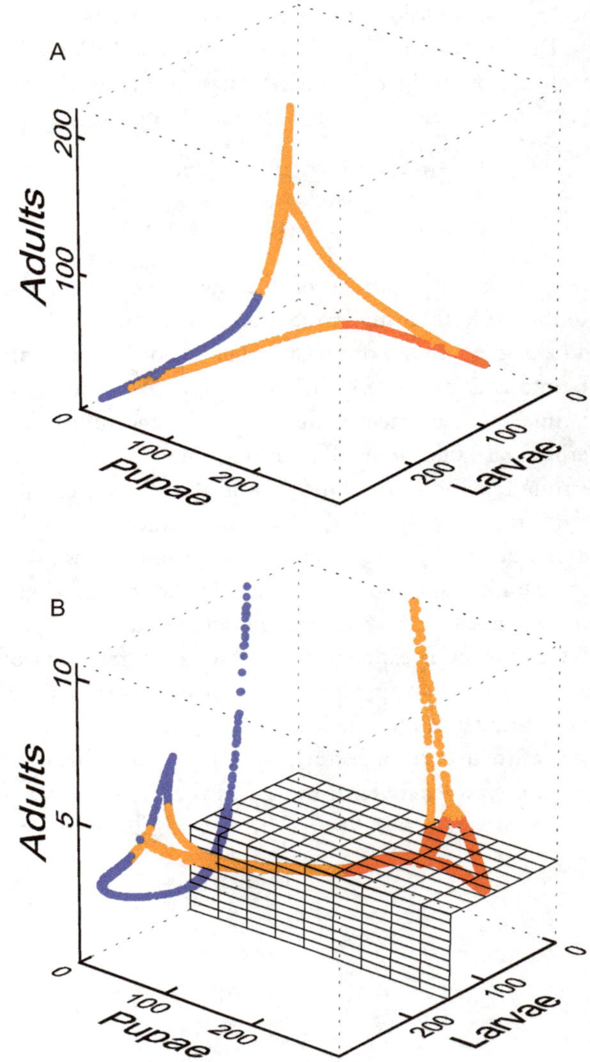

FIGURE 7 Hot spots on the chaotic attractor of the LPA model in phase space. Shown are (A) the full attractor and (B) the boundaries of the in-box and out-box. The points colored in red are the most sensitive to perturbations.

Unstable saddle cycles can be detected in the chaotic oscillations of population data. The LPA model with $c_{pa} = 0.35$ predicts a chaotic strange attractor with an unstable saddle cycle of period 11. Model simulations are strongly influenced by the 11-cycle, and evidence of these cycles appear in the population time series (Fig. 6).

Experimental evidence also exists for the property, characteristic of chaos, of sensitivity to initial conditions. The chaotic attractor of the LPA model with $c_{pa} = 0.35$ has "hot spots," regions of state space where small perturbations have large effects (Fig. 7). In a separate experimental study involving *Tribolium*, an experimental protocol was established where phase space was divided into an in-box region and an out-box region (grid in Fig. 7). The in-box contained the hot regions of the attractor. The experimental protocol established populations with chaotic dynamics. One treatment involved perturbing the populations by adding three adults whenever the LPA values fell within the in-box. A second treatment served as a negative control: populations were perturbed in the same way whenever they fell into the out-box region. Unperturbed populations were maintained as a regular control. As predicted by the model, the in-box perturbations caused a large dampening of the magnitude of the population fluctuations, but no such dampening was exhibited by the out-box treatment or the control (Fig. 8). This experiment provided additional empirical support for the possibility of population chaos.

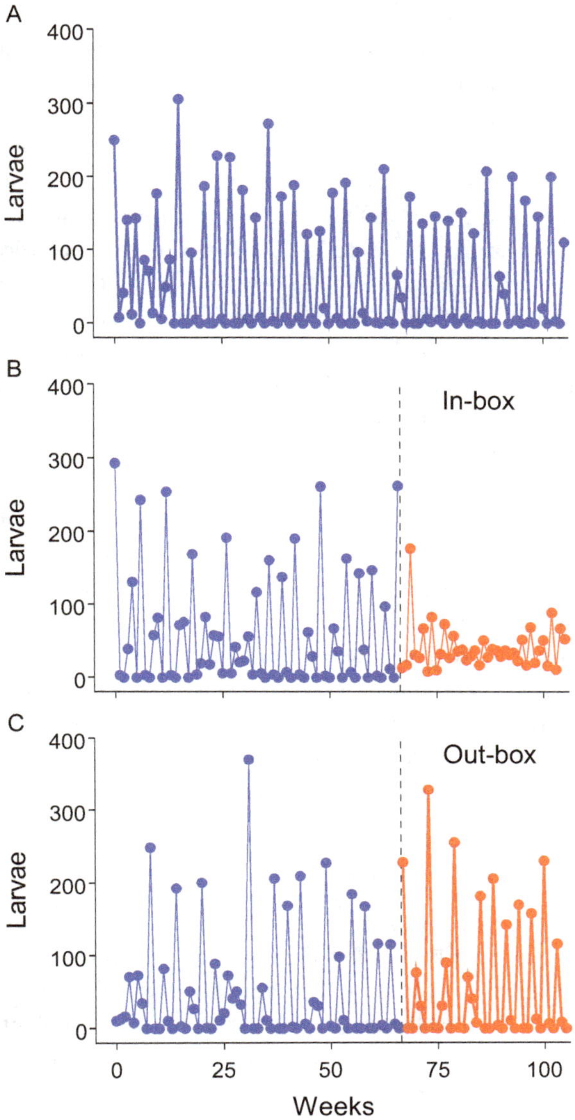

FIGURE 8 Larval numbers from a *Tribolium* experiment investigating sensitivity to initial conditions on a chaotic attractor. (A) The control populations exhibited chaotic fluctuations. (B) The populations perturbed while LPA values were in the in-box showed a large reduction in the magnitude of the oscillations, while (C) the out-box treatment looked much like the controls.

SEE ALSO THE FOLLOWING ARTICLES

Bifurcations / Difference Equations / Metapopulations / Nondimensionalization / Single-Species Population Models / Stability Analysis

FURTHER READING

Costantino, R. F., R. A. Desharnais, J. M. Cushing, B. Dennis, S. M. Henson, and A. A. King. 2005. Nonlinear stochastic population dynamics: the flour beetle *Tribolium* as an effective tool of discovery. In R. A. Desharnais, ed. *Population dynamics and laboratory ecology*. New York: Academic Press.

Cushing, J. M., R. F. Costantino, B. Dennis, R. A. Desharnais, and S. M. Henson. 2003. *Chaos in ecology: experimental nonlinear dynamics*. New York: Academic Press.

Gleick, J. 1988. *Chaos: making a new science*. New York: Viking Penguin.

Hastings, A., C. L. Hom, S. Ellener, P. Turchin, and H. C. J. Godfray. 1993. Chaos in ecology: is Mother Nature a strange attractor? *Annual Review of Ecology and Systematics* 24: 1–33.

Lorenz, E. 1993. *The essence of chaos*. Seattle: University of Washington Press.

May, R. M. 2001. *Stability and complexity in model ecosystems* (*2nd edition with a new introduction*). Princeton: Princeton University Press.

Peitgen, H., H. Jurgens, and D. Saupe. 1992. *Chaos and fractals: new frontiers of science*. New York: Springer-Verlag.

Perry, J. N., R. H. Smith, I. P. Woiwod, and D. R. Morse. 2000. *Chaos in real data: the analysis of non-linear dynamics from short ecological time series*. Dordrecht: The Netherlands: Kluwer Academic Publishers.

Solé, R., and J. Bascompte. 2006. *Self-organization in complex ecosystems*. Princeton: Princeton University Press.

Turchin, P. 2003. *Complex population dynamics*. Princeton: Princeton University Press.

COEVOLUTION

BRIAN D. INOUYE

Florida State University, Tallahassee

Species change over time in response to selection both by the abiotic environment and by other species. When two species interact with each other these interactions can impose selection on their traits, leading to adaptive evolution. Coevolution is reciprocal evolutionary change in two or more species due to their interactions with each other. For example, a flower with a long corolla may cause selection for longer tongues in a pollinator population, and pollinators may in turn cause selection on the shape of a flower for more effective delivery of pollen. Coevolution is also present in competitive and antagonistic interactions, and it may drive patterns of speciation as well as shape the traits present within populations.

OVERVIEW AND HISTORY

In Darwin's description of the mechanism of natural selection and its importance for evolution, his foundational 1859 book *On the Origin of Species*, he recognized that many of the traits we see in species today have been shaped by their interactions with other species. Darwin described species that interact with each other in specialized or obligate ways as being coadapted; however, the term was not restricted to separate species, but was also applied to the parts of an individual that function together (a meaning later adopted for the term coadapted gene complexes, to describe closely interacting genes within a species). It was not until nearly 100 years later, in 1958, that Charles Mode first used the term coevolution to describe the reciprocal evolutionary interactions between plants and pathogens. The genetic basis for plant resistance to certain fungal pathogens had been described in the previous decade, along with the genetic basis for variation in the abilities of different pathogen strains to infect plants, and thus the potential for reciprocal evolutionary interactions between two species was firmly established.

In 1964, Paul Ehrlich and Peter Raven coined a new usage for the word coevolution, describing it as the outcome of close ecological interactions between species. In their widely read and influential paper (see "Further Reading" section), Ehrlich and Raven describe a cyclical scenario whereby selection on plants by insect herbivores drives the evolution of novel chemical defenses against herbivory. The escape from herbivores allows plant populations to expand into new ecological niches and eventually form new species. Finally, to complete one iteration of a coevolutionary cycle, a novel mutation in an insect lineage allows a population to circumvent the novel plant defenses and the insects may undergo their own adaptive radiation to exploit the new plant species. While there is reciprocal evolutionary change in plants and insect herbivores in response to selection by each other, the rounds of selection and adaptation may be separated by large spans of time. Furthermore, the evolution of each group is not necessarily reciprocal, since the insect herbivore lineage that gains the ability to circumvent a plant's novel chemical defense may not be the same as the lineage that drove the evolution of the defense in the first place. Nevertheless, this type of coevolution, linking ecological interactions such as herbivory to species' adaptations and patterns of speciation, became a popular explanation for a wide range of ecological and evolutionary phenomena, and usage soon expanded to describe almost any species that had close interactions.

Daniel Janzen, in 1980, criticized overuse of the term coevolution to describe any groups of species with close ecological interactions, pointing out that actually documenting reciprocal evolutionary change is extremely difficult and rarely attempted. He also defined the term diffuse coevolution to describe the common scenario where multiple species cause selection on each other. This is in contrast to pairwise coevolution, where the evolutionary dynamics of two species depend only on the traits and dynamics of each other. As more attempts to measure reciprocal patterns of selection have been made, and more is learned about the prevalence of indirect trait-mediated ecological interactions and the context dependence of ecological interactions, it appears that most coevolution is likely to be diffuse.

Because of differences in the environment and in community composition from one place to another, populations that live in different parts of a species' geographic range are likely to experience different patterns of selection. This insight led to the development of Thompson's geographic mosaic theory of coevolution in 2005. Across a landscape, species will have strong coevolutionary interactions in some locations but not in others. This mosaic of coevolutionary hot and cold spots contributes to greater genetic diversity within a species and may contribute to patterns of speciation.

Despite the difficulty of documenting simultaneous selection and evolutionary changes in interacting species, coevolution is surely common. The large number of ecologically important species that live with symbiotic species have all undergone coevolution with their symbionts. In fact, whenever species interact—whether via mutualisms, competition, or predation—the necessary ingredients for coevolution (heritable traits, genetic variation, and fitness differences among genotypes) are likely to be present.

MAJOR THEMES AND MODELING APPROACHES IN COEVOLUTION

Species' adaptations to the physical environment will not necessarily cause changes to the environment. For example, a species might adapt to a particular climate and remain well adapted as long as those climate conditions persist. In contrast, a species' adaptation to traits in another species will often provoke reciprocal coevolutionary changes. A major theme in theoretical studies of coevolution is determining the conditions under which coevolution will cause on-going reciprocal changes, as opposed to reaching a static stalemate. This is particularly relevant for models of pairwise reciprocal coevolution, as opposed to diffuse coevolution. Depending on the structure of the mathematical model used to represent coevolution, or sometimes on the values chosen for parameters within a model, any of several types of outcomes are possible. Some models predict coevolution will lead to a single stable equilibrium state, whereas others result in multiple alternative equilibria, systems that cycle through time, or no equilibrium at all. Furthermore, when coevolution does lead to a stable equilibrium state, we wish to know if the solution is one where each species consists of individuals with a single optimal genotype, or a stable mixture of multiple genotypes (genetic polymorphisms). The ability of coevolution to foster genetic diversity both within and among populations may help to explain why natural populations almost always have genetic diversity for important life history traits.

The primary challenge to analyzing mathematical models of coevolution is that the reciprocal nature of the process means that equations describing the evolution of each species are not independent. In other words, because the evolutionary dynamics of one species depends on the current state of a different species, and vice versa, equations to describe the coevolutionary process cannot be written in a way that separates the solution of one from the solution of the other. In some cases, there are solutions to sets of coupled differential or difference equations, but a variety of methods can be used to tackle this challenge when simple solutions are not available. Numerical simulations can usually be obtained, and evolutionary game theory is an approach often used to address difficulties introduced by frequency dependence (i.e., when the mean fitness of one species depends on the relative frequencies of other species).

Coevolution with Diseases and Parasites

Nearly all species are attacked by diseases and parasites. Coevolutionary dynamics are probably best understood for this type of interaction, particularly for cases that have a simple genetic basis. Although many organisms are diploid, and thus can potentially have two different alleles at each locus (and some plants are polyploid, and thus have more than two copies of each gene), most mathematical models of coevolution between a host and a disease or parasite assume for the sake of simplicity that the organisms are haploid. Furthermore, most models only consider one or two genetic loci, although there can be many alleles for each gene. These simplified mathematical models are most appropriate for systems such as coevolution between bacteria and bacteriophage, but they have also been successfully applied to plants and plant pathogens and in some cases when alleles are strictly dominant.

In a gene-for-gene model of coevolution, the presence of a "virulence" allele in a pathogen gives it the ability to attack host individuals, and the presence of a corresponding "resistance" allele in a host allows it to resist infection. This scenario roughly matches the actual genetics of some plants and fungal pathogens and was the basis for the very first mathematical model of coevolution, Charles Mode's 1958 study that coined the word coevolution. In this model and subsequent work, whether the host and pathogen reach a stable equilibrium with a single optimal genotype or a stable polymorphism usually depends on assumptions about how costly it is for the host to be resistant or for the pathogen to be able to attack a host. Higher costs generally lead to a single pathogen genotype and a single host genotype that quickly reach a static equilibrium. However, more complex models that include multiple genetic loci as well as multiple alleles per locus can result in solutions with persistent complex cycles, where different alleles are common at different points in time.

The matching-alleles model of coevolution is at the opposite end of a continuum from the gene-for-gene model. In the matching-alleles model, an appropriate allele at a parasite's locus must match the counterpart allele at a host locus in order to cause an infection; in the absence of a match, the host is completely resistant. This type of model has been used to represent systems with invertebrate parasites that must evade recognition by a host's immune system, such as trematodes. Because matching-allele models make different assumptions about the distributions of host and parasite fitnesses, solutions with persistent cycles are common. Rare alleles may gain a fitness advantage and thus increase in frequency, but alleles inevitably lose their fitness advantage when they become too common. Because this class of models often finds an evolutionary advantage for rare genotypes, matching-allele models have been used to understand the evolution of sexual recombination and the evolution of diploid organisms (having two alleles per locus).

Work in the past decade has developed theoretical frameworks for models that are intermediate hybrids of gene-for-gene and matching-alleles models. These general theoretical frameworks are useful for unifying studies that have applied different assumptions and mathematical approaches to study different types of species. Hybrid models suggest that outcomes of persistent coevolutionary cycles are common and appropriate for a wide range of species. Persistent coevolutionary cycles are often referred to as "Red Queen" dynamics, using a literary reference to the red chess queen in Lewis Carroll's book *Through the Looking Glass*, in which the Queen tells Alice: "Now, here, you see, it takes all the running you can do, to keep in the same place. If you want to get somewhere else, you must run at least twice as fast as that!" By analogy, as fast as a species evolves, other species are coevolving along with it. No species gains an advantage over others, and all must keep coevolving just to maintain their place.

Predator–Prey Arms Races

Species traits like the running speeds of predators and prey or the crushing ability of a crab and the strength of a snail's shell are under the influence of networks of many genes.

Complex traits that are not controlled by a single gene are known as quantitative traits and can, of course, also coevolve. One well-studied example of coevolving quantitative traits is that of a weevil that feeds on the seeds at the center of camellia fruits. As a defense, some populations of camellia plants have evolved fruits with thicker outer layers, so that beetles have difficulty in reaching the seeds. In response, populations of beetles can evolve longer heads that still allow them to chew into the fruits deeply enough to reach the seeds. In some populations, fruits are relatively large and the weevils have heads dramatically longer than the entire rest of the body. Because the genetic simplifications used in models of gene-for-gene or matching-alleles models are not applicable, different conceptual and mathematical approaches have been used to study coevolution of quantitative traits.

Interactions between predators and prey are the most thoroughly studied type of coevolution for quantitative traits, and in the development of this topic the analogy of an arms race has been frequently applied. In a predator–prey arms race, the acquisition of better weapons by a predator caused the evolution of better defended prey, which in turn selects for increased predation ability, and so on. In some cases, the evolution of a predator's arms is literal. For example, Geerat Vermeij has used the extensive marine fossil record to study the coevolution of crab claws and their crushing ability with the coevolution of snail and clam shells. In his book *Evolution and Escalation* (1993), Vermeij argued that this type of predator–prey arms race is one of the dominant trends in the history of the evolution of life, and thus much of the biological diversity we see today has been shaped by coevolution. Some authors have extended this analogy to include coevolution between males and females of the same species, leading to arms races in sexual ornamentation and signaling.

As in the previous section, mathematical models of coevolutionary arms races have focused on whether the predator and prey reach a stalemate, or whether coevolution continues indefinitely or in cycles. Again, the outcome often hinges on how assumptions about the costs of traits are incorporated or the magnitude of the costs (for example, how large is the fitness cost to making a larger diameter camellia fruit or a thicker, better-defended snail shell). A recent development in theoretical approaches to arms races has been to use more realistic representations of the genetic architecture that influences traits. The rate at which a trait can evolve is affected by the amount of additive genetic variation for that trait that is present in a population. Recent models allow the amount of genetic variation to evolve along with the traits themselves, and they can include multiple genetic loci with different genetic constraints and correlations. Complex models that include multiple loci and the evolution of genetic variances as well as trait means can produce coevolutionary cycles, whereas simpler models often progress to a coevolutionary stalemate.

Adaptive Radiations and Diversifying Coevolution

The idea that coevolution can lead to adaptive radiations and bouts of speciation was first raised by Ehrlich and Raven in 1964. Although their paper focused on interactions between butterfly species and the plants eaten by the butterfly larvae, this type of coevolutionary process has now been studied for other kinds of species too. These include other types of insect herbivores attacking plants, hosts and parasites, and plant–pollinator mutualisms. Both verbal theory and simple mathematical models have also extended this approach to the idea that competition for resources, not just interactions between two different trophic levels, can drive diversification. In contrast to models of coevolution between hosts and diseases or parasites, which usually only consider pairwise reciprocal evolution between two species, studies of diversifying coevolution generally consider diffuse coevolution among many species.

Coevolution between competitors can drive diversification if species evolve to use different resources, diminishing the competition between them. One outcome of this type of coevolution is character displacement. This can be seen when populations of two or more species have divergent traits if the populations live in the same geographic location but have more similar traits among populations that do not overlap ranges. While many populations have patterns of traits that are consistent with character divergence and past coevolution, it is extremely difficult to be certain that coevolution has been responsible for what we see in specific cases because this hypothesis rests on historical events leading to the present-day patterns.

Recent mathematical models suggest that coevolution in guilds of mutualists may in fact lead to species sharing similar traits and showing signs of convergence rather than diversifying selection. For example, in a community with multiple plant species with fruits that are dispersed by birds, different plant species may converge upon similar sizes and colors of fruits, and frugivorous birds may coevolve adaptations for finding and consuming these fruits.

THE GEOGRAPHIC MOSAIC THEORY OF COEVOLUTION

With the development and rise in popularity of metapopulation and metacommunity theory, it has become axiomatic that species consist of populations spread out over a geographic range. If populations of a species live in different habitats, individuals in those populations are likely to interact with different competitors, mutualists, predators, and prey. Due the context dependence of ecological interactions, aided by the stochasticity of different mutations arising in different populations, the coevolutionary trajectories of populations in different communities are likely to be divergent. The concept that populations of a species will be involved in different coevolutionary interactions in different parts of a species' range was captured in John Thompson's book *The Geographic Mosaic of Coevolution* (2005).

A coevolutionary geographic mosaic will contribute to biological diversity because populations in some areas will have strong coevolutionary interactions (known as coevolutionary hot spots), whereas in other areas coevolutionary interactions will be weaker (coevolutionary cold spots). Even within hot spots, the process of coevolution may proceed in different ways, leading to novel traits in some populations. The accumulation of novel traits in a population as species adapt to local conditions and to each other can even contribute to the process of speciation. For example, different populations of a species may evolve to specialize on different types of food resources as they coevolve with local competitor and prey populations.

The heuristic concept of a coevolutionary geographic mosaic has recently been extended to include a mosaic of interaction networks, emphasizing the diffuse nature of coevolutionary responses among species in a community. In this view, the structure of community networks (i.e., patterns of species interactions and food webs) will be different from one location to another, but communities might still fall into a smaller number of predictable types or follow simple organizational rules. Whether ecological networks are structured in this way is an open empirical question.

APPLICATION OF COEVOLUTIONARY THEORY

When a species acquires a new defense against the mortality caused by a predator or parasite, the predator can evolve a reciprocal change to counter the new defense. In contrast, if a pest species evolves resistance to the pesticides, herbicides, and antimicrobial drugs developed by humans, the chemical formulation or mode of delivery for the pesticide or drug cannot respond with reciprocal evolution. Because the evolution of resistance is such an important challenge to modern agriculture and medicine, a better understanding of how coevolution functions in natural systems is important.

Although antimicrobial drugs cannot respond to pathogens with reciprocal evolution, some clinical practices are already informed by concepts from coevolution. For example, doctors can control the order in which different drugs are administered to a patient, and the frequency with which the drugs are switched, in an effort to prevent an arms race and manipulate the evolution of resistance by the population of a pathogen within a patient. When new drugs are developed, one question sometimes addressed during trials is whether a pathogen is likely to lose resistance to older drugs as the new drug becomes more common, or whether the drug's mode of action is one that is susceptible to the evolution of simultaneous resistance to multiple drugs.

The central idea of coevolution, that interaction between two entities will cause reciprocal changes in both, has also now spread well beyond its original home in evolutionary ecology. In economics and sociology, the term coevolution is widely used in studies aiming to understanding how conflict and cooperation among groups might be resolved. Although originally developed in economics, game theory, in which the payoff of a strategy depends on what other players of the game are doing, first became widely used in evolutionary models of coevolution. After seeing success in evolutionary biology, the use of game theory has since spread back into economic and sociology studies. In social sciences in general, the term coevolution has become a popular way to describe the manner in which changes in one field, such as a novel technological innovation, lead to changes in a different field, with potential reciprocal influences. Finally, in computer science, the term coevolutionary computation is used to describe extensions of genetic algorithms that allow different "populations" of algorithms to interact in either competitive or cooperative approaches to solving complex problems. The spread of this idea shows that in addition to being a common and important phenomenon shaping diversity in the natural world, coevolution is a useful concept in many human endeavors.

FUTURE DIRECTIONS

Coevolution is a common and widespread phenomenon. However, documenting the reciprocal patterns of selection and evolution in multiple species is very difficult and has only been done carefully for a small set of microbial lab systems and a few field systems. The interdependence of species' fitness values makes it challenging to model complex and realistic coevolutionary scenarios, but researchers have made progress using a wide range of mathematical techniques.

An important future challenge for more tightly integrating theoretical population biology and coevolution is to combine effects of species interactions on population sizes with effects on population genetics. Most genetic models assume that population sizes are nearly infinite, so that complication due to demographic and genetic stochasticity can be ignored. Nonetheless, evolutionary genetic changes can clearly cause changes in population sizes when they affect resource-use efficiency, the ability to resist infections, or the ability to escape predation. In turn, changes in population size can affect patterns of selection on extant genetic variation in a population, since density-dependent natural selection has been found in many species. Some researchers have constructed models of coevolution that include partial differential equations for simultaneous considerations of both population dynamics and genetic dynamics, but general analytical solutions are only available using methods such as separation of scales that assume populations dynamics happen much more quickly than evolutionary changes. Numerical solutions can be found for most well-posed systems of coupled partial differential equations, and they have been used for several studies of coevolution that include factors beyond allele frequencies.

Another promising direction for work on coevolution comes from the rapid rise in availability of genomic data. With the ability to rapidly and cheaply collect data on the genetic structure of populations for multiple species and locations, we will able to refine ideas from the genetic mosaic theory of coevolution using quantitative data. Currently, both the molecular methods for gathering widespread genomic data and the quantitative tools for analyzing these data are in a rapid state of growth. New metrics for incorporating genomic data into coevolution theory still need to be developed.

SEE ALSO THE FOLLOWING ARTICLES

Cooperation, Evolution of / Evolutionarily Stable Strategies / Game Theory / Mutation, Selection, and Genetic Drift / Quantitative Genetics / Sex, Evolution of

FURTHER READING

Ehrlich, P. R., and P. H. Raven. 1964. Butterflies and plants: a study in coevolution. *Evolution* 18: 586–608.

Futuyma, D. J., and M. Slatkin, eds. 1983. *Coevolution.* Sunderland, MA: Sinauer Associates.

Inouye, B. D., and J. R. Stinchcombe. 2001. Relationships between ecological interaction modifications and diffuse coevolution: similarities, differences, and causal links. *Oikos* 95: 353–360.

Janzen, D. H. 1980. When is it coevolution? *Evolution* 34: 611–612.

Mode, Charles J. 1958. A mathematical model for the co-evolution of obligate parasites and their hosts. *Evolution* 12(2): 158–165.

Rausher, M. D. 2001. Co-evolution and plant resistance to natural enemies. *Nature* 411: 857–864.

Thompson, J. N. 2005. *The geographic mosaic of coevolution.* Chicago: University of Chicago Press.

Vermeij, G. J. 1993. *Evolution and escalation.* Princeton: Princeton University Press.

COMPARTMENT MODELS

DONALD L. DEANGELIS

University of Miami, Coral Gables, Florida

Compartment, or compartmental, modeling is a mathematical method of describing flows in a system. Usually the flows are energy or matter, but they can also be flows of information, population numbers, or other quantities. As a general approach, compartment models are used in chemical kinetics, physiology and biomedicine, information sciences, systems biology, epidemiology, social science, and many other fields besides ecology.

THE CONCEPT

A system may be defined a set of entities that interact in specified ways. Ecological systems encompass a broad range; e.g., a leaf, a plant, a food web, or the whole biosphere of the Earth may be viewed as systems. In modeling a phenomenon, the first task of the modeler is to determine what entities and interactions go into the model system. In a compartment model, the system is assumed to be composed of a number of distinct compartments that each have separate pools of energy, biomass, carbon, nutrients, or other entities.

The compartments are linked by transfer functions that describe fluxes of energy, matter, or other entities between compartments and that may depend on the sizes of any of the compartments of the system. In addition, there may be external inputs, or forcing functions, that act on one or more compartments, and there can be losses, or outputs, from compartments to the external world. Both the transfer functions and forcing functions may in general be variable in time and be influenced by environmental conditions.

The first task in compartment modeling is deciding what are the basic compartments of the system and which compartments are connected by flows. Any number of compartments can be used to represent or model a particular system, and they may be connected in many ways to simulate a real system. A three-compartment system is shown in Figure 1A. A compartment is usually

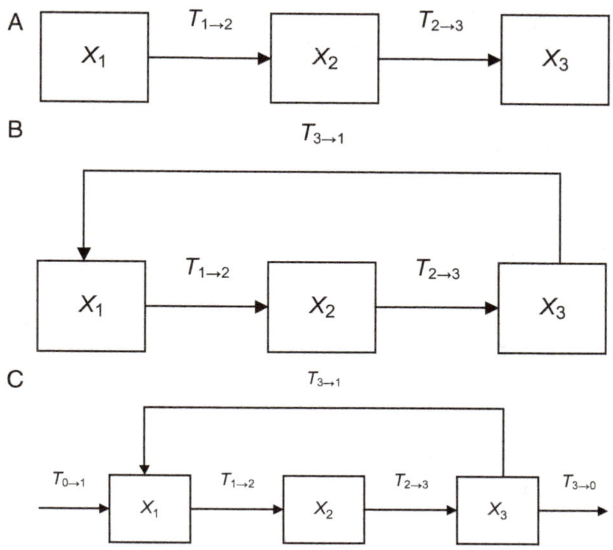

FIGURE 1 Schematics of (A) a closed compartment model with no feedbacks, (B) a closed compartment system with feedback from compartment 3 to compartment 1, and (C) a compartment system that is open to external input and losses.

represented by a box, and transfers of material are represented by arrows. Each arrow is labeled with the corresponding fractional transfer coefficient. This shows a flow of material from compartment 1 to 2 and of 2 to 3. If only the two transfer functions $T_{1\rightarrow 2}$ and $T_{2\rightarrow 3}$ exist, the flow is one way, but there also may be two-way flows between compartments. If a transfer, $T_{3\rightarrow 1}$, is added to Figure 1A, this represents a cycling feedback of material to compartment 1 (Fig. 1B). If there are no connections to the external world, either inputs or outputs, the system is closed. Otherwise it is open (Fig. 1C), and the subscript "0" represents flows either from or to the world external to the system. The configuration of compartments can also be hierarchical; that is, a compartment may be itself composed of internal compartments. The assumption is that any compartment that does not have subcompartments within it is internally homogeneous, meaning it is assumed that the manner in which the material is distributed within the compartment does not affect the flows into and out of the compartment.

The capability of a system to be well approximated by a compartment model depends on properties of the system. The system should be easily partitionable into compartments that are relatively homogeneous internally. This means that transfer rates between compartments should occur on a slower time scale than the rates of mixing of substance within the compartments.

Another basic task in compartment modeling is determining the transfer functions between the compartments. This determination depends on empirical information on the dependence of the transfer rate from i to j, $T_{i\rightarrow j}$. A common assumption in many compartment models is that the transfer from compartment i, the donor compartment, to compartment j, the recipient compartment, is a simple linear function of the size of compartment i. However, in principle, this rate can depend linearly or nonlinearly on the size of either of these compartments, or nonlinearly on some combination of the sizes of both of these compartments or even on other compartments in the system.

MATHEMATICAL FORMULATION AND SOLUTION

A compartment model is typically described by a set of differential equations, with one equation for each compartment. Alternatively, if it is reasonable to assume that there is an inherent discrete time scale upon which transfers between compartments occur, then describing the system as a discrete dynamical system is appropriate. If the flows are linear, the discrete case leads to a matrix model and if stochasticity is included, this leads to a Markov chain. For the three-compartment system above (Fig. 1B), the simplest assumption is usually that the transfer function is linearly dependent on the "donor" compartment; that is, $T_{i\rightarrow j} = k_{ji}X_i(t)$ (note the convention of putting the index of the recipient compartment first in the subscript of k). If all of the transfer functions are linear and donor dependent, the set of equations for Figure 1C has the form

$$\begin{aligned} \frac{dX_1(t)}{dt} &= I_1(t) - k_{21}X_1(t) + k_{13}X_3(t), \\ \frac{dX_2(t)}{dt} &= k_{21}X_1(t) - k_{32}X_2(t), \\ \frac{dX_3(t)}{dt} &= k_{32}X_2(t) - k_{13}X_3(t) - k_{03}X_3(t), \end{aligned} \tag{1}$$

where $T_{0\rightarrow 1}$ has been set to $I_1(t)$. Transfer coefficients denoted k_{ji} express the rate of transfer of matter or energy from compartment i to compartment j. The parameter k_{ji} has dimensions of time^{-1} (e.g., sec^{-1}, day^{-1}, year^{-1}). For example, if $k_{21} = 0.1$ day^{-1}, then a fraction 0.1 of the energy or material in compartment 1 is transferred to compartment 2 in a day. These equations can be written in the more compact matrix form

$$\begin{vmatrix} \dot{X}_1(t) \\ \dot{X}_2(t) \\ \dot{X}_3(t) \end{vmatrix} = \begin{vmatrix} -k_{21} & 0 & k_{13} \\ k_{21} & -k_{32} & 0 \\ 0 & k_{32} & -k_{13} - k_{03} \end{vmatrix} \begin{vmatrix} X_1(t) \\ X_2(t) \\ X_3(t) \end{vmatrix} + \begin{vmatrix} I_1(t) \\ 0 \\ 0 \end{vmatrix}, \tag{2}$$

where the dots represent derivatives with respect to time. This set of linear equations can be solved by computing the eigenvalues from the equation

$$det \begin{vmatrix} -k_{21} - \lambda & 0 & k_{13} \\ k_{21} & -k_{32} - \lambda & 0 \\ 0 & k_{32} & -k_{13} - k_{03} - \lambda \end{vmatrix} \quad (3)$$

$$= \lambda^3 + (k_{13} + k_{03} + k_{21} + k_{32})\lambda^2 + [(k_{21} + k_{32}) \times (k_{13} + k_{03}) + k_{21}k_{32}]\lambda - k_{21}k_{32}k_{03} = 0.$$

This yields three roots, or eigenvalues, λ_1, λ_2, λ_3. The solution has the general form

$$X_i(t) = c_{i1}e^{\lambda_1 t} + c_{i2}e^{\lambda_2 t} + c_{i3}e^{\lambda_3 t} \quad (i = 1,2,3), \quad (4)$$

where the constants c_{ij} are chosen to fit initial conditions; i.e., the values of the $X_i(t)$s at $t = 0$.

In general, the transfer functions between compartments can be nonlinear, and both the recipient compartment and other compartments in the system can influence the rate of flow. This is true in many ecological systems, such as food webs. For example, assume that in Figure 1C $T_{3\to 1} = 0$ and that $T_{1\to 2}$ and $T_{2\to 3}$ depend nonlinearly on both the donor and recipient compartments. The transfer function can be written as $T_{i\to j} = f_{ji}(X_i, X_j)$. Then the above system can be described by the following set of nonlinear differential equations:

$$\frac{dX_1}{dt} = I_1(t) - f_{21}(X_1, X_2),$$
$$\frac{dX_2}{dt} = \eta_2 f_{21}(X_1, X_2) - f_{32}(X_2, X_3), \quad (5)$$
$$\frac{dX_3}{dt} = \eta_3 f_{32}(X_2, X_3) - f_{03}(X_3).$$

In an ecological food web, composed of interacting populations of predators and prey, the function for flow from one compartment (prey) to the next compartment (predator) per unit predator biomass, or f_{ji}/X_j, is called the functional response. The functional response represents how the size of a prey compartment in a food web affects the rate at which its biomass is being consumed per unit predator. Note that in Equation 5 the multiplicative factors, η_2 and η_3 ($\eta_2, \eta_3 < 1$) represent the fact that there are losses to the external world in the movement between compartments, which is necessary if the equations represent trophic interactions, in which biomass (or energy or carbon) are lost in predator prey interactions.

In ecological systems, such as food webs, the transfer functions are frequently nonlinear. However, it is often of interest to assume that the food web is close to steady state equilibrium and is subject to a small perturbation. In Equation 5, assume the perturbation is the introduction of a small amount of tracer, $i_1(t)$ ($<< I_1$, where I_1 is assumed constant here), such as a chemically identical isotope of the substance (e.g., a nutrient) being modeled. Because the system is close to steady state, the transfers can be approximated as being linear. The linear set of equations can be obtained by substituting for each variable:

$$X_i(t) = X_i^* + x_i(t),$$

where $x_i(t) << X_i^*$, where * represents the steady-state solution. Using this substitution, the f_{ji}s can be expanded as Taylor series, and only terms of first order kept (the zeroth-order terms cancel out). The set of equations with only first-order terms is, then,

$$\frac{dx_1(t)}{dt} = i_1(t) - \left(\frac{\partial f_{21}(X_1, X_2)}{\partial X_1}\right)_{X_1^*,X_2^*} x_1(t) - \left(\frac{\partial f_{21}(X_1, X_2)}{\partial X_2}\right)_{X_1^*,X_2^*} x_2(t),$$

$$\frac{dx_2(t)}{dt} = \left(\frac{\eta_2 \partial f_{21}(X_1, X_2)}{\partial X_1}\right)_{X_1^*,X_2^*} x_1(t) + \left(\frac{\eta_2 \partial f_{21}(X_1, X_2)}{\partial X_2} - \frac{\partial f_{32}(X_2, X_3)}{\partial X_2}\right)_{X_1^*,X_2^*,X_3^*} x_2(t) - \left(\frac{\partial f_{32}(X_2, X_3)}{\partial X_3}\right)_{X_2^*,X_3^*} x_3(t), \quad (6)$$

$$\frac{dx_3(t)}{dt} = \left(\frac{\eta_3 \partial f_{32}(X_2, X_2)}{\partial X_2}\right)_{X_2^*,X_3^*} x_2(t) + \left(\frac{\eta_3 \partial f_{32}(X_2, X_3)}{\partial X_3} - \frac{\partial f_{03}(X_3)}{\partial X_3}\right)_{X_2^*,X_3^*} x_3(t).$$

These equations can be used to describe the dynamics of tracers through the steady state system and thus help elucidate the pathways of material fluxes and the effects of the compartments on these fluxes. A tracer represents only a small perturbation to the system, not enough to push it away from the linear dynamic regime. Therefore, linear models can frequently be used to study the dynamics of tracers, such as stable isotopes or radioisotopes, in an ecological system in which the interactions are nonlinear.

Linear or linearized models such as Equation 1 or 6 constitute a very general class of compartment models that are used to study the behavior of ecological models close to equilibrium subjected to small perturbations. However, compartment models are also used to study the nonlinear behavior of ecological systems. Examples of both are considered later.

Compartment models are often applied by ecologists to data from the field, such as measurement of phosphorus in a lake, or radionuclides or a heavy metal in a field or forest food web. The objective is to build an appropriate model of the system in order to make predictions. Usually it isn't feasible to measure rates directly, but only the sizes of several compartments through time. The problem is the inverse problem, that is, the problem of constructing the model from empirical observations.

The inverse problem can have several levels of difficulty. Suppose the all of the compartments in the system under study are known along with the connections between compartments and the system can be approximated as being in steady state. In that case, the problem of constructing the model involves only the estimation of parameter values. Then the problem reduces to using data to estimate values of the eigenvalues from equations like Equation 4 and then determining the transfer coefficients, k_{ji}. It may be possible to do more specific perturbations that allow estimation of some or all of the k_{ij}s directly.

Often, however, all of the possible linkages between compartments are not known, or there may components of the system that are hidden, so that there is only partial knowledge of the system, while the rest is a black box. This problem is called system identification. Methods exist for estimating the structure of a system from data on inputs and outputs to the system. A basic problem involving system identification is determining whether a unique structure exists for the set of data. If the data are insufficient to distinguish between alternative model structures, then more types of data may be needed.

One final general point should be made here regarding compartmental models. Compartment models and continuous-time Markov chain models are related. In the former, flows between compartments are modeled. In the latter, the probabilities of individuals (e.g., molecules) going from one compartment to another (movement between states) are tracked in continuous time. A compartment model can be the basis for following the path of a particle through an ecological system and obtaining a probability distribution for which compartment the particle is in a given time. For example, from our model 1, the transfer coefficient $k_{ji}\Delta t$ can represent the probability of an individual moving from state i to state j during time step Δt, where Δt is chosen sufficiently small that $k_{ji}\Delta t < 1$ for all transfer coefficients. The value $(1 - k_{ji}\Delta t)$ represents the probability of the individual staying in state i during that time step. A Markov chain model can produce a probability distribution of the state of the individual either after a given number of time steps, in which case the system response is approximated in discrete time, or in continuous time in the limit where Δt goes to zero.

APPLICATIONS

Compartment modeling has been employed as a way of representing material flows in ecological systems since at least the time of Alfred. J. Lotka, but its influence grew especially rapidly with the rise of General Systems Theory of Ludwig Bertalanffy and with the progress in high-speed computation. A few applications are described below.

RADIOECOLOGY

One of the first applications of compartment modeling in ecology grew out of a need to understand the potential impact of atomic energy on human health and the environment. It was recognized as early as the 1940s that the effects of radioactive isotopes, released by nuclear activities such as waste from reactors and nuclear weapons testing, could move through ecological systems (food chains) and accumulate within specific ecosystem components such as soil, water, and biota. Surveys of fallout radioactivity in air and food, which began with the atmospheric testing of nuclear bombs in the late 1940s and 1950s, led to recognition that the ultimate fate of long-lived fallout-derived radionuclides was dependent on complex ecological processes within food webs. Compartment modeling of the behavior of radionuclides in ecosystems, described in publications as early as 1948, led to the development of systems analysis methods for the explicit purpose of predicting the movement of radionuclides through environmental pathways to humans. A model of strontium-90 in a small Canadian lake, for example, included compartments for strontium-90 in lake water, bottom sediment, aquatic plants, plankton, minnows, beaver bone, muskrat bone, mink bone, minnows, and perch bone. Further models simulated the uptake of strontium-90 in the human body and concentration within certain tissues, such as bones.

Although the initial development of compartmental models for radionuclides was motivated by the detrimental effects of ionizing radiation in contact with humans and wildlife, the fact that radionuclides could be detected with radiation counters suggested they could be used to as tracers to help determine the pathways of substances through the ecosystem through the development of compartment models. In studies in the 1960s, radionuclides such as cesium-137 were used, as they were somewhat similar in behavior in their movement in an ecosystem to nutrients like potassium and magnesium. Using microcosms in which leaves were labeled with cesium-137, researchers were able estimate the pathways of the radionuclide from leaf litter, through millipedes and other soil invertebrates to the soil, and could estimate leaching and mineralization-immobilization processes in the soil as functions of wetness and temperature. The movement of phosphorus through aquatic microcosms could be delineated using the radioisotope phosphorus-32. The turnover times of phosphorus in different biota and the effects of biota on the movement of phosphorus between the water column and sediments could be explored in detail.

CARBON/ENERGY PATHWAYS IN FOOD WEBS

Compartment models continue to be used in food web ecology to characterize the flows of energy or matter (carbon, nutrients) through the food webs of specific ecosystems. One of the major applications is to marine food webs that include important commercial species. Software products such as Ecopath have been developed to facilitate the construction of models that provide a mass-balance snapshot of biomass flow through large food webs at steady state.

The first step in food web construction is to identify the compartments. Because almost all food webs contain far too many species to attempt to include them all as distinct compartments, groups of functionally similar species (having the same prey and predators) may be lumped into guilds or functional groups and modeled as single compartments, such as "phytoplankton" or "benthic invertebrates." Methods such as radioisotope traces are not available in general for the study of large food webs, and other methods must be relied on. The mathematical heart of developing such models is the formulation of two basic equations regarding biomass, which hold for each compartment. Each compartment consumes biomass from other compartments (consumption), and each compartment produces or assimilates its own biomass from what it consumes (production), and this assimilated biomass may then be exploited by other compartments. The two equations, which represent mass balance, are, if we ignore immigration and emigration to a the local food web,

$$\begin{aligned}\textit{Production} &= \textit{Catch} + \textit{Predation} + \textit{Biomass}\\ &\quad \textit{Accumulation} + \textit{Other Mortality}\\ \textit{Consumption} &= \textit{Production} + \textit{Respiration}\\ &\quad + \textit{Unassimilated Food}\end{aligned}$$

where Catch refers to biomass removed by the fishery, Biomass Accumulation refers to the growth in size of the population, and Unassimilated Food means undigested consumed biomass that is egested and becomes suspended organic matter or detritus. If these quantities could all be estimated, a complete picture of all biomass flows in the system could be constructed. Usually, there is not enough information to allow direct estimations of all of these quantities. However, software such as Ecopath uses a basic set of quantities that can often be estimated based on laboratory and field studies: biomass of the compartment, production/biomass ratio (or total mortality), consumption/biomass ratio, and ecotrophic efficiency. Here, the ecotrophic efficiency expresses the proportion of the production that is used in the system (i.e., it incorporates all production terms apart from the "other mortality"). If three of these four quantities can be estimated for each compartment, then Ecopath can set up a series of linear equations to solve for unknown values, establishing mass balance. Methods of optimization can be used to obtain all of the flows of carbon between compartments. This is a form of inverse modeling. Many such models have been developed for specific marine ecosystems and used to examine the effects of harvesting on the food web.

GLOBAL CYCLES OF MATERIAL

Complex compartment models serve as an important tool in following the global cycles of matter. It is crucially important to understand the fate of CO_2 that enters the atmosphere through fossil fuel burning and interacts with carbon pools in terrestrial and ocean ecosystems. A typical model divides the biosphere into a number of compartments. These represent carbon stored in various forms: as CO_2 in the atmosphere, carbonates in the ocean, and different biomass compartments in the terrestrial and ocean ecosystem. Compartments may include different biomes of the Earth and may also include the carbon stored in key compartments within those biomes, such as soil, woody biomass, roots, foliage, litter, herbivores, and detritivores. An important aspect of modeling the global carbon cycle is distinguishing rapid turnover from long turnover time compartments. The latter are effective for long-term storage. Hence, in the typical distinction of the atmosphere, terrestrial (short-term living, such as foliage and root hairs), detritus, wood, soil, mixed upper layer of ocean, and deep layer of ocean, the soil and the deep ocean compartments are the largest and the slowest turnover. Therefore, most attention is focused on the factors affecting storage in these compartments.

Determining the turnover rates of the compartments of the global carbon cycle is difficult, but tracers can sometimes be used; in particular, the radioisotope carbon-14. Carbon-14 is formed in the atmosphere and decays with a half-life of 5700 years. To examine the turnover time of carbon in the ocean, a simple compartment model with atmosphere, mixed upper layer, and the deep ocean can be used. Both carbon-12 and carbon-14 diffuse from the atmosphere to the ocean, and between the mixed upper layer and the deep ocean—a linear donor-controlled process in both cases. Compartment models for both carbon-12 and carbon-14 are used, with carbon-14 differing in that it decays. The amounts of carbon-12 and carbon-14 can be estimated based on measurements in each of the three compartments. The diffusion constants for both carbon-12 and carbon-14 are the same, but unknown, but the decay coefficient for carbon-14 is known. Steady state conditions for carbon-14 in atmosphere, mixed ocean, and deep ocean provide enough information to solve for the turnover times of carbon in the mixed and deep ocean layers.

EPIDEMIOLOGY

Compartment models can be used to follow the flow of a population through states represented as compartments. The classic Susceptible–Infected–Recovered (SIR) epidemiological models are a type of compartment model. In order to model an epidemic, one divides the population being studied into three pools, or compartments, labeled S, I, and R. The compartment model follows a flow of individuals of a population passing from the susceptible compartment to the infected compartment, and then to the recovered, resistant, compartment. Let $S(t)$ denote the number of individuals who are susceptible to the disease, that is, who are not (yet) infected at time t. $I(t)$ denotes the number of infected individuals, assumed infectious and able to spread the disease by contact with susceptibles. $R(t)$ denotes the number of individuals who have been infected and then removed from the possibility of being infected again or of spreading infection. In the particular model shown in Figure 2, the rate of infections is regarded as nonlinear, dependent on the product of the susceptibles and infecteds, $\gamma S(t)I(t)$. The mortality rate, m_i, of infecteds is assumed different from susceptibles, m_s, and recovereds, m_r. Individuals in all states are assumed to give birth to susceptibles at rates, f_s, f_i, and f_r.

The above types of application are only a small sampling of the uses to which compartment models have been put.

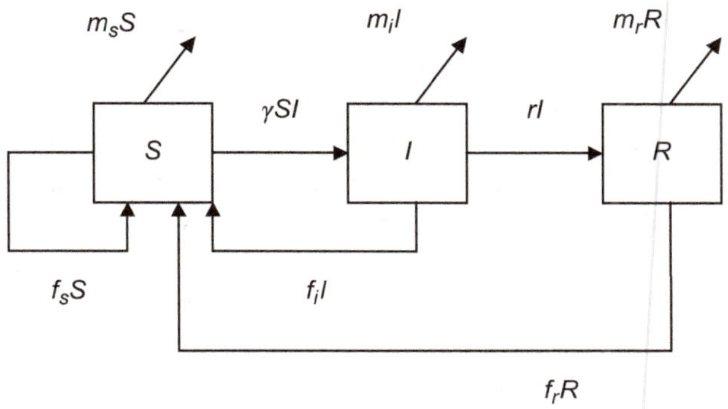

FIGURE 2 A schematic of a Susceptible–Infected–Recovered epidemiological model, showing nonlinear infection rate and mortality and reproductive rates that depend on the current state of individuals.

SEE ALSO THE FOLLOWING ARTICLES

Biogeochemistry and Nutrient Cycles / Ecotoxicology / Epidemiology and Epidemic Modeling / Food Webs / Matrix Models / NPZ Models / SIR Models / Stage Structure

FURTHER READING

Andersson, D. H. 1983. Compartmental modeling and tracer kinetics. *Lecture Notes in Biomathematics* 50. Berlin, Germany: Springer-Verlag.

Godfrey, K. 1983. *Compartmental models and their application.* New York: Academic Press.

Jacquez, J. A. 1972. *Compartmental analysis in biology and medicine.* Amsterdam: Elsevier.

Odum, H. T. 1983. *Systems ecology.* New York: Wiley-Interscience, John Wiley & Sons.

Walter, G. G., and M. Contreras. 1999. *Compartment modeling with networks.* Boston: Birkhäuser.

COMPETITION, APPARENT

SEE APPARENT COMPETITION

COMPUTATIONAL ECOLOGY

STUART H. GAGE

Michigan State University, East Lansing

Computational ecology is the application of computer-enhanced numerical and visual methods to better understand the intricacies of simple and complex ecological systems. Computational ecologists are those who focus their research and teaching resources on application of computer science and mathematics to address ecological issues. Quantitative ecologists are engaged in addressing ecological questions at all scales of biological organization from the individual to the biosphere, including the ecological ramifications of climate change and the environmental consequences of human exploitation of natural resources. Computational ecology quantitatively examines the patterns and processes associated with the interactions of organisms and their environment at both small (organisms, populations) and large scales (landscapes, biosphere). To enable ecologists to address biophysical interactions at multiple levels of complexity, computational tools and methods are increasingly necessary to handle most ecological investigations.

COMPUTATIONAL ECOLOGISTS

E. C. Pielou's 1977 book, *Mathematical Ecology*, is recognized as a pioneering work in computational ecology. Many other authors since then have played an important role in computational ecology, including Louis Gross, Bernie Patten, and C. S. Holling. Today, all ecologists require some computational skills to utilize and apply computational technology to understand the complexity of ecological processes. Others have developed advanced capacity to extend the application of computer-enhanced technology to study ecological processes. The work of Simon Levin and Brian Maurer deals with issues of quantitative biocomplexity, and William Michener has led the way in development of ecological standards. Monica Turner and Michael Goodchild stand as leaders in the advancement of spatial analysis. Many others with primary backgrounds in mathematics and ecology have made significant contributions to computational ecology.

THE STATE OF COMPUTATIONAL ECOLOGY

The state of computational ecology was first assessed by John Helly and colleagues in 1995. This paper addressed the obstacles that impede the progress of ecological research. Notably, several obstacles arose as computational ecologists began to tackle complex problems and large scales of analysis and modeling at the landscape scale or greater. Three primary areas of concern were identified as critical to computational ecology: numerical modeling, data management, and visualization. Prediction and forecasting of ecological change can be accomplished only through a detailed understanding of complex system interactions. Often, the best way to develop an understanding of an ecological process is by codifying computer models to simulate the outcome of interactions between system components. The complexity of ecological interactions, the time over which ecological processes occur, and the spatial extent of the processes have challenged model developers in their quest for increased ecological understanding. Developing complex ecological models can take a long time. Also, inputs to spatially explicit time-series models can be large, complex datasets that must be linked and managed. Such models can produce massive amounts of digital output. Visualization can be the best solution, not only to examine the large data set inputs to the model, but also to examine spatial time series of model output.

Integral to computational ecology is the need for data standards, methods to compare model results, and technologies to enable model component integration. Emergent issues included data sharing incentives, the need for and application of appropriate visualization software systems, and integration of multiscale models across disparate disciplines.

Major computational technological advances have followed Moore's law, but the challenges, educational needs, and priorities are still relevant today as computational ecology matures as an integral aspect of ecology. Laboratories focusing on computational ecology were established in the mid-1990s, including the Computational Ecology and Visualization Laboratory at Michigan State University (CEVL), the Center for Computational Ecology (CCE) at Yale University, and the Institute for Environmental Modeling, University of Tennessee.

HISTORY OF COMPUTING CAPACITY FOR COMPUTATIONAL ECOLOGY

The physical resources necessary to conduct ecological computations have changed dramatically. Prior to the early 1960s, statistical ecological relationships were computed by hand using a pencil and a slide rule. In the early 1960s, mechanical calculators were used. In the mid-1960s, programmable electronic calculators were developed, and advanced programmable calculators became available to ecologists in the late 1960s to automate and program statistical computations. Also in the mid-1960s, mainframe computers became available to computational ecologists for processing larger amounts of ecological information. By the mid-1980s, desktop computers provided a direct, hands-on capacity to apply ecological informatics. Computational ecologists now have access to desktop computers as powerful as yesterday's supercomputers, as well as access to supercomputers that can process billions of computations per second.

APPLICATION OF COMPUTATIONAL ECOLOGY SOFTWARE

Software resources available to computational ecologists in the 1960s were virtually nonexistent, so those engaged in computational ecology developed their own statistical and visualization software. An innovation that advanced computational ecology was to use computers to map the distribution of organisms. One example of early computer mapping software was SYMAP, developed at the Laboratory for Computer Graphics and Spatial Analysis at Harvard University. It enabled ecologists and others to interpolate between samples to produce contour maps of animal and plant distribution based on geographic coordinates. This work led to the development of new software companies (ERDAS Inc., ESRI Inc.) that currently supply commercial geographic and image analysis software. The development of geographic information systems revolutionized ecological thinking and enabled the application of the principles of Landscape Ecology—gaining knowledge from the understanding of ecological processes from patterns observed from above the Earth's surface.

Today, sophisticated software systems are beginning to integrate relational data management with geographic information systems (GIS), Bayesian statistics, image analysis systems, and integrated, programmable statistical systems such as the freeware R. These tools enable the construction of numerical models for addressing ecological issues at all scales of resolution. For example, advances in satellite technology and image analysis software can provide data at multiple resolutions for almost anywhere on Earth and allow computational ecologists to integrate these types of quantitative imagery into models of ecological change. However, all of these software systems require the user to learn and understand the software's capacities, limitations, and intricacies in order to utilize them effectively in ecology. This must lead to a major overhaul of ecological and computational learning at educational institutions.

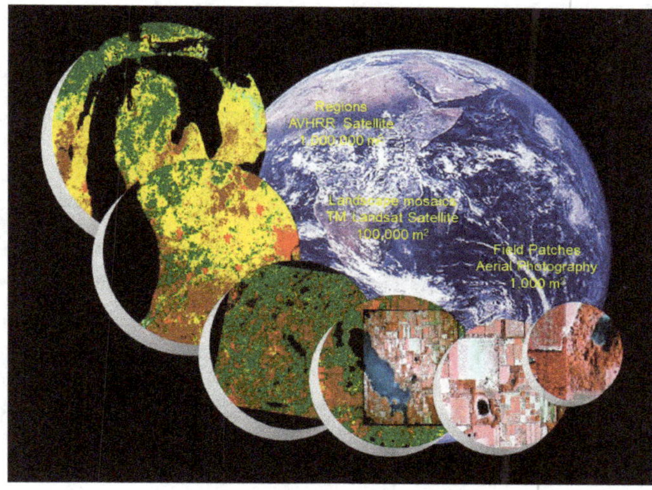

FIGURE 1 Computation and visualization of local to regional scales based on a variety of technologies ranging from aerial photography to satellite imagery.

INTEGRATION HARDWARE AND SOFTWARE

However, such software systems, designed to handle a large set of complex information, are themselves complex and may be designed to operate on specific hardware platforms, requiring additional knowledge of complex hardware and operating systems. In many cases, it is difficult to transition from a desktop computer to a supercomputer even though the scale and speed of the supercomputer may be appropriate for a large-scale computational ecology process.

EXAMPLES IN COMPUTATIONAL ECOLOGY AND VISUALIZATION

Some examples of computational ecology efforts are provided below. The Computational Ecology and Visualization Laboratory created at Michigan State University "promotes the study of temporal and spatial dynamics of features and functions associated with natural and human-dominated landscapes." This laboratory focused on visualization of ecological patterns of interaction at multiple spatial scales by addressing ecological issues, including:

1. Characterizing the flow of organisms at multiple scales illustrating biological and meteorological interactions that govern the movement of organisms in the atmosphere. Figure 1 is an example of integrating spatial scales utilizing different technologies ranging from aerial photography at the landscape scale to low resolution satellite imagery at the regional spatial scale.
2. Examining the role of biodiversity in agriculture by identifying key features in agroecosystems that regulate biotic diversity and the assessment of ecological approaches to manage biological diversity to reduce the need for chemical subsidies.
3. Modeling regional crop production by examining climate interactions in the midwest using a 30-year daily climate database along with an annual crop database and physical attributes (soils, topography) of the region to understand the long-term patterns of agricultural productivity and the interactions with climate variability. Associated with this effort was development of regional models to examine carbon cycling and climate.
4. Modeling the dynamics of land-use change by developing a geographic simulation model using neural network technology to forecast future changes in Michigan's land use based on past and current distribution of land use and to forecast spatial changes in the state of Michigan to the year 2040 for agriculture, forests, build areas, other vegetation, and wetlands. Figure 2 illustrates the application of a model that

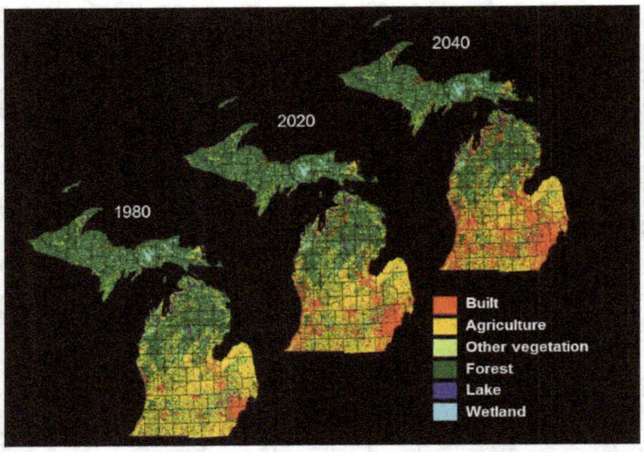

FIGURE 2 Computation and visualization of projected generational change in land use in Michigan (1980–2040) based on a Land Transformation Model using neural network technology.

forecasts the extent of land use change in Michigan over a time period beginning in 1980 until 2040.

5. Development of automated sensors to monitor the sounds of the environments to assess the impact of humans on ecosystems and to classify ecosystems based on their integrity as measured by their acoustic signatures. Associated with this work was development of a digital acoustic archive, pattern recognition analytical tools, and retrieval systems.

FUTURE STATE OF COMPUTATIONAL ECOLOGY

Today, computational ecology is an integral part of many aspects of ecological inquiry at all scales and has evolved into areas such as ecoinformatics. Computational ecology is reemerging from early efforts to establish it as a subfield within ecology. New perspectives are emerging about spatial ecology and the emerging properties that occur when location is added to the data or when objects in multiple dimensions are considered. The computer industry also is establishing a research focus on computational ecology to develop novel computational tools and methods, including grid computing and subsequently cloud computing, that are providing new approaches to ecological analysis and synthesis to predict and mitigate the rapid changes occurring in the Earth's ecosystem services. Large-scale issues such as examining the flow of organisms in the atmosphere will require linking meteorological and biological systems at massive scales. New ecological-sensor technologies have been developed and are being deployed to monitor ecological change. For example, a challenge associated with new sensor deployment is the design and placement of sensors to best meet the ecological objectives of the study, constrained by sensor cost. Geographic models and the use of spatial statistics can assist with the placement of sensors and can compute the number of sensors required to ensure that the study results will be statistically sound. New approaches to statistics are emerging (including Bayesian analyses) that are readily computed with modern software and hardware.

New initiatives such as the National Ecological Observatory Network (NEON) will develop new paradigms for ecological analysis, synthesis, and management of massive sets of automated observations made at high temporal and spatial resolutions. This will again bring computational ecology to the forefront. Due to the national scaling of ecological information collection at NEON sites by autonomous sensors and ecological investigators associated with those sites, issues of data standards and their implication for data sharing are imminent. Much work has been developed to implement spatial data and ecological data and metadata standards. The scientific community must adhere to strict standards associated with data collection to more effectively use the information within multiple contexts including across disciplines and between different hardware and software platforms. The promise of NEON is great, but the generation of knowledge from a vast increase of ecological observations will be proven over time.

Ecoinformatics, or ecological informatics, is emerging as a new area of ecology that has elements similar to those of computational ecology. Ecoinformatics integrates environmental and information science to define entities and natural processes with a language common to both humans and computers. This rapidly developing area in ecology has yet to be fully realized. Supercomputers on the desktop, advances in statistics, new approaches to spatial modeling, autonomous sensors, integrated software systems, and new thinking about monitoring the environment at high temporal and spatial resolution will lead to exciting times in ecology.

SEE ALSO THE FOLLOWING ARTICLES

Bayesian Statistics / Cellular Automata / Ecosystem Services / Geographic Information Systems / Spatial Ecology

FURTHER READING

Beck, M. B., H. Gupta, E. Rastetter, C. Shoemaker, D. Tarboton, R. Butler, D. Edelson, H. Graber, L. Gross, T. Harmon, D. McLaughlin, C. Paola, D. Peters, D. Scavia, L. J. Schnoor, and L. Weber. 2009. Grand challenges of the future for environmental modeling. White Paper, National Science Foundation, Arlington, Virginia.

Helly, J., T. Case, F. Davis, S. Levin, and W. Michener, eds. 1995. The state of computational ecology. Research Paper No. 1. National Center for Ecological Analysis and Synthesis, Santa Barbara, California. http://www.nceas.ucsb.edu/nceas-web/projects/2295/nceas-paper1/.

Isard, S. A., and S. H. Gage. 2000. *Flow of life in the atmosphere: an airscape approach to understanding invasive organisms.* East Lansing, MI: Michigan State University Press.

History of Computing: http://www.computerhope.com/history/.
History of Geographic Information: http://www.gisdevelopment.net/history/1960-1970.htm.
Kemp, K. K., and K. K. Kemp. 2008. *Encyclopedia of geographic information science.* Thousand Oaks, CA: SAGE Publications.
Kruschke. J. K. 2011. *Doing Bayesian data analysis: a tutorial with R and BUGS.* Burlington, MA: Academic Press/Elsevier.
Managing Ecological Data and Information: http://www.ecoinformatics.org.
Pascual, M. 2005. Computational ecology: from the complex to the simple and back. *PLoS Computational Biology* 1(2): e18.
Pielou, E. C. 1977. *Mathematical ecology.* New York: Wiley.
Remote Environmental Assessment Laboratory: http://www.real.msu.edu.
Spatial Data Standards: http://www.fgdc.gov/metadata/geospatial-metadata-standards.
Zhang, W. 2010. *Computational ecology: artificial neural networks and their applications.* Singapore: World Scientific Publishing Company.

COMPUTATION, EVOLUTIONARY

SEE EVOLUTIONARY COMPUTATION

CONSERVATION BIOLOGY

H. REŞIT AKÇAKAYA
Stony Brook University, New York

Conservation biology is the interdisciplinary, applied science of maintaining Earth's biological diversity. Although most closely associated with ecology, it also links other disciplines within the natural sciences (e.g., population genetics, phylogenetics, atmospheric sciences, toxicology, geography, demography), as well as those in the social sciences (e.g., economics, law) and humanities (e.g., philosophy, environmental history). Conservation biologists study the diversity of life on Earth, including the rate and patterns of change in biodiversity, examine the effects of human activities and natural factors that cause loss of species and degradation of ecosystems, and try to find ways to protect and restore biological diversity. Thus, conservation biology has an applied focus, with elements of natural resource management (e.g., forestry, fishery management), and uses methods developed in a variety of applied disciplines. However, in comparison with these resource management disciplines, conservation biology considers much longer time horizons and aims to maximize a larger variety of values (see "Ethics" section, below).

HISTORY

Although the idea and practice of conservation of natural landscapes, populations, and resources has a long history, the science of conservation biology developed mostly in the late twentieth century. Roots of conservation biology include such well-established disciplines as the study of natural history, biogeography, ecology, genetics, and evolutionary biology, as well as applied fields such as forestry, fishery management, and wildlife management. The development of conservation biology as a distinct scientific discipline coincided with and was probably affected by the confluence of several factors and developments, including the accelerating loss of natural ecosystems in the twentieth century, which underlined the necessity of a more inclusive, systematic, and science-based approach to conservation; new developments in diverse fields such as molecular genetics, theoretical ecology, and remote sensing, which provided methodological opportunities; the increasing recognition that human populations and human activities are an integral part of the biosphere; and the increasing public understanding of and support for conservation in the last three decades of the twentieth century. The term is thought to have been used first in the title of a professional meeting in 1978; a subsequent meeting led to the creation of the Society for Conservation Biology in 1985.

BIODIVERSITY

Components of Biodiversity (What to Conserve)

Biodiversity is the variety of life at all levels of organization, from the molecular to the ecosystem. The Convention on Biological Diversity (CBD), an international treaty with the objectives of the conservation and sustainable use of biodiversity and the equitable sharing of the benefits of the utilization of genetic resources, defines biological diversity as "the variability among living organisms from all sources including, inter alia, terrestrial, marine and other aquatic ecosystems and the ecological complexes of which they are part; this includes diversity within species, between species and of ecosystems." Thus, genetic diversity, species diversity, and diversity at higher levels of ecological organization are all considered to be components of biodiversity. In practice, however, the various methods of measuring biodiversity (see below) are most often based on measures at the species level.

Measures and Patterns of Biodiversity

The simplest (and most commonly used) measure of biodiversity is species richness, which is the number of species in a given area (also called alpha-diversity or alpha-richness). Also commonly used are measures such as the number of endemic species (species that only occur in the defined area) and the number or proportion of threatened species (see below). Other measures take into account not only the number of species but also their relative abundances (i.e., the number of individuals of each species found in a given area or counted in samples from a given area) or measure the rate of change or turnover in biodiversity across habitats (beta-diversity) and across larger regions or landscape gradients (gamma-diversity).

A number of biodiversity patterns have been identified. Biodiversity is higher in larger areas; this species–area relationship is commonly approximated as $S = cA^z$, where S is the number of species, A is area, and c and z are constants. Biodiversity is higher in lower latitudes (tropical regions) than in higher latitudes (temperate and polar regions); various hypotheses are offered to explain this pattern. Independent of these, biodiversity is higher in certain areas, sometimes called biodiversity hotspots, which are subject to focused conservation efforts (see "Area-Based Conservation," below).

Indicator Species

Conservation assessments and conservation actions often focus on particular species, both for practical reasons (there are usually not enough resources to study and model each species) and because the status of certain species are thought to provide more information or insight, or their conservation is considered more of a priority. Such species include indicator, sensitive, keystone, dominant, umbrella, threatened, charismatic, and flagship species.

An indicator species is often defined as a species whose presence indicates the presence of a set of other species and whose absence indicates the lack of that set of species. It is also defined as a species that indicates particular environmental conditions, such as certain soil characteristics, or certain human-created abiotic conditions such as air or water pollution. A sensitive species is strongly (and usually negatively) affected by environmental changes, such as modified forestry or agricultural practices, and changes brought about by global climate change (e.g., changing precipitation patterns or fire regimes). Sensitive species serve as an early warning indicator of such changes. A keystone species is a species whose addition to or loss from an ecosystem leads to major changes in abundance or presence of other species. A dominant species contributes a large proportion of the biomass or number of individuals in a given area. An umbrella species, if protected, indirectly causes the protection of several other species (typically because it requires large, undisturbed areas). A threatened species is one that has a high risk of extinction in the near future (see below on identifying threatened species). Finally, a charismatic species has widespread appeal to the general public (sometimes because of national or other social reasons) and therefore may be selected as a flagship species by a conservation organization to represent an environmental cause or a conservation goal.

Ecosystem Services, Ethics, and Values (Why Conserve?)

Two distinct kinds of values are often attributed to biodiversity. Intrinsic (or existence) value is attributed to components of biodiversity such as species, communities, and ecosystems, independent of the preferences or needs of humans. In contrast, utilitarian (or instrumental, or anthropocentric) value is based on the benefits humans receive from these components. These benefits (also known as ecosystem services) include resources (such as food, water, timber, fuel, and pharmaceuticals), regulating services (e.g., flood control, pollination, water purification), cultural values (e.g., scientific, aesthetic, spiritual, and recreational benefits), and supporting services (e.g., soil formation and nutrient cycling). Many ecosystem services and functions are shown to decline with species diversity, pointing out to the value of biodiversity. Some of the ecosystem services and functions can also be valued in economic terms, which is a topic of ecological economics.

As the definition of the field indicates, conservation biology has a specific aim (of maintaining biodiversity) and consequently has been described as mission oriented. Further, there is an assumption of the value of biodiversity, even outright advocacy of it, leading some to label the discipline also as value laden. However, others argue conservation biology is not any more value laden than other applied disciplines such as medicine. Nevertheless, there is often a tradeoff between using resources for conservation vs. for other societal needs, as well as tradeoffs among different conservation goals, leading to debate about who makes the decisions of allocating limited resources and how conservations biologists contribute to these decisions.

METHODS

The theoretical foundations of conservation biology include population dynamics, population genetics, and biogeography, among others. A large variety of experimental, quantitative, spatial, and molecular methods from these fields have been adopted to address questions in conservation biology. In many cases, the specific conservation questions have induced major modifications and additions to these methods, as well as development of new methods.

Inferring Extinctions and Measuring Extinction Rates

A species is extinct when the last individual of that species has died. Declaring a species extinct is often complicated because species are very rare before they go extinct and because false inference of extinction (declaring a species is extinct even though it is extant) may have serious consequences, such as cessation of conservation programs, unavailability of funds, as well as the loss of credibility if the species is rediscovered. Guidelines by IUCN (International Union for Conservation of Nature) require that "exhaustive surveys have been undertaken in all known or likely habitat throughout its historic range, at appropriate times (diurnal, seasonal, annual) and over a timeframe appropriate to its life cycle and life form" in order to list a species as extinct. When data are available, quantitative methods based on time series of observations of the species (i.e., the year of each sighting and the number of years since the last sighting) and time series of survey effort are used to calculate the probability that the species is extinct (note that this is a different probability than that of an extant species going extinct in the future; see "Estimating Extinction Risks," below).

A different but related issue is measuring the overall rate of extinctions (proportion or number of species going extinct per year). For the reasons mentioned above, the number of species listed as extinct likely underestimates the actual number of extinct species. In addition to uncertainty in the number of extinct species and the year that each extinction occurred, there is the uncertainty in the number of described species (species known to science) and the total number of existing species, making any estimate of the rate of extinction uncertain. Nevertheless, various estimates of the overall extinction rate indicate that the current rate of species extinctions, caused mostly by human impacts, is comparable to the extinction rates during the five major mass extinction events observed in the fossil record.

Estimating Extinction Risks

The main goal of conservation biology can also be viewed as preventing extinctions of native species and populations. Thus, assessing the likelihood that a species or a population may go extinct sometime in the future is one of the main methods of conservation biologists. This method is also known as population viability analysis (PVA). Viability is the likelihood that a species (or a population) will remain extant in the future, and PVA is a method for calculating this, either under current conditions or under assumed future changes. Theoretical approaches used for making these assessments include development of stochastic population and metapopulation models that predict a number of measures of population persistence, including risk of population extinction or decline, chance of population recovery, expected population size in the future, and expected time to extinction. A variety of model types are used for assessing extinction risks; the most commonly used models have age or stage structure (e.g., matrix models) and incorporate temporal variability (demographic and environmental stochasticity, including "catastrophes," which are rare events that have large effects on population parameters), density dependence, dispersal among populations, spatial heterogeneity (e.g., survival, fecundity, and other population parameters that vary in space), and spatial correlation (synchrony of temporal fluctuations among populations).

In addition to assessing the vulnerability of populations and species to extinction, PVA is also (perhaps more often) used to evaluate the effectiveness of conservation actions. Such actions include habitat protection or restoration, control of hunting, poaching, and other direct human mortality, reintroduction of species to areas they have been extirpated from, translocation of individuals among existing populations (also known as human-assisted migration or dispersal), and habitat corridors designed to increase connectivity among populations, among others. Because of the rarity and the threatened status of the species involved, and because of the difficulty and cost of experimental study, theoretical approaches such as PVA are often the only methods of systematically analyzing the consequences of such actions and their relative effect on decreasing the extinction risk of the species.

Another important area in which theoretical approaches are used is parameter estimation. Developing a PVA to estimate extinction risks or evaluate conservation actions requires estimating model parameters, often from limited and uncertain data, which may include surveys, censuses, and mark-recapture data (data from resighting or recapture of previously marked or tagged animals). When estimating model parameters such as survival, fecundity, dispersal, or habitat suitability (see below), often a likelihood-based statistical model is used, ideally guided by biological intuition and species-specific life history information. When multiple statistical models are plausible, information theoretic criteria are used to select the best-supported model, or other (e.g., Bayesian) methods are used to find weighted estimates derived from all plausible models. Other model components provide additional complexities; detection and modeling of density dependence, estimating natural variability by removing variance due to measurement error and sampling variability, estimating dispersal rates (based on genetic data, mark-recapture data, behavioral data, and habitat characteristics), and modeling habitat suitability (see below) require sophisticated quantitative methods.

Identifying Threatened and Endangered Species

Many countries have laws for protecting threatened and endangered species (such as the Endangered Species Act in the United States and the Species at Risk Act in Canada), and such laws are some of the strongest legal conservation tools available. These laws require determining which species are threatened and endangered. In addition, international organizations maintain lists of globally threatened species. The most widely known of such global lists is the IUCN Red List of Threatened Species, maintained by the International Union for Conservation of Nature.

The most direct method of determining whether a species is threatened or not is to estimate its risk of extinction with a population viability analysis (PVA), as discussed above. However, for many species the data necessary for carrying out a PVA are not available and would require a long time (and large resources) to collect. Thus, conservation biologists have developed rule-based or score-based systems that use the available information about a species to classify it in one of a few broad categories of threat. The most commonly used system for classifying threatened species is the IUCN Red List Categories and Criteria, which consists of a set of rules based on the principles of population dynamics. These criteria use species-specific information such as generation time, population size, rate of population decline, degree of fragmentation, range area, and area of occupied habitat to classify the species into one of the categories (Fig. 1). These criteria are based on population dynamics and build on concepts such as demographic stochasticity, environmental fluctuations, and the effects of population subdivision.

Monitoring the Status of Biodiversity

In conservation, monitoring is done at various levels, scales, and for various purposes. Specific populations can be monitored with regular censuses or surveys to determine the effects of management practices (e.g., harvest regimes) or conservation measures (e.g., reintroduction). Selected species in an area can be monitored along particular transects or sampling areas to detect early warning signs of sensitive species declining. Large numbers of species can be monitored at regional, continental, or even global scales to uncover general trends in biodiversity.

Examples of large-scale monitoring efforts include the Breeding Bird Survey (BBS) and the Christmas Bird Count in North America, the Living Planet Index (LPI), and the Red List Index (RLI). The BBS is based on annual observations of about 400 breeding bird species at about 4000 routes in North America, some of which have been monitored since the mid 1960s. The LPI is based on the sizes of several thousand vertebrate populations belonging to over 1000 species. The RLI is based on the threat status of selected species. To remove selection bias, the RLI has been calculated for taxonomic groups that are either fully assessed (such as mammals, birds, amphibians, gymnosperms, and corals, in which all species have been assessed according to the IUCN

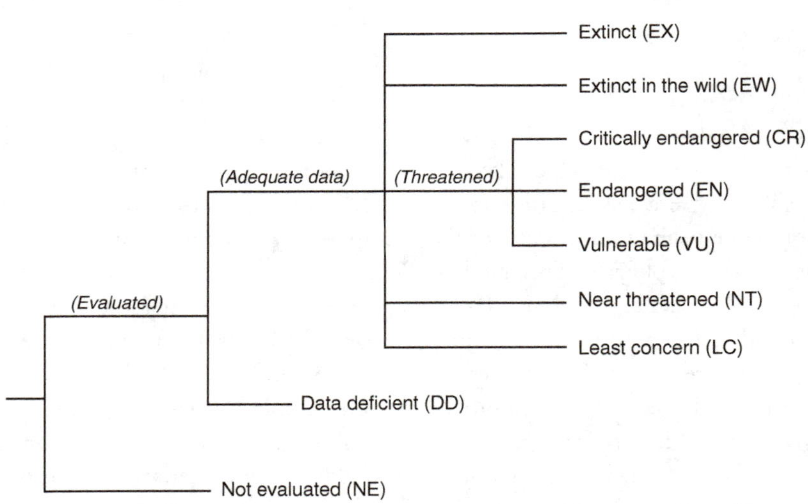

FIGURE 1 Schematic representation of the IUCN Red List classification scheme for placing each species into a category that represents its conservation status. For example, an evaluated species, if there are adequate data, can be placed in one of the threatened categories (Critically endangered, Endangered, or Vulnerable). For more information, see www.iucnredlist.org.

criteria) or sampled. The sampled RLI is based on a randomly selected sample of about 1500 species in each group (reptiles, dragonflies, monocots, freshwater fishes, and freshwater crabs), which are then assessed according to the IUCN criteria.

Area-Based Conservation

As the human population continues to increase and exerts increasing pressure on the remaining natural ecosystems, protection of these ecosystems as nature reserves is becoming one of the most important conservation tools, at least in parts of the world where it is possible to set aside additional land to be protected. Most of the current protected areas had been developed in an ad hoc or opportunistic way—areas with steep slopes, unproductive soils, and high altitudes are often overrepresented in protected areas, whereas areas at lower elevations with flat terrain and productive soils are underrepresented, resulting in underprotection of some ecosystems and the species that make them up. To remedy this situation, conservation biologists attempt to identify priority areas for protection. Variously called hotspots, key biodiversity areas, important (bird/plant/etc.) areas, priority ecoregions, or special protection areas, these areas are intended to provide maximal protection to the largest number of species.

Theoretical approaches to formalize the selection process for these priority areas are known as systematic conservation planning, which aims to select a set of areas that are representative, persistent, and efficient. Representativeness is the need for the selected reserves to represent, or sample, the full variety of biodiversity, ideally at all levels of organization. Persistence (also called resilience) is the need for reserves to promote the long-term survival of the species and other elements of biodiversity contained in them by maintaining natural processes and viable populations and by excluding threats. Efficiency is the need for reserves to achieve representativeness and persistence at a minimum cost (minimum economic impact on the society). Systematic conservation planning uses reserve selection algorithms that rely on concepts such as complementarity and irreplaceability to select the reserves that protect the largest number of conservation targets (e.g., species) at the lowest cost with the maximum potential for long-term persistence. Complementarity is a measure of the extent to which an area contributes unrepresented features (e.g., species that are not in any existing reserve) to an existing set of protected areas. Irreplaceability is a measure of the degree to which an area is needed to meet the objectives of the reserve system (e.g., if an area contains the only known occurrence of a species, then it is completely irreplaceable).

Population-Based Conservation

In many parts of the world that are densely populated by humans, area-based conservation has limited applicability, because most land is privately owned (and either too expensive to acquire for conservation or not for sale) or used intensively by humans. In these regions, conservation involves management of existing reserve areas, habitat restoration (e.g., planting of native species, and providing the abiotic factors for their survival), increasing connectivity of small remnants of natural habitat (e.g., by building habitat corridors), and managing populations of native, and especially threatened, species.

Methods of population management include preventing or at least regulating take (i.e., harvest, poaching, hunting, fishing, and so on), minimizing direct impacts of human activities (such as road mortality), human-assisted dispersal, or migration to counter the effects of fragmentation and isolation of habitat patches, reintroducing species to reestablish extirpated populations, as well as methods of ex situ conservation (discussed below). Population models (and more generally PVA) are often used to find the most effective approaches, as well as to determine the optimal effort for a given conservation method under the existing financial and social constraints and other restrictions. For example, models can be used to determine the maximum level of sustainable harvest that meets a certain viability criterion (e.g., "the probability of a 50% decline is less than 0.01"), to determine the number, age distribution, and source and target populations of translocated animals to maximize the viability of the metapopulation, or to identify the dispersal bottlenecks where the development of a habitat corridor would most benefit the species.

Describing, Identifying, and Mapping Habitat

Many approaches to species conservation require knowing where the species can survive and reproduce (its habitat) and characterizing the species' habitat in terms of its biotic and abiotic components, such as weather, topography, soil, vegetation cover, and so on. Such characterization allows conservation biologists to create a map of areas suitable for the species, thus predicting where a species might be found. Quantitative methods of predicting habitat are varied and are known by a large variety of names (such as habitat modeling, habitat suitability modeling, ecological niche modeling, species distribution modeling, and bioclimatic envelope modeling). Most of these methods use a statistical approach, correlating the occurrence of a species with the environmental conditions at the locations of its occurrence. Others are based on mechanistic models, using information on known physiological tolerances and preferences of species for environmental factors. In both cases, the environmental factors used to predict the species' distribution include a combination of climatic variables (derived from long-term weather data), land-use or land-cover variables (often derived from remote-sensing products such as satellite images), topographic variables (slope, aspect, topographic heterogeneity), and others relevant to the species modeled (e.g., soil types for plants or canopy structure for birds).

It is important to keep in mind that the habitat of a species mapped in this way may not be completely occupied (i.e., there might be areas of seemingly suitable habitat where the species does not occur), because of dispersal limitations, historical constraints, and population demography. For example, populations in suitable habitat may go extinct if the area is very small (and thus the small size of the population makes it prone to demographic stochasticity and Allee effects) or because of exploitation by humans, predation, diseases, and other factors not explicitly included in the statistical model of habitat suitability. Thus, methods of mapping habitat are most useful for conservation of species when they are combined with other quantitative approaches that focus on population dynamics.

Ex Situ Conservation

Ex situ (or off-site) conservation involves approaches to preventing the extinction of a species by maintaining it outside its natural habitat, for example, in zoos, zoological parks, aquariums, botanical gardens, and gene banks (which store live or cryogenically "frozen" samples of seeds, sperm, eggs, or embryos). Ex situ conservation is often considered a last resort, used when the last remaining natural habitat of a species is destroyed or imminently endangered or only a few, scattered individuals of the species remain in the wild. The methods employed in ex situ conservation include captive breeding and maintenance of viable seed banks. With a very small number of individuals of each species available, inbreeding, founder effects, and genetic drift are important concerns for captive populations. Theoretical models of population genetics are used to incorporate these factors into conservation decisions. Combined with the maintenance of accurate pedigrees or studbooks (which identify the parents and record the birth date, sex, and other information of every individual), these approaches inform genetic management techniques such as pairing individuals with lowest kinship and equalizing the reproductive contributions of reproductively capable individuals. Successful captive breeding programs allow reintroduction of species into their native ranges, provided that suitable habitat is available.

THREATS

An important aspect of conservation biology is analyzing and understanding the impact of human activities on biodiversity with the aim of easing, counteracting, or mitigating against these impacts. The threats to biodiversity differ among regions, and the degree of impact of such threats is a function of various factors, including the size and spatial distribution of the human population, per capita resources used, the technologies involved in resource extraction and use, and the characteristics of the natural systems. The threats that are thought to have the most serious impacts on biodiversity, and the methods that conservation biologists use to study them, are discussed below.

Habitat Loss, Degradation, and Fragmentation

Decrease in the quantity (total area) of species' habitats (habitat loss), decrease in habitat quality or suitability (habitat degradation), and division of species' habitats into smaller and often more numerous subdivisions or fragments (habitat fragmentation) are the most prominent threats to biodiversity. Habitat fragmentation (often accompanied by habitat loss and degradation) results in small, isolated populations that are prone to extinction. Such populations are extirpated more often (because of demographic stochasticity, Allee effects, and edge effects), and once extirpated, they are more rarely recolonized from other populations (because of unsuitable areas and other dispersal barriers that separate these isolated populations).

Predicting future habitat loss, degradation, and fragmentation is difficult; in some cases, landscape models can be used to project future patterns of land cover, based on past changes in land cover and information on human population demography and human land use. The effects of projected (or hypothesized) changes in a species' habitat on that species' viability can be estimated using metapopulation models that allow a changing number of populations in time (because fragmentation usually results in populations splitting into larger number of small populations), changing carrying capacities of each population (because habitat loss and degradation usually result in lower habitat suitability, which causes the same area of habitat to support fewer individuals), changing average survival rates and average fecundities (because lower habitat suitability may also cause these average demographic rates to decline in time), and changing dispersal rates (because habitat degradation may cause the landscape between populations to become more hostile, resulting in lower dispersal).

Climate Change

Human use of fossil fuels and widespread deforestation have caused increasing CO_2 concentrations in the Earth's atmosphere, which in turn is causing changes in the Earth's climate that are predicted to continue for centuries. Global climate change is already affecting many species' phenology, ranges, populations, and interactions with other species, and it has even caused or contributed to species extinctions. These impacts will presumably intensify in the coming decades and will affect ever-larger numbers of species. Attempts to predict the impact of global climate change have focused on integrating three different types of models: global circulation (or climate) models that predict future changes in temperature, precipitation, and other climatic variables; habitat (or bioclimatic) models that predict suitable habitat for a species (see "Describing, Identifying, and Mapping Habitat," above); and metapopulation models that predict the viability of the species (see "Estimating Extinction Risks," above). Because of the shifting and fragmentation that occurs in species ranges under climate change, the metapopulation models used for this purpose allow the same types of changes as discussed in the section "Habitat Loss, Degradation, and Fragmentation," above.

Exploitation

Many species are harvested, hunted, or fished for direct human use, often constituting a major source of income, calories, or protein for the human population exploiting them. Both the viability of these species and their long-term, sustainable use for human needs are conservation concerns. Many species are overexploited, which means that their populations cannot be maintained at the present rates of harvest. These include many marine fish species, marine turtles, some mammals and birds taken as bushmeat, and several plant and animal species taken for private collections and aquariums. Information about the population dynamics and life history of these species and data on their rates of exploitation can be used to develop population models to assess the impacts of harvest and to calculate sustainable harvest rates that ensure long-term persistence and sustainable yields. To be accurate, such models need to take into account various types of variability (environmental and demographic stochasticity as well as measurement error or sampling variability), interacting effects of harvest and density dependence on population parameters, and spatial patterns of harvest rates.

Invasive Species and Emerging Diseases

Increasing global commerce and transportation is causing many species to be inadvertently (but sometimes also deliberately) moved out of their natural ranges and released (or introduced) into new environments. Such introductions of nonnative species often fail (because the species is outside the area it evolved), but when they do not fail, the species may become invasive if the new range does not include species that may become its predators or competitors. Invasive species cause both economic and ecological damage, preying on, competing with, and hybridizing with native species, causing or facilitating the spread of diseases, and in some cases causing the extinction of native species.

Pollution

Industrial chemicals, agricultural chemicals (pesticides such as DDT and fertilizers), oil spills, and other human waste products destroy and degrade the habitats of many species. In addition, thermal pollution from power plants change the habitat of some aquatic species, light and noise pollution cause behavioral changes, and atmospheric pollution causes global climate change (see above).

The effects of chemical pollutants are often studied in laboratory experiments and measured in terms of growth, reproduction, and short-term survival of experimental organisms. Extrapolating the results of these short-term, individual-level studies to estimate impacts on natural populations is complicated and rarely done. Such extrapolation requires estimating population level parameters (survival, fecundity), untangling the effects of density dependence and toxicant-caused mortality, and developing theoretical methods that incorporate complex interactions and dynamics caused by spatial variation in pollution.

SEE ALSO THE FOLLOWING ARTICLES

Applied Ecology / Diversity Measures / Ecosystem Services / Ecotoxicology / Population Viability Analysis / Reserve Selection and Conservation Prioritization / Restoration Ecology

FURTHER READING

Groom, M., G. K. Meffe, and C. R. Carroll. 2006. *Principles of conservation biology*, 3rd ed. Sunderland, MA: Sinauer.

Hoffmann, M., C. Hilton-Taylor, and 172 other authors. 2010. The impact of conservation on the status of the world's vertebrates. *Science* 330: 1503–1509.

Hunter, M. L., Jr., and J. P. Gibbs. 2007. *Fundamentals of conservation biology*, 3rd ed. London: Blackwell.

Macdonald, D.W., and K. Service, eds. 2007. *Key topics in conservation biology.* London: Blackwell.
Mace, G.M., N.J. Collar, K.J. Gaston, C. Hilton-Taylor, H.R. Akcakaya, N. Leader-Williams, E.J. Milner-Gulland, S.N. Stuart. 2008. Quantification of extinction risk: IUCN's system for classifying threatened species. *Conservation Biology* 22: 1424–1442.
Primack, R.B. 2006. *Essentials of conservation biology*, 4th ed. Sunderland, MA: Sinauer.

CONSERVATION PRIORITIZATION

SEE RESERVE SELECTION AND CONSERVATION PRIORITIZATION

CONTINENTAL SCALE PATTERNS

BRIAN A. MAURER
Michigan State University, East Lansing

There are a number of intriguingly general patterns in biological diversity across continents, oceans, and island archipelagos. Study of these patterns is the province of macroecology. Recent advances in understanding the mechanics of large species assemblages have lead to a single, albeit complicated, explanation for these patterns. Many patterns can be explained by showing how different sampling protocols emphasize different aspects of the same complex system generated by the mechanics of how individual organisms distribute themselves across a spatially complex environment.

GENERAL CONTINENTAL AND OCEANIC PATTERNS OF SPECIES DIVERSITY

Distance–Decay Relationship

A nearly universal pattern is the relationship between the distance separating two local communities and their similarity in species composition. In nearly every taxon in which this pattern has been documented, there is a monotonic decline in diversity with increasing distance between any pair of communities. This decline in similarity can occur in the absence of any detectable changes in the environment; however, in most cases, there is also a corresponding decline in the similarity in environments inhabited by the communities (Fig. 1). Other than a monotonically decreasing function, there is no particular mathematical form that describes this decline in similarity among communities.

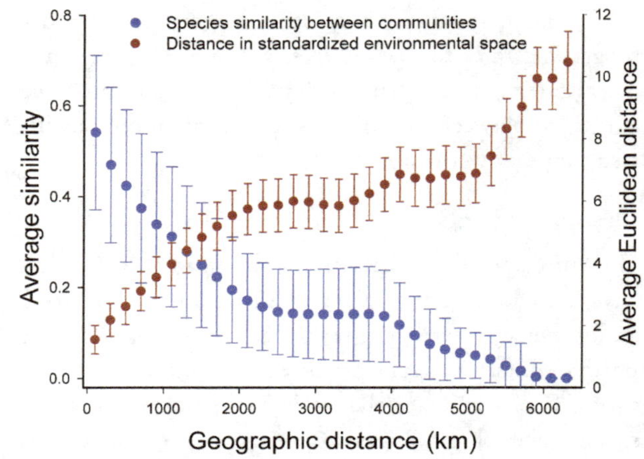

FIGURE 1 Decline in average similarity in terrestrial bird species composition with distance from 1393 surveys done during the breeding season across North America is given in blue. Similarity is defined as the minimum relative frequency of a species on a pair of sites summed across all species. Average similarities in environmental conditions as a function of distance are given in red. Similarity is expressed as the Euclidean distance in a 14-dimensional environmental space defined by standardized values of variables such as temperature, precipitation, and vegetation density. Note that greater Euclidean distance corresponds to lower similarity. Error bars represent two standard deviations.

Species–Area Relationship

It has long been known that as the area of a region sampled increases, so does the number of species. This increase in species number with area sampled follows a law-like pattern that can be approximated by a power relationship, that is, if S is the number of species in a region of area A, then

$$S = cA^z,$$

where z is usually in the range $0 < z < 1$. This pattern has been studied most often on oceanic islands, although it also holds when larger areas are sampled within a continent. For almost any type of habitat that is distributed in geographic space as discrete "islands," the relationship also holds. Even in relatively continuous regions, if samples of different area are arbitrarily selected and surveyed for species number, the species–area relationship is found. Although there are many minor deviations from the power relationship, the monotonic increase in species number with increasing area sampled is a pattern repeated many times in many different geographic regions for many different taxa.

Distribution–Abundance Relationship

Generally, species that are widespread across large expanses of geographic space or across a set of oceanic islands also tend to be the most abundant species wherever they are found. This pattern is greatly modified by a number of factors, most notably by body size, since all else being equal, species of large body size tend to be less common that related species of smaller size. This pattern is strongest when an ensemble of similarly sized habitat patches are considered, but it is also evident when the sizes geographic ranges of a group of related species is compared to the average abundances of those species within their geographic ranges. The latter relationship often shows a greater degree of variability than the former.

A closely related pattern is the observation that species that are locally abundant are also among the most abundant species within a geographic region. There are often species that deviate from this pattern; for example, some species of plants may be common where they are found but may be found at only a few geographic locations. Nevertheless, when their abundance is averaged across a sufficiently large geographic region, such species more closely fit the general pattern.

The positive relationship between distribution and abundance leads to a pattern called nested subsets that is related to the species–area relationship. That is, the species found on the smallest islands (or habitat patches) are only those species that are widespread throughout the ensemble of islands (Table 1). This subset of widespread species is a nonrandom subset of all species found in the ensemble. Conversely, the rarest species are found only on the largest or most productive islands. Ranking islands from largest to smallest reveals a set of nonrandom subsets of species where the subset of species found on all islands is the smallest subset. The subset of species on the largest islands contain all other subsets, with subsets of species on successively smaller islands being nested within subsets defined by larger islands.

Diversity Along Environmental Gradients

A number of patterns have been identified showing regular changes in species diversity across environmental gradients on continents and, recently, across environmental gradients in oceans. The most notable of these is the general increase in species diversity within many taxonomic groups with decreasing latitude. There are many exceptions, however, where species diversity within some taxa increases toward the poles. Proximate explanations for this pattern have focused on trying to relate high species diversity in tropical latitudes to various attributes of tropical locations such as high primary productivity, long-term stability of tropical ecosystems, and greater area of continental masses at tropical latitudes. None of these explanations have received universal support.

A related pattern is the observation that as different measures of primary productivity increase across a continent, species diversity also increases. Productivity has most often been represented by measure of evapotranspiration, a measure that measuring evapotranspiration, which integrates both solar radiation and water availability. Often, it is found that species diversity is highest for some intermediate level of productivity, though the decrease in diversity away from this intermediate level of productivity is asymmetric, being greater as productivity declines away from the diversity peak.

GENERAL EXPLANATION FOR THESE PATTERNS

All of the patterns described above have had numerous explanations posed for them. Very few of these explanations, however, invoke the basic mechanisms

TABLE 1

Nested Subset Pattern of Small Mammal Species in Lamar Cave, Yellowstone National Park, United States

Level	12	3	8	9	10	13	16	6	7	11	14	15	1	5	4	#levels
Marmota flaviventris	1	1	1	1	1	1	1	1	1	1	1	1	1	1	1	15
Neotoma cinerea	1	1	1	1	1	1	1	1	1	1	1	1	1	1	1	15
Microtus spp. = all	1	1	1	1	1	1	1	1	1	1	1	1	1	1	1	15
Spermophilus armatus	1	1	1	1	1	1	1	1	1	1	1	1	1	1	0	14
Sorel spp. = all shrews	1	1	1	1	0	1	1	1	1	0	0	1	0	0	0	9
Lepus townsendii	1	1	1	1	1	1	1	0	0	1	0	0	1	0	0	9
Tamias sp. = all spp	1	1	1	1	1	1	0	1	1	0	1	0	0	0	0	9
Sylvilagus spp. = all	1	1	1	1	0	1	0	1	0	1	1	0	0	0	0	8
*Zapus princeps**	1	0	1	1	1	0	1	0	0	0	0	1	0	0	0	6
Ochotona princeps	1	1	0	0	0	0	0	0	0	0	0	0	0	0	0	2
Mustela erminea	0	0	0	0	1	0	1	0	0	0	0	0	0	0	0	2
Richness	10	9	9	9	8	8	8	7	6	6	6	6	5	4	3	

NOTE: Levels refer to different deposition events within the cave over the past 3000 years. Each level was deposited over a different length of time. Each level represents a "time island." Levels are listed from left to right in descending order of thickness (time span over which deposition occurred). Notice that the levels spanning the shortest time (those on the right) contain the fewest species. Notice also that the most common species tend to be found across all levels. Data provided by E. Hadly.

that must underlie these patterns. Recently, it has become evident that there are several key components that must be included in any explanation. These components include some understanding of how environmental conditions vary across geographic space, how geographical populations of species respond to these conditions, and how these species responses result in the ensembles of species that make up these diversity patterns.

Capacity of the Environment

It is abundantly clear that the environment in which patterns of species richness occurs is spatially complex. Resources available for species are not uniformly distributed across space but exist as multiscaled patches that are fractal-like in structure but also are autocorrelated in space and time. Physical conditions are defined by geological and climatic processes, which result in complex sets of edaphic conditions to which autotrophs respond. Spatial patterns in primary productivity and nutrient availability are created by the responses of species of plants and microbes to these conditions. These complex spatial patterns in autotroph distributions result in complex spatial distributions of resources for consumers.

Geographic Population Dynamics

For a single species, the environment presents a complex mosaic of spatially and temporally autocorrelated resources. Within some subsets of this mosaic, there will be sufficient resources for local populations of the species to maintain themselves through having an excess of births over deaths in the face of environmental pressures (including competition, short-term environmental variation, and the like) For most species, some individuals within favorable portions of the environment will emigrate to other locations. Note that regions of the surrounding environmental mosaic may differ in their permeability to migrating individuals, which can profoundly affect the temporal dynamics of local populations. Groups of populations tied together by migration form metapopulations, which in turn may be aggregated into geographic populations. Patterns of abundance across geographic space are also spatially autocorrelated by the underlying migration dynamics that determine variation in local demography and by the autocorrelated birth–death dynamics resulting from autocorrelated resource distributions. Boundaries of geographic ranges are set when birth and immigration dynamics are not sufficient to offset losses due to death and emigration. Note that geographic range boundaries may fluctuate over time due to temporal changes in resources.

Metacommunity Dynamics

The complex geographic range dynamics of an ensemble of species that share resources within a region of a continent result in regional changes in the relative abundances of species. This ensemble of species has been called a metacommunity. The dynamics of such metacommunities in space and time underlie the spatial patterns of species diversity described above.

Consider first the distance–decay relationship. The decay in similarity of species diversity with increasing distance separating samples taken from a given location is a direct consequence of the spatial autocorrelation of the environment. Species composition within a local community is determined by which species geographic ranges overlap the community. But this occurs because the resources used by each species are available within the spatial boundaries of the community. Communities nearby will have similar resources, resulting in similar species composition. As the similarity in resources decays with increasing distance, range boundaries of species will be crossed, resulting in loss of some species and appearance of others.

Distance–decay relationships can theoretically occur even if species are nearly identical ecologically. Metacommunities that contain ecologically equivalent species are called neutral metacommunities. If local communities are of finite size, then random changes in abundances of different species within local communities (called ecological drift) cause species composition of local communities to diverge from one another over time. This rate of divergence is slowed, however, if species are ecologically different from one another. That is, there is community inertia to spatial and temporal change in species composition in nonneutral metacommunities.

Species–area relationships emerge from metacommunity dynamics because large areas typically cover the geographic range boundaries of more species than smaller areas. Since environmental similarity declines with distance, a large area will cover a greater range of ecological conditions than a smaller area. Hence, in a large area, there is a greater chance on average that a species in the metacommunity will find appropriate ecological conditions and resources in at least some part of the geographic region. This is true for any collection of geographic regions tied together by dispersal among them, including oceanic and habitat islands. In many oceanic islands, migration among islands is very restricted, and the increase in species richness with island area is much more gradual than in a comparable continental region (that is, many island species–area relationships have much smaller exponents than in continental regions).

Distribution–abundance relationships and related patterns are a consequence of asymmetries in ecological attributes among species such that some species are able to find more resources across geographic space than others. Thus, within metacommunities, the relative abundances of species will also be asymmetric, and such asymmetries in relative abundances in metacommunities are preserved even under short-term ecological drift. When a collection of local communities are sampled using sampling units of the same size (thus controlling for area effects), the asymmetries in relative abundances of species within the metacommunity will ensure that some species show up in samples more often than others (Fig. 2). Abundances of species at local scales will also reflect these asymmetries; hence, species that show up more often in samples will also tend to be represented by more individuals within those samples.

When examining species diversity along environmental gradients, the sampled geographic region may represent only a subset of sites contained within the metacommunity of the region. Since there is spatial autocorrelation in environmental conditions, sites that have more resources will typically have more species, and these sites will tend to be clustered together along the gradient. In more extensive samples, such as samples of species diversity with latitude, many metacommunities may be sampled, and the spatial autocorrelation among resources may be very low for regions at different extremes along the gradient. In such cases, the trends in species diversity observed across space must reflect differences in the capacity of the environment to support species.

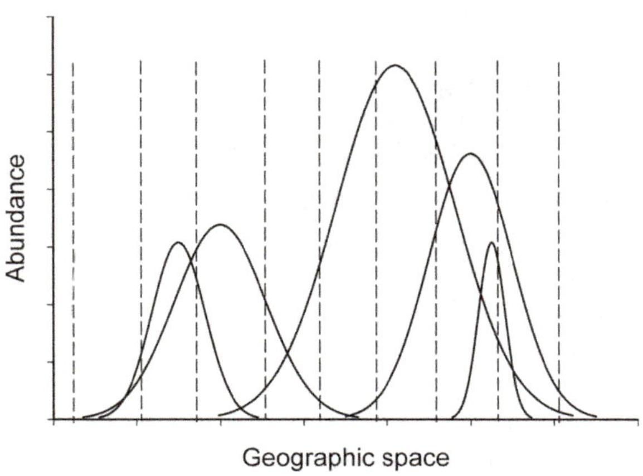

FIGURE 2 A schematic diagram showing how asymmetries in distributions of different species across geographic space result in differences in species diversity among a collection of sampling locations. Each unimodal curve represents the geographic distribution of a single species. Note that some species are found across a greater span of geographic space than others. This would be a consequence of the fact that some species will have attributes that will allow them to use a wider range of environmental conditions than others. The dashed lines represent geographic locations where a survey is taken. Surveys are assumed to be of standard size.

SEE ALSO THE FOLLOWING ARTICLES

Allometry and Growth / Diversity Measures / Metacommunities / Neutral Community Ecology / Species Ranges

FURTHER READING

Blackburn, T. M., and K. J. Gaston, eds. 2003. *Macroecology: concepts and consequences.* Proceedings of the 43rd Annual Symposium of the British Ecological Society (17–19 April 2002). Malden, MA: Blackwell Publishing.

Brown, J. H. 1995. *Macroecology.* Chicago: University of Chicago Press.

Gaston, K. J., and T. M. Blackburn. 2000. *Pattern and process in macroecology.* Malden, MA: Blackwell Science.

Hubbell, S. P. 2001. *The unified neutral theory of biodiversity and biogeography.* Princeton, NJ: Princeton University Press.

Huston, M. A. 1994. *Biological diversity.* Cambridge, UK: Cambridge University Press.

Maurer, B. A. 1999. *Untangling ecological complexity: the macroscopic perspective.* Chicago: University of Chicago Press.

Maurer, B. A. 2009. Spatial patterns of species diversity in terrestrial environments. In S. A. Levin, ed. *Princeton guide to ecology.* Princeton, NJ: Princeton University Press.

Rosenzweig, M. L. 1995. *Species diversity in space and time.* Cambridge, UK: Cambridge University Press.

CONTROL THEORY

SEE OPTIMAL CONTROL THEORY

COOPERATION, EVOLUTION OF

MATTHEW R. ZIMMERMAN, RICHARD MCELREATH, AND PETER J. RICHERSON
University of California, Davis

Cooperation occurs when individuals act together for beneficial results. Most evolutionary theory of cooperation addresses the evolution of altruism, the most difficult type of cooperation to explain. However, observed cooperation can also result from environments favoring other kinds of cooperation, including mutualism and coordination. Theoreticians model the evolution of cooperation by accounting for the effects of cooperative behavior at different levels of analysis, including the genes, the individual, and the group. Different mechanisms have been proposed as ones that encourage cooperative behavior, including limited dispersal, signaling, reciprocity, and biased transmission. All of these mechanisms are based on cooperators assortatively interacting with other cooperators.

TYPES OF COOPERATION

Mutualism

Mutualistic behavior is helping behavior for which the producer's costs are smaller than the producer's benefits. Intraspecific mutualisms lack clear incentives for individuals to withhold cooperation, and so natural selection always favors cooperation over noncooperation. (Community ecologists reserve the term mutualism for interspecies relationships, though here, as in evolutionary game theory generally, we use the term primarily for intraspecific interactions.) To see how mutualisms evolve, imagine a population of wolves that can either pair up to hunt deer cooperatively or hunt deer noncooperatively. (Of course, wolves can cooperatively hunt in larger groups, but the underlying logic of mutualisms holds even if we restrict our analysis to two individuals.) If either of the wolves hunts individually, they cannot bring down large game and go hungry. However, if the wolves hunt together, they can capture a deer and eat three units of meat.

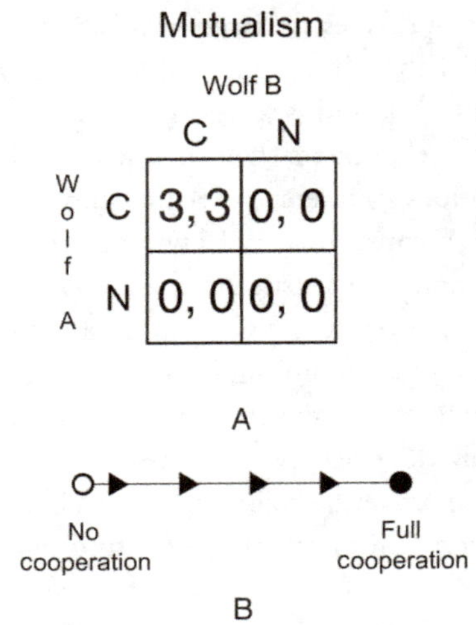

FIGURE 1 (A) A simple game illustrating mutualistic hunting in a hypothetical wolf population. Wolves get meat only if they hunt cooperatively, thus the benefits of cooperation are always positive. (B) The adaptive dynamics of the wolf population when individuals are randomly paired. Because it is always better to cooperate, natural selection increases the frequency of cooperative hunting away from the unstable noncooperative equilibrium to the stable equilibrium at full cooperation.

How would cooperation evolve in a population of wolves made up of cooperative hunters and noncooperative hunters? We will assume that wolves that are more successful hunters are more likely to produce offspring. The potential payoffs to the wolves in this situation are shown in Figure 1A. Both wolves have a choice of cooperative hunting (C) and noncooperative hunting (N). The numbers in the grid show the payoffs for each combination of behaviors, with the payoff to the Wolf A first, followed by the payoff to Wolf B. Notice that a wolf only has a chance at getting meat if it tries to hunt cooperatively. Thus, wolves that hunt cooperatively will tend to produce more offspring and the trait for cooperative hunting will tend to spread in the population.

The line in Figure 1B shows the adaptive dynamics of this wolf population. The arrows represent the direction of natural selection and show that, no matter where the mixture of hunting strategies starts, selection will tend to push the population toward a stable cooperative equilibrium, represented by the solid circle. The open circle at the other end of the line represents a noncooperative equilibrium where there are no cooperative wolves and they cannot reproduce. However, this equilibrium is unstable since any introduction of cooperative wolves, through genetic mutation or migration from other populations, would allow natural selection to push the population to the stable equilibrium.

In some models of mutualisms, since cooperative individuals always achieve greater benefits than noncooperative individuals, the evolution of cooperative behavior does not present much of a puzzle and does not require special explanations for why they persist in a population. However, the origins of mutualisms can be more difficult to explain, as initial benefits may not be sufficient to exceed individual costs. For example, if in a proto-wolf population hunters are mainly solitary, then it stands to reason that the skills necessary for successful cooperative hunting will not exist. Therefore, early cooperative hunters are not assured the large gains that may arise later, once cooperation gets a foothold through other mechanisms. In this way, other types of cooperation, such as coordination (discussed next), can evolve over time into mutualisms, once cooperation is stabilized through other means.

Coordination

In coordination contexts, natural selection favors common behavior, whether it is cooperative or noncooperative. For example, imagine that our population of wolves lives in an environment with both deer and rabbits. Wolves now have two strategies: hunt deer cooperatively, or solitarily hunt rabbits. If a wolf and its partner hunt cooperatively, they manage to catch a deer with three units of meat apiece as before. If, however, one of the wolves hunts deer and the other hunts rabbits, the deer hunter fails to bring down the game, but the rabbit hunter catches two rabbits for two units of meat. However, if both wolves hunt rabbits individually, they are competing for the easy game and only receive one unit of meat each. The payoffs for this situation are shown in Figure 2A.

In this wolf population, the direction of natural selection depends on the current level of cooperative hunting. For example, if all the other wolves in the population are cooperatively hunting deer, a rogue rabbit hunter will have less meat and will therefore be less reproductively successful than the rest of the population. Thus, natural selection works against rabbit hunters. However, if all the other wolves in the population are hunting rabbits, then a lone deer hunter will not find a cooperative hunting partner and starve. Natural selection, in this case, works against the deer hunters.

The line in Figure 2B illustrates the adaptive dynamics of this wolf population. There are two stable equilibria. If deer hunters are more common in the population, natural selection drives the population toward the cooperative deer-hunting equilibrium. If rabbit hunters are more common, natural selection drives the population toward the noncooperative rabbit-hunting equilibrium. These equilibria are stable because the introduction of a nonconforming wolf to the population will have lower payoffs than the rest of the population obtains and thus be outcompeted. Because, in coordination, the equilibrium preferred by natural selection depends on the frequency of behaviors in the population, it is called frequency dependent. (There is also an unstable equilibrium where there are exactly as many deer hunters as rabbit hunters, but this equilibrium will not persist, as natural selection will tend to drive the population away from the center as soon chance factors cause this perfect balance to be broken.)

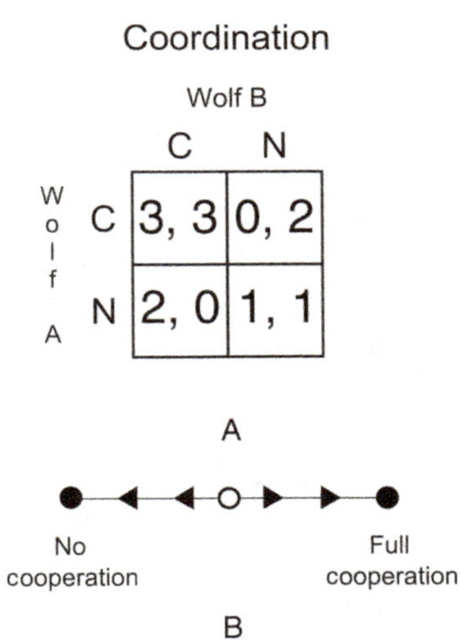

FIGURE 2 (A) A simple game illustrating a coordination scenario in a hypothetical wolf population. When individuals are randomly paired, wolves can expect to get the most meat when they have the most common behavior in the population. (B) The adaptive dynamics of the randomly paired wolf population. When cooperation is common, natural selection drives the frequency of cooperative behavior toward the stable equilibrium at full cooperation. However, when cooperation is rare, natural selection drives the frequency of cooperative behavior toward the stable equilibrium where none of the wolves cooperate. The location of the unstable equilibrium in this model depends on the payoffs in (A) and the assumption that the fitness effect of meat consumption is linear.

This situation shows why coordination can be a more complicated type of cooperation than mutualism. A population of rabbit-hunting wolves would be better off if they just cooperated and hunted deer, in that they would all obtain three units of meat instead of one, but natural selection discourages the population from moving toward the cooperative equilibrium. In resolving this dilemma of coordination, evolutionary theorists focus on equilibrium selection, how a population might move, through the processes of natural selection or drift, from the noncooperative to the cooperative equilibrium. Once a population moves to the cooperative equilibrium, it is less difficult to explain its maintenance. However, our hypothetical wolf population faces a particularly hard case of equilibrium selection since trying to hunt deer without a partner has zero payoff when most wolves are hunting rabbits. It is thus much riskier for a wolf to switch from a noncooperative to a cooperative behavior than the other way around, and the population might spend more time in the noncooperative equilibrium. A population transitioning from the noncooperative equilibrium to the cooperative equilibrium may require mechanisms similar to the evolution of altruism, as described below.

Altruism

A helping behavior is altruistic if the individual fitness costs of the behavior exceed the individual fitness benefits. Altruism is a context in which natural selection on individuals appears to favor noncooperation, even though average fitness across the population is assumed to increase as more individuals cooperate. Altruism is thus a strategic setting that directly opposes the interests of individuals and the interests of groups. Again imagine our hypothetical population of wolves, now living in an environment where rabbits are larger and more abundant. Now, if a wolf hunts rabbit and its partner hunts deer, it gets four units of meat while its partner gets zero. If they both hunt rabbits, they each get two units of meat. If they both hunt deer cooperatively, they each get three units of meat. These payoffs are reflected in Figure 3A. As might be expected, if the benefits of hunting rabbits are higher, then natural selection more strongly favors hunting rabbits over cooperatively hunting deer. In this case, hunting rabbits is always a better strategy for an individual wolf than hunting deer. If a wolf's partner hunts rabbit, also hunting rabbits nets it two units of meat compared to going hungry when hunting deer. If a wolf's partner hunts deer, rabbit hunting nets it four units of meat compared to three for hunting deer. Since hunting rabbits always produces higher payoffs for the individual, natural selection should push the population toward a stable equilibrium of noncooperation, as shown in the line under the payoff matrix in Figure 3B.

Notice that if all wolves in the population hunt rabbits, they all gain two units of meat. But if they all cooperatively hunted deer, they would all have three units. Thus, the equilibrium favored by natural selection is not where the wolf population has the highest total payoff. This type of scenario is the popular model of altruism called a "Prisoner's Dilemma" (named after a hypothetical dilemma faced by prisoners who have to decide whether to rat out their accomplices during police questioning). In the Prisoner's Dilemma (PD), the cooperative strategy is an example of an altruistic act because it requires giving up the individual benefits of hunting rabbits in order to try to achieve the collective benefits of hunting deer.

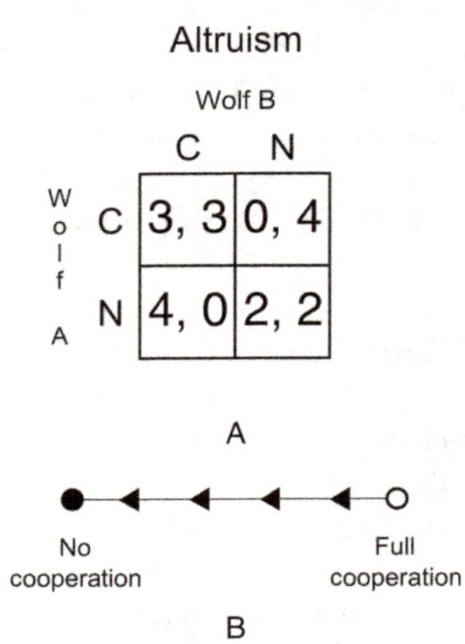

FIGURE 3 (A) A simple game illustrating the difficulty in explaining the evolution of altruistic behavior. Wolves always have a higher payoff when they behave noncooperatively, regardless of what their partner does. (B) The adaptive dynamics in a randomly paired wolf population. Because the individual wolf always gets more meat when it behaves noncooperatively, natural selection drives the frequency of cooperative behavior to zero.

Hunting rabbits is called "defection" because the rabbit hunter does best when it exploits a partner's efforts to hunt cooperatively and forces the cooperative hunter to starve. Starvation, in this scenario, is called the "sucker's payoff."

Because noncooperation appears to be the only stable evolutionary equilibrium in the PD, the conditions under which altruism can evolve in situations that resemble PD are more challenging than trying to understand the evolution of coordination or mutualistic behavior. Because altruism is the most difficult type of cooperation to explain, the evolution of altruism has received by far the most theoretical attention. In fact, many biologists think of altruism as synonymous with cooperation. It is thus important for field ecologists interested in testing theory not to assume that all observed cooperative behavior is altruistic, since this implies evolution has solved a harder problem than it actually has. For example, to understand the evolutionary history of cooperative hunting in our hypothetical wolf population, it would be important to determine and quantify the other hunting options available. However, in practice, quantification of the costs and benefits of animal behavior is often difficult, especially theoretically relevant but unobserved behavior.

Threshold Cooperation

Threshold cooperation is a strategic context in which cooperation is favored when rare but selected against when common. There is a threshold frequency of cooperation at which selection changes direction. In all of the above examples, our hypothetical wolf pack always evolves toward either full cooperation or full noncooperation. This occurs because each of the simple game models favors pure types. But many populations are made up of mixed populations of both cooperators and noncooperators. For a simple model of how this can happen, imagine that to successfully take a deer, our hypothetical wolves need to first find a deer herd. Although both wolves are needed to bring down a deer, they can split up to search more efficiently, and the wolf that first spies a group of deer can signal the other to join in the kill. If they are successful in the kill, the wolves each get a payoff of three units of meat as above. However, one or both of the wolves might shirk their searching duties and, instead, spend the time hunting rabbits and leaving their partner to find the herd. If both wolves shirk searching, they never find deer and each get one unit of meat from hunting rabbits. If only one wolf searches, it is less efficient and it metabolizes, on average, the equivalent of one unit of meat, receiving a net payoff of two units. The other wolf receives one unit of meat from hunting rabbits and three from the deer, for a total of four units. The payoffs for this scenario are in the game matrix in Figure 4A.

This type of game is variously called the hawk–dove or snowdrift game in biology or the game of chicken in economics. These games share the feature of favoring cooperation when cooperation is rare but favoring noncooperation when cooperation is common. In this scenario, if all the wolves are cooperative searchers, then wolves that shirk cooperative searching acquire more meat (four units to two) and thus leave more offspring. So shirking behavior should spread. However, if none of the other wolves search, a wolf that does should acquire more meat than all of the noncooperative wolves except for its partner (two units to one). So searching behavior would spread. If noncooperative shirking spreads when it is rare, and cooperative searching spreads when it is rare, natural selection pushes the population away from homogenous equilibria. Figure 4B shows the adaptive dynamics of this behavior. The population moves to a stable equilibrium where there are both cooperators and noncooperators.

This game can be seen as a mild version of altruism. Behaving cooperatively is mutualistic when cooperation is rare in a population. However, cooperation becomes altruistic when it is already common. The challenge for the theorist is explaining how cooperation might evolve above what is expected at the equilibrium depicted in Figure 4B. The challenge for the empiricist is to determine, given cooperation observed in a population, if it is above the level expected given cooperation's individual payoffs.

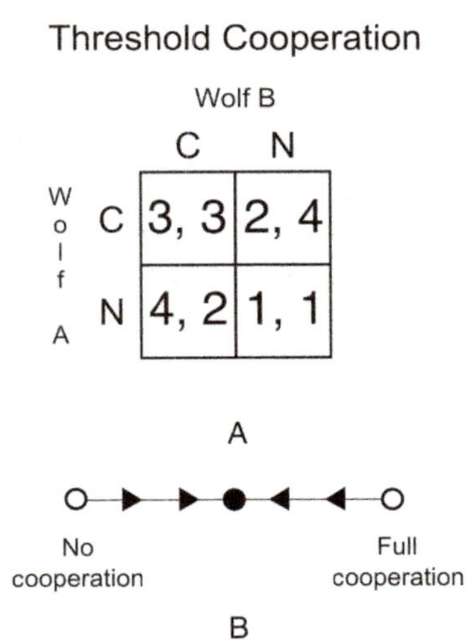

FIGURE 4 (A) A simple game illustrating threshold cooperation. Here wolves have higher payoffs when their behavior is different than the majority of other wolves in the population. (B) In this scenario, when cooperation is rare, natural selection increases its frequency in the population. When cooperation is common, natural selection decreases its frequency in the population. This creates a stable equilibrium of both cooperators and noncooperators in the wolf population. The location of the stable equilibrium in this model depends on the payoffs in (A) and the assumption that the fitness effect of meat consumption is linear.

MODELING THE EVOLUTION OF COOPERATION

When modeling the evolution of cooperation, the modeler must account for all of cooperative behavior's effects on the fitness of possibly many individuals. There are two dominant and mathematically equivalent styles of analysis for accomplishing this objective: inclusive fitness and multilevel selection. The inclusive fitness approach accounts for fitness effects from the perspective of individual genes, focusing on the effects of genes coding for cooperative behavior on related individuals, who may also share copies of these genes. The multilevel selection approach instead distinguishes fitness effects at the individual and the group level—for example, examining how some packs of cooperative wolves might outcompete packs of noncooperative wolves.

Modeling in either of these approaches will yield equivalent answers, as long as the modeler properly accounts for all fitness effects. For example, as long as the modeler is careful, modeling the evolution of cooperation at the gene level will give the same result as modeling cooperation as a tug-of-war between individual and group levels. Thus, old debates concerning the "correct" style of analysis have largely faded in biology. A modeler instead selects styles of analysis that are most useful to understanding a particular question or that are the most mathematically convenient. For example, modeling individual-level selection could be the most useful approach if populations lack social structure and close relatives are widely dispersed. In that situation, the modeler may only need to understand a behavior's effect on the "classical fitness" of an organism, or its direct number of offspring. When these assumptions are relaxed, analyzing fitness at other levels of analysis might be more useful.

Kin selection models adjust individual fitness to be a special sum of the negative effects of behavior on self and the positive effects on relatives. If an individual behaves altruistically, the cost of the behavior decreases its classical fitness. However, the benefits of altruism increase the fitness of others, and if these others are genetically related, they may also have copies of the altruism gene. Since kin share genes from a recent common ancestor, genes favoring altruistic behavior toward kin can increase in the next generation as long as the net benefits of altruism are high enough relative to the cost. For example, if a gene encourages altruistic hunting in wolves, when a wolf altruistically hunts with a sibling who shares the gene, the gene might be more likely to spread to the next generation even if altruistic behavior hurts the fitness of the original wolf. Kin selection models like this are said to consider inclusive fitness, which includes both an organism's classical fitness and the discounted fitness of its kin.

The most famous kin selection model, called Hamilton's Rule, was proposed by and named after biologist William Hamilton. It states that cooperation should spread in a population when

$$rb > c.$$

Here, c is the cost of an altruistic act to the altruistic individual, b is the benefit of the act to a target of the altruistic act, and r is the degree of relatedness between the altruist and the target of altruism. This equation implies that the less related you are to individuals you interact with, the less likely you are to behave cooperatively. Empirical work in understanding altruism in nature has been focused on trying to estimate r, b, and c for specific behaviors and determine how well they conform to Hamilton's rule. For example, a researcher could genetically sample a population of wolves to estimate r and then try to quantify the reproductive costs and benefits of cooperative hunting. With modern genetic techniques, estimating r is often much easier than estimating b or c. Unfortunately, this has led many researchers to ignore b and c when testing Hamilton's Rule, but as we have seen, determining the costs and benefits of cooperation are important for whether organisms are behaving mutualistically, altruistically, or coordinating their behavior.

In contrast, the multilevel selection approach divides the consequences of natural selection into within-group (individual) and between-group levels. For example, wolf packs with more altruistic hunters can achieve higher payoffs, increase in number, and displace wolf packs with less altruistic hunting. However, within a pack, altruistic wolves will have lower fitness than nonaltruistic wolves. Thus, selection at the group level tends to increase altruistic behavior, and selection at the individual level tends to decrease it. Any model of natural selection, whether it is about cooperation or not, can be split up in this way.

Which of these levels, individual or group, dominates depends on the magnitude of fitness effects at each level, as well as on how much variation is present at each level. For example, if all groups are identical in their levels of altruism, then selection at the group level cannot change the frequency of altruistic genes in the population. Similarly, if groups are different in their levels of altruism but individuals within groups are all identical to others within their own groups, selection at the individual level cannot change the frequency of altruism genes.

Since group membership is often a good predictor of relatedness, kin selection models can always be reformulated as multilevel selection models, and vice versa. The choice of model depends primarily on which formulation is easier to construct, is more informative to the research question, is more open to empirical analysis, or better accounts for the relevant mechanisms that encourage cooperation in the population. Note, however, that both approaches, inclusive fitness and multilevel selection, are steady-state approximations of true evolutionary change. They are equivalent, but not exhaustive, ways of modeling the evolution of cooperation.

MECHANISMS OF POSITIVE ASSORTMENT

Much of the effort expended to understand the evolution of altruism has been to specify various mechanisms that favor its evolution or maintenance. A mechanism in this context is a strategy for achieving positive assortment of altruistic strategies. However one chooses to model the evolution of altruism, it cannot evolve unless individuals carrying altruistic genes tend to direct the benefits of altruism to other individuals carrying altruistic genes. For example, imagine that in our pack of wolves, altruistic hunters could see into the genomes of other wolves and were thereby able to pair only with other altruistic hunters. Without the risk of defection, altruistic hunters would have higher payoffs and increased numbers of offspring relative to nonaltruistic wolves interacting with other nonaltruists. While wolves cannot actually peer into one another's genomes, animals have other mechanisms that increase positive assortment.

Limited Dispersal

One mechanism that can generate positive assortment among altruists is when organisms do not disperse far from where they are born. When this occurs, altruists have greater opportunity to interact with related altruists, because kin will share space. For example, wolves may be more likely to live in packs with relatives, because many pups remain in their natal groups. This means that individuals will naturally find themselves in groups with close kin, creating positive assortment, without any need for the ability to know who their relatives are.

However, limited dispersal alone can also enhance competition among relatives and reduce altruism. Understanding the full effects of limited dispersal also requires an understanding of how resources are distributed. If resources are locally limited, then an individual's primary reproductive competitors may also be its group mates. This localized resource competition can cancel or reverse the ability of limited dispersal to favor the evolution of altruism. It is therefore difficult to predict altruistic behavior in organisms based solely on measurements of dispersal. What is required is that altruists positively assort for social behavior but avoid one another when competing for resources.

Signaling

Signaling has also been proposed as a mechanism for generating assortment among altruists. If altruists can signal their altruism to each other, they can coordinate their interaction and avoid nonaltruists. These signals are sometimes called green beards after a hypothetical mechanism by which all altruists grow and display green beards and always help others who display green beards. Altruists can thus recognize and cooperate with each other, creating the required positive assortment. However, green beard–type signals are probably very rare in nature. It is hard for organisms to evolve green beards unless there is a way to keep nonaltruists from copying the signal. If we imagine a population of green beard altruists cooperating with each other, any individual who can display the green beard but avoid costly assistance to other green beards will have high fitness, which will eventually decouple the signal from the altruistic behavior. Because of this problem, altruistic behavior based on these signals are generally considered short-lived phenomena in complex organisms. However, in some simple organisms such as bacteria, there are possible empirical examples of single genes that both create signals and contingently help others producing the signal. Another exception might be the use of socially learned signals, such as human dialect and languages, that are so complex that outsiders cannot easily fake them.

A more commonly discussed signaling mechanism that supports altruistic behavior is kin recognition. Instead of organisms using a universal signal for altruistic behavior, organisms can recognize kin by shared traits and preferentially cooperate with them. While green beard mechanisms attempt to signal the altruism allele itself, kin recognition mechanisms attempt to recognize close kin who may share the altruism allele. If altruistic wolves remember their littermates and their littermates are genetically related to them, then the littermates that interact with one another will positively assort for altruism. There are kin recognition strategies that use uninherited life history events correlated with kinship, such as sharing a mother, and alternative strategies that use biologically inherited markers such as facial similarity. However, current models suggest that kin recognition by inherited markers can be difficult to evolve.

Reciprocity

Reciprocity is a mechanism by which individuals can use a history of past interactions to predict an individual's probability of altruistic behavior. If wolves remember which of the members of their pack hunted altruistically in the past, they may choose to hunt with them in the future. A common framework for modeling reciprocal altruism is the iterative Prisoner's Dilemma (IPD). In the IPD, individuals play the Prisoner's Dilemma repeatedly with the same partner. Individuals are able to remember their past interactions and use that information to decide on their behavior in the next round.

A famous instance using the IPD to understand the evolution of reciprocity was two tournaments run by Robert Axelrod. Axelrod allowed participants from all over the world to submit candidate strategies for playing the IPD. The winning strategy for both tournaments, called Tit-for-Tat (TFT), was also the simplest. The strategy cooperated on the first turn and copied its opponent's behavior in the previous round on any subsequent turn. In other words, it would cooperate as long as its opponent was altruistic and defect as long as its opponent was nonaltruistic. In effect, it was a strategy that used a simple form of reciprocity based on its memory of one previous interaction.

Because of the success of TFT in Axelrod's tournament, many people think of it as the best way to play an IPD. However, later work has confirmed that TFT is not a robust strategy. TFT is not evolutionarily stable and can be reduced in frequency in a population through invasion by other cooperative strategies. For example, in a population made up only of TFT-playing individuals, a strategy that always cooperates (ALLC) will have the same payoffs as TFT, and natural selection will not act against it. In fact, if there is any cost to the cognition and memory of TFT, ALLC will have the advantage. Once ALLC becomes common enough, nonaltruistic strategies can take over the population by exploiting its lack of contingent defection. Thus, TFT is not always a successful strategy in the IPD.

TFT is not alone in its failings, however. In general, there can be no master reciprocity strategy, because many forms of contingent aid can coexist and rise and fall in a population, depending upon the details of the model. What can be said, however, is that when animals can identify individuals and remember past interactions, some form of reciprocity that makes altruism contingent upon past behavior can often evolve. Successful strategies will often be "nice," in the sense that they begin by cooperating in hopes that other individuals are also reciprocal.

Although the IPD model of reciprocity relied on an individual's memory of its past interactions with others, this might not be a useful mechanism in populations where repeated interactions with the same individual are rare. In this case, individuals can obtain similar information about another's probability of altruistic behavior from the other's interactions with third parties. This mechanism for assortment is called indirect reciprocity. A downside to indirect reciprocity as an assortment mechanism is that for it to work as well as regular pairwise reciprocity, it requires that reputation about third parties be as accurate as direct information one obtains from personal experience. This constraint likely limits its evolution to organisms with high-fidelity communication, such as humans.

Culture and Social Learning

Thus far, the mechanisms we have discussed to explain altruistic behavior apply to any inheritance system, and for many species the most important inheritance system is genetic transmission. However, the behavior of some species, especially humans, is heavily influenced by socially learned information, or culture. The mechanisms by which cultural inheritance can produce altruistic behavior are similar to genetic inheritance. However, because culture can spread more quickly than genes to many different individuals, large groups may be much more culturally related than genetically related. This allows for altruism to evolve culturally in much larger groups than is possible with genetic transmission.

This is especially true when cultural transmission is adaptively biased. For example, under many situations it makes sense for individuals to learn the most common behaviors in their group (called a conformist bias). This allows them to quickly adapt to local conditions or contexts where coordination is important. If conformist biases are common in a species, this maintains strong cultural relatedness within a group by discrimination against learning the traits of rare migrants, which strengthens the effect of selection for altruism at the group level.

SEE ALSO THE FOLLOWING ARTICLES

Adaptive Dynamics / Behavioral Ecology / Evolutionarily Stable Strategies / Game Theory

FURTHER READING

Axelrod, R. 1984. *The evolution of cooperation.* New York: Basic Books.

Clutton-Brock, T. 2009. Cooperation between non-kin in animal societies. *Nature* 462: 51–57.

Dugatkin, L. A. 2006. *The altruism equation.* Princeton: Princeton University Press.

Frank, S. A. 1998. *Foundations of social evolution.* Princeton: Princeton University Press.

Lion, S., V. Jansen, and T. Day. 2011. Evolution in structured populations: beyond the kin versus group debate. *Trends in Ecology and Evolution* 26: 193–201.

Maynard Smith, J. 1982. *Evolution and the theory of games.* Cambridge, UK: Cambridge University Press.

Maynard Smith, J., and E. Szathmáry. 1995. *The major transitions in evolution.* Oxford: Oxford University Press.

McElreath, R., and R. Boyd. 2007. *Mathematical models of social evolution.* Chicago: University of Chicago Press.

Richerson, P., and R. Boyd. 2005. *Not by genes alone.* Chicago: University of Chicago Press.

Skyrms, B. 2003. *The stag hunt and the evolution of social structure.* Cambridge, UK: Cambridge University Press.

Sober, E., and D. S. Wilson. 1999. *Unto others.* Cambridge, MA: Harvard University Press.

D

DELAY DIFFERENTIAL EQUATIONS

YANG KUANG
Arizona State University, Tempe

All processes take time to complete. While physical processes such as acceleration and deceleration take little time compared to the times needed to travel most distances, the times involved in biological processes such as gestation and maturation can be substantial when compared to the data-collection times in most population studies. Therefore, it is often imperative to explicitly incorporate these process times into mathematical models of population dynamics. These process times are often called delay times, and the models that incorporate such delay times are referred as delay differential equation (DDE) models.

CONCEPTS AND NOTATION

Recent theoretical and computational advancements in delay differential equations reveal that DDEs are capable of generating rich and plausible dynamics with realistic parameter values. Naturally occurring complex dynamics are often naturally generated by well-formulated DDE models. This is simply due to the fact that a DDE operates on an infinite-dimensional space consisting of continuous functions that accommodate high-dimensional dynamics. For example, the Lotka–Volterra predator–prey model with crowding effect does not produce sustainable oscillatory solutions that describe population cycles, yet the Nicholson's blowflies model can generate rich and complex dynamics.

DDEs are differential equations in which the derivatives of some unknown functions at present time are dependent on the values of the functions at previous times. Mathematically, a general delay differential equation for $x(t) \in R^n$ takes the form

$$\frac{dx(t)}{dt} = f(t, x_t),$$

where $x_t(\theta) = x(t+\theta)$ and $-\tau \leq \theta \leq 0$. Observe that $x_t(\theta)$ with $-\tau \leq \theta \leq 0$ represents a portion of the solution trajectory in a recent past. Here, f is a functional operator that takes a time input and a continuous function $x_t(\theta)$ with $-\tau \leq \theta \leq 0$ and generates a real number $(dx(t)/dt)$ as its output. A well-known example of a delay differential equation is the Hutchinson equation, or the discrete delay logistic equation, $x' = rx(1 - x(t-\tau)/K)$. Some DDEs can be conveniently solved in a stepwise fashion. In fact, the Hutchinson equation can be rewritten as $(\ln x)' = r(1 - x(t-\tau)/K)$, which can be used to solve for x for $0 \leq t \leq \tau$. Some DDEs, such as $x'(t) = rx(t)[1 - a\int_{-\infty}^{0} e^{as}x(t+s)ds/K]$, $a > 0$, are in fact a system of ordinary diffential equations (ODEs) in disguise. This can be seen by letting $y(t) = a\int_{-\infty}^{0} e^{as}x(t+s)ds$ and noticing that $y' = a(x(t) - y(t))$, which yields a system of ODEs $x'(t) = rx(t)(1 - y(t)/K)$; $y' = a(x(t) - y(t))$. Indeed, an integro-differential equation of the form $x'(t) = f(t, x(t)) + \int_{-\infty}^{0} k(s)g(x(t+s))ds$ with initial condition $x_t(\theta)$ where $-\infty < \theta \leq 0$ is equivalent to a system of ODEs with initial condition if k is a linear combination of functions $e^{at}, te^{at}, t^2e^{at}, \ldots, t^m e^{at}$, where a is a real number and m is a positive integer. The method of reducing such a delay differential equation into a system of ODEs is called the linear chain trick.

Numerically solving most delay differential equations or systems is almost as simple as solving ODEs. The popular MATLAB-based dde23 solver developed by Shampine and Thompson for delay differential equations is well tested and user-friendly. Interested readers can find many familiar and informative examples at the website http://www.radford.edu/thompson/webddes/ddetutwhite.html, and more sophisticated users can find additional information at http://www.radford.edu/thompson/webddes/.

As with linear ODEs, stability properties of linear DDEs can be characterized and analyzed by studying their characteristic equations. For example, the characteristic equation for $x'(t) = ax(t) + bx(t - \tau)$ is $\lambda - a - be^{-\lambda\tau} = 0$. The roots λ of the characteristic equation are called characteristic roots. Notice that the root appears in the exponent of the last term in the characteristic equation, causing the characteristic equation to possess an infinite number of roots. However, there are only a finite number of roots located to the right of any vertical line in the complex plane.

SOME CHARACTERISTICS OF DDEs

In most applications of delay differential equations in the sciences, the need for incorporating time delays is often due to the presence of process times or the existence of some stage structures. In engineering applications, such time delays are often modeled via high-dimensional compartment models. In life-science applications, compartmental models can present the additional challenges of estimating some of the involved parameter values. In such cases, low-dimensional delay differential models with fewer parameters can be sensible alternatives.

Since the through-stage survival rate is often a function of such time delays, it is easy to see that these models may involve some delay-dependent parameters. The ubiquitous presence of such parameters often greatly complicates the task of a systematic study of such models. In some special cases, the stability of a given steady state can be determined by the graphs of some functions of time delay that can be expressed explicitly and thus can be depicted. The common scenario is that as time delay increases, stability changes from stable to unstable to stable, implying that a large delay can be stabilizing. This scenario often contradicts the one provided by similar models with only delay-independent parameters.

In addition, a closer look at the cause of a time delay often suggests that the time delay itself maybe dependent on some key model variables. In short, the delays are state dependent. These state-dependent delay differential equations are notoriously difficult to study mathematically. However, they may possess some surprising and more plausible dynamics.

SOME SIMPLE DELAY DIFFERENTIAL EQUATION MODELS

Many consumer species go through two or more life stages as they proceed from birth to death. In order to capture the oscillatory behavior often observed in nature, various models are proposed. They include many difference models and delay differential models. The Hutchinson equation,

$$x' = rx(1 - x(t - \tau)/K), \tag{1}$$

and its variations are among the ones that are most frequently employed in theoretical ecology models. In Equation 1, r is the growth rate, K is the carrying capacity, and τ is a time delay that may have no real biological meaning. Like logistic equations, these models are ad hoc and hence can be misleading. Indeed, they produces artificially complex dynamics such as excessive volatility and huge peak-to-valley ratios (Fig. 1).

On the other hand, if we assume the adults have a constant birth rate of r, the newborns mature in τ units of time, and the mortality rate is proportional to the adult population density, then the following model may be a reasonable model for the adult population:

$$x' = rx(t - \tau)e^{-m\tau} - rx^2/K. \tag{2}$$

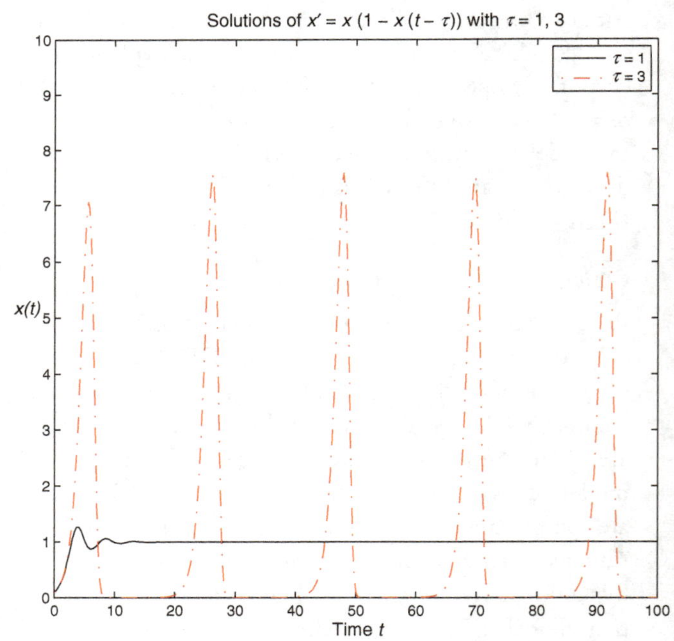

FIGURE 1 Solutions of the Hutchinson equation with delay values of 1 and 3. Note that the peak-to-valley ratio is well over 2000 when the delay is 3.

However, the positive steady state of model 2 is always (globally) stable, similar to the case when the delay is zero. On the other hand, the well-known Nicholson blowflies model,

$$x' = px(t - \tau)e^{-ax(t-\tau)} - mx, \quad (3)$$

exhibits plausible and rich dynamics. In other words, the dynamics of delay differential equations are extremely sensitive to model forms.

To further support the above statement, let us now examine some predator–prey models with age structure. We assume that the prey or the renewable resource, denoted by x, can be modeled by a logistic equation when the consumer is absent. The predators or consumers are divided into two age groups, juveniles and adults, and they are denoted by y_j and y, respectively. We also assume that only adult predators are capable of preying on the prey species and that the juvenile predators live on other resources. We then have the following two-stage predator-prey interaction model:

$$\begin{aligned} x' &= rx(1 - x/K) - yp(x), \\ y' &= be^{-d_j\tau} y(t - \tau)p(x(t - \tau)) - d_a y - my^2. \end{aligned} \quad (4)$$

With the aid of the geometric stability switch criteria that were specifically developed to deal with models with delay-dependent parameters, it can be shown that this model generates increasingly more complex dynamics, as its characteristic equation produces more roots with positive real part when we increase the time delay from 0.25 to 25 (Fig. 2). If we assume that the maturation time delay in population dynamics is determined by the resource uptake, then we may have

$$\int_{-\tau}^{0} p(x(s))ds = M \quad (5)$$

for some positive constant M that measures the resource requirement for a newborn to mature. With this additional reality, solutions of model 4 tend to a steady-state or a limit-cycle dynamics. In addition, the time to approach the limit cycle is much shorter than a typical model without time delay or with constant time delay, suggesting that the more realistic formulation of time delay (Eq. 5) satisfactorily describes the often-observed short duration of transition dynamics in nature (Fig. 3).

Delay differential equation models can be more effective and accurate compared to ordinary differential equation–based models when it is necessary to capture oscillatory dynamics with specific periods and amplitudes. This characteristic has been successfully employed to explain why lemmings often have a 4-year cycle whereas snowshoe hares have a 10-year cycle, and why the putative cycles of the moose–wolf interactions on Isle Royale, Michigan, is 38 years long. In addition, some simple and plausible models with two time delays can generate the

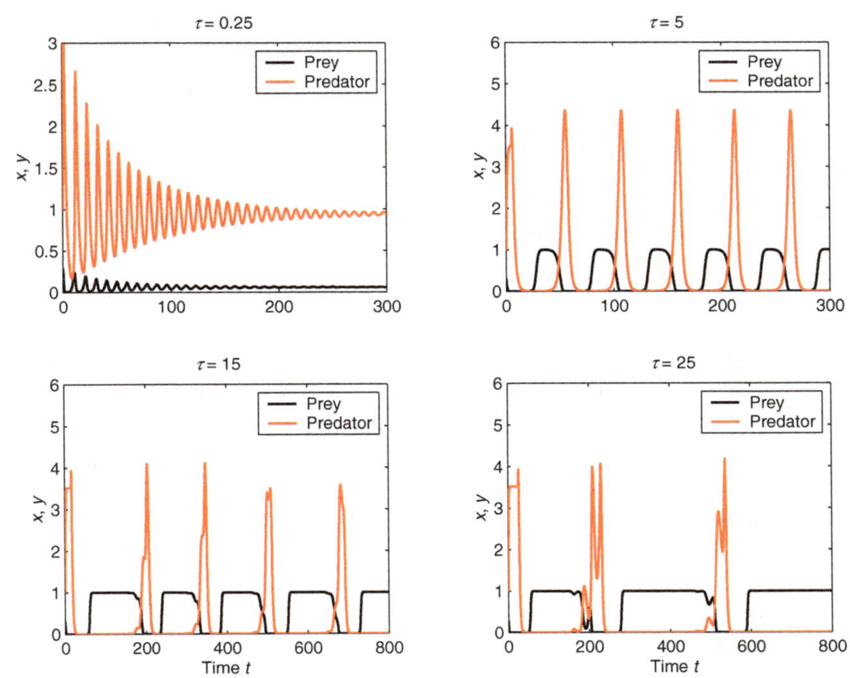

FIGURE 2 A solution of model 4 with $p(x) = px$, where $r = K = 1$, $p = 1$, $b = 10$, $d_j = 0$, $d_a = 0.5$, $m = 0.1$, and τ varies from 0.25 to 25.

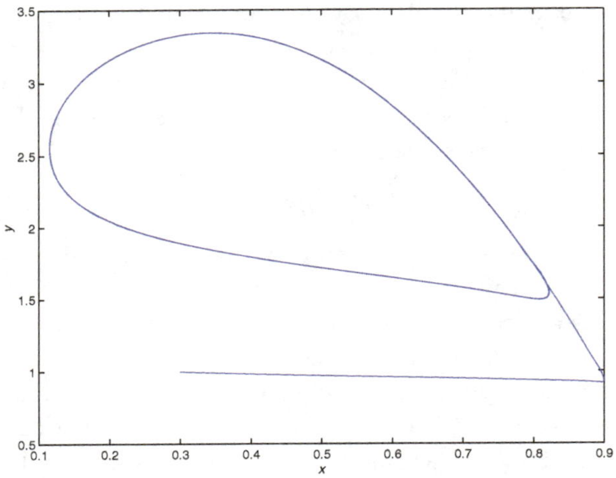

FIGURE 3 Maturation time delay may not generate complex dynamics other than periodic solution. In addition, maturation time delay may significantly cut the transition time from an initial point to an attracting limit cycle.

ubiquitous ultradian insulin secretory oscillations in the human glucose–insulin regulatory system.

SEE ALSO THE FOLLOWING ARTICLES

Difference Equations / Integrodifference Equations / Ordinary Differential Equations / Partial Differential Equations

FURTHER READING

Aiello, W. G., and H. I. Freedman. 1990. A time-delay model of single-species growth with stage structure. *Mathematical Biosciences* 101: 139–153.

Beretta, E., and Y. Kuang. 2002. Geometric stability switch criteria in delay differential systems with delay dependent parameters. *SIAM Journal on Mathematical Analysis* 33: 1144–1165.

Gourley, S. A., and Y. Kuang. 2004. A stage structured predator–prey model and its dependence on maturation delay and death rate. *Journal of Mathematical Biology* 49: 188–200.

Gourley, S. A., and Y. Kuang. 2005. A delay reaction-diffusion model of the spread of bacteriophage infection. *SIAM Journal on Mathematical Analysis* 65: 550–566.

Gurney, W. S. C., S. P. Blythe, and R. M. Nisbet. 1980. Nicholson's blowflies revisited. *Nature* 287: 17–21.

Hale, J. K., and S. M. Verduyn Lunel. 1993. *Introduction to functional differential equations.* New York: Springer-Verlag.

Hutchinson, G. E. 1948. Circular causal systems in ecology. *Annals of the New York Academy of Sciences* 50: 221–246.

Kuang, Y. 1993. *Delay differential equations with applications in population dynamics.* Boston: Academic Press.

Li, J., Y. Kuang, and C. Mason. 2006. Modeling the glucose-insulin regulatory system and ultradian insulin secretory oscillations with two time delays. *Journal of Theoretical Biology* 242: 722–735.

Smith, H. L. 2011. *An introduction to delay differential equations with applications to the life sciences.* Texts in Applied Mathematics. New York: Springer.

DEMOGRAPHIC STOCHASTICITY

SEE STOCHASTICITY, DEMOGRAPHIC

DEMOGRAPHY

CHARLOTTE LEE

Florida State University, Tallahassee

Demography is the study of vital rates, such as mortality and fecundity rates, and their effects on population dynamics. Studies usually focus on how vital rates depend on traits such as age; in ecology, vital rates may include individual growth or shrinkage rates (the latter being most relevant for plants), and the traits on which they depend may include size, developmental stage, or any other state through which individuals transition, including spatial location and environmental state. Commonly investigated population consequences of the vital rates include the population growth rate and the population trait structure (the proportion of individuals in each age, size, stage, or other state class). The predominant tool for the ecological study of demography is the population projection matrix, a discrete-trait approach that is amenable to parameterization using empirical data and which allows analyses of factors such as environmental variation.

OVERVIEW

In almost any population, the rates describing individual activities depend on individual traits, and the activities most likely to have population consequences are the rates at which individuals reproduce, die, and change with respect to rate-determining traits. For example, in an animal population, juveniles may be more likely to die and less likely to reproduce than adults, so differentiating the two developmental stages and understanding the rates at which individuals pass through them is important for understanding overall population death and birth rates. In addition, older or larger animals may survive or reproduce better or worse than younger or smaller ones; better habitat quality in some places or at some times may similarly influence vital rates. Figure 1 illustrates the potential effects of two such traits and hypothetical transitions between trait states. Demography encompasses these and similar processes. Without context, the term is usually taken generally to mean the population dynamics of humans, whose vital rates depend on age and may also depend on socioeconomic status, nationality, or behaviors such as cigarette smoking, for instance. Demography departments in a university or other such organizational units are therefore often interdisciplinary social science units, most often including economists and sociologists.

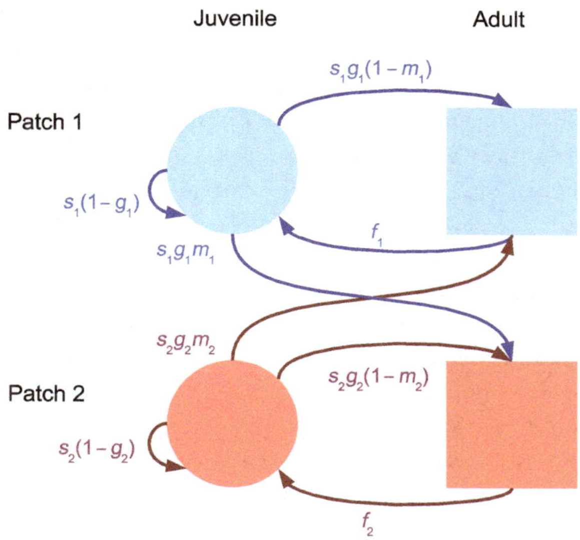

FIGURE 1 Demographic representation of a hypothetical life cycle. This organism has juvenile and adult stages. Juveniles may or may not survive each time interval. If they survive, they may or may not mature into adults (so that the duration of the juvenile stage varies between individuals). A juvenile that matures may establish a territory as an adult in its natal population or migrate to another population to establish territory there. Reproduction as adults results in the production of new juveniles in the population where the adult holds territory. Here juvenile survival is represented by s_1 and s_2, where the subscript denotes the population in habitat patch 1 or 2, and survival rates may or may not differ between patches. Similarly, growth and maturation from juvenile to adult is denoted g_1 and g_2, adult fecundity f_1 and f_2, and migration rate m_1 and m_2. These vital rates combine to yield rates of transition between the trait classes (Juvenile, Patch 1) (Adult, Patch 1), (Juvenile, Patch 2), (Adult, Patch 2). For instance, transition between the first two classes requires survival and growth but no migration at rates obtaining in Patch 1, $s_1g_1(1 - m_1)$, and so on. Other life histories might involve other trait classes and vital/transition rates.

In ecology and evolution, demography is not only a frequent concern of basic ecological investigation but also plays a prominent role in conservation biology and life history evolution.

The many tools for mathematical demographic investigation ("formal demography" in the case of humans) include life table analysis, the McKendrick–von Foerster partial differential equation approach, and integral projection models. But by far the most widely used framework in ecology is the population projection matrix, which is the focus of the rest of this article. Ecologists have found these models amenable to parameterization with data from a straightforward empirical approach: individual organisms are marked and followed to determine their fates at future well-defined time intervals such as 1 year. Mortality, fecundity, and trait-transition rates calculated from these census data are then assembled into a matrix that projects the state of the population through yearly intervals. Standard tools of linear algebra produce the population growth rate and the distribution of individuals among trait classes. As a result, and due also to more sophisticated developments such as are described in the section on special topics, below, the population projection matrix is perhaps the most widely applied theoretical construct in ecology.

ELEMENTS OF THE PROJECTION MATRIX APPROACH

The essential feature of the projection matrix approach to demography is division of a population into discrete trait classes, between which some or all vital rates differ. The definition of trait classes is most straightforward in the cases of organisms having life histories with clearly differentiated developmental stages, such as a plant that has a nonreproductive rosette stage, a flowering stage, and a seed stage. When a continuous variable such as size determines vital rates, discrete trait classes can still be defined, particularly if vital rates cluster in trait space (so that the differences between small individuals and those between large ones in reproductive rate tends to be smaller than the difference between the mean of small and the mean of large individuals, for instance). Although methods developed explicitly for continuous trait variables may be more appropriate in such cases, the matrix approach may be adequate and is popular in part because simple linear algebra tools (described below) yield biologically interpretable information regarding population properties.

One of the most useful aspects of the matrix demographic approach has been its ability to identify trait classes that are particularly important to population dynamics. For instance, projection matrix modeling can enable researchers to conclude that large adults contribute more to a commercially important fish species' population growth than smaller individuals or juvenile stages. This type of application has resulted in widespread use of demographic projection matrices in conservation biology as well as basic ecology, and its use is growing in invasive species control.

The Projection Equation and Population Growth Rate

Once trait classes are defined, one can census a population to determine the number of individuals in each class; these numbers are the elements of the population vector at time t, $\mathbf{n}(t) = \{n_i(t)\}$. At the next census interval, the same individuals are again assigned to classes (including dead individuals), and their reproductive activity is also determined. This information, and any from future census intervals, enables a researcher to calculate the entries of a discrete-time projection matrix, $\mathbf{A} = \{a_{ij}\}$ (Fig. 2),

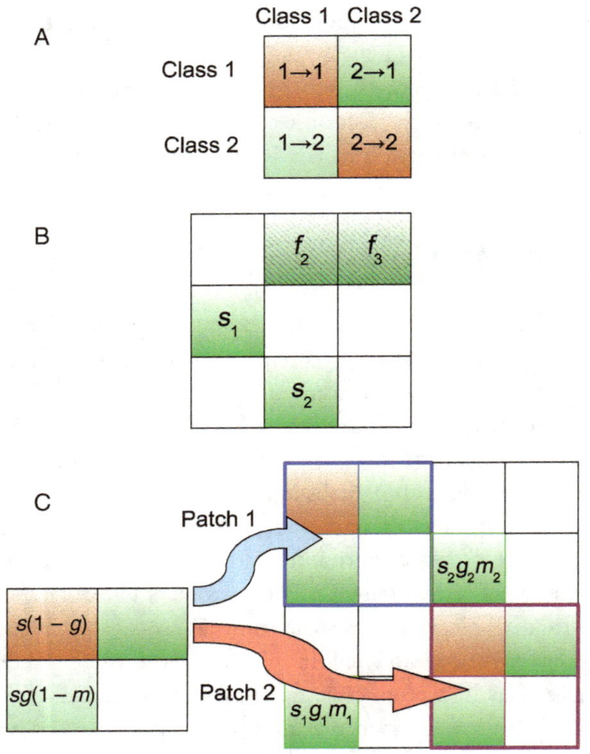

FIGURE 2 Projection matrices project a population forward through time. (A) In general, each element of a projection matrix is the rate at which transitions occur from each trait class to each other trait class, including rates of staying in any given class (red) or changing to a different class (green). (B) A *Leslie matrix* projects a population classified by age through yearly intervals. Staying the same age for more than a year is not possible, and the rate of transition from any age class to the next is the survival rate during that age class (green). Newborn individuals are generated by fecundity (green hatched). In this example, newborns are nonreproductive. (C) A stage-based *Lefkovitch matrix* takes the general form given in part (A), with transition rates corresponding to combinations of the vital rates of a given life history. Here a two-stage matrix expresses the within-patch life history illustrated in Figure 1. A larger matrix with within-patch matrices on the block diagonal and between-patch transition rates on the off-diagonals expresses the complete two-patch life history.

which describes the rates at which individuals in each trait class generate individuals in each trait class. Then the projection equation describing population dynamics is $\mathbf{n}(t+1) = \mathbf{An}(t)$, which states that the projection matrix operates on the population vector in each census interval to produce the population vector in each next census interval. The matrix notation is shorthand for the system of linear equations $n_1(t+1) = a_{11}n_1(t) + a_{12}n_2(t) + \cdots ; n_2(t+1) = a_{21}n_1(t) + a_{22}n_2(t) + \cdots$.

According to standard methods of linear algebra, the projection equation results in a population that grows (or shrinks) at an eventual (asymptotic) rate equal to the dominant eigenvalue of the projection matrix **A**, and the proportion of individuals in each trait class eventually equals the right eigenvector of **A** associated with this eigenvalue. Standard methods and a variety of computer packages make computation of matrix eigenvalues and eigenvectors easy. Thus, it is straightforward to interpret census data in terms of population growth rate and eventual population trait structure. The standard projection equation, however, assumes a constant transition matrix, which is to say that population vital rates do not change from year to year. Furthermore, the projection equation is linear, and therefore assumes that vital rates do not depend on population density (which would result in a nonlinear projection equation). In practice, researchers frequently acknowledge that environmental variation, density dependence, or other factors influence vital rates, but argue that in the absence of evidence to the contrary, it is adequate to assume that these factors are constant in time and are therefore included in empirical estimates of transition matrix elements. Methods do exist for explicitly relaxing many assumptions of the basic projection equation, some of which are described in the section on special topics, below.

Sensitivity and Elasticity

To identify how trait classes contribute to population dynamics, researchers often calculate sensitivities and elasticities of population growth to specific matrix entries. These quantities both involve the local partial derivatives of the asymptotic population growth rate with respect to the matrix entries, which describe how the growth rate would change if, all else being equal, each single matrix entry were slightly changed. Thus, the larger the magnitude of the sensitivity or elasticity with respect to a given matrix element, the more important that element is to determining the population growth rate.

The sensitivity of the growth rate with respect to a matrix element is the raw local partial derivative $s_{ij} = \partial\lambda / \partial a_{ij}$, where a_{ij} is the i, jth element of the projection matrix **A**. A sensitivity is thus a function of a matrix element; the values of the sensitivities evaluated at the observed values of the repsective matrix elements are readily calculated from the right and left eigenvectors of the matrix. The magnitudes of the values of different matrix elements can differ quite dramatically, however, due to their relationship to vital rates with radically different potential distributions (Fig. 2). For instance, survival rates are bounded below by 0 and above by 1, whereas fecundity is a nonnegative number that can range into the hundreds or more, depending on the organism in question. To compare the importance of such very different vital rates to population growth, researchers may use elasticities in addition to sensitivities. The elasticity of the growth rate with respect to a matrix element is the local proportional partial derivative $e_{ij} = (a_{ij}/\lambda)(\partial\lambda/\partial a_{ij}) = \partial \log \lambda / \partial \log a_{ij}$, which describes the proportional change in the asymptotic growth rate given a proportional change in the relevant matrix element. The values of elasticities are readily calculated from the sensitivities.

SPECIAL TOPICS

The basic projection matrix framework described here has been adapted to accommodate important features of biological systems such as spatial or temporal environmental variation, density dependence, or multiple traits determining vital rates. These adaptations typically require modification of the transition matrix or of the projection equation.

Spatial Variability or Other Additional Traits

If several different populations, each having dynamics following the basic projection equation, are connected by migration of individuals between populations, one can describe this process using a large matrix that is the sum of two matrices: a block diagonal matrix where the square matrices on the diagonal are the projection matrices describing demography within each population, and a matrix with elements off the block diagonal describing the contribution, via migration, of individuals in each class in each population to each class in each other population (Fig. 2). This approach can capture the influence of any additional traits that influence demography, not only interpopulation migration and not limited to only one; however, the data requirements for empirical application can quickly become formidable.

Environmental Variability

Most natural populations are subject to environmental variability, including repeated cyclical fluctuation such as arises from seasonality in temperate environments and year-to-year variation such as may be caused by random fluctuations in annual rainfall in any environment. In many such cases, deterministic demographic analysis may be adequate, but methods have been developed to account for such fluctuation explicitly. As in the case of spatial variation, these methods are not limited to variability caused by the weather alone but can capture any source of variability that results in cyclic or random variation in projection matrices.

The projection equation generalizes to accommodate a fluctuating environment by replacement of the constant projection matrix $\mathbf{A}$ with a set of matrices $\mathbf{A}(t)$ that differ between time intervals: $\mathbf{n}(t + 1) = \mathbf{A}(t)\mathbf{n}(t)$. For seasonal or other periodic variation, a finite number of matrix repeats in a fixed sequence. For random variation, the number of matrices may be fixed, assuming that each matrix represents one of a fixed number of possible environmental states. Alternatively, one may sample matrices from an assumed distribution of environmental states; this would result in an infinite pool of potential matrices if the vital rates are assumed to be functions of a continuous variable such as snowpack depth or streamflow. In either case, the population is subject to a random sequence of projection matrices sampled from the pool of potential environmental states. In the absence of concrete hypotheses regarding the generation of the sequence, matrices are frequently sampled to represent an independent and identically distributed environmental process. In the case of recurring important environmental events such as fire or windthrow, a reasonable alternative is to model the environment using a Markov chain, with transitions between environmental states occurring according to an additional transition matrix.

Iterating the projection equation for time-dependent matrices, we see that $\mathbf{n}(t + 2) = \mathbf{A}(t + 1)\mathbf{n}(t + 1) = \mathbf{A}(t + 1)\mathbf{A}(t)\mathbf{n}(t)$; thus, the population vector in any time interval is a function of the sequence of environments the population has experienced, represented by the product of the transition matrices describing those environments. Matrix multiplication is not commutative, so the order of the matrices within the product is important; this represents a biological path dependency where, for instance, two initially identical populations would fare very differently if one experienced alternating good and bad years and the other experienced all bad years followed by all good years (the latter could be extinct before experiencing any good years).

In the case of deterministic, seasonal (or other periodic) environmental variation, the asymptotic population growth rate is given by the dominant eigenvalue of the appropriate product of the time-dependent matrices. When the environment varies randomly, any sequence of environmental states of a given length is a sample of all possible sequences of that same length. Then the average rate over time at which a population grows is a different quantity than the growth rate of the average population, because the latter takes into account all possible environmental sequences. Care must be exercised to evaluate the appropriate quantity; for most applications, the former growth rate is of most interest and is called the stochastic growth rate. It is calculated empirically as the difference between the log of the final population and the log of the initial population, divided by the time interval: $a = (\log N_t - \log N_0)/t$. The analysis of growth rate and its senstivity and elasticity in variable environments is a field of active current research.

SEE ALSO THE FOLLOWING ARTICLES

Age Structure / Individual-Based Ecology / Matrix Models / Stage Structure

FURTHER READING

Boyce, M.S., C.V. Haridas, C.T. Lee, C.L. Boggs, E.M. Bruna, T. Coulson, D. Doak, J.M. Drake, J.M. Gaillard, C.C. Horvitz, S. Kalisz, B.E. Kendall, T. Knight, E.S. Menges, W.F. Morris, C.A. Pfister, and S.D. Tuljapurkar. 2006. Demography in an increasingly variable world. *Trends in Ecology and Evolution* 21: 141–148.

Caswell, H. 2001. *Matrix population models: construction, analysis, and interpretation*. Sunderland, MA: Sinauer.

Crouse, D.T., L.B. Crowder, and H. Caswell. 1987. A stage-based population model for loggerhead sea turtles and implications for conservation. *Ecology* 68: 1412–1423.

Ellner, S.P., and M. Rees. 2006. Integral projection models for species with complex demography. *American Naturalist* 167: 410–428.

Leon, S.J. 2009. *Linear algebra with applications*, 8th ed. Integral projection models for species with complex demography. Upper Saddle River, NJ: Prentice Hall.

Morris, W.F., and D.F. Doak. 2002. *Quantitative conservation biology: theory and practice of population viability analysis*. Sunderland, MA: Sinauer Associates.

Preston, S.H., P. Heuveline, and M. Guillot. 2001. *Demography: measuring and modeling population processes*. Malden, MA: Blackwell.

DIFFERENCE EQUATIONS

JIM M. CUSHING

University of Arizona, Tucson

Difference equations are mathematical formulas that recursively define sequences of numbers (or other mathematical objects such as vectors or functions). They are often employed to model the temporal dynamics of biological populations at discrete time intervals. Such discrete-time models predict future states of a population by describing how a population's state at each point of time depends on its state at previous points in time.

DIFFERENCE EQUATIONS

Difference equations arise in ecology when temporal changes in a population (or a community of populations) are modeled on discrete time intervals. This is in contrast to the use of differential equations, which model population dynamics in continuous time. Discrete-time models are generally motivated by some discrete aspect of the population's biology, its environment, and/or the available data. For example, the dynamics of populations with sharply defined life cycle stages are particularly amenable to description by difference equations. Other examples include abrupt seasonal changes in environmental factors and significant gradients in a spatial habitat. When used in appropriate circumstances, difference equations often have many attractive features as dynamic models when compared to continuous-time models. These features can include a straightforward derivation from biological assumptions, a more transparent parameter identification and estimation, effortless simulation by computers, the ease with which stochastic effects can be incorporated, and a less demanding mathematical background (when compared, for example, to partial differential equations). Indeed, these factors are attractive enough that difference equations are sometimes preferred and used in theoretical and practical studies even in the absence of any clearly defined discrete features in the biological situation.

A scalar difference equation $x(t+1) = f(x(t))$ recursively defines sequences of numbers $x(0)$, $x(1)$, $x(2), \ldots$. (Difference equations can also define sequences of other mathematical objects, such as vectors, functions, matrices, and so on.) For each initial term $x(0)$, the equation determines a unique sequence, which is called a solution of the difference equation. In population dynamics modeling, x denotes a quantitative description of a population (e.g., the number or density of individuals, the amount of biomass or dry weight, and so on). A difference equation serves as a deterministic model that uniquely predicts the state of the population $x = x(t)$ at a sequence of discrete future times $t = 1, 2, 3 \ldots$ from knowledge of the initial state of the population $x(0)$ at time $t = 0$.

In some contexts, it is of interest to describe the population by more than one quantity. For example, so-called structured population models categorize individuals in a population into a finite number of classes (based on, for example, chronological age, body size, life cycle stages, and the like) and the population numbers (or densities) in each class are tracked dynamically in time. In such a case, the state of the population at time t is a vector $x(t)$ consisting of all of these quantities, and $x(t+1) = f(x(t))$ is a vector difference equation that describes how this distribution vector changes in time. This vector difference equation can also be viewed as a system of difference equations for the dynamics of the classes of individuals. Vector (or systems of) difference equations arise in other contexts as well. The vector x can, for example, consist of the population numbers of several interacting species (or structured classes of several species), in which case the vector difference equation describes the dynamics of a multispecies community or ecosystem.

Systems of difference equations also arise when a population's future state depends on population states at several previous times—an assumption that leads to a *multistep difference equation.* Multistep difference equations can be rewritten as a system of one-step difference equations by using lag variables. For example, a two-step difference equation $x(t+1) = f(x(t), x(t-1))$ can be rewritten as the pair $x(t+1) = f(x(t), y(t)), y(t+1) = x(t)$ of two one-step difference equations.

Another type of difference equations arises in circumstances when a population's future state depends not only on the population's current state but also on the current time. In this case, the difference equation $x(t+1) = f(t,x(t))$ is called nonautonomous. For example, a population in a periodically fluctuating environment might be modeled by use of a recursive formula $f(t,x)$ that is mathematically periodic in t, in which case the equation is a periodic difference equation or is said to be periodically forced.

Difference equations in which x at any point in time is a mathematical function, rather than a number or vector, also arise in population modeling. For example, difference equations are used to describe spatial models for the study of population dispersal and diffusion. If discrete habitats are of interest, then one can use a structured model as described above with discrete spatial classes. If a continuous spatial habitat is of interest, then $x = x(t,s)$ is also a function of a continuous spatial variable s (of one or more dimensions), and the difference equation maps $x(t,s)$ to $x(t+1,s)$. One way to do this, for example, is by means of an integrodifference equation

$$x(t+1, s) = \int_S k(s, \sigma) f(t, x(t, \sigma)) d\sigma,$$

where $k(x,\sigma)$ is a dispersal kernel and S is the spatial habitat.

In some difference equation models, the function f contains one or more random variables so as, for example, to account for stochastic effects that influence a population's dynamics. Such stochastic difference equations do not uniquely predict the future states of the population, which as a result need to be described statistically.

THE HISTORY OF DIFFERENCE EQUATIONS

The procedure of computing by means of recursive steps, as represented in modern terms by difference equations, has a long and varied history. Recursive calculations were utilized by mathematicians in ancient Babylonia (as early as 2000 BC), Greece (as early as 450 BC), and India (as early as 200 BC) to solve geometric and number theoretic problems and to approximate irrational numbers, among other applications. Included in these early mathematicians were Pythagoras, Archimedes, and Euclid. Throughout subsequent centuries, numerous scholars used and studied difference equations, including some of the most famous in the history of mathematics: Fibonacci, Leibniz, Newton, Gauss, Euler, Laplace, Lagrange, de Moivre, and Poincaré.

Difference equations are associated with a large variety of problems that arise from a diversity of disciplines, ranging from number theory and geometry to economics and computer science. They arise naturally, for example, when approximations are made in problems from calculus and analysis, such as the approximation of integrals and derivatives. In other circumstances, difference equations arise directly from the problem (such as, for example, in compound interest calculations).

The use of difference equations took on a central role in dynamical system theory after the foundational work, during the late nineteenth and early twentieth centuries, of Henri Poincaré, who utilized difference equations in the study of differential equations. Poincaré introduced what are called return maps, which are in essence discrete time samplings of state variables varying continuously in time. (These maps give rise to a special class of difference equations called diffeomorphisms, which possess the property that they also define a unique sequence in reverse time. Many, if not most, difference equations used in population dynamics, however, do not define unique past histories and hence are not diffeomorphisms.) The use of difference equations as mathematical models of population dynamics in their own right was stimulated by the seminal work of the demographer P. H. Leslie in the 1940 through the 1960s, who popularized the use of linear and nonlinear matrix models for populations structured by chronological age. Together with J. C. Gower, Leslie also utilized difference equations to model multispecies interactions and was (seemingly) the first to use stochastic difference equations in population dynamics.

Difference equations played a prominent role in stimulating interest in chaos and complexity theory after the appearance in the 1970s of a several influential papers by Lord Robert May, who was motivated by the use of difference equations as theoretical models in population dynamics and ecology. Because of this development, difference equations played, and continue to play, a major role in elucidating the fact that complex dynamics (such as chaos) can result from simple (low-dimensional) mechanisms. There is today a large and growing literature that utilizes difference equations as models for both theoretical and applied studies in population dynamics, ecology, and related fields (such as population genetics, evolutionary dynamics, natural resource management, epidemics, and the like).

ANALYSIS OF DIFFERENCE EQUATIONS

The use of difference equations in theoretical population and ecological dynamics has focused primarily on asymptotic dynamics, i.e., the long-term behavior of solutions. The range of a solution $x(t)$ of a difference equation $x(t+1) = f(x(t))$ is a sequence of real numbers (or vectors) called an orbit. The line (or higher-dimensional Euclidean space) in which they lie is called phase space. An attractor is a set of points in phase space that orbits approach as t increases (without bound). The most fundamental candidate for an attractor is an equilibrium, which is a solution (or orbit) that remains unchanged in time. Equilibria are fixed points of $f(x)$, i.e., they are solutions of the equation $x = f(x)$. Another basic candidate for an attractor is a periodic solution (or orbit), often called a *cycle*, which is a solution that visits a set of (different) points sequentially and repeatedly for all t. For example, a two-cycle is a solution that oscillates between two different points (which are fixed points of the composite $f(f(x))$ function). Unlike equilibria and cycles, other types of attractors can be very complicated sets of points in phase space consisting of infinitely many points. The description and classification of complicated attractors is a difficult and technically sophisticated task, one that continues to be a challenge for mathematicians. Strange attractors and chaotic attractors are two well-known types. Strange attractors are distinguished by their geometry and topology (often exhibiting fractal characteristics). Chaotic attractors are characterized by erratic temporal oscillations that are difficult to distinguish from noise. There are numerous technical definitions of chaos, but a hallmark property is that of sensitivity to initial conditions, by which is meant that (arbitrarily) small differences between two initial states lead ultimately to large differences between their solutions (and to oscillations that are uncorrelated).

An equilibrium is stable (technically, locally asymptotically stable) if small perturbations from the equilibrium remain close and ultimately return to the equilibrium. The most basic tool for ascertaining the stability of an equilibrium is the linearization principle. This principle utilizes the eigenvalues of the Jacobian matrix associated with the difference equation. The entries of this matrix are the (partial) derivatives $\partial f_i / \partial x_j$, where f_i is the ith entry in f and x_j is the jth entry in x. For a scalar difference equation, the Jacobian is simply the derivative df/dx. If all eigenvalues of the Jacobian evaluated at an equilibrium are less than 1 in magnitude, then the linearization principle implies the equilibrium is stable. If at least one eigenvalue has magnitude greater than 1, the equilibrium is unstable. If all eigenvalues have magnitude greater than 1, the equilibrium is a repeller. If some have magnitude less than 1 and others greater than 1, the equilibrium is a saddle.

In population models, difference equations usually contain one or more parameters (or coefficients), which represent important biological quantities such as birth and death rates, predation rates, competition modulo, and so on. The stability of an equilibrium usually depends on the numerical values assigned to these model parameters. If stability is lost when a selected parameter is changed, then a bifurcation is said to occur. The selected parameter is called a bifurcation parameter and the value at with the bifurcation occurs is called a bifurcation value.

By the linearization principle, stability is lost by a change in a parameter if the change causes the magnitude of an eigenvalue of the Jacobian (evaluated at the equilibrium) to become greater than 1. Since eigenvalues can be complex, this change can be viewed as an eigenvalue moving from inside to the outside of the unit circle in the complex number plane as the parameter changes. According to (local) bifurcation theory, the resulting asymptotic dynamics depend on how the eigenvalue leaves the unit circle. If it leaves the unit circle through $+1$, then in general an equilibrium bifurcation occurs. This typically results in the creation of new equilibria. Canonical types of equilibrium bifurcations include the saddle-node (blue-sky or tangent) bifurcation, transcritical bifurcation, and pitch-fork bifurcation.

If, on the other hand, an equilibrium destabilization occurs because an eigenvalue leaves the (complex) unit circle through -1, then typically a two-cycle is created. In this case, a period-doubling bifurcation is said to occur.

Finally, if the destabilization occurs because a pair of conjugate eigenvalues leaves the (complex) unit circle at a point other than ± 1, then a more complicated dynamic is created. (This can only occur for a system of two more difference equations.) The Naimark–Sacker (or invariant loop or discrete Hopf) bifurcation theorem implies the creation of a loop in the Euclidean space of the vector x (so-called phase space) that is time invariant (that is to say, a solution starting on the loop remains on the loop for all future time). Orbits on the invariant loop might be, or asymptotically approach, a periodic cycle, in which case period locking is said to occur. Or orbits on the invariant loop might not approach periodic cycles, in which case the motion is aperiodic (which has the appearance of, but is not technically, chaos).

In any of these bifurcation cases, the newly created equilibria, cycles or invariant loops can be stable or unstable. When stable, they might also lose stability upon further changes in the bifurcation parameter. Often a sequential cascade of bifurcation occurs, as the bifurcation parameter varies over a sufficiently large range, which can ultimately result in the appearance of chaotic attractors (such a cascade is called a route to chaos).

The linearization principle for stability and bifurcation analysis is also applicable for other types of difference equations, such as the ingrodifference equations and periodically forced difference equations mentioned in above.

In addition to the fundamental linearization principle, other methods from the theory of dynamical systems are available for the analysis of difference equations, including classic methods such as Lyapunov functions and perturbation methods and modern methods such as persistence theory and the theory of monotone flows, to name a few. The analysis of periodic cycles of a difference equation can be carried out (in principle, but often with difficulty in practice) by performing an equilibrium analysis on the difference equation formed by composition of f. A study of period 2 cycles, for example, becomes a study of equilibria when only every other time step is considered and the difference equation is defined by $f(f(x))$ in place of $f(x)$. On the other hand, an analysis of invariant loops, bifurcating cascades, and chaos is in general difficult, and studies of such complicated dynamical scenarios are usually carried out by computer simulations.

DIFFERENCE EQUATIONS AND POPULATION DYNAMICS

In modeling the dynamics of a population using a difference equation, focus is placed on specifying the function $f(x)$ in the difference equation $x(t+1) = f(x(t))$. This function incorporates the assumptions that the modeler wishes to make concerning how future population states depend on past population states. Basic modeling assumptions about the dynamics of a biological population concern birth and death processes. Other factors such as immigration, emigration, harvesting, predation, and so on can also come into play.

Unstructured Population Dynamics

A basic population model takes into account birth and death processes. If the population at time $t+1$ consists of b newborns per individual alive at time t and of a fraction s of surviving individuals, then (in the absence of any other factors) $f(x) = rx$ is a linear function with $r = b + s$. The resulting linear difference equation, when iterated, predicts exponential decay or growth $x(t) = r^t x(0)$, depending on whether the population growth rate r satisfies $r < 1$ or $r > 1$, respectively. Equivalently, decay or growth occurs if and only if $R_0 < 1$ or $R_0 > 1$ where $R_0 = b/(1-s)$ is the net reproductive number, i.e., the expected number of newborns per individual per lifetime. Nonextinction, but bounded, population dynamics are possible if and only if the growth rate $r = 1$ (or $R_0 = 1$), in which case all solutions are equilibria. Thus, a bifurcation occurs at $r = 1$ (or $R_0 = 1$) as the extinction equilibrium $x = 0$ loses stability. (Mathematically, the equilibria that occur when $r = 1$ constitute a continuum that bifurcates from the extinction state $x = 0$ at $r = 1$ as a function of the parameter r (see Fig. 1).)

If fertility and survivorship depend on population density, a situation referred to as density dependence, then r is a function of x and the resulting nonlinear function $f(x) = r(x)x$ defines a nonlinear difference equation $x(t+1) = r(x(t))x(t)$. The dynamics of nonlinear difference equations are typically more complicated than the exponential dynamics of linear difference equations.

The biomathematics literature contains a plethora of mathematical formulas for use in specifying f (or r) in order to describe biological and ecological mechanisms of interest. These formulas range from relatively simple formulas, designed to capture qualitatively some ecological phenomenon of interest, to more complex formulas aimed at a more accurate quantitative description of biological mechanisms. Two types of formulas frequently used to build models are rational polynomials and exponential

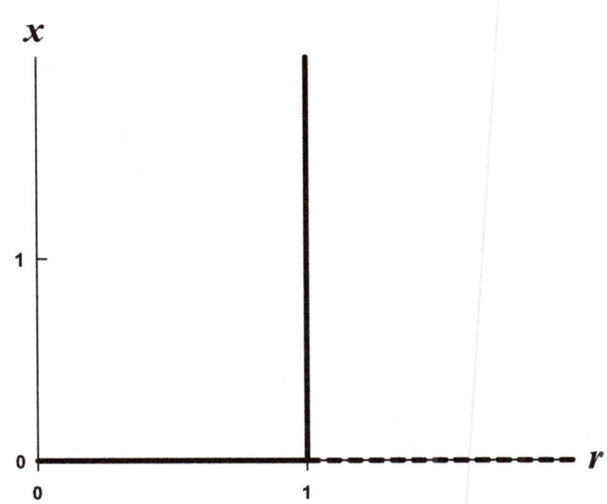

FIGURE 1 The bifurcation diagram for a linear difference equation $x(t+1) = rx(t)$ has a vertical bifurcation at $r = 1$, at which point there is a continuum of equilibria.

functions. The scalar difference equation with the rational polynomial $r(x) = b/(1 + cx)$ is analogous to the famous logistic differential equation (its solutions monotonically approach the equilibrium $K = (b - 1)/c$); it is often called the discrete logistic or the Beverton–Holt equation. The scalar difference equation with the exponential $r(x) = \exp(-cx)$, the so-called Ricker equation, is the iconic example that exhibits a period-doubling cascade to chaos (see below). These two examples represent deleterious density (or compensatory) effects on population growth. The Ricker equation models overcompensatory effects in that, unlike the discrete logistic, $f(x)$ tends to 0 as density x increases without bound.

The linearization principle implies that the behavior of solutions near the extinction state $x = 0$ is the same as that of the linear equation $x(t + 1) = r(0)x(t)$ (which is called the linearization at $x = 0$). Thus, one feature these nonlinear difference equations have in common with linear difference equations is the loss of stability of the extinction state as the inherent population growth rate $r(0)$ (or, equivalently, the inherent net reproductive number R_0) increases through 1. Also similar, it turns out, is the creation (bifurcation) of a continuum of positive equilibrium states as a result of this destabilization, although for nonlinear equations the set (or spectrum) of $r(0)$ values (or $R_0(0)$ values) for which there exists positive equilibria is not in general a single point, as it is for linear equations; instead, it is an interval of values. The stability of the bifurcating positive equilibria, at least near the bifurcation point, depends on direction of the bifurcation. Bifurcation to the right (a forward or supercritical or stable bifurcation), in which positive equilibria near the bifurcation point exist for $r(0) > 1$, produces stable positive equilibria, while bifurcation to the left (a backward or subcritical or unstable bifurcation) produces unstable positive equilibria for $r(0) < 1$ (See Figs. 2 and 3).

A stable bifurcation to the right results from negative feedback effects in which the (per unit) growth rate $r(x)$ decreases with increased population density x (which is the most common assumption in population models). An unstable bifurcation to the left requires positive feedback effects of sufficient magnitude at low population levels, called Allee effects (i.e., $r(x)$ increases with x near 0).

Stable equilibria can lose their stability with sufficient increase in $r(0)$ (or $R_0(0)$). This typically occurs with so-called overcompensatory density effects in which $f(x) = r(x)x$ decreases for large x. This equilibrium destabilization results in a period doubling bifurcation (the creation of oscillatory cycles of period 2). With further increases in $r(0)$ (or $R_0(0)$) these 2-cycles can in turn destabilize and produce 4-cycles, which in turn destabilize and produce 8-cycles, and so on. Even further increases in $r(0)$ (or $R_0(0)$) can ultimately produce complicated chaotic dynamics (see Fig. 4). Seminal papers in the 1970s by R. M. May and by T. Y. Li and J. Yorke stimulated considerable interest on the part of both mathematicians and ecologists in this so-called period doubling route to chaos. Mathematicians have shown this bifurcation scenario to be typical for a broad class of nonlinear scalar

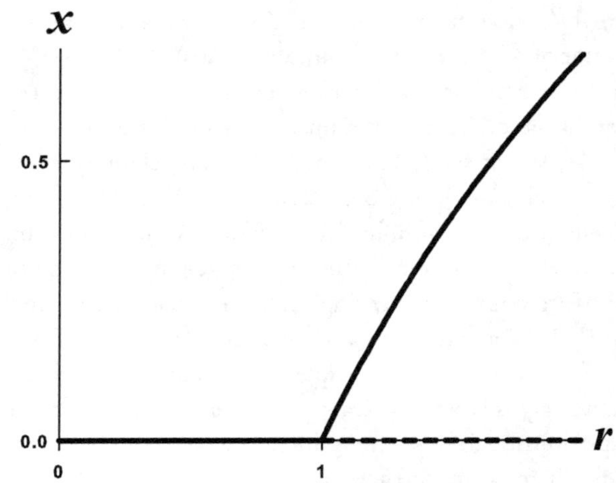

FIGURE 2 The bifurcation diagram for the nonlinear Ricker equation $x(t+1) = rx(t)\exp(-x(t))$ has a supercritical bifurcation at $r = 1$. Solid lines correspond to stable equilibria while dotted line plots correspond to unstable equilibria.

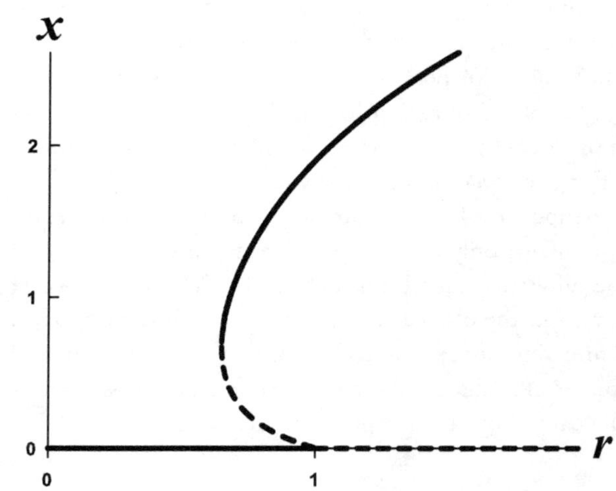

FIGURE 3 The bifurcation diagram for the nonlinear difference equation $x(t+1) = r(1 + cx(t))\,x(t)\exp(-x(t))$, which models an Allee effect when $c > 1$, exhibits a subcritical bifurcation when $c = 3$. The graph bends back to the right at a critical value of r less than 1 (due to negative density effects at high densities), which gives rise to interval of r values associated with two stable equilibria. Solid line plots correspond to stable equilibria while dotted line plots correspond to unstable equilibria.

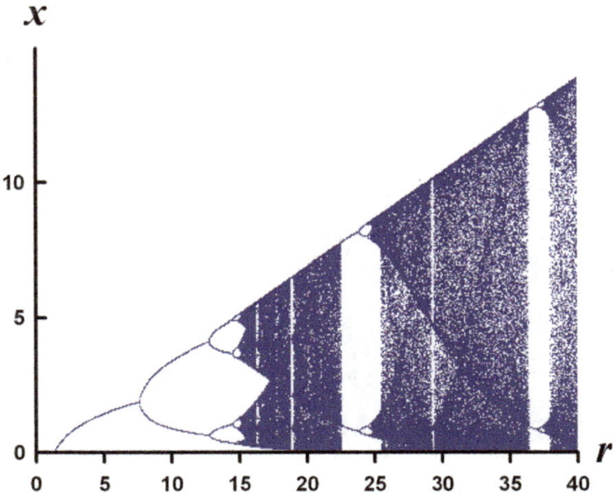

FIGURE 4 The extended bifurcation diagram for the Ricker equation in Fig. 2 shows a period doubling route to chaos.

difference equations. For systems of difference equations, on the other hand, routes to chaos can occur that involve different sequences of bifurcations.

Structured Population Dynamics

If the state of a population is described by a finite number of quantities x_i based on some classification scheme for individual organisms, then $x = \text{col}(x_i)$ is a vector and f is a vector-valued function. For example, if $F = (b_{ij})$ is the matrix of birth rates (b_{ij} is the number of i-class offspring produced per j-class individual) and $T = (s_{ij})$ is the matrix of survival/transition fractions (s_{ij} is the fraction of i-class individuals that survives and moves to class j), then $f(x) = Px$, where the projection matrix is $P = F + T$ is a nonnegative matrix. The resulting difference equation $x(t+1) = Px(t)$ is called a matrix (difference) equation.

The theory of linear matrix equations is a beautiful application of matrix theory in mathematics. If the projection matrix P is irreducible, then it has a positive dominant eigenvalue r, the population's growth rate. (P is irreducible if every class is reachable from every other class, through births or class transitions.) If r strictly dominates all other eigenvalues in magnitude, then the fundamental theorem of demography holds. This theorem states that the population grows or decays asymptotically at rate r and has an asymptotic (sometimes called a stable) normalized distribution.[1] This holds regardless of whether the population goes extinct, $r < 1$ (or $R_0 < 1$), or grows exponentially, $r > 1$ (or $R_0 > 1$). Here, the net reproductive number R_0 is the dominant eigenvalue of the matrix $F(I - T)^{-1}$. This state of affairs corresponds to Fig. 1, where a vertical bifurcation of equilibrium states at $r = 1$ (or $R_0 = 1$).

The general bifurcation alternatives shown in Figures 2 and 3 for scalar difference equations also remain valid for nonlinear matrix models $x(t+1) = P(x(t))x(t)$ with respect to the inherent growth rate $r(0)$ or to the inherent net reproductive number $R_0(0)$ defined by means of the inherent projection matrix $P(0) = F(0) + T(0)$.

The classic example, and historically the first significant application of matrix equations, is the Leslie model based on chronological age (with the time unit equal to the size of the age classes). For m age classes, the projection matrix $P = F + T$ takes the form

$$F = \begin{pmatrix} b_1 & b_2 & \dots & b_{m-1} & b_m \\ 0 & 0 & \dots & 0 & 0 \\ 0 & 0 & \dots & 0 & 0 \\ \vdots & \vdots & & \vdots & \vdots \\ 0 & 0 & \dots & 0 & 0 \end{pmatrix}, \quad T = \begin{pmatrix} 0 & 0 & \dots & 0 & 0 \\ s_1 & 0 & \dots & 0 & 0 \\ 0 & s_2 & \dots & 0 & 0 \\ \vdots & \vdots & & \vdots & \vdots \\ 0 & 0 & \dots & s_{m-1} & 0 \end{pmatrix}.$$

Although no analytic formula is possible in general for the population growth rate r, the formula $R_0 = b_1 + \sum_{i=2}^{m} s_1 s_2 \cdots s_{i-1} b_i$ is available for the net reproductive number.

Multispecies Models

Difference equation models for n interacting species give rise to systems of difference equations. For n interacting species, each species' dynamic equation $x_i(t+1) = r_i x_i(t)$ is coupled with those of other species by prescribing that $r_i = r_i(x_1, \dots, x_n)$ depends on the densities of other species. The mathematical nature of this dependence specifies the ecological relationship among the species (competition, predator–prey, host–parasitoid, mutualism, and so on). In the case of interacting structured species, we have for each species a matrix equation with projection matrices $P = P(x_1, \dots, x_n)$.

For example, consider the dynamics of two (unstructured) species whose growth rates are $r_i = r_i(\Sigma j c_{ij} x_j)$. This class of models is the discrete-time analog of the famous Lotka–Volterra differential equation models for two-species interactions. An example is the Leslie–Gower competition model, in which $r_i = b_i/(1 + \Sigma j c_{ij} x_j)$. This model, as is the case with Lotka–Volterra differential equation models, is based on discrete logistic (or Beverton–Holt) growth for each species in the absence of the other. The

[1] $\lim_{t\to\infty} x(t)/p(t) = v$, where $p(t) = x_1(t) + \cdots + x_n(t)$ is total population size and v is the positive unit eigenvector associated with the dominant eigenvalue r.

Leslie–Gower competition model (which historically was formulated with competing species of insects in mind) has the same dynamic scenarios as the famous Lotka–Volterra competition differential equation model, all of which involve only equilibrium dynamics. Other per capita growth functions, such as the exponential (or Ricker-type) function $r_i = b_i \exp(-\sum jc_{ij}x_j)$, can lead, however, to more complicated nonequilibrium dynamics for competitive interactions. The same is true of difference equation models for predator–prey and host–parasitoid interactions. An example is (a variant of) the Nicholson–Bailey model, in which $r_1 = \exp(r\ (1 - x_1/K) - ax_2)$, $r_2 = (1 - \exp(-ax_2))x_1/x_2$. Difference equation models for basic two-species interactions can serve as the starting point for the building of extended models that include more species and trophic levels, population structure, nonautonomous forcing, spatial environments (e.g., as integrodifference equations), and stochastic effects.

MODELING WITH DIFFERENCE EQUATIONS

Difference equations have found widespread use in population dynamic modeling across numerous disciplines, including population genetics, ecological interactions, evolutionary dynamics, epidemiological models, studies of the spread and dispersal of invasive species and diseases, natural resource management and harvesting, conservation biology, and pest control. Models based on difference equations are used for theoretical explorations into the implications of biological mechanisms and ecological interactions. They are also used, in both natural and experimental settings, to provide quantitative descriptions of observed population data and to make predictions of future population dynamics and how they depend on biological and environmental parameters (e.g., by means of sensitivity analysis).

As population models, difference equations have numerous advantages. The construction and derivation of models is straightforward, and parameters are generally easy to interpret. In comparison with continuous-time models, numerical simulations of difference equations are simple to carry out, the inclusion of stochasticity is easier, and numerous technical issues concerning the existence and nature of solutions are avoided (particularly in comparison to partial differential equation models). Methods for parameter estimation and sensitivity analysis are well developed, and some methods of analysis available for nonlinear difference equations are not yet available for continuous-time models. As population dynamics models, difference equations work best, and are most appropriate, when some important biological characteristics and processes occur (at least approximately) at discrete time intervals or are reasonably well described by discrete categories, such as is the case, for example, for populations whose life histories involve discrete developmental stages.

SEE ALSO THE FOLLOWING ARTICLES

Beverton–Holt Model / Bifurcations / Chaos / Integrodifference Equations / Matrix Models / Ricker Model

FURTHER READING

Caswell, H. 2001. *Matrix population models,* 2nd ed. Sunderland, MA: Sinauer Associates, Inc.

Cushing, J. M., R. F. Costantino, B. Dennis, R. A. Desharnais, and S. M. Henson. 2003. *Chaos in ecology: experimental nonlinear dynamics.* Theoretical Ecology Series 1. San Diego: Academic Press/Elsevier.

Edelstein-Keshet, L. 2005. *Mathematical models in biology.* Classics in Applied Mathematics 46. Philadelphia: SIAM.

Elaydi, S. 2005. *An Introduction to difference equations,* 3rd ed. New York: Springer.

Kot, M. 2001. *Elements of mathematical ecology.* Cambridge, UK: Cambridge University Press.

May, R. M., and G. F. Oster. 1976. Bifurcations and dynamic complexity in simple ecological models. *American Naturalist* 110(974): 573–599.

Otto, S. P., and T. Day. 2007. *A biologist's guide to mathematical modeling in ecology and evolution.* Princeton: Princeton University Press.

DIFFERENTIAL EQUATIONS, DELAY

SEE DELAY DIFFERENTIAL EQUATIONS

DIFFERENTIAL EQUATIONS, ORDINARY

SEE ORDINARY DIFFERENTIAL EQUATIONS

DIFFERENTIAL EQUATIONS, PARTIAL

SEE PARTIAL DIFFERENTIAL EQUATIONS

DISCOUNTING IN BIOECONOMICS

RAM RANJAN

Macquarie University, Sydney, Australia

JASON F. SHOGREN

University of Wyoming, Laramie

Discounting is a concept relating present and future costs and benefits. Since individual and group decisions impact today and the future, discounting exists to relate future streams of costs and benefits in current value terms. All dynamic resource allocation decisions and intertemporal policy evaluations involve discounting.

THE CONCEPT

Discounting plays a central role in bioeconomic analysis and theory. When biological resources generate values that accrue over time, discounting helps maximize the benefits derived by providing guidance over the appropriate rate of extraction. This process involves comparing the growth rate of the resources to the discount rate. In addition, discounting matters in decisions involving optimal preservation and use of endangered resources such as fisheries or forests facing the threat of extinction from overexploitation or natural hazards. Discounting reflects impatience, decreasing marginal values, productivity of a capital stock, and risk preferences. Discounting brings future incomes and costs to their present value—the value as if all the incomes and costs were incurred in the first year. One resource allocation decision's rate of return can be compared to the next best alternative rate of return, i.e., the interest rate earned from a bank. Suppose an individual invests \$100 in a bank, which earns interest. The investment grows at a certain rate (say 10 percent) to become $\$100(1 + 0.1)$ dollars at the end of 1 year, $\$100(1 + 0.1)^2$ dollars at the end of 2 years and $\$100(1 + 0.1)^t$ dollars at the end of t years. This implies if he borrowed A dollars for t years, he would owe $A(1 + 0.1)^t$ dollars at the end of t years. The individual is considering an up-front cost of A dollars from a bank to buy capital that would generate a stream of future benefits. The benefit stream lasts t years, and he would earn B dollars each year. The task is to determine whether he should make this investment. For this venture to be viable, the total sum of income generated has to be larger than the amount borrowed at an interest rate of 10 percent. The money the person would owe the bank would grow to become $A(1 + 0.1)^t$ dollars at the end of t years; his income stream from the investment must sum up to the amount at the end of t years for him to break even.

DISCOUNTING: CONSUMPTION AND UTILITY

Discounting over monetary returns is relatively straightforward. But complications arise when considering factors other than money, such as consumption and utility. Consider a scenario in which our individual has a dollar to spend today or after 1 week. Postponing immediate consumption of a desirable good is a challenge. This desire leads to impatience toward postponing future consumption of the goods. When current consumption is preferred over future consumption, the pure rate of time preference must be considered. First, we derive a continuous version of the discount factor, $df = \frac{1}{(1+r)^t}$. When the time at which money is paid interest upon shrinks from 1 year to instantaneously, the discount factor takes the form $df = \exp(-r \cdot t)$. Now, if the utility derived from consumption at time t is $U(c(t))$, the discounted utility in which the discount factor represents the pure rate of time preference is $U(c(t)) \cdot \exp(-\rho \cdot t)$. Adding the market rate of interest, the discounted utility is

$$U(c(t)) \cdot \exp(-\rho \cdot t) \cdot \exp(-r \cdot t),$$

or

$$U(c(t)) \cdot \exp((-\rho - r) \cdot t).$$

The pure rate of time preference has two components: intergenerational rate of time discounting and intertemporal rate of time discounting. The intergenerational rate involves tradeoffs when an individual or group postpones current consumption for the sake of future generations; the intertemporal rate considers the tradeoffs when an individual postpones current consumption for future consumption.

The choice of intergenerational rate of time discounting is a key issue in public policy analysis, especially within social debates over resource allocation involving biodiversity and resource use. A larger intergenerational rate of time preference leads to a higher discounting of the damages to the future generations and promotes higher consumption today. An example of intertemporal discounting in the ecological context is the management of fisheries. Consider a fishery in which an individual gets a utility out of consuming fish, $U(h)$, given a cost $c(h,s)$ to harvesting the fish. The cost increases as stock declines. The fish grow at a rate given by a logistic growth function as

$$\dot{s} = s \cdot \gamma(1 - s/k) - h,$$

where s is the stock of fish, γ the intrinsic growth rate, h is the harvest rate, and k its carrying capacity. To maximize the present value of the sum of future profits, the marginal utility from consumption $U'(h)$ equals the shadow price of the stock of fish μ:

$$U'(h) - C_h^1(h,s) = \mu.$$

The optimization problem involves maximizing

$$\int_0^\infty (U(h) - c(h,s)) \cdot \exp(-r \cdot t)dt.$$

The marginal utility from consumption is the utility derived from consuming an extra unit of fish, and the shadow price of fish is the cost of changing the stock of fish in the reservoir by an extra unit. The shadow price reflects the forgone utility of all the extra fish this fish would have led to in future if allowed to remain in the reservoir rather than being consumed today. In steady state, the shadow price is

$$\mu = -\frac{\frac{\partial c(h,s)}{\partial s}}{r - \frac{\partial \dot{s}}{\partial s}}.$$

The numerator term on the right-hand side represents the change in the cost of catching fish resulting from a marginal change in fish stock. This term affects the future utility as a decrease in stock could lead to higher catch costs. The term in the denominator has the discount rate in addition to the partial of the growth in stock of fish with respect to a change in its stock. This extra term changes the interpretation of the discount rate. When the growth in the stock of fish is increasing in its own stock, the term $\partial \dot{s}/\partial s$ is positive and works to lower the denominator.

The shadow price can also be interpreted now as the sum of this increased cost of harvesting from a marginal reduction in fish stock from now to infinity. A lower discount rate increases this extra cost, and so does a higher growth rate in the stock of fish. The idea is if the future is valued more by choosing a lower discount rate, reducing a stock of fish marginally is more costly in terms of forgone future utility.

DISCOUNTING: HYSTERESIS AND NONLINEAR FEEDBACKS

Discounting becomes more complicated within complex biological systems. Suppose in the fishery example the individual derives utility by emitting effluents into the lake, which after some level of pollution flips into a degraded state. The level of stock of pollution evolves as

$$\dot{s} = h - \delta \cdot s + \frac{s^{\alpha}}{s^{\alpha} + \beta},$$

where h is the effluent dumped into the lake and δ is the natural rate of purification of the lake. The term $s^{\alpha}/(s^{\alpha} + \beta)$ captures the phenomenon of hysteresis in which the system flips into a degraded state due to this extra feedback effect from the stock of pollutants crossing a certain threshold. The individual maximizes utility is

$$\int_{0}^{\infty} (U(h) - d(s)) \cdot \exp(-\rho \cdot t)dt,$$

where ρ is the discount rate. In a steady state, the shadow price of stock of pollutants is

$$\mu = -\frac{d'(s)}{\rho + \delta - \frac{\partial \frac{s^{\alpha}}{s^{\alpha} + \beta}}{\partial s}}.$$

The denominator has three terms. Term δ augments the discount rate. A higher regenerative capacity of the system removes the stock of pollutants faster and promotes more effluent flow into the lake by lowering the discount rate. The third term in the denominator is the rate of change of the hysteresis function with respect to stock. This term is positive in sign and increasing until the threshold is crossed. The overall implication of this process is to lower the discount rate, but by a low amount in the beginning and by a high amount as the stock of pollution reaches the threshold. This implies the amount of effluents being loaded into the lake must be cut back on once the threshold is neared.

We see that nonlinear effects have a significant effect on how the discount function evolves in a dynamic environment. When the feedback effects are positive, such as through ecosystems exhibiting resilience in the stock of environment, the effect on the shadow price is similar. An example would be of ecosystems becoming more immune to invasion from alien plants after a certain stock level is reached or fisheries becoming less prone to extinction from predator fish invasion once they exceed a certain size. When such stock effects are nonlinear but positive, the shadow price of stock of environment also evolves nonlinearly due to the above third term. The cost of reducing the stock in this case, when it is close to becoming resilient, is high—as a resilient state would mean higher stocks in future or a lower risk of extinction and a higher reward from consumption. The same analysis could be extended to the case when the risks of such changes are nonlinear too.

DISCOUNTING: BEHAVIORAL COMPLEXITIES

Traditionally, modelers use constant discounting to determine present values. Empirical evidence, however, suggests humans and animals do not use a constant discount rate; rather, their implicit discount rate declines as the benefits received move farther and farther into the future—hyperbolic discounting. This empirical evidence has led to a great deal of research, as well as disagreement, on the role of hyperbolic discounting in economic analysis. Most of this work on hyperbolic discounts looks for evidence to either support or reject the use of hyperbolic discounting, whether it is the choice of the correct discount rate or the correct discount model.

Researchers have explored several different models of hyperbolic discounting. A general form of hyperbolic discounting is

$$PV = \frac{p(t)}{\left(1 + \frac{r}{h(t)}\right)^t},$$

where PV is the present value of a future benefit, $p(t)$, collected at time t, r is the annual discount rate, and $h(t)$ is an increasing function of time. Here,

$$\frac{1}{\left(1 + \frac{r}{h(t)}\right)^t}$$

is the discount factor used to discount a future benefit to the current time. If $h(t)$ is equal to 1, this model is the constant discounting model. In constant discounting, the discount factor decreases asymptotically toward zero as we move further into the future—future events are discounted toward zero. If $h(t)$ is instead an increasing function of time, this is hyperbolic discounting, which allows the discount factor to stay larger than in constant discounting—future benefits are not discounted toward zero as quickly. For small values of t, events close to the present, $h(t)$ is small and the discount rate at time t is close to r—for events close to the present, hyperbolic discounting is a fair representation of constant discounting. As we move into the future, t becomes larger, $h(t)$ becomes larger, and the discount rate declines, increasing the discount factor. As events take place in the distant future, the difference between constant discounting and hyperbolic discounting becomes larger. Another model describing the discount rate as a discount weight is $w_t = 1/(1 + r)^{\phi(t)}$, where $\phi(t)$ is a function of how an individual perceives time passing. Note $\phi(t)$ plays the same role as $h(t)$. Changing the form of $\phi(t)$ changes the discount factor so it can be either constant or hyperbolic discounting: if $\phi(t) = t$, the discount factor is constant discounting; if $\phi(t)$ is concave in time, the discount factor is hyperbolic.

SEE ALSO THE FOLLOWING ARTICLES

Ecological Economics / Ecosystem Valuation / Fisheries Ecology / Resilience and Stability

FURTHER READING

Armsworth, P., B. Block, J. Eagle, and J. Roughgarden. 2010. The role of discounting and dynamics in determining the economic efficiency of time-area closures for managing fishery bycatch. *Theoretical Ecology* 5: 1–14.

Frederick, S., G. Loewenstein, and T. O'Donoghue. 2002. Time discounting and time preference: a critical review. *Journal of Economic Literature* XL: 351–401.

Goulder, L., and R. Stavins. 2002. Discounting: an eye on the future. *Nature* 419: 673–674.

Guo, J., C. Hepburn, R. Tol, and D. Anthoff. 2006. Discounting and the social cost of carbon: a closer look at uncertainty. *Environmental Science & Policy* 9: 205–216.

Lind, R., ed. 1982. *Discounting for time and risk in energy policy.* Baltimore, MD: Johns Hopkins University Press.

Weitzman, M. 2001. Gamma discounting. *American Economic Review* 91: 260–271.

DISEASE DYNAMICS

GIULIO DE LEO AND CHELSEA L. WOOD

Hopkins Marine Station of Stanford University, Pacific Grove, California

Disease agents—organisms that live in spatially close and temporally durable associations with hosts that suffer a fitness cost—affect every species on the Earth and include parasitic organisms as various as prions, viruses, bacteria, protozoa, helminths, and arthropods. The changes experienced by populations of pathogens and their hosts are known as disease dynamics, and they can include short- and long-term shifts in abundance, species composition, and distribution. Because diseases are difficult to track and manipulate in nature, the study of disease dynamics has benefited from a variety of mathematical modeling approaches, which have been particularly useful in developing theory for patterns and drivers of disease, as well as strategies for disease control.

OVERVIEW

Every living thing on Earth is affected by one or more disease agents—organisms that live in spatially close and temporally long associations with their hosts and that may also be called parasites, pathogens, or infectious agents. Because the host, by definition, suffers a fitness cost as a result of the association, and because diseases are so ubiquitous, they can be a critical determinant of the structure and function of populations, communities, and ecosystems.

The term *disease dynamics* refers to the short- and long-term changes in number, type, and distribution of diseased individuals in a population that are studied as a branch of the field of epidemiology. The theory of disease dynamics began its development in the first half of the twentieth century, with attempts by Ross and MacDonald to model the dynamics of malaria in humans and mosquitoes, and with work by Kermack and McKendrick to explain the rapid rise and fall in the number of plague cases in the 1665–1666 London outbreak. But only in the early 1980s did this research area become a central component of the epidemiology of infectious diseases. This flowering of interest was mainly due to the seminal work of Anderson and May, who demonstrated that disease dynamics are a powerful lens through which to view public health; specifically, they showed that the study of disease dynamics could explain the patterns of boom and bust that are typical of many infectious diseases, reveal the processes and factors that allow for a pathogenic agent to invade and possibly establish in a host population, and identify conditions under which novel strains arise and outcompete or coexist with dominant strains. Anderson and May's work also showed that the analysis of disease dynamics allows us to derive statistical and modeling tools to predict the potential course of an epidemic and to develop effective disease control strategies (including vaccination, quarantine, and, for animal diseases, culling).

While the study of disease dynamics, as part of epidemiology, can be strongly data-driven and makes use of advanced statistical tools for the analysis of temporal and spatial patterns of disease, the majority of the theoretical work in this field is inherently tied to the use of mathematical models. Models are a very effective approach for formalizing, in a simplified context, alternative hypotheses on disease spread. Moreover, through analytical solutions or computer implementation (necessary for more complex models), mathematical models allow us to simulate the course of the epidemics under alternative intervention and eradication strategies or under any variety of conditions that may be difficult to manipulate or observe in nature.

THE BASIC CONCEPTS

Within-Host Dynamics: The Course of an Infection and Disease Profiles

The course of an infection within a single host is depicted in Figure 1. Soon after the inoculation of a susceptible host with infective propagules at time $t_0 = 0$, the infective agent begins to grow or replicate within its host. In the first phase of infection, the agent may not produce enough propagules for the host to become infective—that is, for the host to be competent to transmit the disease to another individual. The host is thus exposed but not yet infective. The duration of this phase depends upon the

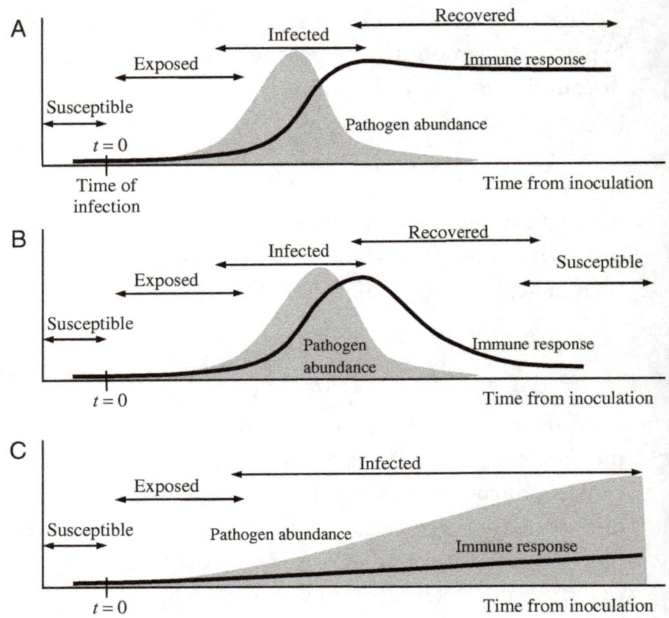

FIGURE 1 A conceptual diagram illustrating the course of infection. Gray shading represents pathogen abundance, and the black line represents host immune response for several levels of host health status: (A) pathogen triggers a permanent immune response; (B) pathogen triggers temporary immune response; (C) long-lasting pathogen for which the host is not able to develop an effective immune response (elaboration from Figure 1.2 in Keeling and Rohani, 2008).

nature of the infective agent, its within-host reproductive rate, the health status of the host, and the host's innate or acquired immunity. The phase can range in length from a few days (as in flu) to years (as in AIDS). If the pathogen is able to defeat the host's initial immunological barriers, the host is then considered infected and infective and begins to shed propagules that can invade other susceptible hosts.

After an initial infection, the host may be able to mount an effective immune response and eliminate the infective agent. The host has thus recovered, and immunity to further infection by the same or related pathogens may be permanent (as in Fig. 1A) or temporary (as in Fig. 1B). For some diseases, such as gonorrhea, AIDS, or infection with parasitic worms, the host might not be able to develop immunity and infection can be long lasting or recurrent (as in Fig. 1C). While there is usually a smooth transition between being exposed and infected or infected and recovered, we can qualitatively divide the stages of infection into simple categories or compartments reflecting the health status of the host: in the cases reported in Figures 1A and 2A, the profile of the disease is referred to as SEIR (susceptible, exposed, infective, recovered).

If immune protection is not conferred by infection or is impermanent, a recovered individual may return to

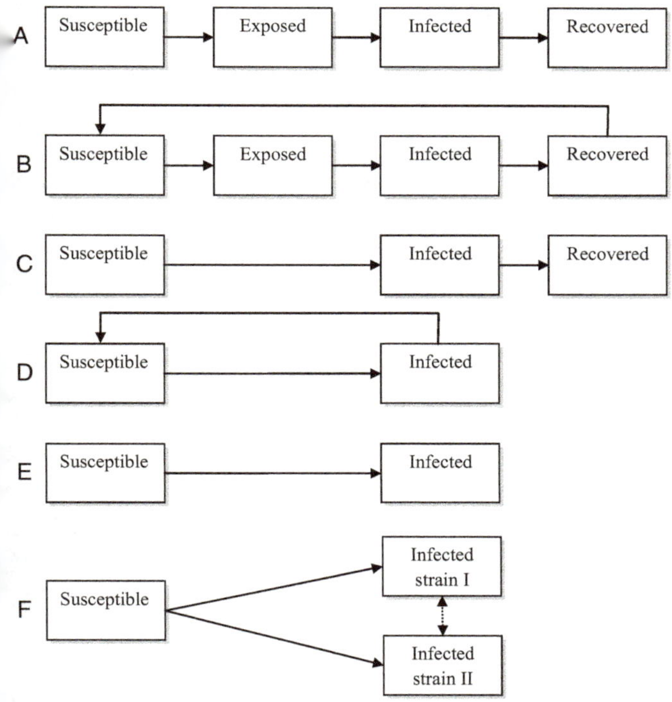

FIGURE 2 A schematic representation of some disease profiles. (A) SEIR with permanent immunity; (B) SEIR with temporary immunity; (C) SEI; (D) SIS; (E) SI; (F) disease profile when host can be infected by two strains of the same pathogen with or without cross-immunity.

the initial susceptible state, as represented in Figure 2B. For some diseases, infection evolves so rapidly that the exposed class can be ignored (Fig. 2C) so as to obtain an SIR (susceptible, infective, recovered) profile. In the case of some diseases, prior infection does not confer immunity, so once the disease has been cleared, the host is again fully susceptible, yielding an SIS profile (susceptible, infective, susceptible; Fig. 2D). For a fatal disease, recovered individuals do not exist, and the disease can be considered to have an SI profile (susceptible, infective; Fig. 2E). In some cases, more than one strain is circulating in the population and either one or the other can infect individuals; additionally, depending on whether there is some degree of cross-protection, an individual infected with one strain may be protected against the other strain (Fig. 2D).

Types of Infective Agents: Microparasites and Macroparasites

Figures 1 and 2 represent only a few of many possible disease profiles. Some infectious diseases may cause a long-lasting state of infection from which the host never recovers, such as AIDS. Yet, with some notable exceptions, the course of infection depicted in Figures 1A and 1B can be considered representative of disease often caused by *microparasites*, or disease agents that replicate within their host and whose offspring infect the same host individual, such that the initial number of infective propagules is decoupled from the pathogenicity of the infection (i.e., in principle, a single infective propagule could produce a full-blown infection). Infection by microparasites is often transient, such as in the case of influenza. In contrast, *macroparasites* replicate within their host, but their offspring are shed from the parent's host and go on to infect other host individuals; thus, pathogenicity is highly dependent upon the number of propagules initially infecting the host (i.e., infection with one propagule will produce low pathogenicity, but infection with many propagules will produce high pathogenicity). While there is substantial overlap in natural history characteristics and taxonomic grouping between microparasites and macroparasites, there are some general patterns. For example, microparasites tend to possess small body size, to produce a strong host immune response, and to be protozoa, bacteria, and viruses, whereas macroparasites tend to be larger, to produce a weak host immune response, and to be helminths and arthropods. Infection with a macroparasite can be long lasting, and the distribution of macroparasites in the host population is often very uneven, with a substantial fraction of hosts harboring few or no parasites and a few hosts harboring the majority of parasites.

The distinction between micro- and macroparasites has important consequences in terms of the conceptual and quantitative description of the resulting disease: in fact, in the case of microparasitic infection, it is generally considered safe to neglect the actual pathogen dynamics and abundance within the host and simply classify hosts as susceptible, exposed, infected, recovered, and temporarily or permanently immune. The dynamics of microparasitic infection are thus usually described through compartmental models as represented in Figure 2, in which only the number or proportion of individuals in each infective class are tracked. In the case of macroparasites, the quantitative aspects of infection (i.e., the number of parasite propagules initially infecting a host) are relevant and the actual distribution of parasites in their host should be taken into account explicitly.

Chronic versus Acute Infections

At the individual level, an infection is defined as chronic if it is long-lasting with respect to the lifetime of the host, as in the case of AIDS and many helminth infections (e.g., schistosomiasis), or recurrent, as in the case of malaria relapses caused by the reemergence of blood-stage parasites from latent stages in the liver. Acute diseases are those that either cause a rapid infection and/or have a short course, such as flu. Whether a disease is acute or chronic is unrelated to its pathogenicity: for example, the common cold virus produces an acute infection of low pathogenicity, and the filarial nematode *Onchocerca volvulus* produces river blindness, a chronic infection of high pathogenicity.

Endemic versus Epidemic Dynamics

At the population level, a disease is said to be endemic when it is maintained locally at a fairly constant (and possibly low) abundance, so that transmission occurs without the need for introduction of infected cases from outside the population. A disease becomes epidemic when it spreads rapidly through a population or among populations, with the increase in infected individuals exceeding typical rates of increase. A pandemic is a global epidemic affecting several continents simultaneously. For example, chickenpox is endemic in the United States, but malaria is not. Every year, there are a few cases of malaria acquired in the United States, but these do not lead to sustained transmission in the population due to the low number of vertebrate reservoirs and mosquito-control efforts.

Endemicity is not a fixed feature of the infective pathogen or a specific disease, but is instead the result of a combination of factors, including the size of the susceptible population and the fraction of individuals with innate or acquired immunity. A single disease can have an initial epidemic phase when it invades a naïve population and may only later become endemic.

An example of endemic disease is reported in Figure 3A for chlamydia in Tennessee. Figures 3B and 3C report the typical boom-and-bust dynamics of plague in London in 1665–1666 and of measles in London in 1955–1956, respectively. On a longer time scale, these diseases show very different dynamics: sporadic and irregular outbreaks for plague (Fig. 3D) and recurrent outbreaks for measles (Fig. 3E). In the case of plague, the disease was not able to maintain itself in the population and required repeated spillover from the wildlife reservoir, i.e., the rodent population. For measles, the dynamics reveal a typical biannual pattern of the epidemics, but the disease does not fade out during interepidemic years.

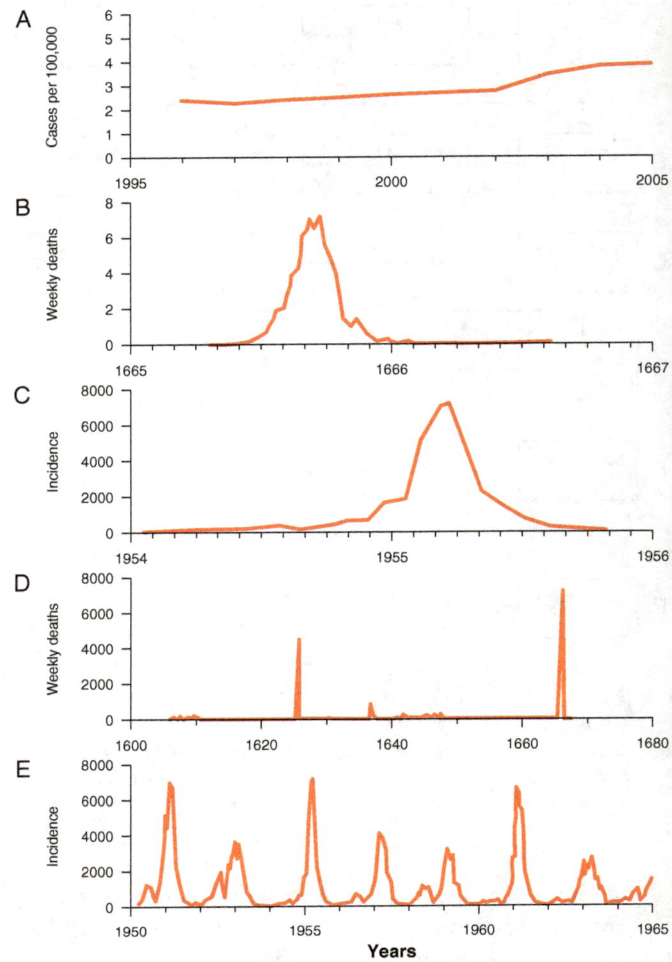

FIGURE 3 (A) Cases of chlamydia per 100,000 in Tennessee; (B) weekly deaths for plague in London in 1665–1666; (C) number of cases of measles in London between 1954 and 1956; (D) weekly deaths for plague in London in the seventeenth century; (E) number of cases of measles in London between 1950 and 1964 (elaboration from Figure 1.2 in Brauer et al., 2008).

TYPES OF TRANSMISSION

Transmission (i.e., the passing of a communicable disease from an infected host individual to a susceptible host individual) is a key process affecting the dynamics of infectious disease. Terms for transmission strategies vary between the epidemiology and parasitology/ecology literatures. In parasitology and ecology, transmission strategies are typically divided into two types: a directly transmitted parasite can complete its entire life cycle using only one host individual, whereas an indirectly transmitted parasite requires multiple host species to complete one life cycle, with each host obligately required for a particular parasite life stage. For example, while *Yersinia pestis*, the plague bacterium, can complete its entire life cycle in one human individual (directly transmitted), *Schistosoma mansoni*, the trematode that causes schistosomiasis, must pass through a larval stage in freshwater snail hosts before infecting a human and reaching adulthood in the human host (indirectly transmitted). Direct transmission may be accomplished through direct contact or close proximity with an infected host (e.g., the common cold), sexual contact (e.g., AIDS), contact with environmental stages of a parasite (e.g., hookworm), contact with infected feces, often through contamination of water or food (e.g., cholera), ingestion of infected tissues (e.g., prion diseases transmitted by cannibalism), or vertical transmission (i.e., transmission from mother to offspring prenatally or during birth). Indirect transmission is often accomplished through trophic transmission (i.e., transmission from prey to predator, as in trichinosis contracted by a human who has consumed infected pork) or through free-living stages of the parasite (e.g., schistosomiasis contracted by a human who contacts a free-living cercarial stage of *Schistosoma mansoni*).

The meaning of these terms differs in the epidemiology literature, where direct transmission is defined as transmission requiring direct contact or close proximity between infected and susceptible individuals, and indirect transmission occurs through resistant propagules shed into the environment. Though the meanings assigned to the terms direct transmission and indirect transmission differ substantially among fields, it is less important to keep track of terminology than it is to note the nature of differences in life histories and transmission strategies among various groups of disease agents, because these differences will strongly influence the structure of models and outcomes of disease outbreaks.

A further distinction is between density-dependent transmission and frequency-dependent transmission. Density-dependent transmission prevails when the contact rate between infected and susceptible hosts (and, hence, the likelihood of transmission) is a function of host density (i.e., number of hosts per unit area). Such a form of transmission assumes that host contact rates are accurately modeled by mass action: that is, hosts intersect (and transmit disease) randomly within a fixed area. Density-dependent transmission is typical of wildlife populations at low density, with transmission rate being a linearly increasing function of population density. Frequency-dependent transmission, in contrast, prevails when the contact rate between infected and susceptible hosts is a function of the proportion or frequency (not the density) of infected hosts in a population and is common for diseases of hosts that do not experience disease-transmitting contact randomly but rather interact selectively with a particular subset of other individuals (e.g., sexually transmitted disease). Frequency-dependent transmission is also common for vector-transmitted diseases; transmission of malaria to a particular human, for example, depends on the prevalence of malaria in biting mosquitoes and on the number of mosquito bites received. Because frequency-dependent transmission does not abate at low densities of the host, diseases that are transmitted in this manner could drive a wildlife population to extinction (i.e., in contrast with a density-dependent disease, whose transmission would slow at low densities of the host due to low contact rates among hosts). The difference between frequency- and density-dependent transmission is thus relevant if the host population size changes.

IMPORTANT EPIDEMIOLOGICAL PARAMETERS FOR DISEASE DYNAMICS

A crucial parameter in the dynamics of infectious disease is the reproductive number, indicated with R_0, which represents the number of secondary infections caused by one infected individual entering a completely susceptible population. In fact, the ability of a pathogen to invade a population can be determined by the R_0 value: if R_0 is greater than 1, the disease is able to invade the population, but if it is less than 1, it is unable to invade the population and will fade out. R_0 is not an intrinsic trait of a pathogenic agent but depends upon many demographic and epidemiological parameters, including host abundance and density, host social structure and the nature of host contact processes, previous exposure and immunity, the ability of the pathogenic agent to bypass host immune defenses, and the type of transmission. As a consequence, the same disease might exhibit a different R_0 in different communities or populations. In general, R_0 increases with the abundance/fraction of susceptible host in the population, the frequency of contacts between infected and susceptible individuals, the probability that such contacts will communicate the disease to a susceptible individual—usually summarized in a single parameter β, called the transmission rate—and the life expectancy G of an infected individual. For fatal or transient infection, G is inversely proportional to the disease-induced mortality rate α (i.e., $G = [\alpha + \mu]^{-1}$, where μ is the average natural mortality rate of a healthy individual). Because direct measures of the transmission rate β are rarely available, an approximation of R_0 can be obtained on the basis of the age at first infection and mean life expectancy of the host; that is, $R_0 = L/A + 1$, where $L = \mu^{-1}$ is the life expectancy of a healthy individual and A is the mean age at infection.

Another important epidemiological parameter is the force of infection λ: the per capita rate at which susceptible individuals become infected. For simple SI models, $\lambda = \beta I$, where I is the number of infected individuals in the population, in the case of density-dependent transmission and variable population size; in the case of frequency-dependent transmission, $\lambda = \beta I/N$, where I/N is the proportion of infected individuals.

SIMPLE MODELS OF DISEASE DYNAMICS

In 1927, Kermack and McKendrick developed a simple but effective compartmental model to describe the dynamics of a nonlethal, infectious disease in a closed population of constant size:

$$\dot{S} = -\beta SI,$$
$$\dot{I} = \beta SI - \gamma I,$$
$$\dot{R} = \gamma I,$$

where S, I, and R are the proportions of susceptible, infected, and recovered individuals in the population ($S + I + R = 1$), the dot on the left-hand side of each equation indicates differentiation with respect to time (i.e., d/dt), β is the transmission rate, and γ is the rate at which infected individuals clear their infections and move into the recovered class. In this case, the reproductive number in an entirely susceptible population is $R_0 = \beta/\gamma$ and, when $R_0 > 1$, the disease exhibits typical boom-and-bust dynamics, as depicted in Figure 4A. The disease eventually fades out and the final number of susceptible individuals is greater than zero, so a proportion of the population never becomes infected.

This system can be modified to account for host turnover in a constant population:

$$\dot{S} = \mu - \beta SI - \mu S,$$
$$\dot{I} = \beta SI - \gamma I - \mu I,$$
$$\dot{R} = \gamma I, - \mu R,$$

where μ is the influx of new susceptible individuals (i.e., the birth rate). The disease does not fade out when $R_0 = \beta/(\gamma + \mu) > 1$, but instead reaches an endemic equilibrium through dumped oscillations, i.e., oscillations that gradually fade out, as shown in Figure 4B.

For wildlife disease, host population is usually variable in time and thus host population dynamics must be explicitly modeled. Anderson et al. (1981) provides a general model for rabies:

$$\dot{S} = \nu S - (\mu + \eta N)S - \beta SI,$$
$$\dot{E} = \beta SI - \sigma I - (\mu + \eta N)I,$$
$$\dot{I} = \sigma E - \alpha I - (\mu + \eta N)I,$$

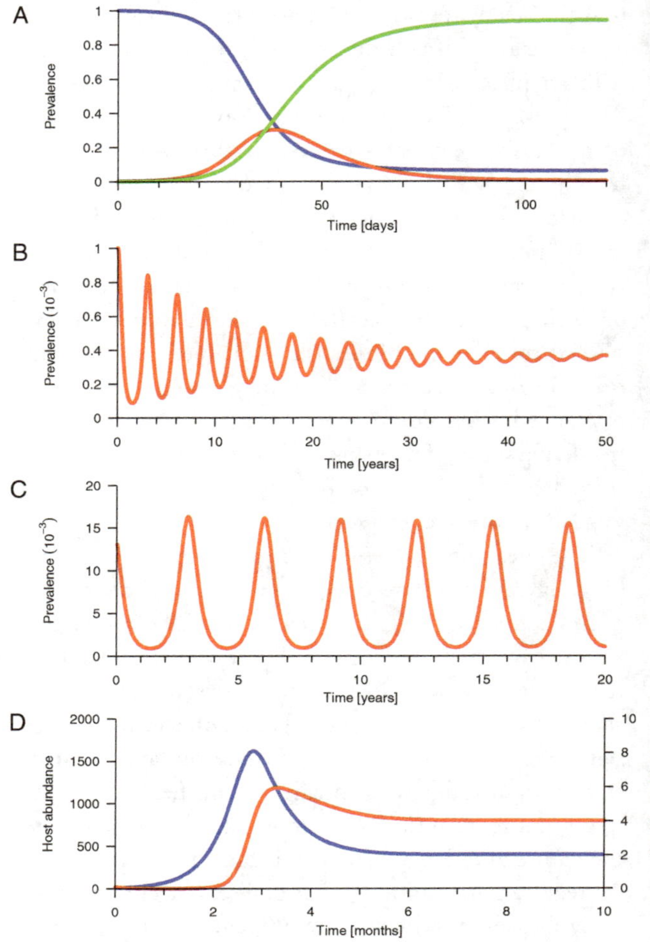

FIGURE 4 (A) Disease dynamics of an SIR model for a nonfatal disease in a closed population of constant size. This figure is plotted assuming that the time spent in the infected class $1/\gamma$ is 10 days and $R_0 = 3$ (thus $\beta = 0.3$ per day). Initial fraction of susceptible (in blue) was set to 0.999, infected (in red) to 0.001, and recovered (in green) to 0. At the end of the epidemic, the fraction of individuals that never got infected is 5.9% of the population. (B) SIR model with host demography and a constant population size. Here, $1/\gamma = 10$ days, the average life expectancy $1/\mu = 70$ years, $R_0 = 10$ (and thus $\beta = 365$ per year). (C) Fraction of infected individuals in SEI model of fox rabies in Europe, parameterized as in Anderson et al. (1981), i.e., $\nu = 1$ per year, $\mu = 0.5$ per year, $K = 8$ individuals per km^2 (and thus $\gamma = 0.063$). The average time spent in the exposed class $1/\sigma$ was set to 30 days and R_0 to 8 (thus $\beta = 80$ per year). (D) The dynamics of the host-macroparasite model of Anderson and May (1978). Here, $a = 3$, $b = 1$, $\alpha = 0.5$, $\mu = 0.1$, $\lambda = 11$, $k = 0.1$. Host density is in blue, and mean parasite burden P/H is in red (i.e., the mean number of parasite per host).

where S, E, and I are the number of susceptible, exposed but not yet infective, and infective individuals, respectively, ν is the natality rate, μ is the natural mortality rate, η is the density-dependent mortality, σ is the rate at which exposed individuals move into the infected class, and α is the disease-induced mortality rate. The population experiences logistic growth when disease-free: $r = \nu - \mu$ is the per capita growth rate at low density and $K = (\nu - \mu) / \eta$ is the disease-free carrying capacity of the population. The disease can invade and establish in the population if $R_0 = \sigma\beta K (\sigma + \nu)^{-1} (\alpha + \nu)^{-1} > 1$. The presence of the exposed class slows down disease transmission, and for large values of R_0 and specific combinations of the model parameters, disease dynamics can be characterized by limit cycles (i.e., undumped oscillations), as shown in Figure 4C.

A different model formulation is required to describe the dynamics of macroparasites like intestinal nematodes. In this case, in fact, the distribution of parasites in the host population can be assumed to be a negative binomial with clumping parameter equal to k (the smaller k, the more aggregated the distribution, with only a few hosts harboring the majority of parasites). Anderson and May showed in 1978 that the dynamics of this host–macroparasite system can be described by the following model:

$$\dot{H} = (b - d)H - \alpha P,$$

$$\dot{P} = \frac{\beta PH}{H_0 + H} - (\mu + d + \alpha) P - \alpha \frac{k+1}{k} \frac{P^2}{H},$$

where H and P are the numbers of host and parasite, respectively, b and d are the host natality and mortality rate, μ is the parasite natural mortality rate, α is the per capita parasite-induced mortality rate, β is the rate at which infective stages are acquired by the host, and H_0 is a parameter equal to η/λ, where η and λ are the natural mortality and the rate of production of free-living stages, respectively. Anderson and May showed that under specific conditions of model parameters, a macroparasite is able to control the population size of an otherwise Malthusian host (i.e., a host that would otherwise increase indefinitely), as depicted in Figure 4D.

MORE COMPLEX DYNAMICS

The simple compartmental models reported above are very useful for understanding disease dynamics from first principles. Yet, the assumption of mass action (i.e., homogeneous mixing of hosts) and the assumption of constant parameters are often too simplistic to accurately predict disease dynamics in the real world. To improve accuracy of model predictions, there are several types of heterogeneities that should be taken into account, such as seasonality, variability in contact rates, stochastic fluctuations, age structure, social structure, spatial dynamics, broad host ranges for parasites (i.e., parasites with multiple host species), diverse parasite faunas for hosts (i.e., hosts with multiple parasite species or parasite strains), and the evolution of virulence.

The Effect of Seasonality

Seasonality can affect several aspects of host demography, especially the contact rate. In wildlife, reproduction is often seasonal and produces a pulse of newborns that lack acquired immunity to many parasites and therefore may substantially increase the pool of susceptible individuals. Natural mortality may also be higher, and immunity lower, during periods of harsh environmental conditions. The contact process is especially likely to exhibit periodic fluctuations. For example, in human populations, the sharp increase in contact rate at the beginning of the school year or in coincidence with seasonal markets or religious festivities can produce a spike in disease. The inclusion of seasonality—described either through smooth sinusoidal, pulse, or stepwise term-time functions—is thus crucial for predicting the course of a disease outbreak, as shown in the case of measles and influenza. Seasonally forced models can exhibit much more complex dynamical patterns that nonforced models, including chaotic behavior and multiyear periodicity. In this case, disease outbreaks may occur every several years, interspersed with long endemic phases at very low prevalence (Fig. 5).

Demographic Stochasticity

The number of infected individuals during the endemic phase of a disease outbreak could be so small that the disease may fade out in the population just by chance, a possibility that cannot usually be accounted for in classical deterministic models. It is thus increasingly common for models of control and eradication measures to provide a description of demographic and epidemiological processes in a stochastic framework. A common way to account for the random nature of transmission events is to use an event-driven approach where the population abundance in each infective class is described by integer numbers and it is assumed that only one individual at a time, rather than fractions of population density, can move from one class to another according to the specific rates of birth, transmission, recovery, and death. This class of models is usually much more computationally intensive, as it requires many replicates to derive average dynamical patterns. The advantage is that such models provide a more realistic way of accounting for chance events and inherent variability in transmission and in other demographic parameters.

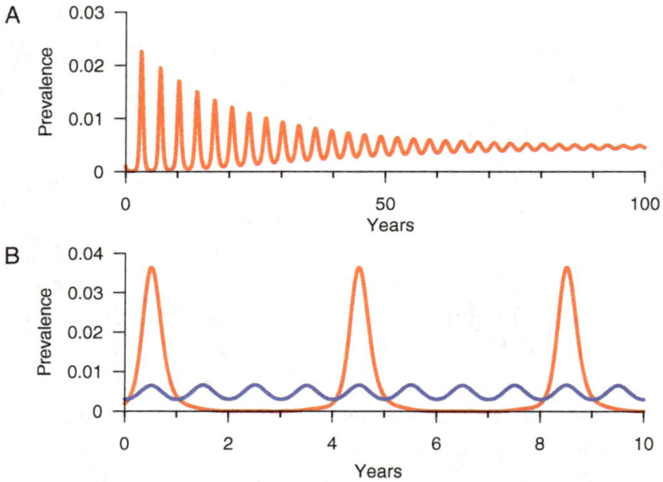

FIGURE 5 (A) Prevalence of infected individuals in an SEI model. Parameters are as in Figure 4D, except for the carrying capacity K = 4 individuals per km^2 (and, consequently, R_o = 4). The model reaches an endemic equilibrium through dumped oscillation. (B) Model parameters are here as in part A, but the transmission rate exhibits regular seasonal fluctuations of 1-year periodicity, i.e., $\beta(t) = \beta_m[1 + \epsilon \sin(2\pi t)]$, where β_m is the same mean transmission rate used in part A) and ϵ = 0.2 is the strength of the seasonal forcing. The two trajectories in red and in blue represent disease prevalence for the same set of parameters but two different initial conditions: in fact, the seasonally forced SEI model may exhibit multiple coexisting attractors for the same set of parameters. Here, the first attractor is the small period-1 epizootic cycle generated by the seasonal forcing function (in blue) and corresponds to the long-term endemic equilibrium of the nonseasonal model (A). The second attractor is a multi-annual cycle characterized by an outbreak occurring every 4 years (in red), i.e., a multiple integer of the period (1 year) of the seasonal forcing function (a phenomenon known as subharmonic resonance).

Age-Structured Models

Age structure is important to consider in models of disease dynamics whenever susceptibility to an infective pathogen, disease-induced mortality, or transmission rate change substantially with the age of the host, as in many wildlife and childhood diseases. For instance, risk of schistosomiasis in Africa varies strongly with age due to differences in behavior among age groups. Overall metazoan parasite burdens increase with age in marine fish and gastropods because such parasites are accumulated over time. The mean age of infection for measles is very low and transmission is highest for pre-school and school children than for younger or older age groups. The inclusion of age structure and age-dependent parameters is thus crucial for a realistic description of disease dynamics and for assessing the optimal age of vaccination.

Social Structure and Patterns of Contact

Another factor that complicates the prediction of disease dynamics is nonhomogeneous mixing, a process that occurs whenever individuals contact one another in a nonrandom manner, such that disease transmission is no longer a function of the random encounters defined by mass action. For example, nonhomogeneous mixing is crucial in the dynamics of sexually transmitted diseases (STDs), where a handful of highly promiscuous individuals (e.g., prostitutes) can have contact rates much higher than the population's average and therefore exert a disproportionate influence on disease transmission. The inclusion of nonhomogeneous mixing is thus crucial for the accurate description of STD dynamics and the dynamics of other diseases influenced by nonrandom patterns of contact.

Spatial Dynamics

A special form of nonhomogeneous mixing occurs when we account explicitly for the spatial distribution of individuals in a population. In plant diseases, for instance, there is a greater chance for airborne disease to infect nearest neighbors, even though occasionally the wind may transport infective propagules far away. Transmission is also a localized process for many wildlife and human diseases. Several methods have been devised to describe the dynamics of such cases, including reaction–diffusion models using partial differential equations, metapopulation models (a collection of weakly connected subpopulations), coupled lattice models (specialized metapopulation models where subpopulations are arranged on a grid and coupling is generally to nearest neighbors only), cellular automata (a lattice of sites with each site generally assumed to hold a single host), individual-based models, and different types of spatial networks (random, scale-free, and small-world; Fig. 6). Each approach has its advantages and limitations and can be more suitable to describe the dynamics in time and space under specific conditions.

Multihost/Multistrain Dynamics

Hosts may be infected by several different pathogens or pathogenic strains that can confer either partial cross-immunity (i.e., exposure to one species or strain confers immunity to multiple species or strains, as for some human influenzas) or, on the contrary, can exacerbate disease virulence, as in the case of the dengue fever, where cross-infections with different strains can cause deadly hemorrhagic fevers. Moreover, there are pathogens that may affect multiple host species, such as *Trichinella spiralis*, which can cause trichinosis in a great number of mammal species. As the majority of infectious diseases are zoonotic—that is, they originate in a nonhuman host and can be transmitted from an animal species to humans—it

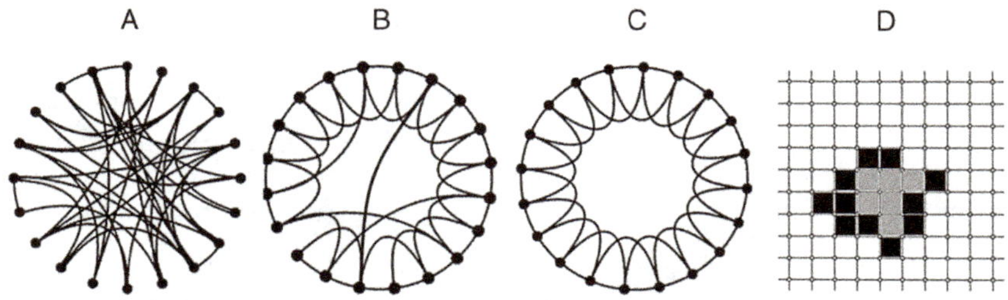

FIGURE 6 Four distinct ways of accounting for nonhomogeneous mixing and spatial dynamics: (A) random network; (B) small-world; (C) regular lattice; (D) cellular automata (an individual per cell, black: infected, gray: recovered). In all four graphs, the average number of contacts per individual is approximately 4.

will be crucial in the future to derive a new generation of models able to provide more realistic descriptions of multihost, multipathogen, and multistrain dynamics and to address a broader set of parasitic life histories.

PRACTICAL USES OF THEORY

The study of disease dynamics has proven to be extremely useful for three main goals of epidemiology and public health: (1) to understand patterns and drivers of disease, (2) to predict future changes in disease, and (3) to develop methods for manipulating and possibly eradicating disease. The most common approaches to disease control include vaccination, reduction of the transmission rate (e.g., through quarantine), and—in the case of wildlife and livestock diseases—culling (i.e., killing animals to reduce their population density). Each of these measures can be entirely effective or utterly useless for manipulating disease dynamics, and their effectiveness can be predicted through the use of the mathematical models we review above. For instance, mathematical models reveal that the proportion of a population that must be vaccinated to eradicate a disease through herd immunity (a form of immunity that occurs when the vaccination of a significant portion of a population provides a measure of protection for individuals who have not developed immunity and are not vaccinated), needs to exceed the threshold value $1 - 1/R_0$. Models also reveal that targeting vaccination toward high-risk groups, such as those that disproportionately influence STD transmission, generally is more effective than vaccination of randomly selected individuals. This type of analysis can save a substantial amount of time, effort, money, and risk by preventing unnecessary vaccinations.

The variety of theoretical approaches available for exploring disease dynamics has allowed scientists to pose many interesting questions about how and why disease changes and what we can do to manipulate these changes. Though many aspects of the epidemiology of infectious disease remain to be explored, research to date has highlighted the value of mathematical models in providing experimental tractability and manipulability that natural disease systems often lack.

SEE ALSO THE FOLLOWING ARTICLES

Cellular Automata / Compartment Models / Epidemiology and Epidemic Modeling / Individual-Based Ecology / Networks, Ecological / Population Ecology / Spatial Models, Stochastic / Spatial Spread

FURTHER READING

Altizer, S., A.P. Dobson, P. Hosseini, P. Hudson, M. Pascual, and P. Rohani. 2006. Seasonality and the dynamics of infectious diseases. *Ecology Letters* 9(4): 467–484.

Anderson, R.M., and R.M. May. 1978. Regulation and Stability of Host-Parasite Population Interactions: I. Regulatory Processes. *Journal of Animal Ecology* 47(1): 219–247.

Anderson, R.M., H.C. Jackson, R.M. May, and A.M Smith. 1981. Population dynamics of fox rabies in Europe. *Nature* 289: 765–771.

Bolzoni, L., M. Gatto, A.P. Dobson, and G.A. De Leo. 2008. Allometric scaling and seasonality in the epidemics of wildlife diseases. *American Naturalist* 172(6): 818–828.

Brauer, F., van den P. Driessche, and J. Wu (Eds.). 2008. *Mathematical Epidemiology*. Berlin: Springer-Verlag.

Diekmann, O., and J.A.P. Heesterbeek. 2000. *Mathematical epidemiology of infectious diseases: model building, analysis and interpretation.* Wiley Series in Mathematical and Computational Biology. Chichester, UK: Wiley.

Ferguson, N.M., D.A. Cummings, S., Cauchemez, C. Fraser, S. Riley, A. Meeyai, S. Iamsirithaworn, and D.S. Burke. 2005. Strategies for containing an emerging influenza pandemic in Southeast Asia. *Nature* 437(7056): 209–214.

Ferrari, M.J., R.F. Grais, A.J.K. Conlan, N. Bharti, O.N. Bjornstad, L.J. Wolfson, P.J. Guerin, A. Djibo, and B.T. Grenfell. 2008. Seasonality, stochasticity and the dynamics of measles in sub-Saharan Africa. *Nature* 451: 679–684.

Keeling, M.J., and P. Rohani. 2008. *Modeling infectious disease in humans and animals.* Princeton: Princeton University Press.

Lloyd-Smith, J.O., D. George, K.M. Pepin, V.E. Pitzer, J. Pulliam, A.P. Dobson, P.J. Hudson, and B.T. Grenfell. 2009. Epidemic dynamics at the human–animal interface. *Science* 326(5958): 1362–1367.

McCallum, H., N. Barlow, and J. Hone. 2001. How should pathogen transmission be modelled? *Trends in Ecology & Evolution* 16(6): 295–300.

Tompkins, D.M., A.M. Dunn, M.J. Smith, and S. Telfer. 2011. Wildlife diseases: from individuals to ecosystems. *Journal of Animal Ecology* 80(1): 19–38.

Truscott, J., T. Garske, I. Chis-Ster, J. Guitan, D. Pfeiffer, et al. 2007. Control of a highly pathogenic H5N1 avian influenza outbreak in the GB poultry flock. *Proceedings of the Royal Society B: Biological Sciences* 274: 2287–2295.

DISPERSAL, ANIMAL

GABRIELA YATES AND MARK S. BOYCE
University of Alberta, Edmonton, Canada

The invasion of Africanized bees from Central America, the reappearance of the endangered red-cockaded woodpecker where it was once extirpated in the pine forests of the southeastern United States, and a Canada lynx originally radiocollared in Colorado found 2000 km north in central Alberta; these instances all speak to the profound and pervasive effects of dispersal. Dispersal is most often defined as a one-way movement of an individual away from its breeding or natal area to another habitat patch. The details of dispersal, such as the minimum distance traveled to be considered dispersal (rather than other movement such as foraging) and the definition of habitat patch, are specific to the species and the research or conservation question. Dispersal can be further broken down into a three-step process: the reaction/decision to begin dispersal, a transient or search phase, and the settlement or habitat-selection phase.

DISPERSAL AND POPULATION BIOLOGY

Dispersal is a building block for population biology because it constitutes a vital rate. It is part of the fundamental parameters (i.e., birth, death, immigration, emigration) that govern population growth. As such, dispersal plays a part in all of the major topics in population biology, including individual fitness, genetics, population dynamics, and the distribution of a species. Lagged dispersal can yield population cycles just like time delays in birth and death processes. On a larger scale, many species experienced a post-glacial expansion after the ice age (colonization/population dynamics). For each species, the expansion of a few individuals who survived the journey (fitness) broadened the species demographic range (distribution). However, because of founder effects, new populations have lower allelic diversity than the original population, whereas populations located near their glacial refugia still maintain higher allelic diversity (genetics).

MAJOR CONCEPTS IN DISPERSAL LITERATURE

The two basic types of dispersal are active dispersal (moving) or passive dispersal (being moved). Passive, or density-independent, dispersal is used by some fish, invertebrates, and other sessile organisms that take advantage of the energy produced by gravity, wind, currents, or other carriers to achieve spread. However, active dispersal requires a reaction or decision-making process on the part of the individual in response to one or more factors.

Stage One: The Decision to Disperse

Active dispersal is driven by a reaction/decision to disperse termed condition-dependent dispersal (sometimes narrowed to density-dependent dispersal). Condition-dependent dispersal is based on the premise that an individual will attempt to match their phenotype or internal condition against prevailing external conditions to maximize their fitness. Individuals with low fitness in their environment will have a propensity to disperse to where they might achieve increased fitness. This theoretical juxtaposition of internal vs. external conditions requires a clear understanding of these two terms. External conditions in condition-dependent dispersal might include both biotic and abiotic factors of the environment: population density, social or demographic structure, resource competition, habitat quality or size, and photoperiod or season. The internal conditions related to dispersal are basic phenotypes like body size, energy or body-fat reserves, and hormones like testosterone or corticosterone, which are further correlated with dispersal timing, survival probability, mating success, and fecundity.

Stages Two and Three: Search and Settlement Phases of Dispersal

Environment not only plays a role in the instigation of dispersal but also is the primary driver in the search and settlement phases. The spatial pattern of environmental features like climate, human disturbance, food resources, cover, predator density, competitor density, and presence of conspecifics is what governs whether an individual is able to escape their previously unfavorable conditions and settle in area where they attain increased fitness. The ability of individuals to disperse over these gradually changing habitats could be what allows the species to survive extreme conditions, such as those brought about by climate change.

Barriers to dispersal can influence both the extent of dispersal and the survival of dispersers in the search and settlement stages. Natural barriers are often species specific, such as impassible mountain ranges or rivers. Artificial barriers can be created by human land use and habitat fragmentation and also are largely species specific. As with any habitat use, barriers to movement also might affect different individuals of the same species to varying degrees. For instance, in many species of mammals adult females are much less willing to cross barriers such as inhospitable habitat when compared to males of the same species. Understanding such individual preferences can be important if conservation efforts are attempting to protect or enhance habitats specific to those individuals, such as female den sites and the like.

METHODS FOR CHARACTERIZING DISPERSAL

Dispersal is commonly treated as a fixed trait during analysis. In some species, such as insects or invasive species, it can be adequately understood as a static function. However, it is clear that for many other species it is both condition dependent and a selective process, rather than a random walk or diffusion process.

Diffusion and Correlated Random Walks

Dispersal is often conceptualized as a diffusion process, using second-order partial differential equations to characterize the spread of a population through space. The seminal work by John G. Skellam in 1951 gave an early example of this diffusion modeling for the spread of muskrats (*Ondatra zibethicus*) after their introduction into central Europe. Random walks (RWs) are related to diffusion models and are based on the idea that animals disperse in random directions (random distribution of turning angles) and distances based on some random distribution of step lengths. However, animals do not display truly random movement or Brownian motion, but rather have correlated walks where the previous direction influences the angle of their next step, i.e., directional persistence. Peter Turchin's foundation text on animal movement describes correlated random walks (CRW) as straight-line step lengths where random turning angles are autocorrelated and, in the case of biased random walks, weighted toward one location or direction. Finally, Lévy flights or Lévy walks characterize random walks that have a heavy-tailed probability distribution. These outliers that define the tails of the distribution often are used to characterize dispersal as a less common but defining and important event for a population. This above group of mathematical models describes the playing out of an underlying process that is ultimately random. CRWs and Lévy walks are modified forms of RWs because they have structure or rules that modify the randomness of movement.

Retrospective Models: Fractal D

Turchin retrospectively characterized details of movement in terms of magnitude and speed, step lengths, directionality, and measures of tortuosity. Although this type

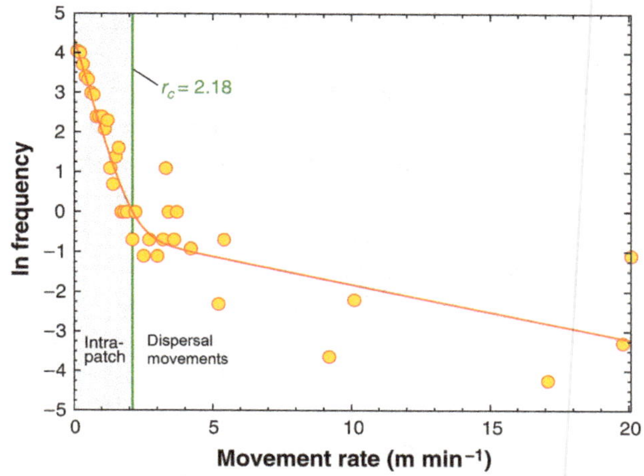

FIGURE 1 The "broken-stick" curve-fitting procedure to calculate a threshold (r_c) based on the $\log_e$ frequency distribution of movement rates for a caribou. Movement rates less than r_c represent foraging movements, whereas movement rates grater than r_c represent dispersal movements (adapted from Johnson et al., 2002, *Journal of Animal Ecology* 71: 225–235.).

of analysis was first done with small organisms and experimental systems (using insects and the like), advances in GPS radiotelemetry technology have allowed for applications with larger spatial and temporal scales. Such retrospective study can reveal switching between behaviors such as foraging vs. relocating motivations during a movement path. Fractal dimension (fractal D) and tortuosity are metrics that have been used to identify behaviors such as foraging vs. dispersal. While tortuosity generally measures the sinuosity of a path, fractal dimension is often preferred to characterize searching behavior because it measures the ability to cover a two-dimensional plane. Greater turning angles and shorter step lengths (higher fractal D) are associated with foraging behavior, and larger step lengths and less tortuosity (lower fractal D) are associated with dispersal. One example is the "broken-stick" curve-fitting procedure that defines a threshold, which separates foraging vs. dispersal movements in caribou (*Rangifer tarandus*; Fig. 1). This retrospective modeling is different than the predictive diffusion-type modeling because it characterizes an underlying nonlinear process that can give clues to behavioral mechanisms.

Statistical Models: Step Selection Functions

A new and promising analysis called a Step Selection Function (SSF) has emerged for characterizing habitats that are likely to support dispersal and other movements. The SSF, first used by Daniel Fortin and colleagues, is an extension of the widely used analysis described in Bryan Manly and colleagues' seminal book on resource selection functions (RSFs). Where an RSF maps the relative probability of habitat selection given an encounter on the landscape, an SSF maps the relative probability of taking a movement step through a landscape. Predictor variables, or covariates, are habitat attributes (e.g., shrub, conifer forest, deciduous forest, and the like). The habitat attributes associated with each step is compared with those of random steps having the same origin but with different lengths and directions. The lengths and turning angles of random steps are drawn from the distributions established from the original complete set of movement observations. Each "used" step is compared with a number (e.g., 20) of random alternative steps that the animal might have taken, to model factors that promote movement across the landscape. The analysis involves entering the candidate covariates into a conditional logistic regression. The covariate coefficients are then used to calculate the SSF typically as $w(\mathbf{x}) = \exp(\beta_1 x_1 + \cdots + \beta_p x_p)$. A set of plausible models (i.e., combinations of the candidate covariates) can be evaluated using Akaike's Information Criterion (AIC) to select the model with the minimum AIC score. By estimating an SSF for longer movements, say as identified by the broken-stick model in Figure 1, one could obtain a model of habitats that an animal selected for dispersal.

Statistical Models: State-Space Models

An alternative retrospective analysis focuses on underlying behavior like switching models but adds another layer of complexity. State–space models characterize the effect of two simultaneous influences on step length: the response of an animal to its immediate surroundings, and also the underlying behavioral state. Such models allow researchers insight into the behavioral state that is associated with observed movement. In some applications, such as research on Yellowstone elk (*Cervus elaphus*), strong autocorrelation terms may even indicate that the underlying behavioral state might have a stronger influence on step length than immediate habitat variables. This type of modeling can sometimes better address biological mechanisms because it considers behavioral states explicitly along with habitat selection, where analyses like SSFs are based strictly on the surrounding environment disregarding differences between dispersers vs. residents, young vs. old, and so on.

GENETICS

Dispersal Maintains Genetic Variability

Genetic variation offers insights into dispersal not only as a measurement tool but also as a selective motivation for dispersal. One of the consequences of dispersal is the maintenance of genetic variability, i.e., increasing the chance that the available genotypes will harbor recombinants preadapted to stochastic events. Without dispersal, genetic drift can shrink the genetic pool down to a puddle, leaving a few genotypes that might be ill equipped to respond to change. Dispersal also can be an avenue to escape local selection against a certain phenotype (and the genotype that produced it). Dispersal also can overcome selection by putting disappearing traits back into a population; for instance, dispersal from unharvested areas can replenish the traits being targeted by hunters (large antler size, body size, and so on).

Dispersal, or the ability to disperse, can itself be selective. Traits associated with dispersal, including morphological (body size, wing length, leg length), behavioral (likelihood to initiate dispersal, duration of dispersal), and physiological (enzymes associated with locomotion) traits have all shown a rapid response to selection. Dispersal can prevent inbreeding depression, or the loss of fitness due to unfavorable homozygosity, by injecting new breeders into the population and increasing the chances of producing diverse heterozygotes and even hybrids. This genetic chess game has the same rules across the board: competition for resources, competition for mates, avoidance of predators and parasites, and inbreeding avoidance. However, species can use very different strategies to win, like sex-specific dispersal. Male-biased dispersal is seen primarily in mammals, whereas female-biased dispersal is more common in birds and some insects.

Genetics as an Analytical Tool

Genetic techniques are rapidly evolving and hold great promise for revealing clues to dispersal and landscape connectivity even in the most elusive species. Genetic analysis has the potential to cover vast spatial and temporal scales. Sample sizes can easily number in the hundreds from samples of blood, hair, or skin and tissue. Researchers are able to produce analyses that show contemporary genetic divergence (e.g., microsatellite analysis) or population histories such as phylogenetic distance and post-glacial expansions (e.g., mitochondrial DNA analysis).

Genetic analysis can document dispersal based on gene flow, or the exchange of alleles between populations; no gene flow means no dispersal. Most studies use genetic analysis as an indirect measure of gene flow because they measure differentiation, which can happen due to a lack of dispersal (but it can be confounded by a number of factors such as selection). Genetic analysis can occasionally measure dispersal directly by identifying an unusual or rare genotype in a source population and then tracking the dispersal of that unique marker into various subpopulations. Similarly, an assignment test can be used to identify dispersers in spatially differentiated populations, a technique that has been used to document gene flow among populations of grizzly bears (*Ursus arctos*).

Indirect Measures of Gene Flow and F_{st}

Indirect measures of dispersal involve estimating parameters like relatedness between mated pairs (inbreeding), gene differentiation among populations (F_{st}), and metapopulation heterozygosity (H). Assumptions behind these measures of gene flow are difficult to verify because many things can influence genetic variability other than dispersal. Genetic analysis requires that the genetic markers are neutral, so that differentiation is not a consequence of natural selection trimming out unfit genes. Populations must be demographically stable (i.e., at equilibrium), rather than experiencing large fluctuations in size such as expanding due to invasion/colonization or influenced by founder effects that temporarily change genetic variability. Mutation rate of the markers must be low or at least known, so that genetic variability can be attributed to gene flow rather than mutation that counters genetic drift. Finally, social barriers that inhibit dispersers from becoming breeders can significantly change the interpretation of genetic results, such as in group-living animals like ants.

Estimates of population subdivision, a possible consequence of no or low dispersal events, are typically done with an F-statistic, such as F_{st}. F_{st} measures the variance in allelic frequencies among populations, standardized by the mean allele frequency (p) at that locus, $F_{st} = \text{var}(p)/p(1 - p)$. Assumptions associated with this estimate are that data were taken from one-time samples, taken from several subpopulations, and meet all of the previously stated assumptions for indirect measures of dispersal. Generally, a low F_{st} could be interpreted as a consequence of high gene flow if assumptions are met.

In addition to proposing the famous F statistic, Sewall Wright also introduced a simple island model of population structure where the number of dispersers (migrants), Nm, entering a population per generation can be estimated from the relationship $F_{st} \approx 1/(1 + 4\,Nm)$. In practice, this dispersal calculation yields unreliable estimates of the number of immigrants, Nm. The island model has several unrealistic assumptions: infinite number of populations, no geographical structure, populations equally likely to receive migrants, and so on. Even if assumptions of the island model were met, other problems exist. Figure 2 characterizes the nonlinear relationship between Nm and F_{st} and shows how the number of immigrants need only be small, as low as one disperser per generation, to maintain genetic variability as measured by F_{st}. Any errors in F_{st} will be amplified if using this relationship to estimate Nm. The enormous confidence intervals associated with low F_{st} (high gene flow) make this estimate highly inaccurate and of little value. Use of F_{st} to estimate Nm has largely been abandoned. Direct measures of dispersal (mark–recapture, band recovery, radiotelemetry, assignment tests) continue to be the most reliable methods for estimating the number of dispersers.

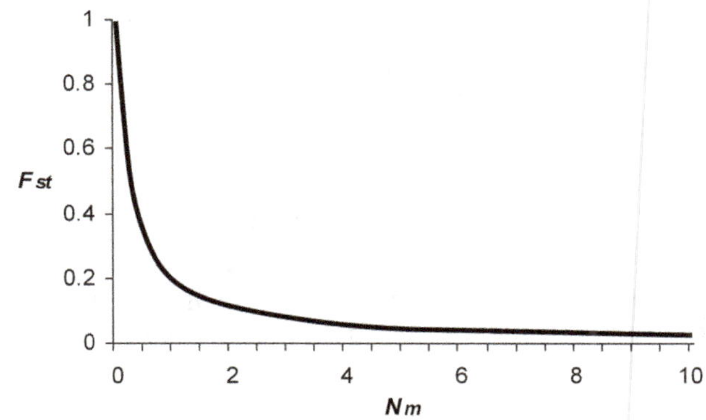

FIGURE 2 The nonlinear function between genetic variability measured by F_{st}, and the number of immigrants, *Nm*, showing how only a few immigrants per generation are required to maintain genetic variability. However, using this relationship to estimate *Nm* is often unreliable because small differences in F_{st} will result in large differences in estimates of *Nm*.

SCALE

Dispersal reciprocally affects and is affected by issues of scale. Dispersal distance often determines the scale at which population processes operate. However, dispersal distance itself and dispersal success are directly dependent on the amount and configuration of habitat, instigating research on the type and scale of habitat fragmentation. Data collection and analyses must be at the biologically relevant grain and extent to obtain reliable knowledge about dispersal.

Dispersal research, like with many landscape-level phenomena, faces difficulties when designing manipulative experiments at the appropriate scale, which can be vast for large mobile species. Some researchers have attempted to use experimental model systems and scale-up conclusions or to apply results from one species to another. Likewise, conservation policy often has little choice but to use information about dispersal from an isolated area to develop policies for an entire region. However, dispersal remains a uniquely species- and location-specific phenomenon and care must be taken when making predictions or extrapolations.

CONSERVATION POLICY APPLICATIONS

Metapopulation theory has emphasized the importance of the rescue effect, where some dispersal among otherwise isolated populations can counter population viability threats such as stochastic extinctions, genetic drift or inbreeding depression, and the loss of community-wide ecological processes. Conservation policy related to dispersal is often focused on slowing biodiversity loss connected with shrinking or fragmented habitats. Other common policy goals include maintaining populations affected by human-caused mortality, such as over-harvesting or biased harvesting. One policy option is the use of reserves as a dispersal source to surrounding vulnerable areas. For instance, strategic placement of unharvested reserves can produce dispersers to replenish the alleles under pressure in adjacent harvested populations, such as reduced antler or body size resulting from selective harvesting.

Another conservation policy related to dispersal is the designation of movement corridors. Corridors have been successful in facilitating dispersal when properly applied, but they can have dangerous costs. Oversimplified notions common in corridor literature, such as the patch–corridor–matrix categorization, should be viewed with caution. In real landscapes, animals encounter gradients of suitability and will use matrix habitat in a number of ways that can compromise or assist successful dispersal (e.g., transients might use matrix as marginal habitat until territory space becomes available in adjacent primary habitat).

CONCLUSION

Dispersal might be among the strongest of population vital rates. Dispersal can happen very rapidly, unlike population growth that is limited by reproductive capacity. When dispersal is allowed to play a major role, population change can dwarf anything that might occur through the balance of birth and death rates. The significance of dispersal is amplified when we see it occurring across sex and age groups. Dispersal has the potential to rapidly adjust distributions in the face of ecological changes, such as the numerous examples of pole-ward or elevation shifts in distributions in response to climate change.

All populations are spatially structured, and in a high proportion of natural populations dispersal is a very powerful force. Dispersal influences all of life, whether it is density dependent (condition-dependent reaction of actively moving organisms) or density independent (passive movement of sessile organisms). Population biologists have discovered particular mechanisms, such as metapopulations, that are tightly linked to dispersal. However, dispersal is relevant to all spatially structured populations, not just metapopulations. As John Wiens states (in Clobert et al., 2001), "dispersal is the glue that binds populations together."

SEE ALSO THE FOLLOWING ARTICLES

Dispersal, Evolution of / Dispersal, Plant / Invasion Biology / Metapopulations / Population Ecology / Species Ranges

FURTHER READING

Bullock, J. M., R. Kenward, and R. Hails. 2002. *Dispersal ecology: the 42nd Symposium of the British Ecological Society.* Oxford: Blackwell.

Clobert, J., E. Danchin, A. A. Dhondt, and J. D. Nichols, eds. 2001. *Dispersal.* Oxford: Oxford University Press.

Fahrig, L., and G. Merriam. 1994. Conservation of fragmented populations. *Conservation Biology* 8: 50–59.

Fortin, D., H. Beyer, M. S. Boyce, D. W. Smith, and J. S. Mao. 2005. Wolves influence elk movements: behavior shapes a trophic cascade in Yellowstone National Park. *Ecology* 86: 1320–1330.

Hilty, J. A., W. Z. Lidicker, and A. M. Merenlender, eds. 2006. *Corridor ecology: the science and practice of linking landscapes for biodiversity conservation.* Washington, DC: Island Press.

Roberts, C. M., J. A. Bohnsack, F. Gell, J. P. Hawkins, and R. Goodridge. 2001. Effects of marine reserves on adjacent fisheries. *Science* 294: 1920–1923.

Soulé, M. E., and J. Terborgh, eds. 1999. *Continental conservation: scientific foundations of regional reserve networks.* Washington, DC: Island Press.

Turchin, P. 1998. *Quantitative analysis of movement: measuring and modeling population redistribution in animals and plants.* Sunderland, MA: Sinauer Associates.

Waser, P. M., and C. Strobeck. 1998. Genetic signatures of interpopulation dispersal. *Trends in Ecology & Evolution* 13: 43–44.

Whitlock, M. C., and D. E. McCauley. 1999. Indirect measures of gene flow and migration. *Heredity* 82: 117–125.

DISPERSAL, EVOLUTION OF

MARISSA L. BASKETT

University of California, Davis

While dispersal encompasses several types of organismal movement, theoretical investigation into the evolution of dispersal focuses primarily on cases where movement has the evolutionary consequence of shifting the distribution of genes and individuals in space. This literature addresses what drives the evolution of dispersal and, given specific evolutionary drivers, what the evolutionary outcome might be for the amount or rate of dispersal.

An understanding of dispersal and its scale is critical to analyzing a variety of dynamics in ecological and evolutionary biology. For example, dispersal determines the distribution and spread of populations, the outcome of certain species interactions in terms of the potential for coexistence and stability, and the potential for evolutionary responses to spatial variation in selection including local adaptation and speciation.

TYPES OF DISPERSAL

Dispersal involving the movement of individuals and genes in space might be categorized as natal dispersal or breeding dispersal. Natal dispersal describes the permanent movement of a given individual from the location of origin to a different location for reproduction. Breeding dispersal describes the movement of a given individual between different instances of reproduction. In the literature on the evolution of dispersal, some use the term "dispersal" interchangeably with "migration" and "movement." Care should be used to clarify these terms, however, since some fields such as wildlife biology refer to migration as potentially a quite different behavior (mass movement with regular periodic—e.g., seasonal—returns to a given site).

MODEL CONSTRUCTION AND THEORETICAL DRIVERS OF THE EVOLUTION OF DISPERSAL

Diverse models of the evolution of dispersal arise based on which selective forces acting on dispersal are incorporated as well as how the evolutionary dynamics, the evolving trait that influences dispersal, and space are represented. The selective forces that might drive the evolution of dispersal fall into two broad categories: interactions with kin and spatiotemporal heterogeneity (Table 1). Each is described in more detail below. In both cases, selection depends on the prevailing dispersal strategy in the population. Specifically, dispersal-driven distributions of a population in space determine the frequency of interactions with kin and the intensity of intraspecific interactions (e.g., competition between individuals of the same species). The selection pressure that an individual experiences with respect to traits that influence dispersal depends on the frequency of different dispersal-related phenotypes in the population.

Evolutionary dynamics models that explicitly account for such frequency dependence utilize game theory. Game theoretical models predict the evolutionary outcome based on strategies that can invade (increase in a population dominated by other strategies) when rare and cannot be invaded by other strategies when common. Because frequency dependence is integral to game theory, the theoretical literature on the evolution of dispersal predominantly employs a game theoretic approach. However, game theory does not provide a mechanistic representation of evolutionary processes. Some dispersal evolution models do incorporate some genetic mechanism through adaptive dynamic and population genetic models. An example is a two-locus population genetic model where one locus determines dispersal and the other determines a fitness-related trait as it might depend on inbreeding or local conditions.

Whether in a game theoretic, adaptive dynamic, or population genetic framework, dispersal can depend on a number of different evolving traits, such as the proportion of offspring dispersing, the rate of individual movement, or the shape of the dispersal kernel (probability density function that describes the distribution of individuals across space in a dispersal event). The choice of trait determines whether dispersal is quantified in terms of the total amount of dispersal and/or the distance over which dispersal takes place. Beyond single traits, some models predict multiple traits as the evolutionary outcome, arising as mixed strategies in a game theoretic approach or polymorphisms in an adaptive dynamic or population genetic approach.

Because dispersal is an inherently spatial process, models incorporate spatial representations, which depend in part on the dispersal trait (Fig. 1). Often space is implicit, as in models that follow the evolution of the proportion of offspring dispersing where all dispersal that occurs

TABLE 1
Summary of general theoretical expectations for factors that influence the evolution of dispersal

Category	Factors that select for dispersal	Factors that select against dispersal
Interaction with kin	Sibling and parent–offspring competition (especially with strong competition and longer-distance dispersal)	
	Inbreeding depression	Outbreeding depression
Spatiotemporal heterogeneity	Temporal heterogeneity (e.g., in habitat quality, population dynamics, or patch occupancy)	Spatial heterogeneity (e.g., in habitat quality; assuming passive dispersal)
Both	Conditional dispersal	Survival or reproductive costs to dispersal

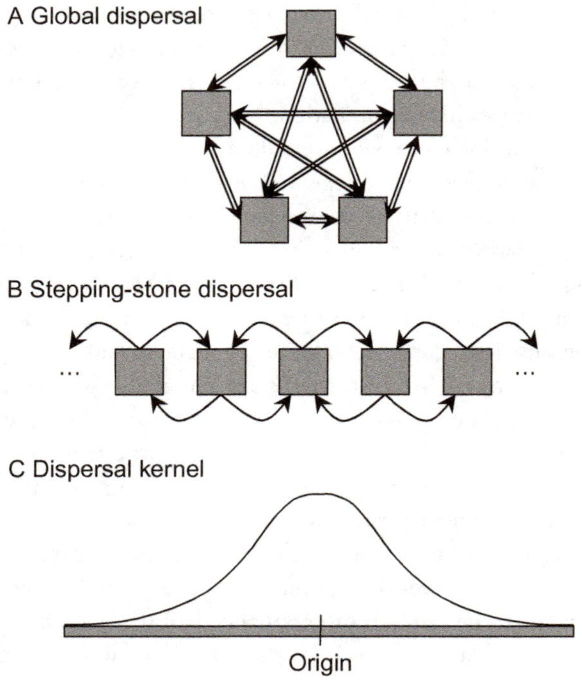

FIGURE 1 Examples of different representations of dispersal in space: (A) global dispersal, where dispersal to any site is equally likely; (B) stepping-stone dispersal, where dispersal is limited to the nearest neighbors; (C) a dispersal kernel, which describes the probability density function of settlement location given a location of origin.

is global (offspring are distributed uniformly across the entire area under consideration; Fig. 1A). More recent models have incorporated spatially explicit dynamics, as in models that follow the evolution of dispersal kernels (Fig. 1C), with results often confirming those from the spatially implicit case. Below is a synopsis of these results, organized by the type of selective force.

Interactions with Kin

In a homogeneous environment, evolution can theoretically converge on high levels of dispersal, even with a cost to dispersing, due to the benefits gained from avoiding competition with kin. With discrete generations, dispersing offspring of a particular individual are less likely to compete with each other. Thus dispersal reduces competition between siblings and increases the potential proportion of patches occupied by that individual's offspring. The foundational paper by Hamilton and May (1977) that explored the evolution of dispersal as a mechanism to avoid kin competition predicts that evolution will converge to a proportion of dispersing offspring $v^* = 1/(2 - p)$, where the dispersal survival probability p accounts for a cost to dispersal. Other models arrive at this result for the evolutionarily predicted proportion of offspring dispersing under the simplifying assumptions originally made by Hamilton and May: one individual per site, global dispersal, no reproductive cost for dispersal, and parthenogenetic reproduction.

Models that relax the simplifying assumptions made by Hamilton and May investigate processes that might drive the evolution of dispersal as it depends on kin competition. The assumption of one individual per site represents the extreme case of strong competition; subsequent models that allow multiple individuals to inhabit each site indicate that the degree of competition is a critical parameter affecting the evolution of dispersal. Selection for dispersal increases with increasing competition (e.g., lower site carrying capacity) because it intensifies the selective force of kin competition. This intensification can be further magnified if more closely related individuals have a greater competitive effect on each other than more distantly related individuals, as opposed to the constant competitive effect independent of relatedness often assumed. The assumption of global dispersal also represents an extreme case, in this instance of long-distance dispersal given that dispersal to any site in the entire landscape is equally likely (Fig. 1A). Selection for dispersal decreases if those dispersing move over shorter distances, as in a stepping-stone model with dispersal constrained to neighboring sites (Fig. 1B). This trend arises because longer-distance dispersal increases the likelihood that dispersing offspring escape kin competition. Relaxing the assumption of discrete generations in Hamilton and May to allow for overlapping generations leads to the potential for both parent–offspring and sibling competition to impact the evolution of dispersal. This additional kin competition enhances selection on dispersal; however, it also raises potential parent–offspring conflict when it is beneficial from the parents' perspective for the offspring to disperse while the offspring might benefit from remaining at the parental site.

In addition to kin competition, mating with kin can favor the evolution of dispersal if inbreeding depression (lower fitness in offspring of related individuals, particularly due to deleterious recessive alleles) is likely to occur. Interactions between individuals with similar genetic makeup can lead to both kin competition and inbreeding depression, which can interact to affect selection on dispersal; the relative influence of each remains a debated topic in the theoretical literature. Just as inbreeding depression can increase selection pressure for dispersal, the potential for outbreeding depression (lower fitness in offspring of individuals from different populations, particularly due to degradation of local adaptation) can create a counteracting selective force against dispersal. Models indicate that the selective balance between inbreeding and outbreeding depression on dispersal depends on the cost of each and the mutation process.

Spatiotemporal Heterogeneity

A consistent conclusion from theoretical investigations into the evolution of dispersal is that spatial heterogeneity alone generally selects against dispersal whereas temporal heterogeneity generally selects for dispersal. Given spatial variation in habitat quality that is constant in time, any individual or its offspring in a higher-quality habitat who disperses will inevitably move to lower-quality habitat. This potential for dispersing individuals to experience decreased habitat quality, in combination with the ability for higher-quality habitat to support larger populations, leads to a prevailing selection against dispersal. However, the inclusion of variation in habitat quality in time as well as space introduces the potential for dispersing individuals to move to a higher quality habitat regardless of the current state of the pre-dispersal location. In this case, dispersal has the potential to reduce crowding and the amount of intraspecific competition that an individual experiences. Finally, dispersal in a spatially and temporally variable environment reduces the variation in survival and/or reproductive success of offspring across time and space, which creates selection pressure for dispersal to evolve as a bet-hedging, or risk-spreading, strategy. Which of these interacting factors (increasing habitat quality, crowding avoidance, bet hedging) is the primary driver for the evolution of dispersal in a temporally heterogeneous environment depends on the model details.

Temporal heterogeneity can stem from a number of sources. Environmental variability can drive temporal heterogeneity in fitness-related processes such as survival or reproduction. Demographic stochasticity and chaotic population dynamics can induce temporal heterogeneity in population densities. All of these potentially affect the evolution of dispersal. Different sources vary in their degree of predictability and amount of autocorrelation (correlation of the varying characteristic with itself across space or time), which can alter theoretical predictions for the level of dispersal. Positive spatial autocorrelation increases selection for dispersal by increasing the probability of moving between favorable habitats given localized dispersal. On the other hand, positive temporal autocorrelation decreases selection for dispersal by decreasing the likelihood that a favorable location will become unfavorable in the near future. More recently, models have begun to explore how the shape of the dispersal kernels that arise from these selective forces depends on the amount of autocorrelation and the type of noise in spatiotemporal heterogeneous habitats.

One specific source of spatiotemporal heterogeneity that has received particular attention is local extinction. Regardless of whether the ecological drivers of local extinction are predictable (e.g., succession) or unpredictable (e.g., stochastic metapopulation dynamics), the resulting creation of empty sites drives temporal variability through the ephemeral or unstable nature of patch occupancy. Selection for dispersal through the opportunity to colonize empty sites was originally postulated in a group selection framework in one of the first models of the evolution of dispersal, by van Valen (1971). Subsequent models have demonstrated that individual-level selection for dispersal can arise under local extinction. For some models with spatiotemporal heterogeneity, individual-level selection on dispersal can lead to the population-level property of an ideal free distribution, a population distribution with equalized fitness across the modeled landscape. Whether an ideal free distribution emerges depends on the model details; for example, it does not arise in the presence of sink habitats.

THE ROLE OF SPECIES INTERACTIONS

While these two general categories of selective forces acting on dispersal focus on interactions with conspecifics in general (e.g., crowding avoidance) and kin in particular (e.g., kin competition, inbreeding avoidance), other potentially important factors that have received much less attention are interspecific interactions such as predation, competition, and mutualism. Interspecific interactions have the potential to influence either category of selective drivers of dispersal. More closely related individuals might attract similar exploiters (parasites or predators) and therefore amplify the selective force of kin competition with apparent competition. In addition, the dynamics of interacting species can induce spatiotemporal heterogeneity similar to how chaotic population dynamics provide a possible source of such heterogeneity. Given the complexity of species interactions, a simple directionality to their influence in terms of increasing or decreasing selection on dispersal is unlikely. For example, heterogeneity in predation risk might select for dispersal, while predation on the dispersal stage might select against dispersal.

CONSTRAINTS AND TRADEOFFS

In addition to the indirect cost of moving to a less favorable habitat, dispersal can incur a direct cost. Many of the above-described models incorporate a cost to dispersal, typically assumed to be constant but in some cases dependent on how dispersal occurs or on the landscape structure. Many models assume lower survival for offspring that disperse compared to those that do not disperse, due to movement through a risky environment or development through a risky stage. The production of dispersal structures, such as seed structures necessary for transport by wind or mobile animals in plant dispersal, incurs some costs. Accordingly, some models assume dispersal induces lower offspring production. Incorporating such costs for dispersal models life history tradeoffs with survivorship or reproductive effort.

Tradeoffs of lower survival or reproduction with dispersal do not universally apply. One example concerns planktonic larval dispersal in nearshore marine systems. In this case, the initial offspring stage is comprised of propagules that have little or no control over their position in the water and therefore primarily disperse according to random drift and currents. Two of the leading hypotheses for the existence of this life history strategy counter the typical survival or reproduction tradeoff assumption about costs of dispersal. First, the hypothesis that offspring that disperse experience less predation because they avoid benthic (bottom-dwelling) grazers is based on the idea that dispersal is a less risky strategy than nondispersal. A second hypothesis is that planktonic propagules with a long feeding stage, and therefore potentially longer-distance dispersal assuming passivity, require less initial investment per propagule and allow the production of more offspring. In this case, dispersal and reproductive output have a positive relationship. However, longer larval duration might confer a survivorship cost just as, in plants, smaller seeds can have greater dispersal capacity but lower survivorship or competitive ability. Models of the evolution of planktonic larval dispersal focus on the offspring size–number tradeoff that emerges from these processes. Model results, based on fitness optimization and game theoretic analyses, often suggest disruptive selection for short and long larval durations. This exemplifies a more general trend for disruptive selection to emerge from models that account for the joint evolution of dispersal and its costs.

In addition to survivorship and reproduction, other life history traits can affect selection on dispersal. Some models have directly explored the joint evolution of dispersal and another relevant life history trait such as seed dormancy. Seed dispersal can provide a bet-hedging adaptation to spatiotemporal heterogeneity by averaging the environment experienced by an individual's offspring in space, while dormancy averages the environment experienced by an individual's offspring in time. Models of spatiotemporal heterogeneity that include dormancy predict that less dispersal will evolve than when dispersal is considered alone.

Beyond life history tradeoffs, local adaptation and dispersal affect each other. Models of local adaptation typically assume dispersal is constant and conclude that dispersal generally impedes local adaptation. This general prediction can break down when dispersal can supply the genetic diversity necessary for adaptation in small or declining populations. The potential for dispersal to lead to movement to a habitat to which an individual is less well adapted drives selection against dispersal analogous to that for movement to a generically lower-quality habitat. Two-locus population genetic models that have one locus for dispersal and a second locus for environment-dependent fitness directly account for the role of local adaptation in the evolution of dispersal. When accounting for interactions between kin, local adaptation is a key potential driver of the outbreeding depression that can counteract inbreeding depression-based selection for dispersal.

CONDITIONAL DISPERSAL

In addition to explaining whether and how much dispersal might evolve, the above-described selective forces can also explain the conditions for the timing of dispersal and which individuals are more likely to disperse. Conditional dispersal generally refers to dispersal as a plastic trait that is dependent on the conditions experienced by a particular individual, such as population density, habitat quality, and physiological status. Models indicate that conditional dispersal typically increases selection for dispersal by reducing the cost of dispersal by constraining it to conditions when it is more likely to be favorable. Habitat-quality-dependent or density-dependent dispersal can constrain dispersal to when individuals reside in a patch of lower quality or with greater crowding than average. Incorporating conditional dispersal into a model can allow for the evolution of dispersal given spatial heterogeneity that is constant in time, which selects against passive dispersal. This theoretical potential for conditional dispersal to increase selection for dispersal depends on the reliability of the signal and the predictability of any heterogeneity. Given their connection to environmental variability and crowding avoidance, habitat-quality-dependent and density-dependent dispersal primarily apply to models where spatiotemporal heterogeneity is the selecting force acting on dispersal. However, dispersal conditional on family size can factor into kin-competition-driven dispersal.

Along with being dependent on exogenous environmental conditions such as habitat quality and population density, dispersal can depend on factors related to individual condition, such as age, sex, and social rank. These types of conditional dispersal not only affect the amount of dispersal that occurs under different conditions but also which individuals disperse. The potential for subordinates in one location to have greater success at another location can bias selection on dispersal due to spatiotemporal heterogeneity with respect to social rank. Such bias provides an individual-level selective force for the idea, originally posited under a group selection framework, that dispersal evolves to reduce crowding. Models indicate that a variety of selective forces, such as inbreeding avoidance, competition for mates, and kin cooperation, can explain sex-biased dispersal. Whether sex-biased dispersal occurs as well as which sex disperses depends on whether the mating system is monogamous, polygynous, or polyandrous and whether mate choice occurs. Finally, models of age-biased dispersal indicate that the selective force(s) acting on dispersal determine whether an optimal age of dispersal exists and, if so, what that age might be. For example, inbreeding-avoidance-driven dispersal tends to select for dispersal before the age at first reproduction, whereas parent–offspring-conflict-driven dispersal tends to select for dispersal at later ages. Similarly, maternal age or condition can factor into the evolution of offspring dispersal by driving variation in the strength of kin competition (e.g., parent–offspring conflict declines with parental age). Thus the individual condition driving dispersal can be that of the parent as well as the dispersing individual.

DATA AND APPLICATIONS

Difficulties in measuring dispersal and its heritability limit the availability of data to test many of the above theoretical predictions. While models can focus on specific selective forces, multiple selective forces are likely to interact in influencing the evolution of dispersal, which can be difficult to disentangle. This challenge particularly applies when the same ecological conditions similarly influence different potential selective forces, such as small local population sizes intensifying both kin competition and demographic stochasticity (a potential driver of spatiotemporal heterogeneity). Given these challenges, empirical studies focus on measuring the costs and benefits of dispersal hypothesized from theory and provide circumstantial evidence for hypothesized drivers of dispersal evolution. Empirical studies frequently find lower survivorship for dispersers compared to nondispersers, as is commonly assumed in the above-described models. However, this is not a universal outcome, and the results can depend on factors such as population density. In addition, a number of empirical studies investigate and demonstrate the influence of environmental and physiological conditions on dispersal, which indicates the potential importance of including plasticity into dispersal evolution models. Finally, some studies reveal empirical patterns consistent with specific theoretical predictions, such as the evolution of flightlessness and reduced plant dispersal on islands, an extreme case of spatial heterogeneity that is constant in time.

More recently, studies of dispersal subsequent to anthropogenic impacts, such as habitat fragmentation and species introductions, take advantage of accidental experiments to test hypotheses about the evolution of dispersal. Such species responses to anthropogenic impacts also demonstrate the importance of understanding dispersal, because dispersal is both critical to and affected by these impacts. Dispersal can influence the potential for population persistence in a fragmented habitat, and habitat fragmentation alters selection on dispersal by artificially increasing deterministic spatial heterogeneity. Dispersal is a key determinant of the spread of invasive species, and species introductions involve novel environments and therefore novel selective forces acting on dispersal. Climate change analogously alters the selective environment a species experiences, and dispersal determines the potential for species range shifts to track climate change. Given the importance of dispersal to species responses to global environmental change and the potential for dispersal evolution to be a part of this response, continued study of evolution of dispersal in theory and in the field will allow deeper understanding of the ecological and evolutionary consequences of global change.

SEE ALSO THE FOLLOWING ARTICLES

Dispersal, Animal / Dispersal, Plant / Game Theory / Movement: From Individuals to Populations / Spatial Ecology

FURTHER READING

Colbert, J., E. Danchin, A.A. Dhondt, and J.D. Nichols, eds. 2001. *Dispersal*. Oxford: Oxford University Press.

Hamilton, W.D., and R.M. May. 1977. Dispersal in stable habitats. *Nature* 269: 578–581.

Johnson, M.L., and M.S. Gaines. 1990. Evolution of dispersal: theoretical models and empirical tests using birds and mammals. *Annual Review of Ecology and Systematics* 21: 449–480.

Levin, S.A., D. Cohen, and A. Hastings. 1984. Dispersal strategies in patchy environments. *Theoretical Population Biology* 26: 165–191.

McPeek, M.A., and R.D. Holt. 1992. The evolution of dispersal in spatially and temporally varying environments. *American Naturalist* 140(6): 1010–1027.

Olivieri, I., Y. Michalakis, and P. Gouyon. 1995. Metapopulation genetics and the evolution of dispersal. *American Naturalist* 146(2): 202–228.

Ronce, O. 2007. How does it feel to be like a rolling stone? ten questions about dispersal evolution. *Annual Review of Ecology, Evolution, and Systematics* 38: 231–253.

Roze, D., and F. Rousset. 2005. Inbreeding depression and the evolution of dispersal rates: a multilocus model. *American Naturalist* 166(5): 708–721.

Travis, J. M. J., and C. Dytham. 1999. Habitat persistence, habitat availability, and the evolution of dispersal. *Proceedings of the Royal Society B: Biological Sciences* 266: 723–728.

van Valen, L. 1971. Group selection and the evolution of dispersal. *Evolution* 25(4): 591–598.

DISPERSAL, PLANT

HELENE C. MULLER-LANDAU

Smithsonian Tropical Research Institute, Panama City, Panama

Plant dispersal is the movement in space of a seed or other unit of plant tissue that is capable of giving rise to one or more reproductive adults. Plant dispersal patterns are highly variable and can be studied and modeled in a variety of ways. Dispersal contributes to plant population and community structure and dynamics, including species coexistence and rates of spread. Human activities are altering plant dispersal patterns and the consequences of dispersal, with important implications for conservation and management.

PLANT DISPERSAL STRATEGIES

The Unit of Dispersal

The unit of dispersal in a plant is referred to as the diaspore. This is may be the seed, the fruit, the spore (in lower plants such as ferns), a vegetative part of the plant that is capable of growing into an adult plant, or even the whole plant. Most plants are sessile as adults and disperse exclusively at the seed or spore stage. However, dispersal of whole plants or viable fragments is the dominant mode of dispersal in aquatic plants. Further, quite a number of herbaceous plant species have special vegetative dispersal organs, termed bulbils. While seeds are the result of sexual reproduction, bulbils and viable fragments result in asexual or clonal propagation. A single plant species may have more than one type of diaspore; for example, an aquatic species may disperse both viable vegetative fragments and seeds.

Modes of Dispersal

Plants can be dispersed by wind, water, and animals, including humans and their conveyances. Diaspores of the same species may be dispersed in multiple ways, either alternatively or in succession. Modes of rare long-distance dispersal may be different from modes of more frequent short-distance dispersal; for example, seeds that are usually dispersed short distances by wind may occasionally be transported long distances by water. Where the same diaspores are often moved two or more ways in succession, the initial movement away from the mother plant to the ground or other substrate is termed primary dispersal, while subsequent movement over the ground or substrate is termed secondary dispersal. Thus, for example, a seed may first experience primary dispersal by wind, and later secondary dispersal by ants.

Many plant species use animals for dispersal. Animals may consume fruits and subsequently defecate or spit out viable seeds; they may move seeds to caches with the intention of consuming them later, yet leave some untouched; and they may unintentionally transport seeds attached to their bodies by, for example, burs. A wide range of animal groups are involved—mammals, birds, reptiles, fish, ants, beetles, snails, earthworms, and more. In many cases, the relationship is a mutualism in which the animal species benefit by consuming plant tissue; in others, the animals receive no benefit and may even incur a cost. Tight dispersal mutualisms between single plant and animal species are the exception: typically, any given plant species will have its diaspores moved by multiple animal species, and any given animal species will be involved in moving diaspores of multiple plant species. However, though many animal species may be involved in dispersing diaspores of a particular plant species, there may be one species that is disproportionately important for the plant, because it moves more diaspores and/or places diaspores in particularly favorable conditions and is thus a more effective disperser, from the plant's perspective. This is reflected in plant traits: the coloration, mode of display, and chemical composition of fruits and seeds may enhance their attractiveness to some animals, and reduce attractiveness to others. For example, the capsaicin in chile peppers deters consumption by mammals, a phenomenon known as directed deterrence, while the red color of the ripe fruits increases visibility to birds.

Plants may instead (or in addition) rely on wind or water for dispersal, or may disperse without any external aid. Many species have seeds, spores, fruits, or even bulbils that are dust-like particles, balloons, plumed, or winged, and thereby adapted for dispersal by wind. Other plants are regularly dispersed by water: this includes aquatic plants and plants of seasonally flooded habitats whose diaspores may float or be transported submerged, as well as terrestrial plants whose seeds are moved by rain wash. Some plant species explosively disperse their diaspores; wind, water, or a passing animal may or may not be needed to provide energy or the trigger for such ballistic dispersal. Finally, there are plant species in which diaspores spread by growth alone (e.g., stalked infructescences curving toward the ground), and those in which the diaspores seem to have no adaptation for dispersal and are said to disperse by gravity or weight alone.

Human activities have provided novel modes of dispersal, and especially long-distance dispersal for many plants. Seeds, viable plant fragments, and whole plants may be moved from place to place deliberately for agriculture, horticulture, or silviculture, or accidentally as hitchhikers on such shipments, on clothing, on vehicle treads, in ballast water, and so on.

MEASURING PLANT DISPERSAL

Two qualitatively different approaches can be used to measure plant dispersal. The Lagrangian approach quantifies the trajectories of individual diaspores by, for example, following marked seeds. The Eulerian approach quantifies the spatial pattern of diaspores after dispersal by, for example, measuring seed arrival in seed traps. Genetic techniques can be applied to either approach and offer the promise of combining the best of both approaches.

The Lagrangian Approach: Tracking Diaspore Movement

Plant dispersal distances and patterns can be evaluated by following the trajectories of individual diaspores. In some cases, the diaspore can be directly observed during dispersal. In other cases, it may be possible to label seeds before dispersal and relocate them afterward. Such labeling can be accomplished with dyes, radioisotopes, metal inserts (permitting relocation with a metal detector), or even radio transmitters. Diaspores may be followed or relocated after natural dispersal from their mother plant or after experimental release or deployment (e.g., previously collected wind-dispersed seeds are dropped by hand, or marked seeds are placed on the ground beneath a fruiting tree).

The advantage of this method is that the source of the seed is known and often also its mode of transport. The key disadvantage is that the observed trajectories are unlikely to be representative of the population of diaspores as a whole. First, some fraction of target diaspores invariably cannot be followed or relocated, and it is likely that these escapees have a different distribution of fates than diaspores that are successfully tracked (long-distance seed dispersal is disproportionately likely to be missed). Second, tracking methods may capture only one mode of dispersal, missing others, especially those associated with secondary dispersal, and thus measurements may fall short of the complete trajectories of the diaspores. Third, experimental releases of diaspores tend to be concentrated in time and space, for logistic reasons, and the dispersal trajectories thus sampled may not be representative of the trajectories in the population as a whole because of temporal and spatial variation in dispersal conditions (e.g., windspeed) as well as failure to correctly mimic the distribution of conditions under which diaspores are naturally dispersed (e.g., seeds naturally released disproportionately when windspeeds are high).

The Eulerian Approach: Quantifying Post-Dispersal Diaspore Distributions

Alternatively, the spatial distribution of diaspores can be mapped following dispersal, and combined with information on the distribution of sources to infer dispersal patterns. This is most commonly done through the use of appropriate seed traps—for example, sticky traps, netting on a frame, or soil in seed trays. Juvenile plants are sometimes mapped instead of diaspores, as they are generally easier to find; however, spatial patterns of juveniles are influenced not only by dispersal but also by spatial variation in establishment success, complicating analyses, and inferences.

The major disadvantage with the Eulerian approach is that the sources of individual diaspores are usually uncertain—except in the extreme case where there is only one source—which introduces corresponding uncertainty into estimates of dispersal patterns. In principle, the key advantage is that the observed patterns usually integrate over complete seasons and include dispersal via all modes. However, many seed traps prevent or restrict secondary dispersal, so that in practice this method often provides information only about primary dispersal. Further, for practical reasons, sampling is unlikely to capture long-distance dispersal to locations far from source plants, and thus, as with the Lagrangian approach, long-distance seed movement is disproportionately likely to be missed.

Genetic Tools

Genetic markers offer additional direct and indirect methods for gaining insight into plant dispersal. The classical indirect method uses measures of spatial genetic structure to estimate long-term effective seed dispersal distances. A key point is that these estimates reflect *effective* dispersal, that is, dispersal that leads to reproductive adults, and thus incorporate the influences of post-dispersal processes, which are likely to vary systematically with dispersal distance. Biparentally inherited markers also reflect the influence of pollen dispersal.

The newer direct genetic method uses high-resolution molecular markers to match seeds (or seedlings) with parents (or specifically mothers). In principle, this could capture the key advantages of the Lagrangian and Eulerian approaches, making it possible to sample representatively from the whole post-dispersal distribution and simultaneously obtain certainty about sources and thus trajectories. In practice, however, the direct method has thus far fallen well short of this promise, and not only because its application continues to be exceptionally time-consuming and expensive. Genotyping errors bias toward overestimation of the number of immigrant diaspores outside the genotyped source pool, while inability to uniquely identify all source trees and inability to distinguish mothers from fathers bias toward underestimation of dispersal distances and immigration. Further, it has only belatedly been recognized that the distribution of dispersal distance estimates obtained with direct genetic methods is not in and of itself an unbiased sample of dispersal distances in the whole population but is contingent upon the distribution of sampling points relative to sources. Appropriate statistical analyses need to be applied to correct for these influences. Direct genetic methods, like all other tools, are imperfect windows on plant dispersal.

MODELING PLANT DISPERSAL

Decomposing Dispersal Patterns

A key concept in analyzing and modeling plant dispersal is the seed shadow, or diaspore shadow—the spatial distribution of diaspores that originated from a single plant (usually in a single season or year). The overall spatial distribution of diaspores in a plant population is the sum of diaspore shadows produced by individual plants. The diaspore shadow of a plant can itself be thought of as the product of its fecundity (number of diaspores produced) and the probability density function for the locations of these diaspores relative to the source plant, termed the dispersal kernel.

The focus of the vast majority of plant dispersal models is the dispersal kernel for a plant species or population, averaged over all individuals. Dispersal kernels are typically defined as functions of distance from the source alone, ignoring other systematic sources of variation in diaspore arrival probability, but this need not be the case. Dispersal by wind is generally anisotropic, reflecting prevailing wind directions, and this pattern can be captured in dispersal kernels that incorporate direction as well as distance. Diaspore arrival often varies systematically with substrate; for example, wind-dispersed seeds are more likely to come to rest in tall grass than on sand, and animal-dispersed seeds are more likely to be deposited in habitats preferred by the disperser. Such substrate-specific deposition generally defies inclusion in simple dispersal kernels that apply for all sources in a population, but can be incorporated into more complex models. It can be particularly important where seeds are disproportionately deposited in favorable habitats, so-called directed dispersal. Finally, focus on the dispersal kernel alone ignores the influences of clumping of diaspore arrival, such as that which results when multiple seeds are dispersed in a group (e.g., an animal disperser defecates multiple seeds in one place). Characterization of the magnitude and scale of this clumping can be important to accurately model the distribution of dispersed diaspores and consequences for recruitment. Clumping can in the simplest case be modeled with a negative binomial distribution around the expectation derived from a dispersal kernel.

Phenomenological Models

Plant dispersal is most often analyzed by fitting purely phenomenological models for dispersal kernels to empirical data. In most cases, the dispersal kernel is a function only of distance from the source. Commonly used empirical models for the dependence upon distance include the exponential, Gaussian, inverse power, lognormal, Student's, Weibull, and the exponential power (exponential function of distance raised to a power). These may be used to describe the dispersal kernel in two dimensions, or the radial integration of this dispersal kernel, i.e., the probability distribution of dispersal distances; this is unfortunately a frequent point of confusion. Many of these functional forms can be motivated in general terms. For example, a Gaussian distribution is expected if diaspores move by Brownian motion for a fixed period of time. More complex phenomenological models may be constructed as mixtures of two distributions, motivated as the sum of two different dispersal processes (e.g., two exponential distributions reflecting frequent short-distance and rare long-distance dispersal).

Empirically, dispersal distance distributions are generally strongly leptokurtic. Thus, among one-parameter models, the (relatively fat-tailed) inverse power generally fits better than the exponential, which generally fits better than the (thin-tailed) Gaussian. Unsurprisingly, models with more parameters tend to more closely fit the data, but they are also more vulnerable to overfitting and have wider confidence intervals on parameter estimates. Thus, two-parameter Student's *t*, lognormal, Weibull, and exponential power models generally outperform the one-parameter exponential, inverse power, and Gaussian models, and may themselves be outperformed by four-parameter mixed models. The lognormal is the only widely used phenomological model in which the highest expected seed density is not (necessarily) at the source, and thus it generally provides the best fit to datasets exhibiting this pattern. Overall, no one phenomenological model outperforms the rest for all datasets; different dispersal datasets are best fit by different models.

Mechanistic Models

An alternative approach is to develop mechanistic models based on an understanding of the underlying dispersal process. Truly mechanistic models can predict diaspore distributions from independently measured characteristics of the dispersal process.

The mechanistic approach is best developed for dispersal by wind. The simplest model of dispersal by wind is the ballistic model, in which dispersal distance is the product of horizontal wind velocity and release height, divided by the fall velocity of the diaspore. This can be used to generate dispersal kernels from empirical measurements of the three parameters and their probability distributions. More complex and realistic mechanistic models can incorporate nonzero vertical wind speeds, systematic variation in windspeed with height above the ground, autocorrelated random variation in wind velocity during flight, differential diaspore release as a function of wind conditions, and more. Most of the resulting models cannot be expressed analytically but must instead be evaluated numerically (with simulations). An exception is the WALD, or inverse Gaussian, derived from simplification of a stochastic model of turbulent transport. The two parameters of this model can be estimated directly from wind statistics, release height, and fall velocity and/or fit based on dispersal data.

Some of the same general approaches used for dispersal by wind have also been applied to dispersal by water. However, overall, there has been little research specifically on plant dispersal by water. It is likely that models of plant diaspore transport in water could usefully borrow from models of the transport in water of flotsam, of animals that possess limited independent movement capacity, and even of certain pollutants.

The biggest challenge in mechanistic models of plant dispersal by animals is the heterogeneity in behavior within and among individuals and species that serve as dispersers of a given plant species. The simplest models compound distributions of the time a diaspore is retained by an animal (e.g., in the gut or on the coat) with distributions of displacement distances as a function of time. Where multiple animal species are involved, and differ in their retention or displacement distributions, a combined kernel can be constructed as the appropriate weighted sum of kernels due to the activities of each species. More realistic models have further incorporated clumped diaspore deposition by animals, differential movement through and deposition in different habitats or substrates, diurnal variation in disperser behavior, spatial variation in disperser behavior depending on resource plant density, and more. Aided by the increasing availability of high-resolution animal movement data and ever-better computing resources, models of animal movement and behavior continue to improve rapidly, promising further improvements in mechanistic models of seed dispersal by animals. At present, however, the utility of these complex models is limited by their requirements for extensive data on the relevant disperser species.

CONSEQUENCES OF DISPERSAL

Dispersal patterns have direct consequences for the fitness of individual plants, as plants whose diaspores are dispersed into relatively more favorable locations will be better represented in subsequent generations. This leads to natural selection on dispersal-related traits. Dispersal also affects population and community patterns, as described below.

Population and Community Structure and Dynamics

Seed dispersal determines the spatial pattern of potential recruitment and thereby can contribute greatly to spatial structure of populations and communities. The probability of successful establishment of a juvenile and successful maturation to adulthood typically varies extensively in space within a plant population. This variation may be related to environmental factors such as temperature and water availability, as well as to purely biotic factors such as proximity to competitors or natural enemies. In many plant species, per-diaspore recruitment probabilities are depressed close to conspecific adults, because of intensified resource competition and/or natural enemy attack in these areas. Spatial patterns among juveniles and adults reflect the combined influences of seed dispersal and post-dispersal processes. The same patterns can arise in multiple ways; for example, clumped distributions of adults may arise from a predominance of short dispersal distributions or from the combination of widespread dispersal and patchy distributions of sites favorable for recruitment, among other possibilities. In general, shorter dispersal distances are expected to increase clustering of individuals, decrease local diversity (of both genes and species), and increase turnover in space (of both genes and species).

Seed dispersal can be critical to understanding the dynamics and structure of populations and communities. For example, in metapopulations (collections of island populations bound loosely together by occasional migration between the islands), the rate of dispersal among island populations is critical to determining patch occupancy, overall species abundance, and persistence in the face of disturbance. Dispersal from favorable to unfavorable habitats may also result in source–sink dynamics, in which species persist in areas in which their per capita population growth rates are negative.

Seed dispersal may also contribute to plant species coexistence. Competition–colonization tradeoffs between the ability to reach potential recruitment sites with diaspores and the ability to succeed following arrival can enable stable species coexistence. Tradeoffs between dispersal and fecundity can contribute to stable coexistence given spatial variation in the density of favorable sites. More generally, variation in dispersal strategies can also contribute to coexistence insofar as it gives each species an advantage in a different set of circumstances. Finally, limited dispersal overall within a community reduces interspecific interactions and thereby slows competitive exclusion and increases opportunities for nonequilibrium coexistence.

Rates of Spread

Plant dispersal patterns are critical to determining the rate of advance of a species in favorable habitats. This is of practical interest for the spread of invasive species and in determining the maximum rate of advance as species respond to changing climates. Where the dispersal of a species is well approximated by random walk and thus by diffusion, the population advances continuously from its periphery, and the asymptotic rate of spread is approximately $2\sqrt{Dr}$, where D is the diffusion coefficient and r is the rate of per capita population growth when rare. This approximation works well for species with Gaussian dispersal. However, many species have dispersal kernels whose tails are fatter than those of the Gaussian, and these can lead to more complicated dynamics of expansion and accelerating rates of spread. Long-distance dispersal events from the tails of these distributions may form satellite patches in advance of the main front that themselves become foci for population expansion. Rates of spread in these cases can be estimated by integrodifference equations, or by simulation.

APPLICATIONS TO CONSERVATION AND MANAGEMENT

Anthropogenic Alteration of Dispersal Patterns

Human activities are changing plant dispersal opportunities, frequencies, and patterns. Intentional and unintentional plant dispersal by humans has increased as diaspores travel in cars, boats, planes, clothing, and more. Hunting and habitat fragmentation have reduced abundances of many animal species even in relatively pristine areas. Densities of some species of animal seed dispersers have thus declined; densities of others may increase in response to the loss or decline of their predators and/or competitors. A number of studies have documented associations of such changes in animal disperser abundance with a decrease in the frequency of fruit and seed removal, changes in seed dispersal distances, and decreases in seedling recruitment. Even where animal dispersers are still present in historic densities, their behavior may be altered in such a way that dispersal services change if, for example, they tend not to cross roads.

Windspeeds, and thus the patterns of seed dispersal by wind, are also affected by human activities. The higher temperatures associated with global warming are expected to increase atmospheric instability and thus the frequency of long distance dispersal by wind. Landscape alteration also affects wind regimes; for example, forest fragments are generally exposed to higher wind speeds than are intact forests. Construction of canals, damming of rivers, constriction of floodplains, and other alterations of water bodies have similarly altered water flows and thus plant dispersal by water.

Anthropogenic Alteration of the Consequences of Dispersal

Human activities are also fundamentally altering the payoffs to dispersal on different scales, and arguably the necessity of dispersal. Habitat fragmentation means that dispersal that takes diaspores outside natural remnants yields nothing for many plant species, whereas historically dispersal to this distance might have been beneficial. At the same time, dispersal that goes far enough to reach another remnant may result in enhanced opportunities relative to the historical norm. Case studies of the evolution of dispersal on islands illuminate the potential consequences—selection for reduced dispersal within remnants, which in the long-term reduces the possibility of colonizing or recolonizing other remnants. Such landscape structure also disproportionately rewards directed dispersal, for example, by birds that fly between remnants.

Global climate change is increasingly making migration a necessity for long-term persistence of many species. Increasing temperatures and shifting rainfall regimes are leading to a growing mismatch between species' current distributions and the climates to which they are best suited. This places a premium on plant dispersal into the newly suitable areas and, indeed, threatens extinction for many species if they fail to disperse. In practice, this often requires dispersing over or around large areas of anthropogenically modified landscapes or through narrow corridors crossing such landscapes. The paleorecord shows that past climate shifts have been accompanied by associated shifts in plant species' ranges, although these have often lagged considerably. Historic climate shifts were accompanied by more extinctions on continents in which east–west mountain ranges barred the way. Unfortunately, anthropogenically modified habitats may for many species prove as much a barrier to dispersal as mountain ranges.

Manipulating Dispersal Opportunities to Promote Conservation

Deliberate measures to preserve, enhance, or inhibit plant dispersal opportunities can constitute valuable tools for conservation and management. Restoration and maintenance of natural densities of animal seed dispersers is an integral part of the conservation of any plant population, community, or ecosystem. Construction of corridors that connect habitat remnants can enable dispersal that enhances short-term population persistence and long-term viability in the face of global change. Habitat restoration and reestablishment of native vegetation can often be speeded through the provision of perches for birds that bring in seeds. Deliberate assisted migration of plant propagules to track climate change should be considered, especially where anthropogenic barriers restrict the possibility for unassisted migration. Finally, the introduction and spread of invasive species can be reduced by measures that restrict the transport of propagules by humans.

SEE ALSO THE FOLLOWING ARTICLES

Dispersal, Animal / Dispersal, Evolution of / Integrodifference Equations / Metapopulations / Restoration Ecology / Spatial Ecology

FURTHER READING

Bullock, J. M., K. Shea, and O. Skarpaas. 2006. Measuring plant dispersal: an introduction to field methods and experimental design. *Plant Ecology* 186: 217–234.

Bullock, J. M., R. E. Kenward, and R. S. Hails, eds. 2002. *Dispersal ecology.* Oxford: Blackwell Science.

Dennis, A. J., E. W. Schupp, R. J. Green, and D. W. Westcott, eds. 2007. *Seed dispersal: theory and its application in a changing world.* Wallingford, UK: CAB International.

Kuparinen, A. 2006. Mechanistic models for wind dispersal. *Trends in Plant Science* 11: 296–301.

Jones, F. A., and H. C. Muller-Landau. 2008. Measuring long-distance seed dispersal in complex natural environments: an evaluation and integration of classical and genetic methods. *Journal of Ecology* 96: 642–652.

Levin, S. A., H. C. Muller-Landau, R. Nathan, and J. Chave. 2003. The ecology and evolution of seed dispersal: a theoretical perspective. *Annual Review of Ecology and Systematics* 34: 575–604.

Levine, J. M., and D. J. Murrell. 2003. The community-level consequences of seed dispersal patterns. *Annual Review of Ecology Evolution and Systematics* 34: 549–574.

Nathan, R., and H. C. Muller-Landau. 2000. Spatial patterns of seed dispersal, their determinants and consequences for recruitment. *Trends in Ecology & Evolution* 15: 278–285.

Schupp, E. W., P. Jordano, and J. M. Gomez. 2010. Seed dispersal effectiveness revisited: a conceptual review. *New Phytologist* 188: 333–353.

Turchin, P. 1998. *Quantitative analysis of movement: measuring and modeling population redistribution in animals and plants.* Sunderland, MA: Sinauer.

DIVERSITY MEASURES

ANNE CHAO

National Tsing Hua University, Hsin-Chu, Taiwan

LOU JOST

Baños, Ecuador

Diversity is a measure of the compositional complexity of an assemblage. One of the fundamental parameters describing ecosystems, it plays a central role in community ecology and conservation biology. Widespread concern about the impact of human activities on ecosystems has made the measurement of diversity an increasingly important topic in recent years.

TRADITIONAL DIVERSITY MEASURES

The simplest and still most popular measure of diversity is just the number of species present in the assemblage. However, this is a very hard number to estimate reliably from small samples, especially in assemblages with many rare species. It also ignores an ecologically important aspect of diversity, the evenness of an assemblage's abundance distribution. If the distribution is dominated by a few species, an organism in the assemblage will seldom interact with the rare species. Therefore, these rare species should not count as much as the dominant species when calculating diversity for ecological comparisons. This observation has led ecologists (and also economists and other scientists studying complex systems of any kind) to develop diversity measures which take species frequencies into account.

There are two approaches to incorporating species frequencies into diversity measures. If the species-abundance distribution is known or can be determined, one or more of the parameters of the distribution function serve as a diversity measure. For example, when a species rank abundances distribution can be described by a log-series distribution, a single parameter, called Fisher's alpha, has often been used as a diversity measure. The parameters of other distributions (particularly the log-normal distribution) have also been used. However, this method gives uninterpretable results when the actual species abundance distribution does not fit the assumed theoretical distribution. This method also does not permit meaningful comparisons of assemblages with different distribution functions (for example, a log-normal assemblage cannot be compared to an assemblage whose abundance distribution follows a geometric series).

A more robust and general nonparametric approach, which makes no assumptions about the mathematical form of the underlying species-abundance distributions, is now the norm in ecology. Ecologists have often borrowed nonparametric measures of compositional complexity (which balance evenness and richness) from other sciences and equated these with biological diversity. The most popular measure of complexity has been the Shannon entropy,

$$H_{Sh} = -\sum_{i=1}^{S} p_i \log p_i, \tag{1}$$

where S is the number of species in the assemblage and the ith species has relative abundance p_i. This gives the uncertainty in the species identity of a randomly chosen individual in the assemblage. Another popular complexity measure is the Gini-Simpson index,

$$H_{GS} = 1 - \sum_{i=1}^{S} p_i^2, \tag{2}$$

which gives the probability that two randomly chosen individuals belong to different species.

However, these two complexity measures do not behave in the same intuitive linear way as species richness. When diversity is high, these measures hardly change their values even after some of the most dramatic ecological events imaginable. They also lead to logical contradictions in conservation biology, because they do not measure a conserved quantity (under a given conservation plan, the proportion of "diversity" lost and the proportion preserved can *both* be 90% or more). Finally, these measures each use different units, so they cannot be compared with each other.

DIVERSITY MEASURES THAT OBEY THE REPLICATION PRINCIPLE

Robert MacArthur solved these problems by converting the complexity measures to "effective number of species" (i.e., the number of equally abundant species that are needed to give the same value of the diversity measure), which use the same units as species richness. Shannon entropy can be converted by taking its exponential, and the Gini–Simpson index can be converted by the formula $1/(1 - H_{GS})$. These converted measures, like species richness itself, have an intuitive property that is implicit in much biological reasoning about diversity. This property, called the replication principle or the doubling property, states that if N equally diverse groups with no species in common are pooled in equal proportions, then the diversity of the pooled groups must be N times the diversity of a single group. Measures that follow the replication principle give logically consistent answers in conservation problems, rather than the self-contradictory answers of the earlier complexity measures. Their linear scale also facilitates interpreting changes in the magnitudes of these measures over time; changes that would seem intuitively large to an ecologist will cause a large change in these measures.

Mark Hill showed that the converted Shannon and Gini–Simpson measures, along with species richness, are members of a continuum of diversity measures called Hill numbers, or effective number of species, defined for $q \neq 1$ as

$${}^{q}D = \left(\sum_{i=1}^{S} p_i^q \right)^{1/(1-q)}. \tag{3a}$$

This measure is undefined for $q = 1$, but its limit as q tends to 1 exists and gives

$${}^{1}D = \lim_{q \to 1} {}^{q}D = \exp\left(-\sum_{i=1}^{S} p_i \log p_i \right) = \exp(H_{Sh}). \tag{3b}$$

The parameter q determines how much the measure discounts rare species. When $q = 0$, the species abundances do not count at all, and species richness is obtained. When $q = 1$, Equation 3b is the exponential of Shannon entropy. This measure weighs species in proportion to their frequency and can be interpreted as the number of "typical species" in the assemblage. When $q = 2$, Equation 3a becomes the inverse of the Simpson concentration and rare species are severely discounted.

The measure ^2D can be interpreted as the number of "relatively abundant species" in the assemblage.

All standard complexity measures can be converted to effective number of species. Since these and all other Hill numbers have the same units as species richness, it is possible to graph them on a single graph as a function of the parameter q. This diversity profile characterizes the species-abundance distribution of an assemblage and provides complete information about its diversity. All Hill numbers obey the replication principle.

Diversity measures that obey the replication principle are directly related to the concept of compositional similarity. If we pool N assemblages in equal proportions, the ratio of the mean single-assemblage diversity to the pooled diversity will vary from unity (indicating complete similarity in composition) to $1/N$ (indicating maximal dissimilarity in composition), as long as the mean single-assemblage diversity is defined properly. This diversity ratio can be normalized onto the unit interval and used as a measure of compositional similarity. Many of the most important similarity measures in ecology, such as the Sørensen, Jaccard, Horn, and Morisita–Horn indices of similarity, and their multiple-assemblage generalizations, are examples of this normalized diversity ratio.

The diversity of an extended region, often called the gamma diversity, can be partitioned into within- and between-assemblage components, usually called alpha and beta diversities, respectively. When all assemblages are assigned equal statistical weights, the beta component of a Hill number is the inverse of the diversity ratio described in the preceding paragraph. Beta diversity is thus directly related to compositional differentiation and gives the effective number of completely distinct assemblages (i.e., assemblage diversity). When the diversity measure is species richness or the exponential of Shannon entropy, this interpretation of beta diversity is valid even when the assemblages are not equally weighted.

The apportionment of regional diversity among assemblages gives clues about the ecological principles determining community composition. In order to test hypotheses about the factors determining community assembly and intercommunity differentiation, biologists need to compare the observed patterns against those that would be produced by purely stochastic effects. The expected values of alpha, beta, and gamma diversities of order 2 (Simpson measures) can be predicted from a purely stochastic "neutral" model of community assembly. This quality makes order 2 measures particularly important in community ecology. Approximate predictions can also be made for order 1 measures.

DIVERSITY MEASURES THAT INCORPORATE SPECIES' DIFFERENCES

Evelyn Pielou was the first to notice that the concept of diversity could be broadened to consider differences among species. All else being equal, an assemblage of phylogenetically or functionally divergent species is more diverse than an assemblage consisting of very similar species. Differences among species can be based on their evolutionary histories (in a form of taxonomy or phylogeny) or by differences in their functional trait values. Diversity measures can be generalized to incorporate these two types of species differences (referred to respectively as phylogenetic diversity and functional diversity).

Evolutionary histories are represented by phylogenetic trees. If the branch lengths are proportional to divergence time, the tree is ultrametric; all branch tips are the same distance from the basal node. If branch lengths are proportional to the number of base changes in a given gene, some branch tips may be farther from the basal node than other branch tips; such trees are non-ultrametric. A Linnaean taxonomic tree can be regarded as a special case of an ultrametric tree.

Most measures incorporating species differences are generalizations of the three classic species-neutral measures: species richness, Shannon entropy, and the Gini–Simpson index. Vane-Wright and colleagues generalized species richness to take into account cladistic diversity (*CD*), based on the total nodes in a taxonomic tree (which is also equal to the total length in the tree if each branch length is assigned to unit length). Faith defined the phylogenetic diversity (*PD*) as the sum of the branch lengths of a phylogeny connecting all species in the target community. These two measures can be regarded as a generalization of species richness (see Table 1).

Rao's quadratic entropy is a generalization of the Gini–Simpson index that takes phylogenetic or other differences among species into account:

$$Q = \sum_{i,j} d_{ij} p_i p_j, \tag{4}$$

where d_{ij} denotes the phylogenetic distance between species i and j, and p_i and p_j denote the relative abundance of species i and j. It gives the mean phylogenetic distance between individuals in the assemblage.

TABLE 1

A summary of diversity measures and their interpretations based on Hill numbers (all satisfy the replication principle)

Diversity order	Traditional diversity measures	Taxonomic diversity measures (L levels)	Phylogenetic diversity measures over T years (Ultrametric)	Phylogenetic diversity measures over $\bar{T}$ mean base changes (Non-ultrametric)	Functional diversity measures over R trait-based distances
$q = 0$	Species richness	CD/L	PD/T	$PD/\bar{T}$	FD/R
$q = 1$	$\exp(H_{Sh})$	$\exp(H_p/L)$	$\exp(H_p/T)$	$\exp(H_p/\bar{T})$	$\exp(H_p/R)$
$q = 2$	$1/[1-(H_{GS})]$	$1/[1-(Q/L)]$	$1/[1-(Q/T)]$	$1/[1-(Q/\bar{T})]$	$1/[1-(Q/R)]$
Diversity or mean diversity of general order q	${}^qD$: Hill numbers (effective number of species)	${}^q\bar{D}(L)$: Mean effective number of cladistic nodes per level	${}^q\bar{D}(T)$: Mean effective number of lineages (or species) over T years	${}^q\bar{D}(\bar{T})$: Mean effective number of lineages (or species) over $\bar{T}$ mean base changes	${}^q\bar{D}(R)$: Mean effective number of functional groups up to R trait-based distances
Related measure		${}^q\bar{D}(L) \times L$: effective number of total cladistic nodes for L levels	${}^q\bar{D}(T) \times T$: effective number of lineage-lengths over T years	${}^q\bar{D}(\bar{T}) \times \bar{T}$: effective number of base changes over $\bar{T}$ mean base changes	${}^q\bar{D}(R) \times R$: effective number of functional distances up to R trait-based distances

H_{Sh}: Shannon entropy; H_{GS}: Gini–Simpson index; CD: cladistic diversity (total number of nodes) by Vane-Wright et al.; PD: phylogenetic diversity (sum of branch lengths) by Faith; FD: functional diversity (sum of trait-based distances) by Petchey and Gaston; Q: quadratic entropy by Rao; H_p: phylogenetic entropy by Allen et al. and Pavoine et al.; $\bar{T}$: mean base change per species for nonultrametric trees; ${}^qD$: Hill numbers (see Eq. 3a); ${}^q\bar{D}(L)$, ${}^q\bar{D}(T)$, ${}^q\bar{D}(\bar{T})$, and ${}^q\bar{D}(R)$: phylogenetic diversity by Chao et al. (see Eq. 6).

Shannon's entropy has also been generalized to take phylogenetic distance into account, yielding the phylogenetic entropy H_p:

$$H_P = -\sum_i L_i a_i \log a_i, \tag{5}$$

where the summation is over all branches, L_i is the length of branch i, and a_i denotes the abundance descending from branch i.

Since Shannon entropy and the Gini–Simpson index do not obey the replication principle, neither do their phylogenetic generalizations. These measures of phylogenetic diversity will therefore have the same interpretational problems as their parent measures. These problems can be avoided by generalizing the Hill numbers, which obey the replication principle. The generalization requires that we specify a parameter T, which is the distance (in units of time or base changes) from the branch tips to a cross section of interest in the tree. The generalization is

$${}^q\bar{D}(T) = \left\{ \sum_{i \in \boldsymbol{B}_T} \frac{L_i}{T} a_i^q \right\}^{1/(1-q)}, \tag{6}$$

where $\boldsymbol{B}_T$ denotes the set of all branches in this time interval $[-T, 0]$, L_i is the length (duration) of branch i in the set $\boldsymbol{B}_T$, and a_i is the total abundance descended from branch i. This ${}^q\bar{D}(T)$ gives the mean effective number of maximally distinct lineages (or species) through T years ago, or the mean diversity of order q over T years.

The diversity of a tree with ${}^q\bar{D}(T) = z$ in the time period $[-T, 0]$ is the same as the diversity of a community consisting of z equally abundant and maximally distinct species with branch length T. The product of ${}^q\bar{D}(T)$ and T quantifies "effective number of lineage-lengths or lineage-years." If $q = 0$, and T is the age of the highest node, this product reduces to Faith's PD.

For nonultrametric trees, let $\boldsymbol{B}_{\bar{T}}$ denote the set of branches connecting all focal species with mean base change $\bar{T}$. Here, $\bar{T} = \sum_{i \in \boldsymbol{B}_{\bar{T}}} L_i a_i$ represents the abundance-weighted mean base change per species. The diversity of a nonultrametric tree with mean evolutionary change $\bar{T}$ is the same as that of an ultrametric tree with time parameter $\bar{T}$. Therefore, the diversity formula for a nonultrametric tree is obtained by replacing T in the ${}^q\bar{D}(T)$ by $\bar{T}$ (see Table 1).

Equation 6 can also describe taxonomic diversity, if the phylogenetic tree is a Linnaean tree with L levels, and each branch is assigned unit length. Equation 6 also describes functional diversity, if a dendrogram can be constructed from a trait-based distance matrix using a clustering scheme.

Both Q and H_p can be transformed into members of this family of measures, and they then satisfy the replication principle (see Table 1) and have the intuitive behavior biologists expect of a diversity. The replication principle can be generalized to phylogenetic or functional diversity: when N maximally distinct trees (no shared nodes during the interval $[-T, 0]$) with equal mean diversities are combined, the mean diversity of the combined tree is N times the mean diversity of any individual tree. This property ensures the intuitive behavior of these generalized diversity measures.

ESTIMATING DIVERSITY FROM SMALL SAMPLES

In practice, the true relative abundances of the species in an assemblage are unknown and must be estimated from small samples. If the sample relative abundances are used directly in the formulas for diversity, the maximum-likelihood estimator of the true diversity is obtained. This number generally underestimates the actual diversity of the population, particularly when sample coverage is low. When coverage is low, it is important to use nearly unbiased estimators of diversity instead of the maximum-likelihood estimator. Unbiased estimators of diversities of order 2 are available, and nearly unbiased estimators of Shannon entropy and its exponential have recently been developed by Chao and Shen. Species richness is much more difficult to estimate than higher-order diversities; at best a lower bound can be estimated. A simple but useful lower bound (which is referred to as the Chao1 estimator in literature) for species richness is

$$S_{Chao1} = D + f_1^2/(2f_2),$$

where D denotes the number of observed species in sample, f_1 denotes the number of singletons and f_2 denotes the number of doubletons. Estimation of phylogenetic or functional diversity from small samples should follow similar principles, but merits more research.

SEE ALSO THE FOLLOWING ARTICLES

Conservation Biology / Neutral Community Ecology / Statistics in Ecology

FURTHER READING

Chao, A. 2005. Species estimation and applications. In S. Kotz, N. Balakrishnan, C. B. Read, and B. Vidakovic, eds. *Encyclopedia of statistical sciences*, 2nd ed. New York: Wiley.

Chao, A., C.-H. Chiu, and L. Jost. 2010. Phylogenetic diversity measures based on Hill numbers. *Philosophical Transactions of the Royal Society B: Biological Sciences* 365: 3599–3609.

Chao, A., and T.-J. Shen. 2010. *SPADE: Species prediction and diversity estimation*. Program and user's guide at http://chao.stat.nthu.edu.tw/softwareCE.html.

Gotelli, N. J., and Colwell, R. K. 2011. Estimating species richness. In A. Magurran and B. McGill, eds. *Biological diversity: frontiers in measurement and assessment*. Oxford: Oxford University Press.

Jost, L. 2006. Entropy and diversity. *Oikos* 113: 363–375.

Jost, L. 2007. Partitioning diversity into independent alpha and beta components. *Ecology* 88: 2427–2439.

Jost, L., and A. Chao, 2012. *Diversity analysis*. London: Taylor and Francis. (In preparation.)

Magurran, A. E. 2004. *Measuring biological diversity*. Oxford: Blackwell Science.

Magurran, A. E., and B. McGill, eds. 2011. *Biological diversity: frontiers in measurement and assessment*. Oxford: Oxford University Press.

DYNAMIC PROGRAMMING

MICHAEL BODE

University of Melbourne, Victoria, Australia

HEDLEY GRANTHAM

University of Queensland, Australia

Dynamic programming is a mathematical optimization method that is widely used in theoretical ecology and conservation to identify a sequence of decisions that will best achieve a given objective. When ecosystem dynamics (potentially including social, political, and economic processes) can be described using a model that is discrete in time and state space, dynamic programming can provide an optimal decision schedule. The technique is most commonly applied to explain the behavior of organisms (particularly their life history strategies) and to plan cost-effective management strategies in conservation and natural resource management. Compared with alternative dynamic optimization methods, dynamic programming is both flexible and robust, can readily incorporate stochasticity, is well suited to computer implementation, and is considered to be conceptually intuitive. On the other hand, the optimal results are generated in a form that can be very difficult to interpret or generalize. Furthermore, models of complex ecosystems can be computationally intractable due to nonlinear growth in the size of the state space.

OPTIMIZATION

In mathematics, optimization is the process by which an agent chooses the best decision from a set of feasible alternatives. It plays a central role in a wide range of ecological, evolutionary, and environmental fields. Optimization both provides normative advice to managers in applied ecology (i.e., how to best manage ecosystems) and offers positive insights into the actions of ecological agents (i.e., understanding why organisms behave in particular ways).

Frequently, agents are required to make a sequence of decisions where the outcome will not be realized until all the decisions have been implemented. Such sequential optimization problems are more difficult to solve than optimizations involving single decisions (static optimization), because actions that appear attractive in the short term may not result in the best long-term outcomes. In these situations, agents are required to undertake dynamic optimization. Dynamic optimization requires the application of more sophisticated analytic methods than those used in static optimization. The two most commonly applied in theoretical ecology are optimal control theory and dynamic programming.

DYNAMIC PROGRAMMING

Dynamic programming was developed in the 1950s by the mathematician Richard Bellman while working at the RAND Corporation. The method advanced the field of dynamic optimization by synthesizing three novel insights. The first was conceptual. Since the development of the calculus of variations in the early eighteenth century, dynamic optimization had been understood to be the search for a continuous control function $u(t)$ that, if applied at each point in time, would maximize a time-integrated objective function. Bellman instead framed the solution to a dynamic optimization problem as a policy: a prescription to undertake a particular action if the agent observes the system to be in a given state. This policy formulation was likely influenced by the economic and political questions being considered at RAND. It makes dynamic programming particularly well suited to applied management questions and helps to explain the method's frequent application to problems in natural resource and environmental management.

Practically, this policy approach requires the optimization problems to be formulated as Markov decision processes, with discretized time and state dimensions. This approach is very robust to different assumptions about the nature of either the control variables or the state dynamics, in contrast to optimal control theory. As time passes, the ecosystem's natural dynamics can lead to changes in the system state, but this evolution is altered by the decision maker's interventions. Their goal is to maximize a well-defined objective function by constructing a policy that defines when and which interventions should be taken. While this optimal decision policy could theoretically be identified by exhaustively considering all the possible combinations of decisions, states, and time periods, the number of candidate solutions makes this practically impossible. Bellman circumvented this problem with two additional insights. One was his principle of optimality:

> An optimal policy has the property that whatever the initial state and initial decisions are, the remaining decisions must constitute an optimal policy with regard to the state resulting from the first decision.

For example, our problem could be to identify the shortest route between two points A and B (Fig. 1). Allow that

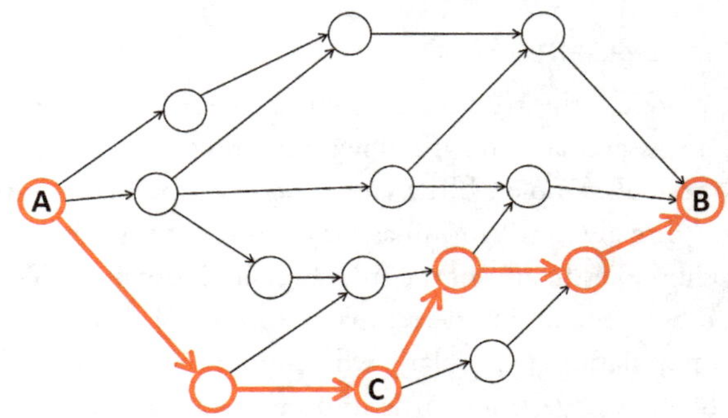

FIGURE 1 Example of Bellman's principle of optimality. If the red arrows denote the optimal path between nodes A and B, then the optimal path between C and B follows the same route.

the red arrows denote the unique optimum. Bellman's principle of optimality states that if we instead started the problem at an intermediate point C on the optimal path, we could achieve the optimal outcome in the reduced problem (i.e., find the shortest route between C and B) by following the path that is optimal for the full problem.

Bellman's principle of optimality allows the final insight—the application of backward induction—to drastically reduce the computational effort required to consider all potentially optimal decision sequences. Instead of trying to determine the optimal policy all at once, backward induction instead solves a much smaller problem in the penultimate time step. It then repeats the backward step, and never again needs to consider those discarded options. This repetitive back-stepping approach, along with the discretized nature of the problem formulation, makes dynamic programming well suited to modern computational methods.

These benefits of dynamic programming are counterbalanced by two primary drawbacks. First, the discretized formulation of the state dynamics leads very rapidly to computational problems when the dimensionality of the state space increases, a problem that Bellman called the "curse of dimensionality." For example, identifying the optimal eradication strategy for a two-species ecosystem is more than twice as computationally intensive as for a single-species ecosystem. As a result, many common ecological optimization problems can be practically unsolvable without drastic simplifications. The second drawback of dynamic programming is the opacity of the optimal solutions generated. The format of the optimal policy—indicating the best decision to take in each state, at each point in time—leads to an unwieldy amount of information. While dynamic programming generates optimal solutions, the method itself offers little insight into what factors determined the form of the solution or how parametric or structural changes to the problem formulation would affect that solution. Creative sensitivity analyses and visualizations can provide a better understanding of the optimal solution's nature, but it can be difficult to confirm that these insights are in fact true.

MATHEMATICAL FORMULATION

A distinction is often made between dynamic programming and stochastic dynamic programming. In the latter, state transition dynamics are not deterministic but follow defined probability distributions. This entry does not distinguish between these two methods, noting that dynamic programming can be considered a limiting case of stochastic dynamic programming where the transition probabilities are either one or zero.

The basic dynamic programming problem has two features. First, the state dynamics of the system are described in discrete time. At each point in time t, the system is in state x_t, and the decision maker has the option of taking an action a_t from a set of alternatives. At the next point in time, the system will have evolved to a new state x_{t+1}, defined as

$$x_{t+1} = f(x_t, a_t, t, w_t), \tag{1}$$

where w_t is a random parameter that allows the uncertain evolution of a stochastic system. Equation 1 shows the traditional formulation of system dynamics in dynamic programming, allowing the state x_t to take continuous, real values. However, when describing or managing ecosystems we are more often interested in dynamics that occur on a finite set of S different system states, for example, the number of individuals in a population. Such state dynamics are most easily understood and expressed using a set of probabilistic state transition matrices. The probability of a system in state x_t transitioning to state x_{t+1}, given that the manager undertakes action a at time t is

$$p_{ij}(a, t) = \mathrm{P}\{x_{t+1} = j \mid x_t = i, a\}. \tag{2}$$

In many cases, these transition matrices remain constant through time (i.e., $p_{ij}(a, t) = p_{ij}(a)$).

The second feature of a dynamic programming problem is an objective, generally the maximization of a reward function. A reward function is some combination of the benefits and costs that the managers derive from being in each state at each point in time, less the cost of the actions necessary to reach those states. A decision maker's objective is to maximize the total reward,

$$\max_{a_{x_t}} \mathrm{E}\left\{R(x_T, T) + \sum_{t=1}^{T-1} R(x_t, a_{x_t}, t)\right\}, \tag{3}$$

by choosing the best intervention a_{x_t} for each system state and time. The objective stated in Equation 3 is a composite of a terminal reward that depends on the final state of the system, and a reward that accumulates at each time step, and it depends on both the system state and the cost of the action taken. Alternative objectives might only include the terminal reward, or only the accumulating reward.

The dynamic programming algorithm introduces a value function defined by the reward function in the final time step,

$$J(x_T, T) = R(x_T, T), \tag{4}$$

and an equation that links the value function in a given time step to the value in the subsequent time step,

$$J(x_t, t) = \max_{a_{x_t}} \left\{R(x_t, a_{x_t}) + \sum_{i=1}^{S} p_{x_t,i}(a_{x_t}, t)\, J(i, t+1)\right\}. \tag{5}$$

With repeated application of Equation 5, Equation 4 can be used to identify the sequence a_t that solves Equation 3.

APPLICATIONS

Ecology and Evolution

A fundamental tenet of the modern evolutionary perspective is that organisms are fitness-maximizing agents—Darwinian individuals whose purpose is to maximize the probability that their genetic material will be successfully passed to the next generation. Based on this understanding, an organism's heritable attributes—both its characteristics and its behavior—are solutions to a constrained fitness-optimization problem. Importantly, an organism's fitness is the result of a large number of evolutionary and behavioral decisions taken over its entire lifetime. Early optimization approaches to behavioral ecology and evolution applied static optimization methods to identify the best decisions to isolated behavioral problems—for example, choosing a maximum foraging distance or clutch size. These approaches were criticized for implicitly assuming that an organism's decisions contributed independently to its fitness. In reality, its decisions are made strongly interdependent by constraints such as energy or time budgets; for example, organisms cannot increase the time they spend foraging without reducing the time available for reproduction. This interdependence even extends across generations, through decisions about maternal investment or offspring size. Dynamic optimization methods such as dynamic programming provide a coherent framework within which to consider the time-integrated and interdependent effect of different decisions on an organism's fitness. The flexible and practical nature of dynamic programming makes it a favorite method of behavioral and evolutionary ecologists.

Dynamic programming was introduced to the broader ecological audience by Mangel and Clark's influential 1988 book *Dynamic Modeling in Behavioral Ecology*. Although dynamic programming has been applied to traditional behavioral ecology questions such as foraging theory, the method's value is most apparent in its application to integrated life history theory, which considers all the events of an organism's life as a single strategy that maximizes its expected number of lifetime surviving offspring. Because life history theory is fundamentally concerned with the integrated effect of multiple sequential decisions, it has found the most use for dynamic optimization. Dynamic programming has been heavily applied to investigations into the optimal amount of parental investment and the seasonal timing of growth and reproduction by annual plants. In the latter example, the strengths of dynamic programming are clearly in evidence when environmental fluctuations influence outcomes. The timing of reproductive investment in plants is one of behavioral ecology's canonical examples, first considered in the 1970s using deterministic optimal control theory. However, because conditions and season lengths vary unpredictably, stochastic dynamic programming provides more accurate insight into optimal life history decisions that must predict and respond to unpredictable changes.

Conservation Management

Conservationists were first exposed to dynamic programming via theories of natural resource management, where the method had been in use since the 1960s. Dynamic programming is particularly common in managing the harvest of forests and waterfowl populations. Fish and wildlife managers also recognized that the feedback-control aspects of dynamic programming were particularly well suited to harvesting in contexts where populations levels are difficult to estimate. Carl Walters' early work on the management of Pacific salmon populations introduced fisheries managers to the method, which he went on to use as the analytic machinery for adaptive management. Active adaptive management in particular displays the strengths of dynamic programming, making use of its ability to consider both uncertainty (as a knowledge state) and stochasticity.

Conservation is a chronically underfunded enterprise. As a consequence, management ambitions must be pursued incrementally, and managers aim to optimize the amount of biodiversity their limited budgets can protect. Dynamic programming, with its focus on sequential decision making and optimality, is therefore an ideal conservation tool and has been applied in a range of conservation contexts. These include ecological fire management, translocation of captive-bred individuals from threatened species, and the search for and eradication of invasive species. Conservation resource allocation theory has applied dynamic programming to many applications, particularly to the problem of land acquisition planning. The example application given below is from this theoretical field.

EXAMPLE APPLICATION

Anecdotally, dynamic programming is often best learned by example, and this entry therefore describes a simple application to a conservation planning problem, using a simplified version of the opportunistic land acquisition model described in McDonald-Madden et al. (2008), an "optimal-stopping" problem.

A conservation organization wants to protect threatened species that are distributed heterogeneously within a privately owned landscape. The species are randomly distributed with a mean density of λ, and thus the probability that a land parcel contains r species is Poisson distributed as $q(r) = \lambda^r e^{-\lambda}/r!$. The organization does not have the power to expropriate land and can only purchase parcels that appear on the open market. This is a problem that many nongovernmental conservation organizations must address when opportunistically protecting land. Each year, the organization receives resources sufficient to purchase a single land parcel; any funds that remain unspent at the end of the period will not roll over to the next. A random parcel becomes available for purchase on the first day of each month; the organization must repeatedly decide whether it wants to purchase the available parcel or wait in the hope that a better option becomes available. Their objective is to maximize the number of species protected in the purchased land parcel.

First, define the states of the system. The ith state of the system is represented by the ordered pair (r_i, B_i), where B_i is a Boolean variable that indicates whether the organization has already purchased a parcel. At the beginning of the year, $B_i = 0$; once the manager purchases a parcel, B_i becomes 1. The variable r_i indicates the number of species present in the parcel currently available for purchase, and it ranges between zero and some maximum species density $R_{\max}$ (e.g., the number of species in a particular parcel cannot exceed the regional species richness). If the state is (4, 0), for example, the manager has not yet purchased a land parcel, but one is currently available for purchase that contains four threatened species. The equation,

$$i = (R_{\max} + 1)B_i + r_i + 1, \tag{6}$$

allocates a unique identification number to each state that allows us to construct probabilistic transition matrices. According to Equation 6, if $R_{\max} = 5$ the states are

$$\begin{aligned} s_1 &:= (0, 0), \\ s_2 &:= (1, 0), \\ s_3 &:= (2, 0), \\ &\vdots \\ s_7 &:= (0, 1), \\ s_8 &:= (1, 1), \\ &\vdots \\ s_{12} &:= (5, 1), \end{aligned} \tag{7}$$

and $N = 2R_{\max} + 2$ is the total number of states in the system.

Each month the manager can take one of two decisions: do not purchase (decision 1) or purchase (decision 2) the parcel currently being offered. If decision 1 is taken and the manager has not yet purchased a parcel ($B_i = 0$), the parcel that becomes available next month will contain r species with probability $\lambda^r e^{-\lambda}/r!$. If the manager already owns a parcel ($B_i = 1$), the state does not change. Thus, the transition matrix for decision 1 is

$$\begin{aligned} p_{ij}(1) &= \mathrm{P}\{x_{t+1} = j \mid x_t = i,\ a(x_t, t) = 1\} \\ &= \begin{cases} \lambda^{r_j} e^{-\lambda}/r_j! & \text{if } B_i = B_j = 0, \\ 1 & \text{if } B_j = 1;\ i = j, \\ 0 & \text{otherwise.} \end{cases} \end{aligned} \tag{8}$$

If the manager does not currently own a land parcel ($B_i = 0$), he can choose to purchase the one currently available (decision 2). However, if a land parcel has already been purchased ($B_i = 1$) and the manager tries to purchase a second, both are lost (e.g., the organization became financially overextended), and the organization is left without any protected species and is unable to purchase another parcel (i.e., state (0, 1)). Thus, the transition matrix for decision 2 is

$$\begin{aligned} p_{ij}(2) &= \mathrm{P}\{x_{t+1} = j \mid x_t = i,\ a(x_t, t) = 2\} \\ &= \begin{cases} 1 & \text{if} \quad B_i = 0, j = i + R_{\max} + 1, \\ 1 & \text{if} \quad B_i = 1, j = R_{\max} + 2, \\ 0 & \text{otherwise.} \end{cases} \end{aligned} \tag{9}$$

The reward function for this problem is very straightforward and depends only on the number of species in the parcel purchased by the organization. The intermediate rewards are zero in Equation 3 (i.e., $R(x_t, a(x, t)t) = 0$ for $t \neq T$). At the end of the year, the conservation organization benefits linearly from the number of species contained in the protected land parcel:

$$R(x_T = i, T) = \begin{cases} 0 & \text{if} \quad B_i = 0, \\ r_i & \text{if} \quad B_i = 1 \end{cases}. \tag{10}$$

With these descriptions of the dynamics, decisions, and rewards, the optimal policy can be calculated. In the final time step, when no actions are possible, $J(x_T, T) = R(x_T, T)$, as described in Equation 4. In the penultimate time step, we follow Equation 5:

$$J(x_{T-1}, T-1) = \max_a \left\{ \sum_{j=1}^{N} p_{x_{T-1}j}(a) R(j, T) \right\}. \tag{11}$$

For example, the value of being in state 1 on December 1st is equal to

$$\begin{aligned} J(1, T-1) = \max_a \{ & p_{11}(a) R(1, T) \\ & + p_{12}(a) R(2, T) \\ & + \cdots p_{1N}(a) R(N, T) \}, \end{aligned} \tag{12}$$

and the optimal decision is the one that maximizes the expected reward in the final time step. Since we know that the reward function is zero if we end the year without purchasing a land parcel (Eq. 8), the optimal decision in the penultimate time step would be to purchase the available parcel, if a parcel has not yet been bought.

The value of being in state 1 on November 1st (i.e., the second-last month) is equal to

$$\begin{aligned} J(1, T-2) = \max_a \{ & p_{11}(a) J(1, T-1) \\ & + p_{12}(a) J(2, T-1) \\ & + \cdots p_{1N}(a) J(N, T-1) \}. \end{aligned} \tag{13}$$

In this case, if the available parcel is purchased, the manager forfeits the opportunity to purchase the parcel that will become available on December 1st. He would therefore only choose to do so if the currently available land parcel contains more threatened species than the average parcel (i.e., if $r_i > \lambda$). Thus, as the backward-induction proceeds, the managers become more selective with the parcels they would be willing to buy, because they have more time (and therefore more options) available to them before they must purchase the December parcel. Figure 2 depicts the solution to the dynamic programming problem: the optimal decision in each month for a manager who has not yet purchased a parcel. The shade of each grid

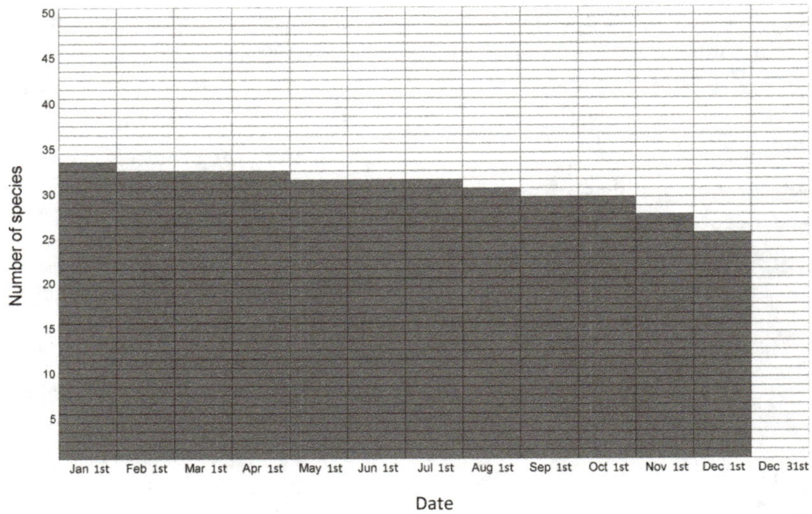

FIGURE 2 The optimal policy for the example problem when the mean species richness is $\lambda = 25$. Shading indicates the optimal decision for the manager to take given the date (on the horizontal axis) and the species richness of the currently available land parcel (on the vertical axis). Gray shading indicates that the optimal decision is not to purchase (i.e., decision 1); white shading indicates that purchase (decision 2) is optimal.

square indicates the optimal decision given the number of species present in the land parcel (indicated on the vertical axis) and the date (indicated on the horizontal axis).

EXTENSIONS

Applications of dynamic programming in ecology have proliferated over the past decade with the method's expansion into conservation applications. Its flexibility and suitability for policy formulation are likely to encourage further use as the method becomes better known and as ever-increasing computational power increases the complexity of problems to which it can be practically applied. For example, computational improvements have recently made possible the dynamic optimization of processes where the state dynamics prove difficult to observe. Although this issue was first addressed in the 1980s, an explicit inclusion of observational uncertainty into the control of dynamic systems was long hampered by increased computational demand of additional information dimensions. Recent years have seen more frequent investigations into partially observable dynamic programming. Dynamic programming is also a field of ongoing research in mathematics, and there is consequently a range of novel variations to the method that have yet to be applied to theoretical ecology (though some have found applications in economics and engineering). These include robust dynamic programming in the presence of structural model uncertainty, and approximation methods for problems whose high dimensionality precludes standard approaches.

SEE ALSO THE FOLLOWING ARTICLES

Conservation Biology / Foraging Behavior / Markov Chains / Optimal Control Theory / Predator–Prey Models

FURTHER READING

Bellman, R. E., and S. E. Dreyfus. 1962. *Applied dynamic programming.* Princeton: Princeton University Press.

Clark, C. W. 1990. *Mathematical bioeconomics,* 2nd ed. New York: Wiley and Sons.

Iwasa, Y. 2000. Dynamic optimization of plant growth. *Evolutionary Ecology Research* 2: 437–455.

Mangel, M., and C. W. Clark. 1988. *Dynamic modeling in behavioral ecology.* Princeton: Princeton University Press.

McDonald-Madden, E., M. Bode, E. T. Game, H. Grantham, and H. P. Possingham. 2008. The need for speed: informed land acquisitions for conservation in a dynamic property market. *Ecology Letters* 7: 1169–1177.

Regan, T. J., M. A. McCarthy, P. W. J. Baxter, F. D. Panetta, and H. P. Possingham. 2006. Optimal eradication: when to stop looking for an invasive plant. *Ecology Letters* 9: 759–766.

Walters, C. 1986. *Adaptive management of renewable resources.* New York: MacMillan.

E

ECOLOGICAL ECONOMICS

SUNNY JARDINE AND JAMES N. SANCHIRICO
University of California, Davis

Ecological economics is the study of the relationship between the economy and the ecosystem to promote the goal of sustainable development. While the field is over 40 years old, there is no agreed-upon definition, because the theories and methodologies of ecological economics draw from a number of social and natural science disciplines. This entry provides background on the emergence of the field as well as several competing interpretations of what constitutes ecological economics.

THE EMERGENCE OF ECOLOGICAL ECONOMICS

The roots of ecological economics are in the theory and methodology of natural resource and environmental economics. However, ecological economics also accepts the existence of a limiting factor to growth, as first described by Thomas Malthus in 1798. Malthus argued that a geometrically growing human population would inevitably surpass the capability of the land to provide for its sustenance. Thus, famine and disease will inevitably result, driving population levels down until the population begins to grow and the cycle repeats itself.

Mainstream environmental and natural resource economists reject Malthus's theory because it does not consider society's ability to develop technological innovations that increase the productive capabilities of the land. In unencumbered markets, they argue, the scarcity of a resource (not just land) will drive up the resource price and provide incentives for people to use the resource more conservatively and explore consumption alternatives. Meanwhile, ecological economists maintain that growth has unavoidable limits in a system with finite resources.

In addition to the differences in underlying assumptions, ecological economics incorporates work from various other social science disciplines in a way that economics traditionally has not done. Finally, ecological economists differ in their view of the economist's role. While resource economists and environmental economists often undertake analyses of resource and pollution problems with the goal of economic efficiency, ecological economists employ methods to discover ways in which to attain the goal of sustainable development, often defined as development that meets the needs of the present without compromising the ability of future generations to meet their own needs.

To better understand the differences and similarities between ecological economics and resource or environmental economics, the following section briefly summarizes the development of economic thought regarding resource and pollution problems. Due to the lack of a unifying paradigm within ecological economics, however, providing a definitive story of its origin and composition is elusive.

For the most part, environmental and resource economics developed alongside each other beginning in the 1920s and 1930s. The two fields differ in the types of problems addressed; while environmental economics focuses on pollution problems, resource economists look at the management of natural resources. Ecological economics is a newer field, beginning in the 1960s. It addresses problems of both types and often contends that environmental and resource economists conduct research too narrow in scope.

Environmental Economics

The origins of environmental economics date back to an 1844 publication by Jules Dupuit entitled "On the Measurement of the Utility of Public Works." In it, Dupuit introduces two concepts that became very influential to the field of environmental economics. First, he argues that if a resource is available that benefits all members of society without the possibility for exclusion, each individual benefits from the contribution of others and therefore does not have the incentives to make contributions equal to their desire for the good. The implication was that resources such as these, which became known as public goods, will be underprovided by the market. A number of environmental goods and services have characteristics of public goods, including air and water quality, forests, and wildlife reserves.

The second concept developed by Dupuit is that of consumer surplus. In Dupuit's example, the benefit an individual consumer receives from having a bridge built is the difference between what the individual is willing to pay for the particular number of times they choose to cross the bridge at a given price and what they are required to pay. The sum of benefits accruing to all the individuals in society from the provision of a public good is the consumer surplus. When the public good is an environmental good or service, consumer surplus can be used as one measure of its value, capturing the tradeoffs individuals in society are willing to make to have the environmental good or service.

Another critical development in environmental economics is found in Arthur Cecil Pigou's 1920 publication, *The Economics of Welfare.* In this volume, Pigou introduces the idea of externalities, which are the byproducts of either production or consumption and characterize the situation where one agent's decision affects the welfare of another agent, as a type of market failure. Externalities can be either desirable or undesirable, and when the externality remains unchecked, there is a divergence between the private costs and benefits of an action and the social costs and benefits of that action.

Pollution as a byproduct of production is the classic example. When a firm faces no costs to polluting a river or the air, the amount of pollution released in the production process will be greater than what is socially desirable. Pigou proposed a pollution tax, where the tax is the price of pollution and is set equal to the damages (at the margin) from the pollution. The tax, therefore, provides the firm with the correct incentives regarding how much pollution to emit. In economic jargon, the tax corrects the market failure.

In 1960, Ronald Coase published "The Problem of Social Cost," in which he proposes the allocation of property rights as another solution to externalities. If property rights to the resource being polluted were appropriated, the owner of the property rights would demand a payment in compensation equal to the damage caused by the pollution. As with a tax, the pollution price the owner demanded would provide the firm with the right incentives and correct the market failure.

Taxation and property rights are the main policy tools proposed by environmental economists and many ecological economists as solutions to environmental externality problems. For example, the discussion of how to reduce greenhouse gas emissions focuses on using a carbon tax or carbon caps, where the caps are a form of property rights that allow the firms to emit a certain amount via emission permits.

Up through the 1960s, externalities were viewed as an exception to the norm. In 1969, Robert Ayres and Allen Kneese published a paper, "Production, Consumption, and Externalities," that challenges this perception. They argue that externalities are the necessary result of *all* acts of production and consumption because these acts produce wastes that, due to the first law of thermodynamics, become externalities with limited environmental capacity for assimilation.

Applied to the natural resources used in the economy, the laws of conservation of mass are known as mass balance or materials balance. While used in engineering, this concept's use in economic modeling had been very limited. Ayres and Kneese show that capturing the entirety of the externalities any economic activity generates requires a model of the entire system or a general equilibrium model adhering to natural laws.

Resource Economics

While welfare economists were busy developing the theory that later evolved into the subfield of environmental economics, other economists were formulating issues of natural resource use as investment problems. This area of research became known as resource economics and is concerned with nonrenewable resources, such as oil and minerals, and renewable resources, such as forests and fisheries.

In 1931, Harold Hotelling published "The Economics of Exhaustible Resources," in which he considers the problem of using the nonrenewable resource over time in a manner that yields the greatest public good. The time dimension to the resource problem, he argues, necessitates setting up a dynamic model and solving for an optimal extraction path given a resource constraint. An example of dynamic optimization applied to resource use is the optimal amount of ore that should be extracted from a mine in each year (the extraction path), given that costs and benefits are associated with extraction and only a finite amount of the mineral is available for extraction (the resource constraint).

Hotelling's work came at a time when resource scarcity was a "new" concern for economic growth and the conservation movement. Interestingly, Hotelling anticipated one of the critiques of his work by the modern conversation movement, which stemmed from the fact that the value of future resources are discounted relative to the present—that is, future pleasures are less valuable than current ones, everything else being equal. Hotelling's response to the potential argument against discounting was to note the link between resources and future pleasures is an uncertain one and that pleasure is also determined by income distribution. He states that economists should be concerned with the value of goods produced from the resource under varying extraction paths and that taxes or other political tools should be used to promote a more equitable income distribution both today and in the future.

Up until the 1950s, most thought on renewable resources was on forestry, dating back to the work of German forester Martin Faustman in 1849. As Hotelling showed for exhaustible resources, Faustman showed that the optimal time to cut a forest was an investment decision that depended on when the benefit was equal to the cost, which included the lost economic value from the future growth of the tree.

Economists became interested in the economic aspects of managing fishery resources starting in the 1950s, when the global demand for fish and a post-WWII boom in vessel catching power were beginning to have an impact on fish populations. At the time, fisheries had virtually no regulations. In works published in the mid-1950s, H. Scott Gordon and Anthony Scott argue that when access to a resource is unrestricted, the resource will be overused and the economic benefits will be dissipated.

The study "The Fishery: The Objectives of Sole Ownership," by Anthony Scott, is thought to be the first economic analysis to incorporate population dynamics to describe the system. As such, Scott's work is a first step in the direction of modeling the human economy as part of a larger ecological system, which later became known as bioeconomics.

In the 1962 publication *Economic Aspects of the Pacific Halibut Fishery*, James Crutchfield and Arnold Zellner incorporate a dynamic population model describing the halibut stock with an economic objective in a dynamic optimization framework. The work by Crutchfield and Zellner can be viewed as the renewable resource counterpart of Hotelling's 1931 paper, creating the foundations for thought on the optimal solution to a dynamic nonrenewable resource problem.

The environmental movement of the 1960s questioned the notion that natural resources should only be valued for the tangible goods and services they generate. In the 1967 paper "Conservation Reconsidered," John Krutilla explores both the use and nonuse values associated with natural resources. The paper introduces the concept of option value, which is associated with the value of having an option to use in the future a resource that is difficult or impossible to replace.

Krutilla uses the Grand Canyon as an example. Option values for the Grand Canyon include the potential value of discovering a species with medicinal uses and the value of having the option to visit this natural wonder even if one does not intend to make the trip (also known as existence value). The key insight of his paper is that at any point in time, the amount of resources required for a given level of production are higher than will be required in the future, as technological innovations continue to relieve pressure on finite resources. In addition, at any point in time, the demand for natural resources will be at its lowest because the prior experience of past generations will generate higher demand by future generations. The combination of these two facts ensures that in the case of a rare and unique natural resource, the benefits to conservation will most likely outweigh the costs.

The intellectual debate surrounding the type of values provided by natural resources was not merely an academic exercise. The United States government had growing demand for benefit–cost analysis of public works programs. The idea of nonuse value raised a new issue; can nonuse value and use value be measured in the same way, or are they fundamentally incommensurable? In 1965, the Water Resources Council was established in part to undertake the task of setting standardized guidelines for benefit–cost analysis. Ultimately, the council decided in favor of measuring market and nonmarket benefits in terms of monetary value, a method that has become standard for environmental and resource economic research. Since the 1960s, resource and environmental economists have developed a number of valuation techniques, including stated preference and revealed preference methods. For example, to examine the tradeoffs an individual is willing to make in order to have access to a national park, survey data could be collected where an individual states their willingness to pay in order to preserve the park, or revealed preference data that captures the time and money expenditures made in visiting the park.

Ecological Economics

In 1966, Kenneth Boulding published "The Economics of the Coming Spaceship Earth," an essay dealing with the philosophy and ethics of consumption in a system with limited resources. Boulding contrasts the "cowboy economy," in which humans view the world as one with unbounded space and resources available for exploitation, with the "spaceman economy," in which humans are aboard a system with finite resources and assimilative capacities.

In the spaceman economy, humans must discover a way of existence that guarantees the maintenance of natural capital stocks and the quality of those stocks. Unlike in the cowboy economy, consumption in and of itself is not a goal of the spaceman economy, and humans should seek to minimize consumption while maintaining quality psychic and physical conditions for themselves. The essay explores the contradictions between viewing consumption as an input to be minimized while recognizing the act of consumption can be enjoyable and provide quality of life. Thus, the work is concerned with human moral obligations to posterity and the ecological system. It sets the stage for ecological economics to combine ethical goals with the types of economic analysis environmental and resource economists have developed.

In "Production, Consumption, and Externalities," Ayres and Kneese cite Boulding's paper as a source of inspiration. Though as resource economists they did not attempt to incorporate the ethical aspects of his arguments, they took the idea of the spaceman economy and modeled it using materials balance as discussed above. The analytical tools Ayres and Kneese developed have become part of the methodological toolkit for environmental, resource, and ecological economists, with greatest impact on this latter group. "Production, Consumption, and Externalities" is one of the most highly cited papers in the ecological economics literature today.

In 1971, Nicholas Georgescu-Roegen published *The Entropy Law and the Economic Process*, in which he argues that the limits to economic growth exist due to energy dissipation as dictated by the second law of thermodynamics. Georgescu-Roegen describes the economy as a system transforming low-entropy resources into high-entropy wastes so that humans can maintain their entropy levels. For example, low-entropy inputs such as water and fuels are used to grow crops for human consumption in order to maintain health, and high-entropy heat wastes are released in the consumption process. Low-entropy materials are Gorgescu-Roegen's limiting factor and differ from the land example, as given by Malthus.

While land can be used for repeated crop cycles, the transformation of a resource from a low-entropy state to one of high entropy can occur only once. Georgescu-Roegen held the opinion that consumption should be limited for ethical reasons because every item produced and consumed in the present period takes away from the consumption and well-being of posterity. The book's focus on the energy costs of an activity versus the monetary costs was very influential in building momentum for developing a new subfield.

In 1987, the International Society for Ecological Economics was formed, and 2 years later the journal *Ecological Economics* had its first publication. Ecological economics arose out of a belief that environmental and economic sustainability should be a priority and that neither ecology nor economics would be able to meet these goals; ecology too often does not consider the human population, and ecological economists have a number of issues with environmental and resource economic research.

In particular, the rejection of the limiting factor to growth has led ecological economists to refer to mainstream economists as technological optimists. Ecological economists also find that discounting future consumption with the market interest rate, as anticipated by Hotelling, is at conflict with their goal of intergenerational equity.

In addition, while many ecological economists employ techniques from mainstream economics to assess the monetary value of nonmarket benefits, such as the existence of a wildlife preserve, for others in this field, incommensurability is an issue mainstream economic theory is unable to overcome.

Finally, while economics seeks to provide an analysis for the most part based on economic efficiency, ecological economists are wary of the divorce of ethics from economic analysis and seek to provide studies with the explicit goal of sustainable development.

ECOLOGICAL ECONOMICS TODAY

Recounting the theoretical foundations of ecological economics is difficult because the field is often characterized by its lack of a unifying theory. Ecological economists' varying attitudes toward economics illustrate this ambiguity. Most ecological economists take theory and methods from economics in a manner described in *The Development of Ecological Economics* (Costanza, Perrings, and Cleveland, 1997):

> Within this pluralistic paradigm traditional disciplinary perspectives are perfectly valid as part of the mix.

> Ecological economics therefore includes some aspects of neoclassical environmental economics, traditional ecology and ecological impact studies, and several other disciplinary perspectives as components, but it also encourages completely new, hopefully more integrated, ways of thinking about the linkages between ecological and economic systems.

However, other scholars in this community reject mainstream economic theory entirely. Paul Christensen articulates this perspective in his article "Driving Forces, Increasing Returns, and Ecological Sustainability" (1991):

> The neoclassical economic theory which dominates resource and environmental analysis and policy is based on atomistic and mechanistic assumptions about individuals, firms, resources, and technologies which are inappropriate to the complex and pervasive physical connectivity of both natural and economic systems.

The lack of general consensus regarding theory and methods defining ecological economics has become a major source of criticism for the field. While it is easy to see how lack of a general theoretical framework can be problematic, others view this as part of the field's vision for itself—a discipline receptive to many alternative perspectives—and the preferred way to approach a problem.

In addition to the ambiguity surrounding what makes up ecological economic theory, there are also various interpretations about what ecological economics is in practice. The following section will describe three different interpretations of ecological economics as a field today. The first describes ecological economics as a transdiscipline, a field that transcends the disciplinary boundaries established in academia today. The next interpretation defines ecological economics as bioeconomics applied to the goal of sustainability. According to a third interpretation, ecological economics is an attempt to incorporate parallel research from other social and natural sciences.

A Transdiscipline

Ecological economics is often described within and outside the field as a transdiscipline. A transdiscipline can be defined as a field of study informing other fields; for example, statistics is a field that develops theory widely used by other disciplines. Alternatively, a transdiscipline is intent on understanding the world free of disciplinary boundaries. A transdisciplinary approach is distinct from an interdisciplinary approach in that interdisciplinary work occurs when researchers from different disciplines approach a problem together, bringing with them the paradigms from their respective fields, while in transdisciplinary work, scholars are not bound by paradigms and the associated tools and methods. It is this second definition that describes ecological economics.

Ecological economics' identity as a transdiscipline presents a tradeoff between the field's enhanced ability to incorporate new ideas and the lack of unity within the field that potentially limits its contributions to academic thought and policy.

Bioeconomics

Another perspective is that ecological economics is bioeconomics research conducted to promote goals of conservation, sustainability, and equity. Anastasios Xepapadeas's 2008 entry for ecological economics in *The New Palgrave Dictionary of Economics* describes ecological economics to be the study of the relation and dynamic evolution between the human economy and ecosystems within which they reside through the use of bioeconomic models.

An Integration of Economics with Other Social and Natural Sciences

The discipline of economics at large has been slow to incorporate qualitative methods in its research, relying heavily on quantitative empirics and mathematical modeling. This identity has provoked criticism, and political scientist Elinor Ostrom's 2009 Nobel Prize in Economics for her research on how institutional structure affects economic outcomes illustrates a growing awareness in the field that other social science disciplines can enrich the understanding of human behavior. Because incorporating theory and research from many disciplines is consistent with the goals of ecological economics, the growing acceptance of this type of work is beneficial to the discipline. Defining ecological economics as open to multiple paradigms perhaps puts it in a better position than resource and environmental economics to bring the richness of knowledge created in the other sciences to economic analysis.

THE FIELD'S STANDING

Forty years after the inception of the field of ecological economics, it remains at the margins in academic circles. To date, the field has had little impact on traditional economics and the paradigm of mainstream economists, and critics lambast its lack of cohesion in theory.

Despite being marginalized, ecological economics has made some significant contributions. First, the very existence of the field effectively generates dialogue between ecologists, economists, and policymakers around important

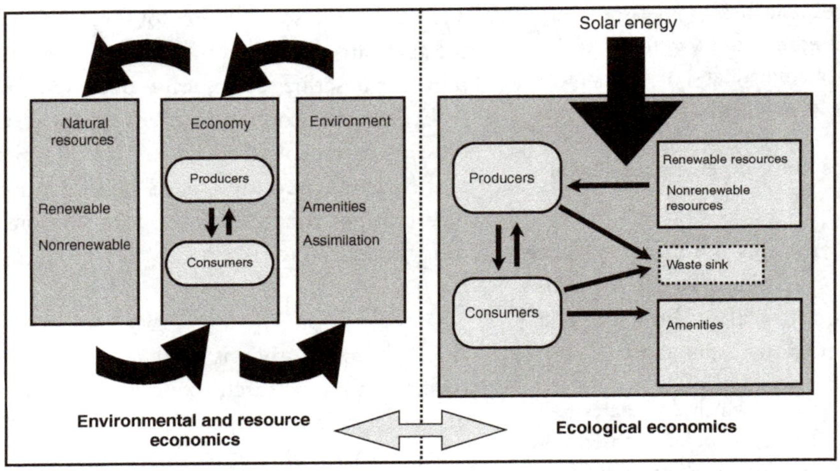

FIGURE 1 Converging of purviews.

issues. Most notably, Herman Daly, Robert Costanza, and Gretchen Daily have brought attention to the value of ecosystem goods and services, or the benefits provided by components of the ecosystem. For example, Costanza and colleagues' 1997 *Nature* article, "The Value of the World's Ecosystem Services and Natural Capital," calculates that global ecosystem services are worth $33 trillion annually. Although highly controversial, the paper did effectively bring economists and ecologists together creating a common language to describe ecosystem functions and services. The paper also increased society's awareness about the potential contribution of ecosystem services to our welfare.

Today, the dichotomy of scope and approach between ecological economics and environmental and resource economics is blurred. Environmental and resource economists are often portrayed as modeling the human economy as separate from the natural resources and environment (Fig. 1, left section), while ecological economists are portrayed as modeling the human economy as entirely contained within the global ecosystem (Fig. 1, right section). In reality, ecological economists do not always model systems; often very focused questions require partial analysis, which seeks to characterize one part of a system. Similarly, environmental and resource economists do model systems; in fact, many methodological innovations in systems modeling come from resource economists like Scott, Ayres, and Kneese. It is easier to base the distinction between the two camps not on approach but on research objectives: the ecological economist's goal of promoting sustainable development versus the environmental or resource economist's goal to provide analyses describing the nature of the tradeoffs in the provision of environmental goods and services.

SEE ALSO THE FOLLOWING ARTICLES

Discounting in Bioeconomics / Ecosystem Valuation / Fisheries Ecology / Harvesting Theory / Optimal Control Theory

FURTHER READING

Boulding, K. E. 1966. The economics of the coming spaceship earth. In H. Jarrett, ed. *Environmental quality in a growing economy.* Baltimore: Johns Hopkins University Press.

Costanza, R., ed. 1991. *Ecological economics: the science and management of sustainability*. New York: Columbia University Press.

Costanza, R., C. Perrings, and C. Cleveland. 1997. *The development of ecological economics.* The International Library of Critical Writings in Economics, Volume 75. Cheltenham, UK: Edward Elgar Publishing Limited.

Georgescu-Roegen, N. 1973. The entropy law and the economic problem. In H. E. Daly, ed. *Economics, ecology, ethics: essays toward a steady-state economy.* San Francisco: W. H. Freeman.

Pearce, D. 2002. An intellectual history of environmental economics. *Annual Review of Energy and the Environment.* 27: 57–81.

Xepapadeas, A. 2008. Ecological economics. In S. Durlaf and L. Blume, eds. *The new Palgrave dictionary of economics*, 2nd ed. New York: Macmillan.

ECOLOGICAL NETWORKS

SEE NETWORKS, ECOLOGICAL

ECOLOGICAL STOICHIOMETRY

SEE STOICHIOMETRY, ECOLOGICAL

ECONOMICS, ECOLOGICAL

SEE ECOLOGICAL ECONOMICS

ECOSYSTEM-BASED MANAGEMENT

SEE MARINE RESERVES AND ECOSYSTEM-BASED MANAGEMENT

ECOSYSTEM ECOLOGY

YIQI LUO, ENSHENG WENG, AND YUANHE YANG

University of Oklahoma, Norman

Ecosystem ecology is a subdiscipline of ecology that focuses on *exchange of energy and materials* between organisms and the environment. The materials that are commonly studied in ecosystem ecology include water, carbon, nitrogen, phosphorus, and other elements that organisms use as nutrients. The source of energy for most ecosystems is solar radiation. In this entry, material cycling and energy exchange are generally described before the carbon cycle is used as an example to illustrate our quantitative and theoretical understanding of ecosystem ecology.

MATERIAL CYCLING AND ENERGY EXCHANGE

Continuous exchanges of materials among compartments of an ecological system form a *cycle of elements*, usually characterized by *fluxes and pools*. Flux is a measure of the amount of materials that flows through a unit area per unit time. Pools store materials in compartments. When one ecological system is delineated with a clear boundary between inside and outside of the system, we also have to consider materials *input* into and *output* out of the system. Cycling of carbon, nitrogen, phosphorus, and other nutrient elements involves both geochemical and biological processes and is thus studied by biogeochemical approaches. In these aspects, ecosystem ecology overlaps with biogeochemistry for studying element cycles among biological and geological compartments. To fully quantify element cycles, we also need to understand *regulations* of flux, pool, input, and output by environmental factors and biogeochemical processes.

Material cycling in ecosystems is usually coupled with energy exchange. Incoming solar radiation, the source of energy for most ecosystems, is partly reflected by ecosystem surfaces and partly absorbed by plants and soil. The absorbed solar radiation is partly used to evaporate water via latent heat flux and partly converted to thermal energy to increase the temperature of ecosystem. The thermal energy is transferred to air via sensible heat flux and to soil by ground heat flux. A very small fraction of the solar radiation is converted to biochemical energy via photosynthesis.

Material and energy exchanges in ecosystem are usually described by mathematical equations. Most of the equations have been incorporated into land process models to quantitatively evaluate responses and feedback of ecosystem material and energy exchanges to global change. In this sense, quantitative analysis of ecosystem dynamics in response to global change is quite advanced. However, most of the land process models are complex and have been rarely analyzed to gain theoretical insights into ecosystem ecology.

Using the carbon cycle as an example of material and energy exchange, the following sections describe each of the major carbon cycle processes followed by their quantitative representations. Those carbon processes include leaf photosynthesis; photosynthesis at canopy, regional, and global scales; carbon transfer, storage, and release in terrestrial ecosystems; dynamics of ecosystem carbon cycling; impacts of disturbances on the carbon cycle; and effects of global change on the carbon cycle. Theoretical principles in ecosystem ecology are also presented.

Leaf Photosynthesis

The terrestrial carbon cycle usually initiates when leaf photosynthesis fixes carbon dioxide from the atmosphere into organic carbon compounds, using the energy from sunlight. Photosynthesis is vital for nearly all life on Earth directly or indirectly as the source of the energy. It is also the source of the carbon in all the organic compounds and regulates levels of oxygen and carbon dioxide in the atmosphere.

Photosynthesis begins with the light reaction when chlorophylls absorb energy from light. The light energy harvested by chlorophylls is partly stored in the form of adenosine triphosphate (ATP) and partly used to remove electrons from water. These electrons are then used in the dark reactions that convert carbon dioxide into organic compounds (i.e., carboxylation) by a sequence of reactions of the Calvin cycle. Carbon dioxide is fixed to ribulose-1, 5-bisphosphate by the enzyme in mesophyll cells, which are exposed directly to the air spaces inside the leaf. Carbon dioxide enters chlorophylls via stomata where water vapor exits the leaf. Stomatal conductance is to measure the rate of carbon dioxide influx into or water efflux from leaf.

Thus, the major processes of photosynthesis at the leaf level include light reaction, carboxylation, and stomata conductance. Those processes can be mathematically represented by the Farquhar model for C_3 plants to calculate gross leaf CO_2 assimilation rate (A, μmol CO_2 m^{-2} s^{-1})

(Farquhar et al., 1980) as

$$A = \min(J_c, J_e) - R_d, \tag{1}$$

where J_c is the rate of carboxylation with CO_2 limitation, J_e is the rate of light electron transport, and R_d is dark respiration. The leaf-level photosynthesis is determined by the one with the lowest rate of the two processes. The rate of carboxylation can be described as

$$J_c = V_m \frac{C_i - \Gamma_*}{C_i + K_C\left(1 + \dfrac{O_x}{K_o}\right)}. \tag{2}$$

The light electron transport process (J_e) is

$$J_e = \frac{\alpha_q \cdot I \cdot J_m}{\sqrt{J_m^2 + \alpha_q^2 \cdot I^2}} \cdot \frac{C_i - \Gamma_*}{4 \cdot (C_i + 2\Gamma_*)}, \tag{3}$$

where C_i is the leaf internal CO_2 concentration (μmol CO_2 mol^{-1}), O_x is oxygen concentration in the air (0.21 mol O_2 mol^{-1}), V_m is the maximum carboxylation rate (μmol CO_2 m^{-2} s^{-1}), Γ_* is CO_2 compensation point without dark respiration (μmol CO_2 mol^{-1}), K_c and K_o are Michaelis–Menten constants for carboxylation and oxygenation, respectively, (μmol CO_2 mol^{-1}), I is absorbed photosynthetically active radiation (PAR, μmol m^{-2} s^{-1}), α_q is quantum efficiency of photon capture (mol mol^{-1} photon), J_m is the maximum electron transport rate (μmol CO_2 m^{-2} s^{-1}). Responses of leaf photosynthesis to leaf internal CO_2 concentration and radiation both follow asymptotic curves (Fig. 1)

The leaf internal CO_2 concentration, C_i, is regulated by stomatal conductance (G_s) and related to leaf photosynthesis by

$$A_n = G_s \cdot (C_a - C_i) \tag{4}$$

and

$$G_s = g_l \cdot \frac{A}{(C_i - \Gamma_*) \cdot \left(1 + \dfrac{D}{D_0}\right)}, \tag{5}$$

where C_a is ambient CO_2 concentration, g_l and D_0 (kPa) are empirical coefficients, and D is vapor pressure deficit (kPa). Stomatal conductance also controls water loss through leaf surface (i.e., transpiration). Since water is almost always a limiting factor for terrestrial plants, there is a tradeoff for absorbing CO_2 and reducing water loss via stomata.

Many of the photosynthetic processes are sensitive to temperature change. The temperature sensitivities of those processes are usually expressed by temperature functions of parameters. Those parameters that are sensitive to temperature change include V_m, Γ_*, K_c, K_o, J_m, and R_d. For one given parameter, which is denoted by P, the temperature is usually expressed by Arrhenius equation as

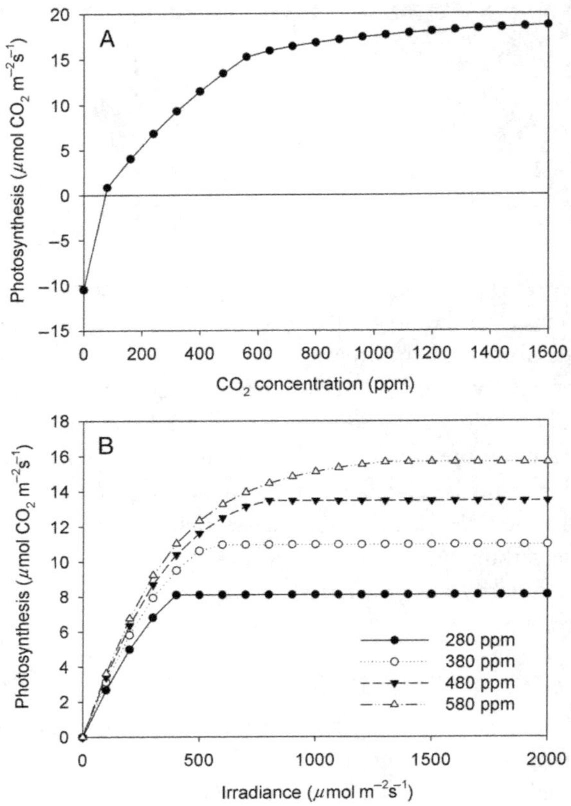

FIGURE 1 Leaf photosynthesis as a function of intercellular CO_2 concentration (A) or irradiance (B).

$$P = P_{25} \cdot \exp\left(\frac{E_p \cdot (T_k - 298)}{R \cdot T_k \cdot 298}\right) \tag{6}$$

where E_p is the activation energy (J mol^{-1}) of a parameter, R is universal gas constant (8.314 J K^{-1} mol^{-1}), T_k is leaf temperature in Kelvin (K), P_{25} is the rate at 25 °C. The temperature sensitivity is sometimes expressed by a peaked function to describe the increase of a process at a low temperature with a peak at an optimal temperature followed by decline (Fig. 2).

Leaf photosynthesis is also affected by nitrogen content in leaves. Leaf photosynthesis is preformed by enzymes, which require nitrogen. Most models use linear equations to relate leaf nitrogen content with maximum caboxylation, maximum electron transport, and dark respiration.

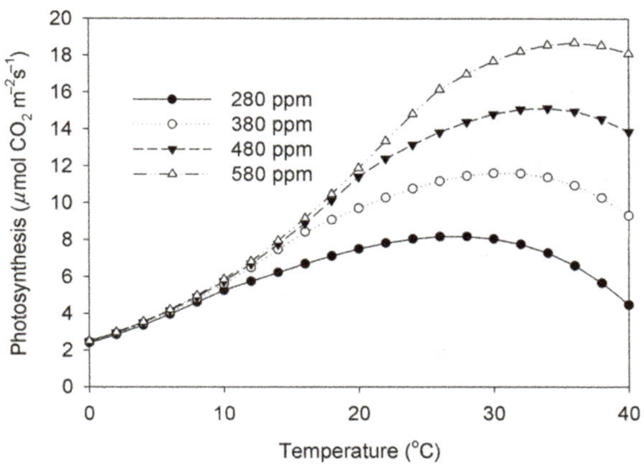

FIGURE 2 Leaf photosynthesis as a function of temperature at different CO_2 concentrations.

PHOTOSYNTHESIS AT CANOPY, REGIONAL, AND GLOBAL SCALES

When leaf photosynthesis is scaled up to the canopy level, the gradients of solar radiation, water vapor pressure, and nitrogen distribution within a canopy should be considered. The penetration of solar radiation through canopies can be described by the Beer's law as

$$I = I_0 \exp(-kL), \tag{7}$$

where I is the radiation at leaf area index L, I_0 is the solar radiation at the top of canopy, and k is light extinction coefficient. Water vapor pressure is different for the leaves within a canopy with those adjacent to bulk air. Canopies can slow down wind speed and decrease boundary layer conductance, leading to changes in the microclimate of leaves in canopies for transpiration. The photosynthetic capability as related to nitrogen concentration of the leaves is different with their positions in a canopy. Usually, nitrogen is distributed in proportion to the distribution of absorbed irradiance in canopy when there are no other limitations.

Many models have been developed to scale up photosynthesis from the leaf to canopy level based on canopy structure and gradients of environmental factors. These models can be categorized into big-leaf (single-layer) models, two-leaf models, and multiple layer models according to how canopy structure is represented and the environmental gradients are treated. The single-layer models take the whole canopy as one "big leaf," by assuming all the leaves in a canopy are the same and have the same water conditions (i.e., the humidity of air in the canopy is the same). The integration of leaf photosynthesis only considers the gradient of solar radiation. The photosynthesis rate (carbon assimilation rate) at canopy level is thus calculated by

$$A_c = A_n \frac{1 - \exp(-kL)}{k}, \tag{8}$$

where A_c is canopy photosynthesis rate, A_n is net photosynthesis rate at leaf level.

The two-leaf models separate leaves into two classes—sunlit and shaded leaves—and thereby simulate photosynthesis in the two classes of leaves individually. The separation of sunlit and shade leaves is based on the structure of canopy and the angles of solar radiation. For the leaves in a canopy, the shade has a linear response to radiation, while the sunlit are often light saturated, and independent on irradiance, which allow averaging of solar radiation in sunlit and shaded leaves separately.

Multilayer models separate a canopy into many layers and calculate water and carbon fluxes at each layer according to its physiological properties and climatic conditions (e.g., solar radiation and water vapor). The distribution of nitrogen in canopies is often optimized for maximizing photosynthesis according to the gradient of solar radiation.

The single-layer models overestimate photosynthesis rate and transpiration. These biases caused by the big-leaf models can be corrected by adding curvature factors or tuning parameters. Single-layer models are appropriate when the details of canopy structure and its microclimate can be ignored, such as when vegetations are taken as a lower boundary of the atmosphere in global circulation models or when canopy structure is relatively simple such as tundra and desert ecosystems. Multilayer models have the flexibility to incorporate the details of canopy environmental and physiological variables. Their complexity demands high computational power for calculations and thus limits their applications at large scales. Two-leaf models can be as accurate as multilayer models but are much simpler. Therefore, they are widely used in current ecosystem and Earth system models.

Leaf and canopy photosynthesis is usually scaled up to estimate regional and global photosynthesis. There are generally two approaches to up-scaling. One is to estimate global photosynthesis from remote sensing data with a light-use efficiency constant by

$$GPP = f_{APAR} PAR \varepsilon^* T_s W_s, \tag{9}$$

where PAR is photosynthetically active radiation estimated from solar radiation and f_{APAR} is the fraction of PAR that is absorbed by leaves. ε^* is the maximum potential light-use efficiency, and T_s and W_s are the temperature and moisture scalars, which are used to reduce the potential light-use efficiency (ε^*) in response to climate conditions. Another approach is to use process-based leaf and canopy photosynthesis

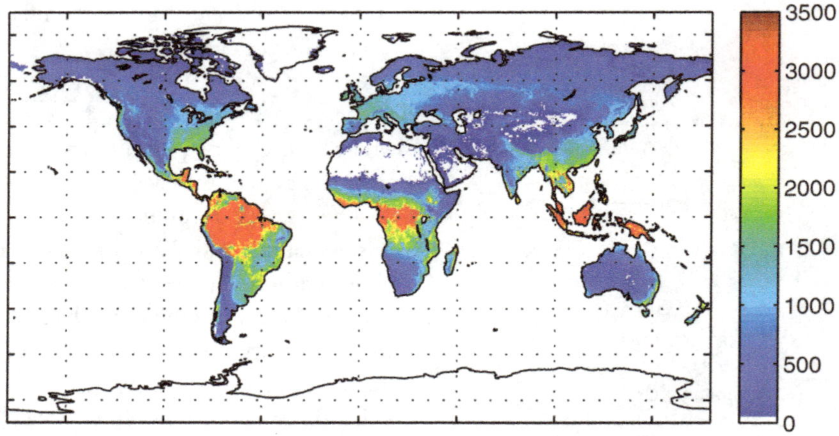

FIGURE 3 Global distribution of photosynthesis (i.e., gross primary production, GPP, Pg C year^{-1}) estimated from spatially explicit approaches. (Adapted from Beer et al., 2010, *Science* 329: 834–838.)

models such as the Farquhar model in combination of vegetation covers measured by remote sensing or simulated by models to estimate regional and global photosynthesis.

At the global scale, photosynthetic organisms convert about 120 teragrams of carbon into organic carbon compounds per year (Fig. 3). Tropical forests accounts for one-third of the global photosynthesis and have the highest photosynthetic rate per unit area. Savannahs account for about one-quarter of the global photosynthesis and are the second most important biome in terms of carbon fixation, largely due to their large area. The rate of energy capture by photosynthesis is approximately 100 terawatts, about six times larger than the power consumption of human civilization.

CARBON TRANSFER, STORAGE, AND RELEASE IN TERRESTRIAL ECOSYSTEMS

Carbohydrate synthesized from photosynthesis is partitioned for plant respiration and the growth of leaf, fine roots, and wood (Fig. 4), with small fractions to root exudates and mycorrhizae. Most large-scale models do not consider root exudation and mycorrhizae, which are usually explored by plant–soil models toward mechanistic understanding. Carbon allocation to plant respiration roughly accounts for 50% of total photosynthesis, with variation from 23% to 83% among different ecosystems. Carbon allocation between aboveground and belowground plant parts reflects the different investment of photosynthate for light harvest and uptake of water and nutrients. It varies in response to different environmental conditions. The optimal partitioning theory predicts that growth-limiting conditions usually lead to greater carbon allocation to those organs that are constrained. For instance, carbon allocation favors leaves if light becomes more limiting and roots if plant growth is greatly limited by nutrient and water availability.

Dead leaves, stems, and roots go to litter pools and become a source of soil organic matter. The rate of litter production is determined by the turnover rates of leaves, stems, and roots. It varies with environment conditions in the short term but equals net primary production in long term. Global litter production ranges from 45 to 55 Pg C yr^{-1}, of which about 20 Pg C yr^{-1} is from aboveground plant

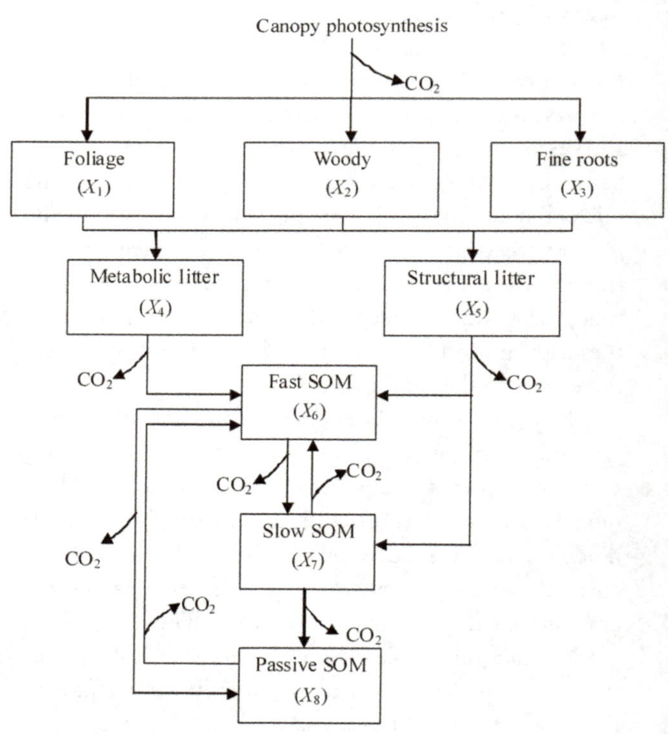

FIGURE 4 Structure of an eight-pool model to illustrate carbon pools and fluxes between them.

parts. The global litter pool is estimated at 50 to 200 Pg C. An additional 75 Pg C is estimated for the coarse woody detritus pool. Global mean steady state turnover times of litter estimated from the pool and production data range from 1.4 to 3.4 years. The mean turnover time is ~5 years for forest and woodland litter and ~13 years for coarse woody detritus. Litter is decomposed by microbes with a part respired to CO_2 and a part converted to soil organic matter.

Soil organic matter is the largest carbon pool of the terrestrial ecosystems. It receives carbon input from plants and litter and release carbon via decomposition and mineralization. Decomposition is a process conducted by microbes and affected by the quality of substrate and environmental conditions. The rate of decomposition is a key factor determining how much carbon can be stored in ecosystems. Any small changes in decomposition rate can lead to substantial changes in the carbon stock of terrestrial ecosystems, therefore affecting CO_2 concentration in atmosphere. Soil organic carbon is usually separately in conceptual models into slow and passive soil carbon pools.

The carbon processes of carbon allocation, plant growth, litter dynamics, and soil organic carbon can be mathematically represented in a matrix form:

$$\begin{cases} \frac{d}{dt}X(t) = \xi ACX(t) + BU(t), \\ X(t=0) = X_0, \end{cases} \tag{10}$$

where $U(t)$ is the photosynthetically fixed carbon and usually estimated by canopy photosynthetic models, B is a vector of partitioning coefficients of the photosynthetically fixed carbon to plant pools (e.g., leaf, root, and woody biomass), $X(t)$ is a vector of carbon pool sizes, X_0 is a vector of initial values of the carbon pools, and A and C are carbon transfer coefficients between plant, litter, and soil pools. ξ is an environmental scalar representing effects of temperature and moisture on the carbon transfer among pools.

For a carbon cycle model as depicted in Figure 4, the vector of partitioning coefficients can be expanded to $B = (b_1 \;\; b_2 \;\; b_3 \;\; 0 \;\; 0 \;\; 0 \;\; 0 \;\; 0)^T$, where b_1, b_2, and b_3 are partitioning coefficients of photosynthetically fixed C into leaf, wood, and root, respectively. $X(t) = (x_1(t), x_2(t), \ldots x_8(t))^T$ is a 8×1 vector describing C pool sizes, A and C are 8×8 matrices describing transfer coefficients and given by

$$A' = \begin{pmatrix} -1 & 0 & 0 & 0 & 0 & 0 & 0 & 0 \\ 0 & -1 & 0 & 0 & 0 & 0 & 0 & 0 \\ 0 & 0 & -1 & 0 & 0 & 0 & 0 & 0 \\ f_{41} & 0 & f_{43} & -1 & 0 & 0 & 0 & 0 \\ f_{51} & f_{52} & f_{53} & 0 & -1 & 0 & 0 & 0 \\ 0 & 0 & 0 & f_{64} & f_{65} & -1 & f_{67} & f_{68} \\ 0 & 0 & 0 & 0 & f_{75} & f_{76} & -1 & 0 \\ 0 & 0 & 0 & 0 & 0 & f_{86} & f_{87} & -1 \end{pmatrix}$$

$$C = diag(c) \tag{11}$$

where f_{ij} is the transfer coefficients from pool j to pool i, $diag(c)$ denotes the diagonal matrix with diagonal components given by elements of vector $c = (c_1, c_2, \ldots, c_8)^T$, and $c_j, (j = 1, 2, \ldots 8)$ represents transfer coefficients (i.e., exit rates of carbon) from the eight carbon pools X_j, $(j = 1, 2, \ldots 8)$. The initial value vector can be expanded to $X_0 = (x_1(0), x_2(0), \ldots x_8(0))^T$.

Equation 10 adequately describes most observed C processes, such as litter decomposition and soil C dynamics. It has been represented in almost all ecosystem models and integrated into Earth system models. The parameters in Equation 10 have recently been estimated from data collected in Duke Forest, North Carolina, with a data assimilation approach (Weng and Luo, 2011, *Ecological Applications*, 21: 1490–1505). Carbon transfer matrix A in Equation 10 is estimated to be

$$A = \begin{pmatrix} -1 & 0 & 0 & 0 & 0 & 0 & 0 & 0 \\ 0 & -1 & 0 & 0 & 0 & 0 & 0 & 0 \\ 0 & 0 & -1 & 0 & 0 & 0 & 0 & 0 \\ 0.9 & 0 & 0.2 & -1 & 0 & 0 & 0 & 0 \\ 0.1 & 1.0 & 0.8 & 0 & -1 & 0 & 0 & 0 \\ 0 & 0 & 0 & 0.45 & 0.275 & -1 & 0.42 & 0.45 \\ 0 & 0 & 0 & 0 & 0.275 & 0.296 & -1 & 0 \\ 0 & 0 & 0 & 0 & 0 & 0.004 & 0.01 & -1 \end{pmatrix}. \tag{12}$$

The values of the eight transfer coefficients in the diagonal matrix, $C = diag(c)$ are

$$C = \begin{pmatrix} 0.00258 & 0 & 0 & 0 & 0 & 0 & 0 & 0 \\ 0 & 0.0000586 & 0 & 0 & 0 & 0 & 0 & 0 \\ 0 & 0 & 0.00239 & 0 & 0 & 0 & 0 & 0 \\ 0 & 0 & 0 & 0.0109 & 0 & 0 & 0 & 0 \\ 0 & 0 & 0 & 0 & 0.00095 & 0 & 0 & 0 \\ 0 & 0 & 0 & 0 & 0 & 0.0105 & 0 & 0 \\ 0 & 0 & 0 & 0 & 0 & 0 & 0.0000995 & 0 \\ 0 & 0 & 0 & 0 & 0 & 0 & 0 & 0.0000115 \end{pmatrix}. \tag{13}$$

The vector of partitioning coefficients B is

$$B = \begin{vmatrix} 0.14 \\ 0.26 \\ 0.14 \\ 0 \\ 0 \\ 0 \\ 0 \\ 0 \end{vmatrix}. \tag{14}$$

$U(t)$ is the C input (GPP) at time t, and its daily average over one year is estimated to be 3.370 g C day^{-1}. The initial values of eight pools are

$$X_0 = \begin{vmatrix} 250 \\ 4145 \\ 192 \\ 93 \\ 545 \\ 146 \\ 1585 \\ 300 \end{vmatrix}. \tag{15}$$

The environmental scalar $\xi(t)$ is a product of temperature and soil moisture response functions as

$$\xi(t) = f_W \cdot f_T, \tag{16}$$

where f_W and f_T are functions of volumetric soil moisture (W) and temperature (T), which are set to be

$$f_W = \min(0.5W, 1.0) \quad \text{and} \tag{17}$$

$$f_T = Q_{10}^{(T-10)/10}, \tag{18}$$

and Q_{10} is a temperature quotient to describe a change in decomposition rate for every 10°C difference in temperature.

PROPERTIES OF ECOSYSTEM CARBON CYCLING PROCESSES

Ecosystem carbon cycle dynamics are dictated by the properties of Equation 10, which is considered to be the governing equation of the carbon cycle in the terrestrial ecosystems. The properties of Equation 10 can be summarized in the following five aspects (see Luo and Weng, 2011). First, photosynthesis is the primary pathway of C entering an ecosystem and described by parameter $U(t)$ in Equation 10. Thus, photosynthesis determines the rate of carbon cycling in an ecosystem. In a barren soil, for example, no carbon flows from photosynthesis to the ecosystem nor does carbon efflux or storage exist. In an ecosystem with a high rate of photosynthesis, carbon efflux is usually high. The rate of photosynthesis in an ecosystem is determined by available light, water, and nutrients, as described above.

Second, carbon in an ecosystem is compartmentalized with clear physical boundaries of different pools of C in leaf, root, wood, litter, and soil. Soil C has been further compartmentalized into conceptual or physically and chemically separable pools in some models to adequately describe its short- and long-term dynamics. Pools are represented by vector $X(t)$ with their initial values by $X(0)$ in Equation 10. Carbon influx into each of the plant pools is determined by partitioning coefficient in vector B times photosynthetic rate $U(t)$. Carbon influx to each of the litter and soil pools is determined by their donor pool sizes times exit rates of carbon from the donor pools (as represented by diagonal matrix C) times transfer coefficients in matrix A. Carbon exiting from each of the pools is described by diagonal matrix C.

Third, each of the pools has a different residence time, which is the inverse of its exit rate described by diagonal matrix C. At the ecosystem scale, the residence time measures the averaged duration of the atoms of C from the entrance via photosynthesis to the exit via respiration from the ecosystem. For individual pools, the residence time measures the averaged duration of the atoms of C from the entrance into the pool to the exit from the pool. Since each atom of C that enters an ecosystem is eventually released back to the atmosphere, residence time is a critical parameter to determine the capacity of ecosystem C storage. The residence times of the eight pools in the Duke Forest are (in units of days)

$$T = \begin{vmatrix} \tau_1 \\ \tau_2 \\ \tau_3 \\ \tau_4 \\ \tau_5 \\ \tau_6 \\ \tau_7 \\ \tau_8 \end{vmatrix} = \begin{vmatrix} 388 \\ 17{,}065 \\ 418 \\ 92 \\ 1053 \\ 95 \\ 10{,}050 \\ 86{,}957 \end{vmatrix}. \tag{19}$$

Thus, residence times are long in the plant wood, slow and passive soil pools. The capacity of an ecosystem to sequester C is proportional to residence times. Thus, an ecosystem sequesters more carbon if more photosynthate is partitioned to pools with long residence times, such as wood and soil. The ecosystem-scale C residence time (τ_E)

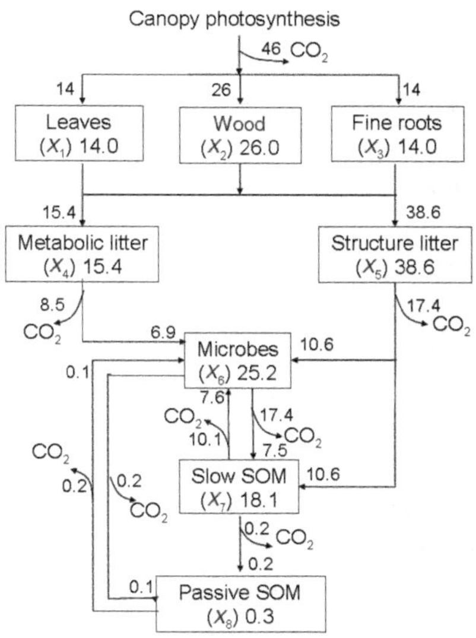

FIGURE 5 Fraction of carbon (×100) that flows through various pathways and is partitioned to the eight pools. The fraction to plant pools is determined by partitioning the coefficient in vector *B* in Equation 10. The fraction to litter and soil pools via each pathway is determined by the transfer coefficient matrix *A*. The values of vector *B* and matrix *A* are estimated from data collected in Duke Forest via data assimilation approach (Weng and Luo 2011, *Ecological Applications*, 21: 1490–1505). The fraction of carbon from photosynthesis is large to plant pools and small to soil pools, particularly to the passive soil carbon pool.

can be computed by at least two methods. One is that τ_E equals total ecosystem C content at equilibrium divided by total carbon influx. The other method is to estimate fractions of carbon entering each of the pools, which are then multiplied by the residence time of each pool. The products are summed up for all the pools. Large fractions of photosynthetically fixed carbon go to plant pools but a very small fraction to the passive soil pool (Fig. 5).

Fourth, C transfers between pools are predominantly controlled by donor pools and not much by recipient pools. C transfer from a plant to litter pool, for example, is dominated by the amount of C in the plant pool (the donor) and not the litter pool (the recipient). Although SOM decomposition is primarily mediated by microorganisms, C transfer among soil pools can be effectively modeled by proportion to donor pool sizes and not to recipient pool sizes. The donor pool–dominated transfer is the primary mechanism leading to convergence of carbon dynamics toward the equilibrium level after disturbances, as discussed below.

Fifth, decomposition of litter and soil organic matter to release CO_2 can be usually represented by a first-order decay function as described by the first term on the right side of Equation 10. Thousands of experimental studies have showed that the first-order decay function can adequately describe the mass remaining of litter with time lapsed from litter and SOM decomposition experiments without many exceptions. The first-order decay function reinforces the property of the donor pool–dominated transfer to drive the C cycle toward equilibrium.

Mathematically, Equation 10 satisfies the Lyapunov stability conditions with negative eigenvalues of the C transfer matrix A. Using estimated parameters from the Duke Forest, the eigenvalues of matrix A in equation 10 are

$$\begin{pmatrix} -0.0106 \\ -0.0000871 \\ -0.0000115 \\ -0.0109 \\ -0.00095 \\ -0.00258 \\ -0.0000586 \\ -0.00239 \end{pmatrix}.$$

According to the conditions of Lyapunov stability of a continuous linear time invariant (CLTI) system, it is stable since all the eigenvalues are negative. Model simulation also shows the convergence of ecosystem carbon storage with varied initial carbon content (Fig. 6). The carbon pool sizes at the converged equilibrium (X_{eq}) are

$$X_{eq} = \begin{pmatrix} 339 \\ 27{,}688 \\ 366 \\ 88 \\ 2536 \\ 113 \\ 10{,}020 \\ 1296 \end{pmatrix}. \tag{20}$$

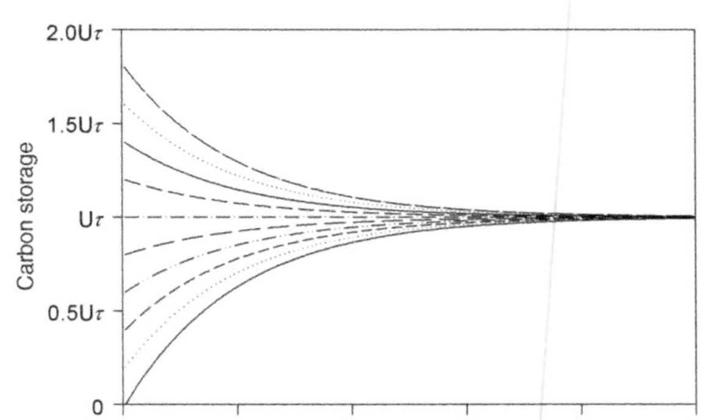

FIGURE 6 Simulated dynamics of carbon content in an ecosystem using different initial values of pools. Symbol *U* represents carbon influx into an ecosystem and τ the ecosystem carbon residence time. The figure illustrates the convergence of carbon storage toward the equilibrium value, which equals the product of the carbon influx *(U)* and residence time (τ).

Empirical evidence from many studies at the ecosystem scale has also shown that C stocks in plant and soil pools recover towards equilibrium during secondary forest succession and grassland restoration after disturbances.

DISTURBANCE EFFECTS ON THE CARBON CYCLE

An ecosystem is subject to frequent natural and anthropogenic disturbances, causing ecosystem carbon cycling processes to be in states away from equilibrium. Disturbances create disequilibrium of the carbon cycle by (1) depleting or adding carbon in pools, (2) decreasing or increasing in canopy photosynthesis, and/or (3) altering carbon residence time via changes in carbon partitioning, transfer, and decomposition.

Anthropogenic land-use conversion from forests and native grasslands to croplands, pastures, and urban areas, for example, not only results in the net release of carbon to the atmosphere, it also reduces ecosystem carbon residence time due to the elimination of carbon pools in plant wood biomass and coarse wood debris, and physical disturbance of long-term soil carbon pools. Nearly 50% of the land surface on the Earth has been used for agriculture and domestic animal grazing, resulting in a net release of 1 to 2 Pg of carbon per year to the atmosphere (Fig. 7).

Fire removes carbon by burning live and dead plants, litter, and sometimes soil carbon in top layers. Fire often reduces ecosystem photosynthetic capacity by removing foliage biomass. It also alters physical and chemical properties of litter and soil organic matter to influence their decomposition so that carbon residence time may be affected. Globally, wildfires burn 3.5 to 4.5 million km^2 of land and emit 2 to 3 petagrams of carbon per year into the atmosphere (Fig. 8). Fire occurs as an episodic event, after which ecosystems usually recover in terms of photosynthetic and respiratory rates, and carbon pools in plant, litter, and soil. Ecosystem carbon processes are also affected by other episodic events like windstorms, insect epidemics, drought, and floods.

Modeling and theoretical analysis of disturbance effects on the ecosystem carbon cycle are still in their infancy. Disturbances are usually treated as prescribed events in an input file to influence biogeochemical processes, vegetation ecophysiology, species composition, age structure, height, and other ecosystem attributes. Most models then simulate recovery of plant growth, litter mass, and soil carbon. Some models consider those recovery processes

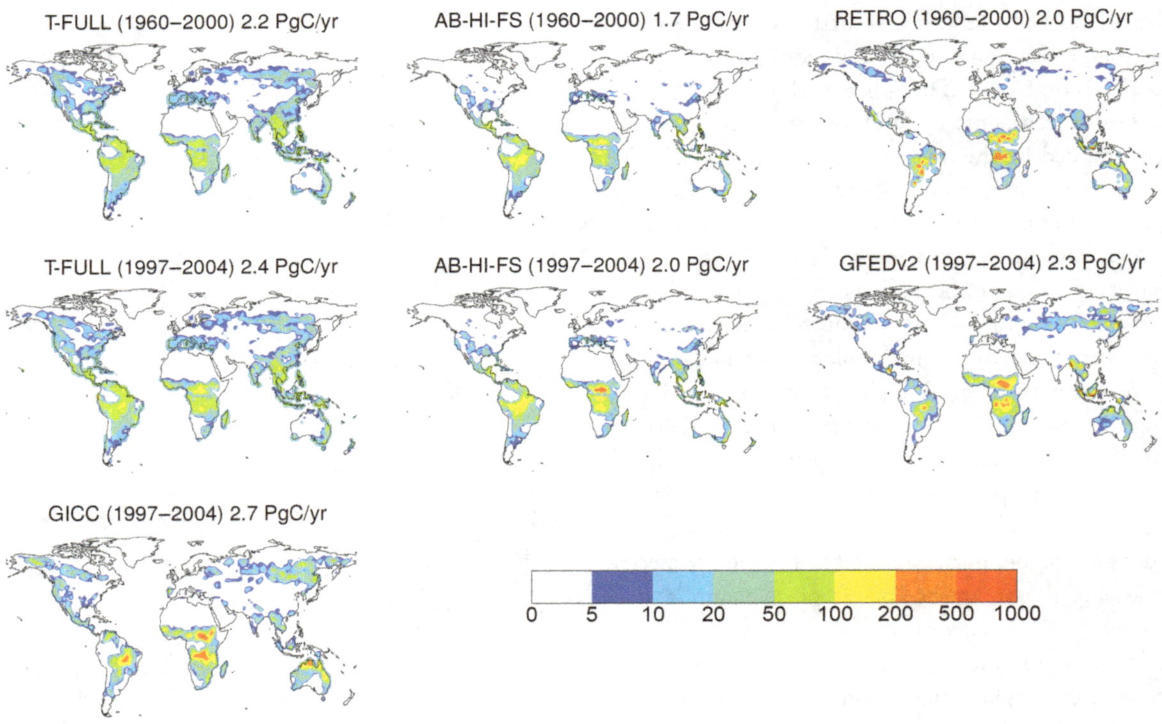

FIGURE 7 Land-use effects on carbon storage. Net emissions, coupling flux, and primary emissions of anthropogenic land cover change (ALCC) accumulated over the given time interval: preindustrial (800–1850), industrial (1850–2000), and future period (2000–2100). Units are Gt C released from each grid cell. (Adapted from Pongratz et al., 2009, *Global Biogeochemical Cycles* 23, GB4001).

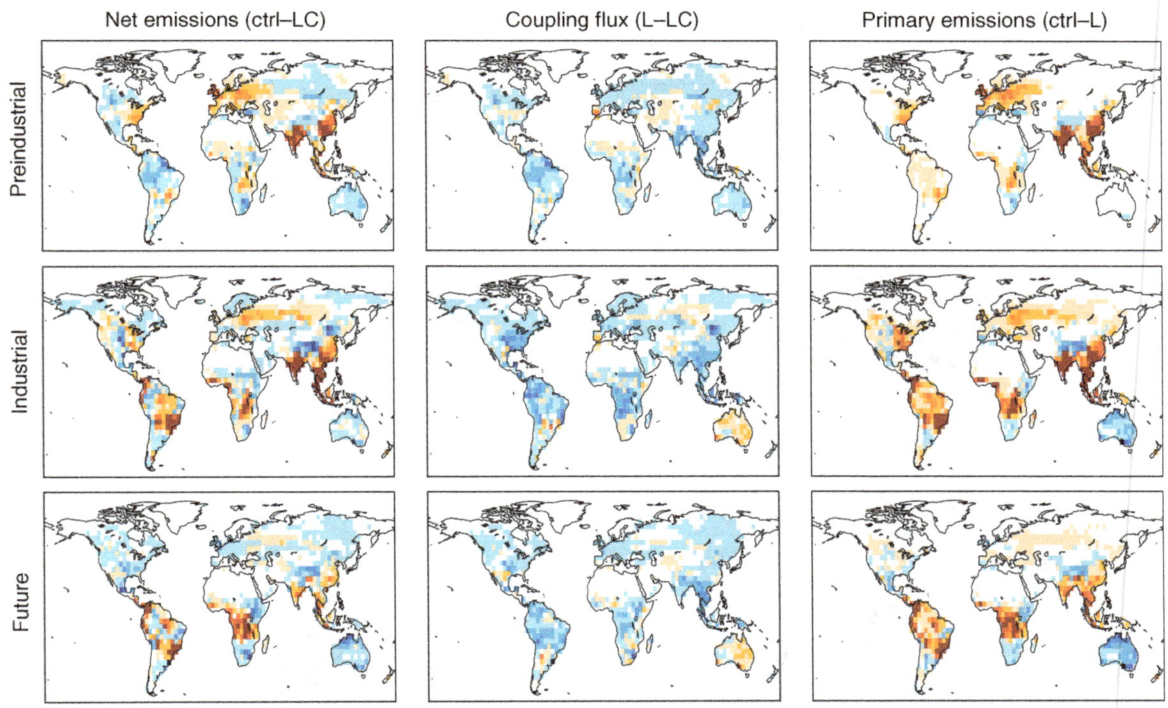

FIGURE 8 Fire effects on carbon sink. Annual mean total (wildfire plus deforestation) fire carbon emissions [g C/m^2/year] compared to emissions reported in other studies, including the fire products GFEDv2, RETRO, and GICC (see the original paper for description of the fire products). The model simulations are averaged over the corresponding observational periods (GFEDv2/GICC: 1997–2004; RETRO: 1960–2000). The numbers in the title of each panel are global mean fire emissions with units of PgC/year. (Adapted from Kloster et al., 2010, *Biogeosciences* 7: 1877–1902).

under the influence of global changes. However, severity of disturbances on ecosystem processes is difficult to model, largely due to the lack of data. The overall net carbon flux from forests to the atmosphere depends on the spatial extent, severity, and heterogeneity of disturbances (e.g., fire suppression, logging, and insect outbreaks). Presently, we have a limited capability of simulating the occurrences and severity of disturbances under climate change.

GLOBAL CHANGE EFFECTS ON THE ECOSYSTEM CARBON CYCLE

Many carbon cycle processes are sensitive to global change factors. For example, leaf photosynthesis is responsive to increasing atmospheric CO_2 concentration as described by the Farquhar model (Eq. 1), which creates a potential for carbon sequestration in plant biomass and soil C pools. However, growing in a high-CO_2 environment, plants may acclimate and adapt to diminish CO_2 effects. Many canopy- and ecosystem-scale processes, such as phenology, as well as nitrogen and water availability, have to be considered when we scale up leaf-level photosynthetic responses to estimate ecosystem-level responses. A meta-analysis showed that carbon and nitrogen contents in the litter and soil pools significantly increased under elevated CO_2 concentration (Fig. 9).

As global warming is happening, land surface temperature increases. While an increase in temperature usually accelerates all physical, chemical, and biological processes of ecosystems, net effects of climate warming on ecosystem carbon balance are extremely variable among ecosystems. Although instantaneous effects of temperature on leaf photosynthesis can be estimated by Equation 6, temperature affects stomatal conductance directly and indirectly via accompanied changes in vapor pressure deficit and water stresses, which further modify photosynthetic responses to climate warming. At the ecosystem scale, additional effects of temperature on photosynthetic C influx over a year is via changes in phenology and the length of the growing season under warmed climate. Similarly complex interactions of multiple processes modify responses of respiration and decomposition of litter and soil organic carbon, although temperature responses of one single processes can be usually modeled by an exponential equation as in Equation 18 or an Arrhenius equation as in Equation 6.

Human activities have also substantially altered the nitrogen cycle. As a consequence, nitrogen fertilization and deposition increase. Increased nitrogen availability usually stimulates photosynthesis and plant growth. But

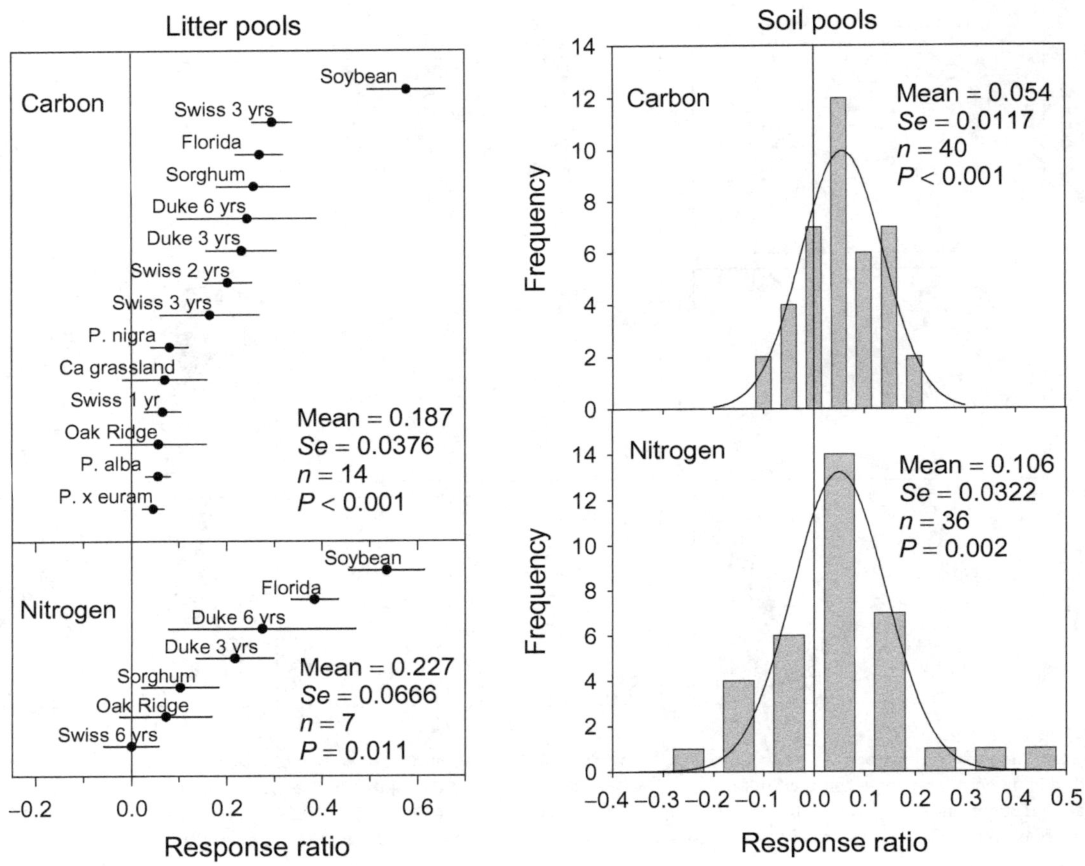

FIGURE 9 Effects of elevated CO_2 on carbon storage in litter and soil pools (Adapted from Luo et al., 2006, *Ecology* 87: 53–63.)

the increased plant growth might not lead to much net C storage in soil partly because litter produced under a high-nitrogen environment decomposes faster than that under a low-nitrogen environment and partly because nitrogen deposition or fertilization stimulate more aboveground than belowground growth, reducing C input into the soil (Fig. 10). Also, litter produced in the aboveground usually contributes much a smaller fraction of carbon than the belowground litter to soil carbon dynamics.

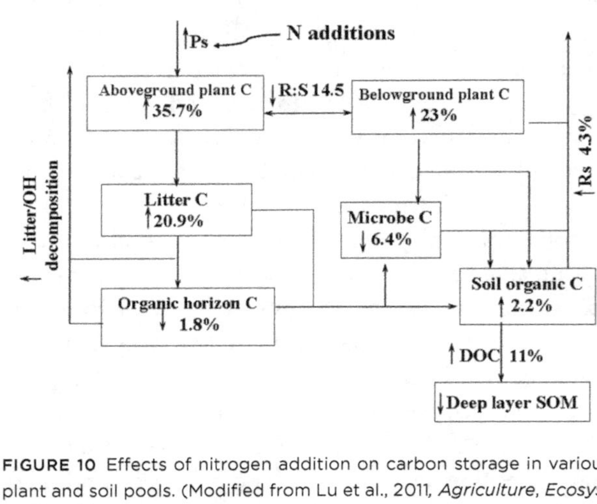

FIGURE 10 Effects of nitrogen addition on carbon storage in various plant and soil pools. (Modified from Lu et al., 2011, *Agriculture, Ecosystems and Environment*, 140: 234–244).

Disturbance frequency, severity, and spatial coverage (collectively called disturbance regimes) are strongly affected by global change. For example, dendrochronological and observational analyses and sedimentary charcoal records have shown tight coupling between fire activities and climate oscillations. During the mid-1980s, a period with unusually warmer springs and longer summer dry seasons, large wildfires in forests occurred more frequently in the western United States. Forest dieback and insect outbreaks usually increase in warm and dry periods. It is challenging to project future disturbance regimes in response to global change so that we can assess their impacts on the ecosystem carbon cycle.

It has long been documented that multiple states of ecosystem equilibrium exist. Natural disturbances, global change, and human intervention may trigger the state changes, resulting in major impacts on the ecosystem carbon cycle. If an ecosystem changes from a high carbon storage capacity (e.g., forest) to a new state with a low

TABLE 1

Dynamic equilibrium and disequilibrium of carbon cycle under four situations

Situation	Equilibrium	Disequilibrium
Global change	An original equilibrium can be defined at a reference condition (e.g., pre-industrial [CO_2]) and a new equilibrium at the given set of changed conditions.	Dynamic disequilibrium occurs as C cycle shifts from the original to new equilibrium. Global change factors gradually change over time, leading to continuous dynamic disequilibrium.
Ecosystem within one disturbance–recovery episode	C cycle is at equilibrium if the ecosystem fully recovers after a disturbance. The equilibrium C storage equals the product of C influx and residence time.	C cycle is at dynamic disequilibrium and an ecosystem sequesters or releases C before the ecosystem fully recovers to the equilibrium level.
Regions with multiple disturbances over time	C cycle is at dynamic equilibrium in a region when the disturbance regime does not shift (i.e., stationary). The realizable C storage under a stationary regime is smaller than that at the equilibrium level.	C cycle is at dynamic disequilibrium and the region sequesters or releases C when the disturbance regime in the region shifts (i.e., nonstationary).
Multiple states	C cycle can be at equilibrium at the original and alternative states.	Dynamic disequilibrium occurs as an ecosystem changes from the original to alternative states.

SOURCE: Adapted from Luo and Weng 2011.

storage capacity, the ecosystem loses carbon. Conversely, a change of an ecosystem from a low to high storage capacity results in a net increase in carbon storage. For example, Amazonian forests are currently the largest tropical forests on Earth, containing 200–300 Pg C in their forests and soils. Some models have predicted that climate change could alter moist convection, leading to a reduction in dry-season rainfall in various parts of Amazonia and triggering positive feedback to state changes of the ecosystems partly to savanna. Permafrost in the high-latitude regions of the northern hemisphere contains nearly 1700 Pg of organic C. Global warming will alter physical, chemical, biological, and ecological states of the permafrost ecosystems in the region. The state change likely results in substantial C loss. State changes among multiple equilibriums can cause destabilization of the ecosystem carbon cycle. We need innovative methods to examine conditions and processes leading to the state changes.

FUTURE CARBON CYCLE DYNAMICS

Future terrestrial carbon cycle dynamics will be still governed by these processes as described by Equations 1 and 10 but strongly regulated by disturbances and global change in several ways (Table 1). First, one disturbance event causes temporal changes in carbon source and sink, followed by recovery. The recovery is driven by converging properties of ecosystem carbon processes. One disturbance event may not have much impact on the long-term carbon cycle unless disturbance regimes shift. Second, shifts in disturbance regimes are usually caused by global change and human intervention. Disturbance regime shifts can result in substantial changes in the long-term carbon cycle over regions. Third, global change can directly alter C influx and residence time, leading to changes in the carbon cycle. Global change can also indirectly affect the carbon cycle via changes in ecosystem structure and disturbance regimes. Fourth, when ecosystem structure changes and disturbance regimes shift, the ecosystem carbon cycle might move to alternative states. State changes among multiple equilibriums can have the most profound impact on future land carbon cycle dynamics, especially if they happen at regions with large carbon reserves at risk.

SEE ALSO THE FOLLOWING ARTICLES

Biogeochemistry and Nutrient Cycles / Environmental Heterogeneity and Plants / Gas and Energy Fluxes across Landscapes / Regime Shifts

FURTHER READING

Bolker, B. M., S. W. Pacala, and W. J. Parton. 1998. Linear analysis of soil decomposition: insights from Century model. *Ecological Applications* 8: 425–439.

Bonan, G. B. 2008. Forests and climate change: forcings, feedbacks, and the climate benefits of forests. *Science* 320: 1444–1446.

Chapin, F. S., III, P. A. Matson, and H. A. Mooney. 2002. *Principles of terrestrial ecosystem ecology.* New York: Springer.

Farquhar, G. D., S. von Caemmerer, and J. A. Berry. 1980. A biochemical model of photosynthetic CO_2 assimilation in leaves of C_3 species. *Planta* 149: 78–90.

Luo, Y. Q., and E. S. Weng. 2011. Dynamic disequilibrium of terrestrial carbon cycle under global change. *Trends in Ecology & Evolution* 26: 96–104.

Parton, W. J., D. S. Schimel, C. V. Cole, and D. S. Ojima. 1987. Analysis of factors controlling soil organic-matter levels in Great-Plains grasslands. *Soil Science Society of America Journal* 51: 1173–1179.

Potter. C. S., J. T. Randerson, C. B. Field, P. A. Matson, P. M. Vitousek, H. A. Mooney, and S. A. Klooster. 1993. Terrestrial ecosystem production: a process model-based on global satellite and surface data. *Global Biogeochemical Cycles* 7: 811–841.

Schlesinger, W. H. 1997. *Biogeochemistry: an analysis of global change,* 2nd ed. San Diego: Academic Press.

Sellers, P. J., R. E. Dickinson, D. A. Randall, A. K. Betts, F. G. Hall, J. A. Berry, G. J. Collatz, A. S. Denning, H. A. Mooney, C. A. Nobre, N. Sato, C. B. Field, and A. Henderson-Sellers. Modeling the exchange of energy, water, and carbon between continents and the atmosphere. *Science* 275: 502–509.

ECOSYSTEM ENGINEERS

KIM CUDDINGTON
University of Waterloo, Ontario, Canada

Ecosystem engineer is a term coined in the mid-1990s by Clive Jones and colleagues to refer to those species that have a large effect on the physical environment. A canonical example is the beaver, which, through the construction of dams, can alter the hydrology of a region, creating new aquatic habitat where none existed previously. Those species that modify the abiotic environment through their actions (allogenic) can be distinguished from those whose physical form may alter the environmental state (autogenic). The key to identifying an ecosystem engineer is the mode of action through which the environmental alteration occurs. An ecosystem engineer modifies the environment via nontrophic interactions. Therefore, consumption of trees by beavers is not engineering, although it may result in a modified environmental state; however, the construction of beaver dams is ecosystem engineering. Other examples of ecosystem engineering include oyster reef influences on estuarine flows and sedimentation, leaf tying by caterpillars, and the actions of earthworms in soils.

CONTROVERSY

There has been controversy over the utility of this concept. Some authors suggest that because all species modify the abiotic environment, ecosystem engineering is not useful for understanding ecological interactions. Others have suggested the term be limited to those scenarios where engineers have large impacts. In response, some ecologists suggest that restricting our attention to cases where there are large effects will undermine our ability to understand when ecosystem engineering is and is not important. Another suggestion is that the term be restricted to those cases where engineers have only a positive impact on their own population growth. Much of this controversy seems to be dependent on confusion about how a concept confers utility. The ubiquity of a mechanism is not necessarily a problem for utility. For example, consumption of resources is a ubiquitous process in ecological systems, but it is nonetheless extremely useful for understanding these systems.

In support of the utility of the ecosystem-engineering concept, one can consider the implications of species modification of abiotic factors for devising management plans for impacted areas. While it is generally realized that human modification of the environment may be necessary for successful restoration efforts (e.g., ripping compacted soil), it is not generally realized that the presence of an engineering species may inhibit the success of such actions or that the introduction of an appropriate engineering species could achieve the same effect with much less expense. In addition, the awareness that very different species with disparate physical impacts may have common functional roles in ecological systems seems an important insight.

Ecosystem engineering and the controversy surrounding it resemble similar ideas and debates of a related concept in the area of evolutionary biology: niche construction. Niche construction refers to habitat creation, which produces the evolutionary environment that affects the engineering species. The idea was developed from an earlier observation by Lewontin that the appropriate description of the interaction of species with their selection environment would include feedback from the species to the environment. Ecosystem engineering is therefore a more general term than niche construction since, although a biotic feedback of some form is required for engineering to occur, the feedback need not be evolutionary nor directed at the engineer species.

EMPIRICAL STUDIES

The study of species effects on the physical environment has a long history, even though the collective term to describe these activities is of recent origin. Early empirical studies of ecosystem engineering date to the 1800s. There are a large number of observational studies of ecosystem engineering that can be roughly divided into works on soil and sediment processes, succession, facilitation and inhibition, and habitat creation.

The study of species effects on soils and sediments has the longest history. Darwin's early work on the effects of earthworms on the physical structure of soils belongs to this genre. Some studies have focused on the impact of burrowing species on erosion and bank stability. More recent work has described the impact of earthworms and other invertebrates on soil or sediment chemistry. For example, burrowing fiddler crabs can cause an increase in soil oxidation-reduction potential. Similar types of work have been conducted for burrowing mammals. In general, research on the effects of invertebrates on soil and sediments is more advanced than any other area of research on ecosystem engineering.

The effects of plant species on soils and sediments have been most frequently studied in the context of succession. Early successional models assumed a species–environment interaction that partially drove changes toward a climax community. The formation of organic soils from abiotic substrates by plants and microbes has been an important assumption for such models.

Another area of active empirical research is that of examining the role of abiotic modification in determining whether species interactions are positive (facilitation) or negative (inhibition). One large area of study is that of the role of "nurse" plants in modifying the local microclimate, which in turn promotes the colonization and survival of other plant species. For example, in arid environments the presence of an established plant reduces the surface temperature under its canopy, and leaf litter can increase soil water retention. These altered conditions benefit seedlings attempting to establish in a harsh desert environment. On the other hand, some vegetative species can inhibit the establishment of other plants through abiotic modification of the environment. Some researchers have found that acidification of the soil through leaf litter decomposition can have a negative impact on other vegetation. These types of impacts are not limited to terrestrial environments. For example, phytoplankton biomass and distribution can alter the thermal stratification of lakes through light interception and reflection.

The most obvious effect of ecosystem engineers is the creation of new habitats, which can alter species diversity and distribution. Phytotelmata, the small aquatic habitats formed by the physical structure of terrestrial plants, are a classic example. Those ecosystem engineers that create habitat may be described as foundation species. This seems particularly appropriate when the created habitats are large, such as the formation of atolls by corals.

It seems clear that all foundation species may be engineers, but probably not all engineers would be classified as foundation species, since it is thought that foundation species act on the environment partially through their large biomass. Using the classification suggested by Jones and colleagues, foundation species and autogenic ecosystem engineers may be synonymous. However, beavers create habitat but are not foundation species: a large number of beavers is not required for large ecosystem impacts. Species with similar impacts may include alligators which create wallows in wetland ecosystems. In recognition of this distinction from foundation species, and of the large impact of these species, allogenic engineers have sometimes been called keystone species. But this may be a confounding of concepts, since most widely recognized keystone species act through predation links (e.g., the sea star *Pisaster*).

Empirical work on habitat creation has, for the most part, focused on the role of engineers in altering the species richness or composition on an area. For example, leaf tiers increase species richness of arthropods on host plants. In the Adirondack region of North America, beavers increase the number of species of herbaceous plants in the riparian zone.

In summary, the empirical study of ecosystem engineers predates the coining of the term; there has been significant work on this topic in areas related to soil processes, succession, facilitation, and habitat creation. These studies make the connection between abiotic modification by particular species and some important ecological concepts such as succession, facilitation, and foundation species.

THEORY OF ECOSYSTEM ENGINEERING

The theoretical analysis of ecosystem engineering is much less developed than the empirical work and has origins scattered through several, seemingly disparate, areas of research. Some early work took the form of conceptual models in which ecosystem engineering played some role (see above comments on this idea in descriptions of succession and facilitation). Although some early theoreticians considered abiotic environment–species interactions, modeling work that has been concerned with the effects of ecosystem engineers per se has been completed very recently.

MODELS OF ECOSYSTEM ENGINEERING: CLIMATE LITERATURE

James Lovelock and coauthors were some of the first authors to generate a model of ecosystem engineering in a specific context. In 1983, Watson and Lovelock presented the "Daisyworld" model to quantify Lovelock's concept of Gaia: the idea that species could produce a beneficial and stable environment. The model includes two species of plants: a black daisy that has an albedo lower than the albedo of the surface of the Earth, and a white daisy that has a higher albedo than the Earth.

The model tracks the percentage of the surface occupied by daisies, where the total albedo of the Earth is determined by the percentage of each kind of daisy, and unoccupied ground. The growth rate of the daisies depends on the physical environment as a parabolic function of local temperature. However, the local temperature experienced by the daisies is both a function of the solar radiation heating the Earth and the albedo of the daisies. Since black daisies absorb more radiation than white daisies, they will be warmer. Planetary temperature therefore also depends on the percentage of black and white daisies.

Watson and Lovelock demonstrate that there is a globally stable planetary temperature for a wide range of solar luminosity that occurs because of modification of surface reflectance by the daisies. The model always shows greater temperature stability with daisies than without them. The Daisyworld models have been investigated for various other scenarios; however, until recently, the majority of this work has appeared in climatology journals rather than in the ecology literature. In a recent contribution, a cellular automata version of Daisyworld that included effects of the Earth's curvature predicted that different latitudes would have different temperatures, and at a critical value of solar luminosity, a desert formed in wide bands across the planet.

MODELS OF ECOSYSTEM ENGINEERING: ECOLOGY LITERATURE

In the ecology literature, pure ecosystem models can be distinguished from ecosystem-engineering models using a scale of species specificity and the inclusion environmental feedbacks. Ecosystem models are generally concerned with energy or material flows related to communities and do not parse out the effects of specific species or species groups. So, for example, global circulation models that do not include the impact of particular species will not be considered in this entry. Also excluded are models that do not include the effects of abiotic feedbacks to biotic dynamics. Note, however, that the abiotic feedback does not necessarily link the engineer species and the modified aspect of the environment, but could instead be a link between an abiotic characteristic and another species in the community, or even a community-wide property (e.g., species richness).

Of models that meet our criteria for describing ecosystem engineering, a distinction can be made between models designed to determine the general properties of ecosystem engineering and those produced to describe the effects of ecosystem engineering in specific contexts. Readers should note that, usually, authors of models designed to predict the effects of species–environment feedbacks for particular systems do not identify the process described as ecosystem engineering.

ENGINEER–ENVIRONMENT MODELS

The first self-identified ecosystem engineer model was produced by Gurney and Lawton in 1993. The authors described the effect of an obligate, allogenic engineering species, E, on the proportion of virgin, V, and modified habitat, H, as

$$\frac{dE}{dt} = rE\left(1 - \frac{E}{H}\right),$$

$$\frac{dH}{dt} = f(V, E)E - \delta H,$$

$$\frac{dV}{dt} = \varepsilon(T - V - H) - f(V, E)E.$$

The model assumes that the total area, T, remains constant and that modified habitat eventually degrades (therefore the amount of degraded habitat would be $T - V - H$). Engineers need modified habitat to survive and convert virgin areas to modified habitat at a rate determined by a function f of virgin habitat available and engineer species density. This habitat degrades at a rate δ but then recovers to virgin status at a rate ε. The model predicts two possible stable states: extinction of the engineers or a positive engineer population. The extinction state is attracting unless the engineer species can modify the virgin habitat at a rate that compensates for degradation. Engineer population density thus determines landscape heterogeneity. In addition, where engineers cooperate (positive feedback), oscillations of landscape state are possible.

Cuddington and Hastings took a different spatial approach to single-species engineer dynamics. They modeled the transient dynamics of an invasive species using an integrodifference formulation where habitat quality and population density were tracked. As compared to the patch approach used by Gurney and Lawton, Cuddington and Hastings assumed that habitat quality would vary smoothly over space as a function of engineer occupancy. This model predicts that standard invasion models could underestimate both spatial spread rates and population densities of engineer species invading suboptimal habitats.

More recently, Cuddington and colleagues developed a simple, general model of ecosystem engineering that eliminates explicit or implicit spatial considerations. Engineer population density, N, and environmental state, E (e.g., temperature), are described in a set of ordinary differential equations as

$$\frac{dN}{dt} = f(N,E) + g(N),$$

$$\frac{dE}{dt} = j(N,E) + k(E),$$

where the function, f, describes that portion of population growth rate which is affected by feedback to the environmental state, g gives the population growth rate that is independent of that particular environmental factor, j

describes modification of the environmental state that is due to the engineering species, and k gives the rate of change of that environmental characteristic in the absence of an engineer species. When the engineering of the abiotic environment is described as a simple linear function of population density that feeds back to the engineer through the environmental influences on the population growth rate, altered equilibrium densities, bistability, or runaway growth of the engineer population is possible.

Collectively, these general models suggest that there are two classes of ecosystem engineers that might be expected to have different dynamics: obligate engineers (extinction is an attracting equilibrium), and nonobligate engineers. For obligate engineers, there is an obvious connection to Allee effects where there is positive feedback from increasing density via the engineering relationship. In this case, more complicated dynamics, such as multiple basins of attraction and stable oscillations, are possible. In addition, it is very clear that the magnitude of engineering impacts depends critically on the environmental context, where highly robust environments would return quickly to their unengineered state. That is, a species that may have significant environmental impacts in one region may have no significant effects in another location.

These general models of ecosystem engineers can be distinguished from models designed to examine the consequences of engineering in specific systems. These models are often spatially explicit descriptions of environmental modification. It is not usually the case that these models are identified as ecosystem engineers models per se.

A large category of models for specific systems explores the impacts of engineering species that occur via the interception of abiotic flows (i.e., water, wind, currents). One of the earliest of these, proposed by Klausmeier in 1999, was a simple mechanistic model to show the effect of vegetation on soil-water infiltration rates, which in turn feeds back to spatial patterns of vegetation. The authors used a partial differential equation model that included the variables of water and plant biomass. Simple linear functions describe the response of plant biomass, n, to increased water, w, and the effect of plant biomass on soil-water infiltration rates as

$$\frac{dw}{dt} = a - w - wn^2 + v\frac{\partial w}{dx},$$

$$\frac{dn}{dt} = wn^2 - mn + \left(\frac{\partial^2}{\partial x^2} + \frac{\partial^2}{\partial y^2}\right),$$

where m is plant biomass loss, a is water input, and v is the speed of water flow downhill. There are two stable equilibria: a nonvegetated state and a vegetated state, which, in the spatial domain, are connected by stripped patterns of vegetation and bare ground. More detailed versions of this type of model have been developed by others, who have found that the emergent spatial arrangement determines the ultimate engineering outcome.

Similar types of models describe the relationship between vegetation and dune stabilization, the development of tidal creeks, fog interception, and wrack accumulation. These models all have a similar advection–diffusion formulation regardless of the abiotic flows described, have been analyzed with similar methods. For example, the authors often make a similar quasi-equilibrium assumption regarding the dynamics of vegetation. This similar formulation and mode of analysis suggests that this body of work should be regarded as tightly related.

ENGINEER-ENVIRONMENT-EVOLUTION MODELS

Authors in the area of niche construction have contributed a series of models that relate ecosystem engineering to gene frequency, although more recently there has been some work that expands this focus in ecological directions. The basic formulation comes from Lewontin (1983), where organism traits, O, and environment state, E, feedbacks are given as

$$\frac{dO}{dt} = f(O,E),$$

$$\frac{dE}{dt} = g(O,E).$$

This formulation is, of course, strictly analogous to the ecological formulation where organism density is tracked rather than trait frequency. In the late 1990s, Laland and colleagues used recursion equations to track the dynamics of traits whose fitness depends on environmental modification and the amount of resources in the environment. The authors find that engineering can lead to the fixation of deleterious alleles and the destabilization of stable polymorphisms.

This evolutionary focus has been expanded to include eco-evolutionary models. In the context of plant–nutrient flow feedbacks, several authors have developed models where plants have the ability to increase the input of inorganic nutrients into the soils, which can lead to partial regulation of soil nutrients. In the quite specific contest of the evolution of flammability, a haploid model with separate loci for flammability and response to fire illustrates that the presence of flammability-enhancing traits can redirect the evolution of other traits though effects on the environment. When this model is extended to the spatial domain, with local processes, flammability may increase in frequency without direct fitness benefit.

ENGINEER-ENVIRONMENT-COMMUNITY MODELS

Models that describe the interaction of engineer species with other species in the environment (either directly or via abiotic feedbacks) are much rarer in the literature. In general, of the few models in this area, the main focus has been on the effects of engineer species on community species richness.

There are few models that describe the effect of specific species interactions with engineers. As an extension to the work by Klausmeier, the role of herbivory on engineering plants in determining the patterns of vegetation in semi-arid systems has been explored. Herbivores with the ability to select foraging locations at a fine scale can shift a system from a stable vegetated state to a stable unvegetated state. In a 2009 publication, Beckage and coauthors explore the facilitation or inhabitation of fire through density-dependent vegetation–fire feedbacks in a grass–forest tree–savanna tree model. The authors find that fire-facilitating savanna trees could engineer landscapes toward savannas. Engineering in aquatic systems has also been explored; researchers combined an individual-based population model and an advection–diffusion model of sulfide flows to describe the relationship between three different engineering groups in deep water vent systems. The tubeworm *Lamellibranchia luynmesi* is thought to release sulfate to the sediments. The sulfate is reduced to sulfide by bacteria that, in turn, are associated with methane oxidizing or hydrocarbon degrading species. Positive feedbacks between the species contribute to the stability of the coexistence in spite of decreasing sulfide levels.

A few theoretical works describe the effects of engineering on aggregate community properties. In 2004, Wright and Jones related ecosystem engineering, primary productivity, and species richness in a conceptual model, where they postulated that engineered patches could have higher or lower resource availability than unengineered patches, and thus alter regional primary productivity and therefore, species richness. In a later mathematical formulation of this idea, using a Lotka–Volterra modeling framework, it was found that species richness will increase with increasing average growth rates caused by engineering, or decrease with ecosystem engineering-enhanced interactions such as interspecific competition. Other authors have combined empirical data with Monte Carlo simulation approaches to conclude that the population dynamics of engineer species could dramatically modify the species richness of a region through habitat patch creation.

It should be noted that there is no necessary relationship between engineering and a beneficial impact on the engineer dynamics, the community characteristics, or ecosystem function. However, there has been much less exploration of any negative impacts of ecosystem engineers on biodiversity, and few models of particular species and specific systems investigate the conditions under which engineering will decrease species richness. Generalizing from previous work, species that decrease the heterogeneity of the landscape might be predicted to decrease species diversity. On the other hand, it does seem possible that new invaders could also increase new habitats that, while increasing total biodiversity, may in fact decrease native biodiversity (e.g., *Spartina alterniflora* on the West Coast).

SEE ALSO THE FOLLOWING ARTICLES

Allee Effects / Belowground Processes / Facilitation / Niche Construction / Succession

FURTHER READING

Beckage, B., W. J. Platt, and L. J. Gross. 2009. Vegetation, fire, and feedbacks: a disturbance-mediated model of savannas. *American Naturalist* 174: 805–818.

Cuddington, K., and A. Hastings. 2004. Invasive engineers. *Ecological Modelling* 178: 335–347.

Cuddington, K., W. G. Wilson, and A. Hastings. 2009. Ecosystem engineers: feedback and population dynamics. *American Naturalist* 173: 488–498.

Gurney, W. S. C., and J. Lawton. 1996. The population dynamics of ecosystem engineers. *Oikos* 76: 273–283.

Jones, C. G., J. H. Lawton, and M. Shackak. 1994. Organisms as ecosystem engineers. *Oikos* 69: 373–386.

Klausmeier, C. A. 1999. Regular and irregular patterns in semiarid vegetation. *Science* 284: 1826–1828.

Laland, K. N., F. J. Odling-Smee, and M. W. Feldman. 1996. The evolutionary consequences of niche construction: A theoretical investigation using two-locus theory. *Journal of Evolutionary Biology* 9: 293–316.

Lewontin, R. C. 1983. Gene, organism and environment. In D. S. Bendall, ed. *Evolution from molecules to men.* Cambridge UK: Cambridge University Press.

Watson A. J., and J. E. Lovelock. 1983. Biological homeostasis of the global environment—the parable of daisyworld. *Tellus Series B: Chemical and Physical Meteorology* 35: 284–289.

Wright, J., W. Gurney, and C. Jones. 2004. Patch dynamics in a landscape modified by ecosystem engineers. *Oikos* 105: 336–348.

ECOSYSTEM SERVICES

FIORENZA MICHELI
Hopkins Marine Center of Stanford University, Pacific Grove, California

ANNE GUERRY
Stanford University, California

Ecosystem services are the "conditions and processes through which natural ecosystems, and the species that make them up, sustain and fulfill human life" (Daily 1997). Simply put, ecosystem services are the benefits people obtain from ecosystems, or the things people want and need from ecosystems.

THE CONCEPT

An awareness that natural ecosystems, in diverse direct and indirect ways, support human societies and needs traces far back in time. Mooney and Ehrlich (in Daily's 1997 book, *Nature's Services*) attribute the origin of the modern concept of ecosystem services to George Perkins Marsh's *Man and Nature*, published in 1864. Nearly 150 years ago, Marsh discussed the links between deforestation, the loss of soil, and degradation of freshwater quality and highlighted the services of waste recycling and pest control provided by natural ecosystems. The link between ecosystem and human well-being is also at the core of the science of ecology. Odum, in his 1959 *Fundamentals of Ecology*, squarely placed human populations as part of ecosystems and described the human use of natural resources through agriculture, forestry, and fisheries as important connections within ecosystems. The term ecosystem services (or its synonyms, such as environmental or ecological services) was used rarely in the scientific literature until the 1990s, when the concept and its applications were brought to the fore (e.g., in Daily's 1997 book and in Costanza and colleagues' 1997 global ecosystem valuation; see below). After the Millennium Ecosystem Assessment (MA) of 2005, the relevance of the concept of ecosystem services to policy and governance was broadly accepted by many, but development of practical guidance on how to apply it was still lacking. Efforts to use the concept in understanding coupled social-ecological systems, improving multiple-objective management, setting up new markets, and designing sustainable management schemes are the new frontier of turning the appealing concept of ecosystem services into practical guidance for decision making.

The MA was an international effort of more than 1300 experts from 95 countries to assess the consequences of ecosystem change for human well-being. It highlighted the diversity of services provided, the range of natural processes and natural–human interactions that underlie them, as well as the fundamental dependence of people on the flow of ecosystem services. The MA classified ecosystem services into four categories: *provisioning services* such as food, water, timber, fiber, genetic resources, and pharmaceuticals; *regulating services* controlling climate, air and water quality, erosion, disease, pests, wastes, and natural hazards; *cultural services* providing recreational, aesthetic, and spiritual benefits; and *supporting services* such as nutrient and water cycling, soil formation, and primary production. The last category, supporting services, are unique because they underpin the provisioning of all the others. A major, alarming finding of the MA was the widespread, global decline and degradation of ecosystem services. According to the MA, approximately 60% of ecosystem services are degraded or used unsustainably, including capture fisheries, air and water purification, and the regulation of regional and local climate, natural hazards, and pests.

Nearly a decade earlier, Costanza and co-authors highlighted the economic value of the services provided by natural ecosystems. They estimated the value of ecosystem services for the entire biosphere to be in the range of US$16–54 trillion per year, with an average of US$33 trillion per year, over twice the global gross national product at the time. The methods and utility of their approach and results have been hotly contested in the literature, but the importance of their work in bringing attention to the concept of valuing ecosystem services is clear.

While useful for highlighting the underappreciation of ecosystem services, estimates of the grand value of ecosystem services do not capture how ecosystem services provision varies depending on the condition of natural ecosystems and human demand, nor do they allow for an evaluation of possible tradeoffs among different services. Recent efforts have focused on developing frameworks and approaches for the direct incorporation of the ecosystem services concept in decision-making contexts (see below). These are based on the ability to explore marginal changes in ecosystem services, or how ecosystem services are likely to change under alternative future scenarios.

The linkages and relationships among ecosystems and the services they provide and support are represented in Figure 1. Ecosystem structure, including the diversity and abundance of species and habitats, produces ecosystem

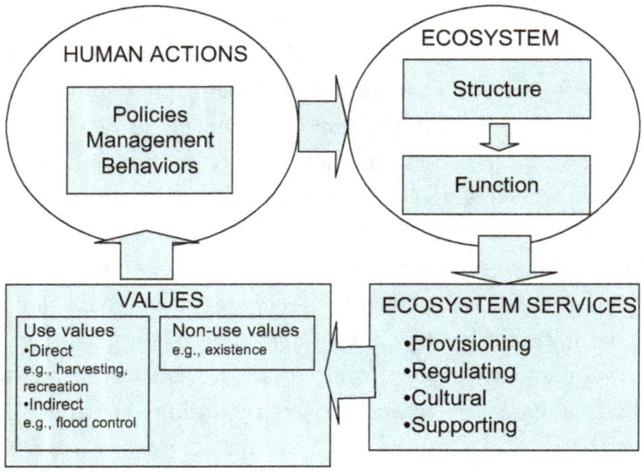

FIGURE 1 Relationships between ecosystem structure and function, services, values, and public policies and human behaviors.

function. Function refers to the suite of processes and interactions occurring within ecosystems, including nutrient cycling and primary and secondary productivity. Ecosystem functions, together with demand from humans, produce ecosystem services. These can be valued in a number of different ways (Fig. 1). Changes in these values can inform behaviors and management actions that can, in turn, affect ecosystem structure and function and their ability to provide ecosystem services.

Tallis and coauthors proposed a three-step framework for clarifying these linkages between ecosystem structure, function, and services and for improving the utility of the ecosystem services concept by concretely identifying what to measure. The structural characteristics and functions of ecosystems provide the supply of ecosystem services (for example, the number of fish in the sea). But the ecosystem services concept is inherently anthropocentric; the supply must be used or enjoyed by humans to be defined as an ecosystem service. Now we are no longer talking about the number of fish in the sea, but rather the number of fish in the net. Finally, we take into account people's preference for services, or ecosystem service values. This may be the price/pound paid for fish in the market or the importance of those fish for nutrition or cultural use.

SERVICES PROVIDED BY DIFFERENT ECOSYSTEMS

As highlighted above, ecosystem services provisioning depends on ecosystem structure and function, and thus varies within and across ecosystems depending on their diversity, dynamics, and condition. Below are presented examples of the rich diversity of ecosystem services provided by forested, freshwater, and marine systems (Fig. 2).

FIGURE 2 (A) Tuscan countryside (Italy) showing a mosaic of olive groves, vegetable and fruit orchards, and oak forests (photograph by F. Micheli); (B) residential, touristic, and industrial developments along the shores of Lake Como (Italy; photograph by F. Micheli); (C) a view of Monterey Bay (California, USA), a marine ecosystem that supports research and education (conducted by the Hopkins Marine Station and Monterey Bay Aquarium, in the picture, and many additional institutions), commercial and sport fisheries, and a thriving marine tourism sector (photograph by Alan Swithenbank).

Forested ecosystems provide a wide range of ecosystem services. For example, forests protect soils, preventing their loss through erosion, and help them retain moisture and store and recycle nutrients. They can regulate water flows and cycling, thereby reducing floods and droughts, and they serve as buffers against the spread of pests and diseases. They modulate climate at local, regional, and global scales through the regulation of rainfall regimes, evaporation, and carbon sequestration in plants and soil. They provide materials such as timber, nonwood products such as cork, rubber, mushrooms, wild fruits and seeds, and wild game. They provide habitat for insects that pollinate important crops. And they provide recreational opportunities such as hiking, camping, and hunting and serve as a source of inspiration and cultural identity. Deforestation and degradation of forest ecosystems has major impacts on provisioning, regulating, cultural, and supporting services on local to global communities.

Similarly, freshwater ecosystems support critical services. Service provision by freshwater ecosystems is disproportionately important, when compared to their extent; only a tiny fraction (<1%) of the Earth's water supply is found in lakes, rivers, wetlands, and underground aquifers. A critical service is water supply for drinking, washing, and other household uses, thermoelectric power generation and other industrial uses, irrigation, and aquaculture. Other provisioning services include the production of fish and shellfish. Regulating, supporting, and cultural services include flood control, pollution dilution and water quality control, transportation, recreation, aesthetic beauty, and cultural heritage.

Marine ecosystems also provide a wealth of services, spanning all four categories identified by the MA. Provisioning services include food from capture fisheries and aquaculture, timber from mangrove forests, and chemicals for cosmetics and food additives. In the future, the oceans may provide energy through biofuels from algae, and wave and tidal energy production. Oceans are key providers of a suite of regulating and supporting services. Among regulating services, most critical are natural hazard and climate regulation and the transformation, detoxification, and sequestration of waste. The oceans provide an estimated 40% of the global net primary productivity and 50–85% of the oxygen within the atmosphere, hold 96.5% of the Earth's water, and are key players in the global water, carbon, oxygen, and nutrient cycles. Also, coastal and marine systems are culturally important: many communities define themselves through a coastal way of life, and coastal tourism is one of the world's most profitable and most rapidly growing industries.

AN ECOSYSTEM SERVICES FRAMEWORK TO GUIDE DECISIONS

An increasing recognition in scientific, management, and governance spheres of the considerable values of ecosystem services to society is beginning to fundamentally change the way in which natural resources are managed. New management and policy appetites for ecosystem service values and scientific advances in conceptual frameworks and tools have begun to inject these values into decision-making processes.

The management of natural resources has often focused on objectives for single sectors and short-term goals. This approach can lead to unintended consequences, including negative impacts on other sectors and the suboptimal delivery of a broad range of ecosystem services. Most decisions that do consider costs reflect only a portion of the full suite of values that nature provides. Also, they tend to consider benefits to particular sectors, rather than to society as a whole. For example, cost–benefit analyses tend to focus on values that are easy to express in economic terms. Nonmarket and nonuse values are ignored or assumed to have lesser value. This one-sidedness can guide decision makers toward actions with negative repercussions for society. For example, mangrove areas are routinely cleared and used for shrimp aquaculture. The high market price of shrimp is the motivation for this action, but the net value of shrimp production (market value minus subsidies and social costs) pales in comparison to the social and ecological benefits that standing mangroves provide. More complete accounting by Barbier and colleagues indicates that intact mangroves provide more benefits for society. A single-sector approach ignores the connections among components of natural and social systems. These connections are often important for the maintenance of ecosystem health, human well-being, and the sector of interest itself.

In many cases, investing in natural capital is more efficient than using built capital to provide desired services. The classic example of the efficiency of natural capital over built capital is the case of the provisioning of New York City's drinking water from the Catskill Mountains. In the 1990s, federal law required water suppliers to filter surface water unless they demonstrated alternative ways of keeping water clean. Managers explored building a filtration plant ($6–$8 billion, excluding annual operating and maintenance costs) or a variety of watershed protection measures (acquisition of land, reduction of contamination, and the like, $1.5 billion). The choice to invest in natural rather than built capital was obvious.

Similarly, ecosystem services are also being used to generate new funding streams for conservation. "Payment for ecosystem service" schemes are being set up around the world—from the United States, to Costa Rica, to Australia, to Colombia. For example, The Nature Conservancy, with many partners, has helped to set up over a dozen "water funds" in the northern Andes. In these programs, water users voluntarily put money into a trust fund that subsidizes conservation projects, improves water quality, and avoids significant water treatment costs. A public–private partnership composed of water users and other stakeholders makes decisions about using the fund to finance conservation activities in the watershed. Such programs are protecting rivers and watersheds, maintaining livelihoods of farmers and ranchers upstream, and helping to provide irrigation, hydropower, and clean drinking water to people—yielding promising results in terms of social, financial, and conservation metrics.

ECOSYSTEM SERVICE VALUATION

The assessment of the value of ecosystem services may involve qualitative and quantitative analyses, from a conceptual depiction of how human activities depend on and affect ecosystems to a quantification of the monetary value of particular services. The goal of these assessments is to link management actions, the development of new markets, and other activities directly to changes in ecosystem conditions and to gain an understanding of how those changes may affect the benefits that various individuals and groups derive from ecosystems. It is also important to note that valuation does not have to be in monetary terms. Ecosystem services can indeed be valued using monetary metrics, but they can also be valued qualitatively by groups of people or using social metrics (such as the number of people displaced by a flood).

One key consideration when attempting to value ecosystem services in monetary terms is to be very clear and consistent about what exactly is being valued. Boyd and Banzhaf propose that ecosystem services are "components of nature, directly enjoyed, consumed, or used to yield human well-being." There are two key elements of this definition. First, they must be directly enjoyed. Thus, "supporting" ecosystem services such as nutrient cycling do not get valued in their own right; rather their value is captured in the total value of the "final" ecosystem services such as clean drinking water and food. Second, ecosystem services are components—things or characteristics—not functions or processes. Nutrient cycling is a process that helps to produce characteristics (e.g., fertile soils) or things (e.g., agricultural crops) that we value. The value of nutrient cycling is captured in the value of the thing, the ecological endpoint. This is a useful framework to be applied when estimating total ecosystem value as it avoids double counting, but there may be other settings, such as monitoring the effectiveness of management practices, where it will be important to measure and value supporting or intermediate services.

Several methodologies have been developed and applied to ecosystem services valuation. Each of these methods has strengths and weaknesses and needs to be carefully matched to the questions being asked, the data available, and other details about the application. Some ecosystem services have explicit prices or are traded in an open market. In these cases, value may be directly obtained from what people pay for a good (e.g., timber, crops, or fish). However, many ecosystem services are not traded in markets. A suite of nonmarket valuation techniques can be used to assign monetary value to these ecosystem services. Revealed-preference approaches use observed behaviors to measure or infer value. For example, travel–cost methods value site-based amenities based on the costs people incur to enjoy them, while hedonic methods derive value from what people are willing to pay for the service through purchases in related markets, such as housing markets (e.g., purchases of properties near natural amenities such as lakes and wilderness areas). Nonconsumptive use values, such as recreation, have often been influential in decision making, and these benefits are sometimes valued in monetary terms using these travel–cost and hedonic methods. In contrast, stated-preference approaches seek to elicit information about values through surveys. Examples include contingent valuation, where people are asked their willingness to pay or accept compensation for change in ecological service (e.g., willingness to pay for cleaner air) and conjoint analysis, where people are asked to rank different service scenarios. Values estimated through stated-preference approaches are generally more contentious than those estimated through revealed-preference methods. A relatively straightforward method for monetary ecosystem services valuation is the cost-based approach. In this approach, the cost of replacing a lost service (e.g., water purification by water treatment plants instead of forested lands or coastal protection by seawalls instead of coastal wetlands) is used to represent the cost of the ecosystem services. Finally, the benefits transfer approach adapts existing ecosystem services valuation (using one of the methods described above) to new contexts that lack data. Avoided damages can also be used—with this approach, the estimated cost of the damage that would have occurred without the service (e.g., avoided flood damage) stands in for the value of the service. The benefits transfer approach can be used to estimate the value of ecosystem services in one place based on the calculated value of those services in another place, but it can also be used to approximate the type, number, and level of services provided. The second, broader application of this method—and its pitfalls—is discussed in the modeling section, below.

The economic approach to measuring benefits has limitations—some arise from the paucity of data necessary to apply otherwise useful approaches, while others arise from approaches that just can't be applied in some instances. First, various types of biophysical and/or economic data needed to understand ecosystem services in a particular place or data on the benefits of ecosystem services to specific groups (versus society as a whole) may be lacking. Second, many ecosystem services cannot be easily (or, in many cases, ever) reflected as monetary values (e.g., spiritual and cultural values). Disputes among different groups may require extensive dialogue and explicit discussion of tradeoffs that will likely be multifaceted rather than measured in a common currency (e.g., disputes between commercial interests, who readily deal with monetary values, and indigenous groups, who do not). Nonmonetary indicators of ecosystem benefits may be better suited to address services that cannot or should not be valued in monetary terms, including spiritual, cultural, or aesthetic values. Interviews, surveys, and other analyses by social scientists can generate evidence about deeply held beliefs of individuals and groups and the benefits they derive from ecosystems. Analysis of voting patterns on public referenda can also shed light on what is important to various constituencies.

While estimates of the monetary value of ecosystem services are useful in some applications, the difficulty or, in some cases, impossibility of establishing monetary value for all services does not diminish the utility of the ecosystem services framework to environmental decision making and as a framework for conveying people's links to and dependence on functioning ecosystems.

MODELING THE FLOW OF ECOSYSTEM SERVICES AND EXAMINING TRADEOFFS

The scientific community has articulated a conceptual framework for moving management toward a more holistic approach that considers the flows of ecosystem services. Today's best available science provides methods, tools, and information for modeling ecosystem services and for multiple-objective planning, decision making, and management. The Millennium Ecosystem Assessment provides a useful treatment of the provisioning of a diverse suite of ecosystem services and their status and trends at a global scale. However, while this is useful contextual information, most resource management decisions are made at local, regional, or national scales. Having information and tools that work at these scales is essential for affecting decision-making processes.

Two primary paradigms for providing ecosystem services information to decision makers have been applied: the benefits transfer approach and the production-function approach. With the benefits transfer approach, estimates of ecosystem services and their values for specific habitats are based on published examples and extrapolated to those habitats in other regions. This approach is simple and can be used when primary data collection is not feasible. However, given the limited set and scale of existing valuation studies, this approach must often be based on the assumption that most (or all) hectares of a given habitat type are of equal value. Information about the rarity, spatial configuration, landscape context, size, and quality of habitats, and about the numbers, practices, and values of nearby people, can be difficult to include. Also, this approach does not address marginal changes. It tallies up the values of services, but it does not provide information about how those values might change with different management actions. Although it has some serious limitations, this approach, when used carefully (i.e., when sites that values are extrapolated to are good matches for sites for which the values were estimated), can provide useful estimations of ecosystem services values.

Process-based, or production-function, approaches model the relationship between the provision of a service and ecological variables. Production-function models showing the relationship between inputs (e.g., fertilizer and labor) and outputs (e.g., crop production) have been used extensively in agriculture, manufacturing, and other sectors of the economy. Similar relationships exist between natural inputs (e.g., structure of oyster reefs) and natural capital or ecosystem services (e.g., protection of the shoreline from erosion). Production-function approaches can use both market prices and nonmarket valuation methods (see "Ecosystem Service Valuation," above) to estimate economic value and show how the monetary value or other value currencies of services are likely to change under different environmental conditions.

There is a long tradition of process-based modeling for single services in marine, freshwater, and terrestrial systems. In marine systems, food-web models (such as EcoPath with EcoSim) and whole ecosystem models (such as Atlantis) have been used to explore management options for the provisioning of seafood. One freshwater example, the Soil and Water Assessment Tool (SWAT), quantifies the impact of land management practices on water yield and water quality (nutrient, sediment, and pesticide loads). In terrestrial systems, relatively simple models like timber yield models in conjunction with inventory and other data allow for the exploration of management and harvesting options and the determination of sustainable timber yields. And more complex models like the CENTURY model can be used to explore plant–soil nutrient cycling to simulate carbon storage and sequestration in different types of ecosystems (e.g., grasslands, agricultural lands, forests).

More recently, efforts to simultaneously model multiple ecosystem services and the tradeoffs or synergies among them have emerged. Because ecosystem services are highly interdependent (e.g., all derived from the same suite of supporting services), attempts to maximize the delivery of a single service often come at the expense of other services. Foley and colleagues (2005) explore the global consequences of local land-use decisions to maximize food, fiber, water, and shelter for more than six billion people. Rodriguez and others (2006) review ecosystem service tradeoffs from the MA and elsewhere and suggest that tradeoff decisions show a preference for provisioning services, often at the expense of others. Although tradeoffs are common, there are also synergies. For example, in the Willamette Basin, Oregon, Nelson and colleagues' comparison of multiple modeled ecosystem services across different land-use/land-cover scenarios demonstrated little evidence of tradeoffs between ecosystem services and biodiversity conservation. Scenarios designed to enhance biodiversity conservation also improved the provisioning of modeled ecosystem services.

A number of efforts to model the flows of multiple ecosystem services have resulted in the development of tools designed to be applicable in various locations. Some approaches use probabilistic (Bayesian) models, machine learning, and pattern recognition (Artificial Intelligence for Ecosystem Services, ARIES); others use a production-function approach to examine how changes in inputs (e.g., land use/land cover) are likely to lead to changes in the delivery of a wide range of ecosystem services (Integrated Valuation of Ecosystem Services and Tradeoffs, InVEST); and still others use spatially explicit simulations of ecosystems and socioeconomic systems to explore valuation and complex tradeoffs among ecosystem services (Multi-scale Integrated Models of Ecosystem Services, MIMES).

Decision-support tools can help make the framework of ecosystem services useful in real-world contexts. For example, the InVEST ecosystem services mapping and modeling tool was used to design ecological function conservation areas, which in part identify development zones that avoid areas of high ecosystem services provision and importance for conservation in Baoxing County, China. The mapping exercise highlighted that development activities are currently planned to occur in areas important for several priority ecosystem services. These developments are now being reconsidered by local government officials in the next iteration of the Land Use Master Plan. Similarly, Kamehameha Schools (KS), the largest private landowner in Hawaii, partnered with the Natural Capital Project (a university–NGO collaboration) to use a quantitative ecosystem services assessment to inform land-use planning for a large tract of land on Oahu where many formerly agricultural lands lay fallow. They examined expected impacts on ecosystem services for three major alternative planning scenarios: (1) returning agricultural lands to sugarcane as a biofuel feedstock, (2) diversified agriculture and forestry, or (3) residential development. The quantified services were carbon storage and water quality, as well as financial return from the land. Cultural services were incorporated qualitatively. An examination of the tradeoffs among the three alternatives helped to prioritize a land use plan involving diversified agriculture and forestry. The results informed KS' decision to rehabilitate irrigation infrastructure and work with local communities to implement a mixed land use plan.

CONCLUSIONS

The concept of ecosystem services has a long history, but as a framework for making natural-resource decisions, it represents a major and critical shift in how natural ecosystems, and the things people receive from them, are viewed. However, the science of ecosystem services is still in the early phases of moving from a conceptual framework to a suite of methodologies for the rigorous application of the concept in real decision contexts, including spatial planning, permitting, and market development.

There are numerous challenges to be faced moving this concept from theory to practice. For example, data on the supply, human demand, and value components of ecosystem services are often scarce. This challenge can be addressed both by new efforts to gather relevant data and by building simple models that can provide first-pass estimation of results for multiple ecosystem services in the majority of cases where data availability does not support the use of more sophisticated methods. Also, there is inherent difficulty in implementing the effective interdisciplinary efforts required to advance the science of ecosystem services—numerous disciplines, from ecology to engineering to economics and beyond, often need not only to talk to one another but to closely collaborate. And while conceptually simple, the extraordinary complexity of the issues and feedbacks involved in ecosystem services cannot be understated.

Despite these challenges, there is a great need for ecosystem services information to feed into decisions. There is a growing commitment by decision makers—from governments to NGOs to the private sector—to include ecosystem services in their day-to-day deliberations, and there are unprecedented opportunities to include ecosystems in economic and social accounting through the concepts and methodologies we have described here. Fortunately, much progress has been made and ecosystem services concepts and tools are currently being used to inform decisions around the world. As future iterations of using ecosystem services in decisions proceed, its scientific underpinnings will certainly increase in sophistication to better support decisions that safeguard the sustainable provisioning of the full range of ecosystem services upon which human lives and societies depend.

SEE ALSO THE FOLLOWING ARTICLES

Conservation Biology / Ecological Economics / Ecosystem Ecology / Ecosystem Valuation / Restoration Ecology

FURTHER READING

Barbier, E. B. 2009. Ecosystems as natural assets. *Foundations and Trends in Microeconomics* 4: 611–681.

Boyd, J. W., and H. S. Banzhaf. 2006. What are ecosystem services? the need for standardized environmental accounting units. Discussion paper. Washington, DC: Resources for the Future. http://www.rff.org/rff/Documents/RFF-DP-06-02.pdf.

Costanza, R., R. d'Arge, R. de Groot, S. Farber, M. Grasso, B. Hannon, K. Limburg, S. Naeem, et al. 1997. The value of the world's ecosystem services and natural capital. *Nature* 387: 253–260.

Daily, G., ed. 1997. *Nature's services: societal dependence on natural ecosystems.* Washington, DC: Island Press.

Daily, G., and K. Ellison. 2002. *The new economy of nature: the quest to make conservation profitable.* Washington, DC: Island Press.

Kareiva, P., G. Daily, T. Ricketts, H. Tallis, and S. Polasky, eds. 2011. *The theory and practice of ecosystem service valuation in conservation.* Oxford: Oxford University Press.

Millennium Ecosystem Assessment (MA). 2005. *Ecosystems and human well-being: synthesis.* Washington, DC: Island Press.

National Research Council. 2005. *Valuing ecosystem services: toward better environmental decision-making.* Washington, DC: The National Academies Press.

Nelson, E., G. Mendoza, J. Regetz, S. Polasky, H. Tallis, D. R. Cameron, K. Chan, G. C. Daily, J. Goldstein, P. M. Kareiva, E. Lonsdorf, R. Naidoo, T. H. Ricketts, and M. R. Shaw. 2009. Modeling multiple ecosystem services, biodiversity conservation, commodity production, and tradeoffs at landscape scales. *Frontiers in Ecology and the Environment* 7: 4–11.

The Economics of Ecosystems and Biodiversity (TEEB). 2010. The economics of ecosystems and biodiversity: mainstreaming the economics of nature: a synthesis of the approach, conclusions and recommendations of TEEB. www.teebweb.org.

ECOSYSTEM VALUATION

STEPHEN POLASKY
University of Minnesota, St. Paul

Humans depend on nature for many things, from basic life support (e.g., food provision and climate regulation) to life-enriching experiences (e.g., appreciation of natural beauty, cultural and spiritual values). The contributions of ecosystems and ecosystem processes to human well-being are called ecosystem services. The value of an ecosystem service is a measure of the relative contribution of that service to human well-being. Interest in ecosystem services and how to document their value to people has surged since the publication of the Millennium Ecosystem Assessment (MA) in 2005. A main goal of the MA was to assess how changes in ecosystems would impact the provision of ecosystem services and how this in turn would impact human well-being. This article discusses issues surrounding the valuation of ecosystem service and describes various methods for assessing the value of ecosystem services.

WHAT IS VALUE?

Value is a term with many meanings. In the context of ecosystem services, value typically refers to a measure of the relative contribution of an ecosystem service to human well-being vis-à-vis other things that contribute to human well-being. For example, if a person is better with improved water quality as compared to having higher income, then the water quality improvement has greater value to the person than does the higher income.

Comparing a change in the provision of an ecosystem service with a change in an alternative measured in money, like income, allows the value of the ecosystem service to be expressed in monetary terms. Money is a convenient common metric in which to measure value, just as CO_2 equivalence is a useful metric to measure the relative contribution of different greenhouse gases to climate change.

Assessing value in a common metric allows aggregation of the value of multiple services into a single value that allows for straightforward comparisons of the value of alternative bundles of ecosystem services. Monetary values can also facilitate communication with policymakers and business leaders trained to think in financial terms.

However, it is not always necessary to measure ecosystem services in monetary values. It can be difficult to get an accurate accounting of the monetary value of some ecosystem services, such as water quality improvements or species conservation. Sometimes it is informative to simply report ecosystem services in biophysical units, such as the degree to which alternative policies will meet water quality standards or conserve species.

Some alternative approaches to measurement of ecosystem services have used concepts such as embodied energy, which measures the total amount of energy required to produce the service, or ecological footprint, which measure the total land area required to produce the service, as a common metric. Such approaches differ from standard approaches to valuation of ecosystem services because they do not directly tie to measures of human well-being and will not be discussed further in this article.[1]

Instrumental versus Intrinsic Value

Some environmental philosophers and conservationists feel strongly that the value of nature is much broader than the value of ecosystem services. Under a standard ecosystem services view, nature has value to the extent that it contributes to human well-being. In other words, nature has instrumental value: nature generates value by providing ecosystem services that contribute to the end goal of improving human well-being. But some object that this places too much emphasis on people and ignores the importance of other species and other elements of nature. An alternative view is that nature has intrinsic value (value in and of itself) regardless of whether or not humans benefit. If nature has intrinsic value, then humans have an obligation to conserve the natural world for its own sake even though doing so might require sacrifices in human well-being. The two approaches to the value of nature—intrinsic value of nature versus ecosystem service (instrumental) value—start from different ethical foundations and generate different rationales for conservation. Like discussions of religion, it can be difficult for proponents of different approaches to see eye to eye. Despite this, however, these two approaches to value often yield similar recommendations about conservation policy in practice because actions taken to conserve nature tend to increase the provision of important ecosystem services, and vice versa.

[1]Readers interested in these approaches can find references in U.S. EPA, 2009, p. 51.

Total versus Marginal Value

We all know that we cannot live without water. We can, however, live without diamonds. Yet the price of diamonds is high and the price for water is typically low. This "diamond–water paradox" aptly illustrates the difference between the concepts of total and marginal value. Because water is essential for life, its total value is high (infinite). But since water is plentiful in many places, the contribution of an additional unit of water—its marginal value (price)—is low. In contrast, diamonds are not essential to life, so they generate relatively low total value. But diamonds are rare, so owning a diamond has relatively high marginal value.

An early influential paper on ecosystem services attempted to estimate the total value of important ecosystem services for the entire Earth (Costanza et al., 1997, *Nature* 387: 253–260). The paper attracted both attention and a storm of criticism. Some of the most pointed criticism focused on what it means to estimate the total value of global ecosystem services. Calculating total value of global ecosystem services requires assessing human well-being with and without ecosystem services. However, if ecosystems provide essential life-support services, then the total value of ecosystem services is infinite: no amount of other goods and services could be traded for the existence of ecosystems to make people as well off.

What is a more interesting and relevant calculation than (infinite) total value is a calculation of the change in the value of ecosystem services with conceivable or realistic changes in environmental conditions. Economists often calculate "marginal value," which measures the change in value with a one-unit change in provision of a good or service. The advantage of marginal values is that they correspond to the notion of "price" (the value of one unit of a good of service). Marginal value calculations also do not require considerations of conditions far different from what currently exists, which is a major advantage in empirical efforts to estimate values.

Marginal values for ecosystem services typically change with level of provision. For example, increasing available water by one unit has low value in areas that already have a lot of water but has high value in dry areas. Therefore, one cannot estimate the value of a large (nonmarginal) change in ecosystem service provision by simply multiplying the marginal value by the change in the level of provision. In the case of nonmarginal changes, the value of ecosystem services needs to be estimated both before and after the change in order to estimate the difference in the value caused by the change.

ECONOMIC METHODS FOR ASSESSING THE VALUE OF ECOSYSTEM SERVICES

Much of the research on the value of ecosystem services has used economic valuation methods. This section describes methods consistent with economic theory for measuring the value of ecosystem services.

Foundations of Economic Valuation

The foundation for value in economics rests on individual subjective preferences. Economists assume that individuals have well-defined preference functions (i.e., an individual knows what he or she likes), which allows the person to rank alternatives in terms what is more or less desirable. Economists impose virtually no restrictions on individual preferences other than consistency: if a person prefers A to B, and B to C, then the person must prefer A to C. Saying that A is preferred to B is equivalent to saying that A is more valuable to the individual than B.

The problem with subjective preferences is that they are a product of the internal mental processes of an individual and therefore not readily observable. However, by assuming that individuals behave rationally, meaning that an individual will choose the most preferred alternative from the set of feasible options, economists can infer how individuals value things based on observed choices among alternatives. So, for example, when an individual chooses to purchase an item for a given price, the assumption of rational choice implies that the individual is "willing to pay" at least the given price to obtain the item. Willingness to pay is a monetary measure of the value of the item to the individual.

Critics of economic approaches to valuation object to the assumptions of rational choice and well-defined preference functions. Psychologists have shown a number of cases in which behavior is inconsistent with postulates of rational choice. For example, changing how alternatives are described (framing) can lead individuals to make different choices even though the consequence of each alternative is not altered. Further, psychologists point to evidence that people construct preferences depending on the circumstances rather than having given well-defined preferences that they use in all circumstances. Based on this evidence, behavioral economists and psychologists argue that individual choice can be an unreliable guide on which to base assessments of value. This critique, though, raises the difficult question of who then gets to decide what is in an individual's best interest if it is not the individual. Other critics argue that environmental protection should trump the desires of individuals in order to assure sustainable outcomes, but this, too, raises the difficult question of who gets to decide.

Market Valuation

Many goods and services are exchanged in markets. By observing how much of a good (or service) an individual chooses to buy at different market prices, one can trace out the individual's willingness-to-pay function, also known as the individual's demand curve. For most goods, individuals have declining marginal utility, meaning that the willingness to pay for an additional unit falls with more units. In principle, a rational individual facing a given market price should purchase a good up to the point where the willingness to pay just matches the price. At the margin then, the price just equals the value to the individual. All units of the good prior to the last unit purchased generate surplus value for the individual because the willingness to pay will exceed the price (consumer surplus). Consumer surplus measures the increase in value to the individual from being able to purchase the good at the market price versus not being able to purchase the good at all. Consumer surplus can be a function of the income of the individual, an effect well known in theory but which can complicate estimation of value in practice.

For society as a whole, the value created by production, exchange, and consumption of a good or service is the sum of consumer surplus for all individuals who buy the good or service plus the profit (producer surplus) for all firms who produce and sell the good or service. Profit is defined as revenue (price times quantity sold) minus production cost.

Market prices and quantities of goods and services bought and sold are observable, and in fact, data on prices and quantities are routinely collected in many countries for a large number of marketed goods and services. With sufficient variation in price and quantity purchased, market demand curves can be estimated. Armed with information about demand and production cost, consumer and producer surplus can be estimated, thereby yielding an estimate of the value to society of the availability of the good or service. Similarly, it is also possible to evaluate the change in value cause by changes in conditions that shifts either willingness to pay or production costs.

Some ecosystem services are directly sold in markets, and other ecosystem services contribute to the production of marketed goods and services. Timber and fish are sold commercially. Pollination services increase the productivity of some marketed agricultural crops, and improved nursery habitat can increase the productivity of commercial fisheries. In cases where an increase in the ecosystem service yields an increase in the output of a marketed commodity with no change in production costs, and the change in provision is small relative to the overall size of the market so that price effects are small, the increase in value can be approximated by simply multiplying the market price by the increase in output. However, analysis of the change in value for cases that cause shifts in market prices or production costs require estimating changes in consumer and producer surplus.

When ecosystem services are inputs into the production of marketed goods and services, such as pollination, the major difficulty typically arises from estimating the contribution of the ecosystem service to the production of the marketed commodity, rather than from assessing value once the provision of the service is known. For example, if habitat conditions are improved, how will this translate to increases in fishery productivity? Answering this question can require detailed understanding of population biology and complex food web dynamics. Once the contribution of the ecosystem service to the provision of the marketed commodity is known, however, information about demand and production cost can generate an estimate of value.

Nonmarket Valuation

If all ecosystem services were tied to goods and services that were traded in markets, the task of generating consistent and comprehensive estimates of value expressed in monetary units would be relatively straightforward and uncontroversial. However, most ecosystem services contribute to the provision of public goods for which markets currently do not exist. With no market, there is no market price and often limited data on provision. Over the past half-century, though, environmental economists have worked to extend methods of valuation developed for marketed goods and services to nonmarket goods and services. These nonmarket valuation methods have been applied to estimate the value of a range of environmental attributes. By now, a large number of nonmarket valuation studies have been done. Coverage, though, is quite uneven, with a large number of studies relevant for some ecosystem services and virtually none for others. Nonmarket valuation techniques can be classified into revealed preference methods and stated preference methods.

REVEALED PREFERENCE METHODS

Revealed preference methods use observed behavior for which environmental attributes play some role in influencing the behavior to infer how people value the environmental attributes.

In the hedonic property price approach, data is collected on property values and variables that should play a role in determining property values. Most hedonic studies focus on single-family homes and use structural characteristics (size of house, size of lot, number of rooms, age), neighborhood characteristics (distance to city center, quality of schools, crime rate), as well as environmental characteristics (local air quality, local water quality, proximity to natural amenities) to explain property values. Estimated coefficients from a multiple regression analysis show the effect on property value from a change in a single attribute, holding other attributes constant. Hedonic property price analyses have been used to estimate how much a property owner would be willing to pay for improvements in local air and water quality, views and access to natural areas, and proximity to wetlands, forests, and other habitats. These studies consistently show positive values for improvements in environmental attributes. Hedonic studies have the advantage of being able to utilize large data bases, often consisting of thousands of home sales that link real estate information on housing characteristics with GIS information about environmental characteristics, though this is more true in urban than rural areas.

Random utility and travel-cost studies use information about visits by individuals to recreation sites to infer the value of these sites to these individuals. These studies collect information about the number of visits to each recreation site by each individual, the cost of visiting various sites, and a range of characteristics of each site, including environmental characteristics, which could influence the desirability of visiting a given site. The willingness-to-pay function is estimated by looking at how visits to sites vary with changes in the travel cost of visiting the site. Shifts in the willingness-to-pay functions with changes in environmental quality of a site indicate the value of environmental improvements at the site. Travel-cost studies have been used extensively to study recreational fishing as well as other forms of outdoor recreation.

A third revealed preference method, averting behavior, uses observations on how much people choose to spend to avoid exposure to environmental degradation in order to estimate the value of environmental improvement. For example, expenditures on water purification in response to reduction in local water quality indicate that having improved water quality is valued at least as much as the cost of purchases of water purification.

An advantage of revealed preference methods is they use data based on observable behavior by individuals. With hedonic property price methods, choices should reflect careful thought, as the purchase of a home is typically the largest purchase a person makes. When they can be applied, revealed preference methods generate useful information about value of ecosystem services. However, revealed preference methods cannot be usefully applied to all ecosystem services. For example, these approaches are unlikely to be of much use in trying to estimate the value of carbon sequestration or endangered species conservation. In addition, behavior reflects the information base of those making decisions. When people are not fully informed, their choices may not accurately reflect their values.

STATED PREFERENCE METHODS

Stated preference methods, also called choice experiments, use survey questions to ask individuals about the choices they would make in hypothetical situations. The most common forms of choice experiments in nonmarket valuation are contingent valuation and conjoint analysis. In contingent valuation, individuals are asked whether they would be willing to pay a specified amount in exchange for a specified increase in some ecosystem service. For example, survey respondents might be asked whether they would be willing to pay $40 per year in order to improve water quality from "fishable" to "swimmable" in a given lake. By varying the amount of payment, an analyst can estimate a willingness-to-pay function comparable to a demand curve for a market good.

In conjoint analysis, choices may vary in a number of dimensions, one of which may be price. Conjoint analysis can ask about tradeoffs that respondents are willing to make among different ecosystem services in addition to asking about willingness to pay.

The main advantage of choice experiments, as compared to revealed preference methods, is that they can be designed to gather information about the value of any ecosystem service. Survey methods also have the advantage of asking directly about the value of ecosystem services rather than having to infer value indirectly via property values, recreational trips, or averting behavior. Choice experiments, however, have been criticized because they ask about hypothetical rather than real choices. Critics claim that people do not necessarily respond in the same fashion when asked a question about a hypothetical choice on a survey as they would when faced with a real decision. Critics also point to evidence of the influence of framing on results. Proponents of choice experiments counter that careful design of survey questions can minimize the influence of hypothetical bias and framing.

Summary

Market and nonmarket valuation methods described above have the advantage of being consistent with the mainstream economic theory of value, which provides a logical and coherent framework for values. These methods have been applied in a wide variety of circumstances to generate estimates of the value of ecosystem services. However, the methods may not apply to all ecosystem services, and there are various methodological and data questions that can reduce reliability of estimates. In addition, these methods may not be easily understood or believed by noneconomists, which reduces their usefulness in providing credible and transparent information about value to decision makers and the general public.

COST-BASED METHODS FOR ASSESSING THE VALUE OF ECOSYSTEM SERVICES

Several methods that focus on estimating costs or damages rather than willingness to pay have been used in recent years to estimate the value of ecosystem services. These methods generate valid estimates of value in some circumstances and are generally easier to explain to decision makers and the general public than are the economic methods described above. However, these methods can be misused, and care should be taken in their application.

Avoided Cost

A common approach for estimating the value of regulating ecosystem services is to calculate how much benefit is provided by moderating disturbances or extreme events and thereby reduce damages. For example, maintaining wetlands or coastal mangroves can reduce inland and coastal flooding. Carbon sequestration in ecosystems has value because it reduces atmospheric carbon levels, thereby reducing the likely impacts of future climate change. Damages, however, are a function of both the severity of the event and human behavior in response to risks. For example, flood damages can be reduced by strictly limiting building in flood zones, and damages from future climate change can be reduced by taking actions that increase adaptive capacity. Avoided cost estimates should take proper account of adaptation in human behavior in response to risk.

Replacement Cost

The replacement cost method attempts to answer the question of what it would cost to provide an ecosystem service by alternative means were it not provided by the ecosystem. For example, what would it cost to provide clean drinking water to a city via a human-engineered water filtration plant rather than having filtration provided for free by an intact functioning ecosystem? Perhaps the most widely cited example of the value of an ecosystem service is the value of the Catskills watershed providing clean drinking water for New York City.

While replacement cost is quite similar to averting behavior and is easily explained to decision makers and the general public, some economists have been reluctant to endorse its use. The main objection is that replacement costs is a measure of cost rather than benefit. In order for replacement cost to be a valid measure of the value of an ecosystem service, it must be the case that a human-engineered solution provides equivalent quality/quantity of the service and total willingness to pay for the service exceeds the cost of providing the services via the human-engineered solution.

Marketable Permit Prices

Cap-and-trade schemes have been established in recent years to cost-effectively limit harmful emissions of various pollutants, including SO_2 and NOx emissions in the United States and CO_2 emissions in Europe. Setting up a market and allowing trade of permits to emit establishes a market price for permits. The permit price reflects the marginal cost of reducing emissions and in this regard is similar to replacement cost measures. However, permit prices reflect the stringency of the cap. Only if the cap has been set such that the marginal cost of reducing emissions equals the marginal benefit of doing so will marketable permit prices be valid measures of the value of benefits. Since this is not generally the case, the use of permit prices in estimating the value of ecosystem services is not recommended.

MAJOR OPEN CHALLENGES IN VALUATION

The art and science of the valuation of ecosystem services is still relatively young, and a number of outstanding challenges remain to be addressed before comprehensive, credible, and transparent estimates of the value of ecosystem services will be routinely available. Evidence on some ecosystem services, particularly those tied fairly directly to marketed commodities or readily identifiable costs, or those readily studied using revealed preference methods, is fairly extensive and growing rapidly. Evidence on other ecosystem services, particularly spiritual or cultural services, or those that involve complex interconnections within ecosystems, is still relatively sparse, and existing estimates often have large error bars. Reducing these error bars will require advances in scientific understanding, greater agreement on proper approaches, and a far greater level of application to build experience and a larger library of comparable results.

An important set of unresolved issues revolves around dynamics of ecosystems and human values. Estimates of the value of ecosystem service are typically based on current conditions. Yet actions taken at present may have wide-ranging and long-lasting effects. Thresholds or other nonlinear dynamics can mean the future will be quite dissimilar from the present. How to value ecosystem services under quite different conditions than exist at present is often unclear. In addition, human preferences change through time, and what our descendants will value may not match what the current generation values.

Another important set of unresolved issues is how the value of ecosystem services should be integrated with issues of poverty alleviation and equity. Should increases in ecosystem services that benefit the poor and disadvantaged be given greater weight than increases in services for those that are relatively well off? Most observers think there should be an explicit focus on the distribution of benefits from ecosystem services, though exactly how this should be done is less clear.

SEE ALSO THE FOLLOWING ARTICLES

Discounting in Bioeconomics / Ecological Economics / Ecosystem Services

FURTHER READING

Champ, P., K. Boyle, and T. Brown, eds. 2003. *A primer on nonmarket valuation.* Dordrecht, Netherlands: Kluwer Academic Publishers.

Daily, G.C., ed. 1997. *Nature's services: societal dependence on natural ecosystems.* Washington, DC: Island Press.

Freeman, A.M., III. 1993. *The measurement of environmental and resource values: theory and methods.* Washington, DC: Resources for the Future.

Millennium Ecosystem Assessment. 2005. *Ecosystems and human well-being: synthesis.* Washington, DC: Island Press.

National Research Council. 2005. *Valuing ecosystem services: toward better environmental decision-making.* Washington, DC: The National Academies Press.

The Economics of Ecosystems and Biodiversity (TEEB). 2010. *The economics of ecosystems and biodiversity: ecological and economic foundations.* London: Earthscan.

U.S. Environmental Protection Agency (U.S. EPA), Science Advisory Board. 2009. *Valuing the protection of ecological systems and services.* EPA-SAB-09-012. Washington, DC: US EPA.

ECOTOXICOLOGY

VALERY FORBES AND PETER CALOW
University of Nebraska, Lincoln

Ecotoxicology is concerned with describing, understanding, and predicting the effects of chemicals used by people, either from natural sources (such as metals) or from synthetic processes (agrochemicals, pharmaceuticals, industrial chemicals), on ecological systems. It provides the basis for ecological risk assessment, predicting likely impacts of chemicals on ecological systems, and is an important contributor to environmental protection legislation.

THE CONCEPT

First defined explicitly by Truhaut in the 1970s, ecotoxicology was specified somewhat broadly as "the study of toxic effects, caused by natural or synthetic pollutants, to the constituents of ecosystems, animal (including human), vegetable and microbial, in an integral context." But the field soon became more focused and sought to study the effects of chemicals on individuals and their systems (biochemistry, physiology, cells) in a few widely tested species. More recently, attention has returned to broader ecological concerns, with a realization that individual-level responses may not translate straightforwardly into population and ecosystem effects. Homeostatic responses might dampen individual responses, or indirect effects (i.e., loss of predators, prey, or competitors through toxic effects leading to changes in interspecies interactions) might serve to magnify them for populations and ecosystems. Thus, extrapolation—e.g., from test species to nontested species, from toxicological effects to ecological impacts, from laboratory to field—is an important element of ecotoxicology.

Organisms can be exposed to chemicals in the environment through various routes, namely, water, air, sediment/soil, and food. An important challenge is to relate chemical concentrations in these environmental compartments to a toxicologically relevant dose. For practical reasons, ecotoxicological tests are most often carried out at high concentrations over short periods of time (acute exposures), and yet chemicals in nature often occur at low concentrations, with organisms exposed for long periods of time (chronic exposures). It is assumed that toxicological responses to acute, high-concentration exposures give some insight into responses to chronic, low-concentration exposures.

At the toxicological level, chemicals can have different modes of action. Some may target very specific metabolic sites or processes (e.g., insecticides that specifically inhibit chitin synthesis and hence molting in arthropods), whereas others may exhibit nonspecific modes of action (e.g., narcotics that impair overall membrane function). Regardless of toxicological mode of action, the effects of chemicals at the ecological level only become relevant when they impact the structure and dynamics of populations, mediated through effects on survival, reproduction, and growth.

Ideally, chemical effects ought to be studied on either natural populations/ecosystems or isolated subsets of them (micro- and mesocosms). But generally, such studies are neither practicable nor desirable. Yet because society obtains great benefits from chemicals in terms of agricultural production, medicine, and general well-being and lifestyle, the pressures to assess chemicals for risks to human health and the environment are enormous. One common tendency, therefore, has been to base risk assessments on worst-case responses from ecotoxicology and to use large uncertainty factors to add a margin of safety to observed effects. In this context, chemicals that are likely to bioaccumulate to high levels or persist for long times in the environment, as well as those that have a high toxicity, are prioritized for management. However, this might lead to overly conservative assessments and hence increased costs to producers and users without commensurate benefits.

THE THEORY

Ecotoxicology is an applied science. It has no definitive theory as such, but instead it relies on theory from environmental chemistry, toxicology, and ecology to develop an understanding of the fate and effects of chemicals in ecological systems.

Theory from environmental chemistry provides a basis for using chemical structure and properties to predict the fate of chemicals in the environment and their uptake by organisms. For example, quantitative structure activity relationships (QSARs) aim to predict chemical behavior from structural attributes. Other models use physico-chemical characteristics to make predictions about reaction products (e.g., chemical speciation) and environmental distribution.

Historically, the most important theoretical concept in toxicology is that the effects of chemicals depend on dose and duration of exposure. There is a substantial body of theory on dose-response and time-response relationships that seeks to provide mechanistic understanding and analytical tools. Typically, the dose-response curves for chemicals are monotonic (continuously increasing or decreasing with increasing dose). However, there are cases for which the dose-response may show a nonmonotonic pattern. For example, some chemicals that act as if they were hormones in organisms can have more effect at low concentration than at high concentration. Similarly, some chemicals seem to show stimulatory effects at low doses but then increasingly adverse effects as dose increases. This is known as hormesis.

Another important issue arising from toxicological theory is how chemicals interact in mixtures to have their effects on organisms. Given that organisms in natural ecosystems are exposed to a complex soup of chemicals from natural and synthetic sources, it is important to have an understanding of possible additive, antagonistic (less-than-additive), or synergistic (more-than-additive) interactions among chemicals. Toxicological models exist to distinguish among these different types of interactions, but mixture toxicity continues to pose challenges for ecotoxicology.

There is now considerable ecological theory that relates individual responses to environmental factors to population structure and dynamics and changes in population dynamics to ecosystem structure and processes. This is increasingly being used to address the extrapolation challenges for ecotoxicology.

At the individual-to-population interface, there are three broad classes of relevant models. The first class, demographic models, describes individuals in terms of their contribution to recruitment and their survivorship. All individuals may be treated as identical (unstructured), or they can be separated into different size or age classes (structured). Spatial structure may also be included, as in metapopulation models, and stochasticity can also be included to represent both demographic and environmental characteristics. The second class of models, energy budget models, are similar to the first in treating all individuals as the same, but they represent the responses of individuals in terms of intakes and outputs of energy that relate to individual growth performance and reproductive output. A third class of models, individual-based models (IBMs), represent each individual in a population as being distinct and describe individual responses with more or less detail.

Species sensitivity distributions aim to bring the effects on different species populations together to predict likely ecosystem effects. They make three key presumptions that are not necessarily valid: (1) that the species included in the distribution are representative of the ecosystem to be protected; (2) that toxicological sensitivity is predictive of ecological sensitivity; and (3) that protecting structure (i.e., species composition) protects ecological function. Ecological theory indicates that the relationship between species composition and ecosystem stability is complex and that some species may be more important than others in both maintaining the structure and the processes within ecosystems. These features have yet to be incorporated into ecotoxicological work except insofar as the precautionary approach of attempting to protect all through protecting the most (toxicologically) sensitive species is applied.

Trophic transfer models and food web models have been used to chart the movement of chemicals through ecological communities, and hence the possibility of bioaccumulation and biomagnification. Food web models have also been useful in demonstrating indirect effects and their consequences for ecosystem structure and processes.

Finally, evolutionary theory predicts that toxic chemicals will act as selection pressures. Classical examples include resistance of pest organisms to pesticides and the evolution of heavy metal–tolerant plants growing in soils contaminated with mine waste. The implication of this adaptation is that some species may become less sensitive to chemicals over time. Another possibility, however, is that selection for tolerance to chemicals may reduce genetic variability and thus the capacity for populations to respond to other forms of stress, whether artificial or natural.

CONCLUSIONS AND FUTURE DEVELOPMENTS

Theoretical models have an important part to play in ecotoxicology, because it is extremely difficult to predict the likely effects of chemicals in complex ecological systems simply on the basis of observation. Ecological models that incorporate mechanistic understanding of the appropriate processes and linkages are particularly useful. An important issue to be addressed in this context is how much complexity needs to be incorporated in models to make them suitably predictive for their application in risk assessment and environmental protection. More generally, effort needs to be put into developing methods for assessing how chemicals impact ecosystem processes that have consequences for valued ecosystem services. Finally, there is as yet little understanding of how evolutionary adaptation should be addressed in risk assessment, and this needs further study.

SEE ALSO THE FOLLOWING ARTICLES

Ecosystem Ecology / Food Webs / Individual-Based Ecology / Metapopulations / Population Ecology / Stochasticity, Demographic / Stochasticity, Environmental

FURTHER READING

Baird, D. J., L. Maltby, P. W. Greig-Smith, and P. E. T. Douben, eds. 1996. *Ecotoxicology: ecological dimensions.* London: Chapman and Hall.

Forbes, V. E., ed. 1999. *Genetics and ecotoxicology.* Philadelphia: Taylor and Francis.

Forbes, V. E., P. Calow, and R. M. Sibly. 2008. The extrapolation problem and how population modeling can help. *Environmental Toxicology and Chemistry* 27: 1987–1994.

Greim, H., and R. Snyder, eds. 2008. *Introduction to toxicology.* Chichester, UK: Wiley.

Luoma, S. N., and P. S. Rainbow. 2008. *Metal contamination in aquatic environments.* Cambridge, UK: Cambridge University Press.

Newman, M. C., ed. 2010. *Fundamentals of ecotoxicology*, 3rd ed. Boca Raton, FL: CRC Press.

Truhaut, R. 1977. *Ecotoxicology: objectives, principles and perspectives. Ecotoxicology and Environmental Safety* 1: 151–173.

Van Leeuwen, C. J., and T. G. Vermeire. 2007. *Risk assessment of chemicals: an introduction*, 2nd ed. Dordrecht, NLD: Springer.

ENERGY BUDGETS

S. A. L. M. KOOIJMAN

Vrije University, Amsterdam, The Netherlands

An energy budget specifies the uptake of energy from the environment by an organism through feeding and digestion and the use of this energy for maintenance, development, growth, and reproduction. A static energy budget represents a kind of snapshot of these fluxes for an individual in a given state, while a dynamic energy budget follows the changes of these fluxes during the life cycle of an organism.

WHY ARE ENERGY BUDGETS IMPORTANT?

Organisms typically grow during their life cycle, and food uptake and maintenance are coupled to the size of the organism. Ultimate size of organisms (i.e., when growth ceases) is controlled by the balance between uptake and utilization of energy due to maintenance and reproduction. Just after the start of maturation, during the embryo stage, organisms typically do not take up food. Food uptake is initiated at a moment called birth (the onset of the juvenile stage). Maturation, a form of metabolic learning, ceases at puberty (the onset of the adult stage), after which energy (metabolite) allocation to reproduction starts. Although not yet very detailed, this natural sequence already structures the underlying processes profoundly. Box 1 summarizes stylized empirical patterns in energy budgets.

Quite a few processes underlying metabolism are tightly interlinked and can only be studied simultaneously, exploiting the law of conservation of energy. Energy (the capacity to do work) is, however, only one aspect; mass, where each of many chemical compounds has its own properties, is another aspect important to understanding what organisms do and can do in terms of feeding and production of offspring and products (such as feces). Energy and mass aspects cannot be separated. Mass aspects are substantially more complex to understand since chemical compounds can be transformed (there are no conservation laws for compounds) and the body can change in chemical composition in response to changes in its nutritional status. We can, however, use conservation laws for chemical elements, (C, H, O, and N being the most important ones) and use a variety of homeostasis concepts to capture what organisms do in terms of mass uptake and use.

Mass aspects impinge various taxa differently. Animals, i.e., organisms that feed on other organisms, acquire the various compounds they need in approximately fixed relative proportions, but bacteria, algae, and plants take up nutrients (e.g., nitrate, phosphate, carbon dioxide) and light from the environment almost independently. However, the various uptake routes are coupled due to the fact that biomass varies in composition in a limited range only; this gives rather complex stoichiometric constraints on uptake. The ecological literature about stoichiometric constraints on production typically simplifies this and follows chemical elements in particular substrates, e.g., nutrients, and considers biomass to be of constant composition.

Most use of energy can only be inferred indirectly from observations of comparisons of performances between individuals of different size and under different feeding conditions. An example is the allocation of energy (or resources) to growth compared to the fixation of these resources in new tissue. The difference is in the overhead costs of growth, which are notoriously difficult to quantify in static energy budgets. Another example is the allocation of energy to maintenance, i.e., the collection of processes using energy that are not directly linked to production of biomass (growth, reproduction). Static energy budgets typically use respiration (the use of oxygen or the production of carbon dioxide or heat) as a quantifier for maintenance. This is overly simplistic, and dynamic energy budgets include, besides maintenance, maturation and overhead costs of assimilation, growth, and reproduction in respiration. Although static energy budget approaches are still popular, this entry focuses only on dynamic energy budgets.

So far, the emphasis has been on individuals because they are the survival machines of life and the target for natural selection. The rules individuals use for the uptake and use of substrates have profound consequences for suborganismal organization of metabolism. Many species are unicellular, so the step to subcellular organization is rather direct. Multicellular species have organization layers between that of the individual and the cell, and allocation to the various tissues and organs has functional aspects.

At the supraorganismal level, populations are groups of interacting individuals of the same species. Apart from other within-population interactions, including food competition, transport of individuals and substrates (or food) through the environment dominates population dynamics. In most situations, seasonal forcing has to be considered, since factors such as temperature and water affect substrate availability and uptake.

At the planetary level, all oxygen in the atmosphere is of biological origin, including the ozone protection shield for UV radiation, and most important greenhouse gases, atmospheric water, and the minor ones (carbon dioxide and methane) are strongly influenced by biota. So the global climate results from complex long-term interactions between biota and the physicochemical environment.

Typical human interests (medicine, agriculture and aquaculture, forestry, sewage treatment, biotechnology) also make intensive use of particular aspects of energy budgets. These considerations illustrate that energy budgets are key to biology and its societal applications.

PRINCIPLES OF DEB THEORY

Dynamic energy budget (DEB) theory is a specific theory that explains each of the patterns listed in Box 1, exploiting the features that all organisms—microorganisms, animals, and plants—have in common through a rather abstract perspective. DEB theory is a set of explicit coherent and consistent assumptions (axioms or hypotheses) that, in combination, specifies a set of models. A modular set-up is exploited, where modules for particular details are only incorporated if the research question requires it. The standard DEB model, in a sense the simplest one in the DEB family, deals with isomorphs, i.e., organisms that don't change shape during growth, with one reserve and one structure feeding on one type

BOX 1. STYLIZED AND EMPIRICAL FACTS

Feeding

- Many species (almost all animals and plants) have an embryo stage that does not feed.
- During starvation, organisms are able to reproduce, grow, and survive for some time.
- At abundant food, the feeding rate is at some maximum, independent of food density.

Growth

- Many species continue to grow after reproduction has started.
- Growth of isomorphic organisms at abundant food is well described by the von Bertalanffy growth curve.
- For different constant food levels, the inverse von Bertalanffy growth rate increases linearly with ultimate length.
- The von Bertalanffy growth rate of different species decreases almost linearly with the maximum body length.
- Fetuses increase in weight is approximately proportional to cubed time.

Reproduction

- Many species (almost all animals and plants) have a juvenile stage that does not reproduce.
- Reproduction increases with size intraspecifically but decreases with size interspecifically.
- A range of constant low-food levels exists at which an individual can survive but not reproduce.

Respiration

- Animal eggs and plant seeds initially hardly use dioxygen.
- The use of dioxygen increases with decreasing mass in embryos and increases with mass in juveniles and adults.
- The use of dioxygen scales approximately with body weight raised to a power close to 0.75.
- Animals show a transient increase in metabolic rate after ingesting food (heat increment of feeding).

Stoichiometry

- The chemical composition of organisms depends on the nutritional status (starved vs. well fed).
- The chemical composition of organisms at constant food density becomes constant during growth.

Energy

- Dissipating heat is a weighted sum of three mass flows: carbon dioxide, dioxygen, and nitrogenous waste.

Aging

- Mean life span typically increases interspecifically with maximum body length in endotherms, but hardly depends on body length in ectotherms.

of food. It applies to animals (i.e., organisms that feed on other organisms). Microalgae need several reserves, and plants also need at least two structures (root and shoot).

Homeostasis Is Key to Life

DEB theory uses five homeostasis concepts to capture what organisms do:

1. *Strong homeostasis* is the strict constancy of the chemical composition of pools of compounds within an organism. This implies stoichiometric constraints on the synthesis of generalized compounds, i.e., mixtures of compounds that do not change in composition. By delineating more and more pools, strong homeostasis becomes less restrictive.
2. *Weak homeostasis* is the constancy of the chemical composition of the individual as a whole as long as substrate availability in the environment remains constant, even when growth continues. This implies constraints on the dynamics of the pools. Weak homeostasis in fact implies strong homeostasis.
3. *Structural homeostasis* is the constancy of the shape of the individual during growth. This implies that surface area is proportional to volume to the power 2/3, a condition referred to as isomorphy (or V2/3-morphy). Isomorphy is assumed in the standard DEB model, but not in DEB models generally.
4. *Thermal homeostasis* is the constancy of body temperature. Endotherms oxidize compounds for heating. Mammals and birds do it "perfectly"; tunas and insects much less so. Homeotherms do not do this, but they make use of spatial differences in temperature to reduce variations in body temperature. Ectotherms (by far the majority of species) have a body temperature (almost) equal to the environmental temperature.
5. *Acquisition homeostasis* is the constancy of the feeding rate, independent of food availability. This is, to some extent, correct for animals near the demand end of the supply–demand spectrum at which organisms can be ranked. Most organisms are near the supply end (see Table 1). Demand systems evolved from supply systems, and developed several adaptations for this while preserving many other properties of supply systems.

TABLE 1

Comparison between supply and demand systems

Supply	Demand
Eat what is available	Eat what is needed
Can handle large range of intake	Can handle small range of intake
Reserve density varies wildly	Reserve density varies little
Low peak metabolic rate	High peak metabolic rate
Open circulatory system	Closed circulatory system
Rather passive, simple behavior	Rather active, complex behavior
Sensors less developed	Sensors well developed
Typically ectothermic	Typically endothermic
Evolutionary original	Evolved from supply systems
Has demand components (maintenance)	Has supply components (max. size depends on food quality)

Reserve Mobilization Drives Metabolism

Six compelling arguments urge us to partition biomass into two compartments: reserve and structure.

1. To include metabolic memory. Think, for instance, of embryos, which develop, grow, and maintain themselves without feeding, or of mother baleen whales, which feed only for half a year and then travel 25,000 km while giving birth to a big calf in tropical waters and feed it daily with 600 liters of milk for several months before resuming feeding in Arctic/Antarctic waters.
2. To smooth out fluctuations in resource availability to make sure that no essential type of resource (nutrient) is temporarily absent.
3. To allow the chemical composition of the individual to depend on the growth rate.
4. To understand why mass fluxes are linear sums of three basic energy fluxes: assimilation, dissipation, and growth (which is basic to indirect calorimetry).
5. To explain observed patterns in respiration and in body-size scaling relationships.
6. To understand how the cell decides on the use of a particular (organic) substrate, as building block or as source of energy.

Most of these arguments arise from Box 1. The reason for being so detailed is because this complicates the theory and its application quite a bit, so there is a need for a careful cost–benefit analysis in composing the theory. The difference between reserve and structure is in their dynamics; only structure requires maintenance, while reserve is synthesized from substrates taken from the environment and used for metabolic purposes. A substantial part of maintenance

BOX 2. THE ASSUMPTIONS THAT SPECIFY THE STANDARD DEB MODEL QUANTITATIVELY

1. The amounts of reserve, structure, and maturity are the primary state variables of the individual; reserve and structure have a constant composition (strong homeostasis), and maturity represents information.
2. Substrate (food) uptake is initiated (birth), and allocation to maturity is redirected to reproduction (puberty) if maturity reaches certain threshold values.
3. Food is converted into reserve and reserve is mobilized at a rate that depends on the state variables only to fuel all other metabolic processes.
4. The embryonic stage initially has a negligibly small amount of structure and maturity (but a substantial amount of reserve). The reserve density at birth equals that of the mother at egg formation (maternal effect). Fetuses develop in the same way as embryos in eggs but at a rate unrestricted by reserve availability.
5. The feeding rate is proportional to the surface area of the individual and the food-handling time is independent of food density.
6. The reserve density *at constant food density* does not depend on the amount of structure (weak homeostasis).
7. Somatic maintenance is proportional to structural volume, but some components (osmosis in aquatic organisms, heating in endotherms) are proportional to structural surface area.
8. Maturity maintenance is proportional to the level of maturity.
9. A fixed fraction of mobilized reserves is allocated to somatic maintenance plus growth, the rest is allocated to maturity maintenance plus maturation or reproduction (the κ-rule).
10. The individual does not change in shape during growth (isomorphism). This assumption applies to the standard DEB model only.
11. Damage-inducing compounds (modified nuclear and mitochondrial DNA) are generated at a rate that is proportional to the reserve mobilization rate; damage-inducing compounds induce themselves at a rate that is proportional to the mobilization rate. Damage-inducing compounds generate damage compounds ("wrong" proteins) at constant rate, which accumulate in the body. The hazard rate is proportional to the density of damage compounds.

relates to the turnover of structure, so compounds in both reserve and structure have a limited life span.

The mobilization of reserve, within the context of DEB theory, is completely quantified by the requirement of weak homeostasis, but the derivation is rather technical. This explains why the assumptions for the standard DEB model (see Box 2) seem to ignore reserve dynamics.

Development and Allocation

Metabolic learning during ontogeny is quantified by the state of maturity—more specifically, by the cumulative investment of reserve in maturity; one can think of the impact of (e.g., hormonal) regulation systems. Maturity does not represent mass, energy, or entropy; it has the formal status of information. Metabolic switches occur when maturity reaches a threshold, e.g., at cell division, birth, puberty, or metamorphosis (in some species). This explains why the age at which these switches occur varies, and also why the body length varies (somewhat) with nutritional conditions. Although body length at stage transitions is typically rather constant, the observation that some taxa (e.g., most bird species) start reproduction only after body weight no longer changes for some time illustrates that state transitions cannot be linked to length.

Allocation is the set of decision rules for the use of mobilized reserve. The simplest situation, capturing quite a few general aspects of growth and development, is that a fixed fraction κ of mobilized reserve is allocated to somatic maintenance and growth (together called soma), and the remaining fraction to maturity maintenance plus maturation or reproduction. Both types of maintenance take priority (demand organization), rendering growth and maturation (or reproduction) into a supply organization. Somatic maintenance comprises the turnover of structure, movement and other forms of behavior, osmotic work (in freshwater), and heating (in endotherms). Maturity maintenance comprises the maintenance of regulation systems and defense (e.g., the immune system). The static generalization of this κ-rule further partitions the allocation to the soma into allocation to body parts (e.g., organs), where each body part receives a fixed fraction of mobilized reserve and the maintenance of that part takes priority over growth. The dynamic generalization releases these fractions and links them to the relative workload of that body part, i.e., its work as a fraction of the maximum work a body part of that size can do. This requires a specification of the relationship between organ size and organ function.

Generalizations of the κ-rule are basic to the concept of heterochrony. They explain, for instance, why relative brain size decreases during ontogeny of animals in view of rapid development in the early stages and why (moderate) use of alcohol leads to enlargement of livers in humans. Another application is in understanding how tumor growth depends on the state of the host (i.e., amounts of reserve and structure) and its response to caloric restriction.

Surface Area–Volume Relationships

Transport, such as food uptake, in three-dimensional space (volume) occurs across a two-dimensional space (surface). Not all of the surface of an individual needs to be involved; it can be a certain fraction (e.g., that of the gut). Since maintenance is linked to (structural) volume, surface area to volume relationships control growth and reproduction.

The simplest situation of no change in shape during growth (structural homeostasis), called isomorphy, holds approximately for most animals. Apart from isomorphs, two special cases of changing shapes repeatedly arise in applications of DEB theory:

(i) V0-morphs, where surface area is proportional to structural volume to the power 0, so it remains constant (biofilms and organisms that increase their structure at the expense of their vacuoles are examples).

(ii) V1-morphs, where surface area is proportional to structural volume to the power 1. Growing filaments and sheets are examples.

Many other cases can be seen as static or dynamic mixtures of these three basic types; rods (most bacteria) are static mixtures of V0- and V1-morphs, plants naturally evolve from V1-morphs, via isomorphs, to V0-morphs during their life cycle, and crusts (e.g., lichens or superindividuals such as forests) from V1- to V0-morphs such that their diameter grows linearly in time at constant substrate. Think also of a population of muskrats, for instance, conceived as a superindividual, that spreads over Europe from individuals released in central Europe. The front of such a population moves at a constant rate for exactly the same reasons the edge of a forest or a lichen in homogeneous space moves at constant rates: it is the scaling of the surface area for the uptake of resource, relative to the volume that requires maintenance.

V1-morphs have the unique feature that the significance of the levels of the individual and the population completely merge; a population of many small V1-morphs behaves identically to that of a few big V1-morphs with equal total structure and reserve. V1-morphs also have no size control as an individual (if they would not reset their size by division); they continue to grow exponentially as long as substrate density remains constant. This argument can also be reversed: if we want to understand population characteristics (such as the maximum specific growth rate) in terms of properties of individuals (such as size at division), we cannot consider them as V1-morphs. The population dynamics of V1-morphs is so much simpler than that of other morphs that it remains attractive to make this simplification. This can, for dividing organisms, be defended mathematically as being a good approximation in quite a few situations.

The scaling of surface area to volume dominates the rate of living processes at all levels of organization. For instance, the production in lakes is typically nutrient limited, and the acquisition of nutrients is via inflowing water. The amounts of water and nutrients are directly linked to the water catchment area of the lake (so a surface), while its effect on (algal) growth is via the concentration of nutrients, which involves the volume of the lake. At the subcellular level, membranes (surfaces) dominate metabolism, while substrate and product concentrations involve cytoplasm volume.

Synthesizing Units and Chemical Transformation Rates

Spatial structure, especially within organisms and cells, complicates the application of the concept of concentration, which implies homogeneous mixing at the molecular level. DEB theory avoids the use of the concept to quantify metabolism using the dynamics of synthesizing units (SUs) at several crucial places. SUs can be conceived as generalized enzymes that generally follow the rules of enzyme kinetics, with two far-reaching modifications: (i) their activity depends on arrival rates of substrates, rather than concentrations of substrates, and (ii) the dissociation of substrate from the SU-substrate complex is assumed to be small (all bound substrate is transformed to products). If substrates are in a homogeneous environment, the arrival rate of substrates can be taken proportional to the substrate concentration on the basis of diffusive transport.

In the simplest situation, as in transformation of one substrate into products, SU- and enzyme-dynamics have the same result. Think, for example, of a feeding individual conceived as an SU that has only two behavioral states: searching for and handling food. Then the Holling type II functional response results, and this has the same mechanistic background as Monod's model for microbial growth, and Michealis–Menten's model for enzyme kinetics. The modeling of fluxes has, however, much larger flexibility, especially if spatial structure matters, and it combines nicely with the concepts of allocation, the partitioning of fluxes. Compounds can be classified as substitutable or complementary, binding to SUs as sequential or parallel; this gives four basic classes from which more complex forms are derived, such as inhibition and co-metabolism.

DEB theory uses SU dynamics in the assimilation module, in the mobilization of reserve and in growth, i.e, the conversion of reserve to structure. SU dynamics is also used if multiple substitutable reserves are present, such as carbohydrates and proteins for maintenance. It can be shown that active excretion of mobilized reserve that is rejected by SUs for growth in multiple reserve systems is unavoidable, with important ecological consequences (e.g., excretion of toxins by algae in eutrophic waters). In its simplest form, DEB theory separates metabolic transformations into three macrochemical reaction equations: (i) assimilation (the conversion of environmental substrate(s) to reserve and products), (ii) dissipation (the conversion of reserve into products) and (iii) growth (the conversion of reserve into structure and products). Substrate, reserve, structure, and products are conceived as generalized compounds; the latter are typically released into the environment and include heat. This explains the success of the method of indirect calorimetry (cf. Box 1), which quantifies heat as a weighted sum of the fluxes of oxygen, carbon dioxide, and nitrogen waste (e.g., ammonia). Since products serve as substrate for other organisms, these three processes, and their coupling, are of fundamental significance for ecosystem dynamics.

When the log of any metabolic rate is plotted against the inverse absolute temperature, a straight line results across a species-specific tolerance range of temperatures; the slope is called the Arrhenius temperature. This Arrhenius relationship can be understood from fundamental principles under simple very idealized conditions, remote from the situation in living organisms. DEB theory treats this relationship empirically only. Outside the temperature-tolerance range, rates are typically lower than expected on the basis of the Arrhenius relationship. At the high-temperature end, the rates are typically a lot less and the individual dies. At the low-temperature end, the individual may send itself into a state of torpor. This situation typically occurs during the bleak season, where substrate availability is low. This deviating behavior can be captured by delineating temperature-dependent transitions of enzymes from an active state and two inactive states (relating low and high temperatures); these transitions again follow the Arrhenius relationship.

Since substrate uptake affects substrate availability, and the Arrhenius temperature is species specific, temperature can have complex effects. Ultimate size (i.e., a state) relates to the ratio of two rates—uptake (food) and utilization (maintenance)—are affected by temperature; food uptake can affect food availability.

If more than one reserve is present, the corresponding assimilation rates might differ in the way they depend on temperature. Therefore, these systems are more flexible than the single-reserve systems. Photon capture hardly depends on temperature, for instance, which implies that carbohydrate content becomes temperature dependent. Algae in Arctic/Antarctic waters have much more starch and/or lipids than those in tropical waters, with consequences for those who feed on them.

STANDARD DEB MODEL

The logical links between substrate, reserve, structure, and maturity in the standard DEB model are given in Figure 1, and the assumptions in Box 2 quantify all fluxes in this figure uniquely and how they change during the life cycle of the individual. Table 2 gives an overview of its primary parameters (excluding the aging module). The 14 parameters fully quantify the 7 processes of feeding, digestion, maintenance, growth, maturation, reproduction, and aging during the full life cycle of the individual. Thus, there are two parameters per process, illustrating the simplicity of this model. Efficiency of fecal production y_{PX} (or the equivalent yield of feces on food y_{PX}) could be added to quantify feces production and the fluxes of

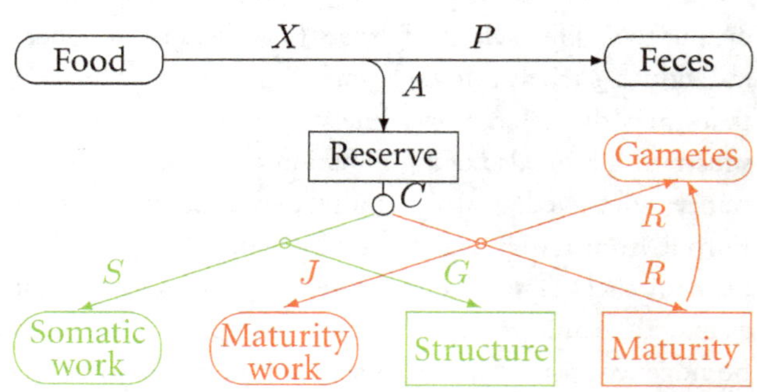

FIGURE 1 Energy fluxes in the standard DEB model. The rounded boxes indicate sources or sinks. Symbols: X, food intake; P, defecation; A, assimilation; C, mobilization; S, somatic maintenance; J, maturity maintenance; G, growth; R, reproduction. The colors illustrate that fluxes G plus S comprise a fixed fraction κ of flux C in the model: the κ-rule for allocation.

TABLE 2

The primary parameters of the standard DEB model in a time-length-energy and a time-length-(dry) mass frame and typical values among species at 20 °C with maximum length $L_m = z\,L_{ref}$ for a dimensionless zoom factor z and $L_{ref} = 1$ cm

Specific searching rate	$\{F_m\}$	$6.51\ cm^{-2}\ d^{-1}$	$\{F_m\}$	$6.51\ cm^{-2}\ d^{-1}$
Assimilation efficiency	κ_X	0.8	y_{EX}	$0.8\ mol\ mol^{-1}$
Max. spec. assimilation rate	$\{p_{Am}\}$	$22.5\ z\ J\ cm^{-2}\ d^{-1}$	$\{J_{EAm}\}$	$0.041\ z\ mmol\ cm^{-2}\ d^{-1}$
Energy conductance	v	$0.02\ cm^{-1}\ d^{-1}$	v	$0.02\ cm^{-1}\ d^{-1}$
Allocation fraction to soma	κ	0.8	κ	0.8
Reproduction efficiency	κ_R	0.95	κ_R	0.95
Volume-spec. som. maint. cost	$[p_M]$	$18\ J\ cm^{-3}\ d^{-1}$	$[J_{EM}]$	$0.033\ mmol\ cm^{-3}\ d^{-1}$
Surface-spec. som. maint. cost	$\{p_T\}$	$0\ J\ cm^{-2}\ d^{-1}$	$\{J_{ET}\}$	$0\ mol\ cm^{-2}\ d^{-1}$
Maturity maint. rate coeff.	k_J	$0.002\ d^{-1}$	k_J	$0.002\ d^{-1}$
Specific costs for structure	$[E_G]$	$2800\ J\ cm^{-3}$	y_{VE}	$0.8\ mol\ mol^{-1}$
Weibull aging acceleration	h_a	$10^{-3}\ z\ d^{-2}$	h_a	$10^{-3}\ z\ d^{-2}$
Gompertz stress coefficient	s_G	0	s_G	0
Maturity at birth	$E_H{}^b$	$275\ z^3\ mJ$	$M_H{}^b$	$500\ z^3\ nmol$
Maturity at puberty	$E_H{}^p$	$166\ z^3\ J$	$M_H{}^p$	$0.3\ z^3\ mmol$

NOTE: The two frames relate to each other via the chemical potential for reserve $\mu_E = 550$ kJ/mol and the volume-specific mass of structure $[M_V] = 4\ mmol/cm^3$. The typical value for the Arrhenius temperature $T_A = 8$ kK.

oxygen and carbon dioxide in association with assimilation. Some parameters vary widely among species (e.g., $E_H{}^b$, $E_H{}^p$, h_a), others much less (e.g., v). The structure of the model could be tested using the effects of toxicants. If the (internal) concentration is sufficiently low, a toxicant affects a single parameter, with particular consequences for energetics. Reactions to perturbations such as acute toxicant input provide strong support for the general structure of the model.

The allocation to reproduction is first accumulated in a reproduction buffer; species-specific buffer-handling rules convert the content of that buffer into gametes. Some species produce offspring as soon as they accumulate enough reserve, and others spawn once a year only. These handling rules, together with the feeding rate, involve the behavioral repertoire, which is notoriously stochastic. Stochasticity has far-reaching consequences for population dynamics.

The methodology of energy budgets is closely related to that of biophysical ecology, a subdiscipline that deals with the details of thermal aspects. Body temperature controls metabolic rates; this temperature is not given but follows from a number of factors. Apart from environmental temperature (which is typically spatially inhomogeneous), metabolic heat (as side-product of metabolic activity) and, for terrestrial organisms, irradiation and water evaporation are elements to consider. Extensions of the standard DEB model that incorporate more detail on nutritional requirements can make the bridge to the geometric framework of nutrition, allowing a detailed quantification of the niche concept. This framework deals with the details of the nutritional needs of organisms and how these change during the life cycle. The needs of young, fast-growing individuals, for instance, differ from those of fully-grown ones, which translates to changes in food preference.

Covariation of Parameter Values and *r/K* Strategies

DEB theory captures the differences between organisms through individual-specific parameter values, allowing for evolutionary responses since parameter values are partly under genetic control. Adaptations involve changes in parameter values. Think, for example, of geographical size variations as adaptations to food availability in the growing season. The Bergmann rule states that the maximum body size of species tends to increase from the equator toward the poles. Toward the poles, seasonality becomes more important, as the harsh season thins populations and so reduces competition during the breeding season. Predictable food levels can to some extent be fixed in parameter values within one species in ways that are more clearly demonstrated interspecifically. The main difference between intraspecific and interspecific parameter variations is the amount of variation.

Notice that structure and reserve are state variables, not parameters, and they vary during the life of an individual, even if its parameters remain constant. Although we can compare an old (large) individual of a small-bodied species with a young (small) individual of a bigger-bodied species, the result can be complex. Even if they are the same size (i.e., length or weight), their metabolisms will differ. For simplicity's sake, we confine interspecies comparisons to fully-grown individuals.

A powerful property of the standard DEB model is that its structure allows us to predict the covariation of parameter values across species without using any empirical argument or new assumption. This is due to three properties:

1. The parameters can be classified into two classes: intensive parameters that only depend on the very local physicochemical suborganismal conditions, and design parameters that depend on the size of the individual (see Table 2).
2. Simple functions of design parameters (typically ratios) are intensive.
3. Maximum length is a function of parameters, of which only one is a design parameter.

The covariation of parameter values is a tendency that is based on physicochemical principles. Species-specific deviations from the mean pattern refer to species-specific adaptations. The better we can characterize this mean pattern, the more we can appreciate the deviations from it and recognize what properties make a particular species special.

Only four parameters in the list in Table 2 depend on size: the surface area–specific assimilation rate, the Weibull aging acceleration, and the maturity levels at birth and puberty. The specific assimilation rate must be proportional to maximum length, because both other parameters that control it—the allocation fraction to soma and the specific somatic maintenance costs—are intensive parameters. The Weibull ageing acceleration is proportional to maximum length, because it is proportional to the specific assimilation rate. Maturity density is intensive, which is easiest to see when the maturity and somatic maintenance rate coefficients are equal, $k_j = [p_M]/[E_G]$, and the maturity density remains constant during growth, so it must be independent of size. Maturity thresholds must, therefore, be proportional to maximum length cubed.

The application of these simple relationships is to express a physiological quantity of interest, such as body weight or respiration rate, as a function of the primary parameters. We know how each of the primary parameters depends on maximum length, so we now can derive how the physiological quantity of interest depends on maximum length.

Feeding rate increases with squared length intraspecifically but with cubed length interspecifically. Reproduction rate increases with size intraspecifically but decreases with size interspecifically. Interestingly enough, respiration rate scales somewhere between a surface and a volume, both intraspecifically and interspecifically, but for very different reasons. Intraspecifically, when we follow a growing individual with food held constant, body weight is proportional to the amount of structure (weak homeostasis), but the allocation to growth declines with size; the overhead costs of growth contribute to respiration. Interspecifically, when we compare fully grown adults and growth plays no role, respiration is dominated by maintenance. Since reserve density increases with maximum length, and maintenance is only used for structure, weight-specific maintenance decreases with maximum length. Reserve density increases with maximum length, because it equals the ratio of the specific assimilation rate and energy conductance. While the specific assimilation rate is extensive, the energy conductance is intensive. The ratio is, therefore, also intensive. The length of the juvenile period tends to increase with body length, as does life span in endotherms, because they typically have a positive Gompertz stress coefficient (around 0.5; see Table 2). Life span in ectotherms, however, hardly depends on body size because for them the Gompertz stress coefficient is small. The consequence is that ectotherms can only evolve large maximum body sizes if they manage to decrease their aging acceleration rate.

The ecological literature talks about *r* and *K* strategists. *r*-strategists are small, abundant, grow fast, have a short generation time and have little parental care. *K*-strategists do the opposite. The symbols *r* and *K* are frequently used for the parameters in the logistic equation and have the interpretation of growth and carrying capacity, respectively. The supposed difference in properties between *r* and *K* strategists mainly follow patterns that DEB theory expects for small-, repectively large-bodies species.

ECOSYSTEM STRUCTURE AND FUNCTION

Ecosystems have a structural aspect (abundance of the various biota) and a functional aspect (nutrient cycling). In a food pyramid at steady state, the top predators mainly suffer mortality from aging and some accidental losses, so their reproduction rate is low to compensate for these small losses. DEB theory expects low reproduction rates for large-bodied species at low food densities. At the bottom of the pyramid, the predation pressure is typically high, rendering losses due to accidents and ageing insignificant, so the reproduction rate must be high to compensate. DEB theory expects high reproduction rates for small-bodied species at high food densities. At the very bottom are nutrients and light as input for the producers that serve as food for the consumers. This gives a natural focus on the processes of nutrient supply and nutrient recycling to fuel the ecosystem; microbial degradation plays a key role. A tree's leaves may last only a year, and without assistance from microorganisms in the soil to release the nutrients locked into the leaf litter, the tree would have a short life span. Leaf litter is a waste product for trees, as are nutrients for the microorganisms—a clear example of syntrophic interactions. Once recognized, such syntrophic interactions can be seen everywhere and represent the dominant form of interaction. DEB theory is specifically designed to deal with such interactions since it specifies product formation quantitatively and uses SU-dynamics to analyze fluxes in a network.

EVOLUTION OF THE INDIVIDUAL AS DYNAMIC SYSTEM

A possible evolutionary scenario of the basic DEB models is presented in Figure 2, where the top row refers to the evolution of prokaryotes, from which the eukaryotes (second row) evolved. The increasing control over the chemical composition of the individual's structure during evolution induces stoichiometric constraints on growth. Since the concentrations of the various complementary substrates fluctuate wildly in the local environment, individuals (here, prokaryotic cells) need to store substrates in reserves to smooth out these fluctuations. The evolution of strong homeostasis might well have been via weak homeostasis.

When uptake became more efficient by using proteins that require turnover, the need to increase the reserve capacity increased; fluctuations in resource availability otherwise combines poorly with a continuous turnover. While homeostasis creates the need for reserves, maintenance enhances it.

Reserves could originally be built up by delaying the processing of internalized substrates, but the need to increase reserve capacity came with the need to temporarily store them in a form that does not create osmotic problems, for otherwise they start to interfere with metabolism. To ensure continuity of the fuelling of maintenance, the payment of maintenance costs was internalized from fluctuating external substrates to much more constant mobilized reserve.

Size control (i.e., the resetting of cell size by division and the control of surface area–volume ratios) boosted population growth but came with the need to install a maturity program. These steps were already taken before the eukaryotes evolved.

After phagocytosis arose in eukaryotes, feeding on other living creatures became popular in one line of development, which coupled the uptake of the various complementary substrates and induced a covariation of reserve densities. This encouraged the animal line of development, where homeostatic needs finally reduced the number of independent reserves to one, and the juvenile stage evolved an embryo and adult stage. The pattern arose with the evolution of mobility, sensors, and a neuronal system to allow for fast information exchange between otherwise rather isolated cells.

Another line of development did not start to feed on living creatures and kept their reserves independent but evolved an increased capacity to cope with changes in the

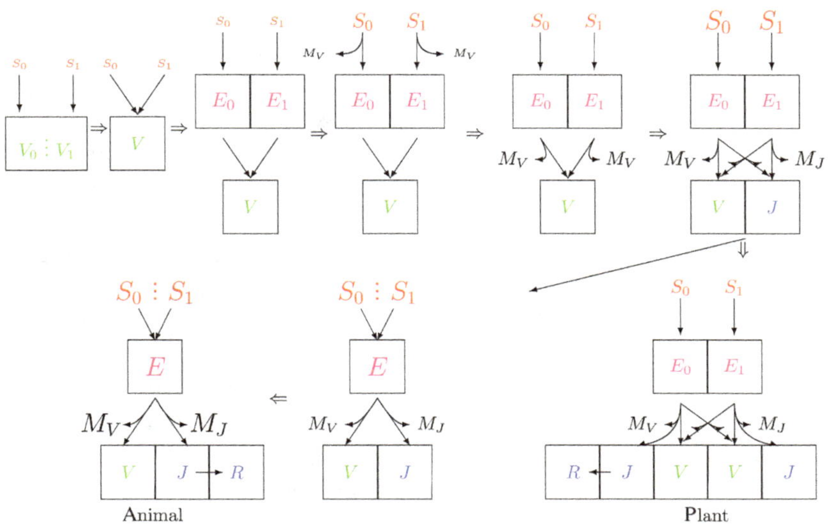

FIGURE 2 Steps in the evolution of the organization of metabolism of organisms. Symbols: *S*, substrate; *E*, reserve; *V*, structure; *J*, maturity; *R*, reproduction; M_V, somatic maintenance; M_J, maturity maintenance. Only two of several possible types of *E* are shown. Font size reflects relative importance. Stacked dots mean loose coupling. The top row shows the development of a prokaryotic system, which bifurcated in a plant and an animal line of development.

local environment: the plant line of development. They partitioned their structure into a root and a shoot, and eventually the use of products (wood) to adapt their shape during growth became feasible. They became masters of the art of torpor to escape bleak periods and evolved a much more open (but slow) mass communication between cells. The embryo/adult stages arose independently for this group.

SEE ALSO THE FOLLOWING ARTICLES

Allometry and Growth / Compartment Models / Ecotoxicology / Individual-Based Ecology / Integrated Whole Organism Physiology / Metabolic Theory of Ecology / Phenotypic Plasticity / Stress and Species Interactions / Stoichiometry, Ecological

FURTHER READING

DEB Research Program of Vrije University. http://www.bio.vu.nl/thb/deb/.

Kearney, M., S. Simpson, D. Raubenheimer, and B. Helmuth. 2010. Modelling the ecological niche from functional traits. *Philosophical Transactions of the Royal Society B: Biological Science* 365: 3469–3483.

Kooijman, S. A. L. M. 2001. Quantitative aspects of metabolic organization; a discussion of concepts. *Philosophical Transactions of the Royal Society B: Biological Science* 356: 331–349.

Kooijman, S. A. L. M. 2010. *Dynamic energy budget theory for metabolic organisation.* Cambridge: Cambridge University Press, 2010.

Sousa, T., T. Domingos, and S. A. L. M. Kooijman. 2008. From empirical patterns to theory: a formal metabolic theory of life. *Philosophical Transactions of the Royal Society B: Biological Science* 363: 2453–2464.

Sterner, R. W., and J. J. Elser. 2002. *Ecological stoichiometry.* Princeton: Princeton University Press.

ENVIRONMENTAL HETEROGENEITY AND PLANTS

GORDON A. FOX
University of South Florida, Tampa

BRUCE E. KENDALL
University of California, Santa Barbara

SUSAN SCHWINNING
Texas State University, San Marcos

Environmental heterogeneity—variation in environmental conditions over space, time, or both—drives much of plant biology. There are a multitude of plant adaptations for tolerating or capitalizing on environmental heterogeneity, and these vary with the scale of the heterogeneity relative to the size and longevity of individuals. It has been difficult to scale up from physiological processes at the level of organs and individuals to that of entire canopies or populations. In both of these problems, the difficulty arises mainly because leaves, roots, whole plants, and indeed entire biomes interact with one another by modulating patterns of heterogeneity, creating dynamic feedbacks so that one cannot simply apply nonlinear averaging. To date, the most successful approaches to heterogeneity at both the ecosystem level and the population level have treated the underlying feedback mechanisms selectively or else as unspecified or stochastic. However, there are important applications, including in the global change context, that require an improved understanding of how physiological processes at the organ level scale up to properties of populations, communities, and ecosystems, and this remains an important challenge.

RESPONSES OF ORGANS AND INDIVIDUALS TO ENVIRONMENTAL VARIATION

Environments vary on spatial scales from the molecular to the global. Most ecological study of temporal variation concerns scales from the diurnal to the generational. On any particular scale, environmental heterogeneity occurs in multiple quantities. Some of these are plant resources—CO_2, water, light, and nutrients—and so by definition can be consumed and locally depleted by individual plants. In turn, resource heterogeneity directly affects rates of resource uptake, and thus growth, reproduction, and survivorship. Heterogeneity also occurs in many quantities that are not plant resources—like temperature, humidity, and wind speed—which cannot be consumed by plants but are also locally affected by the presence of plants. Heterogeneity in these physical parameters affects plant function by modifying metabolic rates, stomatal control, or morphology, and thus resource processing.

At scales smaller than individuals, environmental heterogeneity is typically met by adjusting organ function to local conditions. For example, a forest tree may have specialized "sun leaves" in the top of the canopy (thicker leaves with higher photosynthetic pigment content per area) and "shade leaves" below (thinner leaves with less pigment per area). Another example is that of nutrient "hotspots" in the root zone, leading to highly localized proliferation of fine roots and root hairs. Such local organ adaptations are suitably interpreted by economic analogy; minimizing the cost of resource capture to maximize carbon gain at the whole-plant level.

Heterogeneity at the scale of individuals and populations can produce individuals of very different overall function and morphology, a phenomenon referred to as phenotypic plasticity. For example, root area–leaf area ratios can change dramatically in space or time, to balance the uptake of two or more essential resources (e.g., water and light). In extreme cases, leaves or fine roots may be shed entirely, when their carbon costs no longer justify their upkeep. Leaf phenology is an important factor in predicting the function of ecosystems, and one with worldwide implications for migratory species and climate–vegetation feedbacks. Environmental heterogeneity at the scale of individuals can also produce species or genotype sorting, matching species (or genotypes within species) with the specific environment in which they achieve higher function.

The functional integration of individuals, plasticity, leaf phenology, and species sorting leads to three major challenges for addressing the effects of heterogeneity on plants: understanding how environmental variation at scales smaller than individual plants scale up to whole-plant physiological processes, understanding how physiological processes at the organ and individual levels scale up to the function of ecosystems, and understanding how variation among individuals affects the growth and decline of populations. This entry addresses the latter two of these challenges; the first is considered elsewhere in this volume. As will be seen later, these problems are closely related, as the ecosystem level response to heterogeneity depends on the balance of species serving specific ecosystem functions.

HETEROGENEITY: FROM LEAVES TO ECOSYSTEMS

Heterogeneity impacts at the ecosystem level is a central problem in ecology, as it is plant primary production that determines food availabilities to all heterotrophic organisms, and vegetation patterns that provide habitat, or at least contribute largely to habitat quality. On first consideration, the scaling from organ function to ecosystem function may seem simple enough, as the biochemical machinery of photosynthesis on land and in water is highly uniform across species and environmental conditions, barring the relatively small, and well-documented, variation between the C_3 and C_4 photosynthetic pathways and the more profound photosynthetic variants of a rare group of organisms: purple and green sulfur bacteria.

Almost all organic compounds in the world are produced through the interaction of two processes: the Light Reaction, wherein light is intercepted by pigments and radiation energy transformed to chemical energy producing dioxygen as an end product, and the Carbon ("Dark") Reaction, wherein the carriers of chemical energy (ATP, NADPH) are used to construct organic carbon molecules, $(CH_2O)_n$, from inorganic CO_2. Thus, light (specifically light in the visible wavelength range) and CO_2 are the fundamental plant resources limiting the rate of primary production, but additional macro- and micronutrients are needed for building the machinery of photosynthesis: pigments, enzymes, and electrolytes. Additionally, photosynthesis on land is almost always water-limited, due to the inescapable constraint that uptake of CO_2 from air leads to loss of water vapor. Thus, land plants need to constantly replenish water lost from leaves with water taken up from the ground, necessitating a large investment on nonphotosynthetic biomass: stems and roots.

A key strength of ecosystem ecology is the expression of ecological processes at any scale, from organelles to biomes, in units of common currency—e.g., carbon, water, and nitrogen. However, the challenge of ecosystem ecology lies in the scaling from fundamental processes well understood at one level of organization to another level of organization. The difficulties arising in this context are of two kinds: spatial and temporal averaging of functions that depend nonlinearly on environmental heterogeneity, and, more problematically, feedbacks between environmental heterogeneity and plant function.

These challenges may be exemplified by the relatively simple task of scaling CO_2 exchanges from the leaf to the canopy level. The gas exchange of individual leaves can be described quite accurately by light, CO_2, humidity, temperature, and other types of response functions. However, every leaf in a canopy will also modify the local environment. For example, light interception raises leaf and air temperature and reduces light levels for leaves further below. Concurrently, water vapor evaporation will humidify the air and have a cooling effect on leaf temperature. Changes in light, humidity, and temperature not only change leaf gas exchange levels directly, but consistent exposure to modified environmental conditions will feed back on leaf construction, modifying the very functions that relate gas exchange rates to light, humidity, and temperature.

While one can construct and parameterize simulation models that scale from leaves to canopies, for example, this is much harder to do for natural canopies, which are typically constructed from individuals of mixed species of uneven height in possibly structured arrangements. For some problems, solutions can be found by using different conceptual frameworks for different levels of organization. For example, the water and energy exchanges of whole canopies are more easily described by coupled water and energy budget equations that treat the entire canopy as a uniform surface, where the effects of species composition and canopy structure affect only a few parameters (e.g., albedo and surface roughness) and where empirical functions relate plant-available water in the root zone to evapotranspiration. What is lost by applying different models to different scales is the ability to dynamically link effects of small-scale heterogeneity to plant function at larger scales.

Another important but presently unresolved scaling problem leads from the physiology of individuals to the demographic processes that govern population growth and decline. At the simplest level, growth is constrained by the rate of carbon assimilation. But decomposition analysis of individual growth rates shows that the latter also depend on biomass allocation to leaves versus stems, roots, or reproductive structures, which are controlled by internal physiological and molecular processes interacting with environmental conditions and triggers. Among the most difficult problems is prediction of mortality rates. What level of plant water deficit is lethal? At what point are internal carbon reserves too low to revive a plant from herbivory or fire? When is the resource availability to seedlings too low to maintain positive carbon balance? Where mortality rates have proved to be predictable—in the important case of self-thinning in plant populations—we do not understand it at the physiological level.

Scaling up from the physiology of plant organs has proved to be difficult at both the population and the ecosystem level. But all is not lost: considerable success has been achieved by approaches that rely on a basic physiological understanding but deliberately do not specify many of the mechanistic details. In particular, such approaches have guided much of our understanding of the consequences of heterogeneity at both the biome and population levels.

The fundamental requirements of photosynthesis—chiefly light, water, and temperature above freezing—dominate broad-scale responses of vegetation to environmental heterogeneity, especially the spatial distribution of distinct biomes across the global terrestrial land area. The location and extent of major biomes, from desert scrubland to tropical rainforests depends, perhaps surprisingly, largely on only two major drivers: mean annual temperature and mean annual precipitation. The former is strongly correlated with the length of the growing season and with the necessity for low temperature tolerance in plants. Mean annual precipitation, in conjunction with temperature, determines the degree of drought tolerance required and correlates with fire frequency. This relatively simple state of affairs has made it possible to predict the distribution of major biomes, each represented by one or more characteristic plant types, and to assign biome-specific biogeochemical response characteristics. Biogeochemical and biogeography models take center stage in addressing the timely problem of global vegetation–atmosphere interactions.

Due to fundamental tradeoffs in plant form and function, plant species tend to maximize fitness within a narrow range of environmental conditions. While these conditions can span a large geographic range, plant species inevitably will be replaced by other species at their range limits. Where suites of species are replaced by other suites of species, transition zones are recognized as ecotones. The dynamics of ecotones have recently attracted much research interest, as it is here where range expansion and contraction occurs and where signs of biome redistribution and reorganization may be first recognized.

Even though biome function at the global scale can often be approximated by just one vegetation type, any community contains multiple species with strikingly different forms of adaptation to the local climatic drivers. Ecosystem models operating at a finer grain of temporal and spatial resolution have addressed this complexity by dividing the many coexisting species into a small number of plant functional types. In the community context, plant functional types are recognized as players realizing contrasting adaptive strategies in a game of largely shared environmental constraints. For example, desert ecosystems typically contain evergreen perennials with high drought tolerance, drought-deciduous perennials with somewhat reduced drought tolerance, and wet season ephemerals with no tolerance for drought at all, at least in the vegetative state.

Another way of looking at plant functional types is therefore as a collection of broad ecological niches, and scaling issues emerge in the form of niche interactions. Overyielding—the increased productivity of species mixtures in relation to area-based average yield of monocultures—shows that the integrated function of a species mixture is not merely the sum of its parts. Novel properties at the community scale are the result of complementarity among species in response to environmental heterogeneity.

It is thus no surprise that shifts in functional type composition can have profound effects on ecosystem processes. For example, the global conversion of open grasslands into woodlands for the last 100–150 years is thought to have accelerated carbon sequestration, creating a global sink that may have ameliorated CO_2

accumulation in the atmosphere. In a contrasting example, the "annualization" of sagebrush steppe in the western United States by the invasive species *Bromus tectorum* has introduced a frequent fire cycle that may lead to the irreversible loss of this biome.

There has been much debate as to whether diversity beyond the scale of plant functional types affects ecosystem function. Typically, virtually indistinguishable ecosystem functions are served by many species, suggesting that 1000 species can maintain ecosystem function just as well as 20 species. The counterargument is that uninterrupted ecosystem function requires a degree of functional redundancy, analogous to the fail-safe design of complex machines. It has been suggested that certain types of heterogeneities across space and time subtly favor different species of the same functional type or spread the risk of local extinction between species of similar function, thus stabilizing ecosystem function over space and time. However, this argument has to be weighed against expected increases in extinction risk in smaller populations, and so we now turn to considering the demographic consequences of environmental variation and their effects on population growth rate and extinction risk. The distribution of species and functional types responds to small-scale variation in microclimate or site conditions. For example, community composition often changes between the north- and the south-facing slope of a hill. Species sorting can, but does not always, involve direct positive or negative interactions between individuals. An example where direct interactions do occur is in the "nurse plant" effect, the improvement of seedling survivorship under the canopy of an established perennial. On the other hand, differential germination of seeds in different year types does not involve interactions between species.

A theoretical challenge of recent concern is to understand the roles of dispersal versus site selection. Compared to mobile animals, sessile individual plants have much less control over the site of their establishment or that of their offspring. In reality, it is often difficult to determine whether a spatial association of conspecifics is the result of favorable site conditions, limited dispersal range, or site selection by a dispersal vector like seed caching granivores.

HETEROGENEITY: FROM INDIVIDUALS TO POPULATIONS

Environmental factors varying on the scale of individuals or stands lead to demographic heterogeneity among individuals, and this affects population growth rates and extinction risks.

Interactions between populations, or between individuals within populations, can also cause environmental variation that leads to demographic heterogeneity. For example, the densities of neighboring plants, pathogens, herbivores, or mycorrhizae may vary for plants within a population. Processes like these may also be considered spatial heterogeneity, but it is often useful to consider them separately because their causes and dynamics are quite different from factors like differing parent materials in soils.

Temporal environmental variation can lead to demographic heterogeneity in two different ways. First, different cohorts experience different environmental conditions—for example, weather, fire, plant densities, and pathogens can all vary substantially from year to year. Because these differing conditions occur at different life stages for different cohorts, they often have strong effects on the average demographic performance of individuals. These cohort effects tend to either destabilize inherently stable population dynamics or stabilize inherently unstable dynamics.

There is also an indirect effect of temporal environmental variation. It is well established that in the long term, selection opposes among-year variance in the population growth rate. Much research on mechanisms like seed heteromorphisms has focused on the intriguing possibility that these are bet-hedging adaptations that trade off the mean and variance in population growth rate among years. But if they lead to heterogeneity in survival, mechanisms like these are selectively favorable even without a mean–variance tradeoff, because populations with among-individual heterogeneity in survival have larger long-term growth rates than monomorphic populations with the same mean survival. Thus, many mechanisms that can act as bet hedges are also favored even without a mean–variance tradeoff.

In sum, there are many environmental factors that vary in space or time (or both) and lead to demographic heterogeneity. There are, of course, two further sources of demographic heterogeneity: genetic variation and nongenetic parental effects. These are not caused directly by environmental heterogeneity, but they may be correlated with it. As is well known, genetic variation can be eroded by natural selection, while the other causes of demographic heterogeneity may persist.

The prevalence (and strength) of natural selection in plant populations provides strong evidence that demographic heterogeneity is, indeed, both ubiquitous and important. Natural selection can occur only in populations with demographic heterogeneity. The reverse need not be true, as it is possible for demographic heterogeneity to be random with respect to phenotypes in a population. The evolutionary effects of demographic heterogeneity are well understood; this section considers its effects on population dynamics.

General theory suggests that heterogeneity in survival, especially if an individual retains its relative (dis)advantage throughout its life, can reduce demographic stochasticity, increase the low-density growth rate, and increase the equilibrium density. The latter two results are an effect of cohort selection: as a cohort ages, the individuals with higher mortality risk preferentially die off, leaving increasingly more robust individuals as the population ages. In a population context, this increases the average survival of individuals at the stable age structure, relative to a homogeneous population with the same average survival rate. These general models have not been applied to structured plant population models.

Environmental heterogeneity can cause individuals to grow at different rates, even in the absence of interspecific competition. Herbaceous plants, especially annuals, often grow nearly exponentially (at least while uncrowded) in their pre-reproductive phase. Heterogeneity in the growth constant (called the relative growth rate, RGR) can cause a cohort of initially identical individuals to develop a lognormal size distribution, with the skew increasing with the amount of variance in the RGR. A great deal of effort in the 1970s and 1980s went into trying to understand whether the shape of this distribution provides any information about the intensity or nature of interspecific competition. However, one consequence was largely overlooked: a population with a heterogeneous growth rate will have a larger mean final size than a population of identical individuals with the same mean growth rate. In 1985, Holsinger and Roughgarden incorporated this into a plant population model; if seed production was positively related to size (as it is in many plants), then increasing RGR heterogeneity increases both the low-density growth rate and the equilibrium population density. Such heterogeneity may also be important to include in perennial plant population models as well, but this has not been explored.

The demographic process of reproduction is much more idiosyncratic than survival—What is the distribution of seed size and number? What is the germination rate? How likely is it that a seedling will recruit to the reproductive population?—so there is not, as yet, a general theory about the demographic effects of reproductive heterogeneity. In isolation, its effects are probably usually modest, as reproductive heterogeneity does not change the phenotype distribution in the way that survival heterogeneity does. Nor does it modify mean performance in the way that growth rate heterogeneity does. However, reproductive heterogeneity can exaggerate or mitigate the impacts of those other types of heterogeneity.

Demographic heterogeneity is also likely to have strong impacts on the genetic structure of populations. Natural selection has well-known effects on the genetic composition of populations, and heterogeneous populations will generally have smaller effective population sizes than homogeneous populations. Heterogeneity thus can be expected to contribute to genetic drift as well as to selection. Moreover, insofar as it is caused by spatial environmental heterogeneity, demographic heterogeneity may also have substantial effects on gene flow.

Finally, demographic heterogeneity may have large effects on the coexistence of competing species, through either of two mechanisms. In an influential model, Chesson in 1990 showed that coexistence (or exclusion) can result from the way in which populations' sensitivities to competition covary with their sensitivities to spatial or temporal environmental variation. Using a rather different argument, in 2010 Clark proposed that heterogeneity within populations of coexisting plants can make it possible for intraspecific competition to outweigh interspecific competition, although the particular empirical case for which he proposed this mechanism—trees in the southern Appalachians—has since been disputed.

CONCLUSIONS AND CHALLENGES

Environmental heterogeneity drives plant ecology and diversity. We have a satisfactory understanding of its effects at the organ level for small time scales. However, its effects on larger/longer scales are not additive, because there are feedbacks and interactions of many kinds—back to the environment itself as well as among organs, individuals, and stands. Understanding ecosystem and population-level consequences of environmental heterogeneity has been accomplished by ignoring most of the possible feedbacks and concentrating on specific feedbacks deemed important or instructive. In the case of physiological scaling from leaves to ecosystems, the aim is to accurately track the exchange of elements and energy. In the case of populations, heterogeneity in the function of individuals is taken as given and then the larger-scale, longer-term consequences are explored.

However, there are important problems for which progress may require pursuing the coupling between the ecosystem and population perspectives. At a time of rapidly changing ecosystem function, accelerated species extinction rates, and the real threat of global climate regime change, a pressing issue is the extent to which species diversity and ecosystem productivity and stability are related. Models and experiments exploring this question in particular systems may require at least an approximation of the underlying physiological issues at the population scale, rather than the black box of hypothesized heterogeneity. Conversely, ecosystem science would likely benefit from improved representation of demographic processes, including recruitment, mortality, and dispersal, to predict ecosystem dynamics.

Ecologists recognize that areas like ecophysiology, evolutionary biology, community ecology, and ecosystem ecology are not really separate; in practice, they are, subdisciplines that tend to develop independently of one another. The view outlined in this entry suggests that many large questions in ecology depend on tighter coupling among these lines of inquiry. Such coupling may allow us to make inroads in some of the fundamental problems in ecology.

SEE ALSO THE FOLLOWING ARTICLES

Integrated Whole Organism Physiology / Plant Competition and Canopy Interactions / Phenotypic Plasticity / Population Ecology / Stochasticity, Demographic / Stochasticity, Environmental / Stoichiometry, Ecological

FURTHER READING

Bloom, A. J., F. S. Chapin, and H. A. Mooney. 1985. Resource limitations in plants—an economic analogy. *Annual Review of Ecology and Systematics* 16: 363–392.

Chapin, F. S., III, P. A. Matson, and H. A. Mooney. 2002. *Principles of terrestrial ecosystem ecology.* New York: Springer.

Chesson, P. L. 1990. Geometry, heterogeneity and competition in variable environments. *Philosophical Transactions of the Royal Society B: Biological Sciences* 330: 165–173.

Clark, J. S. 2010. Individuals and the variation needed for high species diversity in forest trees. *Science* 327: 1129–1132.

Holsinger, K., and J. Roughgarden. 1985. A model for the dynamics of an annual plant population. *Theoretical Population Biology* 28: 288–313.

Kendall, B. E., and G. A. Fox. 2002. Variation among individuals and reduced demographic stochasticity. *Conservation Biology* 16: 109.

Koyama, H., and T. Kira. 1956. Intraspecific competition among higher plants. VIII. Frequency distributions of individual plant weight as affected by the interaction between plants. *Journal of the Institute of Polytechnics Osaka City University Series D:* 7: 73.

Loreau, M., S. Naeem, and P. Inchausti, eds. 2002. *Biodiversity and ecosystem function: synthesis and perspectives.* Oxford: Oxford University Press.

Morris, W. F., C. A. Pfister, S. Tuljapurkar, C. V. Haridas, C. L. Boggs, M. S. Boyce, E. M. Bruna, D. R. Church, T. Coulson, D. F. Doak, S. Forsyth, J.-M. Gaillard, C. C. Horvitz, S. Kalisz, B. E. Kendall, T. M. Knight, C. T. Lee, and E. S. Menges. 2008. Longevity can buffer plant and animal populations against changing climatic variability. *Ecology* 89: 19–25.

Smith, T. M., H. H. Shugart, and F. I. Woodward. 1997. Plant functional types: their relevance to ecosystem properties and global change. Cambridge, UK: Cambridge University Press.

ENVIRONMENTAL STOCHASTICITY

SEE STOCHASTICITY, ENVIRONMENTAL

ENVIRONMENTAL TOXICOLOGY

SEE ECOTOXICOLOGY

EPIDEMIOLOGY AND EPIDEMIC MODELING

LISA SATTENSPIEL
University of Missouri, Columbia

Epidemiology is the study of how, when, where, and why diseases occur, with special emphasis on determining strategies for their control. Originally the term referred to the spread and control of the great historical infectious disease epidemics that swept through human populations in Western Europe on a regular basis. The term now relates to the study of the causes, distribution, and control of chronic as well as infectious diseases in humans and other living organisms. Because a full description of all types of epidemiology is beyond the scope of this article, the discussion that follows is limited to infectious disease epidemiology. Readers interested in a broader view of the discipline are encouraged to consult the general epidemiology references listed at the end of this article.

SOME BASIC PRINCIPLES OF EPIDEMIOLOGY

Epidemiology is a population-centered discipline, and as such, it has strong ties to many areas within ecology. A core emphasis in epidemiology is the nature of interactions between the pathogen(s), one or more hosts, and the environment. Therefore, most epidemiological studies consider the biology and behavior of both the hosts experiencing disease and the pathogens that are responsible for the disease, as well as the environmental influences (e.g., climate, nutrition, stress, and the like) that may affect the development and manifestations of the disease.

A number of specialized terms have arisen during the development of epidemiology. The term epidemic is used when the number of cases in a localized area suddenly increases above an expected or normal level for a short time; when this happens worldwide, the term pandemic is often used. When a disease is present in a population at a relatively constant (and often low) level at all times, it is called endemic. The number of cases of a disease present in a specified population over a defined time period is called the prevalence; the number of new cases of a disease in a specified population during a defined time period is called the incidence.

When thinking about infectious disease in an epidemiological framework, it is useful to identify a number of distinct disease-related states that can distinguish members of the population. The most common of these are susceptible—an individual at risk for infection, exposed—an individual who has been infected with a pathogen but does not show symptoms and cannot pass the pathogen on to others, infectious—an individual who is capable of transmitting a pathogen to susceptible individuals, and immune (or recovered)—a formerly infectious individual who is no longer able to pass the pathogen on and is not capable of becoming reinfected.

Epidemiologists also consider the different stages an individual goes through from the time of infection until recovery, which is sometimes referred to as the natural history or the natural progression of an infectious disease. Depending on the questions being addressed, two frameworks have been used to describe these stages, a symptomatic framework and a transmission framework. The symptomatic framework is commonly employed by medical personnel, who tend to focus on ways to control the external manifestations of a disease; the transmission framework is used in epidemiological contexts when the emphasis is on developing methods of control aimed at stopping or lessening transmission of a pathogen within and among populations. The transmission framework is also the most common approach taken in ecological studies of infectious diseases in both humans and other animals.

Figure 1 illustrates the distinctions between these two frameworks. In the symptomatic framework, upon infection an individual enters an incubation period where no symptoms are perceived. Once the individual begins to show symptoms, the incubation period ends and the period of symptomatic illness begins. When symptoms are no longer apparent, the individual is said to have recovered from the disease. In the transmission framework, upon infection the individual enters the latent period. This period continues until the individual is able to transmit the pathogen to a susceptible individual. The onset of infectiousness (end of the latent period) is correlated with the onset of symptoms (end of the incubation period), but the two are not identical. For many diseases, an infected individual can transmit a pathogen before symptoms appear; sometimes the onset of transmission occurs simultaneously with the development of symptoms, and it is even possible for the onset of transmission to occur after symptoms have developed. When an infected individual becomes unable to transmit the pathogen to others, the infectious period ends and the individual enters the immune state. As with the onset of infectiousness, the end of the infectious period can occur before, simultaneous with, or after the cessation of symptoms. In both

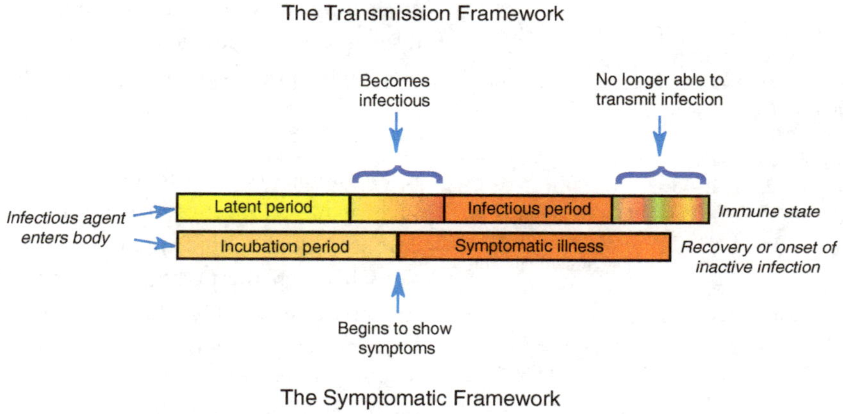

FIGURE 1 Comparison of the transmission and symptomatic frameworks for disease progression stages. The end of the latent period in the transmission framework can occur before, at the same time as, or after the end of the incubation period in the symptomatic framework, depending on the biology of the specific disease. This uncertainty is indicated by the gradient of colors in the box representing when an individual becomes infectious. Similarly, the end of the infectious period can occur before, at the same time as, or after symptoms disappear, so the uncertainty in timing of this event is also indicated by a gradient of colors.

TABLE 1
Modes of transmission of infectious diseases

Mode	Example
Direct transmission	
Fecal–oral	Rotavirus, hepatitis A, *Trichuris suis*
Respiratory	Influenza, measles, equine rhinopneumonitis
Sexual	Syphilis, HIV, bovine genital campylobacteriosis
Vertical (congenital)	Rubella, syphilis, HIV, toxoplasmosis
Direct physical contact	Yaws, diphtheria, chicken pox, herpes simplex
Indirect transmission	
Vector-borne	
Fleas	Bubonic plague, murine typhus fever, dog tapeworm
Flies	Trypanosomiasis, leishmaniasis, anthrax in cattle
Lice	Louse-borne typhus, relapsing fever, trench fever
Mosquitoes	Malaria, West Nile virus, eastern equine encephalitis
Ticks	Rocky Mountain spotted fever, Lyme disease
Other	Chagas disease, blue tongue
Vehicle-borne	
Food-borne	Botulism, salmonellosis, tapeworm
Soil-borne	Hookworm, valley fever, tetanus
Water-borne[a]	Cholera, hepatitis, typhoid fever
Needle sharing[b]	Hepatitis B, HIV
Complex cycles	Dracunculiasis, hydatid disease, whirling disease of fish

[a] Most of these can also be transmitted through contaminated food.
[b] Includes both vaccinations and intravenous drug use.

frameworks, recovered individuals can remain in that state either permanently or temporarily depending upon the nature of immunity to the pathogen involved.

Another basic question emphasized in epidemiology is how pathogens move between hosts, and this behavior is a fundamental aspect of the ecology of the host–pathogen–environment interaction. Infectious disease ecologists and epidemiologists call the mechanism a pathogen uses its mode of transmission, and they have identified several broad classes of transmission mechanisms. These can be divided into (a) directly transmitted pathogens, which spread directly from a host of one species to another host of the same species, and (b) indirectly transmitted pathogens, which either require the aid of an insect vector or inanimate material, or must alternate between one or more distinct host species in order to complete their life cycle. Mechanisms for direct transmission of pathogens include respiratory (pathogens are spread through the air and breathed in by the host), fecal–oral (spread from fecal material to the mouth), sexual (transmission during sexual activities), vertical (transmission from mother to offspring), and direct physical contact. Indirect transmission mechanisms include vector-borne (use of a living organism, usually an arthropod) to carry the pathogen between hosts), vehicle-borne (spread through contact with contaminated food, water, soil, or inanimate objects such as needles), and complex modes of transmission involving multiple host species and often vectors and/or vehicles. Examples of diseases spread by each of these mechanisms are shown in Table 1. Note, however, that many pathogens can be transmitted by multiple mechanisms.

As in other areas of ecology, theoretical questions in epidemiology are addressed using mathematical and computer models. The basic epidemiological principles described here provide many of the necessary building blocks for this task. So how does one translate these concepts and principles into models and use them to enrich epidemiological theory?

MODELING EPIDEMIOLOGICAL PROCESSES

Mathematical models have been used in epidemiology for over 100 years, but they have became much more common since the last few decades of the twentieth century. The vast majority of epidemic models are mathematically based, but the development and use of computer-based models is becoming more frequent. Some of the computer-based models are analogs to mathematical models, but many recent computer-based models do not have an explicit mathematical model underlying their structure.

Some epidemiological models are population based (including most mathematically based models), others are individual based, which, as in other areas of ecology, are being developed with increasing frequency.

Regardless of the basic approach used in developing an epidemiological model, in determining a model's structure, a fairly conventional strategy is used that takes into account the biology and ecology of the host–pathogen–environment interaction. Most epidemic models use a compartment model as their underlying structure. The individuals within a host population are generally divided into groups or compartments according to their disease status. The earliest and simplest epidemiological models considered only two or three such groups—susceptible (S) and infectious (I), with or without recovered (R) individuals. Epidemiological models are commonly referred to by the letters corresponding to the groups that form the heart of their structure; thus, these early models are called SI and SIR models. The SIR model, in particular, is the fundamental basis for a large number of epidemic models, although models including an exposed class (SEIR models) are also common. The choice of which structure to use depends on the underlying biology of the disease being modeled and the questions motivating model development.

As with other kinds of compartment models, epidemiological models of this type must also consider the process by which individuals "flow" from one compartment to the next. Knowledge of the biological, ecological, and social factors influencing pathogen transmission and disease progression must be used to determine these mechanisms, which govern how and when an individual changes disease status (i.e., becomes infectious after being susceptible, or recovers after being infectious).

As an example of this process, consider a simple model of a disease such as measles. Measles is a viral disease that spreads directly from person to person by means of a respiratory mode of transmission. After infection there is a latent period of one to two weeks, followed by an infectious period of about one week. Upon recovery a person becomes permanently immune to further infection. This disease history can be represented by an SEIR model consisting of the four compartments—susceptible, exposed, infectious, and recovered/immune. Transmission of the virus can occur whenever there is suitable contact between an infectious individual and one who is susceptible. In the simplest models, this is assumed to occur at a constant rate per contact. In these models the flow rates from the exposed category to infectious and from infectious to recovered/immune are also assumed to be constant. The entire model can be represented as follows:

$$\frac{dS}{dt} = -\frac{\beta SI}{N}$$

$$\frac{dE}{dt} = \frac{\beta SI}{N} - \sigma E$$

$$\frac{dI}{dt} = \sigma E - \gamma I$$

$$\frac{dR}{dt} = \gamma I$$

where S, E, I, and R are the numbers of individuals in each disease stage, N is the total population size, β is the effective rate at which infectious individuals transmit the pathogen to a susceptible individual with whom there is contact, σ is the rate at which exposed individuals become infectious, and γ is the rate at which infectious individuals recover from the disease. With these constant flow rates, $1/\sigma$ gives the length of the latent period and $1/\gamma$ gives the length of the infectious period. It is also important to note that in this simple formulation, the transmission parameter, β, implicitly includes both the nature of contact between susceptible and infectious individuals and the probability that the pathogen is transmitted given that contact. Note also that the term $\beta SI/N$ gives the number of new infections in a given time unit.

Depending on the questions of interest, a variety of complexities can be added to simple SI, SIR, or SEIR models to make them more realistic. For example, the transmission parameter, β, can be separated into distinct components specific to the contact process and to the risk of transmission given contact. The entire population can also be divided into distinct subgroups corresponding to different communities or risk groups. These two additions facilitate the development of models that are designed to examine how heterogeneity in patterns of contact among subgroups affects the spread of a disease across a population. The issues of how to model mixing patterns between subgroups effectively and the impact of population subdivision have been the focus of a substantial amount of work in the last 2–3 decades. Because of their importance, epidemiological questions related to them and the use of models to address these questions are discussed further below.

Acute infectious diseases such as measles spread very rapidly through a population, and if the population is closed, they quickly die out, largely because infected individuals recover quickly and become immune, so that eventually the number of susceptible and infectious individuals is so small that the chain of transmission cannot be maintained. Most populations are not closed, however, and so many acute diseases tend to cycle through the populations over time. Furthermore, many disease conditions, both infectious and noninfectious, continue to affect an individual over long periods of time. In order to model these long-term situations, it is necessary to add demographic change to the simple epidemic models. This is usually done by incorporating only birth and death terms, but depending on the disease in question, immigration and emigration may also be included. The addition of these demographic processes to a model allows for the replenishment of susceptible individuals, which can effectively keep a chain of infection going for longer periods of time. In addition, if a population being modeled will be followed over several generations or if the relevant disease parameters are known to vary significantly with age, it may be desirable to add age structure to the model. As an example, age-structured models have been used effectively to assess the impact of contact within schools and the timing of school terms on the spread of measles and other infections.

The simple measles model described above is fully deterministic. All parameters are set, and random factors are assumed not to be of importance. This assumption may be reasonable for large populations, but if the population being modeled is small or if the questions of interest center on the impact of potential variation across the population in relevant parameters (e.g., variation in infectious period among individuals), then the model must account for the random effects of these factors. Models that incorporate random effects are called stochastic models, and they can be either partially or fully stochastic. Many mathematically based stochastic models are only partially stochastic because the mathematical analysis of stochastic effects is more complex than the analysis of deterministic models. Computer-based models are more often fully stochastic. In recent years, much research has been devoted to understanding the consequences of stochasticity for the spread of infectious diseases as well as for other ecological phenomena.

Only a small selection of the possible extensions of the basic epidemiological models have been discussed here. Readers interested in learning more about the ways that complexity can be added to epidemiological models are encouraged to consult the epidemic modeling sources listed at the end of this article. It is important to remember, however, that the driving forces for determining how to add complexity to a simple model are the questions that one hopes to address using the model.

SOME FUNDAMENTAL QUESTIONS IN THEORETICAL EPIDEMIOLOGY

As the references listed below suggest, a substantial number of questions have been addressed using epidemiological theory and mathematical and computer modeling, but because of space limitations, this discussion is limited to two of the most significant of these topics: how theory can be used to aid in predicting the risk of a disease outbreak and controlling its spread, and the impact of nonrandom mixing within and among groups on patterns of epidemic spread.

Predicting the Risk of an Epidemic and Controlling Its Spread: R_0

Probably the earliest motivating factor for the development of epidemiological theory is the question of how best to predict whether a pathogen will spread upon introduction into a population and how to control and/or stop its spread if it does gain a foothold. Intuitively, in order for a pathogen to spread, the initial infected individual must be able to pass it on to at least one susceptible individual, thereby ensuring that the chain of infection remains unbroken.

When a pathogen first enters a population, the population consists almost entirely of susceptible individuals. Recall that the number of new infections in simple models is given by the term, $\beta SI/N$, and the average length of the infectious period is $1/\gamma$. Since $S \approx N$, this means that near the beginning of an epidemic, the number of new infections per time unit is approximately βI. Furthermore, the total number of infections caused by all infectious individuals at the beginning of an epidemic can be approximated by $\beta I/\gamma$, the product of the number of new infections per time unit and the length of the infectious period. The number of new infections per time unit caused by a single infectious individual is then given by β/γ. This value has been given the name "basic reproductive number" and it is commonly represented by the symbol R_0. It plays essentially the same role in epidemiology that R.A. Fisher's net reproductive rate plays in models of population growth. If $R_0 > 1$, each infectious individual generates more than one new infection and the epidemic spreads; if $R_0 < 1$, an infectious individual does not generate a sufficient number of new infections and the epidemic dies out; and if $R_0 = 1$, the epidemic is maintained at a constant level of infections.

Because R_0 provides information on whether an epidemic will or will not occur in a fully susceptible population, it is often referred to as the epidemic threshold. It is important to note that prior to the early 1980s, analyses were generally directed at determining a threshold population size rather than R_0. This threshold population size is directly related to R_0 and has a similar interpretation: if the initial size of a susceptible population is larger than the threshold population size, an epidemic can occur; if it is smaller than this threshold, an epidemic will die out; and if it is equal to this threshold, a disease can be maintained at a constant level of infections.

Because of the importance of R_0 in predicting whether an epidemic can potentially occur, one of the first analyses associated with the development of a new epidemic model is the determination of the appropriate formulation of R_0 for that model (specific functional forms are model dependent). This determination is often straightforward for simple models, but it can be quite difficult for more complex models (even some that appear to be relatively simple). Nonetheless, it can usually be shown numerically that a disease model exhibits some kind of threshold behavior during its initial stages.

By definition, R_0 relates to the potential for epidemic spread only at the beginning of an epidemic when the population is fully susceptible, and so technically it is not applicable at other times. While an epidemic is occurring, the number of susceptibles changes and is usually not near N. Thus, a related concept—the general reproductive number, R_t—has been devised to describe the average number of new infections generated by a single infectious individual at any time t during an epidemic. If a population consists of S susceptibles, then under the assumptions of the simplest epidemic models, each infectious individual produces new infections at a rate $\beta S/N$ per time unit. The total number of new infections produced by a single infectious individual is then this rate multiplied by the length of the infectious period, $1/\gamma$. R_t is thus given by the product of R_0 and S/N, the proportion of the population that is still susceptible at time t. Even if $R_0 >> 1$, it is clear that in a closed population, an epidemic will die out once the number of susceptibles declines to a value that causes R_t to fall below 1, even if some susceptibles have escaped infection. Of course, if the number of susceptibles is replenished regularly through births or migration, then it is possible for R_t to remain above 1, ensuring that a disease can be maintained over long periods of time.

The concept of R_0 has taken hold within epidemiology and has proven to be a valuable guide in the development of strategies to consider when faced with a potential or ongoing epidemic. For example, modelers and epidemiologists together have used R_0 and R_t to evaluate the efficacy of vaccination, isolation of infected individuals, and other control efforts in order to prevent or control disease outbreaks such as the 2003 SARS epidemic or avian influenza. Unfortunately, however, it is not always appreciated by epidemiologists that the particular formulations of R_0 and R_t and the estimates that may result from them are highly model dependent. Nonetheless, the concepts are very important and among the most useful contributions that theoretical studies have made to the field of epidemiology.

Population Structure, Contact Patterns, and Nonrandom Mixing

The transmission of a directly transmitted infectious disease can only occur if a susceptible individual has adequate contact with an infectious individual. The simplest epidemic models assume that such contacts occur randomly within a large group of homogeneous individuals. In reality, however, populations are not homogeneous, and contact between susceptible and infectious individuals is clearly nonrandom. One way to make epidemiological models more realistic with regards to the heterogeneity among individuals and nonrandom patterns of contact is to consider a population of distinct individuals, all of whom vary in their personal characteristics. Individuals can then be linked to one another by means of a social network where the individuals form the nodes of the network, and links between individuals indicate that contact between two individuals is possible. This approach is used in some mathematically based epidemiological models and in many computer simulations of epidemic processes. Determining the nature of the contact networks and how those influence disease transmission processes are questions of major interest to epidemiologists. A discussion of networks and their analysis is beyond the scope of this article, but interested readers are encouraged to consult the related entry in this volume and the sources listed below.

A growing number of epidemiological models take the middle ground and incorporate some level of population structure between that of a single homogeneous population and that of a population of completely unique individuals. Essentially what these models do is to divide the population into a number of different subgroups within which individuals are assumed to be homogenous and randomly mixed but between which there is heterogeneity in individual characteristics and nonrandom mixing. For example, to deal with a disease for which the incidence varies markedly with age, a population may be divided into different age classes; to model the geographic spread of an epidemic, a regional population may be divided into different discrete communities. Models of this type are referred to as metapopulation models and are being used widely within many areas of ecology.

In addition to characterizing the groups into which a population is subdivided, models that include heterogeneity among individuals must specify not only how transmission occurs between susceptible and infectious individuals within a group but also how it can occur between individuals from different groups. Recall that in the simple epidemic models described above, the transmission parameter, β, implicitly includes both contact between susceptible and infectious individuals and the probability that contact leads to transmission of the pathogen. In models that focus on nonrandom patterns of contact across distinct subgroups, it is necessary to separate these two processes. The patterns of contact are usually represented by a matrix, sometimes dubbed the "who acquires infection from whom" (WAIFW) matrix, where each element specifies the rate at which individuals from group i come into contact with individuals from group j. Depending on the questions to be asked of the model, the transmission probabilities can be assumed to be constant no matter what type of contact occurs, or they can be made specific to the characteristics of the individuals involved.

Much theoretical work in epidemic modeling has centered on understanding the impact of nonrandom contact patterns on disease transmission, and a variety of simplified patterns of contact have been considered. In a structured setting, the term proportionate mixing is used to describe random mixing across groups. This occurs when individuals have no preference for those with whom they interact and so contacts are made by chance. In this situation, the chance that an individual from group i makes contact with an individual from group j is determined only by the number and activity level of individuals in group j. Assortative mixing occurs when individuals tend to mix more often with others from their own group than expected by chance alone (i.e., under proportionate mixing), while disassortative mixing occurs when within-group contact rates are lower than would be expected under proportionate mixing. Within each of these broad types, a number of different generalized models have been studied. For example, several models developed to examine the spread of HIV considered a type of assortative mixing called preferred mixing, where a portion of contacts are reserved for within-group contacts and the rest are distributed randomly.

In general, it is difficult to find a simple formulation of R_0 in structured populations with nonrandom patterns of contact, but analyses have shown that highly active individuals with many social contacts often contribute disproportionately to transmission. In the context of sexually transmitted diseases, models have shown that these highly active individuals may form a core group within which disease prevalence is high and which can help to maintain circulation of the disease in the population as a whole even though prevalence in the general population is low.

Because of the threat of bioterrorism in the early years of this century and the appearance of new pathogens such as the SARS virus, West Nile virus, and the 2009 strain of H1N1 influenza, structured population models have been used extensively in recent years to help understand and predict the geographic spread of infectious disease epidemics. Detailed data, including airline and other transportation data, are being collected and analyzed to help determine more realistic contact matrices to use in the models. These models have also been used to assess the relative effectiveness of different kinds of control strategies, such as targeted vaccination or prophylaxis, isolation of cases, or culling of infected animals, in order to help slow or limit the worldwide spread of new infectious diseases.

The body of research on structured population models has shown that the nature of mixing assumptions (i.e., who mixes with whom) has strong influences on the timing of introduction of a disease into different groups and the rate of spread of an epidemic within and among groups, as well as the total severity of an epidemic. Heterogeneity in contact and transmission also has important implications for developing efficient strategies to control outbreaks, especially in situations where a small portion of the population, such as a core group, is responsible for a large portion of the active cases. In such a situation, strategies directed at the general population may have little effect overall, while strategies directed at the core group may have a large effect at relatively low cost. Analyses of models incorporating nonrandom mixing make it clear that who mixes with whom can have dramatic effects on patterns of disease spread, and that in situations where these patterns are highly nonrandom, it is essential to include them in order to be able to predict and control epidemic outcomes effectively.

USES AND CONTRIBUTIONS OF EPIDEMIC MODELING

Theoretical epidemiology has provided important insights into the nature of infectious disease outbreaks, and these insights are being used increasingly to find better ways to predict and control the diseases that affect humans and other organisms. Epidemic models have been used hand in hand with more traditional epidemiological methods in the course of responding to several recent epidemics, including the 2001 foot-and-mouth epidemic in the United Kingdom, the 2003 SARS epidemic, the potential spread of avian influenza within humans (which has so far failed to reach epidemic status), and the spread of the 2009 H1N1 epidemic. In each of these cases, modelers were called in to develop sophisticated models for the diseases in question and to explore which of the possible control strategies would be most effective. Much was learned that could be put to use in the emergencies facing the public health community, especially in identifying unique features of spread of each epidemic and the kinds of strategies that might work best given those features.

The modeling activities and results were sometimes controversial, but epidemic models provided an important tool to use in conjunction with traditional epidemiological approaches when faced with the uncertainty of how these newly evolved or reemergent pathogens might have affected their hosts. And as history has shown, pathogens capable of causing major damage to other kinds of organisms are continually evolving, and this, combined with the increasing globalization of all the world's populations, make it essential that a variety of tools and techniques be brought into play in order to control outbreaks of new pathogens. Theoretical epidemiology provides a strong scientific basis for the development of such tools and techniques, and its results can help to direct the use of these tools and techniques toward addressing the fundamental goals of epidemiology—to understand how, when, where, and why diseases occur and how their spread can be controlled.

SEE ALSO THE FOLLOWING ARTICLES

Compartment Models / Disease Dynamics / Individual-Based Ecology / Networks, Ecological / Metapopulations / Single-Species Population Models / SIR Models / Spatial Spread

FURTHER READING

Anderson, R.M., and R.M. May. 1991. *Infectious diseases of humans: dynamics and control.* Oxford: Oxford University Press.

Aschengrau, A., and G.R. Seage, III. 2008. *Essentials of epidemiology in public health*, 2nd ed. Sudbury, MA: Jones and Bartlett.

Diekmann, O., and J.A.P. Heesterbeek. 2000. *Mathematical epidemiology of infectious diseases: model building, analysis and interpretation.* Chichester, UK: John Wiley & Son.

Keeling, M.J., and P. Rohani. 2007. *Modeling infectious diseases in humans and animals.* Princeton: Princeton University Press.

Olekno, W.A. 2008. *Epidemiology: concepts and methods.* Long Grove, IL: Waveland.

Rothman, K.J., S. Greenland, and T.L. Lash. 2008. *Modern epidemiology*, 3rd ed. Philadelphia: Wolters Kluwer.

Sattenspiel, L. (with contributions from Alun Lloyd). 2009. *The geographic spread of infectious diseases: models and applications.* Princeton: Princeton University Press.

Webb, P., C. Bain, and S. Pirozzo. 2005. *Essential epidemiology: an introduction for students and health professionals.* Cambridge, UK: Cambridge University Press.

EVOLUTIONARILY STABLE STRATEGIES

RICHARD McELREATH
University of California, Davis

An evolutionarily stable strategy (ESS; also, evolutionary stable strategy) is a strategy that, if adopted by most individuals in a population, can remain common when challenged by rare alternative strategies. In other words, no rare alternative strategy can increase in frequency when the population adopts an evolutionarily stable strategy. A strategy, used here as in game theory, is a complete plan of action that covers all circumstances. A strategy is an algorithm that produces behavior.

THE CONCEPT

While the formal naming of the concept did not arise until 1973, evolutionary biologists such as Ronald Fisher employed similar logic as far back as the 1930s, notably in the formulation of sex ratio theory. The concept was also important in the development of inclusive fitness theory in the 1960s. John Maynard Smith and George R. Price coined the term evolutionarily stable strategy as a label for strategies that natural selection can maintain at high frequency in a population. Maynard Smith and Price defined an ESS by two conditions. First, a strategy is an ESS if it has higher fitness when interacting with itself than any alternative strategy has when interacting with the ESS. Second, a strategy can also be an ESS if it has higher fitness when interacting with itself than any alternative strategy has when interacting with an alternative.

To clarify these conditions, consider two strategies, labeled X and Y. The average fitness of an individual behaving according to X is determined by which type of individual he or she interacts with. Let $E(X \mid X)$ be the average fitness of an X individual interacting with another X individual. Similarly, let $E(Y \mid X)$ be the average fitness of a Y with an X. Then the strategy X is an ESS provided either

$$E(X|X) > E(Y|X), \quad \text{or} \tag{1}$$

$$E(X|X) = E(Y|X) \quad \text{and} \quad E(X|X) > E(Y|Y). \tag{2}$$

The first condition says that X is an ESS if rare Y individuals do worse than common X individuals do. When Y is rare, it will hardly ever interact with itself, and so only $E(Y|X)$ is relevant for evolutionary stability. The second condition says that an alternative strategy Y can have the same fitness as common X individuals when rare, but X will still be evolutionarily stable, provided that as Y becomes more common, it does worse when interacting with itself than X does when interacting with X.

Together, these conditions identify strategies that can be maintained at high frequency by natural selection, within the context of a game theoretic model. The main use of the ESS concept in biology is to identify possible end points of an evolutionary process. Evolutionary modelers often identify ESS's as a way of generating predictions for the kinds of strategies we might find in animal and plant populations.

The definition of an ESS depends upon a measure of fitness. There are several different fitness measures in evolutionary biology, such as geometric mean fitness used in models of time-varying environments and matrix definitions used in age and class structured populations. The ESS was not proposed with these different definitions in mind, but is often used with them.

AN EXAMPLE: ANIMAL CONFLICT

Maynard Smith and Price's original example concerned fights over indivisible resources, such as nesting sites. When two individuals contest such a resource, imagine that they may adopt one of two strategies. First, they may adopt dangerous fighting tactics, such as using fangs and attacking from behind, that may injure or even kill their opponent. Label this strategy D. Second, they may adopt conventional tactics, such as the wrestling that some snakes engage in, that cannot seriously injure an opponent. Label this strategy C. When two D individuals interact, one of them loses and is injured, while the other gains the value of the resource. Each has a probability 1/2 of winning. When a D meets a C, D always wins, and the C individual escapes uninjured, because it refuses to escalate. Finally, when two C individuals meet, they wrestle harmlessly for some time, until one of them grows tired and loses. Neither is injured, and one chosen at random with probability 1/2 wins the whole resource. To put magnitudes on these benefits and costs, let V be the value of the resource and I be the cost of injury.

With these fitness payoffs in mind, we can ask if either strategy can be an ESS. Consider first C, the conventional fighting strategy. The average fitness of a C who meets another C is $E(C|C) = V/2$. This is because half the time the focal individual wins, receiving V. The average fitness of a rare D in a population of C is $E(D|C) = V$. C can be an ESS if either $E(C|C) > E(D|C)$ or $E(C|C) = E(D|C)$ and $E(C|C) > E(D|D)$. But since $E(C|C) < E(D|C)$, C is never an ESS. A rare D individual always does better.

Now consider whether D can ever be an ESS. $E(D|D) = V/2 - I/2$, because half the time the focal individual wins, getting V, and otherwise loses, subtracting I. $E(C|D) = 0$, because the C individual runs away, losing but avoiding injury. D can be an ESS if $E(D|D) > E(C|D)$, which reduces to $V > I$. If the value of the resource exceeds the cost of injury, then D can be an ESS.

When $V < I$, neither C nor D is an ESS. In this case, it is possible for a mixed strategy, in which individuals play D with some heritable probability and otherwise play C, to be an ESS. In this particular model, the only mixed ESS plays D with probability V/I and C with probability $1 - V/I$. Other examples of mixed ESS's from evolutionary biology include Fisherian sex ratio and optimal dispersal rates.

SOLUTION CONCEPTS

The ESS concept is a refinement of the general equilibrium concept from game theory, used to derive predictions from a model. An ESS is a type of equilibrium, a state of the population at which the frequencies of different strategies do not change over time. There are two basic types of equilibria in the study of simple evolutionary dynamics, stable equilibria and unstable equilibria. In the presence of small numbers of mutations, a population returns to a stable equilibrium. In contrast, a population evolves away from any state identified as an unstable equilibrium, once rare mutants are introduced. Both kinds of equilibria are common in evolutionary models, but only stable equilibria may also be ESSs. A stable equilibrium is also an ESS, provided that at the equilibrium the population contains only one strategy (which may be mixed, as in the animal conflict example).

In economic game theory, no distinction was originally made between these different types of solution concepts, perhaps because economists did not analyze games using evolutionary considerations. Thus, the common Nash equilibrium concept from economics is the same as the general category "equilibrium" in biology, while an ESS is a subset of the "equilibrium" category. Game theory contains many other refinements of the general Nash equilibrium solution concept, most of them being used only in economics. However, the ESS concept is also used in behavioral economics, psychology, and anthropology, in addition to evolutionary biology. The underlying logic of the approach does not require genetic inheritance. And so ESS analyses are used in the study of learning and cultural dynamics, as well.

PROBLEMS WITH THE ESS CONCEPT

The concept of the ESS has a number of issues that may prevent it from adequately describing the evolutionary dynamics of a strategy. Two important limitations include (1) the existence of very many ESS strategies and (2) the possibility that the population can never reach the ESS in the first place.

Very many possible strategies may all be ESSs. In a class of games known as repeated games, it is common to find a large number of strategies that are all evolutionarily stable. In an infinitely repeated game, in fact, there will always be an infinite number of ESSs. This result is often known as the "folk theorem." This multiplicity of possible evolutionary destinations makes ESS analysis unhelpful for making predictions.

Related and equally important is the limitation that the criteria for being an ESS only consider the dynamics once a strategy is common. It may be that a strategy which is an ESS is nevertheless not a good candidate for a long-run evolutionary outcome, because it cannot increase when rare. When this is true, evolutionary dynamics will never lead the strategy to become common in the population. If so, then the fact that it is an ESS is irrelevant; it is an unlikely evolutionary end point.

SEE ALSO THE FOLLOWING ARTICLES

Adaptive Dynamics / Cooperation, Evolution of / Game Theory

FURTHER READING

Fudenberg, D., and E. Maskin. 1986. The folk theorem in repeated games with discounting or with incomplete information. *Econometrica* 50: 533–554.

Maynard Smith, J. 1982. *Evolution and the theory of games*. Cambridge, UK: Cambridge University Press.

Maynard Smith, J., and G. R. Price. 1973. The logic of animal conflict. *Nature* 246: 15–18.

EVOLUTIONARY COMPUTATION

JAMES W. HAEFNER
Utah State University, Logan

Evolutionary computation (EC) is a family of optimization algorithms based on the biological process of evolution by natural selection. In general, the optimization function is a complicated function, typically involving many variables, for which it is desired to find an overall (or global) maximum or minimum value and identify where in the space of variables it attains this maximum or minimum. The problems for which this family is appropriate include those requiring iterative, computational solutions to address large combinatorial problems, model identification, and problems with many local optima.

THE BASIC ALGORITHM

Evolutionary computation attempts to find the global optimum (e.g., a maximum or minimum value) of some complex function through an iterated algorithm that creates many guesses distributed throughout the solution space. In each iteration of the search, the collection of "individuals" are evaluated for their "fitness" in minimizing or maximizing the function. Here, an individual may be thought of as a particular collection of variables in the solution space (i.e., a possible solution to the problem). A subset of the individuals that have high fitness are selected to become the core of the solutions for the following iterations. More formally, the general evolutionary algorithm defines $\mathbf{P}(k)$, a population of N potentially optimal solutions at iteration k, and iterates through the following steps:

1. Initialize $\mathbf{P}(0)$ with random solutions coded in "chromosomes."
2. Evaluate the fitness of each element of the initial $\mathbf{P}(0)$.
3. Recombine elements of the current $\mathbf{P}(k)$ to form a new $\mathbf{P}'(k)$.
4. Mutate $\mathbf{P}'(k)$ to form $\mathbf{P}''(k)$.
5. Evaluate the fitness of $\mathbf{P}''(k)$.
6. Select the best of the $\mathbf{P}''(k)$ to form a new $\mathbf{P}(k)$.
7. Repeat steps 3–7 with $k = k + 1$ until a stopping criterion is met.

As stated here, the method mimics the basic life history biology of an imaginary single-celled organism with sexual reproduction: (a) steps 3–4 are analogous to chromosome reproduction and random mutation to form new population members; (b) step 5 is analogous to determining the reproductive fitness of the phenotype; and (c) step 6 mimics selection to eliminate a portion of the least fit individuals. By retaining a significant number of less fit individuals at iteration *k*, the algorithm explores large domains of the solution space to attempt to locate the global optimum. Like similar computational hill-climbing methods (e.g., steepest descent), by using a finite number of individuals that change in discrete steps (e.g., through mutations), EC only examines a finite subset of the solution space. As a result, there is no guarantee that the global optimum will reside in the subset examined by EC.

There are many variants on this basic algorithm. A key difference is the structure of the "chromosome." Here, a chromosome is the abstract representation of the unknown variables that must be identified. One common implementation uses a literal interpretation of the analogy with a biological chromosome and represents positions along the computational chromosome as bits belonging to "genes." A computational chromosome is composed of a sequence of these genes, each possibly having different numbers of bits, represented as a string. The set of the bits (0s or 1s) constitutes the allele for that gene, where an individual bit is an abstract representation of nucleic acids. As in real biology, this approach requires a method to translate from the genotype to the phenotype (the contribution of the "gene" to the solution). Computational recombination is implemented as exchanges of subsets of the string among the two parents. Mutation occurs by flipping one or more bits.

An advantage of this binary method is that it adheres closely to biology processes and allows an easy connection to the recombination and mutation steps of the algorithm. For some combinatorial problems (e.g., the Traveling Salesperson Problem), the string of bits itself can represent the solution, but for many other problems, the algorithm must translate the bit string to the problem solution. A problem is that if the standard binary coding is used to translate the string to an integer, a single point mutation in the genes can have a large effect on the "phenotype" (the solution), which is contrary to most biological point mutations. For example, a 4-bit string with "0001" has integer value 1, but a mutation at the most signficant bit will alter the phenotype to value 9 ("1001"). This problem is removed by using a Gray code that has the property that each successive integer differs by only a single bit change. Thus, the bit-string representation, while compatible with biological processes, induces computational complexity. Alternatively, some implementations relax the direct connection with biology, representing genes as real numbers, chromosomes as vectors of these numbers, mutations as small perturbations to their values, and recombination as a function applied to the values associated with the parents of new offspring (e.g., the arithmetic average).

MODIFICATIONS

Evolutionary Programming

One of the earliest uses of EC for parameter estimation was by L. J. Fogel, who used evolutionary programming to estimate transition probabilities in finite-state automata. In this approach, genes are coded as real numbers, step 3 is not performed, and for most applications variation in solutions are obtained by random mutation from an *n*-dimensional normal distribution with an iteration dependent standard deviation. Selection (step 6) is performed by a tournament competition in which each solution is compared to *m* randomly chosen alternative solutions and final fitness is determined by the number of tournaments "won" by each solution. Thus, fitness is an index relative to the current set of solutions.

Evolution Strategies

This approach is similar to evolutionary programming but recombination is allowed. Genes are represented as real numbers, mutation occurs by random draws from a probability distribution whose standard deviation varies with iteration, and crossover occurs between genes (or components) when reproducing parents swap values. Selection occurs by comparing parents with their respective mutant offspring.

Genetic Algorithms

In their original formulation, genetic algorithms (GA) used a form of binary coding, had mutation based on bit-flipping, and allowed crossover to occur within genes. Fitness is not restricted to tournament competitions, although they can be used. One of the major differences between a binary coding algorithm and those using real numbers is that substrings consisting of contiguous bits will, if significantly contributing to fitness, persist through several selection episodes due to their resistance to crossover. These "building blocks" (or schemata) then become elements under selection and provide GAs some of their optimization power when they are combined with other building blocks to produce the solution.

GENETIC PROGRAMMING

A large number of problems have solutions that resemble trees. Virtually every mathematical function, computer program, or algorithm can be represented as a tree. This can be visualized using prefix notation in which the operator (e.g., multiplication) is specified first, followed by its two operands. For example, "$y = 3 \times 4$" becomes: "*(3)(4)." This can be visualized as a tree with "*" at the internal node and "3" and "4" being the leaves (terminal nodes) at the end of two branches. Complex operations (or functions) can be given names (e.g., *sine*, for the sine function) with any number of operands according to the requirements of the function. Any mathematical or programming construct (e.g., inequalitites, conditionals, or loops) can be used. Figure 1 is a more complex example that represents a function that computes over a range symbolized by the variable "X" the product of a sine function and the maximum of a constant (0.25) and a negative exponential. In genetic programming (GP), such a tree can be altered by recombination that is performed by swapping branches of parent trees (e.g., swapping the subtree "Angle" with the subtree "Base"). Mutation can be implemented by replacing either internal or terminal nodes with a random, but legimate, subtree (e.g., replacing the subtree "Angle" with a constant).

APPLICATIONS: DATA ANALYSIS VS. THEORY

Most applications of EC are not in ecology or environmental science. A search of *Web of Science* in 2010 over the previous 30 years for the pairs of keywords used above to describe EC and its derivatives revealed 362 of 37,145 hits (1%) were also indexed as "ecology" or "environment." Ecological applications include those concerned with analyzing data and those addressing a specific theoretical question.

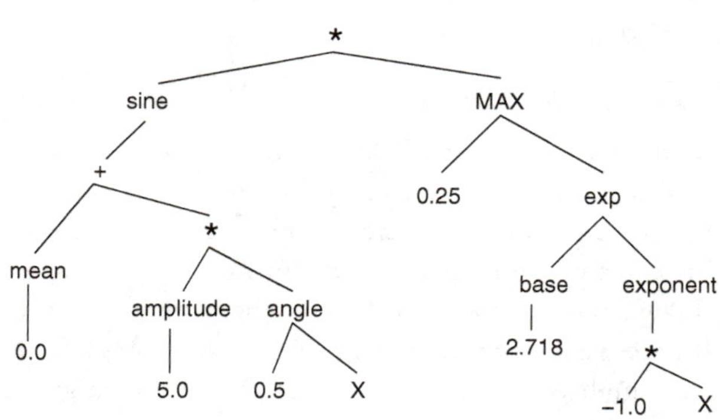

FIGURE 1 A function represented by a tree in prefix notation. Shown is the product of a sine function (with mean, amplitude, and angle variable over a range symbolized by X) and the maximum of a constant and a negative exponential defined over the same range X.

Data Analysis Applications

Data analysis problems solvable by EC include (a) finding the global optimum, (b) estimating parameters in nonlinear equations, (c) solving combinatorial problems, and (d) identifing functions that best describe a data set. High-dimensional parameter estimation with noisy data frequently have multiple local minima, and EC is well suited for the problem of ascertaining appropriate parameter values in complicated ecological models. Symbolic regression (or, model identification) is a major application of GA and GP—for example, modeling species' spatial distributions using *niche models*. The GA-based algorithm GARP (Genetic Algorithm for Rule Set Prediction) is one method to identify niche models. This approach uses a training data set to fit the predictions of species presence and absence at spatial points based on climatic envelopes and other biotic and abiotic variables (i.e., niche variables) in order to define the combination of variables which accurately predicts the distribution of the species. Because the rules frequently have the form of nested conditional statements, they can be represented as trees; consequently, GP-based applications are common. These niche models are applied to reserve design and resource management problems for which an objective is to arrange preserves spatially so as to maximize biodiversity and/or various metrics of resource availability.

Theoretical Ecology Applications

There are fewer examples where EC has significantly affected ecological theory, perhaps because it is difficult to separate EC applied to theoretical ecology from individual-based models (IBMs) that use explicit genetic representations to explore theoretical questions. When IBMs are so applied, they are frequently referred to as *artificial life* (AL) models or genetic-based IBMs (GB-IBM). While using many of the same algorithms as EC, AL, and GB-IBM emphasize the effects of hypothesized genetic mechanisms on genotypic and phenotypic evolution. This modeling approach was pioneered by A. S. Fraser in the 1950s using simulation on the earliest computers.

An example that explicitly uses GA to address a theoretical question is Y. Toquenaga's model of the evolution of life history traits in two competing species of bean beetles. In this system, beetles lay eggs on beans, and the larvae compete by both contest and scramble competition in the bean. Beetles also differ by their rate of development, number of eggs laid per bean, and the depth within the bean at which the larvae forage. Toquenaga and colleagues represented each of the four traits as four genes. Fitness was measured as the number of offspring emerging from the beans, after which the new adults reproduced according to a GA algorithm. Model predictions qualitatively agreed with experimental results: large beans favor scramble competition, small beans favor contest competition, and fitness was maximized if scramble beetles delayed development and used deeper regions of the bean.

GB-IBMs with explicit representations of alleles (i.e., chromosomal structure) are also playing a role in evolutionary ecology theory. These models are evolutionary algorithms without artificial selection imposed by an externally defined objective function (e.g., the minimum of a function). Instead, these models define the fitness function in classical evolutionary terms: net reproductive contribution to future generations.

Michael Doebeli and colleagues used this approach with models that represented chromosomal structure as an undifferentiated sequence of 0s and 1s. Unlike typical GAs (e.g., Toquenaga), the phenotype in Doebeli's models is computed by summing the 1s in the entire sequence. Since the positions of 1s on the chromosome do not influence fitness, recombination is not used, but point mutations do occur. Thus, these models are similar to evolutionary programming. The models demonstrate that this simple representation using competition as the primary ecological process can produce character convergence, sympatric speciation, and the evolution of sexual dimorphic characters.

Another similar, but more abstract, approach builds on the metaphor of organisms as computer programs and in so doing draws indirectly on the concepts of genetic programming applied to adaptive radiation. Following work by Thomas S. Ray in the early 1990s using short computer programs that replicate and compete for computer resources (CPU cycles and memory), Richard Lenski and colleagues showed that the productivity of a digital organism's environment coupled with frequency-dependent selection can permit radiation of bacteria-like organisms to account for maximal biodiversity at intermediate levels of productivity. These results agree qualitatively with Lenski's long-term selection experiments.

In addition to data analysis of trees, GP has been used to identify mathematical functions of a phenotypic trait that optimizes a fitness index. John Koza and Joan Roughgarden used GP to identify a function defining the set of two-dimensional spatial positions at which a lizard should attack a flying insect to maximize energy. The complex solution, while accurate and extensible to other theoretical constraints on lizard behavior and environmental conditions, may not provide the same insights as mathematical analysis of idealized cases that result in simpler equations. A continuing challenge to the use of GP in theoretical ecology is the production of extremely complex functions that are difficult to interpret.

SEE ALSO THE FOLLOWING ARTICLES

Dynamic Programming / Gap Analysis and Presence/Absence Models / Individual-Based Ecology / Mutation, Selection, and Genetic Drift / Quantitative Genetics / Reserve Selection and Conservation Prioritization / Sex, Evolution of / Species Ranges

FURTHER READING

Fogel, L. J., Owens, A. J. and Walsh, M. J. 1966. *Artificial intelligence through simulated evolution.* New York: John Wiley and Sons.

Holland, J. H. 1975. *Adaptation in natural and artificial systems.* Ann Arbor, MI: University of Michigan Press.

Kim, K.-J., and Cho, S.-B. 2006. A comprehensive overview of the applications of artificial life. *Artificial Life* 12: 153–182.

Koza, J. R. 1992. *Genetic programming: on the programming of computers by means of natural selection.* Cambridge, MA: MIT Press.

Olden, J., J. J. Lawler, and N. Poff, N. 2008. Machine learning methods without tears: a primer for ecologists. *The Quarterly Review of Biology* 83: 171–193.

Stockwell, D. 2007. *Niche modeling: predictions from statistical distribution.* Boca Raton, FL: Chapman & Hall/CRC Press.

Toquenaga, Y., M. Ichinose, T. Hoshino, and K. Fugii. 1994. Contest and scramble competitions in an artificial world: genetic analysis with genetic algorithms. In C. Langton, ed. *Artificial life III.* Reading, MA: Addison-Wesley.

F

FACILITATION

MICHAEL W. McCOY, CHRISTINE HOLDREDGE, AND BRIAN R. SILLIMAN

University of Florida, Gainesville

ANDREW H. ALTIERI

Brown University, Providence, Rhode Island

MADS S. THOMSEN

University of Aarhus, Roskilde, Denmark

Facilitation includes direct or indirect interactions between biological entities (i.e., cells, individuals, species, communities, or ecosystems) that benefit at least one participant in the interaction and cause harm to none. Research on ecological facilitation has steadily increased over the past three decades and is now appreciated as a fundamental process in ecology. Facilitation also has many important implications for problems in applied ecology and conservation.

HISTORICAL CONTEXT

Understanding the processes governing species coexistence and community structure is a central goal of community ecology. Historically, most ecological research has focused on the negative effects of abiotic or biotic interactions as the primary drivers of species occurrence and the organization of ecological communities. Consequently, negative ecological interactions, such as competition and predation, contribute disproportionately to the conceptual foundation upon which most ecological theory is built. However, over the past three decades there has been a growing body of literature highlighting the important role of facilitative interactions for population- and community-level processes.

Before discussing recent conceptual advancements leading from facilitation research, it is useful to briefly reflect on the history of ecology and facilitation research. The focus of ecologists on antagonistic species interactions can be traced back even to the publication of *The Origin of Species* in 1859, in which Charles Darwin metaphorically described species as wedges. In this metaphor, ecological space (the resource pie) is divided into a series of wedges that represent species' population or range sizes (i.e., proportion of the pie occupied). The addition of a new species to the resource pie or an increase in the size of the wedge of an existing species must necessarily come at the expense of other species. Indeed, the footprint of this metaphor is imprinted on the conceptual foundations of ecology and lies at the heart of many core ecological concepts, such as the competitive exclusion principle, the niche concept, island biogeography, community assembly, and community invasibility. Interestingly, in addition to setting the stage for research on negative interactions, Darwin also laid the groundwork for studying facilitation by recognizing that reciprocally positive interactions could arise in nature as a result of organisms acting with purely selfish interests. However, his insights did not permeate ecological thought until the mid-twentieth century. Indeed, it is now widely recognized that consideration of facilitation fundamentally changes our understanding of many core ecological concepts and greatly enhances the generality and depth of ecological theory. The stage has been set for rapid development of ecological knowledge as positive interactions, negative interactions, and neutral processes are incorporated into a more sophisticated archetype for ecology.

INTRASPECIFIC FACILITATION, MUTUALISM, AND COMMENSALISM

Although ecologists traditionally invoke facilitation to describe interspecific interactions, facilitation between individuals of the same species (intraspecific facilitation) also plays a key role in driving population and community dynamics. Under some circumstances, organisms can experience positive density dependence whereby individuals living in aggregations have higher growth rates, survivorship, and reproductive output. These benefits arise via a wide variety of mechanisms ranging from the reductions in per capita risk that occur as predator consumption rates saturate at high prey densities (e.g., dilution effects) to the buffering effects of neighbors against harsh physical stressors (e.g., in rocky intertidal zones). Positive density dependence also occurs at low densities via Allee affects where a species' population growth rate rises with increasing density via increased fertilization success and propagule survival.

Facilitative interactions among populations and species are generally categorized as mutualism or commensalism. Mutualism is a specific form of facilitation in which interactions are reciprocally positive for all participants (Fig. 1). Mutualism includes highly transient reciprocally positive interactions as well as interactions that have emerged as a result of long coevolutionary histories between participants (such as plant-pollinator and plant-disperser networks). Commensalism, in contrast, is reserved for cases of facilitation where at least one participant in an interaction is positively affected while others are neither positively nor negatively impacted. Although the literature is replete with examples of mutualisms, there is a relative scarcity of examples of commensalism in nature. Consequently, some ecologists debate the relevance of this term, arguing that species interactions are more likely asymmetrical in strength (i.e., one species exhibits a strong positive response to the other, while the other exhibits a weak positive [or negative] response to the first), rather than being truly commensal in nature.

FIGURE 1 Example of mutualism. In this interaction, coral (*Pocillopora sp.*) provides habitat, shelter, and foods for crabs (*Trapezia rufopunctata.*), which, in turn, provide the coral protection by repelling coral predators. Photograph by Adrian Stier.

THEORETICAL PERSPECTIVES

While there has been much recent interest in facilitation, there is still much to learn about how positive interactions influence population and community dynamics. Theoretical developments on positive interactions have only recently begun to move beyond phenomenological descriptions to identify general mechanisms that drive ecological dynamics. Indeed, a number of recent advancements have illustrated the essential role of positive interactions for key ecological phenomena such as ecological community assembly, determining the geographical distributions of species, maintaining species coexistence, and influencing the diversity and stability of ecological communities.

Simple two-species models have most commonly been used to investigate the effects of positive interactions on species coexistence and to examine the environmental conditions where positive interactions are expected. For example, as early as 1935, Gause and Witt examined two-species Lotka–Volterra models and showed that positive interactions can be destabilizing when they are strong because they create positive feedbacks between species (e.g., mutualisms) that can lead to infinite population growth. When interaction strengths are weak or strongly asymmetrical (e.g., commensalisms), however, positive interactions can have stabilizing effects, especially when they occur in conjunction with external mortality sources such as predation, disturbance, and stressful environments. Recent investigations have shown that incorporating nonlinearities via density dependence in cost–benefit functional responses (i.e., positive effects saturate with increasing population density) has a stabilizing effect in these models. In fact, mutualistic communities characterized by nonlinear functional responses have positive complexity–stability relationships that suggest positive interactions may be important drivers of community resilience as a whole.

Phenomenological models (e.g., simulations and agent-based models) have extended the findings of a large body of empirical research to generate important insight into the conditions where positive interactions are expected to play a key role in species coexistence and persistence. In general, positive interactions are important for promoting species persistence in severe environments (e.g., arctic, salt marsh, and desert ecosystems) and extending the geographic distributions of species along range boundaries. Specifically, positive interactions can expand species range limits by enabling the expansion of the realized niche into more severe environments than would be possible without positive interactions. In these cases, one species makes local environments favorable for colonization by a second species by directly or indirectly enhancing access to resources, dispersal rate, or provision of refuge from competitors, predators, or abiotic stress. Facilitative interactions, however, are often context dependent. Most interactions between species have positive and negative components, and the relative strength of positive and negative effects is often determined by the environmental context of the interaction. For example, mutualism can change to competition along a stress gradient, and so species can exist as facilitators and competitors in different zones of a landscape.

Although models examining the effects of positive interactions on the dynamics of species pairs have provided progressive insights into the conditions where facilitation promotes species coexistence, population stability, and range limits, species rarely interact and coexist in isolation from other species or habitats. Indeed, most natural communities are characterized by complex webs of interacting species that are spatially linked to other communities via dispersal or species movements. It is likely that the effects of facilitation will not be intuitive extensions of two-species models in these multispecies and multipatch assemblages.

Several recent studies have, in fact, begun to examine spatially explicit multispecies metacommunity models that include diverse interspecific interactions and environmental stress gradients. These models reveal some general expectations for the role of facilitation in community assembly and ecosystem diversity and function. For example, poor habitat quality and low spatial connectedness are expected to favor the emergence of highly stable local communities that are strongly facilitative but characterized by low diversity and low productivity. In contrast, high habitat quality and high connectedness among metacommunity patches are expected to promote local communities that are less facilitative and less stable but characterized by high diversity and productivity (Fig. 2).

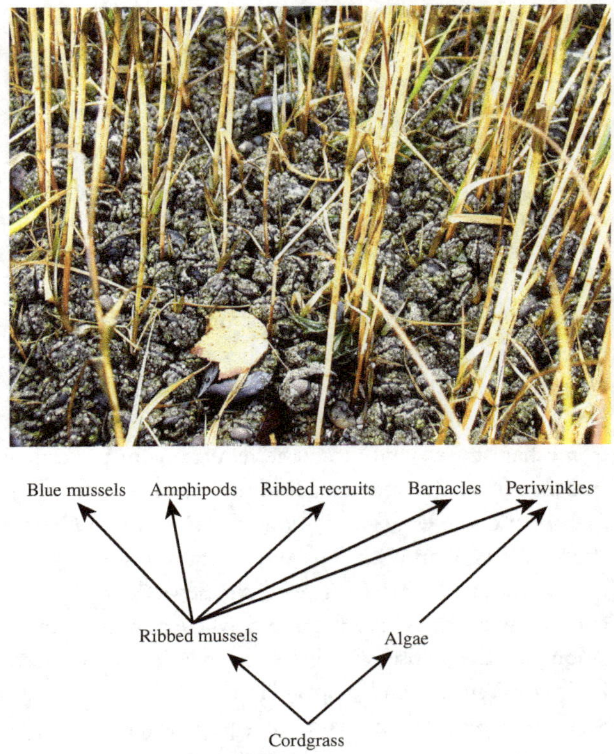

FIGURE 2 Example of a facilitation cascade in cordgrass/mussel bed communities. The establishment of cordgrass intiates a facilitation cascade whereby the establishment of ribbed mussels is facilitated by cordgrass. The synergistic effects of the cordgrass and mussels, in turn, facilitate the establishment of a variety of other taxa, including other species of mussels, barnacles, amphipods, and snails. Photograph by Andrew Altieri.

EMPIRICAL PERSPECTIVES

Foundation Species and Community Facilitation

The structure and dynamics of many widely recognized ecological communities are influenced by facilitation. In fact, many communities are identified by a foundation species that provides or creates the physical structure, conditions, and boundaries of a community (e.g., kelp beds, coral reefs, hardwood forests, mangrove stands, salt marshes, phytotelmata, and so on), which directly or indirectly facilitates a diverse array of species. Though a range of interactions including faciliation, competition, and predation may occur among species in the community, the overall persistence of the community is facilitated by the foundation species. The exact mechanism by which a foundation species facilitates a community varies among habitats. In physically stressful environments, foundation species typically ameliorate environmental stress, whereas in more benign environments they more typically provide refuge from predation. By creating patches of suitable habitat, foundation species influence community structure on a landscape scale and can be important drivers for both local and regional patterns of diversity.

Although community facilitation is often attributed to a single foundation species, it can also be driven by the synergistic interactions between two or more species that in concert provide the foundation for communities. Facilitation of communities via synergistic interactions among foundation species has been explicitly identified in only a few habitats, such as cobble beaches and coral reefs, but could be widespread in habitats that are defined by mixtures of species, such as sea grass meadows. However, like other forms of facilitation, community facilitation is likely context dependent, with the interaction between foundation species changing from synergistic to antagonistic along an environmental gradient. At a landscape scale, this context dependency can lead to abrupt changes in species composition if foundation species exist as faciliators and competitors in different zones of the landscape.

Ecosystem Facilitation

Facilitation is also an important process operating at the highest level of ecological organization—ecosystems. Ecologists have long recognized the importance of fluxes of energy, matter, and organisms for driving ecosystem processes. When this spatial coupling benefits foundation species and their associated species assemblages in a way that increases ecosystem services (e.g., increased stability, productivity, resilience, and so on), then the spatial coupling serves as a form of "ecosystem facilitation." Ecosystem facilitation differs from community facilitation in that it typically occurs along spatial gradients and refers to positive interactions across ecosystem boundaries, whereas community facilitation typically occurs along stress gradients and refers to species facilitation within a single community or ecosystem type. For example, in aquatic systems ecosystem facilitation occurs when plankton produced in the pelagic zone sinks and provides food and energy to support and maintain the benthic ecosystem (benthic–pelagic coupling). Ecosystem facilitation is particularly common between aquatic and terrestrial ecosystems and between productive and unproductive systems. For example, offshore sea grass beds and coral reefs can attenuate storm waves to protect intertidal mangroves and salt marshes that then provide additional energetic protection for inland terrestrial ecosystems. Mangroves and salt marshes can also reciprocally facilitate offshore sea grasses and coral reefs by catching and retaining suspended sediments and nutrients that can drive epiphytic growth and shading that is harmful to seagrass and coral ecosystems.

In these and most other examples of ecosystem facilitation, the positive interactions among ecosystems occur via provision of limiting resources (spatial subsidies of nutrients, organic matter, shelter, habitat to live in) or by reducing environmental stressors (e.g., sediments, toxins, flooding, storm disturbances, abrasion). However, spatial coupling among ecosystems is not always facilitative and can become detrimental in some contexts. For example, although plankton deposition is important for benthic ecosystems, excessive depositions may ultimately disrupt the system by choking filter feeders and stimulating bacterial blooms.

Finally, it is important to note that research on ecosystem facilitation, in contrast to population and community facilitation, is dominated by "correlative studies." Natural ecosystems are typically too large and complicated to manipulate, so the mechanistic basis of ecosystem facilitation is often difficult to identify. Thus, where manipulative experiments cannot be conducted, a weight-of-evidence approach is needed that combines rigorous site selection and data collection criteria, statistical modeling, and natural history.

CONSERVATION APPLICATIONS

Conservation biology is a goal-focused science whose primary objective is to study, protect, and preserve preidentified targets. Historically, conservation targets were specific species of concern, and conservation strategies focused on minimizing negative species and environmental interactions. Stressor-reducing approaches, however, fail to incorporate positive interactions into their designs and are being replaced by strategies that identify and harness positive interactions. Specifically, recent advancements in conservation research have instigated the expansion of conservation targets to include entire ecosystems (i.e., restoration of foundation species) and the functions and services they provide.

The common omission of positive interactions in conservation and restoration approaches has likely resulted in missed opportunities to enhance conservation projects at no increased cost, as many natural synergisms do not emerge from restoration and preservation designs focused on minimizing negative interactions. For example, when restoring coastal marshes and mangroves, traditional planting designs have been plantation style, with all plugs planted at far enough and equal distances from each other to ensure no competition. Experimental work in these systems over the past 20 years, however, has shown that mangrove and marsh plants, when growing in stressful mudflats, grow better in larger clumps and when these clumps are placed closer together. The improved growth stems from the benefits plants receive from the aeration of soils by other nearby plants. These studies also clearly show that these cooperative benefits far outweigh the negative impacts of competition for nutrients between plugs. Thus, by not updating its designs and theoretical approaches to incorporate recent findings that highlight the importance of positive interactions in the success of species under harsh physical conditions, restoration ecology is failing to take advantage of naturally occurring synergisms among species and individuals.

While this wetland restoration example is focused on the importance of incorporating positive interactions at the population level, such synergisms can also occur at the ecosystem level. For example, some approaches to protecting shoreline habitats have already begun to incorporate positive interactions among species, ecosystems, and man-made structures. The overall goal is to maximize positive interactions that surrounding or overlapping ecosystems would naturally provide each other but were lost under the old paradigm of coastal protection (remove all buffers in favor of stronger man-made structures). The combined use of hard structures to fend off flooding and erosion and wetland plant ecosystem restoration can be effective if we identify compatible and complementary aspects of engineering and vegetation adaptation measures. An excellent example comes from Dutch engineering and conservation efforts, where coastal engineers have tried to "build with nature" to increase the resilience of their man-made structures to oceanic disturbance. Levees built to prevent flooding during storms are maintained with a thick grass cover to increase their integrity. In addition, more recent efforts have focused on placing willow trees and marsh grasses just ahead of man-made levees to reduce wave action on and water levels around the protective barrier. The benefits of this type of positive-interaction engineering go well beyond natural ecosystems enhancing the integrity and efficacy of man-made structures, as the services of the planted ecosystems are not limited to this one interaction. For instance, planted marshes likely increase fishery production in surrounding areas, and willows increase carbon sequestration and habitat for local songbirds. Human utilization of the shorelines also increases, as the natural ecosystems planted on top of and around man-made structure provide recreational opportunities.

Despite advances in positive interactions and facilitation theory in ecological research over the past 20 years, the concepts have failed to make a large impact on conservation and restoration ecology. As highlighted in the examples above, incorporating positive interactions into conservation and restoration practices can occur at the organismal, population, community, and ecosystem levels and can reap substantial benefits with little additional investment in resources. Conservation and restoration plans simply need to be modified to explicitly integrate positive interactions. The old paradigm of applying terrestrial forestry and wildlife theory (i.e., minimizing competition and predation on target species) to modern-day conservation and restoration efforts needs to be updated with current ecological theory revealing that positive interactions among species under harsh conditions (i.e., those of stressed targets) are paramount to those species continued existence.

SEE ALSO THE FOLLOWING ARTICLES

Conservation Biology / Metacommunities / Resilience and Stability / Restoration Ecology / Spatial Ecology / Stress and Species Interactions / Succession / Two-Species Competition

FURTHER READING

Altieri, A., B. Silliman, and M. Bertness. 2007. Hierarchical organization via a facilitation cascade in intertidal cordgrass bed communities. *The American Naturalist* 169: 195–207.

Bertness, M., and R. Callaway. 1994. Positive interactions in communities. *Trends in Ecology & Evolution* 9: 191–193.

Bruno, J., J. Stachowicz, and M. Bertness. 2003. Inclusion of facilitation into ecological theory. *Trends in Ecology & Evolution* 18: 119–125.

Stachowicz, J. J. 2001. Mutualism, facilitation, and the structure of ecological communities. *Bioscience* 51: 235–246.

Thomsen, M. S., T. Wernberg, A. H. Altieri, F. Tuya, D. Gulbransen, K. J. McGlathery, M. Holmer, and B. R. Silliman. 2010. Habitat cascades: the conceptual context and global relevance of facilitation cascades via habitat formation and modification. *Integrative and Comparative Biology* 50: 158–175.

FISHERIES ECOLOGY

ELLIOTT LEE HAZEN

Joint Institute for Marine and Atmospheric Research, University of Hawaii

LARRY B. CROWDER

Stanford University, California

Fisheries ecology is the integration of applied and fundamental ecological principles relative to fished species or affected nontarget species (e.g., bycatch). Fish ecology focuses on understanding how fish interact with their environment, but fisheries ecology extends this understanding to interactions with fishers, fishery communities, and the institutions that influence or manage fisher behaviors. Traditional fisheries science has focused on single-species stock assessments and management with the goal of understanding population dynamics and variability. But in the past few decades, scientists and managers have analyzed the effects of fisheries on target and nontarget species and on supporting habitats and food webs and have attempted to quantify ecological linkages (e.g., predator–prey, competition, disturbance) on food web dynamics. Fisheries ecology requires understanding how population variability is influenced by species interactions, environmental fluctuations, and anthropogenic factors.

THE DISCIPLINE

While some researchers view fisheries ecology as basic ecology applied to fisheries management, the complex interactions among individual organisms, trophic levels, and a changing environment have spanned a number of ecological fields. The field of fisheries ecology was described by Magnuson in 1991 as the intersection of ichthyology, fisheries science, and ecology, and also including theory from a variety of ecological disciplines including physiological ecology, behavioral ecology, population ecology, community ecology, ecosystem ecology, and landscape ecology (Fig. 1). More recently, the fields of historical ecology and social-ecological systems have aimed to better define a baseline for change and to reintegrate human dimensions into ecology. These diverse fields highlight the interdisciplinary nature of fisheries ecology but also the interactions at various scales and across individuals, species, communities, and ultimately ecosystems.

Physiological Ecology

The field of physiological ecology initially examined the effects of physical features such as dissolved oxygen (DO), temperature, and salinity on mortality of individual organisms. Similar studies have extended data from individual organisms to identify potential habitat for an entire species using thermal, salinity, and DO preferences and support the concept of environmentally mediated niches.

Behavioral Ecology

Behavioral ecology includes sensory perception and decision making in individual organisms. One particularly applicable example is Pacific salmon that live in saltwater and spawn in freshwater, homing to the same stream where they hatched. By following chemical cues from river outflow, they are able to return to their natal stream to spawn. Decision-making choices such as migration from breeding to feeding grounds and foraging theory such as when to leave one feeding ground for another have been aided by the development of archival tags

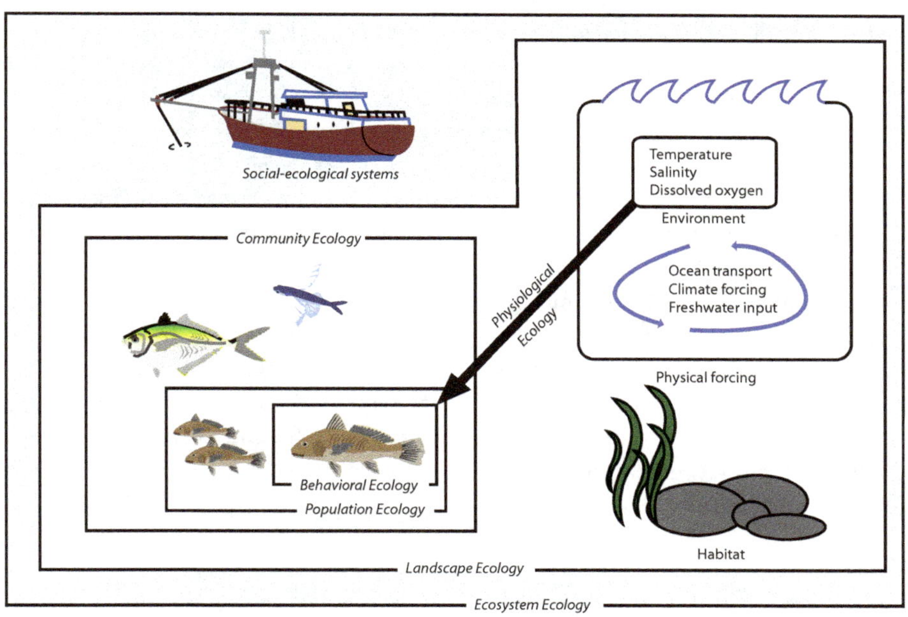

FIGURE 1 Fisheries ecology consists of multiple ecological disciplines (shown in italics) and can focus on the behavior of a single species to the ecosystem and on the economic effects of fishing. *Physiological ecology* focuses on the effects of physical variables in the environment on cell biology and mortality; *behavioral ecology* focuses on the response of individuals or schools of fish to sensory cues; *population ecology* focuses on measuring growth rates, biomass, and age structure of a population; *community ecology* focuses on the interactions among species and how that affects food web structure and population dynamics; *landscape ecology* focuses on how physical features including habitat affect distribution and abundance of one or many species; *ecosystem ecology* examines trophic interactions, physical forcing, and the resiliency of ecosystem states; *social-ecological systems* focuses on the human dimensions of how fishermen affect and are affected by fisheries and the environment; and *historical ecology* focuses on long, often prehistoric time series (not shown).

(e.g., satellite) that can measure movement patterns in otherwise hard-to-study species.

Population Ecology

Population ecology has served as the primary tool of fisheries management since the mid-twentieth century. Since the passing of the Magnuson–Stevens Act in 1976, fishery targets have been set by modeling the maximum sustainable yield (MSY) that keeps a population at an optimal spawning stock biomass (SSB). In unfished populations, biomass would plateau as competition for prey and habitat result in a maximum carrying capacity. The MSY is the peak of the curve at which fisheries stocks are kept at a maximum population growth rate by fishery harvest, although interaction effects make this target difficult to achieve. Life history theory has also been incorporated into management such that individual life stages, particularly those that are vulnerable, can be managed independently. Transition matrices can be used to model probabilities of transition from one life stage to another (e.g., juvenile to spawning adult) as a function of fishery and natural mortality.

Community Ecology

Community ecology focuses on multiple species and their interaction strengths through competition or predation. As predation is a driver of mortality for the prey and fitness for the predator, it has become the dominant model for understanding and modeling species interactions. A classic example of how predation and competition can structure ecosystems is the overfishing of cod (*Gadus morhua*) in the northwest Atlantic, where human predators drove cod to functional extinction. Even with fisheries closures, cod have not recovered as would have been predicted by a single-species model because of competition with other predators in the ecosystem such as dogfish. Because cod were a key interactor in the northwest Atlantic food web, their reduction allowed other species to thrive and occupy their niche, further hampering their recovery. This example highlights both the need to consider multispecies interactions in addition to cod's top predator, human fishers.

Landscape Ecology

While landscape ecology may seem a misnomer for a subdiscipline of fisheries ecology, the focus on patterns in spatial structure and scale have become an important consideration in understanding both species distributions and interactions. Traditional landscape ecology examines two-dimensional distribution in the terrestrial environment. In marine systems, depth becomes an important dimension. Ecological processes are inherently scale dependent both in their interactions and the way they are measured. For example, small zooplankton react to microturbulence in the water at the scales of centimeters, while large fish interact with fronts and eddies at the scale of 100–1000 meters. Moreover, because prey and predator often aggregate at different scales, their interactions can alternate from in and out of phase, dependent on the scale of observation. Understanding these scales of interaction and applying appropriate measurements and models is necessary to move toward spatially explicit ecology and management.

Ecosystem Ecology

Ecosystem ecology has focused on connecting multiple species interactions with physical forcing mechanisms with the goal of understanding how species affect and are affected by the environment. Ecosystem models have used energy flow among trophic levels as a currency to understand ecosystem change. Examples include the effects of direct fishery removal and answering scenario-based questions such as how management policies or environmental change could affect ecosystems of interest. Examples include Ecopath and Atlantis, both of which allow scenarios to propagate through the food web.

Social-Ecological Systems

In the past decade, human dimensions have been recognized as an often-overlooked component of ecosystems, both as drivers of change and as users of ecosystem services. Fishers have been modeled as economic actors, a more complex predator that seeks out prey (fishery landings) with an informed knowledge and cost structure (e.g., license and fuel costs). Economic-based behavior models can directly inform management decisions, creating a more holistic view of fisheries management as managing fishers and the fishery in concert. The future of fisheries ecology will require merging the theoretical with the applied using relevant spatial and temporal frameworks to create interdisciplinary fisheries management tools that focus on the ecosystem from physical processes to human behavior.

CHALLENGES IN FISHERIES ECOLOGY

Direct Effects of Fishing

Fishing has been one of the earliest forcing effects and has had the greatest impact on marine ecosystems. Between 50–90% of top predators have been removed from the world's oceans, including species that are not targeted by the fishery, such as a number of shark species. Many top predators such as elasmobranchs and marine mammals are long lived and have slow reproductive rates, making sustainable harvest near impossible. The longer it takes an individual to become reproductive, the likelihood increases that individuals will be caught before they have a chance to reproduce. As top predator biomass declines, fishing pressure often shifts to lower trophic levels to maintain harvest rates, a process that Daniel Pauly has termed "fishing down marine food webs." Many of the mid-trophic species such as sardines and anchovies are important forage fish that serve as primary prey resources in pelagic food webs. Fisheries targeting mid-trophic-level species can result in user conflicts between fisheries and protected species needs. For example, Steller sea lions (*Eumetopias jubatus*) in the western Aleutian Islands are listed as endangered because they have declined over 40% between 2000 and 2008 following severe stock reductions from fur harvest. In 1999, federal managers closed pollock (*Theragra chalcogramma*) fishing in Steller sea lion critical habitat, and in 2010 have suggested closing Pacific cod (*Gadus macrocephalus*) and Atka mackerel (*Pleurogrammus monopterygius*) because they are a primary prey item of juvenile and adult sea lions.

Fishing pressure leads to direct biomass reduction of target species but can have other direct effects. Because most fisheries have size limits, catching only large fish can have population-level effects. In 2007, Law described the resulting intense selection pressure that can alter the genetic composition of fished populations. Populations mature earlier and at smaller body size in response to years of size-selective fishing. As only large individuals are caught in trawl nets, slow-growing and smaller fish that can escape the mesh have increased reproductive success. In the northwest Atlantic, Atlantic cod mature earlier and at smaller body size in response to years of fishing for larger individuals. Changes in population genetics may not be easily reversible and will require an evolutionary perspective to help long-term protection of species and maximization of yields.

Indirect Effects of Fishing

When fisheries target species that are strong interactors, extreme fishing pressure can result in trophic cascades, regime shifts, or in worst cases, ecosystem collapse. In diverse systems, competitors can proliferate and replace the niche that was made available by species removal. Reduction in apex predators can relieve both competition and predation pressure from mid-level predators resulting in an abundance of mid-trophic species, a process that has been termed mesopredator release. Although mid-tropic-level species are usually less valuable from a fisheries-removal perspective, their release can minimize cascading effects through the food web. In extreme cases, fisheries can trigger regime shifts, particularly in systems

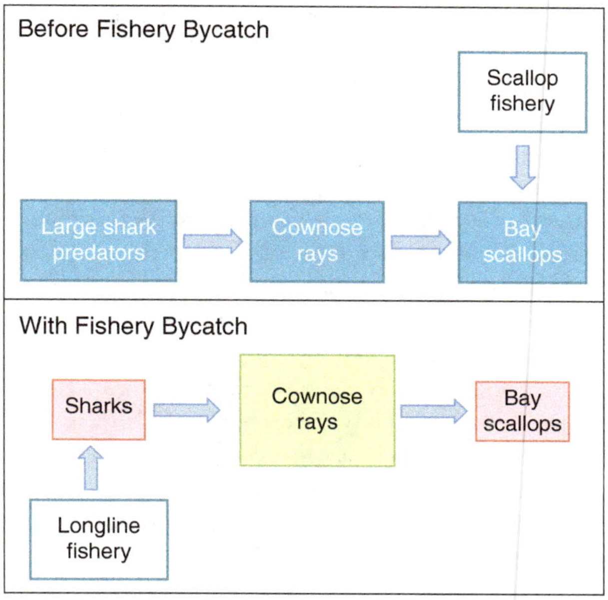

FIGURE 2 Representation of a trophic cascade from reduction of sharks in coastal North Carolina. Each box represents a species or a functional group, with arrows representing predation pressure. As longline fishery effort increased, declines in large sharks resulted in increases in small sharks, skates, and rays. With increased mesopredator abundance, bay scallop abundance crashed, resulting in the closure of the long-standing fishery.

with low functional diversity, when a strong competitor or keystone species is removed from the food web.

Trophic cascades can restructure entire ecosystems and can have unforeseen effects on fisheries landings (Fig. 2). In 2007, Myers and colleagues observed a trophic cascade following the removal of a complex of top predators in the western Atlantic. High bycatch rates in coastal fisheries led to population declines in seven species of coastal sharks of 87–99% from 1970 to 2005. With the functional removal of these top predators, predation pressure was released on smaller elasmobranchs (skates, rays, small sharks) in the mid-Atlantic. During the same period of large shark decline, the smaller elasmobranchs increased at a rate of 1.2 to 23% per year. The cownose ray (*Rhinoptera bonasus*) undergoes a mass migration southward to Florida wintering grounds each autumn and feeds primarily on bivalves in shallow water. With mid-trophic-predator increases such as the cownose ray, bay scallop (*Argopecten irradians*) population numbers crashed, leading to the closure of a century-old fishery in 2004. Predator exclusion experiments confirmed the cownose ray migration increased predation pressure on the bay scallop to the point of reproductive failure and fishery closure.

In addition, when ecosystem engineers are lost, such as algal grazers, food webs can shift from a top-down to bottom-up control. In 2006, at sites where herbivorous grazers were low in abundance, Newman and colleagues found that there was an increase in fleshy algae growth that can cover and smother coral reefs. Instead of top-down control where predators and grazers maintain a coral-dominated ecosystem, the system can become bottom up where algal density is nutrient driven. Although complex food webs are more resilient to change, reduction in diversity leaves them vulnerable to collapse and they can be difficult if not impossible to recover.

In addition to harvesting top predators, overharvesting of important prey resources such as forage fish can decrease prey availability to top predators. If there is no excess production in forage fish, top predators and fisheries can compete for resources. In the highly productive Peruvian upwelling system, sardine (*Sardinops sagax*) and anchovies (*Engraulis ringens*) cycle based on the strength of upwelling, driving the distribution of top predators. During El Niño years, a lack of offshore transport results in warm nutrient-poor water dominating the coastal environment and leading to reduced productivity and stock biomass of forage fish. This reduction of an important forage base echoes through the food web and results in a decline in seabirds and marine mammal predators. Fishery landings also are proportional to anchovy and sardine populations, but the competition between fisheries and top predators can lead to further declines in top predator numbers. These interactions between fishery needs and marine food webs echo the need to manage the ecosystem as a whole.

Bycatch

Bycatch includes all target and nontarget species that are inadvertently killed, injured, or otherwise incapacitated but not retained. If fishers retain nontarget species, they are considered catch. For some marine species such as loggerhead turtles (*Caretta caretta*), bycatch mortality encompasses nearly all of their fishing-related mortality. From a population dynamics perspective, bycatch-associated mortality is indistinct from direct harvest mortality. For bycatch often more than target catch, it is challenging to determine the impacts on marine ecosystems because less information is available regarding the magnitude of removals. Life history characteristics (growth, age at reproduction) can be better predictors of fishing impact than trophic level, and behavioral characteristics of the species may play a major role in their vulnerability, such as birds that are attracted to discarded bycatch or baited longline hooks. In examples where bycatch species are slower growing and later maturing than target catch, their removals can exceed mortality limits and trigger management action before target populations meet quota. This effect can place bycatch taxa at risk of inadvertent overexploitation in even well-managed fisheries.

Seabirds, sea turtles, marine mammals, and chondrichthyans are among the most bycatch sensitive of the long-lived taxa. Although few fisheries target chondricthyans (sharks, rays, and chimaeras), they are common in bycatch and are increasingly retained in many fisheries. Global reported landings have increased steadily since 1984, and numerous studies have found declines in abundance exceeding 95%. For example, shark species are commonly caught as bycatch in the northwest Atlantic longline fishery. In 2003, Baum and colleagues used logbook data to examine rates of population decline in sharks from 1986 to 2000. All of the species caught including coastal and oceanic species showed extreme population declines (60–89%), including localized extinctions such as white sharks (*Carcharodon carcharias*) off the coast of Newfoundland. In some fisheries, juveniles of target species are considered bycatch because when caught below the legal limit, they will be discarded with many injured or already dead. From a fisheries management perspective, the main difference between bycatch and target catch is simply a difference in available data.

Recreational and Artisanal Fisheries

Recreational and commercial fisheries are fundamentally different and present distinct management challenges. Unlike most commercial fisheries, recreational fisheries remain open access with little data available to managers. While most U.S. states issue saltwater fishing licenses, there is generally no mechanism to limit total effort, making it difficult to regulate or reduce total harvest. In many regions, recreational fisheries lack consistent data collection on effort, landings, discards, and expenditures. In the United States, the National Marine Fisheries Service currently uses telephone and intercept surveys to gather data. But the high volume and diffuse nature of angler effort presents sampling challenges, making it difficult for managers to monitor trends in real time. Recreational anglers are also driven by entirely different motivations than their commercial counterparts. Managers often suffer from the mistaken assumption that recreational fisheries are self-regulating and that anglers will exit the fishery if the fishing experience somehow declines in quality. But if anglers are willing to tolerate low catch rates, recreational fishers may be more likely to exploit fisheries to the point of collapse.

The impact of large numbers of small-scale fishers can be rapid and similar in magnitude to commercial fisheries. While artisanal fisheries often operate in a smaller geographic area than their commercial counterparts, the near-shore environment can still suffer from local depletion of fish stocks, habitat damage, and bycatch. In 2007, Peckham and fellow researchers in Baja California Sur, Mexico, found a high degree of overlap between loggerhead sea turtle (*Caretta caretta*) hotspots and small-scale fishery effort. After assessing bycatch rates in bottom-set gillnets and pelagic longlines, two small-scale fisheries had loggerhead turtle bycatch rates (individuals per hook) an order of magnitude greater than industrial longline fisheries. Yearly catch was on the order of 1000 loggerheads, a comparable number to the 1300 loggerheads taken per year in the North Pacific industrial longline fleet. While artisanal fisheries typically do not operate crushing or habitat-altering gear such as trawl nets or dredges, they can damage key structure-forming organisms when walking on coral reefs or using spears or nets. In order to manage global fisheries and the cumulative effects from anthropogenic impacts, better data on artisanal fishing effort, catch rates, and bycatch numbers are needed.

The heterogeneity and complexity of small-scale fisheries provide challenges to assessing their impacts and contrast directly with large commercial-fishing operations. Consequently, mitigation of the impacts of artisanal fisheries proves challenging and may be region and/or gear specific. There is strong evidence for impacts on community structure and trophic relationships, but the general direction and processes do not differ from other sectors of the fishing community. Artisanal fisheries may be somewhat unique with respect to (a) predominance in the inshore zones, (b) multispecies catch and low discard ratios, and consequently (c) low nutrient input from discards to other marine fauna. Better management of post-harvest losses could reduce the demand and subsequent rate of removals. Where artisanal fisheries overlap with intrinsically vulnerable taxa such as sea turtles or marine mammals, a large number of small-scale fishers can rival the impact of commercial fisheries and generate rapid declines.

However, some of the best examples of fisheries management also come from small-scale fisheries where communities manage their marine and terrestrial systems in concert and as a renewable resource rather than a rush to catch fish. These management principles have been incorporated in commercial fisheries where fishers are given "ownership" of the resource on a yearly basis. This approach works both by preventing overcapacity in the fishery and by providing security to have fishing rights in future years and providing a sense of stewardship.

Fishery Recovery

Unfortunately, few examples exist of successful recovery of heavily exploited fishes. A 2004 study by Caddy and Agnew showed that the North Atlantic swordfish recovery was successful because depletion was not excessive (70% of the maximum sustainable yield), recovery time was relatively short (10 years), and the fishing fleets cooperated in management due to their similarities and shared incentives. Otherwise, evidence for recovery of long-lived, slowly maturing top predators is limited and dependent on many factors, both biological and political. Scientists are beginning to agree that the escalating crisis in marine ecosystems is in large part a failure of governance rather than a failure in fishers. Management structures such as catch shares provide a degree of ownership to the fishery and incentives toward preserving future catch rates. Techniques that optimize fishery involvement and consequently incentives toward future sustainability may be a large part of the solution. Many recent assessments and government objectives have called for a transition from managing sectoral activities, including fisheries, toward ecosystem-based management. Environmentalists have sought to implement marine reserves to maintain the structure and function of marine ecosystems. But this too is a sectoral approach. Traditional single-species management has a clearer recovery goal, specifically a certain spawning stock biomass to support future fishing efforts. But it is more difficult to define recovery goals in an ecosystem framework.

THE ROLE OF MANAGEMENT IN FISHERIES ECOLOGY

Traditional Fisheries Management

Single-species management relies on finding the inflection point where fish stock growth slows due to competition for food and habitat at high population sizes. Fisheries quotas are set and divided among commercial license holders for the duration of the fishing season. Once the quota is reached or the season ends, the fishery is closed. Even with appropriate precautionary quotas, overcapacity in fleets and a race to reach fisheries quotas can lead to overfishing. Changes in governance that limit access could both reduce overall fishing pressure and increase catch per unit effort for the remaining fishers.

Although almost 1/3 of the world's stocks have been estimated as overfished, there are examples of successfully managed and sustainably certified fisheries, such as halibut and numerous salmon species in Alaska.

These fisheries have limited bycatch, and the freshwater migration in salmon life history improves data access for stock assessments. However, natural variability in population dynamics, compounding stressors, and economic considerations can add complexity and can hamper successful fisheries management. The lack of sufficient data is still a large problem resulting in many poorly understood and poorly managed fisheries. Fish stocks often span management and governmental boundaries requiring multijurisdictional and international cooperation to adequately manage migratory stocks. Management techniques such as ecosystem-based management and integrated ecosystem assessments aim to incorporate multiple-species interactions and are not limited by economic boundaries.

Catch Shares

Catch shares are a form of single-species management that provides additional incentives to fishers to harvest sustainably. Rather than set an individual quota per license, each fisher or cooperative of fishers purchases a share of the fishery, essentially providing ownership of their quota. Catch shares have been implemented in various forms, including individual transferable quotas (ITQs), fishery cooperatives, and spatial management rights. Catch shares can be successful because they serve as a rights-based management strategy providing fishers with a dependable asset. It is important to note that catch shares are an implementation of management targets and would be compatible with ecosystem approaches.

Ecosystem-Based Management

Ecosystem-based management (EBM) focuses on the entire ecosystem across multiple trophic levels and includes humans. The basic tenet of EBM is to manage in concert how interannual and decadal environmental variability and fishery pressure will affect other species in the ecosystem, adjacent ecosystems when appropriate, and fisheries economics and behavior. In addition to managing biomass, it is important to conserve biodiversity, food web structure, and ecosystem function. The entire ecosystem needs to be managed in an economic context as well, to understand how ecosystem effects will translate to fishers' livelihoods. Because there are many users of the ecosystem, a successful ecosystem approach must include all relevant sectors of society and to be truly interdisciplinary.

Measuring the ecosystem effects of fishing is a daunting task, as many abiotic, ecological, and anthropogenic factors act in synergy. The development of marine food web models (e.g., Ecopath/Ecosim, Atlantis) has allowed fisheries ecologists to assess the anthropogenic impacts on marine ecosystems at temporal and spatial scales that are too large and complex for experimental studies. Because variability in climate forcing can propagate up food webs to affect higher trophic levels, management needs to set targets that incorporate predicted biological responses to climate forcing even though predictions can be difficult. Widespread use of single-species harvesting policies based on maximum sustainable yield can lead to removal of a suite of top predators and extensive loss of ecosystem function. Despite advances in ecosystem approaches, applying robust models to improve ecosystem-based management of commercial fisheries still requires investment in the collection of sufficient biological and environmental data in addition to extensive model-validation exercises.

Marine Spatial Planning

Place-based management and marine spatial planning (MSP) can provide a far more promising approach to implementing ecosystem-based management. Rather than individual sectoral agencies managing their activities everywhere, responsible authorities could collaborate to manage all the human activities in a focused place. These places might align with ecosystem boundaries, socioeconomical boundaries, and/or jurisdictional boundaries. In practice, management always occurs in a delimited space, with processes that cross management boundaries.

The biophysical component of marine ecosystems provides the basic template on which all human activities, including fisheries, occur and that various forms of governance regulate. Approaches to MSP and ocean zoning consider basic ecological concepts so that human activities can be conducted in ways that maintain ecosystem functioning, provide sustainable ecosystem services on which people depend, and maintain resilient ecosystems that can respond to environmental change.

Place-based management of marine ecosystems requires a hierarchy of management practices starting at the most general level with the concept of ecosystem-based management and moving toward the development of an integrated approach that accords priority to the maintenance of healthy, biologically diverse, productive, and resilient ecosystems. The key to success in place-based management of marine ecosystems is to design governance systems that align the incentives of stakeholders, in this case fishermen, with the objectives of management. MSP that fully incorporates the underlying ecosystem template and explicitly integrates the socioeconomic and governance overlays can form the basis for adequate protection of marine ecosystems and the sound use of marine resources, including fisheries.

FISHERIES ECOLOGY MOVING FORWARD

As fisheries management continues the transition to ecosystem-based management spanning physical variability to human dimensions, fisheries ecology will need to become more holistic and interdisciplinary. Recent National Ocean Policy (NOP) reports have tasked managers with developing ecosystem approaches to management that incorporate species interactions and translate to ecosystem services. New tools and research approaches are required that can evaluate management scenarios and their downstream effects on ecosystem function and services.

Fisheries ecology research includes theoretical developments, field-based experiments, and meta-analyses of fisheries data. Research on food web dynamics, particularly under changing climate regimes, will be critical in informing ecosystem models. Field-based experiments that can be built into long time series also allow comparisons of species interactions, abundance, and distribution before and after management decisions or climate regime shifts. These indicators will become critical in detecting ecosystem-wide perturbations, predicting future ecosystem dynamics, and parameterizing mass balance models to predict ecosystem changes. Mass balance models (e.g., Ecopath/Ecosim, Atlantis) are our primary tool for measuring trophic transfer for an entire food web and forecasting change under varied management and climate scenarios. Spatially explicit distribution models (e.g., linear models, generalized additive models, neural networks) can be used to identify biotic and abiotic drivers of distribution and can identify critical spatial scales of distribution and ecological processes. Spatially explicit approaches become necessary for identifying ecological and anthropogenic patterns at relevant scales to inform marine spatial planning.

Ultimately, coastal MSP will require tradeoff-based interaction models incorporating physical forcing up to economics and human behavior. Ecosystem-based management has evolved from an idea into a management framework, yet tools are still being developed to manage entire ecosystems in addition to individual stocks. In 2009, Levin and colleagues proposed integrated ecosystem assessments (IEAs) as a framework to aggregate scientific findings and to inform EBM at various scales and across sectors, although few have been developed to date. More specifically, IEAs provide a five-step pathway to quantitatively assess physical, biological, and socioeconomic factors in concert with defined ecosystem management objectives. This five-step transition from (1) scoping, which involves scientists, stakeholders, and managers identifying key EBM goals, (2) indicator development, which serve as representative proxies for the overall ecosystem state, (3) risk analyses, which assess the risk human activities and natural processes have on indicators, (4) management evaluation, which uses ecosystem models to evaluate outcomes and tradeoffs from proposed management strategies, and (5) ecosystem monitoring and evaluation, which measures the ongoing effectiveness of implemented management at appropriate scales and includes interactions among indicators.

Stakeholders, economists, and social scientists need to play a role in each step of the IEA to ensure the effect of management decisions on fishers and communities remain central goals of ecosystem-based management. More research is desperately needed on recreational and artisanal fisheries, as both target catch and bycatch from these fisheries are often absent in EBM. The field of fisheries ecology is in transition from primarily reductionist single-species approaches to integrated multiple-species responses forming a broader, interconnected framework that crosses multiple disciplines, management sectors, and jurisdictional boundaries to ensure sustainable fisheries for future generations.

SEE ALSO THE FOLLOWING ARTICLES

Beverton–Holt Model / Ecological Economics / Ecosystem Services / Food Webs / Marine Reserves and Ecosystem-Based Management / Population Ecology / Ricker Model

FURTHER READING

Caddy, J. F., and D. J. Agnew. 2004. An overview of recent global experience with recovery plans for depleted marine resources and suggested guidelines for recovery planning. *Reviews in Fish Biology and Fisheries* 14: 43–112.

Hilborn, R. 2007. Reinterpreting the state of fisheries and their management. *Ecosystems* 10: 1362–1369.

Law, R. 2007. Fisheries-induced evolution: present status and future directions. *Marine Ecology Progress Series* 335: 271–277.

Levin, P. S., M. J. Fogarty, S. A. Murawski, and D. Fluharty. 2009. Integrated Ecosystem Assessments: developing the scientific basis for ecosystem-based management of the ocean. *PLoS Biology* 7: 23–28.

Lubchenco, J., and N. Sutley. 2010. Proposed U.S. policy for ocean, coast, and great lakes stewardship. *Science* 328: 1485–1486.

Magnuson, J.T. 1991. Fish and fisheries ecology. *Ecological Applications* 1: 13–26.

Myers, R. A., J. K. Baum, T. D. Shepherd, S. P. Powers, and C. H. Peterson. 2007. Cascading effects of the loss of apex predatory sharks from a coastal ocean. *Science* 315: 1846–1850.

Pauly, D., V. Christensen, J. Dalsgaard, R. Froese, and F. Torres. 1998. Fishing down marine food webs. *Science* 279: 860–863.

Walters, C. J., V. Christensen, S. J. Martell, and J. F. Kitchell. 2005. Possible ecosystem impacts of applying MSY policies from single-species assessment. *ICES Journal of Marine Science* 62: 558–568.

Worm, B., et al. 2009. Rebuilding global fisheries. *Science* 325: 578–585.

FOOD CHAINS AND FOOD WEB MODULES

KEVIN McCANN AND GABRIEL GELLNER
University of Guelph, Ontario, Canada

Ecological systems are enormously complex entities. In order to study them, ecologists have tended to greatly simplify and deconstruct food webs. A result of this simplification has been the development of modular theory—the study of isolated subsystems. Figure 1 shows a common three-species subsystem (the food chain) and a common four-species subsystem (the diamond module) extracted from a whole food web. This entry briefly reviews modular theory with a focus on these two ubiquitous modules. It ends by discussing how the results from this modular theory hope to ultimately contribute to a theory for whole systems.

MODULES AND MOTIFS: TOWARD A THEORY FOR WHOLE SYSTEMS

The term module has been around for some time, although it appears to have had subtly different meanings. Robert Paine first used the term module to describe a subsystem of interacting species; he considered a module as a consumer and its resources that "behave as a functional unit." However, the term did not initially seem to catch on, and it was later reintroduced to ecology in slightly different form by Robert Holt in order to facilitate theoretical development beyond the well-developed theory of single-species and pairwise interactions. Holt defined modules specifically as communities of intermediate complexity beyond the well-studied pairwise interactions but below the diversity found in most natural systems (e.g., combination of 3–6 interacting species).

In a sense, the modular approach discussed by Holt sought to ask if we can extend predator–prey and competition theory in a coherent fashion to small subsystems. Similarly, network theorists—looking for properties in all kinds of nature's complex networks—have simultaneously considered the idea of underlying modular subnetworks that form the architecture for whole webs. Their terminology for this subnetwork topology is motif. Generally, these motifs are defined for two-species interactions (e.g., consumer–resource interactions), three-species interactions (e.g., food chains, exploitative competition, apparent competition, and the like), and beyond. Each motif is an i-species class characterized by the set of all possible interactions within that group. In this case, the single node is of little interest in quantifying network structure since it is everywhere by definition. Nonetheless, ecologists recognize that single nodes (i.e., populations) are dynamically important (e.g., cohort cycles). Further, there is a well-developed dynamic theory for populations in ecology.

In what follows, we will use the most general definition of modules to represent all possible subsystem connections, including the one-node/species case through to the n-node/species cases. In an attempt to delineate the different terminology, Robert Holt has argued that he sees modules "as motifs with muscles." This is reasonable since Holt's modular theory has always sought to understand the implications of the strength of the interactions on the dynamics and persistence of these units. The term module here, therefore, will be used to mean all motifs that include interaction strength.

Network and graph theorists have added to our ability to categorize the structure of real food webs. Network theorists have developed techniques that allow us to rigorously consider which motifs are common—or in the language of network theory, which motifs are over- or underrepresented in nature. Here, as with much of ecology, one compares nature to some underlying null model, and so one must remain cautious about what overrepresentation actually means. Nonetheless, this technique allows researchers to quantify the relative presence of motifs and focus the development of modular theory on common natural subsystems (i.e., overrepresented modules of real webs).

It is natural for ecological theory to move beyond consumer–resource interaction in order to explore common subsystems of food webs. In fact, Robert May, who championed the classical many-species whole food web matrix approach, argued that models of intermediate complexity may be a more direct path to interpreting how food web structure influences population dynamics and stability than matrices.

This entry first briefly reviews consumer–resource theory (hereafter *C–R* theory) before discussing two common higher-order modules, the three-species food chain and the four-species diamond module. The dynamics of the food chain module, which appears to be the most often overrepresented of all three-species modules in food webs, is reviewed, followed by a consideration of the common four-species diamond module, which is, in a sense, simply two food chains

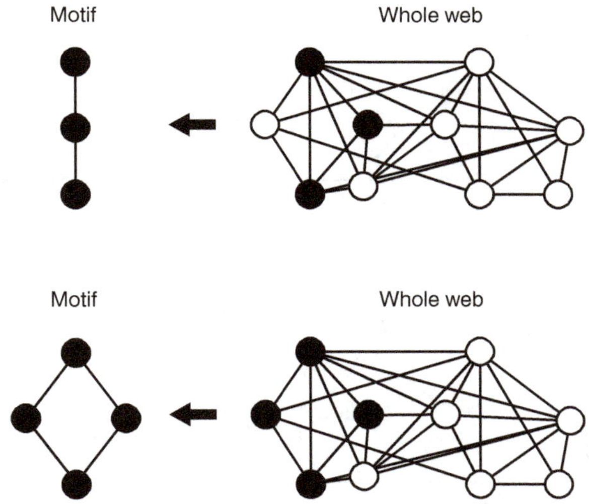

FIGURE 1 Food webs can be deconstructed into subsystems or motifs. Here, two common motifs are shown. A "common" structure means that these motifs appear with greater frequency than expected by pure chance. Different types of networks have different types of ubiquitous structure. It has been found, for example, that information processing networks are different in structure than energy processing-based networks.

concatenated (Fig. 1). This module also appears common in natural systems.

All the while, attention is paid to existing *C–R* theory in order to interpret the dynamic consequences of these modules. Food web modules have the tendency of displaying subsystem signatures such that oscillations can often be attributed to particular underlying *C–R* interactions. Given this, *C–R* theory forms a framework that allows the examination food web module dynamics from an interaction strength perspective.

A FUNDAMENTAL MODULE: CONSUMER-RESOURCE INTERACTIONS

Figure 2 shows the material fluxes that accompany a consumer–resource interaction in nature. There exist fluxes between the consumer and its resource (the interaction itself), there are fluxes into the resource (nutrient or biomass uptake), and there are also fluxes out of the consumer and the resource (e.g., mortality). These *C–R* fluxes occur repeatedly in all the underlying *C–R* modules that ultimately make up whole food webs (e.g., Fig. 1). These fluxes show that consumer–resource theory yields a very general and powerful result when considered from a flux-based interaction strength perspective.

Specifically, there is a tendency for all *C–R* models to produce a destabilizing response to increased production,

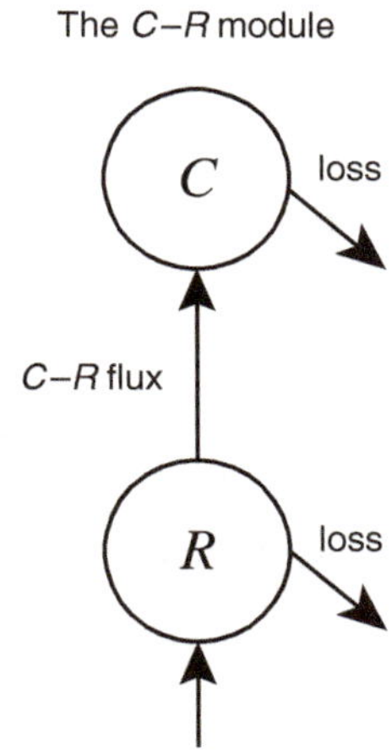

FIGURE 2 The *C-R* interaction module one of the basic building blocks. The basic fluxes are shown: the *C-R* flux (consumption) and the loss rates. Ultimately, flux in and flux out are related to any reasonable metric of direct interaction strengths.

increased interaction rates, or decreases in mortality. Figure 3 shows the dynamic response across a gradient in interaction strength. The gradient follows the ratio of *C–R* flux relative to the consumer loss term. *C–R* models that have high *C–R* flux rates and low mortality loss rates tend to be less stable. "Less stable" warrants some clarification since stability is often assessed in a multitude of ways. Figure 3A expresses this loss of stability as the change from equilibrium dynamics (for low *C–R* flux:loss ratios) to wildly oscillating dynamics at high *C–R* flux:loss ratios. As an example, this result of increased variability with increased flux-based interaction strengths generally occurs for the well-known Rosenzweig–MacArthur model.

Figure 3B, on the other hand, shows a different but very related reduction in stability. Here, solutions remain in equilibrium dynamics, but the return time back to the equilibrium increases as the *C–R* flux rate:loss ratio increases. Part of this increase in return time is due to the fact that the solutions take on oscillatory decays (i.e., eigenvalues become complex). This increased return time also can be seen in that the negative real part of the eigenvalues of the stable equilibria move closer to zero (i.e., weaker attraction to equilibrium). This increased oscillatory decay, and weakened attractor, drive a greater return time to equilibrium and so reduces stability. This destabilization result is qualitatively similar the Rosenzweig–MacArthur model result of Figure 3A. Note that these oscillatory decays can quite readily turn into cycles, or quasi-cycles, in stochastic settings.

There is a large food web literature that employs matrices to examine dynamics. It is of interest to ask whether

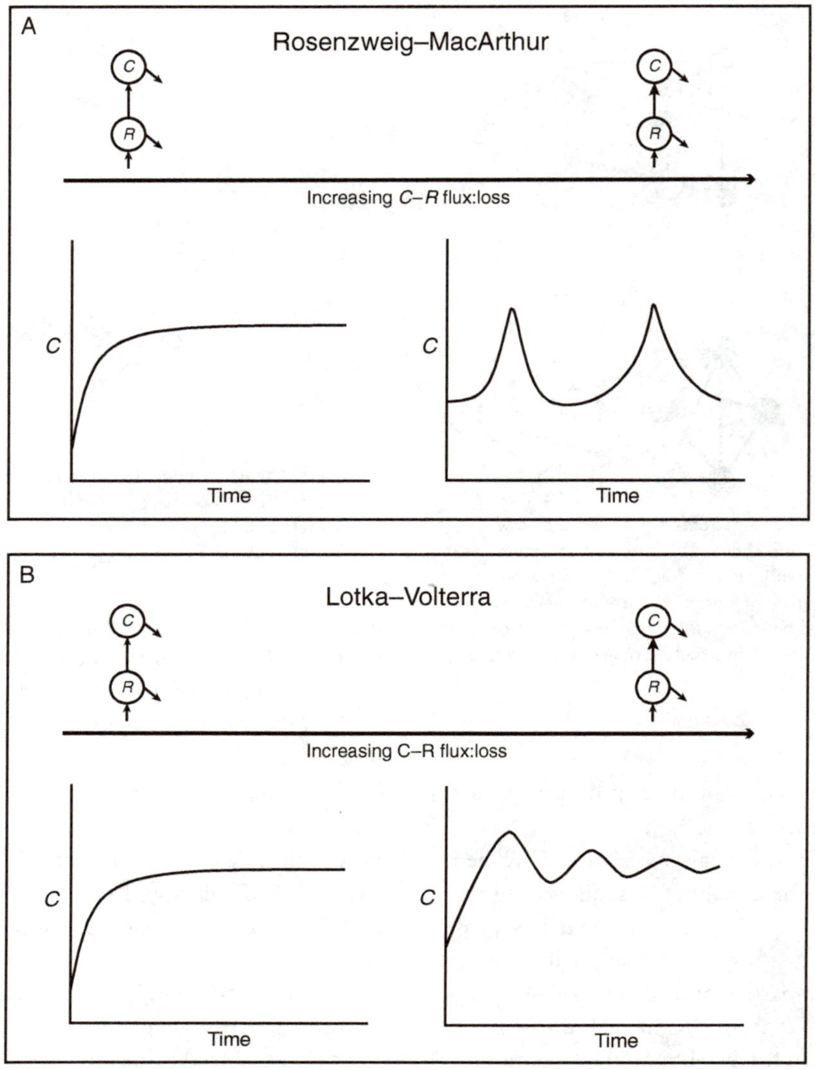

FIGURE 3 The influence of changing interaction strength (here defined as *C–R* flux: *C* loss ratio) on the dynamics of the module. Increasing *C–R* flux:loss ratio destabilizes the interaction by (A) yielding wild oscillations as in the Rosenzweig-MacArthur model, or (B) yielding oscillatory decay as in the Lotka-Volterra model.

the *C–R* matrix approach gives similar answers to the *C–R* theory discussed above. Here, a compatible result is also found since the elements of the matrix can be decomposed in a similar manner (i.e., interaction terms and loss terms). In the *C–R* matrix, indeed for all matrices, the diagonal elements of the matrix are measures of the loss rate, while the off-diagonal entries are measures of the fluxes through the consumer–resource interactions. Analogous to the previous results, increasing the off-diagonal interaction elements relative to the diagonal loss terms tends to decrease the linear stability of the system up to some maximum. This is satisfying, as it means that different mathematical approaches yield similar results.

The above results, taken together, can be stated biologically as follows:

THE PRINCIPLE OF INTERACTION STRENGTH (1)

Any biological mechanism that increases the strength of the flux through a consumer–resource interaction relative to the mortality rate of the consumer ultimately tends to destabilize the interaction.

The degree of this destabilization will necessarily depend on the model formulation, but all models tend to follow this general pattern. This result can be stated differently, and to some advantage for the remaining sections of this entry, as follows:

THE INTERACTION STRENGTH COROLLARY (2)

Any biological mechanism that decreases the strength of the interaction relative to the mortality tends to stabilize that interaction.

The next section turns to modular theory to point out how *C–R* theory explains the dynamics of more speciose food web modules.

FOOD CHAINS: COUPLED *C–R* INTERACTIONS

There exists a lot of theory describing the dynamics of food chains. Researchers, for example, have shown that it is relatively easy to find complex or chaotic dynamics in a common food chain model. Specifically, these researchers have found chaos readily occurrs in the Rosenzweig–MacArthur food chain model when the attack rates were high relative to the mortality rates. This result is strongly related to the interaction strength principle (1) of *C–R* theory stated in the last section. Note that in such a case the underlying interactions that comprise the food chain (i.e., both predator–consumer (*P*–C and *C–R* interactions) would tend to produce cycles if isolated (Fig. 3A).

It is important to realize that chaos often emerges when multiple oscillators interact. Multiple oscillators, once coupled, tend to produce a complex mixture of their underlying frequencies (e.g., Fig. 4A). In the case of the *P–C–R* food chain, there are two underlying oscillators, a predator–consumer oscillator (*P–C*) and the consumer–resource oscillator (*C–R*), which mix to drive complex dynamics (Fig. 4A). The signatures of these underlying oscillators in chaotic attractors are present and visible in the power spectrum of their time series. These results resonate with the interaction strength principle (1) of *C–R* theory stated above.

It becomes interesting to consider what happens when a weaker interaction (i.e., low *C–R* flux:loss ratio) is coupled with a stronger interaction (i.e., high *C–R* flux:loss rate). Figure 4B depicts such a case. Then there is a potentially oscillatory interaction (the *C–R* interaction) and an interaction that would not oscillate in isolation (the *P–C* interaction). This case introduces an example of weak interaction theory. The *C–R* interaction strength corollary (2) reviewed in the previous section allows the interpretation of the results that follow.

Note that in Figure 4B, the stable *P–C* interaction, in effect, adds an additional mortality loss to the consumer (*C*). Thus, the *C* in the *P–C–R* chain experiences more loss than such a consumer would experience in isolation. All else being equal, from the interaction strength corollary (2) it would be expected that any such increase in loss to the consumer would tend to stabilize the underlying

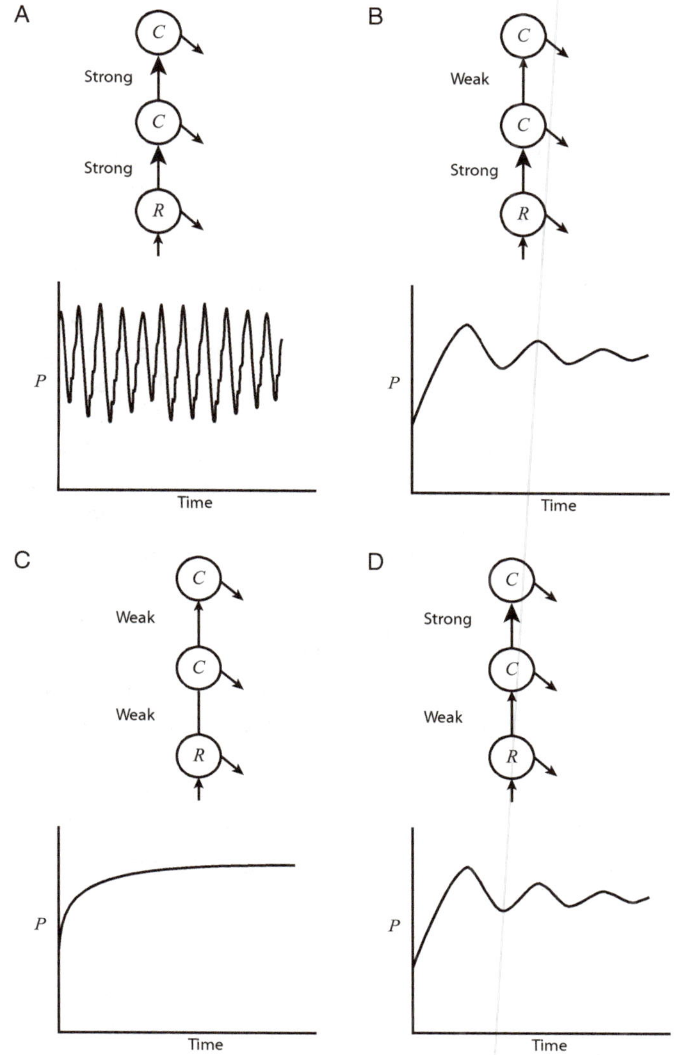

FIGURE 4 The influence of changing interaction strength on the dynamics of the food chain module. (A) Two strong *C-R* interactions couple to produce wild oscillations as in the Rosenzweig-MacArthur model. (B) The weak *P-C* interaction mutes (increases loss) without adding another oscillator and so stabilize the *C-R* interaction. The resultant *P-C-R* food chain dynamic, in this depicted case, is a stable equilibrium, although an oscillatory approach to equilibrium occurs. (C) All interactions are weak and the dynamics are stable. (D) Weak *C-R* interaction nullifies potentially strong *P-C* interaction, yielding stable dynamics.

C–R interaction (Fig. 4B). Indeed, this is what generally happens when a weakly consuming top predator, *P*, taps into a strong *C–R* interaction. The interaction gets less oscillatory—it mutes the *C–R* oscillator—and can even produce stable equilibrium dynamics.

It is worth pointing out that a range of food chain dynamics can be thought of from this very simple, but powerful, perspective. For example, two stable underlying interactions, the *P–C* and the *C–R* interaction, tend to yield stable equilibrium food chain dynamics (Fig. 4C).

Further, and similar to the case discussed in Figure 4B, a weak interaction in the basal $C–R$ can mute a potentially strong $P–C$ interaction (Fig. 4D).

To understand this last result, consider a thought experiment starting with the scenario in which both underlying interactions in the food chain are strong. In other words, imagine starting with the case depicted in Figure 4A. Now, if the strength of the basal $C–R$ interaction is experimentally reduced, the result becomes like the case depicted in Figure 4D. As the lower level $C–R$ interaction is reduced, the amount of energy that reaches the $P–C$ interaction begins to be reduced. As the corollary (2) states, any mechanism that reduces the flux:loss ratio ought to stabilize the underlying interaction. Therefore, one expects the potentially oscillatory $P–C$ interaction to be muted by the reduced $C–R$ interaction strength, and it is (compare Fig. 4A to Fig. 4D).

In summary, a weak $C–R$ interaction means that little actual production reaches the top predator. As such, not enough biomass energy is transmitted up the food chain to drive the potential oscillations in the $P–C$ interaction. In such a case, the weak basal interaction effectively mutes the potentially unstable higher-order food chain interaction. This is again the "interaction strength corollary" operating—a biological mechanism that weakens flux stabilizes a potentially unstable interaction. Properly placed weak interactions therefore can readily inhibit oscillatory dynamics.

Numerous fairly hefty mathematical papers that look at the bifurcation structure of this commonly explored model are consistent with the simple ideas presented here. These papers, however, catalog in more detail dynamical outcomes such as multiple basins of attraction, quasi-periodicity, and chaotic transients. Additionally, researchers are beginning to explore the role of stage structure on food chains, which appear to quite readily produce multiple basins of attraction.

HIGHER-ORDER MODULES: COUPLED FOOD CHAINS AS AN EXAMPLE

The results of the last section appear to work for simple food web modules. That is, multiple strong interactions coupled together readily beget complex dynamics. In addition, weak interactions properly placed can mute potentially oscillatory interactions. However, in more complex food web modules another stabilizing mechanism arises.

As an example, this other stabilizing mechanism arises in cases where predators, P, feed on consumers (C_1 and C_2) competing for a common basal resource (Fig. 5). This module is ubiquitous in real food webs and has become referred to as the diamond module. There is evidence that organismal tradeoffs play a major role in governing the diamond module since tradeoffs frequently manifest around parameters involved with consumption and growth rates. As an example, high tolerance to predation tends to correlate with a lower growth-rate life history.

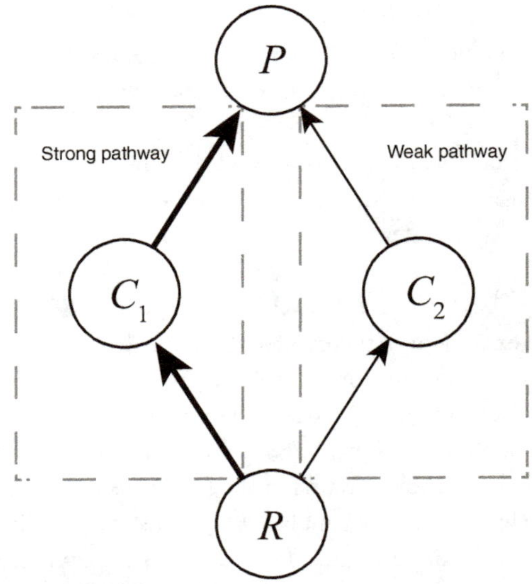

FIGURE 5 The diamond four-species food web module. One pathway is composed of strong interactions and the other, weak interactions due to very general growth-defense tradeoffs. In this food web module, the fast-growing species tends to be highly competitive but also very vulnerable to predation. This tradeoff produces relatively stable modules compared to just the strong food chain model.

This tradeoff occurs because organisms that have adapted to grow rapidly allocate energy to somatic and reproductive effort, with little to no energy allocated into costly physical structures that impede a predators' ability to consume them. Such a rapidly growing consumer is expected to be an extraordinary competitor in situations without a predator. Nonetheless, their ability to produce biomass also makes them vulnerable to predators. Thus, this same consumer species is both productive and potentially strongly consumed by a predator (i.e., the strong chain of Fig. 5)—it is the precise recipe for nonequilibrium food chain dynamics.

On the other hand, an organism that puts much energy into the development of defense structures (e.g., porcupine) may also be expected to grow more slowly and be less competitive when not in the presence of predators. Since predators have less impact on this organism, this slow-growing species will become an excellent competitor when predator densities are high. This species forms the middle node in the weak chain of Figure 5.

This tradeoff, therefore, creates a combination of strong interaction and weak interaction pathways (Fig. 5). In isolation, the strong pathway chain ($P–C_1–R$ in Fig. 5) tends toward instability, while the weak interaction pathway (P–C_2–R in Fig. 5) tends toward heightened stability (Fig. 4C). Together, they blend into a system that has far more stability than the strong interaction pathway alone.

This reduction of the whole system to coupled subsystems allows one to take a simplifying view of this otherwise staggering mathematical problem. If one can find a mechanism that tends to stabilize all the underlying oscillators, then this ought to eliminate the occurrence of oscillatory dynamics in the full system. Similarly, such a mechanism that reduces the amplitude of the underlying oscillators also ought to reduce the amplitude of the dynamics of the full system. Previous work suggest that this type of result may occur frequently.

Toward this effect, one can perform theoretical experiments on a strong focal chain by adding new C–R interactions, one at a time, until the diamond module is created. Figure 6B highlights the numerical experiment of adding exploitative competition (C_2) to a food chain model undergoing chaotic dynamics (Fig. 6A). In this case, C_2, is competitively inferior to C_1, so its ability to persist is mediated by the selective predation of the top predator, P, on C_1. Here, one expects exploitative competition to inhibit the oscillating C_1–R subsystem by deflecting energy away from the strong, potentially excited interaction. It does exactly this, producing oscillatory but muted or well-bounded limit cycles (Fig. 6B). In this particular example, the system still does not reach an equilibrium solution over this range, as the muting potential of the added competitor simply is not capable of deflecting enough energy to cause period-doubling reversals all the way to an equilibrium value.

Now, if one completes the diamond module and allows the top predator to also be a generalist and feed on C_2 weakly, then Figure 6C shows that this final addition has now produced a stable equilibrium dynamic. This final result is not immediately obvious and so a comment is in order. There is another somewhat hidden stabilizing mechanism in this last result. The differential-strength pathways tend to readily produce asynchronous consumer dynamics that the consumer can average over.

As an example, if one perturbs this system by adding additional basal resources, this first tends to increase both of the consumers. They increase synchronously, however; soon this C increase drives a decrease in total resource densities, R, and an increase in total predator densities, P. The system is suddenly quite tricky for consumers to eke out a living in, since there are simultaneously low-resource (high competition) and high-predation conditions.

In this precarious situation, the differential strength pathways produce asynchronous C dynamics. This occurs because as P grows, it starts to consume a lot of C_1, freeing C_2 from the suppressive grips of the superior competitor, C_1. Once freed from competition and yet not strongly consumed by the top predator, P, the weak-pathway consumer, C_2, starts to grow. Thus, one consumer is decreasing and the other consumer is increasing.

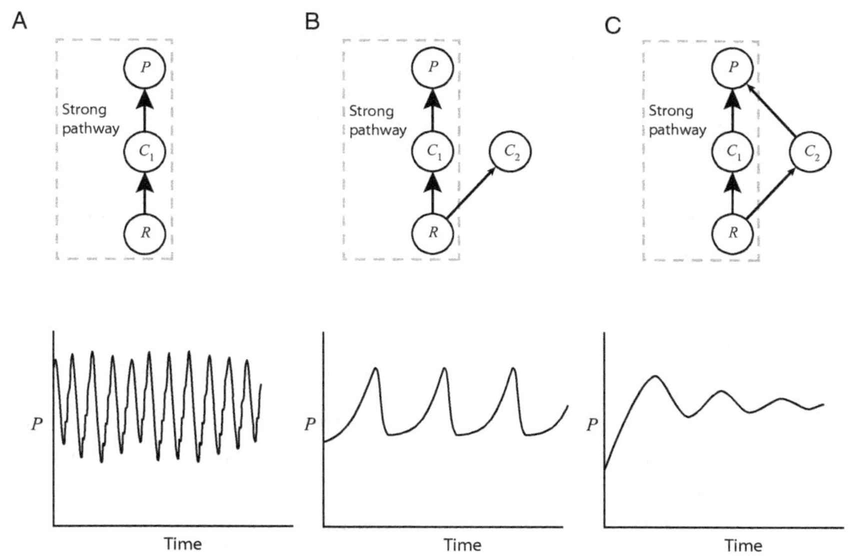

FIGURE 6 Time series for three different food web configurations that ultimately end up as the diamond module with weak and strong pathways. (A) Food chain composed of just strong pathways with complex, highly variable dynamics. (B) One weak competitive interaction added to the food chain mutes the oscillations, leaving a smaller, less complex oscillation. (C) Another weak interaction, completing the diamond module, changes the dynamics to stable equilibrium dynamics.

This out-of-phase consumer dynamic, in a sense, can sum together to give the top predator a relatively stable resource base. That is, when one consumer is low, the other is high, and so the top predator is buffered against the low resource densities of any one prey item. Clearly, this is a much more stabilizing situation than when both consumers are simultaneously low, in which case, the top predator tends to precipitously decline.

In summary, despite the complexity of this system, which includes multiple attractors and numerous bifurcations, the qualitative result remains: relatively weak links, properly placed, simplify and bound the dynamics of food webs. Here, "properly placed" is a result of the underlying biologically expected tradeoffs. On the other hand, strong interactions coupled together are the recipe for chaos and/or species elimination.

DATA AND THEORY

There are accumulating empirical and experimental results that appear consistent with this simple food web module theory. First, it is clear that *C–R* interactions, food chains, and simple food web modules can produce complex oscillatory dynamics in experimental settings. Further, it appears clear that increasing flux through these interactions by increasing nutrients tends to drive more complex dynamics. Thus, strong coupled interactions do indeed drive complex dynamical phenomena in highly controlled experiments.

Experiments have also begun to test whether weak interactions can mute such wild oscillations. Two recent controlled experiments have found that weak interactions in a simple food web module can decrease the variability of the population dynamics. One of these experiments found that some of this stability may also be due to the asynchronous resource dynamics discussed above. Along these lines, an extensive study based on field data from Lake Constance found that edible (strong) and less edible (weak) phytoplankton varied out of phase shortly after a synchronized pulse in zooplankton following spring turnover. Both controlled experiments and field data show signs that weak–strong pathways may produce asynchronous out-of-phase dynamic responses to increases in predator densities.

Finally, a recent analysis of interaction strengths in a Caribbean food web found the cooccurrence of two strong interactions on consecutive levels of food chains occurred less frequently than expected by chance. Thus, real food webs may not commonly produce strongly coupled food chains. Further, where they did find strongly coupled links in food chains there was an overrepresentation of omnivory. While not discussed here, this latter result may imply that omnivory acts in a stabilizing fashion in this case.

SEE ALSO THE FOLLOWING ARTICLES

Bifurcations / Bottom-Up Control / Chaos / Food Webs / Ordinary Differential Equations / Predator–Prey Models / Stability Analysis / Top-Down Control

FURTHER READING

Fussmann, G. F., S. P. Ellner, K. W. Shertzer, and N. G. Hairston. 2000. Crossing the Hopf bifurcation in a live predator–prey system. *Science* 290: 1358–1360.

Hastings, A., and T. Powell. 1991. Chaos in a 3-species food-chain. *Ecology* 72(3): 896–903.

Holt, R. D. 2002. Community modules. In A. C. Gange and V. K. Brown, eds. *Multitrophic interactions in terrestrial ecosystems.* The 36th Symposium of the British Ecological Society. London: Blackwell Science.

McCann, K., and P. Yodzis. 1995. Bifurcation structure of a 3-species food-chain model. *Theoretical Population Biology* 48: 93–125.

McCann, K. S., A. M. Hastings, and G. Huxel. 1998. Weak trophic interactions and the balance of nature. *Nature* 395: 794–798.

Muratori, S., and S. Rinaldi. 1992. Low- and high-frequency oscillations in three dimensional food chain systems. *SIAM Journal on Applied Mathematics* 52: 1688.

Rip, J., K. McCann, D. Lynn, and S. Fawcett. 2010. An experimental test of a fundamental food web motif. *Proceedings of the Royal Society: London B: Biological Sciences* 277: 1743–1749.

FOOD WEBS

AXEL G. ROSSBERG
Queen's University Belfast, United Kingdom

Food webs are the networks formed by the trophic (feeding) interactions between species in ecological communities. Their relevance for population ecology follows directly from the importance of trophic interactions for maintenance and regulation of populations. Food webs are crucial for community ecology whenever trophic interactions control competition, and hence community composition and species richness. Theoretical representations of food webs range from simple directed graphs to complex dynamic models integrating much of theoretical ecological.

STATUS OF THE THEORY

The study of food webs developed from a science that simply records data, through a phase of cataloging and identifying patterns in the data, and then moved toward interpreting data and patterns, first in terms of phenomenological models and later in terms of general ecological mechanisms. A stage in which the theoretical foundations would be settled and their implications studied in depth has not been reached, yet.

By now, a large number of food web models have been developed that combine elements from a certain set of recurring model components with specific original ideas. As these models are becoming more complex, work to systematically evaluate and compare their predictive power is limited by computational issues, differences in model focus, and deficits in the available data. There is currently no standard model for food web structure and/or dynamics. Understanding the roles played by different model elements in generating the patterns found in data remains a major theoretical challenge.

Fisheries management is perhaps the only area where progress has been made in applying complex food web models to real-world problems. While long-term forecasts of the dynamics of interacting, harvested fish populations are limited by high model sensitivity and the problem of food web stability (see below), food web models of fish communities help understanding medium-term changes in abundances, mortalities, and population growth rates.

BASIC ELEMENTS OF FOOD WEB MODELS

Topological and Quantitative Food Webs

Food webs are often represented as in Figure 1, by directed graphs in which species form the nodes and feeding interactions are represented by arrows (trophic links) pointing from resources to consumers, i.e., in the direction of energy flow. By convention, consumers are drawn above their resources whenever possible, so that energy flows upward and top predators are on the top.

Some empirical food webs restrict the set of species included to those that are directly or indirectly eaten by certain consumers (*sink webs*) or those directly or indirectly feeding on certain resources (*source webs*). Theory nearly exclusively considers *community food webs*, implying the idealizations of sharply delineated ecological communities and well-defined sets of member species.

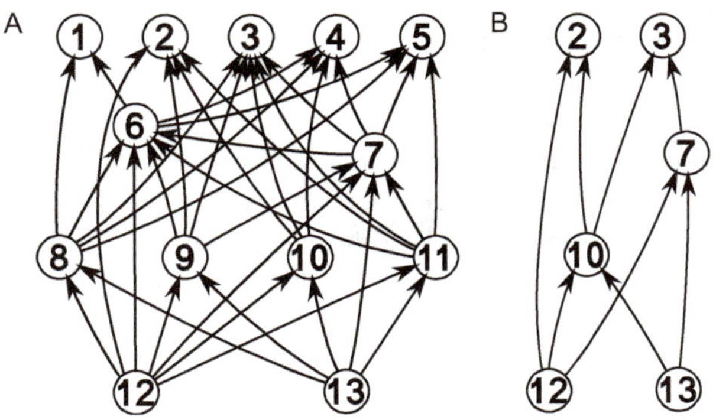

FIGURE 1 The food web of (A) Morton Creek (after G. Wayne Minshall, *Ecology* 48: 139–149, Fig. 3) and (B) the community that persists in the model described in Box 1.

A food web described exclusively by a species list and an adjacency matrices $A_{ij} = 0, 1$ or a corresponding graph is called a *binary food web*. This distinguishes *topological food webs* from *quantitative food webs* that are described by a link-strength matrix C_{ij} and, potentially, species abundances.

Link-Strength Functions and Trophic Niche Space

The trophic link strength C_{ij} between a resource i and a consumer j is often modeled as a function of characteristics of these two species. That is, one assumes that each species i is characterized by a set of vulnerability traits $\mathbf{v}_i$ and a set of foraging traits $\mathbf{f}_i$ such that, with an appropriate choice of the link-strength function $c(\cdot,\cdot)$, trophic link strength is given by $C_{ij} = c(\mathbf{v}_i, \mathbf{f}_j)$. Vulnerability and foraging traits may not be independent. Body size or preferred habitat, for example, affect species in both their roles as resources and as consumers.

The function $c(\cdot,\cdot)$ is often modeled such that for any biologically possible (or technically allowed) combination

TABLE 1
The sets of species lumped into each network node in Figure 1A

Node	Name	Species
1	*Phagocata gracilis*	
2	Decapoda	*Orconectes rusticus rusticus* *Cambarus tenebrosus*
3	Plecoptera	*Isoperla clio* *Isogenus decisus*
4	Megaloptera	*Nigronia fasciata* *Sialis joppa*
5	Pisces	*Semotilus atromaculatus* *Rhinichthys atratulus*
6	*Gammarus minus*	
7	Trichoptera	*Diplectrona modesta* *Rhyacophila parantra*
8	Asellus	*A. brevicaudus* *A. intermedius*
9	Ephemeroptera	*Baetis amplus* *Baetis herodes* *Baetis phoebus* *Epeorus pleuralis* *Centroptilum rufostrigatum* *Pseudocloeon carolina* *Paraleptophlebia moerens*
10	Trichoptera	*Neophylax autumnus* *Glossoma intermedium*
11	Diptera	Tendipedidae *Simulium* sp. Tipulidae *Pericoma* sp. *Dixa* sp. Other
12	Detritus	
13	Diatoms	(Unresolved)

SOURCE: After G. Wayne Minshall, *Ecology*, 48: 139–149.

of vulnerability traits $\mathbf{v}$ there is a specific combination of foraging traits $\mathbf{f}$ that maximizes the interaction strength $c(\mathbf{v}, \mathbf{f})$. Disregarding those $\mathbf{f}$ that do not match any $\mathbf{v}$, one can then identify the set of biologically possible $\mathbf{v}$ with the set of maximizing $\mathbf{f}$. This unified set of trophic traits is called the *trophic niche space*. Under some mild conditions on the convexity of $c(\cdot, \cdot)$ (i.e., that $c(\mathbf{v}, \mathbf{f})$ decreases the more $\mathbf{f}$ differs from the value matching $\mathbf{v}$), satisfied in most models, the interaction strength between two species i and j is then the stronger the closer $\mathbf{f}_j$ is to $\mathbf{v}_i$ in trophic niche space. By fixing some threshold value c_t, one can assign a foraging niche to each consumer j as the set of all $\mathbf{v}$ for which $c(\mathbf{v}, \mathbf{f}_j) > c_t$. The *foraging niche* of a consumer j surrounds $\mathbf{f}_j$ in trophic niche space. Many food web models are defined by combining a link-strength function with rules for assigning trophic traits to species.

An example for a link-strengths function is

$$c(\mathbf{v}, \mathbf{f}) = c_0 \exp\left[v^{(0)} + f^{(0)} - \frac{1}{2} \sum_{i=1}^{D} \sigma_i \left(v^{(i)} - f^{(i)} \right)^2 \right]. \quad (1)$$

In this case, the trophic traits $\mathbf{v}$ and $\mathbf{f}$ are $(D + 1)$ dimensional real-valued vectors with components $v^{(0)}, \ldots, v^{(D)}$ and $f^{(0)}, \ldots, f^{(D)}$, respectively, σ_i equals either $+1$ or -1, and c_0 is a scale constant. Many link-strength functions found in the literature can be brought into this or similar forms. Although mixed sign structures are conceivable as well, theorists generally assume $\sigma_i = +1$ for all i. For a given $\mathbf{v}$ and fixed $f^{(0)}$, trophic link strength is then maximized by matching $f^{(1)}, \ldots, f^{(D)}$ with $v^{(1)}, \ldots, v^{(D)}$. Thus, trophic niche space is here a D-dimensional vector space. The *trophic baseline traits* $v^{(0)}$ and $f^{(0)}$ determine the overall vulnerability of a resource and the overall aggressivity of consumers, respectively.

The idea that trophic link strength depends only on the consumer and the resource is a simplifying idealization (think of prey hiding in bushes). A modification of trophic link strength by a third species has been called a *rheagogy* (based on the Greek *rheô*, "flow," and *agôgeô*, "influence"). Rheagogies can complicate the interpretation of food webs and, when sufficiently strong, qualitatively affect community structure.

Dynamic Food Web Models

Population-dynamical food web models describe how the abundances of all species in a community change over time as a result of trophic interactions. They build on the premise that trophic interactions dominate population dynamics, so that nontrophic effects can be modeled in highly simplified form. Historically, population sizes have been quantified by the number of individuals; recent models generally operate with biomass or bio-energy instead. Minimal ingredients for population-dynamical food web models are appropriately parametrized sub models for (1) functional and numerical responses of consumers to varying resource abundances, (2) nontrophic losses (death and/or metabolic losses), and (3) the population dynamics of basal species. Box 1 provides an example.

BOX 1. A SIMPLE POPULATION-DYNAMICAL FOOD WEB MODEL

This model describes trophic interactions between 13 (trophic) species. The population dynamics of consumer species (i = 1–11) is modeled by a generalized Lotka-Volterra model of the form

$$\frac{dB_i}{dt} = -RB_i + \overbrace{\epsilon \sum_j C_{ji} B_j B_i}^{\text{numerical response}} - \sum_j \overbrace{C_{ij} B_i}^{\text{functional response}} B_j \quad (2)$$

and that of the basal species (i = 12, 13) as

$$\frac{dB_i}{dt} = \overbrace{r(1 - B_i/K) B_i}^{\text{basal dynamics}} - \sum_j \overbrace{C_{ij} B_i}^{\text{functional response}} B_j. \quad (3)$$

With A_{ij} denoting the adjacency matrix corresponding to Morton Creek (Fig. 1A), trophic link strengths are set to pseudo-random values $C_{ij} = A_{ij} \cos(13i + j)^2$, the efficiency for conversion of resource biomass to consumer biomass to $\epsilon = 0.5$, the effective consumer respiration rate to $R = 0.2$, the growth (or replenishment) rate of basal species to $r = 1$, and their carrying capacities to $K = 1$. Simulations are initiated by setting all $B_i = 1$ and continued over 1000 unit times. Species with $B_i < 10^{-10}$ are removed as extinct. The remaining community, shown in Figure 1B, approaches a stable, feasible fixed point.

Food web assembly models describe how the structure of food webs changes as a result of a sequence of invasions and extinctions. Invading species are either chosen from a predetermined species pool (representing a metacommunity) or generated at random. Assembly models in which trophic traits of invading species are determined by randomly modifying traits of resident species are called *evolutionary food web models*. This is often described as "speciations" of resident species in large "mutation" steps. The scheme is perhaps more realistic than this interpretation. The actual speciations can have occurred allopatrically or at distant locations or times. In the focal community, such processes would simply be reflected by invasions of species that are more or less similar to existing ones.

Examples of evolutionary models in the literature cover cases with small and large mutation steps, as well as cases where the decision which species to invade or to extinguish is made depending on population dynamics (populations falling below a given threshold are removed as extinct), depending on other properties of the food web or purely at random (neutral evolution).

PATTERNS AND MECHANISMS

Models and Data

Generic models of complex systems such as food webs are unlikely to reproduce all aspects of empirical data. One therefore has to distinguish between patterns in empirical data, on one hand, and models capable of reproducing and explaining specific sets of patterns, on the other hand. This distinction is blurred when data are characterized in terms of models fitted to them. Yet models fitted to data are useful even on purely theoretical grounds—for example, for identifying the mechanisms by which specific combinations of model elements reproduce particular empirical patterns, or to understand in how far different mechanisms interact under realistic conditions. The discussion here will emphasize mechanisms that relate model elements to patterns, since understanding these should allow us to recombine model elements to balance model complexity against desired descriptive or predictive capabilities.

The sheer amount of data that accurate quantitative descriptions of complete natural food webs would require dashes any hopes that such descriptions will become available in the foreseeable future. Each food web data set is a compromise between accuracy and completeness, and different empiricists will set different priorities. To make food webs manageable, trophic links are often inferred rather than measured, *trophic species* are introduced for special compartments such as "detritus" or to summarize groups of similar species, especially at lower trophic levels, and interactions with neighboring communities are not represented (see, e.g., Table 1, Fig. 1A). Theory interpreting such data has to take these limitations into account.

Size Selectivity

Big fish eat small fish. Patterns of body-size selectivity are evident in food web data, especially for aquatic communities. Models capture this phenomenon by assigning a trophic trait "size" to each species and constraining the relative sizes of resources and consumers in feeding interactions. For models and data that focus on predatory interactions rather than parasites, pathogens, or grazing, the pattern "large eats small" dominates.

Because individuals of many species grow substantially before they mature, the size range of resources fed on by a species as a whole can be large. Pairs of species eating each other's offspring lead to loops in food webs, and species eating their own offspring lead to *cannibalism.* As a simplified representation of these complications, food web models sometimes allow consumers to choose resources among all smaller species, admitting a few exceptions where small eats large. When no exceptions are allowed, food web adjacency matrices can be brought into lower-triangular form by assigning indices to species according to size. One of the earliest topological food web models, the *cascade model* conceived by Cohen, Briand, and Newman in 1989, generates adjacency matrices simply by randomly setting a fraction of matrix entries below the diagonal to 1 and leaving all others 0. The cascade model was shown to generate topologies more similar to empirical webs than other, comparably simple models.

Phylogenetic Constraints

Evolutionarily related species are similar, and similar species tend to have similar resources and consumers. For example, a bird eating a seed is likely to eat other kinds of seeds; and a bird eating an insect is more likely to eat other kinds of insects than to eat seeds because insects are more similar to each other than to seeds. Empiricists are taking these phylogenetic constraints on food web topology for granted when defining trophic species taxonomically (e.g., Table 1). Phylogenetic analyses of food webs confirm this pattern. They also show that consumer sets of species are inherited more strongly than resource sets. For example, granivorous and insectivorous birds may differ in their diets and yet share bird-eating raptors as common consumers.

These phylogenetic constraints can be described by evolutionary food web models in which both vulnerability and foraging traits are inherited, the former stronger than the latter. Such models generate characteristic patterns in food web topologies, that is, patterns evident even without knowledge of the underlying phylogeny. Before considering other mechanisms to explain patterns in food webs, one should therefore always ask first whether these patterns are simply consequences of phylogenetic and size constraints, two empirically well-established facts. Most patterns discussed below are of this kind.

Link-Strength Distributions

Food webs contain many more weak links than strong links, that is, distributions of link strengths are highly skewed toward weak links. This pattern is found independent of whether link strength is measured in terms of absolute biomass or energy flows, or normalized to predator and/or prey abundances. While early studies found link strengths to be exponentially distributed, more recent work points toward power-law or log-normal distributions (a quantity is log-normally distributed if its logarithm is normally distributed).

Log-normal and power-law distributions may be empirically indistinguishable. Assuming log-normal link strength distributions, very strong links are unlikely to be observed because corresponding consumer–resource pairs are rare, and very weak links are unlikely to be observed because they are hard to detect. Thus, only a finite range in link-strength magnitude is empirically accessible, and it is well known that the upper tail of a log-normal distribution can, over a range small compared to the width of the distribution, be approximated by a power law.

Log-normal distributions of trophic link strength arise naturally in models where the link-strength function is a product of many factors,

$$c(\mathbf{v}, \mathbf{f}) = \prod_k c_k(\mathbf{v}, \mathbf{f}), \tag{4}$$

and the factors $c_k(\mathbf{v}, \mathbf{f})$ are all positive and depend differently on consumer and/or resource traits. An example for such a link-strength function is Equation 1 with large D. To the extent that the trophic traits of member species are distributed randomly within a community, $\log c(\mathbf{v}, \mathbf{f})$ is then the sum of many random numbers and can be approximated by a normal distribution. Hence, $c(\mathbf{v}, \mathbf{f})$ is log-normally distributed.

Two empirical patterns—(a) that link strengths are similarly distributed independent of their precise definition, and (b) that resource abundances are bad predictors of diets—are both reproduced by models in which the variance of logarithmic link strengths is large compared to that of logarithmic abundances.

Very weak trophic links are empirically indistinguishable from absent trophic links. Quantitative food web models will therefore in general be more parsimonious if all possible link are modeled as present, but logarithmic link strengths are allowed to vary broadly, such that, as observed, the fraction of links sufficiently strong to be detected (*connectance*) is small. In this view, adjacency matrices A_{ij} derive from link-strength matrices C_{ij}, e.g., by thresholding link strengths. Empirical food web topologies might therefore best be understood by combining a theory for link strengths C_{ij} with an observation model.

Degree Distributions

For directed graphs, one defines the *in-degree* of a node as the number of incoming links, and the *out-degree* as the number of outgoing links. In food web theory, in-degree (number of resources) is also called *generality* of a species, and out-degree (number of consumers) is called its *vulnerability* (not to be confused with vulnerability traits).

Both, in- and out-degrees are distributed much more broadly in food webs than would be expected for links assigned with equal probability to each species pair (which yields binomial distributions). Out-degrees often have an approximately even distribution between zero and some upper limit. This pattern is reproduced by models implementing the "large eats small" rule mentioned above, but otherwise assigning links from consumers to a given resource independently and with equal probability.

Most empirical distributions of in-degrees (and sometimes out-degrees) are approximately exponential (geometric). At least, values near zero tend to be the most frequent and the distributions have long upper tails. This skewed structure can be reproduce by evolutionary models with high heredity of vulnerability traits. The skewed distribution of clade sizes characteristic of phylogenetic trees then leads to the observed skewed distribution of in-degrees, provided consumers tend to forage on a single resource clade. Near-exponential degree distributions are sometimes also found in models without phylogenetic constraints, but the underlying mechanism in this case has not been identified, yet.

For better comparison across food webs, in- and out-degrees can be normalized by the mean out-degree (= mean in-degree = link density). Distributions of normalized degrees are quite similar among food webs and have been hypothesized to follow universal functional forms.

Degree distributions and link-strength distribution are related through the *multivariate distribution* of the strengths of all links from and toward one species. For instance, the hypothesized universality of normalized in-degree distributions implies dependencies among link strengths: a link from one resource to a consumer makes the occurrence of a link of similar strength from another resource to the same consumer more likely (Rossberg, Yanagi, Amemiya, and Itoh, *Journal of Theoretical Biology* 243: 261–272).

Network Motifs

Food webs contain certain types of small connected subgraphs with three or four species more often (others less often) than expected by chance. Using smart randomization algorithms, it can be shown that the prevalence of these *network motifs* is not explained by the degree distributions particular to food webs. Yet it remains unclear how far network motifs are simply consequences of phylogenetic and size constraints, or whether some motifs call for independent explanations.

Intervality

For small food webs, it is often possible to re-index species such that, in each column of the adjacency matrix, the 1s form a contiguous block (consecutive ones property). For larger food webs, this is usually not possible, but the number of 1s one would need to add in the matrix to archive this is much smaller than expected by chance. This phenomenon is called *food web intervality*, with reference to a related concept in graph theory.

Evolutionary food web models with high heredity of vulnerability traits naturally reproduce this phenomenon (see Fig. 2). Its historical interpretation, however, was different. Influenced by an ongoing discourse on ecological niches, Joel E. Cohen, when discovering intervality in 1977, interpreted it as the signature of an approximately one-dimensional trophic niche space. In fact, food webs defined by thresholding link-strength function Equation 1 with $D = 1$ and fixed $v^{(0)}$ will always be perfectly interval. A subsequent search for the trophic trait corresponding to this single dimension, however, remained unsuccessful. Body size, a likely candidate, does not to constrain topology sufficiently to explain the pattern. By now we know that phylogenetic constraints offer a more parsimonious explanation. Yet models constructed around one-dimensional niche spaces are still popular as simple tools to generate interval food webs.

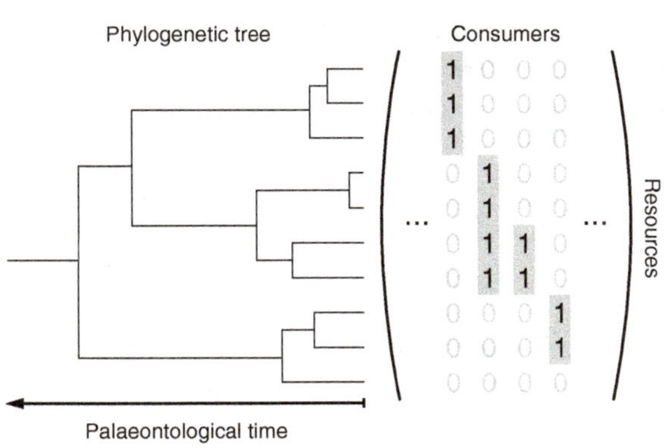

FIGURE 2 The phylogenetic explanation of intervality. Put a food web's member species in an order in which they could appear on the tips of a phylogenetic tree. If consumers feed on groups of species that are similar because they are related, then the 1s in each column of the adjacency matrix (shown in part) will automatically form contiguous blocks (gray boxes). That is, the food web is interval. In practice, food web intervality in large communities is rarely perfect, neither in data nor in models implementing this mechanism *via* phylogenetically correlated vulnerability traits.

Block Structure

By visual inspection, one easily recognizes large rectangular *blocks* of high connectance in empirical adjacency matrices. More precisely, there are pairs of large species sets (P_r, P_c) such that trophic links between resources from P_r and consumers from P_c are much more frequent than trophic links on average. If P_r and P_c are identical, one speaks of a *compartment* structure. Compartments are rare in food web topologies, but they can result from joining food webs of separated habitats—and be modeled as such.

Blocks with nonoverlapping or partially overlapping P_r and P_c arise in models with phylogenetic constraints. If, for example, all members of P_r are closely related and heredity of vulnerability traits is high, then all members of P_r have similar vulnerability traits. Members of P_c are those species with foraging traits located near the vulnerability traits of the members of P_r in niche space. The members of P_c are not necessarily related.

Covariation Patterns

Early food web theory worked with large collections of food web datasets that were each too small to exhibit much structure on their own. Theory therefore focused on the question how simple quantitative properties of topological food webs, such as the number of species, the fraction of top species (species without consumers), the mean length of resource–consumer chains, and so on, covary among datasets. Stochastic food web models reproducing these covariation patterns were sought.

Later, this idea was adapted to smaller collections of larger, high-quality food webs. The problem then amounts to defining a vector of food web properties $\mathbf{x}$ and finding models such that, for each empirical data set i, its properties $\mathbf{x}_i$ are likely to co-occur in the output of a stochastic model for an appropriate set of model parameters $\mathbf{p}_i$. Model-selection methods such as the Akaike Information Criterion have been used to guard against model overparametrization. Applicability of this analysis is limited by its computational cost. In the most advanced application so far, involving 14 food web properties and models with up to 6 free parameters, a million independent model samples had to be generated to fit each of 17 empirical datasets. The model most faithfully reproducing covariation patterns was the *matching model*, which generates food web topologies by combining a large-eats-small rule with phylogenetically correlated trophic traits.

THE PROBLEM OF FOOD WEB STABILITY

What Is the Problem?

Many practical questions that food web theory strives to address relate directly or indirectly to population dynamics. A plausible approach to answering such questions is to construct a quantitative food web for the community in question, to set up equations modeling the population dynamics of the community, and to simulate this system. However, in the majority of cases, one will find this approach to fail: as simulation time proceeds, many species in the community go extinct, despite their observed coexistence in reality.

An example is shown in Figure 1. The empirical food web of Morton Creek is equipped with population dynamics as described in Box 1. When simulating the system, 7 of the 13 trophic species in this food web go extinct. Variations of the parametrization suggested in Box 1 lead to similarly devastating results.

The mathematical concept most adequate for the study of this phenomenon is community permanence. A community is permanent if no population ever drops below an arbitrary but fixed positive threshold, except for initial transients, and all populations remain bounded from above, independent of initial conditions. However, for the sake of simplicity, theorists often consider instead community feasibility (existence of a fixed point where all populations are positive) or the linear stability of feasible fixed points. The three concepts are related: linearly stable communities are typically permanent, and all permanent communities have a feasible fixed point.

What follows is a critical discussion of possible explanations for these difficulties in building realistic, stable population-dynamical food web models and of suggestions that have been made to overcome these. Efforts have been made to cover the most important ideas, yet the list is not complete.

Self-Limitation of Populations

Trophic interactions are not the only factors limiting population growth. For basal species (species not feeding on others), this is obvious. But even consumer population dynamics will be affected by nontrophic factors, such as the spread of diseases or competition for suitable breeding space. In models, such effects are generally captured by self-limiting, density-dependent contributions to the population growth rate (intraspecific competition). In the Jacobian matrix computed at a community fixed point (controlling linear stability; see Stability Analysis), these self-limiting terms lead to negative contributions to the diagonal, which stabilize the system. The magnitude of these diagonal contributions required to achieve linear stability has been used as a measure of system stability (*diagonal strength*).

One finds that substantial intraspecific competition is required to stabilize large food web models by this mechanism alone. The question whether in nature self-limitation of sufficient magnitude regularly occurs appears to be open. In agriculture, population densities much higher than naturally observed can be reached simply by providing sufficient food and protection from natural enemies. Besides, one needs to ask for any mechanism of nontrophic self-limitation whether the same mechanism could also mediate competitive interactions with other, similar species, in which case this mechanism would equally contribute to destabilizing a community as it helps stabilizing it.

Adaptive Foraging

In laboratory studies, consumers that are let to feed on two kinds of resources often feed disproportionally more on the more abundant resource (*adaptive foraging, prey switching*). The reasons can be active behavioral adaptation by the consumer, or passive mechanisms such as shifts in the consumer's feeding grounds. Evolutionary adaptation can have similar effects, too. Independent of the mechanism, the phenomenon leads to a release of rare species from predation pressure. Species that would otherwise go extinct can survive at low abundances. Adaptive foraging therefore stabilizes communities.

However, foraging adaptation appears difficult to measure in the field, leaving the question open how much of this effect should realistically be allowed in models. Besides, current representations of adaptive foraging in food web models often do not allow for switching to become weaker for pairs of very similar resources, an unlikely feature, which effectively disables competitive exclusion through indirect competition.

Sparse Food Webs

In food web models with self-limitation, stability breaks down when the link density becomes too large, unless other, stabilizing model elements are included. With respect to linear stability, this effect was mathematically explained in 1972 by Robert May. Later simulation studies demonstrated a similar constraint for community persistence. It appears as yet unclear if this phenomenon prevails in food web models without self-limitation. Empirically, the question if there is a natural limit on link density is unsettled. A large number of studies indicate that link density increases with food web size, but these studies have been criticised for insufficiently taking account of known biases in empirical data.

Slow Consumers

The larger the typical body mass M of a species, the slower its physiological and ecological dynamics are. Ecologically relevant rate constants of dimension 1/Time, such as consumption rates, respiration rates, or maximal Malthusian growth rates, are known to scale approximately as $M^{-1/4}$ (allometric scaling).

Consumers are often larger than their resources. As a result, ecological processes are slower for consumers than for their resources. When defined appropriately, trophic link strength therefore decreases toward higher trophic levels. This pattern has been shown to stabilize community dynamics and to enhance the likelihood of feasible communities. Randomization of link strengths in model food webs exhibiting this pattern, leaving only the overall distribution of link strengths intact, drastically destabilizes these food webs. Further, it has been shown that for consumer–resource body-mass ratios above 10–100, typical for the observed range of values, this stabilization is particularly efficient.

Stable Trophic Modules

When analyzing certain three-species sub-graphs found in food webs in isolation, i.e., ignoring their interactions with the rest of the web, they are found to be stable (or feasible) more often than expected by chance. It is unlikely that this pattern is simply a consequence of the "slow consumers" pattern described above. Thus, stable trophic modules appear to be an independent phenomenon contributing to food-web stability. This is confirmed by studies directly relating the relative frequency of modules to the stability of random food webs.

Weak Links

Relatively weak trophic links can damp destabilizing oscillations in food webs, that is, when removing these links oscillations become stronger. While this effect is important, it is often overinterpreted to the extent that any weak trophic link would be stabilizing. Simulations by McCann, Hastings, and Huxel (*Nature*, 395: 794–798), for instance, show that a minimum link strength is required. The broader question if specific link-strength distributions contribute to stabilization or destabilization of food webs or emerge from stability constraints requires more research.

Assembly and Evolution

Population-dynamical food webs emerging in assembly models with or without an evolutionary component tend to be substantially larger and more complex than model food webs of the same type obtained by first fixing links and abundances according to some (partially) random algorithm and then simulating dynamics until a persistent community remains. Apparently, assembly selects a set of particularly stable food webs among the set of all possible webs. The particular structural features selected for remain unidentified. Food webs tend to *saturate* in assembly processes; that is, a state is reached in which the perturbations of the community resulting from the invasion of one species lead, on average, to the extinction of one species. There are indications that such saturated states have particular dynamical properties combining fast and slow relaxation processes.

It is plausible to conjecture that natural communities are in such saturated states, too. This would explain the persistent difficulties to reproduce food web stability in models, without offering an immediate solution. Because in saturated communities populations coexist in a delicate ecological balance, models of such communities, unless reproducing population dynamics to very high fidelity, would not be persistent or, after adding stabilizing model elements, be more stable than in reality. Recent observations do indeed indicate that natural communities saturate due to limits to coexistence. However, conclusive evidence linking these observations to community saturation in food web assembly models appears to be missing.

SEE ALSO THE FOLLOWING ARTICLES

Allometry and Growth / Assembly Processes / Food Chains and Food Web Modules / Foraging Behavior / Networks, Ecological / Predator–Prey Models / Stability Analysis / Two-Species Competition

FURTHER READING

Berlow, E. L., A.-M. Neutel, J. E. Cohen, P. C. de Ruiter, B. Ebenman, M. Emmerson, J. W. Fox, V. A. A. Jansen, J. I. Jones, G. D. Kokkoris, D. O. Logofet, A. J. McKane, J. M. Montoya, and O. Petchey. 2004. Interaction strengths in food webs: issues and opportunities. *Journal of Animal Ecology* 73: 585–598.

Bersier, L.-F. 2007. A history of the study of ecological networks. In F. Képès, ed. *Biological networks*. Hackensack, NJ: World Scientific.

Cohen, J. E., F. Briand, and C. M. Newman. 1990. *Community food webs: data and theory.* Berlin: Springer.

de Ruiter, P. C., V. Wolters, J. C. Moore, and K. O. Winemiller. 2005. Food web ecology: playing jenga and beyond. *Science* 309: 68–71.

Drossel, B. 2001. Biological evolution and statistical physics. *Advances in Physics* 50: 209–295.

Loeuille, N. 2010. Consequences of adaptive foraging in diverse communities. *Functional Ecology* 24: 18–27.

May, R. M. 1973. *Stability and complexity in model ecosystems.* Princeton: Princeton University Press.

McKane, A. J. 2004. Evolving complex food webs. *European Physical Journal B:* 38: 287–295.

Montoya, J., S. Pimm, and R. Solé. 2006. Ecological networks and their fragility. *Nature* 442: 259–264.

Pimm, S. 2002. *Food webs.* Chicago: University of Chicago Press. Reprinted with a new foreword.

Williams, R. 2008. Effects of network and dynamical model structure on species persistence in large model food webs. *Theoretical Ecology* 1: 141–151.

Woodward, G., B. Ebenman, M. Emmerson, J. Montoya, J. Olesen, A. Valido, and P. Warren. 2005. Body size in ecological networks. *Trends in Ecology & Evolution* 20: 402–409.

FORAGING BEHAVIOR

THOMAS CARACO

State University of New York, Albany

An animal's survival and reproduction demand that it consume energy and nutrients produced by other organisms. Some animals acquire essential resources in a comparatively simple manner; consider an aquatic filter feeder extracting organic matter from flowing water. Other animals capture resources in a manner requiring a complex series of actions, sometimes involving social relationships; consider a group of lions ambushing, capturing, and then competing for a gazelle. The study of foraging behavior spans diverse questions concerning the mechanisms, evolution, and ecological consequences of animals' food consumption. Foraging theory, more specifically, assumes that natural selection can shape behaviors that directly govern an animal's energy acquisition. Foraging theory has helped advance understanding of the remarkable diversity observed among different species' feeding behavior. Furthermore, models of individual or social foraging can be linked with models for population dynamics to predict stability and complexity at the level of ecological communities.

NUMERICAL AND FUNCTIONAL RESPONSES

A consumer population's birth and death rates, as well as migration rates between populations, may depend on temporal and spatial variation in food availability. Change in a consumer population's density, driven by change in food abundance, is termed the consumer's numerical response to resource density.

Consumers often affect the population dynamics of the biotic resources they exploit. The rate at which consumers deplete food abundance depends on both consumer density and the amount of the food resource eaten per unit time by each consumer. The latter quantity is termed the consumer's functional response to resource density. Properties of the consumer's foraging behavior can directly impact the functional response.

OPTIMAL FORAGING THEORY

Optimal foraging theory (OFT) asks how behaviors governing the acquisition and consumption of resources contribute to survival and reproductive success. OFT offers an understanding of prominent behaviors by evaluating their potential adaptive significance. Characterizing foraging as "optimal" follows from the premise that variation in a behavior influencing an individual's survival or reproduction can be subject to optimizing (i.e., stabilizing) selection. Given this premise, foraging theory commonly invokes mathematical optimization as a metaphor for natural selection. OFT has drawn criticism, largely because the models seldom include intrinsic limitations (e.g., genetic constraints) on phenotypes and their evolution. Foraging theorists reply that they seek general principles linking environments to behavior, predictions independent of any particular organism's mechanistic constraints. Furthermore, any OFT model appreciates that constraints on the forager–environment interaction (constraints often defining the problem) limit the choices or options available to the consumer. OFT has produced quantitative hypotheses about behavior—predictions subject to rejection through experimentation or observation—and the theory has advanced understanding of why certain animals forage as they do.

Behavioral ecologists generally define OFT as the study of solitary, independent foragers and refer to models of interacting foragers as social foraging theory (SFT). For clarity, this distinction is adopted here; some questions concerning social foragers do not apply to solitaries, and methods ordinarily used to solve the two types of models differ.

Model Structure

Models in OFT first identify feasible phenotypes. The set may be discrete (e.g., prey types a predator encounters) or continuous (e.g., the length of time an ambush predator remains at one location). Second, the model specifies limitations intrinsic to the organism (e.g., inability to distinguish prey types) or extrinsic (e.g., time available to feed). As indicated above, OFT stresses the latter constraints. Finally, the model's objective function specifies a quantitative relationship between feasible behavioral phenotypes and a "currency of fitness." That is, the model formalizes the hypothesis that lifetime reproductive success (Darwinian fitness) correlates with a measure of foraging performance, the currency. Maximizing the objective function (or minimizing cost) identifies optimal behavior; predictions are deduced from the model's solution. Testing the predictions asks if the behavior of interest has the functional significance proposed in the model. OFT does not suggest that every trait of a forager is an adaptation.

Diet Breadth

Specialist consumers exploit a narrow range of resources; generalists are less selective. OFT addresses this distinction in a series of models for the prey types (different foods) included in a forager's diet. A basic version, called the contingency model, answers the following question. Given encounter with an item of a recognizable prey type, should the forager consume the item or reject it and search for a more rewarding prey type? The number and identity of the prey types a forager accepts specify its diet breadth.

The contingency model (Fig. 1) assumes that a forager can search for different prey types simultaneously, since prey are intermingled. But the forager discovers only one item per encounter. When an item is accepted, the forager must stop searching and handle the food to extract energy. Prey types can differ in density (and hence encounter rate during search), net energy yield per item, and handling time. The model hypothesizes that a forager's fitness should increase with its average long-term rate of gaining energy. Hence, an optimal diet breadth maximizes the rate of energy gain.

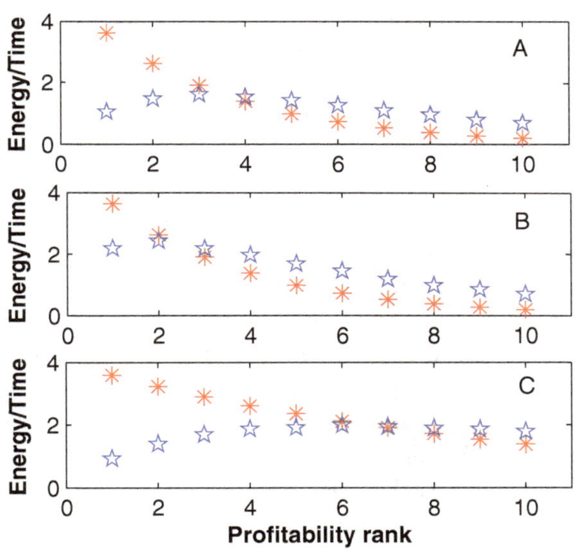

FIGURE 1 The contingency model's optimal diet. Prey types are ranked from highest to lowest profitability, the ratio of net energy yield per item to handling time per item. Profitabilities are indicated by red ✳ symbols. Blue ☆ symbols indicate mean long-term rate of energy gain for a diet composed of the k highest ranked prey types ($k = 1, 2, \ldots, 10$). If the profitability of prey type ($k + 1$) exceeds the rate of gain for a diet composed of the first k prey types, the optimal (rate-maximizing) diet includes prey type ($k + 1$). If the rate of gain for a diet of k prey types exceeds the profitability of prey type ($k + 1$), the optimal diet cannot include prey type ($k + 1$) or any lower-ranked prey. (A) Optimal diet includes first three prey types. Profitability of fourth ranked type is less than optimal diet's rate of gain. (B) Same profitabilities as in Part A, but encounter rates will all prey are increased. Rates of energy gain increase, and optimal diet includes only two highest ranked prey types. (C) Profitabilities decline less rapidly with rank when compared to Part A. Optimal diet includes the six highest ranked prey types.

To find the optimal diet, the model evaluates each prey type by the ratio of its net energy yield per item to handling time per item. This ratio is termed the type's profitability. The model's solution requires that the most profitable type be included in the diet. Adding the prey type ranked second implies that the forager will encounter food it accepts more often while searching. But taking the second type will decrease the mean energy yield per item accepted and/or increase the mean handling time per item. The forager faces a tradeoff between faster prey encounter while searching and reduced mean profitability per item eaten. The model's solution yields a simple rule: the forager should expand its diet if its long-term rate of gaining energy when specializing on the most profitable type is less than the profitability of the second type. If this is true, the expanded diet increases the rate of energy gain. Proceeding from the highest rank in descending order, the profitability of each prey type is compared to the long-term rate of gain for the diet including all types of higher rank, and no others. The first set of prey types where the rate of energy gain exceeds the profitability of the next lower ranked type is the optimal diet. The decision to accept or reject a prey type does not depend on its density, but does depend on the densities of all types of higher profitability. The model predicts that a given prey type is either always accepted or always rejected, and it predicts more specialized diets when profitability decreases steeply across ranks or when densities of the most profitable prey types are increased. Later versions examine diet breadth when prey types are encountered simultaneously, when discriminating prey types imposes a cost, and when profitability of each prey type varies randomly.

Herbivory and Dietary Constraints

The contingency model's assumptions apply to many carnivores, insectivores, and granivores, since search and handling ordinarily are exclusive activities. Furthermore, food consumption by these foragers simultaneously yields both energy and other required nutrients. However, understanding the diverse dietary ecologies of herbivores, animals that consume only (or mostly) green plant material, demands modified approaches.

A generalist mammalian herbivore sees palatable food everywhere in its environment; its diet breadth may have no relationship to searching effort. Variation in the availability of different plants may contribute to the complexity of herbivore diets, but foraging theory for generalist herbivores emphasizes that plant species can vary in digestibility, nutrient profiles, and toxins. Some herbivores include less digestible material in the diet to slow down the rate at which more digestible material passes through the gut. Only digestible material is absorbed and converted to stored energy; too great a rate of passage may reduce the energy extracted from higher-quality food. Some herbivores balance their intake of energy and essential nutrients (crude protein or sodium) by consuming combinations of different plants. Other herbivores' mixed diets may expose them to different anti-herbivore compounds, while averting too great a consumption of any single plant toxin.

OFT models generalist herbivore diets by subjecting fitness maximization to constraints assuring that nutritional, physiological, or ecological criteria are satisfied simultaneously. Some models identify "strategies" of energy maximization or time minimization. Among feasible diets, one may provide the most energy; another diet may be energetically and nutritionally feasible while minimizing foraging time, and so reducing the herbivore's hazard of predation.

Patch Residence Time

For many animals, foraging consists of repeated cycles of travel between food patches and resource extraction within these patches. One of OFT's most enduring results concerns the length of time a forager should remain in a patch, under the hypothesis that selection favors increases in the long-term rate of energy gain. The solution to the patch-residence problem, called the marginal value theorem, has been applied to a number of seemingly different questions in evolutionary ecology.

The patch-residence model assumes an environment containing one or more patch types. The forager knows the mean travel time between patches and recognizes each patch type upon entry. Patch types differ in resource availability; more productive patches yield a greater net energetic gain for fixed residence time. In any type of patch, the forager's energetic gain decelerates as residence time increases. The rate of energy gain is maximal as the forager begins to exploit a patch, and it declines continuously with residence time, due to resource depression. That is, depletion of food (or evasive action by the forager's prey) lowers the rate at which the forager gains energy. Given a reduced rate of gain as residence time increases, the marginal value theorem asks when an optimal forager should leave and travel to the next patch.

For an environment with only one patch type, the model predicts that increased travel time between patches increases the optimal residence time. In an environment with many patch types, a forager that maximizes its long-term rate of energy gain will leave each patch at the same rate of increase in its energy gain within the patch, and that rate equals the long-term gain rate. Energy gain and residence time may differ among patch types, but the *derivative* of energy gain within the patch (the marginal value), with respect to residence time for that type, is identical across patch types for the optimal forager. This result generated remarkable interest among ecologists, and a number of related models followed. Some relax the assumption of an "omniscient" forager. For example, experience within a patch might help the forager discriminate better from worse patches. Other models compare simpler rules for departure; a forager might leave every patch after a fixed residence time elapses, after capturing a fixed number of prey, or as soon as the time since the forager last found food exceeds a critical "giving up" time.

Risk-Sensitivity

In winter, a forager may have only the daylight hours to consume energy fulfilling its 24-hour metabolic demands. During breeding, an individual might have to capture enough prey each day to meet its needs and those of rapidly developing offspring. For these foragers, failure to consume a required amount of energy during a limited period imperils survival or reproduction. If we further assume that energy intake varies randomly among foraging periods, as must often be true, then models for risk-sensitive behavior apply.

Risk-sensitive behavior implies that an individual's preferences respond not only to average benefits but also to the variance in benefits associated with different actions. To demonstrate the idea, consider the "small bird in winter." A forager has T time units available. Total energy intake by time T must exceed the individual's physiological requirement R, or its chance of surviving the nonforaging period is reduced significantly. For simplicity, let the forager choose between two habitats to search for food. Within a habitat, the animal discovers food clumps as a random process. When the forager discovers a clump, the amount of energy available within the clump varies randomly.

Foraging ends at time T; total energy intake is the sum of the amount consumed within each clump discovered.

Under reasonable assumptions, the distribution of energy intake follows a bell-shaped curve. The expected total intake is simply the product of the mean number of clumps discovered and the mean energy available per clump. The variance of the total energy intake increases with the variance of the number of clumps discovered and with the variance in the energy available per clump. The animal behaves as if it knows the mean and variance of energy intake for each habitat.

A plausible currency of fitness is the probability that energy intake fails to exceed the requirement. A risk-sensitive forager should choose its habitat to minimize the probability that its intake is less than or equal to R. If the intake variance is equal for the two habitats, the forager should choose the habitat with the greater mean. If the habitats offer the same mean intake but different variances, the choice is not so simple. If the expected intake exceeds the requirement R (food is plentiful), the forager should choose the lower-variance habitat. However, when the mean intake does not exceed R (so that survival is jeopardized), the forager should choose the higher-variance habitat; the animal should "gamble" when losing energy. When both mean intake and its variance differ between habitats, they interactively govern probabilities of energetic failure and so combine to predict foraging preference.

State-Variable Models

The preceding examples of models in OFT make static predictions. That is, the expression (or choice) of a behavioral phenotype maps directly to a fitness score. More generally, an action may contribute directly to survival and reproduction, or may contribute indirectly by changing the animal's state. The new state and the advance of time together can affect the animal's next action. Feedback between behavior and state continue until a final time (e.g., end of the day) is reached and fitness is scored. State-variable models predict sequences of actions between initial and final times, as a function of state. The models are termed dynamic, rather than static.

Definition of state depends on the question of interest. In foraging theory, state usually refers to the individual's level of energetic reserves. To demonstrate, recall the diet-choice problem, but in a dynamic context. As the foraging period commences, the animal might accept or reject a prey type based on its initial energy reserve. As its reserve grows or decays, and as the time remaining to forage declines, the animal might expand or contract its diet. That is, the predicted diet breadth can vary with state, even if prey densities and profitabilities remain constant. At the end of the period, a hypothesized "terminal reward" function maps the final energy reserve to survival and reproduction. Dynamic state-variable models take the expected value of the terminal reward as currency of fitness; optimal behavior, for given reserve and time remaining to forage, maximizes this expectation.

State-variable models ordinarily require computational solution, so that general predictions are not always apparent. Some interesting applications of state-variable models concern costs of suboptimal behavior. Suppose a forager makes a prey-choice "error" in the middle of the day and then follows the optimal policy until the final time. If the error has little effect on the value of the terminal reward, selective pressure is likely weak. However, if the error induces a significant fitness cost, selective pressure on the state-time combination where the suboptimal choice occurred may be strong.

OFT: Final Comment

The methods used in OFT parallel models in bioeconomics—theory developed to manage ecological resources optimally. Predictions of foraging theory have been applied in anthropology, microeconomics, and psychology. Some ethologists, students of behavioral mechanisms, suggest that OFT offers complex models for simple environments and that animals may use simpler rules that deal efficiently with complex environments. As knowledge concerning the neural bases of decision making increases, a combined functional and mechanistic understanding of foraging may emerge.

SOCIAL FORAGING THEORY

Social foraging implies that the functional consequence of an individual's actions depends on both the individual's behavior and the behavior of other foragers, often competitors for the same resource. Social foraging theory (SFT) models generally rely on methods of game theory. Models for dietary choice or patch departure when individuals forage in groups usually make predictions that differ, at least in detail, from the corresponding model in OFT. To emphasize the distinction between solitary and social foraging, this section reviews some issues that concern social foragers only.

Group Size

Groups ordinarily encounter prey (or food patches) more often than do solitaries, and groups can capture larger prey. Group membership may provide the opportunity to learn locations of food or to acquire a foraging skill via social learning. A group member may be safer from predation than is a solitary forager. But as group size increases, most (or all) group members experience greater competition for food, often leading to aggressive interaction. These benefits and costs, along with mechanisms regulating recruitment/expulsion of group members, govern a foraging group's equilibrium size.

IDEAL FREE DISTRIBUTION

Suppose that a population of identical individuals exploits food occurring only in a small number of patches. Suitability of a patch is given by a constant, representing food density; constancy implies that consumption does not reduce food availability. Increasing the number of consumers occupying a patch decreases the feeding rate of each individual in that patch; consumers interact only through scramble competition.

The ideal free distribution (IFD) predicts consumer density in each patch. "Ideal" implies that an individual knows and chooses that patch where its rate of food consumption is maximal. "Free" implies that a consumer can move between patches without energetic cost or behavioral interference. Foragers move until nothing can be gained by moving elsewhere; the sizes of consumer groups equilibrate when each individual has the same resource-consumption rate. The IFD predicts input matching, where the fraction of consumers in a patch equals the fraction of the total resource available in that patch. That is, the distribution of consumers matches the distribution of resources. The IFD is stable in that an individual switching from one patch to another will reduce its resource consumption, as long as all other consumers do not move.

The IFD has prompted a number of further models. Less than ideal foragers fail to discriminate resource-consumption rates or may learn a patch's suitability only after sampling. Consumers may not be free; travel between patches can be costly. Consumers will not be identical if some forage more efficiently, and interference among individuals will affect the impact of local density on resource-consumption rates. Each of these altered assumptions can predict a different patch-occupation pattern.

AGGREGATION ECONOMIES

In an aggregation economy, benefits of group foraging outweigh costs when groups are small. But as groups become large, competitive interactions eventually increase costs beyond any attainable benefits of foraging socially. Therefore, aggregation-economy models assume that the individual's currency of fitness, as a function of the size of its group, has a single peak. The associated group size is termed, perhaps inappropriately, the optimal group size. Predicted group size depends on how groups form and dissolve, and can depend on genetic relatedness among group members.

If solitaries can freely enter any group where membership increases their fitness, equilibrium group size will likely exceed the optimal size. However, solitaries may hesitate to enter groups of close relatives, if doing so reduces each relative's fitness. If group members collectively accept or repel a solitary trying to join the group, the equilibrium group in the absence of relatedness will be the optimal size. If, however, the solitary is a relative of current group members, the solitary could be admitted.

Social Parasitism

Different members of the same group may choose alternate methods to obtain food. Consider the producer–scrounger distinction, an example of social parasitism. Producers expend effort finding and capturing prey; a producer gets a meal only when it generates a feeding opportunity. Scroungers avoid costs of producing and attempt to exploit every feeding opportunity provided by the group's producers. If all individuals have chosen to produce, the first individual switching to scrounging will have more chances to feed than any other group member. When the scrounger phenotype is rare, its fitness should exceed a producer's fitness. If all individuals have chosen to scrounge, no food is discovered. As long as a producer can obtain a greater-than-average portion of the food it discovers, the first producer will have a fitness exceeding that of a scrounger. When these conditions hold, the frequency of scrounging will equilibrate where each phenotype has the same fitness. The predicted equilibrium will depend on environmental attributes (e.g., prey density) and the model's fitness currency, but in each case the equilibrium frequency of scrounging will qualify as stable.

Scrounging can appear because individuals seek to increase their own food consumption or to reduce their foraging costs. For a given group size, more frequent scrounging (i.e., reduction in the number of producers) reduces total food consumption across group members. Each individual's pursuit of its own advantage means that every group member obtains less resource, a consequence of social parasitism.

FORAGING BEHAVIOR TO POPULATION DYNAMICS

Models of foraging behavior can be written into the growth equations of consumer–resource systems, integrating individual-level processes with the analysis of ecological interactions. Some combined models evaluate consequences of particular foraging preferences or functional responses. Other models assume that foragers respond optimally to varying prey density, to predict effects of adaptive behavior on community stability. The body of results is complex; this section lists only a few prominent lessons.

Suppose that an individual forager's effect on the prey population's growth declines as prey density increases. The consequent decelerating functional response does not tend to reduce density fluctuations in a consumer–resource interaction. However, a sigmoid functional response accelerates at intermediate prey densities, so that the prey mortality imposed by each forager increases with the density of prey. Hence, at some prey densities a sigmoid functional response can stabilize population dynamics. When a consumer population preys on two species, a sigmoid functional response can arise if foragers switch between resources and so concentrate predation on the more common prey. Predator switching can, therefore, stabilize the three-species interaction. When a switching predator prevents one prey species from excluding another competitively, the predator's impact is termed a keystone effect.

Dynamical consequences of foraging preference, and its impact on details of the functional response, have been deduced in analyses of three-species food chains. A resource is exploited by a consumer that, in turn, is exploited by a third species. The third species might be an omnivore (exploiting both the resource and the consumer) or a top predator specializing on the consumer; omnivory should exert the greater stabilizing influence on density fluctuations.

Parasitoids often exploit a host population with a highly clumped spatial distribution; many patches contain few hosts, and some patches contain many hosts. An inefficient forager fails to respond to host spatial heterogeneity, while an optimal forager searches patches with the greatest host density. In models of this interaction, optimal patch use by the parasitoid tends to stabilize the densities of the two species. Finally, consider a predator with access to two prey species of differing profitabilities. Suppose that the contingency model's average rate of energy gain enters the dynamics as a component of both prey mortality rates and the predator's birth rate. The predator always includes the prey of higher profitability in its diet. It adds or drops the second prey as the density of the preferred prey changes, according to the optimal diet's choice criterion. The resulting pattern of prey consumption does not tend to stabilize the dynamics, and it can be destabilizing. In general, adaptive foraging may or may not promote stable ecological interaction; predictions—not surprisingly—depend on model details.

SEE ALSO THE FOLLOWING ARTICLES

Behavioral Ecology / Energy Budgets /
Evolutionarily Stable Strategies / Predator–Prey Models

FURTHER READING

Clark, C. W., and M. Mangel. 2000. *Dynamic state variable models in ecology: methods and applications.* Oxford: Oxford University Press.

Giraldeau, L.-A., and T. Caraco. 2000. *Social foraging theory.* Princeton: Princeton University Press.

Houston, A. I., and J. M. McNamara. 1999. *Models of adaptive behaviour: an approach based on state.* Cambridge, UK: Cambridge University Press.

Stephens, D. W., J. S. Brown, and R. C. Ydenberg, eds. 2007. *Foraging: behavior and ecology.* Chicago: University of Chicago Press.

Stephens, D. W., and A. S. Dunlap. 2008. Foraging. In R. Menzel, ed. *Learning theory and behavior.* Oxford: Elsevier.

Stephens, D. W., and J. R. Krebs. 1986. *Foraging theory.* Princeton: Princeton University Press.

FOREST SIMULATORS

MICHAEL C. DIETZE
University of Illinois, Urbana-Champaign

ANDREW M. LATIMER
University of California, Davis

Forest simulators are computer models used to predict the state and dynamics of a forest. As such, forest simulators are on the more complex end of ecological models, both because of the inherent complexity of forest communities and because these models are typically focused on predicting real assemblages of trees, not abstract "forest vegetation." Diverse motivations have driven the development of forest simulators, but the objectives fall into two general classes: (1) to test and extend ecological theory and (2) to predict responses to management action and environmental change. Increasing scientific concern with climate change and the role of forests in global C and N cycles, together with advances in computational power and modeling, are increasing the importance of forest simulators as predictors of forest responses.

OBJECTIVES OF FOREST SIMULATORS

Forest simulators serve to synthesize our reductionist information about how forests work into a coherent, quantitative framework that can predict mechanistically based on first principles and permit us to verify that inclusion of all the "parts" we study in detail allows us to reconstruct the "whole." In this regard, forest simulation can drive theory by forcing us to codify our assumptions, allowing data–model mismatch to identify false assumptions or understudied processes. A related goal has been to test theoretical predictions about forest dynamics with data from specific systems. Examples include investigating different theories of species coexistence and the roles of disturbance and site history in forest dynamics.

Beyond theory, forest simulators also play an important role in management and policy. A number of applied forest simulators are routinely used to predict growth and yields, such as the U.S. Forest Vegetation Simulator (FVS) and the Canadian Tree and Stand Simulator (TASS). These tend to be far ahead of most ecological models in terms of the diversity of factors they include that impact forest growth, but they also suffer the problem of overparameterization, which leads to high forecast uncertainty. The incredibly high data demands for fully calibrating such models means that they are regularly used with default parameters that may not be appropriate for a given site or situation. In the last few decades, there has also been an explosion of forest simulation research focused on global change issues. The goal here is to make projections that help clarify the potential impacts of global change on forests, such as the change in ecosystem services or the loss of biodiversity, and equally importantly to characterize feedbacks from forests to the climate system via energy, water, and carbon fluxes. These global change applications share the goals of informing policy and management and prioritizing directions for further research. Finally, a more recent application of forest simulators has been in data assimilation, where the goal is to estimate the current state of the forest, rather than some future state, given the constraint of incomplete data. For example, a forest simulator might be used to estimate the structure of a forest that would be compatible with an observed lidar profile and then to make inferences about the likely range of values for other stand properties.

CLASSES OF FOREST SIMULATORS

Forest simulators encompass a wide range of models dealing with different ecological processes and operating across a large range of spatial and temporal scales. While there are exceptions, most forest simulators can be divided into two groups, one that is focused on community ecology and the other on ecosystem ecology. Within the first group, forest simulation is dominated by a class of models generally referred to as gap models because of their origin in simulating forest gap dynamics, the dominant disturbance for many forest types. Gap models originated in the early 1970s with patch-based models such as JABOWA and FORET that accounted for the height-based competition for light among trees of different sizes and species. These models generally predict dynamics driven by growth rate and shade tolerance, with fast-growing but shade-intolerant early successional species giving way over time to slower-growing but shade-tolerant late successional species. The 1990s saw the development of truly spatially explicit, individual-based forest simulators such as SORTIE. In these models, the crowns of individual trees interact with each other in three dimensions and the understory light environment is more heterogeneous, driven by the overlap of the shadows cast by each individual tree. Similarly, in these spatial individual-level models, seed dispersal becomes an explicit two-dimensional process, with models differing as to whether they treat dispersal from a Lagrangian (individual seed, e.g., SORTIE) vs. Eulerian (seed density, e.g., SLIP) viewpoint. In addition to the spatially explicit IBMs, there are also a number of landscape patch models, such as LANDIS, that take a simpler representation of each individual patch but which represent the broader scale interactions of vegetation with the abiotic environment and which are often focused on broad-scale spatial-pattern and disturbance feedbacks.

In contrast with community-focused gap models are ecosystem-focused forest simulation models. These models are focused primarily on fluxes and pools of carbon but may represent other biologically important cycles as well, most commonly water and nitrogen. Forest ecosystem models tend to be much simpler in terms of their representation of interactions among individuals but more complicated in their representation of physiological processes, such as photosynthesis, carbon allocation, and respiration. These models are also more likely to represent belowground processes such as rooting, soil moisture, and soil biogeochemical cycles. There is a much wider range of spatial and temporal scales represented in forest ecosystem models than in forest community models, from individual trees up to the globe and from near instantaneous in time to millennial. That said, the biological processes involved tend to have particular scales they operate at, and thus models are generally built around specific spatial and temporal scales. Indeed, some of the major remaining challenges in forest modeling—both conceptually and computationally—revolve around scaling.

SPATIAL SCALES

Individual Scale

The spatial scales represented explicitly by forest models range from <1 m to global (Fig. 1). At the finest spatial scales are the spatially explicit individual-based models, such as SLIP and SORTIE, that represent the exact location of individual trees and the spatial interactions between trees. The primary focus of these models is competition for light, which is the limiting resource in most forests and which drives interspecific and intraspecific interactions, tree growth patterns, and demography. Fine-scale processes include the 3D representation of light based on ray-tracing algorithms, which are particularly important for capturing the high degree of heterogeneity in the light environment of forest gaps (Fig. 2). Also occurring at a fine scale is crown competition, 2D seed dispersal, and density-dependent interactions in the youngest life history stages (e.g., seed bank, seedlings). The focus of these models has thus far been on autogenic fine-scale heterogeneity, rather than fine-scale exogenous heterogeneity in soils or topography, as a mechanism for promoting coexistence.

Patch Scale

The next spatial scale represented by forest models is the "patch" scale, which is on the order of 10–30m in diameter depending on the model, thus encompassing several to dozens of individuals in what is assumed to be a locally homogeneous, common environment. Patch-based models average over the fine-scale variability of spatially explicit models, and the size of patches are set assuming that every individual within a patch is able to compete with every other and that the mortality of the dominant canopy tree is sufficient to convert a patch to a forest gap. Light within a patch-based model is usually represented by a vertical

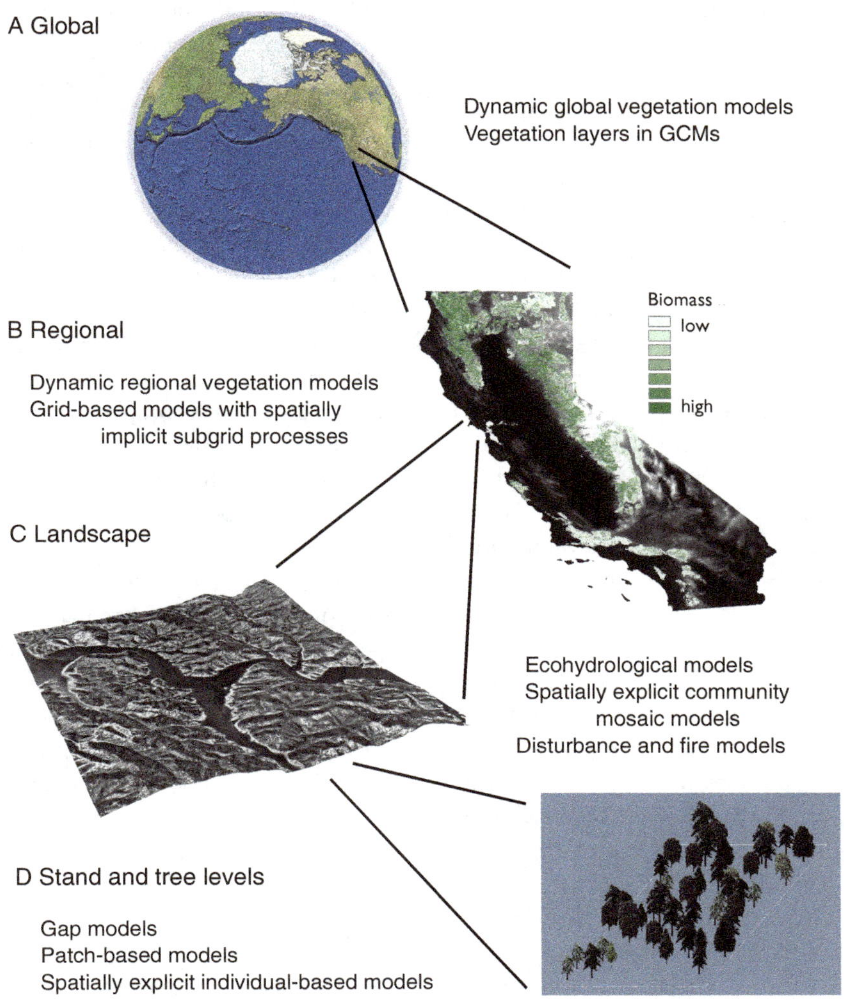

FIGURE 1 Spatial scales addressed by different classes of forest simulators: (A) global vegetation; (B) regional vegetation or forest; (C) landscapes; and (D) forest stands and individual trees.

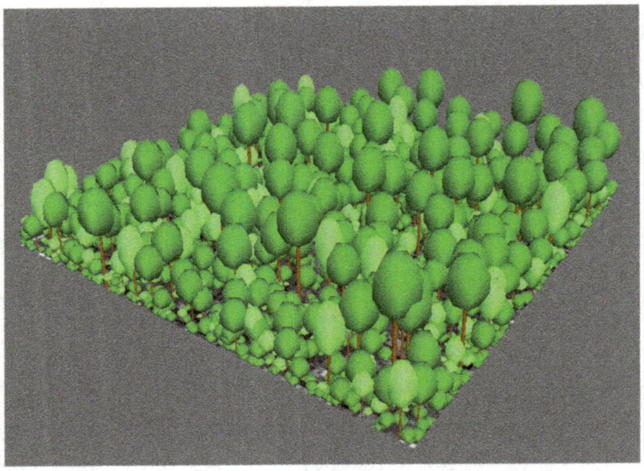

FIGURE 2 Visual representation of forest dynamics in the spatially explicit forest simulator SLIP.

gradient, in which case these models tend to overestimate light levels in gaps, though some models do consider the shadow cast by each patch onto neighboring patches.

The patch is the fundamental scale for a large fraction of models from both the community and ecosystem perspectives, with one key difference: community models are still individual based and thus include multiple trees of multiple sizes and species within a single patch, whereas ecosystem models are based on aggregate carbon pools of a single plant functional type. One ramification of this difference is that while community models will have multiple canopy layers, ecosystem models typically have either a single layer of foliage, often referred to as the "big leaf," or two layers of foliage, representing the functional difference in leaf structural and photosynthetic properties between sun-leaves and shade-leaves. Another key difference in the two modeling approaches at this scale is that because community models are individual based they are focused on the demography of individuals. This means that the fundamental dynamics are conceived in terms of individual demographic responses: the growth rate of a tree based on its size, species, and light environment, the fecundity of individual trees as a function of size and sometimes growth, and the mortality of whole individual trees as a function of growth rate. The inclusion of individual mortality means that almost all forest community models are stochastic, while ecosystem models are almost all deterministic. A consequence of this is that community modelers usually analyze models based on runs with large numbers of patches to average over stochastic dynamics, whereas ecosystem models usually have just one patch in which mortality is simply a deterministic "coarse litter" flux term.

Landscape Scale

The next scale up above patches is the landscape scale (Table 1). The spatial extent of landscapes can vary considerably, from hundreds of meters to tens of kilometers or more. The critical feature of landscape-scale models is not their absolute geographical extent but rather the fact that they account for environmental heterogeneity among patches and aim to provide insight into the effects of such heterogeneity on community or ecosystem dynamics. This heterogeneity can be in terms of the physical template of the landscape itself (e.g., topography, soils, hydrology, microclimate), anthropogenic heterogeneity in the landscape due to land use and fragmentation, or autogenic heterogeneity generated by large-scale disturbances. There is a greater emphasis in landscape modeling on real landscapes rather than on conceptual ones, which are common in gap models that are often focusing on more theoretical questions about the process of succession and community assembly. With this focus on real landscapes also comes a greater emphasis on applied problems and management. A frequent "natural" extent for landscape models is the watershed. Landscape-scale models are most often community focused (e.g., LANDIS, MetaFor), though there are also a number of landscape-scale ecosystem models (e.g., RHESSYS, ForClim), the majority of which are coupled to watershed hydrology models to address ecohydrological questions. Another common feature of landscape models is that there is greater emphasis on spatially contagious processes such as disturbance and dispersal. The most studied of these processes is fire; there are many forest landscape models coupled to fire models that range in complexity from simple "contagious" process models to very detailed mechanistic models of fire spread and intensity (e.g., BEHAVE, FIRE-BGC). While the fundamental unit in landscape models is the patch, the representation of processes within each patch is often simplified compared with patch-scale models. Landscape-scale models are often operating at a spatial scale that encompasses thousands or more patches and necessarily focuses on the distribution of vegetation types and stand ages across patches, rather than the states and dynamics of individual patches. In applying landscape models, users typically assume that they are large enough that the states of constituent patches reach a steady-state distribution (i.e., Watt's patch mosaic) despite the fact that individual patches are far from equilibrium.

Regional to Global Scale

Above the landscape scale are models that take a regional to global perspective on forest dynamics. The questions driving research at this scale primarily surround climate

TABLE 1
Classification of models discussed in the text

Model	Spatial Scale	Temporal Scale	Phenom. or Mechanistic	Descriptive or Predictive	Deterministic or Stochastic	Point or Areal
SLIP (Scaleable Landscape Inference and Prediction)	Individual	Annual	Phenom.	Proscr.	Stochastic	Area
SORTIE	Individual	>Annual	Phenom.	Proscr.	Stochastic	Area
TASS (Tree and Stand Simulator)	Individual	Annual	Phenom.	Proscr.	Stochastic	Area
JABOWA (concatenation of authors Janak, Botkin, Wallis)	Patch	Annual	Phenom.	Proscr.	Stochastic	Point
FORET (Forests of Eastern Tennessee)	Patch	Annual	Phenom.	Proscr.	Stochastic	Area
FVS (Forest Vegetation Simulator)	Patch	>Annual	Phenom.	Proscr.	Stochastic	Point
LANDIS (Forest Landscape Disturbance and Succession)	Landscape	>Annual	Phenom.	Proscr.	Stochastic	Area
MetaFor (Forest Meta-model)	Landscape	Annual	Phenom.	Proscr.	Stochastic	Area
RHESSYS (Regional Hydro-Ecologic Simulation System)	Landscape	Daily	Mech.	Proscr.	Determ.	Area
ForClim (Forests in a changing Climate)	Landscape	Monthly	Phenom.	Proscr.	Stochastic	Point
LPJ-GUESS (Lund-Postdam-Jena General Ecosystem Simulator)	Globe	Daily	Mixed	Proscr.	Stochastic	Avg. point
Hybrid	Globe	Daily	Mech.	Proscr.	Stochastic	Avg. point
ED (Ecosystem Demography)	Globe	Subdaily	Mech.	Proscr.	Determ.	Area
CLM (Community Land Model)	Globe	Subdaily	Mech.	Proscr.	Determ.	Wt. point
Sheffield DGVM (Dynamic Global Vegetation Model)	Globe	Daily	Mech.	Proscr.	Determ.	Wt. point
Orchidee (Organizing C & Hydrology in Dynamic Ecosys.)	Globe	Subdaily	Mech.	Proscr.	Determ.	Wt. point
LPJ (Lund-Potsdam-Jena)	Globe	Daily	Mech.	Proscr.	Determ.	Wt. point
Biome-BGC (Biome BioGeochemical Cycles)	Globe	Daily	Mech.	Proscr.	Determ.	Point
CASA (Carnegie-Ames-Stanford-Approach)	Globe	Monthly	Mech.	Descr.	Determ.	Point

NOTE: Many additional excellent models exist in every category. Spatial scale generally refers to the broadest spatial extent the model is designed to run at, though the individual-based models (IBM) typically function at the scale between patch and landscape. Temporal scale refers to the time step of the model. For point vs. area, wt. point refers to models that have multiple points within a grid cell that are weighted by their proportional area while avg. point refers to models that have multiple stochastic replicates within each grid cell that are averaged. See the text section "Process Representation in Forest Simulators" for discussion of other groupings.

change impacts on the carbon cycle and to a lesser extent on biogeographic/biodiversity issues, though these are usually resolved only to the level of biome or plant functional type rather than to species (e.g., Community Land Model, Sheffield DGVM, Orchidee, LPJ). These models all have an ecosystem component, and only a small subset considers community processes (e.g., ED, LPJ-GUESS). However, there is a growing recognition that disturbance history and successional processes can strongly influence the carbon cycle. These models are typically run on a grid where the grid cells are often much larger in extent than the landscapes in the landscape models. When these models include processes at individual through landscape scales, they must represent them as spatially implicit subgrid processes. For example, in the Ecosystem Demography (ED) model forest stands of different ages are not given spatial locations but are represented by the proportion of the landscape that is in each age class.

Since most global models are based on deterministic ecosystem models, the dynamics of these grid cells are essentially identical to that of a single patch or a weighted average of noninteracting patches representing different plant functional types. Models at this scale include the dynamic global vegetation models (DGVMs) that represent the terrestrial ecosystems in general circulation models (GCMs). While these global models are no longer strictly forest models, almost all originated as forest models (e.g., Forest–BGC evolved into Biome–BGC) and were later modified to incorporate other vegetation types. Because of the emphasis on global change within this research community, there have been a much larger number of model intercomparison projects focused on these models than on other classes of forest models. These include early efforts such as VEMAP (Vegetation/Ecosystem Modeling and Analysis Project) and VEMAP2 focused on the continental United States as well as more recent intercomparisons such as the global scale C4MIP (Coupled Carbon Cycle Climate Model Intercomparison Project), the LBA (Large Scale Biosphere Atmosphere) focused on Amazonia, and the two NACP (North American Carbon Program) intercomparison projects, one focused on the continental scale and the other on site-level comparisons to the Ameriflux network.

TEMPORAL SCALES

Forest models resolve processes that range in temporal scale from the near instantaneous to the millennial. Because the processes involved in community models are essentially demographic, they tend to focus on a narrower range of time scales, from annual to centennial. In contrast, all ecosystem models resolve intra-annual dynamics and some resolve subdaily processes down to a very fine scale. There are two reasons for forest models to resolve processes at subdaily scale. The first reason is to capture the diurnal cycle of photosynthesis using mechanistic photosynthesis models that are driven by instantaneous values of light, temperature, humidity, CO_2, and wind speed. Since these mechanistic models are nonlinear, photosynthesis models operating at a coarser time step either have to make approximations based on an "average" day or use more empirical relationships. The second reason for subdaily modeling is to explicitly resolve the mass and energy budgets of the land surface. These budgets are calculated using a class of process models referred to as land surface models that include a large number of environmental processes beyond the strictly ecological (e.g., boundary layer mixing, snow physics, hydrology, and so on). The primary motivation for including a land surface submodel within a forest ecosystem model is to be able to couple the ecosystem model with an atmospheric model, which requires a lower boundary condition for the land surface. By operating at fine temporal scale and by including atmospheric, vegetation, and hydrological processes, land surface models aim to capture the turbulent mixing and other energy flows that mediate feedbacks among the soil, vegetation, and the atmosphere that are vital to climate projections and to understanding the role of forests in global climate.

At daily to monthly time scales, the processes resolved by forest models are ecophysiological in nature, such as photosynthesis, respiration, carbon allocation, phenology, decomposition, and biogeochemical cycling. Models that have a daily or monthly time scale as their smallest time step typically resolve an explicit mass balance but assume that the energy budget is controlled by some external meteorological driver. At annual to multiannual time steps, forest models typically resolve growth, mortality, reproduction, and disturbance. For most ecosystem models these processes are not resolved explicitly, while for most forest community models this represents the fundamental time step and these processes are the basis for their dynamics. Since most community models ignore intra-annual processes, their calculations for demography are typically based on data-driven empirical relationships rather than physiology. As such, community models are often more constrained to field data, especially with respect to long-term dynamics, but because they generally rely on correlations rather than well-defined mechanisms, they are less suitable for extrapolating responses to novel changes in the environment drivers or novel combinations of environment variables. In contrast, mechanistic ecosystem models are more robust to extrapolation to different conditions, but they often fail to represent long-term dynamics both because they do not include the successional processes that dominate long-term dynamics and because they are often only calibrated to short-term data.

PROCESS REPRESENTATION IN FOREST SIMULATORS

Beyond space and time, forest models can also be classified by how they represent different processes. Below are presented four important contrasts in model dynamics: phenomenological vs. mechanistic, descriptive vs. predictive, stochastic vs. deterministic, and point-based vs. area-based.

Phenomenological vs. Mechanistic

As alluded to above, the phenomenological/statistical versus mechanistic/physiological dichotomy in many ways reflects the community/ecosystem distinction, but it is more useful to view this as a continuum because at some scale of biological organization all our ecological models are phenomenological and within ecosystem models there is a good bit of variability in how different processes are represented. However, the crux of the distinction lies in whether tree growth is based on correlations with environmental variables or on mechanistic representations of NPP/photosynthesis because this distinction largely determines our degree of belief in extrapolating to novel conditions. In principle, other demographic transitions might also be modeled mechanistically, but in fact mechanistic models for mortality simply do not exist, and those for fecundity are rare and difficult to parameterize. The link between growth and productivity is largely one of mass balance—a given amount of net carbon uptake translates into a given amount of growth, and the only real issue is allocation. Mortality, on the other hand, is a complex and multifaceted phenomenon that is often gradual, with many drivers, feedbacks, and lags. Typically, forest gap models assume that mortality is a function of growth rate and disturbance, while in ecosystem models mortality can be as simple as assuming some constant background rate. Beyond mortality and growth, fecundity can be either phenomenological or mechanistic (usually some fixed fraction of NPP), but in either case it is usually poorly constrained to data.

In theory, dispersal can be either phenomenological or mechanistic, though in practice we are unaware of a forest model that has been coupled to a mechanistic dispersal model, but this is bound to happen soon due to their increasing popularity. Mechanistic dispersal models are of varying complexity, but all are fundamentally based on wind speed and seed drag or on movement patterns of animal dispersers. Phenomenological dispersal models, on the other hand, are all based on dispersal kernels, which are probability density functions that give the probability a seed will travel a given radial distance from the parent. Either way, data and theory suggest long-distance dispersal (LDD) is a highly stochastic and inherently unpredictable process. One reason for the use of mechanistic dispersal is that LDD is almost impossible to determine from seed trap data. While the role of LDD in community dynamics is well recognized, its importance for ecosystem responses is less well understood—most large-scale models lack explicit dispersal but instead assume one of two extreme cases that define the endpoints in LDD: (i) new seed is available at all places at all times and thus dispersal is not limiting or (ii) all seed rain is local.

Descriptive vs. Predictive

Another important dichotomy is between models that are descriptive versus predictive. Predictive models attempt to predict biotic responses given a set of initial conditions and meteorological drivers and thus can be run into the future conditioned on meteorological scenarios. Descriptive models, on the other hand, typically require other biotic variables to be specified as drivers. Most commonly these are remotely sensed data, such as LAI, fAPAR, albedo, and the like. Because these models are more constrained by data, they are expected to do a better job of diagnosing unobserved biotic variables. For example, atmospheric inversion models such as the CarbonTracker typically base their continental-scale ecosystem carbon fluxes on descriptive models such as CASA. The tradeoff is that such models cannot be run into the future, and thus climate change forecasts are all based on predictive models.

Stochastic vs. Deterministic

A third contrast is between stochastic and deterministic models. As mentioned above, most ecosystem models are deterministic and most community models are stochastic. This difference is due to mortality and the spatial scales of the models. In fine-scale models, the death of an individual tree is an all-or-nothing event and has a large impact on the microenvironment, and thus these deaths are represented stochastically. In broad-scale models, in contrast, mortality is often modeled as a carbon flux term. Since the fine-scale dynamics of individual tree mortality and gap dynamics are thought to play a large role in overall forest structure and composition, the failure to represent these gaps is one of the main limitations of deterministic ecosystem models at long time scales. The approaches to accommodate this scaling problem can be divided into two categories. First, there are ecosystem models, such as LPJ-GUESS and Hybrid, that are coupled with stochastic gap models and scale up by sampling (i.e., running a large number of replicate stochastic patches). Second, there is the Ecosystem Demography model (ED) and models derived from the ED that treat mortality as a deterministic process and accommodate this by explicitly modeling the distribution of stand ages across the landscape. In essence, mortality is thought of as affecting some fraction of each patch in each year, which is reset to a stand age of 0, while the remainder of the patch does not experience mortality. Simulations with a stochastic version of ED show that the deterministic approach accurately captures the mean of the stochastic version and also is more efficient and tractable, as demonstrated in work by Moorcroft and collaborators (2001).

Point-Based vs. Area-Based

The final contrast considered here is between point-based and area-based models and has to do with how models represent space. Most regional/global models are actually point or patch models that are 0D or 1D (vertically structured) and are simply run on a grid (models represent the nodes on the grid). A few contain spatially implicit subgrid processes, where different patches within a grid cell represent different fractional areas, and thus could be considered to be quasi-area-based. Stand level models are area based in terms of a grid of patches (where each patch truly fills the area allocated to it) or are IBM that are spatially explicit and represent area in 2D or 3D. Landscape models fall in between in that they are explicitly based on a map of polygons or grid cells but these grid cells can start to get too big to represent every tree in them or to safely assume all trees within a cell are interacting. Understanding how a model represents space affects how processes scale in the models, what data can be used to calibrate or test the model, and how we interpret model parameters and model dynamics. For example, a leaf property such as maximum photosynthetic rate means very different things if it is referring to an individual leaf on a tree, the whole forest canopy within a patch, or the aggregate carbon uptake across a 1×1 degree lat/lon grid cell.

DISTURBANCE AND STEADY STATES

A number of disturbances have been included in forest models, the most common being gap phase disturbance, from which gap models derive their name, and fire. Gap phase disturbance can either be autogenic, driven by the mortality of a large adult tree, or externally generated by windthrow or ice storms. As mentioned above, fire models vary enormously in their complexity from simple contagious processes to complex simulations. A number of other disturbances are also included sporadically in different models, such as land-use/land-change, droughts, insects, and pathogens, but overall these have received far less attention than fire and gaps.

One of the reasons the representation of disturbance is so critically important to forest models is that they have such a large impact on if and when an ecosystem reaches steady state. Most community-focused models do not assume that the system is at equilibrium at the start of a run since they are interested in the transient dynamics. Community models often start from bare ground or (less often) from some observed or "typical" composition/structure. That said, community models are often run out to some steady state with a lot of emphasis placed on what that steady state is (despite the fact that there's very little information to judge if the steady state is correct). This is in part a reflection of their conception around questions of long-term coexistence. As the spatial scale increases, more and more models use "steady state" as the initial condition for the computer experiments. This is done even when there is widespread recognition that a particular system is not in steady state (and open debate as to whether any ecosystem ever is in steady state). There are two interconnected reasons for this. First, at broad spatial extents datasets do not exist to serve as the initial conditions. There may be partial information from inventories or remote sensing, but many state variables are unconstrained, especially soil properties such as carbon and nitrogen content. The second reason for a steady-state assumption at a broad scale is that models at these scales generally do not explicitly represent successional dynamics and subgrid (landscape, patch) heterogeneity. Current research into ecological data assimilation is in its infancy, one of its goals is to get around the equilibrium assumption at these scales and to acknowledge the impact of this uncertainty on model predictions. Given what we know about the importance and prevalence of disturbance and transient dynamics in forest community and ecosystem dynamics, this is a vital area of research.

CHALLENGES AND CONCLUSIONS

Forest simulators are likely to continue to play a large role in ecological research for the foreseeable future. Many important basic and applied questions about forest models remain unanswered, and important challenges face model developers. This final section highlights issues believed by the authors to be the most important. In a nutshell, the major challenges for forest simulators are that they are very data intensive, hard to initialize correctly, computationally expensive, lack clear analytical solutions, and face a number of scaling issues, particularly when it comes to bridging the community/ecosystem dichotomy. One unifying characteristic of forest models, whether they are ecosystem- or community-oriented, is that because they are generally aimed at predicting real ecosystems they include a lot of processes and require a large number of parameters. Work by Pacala and colleagues in the 1990s on the SORTIE model was a key turning point in the shift from parameterization of models from "the literature" to being much more data driven and connected to experiments designed with model parameterization as an explicit goal. This is still an ongoing change in perspective, though there is a growing recognition of the importance of formalizing data–model fusion and the propagation of uncertainty through models.

Data for Parameterization and Generalization

One important remaining challenge is to better understand to what extent parameters at one site can be applied to another site. In general, gap model parameters are considered site specific, and for larger models the impact of ecotypic variation is largely unknown. Site-to-site variability is not just a "nuance" parameter for models but has large impacts on our conceptual understanding of how forests work and in testing how general our theories of forest dynamics are. Beyond parameterization, forest models are also data intensive when it comes to initialization and drivers. As discussed in the last section, moving away from simple initial conditions that are either "bare ground" or "steady state" to ones that are based on the current state of specific forests requires large amounts of information. Both community and ecosystem models have so many internal state variables that it is virtually impossible to initialize a model precisely for even a single patch, let alone at broader scales, especially once one acknowledges that empirical measurement error is often nontrivial for many ecological processes (especially belowground dynamics). As ecology moves into a "data rich era" thanks to modern observational technologies (e.g., remote sensing, eddy covariance) and research networks (e.g., NEON, FLUXNET, LTER), these challenges will move from the insurmountable toward the routine as ecologists become more adept at data assimilation and informatics. This is not to say that we won't always be data limited, but that we will be more sophisticated at dealing with the uncertainties. We expect an emerging focus for forest model research will be on determining the quantity, quality, and type of data required to represent and forecast forest dynamics.

Computation

Beyond data, one of the persistent challenges in forest modeling has been computation. While forest simulators have come a long way since the early days of punch cards, the complexity of our models and the scales that we wish to run them on seems likely to continue to outpace Moore's law. In general, forest models are among the most computationally intensive models in ecology. At fine spatial scales, the inclusion of spatially explicit processes can dominate computation (e.g., light and dispersal can be >95% of the computation) and the algorithms involved get disproportionately slower as the spatial scale increases. For broad-scale models, the sheer size of the simulation is usually daunting. For fast time-scale models, such as coupled ecosystem/atmosphere models that include land surface models, the closure of the surface energy budget is computationally expensive and can necessitate complex dynamic numerical integration routines. In all cases, what underlies this computational demand is the fact that forest models lack an analytical solution and thus need to be understood using numerical experiments. The combination of model complexity and lack of a closed-form solution can make forest models difficult to interpret and hampers the ability to reach broad general conclusions. Progress has been made in finding analytical approximations to forest models, and this is an important area of future research and is also closely related to the issues of scaling and crossing the community/ecosystem dichotomy. Partnerships among ecologists, modelers, and mathematicians will be as important as increasing computer power in making these models more useful and interpretable.

Scaling Issues

The frequent dichotomies in the function of forest models (community/ecosystem, annual/diurnal, fine scale/large scale) arise because the processes that affect overall forest dynamics span such a wide range of scales. For computational reasons, it is often impossible to explicitly represent processes important to one class of dynamics (e.g., the emergence of successional dynamics from tree-to-tree competitive interactions) at broader spatial scales. Indeed, individual-based models seem to be limited to a scale of a few km due to the nonlinear scaling of the computation involved as much as the sheer number of trees that need to be tracked. Given that upscaling individual-based models, or even patch-based models, to regional and global scales will effectively never be computationally possible, an important unresolved question is in what ways do the broad-scale ecosystem models lose representative and predictive power by excluding finer-scale processes. These processes are, especially, (a) neighborhood competition and gap dynamics, (b) the importance and persistence of nonequilibrium dynamics, and (c) the landscape-scale effects of interactions with the abiotic environment. We cannot solve this problem by brute-force computation, so it is essential to understand, by extensive model comparison, analytical insight, and large-scale field campaigns, what is lost in scaling and to devise new scaling approaches. As mentioned above, there are already a small number of models (LPJ-GUESS, Hybrid, and ED) that explicitly attempt to integrate ecosystem and community perspectives and to bring together processes operating across a large range of spatial and temporal scales, but these are just the start and many opportunities for innovation remain.

Species and Functional Types

Another challenge in bridging the community/ecosystem dichotomy is that most community models are parameterized around individual species, whereas most broad-scale ecosystem models are built around plant functional types (PFTs). While the use of PFTs is in part driven by the computational demands of representing diversity, it is more often a reflection of the availability of data to accurately parameterize models. This data limitation is only in part a reflection of what trees have been studied but is also a function of what data are available to modelers. Although there are a number of plant trait database initiatives in progress, these databases need to be made more public and there needs to be a greater incentive for field researchers to archive and document data and to deposit it in such databases. Only with such data can modelers and functional ecologists assess how best to summarize species to the level of functional type and whether important dynamics are lost in doing so. There also needs to be more concerted effort on gap-filling research to constrain the processes that drive model uncertainties, such as belowground dynamics.

Prospects for Forest Simulators

Forests structure the ecological dynamics of many ecosystems, influence regional-scale weather patterns, and dominate carbon fluxes from terrestrial vegetation. They are also economically important globally and locally, presenting difficult land management challenges such as those arising from logging and fire policy and enforcement. For these reasons, forest simulators will play an increasingly central role both in forecasting global change and in assessing its impacts on existing forests and management practices. Yet current models, despite their increasing sophistication and power, remain highly data dependent and often make predictions without a robust accounting of uncertainty. Because of this, it remains very difficult to do model intercomparisons and to assess model performance confidently. One of most pressing challenges, accordingly, is the availability and integration of data. There is likely to be rapid progress on this front as large new data sources and computational methods become available and widespread. A second set of challenges lies at the intersection of community and ecosystem models: understanding how competitive spatial dynamics and nonequilibrium successional processes influence ecosystem processes and broad diversity patterns, determining how to scale these processes efficiently, and assessing the adequacy of functional types to bridge between species-level dynamics and ecosystem function. Finally, richer information about the belowground components of ecosystem function—including soil microbial ecology and the role of mycorrhizae in flows of energy and nutrients—are fertile areas of investigation, and belowground dynamics are becoming an important frontier of forest modeling. While these are all very active areas of research, there are no clear answers yet, and progress will depend on collaboration among mathematicians, modelers, and field ecologists.

SEE ALSO THE FOLLOWING ARTICLES

Computational Ecology / Dispersal, Plant / Environmental Heterogeneity and Plants / Gap Analysis and Presence/Absence Models / Gas and Energy Fluxes across Landscapes / Integrated Whole Organism Physiology / Landscape Ecology / Plant Competition and Canopy Interactions / Stoichiometry, Ecological

FURTHER READING

Bugmann, H. 2001. A review of forest gap models. *Climatic Change* 51: 259–305.

Gratzer, G., C.D. Canham, U. Dieckmann, A. Fischer, Y. Iwasa, R. Law, M.J. Lexer, H. Sandmann, T.A. Spies, B.E. Splechtna, and J. Szwagrzyk. 2004. Spatio-temporal development of forests—current trends in field methods and models. *Oikos* 107: 3–15.

Larocque, G., J. Bhatti, R. Boutin, and O. Chertov. 2008. Uncertainty analysis in carbon cycle models of forest ecosystems: research needs and development of a theoretical framework to estimate error propagation. *Ecological Modelling* 219: 400–412.

McMahon, S.M., M.C. Dietze, M.H. Hersh, E.V. Moran, and J.S. Clark. 2009. A predictive framework to understand forest responses to global change. *Annals of the New York Academy of Sciences* 1162: 221–236.

Moorcroft, P. R., G. C. Hurtt, and S. W. Pacala. 2001. A method for scaling vegetation dynamics: the Ecosystem Demography model (ED). *Ecological Monographs* 71: 557–586.

Pacala, S. W., C. D. Canham, J. Saponara, J. A. Silander, Jr., R. K. Kobe, and E. Ribbens. 1993. Forest models defined by field measurements: estimation, error analysis and dynamics. *Ecological Monographs* 66: 1–43.

Perry, G. L. W., and J. D. A. Millington. 2008. Spatial modelling of succession-disturbance dynamics in forest ecosystems: concepts and examples. *Perspectives in Plant Ecology, Evolution and Systematics* 9: 191–210.

Pretzsch H., R. Grote, B. Reineking, T. Rötzer, and S. Seifert. 2008. Models for forest ecosystem management: a European perspective. *Annals of Botany* 101: 1065–1087.

Scheller, R. M., and D. J. Mladenoff. 2007. An ecological classification of forest landscape simulation models: tools and strategies for understanding broad-scale forested ecosystems. *Landscape Ecology* 22: 491–505.

Shugart, H. H., and T. M. Smith. 1996. A review of forest patch models and their application. *Climatic Change* 34: 131–153.

FREQUENTIST STATISTICS

N. THOMPSON HOBBS

Colorado State University, Fort Collins

Frequentist statistics provide a formal way to evaluate ecological theory using observations. Frequentist inference is based on determining the probability of observing particular values of data given a model that describes how the data arise. This probability provides a basis for discarding models that make predictions inconsistent with observations. The probability of the data conditional on a model also forms the foundation for maximum likelihood estimation, which has been the method of choice for estimating the values of parameters in ecological models.

AIMS AND BACKGROUND

Purpose

Ecological theory seeks general explanations for specific phenomena in populations, communities, and ecosystems. Virtually all scientific theory achieves generality by abstraction, by portraying relationships in nature as mathematical models. Models are abstractions that make predictions. Statistical analysis provides a process for evaluating the predictions of models relative to observations, and in so doing provides a way to test ecological theory. Frequentist statistics, also known as classical statistics, have been the prevailing system for statistical inference in ecology for decades.

Textbooks that introduce frequentist statistics usually emphasize methods—how to estimate a parameter, conduct a test, find confidence limits, estimate power, and so on. Because these texts give only brief treatment of the statistical principles behind these methods, students introduced to classical statistics often fail to understand what they are doing when they apply the procedures they learned. They have difficulty extending familiar procedures to unfamiliar problems. They may fail to understand fundamental concepts, for example, the meaning of P values or confidence intervals. This entry will depart from the customary introductory material by describing the principles that underpin frequentist methods. The purpose of the chapter is to provide a conceptual foundation for understanding classical methods and for appreciating their relationship to other inferential approaches.

Models, Hypotheses, and Parameters

A brief treatment of terminology and notation is needed for this foundation. Assume that a hypothesis expressed verbally (abbreviated as H) must be translated into a parameter or parameters (abbreviated as θ) to allow the hypothesis to be evaluated with observations. Thus, a simple hypothesis H = *The mean rate of capture of prey by a predator is 11.5 prey per day* is translated into $\theta = 11.5\ d^{-1}$. A more detailed hypothesis uses a model to explain variation in the parameter in terms of independent variables. In this way, models are mathematical expressions of verbal hypotheses. So, a more detailed, explanatory hypotheses might be expressed as H = *The average number of prey captured by a single predator per day increases asymptotically as prey density increases because searching and handling are mutually exclusive processes.* This verbal hypothesis is expressed mathematically as

$$\mu = \frac{aV}{1 + ahV}, \tag{1}$$

where μ is the average capture rate of prey per predator per day (time^{-1}), a is the search rate (area/time), h is the handling time (time), and V is the density of prey (area^{-1}). Note that is this case, θ is a vector composed of two parameters, a and h.

In the discussions that follow, the terms hypothesis and model will be used interchangeably. Moreover, because a model is composed of parameters, θ will also be used to abbreviate models.

HISTORY

The frequentist approach to statistical inference can be traced to the influential work of J. Newman, E. S. Pearson, and R. A. Fisher during the mid-twentieth century. The seminal work of Newman and Pearson developed statistical procedures for making decisions about two alternative actions based on observations. Their work focused on identifying procedures that had the best operational characteristics for separating these alternatives. Two operational characteristics were important. A test should have a low probability of rejecting a hypothesis that is true and a low probability of failing to reject a null hypothesis that is false. The common scientific practice of rejecting a null hypothesis and accepting an alternative has its roots in these ideas. In contrast, R. A. Fisher was more concerned with the use of observations as evidence for one hypothesis over another. In particular, he developed the use of likelihood as a way to quantify the relative support in data for alternative values of parameters or models. The unifying idea between the work of Newman and Pearson on the one hand and Fisher on the other is that probability is defined in terms of the relative frequency of observations and that probability is objectively verifiable and, as a result, can be used to objectively evaluate scientific hypotheses.

FIRST PRINCIPLES

Probability from Frequency

The term frequentist comes from the definition of probability as the relative frequency of observations chosen randomly from a defined population. Imagine that we have an unknown quantity that represents any possible outcome of an experiment or a sample; we call this quantity a random variable, or Y. Examples of random variables relevant to ecological theory might include the average height of a tree in a forest patch, the number of offspring produced by a bird, the number of species in a habitat, or the biomass of plants produced on agronomic-style plots. The shorthand $\mathbf{y}$ will be used to describe a set of specific observations on the random variable, and the shorthand y_i to represent a single observation. So, if Y is the possible average height of trees, then $y_i = 3.3$ meters is an example of the observed average height. For occasional cases where we can speak of a single observation or a set in the same context, I will use y without a subscript.

The probability that we would observe a particular value, $Y = y_i$, is based on the frequency of that value relative to other values given many repetitions of an experiment or a sample. Simply put, the probability of y_i is the number of times that we observe y_i divided by the total number of observations. The estimated value of the probability of $Y = y_i$ asymptotically approaches the true value as the number of observations approaches infinity. The important message here is that frequentist statistics is based on a specific definition of probability: the relative frequency of observations in an infinite number of repetitions of an experiment or a sample.

Probability Distributions

We treat ecological quantities of interest as random variables because we expect variation when we observe them, variation that arises from many sources—genetic and phenotypic differences among individuals, differences in environmental conditions among sites, errors that arise in our observations, and so on. It follows that if a model predicts no difference in a random variable of interest and if we observe differences in the data, we need to know if the differences we observe should be taken as evidence refuting the model or if these differences would be reasonably expected to arise from natural variation. We use probability distributions to portray variation. Because the idea of a probability distribution and its mathematical description in a density function are central to all statistical analysis, some intuition for probability distributions is developed here.

First, consider the case where data are discrete, which means that integers are the only possible values that we can observe—when we count things, we obtain discrete data. Assume we take all of the y_i in the population and sort them into bins according to their value (Fig. 1). In this case, the bins for sorting the values of y_i are defined by a range of integers (Fig. 1). The relative frequency of values of y_i can be summarized in a histogram, where the heights of the

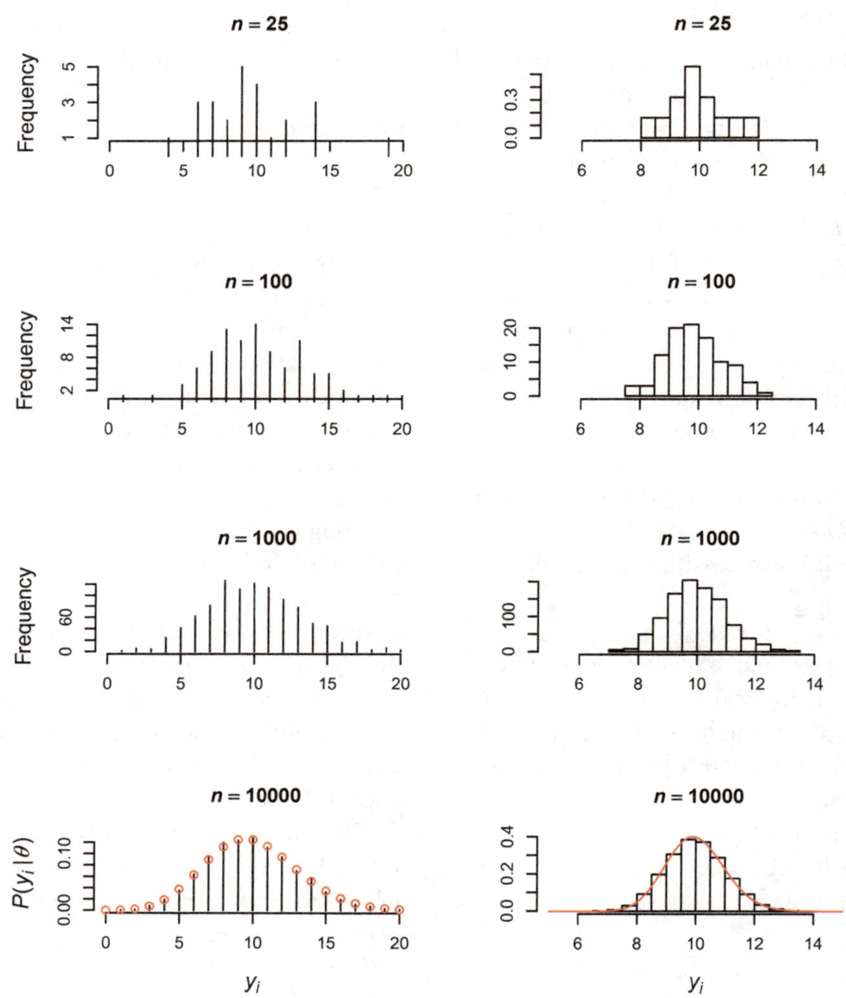

FIGURE 1 Illustration of the relationship between the relative frequency of observations (y_i) drawn from a population with parameters θ and their representation in a probability distribution for discrete (left column) and continuous data (right column). Each panel represents the assignment of n randomly chosen observations from a population to "bins" according to the values observations. The central tendency and shape of the distribution becomes better defined as the number of observations increases, which can be seen in the progression of histograms from the top to the bottom of the figure. As the number of observations approaches infinity, the frequency of each observations divided by the number of observations defines the probability of an observation conditional on the population parameter(s) (bottom row). A discrete density function calculates this probability for discrete data (red circles in lower left panel). A probability density function (red line in lower right panel) calculates the probability density for continuous data. (red solid line in lower right panel).

bars above each integer bin gives the number of observations that we assign to it. If we rescale the height of the bars in the histogram by dividing the number the y_i in each bin by the total number of the observations, then the heights of all of the bars sum to 1 and the height of an individual bar gives the probability the of ith observation, Pr $(Y = y_i)$. As the number of observations gets large (strictly speaking, approaches infinity), then our rescaled histogram defines the probability distribution of the data (Fig. 1).

When data are continuous, our interpretation is somewhat different. Continuous data (for example mass, length, time, temperature) cannot be accurately represented as integers; they must be expressed as real numbers. With continuous data, our binning example works only as the width of the bins becomes infinitely narrow (Fig. 1), causing an important mathematical distinction between probability distributions for discrete and continuous data. When data are discrete, we can talk about the probability that an individual observation, y_i, takes on a given value because the total area of our rescaled histogram (Fig. 1, lower left panel) can be represented as a sum of the heights of the bars multiplied by their width. This sum must = 1. However, when we make our bins "infinitely narrow," we can no longer sum them to find their area. In this case, we must use definite integration to find the area under curve (Fig. 1, lower right panel), which also equals 1. Because we use definite integration to find probability, we cannot talk about the probability of an observation (that is a single value of y_i), because integration is defined only for intervals, not for points. Thus, for continuous data, we can only talk about probability of ranges of values. As a tangible example, we can estimate the probability that the depth of a stream is between 50 and 70 cm [$\Pr(50 \leq y_i \leq 70)$] using the definite integral of the probability distribution between 50 and 70, but we cannot estimate the probability that a stream is 50 cm deep. Instead, for a single observation of continuous data we must talk about probability density, rather than probability, a distinction that will be clarified in the subsequent section.

Density Functions

With an infinite number of observations of our random variable, the relative frequencies of the values of observations define a probability distribution. This distribution can be portrayed mathematically using an equation called a discrete density function for discrete data (also called a probability mass function and a probability function) and a probability density function for continuous data (Fig. 1). These functions can be represented in a general way as

$$\Pr\,(Y = y_i \mid \theta) = f(y_i, \theta), \qquad (2)$$

where the left-hand side of Equation 2 is the probability that we would observe y_i conditional on the value of the parameter θ. Avoiding statistical formalism, "conditional on" means that we fix the value of θ and we seek a value for the probability of a variable y_i, given the specific, fixed value of θ. The right-hand side of Equation 2 is the discrete density function or the probability density function. Equation 2 returns a probability for discrete data and a probability density for continuous data. Probabilities, of course, must be between 0 and 1, but a probability density can take on any value such that the integral of the density function over all values of $y_i = 1$. The probability (or probability density) returned by the function is determined by the value of the observation (y_i) and the parameter(s) (θ).

To illustrate these concepts, consider the Poisson discrete density function, which is often used to describe the probability that we would count a number of objects or events per unit of time or space, given that the average number of counts = θ:

$$\Pr(y_i \mid \theta) = \frac{\theta^{y_i} e^{-\theta}}{y_i!}. \qquad (3)$$

Illustrating this function using the simple capture rate example, above, we hypothesize that the mean capture rate $\theta = 11.5$. We then ask, what is the probability that a predator would capture 15 prey items in a day if the mean capture rate = 11.5 d^{-1}? Thus,

$$\Pr\,(y_i = 15 \mid \theta = 11.5) = \frac{11.5^{15} e^{-11.5}}{15!} = .063. \qquad (4)$$

We conclude that if the mean of the distribution is 11.5 d^{-1}, then we expect to observe 15 prey captured in a day only 6% of the time given many observations.

There are many different density functions that are used to represent the probability distributions of different kinds of observations—the choice of which one to use in a frequentist analysis depends on how the data arise. The most familiar one is the normal distribution, but there are many others. Particularly useful to ecologists are the binomial, multinomial, negative binomial, Student's *t*, *F*, chi-squared, uniform, and gamma distributions.

Parameters and Moments

All discrete density functions and probability density functions have one or more parameters (θ). These parameters determine the specific relationship between the inputs and the outputs of the function. All probability distributions also have moments that describe the central tendency and shape of the distribution. For some density functions, notably the normal and the the Poisson, the parameters are identical with the first two moments (the mean and the variance). However, for all other distributions there is an algebraic relationship between shape parameters and moments, but they are not identical. Translating moments into shape parameters and shape parameters into moments can be accomplished by a method called moment matching that exploits the algebraic relationship between them to solve two equations in two unknowns.

FREQUENTIST INFERENCE BASED ON THE PROBABILITY OF THE DATA

Hypothesis Testing

Statistical hypothesis testing seeks to evaluate scientific hypotheses, which are most often stated as models. Frequentist inference depends on answering the following question: if our hypothesis is true, how probable are the data that we observe? If the observed data have a very low probability under the hypotheses (or model), then the hypothesis can be discarded as false. Notice that this approach to inference sees the hypothesis as fixed while the data are variable.

We gain insight about the hypothesis as follows. We choose a value of θ to represent the hypothesis and choose an appropriate discrete density function or probability density function specifying the probability that a random variable Y takes on a specific value, y_i, conditional on θ. We take an observation (or observations) and use the chosen density function to determine the probability that the random variable would take on values greater than or equal to the observed value. If the data are improbable if the hypothesis were true, we conclude it must be false.

As a simple example, consider a study of survival in a population of birds. We wish to test the hypothesis that the probability that an adult survives from one year to the next is .45, and so we define $\theta = .45$. We will test this hypothesis by observing 100 birds and determining the number that survive a year later. An appropriate probability distribution for this hypothesis is the binomial, which gives the probability of observing a specific number of successes (k = observed number of birds that survive) on a given number of trials (n = number of birds observed) and an overall mean probability of a success (θ = hypothesized survival probability). Thus,

$$\Pr(Y = y_i \mid \theta) = \binom{n}{k} \theta^k (1 - \theta)^{n-k}. \tag{5}$$

We observe that 57 birds survive. Thus, we ask, what is the probability that we would observe 57 birds alive if the true value of the annual survival probability = .45? The probability of obtaining 57 survivals from a sample of 100 assuming that the true survival probability = .45 is

$$\begin{aligned}\Pr(Y = 57 \mid \theta = .45) &= \binom{100}{57} \theta^{57} (1 - \theta)^{100-57} \\ &= .004,\end{aligned} \tag{6}$$

and the probability of observing 57 *or more* survivors is

$$\sum_{n=57}^{100} \binom{100}{n} \theta^n (1 - \theta)^{100-n} = .01. \tag{7}$$

If the true value of the survival probability is .45, and we were to repeat our sample many, many times, we would expect that only 1% of those samples would include 57

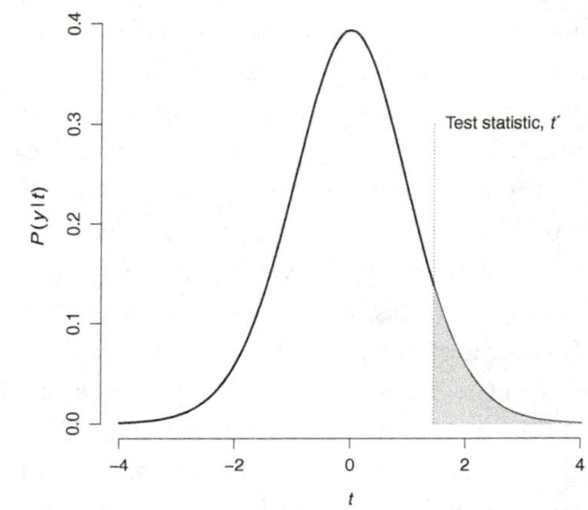

FIGURE 2 Illustration of a test statistic calculated as a function of the data. In this example, we have two data sets, **y** = [11, 16, 18, 24, 16, 8, 32, 8, 6, 13] and **z** = [18, 14, 21, 26, 19, 25, 37, 10, 11, 23]. We want to know if the mean of the distribution from which **z** was drawn exceeds the mean of the distribution of **y**, or, alternatively, if the variation in the distributions is sufficiently great that the observed difference in means could be expected to arise from chance alone. We compose a null hypothesis: "The mean of the population from which **z** was drawn is not greater than the mean of the population from which **y** was drawn." To test this hypothesis, we choose a function of the data, (i.e., a test statistic) that has a known probability distribution under the null hypothesis. Because we have equal sample sizes and the variance of the two samples is the same, we use $t = \frac{\bar{z} - \bar{y}}{\sqrt{(s_y^2 + s_z^2)\frac{1}{n}}}$ where s^2 symbolizes the variance of each sample and n is the sample size. Student's t distribution gives the probability density for given values of t conditional on the degrees of freedom = $2n - 2$. Using the data at hand, the value of t is 1.45. The probability of obtaining a value ≥1.45 if the null hypothesis is true is P = .06 as indicated by the shaded area in the figure. If we used a fixed significance level of α = .05, we would fail to reject the null hypothesis. In this case there would be two possible interpretations: (1) there is no difference between the means or (2) our sample size was not large enough to detect the difference. Alternatively, we could use the value of P as "evidence" against the null hypothesis and could conclude that there is reasonably strong evidence against it—we would expect a value of $t \geq 1.45$ only 6% of the time given many repetitions of the sample if the mean of the distribution of **z** were not greater than the mean of the distribution of **y**.

or more survivors. This result forms evidence that a survival probability of .45 is improbable, allowing us to reject the hypothesized value.

The example above uses the observations directly to test a hypothesis, but more often, we calculate a test statistic as a function of the observations. Probability tells us that any quantity that is a function of a random variable is also a random variable. Exploiting this fact, we let $t = g(y)$ be a function of the observations and let $T = g(Y)$ be the corresponding random variable. We call T a test statistic if we know the probability distribution of t when the hypothesis is true (H) and if larger values of t provide stronger evidence against the hypothesis. We can use a test statistic to evaluate the hypothesis by taking observations on the random variable and calculating a significance level, P, as

$$P = \Pr[T \geq g(y) \mid H]. \tag{8}$$

Assuming many repeated experiments or samples, the significance level gives the probability that we would observe a test statistic more extreme than the one we observed given that the hypothesis is true. If that probability is low, then we can conclude that the hypothesis is false. An example of using a test statistic is given in Figure 2.

Null Hypothesis Testing

The null hypotheses forms a central concept in frequentist statistics, a concept that extends from the argument of logicians that propositions can only be determined to be false; they cannot be proven to be true. "True" statements are those that withstand repeated attempts to show they are false. Thus, if a researcher seeks to determine if there is a difference between two means, he or she seeks to falsify the null hypothesis of no difference. In the Newman–Pearson school, the P value provides a basis for choosing between two actions, which has been translated into the idea of rejecting the null hypothesis and accepting the alternative. Acceptance or rejection of the null hypothesis is accomplished using a critical value, which is the value of T for a set significance level $= \alpha$. If the observed value of T exceeds the critical value, then we reject the hypothesis; if it fails to exceed it, then we fail to reject. The specific value of P matters only in the context of choosing whether to reject the hypothesis. Newman and Pearson viewed α as an operational characteristic of the statistical test, specifically, the probability of making a type I error, rejecting a hypothesis that is true. This level was fixed, and any value of a test statistic associated with $P > \alpha$ provided a bias for falsifying the hypothesis. In contrast, R. A. Fisher's writing about P also included the idea that significance levels provide evidence against the hypothesis and that the smaller the value, the stronger the evidence. Moreover, Fisher also developed the idea of likelihood, which diverged markedly from the approach of Newman and Pearson. (Likelihood is discussed below.)

Confidence Intervals

Thus far, we have discussed the role of observations in testing hypotheses. Observations are also useful for estimating parameters—for example, means, variances, proportions, and rates. Because of the inherent variability among individual observations, we need a way to express uncertainty in parameter estimates, which is accomplished with confidence intervals. The frequentist view of a confidence interval starts with the idea that there is a fixed value of a parameter of interest (θ). Presume we take many samples or run many experiments to estimate θ. We estimate a $1 - \alpha$ confidence interval as a range of values of the random variable of interest $[l(Y), u(Y)]$ such that the range would fail to bracket the true, fixed value of the parameter $(1 - \alpha) \times 100\%$ of the time. More formally,

$$\Pr\,[l(Y) < \theta < u(Y)] = 1 - \alpha. \tag{9}$$

FREQUENTIST INFERENCE BASED ON THE LIKELIHOOD OF THE PARAMETER

Frequentist inference discussed thus far relies on estimating the probability of the data conditional on the parameter. In this framework, the parameter is considered fixed and the data are variable (Fig. 3). Likelihood reverses this relationship by considering the parameter to be variable and the data fixed. In the likelihood framework, we ask what is the likelihood of the parameter given that we have an observation with a particular value?

Definition of Likelihood

Likelihood is defined by the relationship,

$$L\,(\theta \mid y_i) = c \Pr\,(y_i \mid \theta), \tag{10}$$

which simply says that the likelihood of the parameter conditional on the data is proportional to the probability of the data conditional on the parameter. When the data are continuous and the right-hand side of Equation 10 is a probability density function, then likelihood is proportional to probability density. As a result of the relationship in Equation 10, $\Pr(y_i \mid \theta)$ is often referred to as a likelihood function.

Remembering that we express hypotheses as different values for parameters, the difference between probability-and

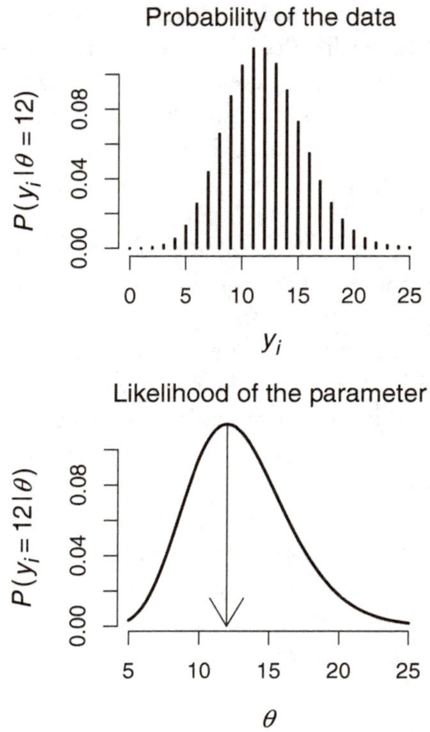

FIGURE 3 Illustration of the difference between a probability distribution (top graph) and a likelihood profile (bottom graph) using a Poisson discrete density function, $\Pr(y_i \mid \theta) = \frac{\theta^{y_i} e^{-\theta}}{y_i!}$. In the probability distribution, we fix the mean ($\theta = 12$) and we vary the value of the data to determine how probable the data are a given value of the mean of the distribution. In the likelihood profile, we fix the value of the data ($y_i = 12$) and we vary the value of the parameter to determine the likelihood of the parameter for a given observed value of the data. The arrow shows the maximum likelihood estimate of the parameter.

likelihood-based inference is as follows. In the probability framework, we assume a value for the parameter and evaluate the probability that the data would arise if the fixed parameter value is correct. In the likelihood framework, we assume we have the data in hand, i.e., they are fixed, and we want to evaluate alternative hypotheses represented as different values of the parameter. Because the purpose of likelihood is to compare the evidence in data for alternative values of parameters (read alternative models, hypotheses; see the section "Strength of Evidence," below), the value of the constant c does not matter; for the purpose of model comparison, we can assume $c = 1$. This assumption is fundamental to likelihood analysis. It means that a single value for likelihood cannot be interpreted; rather, interpretation depends on comparing likelihoods for different hypotheses. Likelihood cannot be used to evaluate a single hypothesis, but rather can only be used to evaluate support in data for one hypothesis relative to another, a core property of likelihood theory called "relativity of evidence." This property is explored further below.

Equation 10 describes the likelihood for a single observation (y_i); how do we estimate the likelihood of sets of observations? The likelihood of n independent observations conditional on the value of the parameter(s) is simply the product of the individual likelihoods,

$$L(\theta \mid \mathbf{y}) = c \prod_{i=1}^{n} \Pr(y_i \mid \theta). \tag{11}$$

The log likelihood of the parameter given multiple observations is obtained using the sum of the individual log likelihoods,

$$\ln L(\theta \mid \mathbf{y}) = ln(c) + \sum_{i=1}^{n} \ln \Pr(y_i \mid \theta). \tag{12}$$

Log likelihoods are often used to estimate parameters because of computational advantages and because of their relationship to other statistical quantities.

Probability Distributions Compared to Likelihood Profiles

The proportionality in Equation 10 might suggest that frequentist inferences based on probability and inferences based on likelihood are similar, but there are fundamental differences between the two approaches. The basis for these differences is revealed by comparing a probability distribution and a likelihood profile (Fig. 3). A probability distribution has values of the data (i.e., $Y = y_i$) on the x-axis, while a likelihood profile has values of the parameter (i.e., θ) on the x-axis. For both approaches, the value on the y-axis is calculated using the function, $f(y_i, \theta)$ (Eq. 2). However, for probability distributions the parameter is held constant in this function and the value of data varies, while in the case of likelihood profiles the data are held constant and we vary the value of the parameter. It is also important to note that the probability distribution can be discrete or continuous, but the likelihood profile is always continuous. A critical difference between the probability distributions and likelihood profiles is that the area under the likelihood profile $\neq 1$, while the area under the probability distribution $= 1$.

Maximum Likelihood

The likelihood profile helps us to understand the concept of maximum likelihood (Fig. 3). The maximum likelihood estimate of the parameter in a model is defined as the value of θ that maximizes $L(\theta \mid \mathbf{y})$, which can be seen graphically as the value of θ at the peak of the likelihood profile (Fig. 3). For simple models, maximum likelihoods can be found using calculus, but more complex models require numerical methods. Maximum likelihood is the prevailing approach to estimation of parameters in the frequentist framework.

Strength of Evidence

To evaluate one model relative to another, we use the ratio of the likelihoods for each model evaluated at the maximum likelihood values of the parameters $(\hat{\theta}_1, \hat{\theta}_2)$:

$$R = \frac{L(\hat{\theta}_1 \mid y)}{L(\hat{\theta}_2 \mid y)}. \tag{13}$$

The quantity R measures the strength of evidence for one model over another. So, presuming that the numerator contains the greater of the two likelihoods, we can say that evidence for model 1 is R times stronger than the evidence for model 2. Although this statement provides a perfectly clear summary of the relative strength of evidence for the two models, it is also possible to calculate a P value (Eq. 8) for the difference between the models with a likelihood ratio test. The likelihood ratio test uses $-2 \ln R$ as a test statistic with the chi-square distribution as a basis for the level of significance. The log of the maximum likelihoods is also used to compare evidence for models using information theoretics, for example, Akaike's information criterion (AIC).

RELATIONSHIPS BETWEEN FREQUENTIST AND BAYESIAN INFERENCE

This section briefly discusses key similarities and differences between the two inferential approaches used to evaluate ecological theory with data, frequentist and Bayesian statistics.

Bayesian inference estimates the posterior distribution of model parameters, i.e., $\Pr(\theta \mid \mathbf{y})$. Bayes' law provides the foundation for this estimation:

$$\Pr(\theta \mid \mathrm{y}) = \frac{\Pr(\mathrm{y} \mid \theta)\Pr(\theta)}{\Pr(\mathrm{y})}. \tag{14}$$

The left-hand side of Equation 14 gives the probability of the parameter conditional on the data, which resembles the likelihood profile (Fig. 3), except that the area under the curve now equals 1. The probability of the parameter (i.e., $\Pr(\theta)$) summarizes the information about the parameter that we knew before conducting an experiment or taking a sample—hence, it is called the prior distribution or prior. The probability of the data [$\Pr(\mathrm{y})$] serves as a normalizing constant, which assures the left-hand-side is a true probability. Because the probability of the data is a constant for any given data set, we can simplify Equation 14 as

$$\Pr(\theta \mid \mathrm{y}) \propto \Pr(\mathrm{y} \mid \theta)\ \Pr(\theta). \tag{15}$$

There are obvious similarities between Equation 15, which defines Bayesian inference, and Equation 10, which defines likelihood. Although it is often said that the difference between frequentist and Bayesian inference is the use of prior knowledge, this statement is not accurate, because it is perfectly feasible to calculate maximum likelihood estimates that include $\Pr(\theta)$. Moreover, Bayesians and frequentists agree that all of the information in the data about the parameter is transmitted through the likelihood function, $\Pr(\mathrm{y} \mid \theta)$.

So, what is the difference between the two approaches? A major divergence is the frequentist and Bayesian view of the parameters and the data. In the frequentist framework, the data are seen as random variables and the parameter is a fixed but unknown constant. In the Bayesian framework, the parameter is viewed as a random variable, which allows calculation of its probability distribution. The practical effect of including the probability of the data in the denominator of Equation 14 is to normalize the likelihood profile, such that the area under the curve integrates to 1. So, computationally, the primary difference between likelihood and Bayesian methods is that likelihood uses maximization to find the parameter that yields the highest probability of the data, while Bayesian methods integrate the likelihood function (or sum for discrete data) over all values of the parameter to obtain a probability distribution of the parameter. Estimates of means of parameters will be identical using the two approaches for simple analyses when priors are uninformative or when identical informative priors are included in maximum likelihood and Bayesian analysis.

A second divergence is the way the two frameworks define probability. As described above, frequentists hold that probability can only be defined in terms of the frequency of an observation relative to a population of potential observations. In contrast, Bayesians view probability as a measure of the state of knowledge or degree of belief about the value of a parameter, a definition that is consistent with the view of parameters as random variables. Historically, this philosophical difference was a cause for heated argument between frequentists and Bayesians. However, it is important to remember that both inferential approaches are abstractions. The idea that the probability of a parameter represents a state of knowledge is no more or less abstract than the idea of probability based on an infinite number of experiments or trials that we never observe.

The contemporary view of the two approaches focuses less on philosophy and more on pragmatism. Modern computational algorithms, particularly Markov chain

Monte Carlo methods, and fast computers have made it feasible to use Bayesian methods to evaluate complex models that heretofore defied analysis using maximum likelihood. If prior, objective information exists, then it can be included in the Bayesian analysis; if it does not, then priors can be made uninformative (with the caveat that there are frequentists who maintain uninformative priors do not exist). When priors are uninformative, Bayesian analysis resembles "normalized likelihood." Many contemporary ecologists use both approaches for evaluating models with data, their choice being guided by the practical requirements of the problem rather than a dogmatic commitment to an inferential philosophy. However, the freedom to match problems to analyses requires a familiarity with fundamental statistical principles. This entry has introduced those principals for frequentist inference.

SEE ALSO THE FOLLOWING ARTICLES

Bayesian Statistics / Information Criteria in Ecology / Markov Chains / Statistics in Ecology

FURTHER READING

Bolker, B. 2008. *Ecological models and data.* In *R.* Princeton: Princeton University Press.

Edwards, E. W. F. 1992. *Likelihood.* Baltimore, MD: Johns Hopkins University Press.

Royall, R. 1997. *Statistical evidence: a likelihood paradigm.* Boca Raton, FL: Chapman and Hall/CRC.

FUNCTIONAL TRAITS OF SPECIES AND INDIVIDUALS

DUNCAN J. IRSCHICK AND CHI-YUN KUO

University of Massachusetts, Amherst

The notion of animal function centers on the idea of biological roles for certain physical traits. Researchers study animal function in several ways, including mechanistic studies of how muscles, soft tissues, and nerves integrate to drive movement and how structural elements (bones, tendons, muscles, etc.) interact to bolster or inhibit animal motion. For example, one basic function of the vertebrate hind limb is movement to facilitate locomotion, such as during walking or running. An essential element for the notion of function is the ability to perform dynamic tasks, such as biting, running, walking, flying, and so on. Such ecological tasks, otherwise known as whole-organism performance capacities, are often essential to the survival of animals, as it enables them to occupy environments, escape predators, and capture prey. Examples of whole-organism performance traits include maximum speed or maximum acceleration incurred during short sprints of locomotion, endurance capacity, or bite force (the maximum amount of force that an animal can produce during biting). In other words, function is a broad term used to describe what the phenotypic trait (e.g., hand, limb) is generally selected and well suited for, whereas performance describes how well an organism can accomplish a certain functional task that is ecologically relevant.

HISTORICAL BACKGROUND

The importance of animal function within the broader field of ecology has emerged slowly over the last 40 or so years. Attempts to integrate animal function into ecology first began when ecologists were attempting to understand why and how species could occupy different niches within ecological communities. Early attempts to address this phenomenon were largely devoid of animal function and relied nearly exclusively on measurements and comparisons of morphological traits (e.g., body size, limb elements). However, during the 1970s and 1980s, the field of ecomorphology arose in an attempt to link the functional traits of animals (originally only morphology, but later changing to measurements of performance) with variation in habitat use. The basic premise of ecomorphology is that variation in morphology and function should be tightly correlated (Fig. 1) with variation in habitat use among species that occupy distinct niches. In some cases, researchers have shown that interspecific morphological and functional differences could enable species to access different resources, supporting the basic tenants of niche theory that such differences have evolved to reduce competition. Therefore, ecomorphology is a paradigm that allows researchers to study how different species have become adapted, or not, to their environments. Over the last 40 years, this field has matured

A *Interspecific* **Morphology → Performance → Habitat use**

B *Interspecific* **Morphology → Performance → Fitness**

FIGURE 1 A heuristic diagram showing relationships between morphology, performance, and either habitat use (panel (A), intraspecific), or fitness (panel (B), interspecific). The arrows do not imply causality directly, but generally mean that variation at each level (e.g., morphology) influences the next level up (e.g., performance) more than the other way around.

and changed, especially in regards to researchers adding in direct measurements of animal function that enable a more complete mapping of morphology → performance → resource use. This trend has been driven by the emergence of field-portable technologies for measuring animal performance, such as portable racetracks, external monitors for measuring speed or metabolic rate, and high-speed video cameras. While the field of ecomorphology has focused primarily on the concept of adaptation among well-defined species, many of the same ideas and approaches have been applied to intraspecific variation to test ideas about adaptation in the context of natural and sexual selection. The idea here is to examine linked variation among individuals between three levels: morphology, performance, and fitness (with fitness substituting for resource use at the intraspecific level; Fig. 1). This practice both tests whether individual variation in morphology is predictive of variation in performance and also whether this morphological variation is adaptive by being under the influence of natural selection. In practice, researchers have not always examined all three components simultaneously, and have mixed and matched various elements (e.g., morphology → performance, performance → fitness, etc.), but this approach is complementary to the interspecific studies. The intraspecific approach relating morphology, performance, and fitness is fundamentally important for ecology because it addresses how environmental and intrinsic factors influence demographic processes of life, death, and reproduction. This explosion of papers examining links between ecology and function at both interspecific and intraspecific levels has allowed evolutionary ecologists to address issues such as how communities are structured, whether members of ecological communities differ in just one aspect (e.g., morphology), or more (morphology, function) aspects, and whether animals use different functional strategies that enable them to occupy different habitats.

INTERSPECIFIC VARIATION: EVOLUTIONARY RADIATIONS

Studies that have linked animal function/performance to ecology among species have done so at various scales. Some have investigated sympatric, or nearly so, groups of closely related species that co-occur within the same community, whereas others have examined broader radiations of species that occur across a wide range of habitats (and therefore could not reasonably be called communities). No studies, to our knowledge, have comprehensively examined multiple taxonomic groups within the same community in terms of examining relationships between function and ecology.

One of the unresolved questions in ecology concerns whether animal species can play different roles within communities. Are species closely bound to the use of a single resource, or can they use many different resources, and therefore become generalists? Functional tests are especially valuable for addressing these questions because they are often used directly to acquire resources (such as in the case of feeding), whereas phenotypic variation alone may be deceiving. The classic (and somewhat dated) view of animal communities is that each species occupies separate niches and uses nonoverlapping resources. While this view has been updated and modified with the addition of new data over the least several decades, recent functional studies have revealed new twists for both how communities and broader radiations perform tasks, and how those tasks can be mapped onto phenotypic variation. Even in the absence of empirical data, however, the expectation of one-to-one (i.e., each species with a distinct phenotype and function occupying a distinct niche) matching is probably unrealistic in most cases for two reasons. First, most phenotypic traits are multifunctional (one-to-many), meaning they can perform several different tasks (consider the human hand, which can perform many different tasks, including gripping objects, throwing, or punching). Second, because of this generality, many different phenotypes might also perform the same function (many-to-one).

Studies with fish and bats both show that there are either weak links between animal function and resource use (bats), or some evidence of many-to-one mapping between jaw morphology and function (fish). Fish jaws present great potential for multifunctionality; they are composed of a complex series of levers (jaw bones, muscles, and soft tissue) that can be moved and used in a variety of ways (e.g., by projecting certain mouthparts, suction, etc.). Fish also feed in myriad ways, ranging from suction feeding (in which no real biting or teeth are required), to durophagy (the crushing of hard prey, such as snails, which typically requires a robust jaw and teeth). Research by Peter Wainwright and his colleagues shows that fish jaws from large and complex coral reef communities, and from larger fish evolutionary radiations, seem to exhibit both many-to-one and one-to-many mapping of fish jaw lever systems (a proxy for functional data on fish jaws) and the kind of prey they consume (e.g., suction feeding, durophagy). In other words, species that consume similar prey can have different jaw morphologies with similar functional capacities. On the other hand, morphologically similar species can differentiate in diet due to functional versatility. The potential implication of this phenomenon is that the most common force driving evolutionary divergence within communities, namely, interspecific competition, is unlikely to result in a one phenotype–one function mapping. This also shows the importance of examining functional data, as simple examination of phenotypic data would lead to the erroneous conclusion that communities are highly ordered and structured, whereas the inclusion of functional data shows more clearly the complex many-to-one or one-to-many patterns. Moreover, these patterns also suggest that morphological differences may have arisen in some cases for reasons other than resource use, although the alternative causes remain opaque. Other studies with bats show that there are also weak linkages between diet and bite force performance, except for highly specialized species, such as those that consume nectar or lap blood. This general lack of connection between function and diet observed in bats and fish is reminiscent of classic ecomorphological studies that showed little correspondence between habitat use and morphological form, such as for some Midwestern bird communities studied by John Wiens and his colleagues. These early studies were instrumental in broadening the approaches of ecomorphologists beyond the well-studied ideas of classic niche segregation theory.

At the other extreme, there seem to exist some systems that show links between morphology, function, and resource use, most notably within relatively simple island systems that likely arose via adaptive radiation. *Anolis* lizards of the Caribbean are notable for their high level of diversity (Fig. 2), both in terms of species numbers (over 400 species in the Caribbean, Central and South America)

FIGURE 2 Images of five different *Anolis* lizard ecomorphs from the Dominican Republic showing the diversity of form in this genus. Caribbean anoline lizards appear show a clear division of matching between morphology, habitat use, and function, suggesting a relatively rare example of "one-to-one" matching. Photographs by Duncan J. Irschick.

as well as behavior and morphology. Caribbean islands are composed of assemblages of anole lizard species of varying numbers (from over 50 in Cuba and Hispaniola to 7 in Jamaica), and each possesses several ecomorphs, which are groups of species that occupy distinct habitats (e.g., tree trunks, crowns of trees) and have corresponding differences in limb, body, and tail dimensions, as well as performance capacities, such as maximum sprint speed on flat surfaces and on rods of varying diameters. This matching of morphology, habitat use, and performance capacities is repeated across islands with remarkable consistency, reinforcing the view that ecological communities can show a one-to-one pattern. Similarly, Darwin's finches (*Geospiza*), of which only 14–15 species exist, show clear relationships between beak morphology (namely, the width and depth of the beak), maximum bite force, and the average size of the prey they consume (e.g., either hard or softer prey), leading to both strong relationships between resource use and function, and also a division of species that access largely nonoverlapping sets of resources. The only caveat to this latter example is that bird beaks are also used for song, a vital behavior employed primarily by males to attract females, and there is evidence that the morphological and kinematic features that promote bird song do not always coincide with features that might be useful for other tasks, such as feeding.

The ecological factors that underlie the formation of communities or that have influenced the broader radiations of animals may explain some of these different outcomes. Two key factors may be the rapidity of the radiation and the simplicity of the ecological context in which species evolve. In cases where evolutionary radiations are rapid and where interspecific competition is intense, likely because of limited resources, then extreme specialization for different roles and a corresponding one-to-one matching can evolve and persist, a scenario that closely fits the profile of Caribbean anoles. By contrast, when ecological conditions are more complex, and the number of potential species interactions increases greatly, and/or if the radiation or community has evolved over a much longer time period, there may be greater evolutionary pressures for species to be opportunistic to survive, therefore creating selection pressures for animals to utilize resources in a myriad of ways and potentially setting the stage for one-to-many or many-to-one mapping.

INTRASPECIFIC VARIATION: METHODOLOGY

Within species, the notion of variation in animal function has often been regarded as a nuisance to be ignored or minimized. Researchers often minimize variation by examining certain sets of animals (e.g., particular age classes) or by gathering large sample sizes such that the standard error around a mean is minimized. One obvious exception to this methodology is the field of behavioral ecology, which seeks to understand the behavior of individuals in relation to intrinsic or extrinsic factors. Whereas reducing variation may be desirable in certain situations, there is an increasing appreciation that intraspecific variation in morphology is often closely tied to variation in function, behavior, and resource use, often in a complex fashion that can affect ecological dynamics. The amount of intraspecific variation in functional traits can be surprisingly large, and many performance traits show a skewed distribution with a few exceptional performers, and a relatively large number of average athletes, although the reasons for this pattern are poorly understood.

Determining whether intraspecific functional variation is "real" (see below) as opposed to random measurement error or simply meaningless variation, is challenging, but two issues are paramount: First, is the trait repeatable? That is, does an animal always run quickly or bite hard if given repeated opportunities to do so? Second, can intraspecific variation in functional capacities be reliably mapped onto morphology, or is there a disconnect among individuals among these and other traits? For example, among individual animals, thicker muscle fibers are usually correlated with the production of greater force, and this means, in general, that variation among individuals in the overall size of their muscles has a direct consequence for an important ability—strength. If no such relationship exists, then variation within each level might carry less value. A third, more rarely applied idea is that variation in both the phenotype and function should have a genetic basis (i.e., be heritable). It is this last ingredient that facilitates evolution via natural selection and therefore sets the table for microevolutionary change.

ANIMAL PERFORMANCE AND FITNESS

Perhaps the most fundamental role of intraspecific functional variation in regards to ecology concerns how it is linked with fitness. Because performance traits are believed to more directly affect fitness compared to morphology per se, there is a prevailing view that natural and sexual selection should be stronger on performance than on morphology, although the available data indicate that the intensity of selection on these two kinds of traits are similar, perhaps because of interrelationships between them.

Do high levels of performance capacity increase the odds for survival and reproductive success? Of 23 studies reviewed by Duncan J. Irschick and Jean-Francois

Le Galliard in 2008, about half showed positive directional selection on performance in relation to survival. In other words, about half the time, high levels of performance increase the odds of survival, whereas the other half of the time, there was no relationship between survival and performance. This supports the view that performance traits play an important role in the demographic dynamics of life and death within ecological communities. Interestingly, stabilizing selection appears to be rare, only being documented in a few isolated cases, a pattern that is similar for morphological traits. Given how frequently environmental conditions can change, thereby altering densities of predators and prey, and therefore the relevant selection pressures, this finding suggests that directional selection may occur in some years and not others, perhaps leading to an on and off "blinking" among years, with a broader pattern of stabilizing selection over microevolutionary time periods. However, there is little data on how variation in performance capacity influences lifetime fitness, or even reproductive success. The data to date suggest that survival selection may not translate into higher reproductive success, perhaps because performance traits appear important primarily in the context of eluding predators or perhaps capturing prey, with few documented links to female choice.

PERFORMANCE TO HABITAT USE

While the correspondence between morphology, performance, and resource use among individuals (intraspecific variation) has been examined less than for among well-defined species or populations, such relationships, at least in theory, should hold. Ecologists have long recognized that individuals of the same species often differ substantially in morphology and can even occupy distinct niches. Some of the various ways that intraspecific variation can occur within a species include ecological dimorphism, ontogenetic shifts, and the presence of behaviorally distinct morphs, often among males.

Many studies have demonstrated both ecological dimorphism and variation among ontogenetic classes in ecology, but few have integrated functional data. In some cases, functional differences among the sexes enables males and females to access different resources, such as in common collared lizards (*Crotaphytus collaris*), in which males have both larger body sizes and relatively larger heads and therefore can bite harder than females. The larger gapes and higher bite forces of male collared lizards enable them to consume larger and bulkier prey than females. Similarly, ontogenetic shifts in morphology and function can enable different size classes to access different resources, such as in sheepshead fish (*Archosargus probatocephalus*), in which oral jaw-crushing force increases with body size across nine ontogenetic classes, enabling larger size classes to consume harder prey, such as bivalves and crabs.

Only a few studies have examined whether morphs within species differ in functional capacities and whether such variation has an impact on their social status, behavior, or ecology. *Uta* lizards form distinct male morphs marked by different throat colors, which correspond to different social types (territorial, mate-guarder, roaming). These morphs differ in circulating levels of testosterone (T), a hormone known to influence vertebrate muscle and certain kinds of fast-twitch performance capacity, a hormonal difference that may enable dominant morphs to more effectively patrol their territories and evict intruders during male conflicts. Other studies with lizard male morphs (*Urosaurus*) similarly show few differences in body shape or performance, suggesting that social behavior, perhaps driven in part by hormones, may be a more important determinant of social status. Not all intraspecific morphs exist as male social classes, and in some cases they may represent the nascent stages of speciation. In African seedcracker finches (*Pyrenestes*) that occur in Cameroon, two morphs exist; the large morphs have larger bills and wider lower mandibles, whereas the small morphs have small bills and narrower lower mandibles. Both feed on sedge seeds, but the large morph feeds more efficiently on harder seeds and the small morph feeds more efficiently on softer seeds, a dietary difference that is reflected in the apparent ability to crush prey and that has arisen and been maintained through disruptive selection.

FUNCTIONAL STRATEGIES

One of the advantages of examining functional traits is the potential for different strategies among individuals within species. Such strategies are analogous to human athletes employing different strategies to win despite some physical constraints. For example, the diminutive (relative to her competitors) Janet Evans (5′6″, 119 lbs) won swimming races by rapidly flapping her arms against the water (the "windmill" method), while much larger female swimmers used the more traditional method of using their long arms to take long and slower strokes. The potential for such strategies in animals is great, but the study of such strategies is in its infancy conceptually and methodologically, and only a few tantalizing examples exist. When bats are forced to fly with an extra load (saline water, harmlessly voided by urination) they do not uniformly respond with the same set of compensatory kinematic (movement) mechanisms, suggesting individual strategies to cope with an ardurous mechanical demand. A recent published study suggested that different strategies can develop in animals exposed to certain environmental conditions. In recent work by Dror Hawlena and colleagues, grasshoppers raised in an environment with predators (spiders) altered their jumping kinematics, which resulted in faster and longer jumps than those raised in a predator-free environment. The difference was solely due to alterations in jumping kinematics without conspicuous morphological change, in contrast to many studies showing morphological modifications that facilitate certain kinds of performance (e.g., burst speed) when animals develop with predators, such as tadpoles in the presence of dragonflies. Animals can also exhibit developmental plasticity in terms of how performance capacity interacts with environmental factors such as food. Research on common lacertid lizards (*Lacerta vivipara*) shows that when juvenile lizards are fed *ad libitum*, those with low endurance catch up and those with high endurance decline. Dietary restriction, paradoxically, allows juvenile lizards with high endurance to retain their locomotor advtange as they mature. In other words, there may be genetic variation both in overall performance proficiency and in how animals respond to environmental cues as they mature.

One promising yet largely unexplored area concerns variation in motivation to perform maximally. Such individual difference in willingness, if proven, would be analogous to "behavioral syndromes" documented in the field of behavioral ecology. This phenomenon refers to consistent individual correlative patterns among multiple behaviors, as when some individuals are bolder than others across all contexts (e.g., being more aggressive toward conspecific individuals, less flighty toward predators, and so on). Similarly, such "performance syndromes" might exist within species, as ontogenetic classes of animals (e.g., juveniles, adults) often show variation in their propensity to run to the limits of their maximum capacities, perhaps because of the differential threat they perceive from predators as a result of vulnerability at different life stages. Interestingly, if such variation in motivation or other manifestations of performance syndromes existed, this might be one reason for a general lack of correspondence between morphology and performance within species. These compelling examples provide suggestive hints as to how animals functionally cope with different environments, but far more data are needed.

SEE ALSO THE FOLLOWING ARTICLES

Behavioral Ecology / Integrated Whole Organism Physiology / Movement: From Individuals to Populations / Phenotypic Plasticity

FURTHER READING

Alexander, R. M. 2003. *Principles of animal locomotion*. Princeton: Princeton University Press.

Bennett, A. F., and R. B. Huey. 1990. Studying the evolution of physiological performance. *Oxford Survey of Evolutionary Biology* 7: 251–284.

Biewener, A. A. 2003. *Animal locomotion*. Oxford: Oxford University Press.

Hawlena, D., K. Holger, E. R. Dufresne, O. J. Schmitz. 2011. Grasshoppers alter jumping biomechanics to enhance escape performance under chronic risk of spider predation. *Functional Ecology* 25: 279–288.

Irschick, D., J. Meyers, J. Husak, and J. Le Galliard. 2008. How does selection operate on whole-organism functional performance capacities? A review and synthesis. *Evolutionary Ecology Research* 10: 177–196.

Podos, J., and S. Nowicki. 2004. Beaks, adaptation, and vocal evolution in Darwin's finches. *BioScience* 54: 501–510.

Vogel, S. A. 1988. Life's devices: the physical world of animals and plants. Princeton: Princeton University Press.

Wainwright, P. C., and S. M. Reilly, eds. 1994. *Ecological morphology: integrative organismal biology*. Chicago: University of Chicago Press.

Wiens, J. A., and J. T. Rotenberry. 1980. Patterns of morphology and ecology in grassland and shrubsteppe bird populations. *Ecological Monographs* 50: 287–308.

G

GAME THEORY

KARL SIGMUND AND CHRISTIAN HILBE

International Institute of Applied Systems Analysis, Laxenburg, Austria

Game theory was developed as a tool for rational decision making. Its basic concepts were later used in evolutionary game theory to describe the evolution of behavioral phenotypes. In the hands of evolutionary biologists, this merger of game theory and population dynamics became an important tool for analyzing frequency-dependent selection and social interaction.

GAME THEORY

Game theory, as originally created by mathematicians and economists, addresses problems confronting decision makers with diverging interests (such as firms competing for a market, staff officers in opposing camps, or players engaged in a parlor game). The "players" have to choose between strategies whose payoff depends on their rivals' strategies. This interdependence leads to mutual outguessing (she thinks that I think that she thinks . . .). There usually is no solution that is unconditionally optimal—i.e., which maximizes a player's utility function—no matter what the coplayers are doing. In contrast to such mutual dependence, monopolists can optimize their budget allocations without having to worry that others will anticipate their decisions. An optimization problem may be fraught with uncertainty, or computationally complex, but what is meant by a solution usually stands beyond doubt. In game theory, this need not be the case. Even in the simple game of "matching pennies" (two players I and II choose independently between two alternatives; I wins if the two agree, and II if they differ), no outcome can leave both players satisfied.

A player can choose between alternative moves, or strategies. Since it is often useful to be unpredictable, a player may also choose a mixed strategy; i.e., opt with specific probabilities for this or that alternative. It can be shown that for any game there exists at least one set of strategies (one for each player) that are best replies to each other (see Box 1). In this case, no player has an incentive to deviate from his or her strategy as long as the other players stick to theirs. This defines a Nash equilibrium. (In the matching pennies game, both players have to choose with probability 1/2 between the two alternatives; as this example shows, Nash equilibria need not exist if mixed strategies are not admitted.)

BOX 1. BEST REPLIES AND NASH EQUILIBRIUM PAIRS

A game between two players I and II can be described by its normal form, which consists of a list of all the strategies $e_1, \ldots, e_n$ and $f_1, \ldots, f_m$ available to player I and player II, respectively, and of their payoff values a_{ij} and b_{ij} obtained when I plays e_i and II plays f_j. A mixed strategy for player I is given by the vector x of the probabilities x_i to use e_i. Since $x + \cdots + x_n = 1$, the vector $x = (x_1, \ldots, x_n)$ is an element of the unit simplex S_n spanned by the vectors of the standard basis in R^n—i.e., the vectors with $x_i = 1$ and $x_j = 0$ for $j \neq i$, which correspond to the pure strategies e_i. If player I uses strategy x and player II uses y, then the payoff for the former is given by the sum of the terms $a_{ij}\, x_i\, y_j$, summed over all i and j, and the payoff for the latter by $b_{ij}\, x\, y_j$. We denote these terms by xAy and xBy, respectively.

The strategy x is said to be a best reply to strategy y if $xAy \geq zAy$ holds for all z in S_n. In this case, player I cannot

BOX 1 *(continued)*.

expect any gain from using a strategy different from x. Similarly, y is a best reply to x if $xBy \geq xBw$ for all w in S_m. A pair of strategies (x, y) is said to be in Nash equilibrium if both conditions are satisfied—i.e., if each strategy is a best reply to the other. In this case, both players have no incentive to deviate unilaterally from their strategy. In the special case of a zero sum game (i.e., when $a_{ij} = -b_{ij}$ holds for all i and j), these strategies are maximin strategies; i.e., each maximizes the minimal payoff and thus guarantees the best security level.

One speaks of a symmetric game if the players have the same sets of strategies and payoff values and thus cannot be distinguished. Formally, this means that $a_{ij} = b_{ji}$ holds for all i and j. In this case, a strategy x is said to be a Nash equilibrium if the symmetric pair (x, x) is a Nash equilibrium pair; i.e., if $zAx \leq xAx$ for all z in S_m.

BOX 2. POPULATION GAMES AND REPLICATOR DYNAMICS

In the simplest formal setup for evolutionary game theory, the e_1 to e_n correspond to different types of individuals in a large, well-mixed population, and the x_i are their relative frequencies (thus, the state of the population is given by x in S_n). The game is assumed to be symmetric. Since an individual of type e_i randomly meets an e_j-individual with probability x_j, and obtains payoff a_{ij} from the interaction, the average payoff for e_i -players is given by $(Ax)_i = a_{i1}x_1 + \cdots + a_{in}x_n$, and the average payoff in the population is given by xAx. The frequencies x_i evolve as a function of time t, according to their success. If one assumes that the per capita growth rate of type e_i is given by the difference between its payoff and the average payoff in the population, one obtains the replicator equation $dx/dt = x_i[(Ax)_i - xAx]$ on the state space S_n. Every Nash equilibrium is a fixed point of the replicator equation, and every stable fixed point is a Nash equilibrium, but the converse statements need not hold.

The notion of a Nash equilibrium satisfies a minimal consistency requirement for the "solution" of a game (since otherwise, at least one player would deviate from it), but it presents a series of pitfalls. Consider, for instance, the following "helping game," where two players have independently to decide whether or not to confer a benefit b to the other player, at a cost c to themselves. If $b > c$, they would both earn $b - c > 0$ by cooperating. But since it is better to defect—i.e., not to incur the cost—each player's best reply, irrespective of the other's decision, is to defect. The unique Nash equilibrium, in the helping game, is thus mutual defection. This game thus displays a "social dilemma": the pursuit of self-interest is self-defeating. In other games, there exist several Nash equilibria, and the choice of the right one can be a tricky issue. A large part of classical game theory deals with equilibrium refinements and equilibrium selection.

EVOLUTIONARY GAME THEORY

In the context of evolutionary biology, the two central concepts of game theory, namely, strategy and payoff, have to be reinterpreted. A strategy is not a deliberate plan of action but an inheritable trait, for instance, a behavioral program. Payoff is not given by a utility scale indicating subjective preferences but by Darwinian fitness, i.e., average reproductive success. The "players" are members of a population, competing for a larger share of descendants. If several variants of a trait occur in a population, then natural selection favors the variants conferring higher fitness. But if the success of the trait is frequency dependent, then an increase of the frequency of variant may lead to a composition of the population for which other variants do better. Similar situations are studied in population ecology. Thus, if prey is abundant, predators increase for a while. But this increase reduces the abundance of prey and therefore leads to a decrease of the predators. Evolutionary game theory can be viewed as the ecology of behavioral programs.

A classical example, which led Maynard Smith to develop evolutionary game theory, is provided by interspecific contests. Assume that there are two behaviorally distinct types: "Hawks" escalate the fight until the injury of one contestant settles the issue, whereas "Doves" stick to some form of conventional display (a pushing match, for instance, where injuries are practically excluded) and give up as soon as the adversary escalates (Fig. 1). If most contestants are Doves, Hawks will be able to settle every conflict in their favor, with a corresponding gain in fitness. Hence, Hawks will spread. If most contestants are Hawks, however, then escalating a conflict will lead with probability one-half to injury. If the object of the fight

		Type of the opponent	
	Hawk-Dove	Hawk	Dove
Type of the player	Hawk	$(G-C)/2$	G
	Dove	0	$G/2$

FIGURE 1 Payoffs for the Hawk-Dove game: If a hawk encounters another hawk, there is an equal chance to win the contest or to get injured, resulting in an expected payoff of $(G - C)/2$. Against doves, a hawk always comes off as the winner, leading to a safe payoff of G. The payoffs for doves are derived analogously.

is not worth the injury, then the Dove trait will spread. No trait is unconditionally better than the other. Hawks can spread only if their frequency is below G/C, where G is the value of the contested object and C is the cost of an injury (both measured in terms of fitness). If their frequency is higher, it will diminish. Oversimplified as it is, this thought experiment shows that heavily armed species, for which the risk of injury is large, are particularly prone to conventional displays, i.e., ritual fighting. This fact had been observed empirically, but before the advent of evolutionary game theory, it was erroneously interpreted as benefiting the "good for the species."

A large number of behavioral traits, but also of morphological or physiological characters, such as the length of antlers or the height of trees, are subject to frequency-dependent selection. Trees invest considerable resources into growth, for instance, because neighboring trees do. To fall behind, in such an "arms race," means to give up a place in the sun. Traits subject to frequency-dependent selection occur in many types of conflicts between two individuals, for instance, concerning territorial disputes (between neighbors), division of parental investment (between male and female), or length of weaning period (between parents and offspring). Moreover, frequency-dependent selection also occurs without antagonistic encounters, as when individuals are "playing the field." The sex ratio is a well-studied example. In the simplest scenarios, the rule is simple: if the sex ratio is biased towards males, it pays to produce daughters, and vice versa. Under specific conditions, however, occurring with inbreeding or local competition for males, the sex ratio may evolve away from 1:1. Other examples of frequency-dependent selection concern the dispersal rate among offspring, the readiness to emit an alarm-call, or the amount of time spent on the lookout for predators.

The evolution of cooperation is one of the best-studied chapters of evolutionary game theory. Traditionally, this is modeled by the helping game described above. If an individual is equally likely to be potential recipient or donor in a given encounter, then a population of cooperators would earn, on average, $b - c > 0$ per interaction and be better off than a population of defectors earning 0. But an individual would always increase its fitness by refusing to help, and hence we should not see cooperation.

Game theorists have encapsulated this social dilemma in the Prisoner's Dilemma (PD) game. In this game, each player can choose between the two strategies C (to cooperate) and D (to defect). Two C players will get a reward R that is higher than the punishment P obtained by two D players. But a D player exploiting a C player obtains a payoff T

		Type of the opponent	
	Prisoner's Dilemma	**Cooperator**	**Defector**
Type of the player	**Cooperator**	R	S
	Defector	T	P

FIGURE 2 Payoffs for the Prisoner's Dilemma (with $T > R > P > S$): Irrespective of the opponent's strategy, it is always better to defect, since $T > R$ and $P > S$. If both players follow this logic, they end up with payoff P instead of R.

(temptation to defect) that is higher than R, and this leaves the C player with the sucker's payoff S, which is lower than P. A rational player will always play D, which is the better move no matter what the coplayer is doing. Two rational players will each end up with payoff P instead of R (Fig. 2).

Many species engage in interactions that seem to be of the Prisoner's Dilemma type. Vampire bats feed each other, monkeys engage in allogrooming, vervet monkeys utter alarm calls, birds join in anti-predator behavior, which includes vigilance and mobbing, guppies and stickleback cooperate in predator inspection, hermaphroditic sea bass alternate as egg-spenders, many species of birds engage in nest helping, and lions participate in cooperative hunting or joint territorial defense. It is difficult, however, to measure the lifetime fitness of free-living animals, and in many cases it remains doubtful whether a given type of encounter is really of the Prisoner's Dilemma type, i.e., satisfies the inequalities $T > R > P > S$. Some of the aforementioned examples could be instances of by-product mutualism, in which both players are best served by cooperating and none is tempted to defect. Other types of encounters (for instance, the Hawk–Dove game) may have the structure of a so-called Chicken game (with $T > R > S > P$), in which the best reply to the coplayer's C is a D, but the best reply to a D is a C. In both cases, cooperation (at least by one partner) is no paradox.

There are several ways in which the Prisoner's Dilemma can be overcome. In general, any form of associative interaction favors cooperation. Such association may be due to kinship, to partner choice, to the ostracism of defectors, or simply to spatial structure and limited dispersal. Indeed, if players can only interact with their nearest neighbors, then clusters of cooperators can grow. This spatial aspect of game theory is likely to operate for many sessile organisms.

Moreover, if interactions of the Prisoner's Dilemma type are repeated between the same two individuals, players can have the option to break up partnerships, or vary the amount of cooperation, depending on past experience. But even without these options, the strategy of

Type of the player	IPD	Type of the opponent: TFT	Type of the opponent: All D
	TFT	R	$(1-w)S+wP$
	All D	$(1-w)T+w$	P

FIGURE 3 Payoffs for the Iterated Prisoner's Dilemma (IPD): When a TFT player meets a coplayer of the same type, both will cooperate mutually, leading to an average payoff of R. Against a coplayer who defects always (All D), a TFT player stops cooperating after the first round and plays D subsequently. If the number of rounds is random and the probability of a further round is w, this results in the payoffs displayed in the matrix.

always defecting is not invariably the best option in the Iterated Prisoner's Dilemma (IPD game). If the probability of a further round is sufficiently high, then even a small amount of conditional cooperators suffices to favor cooperation. The best-known example of such a discriminating strategy is Tit For Tat (TFT). A TFT player cooperates in the first round and from then on always repeats whatever the coplayer did in the previous round (Fig. 3).

The best examples for reciprocation may be found in human societies. Among humans, moreover, reciprocation is often indirect. An act of assistance is returned, not by the recipient, but by a third party. A prerequisite is that players know enough about each other. This condition is likely to hold if groups are close knit and individuals can exchange information about each other.

GAME DYNAMICS

The major new tool of evolutionary game theory consists in using population dynamics. This "technology transfer" from population ecology relies on the assumption that successful traits spread. If there are only two possible types A and B, for instance, then essentially only three scenarios are possible, depending on whether a minority of one type can invade a resident population consisting of the other type only (Fig. 4):

1. A can invade B but B cannot invade A. In this case, the dominant strategy A will always outcompete B. This happens with the Prisoner's Dilemma, if A players defect and B players cooperate.
2. A can invade B and B can invade A. This leads to the coexistence of both types in stable proportions as, for instance, if A are Hawks and B are Doves.
3. No type can invade the other. This is a bi-stable situation; whoever exceeds a certain threshold will outcompete the other. This happens with the Iterated Prisoner's Dilemma if A is TFT and B always defects.

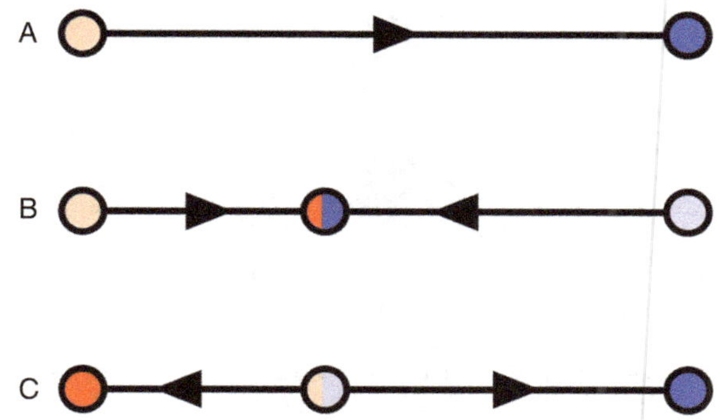

FIGURE 4 Different scenarios for the evolutionary dynamics between two strategies. (A) Dominance: The blue strategy always outcompetes red. Evolution leads to the state in which every individual adopts blue. (B) Coexistence: Red invades blue and blue invades red. Eventually, there is a stable coexistence of both strategies. (C) Bistability: Both red and blue are stable. The eventual outcome depends on the initial population.

With three types A, B, and C, the game dynamics become considerably more complex, in part because "rock–paper–scissors" cycles can occur: A is dominated by B, B by C, and C in turn by A. Several such situations have been documented. In cultures of *E. coli*, for instance, the wild type A can be superseded by a mutant strain B killing the competitors by producing colicin, which acts as a poison. Simultaneously, this mutation produces a protein conferring immunity against the poison to its bearer. A population of type B can be superseded by a further mutant type C that produces the immunity protein but not the colicin (since this poison is inefficient in a population consisting of types B and C). In turn, type C can be invaded and eliminated by type A. Another rock–paper–scissors cycle has been found among males of the lizard *Uta stansburiana*. The three types correspond to inheritable male mating strategies. Type A forms no lasting bonds but looks for sneaky matings, type B lives monogamously and closely guards the female, and C guards a harem of several females, of course less closely.

Depending on the parameters, evolutionary models of rock-paper-scissors games lead either to the stable coexistence of all three strategies or to oscillations with increasing amplitude that lead to the recurrent elimination of the three types (Fig. 5). The competition of male lizards displays the former type of dynamics, and that of *E. coli* bacteria displays the latter.

With four or more types competing, game dynamics can become yet more complex. The frequencies of the different types can keep oscillating in a regular or chaotic fashion. In addition to the dynamics describing

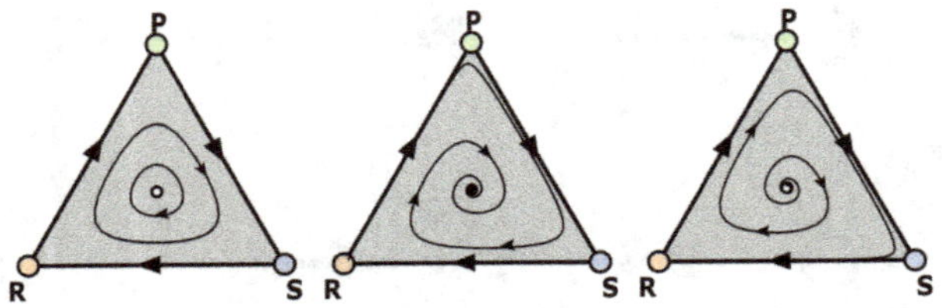

FIGURE 5 Dynamics of the rock-paper-scissors game. Paper beats rock, scissors beats paper, and rock beats scissors. Depending on the exact payoff values, this may result in either closed cycles (left), a stable coexistence of all strategies (middle), or never-ending oscillations (right).

frequency-dependent selection among a given set of types, mutations can produce new types occasionally. This usually proceeds at another time scale. Evolutionary game theory allows studying both short-term and long-term evolution. For the latter, it is often convenient to assume that the transient effects following a random mutation have settled down before the next mutation occurs. As long as the population consists of one type only, this leads to a trait substitution sequence: the fate of a mutant, i.e., its fixation or elimination, is settled before the next mutation occurs. The path of the corresponding "adaptive dynamics" can lead to evolutionarily stable states immune against further invasion or to "branching points" where the population splits up and becomes polymorphic.

Game dynamics can also be used to analyze the interactions between different subpopulations (such as males and females, or territorial owners and intruders). A fast-growing branch of evolutionary game theory deals with structured populations: here, the assumption of random encounters is replaced by that of interaction networks.

Evolutionary game theory deals with phenotypes, and usually assumes that "like begets like." With sexual replication, however, this assumption can fail. Mendelian segregation, pleiotropy, and sexual recombination can lead to situations where more successful types produce less successful variants. In principle, such features can be integrated into models of frequency-dependent selection acting within the gene pool, but this can lead to intractable dynamics. Moreover, arguments from evolutionary game theory can fail, just like optimization arguments from adaptationism, due to genetic constraints. In the absence of specific information on the genotype–phenotype map, however, evolutionary game theory often provides an efficient heuristic tool for understanding frequency-dependent adaptation at the phenotypic level. Moreover, it also proved a suitable tool to describe social learning and cultural evolution.

SEE ALSO THE FOLLOWING ARTICLES

Adaptive Dynamics / Behavioral Ecology / Cooperation, Evolution of / Evolutionarily Stable Strategies / Sex, Evolution of

FURTHER READING

Binmore, K. 2009. *Game theory: a very short introduction.* Oxford: Oxford University Press.

Cressman, R. 2003. *Evolutionary dynamics and extensive form games.* Cambridge, MA: MIT Press.

Dugatkin, L.A. 1997. *Cooperation among animals: an evolutionary perspective.* Oxford: Oxford University Press.

Fudenberg, D., and K. Levine. 1998. *The theory of learning in games.* Cambridge MA: MIT Press.

Hofbauer, J., and K. Sigmund. 1998. *Evolutionary games and population dynamics.* Cambridge, UK: Cambridge University Press.

Maynard Smith, J. 1982. *Evolution and the theory of games.* Cambridge, UK: Cambridge University Press.

Nowak, M. 2006. *Evolutionary dynamics.* Cambridge, MA: Harvard University Press.

Sigmund, K. 2010. *The calculus of selfishness.* Princeton, NJ: Princeton University Press.

Weibull, J.W. 1995. *Evolutionary game theory.* Cambridge, MA: MIT Press.

GAP ANALYSIS AND PRESENCE/ABSENCE MODELS

JOCELYN L. AYCRIGG AND J. MICHAEL SCOTT
University of Idaho, Moscow

Gap analysis is a spatial comparison of biodiversity elements (e.g., species and habitats) within the current network of conservation lands managed primarily for biodiversity protection. The analysis is used to indicate gaps in the conservation network with regard to biodiversity conservation. To accomplish this comparison, spatially detailed maps representing the distribution of vertebrate species are needed. These maps are the result of presence/absence models based on the theory that a species has a high probability of occurring in preferred habitat types within its predicted range. This modeling approach provides an efficient means to conduct biodiversity assessments over large areas and has wide applicability to conservation biology.

BACKGROUND

The concept of gap analysis grew out of the concern for the increasing loss of biodiversity. It was presented as a straightforward and relatively rapid method to assess the distribution and conservation status of multiple components of biodiversity. The ultimate aim was to identify gaps in the current conservation network of lands that could be filled by creating new reserves or simply changing land management practices. This aim later became the goal of the U.S. Geological Survey's Gap Analysis Program, which has assumed the responsibility for producing the data needed for conducting a gap analysis for the entire United States.

To identify the gaps in biodiversity protection, large amounts of spatial data needed to be compiled, created, and analyzed, which advanced theoretical, technological, and spatial data forefronts in conservation biology. One of the forefronts advanced was predicting species distributions across large areas with detail and accuracy. Based on the theory that a species' habitat can predict its presence or absence, empirical presence/absence models for vertebrate species were developed from field surveys and environmental data using a variety of correlative statistical techniques. Often, the resulting models were the first attempt to predict where a species occurred within a state or region.

MODELING APPROACH

Since the inception of gap analysis, presence/absence modeling has expanded tremendously to include a wide range of sophisticated applications of spatial statistics, from ecological niche factor analysis to resource selection functions to occupancy estimation and modeling. However, the basic approach remains to (1) compile species observation records across a species range (i.e., presence/absence data), (2) compile spatial data of pertinent environmental variables, such as habitat types or elevation, (3) relate environmental variables to species observation records and predict the distribution of a species across the landscape (Fig. 1).

Presence/absence models can be put into two general categories of deductive and inductive modeling. Deductive modeling relies on delimiting a species range,

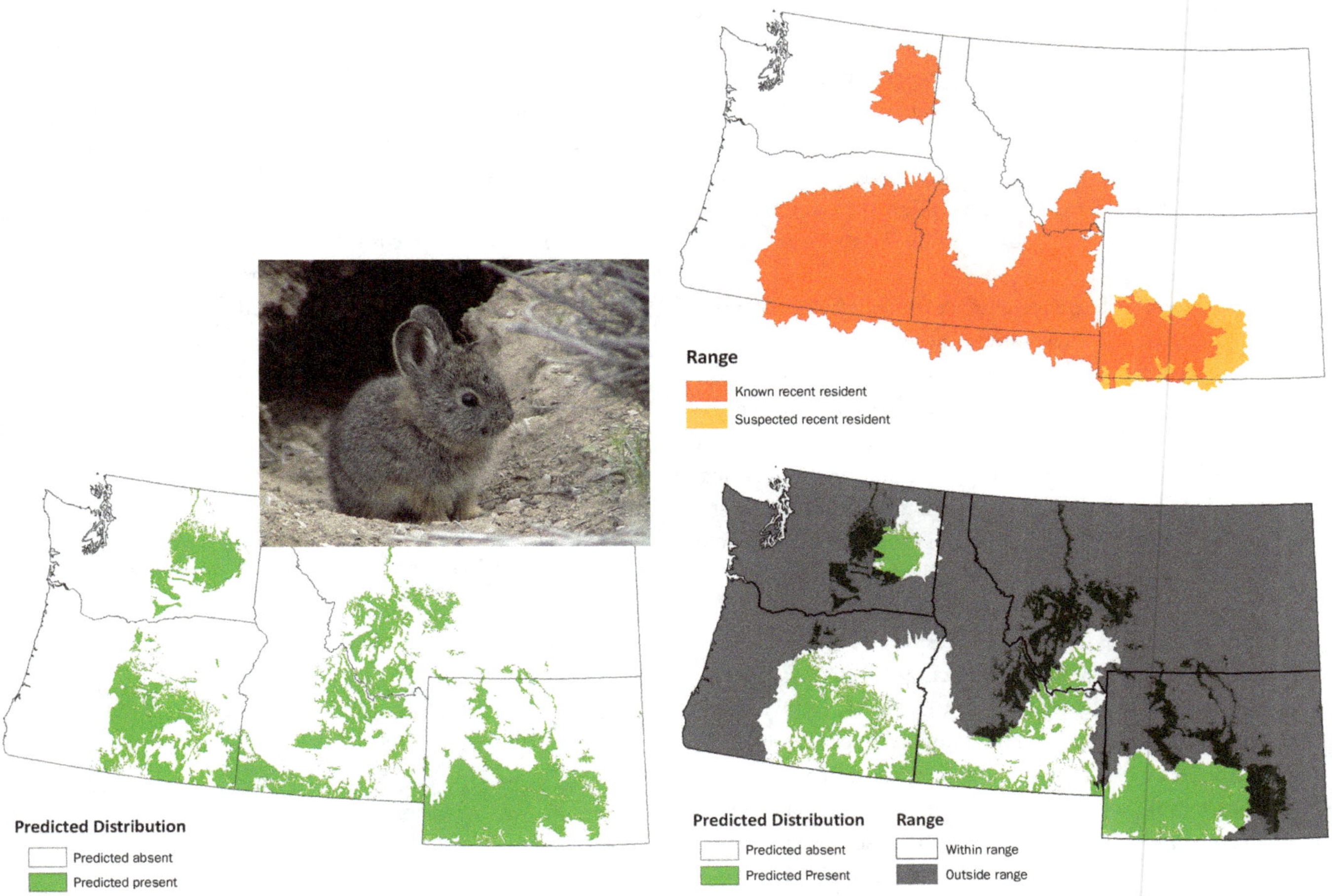

FIGURE 1 Range (upper right) and predicted distribution maps (lower left) for pygmy rabbit (*Brachylagus idahoensis*; upper left) in northwestern United States based on presence/absence modeling. Predicted distribution model is limited to area of range map (lower right)..

establishing species habitat associations based on an exhaustive literature search, and getting input from species' experts to evaluate model predictions. This approach is often criticized as being too subjective and not replicable, but when species observation records are scarce and habitat associations are well established, this approach can be valuable and heuristic. However, overprediction occurs if habitat associations or mapped habitat types are too general. Species that occur in rare or small patch habitats (e.g., wetlands or riparian areas) also are typically overpredicted with this modeling approach, but including higher resolution data can improve the prediction.

Inductive modeling is an alternative approach based on species observation records and spatial environmental variables, such as temperature and elevation. Basically, a species' distribution is predicted based on environmental parameters at known points of occurrence. This is a statistical approach that is more objective and data driven than deductive modeling. There are many statistical tools available ranging from generalized linear models to additive models to classification trees to random forests. Many of these tools are incorporated into software packages such as MaxEnt, BIOCLIM, DOMAIN, and BIOMOD. Inductive modeling works well when both presence and absence data are available and when the species observations are not only abundant but also evenly distributed across a species' range. Underprediction occurs when limited species observation data are available or when the observation data are biased towards specific areas of the species' range.

Limitations to presence/absence modeling include modeling at scales (both resolution and extent) inappropriate for the input data as well as the life history characteristics of a particular species; using functionally irrelevant environmental variables; modeling only the mean response of a species to its environment; ignoring biotic interactions; and lack of testing on model validation and uncertainty. These limitations, however, are currently being researched, and better approaches are continually being proposed.

APPLICATIONS TO CONSERVATION BIOLOGY

Knowledge about factors influencing presence or absence of species, especially at large spatial scales, is invaluable in conservation biology. This knowledge can be used to identify sites with common or threatened and endangered species as well as those of more general conservation and management concern. Areas where a species is absent within the suitable portion of its predicted distribution could potentially be reintroduction sites. Areas with high species richness of species of conservation concern could become priority areas for conservation action. Species conservation can be improved by managing environmental variables, such as habitat, to benefit species occurrence. Not only can gaps in a species distribution be identified using gap analysis, but the causes of those gaps can be determined. On the practical side, species presence/absence models can identify gaps in the species observation data, species-specific life history information, environmental data, and model variables.

FUTURE DIRECTIONS

Many directions can be taken in the future with presence/absence modeling in conservation biology. One vitally important direction is characterizing, reducing, and/or assessing model uncertainty and its influence on conservation decisions. Presence/absence modeling could benefit from expanded techniques in model selection and evaluation with statistics and machine learning. Presence-only data are being used more often, but learning how to deal with biases and evaluating results from presence-only data will be important. The links between presence/absence models and theory need to be strengthened through more development, implementation, and evaluation of models. Models will become more applicable to conservation when theory is used to develop more ecologically relevant predictors. Models could also benefit from including occupancy estimation, biotic interactions, additional ecological processes, and further testing of model results across different temporal and spatial scales. All of these areas for future work as well as development of improved methods of predicting abundance and viability in different parts of their range will advance the theory behind presence/absence models, gap analysis, and conservation biology and the usefulness of model predictions to decision makers.

SEE ALSO THE FOLLOWING ARTICLES

Conservation Biology / Geographic Information Systems / Landscape Ecology / Reserve Selection and Conservation Prioritization / Species Ranges

FURTHER READING

Edwards, T.C., Jr., E.T. Deshler, D. Foster, and G.G. Moisen. 1996. Adequacy of wildlife habitat relation models for estimating spatial distributions of terrestrial vertebrates. *Conservation Biology* 10: 263–270.

Elith, J., and J.R. Leathwick. 2009. Species distribution models: ecological explanations and prediction across space and time. *Annual Review of Ecology, Evolution, and Systematics* 40: 677–697.

Fielding, A.H., and J.F. Bell. 1997. A review of methods for the assessment of prediction errors in conservation presence/absence models. *Environmental Conservation* 24: 38–49.

Franklin, J. 2009. *Mapping species distributions: spatial inference and prediction.* Cambridge, UK: Cambridge University Press.

Graham, C.H., S. Ferrier, F. Huettmann, C. Moritz, and A.T. Peterson. 2004. New developments in museum-based informatics and applications in biodiversity analysis. *Trends in Ecology & Evolution* 19: 497–503.
Guisan, A., and W. Thuiller. 2005. Predicting species distribution: offering more than simple habitat models. *Ecology Letters* 8: 993–1009.
Guisan, A., and N.E. Zimmerman. 2000. Predictive habitat distribution models in ecology. *Ecological Modelling* 135: 147–186.
MacKenzie, D.I., J.D. Nichols, J.A. Royle, K.H. Pollock, L.L. Bailey, and J. E. Hines. 2006. *Occupancy estimation and modeling: inferring patterns and dynamics of species occurrence.* New York: Academic Press.
Manel, S., H.C. Williams, and S.J. Ormerod. 2001. Evaluating presence–absence models in ecology: the need to account for prevalence. *Journal of Applied Ecology* 38: 921–931.
Scott, J.M., F. Davis, B. Csuti, R. Noss, B. Butterfield, C. Groves, H. Anderson, S. Caicco, F. D'Erchia, T. C. Edwards, Jr., J. Ulliman, and R. G. Wright. 1993. *Gap analysis: a geographic approach to protection of biological diversity.* Wildlife Monographs No. 123. *Journal of Wildlife Management* 57(1) supplement.
Scott, J.M., P.J. Heglund, M.L. Morrison, J.B. Haufler, M. G. Raphael, W. A. Wall, and F. B. Samson, eds. 2002. *Predicting species occurrences: issues of accuracy and scale.* Washington DC: Island Press.

GAS AND ENERGY FLUXES ACROSS LANDSCAPES

DENNIS BALDOCCHI
University of California, Berkeley

There is a continuous and invisible stream of gases being transferred into the atmosphere from plants and soils of ecosystems, and vice versa. The suite of gases that have biological origins, or fates, includes water vapor, carbon dioxide, methane, isoprene, monoterpenes, ammonia, nitrous oxide, nitric oxide, hydrogen sulfide, and carbonyl sulfide. Together, these gases exist in trace amounts, as they represent less than 1 percent of the gases in an atmosphere that is dominated by nitrogen (N_2, 78%) and oxygen (O_2, 21%). Yet despite their meager quantity, these trace gases give the atmosphere its signature of life on Earth—these gases serve as either inputs or outputs of the major biogeochemical cycles, they help regulate the atmosphere's climate and chemistry, and they are coupled to the flows of energy into and out of ecosystems.

CONCEPTS

The exchange of trace gases between all living organisms and the atmosphere is a consequence of their necessity to extract energy from their environment. This action occurs because life must comply with two fundamental laws of physics and biology. First, all living things must work to sustain their metabolism, to grow, move, acquire resources, and reproduce. And, second, to perform this work, living organisms need energy. From a thermodynamic perspective the alternate state—death of an organism—is predicated by an inability to acquire and utilize energy.

The sun is the primary source of energy for most of life on Earth; exceptions include chemoautotrophic microbes. Most importantly, sunlight provides the energy that drives photosynthesis. Sunlight also provides biophysical services that enable photosynthesis. Sunlight warms the land and its flora, its heat keeps water in a liquid state over most of the planet, and it evaporates the water that is transpired through the stomata of leaves as they open to facilitate carbon assimilation (Fig. 1).

The amount of sunlight absorbed by a landscape sets the upper limit for the amount of life that can be sustained on an area of land and the trace gas exchange associated with that life. The quantity of light absorbed by a landscape equals the flux density of incident sunlight multiplied by 1 minus surface albedo, or reflectivity. The flux density of solar radiation hitting the Earth at the top of the atmosphere equals 1365 J m^{-2} s^{-1} and sums to 43 GJ m^{-2} y^{-1} when it is integrated over a year. Less sunlight is received at the Earth's surface as a column of sunlight is stretched over a wider area as the angle of the sun, relative to the zenith, declines with hour of the day, day of the year, and latitude. In practice, 1–8 GJ m^{-2} y^{-1} reaches the ground as the sunlight passes through a veiled atmosphere with reflective clouds and aerosols and the sunlight is distributed over the surface of a sphere that revolves on its tilted axis.

The fraction of sunlight that is absorbed by the landscape depends on the optical properties of leaves and the soil and the area of leaves per unit ground area. Individual leaves are relatively dark in the visible wavebands (0.4 to 0.7 microns) and highly reflective in the near infrared wavebands (0.7 to 3 microns). Plants, grouped together, form a canopy consisting of multiple layers of leaves which possess a range of inclination and azimuth angles. This ensemble of leaves is more effective in trapping sunlight than individual leaves and may absorb between 70% and 95% of incident sunlight during the growing season. The amount of light absorbed depends upon how densely plants can group into communities. Tall plants, forming closed canopies, absorb the most light. Conversely, landscapes in dry, sunny regions form open and short canopies and absorb the least light.

The number of individual plants per unit area is constrained by a number of physical laws associated with

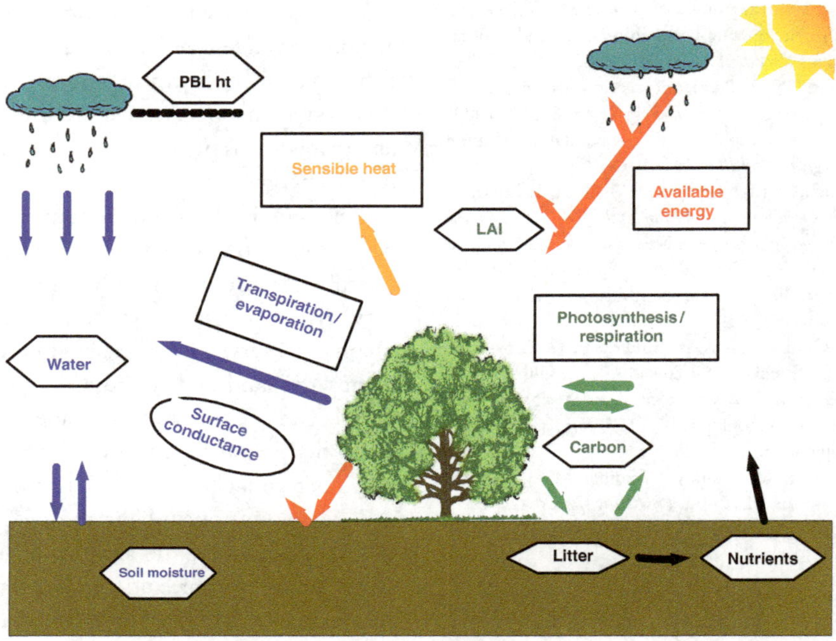

FIGURE 1 Schematic of the processes and fluxes associated with mass and energy exchange between a landscape and the atmosphere.

mass and energy exchange. In practice, the density of plants (plants per unit area) is proportional to the mass of the plants, to the –3/4 power. And the metabolic rates that sustain these individuals scales with mass to the 3/4th power. Because there is only a certain amount of energy available to a unit area of land, it will sustain either a few large individuals or many small individuals, but not many large individuals. Consequently, the metabolism of the landscape on an area basis is invariant with the size of the individual organisms.

In practice, leaf area index is the critical variable for scaling mass and energy exchange between vegetation and the atmosphere, not plant size or density. The theoretical upper limit of leaf area index is associated with the ability of the landscape to establish a vegetated canopy that intercepts over 95% of incident sunlight. The highest values of leaf area index (6 to 10 $m^2\ m^{-2}$) occur in regions with ample rainfall that exceed potential evaporation. In dry, sunny regions, potential evaporation far exceeds rainfall, so landscapes tend to form open canopies and establish canopies with low leaf area indices, below 3 $m^2\ m^{-2}$.

The net radiation (R_{net}) balance supplies the energy that drives sensible heat (H) and latent heat exchange (λE), conductive heat transfer in and out of the soil and vegetation (G) and photosynthesis (P):

$$R_{net} = H + \lambda E + G + P$$
$$= R_{in} - R_{out} + L_{in} - L_{out}. \qquad (1)$$

The radiative energy balance of a landscape is defined as the algebraic sum of incoming solar shortwave radiation (R_{in}) minus outgoing reflected radiation (R_{out}) plus incoming terrestrial longwave radiation (L_{in}) minus terrestrial longwave energy that is radiated by the surface (L_{out}) in proportion to its absolute temperature to the fourth power.

From an ecological perspective, plants "eat" sunlight and by doing so drive the energy cascade that feeds the world's food webs. Solar photons absorbed by chloroplasts in the photosynthetic organs of autotrophs—self-feeders like plants, algae, and cyanobacteria—initiate a series of photochemical reactions in chloroplasts that split water, release oxygen (O_2), and produce electrons. These electrons generate energy-containing biochemical compounds, which are ubiquitous to life—adenosine triphosphate (ATP) and nicotinamide adenine dinucleotide phosphate (NADPH). In turn, ATP and NADPH drive the carbon-reduction cycle that fixes carbon dioxide (CO_2) into the reduced carbon compounds that store chemical energy, including the carbohydrates $(CH_2O)_n$ glucose, sucrose, and fructose.

Theoretically, 8 to 12 photons in the visible band (0.4 to 0.7 mm) of the electromagnetic spectrum are needed to fix one CO_2 molecule, thereby consuming 496 kJ of sunlight per mole of CO_2 fixed. In general, landscapes that form closed canopies with green, well-watered vegetation use less than 2% of incoming sunlight to fix carbon on a given day during the growing season; this value sets an upper limit on how much energy can be extracted from biomass to drive the world's economy. In locales where the growing season is less than 365 days, a lower fraction of incident sunlight is used to produce chemical energy in plants over the course of the year. On a landscape basis, ecosystems assimilate between 100 and 4000 gC m^{-2} y^{-1} via gross photosynthesis.

The oxidation of carbohydrates, fixed by plants, serves as the energy source for the structural growth, maintenance metabolism, and reproduction of plants. Chemical energy stored in the structural material of plants is consumed eventually by plants, aerobic and anaerobic microbes (archaea, bacteria, fungi), herbivores, and carnivores. This acquisition of energy occurs through a set of biologically relevant reduction-oxidation (redox) reactions, leading to another fundamental biophysical law to which living organisms must oblige—the functioning of life requires the uptake and release of a suite of gases (Fig. 2). In general, organisms inhale one set of gases that serve as electron acceptors, oxidants that become reduced (e.g., O_2, H_2, CO_2), and they exhale another set of gases that serve as electron sources, reductants that become oxidized (e.g., CO_2, N_2O, N_2, NO, CH_4, H_2S, N_2). Decomposition, and the eventual mineralization, of dead organic material deposited on the soil by aerobic bacteria, fungi, and soil invertebrates involve the oxidation of complex carbon compounds. Decomposition produces CO_2, organic nitrogen, and ammonium (NH_4^+). Nitrification of ammonium (NH_4+) to nitrite (NO_2^-) or nitrate (NO_3^-) by *Nitrosomanas* and *Nitrobacter* bacteria produces NO or N_2O at the expense of oxygen (O_2) as the terminal electron acceptor. Energy is also extracted from organic matter by anaerobic microbes through such processes as denitrification, fermentation, sulfate reduction, methane fermentation, and ammonium oxidation. A diverse assortment of bacteria and archaea in oxygen-deprived anaerobic microsites, deep in soils or in wetland sediments, use a hierarchy of terminal electron acceptors (NO_3^-, NO_2^-, MnO_2, FeOOH, $SO_4^=$), instead of oxygen, to utilize the energy bound in decaying plant matter and produce gases such as carbon dioxide (CO_2), methane (CH_4), nitrous oxide (N_2O), dinitrogen (N_2), nitric oxide (NO), and hydrogen sulfide (H_2S). Gas exchange also occurs between the vegetation and atmosphere via the evaporative loss of volatile biogenic gases, including ammonia (NH_3), isoprene (C_5H_8), monoterpenes ($C_{10}H_{16}$), and water vapor (H_2O).

The transfer of trace gases between vegetation and the atmosphere is quantified in terms of a flux density, moles per unit area per unit time. Typical magnitudes of the flux densities for a number of ecologically important trace gases rank as follows (Fig. 2): H_2O (mmol m^{-2} s^{-1}); CO_2 (mmol m^{-2} s^{-1}); NO_2, CH_4, C_5H_8 (nmol m^{-2} s^{-1}); H_2S, N_2O (fmol m^{-2} s^{-1}). In the long term, the rates by which gases are assimilated or produced are

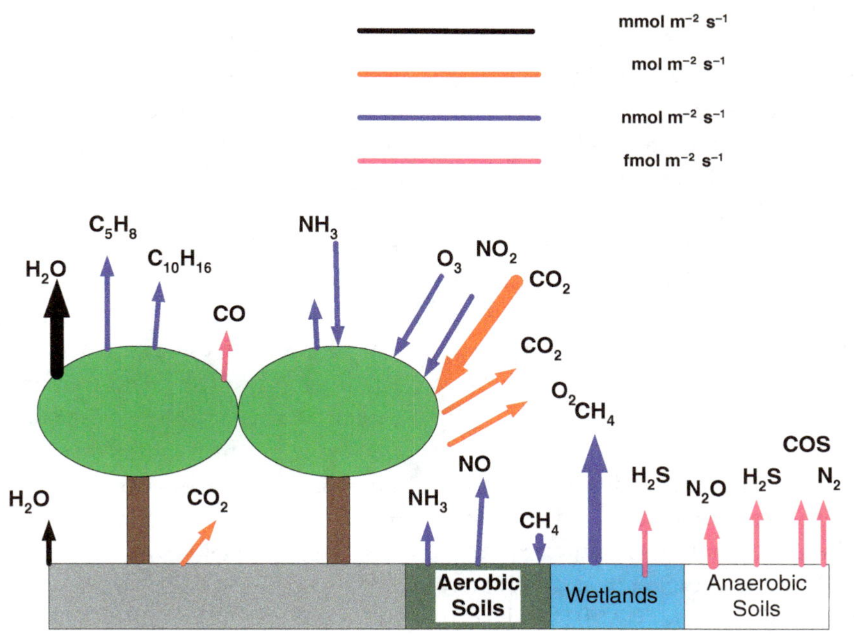

FIGURE 2 Schematic identifying the major trace gas flux, direction, and magnitude between landscapes and the atmosphere. The suite of gases included depends whether the land is vegetated, a wetland, or if the soils are aerobic or anaerobic.

constrained by one another and must follow stoichiometric rules associated with the overall chemical composition of living material, $C_{106}H_{263}O_{110}N_{16}P$. If nutrient limitations occur, the ability of the landscape to sustain rates of mass and energy exchange is impeded and flux densities of trace gas exchange diminish in proportion.

THEORETICAL PRINCIPLES

The law of conservation of mass provides the guidepost for quantifying fluxes of trace gases into and out of the atmosphere and provides a fundamental basis for understanding the range and temporal and spatial variability of these fluxes. In principle, the rate by which the scalar concentration of a trace gas varies with time is a function of the flux divergence acting upon a volume of air. Concentrations of trace gases will remain steady if the rate by which material enters a controlled volume equals that leaving it. Concentrations of trace gases will increase if the flux entering a known volume is greater than that leaving, and vice versa.

Trace gas fluxes occur along a concentration gradient starting at the turbulent atmosphere, through the still and diffusive boundary layers of leaves or the soil, and ultimately through a path that involves a number of biological and physical processes that act to restrict or enhance this transfer. The flux density of a trace gas (F, moles m^{-2} s^{-1}) between a leaf and the atmosphere is typically quantified using an analog to Ohm's law for the flow of electric current, which defines current flow (I) as the ratio between the potential voltage (V) and the resistance against the current (R). With regards to trace gas exchange of a substance like CO_2, the difference in its concentration between the free atmosphere (C_a) and the chloroplast within the mesophyll of the leaf (C_c) is analogous to the potential voltage. The resistances to trace gas exchange along this path include those associated with transfer through the turbulent and diffusive boundary layers of the atmosphere (r_a) and leaves (r_b), respectively, the diffusion through the stomata (r_s) and the mesophyll of the leaf (r_m), and the dissolution in cells.

$$F \approx \frac{C_a - C_c}{r_a + r_b + r_s + r_m}. \qquad (2)$$

The boundary layer resistances are a function of the dimensions of the leaf and the wind speed. The stomatal resistance is a nonlinear function of soil water content, light, temperature, CO_2 concentration, and the photosynthetic pathway (e.g., C_3, C_4, or CAM). Feedbacks between diffusive resistances, enzymatic activity and available energy establish a critical balance between supply and demand. For example, the demand for carbon dioxide by photosynthesis depends upon CO_2 concentration in a saturating manner defined by Michaelis–Menten enzyme kinetics. Subsequently, the demand for a substrate, C, will down-regulate if supply of that substrate cannot keep pace, and vice versa.

FUTURE DIRECTIONS

Experimental information on gas and energy between landscapes and vegetation on ecological time scales has expanded rapidly on several fronts over the past decade. Modern instrumentation, computer hardware, and software enable scientists to measure trace gas fluxes of water vapor, carbon dioxide, and heat between vegetation and the atmosphere routinely, quasi-continuously, and for extended durations (days, years, and decades). For example, energy and trace gas fluxes are being collected from over 500 sites worldwide, spanning most of the world's major biomes and climate spaces, through the FLUXNET project (www.fluxdata.org). Newer advances in laser spectroscopy and proton transfer mass spectrometry are providing a generation of sensors that can measure another assortment of greenhouse gases, including methane, nitrous oxide, and volatile hydrocarbons, with high precision and frequent sampling rates.

There are coincident efforts to use this vast amount of trace gas flux data, in conjunction with satellite-based remote sensing, to evaluate and parameterize the data assimilation models and to use these models to predict trace gas fluxes "everywhere and all of the time."

The next-generation of coupled climate–ecosystem–biogeochemistry models are striving to forecast or hind-cast the exchanges of gas and energy with the atmosphere in a mechanistic manner by considering the birth, growth, competition, and death of plants and how plants and soil respond to environmental conditions. In turn, these models enable us to ask and answer many ecological questions relating to feedback between vegetation and the atmosphere. Examples include the role of land-use change (e.g., deforestation or afforestation) on carbon, water, and energy balances and regional climates; the degree to which species diversity or functional diversity most affects ecosystem metabolism and productivity; and how much energy can be extracted from landscapes to produce biofuels.

SEE ALSO THE FOLLOWING ARTICLES

Biogeochemistry and Nutrient Cycles / Energy Budgets / Landscape Ecology / Metabolic Theory of Ecology

FURTHER READING

Baldocchi, D.D. 1991. Canopy control of trace gas emissions. In T.D. Sharkey, E.A. Holland, and H.A. Mooney, eds. *Trace gas emissions by plants.* San Diego: Academic Press.

Brown, J.H., J.F. Gillooly, A.P. Allen, V.M. Savage, and G.B. West. 2004. Toward a metabolic theory of ecology. *Ecology* 85: 1771–1789.

Burgin, A.J., W.H. Yang, S.K. Hamilton, and W.L. Silver. 2011. Beyond carbon and nitrogen: how the microbial energy economy couples elemental cycles in diverse ecosystems. *Frontiers in Ecology and the Environment* 9(1): 44–52.

Chapin, F.S., P.A. Matson, and H.A. Mooney. 2002. *Principles of terrestrial ecosystem ecology.* Berlin: Springer.

Monteith, J.L., and M.H. Unsworth. 1990. *Principles of environmental physics.* London: Edward Arnold.

GENETIC DRIFT

SEE MUTATION, SELECTION, AND GENETIC DRIFT

GEOGRAPHIC INFORMATION SYSTEMS

MICHAEL F. GOODCHILD

University of California, Santa Barbara

Geographic information systems (GIS) are computer applications designed for the acquisition, storage, analysis, modeling, archiving, and sharing of geographic information—in fact, virtually any conceivable operation on geographic information in digital form. Geographic (or geospatial or georeferenced or geolocated) information, in turn, is defined as information linking locations on or near the Earth's surface to properties present at those locations, and often to specific points or intervals of time at which those properties were present or observed. Geographic information has traditionally been compiled in the form of maps and globes, but the digitization of geographic information beginning in the 1960s opened a host of new, more powerful, and more precise forms of analysis and use.

DEFINITIONS

By the 1980s, GIS had advanced to a readily available set of software tools, some commercial and some (such as GRASS and Idrisi) developed by academic research groups. Landscape ecologists quickly saw the importance and benefit of these tools and began to adopt them as essential software for any research that involves analysis or modeling of the distributions of ecological phenomena over the Earth's surface. A series of conferences in the 1990s (the International Conferences on Integrating GIS and Environmental Modeling) drew attention to the potential of GIS, and especially for the integration of ecological modeling into the multidisciplinary analysis of complex systems.

It is important to understand the significance of the various terms used to describe this area of computer application. While GIS is well defined, its borders inevitably shift, especially in relation to other geospatial technologies, including the Global Positioning System (GPS) and satellite- or aircraft-based remote sensing. At its broadest, the term GIS is used to encompass all computer-based technologies capable of handling geographic information. At its narrowest, however, the term covers only a small subset of this domain and is often associated with the products of the industry-dominant Environmental Systems Research Institute (ESRI), a commercial developer of GIS software. A middle ground will be taken for the purposes of this entry, by excluding GPS and remote sensing but noting their importance as complementary technologies.

A rich set of textbooks on GIS has emerged over the past two decades, including texts that place much of GIS analysis on a strong statistical footing and texts that provide a comprehensive review of analytic techniques based on GIS. The term geographic information science (GIScience) was coined in 1992 to identify the many fundamental issues that arise in the creation and use of GIS, ranging from scale to uncertainty and to the social impacts of GIS technology.

BASIC PRINCIPLES

GIS were originally designed to capture the contents of maps, in part to facilitate editing and production, and in part to support precise measurement of properties of maps such as length and area. The first system clearly identified as a GIS was developed by the Government of Canada, in collaboration with IBM, for the sole purpose of producing statistical summaries of the Canada Land Inventory from maps. It is tempting to explain a GIS as "maps stored in a computer," and the map metaphor remains a powerful model for thinking about GIS, but at the same time it is a powerful constraint on thinking. In the late 1960s, the landscape planner Ian McHarg promoted the idea of maps as co-registered two-dimensional layers of thematic information that could be overlaid and combined in support of complex decision making; for example, the location of a new highway could be planned by overlaying layers representing ecological, hydrological, geological, cost, and social issues. Early GIS simply took these ideas and implemented them in a computing system, with different files corresponding to the different maps or layers of thematic data. This tradition persists today, and it is still easier in GIS to compare the same theme at different locations, because the relevant items of information are stored together in a GIS database, than to compare different themes at the same location.

GIScience recognizes two distinct ways of conceptualizing the geographic world and its representation. In the discrete-object view, the geographic world is likened to an empty tabletop littered with discrete, countable entities, including trees, vehicles, buildings, and animals. Each object is represented as a distinct geometric feature: a point if the object is small and the scale is coarse, a line for objects such as rivers and roads, and an area for objects such as lakes or habitat patches. Each object has an associated set of properties, or attributes, which may be numeric or alphanumeric; in an attribute table, the attributes form the columns and the objects form the rows.

This approach becomes problematic for phenomena that are fundamentally continuous in space, such as rivers or topography, since these must be arbitrarily broken into discrete fragments to be treated in this way. Thus, rivers might have to be broken into reaches, and topography represented by a discrete collection of spot heights.

Topography and many other phenomena are characterized as having defined values at every location on the Earth's surface, and they are better conceptualized using the alternative view of continuous fields. Mathematically, these amount to a mapping from location in two-dimensional space $\mathbf{x}$ to a value z or a class c. A habitat map is better conceptualized this way as a mapping from location to classes of habitat than as a collection of discrete patches (Fig. 1), and, similarly, precipitation is better conceptualized as a continuous field of rainfall. A continuous field must have a single value at every location $\mathbf{x}$, a requirement often termed planar enforcement, whereas discrete objects can overlap, and there are usually empty spaces in a discrete-object representation.

Unfortunately the computer is a discrete machine, and every item of information must ultimately be reduced to a combination of 0s and 1s. Thus, despite the importance of a continuous-field conceptualization, it is still necessary to reduce the representation of a continuous field to a collection of discrete objects. Topography, for example, is represented using one of four approaches: as a collection of heights at points regularly spaced on a rectangular lattice (a digital elevation model); as a collection of spot heights at irregular, user-defined locations; as a collection of digitized isolines or contours; and as a mesh of irregular triangles (a triangulated irregular network). Remotely sensed images capture continuous fields of spectral response but are represented as collections of discrete, rectangular raster

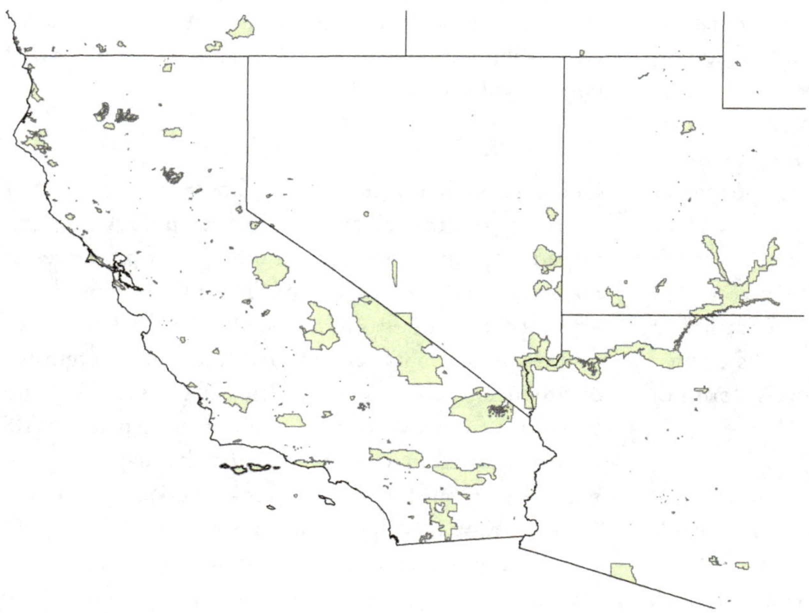

FID	NAME	FCC
8148	Glacier Bay NP and NPRES	D83
8149	Katmai NP and NPRES	D83
8150	Lake Clark NP and NPRES	D83
8151	Kenai Fjords NP	D83
8152	Pu'uhonau O Honnaunau NHP	D83
8153	Hawaii Volcanoes NP	D83
8154	North Cascades NP	D83
8155	Olympic NP	D83
8156	Lake Chelan NRA	D83
8157	Acadia NP	D83
8158	Theodore Roosevelt NP	D83
8159	Badlands NP	D83
8160	Apostle Islands NL	D83
8161	Craters of the Moon NMON	D83
8162	Nez Perce NHP	D83
8163	Voyageurs NP	D83
8164	Lava Beds NMON	D83
8165	Cape Cod NS	D83
8166	Gateway NRA	D83
8167	Sagamore Hill NHS	D83
8168	Indiana Dunes NL	D83
8169	Death Valley NP	D83
8170	Kings Canyon NP	D83

FIGURE 1 Parks of California and contiguous areas, conceptualized as discrete areas in an otherwise empty space, with an extract from the associated attribute table. (Figure created by author from public-domain data.)

cells, or pixels, and many thematic layers are captured as collections of discrete areas. Treating continuous fields in this way means that the constraints imposed by the conceptualization, such as the absence of overlaps, are typically ignored by the system. Thus, a user is free to edit digitized contours so that they cross, the system having no way of knowing that they are actually representative of a continuous, single-valued field that by definition cannot have locations where two or more contours intersect.

The digital representations produced by these options fall into two broad categories: regular arrays, typically rectangular and generally known as rasters; and discrete points, lines, and areas, generally known as vector data. Areas are most often represented as sequences of points (pairs of coordinates) connected by straight lines (polygons), while lines are similarly most often represented as sequences of straight lines termed polylines. Unfortunately the term polygon has come to refer in GIS to any area, whether or not it has straight sides.

One of the strongest arguments for GIS points to its ability to link together information of different types and from different sources based on geographic location. For example, it is possible to link the vegetation cover data shown in Figure 2 with the park data shown in Figure 1, to obtain information about the vegetation cover types found in specific parks. A spatial join is made

FIGURE 2 Natural vegetation classes for California, from the map published by Küchler in 1977. This layer is conceptualized as a continuous field, with exactly one class recorded at every point within the state. (Figure created by author from public-domain data.)

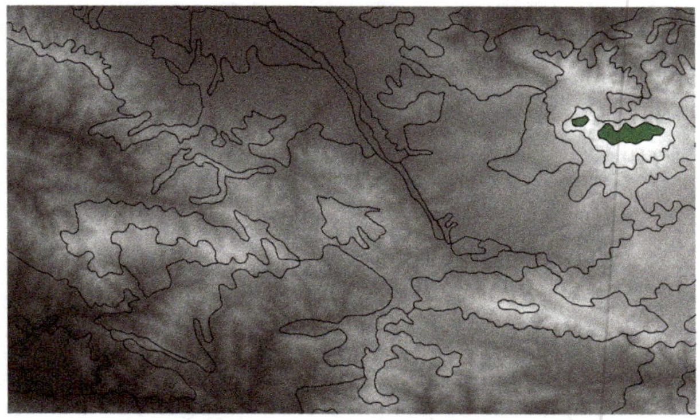

FIGURE 3 Spatial join of the vegetation layer of Figure 2 with elevation data for part of Santa Barbara and neighboring counties. In the attribute table, each polygon now has attributes derived from the elevation data, including minimum, mean, and standard deviation of elevation. The highlighted polygon covers the top of Mt. Pinos and is dominated by white fir.

by overlaying the two data sets and concatenating the respective attributes. Figure 3 shows the results of a spatial join or topological overlay of a digital elevation model with the vegetation cover data of Figure 2. Note that the attribute table now shows both vegetation attributes for each polygon, plus statistics from the elevation data.

It will have become clear from the preceding discussion that most GIS technology takes a planar approach and assumes that the Earth's surface is a flat plane. It is, for example, impossible to lay a raster over the curved surface of the Earth, and the necessary flattening is implicit in most GIS applications. GIS users are thus forced to engage with the complications of map projections and to make assumptions of planarity that become problematic when dealing with large study areas. Moreover the nonspherical nature of the Earth creates additional problems, since many different mathematical approximations to the Earth's shape are in use around the world. Modern GIS technology includes a full set of tools for dealing with these issues, but users should be cognizant of the necessary background in the principles of geodesy and map projections.

Substantial advances have been made in recent years in extending GIS representations to the third spatial dimension and to time. A simple solution to the former problem is to regard elevation as a function of the two horizontal dimensions, an approach often termed 2.5*D*. But truly three-dimensional applications, such as those dealing with the subsurface or with the three-dimensional structure of vegetation, require a more comprehensive approach. The object-oriented paradigm that now dominates GIS makes it straightforward to deal with temporal change, but impediments to full-fledged spatiotemporal

GIS still exist in the lack of a comprehensive set of standard models and analytic techniques and the conceptual difficulty of thinking about information that cannot be visualized in two dimensions. Another research frontier lies in dealing with data on flows, interactions, and other phenomena that are properties of locations taken two rather than one at a time, since these also have traditionally posed problems for cartography.

The topic of uncertainty has generated substantial research interest over the past two decades. A computer is a very precise machine, capable of working to one part in 10^8 in single precision, and one part in 10^{15} in double precision. These precisions far exceed the accuracy that is typical of many GIS applications. For example, using a precision of one part in 10^{15} to record latitude would resolve position to 10^{-8} meters, or a hundredth of a micron, whereas the accuracy of a typical field GPS is in the meter range. The boundaries used to represent habitat patches are infinitely thin, but in reality lines are positioned somewhat arbitrarily within ecotones and are generally not replicable between observers. Most GIS data acquisition is by measurement and subject to inevitable uncertainty. Moreover, many of the terms used to define classes, such as classes of land cover, are fundamentally vague and not replicable. In short, all geographic information is subject to uncertainty and can present at best only a generalized or otherwise abstracted view of the complexity of the real geographic world. Yet the attraction of GIS to many users lies in its precision, the appearance of rigor in its analyses, and the degree to which it provides authority to decisions. Uncertainty has been termed the Achilles heel of GIS.

GIS EVOLUTION

As noted earlier, GIS began in the 1960s with very limited goals. By the late 1970s, however, it was clear that a range of motivations could be satisfied with a single, monolithic software package that recognized several standard types of data, including both raster and vector data. Large markets were developed, initially in forestry management and later in utilities management, mapping, the military and intelligence communities, and academic research. With a foundation of standard database formats, it was possible to add functionality at a great rate, sometimes as additions to the basic software and sometimes through third-party contributions, often from researchers.

The first GIS data formats were entirely single-purpose, but in the late 1970s the relational data model emerged as a preferred standard. ESRI's ARC/INFO, released in the early 1980s, used the INFO relational database management system to handle the tables of a map representation. It did so in a cumbersome way, however, because at that time it was impossible to store the variable number of points required to represent a polyline or polygon in a single cell of a relational table; early versions of ARC/INFO were literally a hybrid of a standard relational database management system and a customized (and still proprietary) format for storage of polyline and polygon coordinate strings. Efforts to move away from this unsatisfactory solution were not successful until the late 1990s and the advent of object-oriented database management, when the increased power of computers, together with advances in database technology, made a unified solution possible.

The popularization of the Internet and the advent of the World Wide Web in the early 1990s had profound effects on the world of GIS. The mainframe solutions of the 1960s and 1970s had given way in the 1980s to minicomputers operating over local area networks, but GIS was still essentially a stand-alone application. But the GIS industry was quick to see the potential of the Internet for sharing data and began to support and advocate the establishment of geolibraries, or libraries of geographic information whose primary search mechanism was based on location. The Open Geospatial Consortium was established in the mid 1990s as a collaborative effort to develop standards for sharing data over the Internet. Today, single points of entry known as geoportals allow the GIS user to search over large numbers of such geolibraries from a single point of entry for data to meet specific needs.

More recently, it has become possible to offer simple GIS functions over the Web using server-based GIS technology. Many agencies and corporations have responded by making their Web assets accessible in map form, and a host of sites now provide simple functionality. Google Maps and Google Earth are popular examples of simple GIS services, offering geolocation, driving directions, and advanced visualization techniques. Moreover, many such services have allowed users to access their programming interfaces, and it is now common to find mashups, or sites that offer services built on two or more such interfaces, through processes akin to the overlays discussed in the previous section. For example, Web users have become used to links that go directly from a hotel Web page to a Google map of the hotel's location. Mashups provide a convenient way for ecologists to produce interactive, Web-based maps of their data.

Another very recent trend that is having enormous impact on the world of GIS is termed volunteered geographic information, a form of crowd sourcing, user-generated Web content, or citizen science. The traditional source of geographic information has been the national mapping agencies that exist in most countries. However, increasing demand for geographic information, the increasing reluctance of governments to fund these activities, and the spectacular drop in the cost and level of expertise required to create maps have led to a phenomenon known a neogeography, a breaking down of the distinction between expert and amateur with respect to mapmaking. Today, a host of projects around the world are relying on volunteers to create and upload geographic information, much of it of ecological significance. Phenological efforts such as Project Budburst are an effective extension of the ideals of citizen science and build on the general trend toward user-generated Web content that is exemplified by blogs and wikis.

GIS IN ECOLOGY

Examples of the ecological significance of GIS have been given at various points in the preceding sections. This section focuses specifically on ecological applications. It would be difficult if not impossible for an ecologist working with phenomena distributed over the Earth's surface to avoid encountering GIS, at least in its broader sense. GPS has made it easy to record the location of any observation, and the base maps needed for field research are commonly available over the Internet. GIS has been adapted to the limited space and other constraints of field technologies, and today it is possible to run a reasonably powerful GIS on a handheld device, either stand-alone or online.

Many GIS applications focus on mapping, through the characterization of the Earth's surface with respect to ecologically important themes. Land cover mapping is often based on remote sensing, because of the latter's continuous coverage, using sensors such as Landsat's Enhanced Thematic Mapper Plus with a spatial resolution of 30 m. Pixels are first classified, and then contiguous pixels of the same class are aggregated to form polygons. The Gap Analysis Project of the U.S. National Biological Information Infrastructure uses such a procedure to classify the U.S. land surface according to habitat suitability for a large number of wildlife and plant species.

GIS has found important applications in the planning of conservation areas for species preservation. Optimization methods have been used to pick the best collection of areas in order to provide sufficient habitat for a sustainable population, and concepts such as corridors to connect islands of habitat have been implemented in GIS-based search procedures.

With the miniaturization of GPS devices, it has become possible to provide space–time tracks of birds and animals and to submit these to GIS-based analysis. New technologies have extended the limits of these approaches by providing devices light enough to be carried by small migratory birds and yet capable of producing useful spatial and temporal accuracies. GIS has been used in conjunction with tracking data to analyze feeding, mating, and migratory behaviors and to parameterize and evaluate agent-based models of individual behavior.

Many GIS functions find useful applications in ecological research. Spatial interpolation methods implemented in GIS allow continuous fields to be interpolated from point samples. Kernel density estimation is a valuable technique for interpolating continuous fields of population density for comparison with edaphic, climatic, and resource factors. Numerous techniques for analyzing patterns of points, patches, and tracks have been developed and implemented on the GIS platform, together with techniques for mapping species ranges from observational data.

SEE ALSO THE FOLLOWING ARTICLES

Computational Ecology / Landscape Ecology / Phylogeography / Spatial Ecology / Spatial Models, Stochastic / Species Ranges

FURTHER READING

Alexander, R., and A. C. Millington, eds. 2000. *Vegetation mapping: from patch to planet.* Chichester, UK: Wiley.

Burrough, P. A., and R. McDonnell. 1998. *Principles of geographical information systems.* New York: Oxford University Press.

De Smith, M. J., M. F. Goodchild, and P. A. Longley. 2009. *Geospatial analysis: a comprehensive guide to principles, techniques and software tools,* 3rd edition. Leicester, UK: Matador.

Haynes-Young, R. H., D. R. Green, and S. Cousins. 1993. *Landscape ecology and geographic information systems.* New York: Taylor and Francis.

Hunsaker, C. T., M. F. Goodchild, M. A. Friedl, and E. J. Case, eds. 2001. Spatial uncertainty in ecology: implications for remote sensing and GIS applications. New York: Springer-Verlag.

Longley, P. A., M. F. Goodchild, D. J. Maguire, and D. W. Rhind. 2005. *Geographic information systems and science,* 2nd ed. Hoboken, NJ: Wiley.

Zhang, J.-X., and M. F. Goodchild. 2002. *Uncertainty in geographical information.* New York: Taylor and Francis.

GROWTH

SEE ALLOMETRY AND GROWTH

H

HARVESTING THEORY

WAYNE MARCUS GETZ

University of California, Berkeley

Mathematical theories of harvesting biological resources can be traced back to Martin Faustmann's 1849 analysis of the optimal interval for the periodic harvesting of cultivated forest stands. Faustmann's work presaged two different pathways that harvesting theory took a hundred years later in the early 1950s. These were R. J. H. Beverton and S. J. Holt's cohort analysis and M. B. Schaefer's maximum sustainable rent analysis. The latter matured largely through the work of C. W. Clark and colleagues in the early to mid-1970s into the field of mathematical bioeconomics. In the late 1970s, cohort theory was linked, as described below, to Leslie matrix modeling theory and extended to include nonlinear stock-recruitment analysis, first begun by W. E. Ricker, and Beverton and Holt in the early 1950s. Today, harvesting theory has been greatly extended beyond these roots to include the uncertainties inherent in the demographic processes of births and deaths, the vagaries of environmental forces impinging on populations, and the incompleteness of biological models, as well as risk analysis and the use of sophisticated social and economic instruments for meeting the competing needs of different stakeholders concerned with the exploitation of any particular biological resources.

HARVESTING PLANTATIONS

Value Function

Plantation or even-aged forest stand management theory can be traced back to Faustmann's 1849 formulation for calculating the optimal rotation period for clear-cutting a stand of trees for timber or pulp, where the stand is then replanted only to be clear-cut again at the end of the next rotation period. The analysis begins with the assumption that a value function exists, which over time represents that net profit $P(t)$ that will be obtained if the stand is clear-cut at time t. It follows immediately after clear-cutting that in the new rotation period $P(0) = 0$, and also that although $P(t)$ may initially be negative for small t (if we clear-cut too soon it may actually cost us more to take this action than the value of the harvest once sold and the costs of replanting the stand may also not be covered) the enterprise will not be worth undertaking unless ultimately $P(t) > 0$.

A point in time may come at which the net profit function has a local maximum value. If we denote this time by t^*, then it follows that $P^* = P(t^*) > P(t)$ for $t \neq t^*$, where we assume that if t^* exists it is unique in satisfying this condition for all $t \in [0, \infty)$. If t^* exists, this implies that the value of the resource starts decaying in quality beyond t^*, e.g., timbers getting knotty with age or wood losing quality as a source of pulp for making paper. If t^* does not exist, this implies $P(t)$ asymptotically approaches P^* as $\rightarrow \infty$. However, rather than considering the value of t that maximizes $P(t)$, we may want to account for an monetary discount rate δ over time by defining the present value function

$$V(t) = P(t)e^{-\delta t}.$$

To find the time t_δ at which $V(t)$ has its maximum we solve for t in the equation

$$V'(t) = 0 \left(V'(t) \equiv \frac{dV}{dt} \right);$$

i.e.,

$$\begin{aligned} V'(t) = 0 &\Rightarrow P'(t_\delta)e^{-\delta t_\delta} + P(t_\delta)(-\delta)e^{-\delta t_\delta} = 0 \\ &\Rightarrow P'(t_\delta) = \delta P(t_\delta). \end{aligned}$$

In order for t_δ to maximize rather than minimize $V(t)$, we require $P(t_\delta) > 0$, from which it follows that since $\delta > 0$ we must have $P'(t_\delta) > 0$. Hence, as illustrated in Figure 1, $0 < t_\delta < t^*$ and $P(t_\delta) < P^*$.

Optimal Rotation Period

Since a harvested stand can be replanted and re-harvested again some time in the future, the stand has some intrinsic value immediately after harvesting. The quantity $V(t_\delta)$ solved for above only finds the present value for a single rotation period: it ignores the fact that the stand can be replanted and the process of growth and clear-cutting repeated. If we now assume that we harvest the stand every T periods of time and that in each re-growth period, the value of the stand is given by the same value function P (i.e., no exhaustion of soil nutrients occurs and the climate repeats itself each rotation period); that is, $P(nT + t) = P(t)$ for all $n = 1, 2, 3, \ldots$, and $t \in [0, T]$ (note we include $t = T$ but not $t = 0$ to avoid assigning two different values to $P(t)$ at points nT, $n = 1, 2, 3, \ldots$; cf. Figure 1); then we can define the present value function for the infinitely repeated clear-cutting operation as

$$V_\infty(T) = \sum_{n=1}^{\infty} P(nT)e^{-\delta nT}.$$

From the periodicity of $P(t)$ in T, and the fact that $\sum_{n=1}^{\infty} x^n = x/(1 - x)$ when $0 \leq x < 1$, it follows that

$$V_\infty(T) = P(T)/(e^{\delta T} - 1). \quad (1)$$

To find the optimal rotation period T_δ that maximizes $V_\infty(T)$, we need to solve for T in the equation $V'_\infty(T) = 0$; i.e.,

$$\begin{aligned} V'_\infty(t) = 0 &\Rightarrow \frac{P'(T_\delta)}{e^{\delta T_\delta} - 1} - \frac{\delta P(T_\delta)}{(e^{\delta T_\delta} - 1)^2} = 0 \\ &\Rightarrow P'(T_\delta) = \frac{\delta e^{\delta T_\delta} P(T_\delta)}{e^{\delta T_\delta} - 1}. \end{aligned} \quad (2)$$

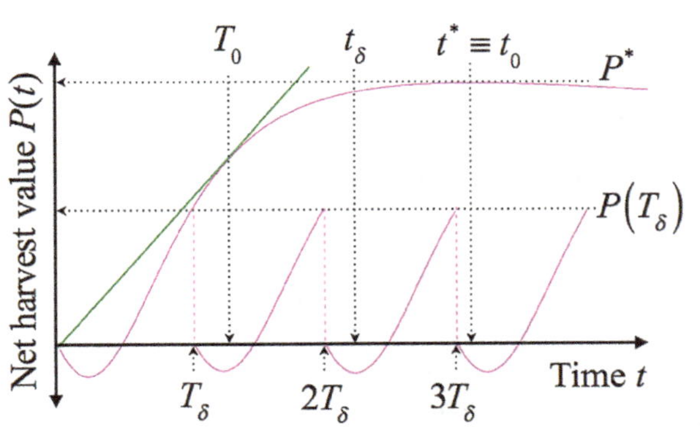

FIGURE 1 A graphical summary of Faustmann Optimal Rotation Period analysis.

Making use of the relationship $\frac{k}{k-1} = 1 + \frac{1}{k-1}$ and of Equation 1, Equation 2 reduces to

$$P'(t_\delta) = \delta P(T_\delta)\left(1 + \frac{1}{e^{\delta T_\delta} - 1}\right) = \delta(P(T_\delta) + V_\infty(T_\delta)). \quad (3)$$

Since $P(T_\delta)$ is the nondiscounted value of the harvest at the optimal rotation period T_δ and $V_\infty(T_\delta)$ is the stand's present value at the start of any rotation period before planting has occurred, Equation 3 represents the principle (Faustmann) that the optimal rotation period occurs when the marginal value of the harvest is equal to the discount rate multiplied by the value of harvest and of the stand itself immediately after it has been clear-cut.

Finally, we consider what happens as the discount rate goes to 0 in Equation 3. In this case, we obtain $P'(T_0) = P(T_0)\left(\lim_{\delta \to 0} \frac{\delta e^{\delta T_\delta}}{e^{\delta T_\delta} - 1}\right) = \frac{P(T_0)}{T_0}$, which implies that when the discount rate is 0, the optimal rotation period maximizes the rate at which revenue is accumulated over time (Fig. 1).

MATHEMATICAL BIOECONOMICS

Harvesting Formulation

Early harvesting theory focused on modeling the dynamics of a population biomass variable $x(t) \geq 0$ by a single ordinary differential equation using a function $F(x)$ that describes that growth rate of biomass in terms of the biomass itself, and a function $h(v,x)$ that describes the rate at which biomass is removed from the population as a function of the population's biomass and of a harvesting effort variable $v(t)$:

$$\frac{dx}{dt} = F(x) - h(v,x) \quad \text{with} \quad x(0) = x_0 > 0. \quad (4)$$

The canonical approach developed by Schaefer assumes that (i) $F(x) = rx(1 - x/K)$ (i.e., the logistic equation) where r is the so-called intrinsic growth rate and K is the so-called environmental carrying capacity, and (ii) $h(v,x) = qvx$ (i.e., the mass-action principle) where q is the so-called catchability coefficient. The theory, as developed later by Colin Clark, uses general functions $F(x)$ and $h(x,v)$. We present a mixture of both the specific and general, but in all cases use a hat to denote equilibrium solution pairs $(\hat{v},\hat{x})$ that satisfy the equilibrium equation $F(\hat{x}) = h(\hat{v},\hat{x})$ associated with Equation 4.

By definition, since $h(x,v)$ in Equation 4 is the rate at which biomass is extracted from the population, the biomass yield Y rate (i.e., yield per unit time) harvested from the population over any interval of time $[0,T]$ is

$$Y = \frac{1}{T}\int_0^T h(v(t), x(t))dt. \quad (5)$$

This definition, of course, only holds provided we define $h(v,0) = 0$ for any value of v, since we cannot continue to extract biomass from a population that has been driven to extinction! Although, from a biological perspective, yield is a focal quantity for the development of a harvesting theory, from a resource economics point of view, the net rent, or rate of return, is a more salient quantity; and the present value V of a resource, a quantity that takes account of both net rent and the discount rate is yet more salient. Per unit of time if p is the value of a unit of harvest and c is the cost of applying a unit of harvesting effort v, then the net revenue rate (or rent) is $ph(v, x) - cv$. Thus, if $\delta > 0$ is the discount rate, then the present value is

$$V_\delta = \int_0^\infty e^{-\delta t}\,(ph(v(t),x(t)) - cv(t))dt. \qquad (6)$$

We note that V_δ is defined for all $\delta > 0$ but not for $\delta = 0$, since the integral is then infinite.

Equilibrium Yields

Whenever the harvesting rate is selected to satisfy $F(x) = h$, then, from Equation 4, $dx/dt = 0$ and the population is at some equilibrium value $\hat{x}_h$ (the subscript h to remind us that this value depends on our choice of h).

If h is a constant (i.e., independent of v or x), then in the case of logistic growth $\hat{x}_h$ is the solution to the equation $h = rx(1 - x/K)$. This quadratic equation in x has real positive solutions

$$\hat{x}_{h\pm} = \frac{K}{2} \pm \frac{\sqrt{r^2K^2 - 4hrK}}{2r} \quad \text{provided} \quad 4h \le rK. \qquad (7)$$

The value of $\hat{x}_h$ that corresponds to $h = rK/4$ is given by Equation 7, where $\hat{x}_{h-} = \hat{x}_{h+} = K/2$ is the value of x where the function $F(x) = rx(1 - x/K)$ has its maximum value $F\left(\frac{K}{2}\right) = \frac{rK}{4}$. We designate this value of h as $h_{\mathrm{MSY}} = \frac{rK}{4}$, since it is the largest value of h for which an equilibrium exists: as illustrated in Figure 2A, when $h > rK/4$ no equilibrium solutions are possible and the population is driven to 0 because for these values of h we have $dx/dt < 0$ for all x, which implies that the yield cannot be sustained over time.

For constant $h \le rK/4$, it follows from Equation 5 that the corresponding equilibrium yield, or so-called sustainable yield is $Y_h = \lim_{T\to\infty} \frac{h}{T}\int_0^T dt = h$, which also has a maximum $h_{\mathrm{MSY}} = h_{\mathrm{MSY}}$. For the case of logistic growth the maximum yield that can be sustained per unit time occurs at $h_{\mathrm{MSY}} = rK/4$ (Fig. 2A), which implies

$$\text{Logistic maximum sustainable yield:} \quad Y_{\mathrm{MSY}} = rK/4. \qquad (8)$$

Gordon-Schaefer Theory

If $h = qvx$ then, from Equation 4, $dx/dt = 0$ implies for the logistic case that the equilibrium value $\hat{x}$ satisfies the equation $rx(1 - x/K) = q\hat{x}\hat{v}$ so that either $\hat{x} = 0$ or $\hat{x} = K\left(1 - \frac{q\hat{v}}{r}\right)$. The latter only holds (i.e., satisfies $\hat{x} \ge 0$) for $\hat{v} \le r/q$. In this case, the sustainable yield per unit time is

$$\hat{Y} = Y(\hat{v}) = \lim_{T\to\infty} \frac{1}{T}\int_0^T q\hat{x}\hat{v}dt = q\hat{v}K\left(1 - \frac{q\hat{v}}{r}\right). \qquad (9)$$

Again, as in Equation 5, this has a maximum value $Y_{\mathrm{MSY}} = \hat{Y}(r/(2q)) = rK/4$, with corresponding harvest effort level $v_{\mathrm{MSY}} = r/(2q)$ (Fig. 2A). Thus, as $\hat{v}$ varies from 0 to r/q, the sustainable yield per unit time is the quadratic $\hat{Y} = q\hat{v}K\left(1 - \frac{q\hat{v}}{r}\right)$ in $\hat{v}$: it starts out at $\hat{Y} = 0$ when $\hat{v} = 0$, increases to a maximum value at $\hat{Y} = \frac{rK}{4}$ at $\hat{v} = r/(2q)$, and then decreases back to $\hat{Y} = 0$ at $\hat{v} = r/q$. For $\hat{v} > r/q$, no equilibrium $\hat{x} > 0$ (such as v_2 in Fig. 2A)

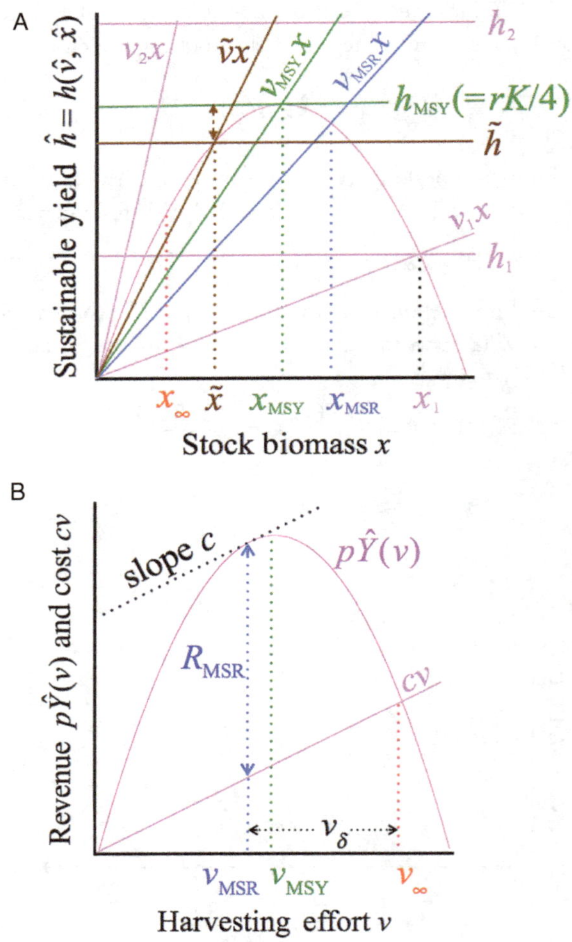

FIGURE 2 A graphical summary of Gordon-Schaefer Sustainable Yield (A) and Sustainable Rent (B) analyses.

exists, and hence the sustainable yield is 0. (Note if we formally apply Equation 9 when $\hat{v} > r/q$ we obtain $\hat{Y} < 0$.)

Recalling in Equation 6 that $p > 0$ is price per unit harvest and $c \geq 0$ is cost per unit harvest effort v, then under equilibrium conditions the sustainable revenue is $R(\hat{v}) = pY(\hat{v}) - c\hat{v}$, which is maximized at the value of $\hat{v}$ for which $\frac{d}{dv}(pY(\hat{v}) - c\hat{v}) = 0 \Rightarrow \hat{Y}' = \frac{c}{p}$ (Fig. 2B). This value of $\hat{v}$ is called v_{MSR} (MSR stands for maximum sustainable rent), and the corresponding sustainable yield and rent are referred to as Y_{MSR} and R_{MSR}, respectively (Fig. 2B). Since $c > 0$, it is easily shown, as we see from Figure 2B, that $v_{MSR} < v_{MSY}$ and that the corresponding population equilibrium biomass values satisfy $x_{MSY} < x_{MSR}$ (Fig. 2A). The value of $\hat{v}$ that solves the equation $pY(\hat{v}) = c\hat{v}$ is called the bionomic equilibrium (Fig. 2B), denoted by the equilibrium pair (v_∞, x_∞) (the reason for the infinity subscript will become apparent below). As argued by H. S. Gordon in the early 1950s, it is the effort level value that unregulated harvesting should equilibrate at, under so-called open-access conditions of exploitation. Extending the above inequalities, we see from the graph that $v_{MSR} < v_{MSY} < v_\infty$, which implies $x_{MSY} > x_{MSR} > x_\infty$.

Nonequilibrium Harvesting

Although the optimal v_{MSY} and v_{MSR} solutions can be generated for resource populations that are reasonably well modeled by a differential equation, such as Equation 4, solutions cannot be implemented directly, because the population is unlikely to initially satisfy $x_0 = x_{MSY}$ or $x_0 = x_{MSR}$, respectively. In unexploited populations, the initial biomass level x_0 of the population is generally twice as large as x_{MSY}. The strategy that optimizes V_δ defined in Equation 6 is to drive the population biomass abundance variable $x(t)$ as rapidly as possible to a value x_δ, that is, the equilibrium value corresponding at sustainable harvesting level v_δ, where singular optimal control theory can be used to solve for the equilibrium solution pair (v_δ, x_δ) (e.g., Clark, 2010). This involves setting $v(t) = v_{max}$ whenever $x(t) > x_\delta$ (which only does the job if $v_{max} > v_\delta$) and $v(t) = 0$ whenever $x(t) < x_\delta$. Further, it can also be demonstrated that $v_{MSY} < v_\delta < v_\infty$ and $x_{MSY} > x_\delta > x_\infty$, and that as $\delta \to 0$ the equilibrium pair $(v_\delta, x_\delta) \to (v_{MSR}, x_{MSR})$ and that as $\delta \to \infty$ the equilibrium pair $(v_\delta, x_\delta) \to (v_\infty, x_\infty)$, respectively.

Equation 4 is only a crude model of the many processes that impact the biomass dynamics of a harvested biological resource. In practice, we expect environmental events such as unusual weather patterns or epidemics to periodically perturb a resource away from its equilibrium. In this case, as discussed in the "Stochasticity" section, below, the best approach is to regard harvesting from a stochastic systems point of view and to avoid strategies based on selecting h_{MSY} or h_{MSR} rather v_{MSY} or v_{MSR}.

Investing in Effort

There are many ways to extend Gordon–Schaefer theory to include multiple interacting biological populations—which takes us into the realm of multispecies harvesting or the consideration of the deeper economic issues relating to fixed versus operating costs and the elasticity of capital needed to invest to increase harvesting effort (e.g., buy more boats) or disinvest to decrease harvesting effort (e.g., sell boats or divert boats to activities other than harvesting the population being managed). One obvious extension is to assume that the effort level $v(t)$ itself satisfies a differential equation in which investment or disinvestment in $v(t)$ is proportional to the rate at which the rent $ph(v(t), x(t)) - cv(t)$ accumulates or is lost, respectively, depending on whether this rent is positive or negative. In this case,

$$\frac{dv}{dt} = a(ph(v,x) - cv),$$

where the constant $a > 0$ reflects the elasticity of capital in the harvesting enterprise.

For the case $h(v,x) = qvx$, we see Equation 4 together with the above equation have the form of the Lotka–Volterra prey–predator equations that in the logistic case $F(x) = rx(1 - x/K)$ cause the trajectory pair $(v(t), x(t))$ to asymptotically approach an equilibrium pair $(\hat{v}, \hat{x})$ that satisfies the two equations

$$rx(1 - x/K) - qvx = 0 \quad \text{and} \quad pqvx - cv = 0.$$

This model is dynamically analogous to the classic Lotka–Volterra prey–predator model with logistic growth for the prey and mass-action extraction of prey by predators.

COHORT ANALYSIS

Cohort Model

Cohort analysis, developed by Beverton and Holt in the early 1950s, follows the biomass abundance $x(t)$ of a single cohort of individuals recruited at time $t = 0$ to a population of individuals that by virtue of their life history stage and location are vulnerable to harvesting from the point of recruitment onward. This biomass $x(t)$ is computed by knowing at time t the actual number $n(t)$ of individuals in the cohort and the average weight $w(t)$

of each individual, i.e., $x(t) = n(t)w(t)$. By the chain rule of calculus, the differential equation for the biomass abundance of the population at time t is thus

$$\frac{dx}{dt} = w(t)\frac{dn}{dt} + n(t)\frac{dw}{dt} \quad \text{with} \quad x(0) = n_0 w(0), \quad (10)$$

where n_0 is the number of individuals recruited to the population and $w(0)$ is a shifted weight-at-age function so that $w(t)$ represents the weight of an individual t units of time after recruitment (i.e., if an individual is recruited at age a at time 0, then at time t its age is $a + t$ and weight is $w(t)$).

If $x(t)$ is the solution to Equation 10, then a necessary condition for $x(t)$ to have its maximum biomass at $t = t^*$ is for $x'(t)^* = 0$, where we use the "prime" notation $x'(t)$ to represent the derivative of $x(t)$ that is itself a function of t. With this notation, Equation 10 implies that t^* is the solution to

$$\frac{1}{n(t^*)}n'(t^*) = -\frac{1}{w(t^*)}w'(t^*).$$

In Beverton and Holt's treatment, and almost all other treatments, it is assumed that $n(t)$ experiences a constant exponential decay rate $\alpha > 0$ over time, which implies

$$\frac{1}{n(t)}n'(t) = -\alpha \Rightarrow n(t) = n_0 e^{-\alpha t} \quad (11)$$

and hence

$$\frac{1}{w(t^*)}w'(t^*) = \alpha. \quad (12)$$

Typically, $w(t)$ is a function, such as the Bertalanffy growth function $w(t) = w_{max}(1 - be^{-ct})^3$ (where $w_{max} > 0$ is an upper bound, $0 < b < 1$ implies $w(0) = w_{max}(1 - b)$, and $c > 0$ controls the asymptotic rate of approach of $w(t)$ to w_{max}), that has the following properties: (i) $w(t) \geq 0$ is an increasing function of time satisfying $0 < w(0) < w(t) < w_{max}$ for all $t > 0$, and (ii) the proportional rate of increase in $w(t)$ is itself a decreasing function of time (this implies $w'(t)/w(t)$ is decreasing over time). Under these assumptions if $w'(0)/w(0) < \alpha$ then Equation 12 has a solution $t^* > 0$, in which case the biomass of the cohort does indeed have a maximum at t^*. Hence, if Equation 10 represents the biomass trajectory of a cohort of organisms being reared for consumption, whether fish in an aquaculture facility or buffalo on a game ranch, then t^* is the optimal time to harvest (or slaughter the cohort) if biomass maximization is the only criterion. However, other economic criteria come into play such as cost and discount rates, as we have seen in the "Harvesting Plantations" section, above, and explore further in the "Cohort Analysis" section, below.

Yield-per-Recruit Computation

If t^* is the solution to Equation 11 it follows that the maximum biomass that can be extracted from a cohort of n_0 individuals recruited at time 0 is

$$Y_{max} = x^* = w(t^*)n_0 e^{-\alpha t^*}. \quad (13)$$

However, this assumes that we are able to harvest all individuals at the instant of time t^*. This may be possible in a small aquaculture or ranching facility, but if the cohort represents individuals in an ocean fishery then this is an unrealistic assumption. In their canonical analysis, Beverton and Holt assumed that from a critical time $t_e \geq 0$ onward it is possible to subject the population to a harvesting intensity v that induces a constant harvest-mortality rate $h = qvx$, as was assumed in deriving the sustainable yield expression in Equation 9. In this case, the biomass in the population is given by (cf. Eq. 11)

$$x(t) = \begin{cases} w(t)n_0 e^{-\alpha t} & \text{for} \quad 0 \leq t \leq t_c, \\ w(t)n_0 e^{vt_c-(\alpha+v)t} & \text{for} \quad t \geq t_c, \end{cases} \quad (14)$$

and the biomass harvested over the life time of the cohort is

$$Y(t_c, v) = qv\int_{t_c}^{\infty} x(t)dt = qvn_0 e^{vt_c}\int_{t_c}^{\infty} w(t)e^{-(\alpha+v)t}\,dt. \quad (15)$$

If we know $w(t)$ then we can integrate Equation 15 and calculate the yield-per-recruit $Y(t_c,v)/n_0$ as a function of the age t_c at which the individuals are first subject to harvesting intensity v. For any yield-per-recruit function, we can define the so-called eumetric and cacometric yield per recruit functions as

$$y_{eumetric}(v) = \max_{t_c} Y(t_c, v)/n_0 \quad \text{and}$$
$$y_{cacometric}(t_c) = \max_v Y(t_c,v)/n_0.$$

It is worth noting from Equation 13 that both $y_{eumetric}(v)$ and $y_{cacometric}(t_c)$ are bound above by $w(t^*)e^{-\alpha t^*}$, which is only achieved in the limit for $t_c = t^*$ as $v \to \infty$.

The yield-per-recruit $Y(t_c, v)/n_0$ represents that actual biomass we can expect to get from each individual in a cohort that over the total life time of the cohort is subject to the harvesting policy represented by the pair of values (t_c, v), where v represents the harvesting intensity level that all individuals are subject to from age $t_c + a$ onward, where a is their age at recruitment (since fish have larval stages feeding in nursery areas during part or all of their first year of life, a is typically 1 for many marine fisheries, as is the case for pacific salmon). Now consider a population that consists of a mixture of cohorts such that at the beginning of each unit of time (i.e., annual if units are years) a new cohort of n_0 individuals enters the pool. If individuals in this pool are subject to harvesting with knife-edge selectivity—i.e., although a harvesting intensity v is applied to the whole pool only those individuals who have been in the pool for t_c units of time or longer are subject to harvesting—then $Y(t_c,v)$ also represents the yield that can be sustained from the pool as a whole in each unit of time under equilibrium conditions. This assumes the specified recruitment and harvesting conditions apply to all cohorts, in the pool that have been recruited going back infinitely far in the past. If units are years, then $Y(t_c,v)$ is the annual sustainable yield for this so-called dynamic pool resource in which n_0 individuals are newly recruited to the pool in each time period.

INCLUSION OF AGE STRUCTURE

Leslie Model

An important category of harvesting problems relate to selecting individuals of particular sizes or ages from a population structured into size or age classes. Examples of these are selecting trophy animals by age for recreational hunting or trees by size class for timber. Most of the theory has been developed in the context of the Leslie matrix age-structure model and its extension to include size or other stage-specific classes, where the number of individuals in each of the n stage classes is represented by a vector $\mathbf{x} = (x_1, \ldots, x_n)^T$ (where superscript T denotes vector transpose so that the row is actually a column). We stress that in this section the variables x_i represent numbers density rather than biomass density, as has been the case in the previous sections.

The Leslie model is formulated as a discrete-time matrix-iteration process involving an age-structure vector $\mathbf{x}(t)$ and projection matrix A. The elements x_i, $i = 1, \ldots, n-1$, of the vector $\mathbf{x}$ represents the number of individuals in the consecutive age classes 1 to $n - 1$, though the final class x_n may represent individuals of age n and older. The Leslie matrix A, depicted in Equation 16 below, is a sparse matrix, with all entries 0 except for the following:

i. Elements $r_i \geq 1$, $i = 1, \ldots, n$, (typically $r_n > 0$) in the first row representing the number of individuals recruited to the first age class at the start of the current time period as a consequence of reproduction by individuals (typically females) in the ith age class at the start of the previous time period.

ii. Entries $s_i > 0$, $i = 1, \ldots, n-1$, down the lower diagonal representing the proportion of individuals in the ith age class at the start of the previous time period that survive to become individuals in the $(i + 1)$th age class at the start of the current time period.

iii. An entry $s_n \geq 0$ in the last position of the last row, representing the proportion of individuals in the nth age class at the start of the previous time period that survive to remain in the nth age class at the start of the current time period.

With these definitions, the Leslie matrix A has the form

$$A = \begin{pmatrix} r_1 & r_2 & \cdots & r_{n-2} & r_{n-1} & r_n \\ s_1 & 0 & \cdots & 0 & 0 & 0 \\ 0 & s_2 & \ddots & 0 & 0 & 0 \\ \vdots & \vdots & \ddots & \ddots & \vdots & \vdots \\ 0 & 0 & \cdots & s_{n-2} & 0 & 0 \\ 0 & 0 & \cdots & 0 & s_{n-1} & s_n \end{pmatrix}, \tag{16}$$

and the Leslie model is given

$$\mathbf{x}(t+1) = A\mathbf{x}(t) \quad \text{with} \quad \mathbf{x}(0) = \mathbf{x}_0. \tag{17}$$

Linear Harvesting Formulation

At the end of each time period—that is, after we have computed how many individuals survived but have not yet changed their current age designation to the next age class—we can image that we harvest u_i, individuals from the ith age class. In this case, if $\mathbf{u} = (u_1, \ldots, u_n)^T$ is the vector of numbers harvested by age class then the Leslie matrix model generalizes to

$$\mathbf{x}(t+1) = A\mathbf{x}(t) - \mathbf{u}(t), \tag{18}$$

where, for consistency, we require $\mathbf{u}(t) \leq A\mathbf{x}(t)$, because we cannot harvest more individuals than there are in the population. If $w_i \geq 0$, the ith element of a vector of weights $\mathbf{w} = (w_1, \ldots, w_n)^T$ represents the value of a harvested individual from age class i, then the inner product $\mathbf{w}^T\mathbf{u}$ is the total value of the harvest over the period of concern so that the total discounted value of the harvest for all time in the future for a given constant discounting rate δ is

$$V_\delta = \sum_{t=1}^{\infty} e^{-\delta t}\, \mathbf{w}(t)^T \mathbf{u}(t). \tag{19}$$

If, instead of characterizing the harvest in terms of numbers $\mathbf{u}$, we characterize it in terms of a diagonal matrix $H(t)$, with diagonal elements $h_i(t)$ representing the proportion of the ith age class of the population $A\mathbf{x}(t)$ that is available once survival has been taken into account (all nondiagonal elements of H are 0), then we have $\mathbf{u}(t) = HA\mathbf{x}(t)$. This leads to the following optimization problem.

Present-value maximization:

$$\max_{0 \le h_i(t) \le 1,\, i=1,\ldots,n} \quad V_\delta = \sum_{t=1}^{\infty} e^{-\delta t}\mathbf{w}(t)^T H(t)A\mathbf{x}(t)$$

subject to the constraints (where I is the identity matrix)

$$\mathbf{x}(t+1) + (I - H(t))A\mathbf{x}(t) \quad \text{and} \quad \mathbf{x}(t) \ge 0$$
$$\text{for all } t = 1, 2, 3, \ldots.$$

If, in addition, we require the population to be at an equilibrium then the present value formulation reduces to

Sustainable present-value maximization:

$$\max_{0 \le h_i \le 1,\, i=1,\ldots,n} \quad V_\delta = \frac{\mathbf{w}^T HA\mathbf{x}}{e^\delta - 1}.$$

subject to the constraints

$$\mathbf{x} = (I - H)A\mathbf{x} \quad \text{and} \quad \mathbf{x} \ge \mathbf{0} \quad \text{for all } t = 1, 2, 3, \ldots.$$

Although the general present-value maximization problem is an infinite linear programming problem, the sustainable present-value version is finite and easily solvable for any applied problem once values are known for the population projection matrix A, whether it has the Leslie structure given in Equation 17 or the more general Lefkovitch structure in which the diagonal elements are no longer 0 but allow for partial transitions in each time step from what are now stage classes i (e.g., size or life history stage) rather than strict age classes.

Before moving on to discuss nonlinear extensions to the above maximization problems, it will be useful to remind ourselves of the Perron–Frobenius theorem, which relates to the eigenvalues and eigenvectors of the matrix A and has implications for the solution to the discrete-time system of linear equations represented by Equation 17. First, for any matrix A we can order its n eigenvalues λ_i of A (not all necessarily distinct values) according to their absolute values to satisfy $|\lambda_1| \ge \ldots \ge |\lambda_n| \ge 0$. The Perron–Frobenius theorem states that if the elements of A are nonnegative (which is true of the Leslie matrix Equation 16) and we can find an integer $k > 0$ such that all the elements of A^k are positive, then the matrix A is said to be a nonnegative primitive matrix and it has a positive dominant eigenvalue; i.e., with our above ordering, we have $\lambda_1 > |\lambda_2|$ called the Perron root. If A is a nonnegative primitive matrix then all solutions to Equation 17 either grow without bound when $\lambda_1 > 1$ or decay to extinction when $\lambda_1 < 1$, unless the dominant eigenvalue A fortuitously satisfies $\lambda_1 = 1$. In addition, in the absence of harvesting, as time progresses each stage class asymptotically satisfies the equation

$$x_i(t+1) = \lambda_1 x_i(t). \tag{20}$$

Thus, only if $\lambda_1 > 1$ will the above sustainable present-value maximization problem have a solution: in this case the constraint equation in the formulation implies that elements h_i of H are selected so that the dominant eigenvalue $\lambda_1(H)$ of the matrix $(I - H)A$ (i.e., $\lambda_1 = \lambda_1(0) > 1$ above is reduced with positive harvesting to $\lambda_1(H) = 1$ when one or more of the diagonal elements of H is nonzero).

Nonlinear Harvesting Formulation

In the Leslie model, one approach is to regard r_i as composed of two parts, the number of individuals b_i born per individual in age class i multiplied by the proportion s_0 of these newborns that make it to the first age class by the end of the time interval. This approach assumes reproduction occurs just after reassigning individuals to their new age class. Thus, we have $r_i = s_0 b_i$ so that the first equation in the Leslie matrix model can be written as

$$x_1(t+1) = s_0 \sum_{i=1}^{n} b_i x_i(t) = s_0 \mathbf{b}^T \mathbf{x}(t), \tag{21}$$

where $\mathbf{b} = (b_1, \ldots, b_n)^T$ and the quantity

$$z(t) = \mathbf{b}^T \mathbf{x}(t)$$

can be thought of as some kind of birth or egg index. Instead of the new recruits x_1 to the population being a linear function of z, several different nonlinear functions have been proposed, the two most important involving two parameters s_0 and $\beta > 0$:

Beverton and Holt recruitment:

$$x_1(t+1) = \frac{s_0 z(t)}{1 + \beta z(t)}$$

and

Ricker recruitment:

$$x_1(t+1) = s_0 z(t) e^{-\beta z(t)}.$$

Both of these are special cases of the more general DeRiso–Schnute stock recruitment function the involves and additional parameter γ that can be both positive and negative:

DeRiso–Schnute recruitment:

$$x_1(t+1) = s_0 z(t)(1 - \beta\gamma z(t))^{1/\gamma}.$$

The remaining equations for updating $x_i(t+1)$, $i = 2, \ldots, n$, can then either be the same as those in the Leslie model or the survivorship proportions s_i, rather than being regarded as constants, can themselves be made functions of the number or density of individuals in their own and other age classes. Typically, we can expect survivorship to decrease with increasing numbers of individuals in the various age classes if intraspecific competition affects survivorship. However, in the case where the only nonlinearity in the population growth process is the recruitment process, the following result holds under the assumption that each age class can be harvested independently of all other age classes. The maximum sustainable yield—also known as the ultimate sustainable yield because of the assumption of independent harvesting of each age class—involves harvesting at most two age classes: (i) first the age class containing the age at which the unharvested cohort attains its maximum biomass (through the tradeoff of losses of individuals to natural mortality and gains due to increasing weight-with-age of each individual, as presented in the "Cohort Analysis" section, above) is harvested, but only completely if the particular age cohort is sufficiently reproductively mature to have allowed the population to replace itself in a sustainable way prior to harvesting. (ii) If the first age class does not satisfy this latter condition, then it can be harvested only partially or not at all, with complete harvesting of an older age class implemented once reproductive levels needed for sustainability have been accomplished.

SYNTHESIS OF FORMULATIONS

The primary difference between plantation harvesting theory presented in the first section of this entry versus the dynamics harvesting theory presented in the second and fourth sections, above, is that harvesting occurs at a single point in time in the plantation analysis and hence does not require a dynamic model to evaluate its influence on the resources through time. The basic notions of harvest value and its transformation into a present-value integral are introduced in the "Harvesting Plantations" section and carried over to varying degrees of refinement in subsequent sections. Further, the one-period optimal rotation plantation harvesting problem presented represents the limiting solution to the cohort harvesting analysis for the case where the harvesting effort variable v approaches infinity: in the limit, the rate of biomass removal at each point in time accumulates to an actual biomass quantity removed at the single critical point t_c in the cohort analysis. Further, the plantation harvesting theory is an even-aged stand management theory that in the context of harvesting age-structured populations is generalized to an uneven-aged stand management theory, since in the Leslie matrix model each of the age-classes can be harvested separately at each time period and the impact modeled using Equation 18.

In the "Mathematical Bioeconomics" section, we used a continuous differential equation to develop MSY and MSR solutions for biological resources described purely in terms of the dynamics of their biomass abundance in a common pool over time. In the "Cohort Analysis" section, we continued to use differential equations to model biomass abundance, but here we took cognizance of the fact that individuals in the population increase in size and, hence, value as they age. However, in this analysis we only focused on one cohort at a time, making the link to other cohorts under the assumption of all cohorts being in the same pool under equilibrium conditions. With the subsequent inclusion of age structure, we assumed that we could differentially harvest various cohorts using a difference equation (i.e., discrete-time) Leslie matrix model. Here, we bring all these models together in a formulation that has provided the backbone for fisheries harvesting analysis of the past 40 years.

Harvesting and Survival

When written out in detail, the transition of the ith age class at time t to the $(i + 1)$th age class at time $t + 1$ in the Leslie matrix model has the form

$$\begin{aligned} x_{i+1}(t+1) &= s_i(v(t))x_i(t) \quad i = 1, \ldots, n-2, \\ x_n(t+1) &= s_{n-1}(v(t))x_{n-1}(t) + s_n(v(t))x_n(t), \end{aligned} \tag{22}$$

where we now explicitly indicate that the proportion s_t that survive in a harvested population is a function of the harvesting effort level $v(t)$. Over the interval $[t, t+1)$, the number of individuals at any time $t + \tau$ where $\tau \in [0, 1)$ for constant harvesting v is given by the equation

$$\begin{aligned} &\frac{dx_i}{d\tau} = -(\alpha_i + q_i v)\, x_i \text{ with } x_i(t) \text{ specified} \\ &\Rightarrow x_i\,(t+\tau) = x_i(t)e^{-(\alpha_i + q_i v)\tau} \text{ for } \tau \in [0, 1) \end{aligned} \tag{23}$$

where α_i and $q_i v$ are age-specific natural and harvesting mortality rates, respectively. Note that the differential equation includes the point t at the start of each interval but not the point $t + 1$ at the end of each interval, since at $t + 1$ we change the name of the variable to indicate that the individuals this variable represents have aged by 1 year. Also note that Equation 23 is easily generalized to harvesting the age classes over only part of each time interval for the case where harvesting is limited to a fishing season that is a subset of the full interval.

If we now specify that $\lim_{\tau \to 1} x_i(t + \tau) = x_{i+1}(t + 1)$, then the first equation in Equation 22 and the second part of Equation 23 are the same equation under the identity

$$s_i(v) = e^{-(\alpha_i + q_i v)}. \tag{24}$$

Using this identity for the survival elements in the Leslie matrix model and replacing the linear Leslie relationship given by Equation 19 with a nonlinear relationships

$$x_1(t + 1) = f(z(t)) \text{ where } z = \mathbf{b}^T \mathbf{x}, \qquad (25)$$

where $f(z)$ is the Beverton and Holt, Ricker, or a more general stock-recruitment function, we have generalized Leslie and cohort harvesting theory in following way: (i) both natural mortality and catchability are now age dependent, with the latter replacing the knife-edge selectivity assumption (which is equivalent to $q = 0$ for $t < t_c$ and $q > 0$ for $t \geq t_c$); (ii) Equation 25 provides a link between the current population level represented by the vector $\mathbf{x}(t)$ with the next time period's recruitment level; and (iii) we have dispensed with the equilibrium assumption that we need to link cohort to dynamic pool harvesting theory: Equations 22, 24, and 25 provide a full dynamic pool description the allows yields to be calculated under nonequilibrium conditions for a general time-varying harvesting effort level $v(t)$.

Yields

From Equation 23, we see that the rate at which biomass is harvested from the ith age class, if we assume individuals in this age-class on average weigh w_i, is $w_i q_i x_i(t + \tau)v(t + \tau)$ for $\tau \in [t, t + 1]$. Hence the total yield obtained from the population over this time period is

$$Y_t = \int_0^1 \left(\sum_{i=1}^{n} q_i w_i x_i(t + \tau) v(t + \tau) \right) d\tau$$

$$= \sum_{i=1}^{n} q_i w_i \left(\int_0^1 x_i(t + \tau) v(t + \tau) d\tau \right). \qquad (26)$$

If v is constant and the population is at an equilibrium $\mathbf{x}(t) = \mathbf{x}_v = (x_{1v}, \ldots, x_{nv})^T$ for $t = 0, 1, 2, \ldots$, then it follows from Equation 23 that the yield in Equation 26 can be expressed as

$$Y(v, \mathbf{q}) = \int_0^1 \left(\sum_{i=1}^{n} q_i w_i x_i(t + \tau) v(t + \tau) \right) d\tau$$

$$= \sum_{i=1}^{n} \frac{q_i w_i v x_{iv} \left(1 - e^{-(\alpha_i + q_i v)}\right)}{\alpha_i + q_i v}. \qquad (27)$$

Note, as in Equation 15, that the yield given by Equation 27 depends on two kinds of instruments: the fishing effort v and the catchability coefficient vector $\mathbf{q}$, where in Equation 15 the catchability process is completely characterized by the time t_c at which harvesting switches from being "off" to being "on." The catchability vector $\mathbf{q}$ in Equation 27 is a reflection of the hardware or gear used to harvest the population—e.g., the mesh size of nets or the size of hooks used to catch fish. For a given gear type with corresponding $\mathbf{q}$, one can then find the harvesting effort v that solves the problem $Y^*(\mathbf{q}) = \max_{v \geq 0} Y(v, \mathbf{q})$. One can also compare $Y^*(\mathbf{q}_1)$ and $Y^*(\mathbf{q}_2)$ where $\mathbf{q}_1$ and $\mathbf{q}_2$ represent two different types of gear, e.g., nets of different mesh sizes.

As in the case of the Gordon–Schaefer theory, if we look at the problem of maximizing revenue rather than yield per se, we can include discounting as well as evaluate the economic efficiencies of different gear types. Such analyses are conducted by resource economists, particularly in the context of regulating resources, especially fisheries, to protect them from overexploitation.

Ad Hoc Approaches

The logistic equation $\frac{dx}{dt} = rx(1 - x/K)$ can be integrated to obtain a discrete time equation for $x_{t+1} = x(t + 1)$ in terms of given value $x_t = x(t)$:

$$x_{t+1} = \frac{Ke^r}{K + x_t(e^r - 1)\, x_t}.$$

For $x_t << K$, this formula implies $x_{t+1} \approx e^r x_t$. If we now compare this expression to Equation 20, we see that we can identify e^r with λ_1, i.e.,

$$\lambda_1 = e^r \qquad (28)$$

in terms of the rate at which both the logistic and Leslie models predict that a population will grow when far from their carrying capacity and when age structure is close to its stable age distribution.

If the unharvested population has a carrying capacity K, as in the Gordon–Schaefer theory presented in the "Mathematical Bioeconomics" section, and the maximum growth rate of an age-structured population when density is not a dampening factor is represented by the dominant eigenvalue λ_1 (i.e., under conditions when population growth can be modeled by a Leslie-type model as presented in the previous section), then Robinson and Redford in their analysis of sustainably harvesting wildlife for *bushmeat* recommend that the maximum sustainable yield is best estimated by following the adhoc formulation:

Robinson–Redford MSY rule of thumb:

$$Y_{\text{MSY}} = 0.6c(\lambda_1 - 1)K, \qquad (29)$$

where c is a correction factor based on the maximum life span L of individuals in the population. Robinson and Redford suggested the following values for c: $c = 0.2$ if $L \geq 10$ years, $c = 0.4$ if $5 \leq L < 10$ years, and $c = 0.6$ if $L < 5$ years. Of course, this approach is ad hoc, but if we compare Robinson–Redford with the logistic MSY specified by Equation 8 and make use of identity Equation 28 in the case $r << 1$—i.e., using the approximation $(e^r - 1) \approx r$—we see that Robinson and Redford (Eq. 29) estimate the MSY yield to be $0.6c/0.25 = 2.4c$ times the logistic MSY yield (Eq. 8). If $c = 0.2$, this is half the logistic-based prediction; if $c = 0.4$, it is the same; and if $c = 0.6$, it is 1.5 times the logistic-based prediction. The latter appears to be a serious discrepancy that could well lead to overexploitation of many species of small animals if the Robinson–Redford formula is used in preference to the logistic-based formula for animals with maximum life span of 5 years or less.

STOCHASTIC EXTENSIONS AND ADAPTIVE HARVESTING

Sources of Stochasticity

In the material presented so far, a variety of population models and resource analysis frameworks have been presented for estimating optimal harvesting strategies in the context of maximizing equilibrium yields (MSY solutions), equilibrium rents (MSR solutions), and present values of resources harvested for all time into the future. All models and frameworks have been deterministic when the real world is subject to considerable variation and noise arising from (i) demographic stochasticity, which is the inherent randomness associated with population birth and death processes; (ii) environmental stochasticity, which is the influence of variation in the peaks and troughs of seasonal and other longer-term cycles relating to rainfall, temperature, disease, and other influences extrinsic to the harvested population; (iii) observational errors relating to measurements we make in assessing the state of the population, such as current abundance and age structure of the population, or even the size or age structure of the harvest; and (iv) processes errors relating to the fact that the models used in the analyses do not describe in full detail all the ecological and socioeconomic factors influencing both population and market processes.

Expected Returns

The various models that have been presented in the first five sections of this entry can be augmented to include the different types of uncertainties listed above. Such stochastic models can then be used to obtain estimates of the expected values of sustainable yields, although in stochastic settings a nonzero probability exists that the population may go extinct. Thus, stochastic models generally make two types of predictions in the context of the harvesting policies at hand: (i) the probability that the population will go extinct during the period of management, and (ii) the distribution of population trajectories that persist over the period of interest, given that the population does not go extinct during the period of management. For stochastic systems, an equilibrium concept is inappropriate since the population is always being perturbed in one way or another. However, long-run analyses can be undertaken for situations where the same harvesting policies are repeated year after year and the environment is stationary (i.e., each year the state of the environment is governed by a distribution where the distribution itself does not change over time). In this case, the distribution of the population state over the long run, conditioned on the fact that the population does not go extinct, is often stationary, and, when such a conditional stationary distribution exits, the distribution is called quasi-stationary. Such quasi-stationary distributions can be used to estimate expected sustainable yields with the maximum of these called the maximum expected sustainable yield (MESY). Of course, the concept could be applied to rent rather than yield to obtain estimates of MESR population, harvest, and effort levels. In nonstationary situations, the expected present value can also be calculated.

In addition to carrying out analyses to obtain MESY and MESR solutions, one can also ascribe a cost to the risk of extinction or loss of individuals set aside to let the resource gain in value before harvesting—such as in the optimal stand rotation formulation, or in cohort analysis in waiting for the optimal age or size at which to harvest individuals. Accounting for this type of risk is akin to the process of discounting the value of revenues expected in the future, since we need to account for the fact that the individuals we plan to harvest may die of disease, be harvested by others, or destroyed by some catastrophe, such as a forest fire, before they can be harvested as planned. In this case, if we assess the risk rate to be ρ—which implies a probability $e^{-\rho t}$ that an individual "banked" for future harvest is no longer available at the planned time of harvest—then augmenting the discounting rate δ by the risks ρ obtains the augmented discount rate $\delta + \rho$. For the optimal rotation period analysis carried out in the Harvesting Plantations section, the Faustmann formula still holds, but now applying the augmented discount rate to obtain (cf. Eq. 1)

$$R'(T_{\delta+\rho}) = \frac{(\delta + \rho)R(T_{\delta+\rho})}{\left(1 - e^{-(\delta+\rho)T_{\delta+\rho}}\right)}.$$

This same principle can be applied to other resource present value calculations when a risk exists that individuals being banked for harvest may disappear before harvesting can be implemented.

Adaptive Management, Fixed Escapement, and Reserves

Another way of dealing with uncertain resources when we know that our model projections are not very accurate is to design harvesting strategies that are self-correcting. It has also been proposed under the rubric of adaptive management under uncertainty that we need to actively evaluate how a resource responds to various harvesting regimes in an endeavor to better characterize the response of the response to harvesting, thereby gaining information that might lead to improved design of long-term harvesting strategies. In such cases, it is not completely adequate to characterize harvesting, as we did in plantation and age-structured harvesting formulations, purely in the context of rates h of removing individuals (or proportions of age classes) or rates v of applying a particular effort level. In these formulations, we assumed that h and v are related through stock levels x by some function $h = h(v,x)$, which might apply at the whole population level or at the individual age-class level. Specifying h is equivalent to setting a particular biomass or numbers harvest quota for each time period, while specifying v is equivalent to regulating the number of human or boat days, hook or snare densities per unit time, or other measure of effort rates that can be applied to harvesting the population in a given time period. However, in stochastic systems we should also be keeping track of the abundance variable x to make sure our harvesting strategies are not going awry.

In stochastic systems, the application of a particular equilibrium harvest rate $\hat{h} = h(\hat{v},\hat{x})$ will not maintain a population at its corresponding equilibrium level $\hat{x}$ for all time. However, we can ask if it is better to apply the optimal $h_\delta = h(v_\delta,x_\delta)$ or to apply v_δ in the expectation that $x(t)$ will frequently exhibit small perturbations from x_δ, with possible large perturbations happening less often. In the face of small perturbations, it turns out that it may be better to apply a constant harvesting effort v_δ (e.g., number of boats days in a season, number of hooks per unit time on long lines, hunter hours in a season) rather than a constant harvest rate h_δ (total catch biomass per season or number of trophy animal permits per season). The reason for this is very clear in the case that δ has a value that results in $v_\delta \approx v_{MSY}$. In this case, if $x(t)$ is perturbed to some $\tilde{x}$, which necessarily satisfies $\tilde{x} < x_{MSY}$ because x_{MSY} maximizes the growth function $F(x)$ (quadratic-shaped function in Figure 2A), then following from the fact that $h_{MSY} > \tilde{h} = h(v_{MSY}, \tilde{x})$ (brown lines and variables in Figure 2A) since $h(v,x)$ is always an increasing function of x, we have that the lower harvest level $\tilde{h}$ allows $x(t)$ to return to equilibrium because $\tilde{h} < F(\tilde{x})$, while the higher harvest level h_{MSY} will drive the population to zero unless a second perturbation acts to restore $x(t)$ to a level at or above x_{MSY}.

Specifying effort rather than actual harvest rates, however, can be specious in that the harvest is more easily monitored, while actual effort is difficult to assess. This is particularly true in the face of improving harvesting technologies. In fisheries, for example, improved sonar devices make boats much more efficient than in the past in allocating shoals of fish so that one boat day in an efficient fleet of boats can lead to the harvest of many more fish than in an inefficient fleet of boats. Thus, although specifying effort is a more robust way to manage a resource to ensure sustainability, because harvest rates drop with stock levels for the same level of effort, it is far easier to monitor compliance of harvesting policies if actual quotas are specified and then take rates (number of animals brought to market or tons of fish landed at docks) assessed.

One way to combine the robustness of the negative feedback aspect of effort level policies (i.e., actual harvest rates drop with a decrease in stock for specified effort v because $h(v,x)$ is a decreasing function of increasing x) is to monitor catch-per-unit–effort (CPUE) and then reduce harvest when CPUE levels drop below values that correspond to MESR solutions. A class of harvesting strategies that meets this criterion is called fixed escapement strategies. These strategies are based on using age-or stage-structure models to calculate the optimal population abundance after harvesting has been implemented in each time period, instead of using the optimal harvest rate or optimal effort level to generate the desired harvest policies. Then, instead of assuming the population is going to be at a particular optimal abundance level, which we denote by the stage-structure vector $\mathbf{x}_{MSR}$ rate, we estimate the actual population level $\mathbf{x}^o(t)$ (subscript "o" for observed) and then adjust our harvesting quota in the next time interval to either mitigate or take advantage of the fact that particular age classes in $\mathbf{x}^o(t)$ may respectively be below or above $\mathbf{x}_{MSR}$ levels. Such fixed escapement policies have been found to be more robust in protecting the harvested population than effort- or quota-based harvesting policies in highly variable fisheries, and the real difference between these approaches depends on the level of variability that can be tolerated from one time period to the next in harvest levels versus effort levels versus stock levels.

Finally, management policies that involve setting aside terrestrial and marine areas as reserves, where resources are protected from harvesting, are gaining increasing attention. Analyses of the efficacy of such policies require that the spatial heterogeneity of resources be taken into account and consideration be given to the movement of individuals across space. The goal is to protect populations in selected areas as a natural or managed source (e.g., a case of the latter is the augmentation of salmon stocks from hatcheries) that then become a replenishing stock for harvesting populations that may possibly not be sustainably harvested without such replenishment. Clearly, the complexities involved in such analyses require rather specific information on the biology and geographical distributions of the species involved.

SEE ALSO THE FOLLOWING ARTICLES

Age Structure / Beverton–Holt Model / Discounting in Bioeconomics / Ecological Economics / Fisheries Ecology / Reserve Selection and Conservation Prioritization / Ricker Model / Stochasticity

FURTHER READING

Clark, C. W. 2006. *The worldwide crises in fisheries.* New York: Cambridge University Press.

Clark, C. W. 2010. *Mathematical bioeconomics: the mathematics of conservation,* 3rd ed. New York: John Wiley & Sons.

Getz, W. M., and R. G. Haight. 1989. *Population harvesting: demographic models of fish, forest, and animal resources.* Princeton: Princeton University Press.

Robinson, J. G., and E. L. Bennett, eds. 2000. *Hunting for sustainability in tropical forests.* New York: Columbia University Press.

Walters, C. J. 1986. *Adaptive management of renewable resources.* New York: Macmillan.

HETEROGENEITY, ENVIRONMENTAL

SEE ENVIRONMENAL HETEROGENEITY AND PLANTS

HYDRODYNAMICS

JOHN L. LARGIER

Bodega Marine Laboratory of UC Davis, Bodega Bay, California

Hydrodynamics is a branch of fluid dynamics concerned with the flow of water. It presents theoretical and empirical relations between dynamical properties of a fluid such as velocity, pressure, and density. In this entry, attention is restricted to landscape-scale water flow in the coastal ocean and its importance in transporting material between organisms, habitats, and ecological communities—where coastal ocean includes bays, estuaries, and nearshore and shelf waters. Comparable flow phenomena can be expected in freshwater systems, including lakes and rivers, but the specific relevance to ecology varies. Water motion is critical to aquatic ecology, which is defined by an immersion in water. Not only does water move, but it has a large capacity to transport material and energy due to special properties, such as high heat capacity, high density, and high capacity for solutes. The properties and motion of water have a first-order effect on both form and function at all scales of organization (ecosystems, communities, populations, individuals).

BASIC PRINCIPLES

The Dynamics of Fluid Motion Are Described by the Equations of Motion

The equations of motion are obtained from two principles: (i) the conservation of linear momentum (force balance), as outlined in Newton's second law that force equals the product of mass and acceleration; and (ii) the conservation of mass, or the principle of continuity. Water motion is driven by stress on the boundaries where it meets air or earth (e.g., surface wind stress) and by pressure gradients due to gradients in water level or gradients in the height of isopycnals (lines of equal density). In addition, larger scale motions are influenced by the rotation of the Earth. As fluid is a continuous material, the force balance is expressed per volume, so that mass is expressed as density. In any of the three dimensions of space, forces are balanced by acceleration, e.g., for the horizontal x-dimension:

$$\partial_t u + u\partial_x u + v\partial_y u + w\partial_z u = \frac{1}{\rho}\partial_x P + fv + \frac{1}{\rho}F_x,$$

where u represents velocity in the x-direction (v and w are velocities in y- and z-directions, respectively; note that in the vertical z-dimension, pressure gradient is balanced by gravity). Acceleration $d_t u$ is expanded to the sum of local acceleration $\partial_t u$ plus field acceleration $d_t u = \partial_t u + u\partial_x u + v\partial_y u + w\partial_z u$. Further, P represents pressure, ρ represents density (related to water temperature and salinity), $f = 2\pi\,\sin(\text{latitude})/12\,\text{hr}$ represents rotation (Coriolis parameter), and F_x represents friction forces due to the viscous drag of surrounding water or boundaries (e.g., wind stress at the surface). Recognizing that pressure is due to gravity pulling water down on itself and thus $P = \int_{-z}^{\eta(x)} \rho(x,z)\, gdz$, expansion of the pressure-gradient forcing term yields two pressure gradient terms, due to differences in level of the free surface $\partial_x \eta$ and differences in water density $\partial_x \rho$ ("barotropic" and "baroclinic," respectively). Viscous drag is separated into that due to horizontal shear $A_h\partial_y u$ and that due to vertical shear $A_z\partial_z u$—and the net force

FIGURE 1 Tidal jet outflow from Bodega Harbor, showing the development of eddies on multiple scales owing to the strong velocity shear between fast-flowing turbid outflow and darker blue receiving waters in the bay. Photograph by John Largier.

F_x is due to the divergence in this stress, that is, $\partial_y(A_h\partial_y u)$. In addition to the balance of forces in three dimensions, a fourth equation is obtained from the principle of mass conservation (i.e., for an incompressible fluid, inflow to a finite control volume must equal outflow): $\partial_x u + \partial_y v + \partial_z w = 0$. While these four equations prescribe how water moves, analytical solutions for u, v, and w are obtained only for special conditions (e.g., steady geostrophic flow), and the complex motion of water increasingly is simulated in computer models that resolve time-varying velocity, water level, and density fields (i.e., computational fluid dynamics, or CFD).

Mixing Is Fundamental in Fluids

Mixing is due to small-scale quasi-stochastic movements, spreading and diluting concentrations of momentum, heat, salt, nutrients, dissolved oxygen, plankton, suspended particles, and more. These small-scale fluctuations in motion are due to the viscosity of water, such that it does not flow easily past itself but rather forms whirls and swirls (Fig. 1, Fig. 2). Turbulence results, with unsteady vortices or eddies of many different sizes forming and interacting with each other. Energy cascades from shear in large-scale flow, through energetic large-scale eddies, to smaller and smaller eddies, until eventually eddies are so small that molecular viscosity is important and the remaining energy is lost (and molecular diffusivity removes remaining small-scale gradients in solute concentrations). Turbulence plays a key role in how fluid flow adjusts to the drag of a boundary or to the drag of adjacent fluid moving at a different speed (i.e., velocity shear). Quasi-random turbulent motions toward and away from a boundary (or zone of faster or slower flow) carry different amounts of momentum, with a bias for flows toward a nonmoving boundary transporting faster moving fluid, and vice versa. The net result is a diffusion of momentum to the boundary (or across a shear zone) due to turbulent eddies and described by an eddy viscosity A_h or A_z that relates momentum transfer (or stress) to velocity shear via a flux-gradient model $\tau_x = \rho A_h \partial_y u$. Further, the shear stress τ_x imposed by the boundary is expected to be uniform through a thin boundary layer, with a decrease in eddy viscosity A_h toward the boundary (as the size of eddying motions is limited by proximity to boundary) accompanied by an increase in shear $\partial_y u$ described by the logarithmic boundary layer profile $u(y) = u_*/\kappa \cdot \log(y/y_0)$. Similarly, an eddy diffusivity K_h is used to relate the flux q of a fluid-borne property to the gradient in concentration $\partial_y c$ so that $q = -K_h \partial_y c$ (and K_z used for vertical fluxes).

As Water Moves, It Transports Material

Water motion is important for connectivity in aquatic systems, transporting material between organisms, populations, habitats, and communities. Whether suspended particles or solute, the dispersion of material may be represented by a random-walk model in which the fate of a group of particles is determined by following the position x of each particle (and likewise y and z dimensions):

$$x_{N+1} = x_N + \Delta t(U + u_N),$$

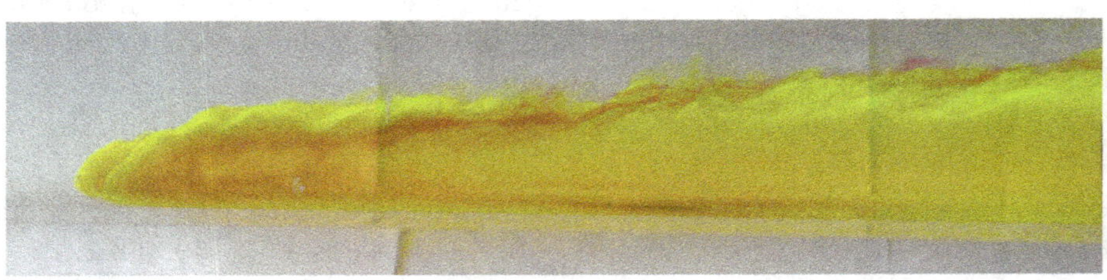

FIGURE 2 A laboratory-scale example of dense water (dyed yellow-orange) intruding underneath clear low-density water. The sloping upper surface of the dyed waters represents a sloping pycnocline, and thus induces a baroclinic pressure gradient in the lower layer (to the left). The dense water is dyed evenly with fluorescein, while purple streaks are entrained into the turbulent dense-water intrusion from potassium permagnate crystals resting on the bottom boundary. Photograph by John Largier.

where N and $N + 1$ represent subsequent time steps, U mean velocity, and u a fluctuating velocity randomly assigned but changing on a time scale given by the observed decorrelation time scale of the flow. To properly evaluate this model, one requires a complete knowledge of the flow field.

Transport and mixing can also be represented by an advection–diffusion model, which is an expression of the conservation of mass. Complex transport patterns due to many scales of motion are reduced to two terms: "advection" describes and represents the mass flux due to transport $q_a = uc$ associated with ordered large-scale motion $u(x, y, z, t)$, while "diffusion" describes and represents the flux due to mixing $q_d = -K_h\partial_x c$ associated with smaller-scale quasi-random motions with zero-mean transport (Fig. 3). Note that eddy diffusivity K_h is often used to represent fluctuating motions at scales much larger than turbulent eddies, if the mean is zero and the time scale of fluctuations is much less than the time scale of interest (e.g., tidal diffusion in the study of seasonality). Diffusive flux can be measured directly through high-frequency observations of concentration and velocity, so that one can obtain zero-mean fluctuating values $u'(t) = u(t) - \bar{u}$ and $c'(t) = c(t) - \bar{c}$ and calculate the flux from the covariance $\langle u'c' \rangle$, where brackets $\langle\ \rangle$ represent a mean over the time scale of interest. The decision as to which scales to resolve as advection is a subjective one (and at times expedient), based on the problem, but the appropriate separation of effects rests on a separation of scales (i.e., mixing effects are due to active motions on scales that are much less than the scale of the circulation). Combining this with a reaction term, the rate of change of the concentration c of material in a control volume is due to net growth within that volume λ plus divergence in advective fluxes or diffusive fluxes in three dimensions. Restricting attention to one dimension, one can write out the advection–diffusion reaction equation as

$$\partial_t c = -\partial_x(u_c - K_x\partial_x c) + \lambda.$$

Of course, λ may be positive or negative (growth or decay) and represents ecological/biogeochemical dynamics—often with the rate dependent on the concentration of another constituent (thus coupling this mass balance to that of another). Eddy diffusivity K_h represents the nature of the eddying flow and can be expected to be isotropic (equal strength in all horizontal directions) and of the same value for all constituents. However, due to stratification and the proximity of top and bottom boundaries, vortices are typically more constrained in the vertical, and thus vertical eddy diffusivity K_z is typically smaller (Fig. 2).

FIGURE 3 Two superimposed photographs of a patch of fluorescein dye dispersing in nearshore waters off La Jolla Shores (southern California). The box in the bottom left corner is the initial photograph taken 19 minutes before the main photograph—photographs are aligned. In the lower panel, a bold arrow indicates the net movement of the center of mass of the patch, which represents "advection." This is the common flow that all dye-marked water parcels experience (the average of flow velocities over the time and space in which this patch dispersed). The center of the patch moved 40 m in 19 minutes, giving an advection $u \sim 0.035\ \mathrm{ms}^{-1}$. Also shown are two ellipses that indicate the concomitant spreading out of the patch, and dilution of dye concentration, which represents "diffusion." This is the net effect of flow fluctuations (deviations from the common/average flow)—each water parcel experiences a unique set of fluctuating velocities and thus experiences a parcel-specific displacement. The patch spreads more rapidly in the direction of advection due to shear dispersion (mixing vertically across shear), so that one differentiates between streamwise mixing and cross-stream mixing, calculating each from $K = 0.5d_t\,\sigma^2$ where σ is the patch length scale. The streamwise patch length increased from 15 m to 60 m in 19 minutes so that $K_s \sim 1.5\ \mathrm{m^2s^{-1}}$ while cross-stream patch length increased from 5 m to 15 m so that $K_c \sim 0.09\ \mathrm{m^2s^{-1}}$. As K_s includes shear dispersion effects, not just eddy diffusion effects, this is better referred to as a dispersion coefficient. The cross-stream K_c provides a better estimate of the true eddy diffusion effect. Photographs by Linden Clarke.

In many ecological problems, advection–diffusion is a valuable conceptual model, as it is used to track and account for the rate of accumulation, delivery, or removal of waterborne material. However, simplifications need caution, e.g., the reduction of the complexity of flow fluctuations to eddy diffusivity is not valid when exploring shorter time periods during which the collection of displacements does not approach a normal distribution. Further, it is difficult to properly account for particle motions in the advection–diffusion approach. While swimming, settling, and floating motions that result in particles crossing streamlines may be represented by an additional term $u_s c$ representing the flux due to typical particle speeds u_s, this results in a disconnect between the mass balance for the material of interest and the mass balance for water itself (which requires $\partial_x u + \partial_y v + \partial_z w = 0$ without u_s). Given the importance of particle motions in allowing for accumulations (i.e., concentration increase), it is best to explore this phenomenon with a particle-specific Lagrangian approach like the random walk.

COASTAL OCEANOGRAPHY

Amid the complexity of flows in coastal waters, one can discern characteristic flow patterns in response to specific forcing due to winds, buoyancy, or waves—and in each type of flow system one sees characteristic ecological structures and processes.

Wind Forcing

Where wind stress acts on the ocean, it drags along surface waters—and where winds persist for more than an inertial period (about a day), the rotation of the Earth is felt and this surface boundary layer turns to the right in the northern hemisphere (and left in the southern hemisphere). This is called the Ekman layer, and Ekman transport is given by $\tau/\rho f$ (actually transport per unit length of wind stress). Along mid-latitude west coasts of major continents, prevailing equatorward winds drive offshore Ekman transport and surface divergence at the coast, which is compensated by vertical flows, known as coastal upwelling (Fig. 4). Cold deep waters rise to the surface,

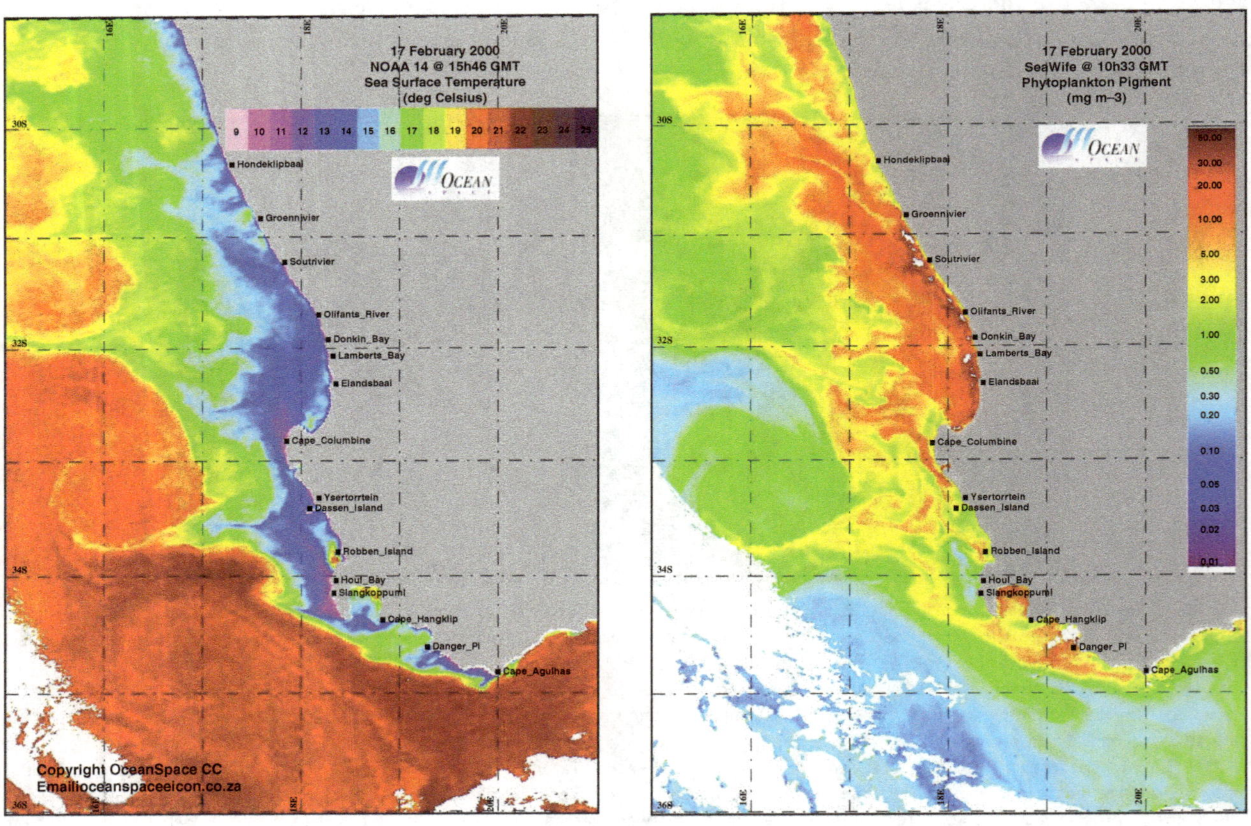

FIGURE 4 Satellite-derived sea-surface temperature (A) and chlorophyll concentration (B) off the west coast of South Africa, 17 February 2000 (data from AVHRR and SeaWiFS sensors). High concentrations of phytoplankton are evident as yellow/red, streaming north and offshore away from upwelling centers at Cape Point, Cape Columbine, and Cape Agulhas (marked by nearshore water temperatures below 10 °C—pink colors). Data courtesy of Scarla Weeks (Ocean Space).

and the sea level drops along the coast. In turn, the pressure gradient toward the coast drives a strong geostrophic coastal upwelling jet (alongshore flow over continental shelf) equatorward with speed $v = \frac{1}{\rho f}\partial_x P = \frac{g}{f}\partial_x \eta$, where x now denotes the cross-shore direction (oceanography convention).

Buoyancy Forcing

Buoyancy forcing occurs where freshwater flows into the ocean (Fig. 5) or where high-salinity waters intrude into an estuary (Fig. 2). In a surface plume, the surface is slightly elevated at the source, sloping down away from the source such that a barotropic pressure gradient pushes water away from the source. However, isopycnals (lines of equal density) slope upward underneath the low-density water, resulting in a baroclinic pressure gradient that pushes denser waters at depth toward the source. The result is an exchange flow, with low-salinity waters flowing seaward as a thin surface plume over the ambient higher-salinity seawater (Fig. 5). A similar buoyancy-driven circulation accounts for the intrusion of dense salty waters into estuaries (Fig. 2), known as estuarine circulation: the seaward-sloping free surface accounts for a seaward pressure gradient in the low-salinity surface layer, while the landward-sloping interface between surface and deeper dense waters accounts for a landward pressure gradient in the high-salinity waters beneath the interface. Meanwhile, turbulence in the surface layer outflow entrains deeper waters upward into the surface layer and completes the conveyor-belt-like circulation that so effectively renews waters in estuaries. Finally, discontinuities at the landward end of the near-bottom salinity intrusion into the estuary (Fig. 2) and at the seaward end of the near-surface plume (Fig. 5) are known as fronts—sudden changes in water properties accompanied by vigorous small-scale vertical circulation. In a surface front (Fig. 5), there is a convergent flow at the surface and a subduction of both water types, descending obliquely underneath the buoyant surface plume. Where biotic or abiotic particles have an affinity for the surface (owing to swimming or floating), they will remain at the surface or soon return there as the water moves down and away from the front—these particles then will be swept back into the convergence where they will accumulate with other particles that were transported to the front earlier or later.

FIGURE 5 A plume of low-salinity turbid water spreads seaward from the mouth of the Russian River (northern California, USA). A sharp line between brown plume and green offshore waters represents a density front, which can accumulate material with surface affinity (e.g., driftwood and plankton). In the foreground, the powerful mixing effect of breaking waves overcomes the density difference between water types and mixes the brown and green colors. Photograph by John Largier.

Wave and Tide Forcing

Propagating waves radiate away from locations where the sea surface is perturbed; most familiar are wind-driven surface gravity waves that are seen as waves breaking on the shore. Breaking occurs mostly near-surface, imposing a radiation stress and imparting an onshore flow that is balanced by offshore flow at depth (downwelling vertical circulation). Ubiquitous and incessant like waves, tides are forced by gravitational forces due to sun and moon—these appear as a slow rise and fall of the sea level along the coast and strong topographic jets (Fig. 1). Waves and tides transfer energy to shorelines, dominating currents in shallow waters and bays—as well as accounting for high levels of turbulence. Below the surface, internal waves propagate on isopycnals (lines of equal density). As for surface waves, these internal waves dissipate their energy in shallow coastal waters, accounting for strong vertical mixing and tidal or high-frequency pulses of deeper, cold, and nutrient-rich waters into the nearshore.

TRANSPORT AND MIXING IN MARINE ECOLOGY

Hydrodynamics influence drifting organisms, swimming organisms, and fixed organisms at many scales, from individual interactions with the fluid, to predator–prey interactions, dispersal of propagules across populations, and global biogeography. A few examples of the ecological importance of water motion are discussed here.

Accumulation of Plankton

Water motion interacts with population growth to control the concentration of plankton at a given place. The advection–diffusion model is well suited to an exploration of changes in this concentration. Four terms control concentration change—advection, diffusion, growth, decay—but only the growth term explains increase in water-borne material. Advection moves plankton around but does not change concentration, while diffusion always acts to reduce concentrations, as does the decay term. An intriguing interplay between growth and diffusion arises owing to the dependence of eddy diffusivity K_h on plankton patch size—in small patches, growth may dominate and the patch population increases, as does the patch size, but above a critical patch size the gain of plankton through growth is dominated by the loss of plankton through larger eddies mixing outward across the edges of the patch. The dependence of K_h on patch size L is known as the 4/3-rule: $K_h \sim a \cdot L^{4/3}$, where a is a proportionality constant. In the ocean, this characteristic scale of phytoplankton patches is 1–10 km.

Advected plumes of phytoplankton are evident in satellite imagery, streaming away from locations where nutrients upwell into the euphotic zone (Fig. 4). Concentration increases as the newly upwelled waters are advected offshore and the phytoplankton population develops in near-surface, nutrient-rich, and illuminated surface waters. Given a time scale of 3–7 days for diatom blooms to deplete nutrients in the surface layer, and flow speeds of 0.1–0.7 m/s, coherent plumes are observed on scales of 10s to 100s of kilometers. Although primarily an interplay between advection and growth, neither is constant, as currents change with variable winds and growth decreases with limiting light or nutrients; further, predation and mixing (horizontal and vertical) become important as the bloom develops.

Recently, attention has focused on the interplay of phytoplankton population growth and vertical dispersion and how this leads to thin layers of high concentration. In the pycnocline, vertical eddy diffusivity K_z is weak, and growth can exceed mixing, but elsewhere mixing is stronger and often growth is weaker due to either limited light (at depth) or limited nutrients (near surface). Where phytoplankton have some swimming ability (e.g., dinoflagellates), they can forage into the weakly turbulent high-nutrient zone in the lower pycnocline as well as into higher light levels in the upper pycnocline. Mixing between the nutrient zone in the lower pycnocline and the deeper high-nutrient waters is sufficient to top up the nutrients at a rate comparable with uptake by plankton foraging out of the thin layer. Further, these vertical motions across the strong shear centered on the pycnocline result in enhanced shear dispersion, which spreads developing thin-layer blooms over large horizontal scales—so that these layers may be just a meter thick, yet kilometers in extent. Weakly swimming primary consumers will also accumulate in these thin layers.

Accumulation can also occur without growth if particles are not perfectly water following and exhibit some independent behavior (e.g., sinking, swimming, floating, rafting, attaching). They can then cross streamlines (i.e., follow a trajectory that is not exactly the same as the trajectory of a parcel of water), which introduces a different nonconservative effect and can yield an increase in concentration. The most common scenario is an accumulation of plankton where waters move downward or upward away from a boundary while particles move vertically through water to remain near to that boundary (even weak particle motion can be important in the shorter vertical distances typical of ocean flows). A classic example is accumulation at surface fronts (Fig. 5), as discussed above. Where different taxa have different upward motility, they will be accumulated different distances from the surface front. Comparable accumulation also occurs in vertical circulation associated with wind-driven coastal upwelling or downwelling, buoyancy-driven circulation (estuaries, salinity intrusion, estuarine turbidity maximum), internal tides and internal waves, and wave-driven downwelling nearshore. Where these vertical flow structures are associated with fixed topography, the flow-induced accumulation also acts as a retention mechanism, holding plankton in a given location, which may be critical in dispersal of propagules or pelagic–benthic trophic coupling (see below). The ecological importance of these flow-associated accumulations of plankton propagates trophically through weakly swimming taxa (that may find an accumulation of prey through a combination of dispersion luck and active foraging behavior) to higher-trophic-level taxa that are strong swimmers or fliers and actively seek and find prey accumulations. In each case, trophic efficiency is much improved by the availability of food at high concentrations—at times a necessary condition for foraging to yield a positive energy balance.

Dispersal of Early Life Stages

Water movement has a first-order effect on the distribution and success of populations in which early life stages are planktonic, which includes many benthic invertebrates and reef fish. Understanding dispersion of planktonic propagules requires one to connect the origin and destination of specific particles. A dispersion matrix (often called a connectivity or dispersal matrix) represents the likelihood of propagules moving between defined adult habitat patches (Fig. 6), either naturally fragmented or resulting from marine protected area networks. With a focus on population connectivity, only successful recruits matter, and the general patterns represented by

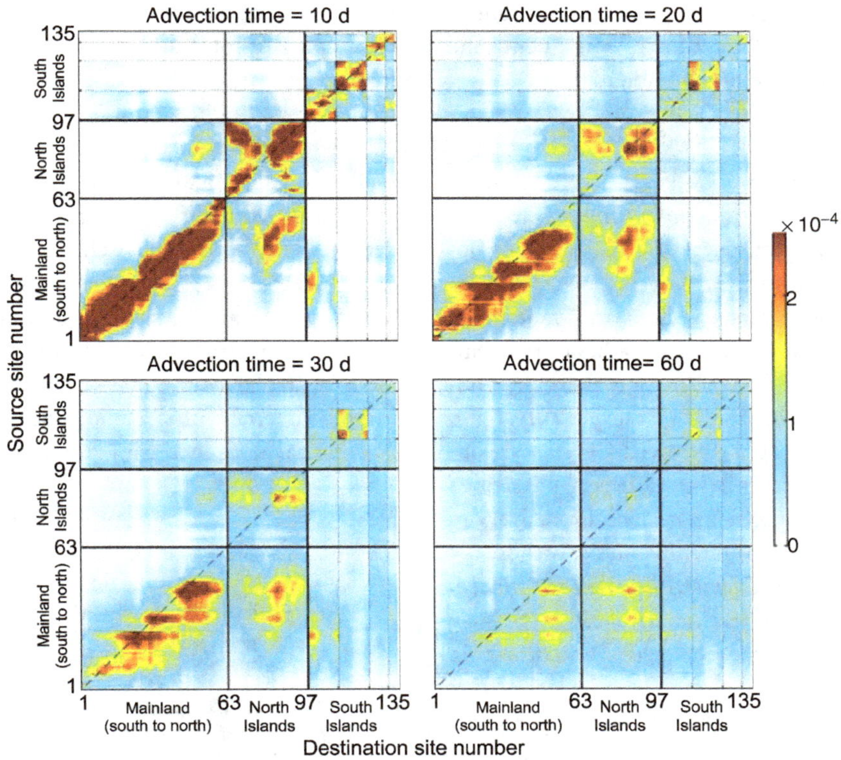

FIGURE 6 A dispersal matrix generated by Mitarai et al. (2009, *J. Geophys. Res.* 114, C10026, doi:10.1029/2008JC005166) from a numerical model of circulation in ocean waters off southern California. The *y*-axes represent 135 source locations on the mainland and islands, and the *x*-axes represent the same 135 sites at which competent larvae may be delivered after defined periods in the plankton (10, 20, 30, and 60 days—here called "advection time" as the model resolves motion at small scale and thus considers all relevant transport to be advection). For each source-destination grid point, a color represents the proportion of particles released at the source that arrive at the destination (high values are order 10^{-4}, or 1-in-10,000). Highest values are along the matrix diagonal, indicating that strongest settlement is near-local.

central tendency statistics (mean, median, or mode) may be misleading. Considering dispersal for a given year or a limited number of years (dispersal events), a particle-specific Lagrangian approach is required, as dispersal outcomes can vary markedly from year to year—both in terms of where settlement is high and also where new recruits were spawned. This particle-tracking approach is normally implemented through numerical models of the time-varying, three-dimensional flow field (Fig. 6). In spite of recent increases in the resolution of flow structure in models, model-generated dispersal outcomes are limited by incomplete knowledge of (i) the behavior of meroplankton, specifically ontogenetic or diel vertical movements, (ii) spatial patterns in mortality, and (iii) small-scale larvae–fluid interactions, specifically during spawning and settlement in adult habitat.

As one explores longer times, aggregating over a diversity of years and dispersal outcomes, the advection–diffusion approach becomes more valid. Thus, it is appropriate for the study of trends in metapopulations and the study of gene exchange and population evolution. Advection–diffusion models yield Gaussian-like dispersal patterns emanating from a single origin, with advection accounting for offset of the Gaussian from the origin and diffusion accounting for the Gaussian width. The effects of shear, accumulation, and retention can be included in higher-order Gaussian parameters, like skewness and kurtosis.

The problem of population dispersal through meroplankton remains only partially resolved, representing an intriguing interplay of biota with fluid dynamics. To date, studies have focused more on dispersion (transport and mixing of particles) than on dispersal (successful exchange of propagules from one habitat to another, which requires an understanding of ecology in the plankton, as well as spawning and settlement behaviors). Further, for population ecology, it is connectivity that is critical, which includes post-settlement survival and reproduction success (closing the life cycle from one generation to the next). While many of the emerging issues are more biological, fluid-dynamic challenges include (i) better resolution of nearshore flow, e.g., coastal boundary layer,

(ii) the influence of wave-driven processes on contact rate, olfactory cues, and disruption of settlement, and (iii) fine-scale fluid flow over adult habitat.

Trophic Coupling of Pelagic and Benthic Habitats

Understanding the trophic coupling between offshore waters and nearshore benthic environments requires knowledge of nearshore transport patterns, including opportunities for accumulation of biogenic material prior to delivery to benthic habitats. Waterborne material is important in terms of allocthanous subsidies of nutrients or particulate organic matter (POM) to benthic communities, habitat determination (e.g., temperature, light, salinity, dissolved oxygen, pH, dissolved CO_2), and contaminant exposure. Concepts in larval dispersal described above are similar for the problem of plankton delivery to the shore, specifically net onshore flux and nearshore accumulation associated with vertical circulation. Indeed, in the absence of nearshore vertical circulation, one may expect a much weaker link between pelagic and nearshore benthic habitats.

In the case of solutes (and the smallest nonaggregated particles), material is fixed in the water and must move with it (Fig. 3). The motion that delivers to the shore will also remove from the shore, so that waterborne material can only have impact if it is taken up rapidly as water flows past organisms (e.g., filter feeders). The interplay between landscape-scale delivery mechanisms, organism-scale boundary layers, and organism physiology control the rate of uptake—and any one can be limiting. Turbulence due to waves on the shore can increase fluxes across organism or substrate boundary layers, but strong breaking waves can stall vertical circulation and renewal of nearshore waters. For example, while internal wave runup can inject high-nutrient (or high-POM) water into the nearshore, this stratified intrusion is likely to stall in the outer surf zone owing to strong wave breaking and mixing. For the same reason, polluted inflows may be retained in the surf zone, exposing disproportionately long stretches of shoreline to pathogens and toxic material. In estuaries with variable depth, stratification can trap material, leading to accumulation of undesirable effects such as the depletion of oxygen.

Connectivity in coastal waters remains poorly studied. Given patterns of transport and mixing, what is an optimal arrangement of habitats? Material produced in one habitat (e.g., detritus from kelp forests) can be a critical subsidy for uptake in a habitat with high demand for particulate organic matter (e.g., mussel bed). Do observed landscapes represent a mosaic of habitat types that reflects underlying trophic exchange due to water motions in this area? As society's interest trends from marine protected areas to marine spatial planning, science interest is likely to trend from dispersion of propagules to a more inclusive analysis and understanding of the dispersion of trophic material within a landscape of coastal ocean habitats that includes increasing human activity.

SEE ALSO THE FOLLOWING ARTICLES

Dispersal, Evolution of / Marine Reserves and Ecosystem-Based Management / NPZ Models / Ocean Circulation, Dynamics of / Partial Differential Equations / Reaction–Diffusion Models

FURTHER READING

Denny, M. W., and B. Gaylord. 2010. Marine ecomechanics. *Annual Review of Marine Science* 2: 89–114.

Franks, P. J. S. 1992. Sink or swim: accumulation of biomass at fronts. *Marine Ecology Progress Series* 82: 1–12.

Haidvogel, D. B., and A. Beckmann. 1999. *Numerical ocean circulation modeling*. London, UK: Imperial College Press.

Hearn, C. J. 2008. *The dynamics of coastal models*. Cambridge, UK: Cambridge University Press.

Jumars, P. A. 1993. *Concepts in biological oceanography*. Oxford, UK: Oxford University Press.

Kundu, P. K., and I. M. Cohen, 2004. *Fluid Mechanics*, 3rd ed. Burlington, MA: Elsevier/Academic Press.

Largier, J. L. 2003. Considerations in estimating larval dispersal distances from oceanographic data. *Ecological Applications* 13(1): S71–89.

Mann, K. H., and J. R. N. Lazier. 2006. *Dynamics of marine ecosystems: biological–physical interactions in the oceans*. Malden, MA: Blackwell Publishing.

Okubo, A., and Levin, S. 2001. *Diffusion and ecological problems: modern perspectives*, 2nd ed. New York, NY: Springer-Verlag.

Siegel, D. A., B. P. Kinlan, B. Gaylord, and S. D. Gaines. 2003. Lagrangian descriptions of marine larval dispersion. *Marine Ecology Progress Series* 260: 83–96.

Thorpe, S. A. 2005. *The turbulent ocean*. Cambridge, UK: Cambridge University Press.

Valle-Levinson, A. 2010. *Contemporary issues in estuarine physics*. Cambridge, UK: Cambridge University Press.

Vogel, S. 1981. *Life in moving fluids: the physical biology of flow*. Princeton, NJ: Princeton University Press.

INDIVIDUAL-BASED ECOLOGY

STEVEN F. RAILSBACK

Lang, Railsback & Associates, Arcata, California

VOLKER GRIMM

Helmholtz Centre for Environmental Research, UFZ, Leipzig, Germany

Individual-based ecology (IBE) refers to theoretical and applied ecology conducted in ways that recognize that individuals are important. IBE addresses how the dynamics of populations, communities, and ecosystems are affected by characteristics and behaviors of the individual organisms that make up these systems. The main tools of IBE are individual-based models (IBMs), simulation models that represent ecological systems as collections of explicit individuals and their environment. Theory in IBE consists of models of *individual* characteristics—especially adaptive behaviors—that have proven useful for explaining *system*-level phenomena, by testing them against empirical information. This kind of theory is unique in two ways. First, it is across-scales: the aim is not to merely seek good models of individual behavior but models of individual behaviors that explain system dynamics. Second, theory is readily tested against many kinds of observations so that its definition more closely resembles that of theory in physical sciences: models that are "general" in the sense of solving many real-world problems.

THE MOTIVATION FOR INDIVIDUAL-BASED ECOLOGY

IBE arose as an attempt to bring more of the complexities of real ecosystems into ecological modeling and theory. It typically combines empirical knowledge and research at both the individual and system levels, relatively simple models of individual-level processes, and computer simulation of the system as a collection of individuals. The rising popularity and importance of IBE is often attributed to the availability of computing power, but it also results from increasing awareness of how important individual behavior and environmental effects can be.

IBMs, the main tool of IBE, are used if one or more of the following individual-level aspects are considered important for explaining system-level phenomena: heterogeneity among individuals, local interactions, and adaptive behavior—individuals adapting their behavior to the current states of themselves and their biotic and abiotic environment to increase their potential fitness. In ecology, IBMs have originally focused on individuals' heterogeneity and local interactions, whereas in models of human systems the focus has been on adaptive behavior and the term agent-based model (ABM) is often used. Over the last ten years, however, differences between IBM and ABM are fading away so that both terms can be, and should be, used interchangeably.

Ecological problems over a wide range of complexity have been addressed with IBE. Simple problems have included understanding how birds in nesting colonies synchronize their breeding and how the ability of a fish school to migrate successfully depends on how many of its members are older and experienced with migration versus young and inexperienced. These problems were addressed using simple IBMs containing only one simple individual behavior. At the other extreme, problems such as predicting how trout populations respond to changes in river flow, temperature, and channel shape, and how loss of intertidal habitat affects shorebird overwinter survival, have been addressed with complex IBMs that include multiple adaptive behaviors and detailed representation of individual physiology and the environment.

Currently, the majority of IBMs in ecology are between these two extremes; particularly in plant ecology, most IBMs still focus on individual heterogeneity and local interactions. Over the last ten years, however, the number of IBMs that include adaptive behavior has increased rapidly, indicating that adaptive behavior is increasingly recognized as key to explaining many system-level phenomena. Thus, IBMs can integrate behavioral ecology, which focuses on evolutionary explanations but usually ignores the individuals' systems, and system-level ecology that focuses on demographic rates or fluxes of nutrients and energy but ignores adaptive behavior. From this integration, IBE emerges as a new and promising branch of ecological research.

TYPICAL STEPS IN INDIVIDUAL-BASED ECOLOGICAL RESEARCH

This section outlines a typical research program of IBE. This program includes a cycle of developing, testing, and applying theory for how the individual and system levels affect each other. The program relies on a strategy called pattern-oriented modeling, which is critical for the two most fundamental challenges of IBE: designing an IBM that has the right mechanisms and just enough complexity to meet its purpose, and developing theory for the individual characteristics that drive the system dynamics of interest.

1. Identify a Problem in Which Individuals Are Important

All research should start by clearly defining a well-bounded problem to address. Stating the problem for an individual-based study may seem trivial because IBE is typically chosen because it seems the only way to address a problem that is already the research focus. The researcher tries IBE because individuals seem too important for simpler methods to work. But an important second part of this step is hypothesizing *how* individuals are important. What individual-level characteristics seem essential to the system's dynamics? These characteristics typically are individual variation in particular variables, adaptive behaviors, and ways that individuals interact with each other or their environment.

2. Identify Characteristic Patterns of the System and Individual-Level Processes

This is the critical step in pattern-oriented modeling: identifying, from empirical knowledge of the system being studied, a set of observed patterns that characterize the system's internal workings. Ideally, these patterns are unambiguous and well-understood ways that the system or its individuals behave as a consequence of the individual-level processes identified in step 1. Example patterns include ways that individuals behave in response to some change in their environment or population, ways that the population responds to several different external forces, and characteristic spatial and temporal distributions. These patterns will be used for the model's design, verification, and calibration. For this purpose, a variety of diverse patterns—at both individual and system levels, in response to several kinds of stimuli, in different state variables—that are qualitative and even weak are often more useful than one or two strong quantitative patterns. IBE is thus based on multiple-criteria assessment of models: instead of fitting a model to just one pattern, which is likely insufficient to capture the internal workings of the real system, multiple criteria—patterns—are used to design and test the model.

3. Design and Implement an Initial IBM

Now there is sufficient information to build a pilot version of an IBM. The model needs to simulate the system and how its dynamics emerge from its individuals, but with no more detail and complexity than necessary. What is necessary? The IBM should only include the structures and processes of the system and individuals that appear necessary to (a) solve the problem defined in step 1 and (b) make it *possible* for the patterns identified in step 2 to emerge—while being careful not to accidentally *force* the model to reproduce those patterns. (Further guidance for developing IBMs is provided in the section "A Conceptual Framework for IBMs," below.)

4. Hypothesize Theory for Individual Traits

The initial IBM developed in step 3 will have initial traits for the individual characteristics that were identified in step 1 as important. (By "trait" we mean a submodel of the IBM that represents a single specific individual-level process.) For some well-understood characteristics of widely studied species, the literature can provide reliable traits that need no further testing; examples might include bioenergetic models of how growth depends on food intake and temperature, and foraging models that represent how food intake varies with individual size, food availability, and behavior. But, especially in these early years of IBE, there will often not be reliable traits available for some characteristics, especially adaptive behavior. Hence, in this step we need to hypothesize one or several traits for the key individual characteristics or behaviors for which we lack existing submodels.

To "hypothesize theory" means to propose several alternative traits for a key individual characteristic; here, the focus is on individual adaptive behavior. Traits for adaptive behaviors should not be hypothesized by making them up with no reference to existing behavioral theory or empirical knowledge. Instead, this step should include developing a good understanding of the organisms and behaviors (for which models of completely different taxa may be useful) from the literature and empirical research, and attempting to apply existing theory of behavioral ecology. The result should be a set of traits, varying perhaps in their complexity or fundamental assumptions, implemented as alternative submodels in the IBM and ready to test in simulation experiments.

5. Test Theory by How Well It Reproduces Patterns in the IMB

Now the job is to test the traits hypothesized in step 4 to determine which provide the best theory for how system dynamics arise from the individual characteristic. This is done by multiple-criteria assessment, i.e., by seeing how well the IBM reproduces the characteristic patterns (step 2) when it uses each alternative trait. This step can also include cycles through steps 2–4, revising and refining traits and finding additional characteristic patterns that are better able to distinguish among more- and less-useful traits.

When these simulation experiments demonstrate a hypothesized trait's ability to make the IBM reproduce a diversity of patterns observed in the real system, they make a persuasive argument that the trait is a useful, general theory for the individual characteristic from which system dynamics emerge.

6. Use Completed Model and Theory to Simulate and Analyze the Problem

A major purpose of any model is to understand the internal organization of a system of interest. Mechanistic understanding is not only of basic scientific interest but also the basis for solving applied problems, for example, ranking management alternatives. In IBE, a first step for achieving "understanding" is identifying the mechanisms that generate the observed patterns we chose to characterize the system. Once we have found the right theories of individual behavior to reproduce these patterns, detective work is required to understand the relative importance of the mechanisms represented in the model. For this, simulation models are analyzed like real systems: by performing controlled (simulation) experiments. "Controlled" means that the model is deliberately made unrealistic again, by turning certain mechanisms off, by reducing heterogeneity among individuals and in the environment, by studying scenarios that never could occur in reality, and so forth (Fig. 1).

To evaluate these experiments, one or a few currencies, or outputs, are needed that allow us to characterize the output of a simulation run by just one, or a few, numbers. Examples include extinction time; mean, variance, and skewness of size distributions; variability in abundance of total biomass; the occurrence of population cycles or outbreaks; time for recovery after a disturbance; primary production over a certain time period; and measures of

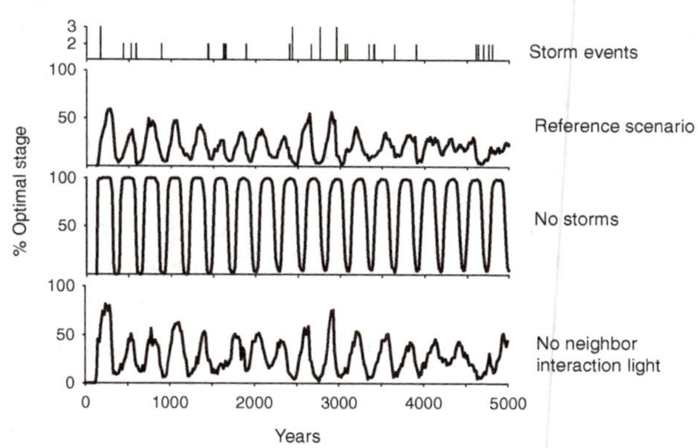

FIGURE 1 Example of analyzing unrealistic scenarios in a realistic individual-based model. The model BEFORE explores drivers of spatiotemporal dynamics in mid-European natural beech forests (Rademacher et al., 2004, *Forest Ecology and Management* 194: 349–368). The model's local scale is defined by grid cells of about 15×15 m^2. Canopy and subcanopy trees are represented individually. The model includes storm events of varying strengths. The reference scenario shows oscillations in the percentage of the forest's area in the "optimal stage," characterized by a closed canopy and almost no understory. If storm events are turned off ("no storm"), forest structure and dynamics are almost completely synchronized. What creates this synchronization? Light falling through local canopy gaps induces growth and reduces mortality not only in the gap's grid cell but also in its neighborhood. If this type of interaction among neighbor cells is turned off, but storm events turned on ("no neighbor interaction light"), dynamics are similar to the reference scenario.

spatial patterns (for example, Ripley's L for the spatial distribution of plants). Evaluation of simulation experiments can also use statistical tools such as general linearized models or analysis of variance.

7. Test Sensitivity and Robustness of Conclusions

Every ecological model includes parameters that are uncertain. Sensitivity analysis is thus part and parcel of model analysis in IBE: the sensitivity of model outputs to variations in model parameters is explored. If a model is overly sensitive to parameters that are also overly uncertain, resources available for empirical work should go into reducing uncertainty in those parameters. Sensitivity also indicates which processes are more important than others.

Computational limitations used to restrict sensitivity analysis of many IBMs to local analyses where only one parameter was varied a little at a time. This limitation has largely disappeared with hardware now available, but subsampling of parameter values (e.g., Latin hypercube sampling) is often required to keep the number of parameter sets to be examined manageable. The complexity of analyzing and understanding results remains a potential limitation for sensitivity analysis of some IBMs.

Robustness analysis refers to exploring not only the sensitivity, or robustness, of model behavior to parameter values but also to variation in model structure. In particular, the full model can be simplified in a systematic way to see what elements of it are really needed for the model's purpose. This is important because model construction usually is path dependent. It can be a long and complicated process, starting from a conceptual model, to construct a full IBM that captures a set of observed patterns. In this process, the focus necessarily is on construction, i.e., adding of new elements, alternative submodels, and so on. This should be followed by a phase of deliberate "deconstruction" that not only helps understand how important the model's elements are but also helps find the most parsimonious model that still answers the original question.

A CONCEPTUAL FRAMEWORK FOR IBMs

One of the biggest limits to progress in IBE has been the lack of a clear conceptual framework: the kind of guide to thinking about, organizing, and communicating models that differential equations provides for classical theory and that matrix models and Bayesian and frequentist statistics bring to other kinds of ecological analysis. Simulation modeling does not impose such a framework, leaving us seemingly free to do whatever we want. However, complexity scientists and experienced users of IBMs and ABMs have developed a list of important design concepts (Table 1) to provide a conceptual framework. Thinking about the key questions of each concept helps us design models that are useful, simple, and theoretically sound, and documenting how the concepts were addressed helps make models seem less ad hoc and easier to understand.

TOOLS FOR INDIVIDUAL-BASED ECOLOGY

One serious limitation to IBE is the need for tools and skills that many ecologists are not exposed to in their education. Primary among these is simulation software design. Especially in the early years of IBMs, few ecologists had the skills and experience to do a good job on tasks such as selecting appropriate software platforms, designing and writing programs, testing and documenting their programs, and providing interfaces allowing the model's dynamics to be observed. Fortunately, a great deal of progress has been made on software tools that make IBMs much more accessible to ecologists. One particular platform, NetLogo, has evolved into a powerful tool widely used by scientists. NetLogo's documentation and learning materials are extensive and very professional, and there are now textbooks on scientific individual-based modeling with NetLogo. Other platforms are more flexible but pose a much steeper learning curve. If combined with a recent standard format for describing IBMs, the so-called ODD protocol (Overview, Design concepts, Details), NetLogo can provide a "lingua franca" for individual-based and agent-based modeling.

IMPLICATIONS FOR FIELD AND LABORATORY RESEARCH

What does using IBE mean for the empirical research that ecologists do to support theory? The most obvious answer to this question is that individual-based models and theory are typically much more closely tied to real systems and problems than classical theoretical ecology and address both individual and system levels. Hence, IBE uses a much wider range of empirical information. Many other approaches to ecology also make extensive use of empirical data and information; how would you design field or laboratory research differently if you planned to use IBE instead of perhaps statistical or matrix modeling?

First, IBMs are generally more mechanistic than other ecological models, so traditional study designs that focus on measuring abundance (or biomass, or density) and contrasting discrete treatments tend to be less useful. Instead, for IBE it is more important to quantify

TABLE 1
Design concepts for individual-based models

Concept	Key Questions
Basic principles	1. What general concepts, theories, hypotheses, or modeling approach underlie the model's design? How is the model related to previous theory or thinking about the problem it addresses? 2. How were these principles incorporated in the model? Does the model implement the principles, or address them as a study topic such as by evaluating them?
Emergence	3. What are the model's key outputs and results? Do they emerge from adaptive behavior of individuals and interactions among individuals and their environment, or are some of them imposed by rules that force the model to produce certain results?
Adaptation	4. What adaptive behaviors do individuals have? In what ways do they respond to changes in their environment and themselves? What decisions do they make? 5. How are these behaviors modeled? Do adaptive traits assume individuals choose among alternatives by explicitly seeking to increase some specific fitness objective ("direct objective-seeking"), or do traits simply force individuals to reproduce behavior patterns observe in real systems ("indirect objective-seeking")?
Objectives	6. For adaptive traits modeled as direct objective seeking, what measure of individual fitness is used to rate decision alternatives? This objective measure is the individual's internal model of how it would benefit from each choice it might make. What elements of future fitness are in the objective measure (e.g., survival to a future reproductive period, reproductive output)? How does the objective measure represent processes (e.g., mortality, feeding, and growth) that link adaptive behaviors to important variables of the individuals and their environment? 7. How were the variables and mechanisms in the objective measure chosen considering the model's purpose and the real system it represents? How is the individual's current internal state considered in modeling decisions? Does the objective measure change as the individual changes?
Learning	8. Do individuals change their adaptive traits over time as a consequence of their experience? How?
Prediction	9. Do individuals predict, explicitly or implicitly, future conditions (environmental and internal) in their adaptive traits? 10. How does simulated prediction make use of mechanisms such as memory, learning, or environmental cues? Or is prediction "tacit": only implied in simple adaptive traits?
Sensing	11. What variables of their environment and themselves are individuals assumed to sense and, therefore, consider in their behavior? 12. Are sensing mechanisms modeled explicitly, or are individuals instead assumed simply to "know" some variables? 13. With what accuracy or uncertainty are individuals assumed to sense which variables? Over what distances?
Interaction	14. How do the model's individuals interact? Do they interact directly with each other (e.g., by eating or fighting with each other)? Or is interaction mediated, such as via competition for a resource? 15. With which other individuals does an individual interact? 16. What real kinds of interaction were the model's interactions based on? At what spatial and temporal scales do they occur?
Stochasticity	17. How are stochastic processes used in the model and why? Example uses are setting initial values of variables, to make some processes variable without having to represent the causes of variability, and reproducing observed behaviors by using empirically determined probabilities.
Collectives	18. Are collectives—social groups or other aggregations of individuals that affect the state or behavior of member individuals and are affected by their members—represented in the model? 19. If collectives are represented, do they emerge from the traits of individuals, or are individuals given traits that impose the formation of collectives? Or are the collectives modeled as another type of "individual" with its own traits and state variables?
Observation	20. What outputs must be observed to understand the model's internal dynamics as well as its system-level behavior? What tools (graphics, file output, data on individuals, etc.) are needed to obtain these outputs? 21. What outputs are needed to test the model and, finally, to solve the problem the model was designed for?

relationships—how the system changes as this or that input is increased, how individuals change their behavior as predation risk changes, and so on. Research designed to support IBE also puts more emphasis on understanding processes and interactions, so it is often more important to observe dynamics and patterns than to quantify averages or other static variables. Perturbed systems may be more useful to study than stable or undisturbed ones.

Second, it is clearly important to study individuals (and "collectives" such as family and social groups), but we must remember that individual-based ecology is not behavioral ecology and that an IBM is a population model, not a model of individuals. To build IBMs we do not need, or even want, a complete model of individual behavior or a complete understanding of all the ways organisms adapt. Instead, we want traits that capture the essential parts of adaptive behavior while otherwise being as simple as possible. However, we need traits that work in realistic contexts, and IBMs often include more of the real world's complexities than highly simplified laboratory systems do. That means we often need a trait that captures the essence of how real organisms solve a more complicated problem than classical models of behavioral ecology address. For example, a relatively simple foraging model—e.g., how an elk selects among habitat patches that vary only in (constant) food availability by optimizing growth—will not be useful in an IBM of a population of elk competing for and depleting these patches every day while also avoiding wolves or hunters.

Finally, many (but not all) IBMs are full life-cycle models that represent population dynamics over many generations. These models should capture the essence of the basic elements of individual fitness: What determines how much the individuals get to eat, what determines who eats the individuals, and what determines reproductive success? Field ecology tends to focus on reproductive processes and feeding, but there is often less study of the other major driver of abundance, mortality. Empirical research on predation risks and rates is often particularly valuable.

IMPLICATIONS FOR THEORETICAL ECOLOGY

Classical theoretical ecology, which is based on analytically formulated models, strives for general theories that apply to a wide range of systems and situations. Early models in this field, such as Lotka–Volterra models of interspecific competition or food webs, are thus highly abstract and ignore virtually every detail in structure and mechanism of real systems. The price for this is that such models can be hard, or impossible, to test. They provide insights in the sense of logical relationships between model entities, but the relevance of such insights for understanding real systems often remains unclear.

IBMs, in particular those developed according to the research program described above, are still much simpler than real systems but nevertheless tend to be considerably more complex than simple mathematical models. Many IBMs have more than ten parameters. They would thus contain too many degrees of freedom in parameters (and model structure) to lead to any robust insights—unless tied to specific systems for which there are data and patterns that allow us to fix parameter values.

Still, IBMs should be designed to be generic in that they do not represent a specific system or site, but capture overall, robust features that are observed over a wide range of conditions. For example, a savanna model might originally be designed for a specific semi-arid region. Once tested and understood, however, its scope could be extended to semi-arid regions in general. Systematic analyses of patterns in regions with higher rainfall might even allow the model to be adapted to those regions.

An important tool for making IBMs as generic as possible is robustness analysis, which is not yet routinely performed, perhaps because many IBMs are developed for solving applied problems rather than for contributing to theoretical ecology. Therefore, an important role of classical theoretical ecology in IBE is to provide theoretical questions and general concepts that are explored, very much as in virtual laboratories, in well-tested IBMs. In this way, the limitations of models from both classical and individual-based theoretical ecology can be overcome. Success in IBE thus does not only require knowledge in simulation but also in classical theoretical ecology.

CONCLUSIONS

Individual-based ecology is, essentially, an adaptation to ecology of ways to think about and model complex systems long used in the physical sciences. Many very accurate and robust models in physics, chemistry, and engineering are built by assembling simplified models of the system's individual components—subatomic particles, molecules, resistors and transistors, beams or pipes—and how these components interact with each other. The difference in ecology is that the individuals have adaptive behavior, which means their behavior is more complex, but it also means that evolutionary fitness provides a strong guide to modeling behavior. Ecologists have actually been leaders among the complex sciences in the development and use of individual-based approaches.

We use IBE and IBMs when we realize that more traditional theory and techniques will not work because characteristics of individuals are too important to how the system behaves. But we must remember that IBE is not just about individuals and behavior, and it is not just simulation modeling; it is a theoretical approach to understanding how the dynamics of systems emerge from the characteristics of individuals.

IBE is a particularly fertile research field because it provides a new way of looking at the systems we already know much about as well as the ability to address many new questions. Even in well-studied systems, there is still little theory relating individual characteristics and system dynamics. Now that we have a proven strategy (pattern-oriented modeling), a conceptual framework, and powerful software tools, progress is likely to be rapid and exciting.

SEE ALSO THE FOLLOWING ARTICLES

Adaptive Behavior and Vigilance / Applied Ecology / Computational Ecology / Environmental Heterogeneity and Plants / Forest Simulators / Population Ecology

FURTHER READING

Auyang, S. Y. 1998. *Foundations of complex-system theories in economics, evolutionary biology, and statistical physics*. New York: Cambridge University Press.

DeAngelis, D. L., and W. M. Mooij. 2005. Individual-based modelling of ecological and evolutionary processes. *Annual Reviews of Ecology, Evolution, and Systematics* 36: 147–168.

Grimm, V., and Railsback, S. F. 2005. *Individual-based modeling and ecology*. Princeton: Princeton University Press.

Grimm, V., E. Revilla, U. Berger, F. Jeltsch, W. M. Mooij, S. F. Railsback, H.-H. Thulke, J. Weiner, T. Wiegand, and D. L. DeAngelis. 2005. Pattern-oriented modeling of agent-based complex systems: lessons from ecology. *Science* 310: 987–991.

Grimm, V., U. Berger, F. Bastiansen, S. Eliassen, V. Ginot, J. Giske, J. Goss-Custard, T. Grand, S. K. Heinz, G. Huse, A. Huth, J. U. Jepsen, C. Jørgensen, W. M. Mooij, B. Müller, G. Pe'er, C. Piou, S. F. Railsback, A. M. Robbins, M. M. Robbins, E. Rossmanith, N. Rüger, E. Strand, S. Souissi, R. A. Stillman, R. Vabø, U. Visser, and D. L. DeAngelis. 2006. A standard protocol for describing individual-based and agent-based models. *Ecological Modelling* 198: 115–126.

Grimm, V., U. Berger, D. L. DeAngelis, G. Polhill, J. Giske, and S. F. Railsback. 2010. The ODD protocol: a review and first update. *Ecological Modelling* 221: 2760–2768.

Huston, M., D. L. DeAngelis, and W. Post. 1988. New computer models unify ecological theory. *BioScience* 38: 682–691.

Railsback, S. F., and V. Grimm. 2012. *Agent-based and individual-based modeling: a practical introduction*. Princeton: Princeton University Press.

Stillman, R. A., and J. D. Goss-Custard. 2010. Individual-based ecology of coastal birds. *Biological Reviews* 85: 413–434.

Wilensky, U. 1999. NetLogo. http://ccl.northwestern.edu/netlogo/. Center for Connected Learning and Computer-Based Modeling, Northwestern University, Evanston, IL.

INFORMATION CRITERIA IN ECOLOGY

SUBHASH R. LELE

University of Alberta, Edmonton, Canada

MARK L. TAPER

Montana State University, Bozeman

Science progresses by successively replacing existing models with new models that approximate nature's underlying reality with greater accuracy and precision. Given a collection of models, information criteria guide the selection of the best model from that class. These techniques try to balance bias in prediction and variability in estimation. Different definitions of "best" lead to different model selection criteria.

THE CONCEPT

Scientific inference is about distinguishing between different mechanisms. To think about the real world, scientists must construct some sort of simplified representation of possible mechanisms. These representations are models and they range from conceptual/verbal models to huge simulation models with thousands of parameters. Science makes progress by successively replacing existing models with new models of increasingly accurate approximation to nature's underlying reality. Model choice, thus, has immense influence on scientific understanding and decision-making capabilities.

Reaching back all the way to Aristotle, the principle of parsimony is the oldest method of choosing a scientific theory. Commonly known as Ockham's razor, one statement is: "It is futile to do with more things that which can be done with fewer." While no one can deny the aesthetic pleasure that a theoretician feels when proposing an elegant solution, we do not see simplicity as an intrinsic value in science. However, simplicity does have several important instrumental values for science. First, simple models are more precisely estimable than complex models, and second, simple models are generally more understandable than complex models. These values enhance the two primary functions of models in science: to make predictions and to give explanations.

Unfortunately, while the estimates of simple models are less variable than those of more complex models, for every simple model, a more complex model can be found that is less biased for prediction. Thus, for the predictive use of models, the value of estimation precision that comes from simplicity is offset with a cost due to reduction in prediction accuracy. Similarly, for the explanatory use of models, the benefit of comprehensibility that comes with model simplicity is offset with a cost of lack of comprehensiveness.

The family of model selection techniques known as information criteria try to balance bias in prediction and variability in estimation in such a fashion that they lead to the "best" model. It is clear that what is chosen as the best model depends on the relative importance given to these two components. Hirotsugu Akaike first proposed an information criterion for model selection in 1973. The highly influential book by K. P. Burnham and D. R. Anderson in 2002 popularized their use in ecology. It is not an exaggeration to say that information criteria have profoundly influenced statistical thinking in ecology. Changes continue as the deep implications of a switch from a hypothesis refutation mode of inference to a model selection (model comparison) mode of inference are realized.

LOGIC OF MODEL SELECTION

At least in the ecological literature, model selection is considered as something different than either hypothesis testing or parameter estimation. In the following, we show that the logical basis behind parameter estimation, testing of hypothesis and model selection has a commonality. Let vector y denote the data set in hand. In the following let $t(y)$ denote the true, but unknown, distribution from which the data arises. It is assumed that such a distribution exists.

1. Maximum likelihood estimation and the Kullback–Leibler divergence Let $f(y;\theta)$; $\theta \in \Theta$ denote the class of models under consideration. For example, if the data are survival times, one may consider fitting an exponential distribution with mean $\theta \in (0,\infty)$. Although commonly this is considered as fitting an exponential model, this is, in fact, a collection of models, with each different value of the parameter θ representing a different model. The likelihood function calculates how probable the observed data are under different parameter values. In this function, data are fixed and the parameter values are changing. If the observed data are more probable under one parameter value (θ_1) than another (θ_2), then parameter value θ_1 is considered better supported than θ_2. The maximum likelihood estimator is that value of θ which is better supported than any other value in the parameter space. Because this is a single value, it is called a point estimator. Instead of providing the single best value, if we provide a set of values of the parameters that could also have been supported reasonably well with the data, such a set is called an interval estimator.

Similar to the distance between any two points on a plane, one can define a distance between the data and a statistical model. Although distance from point A to point B is the same as the distance from point B to point A, the metrics of difference between two statistical distributions are not always symmetric. In this case, we call such metrics divergence measures. In the divergence measure approach, the best supported value as described above is that value of θ (or, model) that is closest to the observed data.

It is well known that the maximum likelihood estimator of θ, denoted by $\hat{\theta}$, corresponds to the model from the collection $f(y;\theta)$; $\theta \in \Theta$ that is closest to the true model $t(y)$ in terms of the Kullback–Leibler divergence. This statement holds even if the true model is not a member of the collection. Other measures, such as the Hellinger divergence, are possible and lead to other ways to select the best-fitting model from the collection. Thus, the problem of parameter estimation is, at its core, a model-selection problem. We select the best-fitting model in terms of some divergence measure. In statistics, point estimate provides the best-fitting parameter (model), and interval estimate provides the set of parameters/models that are not strongly differentiable from the best-fitting parameter (model). The concept of sampling distribution is used to attach probabilities to these sets (coverage probability). The ideas of point and interval estimation for parameters generalize to the model-selection problem in a very straightforward fashion.

2. Nested models and the likelihood ratio test Let us generalize the situation to the case where the model space is somewhat more complex. Suppose we consider the collection of models to be that of the gamma distribution with parameters (α,β). It is known that the exponential model collection is a subset of the gamma model collection where $\alpha = 1$. The exponential model is said to be "nested" within the gamma model. Clearly, if we simply decide to choose a model that is closest according to the KL or any other divergence measure, we will always choose a complex model (gamma model) because such a model fits at least as well as any of its particular cases (exponential model). The best-fitting exponential model cannot have a greater likelihood value than the best-fitting gamma model. Now we need to decide when a complex model is worth its complexity. If the model collections are nested, the answer is available in terms of the likelihood ratio test (LRT). The likelihood ratio test first calculates the difference between support for the best-fitting values under two competing model collections; in our example the value of the likelihood function for the best-fitting exponential and the best-fitting gamma model and then tests whether that difference is statistically significant or not. In this formulation, the simple model (exponential) and the complex model (gamma) are a priori treated asymmetrically. The simple model corresponds to the null hypothesis, and a complex model is selected only if the LRT statistic is small or equivalently if its p-value is small. Under the nested structure, this p-value can be computed explicitly using the chi-square distribution. After the model is selected in this fashion, however, the confidence intervals for the parameters are computed only under the selected model, either the exponential or the gamma. Such confidence intervals do not account for uncertainty in model selection.

3. Nonnested model collections and the Akaike Information Criterion (AIC) Now suppose we want to fit either a gamma model collection or a log-normal model collection to the survival data. Because one model collection is not a particular case of the other, these model collections are not nested. In this situation, it is unclear which model collection should be considered preferable or be the null model collection. Thus, the concept of the null distribution of the likelihood ratio statistics does not apply. Hence, the LRT cannot be applied. (If one has an a priori preference for one of the model collections, we can use the p-value logic by considering parametric bootstrap under that collection to compute the null distribution.) Akaike in 1973 suggested selecting the model that minimizes the quantity, now called the Akaike Information Criterion (AIC), namely, AIC $= -2\log L(y;\hat{\theta}) + 2p$ where p denotes the number of parameters in the model.

Notice that the maximum likelihood estimation approach chooses the parameter (model) that minimizes the AIC. If the model collections are nested, AIC simply chooses that model collection that has observed chi-square value farthest from its theoretical mean under the null model space. Under the LRT formulation, this has the smallest p-value. Thus, the two concepts lead to similar inferences.

The difference $\mathrm{KL}(f,t)-\mathrm{KL}(h,t)$ tells us between the two models, f and h, which one is the closest to the true model t. The difference of AIC values (ΔAIC) has been justified as a model selection criterion by showing that it is an estimator of the difference between two KL divergences. If ΔAIC is an estimator of this difference, some natural questions arise. Is this estimator consistent? That is, as the sample size increases would we select the closest model? Is this an unbiased estimator? If not, can we reduce the bias? In general, can we construct better estimators of this difference than the ΔAIC?

The AIC is biased, tending to choose more complex models than the true model, at least for small sample sizes. This has led to many different information criteria, all with the same basic form as the AIC; that is, IC $= -2\log L(y;\hat{\theta}) +$ *complexity penalty*. For example, the Schwarz information criterion (SIC), also known as the Bayesian information criterion (BIC), and the Hurvich and Tsai criterion (AICc) correct for this by using penalty functions that depends on the sample size ($\ln(n)p$ and $2p(n/(n-p-1))$, respectively). The extended information criterion (EIC), on the other hand, corrects for the bias using bootstrap techniques.

The comparison of models with information criteria does not employ sharp cutoffs as in significance testing. There are ad hoc but generally accepted guidelines for interpreting ΔICs. If the ΔIC is less than 2, there is only weak evidence that the observed ranking of the models is correct. If $2 < \Delta\mathrm{AIC} < 4$ or $\Delta\mathrm{AIC} > 4$, the evidence for a correct model ranking is considered strong and very strong, respectively.

In regression analysis, covariate (model) selection is usually conducted by minimizing prediction error. This takes into account both the variance of the parameter estimates and the prediction bias. The AIC has been shown to be asymptotically equivalent to cross-validation prediction error at least for linear models. Generalizations of this idea are discussed by Bozdogan in an excellent 1987 review article in *Psychometrika*. Different weighting for prediction accuracy and estimation accuracy has not been explored in detail so far.

Given the close relationship between parameter estimation and model selection, it seems reasonable to consider unifying model estimation and selection by simply maximizing a penalized likelihood where the likelihood is penalized by model complexity. It is not always sensible to characterize the complexity of a model simply based on counting the number of parameters. In many ecological models, the parameters are highly dependent on each other, and the realized dimension of the parameter space is not always equal to the count of the parameters because of nonidentifiability. If the models are nearly nonidentifiable, the parameter space effectively, although not exactly, is of somewhat lower dimension. A model selection criterion called ICOMP, proposed by Bozdogan in 2000, selects the model that minimizes the quantity

$$\mathrm{ICOMP} = -2\log L(y;\hat{\theta}) + p \cdot \log\left[\frac{tr\left(\Sigma(\hat{\theta})\right)}{p}\right] - \log\left(\Sigma(\hat{\theta})\right),$$

where $\Sigma(\hat{\theta})$ denotes covariance matrix. Here, ICOMP stands for information complexity. The model complexity is characterized by the difference between the putative number of parameters and the effective number of parameters. This penalty can be made scale-invariant by considering correlation instead of the covariance matrix.

AN EXAMPLE APPLICATION

To illustrate the use of information criteria for model selection, we revisit the single-species population growth data from Gause's (1934) laboratory experiments with *Paramecium aurelia* with interest in the following questions: does the population exhibit density-dependent population growth, and if so, what is the form of density dependence? The observed growth rate for a population is calculated as $r_t = \ln(N_{t+1}/N_t)$. By definition, the growth rate of a population with density dependence is a function of population size, N_t (Fig. 1). Consequently, we model the population's dynamics by $r_t = g(N_t, \underline{\theta}) + v_t(\sigma)$, where g is a deterministic growth function, $\underline{\theta}$ is a vector of parameters, $v_t(\sigma)$ is an independent random

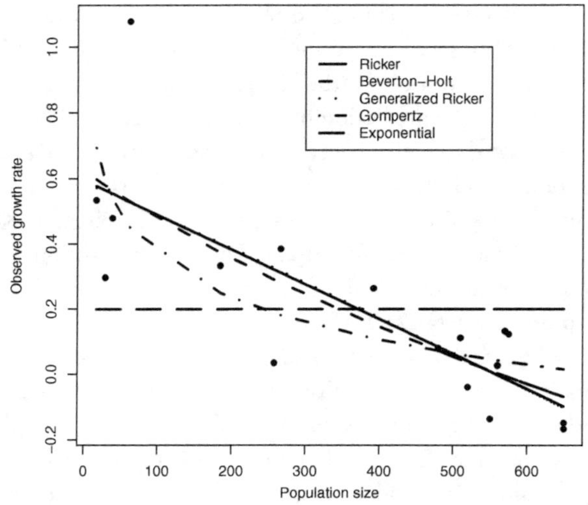

FIGURE 1 Observed population growth rate plotted population size. The lines are expected growth rates for five fitted growth models. The data (available online) are the first of three replicate time series for *Paramecium aurelia* given in G. F. Gause, 1934, *The Struggle for Existence*, Appendix 1, Table 3.

normally distributed environmental shock to the growth rate with mean 0 and standard deviation σ.

We use a suite of common population growth models:

Ricker: $g(N_t, \underline{\theta}) = r_i(1 - N_t/K)$
generalized Ricker: $g(N_t, \underline{\theta}) = r_i(1 - (N_t/K)^{\gamma})$
Beverton–Holt: $g(N_t, \underline{\theta}) = r_iK/(K + r_iN_t - N_t)$
Gompertz: $g(N_t, \underline{\theta}) = a(1 - \ln(N_t/K))$
density-independent exponential growth model: $g(N_t, \underline{\theta}) = r_i$.

These models have been parameterized in as similar a fashion as possible. K represents the equilibrium population size, and r_i is the intrinsic growth rate, or limit to growth rate as N_t approaches zero. In the Gompertz model, the parameter a also scales growth rate but is not quite the same thing as r_i, because in this model growth rate is mathematically undefined at zero.

The log-likelihood function for all of these models is

$$\log L(r_t, N_t; \underline{\theta}, \sigma) = \frac{\sum_{t=0}^{T-2} (g(N_t, \underline{\theta}) - r_t)^2}{2\sigma^2} - \frac{(T-1)\log(2\pi\sigma^2)}{2},$$

where T is the total number of population sizes observed. For the construction of information criteria, the number of parameters, p, is the length of the vector $\underline{\theta} + 1$, the addition of 1 for the parameter σ.

Table 1 is typical of the tables produced in information criteria analysis. It contains the log-likelihoods, the number of parameters, and, for several common criteria, the IC and ΔIC values. From this table one can make a number of observations that are very useful in framing our thinking about our driving questions. (1) The Ricker model is nested within the generalized Ricker, and the exponential within the Ricker. As dictated by theory, the generalized Ricker has the highest log-likelihood among these three models, but it is not the best model according to the information criteria. (2) Different information criteria favor different models with different degrees of strength. Both the SIC and the AICc indicate strong evidence that the generalized Ricker is not the best model. The evidence from the AIC is more equivocal than that registered by the other two criteria. This is an example of the tendency of the AIC to over-fit. Although not the case in this example, the rank order for some models can change between different criteria. (3) The Ricker model has the lowest IC value, but the difference with the Beverton–Holt model is small; thus the evidence that the Ricker model is superior to the Beverton–Holt is very weak, and both models should be considered for prediction and interpretation. (4) There are three classes of nonnestable models in this problem. Classical likelihood ratio tests do not compare across model families, and thus an information criterion based analysis allows a richer probing

TABLE 1
Information criteria values for 5 common population growth models fitted to Gause's 1934 *Paramecium aurelia* data

Model	Log Likelihood	# Parameters	AIC	AICc	SIC	ΔAIC	ΔAICc	ΔSIC
Ricker	4.90	3	−3.80	−1.96	−1.30	0.00	0.00	0.00
Beverton–Holt	4.82	3	−3.63	−1.79	−1.13	0.17	0.17	0.17
Generalized Ricker	4.91	4	−1.81	1.52	1.52	1.99	3.48	2.83
Gompertz	2.66	3	0.68	2.53	3.18	4.48	4.48	4.48
Exponential	−3.72	2	11.40	12.30	13.10	15.20	14.20	14.40

NOTE: ΔIC values are given relative to the best (minimum IC) model.

of nature. In this case, we see that the Beverton–Holt model is essentially indistinguishable in merit from the Ricker, at least on the basis of this data. (5) The ΔIC values are all >14 for the exponential model, confirming quantitatively what is visually obvious from the figure, that it is essentially impossible that *P. aurelia* is growing in a density-independent fashion under the conditions of Gause's experiment.

CURRENT DIRECTIONS AND FUTURE TOPICS

Model Selection for Hierarchical Models

These models form an important class of models that are useful in describing many ecological datasets. Unfortunately, the model selection approach for this class of models is still in its infancy. A few papers in this direction are beginning to appear in the literature, but with no clear consensus. The major problem with these models is that counting the number of parameters is unclear. Should the random effects be counted as parameters, or should only the variance components be counted in the number of parameters? The most commonly used criterion is the divergence information criterion (DIC). However, there are various studies that indicate that the DIC might not be an effective model selection criterion. Recently, Vaida and his colleagues have used conditional AIC, and Yafune and his colleagues have used EIC, as model selection criteria for mixed models. However, it is unknown how effective these are in actually choosing a hierarchical model. We suspect that Bozdogan's ICOMP criterion might be a good way to select mixed and hierarchical models because it sidesteps counting the number of parameters and focuses on the structure of the parameter estimate variance/covariance matrix for its complexity penalty.

Confidence Set for the Models

Using information criteria, one can provide a set of model spaces that may not be strongly distinguishable from each other, e.g., ΔAIC value less than a pre-decided value such as 2. Thus, one approach is to provide a confidence set. This is an interesting idea, although because there is no probabilistic structure across different model collections, there cannot be the concept of error probabilities associated with such a set. Under the Bayesian formulation of model selection, such a super-probability structure is assumed through the prior distribution, but it faces exactly the same difficult philosophical issues that any Bayesian analysis faces.

Model Averaging

An argument has been made that if there are many models that are not strongly distinguishable, one should consider a weighted average model with weights dependent on the AIC differences. However, the meaning of such an average model is not apparent. Because there is no super-probabilistic structure across model spaces, we cannot see any statistical justification for model averaging. Such a super-probabilistic structure can exist if we are willing to assume that there are different mechanisms underlying the observations and each mechanism is at work only a certain proportion of times. That is to say that the true model is in fact a mixture model with mixture proportions corresponding to the proportion of times a mechanism is applicable. In practice, model averaging is conducted in a statistically flawed fashion by averaging the parameter estimates or by holding one of the parameters equal to zero. These ideas do not seem to have any strong statistical foundation. Recalling the fundamental unity between parameter estimation and model identification discussed earlier, just as "mean likelihood estimator" has no statistical justification, there appears to be no real justification for the "mean model" either.

CONCLUSIONS

Information criteria have been beneficial to the practice of ecology because they promote the consideration of many models simultaneously and thus go to the heart of science in a more efficient and, importantly, less biased fashion. Information criteria do a good job of automatically matching model complexity with the available information in the data. However, it is important to realize that this matching is conditional on the set of models considered and is necessarily provisional. Further, as more data accumulate one should expect the "best" model to change. Perhaps, instead of trying to choose the "true" model, we should consider choosing a model that will answer the scientific questions of interest over a range of data conditions.

SEE ALSO THE FOLLOWING ARTICLES

Bayesian Statistics / Beverton–Holt Model / Frequentist Statistics / Model Fitting / Ricker Model / Statistics in Ecology

FURTHER READING

Akaike, H. 1973. Information theory and an extension of the maximum likelihood principle. Second International Symposium on Information Theory (B.N. Petrov and F. Csaki, eds.). Budapest: Akademiai Kiado.

Burnham, K.P., and D.R. Anderson. 2002. *Model selection and multimodel inference: a practical information-theoretic approach.* 2nd ed. New York: Springer-Verlag.

Claeskens, G., and N. L. Hjort. 2008. *Model selection and model averaging.* Cambridge, UK: Cambridge University Press.

Myung, I. J., M. R. Forster, and M. W. Browne. 2000. Special issue on model selection. *Journal of Mathematical Psychology* 44(1): 1–231.

Taper, M. L., and S. R. Lele, eds. 2004. *The nature of scientific evidence: statistical, philosophical and empirical Considerations.* Chicago: University of Chicago Press.

INTEGRATED WHOLE ORGANISM PHYSIOLOGY

ARNOLD J. BLOOM

University of California, Davis

Biologists—to paraphrase Justice Potter Stewart in *Jacobellis v. Ohio*—may know an organism when they see it, but most readers will benefit from a more precise definition. A simple definition is that an organism is an independent living unit that persists for a substantial period of time. Persistence through time may require an organism to reproduce. The following introduces the characteristics of organisms and the physiological mechanisms that many organisms employ to meet their need for independence, persistence, and reproduction.

CHARACTERISTICS OF ORGANISMS

Independence

An organism is self-sufficient in terms of (a) responses to external stimuli, (b) synthesis and organization of organic compounds through a biochemical metabolism, and (c) growth and development. Below are some examples of species that may lack sufficient independence to classify as organisms.

Viruses are usually not considered organisms, because they fail to satisfy the last two criteria for self-sufficiency because their metabolism, growth, and development depend on their host.

Obligate intracellular parasites such as the bacterium *Rickettsia* that causes typhus in humans and the protozoa *Plasmodium* that causes malaria are also less than organisms: although these parasites have their own metabolism, their growth and development depend on their hosts.

Obligate endosymbionts include the fungi that join with green algae or cyanobacteria to form lichens, the giant amoeba *Pelomyxa* that lacks mitochondria but has aerobic bacteria that carry out a similar role, and the marine worms of the genera *Olavius* or *Inanidrillus* that obtain their nutrition from the chemoautotrophic bacteria that fill their bodies. These meet the criteria for independence only together with their associated symbiont.

Some species are found only in colonies. For example, species in the genus *Volvox* are considered to be intermediaries between single-celled and multicellular organisms, where the colony assumes the status of an organism.

Persistence

Expanding the criteria for what constitutes an organism to include persistence for a substantial period of time can increase the ambiguity. Parts of some organisms may operate independently until they run out of resources. Here are two examples.

Some members of symbiotic associations may remain viable as independent units in the short term, but not in the long term. For example, corals often are associated with endosymbiotic photosynthetic protozoa called zooxanthellae. Under environmental stress such as high temperatures, corals expel their zooxanthellae and become colorless as they reveal the white of their calcium-carbonate skeletons, a process known as coral bleaching. Bleached corals usually expire within months unless the zooxanthellae reestablish.

Plant roots slough off individual cells from their root cap that remain intact for weeks in the surrounding soil. These so-called border cells secrete materials that attract beneficial soil microbes to the rhizosphere but deter pathogenic ones. Nonetheless, few would consider border cells to be organisms.

Reproduction

Persistence of an organism through time eventually entails reproduction, be it sexual or asexual. Several other entries in this volume focus on the theory of reproduction. Reproduction may be as straightforward as simple mitotic division or as complicated as an elaborate sexual life cycle. The various life stages of an organism may bear little resemblance to one another.

Slime molds of the genus *Dictyostelium,* for example, spend part of their life cycle as cellular amoebae that divide through mitosis. When food becomes limiting, they aggregate into a multicellular assembly, called a pseudoplasmodium or slug. The slug has a definite front and back orientation, responds as a unit to external stimuli such as light and temperature gradients, and has the ability to migrate. Under the correct conditions, the slug forms a fruiting body with a stalk supporting one or more balls of spores that it disperses widely. These spores, inactive cells protected by environmentally resistant cell walls, develop into new cellular amoebae upon exposure to food.

Sexual reproduction usually involves haploid and diploid life stages that are distinct in form and function. In animals such as ourselves, the diploid stage dominates in size and longevity, whereas in bryophytes (mosses, liverworts, and hornworts) the haploid dominates, and in many multicellular algae, the haploid and diploid stages are similar in size and shape. In some sense, both the haploid and diploid stages are parts of one organism.

Quaking aspen (*Populus tremuloides*) reproduces asexually via a process called suckering in which an individual stem can send out lateral roots that send up other erect stems that look just like individual trees. A colony of a single male aspen that contains more than 47,000 stems interconnected by a common root system covers 43 hectares in south-central Utah and may be over 80,000 years in age (Fig. 1).

Some eusocial species—prominently insects in the orders Hymenoptera (e.g., bees, ants, and wasps) and Isoptera (e.g., termites)—form colonies in which only one female is fertile and the vast majority of the colony are sterile workers or soldiers that build, maintain, and protect the colony.

This brings us to the distinction between genet and ramet that is common in plant population biology: a genet is a group of genetically nearly identical individuals that originate from the reproduction of a single ancestor, whereas a ramet is each separate individual from such a population. For instance, the colony of aspens or bees would be the genet, whereas the stem of an aspen or the worker bee would be the ramet. From a genetic perspective, the genet is the organism, but from a physiological perspective, the ramet is the organism. The following focuses on physiology and, thus, treats the ramet as the organism.

FIGURE 1 Aspen grove composed of a single male aspen in Fishlake National Forest in south-central Utah. Photograph courtesy of the U.S. Forestry Service.

REQUIREMENTS OF ORGANISMS

To survive, organisms need to acquire resources from their environment, reproduce, and sustain their well-being.

Resources

Resources vital to living organisms fall into three categories: energy (carbon or sunlight), water, and nutrients (Fig. 2). Organisms may expend stores of one resource to acquire additional amounts of another resource if this second resource limits their growth. For example, plants will allocate more carbohydrates (energy) on root growth when water or nutrients are scarce. Conversely, plants may lose more water to transpiration or allocate more nitrogen to their photosynthetic apparatus when sunlight (energy) is limiting.

ENERGY

Many structures and functions within organisms are viable only in a chemical environment that is far removed from chemical equilibrium. To sustain this nonequilibrium state, organisms require a selective barrier between themselves and their surroundings and the continual input of energy. Autotrophic organisms either convert sunlight into chemical energy or oxidize inorganic compounds such as ferrous iron, ammonia, sulfite, or hydrogen to generate energy. In contrast, heterotrophic organisms ingest and then catabolize organic compounds synthesized by other organisms to generate energy. One approach for accounting for these energy expenditures is to trace the flow of carbon through an organism.

WATER

Life on Earth is based on chemical reactions conducted in a well-defined aqueous medium. Aquatic organisms must maintain an appropriate internal environment that is usually distinct from the surrounding bodies of water. Terrestrial organisms, although they may have better access

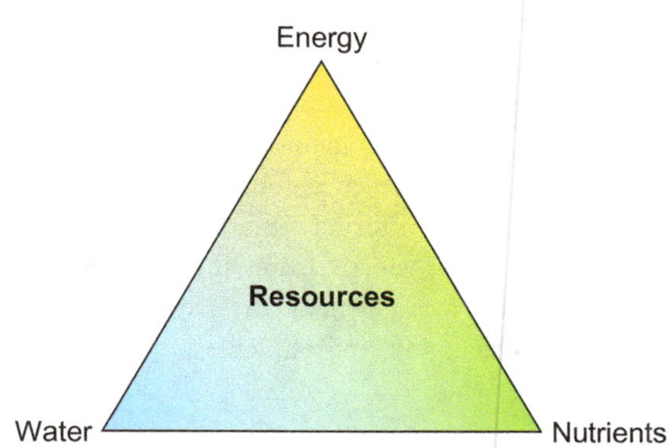

FIGURE 2 Resource Triangle: three categories of natural resources that organisms must obtain from their environment.

to energy (less attenuated sunlight and higher concentrations of carbon dioxide), have the additional challenge of water acquisition and retention under conditions that are often extremely desiccating.

NUTRIENTS

The materials that constitute an organism are composed of hydrogen and oxygen derived from water, carbon from CO_2 in the atmosphere, mineral elements such as nitrogen from ions dissolved in the medium, and organic compounds from what it ingests. The concentrations of nutrients within an organism usually deviate strongly from the concentrations in its surroundings. Therefore, an organism selectively acquires or excludes nutrients.

Reproduction

Reproduction affords organisms an opportunity not only to replicate themselves but also to renew, protect, distribute, and adapt themselves. Most cells demonstrate a limited ability to divide because of telomere shortening or more general wear and tear; the passage through ontogeny tends to reset the cellular clock and renew the species. In many organisms, one stage of the life cycle associated with reproduction—for example, seeds, spores, and pupae—is more resistant to stress than others and protects the species from adverse conditions. Moreover, in many organisms, one stage of the life cycle associated with reproduction—for example, seeds, spores, pollen, and sexually active adult—is more mobile than others and promotes the geographical distribution of the species. Finally, genetic variation or morphological reallocation during reproduction is the basis of adaptation to a changing environment. Reproduction, as discussed in the previous sections, may include sexual and asexual mechanisms.

SEXUAL

The main two processes in sexual reproduction are meiosis, a type of cell division in which the number of chromosomes per cell halve, and fertilization, in which two cells fuse and restore the original number of chromosomes per cell. During meiosis, homologous recombination can occur, in which chromosome pairs exchange DNA. Offspring from sexual reproduction, thus, receive a unique mixture of genetic material from their parents. This results in greater genetic diversity that promotes adaptation to changing environments.

ASEXUAL

In asexual reproduction, a single parent produces offspring via a number of mechanisms, including suckering (introduced in the previous discussion of aspen). Although asexual reproduction does not increase genetic diversity, it still serves to replicate, renew, protect, distribute, and reallocate resources in an organism. This benefits fast growing populations in stable environments where rapid adaptation may be less imperative.

Well-Being

Well-being for an organism entails maintaining a healthy and comfortable lifestyle. This implies that an organism avoids biotic and abiotic stresses that significantly inhibit its productivity. Biotic stresses include competitors and pathogens. Abiotic stresses include resource deficiencies and exposure to extreme environmental parameters such as high or low temperatures, drought or flooding, and salinity or acidity. Seldom is any organism stress-free, but an organism can minimize the amount of stress it encounters through continual space- and time-management and appropriate resource allocation.

PHYSIOLOGICAL MECHANISMS

Biologists have thoroughly examined the major mechanisms that most organisms employ to acquire resources, reproduce, and sustain well-being, but information about variations on these themes or about novel mechanisms are still sparse. For example, something as fundamental as the genetic code has variations whereby different organisms and even the same organism may translate one codon of three bases into different amino acids. Most organisms also have many variants of a single enzyme (isozymes) that perform similar functions but under different optimal conditions, different stages of development, or different compartments. How such variation optimizes performance and thus fitness of an organism is still an area of active research.

Often underappreciated is the role of compartmentalization via a selectively permeable membrane. Compartmentalization separates an organism from its surroundings and sustains the highly organized internal environment necessary for its survival. Compartmentalization into tissues and cells in multicellular organisms allows specialization. Compartmentalization into organelles within a cell contains highly reactive and potentially dangerous materials in specialized structures such as mitochondria and chloroplasts, removes wastes in peroxisomes and vacuoles, and packages materials for excretion in Golgi apparati.

Consider that in *Hemophilus influenzae,* the first prokaryote for which the entire genome was sequenced, more than 10% of the 1743 genes code for transport proteins to move materials between compartments. Similarly, as many as one-third of the 6000 genes in yeast code for transport proteins. Somewhere between 15% and 39% of the human proteome are membrane proteins. That such a large proportion of the genome or proteome serves for transport and membrane functions attests to the importance of compartmentalization.

Energy Mechanisms

AEROBIC RESPIRATION AND EVOLUTION

Our solar system, including planet Earth, began to coalesce about 4.5 Ga (billion years ago). When the surface of primitive Earth cooled below the boiling point of water (100 °C) about 3.8 Ga, the atmosphere consisted primarily of gases released from volcanoes including high concentrations of carbon dioxide (CO_2), carbon monoxide (CO), water vapor (H_2O), dinitrogen (N_2), and hydrogen chloride (HCl), and small amounts of methane (CH_4). Soon appeared the first organisms on Earth, chemotrophs, which generate chemical energy from the removal of electrons from methane, hydrogen sulfide, elemental sulfur, ferrous iron, molecular hydrogen, and ammonia.

Within a relatively short time, as early as 3.6 Ga, photosynthetic bacteria (cyanobacteria) were present to such an extent that the oxygen (O_2) released during photosynthesis rusted (oxidized) iron near the Earth's surface, thereby producing layers of rock called *banded iron formations.* These cyanobacteria use solar energy to convert a low-energy carbon in CO_2 into a high-energy carbon in carbohydrate (CH_2O) via a mechanism that splits water (H_2O) and generates dioxygen (O_2):

$$CO_2 + H_2O + \text{light} \rightarrow CH_2O + O_2.$$

Cyanobacteria proliferated widely during the next billion years, especially after 2.7 Ga. Their photosynthesis depleted CO_2 in the atmosphere to below 10% and released sufficient O_2 to exhaust minerals such as ferrous iron and elemental sulfur near the Earth's surface and CH_4 in the atmosphere.

Sometime between 2.45 and 2.22 billion years ago, atmospheric O_2 concentrations jumped from less than 0.02% to around 3% in what has been termed the Great Oxidation Event. This had major biological repercussions. Many simple life forms, which had theretofore dominated Earth, could not tolerate exposure to high concentrations of O_2. Today, descendants of such organisms are limited to the few remaining hypoxic environments on Earth such as bogs.

Eukaryotic organisms appeared after the height of the Great Oxidation Event. This was no coincidence. Cells of eukaryotes contain mitochondria, peroxisomes, and chloroplasts that isolate O_2 reactions. Aerobic respiration, O_2-based catabolism of organic compounds, generates far more energy than its anaerobic alternatives. In aerobic respiration, CH_2O reacts with O_2:

$$CH_2O + O_2 \rightarrow CO_2 + H_2O + \text{chemical energy (more than 5 ATP per carbon).}$$

One pathway of anaerobic respiration involves the breakdown of CH_2O to ethanol:

$$3CH_2O \rightarrow C_2H_5OH + CO_2 + \text{chemical energy (less than 1 ATP per carbon).}$$

The energy bonanza of aerobic respiration fostered the development of more complex organisms. Additionally, higher O_2 levels gave rise to an ozone (O_3) layer in the upper atmosphere that screened out ultraviolet radiation harmful to life; eukaryotes could complete relatively long life cycles without extensive ultraviolet damage.

At around 0.57 Ga, Earth's climate became warm, photosynthetic organisms proliferated, and atmospheric O_2 concentrations climbed to about 15%. Multicellular organisms appeared. Again this was not a coincidence. At higher O_2 concentrations, parts of organisms could be buried deeper within the organism and still receive sufficient O_2 to conduct aerobic respiration. Also, the thicker ozone (O_3) layer in the upper atmosphere afforded greater UV protection, and organisms could complete still longer life cycles.

ENERGY COMPOUNDS

Nearly all organisms use ATP, NADH, and other phosphorylated nucleotides as high-energy transfer agents. These compounds are too volatile, however, for large-scale energy storage or for transport across biological membranes. Consequently, most organisms employ a mix of carbohydrates, lipids, and organic acids for these purposes. For example, organisms shuttle reducing power among organelles or tissues in the form of malic acid because it is stable but readily converts to oxaloacetate generating NADH where and when needed.

Mechanisms for Water and Nutrients

Physiological mechanisms for acquiring and retaining water are often identical or, at least, similar to those involved in acquiring and retaining nutrients. Physical processes such as diffusion, mass flow, adsorption, and capillary action can be involved, but are not sufficient because they lack the specificity to maintain the nonequilibrium conditions of life. More important is active ion transport, the process through which organisms expend metabolic energy to translocate particular ions between compartments and thereby maintain an appropriate water balance.

Active transport includes primary and secondary active transport. In primary active transport, an organism expends energy to alter the conformation of a transporter that moves an ion (e.g., a proton pump driven by the hydrolysis of ATP). In secondary active transport, an organism expends energy to produce a local deformation of the electrochemical gradient between compartments, and this deformation drives the movement of an ion through a cotransporter or channel (e.g., potassium-chloride cotransport). Water flows between compartments according to its electrochemical gradient because biological membranes contain multiple water channels (aquaporins) that usually keep water permeability through membranes high.

Even if an organism has the right equipment in terms of transport mechanisms, its acquisition of water and nutrients may be unsuccessful unless it is in the right place at the right time. Organisms must position themselves to optimize their resource acquisition. This can also be considered to be optimal foraging.

For example, oceans are generally nutrient poor. When marine organisms eliminate wastes or when they die, the nutrients that they contain eventually sink and collect in deep waters. Ocean currents at high latitudes turn over deep waters, bringing their nutrients to the surface. These upwellings promote algal blooms, striking increases in algal populations that support large swarms of krill. Baleen whales spend extended periods in the high latitude oceans feeding on these swarms.

Plant roots proliferate in soil patches that are relatively rich in nutrients or water (Fig. 3). This proliferation results from additional resources being available for root growth as well as from chemotropism, the reception of and response to chemical signals. Thus, plants use directional growth to optimize resource foraging.

Reproductive Mechanisms

Physiological mechanisms for reproduction, although diverse, have several elements in common. First, there are mechanisms for duplicating the genetic material; this involves DNA replication and mitotic or meiotic cell divisions. Second, there are mechanisms for provisioning

FIGURE 3 A 20-day-old barley plant grown in a hydroponic system that supplied the middle root zone with a nutrient solution containing 1.0 mM nitrate; the top and bottom received 0.01 mM nitrate (Drew and Saker, 1975).

the offspring with enough supplies to support them until they can become self-sufficient. Third, there are mechanisms for protecting the offspring from the cold, cruel world until they can better defend themselves; nonetheless, offspring mortality is usually much greater than that of any other life stage.

Mechanisms for Well-Being

Organisms have a variety of mechanisms to distinguish and protect themselves from others who might threaten their health. The first line of defense is to keep out of harm's way, that is, to recognize and avoid dangerous situations if at all possible. The second line may be a mechanical barrier to the external world such as leaf cells with a waxy cuticle, skin cells of mammals, and insect exoskeleton. Once an organism recognizes that an invader has breeched its mechanical barrier, the organism may unleash mechanical, chemical, and biological warfare. Blood, sweat, and tears as well as coughing, sneezing, and urinating may forcibly expel pathogens. Mucus or slime may entangle microorganisms. Chemical warfare includes secretion of a variety of bactericidal and fungicidal proteins. Immune responses, a form of biological warfare, range from the production of leukocytes to apoptosis.

Finally, a comfortable lifestyle depends on the ability of an organism to anticipate environmental changes and behave proactively. Circadian rhythms found in most organisms serve this purpose. Also common is orientation to environmental cues such as light, smells, magnetic fields, temperature, and carbon dioxide concentration. Organisms avoid discomfort through movement and may transition to another phase of their life cycle when moving proves inadequate. The reader, as a theoretical organism, will hopefully feel comfortable with the general tendencies outlined here, or for their own well-being, will move to other parts of the book or transition to other phases of their life cycle.

INTEGRATION OF ORGANISMAL FUNCTIONS

An organism may be greater than the sum of its parts, but even enumerating all the parts of an organism proves daunting. Walter Elsasser argued that the structural complexity of even a single living cell is beyond the power of any imaginable system to compute. Still, models of organismal growth and development based on submodels of organismal functions already provide useful approximations for management of diverse endeavors including crop production, fisheries, public health, and wildlife conservation and restoration. A greater understanding of organismal functions and their interactions will undoubtedly lead to better approximations.

SEE ALSO THE FOLLOWING ARTICLES

Energy Budgets / Functional Traits of Species and Individuals / Movement: From Individuals to Populations / Mutation, Selection, and Genetic Drift / Sex, Evolution of / Stress and Species Interactions / Transport in Individuals

FURTHER READING

Bloom, A. J. 2010. *Global climate change: convergence of disciplines.* Sunderland, MA: Sinauer Associates.

Bloom, A. J., F. S. Chapin, III, and H. A. Mooney. 1985. Resource limitation in plants: an economic analogy. *Annual Review of Ecology and Systematics* 16: 363–392.

Bullock, J. M., R. Kenward, and R. Hails, eds. 2002. *Dispersal ecology: the 42nd Symposium of the British Ecological Society held at the University of Reading, 2–5 April 2001.* Malden, MA: Blackwell Publishers.

Epstein, E., and A. J. Bloom. 2005. *Mineral nutrition of plants: principles and perspectives,* 2nd ed. Sunderland, MA: Sinauer Associates.

Harper, J. L. 1977. *Population biology of plants.* London: Academic Press.

Hawes, M. C., L. A. Brigham, F. Wen, H. H. Woo, and Z. Zhu. 1998. Function of root border cells in plant health: pioneers in the rhizosphere. *Annual Review of Phytopathology* 36: 311–327.

Jones, S. E., and J. T. Lennon. 2010. Dormancy contributes to the maintenance of microbial diversity. *Proceedings of the National Academy of Sciences of the United States of America* 107: 5881–5886 (DOI:DOI 10.1073/pnas.0912765107).

Mitton, J. B., and M. C. Grant. 1996. Genetic variation and the natural history of quaking aspen. *Bioscience* 46: 25–31.

Morowitz, H. J. 1970. *Entropy for biologists: an introduction to thermodynamics.* New York: Academic Press.

Somero, G. N., C. B. Osmond, and L. Bolis. 1992. *Water and life: comparative analysis of water relationships at the organismic, cellular, and molecular levels.* New York: Springer-Verlag.

INTEGRODIFFERENCE EQUATIONS

MARK KOT
University of Washington, Seattle

MARK A. LEWIS
University of Alberta, Edmonton, Canada

MICHAEL G. NEUBERT
Woods Hole Oceanographic Institution, Massachusetts

Many species have distinct growth and dispersal stages. Some annuals, for example, germinate in spring, flower in summer, and spread their seeds in autumn. Can we predict the growth and spread of such species? One approach is to use integrodifference equations (IDEs). IDEs are nonlocal models that treat time as discrete, space as continuous, and state variables, such as population density, as continuous.

FORMULATION

A deterministic IDE for a single, unstructured population living on a one-dimensional spatial habitat is typically written

$$n_{t+1}(x) = \int_{\Omega} k(x, y)\, f(n_t(y))\, dy. \qquad (1)$$

The state variable $n_t(x)$ is the population density at location x and time t. Space is continuous, but time is discrete. Thus, x and y are real numbers, but t is an integer. Ω is the spatial domain for the population and the model. This domain may be finite, e.g., the interval 0 to 1. For some problems, however, it is convenient to take the domain and the limits of integration as infinite, $\Omega = (-\infty, +\infty)$.

Equation 1 maps the density of one generation to the density in the next generation in two distinct stages (Fig. 1). During the first or sedentary stage, individuals may grow, reproduce, or die. At each point x, the local population, of density $n_t(x)$, produces propagules of density $f(n_t(x))$.

During the second or dispersal stage, the propagules move. The dispersal kernel, $k(x, y)$, describes this movement. For a fixed source y, the kernel is a probability density function for the destination, x, of the propagules.

In general, the dispersal kernel, $k(x, y)$, can depend upon the source y and the destination x in some complicated way. To simplify matters, ecologists often assume a difference kernel,

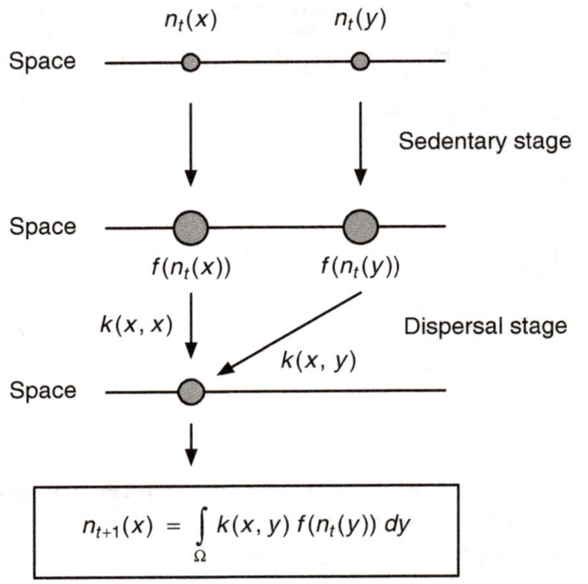

FIGURE 1 Stages of an integrodifference equation. During the sedentary stage, individuals may grow, reproduce, or die, in place, leaving propagules. The sedentary stage is described by a simple growth function or recruitment curve that maps $n_t(x)$ into $f(n_t(x))$. During the dispersal stage, the propagules disseminate. For a fixed source y, the dispersal kernel $k(x, y)$ is the probability density function for the destination x of the propagules.

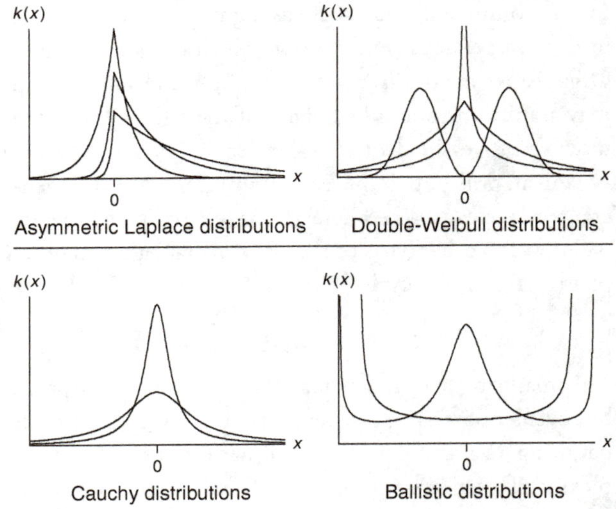

FIGURE 2 Disperal kernels. Difference kernels come in a variety of shapes and sizes. In each subfigure, a family of dispersal kernels was plotted for two or more pairs of shape and scale parameters. Many other dispersal kernels have also been observed and used.

$$k(x, y) = k(x - y), \tag{2}$$

so that the integrodifference equation,

$$n_{t+1}(x) = \int_\Omega k(x - y) f(n_t(y))\, dy, \tag{3}$$

is built around a convolution integral. A difference kernel is the probability density function for the displacement of the propagules.

Dispersal kernels are frequently estimated from seed shadows, plant-disease dispersal gradients, mark-recapture studies, and other empirical data; they may also be derived from first principles. Because dispersal kernels assume a wide variety of shapes (Fig. 2), IDEs can accommodate many dispersal mechanisms.

Additional spatial dimensions, population age or stage structure, and interspecific interactions are easily incorporated into the IDE framework.

HISTORY

The study of IDEs is closely related to that of random walks and, more specifically, to that of branching random walks. Integrodifference equations are also implicit in the mid-twentieth-century writing of John Gordon Skellam. Skellam formulated the critical patch size for a population in two dimensions as an integral equation. This integral equation may be thought of as the steady-state equation for an integrodifference equation.

IDEs first appeared as explicit biological models in population genetics. In 1973, Montgomery Slatkin used an integrodifference equation to study clines. He also introduced an elegant method for characterizing the equilibria of integrodifference equations. Several years later, Hans Weinberger and Roger Lui introduced IDEs more formally and developed methods for determining the wave speeds of spreading populations.

Integrodifference equations entered ecology in the 1980s and 1990s. IDEs quickly proved useful for modeling the spread of organisms with leptokurtic or heavy-tailed dispersal kernels and they gained popularity as one way of resolving Reid's paradox of rapid plant migration.

Ecologists have used integrodifference equations to study the effects of age and stage structure, Allee effects, biological invasions, chaos, competition, epidemics, the effects of fluctuating environments, long-term transients, pattern formation, persistence, pest and weed control, phytopathology, resource-dependent growth and dispersal, and many other phenomena. IDEs have been applied to many organisms, ranging from bacteria to birds. Theory developed for IDEs has also shed light on the behavior of other types of models, including cellular automata and coupled map lattices.

CORE PROBLEMS

Much research on integrodifference equations focuses on a few key problems. These problems resemble those that arise in the study of reaction–diffusion equations. Indeed, a major goal of the study of IDEs is to understand how the different assumptions that underlie reaction–diffusion models and IDEs lead to different behaviors and outcomes.

Population Persistence

What is the critical patch size for a population? That is, how much habitat does a population need to persist, given that some individuals leave due to dispersal and perish? Theoretical ecologists often analyze this problem by studying the stability of the equilibria of the integrodifference equation

$$n_{t+1}(x) = \int_{-L/2}^{+L/2} k(x-y)\, f(n_t(y))dy \qquad (4)$$

on a finite domain of length L. For a population that can grow at low densities on an infinite domain, the critical patch size L_c depends on the eigenvalues that determine the stability of the trivial (zero) equilibrium for the finite domain. For small patches, $L < L_c$, these eigenvalues are all less than 1 in magnitude and the population dies out. For large patches, $L > L_c$, at least one eigenvalue is bigger than 1 and the population persists. For a given recruitment curve, different dispersal kernels generate different critical patch sizes.

Range Shifts

Can a population keep pace with climate-induced range shifts? This problem is closely related to the critical patch size problem except that the patch now moves because of climate change. In the simplest case, one studies an integrodiifference equation,

$$n_{t+1}(x) = \int_{-L/2+ct}^{+L/2+ct} k(x-y)\, f(n_t(y))dy, \qquad (5)$$

in which the the patch moves with constant range-shift speed c. A population that persists when the patch is fixed may die if c is too large. A population's response to a given range-shift speed depends on its growth function and its dispersal kernel.

Spread Rates

How quickly does an introduced or an invading population spread? Ecologists have gained insight into this problem by considering IDEs on an infinite domain,

$$n_{t+1}(x) = \int_{-\infty}^{+\infty} k(x-y)\, f(n_t(y))dy. \qquad (6)$$

IDEs can, like reaction–diffusion equations, generate constant-speed traveling waves. In the simplest cases, wave speeds depend on the net reproductive rate of the population and on the moment generating function of the dispersal kernel. If, however, an IDE has a dispersal kernel that is heavy-tailed, it may instead generate accelerating invasions. This result has led to keen interest in long-distance dispersal in ecology.

Pattern Formation

Can spatial patterns in population density arise from trophic interactions and dispersal in homogeneous environments? Consider, for example, an IDE predator–prey system of the form

$$n_{t+1}(x) = \int_{-\infty}^{+\infty} k_1(x-y)\, f(n_t(y), p_t(y))\, dy, \qquad (7a)$$

$$p_{t+1}(x) = \int_{-\infty}^{+\infty} k_2(x-y)\, g(n_t(y), p_t(y))\, dy, \qquad (7b)$$

where $n_t(x)$ is the density of the prey and $p_t(x)$ is the density of the predator. Under what circumstances do differences in the dispersal of the predator and its prey destabilize spatially uniform solutions? In general, integrodifference equations exhibit dispersal-driven instability under a broader set of ecological conditions than do reaction–diffusion models.

Stage Structure

How do age or stage structure, in growth and in dispersal, affect the answers to the above questions? Stage structure is added to a simple IDE by replacing Equation 1 with the more general model

$$\mathbf{n}(x, t+1) = \int_{\Omega} [\mathbf{K}(x-y) \circ \mathbf{B}(\mathbf{n}(y,t))]\mathbf{n}(y,t)\, dy. \qquad (8)$$

Here, $\mathbf{n}(x, t+1)$ is a vector of population-stage densities. We now use matrices to describe the transitions and transformations that occur during the growth and dispersal intervals: $\mathbf{B}(\mathbf{n}(y, t))$ is a density-dependent population projection matrix, $\mathbf{K}(x-y)$ is a matrix of stage-specific dispersal kernels, and "$\circ$" is the Hadamard or element-by-element matrix product. It appears that differences in life cycles can indeed have a profound effect on species' abilities to persist, shift their ranges, invade, and interact with other species.

In summary, IDEs shed light on many fascinating questions concerning the spatial dynamics of populations and can be formulated to include key aspects of nonlinear growth, dispersal, community interactions, and stage structure.

SEE ALSO THE FOLLOWING ARTICLES

Demography / Dispersal, Plant / Invasion Biology / Reaction–Diffusion Models / Spatial Spread / Stage Structure

FURTHER READING

Clark, J. S. 1998. Why trees migrate so fast: confronting theory with dispersal biology and the paleorecord. *American Naturalist* 152: 204–224.

Clark, J. S., C. Fastie, G. Hurtt, S. T. Jackson, C. Johnson, G. A. King, M. Lewis, J. Lynch, S. Pacala, C. Prentice, E. W. Schupp, T. Web III, and P. Wycko. 1998. Reid's paradox of rapid plant migration: dispersal theory and interpretation of paleoecological records. *Bioscience* 48: 12–24.

Hart, D. R., and R. H. Gardner. 1997. A spatial model for the spread of invading organisms subject to competition. *Journal of Mathematical Biology* 35: 935–948.

Kot, M., and W. M. Schaffer. 1986. Discrete-time growth–dispersal models. *Mathematical Biosciences* 80: 109–136.

Kot, M., M. A. Lewis, and P. van den Driessche. 1996. Dispersal data and the spread of invading organisms. *Ecology* 77: 2027–2042.

Lee, C. T., M. F. Hoopes, J., Diehl, W. Gilliland, G. Huxel, E. V. Leaver, K. McCann, J. Umbanhowar, and A. Mogilner. 2001. Non-local concepts and models in biology. *Journal of Theoretical Biology* 210: 201–219.

Lewis, M. A., M. G. Neubert, H. Caswell, J. Clark, K. Shea. 2006. A guide to calculating discrete-time invasion rates from data. In M. W. Cadotte, S. M. McMahon, and T. Fukami, eds. *Conceptual ecology and invasions biology: reciprocal approaches to nature.* Dordrecht, Netherlands Kluwer.

Neubert, M. G., and H. Caswell. 2000. Demography and dispersal: calculation and sensitivity analysis of invasion speed for structured populations. *Ecology* 81: 1613–1628.

Neubert, M. G., M. Kot, and M. A. Lewis. 1995. Dispersal and pattern-formation in a discrete-time predator–prey model. *Theoretical Population Biology* 48: 7–43.

Veit, R. R., and M. A. Lewis. 1996. Dispersal, population growth, and the Allee effect: dynamics of the house finch invasion of eastern North America. *American Naturalist* 148: 255–274.

INVASION BIOLOGY

MARK A. LEWIS

University of Alberta, Edmonton, Canada

CHRISTOPHER L. JERDE

University of Notre Dame, Indiana

Exotic or alien species come from foreign locations. Humans introduce most, but not all, exotic species. Some of these are beneficial to humans, such as crops. However, others have harmful effects. If an exotic species manages to survive, reproduce, spread, and finally harm a new environment, it is called *invasive*. The impacts of invasions are varied. Invaders can change habitats and ecosystem dynamics, crowd out native species, or damage human activities. Invasive species rank only second to habitat destruction as a threat to biodiversity. One estimate of the direct and indirect economic impacts of invasive species is $137 billion per year in the United States alone. The process of invasion, while dramatic, occurs in the context of dispersal of individuals, population growth, perturbations of biological communities, and changes to the social value and economic benefits of ecosystems. Invasion biology provides a new lens through which we can observe and understand population and community ecology.

HYPOTHESES TO EXPLAIN BIOLOGICAL INVASIONS

Why do some invasive species have the ability to thrive in a new environment? Numerous hypotheses have been put forward. The simplest hypothesis is that successful invaders may have attributes that make them competitively superior to the native communities they invade. They are inherently more successful as colonizers, competitors, or predators (inherent superiority hypothesis). A well-known case is the brown tree snake, a very effective predator of birds, which effectively eliminated 10 out of 11 native species of birds when introduced to Guam. Alternatively, the invaders may have "novel weapons" that are unfamiliar to the native species, weapons that make them more effective than natives at fighting off predators or catching prey (novel weapons hypothesis). Our understanding of novel weapons comes from plant species that produce allelopathic chemicals that prevent or inhibit the growth of other species. Also, preadaption may make an invader superior when it is introduced to agricultural or disturbed habitats (preadaption/disturbance hypothesis). For example, species that are long associated with human disturbance or agriculture may thrive when introduced to newly disturbed habitats. This is the case for some birds, such as house sparrows, and for many European species of weeds that can thrive when released into agricultural habitats of the new world.

Shifting the focus to community attributes, successful invasion may depend on who is and, importantly, who is not in the new environment encountered by invaders. If there are no species occupying similar niches, the invader may use unexploited resources, establishing itself in a new niche (empty niche and fluctuating resources hypotheses). The invasion process may also filter out certain parasites, diseases, or predators. These natural enemies, at least initially, are unlikely to make the same journey as the invader. Released from these enemies, the invader can then be a more effective competitor against the native species, which must also contend with their own native enemies. This is known as the enemy release hypothesis.

Successful establishment of an invader depends largely on the incipient interactions upon introduction to a new environment. Strong competitive interactions with native species can reduce the chances of invasion success and limit invader impacts. Hence, the biotic resistance hypothesis states that simpler communities, such as those found in disturbed habitats, islands, or isolated desert springs, will have fewer competitors to resist the invader and will be more susceptible to invasion and its environmental impacts. Islands have borne the brunt of many invasion processes, providing evidence for this hypothesis. By way of contrast, the invasional meltdown hypothesis focuses on mutualistic interactions between successive invaders. The more often a community is invaded, the more susceptible it is to further invasion. Many species are dependent upon mutualists at a crucial life history stage. For example, plants may require specialized pollinators, or seeds may require specialized birds to disperse their seeds. There is evidence that mutualisms facilitate invasions, and some introduced species have been known to remain latent as invaders until their mutualistic counterpart is introduced.

Evolutionary pressures can change when invasion occurs. For example, when an invader is released from enemies, an evolutionary loss of the invader's defenses is possible. The evolution of increased competitive ability hypothesis is a corollary of the enemy release hypothesis. It states that such an evolutionary loss of defenses allows for resources to be directed toward growth and reproduction or toward other factors that influence fitness of the invader. The end result is a more competitive invader. The invader may also hybridize with native species. Although this may result in a temporary decrease in fitness, it could confer long-term benefits, such as the ability to cope with local environmental conditions. The hybridization hypothesis states that hybridization can increase the invasiveness of exotic species by generating genetic variability and hybrid vigor. Finally, founder events can play a key role in the evolutionary development of an invader. Because few individuals are actually introduced into the new environment, they may be, by chance, different from their parent population. The founder hypothesis states that successive breeding of the small initial population could lead to rapid evolution, or even speciation. This is the case, presumably, with the North American invasion of Argentine ants.

MODELING THE DYNAMICS OF BIOLOGICAL INVASIONS

Number of Invaders over Time

In many regions, invader richness appears to be accelerating with time. By itself, this may not reflect increased introduction rates or invasion success. One approach to understanding the process is to create a null model to identify the expected trend in invasion records over time. This null model assumes that the rates of introduction and invasion success are constant. The model can give rise to increasing rates of discovery, because it assumes that the discovery process itself is imperfect and it may take many years to discover a new invader. Such models have been used to understand observations of increasing global invasion rates in freshwater, marine, and terrestrial environments. Inferences resulting from null and alternative models of invader richness are critical for evaluating the effectiveness of economic and policy changes (see also the section "Bioeconomics," below).

Population Spread Models

Early invasion models assumed homogeneous environments in which individual organisms reproduced via logistic growth and dispersed via diffusion. These models yielded a compact formula for the rate of spread of an invader.

Work by Fisher (1937) on the rate of spatial spread of an advantageous gene forms the basis of our modern-day understanding of invasive spread. The frequency of an advantageous allele p in a homogeneous environment is modeled to change at a rate defined by

$$\underbrace{\frac{\partial p}{\partial t}}_{\text{rate of change of allele density}} = \underbrace{mpq}_{\text{production of new alleles}} + \underbrace{D\frac{\partial^2 p}{\partial x^2}}_{\text{diffusive spread}}. \quad (1)$$

Here, x is a spatial location, t is time, m is the selection coefficient favoring the allele $p(x, t)$, and $q = 1 - p$ is the frequency of the wild type allele. The diffusion coefficient D describes random (Brownian) movement of the mutant allele through the landscape and can be understood in terms of the mean squared displacement per unit time (MSD) by $D = \text{MSD}/2$.

Fisher's equation, recast in an ecological context, involves logistic growth and random motion and is given by

$$\underbrace{\frac{\partial n}{\partial t}}_{\text{rate of change of local density}} = \underbrace{rn\left(1 - \frac{n}{K}\right)}_{\text{logistic growth}} + \underbrace{\frac{\partial}{\partial x}\left(D\frac{\partial n}{\partial x}\right)}_{\text{diffusive spread}}, \quad (2)$$

where $n(x, t)$ describes the population density as a function of space and time, r is the intrinsic growth rate, and K is the carrying capacity. This equation, or variants of it, has been applied widely to understand the rate of spatial spread of invaders (Fig. 1). Some of the earliest work in this area was by Skellam (1951), who considered

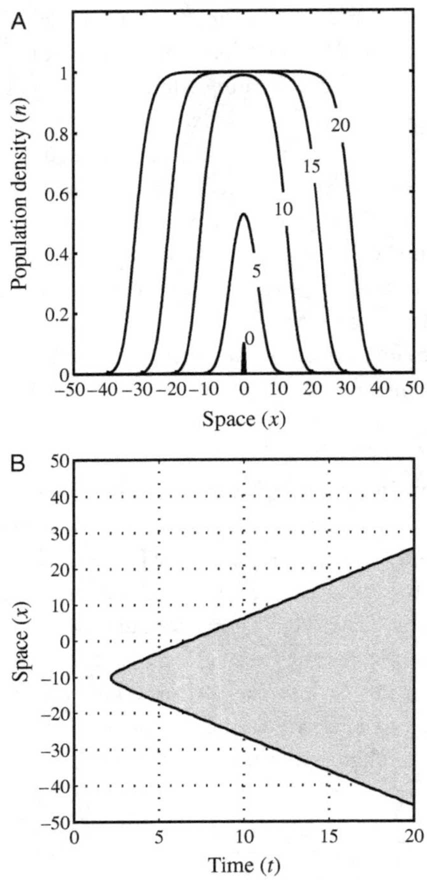

FIGURE 1 A typical solution of Fisher's equation; (A) The solution is plotted for equally spaced time intervals. (B) The gray area denotes the region in space where the population is larger than a threshold level $n = 0.5$. The boundaries of this area have slopes equal to the spread rate $c^* = 2\sqrt{(rD)}$. Here, $r = D = K = 1$. From Neubert and Parker (2004).

the spread of invasive species. The spread rate formula provides a straightforward prediction regarding invasiveness that involves only two key parameters that can be measured from life history tables (r) and mark–recapture data (D). This formula states that locally introduced populations will eventually spread at rate $c^* = 2\sqrt{(rD)}$. This compact formula has been used to assess spread rates of invasive flora and fauna including weeds, sea otters, butterflies, birds, and some insect pests.

One appealing mathematical feature of Fisher's equation (Eq. 2) is that the spread rate is linearly determined. That is, the spread rate remains unchanged if the logistic growth term $f(n) = rn(1 - n/K)$ is replaced with its linear analog, $f(n) = rn$, which includes no density dependence. This feature is lost, however, when populations exhibit Allee effects, or positive density dependence at low densities. Under these circumstances, the linearized equation no longer gives the spread rate, and further analysis is needed. Allee effects have a net effect of slowing spread rates.

In many cases, Fisher's spread rate prediction $c^* = 2\sqrt{(rD)}$ is close to empirically observed spread rates. However, it seriously underestimates the spread rate for species that use nondiffusive jump dispersal. Cases where the formula underestimates the spread rate include beetles, gypsy moth, and starlings. A famous example comes from Skellam's analysis of the spread rate of oak trees recolonizing the UK after the last ice age. Here, Fisher's spread rate underestimates the historically observed spread rate, giving rise to so-called Reid's Paradox of rapid plant invasion. To explain this paradox, rare, long-distance, animal-mediated dispersal of acorns must be included in the model.

Nondiffusive long-distance dispersal can increase spread rates. Diffusive movement implicitly assumes that, in a fixed time t, individuals disperse via a Gaussian dispersal kernel with variance $2Dt$ (one dimension) or $4Dt$ (two dimensions). However, biological dispersal kernels are often leptokurtic, with more short- and long-range dispersers and fewer midrange dispersers. The resulting patterns of dispersal can be described by stratified diffusion and, when growth is included, by scattered colony models. The net effect is an increase in the spread rate.

Equations describing invasive spread via long-distance dispersal typically involve integrodifference equations (discrete time and continuous space) or related integrodifferential equations (continuous time and continuous space). The integrodifference equation gives a spread rate of

$$c^* = \min_{x>0} \frac{1}{s} \log(RM(s)), \qquad (3)$$

where R is the geometric growth rate of the population and $M(s) = \int \exp(su)k(u)du$ is the moment generating function of the dispersal kernel $k(x)$. A choice of $r = \log(R)$ and $k(x)$ as a Gaussian with variance $2D$ regains Fisher's spread rate $c^* = 2\sqrt{(rD)}$. However, leptokurtic dispersal kernels can greatly increase the spread rate (Fig. 2).

Spatial heterogeneity in the local environmental conditions can impact spread rates. Here, fine scale spatial variations in the intrinsic growth rate $r(x)$ and the diffusion coefficient $D(x)$ affect the spread rate differently. For example, the simple case with alternating good and bad patches in Fisher's equation (Eq. 2) yields $c^* = 2\sqrt{(<r>_A<D>_H)}$, where $<r>_A$ is the arithmetic average of growth rates between good and bad patches and $<D>_H$ is the harmonic average of diffusion rates between good and bad patches.

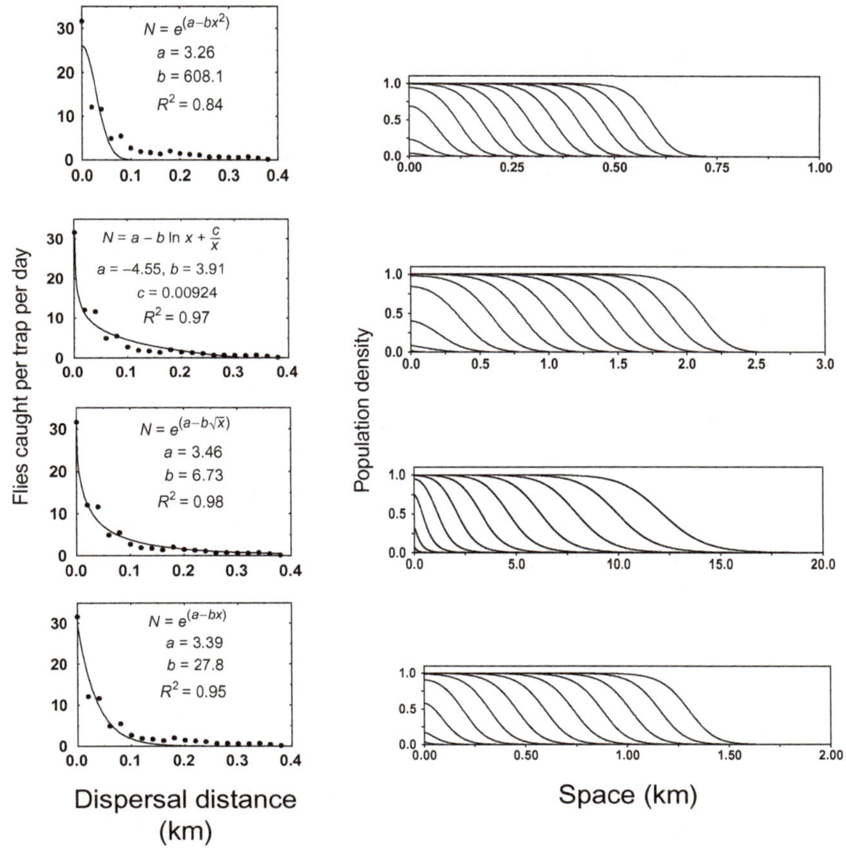

FIGURE 2 Fitted functions to *D. pseudoobscura* dispersal data provide ingredients for an integrodifference model for insect spread. It was assumed that dispersal was equally likely in both directions, so the dispersal kernels were $k(x) = (N(x) + N(-x))/2$, where N is the fitted function in the left panel. Simulations of the integrodifference equations assume Beverton–Holt population dynamics with a geometric growth rate of 10 and a carrying capacity scaled to 1. Note the different spreading speeds. Based on Kot et al. (1996).

Multispecies interactions can seriously complicate invasive spread. Typical interactions include competition, mutualism, or predation. Multispecies spread can involve either both species spreading in the same direction or the spread of one species into another. When spread rates are linearly determined, they can be calculated, but in some cases (such as unequal competition coupled to unequal diffusion) rates may not be linearly determined and are difficult to calculate. The analysis of multispecies population spread is an ongoing area of mathematical research.

Gravity Models

There is interest in modeling the anthropogenic dispersal of invaders to explain landscape patterns of invasion and to identify locations at risk of becoming invaded. There is strong evidence that the likelihood of invasive establishment is positively correlated to the inward flow of invaders (propagule pressure). In contrast to diffusion models, which are useful across continuous space, gravity models are used to estimate the flow of invaders to each node or discrete patch on a network. They are called gravity models due to a superficial resemblance to Newton's universal law of gravitation. Gravity models are commonly applied to the estimation of propagule pressure for aquatic invaders, such as spiny water flea, zebra mussels, or Eurasian watermilfoil on lake networks that are connected by recreational boaters. Other applications include disease transfer between cities and economic flow between commerce regions via movement of people.

Direct measurement of propagule pressure is logistically difficult if estimates are needed for each connection in a network. With a network of n lakes, then there are $n(n - 1)/2$ pairs of lakes and $n(n - 1)$ potential flows between lakes, as flow can be different in each direction. When n is large (say, several hundred), the number of potential flows between lakes becomes enormous, and it is difficult to acquire sufficient data to estimate invader movement. This is when a gravity model is needed. Gravity models can estimate surrogate measures of flow between patches, such as the number of boats moving between lakes, to build gravity scores that are then assumed to be proportional to the propagule pressure of invasive species being introduced.

The gravity model is a phenomenological formulation that relates the flow of boaters from lake i to lake j to empirically measured quantities, such as lake area and road distance between lakes. A secondary model is needed to relate the propagule pressure (measured in invaders per unit time) to the flow of boaters that could carry the invaders (measured in boaters per unit time). Different types of gravity models exist, based upon the level of information known.

In the production-constrained gravity model, the flow of boaters traveling from lake i to lake j (T_{ij}) is given as a function of flow of boaters leaving lake i (O_i), the attractiveness of lake j (W_j), the distance between lake i and lake j (c_{ij}), and an exponent α:

$$T_{ij} = \frac{O_i W_j c_{ij}^{-\alpha}}{\sum_{j=1}^{n} W_j c_{ij}^{-\alpha}}. \tag{4}$$

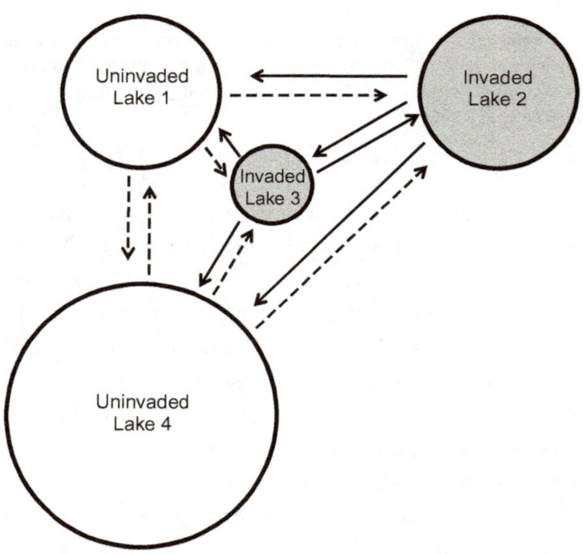

FIGURE 3. Hypothetical lake network. Solid lines show the connections between lakes that can deliver invasive species to uninvaded lakes, and dashed lines show the connections from lakes without invaders present. Gravity scores are a sum of the composite weighting of the attractiveness (commonly area) and the distance from invaded lakes as a measure of invasion risk. As such, an invaded large lake (lake 2) some distance away from an uninvaded lake (lake 4) may contribute as much to the gravity score as a nearby, small, invaded lake (lake 3). The larger the gravity score, the greater the propagule pressure and invasion risk.

Other versions of gravity models can be formulated, depending on the amount of information available on propagule movement.

Gravity models have been used successfully to predict invasion sequences in lake networks and, at larger scales, to predict long-distance invasions, such as expansion of zebra mussels into the western United States. Figure 3 shows one hypothetical network of lakes. Summing the T_{ij} from invaded lakes provides a gravity score, which is assumed to be proportional to propagule pressure. Lakes with relatively large gravity scores are at greater risk for invasion than lakes with smaller gravity scores. From the example, if attractiveness W_j (area of the circles in Fig. 3) were the main determinant of the gravity score, lake 4 would be at greater risk of invasion than lake 1. In contrast, if distance c_{ij} (length of the arrows in Fig. 3) were the main determinant, then lake 1 would be at greater risk of being invaded.

RISK ASSESSMENT FOR BIOLOGICAL INVASIONS

Hierarchical Models for Invasion

A simple theory for invasive plants, suggested by Williamson and Fitter in 1996, is the "tens rule." This posits that, on average, one in ten imported species escapes to survive in the wild, one in ten of these surviving species will reproduce to become self-sustaining, and one in ten of these self-sustaining species will spread and become a pest. This is simply a rule of thumb and, like all theories, is an approximation to reality. However, the prediction is that about 1 in 1000 introduced plant species will become a true invader.

This is a simple version of what is called a *hierarchical model* for invasion: each step in the invasion process is reliant on the previous steps. Hierarchical models can operate at the level of the population or the level of the individual. The usefulness of hierarchical modeling comes from reducing the larger process to simple subprocesses.

A general hierarchical model for invasion that operates at the level of individuals can be used to understand the process of invader transport and establishment. When transoceanic invaders are carried in the ballast water of ships from a previously invaded port to a previously uninvaded port, there is a source pool of N_s individuals at the previously invaded port, a dispersal pool of N_{dp} individuals being transported by ships, a destination pool of N_d individuals that are introduced at the previously uninvaded port, and, finally, an established pool of N_E individuals in the new port (Fig. 4).

The hierarchical model assumes that transitions from one step to the next are probabilistic. A simple version of

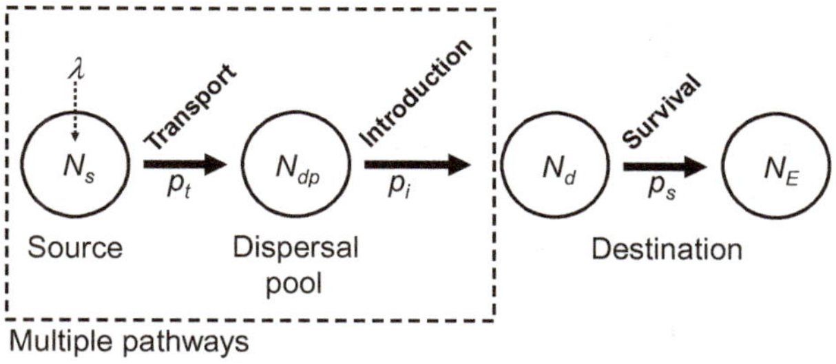

FIGURE 4 Flow diagram of the invasion process from source to established population. Figure from Jerde and Lewis (2007) and used with authors' permission (© 2007 by The University of Chicago).

the model assumes individuals act independently of one another so that

$$\begin{aligned} N_E \mid N_d &\sim \text{binomial}(N_d, p_s), \\ N_d \mid N_{dp} &\sim \text{binomial}(N_{dp}, p_i), \\ N_{dp} \mid N_s &\sim \text{binomial}(N_s, p_t). \end{aligned} \tag{5}$$

Each random variable has a distribution with a parameter that comes from another random variable. This is what makes the model hierarchical. An assumption that the source population fluctuates according to a Poisson random variable with intensity λ completes the model with

$$N_s \sim \text{Poisson}(\lambda). \tag{6}$$

The rules for hierarchical models simplify this to yield the number of surviving as invaders to a simple Poisson random variable

$$N_E \sim \text{Poisson}(\lambda p_t p_i p_s). \tag{7}$$

The process from source to destination represents one unique pathway of introduction. However, when there are many possible pathways (e.g., multiple ships and ports), the number of establishing individuals, summed across all n possible pathways, is

$$\begin{aligned} \sum N_E &\sim \sum_{k=1}^{n} \text{Poisson}(\lambda_k p_{tk} p_{ik} p_{sk}) \\ &= \text{Poisson}\left(\sum_{k=1}^{n} \lambda_k p_{tk} p_{ik} p_{sk} \right). \end{aligned} \tag{8}$$

This final formula allows us to calculate probabilities associated with having a certain number of establishing individuals in a port, based on our understanding of the underlying stochastic processes.

Managers interested in preventing invasion typically focus on methods to reduce propagule pressure N_d. When freshwater transoceanic invaders are carried in the ballast water of ships, one method to reduce propagule pressure is to reduce the probability of surviving the voyage p_t. This can be achieved by exchanging ballast water at sea, long before entering port. The timing of such ballast water exchange can be optimized using dynamical models for the growth and mortality of the invasive species under different salinities.

Hierarchical models become more complex when individuals interact, giving rise to population thresholds, such as those for establishment. When these Allee effects are present, it may be that many small or diffuse releases of infected ballast water are preferable to large or concentrated releases. Small releases may then soon drop below the Allee threshold and no longer be such a threat.

Quantitative Trait-Based Risk Assessment

Many species normally found in the confined environments of agriculture, pet trade, aquarium trade, or ornamental trade are potential invaders if they escape into the wild. Thus, when assessing risk for these species, policymakers must balance benefits in trade against potential damages that would be incurred if the species were to become feral and then invasive. Preventing the arrival of a potential invader is a sure-fire method for avoiding the associated costs and damages. Thus, effective management needs reliable methods for assessing whether it is worth preventing the trade in a particular species so as to reduce invasion risk. As discussed earlier, while scientists have underlying hypotheses on what makes a good invader, the evidence supporting these hypotheses is not always clear. However, trait identification has been formalized as a quantitative tool for predicting invader impact. As such, predictions can be used to motivate and formulate policy and guide management.

The simplest type of risk assessment takes the form of a questionnaire, such as the Australian Weed Risk Assessment (WRA). The WRA uses answers to 49 questions to predict whether a species has a high risk of becoming invasive. This assessment is based on the properties of 139 known "serious invaders," 147 "moderate invaders," and 84 "noninvader" plants. A species that scores a high score and is commonly associated with invaders or moderate invaders is banned from importation in order to protect against potential damages. A species with a score that is low and is commonly associated with known noninvaders will be approved for import so that society can reap the economic benefits from trade in the organisms.

An alternative to a questionnaire-based risk assessment is quantitative risk assessment. This involves evaluation of criteria by statistical models or by machine learning. Statistical models include logistic regression and discriminant analysis. These analyses find the best possible combination of trait data and weighting of characteristics to discriminate between those species that have and have not invaded. Categorical and Regression Tree (CART) analysis takes a

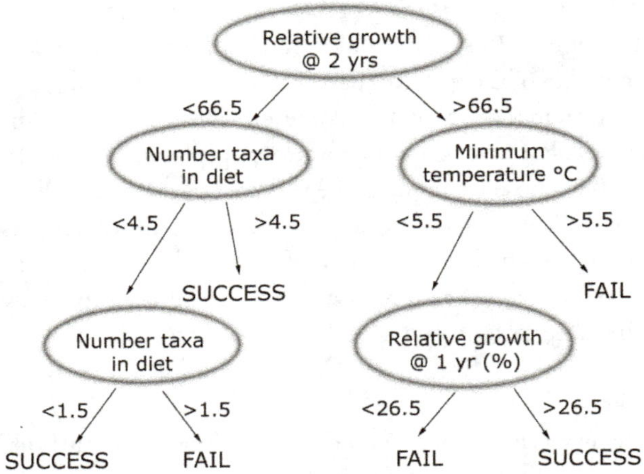

FIGURE 5 Categorical and Regression Tree (CART)-based risk assessment for fish introductions in the Great Lakes. Based on Kollar and Lodge (2002).

machine learning approach. By making successive splits in the predictor variables, the CART method maximizes within-group homogeneity of the resulting groups produced. The process continues until a user-defined limit is reached, such as the minimum number of species at each node. Kolar and Lodge used CART to assess risk for fish introductions in the Great Lakes. They showed that only 4 of 25 species characteristics are needed to correctly classify 42 of 45 introduced species as either established or not (Fig. 5).

Environmental Niche Modeling

Environmental niche modeling attempts to determine locations that are susceptible to invasion by a given species. This is achieved by matching environmental conditions to species presence/absence in regions where the species is native and then extrapolating these results to the new environment. A variety of statistical and machine learning methods are used. Ideally, environmental niche modeling would use both presence and absence data for the species, although absence data typically is difficult to come by.

Environmental niche models have been applied effectively to predict locations where invaders may be successful. The Chinese mitten crab, for example, has invaded European rivers and has recently invaded the San Francisco and Chesapeake Bays. An environmental niche model predicts its spread to other North American coastal regions, based on presence–absence data for Chinese mitten crab in its native Asian range and its introduced European range, as well as climate data (mean, minimum, and maximum air temperature, frost frequency, precipitation, wet day index), hydrological data (river discharge and river temperature), and oceanographic data (spring ocean temperature).

THEORETICAL CONNECTIONS

Bioeconomics of Biological Invasions

While mathematical bioeconomics is an established field, the application of bioeconomics to invasive species is recent. However, bioeconomics can be used to formulate methods to control invaders. A key management tool to combat invaders is the application of control measures. These take several forms. If a potential invader has not yet been established, control measures focus on reducing the chance of establishment. For established invaders, control measures can be broadly grouped into prevention, which attempts to slow or stem the spread, and eradication, which attempts to remove the invader. Analysis of the kind, amount, place, and timing of control requires us to understand the costs and benefits. If the impacts of having the invader present, the costs of control, and the benefits of control can be measured in monetary value, the problem of control can be couched in bioeconomic terms using mathematical equations.

Classical bioeconomic models track the invader abundance $u(t)$ as a function of time t and the level of control $x(t)$. A simple model using a differential equation is

$$\frac{du}{dt} = f(u(t), x(t)). \tag{9}$$

The initial abundance of invader is given by $u(0) = u_0$. Application of the control measure incurs a cost, but reduces invader abundance, so that f is a decreasing function of x. Decreased invader abundance means savings due to reduced damages. The bioeconomic problem is to find the optimal time trajectory for the control measures $x(t)$ so as minimize the present value of all costs to the system

$$\min_{\substack{x(t) \\ 0<t<\infty}} = \int_0^\infty e^{-rt} C(u(t), x(t))dt, \tag{10}$$

where $C(u(t), x(t))$ is the cost per unit time for having an invader at level $u(t)$ and control at level $x(t)$ and r is the discount rate on future costs. A solution to a problem of this form can be calculated using optimal control theory. Alternative formulations employ stochastic dynamic programming for solution.

Bioeconomic theory can also be used to calculate the optimal investment needed to combat the spread of organisms such as zebra mussels into previously uninvaded lakes. Here it is possible to show that high levels of investment are needed early on in the invasion process, before the species attains a solid foothold. The theory can also be used to calculate optimal management actions so as to reduce overall costs. Management actions include education of the public, new regulations or rules regarding shipment or trade in species, or the formation of dynamic "barrier zones" so as to reduce the spread of an established invader. Thus, the bioeconomic modeling suggests how, as a society, we can best deal with the economics of invasions.

Biological Control

Biological control agents are nonnative organisms released into the natural environment to control pest species. In other words, the release of control agents is an artificial "invasion," sponsored by managers, in hopes of beneficial side effects. Links between invasion biology and biological control provide an area of growing interest. Shea and Possingham (2000) suggested rules of thumb to guide releases. A few large releases of control agents constitute the optimal strategy when there is low probability of establishment, while many small releases become optimal as the probability of establishment increases. Mixed strategies allow managers to learn how innoculum size influences colonization success.

A simple model for control uses spread rates, based on Fisher's equation, to track the spread of the pest species as well as that of the control agent as it moves into an established pest population. If the control agent (predator) can spread faster than the pest (prey), then there is the possibility of catch-up and control. If the pest has an Allee effect, then the control agent can actually reverse the spread of the pest, potentially driving it to extinction.

Communities, Succession, and Species Assembly

Community dynamics are susceptible to novel interactions from invasive species. Theoretical and empirical studies have demonstrated that when species invade, local diversity may drop. At the same time, regional- or landscape-level diversity may increase, simply because invaders increase the pool of species present. New communities can then develop from this larger species pool. Thus, when invaders are present, community end states, or assemblages of communities that are resistant to further invasion, will differ between invaded and uninvaded regions.

SEE ALSO THE FOLLOWING ARTICLES

Allee Effects / Ecological Economics / Integrodifference Equations / Optimal Control Theory / Partial Differential Equations / Reaction–Diffusion Models / Spatial Spread

FURTHER READING

Elton, C. S. 1958. *The ecology of invasions by animals and plants.* London: Methuen.

Hastings, A., K. Cuddington, K. F. Davies, C. J. Dugaw, S. Elmendorf, A. Freestone, S. Harrison, M. Holland, J. Lambrinos, U. Malvadkar, B. A. Melbourne, K. Moore, C. Taylor, and D. Thomson. 2005. The spatial spread of invasions: new developments in theory and evidence. *Ecology Letters* 8: 91–101.

Jerde, C., and M. A. Lewis. 2007. Waiting for invasions: a framework for the arrival of non-indigenous species. *American Naturalist* 170: 1–9.

Keller, R. P., D. M. Lodge, M. A. Lewis, and J. F. Shogren. 2009. *Bioeconomics of invasive species: integrating ecology, economics, policy, and management.* New York: Oxford University Press.

Kolar, C. S., and D. M. Lodge. 2001. Progress in invasion biology: predicting invaders. *Trends in Ecology & Evolution* 16: 199–204.

Kot, M., M. A. Lewis, and P. van den Driessche. 1996. Dispersal data and the spread of invading organisms. *Ecology* 77: 2027–2042.

Moyle, P. B., and T. Light. 1996. Biological invasions of fresh water: empirical rules and assembly theory. *Biological Conservation* 78: 149–161.

Neubert, M. G., and I. M. Parker. 2004. Projecting spread rates for invasive species. *Risk Analysis* 24: 817–831.

Shigesada, N., and K. Kawasaki. 1997. *Biological invasions: theory and practice.* Oxford: Oxford University Press.

Williamson, M. H. 1996. *Biological Invasions.* London: Springer.

L

LANDSCAPE ECOLOGY

JIANGUO WU

Arizona State University, Tempe

Spatial heterogeneity is ubiquitous in all ecological systems, underlining the significance of the pattern–process relationship and the scale of observation and analysis. Landscape ecology focuses on the relationship between spatial pattern and ecological processes on multiple scales. On the one hand, it represents a spatially explicit perspective on ecological phenomena. On the other hand, it is a highly interdisciplinary field that integrates biophysical and socioeconomic perspectives to understand and improve the ecology and sustainability of landscapes. Landscape ecology is still rapidly evolving, with a diversity of emerging ideas and a plurality of methods and applications.

DEFINING LANDSCAPE ECOLOGY

Landscapes are spatially heterogeneous geographic areas characterized by diverse interacting patches or ecosystems, ranging from relatively natural terrestrial and aquatic systems such as forests, grasslands, and lakes to human-dominated environments including agricultural and urban settings (Fig. 1). Landscape is an ecological criterion whose essence is not its absolute spatial scale but rather its heterogeneity relevant to a particular research question. As such, the "landscape" view is equally applicable to aquatic systems. This multiple-scale concept of landscape is more appropriate because it accommodates the scale multiplicity of patterns and processes occurring in real landscapes, and because it facilitates theoretical and methodological developments by recognizing the importance of micro-, meso-, macro-, and cross-scale approaches.

The term landscape ecology was coined in 1939 by the German geographer Carl Troll, who was inspired by the spatial patterning of landscapes revealed in aerial photographs and the ecosystem concept developed in 1935 by the British ecologist Arthur Tansley. Troll originally defined landscape ecology as the study of the relationship between biological communities and their environment in a landscape mosaic. Today, landscape ecology is widely recognized as the science of studying and improving the relationship between spatial pattern and ecological processes on a multitude of scales and organizational levels. Heterogeneity, scale, pattern–process relationships, disturbance, hierarchy, and sustainability are among the key concepts in contemporary landscape ecology. Landscape ecological studies typically involve the use of geospatial data from various sources (e.g., field survey, aerial photography, and remote sensing) and spatial analysis of different kinds (e.g., pattern indices and spatial statistics). The intellectual thrust of this highly interdisciplinary field is to understand the causes, mechanisms, and consequences of spatial heterogeneity in landscapes.

Heterogeneity refers to the spatial variation of the composition and configuration of landscape, which often manifests itself in the form of patchiness and gradient. In landscape ecology, scale usually refers to grain (the finest spatial or temporal resolution of a dataset) and/or extent (the total study area or duration). When heterogeneity becomes the focus of study, scale matters inevitably because the characterization and understanding of heterogeneity are scale dependent. Landscape

FIGURE 1 Landscapes of the real world. The study objects of landscape ecology range from natural, to agricultural, to urban landscapes. Not only may they be dominated by different vegetation types (e.g., forests, grasslands, and deserts), but they may also have either a terrestrial or an aquatic matrix (e.g., a lakescape, seascape, or oceanscape). Photographs by J. Wu.

pattern involves both the composition of landscape elements and their spatial arrangement, and the relationship between pattern and process also varies with scale. Disturbance—a temporally discrete natural or anthropogenic event that directly damages ecosystem structure—is a primary source of spatial heterogeneity or pattern. Like pattern and process, disturbance is also scale dependent—meaning that the kind, intensity, and consequences of disturbance will change with scale in space across a landscape. This scale multiplicity of patterns and processes frequently results in the hierarchical structure of landscapes—that is, landscapes are spatially nested patches of different size, content, and history. The goal of landscape ecology is not only to understand the relationship between spatial pattern and ecological processes but also to achieve the sustainability of landscapes. Landscape sustainability is the long-term ability of a landscape to support biodiversity and ecosystem processes and provide ecosystem services in face of various disturbances.

EVOLVING PERSPECTIVES

Two dominant perspectives in landscape ecology are commonly compared and contrasted: the European perspective and the North American perspective. The European perspective traditionally has been more humanistic and holistic in that it emphasizes a society-centered view that promotes place-based and solution-driven research. It has focused on landscape mapping, evaluation, conservation, planning, design, and management. In contrast, the North American approach is more biophysical and analytical in that it has been dominated by a biological ecology–centered view that is driven primarily by scientific questions. It has had a distinct emphasis on the effects of spatial pattern on population and ecosystem processes in a heterogeneous area. This research emphasis is practically motivated by the fact that previously contiguous landscapes have rapidly been replaced by a patchwork of diverse land uses (landscape fragmentation) and conceptually linked to the theory of island biogeography developed in the 1960s and the perspective of patch dynamics that began to take shape in the 1970s.

However, this dichotomy most definitely oversimplifies the reality because such geographic division conceals the diverse and continuously evolving perspectives within each region. In fact, many ecologists in North America have recognized the importance of humans in shaping landscapes for several decades (especially since the dust bowl in the 1930s). Although humans and their activities have been treated only as one of many factors interacting with spatial heterogeneity, more integrative studies have been emerging rapidly in the past few decades with the surging interest in urban ecology and sustainability science in North America. On the other hand, the perspective of spatial heterogeneity has increasingly been recognized by landscape ecologists in Europe and the rest of the world. Thus, the current development of landscape ecology around the world suggests a transition from a stage of diversification to one of consolidation (if not unification) of key ideas and approaches.

Both the European and North American perspectives are essential to the development of landscape ecology as a truly interdisciplinary science. To move landscape ecology forward, however, one of the major challenges is to develop comprehensive and operational theories to unite the biophysical and holistic perspectives. Viewing landscapes as complex adaptive systems (CAS) or coupled human–environmental systems provides new opportunities toward this end. In addition, spatial resilience—which explores how the spatial configuration of landscape elements affects landscape sustainability—may also serve as a nexus to integrate a number of key concepts, including diversity, heterogeneity, pattern and process, disturbance, scale, landscape connectivity, and sustainability.

KEY RESEARCH TOPICS

Although landscape ecology is an extremely diverse field (Fig. 2), a set of key research areas can be identified. These include quantifying landscape pattern and its ecological effects; the mechanisms of flows of organisms, energy, and materials in landscape mosaics; behavioral landscape ecology, which focuses on how the behavior of organisms

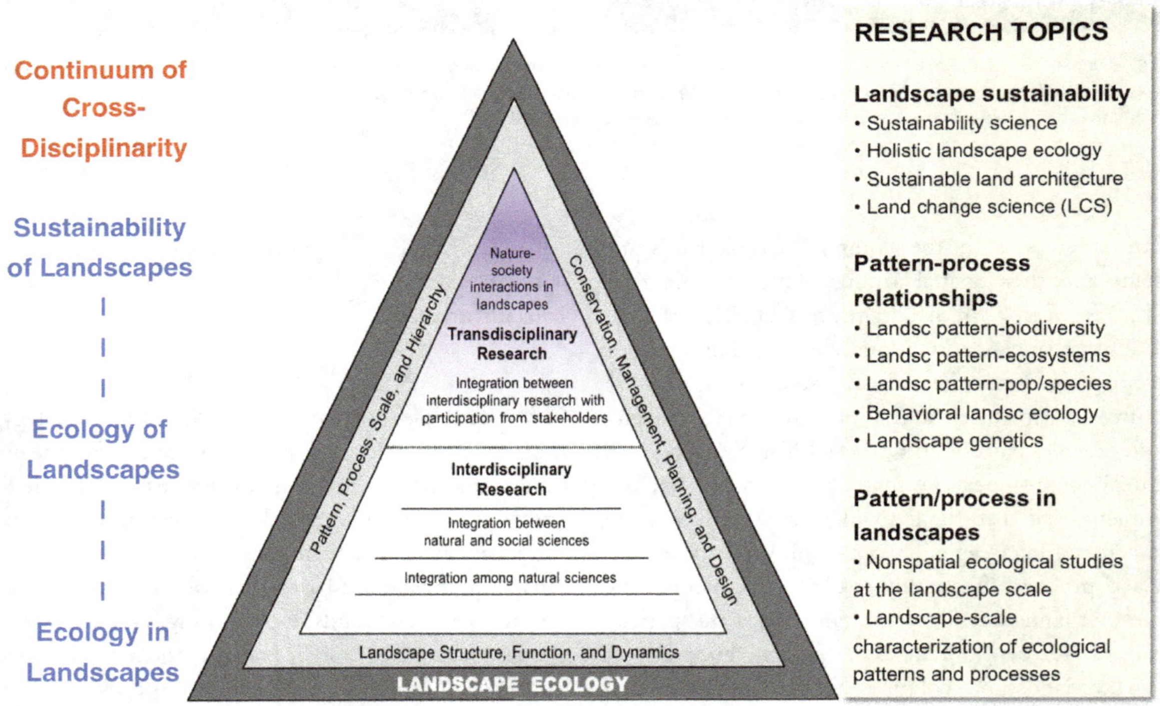

FIGURE 2 A hierarchical and pluralistic view of landscape ecology (modified from Wu, 2006). "Hierarchical" refers to the multiplicity of organizational levels, spatiotemporal scales, and degrees of cross-disciplinarity in landscape ecological research. "Pluralistic" emphasizes the values of different perspectives and methods in landscape ecology derived from its diverse origins and goals.

interacts with landscape structure; landscape genetics, which aims to understand how landscape heterogeneity affects population genetics; causes and consequences of land use and land cover change; spatial scaling, which deals with translation of information across heterogeneous landscapes; and optimization of landscape pattern for conservation or sustainability. A number of theoretical challenges exist in the study of these key topics. These challenges hinge on the spatialization of processes of interest—i.e., explicitly describing where processes take place and how they relate to each other in space. Mathematically, spatialization introduces heterogeneity, nonlinearity, and delays into models.

Thus, spatial explicitness is a salient characteristic of landscape ecological studies. Metapopulations and metacommunities have been fair game in landscape ecology. In general, however, a landscape mosaic approach considers more than the network of patches. For example, in contrast with metapopulations, landscape populations emphasize not only the dynamics of, and interactions between, local populations but also the effects of the heterogeneity of the landscape matrix. Thus, landscape population models explicitly consider the size, shape, and spatial arrangement of all habitat and nonhabitat elements. Also, the landscape population approach allows for explicit examination of how idiosyncratic features of habitat patches and the landscape matrix affect the dispersal of organisms or propagules. In general, the theory of island biogeography, metapopulation theory, and most population viability analysis (PVA) models focus mainly on the islands in a homogeneous matrix, whereas the landscape population approach explicitly considers all landscape elements and their spatial configuration in relation to population dynamics across a heterogeneous geographic area.

In the study of ecosystem processes in a heterogeneous area, the landscape mosaic approach is characterized by the explicit consideration of the effects of spatial heterogeneity, lateral flows, and scale on the pools and fluxes of energy and matter within an ecosystem and across a fragmented landscape. While models of ecosystem processes have been well developed in the past several decades, landscape-scale ecosystem models are still in their infancy. The primary difference between these two kinds of models lies in the fact that landscape models explicitly account for the locations of pools and rates, and in many cases multiple interactive ecosystems are considered together. The landscape approach to ecosystem dynamics promotes the use of remote sensing and GIS in dealing with spatial heterogeneity and scaling in addition to more traditional methods of measuring pools and fluxes commonly used in ecosystem ecology. It also integrates the pattern-based horizontal methods of landscape ecology with the process-based vertical methods of ecosystem ecology and promotes the coupling between the organism-centered population perspective and the flux-centered ecosystem perspective.

FUTURE DIRECTIONS

Emphasis on heterogeneity raises questions of the relationship between pattern and process. Heterogeneity is about structural and functional patterns that deviate from uniform and random arrangements. It is this pervasively common nonhomogeneous characteristic that makes spatial patterns ecologically important. Thus, studying pattern without getting to process is superficial, and understanding process without reference to pattern is incomplete. Emphasis on heterogeneity also makes scale a critically important issue because heterogeneity, as well as the relationship between pattern and process, may vary as the scale of observation or analysis is changed. Thus, whenever heterogeneity is emphasized, spatial structures, underlying processes, and scale inevitably become essential objects of study. From this perspective, landscape ecology is a science of heterogeneity and scale. On the other hand, with increasing human dominance in the biosphere, emphasis on broad spatial scales makes it inevitable to deal with humans and their activities. As a consequence, humanistic and holistic perspectives have been and will continue to be central in landscape ecological research.

Various effects of the compositional diversity and spatial configuration of landscape elements have been well documented, and a great number of landscape metrics (synoptic measures of landscape pattern) and spatial analysis methods have been developed in the past decades. The greatest challenge, however, is to relate the measures of spatial pattern to the processes and properties of biodiversity and ecosystem functioning. To address this challenge, well-designed field-based observational and experimental studies are indispensable, and remote sensing techniques, geographic information systems (GIS), spatial statistics, and simulation modeling are also necessary. Landscape ecology is leading the way in developing the theory and methods of scaling that are essential to all natural and social sciences. However, many challenges still remain, including establishing scaling relations for a variety of landscape patterns and processes as well as integrating ecological and socioeconomic dimensions in a coherent scaling framework. Massive data collection efforts (e.g., NEON) are expected to provide both opportunities and challenges to landscape ecology to develop and test models and theories using detailed data spanning from local to regional scales. Mathematical theory and modeling are critically important for the development of a science of spatial scale.

Overall, landscape ecology is expected to provide not only the scientific understanding of the structure and functioning of various landscapes but also the pragmatic guidelines and tools with which resilience and sustainability can be created and maintained for the ever-changing landscapes (Fig. 2). The rapid developments and advances in landscape ecology are best reflected in the pages of the flagship journal of the field, *Landscape Ecology* (http://www.springeronline.com/journal/10980/).

SEE ALSO THE FOLLOWING ARTICLES

Adaptive Landscapes / Computational Ecology / Geographic Information Systems / Metacommunities / Metapopulations / Population Viability Analysis / Spatial Ecology / Urban Ecology

FURTHER READING

Forman, R. T. T. 1995. *Land mosaics: the ecology of landscapes and regions.* Cambridge: Cambridge University Press.

Forman, R. T. T., and M. Godron. 1986. *Landscape ecology.* New York: John Wiley & Sons.

Naveh, Z., and A. S. Lieberman. 1994. *Landscape ecology: theory and application.* New York: Springer.

Pickett, S. T. A., and M. L. Cadenasso. 1995. Landscape ecology: spatial heterogeneity in ecological systems. *Science* 269: 331–334.

Turner, M. G. 2005. Landscape ecology: what is the state of the science? *Annual Review of Ecology and Systematics* 36: 319–344.

Turner, M. G., and R. H. Gardner. 1991. *Quantitative methods in landscape ecology: the analysis and interpretation of landscape heterogeneity.* New York: Springer-Verlag.

Turner, M. G., R. H. Gardner, and R. V. O'Neill. 2001. *Landscape ecology in theory and practice: pattern and process.* New York: Springer.

Wiens, J., and M. Moss, eds. 2005. *Issues and perspectives in landscape ecology.* Cambridge: Cambridge University Press.

Wu, J., and O. L. Loucks. 1995. From balance-of-nature to hierarchical patch dynamics: a paradigm shift in ecology. *Quarterly Review of Biology* 70: 439–466.

Wu, J., and R. Hobbs, eds. 2007. *Key topics in landscape ecology.* Cambridge: Cambridge University Press.

M

MARINE RESERVES AND ECOSYSTEM-BASED MANAGEMENT

LEAH R. GERBER AND TARA GANCOS CRAWFORD

Arizona State University, Tempe

BENJAMIN HALPERN

National Center for Ecological Analysis & Synthesis, Santa Barbara, California

Marine reserves protect marine biodiversity from direct human impacts, primarily fishing, by providing spatial refuge to marine organisms. Most oceans, however, are affected by indirect threats as well, such as pollution and climate change, which cannot be effectively mitigated by restricting human access. Recently, there has been growing interest in ecosystem-based management (EBM) as a marine conservation strategy, because it aims to protect biodiversity and enhance ecosystem resilience by integrating management across sectors and addressing cumulative impacts of diverse human activities. While there are burgeoning bodies of literature on both marine reserve theory and EBM, there is little integration between them. In this entry, the two management approaches are compared and contrasted, and opportunities for synergy are suggested.

MARINE RESERVE THEORY

Concept

Marine reserves, or no-take areas, are management tools intended to conserve marine biodiversity and recover exploited species, primarily commercially harvested fish stocks, by providing a spatial refuge in which access is limited and extractive activities are restricted. Reserves are a special type of marine protected area (MPA)—a broad classification of areas of the ocean set aside for conservation purposes, which range from totally protected to zones that allow various levels of use. Unlike standard fisheries regulations, which apply to single species and stocks, the marine reserve approach is more similar in spirit to terrestrial nature reserves as it protects resident species and habitats. It differs from terrestrial reserves, however, in that terrestrial reserves typically do not harbor heavily harvested species or species that regularly cross reserve boundaries (because terrestrial reserve boundaries often represent sharp contrasts in habitat).

The design of a given marine reserve depends upon specific management goals, which may include biodiversity conservation, fishery management, recreation, aesthetics, intrinsic value, and research or educational opportunities. Reserve designs that meet one management goal may not sufficiently address others. For example, the justification for establishing a reserve for marine mammals might be to provide refuge from undesirable human activities such as shooting, entanglement, oil spills, waste dumping, fishing, noise disturbance, or ship traffic. In other cases, a reserve might be designed to promote recovery of a declining population of a commercially harvested fish species such that harvest can be sustained. Ideally, reserves should include a comprehensive and representative distribution of habitats and encompass sufficient area to support viable target populations.

A majority of marine reserves address fishery management concerns. From these, most studies conclude that reserves increase fishery yields when populations would otherwise be overfished. They help recover overexploited fish stocks directly by providing a respite from harvest and indirectly through habitat protection, trophic cascades, and community-wide changes that can increase productivity. Reserves are less effective for species with high rates of juvenile and adult movement, such as wide-ranging marine mammals and sharks, and they do little to address threats that originate beyond reserve boundaries.

State of Practice

Although the idea of setting aside areas where fishing is prohibited to facilitate population recovery has been around for centuries, the first marine reserves were formally established in the 1980s. Today, there are more than 200 reserves worldwide; however, as a whole, they cover only a fraction of the world's oceans (i.e., less than 13,900 square miles, or 0.01%), and many individual reserves cover less than 1.5 square miles. The science of marine reserves for marine conservation is also relatively new, the vast majority of literature on marine reserve theory having been published since 1992. As the empirical basis for reserves advances, rules of thumb have been developed to facilitate achievement of various conservation objectives, but not all objectives can be achieved simultaneously. For example, conservation benefits offered by individual reserves are not enough to protect the full spectrum of marine species because the extent of habitat covered by a single reserve is often minuscule relative to the geographic distribution of particular target populations.

While there is empirical evidence that marine reserves may increase population vital rates for some fish and invertebrates, data are needed to determine whether reserves reduce mortality for long-lived vertebrates. Reserves may benefit these species when established to protect foraging areas surrounding breeding grounds (e.g., seabird colonies), productive oceanic habitats (e.g., seamounts), and migration routes (e.g., frontal systems).

More work also remains with regards to integrating social and ecological objectives into marine reserve design processes. When the human component has been considered, it has usually been with regards to impacts of and on fisheries (i.e., the tradeoff between conservation and fisheries values). Broader social and economic considerations are important because short-term biological benefits may wane if social issues are neglected (e.g., if there is narrow participation in management, economic benefits are unevenly distributed, and/or mechanisms for conflict resolution are absent).

Further, while reserves are generally planned and evaluated using biophysical metrics, biological changes due to reserve establishment result from changing human use of designated areas. It is increasingly recognized that planning and monitoring marine reserves based solely on biological criteria may lead to erroneous management decisions, and that a marine reserve will have no effect unless it alters human behavior, as demonstrated by reserves worldwide that are unable to meet their conservation goals due to poaching and other noncompliant activities. The success of marine reserves depends on the ability of management to motivate a shift in human behavior, and the nature of the resulting behavioral shift is of consequence.

Behavioral responses to spatial or temporal closures can be unpredictable, diluting or eliminating the intended benefits of policy action. Because factors such as stakeholder support and changes in fishing behavior due to a reserve are pivotal in determining reserve success, there have been increasing calls for marine reserve implementation guided by economic incentives and an understanding of local social dynamics. In particular, empirical studies of fisher behavior are needed to integrate biological monitoring programs with social systems.

Strategies and Tools

The underlying physical and biological heterogeneity of seascapes makes designing marine reserves and networks of reserves complex. Network design relies on some understanding of patterns of population dispersal and connectivity, and one of the most daunting theoretical challenges facing studies of the impacts of marine reserves relates to understanding the influence of dispersal among them. The prominence of marine life histories that involve planktonic larvae means dispersal of juveniles is a key component of connectivity among marine populations. Because it is not possible to quantify the dynamics of all components of marine ecosystems, it is important to identify simplified assumptions to guide marine reserve design. Some models have explicitly considered the dispersal of larvae along a coastline; however, most models have simplified the problem by considering plankton as a well-mixed larval pool (i.e., planktonic larvae produced along a continuous coastline) or as coming from discrete sites and entering a common larval pool that is then redistributed equally among adult populations.

It has been suggested that alternative reserve designs be compared with demographic and ecological models before reserves are established as experiments within adaptive management frameworks. Because life history information is lacking for many marine populations, categorizing life histories according to their response to changes in stage-specific mortality may provide a useful framework for considering conservation alternatives. More recent modeling approaches take into account other life histories, such as long-lived vertebrates that do not disperse as larvae.

While the efficacy of reserves for protecting target species with high rates of juvenile or adult mobility, whole communities, and ecosystem processes remains unclear, emerging tools for designing and evaluating marine reserves show promise with regard to incorporating social and biological information MARXAN, a free reserve selection software, is a common tool used to select reserve sites. It is most famous for its use in the creation of the Great Barrier Reef marine reserve network in Queensland, Australia. MARXAN chooses sites that minimize economic losses while achieving adequate representation of conservation features by selecting areas that achieve conservation goals while reducing the overall size of the reserve system. For example, using MARXAN, planners in south Australia state waters explicitly considered the heterogeneous distribution of catch values for rock lobster across the planning region and assigned an integrated measure of each planning unit's area and commercial value. Then they explored different design scenarios and evaluated each planning unit's ability to address both economic and conservation objectives. With this strategy, they were able to minimize economic costs without compromising conservation goals or spatial design requirements.

MarineMap is another, yet slightly different, marine reserve planning tool. It is a free, web-based decision support tool that enables different reserve network arrays to be evaluated by resource managers, scientists, stakeholders, and the public. A critical tool in California's Marine Life Protection Act-Initiative (2004), which is establishing a state-wide network of marine protected areas, MarineMap provides access to the data, methods, and analyses scientist use to plan and evaluate marine protected areas. Its user-friendly format facilitates stakeholder engagement by allowing users to visualize the planning area, which is currently limited to the California coast, in terms of social and ecological attributes. It also makes it possible for them to create networks of prospective MPAs, assign objectives and regulations to each area within the network, and evaluate them according to scientific guidelines and social and economic impacts.

ECOSYSTEM-BASED MANAGEMENT

Concept

Ecosystem-based management emerged from the growing recognition that fragmented, single-species, and sector management approaches inadequately address the multitude of growing threats to marine systems. EBM, in contrast to marine reserves, strives to mitigate diverse threats by holistically considering the entire ecosystem, including humans. It is a dynamic, integrated approach to management based on five key principles that play out differently in each context.

First, management focuses on enhancing ecosystem resilience and sustaining ecosystem services that promote human well-being. Management objectives are related to ecological, social, and institutional goals, and humans are considered part of the system as beneficiaries of services as well as drivers of change. Second, management is carried out at an ecologically relevant scale, delineated by natural boundaries that are understood to be porous as people and resources readily move across them. Further, the scope of EBM corresponds with the ecosystem service(s) of interest, which theoretically minimizes the number of factors outside the system that need to be considered. Third, cumulative impacts of human activities are explicitly considered because they influence ecosystem services across space and time. In addition, maximizing particular ecosystem services inherently limits others; hence, tradeoffs are evaluated and made between different sectors and stakeholder objectives. Fourth, integrated management involving collaboration and coordination among a broad set of stakeholders is encouraged to ensure different priorities are considered, cumulative impacts and tradeoffs are identified, and different knowledge sources and areas of expertise have an opportunity to contribute to the management process. Fifth, because management decisions are made with incomplete information, they are made using the precautionary principle and implemented such that management is carried out as an experiment that is monitored and evaluated. The results of which are used in later planning and decision-making processes.

At its core, EBM acknowledges connections that link and affect all parts of the ecosystem, including interactions between target and nontarget ecosystem services, as well as social and ecological systems across different spatial and temporal scales. In addition, maintaining and enhancing ecosystem structure and functions is often a fundamental goal of EBM that helps ensure ecosystem processes responsible for service production

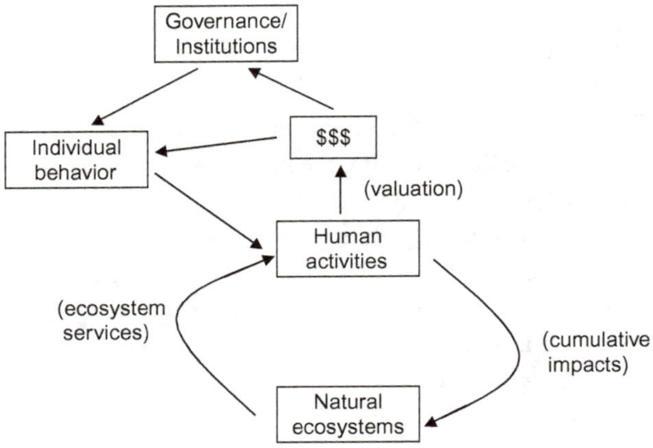

FIGURE 1 Schematic of the coupled social-ecological system that forms the basis for EBM. Human institutions (top box) regulate human behavior, both of which are influenced by how much money people can and do make from ecosystem services ($$$ box). Human activities are driven by the desire for and delivery of ecosystem services from natural systems, while simultaneously impacting those systems through those activities.

are sustained. EBM requires an understanding of the processes that drive social-ecological systems and affect their resilience, how people impact those processes, if and how people benefit from those processes, and the synergies and interactions among different drivers of change (Fig. 1).

State of Practice

An EBM perspective has been applied to terrestrial contexts for more than 30 years. Only in the last decade has it been adopted in coastal and marine systems, but it is now being pursued at sites around the world, ranging in scope from local to international (Fig. 2). Just as the theory of EBM was built upon tenets of other natural resource management approaches, such as integrated coastal zone management, the practice of EBM has capitalized on previous and existing conservation and management efforts at each site.

FIGURE 2 Picture of Morro Bay, California, where the San Luis Obispo Science and Ecosystem Alliance (SLOSEA), an ecosystem-based management initiative, is located. Photograph courtesy of Tara Gancos Crawford.

Context appears to strongly influence the nature of the EBM effort at each site, as no two EBM initiatives are the same. Management objectives, timelines, previous management regimes, and the nature and scope of stakeholder engagement influence the trajectory of EBM, but productive paths forward can be achieved from many different starting points.

While making progress, many EBM sites are not yet implementing comprehensive EBM as it is theoretically articulated. There are several reasons for this. As interdisciplinary, multi-stakeholder efforts, EBM planning processes are complicated, time-consuming, and expensive. Few sites have progressed beyond initial stages. Fewer have begun implementation or undergone formal evaluation and completed a cycle of adaptive management. In addition, information is still accumulating with regards to the production, distribution, and values of many ecosystem services beyond fisheries, and few dynamic models have been able to couple socioeconomic and ecological systems, and the interactions of multiple stressors on these systems. Furthermore, uncertainty about ecosystem variability and change, particularly in the face of climate change, limits predictions of management outcomes. The preexisting, fragmented regulatory landscape is also an impediment to EBM. Each of these issues is being addressed in unique ways among different sites, and solutions are slowly forthcoming. While such knowledge and empirical support for the practice of EBM accumulates, there is growing interest among policymakers and resource managers in adopting this type of management as demonstrated, for example, by President Barack Obama's 2010 Executive Order, *Stewardship of the Ocean, Our Coasts, and the Great Lakes.*

Strategies and Tools

Because EBM is an integrated approach that engages diverse stakeholders and tries to coordinate management of all activities within a particular place, a myriad of strategies and tools must be employed, including decision-support tools, modeling and analysis tools, and conceptual models and visualization tools. Decision-support tools facilitate general decisions for particular sectors or processes. Such tools include those that facilitate conservation and restoration site selection, ocean zoning and coastal zone management, fisheries management, hazard assessment and resiliency planning, and land use planning. Modeling and analysis tools enable practitioners to analyze the ecosystem and the processes therein and to assess possible impacts of particular activities and management actions. These tools include models for socioeconomic processes, watersheds, estuaries and marine ecosystems, oceanographic flows and dispersal patterns, habitat suitability and species distribution, and geographic information systems (GIS). In addition, EBM initiatives have made use of data collection, processing, and management tools; stakeholder engagement and outreach tools; conceptual models and visualization tools; and project management, monitoring, and assessment tools. A comprehensive and searchable listing of EBM tools can be found at www.ebmtools.org. Because each EBM initiative is unique, some tools are more relevant than others in different contexts. We will briefly describe some of EBM's most prominent tools below.

EBM theory requires models that capture the complex nature of entire ecosystems. The dynamics and interactions of species and habitats in a system and the roles of environmental drivers in modifying those interactions must be represented, as best as possible. To date, the best examples of ecosystem models, neither of which were originally designed with EBM in mind and both of which explicitly address anthropogenic disturbances to ecosystems, are Ecopath with Ecosim (EwE) and Atlantis. The strengths and weaknesses of these models, along with other relevant modeling frameworks, have been reviewed elsewhere.

EwE is a free ecosystem-modeling software suite with three key components: Ecopath, a fixed, mass-balanced depiction of the system; Ecosim, a time-dynamic simulation module that enables investigation of policy alternatives; and Ecospace, a spatially and temporally dynamic module that makes it possible for users to investigate the effects of protected areas in particular places (Ecopath with Ecosim). EwE is one of the most user-friendly and least data-intensive whole ecosystem models. The Atlantis model couples food web models and oceanographic models of water movement and biogeochemical processes to simulate whole-system dynamics. It enables practitioners to investigate appropriate strategic management options for regional fisheries and the strength of indicators for fisheries' ecological impacts, among other things.

These models hold great promise for advancing the practice of EBM; however, they neglect a more complete accounting of how the full range of human activities and socioeconomic and institutional dynamics modifies the species models within the system "boxes," and how changes in the whole system translate into changes in services delivered from the system (beyond fishing, which remains the focus of both models). The human impact gap can be addressed by calculating the cumulative impact of human activities based on habitat-specific vulnerability weights and information on the intensity and distribution of key human drivers of ecosystem change. This approach allows one to assess the level of degradation of a system while avoiding the overwhelming number of parameters and assumptions that would be necessary if human activities and their impacts were added as parameters to species population models. However, this does not help translate that estimate of ocean condition into consequences for people.

Efforts are also underway to assess values of ecosystem services and model their production and distribution, and work is being done to integrate those efforts with the ecosystem models and decision-support tools discussed above. Projects that are working to pull all of these pieces together to bolster our understanding of coupled social-ecological systems and their management include the Integrated Valuation of Ecosystem Services and Tradeoffs modeling suite (InVEST), the Multiscale Integrated Model of the Earth System's Ecological Services (MIMES), and the Artificial Intelligence for Ecosystem Services (ARIES) applications.

InVEST demonstrates the delivery, allocation, and economic importance of ecosystem services presently and in the future, and makes it possible for users to visualize the consequences of alternative decisions and identify tradeoffs and congruence among social, economic, and environmental benefits. MIMES is a collection of models that assesses the contribution of ecosystem services by quantifying the effects of changing environmental conditions resulting from changes in land use. ARIES is a globally accessible Internet-based technology that enables users to conduct rapid ecosystem service assessment and valuation to facilitate decisions. ARIES makes it possible to discover, explore, and measure environmental assets in a geographical area and understand the factors influencing their values as determined by the priorities and requirements of its users. Key differences between these efforts are that InVEST models individual services separately using models that range from simple to complex and then stitches those pieces together, whereas MIMES and ARIES model comprehensive systems simultaneously. ARIES uses a Bayesian approach to parameter estimation.

INTEGRATED MARINE RESERVE THEORY AND ECOSYSTEM-BASED MANAGEMENT

Commonalities

Marine reserve theory and ecosystem-based management have many commonalities (Table 1). Both are spatially focused measures that seek to sustain marine biodiversity and enhance resilience of marine communities. Neither is a management goal itself per se. Instead, they are means through which particular management goals such as fisheries yield improvement or biodiversity conservation are achieved. They share a number of best practices, such as implementing management measures in an adaptive manner and encouraging co-management and/or collaborative management of resources among agencies and user groups. Success of both approaches depends on social dynamics and understanding the role of people in marine ecosystems as drivers of positive and negative change, and both approaches are learning by doing. Best practices and factors that contribute to or inhibit success are being learned from "in the water" implementation among pioneering sites.

Both theories emerged within the fields of marine biology and ecology, where connections are a prominent, fundamental concept. This includes connections among land and sea, species, habitats, stressors, ecosystem structure and functions, and knowledge and uncertainty. Such linkages and interdependences influence management outcomes and must be understood to ensure informed management decisions are made. The literature from both bodies of theory recognizes the importance of interactions and feedbacks across different spatial scales as important influences on ecosystem resilience.

TABLE 1
Commonalities and differences between marine reserve theory and ecosystem-based management

Commonalities	Differences
Spatially-focused management	Abundance and focus of objectives
Conservation interest	Boundaries
Best practices	Spatial scope
Success dependent on social context and responses	Empirical basis
Learning by doing	Clarity of definition
Origin in marine biology and ecology	
Importance of connectivity	
Knowledge gaps	

In seeking to acknowledge connections among resources and habitats, marine reserve theorists are championing networks of reserves. Reserve networks amplify individual reserve success by accommodating population connectivity and providing aggregate and emergent benefits that enhance the likelihood of conservation success for a broader range of species. With a special interest in resilience science, EBM addresses connectivity by highlighting the nested nature of EBM efforts within broader seascapes and stressing the importance of coordination and integration of top-down and bottom-up management approaches. Additionally, literature from both areas cautions against a "one size fits all" prescription. Instead, they encourage that the extent and location of reserves and EBM efforts correspond with management objectives, acknowledge site-specific conditions, and are appropriate for the local socioeconomic and institutional context.

Marine reserves and EBM also share a number of key knowledge gaps. Both need a better understanding of the feedbacks between human activities, ecosystem conditions, and management actions to better anticipate ecosystem changes and management outcomes. Such understanding may be derived from additional research as well as refinement of existing ecosystem models and tools. In addition, both sets of theory acknowledge that management boundaries are porous and resources and habitats are linked across these borders, but the nature and extent of these linkages, and their influence on management success, are poorly understood. In addition, models developed for both approaches have disproportionately focused on fisheries activities and need to be expanded to include additional ecosystem threats and benefits. Finally, both research communities are actively exploring and developing mechanisms for comprehensively evaluating, and fairly addressing, desires for use and protection of marine resources.

Differences

In a number of ways, the scope of EBM goes beyond what is necessary to achieve effective marine reserves, but the distinction between the two is often unclear, as reserves are sometimes considered ecosystem-based approaches. The most apparent theoretical difference between the two is that marine reserves are primarily focused on enhancing resource protection whereas EBM embodies a broader set of objectives that include use and conservation. Additionally, reserves are often delineated by arbitrary borders, whereas EBM defines the management

FIGURE 3 A diver studies reef fish communities in the Great Barrier Reef marine reserve in Australia. Photograph courtesy of Leah Gerber.

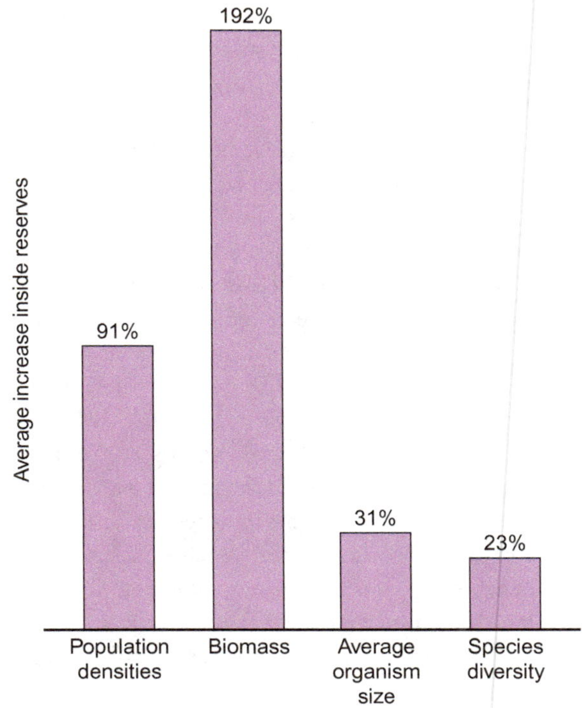

FIGURE 4 Summary of a meta-analysis of empirical work on the impacts of marine reserves on several biological measures (density, biomass, size of organisms, and diversity). Results for 89 studies show that all four biological measures are significantly higher inside reserves compared to outside (or after reserve establishment vs. before).

area by ecological boundaries. Further, while there is not a particular scale at which EBM should be implemented, EBM efforts currently underway include a larger set of broad-scale applications. Marine reserves are moving in that direction by increasing reserve size and designing marine reserve networks (Fig. 3).

EBM and marine reserves also differ in the extent of their empirical basis. Having been around longer than EBM, marine reserve theory is based on greater empirical evidence (Fig. 4). Formal marine reserves have existed for several decades, and their experiences have informed the body of literature on the subject. EBM initiatives, in contrast, are more novel and many have yet to reach implementation stages. In fact, few examples of comprehensive EBM exist. EBM literature is presently dominated by theoretical assertions and discussions regarding how to approach management in this idealized way. It is less informed by the practice of EBM, but efforts are underway to overcome this limitation.

Another key distinction between the two is clarity of definition: the definition of a marine reserve is unambiguous, and it is easy to classify particular protected areas as such, whereas the definition of EBM is more versatile. It is harder to identify sites as EBM since few are enacting all of the principles described in the scientific literature. This may be a consequence of the novelty of this approach or a function of the way contextual differences influence the nature and trajectory of EBM initiatives. On the other hand, it is easy for sites to label themselves as EBM initiatives since there are no formal criteria for EBM classification and the concept's application is intended to be adaptable to local needs and desires.

SYNERGY AS NESTED APPROACHES TO MARINE CONSERVATION

In many respects, EBM and marine reserve theory are advancing in parallel, with overlap among ideas and opportunities for mutual learning. Work in each area complements the other. Advances in our understanding of marine ecosystems and further development of the models and tools discussed above will improve management and contribute to the success of both approaches. Improvements in models that capture the complex nature of marine ecosystems and incorporate multiple stressors into population and trophic interaction models may be accomplished by parameterizing these models using variables that alter growth, mortality, reproduction, and movement and are based on how various stressors affect ecosystem processes.

Synergy between these two approaches to marine conservation and management arises from use of marine reserves as tools in the marine ecosystem–based management portfolio. Within reserves and reserve networks, conservation objectives may be achieved as resident target species and habitats are protected from direct human impacts. However, reserves are incomplete management tools for many species, including highly migratory species, and they may not deliver long-term conservation benefits or be effective at reducing indirect threats that originate beyond their boundaries, such as pollution and climate change and contribute to species declines. Establishing reserves in the context of EBM, which coordinates human activities across sectors and mitigates cumulative impacts throughout the marine matrix, promises to reduce peripheral threats to in-reserve success, thus facilitating broader conservation benefits. As evidence of these benefits is generated among pioneering sites, managers must quantify them such that they can inform future practices. When implemented as nested strategies, these approaches have the greatest potential to enhance and sustain marine resources and ecosystem services. Enacted concurrently, marine reserves and ecosystem-based management can address stakeholder preferences for use and protection across the seascape.

SEE ALSO THE FOLLOWING ARTICLES

Applied Ecology / Conservation Biology / Ecosystem Services / Fisheries Ecology / Ocean Circulation, Dynamics of / Reserve Selection and Conservation Prioritization / Resilience and Stability

FURTHER READING

Botsford, L.W., F. Micheli, and A. Hastings. 2003. Principles for the design of marine reserves. *Ecological Applications* 13: S25–S31.

Environmental Law Institute. 2009. *Ocean and coastal ecosystem-based management: implementation handbook.* Washington, DC: Environmental Law Institute.

Gaines, S.D., J. Lubchenco, S.R. Palumbi, and M.N. Dethier. 2001. Scientific consensus statement on marine reserves and marine protected areas. Signed by 161 leading marine scientists and experts on marine reserves and published by the National Center for Ecological Analysis and Synthesis (NCEAS).

Gerber, L.R., L.W. Botsford, A. Hastings, H.P. Possingham, S.D. Gaines, S.R. Palumbi, and S. Andelman. 2003. Population models for marine reserve design: a retrospective and prospective synthesis. *Ecological Applications* 13: S47–S64.

Guerry, A.D. 2005. Icarus and Daedalus: conceptual and tactical lessons for marine ecosystem-based management. *Frontiers in Ecology and the Environment* 3: 202–211.

Halpern, B.S. 2003. The impact of marine reserves: do reserves work and does reserve size matter? *Ecological Applications* 13: 117–137.

Halpern, B.S., K.L. McLeod, A.A. Rosenberg, and L.B. Crowder. 2008. Managing for cumulative impacts in ecosystem-based management through ocean zoning. *Ocean and Coastal Management* 51: 203–211.

Lubchenco, J., S.R. Palumbi, S.D. Gaines, and S. Andelman. 2003. Plugging a hole in the ocean: the emerging science of marine reserves. *Ecological Applications* 13: S3–S7.

McLeod, K., and H. Leslie, eds. 2009. *Ecosystem-based management for the oceans.* Washington, DC: Island Press.

McLeod, K.L., J. Lubchenco, S.R. Palumbi, and A.A. Rosenberg. 2005. Scientific consensus statement on marine ecosystem-based management. Signed by 221 academic scientists and policy experts with relevant expertise and published by the Communication Partnership for Science and the Sea (COMPASS).

MARKOV CHAINS

LOUIS J. GROSS

University of Tennessee, Knoxville

Markov chains are a mathematical tool applied to problems with components that cannot be determined with certainty. They are used in many biological situations when it is feasible to ignore the past history of the situation, given the present. That is, the future state of the system is possible to project based upon your current knowledge of the system state, so the previous states do not matter. The system description is defined by a set of state variables that are specified by a probability distribution. Examples include projecting future population sizes, genotype frequency, and individual behaviors from past observations. The utility of Markov chains arises due to the relatively small number of parameters required to project future states, the ease with which these parameters can be estimated from observations or hypothesized from theory, and the availability of general mathematical results that provide insight concerning long-term behavior and other biologically relevant properties of the system.

STOCHASTIC PROCESSES

The majority of applications of Markov chains in biology concern a system characterized by a single state, or a collection of states, that is a function of time. Thus, there are measurements $X(t)$ if the observations are made continuously in time, or X_n if the observations are made at discrete time periods, $n = 1, 2, \ldots, N$. Here, $X(t)$ and X_n are random variables, described by some underlying probability distribution $F_t(x) = P(X(t) \leq x)$ or $f_t(k) = P(X(t) = k)$, where the first situation applies if the measurement is of a continuous variable such as height, mass, or population density, and the second is for the case in which the measurement can take on only discrete values, such as counting numbers of individuals or observing a behavioral state such as resting, active, or feeding. If observations involve several components at each time, this leads to a random vector $\mathbf{X}(t)$ or $\mathbf{X}_n$ where the length of each vector is the number of components measured. For example the elements of $\mathbf{X}(t) = (X_1(t), \ldots, X_m(t))$ could represent the population sizes at time t in each of m patches, or the behavioral states at time t of m individuals. The associated joint probability distributions are

$$F_t(x_1, \ldots, x_m) = P(X_1(t) \leq x_1, \ldots, X_m(t) \leq x_m)$$
$$\text{or} \quad f_t(k_1, \ldots, k_m) = P(X_1(t) = k_1, \ldots, X_m(t) = k_m).$$

In the above, there is a collection of random variables indexed by some set. In our case, the index set is time, either discrete or continuous. Such a collection of random variables is called a stochastic process, and such a process is considered well defined if all joint distributions of the process, $F_{t_1,\ldots,t_m}(x_1, \ldots, x_m) = P(X(t_1) \leq x_1, \ldots, X(t_m) \leq x_m)$ are known for any sequences of times $t_1, \ldots, t_m$ and state values $x_1, \ldots, x_m$. If the random variables take on discrete values, the process is well-defined if all joint distributions $f_{t_1,\ldots,t_m}(k_1, \ldots, k_m) = P(X(t_1) = k_1, \ldots, X(t_m) = k_m)$ are known for any sequences of times $t_1, \ldots, t_m$ and state values $k_1, \ldots, k_m$. A similar requirement applies if the random variables are vectors.

DEFINITION

A Markov chain is a particular type of stochastic process in which the states of the random variables take on only a finite or countable number of values (e.g., our observations involve a set of discrete values such as population numbers, numbers of alleles, or behavioral states) and the process possesses the Markov property. A process is Markov if, given the present, the future is independent of the past. This is a form of history independence—given that you have a measurement of the current state of the system, where the process goes in the future does not depend on how it arrived at the current state. Mathematically, for any sequence of times $t_1 < t_2 < \cdots < t_{m-1} < t_m$ and any possible state values $k_1, \cdots, k_m$, then $P(X(t_m) = k_m \mid X(t_1) = k_1, \ldots, X(t_{m-1}) = k_{m-1}) = P(X(t_m) = k_m \mid X(t_{m-1}) = k_{m-1})$, which are conditional probabilities. The | symbol should be read "given." Knowledge of the history of the process before the most recent time for which you have information (t_{m-1}) provides no additional information about the value of the process at time t_m than you have from knowledge of the process at the most recent time (t_{m-1}). The term on the right-hand side here gives the chance of going to a particular state k_m at time t_m given that the process was in state k_{m-1} at time t_{m-1}. This is called a transition probability for the Markov chain.

Specifying a Markov chain requires the following: (i) defining the possible states of the process (e.g., all possible observations such as individual behaviors); (ii) giving the probability distribution for the initial state of the system (if the process starts in a single fixed state, such as an animal sleeping, then this probability distribution assigns all probability to that single state); and (iii) determining all possible transition probabilities for the system (e.g., giving $P(X(t) = j \mid X(s) = i)$ for all possible times $t > s$ and states i, j). A continuous-time Markov chain has states that are followed continuously in time, whereas a discrete-time Markov chain has states observed only at discrete times. A Markov chain has homogeneous transition probabilities if the transition probabilities only depend upon the time difference between the two observations, so that $P(X(t) = j \mid X(s) = i)$ depends upon $t - s$ and the states i, j and not the exact times t and s. The vast majority of Markov chain models assume homogeneous transition probabilities, though this assumption may not be appropriate if seasonal or diurnal variations affect the transition probabilities.

In practice, Markov chain transition probabilities are determined either from assumptions of a particular model or from estimates derived from data. In the latter case, a Markov chain can be thought of as an elaborate curve-fit in which data are used to specify the statistics of future states of the system, and the estimates are constrained by the time periods between observations. It is usual to organize the transition probabilities in a square matrix $\mathbf{P}$ with the element in the ith row and jth column of this matrix giving the transition probability $p_{i,j}$ of going from state j to state i in one time step. To properly define this matrix, the states must be ordered in some manner, which is trivial if the states represent ordinal values such as numbers of individuals but is arbitrary if the states represent discrete characters such as behavioral states. Here, we are defining the matrix elements to give the probabilities of going from the column to the row state, while some applications define the matrix as the transpose of this—the elements give the transition probabilities from the row to the column state. So be aware that either approach may be used.

EXAMPLE—BEHAVIORAL SEQUENCES

Suppose observations of an individual in a zoo enclosure are taken every minute for an hour and the individual's state is listed as Resting (R), Eating (E), or Grooming

(G) at the time of each observation. The sequence for an individual could be

RRRREEEEGEGGRRREEEGGGGRRRRRR
GGGGRRRRRRRRREEEEGGGGGRRRRRRRRR.

A variety of statistical methods could be applied to summarize this sequence, including a simple bar graph of the amount of time in each state, but a Markov chain approach is appropriate when the objective is to characterize the dynamics of states (e.g., what behavior is likely to occur in the next time period). There are two methods to associate a Markov chain with this sequence: (i) the random variables X_n for the Markov chain give the state at each minute n, or (ii) the random variables X_i give the state at each transition where i represents the ith transition from a state to a different state. The second of these methods is often called the embedded Markov chain arising from the time sequence. Note that the second method does not include an inherent time scale, although the Markov chain developed this way can be connected to probability distributions for the time to make each transition, giving a Markov Renewal Process.

In the above, note that there are 32 time periods in the R state, 12 time periods in the E state, and 16 time periods in the G state. For times when the animal is in the R state, there are 27 periods in which it stays in state R the next time period, 3 times it transitions to state E, and one time it transitions to state G. When the animal is in the E state, there are 8 times it stays in the E state, 4 times it transitions to the G state, and no transitions to the R state. When the animal is in the G state, there 11 times it stays in the G state, 1 time it transitions to E, and 4 times it transitions to the R state. From this we can construct the transition matrix P where the states are ordered R, E, G for the rows and columns

$$P = \begin{pmatrix} \frac{27}{31} & 0 & \frac{4}{16} \\ \frac{3}{31} & \frac{8}{12} & \frac{1}{16} \\ \frac{1}{31} & \frac{4}{12} & \frac{11}{16} \end{pmatrix}.$$

Note that the columns each sum to one and that although the observation series has 32 time periods in state R, there are only 31 possible transitions since the initial observation is state R.

To construct the chain accounting for transitions to a different state, we only consider times at which transitions occur so the transition matrix is

$$P_e = \begin{pmatrix} 0 & 0 & \frac{4}{5} \\ \frac{3}{4} & 0 & \frac{1}{5} \\ \frac{1}{4} & 1 & 0 \end{pmatrix}.$$

To finish specifying the Markov chain, we need to determine an initial condition that is a probability distribution across the states, (e.g., a vector which sums to 1 and gives the probability of initially being in the R, E, and G states). If each observation set began with animals at rest, then this vector would be $(1, 0, 0)$.

The Markov chain can be used in a variety of ways. It gives us a direct Markov to project the state at a future time given that an animal is in a particular state initially, allows us to find the long-term fraction of time an animal is in each state, and it provides a way to compare the behaviors of different animals (e.g., by age, gender, social status, and so on) or of the same animal in different periods (e.g., diurnal or seasonal).

BASIC PROPERTIES

Projecting the Markov Chain

Given a transition matrix P and a probability distribution for the current state of the Markov chain, expressed as a column vector **v** with elements giving the probability that the Markov chain is in each of the possible states, then the probability distribution for the states of the Markov chain in the next time period is the product P**v**. If the transition matrix is for an embedded chain, then this gives the probability distribution for the next state following a transition. If the initial probability distribution for the state of the chain is $\boldsymbol{v}_0$, then the probability distribution at time period n is $\boldsymbol{v}_n = P^n \boldsymbol{v}_0$, where P^n is the nth power of the transition matrix. For the behavioral sequence example above, given that the animal starts in state R, then $\boldsymbol{v}_0 = (1, 0, 0)^T$ (here, the T means transpose so this is a column vector), and taking the matrix product, we find $\boldsymbol{v}_5 = (.594, .202, .204)^T$, $\boldsymbol{v}_{10} = (.531, .205, .264)^T$, $\boldsymbol{v}_{100} = (.526, .203, .271)^T$, and $\boldsymbol{v}_{200} = (.526, .203, .271)^T$. So after a long time, the probability of being in each state approaches a particular value.

Limiting Distribution

As illustrated above for the behavioral sequence example, in many cases the long-term probability that the Markov chain is in any state approaches a single value so that $\lim_{n\to\infty} \boldsymbol{v}_n = \boldsymbol{v}_\infty$. Under quite general conditions, it is possible to show that this limiting distribution is stationary (e.g., if the Markov chain starts with this distribution of states, then it keeps this distribution), does not depend upon the initial distribution $\boldsymbol{v}_0$, and can be found simply as the eigenvector π associated with the eigenvalue 1 for the transition matrix P, so $\pi = P\,\pi$, where π is a column vector giving the stationary distribution. The situations in which this general limiting distribution does not exist occur in two main situations: (i) the Markov chain has an underlying periodicity, and (ii) the Markov chain has groupings of states in which it is possible for the process to be caught within a subset of the states. Whether either of these two situations arises can be determined from analyzing the transition matrix P. In case (i), there will be a periodicity in the nonzero entries of P, and in case (ii) there will be states or collections of states that are "absorbing" in that once the process is in this state, it never leaves. For a behavioral sequence, death is an absorbing state, and if i is an absorbing state, then $p_{i,i} = 1$.

An absorbing state in a simple population model is the zero state, with the population extinct, assuming no immigration occurs. In a Markov chain with an absorbing state, properties of interest include how long it takes to be absorbed (e.g., finding the distribution of time to extinction), and what the distribution of states are in the case that extinction has not yet occurred. Under very general circumstances, it is possible to determine both of these for a Markov chain, as well as to find the probability that the Markov chain has reached some value (e.g., that the population size has crossed some threshold, which is called level-crossing for stochastic processes). In general, for a Markov chain with an absorbing state (e.g., extinction), it is possible to find the quasi-stationary distribution conditional on nonextinction, which is the long-term probability that the Markov chain is in a state, conditional on it not yet going extinct.

Ergodicity

One way of thinking about a Markov chain is to consider different samples through time, which could be viewed as tracking the behaviors of different individuals, or the population sizes on different identical islands. For each individual or island followed, the Markov chain gives a function that characterizes the time-course of that particular island or individual, called a trajectory or sample path of the Markov chain (Fig. 1). If we look across all possible sample paths, which can be thought of as looking at a very large number of islands, at any particular time t and count

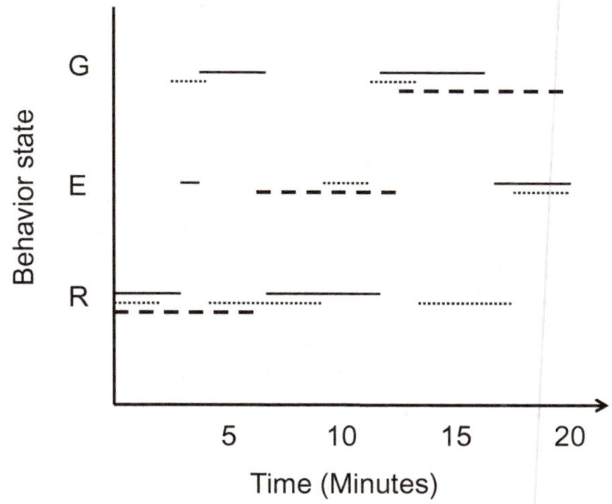

FIGURE 1 Illustration of three sample paths from the behavioral sequence Markov chain for three states (Resting, Eating, Grooming) for 20 minutes, with R as the initial state. The sample paths jump between states at the time of transition to a new state.

up what fraction of all the sample paths take on a particular value k at that time, this is the probability that the process at that time has that value, $P(X(t) = k)$.

A stochastic process is ergodic if the statistical properties over time of the process converge to limiting values, the simplest example being convergence of the mean value. This is important in practice because it allows one to consider measurements along a single sample path (e.g., a single island) to provide information about all possible sample paths after a long time. In the island case, this would allow us to follow the population size over time on a single island, take its mean, and this would be a good estimate of the long-term mean population size across *all* islands. A Markov chain that has a stationary distribution is ergodic, so we can find the long-term distribution of the Markov chain by following a single trajectory. If it followed a Markov chain, then the long-term distribution of population sizes across all islands could be estimated by following a single island over a long time period.

APPLICATIONS

Succession and Landscape Change

A classic application of Markov chains is to follow the dynamics of community composition, including the vegetation succession process. The states of the system are typically characterized by vegetation states such as bare soil, grasses, shrubs, pines, hardwoods, and so on. The transition matrix P can then be estimated from historical analysis of landscape data from which one can obtain the long-term stationary distribution, which in a simple model might be a climax forest state. The availability of Google Earth time sequences of aerial photos and satellite images allows one to directly estimate transition probabilities for landscapes that incorporate human structures and project the transition of agricultural to developed land. It is feasible to link the Markov chain model to one that incorporates disturbances such as fire, harvesting, hurricanes, or disease to evaluate the impacts on landscape structure of alternative patterns and timing of these disturbances. If aspects of this are under human control, such as harvest or fire management, then bioeconomic models can be developed in which there is an objective function with costs and benefits associated with the controls, and constraints on the allowable controls.

Markov Chain Monte Carlo Methods

Monte Carlo methods are numerical ways to estimate some quantity, including probabilities, by repeating numerical experiments a large number of times. These are done through the calculation of pseudorandom numbers (e.g., computed to mimic the numbers arising from a given probability distribution, called "pseudo" since they come from a numerical method that is not perfect). For example, Markov chain models are regularly used in population genetics to analyze the impact of different assumptions about selection, mutation, migration, and the like, on allele frequencies. The states of the Markov chain are the possible allele frequencies and for a single-locus, diploid population with N individuals; the allele frequencies are $0, 1/2N, 2/2N, \ldots, (2N-1)/2N, 1$, where the state being 1 corresponds to fixation of the allele. To calculate the first time to fixation given an initial allele frequency, simulate the Markov chain by choosing pseudorandom numbers from the transition probabilities from the initial state and then each succeeding state and counting the time steps until the Markov chain reaches state 1, then repeating this many times. This gives a list of times to fixation from which you can construct a histogram and the distribution of time to fixation. Note that this assumes there is mutation or some other force that does not absorb the population in the 0 frequency state, or else the fixation time could be infinite.

In the situation for which the quantity you wish to estimate arises from a probability distribution that is the limiting distribution of a Markov chain, it is then feasible to simulate the Markov chain using a Monte Carlo method as a way to estimate the quantity or distribution of interest. This is called Markov Chain Monte Carlo (MCMC), which has become particularly useful due to the feasibility of computing large numbers of sample paths for Markov chains and due to the fact that many models arising in Bayesian statistical analysis can be shown to have a posterior distribution that is the same as an appropriately chosen Markov chain's stationary distribution. Since Bayesian methods rarely have an explicit form for the posterior distribution, these can be estimated by numerically simulating a Markov chain, obtaining its stationary distribution, and calculating the Bayesian estimate from this stationary distribution. Doing this effectively requires determining when the Markov chain has reached its limiting distribution (the burn-in period for the MCMC) and whether the statistical properties calculated from the MCMC are appropriate estimators for the desired Bayes posterior (the variance estimation problem).

SEE ALSO THE FOLLOWING ARTICLES

Bayesian Statistics / Behavioral Ecology / Matrix Models / Stochasticity / Succession

FURTHER READING

Allen, L. J. S. 2011. *An introduction to stochastic processes with applications to biology*, 2nd ed. Boca Raton, FL: CRC Press.

Baltzter, H. 2000. Markov chain models for vegetation dynamics. *Ecological Modelling* 126: 139–154.

Caswell, H. 2001. *Matrix population models.* Sunderland, MA: Sinauer.

Clark, C. W., and M. Mangel. 2000. *Dynamic state variable models in ecology: methods and applications.* Oxford: Oxford University Press.

Clark, J. S., and A. E. Gelfand, eds. 2006. *Hierarchical modelling for the environmental sciences: statistical methods and applications.* Oxford: Oxford University Press.

Hill, M. F., J. D. Witman, and H. Caswell. 2004. Markov chain analysis of succession in a rocky subtidal community. *American Naturalist* 164: 46–61.

Karlin, S., and H. M. Taylor. 1975. *A first course in stochastic processes*, 2nd ed. San Diego, CA: Academic Press.

Mangel, M., and C. W. Clark. 1988. *Dynamic modeling in behavioral ecology.* Princeton: Princeton University Press.

MATING BEHAVIOR

PATRICIA ADAIR GOWATY
University of California, Los Angeles

Whether an individual animal accepts or rejects copulations with a potential mate is the first mating behavior upon which rests all other reproductive decisions (e.g., to ejaculate or not, to nurture or kill donated sperm, to lay eggs or give birth or not, to raise offspring or not, to provide resources to offspring or not, to differentially invest in some offspring or not, and so on). In sexually reproducing species, mating is a necessary preliminary to reproductive success, the most important component of Darwinian fitness.

QUESTIONS ADDRESSED BY THEORIES OF MATING BEHAVIOR

Mating models may be intuitive qualitative scenarios, arguments from correlation inferred from past observations, deductions from simulations (numerical experiments), or quantitative analytical solutions from ad hoc or first-principle assumptions. Mating models address a variety of related problems: When will this or that individual mate at random or be "choosy"? How many times in its lifetime or during a breeding season will an individual mate? How many different mates will an individual have during a breeding season or over a lifetime? What is the mean number of mates per sex? What is the within-sex variance in number of mates? Does the number of mates affect an individual's number of offspring and, if so, how? Do some individuals accept more potential mates than other individuals? Do some individuals reject more potential mates than other individuals? Do traits vary between accepted and rejected potential mates?

Most mating behavior theory starts from qualitative assumptions about the origins of sex differences that explain why females are choosy (when they are) and males are indiscriminate (when they are), that is, most mating theory assumes sex differences to predict further sex differences. The resulting theories are essentialistic (neither assuming nor predicting within-sex, between-individual variation), strictly binary (about males and females implying no gender variation), only about heterosexual mating, and often produce inexact predictions (as in "more than" and "less than"). At the other end of the continuum, answers start with fundamental first-principle assumptions about chance and deterministic (e.g., those from selection) forces affecting variation in quantifiable parameters, which predict quantitative variation in the mating behavior of individuals, sometimes independent of their sex.

QUALITATIVE MODELS

The qualitative models begin with Darwin's insights about natural and sexual selection (including narrow-sense sexual selection only among males and broad-sense sexual selection). Darwin's experiences led him to generalize that males were usually profligate, readily mating with any potential mate, and "ardently assertive" about mating, while females were choosy, "coy," passive, and reticent about mating. Darwin said that because males competed among themselves and females were choosy about with whom they mated, males evolved extravagant traits; however, Darwin did not explain why the sexes seemed so different. Other investigators who, like Darwin, believed Victorian sex stereotypes, explained sex differences in choosy and indiscriminate mating behavior as due to selection acting through the relative costs of reproduction for males and females. The intuitive argument provided by George Williams almost 100 years after Darwin was that, for example, in a female mammal compared to a male mammal, any copulation she accepted could lead to extraordinary costs of extended gestations, later periods of lactation, and further offspring dependence, while males suffered only the costs of the copulation. Thus, the hypothesis went that selection favored different genes in each sex: those in males for indiscriminate mating and those in females for choosy mating. This intuitive argument was justified from anisogamy theory, which posited that disruptive selection acting on gamete size favored two distinct gamete sizes: large, sessile, resource-accruing eggs and small, mobile sperm that competed among themselves for access to eggs. Geoff Parker then argued from correlation that "the sexes are what they are because of the size of the gametes they carry." Many exceptions to the anisogamy argument exist: choosy males and profligate females occur in many species with typical gamete size asymmetries, and in some species, like *Drosophila hydei,* there are minor or nonexistence differences in size of female and male gametes. Thus, other explanations or at least more nuanced explanations of perceived sex differences in mating behavior appeared, one of which was the plausible selection argument of Robert L. Trivers' parental investment (PI) theory.

PI theory holds that the sex exhibiting higher parental investment, independent of that sex's gamete size, will be choosier than the sex with lower parental investment. PI theory predicted that when males invest more in offspring then females, males are choosier than females. Trivers bolstered PI theory using an experimental 1948 study on intrasexual selection on fruit flies (*Drosophila melanogaster*), in which Angus Bateman, paraphrasing Darwin, concluded that the higher fitness variance of the experimental male flies was due to "discriminating females and ardent, competitive males." Trivers thus predicted that the competitiveness of males and the choosiness of females produced higher variance in number of mates and fitness for the sex with the lower parental investment. Easy to understand and highly intuitive, these ideas—known collectively as the Bateman–Trivers paradigm—assert that (a) males are aggressive to each other and eager to mate whereas females are passive and discriminating, which results in (b) the reproductive success of males being more variable than that of females and (c) mating with multiple partners having a greater effect on male fitness than female fitness.

The Bateman–Trivers paradigm fueled more than 30 years of empirical research, which ultimately provided the data for its rejection: (1) As Sarah Hrdy pointed out in 1981 in *The Woman That Never Evolved*, even in species with female-biased PI, female–female aggression and competition are common and sometimes dramatic. In experimental and naturalistic field studies, females often mate at random and males are often choosy. In common fruit flies—*Drosophila melanogaster*, with typical asymmetries in gamete sizes; in *D. pseudoobscura*, the fruit fly with the largest female-biased asymmetry in gamete sizes; and in *D. hydei*, with negligible gamete size asymmetries—female ardor is not statistically different from male ardor and on occasion exceeds male ardor. In common chimpanzees and in bonobos (humans' closest relatives), which have much higher female PI than male PI, female enthusiasm for sex is notable. In *D. pseudoobscura* and *Mus musculus*, species with female-biased PI, males are choosy and prefer to mate with some females more than others, and males gain fitness benefits similar to those choosy females gain in those same species. In hundreds of studies in the wild, females and males are more flexible—able to change their behavior much more rapidly—than the between-generation selection usually associated with the Bateman–Trivers assertions. (2) In birds and flies, female reproductive success often varies as much as male reproductive success independent of the direction of bias in PI. In fact, it has been easier to demonstrate that females in socially monogamous bird species mate with multiple males than to demonstrate that males of those species mate with multiple females. (3) In experimental studies, females enhance the viability of their offspring and the number of their offspring that reach reproductive age when they mate with more than one male compared to when they mate with only one male. Furthermore, reanalysis of Bateman's data showed that female fitness, like male fitness, increases with females' number of mates, and in some of Bateman's trials, this reanalysis revealed that there were no statistical sex differences in the slope describing the relationship between number of mates and the number of adult offspring (fitness).

The research that the Bateman–Trivers paradigm inspired revealed more as well. The mating behavior of males and females is often similar, making it sometimes very hard to tell males and females apart; mating behavior in nonhuman animals sometimes involves competitive interactions between more than two genders; mating between same-sex individuals (homosexuality) is far more common than previously imagined, and the functions of sexual behavior in many nonhuman animals includes more than reproduction alone. As a result of the mismatch between the predictions of classic theory and observations, the intuitive and qualitative Bateman–Trivers paradigm is losing sway, and newer, quantitative mating theories about flexibility among individuals, not sexes, are appearing. Some are focused on within-sex flexibility and sometimes within- and between-individual flexibility.

THEOREMS PREDICTING FITNESS MEANS AND FITNESS VARIANCES

Inspired by Sutherland's challenge to the Bateman–Trivers assertions that chance could produce variances in mating not different from Bateman's data, Hubbell and Johnson produced an analytical solution to the key parameter in almost all mating theories: lifetime variance in mating success:

$$\text{Variance} = \frac{es^2[1 - s + es\,(1 - s + s^{o+1})]}{[1 - s + es\,(1 - s^{o+1})]^2}. \qquad (1)$$

Equation 1 predicts variance in number of matings without any assumption of the adaptive significance of mating, assumes stochastic demography from variation in encounters (e), survival (s), and the duration of postmating nonreceptive periods (o) of individuals, and is a useful statistical tool for partitioning variances into components representing the opportunity for selection: As Sutherland and then Snyder and Gowaty showed, most of the variances that Bateman previously attributed to sexual selection in *Drosophila melanogaster* are not statistically different from variances calculated under assumptions of demographic stochasticity. The empirical comparison of observed and stochastic variances raises questions about the force of selection in many studies in which investigators ignored the effects of demographic stochasticity on "fitness" variances.

The insight that stochastic demography can produce variances in supposed components of fitness similar to those produced by within-sex competition had another, also overlooked, implication: it turned around the arrow of causation in arguments about mating behavior and fitness. Under the Bateman–Trivers formulation, sex differences in the cost of reproduction favored sex differences in choosy and indiscriminate mating behavior, which in turn, it was said, produced sex differences in variances in number of matings. However, as Gowaty and Hubbell realized, it is also true that stochastic demography in the absence of mating competition can produce variances in both number of matings and number of mates that then are an ecological constraint with selective force on individuals that can induce the expression of flexible choosy and indiscriminate behavior of individuals. Thus, it is possible that fitness variances preceded, rather than followed, the evolution of sex differences in mating behavior. Theoretically, it is also true that the arrows of causation can go in both directions simultaneously, so that the task at hand is more complicated than the Bateman–Trivers paradigm alone argues what we observe in nature is a correlation. What now is needed is careful experimental manipulation to tease out the mechanisms and the direction of causation between variances in numbers of mates and behavior.

MODELS OF SEX DIFFERENCES IN MATING RATE

Despite the questionable status of the Bateman–Trivers paradigm, many recent hypotheses of mating still begin with the Bateman–Trivers assertions. From these assumptions of preexisting sex differences, theorists then predict further sex differences in tendencies to mate at random (indiscriminately) or to sometimes reject potential mates. There are now scores of such models, sometimes, as H. Kokko has noted, bewilderingly convoluted in their assumptions and predictions, making it difficult for empiricists to know how to proceed in testing them. These models address questions such as which sex should be choosier, how much should each sex invest in offspring care, and so on. Many of these models attempt to predict "the direction of sexual selection" or the sources of variation on relative mating rates of males and females. The operational sex ratio and differences in the potential reproductive rate of males and females have been important in such models, but many suffer from the weakness that they do not constrain actual mating rates in the two sexes to be equal when the adult sex ratio is 1:1, a requirement noted by the geneticist R. A. Fisher when he said that every offspring in sexual species has a mother and a father. Models of H. Kokko and her associates, which dominate this subfield of mating behavior theory, do not have this problem, because they factor in the relative amounts of time that individuals of both sexes spend in "time-out" periods when they are dealing with the consequences of a prior mating and are not receptive to additional matings. If these time-out periods differ between the sexes, this can produce sex differences in mating behavior, a conclusion reached much earlier by the demographic approaches of Sutherland, Hubbell and Johnson, and Clutton-Brock and Parker. Kokko's models also consider, usually in different models, relative mortality rates of the sexes, relative parental investment, sex ratio at maturation, and the mortality cost of reproduction. A conclusion is that the mortality cost of a single breeding attempt is more determinative of sex differences in mating behavior than is variation in the operational sex ratio or sex differences in potential reproductive rate, a conclusion similar to those of much simpler models. However, to make the prediction of a sex difference, Kokko's model assumes a sex difference in the mortality cost of reproduction. A further limitation of all models that deal with relative mating rates in the sexes—virtually all of which are formulated as continuous-time differential equations—is that they predict only mean mating rates; thus, they do not provide a mathematical framework for predicting the within-sex variance in mating rate, the key to understanding the force of selection in mating behavior. Therefore, in these models, sex differences in mating behavior do not arise as a consequence of selection operating on variable individuals of either sex experiencing different ecological and social constraints.

MODELS OF INDIVIDUALS: WHOM TO ACCEPT AND WHOM TO REJECT

Building on the certainty that demographic forces (individuals enter or leave the population, die or are born, or enter post-mating nonreceptive stages) affect availability of mates, theorists realized that a fundamental, first-principle concern could organize novel solutions to questions about who should accept or reject whom. Lifetimes are finite, so the tempo of individual acceptances and rejections must profoundly affect the relative reproductive output (fitness) of individuals. Finite lifetimes allowed theorists to state a synthetic theory, bringing together threads of earlier and later mating theories, of how individuals adaptively modify their behavior even moment-to-moment to enhance fitness. Selection should act so that individuals are sensitive to indicators of how much time they have and thus should make reproductive decisions by trading off time left with fitness gains and losses. This logic says that encounter rates with potential mates can appear to "speed up or slow down time"; thus, when one's encounter rate with potential mates is high, an individual's time available for mating is greater than when encounter rates are low, so that adaptively flexible individuals will reject more potential mates. When encounter rates are low, so that individuals must spend more time searching for, advertising for, or waiting to encounter potential mates, individuals have less time available for mating and reproduction, and the time models predict they will accept more potential mates. When predators, parasites, and pathogens threaten individuals' lives so that their likelihood of survival is reduced, individuals may modify their behavior to avoid predators, which may also lower their encounters with potential mates. Both effects of predation risk reduce time for mating, and the time models predict that previously discriminating individuals thus will more often mate at random. The synthetic quantitative time model is the switch point theorem of Gowaty and Hubbell, which proved that selection is such that individuals who trade off fitness gains and losses of a given mating against time available for mating have significantly higher reproductive success than individuals of either sex who were always indiscriminate or who were always choosy. It showed that, for species with female biases in parental investment, males who are always indiscriminate or females who are always choosy will have lower fitness than individuals who switch their behavior to accept or reject potential mates with time available for mating.

The switch point theorem comes from an absorbing Markov chain model in which individuals move through states with some probability, thus allowing one to compute the mean and variances in the number of times individuals pass through each stage in the model. The stages include an absorbing state of death from which individuals cannot escape, a stage during which they are receptive to mating, a mating stage in which they encounter and mate with specified potential mates in the population (a stage for each potential mate), and a period after mating called latency or time out when individuals are not receptive to further mating. The mating model assumes that individuals gain information about relative fitness rewards of mating with alternative potential mates during development before individuals enter first receptivity to mating; that individuals thus can assess instantaneously on encountering a potential mate the likely fitness rewards of mating with the encountered potential mate; and that with this information focal individuals rank potential mates from best at rank 1 to worse at rank *n,* where *n* equals the number of potential mates in the population. This approach provides answers to questions about real-time evolutionary dynamics and inferences about the evolution of mating. If an investigator knows or can estimate (i) the number of potential mates in the population *n*, (ii) the encounter probability, *e*, with potential mates, (iii) the likelihood of survival, *s*, of the focal, (iv) the duration of, *o*, and (v) the distribution of fitness under the assumption of random mating, *w*-distribution (Fig. 1), one can then calculate the mean and the variance

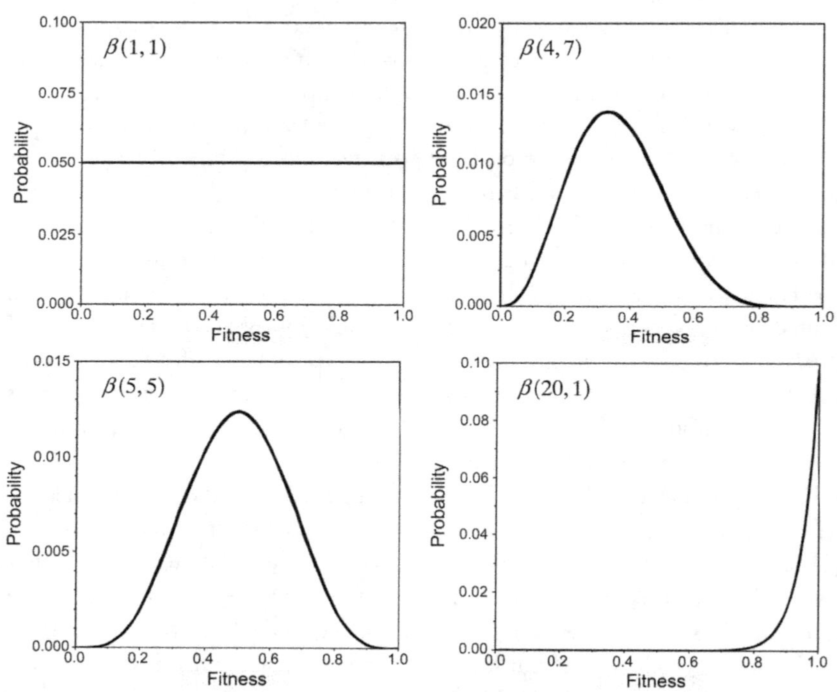

FIGURE 1 Examples of fitness distributions plotted using the two shape parameters (upper left corner of each graph) of the beta distribution. *w*-distribution, a parameter of the switch point theorem, is the distribution of fitness in a population under random mating or of the matrix of all females mated to all males and all males mated to all females without carry-over effects. *w*-distributions have seldom been empirically estimated; nonetheless, it is clear that the shape of the *w*-distribution can have profound effects on the reproductive decisions of individuals. Consider panel $\beta(20,1)$, which shows an extremely right-skewed *w*-distribution. When most of the fitnesses that would be conferred are very high, the switch point theorem predicts that adaptively flexible individuals accept potential mates as they are encountered. When *w*-distributions are uniform $\beta(1,1)$, there is much more variation in the fitnesses that would be conferred than under $\beta(20,1)$, so that the switch point theorem predicts that adaptively flexible individuals much more frequently reject potential mates.

in fitness among individuals, among individuals of different sexes, or in different populations and predict the behavior of individuals. Among the novelties of the switch point theorem is the distribution of fitness conferred, which is the distribution of fitness from random mating. This distribution links the effects of stochastic demography to fitness-enhancing mating behavior of individuals. The shapes of *w*-distributions have an important effect on the behavior of individuals, regardless of their sex. For example, if the *w*-distribution (by convention from 0 to 1) is highly skewed, say, to the right, so that the vast majority of matings will yield high fitness, most individuals will gain higher fitness by mating with potential mates as they are encountered. If the distribution is flat (uniform), the payout from rejecting some potential mates may be substantial. Figure 2 contains a very few examples of switch points for individuals experiencing different values of the inducing parameters, *e*, *s*, *o*, *n*, and *w*-distribution. A sensitivity analysis of the changes in fitness given changes in *e*, *s*, and *o* indicated that *s* has the most exaggerated effect, a result similar to empirical analyses of the major effects on number of mates. With the analytical solution, one can predict individual behavior (who they accept and reject and when) and important values of fitness components (individual mating success, individual number of mates, the relative number of their offspring). The switch point theorem makes these predictions without prior information about the morphological or physiological cues that may (or may not) indicate fitness payouts for a focal who mates with a given potential mate. Then investigators might use the analytical rules of the switch point theorem in the context of numerical experiments run under stochastic effects on *e*, *s*, *o*, and *n* to infer within-sex variances or other population-level phenomena associated with stochastic effects. One can then also partition observed within-sex variances into components due to demographic stochasticity and other effects (such as deterministic effects of selection). One can thus predict, both within and between populations within-sex means and variances in behavior, numbers of mates, and fitness under stochastic demography and deterministic forces. Because demographic stochasticity is the basis for the switch point theorem, it is a stepping-stone from individuals to populations. It may also inform other models explaining the evolution of traits thought to be important in mate choice (see below). Most important is the point that even large variances in number of mates or fitness do not necessarily indicate selection or even the opportunity for selection.

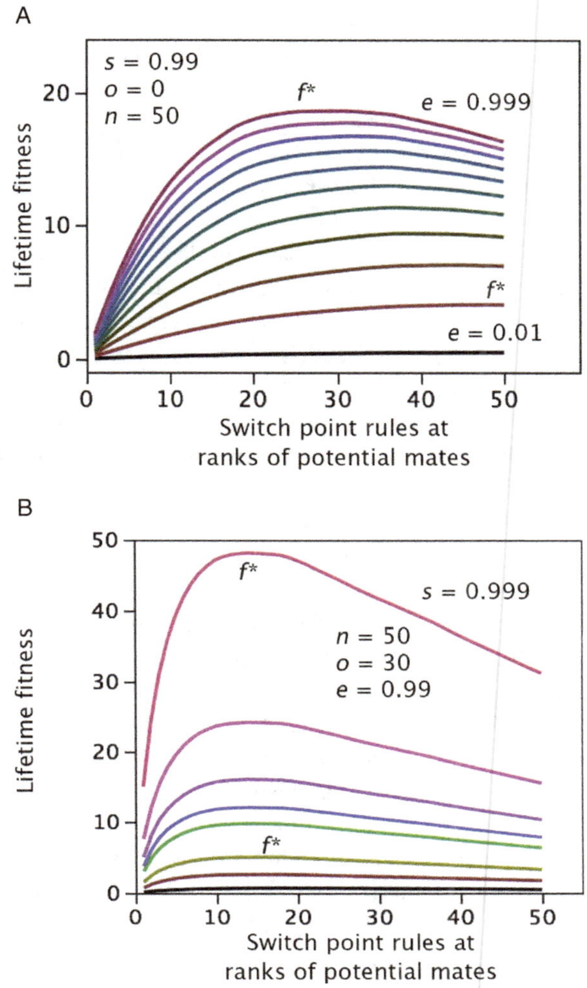

FIGURE 2 The switch point rule that maximizes the fitness of individuals, under specified *e*, *s*, *o*, *n*, and *w*-distribution, is at the hump or peak (f^*) of each curve on a graph. Read the switch point rule as "accept up to rank *x*, reject from $n - (x)$." (A) Graph of switch point rules for virgins (for whom $o = 0$), when *s* is high, $n = 50$, experiencing a uniform *w*-distribution (Beta 1,1) while *e* varied, with values from the top line and descending being $e = 0.999$, 0.9, 0.8, 0.7, 0.6, 0.5, 0.4, 0.3, 0.2, 0.1, 0.01. f^* is indicated for only two values of *e*: $e = 0.999$ (top curve) and $e = 0.1$ (next to bottom curve). The adaptively flexible individuals experiencing given *e*, *s*, *o*, *n*, and *w*-distribution reject more potential mates when *e* is high, and they reject more potential mates when *e* is lower. (B) Variable values of *s* beginning with the line at the top of the graph are $s = .999$, 0.998, 0.997, 0.996, 0.995, 0.99, and the bottom line being $s = 0.9$. When $s = 0.9$, an individual has $1/1 - s$ time units remaining, and thus it may have only 10 hours to live in those species in which the shortest nonreceptivity period after mating is 1 hour; therefore, it is not surprising to see that such individuals have no switch point but will accept all potential mates as they are encountered.

MATING MODELS AND THE EVOLUTION OF SEX ROLES

The mating models that assume sex differences seldom predict sex similarities, as the sex-differences assumption precludes study of sex similarities. Mating models that assume sex differences are necessarily silent about long-standing debates revolving around three hypotheses for the origins of sex differences in mating behavior: (1) Ancient selection pressures resulted in fixed, genetically determined sex differences. (2) Social and ecological circumstances of individuals induce arbitrary (not fitness-related) sex differences. (3) Demographic stochasticity and selection together favored adaptively flexible individuals able to modify their mating behavior moment to moment as their ecological and social circumstances changed. Not until the emergence of models that began with the effects of demographic stochastic effects on individuals' time available for mating were simultaneous experimental tests using strong inference to sort among the three hypotheses possible. Here is a way to proceed: Estimate the shape of the w-distribution. Systematically control for encounter rates and survival rates of focal subjects. Take virgins, who, according to the assumptions of the switch point theorem, have few or no intrinsically determined sex differences. Note their sex/gender and evaluate experimentally the number of potential mates the subjects accept and reject. Hypothesis (3) will be rejected if virgin males and females show systematic differences in the proportion of potential mates accepted and rejected. Hypothesis 2 will be rejected if individual males with high encounter rates, high survival rates, and experiencing a uniform w-distribution always accept potential mates. Hypothesis 1 will be rejected if individual females and males trade off time and fitness in experimental mating tests.

MATING BIASES AND THE EVOLUTION OF TRAITS

The theoretical literature on the evolution of traits is vast, often contentious, and not the subject of this article. However, mating models inform some of the debates about the evolution of traits, and, thus, the following brief synopsis is included only as an indicator that a major use of mating models is to explain the evolution of traits.

Darwin's most mature formulation of sexual selection published in 1871 was a "broad sense definition" that contrasted with Darwin's earliest published definition, which was a "narrow sense" definition. Darwin's book on sexual selection was a response to critics who argued that traits that reduced the survival of their bearers could not evolve by natural selection. Ever since, investigators have sought the details of mechanisms by which the fancy, bizarre, elaborate traits of males can evolve. The hypotheses include Fisher's idea that female mate preferences are arbitrary (not fitness enhancing), perhaps arising from sensory biases of females (e.g., females might prefer males who resemble a favorite food, as seems to happen in guppies). If there is a genetic correlation between female sensory biases (preferences) and a fancy male trait so that sons inherit the trait and daughters the preference for it, Fisher's idea predicts that variation among males in number of mates positively correlates with variation in the male trait. One of the effects of Fisher's long reach has been that most investigators find it nonsensical to ask questions about the evolution of female preferences without specific reference to obvious traits in males, and very few investigators ever ask about male choice of females (perhaps because females so seldom have fancy traits). This research bias has resulted in a strong asymmetry in what we know about males and about females.

Perhaps the most famous of the trait-based hypotheses is that of Hamilton and Zuk. They argued that fancy male traits indicated their bearers possessed "good genes" that protected them from pathogens and parasites, i.e., that males were healthy. If one assumes that the pathogens and parasites of the offspring generation are no different from the parent generation, as Hamilton and Zuk assumed, healthy fathers would have healthy offspring so that females who preferentially mate with males with fancy traits would earn an indirect fitness payout through their production of healthy offspring. However, if the pathogen/parasite populations are different between generations, as Hamilton and Zuk also pointed out, the fancy traits of males will not predict their likelihood of having healthy offspring. Many empirical studies support the assumption that males with the fanciest traits are healthier than males with less fancy traits. However, as D. Clayton reviewed in 1991 and Rick Prum reviewed recently, very few empirical studies have supported the link between fancy male traits and healthier offspring; thus, the current best guess is that nature most often violates the assumption of no evolution of the pathogens between parent and offspring generations. Indeed, because pathogens have shorter generation times than their hosts, pathogens evolve more rapidly than their hosts, so that the Red Queen's challenge to parents to produce healthier offspring with a high likelihood of remaining healthy is potentially difficult, because simple genetic considerations ensure that most offspring will have the ability to fight off the pathogens of the parental generation, not the pathogens of their own generation.

The foregoing suggested an alternative idea that prospective parents choose mates based not on the extravagance of a trait that correlates with the bearer's health but with other information that indicates the likelihood that a chooser's immune-coding genes are complementary (different from) to those of potential mates they prefer. This idea suggested that a primary sensory modality for mate preferences is olfactory, as has been shown in mice and suspected in insects and fish, rather than the visual modality of fancy plumage in birds or weird horns and antlers of mammals, or even size. Experimental trials designed to inform questions about the fitness payouts of preferences irrespective of variation in fancy traits showed that females *and males* (even in species with female-biased PI) prefer some potential mates rather than others. Furthermore, when choosers of either sex in flies, mice, mallards, and fish were experimentally paired with the potential mate they preferred versus the one they did not prefer, offspring viability and the number of offspring that reached reproductive age were significantly higher for choosers paired with their preferred partners. These interesting results further challenge the usual hypotheses for the evolution of fancy traits, because fancy traits were unnecessary for preferences (which even males express for unornamented females).

Joan Roughgarden entered the fray over sexually selected trait evolution with an alternative hypothesis: elaborate fancy male traits were "signs of within-sex inclusion." The hypothesis says that males without fancy badges or with more poorly developed badges suffer lower survival because of between-male aggression associated with the dynamics of exclusion. Thus, Roughgarden argues that the most important component of fitness acting on fancy male traits is not variance in number of mates but variance in survival of males. This turned out not to have been a new idea in that others cast similar qualitative theory two decades earlier and data in support of it have accumulated from naturalistic studies ever since. Nevertheless, the idea that male–male traits of inclusion affect between-male survival remains an interesting and testable potential explanation for some extravagant traits.

Another testable hypothesis comes from the switch point theorem: it predicts that individuals of either sex with fancier traits have higher numbers of encounters so that they are able to reject more potential mates, ensuring that they mate preferentially with potential mates with whom they are likely to gain highest fitness. This hypothesis predicts that fancy males, for example, encounter more potential mates and therefore are choosier than those less fancy and as a result reject females more often than males with not-so-fancy traits. The switch point theorem says that any trait that enhances individual probabilities of encounter, including nonfancy but convenient traits, like, e.g., higher metabolic rates, will work as just described. The switch point theorem's prediction suggests that fancy traits may be one end of a continuum of plain to fancy traits that solve similar problems.

These alternative hypotheses for fancy male trait evolution are not mutually exclusive, and thus a consensus conclusion about their adaptive significance—that may or may not result in within-sex mating variance—appears to remain a theoretical and empirical problem for the future.

SEE ALSO THE FOLLOWING ARTICLES

Adaptive Behavior and Vigilance / Allee Effects / Behavioral Ecology / Mutation, Selection, and Genetic Drift / Sex, Evolution of / Stochasticity, Demographic

FURTHER READING

Clutton-Brock, T.H., and G.A. Parker. 1992. Potential reproductive rates and the operation of sexual selection. *The Quarterly Review of Biology* 67: 437–456.

Darwin, C. 1871. *The descent of man and selection in relation to sex.* London: John Murray.

Fisher, R.A. 1930. *The genetical theory of natural selection.* Oxford: Endon Press, Oxford.

Gowaty, P.A., and S.P. Hubbell. 2009. Reproductive decisions under ecological constraints: it's about time. *Proceedings of the National Academy of Sciences of the United States of America* 106: 10017–10024.

Hamilton, W.D., and M. Zuk. 1982. Heritable true fitness and bright birds: a role for parasites. *Science* 218: 384–387.

Hubbell, S.P., and L.K. Johnson. 1987. Environmental variance in lifetime mating success, mate choice, and sexual selection. *American Naturalist* 130: 91–112.

Parker, G.A., R.R. Baker, and V.G.F. Smith. 1972. The origin and evolution of gamete dimorphism and the male-female phenomenon. *Journal of Theoretical Biology* 36: 529–553.

Snyder, B.F., and P.A. Gowaty. 2007. A reappraisal of Bateman's classic study of intrasexual selection. *Evolution* 61: 2457–2468.

Sutherland, W.J. 1985. Chance can produce a sex difference in variance in mating success and account for Bateman's data. *Animal Behaviour* 33: 1349–1352.

Trivers, R.L. 1972. Parental investment and sexual selection. In B. Campbell, ed. *Sexual selection and the descent of man.* Chicago: Aldine.

MATRIX MODELS

EELKE JONGEJANS AND HANS DE KROON

Radboud University Nijmegen, The Netherlands

Transition matrix population models quantify all the ways (through survival and reproduction) in which individuals contribute to the size of the population after one time step. Matrix models thus represent the life cycle of individuals and can be used to investigate the dynamics of a population. Several analytical characteristics of transition matrices have clear biological interpretations, making matrix models simple and insightful models for both fundamental and applied population studies.

THE CONCEPT

Individual plants and animals are born and survive until a certain age, and in the meantime they change in size and appearance and may reproduce sexually and/or asexually. Together with migration, these demographic processes determine whether local populations of individuals grow or decline in size. Quantifying all the demographic rates involved allows a researcher to study the dynamics of population size through time. Models of population dynamics can not only be used to answer fundamental questions like "Which demographic process contributes most to population growth?" but also to important applied questions like "How effective will various management options be in controlling an invasive population?" or "How likely is a population of an endangered species to go extinct over time?" At first, such fundamental and applied questions were tackled by summarizing the fate (e.g., survival and number of offspring) of cohorts of individuals in tables or flow charts. Later, the important step was made from flow charts to life cycle graphs and transition matrix population models.

Transition matrices contain exactly the same information as life cycle graphs, but they are organized in matrix form. The two life cycle examples in Figures 1 and 2 illustrate the basic idea of matrix models. The purpose

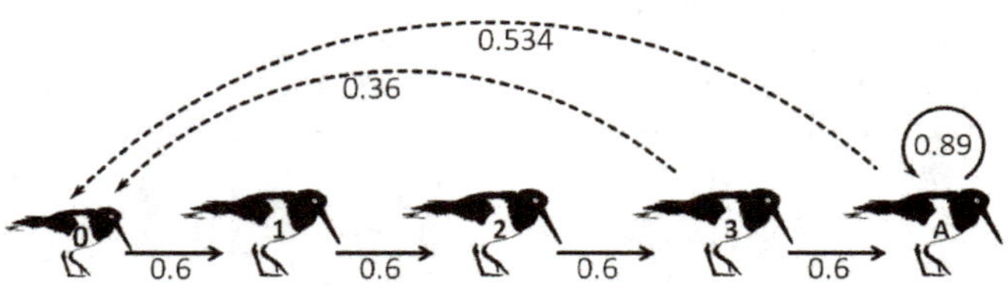

$$\begin{pmatrix} fledglings_{t+1} \\ one_{t+1} \\ two_{t+1} \\ three_{t+1} \\ adults_{t+1} \end{pmatrix} = \begin{pmatrix} 0 & 0 & 0 & 0.36 & 0.534 \\ 0.6 & 0 & 0 & 0 & 0 \\ 0 & 0.6 & 0 & 0 & 0 \\ 0 & 0 & 0.6 & 0 & 0 \\ 0 & 0 & 0 & 0.6 & 0.89 \end{pmatrix} \begin{pmatrix} fledglings_{t} \\ one_{t} \\ two_{t} \\ three_{t} \\ adults_{t} \end{pmatrix}$$

FIGURE 1 Life cycle graph and corresponding transition matrix of an oystercatcher (*Haematopus ostralegus*) population in Wales, with post-breeding census and age-based classes (data from C. Klok et al., 2009, *Animal Biology* 59: 127–144; drawings by N. Roodbergen). Only females are modeled. The solid arrows and accompanying numbers are annual survival rates, whereas the dashed arrows represent the average number of fledglings produced after 1 year. The projected population growth rate (λ) is 0.974, meaning that the population is projected to decline with 2.6% per year if these model parameters remain unchanged over time.

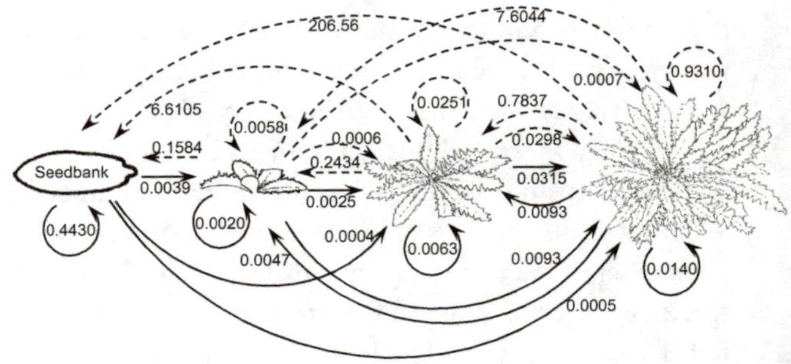

$$\begin{pmatrix} seedbank_{t+1} \\ small_{t+1} \\ medium_{t+1} \\ large_{t+1} \end{pmatrix} = \begin{pmatrix} 0.4430+0 & 0+0.1584 & 0+6.6105 & 0+206.56 \\ 0.0039+0 & 0.0020+0.0058 & 0.0000+0.2434 & 0.0047+7.6044 \\ 0.0004+0 & 0.0025+0.0006 & 0.0063+0.0251 & 0.0093+0.7837 \\ 0.0005+0 & 0.0093+0.0007 & 0.0315+0.0298 & 0.0140+0.9310 \end{pmatrix} \begin{pmatrix} seedbank_{t} \\ small_{t} \\ medium_{t} \\ large_{t} \end{pmatrix}$$

FIGURE 2 Life cycle graph and corresponding transition matrix of a nodding thistle (*Carduus nutans*) population in Australia, with post-breeding census and stage classification based on developmental stage (seed or rosette) and rosette size (data from K. Shea et al., 2010, *Ecological Applications* 20: 1148–1161). The solid arrows represent survival/growth and the accompanying rates give the proportion of individuals that survive and move to a particular class. The dashed arrows involve production (via seed) of new individuals after one year. Transition and contribution rates >0.0010 are left out of the life cycle graph. The elements of the accompanying 4 × 4 matrix model are often a sum of a survival transition and a reproduction contribution. The projected population growth rate (λ) is 1.207 (based on the rounded matrix elements shown here), projecting an increase in population size of 20.7% per year.

is to bring together all the demographic processes (at a single location, while migration is often ignored). The first thing to note in these examples is that individuals are classified in different groups, based on, for example, their age, developmental stage, or size. Secondly, the arrows between these classes of individuals. Each arrow represents the contribution of an average individual in one class to the number of individuals in a particular class one time step later. Often this time step is 1 year (as assumed in the remainder of this entry), but other transition periods are possible as well. There are two ways in which an individual can contribute to the number of individuals in the next year: either through surviving (solid arrows; including age and size changes) or through reproducing (dashed arrows).

Once all transition arrows are quantified, and if the initial number of individuals in each class is known, the population size after one time step can be calculated by multiplying each arrow with its corresponding initial group size. This multiplication can be made easier by rearranging the transition arrows into a matrix, as in Figures 1 and 2. In a matrix, the transition rates are arranged so that a column contains all contributions *by* an average individual in a particular class, while a row contains all contributions *toward* the number of individuals in a particular class after one time step. Multiplication of the transition matrix with a vector of the initial number of individuals per class immediately results in the population vector after one time step:

$$\begin{pmatrix} fledglings_{t+1} \\ one_{t+1} \\ two_{t+1} \\ three_{t+1} \\ adults_{t+1} \end{pmatrix} = \begin{pmatrix} 0 & 0 & 0 & 0.36 & 0.534 \\ 0.6 & 0 & 0 & 0 & 0 \\ 0 & 0.6 & 0 & 0 & 0 \\ 0 & 0 & 0.6 & 0 & 0 \\ 0 & 0 & 0 & 0.6 & 0.89 \end{pmatrix} \begin{pmatrix} 100 \\ 200 \\ 300 \\ 400 \\ 500 \end{pmatrix}$$

$$= \begin{pmatrix} 0.36 * 400 + 0.534 * 500 \\ 0.6 * 100 \\ 0.6 * 200 \\ 0.6 * 300 \\ 0.6 * 400 + 0.89 * 500 \end{pmatrix} = \begin{pmatrix} 411 \\ 60 \\ 120 \\ 180 \\ 685 \end{pmatrix}.$$

If the initial population size was 1500 birds (including 100 fledglings and 500 adults), the transition matrix projects 1456 birds after 1 year. Multiplication of the population vector with the same transition matrix for ten successive time steps results in Figure 3.

Transition matrices are useful not only for matrix-vector multiplication but also because matrix algebra can be applied. Many of the analytical properties of matrices have clear biological meaning. One of these, the dominant eigenvalue of a matrix, for instance, gives the asymptotic growth rate of a population (see Box 1). This is the per capita growth rate that is eventually reached when the same transition matrix is used over many time steps to project what will happen to population size in the long run. In population studies, the projected population growth rate is often symbolized by λ, and λ values above 1 indicate that a population will increase in size if the demographic rates in the transition matrix remain the same. Values of λ between 0 and 1 signal decreases in population size, as with the oystercatcher example in Figure 3.

Not only does the proportional increase (or decrease) in population size reach an asymptotic value after repeatedly multiplying the same matrix many times to a population vector, but the ratios between the number of individuals in the various classes also become fixed (see Fig. 3): all these numbers eventually also change by a factor λ each time step. The asymptotically fixed proportions of the total population size in each of the classes are called the *stable stage distribution* and are given analytically by the right eigenvector (another property of the transition matrix) that corresponds to the dominant eigenvalue (see Box 1). The left eigenvector is also informative: it gives the reproductive values for each of the classes, i.e., the number of offspring produced during the lifetime of an average individual starting in that class.

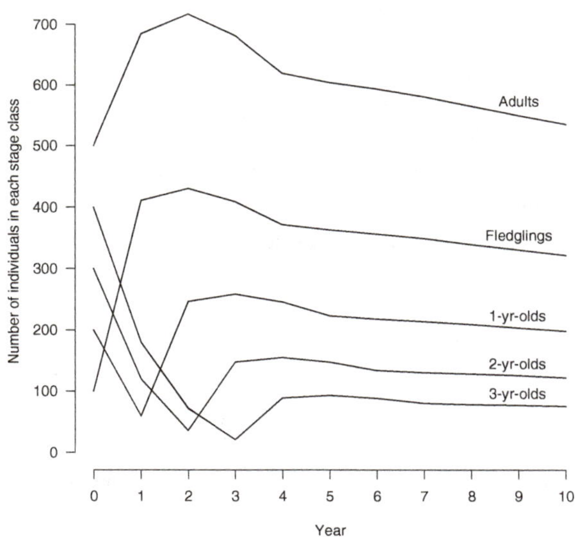

FIGURE 3 Trajectory of the number of oystercatchers in each of the five age classes over ten years. The initial population in year 0 consisted of 100 fledglings, 200 1-year-olds, 300 2-year-olds, 400 3-year-olds, and 500 adults. Each time step the population vector was multiplied by the transition matrix depicted in Figure 1. After some years with transient dynamics, the distribution of the number of birds over the classes stabilizes: eventually both the total population size and the number of individuals in each class declines by the same factor, $\lambda = 0.974$.

BOX 1. EXTRACTING EIGENVALUES AND EIGENVECTORS FROM TRANSITION MATRICES

Eigenvalues and eigenvectors are characteristic properties of transition matrices. Here is a short explanation of how these can be determined for a simple 2 × 2 transition matrix:

$$A = \begin{pmatrix} 0.1 & 2 \\ 0.3 & 0.5 \end{pmatrix}.$$

As shown in Figure 3, when the population vector is multiplied with the same transition matrix each time step, any initial population will eventually grow in size with a certain asymptotic population growth rate. This asymptotic population growth rate is equal to the dominant eigenvalue (λ) of the transition matrix. The right eigenvectors (w) associated with the dominant eigenvalue are population vectors of which the proportional distribution of individuals over the stage classes does not change after multiplication with the transition matrix, e.g.:

$$\begin{pmatrix} 0.1 & 2 \\ 0.3 & 0.5 \end{pmatrix} \begin{pmatrix} 20 \\ 10 \end{pmatrix} = \begin{pmatrix} 0.1 * 20 + 2 * 10 \\ 0.3 * 20 + 0.5 * 10 \end{pmatrix} = \begin{pmatrix} 22 \\ 11 \end{pmatrix}.$$

BOX 1 *(continued).*

This 67%:33% distribution of the individuals over the two stage classes was the same before and after the matrix multiplication, and is called the stable stage distribution. We can also see that λ must be 1.1. But how can we derive these dominant eigenvalues and eigenvectors? The first thing to realize is that for a population with a stable structure multiplication with transition matrix **A** gives the same result as multiplication with λ, the asymptotic population growth rate: $\mathbf{A}w = \lambda w$. This can be rewritten as $\mathbf{A}w - \lambda w = 0$ or $(\mathbf{A} - \lambda I)w = 0$, where I is an identity matrix with ones on the diagonal and zeros elsewhere. This set of linear equations can be solved for λ by using the determinant: $\det(\mathbf{A} - \lambda I) = 0$. For our example, this works out as follows:

$$\det\left(\begin{pmatrix} 0.1 & 2 \\ 0.3 & 0.5 \end{pmatrix} - \begin{pmatrix} \lambda & 0 \\ 0 & \lambda \end{pmatrix}\right) = 0,$$

$$\det\begin{pmatrix} 0.1 - \lambda & 2 \\ 0.3 & 0.5 - \lambda \end{pmatrix} = 0,$$

$$(0.1 - \lambda)(0.5 - \lambda) - (0.3)(2) = 0,$$

$$\lambda^2 - 0.6\lambda - 0.55 = 0.$$

Of the solutions for λ, -0.5 and 1.1, the latter has the highest absolute value and will determine the asymptotic population growth rate. The stable stage distribution (w) can now be found by substituting $\lambda = 1.1$ in $(\mathbf{A} - \lambda I)w = 0$:

$$\begin{pmatrix} 0.1 - 1.1 & 2 \\ 0.3 & 0.5 - 1.1 \end{pmatrix}\begin{pmatrix} n_1 \\ n_2 \end{pmatrix} = \begin{pmatrix} 0 \\ 0 \end{pmatrix}.$$

Suppose that n_2 is 1, then the second equation becomes $0.3n_1 - 0.6 = 0$, leading to $n_1 = 2$. This shows that the stable ratio $n_1{:}n_2$ is 2:1, or 67%:33%.

Such matrix calculations become increasingly complex with larger matrix dimension, but can easily be done with mathematics-oriented software packages (e.g., Matlab). However, most calculations can also be done with, for example, the PopTools add-in in Excel or with the popbio package in R.

MODEL PARAMETERIZATION

How is a matrix model parameterized from raw demographic field data? First it must be decided whether to classify individuals by age or by some other classification based on, for example, age, size, or developmental or reproductive status. Other important decisions are whether all individuals are included in the model (or, e.g., only females as in the oystercatcher model) and at what time of the year the population is censused. The latter is important for the stage classification as in a post-breeding census newborn individuals can be a class, whereas in a pre-breeding census the youngest individuals are almost 1 year old already.

In age-based matrices in which the individuals move up one class every time step, survival and growth are indistinguishable and entered in the subdiagonal of the matrix (Fig. 1). Often, as in the oystercatcher example, age-specific classes are combined with age-based stage classes like "Adults" in which individuals above a certain age are pooled. When classification is based on size as in the nodding thistle example, many more transitions between stages are possible (Fig. 2). These represent life histories in which it is possible for individuals to skip classes, remain in the same class ("stasis"), or regress to a previous class (e.g., shrinkage in plants). The nonreproductive elements of these matrices can be seen as products of two so-called vital rates: first, the survival rate of individuals in a class until the next census (irrespective of the classes to which individuals move) and, second, the "growth" rates—next-year's distribution of individuals from a class over all classes in the model (conditional to survival).

When calculating reproduction rates, it is important to remember the time step of the model: how many new offspring are alive 1 year later per average individual in a class now. The estimation of reproduction matrix elements therefore often involves multiplying different vital rates. In the oystercatcher model, the reproduction matrix elements consist of the product of the survival probability of a bird until the next year and the average number of (female) fledglings per surviving bird. The order and time span of these vital rates also depend on the choice of the census moment (e.g., after the reproduction pulse as in the oystercatcher and nodding thistle cases). In some cases, it might also be possible to produce different types of offspring—for example, small versus large seedlings, or sexually versus asexually produced offspring.

Because each matrix model represents a species- or even population-specific life history, and because of different model construction choices made by researchers, it can be hard to compare models between studies. Attempts have been made, therefore, to categorize matrix elements into groups representing, for example, reproduction, survival, and growth. The problem, however, with such matrix element classifications is that the nonreproductive elements in a column are not independent: their sum is bound between 0 and 1 (i.e., between 0% and 100% survival), as in the first column of the thistle model: when a high proportion of seeds in the seed bank remain dormant, the transition rates from the seed bank to the rosette stages cannot be very high. For comparative demography, it is therefore more useful to focus on the vital rates that were used to construct the matrix elements. The survival and growth rates described in the previous paragraphs do vary independently of one another. An additional advantage of analyses at the level of vital rates is that these vital rates have clearer biological meaning than matrix elements that are often constructs of multiple vital rates.

SENSITIVITY AND ELASTICITY

Once a matrix model has been constructed, one might wonder what would happen if one of the parameters in the model is changed. For instance, how much does λ increase if 0.01 is added to matrix element a_{21} (the transition from the seed bank to small rosettes; 0.0039 + 0.01 = 0.0139) in the nodding thistle model? Calculation of λ before and after the change shows that the projected population growth rate increases from 1.207 to 1.225. This change in λ not only depends on the magnitude of the change in the matrix element but also on the sensitivity of λ to changes in that matrix element. This sensitivity ($\delta\lambda/\delta a_{ij}$) to small changes in a_{ij} can be calculated analytically using the left and right eigenvectors associated with λ. When comparing the λ-sensitivity values for all matrix elements, one can find out in what element a certain increase has the biggest impact on λ.

However, a 0.01 increase in a survival matrix element is hard to compare to a 0.01 increase in a reproduction matrix element, because the latter is not bound between 0 and 1 and can sometimes take high values. Increasing matrix element a_{14} (number of seeds in the seed bank in the next year produced by an average large rosette) by 0.01 from 206.56 to 206.57 does not have a noticeable effect on λ. Therefore, for comparison between matrix elements it can be more insightful to look at the impact of proportional changes in elements: by what percentage does λ change if a matrix element is changed by a certain percentage? This proportional sensitivity is termed elasticity—$(\delta \log \lambda)/(\delta \log a_{ij}) = (a_{ij}\,\delta\lambda)/(\lambda\,\delta a_{ij})$. It turns out that the reproduction element a_{14} has a 16-fold higher λ-elasticity value as survival element a_{21} (0.0950 vs. 0.0059).

Since λ-elasticity values of all elements in a matrix sum to 1, they can also be considered as a measure of how much a matrix element contributes to λ, relative to the contributions of other elements. Quantifying the relative importance of different types of matrix elements to λ—elasticities—are therefore also very useful in comparative demography: different populations and species can be compared with respect to how much, say, survival or reproduction contributes to population growth. The reproduction transitions (i.e., dashed arrows in Figure 1) of the oystercatcher matrix model have a λ-elasticity total of 0.07, which is much lower than that of the nodding thistle model (elasticity total of the dashed arrows in Figure 2 is 0.78). This places the nodding thistle near the fast end of a slow–fast continuum of species, and the oystercatcher on the slow side. A cautionary note for comparison of elasticity patterns between species and population is required though: elasticity values can also be correlated to matrix dimension and to λ itself. A matrix with a high λ will likely have higher elasticity values for reproduction elements than a matrix for the same species with a low λ.

Elasticity values can also be used to quantify the relative contributions of different life history loops within the life cycle of a species—the pathways that individuals take in the course of their lifetime. In the oystercatcher example, for instance, three unique loops can be discerned: a 4-year loop of new fledglings surviving for 4 years before producing new fledglings themselves, a similar 5-year loop, and a "self-loop" of surviving adults remaining adults (Fig. 4). Each of these loops has one or more transitions (arrows) unique for that loop. Since all transitions within a loop are equally important for the joint contribution of that loop to λ, each transition within a loop has the same elasticity value: all transitions in a loop have the same elasticity value as the unique transitions in the loop. The total loop elasticity is equal to the number of transitions in the loop times the elasticity values of a unique transition in that loop. In the oystercatcher example, the loop elasticities are 0.024, 0.313, and 0.663, respectively, which again adds up to 1. In other cases, such as that of the nodding thistle in Figure 2, it is harder to discern all unique loops by hand. Fortunately, algorithms have been developed to do this automatically.

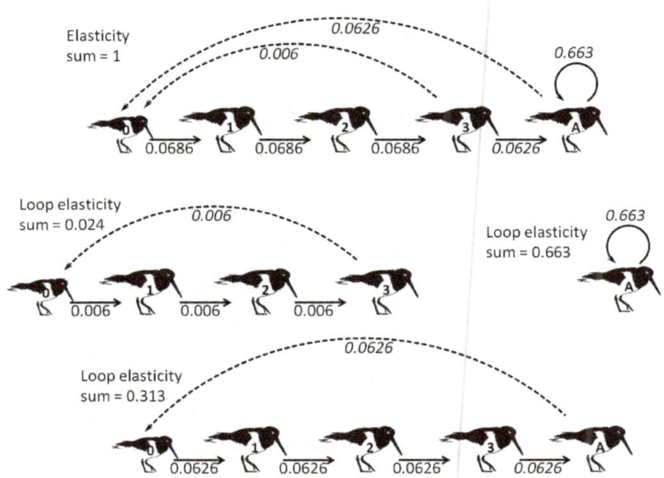

FIGURE 4 λ-elasticity values of transition in the oystercatcher example. The life cycle graph contains three loops, each with at least one unique (italicized) transition. The unrounded elasticity values of all matrix elements always add up to 1. The shown elasticity values are rounded to add up to exactly 1 for clarity. Drawings by N. Roodbergen.

Sensitivity and elasticity values can also be calculated for lower-level parameters such as vital rates. Vital rate elasticities, however, most often add up to more than 1 because single matrix elements can be the products of multiple vital rates. Therefore, the sum of the vital rate elasticities depends on the vital rate construction of the matrix model. This is not a problem for comparisons between populations or years of the same species studied with the same model. But for comparisons between different types of models, this is a problem. The best solution for cross-study comparisons seems to be to cast the different models of these studies into the same general vital rate structure and then compare vital rate elasticities across studies.

So far, this entry has focused on the (relative) sensitivity of λ to small changes in model parameters. However, similar sensitivities can be calculated for any other model output, such as the stable stage distribution or reproductive values. It is also important to realize that these sensitivity and elasticity values are characteristics of the matrix that is being studied. As soon as the transition matrix is changed considerably, the sensitivity values will also change. Although these "local" sensitivities will give a fair indication of what will happen to, say, λ when larger perturbations occur, it is better to study in detail the often nonlinear response of λ to a model parameter along a large range of values of that parameter.

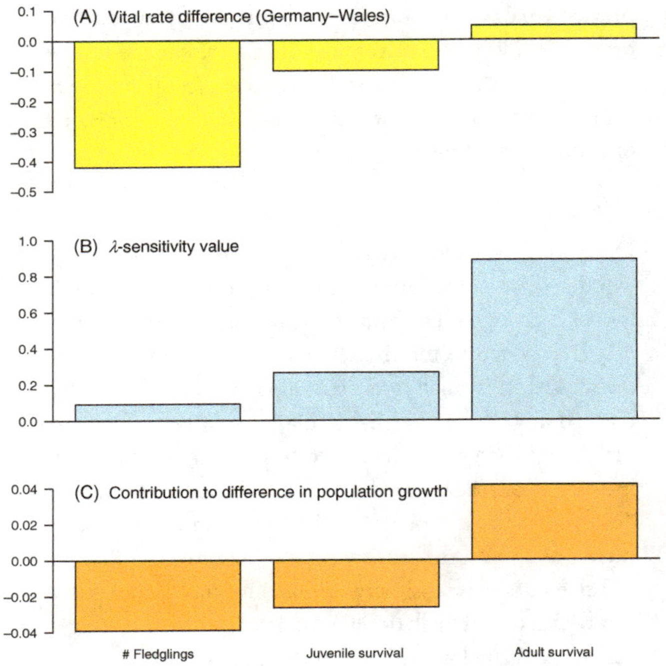

FIGURE 5 Steps of a decomposition (a fixed-effect life table response experiment) of the difference in λ between a German (λ = 0.950) and a Welsh (λ = 0.974) oystercatcher population. The difference between the vital rates of the two populations are plotted in (A). The number of fledglings was especially lower in Germany, whereas adult survival was slightly higher compared to Wales. (B) shows the λ-sensitivity values of the three vital rates (calculated with a reference matrix containing the mean vital rates across the two populations). The projected population growth rate (λ) is especially sensitive to changes in adult survival. Multiplying the vital rate difference (A) and the λ-sensitivity (B) for each vital rate results in a linear approximation of the contributions to the difference in λ (C). The overall effect is negative (lower λ in Germany) due to smaller clutch size and lower juvenile survival but is partly buffered by a positive effect of increased adult survival. The approximation fits well: the sum of the three contributions (−0.0239) is very close to the difference in λ (-0.0240).

VARIANCE DECOMPOSITION

"Why is the projected population growth rate of one studied population higher than that of another?" is a common research question. Obviously some of the model parameter estimates were different, but how much does each of the parameter differences contribute to the difference in λ between the populations? To answer such questions, variance decomposition techniques have been developed, generally referred to as life table response experiments, or LTREs. The basic idea of LTREs is that the contribution of a parameter deviation is approximated by multiplying that deviation and the λ-sensitivity of that parameter. As discussed above, this is a rough approximation since sensitivity values are "local" characteristics that often change nonlinearly with changes in any model parameter. Still, the linear approximation works reasonably well: adding up the contributions of the deviations in all parameters often results in a value that is within a ±1% range around the actual difference in λ. For instance, when the Welsh oystercatcher population is contrasted with a German population (Fig. 5), the decomposition into contributions by vital rate differences fits very well (−0.01%). Model fit may deteriorate when the life histories that are being compared in the matrix models are increasingly different. In such cases, the linear approximation is no longer satisfactory, and researchers might want to use second derivatives or the exact relationships between λ and each model parameter.

This so-called fixed-effect LTRE decomposes the deviation in λ for each separate matrix (representing, for example, a population or a treatment) compared to a chosen reference matrix. In some cases, however, one might not be as interested in how each separate site or year deviates in the contributions to λ but rather in how vital rate variation contributed to the overall variance in λ. In such cases, a random-effect LTRE is more appropriate. In random-effect LTREs, the variance of λ across years is decomposed not only into contributions of the variances of each vital rate but also into contributions of the covariances among vital rates.

TRANSIENT DYNAMICS

Matrix analyses based on the projected population growth rate (λ) or other asymptotic matrix properties assume that these are informative when studying the dynamics of a population and how it will respond to, say, management or environmental change. However, as can be seen in Figure 3, it takes some years before these asymptotic dynamics are reached. Furthermore, the realized population growth depends not only on λ but also on the stage distribution a population happens to be in. The oystercatcher population simulated in Figure 3 is first decreasing, then increasing again, before steadily decreasing, while λ (0.974) is constant throughout.

The asymptotic properties can still inform researchers about how quickly asymptotic population behavior is expected to be reached. The ratio between the dominant eigenvalue and the second-highest eigenvalue of a transition matrix is called the damping ratio. High damping ratios tell you that the dominant stable stage distribution is reached fairly soon.

Recently, researchers have become increasingly interested in transient dynamics rather than asymptotic dynamics. The general notion is that transient dynamics (e.g., the first few years in Figure 3) are more relevant for what happens in real populations that are never in a stable state due to stochastic dynamics or disturbances. A wide arsenal of transient indices has been developed to study transient dynamics in response to all kinds of disturbances and across a wide range of life histories.

STOCHASTIC DYNAMICS

As we have seen, it takes some years with transient dynamics before asymptotic dynamics are reached. In the real world, however, environmental conditions (e.g., climate, competitors, predators) do not remain constant: every year is going to be different, resulting in different vital rates each year. As a consequence, populations are pulled toward different "stable" stage distributions each year. Population dynamics are therefore inherently stochastic in nature. Transition matrices can also be used to study stochastic dynamics.

Year-to-year variation in vital rates can be included in stochastic analyses in different ways. In simulations, vital rate values can be drawn each simulated year from measured or assumed probability distributions of vital rate values. When using random draws for each vital rate, extra care needs to be taken to make sure that the resulting transition matrix does not include biological impossibilities like survival rates larger than 100%. Some researchers therefore prefer another method of simulating population dynamics over long time spans: each year, a single transition matrix is randomly chosen to multiply the current population vector with (Fig. 6). The advantage of this method is that all vital rate combinations are biologically realistic.

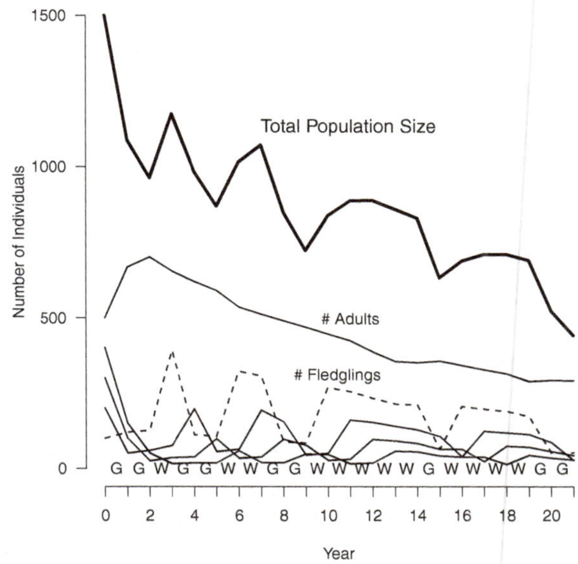

FIGURE 6 A stochastic simulation of an oystercatcher population. The initial population of 1500 is the same as in Figure 3. At each simulated year the population vector (with the number of individuals in each of the five stage classes) is multiplied with a randomly chosen transition matrix: each of two matrices had a 50% chance to be drawn each year. Ideally a pool of matrices representing observed data from a number of different years in the same population is used for such stochastic simulations. For illustrative purposes only, we here use matrices from two different populations: W = Wales and G = Germany (see also Fig. 5). The letters in the graph indicate which transition matrix was randomly drawn and applied each year in this simulation. Although both matrices W and G project asymptotic population declines, the simulated population is actually growing in some years due to changes in the distribution of individuals over the age classes.

From stochastic simulations like those in Figure 6, one can calculate the stochastic population growth rate, λ_s. To get a proper estimation of λ_s, it is important to remove the initial transient dynamics caused by the arbitrarily chosen initial population structure. Elasticity analyses can also be done for λ_s. At the level of matrix elements, stochastic elasticities again sum to 1. Stochastic elasticities, however, can be further split into an elasticity value of the mean of a matrix element and an elasticity value of the variance of that matrix element. Since a small increase in the variance of a matrix element can either increase or decrease stochastic population growth, λ_s-elasticity values of matrix element variances can be both positive and negative. Increases in matrix element means always have a positive effect on λ_s, and these elasticity values are therefore always positive. Such stochastic elasticities can also be used to study whether there are relationships between the amount of variation in vital rates and how much vital rate variance contributes to λ_s. Several such stochastic variance decomposition techniques (SLTREs) have recently been developed.

Population viability analyses often use stochastic simulations to estimate quasi-extinction risks. By repeating the stochastic simulation of Figure 6 many times, we can calculate how many years it takes for this hypothetical population of 1500 oystercatchers to become smaller than 50 birds: 70 years (±3.2 standard deviation). This is, of course, not a prediction but a projection of what would happen if the two transition matrices used in the simulation each have a 50% chance of occurrence each year. Ideally, large numbers of annual matrices from the same population are used. If correlations with climatic drivers can be made, stochastic simulation can then be used to study how the quasi-extinction risk chances with altered occurrence of, for example, good or bad years.

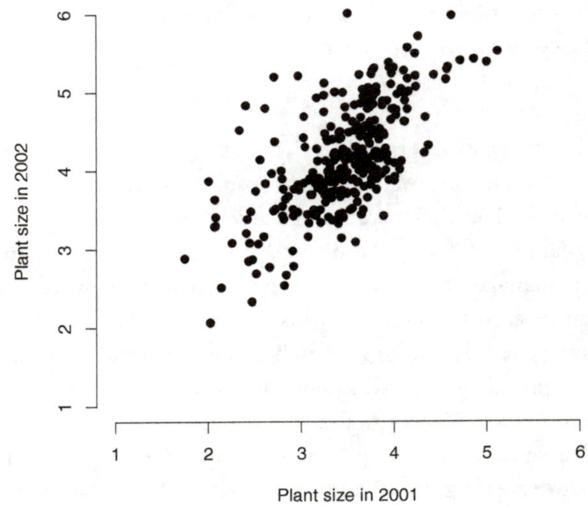

FIGURE 7 Relationship between plant size and the size of the same plants 1 year later in a devil's bit scabious (*Succisa pratensis*) population (data from E. Jongejans and H. de Kroon, 2005, *Journal of Ecology* 93: 681–692). No distinct size classes are apparent, and there is large variation in the size of individuals that were the same size the year before. In this case, plant size is the natural log of the product of the number of leaves and maximum leaf length.

INTEGRAL PROJECTION MODELS

One problem with size-based stage classifications is that it can be hard to find clear borders between one size group and the next: the sizes of individuals in a population often follow a continuous distribution, and vital rates mostly show gradual change with size (e.g., Fig. 7). Furthermore, the choice of the number of size classes is not without consequences: for example, matrix dimension can have profound impact on the projected population growth rate (λ). This is especially the case in long-lived, slow-growing species like trees: if there are only a few size classes, progression to the next class will be rare. However, according to such a model, it would still be possible for some individuals to reach the size of a large reproductive tree in only a few years. Such biologically unrealistic "shortcuts" in the life history increase λ disproportionally. However, when the variation in growth is relatively large compared to the range of observed sizes of individuals (as can be seen in Fig. 7), lower numbers (e.g., ≥7) of size classes will suffice.

Researchers have tried to deal with the problem of seemingly arbitrary size classification by using an algorithm developed by Moloney. This algorithm aims to optimize the number of distinctive classes taking the sample and distribution errors into account. Another method is to use statistical tests to find size classes with significantly different vital rates. However, these methods are strongly driven by sample size rather than based solely on the species' life history. An elegant solution was introduced in 2000: no size classification at all. In their integral projection models (IPMs), vital rates are continuous functions of size, functions that can be copied directly from statistical regression analyses. An additional advantage is that often fewer parameters need to be fitted compared to matrix models with different vital rates for each size class. Though IPMs contain continuous functions, it is still necessary to turn them into sufficiently large matrices before matrix algebra can be applied. In this way, IPMs combine the best of two worlds: parameterization and interpretation along continuous state variables, and the analytical toolbox of discrete matrix models.

IPMs are becoming increasingly popular in studies with continuous state variables (e.g., size), and all kind of combinations of discrete classes (e.g., seed banks) and continuous state variables are possible. An interesting development is to make vital rates not only functions of size but also of time within years. In some cases, for instance, winter and summer survival are modeled as separate vital rates, and if sufficient data are available, survival rates could be continuous functions of size and time between annual census days. Future studies will have to show the added value of adding such detail to projection models.

For comparative demography, IPMs may also prove to be very useful because they solve the problem that different matrix studies use a range of matrix dimensions. However, in comparative studies based on IPMs, decisions still will have to be made about how to compare the various vital rate functions and definitions between studies.

IPMs are also useful for studying the population-level impact of environmental factors like soil conditions, climate, management, and biotic interactions. These explanatory factors can be included in the statistical regression analyses of the vital rates. In such hierarchical population models, the projected population growth rate becomes a function of environmental drivers. For instance, one can then answer research questions like "How much is population size affected by cold winters?" and "How much of that response is due to changes in survival, growth, or reproduction?" Population size itself can also be included as a driver of vital rates. The asymptotic behavior of such density-dependent models is somewhat different from that of models in which density-dependence is not included explicitly. Fortunately, analytical tools for density-dependent matrix models have recently been developed.

SEE ALSO THE FOLLOWING ARTICLES

Age Structure / Conservation Biology / Demography / Population Ecology / Population Viability Analysis / Spatial Spread / Stage Structure / Stochasticity, Environmental

FURTHER READING

Caswell, H. 2001. *Matrix population models construction, analysis and interpretation.* Sunderland, MA: Sinauer.

Caswell, H. 2009. Sensitivity and elasticity of density-dependent population models. *Journal of Difference Equations and Applications* 15: 349–369.

Ellner, S.P., and J. Guckenheimer. 2006. *Dynamic models in biology.* Princeton: Princeton University Press.

Ellner, S.P., and M. Rees. 2006. Integral projection models for species with complex demography. *American Naturalist* 167: 410–428.

Heppell, S., C. Pfister, and H. de Kroon, eds. 2000. Elasticity analysis in population biology: methods and applications. Special feature. *Journal of Ecology* 81: 605–708.

Jongejans, E., H. Huber, and H. de Kroon. 2010. Scaling up phenotypic plasticity with hierarchical population models. *Evolutionary Ecology* 24: 585–599.

Morris, W.F., and D.F. Doak. 2002. *Quantitative conservation biology: theory and practice of population viability analysis.* Sunderland, MA: Sinauer.

Salguero-Gómez, R., and H. de Kroon, eds. 2010. Advances in plant demography using matrix models. Special feature. *Journal of Ecology* 98: 250–355.

Tuljapurkar, S. 1990. *Population dynamics in variable environments.* New York: Springer-Verlag.

META-ANALYSIS

MICHAEL D. JENNIONS

Australian National University, Canberra

KERRIE MENGERSEN

Queensland University of Technology, Australia

Meta-analysis is a statistical approach to analyze the combined results from two or more empirical studies that each test for the existence of the same relationship. The putative relationship could be a causal one that is investigated experimentally (e.g., Does removal of predators reduce herbivore diversity?) or a correlational one (e.g., Does body size increase with latitude?). If the studies analyzed are an unbiased, representative sample of the available studies, the meta-analysis provides a systematic review of the study question. In this broader sense, a meta-analysis is a review that interprets the information in the scientific literature with the same statistical rigor that we now consider obligatory when testing hypotheses in primary empirical studies.

THE BASIC STATISTICS

Meta-analysis involves a set of statistical procedures that allow for the combination of statistical results from different studies. The process starts by converting the results of each study into a measure of the strength of the relationship being tested. Each of these effect sizes has an associated sampling variance that declines with an increase in the number of independent data points in the study. The more data we have, the more precise the estimate. The most commonly used effect sizes are unitless measures that include correlation coefficients, standardized differences between control and treatment means (standardized by dividing by a pooled estimate of the standard deviations from the two group means), and ratios (e.g., risk or odds ratio). All ecologists are familiar with calculating probability when using frequentist statistics to test a null hypothesis. This task is equivalent to determining whether the confidence interval for the effect size includes the null value (i.e., $P < 0.05$ if the 95% CI excludes the null value). An effect size can be described as a sample size–corrected P-value. For a given P-value, the smaller the sample size the larger the estimated effect size must be because less data increases the associated variance of the estimate. Greater variance widens the 95% confidence interval, which is then more likely to include the null value, unless the mean is further away from the null value (i.e., the absolute effect size is larger).

Meta-analysts are interested in exploring the distribution of a population of effect size rather than focusing on the results of any single study. Thinking in terms of effect sizes is an important by-product of conducting meta-analyses. It leads to a more sophisticated statistical approach to the scientific literature and decreases the risk of mistaken interpretation of raw *P*-values. For example, it becomes obvious that two studies can yield the same effect size estimate but have very different *P*-values due to differences in the width of the associated confidence intervals (i.e., due to sample size differences); or that two studies can yield very different estimates of the mean effect and might, or might not, have very different *P*-values. In neither case is there cause to view study findings as conflicting simply because one is significant and the other nonsignificant. The conclusion that the studies disagree is only appropriate if the confidence intervals for the effect sizes do not overlap. Even then, given a large population of effect sizes, we would expect some of the estimated means to differ from each other by chance alone.

The simplest use of a meta-analysis is to estimate the mean effect size for a putative relationship and calculate the likelihood that it differs from a value of interest. This is often a test of whether the observed relationship differs from the null expectation, performed by examining the population of estimates obtained from the analyzed studies. For example, does the 95% confidence interval for the mean effect size r differ from zero? The mean is calculated after weighting each study by the inverse of its variance, which gives greater weighting to larger studies with smaller variances that produce more precise effect size estimates. Crucially, and especially important in ecology where studies often differ in their details, a range of follow-up questions can then be asked about the observed variation in effect sizes among studies. Do the estimates vary more than expected by chance (i.e., due to sample size–based sampling error)? If so, one can test whether certain factors account for some of the heterogeneity among studies. For example, do effect sizes differ between studies conducted in marine and terrestrial ecosystems, or between studies using different methodologies? If so, one can formulate new hypotheses that can be tested with additional manipulative experimental studies.

A BRIEF HISTORY OF META-ANALYSIS

Meta-analysis has revolutionized the social and, especially, the medical sciences. In the late 1970s, health practitioners became aware that it was risky to rely on the outcome of any single study to determine the efficacy of a drug or medical intervention. Even the most well-designed, large-scale study can produce results that inaccurately estimate the general effect of a treatment in the broader population due to unique features of the study population or specific details of the experimental design. For ecologists, the danger of inferring from a single ecosystem or species what will happen in other contexts should be immediately apparent. Using information from multiple studies is always likely to improve our ability to estimate the average effect. Similarly, it became clear to medical and social scientists that when treatment effects are small, many studies will fail to report a statistically significant effect due to the low statistical power imposed by logistically feasible sample sizes. Again, meta-analysis of the combined evidence from the available studies, most of which might be modest in size, allows one to test whether there is a consistent trend for a given treatment to produce the predicted effect. It is important to note that this does not involve tallying how many studies have or have not produced a significant result in the required direction. Vote counting based on whether a given probability threshold is exceeded yields an overly simplistic form of meta-analysis with very low statistical power. It is not recommended. All modern meta-analyses treat effect sizes as continuous variables.

The use of meta-analysis is now standard practice in medicine. It can even be a legal requirement before a new clinical trial is authorized, or be a prerequisite component of funding applications. The "evidence-based medicine" approach has superseded earlier reliance on narrative reviews that placed excessive reliance on expert opinion and the subjective weighting (often based on unstated criteria) these experts gave to different studies. The history of meta-analysis in medicine is instructive about the potential rewards of greater use of meta-analysis by ecologists. For example, there are several well-known cases where additional clinical trials were initiated resulting in unnecessary deaths. At the time these trials started, the available studies already showed the efficacy of the treatment. Conducting a meta-analysis as each new study was completed (cumulative meta-analysis) would have reveal a substantial treatment effect many years prior to trials being terminated. Ecologists have limited resources, so it would seem judicious to know when a relationship has been sufficiently well established so that a new topic can be tackled or future studies modified to ask more subtle questions about variation in the focal relationship.

The first ecological meta-analyses were published in the early 1990s. The use of meta-analysis grew rapidly after a 1995 article aimed at ecologists by Arnqvist and Wooster, and a collection of papers on meta-analysis in *Ecology* in 1999. Just over a decade later, the use of meta-analysis now occurs in almost every ecological discipline, including applied, evolutionary, population, and community ecology. To date, no manual on how to conduct a meta-analysis has been produced for ecologists, but a handbook is due in 2012.

The widespread use of meta-analysis by ecologists is particularly pressing for several reasons. First, most ecological studies have low statistical power. This makes it dangerous to rely on the repeated occurrence of significant results in individual studies to infer that a relationship exists. This is especially true for questions that require replication at high levels to avoid pseudoreplication (e.g., where distinct habitat patches or different river systems are the unit of replication) or where measurement error is severe (e.g., it is difficult to measure the feeding rate or diet of wild animals with great accuracy). Second, many factors interact and contribute to ecological relationships. The presence/strength of uncontrolled and often unmeasured factors differs among studies so that even precise replication of an experiment manipulating a single focal factor will produce a range of effect sizes depending on when and where it is conducted. More generally, effect sizes are often small in ecology, so that, again, individual studies have little statistical power. Third, there are genuine biological differences between ecosystems, species, and populations. The use of meta-analysis to test for significant variation in effect sizes can allow one to identify larger-scale patterns that might otherwise go unnoticed. This can lead to formulation of interesting new hypotheses.

CRITICISM OF META-ANALYSIS

Some ecologists have criticized the use of meta-analysis. Their critiques largely echo objections that initially arose in the medical and social sciences. There are three concerns that are worth highlighting. First, it is claimed that the published literature is biased (e.g., that it is easier to publish significant results), so that a meta-analysis will overestimate the true effect size. This is possibly true—although the available evidence suggests it is not a major problem in ecology. Note, however, that the same problem arises when conducting a traditional narrative review, even though publication bias is rarely mentioned in such reviews. In fact, direct investigation of publication bias in ecology has been conducted almost exclusively by researchers that conduct meta-analyses. One advantage a meta-analysis has over a narrative review is the emphasis placed on providing a detailed protocol of the methods by which studies were located. For example, did the reviewer seek out unpublished work or only look at English-language journals? The meta-analyst tries to conduct a systematic review rather than highlight a subset of papers that she or he thinks are worthy of special attention or illustrative of wider patterns. The meta-analyst's toolkit also includes statistical tests to detect potential publication bias and assess how strong such a bias must be to negate the reported findings. For example, how many negative studies would need to lie undetected to obliterate a reported significant mean effect size based on the currently located studies?

Second, it is claimed that meta-analysis is inappropriate because it combines "apples and oranges." Each species is by definition unique, so how can it be appropriate to talk about the average effect of, say, distance from the parent tree on seedling survival? A quick retort is that apples and oranges are both fruit and that it is possible, and desirable, to make generalized statements about fruit (such as the average weight) even though no single species might have the described mean properties. All science requires the willingness to generalize. Ultimately, the diversity of studies included in a meta-analysis reflects the extent to which the analyst wishes to generalize beyond the limited conclusions that can be draw from a primary study conducted at a single point in space and time. Again, it is noteworthy that most ecologists are already comfortable informally combining results from studies that involve different species or systems. Indeed, the use of citations in primary papers is implicitly an attempt to place a study in a wider context (i.e., did other studies report the same trend?). Finally, it is noteworthy that in a meta-analysis either fixed or random effect models can be used to estimate mean effects. The former assume that all variation in effect sizes among studies is due to sampling error. The latter assume a true underlying effect that also varies among studies. Both the random variance among studies and the variance due to sampling error are calculated. There is no need to assume that the effect size will be identical in every study when conducting a meta-analysis, and most ecologists use random effects models.

Third, there is concern that meta-analysis leads to the uncritical combination of studies that vary widely in quality. In fact, the strict systematic review protocols used to identify studies for inclusion in a meta-analysis can exclude "poor" studies that do not meet quality or bias criteria *if* these standards can be clearly defined. The real problem is that in a narrative review when studies are ignored or dismissed, no objective criteria are provided and the decision is usually ad hoc. In contrast, the more objective approach underlying a well-conducted meta-analysis allows the reader to see the decisions that were made prior to synthesizing the literature. Indeed, the data is often presented in sufficiently detail that a reader can even reanalyze it using his/her own inclusion criteria.

THE PATH AHEAD

Systematic reviews and meta-analyses will become the default option for anyone conducting a literature review on a quantitative ecological question. The basic statistical techniques are in place, but more sophisticated and/or alternate approaches are being developed. For example, some statisticians take a Bayesian approach to meta-analysis, and new multivariate models are appearing. A key concern for ecologists is obtaining better access to models that can handle nonindependence and account for the inevitable spatial, temporal, and/or phylogenetic correlations among studies that arise in ecology. In practice, the uptake of more-sophisticated techniques will depend on the speed with which user-friendly, point-and-click software is developed that does not require advanced programming skills or writing code to run models. Greater emphasis on effect sizes rather than *P*-values is, arguably, the most immediate benefit of wider use of meta-analysis. Many current arguments among ecologists about differences between specific studies are likely to dissipate once the population of effect sizes is considered and the relative position of these focal studies among a continuum of findings is taken into account.

SEE ALSO THE FOLLOWING ARTICLES

Bayesian Statistics / Frequentist Statistics / Information Criteria in Ecology / Model Fitting / Statistics in Ecology / Stochasticity

FURTHER READING

Borenstein, M., L.V. Hedges, J.P.T. Higgins, and H.R. Rothstein, eds. 2009. *Introduction to meta-analysis.* Chichester, UK: John Wiley & Sons.

Cooper, H., L.V. Hedges, and J. Valentine, eds. 2009. *The handbook of research synthesis.* New York: Russell Sage.

Jennions, M.D., and A.P. Møller. 2002. Publication bias in ecology and evolution: an empirical assessment using the "trim and fill" method. *Biological Reviews of the Cambridge Philosophical Society* 77: 211–222.

Koricheva, J., J. Gurevitch, and K. Mengersen, eds. 2012. *Handbook of meta-analysis in ecology and evolution.* Princeton: Princeton University Press.

Nakagawa, S., and I.C. Cuthill. 2007. Effect size, confidence intervals and statistical significance: a practical guide for biologists. *Biological Reviews of the Cambridge Philosophical Society* 82: 591–605.

METABOLIC THEORY OF ECOLOGY

JAMES F. GILLOOLY AND APRIL HAYWARD

University of Florida, Gainesville

MELANIE E. MOSES

University of New Mexico, Albuquerque

The metabolic theory of ecology (MTE) aims to utilize our knowledge of metabolism to link the structure and function of ecological systems across scales of biological organization. The theory consists of a series of interrelated mathematical models that fall into two classes. The first class aims to better understand the mechanisms underlying the body size- and temperature-dependence of metabolism at the level of cells and individuals. The second class explores the consequences of these size- and temperature-dependencies at different levels of biological organization, from cells to ecosystems, by invoking simple principles of mass and energy balance. For both classes, the models are intended to be sufficiently general so as to make first-order predictions across diverse species (e.g., plants, animals, microbes) and environments (e.g., aquatic, terrestrial). Together, the two classes of models in MTE represent an attempt to provide a general, synthetic theory for ecology that predicts aspects of the structure and function of populations, communities, and ecosystems based on the structure and function of individuals that comprise these systems. In all MTE models, energy is taken to be the fundamental biological currency.

METABOLISM AS A LINK AMONG BIOLOGICAL LEVELS OF ORGANIZATION

Biology in general, and ecology in particular, has long sought to link different levels of biological organization. Progress toward linking properties of cells and individuals to properties of communities and ecosystems provides a more synthetic view of ecological systems. At present, though, the study of populations, communities, and ecosystems in ecology is often distinct from the study of other levels of biological organization and thus from other subdisciplines focused on these levels (e.g., physiology, cell and molecular biology). This makes it difficult to discern idiosyncratic features of particular ecological systems from their more general features. It also makes it more difficult for the field of ecology to develop general theory that incorporates well-established laws of physics, chemistry, and biology. Note, however, that progress continues to be made on this front. Since ecology's inception, many subdisciplines have emerged with this general goal in mind, subdisciplines such as physiological and biophysical ecology. However, in many cases, the progress in these subdisciplines has not been incorporated into studies of communities and ecosystems. In more recent years, emerging subdisciplines such as ecological genomics provide yet another exciting opportunity to link genetics to ecosystem function.

In all cases, linking phenomena at different levels of biological organization requires addressing the vast differences in spatial and temporal scales that occur across levels. From unicellular organisms to giant sequoias, the life spans of species vary by nearly 10 orders of magnitude, and body mass varies by nearly 20 orders of magnitude. From a theoretical perspective, this has made the challenge of linking levels in ecology a daunting task!

Metabolism provides an ideal starting point for establishing conceptual and quantitative linkages across levels of organization. Metabolic rate, which is the rate at which an organism takes up, transforms, and uses energy for survival, growth, and reproduction, is fundamental to the structure and function of life at all levels. Moreover, there are at least three special features of metabolism that make it well suited for linking ecological pattern and process to those at smaller scales. First, many cellular-level attributes of organisms, such as organelle density and cell lifespan, vary in proportion to metabolic rate across the diversity of life. Second, it has been known for over a century that metabolic rate varies predictably with body size and temperature. Together, this means that just by knowing the size and temperature of an animal one can estimate its metabolic rate. Third, metabolic rate, by definition, is the process that links the physiology of organisms to the ecology of their environments. So, in many respects, metabolic rate represents the nexus between cell structure and function and the physiology and ecology of individuals.

BACKGROUND

Nearly a century ago, Max Kleiber observed that the metabolic rate of a mammal scales with body mass (M) as a power law of the form $y = aM^b$, where b is approximately 3/4. This relationship has since been observed in many groups of plants and animals. The fact that the exponent is less than 1 indicates that metabolism increases less than proportionally with body mass. For example, a large 100-kg human is 10,000 times bigger than a small 10-g mouse, but its metabolism is only $(10{,}000)^{3/4}$, or 1000, times greater. Notably, the 3/4-power exponent of whole-organism metabolic rate also indicates that mass-specific metabolic rate, or the metabolic rate per gram of tissue, decreases with increasing body mass to the $-1/4$ power. As such, larger animals, pound for pound, use less energy than smaller animals.

Over the years, many theoretical models have proposed to explain the characteristic 3/4-power scaling of metabolic rate with body mass. The explanation that has received the most attention recently is that of West, Brown, and Enquist, who proposed that the 3/4 exponent arises from maximizing energy delivery though the networks that deliver metabolic substrates to tissues, such as the vertebrate arterial network or the network of xylem and phloem that connects leaves to roots in trees. While the derivation of the model is mathematically complex, it is based on simple physical principles that apply across a wide variety of networks. In essence, the theory shows that networks have diminishing returns—that is, the ability of the network to supply energy to cells does not keep pace with the size of the network and the mass of the organism. Instead, physics and geometry constrain the rate of energy supply to be proportional to $M^{3/4}$. To be sure, branching networks such as those described by West, Brown, and Enquist, where a small number of large vessels branch into successively more and smaller vessels, are common in biology. Since the development of this model, alternative network models based on different assumptions have been proposed, and the question of whether more general properties of resource distribution networks could explain the scaling of metabolic rate continues to be investigated. Still, as the scientific debate improves our understanding of the mechanisms that *cause* the relationship between metabolism and body size, much progress is being made in understanding the *consequences* of that relationship.

At about the same time as Kleiber's seminal work, the classic work of Krogh, Arrhenius, and Boltzmann showed that metabolic rate also increases exponentially with temperature (Fig. 1). This relationship is well described by the Boltzmann–Arrhenius factor, $e^{-E/kT}$, where E is the average activation energy of the respiratory complex, k is Boltzmann's constant, and T is absolute temperature. Given that E is approximately 0.65 eV, this equates to about a 2.5-fold increase in metabolic rate for every 10 °C

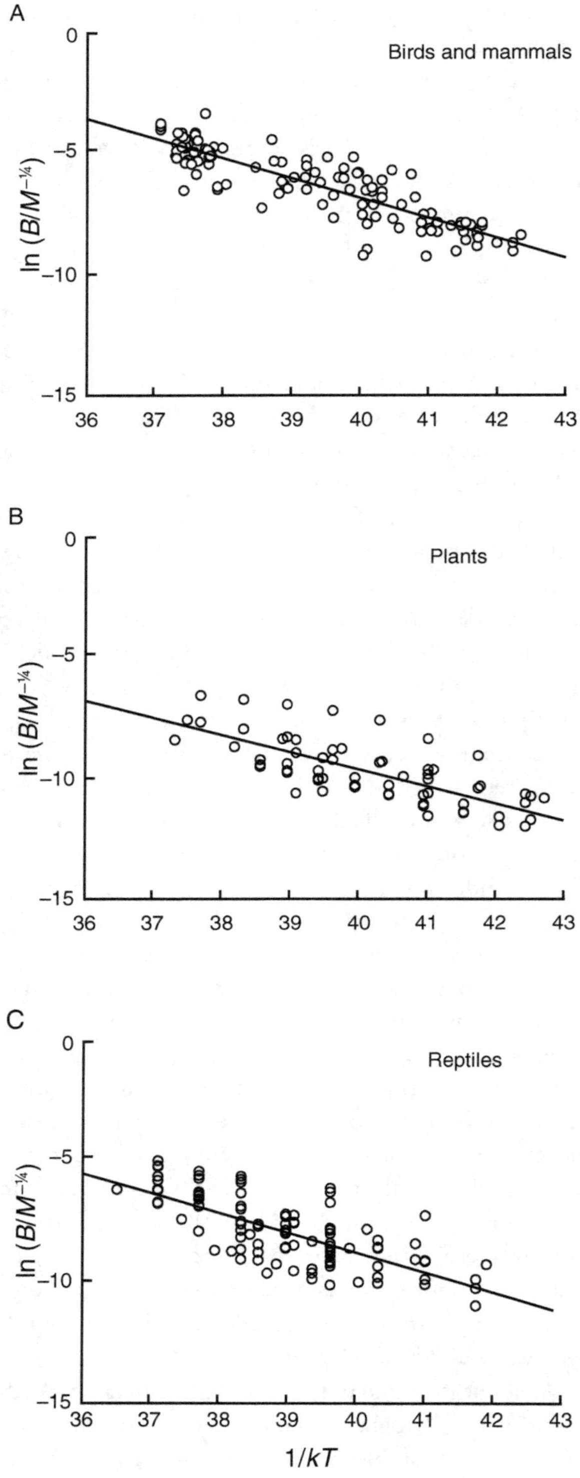

FIGURE 1 Mass-corrected metabolic rate as a function of temperature for (A) birds and mammals ($\ln(y) = -25.90 - 0.82x$, $r^2 = 0.79$, $p < 0.0001$); (B) plants ($\ln(y) = 18.70 - 0.70x$, $r^2 = 0.57$, $p < 0.0001$); and (c) reptiles ($\ln(y) = 23.62 - 0.81x$, $r^2 = 0.52$, $p < 0.0001$). Together, body mass and temperature explain a substantial portion of the variation in metabolic rate across the diversity of life. Data from Gillooly et al. (2001, *Science* 293: 2248–2251).

increase in temperature (this is equivalent to what physiologists refer to as a "Q_{10}" of 2.5).

It is advantageous to express the temperature dependence of metabolism by the Boltzmann–Arrhenius factor rather than the more traditional Q_{10} because it emphasizes that the temperature dependence arises from the basic physics governing biochemical reactions. This formulation also circumvents the temperature dependence of Q_{10} factors across larger temperature ranges. Additionally, it provides a simple equation that relates whole-organism metabolic rate to both body size and temperature as

$$B = b_o M^{3/4} e^{-E/kT}, \tag{1}$$

where b_o is a normalization constant that varies among taxonomic groups, T is temperature, and E and k are constants representing the temperature dependence of biochemical reactions. This expression simply combines Kleiber's power law with Boltzmann's exponential temperature dependence (see Box 1). In one sense, the simplicity of Equation 1 is surprising since metabolism is a very complex biological process that entails thousands of chemical reactions. In another sense, though, it should not be surprising, since metabolic pathways are largely conserved across very different taxa, so the temperature dependence of all of the chemical reactions of metabolism are, on average, very similar.

Still, the body size dependence and temperature dependence of metabolism, and the mechanisms that may generate these relations, remains a contentious topic of debate. With respect to the temperature dependence, some have argued that there is no characteristic temperature dependence and/or that such a dependence cannot be attributed to the biochemical kinetics of metabolic reactions. Similarly, with respect to the size dependence, some have argued that there is no characteristic exponent or mechanistic explanation. Even so, Equation 1 provides a useful functional form for linking different levels of biological organization. Equation 1 captures how size and temperature jointly influence metabolic rate: body size determines the rate that metabolic substrates can be delivered to cells, and temperature determines the rate that those substrates can be converted into energy by metabolic reactions. Because mass varies so widely and biochemical reactions are so sensitive to temperature, Equation 1 captures an enormous amount of variation in metabolic rate. As we explain below, the size and temperature dependence of metabolic rate has important implications for understanding the structure and function of ecological systems.

BOX 1. METABOLIC RATE IS A FUNCTION OF BODY MASS AND TEMPERATURE

Metabolic rate can be described as a function of (a) temperature and (b) body mass. Here, B is the metabolic rate of an organism, b_o is a normalization constant that determines the intercept of the relationship, M is body mass, E is the average activation energy of metabolism (0.6–0.7 eV), k is Boltzmann's constant (8.62×10^{-5} eV K^{-1}), and T is absolute temperature. Specifically, (a) since whole-organism metabolic rate scales with body mass to the power of 3/4, mass-specific metabolic rate B/M—metabolic rate per gram of tissue—scales as a −1/4 power with body mass. Thus, the natural logarithm of temperature-corrected mass-specific metabolic rate scales as a function of body mass to the power of −1/4. (b) The natural logarithm of mass-corrected metabolic rate is negatively related to the inverse of temperature, meaning that metabolic rate is faster at higher temperatures. The slope of the line reflects the average activation energy of the metabolic reactions involved in heterotrophic respiration (~0.65 eV).

$$B = b_o \cdot M^{-1/4} \cdot e^{-E/kT}$$

A $\ln(B/M \cdot e^{E/kT}) = -1/4 \ln M + \ln(b_o)$ B $\ln(B/M^{-1/4}) = -E\,(1/kT) + \ln(b_o)$

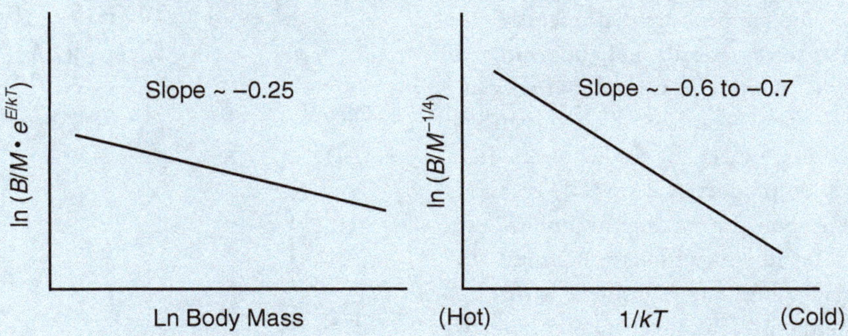

MTE AND INDIVIDUAL-LEVEL ECOLOGY

MTE can help us predict the rates of nutrient cycling, as well as rates of survival, growth, and reproduction of individuals. In terms of nutrient cycling, MTE predicts that the flux, storage, and turnover of elements in organisms is largely governed by individual metabolism. Thus, MTE proposes that these should have predictable body size dependencies and temperature dependencies. Specifically, MTE predicts that rates of biomass uptake, use, and excretion show the same quarter-power body mass scaling and the same temperature dependence as metabolic rate, at steady state.

A compilation of data on biomass production rates for a diversity of plants and animals support this hypothesis (Fig. 2). In Figure 2, body mass and temperature account for 99% percent of the variation in biomass production across species, and the slope of the relationship is nearly identical to the predicted value of 3/4. Other studies have shown similar scaling relationships for uptake and excretion rates in organisms. Moreover, for both plants and animals, MTE quantitatively predicts how biomass is allocated among different tissues (e.g., leaves, roots, and shoots). For example, the number of nitrogen-rich chloroplasts, phosphorous-rich ribosomes, and even the number of leaves in a tree scale in proportion to metabolic rate. In terms of survival, growth, and reproduction, MTE proposes that differences in body size and temperature largely govern metabolic rate, and metabolic rate in turn governs these basic rates. Thus, MTE predicts that

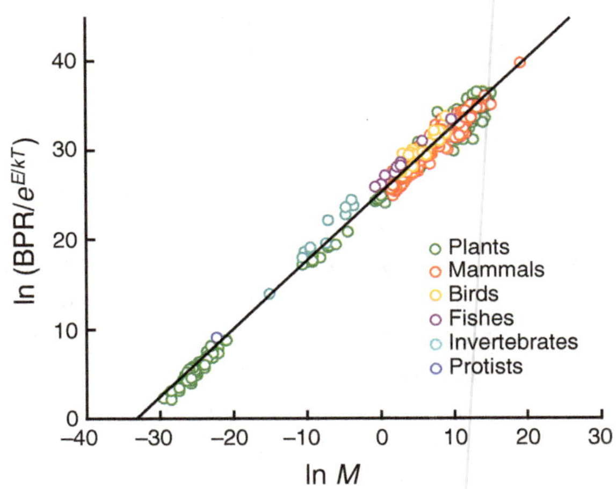

FIGURE 2 The mass-dependence of temperature-corrected biomass production rates (g • $year^{-1}$). As predicted by MTE, biomass production rates scale as a 3/4-power function of body mass: $\ln(y) = 25.25 + 0.76 \ln(x)$, $r^2 = 0.99$, $p < 0.0001$. Data from Ernest et al. (2003, *Ecology Letters* 6: 990–9959).

biological rates, like rates of growth, reproduction, or excretion, are proportional to mass-specific metabolic rate. The implications of this for growth are addressed in detail elsewhere in this volume. Consequently, according to MTE, biological rates are expected to show the following proportionality:

$$\text{Biological Rates} \propto M^{-1/4}\, e^{-E/kT}. \tag{2}$$

Times are expected to be inversely proportional to rates:

$$\text{Biological Times} \propto M^{1/4}\, e^{E/kT}. \tag{3}$$

As an example, Figure 3 shows that life spans of both plants and animals living in their natural environments are predicted by the size- and temperature-dependence of metabolism. Figure 3 also shows that size and temperature explain almost half of the variation in natural mortality rates (i.e., the inverse of life span) regardless of whether mortality is caused by predation, disease, or abiotic forces. Interestingly, life span scaling proportional to $M^{1/4}$ combined with mass-specific metabolic rate scaling in proportion to $M^{-1/4}$ implies that the amount of energy expended in a lifetime is approximately invariant, consistent with Pearl's Rate of Living Hypothesis. This invariance implies that if an ectotherm is living in an environment that is 10°C warmer than another ectotherm of the same size, it will expend energy at a rate that is approximately 2.5-fold higher than the colder animal, but it will also die 2.5-fold sooner. But life span is just one of the many demographic times that show this relationship to mass and temperature. Others include egg hatching time and time to first reproduction, gestation period, and age at weaning.

The metabolic basis of demographic rates and times also provides some insight into fitness. For example, we have learned from MTE that, integrated over an entire life span, reproductive effort, measured as the fraction of mass or metabolism allocated to reproduction, is size invariant for diverse vertebrates, including lizards and mammals. This is a consequence of reproductive rates ($\sim M^{-1/4}$) and reproductive times ($\sim M^{1/4}$) scaling in opposite directions with body mass. This means that, on average, total lifetime reproductive output, a common measure of fitness, does not vary with body mass. Even more interestingly, combining MTE with a life history optimization model predicts the actual amount of reproductive effort as the inverse of the metabolic scaling exponent. Thus, knowing the metabolic exponent allows one to accurately predict the amount of reproductive effort made by mammals (including humans in hunter-gatherer societies) and lizards. More generally, this example demonstrates how allometric theory can be combined with life history theory to better understand fitness. Other recent work has suggested that the temperature dependence of r_{max} predicted by MTE, which is discussed in more detail below, indicates that "hotter is better" in terms of fitness, since warmer temperatures result in higher reproductive rates (another measure of fitness). It is important to point out that MTE's models at the individual level, and at all higher levels considered here, generally assume some steady state in which consumption rates match excretion rates, which makes the mathematics more tractable. In principle, this

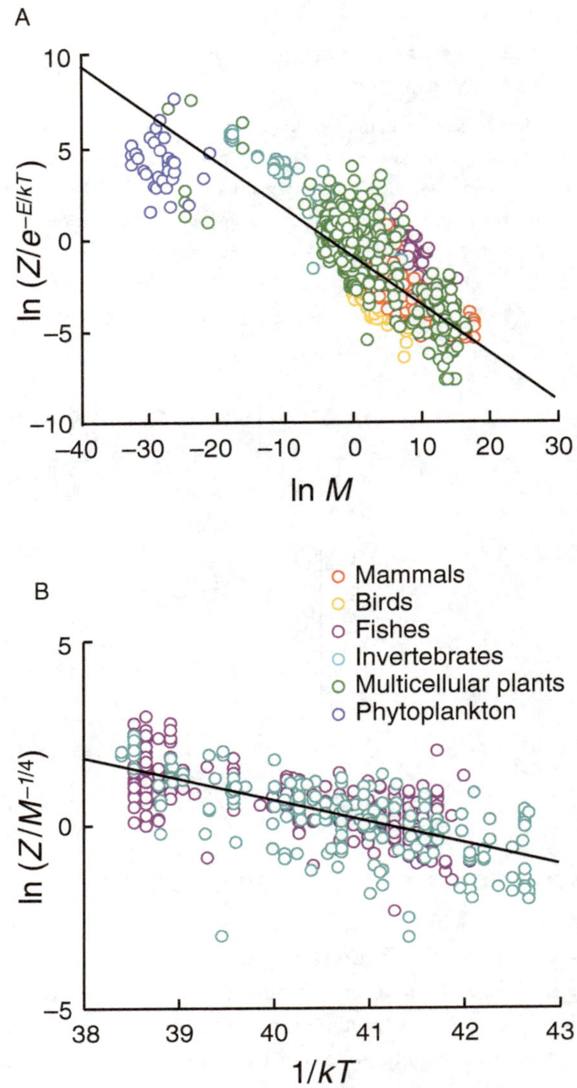

FIGURE 3 The mass- and temperature-dependence of mortality rate (individuals • individual^{-1} • year^{-1}). (A) Temperature-corrected mortality rate as a function of body mass: $\ln(y) = -0.99 - 0.26 \ln(x)$, $r^2 = 0.59$, $p < 0.0001$. (B) Mass-corrected mortality rate as a function of temperature: $\ln(y) = 23.47 - 0.57 \ln(x)$, $r^2 = 0.46$, $p < 0.0001$. Both relationships conform to the predictions of MTE. Data from McCoy and Gillooly (2008, *Ecology Letters* 11: 710–716).

assumption can be relaxed, and when it is, the results are often surprising. For example, predicting the amount of food required by growing animals requires relaxing the steady state assumption because growing animals consume more than they excrete. MTE predicts that food consumption over ontogeny is not a simple power law of mass. This is because the rates of food consumption required to fuel growth and maintenance are described by different power laws, and the sum of these does not yield a power law.

MTE AND POPULATION-LEVEL ECOLOGY

The dynamics of populations in space and time have long been a primary focus of both ecology and evolution. Population dynamics are key for understanding how organisms inhabit, interact, and ultimately coexist and evolve in their natural environments. But these dynamics are often quite complex because they are impacted by a large array of abiotic and biotic factors. Historically, two of the most useful measures of populations to ecologists have been the intrinsic rate of maximum growth, r_{max}, and population density at carrying capacity, K. The term r_{max} describes the potential of a population to grow given unlimited resources. It describes a population's ability to recover from disturbance, to expand into newly available habitats, and to compete with other species for resources. Carrying capacity is conceptually useful because it speaks to the maximum density of individuals an environment can support given some finite supply of resources.

Despite the complexity of population dynamics, the demographic rates of individuals that drive r_{max} and K are tightly linked to metabolic rate. MTE predicts the intrinsic rate of population increase, r_{max}, shows the same size dependence and temperature dependence as mass-specific metabolic rate such that

$$r_{max} \propto M^{-1/4}\, e^{-E/kT}. \qquad (4)$$

Note that this predicted size dependence and temperature dependence of r_{max} is the same as that for mortality rates above, but only under the assumption of steady state. In reality, the extent to which populations experience a steady state is debatable since many factors in natural systems are constantly changing (e.g., resource availability).

MTE can also make predictions about carrying capacity, K, by assuming that the number of individuals an area with finite resources (R) can support depends on the rates of resource use of the individuals in that area. This is because these rates are governed by metabolic rate. Specifically, MTE predicts

$$K \propto [R]\, M^{-3/4}\, e^{E/kT}. \qquad (5)$$

This equation is based on the assumption that resource consumption rates scale in proportion to whole-organism metabolic rate and that the total consumption rate by the individuals in the area matches resource supply rate. Empirically, there is some support for these predictions. John Damuth was the first to observe that $K \propto M^{-3/4}$ in mammals, and it has since been shown to hold for both terrestrial and marine populations. Interestingly, this implies that the amount of energy used by whole populations is approximately invariant with body mass (i.e., $\propto M^0$) since the number of individuals $\propto M^{-3/4}$ but the metabolic rate of an individual $\propto M^{3/4}$. In other words, on average there is "energetic equivalence" among populations of different species such that a hectare of small herbs and a hectare of large trees will use the same amount of energy given the same amount of resources.

Perhaps more intriguingly, MTE's model for carrying capacity (Eq. 6) also predicts that population density should decrease with increasing temperature. There is at least some evidence to suggest it is the case. In one study, after correcting for temperature, ectotherm and endotherm abundance showed the same relationship with body mass, with a slope very close to the predicted value of $-3/4$.

MTE AND COMMUNITY AND ECOSYSTEM ECOLOGY

Communities and ecosystems undergo dynamic shifts in species richness, abundance, energy, and biomass as a consequence of the complex structure and dynamics in multispecies assemblages. MTE can contribute to our understanding of these shifts and the species interactions that underlie them. At the community level, MTE can provide insights into interspecific interactions, including competition and predation, and inform our understanding of food web structure and dynamics. At the ecosystem level, MTE holds promise for predicting the flux, storage, and turnover of energy and elements among different compartments or functional groups in ecosystems. This is because the rates of movement, consumption, excretion, and population growth and death that govern these dynamics among biota are all constrained by individual metabolism. Community ecology typically focuses on changes in species and abundance, and ecosystem ecology typically focuses on changes in energy and elements. MTE makes testable predictions about both. Here, we describe what MTE predicts about interactions between species pairs and about the flux, storage, and turnover of individuals, energy, and biomass in multispecies systems. In doing so, we leave out a growing literature that addresses how MTE can be applied to our understanding of the origin and maintenance of biodiversity at broad spatial scales.

The study of species interactions in ecology has largely focused on the effects of competition or predation on species coexistence since the classical studies of Lotka and Volterra. MTE reminds ecologists that the nature of these interactions will depend on the metabolic rates of the species involved. For example, competition between two species of the same size consuming the same resource should increase exponentially with temperature in the same way as metabolic rate. Moreover, rates of predation of the same-sized prey should increase as a power law with predator body mass. MTE's predictions about species interactions have proven useful in a number of contexts. For example, a reexamination of Park's classic competitive exclusion experiments with flour beetles shows that across three temperatures, time to competitive exclusion declined exponentially with increasing temperature with an activation energy of approximately 0.64 eV, nearly identical to the temperature dependence of metabolism. Since these experiments, a number of other interaction rates and times, including rates of parasitism, herbivory, and predatory attack, have been shown to exhibit temperature and/or body size relations that are generally consistent with those predicted by MTE. Furthermore, the size dependence and temperature dependence of metabolism have been explicitly incorporated into mathematical models to yield novel predictions about consumer–resource dynamics and the interaction strength of species in communities.

The study of trophic dynamics in community and ecosystem ecology has largely focused on how energy, biomass, or abundance is partitioned among multiple species occupying different trophic levels. MTE can help here, too, since body size, temperature, and resource supply determine the magnitudes of stores and rates of flux within and between compartments of multispecies assemblages or functional groups such as primary producers, herbivores, predators, and detritivores. By assuming that the total energy or biomass being fluxed, turned over, or stored in a given biological compartment is simply the sum of those of the constituent individuals, MTE links these dynamics to individual metabolism by employing mass and energy balance. In community ecology, this approach has been used to examine the dynamics, stability, and topology of food webs. In ecosystem ecology, this approach has proven useful for predicting the flux, storage, and turnover of elements such as carbon, nitrogen, or phosphorus. For example, MTE correctly

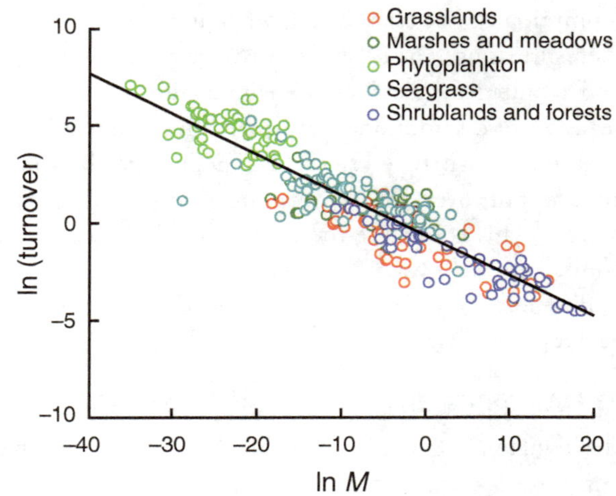

FIGURE 4 The mass-dependence of annual carbon turnover rate (year^{-1}) in aquatic and terrestrial plant communities, where mass is average plant size (g dry mass). The slope of the line is close to the predicted value of −1/4: $\ln(y) = -0.66 - 0.21 \ln(x)$, $r^2 = 0.84$, $p <$ 0.0001. Data from Allen et al. (2005, *Functional Ecology* 19: 202–213).

predicts that the logarithm of carbon turnover rate in different ecosystems would decrease with the logarithm of average plant size in those systems with a slope of −0.25 (Fig. 4). Figure 4 shows that the prediction is very close to the observed relationship for both aquatic and terrestrial ecosystems. MTE also correctly predicts that root decay rates from different sites should show the same temperature dependence as heterotrophic respiration and that variation about this line should be due to differences in nutrient availability (Fig. 5). But much remains to be done to fully incorporate the conceptual and quantitative models of MTE into community and ecosystem ecology. And we are just scratching the surface of what's already been accomplished.

CONCLUSIONS AND FUTURE DIRECTIONS

MTE continues to lend insights into new areas of ecology, and the theoretical foundations of MTE continue to evolve. At present, MTE arguably stands alone as a theory that can quantitatively link the vast differences in scale that occur in space and time across levels of biological organization. It is also the only general ecological theory that conceptually and quantitatively links the structure and function of cells to the structure and function of populations, communities, and ecosystems. This entry has presented just a few simplified examples of the models that comprise the theory here and has largely avoided most of the mathematical derivations. In recent years, the theory has been extended to address various

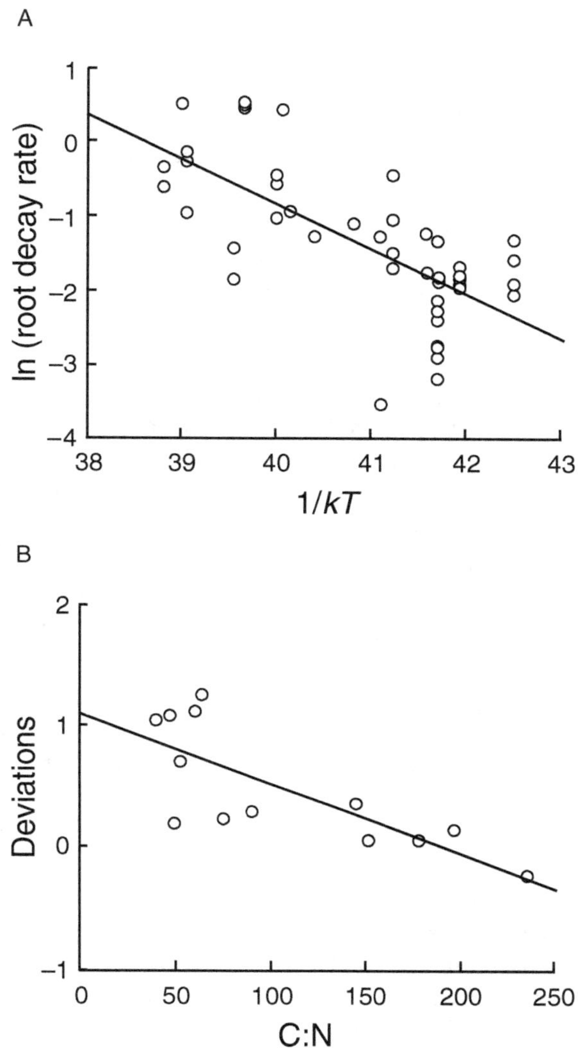

FIGURE 5 The temperature-dependence of short-term root decay rate (d^{-1}). (A) The temperature-dependence of root decay rate falls close to the predicted value of −0.65: $\ln(y) = 23.38 - 0.60 \ln(x)$, $r^2 = 0.48$, $p < 0.0001$. (B) Remaining variation in root decay rates (i.e., residual values from panel (A) are well explained by the ratio of carbon to nitrogen: $y = 1.11 - 0.01x$, $r^2 = 0.58$, $p < 0.01$. Thus, as predicted by MTE, root decay rate increases exponentially with temperature, and much of the remaining variation can be explained by differences in nutrient availability. Data from Silver and Miya (2001, *Oecologia* 129: 407–419).

questions related to cell structure and function, disease dynamics, evolutionary processes, and even phenomena in nonbiological complex systems such as cities and computer networks.

As MTE matures, its strengths and limitations are being revealed, and new and revised models are emerging that provide more clarity into the rules that govern biological systems. For example, with respect to the mechanisms governing the mass dependence of metabolism, recent work has shown that certain assumptions of the WBE model are incorrect, and new, alternative explanations have been proposed. Likewise, recent work has shown that prokaryote metabolism scales superlinearly with body size. This implies that biological rates, including population growth rates, are faster in larger prokaryotes, which is also observed empirically. This relationship appears to contradict the predictions of MTE, but it may be an exception that proves the rule—single cells that lack vascular networks do not show the 3/4-power scaling between metabolism and mass that is predicted to arise from the properties of such networks. This lends support to the hypothesis that distribution networks generate the 3/4 power exponent. Thus, MTE continues to inform, and be informed by, empirical evaluations of model predictions and assumptions.

MTE will continue to be merged with other theories and models designed to address questions outside the domain of MTE. In many cases, this will lead to a more synthetic view of ecological systems and models with greater predictive power. In just the past 5 years, MTE has been successfully combined with Kimura's neutral theory, Hubbell's neutral theory of biodiversity, resource limitation theory, life history theory, and food web theory. MTE will also continue to be applied to make predictions in new areas of biology, and at least three hold particular promise. First, linking MTE to the sensory systems organisms use to exchange information could lead to a more general understanding of the spatial and temporal scales over which species perceive and respond to their environments. This could help us better understand and predict the nature of species interactions. For example, a recent study of animal acoustic communication showed that many of the basic features of animal calls were predictable based on the size dependence and temperature dependence of metabolism. Second, extensions of MTE may help us understand to what extent biological rules of energy use govern the structure and function of whole societies—including human societies. For example, it has been shown that both modern and hunter-gatherer fertility and population growth rates can be predicted using the basic principles of MTE. Third, MTE holds promise for understanding the biological consequences of environmental change, including global warming. A number of recent studies have used MTE to predict how individuals, populations, communities, and ecosystems will be affected by anthropogenic changes in environmental temperature, resource supply, or the size structure of populations. For example, in the past few years, MTE has been used to predict how mass-specific metabolism, producer–consumer biomass ratios, food web stability, and ocean carbon balance will change in response to global warming. These and other examples point to the promise of MTE for linking species to ecosystems, genes to phenotypes, and ultimately for linking subdisciplines of biology such as physiology and behavior to ecology and evolution.

SEE ALSO THE FOLLOWING ARTICLES

Allometry and Growth / Biogeochemistry and Nutrient Cycles / Demography / Ecosystem Ecology / Energy Budgets / Gas and Energy Fluxes across Landscapes / Individual-Based Ecology

FURTHER READING

Allen, A. P., and J. F. Gillooly. 2009. Towards an integration of ecological stoichiometry and the metabolic theory of ecology to better understand nutrient cycling. *Ecology Letters* 12: 369–384.

Banavar, J. R., M. E. Moses, J. H. Brown, J. Damuth, A. Rinaldo, R. M. Sibly, and A. Maritan. 2010. A general basis for quarter power scaling in animals. *Proceedings of the National Academy of Sciences* 107: 15816–15820.

Brown, J. H., J. F., Gillooly, A. P. Allen, V. M. Savage, and G. B. West. 2004. Toward a metabolic theory of ecology. *Ecology* 85: 1771–1789.

Gillooly, J. F., J. H. Brown, G. B. West, V. M. Savage, and E. L. Charnov. 2001. Effects of size and temperature on metabolic rate. *Science* 293: 2248–2251.

Hou, C., W. Zuo, M. E. Moses, J. H. Brown, and G. B. West. 2008. Energy Uptake and Allocation During Ontogeny. *Science* 332(5902): 736–739. DOI:10.1126/science.1162302.

Peters, R. H. 1983. *The ecological importance of body size.* Cambridge, MA: Cambridge University Press.

Price, C. A., J. F. Gillooly, A. P. Allen, J. S. Weitz, and K. J. Niklas. 2010. The metabolic theory of ecology: prospects and challenges for plant biology. *The New Phytologist* 188(3): 1–15.

Savage, V. M., J. H. Brown, J. F. Gillooly, W. H. Woodruff, G. B. West, A. P. Allen, B. J. Enquist, and J. H. Brown. 2004. The predominance of quarter power scaling in biology. *Functional Ecology* 18: 257–282.

West, G. B., J. H. Brown, and B. J. Enquist. 1997. A general model for the origin of allometric scaling laws in biology. *Science* 276: 122–126.

METACOMMUNITIES

MARCEL HOLYOAK

University of California, Davis

JAMIE M. KNEITEL

California State University, Sacramento

Metacommunities are groups of ecological communities that are connected together through movement of potentially interacting species. The concept brings together a local scale, which represents the spatial scale at which a community exists, and a regional or metacommunity scale at which local communities are connected through dispersal. Metacommunities offer the fullest representation to date of how local communities are coupled: persistent local communities can contribute to regional species diversity, and local communities are, in turn, augmented by immigration from other local communities within the metacommunity. The concept of a metacommunity builds on a rich tradition of metapopulation theory, community ecology, and island biogeography, and it is relevant both to explaining spatial patterns of biodiversity and managing this biodiversity.

CORE CONCEPTS

Exploring planet Earth, it is striking how biodiversity is patchily distributed. The obvious examples include freshwater ponds, marine coral reefs, and terrestrial habitats with unique soils (e.g., serpentine), but even within habitats that appear uniform, species and individuals are often highly aggregated. Frequently, we can identify resources (food, water, and so on) as proximate causes for accumulations of large numbers of individuals and species in a local community, but sometimes the causes of such patterns are unclear. For example, the dispersal of individuals among local communities can obscure or enhance the effects of resources on species diversity. The concept of a metacommunity grew up in the 1990s and most actively since 2000 as a means of considering how local- and regional-scale spatial factors influence ecological communities and contribute to spatial biodiversity patterns. While a metacommunity is most easily visualized as consisting of local communities inhabiting habitat fragments such as terrestrial islands, freshwater ponds, or coral atolls, the concept has also been applied to explain patterns of biodiversity in more spatially continuous environments such as oceans or tropical rainforests. The relevant spatial scale for understanding a metacommunity is contingent on the scale over which individuals of species readily migrate (disperse); this can encompass areas that are identifiable as local communities and was termed "the mesoscale" by Robert Holt.

The central concepts behind metacommunities can be divided into four categories or perspectives, termed Neutral Community Dynamics, Patch Dynamics, Species Sorting, and Mass Effects. These were described in a key paper by Leibold et al. (2004. *Ecology Letters* 7: 601–613) and further elaborated in an edited book about metacommunities. A common starting point for each of these perspectives is to explain the mechanisms that allow large numbers of species to coexist in a region (a metacommunity) without a single dominant species driving most of the others to extinction throughout the entire region. The perspectives are summarized in Figure 1.

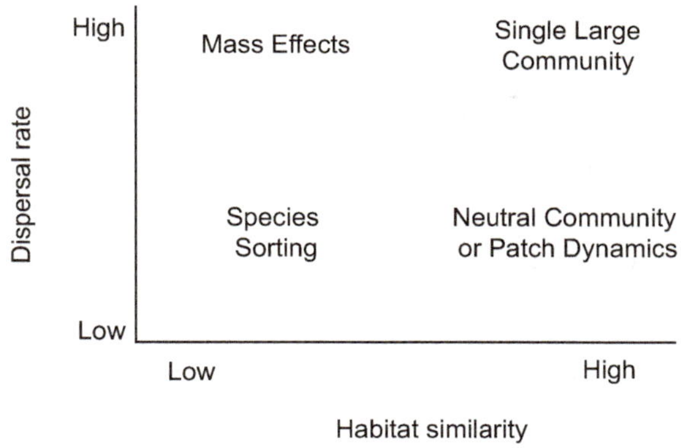

FIGURE 1 A categorization of the four metacommunity perspectives based on dispersal among local communities and the degree of habitat heterogeneity. Patch Dynamics and Neutral Community Dynamics can further be separated by whether species show variation in traits and tradeoffs in factors like their competition and colonization ability.

Neutral Community Dynamics

Neutral community models assume that individuals of all species are, on average, equivalent in their birth, death, and movement rates; that is, all individuals are drawn from identical random distributions. Further, all species are considered to be identical in their ability to compete for resources, and the habitat is effectively uniform (homogeneous) since it does not influence fitness. Neutral models are usually simulated on a grid of spatial locations, each of which can hold a single individual, and areas of space are defined that represent local communities while the larger region represents the metacommunity. Simulations are started with a pool of individuals of a number of species. In each timestep (generation) individuals randomly give birth to new individuals, offspring disperse, and then compete with individuals occupying the same *loci*. Pairwise interactions among individuals that would occupy the same *loci* lead to a random winner that is independent of species identity. Without an input of new species, species diversity would gradually decrease, eventually leaving only a single arbitrary species. An arbitrary species initially becomes numerically dominant in a local community, and eventually in the entire metacommunity.

Simple neutral community models were first described by Stephen Hubbell in a 1979 paper, and then in more complex (spatially explicit) form in an important book in 2001. In its wake, many variations of neutral community models have also been developed. In Hubbell's (2001) model, new species are added through a process of *speciation,* whereas other neutral community models are simulated for shorter periods of time. Input of new species through speciation and loss of species through extinction causes *ecological drift* in both levels of diversity and species composition. The concept of ecological drift is analogous to genetic drift, which is a change in the relative frequency at which a gene variant (allele) occurs in a population due to random sampling (a random sample of alleles of the parents arrives in offspring, and then offspring are subject to random survival and reproductive success). Importantly, Hubbell added to earlier models that species had limited dispersal ability, such that they could not freely disperse across an entire region within a single generation. Such dispersal limitation causes individuals of a species to aggregate in space, which prolongs times to extinction for individual species by increasing the frequency of conspecific encounters (and hence intraspecific competition) and reducing the frequency of heterospecific encounters (and hence interspecific competition). This shift toward increased intraspecific competition over and above interspecific competition slows extinction, just as it does in Lotka–Volterra competition models. Evolution is limited to speciation because any local adaptation would imply differences in fitness, which are assumed away in neutral community models.

Dispersal-Limited Patch Dynamics

The patch-dynamics perspective frequently represents an extension of Levins' metapopulation model consisting of "a population of populations." One such model, for multiple competing species, was developed by Levins and Culver in 1971. As in neutral community models, it is assumed that habitat is uniform (homogeneous) across space, and species are assumed to be limited in their dispersal abilities. However, unlike neutral community models, patch dynamics models assume that there is a tradeoff among species, such that species that are good colonists are poor competitors, and vice versa. The competition–colonization ability tradeoff is critical for coexistence of many species. Species can survive by being good colonists of vacant habitat, by being able to withstand competition within local patches, or by some mixture of these strategies. If randomness is introduced into the competition–colonization tradeoff such that species have variation from a strict tradeoff (where there is an exact hierarchy of increasing colonization ability) with decreasing competitive ability then both local and regional species diversity is reduced.

Models with competition–colonization tradeoffs can also be used to consider the effect of a uniform change in dispersal rates (while still maintaining a competition–colonization tradeoff). This might represent habitat patches that are isolated from one another to a greater or lesser degree. Under such conditions, both local and regional species diversity is expected to be greatest at intermediate levels of dispersal. Dispersal being too low causes colonization rates to be low compared to local extinction rates, and consequently diversity is low and species that are the best colonists are favored relative to species that are poor colonists but strong competitors. If dispersal is uniformly high, then the competitive dominants are likely to reach all patches and drive inferior competitors (good colonists) to extinction, thereby reducing diversity locally and regionally. Therefore, we expect to see a preponderance of competitively dominant species with uniformly high dispersal. Analogous effects of the overall dispersal rate are also seen in metapopulation models of specialist predators and their prey, with the likelihood of persistence of both species being highest at intermediate levels of dispersal of both species.

With patch dynamics, we might expect there to be good potential for evolution to occur given that dispersal is limited and so gene flow does not swamp selection pressures that promote adaptation to local conditions. However, evolution is, in fact, likely to be limited because local extinction is likely to be sufficiently frequent to cause a loss of individuals that are adapted to local conditions. Instead, it may occur at a larger regional spatial scale, at which among-species tradeoffs emerge. Such tradeoffs might arise if there was extinction of species that did not fit the tradeoff.

SPECIES SORTING

The third perspective, species sorting, is a spatial extension of classic niche theory. It proposes that habitat is heterogeneous across space, either through gradients or habitat patches of different kinds. Different species are matched with different habitat types (or points along a gradient) in which they maintain high fitness, and these species may, in turn, support or permit certain other species to coexist with them. Species specialize in these "matched" habitats, which create refuges for species that generally have low fitness in other habitat types. Dispersal is sufficiently frequent to allow species to colonize habitat and replace occasional extinction from matched habitats but not sufficient to create a spillover of individuals (and species) into nonmatched habitats. Spatial turnover of species is strongly associated with turnover in habitat types. Evolution is expected to reinforce this coupling between habitat types and the occurrence of particular species, because gene flow is relatively low and there would likely be individual fitness advantages of specialization to particular habitat types.

Mass Effects

The final of the four perspectives builds on ideas about source–sink dynamics and mass effects. It is also a logical extension of species sorting dynamics but with an emphasis on dispersal from matched into nonmatched habitats. In source–sink dynamics, species maintain finite growth rates >1 in source populations, but finite growth rates are <1 in sink habitats, and immigration is required to prevent local extinction. Sink populations are rescued from extinction by immigration from source populations, raising population size in a "rescue effect." Mass effects are a multispecies extension of rescue effects where the emphasis is on the difference in size (mass) of local populations, which creates a flux of individuals from large to small populations. Whereas species sorting creates a strong correspondence between habitat types and species composition, in the mass effects perspective this correspondence is blurred by spillover of species across habitat types. This spillover may also limit the potential for local adaptation of species.

Empirical Support for Metacommunity Dynamics

There is a wide range of support for various aspects of the four perspectives listed above and the importance of spatial dynamics in ecological communities more generally. The most general evidence for metacommunity dynamics comes from studies that use multivariate statistics (e.g., nonmetric multidimensional scaling) to test whether differences in species composition among localities can best be explained by differences in habitat or by distance. If the variation in species composition is found to be due to habitat characteristics independent of distance between sites, this would be consistent with species sorting (Table 1). Distance is a proxy for limited dispersal because if dispersal was sufficient we would expect the species pool to be well mixed and for there not to be local variation in species composition. Therefore, if distance is important but not habitat type in explaining variation in species composition, this would be consistent with either patch dynamics or neutral community dynamics, both of which assume dispersal is limited. Conversely, a mix of distance and habitat variables being able to explain spatial variation in species composition is consistent with a mass effects perspective, where habitat filters species composition and spillover from one habitat to another is distance dependent. An analysis of

TABLE 1
The types of metacommunity dynamics indicated by an analysis of 158 published datasets

Metacommunity type	Environment	Distance	Distance independent of environment	Environment independent of space	Space by environment interaction	Percentage of datasets
Neutral Community	no	no	yes	no	no	8%*
Patch Dynamics	no	no	yes	no	no	8%*
Species Sorting	either	either	no	yes	no	44%
Mass Effects	either	either	no	no	yes	29%

* Indicates that Patch and Neutral Dynamics could not be distinguished between in this analysis and jointly had a frequency of 8% of datasets.

SOURCE: Cottenie, 2005. *Ecology Letters* 8: 1175–1182. Columns indicate whether variation in species composition could be explained by environment (habitat) factors, distance (indicating dispersal) or a combination of these things. A further 12% of datasets could not be uniquely associated with these categorizations of perspectives, and 7% had no statistically significant explanatory components for species composition.

158 published datasets by Karl Cottenie in 2005 from a variety of community types, broad habitat categories, and types of organisms found support for species sorting in 44% of datasets, mass effects in 29%, and neutral community or patch dynamics in only 8% of datasets. A further 12% of datasets could not be uniquely associated with these categorizations of perspectives, and 7% had no statistically significant explanatory components for species composition. Hence, some kind of metacommunity dynamics were indicated by 81% of datasets in Cottenie's analysis.

Studies within an edited volume on metacommunities (Holyoak et al. 2005) reported some degree of species sorting in most systems examined. In addition, an unusually detailed study of specialist (endemic) plant diversity in 109 local serpentine sites in 78 regions by Harrison and colleagues in 2006 found strong support for species sorting. A total of 66% of variation in local species richness and 73% for regional richness were accounted for by local and regional environmental variables. Like species sorting, most variation in local richness was accounted for by local variables (e.g., soil, rock and vegetation) rather than by spatial variables. The occurrence of mass effects is best demonstrated by a simultaneous manipulation or examination of habitat type and dispersal. Studies in the edited volume on metacommunities by Holyoak and colleagues indicated that mass effects were less common than species sorting but more common than patch or neutral dynamics. The strongest evidence came in 2003 from a study by Karl Cottenie of interconnected ponds, where habitats differed depending on the presence of fish that could not readily move between ponds whereas zooplankton could and were readily structured by a combination of habitat type (fish/no fish) and dispersal.

Evidence for patch dynamics is more limited, which in part shows that it is difficult to rule out alternative causes. A review of neutral community dynamics by Brian McGill and colleagues overwhelmingly found that studies with the weakest evidence supported the occurrence of this kind of dynamics, and studies with more information supported alternative models and perspectives. However, other studies find that these dynamics may be contingent on the ecological background, including trophic level and the presence of disturbances. Nonetheless, the roles of evolution, limited dispersal, transient coexistence, and stochasticity identified in neutral community models are likely important in a wide variety of systems, and it is likely that each of these factors contributes on some level to species diversity.

LIMITATIONS OF THE FOUR METACOMMUNITY PERSPECTIVES

Metacommunity ideas convey a broad way of thinking about the spatial dynamics of ecological communities. They were not intended to imply a simple local vs. regional dynamics split of communities, but rather maintain that all spatial scales are relevant for community patterns. Further, they were not intended to negate the temporal structuring of communities, by succession or the storage effect (persistence through seed banks).

Implicit in the four metacommunity perspectives is that metacommunities are somewhat at equilibrium with the environment. Conversely, the idea that disturbances such as fires or floods alter species diversity and composition is an old one in community ecology. More recent work recognizes that disturbance interacts with the mechanisms that maintain species diversity and that species may become adapted to disturbance regimes. Species may become differentiated by responding in different ways to disturbance, in an analogous way to species undergoing a life history tradeoff. Hence, disturbance provides a niche axis that may enhance persistence and coexistence of species, just as habitat type does in species sorting or as a competition–colonization tradeoff does in patch dynamics. Alternatively, disturbance may simply add to the rate of local extinction of species in a straightforward way in patch dynamics.

Most community ecology studies, including metacommunities, focus on species diversity of a single trophic level, but communities are much more complex. Food web structure represents a common form of complexity, and the spatial dynamics of food webs is only just beginning to be explored by ecologists. A variety of models of simple food web modules describe how both predators (top-down) and nutrients (bottom-up) can influence food webs. Connecting even simple food chains together across habitat patches shows that species mobility across patches can have strong influences on the trophic levels both above and below these species in ways that we would not necessarily expect based on theories from isolated communities. Empirical studies also show that food webs may undergo temporary contractions and expansions spatially in response to the spatial movement of predator or prey species, such as outbreaks of pest insects.

CONCLUSIONS

Metacommunities represent a flexible and dynamic set of ideas about how spatial dynamics structure ecological communities. The literature on metacommunities to date can be divided into two main topics. Several hundred published studies test neutral community models vs. niche models, often without specifying the precise form of niche differences or distinguishing alternative hypotheses. In contrast, there are several hundred studies that develop and/or test all four of the metacommunity perspectives described above. The latter include more detailed descriptions of the way in which dispersal, habitat factors, and species traits combine to maintain species diversity. Species sorting seems to be overwhelmingly the most common kind of spatial community dynamic in nature, followed by mass effects, although it is hard to clearly demonstrate both patch dynamics and neutral dynamics. The implications of metacommunity dynamics for food webs, evolution, and complexities such as disturbance are just beginning to be explored.

SEE ALSO THE FOLLOWING ARTICLES

Assembly Processes / Food Chains and Food Web Modules / Food Webs / Metapopulations / Neutral Community Ecology / Spatial Ecology / Spatial Models, Stochastic

FURTHER READING

Cadotte, M.W. 2006. Dispersal and species diversity: a meta-analysis. *American Naturalist* 167: 913–924.

Chave, J. 2004. Neutral theory and community ecology. *Ecology Letters* 7: 241–253.

France, K.E., and J.E. Duffy. 2006. Diversity and dispersal interactively affect predictability of ecosystem function. *Nature* 441: 1139–1143.

Holyoak, M., M.A. Leibold, and R.D. Holt, eds. 2005. *Metacommunities: spatial dynamics and ecological communities.* Chicago: University of Chicago Press.

Hubbell, S.P. 2001. *The unified neutral theory of biodiversity and biogeography.* Princeton: Princeton University Press.

Leibold, M.A., M. Holyoak, N. Mouquet, P. Amarasekare, J.M. Chase, M.F. Hoopes, R.D. Holt, J.B. Shurin, R. Law, D. Tilman, M. Loreau, and A. Gonzalez. 2004. The metacommunity concept: a framework for multi-scale community ecology. *Ecology Letters* 7: 601–613.

Urban, M.C., M.A. Leibold, P. Amarasekare, L. De Meester, R. Gomulkiewicz, M.E. Hochberg, C.A. Klausmeier, N. Loeuille, C. de Mazancourt, J. Norberg, J.H. Pantel, S.Y. Strauss, M. Vellend, and M.J. Wade. 2008. The evolutionary ecology of metacommunities. *Trends in Ecology & Evolution* 23: 311–317.

METAPOPULATIONS

ILKKA HANSKI
University of Helsinki, Finland

Most landscapes are complex mosaics of many kinds of habitat. For a particular species, only some habitat types provide the necessary resources for population growth, and the remaining landscape, often called the (landscape) matrix, can only be traversed by dispersing individuals. Often, the suitable habitat occurs in discrete patches, which comprise a network at the landscape level. Individual habitat patches may be occupied by a local population, but many patches are temporarily unoccupied because the population went extinct in the past and a new one has not yet been established. The set of local populations inhabiting a patch network is called a metapopulation. In other cases, the habitat does not consist of discrete patches, but even then habitat quality is likely to vary from one place to another. Habitat heterogeneity tends to cause a more or less fragmented population structure, and such spatially structured populations may be called metapopulations.

METAPOPULATION PATTERNS AND PROCESSES

Metapopulation biology addresses the ecological, genetic, and evolutionary processes that occur in metapopulations. For instance, in a highly fragmented landscape all local populations may be so small that they all have a high risk of extinction, yet the metapopulation as a whole may persist if new local populations are established by dispersing individuals fast enough to compensate for local extinctions. Metapopulation structure and extinction–colonization dynamics may greatly influence the maintenance of genetic diversity and the course of evolutionary changes. Metapopulation processes play a role in the dynamics of most species, because most landscapes are spatially more or less heterogeneous and many comprise networks of discrete habitat patches. Human land use tends to increase fragmentation of natural habitats, and hence metapopulation processes are particularly important in many human-dominated landscapes.

Many Kinds of Metapopulations

Researchers have proposed many classifications of metapopulations, which serve a purpose in facilitating communication, but it should be recognized that in reality there exists a continuum of spatial population structures rather than discrete types. The following terms are often used.

Classic metapopulations consist of many small or relatively small local populations occupying networks of habitat patches. Small local populations have high risk of extinction, and hence long-term persistence can only occur at the metapopulation level, in a balance between local extinctions and re-colonizations. As an example, Figure 1 shows the large patch network inhabited by a classic metapopulation of the Glanville fritillary butterfly.

Mainland–island metapopulations include one or more populations that are so large and live in sufficiently great expanses of habitat that they have a negligible risk of extinction. These populations, called mainland populations, are stable sources of dispersers to other populations

FIGURE 1 A large network of ca. 4000 small dry meadows (marked with red) inhabited by a classic metapopulation of the Glanville fritillary butterfly (*Melitaea cinxia*) in the Åland Islands in Finland. Figure 4 gives an example of how the spatial structure of the landscape influences the occurrence of the butterfly in this network.

in smaller habitat patches (island populations). The MacArthur–Wilson model of island biogeography is an extension of the single-species mainland-island metapopulation model to a community of many species.

Source–sink metapopulations include local populations that inhabit low-quality habitat patches and would therefore have negative growth rate in the absence of immigration (sink populations), and local populations inhabiting high-quality patches in which the respective populations have positive growth rate (source populations).

Nonequilibrium metapopulations are similar to classic metapopulations but there is no balance between extinctions and recolonizations, typically because the environment has recently changed and the extinction rate has increased, the recolonization rate has decreased, or both.

Dispersal and Connectivity

Three processes are fundamental to metapopulation dynamics—dispersal, colonization of currently unoccupied habitat patches, and local extinction. Dispersal has several components: emigration, departure of individuals from their current population; movement through the landscape matrix; and immigration, arrival at new populations or at empty habitat patches. All three components depend on the traits of the species and on the characteristics of the habitat and the landscape, but they may also depend on the state of the population, for instance on population density.

Dispersal may influence local population dynamics. In the case of very small populations, a high rate of emigration may reduce population size and thereby increase the risk of extinction. Conversely, immigration may enhance population size sufficiently to reduce the risk of extinction, which is especially important in sink populations. Immigration to a currently unoccupied habitat patch is particularly significant in potentially leading to the establishment of a new local population. From the genetic viewpoint, two extreme forms of immigration and gene flow have been distinguished, the migrant-pool model, in which the dispersers are drawn randomly from the metapopulation, and the propagule-pool model, in which all immigrants to a patch originate from the same source population. The latter is likely to reduce genetic variation in the metapopulation.

Colonization of a currently unoccupied habitat patch is more likely the greater the connectivity of the patch. Connectivity is best defined as the expected number of individuals arriving per unit time at the focal patch.

Connectivity increases with the number of populations (sources of dispersers) in the neighborhood of the focal patch, with decreasing distances to the source populations (making successful dispersal more likely), and with increasing sizes of the source populations (larger populations send out more dispersers). A measure of connectivity for patch i that takes all these factors into account is defined as

$$S_i = A_i^{z_{im}} \sum_{j \neq i} O_j A_j^{z_{em}} e^{-\alpha d_{ij}}. \tag{1}$$

Here, A_j is the area and O_j is the occupancy of patch j (1 for occupied, 0 for unoccupied patches), d_{ij} is the distance between patches i and j, $1/\alpha$ is species-specific average dispersal distance, and z_{im} and z_{em} describe the scaling of immigration and emigration with patch area. This formula assumes that the sizes of source populations are proportional to the respective patch areas; if information on the actual population sizes N_j is available, the surrogate $O_jA_j^{z_{em}}$ in Equation 1 may be replaced by N_j.

Stochasticity and Local Extinction

Metapopulation dynamics are influenced by four kinds of stochasticity (types of random events): demographic and environmental stochasticity affect each local population, while extinction–colonization and regional stochasticity affect the entire metapopulation. Both local dynamics and metapopulation dynamics are inherently stochastic, because births and deaths in local populations are random events (leading to demographic stochasticity) and so are population extinctions and recolonizations in a metapopulation (extinction–colonization stochasticity). Environmental stochasticity refers to correlated temporal variation in birth and death rates among individuals in local populations, while regional stochasticity refers to correlated extinction and recolonization events in metapopulations. Metapopulations are typically affected by regional stochasticity, because the processes generating environmental stochasticity, including temporally varying weather conditions, are usually spatially correlated.

All local populations have smaller or greater risk of extinction due to demographic and environmental stochasticities, while the extinction of an entire metapopulation is increased by extinction–colonization and regional stochasticities. Additionally, local populations may go extinct for many other reasons. Habitat quality may turn unsuitable for natural reasons (e.g., succession) or because of human land use. Natural enemies and competitors may increase the risk of extinction. Small local populations often suffer from inbreeding depression, which may be strong enough to increase the risk of local extinction. Typically—and regardless of the actual mechanism of extinction—the smaller the population size, the greater the risk of extinction. As small habitat patches typically have smaller populations than large ones, populations living in small patches have generally higher risk of extinction than populations living in large patches.

Population Turnover and the Incidence of Occupancy

Population turnover refers to extinctions of current populations and establishment of new populations via dispersing individuals at currently unoccupied habitat patches. Assuming that a local population in patch i has a constant probability of going extinct, denoted by E_i, and that patch i, if unoccupied, has a constant probability of becoming recolonized, denoted by C_i, the state of patch i, whether occupied or not, is determined by a stochastic process (Markov chain) with the stationary (time-invariant) probability of occupancy given by

$$J_i = \frac{C_i}{C_i + E_i}. \tag{2}$$

J_i is called the incidence of occupancy. This formula helps explain the common metapopulation patterns of increasing probability of patch occupancy with increasing patch area (which typically decreases extinction probability E_i and hence increases J_i) and with decreasing isolation (which typically increases colonization probability C_i and hence increases J_i). Figure 2 gives two examples.

It can be demonstrated mathematically that all metapopulations with population turnover caused by extinctions and recolonizations will eventually go extinct: it is a certainty that given enough time a sufficiently long run of extinctions will arise by "bad luck" and extirpate the metapopulation. However, time to extinction can be very long for large metapopulations inhabiting large patch networks (Fig. 3), and the metapopulation settles for a long time to a stochastic quasi-equilibrium, in which there is variation but no systematic change in the number of local populations.

Source and Sink Populations

Populations may occur in low-quality sink habitats if there is sufficient dispersal from other populations living in high-quality source habitats to boost population growth rate in sink habitats. Therefore, the presence of a species in a particular habitat patch does not suffice to demonstrate that the habitat is of sufficient quality to support a viable population. Conversely, a local population may be

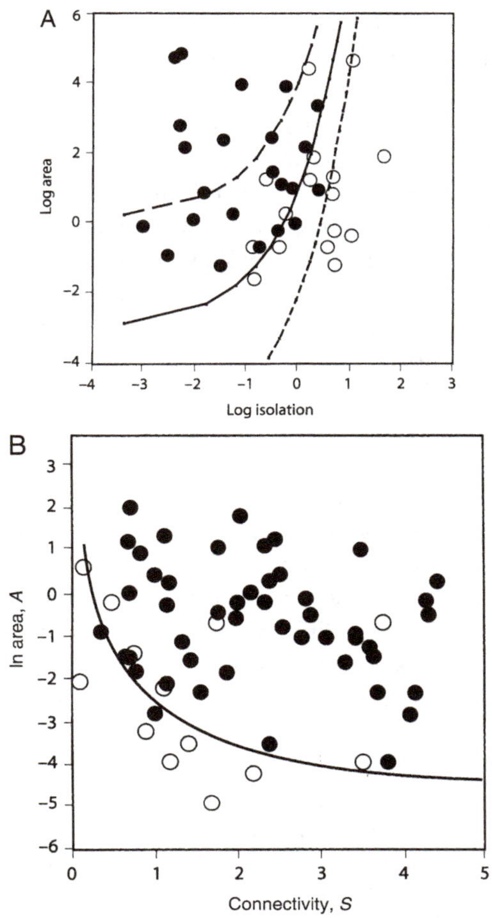

FIGURE 2 Two examples of the common metapopulation pattern of increasing incidence (probability) of habitat patch occupancy with increasing patch area and connectivity. Black dots represent occupied, open circles unoccupied habitat patches at the time of sampling. (A) Mainland–island metapopulation of the shrew *Sorex cinereus* on islands in North America. Isolation, which increases with decreasing connectivity, is here measured by distance to the mainland. The lines indicate the combinations of area and isolation for which the predicted incidence of occupany is greater than 0.1, 0.5, and 0.9, respectively (from I. Hanski, 1993, Dynamics of small mammals on islands, *Ecography* 16: 372–375). (B) Classic metapopulation of the silver-spotted skipper butterfly (*Hesperia comma*) on dry meadows in southern England. The line indicates the combinations of area and connectivity above which the predicted incidence of occupancy is greater than 0.5 (from I. Hanski, 1994, A practical model of metapopulation dynamics, *Journal of Animal Ecology* 63: 151–162).

absent from a habitat patch that is perfectly suitable for population growth, when a local population happened to go extinct for reasons unrelated to habitat quality.

In a temporally varying environment, sink populations may counterintuitively enhance metapopulation persistence. This may happen when source populations exhibit large fluctuations leading to a high risk of extinction. The habitat patches supporting such sources may become recolonized by dispersal from sinks, assuming this happens before the sink populations have declined to extinction. In general, dispersal among local populations that fluctuate relatively independently of each other (weak regional stochasticity) enhances metapopulation growth rate. This happens because when a population has increased in size in a good year and the offspring are spread among many independently fluctuating populations, subsequent bad years will not hit them all simultaneously. It can be shown that this spreading-of-risk effect of dispersal may be so great that it allows a metapopulation consisting of sink populations only to persist without any sources.

LONG-TERM VIABILITY OF METAPOPULATIONS

Mathematical models are used to describe, analyze, and predict the dynamics of metapopulations living in fragmented landscapes. A wide range of models can be constructed differing in the forms of stochasticity they include, in whether change in metapopulation size occurs continuously or in discrete time intervals, in how many local populations the metapopulation maximally consists of, in how the structure of the landscape is represented, and so forth. A minimal metapopulation includes two local populations connected by dispersal. At the other extreme, assuming infinitely many habitat patches leads to a particularly simple description of the classic metapopulation, which will be discussed next.

The Levins Model and the Extinction Threshold

The Levins model has special significance for metapopulation ecology, as it was with this model that the

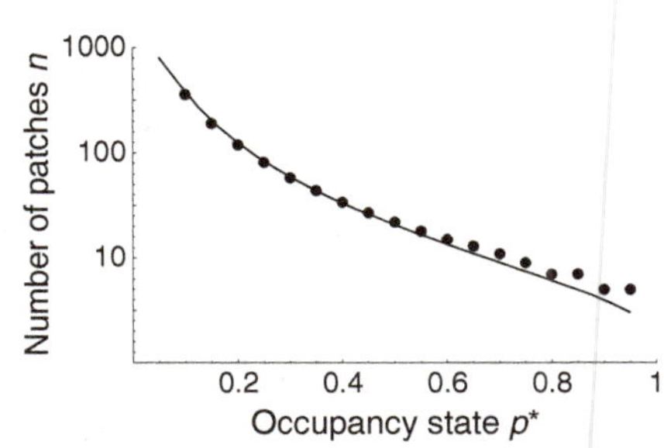

FIGURE 3 The number of habitat patches n needed to make the mean time to metapopulation extinction T at least 100 times longer than the mean time to local extinction in the stochastic Levins model. The dots show the exact result based on the stochastic logistic model, the line is based on the following diffusion approximation:

$$T = \sqrt{\frac{2\pi}{n}}\,\frac{e^{-(n-1)p^*}}{p^{*2}(1-p^*)^{n-1}},$$

p^* is the size of the metapopulation at quasi-equilibrium (from O. Ovaskainen, and I. Hanski, 2003, Extinction threshold in metapopulation models, *Annales Zoologici Fennici* 40: 81–97).

American biologist Richard Levins introduced the metapopulation concept in 1969. The Levins model captures the essence of the classic metapopulation concept—that a species may persist at the network level in a balance between stochastic local extinctions and recolonization of currently unoccupied patches. For mathematical convenience, the model assumes an infinitely large network of identical patches, which have two possible states, occupied or empty. The state of the entire metapopulation can be described by the fraction of currently occupied patches, denoted by p, which varies between 0 and 1.

Assuming that each local population has the same risk of extinction, and that each population contributes equally to the rate of recolonization, the rate of change in the size of the metapopulation is given by

$$\frac{dp}{dt} = cp(1 - p) - ep, \qquad (3)$$

where c and e are colonization and extinction rate parameters. This model ignores stochasticity, but it is a good approximation of the corresponding stochastic model for a large metapopulation inhabiting a large patch network and for species with long-range dispersal. The Levins model is structurally identical with the logistic model of population growth, which can be seen by rewriting Equation 3 as

$$\frac{dp}{dt} = (c - e)p\left(1 - \frac{p}{1 - e/c}\right); \qquad (4)$$

$c - e$ gives the rate of metapopulation growth when it is small, and $1 - e/c$ is the equilibrium metapopulation size (carrying capacity), denoted by p^*. The ratio c/e defines the basic reproductive number R_0 in the Levins model. A species can increase in a patch network from low occupancy if $R_0 = c/e > 1$. This condition defines the extinction threshold in metapopulation dynamics. In reality, in a finite patch network a metapopulation may go extinct because of extinction–colonization stochasticity even if the threshold condition $c/e > 1$ is satisfied. Using a diffusion approximation to analyze the stochastic Levins model, the mean time to extinction T can be calculated as a function of the number of habitat patches (n) and the size of the metapopulation at quasi-equilibrium (p^*). Figure 3 shows the number of patches that the network must have to make T at least 100 times as long as the expected lifetime of a single local population. For metapopulations with large p^*, a modest network of $n = 10$ patches is sufficient to allow long-term persistence, but for rare species (say, $p^* < 0.2$) a large network of $n > 100$ is needed for long-term persistence.

The stochastic Levins model includes extinction–colonization stochasticity but no regional stochasticity, which leads to correlated extinctions and colonizations. In the presence of regional stochasticity, the mean time to metapopulation extinction does not increase exponentially with increasing n as in Figure 3 but as a power function of n, the power decreasing with increasing correlation in extinction and recolonization rates, which reduces long-term viability. This result is analogous to the well-known effects of demographic and environmental stochasticities on the lifetime of single local populations.

Spatially Realistic Metapopulation Models

There is no description of landscape structure in the Levins model; hence, it is not possible to investigate with it the consequences of habitat loss and fragmentation. Real metapopulations live in patch networks with a finite number of patches, the patches are of varying size and quality, and different patches have different connectivities, which affects the rates of dispersal and recolonization. These considerations have been incorporated into the spatially realistic metapopulation model. The key idea is to model the effects of habitat patch area, quality, and connectivity on the processes of local extinction and recolonization. Generally, the extinction risk decreases with increasing patch area, because large patches tend to have large populations with a small risk of extinction, and the colonization rate increases with connectivity to existing populations.

The theory provides a measure to describe the capacity of an entire patch network to support a metapopulation, denoted by λ_M and called the metapopulation capacity of the fragmented landscape. Mathematically, λ_M is the leading eigenvalue of a "landscape" matrix, which is constructed with assumptions about how habitat patch areas and connectivities influence extinctions and recolonizations. The model is closely related to other matrix models in population ecology. The size of the metapopulation at equilibrium is given by

$$p_\lambda^* = 1 - e/(c\lambda_M), \qquad (5)$$

which is similar to the equilibrium in the Levins model, but with the difference that metapopulation equilibrium now depends on metapopulation capacity and metapopulation size p_λ is measured by a weighted average of patch occupancy probabilities. The threshold condition for metapopulation persistence is given by

$$\lambda_M > e/c. \qquad (6)$$

In words, metapopulation capacity has to exceed a threshold value, which is set by the extinction proneness (e) and colonization capacity (c) of the species, for

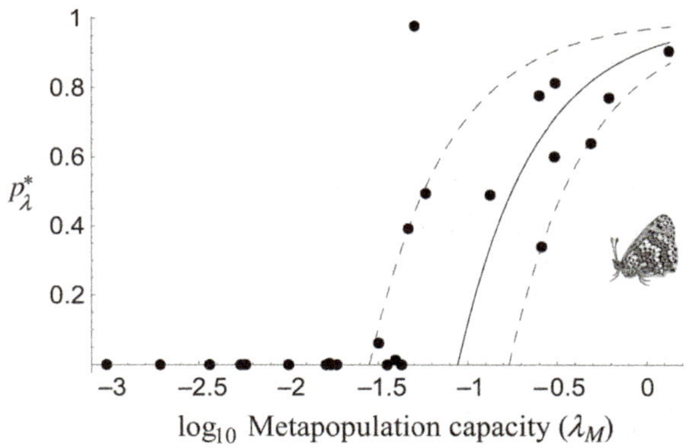

FIGURE 4 Metapopulation size of the Glanville fritillary butterfly (*Melitaea cinxia*) as a function of the metapopulation capacity λ_M in 25 habitat patch networks. The vertical axis shows the size of the metapopulation based on a survey of habitat patch occupancy in 1 year. The empirical data have been fitted by a spatially realistic model. The result provides a clear-cut example of the extinction threshold (from Hanski and Ovaskainen, 2000).

long-term persistence. To compute λ_M for a particular landscape, one needs to know the range of dispersal of the focal species, which sets the spatial scale for calculating connectivity (parameter α in Equation 1), and the areas and spatial locations of the habitat patches. The metapopulation capacity can be used to rank different fragmented landscapes in terms of their capacity to support a viable metapopulation: the larger the value of λ_M, the better the landscape. Figure 4 gives an example, in which metapopulation capacity explains well the size of butterfly metapopulations in dissimilar patch networks. The spatially realistic metapopulation theory facilitates the conceptual integration of metapopulation ecology and landscape ecology.

METAPOPULATIONS IN CHANGING ENVIRONMENTS

A fundamental question about metapopulation dynamics concerns long-term viability, which has great significance for the conservation of biodiversity in fragmented landscapes. A patch network will not support a viable metapopulation unless the extinction threshold is exceeded, and even if it is, a metapopulation may go extinct for stochastic reasons. Long-term viability is further reduced by environmental change.

Ephemeral Habitat Patches

Innumerable species of fungi, plants, and animals live in ephemeral habitats such as decaying wood. A dead tree trunk may be viewed as a habitat patch for local populations of such organisms. The trunk is not permanent, however, largely due to the action of the organisms themselves, and local populations necessarily go extinct at some point. Parasites living in a host individual can be similarly considered as comprising a local population, which necessarily goes extinct when the host individual dies. This example reflects fundamental similarities between metapopulation biology and epidemiology.

Regular disappearance of habitat patches increases extinctions, but the metapopulation may still persist in a stochastic quasi-equilibrium. Equation 2 may be extended to include the disappearance of habitat patches and the appearance of new ones:

$$J_i = \frac{C_i - [C_i(1 - C_i - E_i)^{age}]}{C_i + E_i}, \tag{7}$$

where *age* is the age of patch *i*. Following its appearance, a new patch is initially unoccupied, and hence $J_i = 0$ when $age = 0$. When the patch becomes older, the incidence of occupancy approaches the equilibrium ($C_i/[C_i + E_i]$) given by Equation 2 and which is determined by the extinction–colonization dynamics of the species. The precise trajectory is given by Equation 7, where the term in square brackets declines from C_i when $age = 0$ to zero as *age* becomes large and when Equation 2 is recovered.

Transient Dynamics and Extinction Debt

Human land use often causes the loss and fragmentation of the habitat for many other species. Following the change in landscape structure, it takes some time before the metapopulation has reached the new quasi-equilibrium, which may be metapopulation extinction. Considering a community of species, the term extinction debt refers to situations in which, following habitat loss and fragmentation, the threshold condition is not met for some species, but these species have not yet gone extinct because they respond relatively slowly to environmental change. More precisely, the extinction debt is the number of extant species that are predicted to go extinct, sooner or later, because the threshold condition for long-term persistence is not satisfied for them following habitat loss and fragmentation.

How long does it take before the metapopulation has reached the new quasi-equilibrium following a change in the environment? The length of the transient period is longer when the change in landscape structure is greater, when the rate of population turnover (extinctions and recolonizations) is lower, and when the new quasi-equilibrium following environmental change is located close to the extinction threshold (Fig. 5). The latter result has important implications for conservation. Species that have become endangered due to recent changes in landscape structure are located, by definition, close to their

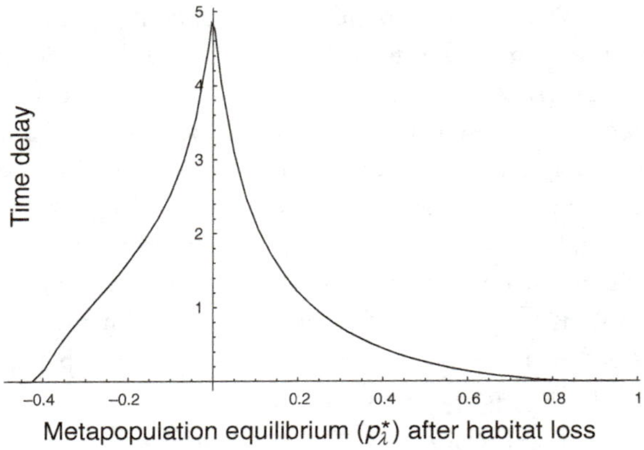

FIGURE 5 The length of the time delay in metapopulation response (vertical axis; relative time units) in relation to the new equilibrium size of the metapopulation following a change in landscape structure. Note that the time delay is especially long when the new equilibrium is close to the extinction threshold (zero metapopulation size; negative sizes correspond to metapopulation extinction; from O. Ovaskainen, and I. Hanski, 2002, Transient dynamics in metapopulation response to perturbation, *Theoretical Population Biology* 61: 285–295).

extinction threshold, and hence the length of the transient period in their response to environmental change is predicted to be long. This means that we are likely to underestimate the level of threat to endangered species, because many of them do not occur presently at quasi-equilibrium with respect to the current landscape structure but are slowly declining due to past habitat loss and fragmentation. On the positive side, long transient time in metapopulation dynamics following environmental change gives us humans more time to do something to reverse the trend.

EVOLUTION IN METAPOPULATIONS

The hierarchical structure of metapopulations, from individuals to local populations to the entire metapopulation, has implications for evolutionary dynamics. In addition to natural selection occurring within local populations, different selection pressures may influence the fitness of individuals during dispersal and at colonization. Individuals that disperse from their natal population and succeed in establishing new local populations are likely to comprise a nonrandom group of all individuals in the metapopulation. Particular phenotypes and genotypes may persist in the metapopulation due to their superior performance in dispersal and colonization even if they would be selected against within local populations. This is often called the metapopulation effect.

The most obvious example relates to emigration rate and dispersal capacity. The most dispersive individuals are selected against locally because their local reproductive success is reduced by early departure and possibly by other factors, such as life history tradeoff between dispersal capacity and fecundity. However, these individuals may find favorable habitat elsewhere, which increases their fitness in the metapopulation. Which particular phenotypes and genotypes are favored in a particular metapopulation depends on many factors. Local competition for resources and competition with relatives for mating opportunities selects for more dispersal, and so does temporal variation in fitness among populations, but mortality during dispersal selects against dispersal. Because of the many opposing selection pressures, habitat fragmentation may select for more or less dispersive individuals depending on particular circumstances.

METAPOPULATIONS AND CONSERVATION

Loss and fragmentation of natural habitats is the most important reason for the current catastrophically high rate of loss of biodiversity on Earth. The amount of habitat matters because long-term viability of populations and metapopulations depends on, among other factors, the environmental carrying capacity, which is typically positively related to the total amount of habitat. Additionally, the spatial configuration of habitat may influence metapopulation viability, because most species have limited dispersal range and hence not all habitat in a highly fragmented landscape is readily accessible, giving rise to the extinction threshold.

Habitat Loss and Fragmentation

Metapopulation models have been used to address the population dynamic consequences of habitat loss and fragmentation. In the Levins model, where there is no description of landscape structure, habitat loss has been modeled by assuming that a fraction $1 - h$ of the patches becomes unsuitable for reproduction, while fraction h remains suitable. Such habitat loss reduces the colonization rate to $cp(h - p)$. The species persists in a patch network if h exceeds the threshold value e/c. At equilibrium, the fraction of suitable but unoccupied patches ($h - p^*$) is constant and equals the amount of habitat at the extinction threshold ($h = e/c$).

The spatially realistic metapopulation model combines the metapopulation notion of the Levins model with a description of the spatial distribution of habitat in a fragmented landscape. In this model, the metapopulation capacity λ_M replaces the fraction of suitable patches h in the Levins model, and the threshold condition for metapopulation persistence is given by $\lambda_M > e/c$. λ_M takes into account not only the amount of habitat in the landscape but also how the remaining habitat is distributed among the individual habitat patches and how the spatial configuration of habitat influences extinction and recolonization rates and hence metapopulation viability. The metapopulation capacity can be computed for multiple landscapes and their relative capacities to support viable metapopulations can be compared: the greater the value of λ_M, the more favorable the landscape is for the particular species (Fig. 4).

Reserve Selection

Setting aside a sufficient amount of habitat as reserves is essential for conservation of biodiversity. Reserve selection should be made in such a manner that a given amount of resources for conservation makes a maximal contribution toward maintaining biodiversity. In the past, making the optimal selection of reserves out of a larger number of potential localities was typically based on their current species richness and composition, without any consideration for the long-term viability of the species in the reserves. More appropriately, we should ask the question which selection of reserves maintains the largest number of species to the future, taking into account the temporal and spatial dynamics of the species and the predicted changes in climate and land use. Metapopulation models can be incorporated into analyses that aim at providing answers to such questions.

CONCLUSION

Metapopulations are assemblages of local populations inhabiting networks of habitat patches in fragmented landscapes. The local populations have more or less independent dynamics due to their isolation, but complete independence is prevented by large-scale fluctuations in environmental conditions leading to regional stochasticity and by dispersal, which occurs at a spatial scale characteristic for each species. Metapopulation models are used to describe, analyze, and predict the dynamics of metapopulations. Important questions include the conditions under which metapopulations may persist in particular patch networks and for how long, how landscape structure influences metapopulation persistence, and the response of metapopulations to changing landscape structure. Metapopulation dynamics in highly fragmented landscapes involve an extinction threshold, a critical amount and spatial configuration of habitat that is necessary for long-term persistence of the metapopulation.

SEE ALSO THE FOLLOWING ARTICLES

Conservation Biology / Dispersal, Animal / Landscape Ecology / Metacommunities / Reserve Selection and Conservation Prioritization / Spatial Ecology / Stochasticity, Demographic / Stochasticity, Environmental

FURTHER READING

Hanski, I. 1998. Metapopulation dynamics. *Nature* 396: 41–49.

Hanski, I. 1999. *Metapopulation ecology*. Oxford: Oxford University Press.

Hanski, I., and O. E. Gaggiotti, eds. 2004. *Ecology, genetics, and evolution of metapopulations*. Amsterdam: Elsevier.

Hanski, I., and O. Ovaskainen. 2000. The metapopulation capacity of a fragmented landscape. *Nature* 404: 755–758.

Hanski, I., and O. Ovaskainen. 2002. Extinction debt at extinction threshold. *Conservation Biology* 16: 666–673.

Hastings, A., and S. Harrison. 1994. Metapopulation dynamics and genetics. *Annual Review of Ecology and Systematics* 25: 167–188.

Tilman, D., R. M. May, C. L. Lehman, and M. A. Nowak. 1994. Habitat destruction and the extinction debt. *Nature* 371: 65–66.

Verheyen, K., M. Vellend, H. Van Calster, G. Peterken, and M. Hermy. 2004. Metapopulation dynamics in changing landscapes: a new spatially realistic model for forest plants. *Ecology* 85: 3302–3312.

MICROBIAL COMMUNITIES

THOMAS G. PLATT, PETER C. ZEE,
KEENAN M. L. MACK, AND JAMES D. BEVER

Indiana University, Bloomington

Microorganisms are ubiquitous, live in diverse and dense communities, and exhibit tremendous physiological, metabolic, and phylogenetic diversity. Microbial communities are the seat of rampant inter- and intraspecific interactions—including competition, facilitation, cooperation, and predation—that drive complex community dynamics. Moreover, many microbes have direct consequences on plants, animals, and their populations through symbiotic associations that range from mutualistic to parasitic. Microbial communities also feature prominently in many important global processes by helping drive biogeochemical transformations, including those involved in nitrogen cycling and biodegradation. Motivated by the widespread importance of these communities, microbial community ecology seeks to understand how biotic and abiotic interactions influence the distribution, abundance, diversity, and function of microbial assemblages.

THE IMPORTANCE OF MICROBIAL ECOLOGY

Advances in molecular sequencing techniques have greatly facilitated efforts to characterize the diversity of microbes, revealing a staggering array of metabolic, physiological, and species diversity at spatial scales ranging from global down to the cubic centimeter. Despite their small size, microorganisms are abundant and ubiquitously distributed, playing an essential role in all of the Earth's biogeochemical cycles. Moreover, interactions between microorganisms and macroorganisms can have important fitness effects on both parties, which translate into controlling roles in their population dynamics. In some cases, such as the mutualism between legumes and nitrogen-fixing bacteria, these associations involve benefits for both parties. In contrast, other macroorganism–microbe interactions involve microbes acting as rapacious pathogens capitalizing on the antagonistic exploitation of host tissues, as exemplified by any number of bacterial human pathogens.

While the study of macroorganismal community ecology has been relatively well integrated with theory, conceptual models have a more uneven history of integration with the studies of microbial communities. Research in microbial ecology often focuses on describing the distribution, functional role, and diversity of microbial communities—a considerable challenge in light of the intrinsic difficulties of studying rapidly changing, microscopic organisms. This work has often proceeded without the benefit of a theoretical foundation that can motivate, organize, and integrate the wealth of information that advances with molecular techniques are now making possible. Consequently, our understanding of the mechanistic processes determining the structure and dynamics of natural microbial communities remains poorly developed and will only be achieved through the integration of ecological theory with observational and experimental data on microbial communities.

There are, however, important successes that illustrate the utility of applying ecological theory to microorganisms. In the 1930s, G. F. Gause recognized the potential of using microbial systems to test the ecological theory of species interactions independently developed by Alfred Lotka and Vito Volterra. In a series of renowned laboratory experiments, Gause demonstrated that simple models could predict the qualitative dynamics of pairwise competitive and predator–prey interactions between microbes. This tradition of using microorganisms to advance and test ecological theory continued during the 1940s with Jacques Monod, who recognized that the growth of bacterial populations was often poorly described by the logistic equation. This led him to adapt the Michaelis–Menten equation of enzyme kinetics to incorporate the importance of resource availability to population growth. This work laid the foundation of David Tilman's revolutionary predictive models of mechanistic resource competition, which he empirically tested using microbial algae competitors.

INTRA- AND INTERSPECIFIC INTERACTIONS AMONG MICROBES

Microorganisms engage in a wide variety of interactions. These range from negative interactions like exploitative and interference competition to positive ones like cooperation and facilitation. These interactions have important consequences shaping the composition and temporal dynamics of microbial assemblages.

Resources and Microbial Communities

The harvest and consumption of resources from the environment is essential to the population growth of all organisms. Microbial communities are typically very diverse, but because of similar needs, neighboring microbes compete for access to resources that are in limited supply. These competitive interactions can have important impacts, shaping the composition of microbial communities. Building on the seminal work of Tilman (1982) and Monod (1949), models examining the process of resource competition describe how consumer populations grow as well as how resources change with this growth. The growth of resource consumer populations is determined by the relative balance of the birth and death rates of the consumer population:

$$\frac{dN_i}{dt} = N_i\left[B_i\left(e_i^{j=1}, h_i^{j=1}, u_i^{j=1}, e_i^{j=2}, \ldots\right) - D_i\right]$$

where N_i is the population size, B_i is the per capita birth rate, and D_i is the per capita death rate of the ith consumer. Importantly, the per capita birth rate of each consumer depends on its encounter rate (e), handling time (h), and utilization efficiency (u) of each resource it consumes. The consumer's harvest of the resource, in turn, influences the available concentration of the resource in the environment:

$$\frac{dR^j}{dt} = S^j - \sum_i^n C_i(N_i, e_i^j, h_i^j, R^j)$$

where S^j is the supply rate of the jth resource and C_i is the rate at which the ith consumer removes the resource from the environment. This consumption rate depends on the population size, encounter rate, and handling time of that consumer for the resource as well as the concentration of the resource in the environment.

These resource–consumer models were first applied to and are well suited to well-mixed microbial communities wherein resources flow in and out of the competitive arena. The utility of these models stems from their ability to predict the outcome of competition based on each competitor's population growth in response to resources. The most basic prediction is that one consumer will displace another if it is able to consume resources below the minimal resource needs of the other consumer. Consequently, when consumers are competing for only one resource, the superior competitor is predicted to be the one able to subsist on the lowest concentration of the resource—a result termed the R^* rule, in reference to the equilibrium concentration of the resource for each competitor.

The same basic rationale also applies to resource–consumer dynamics when consumers compete for multiple resources. These multiple resources can be substitutable (like ammonia and nitrate forms of nitrogen) or essential but unsubstitutable resources (such as glucose and ammonia). With multiple resources, tradeoffs associated with different resources can allow for coexistence of different consumers. This is because the advantage that a consumer has with one resource can allow it to persist despite a disadvantage associated with consumption of a second resource, where neither competitor can drive the resource levels below the minimal needs of the other competitor (Fig. 1). Such resource-use tradeoffs are likely important to the maintenance of high levels of local diversity within microbial communities.

In addition to these negative competitive interactions, bacterial communities are also rife with positive interactions mediated by resources that can help promote the

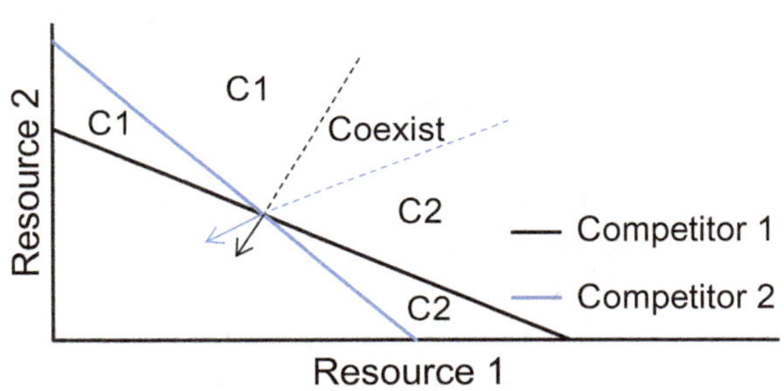

FIGURE 1 Resource-consumer models of competition predict the outcome of competition based on resource and population dynamics. One competitor is predicted to displace another if it is able to drive resource levels below the minimal needs of the competitor. These models can be applied to a variety of situations including competition for substitutable resources. In these models, coexistence between competitors is possible if there are tradeoffs associated with the different resources, such that each competitor can subsist on a lower concentration of one resource than the other. In order for stable coexistence to occur, at least one of the competitors must consume relatively more of the resource for which it is a superior competitor (i.e., the one for which it has a lower R^* when compared to the other competitor) than it does the other resource. For simplicity, competitive scenarios allowing for stable coexistence are presented. Tilman (1982) provides more detailed analysis of this and other possible competitive scenarios. Solid lines represent the zero-net-growth-isoclines, vectors represent the consumption vectors, and the predicted outcome of competition is given for each region of the environment resource supply parameter space.

coexistence of microbes. One such prevalent interaction occurs when one microbe produces a substrate that facilitates the growth of another microbe. In some cases, the cross-feeding substrate is merely a by-product of the metabolism of the facilitating strain such that the interaction is analogous to plants metabolizing carbon dioxide released by neighboring animals. However, some cross-feeding interactions are much more intimate. This is particularly the case with syntrophic associations involving mutualistic partners that can be distantly related. Interactions between syntrophic bacteria can even be obligate and reciprocal such that both species depend on the activity of the other in order to function. Explaining the maintenance of such mutualisms poses a challenge to evolutionary theory in that, as with other cooperative systems, uncooperative freeloaders can benefit from the actions of cooperating individuals.

Microbial Cooperation

Cooperation, in which the costly action of one individual benefits others, is widespread among microorganisms. Such behaviors feature prominently in many aspects of microbial communities, including nutrient acquisition, predation, host interactions, motility, and metabolism. Despite the ubiquity of cooperation in the microbial world, the potential for freeloading challenges its stability. Noncooperative individuals should have a competitive advantage over cooperative individuals because they avoid the costs of cooperation but can benefit from the actions of others. Thus, the maintenance of cooperative phenotypes requires that interactions are genetically structured so that cooperative individuals tend to interact with each other, such that cooperative groups are more productive than less cooperative groups.

Importantly, not all positive interactions among microbes constitute cooperation. As previously discussed, many instances of cross-feeding involve one species capitalizing on a waste product of another species. Although positive, such facilitation is unlikely to constitute cooperation. Cooperative cross-feeding is possible, but it requires that the production of the cross-feeding substrate is costly and benefiting the syntrophic partner somehow benefits the producer.

Antagonistic Interactions

Exploiter–victim interactions are a ubiquitous feature of microbial communities, with predatory and parasitic interactions between microbes being empirically documented across a range of taxa. The interaction between phage and bacteria is one of the classic examples of a microbial exploiter–victim relationship. Bacteriophages are obligate viral parasites that infect bacterial host cells. Following infection, some phages immediately replicate within and then emerge from the cell, killing the bacterial cell in the process. In contrast, the genomes of other phages integrate into the host bacterial genome until environmental conditions stimulate viral replication and eventual host cell lysis. As in other host–parasite systems, the phage–bacteria interaction can lead to cyclical population dynamics.

Many microorganisms produce toxic factors that antagonize neighboring microbes. These toxins are tremendously diverse, varying in their structure, mode of action, and killing range. Many of the best-characterized bacteriocins antagonize a relatively narrow range of microorganisms that tend to be closely related to the toxin-producing strain. Because of this, many bacteriocins are thought to mediate intraspecific interference competition. However, other bacteriocins inhibit the growth of a much broader range of microbes and in some cases even seem to specifically inhibit distantly related competitors, suggesting that toxin-mediated interference competition also impacts community ecology. Because of this, bacteriocins are able to mediate intraguild predation wherein toxin-producing strains are able to kill susceptible competitors and catabolize the resources stored in their bodies.

The resources liberated upon cell death caused by bacteriocins become a public good, potentially available to any cells that survive the toxin such that bacteriocin production is a sort of cooperative action. When the environment is well mixed, resistant strains that do not produce the toxin are able to benefit from the interference or death of susceptible competitors caused by toxin producing strains. However, spatially structured environments restrict this possibility because toxin-producing strains tend to have greater access to the benefits of toxin production due to their proximity to the lysed-susceptible cells. For this reason, toxin-producing strains tend to be more successful in spatially structured environments.

Surface-Associated Biofilm Communities

Natural microbial communities are often associated with surfaces. Such biofilms are ubiquitous, ranging from familiar examples like dental plaques to more exotic ones like marine stromatolites. These communities are composed of remarkably diverse and dense aggregates of cells embedded in a matrix of extracellular polymeric substances. Biofilms are thought to be the natural context for many forms of microbial interactions, including pathogenesis, horizontal gene transfer, cross-feeding, and resource competition. The maturation of biofilms establishes resource and chemical gradients, as well as species and genetic structure, that are likely to have important consequences on the dynamics of surface-associated microbial communities.

COEXISTENCE AND DIVERSITY

Microbial communities harbor a tremendous amount of diversity, with important interactions occurring between both closely and distantly related taxa. These interactions, in concert with abiotic environmental factors, help shape the complex dynamics of these communities. Integrating observations of the dynamics and distribution of microbial communities with ecological theory is essential to developing and understanding the mechanisms underlying the maintenance of the local and global diversity of microbial communities.

Microbial Resource Partitioning

The process of resource partitioning and specialization has long been a centerpiece of thinking about coexistence in ecological communities. As in macroorganisms, microbial communities likely face more than one resource simultaneously and can thus partition the resources in the system through specialization. Because of their large population sizes, rapid generation times, and limited dispersal, it is conceivable that microbial communities are able to partition resources faster than their macroorganism counterparts through rapid response to selection imposed by new environments. This rapid diversification and specialization has been demonstrated in numerous laboratory mesocosms. In these experiments, one strain rapidly diversifies into several novel genetic variants, each specialized to utilize different aspects of the environment.

The storage effect can also contribute to the maintenance of diversity in ecological communities in heterogeneous environments. This hypothesis posits that each species may have a competitive advantage in some environmental conditions while being competitively inferior in others. When experiencing unfavorable environmental conditions, species are able to subvert this disadvantage by going into a dormant state. In this way, multiple species are able to temporally partition the environment such that each species is most active when the environment suits them. A wide variety of both bacterial and fungal microbes form dormant, metabolically quiescent spores during their life cycle. In addition to sporulating species, many nonsporulating bacteria have reversible metabolically dormant states where cell division is ceased. These spores can survive through environmental changes unsuitable to growth and then germinate in periods with favorable conditions. Microbes typically enter these dormant life-history stages during periods of poor environmental conditions such as resource limitation or physical stress.

COMPLEX TROPHIC INTERACTIONS

Many microbial interactions blur the boundaries of trophic levels. As discussed above, bacteriocins can mediate intraguild predation while cross-feeding interactions present widespread potential for resource-mediated facilitation of one competitor by another. The prevalence of these types of interactions shapes the complex trophic interactions of microbial communities and allows for unusual avenues to coexistence and diversity beyond resource use tradeoffs (Fig. 1). Understanding these complex relationships and their consequences for the dynamics and composition of microbial communities presents a challenge for theoretical ecology.

MICROBES AND HOST DYNAMICS

Microbial communities associated with plants and animals can have strong effects on the fitness and dynamics of their hosts. Standard Susceptible–Infected–Resistant (SIR) models of disease dynamics succeed in capturing important aspects of the consequences of transmissible pathogens. These models, however, do not generally represent the dynamics of microbial communities within hosts. With multiple infections, the community dynamics within hosts could result in reduced or increased virulence, with the outcome being determined by factors such as the costs of virulence and the spatial structure of the microbial interactions. For pathogens that persist for significant portions of their life history outside of their hosts, such as the causative agent of cholera, overall disease incidence will partially depend upon the competitive interactions in the environment, another dimension not represented in SIR models. Models including dynamics within the environmental reservoir reveal the potential for disease dynamics being triggered by physical forces such as climate change.

The dynamics of diverse microbial communities, such as those occurring in association with plant roots within the soil, can be difficult to characterize. Despite this, the net effect of host-specific differentiation of microbial communities can be integrated into macroecology dynamics through models of microbial feedback on host fitness. Microbial community feedback involves two steps: first, the density and/or composition of the soil microbial community changes in response to composition of the plant community, and second, the

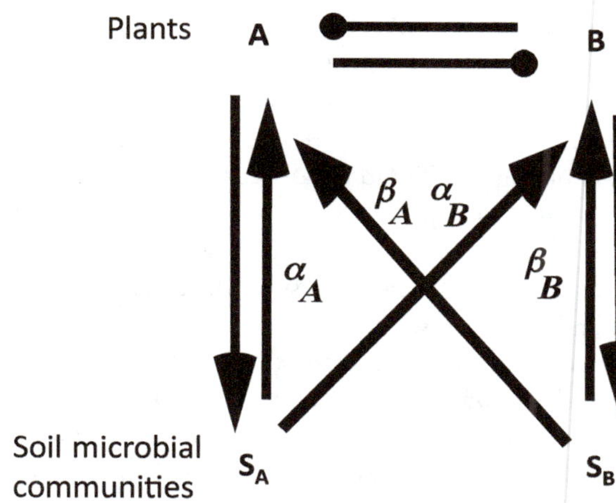

FIGURE 2 Soil microbial feedback. A conceptual representation of soil community feedback (modified from Bever et al., 2010). The presence of plant A can cause a change in the composition of the soil community, represented by S_A. This change in the soil community can directly alter the population growth rate of species A (represented by α_A) and it can alter the growth rate (α_B) of competing plant species B (with negative effect represented by the club symbol). Similarly, the presence of plant B can cause a change in the composition of the soil community (S_B) which can directly feedback (β_B) on the population growth rate of plant B or indirectly feedback on the growth rate of plant B through changes in the growth rate (β_A) of competing plant A. The net effect of soil community dynamics on plant species coexistence is a function of the sign and magnitude of an interaction coefficient $= \alpha_A - \alpha_B - \beta_A + \beta_B$, which represents the net pairwise feedback.

change in composition alters the relative growth rates of individual plant species (Fig. 2). Changes in microbial composition that increase the relative performance of the locally abundant plant species generate a positive feedback dynamic that would lead to loss of diversity at a local scale. Conversely, changes in microbial composition that decrease the relative performance of the locally abundant plant species generate a negative feedback that can contribute to plant species coexistence. As plant–microbe interactions likely occur at a local scale, feedbacks are commonly measured at the scale of individual plants. Recent work suggests that microbially mediated negative feedback predominates in terrestrial plant communities and drives plant species coexistence and relative abundance patterns.

APPLICATION OF ECOLOGICAL THEORY TO MICROBIAL COMMUNITIES

The development of ecological theory for microbial communities is likely to proceed through the adoption and modification of established theory developed for plants and animals as well as the development of novel theory. Microbial communities differ from macroorganismal communities in a number of important aspects, including the dramatic diversity that they exhibit at both local and global spatial scales. Microorganisms often have large population sizes and relatively rapid generation times, facilitating their ability to rapidly adapt to their local environment. Consequently, the ecological and evolutionary time scales of microbial systems are more likely to converge than for macroorganisms. The clonal reproduction and limited dispersal of many microbes result in a high degree of spatial structure that can have important consequences on their adaptation to the local environment and their interactions with one another. Further, although horizontal gene transfer is relatively rare in macroorganisms, it is common among many microbes, particularly among bacteria. The horizontal acquisition of novel genetic systems can further facilitate the ability of microbes to adapt to their local environment. The rapid spread of antibiotic-resistant human pathogens driven by strong selection and horizontal gene transfer typifies this and poses a significant global public health concern.

The extreme diversity, varied interactions, spatial structure, and rapid evolution occurring within microbial communities pose significant challenges to the development of theoretical frameworks describing these communities. Thus, in light of their importance to natural ecosystems, the development, integration, and application of ecological theory for microbial communities poses an important opportunity for ecologists and microbiologists.

SEE ALSO THE FOLLOWING ARTICLES

Belowground Processes / Cooperation, Evolution of / Disease Dynamics / Facilitation / SIR Models / Storage Effect / Two-Species Competition

FURTHER READING

Bever, J. D., I. A. Dickie, E. Facelli, J. M. Facelli, J. Klironomos, M. Moora, M. C. Rillig, W. D. Stock, M. Tibbett, and M. Zobel. 2010. Rooting theories of plant community ecology in microbial interactions. *Trends in Ecology & Evolution* 25: 468–478.

Grover, J. P. 1997. *Resource competition.* London: Chapman & Hall.

Hall-Stoodley, L., J. W. Costerton, and P. Stoodley. 2004. Bacterial biofilms: from the natural environment to infectious diseases. *Nature Reviews Microbiology* 2: 95–108.

Monod, J. 1949. The growth of bacterial cultures. *Annual Review of Microbiology* 3:371–394.

Prosser, J. I., B. J. M. Bohannan, T. P. Curtis, R. J. Ellis, M. K. Firestone, R. P. Freckleton, J. L. Green, L. E. Green, K. Killham, J. J. Lennon, A. M. Osborn, M. Solan, C. J. van der Gast, and J. P. W. Young. 2007. The role of ecological theory in microbial ecology. *Nature Reviews Microbiology* 5: 384–392.

Tilman, D. 1982. *Resource competition and community structure.* Princeton, NJ: Princeton University Press.

West, S. A., S. P. Diggle, A. Buckling, A. Gardner, and A. S. Griffins. 2007. The social lives of microbes. *Annual Review of Ecology, Evolution, and Systematics* 38: 53–77.

MODEL FITTING

PERRY DE VALPINE

University of California, Berkeley

Model fitting is the task of estimating parameters of a mathematical model from data. This task encompasses determination of both the single parameter values for which the model can best match the data and also ranges of parameter values that characterize uncertainty about the best values. The single best parameter values are called the best fit parameters, point estimates, or simply parameter estimates. Justification and improvement of methods for model fitting for all manner of data and models are major goals of statistical research. Model fitting and the related statistical inferences can be viewed as the formal process of treating each model as a hypothesis and confronting the hypothesis with data.

HOW MODEL FITTING WORKS

Model Structure

Model fitting cannot begin without specification of a model structure (but see "nonparametric methods," below) with one or more parameters that must be estimated. The model structure includes deterministic and stochastic (random) components. The deterministic components often stem from ecological theory, while the stochastic components are needed to accommodate realistic variation in data, although they may stem from ecological theory in some cases. The concepts of deterministic and stochastic parts of a model can apply to models for all kinds of data, including for spatial and/or temporal processes or static relationships among variables, and for observational or experimental data.

DETERMINISTIC MODEL STRUCTURE

The deterministic part of the model structure can explain systematic patterns or trends in data. For example, in linear regression—determination of the best linear relationship between paired X and Y values—the deterministic part of the model structure is $Y = a + bX$. The unknown intercept, a, and slope, b, must be estimated from the data. Another example is a Ricker model for time-series data of population size. In the Ricker model, the expected population size $N(t)$ at time t is a function of size at the previous time, $N(t - 1)$, using the function $N(t) = N(t - 1)\exp[r - cN(t - 1)]$, which has unknown parameters r and c to be estimated. The deterministic part of the model predicts expected data values from different explanatory data variables (e.g., temperature experienced by a population) and/or from the same type of data values (e.g., population size) at other times or locations, but it does not describe realistic variation anticipated in data.

STOCHASTIC (RANDOM) MODEL STRUCTURE

Ecological data typically display a great deal of variation around any trends. The stochastic or random part of a model describes the probability distribution(s) of such variation around the systematic patterns (the deterministic model structure). The most common choice is for each data value to have a normally distributed discrepancy (also called error, noise, or deviation) from the model's predicted value. For example, in linear regression, the Y data values often follow a normal distribution centered around the line $a + bX$. In the Ricker model, the stochastic model structure typically allows $N(t)$ to be impacted on a log scale by normally distributed environmental randomness, i.e., environmental stochasticity. (See below for incorporating measurement inaccuracies together with environmental stochasticity.)

Some types of data call for nonnormal distributions for the stochastic part of a model. For example, count data often follow a Poisson distribution. Survival or other binary data follow a binomial distribution. Data on the time until some event—e.g., maturation or death—follow distributions such as Weibull, gamma, or log-normal.

Not all model-fitting methods require the stochastic components of a model to be explicitly defined. In some cases, such as least squares or generalized estimating equations, the measure of fit between a model and data is calculated in a way that incorporates variation in data implicitly or indirectly rather than with explicit probability distributions. In either case, the stochastic part of a model may include additional parameters to be estimated. For example, the variance of a normal distribution is usually unknown and can be estimated.

Calculation of Model Fit for Candidate Parameters

The next step is to determine how to measure the quality of the model's fit to the data for any candidate values of the unknown parameters. A measure of model fit is a mathematical statement about how deviations between the model and data should be weighted as relatively better or worse. For example, the sum of squares used in linear regression measures the fit of the model with specific values of a and b as the sum of the squared differences between the observed and predicted (model) values of Y for each value of X.

The most widely applicable and theoretically justified measure of model fit is the probability (or probability density) of the data given the model with specific parameters; this is known as the likelihood. Calculation of likelihood values explicitly uses the probability distributions in the stochastic part of the model. For example, in the Ricker model, the probability of observing $N(t)$ given $N(t - 1)$ uses the normal distribution probability for the log ratio between $N(t)$ and the value predicted from $N(t - 1)$ with some particular parameter values. Due to the form of the normal distribution function, sums of squared discrepancies yield an equivalent measure of model fit as normal distribution likelihoods for parameters such as a and b or r and c in the examples here. For some measures of model fit (e.g., sum of squares), lower values indicate better fits, while for others (e.g., likelihood), higher values indicate better fits.

Optimization of Model Fit

The best-fit parameters are those for which the measure of model fit is as good as possible. In simple cases, equations for the best fit parameters can be derived using calculus and solved using algebra. More generally, a computer program determines the best fit parameters using an optimization algorithm. An optimization algorithm searches efficiently around possible parameter values to find the ones that optimize (maximize or minimize) the measure of model fit (called the objective function in the jargon of optimization methods). For some models, including hierarchical models that describe data where some or all model discrepancies are not independent from each other, calculating model fits is not straightforward. Optimization of the sum of squares is called least squares estimation, and optimization of the likelihood is called maximum likelihood estimation.

An exception to the optimization step in model fitting is often found in Bayesian methods. In Bayesian methods, the measure of model fit is the posterior distribution, which is proportional to the likelihood multiplied by a prior distribution for parameters. The optimal fit corresponds to the highest density of the posterior distribution, but Bayesian analysis often instead uses an algorithm such as Markov chain Monte Carlo to sample parameter values from the posterior distribution. This means that the full range of parameters that fit the data well or poorly are explored by the algorithm. Results are summarized from the posterior sample using information such as averages, 95% credible intervals, or other summaries that may or may not include highest posterior density parameters.

Assessment of Uncertainty in Estimated Parameters

If the slope parameter, b, in a linear regression is estimated to be 0.2, one does not yet know how to judge the evidence that X and Y are in fact related. If the range of uncertainty extends 0.6 on either side of the estimate—i.e., from −0.4 to 0.8—then the relationship could quite plausibly be 0 or negative. On the other hand, if the range of uncertainty is from 0.1 to 0.3, one might reach a conclusion that they are in fact related, although with an imperfectly known slope.

Assessments of uncertainty take two common formats, following the two major paradigms of statistical inference: frequentist and Bayesian. Frequentist 95% (or other %) confidence intervals give a parameter range that will include the correct parameters in 95% of studies. For complex model-fitting problems, confidence intervals are typically approximate—because exact derivation requires knowledge of the correct model and parameters, which are unknown—but they are increasingly accurate in 95% coverage as the amount of data increases. Confidence intervals are often calculated from estimates of the standard error, which is the standard deviation of estimates of the parameter that would be obtained if the study was hypothetically repeated. Bayesian 95% credible intervals give a parameter range including 95% of the posterior distribution. The relative strengths, weaknesses, and scientific philosophies of frequentist and Bayesian inference are topics of debate.

USES OF MODEL FITTING

Evaluation and Formulation of Ecological Theories

The need to estimate parameters of a mathematical model from data arises in many situations that may be loosely or tightly tied to specific ecological theories. For example, linear regression of log metabolic rate as a function of log body mass might—if there were no theoretical models for the relationship between these traits—be used to ask whether they are significantly related, how strong the relationship is, and whether it is linear. In fact, this relationship is the subject of theoretical modeling, and linear regression is used to evaluate specific hypotheses about the value of the slope, an example of tight coupling to ecological theories. Different labels have been used to describe this spectrum, such as heuristic or phenomenological vs. mechanistic models and statistical vs. biological models.

Another example is estimation of population dynamics models for time-series data of population size. A model loosely tied to specific theories might simply allow any smooth function to relate log population size at one time to the previous time. (Such relationships have been estimated with nonparametric methods such as kernel smoothing.) Results from such a generic model might be used to ask whether there is any clear relationship at all, whether the relationship is nonlinear (density dependent), and how well the relationship explains the data. A model tightly tied to a specific theory might be a particular type of model, such as a Nicholson–Bailey model if the data include population sizes of both a host and parasitoid species. Then the question may be whether this specific model adequately explains the data and if so what its estimated parameters (and their uncertainties) are. The approach of model fitting stands in contrast to the approach of trying to measure each demographic rate separately and combine them into a model for prediction. Similar examples could be considered for many other topics of ecological theory.

Fitting Models of the Data-Sampling Process to Estimate Ecological States or Rates

The above examples emphasize models of ecological processes or relationships. In many situations, the model describes the data-sampling process, and the goal is to estimate one or more ecological states or rates. For example, capture–mark–recapture (CMR) studies of wildlife populations involve capturing and marking (or identifying) many individuals, releasing them, and attempting to recapture them at a later time. A goal of CMR studies is to estimate the survival rate, which can in turn be used as part of a matrix model, to investigate patterns of survival rates relative to other factors, or for other applications. However, in order to estimate survival rate, the efficacy of capturing and recapturing individual animals must be estimated. Study designs and estimation methods for such data can include many types of complexity so that unbiased estimates of survival rates and accurate assessments of uncertainty can be obtained. Another situation in which modeling and model fitting focus on the sampling process is for estimating population size from some sampling design. For example, in distance sampling, data such as avian point counts and their distances can be fit to models of the probability of sighting a bird as a function of distance in order to estimate population size.

Relation Between Model Fitting and Model Selection

Model selection is the task of choosing among different deterministic model structures, and/or different candidate explanatory variables, to find the ones that could best be used to predict hypothetical future data. This requires that overfitting be avoided (see below) and is often accomplished with information theoretic methods such as the Akaike Information Criterion (AIC). AIC is defined using the maximum likelihood value penalized by the number of parameters estimated. Thus, model fitting for each candidate model (or set of explanatory variables) is a step in the process of model selection.

MORE COMPLICATED DETERMINISTIC AND STOCHASTIC MODEL STRUCTURES

The opening section, "How Model Fitting Works," introduced the basic concepts of model fitting. More complicated types of deterministic and stochastic model structures lead to more complicated model-fitting challenges. Deterministic model structures can become more complicated by allowing nonparametric relationships between variables or by using a more mathematically difficult model. Stochastic model structures can be more complicated by explicitly incorporating multiple sources of variation in ecological processes and/or data sampling.

Parametric, Nonparametric, and Semi-Parametric Models

The examples above of linear regression and a Ricker model for a time-series of population abundance are parametric models because they are formulated with specific parameters to be estimated. Nonparametric models can take a great variety of shapes to smoothly characterize relationships among variables. For example, a smoothing spline or a kernel smoother to relate paired X and Y values are not limited to linear, quadratic, exponential, or other specific type of functions but rather can relate the two variables in much more flexible ways. Mathematically, nonparametric models may have parameters to be estimated—e.g., a cubic smoothing spline is composed of many short cubic polynomials connected with continuous first and second derivatives—but their purpose is to allow in effect arbitrary shapes to explain the data. For nonparametric models, determination of appropriate model complexity, commonly measured as total curvature, is a vital part of the model-fitting process. Methods to accomplish this include generalized cross-validation, restricted maximum likelihood (REML), and AIC. Semiparametric models are models that have some parametric relationships and some nonparametric relationships among variables.

Mathematically Challenging Deterministic Model Structures

In some cases, a model stemming from ecological theory involves mathematically challenging calculations that render model fitting more difficult. For example, models of temporal dynamics formulated as continuous time nonlinear differential equations, such as some models of disease dynamics, require numerical integration to calculate predicted model dynamics. This means that the numerical integration must be repeated over and over for different candidate parameter values to find the optimal fit. Moreover, incorporating random variation in the disease dynamics is particularly challenging for a continuous-time model.

Multiple Sources of Variation and Hierarchical Models

Multiple sources of variation can arise for many kinds of models and data. For example, in linear regression or analysis of variance, data blocks refer to groups of data measured from the same part of a field, or the same individual organism, or at the same time, or with some other commonality. Data from one block may be more similar to each other than they are to data from different blocks. A common stochastic modeling scheme to accommodate this is to assume that there is a shared random deviation—a random effect—associated with the block, and therefore with any samples from that block. In addition, each sample has its own random deviation that is combined with the block effect. Therefore, there are two sources of random variation, and they cannot be perfectly disentangled from the data. Random effects are a mathematical device by which the model can predict that some data values will be more similar than others, and the degree of this pattern is represented by additional parameters that must be estimated.

The concept of multiple sources of random variation that can only be observed in unknown combinations is also used in temporal and/or spatial models. In such models, two important sources of variation are considered to be process and sampling variation. Process variation represents all of the random effects on how the system, such as a population or an ecosystem, changes through time or across space. Sampling variation represents deviations between the estimated values of the system and its actual values.

The approach of formulating models that combine multiple sources of variation has been labeled differently for different types of models and in different fields of application, including random effects models, mixture or mixed models, generalized linear mixed models, latent variable models, shared frailty models for survival analysis, state-space models for time-series, and random field models for spatial data. The term hierarchical models now encompasses all of these situations.

INCORPORATING MULTIPLE SOURCES OF VARIATION IN MODEL FITTING

Multiple sources of variation make it harder to say how well a model with particular parameters fits the data. For example, with random effects for blocks, all the data within a block share the same random effect for that block, but the random effect cannot be directly measured. This means that the discrepancies between model predictions and the data from one block should not be treated as if they are independent. In temporal and/or spatial models, the random effects together with the sampling variation determine relationships and correlations among data through time and/or space. A variety of model-fitting methods have been developed for hierarchical models, of which several are summarized here.

To accomplish maximum likelihood estimation for a hierarchical model, all possible values of the unknown random effects must be considered. Mathematically, this corresponds to integrating over the distributions (whose parameters are among those to be estimated) of the random effects to obtain the marginal probability of the data (or, technically, probability density for data measured on a continuous scale). Except for special cases such as linear models with normally distributed random effects, this integration is accomplished computationally and/or with mathematical approximations.

Bayesian methods are especially popular for hierarchical models because the computational methods (such as Markov chain Monte Carlo) used to explore the posterior distribution can naturally accommodate multiple sources of variation. For linear models with normally distributed random effects, restricted maximum likelihood (REML) gives better estimates of variances of random effects and sampling variation, known as variance components, than does maximum likelihood. Another approach is maximum penalized likelihood—sometimes called errors-in-variables or total maximum likelihood—in which specific random effects values are estimated along with model parameters; this approach is practical but can have undesirable properties such as biased parameter estimates (see below). Quasi-likelihood (which can also be incorporated in penalized quasi-likelihood) allows additional variation in data beyond that accommodated in some sampling distributions, such as Poisson and binomial, using a mathematical adjustment to the likelihood equations that may not correspond to an explicitly defined distribution. Generalized estimating equations do not involve explicit assumptions about the random effects distributions but rather allow a general pattern of correlations among data following the sampling design (e.g., blocks) together with general samping distributions. The "method of moments" (which can be used for simpler models as well) involves finding parameters that best match the predicted and observed moments, which are mathematical values such as means, variances, covariances, auto-covariances, skews, and so on. In summary, for models with more complex stochastic components, there is a more complicated range of model-fitting methods.

PROPERTIES OF ESTIMATION METHODS

In frequentist statistical reasoning, properties of estimation methods are defined in terms of how they would perform for many hypothetical data sets. The concept of hypothetical data sets is that if a study was repeated again—and/or an ecological process unfolded again—one would obtain different data sampled from the same distributions as the actual data. For example, a study of phytoplankton density in a lake using 20 random samples could hypothetically be repeated, yielding 20 different samples. Some studies require a more abstract notion of hypothetical data. For example, if the data are a 20-year time series of phytoplankton abundance, then hypothetical data would involve different patterns of population change according to different outcomes of random variation such as abiotic processes or species interactions. In this sense, "random" takes on a statistical meaning of processes for which one has no explanatory data and are therefore represented mathematically by a distribution of unpredictable outcomes.

Bias, Variance, Mean Squared Error, and Coverage Probability

It is easiest to define properties of estimation methods for one parameter at a time, although the definitions can be extended to multiple parameters together. The theoretical bias of an estimation method is the average difference between estimated and actual parameter values, where the average is over many hypothetical data sets. For example, a bias of 0.1 means there is a systematic tendency to overestimate a parameter by 0.1. The theoretical variance of an estimation method is the average squared difference between estimated parameter values and the average estimate. In other words, estimation variance measures typical deviations not around the actual parameter values but around the average estimate of them. The mean squared error (MSE) is the average squared difference between estimated and actual parameter values, which equals the bias squared plus the variance. The root mean squared error (RMSE) is the square root of MSE, which has the same units as the parameters.

The coverage probability of a confidence interval or credible interval is the actual probability that it contains the correct parameters, which may differ from the nominal (claimed) probability, typically 95%. Good model-fitting methods have as low RMSE as possible and as accurate coverage probability as possible.

Bias-Variance Tradeoffs

Choices of models and model-fitting methods can often be conceptualized in terms of a tradeoff between the bias and variance of parameter estimates—and hence of predictions. For example, a common situation is that researchers have data on many candidate explanatory variables for some response variable. A model that includes all of the explanatory variables may have a small ratio of data to parameters and so yield highly variable parameter estimates. This means that if a new data set of the same size was acquired, the estimated parameters might be very different due to sampling variation. However, *on average,* the estimated parameters of a complete model will be centered around the true parameters. In this situation, there would be large variance but small bias in parameter estimates.

On the other hand, a model that omits some explanatory variables runs the risk of missing real scientific relationships but has the benefit of smaller estimation variance. For example, if two explanatory variables are correlated (called collinear in this context) but one is omitted, then the parameter estimated for the included variable will be biased because it compensates mathematically for the omitted variable, i.e., it takes on the explanatory role of the omitted variable. However, using fewer explanatory variables reduces estimation variance, which may reduce mean squared error if any resulting bias is small. Thus, it is possible to have estimation methods with high bias and low variance (always giving nearly the same incorrect estimate) or with low bias and high variance (giving highly variable estimates that are on average correct). Model-selection methods attempt to find the right balance on this spectrum. Similar tradeoffs arise concerning the complexity of curves or other relationships to be included in a model relating multiple variables.

Overfitting

Overfitting refers to drawing too-strong conclusions or estimating a too-complex model relative to the amount and/or quality of data. Other terms used for overfitting include "data dredging" and "statistical fishing expedition." One type of overfitting arises from including too many explanatory variables relative to the amount of data. Then it can be highly likely that some explanatory variables spuriously appear to be important—even according to statistical significance tests—simply because so many were tried. This is a multiple-testing problem and can be formally evaluated as part of an analysis. Another type of overfitting arises from allowing a curve (or other type of relationship) to take a shape too complex relative to the meaningful patterns in the data. For example, a nonparametric spline curve (see above) can take nearly any shape. If the curve is allowed to be too wiggly, then it will pass close to each data value, in effect estimating a relationship that is in fact due to noise in the data.

The hallmark of overfitting is increased prediction error. This means that if a researcher could obtain one new value of the response variable and any associated explanatory variables, the model's prediction for the response variable, based on parameters estimated from the previous data, would be inaccurate because the estimated parameters are too closely tied to the previous data. An overfit model appears to match the data more strongly than simpler models, but it yields less accurate predictions because it has represented random noise as real patterns. The same concept of prediction error can be used in a temporal or spatial model for new data values in time or space, respectively. Even when new data will not actually be acquired, the concept of prediction error for new data provides a line of theoretical reasoning used to develop methods to avoid overfitting. Cross-validation and information-theoretic methods such as AIC are two approaches to avoid overfitting; both do so by attempting to reduce prediction error.

Identifiability

Identifiability refers to the ability to estimate multiple parameters from the same data. For example, if one has data on the number of adults of some species in replicated enclosures at one time and the number of surviving offspring at a later time, then the rates of offspring production and offspring survival are not identifiable. Only their product, the rate of surviving offspring production, can be estimated. This example illustrates mathematical nonidentifiability: any combination of offspring production and survival that yields the same product will mathematically have the same fit to the data. In practice, parameters that are mathematically identifiable may nevertheless be only weakly identifiable from limited data. This occurs when the data allow a wide range of parameter combinations to give similar fits. For example, in multiple linear regression when two or more explanatory variables are highly correlated (collinear), a wide range of combinations of their predictive roles can yield close but not identical fits to the data.

SUMMARY

Model-fitting methods are used to estimate parameters of mathematical models from data. Model fitting involves specifying the types of systematic relationships and patterns of variation in the data, defining how to measure the quality of fit between model and data, and estimating parameters by optimizing the fit. Statistical methods to characterize the uncertainty in estimated parameters are an integral part of model fitting. Model fitting is used to formulate and test ecological theories and to estimate states or rates of ecological systems. More advanced model-fitting methods are used for estimation when there are multiple sources of variation reflected in the data or when a more detailed or flexible model structure is of scientific interest.

SEE ALSO THE FOLLOWING ARTICLES

Bayesian Statistics / Frequentist Statistics / Information Criteria in Ecology / Statistics in Ecology / Stochasticity

FURTHER READING

Bolker, B. 2008. *Ecological models and data in R.* Princeton: Princeton University Press.

Cressie, N., C. Calder, J. Clark, J. Hoef, and C. Wikle. 2009. Accounting for uncertainty in ecological analysis: the strengths and limitations of hierarchical statistical modeling. *Ecological Applications* 19: 553–570.

Hilborn, R., and M. Mangel. 1997. *The ecological detective: confronting models with data.* Princeton: Princeton University Press.

Schabenberger, O., and C. A. Gotway. 2005. *Statistical methods for spatial data analysis.* Boca Raton, FL: Chapman & Hall/CRC.

Shumway, R. H., and D. S. Stoffer. 2006. *Time series analysis and its applications: with R examples,* 2nd ed. New York: Springer.

Zuur, A. F., E. N. Ieno, and G. M. Smith. 2007. *Analysing ecological data,* 1st ed. New York: Springer.

MOVEMENT: FROM INDIVIDUALS TO POPULATIONS

PAUL R. MOORCROFT

Harvard University, Cambridge, Massachusetts

Organisms move for a variety of reasons, including to acquire resources such as food, mates, and shelter, and to avoid competitors and predators. Movement is thus a central component of the ecology of animals and other mobile organisms that determines the nature and scale of their interactions with the physical environment and the nature and scale of their interactions with conspecifics and individuals in other populations.

QUANTIFYING MOVEMENT

Prior to the 1950s, information on the movement of animals in the wild came from either an observer directly recording an animal's successive spatial locations or by inferring movement from animal tracks. The development in the 1950s of radio telemetry, in which animals are outfitted with a radio transmitter that can be used to triangulate their position, transformed the study of animal movement, allowing for the routine and systematic collection of information on animal locations in the wild. Global Positioning System (GPS)–based telemetry systems are now also being used, which generally yield more accurate and more frequent information on an animal's location and do not require the animal to be within range of a receiving antenna in order to determine its location. While radio telemetry datasets are typically comprised of hundreds to thousands of relocations, datasets generated with GPS telemetry typically comprise tens to hundreds of thousands of relocations per animal collected over periods ranging from a few months to several years and thus are able to provide highly detailed pictures of how animals move around their landscape (Fig. 1).

MATHEMATICAL APPROACHES TO MODELING ANIMAL MOVEMENT

The mathematical analysis of animal movement is rooted in the random walk models of statistical physics, in which the motion of an animal is represented in a manner analogous to that of a particle, consisting of sequences of movements of different lengths, directions, and turning frequencies (Fig. 2). The term random walk embodies the

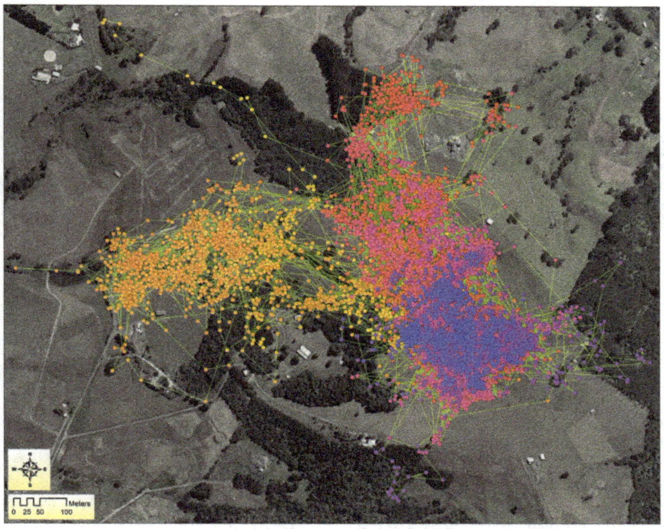

FIGURE 1 Example of a modern-day GPS-telemetry dataset collected for common brushtail possums (*Trichosurus vulpecula*) by Todd Dennis and colleagues. The complete dataset consists of > 140,000 relocations collected at 5-15 minute intervals over two-year period. The figure shows 13,000 relocations for a single individual. The color of each relocation indicates the time of relocation. SOURCE: Todd Dennis (unpublished data).

fact that, at the scale of the individual, movement is a stochastic process, which can be characterized in terms of a redistribution kernel $k(\mathbf{x}, \mathbf{x}', \mathrm{t}, t)$ where $k(\mathbf{x}, \mathbf{x}', \mathrm{t}, t)\, d\mathbf{x}'\, d\mathbf{x}$ specifies probability of an animal moving from any small rectangle $d\mathbf{x}'$ located at $\mathbf{x}'$ at time t to a small rectangle $d\mathbf{x}$ located at $\mathbf{x}$ in a specified time interval. The redistribution kernel is often formulated in terms of a composite of a distribution of distances moved and a distribution of movement directions per time interval (Fig. 2A). Alternatively, an individual's movement can be characterized in terms of distributions of movement velocities and turning angles. Correlated random walk models were first applied to study the movements of cells and microorganisms such as bacteria and slime molds; however, they are increasingly being used to study the movement of more complex organisms such as insects, fish, birds, and mammals.

While individual-based descriptions of an animal movement behavior are a natural way to conceptualize and characterize animal motion, from an ecological perspective it is useful to be able to translate such individual-centered (Lagrangian) descriptions of an animal's movements into a corresponding place-centered (Eulerian) description of the resulting intensity of space use in different areas. This translation is desirable because explaining the spatial distributions of animals on landscapes is one of the central underlying themes of ecology.

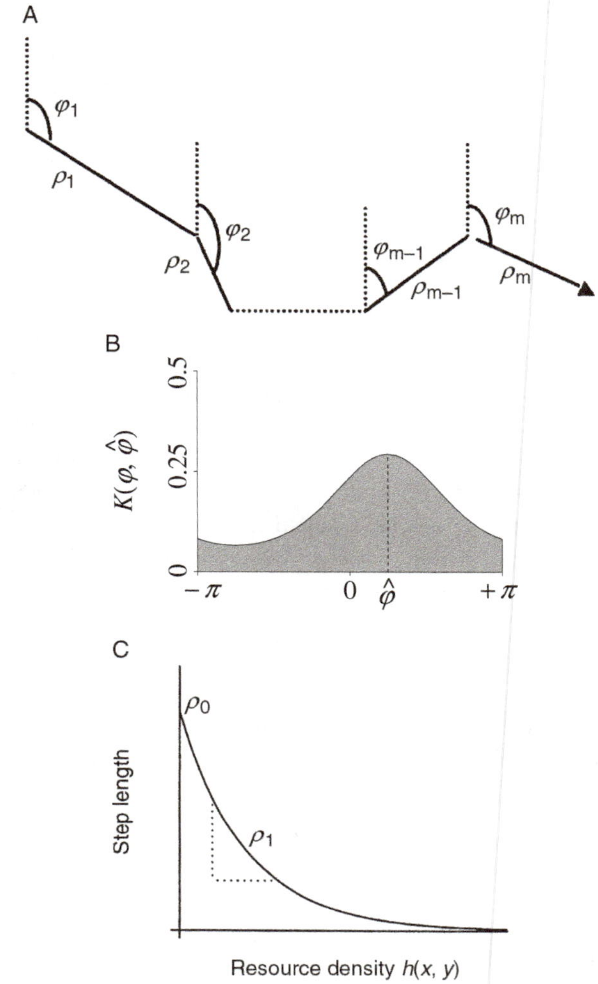

FIGURE 2 (A) Schematic illustrating the underlying model of individual movement behavior that underpins a mechanistic home range model. The movement trajectory of individuals is characterized as a stochastic movement process, defined in terms of sequences of movements between successive relocations ($i = 1, \ldots, m$) of distance ρ_i and directions φ_i drawn from statistical distributions of these quantities that are influenced by relevant factors affecting the movement behavior of individuals. (B) and (C) Two qualitatively different kinds of movement responses individuals can display in response to external or internal stimuli: (B) In tactic movement responses, individuals bias their direction of movement in response to an external or internal stimulus, yielding nonuniform distribution of movement directions $K(\varphi)$. This gives rise a nonzero advection term in the equation for space use (Eq. 3) whose magnitude is governed by the degree of nonuniformity in the distribution of movement directions and whose direction corresponds to mode of the distribution ϕ. (C) In kinetic movement responses, the mean step length of individuals changes in response to an internal or external stimulus. In the example shown, the stimulus is resource density $h(x, y)$, ρ_0 is the mean distance between successive relocations in the absence of resources and r_1 governs the rate at which the mean step length of individuals decreases with increasing resource density $h(x, y)$.

Arguably, the simplest movement model is the case of pure random motion, in which an individual's directions, magnitudes, and frequency of movements are uninfluenced by its environment or by the presence of other individuals. The expected pattern of space use for an individual moving in a purely random manner can, under quite general conditions, be described by the solution of a corresponding diffusion equation specified in the appropriate number of space dimensions. For example, for an animal moving randomly on a one-dimensional landscape x, its expected pattern of space use can be described by the following partial differential equation (PDE):

$$\frac{\partial u(x,t)}{\partial t} = d\frac{\partial^2 u}{\partial x^2}, \quad (1)$$

where $u(x,t)$ is the expected pattern of space use, expressed in terms of a dynamically evolving probability density function for a single individual, or a population density function for multiple individuals. The parameter d of Equation 1 reflects statistical properties of the underlying redistribution kernel describing the individual's movement behavior. Specifically,

$$d = \lim_{\tau \to 0} \int (x - x')^2 k(x, x', \tau, t)\, dx', \quad (2)$$

i.e., the second moment of the animal's redistribution kernel as the time steps t become vanishingly small. The corresponding equations for an animal moving randomly on a two-dimensional $\mathbf{x} = (x,y)$ landscape are

$$\frac{\partial u}{\partial t}(\mathbf{x},t) = d\nabla^2[u(\mathbf{x},t)]$$

where

$$\nabla = \left(\frac{\partial}{\partial x}, \frac{\partial}{\partial y}\right), \quad (3)$$

and

$$d = \lim_{\tau \to 0} \frac{1}{\tau}\int |\mathbf{x} - \mathbf{x}'|^2 k(\mathbf{x}, \mathbf{x}', \tau, t)\, d\mathbf{x}'. \quad (4)$$

Numerous analytic insights can be derived from Equations 1 and 3, including that the mean-squared displacement of the individual from its original starting position increases linearly as function of time and that in a bounded environment the long-term equilibrium pattern of space use by randomly moving individuals will be spatially uniform.

Many studies have shown, however, that animals typically do not move in a random manner. Instead, their movements are influenced by a variety of factors, including the characteristics of their environment (such as resources, terrain, and habitat), by their internal physiological or behavioral state (e.g., hunger, and fear), and by the presence of other individuals (e.g., conspecifics, prey, predators, and competitors). These influences can be incorporated into correlated random walk models of animal movement through appropriate specification of how the animal's redistribution kernel is modified by variables describing its external environment, its internal physiological or behavioral state, or the presence of other individuals.

As for the case of random motion, individual-based movement models incorporating behavioral and ecological responses can often be translated into corresponding partial differential equations (PDE) describing the expected pattern of space use. These take the form of an Advection–Diffusion Equation (ADE):

$$\frac{\partial u}{\partial t}(\mathbf{x},t) = \nabla^2[d(\mathbf{x},t)\,u(\mathbf{x},t)] - \nabla \cdot [\mathbf{c}(\mathbf{x},t)\,u(\mathbf{x},t)]. \quad (5)$$

As in Equations 1 and 3, $u(\mathbf{x},t)$ is a time-evolving density function describing the intensity of space use in different areas. The first term is a diffusion term whose magnitude $d(\mathbf{x},t)$ reflects the random component of the animal's motion that arises because of the probabilistic nature of underlying movement rules, while the second term in Equation 5 is an advection term whose direction and magnitude are specified by the vector $\mathbf{c}(\mathbf{x},t)$ that reflects the directed component of the animal's motion arising from its responses to its environment, its internal state, and the presence of other individuals. Specifically,

$$\mathbf{c}(\mathbf{x},t) = \lim_{\tau \to 0} \frac{1}{\tau}\int_{-\infty}^{\infty} (\mathbf{x} - \mathbf{x}')k(\mathbf{x}, \mathbf{x}', \tau, t)\, d\mathbf{x}', \quad (6)$$

i.e., the first moment of the redistribution kernel as the time step t becomes vanishingly small.

INCORPORATING ECOLOGICAL RESPONSES INTO MODELS OF ANIMAL MOVEMENT

The analytic and numerical tractability of ADEs such as Equation 5 has yielded numerous insights regarding the consequences of movement behavior for the spatial distributions of animals on landscapes. The sections below describe in more detail how different aspects of species' ecology can be incorporated into models of animal movement and used to understand their effects on resulting patterns of space use.

Responses to Resources

Effective search strategies should direct animals toward regions of increased resource availability. From a mathematical standpoint, it is useful to distinguish between two qualitatively different kinds of movement response that animals can exhibit. The first are so-called *tactic* movement responses, in which animals change their distribution of movement directions in response to spatial variation in a quantity such as resource availability (Fig. 2B). A classic example of such a tactic movement response is the chemotaxis of bacteria *Escherichia coli* in response to gradients in glucose concentration.

The second are so-called *kinetic* movement responses, in which organisms alter the distance they move between turns or the frequency at which they change their direction of movement (Fig. 2C). Kinetic movement responses often underlie the "area-restricted search" behavior exhibited by many species of animals when foraging. A well-studied example is the foraging behavior of ladybird beetles feeding on aphids, in which the beetles turn more frequently in areas of high aphid density.

Animals can exhibit similar tactic and kinetic movement responses to other resources, such as water, shelter, or mates. Tactic and kinetic movement responses to spatial gradients in resource availability yields corresponding advection terms in the resulting equations describing the expected pattern of space use by individuals. It is these directed components of the animal's motion that result in animals concentrating in areas of high resource availability.

Responses to Internal State Variables

Many responses to the external environment are mediated through changes in the internal physiological and behavioral state of animals. For example, in the case of ladybird foraging described above, the decline in ladybird movement distances in response to increasing prey density is mediated by their level of satiation (gut fullness). In a similar manner, the movement behavior of many prey species appears to be influenced by their current level of fear of being subject to predation.

An internal state variable that has received a lot of recent theoretical attention is memory. In particular, memory is thought to play a key role in the formation and development of animal home ranges. The implications of memory for animal space use has been explored using so-called *self-attracting* random walk models, in which individuals develop and maintain memories of places that they have visited and display an increased probability of moving toward previously visited locations. These analyses have shown that, under quite general conditions, movement models of this kind that incorporate memory result in individuals developing characteristic stable or quasi-stable home ranges for individuals.

Responses to Conspecifics

In addition to resources and internal state variables, the movements of individuals are often also influenced by conspecifics. In some species, individuals exhibit avoidance responses to the presence of other individuals that result in spacing out of animals across a landscape. These responses can arise as the result of direct aggressive interactions between individuals, or indirectly through signaling cues such as singing in birds or scent marking in mammals. For example, an analysis of space use by coyotes in Yellowstone utilized a individual-based movement model incorporating a conspecific avoidance term, in which individuals displayed a tactic movement response (Fig. 2B), biasing their movements toward their den site following encounters with the scent marks of other individuals. This conspecific-avoidance response was linked to a simple model of scent marking, in which animals scent marked as they moved and increased their marking rate when they encountered scent marks deposited by individuals in other groups. When combined with a kinetic foraging response to food availability (Fig. 2C) in which the animals moved shorter distances per unit time in areas with higher densities of small mammals, this movement model was able to capture the observed patterns of space use by the different coyote packs in the study region (Fig. 3).

In other species, individuals exhibit aggregative responses to the presence of conspecifics. These forms of movement response underlie bacterial congregations, insect swarms, fish schools, bird flocks, and mammal herds. The movement behaviors that underlie the dynamics of animal aggregations have been a subject of considerable theoretical interest. A classic analysis in the 1970s by Okubo and colleagues documented the movement behavior of individuals in a swarm of midges quantifying the patterns of changes in position, velocity, and acceleration (Fig. 4A). Their analysis showed that individuals at the edges of swarm moved toward the center, with their movement speed decelerating as they approached the center of the swarm.

A recent theoretical analysis of has shown how the patterns of acceleration observed in Okubo's midge study can be characterized by the following equation:

$$\frac{\partial \mathbf{c}}{\partial t}(\mathbf{x}, t) = \kappa\left(1 - \frac{u}{u_c}\right)\nabla u + \delta_v \nabla^2 \cdot [\mathbf{c}(\mathbf{x}, t)], \qquad (7)$$

where $\mathbf{c}(\mathbf{x}, t)$ is the velocity of the individual at location $\mathbf{x}$. The first term in Equation 7 specifies that below a specified threshold population density u_c, an individual's rate of acceleration is positively related to the spatial gradient in population density ∇u, but above this threshold the relationship between acceleration and population density becomes negative, indicating repulsion. The parameter κ determines the strength of the aggregative and repulsive responses of the individuals as population density changes. The second term of Equation 7 is a diffusion term that specifies how velocity dissipates over time.

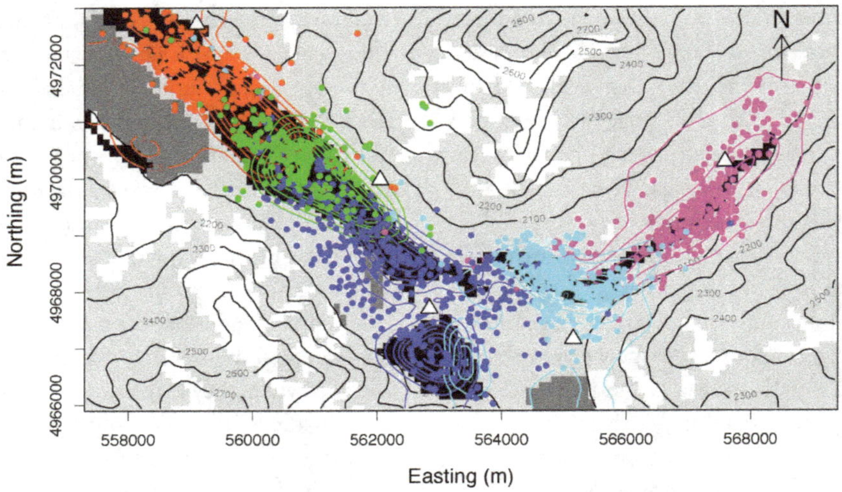

FIGURE 3 Colored contour lines showing fit of a mechanistic home range model to relocations (filled circles) obtained from five adjacent coyote packs in Lamar Valley, Yellowstone National Park. As described in the text, the PA+CA mechanistic home range model used in this study incorporates a foraging response to small mammal prey availability plus a conspecific avoidance response to the scent marks of individuals in neighboring packs. Also shown are the home range centers for each of the packs (triangles), and the grayscale background indicates small mammal prey density (kg ha^{-1}) across the landscape (Moorcroft and Lewis, 2006).

As seen in Figure 4A, Equation 7 captures the spatial patterns of acceleration observed in Okubo's study. Equation 7 can be combined with Equation 5 in order to predict the observed pattern of space use resulting from these movement responses. Analysis of Equations 5 and 7 indicates that below a threshold value of the aggregation parameter κ no clustering occurs, but above this value the population forms clusters with the degree of clustering increasing as values of the aggregation parameter increases.

Species Interactions

The way in which animals move has important consequences for their interactions with other species. Analyses of the consequences of individual movement behavior for species interactions have principally focused on two interrelated issues: the role of movement in pattern formation, and its consequences for the population dynamics of the interacting species.

One important consequence of animal movement is its effect on the rate at which individuals encounter prey items and thus the functional responses of predators to their prey. As discussed elsewhere in this volume, predator functional responses are an important factor influencing the dynamics of predator–prey interactions. In the original derivation of predator functional responses, by Holling in 1959 assumed that predators search a constant area per unit time, an assumption consistent with pure directed motion by a predator. Mathematically, this corresponds to a nonzero advection term in Equation 2, but with no diffusion term (i.e., $d = 0$). For an animal foraging on static, randomly distributed prey items, such pure directed motion results in a linear relationship between prey density and the rate at which predators encounter prey while searching, yielding a linear (type I) functional response when there is no handling time for each prey item consumed and a saturating (type II) functional response if there is a finite handling time per prey item consumed.

As discussed earlier, the probabilistic nature of individual movement behavior means, however, that animal motion usually contains random as well as directed components of motion. A recent analysis explored the implications of this for predator functional responses. When a predator moves randomly, the area searched by the individual does not increase linearly with search time, but instead increases as the square root of the time spent searching. Analysis shows that, again for a predator foraging on static, randomly distributed prey items, this results in a quadratic relationship between prey density and the rate at which predators encounter prey while searching. If there is negligible handing time per prey item, this yields a quadratic functional response of the predator to prey density, and a sigmoidal (type III) functional response if there is a nonzero handling time per prey item. A significant implication of this result is that animals that have a finite handling time per prey item can have type III functional responses even in the absence of learning by the predator or the existence of prey refuges. More generally, the above result implies that the functional responses of animals are likely to vary between a type II and type III

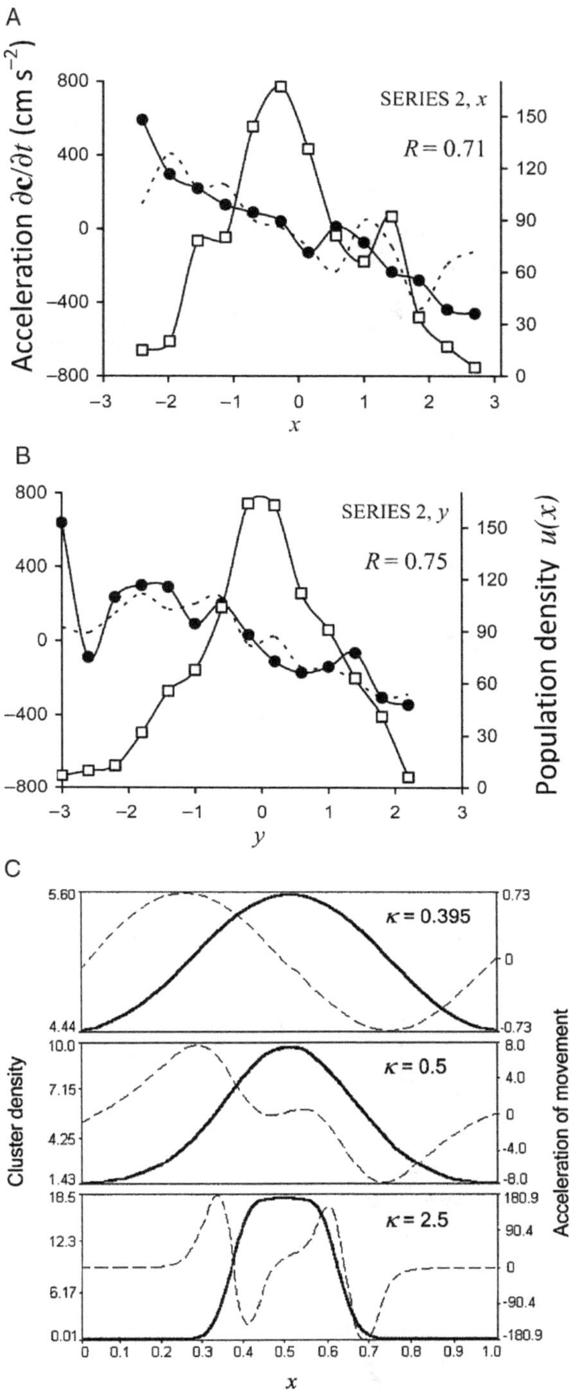

FIGURE 4 (A) and (B) Mean values of acceleration (•) and the density of individuals (▫) within a swarm of midges projected onto horizontal plane (x,y). (A) Acceleration in the x-direction; (B) acceleration in the y-direction. The x and y coordinates are centered on the swarm center and are measured in units of the standard deviation of midge position. Dashed curves show the best fit of Equation 3 to the observed acceleration rates, and R is the multiple correlation coefficient. (C) One-dimensional solutions of Equations 2–3 on finite enclosed domain. Increasing the taxis coefficient κ strengthens the degree of aggregation. The solid line is the density distribution, and the dashed line is the acceleration. From Tyutyunov et al., 2004.

functional response depending on the relative strength of the directed and random components of their motion as they move across a landscape.

Another recent analysis explored the consequences of predator taxis for the pattern and dynamics of predator–prey interactions using the following system of equations:

$$\frac{\partial N}{\partial t} = rN\left(1 - \frac{N}{K}\right) - \frac{aNP}{1 + ahN} + \delta_N \nabla^2 N,$$
$$\frac{\partial P}{\partial t} = e\frac{aNP}{1 + ahN}\mu P - \nabla \cdot (P\mathbf{c}) + \delta_N \nabla^2 P, \quad \text{(8a–c)}$$
$$\frac{\partial \mathbf{c}}{\partial t} = \kappa \nabla N + \delta_c \nabla^2 \mathbf{c}.$$

Equations 8a–c are modified versions of the classic Rosenzweig–MacArthur predator–prey model that describes a prey population (N), which grows logistically in the absence of a predator (P), and a predator population that consumes the prey with a saturating type II functional response and dies at a constant, density-independent rate. The parameters r, K, a, h, e, and m are, respectively, the pest reproduction rate, the pest carrying capacity, the search efficiency, the conversion efficiency, and the predator mortality rate.

Equations 8a–c are modified from the original nonspatial Rosenzweig–MacArthur model through inclusion of spatial movement terms for the prey and the predator. As indicated by the diffusion term in prey density equation Equation 8a, individuals in the prey population are assumed to move at random, with the parameter d_N reflecting their mean-squared displacement per unit time. The predators, in contrast, exhibit a tactic movement response in which their acceleration is proportional to the spatial gradient in prey density. The predator density equation (Eq. 8b) thus has advection and diffusion terms, with the diffusion parameter d_p reflecting the random component of the predator's motion and the advection term $\mathbf{c}(\mathbf{x}, t)$ being given by the solution of Equation 8c, in which the parameter k controls the strength of the predator's movement response to prey density ∇N and the parameter d_c governs the rate at which their directed movement velocity dissipates over time. Note that Equation 8c has a very similar form to Equation 7, except that the predators are responding to the spatial distribution of prey, as opposed to the spatial distribution of their conspecifics, and do not exhibit repulsion at high densities of prey.

In the absence of movement terms, Equations 8a and b correspond to the original Rosenzweig–MacArthur predator–prey model, which, depending on the values of the demographic and functional response parameters, exhibits a stable equilibrium, a stable limit cycle, or unstable oscillations. The curve labeled C_0 in Figure 5A shows the dynamics of the prey population for a case in which the model in the absence of predator and prey movement exhibits a stable limit cycle, with the prey population undergoing large, infrequent oscillations in abundance.

The incorporation of movement terms into Equations 8a–c significantly modifies the patterns and dynamics of the predator–prey interaction. The curve labeled R_k in Figures 5A and 5B shows the resulting dynamics of the predator and prey populations when the prey's movement rate is relatively slow (i.e., low d_N), and the response of the predator to the gradient in prey density is relatively weak (i.e., low value of κ in Equation 8c). As seen in Figure 5A, at the local scale, the movement responses of the predator and prey results in more frequent and smaller amplitude oscillations, indicating that the predator is now more effective in regulating the prey's abundance. At the population scale, however, the system still exhibits a stable limit cycle (Fig. 5B). Figures 5C and 5D show the dynamics for a higher value of κ in Equation 8c, indicating that predator movements now respond more strongly to prey density. This destabilizes the predator–prey interaction, and the prey population now exhibits unstable, chaotic dynamics at both the local scale (Fig. 5C) and at the population level (Fig. 5D).

By comparison, there have been relatively few theoretical investigations looking at the effects of movement behavior on the dynamics of other forms of species interactions, such as interspecific competition and mutualisms. One exception is an analysis of competitive interactions between wolf and coyote populations, which showed that coyote avoidance responses to wolves, in conjunction with intraspecific avoidance responses between wolf packs, can lead to spatial segregation of the two species.

MATHEMATICAL CONSIDERATIONS

The derivation of Equations 1, 3, and 5 from the redistribution kernel k relies on taking the classic diffusion limit, which assumes that the second moment of the redistribution kernel scales with the time step $\tilde{\tau}$. Technically, this is not realistic, because on the finest time scales it implies an infinite speed of movement by individuals. In essence, then, when taking the classic diffusion limit one is approximating the fine-scale movement behavior of the individual with a movement process that has similar statistics to the actual movement behavior on the time scales at which movement is observed (i.e., a distribution kernel with similar means and variances), but on time scales shorter than this has different statistical properties than the observed movement process.

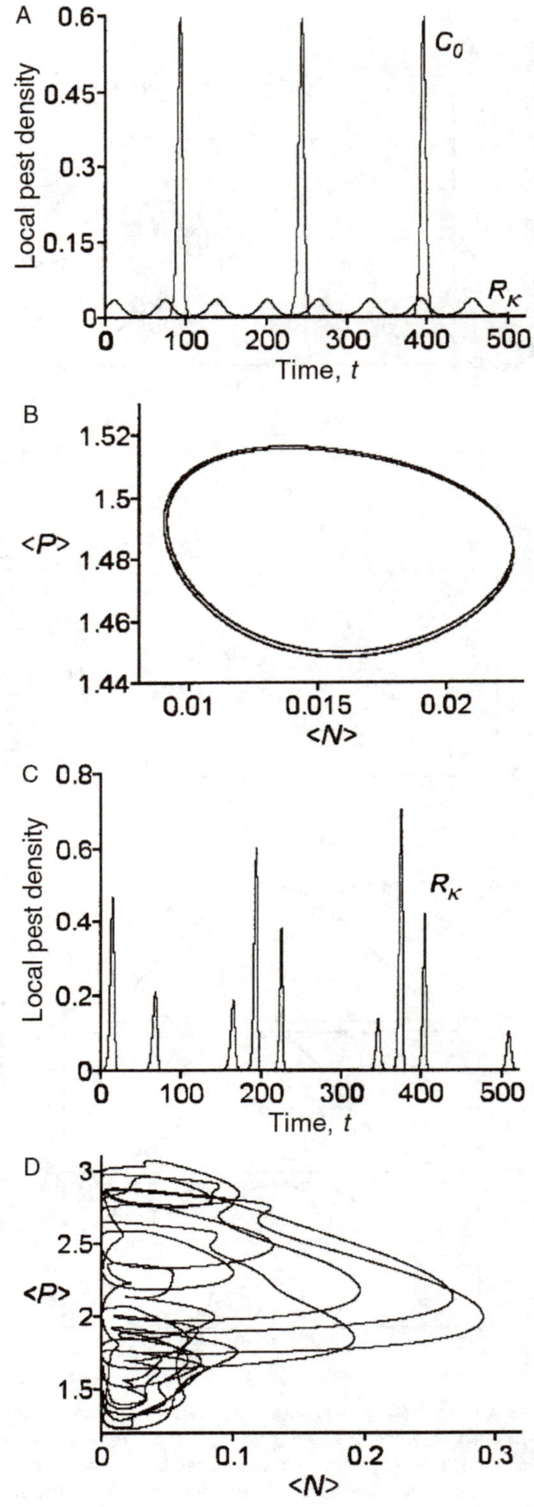

FIGURE 5 Panels (A) and (C) show local-scale fluctuations of the pest density in R_κ in comparison with the homogeneous dynamics C_0; panels (B) and (D) show phase trajectories of the average population density of the prey <N> and predator <P> predicted by Equation 5; (A) and (B) show the case where $\kappa = 0.5$ in which R_κ exhibits a limit cycle; (C) and (D) $\kappa = 1.5$, in which R_κ is chaotic. From Sapoukhina et al., 2003.

A second issue arises if an animal's distribution of movement distances has a sufficiently fat tail that it does not have a finite variance. Its fine-scale movement behavior then corresponds to a so-called Levy walk, which from a theoretical standpoint presents problems because it is "super-diffusive" and thus is not compatible with taking the classic diffusion limit to obtain equations such as Equations 1, 3, and 5 for the expected pattern of space use for the animal. Analyses of some animal movement trajectories have implied that their distributions of movement distances conform to a Levy walk; however, this conclusion remains controversial because distributions of movement distances similar to that of Levy walks can arise when animals move in spatially heterogeneous environments and their distributions of movement distances are significantly different between the different environments.

The adoption of alternative theoretical frameworks for analyzing animal movement behavior can address some of the above theoretical limitations. In particular, Markov operator and semi-group theory are useful approaches for analyzing the implications of animal movements for patterns of animal space use. While more challenging mathematically, these methods avoid some of the restrictions that arise in deriving PDE-based descriptions of space use from stochastic, individual-based movement models.

SEE ALSO THE FOLLOWING ARTICLES

Facilitation / Foraging Behavior / Functional Traits of Species and Individuals / Individual-Based Ecology / Predator–Prey Models / Reaction-Diffusion Models / Spatial Spread / Species Ranges

FURTHER READING

Grunbaum, D. 1998. Using spatially explicit models to characterize foraging performance in heterogeneous landscapes. *American Naturalist* 151: 97–115.

McKensie, H. W., M. A. Lewis, and E. H. Merrill. 2009. First passage time analysis of animal movement and insights into the functional response. *Bulletin of Mathematical Biology* 71: 107–129.

Moorcroft, P. R., and M. A. Lewis. 2006. *Mechanistic home range analysis.* Princeton Monographs in Population Biology. Princeton: Princeton University Press.

Okubo, A., and S. A. Levin. 2001. *Diffusion and ecological problems: modern perspectives,* 2nd ed. Volume 14 of Interdisciplinary Applied Mathematics. Basel, Germany: Springer-Verlag.

Sapoukina, N., Y. Tyutyunov, and R. Arditi. 2003. The role of prey taxis in biological control: a spatial theoretical model. *American Naturalist* 162: 61–76.

Turchin, P. 1998. *Quantitative analysis of movement: measuring and modeling population redistribution in animals and plants.* Sunderland, MA: Sinauer.

Tyutyunov, Y., I. Senina, and R. Arditi. 2004. Clustering due to acceleration in the response to population gradient: a simple self-organization model. *American Naturalist* 164: 722–735.

MUTATION, SELECTION, AND GENETIC DRIFT

BRIAN CHARLESWORTH

University of Edinburgh, United Kingdom

Evolutionary change requires variation, which is ultimately generated by new mutations. But these originate as rare variants within large natural populations. Natural selection and random genetic drift are the main evolutionary factors that cause them to rise in frequency and spread through a population or species. Theoretical and empirical knowledge of these population-level processes is fundamental for understanding the mechanism of evolution.

MUTATION

A mutation is defined as any alteration in the genetic material, which can occur in any cell of a multicellular organism. Mutations that are transmitted from parent to offspring (germline mutations) provide the raw material for evolution. These mainly involve changes in the sequences of nucleotides in the DNA or RNA molecules that constitute the genome of an organism. In addition, the expression of a particular gene in a given cell of an individual can be modified by signals from other cells or by environmental factors. Such changes can be transmitted to the cell's descendants and are essential for the differentiation of cell types during development. These "epigenetic" changes are rarely passed between generations; in most known examples of transgenerational transmission, they are less stable than mutational changes in nucleic acid sequences. If they are stably inherited, they must obey Mendelian rules, and hence they can be treated in a similar way to sequence variants.

Transmission of Epigenetic Changes

Some biologists have recently argued that epigenetic effects can cause the inheritance of phenotypic changes that are acquired during the life of an individual, reviving a Lamarckian view of evolution. Much evidence from classical genetics shows, however, that this process is unlikely to be important. One type of evidence comes from the examination of variation within inbred lines, made by many generations of matings between close relatives, which creates individuals that are genetically nearly identical. In contrast to what is found for genetically mixed populations (see below), it is usually found that there is no detectable correlation between parent and offspring within an inbred line, and selection is correspondingly ineffective. This brings out the distinction between the individual's measurable phenotype (the value of a trait of interest) and its genotype. Phenotypes are controlled by the joint effects of genetic factors (that can be transmitted to the offspring) and nongenetic factors (that are usually not transmissible).

For selection on a genetically uniform stock to be effective, enough time must therefore have elapsed for new genetic variants to be generated by mutation. Measurements over many generation of variability within inbred lines for quantitative traits such as body size or fertility show that the rate of increase of the variance in a trait caused by mutation is of the order of 10^{-3}, when measured relative to the nongenetic variance within the line.

Further evidence is provided by selection experiments that are based on choosing the relatives of individuals with a desired characteristic. These have been found to be effective in several different systems, the classic example being antibiotic resistance in bacteria. Since the relatives of the individuals that are scored by the experimenter have not been exposed to any agency that could have induced the phenotype that is the object of selection, but do have genotypes in common with them, the genetic variability utilized by selection must exist independently of any component of the phenotype acquired by the individual during its life. The differences between castes of social insects are a natural example of traits that have evolved in this way, as was pointed out by Darwin in *The Origin of Species.*

The Randomness of Mutations

These observations show that "random" mutations that are stably transmitted over many millions of generations are the primary source of the raw material of evolution. Randomness does not mean that mutations can have any conceivable effect on the phenotype. As pointed out by H. J. Muller in 1949, the effects of mutations are "conditioned by the entire developmental and physiological system resulting from the action of all the other genes already present." According to Muller, randomness simply means that "there is no relation between the kind of natural circumstances (e.g., climate or prevailing physiological state) under which mutations arise and the direction of phenotypic change which these mutations result in; still less is there any adaptive relation between them in the sense of the phenotypic changes tending to be more useful under the circumstances that prevailed when they arose."

The Nature and Frequency of Mutations

It should also be noted that the phenotypic effects of mutations vary widely, from none, through an almost undetectably small effect on a quantitative trait, to a very large, qualitative change in phenotype. A large fraction of the DNA sequence of a higher organism has little or no functional significance, so that many mutations have no effect on the phenotype of their carriers. At the other end of the spectrum, a change to a single base pair in a gene that codes for a protein that is essential for survival can be fatal.

The most frequent mutations (in terms of their rate of occurrence per generation) are nucleotide substitutions—changes from one base pair at a nucleotide site to an alternative base pair. Small deletions and insertions of sets of nucleotides, as well as insertions of mobile genetic elements, are also relatively common. Larger-scale deletions and duplications of segments of chromosomes, duplications and deletions of entire chromosomes or of the entire genome, and rearrangements of the chromosomal material such as inversions and translocations also occur, albeit at much lower rates.

Rates of spontaneous mutation in organisms with DNA genomes are extremely low, due to proteins that correct errors in the replication of DNA sequences. The rates of nucleotide substitutions per base pair per cell generation in DNA-based microbes are of the order of 10^{-10}. Similar values apply to mutation rates per cell division in multicellular organisms such as *Drosophila* and humans, but the rate per organismal generation is as much as two orders of magnitude higher, owing to the many cell divisions that occur during the production of germ cells. This yields a mutation rate per base pair per generation (u) of about 3×10^{-9} in *Drosophila* and about 1×10^{-8} in humans. Since the coding sequences of genes in these organisms on average contain 1500 base pairs, the rate of mutation per gene to a new allelic form with a detectable phenotypic effect is substantially higher—on the order of 10^{-5} per gene per generation. The large number of base pairs in the genomes of higher organisms means that each newly formed individual may carry more than one mutation somewhere in its genome—over 50 on average in the case of humans. Mutation rates per base pair in mitochondrial genomes, and in viruses with RNA genomes, are often much higher than the values per base pair in DNA genomes, due to a lack of repair mechanisms.

The generally low rate of mutation at the nucleotide level, even for nucleotide substitutions, means that the time scale of changes in the frequencies of the four alternative states at a nucleotide site due to mutation pressure is of the order of 100 million generations or more. This implies that mutation at this level is a weak force and can easily be counteracted by other evolutionary forces. Mutation is therefore important mainly as a source of new variation, but not as a directional factor in evolution, as was argued forcefully by R. A. Fisher. Nevertheless, as mentioned above, increases in quantitative trait variability due to mutation are detectable over relatively short time scales, reflecting the fact that such traits are affected by many different genes. Since a functionally significant trait must have been adjusted by natural selection to be close to its optimal value, most phenotypic changes caused by mutations are harmful to the organism. In consequence, the mean values of fitness-related traits such as viability and fecundity tend to decline under mutation pressure when selection is relaxed by experimental manipulation. In the wild, environmental changes in the environment that remove selection on a trait may lead to its decline under the pressure of mutation, as is probably true of the loss of vision by cave-dwelling animals.

Mutations enter a population each generation in numbers that are determined by the product of the number of breeding individuals (N) and the rate of mutation. In a species with a population size of even a few thousand, many thousands of new mutations arise each generation. But each of these will be present initially as a single copy or a handful of copies. Other evolutionary forces are therefore required to cause the spread of a new mutation to a high frequency over the time scale required to explain evolutionary change.

SELECTION

Natural selection is one such force, and the only one that can explain adaptive evolution. The elementary process involved in evolution by natural selection is a change in the relative frequencies within a population of alternative variants at a particular location in the genome, often involving two different alternative base pairs at a single nucleotide site. Many quantitative traits of ecological significance are controlled by genetic variants at several locations within the genome, and changes in the mean values of such traits reflect the net effects of changes in the frequencies of individual variants at all of these locations.

Haploid Populations

For this reason, much attention has been given to understanding the effects of selection on a pair of alternative variants, or alleles, which can be referred to as A_1 and A_2, where A_2 is the new form produced by a mutation. Let the frequencies of A_1 and A_2 at the start of a generation be p and q, respectively, where $p + q = 1$. The problem is to calculate the value of p and q in the next generation, given a specification of the relative fitnesses of the different genotypes. This is most simply done in the case of a haploid organism, such as a bacterium, in which the two alleles correspond directly to the two possible genotypes. For simplicity, discrete generations are assumed, so that individuals are born into the population, survive to adulthood, reproduce, and then die.

Selection acting on survival can be modeled by assigning different probabilities of survival to adulthood to the two different genotypes. The survival probability of genotype A_i is denoted by w_i ($i = 1$ or 2). In the absence of fertility differences, this represents the fitness of the genotype. If selection acts on fertility differences, the w_i values also take account of each genotype's number of offspring.

Only the ratio w_2/w_1 is important for determining the rate of spread of A_2 from a given initial frequency. This ratio can be written as $1 + s$, where s is the "selection coefficient"; s is positive if A_2 is fitter than A_1, and negative if it is less fit. In a very large population, where we can neglect the effects of random sampling on allele frequencies (considered later in the section on genetic drift), A_2 will be eliminated if $s < 0$ and will spread through the entire population to approach fixation if $s > 0$. Let t be the time taken for A_2 to spread from being present as a single copy (with $q = 1/N$) to almost completely replacing A_1. If s is relatively small, of the order of 1% or less ("weak selection"), t is approximately equal to the natural logarithm of N, multiplied by $2/s$.

This means that t is insensitive to the population size and is equal to a relatively small multiple of $1/s$; e.g., with $N = 10^6$, $t = 28/s$ generations, and with $N = 10^9$, $t = 41/s$. This result, derived by J. B. S. Haldane in 1924, is fundamentally important: it shows that selection can cause favorable mutations to spread through very large populations in a relatively short time as far as evolution is concerned, regardless of their initial frequencies. This firmly establishes selection as a potent factor in evolution.

Diploid Populations

Similar conclusions hold with diploid inheritance, where each individual receives one genome copy from its mother and another copy from its father. There are two main differences from the haploid case. First, if a favorable mutation A_2 is recessive in its effect on fitness (i.e., individuals with both A_1 and A_2 have the same fitness as individuals who carry only A_1), the rate of spread of A_2 will be very small if individuals mate randomly, since A_2 is initially mostly found in individuals that carry A_1, i.e., as A_1A_2 heterozygotes. This explains why cases of recent adaptive evolution in large, randomly mating diploid populations that are associated with the spread of single mutations, such as insecticide resistance, usually involve mutant alleles that have at least some effect in heterozygotes—recessive mutations lag far behind the others in their rate of increase in frequency ("Haldane's sieve"). This contrasts with the fact that most mutations with detectable phenotypic effects are recessive, and it sheds light on the much-debated question of whether the nature of evolutionary change is strongly constrained by the rules of developmental and functional biology. The recessivity of most mutations probably arises from the inherent properties of metabolic and developmental pathways. But the data show that selection seizes on the infrequent nonrecessive mutations. In this case, natural selection is the major factor causing the observed pattern.

Balanced Polymorphisms

The other major distinguishing feature of diploid inheritance is that variability can be actively maintained in the population when there is heterozygote advantage—i.e., the fitness of A_1A_2 individuals is higher than that of A_1A_1 and A_2A_2 individuals—discovered by Fisher in 1922. This results in a stable equilibrium of allele frequencies in a very large population, or a "balanced polymorphism." The classic example of this is "sickle cell" hemoglobin, found in some human populations. Heterozygotes for the A allele (coding for normal hemoglobin) and the S mutation in the β-globin gene are protected against malaria compared with AA individuals, but SS individuals suffer from near-lethal sickle cell anemia, due to the clogging of blood vessels by misshapen red blood cells.

Balanced polymorphisms can be maintained in several other ways, some of which act equally well in haploids and diploids. One of the most biologically important of these is negatively frequency dependent selection, whereby the fitness of a genotype declines as it becomes more common in the population. The ecological relations between hosts and parasites give rise to one form of frequency dependence. The transmission of parasites among hosts, and hence the survival of a parasite population, requires encounters between infected host individuals and susceptible uninfected host individuals. Assume that the population sizes of the host and its parasite have come into equilibrium, so that the numbers of each are stable. A new mutation arises in the host, which confers greater resistance to the parasite, and starts to spread. As the mutation becomes more common, the parasites find fewer new hosts to infect, and so their abundance decreases. The chance that nonresistant hosts acquire new infections also declines, reducing the advantage of resistance. If resistance comes with a cost, so that the fitness of resistant individuals (in the absence of the parasites) is lower than that of nonresistant individuals, there may be a polymorphic equilibrium, with resistant and nonresistant hosts coexisting.

The parasite may also experience frequency-dependent selection, if there is genetic variation in resistance to the parasite. This type of interaction between host and parasite genotypes can give rise to cycles of allele frequency changes in both host and parasite. Relations between plant hosts and their pathogens often involve interactions of the kind assumed in this model, and there is evidence that these have sometimes resulted in the long-term maintenance of variability.

Other Ways of Maintaining Variability

Variability can also be maintained by both temporal and spatial variation in the relative fitnesses of genotypes. It can also be maintained within a local population by migration bringing in alleles that are locally disfavored by selection. There is much evidence for geographic patterns in allele frequencies and in the means of quantitative traits, reflecting spatial variation in selection pressures. Another major source of variability is mutation–selection balance, where the constant production of deleterious mutations in a gene comes into equilibrium with the rate at which selection eliminates them. A sizeable fraction of the variation in quantitative traits, and in fitness itself, probably comes from this source.

As mentioned above, much evolutionary change of ecological significance involves quantitative traits, affected by genetic variants at many different sites in the genome. Both theory and experiment show that, by combining together variants at different sites that were initially present at relatively low frequencies in the population, selection can produce phenotypes and combinations of phenotypes that are far outside the range of variability in the initial population. But this depends critically on the occurrence of sexual reproduction and genetic recombination between the sites involved, which allow sites to evolve more or less independently of each other. In the absence of recombination, this is not possible, and the response to selection is slowed down by the need to wait for new mutations.

GENETIC DRIFT

There is overwhelming evidence that natural selection explains the adaptive evolution of phenotypes, from the sequences of proteins to morphology and behavior. However, among-species comparisons of DNA sequences show that the fastest rates of evolution usually occur at sites in the genome that have little or no functional significance and are unlikely to have evolutionarily significant effects on fitness, i.e., they are selectively neutral or nearly neutral. This aspect of evolution reflects the effects of random genetic drift—random changes in variant frequencies caused by the sampling of the genetic composition of populations, which occurs when each new generation is formed. This process was first investigated theoretically by Fisher and by Wright in the 1920s. Neutral or nearly neutral variants provide an invaluable tool for studying evolutionary relationships among species, and the extent of genetic divergence among populations of the same species.

Wright–Fisher Populations

The simplest way to think about drift is in terms of the Wright–Fisher model. Imagine a randomly mating population consisting of N diploid, hermaphroditic individuals, counted at the time of breeding and reproducing with discrete generations. New individuals are formed each generation by random sampling from a large pool of gametes produced by the parents. Each parent thus has an equal probability of contributing to an individual of the next generation. If the current frequencies of the selectively neutral alleles A_1 and A_2 are p and q, in the new generation there will be a binomial probability distribution of the new frequency of A_2, centered around q, with variance $pq/(2N)$. If the new frequency is $q + \delta q$, in the next generation the distribution will be centered around $q + \delta q$, with variance $(p - \delta q)(q + \delta q)/(2N)$, and so on.

This process continues generation after generation, until one or other allele is fixed in the population. There are two consequences of genetic drift (Fig. 1).

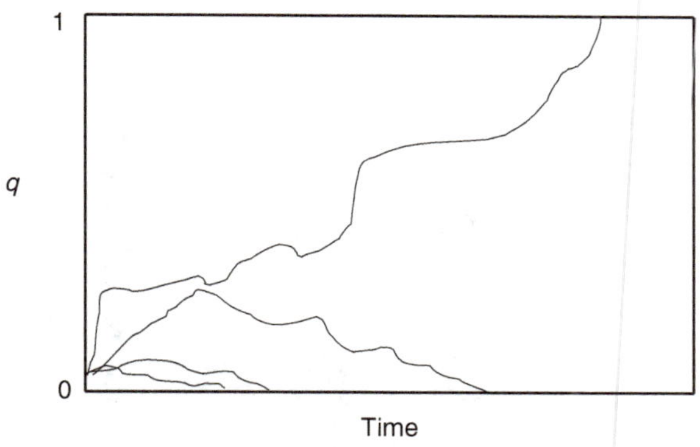

FIGURE 1 The lines indicate the trajectories over time of the frequencies (p) of selectively neutral mutations arising independently in different parts of the genome. These reflect changes caused by random genetic drift. The ultimate fate of each mutation is loss or fixation, but each mutation follows a unique trajectory, so that their frequencies are quite different at a given time.

1. A population of finite size eventually becomes genetically uniform, in the absence of mutations. The time scale for this to occur is of the order of $2N$ generations.
2. The frequencies of variants in isolated populations, such as two species derived from the same common ancestor, diverge over time, because independent replicates of a population with the same initial state arrive at different variant frequencies by chance.

Mutation and Genetic Drift

Examination of homologous DNA sequences from different individuals of a population shows that there is often substantial variability at nucleotide sites that are expected to be close to selective neutrality—notably, silent sites, where mutations do not affect the sequences of proteins. This can be explained by the fact that there is an influx each generation of new mutations, which eventually comes into balance with their loss or fixation by drift. The extent of such variability can be measured by the nucleotide site diversity, π: the frequency per nucleotide site at which a pair of homologous sequences sampled from a population differ in state. If $4Nu << 1$, then the equilibrium value of π is $4Nu$, where u is the mutation rate per nucleotide.

The Coalescent Process

An alternative way of looking at drift is in terms of the coalescent process. If we sample a set of allelic DNA sequences from a population, we can determine the probability that a pair of alleles are derived from the same ancestral allele in the previous generation, i.e., they

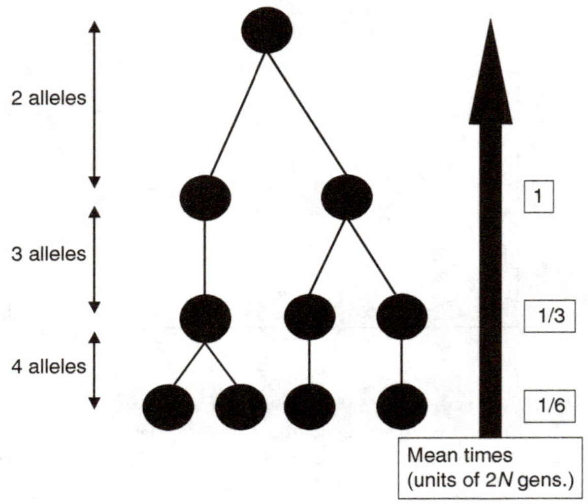

FIGURE 2 The coalescent process for four allelic sequences sampled from a Wright–Fisher population (filled circles at the bottom of the diagram). The left-hand pair is derived from a common ancestral allele that was present in the population at some time in the past (the first coalescent event); the right-hand pair is derived from another ancestral allele further back in time (the second coalescent event). These two ancestral alleles are derived from a common ancestor even further back in time (third coalescent event). The numbers of distinct alleles remaining at each of these times are indicated at the left of the figure. The numbers on the right indicate the mean times to each coalescent event, in units of $2N$ generations. In general, for i alleles, the mean time to a coalescent event is $4N/i(i - 1)$ generations, which reduces to $2N$ generations for a single pair of alleles.

coalesce into a single allele (Fig. 2). This is equal to $1/(2N)$ for a Wright–Fisher population. The time to coalescence of a pair of alleles has a mean and standard deviation of $2N$ generations. This approach provides a powerful means of determining the expected properties of samples from populations, as well as for testing evolutionary hypotheses.

Effective Population Size

Most populations do not match the assumptions of the Wright–Fisher model. The rate of genetic drift can then be represented by the "effective population size," N_e, where N_e replaces N in the relevant formulae for the rate of drift, and can be calculated from knowledge of the breeding structure of the population. A complication is that different formulae for N_e may apply to different aspects of drift (e.g., rate of increase in variance versus loss of variability), especially if the population is not constant in size. In most situations, N_e is much less than N; e.g., with different numbers of breeding males and females, it is close to the number of individuals of the rarer sex, and with varying numbers over time it is strongly affected by the smallest population size in the series. In humans, the mean value of π for silent mutations is about 10^{-3}; with a mutation rate of 1×10^{-8}, the formula $\pi = 4N_e u$ implies that $N_e = 25{,}000$. This reflects the fact that most existing neutral variation is of very ancient origin and that the size of the ancestral human population was very small until recently. In contrast, species of insects such as *Drosophila*, and many plants, have effective sizes of a million or more at the level of the whole species.

Selection vs. Genetic Drift

A major question in modern evolutionary genetics is the extent to which evolutionary change in DNA and protein sequences is controlled by mutation and drift versus selection. One way of modeling the effectiveness of selection versus drift is to determine the probability of fixation (Q) of a single copy of a new mutation, i.e., the chance that it survives in the population and eventually spreads all the way through it. This can be done using diffusion equations, which describe the change over time in the probability distribution of allele frequency. For the haploid model of weak directional selection described above, we have $Q \approx 2(N_e s/N)/[1 - \exp(-4N_e s)]$.

For a favorable mutation ($s > 0$), $Q \approx 2(N_e s/N)$ when $N_e s > 1$; with $N_e s << 1$, Q approaches $1/(N)$, the value for a neutral mutation. For a deleterious mutation ($s < 0$), Q is only slightly less than $1/(N)$ when $-N_e s << 1$, but it becomes close to zero when $-N_e s > 1$. A weakly selected, favorable mutation thus has an enhanced chance of loss when $N_e s$ is small and has a fixation probability of $2s$ in a large Wright–Fisher population, bringing out the point that favorable mutations have a significant chance of loss even in very large populations. Deleterious mutations have almost the same chance of fixation as neutral ones when the magnitude of $N_e s$ is small. Similar results apply to a randomly mating diploid population, where s is now the selection coefficient for heterozygous carriers of the mutation. The fact that N_e is usually more than 10,000, even for a species such as humans with a low effective size, shows that selection coefficients as small as 10^{-4} can be important in influencing evolution, as was strongly emphasized by Fisher.

The rate of DNA sequence evolution can be determined as follows. Each generation, a mean number of $2Nu$ new mutations enters a diploid population at a given nucleotide site. The chance that any one of these becomes fixed is Q, so that the mean number of mutations destined for fixation is $K = 2NuQ$. This is equivalent to the rate at which mutations become substituted over evolutionary time, which can be measured by comparing the sequence divergence between species with known dates of separation. For neutral mutations, this reduces to $K = u$, Kimura's well-known formula for the rate of neutral sequence evolution. Deleterious mutations have $K < u$, and favorable mutations have $K > u$. These results form the basis for comparative studies of DNA sequences, especially the justification for the observation of an approximately constant rate of sequence evolution—the "molecular clock."

Powerful tools for answering the question of the importance of selection versus mutation and drift in controlling sequence evolution are provided by applying predictions from diffusion equations and coalescent theory to data on variability within species and between-species differences. These methods are providing evidence for an important role for natural selection as well as drift in the evolution of sequence differences between species. For example, current estimates suggest that about 50% of the protein sequences differences among related *Drosophila* species are the result of the fixation of favorable mutations. The use of these methods also suggests that there are situations in which mutations have selection coefficients of the order of $1/N_e$, but that nevertheless selection has significantly influenced the outcome of evolution. This is true, for example, of alternative variants that correspond to the same amino acid in a protein sequence and of many mutations in noncoding sequences with functional significance.

CONCLUSIONS

Over the past century, classical and molecular genetics have provided increasingly refined knowledge of the nature of the genetic material and the mechanism of inheritance. This has provided an essential underpinning for our understanding of the forces that cause evolutionary change—population genetics theory relates genetic mechanisms to these forces. The theory provides important insight into the ways in which variability can be maintained in natural and human populations, and in how this variability can be transformed into differences between populations in space and time. It has generated a set of tools for making inferences about the causes of observed patterns of variation, for testing for the effects of selection at the DNA sequence level, and for interpreting the data of molecular evolution, as well as providing an essential basis for understanding evolution at the level of phenotypes. The advent of comparisons of multiple genome sequences from single species provides the ultimate source of data for the application of population genetics principles to questions such as the prevalence of natural selection on coding and noncoding sequences, and for relating phenotype to genotype.

SEE ALSO THE FOLLOWING ARTICLES

Adaptive Dynamics / Coevolution / Game Theory / Integrated Whole Organism Physiology / Quantitative Genetics / Sex, Evolution of

FURTHER READING

Charlesworth, B., and Charlesworth, D. 2010. *Elements of evolutionary genetics.* Greenwood Village, CO: Roberts and Company.

Falconer, D. S., and T. F. C. Mackay. 1996. *An introduction to quantitative genetics,* 4th ed. London: Longman.

Fisher, R.A. 1930. *The genetical theory of natural selection.* Oxford: Oxford University Press. (Reissued 1999 as a Complete Variorum Edition, ed. J.H. Bennett.)

Haldane, J.B.S. 1932. *The causes of evolution.* London: Longmans, Green and Company. (Reprinted 1966 by Cornell University Press, Ithaca, NY.)

Kimura, M. 1983. *The neutral theory of molecular evolution.* Cambridge: Cambridge University Press.

Muller, H.J. 1949. Reintegration of the symposium on genetics, paleontology, and evolution. In G.L. Jepsen, G.G. Simpson, and E. Mayr, eds. *Genetics, paleontology and evolution.* Princeton: Princeton University Press.

Wright, S. 1931. Evolution in Mendelian populations. *Genetics* 16: 97–159.

N

NETWORKS, ECOLOGICAL

ANNA EKLÖF AND STEFANO ALLESINA
University of Chicago, Illinois

Ecological networks are abstract representations of nature describing species diversity, trophic (i.e., feeding) and nontrophic (e.g., facilitation, mutualism) relationships between species, and flows of energy and nutrients or individuals within an ecosystem. Traditionally, these features have been studied separately, such that each network describes just one type of interaction. Combining different types of interactions is more challenging, and there is a growing body of literature and methods attempting to tackle this problem.

GENERAL NETWORK PROPERTIES

The search for unifying principles that give rise to the structure of ecological networks dates back to the 1970s. Networks are characterized by the components—the nodes—and their interactions—the edges (or links or arrows)—connecting them. From this information, the network structure can be derived. Links can either be undirected (a link from node *A* to node *B* implies a symmetric link back from *B* to *A*) or directed (the connection from *A* to *B* can differ from that from *B* to *A*). The links can be binary (representing their presence or absence only) or weighted (i.e., have a measure of their strength). In network analysis, the focus is the topological structure and functional relationships in the network. A path in a network is a route between different nodes where each node and link are visited only once. The length of such a path is the sum (weighted or unweighted) of all the links visited. The shortest path between any two nodes is a frequently used measure in network theory. The longest of all the shortest paths is known as the diameter of the network. The average of all the shortest paths in a network is called the characteristic path length. Because of the large number of nodes and the even larger number of edges that combine to form such an intricate structure, ecological networks are a prominent example of complex systems.

Networks are potentially difficult to understand because of their structural complexity, their dynamic nature (the number of nodes and edges frequently change through time), the diversity of types of links and nodes, the presence of nonlinear dynamics in the relationships between nodes, and the fact that various network properties can often influence each other. Three basic properties are often used to characterize a given ecological network: the number of nodes (S, species richness), the number of links (L), and the connectance (the fraction of realized connections, L/S^2). Network connectance has been shown to be an important descriptor of how sensitive the network is to disturbances such as the removal of nodes (robustness of the network) and the stability of the underlying population dynamics. Other characteristics that are commonly described for ecological networks are the proportion of nodes belonging to a specific group or guild, intervality, modularity, and nestedness. Presented below is a brief summary of these different properties.

WHY STUDY ECOLOGICAL NETWORKS?

There is no doubt that humans affect ecosystems in many different ways and at large scales. A major challenge in ecology is forecasting the effect of these disturbances on ecosystems. In autoecology the focus is on a single species or a few species within an ecosystem. In the network approach, the focus is upon understanding the structure and the functioning of the whole ecosystem with all its interacting parts. In fact, species are not isolated; rather, they interact with each other, forming intricate networks (Darwin's "entangled bank"). Also, even though some disturbances can be considered local, their effects are often global, and this highlights the importance of putting single species into a community context. To determine the effect of disturbances, a community approach is needed. However, the complexity of ecosystems is often perceived to be an insurmountable problem for understanding their function.

Hutchinson pioneered the network approach when he championed a community-based analysis of ecosystems' complexity: "In any study of evolutionary ecology, food relations appear as one of the most important aspects of the system of animate nature. There is quite obviously much more to living communities than the raw dictum 'eat or be eaten,' but in order to understand the higher intricacies of any ecological system, it is most easy to start from this crudely simple point of view" [*The American Naturalist*, 870: 145–159 (1959)]. Networks (in this case, food webs) explicitly take this "crudely simple point of view" and represent simple (trophic or nontrophic) relations among species. In most ecological networks, each node represents a species, and two nodes are connected by an edge if they interact (Fig. 1). There are many different types of interspecific interactions: predation/consumption (positive effect in on the consumer, negative on the resource), competition/interference (negative effect on both competitors), and mutualism (positive effect for both species), as well as other more complicated and indirect interactions (e.g., the presence of one species could modify the interaction between other species). In ecosystems, all these different types of interactions are present, but in network theory they are usually handled separately; i.e., analysis is performed on either predator–prey networks (food webs), competitive networks, or mutualistic networks. Some of these networks are unipartite (i.e., just one type of node, all of which can potentially interact with each other). A prominent example of unipartite ecological networks are food webs, in which all species can potentially consume

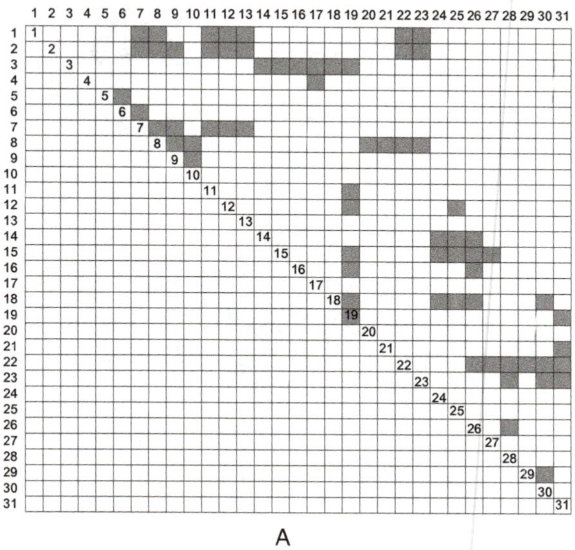

A

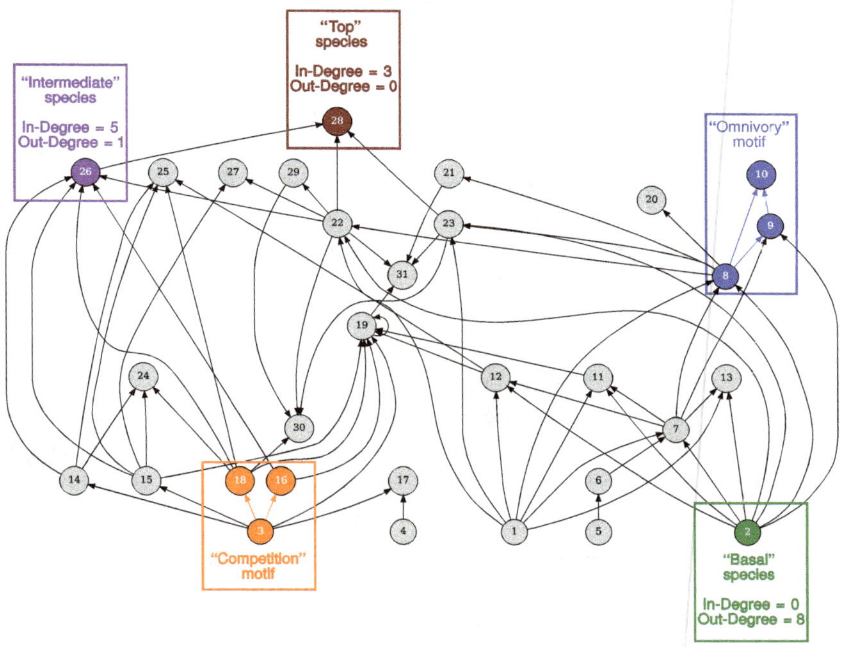

B

FIGURE 1 Example of a food web. (A) Adjacency matrix. In this matrix, each coefficient represents the presence (black square) or absence (white square) between the resources (rows) and the consumers (columns). Each number stands for a species or a group of closely related species. (B) Corresponding network. For specific nodes, the In-Degree (number of incoming connections) and Out-Degree (number of outgoing connections) are reported. Degrees define specific types of species: top species have no outgoing connections, basal species have no incoming connections, and intermediate species have both incoming and outgoing connections. Two typical "network motifs" are also reported: in orange, a "competition motif," in which two species compete for the same resource; in blue, an "omnivory motif," in which species 10 consumes both species 9 and its resource, species 8.

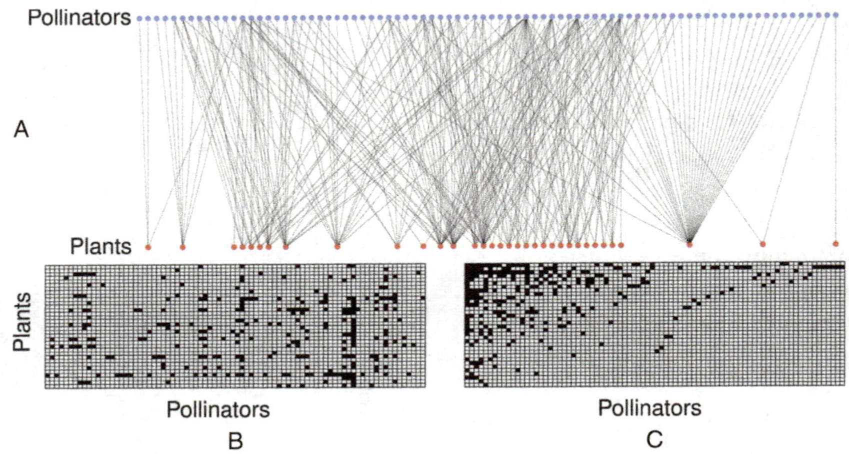

FIGURE 2 Example of a bipartite ecological network. (A) Pollination network. The blue nodes are pollinators, and the red nodes are the plants interacting with them. In this bipartite network, blue nodes are connected to red nodes, but not to blue nodes. Similarly, there is no connection between red nodes. (B) Rectangular incidence matrix. Because interactions are symmetric (implies), the squared adjacency matrix would be symmetric. Because each type of node can interact only with nodes belonging to the other type, it is possible to summarize all the information in a (much smaller) incidence (rectangular) matrix. (C) The same incidence matrix can be rearranged to display the meaning of nestedness. In this matrix, moving from top to bottom, each plant tends to interact with a subset of the pollinators of the plant above in the order. From left to right, the same happens for pollinators.

every other species (Fig. 1). Other networks are bipartite, in which there are two types of nodes (e.g., plants and herbivores) with interactions (e.g., feeding) happening only between nodes belonging to different classes. Examples of bipartite networks are host–parasite, plant–pollinator, and plant–seed dispersal networks (Fig. 2).

ECOLOGICAL NETWORKS

Generally speaking, ecological networks can be divided into three different categories: consumer–resource (food webs, host–parasite/plant–herbivore networks), mutualistic (plant–pollinator/plant–seed dispersal networks) and competitive networks (usually with nodes belonging to a single guild, e.g., plants). These networks can be described at three different levels: topological, quantitative, and dynamical. Topological networks are binary in that they describe only the presence/absence of interactions among species. Quantitative networks add to the topology a weight for each link. This value can describe the frequency or the strength of interaction. Finally, dynamical networks represent not only the species and the topology/strength of interactions but also the underlying population dynamics. From this perspective, the various descriptions can be "coarse-grained" hierarchically: from dynamical networks, we can infer the weight for each connection. Making this weight 0 or 1, we obtain topological networks. Dynamical networks can capture more features compared to just topological networks, such as indirect effects and bottom-up trophic cascades. However, data collection and analysis are usually more difficult because we need to fully parametrize very complex models (e.g., systems of differential or difference equations) that can contain thousands of parameters for which we need either empirical estimates or fitted values. Quantitative networks, on the other hand, contain important information on the dynamics and, at the same time, can be much easier to derive from empirical measurements.

Because topological networks are much easier to assemble from empirical data, the bulk of published ecological networks are described by binary matrices with equal numbers of rows and columns where the species are listed in the same order in both dimensions (adjacency matrices) (Fig. 1). By convention, in food webs the rows represent the resources and the columns the consumers. A 0 indicates that there is no interaction between the species on that specific row and column, respectively, and a 1 represents interaction between the species. Whenever the interactions are symmetric, or when the networks are bipartite, rectangular matrices are typically used (e.g., where the rows are plants and the columns are their pollinators; Fig. 2).

TOPOLOGICAL DESCRIPTIONS OF NETWORKS

Topological properties can be divided into local and global. Local properties relate each node and its interactions to those of its closest neighbors. The most basic local property is the number of links belonging to a node (degree). These links can be further divided into incoming (in-degree) and outgoing (out-degree) connections. For example, each producer in a food web will be characterized by an in-degree of 0, while a top predator would have 0 out-degree (Fig. 1). For each species/node, the degree defines how much of a generalist or specialist a species is, both in its role as a consumer and in that as a resource. In food web literature, *generality* is the number of resources of a given species, while *vulnerability* is the number of consumers a species has.

Moving from the single-node perspective toward whole-network properties, the next step is represented by network motifs. Network motifs are unique *n*-species subgraphs. These can be thought as building blocks for the whole network (Fig. 1). By analyzing the relative frequency of the various motifs, we find insights into the organization of the network. For example, in food webs certain motifs are overrepresented and others are underrepresented. Overrepresented motifs are those believed to confer dynamical persistence to networks: a possible explanation is that networks containing many destabilizing motifs would die off quickly, such that a process of selection would be shaping biological networks. Also, given the rarity of many motifs, one can conclude that a handful of motifs constitute the backbone of food web topology. Another quasi-local property is the clustering coefficient, which measures how likely it is to find interactions between the neighbors of a node (i.e., the probability of interactions between nodes that are connected to a focal node).

Global network properties describe the structure of the entire network. Examples of global properties are nestedness and modularity (Fig. 2, Fig. 3). Nestedness, originally defined in the context of island biogeography, measures patterns in how generalists and specialists interact in bipartite networks. In a perfectly nested network, we can find an ordering of the species from least to most general, such that interactions form "proper subsets." For example, in a perfectly nested plant–pollinator network, the most specialized plant interacts with a subset of the animals interacting with the next-to-last specialist plant, and so forth. In nested communities, a specialist plant would therefore interact prominently with generalist insects, and vice versa.

Modularity (Fig. 3) describes to what extent the network can be divided into nonoverlapping groups of highly interacting species. A module is a set of species that have a disproportionate number of within-module connections compared to among modules. For ecological networks, it is debated whether modules are a common characteristic or not. In networks that contain several habitat types,

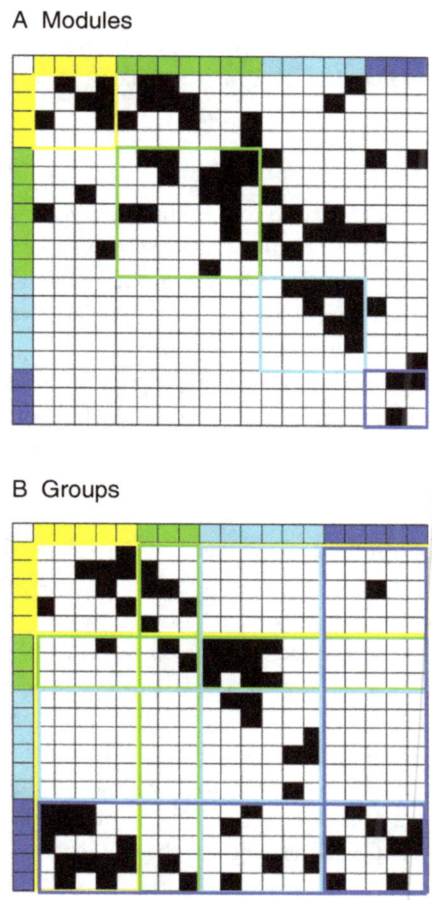

FIGURE 3 Modules and groups in food webs. (A) Division of the species into modules. Species are divided into four modules (different colors) of highly interacting species. The density of within-module interactions (blocks lying on the main diagonal of the matrix—colored squares) is much higher than the density of between-module interactions (the rest of the matrix). (B) Division of the species into "trophic groups." Each network is divided into four groups of species with characteristic patterns of interaction. For example, "green" species tend to consume "yellow" species and be consumed by "cyan" species.

modules should be more common because species living in the same habitat are more likely to interact with one another compared to species from different habitats. For example, in a stratified lake we expect pelagic species to predominantly interact with other pelagic species, while benthic species would predominantly interact with other benthic species. In ecological networks describing a single habitat, other types of groupings are also possible. For example, trophic guilds (such as producers, herbivores, and carnivores) have defined patterns of interaction (e.g., herbivores exclusively consume plants and are, in turn, consumed by carnivores and omnivores). These different types of groups, and unsupervised algorithms to detect them in networks, have been the focus of recent literature. The division of species into nonoverlapping groups may reveal important ecological roles in food webs (Fig. 3).

QUANTITATIVE DESCRIPTION OF NETWORKS

Starting with Odum, and then with Ulanowicz and Patten, network theory branched toward ecosystem ecology. Ecological network analysis (ENA) aims for a quantitative description of nutrient flows within ecosystems. Even in the 1970s, ENA provided quite complex methods, largely based on linear algebra, to assess relationships between "compartments" (nodes in a network, that represent species—and groups of species—as well as nutrient pools and detritus compartments). While food web theory and mutualistic networks have only recently started to include quantitative information on the connections between species, from the very beginning ENA aimed at the quantification of energy and matter flows between any two compartments. Moreover, inputs and outputs to/from ecosystems are also considered in ENA, in order to achieve a complete quantitative picture of the flows within the ecosystem. Finally, arguments based on thermodynamic principles and information theory have been proposed to characterize ecosystems' succession. In particular, Ulanowicz proposed the measure of Ascendency, a quantity accounting for the size and maturity of an ecosystem.

Other key elements of ENA include detailed analysis of indirect effects (e.g., how each species depend on any other for its energy) and of trophic levels. In particular, within ENA the original description of trophic level into producers, primary consumers, secondary consumers, and so forth, has been extended to include a fractionary trophic level. For example, a species consuming a producer (with one-third of the flow to the species) and a herbivore (with two-thirds of the flow) would have a trophic level of $2 \times 1/3 + 3 \times 2/3 = 2.66$. Cycles in networks (largely overlooked in food web theory) are central to ENA and include a complete decomposition of the network into basic cycles. The prominence of cycles is due to the inclusion of nonliving nodes (e.g., detritus), which are responsible for nutrient recycling. Finally, a large body of literature in ENA is devoted to system-wide properties, including the construction of information theories based indices to measure the growth and development of ecosystems.

Recently, the importance of quantifying connection has been rediscovered in many branches of science that make use of networks. Including quantification requires a general rethinking of the methods and algorithms used for their analysis.

DYNAMICAL DESCRIPTIONS OF NETWORKS

Static network representations provide a large amount of information on the behavior of ecological systems. However, network topology can be complemented with and informed by population dynamics. This addition is necessary in order to answer some fundamental questions regarding ecological systems. For example, the study of processes driven by indirect effects such as trophic cascades requires a full specification of the underlying dynamical model. In the same spirit, the effect of species extinction can be assessed in a bottom-up way using topological descriptions (i.e., a species that remain following the extinction of its prey remains without exploitable resources will go extinct as well), but more complex top-down approaches (i.e., lack of regulation due to the extinction of keystone predators) do require a fully specified dynamical system.

Because of the intrinsic difficulties in parameterizing large dynamical systems, the study of dynamical networks has so far mostly been confined to the study of computer-generated food webs or to the parameterization of highly idealized models from empirical data.

TYPES OF ECOLOGICAL NETWORKS

Food Webs

Food webs have been described as networks since the early days of ecology. However, to date food web research has produced very few indisputable trends and empirical laws. It should be noted that the quality of published data has increased dramatically over the last decade, so that sufficient statistical power and a sufficiently large body of networks have recently been made available. Food webs are schematic descriptions of who eats whom in ecological systems; i.e., the focus is on consumption between species spanning different trophic levels. Food webs across different habitats and geographical locations have been shown to contain remarkable regularities. The search for common descriptors of food webs attempts to provide cogent ecological patterns on the distribution of species and their interactions. From the first generation of food web analysis, largely based on networks reconstructed from the literature, it was suggested that food webs contained scale-invariant properties, i.e., properties that did not change with increasing species diversity. The most important scale-invariant properties that were suggested were that (a) the number of trophic levels rarely exceeds five, (b) the link density is constant (L/S), and (c) the fraction of species at different trophic levels and the fraction of links between different trophic levels are constant. However, analysis of more detailed data showed that at least the latter two properties do not generally hold.

There have been several attempts at constructing food webs with the same network characteristics found in natural systems. The first example was the "cascade model" in the early 1980s. The model orders all species in a food web along a single dimension. This is done by assigning a niche value (typically a real number between 0 and 1) to each species. Networks are generated by connecting each species to consumers that are ranked above the species with probability $p = 2CS/(S - 1)$. Thus, species cannot consume species with higher rank. Therefore, networks generated by the model cannot contain trophic cycles or cannibalism. The cascade model was the first to implicitly assume that food webs are governed by some (unspecified) general rule rather than being random assemblages of interacting species.

In order to address the limitation of the cascade model, the "niche model" was developed. The niche model also orders species along a single dimension and assigns each species a niche value. The difference with the cascade model lies in the way feeding links are distributed: each species is in fact given a niche range, and the consumers prey upon all the species falling within their respective niche ranges. The model produces food webs that are completely interval; i.e., there are no gaps in the consumers diet: each predator consumes resources that are adjacent in the niche ordering. Food webs rarely show complete intervality, and several models were developed to solve this problem. Recently, the research on simple models for food web structure has seen an increase in the number of models proposed as well as in the statistical sophistication of the methods used for model selection, making it one of the more active areas of research in food webs.

Another controversial characteristic of food webs is the degree of presence or absence of compartments, also called modules (Fig. 3). A compartment is a subgroup of species interacting strongly with each other but only weakly with other subgroups. It has been argued that this kind of organization would increase food web stability and resistance to disturbances: perturbations are likely to remain within a specific module and not impact the network as a whole. Data resolution, however, affects the detection of compartments: highly resolved and weighted data is more likely to find compartments, which typically represent different biotic habitats, such as the benthic and pelagic components of an aquatic system. However, how common compartments are in food webs and the role they play for stability have yet to be completely characterized.

Parasitic Networks

A large number of species in natural ecosystems are parasites. Despite this fact, parasites have been almost completely disregarded in ecological networks; only recently have ecologists started to include parasites in food webs.

Although parasites have been historically neglected in the food web literature, there have been networks centered on parasites. These are commonly referred as parasite–host networks or parasitoid–host network. These networks, even though based on consumer–resource dynamics, share similar features with mutualistic networks. In fact, they are bipartite and display a high degree of nestedness and modularity.

Ecologists face several challenges when including parasites in ecological networks. First, parasitism is not equivalent to predation. In fact, the classification of a consumer–resource interaction needs to be revised in order to accommodate parasitism: for example, parasites that do not kill their hosts will have different dynamical effects from parasitoids, which kill hosts, and these will be different from predators who ingest their prey. Also, parasites can influence several ecologically relevant traits of their hosts (e.g., fecundity, behavior). Another major challenge is the inclusion of multiple life stages with specific consumers and resources. Complex life cycles break the bijection "one species one node" that characterizes most ecological networks. Solving the complex life-cycle problems could be of great importance for other classes of consumers (such as most amphibians, many insects) for which different stages or size classes have different diets. Parasites typically live inside or in close poximity with their host (intimacy). This raises the question of incidental predation: when a predator consumes an infected prey, does it also consume its parasites? Although consumption may seem detrimental to the parasite, many organisms use incidental predation to move between hosts for the different life stages. Therefore, being consumed could be advantageous to some parasites, although it is typically disadvantageous for free-living species.

Many models of food web structure assume (explicitly or implicitly) that there is a tendency for an optimal ratio between the size of predators and prey, with predators typically being larger than their prey. This logic, clearly inspired by marine ecosystems (in which predators must ingest their prey), has to be reversed for parasites. Newly proposed models have different "attachment rules" for parasites and free-living species that take this into account.

What are the consequences of including parasites in food webs? Recent studies showed that inclusion of parasites would change both the topology and dynamics of the network. The topological changes included an increase in the connectance and the number of specialists (as many parasites are highly specialized). This suggests a possible change in the robustness of the networks as well, since both network characteristics affect the risk of secondary extinctions. The idea of network robustness should also be amended to include life stages. Moreover, equations and functional forms should be dramatically altered to account for parasites. For example, parasites may have metabolic scaling coefficients that differ from other species due to the parasite–host body-size ratio. Also, the body-size ratio itself might have a destabilizing effect on network dynamics since a positive relationship has been shown to have a stabilizing effect. Stability may increase when consumers are larger than their prey, but the addition of parasites would increase the number of interactions having the opposite relationship, which may have a destabilizing effect. All these challenges need to be solved to include parasites in networks.

Competitive Networks

In ecological communities, species compete for resources. Niche theories assume that similar species are going to compete strongly, possibly leading to competitive exclusion. A special type of network arising in competitive studies is the "tournament." In a tournament, for any couple of species, the inferior competitor is connected to the superior competitor. Studies on these networks showed how the network properties (especially the structure of the cycles) can predict the coexistence levels in the community. Typically, competitive networks represent a single trophic level.

Mutualistic Networks

Mutualism is defined as any relationship between species where all species benefit from the interaction. Mutualism is a common phenomenon in all ecosystems and has been shown to be important for ecosystem persistence and functioning. Examples of mutualistic networks include pollination and seed dispersal networks. In these networks, the plant benefits from the services provided by the animals (pollination, seed dispersal) and the animal is rewarded with pollen, nectar, or fruit. Mutualism is also believed to be a major driving force for coevolution among species and is therefore an important process for the maintenance of biodiversity.

Mutualistic networks have well-defined structures, and several patterns have been found in a vast array of datasets encompassing highly diverse ecosystems and varying degrees of complexity. One of these common properties is the heterogeneity in the number of connections per species. In mutualistic networks, most of the species have few connections (specialists), while few of the species have a large number of connections (generalists). The frequency distribution for the number for links is therefore wide, but the upper limit is truncated compared to other types of broad-scale networks. This is due to the fact that in mutualistic networks, as in every ecological network, there are physical constraints on the number of interaction per species (e.g., size of the corolla or the shape of specialized mouth parts of insects). These constraints limit the number of possible interactions, i.e., not all species can interact with all other species. Also, the strength of interaction between species in mutualistic networks is distributed in a heterogeneous way: the majority of links are weak, but few are very strong. Finally, there seems to be an asymmetric relationship between interacting species: if species *A* interacts very strongly with species *B*, then *B* is likely to interact weakly with *A*. The heterogeneous architecture of mutualistic networks makes them robust to random removals of nodes but sensitive to loss of specific nodes (such as the most connected ones).

Most networks describing mutualistic interactions show a nested pattern (Fig. 2). As explained earlier, this means that specialist species tend to interact more with generalist species than with other specialists. This nested structure, due to asymmetric specialization, makes the system more robust and increases the probability of persistence for specialists. Recently, a connection has been drawn between the degree of nestedness and the dynamical properties of the networks.

Another feature in mutualistic networks is represented by compartments or modules (Fig. 3). In mutualistic networks, the compartmentalization is less debated than in food webs, and compartments are suggested to be their basic building blocks. Modularity has been observed regardless the type of mutualism and habitat. Because of the specific matching of traits between, for example, pollinators and their corresponding plants, modules have been suggested to represent coevolutionary units. They would thereby provide important information about the evolution and functional diversity of mutualistic networks. Species in compartmentalized networks play different roles depending on their patterns of interactions: hubs are species that are highly linked in their own compartment, while connectors are species that link different modules to one another. Disturbance to different types of nodes would impact the network in different ways.

Spatial Networks

All ecological interactions take place in a spatial context, and many of the disturbances to ecosystems have a spatial component. For example, habitat fragmentation and destruction is currently the largest cause of species' extinction. Analysis of these problems requires a spatial representation of ecosystems. A landscape often consists of patchily distributed suitable habitats surrounded by nonsuitable areas. In a network representation of a landscape, the nodes are habitat patches or local populations, and the links describe the interaction between patches due to the dispersal, nutrient flows, or other processes. Links can be either bidirectional or directional, i.e., either the dispersal happens with the same strength between the two patches or the relationship is nonsymmetrical, meaning the dispersal is larger or more frequent in one direction compared to the other. The links can be binary or a value accounting for the geographic or functional distance between patches, for example. Functional distances reflect the resistance species are facing when dispersing in the unsuitable matrix. In some applications, the distances are relative measures truncated to the maximum dispersal distance for the species of interest. Also, the nodes (here the habitat patches) can be represented as weighted or unweighted. In an unweighted representation, the species are either present or not. In a weighted representation, the habitat patches are described according to the number of individuals present in the patch. Often, there is a relationship between patch size and species abundance. Depending on the growth rate of the species in a particular patch, these can be classified as sinks or sources. In sink patches, populations have negative growth rate, and the population is only sustained by immigration. In source patches, populations have positive growth rates and can support surrounding patches through dispersal. Most analyses have been performed on metapopulations (local populations of a species connected by dispersal). More recently, theory has been developed to include the metacommunity concept (local communities connected by dispersal).

In the network perspective, the total landscape connectivity is central. For individual patches, their degree of connectivity and centrality are factors describing them. The landscape connectivity can be described in different ways, depending on the assumptions made about the dispersal capacity of the species. Different measures can give different values for network connectivity. A useful illustration is to describe the same landscape with graphical representations based on the different measures to show how the connectivity changes with the assumptions. The centrality of a particular patch describes the ability of the patch to influence other patches in the network. The network perspective of local ecological communities and the network perspective of the landscape can be combined to form a network of networks.

CHALLENGES AND LIMITATIONS FOR THE NETWORK APPROACH

Several challenging questions have been successfully answered, and it is clear that the network perspective is a useful, and sometimes necessary, tool for analyzing ecosystems. However, there are several limitations and challenges for the future development of the discipline.

The first challenge is the quality of the empirical data. Although the data available today is much more detailed and of higher quality than those used for the initial studies of food webs, there are few datasets that have an even resolution for all trophic levels. Usually, the resolution increases with trophic level and size, with vertebrates being overrepresented and other organisms being either neglected or lumped coarse grained into groups.

With increasing data quality, we can study networks containing thousands of nodes and tens of thousands of connections. However, at this end of the data-quality spectrum new problems arise. In fact, many of the computational techniques used to analyze ecological networks do not scale well with size, such that that the complete analysis of large networks is computationally prohibitive. It is likely that the same network can be analyzed at different scales, from coarse grained to fine grained. Different levels of resolution will require different tools.

A major challenge is to combine different types of interactions in one network. Recently, food webs in which parasitic and mutualistic links have been included have started appearing in the literature. The combination of different types of interaction would give a holistic view of ecological networks, although it would increase their complexity. There is also a need to account for other types of trophic pathways such as flows of nutrients from detritus, allochthonous inputs, and life history interactions.

The network perspective powers the opportunity to take the whole ecological community into account. This is required for the understanding of ecological functions and systems response to different kinds of perturbations. To fully understand the complexities of ecosystem functioning, we must address not only the species individually but also the interactions and interdependences between species.

SEE ALSO THE FOLLOWING ARTICLES

Bottom-Up Control / Compartment Models / Ecosystem Ecology / Food Webs / Metapopulations / Spatial Ecology / Top-Down Control

FURTHER READING

Allesina, S., D. Alonso, and M. Pascual. 2008. A general model for food web structure. *Science* 320: 658–661.

Bascompte, J. 2009. Mutualistic networks. *Frontiers in Ecology and the Environment* 7(8): 429–436.

Besier, L.-F. 2007. A history of the study of ecological networks. In F. Képès, ed. *Biological networks.* Tuck Link, Singapore: World Scientific Publishing.

Dunne, J. A., and M. Pascual. 2006. *Ecological networks: linking structure to dynamics in food webs.* New York: Oxford University Press.

Henson, K. S. E., P. G. Craze, and J. Memmott. 2009. The restoration of parasites, parasitoids, and pathogens to heathland communities. *Ecology* 90: 1840–1851.

Krause, A. E., K. A. Frank, D. M. Mason, R. E. Ulanowicz, and W. W. Taylor. 2003. Compartments revealed in food-web structure. *Nature* 426: 282–285.

Lafferty, K. D., S. Allesina, M. Arim, C. J. Briggs, G. De Leo, A. P. Dobson, J. A. Dunne, P. T. J. Johnson, A. M. Kuris, D. J. Marcogliese, N. D. Meartinez, J. Memmott, J. P. Marquet. J. P. McLaughlin, E. A. Moredecai, M. Pascual, R. Poulin, and D. W. Thieltges. 2008. Parasites in food web: the ultimate missing link. *Ecology Letters* 11: 533–546.

Laird, R. A., and B. S. Schamp. 2009. Species coexistence, intransitivity, and topological variation in competitive tournaments. *Journal of Theoretical Biology* 265: 90–95.

Ulanowicz, R. E. 2004. Quantitative methods for ecological network analysis. *Computational Biology and Chemistry* 28: 321–339.

Urban, D. L., E. S. Minor, E. A. Treml, and R. S. Schick. 2009. Graph models of habitat mosaics. *Ecology Letters* 12: 260–273.

NEUTRAL COMMUNITY ECOLOGY

STEPHEN P. HUBBELL

University of California, Los Angeles

Neutral community ecology studies the properties of model communities of trophically similar species under the simplifying assumption that, to a first approximation, species in the community are identical on a per capita basis in their vital rates. The theory is both stochastic and mechanistic in the sense that it embodies the familiar processes in population biology of birth and death, immigration and emigration, as well as speciation and extinction. Because the theory is neutral, it makes few assumptions, but it nevertheless makes many novel and interesting predictions about static and dynamic patterns at multiple spatial and temporal scales, from patterns of species distribution and abundance at local population and community scales, to patterns of biodiversity on landscape to biogeographic scales. Neutral community ecology comes with a quantitative sampling theory for testing its predictions, so it provides a powerful tool and null model. Neutral theory has generated a lively ongoing discussion in community ecology on the relative importance of random dispersal and drift versus deterministic factors such as niche differentiation in controlling the species composition and dynamics of ecological communities.

THE CONCEPT

Origins

Neutral community ecology had its first major expression in MacArthur and Wilson's theory of island biogeography, which hypothesized that the number of species on islands was a dynamic equilibrium between the immigration of species from a mainland source area, and their extinction after their arrival on the island. Neutral community ecology is a generalization of the theory of island biogeography. It differs from the original theory in two major ways. First it adds speciation, which was missing from the theory of island biogeography. Second, it makes the neutrality assumption at the level of individuals rather than at the level of species. This distinction allows neutral community ecology to predict relative species abundance, not just the number of species. It is not widely appreciated that the theory of island biogeography is neutral. In its classical graphical representation of the immigration–extinction equilibrium (Fig. 1), species are unnamed and interchangeable. The theory does not specify which species are present at equilibrium, and the theory predicts a turnover of species at the equilibrium level of island diversity. However, the biological level of the neutrality assumption does matter to the shapes of the immigration and extinction curves. If one makes the neutrality assumption at the level of species, one obtains the graph in Figure 1A. However, instead if one makes the neutrality assumption at the level of individuals, the result is the graph in Figure 1B. The curves in Figure 1B are asymmetric because it only takes few individuals to colonize an island, but all the island individuals of a species must die for a species to count as extinct. Note that neutral theory also generates curved immigration

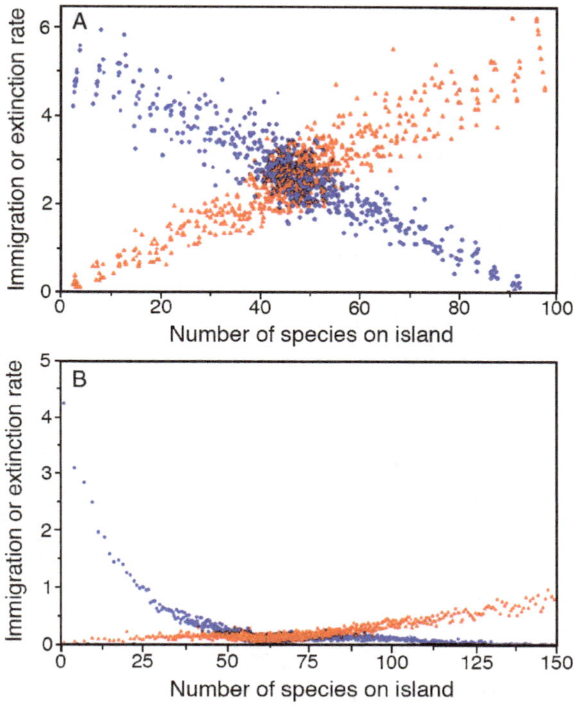

FIGURE 1 Effect of the level of the neutrality assumption on the shapes of the immigration and extinction curves in the graphical model of island biogeography theory. Blue circles are immigration and red triangles are extinctions for an ensemble of 10 simulations of a neutral island community receive immigrants from a mainland. (A) Making the neutrality assumption at the species level results in linear immigration and extinction curves. (B) Making the neutrality assumption at the individual level results in asymmetric and curved immigration and extinction curves. The immigration curve is higher because a single individual can count as a new species immigrant, but an extinction requires the elimination of all individuals of a species from the island. This result shows that curvilinear immigration and extinction curves are predicted by neutrality, not just by the hypothesis of negative species interactions. Figure from Hubbell (2009).

and extinction curves, showing that biotic interactions do not need to be invoked to explain curved immigration and extinction lines.

Brief History

In the 1970s, Watterson and, independently, Caswell took the first steps to develop neutral community ecology, particularly focusing on patterns of relative species abundance. Borrowing approaches used in neutral theory in population genetics, they studied model species undergoing a random walk in abundance, starting from an initial immigration event, analogous to the changes in abundance of a neutral allele undergoing genetic drift. Both Watterson and Caswell noted that the resulting distribution of relative species abundance was Fisher's log series distribution. However, the significance of this result and the connection of Fisher's log series to neutral theory were not generally appreciated. In 1943 Fisher, the renowned evolutionary theorist and statistician, and two entomologists, Corbet and Williams, published a seminal paper in which they described observed patterns of relative species abundance in collections of moths and butterflies using a then new distribution, the log series. The log series is a two-parameter distribution in which the expected number of species $\phi(n)$ having n individuals is given by

$$\phi(n) = \alpha\frac{x^n}{n}, \qquad (1)$$

where α is a fitted diversity parameter and x is a parameter close to but less than unity (if $x > 1$, the log series does not converge). Since then, Fisher's α, as parameter α is now known, has become one of the most widely used measures of species diversity because empirically its value is very stable in the face of increasing sample sizes of individuals drawn from communities. The reasons for the constancy of Fisher's α and the biological interpretation of parameters α and x of the log series would not become known until the development of neutral theory in community ecology.

In the 1970s, in addition to neutral studies of relative species abundance, there were also pioneering studies of neutral models of speciation and phylogeny, by Raup, Gould, Schopf, and Simberloff. However, the connection of this work to neutral community ecology also went unrecognized until much later. The relevance of this work is that Raup and his colleagues took a demographic approach to phylogeny, modeling monophyletic clades evolving by a stochastic birth–death branching process, based on the much earlier work by Yule. They were pursuing the question of whether random birth–death processes could reproduce observed patterns of phylogenetic diversification in the fossil record. In particular, they were interested in whether patterns of punctuated equilibria, that is, long periods of relative stability in diversity interrupted by relatively short bursts of rapid speciation would result from the models. Lineages in their models were assigned birth (speciation) and death (extinction) probabilities. The general outcome of these models was exponential growth in the number of descendant lineages when the probability of birth exceeded the probability of death, and slower growth if all extinct lineages were pruned out of the phylogenetic trees. In no case, however, did the models yield patterns of punctuated equilibria postulated by Eldridge and Gould. In the 1990s, theoretical studies of neutral phylogenetic models by Nee and colleagues reached similar conclusions. Later work in neutral community ecology would show, however, that these conclusions were premature.

In the 1980s, development of neutral theory in ecology slowed because neutral models were criticized for their lack of realism and biological content. However, a limitation of these early models is that they were not dynamical models constructed from fundamental processes in population biology but were simply null models for assembling random communities by one or another sampling protocol. Typically they assigned equal probability of being sampled to each species, irrespective of its abundance or probability of dispersal. Some of these models had statistical and other conceptual problems. They were largely forgotten with the rise in the 1980s of quantitative, resource-based competition theory, which promised both a mechanistic and deterministic explanation for the assembly of competing species into communities, based on niche-differentiation for exploiting limiting resources. This led to a hiatus in the further development of neutral community theory for nearly two decades.

The Unified Neutral Theory

In the mid- to late 1990s, Stephen Hubbell and Graham Bell independently rediscovered and extended the framework of neutral theory in ecology. Hubbell's and Bell's versions of neutral theory differed in several ways. In particular, Hubbell's version (Fig. 2) built on the theory of island biogeography, and emphasized dispersal limitation, whereas Bell's did not. Also Hubbell's version recognized two spatial scales as in island biogeography theory, with a mainland source area, called the metacommunity in Hubbell's theory, and a local (or island) community that draws immigrants from the metacommunity, which replaces the source area concept in the theory of island biogeography. The metacommunity is the biogeographic unit in which species originate and spend their entire evolutionary lifespans. Bell's version made no such distinction. Another difference was that species did not interact in Bell's version, whereas in Hubbell's version species competed diffusely and community dynamics was a zero-sum game, such that increases in one species were matched by collective decreases in all other species. Hubbell's species are trophically similar, whereas species in Bell's version need not be, so long as the species were demographically equivalent.

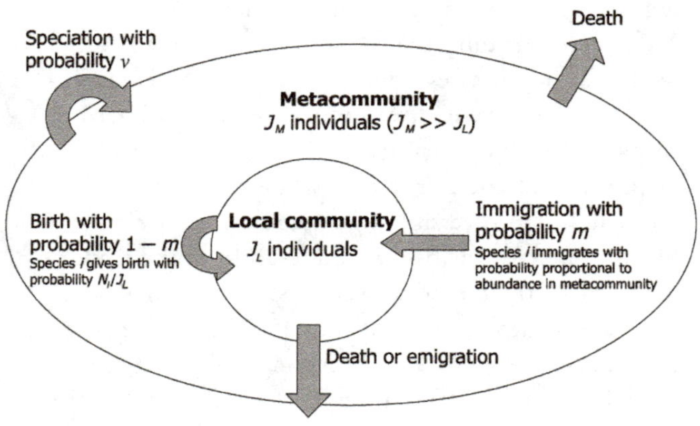

FIGURE 2 Schematic of the implicit-space neutral theory proposed by Hubbell (2001). A local community of J_L individuals of all species is imbedded in a much larger metacommunity of size J_M, and receives immigrants from the metacommunity. The probability that a local death is replaced by an immigrant is m. In the metacommunity, there is a slow rate of addition of new species at rate ν. Figure from Etienne and Alonso (2007).

Parameters of Neutral Theory

Neutral theory in ecology, befitting its name, has only a few free parameters. A free parameter is a number that cannot be derived from the theory itself and whose value must be provided from some other source of information. The free parameters in neutral theory (Fig. 2) are (i) the size of a local community, J_L, defined as the number of individuals of all species in the local community; (ii) the size of the metacommunity, J_M, defined as the sum of the sizes $J_M = \sum J_L$, of all the local communities in the metacommunity; (iii) the probability of immigration, m, the probability that an individual dying in a local community is replaced by an immigrant from the metacommunity; (iv) b/d, the ratio of the average per capita birth rate to the average per capita death rate in the metacommunity; and (v) the probability of a new species arising per birth, ν.

Mathematical Formulations of Neutral Theory

Neutral theory has been formulated in several ways. It can treat space either implicitly or explicitly. The classical island–mainland problem of the theory of island biogeography treated space implicitly, and most of the analytical results on neutral theory come from the implicit space versions. Recently, however, there have been analytical advances in the continuous space versions, particularly in the study of species–area relationships (see below). Also, it is possible to study community drift in model populations of discrete individuals, using Markov chain or "master equation" models, or to model population sizes as continuous variables, making it possible to use a variety of well-known mathematical tools such as Komolgorov and Fokker–Planck equations. There are also quasi-neutral models in which all species are "symmetric" but in which stabilizing mechanisms can operate, such as density dependence and frequency dependence, but mechanisms that operate identically in all species. Some researchers prefer to limit neutrality to discussions of nonstabilizing drift and dispersal and consider communities of symmetric species as a separate class of models. Recent theoretical work has proven that Hubbell's model with zero-sum dynamics and the master equation approach modeling noninteracting species are asymptotically identical as community size increases.

RECENT DEVELOPMENTS IN NEUTRAL THEORY

Since the publication of Hubbell's book in 2001 on the unified neutral theory of biodiversity and biogeography, there have been many major developments of the theory, including many new analytical solutions to problems that previously could be studied only by simulation in numerical experiments. The more salient of these developments are in five major research topical areas: relative species abundance, dispersal, the dynamics of communities, species–area relationships and β diversity, and speciation and phylogeny. It is useful to begin at the largest spatiotemporal scale and discuss advances in understanding speciation and phylogeny.

Speciation and Phylogeny

One of the principal achievements of neutral theory has been to connect speciation with macroecological patterns of species abundance and diversity. Under a model in which new species arise from lineages founded by single individuals (point mutation speciation), the expected steady-state distribution of relative species abundance in the metacommunity is Fisher's log series. The neutral theory also provides a biological interpretation of the parameters of the log series. In neutral theory, Fisher's α is called θ and is a fundamental biodiversity number equal to

$$\theta = \frac{v}{1 - v}(J_M - 1), \qquad (2)$$

where v is the per capita speciation rate and J_M is the size of the metacommunity. Note that the speciation rate v is generally a very small number and J_M is very large, so that $\theta \approx vJ_M$. An alternate derivation with slightly different assumptions (the so-called Fisher–Wright model) leads to a value of $\theta = 2vJ_M$ (for details, see Etienne and Alonso, 2007), but the important point is that either value explains why Fisher's α (or θ) is so stable in the face of increasing sample size: α is proportional to the product of two very stable numbers, one very large (the size of the metacommunity), and the other very small (the average per capita speciation rate). The biodiversity number θ can vary in theory from a minimum of zero (one species in the entire metacommunity) to positive infinity (every individual is a new and different species). Actual values of θ vary from less than unity (e.g., species-poor boreal tree communities) to about 743 (estimated for the entire tree flora of the Amazon Basin).

Neutral theory also reveals the biological meaning of parameter x of the log series (Eq. 1) and why it is very near but slightly less than unity. Parameter x is the ratio of the average per capita birth rate in the metacommunity to the average per capita death rate, b/d. This ratio is very close to but less than unity because all species eventually go extinct. When one adds per capita speciation to the per capita birth rate, species diversity in the metacommunity is in mass balance between speciation and extinction, and the ratio is unity: $(b + v)/d = 1$.

Neutral theory provides insights into phylogeny and the evolution of diversity. Previous neutral models of phylogeny were pure birth–death branching processes in which lineages were assigned birth and death rates, but the effect of the lineage abundance on the number of daughter lineages was ignored. In neutral theory, more abundant lineages have longer evolutionary lifespans and more births, so abundant lineages have many more opportunities for generating new species than do rare lineages. The variation in lineage abundance thus generates patterns of speciation more concentrated in some lineages than others, so an observation of clumped speciation cannot be taken as prima facie evidence for the operation of nonneutral phylogenic processes.

Ricklefs, and independently Gavrilets, have raised objections to the point mutation speciation model on the grounds that it generates too many rare and short-lived species. However, this model is the only one that generates log series relative abundance patterns, and, remarkably, it gives the best fit to data on species abundance of a number of potentially more realistic speciation models, including random fission speciation, in which a parent lineage is randomly cleaved into two daughter lineages. Rosindell and colleagues have suggested a resolution of this issue by modeling incipient speciation as a protracted process, and Allen and Savage have suggested that there is a minimum size threshold for new species. Both approaches bring speciation rates and species life span estimates from neutral theory into closer agreement with data. Allen and Savage's approach, which models population size as a continuous variable, not discrete, also enables speciation rates to be calibrated in units of absolute time, rather than in units of generational time, as is the case in Hubbell's original theory.

Interest is growing in applying a genealogical approach to modeling neutral communities, in which the genealogical histories of lineages are studied spatially explicitly (see Etienne and Olff, 2004; de Aguiar et al., 2009). This work finds that there are different area scaling relationships describing species diversity and genetic diversity in neutral model communities.

Relative Species Abundance

Fisher's log series distribution is the expected distribution of relative species abundance in the metacommunity under point mutation speciation. But what is the expected distribution of relative species abundance in local communities that receive immigrants from the metacommunity under dispersal limitation? This is the classical problem in island biogeography theory. In 2001, Hubbell showed that the distribution of relative species abundance in local communities is affected by the immigration rate. When the immigration rate is low, steady-state species richness in the local community is lower, common species become more common, and rare species rarer. This finding revealed that local patterns of commonness and rarity were not static but were dynamic and functions of dispersal rates. It also showed that the local community pattern of relative species abundance was not the log series but a distribution in which rare species were less abundant than a random sample of the metacommunity log series would produce. The effect of dispersal limitation of the local abundance of rare species is illustrated in Figure 3, for the tree community in a 50 ha mapped plot of tropical forest on Barro Colorado island in Panama.

However, Hubbell's results for the local community were obtained by simulation. An error-free analytical derivation of the distribution of local species abundance under dispersal limitation took several attempts and years to complete (history summarized in Etienne and Aloso, 2007). There is now a full sampling theory for neutral relative species abundance, and the biodiversity number θ and the immigration rate m can be jointly estimated by maximum likelihood methods. Figure 4 shows the log-likelihood surface for the combination of parameters (θ, m) that yielded the parameter values used in calculating the predicted distribution of relative species abundance for BCI trees in Figure 3.

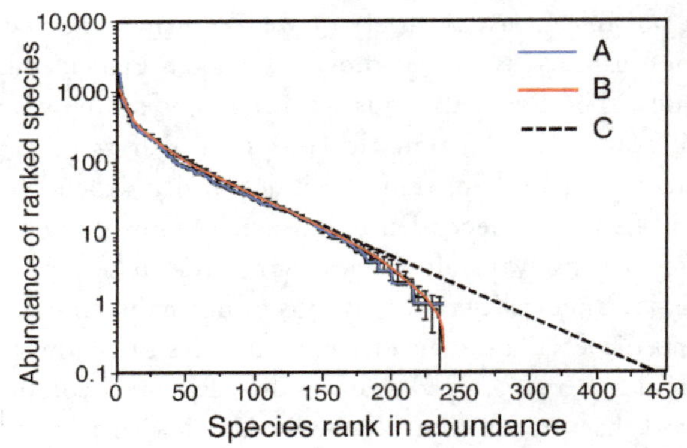

FIGURE 3 Distribution of relative tree species abundance in a 50-ha mapped plot on Barro Colorado Island (BCI), Panama, in the 1982 census, and the fit of neutral theory to the data. Data are for 20,541 trees >10 cm in trunk diameter (dbh) of 253 species. The *x*-axis is species rank in abundance from the commonest species in the rank-1 position at the left, to the rarest species at the right. The *y*-axis is the abundance of the ranked species on a log scale. (A) The blue line is the observed species abundance data. (B) The red line is the maximum likelihood fit of neutral theory to the data, which is obtained for a value of the biodiversity number θ = 50 and an immigration probability m = 0.1 (See Fig. 4). The vertical black bars are ±1 standard deviation of the expected abundance. (C) The dotted black line is the curve expected from a random sample of the metacommunity log series with θ = 50, but no dispersal limitation (m = 1). Note that dispersal limitation results in a lowered local abundance of rare species relative to their expected abundance in a random sample of the metacommunity. Modified from Hubbell (2001).

On local scales, within the body of literature on neutral theory inclusive of symmetric neutral models, two distinct mechanisms have been shown to be consistent with static patterns of relative tree species abundance, at least in tropical forest tree communities. One mechanism is dispersal limitation, the mechanism of Hubbell's original model. The other mechanism is symmetric density dependence, in which species in the community experience the same density dependence and rare-species advantage, when they are at equivalent abundance. In the case of tropical forests, the quality of fit of these two models cannot be distinguished, nor are they mutually exclusive mechanisms. However, in limited tests done to date, the dispersal limitation version has performed much better than the density-dependent version in characterizing the dynamics of species-rich tropical forests—changes in tree species composition over time (see Hubbell, 2009, Condit et al, in press).

Critics of neutral theory have pointed out that patterns of relative species abundance consistent with neutral theory do not in fact prove that neutrality is the mechanistic explanation for them (see Purves and Pacala, 2004). In almost all cases, nonneutral

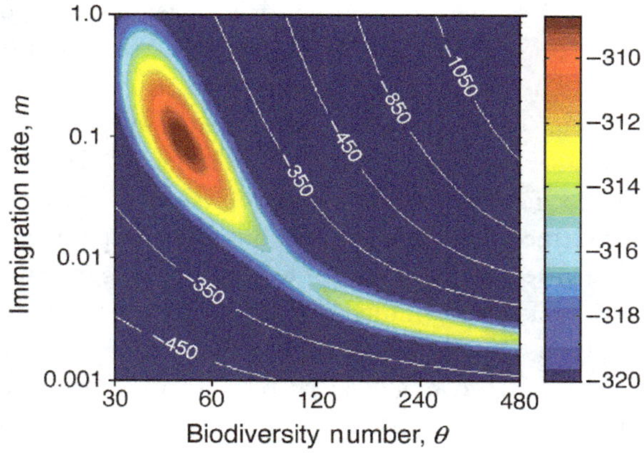

FIGURE 4 Estimating the biodiversity number θ and immigration rate m from the relative species abundance data in Figure 3 for the tree community in the Barro Colorado Island forest plot, using the sampling formulae of neutral theory. The figure shows the log likelihood surface of the (θ, m) parameter combination. The color bar at the right shows the log likelihoods greater than −320. Dark blue areas represent log-likelihood values that are lower than −320 (white contour lines). One of the challenges in estimating θ and m is that there is a hyperbolic interaction between θ and m in the log-likelihood surface. Note that there is both a local and a global maximum. However, in the case of BCI, there is good resolution of the global maximum at $\theta = 50$ and $m = 0.1$. After Etienne and Alonso (2007), color version courtesy of Rampal Etiennne.

explanations require more complex models often with many species-specific parameters, and the burden of proof that greater complexity is necessary rests on these models. The degree to which ecological detail is required also depends on spatial and temporal scale. Chisholm and Pacala (2010) have demonstrated that on large scales, niche-based models and neutral models are convergent and make identical predications.

Dispersal

Understanding dispersal and immigration in neutral communities has also advanced in the past decade. One advance was to recognize that, in addition to a fundamental biodiversity number, there is also a fundamental dispersal number I, which standardizes immigration rates as a function of m and the size of the local community, J_L:

$$I = \frac{m}{1 - m}(J_L - 1). \tag{3}$$

However, Equation 3 still does not provide a clear biological interpretation of parameter m of neutral theory. It has been difficult to compare estimates of m from different communities for two reasons, first because of a lack of direct data on dispersal, and second, because the linkage between the implicit and explicit-space versions of neutral theory remained to be clarified. Recently, Chisholm and Lichstein (2009) have derived a good approximation of m that takes into account the perimeter P and area A occupied by the local community and mean dispersal distance d. The relationship is given by the simple formula

$$m \approx \frac{Pd}{\pi A}. \tag{4}$$

Species–Area Relationships and β Diversity

A significant development in neutral theory is the discovery of an analytical solution to the oldest and most celebrated problem in biogeography and ecology: the species–area relationship. Previously, the full species–area relationship under neutrality was known only from simulations, and it was conjectured to be triphasic on a plot of log number of species against log area. At small spatial scales, the shape of the curve is sensitive to the sampling of the local abundances of species. At intermediate scales, the curve becomes log–log linear, responding to the logarithmic phase resulting from sampling the log series in the metacommunity. At very large spatial scales, the curve inflects upward to a limiting slope of unity. The limiting slope is approached as the correlation length of a single metacommunity is exceeded, when one passes into dynamically uncoupled biogeographic realms. Nine years after Hubbell's book, O'Dwyer and Green (2010) found an analytical solution to the species–area relationship under neutrality in an explicitly spatial model, making use of partition functions from quantum field theory. Their analytical result confirmed the triphasic species–area curve. The shape of the curve depends on the fundamental biodiversity number, θ_s, redefined on a per unit area basis, on σ, the mean distance traveled in dispersal events, and on the quantity $(d/b) - 1$, where, as before, b and d are the average per capita birth and death rates. As a corollary, O'Dwyer and Green also derived an analytical expression under neutrality for a measure of β diversity—the turnover of species across landscapes and geographic regions.

Dynamics of Communities

Although most of the developments in neutral theory have focused on predictions of static patterns of species abundance and diversity, a paper by Azaele et al. (2006) has made a significant advance in understanding the nonequilibrium dynamics of the species composition of communities under drift by demographic stochasticity. The authors were able to calculate a characteristic time scale for a well-studied 50-ha plot of tropical rain forest (Barro Colorado Island (BCI), Panama), which was on the order of 3500 yr based on empirical data on birth and death rates. This allowed them to calculate the residual

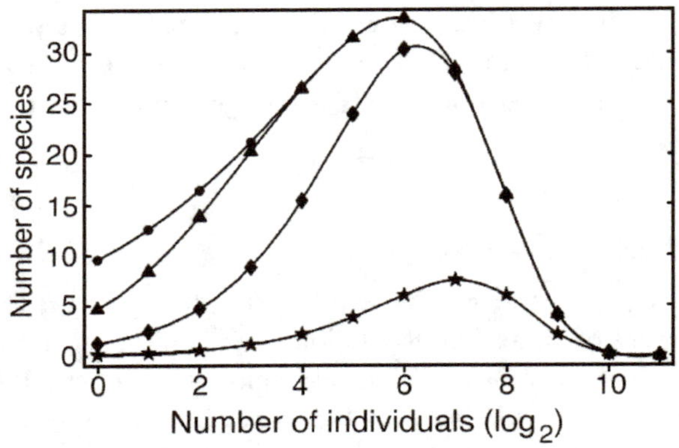

FIGURE 5 Neutral theory predictions of the future dynamics of the Barro Colorado Island tree community over the next 10,000 years. The curves show the predicted changes in the distribution of the relative abundances of those tree species originally present in the first census of the BCI plot. No trees that immigrate to the plot after the initial census are included. Curve with circles: distribution at $t = 0$ (1982). Curve with triangles: $t = 100$ yr (residual abundances after a century). Curve with diamonds: $t = 1000$ yr (residual abundances after a millennium). Curve with stars: $t = 10{,}000$ yr (residual abundances of original tree species after 10 millennia). Note that initially more common species have a higher probability of remaining than initially rare species. These curves were calculated from the estimates of θ and m given in Figures 3 and 4, and additional estimate of the average per capita birth rate to death rate ratio b/d observed for trees in the BCI plot.

abundances of the tree species originally present after *t* yr had elapsed, not including species entering the plot after the start (Fig. 5). Note that there is a preponderance of species remaining after 10,000 years that were initially among the most common species in the plot.

Other Advances

Many other problems in community ecology are currently under study in the context of neutral theory. One of the areas of intense activity is to connect neutral theory with metabolic scaling theory, and a few papers have appeared that are addressing how to put body size variation into neutral theory. Other efforts are being devoted to exploring connections with population genetics and exploiting preexisting mathematical tools in population genetics in neutral community ecology. Recent papers have been exploring neutral models of habitat heterogeneity, productivity patterns, and conservation problems as well (for example, Allouche and Kadmon, 2009).

THE USE AND FUTURE OF NEUTRAL THEORY IN COMMUNITY ECOLOGY

Neutral theory in community ecology is here to stay because it provides a powerful set of statistical tools for testing a diverse array of null hypotheses about the species composition, abundance patterns, and dynamics of ecological communities on multiple spatial and temporal scales. The degree to which different communities are governed by nearly neutral processes is still an open question. There have been many attempts to test neutral theory, and in many communities the theory's assumptions are not well met. In other communities, however, neutral theory's predictions are often good approximations. These reasonable approximations suggest that many ecological communities in nature are subject to significant demographic and environmental stochasticity and drift, just as originally postulated in the theory of island biogeography. Moreover, a failure of the theory to fit data from one or another community does not refute neutral theory, as some authors have claimed. Such claims are like asserting that "selection refutes the Hardy–Weinberg equilibrium." Neutral theory is a set of theorems that follow inevitably from its assumptions. Rejection of neutrality means rejection of the applicability of one or more of these assumptions in particular cases. Moreover, rejection of a neutral explanation does not constitute support for specific nonneutral alternative hypotheses. A major reason for the popularity of neutral theory is that, unlike most contemporary theories in community ecology, it has a sampling theory and can actually be tested. It is unclear how many theories in community ecology would fail as frequently or even more often if they provided a sampling theory for testing them. One measure of good theory is to fail in informative ways. In the recognition that all theory is an approximation of nature, the appropriate use of neutral theory is as a tool for theory improvement and advancement of community ecology.

SEE ALSO THE FOLLOWING ARTICLES

Diversity Measures / Information Criteria in Ecology / Metacommunities / Phylogeography / Spatial Ecology / Stochasticity, Demographic

FURTHER READING

Allen, A. P., and V. M. Savage. 2007. Setting the absolute tempo of biodiversity dynamics. *Ecology Letters* 10: 637–646.

Allouche, O., and R. Kadmon. 2009. A general framework for neutral models of community dynamics. *Ecology Letters* 12: 1287–1297.

Alonso, D., R. S. Etienne, and A. J. McKane. 2006. The merits of neutral theory. *Trends in Ecology & Evolution* 8: 451–457.

Azaele, S., S. Pigolotti, J. R. Banavar, and A. Maritan. 2006. Dynamical evolution of ecosystems. *Nature* 444: 926–928.

Bell, G. 2001. Neutral macroecology. *Science* 293: 2413–2418.

Chave, J. 2004. Neutral theory and community ecology. *Ecology Letters* 7: 241–253.

Chisholm, R. A., and J. W. Lichstein. 2009. Linking dispersal, immigration, and scale in the neutral theory of biodiversity. *Ecology Letters* 12: 1385–1393.

Chisholm, R. A., and S. W. Pacala. 2010. *Proceedings of the National Academy of Sciences USA* 107: 15821–15825.

De Aguiar, M.A.M., M. Baranger, E.M. Baptestini, and Y. Bar-Yam. 2009. Global patterns of speciation and diversity. *Nature* 360: 384–387.
Etienne, R. S., and H. Olff. 2004. A novel genealogical approach to neutral biodiversity theory. *Ecology Letters* 7: 170–175.
Etienne, R.S., and D. Alonso. 2007. Neutral community theory: How stochasticity and dispersal-limitation can explain species coexistence. *Journal of Statistical Physics* 128: 485–510.
Hubbell, S.P. 2001. *The unified neutral theory of biodiversity and biogeography.* Princeton: Princeton University Press.
Hubbell, S.P. 2009. Neutral theory and the theory of island biogeography. In J. Losos and R. E Ricklefs, eds. *The theory of island biogeography revisited.* Princeton: Princeton University Press.
O'Dwyer, J.P., and J.L. Green. 2010. Field theory for biogeography: a spatially explicit model for predicting patterns of biodiversity. *Ecology Letters* 13: 87–95.
Purves, D.W., and S.W. Pacala. 2004. Ecological drive in niche-structured communities: neutral pattern does not imply neutral process. In D. Burslem, ed. *Biological interactions in the tropics.* Cambridge, UK: Cambridge University Press.
Volkov, I, J.R. Banavar, S.P. Hubbell, and A. Maritan. 2003. Neutral theory and relative species abundance in ecology. *Nature* 424: 1035–1037.

NICHE CONSTRUCTION

JOHN ODLING-SMEE
University of Oxford, United Kingdom

Niche construction theory (NCT) is a branch of evolutionary biology that focuses on the capacity of organisms to modify their environments through their metabolisms, activities, and choices. The term also includes migration, dispersal, and habitat selection—when organisms relocate in space and encounter different selective environments by doing so. The defining characteristic of niche construction, however, is not the modification of environments per se but the modification of natural selection by niche-constructing organisms and the inheritance of previously modified selection pressures by descendent organisms via a second inheritance system called ecological inheritance. Niche construction is therefore a source of feedback in evolution that depends on the co-direction of evolutionary processes by the activities of organisms.

NICHE CONSTRUCTION

There are innumerable examples of niche construction familiar to ecologists. Animals manufacture nests, burrows, webs, and pupal cases; plants modify fire regimes, levels of atmospheric gases, and nutrient cycles; fungi decompose organic matter; and bacteria fix nutrients. Niche construction has often been modeled by theorists. For example, components of niche construction are implicit in many standard evolutionary models of frequency and density-dependent selection; habitat selection; maternal inheritance; indirect genetic effects and coevolution. Other models have explicitly investigated niche construction and ecological inheritance. To date, all the latter have found niche construction to be consequential because it changes the dynamics of evolution.

WHY NICHE CONSTRUCTION THEORY IS DIFFERENT

In 1983, Richard Lewontin, who was an early advocate of niche construction, summarized the differences between standard evolutionary theory (SET) and niche construction theory (NCT) in two pairs of coupled differential equations. His first pair summarizes SET:

$$dO/dt = f(O,E), \tag{1a}$$

$$dE/dt = g(E). \tag{1b}$$

In Equation 1a, evolutionary change in organisms, *dO/dt*, depends on both organisms' states, *O*, and environmental states, *E*. In Equation 1b, however, environmental change, *dE/dt*, depends exclusively on environmental states. With many caveats and complications, organisms are not treated as the cause of any evolutionarily significant changes in their environments.

A second pair summarizes NCT:

$$dO/dt = f(O,E), \tag{2a}$$

$$dE/dt = g(O,E). \tag{2b}$$

Change in organisms, dO/dt, is again assumed to depend on both organisms' states and environmental states (Eq. 2a), but environmental change, dE/dt, is now assumed to depend on both environment states and the niche-constructing activities of organisms (Eq. 2b).

Figure 1 illustrates the difference graphically. In SET, the evolution of organisms is directed solely by natural selection pressures in environments (Fig. 1A). Selective environments, *E*, act on populations of diverse phenotypes and influence which individuals survive and reproduce and pass on their genes to the next generation via a single inheritance system, genetic inheritance. In contrast, in NCT (Fig. 1B) the evolution of organisms is directed by both natural selection and niche construction. The transmission of genes by ancestral organisms to their descendants is influenced by natural selection. However, selected habitats, modified habitats, and

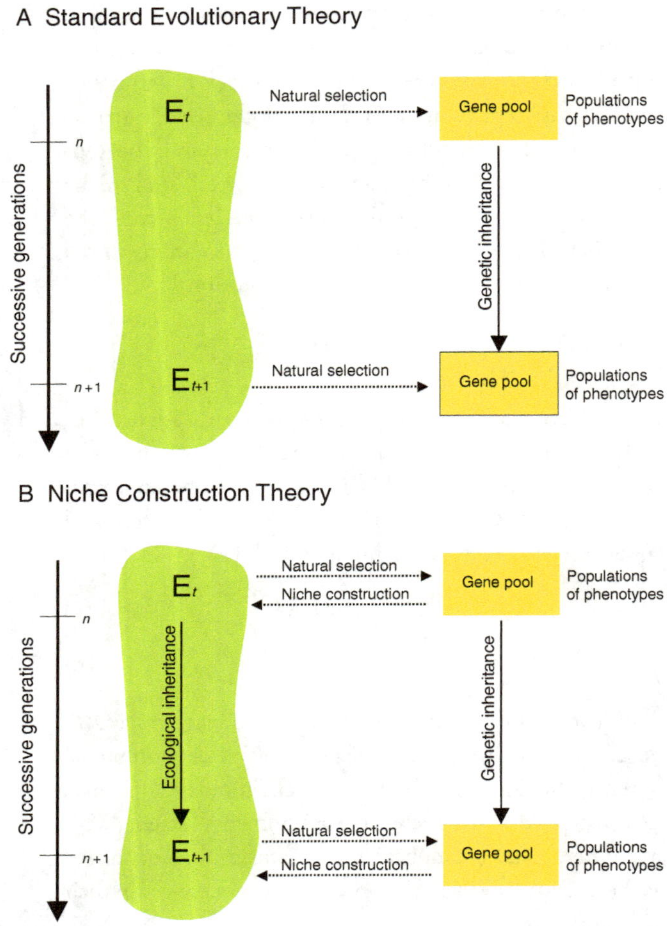

FIGURE 1 (A) Standard evolutionary theory (SET). (B) Niche construction theory (NCT). (Figure based on Odling-Smee et al., 2003: 14, Fig 1.3.)

modified sources of natural selection in those habitats are also transmitted by niche-constructing organisms to their descendants by ecological inheritance. In NCT, the selective environments of organisms are partly determined by independent sources of natural selection—for instance, by climate or physical and chemical events. But they are also partly determined by what niche-constructing organisms do, or previously did, to their own, and each others' environments.

HISTORICAL BACKGROUND

Niche construction is an obvious process, so why has it been marginalized by evolutionary biologists for so long? The answer is probably hidden in a seldom reconsidered foundation assumption of SET concerning the role of environments in evolution.

Philosopher Peter Godfrey-Smith drew attention to the problem by describing SET as an "externalist" theory of evolution because it uses the external properties of environments as its sole explanatory reference device. SET seeks to explain the adaptations of organisms exclusively in terms of natural selection pressures in their external environments (Fig. 1A). This approach owes more to Newton than to Darwin. A giveaway is that natural selection used to be called a "force," a description John Endler once called "a vague and most improper analogy with physics." It implies that a "force," natural selection, acts on an "object," a population that "reacts" by evolving. The analogy with Newton's $F = ma$ is clear.

The vital point obscured by this approach is that organisms are active as well as reactive. Organisms have to be active to stay alive and reproduce. Minimally, all organisms must gain resources from their external environments by genetically informed, nonrandom, fuel-consuming work, and they must return detritus to their environments. Organisms are therefore bound to perturb their environments, constructively and destructively (just change the sign), to stay alive and reproduce. It follows that organisms resemble programmed fuel-consuming "engines" more closely than passive abiotic Newtonian object, and are better described by the laws of thermodynamics. Darwin's discovery of natural selection largely preceded the discovery of the laws of thermodynamics, so SET got stuck with Newton.

Given that organisms must perturb their environments, eventually they are bound to modify some natural selection pressures in their environments too. This point is captured by Lewontin's second pair of equations. In effect, Lewontin couples a second "causal arrow" in evolution, niche construction in Equation 2b, to Darwin's first "causal arrow," natural selection, in Equation 2a.

Now it is possible to see why SET neglects niche construction. Darwin's first causal arrow, natural selection, is compatible with SET's externalist assumption because it points in the "right" direction, from environments to organisms. It is therefore straightforward to describe how natural selection "causes" adaptations in organisms. However, the second causal arrow, niche construction, points in the "wrong" direction, from organisms back to their environments. That makes it difficult or impossible for SET to describe the feedback from niche construction to natural selection as co-causal. Instead, SET is bound to explain away all observed instances of niche construction as nothing but phenotypic, or possibly extended phenotypic, consequences of prior natural selection. SET can recognize niche construction as a product of evolution, but it cannot recognize it as a cause.

NCT overcomes this obstacle by changing evolutionary theory's reference device. Instead of describing the evolution of organisms relative to external environments, it describes evolution relative to niches:

$$N(t) = h[OE]. \quad (3)$$

Equation 3 defines a niche, $N(t)$, in terms of an interactive, two-way $[OE]$ relationship between a population of organisms O and its environment E at time t. The $[OE]$ relationship changes over time, and its dynamics are driven by both natural selection and niche construction. The $[OE]$ relationship itself is neutral. It does not impose any theoretical bias either in favor of natural selection and against niche construction, nor vice versa. Instead, it treats natural selection and niche construction as reciprocal causal processes in evolution. In NCT, the adaptations of organisms are not just products of natural selection. They are products of natural selection and niche construction.

SUBSTITUTING NCT FOR SET IN ECOLOGY

The main difference NCT makes to ecology is that it changes the relationship between Hutchinson's "ecological theatre" and the "evolutionary play." The "theatre" continues to influence the "play," but NCT describes how the "players" are constantly reconstructing the "theatre." Post and Palkovacs (2009) revisited this relationship in the light of NCT, and they suggested a novel theoretical framework for handling it. Their key proposal is to divide niche construction into two subprocesses. They argue that the impacts of niche-constructing organisms on environments need to be considered separately from the subsequent evolutionary responses of populations to organism-modified environments because these two subprocesses operate independently.

The first subprocess is equivalent to one many ecologists now call ecosystem engineering, although possibly more general because it includes all the by-products of living, for example, eating, excreting, nutrient uptake, and mineralization, as well as such obvious "constructs" as beaver dams. The principal ecological consequence of this subprocess is that it can establish engineering webs that exercise additional controlling influences over energy and matter flows and trophic interactions in ecosystems.

The second subprocess cuts in when the modification of environmental components by the first subprocess translates into nontrivial ecological inheritances, in the form of modified selection pressures for evolving populations, and causes at least one population to evolve further. One reason why Post and Palkovacs want to separate the two subprocesses is because niche-constructing populations are not restricted to influencing their own evolution. They frequently influence the evolution of other populations, too. It is therefore necessary to distinguish between environment-altering populations and recipient populations that respond to altered environments, especially when these populations are not the same.

This was originally done by introducing the concept of environmentally mediated genotypic associations, or EMGAs. EMGAs connect environment-altering traits expressed by genotypes in niche-constructing populations to recipient genotypes in recipient populations via a shared external environment. Each EMGA works by providing an evolutionarily significant "bridge" in an external environment between a genotype in a niche-constructing population and a genotype in a recipient population in the form of a modified natural selection pressure. Genotypes connected by EMGAs may either be in the same single population or in different populations. Also, the environmental "bridge" that connects them may either be biotic or abiotic, or some combination of both. To make the idea more concrete, suppose a population of earthworms changes an abiotic resource in the soil in a way that modifies the selection on a population of plants growing in the soil, plus its own selection. Via different EMGAs, the worms could subsequently affect the evolution of both the plants and themselves by feed-forward and feedback loops.

Assuming that naturally selected genotypes carry semantic information between generations in the form of uncertainty-reducing signals relative to specific selection pressures, then EMGAs must also establish information-carrying communication channels between coevolving populations in communities and in ecosystems. Translated into the language of information theory, a single EMGA connects a genotype in a "transmitting" population to a genotype in a "receiving" population, through a channel comprising some "constructed" component of an environment. The "signal" is then "received" by the recipient population in the form of an altered selection pressure in its ecological inheritance, and it may then cause the recipient population to evolve further.

Genetic inheritance and ecological inheritance, however, do not transmit semantic information between generations in the same way. Selected DNA sequences carry uncertainty-reducing information about specific natural selection pressures in environments between generations. If a natural selection pressure is modified by niche construction, and if it is also transmitted between generations by ecological inheritance, it must change the capacity of any relevant DNA sequence to reduce uncertainty. It will therefore change the semantic value, or fitness of any DNA sequence that relates to that selection pressure, by changing its capacity to reduce uncertainty. The degree to which its semantic value changes is also potentially measurable. Minimally, it could be measured by a change in the value of a modified selection coefficient. In an experimental situation in which niche construction is artificially manipulated, whatever change is caused by the artificial manipulation, and consequently in a population's ecological inheritance, is logically equivalent to indirect artificial selection. The change itself can be either positive or negative, depending on the precise nature of the niche construction. Thus, semantic information transmitted between generations by ecological inheritance is the complement of semantic information transmitted by genetic inheritance.

THE IMPACT OF NICHE CONSTRUCTION IN ECOSYSTEMS

In actual ecosystems, niche construction's second subprocess is likely to generate multiple EMGAs, connecting the genetics of multiple niche-constructing populations to the genetics of multiple recipient populations, and to establish rich networks of evolutionarily significant communication channels between populations. They can be called "semantic information nets" in contrast to the "engineering webs" produced by niche construction's first subprocess. We already know that "engineering webs" need to be added to the list of ecological processes that influence and regulate ecosystem processes. Do EMGA-based semantic information nets also need to be added to the same list? With caveats, NCT suggests they do.

The caveats include the point that not all genes in contemporary populations are equally susceptible to changes in external selection. For example, some genes belong to ancient, extremely stable, "kernel" developmental gene-regulatory networks that can no longer be changed by external natural selection, but can only be rearranged by internal selection in genomes. The caveats also refer to very different rates of evolution in different populations and to wide differences in spatial and temporal environmental scales. NCT nevertheless suggests several ways in which EMGA-based semantic information nets should affect how ecosystems work. For instance, partly because of its emphasis on the indirect coevolution of populations via intermediate abiota, as well as via biota, NCT expects far more coevolution in ecosystems than SET anticipates.

THE PERPETUAL RECONSTRUCTION OF ECOSYSTEMS

Semantic information nets should also change the relationship between Hutchinson's "ecological theatre and the evolutionary play." According to NCT, to greatly varying extents, all organisms niche construct. All populations are also recipient populations relative to at least some selection pressures previously altered by niche-constructing populations. Therefore, even if the niche-constructing activities of many populations are discounted as neither ecologically nor evolutionarily significant, this dual role of all populations should ensure the constant reconstruction of ecosystems and the recursive evolution of their populations.

If niche-constructing populations modify natural selection for other populations, and if some of the recipient traits the modified selection affects in the recipient populations are themselves niche-constructing traits, it follows that the niche-constructing activities of some populations will change the niche-constructing activities of other populations in ecosystems. If the recipient populations later niche construct in new ways, they too will be likely to reconstruct one or more components of their ecosystems, sometimes in innovative ways. They may therefore modify one or more natural selection pressures for themselves, or for other populations in their ecosystem, and so on, ad infinitum. EMGA-based semantic information nets should therefore ensure the perpetual reconstruction of ecosystems as a consequence of the first subprocess of niche construction and ensure the perpetual evolution of populations as a consequence of the second subprocess, regardless of all other sources of change in ecosystems. So Hutchinson's "players" should constantly reconstruct their "theatre," and the reconstructed theatre should constantly select for further evolutionary changes in its "players."

CONCLUSION

The full implications of these mechanisms for ecosystems are beyond the scope of this entry. There could be many. Post and Palkovacs touch on some of them by stressing the importance of eco-evo feedback cycles in ecosystems that have the potential to alter the direction of evolution and strongly modify the role of species in ecosystems.

They also provide examples of the small number of eco-evo feedback cycles that have already been studied, while admitting that as yet there is not a single system where the evidence of dynamic feedbacks is without gaps. Future work in this area is therefore likely to demand even closer cooperation between ecologists and evolutionary biologists, and probably geneticists and molecular biologists too. But we also need to ask the right questions. The role of NCT is to encourage diverse scientists to ask those questions.

SEE ALSO THE FOLLOWING ARTICLES

Coevolution / Ecosystem Engineers / Evolutionarily Stable Strategies / Metacommunities / Mutation, Selection, and Genetic Drift / Neutral Community Ecology / Niche Overlap / Succession

FURTHER READING

Cuddington, K., J.E. Byers, W. G. Wilson, and A. Hastings. 2007. *Ecosystem engineers: plants to protists.* Burlington: Academic Press.

Hutchinson, G.E. 1957. Concluding remarks. *Cold Spring Harbor Symposia on Quantitative Biology* 22: 415–427.

Lewontin, R.C. 1983. Gene, organism, and environment. In D.S. Bendall, ed. *Evolution from molecules to men.* Cambridge, UK: Cambridge University Press.

Odling-Smee, F. J., K.N. Laland, and M. W. Feldman. 2003. *Niche construction: the neglected process in evolution.* Princeton: Princeton University Press.

Pigliucci, M., and G.B. Muller. 2010. *Evolution: the extended synthesis.* Cambridge, MA: MIT Press.

Post, D.M., and E.P. Palkovacs. 2009. Eco-evolutionary feedbacks in community and ecosystem ecology: interactions between the ecological theatre and the evolutionary play. *Philosophical Transactions of the Royal Society B: Biological Sciences* 364: 1629–1640.

NICHE OVERLAP

HOWARD V. CORNELL

University of California, Davis

Niche overlap refers to the partial or complete sharing of resources or other ecological factors (predators, foraging space, soil type, and so on) by two or more species. For example, warblers in a woodlot might all feed on insects and thus overlap in their diets, or plants in a meadow might all overlap in their need for light. Niche overlap is an important concept in community ecology because it is expected to determine how many and which species can coexist in a community. Interest in niche overlap began with the competitive exclusion principle, which states that two species using identical resources and/or environments cannot coexist. The niche concept developed as a way to describe the range of environments and resources required for the persistence of individual species. The competitive exclusion principle was then reformulated using niche terminology; two species overlapping completely in their niches cannot coexist. Theory then developed to link limits to niche overlap with limits to the number of coexisting species.

THE NICHE CONCEPT

G. E. Hutchinson formulated the first sophisticated quantitative niche concept. He envisioned a set of n intersecting axes, one for each environmental factor acting on an organism, defining what he termed the "n-dimensional hyperspace." The organism's niche was then an "n-dimensional hypervolume" demarcated by its fitness along each dimension (Fig. 1). This "fundamental niche" could be reduced to a smaller "realized niche" by the presence of other competing species that excluded the focal species from some subset of its hypervolume. Hutchinson's concept of the realized niche did not allow for niche overlap, because species are viewed as competing for single values of an environmental factor at each position in the niche hyperspace, and the competitive exclusion principle holds that one species will always exclude the other in such conditions.

A subsequent niche concept altered Hutchinson's original formulation from fitness limits at various levels

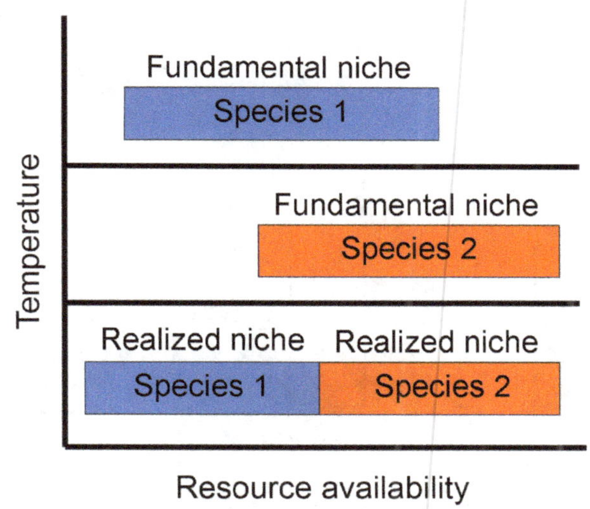

FIGURE 1 Hutchinson's niche concept in two dimensions representing temperature and resource availability. The axes define the niche space, and the colored area defines the boundaries of the niche beyond which the fitness of the species falls to zero. The fundamental niche defines these boundaries in the absence of competitors, and the realized niche defines them when competitors occur in the same community.

of resource availability (e.g., density of food of a given size, nutrient concentration, and so on) or values of other environmental factors (e.g., temperature, salinity, and so on) along a niche dimension to the utilization distribution of resources or environments of a given type (MacArthur and Levins 1967, *American Naturalist* 101: 377–385). The utilization distribution is the fractional use of resources or environments arranged along a niche dimension, such as the proportion of foods of different size that are eaten or the proportion of time spent in different microhabitats. The range of environmental values used became niche breadth, which substituted for Hutchinson's fundamental niche. In its simplest form, the utilization distribution is depicted as a bell curve with a fixed niche breadth, w, and position along a single axis. With the concepts of the utilization distribution and niche breadth, the possibility of niche overlap arose and can be depicted as intersecting utilization distributions with their peaks d distance apart (Fig. 2). Hutchinson's original theory allowed for niche partitioning over multiple dimensions in niche hyperspace but utilization distribution theory focused mainly on overlap in one dimension, probably for ease of analysis. Analyses in multiple dimensions are possible but suffer from complications (see "Measuring Niche Overlap," below). The concept of niche overlap recognizes that species comprise populations of individuals that can vary in their niche utilization. Two species populations may thus interact with each other for resources that are represented by a distribution of values (e.g., frequencies of foods of given sizes), rather than single values (e.g., density of foods of a given size) as suggested by Hutchinson's formulation.

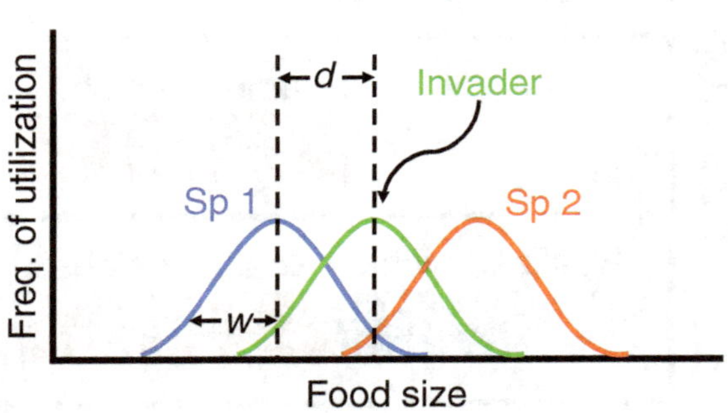

FIGURE 2 Utilization distributions for three species along a single niche dimension such as food size. The standard error of each distribution (w) defines niche breadth, and niche overlap is defined by the distance (d) between the mean utilization frequencies. The ability of a third species to invade between the other two depends upon the niche overlap and therefore the competition coefficients of species 1 and 2. The carrying capacities (K) of all species are assumed to be equal in this model.

MEASURING NICHE OVERLAP

Measures of niche overlap generally quantify the similarity of resource use by different species. Most measure the proportion of resources used by each species (overlap in resource utilization) along a single niche dimension (e.g., prey size, prey species, or microhabitat use). One popular version, first proposed by Eric Pianka, is

$$O_{jk} = O_{kj} = \frac{\sum_i^n p_{ij}p_{ik}}{\sqrt{\sum_i^n p_{ij}^2 \sum_i^n p_{ik}^2}}$$

where overlaps (O's) are symmetrical and p_{ij} and p_{ik} are the proportions of the ith resource used by the jth and kth species. In many theoretical treatments, overlap is measured simply as the ratio d/w, as in Figure 2, but this measure is sensitive to the shape of the utilization distribution and should not be used when the actual distribution is unknown or when species are known to have differently shaped distributions.

Measuring overlap along more than one niche dimension, such as food size plus microhabitat, is possible but difficult. Ideally, overlaps should be measured as the simultaneous proportional overlaps of resource use along each separate dimension (e.g., the proportion of food of a given size utilized in a particular microhabitat by each species). Failing that, if niche dimensions are totally independent (e.g., foods of a given size occur equally in all microhabitats), then multidimensional overlaps can be estimated as the product of overlaps along each separate dimension. If niche dimensions are totally dependent (e.g., food types of a given size only occur in one microhabitat), the overlaps need to be estimated as the mean overlap along all dimensions. Since niche dimensions are unlikely to be at these extremes, both the multiplicative and additive estimates can be inaccurate.

NICHE OVERLAP AND COMPETITION

The Lotka–Volterra Model

Studies of niche overlap are generally concerned with the outcome of species interactions, particularly competition, although the theory can also be applied to predation, parasitism, disease, mutualism, and so on. In their seminal 1967 paper, MacArthur and Levins used utilization distributions to derive competition coefficients from a measure of niche overlap. Competition coefficients are key parameters in the Lotka–Volterra equations that provide the foundation for much competition theory. The equations for competition between two species parameterized as they are usually found in textbooks are

$$\frac{dN_1}{N_1dt} = r_1\frac{(K_1 - N_1 - \alpha_{21}N_2)}{K_1}$$
$$\frac{dN_2}{N_2dt} = r_2\frac{(K_2 - N_2 - \alpha_{12}N_1)}{K_2} \quad (1)$$

where N_1 and N_2 are the population densities of species 1 and 2, r_1 and r_2 are their per capita population growth rates at negligible population densities, K_1 and K_2 are their carrying capacities (equilibrium population sizes in the absence of competitors when resources are limiting), and α_{21} and α_{12} are the interspecific competition coefficients (the intraspecific coefficients are implied and are always assumed to be 1 in this model). Although α_{21} and α_{12} can take different values, for simplicity they will be assumed to be equal in all that follows. The equations express the per capita growth rates $\frac{dN_i}{N_idt}$ as a function of the density-dependent effects of adding individuals of each of the species to the competitive arena.

The terms "niche" and "niche overlap" were never used in the original formulation of the theory. However, the competition coefficients can be thought of as expressing niche overlap in terms of the effect of adding one individual of a competitor species to a competitive arena relative to the effect of one individual of the focal species on the focal species' per capita growth rate. In other words, they summarize the relative strength of interspecific vs. intraspecific competition. High intraspecific relative to interspecific competition has a stabilizing effect on coexistence because each species limits its own growth rate more than it does the competitor's. When resources are limiting, this effect is expressed as a proportional reduction in the focal species' equilibrium population density relative to its equilibrium density in the absence of the competitor (K). In terms of utilization distributions, the broader the overlap in utilization, the higher the niche overlap, the higher the values of alpha, and the stronger is interspecific relative to intraspecific competition.

The carrying capacities K_1 and K_2 are also important in the model. Their values are jointly determined by r_1 and r_2 and the strength of intraspecific density dependence a such that $K_i = \frac{r_i}{a_i}$. They are sometimes thought of as the competitive abilities of the two species. The higher the value of r and the lower the value of a, the weaker the intraspecific density dependence, and the higher will be the value of K. High values of K indicate high growth rates, efficient use of resources, and, consequently, high competitive ability. Whereas the α's indicate the *relative* strength of interspecific vs. intraspecific competition, the K's indicate the strength of competition per se. Values of the α's and K's must both be specified to determine the competitive outcome for the two species.

Coexistence and Limiting Similarity

According to the competitive exclusion principle, complete niche overlap leads to competitive exclusion of one species because it is unlikely that both species will be equally good at utilizing that niche. For example, one species may use limiting resources more efficiently than the other. It follows that coexisting species must differ in their niches, but how different do they need to be? The competitive exclusion principle is silent on this issue. The Lotka–Volterra model predicts that with the right selection of parameter values, niche overlap, although it cannot be complete, can be nearly so and species can still coexist (Fig. 3). Coexistence is possible if at equilibrium, each species can increase when rare. This occurs when $\alpha_{21}\alpha_{12} < 1$ (niche overlap is incomplete and therefore intraspecific competition > interspecific competition) and $\alpha_{21} < \frac{K_1}{K_2} < \frac{1}{\alpha_{12}}$ (competitive abilities are sufficiently similar that they don't overwhelm the stabilizing effects of the first condition). If niches overlap completely ($\alpha_{21}\alpha_{12} = 1$), there is no stabilizing effect and only the

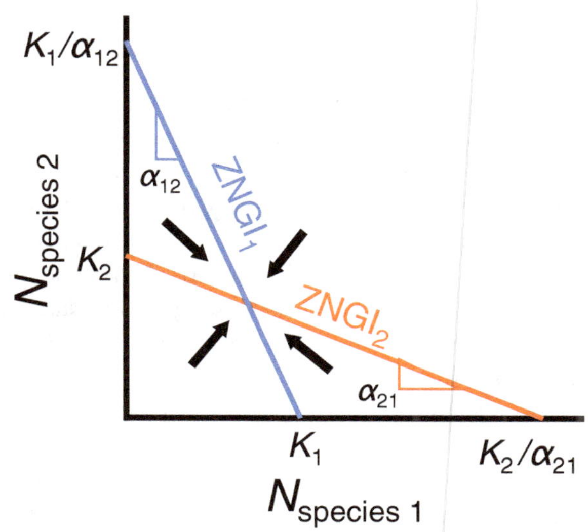

FIGURE 3 Conditions for coexistence in the Lotka–Volterra model depicted on a state-space graph. The axes define the densities of two competing species and the sloping lines are the zero net growth isoclines (ZNGI) representing mixtures of the two species where the population growth rate of each is 0. Where the isoclines cross, the populations are at a stable equilibrium and the species coexist. The slopes (α's) and intercepts of the isoclines (K's) define the conditions for coexistence.

species with the highest K can increase when rare and will exclude the other.

Niches in nature are often quite different, suggesting that there may be finite limits to niche overlap. An early example was Hutchinson's observation that size ratios (linear dimensions) between similar competing consumer species often approached 1.3, implying limits to overlap in food size utilization. The idea was subsequently explored theoretically, and early models indeed predicted a "limiting similarity" to niches necessary for coexistence. The first such exploration by MacArthur and Levins in 1967 was a deterministic model addressing the question, "When resources are limiting, how much must identically shaped utilization distributions of two species be separated in order to allow a third to invade between them?" (Fig. 2). The carrying capacities of all three species are assumed to be identical. The model predicted that the distance needed to be slightly greater than the niche breadths ($d/w > 1$) before invasion could occur. In a 1989 book chapter, Tom Schoener reviews subsequent analyses that explore the effects of more than three species and more than one niche dimension on limiting similarity. Multispecies analyses showed that the limit increases with the number of species, because low-intensity "diffuse competition" with several species adds up to the equivalent of more intense competition with a single species. Little theory has been devoted to multiple niche dimensions, but empirical results suggest patterns of complementarity where high overlap along some dimensions are accompanied by lower overlap along other dimensions in order to allow coexistence.

MacArthur and Levins' model was highly influential and stimulated many subsequent modeling efforts using the same framework. An early modification of the original model by Robert May relaxed the assumption of equal carrying capacities and the limit to similarity disappeared. Later models by Robert MacArthur and Robert May incorporating stochasticity to simulate environmental variability at first appeared to rescue limiting similarity, but, ultimately, Michael Turelli showed in a series of papers that the results for environmental variability were similar to those for the deterministic model. The general conclusion is that for this class of models, although some conditions can produce limits to similarity, a general limit to similarity does not exist. One probable reason is that the utilization distribution framework is incomplete. It focuses primarily on measures of niche overlap and thus considers only the relative strength of intraspecific vs. interspecific competition when predicting coexistence.

But the Lotka–Volterra equations indicate that information on relative competitive abilities (K's) is also required to predict coexistence. Once both criteria are considered, limiting similarity is rescued, but it is not general as predicted by the early models. The more different the competitive abilities of the two species, the more different niches have to be to stabilize the interaction and to allow coexistence. In a 1983 review, Peter Abrams points out that since different groups of coexisting species will have different competitive abilities, the degree of limiting similarity required will vary from community to community. This conclusion explains why evidence for limiting similarity is not consistent from community to community.

Other Models

Although the L–V equations have had tremendous heuristic value and have underpinned much of the early theory of niche overlap, their parameters are not easily measured in ecological systems. Other models have the same qualitative predictions as the L–V models but are more intuitive representations of niche overlap and competitive ability. As a result, modern niche overlap theory has placed more stress on these models.

RESOURCE CONSUMER THEORY

Niche overlap is more easily related to ecological mechanism within the framework of resource consumer theory where interactions are mediated through shared resources. Resource consumer theory, which is authoritatively discussed in the 1983 book *Ecological Niches* by Chase and Leibold, expresses competitive outcomes in terms of resource requirements, the supply rate of resources, and the impacts of consumers on resource levels. Resource requirements are measured as the rate of increase in the per capita growth rate of species i with increases in resource availability R:

$$\frac{dN_i}{N_i dt} = f_i(R) - m_i, \tag{2}$$

where $f_i(R)$ is some increasing function of resource availability and m_i is resource-independent mortality.

This formulation is very similar to a Hutchinsonian niche axis, which represents fitness as a function of resource availability or other environmental variables. Two consumer species whose per capita growth rates respond differently to increases in a resource's availability differ in their resource requirements. A consumer's impact on a resource is its ability to lower its concentration; the higher the per capita rate of resource consumption, the higher the impact. At equilibrium, the resource supply rate matches the consumption rate and R is the

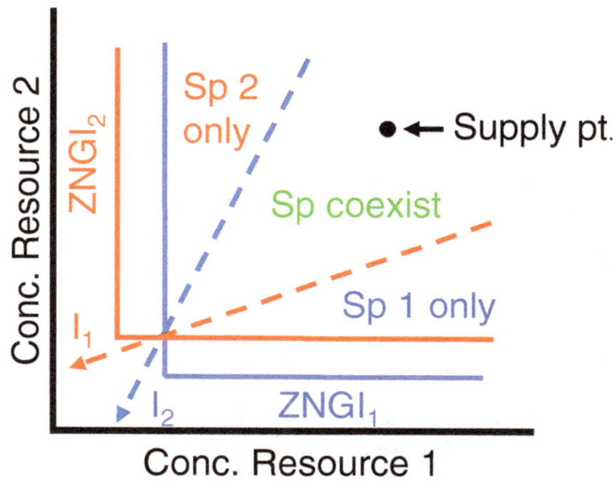

FIGURE 4 Conditions for coexistence according to resource consumer theory. The axes define the availability of two resources (nonsubstitutable in this case) and the solid lines represent zero net growth isoclines (ZNGI) for two species at various resource ratios. The dashed lines represent the impacts of the two species on the two resources. The supply rates of the two resources are indicated by the black dot. Supply rates must fall within the indicated region for coexistence. A stable equilibrium occurs where the isoclines intersect.

minimum resource concentration necessary to maintain the consumer. Coexistence at equilibrium is possible when there are two limiting resources which are supplied at sufficient rates, when each consumer differs in its most limiting resource in the absence of other species and has the greatest impact on its most limiting resource (Fig. 4). This is true even if each consumer utilizes some of the other consumer's limiting resource, that is, even if niches overlap somewhat. Resource requirements and impacts are roughly equivalent to the K's and α's in the Lotka–Volterra model and thus represent competitive abilities and niche overlap respectively. The greater the difference in resource requirements, the greater the difference in impacts required to stabilize the interaction and to allow coexistence. The advantage of this mechanistic framework over the Lotka–Volterra equations is that if the limiting resources are known, they can be measured and coexistence predicted. Many experiments have been done which confirm the theory.

GENERALIZED COMPETITION THEORY

In a paper published in 2000, Peter Chesson proposed a generalized model of competition based around the ability of species populations to increase when rare, which he labels $\bar{r}_i$. Unlike the Lotka–Volterra and resource consumer models, it is not strictly an equilibrium model but shares certain features with them. The variable $\bar{r}_i$ is the per capita growth rate $\frac{dN_i}{N_i dt}$ at low population density. The model can be written in a simple form for a species i which invades a region where species j is resident and at its carrying capacity:

$$\bar{r}_i = a_i(k_i - k_j) + a_i(1 - \rho)k_j, \tag{3}$$

where a_i expresses the strength of density dependence in reducing the invader's per capita growth rate, the k's are fitnesses of species I and j, and ρ is a measure of niche overlap. The first term of the equation expresses average fitness differences of the two species, and the second term is a stabilizing component expressing niche differences. Fitness differences measure the density-dependent per capita growth potential in a given environment. They are roughly equivalent to the differences in competitive ability (K's) in the L–V model and differences in resource requirements in the resource consumer model. The stabilizing term measures the relative impact of intraspecific vs. interspecific competition in terms of niche differences. When the former exceeds the latter, each species inhibits its own per capita growth rate more than that of the competitor's per capita growth rate. The stabilizing term is equivalent to the product alphas in the L–V model and differences in impacts in the resource consumer model.

The stabilizing component compensates for fitness differences between the two species (Fig. 5). If there is no stabilizing component, then the species with the highest average fitness wins in competition. Two species can coexist if the value of the stabilizing term exceeds the value of the fitness difference term; that is, niche differences must

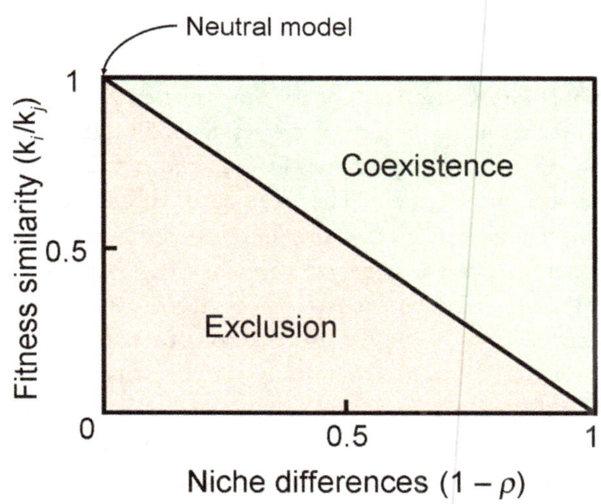

FIGURE 5 Conditions for coexistence of two species according to Equation 3. Fitness similarity and niche differences are represented on the y- and x-axes, respectively. The plotted line defines the boundary between conditions for coexistence and exclusion. The plot shows that when niche differences are greater, stable coexistence is possible even when fitness differences are large. The case where fitnesses are identical and niche differences are 0 represents Hubbell's neutral model.

exceed fitness differences. As for the previous models, the greater the fitness differences, the greater the niche differences required for coexistence. The advantage of this theoretical framework is that its parameters can be measured by observation and experiment in natural systems and the specific limiting factors need not be known. The theory predicts that if niche differences are important for coexistence, then per capita growth rates for a focal species should increase when that species becomes rare (as compared with a situation where the total density is the same, but the frequency of the focal species is higher); that is, growth rates are negatively frequency dependent for a given total density. Also, if niche differences are eliminated—that is, if population growth rates are calculated and experimentally imposed to simulate the condition where each species limits itself and its competitors equally making population growth rates independent of species relative abundance (intraspecific competition = interspecific competition)—the species with the highest fitness should increase in relative abundance. These predictions were borne out in an experimental study of annual grassland species by Jonathan Levine and colleagues, supporting the importance of niche differences for coexistence in this system.

CHARACTER DISPLACEMENT

Niche models employing utilization distributions are purely ecological with no evolutionary component, so niche position and breadth are fixed. But if resources are limiting, competition can select for divergence along niche axes. Phenotypes in the zone of niche overlap are disfavored by interspecific competition so that genes for nonoverlapping phenotypes will spread in each species population. Over evolutionary time, niche overlap (and sometimes niche breadth) decreases. Such divergence was commented upon by Charles Darwin and is today called character displacement. The characters involved are often morphological (e.g., jaw size and structure), which indirectly indicate divergence in ecology (e.g., food size). Character displacement is often observed between two species whose geographic distributions overlap in some places but not others. If divergence is observed in the zone of overlap but not elsewhere, character displacement is indicated (Fig. 6). Lack of divergence outside the zone of overlap has sometimes been called character release. In the case of character displacement, low levels of present-day niche overlap result in little present-day competition but were the result of past competition. This has sometimes been called the "ghost of competition past."

Early models allowing for dynamic niche positions and breadths through evolutionary time generally predict larger limiting similarities consistent with character displacement. These early models considered only niche differences. More recent models predict that selection should not only increase niche differences but also decrease fitness differences. Other recent models with added realism often predict that characters will converge rather than diverge under selection. In one interesting case examined by Sheffer and van Ness in 2006, simulations using the utilization distribution framework for multiple evolving species have shown that the assemblage will organize itself into groups of species with high niche overlap distantly separated from other such groups, even when the resource axis is continuous. In his 2007 book, Robert May proposed a likely reason: species will tend to avoid competition with two species simultaneously and so will evolve to be more similar to one of the two adjacent to it.

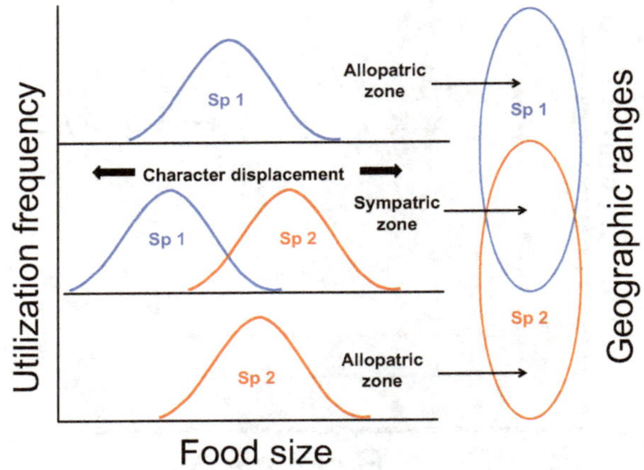

FIGURE 6 Character displacement of food size for two species with similar resource requirements. Food size utilizations are similar where each species occurs by itself (allopatric zone), but they diverge under selection where the geographic distributions overlap (sympatric zone).

Character displacement is now well established and provides one possible mechanism for the occurrence of limiting similarity. However, limiting similarity does not necessarily indicate that niche differences evolved where both species now coexist. Differences could have evolved earlier and elsewhere and then become evolutionarily static because of conflicting selection in different parts of a species' geographic distribution. Later, the two species were able to coexist when conditions allowed one to invade the range of the other. Daniel Janzen termed this phenomenon "ecological fitting," and in such a case, static ecological models are more appropriate for exploring niche overlap.

NICHE OVERLAP AND DIVERSITY

The concept of niche overlap developed in parallel with efforts to explain the number of species that can coexist in an ecological community. If there are limits to similarity

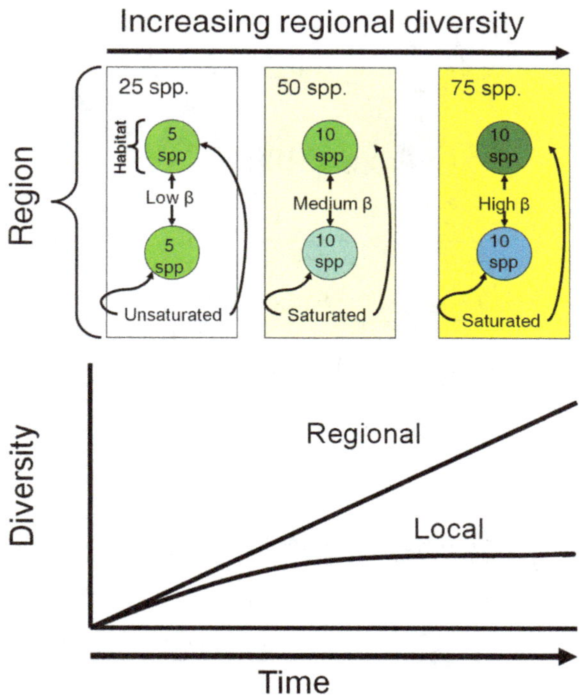

FIGURE 7 Classical view of the way in which niche space limits diversity in local communities. Diversity in a region increases through time and local diversity also increases for a time, but then niche spaces become saturated due to limiting similarity. Regional diversity can continue to increase, but only if composition among communities becomes more different. β measures the degree of dissimilarity among communities.

at competitive equilibrium, then such a number is limited by the size of the niche space; the lower the niche overlap, the fewer the species that can fit into a community. In the classical view exemplified by Robert MacArthur in 1965, diversity within a geographic region increases through evolutionary time via speciation and dispersal, but communities quickly become saturated with species due to liming similarity in finite niche space. Further increases in regional diversity are only possible because differences in species composition among habitats within the region increase (Fig. 7). Evidence for niche differences among species is persuasive, but documenting the relevance of niche differences to species diversity has been problematical.

One way to test whether limiting similarity constrains the number of coexisting species is to determine whether niche differences in the community are more even than in a random selection of species from the regional pool. Such a pattern has been called "community-wide character displacement," although it can be generated by ecological fitting as well as evolutionary character displacement. Data supporting community-wide character displacement are inconsistent; coexisting species were sometimes more different than random, sometimes more similar than random, and sometimes not different from random. It remains controversial whether niche differences are required for species to coexist locally. There are many cases where species coexist even though there appear to be few limiting factors and thus few opportunities for reducing niche overlap. For example, communities of tropical forest trees are rich even though they seem to compete for only a few limiting resources (water, light, a modest number of nutrients). How can such contradictory results be explained? Some possibilities are discussed in the following sections.

NICHE OVERLAP IN NONEQUILIBRIUM CONDITIONS

Classical niche overlap models assume that species assemblages normally exist at competitive equilibrium. In light of strong evidence for population fluctuations generated either by internal dynamics (e.g., time lags, limit cycles, and the like) or via external environmental fluctuations, competitive equilibrium is often not reached. In nonequilibrium conditions, resources are no longer limiting, species are no longer competing and competitive exclusion is avoided even when niches broadly overlap. Under these circumstances, niche overlap may have little relationship or even an inverse relationship to competitive intensity: the greater the niche overlap, the less intense is competition.

The idea that population fluctuations can enhance coexistence goes back at least to Hutchinson's 1961 "paradox of the plankton." Hutchinson asked how numerous plankton species could coexist in a lake since there were only a few well-mixed limiting resources. He proposed that climatic- and weather-generated population fluctuations kept densities below their competitive equilibria. Early modeling efforts showed that niche overlap could indeed be complete and coexistence was still possible as long as resource availability fluctuated. However, tradeoffs in resource utilization efficiency at different resource densities are required for stable coexistence. Indeed, a large body of theory confirms that species cannot coexist stably without such tradeoffs.

For example, there might be tradeoffs in growth rates in different seasons or tradeoffs in competitive ability without fluctuations and the ability to deal with variability. That is, one species might be more efficient at using resources when there are no fluctuations but cannot recover quickly after an environmentally unfavorable period, whereas the other species can recover quickly but cannot use resources as efficiently. These differences can be thought of as niche differences or stabilizing factors; again, intraspecific competition must in some sense exceed interspecific competition. There may not be limiting similarity in resource use per se, but there has to be some limit to similarity with respect to such tradeoffs or coexistence is not possible.

Chesson's temporal storage effect, an aspect of generalized competition theory, emphasizes the importance of tradeoffs to coexistence. If the environment fluctuates through time, if intraspecific competition increases with environmental quality, and if each species has a different optimal environment, then conditions exist where each species can increase when rare and coexistence is possible. However, population decline for each species in bad years must be buffered by some "storage mechanism" such as long-lived life stages, dormancy, or hibernation. In good years, a focal species experiences strong intraspecific competition. In bad years, it experiences a combination of poor conditions and high interspecific competition from other species for which these are good years. If, however, the focal species has a storage mechanism, it can buffer the effects of interspecific competition by, for example, becoming dormant. As a result, intraspecific competition in good years exceeds interspecific competition in bad years and the species can coexist (Fig. 8).

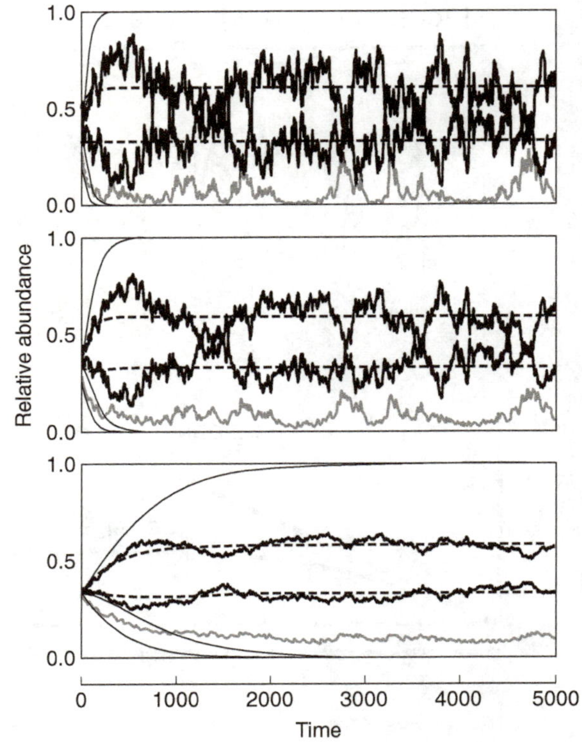

FIGURE 8 The role of environmental variation and the temporal storage effect on the coexistence of three species in a lottery competition model where species are competing for space. The smooth solid lines represent coexistence in a three-species Lotka–Volterra model modified for spatial competition for comparison. The dashed lines represent lottery competition with a temporal storage effect due to long-lived adults but without environmental variation. One species outcompetes the other two under these conditions. When environmental fluctuations are added, coexistence is possible and is enhanced by long-lived adults which smooth fluctuations in population density such that the equilibrium densities more precisely match the predictions of the equivalent Lotka–Volterra model. Coexistence requires that variances of population responses to environmental fluctuations, which are equivalent to intra-specific competition, exceed covariances, which are equivalent to inter-specific competition. SOURCE: Chesson, 2000, *Annual Review of Ecology and Systematics* 31: 343–366.

NICHE OVERLAP IN SPACE AND ENVIRONMENTAL HETEROGENEITY

Tradeoffs can occur spatially as well as temporally. For example, if only close neighbors compete and if inferior competitors are better dispersers (competition–colonization tradeoffs), then coexistence is possible even with broad niche overlap and identical environmental conditions from place to place. One example might be that clonal plants are superior competitors but inferior dispersers. They compete better because the parent can provide resources for the developing clone, but they can only disperse locally by fission. Conversely, seed plants can disperse seeds widely but are poorer competitors. Better dispersers can find unoccupied sites before they are excluded from others by the slower dispersing superior competitor. The better dispersal ability of the inferior competitor equalizes the fitnesses between the two species, and stable coexistence is possible even with broad niche overlap. The model assumes that the superior competitor cannot hold a site indefinitely but must be driven extinct periodically by disturbances.

Spatial heterogeneity in environmental conditions can also result in coexistence. Generalized competition theory is again informative here, but the storage effect in this framework is due to spatial rather than temporal environmental variation. For example, if there are two limiting resources that occur at different ratios from place to place and each species does better on one than the other, it can be shown that both can stably coexist via spatial segregation. In such a case, strong competition varies inversely with niche overlap, just the opposite pattern that would be predicted from models that assume resources are spatially well mixed. Niche conditions that vary spatially are sometimes called spatial niches. Similar to the temporal storage effect, coexistence requires greater competition in good vs. bad environments, different optimal environments for each species, and buffered population growth. Buffered population growth does not require a special storage mechanism but happens naturally because bad environments affect only the part of the population in the bad location and not the population as a whole. Better growth in good locations thus counteracts poor growth in the bad locations. Other trophic levels such as predators, parasites, or diseases can also enhance coexistence if each species responds differently to natural enemies.

In general, Hutchinson's paradox is now explainable by many models that add realism in the form of spatial heterogeneity and structure, temporal fluctuations, competition–colonization tradeoffs, and other trophic levels (e.g., predators). These models have shown that an infinite number of species can coexist locally in spite of broad niche overlaps. Some kind of tradeoff is essential for coexistence in all cases, and there is always a limit to similarity with respect to these tradeoffs.

Such results do not necessarily indicate that local diversity is unlimited. Most obviously, diversity cannot exceed the size of the pool of species available to colonize and persist in a local area. The factors that determine pool size (e.g., speciation, extinction, long-range dispersal, local adaptation) rather than ecological factors would then be setting local diversity. Less obviously, even though there is no theoretical limit on the number of coexisting species set by competition, there may be limits on the number of possible tradeoffs. Moreover, some newer models (e.g., stochastic niche theory) predict that addition of species becomes more difficult as local diversity increases, even though there is no hard limit to diversity.

NICHES VS. NEUTRALITY

Classical equilibrium models of niche theory predict that niches can overlap, but not completely or competitive exclusion will result. Stephen Hubbell's unified neutral theory of biogeography holds that niches can overlap completely and species can still persist together in ways that superficially resemble stable coexistence. Neutral theory is a special case of generalized competition theory where fitnesses and niches are identical. Stabilizing factors are thus absent and all species have equal competitive abilities. Under these conditions, competitive exclusion can be forestalled indefinitely because in assemblages of realistic sizes, exceedingly long time periods are required for identical species to be driven extinct via demographic stochasticity. The time periods can be so long that eventual local extinction can be balanced by speciation and immigration from the regional species pool. The theory was motivated by the observation that tropical tree species seem to overlap broadly along a limited number of niche dimensions (light availability, water availability, availability of a limited number of nutrients) and thus opportunities for niche differentiation were restricted.

Neutral theory explains some diversity and abundance patterns surprisingly well but not others. It is also highly sensitive to the assumption that all species have identical niches; if they differ even by a little, competitive exclusion will occur. Moreover, it cannot explain why niche differences are so ubiquitous, why even tropical forests show certain regularities in community structure over large distances, or why some communities return to equilibrium when they are perturbed; the latter is easily explained by niche theory. In the case of tropical forests, it may be difficult to find relevant niche differences because the number of niche dimensions studied is so small (three or four). In a 2008 paper, Jim Clark reminds us that real niche space has many niche dimensions, most of which are unknown and it is this high dimensionality that may explain coexistence in species-rich communities. Thus, niches may still be important for coexistence in real systems even if neutral models can predict some natural patterns. The controversy might be resolved by testing for stabilizing factors and fitness differences among species in the assemblage that would support the idea that niches are important without actually identifying the critical niche dimensions.

SEE ALSO THE FOLLOWING ARTICLES

Apparent Competition / Diversity Measures / Neutral Community Ecology / Niche Construction / Stochasticity, Demographic / Storage Effect / Two-Species Competition

FURTHER READING

Abrams, P. 1983. The theory of limiting similarity. *Annual Review of Ecology and Systematics* 14: 359–376.

Adler, P. B., J. HilleRisLambers, and J. M. Levine. 2007. A niche for neutrality. *Ecology Letters* 10: 95–104.

Chase, J. M., and M. A. Leibold. 2003. *Ecological niches: linking classical and contemporary approaches.* Chicago: University of Chicago Press.

Chesson, P. 2000. Mechanisms of maintenance of species diversity. *Annual Review of Ecology and Systematics* 31: 343–366.

Clark, J. S. 2008. Beyond neutral science. *Trends in Ecology & Evolution* 24: 8–15.

Hutchinson, G. E. 1978. *An Introduction to population ecology.* New Haven: Yale University Press.

May, R., and A. McLean, eds. 2007. *Theoretical Ecology: principles and applications,* 3rd ed. Oxford: Oxford University Press.

Pianka, E. R. 1974. Niche overlap and diffuse competition. *Proceedings of the National Academy of Sciences (USA)* 71: 2141–2145.

Schoener, T. W. 1989. The ecological niche. In J. M Cherrett, ed. *Ecological concepts: the contribution of ecology to and understanding of the natural world.* Oxford: Blackwell Scientific Publications.

Schoener, T. W. 2009. The niche. In S. Levin, ed. *Princeton guide to ecology.* Princeton: Princeton University Press.

Tilman, D. 2004. Niche tradeoffs, neutrality, and community structure: a stochastic theory of resource competition, invasion, and community assembly. *Proceedings of the National Academy of Science (USA)* 101: 10854–10861.

NICHOLSON–BAILEY HOST PARASITOID MODEL

CHERYL J. BRIGGS

University of California, Santa Barbara

Parasitoids are species (usually insects in the orders Hymenoptera or Diptera) that lay their eggs on or in the body of individuals of another species (their host). The juvenile parasitoid uses its host for food as it develops, usually killing the host in the process. Parasitoids are frequently used as natural enemies in the biological control of insect pests. In classical biological control, the goal is the establishment of a long-term, persistent interaction between pest and parasitoid, with the parasitoid suppressing the pest density and maintaining it at a level below an economically tolerable threshold. Much of the theory on host–parasitoid interactions, therefore, has focused on determining what mechanisms can lead to a stable equilibrium with a low host density. In 1935, A. J. Nicholson and V. A. Bailey developed a model to describe the interaction between an insect host attacked by a specialist parasitoid. The equilibrium in the Nicholson–Bailey model is unstable, with host and parasitoid population trajectories undergoing oscillations with an amplitude that increases through time. Many modifications of the model have been investigated to attempt to explain the stability of real host–parasitoid systems.

PARASITOID LIFE HISTORIES

Parasitoids display an amazing range of behaviors and life history strategies, and many of these behaviors and strategies have been incorporated into models. For example, some parasitoid lay a single egg on each host that it attacks; others lay multiple eggs per host. Some parasitoids kill their host immediately upon attack (idiobionts); others allow the host to continue to feed and grow for some time before eventually killing their host (koinobionts). Hymenopteran parasitoids have a haplo-diploid genetic system, with males resulting from unfertilized eggs and females from fertilized eggs, and in many species the female can choose the sex of her offspring on a particular host.

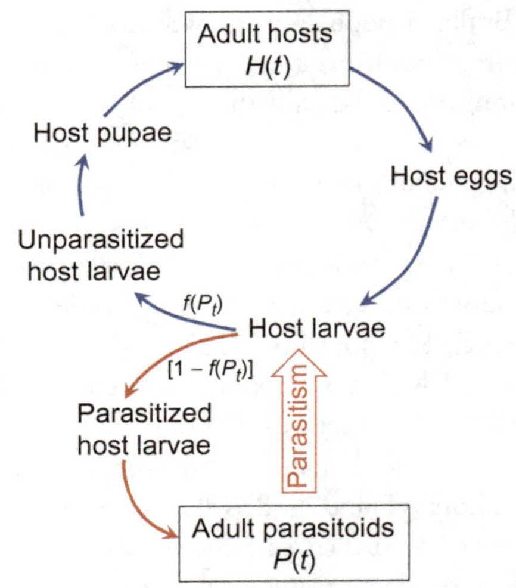

FIGURE 1 Diagram of the hypothetical host and parasitoid life cycles assumed by the Nicholson–Bailey model.

The Nicholson–Bailey model is a discrete-time model that best describes insect host–parasitoid interactions in temperate regions (Fig. 1) in which both host and parasitoid have a single, nonoverlapping generation per year (the term "parasitoid" was not in common usage in 1935, so Nicholson and Bailey instead used the term "entomophagous parasite"). For example, it describes a life cycle in which the adult host emerges in the spring from overwintering pupae, lays her eggs, and then dies. The host is vulnerable to attack by the parasitoid during only a limited portion of the juvenile host development (e.g., the larval stage), and the adult female parasitoid usually has a relatively short life span (days, weeks, or months) in which to search and attack host prior to death.

EQUATIONS

The equations describing the Nicholson–Bailey host parasitoid model are

$$\begin{aligned} H_{t+1} &= RH_t f(N_t, P_t) \\ P_{t+1} &= cH_t \left[1 - f(N_t, P_t)\right] \end{aligned} \qquad (1)$$

where H_t represents the density of female adult hosts in generation t and P_t represents the density of female adult parasitoids in generation t. Each adult female host produces F eggs that develop into larvae, which are vulnerable to parasitism by the adult female parasitoids. $f(N_t, P_t)$ is a function that describes the fraction of larval hosts that escape parasitism in generation t. A fraction s_H of the hosts that escape parasitism survives all of the other sources of mortality during host development, and m is the female fraction of surviving hosts. Therefore,

$R = mFs_H$ is the net reproductive rate of the host in the absence of the parasitoid, and if $R > 1$, the host population density will grow geometrically if the parasitoid is not present. $[1 - f(P_t)]$ is the fraction of larval hosts that do not escape parasitism. If the parasitoid lays e female eggs in each of these larval hosts, and a fraction s_P of the developing juvenile parasitoids survive to adulthood, then $c = eFs_P$. The Nicholson–Bailey model assumes that the only process limiting the host population is attack by the parasitoid and that the parasitoid's population is limited only by the single host species. No intraspecific competition or density dependence is included in the equation for either species.

The Nicholson–Bailey model assumes that each female parasitoid searches for hosts at random and can search over an area a in her lifetime, where a is the per capita searching efficiency, or Nicholson's "area of discovery." Each parasitoid can lay an unlimited number of eggs, such that as the density of hosts increases, the number of hosts attacked per parasitoid can continue to increase. However, as the density of parasitoids increases, an increasing fraction of hosts will be encountered more than once. The model assumes that repeat encounters with a host neither increase nor decrease the number of parasitoid offspring resulting from the host (that is, there is no successful superparasitism and repeat attack does not alter the survival of the juvenile parasitoids within the host) and that the parasitoid wastes no time or eggs by reencountering a host. These assumptions result in the fraction of hosts escaping parasitism being unaffected by host density but decreasing exponentially with increasing parasitoid density $f(N_t, P_t) = f(P_t) = \exp(-aP_t)$. This is the zero term (the fraction attacked zero times) of a Poisson distribution with rate aP_t.

EQUILIBRIUM AND STABILITY (OR LACK THEREOF)

The equilibrium host and parasitoid densities for the Nicholson–Bailey model are

$$\text{Host equilibrium:} \quad H^* = R\ln(R)/[ac(1 - R)],$$

$$\text{Parasitoid equilibrium:} \quad P^* = \ln(R)/a.$$

A positive host–parasitoid equilibrium exists ($H^* > 0$, $P^* > 0$) whenever the host net reproductive rate $R > 1$, that is, whenever the host population would increase in the absence of the parasitoid. The equilibrium host density decreases as the number of parasitoid eggs per host increases, and both host and parasitoid densities decrease with increasing parasitoid attack rate.

The host–parasitoid equilibrium is always unstable. The host and parasitoid dynamics are characterized by oscillations that increase in magnitude through time, with the parasitoid trajectory lagging behind that of the host (Fig. 2A). The diverging oscillations inevitably result in either the parasitoid population reaching vanishingly small densities followed by the host population growing without bounds, or the host population approaching zero followed by effective extinction of the parasitoid population. In this model, long-term persistence of the host–parasitoid system is not possible.

The Nicholson–Bailey model is the discrete-time analog of the Lotka–Volterra predator–prey model, making the same assumptions about density independence in all of the population parameters and about random searching by the consumer. The only difference is that the discrete-time structure in the Nicholson–Bailey model introduces a time lag in the dynamics.

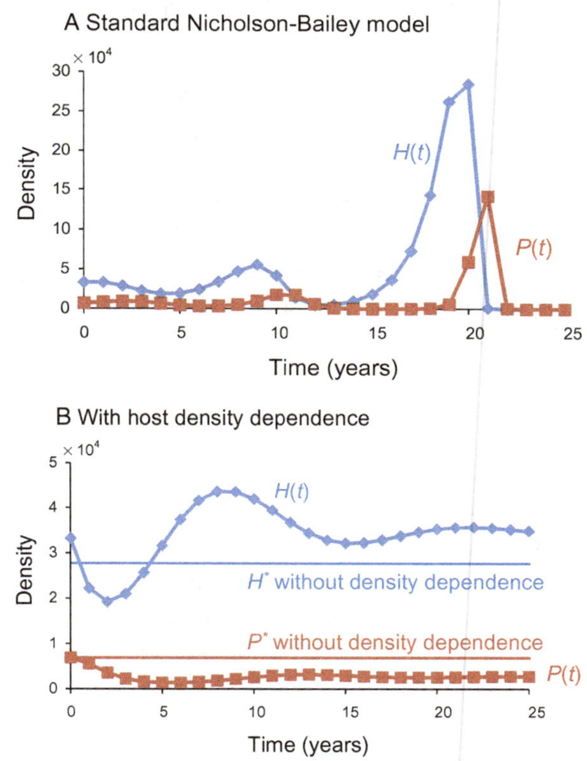

FIGURE 2 Examples of host-parasitoid dynamics predicted by the Nicholson-Bailey host-parasitoid model and its variants. (A) Standard Nicholson-Bailey model (Eq. 1) produces diverging oscillations. $R = 2$, $c = 0.5$, $a = 1 \times 10^{-4}$, with initial conditions: $H(0) = 1.2\ H^*$, $P(0) = P^*$. (B) Model with added host density-dependent fecundity (Eq. 2) can lead to a stable host-parasitoid equilibrium. $g(H_t) = (1 + \alpha H_t)^{-b}$, with $\alpha = 1 \times 10^{-5}$, $b = 1.4$, and all other parameters and initial conditions as in (A). Solid lines show the equilibrium host and parasitoid densities for the standard Nicholson-Bailey model without host density dependence ($g(H_t) = 1$).

This time lag has a destabilizing effect, which converts the neutrally stable equilibrium of the Lotka–Volterra model into the diverging oscillations of the Nicholson–Bailey model.

MODIFICATIONS SEARCHING FOR STABILITY

Parasitoid–host interactions do persist in the real world, and such systems are often remarkably stable, suggesting that the Nicholson–Bailey model is lacking or misrepresenting some important feature or mechanism present in real systems. This disparity between the predictions of the Nicholson–Bailey model and the persistence of real systems has spawned a large body of literature involving various modifications of the Nicholson–Bailey model incorporating either hypothetical processes or features observed in real parasitoid–host systems. Two of the most commonly added stabilizing features are described here. Many other mechanisms have been investigated, most notably dispersal and spatial structure (metapopulation dynamics) and various forms of host dependence in the parasitoid functional response.

Host Density Dependence

Not surprisingly, the host–parasitoid equilibrium can be stabilized by the addition of density dependence in the rate of increase. Many model variants have been investigated that differ in their assumptions about whether the density dependence affects host fecundity or survival, in a stage before or after parasitoid attack, and in the form of the density dependence.

For example, if density dependence occurs in host fecundity, then the equations become

$$\begin{aligned} H_{t+1} &= RH_t\, g(H_t) f(N_t, P_t), \\ P_{t+1} &= cH_t\, g(H_t)[1 - f(N_t, P_t)], \end{aligned} \tag{2}$$

where $g(H_t)$ is a function (with values between 0 and 1) describing the reduction in host fecundity due to density dependence. Many forms of $g(H_t)$ have been used, including, for example, an overcompensating Ricker function $g(H_t) = \exp(-\alpha H_t)$, or a flexible function $g(H_t) = (1 + \alpha H_t)^{-b}$, which can represent overcompensating ($b > 1$), perfectly compensating ($b = 1$), or undercompensating ($b < 1$) density dependence.

Figure 2B illustrates that the addition of host density dependence can stabilize the host–parasitoid equilibrium, but it also leads to an increase in the host equilibrium density. This is a general property of many stabilizing mechanisms in consumer–resource systems: increased stability also leads to an increase in the host (resource) equilibrium. If the host density dependence is too strong, the parasitoid cannot persist and the host population dynamics are determined by the form of the host density dependence (e.g., stable equilibrium, cycles, or chaos).

Skewed Risk of Parasitoid Attack

One of the most common modifications of the Nicholson–Bailey model is to alter the function $f(P_t)$, which describes the fraction of hosts that escapes parasitism, by replacing the zero term of a Poisson distribution $f(P_t) = \exp(-aP_t)$ with the zero term of a negative binomial distribution $f(P_t) = \left[1 + \frac{aP_t}{k}\right]^{-k}$. This model has been used to describe a number of different situations. Robert May derived this function for a situation in which hosts occur in patches, and within each generation the parasitoids are distributed across the patches according to a gamma distribution, but within each patch the parasitoids search and encounter hosts at random (as in the Nicholson–Bailey model; May 1978, *Journal of Animal Ecology* 50: 855–866). The parameter k is the clumping parameter of the negative binomial distribution. As k is reduced, the parasitoids are more and more aggregated, such that the parasitoid attacks are increasingly concentrated on a small fraction of the host population. As $k \to \infty$, the Poisson distribution, and the standard Nicholson–Bailey model, are recovered.

This model has usually been used to describe aggregation of parasitoid attack, independent of host density, but it can also describe other sources of variability in host susceptibility, e.g., due to differences in host defenses, host location, or length of time the host is exposed to attack.

The equations are

$$\begin{aligned} H_{t+1} &= RH_t\left[1 + \frac{aP_t}{k}\right]^{-k}, \\ P_{t+1} &= cH_t\left\{1 - \left[1 + \frac{aP_t}{k}\right]^{-k}\right\}. \end{aligned} \tag{3}$$

The equilibrium is stable if $k < 1$. Stability of the equilibrium in this model is determined by the coefficient of variation (CV) of the distribution of parasitoids across patches, and the criterion of $k < 1$ for a stable equilibrium translates into CV > 1 (or $\text{CV}^2 > 1$). Stability occurs when there is a strongly skewed distribution of attacks among hosts, with the parasitoids concentrating their attacks on a fraction of the host population. This results in density dependence in the parasitoid attack rate, with each parasitoid becoming less efficient as the parasitoid density increases. As k is reduced, and the parasitoids increasingly concentrate their attacks on a smaller fraction of the host population, the equilibrium host density increases, because an increasing fraction of the host population escapes parasitism. Therefore, once again, there is a tradeoff between stability and suppression of the host population.

A recent investigation of a general model of variability of risk across the host population (Singh et al., 2009, *Ecology* 90: 1679–1686) showed that the $CV^2 > 1$ rule holds only when the risk of attack follows a gamma distribution, or when the host net reproductive rate, R, is close to 1. They found that the shape of the distribution of risk is more important in determining the stability of the host–parasitoid equilibrium than is the variability in risk. In particular, for high values of R, stability requires a distribution with a mode of zero (i.e., the most common risk of attack is zero), and that declines sharply away from zero. Thus, stability of the host–parasitoid equilibrium resulting from variability in risk is associated with weak suppression of the host population.

INCLUSION OF WITHIN-GENERATION DYNAMICS

There have been numerous modifications of the function that describes the fraction of hosts that escape parasitism, $f(N_t, P_t)$, to include various behaviors of parasitoids as they search for hosts. One major limitation of investigating such behaviors in a discrete-time model is that the equations are based on the densities at only a single point each generation, even though the behaviors of the parasitoids in searching and attacking hosts will alter the density of the host within the generation. The function $f(N_t, P_t)$ needs to capture the net effects of these changes in density integrated over the entire generation. The form of this function used in models is frequently chosen in a phenomenological way. A number of authors have used hybrid discrete/continuous-time models or semi-discrete models to represent the discrete, nonoverlapping generations of temperate insect systems, but continuous depletion of hosts within each generation.

SEE ALSO THE FOLLOWING ARTICLES

Difference Equations / Population Ecology / Predator–Prey Models / Stability Analysis / Top-Down Control

FURTHER READING

Godfray, H. C. J. 1994. *Parasitoids: behavioral and evolutionary ecology.* Princeton: Princeton University Press.

Hassell, M. P. 1978. *The dynamics of arthropod predator-prey systems.* Princeton: Princeton University Press.

Hassell, M. P. 2000. *The spatial and temporal dynamics of host-parasitoid interactions.* Oxford: Oxford University Press.

May, R. M. 1978. Host parasitoid systems in patchy environments: a phenomenological model. *Journal of Animal Ecology* 47: 833–844.

Murdoch, W. W., C. J. Briggs, and R. M. Nisbet. 2003. Consumer-resource dynamics. Princeton: Princeton University Press.

Nicholson, A. J., and V. A. Bailey. 1935. The balance of animal populations—Part I. *Proceedings of the Zoological Society of London* 105: 551–598.

Singh, A., W. W. Murdoch, and R. M. Nisbet. 2009. Skewed attacks, stability, and host suppression. *Ecology* 90: 1679–1686.

Singh, A., and R. M. Nisbet. 2007. Semi-discrete host-parasitoid models. *Journal of Theoretical Biology* 247: 733–742.

NONDIMENSIONALIZATION

ROGER M. NISBET

University of California, Santa Barbara

Most quantities of interest in ecology have units. Yet ecological models describe processes that are unaffected by the units in which individual scientists choose to measure or describe them. Recognizing this provides the theorist with tools that can simplify mathematical analyses and provide useful insight on ecological dynamics.

UNITS AND DIMENSIONS IN ECOLOGICAL MODELS

Ecological models typically describe relationships between quantities with *units*, such as population density[1] (# m^{-2} or # m^{-3}), irradiance (W m^{-2}), nutrient concentration (mol L^{-1}), time (hours or days). Any meaningful equation describing ecological processes should have consistent units. We can only add or subtract quantities with the same units, and the quantities on the left and right sides of an equation should have identical units.

The fundamental quantities in a model, from whose units all others are derived, are often called dimensions. The word dimension in this context has a completely different meaning from its use in geometry or in describing a dynamical system where, for example, a system of N first-order differential equations is described as N-dimensional. In classical physical systems, there are only four fundamental dimensions (mass, length, time, and electric charge) from which all others are derived, but in ecological models, a larger number of effectively independent dimensions may be invoked—for example,

[1] In this entry the symbol # represents "number" as in "number of individuals" or "number of food items."

masses of different elements or compounds, population of different species, and so on. Treating population number of a species as a dimension may seem surprising, but it simply reflects the well-worn adage that we cannot "add apples and oranges."

Models sometimes involve dimensionless quantities. These may be pure numbers (e.g., the number of eggs in a clutch) or combinations of quantities where the units cancel out. An obvious example of the latter would be the quantity π, which is the ratio of two lengths. Much more complicated examples occur in models involving fluid dynamics. For example, formulae for the rate of nutrient absorption by benthic organisms in a river or for CO_2 uptake by a leaf in the wind involve dimensionless quantities whose calculation involves sophisticated fluid dynamics.

In many theoretical ecology papers, the units for model parameters are unfortunately not stated explicitly, but they can be inferred from the model equations. For example, the type II functional response of a terrestrial animal relating feeding rate I (# food items/time) to food density F (# food items/area) in the environment can be written in the form

$$I = \frac{aF}{H + F}. \qquad (1)$$

This equation has two parameters: a and H. The units of H and F must be the same as they are added; thus, H has units of food items/area. Since the ratio $\frac{F}{H+F}$ is dimensionless, the parameter a must have the same units as I, implying that a has units # food items/time.

Particular care with units is required when using mathematical functions such as powers, exponentials, or logarithms. The only safe way to ensure consistency is to recognize that only dimensionless quantities may be meaningfully raised to noninteger powers or used as the argument for logarithms or exponentials. In practice, many authors ignore this requirement, especially with equations obtained by fitting empirical data, so the theorist commonly has to rewrite the equations in the appropriate form. As an example, consider the commonly invoked empirical allometric relationship describing the interspecific variation in respiration rate R with body mass M:

$$R = bM^q, \qquad (2)$$

where the exponent q is dimensionless. Such a relationship may have been derived by plotting R against log M and fitting a straight line; the slope q and b would be derived from the intercept. Suppose the parameters were estimated from data where R was quoted in units mol O^2 min^{-1} and M in g. Then suppose we want to convert to a form where M is expressed in kg. The safe way to do this is to be sufficiently disciplined to only allow dimensionless quantities to be raised to powers (or used as the argument for logarithms or exponentials). We define M_s to be the unit used to derive Equation 2 and rewrite Equation 2 in the form

$$R = b\left(\frac{M}{M_s}\right)^q. \qquad (3)$$

The constant b now has the same units as R and conversion to other units if required for modeling is straightforward.

NONDIMENSIONALIZATION

The relationships between quantities in a real ecological system cannot be affected by the units chosen by ecologists to characterize these quantities. We may write down a prey–predator equation describing the interaction of voles and weasels in which population densities are expressed in # km^{-2} and time is expressed in days, but the interactions between real voles and weasels in nature do not recognize our choice of units.

Since nature does not "know" our choice of units, it should be possible to express the defining equations of an ecological model in a form that does not involve quantities with units. Turning equations into such a form is often called *nondimensionalization*. Although in principle this often seems a simple exercise, it is remarkably easy to make subtle errors through taking short cuts. However, nondimensionalization can be performed in a safe and rigorous, though sometimes tedious, way by following steps similar to those used with the allometric equation above. We define a scale, or base unit, for each quantity in the model equations and then use the (dimensionless) ratio of each variable to its chosen scale in our dynamic description.

The process is more easily understood through an example involving a differential equation. Consider the von Bertalanffy equation commonly used to describe the growth of a fish. If $L(t)$ denotes the length of the fish at time t, then growth is described by the differential equation

$$\frac{dL}{dt} = r_B(L_\infty - L). \qquad (4)$$

This equation involves two parameters: L_∞ is the maximum possible length for the fish, and r_B is a parameter (known as the von Betalanffy growth rate) with dimension $time^{-1}$. To solve this equation, we need to specify three quantities and their units: the two parameters and the initial length $L(0)$.

To nondimensionalize, we formally introduce two further quantities, L_s and t_s, the *scales* of length and time, respectively. Choosing these scales comes later—at this point we are simply doing algebra. We define dimensionless variables

$$l = L/L_s \text{ and } \tau = t/t_s \text{ implying } L = L_s l \text{ and } t = t_s\tau. \quad (5)$$

Substitution in Equation 4 followed by a few steps of algebra leads to a differential equation for the dimensionless variable:

$$\frac{dl}{d\tau} = (r_b t_s)\left(\left(\frac{L_\infty}{L_s}\right) - l\right). \quad (6)$$

A solution of the dimensionless Equation 6 requires specification of three dimensionless quantities $r_b t_s, L_\infty/L_s$ and $L(0)/L_s$.

This algebraic manipulation might appear to have made analysis more, rather than less, complicated. The "black art" of nondimensionalization lies in recognizing that we are free to choose any values we like for the scales. Thus, if we choose

$$t_s = 1/r_B \quad \text{and} \quad L_s = L_\infty, \quad (7)$$

Equation 6 collapses to the simple form

$$\frac{dl}{d\tau} = 1 - l. \quad (8)$$

The solution of Equation 8 is determined by *only one* dimensionless quantity: $l(0) = L(0)/L_\infty$. So an ecologist wanting to understand the mathematical aspects of von Bertalanffy growth need only study how the solution of Equation 8 depends on initial conditions.

The simplification is not mathematical magic. Algebraic manipulation took us as far as Equation 6, but the reduction from 3 to 1 in the number of quantities controlling the dynamics involved judicious choice of scales. The simplification arose because the choice (Eq. 5) reflected in some sense "natural" scales for the problem. The length scale is the maximum size for a particular fish, and the time scale is related to the speed at which the fish's length approaches the maximum value.

HOW MUCH SIMPLIFICATION CAN BE ACHIEVED?

The analysis of von Bertalannfy growth illustrates well the mechanics of nondimensionalization, but the model is too parameter-sparse to give insight on the level of simplification that can be achieved. To illustrate this, now consider a slightly more complex model, the Rosenzweig–MacArthur model of prey–predator interactions. In standard (dimensional) form, this model has two differential equations describing the changes in prey (F) and predator (C) populations:

$$\begin{aligned}\frac{dF}{dt} &= rF\left(1 - \frac{F}{K}\right) - \frac{aFC}{H+F}\\ \frac{dC}{dt} &= \frac{\varepsilon aFC}{H+F} - \delta C.\end{aligned} \quad (9)$$

The model has six parameters: r (time^{-1}) and K (# prey) are parameters in the logistic equation, a (# prey/time/# predators)) and H (# prey) are parameters in the type II functional response (see Eq. 1), ε (# predators/# prey) is a conversion efficiency, and δ (time^{-1}) is the per capita death rate of the predators. There is no unique way to choose any of the scales, as is evident from the fact that three of the parameters involve time. One possible choice is to choose scales that emphasize the prey population, by setting

$$t_s = r^{-1}, \qquad F_s = K, \qquad C_s = \varepsilon K. \quad (10)$$

Here, the time scale is set by the maximum per capita growth rate of the prey (to be precise, it is around 1.44 times the doubling time of a small prey population in the absence of predators), and the scale of prey population is its carrying capacity. The scale for predators takes account of the conversion efficiency of prey to new predators, with C_s representing the number of new predators produced by consuming K prey.

We define dimensionless variables

$$\tau = t/t_s, \qquad f = F/F_s, \qquad c = C/C_s \quad (11)$$

in terms of which Equation 9 can be shown, with a little laborious algebra, to become

$$\begin{aligned}\frac{df}{d\tau} &= f(1-f) - \frac{Pfc}{Q+f}\\ \frac{dc}{d\tau} &= \frac{Pfc}{Q+f} - Rc\end{aligned} \quad \text{with}\begin{cases} P = \frac{\varepsilon a}{r} \\ Q = \frac{F_h}{K}. \\ R = \frac{\delta}{r}\end{cases} \quad (12)$$

In this example, it thus turns out that a model with six parameters and three dimensions (time, # food items, # consumers) can be expressed as a dimensionless system with three controlling parameter groups (P, Q, R). This is a particular case of a fairly general property; broadly speaking, we expect that in deterministic models with m parameters and k independent dimensions, it will commonly be possible to obtain a dimensionless model with $m - k$ dimensionless parameter groups. This is *not* a completely general result—there are restrictions, stated explicitly for physical systems in Buckingham's Π theorem—but it is a valuable rule of thumb.

RETURNING TO DIMENSIONAL FORMS

There is one important payoff from the systematic, some might say pedantic, approach to nondimensionalization advocated in this article. This is the ease with which it is possible to work back to quantities with units, as is normally required when comparing model predictions with empirical data. The value of any state variable is simply the product of the dimensionless quantity and the scale.

DIMENSIONAL ANALYSIS AS AN AID TO INTUITION

For models with a small number of parameters, there is a very limited choice of natural scales. In such models, insight into possible dynamics can be deduced from the dimensionless equations without solving them explicitly. Again we illustrate through an example: the Fisher equation that has been used to describe biological invasions. The model assumes local logistic population growth and random (diffusive) movement. In one dimension, this leads to the partial differential equation

$$\frac{\partial N}{\partial t} = rN\left(1 - \frac{N}{K}\right) + D\frac{\partial^2 N}{\partial x^2}, \tag{13}$$

where N denotes population density (#/length), r (time^{-1}) and K (#/length) are respectively intrinsic growth rate and carrying capacity, and D (length^2/time) is a diffusion coefficient. We choose base units for population density, length, and time, convenient choices being

$$t_s = r^{-1}, \qquad x_s = \sqrt{\frac{D}{r}}, \qquad N_s = K, \tag{14}$$

and we define dimensionless variables

$$n = N/N_s, \qquad u = x/x_s, \qquad \tau = t/t_s. \tag{15}$$

The nondimensional form of the Fisher equation is then

$$\frac{\partial u}{\partial \tau} = u(1 - u) + \frac{\partial^2 u}{\partial u^2}, \tag{16}$$

which, like our equation for von Bertallanfy growth, has no parameters. Its solution is determined solely by the scaled initial population distribution.

Suppose we are interested in "traveling wave" solutions to the equation describing an invasion advancing at a constant speed that is independent of initial conditions. As Equation 16 has no additional parameters, the nondimensional invasion speed must be some pure number that we call A. In dimensional form, the reasoning of the section "Returning to Dimensional Forms," above, tells us that the invasion speed v must take the form

$$v = A\frac{x_s}{t_s} = Ar\sqrt{\frac{D}{r}} = A\sqrt{rD}. \tag{17}$$

Thus, dimensional considerations alone reveal how the model parameters impact the invasion speed, and in particular that it is unaffected by the invader's carrying capacity.

A detailed analysis of the model is required to evaluate the constant (it turns out that $A = 2$), but for many ecological questions, knowledge of this constant is unnecessary. For example, if environmental change increased the intrinsic growth rate of an invasive species by 30%, the asymptotic invasion speed would increase by a factor $\sqrt{1.3} \approx 1.14$, i.e., by 14%.

CAUTIONARY REMARKS

The preceding examples demonstrate a systematic approach to nondimensionalization and some of the benefits that ensue. Yet there are also potential pitfalls, both technical and ecological.

Technical

The above examples involve dynamic models in continuous time. Many ecological models are couched in a discrete time framework where an update rule relates the state of a system at one time to its state one time step later. The update interval is frequently one unit, thus already setting the natural "scale" of time. Consequently, considerable care is required in relating parameters in discrete- and continuous-time representations of the same process. The problem is sometimes exacerbated by sloppy use of the word "rate" in describing discrete-time models.

Related to the above problem is an issue beyond the scope of this article—subtleties in nondimensionalizing stochastic models, especially those involving so-called white noise. Discrete-time stochastic models frequently include a random component whose mean and variance are specified and whose values at successive time steps are uncorrelated. The corresponding continuous-time models are derived, in principle, by considering the limit of an infinitesimal time step, but there are many important mathematical complications. Again, the literature often contains some unfortunate terminology; variances of white noise terms in continuous-time models are likely to have dimensions involving time in addition to the quantity that is varying randomly.

Ecological

Many applications of ecological theory focus on the effects of environmental change modeled as changes in parameter values in a dynamic model. For example, sublethal effects of contaminants may lead to changes in both parameters in the von Bertalanffy Equation 4 and hence in the choice of scales. When the natural scales in a model are changing over time, nondimensionalization may obfuscate rather than simplify interpretation of model dynamics.

In summary, care is required when working with nondimensionalized models, and the approach is not universally useful. Nevertheless, nondimensionalization is a powerful technique that frequently exposes key properties of dynamic models.

OTHER APPROACHES

The approach in this entry follows Gurney and Nisbet (1998). Okubo (1980) has deeper discussion of dimensional analysis and nondimensionalization in spatially explicit models. Buckingham's theorem is discussed in many engineering texts, e.g., Pankhurst (1964). There is a lengthy introduction to stochastic models in continuous time that pays particular attention to units and dimensions in Chapters 6 and 7 of Nisbet and Gurney (2003). Segel (1972) offers a lucid, mathematically oriented introduction to dimensional analysis and scaling that shows how nondimensionalization can facilitate intuitively motivated approximations to solutions of dynamic equations.

SEE ALSO THE FOLLOWING ARTICLES

Allometry and Growth / Hydrodynamics / Ordinary Differential Equations / Predator–Prey Models / Stability Analysis

REFERENCES

Gurney, W. S. C., and R.M. Nisbet. 1998. *Ecological dynamics.* New York: Oxford University Press.

Nisbet, R.M., and W. S. C. Gurney. 2003. *Modeling fluctuating populations.* Princeton: Blackburn Press.

Okubo, A. 1980. *Diffusion and ecological problems: mathematical models.* Berlin: Springer-Verlag.

Pankhurst, R.C. 1964. *Dimensional analysis and scale factors.* London: Chapman and Hall.

Segel, L.A. 1972. Simplification and scaling. *SIAM Review* 14(4): 547–571.

NPZ MODELS

PETER J. S. FRANKS

University of California, San Diego

NPZ (nutrient–phytoplankton–zooplankton) models are a fundamental tool to explore and understand the dynamics of marine planktonic ecosystems. Usually using nitrogen as a measure of the concentrations of the state variables, these models vary in complexity from simple three-compartment (NPZ) models to multi-compartment models representing different size classes of plankton, different functional types of plankton, and/or different pools of organic and inorganic nutrients. NPZ models are often coupled to models of the physics of the ocean to gain understanding into the potential responses of planktonic ecosystems to physical forcings such as wind-driven upwelling, eddies, and ocean-basin scale circulations. Because biological dynamics exercise significant controls on nutrient cycling in the ocean, NPZ models often form the nucleus of biogeochemical models used to better quantify the vertical and horizontal fluxes of important elements in the ocean such as nitrogen, iron, silica, phosphorus, and oxygen. New applications of NPZ models are providing insights into the physiological, ecological, and environmental controls on the biogeographic distributions of planktonic taxa in the ocean.

TERMINOLOGY

NPZ models are mathematical representations of hypotheses concerning the processes controlling planktonic ecosystems. The model equations contain terms representing the various dynamics (e.g., growth, grazing, mortality, and so on) that would alter the state variable (dissolved nutrient, phytoplankton, zooplankton, and so on). A general NPZ model with the three state variables nutrients (N), phytoplankton (P) and zooplankton (Z) might appear as

$$\begin{aligned}\frac{dP}{dt} &= f(I)g(N)P - h(P)Z - i(P)P,\\ \frac{dZ}{dt} &= \gamma h(P)Z - j(Z)Z, \qquad (1)\\ \frac{dN}{dt} &= -f(I)g(N)P + (1-\gamma)h(P)Z + i(P)P + j(Z)Z.\end{aligned}$$

The state variables are linked by the transfer functions $f(I)$, $g(N)$, $h(P)$, $i(P)$, and $j(Z)$. These transfer functions represent the functional responses of phytoplankton growth to irradiance, $f(I)$, and nutrients, $g(N)$; losses of phytoplankton to zooplankton grazing, $h(P)$, and respiration/mortality, $i(P)$; and losses of zooplankton to respiration/mortality, $j(Z)$. The shapes of the transfer functions will be determined by their mathematical form and the parameters of these functional forms. These parameters will include values such as the maximum phytoplankton growth rate and the maximum zooplankton ingestion rate. The parameter γ represents the fraction of ingested food that is assimilated by the zooplankton. In this particular case, the sum of dN/dt, dP/dt, and dZ/dt is zero, indicating that the total amount of nutrient in the system, $N + P + Z$ is conserved.

HISTORY

NPZ models originated as modifications of Lotka–Volterra-type models, but they used nitrogen as the measure of biomass in the state variables rather than numbers of organisms. Gordon Riley was among the first to apply NPZ models to oceanic systems, studying the controls on phytoplankton and zooplankton cycles on Georges Bank in the Northwest Atlantic. Riley's analyses were limited by the computing power available to him—it was a considerable effort to integrate nonlinear differential equations back in the 1940s. John Steele was one of the more important contributors to the formulation and analysis of NPZ models between 1960 and 1990; his collaboration with Bruce Frost in 1977 produced what is arguably the best-parameterized NPZ model that exists today. Much of the recent work on NPZ models has been concentrated on choosing the appropriate level of biological aggregation for the models: can the system being studied be represented by one phytoplankter and one zooplankter, or must the dynamics include separate representations of two or more phytoplankton types (distinguished by size or nutrient requirements), zooplankton types (distinguished by their food preferences, life histories, or ingestion rates), dissolved nutrient types (various oxidation states of inorganic nitrogen, organic nitrogen, and other potentially limiting nutrients such as iron or silicate), and detrital pools (labile and nonlabile particulate material, potentially of various sizes)? An important tradeoff that must always be considered is that the more detailed the model, the more parameters that must be specified and the less likely there will be data to parameterize and test the model.

CONSTRUCTION

The first step in building an NPZ model is choosing the state variables. This choice will usually be determined by the question being asked. For some questions, a simple three-compartment model would suffice. Other questions may require a distinction between different phytoplankton or zooplankton types, or the inclusion of heterotrophic bacteria or detrital pools (Fig. 1). Planktonic types are often distinguished by taxa (e.g., cyanobacteria, dinoflagellates, diatoms, coccolithophorids, all of which have different nutrient requirements) or size (e.g., protistan microzooplankton vs. metazoan crustaceans). For many biogeochemical questions it is necessary to distinguish several dissolved nutrient types (e.g., nitrate, ammonium, silicate, iron) to reproduce the observed patterns of elemental fluxes.

Once the state variables have been chosen, it is necessary to decide how they are linked to each other. These linkages are the transfer functions, and their shape and connectivity determines the behavior of the NPZ model. Many biological processes such as nutrient uptake rate and grazing rate tend to show a nonlinear saturating response

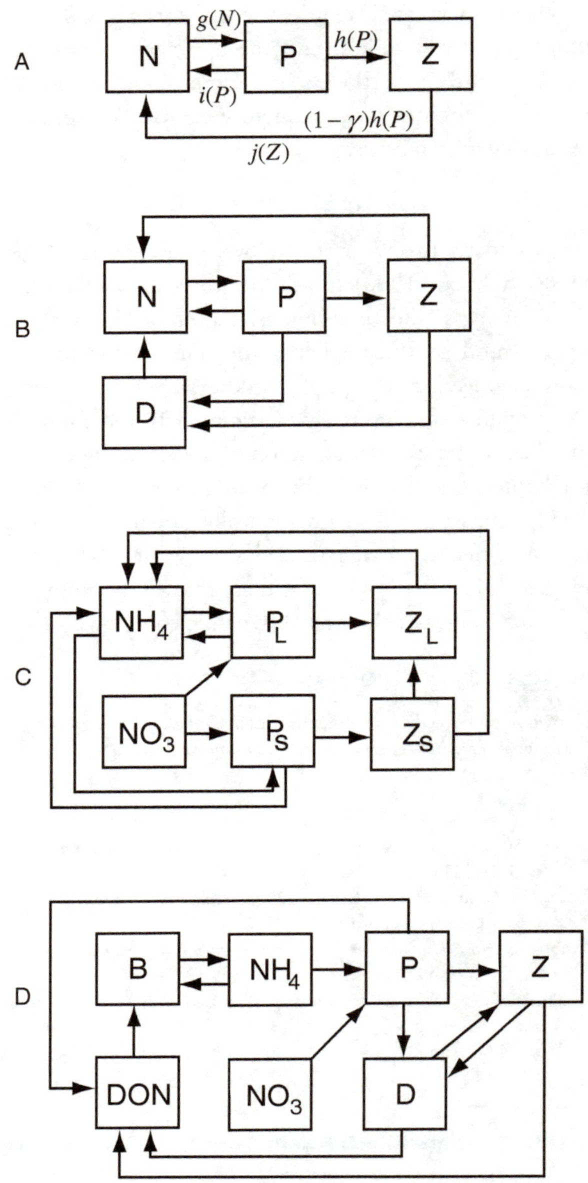

FIGURE 1 Some possible structures of NPZ models. (A) The basic nutrient–phytoplankton–zooplankton (NPZ) model. The functional forms correspond to Equation 1. (B) An NPZ model with a detritus (D) state variable. (C) A simple size-structured NPZ model with two dissolved nutrients (ammonium and nitrate), large and small phytoplankton, and large and small zooplankton. (D) A multicompartment NPZ model that includes detritus, bacteria (B), and dissolved organic nitrogen (DON), as well as two dissolved nutrients.

of the rate to increases in the "substrate" (dissolved nutrient or prey) concentration. Common examples of these saturating responses include Michaelis–Menten enzyme kinetics, and the Ivlev, Holling Type II, and Holling Type III grazing functions (Table 1). Some processes that are commonly included in models, such as phytoplankton death/respiration ($i(P)$ and $j(Z)$ in Equation 1) are not

TABLE 1

Examples of functional forms used in many NPZ models (see Eq. 1 for the appropriate transfer function)

Functional form	Description
Some functional forms for $f(I)$, the phytoplankton response to irradiance I. In some implementations, a parameter P_{max}, the maximal photosynthetic rate, will be multiplied by the functional forms below. Parameters: I_o	
$\frac{I}{I_o}$	Linear response
$\frac{I}{I_o + I}$	Saturating response
$1 - \exp\left(-\frac{I}{I_o}\right)$	Saturating response
$\tanh\left(-\frac{I}{I_o}\right)$	Saturating response
$\frac{I}{I_o}\exp\left(1 - \frac{I}{I_o}\right)$	Saturating and photo-inhibiting response. Parameter I_o determines the irradiance at the photosynthesis maximum.
Some common functional forms for $g(N)$, the phytoplankton nutrient uptake. Parameters: V_m, the maximum uptake rate; k_s, the half saturation constant, k_Q, the minimum cell nutrient quota.	
$\frac{V_m}{K_s + N}$	Michaelis–Menten uptake: saturating response. Two parameters, V_m and k_s.
$V_m \min(\mu_1, \mu_m)$	Uptake rate determined by the process most limiting to growth (μ): light or nutrients. Potential growth rates usually calculated using Michaelis–Menten uptake, and a functional form from Table 1.
$V_m\left(1 - \frac{k_Q}{Q}\right)\frac{dQ}{dt} = V_m I - \mu_m(Q - k_Q)I$	Luxury uptake (Droop cell quota model): nutrients stored in an internal pool Q, then used up through growth. Requires an equation for nutrient uptake from Q. Minimum cell quota for Q is k_Q.
Some of the functional forms used for $h(P)$, the zooplankton grazing on phytoplankton. Parameters: R_m, the maximum grazing rate; P_o, a grazing threshold; λ, sets the slope of the grazing curve. Note that the units for the parameters R_m and λ are not the same in every case.	
$R_m P$	Linear
$\min[cP, R_m]$	Bilinear with saturation at R_m
$\frac{R_m(P - P_o)}{\lambda + P - P_o}$	Saturating, with lower feeding threshold P_o
$\frac{R_m P^n}{\lambda + P^n}, n = 1,2$	Saturating, with curvature determined by n
$R_m[1 - \exp(-\lambda P)]$	Saturating (Ivlev)
$R_m[1 - \exp(-\lambda(P - P_o))]$	Saturating with feeding threshold P_o
$R_m \lambda P[1 - \exp(-\lambda P)]$	Acclimating to ambient food—relatively linear at high P
Some of the functional forms for $i(P)$, the phytoplankton death rate, and $j(Z)$, the zooplankton death rate. Parameters: ε, δ, sets the rate (different units for different formulations).	
Form of $i(P)$	Description
ε	Constant fraction of P
εP	Quadratic (nonlinear)—density-dependent
Form of $j(Z)$	Description
δ	Constant fraction of Z
δZ	Quadratic (nonlinear)—density-dependent
$\frac{\delta Z}{b + Z}$	Nonlinear, density-dependent but saturating rate (constant fraction of Z) at high zooplankton densities

well defined physiologically or ecologically, and the choice of the transfer functional form tends to be quite arbitrary.

A complication in choosing the transfer functions is determining how they will interact. For example, phytoplankton growth is a function of irradiance and dissolved nutrients. Which one determines the actual growth rate? Two methods are generally employed for these interactions: (1) the minimum of the two growth rates determines the final rate, or (2) the rates are multiplicative. The first method tends to be more physiologically accurate, but it is somewhat intractable when attempting to obtain analytical solutions to the equations. The second method is analytically tractable, but it has the disadvantage that the final realized growth rate will almost always be less than the maximum possible. For example, if irradiance $f(I)$ allows the phytoplankton to grow at 90% of their maximum, and nutrients $g(N)$ also allow them to grow at 90% of their maximum, the realized growth rate will be 90% of the maximum using method 1 ($min\{f(I), g(N)\} = 0.90$), but only 81% using method 2 ($f(I)g(N) = 0.81$).

The choice of transfer functions completely determines the potential range of behaviors of the model. This includes whether the model has stable steady states, limit cycles, bifurcations, and/or chaotic fluctuations. Whether the model exhibits any particular behavior is determined in large part (though not exclusively) by the model parameters. Choosing the values for the parameters of an NPZ model is a fundamentally important task, since the model results depend so heavily on the values of these numbers. Unfortunately, model parameterization is often done by adjusting the parameters so that the model fits the data, rather than choosing "realistic" values, a priori. To some extent, this is a function of the lack of data to constrain the parameter values, though sometimes the choice of parameters reflects a lack of understanding of what the model is actually modeling. Many of the transfer functions to be parameterized do not represent individual physiological processes (by which they are usually justified), but rather community-level aggregate responses, and are therefore difficult to parameterize rationally: there are often no measurements quantifying the dynamics included in the models.

STABILITY

NPZ models are known for their pronounced temporal oscillations for certain functional forms and parameter ranges. The presence of these oscillations is strongly dependent upon the shape of the grazing function $h(P)$ and the zooplankton loss term $j(Z)$. Using linear stability analyses it can be shown that grazing functions that have a feeding threshold (grazing is small or zero until $P > P_o$) or are concave upward (in contrast to most of the conventional functional forms; Table 1) will contribute to NPZ models being stable. A zooplankton loss term that is quadratic in Z (Table 1) may also aid in model stability.

Whether the oscillations exhibited by NPZ models are realistic or not is debatable and may depend largely on what is being modeled. For example, the chlorophyll *a* concentration in the ocean (representative of phytoplankton biomass P) is relatively stable, varying by $<2\times$ in much of the ocean over a year. However, the P state variable in many NPZ models often varies by many orders of magnitude over tens of days. If the NPZ model is being used to reproduce fluctuations in the total phytoplanktonic biomass, these model fluctuations are quite unrealistic. However, it is possible that certain taxa within the phytoplankton community will show these fluctuations. In the ocean, these changes in biomass of a given taxon must be offset by compensating changes of other taxa to keep the total biomass relatively constant. The processes underlying this apparent functional stability of the planktonic ecosystem are very poorly understood.

APPLICATION

NPZ models are generally used for theoretical/process studies or hindcasts/simulations. Theoretical/process studies are used to explore the effects of specific dynamics on the ecosystem response. Examples of such studies include investigations of different grazing or mortality formulations, the spatial or temporal scales of plankton variability in response to various idealized flow fields or forcings, and the effects of different model structures on the resulting dynamics. NPZ models have also been used for hindcast/simulations of measureable systems. Hindcasts of existing data have been a particularly useful approach for teasing out the possible dynamics that generated the variability observed in the data. It is important when doing hindcasts to keep some data in reserve (independent) for testing the model. A model that has been tuned to fit a particular data set may easily give the "right" answer (i.e., fit the data well) for the wrong reasons (e.g., an incorrect dynamic balance of terms).

The fact that the planktonic ecosystem is embedded in an ever-moving fluid offers particular challenges relative to the fairly stationary organisms in much of terrestrial ecology. To include spatial variability in an ocean ecosystem model it is necessary to specify the velocities (u, v, w) that move (advect) material in (x, y, z) space, and the diffusivities (κ_x, κ_y, κ_z) that mix (diffuse) material across their gradients. The biological model is coupled to the physical advection–diffusion model, allowing the flow fields to move the biological components while they interact with each other. For example, from Equation 1, a physical-biological model describing the dynamics causing local changes in phytoplankton (P) would be

$$\frac{\partial P}{\partial t} = -u\frac{\partial P}{\partial x} - v\frac{\partial P}{\partial y} - w\frac{\partial P}{\partial z} + \kappa_x\frac{\partial^2 P}{\partial x^2} + \kappa_y\frac{\partial^2 P}{\partial y^2} + \kappa_z\frac{\partial^2 P}{\partial z^2} + f(I)g(N)P - h(P)Z - i(P)P. \quad (2)$$

There would be similar equations for the zooplankton, nutrients, and any other state variables. The first three terms on the right-hand side of Equation 2 describe the advection of phytoplankton. The second three terms describe the diffusion of phytoplankton, while the last three terms describe the biological dynamics. To run the model, it is necessary to specify the velocities and diffusivities in space and time. This is a field unto itself and beyond the scope of this entry.

FUTURE

There are several interesting and important directions of active research using the basic NPZ model structure. One particularly promising research avenue is the use of data assimilation and model skill assessment to better utilize the data in the formulation, parameterization, and testing of models. In data assimilation, measurements (e.g., N, P, or Z) can be introduced into the model during its integration to keep the model close to "reality." A number of different techniques exist for drawing the model state variable or rate closer to a measured value. Some considerations in applying this process are the time or space scale over which the model adjusts to the new information and whether mass (nutrient) is conserved during the process. Data assimilation can also be used in an iterative manner to objectively determine parameter values for the model to obtain the best fit with the data. This can be a powerful technique for identifying deficiencies in the model parameterization and formulation.

A fundamental question of any hindcast or simulation is how well (statistically) the model reproduces the data—its "skill." The development of techniques for assessing model skill is an area of active research. Metrics of model skill will usually consider the number of model parameters and the errors of the data. One important result of many of these studies is that, while more data will allow a better-constrained model, certain kinds of data—particularly rates—provide a stronger constraint for choosing the "best" model.

There has been a recent resurgence in the development and application of size-structured NPZ models. While size is only one of many potential criteria by which to subdivide communities, many fundamental physiological and ecological processes scale smoothly with the size of organisms. This "allometric" scaling of processes such as growth rate, respiration rate, or ingestion rate with size allows the inclusion of size dependence while minimizing the number of new parameters.

The larger the organism, the more rare it is in the ocean. The NPZ models presented here are continuum models: individual organisms are not resolved. In general, NPZ models are good at describing planktonic dynamics but poor at describing the dynamics of larger organisms (which are typically at higher trophic levels) such as euphausiids, gelatinous organisms, and fish. These larger organisms often have complex life histories spanning a range of sizes and can be strong swimmers. Because of this, the larger marine organisms are fruitfully modeled using an individual-based approach. These individual-based models (IBMs) can be formulated and parameterized for a particular species and then coupled to the continuum NPZ planktonic ecosystem model. This coupling can be either one-way (the NPZ model provides the environment for the IBM organism to feed in, but the IBM does not affect the NPZ environment) or two-way (the IBM organism affects the local NPZ environment). The main challenge in creating such models is acquiring sufficient data to formulate and parameterize the IBM.

A recent introduction to the NPZ model realm is the 2007 so-called Darwin model of Mick Follows and colleagues. In the Darwin model, dozens of phytoplankton "species" are specified. For each species, the parameters for functional forms such as the response to light or nutrients—$f(I)$ or $g(N)$—are chosen from a probability distribution. These species are then allowed to compete for resources in a physical circulation model. After many model years, the properties of the dominant surviving species are compared to species found in the oceans. Models such as this can help elucidate the physical, chemical, and biological dynamics that control biodiversity and biogeographic distribution patterns in the oceans.

SEE ALSO THE FOLLOWING ARTICLES

Allometry and Growth / Compartment Models / Stoichiometry, Ecological / Individual-Based Ecology / Microbial Communities / Model Fitting / Ocean Circulation, Dynamics of / Stability Analysis

FURTHER READING

Armstrong, R. A. 1999. Stable model structures for representing biogeochemical diversity and size spectra in plankton communities. *Journal of Plankton Research* 21: 445–464.

Baird, M. A., and I. M. Suthers. 2007. A size-resolved pelagic ecosystem model. *Ecological Modelling* 203: 185–203.

Follows, M. J., S. Dutkiewicz, S. Grant, and S. W. Chisholm. 2007. Emergent biogeography of microbial communities in a model ocean. *Science* 315: 1843–1846.

Franks, P. J. S. 2002. NPZ models of plankton dynamics: their construction, coupling to physics, and application. *Journal of Oceanography* 58: 379–387.

Journal of Marine Systems 76(1–2). 2009. Special issue on Skill assessment for coupled biological/physical models of marine systems.

Kishi, M., et al. 2007. NEMURO—a lower trophic level model for the North Pacific marine ecosystem. *Ecological Modelling* 202: 12–25.

Riley, G. A. 1946. Factors controlling phytoplankton populations on Georges Bank. *Journal of Marine Research* 6: 54–73.

Steele, J. H., and B. W. Frost. 1977. The structure of plankton communities. *Philosophical Transactions of the Royal Society of London B: Biological Sciences* 280: 485–534.

NUTRIENT CYCLES

SEE BIOGEOCHEMISTRY AND NUTRIENT CYCLES

O

OCEAN CIRCULATION, DYNAMICS OF

CHRISTOPHER A. EDWARDS

University of California, Santa Cruz

Ocean circulation describes the amplitude and pathways of fluid transport within the world's oceans. It is responsible for the transport of mass and heat, chemical constituents, and biological organisms throughout the ocean basins. Oceanic motion contributes significantly to the planet's meridional heat transport and helps define oceanic regions fertile for biological growth. In the time mean, a general structure to the circulation exists and can be described phenomenologically and dynamically. For example, the Gulf Stream is a very well-known feature in the western North Atlantic Ocean. This ocean current is among the fastest on Earth and is part of the great subtropical gyre of the North Atlantic; it is influenced by the wind stress curl over the North Atlantic and by frictional processes near the ocean boundary. Less well-known but entirely analogous currents and gyres are also found in the North Pacific and in each of the southern hemisphere oceans. This entry describes the underlying dynamics that produce these features and relates the dominant physical processes to observed distributions of chlorophyll-a as a proxy for phytoplankton in the near-surface ocean.

DYNAMICAL CONSIDERATIONS

The oceans reside within basins bounded from below and laterally by largely insulating, solid material. Although some regions are locally influenced by fluxes of heat and other properties through vents in the ocean floor, these fluxes do not contribute meaningfully to the large-scale circulation and are hereafter neglected. Instead, heat, freshwater, and momentum are predominantly exchanged at the ocean surface. Heat and freshwater fluxes are able to modify surface ocean temperature and salinity and, in places where unstable water columns result, drive vertical convection, which locally mixes a portion of the water column. However, energetic, large-scale horizontal flows cannot be driven by heat and freshwater fluxes at the upper boundary of a fluid. Rather, the general circulation of the ocean, particularly within the upper several hundred meters, is mechanically driven by atmospheric winds and significantly affected by the Earth's rotation. The theoretical foundation of this wind-driven circulation is outlined below.

Wind Stress

In nature, atmospheric fields vary over an enormous range of time scales (e.g., from transient motion associated with cumulous convection to interannual and interdecadal variability of large-scale indices), yet horizontal winds have characteristic amplitudes that allow their effect on ocean currents to be estimated. A typical wind speed of 10 m s^{-1} is much greater than even the swiftest ocean velocities ($\sim$2 m s^{-1}), and most ocean currents are much slower. Friction with this relatively motionless surface causes atmospheric winds to drop in amplitude close to the boundary, exerting a stress, a horizontal force per unit area, on the fluid beneath. This stress exchanges momentum between the ocean and atmosphere and mechanically drives the large-scale circulation. Although it is the vertical gradient of the horizontal velocity at the ocean surface that determines the stress (τ), this quantity is usually estimated using a formula based on the more easily observable horizontal wind speed 10 m above the ocean surface (U_{10}). It is given by the expression

$$\tau = \rho_a C_D U_{10}^2, \qquad (1)$$

where ρ_a is the air density and C_D is the drag coefficient (which takes values of about 1 or 2×10^{-3} depending on atmospheric conditions). For typical wind stress, the low molecular viscosity of water ($10^{-5}\text{m}^2\text{s}^{-1}$) leads to large shear just beneath the ocean surface and, even under steady forcing conditions, a breakdown of oceanic motion into complex, time-dependent, three-dimensional turbulence. The detailed description of this turbulence is involved, but one broader effect is to mix ocean properties such as temperature and salinity vertically between the surface and some depth, where the stability of the underlying ocean stratification prevents deeper turbulent mixing.

Ekman Dynamics

In the late nineteenth century, Norwegian oceanographer Fridtjof Nansen noted on seagoing expeditions that icebergs drift at an angle to the downwind direction. These observations led to a theoretical study of wind stress–driven ocean currents under the influence of rotation by V. Walfrid Ekman, a graduate student at the University of Uppsala. In 1905, Ekman described currents that result from steady wind stress at the surface over an infinitely deep and horizontally infinite and uniform ocean. Ekman's key dynamical insight was to incorporate the Coriolis force into his simplified model. The Coriolis force is a pseudo-force that results from the change in coordinate system from an inertial (nonaccelerating) frame to one that is rotating (accelerating). We live on a rotating planet and observe atmospheric and oceanic phenomena relative to continents and lines of longitude that rotate with the planet. For motions that change appreciably over time scales longer than a rotation period (~1 day), the Coriolis force typically plays a significant quantitative role. Acceleration by the Coriolis force is equal to the product of the Coriolis parameter and the fluid velocity. The Coriolis parameter is given by

$$f = 2\Omega \sin\phi, \qquad (2)$$

where Ω is the angular velocity of the Earth ($\Omega = 2\pi$ radians per sidereal day $= 7.29 \times 10^{-5}\text{s}^{-1}$) and ϕ is the latitude. For a given fluid velocity in any rotating system, acceleration by the Coriolis force is proportional to the rotation rate. The Coriolis parameter varies with latitude because the projection of the Earth's rotation vector, which is directed from the planet's center toward the North Pole, on the local vertical axis changes with latitude. The magnitude of f is greatest at the North and South poles and vanishes at the equator; its sign is positive in the northern hemisphere and negative in the southern hemisphere, leading to a Coriolis force directed to the right of the fluid motion in the northern hemisphere and to the left south of the equator.

In Ekman's model, a steady-state solution occurs when accelerations balance. Parcels of fluid near the surface experience acceleration by wind stress, the Coriolis force, and a viscous force associated with adjacent layers of fluid moving at different velocities. Ekman showed that these three forces balance when the surface velocity is directed 45 degrees to the right of the wind stress in the northern hemisphere (and to the left in the southern hemisphere). The theory also predicted that currents weaken and spiral with depth over a characteristic vertical scale,

$$D_{Ek} = \sqrt{\frac{2A_V}{|f|}}, \qquad (3)$$

where A_V is a mixing coefficient that is analogous to, but considerably larger than, molecular viscosity. The mixing coefficient, also known as an eddy coefficient, results from relating Fick's law for momentum flux down the time-mean gradient to complex turbulent motions. This concept is an oversimplification of mixing by turbulent eddies, but it captures important elements and appears widely in governing equations used in meteorology and oceanography. The efficiency of turbulent eddies to mix fluid properties is reflected by the large values used for A_V (typically $10^{-4}\text{m}^2\text{s}^{-1} - 10^{-1}\text{m}^2\text{s}^{-1}$ as compared to $\sim 10^{-6}\text{m}^2\text{s}^{-1}$ for molecular viscosity). One can develop intuition for the depth scale of Ekman dynamics by introducing numbers relevant to the ocean. At 40N latitude and for $A_V = 0.01\text{m}^2\text{s}^{-1}$, the Ekman depth D_{Ek} is 14.6 m. Changing A_V to $0.1\text{m}^2\text{s}^{-1}$, appropriate for more vigorous turbulence, leads to a D_{Ek} of 46 m. Velocities decay exponentially with an *e*-folding scale of D_{Ek}, and therefore predicted currents become vanishingly small for depths greater than $\sim 3D_{Ek}$, about 45 m and 140 m in these examples.

While the simplicity of the model is appealing, D_{Ek} varies considerably with the amplitude of the mixing coefficient, which is not easily measured over broad areas of the ocean. Furthermore, the detailed structure of the predicted currents relies on the vertical profile of the mixing coefficient. In general, A_V is a complicated function of three-dimensional space and very time dependent, often on time scales shorter than a day. Despite these issues, a very important and robust result of the theory is obtained by vertically integrating the velocity over the water column. The resulting volume transport, called Ekman transport, is determined solely by the local wind stress, the Coriolis parameter, and the fluid density,

$$M_{Ek} = \frac{\tau}{\rho_0 f}. \tag{4}$$

Furthermore, it is directed 90 degrees to the right of the stress in the northern hemisphere and 90 degrees to the left in the southern hemisphere. This fundamental result describes the net transport directly driven by wind stress and is independent of the detailed structure and time dependence of oceanic turbulence.

Vorticity and Sverdrup Dynamics

In 1947, decades following Ekman's theory for surface mixed-layer motion, H. Sverdrup, director of Scripps Institution of Oceanography in California, added to the foundation for understanding large-scale circulation. Ekman's theory described motion only within a thin layer of fluid very near to the ocean surface (compare 140 m or less with the full ocean depth of approximately 4000 m over most of the ocean basins), and it assumed a horizontally uniform ocean and wind field. In nature, time-averaged, atmospheric winds compose well-defined spatial patterns over the ocean (Fig. 1). Well known, for example, are low-latitude trade winds that blow from east to west. At middle latitudes, surface winds are predominantly westerly (one faces west to look into the oncoming breeze), and these westerlies weaken poleward. Although seasonal shifts in the amplitude and latitude of maximum intensity occur and weather systems transiently disrupt the time-mean flow, this general pattern of westward stress at low latitudes and eastward stress at middle latitudes is quite robust. Meridional components to the mean stress also occur, particularly near continental boundaries, and their influence will be discussed at more depth later in this entry. Sverdrup examined how this broad structure of largely zonal winds drive large-scale flow in the ocean. Large-scale is defined in terms of several nondimensional numbers that characterize the fluid flow. Among the most critical is the Rossby number, $R_0 = \frac{U}{fL}$, where L and U are length and velocity scales characteristic of the motion under consideration. Large-scale flows have Rossby numbers much less than 1. For oceanic motion, a typical velocity scale is 1 cm s^{-1}, $L \simeq 5000$ km, and $f \simeq 10^{-4}\ s^{-1}$, which together give $R_0 < 10^{-4}$. A more complete discussion of constraints on other nondimensional numbers critical for Sverdrup theory to be valid can be found in various textbooks on geophysical fluid dynamics.

By taking the curl of the horizontal momentum equations and integrating over the full ocean depth, Sverdrup developed a governing equation for the vertical component of vorticity. In a nonrotating system, fluid parcel vorticity is equal to the curl of its velocity: $\vec{\omega} = \nabla \times \vec{u}$, where $\vec{u} = (u, v, w)$ is velocity in the (eastward, northward, upward) directions, ∇ is the gradient operator, and the arrow indicates a vector quantity. Curl is a mathematical operation on a vector field, but curl of fluid velocity can be understood in a straightforward manner. A simple and common example is water in a bathroom sink. Often one can observe water rotating slowly around the sink as it drains. Generally, a floating object placed in the basin would translate around the basin and it would spin, indicating a curl to the velocity field. The sign of the curl is identified by curling the fingers of one's right hand in the direction of the spin. In a conventional coordinate system, thumbs pointing upward indicate positive vorticity and thumbs pointing downward indicate negative values. Vorticity is a property of a fluid parcel (like

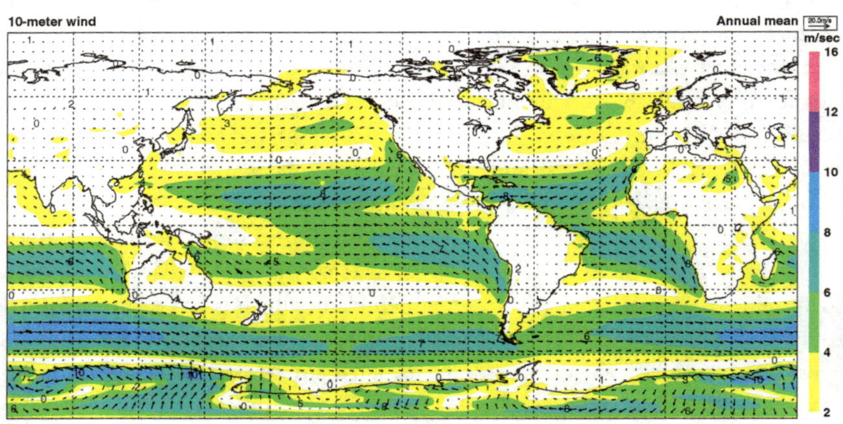

FIGURE 1 Annual mean vector wind with isotachs from the European Centre for Medium-Range Weather Forecasts ERA-40 Atlas. From Kållberg et al., 2005, ERA-40 Project Report Series 19, ECMWF.

temperature, salinity, or momentum) that obeys a conservation law. A fluid element's vorticity can be altered when forced by an external torque. In a rotating system, the concept of vorticity is complicated by the fact that the fluid's total vorticity is partitioned into the spin of the fluid relative to the rotating frame and the rotation of the frame itself. Consider, for example, water sitting motionless relative to a container placed at the center of a spinning merry-go-round. In the rotating frame, the water has no motion and therefore no spin. However, when viewed in an inertial frame from above, the fluid is spinning at the rate of the merry-go-round. Thus, total (or absolute) vorticity of the fluid is equal to the sum of vorticity associated with the rotating reference frame and vorticity relative to that frame. Objects on a spinning planet exist within a rotating frame of reference whose vorticity changes with latitude. It is most positive at the North Pole where the Earth's rotation vector is directed parallel to the local vertical, most negative at the South Pole where they are anti-parallel, and it is zero at the equator where the local vertical is perpendicular to the Earth's rotation axis. Quantitatively, fluid elements on Earth possess planetary vorticity equal to the Coriolis parameter, f (Eq. 2).

Sverdrup treated the ocean as frictionless and showed that for large-scale motion the resulting vorticity equation can be approximated by a dominant balance of two terms, expressed

$$\beta V_S = curl\left(\frac{\vec{\tau}}{\rho_0}\right). \tag{5}$$

In this entry, the word *curl* denotes only the vertical component of the full three-dimensional vector: $curl(\vec{v}) = \hat{k} \cdot \nabla \times \vec{v}$, where $\hat{k}$ is a unit vector in the vertical direction. In Equation 5, V_S is called the Sverdrup transport and is the vertically integrated, meridional velocity. Also, ρ_0 is a reference density for seawater (e.g., 1027 kg m^{-3}), and the stress $\vec{\tau} = (\tau^{east}, \tau^{north})$ is now a vector. The symbol β represents the meridional gradient of the planetary vorticity,

$$\beta = \frac{1}{R_e}\frac{df}{d\phi}, \tag{6}$$

where R_e denotes the radius of Earth. Unlike the Coriolis parameter f, β is always positive and takes its largest value ($\sim 2.28 \times 10^{-11}\text{m}^{-1}\text{s}^{-1}$) at the equator. Equation 5 states that the net meridional transport at any location is proportional to the local curl of the wind stress. Stated differently, winds with nonzero curl impose a torque (vorticity flux) on the water column beneath, and the result is oceanic transport to a different latitude. This result is easily contrasted with that for a fluid at rest in a nonrotating system, where a vorticity flux would initiate a change in relative vorticity and the fluid would begin to spin. In a rotating system *and* one with a gradient in planetary vorticity (such as a spinning planet), a vorticity flux alters the total vorticity of the fluid, but at these large scales the total vorticity is partitioned mostly in the planetary component. Thus, a negative vorticity flux into the ocean by a negative wind stress curl induces transport south, and vice versa.

Sverdrup analyzed the spatial structure of atmospheric winds to identify regions of nonzero curl (Fig. 1). The characteristic wind patterns over the ocean with easterlies at low latitudes and westerlies at middle latitudes yields large regions with nonzero curl and therefore a net vorticity flux across the ocean surface. As an example, consider the region between 10N and 45N at 180E over the central North Pacific Ocean. The dominant wind direction is zonal (east–west) and $curl(\vec{\tau}) \cong -\frac{1}{R_e}\frac{\partial \tau^{east}}{\partial \phi}$ is less than zero. Thus, subtropical latitudes within the northern hemisphere experience a negative vorticity flux into the ocean. The resultant transport must be southward according to Sverdrup theory. Conversely, subtropical latitudes in the southern hemisphere experience a positive vorticity flux into the ocean with an expected northward oceanic transport. Equatorward transport at these latitudes in both hemispheres results from the symmetry in wind stress structure across the equator.

Ekman's theory argued that directly wind-driven motion must occur only in a narrow layer of fluid near to the ocean surface, whereas Sverdrup theory predicts net vertically integrated transport over the full water column. How then can Sverdrup theory and Ekman theory be related? Ekman theory must be considered in the context of spatially varying stress. For example, along 180E Ekman transport at 10N is driven by trade winds and therefore largely directed to the north. At 45N, Ekman transport from westerlies is directed predominantly southward. Between these two latitudes mean stress varies monotonically as does the resulting Ekman transport. As a result, fluid parcels within the Ekman layer converge at subtropical latitudes (between the tropics and the latitude of maximum westerly wind stress) and at most longitudes within the basin, exceptions being near continental boundaries, which will be discussed later. To a very good approximation water is incompressible, and no net convergence in a three-dimensional sense can occur. Thus, fluid converging horizontally within the Ekman layer must be exported vertically to greater depth in the ocean. Downward vertical velocity at the base of the Ekman layer is referred to as Ekman pumping because it pumps fluid from the Ekman layer into the water column below. In places such as higher latitude subpolar regions, the structure of wind stress is reversed, a divergence of fluid parcels occurs in the Ekman layer, and fluid from depth must be supplied to the surface layer. Upward vertical velocity at the base of the Ekman layer is referred to as Ekman suction.

The vertical velocity at the base of the Ekman layer is expressed mathematically by

$$w_E = curl\left(\frac{\vec{\tau}}{\rho_0 f}\right), \tag{7}$$

and this motion drives large-scale circulation well beneath the surface. This vertical velocity causes either a stretching ($w_E > 0$) or compression ($w_E < 0$) of deep fluid columns. Through mass conservation, stretching causes a lateral contraction in fluid columns leading to an increase in total vorticity. In a nonrotating system, laterally contracted fluid columns rotate faster just as a figure skater's rotation rate increases with arms drawn in. Conversely, a skater rotates more slowly with arms laterally outstretched; fluid columns that are compressed vertically expand laterally and their total vorticity decreases. For large-scale motion in a rotating system, changes in total vorticity occur through changes in the planetary component and not relative vorticity. The annual average wind pattern indicates that $w_E < 0$ at subtropical latitudes (Fig. 1). At these latitudes, Ekman pumping compresses fluid columns, reducing their absolute vorticity and driving equatorward motion. Though expressed in very different physical terms, this description is fully consistent with the Sverdrup theory outlined above.

The important and nonintuitive result is that while local wind stress at any location may be directed east or west (or north or south), the depth-averaged meridional motion beneath is determined by its curl. In subtropical latitudes, meridional Sverdrup transport is equatorward and at subpolar latitudes, where the wind stress curl has the opposite sign, it is poleward.

WESTERN INTENSIFICATION

Sverdrup is credited with deriving the powerful relationship between meridional motion and wind stress curl, but his theory was not able to predict the zonal component of flow in the basins. That final key arrived with the description of western intensification of the great gyre system. It has been well known for centuries that currents on the western edges of ocean basins are significantly stronger and more narrow than currents along eastern boundaries. Examples are the Gulf Stream in the Atlantic Ocean and the Kuroshio in the Pacific. Dynamical interpretation for this asymmetry in the boundary currents was provided by H. Stommel in 1948 and then W. Munk in 1950. These two theories built on the earlier model of Sverdrup by including friction in the momentum equations. Stommel considered acceleration by bottom friction in which drag is proportional to velocity,

$$\vec{\mathcal{F}} = -r\vec{U}, \tag{8}$$

where r is a constant and $\vec{U} = (U, V)$ denotes the vertically integrated horizontal velocity vector; Munk's solution included lateral friction using a mixing coefficient similar to A_V described above but parameterizing horizontal momentum exchanges by large scale eddies:

$$\vec{\mathcal{F}} = A_H \nabla^2 \vec{U}, \tag{9}$$

where ∇^2 denotes the Laplacian.

Despite including friction in the governing equations, both Stommel and Munk showed that its influence away from lateral boundaries was negligible and thus supported Sverdrup theory over most of the ocean. However, they found that Equation 5 was invalid near the western boundary. In this region of the ocean, a different vorticity balance was dominant:

$$\beta V_B = curl(\vec{\mathcal{F}}). \tag{10}$$

Here, meridional transport on the left-hand side is driven by a frictional vorticity flux on the right-hand side. Equation 10 does not require that surface forcing vanish. Rather, the fluid self-organizes its velocity and length scales such that the frictional vorticity flux overwhelms that due to wind stress curl within this boundary layer. Conceptually, boundary layer velocities are much greater than interior velocities $\left(|V_B| \gg |V_S|\right)$, and cross-current length scales in the boundary layer are much smaller than those of the basin ($L_B \ll L$). Higher velocities and smaller length scales combine to increase the frictional effect in this region.

When integrated laterally across the boundary layer, the net fluid transport is equal and opposite to the net transport of the rest of the basin driven by wind stress curl. The frictional boundary current closes the gyre circulation, and it does so only along the oceanic western boundary. What explains this east–west asymmetry? The answer is that the sign of the vorticity flux by friction can only drive motion opposite to the interior flow along the western boundary. Intuition on this admittedly subtle point can be developed by considering the sign of the vorticity flux in either an eastern or western boundary current, within a northern hemisphere subtropical gyre with southward interior transport. To conserve mass and close the circulation, the boundary current must transport fluid to the north. For simplicity, consider the vorticity balance using Stommel's model,

$$\beta V_B = -r\zeta, \tag{11}$$

where $\zeta = curl(\overrightarrow{U})$ is the relative vorticity of the depth-integrated flow. Ensuring $V_B > 0$ requires $\zeta < 0$. Figure 2 plots meridional transport as a function of zonal position within a basin for two possible boundary current scenarios in the Stommel model. In both cases, transport is broadly southward over much of the domain, consistent with Sverdrup transport driven by wind stress curl. In Figure 2A, a possible boundary current returning mass to the north is drawn adjacent to the eastern boundary. In Figure 2B, the mirror image is drawn with the boundary current at the western basin edge. Near the boundaries, the meridional velocity is much greater than any zonal velocity (not drawn), and the zonal scale of variation over which the boundary layer velocity changes is much smaller than its meridional scale. As a result, vorticity can be approximated by $\zeta \cong \frac{\partial V}{\partial x}$. From the profiles in Figure 2, it is clear that $\zeta < 0$ only in the case of a western boundary current. An eastern boundary current has positive vorticity, which yields greater southward (and not northward) motion according to Equation 11. Thus, friction in an eastern boundary current cannot supply the needed positive vorticity flux to drive fluid columns northward. Though this discussion has been based on friction as represented in the Stommel model, an extremely similar argument results within the Munk model as well.

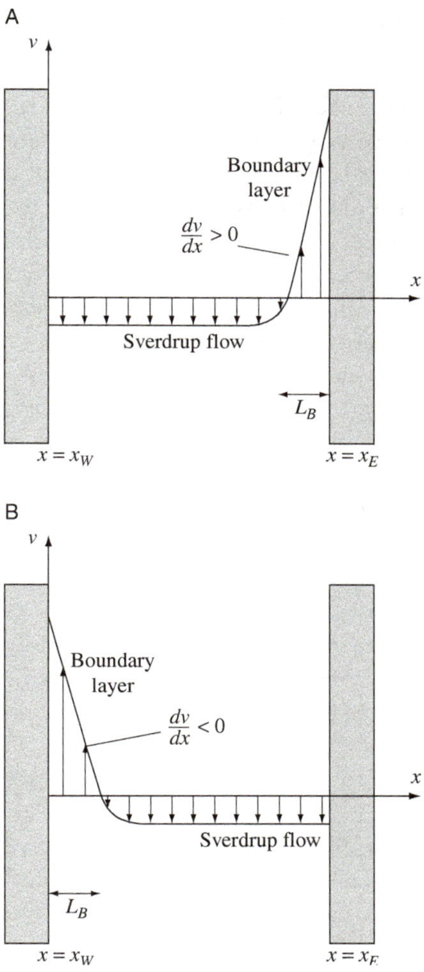

FIGURE 2 Two possible meridional velocity profiles across an idealized subtropical gyre with meridional boundaries at x_W and x_E. In both cases, weak southward motion occurs over most of the domain (Sverdrup flow). Intense northward motion occurs in a boundary layer adjacent to either the (A) eastern boundary or (B) western boundary. From *Introduction to Geophysical Fluid Dynamics,* 2nd ed., Vol. 98, B. Cushman-Roisin and J.-M. Beckers. Copyright Elsevier 2011.

OBSERVATIONS OF THE GENERAL SURFACE CIRCULATION

Dynamical principles outlined above help to interpret and explain the general structure of the circulation as it can be observed. Qualitative and quantitative descriptions of this large-scale motion have resulted from a wide range of observational techniques. Ocean currents, for example, can be measured by ship drift, the difference between a ship's heading and its actual path. Drifters that float but have no wind component can provide similar information. Drifters are deployed for scientific purposes and have complex electronics that record or transmit their position; occasionally, shipping accidents at sea release floating objects that can be tracked by observers on beaches. Current meters, like anemometers measuring wind, can be attached to moorings that sit for months or years at one depth in one location of the ocean.

Ocean currents can also be estimated from an observed pressure field using the geostrophic approximation. For large-scale oceanic motion characterized by a small Rossby number, the horizontal equations of motion can be reduced to a leading order balance between Coriolis accelerations and accelerations due to a pressure gradient. Near the ocean surface, this geostrophic relationship can be combined with a statement of hydrostatic balance to yield expressions for the near-surface circulation. Neglecting (small) changes in atmospheric pressure and near surface ocean density over broad regions of the ocean, the geostrophic relation can be expressed in Cartesian coordinates for simplicity:

$$-fv = -g\frac{\partial \eta}{\partial x}, \tag{12}$$

$$fu = -g\frac{\partial \eta}{\partial y}. \tag{13}$$

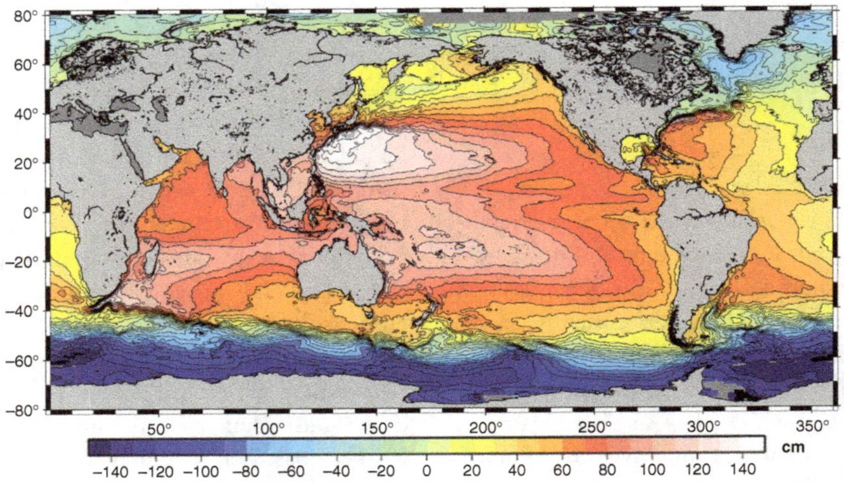

FIGURE 3 Estimate of mean dynamic topography (MDT CNES-CLS09) showing time-averaged sea surface height relative to the geoid. MDT_CNES-CLS09 was produced by CLS Space Oceanography Division and distributed by Aviso, with support from CNES (http://www.aviso.oceanobs.com/).

Here, $g = 9.81\text{ms}^{-2}$ is the acceleration due to gravity, and $\eta(x, y)$ represents the deviation of the ocean surface from its form if the ocean were motionless and unforced.

Historically, oceanographers could use the geostrophic and hydrostatic relations along with shipboard measurements of in situ temperature and salinity (and thus density) to estimate the shape of the ocean surface relative to a deeper level (often chosen to be 500 m) assumed to have zero horizontal velocity. However, since 1992, satellites with remarkably accurate (~2 cm) altimeters have orbited the planet allowing unprecedented, direct measurements of the ocean surface elevation and estimates Earth's gravitational field (needed to determine the equipotential surface called the geoid) to yield η. A recent estimate of the time-mean sea surface (known as mean dynamic topography) is given in Figure 3. This map shows that the ocean surface has organized hills and valleys in each ocean basin. Total sea level variation over the globe is about 3 m, with smaller changes occurring across limited regions. The geostrophic equations allow interpretation of Figure 3 in terms of the surface circulation. Equations 12 and 13 indicate that geostrophic currents flow perpendicular to sea surface gradients. Stated differently, geostrophic flow is directed parallel to isolines of sea surface topography (shown as dark lines in Figure 3).

Broadly speaking, Figure 3 reveals the sea surface signature of the oceans' great gyres. Subtropical regions between 10 and 30 degrees north or south latitude exhibit pronounced sea surface highs around which the ocean circulates. The direction of flow can be determined by Equations 12 and 13. For example, the sea surface slopes downward to the east in the central North Atlantic ocean between about 20N and 40N. According to Equation 12, acceleration due to the surface pressure gradient can be balanced only by a southward velocity, consistent with the meridional transport predicted by Sverdrup as discussed above. The same general southward velocity is visible also across most of the subtropical North Pacific Ocean. Working around an isoline of surface topography, it is evident that northern hemisphere subtropical gyres circulate clockwise. Similar scrutiny in the southern hemisphere reveals gyres that rotate counterclockwise. Oceanographers refer to subtropical gyres as circulating cyclonically in both hemispheres because the projection of the gyre vorticity vector on the Earth's rotation vector is positive both north and south of the equator.

At higher latitudes in the northern hemisphere, local sea surface valleys are found in Figure 3. For example, a local depression in sea surface is found just south of Greenland in the Atlantic ocean and southeast of the Kamchatka Peninsula in the North Pacific. Following isolines in these regions indicates counterclockwise circulation. These anticyclonic gyres are also referred to as subpolar gyres.

Well-known, major ocean currents are also visible in Figure 3. Geostrophic currents are proportional the sea surface gradient, and therefore narrowly separated isolines of sea surface topography indicate faster ocean currents. Indeed, for a given latitude, the net transport between two isolines of sea surface height remains the same whether their separation is wide or narrow. Excellent examples of swift currents are the Gulf Stream in the western subtropical North Atlantic and the Kuroshio Current in the western subtropical North Pacific ocean. Natural manifestations of the idealized western boundary currents of Stommel and Munk, these features hug the continental slope boundary before entering the interior of their respective ocean basins. That each of these currents transports roughly the same volume of fluid northward as the rest of the gyre circulation transports southward reveals the stark asymmetry (i.e., western intensification) of the gyre circulation. An example western boundary current in the southern hemisphere is the Agulhas Current, which carries warm water southward in the Indian Ocean along the eastern coast of South Africa until it reverses suddenly (retroflects) near 40S as the continental boundary veers back north. Just south of this retroflection region and within the Southern Ocean is the Antarctic Circumpolar Current, the only ocean current to circumnavigate the globe.

ECOLOGICAL IMPLICATIONS

The description of the general circulation and observations of the great gyres provides a natural framework to consider ecological processes active in the ocean and the resulting biological distributions on these large scales. Ocean circulation influences organisms through advection, but more importantly, mean currents transport nutrients. The vast majority of primary production in the ocean is carried out by photosynthesizing organisms (phytoplankton) in the well-lit, upper 100 m of the water column. The remains of these organisms and their consumers ultimately sink, and bacterial processes remineralize organic tissue well below this euphotic zone. As a result, nutrient profiles in the ocean typically exhibit low values in the euphotic zone and a nutricline just beneath with nutrient levels increasing to a maximum 1000 meters or more beneath the surface and maintaining high values to the ocean bottom.

In places where mean vertical velocities are downward, nutrients are exported from the euphotic zone to depth, and biological production is low. Such a description applies to subtropical gyres. In these regions, Ekman-transport convergence due to wind stress curl drives Ekman pumping at the base of the mixed layer. This vertical velocity transports material away from the surface and into the main thermocline beneath. As a result, subtropical gyres are also referred to as oligotrophic gyres, generally exhibiting very low primary production and maintaining low concentrations of phytoplankton near the surface. In contrast, subpolar gyres are characterized by opposite wind stress curl that drives Ekman suction. Upward velocity draws fluid and nutrients from beneath the surface mixed layer toward the surface and into the euphotic zone. Thus, subpolar gyres exhibit relatively higher levels of production and maintain higher stocks of phytoplankton than their subtropical counterparts.

Historically, observations from shipboard samples easily identified differences in phytoplankton concentration and community across gyre boundaries. Today, satellite sensors detect ocean color, measuring intensity at multiple wavelengths that can be inverted to estimate pigment concentration, such as chlorophyll-a. The broad coverage and high resolution of the satellite data revolutionizes our perspective on phytoplankton distributions. An example is shown in Figure 4, providing an estimate of the time-mean chlorophyll-a concentration from 8 years of color data collected by the the Moderate Resolution Imaging Spectroradiometer (MODIS). A clear correspondence between features in Figure 4 and Figure 3 is evident. Large swaths of blue and purple indicate low standing stocks of phytoplankton within all subtropical gyres. Higher levels

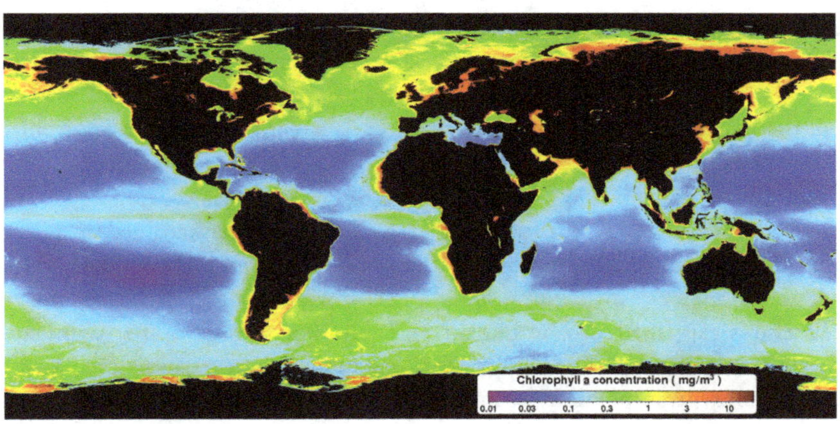

FIGURE 4 Estimate of 2002–2010 mean, near surface chl-a concentration (mg/m^3) from the Aqua MODIS color sensor. A logarithmic scale is used. Feldman, G. C., C. R. McClain, Ocean Color Web, Aqua MODIS Reprocessing 2009, NASA Goddard Space Flight Center. Eds. Kuring, N., Bailey, S. W. October 17, 2010 (http://oceancolor.gsfc.nasa.gov/).

of chlorophyll-a are easily found in subpolar regions of the North Atlantic and North Pacific oceans.

EQUATORIAL AND COASTAL UPWELLING

Figure 4 details many other features in the chlorophyll-a distribution that can be considered in the context of the general circulation using ideas presented above. For example, relatively higher concentrations of chlorophyll-a are visible along the equator in both the Atlantic and Pacific oceans, bisecting oligotrophic subtropical regions in each hemisphere. Generally elevated levels at the equator can be understood by considering low-latitude wind stress. As previously discussed, trade winds are easterly with a resulting westward stress applied to the ocean surface. Although the Coriolis force vanishes at the equator where f goes to zero, this influence is distinctly nonzero a few degrees to the north or south. In the northern hemisphere, trade winds drive northward Ekman transport; in the southern hemisphere, it is to the south. Ekman transport divergence straddles the equator, with deeper, nutrient-rich water required for mass conservation. It is this subsurface upwelling that enables locally enhanced production and phytoplankton standing stock along and near to the equator. To be clear, equatorial circulation is three-dimensional and more complex than this brief description alone. Upwelled fluid derives from deeper source regions to the west via a shoaling undercurrent, and this complexity is reflected in part by the east–west asymmetry of the chlorophyll concentrations within each basin and visible in Figure 4. Though added complexity suggests a more circuitous pathway for upwelled water, it does not significantly alter the fact that subsurface, nutrient-rich water rises to the surface along the equator to supply the Ekman transport divergence and yield locally high levels of primary production and phytoplankton concentrations.

A related process occurs near oceanic eastern boundaries, where Figure 4 also reveals elevated levels of chlorophyll-a. Two examples from the Pacific Ocean are the U.S. west coast in the northern hemisphere and the Peru/Chile coasts in the southern hemisphere. Similarly, in the Atlantic Ocean, South African and Namibian coasts in the southern hemisphere, and north African coast in the northern hemisphere exhibit chlorophyll-a concentrations well above open ocean levels at similar latitudes. This increase is explained also by Ekman transport divergence and nutrient-rich upwelling. In these eastern boundary regions, atmospheric winds veer equatorward on a seaonal basis, though the structure is also evident in the annual average fields of Figure 1. Equatorward wind stress in either hemisphere drives westward Ekman transport. Cross-shore transport must vanish at the coast, resulting in a surface layer divergence. As at the equator, cold water from one to a few hundred meters depth supplies the Ekman divergence, and with it, bountiful nitrate, silicate, and phosphate for primary production. Eastern boundary regions also have complex, multidimensional flow including undercurrents, but as at the equator, atmospheric wind stress is the primary driver for regional circulation patterns that transport nutrients into the euphotic zone with a dramatic ecosystem response.

FINAL WORDS

This discussion has purposely focused on the surface circulation so easily observed through the sea surface topography and quite relevant for many ocean ecosystems. Subsurface variations in temperature and salinity alter pressure gradients and the resulting circulation. The depth to which the surface description of circulation applies varies significantly across the ocean but generally extends between a few hundred to several hundred meters below the surface, and in some places considerably deeper. Theoretical advances have extended the simplified models of Sverdrup, Stommel, and Munk to include and even explain upper ocean stratification and the presence of a main thermocline beneath the subtropical gyres. Other theoretical studies have investigated dynamics of the abyssal ocean that resides beneath the main thermocline and contributes meaningfully to the meridional overturning circulation.

In addition, this entry has addressed only aspects of the steady circulation. Just like the atmosphere, the ocean is rich with complex, time-dependent motion superposed on the time-mean flow. Variable forcing by the atmosphere, interactions between currents and topography, and internal instabilities of the fluid motion itself result in ocean basins filled with eddies having intermediate scale (called the ocean mesoscale), smaller than the basin and much larger than small scale turbulence. Associated with mesoscale activity is vertical motion that also influences ocean ecosystems.

This review has outlined the foundations of large-scale circulation theory. Driven by largely zonal winds that reverse with latitude, the circulation divides into massive counterrotating gyres that dominate ocean basins. Circulation transports heat and salt and importantly also establishes through vertical transport nutrient environments that support a range of phytoplankton and larger ecological communities across the world ocean.

SEE ALSO THE FOLLOWING ARTICLES

Biogeochemistry and Nutrient Cycles / Hydrodynamics / Microbial Communities / NPZ Models / Partial Differential Equations

FURTHER READING

Pedlosky, J. 1996. *Ocean circulation theory.* Berlin: Springer-Verlag.

Salmon, R. 1998. *Lectures on geophysical fluid dynamics.* Oxford: Oxford University Press.

Vallis, G.K. 2006. *Atmospheric and oceanic fluid dynamics.* Cambridge, UK: Cambridge University Press.

Veronis, G. 1981. Dynamics of large-scale ocean circulation. In B.A. Warren and C. Wunsch, eds. *Evolution of physical oceanography.* Cambridge, MA: MIT Press.

Wunsch, C., and R. Ferrari. 2004. Vertical mixing, energy, and the general circulation of the oceans. *Annual Review of Fluid Mechanics* 36: 281–314.

OPTIMAL CONTROL THEORY

HIEN T. TRAN

North Carolina State University, Raleigh

Optimal control theory, a mathematical theory for modifying the behavior of dynamical systems through control of system inputs, is ubiquitous in both science and engineering. The history of optimal control theory dates back to 1696 when Johann Bernoulli challenged his contemporaries with the brachistochrone problem: Given two points A and B in a vertical plane, find the path AMB of the movable point M that, starting from A and under the influence of constant gravity and assuming no friction, arrives at B in the shortest possible time. This problem can be solved using the machineries from the calculus of variations. The era of modern optimal control theory began in the 1950s with the development of dynamic programming and Pontryagin's minimum principle. The theory is important in both behavioral ecology and renewable natural resource management.

THE CONCEPT

The formulation of an optimal control problem consists essentially of the following three components: (i) a mathematical description (or model) of the process to be controlled, (ii) a statement of the biological or physical constraints, and (iii) a specification of a performance criterion or objective function. A nontrivial part of any control problem is modeling the process. The goal is to obtain the simplest mathematical description that can adequately provide predictions of the biological system response to all anticipated inputs. In its standard formulation, optimal control theory models system dynamics by means of a differential equation,

$$\frac{dx}{dt} = f(x(t), u(t), t),\ 0 \leq t \leq T,$$

where the initial condition $x(0)$ is given and $x(t)$ denotes the state variable and $u(t)$ the control, input, or decision variable. In general, both the state and control may be multidimensional. For an illustration, let us consider the modeling process of how social insects (in particular, a population of bees) determine the formation of their society. To this end, we introduce the following variables: state variable $x(t) = [w(t), q(t)]^T$ denotes the number of workers $w(t)$ and queens $q(t)$ at time t, and the control $u(t)$ the fraction of colony effort committed to increase work force. A simplified model for the worker dynamics is given by

$$\frac{dw(t)}{dt} = -\mu w(t) + as(t)u(t)w(t),$$

with the initial condition $w(0) = w_0$. Here, μ is a constant death rate, a is a constant, and $s(t)$ is the time-dependent rate at which each worker contributes to the bee economy. Similarly, a model for the queen population can be formulated as

$$\frac{dq(t)}{dt} = -\beta q(t) + b(1 - u(t))s(t)w(t),$$

with the initial condition $q(0) = q_0$. The constant β denotes the death rate, and b is another constant. After we have developed a mathematical model, the next step is to define biological constraints on the state and control values. For this example, the only constraint that we need to impose is on the control variable

$$0 \leq u(t) \leq 1.$$

It is noted that a control history that satisfies the control constraint during the entire time interval $[0, T]$ is called an *admissible control.* Admissibility is an important concept, because it reduces the range of values that can be assumed by the controls. That is, rather than consider all control histories to see which are best (according to some performance criterion to be defined later), we investigate only those controls that are admissible. Finally, in order to evaluate the performance of a system quantitatively, the modeler needs to select a performance measure, a specific payoff, or a reward criterion. An optimal control is defined as one that minimizes (or maximizes) the performance measure. In general, the performance measure is expressed as an integral

$$J = \int_0^T L(x(t), u(t), t)dt + h(x(T), T).$$

The function h is called the terminal cost function. For example, for the social insects problem described above, a goal (or rather the bees' goal) is to maximize

$$J = q(T),$$

the number of queens at time T. In this performance measure, $L \equiv 0$ and $h(x(T), T) = q(T)$. This optimal control problem is known as the Mayer problem. On the other hand, if the terminal condition $h(x(T), T) = 0$, it is called the Lagrange problem.

Even though optimal control theory was originally designed for engineering applications, it is now at the center of many fields, including behavioral ecology and the management of renewable resources. Behavioral ecology began in the late 1960s with the development of optimal foraging theory and models. Optimal foraging theory is a branch in ecology based on the study of foraging behavior of animals in response to the environment in which the animal lives. This includes the animal's decision of which types of food to eat (i.e., optimal diet), which patch types to feed in (i.e., optimal patch choice), and optimal time allocation to different patches and optimal patterns and movements. In the 35 years since, this basic theory has developed and successfully adapted to mating tactics and fertility decisions, conservation biology, children's foraging, life history, and resource intensification and distribution. In addition, the management of renewable resources can also be placed in the dynamical and optimal control context. For example, in the problem of optimal fishery policy, a size-structured population model has been used to model the fish population size density $n(t,x)$ as a function of both time t and size x and is given by

$$\frac{\partial n(t,x)}{\partial t} = \begin{cases} -\frac{\partial}{\partial x}\left[g(t,x)n(t,x)\right] - [d(x) + u(t)]n(t,x), & x_l \le x \le x_u, \\ -\frac{\partial}{\partial x}\left[g(t,x)n(t,x)\right] - d(x)n(t,x), & x < x_l, x > x_u, \end{cases}$$

with initial and boundary conditions

$$n(0,x) = n_0(x),$$
$$g(t,x_0)n(t,x_0) = R(n(t,\cdot),t).$$

Here, $g(t,x)$ denotes the growth rate, $d(x)$ is the size-dependent fractional death rate, x_0 is the size at which the young fishes enter the population, and R is the recruitment rate. Finally, the control function $u(t)$ is the fishing effort, which is defined to be the fractional mortality rate per unit of time at time t. The optimal fishery harvest problem is the Lagrange problem of finding the functions $u(t)$, lower fish size limit $x_l(t)$, and upper fish size limit $x_u(t)$ to maximize the resource level given by

$$J = \int_0^T e^{-\delta t} u(t) \left[\int_{x_l(t)}^{x_u(t)} p(t,x)n(t,x)dx - K \right] dt,$$

where $p(t,x)$ is the market value of an individual fish, K is a constant denoting fishery cost per unit fishing mortality rate, and δ is the discount rate. In the above expression, the term

$$J(t) = u(t) \int_{x_l(t)}^{x_u(t)} p(t,x)n(t,x)dx - u(t)K$$

denotes the difference between the rate at which revenues are generated due to fishing and the rate at which revenues are spent through fishing. Since fishing effort is naturally limited, a control constraint is usually imposed on $u(t)$ as

$$0 < u(t) \le u_{\max}.$$

At this point, several comments are in order. First, for nonlinear dynamical problems we may not know in advance that an optimal control exists. Since existence theorems are, in general, difficult to obtain, one in most cases attempts to find an optimal control rather than try to prove that one exists. Second, even if an optimal control exists, it might not be unique. Even though nonuniqueness complicates numerical procedures, it allows the possibility of choosing among several possible controller configurations. This can be very helpful, as the modeler can then consider other factors, such as revenue, crop harvest, timing, and so on, which may not have been included originally in the performance measure. Third, by an optimal solution we mean an optimal control $u^*(t)$ and a corresponding optimal state dynamics $x^*(t)$ such that

$$J^* = \int_0^T L(x^*(t), u^*(t), t)dt + h(x^*(T), T)$$
$$\le \int_0^T L(x(t), u(t), t)dt + h(x(T), T)$$

for all admissible controls $u(t)$ and corresponding admissible state dynamics $x(t)$. The above inequality simply states that an optimal control and its corresponding state dynamics cause the performance measure to have a value smaller or equal to the performance measure for *any other* admissible control and state dynamics. Thus, we are seeking the global minimum of J, not just local minima. Of course, if the number of local minima is finite, one way to find the global minimum is to simply pick out the local minimum that yields the smallest value for the performance measure.

FORMS OF THE OPTIMAL CONTROL

If the optimal control is a function of the state variable, that is,

$$u^*(t) = g(x(t), t),$$

then the optimal control is said to be in closed-loop (feedback) form. Another class of control, called open-loop (nonfeedback) controller, is used to describe the functional relationship

$$u^*(t) = g(x(0), t).$$

That is, the optimal control is determined as a function of time for a specified initial state value $x(0)$. Hence, for an open-loop controller, the optimal control is computed based on the performance measure and all available a priori knowledge about the process. The control is in no way influenced by the state $x(\cdot)$ and thus if unexpected disturbances act upon the process or there are changes in operating conditions, the controlled process will not behave precisely as desired. On the other hand, for a closed-loop controller there is a feedback of process information to the controller. Thus, the controlled process is better to adapt to changes in the process parameters or to unexpected disturbances. Any self-regulating natural process involves some forms of feedback. A well-known example in ecology is the oscillation of the population of snowshoe hares dues to predation from lynxes. Figure 1 depicts the conceptual difference between an open-loop optimal control and a closed-loop optimal control.

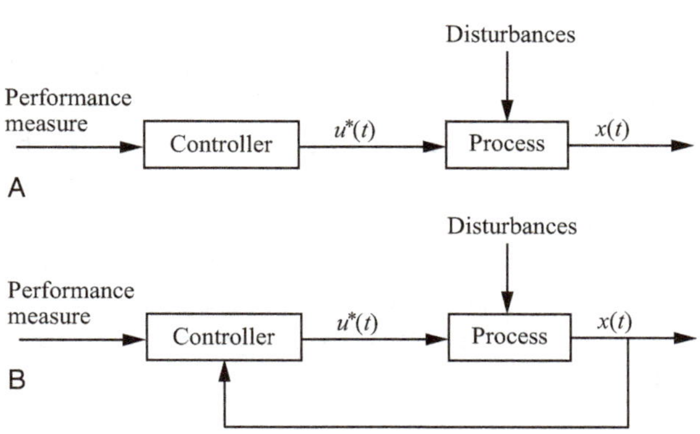

FIGURE 1 (A) Open-loop optimal control. (B) Closed-loop optimal control.

OPTIMAL SOLUTION METHODOLOGIES

Solutions to the optimal control problems can be determined from a collection of necessary conditions, known as the Pontryagin's minimum principle. However, unless the optimal control problem is simple (e.g., low-dimensional, linear process, and the like), the solution must be approximated by numerical procedures. Indeed, over the past two decades the subject of optimal control has transitioned from theory to computation resulting in a variety of numerical methods and corresponding software implementations of these methodologies. The vast majority of software implementations of optimal control today are those that involve the transformation of a continuous-time optimal control problem into a discrete nonlinear programming problem, which can then be solved by one of a variety of well-known nonlinear programming algorithms. One such implementation, which has shown great potential for solving optimal control problems, is a particular Matlab implementation called GPOPS (which stands for "General Pseudospectral Optimal Control Software"). GPOPS is an open-source Matlab software that utilizes the Gauss pseudospectral method. It has been well tested against several well-known and commercially available software packages, including PROPT and SOCS (http://www.gpops.org/index.html).

Each of the above mentioned software packages determines an open-loop optimal control, that is, the optimal control history associated with a specified set of initial conditions. One approach to extend open-loop controls to closed-loop controls is via an interpolation over the state space. That is, this methodology uses techniques for finding an open-loop control for the process with a chosen initial condition $x(0)$. This is then repeated for many values of $x(0)$. At each point, the initial control that is found is considered the control value at $x(0)$. Then an interpolation is done over these initial points to form a closed-loop (feedback) control $u(x)$. The use of the interpolation makes this method very time consuming, given that the open-loop control must be found for many different initial states. However, all of the expensive open-loop control computation can be done offline, so the actual control of the system can still be done relatively quickly. Also, there are no restrictions on the types of problems that can be solved by this method, and the process of increasing the accuracy of the optimal feedback control by adding more interpolation points is also straightforward.

An important class of optimal control problems, which has been extensively studied, is the so-called linear quadratic regulator (LQR) problem. In LQR problems, as the name implies, the process is described by linear differential equations, $\dot{x}(t) = Ax(t) + Bu(t)$, and the performance measure is quadratic in both the state and control, $J = \int_0^\infty (x^TQx + u^TRu)dt$, where the matrices Q and R are symmetric positive semi-definite and symmetric positive definite, respectively. If one assumes full state knowledge, then the optimal feedback control is a linear state feedback law, $u^*(x) = -R^{-1}B^TPx(t)$, where P is the solution to the matrix Riccati equation $PA + A^TP - PBR^{-1}B^TP + Q = 0$. The success of this LQR problem in many applications is due to the successful development of robust and efficient algorithms for solving the Riccati equation. For the nonlinear case, $\dot{x}(t) = f(x(t)) + Bu(x(t))$, the optimal feedback control is known to be of the form

$$u^*(x) = -\frac{1}{2}R^{-1}B^TV_x(x(t)),$$

where the function V is the solution to the Hamilton–Jacobi–Bellman (HJB) equation,

$$V_x^T(x)f(x) - \frac{1}{4}V_x^T(x)BR^{-1}B^TV_x^T(x) + x^TQx = 0.$$

The HJB equation provides the solution to the optimal control problem for general nonlinear systems; however, it is very difficult to solve analytically for any but the simplest problems. This has led to many methodologies being proposed in the literature for ways to approximately obtain the solution to the HJB equation as well as obtain a suboptimal feedback controller for general nonlinear dynamical processes. One possibility is to construct the feedback control as a power series, either by separating out the nonlinearities in the system into a power series or by introducing a temporary variable and expanding around it. Then the first few terms in the series for the control are found by various techniques. This idea is based on considering the system as a perturbation of a linear system, with the control being an extension of the linear control (the first term of the power series is obtained by solving the Riccati equation for the solution of the linearized system). This method has some limitations, as it is not possible to increase the accuracy of the approximation by adding more terms to the power series without the tedious process of solving for the polynomial of the appropriate order. However, this method is very effective for problems with one order of nonlinearity (whether this order is quadratic, cubic, or higher), a category including many biological problems of great interest. Another approach is through successive approximation, where an iterative process is used to find a sequence of approximations approaching the solution of the HJB equation. This is done by solving a sequence of generalized Hamilton–Jacobi–Bellman (GHJB) equations. A more concrete technique for finding the desired solution is to use a Galerkin procedure to approximate the solution to the GHJB equation. The successive Galerkin approximation method is applicable to a larger class of problems, but it is more computationally intensive and, being an iterative method, its convergence depends on the chosen initial iterate.

More recently, other methods have been proposed including the state-dependent Riccati equation (SDRE), which is an extension of the Riccati equation to nonlinear dynamical systems. That is, the idea behind the SDRE method is to parallel the use of the Riccati equation for linear problems by rewriting the nonlinear function in the state-dependent coefficient form $f(x) = A(x)x$. Note that, in general, $A(x)$ is unique only if x is scalar. For the multivariable case, different choices of $A(x)$ will result in different closed-loop controls. As an illustrative example, consider $f(x) = [x_2, x_1^3]^T$. An obvious parameterization is

$$A_1(x) = \begin{bmatrix} 0 & 1 \\ x_1^2 & 0 \end{bmatrix}.$$

However, we can find another state-dependent parameterization

$$A_2(x) = \begin{bmatrix} x_2/x_1 & 0 \\ x_1^2 + x_2 & -x_1 \end{bmatrix}.$$

Because of the many available parameterizations in the multivariable case, as we design the controller we must choose the one parameterization that is most appropriate for the process and control objectives of interest.

With the nonlinear function f rewritten in this way, the state-dependent Riccati equation is of the form

$$P(x)A(x) + A^T(x)P(x) - P(x)BR^{-1}B^TP(x) + Q = 0,$$

and the suboptimal feedback control is given by $u^*(x) = -R^{-1}B^TP(x)x$. We note that the Riccati solution $P(x)$ is state dependent and is not as easy to find as for the constant coefficient case, except for simple problems with certain structures. One proposed method is to use the Taylor series expansion by rewriting the matrix $A(x)$ as a sum of a constant matrix A_0 and a state-dependent incremental matrix $\Delta A(x)$, $A(x) = A_0 + \varepsilon\Delta A(x)$, and to expand the state-dependent Riccati solution $P(x)$ as a power series in ε. However, the methodology of Taylor series approximations works only for processes with constant control coefficients (that is, B is not dependent on the state x) and is only effective locally. For more complex systems, another method, which is similar in spirit to the approach discussed above to extend open-loop control to closed-loop control, involves varying the state over the domain of interest, solving for and storing the control $u(x)$ or the SDRE solution $P(x)$ in a grid, and interpolating over the stored solutions to approximate the suboptimal feedback control.

SEE ALSO THE FOLLOWING ARTICLES

Behavioral Ecology / Computational Ecology / Conservation Biology / Dynamic Programming / Fisheries Ecology / Foraging Behavior / Harvesting Theory

FURTHER READING

Anderson, B. D. O., and J. B. Moore. 1989. *Optimal control: linear quadratic methods*. Englewood Cliffs, NJ: Prentice-Hall.

Banks, H. T., S. C. Beeler, and H. T. Tran. 2000. Feedback control methodologies for nonlinear systems. *Journal of Optimization Theory and Applications* 107: 1–33.

Banks, H. T., B. M. Lewis, and H. T. Tran. 2007. Nonlinear feedback controllers and compensators: a state-dependent Riccati equation approach. *Journal of Computational Optimization and Applications* 37: 177–218.

Bellman, R. E. 1957. *Dynamic programming*. Princeton: Princeton University Press.

Bertsekas, D. P. 2000. *Dynamic programming and optimal control*. Nashua, NH: Athena Scientific.

Botsford, L. W. 1981. Optimal fishery policy for size-specific, density-dependent population models. *Journal of Mathematical Biology* 12: 265–293.

Clark, C. 2010. *Mathematical bioeconomics: the mathematics of conservation*. Hoboken, NJ: Wiley.

Kirk, D. 1998. *Optimal control theory: an introduction*. Mineola, NY: Dover Publications.

Mangel, M., and C. W. Clark. 1988. *Dynamic modeling in behavioral ecology*. Princeton: Princeton University Press.

Pontryagin, L. S., V. G. Boltyanskii, R. V. Gamkrelidze, and E. F. Mishchenko. 1962. *The mathematical theory of optimal processes*. New York: Interscience Publishers.

ORDINARY DIFFERENTIAL EQUATIONS

SEBASTIAN J. SCHREIBER

University of California, Davis

Since their Newtonian inception, differential equations have been a fundamental tool for modeling the natural world. As the name suggests, these equations involve the derivatives of dependent variables (e.g., viral load, species densities, genotypic frequencies) with respect to independent variables (e.g., time, space). When the independent variable is scalar, the differential equation is called ordinary. Far from ordinary, these equations have provided key insights into catastrophic shifts in ecosystems, dynamics of disease outbreaks, mechanisms maintaining biodiversity, and stabilizing forces in food webs.

BASIC DEFINITIONS AND NOTATION

Before getting caught up in definitions, let us consider two classical ecological examples of ordinary differential equations. The logistic equation describes population growth rate of a species with density x with respect to time t

$$\frac{dx}{dt} = rx(1 - x/K), \tag{1}$$

where r is the intrinsic rate of growth of the population and K is its carrying capacity. The left-hand side of this equation is the derivative of x with respect to time t and corresponds to the population growth rate. The right-hand side describes how the population growth rate depends on the current population density. For this equation, one might be interested in understanding how the species' density changes in time and how these temporal changes depend on its initial density. Often, ecological systems involve many "moving parts" with multiple types of interacting individuals, in which case describing their dynamics involves systems of ordinary differential equations. A classical example of this type is the Lotka–Volterra predator–prey equations that describe the dynamics between a prey with density x_1 and its predator with denisty x_2:

$$\begin{aligned}\frac{dx_1}{dt} &= rx_1 - ax_1x_2,\\ \frac{dx_2}{dt} &= cax_1x_2 - mx_2,\end{aligned} \tag{2}$$

where a, c, and d are the predator's attack rate, conversion efficiency, and per capita mortality rate, respectively. For these more complicated equations, one might ask, "Do these equations have well-defined dynamical behaviors?" and "How complicated can these behaviors be?" This entry addresses the answers to these and many other questions.

A system of ordinary differential equations involves a finite number of state variables, say, $x_1, x_2, \ldots, x_n$, possibly representing densities of interacting species, abundances of different age or size classes, frequencies of behavioral strategies, or availability of abiotic factors such as light, water, or temperature. These equations assume the state variables vary smoothly in time. If the instantaneous rates of change of the state variables are given by functions $f_1(x_1, \ldots, x_n), \ldots, f_n(x_1, \ldots, x_n)$ of the dependent variables, then one arrives at a system of autonomous, first-order ordinary differential equations:

$$\begin{aligned}\frac{dx_1}{dt} &= f_1(x_1, \ldots, x_n),\\ \frac{dx_2}{dt} &= f_2(x_1, \ldots, x_n),\\ &\vdots\\ \frac{dx_n}{dt} &= f_n(x_1, \ldots, x_n),\end{aligned} \tag{3}$$

where t denotes time. Here, *first-order* refers to the fact that only first-order derivatives appear in the equations. *Autonomous* refers to the fact that the rates of change $f_1, \ldots, f_n$ do not depend on t. Equation 3 can account for higher-order time derivatives or time dependence by introducing extra state variables as discussed later.

A simple example of this type of differential equation is the logistic equation.

Equation 3 can be written more succinctly in vector notation. Let $\mathbf{x} = (x_1, x_2, \ldots, x_n)$ be the vector of state variables and $\mathbf{f}(x) = (f_1(\mathbf{x}), f_2(\mathbf{x}), \ldots, f_n(\mathbf{x}))$ be the vector of their corresponding rates of change. Then Equation 3 simplifies to the vector-valued differential equation:

$$\frac{d\mathbf{x}}{dt} = \mathbf{f}(\mathbf{x}). \tag{4}$$

EXISTENCE, UNIQUENESS, AND GEOMETRY OF SOLUTIONS

After writing down a differential equation model of an ecological system, a modeler wants to know how the variables of interest change in time. For instance, how are densities of the prey and predator changing relative to one another? Moreover, how does this dynamic depend on the initial densities of both populations? To answer these types of questions, one needs to find solutions to the differential equation. A solution to Equation 4 is a (vector-valued) function of time $\mathbf{x}(t)$ such that when $\mathbf{x}(t)$ is plugged into both sides of the equation, both sides are equal to one another i.e., $\mathbf{x}'(t) = \mathbf{f}(\mathbf{x}(t))$ where $'$ denotes a derivative with respect to time. If $\mathbf{x}(t)$ is a solution and the initial state of the system is given by $\mathbf{x}(0) = \mathbf{x}_0$, then $\mathbf{x}(t)$ is a solution to the initial value problem:

$$\frac{d\mathbf{x}}{dt} = \mathbf{f}(\mathbf{x}) \qquad \mathbf{x}(0) = \mathbf{x}_0. \tag{5}$$

Verifying a function $\mathbf{x}(t)$ is a solution is straight-forward: plug $\mathbf{x}(t)$ in both parts of the equation and verify both sides are equal.

Consider, for example, the simplest model of population growth in which the per-capita growth rate r of the population is constant in time. If $\mathbf{x} = x$ is the population density, then

$$\frac{dx}{dt} = rx \qquad x(0) = x_0.$$

The solution of this initial value problem is $x(t) = x_0 e^{rt}$. Indeed, $x'(t) = r x_0 e^{rt} = rx(t)$, and $x(0) = x_0$. Intuitively, when the per capita rate r is positive and x_0 is positive, the population exhibits unbounded exponential growth. When r is negative, the population declines exponentially to extinction. However, extinction is only achieved in the infinite time horizon. When the per capita growth rate or the initial population density is zero, the population density remains constant for all time. This latter type of solution is known as an equilibrium or steady-state solution.

In general, an equilibrium for Equation 4 is a state $\mathbf{x} = \mathbf{x}^*$ such that the rates of change are zero at this state: $\mathbf{f}(\mathbf{x}^*) = 0$. The solution corresponding to an equilibrium $\mathbf{x} = \mathbf{x}^*$ is the constant solution $\mathbf{x}(t) = \mathbf{x}^*$ for all t. Indeed, plugging $\mathbf{x}(t) = \mathbf{x}^*$ into the left and right side of equation (3) yields $\mathbf{x}'(t) = 0$ and $\mathbf{f}(\mathbf{x}^*) = 0$. For example, for the logistic equation $x = 0$ is the "no-cats, no-kittens" equilibrium and $x = K$ is the equilibrium corresponding to the carrying capacity of the population.

A fundamental question is, "When do solutions to the initial value problem exist?" After all, if there is no solution, there is no point in looking for one. In 1890, the Italian mathematician Giuseppe Peano proved that solutions to Equation 5 exist whenever $\mathbf{f}(\mathbf{x})$ is continuous. This continuity assumption is met for many models (in particular, the logistic equation and the Lotka–Volterra equations), but not all. For example, population models including optimal behavior can exhibit discontinuities at population states where multiple behaviors are optimal. When these discontinuities occur, what constitutes the dynamic of the equations needs to be defined by the modeler.

Given a solution to the initial value problem exists, one has to wonder whether it is unique. After all, if it is not unique, one may never know if all solutions have been uncovered and which, if any, are biologically relevant. Provided that $\mathbf{f}(\mathbf{x})$ is differentiable, French mathematician Émile Picard proved that solutions are unique. When $\mathbf{f}(\mathbf{x})$ is not differentiable, biologically plausible things can still happen. For example, if $\mathbf{x} = x$ is the density of a declining population whose growth rate is $\frac{dx}{dt} = -\sqrt{x}$, then the growth rate is not differentiable at $x = 0$. This equation yields an infinite number of solutions in which the population has gone extinct by some specified time (Fig. 1A). Despite the model being deterministic, knowing the population is extinct now doesn't tell you when it went extinct. This biologically realistic feature does not occur in

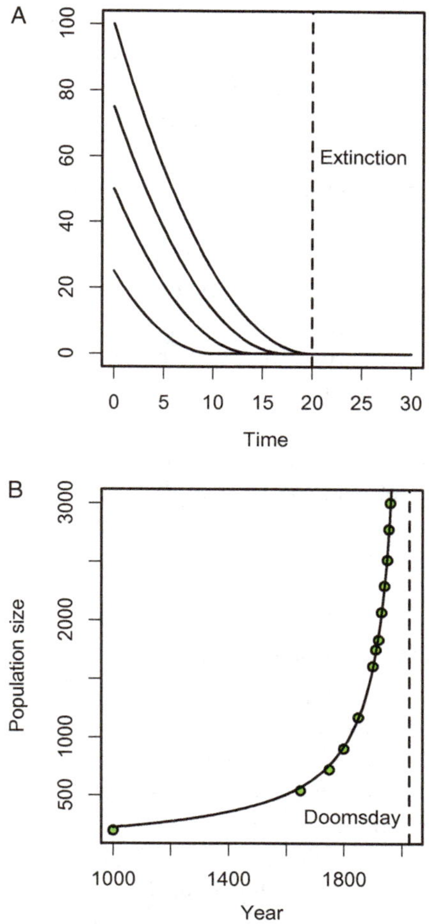

FIGURE 1 Population extinction and blow-up in finite time. In (A), solutions to $dx/dt = -\sqrt{x}$ that all satisfy $x(20) = 0$; i.e., there are multiple population trajectories for which the population is extinct at time 20. In (B), data on human population growth until 1960 (green circles) and the best fitting (in the sense of least squares) solution of $dx/dt = rx^{1+b}$ that blow up in finite time in year 2026.

models where **f**(**x**) is differentiable. While **f**(**x**) is differentiable for most ecological models, there are important exceptions such as predator–prey models with a ratio-dependent functional response (e.g., $a(x_1/x_2)/(1 + ahx_1/x_2)$ where x_1/x_2 is the ratio of prey density to predator density), epidemiological models with frequency-dependent transmission rates, or models incorporating optimal behavior.

Even if solutions to the initial value problem exist and are unique, they might not be defined for all time. An important example is superexponential population growth where the rate of change of the population density x is $\frac{dx}{dt} = rx^{1+b}$ where $b > 0$ and $r > 0$. This model with $b \approx 1$ was the basis for the 1960s prediction of "Doomsday: Friday, 13 November, 2026." Indeed, for $b = 1$, this equation can be solved by separating variables and integrating, which yields $x(t) = 1/(1/x_0 - rt)$ where x_0 is the initial population density. Amazingly, this solution approaches infinity as t approaches $t = 1/(x_0 r)$. This phenomenon is called blow-up and occurs in models of combustion and other runaway processes. In this ecological context, the time of blow-up $t = 1/(x_0 r)$ corresponds to doomsday. In the words of a Pogo cartoon, at this time "everyone gets squeezed to death" (Fig. 1B).

Provided **f**(**x**) is differentiable, solutions to Equation 5 exist for all time if the system is dissipative: there is a bounded region that all solutions eventually enter and remain in for all time. Since the world is finite and can only sustain a finite density of individuals, all ecologically realistic models should be dissipative, i.e., there populations densities eventually are bounded by a constant independent of initial conditions. Verifying dissipativeness can be challenging, and may be violated in unexpected ways. For example, in Lotka–Volterra type models of mutualistic interactions, population densities can blow up in finite time despite each species being self-limited. In the words of Robert May, "For mutualisms that are sufficiently strong, these simple models lead to both populations undergoing unbounded exponential growth, in an orgy of reciprocal benefaction."

Geometrically, solutions trace out curves in the state space, the set of all possible states. For example, for models of n interacting species, the state space consists of all vectors $\mathbf{x} = (x_1, \ldots, x_n)$ whose components are nonnegative. Alternatively, in metacommunity models where each patch can be in one of n different states (e.g., a particular community of species is occupying the patch), the state space consists of all distributions on the n states: nonnegative $\mathbf{x} = (x_1, \ldots, x_n)$ such that $x_1 + \cdots + x_n = 1$. Solutions $\mathbf{x}(t)$ are characterized by being curves whose tangent vectors are given by the right hand side $\mathbf{f}(\mathbf{x}(t))$ of the differential equation (Fig. 2). This geometric interpretation

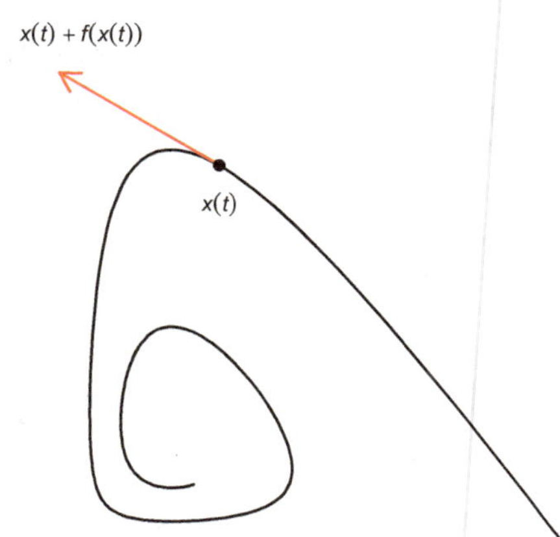

FIGURE 2 A solution $\mathbf{x}(t)$ of a differential equation plotted in its state space. At each point along the solution, its tangent vector is given by $\mathbf{f}(\mathbf{x}(t))$.

of solutions is extremely useful for understanding their qualitative behavior as discussed further below.

LINEAR EQUATIONS

In the absence of density-dependent or frequency-dependent feedbacks, ecological dynamics can be described by systems of linear differential equations. These linear models, just like their discrete-time matrix model counterparts, are particularly useful for describing populations structured in space, size, or age. For these models, the rates of change are linear functions of the state variables:

$$f_i(\mathbf{x}) = a_{i1}x_1 + a_{i2}x_2 + \cdots + a_{in}x_n.$$

If A is the matrix whose ijth entry is a_{ij}, then these linear differential equations can be written more simply as

$$\frac{d\mathbf{x}}{dt} = A\mathbf{x},$$

where $A\mathbf{x}$ denotes matrix multiplication of A and $\mathbf{x}$.

Remarkably, as in the scalar case, the general solution for these equations can be written down explicitly as $\mathbf{x}(t) = \exp(At)\mathbf{x}(0)$, where $\exp(A)$ denotes matrix exponentiation of A and $\mathbf{x}(0)$ is the initial state of the system. As in the case of the exponential function, matrix exponentiation is given by an infinite power series, $\exp(A) = I + A + A^2/2! + A^3/3! + \cdots$, and can be carried out easily in most numerical software packages and program languages.

For the typical matrix A (i.e., the eigenvalues all have nonzero real part), the behavior of the matrix Equation 4 can be classified into two types (for $n = 2$, see Fig. 3). Analogous to the exponential growth model with a negative per-capita growth rate, all solutions of Equation 4 decay exponentially to zero if all eigenvalues of A have a negative real part. In this case, the origin is stable: solutions starting near the origin asymptotically approach the origin. For structured populations, this behavior corresponds to a deterministic asymptotic decline to extinction. Alternatively, all (or almost all) nonzero solutions of Equation 4 grow exponentially when one the eigenvalues of A has a positive real part. In particular, in this case, the origin is unstable: small perturbations away from the origin result in solutions moving further away from the origin. For structured populations, this behavior corresponds to unbounded exponential growth of the entire population. When the eigenvalues of A have imaginary parts, the dynamics can exhibit dampened oscillations if the corresponding real parts of the eigenvalues are negative or undampened oscillations if the corresponding real parts are positive.

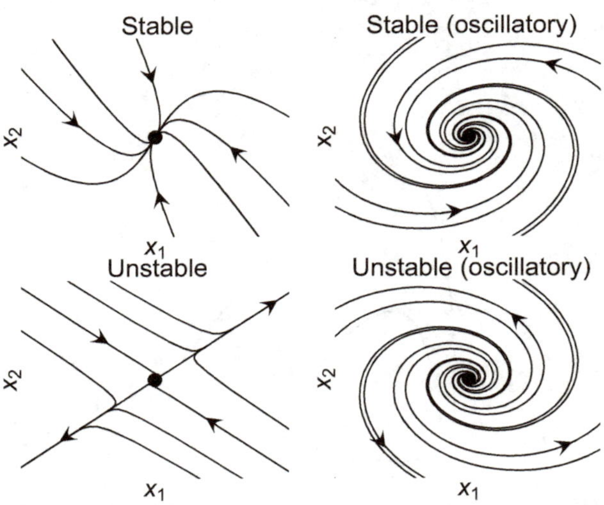

FIGURE 3 Different dynamical behaviors in the phase plane for two-dimensional linear equations.

NONLINEAR EQUATIONS

All natural populations exhibit some degree of density dependence or frequency dependence. Therefore, more realistic models are necessarily nonlinear, e.g., the per capita growth rates are no-longer constant. While it is possible to solve special nonlinear equations (e.g., the logistic equation or the doomsday equation), in general it is impossible to find explicit solutions. In the words of Henri Poincaré, a French mathematician and the founder of modern dynamical systems theory, "Formerly an equation was not considered solved except when the solution was expressed by means of a finite number of known functions; but that is possible scarcely once in a hundred times. What we can always do, or rather what we may always try to do, is to solve the problem qualitatively, so to speak—that is, to find the general shape of the curve which the unknown function represents." Poincaré's suggestion to find "the general shape of the curve" is the genesis of the qualitative theory of differential equations. Instead of finding or approximating the elusive solutions of nonlinear differential equations, this theory examines whether solutions tend to increase or decrease, oscillate or not, or exhibit more complicated long-term behaviors. To achieve these goals, a diversity of techniques have been developed including linearization to understand qualitative behavior of solutions near a well-understood solution such as an equilibrium, bifurcation theory to understand how long-term behaviors change (i.e., bifurcate) as one varies parameters of the system, and ergodic theory to understand the long-term statistical behavior of solutions.

The number of possible asymptotic behaviors exhibited by nonlinear differential equations increases with the dimensional of the system. Some of the highlights of this increasing complexity are as follows.

ONE-DIMENSIONAL SYSTEMS

When there is only one state variable, say, population density, the solutions of Equation 3 are monotonic: either the population density remains constant for all time (i.e., the population is at equilibrium), the population density constantly increases, or the population density constantly decreases. If the population dynamics are bounded in time, then increasing or decreasing solutions always asymptotically approach an equilibrium state. All three of these behaviors can be observed in the logistic model of population growth (Fig. 4A). For solutions initiated below the carrying capacity, the population density increases and asymptotically approaches the carrying capacity. For populations starting at the carrying capacity, their density remains constant for all time. For population starting above the carrying capacity, overcrowding results in the population density decreasing to the carrying capacity.

Despite the simplicity of the short-term and long-term behaviors of one-dimensional models, they can produce surprising predictions due to the appearance and disappearance of equilibria as parameters vary. For example, consider the logistic model with constant harvesting at rate h, $dx/dt = rx(1 - x/K) - h$. For harvesting rates below a critical threshold $h = rK/2$, the population can persist at an equilibrium supporting a density greater than half of the carrying capacity K. For harvesting rates above this threshold, the population goes deterministically extinct in finite time. Hence, increasing harvesting rates above this critical threshold cause sudden population disappearances as population crash from densities greater than $K/2$ to extinction.

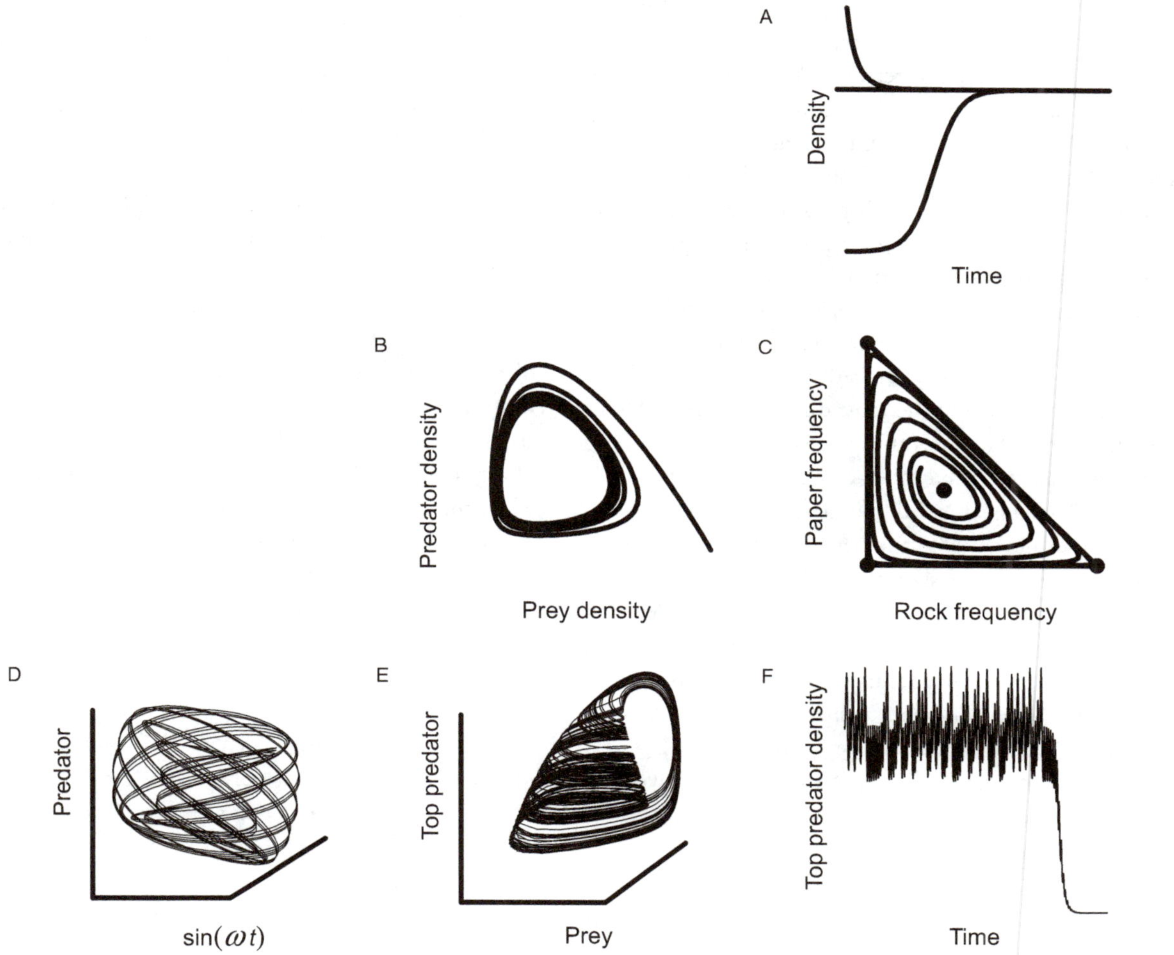

FIGURE 4 Asymptotic behaviors of nonlinear differential equations. For one-dimensional equations such as the logistic equation (A), solutions plotted against time are monotonic and asymptotically approach an equilibrium. For two-dimensional systems, two new asymptotic behaviors are introduced in the phase plane: periodic motions as in predator–prey models (B) and heteroclinic cycles as in rock-paper-scissors population games (C). For three-dimensional systems, the dynamics become infinitely enriched. This enrichment includes quasi-periodic motions of periodically forced predator-prey interactions (D), and chaotic motions (E) and chaotic transients (F) of tritrophic interactions.

TWO-DIMENSIONAL SYSTEMS

Nonlinear feedbacks between two state variables introduces two new asymptotic behaviors for solutions: periodicity and heteroclinic cycles.

Periodicity occurs when the system supports a periodic solution: there exists a period $T > 0$ such that $\mathbf{x}(t + T) = \mathbf{x}(t)$ for all time t. The classic ecological example of periodic behavior is logistic Holling model of predator–prey interactions. For this model, the prey exhibit logistic growth in the absence of the predator and the predator has a saturating functional response. If x_1 and x_2 denote the prey and predator densities, respectively, then their dynamics are given by

$$\frac{dx_1}{dt} = rx_1(1 - x_1/K) - \frac{ax_1}{1 + ahx_1},$$
$$\frac{dx_2}{dt} = \frac{cax_1}{1 + ahx_1} - mx_2,$$

where and a, h, c, m denote the predator's searching efficiency, handling time, conversion efficiency, and per capita mortality rate, respectively. When prey's carrying capacity is sufficiently large, the equilibrium supporting the predator and prey becomes unstable and there is a periodic solution supporting both species (Fig. 4B). For this model, this periodic solution is globally stable (almost all solutions approach it asymptotically) and unique. However, for other forms of functional responses, there can be multiple periodic solutions. Determining the multiplicity of periodic solutions is a tricky affair. Unlike equilibria, there is no simple algebraic procedure to solve for periodic solution. In fact, one of the most difficult open problems in mathematics, Hilbert's 16th problem, involves finding upper bounds to the number of possible periodic solutions to two-dimensional differential equations.

The other new asymptotic behavior introduced by the second dimension corresponds to solutions approaching a heteroclinic cycle: equilibria connected in a cyclic fashion by other solutions of the differential equations. A classic example of this behavior occurs for three competing strategies playing an evolutionary game of rock-paper-scissors (Fig. 4C). If x_1, x_2, x_3 denote the frequencies of the rock, paper, and scissors strategies, respectively, then this dynamic can be described by a replicator equation of the form:

$$\frac{dx_1}{dt} = x_1(bx_3 - cx_2 - W(\mathbf{x})),$$
$$\frac{dx_2}{dt} = x_2(bx_1 - cx_3 - W(\mathbf{x})), \qquad (6)$$
$$x_3 = 1 - x_1 - x_2,$$

where b is the benefit a dominant strategy receives in a pairwise interaction, c is the cost the subordinate strategy plays, and $W(\mathbf{x}) = (b - c)(x_1x_2 + x_1x_3 + x_2x_3)$ is the population average fitness. For this evolutionary game, pairwise interactions always result in the dominant strategy (e.g., rock) displacing the subordinate strategy (e.g., scissors). These pairwise displacements create the heteroclinic cycle between the pure strategy equilibria: rock beats scissors, paper beats rock, and scissors beats paper. If costs are greater than benefits ($c > b$), then most solutions initially supporting all strategies asymptotically approach this heteroclinic cycle (Fig. 4C). As the solutions wrap around this heteroclinic cycle, they spend a longer and longer time near each of the pure strategy equilibria. Moreover, as they spend longer times near pure strategy equilibria, the frequencies of the remaining strategies get lower and lower. Biologically, this dynamic implies that all but one of the strategies are lost in the long run. These heteroclinic cycles arise quite naturally in a variety of ecological models in higher dimensions.

THREE DIMENSIONS AND HIGHER

In three dimensions or higher, many new dynamical phenomena arise, and, unlike planar systems, there is no simple classification of all these phenomena. Three new behaviors that cannot occur in lower dimensions are quasi-periodic motions, chaos, and chaotic transients.

Quasi-periodic motions (Fig. 4D), roughly, correspond to solutions that exhibit multiple, incommensurable frequencies; i.e., the ratio of frequencies is irrational. When there are two frequencies, one can visualize quasi-periodic motions as a curve on a torus (i.e., the surface of a donut) that wraps around the torus without ever returning to the same point. These quasi-periodic motions arise quite naturally when there is periodic forcing of a system with biotically generated periodic behavior or when there are periodic subsystems that interact with one another as discussed below.

One of the most astonishing behaviors exhibited by ordinary differential equations with three or more state variables is chaotic behavior: complex dynamics that make long-term predictions about the state of the system difficult by exponentially amplifying the smallest uncertainty about the current state of the system. This chaotic behavior is exhibited in models of tritrophic interactions. If x_1, x_2, and x_3 are the densities of the prey, predator, and top predator, the prey exhibit logistic growth, and the predators have saturating functional responses, then a classical model of these interactions is

$$\frac{dx_1}{dt} = rx_1(1 - x_1/K) - \frac{a_2x_1x_2}{1 + a_2h_2x_1},$$
$$\frac{dx_2}{dt} = \frac{c_2a_2x_1x_2}{1 + a_2h_2x_1} - m_2x_2 - \frac{a_3x_2x_3}{1 + a_3h_3x_2}, \qquad (7)$$
$$\frac{dx_3}{dt} = \frac{c_3a_3x_2x_3}{1 + a_3h_3x_2} - m_3x_3,$$

where a_i, h_i, c_i, m_i are searching efficiencies, handling times, conversion efficiencies, and per capita mortality rates of the predator and top predator, respectively. Alan Hastings and Thomas Powell showed that this deceptively simple system exhibits chaotic motions (Fig. 4E) for biologically plausible parameter values. These chaotic motions trace out a tea cup in the three-dimensional state space: the predator and prey exhibit oscillatory dynamics that dampen as they wind up the tea cup, a reduction in the predator's density causes the top predator's density to crash down the handle of the tea cup, and the oscillatory path up the tea cup reinitiates.

Chaotic motions may be unstable; i.e., population trajectories initiated near a chaotic motion ultimately move away from this chaotic motion. However, unstable chaotic motions can "trap" nearby solutions in their topologically complex maze for exceptionally long periods of time. When this occurs, the system exhibits chaotic transients before switching to its final asymptotic behavior. For example, Kevin McCann and Peter Yodzis showed that the tritrophic model (Eq. 7) can exhibit long-term chaotic transients supporting all three species, but ultimately the top predator is lost as the dynamics simplify to a predator–prey oscillation (Fig. 4F). This outcome is somewhat shocking as it implies that even without any additional disturbances, a population that appears to be persisting heartily today may be deterministically doomed to extinction in the near future.

One startling way to generate chaotic dynamics and quasi-periodic motions is to increase dimensions by spatially extending models. For ordinary differential equations, this extension can be achieved by modeling space as a finite number of discrete patches and coupling local interactions by dispersal. These spatial extended models can generate chaos even when the local interactions do not. Figure 5 illustrates this phenomena for a spatially extended version of the rock-paper-scissor model (Eq. 6).

COMPETITIVE AND COOPERATIVE SYSTEMS

While there is no complete qualitative theory for higher-dimensional systems, there are two important classes of systems for which there is an exception. The first class are cooperative systems, which are characterized by all direct feedback between pairs of state variables being positive: $\frac{\partial f_i(\mathbf{x})}{\partial x_j} > 0$ for all $i \neq j$. As the name suggests, this assumption is meet for some models of mutualistic interactions. For these models, the American mathematician Morris Hirsch has shown that most solutions converge to one of possibly several equilibria. Hence, the long-term dynamics are typically quite simple. The second class are competitive systems, which are characterized by all direct feedbacks between pairs of state variables being negative: $\frac{\partial f_i(\mathbf{x})}{\partial x_j} < 0$ for all $i \neq j$. As their name suggests, this assumption is met for many models of competing species. Remarkably, Hirsch has shown that-the dynamics of an n-dimensional competitive system can be projected onto an $(n - 1)$-dimensional system. Hence, while models of three competing species can exhibit periodic motions and heteroclinic cycles, they cannot exhibit chaotic or quasi-periodic motions.

NONAUTONOMOUS EQUATIONS

In the natural world, environmental conditions rarely remain static over time. Consequently, it is valuable to consider models where the rates of change $\mathbf{f}(\mathbf{x}, t)$ depend on time t. A trick to doing this and retaining the sheep's clothing of autonomy of Equation 4 is to introduce t as an extra state variable, say, $x_{n+1} = t$, that satisfies the trivial equation $dx_{n+1}/dt = 1$. Thus, a nonautonomous system can be viewed as an autonomous system with an extra dimension. An important special case is periodically forced systems: the rates of change $\mathbf{f}(\mathbf{x}, t)$ are periodic in t. Intuitively, one-dimensional periodically forced systems can exhibit periodic solutions. For example, periodically varying the carrying capacity $K(t)$ in the logistic equation results in all nonzero solutions approaching a unique periodic motion. Due to intrinsic lags of population growth, this periodic motion does not perfectly track $K(t)$. In contrast, two-dimensional periodically forced systems can exhibit the dynamic complexity of three-dimensional autonomous systems. For example, periodically forcing the carrying capacity of the prey in a predator–prey system can generate quasi-periodic motions (Fig. 3D) and chaos.

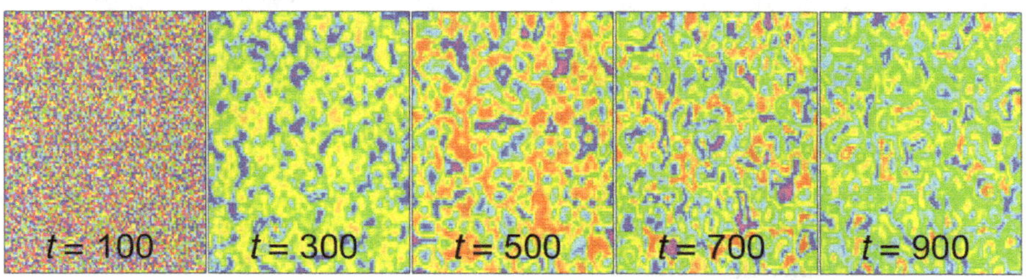

FIGURE 5 Spatial-temporal chaos in a spatial version of the rock-paper-scissors game on a 100 by 100 grid with dispersal to nearest neighbors.

NUMERICAL CONSIDERATIONS

When the chips are down and analysis is not possible, one can close one's door, turn one's computer on, and compute with a fury. Over the past 50 or so years, numerical analysts have developed methods to accurately and efficiently approximate solutions of ordinary differential equations. The simplest method is Euler's method, in which you approximate dx/dt with the difference quotient $(\mathbf{x}(t+h) - \mathbf{x}(t))/h$ where h is a time step. This approximation yields the difference equation

$$\mathbf{x}(t+h) = \mathbf{x}(t) + h\mathbf{f}(\mathbf{x}(t))$$

that can be applied iteratively to approximation solutions to Equation 4. While this method has been used by many theoreticians due to its simple and intuitive appeal, it is generally inefficient for two reasons: errors are of order h^2, and time steps do not dynamically adjust to the magnitude of $\mathbf{f}(\mathbf{x})$; e.g., one can take larger time steps when $\mathbf{f}(\mathbf{x})$ is small. Higher-order methods (e.g., fourth-order Runge–Kutta) with adaptive time steps are available in standard libraries for computing languages like C or Fortran, and computational software like Matlab, R, Maple, and Mathematica. When using these advanced libraries, however, it is important to find implementations that preserve basic structural aspects of the models, e.g., preserve nonnegativity of population densities.

PHILOSOPHICAL CONSIDERATIONS

When is it appropriate to model ecological dynamics with differential equations? This question can be answered indirectly by examining two defining features of differential equations.

First and foremost, differential equations are deterministic. In the words of Henri Poincaré, "If we knew exactly the laws of nature and the situation of the universe at the initial moment, we could predict exactly the situation of the same universe at a succeeding moment." Differential equations cannot account for populations being finite collections of interacting individuals and the associated uncertainty of the individual demography. However, when population abundances are sufficiently large, the American mathematician Thomas Kurtz showed that differential equations provide good approximations (over finite time horizons) to individual-based stochastic models. For instance, consider the Levins' metapopulation model for which the fraction of occupied patches p is modeled by $dp/dt = cp(1-p) - ep$ with per capita extinction and colonization rates e and c. Implicit in Levins' model formulation is that there are an infinite number of patches. This idealization does a good job approximating stochastic models with a finite number of patches whenever the number of patches is sufficiently large (Fig. 6).

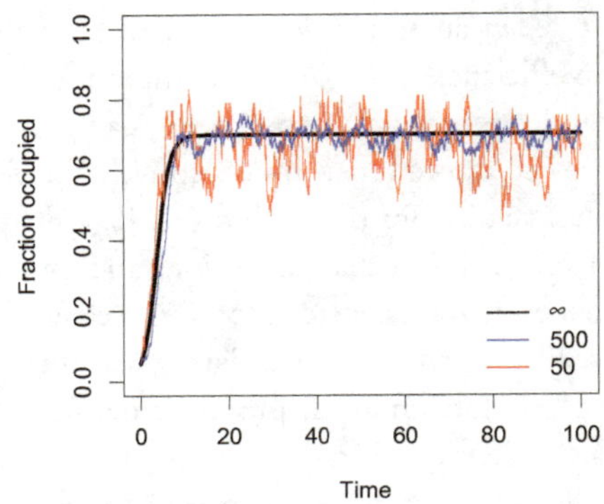

FIGURE 6 Deterministic approximation of metapopulation dynamics. A solution for the Levin's metapopulation model $dp/dt = cp(1-p) - ep$ is plotted in black. Sample trajectories for the stochastic counterpart of this model with 50 and 500 patches are plotted in red and blue, respectively.

A second defining feature of differential equations is that they assume all demographic processes occur continuously in time. For autonomous differential equations, this assumption is quite strong and implies that the populations exhibit overlapping generations. However, by adding time dependence into differential equations, one can account for temporally concentrated demographic events such as synchronized reproductive or migratory events. Classically, populations with these concentrated demographic events have been modeled by difference equations. For populations exhibiting a combination of discrete and continuous demographic processes, there exist impulsive differential equations that lie at the crossroads of difference and differential equations. Despite having a long mathematical history, the impulsive differential equation models have only recently entered the theoretical ecology literature. Whether these hybrid models will overtake the literature remains to be seen.

SEE ALSO THE FOLLOWING ARTICLES

Bifurcations / Chaos / Difference Equations / Food Chains and Food Web Modules / Matrix Models / Nondimensionalization / Phase Plane Analysis / Two-Species Competition

FURTHER READING

Hofbauer, J., and K. Sigmund. 1998. *Evolutionary games and population dynamics*. Cambridge, UK: Cambridge University Press.

Katok, A., and B. Hasselblatt. 1997. *Introduction to the modern theory of dynamical systems*. Cambridge, UK: Cambridge University Press.

Kurtz, T. G. 1981. *Approximation of population processes*. Philadelphia, PA: Society for Industrial and Applied Mathematics.

Perko, L. 2001. *Differential equations and dynamical systems*. Berlin: Springer-Verlag.

Smith, H.L. 1995. *Monotone dynamical systems: an introduction to the theory of competitive and cooperative systems*. Providence, RI: American Mathematical Society.

Strogatz, S.H. 1994. *Nonlinear dynamics and chaos: with applications to physics, biology, chemistry, and engineering*. Reading, MA: Addison-Wesley.

P

PAIR APPROXIMATIONS

JOSHUA L. PAYNE

Dartmouth College, Lebanon, New Hampshire

Pair approximations are analytical techniques for estimating the dynamics and equilibrium properties of network-based models. As the name implies, pair approximations capture the dynamics of the states of neighboring pairs of vertices in a network, as opposed to the dynamics of individual vertex states. These methods have been successfully applied to a variety of network-based ecological and evolutionary models, ranging from the evolution of cooperation to the spread of infectious disease.

MODELING INTERACTIONS

Many ecological processes occur on spatial scales that are much smaller than the entire geographic range of a population. To model such local interactions, populations are often represented as networks, where vertices denote individuals and edges denote their interactions. Finding exact analytical solutions to models of dynamical processes on networks is often exceedingly difficult. Pair approximations are commonly used to overcome these difficulties. Instead of providing an exact solution, these methods use differential equations to approximate the rates of change in the states of connected pairs of vertices, allowing for an estimation of the model's dynamics and equilibrium properties.

THE *SIS* MODEL OF DISEASE SPREAD

Pair approximations will be presented in the context of the classical Susceptible–Infected–Susceptible (*SIS*) model of disease spread, in order to provide a concrete example of their application. In the *SIS* model, a population of N individuals is compartmentalized into two discrete states: susceptible (S) and infected (I). A susceptible individual does not have the disease but is vulnerable to it, whereas an infected individual has the disease and the potential to pass it on.

In the original formulation of this model, which is commonly referred to as a mass-action model, individuals are assumed to come in contact with one another randomly, at rate β individuals per unit time. Letting $[S]$ and $[I]$ denote the number of susceptible and infected individuals in the population, the dynamics of disease spread ($\frac{d[I]}{dt}$) can be simply described with the following differential equation

$$\frac{d[I]}{dt} = \beta\frac{[S]}{N}[I] - g[I]. \tag{1}$$

The rate of change in the number of infected individuals reflects a balance between infection and recovery events. $[I]$ increases at rate $\beta\frac{[S]}{N}[I]$, because each of the $[I]$ infected individuals come into contact with β individuals per unit time, of which $\frac{[S]}{N}$ are susceptible. $[I]$ decreases at rate $g[I]$, as infected individuals recover and return to the susceptible state. The ratio between the contact rate β and the recovery rate g is known as the basic reproductive ratio $R_0 = \beta/g$, which determines whether or not a disease will spread throughout a population. Specifically, if $R_0 > 1$ the contagion will spread because each infected individual transmits the disease to, on average, more than one susceptible individual.

PAIR APPROXIMATIONS

The assumption that individuals encounter one another at random is an oversimplification of the interaction patterns of natural systems; individuals typically encounter

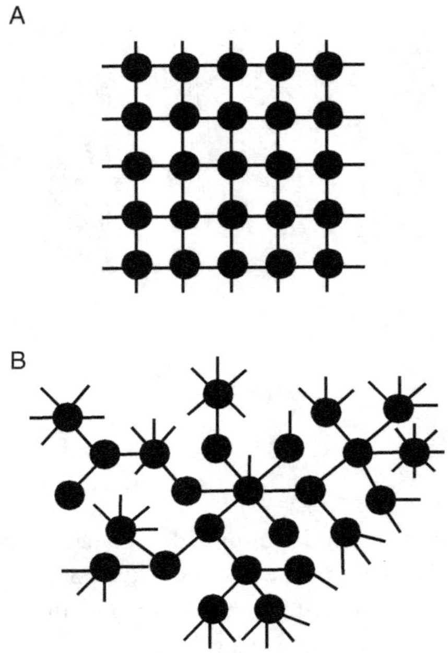

FIGURE 1 Two examples of commonly used interaction networks. (A) Lattice interaction network where each vertex is connected to its nearest neighbors. (B) Random interaction network, with the same average number of edges per vertex as in (A). For illustration, only a small portion (N = 25) of the entire network is depicted. Dangling edges denote connections to vertices that are not shown.

only a small fraction of the total population in their lifetime, and these interactions are usually not random. Such structured interactions are often captured using networks, where individuals are represented as vertices and interindividual interactions are represented as edges (Fig. 1). When a disease spreads throughout such a structured population, its dynamics deviate from those provided by Equation 1, because the assumption of random interaction is violated. The probability of an individual contracting the disease depends on whether or not any of its neighbors in the network are infected. Thus, correlations exist between the states of connected pairs of vertices.

Pair approximations use differential equations to explicitly track these correlations. They were first applied to epidemiological models by Matt Keeling (1999), and we will use his approach and notation. Consider a population structured on a network where every vertex has k edges, and disease transmissibility across an edge is given by $\tau = \beta/k$. Let $[SI]$ denote the number of pairs of connected vertices where one vertex is susceptible and the other is infected (the terms $[SS]$ and $[II]$ are similarly defined, but are counted twice; i.e., these quantities are always even), and let $[SSI]$ denote any connected three-vertex configuration (referred to as a triplet) where the first two vertices are susceptible and the last is infected ($[ISI]$ is similarly defined).

To estimate the dynamics of disease spread, we need to monitor the coupled dynamics of four quantities: $[SS]$, $[SI]$, $[IS]$, and $[II]$. However, we can exploit both symmetry ($[SI] = [IS]$) and redundancy ($[II] = Nk - [SS] - 2[SI]$) so that we only have to monitor the rate of change in $[SI]$ and $[SS]$. The quantity $[SI]$ can change in five ways: It can increase if

(i) one of the vertices in an II pair reverts back to the susceptible state ($II \Rightarrow SI$), which occurs at rate g, or

(ii) one of the vertices in an SS pair contracts the disease from an infected individual outside the pair ($SS \Rightarrow SI$), which occurs at rate τ.

It can decrease if

(iii) an infected vertex in an SI pair reverts back to the susceptible state ($SI \Rightarrow SS$), which occurs at rate g,

(iv) an infected vertex in an SI pair transmits the disease to the susceptible vertex ($SI \Rightarrow II$), which occurs at rate τ, or

(v) a susceptible individual in an SI pair contracts the disease from an infected individual outside the pair ($SI \Rightarrow II$), which occurs at rate τ.

Conditions (ii) and (v) both require information about a vertex state outside of the SI pair. Specifically, condition (ii) can only occur if an SS pair is part of an SSI triplet, and condition (v) can only occur if an SI pair is part of an ISI triplet.

Thus, the rates of change in the number of pairs depend upon the numbers of configurations larger than pairs, and this information is not available. These higher-order quantities are approximated by assuming that the vertices at the opposing ends of a triplet are independent of one another (i.e., triplets form linear chains, not triangles). Under this assumption, $[SSI]$ can be approximated as

$$[SSI] = \frac{(k-1)\,[SS][SI]}{\Sigma_{X \in S,I}\,[SX]} = \frac{(k-1)}{k}\frac{[SS][SI]}{[S]}, \quad (2)$$

where the last equality is valid because the number of singles (e.g., $[S]$) can always be recovered from the number of pairs,

$$[S] = \frac{1}{k}\sum_{X \in S,I} [SX]. \quad (3)$$

The approximation of higher-level quantities from their lower-level counterparts is referred to as "closing" the system, and the name "pair approximation" comes from the fact that this system is closed at the level of pairs.

Using Equations 2 and 3 and the transition rules (i)–(v), we can now describe the rate of change in $[SI]$ as

$$\frac{d[SI]}{dt} = \overbrace{g[II]}^{(i)} + \overbrace{\tau[SSI]}^{(ii)} - \underbrace{g[SI]}_{(iii)} - \underbrace{\tau[SI]}_{(iv)} - \underbrace{\tau[ISI]}_{(v)}. \quad (4)$$

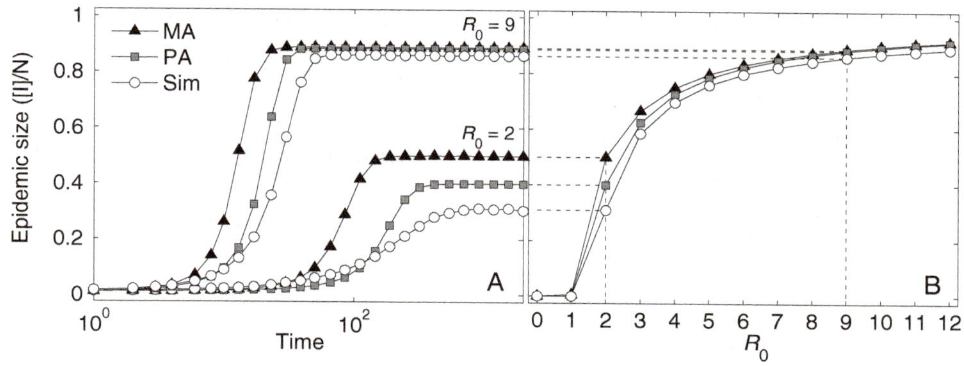

FIGURE 2 (A) Dynamics and (B) equilibrium size of epidemic outbreaks as estimated by the mass-action model (MA, Eq. 1), the pair approximation (PA, Eqs. 4 and 5), and as observed via direct simulation (Sim) on a 10 × 10 square lattice with nearest-neighbor interactions ($k = 4$). Simulation results correspond to 1000 independent replications for each value of the reproductive ratio $R_0 = \beta/g$. In (A), the x-axis is logarithmically scaled.

Similarly, we can describe the rate of change in $[SS]$ as

$$\frac{d[SS]}{dt} = 2g[SI] - 2\tau[SSI], \qquad (5)$$

where the first term on the right-hand side of the equation captures the infected individuals in SI pairs reverting back to susceptibility, and the second term captures SS pairs changing to SI pairs due to their involvement in SSI triplets. These quantities are doubled to ensure that $[SS]$ is counted twice, which is required by Equation 3.

COMPARING THEORY AND DATA

In Figure 2, we depict the dynamics and equilibrium conditions of disease spread estimated by the mass-action model (Eq. 1) and the pair approximation (Eqs. 4 and 5), and as observed through direct simulation on a square lattice with nearest-neighbor interactions (Fig. 1A). As expected, the mass-action model consistently overpredicts the rate of disease spread, relative to the simulations (Fig. 2A). The pair approximation offers considerable improvement, estimating a slower rate of spread than the mass-action model, though not quite as slow as that observed through simulation.

The dynamics estimated by the pair approximation are therefore not completely accurate. However, the equilibrium outbreak size is generally well captured, and for low R_0, the pair approximation provides a more accurate estimation of equilibrium outbreak size than the mass-action model (Fig. 2B).

DISCUSSION

By tracking the correlations between the states of connected vertices, pair approximations provide a more accurate description of the dynamics and equilibrium conditions of disease spread through structured populations than the mass-action model. The discrepancies observed between the pair approximation and the simulation data result from the violation of one principal assumption of the closure method used in the pair approximation: that the underlying contact network is perfectly branching (i.e., possesses no loops, as in Fig. 1B). In that case, it is accurate to assume that the distant ends of triplets are completely independent of one another, as is done in Equation 2. However, in the lattice network considered herein (Fig. 1A), loops abound, and they considerably impact the rate of disease spread.

The accuracy of the pair approximation in predicting the pre-equilibrium dynamics of disease outbreaks can be improved considerably by taking into account some of the topological features of the network. For example, the proportion of triplets that form closed triangles can be incorporated into the closure method (Eq. 2), or the ratio of the local neighborhood size to the underlying lattice size can be used to parameterize the differential equations that describe the rate of disease spread (Eqs. 4 and 5). Accuracy can also be improved by explicitly tracking the dynamics of higher-order motifs, such as triplets. However, the required number of differential equations grows exponentially with the size of the motifs being tracked, which is why the system is usually closed at the level of pairs.

AREAS OF APPLICATION

The pair approximation is a versatile technique that has been applied to many network-based models of ecological and evolutionary processes. For example, it has been used to derive explicit conditions for species invasions in viscous populations, particularly in the context of vegetation dynamics. The pair approximation has also been used to derive simple rules for the evolution of cooperative behavior in social dilemmas and to estimate the equilibrium proportion of cooperators in various evolutionary games. As discussed in this entry, the pair approximation has also been applied to epidemiology, capturing the rate of disease spread and the final epidemic size in structured populations.

SEE ALSO THE FOLLOWING ARTICLES

Adaptive Dynamics / Cooperation, Evolution of / Disease Dynamics / Epidemiology and Epidemic Modeling / Networks, Ecological

FURTHER READING

Keeling, M. J. 1999. The effects of local spatial structure on epidemiological invasions. *Proceedings of the Royal Society London B: Biological Sciences* 266: 859–8767.

Keeling, M. J., and K. T. D. Eames. 2005. Networks and epidemic models. *Journal of the Royal Society Interface* 2: 295–307.

Newman, M. E. J. 2010. *Networks: an introduction.* Oxford: Oxford University Press.

Payne, J. L., and M. J. Eppstein. 2009. Pair approximations of takeover dynamics in regular population structures. *Evolutionary Computation* 17: 203–229.

Petermann, T., and P. De Los Rios. 2004. Cluster approximations for epidemic processes: a systematic description of correlations beyond the pair level. *Journal of Theoretical Biology* 229: 1–11.

Satō, K., and Y. Iwasa. 2000. Pair approximations for lattice-based ecological models. In U. Dieckmann, R. Law, and J. A. J. Metz, eds. *The geometry of ecological interactions: simplifying spatial complexity.* Cambridge, UK: Cambridge University Press.

van Baalen, M. 2000. Pair approximations for different spatial geometries. In U. Dieckmann, R. Law, and J. A. J. Metz, eds. *The geometry of ecological interactions: simplifying spatial complexity.* Cambridge, UK: Cambridge University Press.

PARTIAL DIFFERENTIAL EQUATIONS

NICHOLAS F. BRITTON

University of Bath, United Kingdom

There are extensive applications of partial differential equations (PDEs) in ecology, covering all the main aspects of the science. Questions about the distribution and abundance of organisms may involve PDEs for the population density of various species depending on time and space, for example, in a process of ecological succession. Questions about the movement of materials and energy through living communities may involve PDEs for the concentration of particular chemicals, again depending on time and space. The age structure of a population may change over time according to a PDE, and it may be crucial in considering life processes. The genotypic structure of a population or of a community, which may with time move through some trait space according to a PDE, may be analyzed to explain adaptations and coadaptations. PDEs in science, and in ecology in particular, are often mathematical expressions of conservation laws (such as the law of conservation of mass) and are therefore based on a sound conceptual foundation. Modeling a single population in isolation will generally lead to a single PDE, while models of interacting populations or the interaction of a population with an abiotic resource will lead to systems of coupled PDEs.

SCOPE OF PDE MODELS

A PDE model is not appropriate unless the independent variables are continuous. Time is essentially continuous, but an insect population with nonoverlapping generations is often modeled in discrete time, and although space is essentially continuous, a population in a patchy habitat may be modeled as occupying discrete space. Insects pass through discrete stages in their life history, in a model of a disease the host population is often divided into a finite number of classes, and the simplest population-genetic models consider a finite number of genotypes. Such populations cannot be described using PDEs in the time, space, or structure variables, respectively.

A PDE model is also inappropriate if the dependent variables are not continuous functions of the independent variables, or at least if they may not be approximated as continuous functions. For example, phosphate uptake by phytoplankton in the ocean depends on the phosphate concentration u, which varies as a function of space $\mathbf{x} = (x, y, z)$ and time t. Concentration may be defined as amount of substance per unit volume. The concentration of phosphate in a volume of water containing a point $\mathbf{x}$ is therefore defined, but its concentration *at* the point $\mathbf{x}$ is not. We circumvent this problem by using a continuum approximation: we consider the concentration of phosphate at a point to be its concentration in a volume containing the point where the volume is small compared to the phytoplankton, but large enough that the discrete nature of phosphate ions does not have to be taken into account. This separation in spatial scales is crucial to the approximation. We may define the population density of the phytoplankton similarly, either as the number of individual phytoplankton or as the biomass of phytoplankton per unit volume of water, but it may be that the nature of phytoplankton as discrete organisms plays an essential role in determining the behavior under investigation. If this is so, a PDE model is inappropriate and an individual-based model is required.

In this article we shall consider PDE models for biological populations and abiotic resources distributed in space and moving in time, which involves macroscopic theories for their motion, and PDE models for structured populations.

MOTION OF POPULATIONS AND RESOURCES

Conservation Laws

In PDE models of the motion of populations and resources, the independent variables are time and space and the dependent variables are population densities and concentrations. Consider a population, in some region of interest Ω, whose density is defined as the number of individuals in the population per unit volume. (It is straightforward to consider instead numbers per unit area if appropriate.) PDE models are derived from laws of conservation of matter (expressed as (bio)mass or, in this case, as numbers of individuals). It is important that we can apply this law to any region V with surface S completely contained within Ω. It states that the rate of change of the number of individuals in V is equal to the rate of birth of individuals in V, minus the rate of death of individuals in V, plus the rate of immigration across S into V, minus the rate of emigration across S out of V. (Obvious changes are required if the population is modeled in terms of biomass or if we are interested instead in the concentration of an abiotic resource.) Models for birth and death are analogous to those for ordinary-differential-equation (ODE) models, and we shall restrict discussion in this article to models for motion, and hence for immigration and emigration.

Microscopic and Macroscopic Descriptions of Motion

The motion of populations and resources often has some random component to it. In analyzing the availability of phosphate to phytoplankton, for example, we might be concerned with molecular diffusion, the random motion of the phosphate ions as they are buffeted by molecules of water. The water may be flowing smoothly (so-called laminar flow), and the phosphate is then also carried along by the current, a process known as advection; alternatively, the water may be turbulent, flowing in an irregular and stochastic fashion with eddies at many length scales, so that the phosphate ions undergo turbulent diffusion. A microscopic theory of random particle motion is a description of the statistical properties of this motion for a particle or ensemble of particles, a problem investigated by Einstein. We shall consider instead a macroscopic theory, a description of particle motion in terms of bulk properties such as the velocity field of the fluid (which describes its rate and direction of motion), the particle flux (which describes the rate and direction of motion of the particles), and the concentration or particle density. Advection and turbulent diffusion are inherently macroscopic, as they depend on the fluid flow.

The Concept of Flux

Consider a fluid, which may itself be in motion, containing particles moving within it, and occupying a region Ω. We can think again of phosphate ions or phytoplankton in the ocean. The particle flux at a point $(\mathbf{x}, t)$ within Ω, denoted by $\mathbf{J}(\mathbf{x}, t)$, is defined as follows. First, let $\mathbf{J}$ have magnitude $J = |\mathbf{J}|$ and direction $\mathbf{m} = \mathbf{J}/J$, so $\mathbf{J} = J\mathbf{m}$. Define $\mathbf{m}$ to be the direction of net flow, and define J by placing an infinitesimal test surface of area dS and normal $\mathbf{m}$ within Ω at $(\mathbf{x}, t)$; then JdS is the net number of particles crossing the test surface in the (positive) $\mathbf{m}$-direction per unit time, the so-called current across the surface.

This concept of flux allows us to quantify the rate of net migration out of an arbitrary region V with surface S contained within Ω, as an integral of the outward-pointing normal component of flux over the surface of V, $\int_S \mathbf{J} \cdot \mathbf{n} dS$. Under certain technical conditions, the divergence theorem allows us to convert this surface integral to a volume integral, $\int_V \nabla \cdot \mathbf{J} dV$. We can then make an argument based on conservation of matter in the arbitrary volume V to give

$$\frac{\partial u}{\partial t} + \nabla \cdot \mathbf{J} - f = 0, \tag{1}$$

where f is the net source density, the net number of particles created per unit time and per unit volume. This PDE in time and space is the equation of conservation of matter with a source term and is valid within Ω. It remains to model the flux $\mathbf{J}$ in each case of interest.

Passive Motion

First, consider the flux due to advection with specified velocity $\mathbf{v}$. The particles are moving in the direction of $\mathbf{v}$, and the rate at which they cross an infinitesimal test surface of area dS placed perpendicular to the flow (the current across the surface) is $|\mathbf{v}|udS$, where u is the concentration (particles per unit volume). Thus, $\mathbf{J}_{\text{adv}} = \mathbf{v}u$. The advection equation with a source term is

$$\frac{\partial u}{\partial t} = -\nabla \cdot \mathbf{J}_{\text{adv}} + f = -\nabla \cdot (\mathbf{v}u) + f.$$

Pure advection equations are not common in ecology. They may be tackled analytically by the method of characteristics, but their numerical treatment is not easy. Second, consider the flux due to molecular diffusion. Empirically, the net flow of particles is down the concentration gradient and proportional to its magnitude, so $\mathbf{J}_{\text{diff}} = -D\nabla u$, where D is the scalar diffusivity or diffusion coefficient. This mathematical model of diffusive flux is known as Fick's law. The diffusion equation with a source term is

$$\frac{\partial u}{\partial t} = -\nabla \cdot \mathbf{J}_{\text{diff}} + f = \nabla \cdot (D\nabla u) + f.$$

Diffusion equations may be tackled analytically by the method of fundamental solutions or the method of separation of variables. Numerically, the method of lines is often a good choice. This involves discretizing the equations in space to obtain a system of coupled ordinary differential equations (ODEs) in time, and writing or more likely using a built-in routine to solve these. If we have both advection and diffusion, then $\mathbf{J} = \mathbf{J}_{\text{adv}} + \mathbf{J}_{\text{diff}}$, and the advection–diffusion equation with a source term is

$$\frac{\partial u}{\partial t} = -\nabla \cdot (\mathbf{v}u) + \nabla \cdot (D\nabla u) + f.$$

In turbulent diffusion, the velocity field in the fluid has random elements to it, and we can do no better than model the mean population density over many realizations of the turbulent flow in which the macroscopic conditions creating the turbulence do not vary. A standard model for the flux is then given by $\mathbf{J}_{\text{turb}} = \bar{\mathbf{v}}\bar{u} - \mathrm{K}\nabla\bar{u}$, where bars denote means and $K = \text{diag}\,(K_1, K_2, K_3)$ is the so-called turbulent diffusion or eddy diffusion matrix, a diagonal matrix with the horizontal diffusion coefficients K_1 and K_2 generally many orders of magnitude greater than the vertical diffusion coefficient K_3. The turbulent diffusion equation becomes

$$\frac{\partial \bar{u}}{\partial t} = -\nabla \cdot \mathbf{J}_{\text{turb}} + f = -\nabla \cdot (\bar{\mathbf{v}}\bar{u}) + \nabla \cdot (K\nabla\bar{u}) + f.$$

It is to be solved with the mean velocity field $\bar{\mathbf{v}}$ and the turbulent diffusion matrix K specified by the turbulent motion of the fluid.

Active Motion

Biological organisms not only move at random but sense their environment and respond to it. The response often involves a *taxis*, a movement toward or away from an external stimulus. Some examples of these taxes are chemotaxis, a response to a chemical gradient, phototaxis, a response to a light source, geotaxis, a response to a gravitational field, galvanotaxis, a response to an electric field, and haptotaxis, a response to an adhesive gradient. Tactic responses to conspecifics or to predators and prey are also observed. One of the most important of these taxes is chemotaxis. The chemical may be produced by conspecifics, as when ants follow pheromone trails or when bacteria or slime moulds aggregate. The chemotactic response may be positive, leading to motion up the chemical gradient, or negative. Let us consider bacteria with population density u responding to a chemical with concentration c. The most widely used model for the chemotactic flux of the bacteria is $\mathbf{J}_{\text{chemo}} = \chi u\nabla c$, where χ is a constant of proportionality, known as the chemotactic coefficient or chemotactic sensitivity. It is positive if the bacteria are attracted to the chemical, negative if they are repelled. If c is specified, then this leads to a PDE for the bacteria, but it is often the case that the chemical is being produced by the bacteria. This leads to a system of coupled PDEs typically of the form

$$\frac{\partial u}{\partial t} = f(u) - \nabla \cdot (\chi u\nabla c) + \nabla \cdot (D_u\nabla u),$$

$$\frac{\partial c}{\partial t} = g(u,c) + \nabla \cdot (D_c\nabla c),$$

where f is the net rate at which bacteria are produced and g is the net rate at which the chemical is produced. We are now straying into the area of reaction–diffusion equations, which are covered in a separate article.

Initial and Boundary Conditions

It is clear from the derivation of the PDE models above that we shall need additional conditions to solve the problems. The derivation gave the rate of change of the population size in a region V, and hence we require that the population density at some initial time t_0, the initial condition, be given. Fluxes across surfaces S were considered on the understanding that they were strictly within Ω and that the models were only valid within this region of interest. The conditions *on* the boundary Σ of Ω must be considered separately. It is often the case in ecological models that Ω is the region occupied by the population and there is no flux into or out of it; in that case, the condition on the boundary is given as $\mathbf{J} \cdot \mathbf{n} = 0$ on Σ, where $\mathbf{n}$ is the outward-pointing normal, the so-called zero-flux (or Neumann) boundary condition. An alternative is the zero-density (or Dirichlet) boundary condition $u = 0$ on Σ.

STRUCTURED POPULATIONS

Populations with Age Structure

In the simplest models of population dynamics, age structure is neglected. This can be justified if, despite this neglect, an acceptable model for the expected numbers of births and deaths may be derived. However, the numbers of births and deaths usually depend on the age structure of the population, and this structure must then be taken into account explicitly. In modeling spatially structured populations, a continuum approximation must be made to define the population density (with respect to space) *at* a point **x**. A similar approximation must be made here to define the population density $u(a, t)$ (with respect to age) at age a and time t. We do so by counting the individuals in an age interval including a and dividing by the length of the interval, ensuring that the interval is small compared to the time scale of vital processes but large enough that the discrete nature of the population does not have to be taken into account. A PDE model may then be derived whose independent variables are time and age and whose dependent variable is population density (with respect to age).

In modeling a population of sexual organisms, it is simplest to account for females only. Let $u(a, t)$ be the density of females of age a at time t. Taking into account aging and death, a conservation argument on a cohort leads to

$$\frac{\partial u}{\partial a} + \frac{\partial u}{\partial t} = -du, \tag{2}$$

where $d(a, t)$ is the mortality rate at age a and time t, so that an individual of age a at time t has probability $d(a, t)dt$ of dying in the next infinitesimal interval of time of length dt. This is McKendrick's PDE. It is to be solved for $0 \leq a \leq \omega$, $t \geq 0$, where ω is the greatest possible age in the population, with initial conditions given at $t = 0$,

$$u(a, 0) = u_0(a).$$

Births enter as a boundary condition at age zero,

$$b(t) = u(0, t) = \int_{\alpha}^{\beta} u(a, t)m(a, t)da,$$

where m is the maternity function (so that the expected number of offspring produced by a female of age a in the infinitesimal interval $(t, t + dt)$ is given by $m(a, t)dt$), and α and β are the youngest and oldest ages for childbearing.

This problem may be analyzed by the method of characteristics, which are the lines $t = a + c$, for constant c. Numerically, the simplest approach is to discretize using equal intervals of length h in t and a, so that the numerical solution proceeds along the characteristics. Then Equation 2 is approximated by $u_{i,j} - u_{i-1,j-1} = -d_{i,j}u_{i,j}h$, or $u_{i,j} = u_{i-1,j-1}/(1 + d_{i,j}h)$, using an obvious notation. In the stationary case, where conditions are constant over time, we replace $d(a, t)$ by $d(a)$ and $m(a, t)$ by $m(a)$. Then the solution may be found explicitly, and is given by

$$u(a, t) = \begin{cases} u_0(a - t)l(a)/l(a - t) & \text{for } a > t, \\ b(t - a)l(a) & \text{for } a < t, \end{cases}$$

where the probability $l(a)$ of surviving to age a is given by

$$l(a) = \exp\left(- \int_0^a d(b)db \right).$$

The expression for $a > t$ represents those who were initially of age $a - t$ and who then survived from age $a - t$ to age a in time t, and the expression for $a < t$ represents those who were born at time $t - a$ and then survived to reach age a at time t. The population eventually grows exponentially at rate r, Lotka's intrinsic rate of natural increase, where

$$\int_0^{\omega} e^{-ra} l(a)m(a)da = 1.$$

Note that there are no density-dependent effects in this model (which is why we should expect exponential growth). The total population size $U(t)$ at time t is given by $U(t) = \int_0^{\omega} u(a, t)da$. If the mortality rate d depends on U (or on some age-weighted version of it), then the PDE of Equation 2 becomes an integro-PDE, involving integrals and partial derivatives, taking the problem beyond the scope of this article and making its solution much more difficult. A similar situation occurs in modeling infectious diseases in age-structured populations, where the force of infection generally depends on an age-weighted integral of infectious individuals. The problem then involves a set of coupled integro-PDEs for the susceptible, infectious, and immune population densities.

Populations with Size Structure

It may be the size rather than the age of an individual in a population that determines its mortality rate and maternity function. In that case, we define population density $u(s, t)$ in terms of size rather than age, by counting the number of individuals in a size interval rather than an age interval and dividing by the length of the interval. A conservation argument leads to

$$\frac{\partial(gu)}{\partial s} + \frac{\partial u}{\partial t} = -du,$$

where $g(s, t)$ is the growth rate and $d(s, t)$ the mortality rate for individuals of size s at time t. This is to be solved with an initial condition $u(s, 0) = u_0(s)$ and a birth condition $g(s_0, t)u(s_0, t) = \int_{s_2}^{s_1} m(s, t)u(s, t)ds$, where s_0 is size at birth and s_1 and s_2 are the smallest and largest size at reproduction. It may be generalized (i) if newborns are not all of the same size and (ii) to populations with both age and size structure.

CONCLUDING REMARKS

This article only scratches the surface of applications of PDEs in ecology. It does not include PDEs that arise from stochastic processes, e.g., from diffusion approximations to Markov chain models in population genetics, which are based on the theory of stochastic processes and are consequently beyond its scope. It does, however, cover the main areas of application of deterministic models, to the continuous-time dynamics of populations distributed in continous space or populations structured by a continuous variable such as age or size, and where, necessarily, a continuum approximation to population density is appropriate.

SEE ALSO THE FOLLOWING ARTICLES

Age Structure / Dispersal, Animal / Hydrodynamics / Movement: From Individuals to Populations / Ordinary Differential Equations / Reaction–Diffusion Models / Spatial Spread

FURTHER READING

Britton, N. F. 2003. *Essential mathematical biology.* London: Springer-Verlag.

Cushing, J. M. 1998. *An introduction to structured population dynamics.* Philadelphia: Society for Industrial and Applied Mathematics.

Okubo, A., and S. A. Levin. 2002. *Difffusion and ecological problems: modern perspectives.* New York: Springer-Verlag.

PHASE PLANE ANALYSIS

SHANDELLE M. HENSON

Andrews University, Berrien Springs, Michigan

Phase plane analysis is a tool used in ecology to analyze dynamical systems having two dependent variables (state variables)—for example, predator–prey systems. Phase space is the space spanned by the dependent variables. This entry focuses on the phase space analysis of first-order systems of ordinary differential equations (ODEs); however, phase space is a useful concept for other dynamical systems as well. For one- and two-dimensional systems of ODEs, phase space configurations are catalogued according to the nature of easily computable quantities called eigenvalues. Researchers therefore can analyze the dynamics of a system without solving the equations. A standard application of phase plane analysis in theoretical ecology arises in the study of Lotka–Volterra community models.

AUTONOMOUS SYSTEMS

The idea of phase space is most useful when applied to autonomous systems, that is, systems in which the rates of change do not depend explicitly on time. Phase space is spanned by the axes for the state variables x_1, x_2, . . . , x_n of the dynamical system. The n-tuple at each time t is a point in phase space. The set of such points forms a trajectory (orbit) in phase space with a time arrow direction.

Under certain technical conditions, the fundamental existence and uniqueness theorem guarantees that every point in phase space has a unique solution passing through it. Consequently, two different trajectories in phase space cannot intersect. This fact is fundamental to phase space analysis. If trajectories are confined to a line or plane, as they are in one-dimensional or two-dimensional systems, the possible configurations are limited and easily catalogued. In three or more dimensions, however, trajectories can create complicated tangles in phase space. Complex dynamics, including chaos, can occur only in systems of dimension 3 and higher. (Discrete-time systems, however, can have complex behavior in one and two dimensions because the lines connecting successive points for two different solutions can cross. The simplest nonlinear one-dimensional discrete-time system, the quadratic map $x_{t+1} = rx_t(1 - x_t)$, produces chaotic behavior.)

THE PHASE LINE

Although phase plane analysis refers to systems with two dependent variables, it is illuminating to begin with one-dimensional systems of the form

$$\frac{dx}{dt} = f(x). \tag{1}$$

These have monotone dynamics; in particular, every solution is either constant, strictly increasing, or strictly decreasing. Constant solutions are called equilibria; they are the values of x that make $dx/dt = 0$ for all t, that is, they are the roots of the equation $f(x) = 0$. Phase space is the x-axis; it is called the phase line.

In general, phase space analysis consists of two steps. First, one analyzes the possible behaviors of linear systems, and then one analyzes nonlinear systems using the technique of linearization.

Linear Theory

The linear homogeneous equation

$$\frac{dx}{dt} = px, \tag{2}$$

in which p is constant, has solutions of the form $x = ce^{pt}$. The value $x_e = 0$ is an equilibrium (unique if $p \neq 0$) and there are three possible phase line portraits, depending on the sign of p.

1. $p < 0$ (Sink). If $x > 0$, then $dx/dt < 0$ and so $x(t)$ is decreasing; if $x < 0$, then $x(t)$ is increasing. The phase line portrait is called a sink, and $x_e = 0$ is asymptotically stable (Fig. 1A).

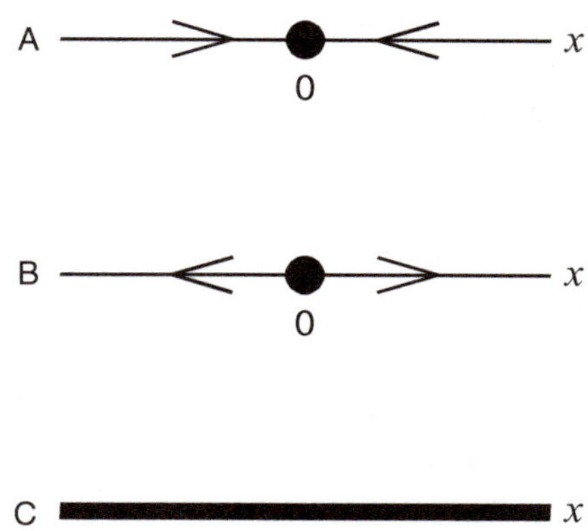

FIGURE 1 Possible phase line portraits of linear homogeneous ODEs with constant coefficient p. (A) If $p < 0$, the origin is a sink. (B) If $p > 0$, the origin is a source. (C) If $p = 0$, every point on the phase line is a neutrally stable equilibrium.

2. $p > 0$ (Source). The arrows are reversed, and $x_e = 0$ is unstable (Fig. 1B).
3. $p = 0$. In this case, all values of x are equilibrium solutions, and the phase line portrait is a solid line of neutrally stable equilibria (Fig. 1C).

The first two cases ($p \neq 0$) are called hyperbolic. Hyperbolicity is important for the linearization of nonlinear systems.

Linearization

The phase line portrait for a nonlinear ODE of the form of Equation 1 is obtained by three steps. First, find the equilibria, that is, the roots of $f(x) = 0$. Second, at each equilibrium x_e, linearize f about x_e and translate the equilibrium to the origin to obtain a linearized ODE of the form of Equation 2. Third, if x_e is hyperbolic, the local phase line portrait (about x_e) can be inferred from the phase line portrait of the linearized ODE. The second of these steps is accomplished as follows.

At an equilibrium x_e, the tangent line approximation of f is

$$f(x) \approx \lambda(x - x_e) \text{ for } x \approx x_e, \tag{3}$$

where the eigenvalue $\lambda = f'(x_e)$ is the slope of f at the equilibrium x_e. Locally, then, near the equilibrium, Equation 1 is approximated by the ODE

$$\frac{dx}{dt} = \lambda(x - x_e). \tag{4}$$

The change of variable $u = x - x_e$ translates the equilibrium to the origin and obtains the linearized ODE

$$\frac{du}{dt} = \lambda u, \tag{5}$$

which has the form of Equation 2. The local phase line portrait depends on the sign of the eigenvalue: if $\lambda < 0$, then x_e is a sink (locally asymptotically stable); if $\lambda > 0$, then x_e is a source (unstable); and if $\lambda = 0$, then x_e is nonhyperbolic and linearization does not apply. In the latter case, the phase line portrait depends on higher-order terms.

EXAMPLE

The logistic model

$$\frac{dx}{dt} = rx\left(1 - \frac{x}{K}\right); \quad r, K > 0, \tag{6}$$

has two equilibria: $x_e = 0$ and $x_e = K$. Here, $f'(x) = r(1 - x/K) - rx/K$ and so the eigenvalues corresponding to the equilibria are $\lambda = r > 0$ and $\lambda = -r < 0$, respectively. Thus, $x_e = 0$ is a source and $x_e = K$ is a sink (Figs. 2A, B).

Phase line portraits of the form $\rightarrow x_e \rightarrow$ and $\leftarrow x_e \leftarrow$ are called shunts. Despite the terminology, solutions are not "shunted" across the equilibrium; rather, they asymptotically approach the equilibrium from the stable side but do not reach it in finite time. Shunts are unstable. Although all shunts are nonhyperbolic, it is not the case that all nonhyperbolic equilibria are shunts. They also can be sinks or sources (Fig. 2C).

PHASE PLANE: EQUILIBRIA

Phase space for two-dimensional systems of ODEs

$$\begin{cases} \dfrac{dx}{dt} = f(x,y) \\ \dfrac{dy}{dt} = g(x,y) \end{cases} \tag{7}$$

is the plane spanned by the x- and y-axes. Solutions $(x(t), y(t))^T$ are depicted by parametric curves in the

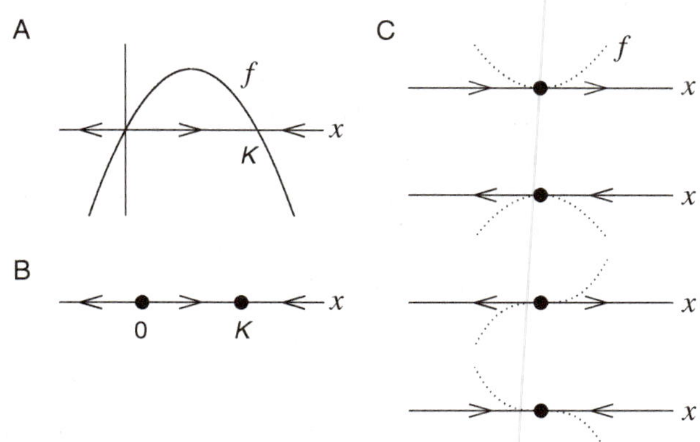

FIGURE 2 Constructing the phase line portrait for a nonlinear ODE. (A) The graph of f vs. x for the logistic model shows the intervals on which $x(t)$ is increasing and decreasing (arrows). (B) In the logistic model $x_e = 0$ is a source and $x_e = K$ is a sink. (C) Nonhyperbolic equilibria can be shunts to the right or left, sources, or sinks.

phase plane, parameterized by time t, where the superscript T denotes the matrix transpose. Equilibria are pairs $(x_e, y_e)^T$ for which $f(x_e, y_e) = 0 = g(x_e, y_e)$. Solutions, unlike those of one-dimensional equations, are not necessarily monotone; in particular, they can oscillate. Oscillating solutions give rise to trajectories in the phase plane that rotate periodically around a closed solution curve or spiral toward or away from such a closed curve or equilibrium point.

Linear Theory

Autonomous linear homogeneous systems have the form

$$\begin{cases} \dfrac{dx}{dt} = ax + by \\ \dfrac{dy}{dt} = cx + dy \end{cases} \tag{8}$$

in which a, b, c, and d are constants. The origin $(0, 0)^T$ is an equilibrium, and it is unique as long as

$$\begin{vmatrix} a & b \\ c & d \end{vmatrix} \neq 0, \tag{9}$$

where $|\cdot|$ denotes the determinant. Substitution of an Ansatz solution

$$\begin{cases} x = v_1 e^{\lambda t} \\ y = v_2 e^{\lambda t} \end{cases} \tag{10}$$

into Equation 8, where the v_i, λ are (possibly complex) constants, yields

$$\begin{cases} 0 = (a - \lambda)v_1 + bv_2 \\ 0 = cv_1 + (d - \lambda)v_2 \end{cases}. \tag{11}$$

The system of algebraic equations in Equation 11 has a nontrivial solution (v_1, v_2) if and only if its coefficients satisfy

$$\begin{vmatrix} a - \lambda & b \\ c & d - \lambda \end{vmatrix} = 0, \tag{12}$$

that is, if and only if

$$\lambda^2 - (a + d)\lambda + (ad - bc) = 0. \tag{13}$$

The eigenvalues are the two roots λ_1, λ_2 of the characteristic Equation 13. For each value of λ, corresponding values of v_1 and v_2 can be determined from Equation 11. The vector $(v_1, v_2)^T$ is the eigenvector belonging to eigenvalue λ, and the associated eigensolution, given in vector form, is

$$\begin{pmatrix} x \\ y \end{pmatrix} = \begin{pmatrix} v_1 \\ v_2 \end{pmatrix} e^{\lambda t}. \tag{14}$$

If $\lambda_1 \neq \lambda_2$, then the general solution of Equation 8 is a linear combination of the two corresponding eigensolutions

$$\begin{pmatrix} x \\ y \end{pmatrix} = c_1 \begin{pmatrix} v_{11} \\ v_{21} \end{pmatrix} e^{\lambda_1 t} + c_2 \begin{pmatrix} v_{12} \\ v_{22} \end{pmatrix} e^{\lambda_2 t} \tag{15}$$

with six possible "generic" phase plane portraits:

1. $\lambda_1 < \lambda_2 < 0$ (Asymptotically stable node). All solutions approach the origin. The orbits (ranges) of the two eigensolutions form two stable manifolds, each of which is a line through the origin in the direction of the corresponding eigenvector. The origin is called an asymptotically stable node (Fig. 3A).
2. $0 < \lambda_1 < \lambda_2$ (Unstable node). All nontrivial solutions grow without bound. The orbits of the eigensolutions are unstable manifolds. The origin is called an unstable node (Fig. 3B).
3. $\lambda_1 < 0 < \lambda_2$ (Saddle). The manifold for the eigensolution with $\lambda_1 < 0$ is stable; the other is unstable. Solutions on the stable manifold approach the origin; all other solutions eventually grow without bound. Initial values close to the stable manifold give rise to trajectories along hyperbolic paths that first approach the origin in the direction of the stable manifold and then leave the origin in the direction of the unstable manifold. The origin is called an (unstable) saddle (Fig. 3C).
4. $\lambda_{1,2} = \pm\beta i$ (Center). The exponential $e^{\lambda t} = e^{\pm\beta i t}$ in the eigensolution of Equation 14 can be written trigonometrically as $\cos \beta t \pm i \sin \beta t$. Solution 15 is reduced to a complex solution whose real and imaginary parts are two independent, real periodic functions of sine and cosine. The linear combination of these two real solutions forms the (periodic) general solution. In the phase plane, solutions are nested closed curves containing the origin. The direction of rotation is determined by the signs of the coefficients in Equation 8. The origin is called a (neutrally stable) center (Fig. 3D).
5. $\lambda_{1,2} = \alpha \pm \beta i$, with $\alpha < 0$ (Asymptotically stable spiral). In this case, $e^{\lambda t} = e^{\alpha t}e^{\pm\beta i t} = e^{\alpha t}(\cos \beta t \pm i \sin \beta t)$, so solutions oscillate with decaying amplitude $e^{\alpha t}$. In the phase plane, trajectories spiral toward the origin. The origin is called an asymptotically stable spiral (Fig. 3E).
6. $\lambda_{1,2} = \alpha \pm \beta i$, with $\alpha > 0$ (Unstable spiral). Solutions oscillate with increasing amplitude; hence, trajectories in phase plane spiral away from the origin. The origin is called an unstable spiral (Fig. 3F).

An equilibrium is called hyperbolic if its eigenvalues have nonzero real parts. Hence, the origin is hyperbolic in each of the above cases except for case 4; centers are nonhyperbolic.

If $\lambda_1 = 0$ and $\lambda_2 \neq 0$, there are infinitely many equilibria, and the general solution of Equation 8 has the form

$$\begin{pmatrix} x \\ y \end{pmatrix} = c_1 \begin{pmatrix} v_{11} \\ v_{21} \end{pmatrix} + c_2 \begin{pmatrix} v_{12} \\ v_{22} \end{pmatrix} e^{\lambda_2 t}. \tag{16}$$

Phase portraits have the following types:

1. $\lambda_1 = 0$ and $\lambda_2 < 0$. There is a line of neutrally stable equilibria passing through the origin with slope v_{21}/v_{11}.

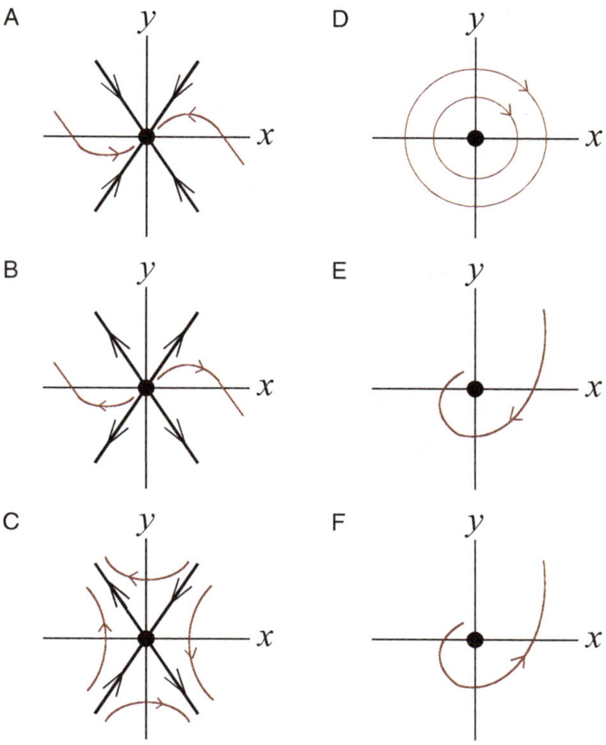

FIGURE 3 The six generic phase plane portraits for linear homogeneous systems with constant coefficients. (A) $\lambda_1 < \lambda_2 < 0$; asymptotically stable node. (B) $0 < \lambda_1 < \lambda_2$; unstable node. (C) $\lambda_1 < 0 < \lambda_2$; saddle. (D) $\lambda_{1,2} = \pm\beta i$; center. (E) $\lambda_{1,2} = \alpha \pm \beta i$, with $\alpha < 0$; asymptotically stable spiral. (F) $\lambda_{1,2} = \alpha \pm \beta i$, with $\alpha > 0$; unstable spiral.

All other solutions approach this line, with trajectories parallel to the stable manifold with slope v_{22}/v_{12} (Fig. 4A).

2. $\lambda_1 = 0$ and $\lambda_2 > 0$. Same as case 1, except the arrows are reversed. All equilibria are unstable (Fig. 4B).

If $\lambda_1 = \lambda_2 = \lambda$, there are two cases of general solutions. For the special case $a = d$ and $c = b = 0$, the general solution is

$$\begin{pmatrix} x \\ y \end{pmatrix} = \begin{pmatrix} c_1 \\ c_2 \end{pmatrix} e^{\lambda t}. \tag{17}$$

Otherwise, the general solution is

$$\begin{pmatrix} x \\ y \end{pmatrix} = c_1 \begin{pmatrix} v_1 \\ v_2 \end{pmatrix} e^{\lambda t} + c_2 \begin{pmatrix} w_1 + v_1 t \\ w_2 + v_2 t \end{pmatrix} e^{\lambda t}, \tag{18}$$

where v_1 and v_2 satisfy Equation 11 and w_1 and w_2 satisfy

$$\begin{cases} v_1 = (a - \lambda) w_1 + b w_2 \\ v_2 = c w_1 + (d - \lambda) w_2 \end{cases}. \tag{19}$$

The phase portrait types are as follows:

1. $\lambda_1 = \lambda_2 < 0$ (Asymptotically stable improper node). In general, trajectories approach the origin tangentially to the single stable manifold (Fig. 4C). If $a = d$ and $c = b = 0$, the trajectories approach the origin radially (asymptotically stable star; Fig. 4D).
2. $\lambda_1 = \lambda_2 > 0$ (Unstable improper node). Like the previous case, except arrows are reversed (Fig. 4E). If $a = d$ and $c = b = 0$, the trajectories leave the origin radially (unstable star, not shown; like Figure 4D except arrows are reversed).
3. $\lambda_1 = \lambda_2 = 0$. In general, there is a line of equilibria through the origin; trajectories are parallel to this line. All the equilibria are unstable (Fig. 4F). If $a = d$ and $c = b = 0$, every point in the plane is a neutrally stable equilibrium (not shown).

Nonlinear Theory

The phase plane portrait for a nonlinear system (Eq. 7) is obtained by three steps. First, find the equilibria. Second, at each equilibrium $(x_e, y_e)^T$, linearize f and g and translate the equilibrium to the origin to obtain a linearized ODE of the form of Equation 8. Third, for hyperbolic equilibria, the local stability and local topology of the phase plane portrait can be inferred from the phase plane

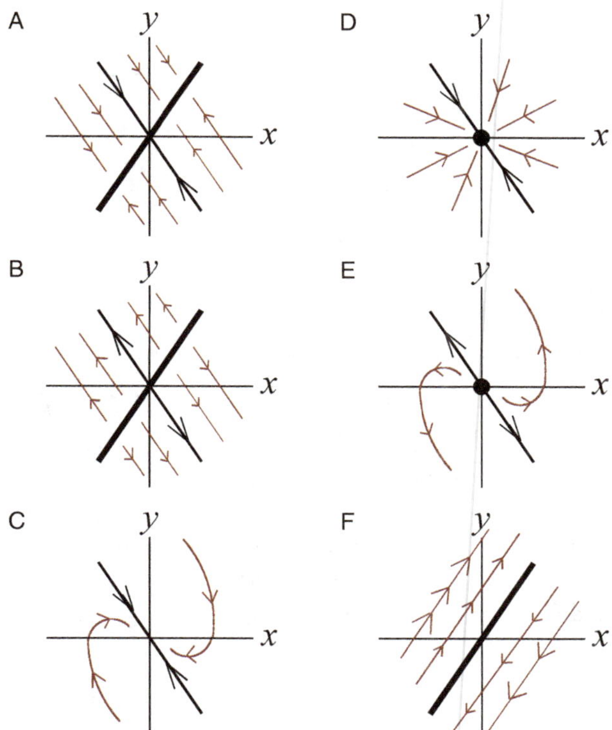

FIGURE 4 Nongeneric phase plane portraits. (A) $\lambda_1 = 0$ and $\lambda_2 < 0$. (B) $\lambda_1 = 0$ and $\lambda_2 > 0$. (C) $\lambda_1 = \lambda_2 < 0$, general case, stable improper node. (D) $\lambda_1 = \lambda_2 < 0$ with $a = d$ and $c = b = 0$, stable star. (E) $\lambda_1 = \lambda_2 > 0$, general case, unstable improper node. Case with $a = d$ and $c = b = 0$ (unstable star) not shown; as in Figure 4D except with reversed arrows. (F) $\lambda_1 = \lambda_2 = 0$, general case. Case with $a = d$ and $c = b = 0$ not shown; every point in the plane is a neutrally stable equilibrium.

portrait of the linearized system by the linearization theorem and the Hartman–Grobman theorem.

The second step above involves replacing functions f and g in Equation 7 with their tangent plane approximations at equilibrium $(x_e, y_e)^T$ and then making the change of variables $u = x - x_e$ and $w = y - y_e$, which translates the system to the origin. This yields the linearized system

$$\begin{cases} \dfrac{du}{dt} = \dfrac{\partial f}{\partial x}(x_e,y_e)u + \dfrac{\partial f}{\partial y}(x_e,y_e)w \\ \dfrac{dw}{dt} = \dfrac{\partial g}{\partial x}(x_e,y_e)u + \dfrac{\partial g}{\partial y}(x_e,y_e)w \end{cases}, \tag{20}$$

which has the form of Equation 8. The Jacobian is defined to be the matrix

$$J = \begin{pmatrix} \dfrac{\partial f}{\partial x} & \dfrac{\partial f}{\partial y} \\ \dfrac{\partial g}{\partial x} & \dfrac{\partial g}{\partial y} \end{pmatrix}, \tag{21}$$

and the eigenvalues are the roots of the equation $|J(x_e, y_e) - \lambda I| = 0$, where I is the identity matrix. That is, the eigenvalues are the roots of the characteristic equation

$$\begin{vmatrix} \dfrac{\partial f}{\partial x}(x_e,y_e) - \lambda & \dfrac{\partial f}{\partial y}(x_e,y_e) \\ \dfrac{\partial g}{\partial x}(x_e,y_e) & \dfrac{\partial g}{\partial y}(x_e,y_e) - \lambda \end{vmatrix} = 0. \tag{22}$$

EXAMPLE

Consider the nonlinear system

$$\begin{cases} x' = 2x + y^2 - 2 \\ y' = x - y - 1 \end{cases}. \tag{23}$$

The equilibrium equations are

$$\begin{cases} 0 = 2x + y^2 - 2 \\ 0 = x - y - 1 \end{cases}. \tag{24}$$

The second equation in Equation 24 yields $x = y + 1$, and so the first equation in Equation 24 implies $0 = y(2 + y)$. Hence, $y = 0$ or $y = -2$. If $y = 0$, then $x = 1$; if $y = -2$, then $x = -1$. The equilibria are therefore $(1, 0)^T$ and $(-1, -2)^T$. From Equation 21, the Jacobian is

$$J = \begin{pmatrix} 2 & 2y \\ 1 & -1 \end{pmatrix}. \tag{25}$$

At equilibrium $(1, 0)^T$, the eigenvalues are determined by

$$\begin{vmatrix} 2 - \lambda & 0 \\ 1 & -1 - \lambda \end{vmatrix} = 0, \tag{26}$$

which yields $\lambda = 2, -1$. Since the eigenvalues have nonzero real parts, the equilibrium is hyperbolic, and the linearization theorem applies. The eigenvalues have opposite sign, so the equilibrium $(1, 0)^T$ is a saddle.

To determine the local topology of the saddle, note that the linearization at $(1, 0)^T$ is, from Equation 20,

$$\begin{cases} u' = 2u \\ w' = u - w \end{cases}. \tag{27}$$

For $\lambda = 2$, the eigenvector equation, Equation 11, for the linearized system 27 is

$$\begin{cases} 0 = 0v_1 + 0v_2 \\ 0 = 1v_1 - 3v_2 \end{cases}, \tag{28}$$

which implies $v_1 = 3v_2$. Setting $v_2 = 1$ yields the eigenvector $(3, 1)^T$. Hence, the unstable manifold (associated with the positive eigenvalue $\lambda = 2$) for the linearization 27 is the line through the origin with slope 1/3 (Fig. 5A). Thus, the unstable manifold for the nonlinear saddle $(1, 0)^T$ is, locally to first order, the line through (1, 0) with slope 1/3 (Fig. 5B).

For $\lambda = -1$, the eigenvector equation is

$$\begin{cases} 0 = 3v_1 + 0v_2 \\ 0 = 1v_1 + 0v_2 \end{cases}. \tag{29}$$

Therefore, $v_1 = 0$; however, v_2 can have any value. Setting $v_2 = 1$ yields the eigenvector $(0, 1)^T$. Hence, the stable manifold (associated with the negative eigenvalue $\lambda = -1$) for the linearization 27 is the y-axis (Fig. 5A). Thus, the stable manifold for the nonlinear saddle $(1, 0)^T$ is, locally to first order, the vertical line through (1, 0) (Fig. 5B).

At equilibrium $(-1, -2)^T$ the eigenvalues are determined by

$$\begin{vmatrix} 2 - \lambda & -4 \\ 1 & -1 - \lambda \end{vmatrix} = 0, \tag{30}$$

which yields the characteristic equation $\lambda^2 - \lambda + 2 = 0$, with eigenvalues $\lambda = 1/2 \pm i\sqrt{7}/2$. Since the eigenvalues have nonzero real part, the equilibrium is hyperbolic, and the linearization theorem applies. Since the eigenvalues are complex with positive real part, equilibrium $(-1, -2)^T$ is an unstable spiral. To determine the direction of rotation, consider the linearization at $(-1, -2)^T$, which is, from Equation 20,

$$\begin{cases} u' = 2u - 4w \\ w' = u - w \end{cases}. \tag{31}$$

Choose a test point, for example (0, 1). At this point, Equation 31 implies that $u' = -4$ and $w' = -1$; hence, both u and w are decreasing in time, which implies a counterclockwise spiral away from the origin in the linear system (Fig. 5A). Thus, the equilibrium

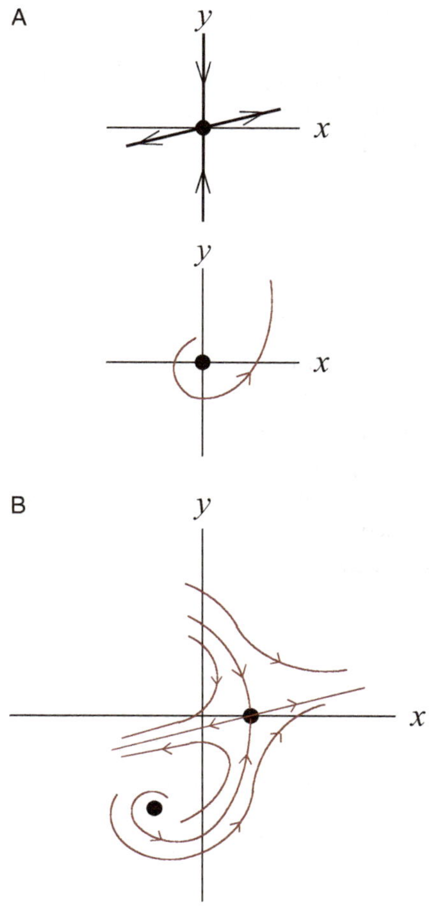

FIGURE 5 Constructing a nonlinear phase plane portrait. (A) The phase plane portraits for the linearizations about the two equilibria, translated to the origin. (B) The nonlinear phase plane portrait.

$(-1, -2)^T$ of the nonlinear system is also (locally) an unstable spiral with counterclockwise rotation (Fig. 5B).

The complete nonlinear phase portrait is shown in Figure 5B.

PHASE PLANE: LIMIT CYCLES, CYCLE CHAINS, AND BIFURCATIONS

Periodic solutions of two-dimensional systems are associated with closed-loop orbits, called cycles, in the phase plane. If the cycle attracts nearby orbits, it is called a limit cycle. Cycles always surround at least one equilibrium. The Lotka–Volterra predator–prey model produces a well-known example of a cycle in ecology.

A bifurcation is an abrupt change in phase portrait type that occurs as a parameter is tuned through a critical value. The following example illustrates both limit cycles and bifurcations.

EXAMPLE

Consider the nonlinear system

$$\begin{cases} x' = y - x^3 + ax \\ y' = -x \end{cases} \tag{32}$$

in which a is a real constant, close to zero, called a parameter. The unique equilibrium is $(0, 0)^T$ and the Jacobian is

$$J = \begin{pmatrix} -3x^2 + a & 1 \\ -1 & 0 \end{pmatrix}; J(0, 0) = \begin{pmatrix} a & 1 \\ -1 & 0 \end{pmatrix}. \tag{33}$$

The characteristic equation is $\lambda^2 - a\lambda + 1 = 0$, and the eigenvalues are

$$\lambda = \frac{a}{2} \pm \frac{i\sqrt{4 - a^2}}{2}. \tag{34}$$

If $a < 0$, the equilibrium is hyperbolic and the nonlinear system has an asymptotically stable spiral at the origin. If $a = 0$, the real parts of the eigenvalues are zero, so the equilibrium is nonhyperbolic and the linearization theorem does not apply. Although the linearized system has a center at the origin, it can be shown that the nonlinear system has an asymptotically stable spiral at the origin. If $a > 0$, the equilibrium is hyperbolic and the nonlinear system has an unstable spiral at the origin.

If the parameter a is "tuned" from $a < 0$ through $a = 0$ to values of $a > 0$, the origin changes (locally) from a stable spiral to an unstable spiral; therefore this system has a bifurcation at $a = 0$. For small $a > 0$, local trajectories spiral away from the origin; however, trajectories further away from the origin are still spiraling toward the origin. The trajectories "spiraling out" and those "spiraling in" approach a closed periodic solution—a limit cycle (Fig. 6A). As a is tuned from negative to positive values, the limit cycle is born at the origin when $a = 0$ and grows in radius as a increases.

Another type of configuration in the phase plane is the cycle chain. This occurs when multiple equilibria are connected by heteroclinic orbits (orbits that connect two different equilibria) in a loop structure (Fig. 6B). Oscillatory solutions may approach a cycle chain, suggesting the erroneous conclusion that they are approaching a limit cycle. The cycle chain, however, is not an orbit.

The Poincaré–Bendixson theorem implies that bounded orbits in the phase plane must approach equilibria, cycles, or sets of equilibria connected by heteroclinic and homoclinic orbits.

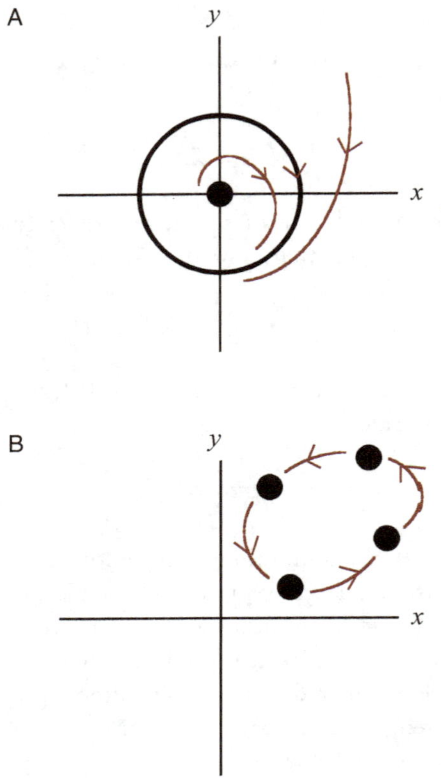

FIGURE 6 Other types of nonlinear phase plane configurations. (A) Limit cycle. (B) Cycle chain.

PHASE PLANE: NULLCLINES AND THE LOTKA-VOLTERRA MODELS

Nullcline (or isocline) analysis is a type of phase plane analysis used to locate equilibria and determine the direction of the vector field in the phase plane. It is particularly useful when the computations of equilibria and eigenvalues are complicated by the presence of many parameters. A nullcline for state variable x is the curve in the phase plane along which $dx/dt = 0$. Nullclines for x and y intersect at equilibria. Nullcline analysis is used to analyze the Lotka–Volterra models in ecology.

Lotka-Volterra Competition

The model is

$$\begin{cases} x' = x(r_1 - a_{11}x - a_{12}y) \\ y' = y(r_2 - a_{21}x - a_{22}y) \end{cases} \tag{35}$$

with positive coefficients. The nullclines for x are the lines $x = 0$ and $y = r_1/a_{12} - a_{11}x/a_{12}$. Along these two lines the change in x is zero, that is, the vector field is vertical. The nullclines for y are the lines $y = 0$ and $y = r_2/a_{22} - a_{21}x/a_{22}$. Along these lines the vector field is horizontal. Figure 7 illustrates the four possible configurations. In two of the cases, one species always wins (Figs. 7A, B). In the third case, there is a coexistence equilibrium, but it is a saddle and the winner depends on the initial condition (Fig. 7C). In the fourth case, the coexistence equilibrium is a stable node and the species coexist (Fig. 7D).

Lotka-Volterra Cooperation

The model is

$$\begin{cases} x' = x(r_1 - a_{11}x + a_{12}y) \\ y' = y(r_2 + a_{21}x - a_{22}y) \end{cases} \tag{36}$$

with positive coefficients. The nullclines for x are the lines $x = 0$ and $y = -r_1/a_{12} + a_{11}x/a_{12}$, and the nullclines for y are the lines $y = 0$ and $y = r_2/a_{22} + a_{21}x/a_{22}$. Figure 8 shows the two possible phase plane configurations. In the first case (Fig. 8A), the cooperators coexist. Robert May famously referred to the second case (Fig. 8B) as an "orgy of mutual benefaction."

Lotka-Volterra Predator-Prey

The model is

$$\begin{cases} x' = x(r_1 - a_{12}y) \\ y' = y(-r_2 + a_{21}x) \end{cases} \tag{37}$$

with positive coefficients, where x is the prey density and y is the predator density. The nullclines for x are the lines $x = 0$ and $y = r_1/a_{12}$, and the nullclines for y are the lines $y = 0$ and $x = r_2/a_{21}$. It is not clear from the nullclines (Fig. 9) whether the coexistence equilibrium is a spiral or a nonlinear center. It is possible to prove, however, that the orbits are closed curves. When the periodic solutions are graphed as time series, they exhibit the well-known predator–prey oscillations.

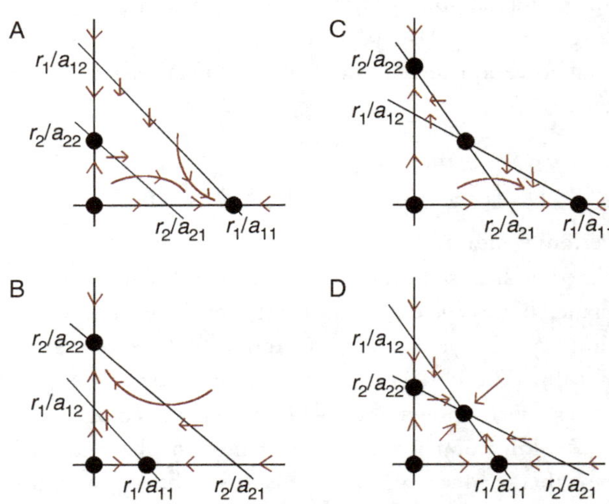

FIGURE 7 Lotka-Volterra competition dynamics. (A) Species x wins. (B) Species y wins. (C) Winner depends on the initial condition. (D) Competitive coexistence.

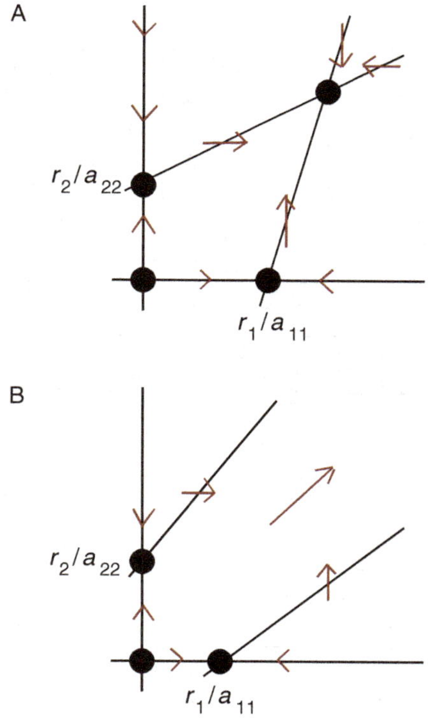

FIGURE 8 Lotka–Volterra cooperation dynamics. (A) Coexistence. (B) An "orgy of mutual benefaction."

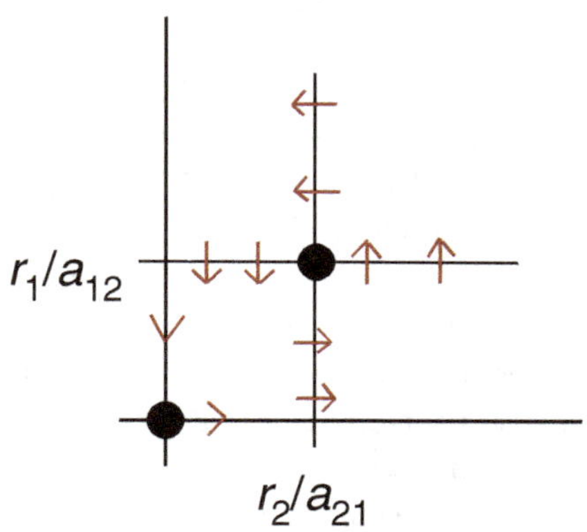

FIGURE 9 Lotka–Volterra predator–prey dynamics exhibit periodic solutions around a neutrally stable coexistence equilibrium.

SEE ALSO THE FOLLOWING ARTICLES

Bifurcations / Chaos / Difference Equations / Ordinary Differential Equations / Predator–Prey Models / Stability Analysis

FURTHER READING

Boyce, W. E., and R. C. DiPrima. 2009. *Elementary differential equations and boundary value problems,* 9th ed. Hoboken, NJ: John Wiley & Sons.

Cushing, J. M. 2004. *Differential equations: an applied approach.* Upper Saddle River, NJ: Pearson–Prentice Hall.

Hirsch, M. W., and S. Smale. 1974. *Differential equations, dynamical systems, and linear algebra.* San Diego: Academic Press.

Hirsch, M. W., S. Smale, and R. L. Devaney. 2004. *Differential equations, dynamical systems, and an introduction to chaos,* 2nd ed. San Diego: Academic Press.

Strogatz, S. H. 1994. *Nonlinear dynamics and chaos: with applications to physics, biology, chemistry, and engineering.* Cambridge, MA: Westview Press.

PHENOTYPIC PLASTICITY

MARIO PINEDA-KRCH

University of Alberta, Edmonton, Canada

Phenotypic plasticity is the ability of a single genotype to express different phenotypes in response to environmental conditions, e.g., alternative morphologies, physiological states, or behavior. Such plasticity can be expressed as several highly distinct phenotypes or as a continuous norm of reaction describing the functional interrelationship of a range of environments to a range of phenotypes.

THE CONCEPT

A water flea sensing the presence of predators changes its morphology by elongating its head and extending its tail spine. A plant experiencing competition for light adjust its growth patterns by elongating its stem. A conifer facing the threat of bark boring beetles alters the chemical composition of its resin by modifying the biochemical pathways producing defensive compounds. The arm of the blacksmith, wielding an incessantly heavy hammer, becomes reshaped as the strength and size of its muscles increase. Phenotypic plasticity—the remarkable ability of a single genotype to express multiple distinct phenotypes when exposed to different environments—permeates the living world and is at the center of the age-old question of nature versus nurture. An understanding of how nature (genetics) and nurture (environment) interact to yield the organisms (phenotypes) is not merely of interest to scientists but directly influences our social norms, cultural practices, and daily lives. Historically, the debate of how traits are shaped by nature and nurture has figured prominently in social, political, and philosophical agendas, often with racist and eugenics undertones. For example, the Nazi regime pursued an agenda based on the concept of human nature (phenotype) defined by race (genetics), while the Communist regime in the former Soviet Union defined human identity (phenotype) as subject to social structures (environment). Although these ideologies misconstrued science in an attempt to justify their eugenics programs, they provide sobering lessons of the far reach and tenuous role phenotypic plasticity has had historically.

A BRIEF HISTORY OF PHENOTYPIC PLASTICITY

Today, the study of phenotypic plasticity is central for our understanding of how genetics and environment interact to produce the phenotypic diversity of organisms we see around us and figures prominently in many scientific disciplines. The concept did not, however, always enjoy such prominent position. Following its introduction onto the scientific stage in the early twentieth century, it received virtually no attention within the scientific community in the West over the next six decades. On the contrary, in genetics phenotypic plasticity was considered an annoying interference making experimental results difficult to interpret, while in the agricultural sciences it was taken to imply lack of adaptation making crop yields unreliable (and because of this it is well documented in the agricultural literature). As a result, substantial efforts were directed toward reducing plastic responses by selecting for phenotypes that were genetically stable. It was not until the second half of the twentieth century that geneticists and evolutionary biologists started to fully appreciate phenotypic plasticity as something more than just an aberration and nuisance. Over the past few decades, the study of phenotypic plasticity has expanded significantly to include a wide range of novel approaches addressing fundamental questions at the interface of genetics, developmental biology, physiology, ecology, and evolution. These approaches include state of the art quantitative and molecular genetics techniques as well as theoretical modeling of the evolution of phenotypic plasticity.

The concept of phenotypic plasticity was introduced in 1909 by the German zoologist Richard Woltereck, simultaneously with the then newly coined concepts of gene, genotype, and phenotype. The first time the word

FIGURE 1 Two genetically identical individuals of the water flea, *Daphnia lumholtzi*. The individual on the left was exposed to chemical cues from a predatory fish, and the individual to the right was not. The pointed anterior prolongation, or helmet, on the head and the extended tail spine provide protection predators locating its prey by sight. The development of the helmeted phenotype confers a higher metabolic cost than the smaller rounded head, resulting in smaller brood volumes. The two phenotypes were initially believed to be a separate species. Scanning electron micrograph courtesy of C. Laforsch and R. Tollrian.

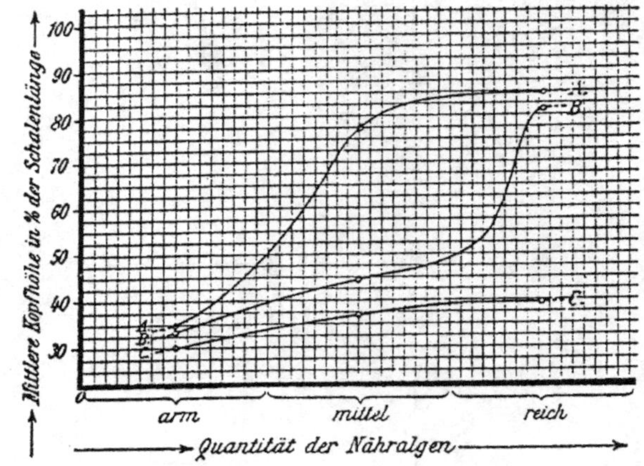

FIGURE 2 The response curves obtained by Woltereck when rearing three pure lines (A, B, and C) of a freshwater *Daphnia (Hyalodaphnia cucullata)* on different levels of resources. Horizontal axis: resource level (poor, intermediate, and enriched). Vertical axis: relative head height. Woltereck expected the phenotypic response of the different lines to vary with resources availability; he did not expect, however, that the phenotypic response of each genotype under the same environmental conditions would be different. Of the three lines examined for helmet height, one was nearly indifferent to changes in resources (C), another responded to a change from intermediate to abundant resources (B), whereas the third responded to a change from poor to intermediate resources (A). Figure from Woltereck, 1909, *Verhandlungen der Deutschen Zoologischen Gesellschaft* 19: 110–173.

plasticity was used to describe the effects of environment on the phenotype was not until 1914, however, by the Swedish geneticist Herman Nilsson-Ehle. Woltereck studied parthenogenetic water fleas, *Daphnia,* that undergo seasonal variation in body form, so-called cyclomorphosis (meaning "temporal cyclic morphological changes"). During the summer the crest of the head extends forward as a thin transparent helmet, and during the winter the helmet is reduced and the head assumes a round form (Fig. 1). When Woltereck experimentally examined the influence of resource availability on the development of head size in *Daphnia* from three different German lakes, he found that the phenotypic response to the same environmental change was different in the strains from the various lakes (Fig. 2). Today, we know that the helmeted morphology is a predator-avoidance strategy that can be cued either by environmental signals correlated with predator abundance or by direct predator exposure.

Woltereck coined the term *Reaktionsnorm* for the relationship between the complete set of phenotypic curves obtained from a particular quantitative trait. This is a slightly different definition of the concept than what is used today (see the section "The Role of Phenotypic Plasticity in Evolutionary Diversification," below). The reaction norm introduced the idea that the genotype was less a deterministic force than an enabling agent when it came to expressing phenotypes. Unfortunately Woltereck's ideas threatened not only to deprive genetics of its major new tool for targeting genetic entities—the then newly introduced concepts of gene and genotype—but could also be interpreted as supporting the notion of inheritance of acquired characters which had been ruled out by the German evolutionary biologist August Weismann a few decades earlier. As a result, over the following decades the concept was largely ignored in the West. In contrast, the reaction norm emerged as an important conceptual tool in Soviet genetics in the 1920s where it was used primarily to deflate claims of the inheritance of acquired characteristics. Starting in 1926, Ukrainian-born Theodosius Dobzhansky, one of the most prominent geneticists and evolutionary biologists of the twentieth century, started advocating that the reaction norm should to be treated as a Mendelian unit of inheritance and that what changes in evolution is the reaction norm of the organism to the environment. In 1927, Dobzhansky emigrated from the Soviet Union to the United States, effectively repatriating the reaction norm concept to the West and ultimately ensuring its permanent place in Western genetics. In his classical 1937 book, *Genetics and the Origin of Species,* he reintroduced the reaction norm to the Western scientific world with these words: "One must constantly bear in mind the elementary consideration which is all too frequently lost sight of in the writings of some biologists; what is inherited in a living being is not this or that morphological character, but a definite norm of reaction to environmental stimuli." Despite this, the number of studies focusing on unraveling the mechanisms and genetic basis of this norm of reaction to environmental stimuli continued to languish. It was not until 1965, when the British evolutionary biologist Anthony Bradshaw published his influential review *Evolutionary Significance of Phenotypic Plasticity in Plants* that the renaissance of the concept in the West begun in earnest. Bradshaw's two key points were that environmental effects on the phenotype were as important as genetic effects (rather than simply inconvenient errors) and that these effects were themselves under genetic control and could therefore evolve. His review was the starting point of a large number of subsequent studies on phenotypic plasticity and remains today one of the key references in the field.

MANIFESTATIONS OF PHENOTYPIC PLASTICITY

The amount by which the expression of an individual genotype is changed by different environments is a measure of the plasticity of the phenotype. Phenotypes having a low plastic response exhibit constraint variation around one or more modes as a result of the developmental system buffering the effect of phenotypic perturbations. The concept of a buffered phenotypic response, or canalized development, was introduced by the British developmental biologist Conrad Waddington in 1957 and is referred to as canalization. Canalized phenotypes may have a zone of canalization within which considerable environmental variation results in a limited or no response in the phenotype. The edges of this zone mark environmental thresholds at which the plastic response increases. A classical example of a canalized phenotype is the number of scutellar bristles in *Drosophila* in response to temperature change.

While some traits vary continuously in response to the intensity of the environmental stimulus, e.g., head height in *Daphnia* in response to chemical cues from predators, penis size in barnacles in response to wave action exposure, and the circumference of the blacksmith's arm in response to years of forging, many other traits exhibit discrete responses. A discrete phenotypic response, also referred to as polyphenism, exhibits a series of discrete forms, often only two, with no intermediates. Historically, polyphenism was believed to be rare due to, among others reasons, misclassification of alternative phenotypes as being genetically distinct and preconceptions about the developmental and evolutionary difficulties associated with obtaining alternative and discrete phenotypes. Today it is, however, abundantly clear that polyphenism is not only taxonomically widespread in nature but is also responsible for some of the most spectacular examples of plastic phenotypes. A few examples are leaf shape in semi-aquatic plants in response small-scale spatial heterogeneity, omnivorous filter feeding or carnivorous tadpole phenotypes in several species of spadefoot toads, horned or hornless male phenotypes in some dung beetle species, castes of social insects, the solitary and gregarious phases of migratory locusts, and the winged and wingless forms of aphids.

Nowhere does phenotypic plasticity play a more important role than in sessile organisms such as in plants and animals permanently attached to a substrate. Most animals are able to meet environmental challenges by behavioral responses, e.g., actively searching for resources and mates while avoiding unfavorable environments. Sessile organisms, on the other hand, are only able to respond to environmentally adverse conditions by being phenotypically plastic. One of the classical and most remarkable types of plasticity is found in aquatic plants growing in shallow water habitats, e.g., the water crowfoot plant. Here, three different environments occur in close proximity: under water, at the water surface, and in the air above the water. Many plants in this habitat exhibit phenotypic plasticity for leaf shape (heterophylly) with finely dissected leaves produced in the submerged habitat and entire or lobed floating leaves produced at the surface. Another, more recently described example of plasticity in a sessile organisms is found in intertidal barnacles. As one of few sessile animals to copulate, most barnacles reproduce by extending their penises to find and fertilize distant mates. The barnacles face a tradeoff, however, between having a long penis capable of reaching more mates and controlling ever-longer penises in turbulent water flow. To address this tradeoff, intertidal barnacles have evolved a capacity for phenotypic plasticity in penis size and shape to suit local hydrodynamic conditions where penises from wave-exposed shores are shorter and stouter than those from wave-protected shores.

In mobile organisms, phenotypic plasticity plays a particularly important role in interspecific interactions. When individuals of two species interact, they can adjust their phenotypes in response to their respective partner. A classical example of this are induced morphological defences in prey in response to exposure to predators. Cyclomorphosis in Woltereck's Daphnia is a classical example, but it also occurs in vertebrates. For example, in laboratory experiments exposure to northern pike (predator) induces a phenotypic response in crucian carp (prey) by increase its body depth. This plastic response reduced predation pressure on deep-bodied carps, as they are more difficult to handle for the gape-limited pike.

THE REACTION NORM CONCEPT

The reaction norm is a central concept for describing the functional relationship between a single genotype and the set of quantitative phenotypes, i.e., phenotypes that vary on a continuous scale (e.g., body size or viability), it produces in different environments. The environment can refer to abiotic factors, e.g., temperature, light, or nutritional availability, or biotic factors, e.g., interactions with prey or predators or conspecifics. The reaction norm embodies the fundamental notion that the phenotype has to be described dynamically, in a variety of environments, rather than a one-to-one relationship between genotype and phenotype. In other words, it only makes sense to talk about a genotype's phenotype in the context of a specific environment. This definition of the reaction norm differs from Woltereck's usage (see the section "A Brief History of Phenotypic Plasticity," above). Woltereck referred to the functions representing the response of a given genotype to the environmental variable as "phenotypic curves" (i.e., the current definition of reaction norm), while he used the term *Reaktionsnorm* to describe all the potential phenotypic curves for a particular quantitive trait, i.e., encompassing all possible genotypes.

Reaction norms are commonly visualized as functions describing the response of a genotype to a quantitative environmental gradient on the *x*-axis and the corresponding phenotype on the *y*-axis. This representation provides an explicit association of a given phenotype with a particular environment in which it is expressed and a connection between the environmental and phenotypic distribution in the population. Reaction norms for a group of genotypes are typically plotted together (the way Woltereck did it in Figure 2) to represent the pattern of genotypic and phenotypic variance within and across environments. If the reaction norms are parallel, different genotypes have the same response to a given environmental change (Fig. 3A). Typically, however, different genotypes respond differently, resulting in reaction norms with different slopes (Fig. 3B) or norms crossing each other (Fig. 3C). These type of interactions between genotype and the environment in shaping the phenotype is referred to as genotype-by-environment interaction and has important theoretical and applied consequences. For example, a specific difference of environment may have a greater effect on some genotypes than others, resulting in different range and variance of phenotypic distributions. Specifically, a steep reaction norm (representing a more plastic phenotype) transforms a unimodal distribution of environmental conditions into a phenotypic distribution that is flattened out, while a flat reaction norm (representing a less plastic, i.e., more canalized, phenotype) will compress the phenotypic distribution (Fig. 3B). Crossing reaction norms exhibit particularly strong genotype-by-environment interactions and have important effects on the amount of genetic variation perceived to be present in a population. In the region where the reaction norms cross, the blurring of phenotypic differences among genotypes makes it difficult to assign phenotypes unambiguously to genotypes, while phenotypes on either side of the crossing point have a reversed ranking (Fig. 3C). As a result, under directional selection for higher phenotypic values, one genotype would be favored to the right of the crossing point and another to the left. The reaction norms in Figure 3 are, of course, very simplified. In natural populations, reaction norms do not have to be linear. For example, curved reaction norms can, in addition to changing the range and variance, also change the shape of the phenotypic distribution.

Genotype-by-environment interactions have important applied ramifications when individuals of a particular genotype are reared in different environments. For example, a breed of livestock may be reared under different conditions on different farms or crop varieties are grown in different seasons, at different places, and under different growing conditions making it difficult to achieve predictable and consistent yields. Typically,

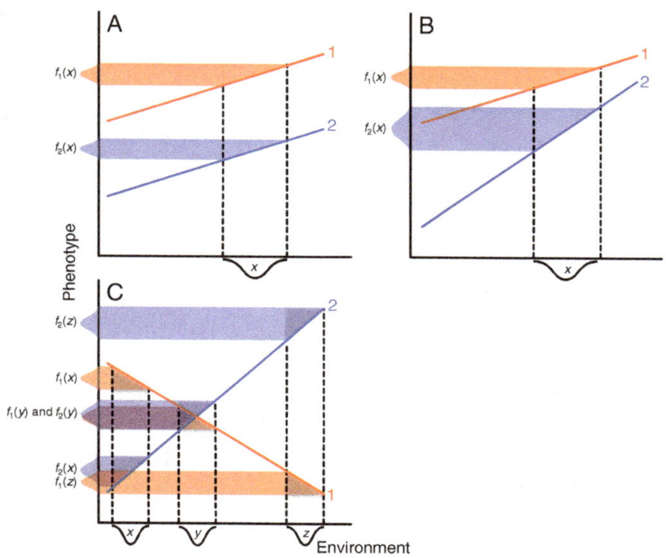

FIGURE 3 Effects of genotype and environment on the distribution of phenotypes in a series of hypothetical reaction norms. A continuous environmental gradient (e.g., temperature) is shown on the horizontal axis, with the distribution of environments experienced by a population indicated (*x*, *y*, and *z*). The blue and red lines represent the reaction norms of different genotypes, with the resulting distribution of phenotypes given on the *y*-axis; e.g., $f_1(x)$ is the distribution of phenotypes for genotype 1 in the environmental distribution *x*. (A) In the absence of genotype-by-environment interaction, the reactions are parallel and, although different in phenotype ($f_1(x) \neq f_2(x)$), respond similarly to differences in the environment. Note how the variance of the phenotypic distributions differs from the variance of the environmental distribution. (B) In the presence of genotype-by-environment interaction, manifested by the reaction norms having different slopes, the two genotypes respond differently to the environment. Note how the variance of $f_1(x)$ now is different from the variance of $f_2(x)$. In both (A) and (B), the distribution of phenotypes is bimodal, and most of the variation is genetic because the genotypes differ substantially in their phenotypes. (C) Crossing reaction norms are an especially strong case of genotype-by-environment interaction. Around the crossing point of the norms, the genotypes are indistinguishable in the phenotypic mixture (i.e., $f_1(y) \approx f_2(y)$), giving an appearance of less genetic diversity than what is actually present in the population. Away from the crossing point, the individual genotypes can be identified in the bimodal phenotypic distribution but with a reversed ranking on either side of the crossing point. For example, while genotype 1 is superior in environment *x*, $f_1(x) > f_2(x)$, genotype 2 is ranked higher in environment *z*, i.e., $f_1(z) < f_2(z)$.

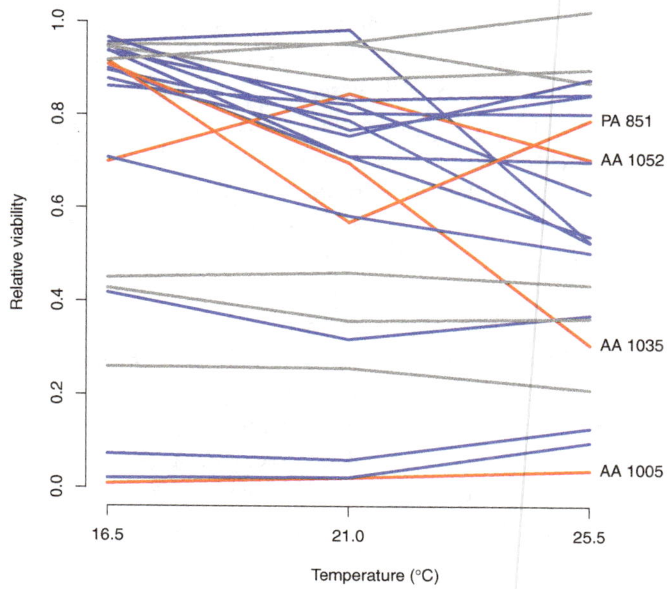

FIGURE 4. Reaction norms for larval viability in a natural population of *Drosophila pseudoobscura*. Each line is the reaction norm for the relative larval viability at three different temperatures for a fourth chromosome homozygote. Most genotypes have a variable phenotype with qualitatively different environmental sensitivities and substantial genotype-by-environment interaction. For example, 6 of the 23 genotypes show no significant environmental sensitivity (gray). Other genotypes exhibit poor viability for all temperature regimes (AA 1005), while some genotypes have a high viability at low temperatures but deteriorate with increasing temperature (AA 1035). In contrast, genotype AA 1052 exhibits the highest viability at intermediate temperatures, while PA 851 shows the opposite trend with higher viability for marginal temperatures. Data from Dobzhansky and Spassky 1944, *Genetics* 29: 270–290.

natural populations exhibit highly heterogeneous patterns in the reaction norms, with strong genotype-by-environment interactions and wildly different environmental sensitivities in different genotypes (Fig. 4).

THE ROLE OF PHENOTYPIC PLASTICITY IN EVOLUTIONARY DIVERSIFICATION

Although phenotypic plasticity was once thought to result from developmental accidents, we now know that environmentally induced phenotypic variation can be selectively advantageous. Hence, phenotypic plasticity has come to be viewed as a trait subject to selection, just like any phenotypic character, and can undergo evolution and be adaptive if fitness is increased by having a plastic phenotype. While genetic variation for plasticity and selection for plastic responses has been documented in natural populations, costs and limits impose constraints on the evolution of plastic responses. Ecological costs of plasticity are defined in terms of a reduction of fitness of plastic individuals compared with less plastic individuals. For example, an imperfect match between a phenotype and the environment resulting in a low fitness results in an ecological cost that is realized in a particular set of environments. Another common cost associated with a plastic phenotype is the energetic cost of maintaining a facultative phenotype. For example, ethylene growth response in plants requires the response of an ethylene receptor protein on the cell membranes of plants. If development would be insensitive to ethylene, the energetic cost associated with the production of the receptor proteins would not be incurred.

The success of a plastic response largely depends on the predictability of the environment. Lags in the response to environmental changes or unpredictable environmental changes can impose significant ecological costs compared with that experienced by fixed genotypes. Using theoretical models, it has been shown that an increased ability for plastic responses is less likely to evolve in stable environments, while plastic phenotypes tend to be selectively favored in environments, that are intrinsically variable in space and time, e.g., due to species interactions and in fluctuating environments. The time scale over which the environment is changing relative to the average life span of individuals in the population is one important aspect determining the evolutionary outcome. For example, if the duration of an environmental regime is less than the average generation time of individuals, the population cannot easily respond by adaptation in a fixed (nonplastic) genotype. Under this scenario, individuals having genetic variation for phenotypic plasticity will be favored. Whether the population will adapt by increasing its capacity for phenotypic plasticity ultimately depends on the costs and physiological limits associated with the plastic phenotype. Finally, it is worth noting that already in 1881 (in a letter to Karl Semper) Charles Darwin, clearly before his time, was able to envision the possibility of plastic phenotypes evolving in response to environmental changes: "I speculate whether a species very liable to repeated and great changes of conditions might not assume a fluctuating condition ready to be adapted to either condition."

SEE ALSO THE FOLLOWING ARTICLES

Environmental Heterogeneity and Plants / Evolution of Dispersal / Integrated Whole Organism Physiology / Mutation, Selection and Genetic Drift / Quantitative Genetics

FURTHER READING

Agrawal, A. A. 2001. Phenotypic plasticity in the interactions amd evolution of species. *Science* 294: 321–326.

deWitt, T. J., and S. M. Scheiner, eds. 2004. *Phenotypic plasticity: functional and conceptual approaches.* New York: Oxford University Press.

Miner, B. G., S. E. Sultan, S. G. Morgan, D. K. Padilla, and R. A. Releya. 2005. Ecological consequences of phenotypic plasticity. *Trends in Ecology and Evolution* 20: 685–692.

Pigliucci, M. 2001. *Phenotypic plasticity: beyond nature and nurture.* Syntheses in Ecology and Evolution. Baltimore, MD: Johns Hopkins University Press.

Pigliucci, M. 2005. Evolution of phenotypic plasticity: where are we going now? *Trends in Ecology and Evolution* 20: 481–486.

Pfennig, D. W., M. A. Wund, E. C. Snell-Rood, T. Cruickshank, C. D. Schlichting, and P. Moczek. 2010. Phenotypic plasticity's impact on diversification and speciation. *Trends in Ecology and Evolution* 25: 459–467.

Price, T. D., A. Qvarnström, and D. E. Irwin. 2003. The role of phenotypic plasticity in driving genetic evolution. *Proceedings of the Royal Society London B: Biological Sciences* 270: 1433–1440.

Tollrian, R., and C. D. Harvell, eds. 1998. *The ecology and evolution of inducible defenses.* Princeton: Princeton University Press.

West-Eberhard, M-J. 2003. *Developmental plasticity and evolution.* New York: Oxford University Press.

PHYLOGENETIC RECONSTRUCTION

BRIAN C. O'MEARA

University of Tennessee, Knoxville

A phylogeny is a depiction of the evolutionary history of a set of organisms. Typically, this is a branching diagram showing relationships between species, but phylogenies can be drawn for individual genes, for populations, or for other entities.

WHAT DO PHYLOGENIES MEAN?

A phylogeny represents a history of populations. Take the example of Figure 1. Starting at the bottom (root) of the tree, one population splits into two. The population on the left speciates again, but one of the descendant species eventually goes extinct without leaving any descendants. Various other processes occur: population sizes (width of the tree's branches) vary, speciation happens through a gradual rather than instant reduction of gene flow, populations develop and lose subdivision, one species forms as a hybrid of two other species, a few genes introgress from one species to another, and so forth. The history of genes evolving within these populations may be even more complex, with selective sweeps, ancestral polymorphisms persisting across speciation events, gene copies being duplicated and lost within the genome, and recombination shuffling histories within and between genes. All of this complex history is typically summarized by a figure like that of Figure 1, with most of the complex history abstracted away to leave only a simplified history of populations.

It is important to interpret phylogenies correctly. Chimpanzees (*Pan troglodytes*) and bonobos (*Pan paniscus*)

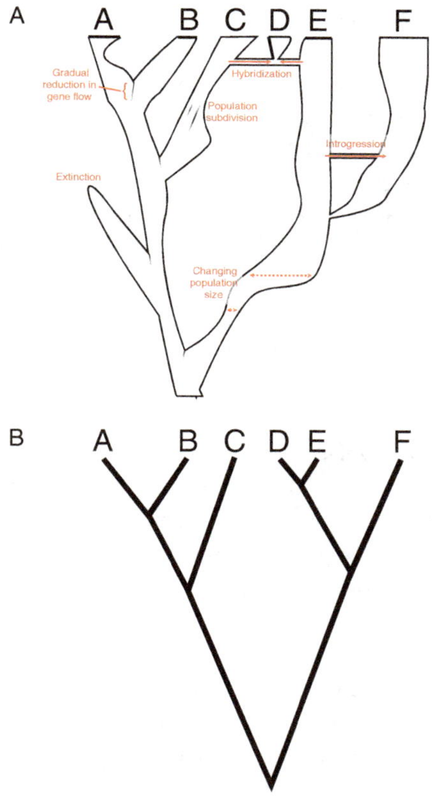

FIGURE 1 An actual, messy population history (A) and its representation as a phylogenetic tree (B). Note that reticulation events are not typically reflected on a phylogeny.

descended from the same parent species and are each other's closest relatives. Their parent species shared a common ancestor with humans (*Homo sapiens*) several million years ago. A phylogenetic tree shows these relationships (see Fig. 2). For a pair of species, one can work down the tree to find where their lineages connect, which is their most recent common ancestor. All the taxa descended from a point on the tree form a clade and are all more closely related to each other than they are to any organism not in the clade. It is important to note that a phylogeny gives information about these nestings, and perhaps information about the timing of splits. A common mistake in interpreting trees is using the order of labels on the tree to make inferences about history of evolution. For example, in Figure 2, moss appears at one end of the list of taxon labels on the tree, while a whale appears in the middle of this set of labels. This does not mean that mosses are more advanced in any way than whales, or that the ancestors of mosses went through a morphology similar to that of a whale on the way to evolving moss form. Similarly, the fact that the names of lampreys and lobsters occur next to each other does not mean that they are especially related to each other (each has far closer relatives). This may seem obvious in Figure 2, but

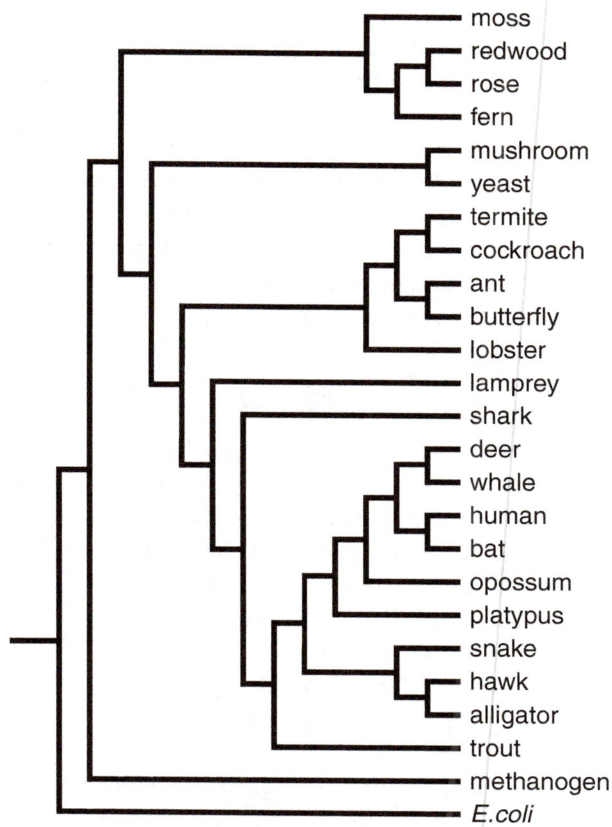

FIGURE 2 A sample tree of life (with extensive pruning of taxa). Note that humans appear in the middle of the list of taxa, not at either of the ends. One common misperception is that phylogenetic trees track advancement from primitive to advanced organisms. In fact, they just show nestings of organisms sharing common ancestors.

it is easy to revert to progressive thinking when presented with the typical textbook metazoan phylogeny: sponges on one side, then cnidaria, various worm-like taxa, arthropods, echinoderms, and finally vertebrates. The nestings provide information (vertebrates are more closely related to echinoderms than to jellyfish), but the relative order does not. Any node may be rotated to return an equally valid tree.

Utility of Phylogenies

Given the extremely simplified nature of a phylogeny relative to the actual evolutionary process, what is the point of reconstructing one? First, phylogenies provide basic information about relatedness of organisms: it is worth knowing that the closest relatives to fungi are animals, not plants, or that alligators are more closely related to birds than to iguanas. Second, knowledge of relatedness can be used to control for the nonindependence of species in ecological or evolutionary studies. Third, and perhaps most compelling, with a phylogeny questions can be answered that are otherwise difficult or impossible to address. Phylogenies have been used to estimate behaviors in ancestral species, investigate factors affecting diversification, examine coevolution between hosts and parasites, predict which species are most at risk for extinction, estimate biogeographic connections, match juveniles with adults, track viral evolution, test community assembly theories, delimit species, and more. This entry focuses on reconstructing phylogenies, but the "Further Reading" section and other entries in this volume provide information on some of their many uses.

DATA

Phylogenies are built from data. The data used could be restriction site polymorphisms, morphological traits such as placement of bristles or shapes of cell walls, order of genes along a chromosome, amino acid sequences, or other information. Morphological traits were most important historically and remain important in constructing phylogenies containing extinct organisms or when placing fossils on the tree for time calibrations. In neontology, however, DNA sequence data is clearly the most common data type and is becoming even more abundant.

To use sequence data in most phylogenetic reconstruction techniques, it must first be aligned: does this particular A for this taxon correspond to the first or second site in the sequence of another taxon? In the case of the sections of protein-coding genes that are transcribed and then translated to proteins (exons), this is relatively straightforward: such sequences rarely have the insertions or deletions of bases (indels) that make alignment difficult since deletions or insertions (other than in multiples of three nucleotides) will induce a frame-shift mutation, changing the protein's amino acids and perhaps reducing or eliminating the protein's functionality. However, for other parts of the genome, insertion or deletion of one to many bases may have little to no detrimental effect, and so these mutations may persist across evolutionary time. There are three main approaches to doing alignment: (1) manual editing, (2) automated tree-free alignment (other than perhaps a guide tree), and (3) joint estimation of the tree and the alignment. The second approach, perhaps combined with subsequent manual adjustment, is most frequently used.

OPTIMALITY CRITERIA

Phylogeny reconstruction is a contentious area, most dramatically when it comes to optimality criterion used for the reconstruction. Some of the debates regard practicality (which criterion is fastest to compute; which ones are least likely to return an incorrect population history) and are amenable to analysis. Other debates regard scientific or statistical philosophy (should prior beliefs be incorporated in analyses?; does use of an explicit model mean an approach is not possible to falsify sufficiently to qualify as a proper scientific method?) and are not readily resolvable through data or analyses. Figure 3 shows the frequency of use of different criteria through time, as well as program usage and the growth of phylogenetics as a field.

Parsimony

Parsimony is simply choosing the tree that minimizes the required number of changes. For example, in Figure 2, mosses, redwoods, roses, and ferns all have chloroplasts. Based on that phylogeny, one would infer that usage of chloroplasts evolved once and was maintained. This is more parsimonious than a tree that would scatter those four taxa within the group of mammals, which would require multiple origins or losses of chloroplasts to explain the evolution of that trait.

Distance

Distance methods typically use a matrix of pairwise distances between taxa (calculated in a variety of ways) and one of several algorithms to reconstruct a tree. Neighbor-joining is a common distance algorithm.

Likelihood

Likelihood methods use an explicit model of evolution and select the tree and model parameters that maximize the probability of the observed data. Most models apply the same parameters to all sites in a sequence, but advances in the past few years allow different models for different sites or even different taxa, at the cost of increased model complexity.

Bayesian Approach

Bayesian approaches use the same set of models as likelihood approaches but seek to return the tree with maximum posterior probability (in contrast to the tree that maximizes the probability of the data). This is done using Bayes' rule:

$$P(\text{Tree \& Model} \mid \text{Data}) = \frac{P(\text{Data} \mid \text{Tree \& Model}) \times P(\text{Tree \& Model})}{P(\text{Data})},$$

where the "Tree" is the phylogeny with branch lengths and the "Model" is the set of parameter values (such as substitution rates). $P(\text{Data} \mid \text{Tree \& Model})$ is just the likelihood of the data. $P(\text{Tree \& Model})$ is the prior probability of the tree and model parameter values. $P(\text{Data})$ is the total probability of the data over all possible trees and

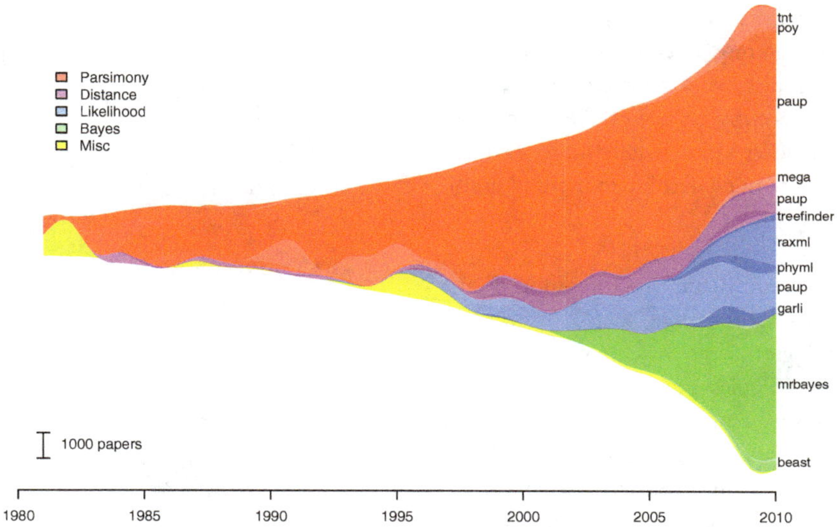

FIGURE 3. The growth of phylogenetics as a whole and of reconstruction algorithms and programs used, as a function of time. In each time interval, the total thickness of the plot represents the number of papers published in that year where the ISI Web of Science topic included "phylogen*" (note that this is not on a log scale). Each tree inference method is indicated by color, with different programs implementing that method indicated by a different intensity of that color. Abundances were based on information in TreeBase.org. Over 45 different computer programs were used to make tree inferences, but only the most frequently used are labeled. Note that some programs, such as PAUP, were used with different inference methods: each unique combination of program and method was counted independently. Information for 2010 is projected based on information available in July 2010. Important features to note are the overall growth in the number of papers using phylogenetics (going from 1507 in 1981 to 18,202 in 2009) and the increasing frequency of use of likelihood and Bayesian methods. One important caveat to the relative abundances of different methods and software is that the source database, TreeBase, primarily contains information on analyses performed in evolutionary biology journals and so underrepresents the techniques used in phylogenetics in fields such as molecular biology.

parameter values. This is quite impractical to calculate analytically. Markov chain Monte Carlo (MCMC) and related techniques can be used instead, directly returning the desired $P(\text{Tree \& Model} \mid \text{Data})$ once priors are specified and the Markov chain has run sufficiently long and with enough mixing.

Gene Tree–Species Tree

It has long been known that the history of genes may differ from the history of populations, or as it is commonly phrased, the gene tree may not match the species tree. This can be due to factors like introgression or hybridization, but even when species are completely reproductively isolated after speciation, lack of coalescence of all gene copies within a species may lead to ancestral polymorphism persisting and then being segregated later. Gene duplication and loss may also lead to difficulties. An algorithm to find a species tree taking into account gene tree conflict as part of a parsimony cost was developed over 30 years ago by Goodman and colleagues and has been used infrequently since then. However, until recently, most phylogenetic studies operated under the assumption that the gene and species trees generally would match and either used concatenated data from multiple genes to try to recover the dominant signal or would run tests to determine whether genes significantly disagreed over the phylogeny. Recent advances have resulted in both likelihood and Bayesian methods that account for this potential conflict. Some require gene trees to be inferred first and then species trees are created, while others go directly from raw sequence data to species trees, integrating over possible gene trees.

Supertrees

A different way to return trees is not from a set of input characters but from a set of input trees. For example, an ecologist may want to examine the phylogenetic diversity of two communities but cannot find a single tree that encompasses all the species needed to be included. However, there may be trees for various clades (monocots, conifers, eudicots) that collectively have all the necessary species, as well as a larger tree of land plants that has representatives from all these groups but not all the species. One common approach is to manually stitch together trees, using personal judgment in the case of conflict. Another approach is to convert each input tree to character data and then do a parsimony search on this tree-derived character data.

Trends

Figure 3 shows both the growth of the number of papers utilizing phylogenetics and the types of analyses and programs used in these papers. Note that of all the methods for inferring a tree, parsimony, likelihood, and Bayesian approaches are most popular, with the latter two methods growing in popularity recently (many of the papers still using parsimony use it along with a likelihood or Bayesian search to test robustness of tree inference to method used).

SEARCH STRATEGIES

Finding the "best tree" is a difficult process—indeed, one issue is agreeing on criteria for what "best" means. The sheer number of topologies grows extremely quickly. The number of possible topologies grows double-factorially ($3 \times 5 \times 7 \times \cdots$) with the number of taxa, which is much faster than exponential growth (i.e., $3 \times 3 \times 3 \times \cdots$.). For four taxa, there are just three unrooted bifurcating topologies, but for ten taxa, there are 2,027,025 such topologies. Besides the space of possible trees growing large very quickly, under many criteria the problem of finding the optimal tree has been shown to be one of a class of problems known as NP-hard in computer science, meaning (among other things) that there are no known algorithms for their solution that scale well (polynomially) with the problem size. There are algorithms that scale exponentially with problem size and are guaranteed to return the optimal tree under a given criterion, but in practice these are too slow for most studies, and thus heuristic methods that tend to get solutions near or at the actual solution (but with no guarantee of getting the optimal solution) are used instead (see Fig. 4). Creating better ways to do this search is an active area of development.

Bayesian approaches typically seek to estimate the posterior probabilities for all trees (and combinations of parameter values). Given that there are easily millions of trees even for small numbers of taxa, this initially appears like an insurmountable difficulty. However, Markov chain Monte Carlo techniques can be used to estimate the posterior probability distribution. A Markov chain is constructed that, when performing properly, moves across tree and parameter space in such a way that the amount of time spent in each tree or parameter value is proportional to its posterior probability.

Potential Problems

Numerous issues arise in getting the right tree due in part to the huge search space. Without approaches guaranteed to return the optimal tree under a given criterion in a feasible amount of time, heuristic methods are used. A more troubling issue is when the best tree, under some optimality criterion, differs from the true population tree that is the goal of phylogenetics, even when the amount of data grows infinitely large. In this situation, the method is called inconsistent. This typically happens when the assumptions of a model are violated. The most famous case of this is "long branch attraction." In this, two long branches have enough convergent changes that putting the branches together in a clade improves the tree score enough (by converting a pair of convergent changes into what is inferred to be one change inherited by both descendants, at the cost of splitting some valid single changes into two changes) that methods seeking to minimize change or evolutionary distance may be misled. Methods have been found to be inconsistent when they assume a single rate of evolution per branch but in actuality different sets of characters may have different rates on the same branch.

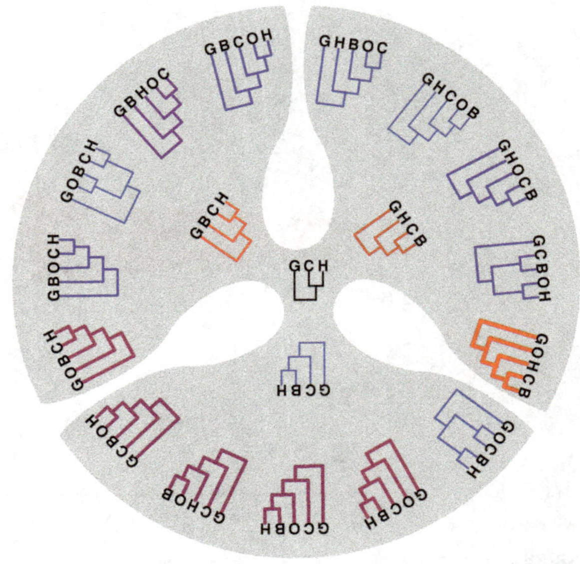

FIGURE 4 Search across a space of trees for orangutan (O), gorilla (G), human (H), chimpanzee (C), and bonobo (B). The analyses used sequences of the trace amine-associated receptor 4 (TAAR4) gene downloaded from GenBank. Branch width and red coloration represent tree score under a likelihood HKY+gamma model (red trees with thick branches are best). This figure illustrates a search technique known as sequential taxon addition. To find the best tree for N taxa, a tree for three taxa is created, and then the best position for a fourth taxon on this tree is found. The 4-taxon tree resulting from that is used to find the position of the fifth taxon to be optimized, and so on up to N taxa. In this empirical example, doing that strategy would result in finding a suboptimal tree: the best tree in the 4-taxon stage does not lead to the best 5-taxon tree, only the second-best tree. However, that strategy dramatically reduces the search space: rather than examining $3 \times 5 \times 7 \times \cdots \times 2N - 3$ trees in an exhaustive search, this only requires examining $3 + 5 + 7 + \cdots + 2N - 3$ trees. For just 15 taxa, if the stepwise addition search took 1 second, the exhaustive search would take over 34,000 years on the same computer (over 10^{12} times as many trees to examine). There are faster exact algorithms than exhaustive search, and in practice, algorithms in addition to stepwise addition would be used, but this small but real example suggests why heuristic strategies are important and also how they may fail to find the optimal tree.

MODELS

Some phylogenetic approaches do not use an explicit model (i.e., parsimony), but many others do. The development of these models has been a major focus in phylogenetics, especially given the concerns about incorrect models leading to inconsistent results.

Nucleotide Transition Matrix

Nucleotides typically have four states. Nucleotide models thus have a 4 × 4 matrix of instantaneous substitution rates to and from any pair of nucleotides. In the most unrestricted model, each entry off the diagonal can be independent (the diagonal entries, which are the rates between a state and itself, are determined by the three rates from a given state to another state). This unrestricted model with 12 free parameters is rarely used. Instead, it is usually simplified so that the rate going from state X to state Y is the same as the reverse rate, so that the model becomes the general time-reversible (GTR) model with 6 free parameters. A further simplification is to make all transitions (just A↔G and C↔T) have one rate and allow transversions to have a second rate. This is far simpler (two free parameters) but has some biological realism, as the two kinds of changes differ in how frequently a base change will result in an amino acid change. The two common parameterizations of these models are known as HKY85 and F84. The simplest restricted model is the Jukes–Cantor (JC) model, which sets all the substitution rates to be equal and so has just one free parameter. There are also many other possible ways of restricting the 12-parameter unrestricted model to simpler models. Note that state frequency at the root of the tree is also important in these models, and that this can be estimated as part of the overall model, equilibrium frequencies from the substitution matrix can be used, or the frequencies can be fixed at empirical or equal values.

Amino Acid Transition Matrix

There are 20 amino acids used in most organisms, leading to a 20 × 20 matrix of instantaneous substitution rates. This would mean 380 free parameters in the most unrestricted model, and 190 free parameters in the time-reversible model. More common than estimating these for a given analysis is to use one of a set of standard fixed substitution rate models (such as DAYHOFF, JTT, WAG, mtREV, cpREV, BLOSSUM, PAM, and others).

Codon Models

Amino acids (20 states) are coded for in DNA (4 states) using codons (3 DNA bases each, so $4^3 = 64$ possible codons). In some ways, a codon model is better than either a single nucleotide or amino acid model: it keeps more of the information in the raw DNA sequence than an amino acid model (redundancies in the code mean that converting DNA to amino acids can mask some of the underlying variation) while potentially having more biological realism than a single nucleotide model.

Alignment Models

DNA evolves through substitutions, but not only substitutions: individual bases or longer sections can also be duplicated, deleted, or rearranged. It is these processes that make alignment of DNA a difficult problem. However, they may also be important to include in phylogenetic models. There are now some models that can take insertions and deletions into account.

Rate Heterogeneity

All the models discussed above assume that each site in the sequence (or each codon) has the same rate. This is biologically unrealistic: different positions in the genome will evolve at different rates, due to different amounts of selection (consider exons vs. introns), different mutation rates, or other heterogeneous processes. There may also be heterogeneity in rate at a given site across a tree. There are many models that seek to include some of this heterogeneity. One set of models calculates the probability of the data at each site under a set of different discrete rates and takes as the probability of the data at the site the sum of these probabilities under different rates (note that the sum is taken, not the max). One such model is known as the discrete-gamma model: a single parameter gamma distribution (shape and rate parameters set equal) is divided into a preselected number of rate categories. At each site, the probability of the observed data under the mean rate in each of these categories is calculated and then summed to get the probability of the data. A similar model, the invariant sites model, uses two rate categories, one with a rate of zero and the other with a nonzero rate. Another approach is to use site-specific rate models: all sites may have the same transition matrix, but first, second, and third codon positions may all have different estimated rates.

Partitioning

An extension of site-specific models are models that apply different substitution models to different sets of sites. For example, one could apply a GTR model to one gene and an HKY model to another gene. Even with the increased complexity (number of free parameters) this causes, it is often a better fitting model than one that applies the same model, even one with rate heterogeneity, to all sites.

Model Choice

A wide variety of models are available, but choosing the wrong model may lead to inconsistent results. For users of maximum likelihood inference, the general approach is to use the Akaike Information Criterion (AIC) to select between models. Model parameters (such as transition rates) are sometimes estimated on an initial tree and then fixed for subsequent searches in order to accelerate searches For Bayesian analyses, AIC scores may also sometimes be used, but so are Bayes factors (the ratio of integrated likelihoods between models). Reversible jump Markov Chain Monte Carlo methods, where a Bayesian MCMC run can move between models of differing complexities, are now available for use.

CLOCKING

Branch lengths on phylogenies typically reflect amount of change. For many uses of phylogenies, such as investigating character evolution, estimating diversification rates, or examining when certain groups originated, it may be more useful to have branch lengths in units of time. The simplest approach to calibrating a phylogeny is to fix one node at a known divergence time and then to constrain the model such that all taxa have the same rate of evolution (so that, for a set of extant taxa, the root to tip length for all taxa is the same). However, this ignores potential issues with calibration uncertainty, and in many cases the data fit to a clock model is significantly worse than fit to a nonclock model. There have been various approaches to relax the clock model while still allowing calibration. One approach uses a fully Bayesian approach allowing local rates on branches and prior distributions on calibration ages rather than fixed ages. A different approach, penalized likelihood, takes a tree and possibly calibrations or bounds as input and uses an approach that estimates correlated but variable rates of evolution on a tree.

CONFIDENCE

A phylogeny reconstruction typically returns a tree (but see below for the Bayesian case). But how does a researcher know the quality of this tree—is the tree strongly supported by the data? Are there certain parts that are good, but others that are uncertain? Though various approaches have been advocated (decay indices, jackknifing, parametric bootstrapping, and others), basically two approaches are used.

Nonparametric Bootstrapping

An estimate (in phylogenetics, an inferred tree) is based on a finite sample of data. A way to evaluate the variability in that estimate would be to get new samples of data from the same underlying distribution and construct new estimates and then examine how much they differed from each other. Bootstrapping applies this idea, but as the underlying distribution is unknown, it uses the available estimate for that distribution, the raw dataset itself, and draws new pseudoreplicate datasets from that (sampling with replacement). Each pseudoreplicate dataset is then analyzed in the same way as the original dataset and the trees returned recorded. If the data weakly support a given tree inference (such as a close relationship between primates and rodents), it will only appear in a subset of the reconstructions. This approach can be used with all methods of phylogenetic reconstruction.

Posterior Probabilities

A great appeal of Bayesian approaches is that rather than return the probability of the data given a model (the likelihood), they return the probability of the model given the data. In the phylogenetics case, after a Bayesian MCMC run is done, the saved set of trees and parameters sampled during the search can be used to estimate the posterior probabilities of their values: a topology found twice as often as another has twice the posterior probability.

Bootstrap/Bayes Comparison

Results from bootstrap and Bayesian runs are generally summarized in the same way: for a given branch on the tree, the proportion of trees in the final sample (bootstrap reconstructions or Bayesian samples) that have that branch is recorded and represents the score or posterior probability. These proportions have the same range (0–1) and can be (and often are) plotted on the same tree, so they are often compared. However, it is important to note that they are measuring different things. Bootstrapping is measuring the sensitivity of point estimates, whereas a Bayesian approach is attempting to estimate the posterior probabilities of all trees. For example, suppose the dataset is

Species_1: AGGGGGGGGGGGGGGGGGGGGG...

Species_2: AGGGGGGGGGGGGGGGGGGGGG...

Species_3:TGGGGGGGGGGGGGGGGGGGGG...

Species_4:TGGGGGGGGGGGGGGGGGGGGG...

With just one variable site, the best tree under all phylogenetic methods would show species 1 and 2 separated by a branch from species 3 and 4. However, bootstrapping and Bayesian approaches would return very different estimates for the support for that branch. As the number of invariant sites increases, the proportion of bootstrap replicates that include the only variable site becomes slightly less than two-thirds. Assuming that in the cases with only invariant sites each of the three possible trees is returned with equal probabilities, this means that the branch will get bootstrap support of approximately 80%. However, in the Bayesian case, all the data are always available. As the number of invariant sites increases, the overall estimated rate of evolution decreases, and the possibility that the one variable site represents multiple changes rather than one change drops, so the posterior probability of inferred branch increases. The Bayesian approach would say that given the dataset and prior beliefs, this tree has ~100% of the posterior probability.

GETTING A PHYLOGENY

Phylogenetic tree reconstruction can take a fair bit of time and computing resources. Fortunately, there is an expanding array of resources for doing this. Downloading existing trees is made easier by the online repository treebase.org. There are ongoing research projects that return current best estimates of phylogenetic trees for various groups (searching online for Phylota, Phylomatic, or Mor will return some of these). There are also online services that will run phylogenetic analyses on supercomputers for free: the CIPRES web portal is one such service, though there are many others.

Given continuing improvements in search algorithms and computer hardware, it is generally feasible to do small- to moderate-sized analyses on personal computers. Much of the software in this area is free and open source.

SEE ALSO THE FOLLOWING ARTICLES

Bayesian Statistics / Information Criteria in Ecology / Markov Chains / Model Fitting / Phylogeography

FURTHER READING

Felsenstein, J. 2004. *Inferring phylogenies.* Sunderland, MA: Sinauer Associates.

Huelsenbeck, J. P., F. Ronquist, R. Nielsen, and J. P. Bollback. 2001. Bayesian inference of phylogeny and its impact on evolutionary biology. *Science* 294: 2310–2314.

Kelchner, S. A., and M. A. Thomas. 2007. Model use in phylogenetics: nine key questions. *Trends in Ecology and Evolution* 22: 87–94.

Rannala, B., and Z. Yang. 2008. Phylogenetic inference using whole genomes. *Annual Review of Genomics and Human Genetics* 9: 217–231.

Semple, C., and M. Steel 2003. *Phylogenetics.* Oxford: Oxford University Press.

Yang, Z. 2006. *Computational molecular evolution.* Oxford: Oxford University Press.

PHYLOGEOGRAPHY

SCOTT V. EDWARDS, SUSAN E. CAMERON DEVITT, AND MATTHEW K. FUJITA

Harvard University, Cambridge, Massachusetts

Phylogeography is the study of the geography and evolution of genetic lineages in natural populations. Its main goals are to elucidate the recent phylogenetic and demographic history of species and the population processes and responses to environmental change that have resulted in species' current geographic distributions. As its name implies, phylogeography shares methodologies and perspectives with phylogenetics—interpreting genetic variation specifically in terms of genealogical or phylogenetic trees—and biogeography. The dominant perspective views phylogeography as one subdiscipline, along with ecological biogeography, of the overarching field of biogeography. Phylogeography also shares many methods and analytical tools with population genetics.

HISTORY AND SCOPE OF PHYLOGEOGRAPHY

The term *phylogeography* was coined by John Avise and colleagues in a seminal paper in 1987, although the principles and methods of the approach extend to as early as 1979. This 1987 paper and much of the field until the early 1990s was focused on trees of mitochondrial DNA lineages (see below) and the interpretation of genetic lineages in broad qualitative categories of population histories, such as population subdivision, population expansion, and intergradation of formerly separated populations. Since then, the field has embraced analysis of genetic variation in the nuclear genome and has shifted to a focus on populations harboring genetic lineages, rather than the lineages per se. As the field grapples with this transition, there is still controversy about how to translate genetic patterns into demographic scenarios and how to accommodate the ubiquitous variation in genealogical patterns from gene to gene. Some researchers prefer heuristic methods that provide stepwise inference keys to interpret patterns in gene trees, whereas others prefer ensemble methods that accommodate stochastic variation from gene to gene and statistically model populations as collections of genetic lineages. Much of the challenge stems from the often poor resolution provided by trees of individual nuclear genes within species, as well as the variation from gene to gene owing to the stochasticity of the coalescent process, the processes by which genetic lineages in extant populations coalesce backward in time to ever fewer ancestral alleles until a single ancestral allele is reached, creating a gene tree. How to combine information from different genes into a single scenario of population history is a major current goal of phylogeographic research.

The original focus of phylogeography on mtDNA and the visual, often qualitative interpretation of gene trees has evolved into a diverse discipline with many molecular and statistical approaches that are well grounded in the framework of hypothesis testing. Most prominent has been the incorporation of nuclear markers in addition to mtDNA as tools in phylogeography. Although mtDNA is still widely and effectively used in the phylogeographic study of animals, particularly vertebrates, this genome has proven less useful in the phylogeographic study of plants, fungi, and other eukaryotes. Similarly, chloroplast DNA evolves too slowly to be of widespread use as a phylogeographic marker in plants, although some progress has been made. For these reasons, nuclear markers (or chromosome or plasmid markers in bacteria and archaea) provide the only truly universal markers for phylogeography. A view is emerging in which the multiplicity of markers is essential for providing robust estimates of population history and demographic processes. This view is not universally held, particularly for phylogeographers working on vertebrates and for whom animal mtDNA provides a powerful marker of phylogeographic history. Even in these situations, though, inference about the history of populations in which gene trees and genetic variation are embedded necessarily require multiple markers—whether for delimiting species, estimating gene flow, or attempting to detect population growth or other processes.

The role of the gene tree per se is shifting in phylogeographic research. Whereas robust gene trees estimated from molecular data have been a mainstay of phylogeographic research, with many inferences drawn directly from inspection of gene trees, other aspects of molecular data, such as allele frequencies and the site-frequency spectrum (the distribution of frequencies of single nucleotide polymorphisms, or SNPs, within a population), are now becoming key tools of inference and estimation. This shift is again the result of the increasing focus on population parameters as goals of phylogeographic estimates, rather than gene trees per se and inferences from them (Figs. 1A–D). The increasing use of microsatellite (simple sequence repeat) markers and SNPs in phylogeography, from which meaningful gene trees are challenging to infer, is another sign that the dominance of gene trees in phylogeographic inference is being broadened to include other types of markers and statistics.

In addition to shifts in the types of molecular data used for phylogeography since its inception, there have been concomitant shifts in the analytical and statistical tools employed. For example, statistical phylogeography, a term coined by Lacey Knowles and Wayne Maddison, has emerged as a perspective that advocates comparison of competing historical hypotheses to explain a given phylogeographic dataset and evaluation of those hypotheses through statistical tests such as likelihood or Bayesian methods or computer simulation. Statistical phylogeography, and associated methods such as approximate Bayesian computation (ABC), emphasize the comparison of statistics observed in collected data to null distributions of those statistics derived from computer simulation from a priori population models. Some examples of simple population models for a single population are given in Figure 2. Many statistical methods will use one or more summary statistics derived from sequence data and gene trees, or simulations of sequences and trees from population models, to estimate population histories such as stability, exponential growth, or abrupt growth (Fig. 2), or to rule out alternative scenarios. In addition to model evaluation, direct inference of population genetic parameters such as gene flow, effective population size, and population growth rate are possible using a variety of likelihood and Bayesian inference-based methods, usually employing coalescent models. Both types of methods allow quantitative estimation of population parameters, which is a major advance over earlier more descriptive approaches in phylogeography (Figs. 1A–D).

Increasingly, traditional phylogeographic study using molecular markers is being complemented by quantification of the physical and biotic environment. Although mapping the distribution of lineages has long been a key component of phylogeography, with the recent increasing availability of high-resolution environmental data and large, publicly available databases of georeferenced species locality data, it is possible to quantify geographic and spatial relationships within a Geographic Information System (GIS). The ability to map a taxon's environmental and geographic range to test scenarios of diversification is a key innovation for phylogeography in recent years. Environmental niche nodels are increasingly applied to questions in statistical phylogeography because they permit testing of spatially explicit biogeographic hypotheses. These models combine spatial environmental data (usually climate) with georeferenced species locality data in a mathematical model. The model can then be used to predict, i.e., map, the species' current distribution,

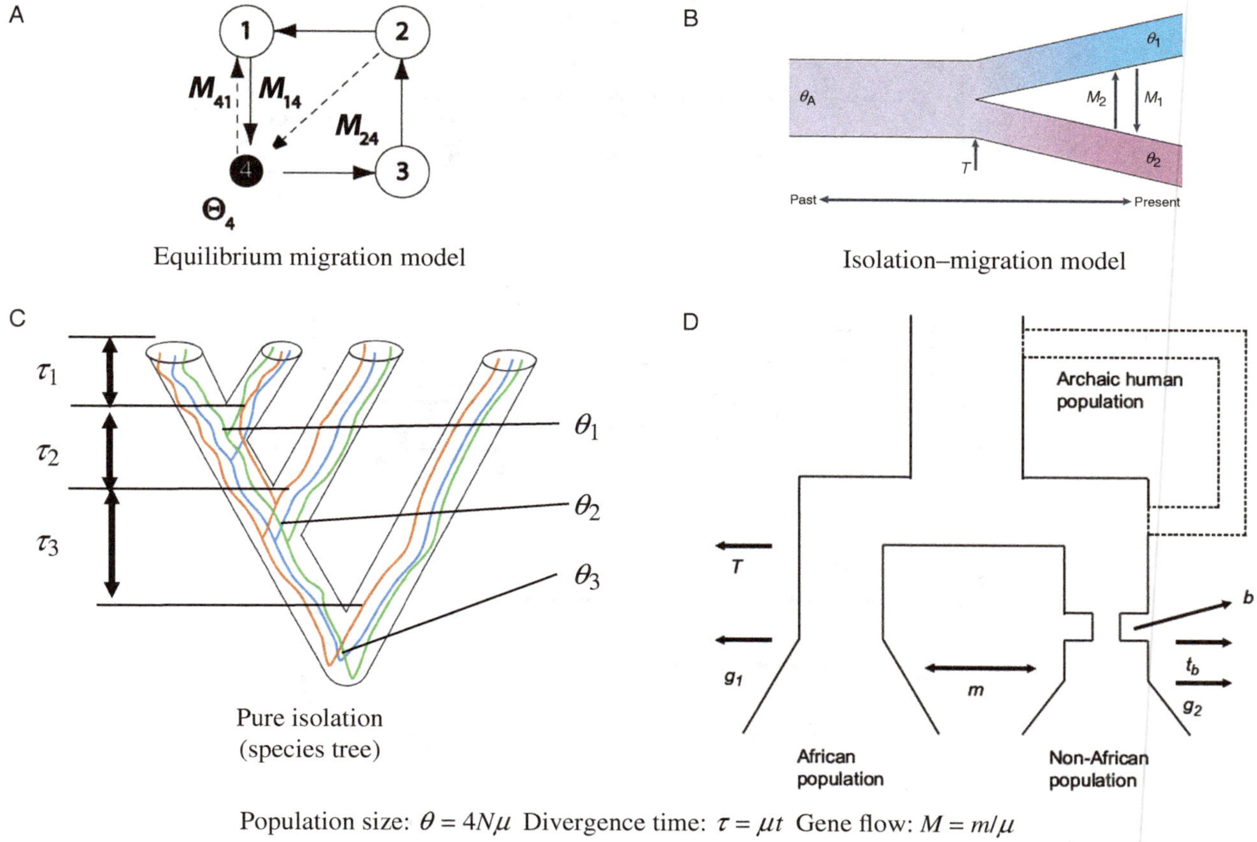

FIGURE 1 Phylogeographic models. These models are standard in modern phylogeographic research and all require data from multiple genetic loci to estimate the parameters indicated. (A) General migration matrix model; (B) Isolation—migration model; (C) Phylogenetic model involving successive isolation events with no gene flow; (D) An example of a complex demographic model such as is used to simulate data sets for comparison with observed data. Demographic parameters indicated include population divergence times (T or t; in (D), t_b is time of bottleneck b); effective population sizes of current (θ) or ancestral (θ_A); migration rates scaled by the mutation rate, $M = m/\mu$, in panel (A); here, number subscripts ab indicate number of migrants leaving population a and entering population b; or m (migration rate per individual per generation) in panel (D); and population growth, g (panel (D)). Figures used with permission as follows: A, from P. Beerli, 2006, *Bioinformatics* 22: 341–345; B, J. Hey, and C. A. Machado, 2003, *Nature Reviews Genetics* 4: 535–543; D, J. D. Wall, K. E. Lohmueller, and V. Plagnol, 2009. *Molecular Biology and Evolution* 26: 1823–1827.

and if historical climate data or models are available, the species' historical distribution as well. Other scenarios of diversification can be generated within a GIS and tested as well, including presence or absence of barriers to dispersal, migration pathways, historical demography, and the presence of contact zones.

MOLECULAR TECHNIQUES AND PHYLOGEOGRAPHY

At the heart of any phylogeographic study are the genetic data, and over the past few decades improvements and methodological transitions in data collection have allowed greater precision and more detailed and complex historical demographic inferences. The overwhelming majority of datasets include a quantification of diversity in DNA sampled from individuals across the distribution of a species. The first method that applied quantification of DNA diversity in phylogeography—indeed the first method able to yield true gene trees within populations—was restriction fragment length polymorphisms (RFLP), in which an endonuclease (restriction enzyme) fragments DNA molecules at particular motifs in the genome (restriction sites). Patterns of fragment lengths, as assayed via gel electrophoresis, represent measures of diversity within a species; similar fragment patterns among individuals imply species lacking significant diversity, while mixtures of fragment patterns suggest significant DNA diversity within species. Crucially, these fragment patterns can be used to infer gene trees by any number of phylogenetic methods, and these gene trees can provide the basis of phylogeographic inference.

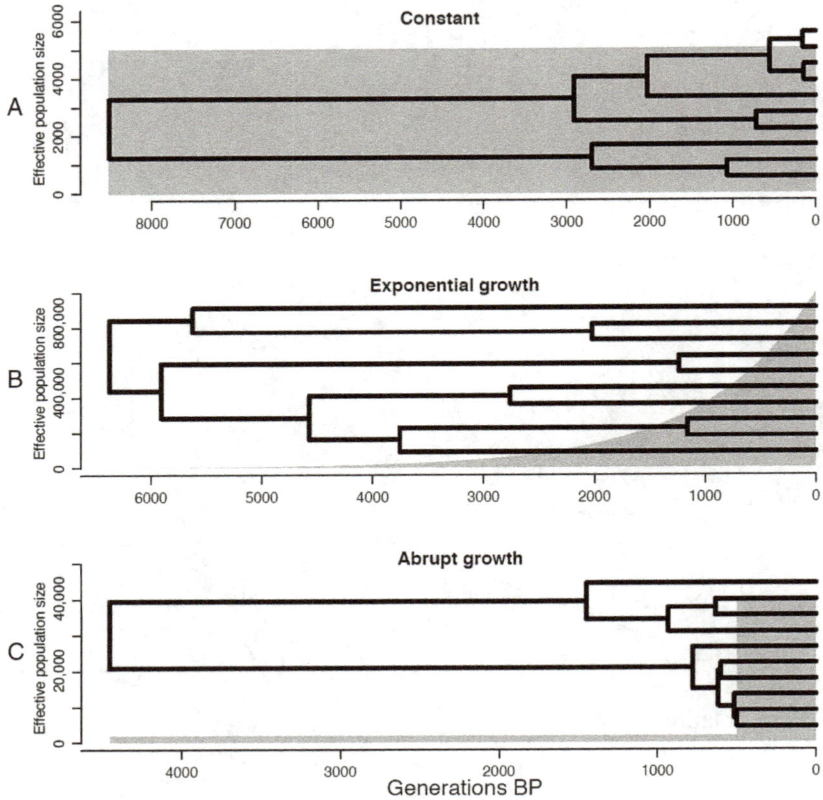

FIGURE 2 The effect of population expansions on gene trees. In each panel the population growth trajectory is depicted in gray, and the scale at the bottom is in generations before the present. Panel (A) shows a typical coalescent tree in a population of constant size. Panel (B) shows the effect of exponential growth, resulting in a protracted gene tree. Panel (C) shows the ability of an abrupt increase in population size to compress coalescent events into a small slice of time. Note the slight difference in scale of the *x*-axis of each panel. Simulations were performed by Christian Anderson in serial simcoal (C. N. K. Anderson, U. Ramakrishnan, Y. L. Chan, and E. A. Hadly, 2005, *Bioinformatics* 21: 1733–1734) so as to result in gene trees of roughly similar depth.

Mitochondrial DNA

Since its inception, the workhorse for phylogeography has been mitochondrial DNA (mtDNA), and for several good reasons. For instance, the mitochondrial genome can be separated from the nuclear genome by standard centrifugation procedures. In addition, because of its small size (~14–17 kb for a typical vertebrate mitochondrial genome), it is possible to measure the diversity of the entire genome—for example, by RFLP or DNA sequencing methods. However, it is the evolutionary properties of mtDNA, at least in animals, that make it a desirable genetic marker, and today, despite the increasing frequency of multilocus studies, mtDNA continues to be an important contribution to phylogeography. First, mtDNA is inherited maternally in the vast majority of animals. Barring significant paternal inheritance (called paternal leakage), this attribute provides a single gene history. This enables tracking the population's history owing to its small effective population size, which allows allelic lineages to have ancestors within their respective populations more frequently than for nuclear genes, as these have a larger effective population size. Second, mtDNA lacks significant recombination, which can complicate phylogeographic inferences. Third, animal mtDNA evolves quickly, perhaps an order of magnitude more quickly than a typical nuclear marker, as a result of a higher mutation rate (probably due to exposure to radicals as a byproduct of respiration in the mitochondrion).

The importance of mtDNA to phylogeography accelerated as DNA sequencing gained momentum in evolutionary studies in the 1990s, primarily as an outcome of the ease of obtaining mtDNA sequences compared to nuclear sequences from nonmodel organisms. The Sanger method for sequencing using polymerase chain reaction (PCR) amplification provides targeted markers as the primary method of obtaining homologous sequences from individuals. The availability of PCR primers that permitted PCR amplification in diverse organisms helped maintain and propel the importance of mtDNA in evolutionary studies. Along with DNA sequence data came increased use of advanced algorithms for inferring genes trees (e.g., maximum likelihood and Bayesian phylogenetics) and the development of user-friendly analytical tools based on coalescent theory (see the section "Statistical Models and the Future of Phylogeography," below).

Sequence-Based Nuclear Markers

The reliance on DNA sequencing at the population level continues today, and the pitfalls of single-marker inferences have resulted in a shift in focus to multilocus analyses of unlinked markers from the nuclear genome. The history of genes within populations and species may not be the same for each gene; mtDNA may have a different history than one or more nuclear genes, and these histories may not reflect the lineage history of the organisms. These historical discrepancies arise for several reasons, including recombination between divergent alleles and migration between populations, but the most pervasive reason for discordant histories is incomplete lineage sorting. Alleles that arise in ancestral populations can sort in descendant populations and divergences such that gene history and lineage history do not match; in the absence of selection, this incomplete lineage sorting arises stochastically as a result of genetic drift.

Every marker—nuclear and mitochondrial—can exhibit incomplete lineage sorting with respect to lineage history, and as a result inferences based on a single marker can provide a biased interpretation of phylogeographic patterns. Recent research on a variety of organisms from humans to plants has shown that discordance between gene tree topologies can be common in intraspecific studies, as well as in studies between closely related species. For this reason, there has been a shift away from using gene tree topologies as the primary inference tool and instead a greater focus on inferring demographic parameters either using other aspects of genetic variation (such as the frequency spectrum of SNPs) or by statistically integrating across possible gene trees. This shift in focus reduces the concern of incomplete lineage sorting and reduces gene tree topologies to statistical "nuisance" parameters. Developing nuclear markers for nonmodel organisms has become routine and generally involves building genomic resources, such as genomic or cDNA libraries, from which "anonymous" loci are characterized for PCR amplification and sequencing.

Microsatellites

Another class of marker often used for phylogeography are simple sequence repeats (SSRs), or microsatellites. SSRs may be the most popular class of marker for phylogeography, in part because of their extraordinary variability, which allows fine-scale discrimination between populations. Despite this, these markers are analyzed as fragments, not as sequence data, and are therefore amenable to fewer and less powerful analytical tools. The mutation processes by which they evolve are challenging to model, and hence SSRs can be challenging to use to infer organismal history. Finally, SSRs are genomically anomalous due to their hypervariability, and hence they do not represent the rate of evolution across the genome generally, making them less useful for inferences about the genome as a whole. This hypervariability, although a perceived advantage, can compromise some indexes of population differentiation, such as F_{st}. As the development of SNP and sequencing markers becomes easier with advances in massively parallel sequencing, microsatellites will likely lose their popularity to sequence-based markers in phylogeography. Nonetheless, due to their hypervariablity, they will continue to play an important role in studying co-ancestry of individuals within populations, parentage, and short-range dispersal.

EMPIRICAL RESULTS OF PHYLOGEOGRAPHY

Because it is a young science, phylogeography has necessarily been used to test sometimes decades-old hypotheses developed in the pre-molecular era of biology. For example, several of the most important patterns that have emerged from phylogeography studies include the existence of refugia during the glacial cycles of the Pliocene and Pleistocene, the formation of hybrid zones as lineages expanded from disparate refugia, and the role of geographic barriers to population structure. Although refugia had long been hypothesized throughout the globe, and had been tested with nongenealogical molecular markers such as allozymes, phylogeographic methods vividly and often graphically capture the geological, climatic, and genetic processes driving the response of populations to glacial processes. Two key phylogeographic signatures of past refugia include the greater genetic diversity typically found in unglaciated regions as compared to formerly glaciated regions, and signatures of population expansion realized as a species rapidly expands from refugia into unoccupied regions as glaciers retreated. Recently, GIS tools have been used to directly infer the location of Pleistocene refugia by modeling species' distributions in the past. Jointly using genetic and GIS tools provides a powerful approach for phylogeographic inference. This section reviews the comparative phylogeographic inferences from three well-studied regions—Southeast United States, Southern Europe, and Australia—as well as the demographic history of humans.

Southeast U.S.

The Southeast of the United States has a special historical relevance to phylogeography because the pocket gophers in this area were the subject of one of the first phylogeographic studies by John Avise and colleagues. Rather than identifying refugial areas, these first studies discovered hierarchical population structure that often corresponded to drainage systems, with major phylogeographic breaks between the Gulf and Atlantic basins. Subsequent to this early study, Avise and colleagues conducted numerous phylogeographic surveys of marine and freshwater species distributed on either side of the Florida peninsula, primarily using mtDNA (Fig. 3). Invariably, these studies showed a strong phylogeographic break corresponding to the Florida peninsula and demarcating populations inhabiting the Gulf of Mexico and the Atlantic Ocean. The geographic consistency of these breaks made this area a classic case of phylogeographic vicariance. Together these studies represent probably the earliest example of the field of comparative phylogeography, in which the genealogical histories of multiple species inhabiting similar geographic ranges are compared. However, the coalescence time of mitochondrial lineages spanning the Florida peninsula varied markedly from species to species, suggesting the possibility of recurrent cycles of isolation across the peninsula. Alternatively, because gene trees inherently are influenced by the stochastic nature of the coalescent process and genetic drift, variation in coalescence times of mtDNA could have also been caused by variation in the sizes of populations of common ancestors of the Gulf and Atlantic regions. Discriminating between the hypotheses of contemporaneous or recurrent isolation is more easily tackled with multiple genetic markers, the genealogical patterns among which can then be used to discriminate between varying ancestral population size and varying times of isolation as explanations for the observed genetic patterns.

Southern Europe

Phylogeographic research has helped shed light on the role of Southern Europe and the Balkans in refugial isolation of diverse species. Phylogeographic structure across a variety of taxa, including insects, mammals, reptiles, amphibians, birds, and plants, has identified three major refugial areas in Europe corresponding to the southern peninsulas: Iberia, Italy, and the Balkans. Post-glacial expansion northward from these refugia established contemporary distributions. Lineages that occupied more than one refugia often formed hybrid zones where the leading edges of their expanding distributions meet. A

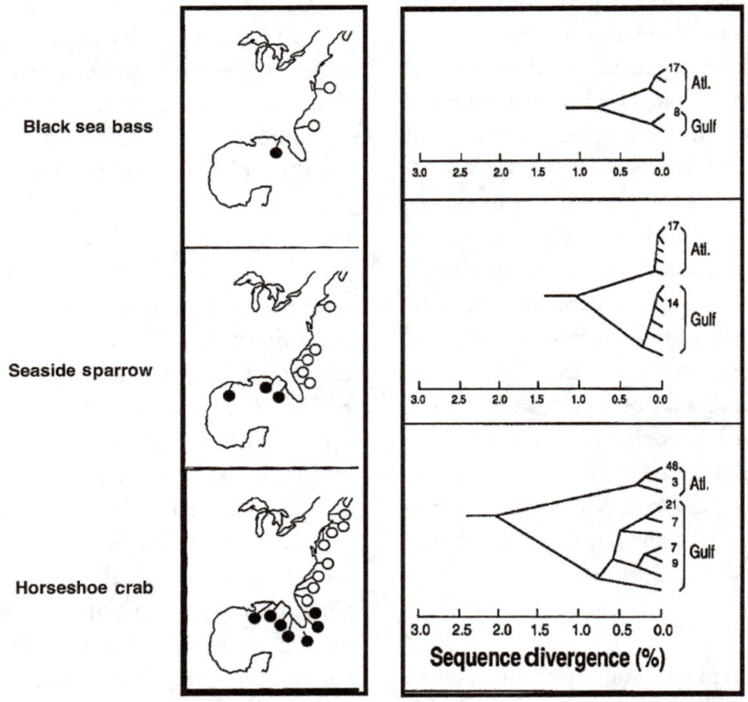

FIGURE 3 Examples of mitochondrial gene trees across the Florida peninsula. Each gene tree has a similar geographic pattern but a unique temporal depth, suggesting recurrent bouts of isolation of populations to the east and west of the peninsula. Adapted from J. C. Avise, 1994, *Molecular Markers, Natural History and Evolution* (New York: Chapman and Hall).

good example of these phenomena is the European rabbit, which shows microdifferentiation within the Iberian peninsula and which has been shown through phylogeographic surveys to be expanding from refugia towards the center of the peninsula. Hybridization between formerly isolated and detectably differentiated populations is now occurring. On a broader geographic scale, the genetic signatures of expansion from refugia in Europe and the subsequent intergradation of populations are among the best characterized phylogeograhic patterns thus far. Mammalian populations in North America also exhibit the classic genetic signatures of population expansion as a result of receding glaciers.

Australia

Other examples of applications of phylogeography to whole faunas come from Australia. The climate history in Australia was different than in the Northern Hemisphere; rather than experiencing repeated glaciation during the glacial cycles, Australia went through waves of intense aridification during the Pliocene and Pleistocene. Once largely tropical, Australia today contains a large arid interior, with few remaining pockets of rainforest around the periphery of the continent, the largest and best known of which are the Australian Wet Tropics (AWT) in northeastern Queensland. Numerous mtDNA and multilocus surveys of vertebrates and invertebrates have been conducted by Craig Moritz and colleagues and have identified the Black Mountain corridor as a site of major genetic breaks in multiple species (Fig. 4). Like the breaks around the Florida peninsula, the breaks around the Black Mountain corridor vary in their coalescence time, with a similar range of potential hypotheses to explain the pattern. In general these surveys, often including both nuclear and mitochondrial markers, do in fact imply a range of divergence times, sometimes exhibiting taxon-specificity (for example, birds tend to have shallower divergences than reptiles and amphibians). The AWT studies have also identified regions that harbor enhanced genetic diversity, and these regions tend to map to areas that have been identified by GIS as being climatically stable over the last 20,000 years. Indeed phylogeographic studies of AWT, as well as similar studies of amphibians in the Atlantic rainforest of Brazil, show a great correspondence between level of genetic diversity, strongly monophyletic clades in gene trees, and regions possessing high climatic stability. Such studies are useful not only for fleshing out the likely drivers of geographic patterns but also for identifying areas of high conservation concern.

Phylogeographic surveys of Australian birds, bats, and other vertebrates have been instrumental in identifying

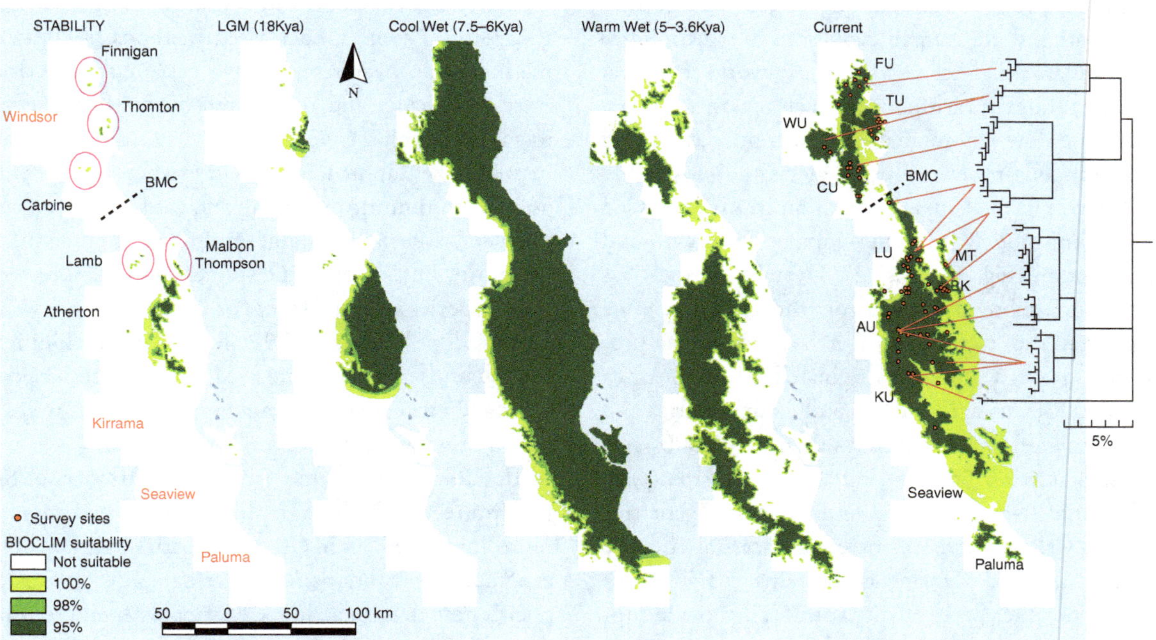

FIGURE 4 Analysis of phylogeographic data using habitat modeling from geographic information systems (GIS). The analysis focuses on a snail (*Gnarosophia bellendenkerensis*) endemic to the wet tropics of northeast Australia. Each panel shows actual or inferred extent of preferred habitat, with light green regions indicating use of 100% of climate parameter values and darker green regions indicating smaller percentages of climate parameters. Leftmost panel indicates areas of habitat stability over the 18,000 years modeled in this study. From Hugall et al., 2002.

phylogeographic breaks in other parts of Australia, particularly across the Carpentarian Barrier, a savannah region just south of the Gulf of Carpentaria that separates the faunas of Cape York from those of the Top End (north-central) and Kimberly (northwest) regions. This barrier, affecting species inhabiting predominantly eucalypt woodland and mangrove habitats, provides a useful counterpoint to the AWT studies, yet the extent to which faunas of different habitats and regions have diversified on different time scales in the two areas is still not known. Areas of phylogeographic diversification in Australia often map well to so-called areas of endemism identified through phylogenetic analyses of populations based on taxonomic and morphological analyses.

PHYLOGEOGRAPHIC PATTERNS IN HUMAN POPULATIONS

Human populations are just as amenable to phylogeographic research as are those of nonmodel species. Since the sequencing of the human genome in 2001, it is no surprise that the enormous genetic resources available for humans have allowed researchers on human population history to lead the way in phylogeographic research. Early research utilizing mtDNA suggested rejection of the "Candelabra" model of human evolution, which posited that human races diverged from one another about 1 million years ago and remained distinct since then. Similarly, mtDNA studies questioned the "multiregional hypothesis" for human evolution, in which *Homo erectus* populated distinct regions of the world ~2 mya, with modern humans arising from each region while occasional gene flow maintained genetic connectivity. The fact that humans are relatively panmictic and most likely recently derive from a single population in Africa has led to a variety of models involving population growth and rapid migration (Fig. 1D, Fig. 2). Overall, the mtDNA data and much subsequent data from the nuclear genome has supported the "out-of-Africa" hypothesis, which proposes that modern humans originated in East Africa about 200,000 years ago, and expanded out of Africa and populated the rest of the world less than ~100,000 years ago, displacing *H. erectus* and *H. neanderthalensis*. Mitochondrial, Y-chromosome, and autosomal gene tree analyses, as well as inferences of small ancestral effective population sizes and a greater genetic diversity in Africa (compared to other parts of the world), all provide support for the out-of-Africa hypothesis.

While there is a general acceptance of the out-of-Africa hypothesis, the multiregional hypothesis is sometimes supported by multilocus studies. Variation in coalescent times in genes have led some researchers to conclude that humans expanded out of Africa multiple times, though these patterns may just reflect the inherent stochasticity of the coalescent process. Complicating the picture is the fact that many human genes exhibit evidence of natural selection, and some show surprisingly deep ancestries in excess of 2 mya, a time deemed too deep to be explained by coalescent stochasticity. Such data, as well as the recent sequencing of a Neanderthal genome, lend support to scenarios involving gene flow between Neanderthals and Europeans after they migrated out of Africa, although this scenario is by no means settled and will likely be contentious for years to come (Fig. 1D).

STATISTICAL MODELS AND THE FUTURE OF PHYLOGEOGRAPHY

Currently, there is a heated debate about the relative merits of inference key approaches to estimating demographic parameters and model-based methods that directly estimate demographic parameters or that use simulation to produce a range of scenarios with which the data are compatible. Inference key approaches to phylogeography run the risk of ignoring the large stochastic component of variability in gene trees and genetic variation as a result of the coalescent process. Model-based methods, though generally viewed as statistically more rigorous, can present a small range of relatively simple models that real populations almost invariably violate. In addition, some model-based methods, such as ABC methods, allow the researcher to explore models of arbitrary complexity and therefore achieve a modest match to real situations. Given the high false discovery rate of inference key approaches, on the whole the phylogeographic community is moving toward model-based approaches, especially as large multilocus datasets become the norm. Phylogeographic studies in humans and other model species such as *Drosophila* and *Arabidopsis* were the first to adopt large-scale SNP analyses in which hundreds of thousands of markers are used to finely delimit populations and estimate past demographic events as well as the action of natural selection on the genome.

In addition to routine use of large numbers of SNPs, the future of phylogeographic research becoming ever more integrated with GIS and spatial analyses. Phylogeographic software packages are being developed to explicitly test diversification scenarios with environmental data. Such methods can simulate migration based upon probability of dispersal through different habitat types using the coalescent, quantify effective distances between populations, estimate historical gene flow and locations of ancestral populations using likelihood, and visualize phylogenetic trees within a GIS framework. As analytical methods improve and spatial data become more available, visualization and spatial analysis within a GIS will become the standard within phylogeography.

SEE ALSO THE FOLLOWING ARTICLES

Demography / Geographic Information Systems / Mutation, Selection, and Genetic Drift / Phylogenetic Reconstruction / Spatial Spread / Stochasticity

FURTHER READING

Avise, J.C. 2000. *Phylogeography: the history and formation of species.* Cambridge, MA: Harvard University Press.

Avise, J.C., J. Arnold, R.M. Ball, E. Bermingham, T. Lamb, J.E. Neigel, C.A. Reeb et al. 1987. Intraspecific phylogeography: the mitochondrial DNA bridge between population genetics and systematics. *Annual Review of Ecology and Systematics* 18: 489–522.

Balakrishnan, C.N., J. Lee, and S.V. Edwards. 2010. Phylogeography and phylogenetics in the nuclear age. In P. R. Grant, and B. R. Grant, eds. *In search of the causes of evolution: from field observations to mechanisms.* Princeton: Princeton University Press.

Brito, P., and S. Edwards. 2009. Multilocus phylogeography and phylogenetics using sequence-based markers. *Genetica* 135: 439–455.

Hewitt, G. 2000. The genetic legacy of the Quaternary ice ages. *Nature* 405: 907–913.

Hey, J., and C.A. Machado. 2003. The study of structured populations—new hope for a difficult and divided science. *Nature Reviews Genetics* 4: 535–543.

Hugall, A., C. Moritz, A. Moussalli, and J. Stanisic. 2002. Reconciling paleodistribution models and comparative phylogeography in the wet tropics rainforest land snail Gnarosophia bellendenkerensis (Brazier 1875). *Proceedings of the National Academy of Sciences* (USA) 99: 6112–6117.

Knowles, L. L. 2009. Statistical phylogeography. *Annual Review of Ecology Evolution and Systematics* 40: 593–612.

Nielsen, R., I. Hellmann, M. Hubisz, C. Bustamante, and A. G. Clark. 2007. Recent and ongoing selection in the human genome. *Nature Reviews Genetics* 8: 857–868.

PLANT COMPETITION AND CANOPY INTERACTIONS

E. DAVID FORD

University of Washington, Seattle

A foliage canopy is a spatially continuous cover of foliage and its support structures of stems, trunks, and/or branches formed collectively by adjacent plants. Foliage is generally contiguous, and gaps that may arise due to natural mortality of whole plants or their parts, or as the result of natural disturbances, are filled by growth from existing or regenerating plants. Plant canopies exchange heat, mass, and water with the atmosphere, provide habitat for diverse organisms, and produce food and timber. The essential organizational feature in canopy formation is that plants live closely together and so are in symbioses of different types that affect canopy development and stability. These include competition, where some plants benefit at the expense of others; commensalism, where some individuals benefit while neither harming nor helping others; and mutualism, where all individuals benefit from the association. In natural plant communities, all of these interactions may occur simultaneously. This entry considers three lines of development of the theory: (i) interactions between whole plants with an emphasis on competition where theory development has attempted to explain results of empirical observations of population processes; (ii) development of microclimate from an ecological perspective with an emphasis on penetration of light into foliage canopies and its effect on production; (iii) construction of functional–structural models of plant growth and interactions in canopies where an attempt is made to integrate interactions between physiology, morphology, and climate.

INTERACTIONS BETWEEN WHOLE PLANTS

Much research into biological interactions in canopies has been developed using single-species plant communities with a focus on competition, and development of theory has generally been at the level of the whole plant; however, questions are increasingly posed that require study at the level of leaves and shoots. Competition is the most-studied interaction between plants growing closely together and forming canopies. Whether explicitly or implicitly, most scientists assume that competition occurs for environmental factors such as light water or nutrients.

Competition is considered in terms of reduction by one plant of environmental factors that could benefit growth of another, rather than through allelopathy. Many models assume that competition is for space and, in effect, use space as a surrogate for collective resources. With few exceptions, theories about competition and models associated with them have been developed to explain observed patterns in data from plant communities in which competition is assumed to have taken place. These patterns are of two major types: (a) relationships between mean individual plant weight and the density of plants (numbers × area^{-1}) during self-thinning; (b) structure of plant stands as defined by the relative sizes of individuals and in some cases the relative spatial distribution of plants of different sizes including dead plants.

Self-Thinning

In 1963, Yoda and colleagues studied the change in individual plant mass and plant density as some plants die while others increase in size during competition. From the results of empirical experiments, they propose that in single-species plant stands undergoing competition, a linear relationship exists,

$$\log \overline{m} = \gamma \log N + \log K, \qquad (1)$$

where $\log\overline{m}$ is average individual plant mass in grams, N is plant density in individuals per square meter, $\gamma = -\frac{3}{2}$, and K is a species specific constant. However, it is more appropriate to relate direct quantities such as B, stand biomass, and N, rather than the ratio, $\overline{m} = B/N$ with a component of the ratio itself, i.e., N, so the self-thinning relationship is defined as

$$\log B = \beta \log N + \log K, \qquad (2)$$

where B is stand biomass density or yield in grams $\times$ m^{-2}, and N is plant density in individuals per square meter, and β and K are constants (Fig. 1).

$\beta = -\frac{1}{2}$, is equivalent to $\gamma = -\frac{3}{2}$. Yoda and colleagues developed a model to explain the self-thinning relationship with $\gamma = -\frac{3}{2}$. This assumes that there is a finite amount of two-dimensional space that remains completely allocated throughout self-thinning and that stand density is inversely proportional to the mean projected crown area. Three dimensions are considered: R, the radius of the exclusive ground area occupied by an average plant; H, plant height; and V, volume of the plant. Yoda and colleagues assume that plants grow with geometric similarity whereby the structure of the plant remains in constant proportions as it grows. The assumption of geometric similarity for plant growth during self-thinning results in the values of $\gamma = -\frac{3}{2}$, $\beta = -\frac{1}{2}$.

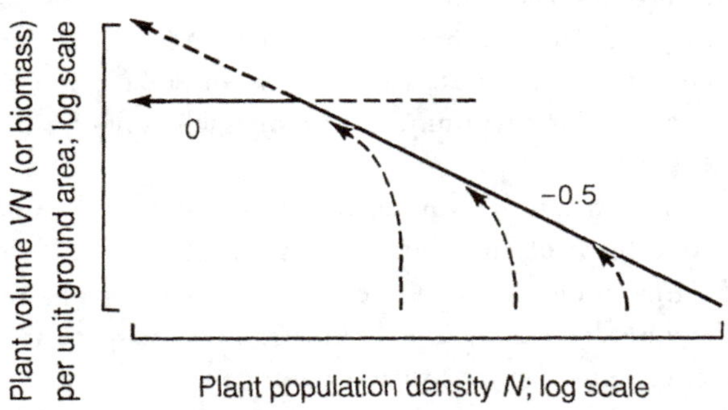

FIGURE 1 Self-thinning in relation to total stand biomass (Eq. 2). Over time, total stand biomass increases while number of plants declines due to self-thinning in response to competition for light. The −0.5 slope of the solid line is the trajectory with the assumption of geometric similarity in the relationship of mean individual plant weight (or plant volume) to mean space occupied. Plant populations with different initial densities, represented by broken curved lines, are projected to join the self-thinning line at different points but thereafter self-thin along the same −0.5 gradient. Initially biomass increases without self-thinning mortality and the curved segment represents a gradual onset of mortality. The segment of the solid line with slope = 0 represents the condition of a constant biomass but some continuing decrease in population density. (After Norberg, 1988.)

In 1987, Weller examined a large number of datasets and found that while some values of β were not significantly different from -0.5, others were significantly greater or significantly less, particularly for trees, e.g., -2.56 for angiosperm trees. He found relationships for β with shade tolerance: for angiosperms, more shade tolerant trees tend to have steeper more negative thinning slopes, while more tolerant gymnosperms have shallower thinning slopes. Other investigators have found γ varying with soil nutrient conditions and site index.

Norberg, in 1988, proposed that geometric similarity does not hold for trees. Longer branches and taller trunks are proportionally thicker than shorter ones. The disproportionate increase in diameter maintains the structural support required by a tree, referred to as elastic similarity. Norberg argues that both branch and stem wood must be considered because branch length determines crown area and so the self-thinning relationship. Norberg considers branch diameter of the largest branch, b, proportional to trunk diameter, t, throughout growth (Fig. 2) so that they have the same relative growth rate.

Norberg derives an equation for self-thinning when plants grow with geometric similarity of

$$N = k_2R^{-2} = k_2(0.5t + k_3t^{2/3})^{-2}, \qquad (3)$$

where k_2 and k_3 are constants. With the assumption of elastic similarity, self-thinning proceeds along a curve rather than a straight line. When trunk diameter is small and population density is high, then β approaches a limit of -1. As trunk diameter increases and population density decreases, the gradient increases to a theoretical maximum of $\beta = -0.33$ (Fig. 3).

Generally, plants without substantial support structures self-thin with gradients close to $\beta = -0.5$, but trees have steeper, more negative gradients. Models based on the mean plant following elastic similarity in its growth provide some insight into the reasons for this—that the weight per unit area of ground occupied increases to a large amount for trees. However, other features than the quotient of plant weight per unit space occupied may affect the course of the self-thinning relationship. Two features are of particular biological interest: the capacity to exert a competitive influence per unit of biomass, such

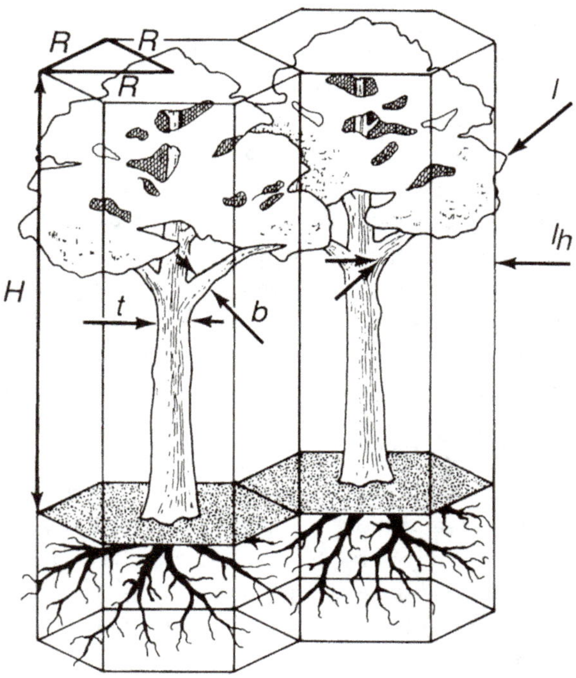

FIGURE 2 The geometric representation of trees as used in calculation of self-thinning assuming geometric similarity. Plants are assumed identical: H, height; t, trunk diameter; b, branch diameter; l, branch length; l_h, horizontal projection of branch length. As trees grow, their height and horizontal projection area they occupy increases. The amount of support tissue in trunk and branches increases disproportionally which ensures the support of the tree. (After Norberg, 1988.)

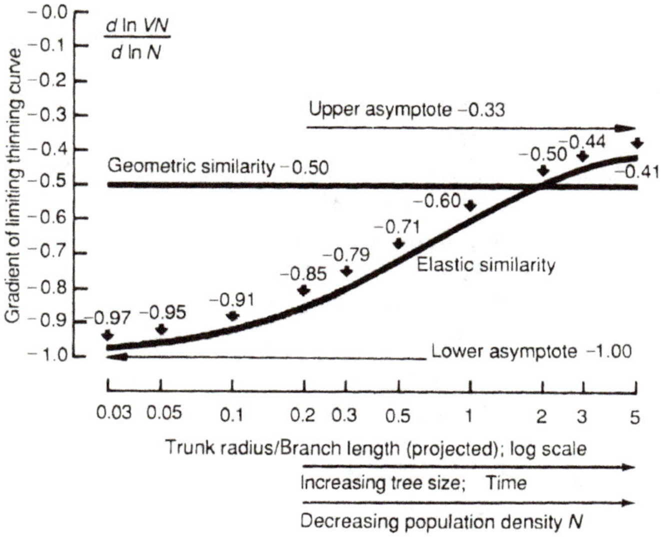

FIGURE 3 Gradient of the self-thinning curve in relation to increasing plant size represented as t/l_h, trunk radius/projected branch length (see Fig. 2). For geometric similarity plant size is maintained in a constant proportionality to space occupied and self-thinning proceeds with a constant gradient of $\beta = -0.5$. For elastic similarity the ratio of plant weight to space occupied changes as trees grow resulting in a curvilinear progression of the gradient of self-thinning curve.

as the rate of lateral growth, and the ability to withstand being shaded and delay mortality, such as through shade tolerance. Self-thinning models based on mean plants are unsuitable for analyzing these features because they represent stand development simply as a response to mortality with no functional representation of how stand development causes mortality.

Environmental factors also affect the rate of self-thinning. From experiments using different nutrient levels, Morris and Myerscough show that self-thinning proceeds fastest in stands grown at high nutrient supply so that competition at a common time is most intense in these stands. Plants from stands grown at lower nutrient levels tend to have less radial extension of the canopy for a given height and to be shorter and have less radial extension for a given shoot weight. When comparisons are made on the basis of the numbers of plants achieving the same mean plant weight, competition appears to be more intense in the stands grown at low nutrient levels.

The self-thinning relationship with $\gamma = -\frac{3}{2}$ was once considered a candidate for a universal law of ecology. Substantial empirical and theoretical research now argues against this, and calculations of the relationship between a plant's weight and the space it occupies suggest variability in the plant weight–plant spacing relationship. Geometric and elastic similarity are both approximations to the relationships that exist between plant form and function as plants grow. Niklas defines different types of relationships that can exist and how they may have evolved, but these relationships have not been incorporated into competition theory.

Structure of Single-Species Populations

Empirical analysis of population structures has led to models of individual plant–plant interactions usually based on the implicit assumption that competition is for light. Two features of population structure are important: the frequency distribution of plant size and the spatial distribution of surviving and/or large individuals. Both features change over time as competition takes place.

Distributions of plant weight in populations where competition has occurred are right skewed, with many smaller than large plants. Such distributions have often been referred to as log-normal, but a Box–Cox transformation may show that, although skewed, they are not strictly log-normal. The Gini coefficient can be used as a general indicator of skewness.

A variant of such skewed distributions is that there is a secondary maximum of larger-size individuals so that distributions are bimodal. This secondary mode is more readily seen in bivariate distributions of height and

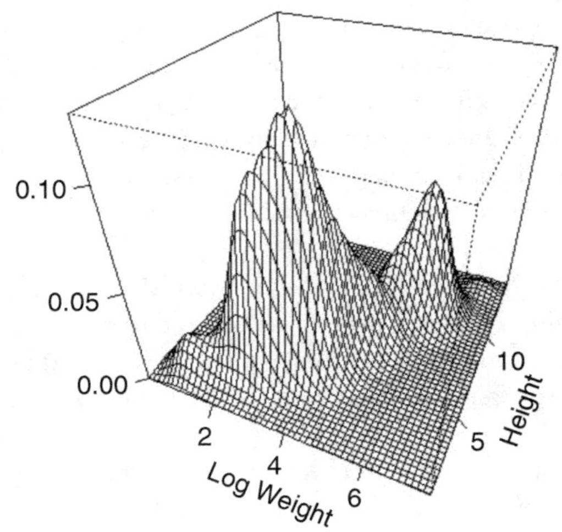

FIGURE 4 Kernel density estimate of the relationship between log plant weight (mg) and plant height (cm) for the annual plant *Tagetes patula* after 56 days grown at 5 cm initial spacing (462 plants·m^{-2}), showing a bimodal distribution of plant size with a smaller number of plants in the large size mode. At this time 36% of plants had died.

weight (Fig. 4) than in univariate distributions. Where individual plants have been remeasured over successive intervals, particularly, though not exclusively, for stands of trees, large plants have the highest relative growth rates and mortality is confined to small-sized plants.

Competition involves spatial interactions between neighboring plants. For individual plants growing in a community, two important basic features of community structure are spatial regularity and spatial clumping, which, when the spatial location of all individual plants is known, can be defined using the cumulative frequency distribution of nearest-neighbor distances. The G statistic, which uses nearest-neighbor distances, is calculated and examined for spatial regularity or clumping. G is calculated as an increasing distance t:

$$G(t) \approx \frac{1}{n}\sum_{i=1}^{n} [I(d_i < t)], \tag{4}$$

where n is the number of plants in the given pattern, d_i is the distance from plant i to its nearest neighbor, and $I(\,)$ is an indicator function that equals 1 if the argument is true and 0 otherwise.

A pattern of spatial regularity of survivors has been found in dense, even-aged natural stands where initial spatial distributions were clumped, and in experimental plantings large plants have been found to be spatially separated but surrounded by near neighbors of smaller plants. Many investigators use the K statistic, based on the cumulative distribution of all plant–plant distances rather than n.n.d., but this is not as powerful as the G statistic for detecting the basic features of clumping and regularity. Comparison of empirical distributions against those from complete spatial randomness is a useful first guide, but comparison against patterns generated from more informative null models is to be preferred.

Spatial regularity is an important result of competition because it implies that the result of competition does not reflect genetic differences, which most likely occur at random through the population, but is the result of a continuing spatial equalization process. New foliage grows upward and outward from large plants. Those large individuals that are closer to more large neighbors are more likely to have a relatively greater decline in their relative growth rate due to shading than large plants with fewer large neighbors. This means that canopy competition is a stochastic process. For any individual plant, its outcome in canopy competition is in part indeterminate. Larger size is more likely to result in success in competition, but the number and size of competing individuals also affects the outcome and can be considered at least in part as determined by a series of chance events.

Models of competition between whole plants represent the process as a spatial interaction. Individuals compete for space, but most frequently in two rather than three dimensions. Models have attempted to define how the size of the individual should be represented so that its influence on neighbors can be calculated. Three general frameworks have been most used: fixed-radius neighborhood models, zone of influence models, and field of neighborhood models.

In fixed-radius neighborhood models, each individual plant is represented by a circle, and other individuals within that circle interact with it. The strength of interaction may be estimated using regressions between individuals, e.g., for trees in diameter or basal area, and the growth plants make over a time interval. Models of this type using competition indices for different species are used to examine how spatial structure of an annual plant community may affect community dynamics over time.

In zone of influence models, plants are also represented by a circle, but this is taken to be the area over which a plant may obtain resources for growth, e.g., light. If zones of influence overlap, then competition is considered to occur. The size of the circle is usually related to some measure of plant size.

A fundamental statement for rules of interaction between zones of influence is given by Gates and colleagues (Fig. 5). Plants are represented as horizontal disks that may overlap. Resources that a plant will use to grow are calculated according to its area, and the question is how to partition the area of overlap. Gates and colleagues

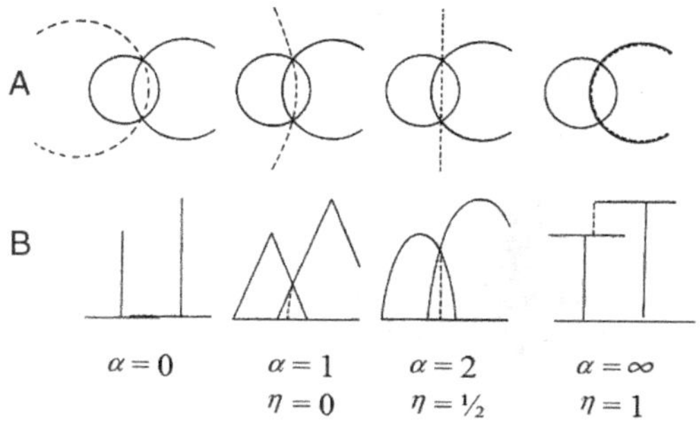

FIGURE 5 (A) Representation of interaction between two plants, larger on the right, represented as circles and with partition of their common area by a boundary curve (broken line) given by Equation 3 with different values of α. (B) The corresponding surfaces along with curve of intersection represented by broken line. (Redrawn from Gates et al., 1979.)

prove that the partition can be specified by a boundary curve between the two disks of the form

$$R_x^\alpha - |r - x|^\alpha = R_y^\alpha - |r - y|^\alpha, \quad \alpha \geq 1, \qquad (5)$$

where R_x, R_y are the radii of the disks, x, y are their centers, and r is a point on the boundary. As α is varied, the partition varies. For $\alpha = 1$, a hyperbolic boundary gives equal areas, for $\alpha = 2$ it gives division along a chord, and for $\alpha = \infty$, it gives all resources going to the larger disc. α is related to the degree of domination of the larger plant over the smaller. Gates and colleagues give a measure of domination as the ratio of distance from plant centers to the partition (Fig. 6). They define a domination index $\eta = 1 - \frac{1}{\alpha}$ for $\alpha \geq 1$. Thus, $0 \leq \eta \leq 1$, while $\eta = 0$ gives unbiased competition (hyperbolae division of overlap), $\eta = \frac{1}{2}$ gives the common chord and favors larger discs, and $\eta = 1$ gives total domination to larger discs. In a 1979 study, Gates and colleagues showed that the partitions can be considered as projections onto the plane of surfaces with varying shapes (Fig. 5B) and point out similarities between types of intersection of crown types.

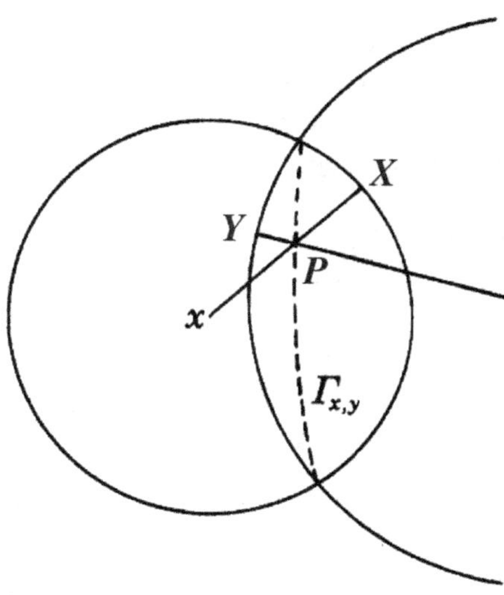

FIGURE 6 For two plants represented as circles with centers at x and y and a partition of the union of the disks represented by $\Gamma_{x,y}$, then PX/PY is a measure of the dominance of the larger disk, centered at y, over the smaller disk. (Redrawn from Gates et al., 1979.)

Field of neighborhood models are an extension of zone of influence models in which each individual plant exerts a field of neighborhood (Fig. 7). Summation of the fields represents how the strength of interaction varies over space, taking into account all potential neighborhood plants. Interactions for different environmental variables can be represented. This approach has been used to study the interactions between species that exert different competition intensities that can be represented by different fields.

Both zone of influence models and field of neighborhood models have potential to answer questions about the type of competitive interaction occurring between individuals—for example, as specified by η. However, the models need to be tested against empirical data and there are differences between researchers in what type of data should be used and which patterns in the data it is important to explain. For example, Gates and colleagues use the union of discs model with $\alpha = 2$ to represent one-sided competition and through a series of pairwise encounters simulated bimodal size distributions effectively. Weiner refers to competition as asymmetric, equivalent to $0 \leq \eta \leq 1$, and has sought support for this approach from the occurrence of right-skewed distributions. However, the shape of frequency distributions of plant size is not constant as competition occurs and an important task for modeling is to define how it changes. From the perspective of modeling a population of plants such changes might be described by changes in the distribution of plant relative growth rate in relation to relative plant size. There is the possibility that η changes both between plants within a stand and over time.

We do not have a comprehensive theory for how plants compete. To date most effort has been toward explaining how a limited set of population features arise in specific examples. In general, we can expect that as a result of competition large plants and/or survivors are spatially evenly distributed, but exceptions to this are likely to occur where initial spatial distribution is strongly clumped. Similarly, we expect that plant size frequency distributions become right skewed as competition occurs, but different measures of plant size, particularly height and weight, have different distributions that change over time during competition.

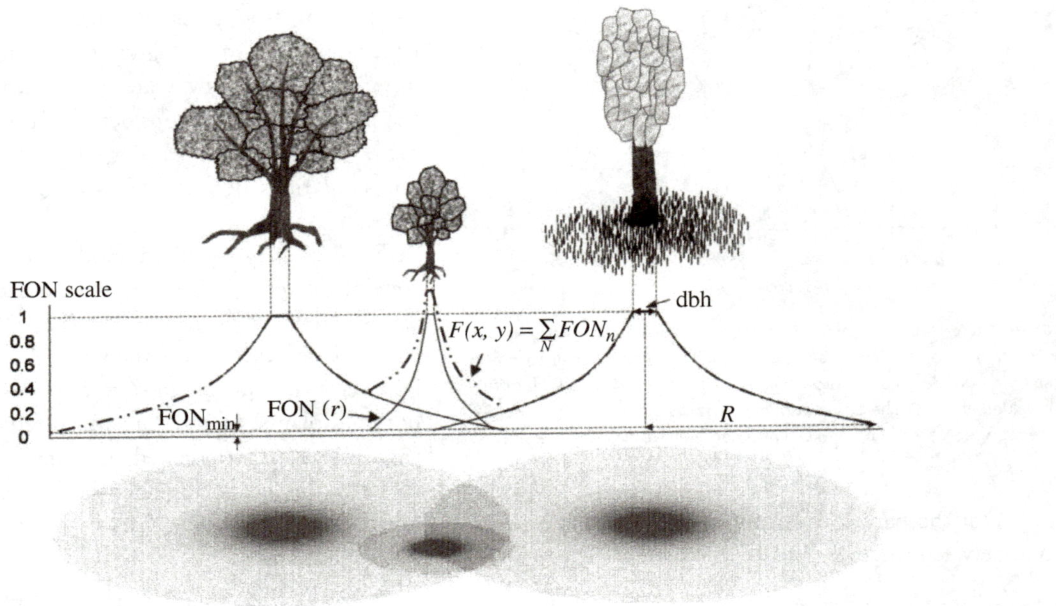

FIGURE 7 Diagrammatic representation of the field-of-neighborhood approach to plant competition. The radius of the zone of influence, *R*, is related to the diameter at breast height (dbh) of the tree. The field of neighborhood (FON) of a tree indicates the strength of competition the tree exerts within its zone of influence and represented here as declining exponentially. The aggregate field of influence, represented for the area around the central tree as the uppermost line, is the sum of all individual fields of influence (Berger and Hildenbrandt, 2000).

Fundamental questions such as "Does competition affect stand productivity?" and "Do differences in plant form and function affect the competition process?" have not been investigated in detail, and no coherent theory with wide explanatory power has been developed. Progress requires developments in two directions: (i) More precise definition of population structure. Analysis of size frequency distributions needs to be joined with spatial distributions of plants of different sizes. (ii) Analysis of differences between species and conditions of growth in both the results of competition and how it occurs. Progress in both directions may require movement away from considering competition as interaction between individual plants and analyzing the detailed interactions and responses between foliage, stems, and branches in canopies.

THE CANOPY MICROCLIMATE

In growing together and forming a canopy, plants modify the environment and a microclimate develops with distinct vertical and horizontal gradients of light, temperature, and humidity. The structure of foliage canopies and their physiological activity, particularly radiation absorption and reflection and evapotranspiration, determine rates of momentum, heat, and water vapor transfer between the canopy and the atmosphere. These rates of transfer, and their variation over time, result in vertical and horizontal gradients of light, wind speed, temperature, and humidity that define a microclimate

Differences in structure result in differences in the amount of light reaching to different parts of the canopy, variation in conditions that affect growth and how well the canopy is ventilated, which affects gas exchange. Much empirical research and development of microclimate models associated with foliage canopies has been stimulated by the objective of predicting biomass production with a focus on defining the light microclimate. In general, this work has concentrated on describing variation with depth in the canopy, whereas research into competition has concentrated more on horizontal interactions between plants.

A dominant theory for the attenuation of light in foliage canopies is based on the Beer–Lambert law of optics relating absorption of light to the properties of the material through which it is traveling. This was applied to foliage canopies by Monsi and Saeki in 1953 to calculate canopy light attenuation approximated as

$$I = I_0 e^{-kF}, \tag{6}$$

where I_0 is photon flux density on a horizontal plane above the top of the canopy, F is the cumulative leaf area index from the top of the canopy to the depth where I is measured, and k is the extinction coefficient.

Equation 6 has been used extensively in developing models of whole canopy photosynthesis. Difficulties in this application arise because within a canopy leaves vary systematically in their angle and absorptive properties. This is seen in systematic variation of k within canopies, and models tend to calculate photosynthesis in successive layers each having specific values of k.

FUNCTIONAL–STRUCTURAL PLANT MODELS

Increase in computer power and its availability is enabling increasingly detailed modeling of relationships between plant structure and function, presenting the possibility of defining how these features affect interactions with the environment and change biological processes. In 2005, functional–structural plant models (FSPM) were distinguished as a distinct class of model by Godin and Sinoquet. FSRM were initiated in response to the perceived limitation of existing models of forest growth both in what they could predict and their representation of growth processes. FSPM are simulation models, and plants can be represented as connected sequences of stems, foliage, and roots. For example, an FSPM for an annual plant might have a class of objects called *leaf* that have properties of photosynthetic rate, import and export of carbon compounds, and growth rate, all determined by environmental conditions. The FSPM could simulate plant growth as leaves are produced, grow, and senesce.

FSPM provide two advantages to the study of plant competition and canopy interactions. First, while continuing to represent complete plants, the interactions between neighbors can be represented more effectively, as extending foliage and branches, than when plants are considered as circles. For example, individual plant crowns are known to become irregular in shape when competition occurs, and this is an important consequence of plant plasticity and the details of how morphological and physiological characteristics affect competition can be studied.

A second potential advantage of FSPM is that they make it possible to represent direct interaction between foliage and its environment using such techniques as ray tracing algorithms to calculate the light environment associated with individual foliage elements.

Progress has been made in both these directions but much work in FSPM has focused on constructing models examining how components in models should be organized. Open-source software platforms OpenAlea (http://openalea.gforge.inria.fr/dokuwiki/doku.php?id=openalea) and GroImp (http://sourceforge.net/projects/groimp/) have been constructed. The full value of such models for developing ecological theory about competition and canopy interactions has yet to be realized.

SEE ALSO THE FOLLOWING ARTICLES

Allometry and Growth / Environmental Heterogeneity and Plants / Individual-Based Ecology / Integrated Whole Organism Physiology / Phenotypic Plasticity / Single-Species Population Models / Spatial Ecology

FURTHER READING

Berger, U., and H. Hildenbrandt. 2000. A new approach to spatially explicit modeling of forest dynamics: spacing, ageing and neighbourhood competition of mangrove trees. *Ecological Modelling* 132: 287–302.

Gates, D.J., A.J. O'Connor, and M. Westcott. 1979. Partitioning the union of disks in plant competition models. *Proceedings of the Royal Society London Series A: Mathematical Physical Sciences and Enginering Sciences* 367: 59–79.

Godin, C., and H. Sinoquet. 2005. Functional–structural plant modelling. *New Phytologist* 166: 705–708.

Hirose, T. 2005. Development of the Monsei–Saeki theory on canopy structure and function. *Annals of Botany* 95: 483–494.

Morris, E.C., and P.J. Myerscough. 1991. Self-thinning and competition intensity over a gradient of nutrient availability. *Journal of Ecology* 79: 903–923.

Niklas, K.J. 1992. *Plant biomechanics: an engineering approach to plant form and function.* Chicago: University of Chicago Press.

Norberg, R.A. 1988. Theory of growth geometry of plants and self-thinning of plant populations: geometric similarity, elastic similarity, and different growth modes of plant parts. *American Naturalist* 131: 220–256.

Weiner, J. 1990. Asymmetric competition in plant populations. *Trends in Ecology and Evolution* 5: 360–354.

Weiner, J., and O.T. Solbrig. 1984. The meaning and measurement of size hierarchies in plant populations. *Oecologia* 61: 334–336.

Weller, D.E. 1987. A reevaluation of the 3/2 power rule of plant self-thinning. *Ecological Monographs* 57: 23–43.

Yoda, K., T. Kira, H. Ogawa, and H. Hozumi. 1963. Self-thinning in overcrowded pure stands under cultivated and natural conditions. *Journal of the Institute Polytechnics, Osaka City University,* Series D 14: 107–129.

PLANT DISPERSAL

SEE DISPERSAL, PLANT

POPULATION ECOLOGY

MICHAEL B. BONSALL AND CLAIRE DOOLEY

University of Oxford, United Kingdom

Population ecology is the study of the change in species distributions and abundances over time and across space. Since its emergence within the science of ecology (during the 1920s), this quantitative discipline relies heavily on mathematics to formalize concepts, test hypotheses, and substantiate arguments. Four key processes—birth, death, immigration, and emigration—determine the patterns of change in the temporal and spatial distribution of species. These changes in numbers of a population ($N_{t+1} - N_t$) are often expressed as

$$N_{t+1} - N_t = b_t - d_t - i_t - e_t, \tag{1}$$

where b_t is the number of births, d_t is the number of deaths, i_t is the immigration rate, and e_t is the emigration rate. This is the fundamental equation of population ecology. Although a simple expression, it masks a wealth of biological detail that is essential to understanding patterns in the distribution and abundance of species.

MALTHUSIAN DYNAMICS

In 1798, Thomas Malthus, an English vicar in a parish near Albury, Surrey, in the UK, published a treatise entitled "On the Principle of Population." In this essay, he argued that populations increase through a geometric progression (i.e., as 2^t), whereas the availability of resources that a population depended upon would only increase linearly. Malthus suggested that population expansion would outstrip food supply and this geometric increase in populations would lead to mass starvations. Malthus believed that population growth would only be held in check by positive checks (that raised death rate) or preventatitive checks (that lowered birth rate).

For population ecology, Malthus highlighted two key concepts: the idea that in the presence of unlimited resources population growth would be unbounded (this is essentially exponential population growth) and that resources are limited and would impose an upper limit to growth rate and hence population size (we now refer to this as the population carrying capacity).

Appreciating the link between the rate of increase and the observation of exponential growth is crucial for population ecology. This simple process holds a whole host of ecological implications, not least of which centers on the factors than ensure the intrinsic rate of increase is constant. That is, resources are supplied at a constant and unwavering rate. A simple expression for the dynamics of exponential growth (Fig. 1A) is

$$\frac{dN}{dt} = rN(t), \tag{2}$$

where N is population abundance and r is the intrinsic rate of increase. However, if the intrinsic rate of increase is directly proportional to a (the most limiting) resource, R, such that $r = aR$, then the size of the population at any point in time will depend on $N(t) = N(0)\exp(aR)$, and the dynamics of the population will show alternatives to simple exponential growth. This phenomenon of resource limitation affecting dynamics is critically important in population ecology through the concept of density dependence.

SINGLE-SPECIES DYNAMICS AND DENSITY DEPENDENCE

Forty-seven years after the first appearance of Malthus's essay, Pierre-Francois Verhulst formalized the ideas of

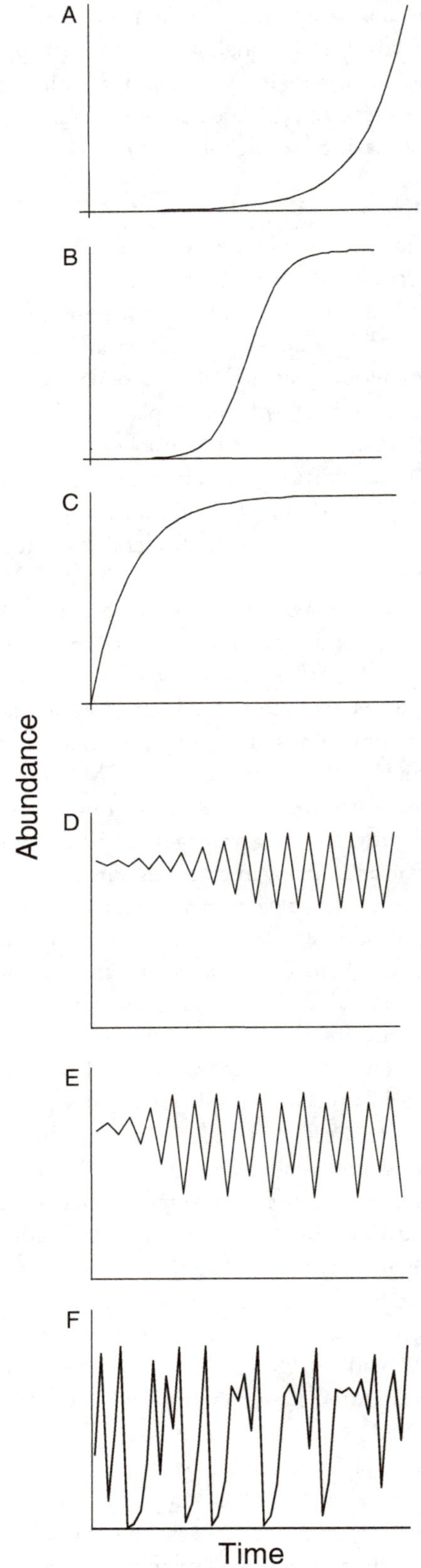

FIGURE 1 Range of population dynamics. Populations may exhibit (A) exponential growth, (B) logistic growth to a carrying capacity, (C) stable equilibrium dynamics, (D) 2-point limit cycles, (E) 4-point limit cycles, or (F) chaotic dynamics.

population growth and limitation. Verhulst reasoned that when populations were small and resources abundant, then population growth would be exponential. As populations become limited, rate of increase slows and populations would reach a carrying capacity. By assuming that the rate of population growth declines linearly as population size increases, Verhulst generated a sigmoid curve for the size of a population through time. This is known as the logistic equation (Fig. 1B). This logistic model (also introduced by Raymond Pearl and Lowell Read) is

$$\frac{dN}{dt} = rN\left(\frac{K-N}{K}\right). \tag{3}$$

This differential equation of the change in population size (N) through time (t) introduces the notion of density dependence (through a fixed carrying capacity, K), whereby an increase in mortality or decrease in fertility occurs as population size increases. At low population density, the rate of change of population is essentially r. As the population increases, the net change in population size is

$$\frac{1}{N}\frac{dN}{dt} = r - \frac{rN}{K}. \tag{4}$$

When $1/N\, dN/dt = 0$, $N = K$ and this is the equilibrium state of the population (Fig. 1C) and increases in population density lead to linear decline in (per capita) reproductive rate under this logistic model. The idea of density dependence introduces the processes of negative feedback, and although in the simplest case these can lead population growth to an equilibrium, alternative expressions (more consistent with data) can generate a range of population dynamics (Fig. 1).

Chaos

Perhaps one of the simplest ways in which to encapsulate the idea of density dependence and negative feedback processes is

$$N_{t+1} = \lambda N_t(1 - N_t), \tag{5}$$

where λ is the population growth rate. This simple model was introduced by Robert May to show how simple population models can display a range of dynamical behaviors from stable population equilibrium through to limit cycles and chaos. These sorts of dynamics are well captured in a bifurcation diagram (Fig. 2). Alternative mathematical formulations of this quadratic model show similar phenomena. For instance, a discrete-time equivalent of the logistic model,

$$N_{t+1} = N_t \exp(r - (N_t)/K) \tag{6}$$

where r is the intrinsic rate of population increase and K is the population carrying capacity, reveals a wide rep-

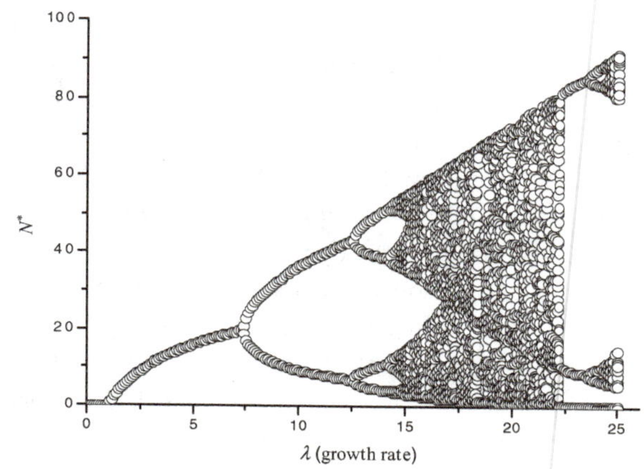

FIGURE 2 Bifurcation diagram illustrating population dynamic patterns (for Eq. 6) as population growth rate increases. Population dynamic patterns are simulated for a large number of time steps, and the asymptotic dynamics (stable, limit cycle, chaos) are illustrated.

ertoire of population dynamics. As the growth rate (r) of the population increases, the dynamics change from stable equilibrium ($r < 2$) through two-point stable limit cycles ($2 < r < 2.526$) and into highly periodic and eventually aperiodic (chaotic) dynamics.

The fundamental characteristic of chaotic population dynamics is extreme sensitivity to initial conditions. That is, two populations that differ slightly (but potentially imperceptibly) in initial starting densities will eventually show completely different population trajectories through time. Even though these dynamics may be indistinguishable from random noise, they are completely described by a deterministic process fluctuating between fixed bounds in population size.

Time Lags

Time lags, or delays in the effects of births and deaths on changes in population numbers, are referred to as delayed density dependence. Delays in the key population processes can have a profound influence on populations and lead to a range of population dynamics. Delayed density dependence can lead to the observations of other types of population dynamics such as limit cycles or chaos (Fig. 1). Depending on the strength of the feedback from different past generations can have important implications for the stability and dynamics of single species interactions. If the feedback from the past generation is small, then the model follows the classic single-species dynamics (e.g., Eq. 5). However, for large contributions of past generations the population dynamics can show quasiperiodic oscillations, and under certain combinations of direct and delayed feedback the dynamics can be highly erratic.

In intraspecific interactions, delayed density dependence may arise through maternal effects. That is, the fecundity of females and the survival of immatures to adulthood are functions of individual quality. However, individual quality is a function of maternal quality, and these maternal effects introduce delays into populations that can give rise to cyclic dynamics.

As noted, variation or pulses in resource supply can affect the deterministically predicted dynamics in single-species interactions and can lead to effects consistent with time lags. In a simple ecological system with direct density dependence (Eq. 5), λ is constant and the system equilibrates at $N^* = \frac{\lambda - 1}{\lambda}$. However, if it is assumed that reproductive output is proportional to resource availability (R_t) at time t such that $\lambda = aR_t$, then the equilibrium is better expressed as $N^* = \frac{(aR - 1)}{aR}$. A pulse of resource of size ε at time t will perturb the population and increase it in size in the next generation ($t + 1$) by εN. In the subsequent generation ($t + 2$), the size of the population will be determined by $N_2 = \lambda \varepsilon N^* (1 - \varepsilon N^*)$. As such, the density of the consumers following a resource pulse can lead to an overshoot of the equilibrium (carrying capacity), which then can lead substantial feedback effects in future generations. These time-lagged effects have important consequences for population persistence and extinction.

Age Structure

In contrast to the simple population models described so far, the dynamics of many species are underpinned by age-structure effects. That is, a population consists of cohorts of individuals born at a range of different points in time and whose probability of reproduction and/or survival is dependent on their age. If each age class is of equal duration then the age-structure effects can be described by a simple discrete-time matrix model:

$$\mathbf{n}_{t+1} = \mathbf{A} n_t \tag{7}$$

where **n** is a vector of numbers of individuals in different age classes in the population and **A** is a projection matrix of the probability of survival and the reproductive capacity of individuals in each age class. In the absence of density dependence acting on survival and/or reproduction, these sorts of population models predict that abundance will either increase or decrease exponentially depending on whether the overall population growth rate is greater or less than 1. However, in comparison to the simple models of population growth, age-structured populations eventually reach a stable age distribution. That is, independent of initial numbers in each of the age classes, a limiting distribution will be obtained in which the proportion of individuals in each class remains constant.

Confronting Theory with Data

Density dependence is the central concept in population ecology, and the time series collected over ecologically relevant spatial and temporal scales provide a cornerstone against which hypotheses on population dynamics processes such as density dependence can be evaluated. Understanding the processes of population regulation, limitation, and growth around an equilibrium point depends on determining the correlation between population densities at different time points. The strength of this correlation is affected by both variety of biotic and abiotic processes and our ability to measure accurately the abundance of a population at a particular temporal or spatial scale.

It is therefore important to examine ecological dynamics in terms of both the density-dependent (endogenous, deterministic) and density-independent (exogenous, noise) processes using appropriate statistical methodologies (Fig. 3). Given the nonlinear nature of ecological interactions, approaches for analyzing population ecological observations have been well-developed and can be broadly divided into four stages of analysis:

1. Define an appropriate population model or set of population models. The appropriate population model is often unknown, and hence a set of alternative candidate population models is required. A carefully chosen set of models is essential to reach appropriate inferences from the analysis of population ecological observations. This candidate set of ecological models is unlikely to be a nested set of models and appropriate fitting and assessment criteria need to be thoroughly evaluated.
2. Set appropriate assumptions about the stochasticity acting on the population. Stochastic effects on population dynamics can occur in two ways: through environmental or demographic effects. Environmental stochasticity refers to randomness imposed on populations by the environment—these are the classic density-independent processes that lead to fluctuations in overall population abundance. Demographic stochasticity depends on the intrinsic uncertainty associated within an individual's fecundity, dispersal, and/or survival and is often thought to be most influential in small populations.
3. Obtain, under appropriate statistical assumptions, estimates for the parameters in the population model. One appropriate statistical framework for

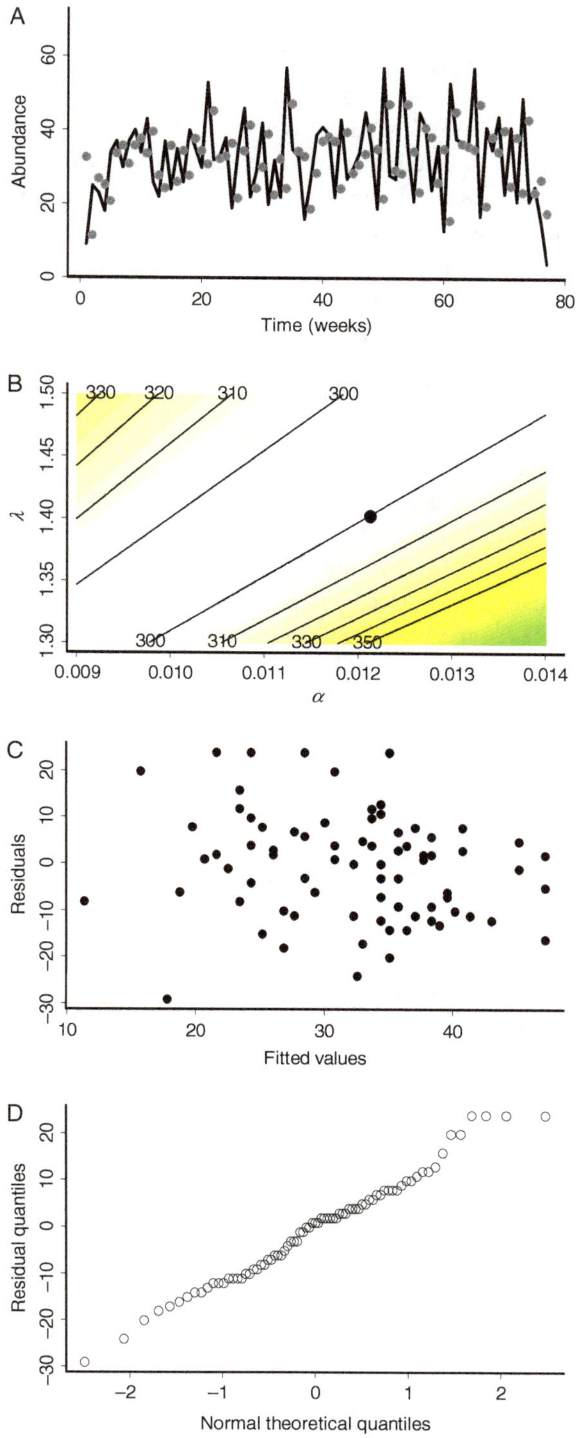

FIGURE 3 Model fitting approach using bruchid beetle (*Callosobruchus chinensis*) population data and Skellam's population model. (A) Population dynamics (solid line) and one-step-ahead predictions for Skellam's model ($N_{t+1} = \lambda N_t(1 + \alpha N_t)^{-1}$) with different assumptions about the noise (Environmental—Gaussian: AIC = 598.812 (**P** = [σ = 11.366, λ = 1.402, α = 0.012]); Demographic—Poisson: AIC = 708.739 (**P** = [λ = 1.089, α = 0.002]); Demographic—Negative Binomial: AIC = 603.586 (**P** = [k = 9.545, λ = 5.908 α = 0.149]). (B) Likelihood surface with likelihood contours (solid lines) and maximum likelihood parameters (λ,α) from Skellam's model with Gaussian errors (solid point). (C) Residuals versus model predicted (fitted) values. (D) Normal quantile plot.

estimating model parameters is likelihood. Likelihood is the measure that a set of observed data is generated by an underlying model and is used to maximize the estimate of an unknown parameter (or set of parameters). Based on the statistical principle that successive points in a time series are likely to be correlated, an appropriate general likelihood ($L(\mathbf{P},n)$) framework is based on a conditional model:

$$L(\mathbf{P}, n) = g_{N_1}(\overline{n}; \mathbf{P}) \prod_{i=2}^{T} g_{N_i|N_{i-1}}(n_i | n_{i-1}; \mathbf{P}), \qquad (8)$$

where $g_{N_1}(n_1;\mathbf{P})$ is the probability density of N based on n observations and an unknown parameter set **P**. The term $g_{N_1}(\overline{n}; \mathbf{P})$ is the likelihood of the first observation conditioned on the mean of the model (essentially the equilibrium expression), and the following term $\prod_{i=2}^{T} g_{N_i|N_{i-1}}(n_i | n_{i-1}; \mathbf{P})$ is the product of the second and subsequent values conditioned on the previous expected value determined from the population model.

4. Assess the goodness of fit of the model to the data. Critically choosing between models descriptions of the underlying population dynamics, noise, and correlation structures needs a rigorous model selection scheme. The use of information theoretic criteria, such as the Akaike Information Criteria (AIC), allows appropriate estimators based on a likelihood framework to be used in nonnested model selection and model criticism. While inference on the most appropriate model is standard practice, the use of estimators such as AIC allows alternative models to be evaluated, evidence ratios computed, and the strength of different hypotheses underpinning the population ecological process to be critically distinguished.

COMPETITION

Species do not exist in isolation: species engage in antagonistic and synergistic interactions that affect their distribution and abundance. Pure single-species dynamics are unlikely to be common, and extending ecological theory to explore and understand species interactions has provided novel insights into the dynamics of ecological systems. One of the simpler extensions of a single-species framework is to explore the dynamical outcomes when two different species interact and compete for a common, shared resource. This concept of interspecific competition is widespread through both ecological and evolutionary dynamics.

A population model of interspecific competition based on the logistic framework (Eq. 3) was originally derived by Alfred Lotka and Vito Volterra. Population growth of any one species is inhibited by both intraspecific and interspecific processes. The Lotka–Volterra model for the competitive interactions between two species is

$$\frac{dN_i}{dt} = r_i N_i \left(\frac{K_i - N_i - \alpha_{ij} N_j}{K_i} \right), \quad (9)$$

where N_i is the density of the ith species, r_i its intrinsic rate of increase, K_i is its carrying capacity, α_{ij} is the competition coefficient that determines the per capita effect of species $j(N_j)$ on species i. If 50 individuals of species j have the same effect as one individual of species i, then the total competitive effect on species is $(N_i + N_j)/50$. This competition coefficient expresses the interspecific effects: if $\alpha_{ij} < 1$, then the jth species has a weaker effect on species i than species i has on itself. To determine the dynamical outcome of competition, that is, whether there is exclusion or coexistence of species, requires understanding the solutions to Equation 9. If the rate of change (dN_i/dt) of each species is zero, then for two species Equation 9 reduces to

$$N_i = K_i - \alpha_{ij} N_j, \quad (10)$$

$$N_j = K_j - \alpha_{ji} N_i. \quad (11)$$

These linear expressions define zero net isoclines that separate regions of positive and negative population growth. Below a linear isocline, a population increases $(dN_i/dt > 0)$ while above it the population declines $(dN_i/dt < 0)$. Combining isoclines from two species allows the interaction between species to be determined (Fig. 4). Determining how isoclines intersect in phase space allows four different types of dynamical behavior to be identified: exclusion of species i, exclusion of species j, stable coexistence of both species, and unstable coexistence of both species.

One limitation immediately obvious from the simple Lotka–Volterra formalism for competition is that any interspecific effects are described in the absence of underlying resource dynamics. When several species compete for the same limiting resource, supply and uptake rates of these resources will be important determinants of the outcome of competition. David Tilman illustrated the population-ecological consequences of mechanistic resource competition using an explicit resource–consumer model for competition between species that use the same resource (R_i):

$$\frac{dN_i}{dt} = N_i \left(\frac{r_i - R}{R + k_i} - m_i \right), \quad (12)$$

$$\frac{dR}{dt} = a(S - R) - \sum_{i=1}^{n} \left(\frac{\frac{dN_i}{dt} + m_i N_i}{Y_i} \right), \quad (13)$$

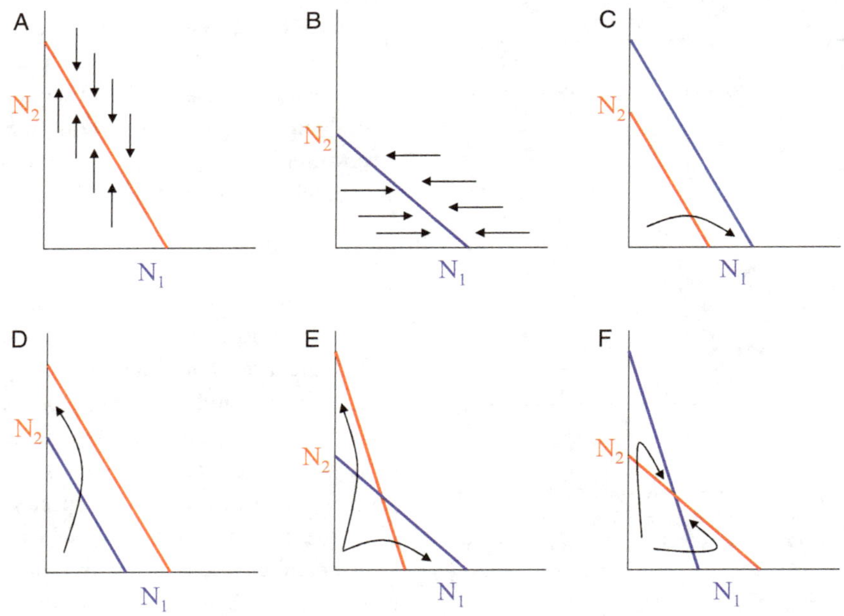

FIGURE 4 Phase-plane portraits of interspecific competition. Phase-planes are the plot of the density of species 1 versus the density of species 2. Below the isoclines, a species increases; above the isocline, a species decreases. (A) The direction of increase and decrease for species 1 (N_1). (B) The direction of increase and decrease for species 2 (N_2). (C)–(F) The four outcomes of interpecific competition: (C) exclusion of species 2 (N_2), (D) exclusion of species 1 (N_1), (E) unstable equilibrium with exclusion based on initial conditions, and (F) stable coexistence of both species. Coexistence (F) is favored only when a species limits itself more than its competitors.

where N_i is the density, r_i is the growth rate, m_i is the death rate of the ith competition, and k_i is the saturation constant. S is the amount of resource supplied at rate a, and Y_i is the number of individuals of species i produced per unit of resource. At equilibrium ($dN_i/dt = 0$ and $dR/dt = 0$), the resource necessary for the ith species to persist (R_i^*) is

$$R_i^* = \left(\frac{k_i m_i}{(r_i - m_i)}\right). \quad (14)$$

The outcome of interspecific competition is determined from these resource dynamics. The species that wins out in competition is the one with the lowest R^* value, as this species requires the least amount of the limiting resource to maintain positive population growth. This species will competitively displace all others and two species can only coexist on a single limiting resource if their R^* values are equal.

Extending this idea of resource supply and exploitation to multiple resources provides predictions on the population ecology of multiple-species competition, the organization of ecological communities, and diet specialization. Resources can be broadly classified into essential resources and substitutable resources. Essential resources are those that must be taken together as each resource provides a different but necessary requirement. Substitutable resources are sets of resources that satisfy the same essential requirement. Other resource classifications have been defined and extended to considered types such as antagonistic resources and switching resources.

For all these sorts of resource types, the general prediction is that for coexistence each species must consume relatively greater amounts of the one resource that limits its growth rate. The types of resource also determine patterns of species abundance and distribution. For instance, multiple species might be expected to coexist when competing for essential resources in a heterogeneous habitat, while species that use antagonistic resources might be excluded in similarly heterogeneous habitats. Understanding this outcome of interspecific competition under multiple resource dynamics also has important implications particular for the population ecology of resource specialism and generalism.

RESOURCE–CONSUMER INTERACTIONS

Trophic interactions whereby a consumer eats a resource are a second class of interactions involving two (or more) species. These enemy–victim interactions have received considerable theoretical attention and broadly fall under the classification of predation. In general, four types of predation have been identified: (i) herbivory (where the consumer eats a plant), (ii) carnivory (where the consumer eats a herbivore or another carnivore), (iii) parasitism (where a parasite attacks or infects a host), and (iv) cannibalism (where the consumer eats individuals of its own species).

A major class of predator–prey interactions that has been the focus of intense research effort is that of host–parasitoid interactions. Parasitoids are insects in the orders Hymenoptera, Diptera, or Coleoptera that attack other arthropods by laying eggs on, in, or near their host. A juvenile parasitoid develops using the host resource, eventually killing its host. The population dynamics of these sorts of association have been shown to be fascinatingly complex with a rich array of biological and ecological processes being important in the dynamics of host–parasitoid interactions. For instance, the role of host refuges (where a proportion of hosts are protected from parasitism), intraspecific host competition, the effects of patchy environments, and the foraging activities of the parasitoid are critically important to the persistence of these sorts of resource–consumer interactions.

A second class of resource–consumer interactions involves the association between hosts and true parasites. It has long been known that the population dynamics of a host can be influenced by a wide variety of different parasites. A key concept in host–parasite theory and epidemiology is the basic reproductive ratio. This determines the (deterministic) potential for parasitic infections to spread and is defined as the number of secondary infections resulting from a single infection in a population of susceptible individuals. Simply, if this ratio is greater than 1, then an infection will spread. If the ratio is less than 1, the infection will not persist and will fade out. Understanding the processes of parasite transmission and spread is critical to the population ecology and hence epidemiology of this type of predator–prey interaction.

Unlike predators, the principal effect of herbivory is often to reduce plant vigor (e.g., numbers of seeds) and reproductive output rather than inflict outright mortality. Thus, in plant–herbivore interactions it is important to distinguish between effects on resource performance and effects on resource dynamics. The focus then is on the amount of herbivory per plant rather than the direct impact of herbivores on plant mortality. This can lead to novel predictions about the role of herbivores on the dynamics, distribution, and abundance of plants. For instance, it is predicted that plants have more impact on the distribution and abundance of specialist herbivores than herbivores have on the distribution and abundance of their resource, herbivores are top-down regulated (as the world is green and resources are abundant), and the ecological processes associated with herbivory in the same system vary from place to place.

Cannibalism is an important class of predation that has important implications for the population dynamics of a range of different species. For instance, in nonseasonal organisms such as the stored product moth, *Plodia interpunctella*, a commonly reported type of population dynamic is generation cycles. These cyclic fluctuations are of a period equal to the generation time of the organism. In *Plodia*, egg and larval cannibalism are common and when coupled with differential competitive effects among different aged larval groups (that is, older larvae having a disproportionate effect on smaller larvae), population models predict that aseasonal cyclic dynamics and, in particular, generation cycles are likely to occur. Similar effects of cannibalism in the flour beetle *Tribolium castaneum* have also been shown to lead to a range of population dynamics from stable dynamics through to chaotic fluctuations.

While each of these forms of predation has specific caveats, the general form of predation can be described with a simple set of coupled (differential) equations. If R denotes the prey density and C the density of predators (parasites or cannibals), then the deterministic theoretical framework for predation is

$$\frac{dR}{dt} = rR - aRC, \tag{15}$$

$$\frac{dC}{dt} = C(bR - d), \tag{16}$$

where r is the prey growth rate, a is the predator attack rate, b is the conversion of prey biomass into predator biomass, and d is the predator death rate. Prey increase in abundance through births and decline (simply) due to predation. In the absence of predators, prey increase exponentially. Predator abundance increases through the conversion of prey and declines through predator death. In the absence of prey, predators' decline is exponential. The predator–prey interaction given in Equations 15 and 16 can be rewritten as to express the changes in prey with respect to changes in predator abundance:

$$\frac{dR}{dC} = \frac{rR - aRC}{C(bR - d)}, \tag{17}$$

which can be simplified to

$$\left[b - \frac{d}{R}\right]dR = \left[\frac{r}{C} - a\right]dC. \tag{18}$$

Following integration, Equation 18 becomes

$$bR - d\ln(R) - r\ln(C) + aC = \text{constant}. \tag{19}$$

This expression (Eq. 19) represents a set of closed curves. This solution to the predator–prey interaction (described by Eqs. 15 and 16) is determined simply by the initial densities of predators and prey. Further interpretation of the predator–prey interaction can be gained from phase-plane graphical analyses around the predator and prey equilibrium (Fig. 5). At steady state, the abundance of prey is $R^* = \frac{d}{c}$ and predators $C^* = \frac{r}{a}$. Prey will increase below and decrease above $\frac{d}{c}$ for all predator densities. Similarly, predators will increase below and

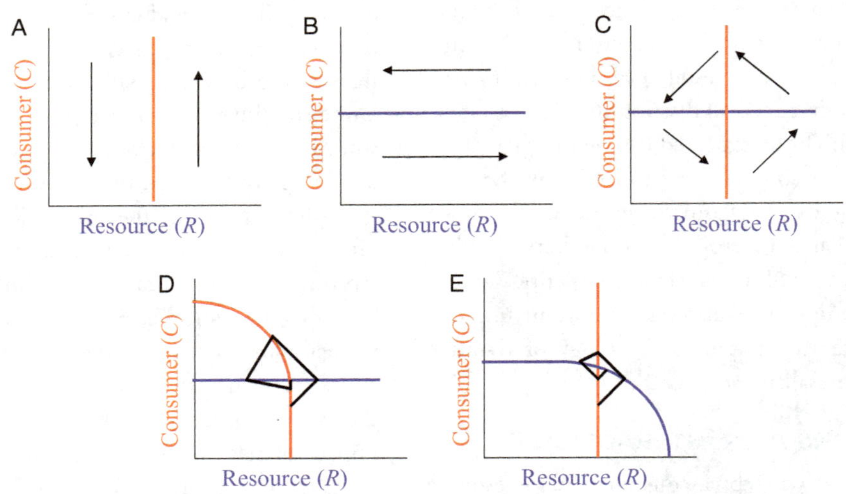

FIGURE 5 Phase plane graphs depicting the interaction between a prey species (*R*) and predator species (*C*). (A) Predator Isocline: when prey are abundant, predators increase; when prey are scarce, predators decline. (B) Prey Isocline: when predators are abundant, prey decline; when predators are scarce, prey increase. (C) Combined isoclines: these show the anticlockwise dynamics and graphical solution to the predator–prey model (Eqs. 15–16). (D) Inclusion of predator density dependence leads to local (but not global) stability. (E) Inclusion of prey density dependence again leads to local (but not global) stability.

decrease above $\frac{r}{a}$. This is illustrated as phase-plane graphs (Figs. 5A, B). Combining these two phase-plane graphs and resolving the trajectories for the growth of each population in each of the quadrants allows the dynamics of the predator–prey interaction to be analyzed. As noted (Eq. 19), this is a closed curve, so the dynamics of the predator–prey interaction are cyclic with changes in predator abundance lagging behind that of changes in prey abundance (by $\frac{1}{4}$ of a cycle). If perturbed by increases or decreases in initial densities, the general pattern of cyclic dynamics is unaltered (although the amplitude is affected). These are known as neutral cycles.

The graphical analysis can be extended to explore other ecological processes such as density dependence. Density dependence on predators and/or prey has crucial effects on the interaction and can lead to stable dynamics (Fig. 5). However, two important conclusions emerge. First, density dependence acting on prey or predator population leads to the same observed phenomenon: stable dynamics. Simple observation of a predator–prey time series may not necessarily enable the identification of the key mechanism of stability. Second, density dependence is a necessary but not sufficient condition for stability. This ecological process only favors local stability in the neighborhood of the equilibrium. A predator–prey interaction with density dependence is not globally stable, as large perturbations can drive both predators and prey extinct.

MULTISPECIES INTERACTIONS

The organization of ecological assemblages is influenced by the biotic processes of both competition and predation. The original premise, based on Darwin's idea of an entangled bank, was that competition was the driving force in shaping the difference between a fundamental and realized niche and hence shaping ecological assemblages. While competitive interactions in all their manifestations (e.g., nonlinear, asymmetric, diffuse) can have profound effects on the structure of species communities, the role of predation is equally important. For instance, Robert Paine demonstrated that the removal of the predatory starfish *Pisaster ochraceus* from intertidal assemblages along the Pacific Northwest coast led to dramatic alterations in the structure of the invertebrate assemblage. In the presence of the predator, the prey assemblage was made up of 15 species. After the removal of *Pisaster*, 7 species were lost through intense interspecific competition. Predation acted to alleviate competition and favor the coexistence of several species that would otherwise have been excluded by the dominant competitor.

The role of predation in shaping ecological assemblages can be even more subtle. If two species do not compete or interact for resources but do share a common predator, then they can still negatively affect one another. Consider a single prey–single predator interaction in which a second prey species invades. Although this species feeds on a different resource, it is susceptible to predation. The population ecological consequences can be illustrated by extending the predator–prey interaction (Eqs. 15 and 16):

$$\frac{dR}{dt} = r_i R_i - a_i R_i C, \tag{20}$$

$$\frac{dC}{dt} = -dC + C\sum_i b_i R_i, \tag{21}$$

where now r_i is the growth rate of ith prey species, a_i is the predator attack rate on the ith prey species, and b_i is the conversion of biomass of the ith prey species into predator biomass; d is the predator death rate. This invasion of a second prey species occurs if when rare the per capita rate of change in abundance of this species (say, R_2) is greater than zero:

$$\frac{1}{R_2}\frac{dR_2}{dt} > 0. \tag{22}$$

More explicitly this per capita change is

$$\frac{1}{R_2}\frac{dR_2}{dt} = r_2 - a_2 C_1^*, \tag{23}$$

$$r_2 - a_2 > 0, \tag{24}$$

$$\frac{r_2}{a_2} > C_1^*, \tag{25}$$

where C_1^* is the equilibrial abundance of predators in the presence of the first prey species ($C_1^* = r_1/a_1$). By similar arguments, the invasion of the first prey species into a predator–prey interaction where prey species 2 is already established occurs if

$$\frac{r_1}{a_1} > C_2^* \tag{26}$$

and $C_2^* = r_2/a_2$. These two inequalities (Eqs. 25 and 26) cannot both hold simultaneously, and hence one prey species will be lost due to the action of the shared predator. In the absence of other ecological processes (such as density dependence, refuges, temporal heterogeneity), the species that is excluded is the one that suffers the higher rate of predation and/or has the lower population growth rate. This ecological process is known as apparent competition and leads to a state of dynamically driven diet specialization.

Apparent competition can have a profound effect on the population ecology of multispecies interactions. Shared predator effects have been accredited to the exclusion of red squirrels by gray squirrels mediated through a virus in the UK, the replacement of one species of leafhopper by another via shared parasitoids in California vineyards, and the structure of host–parasitoid assemblages. If apparent competition is a prevalent process in species assemblages, then it is enigmatic how complex interactions persist. While a number of mechanisms such as intraspecific competition, refuges, metapopulations, and predator behaviors have been suggested as potential mechanisms acting to reduce the influence of apparent competition and favor coexistence, as with direct interspecific competition, it is simply necessary for the intraspecific effects acting within species to outweigh those acting between species.

More complex communities can show a range of population dynamic behavior. Population models in which prey species do not compete for the same resources but have specialist and generalist predators have a range of population behaviors from stable interactions between all members of the assemblage to situations in which the assemblage is not persistent but repeatedly invaded by all species. This sort of pattern of repeated invasion and nonpersistence of any one strategy is familiar in the classic "rock–paper–scissors" game where *rock* beats *scissors* beats *paper* beats *rock*. It has important (but often neglected) implications for understanding the ecological processes that structure multispecies communities. For example, extending the ideas of competition to multiple species

$$\frac{dN_i}{dt} = r_i N_i \left(1 - \sum_{i=1}^{n} \alpha_{ij} N_j\right), \qquad (27)$$

where N_i is the abundance of the *i*th species, r_i is the growth rate of the *i*th species, and α_{ij} is the competitive effect of species *j* on species *i*. The ideas for two species are well known (see the section "Competition," above). For three species, the possible equilibrium solutions are all species extinct, the three individual single population equilibria, the three 2-species population equilibria, and the full 3-species equilibrium. Under particular conditions (leading to a periodic limit cycle defined by a closed manifold in three-dimensional population space), the population trajectories in phase space lead to the repeated invasion, growth to equilibrium, and collapse of each of the single populations. This is a heteroclinic cyclic system in which the output from one cycle is connected to the input of another cycle, and as with other cyclic systems (e.g., predator–prey interactions), these heteroclinic systems can be stable or unstable. If stable, the system gradually approaches the points involved, but if unstable, the system oscillates away from these points.

Extending this idea to include natural enemies has important implications for these sorts of ecological processes. Resource–consumer interactions (as noted) are inherently cyclic ecological systems. Coupled with competition for resources (or through apparent competition) the idea of a heteroclinic system where the outflow from one point is connected to the inflow of another point (and vice versa) drives the complexity of ecological system. Our basic concepts of population ecology can be brought to bear on multispecies interactions but lead to complexities and challenges in understanding the nonlinear dynamics associated with these sorts of ecological processes.

SPATIAL ECOLOGY

In the simple equation of population change (Eq. 1), the dynamics of any population are influenced by spatial processes through immigration and emigration. This role of spatial scale and spatial structure has important implications for ecological interactions and is very well represented by the specialized interaction between microparasitic diseases and their hosts, and in particular the dynamics of measles in the UK. Aggregated dynamics from urban and rural regions show the seasonal, biennial dynamical pattern of measles epidemics in urban places coming ahead of rural ones. This difference in the timing of these epidemics is due to the strength of local coupling between urban and rural places. Patterns of spatial synchrony are related to population size: the number of cases in larger cities (large population size) tends to be negatively correlated, whereas there is no correlation in case reports between rural places with small population sizes. This regional heterogeneity leads to hierarchical epidemic patterns. In the small rural places, the infection fades out in epidemic troughs, although this is clearly dependent on the degree of coupling to larger urban places. These ecological processes give rise to a range of population dynamical patterns such that larger cities show regular biennial cycles, whereas in small towns disease dynamics are strongly influenced by stochastic effects. The dynamics of measles epidemics are a balance between nonlinear epidemic forces, demographic noise, environmental forcing, and how these processes all scale with host population size.

The key ecological process in these sorts of spatial interactions is the dispersal of individuals. This movement of individuals among places or patches not only affects the local population processes but is the important driver of the distribution and abundance of groups of populations at the regional scale. One of the simplest ways to envisage this idea of movement and spatial population ecology is through the concept of a metapopulation, that is, a population of populations linked by dispersal events.

Metapopulations

The key processes in metapopulation dynamics are those of colonization and extinction, and four central conditions define a metapopulation. First, populations breed in discrete patches of habitat. Second, populations have a high risk of extinction. Third, recolonization occurs through a "rescue effect." And finally, local populations are asynchronous in their dynamics.

The simplest characterization of spatial ecological processes is to ignore the full ecological complexities and focus on the proportion of a regional landscape occupied by a species. This is often expressed in terms of a single-species model developed by Richard Levins in 1969. The Levins metapopulation model excludes local population dynamics and is instead based on the balance of extinctions and colonizations of discrete habitat patches and has the form

$$\frac{dP}{dt} = cP(1 - P) - eP, \tag{28}$$

where c represents colonization rate and e represents extinction rate. At equilibrium, the patch occupancy (P_{eq}) is

$$P_{eq} = 1 - (e/c). \tag{29}$$

A metapopulation with a colonization rate larger than its extinction rate will result in positive equilibrium patch occupancy. If the two rates are equal or if extinction rate is greater than colonization rate, i.e., $e \geq c$, P_{eq} will be zero and the metapopulation will have gone extinct. The Levins model treats all patches as being the same. The patches have equal colonization rates, equal extinction rates, and are of equally large area. Individual distances between patches are not considered, and therefore each pair of patches is connected to the same extent.

In fragmented landscapes, habitat suitable for supporting any particular specialist species is often found in discrete patches. Such landscapes exhibit an array of separate local populations that are connected by migrating individuals moving between the local populations. Local population dynamics are often described in terms of changes in population size (e.g., Eq. 5) and metapopulations can describe dynamics in similar fashion, albeit at a larger spatial scale.

For instance, discrete-time logistic models used to illustrate local population dynamics can be coupled in such a way that migration is included in a spatially implicit manner. A coupled-logistic map for a two-population system takes the form

$$x_{t+1} = (1 - \delta)f(x_t) + \delta f(y_t), \tag{30}$$

$$y_{t+1} = (1 - \delta)f(y_t) + \delta f(x_t), \tag{31}$$

where $f(x) = \beta x(1 - x)$ and $f(y) = \beta y(1 - y)$, and x_t and y_t represent the number of individuals occupying patches x and y, respectively, at time interval t; δ describes the diffusive ability of individuals between patches, and β is the local population growth rate. Environmental homogeneity is assumed, and therefore δ is the same for both patches. Studies on these sorts of coupled-logistic maps have highlighted the stabilizing and synchronizing effect of migration on local population dynamics. However, a two-population system is an unrealistic scenario, and coupled maps have been developed and extended to multiple patches in lattices and grids.

Coupled-map lattices transfer the concepts encapsulated by the coupled-logistic map method to a much larger metapopulation system. While these sorts of spatial models are useful for theoretical investigation into the effects of migration on local population dynamics, they provide limited scope when formulating predictions about metapopulation persistence for naturally occurring metapopulations. For instance, it is rarely feasible to obtain accurate data on local population dynamics (see the section "Confronting Metapopulation Theory with Data," below), and therefore a model that can utilize easily obtainable data might be more beneficial for understanding spatial ecological processes such as metapopulation persistence. One example of data that can be collected and collated is species occupancy of a metapopulation's constituent patches. Patches are recorded as being either occupied or empty and, under appropriate statistical assumptions, can reveal patterns in metapopulation extinction and colonization.

Confronting Metapopulation Theory with Data

These limitations in metapopulation theory need to be addressed in order to allow suitable tests of theory and concepts. One development has been to use spatially realistic models to develop more accurate predictions about (meta)population persistence. Ilkka Hanski developed a patch-specific model known as the Incidence Function Model where each patch is assigned an area, and distance between patches is explicitly expressed. Both of these spatial characteristics can be measured. The model scales risk of patch extinction (E_i) with patch area:

$$E_i = \frac{\mu}{A_i^{\chi}} \tag{32}$$

where A_i is the area of patch i, μ describes the extinction probability of a patch of unit size, and χ defines the relationship between extinction risk and patch area. Strength of patch colonization is related to patch area and distance between patches by

$$C_i = f\,[S_i(t)], \tag{33}$$

where $S_i(t) = \Sigma_{j \neq i} O_j(t) \exp[-\alpha d_{ij}] A_j^b$, in which $S_i(t)$ represents the connectivity of patch i, d_{ij} is the distance between patches i and j, and A_j is the area of patch j. $O_j(t) = 1$ for occupied patches and $O_j(t) = 0$ for empty patches; α describes the distribution of dispersal distances and will be both species and habitat type specific and b describes the degree to which patch area influences connectivity. Estimation of the parameters is achieved using patch occupancy data collected over several years. Once a set of parameters have been derived for a patch network, extinction–colonization dynamics can be simulated to calculate the probability of extinction for each patch as well as the whole metapopulation.

Metacommunities

Extending the concept of metapopulations to interacting species and communities leads to the concept of a metacommunity (set of local communities linked by the dispersal of multiple potentially interacting species). Metacommunities scale local population and community processes to a regional scale and introduce novel ways to think about species interactions, diversity, and distributions.

At this regional scale, dispersal can regulate community assembly. If dispersal rates are low, then colonization will be the process determining community structure. If dispersal is high, then effects that modify both species abundances and interactions will be predominant. Diversity will be influence by region-wide effects on limiting similarity and character displacement. Character displacement making species less similar (at the local scale) is often heralded as evidence for the effects of interspecific competition. Ecological processes associated with tradeoffs among species traits, patterns of resource utilization between species in metacommunities, and regional-level environmental heterogeneities (e.g., productivity gradients) challenge this perspective and suggest local species assemblages might be more similar than expected.

SEE ALSO THE FOLLOWING ARTICLES

Age Structure / Apparent Competition / Cannibalism / Chaos / Metapopulations / Predator–Prey Models / Single-Species Population Models / Two-Species Competition

FURTHER READING

Begon, M., C. R. Townsend, and J. L. Harper. 2005. *Ecology: from individuals to ecosystems,* 4th ed. Oxford: Wiley-Blackwell.

Chesson, P. 2000. Mechanisms of maintenance of species diversity. *Annual Review of Ecology and Systematics* 31: 343–366.

Grenfell, B. T., O. N. Bjornstad, and J. Kappey. 2001. Travelling waves and spatial hierarchies in measles epidemics. *Nature* 414: 716–723.

Hassell, M. P. 2000. *The spatial and temporal dynamics of host-parasitoid interactions.* Oxford: Oxford University Press.

Hanski, I. 1999. *Metapopulation ecology.* Oxford, UK: Oxford University Press.

Holt, R. D., and J. H. Lawton. 1994. The ecological consequences of shared natural enemies. *Annual Review of Ecology and Systematics* 25: 495–520.

Malthus, T. R. 1798. *An essay on the principle of population.* London: John Murray.

May, R. M. 1976. Simple mathematical models with very complicated dynamics. *Nature* 261: 459–467.

Tilman, D. 1982. *Resource competition and community structure.* Princeton: Princeton University Press.

POPULATION VIABILITY ANALYSIS

WILLIAM F. MORRIS

Duke University, Durham, North Carolina

Population viability analysis (PVA) is the use of quantitative methods (mathematical models and computer simulations) to assess extinction risk and guide management of populations of threatened or endangered species. PVA is to theoretical ecology as engineering is to physics. Just as engineers apply physical laws to construction materials when designing a building, PVA practitioners apply the theoretical population models described throughout this book to field data on rare species to assess population viability.

USES OF POPULATION VIABILITY ANALYSIS IN CONSERVATION

A quantitative index of population viability can be put to many important uses in conservation management, including the following:

1. Deciding which of several populations is most worth saving (has the lowest extinction risk) or is most in need of intervention (has the highest extinction risk)
2. Identifying which of multiple threats to a population are the most worrisome and in need of mitigation
3. Assessing which life stages or vital rates (survival, growth, and reproduction) are the most profitable targets for management efforts
4. Quantifying how adding new populations (for example, by a combination of habitat restoration and translocation) or linking populations with dispersal corridors would reduce extinction risk for a metapopulation or an entire species.

PVA involves an interplay between theory and data, as the availability of data dictates which of the many potential causes of extinction can be profitably included in the model; there is little value in including, for example, inbreeding depression in the viability model if no data on the magnitude of inbreeding depression exist for the population whose viability is to be assessed. The following sections review factors that may influence population viability.

UNDERLYING CAUSES OF POPULATION EXTINCTION

Perhaps the most common reason that once-healthy populations become endangered is that the vital rates deteriorate due to trends in environmental conditions (such as climate change, invasive species, habitat fragmentation, pollution, and the like) that often result from human activities. Unfortunately, incorporating vital rate trends into viability assessments is often precluded by the lack of long-term data needed to identify those trends. As discussed in the following section, environmental stochasticity, or variability in environmental conditions from year to year and the variability in vital rates that results, also contributes to extinction risk. Whereas environmental stochasticity has a more or less equal effect on all individuals in the population (or at least in a given life stage), random variation in the fate of individuals under the same environmental conditions—so-called demographic stochasticity—is a potentially powerful cause of extinction in very small populations. Changes in vital rates as population density changes also influence viability. Declines in vital rates as density increases (negative density dependence) impose a ceiling that keeps a population closer to extinction, but they also mean that vital rates improve as density declines, buffering against extinction. In contrast, an increase in vital rates as density increases from low levels (positive density dependence, or Allee effects) means populations will do even worse as their density declines. Genetic effects in small populations (such as greater inbreeding leading to inbreeding depression, and enhanced genetic drift leading to loss of beneficial alleles) can elevate extinction risk. Finally, in metapopulations (sets of local populations), dispersal of individuals between populations (which may facilitate reestablishment of extinct local populations) and fluctuations in vital rates that are shared among populations (for example, due to large-scale climate variation) are important factors to consider in assessing metapopulation viability.

THE IMPORTANCE OF ENVIRONMENTAL STOCHASTICITY FOR POPULATION VIABILITY: A SIMPLE MODEL

Of the factors listed in the preceding section, environmental stochasticity is perhaps most often included in viability models. Here, the simplest stochastic population model is used to illustrate how environmental stochasticity affects extinction risk, employing two different metrics of population viability: the long-term stochastic population growth rate and the probability of quasi-extinction. The text then describes how this simple model can be fit to field data.

Let N_t be the density of all individuals in a population at census t, b_t be the number of new individuals to which each of the N_t individuals gives birth and that survive to census $t + 1$, and d_t be the fraction of the N_t individuals that themselves die before census $t + 1$. Therefore,

$$N_{t+1} = [b_t + (1 - d_t)]N_t = [1 + (b_t - d_t)]N_t. \quad (1)$$

The last expression shows that if births exceed deaths on a per capita basis between the two censuses, the population will grow ($N_{t+1} > N_t$). Defining $\lambda_t = 1 + b_t - d_t$ as the annual population growth rate in year t, $\lambda_t > 1$ implies population growth and $\lambda_t < 1$ implies population decline over the intercensus interval. Variation in birth and death rates among intervals causes the per capita population growth rate λ to vary over time, and the variance in λ reflects the magnitude of environmental stochasticity.

Imagine that λ follows a log-normal distribution (which cannot be negative, as is true of a population growth rate) with an arithmetic mean of $\bar{\lambda} = 0.95$. Thus, in an average year, the population declines, as is unfortunately the case for many endangered species. Figure 1A shows two different distributions of λ with this same arithmetic mean, but one of which (black) has a variance of 0.002 and the other (blue) has a variance ten times larger. Now imagine starting two populations with 1000 individuals and projecting their sizes over time, each year drawing a new value of λ from the two distributions. Each line in Figures 1B and 1C shows different trajectories these populations might follow over 50 years. Not surprisingly, the trajectories spread out more rapidly when the year-to-year variance in λ is higher. But it is also apparent that for both populations, the median of the trajectories in each future year is well predicted by the heavy black line, and that line declines more rapidly when environmental stochasticity is higher (population B). The heavy line is produced by the recursion equation $N_{t+1} = \lambda_s N_t$, where the subscript "s" indicates that λ_s is the long-term "stochastic" growth rate of the population. λ_s is a first index of population viability as, all else being equal, a population will face extinction sooner when λ_s is small. Importantly, λ_s is not the arithmetic mean of the annual population growth rates (which, after all, is the

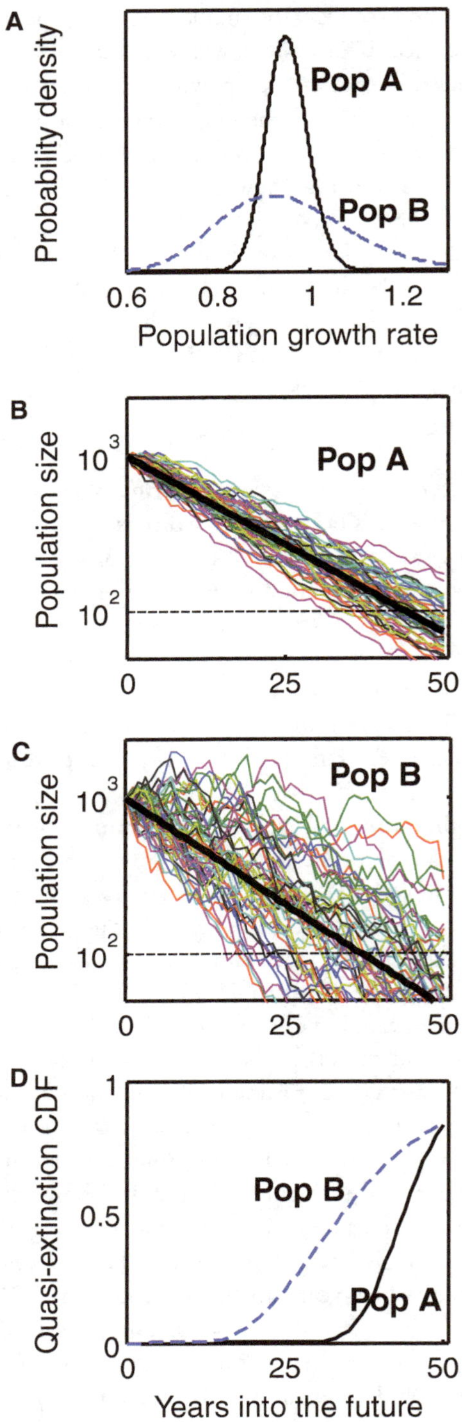

FIGURE 1 Influence of environmental stochasticity on extinction risk in a simple count-based population model. (A) The annual population growth rate varies more from year to year in population A than in population B, although the arithmetic mean growth rate equals 0.95 in both. (B) and (C) If both populations start at the same size (1000 individuals), population B, with greater environmental stochasticity, is more likely to hit a quasi-extinction threshold of 100 individuals (dashed line) sooner; note the logarithmic *y*-axis. (D) The quasi-extinction cumulative distribution functions for the two populations.

same in Figures 1B and 1C), but rather it is the geometric mean, which for a long sequence of annual λ values (for years 1 to t) can be computed as

$$\lambda_s = \sqrt[t]{\lambda_1\lambda_2\lambda_3 \cdots \lambda_{t-2}\lambda_{t-1}\lambda_t} \approx \bar{\lambda}\exp\{-\mathrm{Var}(\lambda)/(2\,\bar{\lambda}^2)\}, \quad (2)$$

where $\bar{\lambda}$ and Var(λ) are the arithmetic mean and variance of the annual population growth rate. The approximation in Equation 2 shows that as environmental stochasticity (represented by Var(λ)), and preceded by a minus sign in the equation) increases, the stochastic population growth rate decreases, clearly indicating that high levels of environmental stochasticity threaten population viability.

A second index of population viability can be obtained from this model by specifying a "quasi-extinction threshold" (QET). Typically, a threshold substantially above zero is chosen, to indicate that a population has declined to a lower level before it has actual gone extinct and to prevent the population from reaching the small sizes at which its viability is increasingly threatened by demographic stochasticity and Allee effects. Once a QET is determined (typically based on both biological and political grounds), the probability that the population hits the QET at any time between now and a specified future time is used as a second index of viability. A plot of this probability as a function of time into the future is called the quasi-extinction cumulative distribution function (or CDF). For any population model, the quasi-extinction CDF can be obtained by computer simulation, but for the simple model in Equation 1 an equation approximating the CDF is often used, which comes from making an analogy between the process of stochastic population growth and the physical process of diffusion. The quasi-extinction CDFs from the diffusion approximation (Fig. 1D) show that the probability the population will decline from an initial size of 1000 to a QET of 100 increases more rapidly (at least initially) for the population experiencing a higher level of environmental stochasticity.

This simple example illustrates several components common to most PVAs: the importance of environmental stochasticity for population viability, the use of multiple viability metrics, and the usefulness of specifying a quasi-extinction threshold. Importantly, to actually use this basic model to assess viability for a threatened population, it would first be necessary to use field data to estimate the parameters of the model, namely, the mean and variance of the annual population growth rate (or the log growth rate if a diffusion approximation is used to compute the quasi-extinction CDF). As the model only tracks the total number of individuals in the population (and not, for example, the number of juveniles and adults separately), a series of annual censuses of the total size of the population can be used to estimate the mean and variance of the population growth rate. Specifically, if $\hat{N}_t$ is the estimate of the size of the population in year t, an estimate of the growth rate in year t is the ratio of successive counts: $\hat{\lambda}_t = \hat{N}_{t+1}/\hat{N}_t$. The mean and variance of the $\hat{\lambda}_t$'s across all years of the census then provide estimates of the model parameters.

Although the preceding description makes it sound fairly simple, the process of estimating parameters and assessing viability using a series of population counts faces a number of challenges in practice. First, if the censuses are not evenly spaced in time, the $\hat{\lambda}_t$'s are not comparable; fortunately, a regression method is available to overcome this difficulty. Second, estimating a single mean and variance from the series of censuses assumes that birth and death rates do not change as population density changes. With enough data, density-dependent population models could be fit to the set of $\hat{\lambda}_t$'s, but sufficiently long time series are often not available for rare species. Third, the census counts may contain errors (e.g., they may miss individuals that are hard to see), and in cases in which they are not exhaustive counts of the entire population but instead are estimates (e.g., counts from a set of plots extrapolated to the entire area the population occupies), they are subjected to sampling variation. That is, repeated sets of plots would not give exactly the same estimate of total population size, even if the counts within plots are completely accurate. When the $\hat{N}_t$'s are estimates, the variance of the $\hat{\lambda}_t$'s will include a component due to environmental stochasticity (which is needed to assess viability) and a component due to the sampling variance of the population size estimates, which does not actually influence population viability but does affect the ability to estimate viability accurately. Regressing the variance between population counts separated by different numbers of years against the number of years elapsed is one way to separate the variance in population growth due to environmental stochasticity from the sampling variance. A fourth challenge is that, even if population size, and thus growth rate, is measured accurately each year, a limited sequence of censuses will provide only an *estimate* of the true mean and variance of the population growth rate. When environmental stochasticity is high, the estimate of the variance in particular will be imprecise, because extreme values of the growth rate are likely to be missed. This imprecision places limits on how far into the future extinction risk can be accurately predicted using a given sequence of censuses.

TARGETING LIFE STAGES OR DEMOGRAPHIC PROCESSES USING STRUCTURED POPULATION MODELS

With the caveats just discussed, an advantage of the simple model (Eq. 1) is that it can be fit to data on total population counts, which are often easier to collect than detailed demographic data on individuals in different life stages (for example, newborns, juveniles, and adults, or individuals in different size classes). However, another aim of a PVA model is to identify critical life stages or demographic processes that are the most promising management targets. To do so, a so-called structured model is needed that accounts for the contributions of different life stages to population viability, as are demographic data to estimate the parameters of the model.

The type of structured population model most commonly used in PVA is the projection matrix model, which divides individuals into discrete life stages that may represent age groups, size classes, or developmental stages. For example, to construct a projection matrix for the endangered loggerhead sea turtle (*Caretta caretta*), which nests on beaches and is threatened by trampling of nests by humans and drowning of turtles in fishing nests, individuals are first classified into five stages (in the order in which they appear in the life cycle): eggs and hatchlings (E/H), small (SJ) and large (LJ) juveniles (nonreproductive), novice breeders (NB), and fully reproductive adults (A). The following matrix describes the mean annual rates of transition between these 5 stages estimated for a loggerhead population in the southeastern United States:

Life stage in year $t+1$:	Life stage in year t:					
	E/H	SJ	LJ	NB	A	
E/H	0	0	0	4.665	61.896	
SJ	0.675	0.703	0	0	0	
LJ	0	0.047	0.657	0	0	(3)
NB	0	0	0.019	0.682	0	
A	0	0	0	0.061	0.809	

As indicated by the row and column labels, entries in the matrix represent the per capita contribution of each life stage this year to each life stage next year. The first row in the matrix thus represents reproduction (i.e., production of new eggs/hatchlings by novice breeders and adults), diagonal elements represent survival without advancement to the next life stage, and subdiagonal elements represent survival and advancement. Matrix elements sometimes represent combinations of underlying vital rates (for example, reproduction often combines breeding probability, number of offspring if breeding occurs, and probability of offspring survival to the next census).

The structure of the population at each census is represented by a column vector with as many rows as there are life stages and with entries representing the current abundances of individuals in each stage. Just as Equation 1 generates next year's population size by multiplying this year's population size by this year's population growth rate, in a structured model next year's population vector is predicted by multiplying the projection matrix by this year's population vector. While the matrix in expression 3 shows the mean demographic rates, a multi-year demographic study may yield an estimate of each vital rate (and thus each matrix element) for each year. In this case, environmental stochasticity can be incorporated into this structured population model by choosing different values of each vital rate or entire matrices each year before projecting the population vector forward in time. Now, the long-term stochastic growth rate of the total population size (the sum of the population vector) will depend on the means and variances of each of the vital rates, but also on how pairs of vital rates covary. But importantly for conservation management, it will also depend on the impact that each vital rate has on the population growth rate in an average year, which is the subject of sensitivity analysis, covered in greater detail in other entries in this book but summarized here in the context of PVA.

If the same projection matrix is used every year, the population growth rate converges to the dominant eigenvalue λ_1 and the population vector converges to the dominant right eigenvector of the matrix (eigenvalues and eigenvectors are easily computed using matrix-based computer languages such as MATLAB and R). A sensitivity is simply the partial derivative of λ_1 with respect to a matrix element, holding all other matrix elements at their current values. For example, if Equation 3 is called the matrix **A**, the annual survival rates of eggs/hatchlings and of adults are represented by matrix elements $a_{21} = 0.675$ and $a_{55} = 0.809$, respectively (where a_{ij} is the element in row *i* and column *j* of matrix **A**). The sensitivities of λ_1 to these two matrix elements are simply $\partial\lambda_1/\partial\alpha_{21}$ and $\partial\lambda_1/\partial\alpha_{55}$, respectively, both computed keeping the values of all matrix elements as shown in Equation 3. These two sensitivities are particularly useful for guiding management of declining loggerhead sea turtle populations, as is now demonstrated.

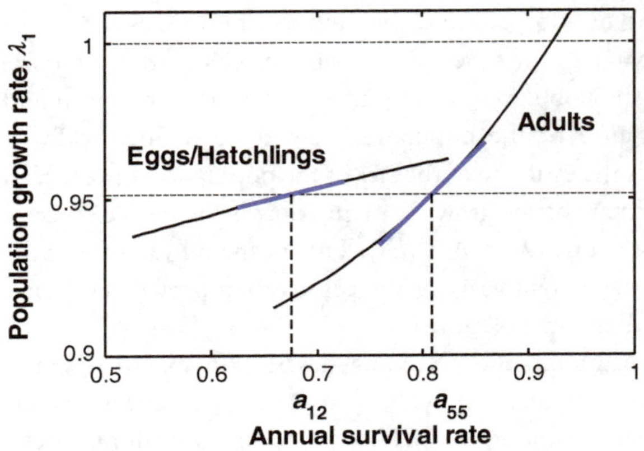

FIGURE 2 Predicted population growth rate (dominant eigenvalue of the projection matrix) as a function of the survival rates of eggs/hatchlings and adult loggerhead sea turtles. a_{21} and a_{55} are the values of egg/hatching and adult survival rate (respectively) in the matrix in the text (Eq. 3), and the slopes of the blue lines are the sensitivities.

One management strategy is to reduce egg mortality by fencing turtle nests on beaches and transferring hatchlings to the sea when they emerge from the nests (predation on beaches is often high), which would increase a_{21}. Another strategy would be to install devices that would eject adult turtles caught in fishing nets, which would increase a_{55}. (These devices would also increase survival of the middle three life stages that are also vulnerable to drowning in nets, but assuming only a_{55} would increase is conservative.) Figure 2 shows the population growth rate as a function of both a_{21} and a_{55}. The dashed horizontal line at $\lambda_1 = 0.952$ is the population growth rate predicted by the projection matrix in Equation 3; because this is less than 1, the loggerhead population is predicted to decline. The goal of management is to elevate λ_1 to 1 (the upper horizontal line) or higher. The solid black lines show the effects of changing either egg/hatchling or adult survival rate, keeping all other matrix elements as shown in Equation 3. Both lines cross the line $\lambda_1 = 0.952$ when the respective survival rate equals the value in Equation 3, and increasing both rates increases λ_1 (linearly for egg/hatchling survival but in an accelerating fashion for adult survival). However, increasing adult survival has a much more pronounced effect on population growth. The blue lines are tangents to the solid lines at the current values of the survival rates, thus they have the same slope as the solid lines and that slope is the sensitivity; population growth is much more sensitive to adult survival than it is to egg/hatchling survival. The two sensitivities indicate that a given increase in adult survival would be much more beneficial than the same amount of increase in egg/hatchling survival and point to installing anti-drowning devices in fishing nets as a more promising management alternative than nest protection. By extrapolating the black or blue lines to the maximum possible survival rate of 1, it is easily seen that even protecting every egg and hatchling would be unlikely to rescue the population, while less dramatic change in adult survival could do so. This example illustrates that an important aim of PVA is not merely to quantify extinction risk but also to help find ways to reduce it.

If multiyear demographic data were available, a sensitivity analysis could also be performed for a stochastic projection matrix model, only now it would be possible to evaluate how changing both the means and the variances (e.g., by ameliorating extreme environmental conditions) of different vital rates would alter the long-term stochastic growth rate of the population. By first defining a quasi-extinction threshold (for total population size or the abundance of key life stages), a quasi-extinction CDF could then be computed, as for the count-based model (Fig. 1D), and with data over a sufficient range of densities, density-dependent functions for the different vital rates could be estimated to incorporate density dependence into the viability assessment. Finally, demographic stochasticity could be included by viewing the matrix as containing average values but drawing random numbers around these averages to decide whether each individual in each life stage survives, grows, and reproduces.

FURTHER EXTENSIONS

The discussion here has taken a single population perspective. If a threatened species exists in more than one location, then a local population that has gone extinct can be replaced if dispersers from other populations successfully recolonize the site. Metapopulation models covered elsewhere in this volume are useful in assessing metapopulation viability (the likelihood that at least some local populations avoid extinction). Such models range from versions of the simple count-based model (Eq. 1), but with multiple populations linked by migration, to structured models with multiple populations, to detailed, spatially explicit computer simulation models that track virtual individuals moving across heterogeneous landscapes composed of breeding and nonbreeding habitat. Parameterizing these highly detailed models requires a large amount of data, so they have been developed for relatively few rare species.

Finally, PVA and management can be profitably combined in an iterative process called "population viability management." First, a PVA model fitted to field data is used to quantify how vital rates and population size influence extinction risk. Then ongoing threats and their effects on vital rates are simulated, as are monitoring "data" measured with different degrees of uncertainty. The monitoring "data" then trigger management activities, which feed back to the vital rates, and finally the PVA model is used to assess extinction risk. This approach can identify a suite of potential monitoring and management programs that would best enhance population viability. As more data are collected, this entire process can be repeated, leading to both a better PVA and better monitoring and management, through an ongoing dialogue between theory and data.

SEE ALSO THE FOLLOWING ARTICLES

Allee Effects / Birth–Death Models / Conservation Biology / Demography / Matrix Models / Metapopulations / Model Fitting / Stochasticity, Environmental

FURTHER READING

Bakker, V. J., and D. F. Doak. 2009. Population viability management: ecological standards to guide adaptive management for rare species. *Frontiers in Ecology and the Environment* 7: 158–165.

Beissinger, S. R., and D. R. McCullough, eds. 2002. *Population viability analysis*. Chicago: University of Chicago Press.

Caswell, H. 2001. *Matrix population models: construction, analysis, and interpretation*, 2nd ed. Sunderland, MA: Sinauer Associates.

Ellner, S. P., and E. E. Holmes. 2008. Commentary on Holmes et al. (2007): resolving the debate on when extinction risk is predictable. *Ecology Letters* 11: E1–E5.

Morris, W. F., and D. F. Doak. 2002. *Quantitative conservation biology: theory and practice of population viability analysis*. Sunderland, MA: Sinauer Associates.

PREDATOR–PREY MODELS

PETER A. ABRAMS

University of Toronto, Ontario, Canada

Predator–prey interactions are arguably the most important type of interaction between species in ecological communities. The eating relationship that is the focus of predator–prey systems forms a key element of many other interactions between species. Eating is a necessity for all heterotrophic organisms. When the foods are themselves living organisms, the interaction between consumer and food is predation. Competing species are often predators that share common prey, and both mutualisms and a wide variety of other interactions involve consumption of foods. It is therefore not surprising that mathematical models of predator–prey interactions are a central component of theoretical ecology.

COMPONENTS OF PREDATOR–PREY MODELS

Ecological theory requires simplification of nature. Every individual in any predator or prey population is unique, and almost all populations have different spatial distributions. Real predators usually consume several types of prey, and prey are consumed by several classes of predator. The prey themselves are usually consumers of resources, which in turn may be living organisms, making the prey a predator in its own right. Theory, of necessity, must begin with simpler cases in which only those elements that are always present in a predator–prey interaction are represented in the mathematical description. Thus, this section focuses on a single predator population that consumes a single prey population. Each population is homogeneous, in that individuals within each species are indistinguishable in their effects on, and how they are affected by, other individuals and by the environment. Furthermore, the growth of the prey in the absence of the predator depends only on prey population size. Theory must also provide guidance about the limits of applicability of such simple models, how they can be elaborated, and what consequences these elaborations may have. Three key elements form the backbone of all predator–prey models: the prey growth function, the predator's functional response, and the predator's numerical response.

Because the prey is assumed to be a biological species, there must be some description of how its population changes over time in the absence of predators. This is generally provided by a single-species population growth model. Any model must also describe how the predator changes the prey's population growth. Predators may have a variety of effects on prey, but the effect that is present in 100% of predator–prey systems is that they prey's population is reduced (at least temporarily) by individuals, or parts thereof, being eaten. The number of prey eaten per unit time by an average predator is known as the predator's functional response to that prey. The loss of prey individuals or biomass per unit time is given by the number of predators multiplied by their functional response. The consumed prey must contribute to predator population growth, and the numerical response (the third and final component) describes this relationship. These three components are also the backbone of any consumer–resource system. Plants can be regarded as having functional responses to light, water, and mineral nutrients, and the intake rates of these different essential resources combine to determine the plant's demographic (numerical) response. Predator–prey models represent a special form of consumer–resource models in which the resource increases in abundance by reproducing itself. In contrast, the resources used by plants (and other nonpredatory consumers such as detritivores) are introduced by processes that do not depend on the abundance of resource that are already in the system.

No population is truly homogeneous. Size, age, and location are some of the most common characteristics that distinguish individuals and affect their interactions with predators, prey, or both. An accurate description of a heterogeneous population requires different functional and numerical responses for the different classes of predators and prey. Individuals may appear to be very similar to each other, but invisible differences that alter individual functional and/or numerical responses can alter the effective form of a population-level response. The concept that a response can be expressed as a function of the population size of a prey species also implies that the population is well mixed, so that any individual prey can be encountered by the predator during the next time interval. In reality, organisms have spatial locations that cannot change infinitely rapidly. The result is that an individual predator can deplete its prey locally and starve in spite of an abundance of prey elsewhere. This may require an accounting of the populations by spatial location to fully describe the dynamics. The problematic assumption of homogeneity in the simple models again arises when extensions to the models are discussed below.

FOUNDATIONAL MODELS OF PREDATOR–PREY INTERACTIONS

The Lotka–Volterra Predator–Prey Model

Alfred J. Lotka and Vito Volterra independently proposed the simplest possible model for the dynamics of a predator–prey system in the 1920s. This model is included in almost all ecology textbooks. It is an ordinary differential equation model, which means that the rates of change in the sizes of the two interacting populations are determined by their current population sizes and a number of parameters that reflect environmental characteristics. Lotka and Volterra assumed that prey have constant per-individual birth and death rates, resulting in exponential prey growth in the absence of predators. An average predator eats a number of prey per unit time that is directly proportional to the population size of the prey (i.e., a linear functional response). Finally, the predator has a constant per capita loss rate and a birth rate that is directly proportional to the amount eaten. At the time these models were developed there was essentially no empirical work to guide model formulation, so the forms of the relationships were motivated by mathematical simplicity.

The Lotka–Volterra model can be written as follows, where P is the predator population size and N is the prey population size. Population sizes can be measured either as the biomass or number of individuals, and are usually expressed as a density per unit area rather than total numbers. The left-hand sides of the following equations are standard symbols for the rate of change of a variable with time, and these are quantified by the expressions on the right-hand sides of the equations:

$$\frac{dN}{dt} = rN - cPN,$$
$$\frac{dP}{dt} = P(bcN - d). \qquad (1)$$

There are four constants in this model: the intrinsic growth rate of the prey, r; the slope of the predator's functional response, c; the slope of the predator's numerical response, b; and the per capita loss rate of predators, d. Usually b is interpreted as an efficiency of converting consumed prey into new predator biomass, and d is the per capita loss of biomass due to the combined impacts of mortality and metabolism. Slightly different interpretations of the constants follow from different methods of defining population size, and appropriate choice of units to measure the two variables can reduce the number of constants to three. This is one of the few predator–prey models for which we can derive an expression giving the exact future population sizes given any pair of initial population sizes. Unfortunately, the dynamics predicted by this model have never been documented in the field or the laboratory. It predicts cycles whose amplitude depends on the difference between initial densities and equilibrium densities (which are $N = d/(bc)$ and $P = r/c$). The populations have no tendency to move either toward or away from equilibrium. Robert May, among others, has criticized the use of the Lotka–Volterra model in any predictive role because of this feature. There are other predictions of the model that also seem inconsistent with the everyday experience of ecologists; for example, the final predator population (either the equilibrium or the average over the course of a cycle) is unaffected by increasing its mortality rate, and the same is true of the prey.

There are two reasons why the Lotka–Volterra model is included in most textbooks in spite of these shortcomings. First, it exemplifies a widespread mechanism by which predator–prey interactions lead to population cycles in both species. Secondly, it provides a framework for constructing more realistic models. There is now a good deal of evidence regarding the nature of functional responses of predators and density-dependent growth of prey that can provide alternatives to the exponential growth and linear functional response in this bare-bones model.

The Rosenzweig-MacArthur Model

The best-known of this "second generation" of models is what is now usually referred to as the Rosenzweig–MacArthur model, after Michael Rosenzweig and Robert MacArthur's 1963 article in the *American Naturalist.* Although that paper used a graphical approach, it did identify the prey's density dependence and the shape of the predator's functional response as factors that affected the stability of a predator–prey system. Later work by Rosenzweig explored a set of fully specified models, one of which he used Crawford Holling's (1959) type II predator functional response (see below) and logistic prey growth for the prey dynamics. The type II response increases at a decreasing rate with higher prey population density because catching and processing abundant prey reduces the time available for the predators to catch new victims. This has a destabilizing effect because greater prey population density reduces their per capita risk, which reduces the force acting to bring a high prey population size back to the equilibrium. The full model has the following form:

$$\frac{dN}{dt} = rN\left[1 - \frac{N}{K}\right] - \frac{cPN}{1 + chN},$$
$$\frac{dP}{dt} = P\left(\frac{bcN}{1 + chN} - d\right), \qquad (2)$$

where the parameters are identical to those in the Lotka–Volterra model except for two additions. Handling time, h, represents the amount of time the predator requires for handling and processing a single resource item. Handling by definition is time that cannot be spent searching for or attacking additional prey. In practice, digestion is likely to be the largest component of handling time in most species. The second new parameter is the carrying capacity of the resource, K, which represents its equilibrium density in the absence of the predator. The population dynamics predicted by the model may be cycles or an approach to a stable equilibrium point. The former is likely when handling time is relatively large, predator loss rate is small, and prey carrying capacity is large. Stability is likely under the opposite conditions. Figure 1 provides an example of the cycles that can arise in this model. The two panels show population dynamics for the same predator–prey pair, but in the lower panel the prey has an equilibrium population in the absence of predation that is 2.5 times larger than in the upper panel. The period of the cycles is clearly longer and its amplitude higher in the lower panel than in the upper one. This increase in cycle amplitude was the basis of Rosenzweig's prediction of a "paradox of enrichment," under which periodic very low populations that were caused by enrichment could be

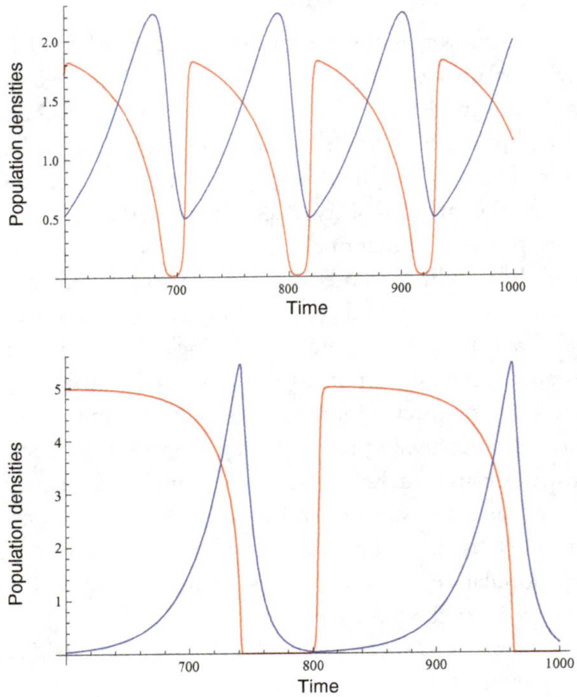

FIGURE 1 Two examples of predator-prey dynamics based on the Rosenzweig-MacArthur model. The red line gives the trajectory of prey population density, and the blue line gives predator population density. The bottom panel shows the impact of increasing the prey's equilibrium density (when alone; i.e., *K*) by a factor of 2.5.

sufficiently extreme to produce extinction of the predator, prey, or both. The Rosenzweig–MacArthur model has been used to uncover a variety of other nonobvious consequences of predator–prey interactions. The type II functional response in this model has the potential to lead to an increase in predator population size with an increase in its own mortality rate. It can cause evolutionary cycles in anti-predator traits in the prey, and it has even been suggested to lead to evolutionary extinction under some conditions.

Although it uses the most commonly assumed model of density dependence together with the most commonly documented functional response, the Rosenzweig–MacArthur model certainly cannot describe all predator–prey systems. Furthermore, the dynamics it predicts differ significantly from the dynamics observed in the best-studied natural and laboratory predator–prey systems. One of its main failings is that, for laboratory systems that have persistent cycles, the model predicts too high an amplitude of oscillation and also predicts that the amplitude is too sensitive to parameters that influence stability. However, in some laboratory predator–prey systems, the observed dynamics have been cycles of ever-increasing amplitude that end with extinction. This was the case for one of the original laboratory studies of predation, the experiments by the Russian ecologist Georgii Gause in the 1930s on the protozoan predator–prey pair, *Didinium* and *Paramecium.* This outcome is never predicted by the model.

ARE THESE MODELS SUFFICIENT TO DESCRIBE MOST PREDATION?

The short answer to this question is no. These models were probably both designed with a typical carnivore in mind. It is not surprising that some interactions in which organisms consume others are so different as to require a very different type of model. The range of interactions that can reasonably be approximated by models that are similar to the Rosenzweig–MacArthur equations is still being debated. It is likely that interactions between large herbivores and plants can be described reasonably well with some small changes, but parasites and parasitoids often require a different framework.

Different Types of Predation

The standard models were developed to represent predator and prey that have independent lives and prey individuals that do not survive consumption. However, most definitions of predation also include interactions in which only part of the prey is eaten, and interactions in which the predators (or some life stage of the predators) lives on or inside of the prey. Partial consumption is probably the rule rather than the exception for invertebrate herbivores eating plants. Parasitoids and disease organisms live on or in their host, which often implies that they die when the host dies. The transmission of individuals to another host is usually more crucial in predicting the fate of a disease than is the total number of individuals of the disease organism in all hosts. Partial consumption is often consistent with standard models when the variables are interpreted as biomass. However, complications can arise when plants are so well adapted to being eaten that they suffer reduced fitness when they are not consumed. Host occupation makes the host similar to a habitat patch in a system with free-living predators, and host-disease models have many similarities with models of predators and prey distributed across semi-isolated habitat patches. An important feature of such models is that the fraction of patches occupied is often more important than is the average number of individuals within a patch if one wants to predict the future course of the entire predator or disease population. It is also important to bear in mind that the predator and prey can be different individuals of the same species, which represents cannibalism. Cannibalism does not alter the concept of the functional response, but it requires that different components of the numerical response (e.g., adult birth rate and juvenile survival) change in opposite directions in response to interactions.

Elaborations of the Basic Models

There are two major sets of extensions to the foundational models described above, which predict the dynamics of a single predator and single prey species. One is to maintain the framework of a simple dynamical systems model with two variables but incorporate different growth functions, functional responses, and numerical responses. Included in this first set is the presence of temporal variation in parameters affecting one or more of the three basic functional components. The second group of modifications abandons the assumption that both predator and prey populations can be described by the single variable, population size. The properties of individuals in one or both species are likely to change over time, and almost all populations consist of individuals that differ in size, age, developmental stage, or all three. Perhaps most important in this second group of extensions is the presence of spatial heterogeneity, so that populations are subdivided based on spatial location. Extensions to systems with more species are treated in the "More Than Two Species" section, below.

ALTERNATIVE FORMS FOR THE THREE COMPONENT FUNCTIONS

It is only possible to give a brief overview of some of the most commonly used alternative forms for predator functional and numerical responses and growth functions in a 1-predator–1-prey system. Some of the important alternatives are:

1. *Direct negative effects of predator density on the predator's functional or numerical response.* Direct negative effects of the predator on its own birth and/or death rates arise because food is not the sole determinant of these rates. There are often limited quantities of the best hunting or nesting sites or general-purpose territories for the predator. As a result, the predator's functional and numerical responses are both likely to be declining functions of predator population density. Both of these effects generally increase the chance that the predator–prey system is stable. However, they also make it possible to have alternative equilibria, where the predator is primarily limited either directly by its own population or indirectly by a shortage of prey. Adaptive defense by prey that increases with predation risk also has the effect of making the predator's response predator dependent; i.e., at any particular prey population density, the predator's consumption rate decreases with increasing predator density—this is discussed in more detail in point 5, below.

2. *Functional responses can have a wide variety of functional forms even when they only depend upon prey density.* Crawford S. Holling, who laid the groundwork for our current understanding of functional responses in the 1950s and 1960s, proposed four types. The linear and convex forms discussed under the Lotka–Volterra and Rosenzweig–MacArthur models were called type I and type II responses. However, Holling suggested that responses could also be sigmoid (type III) or unimodal (type IV). Figure 2 illustrates the shapes of these four classes of responses. Sigmoid responses could come about when prey use refuges adaptively and therefore are not exposed to much predation until their population size exceeds some threshold. Type IV responses have been observed in a number of systems, but the underlying mechanism is usually not known. Such responses have been predicted to arise as an adaptive response to toxins in foods. Type IV responses could also arise from the distracting effect of many targets when prey are abundant. This set of four responses does not exhaust the range of possible shapes; adaptive foraging can produce a variety of more complicated forms in theory, but few if any have been documented in field or laboratory settings. Even the relatively simple alternatives, such as type III and type IV functional responses, can generate alternative states and change the stability properties of those states in some models.

3. *Numerical responses with nonlinear shapes.* The vast majority of models assume linear numerical responses

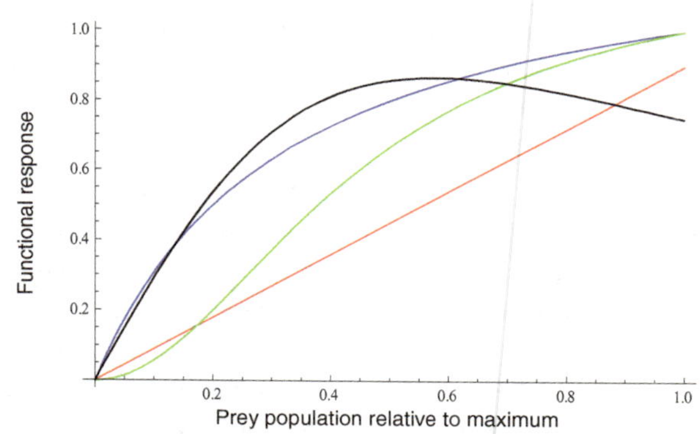

FIGURE 2 Examples of the four classes of functional responses defined by Crawford Holling. Each line describes the intake rate of prey per unit time by an average predator individual as a function of prey population size. The red line is a type I (linear) response, the blue line is type II, the green line is type III, and the black line is type IV.

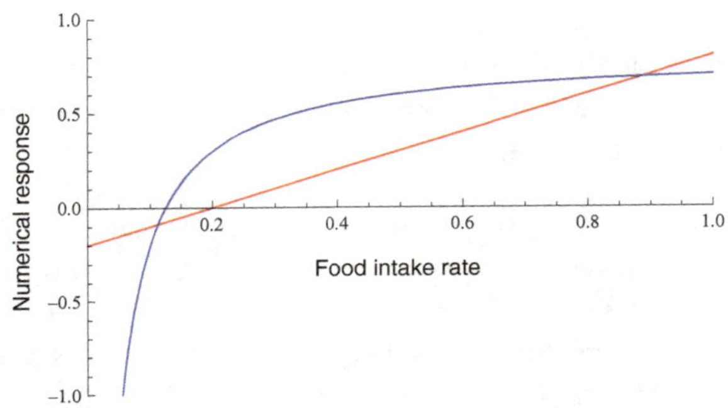

FIGURE 3 Two types of numerical response of a predator. Each describes the difference between the per-individual birth rate and the per-individual death rate of the predator as a function of its intake rate of food. The red line is the traditional linear numerical response, and the blue line is a convex response which more accurately reflects the impact of starvation.

without any empirical justification, usually assuming a constant per capita death rate and a birth rate directly proportional to food consumption. However, these are inconsistent with what is known about starvation; death rates can become very high in the absence of food. Starvation is a lagged effect of lack of food over a finite period of time. However, within the framework of instantaneous effects assumed by the standard models, starvation is best represented by making the numerical response a convex function of food intake. Per capita growth then increases with food intake rate at a decelerating rate from a large negative value at zero intake. Figure 3 compares this convex response to the traditional linear numerical response. Convex numerical responses can alter predator–prey dynamics considerably, particularly when there are cycles. The form of the numerical response has also been shown to play a major role in how organisms should alter foraging effort when experiencing predation risk. Convex numerical responses favor decreased foraging when more food is available. It is surprising that so little effort has been directed toward determining numerical response shape. The extent to which the assumption of a linear numerical response dominates the theoretical literature is also surprising.

(4) *Periodic (e.g., seasonal) variation in the parameters of predator and/or prey growth functions.* For example, many organisms become inactive when it is too hot, too cold, or too dry. Mortality rates vary stochastically. Several authors in the 1980s and 1990s showed that seasonal forcing of simple predator–prey models having some cyclic tendency in the absence of forcing could produce a variety of complicated cycles or chaotic dynamics. It is not known whether these complicated dynamics occur in real systems where the organisms have life history traits specifically adapted to the prevailing form of seasonality.

(5) *Interactions may include adaptive foraging and defensive behavior by the predator, prey, or both.* Population ecology has been relatively slow to incorporate adaptive behavior into models. Switching behavior is preferential foraging on more abundant prey types, and it was first recognized as being important for predator–prey interactions by William Murdoch in the late 1960s. Switching is adaptive when different types of foraging behavior are best for catching different prey. However, even the adaptive dropping of low-quality prey because the time required to catch and process them did not make up for their low nutritional value was not incorporated into any models in population ecology until the mid-1980s, a decade after a large behavioral literature had developed on that topic. The adjustment of foraging behavior due to predation risk did not begin to be incorporated into foraging theory until the 1980s, but it is now recognized as an important determinant of the shapes of functional responses and of the impacts of predators on lower trophic levels. Some aspects of adaptive behavior may be approximated using previously proposed alternative functions. For example, if prey respond to the presence of more predators by greater use of a refuge, then, as noted under point 2, this is effectively a predator-dependent functional response (i.e., the predator's functional response decreases as its own density increases). However, adaptive defense is also likely to lead to changes in the nature of density dependent population growth in the prey, something that is not represented in the set of commonly used alternative functions for standard models. Density dependence changes because it is affected by the prey's activity and resource use, both of which are likely to decline when the prey are hiding from predators.

It is likely that each of the five factors listed above has a significant impact on population dynamics in a large fraction of predator–prey systems; for several of the factors, that fraction may be close to unity. It is not yet clear how deeply these factors will change predator–prey theory. Because each one alone greatly expands the range of potential models, theorists frequently ignore all in the name of simplicity. Future theoretical research should at least try to find out what sorts of biases this continued focus on simplicity is likely to introduce into our understanding of the natural world.

REPRESENTING POPULATION SUBDIVISION

While the above section suggested a large number of new predator–prey models, it was still confined to a framework in which populations were homogeneous and the entire interaction was defined by three functions. The other set of expansions of predator–prey theory consists of models in which each of the species (predator and prey) consists of two or more classes that differ in their demographic properties. In other words, different prey classes differ in their population dynamics and/or their vulnerability to predators, and predator classes differ in their numerical responses and/or their functional responses to one or more classes of prey. The most common subdivisions considered are those based on age, size, sex, and spatial location. More detailed models might consider such features as previous learning or genotype. Such subdivisions do not necessarily pose a major challenge to the standard framework, because the three component functions may simply be defined using numbers of individuals in each subdivision as the variables. If the system is spatially heterogeneous, consisting of a number of habitat patches connected by very low rates of migration, the dynamics within a patch may be represented by the standard model, so all that needs to be added is a description of movement rates between patches.

Unfortunately (for those who wish for simplicity), there are a number of aspects of subdivision that are not so easily merged with the standard model. Many of the properties that differ between individuals are continuous rather than discrete. Representing this continuity requires either partial differential equations or ordinary differential or difference equations having large numbers of variables. Either approach puts more limits on what can be achieved by mathematical analysis. The other problem is that these sorts of subdivisions vastly increase the number of potential models. There are an infinite variety of size/age/stage structures in different populations, and spatial heterogeneity consists of variation in many environmental variables with different effects on different ecological processes. Generally applicable results are very hard to come by. However, it is possible to use simple models with two or three classes to determine what types of population structure are most likely to alter predictions based on models with homogeneous populations. Results so far suggest that many of the predictions of models with homogeneous populations will change significantly with many types of population structure. The "paradox of enrichment," mentioned above, is a case in point. Assume there are two patches, one of which provides the prey with good protection from the predator. If prey move adaptively between patches, then enrichment of both patches often stabilizes the system, which is the opposite of the destabilization that occurs with no spatial structure. The basic reason is that the refuge, with low predation rates and stable dynamics, constitutes a larger fraction of the prey population when the two-patch system is enriched. Models of homogeneous predator and prey populations predict that when instability is possible, it is most likely to occur at low mortality rates, but that is not the case for models with simple juvenile/adult structure in the prey population. Exceptions to the paradox of enrichment have also been found for models with continuous age or spatial structure.

MORE THAN TWO SPECIES: FROM PREDATOR–PREY MODELS TO FOOD WEBS

The tendency of theorists to concentrate on the simplest cases has probably led to an overemphasis on specialist predators. Few true specialists exist in natural communities; even predators that consume only a single species of prey can attack a range of different phenotypes within that species. Even nearly ubiquitous aspects of population heterogeneity, such as greater vulnerability of older prey individuals, have been shown to have major impacts on dynamics. Similarly, most potential prey species are attacked by more than a single predator species. To be used in management or conservation, predator–prey models must be extendable to multispecies interactions. This raises the issues of how to construct such multispecies models, and of how the presence of multiple species is likely to alter the understanding that has been built from 1-predator–1-prey models.

Multispecies models grow out of consideration of prey growth and predator functional and numerical responses. As is true of simpler models, it is important to determine whether each population must be subdivided into functionally distinct classes in representing the interaction. The main impact of the presence of multiple prey species on a given prey's population growth function is the possibility of both resource competition and apparent competition between prey species. However, if prey density dependence is altered because of adaptive balancing of foraging and risk, there may be a variety of more complicated indirect effects of different prey species on each other's growth functions. The presence of other prey is likely to alter the functional response of any given predator to a given prey. Both positive and negative effects of the density of prey B on a predator's functional response to prey A are possible. The best-known impact is that handling prey B reduces the time available to search for prey A. However, the presence of B, by raising total density, may increase (or in some circumstances decrease) a predator's foraging effort. If A and B require different types of predator search behavior, then an abundance of B may reduce the predator's use of search behavior that is more likely to find A, producing switching behavior. Nutritional interactions between A and B may also alter the form of the functional responses. For example, if there are different, nutritionally essential nutrients in the two prey, an adaptive predator is expected to maximize its effort to obtain the most limiting nutrient. This can produce greater per individual predation rates as a prey becomes rarer (anti-switching), a highly destabilizing circumstance.

The discussion of nutritional interactions in the context of a functional response raises the issue of how numerical responses should be represented when several different prey species contribute different nutrients to the predator's diet. The standard model, in which the numerical response is a weighted (by net caloric value) sum of consumption rates, does not work here. One limiting case is Leibig's "law of the minimum," where only the nutrient that is consumed in the smallest amount relative to requirements influences demographic rates. The norm is likely to be less extreme, with a given food having a larger effect on demographic rates when other food types are being consumed at a high enough rate. Again, there has been little work on this important topic.

WHAT ARE THE APPLICATIONS OF PREDATOR–PREY THEORY?

Many of the other entries in this volume treat direct or indirect applications of predator–prey theory. Here is a very incomplete list of how predator–prey models have been applied in ecology. It is clear that predator–prey models (and the slightly broader category of consumer–resource models) are the building blocks of food web models. A variety of indirect effects, including exploitative and apparent competition, are driven by predation. A good deal of our current understanding of competition follows from Robert MacArthur's 40-year-old analysis of the dynamics of two predators that share a set of many prey species, all of which have logistic growth. Some evolutionary biologists regard predation as the prime driver of evolutionary change. The need to find prey and/or evade predators has had large impacts on many traits and has probably resulted in some of the progressive features in evolutionary history. Much of human exploitation of natural populations has been based on single-species growth models. However, it is likely that the majority of species to which these models are applied are either predators, prey, or both. Predator–prey theory therefore has much to tell us about when different single species models might be most appropriate, or whether any single-species model can adequately represent the dynamics of either a predator or a prey species.

SEE ALSO THE FOLLOWING ARTICLES

Adaptive Behavior and Vigilance / Disease Dynamics / Food Chains and Food Web Modules / Food Webs / Foraging Behavior / Metacommunities / Nicholson–Bailey Host Parasitoid Model / Single-Species Population Models

FURTHER READING

Abrams, P. A., and L. R. Ginzburg. 2000. The nature of predation: prey dependent, ratio dependent, or neither. *Trends in Ecology and Evolution* 15: 337–341.

Barbosa, P., and I. Castellanos, eds. 2004. *Ecology of predator–prey interactions.* New York: Oxford University Press.

Case, T. J. 2000. *An illustrated guide to theoretical ecology.* New York: Oxford University Press.

Jeschke, J. M., M. Kopp, and R. Tollrian. 2004. Consumer-food systems: why type-1 functional responses are exclusive to filter-feeders. *Biological Reviews of the Cambridge Philosophical Society* 79: 337–349.

Murdoch, W. W., C. L. Briggs, and R. M. Nisbet. 2003. *Consumer-resource dynamics.* Princeton: Princeton University Press.

Polis, G. A., and K. O. Winemiller, eds. 1996. *Food webs: integration of patterns and dynamics.* New York: Chapman and Hall.

Stephens, D. W., J. S. Brown, and R. C. Ydenberg. 2006. *Foraging: behavior and ecology.* Chicago: University of Chicago Press.

PRESENCE/ABSENCE MODELS

SEE GAP ANALYSIS AND PRESENCE/ ABSENCE MODELS

PROGRAMMING, DYNAMIC

SEE DYNAMIC PROGRAMMING

Q

QUANTITATIVE GENETICS

PAUL DAVID WILLIAMS

University of California, Davis

Quantitative genetics is the body of theory and methods used to study the inheritance of organismal traits that exhibit continuous variation. This variation is usually determined by contributions from alleles at many loci, each of small effect, as well as contributions from environmental sources. The statistical techniques of quantitative genetics have been used for decades to inform agricultural breeding practices and predict the response to artificially imposed selection. The same formalism has been utilized to develop models of quantitative trait evolution and to reconstruct the historical force of selection acting on traits. More recent advances, involving the identification of quantitative trait loci (QTL), offer the possibility of a more comprehensive understanding of the genetic architecture underlying complex traits. In turn, this knowledge can be utilized in the study of ecologically relevant traits in extant, natural populations to understand the roles they play in shaping ecological and evolutionary contexts.

DEFINITIONS AND HISTORICAL DEVELOPMENT

Classical population genetics studies traits that are conditioned by a small number of loci, each with large effects, whose genetic basis can be inferred by observing the discrete phenotypic classes resulting from individuals segregating into Mendelian ratios, as exemplified by the pea traits studied by Mendel in his seminal studies of inheritance. In contrast, many of the traits studied by researchers in fields as diverse as agriculture, medicine, and evolutionary biology, such as height, weight, milk yield, and risk of disease, are not only conditioned by alleles at many loci but by environmental effects as well, like nutritional status or history of exposure to disease risk factors. In this case, trait values of different individuals differ in degree rather than kind, and fail to sort into clearly demarked qualitative classes, like big/small or yellow/green. Instead, distinguishing between individuals requires that a quantitative measurement be made. These traits presented a conundrum with respect to understanding the laws governing their inheritance, as continuous trait distributions appeared to be at odds with the discrete distributions expected by Mendelian genetics.

The British statistician Ronald Fisher is generally credited with the formal establishment of quantitative genetics for his 1918 treatise that consolidated variation in quantitative characters within the Mendelian formalism of inheritance. Further seminal contributions to the foundations of the field were made by the geneticists J. B. S. Haldane and Sewell Wright. Fisher's work introduced the concept of partitioning variance components by analysis of variance (ANOVA) and made extensive use of the statistical concepts of regression and correlation developed by the British biometricians Francis Galton and Karl Pearson at the turn of the nineteenth century. This approach allows for a very general development of the theory in terms of statistical quantities obtained from phenotypic observations of populations, without the need to discuss more detailed and genetically descriptive concepts from population genetics, like allele frequencies.

CONCEPTS

The statistical approach to quantitative genetics is chiefly concerned with quantifying the degree to which relatives resemble one another in phenotype and predicting how much a population's mean phenotypic value will change given a particular selective regime. The purpose of the first of these tasks is to obtain an estimate of the degree to which phenotype depends on inheritance. Inheritance, in turn, depends on a quantity termed "breeding value," defined as the value of an individual's trait as measured by the mean value of the same trait in his/her offspring. Once this is achieved, the result can be used to provide the framework for meeting the second goal.

Before addressing either of these aims, a model of phenotype determination must first be developed. This involves partitioning an individual's phenotypic value for some quantitative trait into a sum of genetic and environmental contributions. An individual is thus imagined to have an intrinsically determined phenotype, P, that depends solely on its genotypic value, G, and a deviation from that intrinsic value generated by environmental sources, E, so that

$$P = G + E. \tag{1}$$

By convention, individual values for each of these components are expressed as deviations from the mean value of the population as a whole, so that the expected values of each of the terms appearing in Equation 1 are zero. The genotypic influence on phenotype determination is typically subdivided into additive (A), dominance (D), and epistatic (I) components. Additive effects are defined as those attributable to the breeding value of an individual. Dominance effects are deviations from the expected phenotype due to breeding value that arise due to interactions between alleles within a given locus. Similarly, epistatic contributions are deviations caused by between-loci interactions. Thus, phenotype can be represented as being generated by the collective additive effects of various genetic and environmental sources

$$P = A + D + I + E. \tag{2}$$

Broad and Narrow Sense Heritabilities

Since all terms on the right side of Equation 2 are uncorrelated with one another, the total phenotypic trait variance in a population is given by

$$\text{var}(P) = \text{var}(A) + \text{var}(D) + \text{var}(I) + \text{var}(E). \tag{3}$$

The relative contribution of genetic factors to the total amount of variance in a trait, defined as the broad-sense heritability, is given by $H^2 = \text{var}(G)/\text{var}(P)$. This quantity thus provides a measure of the degree to which genetics contribute to phenotype determination. From the perspective of selective breeding or predicting evolutionary change, a more important parameter is the narrow-sense heritability (hereafter simply "heritability"),

$$h^2 = \frac{\text{var}(A)}{\text{var}(P)}. \tag{4}$$

As the ratio of variance in breeding values (termed the additive genetic variance) to the total phenotypic variance, h^2 measures the component of the phenotypic value that is determined by the inheritance of parental alleles and hence quantifies the degree to which selection can change the phenotypic distribution of a population. Heritability occupies a central role in quantitative genetics due to the key role it plays in fashioning the theory with a predictive framework. Estimating this parameter (and its generalizations; see next section) is thus of fundamental importance in many applied fields utilizing quantitative genetic methods, such as those engaged in maximizing crop and livestock yields. Calculation of heritabilities involves disentangling the relative contributions of genetic and environmental components to the expression of a trait, which can be achieved through observations and measurements of the phenotypic covariance between parent and offspring (or between other relatives') traits.

Resemblance Between Relatives

Estimates of heritability can be obtained by observations of the relationship between phenotypic trait values in relatives, most commonly between offspring and one parent, offspring and midparent (the average of both parental phenotypes), half siblings, or full siblings. An implicit assumption in this development, generally supported by empirical findings, is that the joint distribution of the relatives' traits is multivariate normal. The expectation of Gaussian trait distributions for quantitative characters can also be formally derived as a consequence of the central limit theorem from probability theory. Roughly, this states that the mean value of a large number of independent random variables (here, the mean effect of alleles at many independent loci) will have an approximately normal distribution. In any case, the effect of this normality assumption is that the relationship between the two variables can be described by a linear regression of the form $P_2 = aP_1 + b$, where P_1 and P_2 are the relatives' phenotypic values, and the slope term a, is calculated as

$$a = \frac{\mathrm{cov}(P_1, P_2)}{\mathrm{var}(P_1)}. \qquad (5)$$

Since the covariance between the environmental deviations E_1 and E_2 can presumably be minimized with a sufficiently well-controlled breeding design, all phenotypic covariance is due to genetic terms. Indeed by Equation 2 and the properties of covariance, the numerator of Equation 5 can be expanded (noting the covariance terms between distinct genetic terms are zero by definition) to give $\mathrm{cov}(A_1, A_2) + \mathrm{cov}(D_1, D_2) + \mathrm{cov}(I_1, I_2)$. Epistatic interaction covariances are typically small relative to the contributions of additive genetic and dominance deviation covariances, and so can often be neglected (see Falconer and Mackay (1996) in the "Further Reading" section for a more complete discussion). By a slight re-expression of terms, the covariance term in Equation 5 can be written as

$$\mathrm{cov}(P_1, P_2) = r\,\mathrm{var}(A) + u\,\mathrm{var}(D), \qquad (6)$$

where the coefficient of relatedness, r, gives the probability that alleles drawn at random from the related pair are identical by descent, and u is the probability that genotypes drawn at random from the related pair are identical by descent.

For example, a common relationship to use for heritability calculations is between offspring (so that $P_2 = P_O$ where O indicates offspring value) and midparent, defined as the average of the parental phenotypic values, that is, $P_1 = \frac{1}{2}(P_{1m} + P_{1f})$, where m and f stand for mother and father values, respectively. In this case, we have that $r = \frac{1}{2}$ and $u = 0$, giving $\mathrm{cov}(P_1, P_O) = \mathrm{var}(A)/2$. Moreover, assuming that phenotypic variance is the same for both sexes, $\mathrm{var}(P_{1m}) = \mathrm{var}(P_{1f}) = \mathrm{var}(P_1)$, we can calculate $\mathrm{var}(P_1) = \mathrm{var}(P)/2$, so that by Equation 5 $a = h^2$, and heritability is simply given by the regression of offspring phenotypic value on the midparent value. When using other pairs of relatives to calculate the phenotypic covariance in Equation 5, some account must be made for dominance effects if the relatives can share genotypes by descent (see Table 1).

By such methods, estimates of the heritabilities of many traits have been obtained (Table 2). An interesting pattern that emerges is the natural grouping of various types of traits based on their heritabilities. In particular, if ordered with respect to increasing average heritability, life history traits (that is, traits linked to the timing of growth, reproduction, and survival) rank the lowest,

TABLE 1

Coefficients of the variance components in the calculation of the covariance between relatives

Relationship	Coefficient r (of V_A)	u (of V_D)
Morphological	1	1
Parent–offspring	½	0
Full sibling	½	¼
Half sibling	¼	0
Grandparent–offspring	¼	0
Uncle (aunt): nephew (niece)	¼	0
Great grandparent–offspring	⅛	0
First cousin	⅛	0

TABLE 2

Heritability calculations for various traits and organisms. Many of these were estimated in field populations

Trait	Organism	h^2	Reference
Morphological			
Orange spot size	Guppy	1.08	House, 1992
Bill length	Darwin's finch	0.65	Boag, 1983
Male body size	Seaweed fly	0.70	Wilcokson et al., 1995
Corolla width	Scarlet gilia	0.29	Campbell, 1996
Shell size	Land snail	0.81	Murray and Clarke, 1968
Behavioral			
Chemoreceptive response	Garter snake	0.32	Arnold, 1981
Flight duration	Milkweed bug	0.20	Caldwell and Hegmann, 1969
Pecking behavior	Japanese quail	0.75	Nol et al., 1996
Life history			
Female fecundity	Red deer	0.46	Kruuk et al., 2000
Growth rate	Meadow vole	0.54	Boonstra and Boag, 1987
Male lifespan	Collared flycatcher	0.15	Merila and Sheldon, 2000
Development time	Cricket	0.32	Simons and Roff, 1994

followed by behavioral, then morphological traits. One hypothesis to explain this pattern posits that, since life history traits are closely tied to fitness, they have historically experienced the strongest directional selection and hence have the lowest additive genetic variance. Other research, however, indicates that life history traits typically have higher additive variance than other traits. This suggests that, for traits with large fitness consequences, environmental, epistatic, and dominance variances are, on average, greater than for traits less relevant to fitness.

The Breeder's Equation

After obtaining an estimation of a trait's heritability, it can be used to make predictions about the effects of selection. In particular, the relationship

$$R = h^2 S, \tag{7}$$

known as the Breeder's equation, has been the principal tool of generations of plant and animal breeders charged with the selective improvement of animal and crop species. Here, R, the response to selection, is the difference between the mean trait value in the offspring and parental generations, and S, the selection differential, is the difference between the mean trait value after and before selection, respectively, in the parental generation. It thus relates change in the mean trait value across generations (R), to within-generation selection (S) and a measure of the across-generation transmission fidelity (h^2; Figure 1).

Selection Intensity

As long as the trait of interest is normally distributed in the parental generation and selection is by truncation (all individuals beyond a certain phenotypic value are selected), an alternative way of writing the Breeder's equation, useful for some applications, is in terms of the selection differential standardized by the standard deviation of the phenotypic distribution, $S/\sqrt{\text{var}(P)}$. This quantity is termed the selection intensity, $i(p)$, and the response to selection can be written as

$$R = h^2 i(p)\sqrt{\text{var}(P)}. \tag{8}$$

With the above assumptions of distribution normality and truncation selection, selection intensity depends only on the parameter, p, which is the proportion of individuals of the total population selected to form the parental generation (see Figs. 1 and 2).

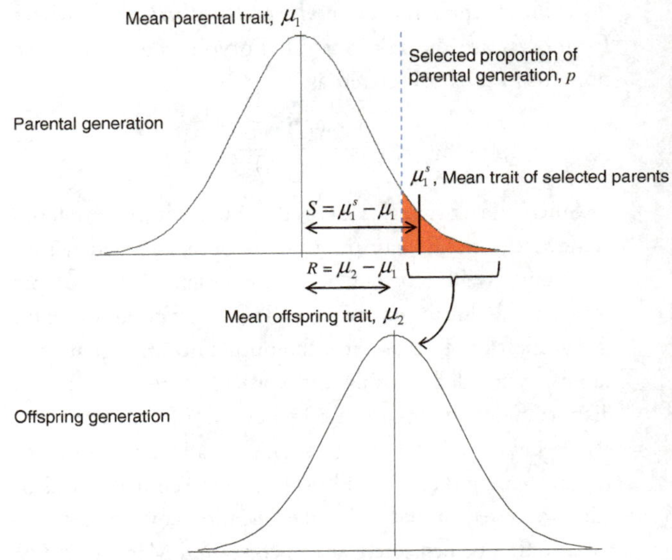

FIGURE 1 Representation of the response to selection. Trait values in the parental and offspring generations are normally distributed. A proportion, p, of individuals from the parental generation is selected as parents. The difference between the mean trait value in the selected and total population is the selection differential, $S = \mu_1^s - \mu_1$. The response to selection, $R = \mu_2 - \mu_1$, is the difference between the mean trait value in the offspring generation and the whole parental generation.

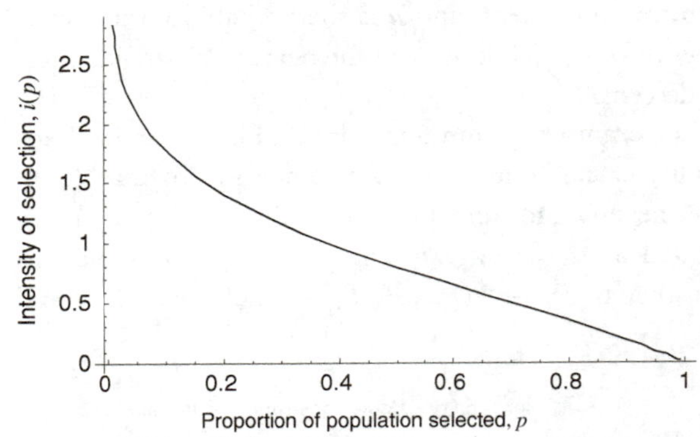

FIGURE 2 Selection intensity, (), as a function of proportion of the parental generation selected, , via truncation selection.

MULTIVARIATE QUANTITATIVE GENETICS

The relative simplicity of the single-trait Breeder's equation contrasts with the empirical reality that organisms are composed of suites of traits, all of which can potentially covary with one another. For example, bigger-than-average individuals in a population are likely to be bigger than average in a number of traits, like hands and feet. Therefore, direct selection of large-handed individuals for the parental generation also indirectly selects for large feet. If this correlation has an additive genetic basis (due, perhaps, to pleiotropic genetic effects), then an evolutionary response in foot size will result.

In light of this possibility, Russell Lande developed the mathematical machinery required to account for multivariate trait relationships. Matrix algebra provides a particularly convenient notation for extending the single-trait Breeder's equation to a multitrait setting. Letting **R** and **S** represent vectors whose entries are the response and selection differentials, respectively, for each trait under consideration, the multivariate Breeder's equation is given by

$$\mathbf{R} = \mathbf{G}\mathbf{P}^{-1}\mathbf{S}. \tag{9}$$

Here, **G** and **P** are (resp.) the additive genetic and phenotypic covariance matrices, whose diagonal entries are the additive genetic (resp. phenotypic) variances for each trait, with off-diagonal entries representing between-trait additive genetic (resp. phenotypic) covariances. In conjunction with the assumption of multivariate normality of the parent–offspring phenotypic variance–covariance distribution, the entries of **G** and **P** can be estimated by the techniques of multiple linear regression.

Despite the structural correspondence between Equation 9 and the single-trait Breeder's equation, made evident by realizing that Equation 7 can be written as $R = \text{var}(A)\text{var}(P)^{-1}S$, some important differences exist. Chief among these is the possibility for counterintuitive trait evolution in the multivariate setting. Since **G** and **P** take into account covariances between different organismal traits, evolutionary responses will generally be determined by the influence of both direct and indirect selection, which can be in opposing directions when some covariances are negative. When this tension is resolved in favor of the indirect forces, evolutionary change in the trait will proceed in the direction opposite to that implied by the direct component of selection. In contrast, since heritability (being the ratio of two variances) is always positive, phenotypic evolution is always predicted to occur in the direction dictated by direct selection on the trait of interest in the case of single-trait selection. The multitrait perspective thus demonstrates how phenotypic traits are generally not optimized with respect to direct selection but must evolve according to the constraints embodied in the additive genetic covariance relationships between traits as well. Moreover, it also shows how unexpected responses to selection on focal traits can result from hidden covariances with other, unmeasured traits.

In more recent work, Mark Kirkpatrick, Richard Gomulkiewicz, and other researchers have extended the single- and multivariate-trait formalisms to include the study of infinite-dimensional quantitative traits. Just as single and multiple quantitative traits can be represented as scalars and vectors, respectively, the objects of study in the infinite-dimensional setting are quantitative traits that vary with respect to some underlying continuous parameter. As such, they are infinite-dimensional, or function-valued, traits. For example, environmentally plastic quantitative traits, like leaf thickness, have reaction norms that describe how the trait changes as a function of temperature, elevation, or some other factor. Similarly, state-dependent traits, such as resource allocation strategies, may vary based on structuring factors like age or size. While the mathematics required for modeling the evolution of these types of quantitative traits is considerably more sophisticated than that for scalar or vector-valued traits, the expansion of the basic formalism to include function-valued traits could provide valuable tools in the study of their evolution.

EVOLUTIONARY BIOLOGY AND QUANTITATIVE GENETICS

As well as being useful tools in designing selection regimes in applied fields like agriculture and livestock breeding, the theoretical developments of quantitative genetics, like the Breeder's equation, have also been used to investigate various types of questions in evolutionary biology. One particular area where these ideas have been usefully employed is in the reconstruction of the historical action of the selective forces that have shaped some trait of interest.

As an example, under the hypothesis that a trait has been under the influence of directional selection (and hence subject to the action of truncation selection), and given an estimate of its heritability, along with measurements of the change in the trait value and the phenotypic standard deviation over evolutionary time, the single-trait Breeder's equation (Eq. 7) can be rearranged to give

$$i(p) = \frac{R}{h^2\sqrt{\text{var}(P)}}. \tag{10}$$

Solving for the right-hand side, the value of p giving this equality can be obtained analytically (Fig. 2); $1 - p$ is then the proportion of potential parents each generation that must be selectively removed from the breeding pool in order to achieve the calculated level of selection intensity.

Modeling in Evolutionary Ecology

Evolutionary ecology studies take the perspective that evolutionary history shapes ecological interactions both within and between species, while a species' ecological context generates the selective forces that determine the evolutionary course of populations. Evolutionary quantitative genetics modeling is a very active area of inquiry and an important source of theory and predictions for fields as diverse as life history theory, the evolution of behavior, sexual selection, and interspecific interactions.

The modeling framework supposes that, as usual, the quantitative trait(s) being modeled are normally distributed. A further implicit assumption is that variation-promoting processes, like mutation and recombination, have reached some dynamic balance with the variation-eroding process of selection, such that additive genetic variance/covariances remain approximately constant over evolutionary time. Since the trait(s) are assumed normal, the phenotypic distribution is completely specified by its mean and variance. By making an additional assumption that the variance remains largely unchanged, to track evolutionary change it is sufficient to follow the change in the trait mean(s) only.

Work along these lines, initiated by Russell Lande in the mid-1970's, focuses largely on a reparameterization of the Breeder's equation (Eq. 7 and Eq. 9) to express it in a way that is more applicable to evolutionary problems. Evolution is predicated on differences in the fitness of individuals with different trait values, so it is natural that this reparameterization be in the currency of fitness. Since the selection differential S (or $\mathbf{S}$ in the multivariate case) is the difference in the mean trait value before and after selection in the parental generation, it can be written as $\overline{P}_s - \overline{P} = \sum_i P_i \phi_i \frac{W_i}{\overline{W}} - \sum_i P_i \phi_i$, where P_i is the phenotypic value of an individual with the ith trait type, ϕ_i is the proportion of individuals with that trait type, W_i is their fitness and the summation is over all trait types. Since $\overline{W} = \sum_i W_i \phi_i$, it follows that $S = \mathrm{cov}(P,W)/\overline{W}$, the covariance between the phenotypic trait value and fitness divided by mean fitness.

The Breeder's equation can therefore be expressed as

$$R = \frac{\mathrm{var}(A)}{\overline{W}}\beta, \tag{11}$$

where $\beta = \mathrm{cov}(P,W)/\mathrm{var}(P)$ is the slope of the least-squares linear regression of fitness on phenotypic value. Under the assumptions that the trait distribution is Gaussian and that fitness is independent of the mean trait value, $\overline{P}$, (i.e., fitness is frequency independent), Lande showed that β can be written as the derivative of mean fitness with respect to the mean trait value, or $\beta = d\overline{W}/d\overline{P}$. As this quantity can be difficult to calculate in practice, an additional assumption of small variance in the phenotypic trait distribution is included, allowing for β to be approximated as the derivative of fitness with respect to the trait value, evaluated at the mean trait value, or

$$R = \frac{\mathrm{var}(A)}{\overline{W}} \frac{\partial W}{\partial P}\bigg|_{P=\overline{P}}. \tag{12}$$

Similarly, with selection acting on multiple traits, $P^1, P^2, \ldots,$ P^n, the multivariate version of Equation 12 is written as

$$\mathbf{R} = \frac{\mathbf{G}}{\overline{W}} \nabla W\bigg|_{P^1=\overline{P}^1, P^2=\overline{P}^2, \ldots, P^n=\overline{P}^n}, \tag{13}$$

where ∇W is the vector $= \partial W/\partial P^1, \partial W/\partial P^2, \ldots,$ $\partial W/\partial P^n$. Importantly, from an evolutionary ecology perspective, expressions 12 and 13 hold even when fitness is frequency-dependent, which allows for the modeling a broader class of evolutionary scenarios.

While this formalism has been very influential and has spawned many quantitative genetic models exploring diverse evolutionary phenomena, there are some important caveats regarding the many assumptions required to arrive at expressions 12 and 13. Among the most serious is the assumed constancy of the additive genetic variances (and covariances in the multivariate case). It has long been known that the Breeder's equation is typically only accurate for one or a few generations, as selection tends to erode additive genetic variation. This concern has been addressed somewhat with investigations of the evolution of the $\mathbf{G}$ matrix.

QUANTITATIVE TRAIT LOCI

The classical development of quantitative genetics proceeded largely in the absence of knowledge of the genetics underlying complex phenotypes, relying instead on population-level statistical descriptors. While the methods of traditional Mendelian genetics cannot be used to study the genetic basis of these phenotypes, recent attempts to identify of the genetic determinants of these phenotypes through the study of quantitative trait loci (QTLs) promises to further unify population and quantitative genetics.

QTL analysis aims to identify localized chromosomal elements, as well as their interactions, that are responsible for the determination of a quantitative trait by identifying linkages or associations between a genetic marker and a region that influences a quantitative trait. This is typically achieved by first identifying multiple strains of an organism that are genotypically distinct with respect to the trait of interest to act as parental lines. Since these strains need not be phenotypically variable, a genetic marker to distinguish between them is also required. Because of their selective neutrality, molecular markers, including single nucleotide polymorphisms (SNPs), restriction fragment length polymorphisms (RFLPs) microsatellites, and transposable element positions, they are particularly well suited for this purpose. Heterozygote offspring are then produced by crossing parental strains, which are then either backcrossed with parental strains or intercrossed with themselves. These individuals are scored for their phenotypic value of the trait of interest, as well as for the genetic markers. Markers that are genetically linked to QTLs affecting the trait will typically segregate with them, allowing for the building of a map that indicates how much of the total variation in a trait's expression is accounted for by each of these chromosomal regions.

The Gaussian phenotypic distribution demonstrated by many quantitative traits is consistent with the action of many contributing loci but, from a statistical perspective, relatively few contributing loci are required in order for approximate trait normality to result. Distinguishing between these two possibilities is therefore an empirical matter that can be investigated with QTL analysis. Early investigations of this matter seemed to indicate that the latter case typically holds. However, later studies utilizing more precise techniques offer support that the former situation is the norm. Such studies indicate that hundreds of loci can have demonstrable phenotypic effects on traits like yield (corn: 276 QTLs), weight (mouse: 137 QTLs), and bristle number (fruit fly: 130 QTLs).

Since the action of selection would presumably rapidly diminish the variation present at a small number of loci, it would be difficult to reconcile the empirically confirmed finding that many quantitative fitness traits display significant amounts of additive genetic variation with a model of genetic determination that included few loci. These kinds of observations thus offer some important insights into the factors that maintain variation in quantitative traits. Moreover, by aiding in the identification of the genetic basis of complex traits, QTL analyses also enable researchers to investigate questions of fine-grained genetic architecture, including how alleles at multiple loci interact to influence quantitative trait variation.

ECOLOGICAL GENETICS

The application of quantitative genetic tools to study questions regarding the evolution of fitness-determining traits that influence the distribution and abundance of organisms is broadly defined as falling within the realm of ecological genetics. The field is generally held to have been founded by the British geneticist E. B. Ford, for his pioneering work on genetic polymorphisms in natural populations and culminating in the publication of his 1964 classic treatise "Ecological Genetics." The work of Theodosius Dobzhansky on chromosome polymorphisms in *Drosophila* species, which helped to confirm the importance of the action of selection in shaping natural populations, also played an important role in the establishment of the discipline.

Early examples of ecological genetic research primarily focused on single populations and traits with simple underlying genetics. A classic example is the work of Bernard Kettlewell (a student of Ford's), who experimentally demonstrated the importance of predation as the selective force favoring the evolution of industrial melanism in the peppered moth, *Biston betularia*. Modern ecological genetic studies are often concerned with evolutionary issues regarding traits with complex genetics. Research in this area includes important applied questions in such fields as evolutionary epidemiological studies of the evolution and spread of pathogen resistance to antibiotics and conservation biology investigations of species invasions and extinctions.

As fitness traits are typically quantitative, modern quantitative genetic techniques, like QTL methods, have become indispensible instruments in the ecological geneticist's toolbox. For example, QTL studies have been employed to investigate the process of population divergence by elucidating the genetics of male calling song in multiple species of Hawaiian cricket, study loci relevant to the reproductive success of invasive species, and quantify gene flow between wild and domestic crops. QTL analysis is also heavily used in the closely allied disciplines of functional and population genomics, the former being concerned with the determination of the fitness and whole organism functional consequences of allelic variation, the latter with the identification of loci controlling ecologically relevant traits. Information gleaned from such studies promises to greatly enable and augment ecological genetic studies.

SEE ALSO THE FOLLOWING ARTICLES

Adaptive Dynamics / Evolutionary Computation / Frequentist Statistics / Mutation, Selection, and Genetic Drift / Niche Construction / Phenotypic Plasticity

FURTHER READING

Abrams, P. A. 2001. Modelling the adaptive dynamics of traits involved in inter- and intraspecific interactions: an assessment of three methods. *Ecology Letters* 4: 166–175.

Conner, J. K., and D. A. Hartl. 2004. *A primer of ecological genetics*. Sunderland, MA: Sinauer Associates.

Falconer, D. S., and T. F. C. Mackay. 1996. *Introduction to quantitative genetics,* 4th ed. Harlow, Essex, UK: Addison Wesley Longman.

Gillespie, J. H. 2004. *Population genetics: a concise guide,* 2nd ed. Baltimore: Johns Hopkins University Press.

Hartl, D. L., and A. G. Clark. 1997. *Principles of population genetics,* 3rd ed. Sunderland, MA: Sinauer Associates.

Lande, R. 1976. The maintenance of genetic variability by mutation in a polygenic character with linked loci. *Genetical Research* 26: 221–235.

Lande, R. 1979. Quantitative-genetic analysis of multivariate evolution applied to brain: body size allometry. *Evolution* 33: 402–416.

Lynch, M., and B. Walsh. 1998. *Genetics and analysis of quantitative traits.* Sunderland, MA: Sinauer Associates.

Rice, S. H. 2004. *Evolutionary theory: mathematical and conceptual foundations.* Sunderland, MA: Sinauer Associates.

R

REACTION–DIFFUSION MODELS

CHRIS COSNER

University of Miami, Coral Gables, Florida

Reaction–diffusion models are spatially explicit models for the dispersal, population dynamics, and interactions of organisms. They are deterministic and treat time and space as continuous variables, as opposed to integro-difference models, discrete diffusion models, interacting particle systems, cellular automata, and metapopulation models. They use partial differential equations or systems of such equations to describe how population densities vary in time and space. Reaction–diffusion models are used to understand various spatial phenomena in ecology. Specifically, they are used to characterize the minimal size that habitat patches must have to support populations, the speed of biological invasions, and the spontaneous formation of patterns in homogeneous environments.

THE MODELS

Diffusion

Physical diffusion refers to the random movement of small particles dissolved or suspended in a fluid that is caused by collisions with molecules of the fluid. Various other movement processes that arise from many small random steps can be described mathematically in the same terms as physical diffusion. In ecology, diffusion models for the dispersal of organisms can arise from physical processes similar to diffusion, such as small-scale turbulence in aquatic environments, or from suitable scaling of the results of individual movements described in terms of unbiased, uncorrelated random walks. As a simple example, let x denote location in a single space dimension and let t denote time. Suppose that a population consists of a large number of individuals and that at each time step Δt each individual moves a distance Δx with probability ½ of going to the left and probability ½ of going to the right. In the limit as Δx and Δt approach zero, with the diffusive scaling $(\Delta x)^2/2\Delta t = D$, the expected density $u(x, t)$ of the population will satisfy the equation

$$\frac{\partial u}{\partial t} = D\frac{\partial^2 u}{\partial x^2}.$$

This is the one-dimensional case of the diffusion equation, which is mathematically identical to the heat equation. If a population of u_0 individuals is initially concentrated at a point $x = x_0$, then the diffusion equation predicts that at any time $t_0 > 0$ the density of the population will be given by $u_0 N(x, x_0, 2Dt_0)$, where $N(x, x_0, 2Dt_0)$ is the normal (i.e., Gaussian) distribution with mean x_0 and variance $2Dt_0$. In two and three dimensions, the diffusion equation is given by

$$\frac{\partial u}{\partial t} = D\left(\frac{\partial^2 u}{\partial x^2} + \frac{\partial^2 u}{dy^2}\right)$$

$$\text{and}\quad \frac{\partial u}{\partial t} = D\left(\frac{\partial^2 u}{\partial x^2} + \frac{\partial^2 u}{\partial y^2} + \frac{\partial^2 u}{\partial z^2}\right),$$

respectively. Diffusion models can be derived as limits of various random walks and related processes, but the diffusive scaling is crucial. For diffusion in n dimensions with diffusion coefficient D, the mean squared distance traveled by an individual in unit time is equal to $2Dn$. Thus, D can be measured empirically by using mark–recapture experiments. In the absence of other effects, diffusion tends to reduce density at high points and increase it at low points, causing the spatial distribution of the population to become more uniform.

If space is viewed as a lattice of points rather than as a continuum, and individuals are assumed to move to or from neighboring points at a rate proportional to the difference in densities at those points, then the way that dispersal affects the population densities at lattice points is described by a discrete diffusion equation. Suppose that the lattice is the set of integer points on a line and that $u_i(t)$ is the density at point i and time t. A discrete diffusion model would take the form

$$\frac{\partial u_i}{\partial t} = D(u_{i+1} - 2u_1 + u_{i+1}) \text{ for each integer } i.$$

Discrete diffusion models are generally analogous to diffusion models except they treat space as discrete. They have many of the same properties and support many of the same phenomena as reaction–diffusion models. They are distinct from patch occupancy models, which also treat space as discrete, but which describe the probabilities that patches are occupied rather than explicitly describing population densities. If space is viewed as continuous but time is viewed as discrete, integrodifference models can be used to describe dispersal. These models also have many features in common with reaction–diffusion models, but they can incorporate more general types of dispersal.

If a diffusing population inhabits a region with boundaries, then to uniquely determine the future population density from its initial density requires a specification of what individuals do when they reach a boundary. Mathematically, such specifications of behavior at boundaries are called boundary conditions. Suppose that a diffusing population with density u inhabits a region Ω, and let $\partial\Omega$ denote the boundary of Ω. If all individuals that reach the boundary $\partial\Omega$ leave Ω and do not return, the boundary condition is that $u = 0$ on $\partial\Omega$. If all individuals that reach the boundary return to Ω, then the boundary condition is that the derivative of u in the direction perpendicular to the boundary is zero on $\partial\Omega$. This is typically written as $\partial u/\partial n = 0$ on $\partial\Omega$, where n represents the outward-pointing unit normal vector on $\partial\Omega$. If some individuals reaching $\partial\Omega$ leave but others return, the boundary condition is $\partial u/\partial n + \beta u = 0$ on $\partial\Omega$ for some $\beta > 0$.

The diffusion equation can be generalized to include directed motion such as advection or taxis. Advection is described by terms involving the first derivatives of density with respect to the spatial variables. In one space dimension, a model with diffusion and constant advection to the right with advection speed A would take the form

$$\frac{\partial u}{\partial t} = D\frac{\partial^2 u}{\partial x^2} - A\frac{\partial u}{\partial x}.$$

The diffusion rate and the coefficients of advection terms can be allowed to depend on time, location, abiotic factors, or densities of conspecifics or other populations. However, in all of the more general dispersal processes that can be naturally incorporated into diffusion equations, the movements of individuals are local in the sense that over any short period of time an individual moves a short distance, as opposed to moving by discrete long-distance jumps. If the diffusion rate depends on location then in a single space dimension the diffusion equation should be written as either

$$\frac{\partial u}{\partial t} = \frac{\partial}{\partial x}\left(D(x)\frac{\partial u}{\partial x}\right) \text{ or as } \frac{\partial u}{\partial t} = \frac{\partial^2}{\partial x^2}(D(x)u).$$

Roughly speaking, the first form would arise as a limit of a random walk where the probability of an individual moving from x to $x + \Delta x$ in a given time step does not depend on conditions at the departure or arrival point, so that it is approximately constant at the scale of Δx. That would be the case for physical diffusion in a heterogeneous medium. The second form would arise if the probability of an individual moving depends on conditions at the departure point and might vary significantly at the scale of Δx. That could be the case for organisms that adjust their movement rate on the basis of fine-scale environmental variations. The reason why those forms should be used rather than the form

$$\frac{\partial u}{\partial t} = D(x)\frac{\partial^2 u}{\partial x^2}$$

is because that form may give the unrealistic prediction that the total population (that is, the integral of the density u over the spatial domain Ω) will change due to dispersal even if there are no population dynamics and no individuals enter or leave Ω.

Another way of deriving and interpreting equations involving advection and diffusion is based on the theory of stochastic processes. A stochastic process is a process that is described by a random variable that depends on time. Stochastic processes include random walks in space, but they also arise in many other contexts, including demographic models where births and death occur at random according to some distribution that may depend on time and/or population density. In the context of models for the dispersal of organisms, the random variable would typically represent the location $x(t)$ of an individual at a given time t. A fundamental type of stochastic process in continuous time is the Wiener process, where the random variable $W(t)$ has the following properties: $W(0) = 0$, the increment $W(t) - W(s)$ is normally distributed with mean 0 and variance $t - s$ for any s and t with $t > s$, and increments for nonoverlapping intervals are independent. The Wiener process arises as a description of Brownian motion. Since the increments of $W(t)$ are normally distributed, the probability density function for $W(t)$ satisfies a diffusion equation. More general stochastic processes can be specified by a stochastic differential equation, sometimes called the Langevin equation, for the random variable $x(t)$:

$$dx(t) = a(x, t)\, dt + b(x, t) dW(t).$$

In the Langevin equation, $a(x, t)$ represents any deterministic aspects of the process, such as advection in the case of spatial models, while $b(x, t)$ represents random aspects such as diffusion. Mathematically, $a(x, t)$ represents the expected rate of change in $x(t)$ and $b(x, t)^2$ represents the variance. The probability density function for the random variable $x(t)$ will satisfy an advection–diffusion equation known as a forward Kolmogorov or Fokker–Planck equation. To be precise, if $p((x, t)|(y, s))$ is the density function for the conditional probability of $x(t)$ given that $x(s) = y$ for some $s < t$, then p satisfies the advection—diffusion equation

$$\frac{\partial p}{\partial t} = -\frac{\partial}{\partial x}[a(x, t)p] + \frac{1}{2}\frac{\partial^2}{\partial x^2}[b(x, t)^2 p].$$

Advection–diffusion equations can also approximate stochastic processes in discrete time. One approach to making such approximations is to describe the process by using a discrete time version of the Langevin equation:

$$\Delta X = a(x, t) + b(x, t)\varepsilon(t),$$

where t takes integer values, $\varepsilon(t)$ is a random variable with mean 0 and variance 1, and $\varepsilon(t)$, $\varepsilon(s)$ are uncorrelated for $s \neq t$. That equation is approximated by the same advection–diffusion as shown above for the corresponding continuous time process. Interpreting advection–diffusion equations in terms of stochastic processes, and vice versa, can facilitate the calculation of various quantities of interest.

Reaction and Diffusion

In reaction–diffusion models in ecology, "reaction" generally refers to terms describing population dynamics or species interactions. These are usually taken from standard population models, such as the logistic equation for a single population or Lotka–Volterra, Rosenzweig–MacArthur, or other models for interacting populations. In the reaction–diffusion setting, the coefficients in the models for population dynamics and species interactions may depend on location or time. Reaction–diffusion models incorporate the assumptions that continuous time models are suitable for the underlying population dynamics and that diffusive dispersal and population dynamics occur on roughly comparable timescales. To completely specify a reaction–diffusion model requires specifying the diffusion coefficient and reaction terms for each population, the underlying spatial region, the initial densities of the populations in the model, and boundary conditions if the spatial region has a boundary. The analysis of reaction–diffusion models is similar in many respects to the analysis of systems of ordinary differential equations. It typically involves methods from dynamical systems theory such as bifurcation analysis combined with methods from the classical theory of ordinary and partial differential equations.

A typical prototype for reaction–diffusion models is the diffusive logistic equation. A version of that equation was introduced in the context of population genetics by R. A. Fisher in the 1930s. It was introduced in theoretical ecology by J. G. Skellam in the 1950s. In the standard notation where r represents the local per capita population growth rate without crowding effects and K represents the carrying capacity, the full model in two space dimensions for the population density $u(x,y,t)$ would take the form

$$\frac{\partial u}{\partial t} = D\left(\frac{\partial^2 u}{\partial x^2} + \frac{\partial^2 u}{\partial y^2}\right) + r\left(1 - \frac{u}{K}\right)u \text{ for } (x, y)$$

in Ω and $t > 0$,

with an initial condition

$$u(x, y, 0) = u_0(x, y) \text{ for } (x, y) \text{ in } \Omega,$$

and if Ω has a boundary $\partial\Omega$, a boundary condition such as $u(x, y, t) = 0$ or $\partial u/\partial n + \beta u = 0$ for (x, y) in $\partial\Omega$ and $t > 0$.

This type of model can be extended by allowing spatial heterogeneity in the diffusion rate and population dynamics, but some care must be taken with the way the equation is expressed. In particular, if r can change sign then the usual form of the logistic equation is not correct, since when r is negative it would predict local population growth if the population exceeds the local carrying capacity, which is not a reasonable prediction. An appropriate form is

$$\frac{\partial u}{\partial t} = \frac{\partial}{\partial x}\left(D_1(x, y, t)\,\frac{\partial u}{\partial x}\right) + \frac{\partial}{\partial y}\left(D_2(x, y, t)\,\frac{\partial u}{\partial y}\right) + (a(x, y, t) - b(x, y, t)u)u$$

where D_1, D_2, and b must be positive but a could change sign. Other forms of population dynamics are also possible; for example, population growth terms incorporating an Allee effect could be used instead of logistic growth.

The Lotka–Volterra competition model with diffusion is a representative example of a reaction–diffusion model for interacting populations. If u and v are the densities of the two competing populations, the model (again, in two space dimensions) takes the form

$$\frac{\partial u}{\partial t} = D\left(\frac{\partial^2 u}{\partial x^2} + \frac{\partial^2 u}{\partial y^2}\right) + (a - bu - cv)u,$$

$$\frac{\partial v}{\partial t} = d\left(\frac{\partial^2 v}{\partial x^2} + \frac{\partial^2 v}{\partial y^2}\right) + (f - gu - hv)v,$$

where D, d, a, b, c, f, g, and h are positive constants. As before, D and d represent diffusion rates, a and f are local per capita population growth rates, and b and h describe logistic self-limitation. The coefficients c and g describe the strength of competition. Again, the equations must be supplemented with initial and boundary conditions.

POPULATION DYNAMICS, DIFFUSION, AND CRITICAL PATCH SIZE

If a population inhabits a patch containing both source and sink habitats, so that the intrinsic population growth rate is positive in some places and negative in others, then population models without dispersal would generally predict that the population would persist in source habitats where the growth rate is positive and become extinct in sink habitats where the growth rate is negative. As long as the population growth rate is positive somewhere, a sedentary population typically will be predicted to persist but it will be restricted to source habitats. If the organisms in the population diffuse, then some of them will move out of source habitats, because diffusive movement is random. For some boundary conditions, some individuals will also leave the patch and not return. These effects can cause extinction if the diffusion rate is too high or the environment is too small. The observation that reaction–diffusion models can predict a minimum patch size needed for population persistence was made independently by J. G. Skellam and by H. Kierstead and L. B. Slobodkin in the early 1950s.

Reaction–diffusion models provide criteria for the persistence of populations in bounded habitat patches by synthesizing the effects of spatial heterogeneity, boundary conditions, dispersal, and population growth. In the case where the local rate of population growth or decline is density independent and the coefficients of the model are constant or periodic in time, the effective per capita growth rate of the population is characterized by a single quantity called the principal eigenvalue of the differential operator describing diffusion and the density-independent aspects of local population dynamics. This eigenvalue plays a role that is analogous to that of the principal eigenvalue of the population projection matrix in matrix models for age or stage structured populations. In the one-dimensional model

$$\frac{\partial u}{\partial t} = \frac{\partial}{\partial x}\left(D(x)\,\frac{\partial u}{\partial x}\right) + r(x)u \text{ for } 0 < x < L,\ t > 0,$$

$$\text{with } u(0, t) = u(L, t) = 0,$$

the boundary of the interval $(0, L)$ acts as a sink. If there are regions where $r < 0$, those act a sinks as well. Regions where $r > 0$ act as sources. The eigenvalue problem associated with the model is given by

$$\frac{\partial}{\partial x}\left(D(x)\,\frac{\partial \phi}{\partial x}\right) + r(x)\phi = \sigma\phi$$

for $0 < x < L$, with $\phi(0) = \phi(L) = 0$.

The eigenvalues are the values of σ for which there is a nonzero solution $\phi(x)$. The principal eigenvalue σ_1 is the largest of those eigenvalues. If $\sigma_1 > 0$, the population grows, but if $\sigma_1 < 0$, it declines toward extinction. In the case that D and r are constants, then the principal eigenvalue for this problem can be computed to be $\sigma_1 = r - D\pi/L^2$. Thus, for a prediction of population growth, it is necessary that $L > \sqrt{D\pi/r}$, which gives the minimum patch size needed to support a population. Alternatively, the condition can be viewed as an upper bound on the diffusion rate or a lower bound on the population growth rate needed for persistence. Related results hold for more general models. For regions in two or three space dimensions, the principal eigenvalue of the analogous model is sensitive to the amount of core area in regions of favorable habitat, but it is not as sensitive to the perimeter/area or surface area/volume ratio. If r and D are positive constants so that the only sink in the model is the boundary, then the principal eigenvalue for a long narrow rectangle will be smaller than the one for a square of the same area. That follows from the fact that the square has more core area because a larger fraction of its area is far from the boundary. On the other hand, an ordinary square and a square where the boundary is very irregular at a small scale will typically produce similar principal eigenvalues, even though the latter may have a much larger perimeter/area ratio than the former, because they will have core areas of similar sizes.

In density-dependent models, the stability or instability of equilibria can often be analyzed in terms of the principal eigenvalues of the corresponding linearized model. For example, for the diffusive logistic equation

$$\frac{\partial u}{\partial t} = \frac{\partial}{\partial x}\left(D(x)\,\frac{\partial u}{\partial x}\right) + r(x)u - b(x)u^2$$

for $0 < x < L$, $t > 0$, with $u(0, t) = u(L, t) = 0$, the linearization at the equilibrium $u = 0$ is precisely the density-independent (i.e., linear) model shown in the previous paragraph. If the principal eigenvalue σ_1 for that linear model is positive, then the equilibrium $u = 0$ is unstable. In that case, the logistic model has a unique positive equilibrium that is globally stable relative to positive solutions. If σ_1 is negative, then $u = 0$ is stable and all positive solutions of the logistic model approach zero as t approaches infinity. More generally, the dynamics of a reaction–diffusion system on a bounded spatial region often can be studied by using stability analysis based on principal eigenvalues together with bifurcation theory, persistence theory, and other ideas from the theory of dynamical systems.

TRAVELING WAVES AND SPREADING SPEEDS

Reaction–diffusion models on unbounded spatial domains can support traveling waves, that is, solutions that move by translation in space but maintain a fixed shape. The existence of traveling waves in reaction–diffusion models was first noted by R. A. Fisher in the 1930s for a model in population genetics that is mathematically equivalent to a diffusive logistic equation. In the context of ecology, traveling waves describe how a species would spread throughout a region where it was not previously present, so they are used to model biological invasions. Reaction–diffusion models often predict that a population initially inhabiting a bounded region of an unbounded domain will have a fixed rate of spread even if it does not form and maintain a traveling wave with fixed shape. For a population to have a fixed rate of spread c^* means that an observer moving faster than c^* would outrun the spreading population but the population would outrun an observer moving slower than c^*.

Mathematically, the population density corresponding to a traveling wave is described by $u(x, t) = U(x \pm ct)$ where c is the speed of propagation and the function U describes the shape of the wave. Generally $U(z)$ approaches equilibria for the population dynamics of the model as z approaches $\pm\infty$. For the logistic equation

$$\frac{\partial u}{\partial t} = D\frac{\partial^2 u}{\partial x^2} + r\left(1 - \frac{u}{K}\right)u \text{ for } -\infty < x < \infty,$$

there are traveling waves of all speeds c greater than or equal to $2\sqrt{Dr}$. Those waves always move so that as time passes the population density increases. For a rightward moving wave $U(x - ct)$, the function $U(z)$ satisfies $U(z) \to K$ as $z \to -\infty$ and $U(z) \to 0$ as $z \to \infty$. The spread rate c^* is $2\sqrt{Dr}$. The case where the logistic nonlinearity is replaced with any smooth function $f(u)$ with $f(0) = f(K) = 0$, $f(u) > 0$ for $0 < u < K$, $f'(0) > 0$, $f'(K) < 0$, and $f(u)/u$ decreasing is similar, with $c^* = 2\sqrt{Df'(0)}$. The case where the population dynamics have an Allee effect (that is, $f(u)/u$ is increasing for some values of u) is different. The spread rate and minimum wave speed are generally not determined by $f'(0)$. In the case of a strong Allee effect, where $f(0) = f(K) = 0, f(a) = 0$ for some a with $0 < a < K, f'(0) < 0, f(u) < 0$ for $0 < u < a, f'(a) > 0$, $f(0) > 0$ for $a < u <$ K, and $f'(K) < 0$, there is a unique speed for traveling waves, but it is not given by $f'(0)$. The waves will move so that population density at each point increases toward K as time passes if the integral of $f(u)$ from 0 to K is positive but will move so that the density decreases toward 0 as time passes if the integral of $f(u)$ from 0 to K is negative. If the integral of $f(u)$ from 0 to K is zero, then the model will support standing waves. Traveling waves can occur in reaction–diffusion systems describing interacting populations as well as single-species models. Discrete diffusion models and integrodifference models also can support traveling waves.

PATTERN FORMATION

In a single equation, diffusion is generally a stabilizing mechanism, but in systems of reaction–diffusion equations diffusion can sometimes cause spatially constant equilibria to become unstable so that spatial patterns form. This was observed by A. M. Turing in the 1950s in his studies of the mechanisms of morphogenesis. In ecology, reaction–diffusion systems have been proposed as models for the formation of spatial patterns in plankton density, the distribution of vegetation in water-limited systems, and the distributions of predators and their prey or parasitoids and their hosts. The most commonly studied scenario that can lead to pattern formation occurs in a system of two interacting species (chemical, biological, or other) where one species functions as an activator in the sense of increasing the rate of production of the other species, the second species acts as an inhibitor on the first, and the inhibitor diffuses more rapidly that the activator. In the ecological context, the activator is often a resource and the inhibitor is often a consumer.

The mechanism for pattern formation proposed by Turing can occur in linear reaction–diffusion systems of the sort that arise as linearizations of reaction–diffusion models with two or more components. Suppose that a system is linearized at a positive spatially constant equilibrium and that the resulting linear system with the equilibrium shifted to (0, 0) is

$$\frac{\partial u}{\partial t} = \frac{\partial^2 u}{\partial x^2} + au + bv,$$

$$\frac{\partial v}{\partial t} = D\frac{\partial^2 v}{\partial x^2} + cu + dv,$$

for $0 < x < L$ with reflecting boundary conditions $\partial u/\partial x = \partial v/\partial x = 0$ for $x = 0, L$. (The reflecting boundary conditions are chosen so that the original system can have positive spatially constant equilibria.) The equilibrium (0, 0) in the corresponding nondiffusive system

$$\frac{\partial u}{\partial t} = au + bv,$$

$$\frac{\partial v}{\partial t} = cu + dv$$

is stable if $a + d < 0$ and $ad - bc > 0$. Solutions of the ordinary differential equations are also spatially constant solutions of the reaction–diffusion system. In particular, (0, 0) is still an equilibrium for the reaction–diffusion system. To understand diffusion-induced instabilities in this system, note that if $w(x) = \cos(n\pi x/L)$ then $d^2w/dx^2 = -\lambda w$ where $\lambda = n^2\pi^2/L^2$, and w satisfies the boundary conditions $\partial w/\partial x = 0$ for $x = 0, L$, so that $(u, v) = (U(t)w(x), V(t)w(x))$ is a solution of the reaction–diffusion system provided U and V satisfy

$$\frac{dU}{dt} = (a - \lambda)U + bV,$$

$$\frac{dV}{dt} = cU + (d - D\lambda)V.$$

If (0, 0) is unstable in this new system of ordinary differential equations, then the reaction–diffusion system will have spatially varying solutions $(U(t)w(x), V(t)w(x))$ that increase in magnitude over time, forming a pattern. For this new system, (0, 0) is unstable if $(a - \lambda)(d - D\lambda) - bc < 0$, which will be possible for some values of λ if the quadratic $D\lambda^2 - (Da + d)\lambda + ad - bc = 0$ has positive real roots, which in turn will be true if $Da + d > 0$. If $Da + d > 0$, then since λ can be any positive number (depending on L and n), there will be instability leading to pattern formation for some values of L and n. The size of the spatial domain affects the possible values of λ and hence affects the existence and nature of emerging patterns. Since $a + d < 0$ is needed for stability of (0, 0) in the nondiffusive system, $Da + d > 0$ is possible only if a and d have opposite signs and $D \neq 1$. Since $ad - bc > 0$ is also necessary for (0, 0) to be stable in the nondiffusive model, that forces b and c to have opposite signs as well. Thus, one component must act as an activator and one must act as an inhibitor for the Turing mechanism to induce a diffusion-driven instability. One possible case would be to have $a > 0, b < 0, c > 0, d < 0$ so that u represents an activator ($c > 0$) and v an inhibitor ($b < 0$). That situation could occur in a predator–prey model where u represents the prey and v represents the predator. Pattern formation would then require the predator to disperse more rapidly than the prey ($D > 1$). In this case, the prey would have an Allee effect ($a > 0$) and the predator would have some sort of self-limitation ($d < 0$). Other cases are possible. Turing instabilities are not the only possible source of pattern formation in reaction–diffusion systems. Pattern formation also can occur in discrete diffusion models and cellular automata. The general question of pattern formation is a topic of current research in theoretical ecology and applied mathematics.

SEE ALSO THE FOLLOWING ARTICLES

Integrodifference Equations / Invasion Biology / Movement: From Individuals to Populations / Partial Differential Equations / Spatial Spread

FURTHER READING

Cantrell, R. S., and C. Cosner. 2003. *Spatial ecology via reaction-diffusion equations.* Chichester, UK: Wiley.

Cosner, C. 2008. Reaction–diffusion equations and ecological modeling. In A. Friedman, ed. *Tutorials in mathematical biosciences IV.* Berlin: Springer.

Grindrod, P. 1996. *The theory and applications of reaction–diffusion equations: patterns and waves,* 2nd ed. Oxford: Oxford University Press.

Lande, R., S. Engen, and B.-E. Sæther. 2003. *Stochastic population dynamics in ecology and conservation.* Oxford: Oxford University Press.

Murray, J. D. 2003. *Mathematical biology II: spatial models and biomedical applications.* Berlin: Springer.

Okubo, A., and S. A. Levin. 2001. *Diffusion and ecological problems,* 2nd ed. Berlin: Springer.

Shigesada, N., and K. Kawasaki. 1997. *Biological invasions: theory and practice.* Oxford: Oxford University Press.

Turchin, P. 1998. *Quantitative analysis of movement.* Sunderland, MA: Sinauer.

REGIME SHIFTS

REINETTE BIGGS, THORSTEN BLENCKNER, CARL FOLKE, LINE GORDON, ALBERT NORSTRÖM, MAGNUS NYSTRÖM, AND GARRY PETERSON

Stockholm Resilience Centre, Stockholm University, Sweden

Regime shifts are large, abrupt, persistent changes in the structure and function of ecosystems. Regime shifts have been empirically documented in a variety of terrestrial and aquatic systems and studied in mathematical models. Understanding of regime shifts is important for ecosystem management, as such shifts may have substantial impacts on human economies and societies and are often difficult to anticipate and costly to reverse.

OVERVIEW

Sudden, large, long-lasting shifts in ecosystem structure and function have been documented in a variety of terrestrial and aquatic ecosystems. For example, diverse coral reefs that have existed for hundreds of years may become overgrown by fleshy algae within the space of a few years and then remain algae-dominated for decades (Box 1). Another more recent example is the shift in the ecosystem dynamics in the North Pacific around 1989 probably caused by climate (Fig. 1A). Similar shifts have been documented at large spatial and temporal scales. For instance, about 5500 years ago the Sahara region in Africa abruptly shifted from a moist, vegetated region to the desert state we know today (Fig. 1B).

Empirical observations of regime shifts are supported by dynamical systems theory, a branch of mathematics that studies the behavior of complex systems. Mathematical models show that many complex systems tend to

BOX 1. REGIME SHIFTS IN CORAL REEF ECOSYSTEMS

Theoretical, experimental modeling and observational work on coral reefs has provided some of the best insights into the dynamics of regime shifts in human-influenced ecosystems. A considerable proportion of the world's coral reefs are rapidly losing live coral cover. This is often associated with reefs undergoing regime shifts where they become dominated by macroalgae, sponges, soft corals, sea anemones, and sea urchins. The archetypical example of a coral reef regime shift, where the system shifts from coral dominance (A) to macroalgal dominance (B), is almost always preceded by the loss of macroalgal consumers (herbivorous fish or sea urchins), due to overexploitation and disease. These changes precipitate the large macroalgal blooms that occur when a proximal trigger, such as mass bleaching and severe hurricanes, kills off the living coral and opens up space for algal colonization.

Coral regime shifts can be problematic for managers and resource users as the new, undesirable states can be "locked" in place by strong feedbacks. For example, as alternate benthic dominants colonize and proliferate, the frequency of their interactions with the remaining coral colonies increases, affecting coral growth rates, impeding coral settlement, and hampering recovery of coral tissues. Bolstering the resilience of coral reefs should involve breaking such ecological feedbacks as they start to emerge, by promoting herbivore abundances and improving water quality. It will also require creating governance structures that support the empowerment of reef resource users as stewards of reef resilience.

B. Christensen/Azote

J. Lokrantz/Azote

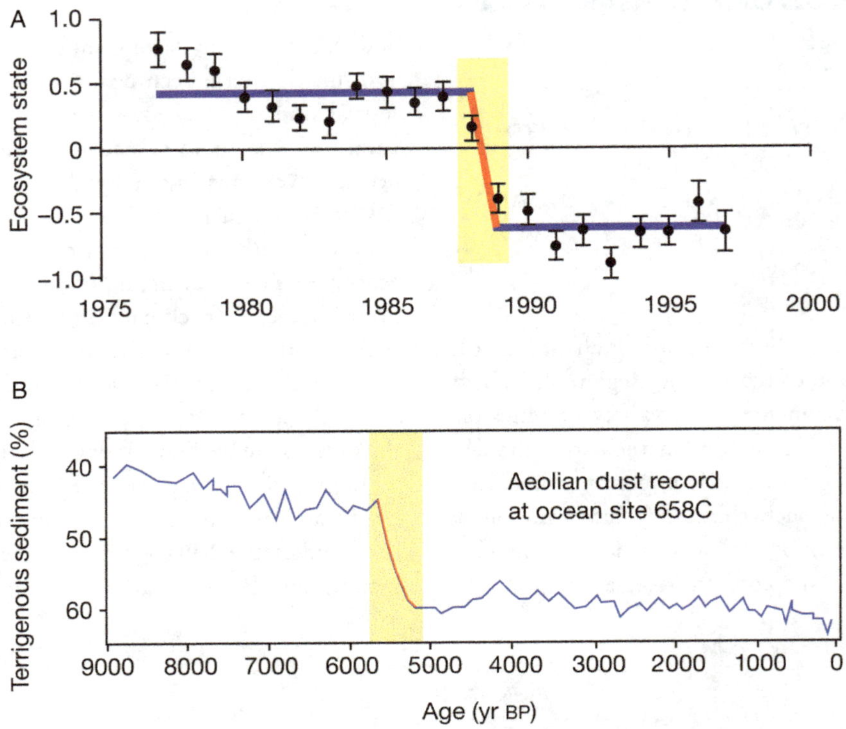

FIGURE 1 Regime shifts are characterized by large, abrupt, persistent changes in ecosystem structure and dynamics. For example, an abrupt shift in climatic and biological conditions was observed in the Pacific Ocean in 1989 (A). Regime shifts are also observed on much longer time scales; for example, an abrupt change in the climate and vegetation of the Sahara occurred about 5500 years ago, reflected in the contribution of land-based dust in the oceanic sediment (B). Figures from Scheffer et al., 2001.

organize and fluctuate around several equilibrium points or attractors. The zone of fluctuation around a particular attractor is referred to as the domain of attraction for that point. Different domains of attraction represent potential regimes in which a system may find itself at a particular point in time, and can be mathematically represented by a stability landscape. In terms of this theory, a regime shift entails the shift of a system from one basin of attraction to another when a critical threshold or tipping point is exceeded (Fig. 2).

Importantly, both theory and observation emphasize that a particular regime (e.g., a coral-dominated reef) is not a stable, unchanging condition. Rather, a particular regime is characterized by dynamic fluctuations of the ecosystem around a specific attractor. A regime shift therefore does not simply involve a large change in an ecosystem (which could occur even if the system remains in the same regime), but rather an abrupt, persistent shift in the structure and dynamics of the system. In most cases, the change in structure and function involves a change in the internal feedbacks of the system. As illustrated in Figure 1, "abrupt" and "persistent" are relative terms indicating that the time period over which the regime shift occurs is much shorter than the duration of each regime.

Regime shifts are an active area of ecological research because they often have large impacts on human economies and societies. For example, the collapse of the Newfoundland cod fishery in Canada in the early 1990s

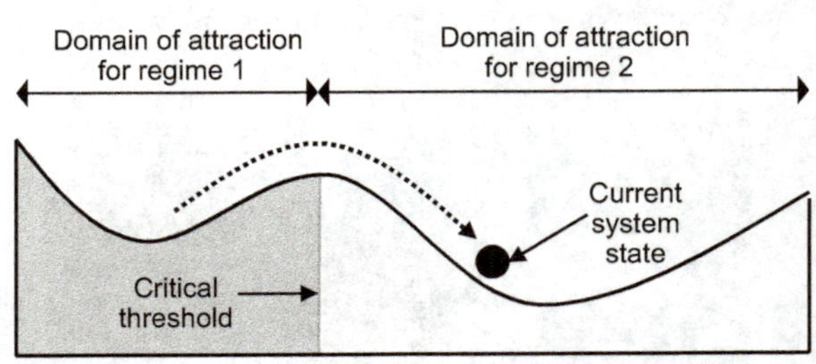

FIGURE 2 Different regimes can be mathematically represented by a stability landscape with different domains of attraction (valleys). A regime shift entails a shift in the current system state (represented as a ball) from one domain of attraction to another. While in a particular regime, the system does not remain stable but fluctuates dynamically around the equilibrium point for that regime.

directly affected the livelihoods of about 35,000 fishers and fish-plant workers, and led to a decline of over $200 million dollars per annum in local revenue from cod landings. In addition it appears that the likelihood of some regime shifts is increasing due to extensive human impacts on the biosphere. The impacts of regime shifts are exacerbated by the fact that such shifts are notoriously difficult to predict. Better understanding of regime shifts—especially their drivers, dynamics and impacts—is therefore important for informing development and environmental management policy in the coming decades.

Historical Context and Related Terms

The regime shift concept has its roots in catastrophe theory, an area of dynamical systems theory that analyzes abrupt changes in system behavior. The concept has been applied and further developed in ecology and physical oceanography, amongst many other fields. In ecology, the concept of regime shifts is closely related to that of resilience. In a seminal paper in 1973, Holling defined resilience as the ability of a system to persist in a particular domain of attraction rather than being pushed into a different domain—i.e., the ability to withstand a regime shift. A similar concept in fisheries dates back to Isaacs (1976). While fisheries and traditional population and community ecologists have focused on the role of exogenous processes (such as climate) in triggering regime shifts, ecosystem ecologists have emphasized the role of slow changes in underlying system variables and internal feedbacks.

The work by Holling and other ecologists in the 1960s and 1970s led to research on what were termed *alternative stable states* or *multiple equilibria*. The concept of multiple equilibria was explored in a wide variety of ecosystems, mostly using computer models. However, following criticism by Connell and Sousa in the early 1980s, the rate of research in this area slowed considerably. Connell and Sousa suggested that demonstrating alternative stable states requires long-term data and the removal of human effects. While their demand for more empirical evidence was valid, the condition that human impact be removed was not very useful given the key role of human effects in many regime shifts. In the 1990s a reinvestigation of the concept began and started providing convincing evidence of regime shifts in, for instance, kelp forests, shallow lakes, and drylands. This work was summarized and formalized in a series of influential papers in the early 2000s that reignited interest in this area.

One difficulty with the early work on alternative stable states is that the terminology was confusing. In particular, the term *state* had multiple meanings. The concept of *system state* usually describes the condition of a system at a particular point in time in terms of specific variables such as pH or population size. However, the term *alternative stable states* refers to different regions of state space–i.e., different domains of attraction. It therefore leads to confusion between the state of the system at a particular point in time and the potential domains of attraction. In addition, the term alternative stable states created the impression that the system remains "stable" within a particular domain of attraction, whereas in reality the system may fluctuate substantially within a particular regime. To avoid such confusion, many ecologists now prefer to use the term alternative dynamic regimes to describe the different domains of attraction and emphasize the fact that the system state may vary substantially within a particular regime.

EVIDENCE FOR REGIME SHIFTS

Our understanding of regime shifts builds on a growing body of evidence from observational, modeling, and experimental work (Table 1). Observations of sharp, persistent shifts in long-term data series, such as those illustrated in Figure 1, have provided some of the strongest arguments that regime-shifts occur. Particularly well-studied examples of regime shifts include the shift from clear water to eutrophic conditions in lakes driven by the inflow of phosphorus-rich water from agricultural fertilizers; the shift in semi-arid grasslands from drought-tolerant grasses to woody plants related to gradual increases in livestock grazing; soil salinization associated with the clearing of woody vegetation and irrigation in large parts of semi-arid Australia; and the shift from diverse coral reef systems to ones dominated by macroalgae. More controversially, work on pelagic marine ecosystems has shown how some fish (e.g., cod) and plankton communities exhibit persistent jumps in time series as a response to changes in fishing pressure, eutrophication, and/or climatic changes.

Experimental evidence of regime shifts includes results obtained from field manipulations (e.g., temporary reduction in fish biomass) and small-scale low-diversity systems (e.g., microcosms). Evidence based on full-scale (ecosystem) experiments is more limited due to logistic and ethical problems associated with conducting experiments on this scale. However, regime shifts have been experimentally induced in shallow freshwater lakes through the targeted removal of fish or the addition of nutrients, yielding much valuable understanding.

TABLE 1
Well-known examples of regime shifts and the supporting sources of evidence

Regime shift	Regime A	Regime B	Impacts of shift from A to B	Evidence	Source of evidence
Freshwater eutrophication	Noneutrophic	Eutrophic	Reduced access to recreation, reduced drinking water quality, risk of fish loss	Strong	Observations, experiments, models
Bush encroachment	Open grassland	Closed woodland	Reduced grazing for cattle, reduced mobility, increased fuelwood	Medium	Observations, experiments, models
Soil salinization	High productivity	Low productivity	Yield declines, salt damage to infrastructure and ecosystems, contamination of drinking water	Strong	Observations, experiments, models
Coral reef degradation	Diverse coral reef	Reef dominated by macroalgae	Reduced tourism, fisheries, biodiversity	Strong	Observations, experiments, models
Coastal hypoxia	Nonhypoxic	Hypoxic	Fishery decline, loss of marine biodiversity, toxic algae	Strong	Observations, models
River channel position	Old channel	New channel	Damage to trade and infrastructure	Strong	Observations, models
Vegetation patchiness	Spatial pattern	No spatial pattern	Productivity declines, erosion	Medium	Observations, experiments, models
Wet savanna–dry savanna	Wet savanna	Dry savanna or desert	Loss of productivity, yield declines, droughts/dry spells	Medium	Models
Cloud forest	Cloud forest	Woodland	Loss of productivity, reduced runoff, biodiversity loss	Medium	Observations, models

NOTE: From Gordon et al. (2008) and Scheffer et al. (2000). Other examples of regime shifts and their impacts on ecosystems and human well-being can be found at www.regimeshifts.org.

Models have been useful in highlighting the potential existence of critical thresholds and multiple equilibria in a range of marine and terrestrial ecosystems. Model simulations are particularly important for investigating regime shifts at large regional to continental scales where experiments are impossible, and the social and ecological impacts of regime shifts can be huge. Models are also useful at smaller scales because they are often stripped of many of the confounding temporal and spatial dynamics that occur naturally in ecosystems. The dynamics underlying regime shifts are therefore usually clearer and easier to investigate and understand in models than in the real world.

HOW REGIME SHIFTS WORK

Modeling, observation, and experimental work have shown that regime shifts typically result from a combination of an external shock, such as a storm or fire, and gradual changes in underlying drivers and internal feedbacks (Fig. 3). While the impacts of a regime shift are usually highly visible, changes in the risk of a regime shift (i.e., changes in system resilience) often go unnoticed. This is because gradual changes in underlying drivers and internal feedbacks that move a system closer to or further away from a critical threshold usually have no visible impact on the system state until the point at which a regime shift is triggered. However, once a system is close to a threshold, a regime shift can be precipitated by even a small shock to the system. Due

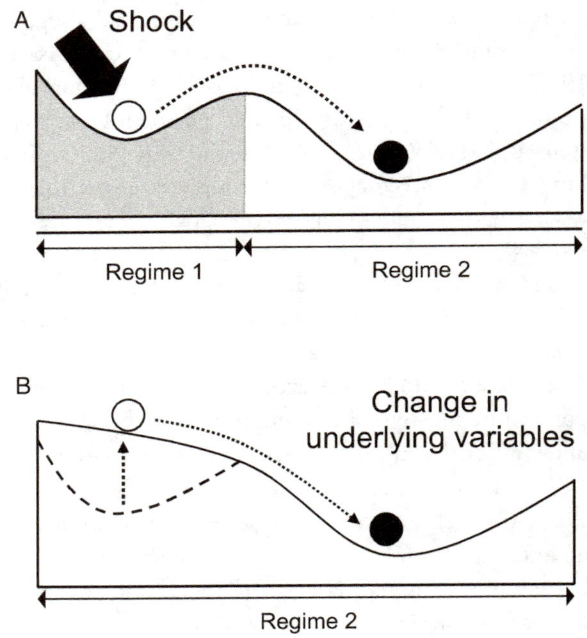

FIGURE 3 Regime shifts are usually due to a combination of (A) a shock such as a drought or flood, and (B) slow changes in underlying variables and internal feedbacks that change the domains of attraction (or resilience) of the different regimes. Gradual changes in underlying variables can lead to the disappearance of some domains of attraction, or to the appearance of new domains of attraction that did not previously exist. The critical thresholds that separate different regimes are typically determined by multiple underlying variables, rather than a single variable.

to these dynamics, regime shifts are often experienced as a complete surprise to the people living in or managing the ecosystem.

A good example is the increased risk of soil salinization associated with rising groundwater table levels in semi-arid regions such as Australia. Rising water tables often go unnoticed since they have little or no impact on the production of agricultural crops until the water table reaches about 2 m below the surface. At this point, capillary action rapidly pulls the water to the soil surface, bringing with it dissolved salts that lead to salinization of the topsoil and dramatically reduce crop growth. Once the water table is close to the 2 m threshold, the shift to the saline regime can be triggered by even a relatively small rainfall event.

Changes in the relative strength and balance of feedbacks in a system are central to understanding regime shifts. All complex systems contain both damping (also known as negative or balancing) and amplifying (also known as positive or reinforcing) feedback loops. Usually damping feedbacks dominate and keep the system within a particular domain of attraction. However, these damping feedbacks may be overwhelmed if there is a particularly large shock to the system, or if a slow variable (e.g., gradual loss of habitat) erodes the strength of the damping feedbacks. Amplifying feedbacks may then come to dominate and drive the system across a threshold into an alternate regime where a different set of damping feedbacks dominates. It is this ability for different combinations of feedbacks to dominate and structure a system that creates the possibility for different regimes and explains the abruptness of regime shifts.

An illustrative example is the rapid shift from clear water to turbid, eutrophic conditions that occurs in shallow lakes when nutrient input levels (especially phosphorus) exceed those which can be absorbed by rooted aquatic plants (Fig. 4). When this threshold is exceeded, the excess nutrients in the water lead to dense growth of planktonic algae. The algae reduce light penetration, leading to the death of the rooted vegetation that stabilizes sediments on the lake floor. This in turn results in resuspension of nutrients that have been trapped in the sediments by the rooted plants, further increasing algal growth, and creating an amplifying feedback. Even if the external input of nutrients is then reduced, the turbid, algal-dominated state is maintained through constant recycling of nutrients from the lake sediment. In order to return to the clear water state, nutrient inputs usually have to be reduced significantly below the threshold level at which the system originally shifted to the eutrophic state. This

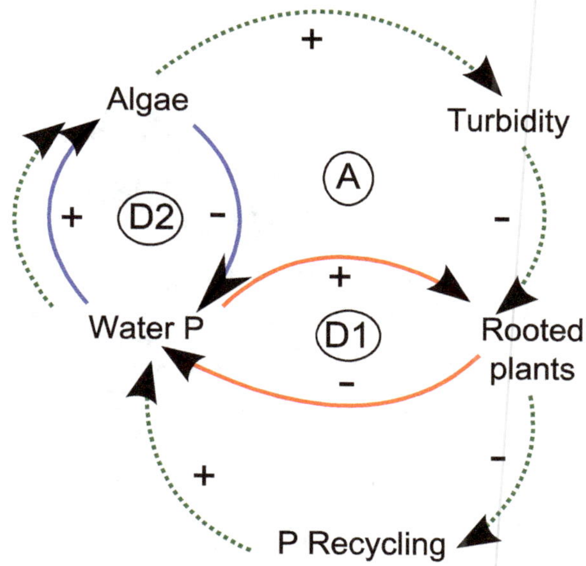

FIGURE 4 Changes in the strength and balance between competing feedback loops in a shallow lake shift the system from one domain of attraction to another. The clear water regime is characterized by the dominance of feedback D1 (shown in red). As phosphorus levels increase, D1 weakens, and the amplifying feedback A (green) starts to dominate, driving the system into a eutrophic state limited by feedback D2 (blue).

phenomenon, where the critical threshold that triggers the shift from regime A to B differs from the threshold at which the system shifts from regime B to A, is known as hysteresis and characterizes many regime shifts.

DETECTING AND PREDICTING REGIME SHIFTS

Determining whether an observed ecosystem change represents a regime shift is often difficult, particularly in highly variable systems. Identifying and defining regimes and regime shifts requires a clear definition of the system being investigated, including the focal system variables of interest and the characteristic temporal and spatial scales of the key processes underlying the dynamics of the focal variables. For instance, the appropriate spatial and temporal scale of the data required to identify regime shifts in coral reef ecosystems will differ if the focal variable is water chemistry as opposed to fish diversity. Similarly, identifying regime shifts in soil structure can be done using data that span only a few years and were collected at the field scale, while the scale of the processes underlying the shift from wet to dry savanna requires regional scale data that spans decades to centuries (Fig. 5).

Once data at the appropriate spatial and temporal scale has been gathered, regime shifts can be detected in several ways. Large, persistent jumps in time series data can sometimes be detected by eye (e.g., Fig. 1B). However,

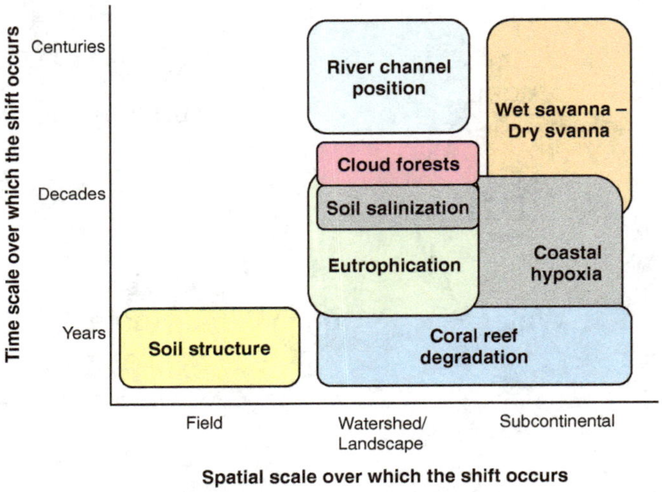

FIGURE 5 Regime shifts operate at different spatial and temporal scales. Identifying and defining regime shifts therefore requires (i) deciding on the focal system variables of interest, and (ii) data at appropriate spatial and temporal scales that capture the key processes underlying the focal system variables. Modified from Gordon et al., 2008.

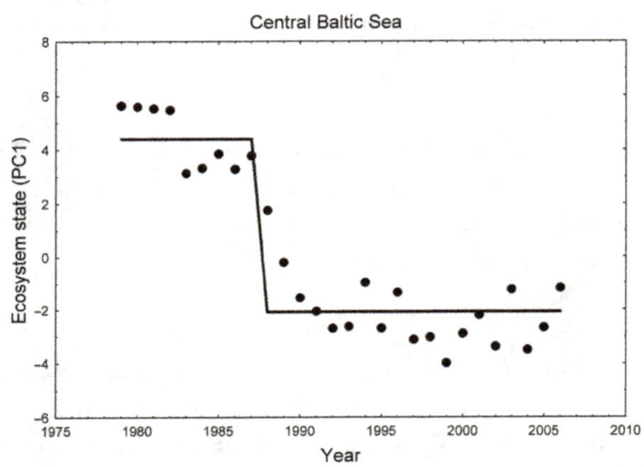

FIGURE 6 Statistical techniques such as PCA and *t*-tests can be used to detect regime shifts. For instance, in an analysis of the central Baltic Sea, all abiotic and biotic variables were included in a PCA analysis. The first component was used as an indicator of the ecosystem state, suggesting a regime shift in 1987. The significance of this shift was then tested using a *t*-test.

especially in highly variable systems, it is often unclear whether there is a step change in the time series, and if so, whether this represents a regime shift or not (Fig. 6). Sophisticated statistical techniques are therefore often required to identify regime shifts. As regime shifts typically occur at the ecosystem level, several variables usually need to be combined and aggregated to detect regime shifts. The most commonly used tools for detecting regime shifts are ordination methods, such as principal component analysis (PCA), which can compress a large number of correlated time series into a small number of uncorrelated ones.

In searching for regime shifts there is always a risk of detecting spurious shifts that are in fact just random fluctuation. There are several approaches to checking whether a change in a time series represents a regime shift or not. If several independent methods (e.g., PCA, chronological clustering) detect the same regime shift, the case for a shift is strengthened. In addition, standard statistical techniques (e.g., *t*-test) can be used to test if there is a significant change in the mean before or after a threshold event (e.g., the introduction of an alien species). Where the change point is unknown, sequential *t*-tests can be used to test for a regime shift at every point in time (with appropriate adjustments for multiple testing). Besides statistical techniques, an understanding of the processes underlying an observed change in a set of ecological variables, especially an understanding of the key system feedbacks, are important for deciding whether an observed change represents a regime shift or not. If feedback mechanisms that reinforce and maintain the different regimes can be identified, the shift is almost certainly a regime shift. As discussed earlier, the way in which such feedbacks operate are often best investigated with simple dynamic systems models. Carefully planned experiments can also be invaluable in uncovering and understanding system feedbacks. In general, identifying and detecting regime shifts often requires several independent lines of inquiry.

Recently there has been substantial interest in improving our ability to predict regime shifts before they happen, in order to for instance avert undesirable shifts or prepare for shifts that cannot be averted. There are two broad approaches to predicting regime shifts. If the mechanisms underlying a particular regime shift are understood, it may be possible to monitor the driving variables directly and provide warning when they approach levels at which a regime shift becomes likely. For example, we now have a good understanding of the mechanisms that lead to freshwater eutrophication and the range of phosphorus input levels at which a shift to eutrophic conditions becomes more likely. Importantly, because the critical threshold is typically a function of several variables (e.g., soil type, local rainfall), the threshold usually varies over time and space. The exact threshold can therefore seldom be predicted, but it is usually possible to define a "tipping zone"—a range of values of key driving variables where a regime shift becomes likely.

In many cases, however, the processes leading to regime shifts are only vaguely understood. In these cases,

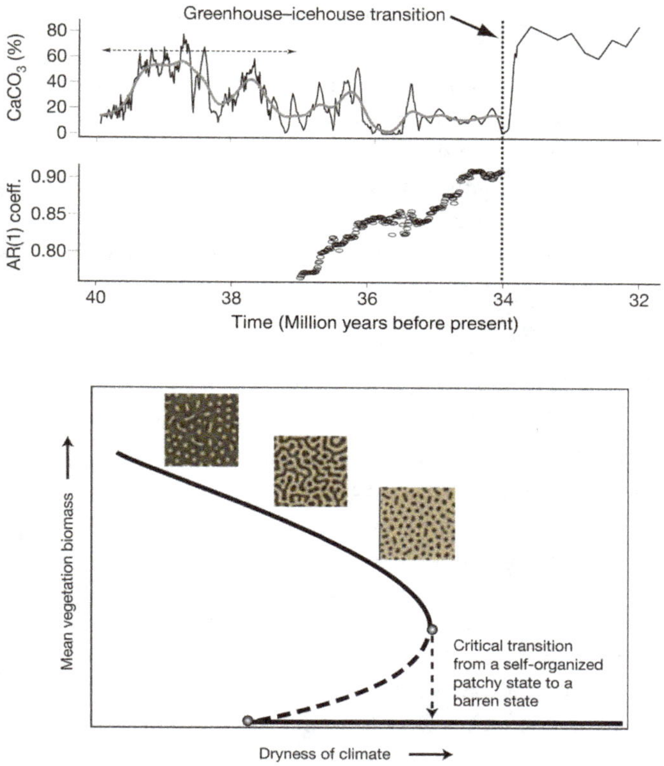

FIGURE 7 Statistically based indicators can provide early warning of regime shifts. (A) Such indicators can be derived from time series data, and show up for example as an increase in the autoregression coefficient as a regime shift is approached. (B) Early warning indicators can also take the form of a predictable sequence of self-organized spatial patterns as a critical threshold is approached. From Scheffer et al., 2009.

a second approach relying on statistically based generic early warning indicators can be helpful. It has been theoretically shown that as a critical threshold is approached in any complex system, marked and predictable changes occur which can be captured by statistics such as the autocorrelation and variance of system variables subject to regime shifts. Such changes occur not only in time series data, but can also be found in spatial data (Fig. 7). Because these early warning indicators are based on generic changes in system behavior that precede regime shifts driven by gradual changes in underlying variables and internal feedbacks, they can potentially be used to detect new or unknown thresholds. Such indicators are thus particularly relevant in the context of the novel and extensive changes occurring in the biosphere today.

USING REGIME SHIFT CONCEPTS IN MANAGEMENT

Management of regime shifts in ecosystems deals with both avoidance and reversal of regime shifts. While our ability to predict ecological regime shifts is improving, recent work suggests that unless the underlying mechanisms are well understood, the current set of generic early warning indicators will normally not provide sufficient warning to take action to avoid a regime shift. Another approach to avoiding undesirable regime shifts is therefore to manage for increased resilience. This approach emphasizes the role of diversity in sustaining the capacity of an ecosystem to cope with shocks and stay within its domain of attraction.

At the species level, functional groups, i.e., guilds of species performing similar ecological functions, provide the link between biodiversity and resilience. Two properties of functional groups are important in this context: response diversity and redundancy. Response diversity is the diversity of responses to environmental change among species within a functional group. Redundancy describes the capacity among species within a functional group to functionally replace each other. Management strategies that bolster these properties help prevent regime shifts by ensuring that critical system feedbacks are maintained in the face of unexpected shocks and disturbances to the system.

Similarly, spatial heterogeneity of land- and seascapes is a critical for maintaining resilience at the landscape level, and spatial analysis can be used to identify particularly vulnerable or important regions. Due to the variation in ecological processes across a landscape some locations will be more vulnerable to regime shifts than others. By identifying locations where processes that maintain an existing regime are weak, managers can identify where regime shifts are more likely to occur. For example, it has been shown that moist savanna systems can exhibit regime shifts between savanna and forest, while drier savanna systems cannot. The potential extent of bush encroachment in drier areas is therefore limited, and potentially less of a concern than in wetter regions. Identifying critical areas or hotspots in relation to the risk of regime shifts can therefore enable managers to focus their rehabilitation or monitoring efforts where they are likely to have the greatest effect. As another example, research on eutrophication has shown that relatively small areas in watersheds that combine high soil phosphorus concentrations and high runoff potential are disproportionately responsible for the majority of phosphorus runoff into lakes. The risk of a regime shift can thus be most effectively reduced by focusing especially on these critical source areas.

Unfortunately, an increasing number of ecosystems are already in degraded or undesirable regimes from a human perspective. Successfully steering them toward more productive or desirable regimes often requires reversing previous regime shifts (Box 1). This task is complicated by self-reinforcing feedbacks associated with the current degraded regimes. Management may have to actively engage in breaking these feedbacks (i.e., eroding resilience of the undesirable regime) in order to enable a reverse shift to a more desirable regime. In this context, it has been proposed that large-scale pulse disturbances could be viewed as windows of opportunities for breaking feedbacks. For example, periods of high precipitation associated with El Niño events can, in conjunction with herbivore exclusion, induce vegetation recovery in degraded arid ecosystems that may be difficult or impossible under other conditions. Clearly, using such natural pulse events to shift ecosystems back to more desired regimes demands a good understanding of the mechanisms underlying the regime shift and a highly responsive governance system that can react rapidly when an opportunity presents itself.

RESEARCH FRONTIERS

The regime shift concept has become increasingly developed over the past several decades. Much of the conceptual work has, however, been derived from mathematical models, and developing ways to apply regime shift ideas to understanding and managing change in real ecosystems, and to compare change across different ecosystem types, is a key research need. As theory moves into practice there are a number of key research frontiers that need to be addressed.

First, there is a need for practical and operational definitions of the regime shift concept. This is necessary to provide guidance on how to identify and conceptualize regime shifts in actual ecosystem management settings, at time scales that are relevant to human societies. One attempt at such an operational definition is being implemented in the *Regime Shifts Database*, where regime shifts are defined as any large, abrupt, persistent change in critical ecosystem services that last long enough (usually 3–5 years or more) to have significant impacts on human economies and societies.

Second, there is a need for practical methods to detect regime shifts in heterogeneous environments experiencing directional change and having limited data. Many of the current statistical approaches have been developed in econometrics and climate research and require data of 100 time steps or more, while most ecological time series usually only have 20–40 time steps (usually years). There is a similar need to develop practical methods that can provide early warning of regime shifts, with sufficient lead time that action to avert an undesirable shift is possible.

Third, there is a need for improved knowledge and tools for managing regime shifts. In particular, we need a better understanding of the mechanisms underlying different regime shifts, their key drivers, impacts and potential intervention strategies. This can only be achieved through dialogue between science and ecosystem managers. Such knowledge will enable us to better assess which places on Earth are particularly vulnerable to specific regime shifts, and the conditions under which the likelihood of different regime shifts increases, in order to better direct management efforts. In addition, it will enable us to better assess the potential risks and costs of regime shifts in order to decide when it makes sense to take action to avert or precipitate a regime shift. To enable managers to bolster or break feedback processes to avert or precipitate regime shifts we need a better understanding of how different feedbacks operate and interact in space and time.

Finally, in a connected world we need to better understand how different types of regime shift are connected to one another. We know that there are cross-scale connections, where local and regional regime shifts interact. For example, land degradation regime shifts can interact with climatic regime shifts. There are also horizontal connections, where a regime shift in one ecosystem can trigger a shift in a connected ecosystem, such as when freshwater eutrophication triggers coastal hypoxia downstream. Identifying the processes that connect different regime shifts and mediate the strength of these connections is vital for understanding the ability of ecosystems to respond to global changes such as climate change and increased habitat conversion.

SEE ALSO THE FOLLOWING ARTICLES

Bifurcations / Continental Scale Patterns / Ecosystem Ecology / Phase Plane Analysis / Resilience and Stability / Stability Analysis / Stress and Species Interactions

FURTHER READING

Andersen, T., J. Carstensen, E. Hernández-García, and C. M. Duarte. 2009. Ecological thresholds and regime shifts: approaches to identification. *Trends in Ecology & Evolution* 24: 49–57.

Beisner, B. E., D. T. Haydon, and K. Cuddington. 2003. Alternative stable states in ecology. *Frontiers in Ecology and Environment* 1: 376–382.

Carpenter, S. R. 2003. *Regime shifts in lake ecosystems: pattern and variation.* Oldendorf/Luhe, Germany: International Ecology Institute.

Folke, C., S. R. Carpenter, B. H. Walker, M. Scheffer, T. Elmqvist, L. H. Gunderson, and C. S. Holling. 2004. Regime shifts, resilience and biodiversity in ecosystem management. *Annual Review of Ecology, Evolution and Systematics* 35: 557–581.

Gordon, L. J., G. D. Peterson, and E. M. Bennett. 2008. Agricultural modifications of hydrological flows create ecological surprises. *Trends in Ecology & Evolution* 23: 211–219.
Holling, C. S. 1973. Resilience and stability of ecological systems. *Annual Review of Ecology and Systematics* 4: 1–23.
Regime Shifts Database. www.regimeshifts.org.
Scheffer, M. 2009. *Critical transitions in nature and society*. Princeton: Princeton University Press.
Scheffer, M., J. Bascompte, W. A. Brock, V. Brovkin, S. R. Carpenter, V. Dakos, H. Held, E. H. van Nes, M. Rietkerk, and G. Sugihara. 2009. Early warning signals for critical transitions. *Nature* 461: 53–59.
Suding, K. N., K. L. Gross, and G. R. Houseman. 2004. Alternative states and positive feedbacks in restoration ecology. *Trends in Ecology & Evolution* 19: 46–53.

RESERVE SELECTION AND CONSERVATION PRIORITIZATION

ATTE MOILANEN
University of Helsinki, Finland

Reserve selection is a decision analysis method used in conservation biology. Assume you have a landscape with different species occurring in different parts of it. Which parts should be selected into a reserve network so that biodiversity conservation objectives are achieved most efficiently? This is the question most traditionally asked by reserve selection. Spatial conservation prioritization techniques address more general questions about the spatial allocation of conservation action.

EVALUATING RESERVE NETWORKS

Habitat loss, overexploitation of populations, pollution, and climate change are among the processes contributing to a global deterioration of natural habitats and ecosystems. Anthropogenic land use pressures are, if anything, increasing. Consequently, it makes sense to use effectively the limited resources available for conservation. In addition to the near-optimal design of entire reserve networks, variants of reserve selection methods can also answer the following questions: What is the optimal expansion of a reserve network? Where should alternative land uses be placed so that they disturb biodiversity as little as possible? How good is the present reserve network compared to what it could have been if it had been systematically developed from the beginning? In other words, is the present reserve network good and balanced?

TRADITIONAL RESERVE SELECTION

Biodiversity Features and Surrogacy

Most often reserve selection operates on information about the occurrences of species across the landscape. But species occurrence is not the only quantity relevant for conservation. The term biodiversity feature, or just feature in short, is used when the general applicability of an analysis is emphasized. Different types of biodiversity features include, for example, species, habitat types, ecological communities, genes and alleles, environmental conditions and ecosystem services—all these can be used as data for reserve selection.

Biodiversity features can act as surrogates to other features. There will rarely if ever be data available for every feature that can occur in the landscape. This is easily illustrated by the tropics, where a large fraction of insect species have not yet even been described let alone had their distributions surveyed. Doing reserve selection, we need to assume that the features for which there are data act as surrogates for all other biodiversity features. Different types of surrogates have been studied a lot, with the conclusion that they either work or don't, i.e., their effectiveness most often cannot be guaranteed beforehand. For example, basing reserve selection on the distributions of birds could not guarantee a good coverage for insects or fish. Consequently, one should keep in mind the concept of surrogacy and that the usefulness of a reserve selection analysis is limited by the amount and quality of information about features and their distributions.

Reserve Selection

In literature, reserve selection and closely related analyses have been called by many names, including site selection, area selection, reserve design, reserve network design, and spatial optimization. Most traditionally, reserve selection concerns the optimal selection of areas that would jointly make up a reserve network. There are two common variants of the reserve selection problem, minimum set coverage and maximum coverage:

(1) Minimum set coverage: find the least expensive set of sites that satisfies given targets for all features.

(2) Maximum coverage: use all resources (money) so that as many targets as possible are covered.

TABLE 1

Minimum set and maximum coverage planning illustrated

		Species A–F, present (1) or not (0)					
Site	Cost	A	B	C	D	E	F
1	5	1	1	1	1	0	0
2	2	1	1	1	0	0	0
3	2	0	0	0	0	1	1
4	6	1	0	0	1	0	0

NOTE: The minimum set solution, for a target of 1 population per species, is sites 1 and 3 with cost 7. The maximum coverage solution with resource 2 is site 2 (three targets covered), and with resource 4, sites 2 and 3 (five targets covered).

Both of these problems are concerned about targets and resources (~ cost, land area). Targets are feature-specific goals that have been set for what would be an adequate representation (amount of occurrences) of the feature in the reserve network. Adequacy may be conceptually based on estimated persistence of features: targets need to be high enough so that the long-term persistence of each feature can reasonably be assumed. Target setting may be a complicated task in itself. The minimum set and maximum coverage problems are illustrated by Table 1.

Technically, an optimization problem consists of an objective and constraints. The objective in minimum set coverage is minimization of cost while the objective for maximum coverage is maximization of the number of targets covered. The constraints in minimum set coverage are the species-specific target levels (which must be equaled or exceeded); the single constraint for maximum coverage is limited money. The minimum set and maximum coverage problems can be solved using exact optimization techniques such as integer programming, or using heuristic iterative local search methods (hill climbers) or stochastic optimization (simulated annealing, genetic algorithms, tabu search). Descriptions of many such solution methods can be found from literature.

Note that reserve selection is applicable not only to the selection of reserves for the establishment of conservation area networks, but to any binary conservation decision problem. One can either protect or not, maintain habitat or not, or restore habitat or not. The contrast is between the doing (selected) and not doing (not selected). For example, in the case of habitat maintenance one would enter data as the estimated occurrences of features assuming habitat maintenance is done in each area. The cost of an area would not be an acquisition cost, but cost of habitat maintenance operations there. Nevertheless, one should realize that the traditional reserve selection framework remains limited in scope, due to the binary choice (selected or not), because of the reliance on target levels, and because the typical problem formulation may involve unrealistic assumptions. For example, independence between sites is commonly implied, but such an assumption is invalid when sites are small and linked by spatial population dynamics.

Systematic Conservation Planning

No discussion about reserve selection is complete without mention of systematic conservation planning. This influential framework was proposed in 2000 by Margules and Pressey, and it has been applied many times since. Reserve selection can be applied as a component of the systematic conservation planning framework.

In a nutshell, systematic conservation planning is an operational model for conservation that includes a series of steps that should be taken when doing conservation planning. These steps include: (1) survey of existing reserves, (2) acquisition of data about biodiversity in the area, (3) specification of adequate conservation targets, (4) solution of the reserve selection optimization problem, (5) negotiating the actual implementation of conservation action on the ground, and (6) monitoring the success of conservation action. While reserve selection is a specific kind of decision analysis, SCP includes broader ecological and sociopolitical considerations. What is adequate for a species? What sociopolitical concerns need to be addressed? What follows after the reserve network has been established?

Dynamic Landscapes, Representation and Retention

There is a critical assumption hidden in the framework of reserve selection: all focus is on what should be in the reserve network. What are the representation levels of features in the network? Are the targets covered? What is the least expensive network that covers all targets? The blind spot of such analysis is that ideally we should be interested in all features and how they occur and persist across the entire landscape, not only in reserves. Considering only biodiversity in reserves is relevant when land use pressures are so high that anything outside reserves will be lost. However, for large regions of the world, the landscape as a whole supports significant biodiversity—persistence cannot be attributed to conservation areas and their networks only. When we are interested in the entire landscape, we talk about retention.

The distinction between representation in reserves and retention in the landscape is fundamentally critical, both conceptually and methodologically. Conceptually, looking at retention takes one to a whole-landscape view where processes that influence the landscape become important. Frequently these processes are called threats in the context of conservation; what are the anthropogenic processes ongoing across the landscape? Which habitats would most likely be (negatively) impacted? Are there regional differences in threats; are different habitats lost at different rates? Which threats can be stopped or reduced by local conservation action (e.g., land clearing, water pollution), and which cannot (e.g., climate change)?

Some variants of retention-based selection problems can be addressed using traditional reserve selection methods. This requires developing analysis inputs in a particular manner. With typical reserve selection, the selected sites would all jointly efficiently cover many biodiversity features. Selected sites are likely to have high richness or unique features or both. When changing focus to retention, a site may be wonderful with respect to biodiversity content but it still can be left without conservation action. This can happen if the degradation rate of the site is estimated to be effectively zero—if nothing negative happens in the site irrespective of whether it is protected, then what is the point of conservation action there? Thus, focus switches to a site-specific combination of biodiversity content and expected loss in the absence of action. Action is needed most urgently when both biodiversity content and loss rates are high. If the loss rate is comparatively low, action in the site is not urgent.

Ability to operate on retention relies on the existence of information about threatening landscape-level processes. Such information may be difficult to obtain. Further information about this topic can be found under the terms threats and vulnerability, scheduling of conservation action, sequential reserve design or dynamic reserve selection. Overall, a focus on retention and the entire landscape suggests a broader view to spatial allocation of conservation resources, as described next.

SPATIAL CONSERVATION PRIORITIZATION: A GENERAL FRAMEWORK

From Ecology to Conservation Value to Conservation Action

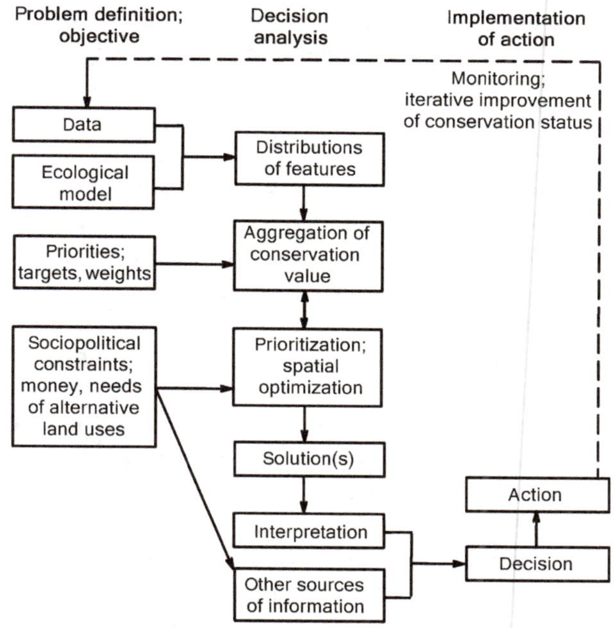

FIGURE 1 A schematic framework for spatial conservation prioritization.

Figure 1 describes a schematic framework for spatial conservation prioritization. We start from an ecologically based model which describes how our actions influence the distributions and local occurrence densities of biodiversity features. Targets or weights are assigned to features to model the priorities given to them, and based on the ecological model a measure of conservation value can be devised. This measure can be used for evaluating individual sites or sets of conservation actions that are done across the landscape. Decision analysis, frequently in the form of spatial optimization, is then applied to find a solution for the near-optimal allocation of conservation action. Typically, a combination of quantitative analysis, local expert knowledge and stakeholder involvement would ultimately decide the real on-the-ground action. This process could be repeated after new data about the environment is obtained, when the landscape has changed since previous analysis, when objectives change or when new resources become available. Conservation typically is not a once-off exercise but more commonly an iterative ongoing process. Components of this framework are elaborated below.

Methodologically, conservation prioritization is a multidisciplinary science. The ecological model uses methods from spatial ecology, landscape ecology, metapopulation ecology, restoration ecology, and statistical ecology. Conservation biology is the conceptual basis of evaluation of conservation value. The decision model belongs to the realm of applied mathematics, operations research, optimization, and decision analysis. Social sciences are influential when going from quantitative analysis to on-the-ground conservation action. With influences from many sciences, the terminology of the field is not most easily accessible. Some key concepts are enclosed in the glossary. Others often encountered in literature include adequacy, complementarity, comprehensiveness, flexibility, irreplaceability, replacement cost, and selection frequency, plus a variety of further concepts from the fields of optimization and decision analysis.

The Ecological Model

What is the distribution of a feature? In a sense, all spatial allocation of conservation resources depends, explicitly or implicitly, on an underlying ecologically based understanding of where biodiversity features occur. At its simplest, the ecological model is an observation model. The landscape is surveyed and features are assumed to occur and persist at locations where they have been observed—historically, the most common occurrence data used in reserve selection has been a sites × species matrix of species presence/absence. This approach may be sufficient when individual sites (reserve areas) are so large that they can plausibly support long-term persistence of features locally. But basing reserve selection on observation is difficult in the general case. Management decisions would commonly be done at resolutions of hectares or square kilometers, and the effort of accurately surveying everything at such resolutions is likely to be prohibitive.

Figure 2 shows a progression of increasingly complex models for the distributions of biodiversity features. The next level of complication is static prediction of pattern of occupancy (presence–absence, density, abundance), done for example using statistical habitat models to explain occurrences by environmental data. Here, a critical assumption is that data on environmental factors (rainfall, temperature, elevation, ground cover, . . .) are available from all across the landscape, making it possible to model and extrapolate occurrence to locations where direct observations do not exist. Typically, data on environmental factors would come from remote sensing and GIS or from long-term observational surveys. Effects of population dynamical connectivity can be included in these models via the inclusion of various neighborhood measures, such as coverage of forest within a 2-km radius around the focal site, into the set of explanatory variables.

Then it is possible to go from static pattern to process. The time dimension and ecological and/or anthropogenic processes may be brought in to predict the dynamics of the landscape and biodiversity features in detail. A variety of modeling approaches could be utilized at this stage, including landscape simulators, climate change models, metapopulation models and spatial PVAs. Some notion of dynamics is compulsory if one is working with retention in the landscape rather than with representation in reserves; there must be some understanding of what is likely to happen across the landscape with and without specific conservation action.

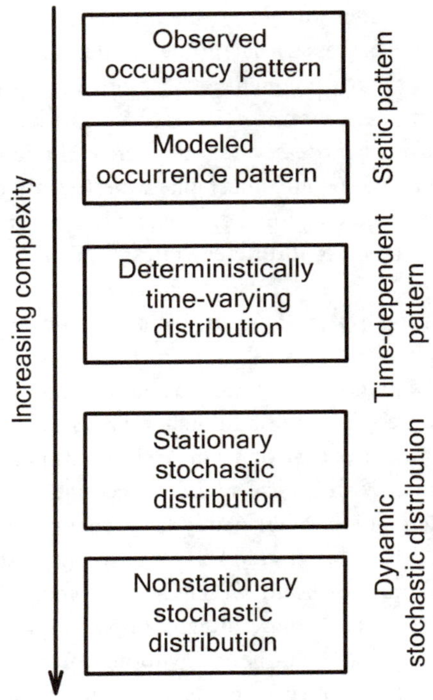

FIGURE 2 Types of ecological models with increasing complexity, used here for the estimation of occurrence patterns of species or other biodiversity features.

Spatial conservation prioritization is ideally based on a broad range of biodiversity features that act as a plausible surrogate to biodiversity as a whole. This implies use of many features, including, for example, species across many taxa and information about habitat types and human impacts on the landscape. The need to model many features presents a problem to the ecological model, which becomes progressively more difficult to parameterize as it becomes more complex and data hungry. There are no fixed rules here; the choice of modeling framework(s) would depend on the objective of the analysis, and the availability of data and modeling skills. In any case, dealing with prediction uncertainty is important when analysis is based on complex models.

From Ecology to Conservation Value: Objectives, Targets, and Weights

Either direct observation or a more complicated ecological model produces an estimate of where biodiversity features would occur, and by what density, assuming a set of conservation actions are taken. While a good start, this is

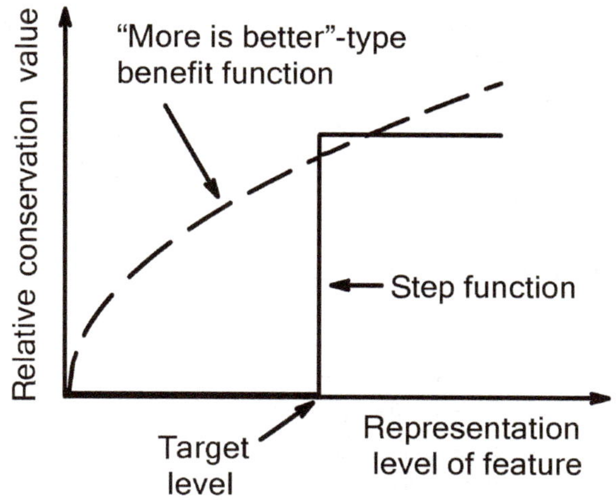

FIGURE 3 Benefit functions. A benefit function is a simple way to convert feature representation to conservation value. A target is implemented by a benefit function that is a step function. Continuously increasing benefit functions model the conception that increasing representation should result in increasing conservation value.

not sufficient for determining where conservation action should be taken. What do you prefer, a species-rich site that has occurrences of 20 species or a species-poor site that has the occurrence of one unique species? When everything cannot be had for conservation, there will inevitably be tradeoffs between features; having a lot for one may imply less for another.

The characteristics of an acceptably good solution may be defined via targets (for target-based planning) or the balance between features may be influenced via weights (in utility-based planning and prioritization). A multitude of factors can be accounted for when deciding feature targets or weights. Such include the international conservation status of the feature, whether the feature is a known surrogate for many other features, how much of the feature has been lost relative to historical distribution, expected threats to remaining occurrences, population trend, economic value, productivity of habitat type, taxonomic uniqueness of species, etc. While the simplest approach is that all features are treated equally (same target/weight to all), substantial effort can be put into developing more sensible targets or weightings.

Going further from targets, benefit functions can be used for converting representation to conservation value (Fig. 3). Typically, a benefit function would assume that increasing representation translates to increasing conservation value—with biodiversity, more is better. The weight given to a feature would influence the magnitude of the respective benefit function. Conceptually, the aggregation of conservation value can adopt a structure nested like the following:

> Aggregate value across all features j accounting for weights and benefit functions
>> Aggregate across time steps t for feature j
>>> Aggregate across sites i (the landscape), for feature j at time t
>>>> Occurrence for feature j at time t in site i depends on habitat quality and conservation action taken at site i and its neighborhood before time t

Written explicitly, it becomes apparent that a balance between features and a balance between now and the future needs to be considered. While most applications would not use a structure as complicated as above (for example, the time dimension is frequently dropped), it is relevant that the initial condition of the landscape, actions, sites, features, time, and interactions between these factors all are conceptually relevant for the distributions of features and thus conservation value.

Finding a Solution: The Decision Model and Different Planning Modes

The decision model specifies the objective of analysis and the structure of the optimization problem. This entry has already described two common decision models, the minimum set coverage model and maximum coverage. Yet another decision model is maximum utility: "maximize conservation value, as defined by an ecologically based model of conservation value, subject to constraints in cost and possibly other factors." The difference between maximum coverage and maximum utility is that maximum utility is more general because it does not require specification of targets for features. Interactions between features or sites, landscape dynamics, and other such complications fit naturally into the maximum utility framework. These decision models are based on a binary choice of "select or not."

The decision model and consequent analysis become significantly more complex if more than two alternative actions (choices) are allowed for each site. What is good for one feature may be bad for another, and a complicated balancing of tradeoffs between features is needed as a consequence. This multi-action decision problem answers a question of optimal allocation of multiple alternative conservation actions across the landscape.

The mode of planning is another consideration. So far we have been discussing what can be called "optimal plan generation." Minimum set, maximum coverage, and maximum utility all aim at providing an optimal plan for spatial resource allocation. However, there are two other common planning modes that are relevant for real-world decision making: scenario evaluation and priority ranking. Scenario evaluation simply means that alternative scenarios (that have been developed by someone else) are evaluated according to the ecological model—there is no free choice in the allocation of action between sites. The other is priority ranking, where each area (or action) is assigned a priority. Priority ranking can support selection of areas for action, but it can also be used, e.g., for the targeting of incentive funds, or for the allocation of intensive economical activity to areas with low conservation priority.

When constructing the ecological model, the model of conservation value and the decision model, one can keep in mind the concept of generalized complementarity. According to this concept, conservation benefits of all conservation actions across the landscape should be evaluated jointly, accounting for long-term consequences of interactions between actions. These interactions include spatial population-dynamical interactions between areas that are close to each other, tradeoffs between species—what is good for one species may be bad for another, and interactions via resource use—money spent on one action may be away from alternatives. Overall, a profitable balance between the multitude of considerations is sought.

Costs and Alternative Land Uses

The ecological model is the basis of evaluation of conservation value—it defines the benefits. On the other side of the balance are costs. Conservation may involve different kinds of costs, which can in fact be rather dominant for the outcome. When the objective of optimization is changed to the form (conservation value)/(cost of actions), the analysis is framed in terms of cost-efficiency. The optimal analysis will be one that maximizes the ratio of benefits to costs, subject to constraints. This formulation is relevant when, for example, the acquisition or maintenance cost of a reserve network needs to be accounted for.

Opportunity costs are indirect costs that impact stakeholders. Often, land would be valuable for other uses than conservation as well: tourism, urban development, agriculture, forestry, etc. Thus, conservation action may cause a cost of lost opportunity to some other party. Such costs can be accounted for in optimization using a structure (conservation value) − (multiple opportunity costs). When multiple antagonistic land uses are included in the same analysis, typically some kind of weighting between the different objectives would be utilized to influence the balancing of objectives.

When costs of different areas vary widely, it should be recognized that cost differences may be driving the solution. Therefore, it may be advisable to always also do a biodiversity-only analysis without (opportunity) costs so as to gain understanding of what is lost when costs are entered into the analysis.

Analysis and Software

When data have been collected, and an ecologically based model of conservation value defined, it is time to do a quantitative computational decision analysis that will suggest a solution for the spatial allocation of conservation resources.

There are several software packages intended for such tasks, including MARXAN, Zonation, ConsNet, C-Plan, and Worldmap, plus a number of implementations using generic linear (integer) programming software. The capabilities of these software differ in how complex ecological models they can handle, what kind of data they analyze (grids or polygons), which decision models they support, and how large datasets they can in fact analyze. Many of these approaches are intended primarily for target-based planning, while some do both targets and more general prioritization.

Further information of these software can be found from literature, but there is one general issue relevant for the interpretation of literature: avoid confounding objective, solution algorithm, implementation of algorithm, and software package (Fig. 4). It is not uncommon to see wording like "the solution was found by application of [insert your favorite software name]." A more explicit description of what was done is "a reserve selection problem of type [minimum set, maximum coverage,

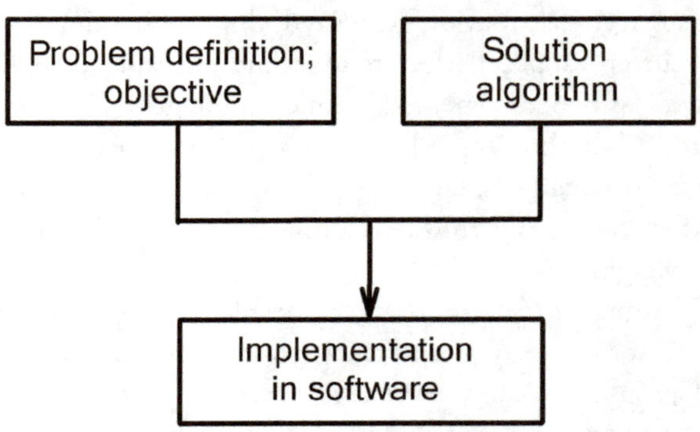

FIGURE 4 The relationship between an objective, an algorithm, and the implementation of an algorithm in a software. These are all conceptually different things even when analyses frequently are identified by the name of the spatial planning software.

maximum utility, . . .] was solved using a [exact, heuristic, stochastic global search] solution algorithm implemented in the software [insert name]." The suitability of an algorithm to solve a specific problem varies. The efficiency of an algorithm implementation in a software varies. How close to optimal a software implementation can get varies. The ease of use of software implementations varies. When choosing software, it may be helpful to keep in mind such considerations.

Dealing with Uncertainty

The world is very complex. Our understanding of the ecology of individual species is limited, and indeed, there are millions of species that have not even been described. Unpredictable international politics may influence nature globally, via, e.g., climate change. Consequently, ecological data and conservation decision analysis are pervaded with uncertainties. Nevertheless, national-level administrative bodies make on a daily basis decisions that impact land use. Despite uncertainties, conservation decision analysis needs to provide output that can support decision making while helping conservation of biodiversity.

There are numerous frameworks that support uncertainty analysis: Bayesian statistics, robust optimization, risk analysis, interval arithmetic, information gap decision theory and so on. For most conservation decision analysis, trying to deal with first-order influences of uncertainty is enough. When considering actions in sites one may succeed in simplifying the decision problem by considering four types of conservation actions/areas:

(1) Most desirable for conservation are areas and/or actions that are good with high certainty (e.g., pristine quality site that conservation action stops from being lost).

(2) Least desirable for conservation are areas and/or actions that are poor with high certainty (e.g., a parking lot, highly unreliable eradication of invasive species, etc.).

(3) If an area/action is evaluated as good, but with uncertainty, then there is scope for disappointments. The possibility of getting less than expected can be controlled using robustness analysis. For example, a rare species may have been observed in the area several years ago. Our assumption of high conservation value could thus be dependent on an uncertain assumption about the continued presence of the species there. In general, external threats, anthropogenic or natural, can cause uncertainty about the future conservation value of a site.

(4) An area may have apparently low value, but if this information is uncertain, there could be potential for positive surprises. For example, an area could be remote and poorly surveyed and thus hard evidence of conservation value would be lacking. But any surprises should be positive, in a sense elevating the value of the site.

Of the alternatives above, choices of type (1) can be taken with little risk, choices of type (2) should be avoided, and choices of type (3) and (4) might benefit from an examination of uncertainties and their likely impact to the achievement of objectives.

STATUS OF RESEARCH AND EXPECTED DEVELOPMENTS

Ecologically based conservation decision analysis (including reserve selection and conservation prioritization) is a broad and active field of research. An overview of past and present research gives a general idea of where the field is going and where developments can be expected (Fig. 5). In the author's opinion, the decision model, including the role of costs and opportunity costs, is the component that has been most comprehensively researched. Implementation always follows conceptual understanding, and not all analyses are publicly available and operational for real large datasets. Nevertheless, significant implementations and software also are available. Quickly improving availability of remote sensing data is further increasing the need for large-scale high-resolution applications.

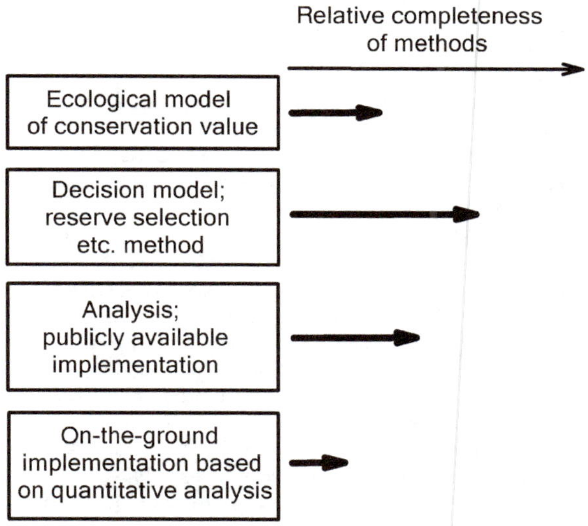

FIGURE 5 Past research and expected developments. The figure shows how complete, relatively, research around the major components of conservation prioritization appears, based on a view of published literature.

The ecological model is somewhat less developed. Most of the reserve selection research between 1985 and early 2000s was (implicitly) using very simple observation models for ecology and conservation value—focus of research was on developing the decision analysis techniques. Lately, there has been increasing effort in analysis based on environmental factors, community level analysis, ecosystem services, or ecological processes. Understanding threats and landscape dynamics has become critical for all retention-based analyses. Climate change has attracted massive attention, but a final conclusion about how it could be systematically accounted for in reserve selection is missing. In summary, more complicated ecologically based models of conservation value are developed actively. This poses new challenges to the decision analysis part, as structurally more complicated models will benefit from more advanced solution methods.

Finally, conservation decision analysis needs to provide information that holds utility for real-world decision making. Of the four main components depicted in Figure 5, the presently least developed part is systematic integration within sociopolitical and governance frameworks.

SEE ALSO THE FOLLOWING ARTICLES

Computational Ecology / Conservation Biology / Ecological Economics / Ecosystem Services / Geographic Information Systems / Landscape Ecology / Population Viability Analysis / Restoration Ecology

FURTHER READING

Ferrier, S., and M. Drielsma. 2010. Synthesis of pattern and process in biodiversity conservation assessment: a flexible whole-landscape modelling framework. *Diversity and Distributions* 16: 386–402.

Knight, A. T., R. M. Cowling and B. R. Campbell. 2006. An operational model for implementing conservation action. *Conservation Biology* 20: 408–419.

Margules, C. R., and S. Sarkar. 2007. *Systematic conservation planning.* Cambridge, UK: Cambridge University Press.

Moilanen, A., K. A. Wilson, and H. P. Possingham, eds. 2009. *Spatial conservation prioritization: quantitative methods and computational tools.* Oxford: Oxford University Press.

Nelson, E., G. Mendoza, J. Regetz, S. Polasky, H. Tallis, D. R. Cameron, K. M. A. Chan, G. C. Daily, J. Goldstein, P. M. Kareiva, E. Lonsdorf, R. Naidoo, R. H. Ricketts, and M. R. Shaw. 2009. Modeling multiple ecosystem services, biodiversity conservation, commodity production, and tradeoffs at landscape scales. *Frontiers in Ecology and the Environment* 7: 4–11.

Pressey, R. L., M. Cabeza, M. E. Watts, R. M. Cowling, and K. A. Wilson. 2007. Conservation planning in a changing world. *Trends in Ecology & Evolution* 22: 583–592.

Walker, S., A. L. Brower, R. T. Stephens, and G. L. William. 2009. Why bartering biodiversity fails. *Conservation Letters* 2: 149–157.

Wilson, K. A., E. C. Underwood, S. A. Morrison, K. R. Klausmeyer, W. W. Murdoch, B. Reyers, G. Wardell-Johnson, P. A. Marquet, P. W. Rundel, M. F. McBride, R. L. Pressey, M. Bode, J. M. Hoekstra, S. J. Andelman, M. Looker, C. Rondonini, M. R. Shaw, and H. P. Possingham. 2007. Conserving biodiversity efficiently: what to do, where, and when. *PLOS Biology* 5: 1850–1861.

RESILIENCE AND STABILITY

MICHIO KONDOH
Ryukoku University, Otsu, Japan

Resilience and stability are dynamics-related properties of ecological systems. Resilience refers to the capacity of such systems to maintain internal self-organizing processes against external perturbations, while stability indicates the ability of ecological systems to maintain equilibrium. These concepts are not only relevant to fundamental ecological issues but also strongly reflect the management strategies of ecosystems.

THE CONCEPT

The structure and function of ecological systems may be disturbed by external perturbations, which may vary in magnitude, type, and spatiotemporal scale. Population density, community composition, and ecosystem functioning are all ecological properties that may be affected by external perturbations. Disturbance caused by a perturbation varies both qualitatively and quantitatively. Furthermore, the recovery process may also be subject to variability. For example, the disturbance may either remain within the system for an extended period or may eventually be completely absorbed. Resilience and stability are theoretical concepts that describe how a system responds to an external perturbation. In the classical view, an ecological system is often assumed to be at a state of equilibrium and thus, when considering a consequence of perturbation, the behavior of the system near the equilibrium state is focused on. A system is regarded as being stable if the focal state variable (i.e., population density) resists departure from equilibrium and is not subject to temporal fluctuation. These stability indices include local (i.e., asymptotic) stability and engineering resilience (often referred to as simply resilience by the early 1990s). The former type represents the qualitative property, whereby the system state always returns to the original equilibrium state following a sufficiently small disturbance. The latter type is a quantitative index, which measures the rate of return to the pre-disturbance state. These stability indices are mathematically tractable and have contributed extensively to our understanding of ecological dynamics at population, community, and ecosystem levels. However, these stability indices only focus on the system behavior around the equilibrium state, and thus they cannot be used when dynamics far from the equilibrium state are of interest. For example, the stability indices exclude periodic and chaotic solutions, which are consistent with the persistence of the system. Furthermore, the stability indices may not be viable when the system has multiple steady states. If the disturbance is sufficiently small, the system may return to the original state (which may be stable or not) driven by internal self-organizing processes. Yet this may be no longer true if there is a threshold in perturbation strength above which transitions among alternative states are caused. This happens, for example, when the perturbation alters the internal process that drives state dynamics, or when the disturbance brings the system state out of the attracting domain of the pre-disturbance state. To capture the transitional feature of ecological systems induced by external perturbations, in 1973 Crawford S. Holling introduced the term "(ecological) resilience" (hereafter termed "resilience"), which is measured by the magnitude of disturbance that may be tolerated before the system moves into an alternative state. Note that stability does not beget resilience, and vice versa. While some globally stable systems return to a unique equilibrium after any magnitude of disturbance (i.e., are perfectly ecologically resilient), other systems may be perfectly resilient and nonstable at the same time (e.g., a system where any initial conditions converge to a unique periodic solution).

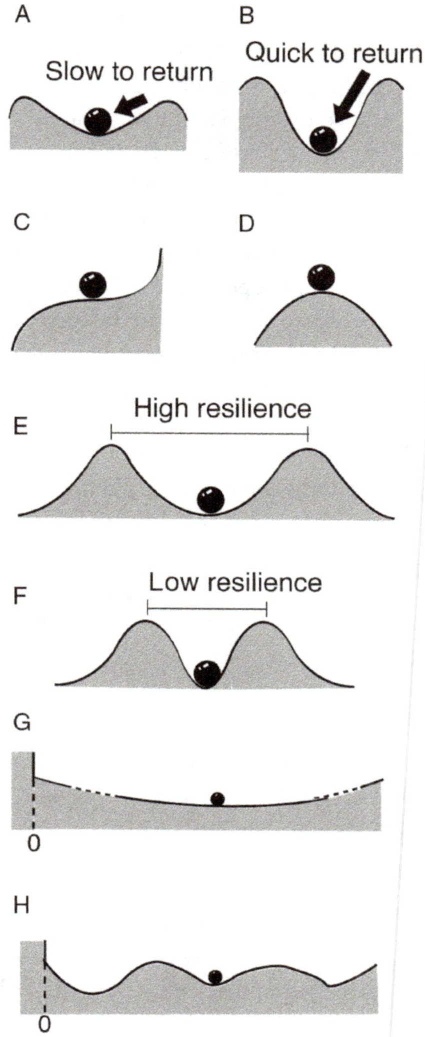

FIGURE 1 "Ball and cup" heuristic models for the "stability" and "resilience" concepts. (A, B) Locally stable equilibrium states, in which the engineering resilience of A is lower than that of B. (C, D) Locally unstable equilibrium states. (E, F) Ecological resilience, represented by the width of cup top (horizontal bars), is higher in E than F. (G) Globally stable equilibrium state. (H) Persistence and permanence.

HEURISTIC AND MATHEMATICAL MODELS

Heuristic Model

The "ball and cup" heuristic is often used to represent the basic concepts of resilience and stability. For example, if a ball on a one-dimensional lagged landscape is considered (Fig. 1), the horizontal position of the ball represents the system state and the landscape determines the direction that the ball moves.

Stability indices characterize the behavior of the ball around the equilibrium state. In the absence of external perturbation the ball may remain stationary (i.e., equilibrium state) at any position where the landscape is locally flat, which may be at the bottom of the cup (Figs. 1A, B), at the top of the shoulder (Fig. 1C), or even at the top of the hill if the ball is precisely positioned (Fig. 1D). Yet the consequence of a slight departure from the original position is completely different between cases. When the ball is at the top or shoulder of the hill, a slight departure may move the ball permanently away from the original position. On the other hand, a ball placed at the bottom of the cup will return to the original pre-disturbance position. Thus, the bottom of cup and the former two positions represent locally stable and locally unstable equilibrium states, respectively. Another stability index, engineering resilience, indicates how quickly the ball returns to the original position after a slight displacement, and is determined by the slopes around the equilibrium state.

While it is only the local landscape that determines the stability of the equilibrium state, resilience is determined by the regional landscape that may include neighboring cups. When the landscape has other cups in addition to the original, a ball moved sufficiently far from the original position would be attracted to another cup and would not return. Ecological resilience represents the distance of how far the ball should be moved for this to happen and is measured by the width at the top of the cup (Figs. 1E, F). More resilient systems may experience a larger magnitude of disturbance before transition to an alternative state. When there is only one cup in the entire landscape (i.e., the top of the cup is infinitely large), the ball will return to the original position after any magnitude of disturbance, meaning that the system is completely resilient. Such an equilibrium state is referred to as globally stable. Persistence and permanence are other indices that are used in the community-ecology context and which take the global landscape into account by focusing on whether the system loses a population following disturbance. Note that a persistent or permanent (Fig. 1H) community may have more than one equilibrium state, and thus may not be completely ecologically resilient.

Mathematical Model

Clearer insight into the concepts of resilience and stability is obtained by using a formal mathematical model. Consider population dynamics governed by the following time-continuous model:

$$\frac{dX}{dt} = [\text{birth}] - [\text{death}] = R(X) - M(X). \quad (1)$$

The state variable is population density, X. Birth, $R(X)$, and death processes, $M(X)$, are the functions of X. Assume (i) that R is a convex function of X, (ii) that M linearly increases with increasing X, and (iii) that R and M cross three times, as shown in Figure 2. The three intersections, where it holds that $R(X^*) = M(X^*)$ ($dX/dt = 0$), correspond to the three equilibrium states (X_0, X_1, and X_2). The population is extinct at X_0 ($X^* = 0$).

Two equilibrium states, X_0 and X_2, are locally stable, as the dynamics immediately around the equilibrium brings the population density back to equilibrium. In other words, the net population growth, $R - M$, takes a negative value ($dX/dt < 0$; the population density decreases) when X is slightly larger than the equilibrium density, while taking a positive value ($dX/dt > 0$; the population density increases) when X is slightly smaller than the equilibrium density. X_1 is a locally unstable equilibrium, as the net population growth has a sign that brings the population density away from equilibrium ($dX/dt > 0$ for $X > X^*$ and $dX/dt < 0$ for $X < X^*$).

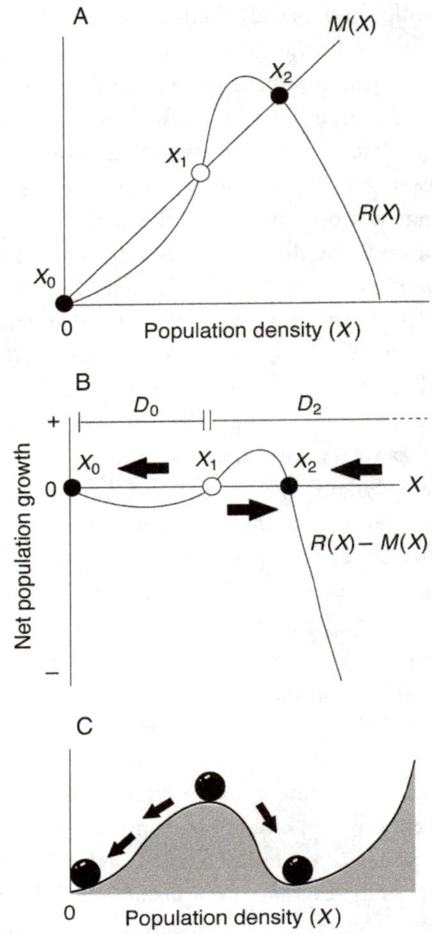

FIGURE 2 The population dynamics model. (A) The dynamics are driven by birth (R) and death (M) processes. Birth and death are balanced at the equilibrium states, X_0, X_1, and X_2, among which X_0 and X_2 are locally stable (filled circle) while X_1 is locally unstable (open circle). (B) Net population growth, $R - M$, determines the population dynamics. The arrows indicate the direction that the system state moves. D_0 and D_2 are the range of attracting domain for the locally stable equilibrium states, X_0 and X_2. (C) The heuristic model for the population dynamics.

The set of the initial state that leads to the specific stable state, X_i, is called an attracting domain of X_i. The system state may return to the pre-disturbance state as long as the magnitude of departure is within the range of the attracting domain. For example, any initial states within the range $(0, X_1)$ eventually lead to equilibrium state X_0, suggesting that the range D_0 $(0, X_1)$ is the attracting domain of equilibrium state X_0. Population density will return to the pre-disturbance equilibrium, X_0, if the disturbance caused by the perturbation is within the attracting domain, D_0. Similarly, the attracting domain of X_2 is determined as D_2 (X_1, ∞). No initial state leads to equilibrium state X_1, and thus X_1 has no attracting domain. Resilience is defined for each state as the size of the attracting domain.

BIOLOGICAL, TEMPORAL, AND SPATIAL SCALES

An ecological system is characterized by scale-specific structure and function, which are shaped by processes and mechanisms that simultaneously act at different scales. Therefore, either stability or resilience is also specific to the biological, spatial, and temporal scales that are of interest.

First, the biological scale is important. For example, in a community comprising multiple interacting species, we may define stability/resilience either at the population level (e.g., population density as a state variable), the community level (e.g., species richness, food-chain length), or the ecosystem level (e.g., ecosystem functions, such as productivity). It is important that stability/resilience at a biological level does not necessarily represent stability/resilience at another biological level. For example, studies with respect to the effect of species richness on ecosystem functioning have revealed that increased species richness may reduce the temporal variability of ecosystem functioning, while inducing larger fluctuation at the population level. This observation indicates that instability at the population level accompanies stability at the ecosystem level. In addition, resilience is dependent on biological scale. Assume that a community of two competing species, A and B, has two stationary states with both species present, with one state dominated by species A and the other by species B. If species number is selected as a state variable, this system does not lose a species and is thus completely resilient. However, if resilience is evaluated by the population densities of both species A and B, the system is not completely resilient in the context that external perturbation may cause a transition from one state (i.e., dominance of species A) to the other (i.e., dominance of species B).

Second, scale dependence is also observed for space and time, since the process that drives population dynamics varies depending on these scales. A good example is provided by considering a metapopulation that contains a number of local populations connected by weak immigrations. Time scale dependence is observed at individual local populations. Each local population may disappear due to local extinction and, therefore, is not resilient in a short time scale. However, in the presence of migration between the local habitats, the extinct population would be eventually "rescued" by immigration from other local populations. This suggests that a local population is resilient in the long time scale of immigration. Similarly, clear spatial scale dependence occurs in a metapopulation. Theory suggests that in a metapopulation there is a proportion of local habitat at equilibrium that is inhabited by the species. This implies that total population density may be almost constant (stable) at the large scale of a metapopulation, while local population density may largely fluctuate due to local extinction and recolonization.

EXTERNAL PERTURBATION

External perturbation affects ecological systems in two different ways (Fig. 3): (1) disturbance to the state variable and (2) alternation of the internal process that drives system dynamics. These two types of external perturbations are explained here by using the example of a biological community containing multiple species. The state variable is population density, which is driven by birth and death processes.

Disturbance of the State Variable

An external perturbation may directly change the state variable (Fig. 3A). For example, the introduction or removal of individuals, or a whole population, brings

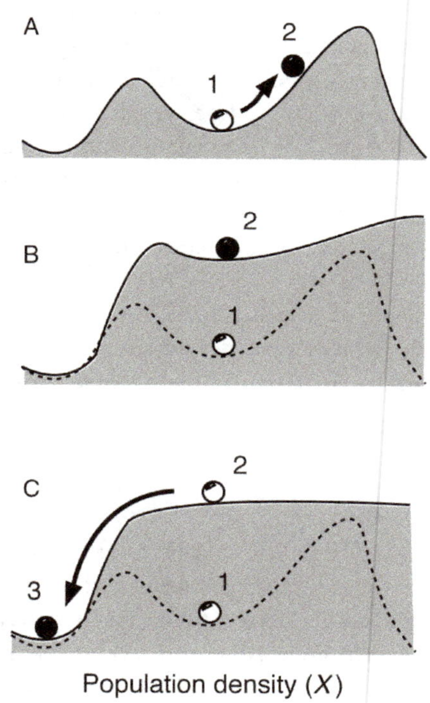

FIGURE 3 Disturbance caused by an external perturbation depicted by the heuristic cup-and-ball model. (A) A direct disturbance to population density (X, state variable). (B) A disturbance to internal process. (C) A disturbance to the internal process, which causes a change in population density. The numbers at the balls represent the time sequence (1 for the initial state).

the system state away from the original one by directly affecting population density or species composition. Biological invasion, which adds a population that has not been present in the system, and intensive exploitation, which removes individuals from the community, are considered to be the major perturbation through which human activity directly affects the biological community.

Alternation of Internal Processes

A perturbation may alter the internal process that drives the dynamics of the state variable, and it may thus affect the stability or resilience of the system. In the case of population dynamics, this perturbation includes the direct or indirect alternation of major ecological processes, such as birth–death processes, immigration, and emigration. A perturbation may alter the state variable and internal processes at the same time. This happens, for example, when the alternation of internal processes in turn cause a change to the state variable (Figs. 3B, C; Fig. 4). "Regime shift" is a typical case, where there is a gradual change in internal processes caused by external perturbations, which leads to an abrupt and nonlinear change in the state variable.

Noting that an external perturbation may cause a range of changes to the system, we may define another type of system property, resistance, which expresses how resistant a system is to external perturbation. Resistance is measured by the magnitude of change in the state variable or internal processes, which is caused by a unit magnitude of external perturbation. When the system is more resistant, it means that a larger perturbation would be required to cause a given magnitude of disturbance to the system. Whether a system may persist against an external perturbation or not would be determined by the complementary effect of resistance and resilience.

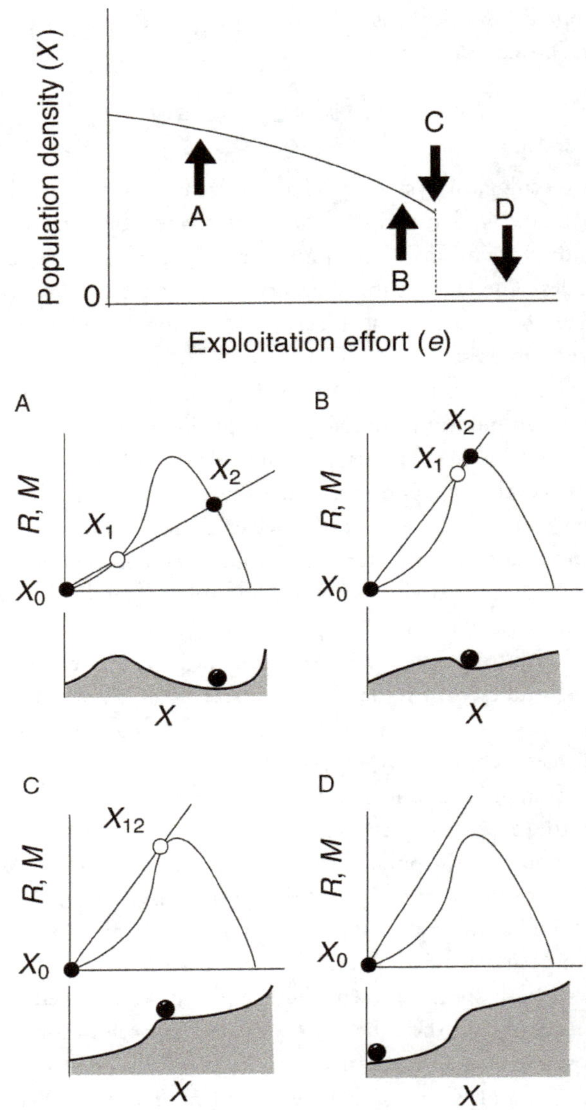

FIGURE 4 The response of population density, X, to increased exploitation effort (e). This is the same model as in Figure 2, except that mortality is given as $M = eX$. The population density, stability and resilience change with increasing exploitation effort. (A) Initial state; assume that equilibrium X_2 is realized. Increase in effort, e, leads to an increase in yield, M, in this phase. (B) The yield, M, is maximized; note that resilience is considerably lower than in (A). A further increase in effort will reduce the yield. (C) The threshold where local stability and resilience is lost for the equilibrium population density (>0). (D) The system has only one equilibrium in which the population is extinct.

MANAGEMENT IMPLICATIONS

Resilience and stability concepts correspond to two alternative views about ecosystems, which yield contrasting approaches to ecosystem management. The "stability" perspective is that an ecosystem is in the unique equilibrium state and, when disturbed, is always able to return to the pre-disturbance state. In this perspective, the alternative stable state would not be taken into account. Hence, the internal process that drives system dynamics is unchanged by external forces. Therefore, the system behavior is predictable and controllable, once an understanding of the process or relationships within the present system is obtained. When the equilibrium view was applied to the management of renewable natural resources, such as food and timber, it contributed toward building an optimal control strategy of the objective system. A typical example is the policy of maximum sustainable yield, which aims to maximize the yield of natural resources obtained from the ecosystem when at equilibrium. In contrast, the "resilience" perspective emphasizes that the system may have alternative states in which an external perturbation may result in a transition between the states or alter the internal process qualitatively. With this view, human activity may reduce system resilience to cause a qualitative change in the system state. The resilience perspective further suggests that an effort to "optimize" the system state may reduce system resilience and lead to an abrupt shift to an unfavorable state. This phenomenon is demonstrated by using the mathematical model described earlier in this entry (Fig. 4). Let us say that natural mortality is negligible and that loss due to death, M, is mainly due to human exploitation, that is, $M = eX$, where e is exploitation rate. As exploitation mortality, M, or the total yield, is equal to fecundity, R, under the assumption of equilibrium, it is clear that the yield, eX, is maximized (i.e., maximum sustainable yield) when the mortality curve intercepts the fecundity curve at maximum fecundity. Noting that population density at the optimal yield is very close to the unstable equilibrium state, X_1, it suggests that under "optimal" resource management a small disturbance may bring the system to the alternative state of population extinction, X_0. This indicates that optimal control under the stability perspective could lead to a "less-resilient" system.

SEE ALSO THE FOLLOWING ARTICLES

Bifurcations / Ecosystem Ecology / Metacommunities / Regime Shifts / Stability Analysis / Stress and Species Interactions

FURTHER READING

Carpenter, S. R., B. H. Walker, J. M. Anderies, and N. Abel. 2001. From metaphor to measurement: resilience of what to what? *Ecosystems* 4: 765–781.

Connell, H. J., and W. P. Sousa. 1983. On the evidence needed to judge ecological stability of persistence. *American Naturalist* 121: 789–825.

Holling, C. S. 1973. Resilience and stability of ecological systems. *Annual Review of Ecology and Systematics* 4: 1–23.

Lewontin, R. C. 1969. The meaning of stability. *Brookhaven Symposium on Biology* 22: 13–23.

Ludwig, D., D. D. Jones, and C. S. Holling. 1978. Qualitative analysis of insect outbreak systems: the spruce budworm and forest. *Journal of Animal Ecology* 47: 315–332.

May, R. M. 1977. Thresholds and breakpoints in ecosystems with a multiplicity of stable states. *Nature* 269: 471–477.

Sutherland, J. P. 1974. Multiple stable points in natural communities. *American Naturalist* 108: 859–873.

RESTORATION ECOLOGY

RICHARD J. HALL

University of Georgia, Athens

Ecological restoration is the process of repairing damage caused by humans to the diversity and dynamics of indigenous ecosystems. Examples of restoration include the removal of harmful nonnative species and repair of residual damage left by these species, conservation or reintroduction of native species crucial to healthy ecosystem function, or returning natural patterns of disturbance and succession to a landscape (e.g., through fire management). Since the success of these interventions is frequently limited by financial, environmental, and time constraints, a body of theory has been developed to optimize the efficiency and effectiveness of restoration strategies.

ECOLOGICAL THEORY AND RESTORATION

Since the establishment of restoration ecology as a recognized subdiscipline of ecology, there have been urgent calls for the development of a theoretical framework to guide practitioners. A growing body of literature has acknowledged the importance of general ecological theory in restoration; for example, knowledge of habitat succession, movement of species across the boundaries of the restored area, and the existence of alternate stable states of communities is necessary for defining realistic endpoints for restoration. This article focuses on the quantitative approaches used to formalize restoration problems, starting with the mathematical formulation of the problem in terms of an objective function and constraints, illustrated with some simple examples, and a review of some solution methods along with a brief discussion of their relative merits and drawbacks.

DEFINING AN OPTIMAL RESTORATION PROBLEM

In order to efficiently deploy limited resources to repair a damaged ecosystem, it is vital to frame the restoration problem in terms of measurable quantities relating to the underlying biology and proposed restoration actions and to devise mathematical relations describing their interaction. It is first necessary to quantify the state variables, the set of management units to be worked with, the set of possible restoration actions and their costs, and the range of values these variables can take. These will then be used to define the objective function, a quantity to be maximized or minimized whose value reflects the success of the restoration activity, and the constraints, a set of biological or financial restrictions limiting the scale of restoration. Depending on the form of the objective function and constraints, an appropriate solution technique is chosen and used to compute the optimal set of restoration actions. This process is summarized in Figure 1.

State Variables

Let $s(t) = s_1(1), \ldots, s_n(t)$ be the set of n state variables of interest at time t. These may represent a set of candidate sites, ecosystem services or species for restoration, or different life stages of a single focal species. The s_i may take binary values (e.g., 0 = unoccupied, 1 = occupied by the target species), a discrete set of integer values (e.g., scores of 1–5 for habitat quality ranging from degraded to pristine, species richness), or vary continuously (e.g., species density, area occupied by a given species).

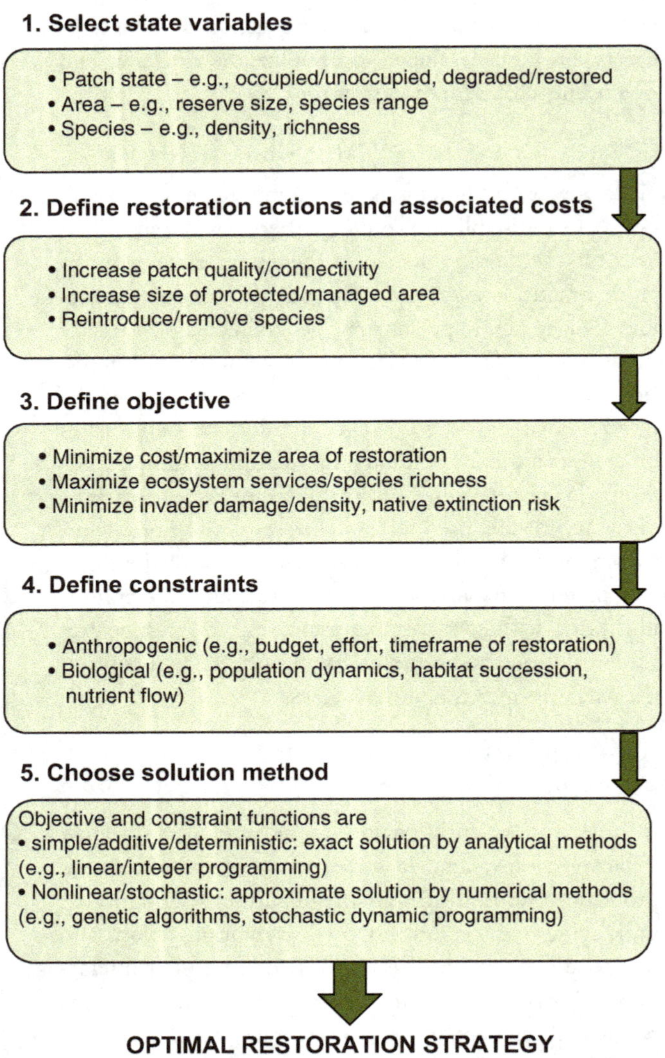

FIGURE 1 Schematic depicting a five-step process for defining an ecological restoration problem in a form amenable to mathematical or computational solution.

Restoration Actions

Let $r(t) = r_1(t), \ldots, r_n(t)$ be the set of restoration actions affecting each of the n state variables above, and undertaken at time t. These actions must be measured in (or transformed to) the same units as the state variables. For the sake of illustration, assume that restoration will take place at discrete time units $t = 1, \ldots, t_{max}$, where t_{max} is the time horizon. At each time step the relationship between the dynamics of the ith state variable pre- and post-restoration, and the restoration action undertaken is described by a function f_i, such that

$$s_i(t+1) = f_i(s_i(t), r_i(t)). \qquad (1)$$

If the outcome of the restoration activity is uncertain, the dynamics are instead described by a set of conditional probabilities that state variable s_i takes on the value $s_i new$ at the next time step, given its value at the previous time step and the action taken,

$$P[s_i(t+1) = s_i\, new \mid s_i(t), r_i(t)]. \qquad (2)$$

Objective Functions

The objective function can take a variety of forms, depending on the specific problem. In many cases, it is expressed as a weighted sum of the state variables or restoration actions at the final time step, or summed over all time steps. Two illustrative examples of objective functions are given below.

EXAMPLE 1: MAXIMIZE BIODIVERSITY IN A MANAGED HABITAT

Suppose the state variables s_i represent the presence or absence of indigenous species i, but that some species are of higher conservation concern than others, as measured by a weighting index w_i. The objective of optimization may be to maximize biodiversity, as measured by a weighted sum of the total number of species present following restoration:

$$\sum_{i=1}^{n} w_i s_i(t_{max} + 1). \qquad (3)$$

EXAMPLE 2: ERADICATE INVADER AT MINIMUM COST

Suppose instead that the s_i represent the area occupied by the ith life stage (e.g., seedling, juvenile, adult) of an invasive species, that the corresponding restoration activities r_i denote the area of stage i removed, and that the aim of restoration is to minimize the total cost of eradication. If the cost of removing one unit of stage i of the invader is c_i, and the economic discount rate is d, the objective function to be minimized is

$$\sum_{t=1}^{t_{\max}} e^{-dt}\left(\sum_{i=1}^{n} c_i r_i(t)\right). \tag{4}$$

Example 1 is a simple illustration of a more general class of objective functions that seek to maximize the total *utility* from a set of biodiversity *assets* (e.g., species, habitat types, ecosystem processes). Example 2 typifies a class if objective functions aiming to minimize the restoration *effort* (e.g., total restoration cost, total number of sites restored) needed to meet a given restoration goal (e.g., invader eradication, all native species represented in at least one reserve in a network). For damaged ecosystems, an appropriate objective function may aim to minimize both the cumulative effects of past damage and future risks to ecosystem health. For example, some invasive species leave lasting damage through changes to the physical or chemical environment that slow or prevent natural recovery of the invaded community, and they may recolonize restored areas if eradication is impossible during the time frame of restoration. In this case, an appropriate objective may be to minimize the sum of residual damage (a function of the cumulative amount of the invader removed) and future colonization risk (a function of the density of the invader remaining at the end of the restoration period). If uncertainty is incorporated explicitly into the model formulation, the aim of the restoration will be to optimize the expected value of the objective function.

Constraints

Many restoration problems are subject to budget constraints. If c_i is the per unit cost of restoration activity r_i, and $C(t)$ is the total budget available at time step t, then at each time step the budget constraint can be expressed

$$\sum_{i=1}^{n} c_i r_i(t) \leq C(t). \tag{5}$$

Frequently, the units of the state variables and the restoration actions (e.g., area, population density) are constrained to be positive, yielding the biological constraints

$$s_i,\ r_i \geq 0 \text{ for all } i. \tag{6}$$

Finally, if the objective is to minimize the cost or effort required to achieve a given outcome, this outcome appears as a constraint. In the case of the invader eradication at minimum cost (example ii), the invader density at all n life stages must be exactly zero after the final round of restoration, i.e.,

$$s_i(t_{\max} + 1) = 0 \text{ for all } i. \tag{7}$$

SOLUTION METHODS

There are many potential solution methods for optimization problems, each with advantages and disadvantages. Computational limitations dictate that simplifications are necessary, either in describing the underlying biology of the system, the functional form of the objective and constraints, the time horizon of restoration, or the number of species, patches, or states considered; where these simplifications are made depends on the actual system being considered for restoration and the judgment of the user. A brief survey of solution methods, and their relative strengths and weaknesses, follows.

If the inherent variability in the biology, environmental conditions, and success of a planned restoration action is relatively low, a deterministic modeling framework may be appropriate, whereby the dynamics of the system can be described by systems of (continuous-time) differential or (discrete-time) difference equations. If, in addition, the expressions describing the dynamics, the objective function, and constraints are sufficiently simple, these systems can be solved exactly and rapidly for the optimal restoration strategy. Consider the example of a colonizing invasive species whose population dynamics are effectively density independent, where the objective is to minimize population size at the end of the restoration period, and removal costs are directly proportional to the number of individuals removed. If removal occurs at discrete time steps the system can be solved using linear programming, or if removal occurs continuously the calculus of variations may be the appropriate solution method. While real-life restoration problems are rarely this simple, the fast, efficient, and exact solution of the simplified system may yield key insights by allowing longer time frames and exhaustive sensitivity analyses to be conducted that may not be possible using more computationally intensive solution methods.

Unfortunately many restoration problems cannot be described in such simple terms, and in particular, accounting for uncertain outcomes explicitly in the formulation of the optimization problem is desirable. In this case, a suite of approximate numerical solutions can be determined by computational methods including stochastic dynamic programming, simulated annealing, and genetic algorithms. The chief disadvantage of these methods is the exponentially rapid increase in computation time that comes with increasing either the time horizon of restoration or the number of states considered (e.g., patches, species, population sizes). Additionally, it is difficult to be sure that the algorithm employed is converging on the "true" optimal restoration strategy without undertaking time-consuming sensitivity analyses.

OPTIMAL STRATEGY

When the optimal strategy has been computed, the results will be output as the set of restoration actions $r_{opt} = r_{opt}(1), \ldots, r_{opt}(t_{max})$ yielding the optimal value of the objective function. In many cases, the optimal strategy at any given time step is to prioritize restoration activity on a single state variable. However, unlike ad hoc methods where the sequence of restoration events is arbitrarily predefined, the optimal strategy often switches which state variable is the target of restoration. The timing of this switch is dictated by how much future damage is valued relative to immediate outcomes by the restoration planner as the end of the restoration period is approached. In spite of the challenges involved in defining a parsimonious model of a restoration problem and choosing an appropriate solution method, optimal restoration strategies are usually more efficient and effective than ad hoc or fixed management strategies, and they also provide a valuable baseline analysis for the feasibility of a proposed restoration effort before costly field trials are conducted.

SEE ALSO THE FOLLOWING ARTICLES

Conservation Biology / Diversity Measures / Dynamic Programming / Ecological Economics / Ecosystem Services / Invasion Biology / Optimal Control Theory / Reserve Selection and Conservation Prioritization

FURTHER READING

Choi, Y. D. 2004. Theories for ecological restoration in changing environment: toward "futuristic" restoration. *Ecological Research* 19: 75–81.

Cipollini, K. A., A. L. Maruyama, and C. L. Zimmerman. 2005. Planning for restoration: a decision analysis approach to prioritization. *Restoration Ecology* 13: 460–470.

Epanchin-Niell, R. S., and A. Hastings. 2010. Controlling established invaders: integrating economics and spread dynamics to determine optimal management. *Ecology Letters* 13: 528–541.

Hall, R. J., and A. Hastings. 2007. Minimizing invader impacts: striking the right balance between removal and restoration. *Journal of Theoretical Biology* 249: 437–444.

Hobbs, R. J., and D. A. Norton. 1996. Towards a conceptual framework for restoration ecology. *Restoration Ecology* 4: 93–110.

McBride, M. F., K. A. Wilson, J. Burger, Y. Fang, M. Lulow, D. Olson, M. O'Connell, and H. P. Possingham. 2010. Mathematical problem definition for ecological restoration planning. *Ecological Modeling* 221: 2243–2250.

McCarthy, M. A., C. J. Thompson, C. Hauser, M. A. Burgman, H. P. Possingham, M. L. Moir, T. Tiensin, and M. Gilbert. 2010. Resource allocation for efficient environmental management. *Ecology Letters* 13: 1280–1289.

Westphal, M. I., M. Pickett, W. M. Getz, and H. P. Possingham. 2003. The use of stochastic dynamic programming in optimal landscape reconstruction for metapopulations. *Ecological Applications* 13: 543–555.

Williams, J. C., C. S. ReVelle, and S. A. Levin. 2004. Using mathematical optimization models to design nature reserves. *Frontiers in Ecology and the Environment* 2: 98–105.

RICKER MODEL

ERIC P. BJORKSTEDT

Southwest Fisheries Science Center, Santa Cruz, California

The Ricker model is one of the simplest possible models for characterizing the density-dependent dynamics of a population. Since its early development by W. E. Ricker (for whom the model is named) and others, it has been used to describe the relation between stock and recruitment that forms one of the foundations of the population models on which quantitative fisheries management rests. Moreover, ever since May and colleagues focused ecologists' attention on the ability of simple models to describe highly complex population dynamics, the Ricker model has served as both touchstone and tool for theoretical and applied ecologists.

ECOLOGICAL MOTIVATION

The Ricker model is one of several simple models for describing density-dependent population dynamics for a single species in terms of changes in abundance observed at discrete intervals. It is well suited for studies of species for which annual (or generational) estimates of abundance adequately characterize the dynamics of populations and as a model for transitions among life history stages, such as production of offspring or survival from one stage to the next. The latter application is common practice in fisheries science, where the Ricker model is often used to relate production of recruits (young fish that survive to join the population, or "stock") to some index of stock size, such as abundance, total biomass, or total spawning potential of adult fish, as part of a comprehensive population model. With few exceptions (such as some semelparous salmonids and short-lived crustaceans) the age structure in fish populations dampens fluctuations in biomass due to variable recruitment.

To provide ecological context for the remaining discussion, let us first briefly examine the sort of ecological mechanisms that can generate dynamics described by the Ricker model (or one of its analogues). The Ricker model describes dynamics that exhibit overcompensatory density dependence, which is characterized by a transition from increasing to decreasing recruitment as population size increases. Ecological mechanisms that can generate such dynamics may be broadly classified as either direct inhibition by the existing population of recruitment from a new cohort or scramble competition within a population.

Examples of direct inhibition include cannibalism (e.g., cannibalism of eggs and larvae by adult planktivorous fishes or of newly settled juveniles by adult demersal fishes) and preemption of resources (e.g., space for planktonic larvae to settle into benthic habitats). In these cases, the total number of offspring produced and the rate at which they die prior to recruitment both increase with adult abundance. The proportion of offspring that survive is determined by integrating mortality over time and so declines steeply with increasing population size. Recruitment is the product of the number of offspring produced and the fraction that survive. Thus, as density-dependent effects on survival strengthen with increasing population size, recruitment will be seen to decline once population size becomes sufficiently large (Fig. 1).

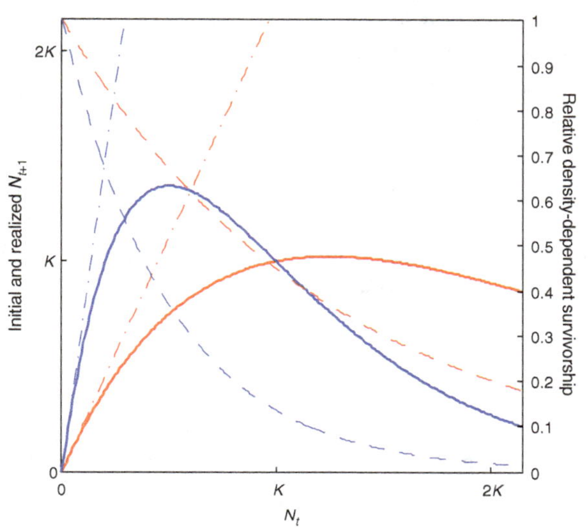

FIGURE 1 Illustration of spawner-recruit curves for a semelparous population generated by overcompensatory density dependence as a result of the combined effect of opposing responses of cohort size (dash-dot lines) and survivorship (dashed lines) to initial population size.

Scramble competition, whether it involves all individuals in a population or occurs at relatively local scales among neighbors, describes scenarios in which individuals compete by consuming resources and thereby reducing the resources available to others. For example, competition among juveniles for food can affect success in completing critical life history stages (e.g., achieving metamorphosis before the juvenile habitat deteriorates, growing out of a size range susceptible to high predation rates). As a general consequence of scramble competition, each additional individual reduces all individuals' (average) share of resources. As a consequence, at sufficiently high abundance, an additional individual does not itself contribute sufficiently to production to offset the cumulative effects of this loss across all individuals in the population. Again, this causes total production to decline at sufficiently large population sizes.

A common characteristic of these ecological mechanisms is that the strength of density-dependent control is a function of the size of the population at the start of the period over which recruitment is determined. This means that density dependence has a delayed, rather than instantaneous, effect on the population, which is a necessary condition for a population to exhibit overcompensatory density dependence. For this reason, the Ricker model is sometimes referred to as exhibiting "stock dependence" to distinguish it from models, such as the Beverton–Holt model, in which within-interval dynamics depend on population size as it changes over time and thus allow the population to exhibit a compensatory response to density-dependent attrition.

Before proceeding, we must recognize that the Ricker model is not a universal consequence of direct inhibition or scramble competition. Rather, its derivation as a description of population dynamics depends on particular assumptions regarding how the strength of density-dependent population control is related to abundance (see Box 1).

MATHEMATICAL EXPRESSION

The Ricker model is in a class of equations known as first-order difference equations that have the general form $N_{t+1} = f(N_t)$ and which describe how a quantity—in this case, abundance—changes (recursively) over discrete periods of time. It can be written in any of several forms, each of which can be simplified to obtain $f(N) = aNe^{-bN}$, where a is a (positive) slope that represents maximum per capita productivity and b governs how strongly density-dependent mechanisms reduce per capita productivity with increasing

BOX 1. EXAMPLE DERIVATIONS OF THE RICKER MODEL

DIRECT INHIBITION

Let us consider recruitment in a species of which each adult produces f offspring (the initial size of the cohort is fN) that then die at a constant rate, Z. Assume that this mortality rate can be partitioned into a density-independent component (c) and a density-dependent component that increases linearly (with slope b) with N (and that N does not change appreciably over the period when recruitment is determined). Integrated from t to $t + 1$, this yields $fNe^{-(c+bN)}$ survivors that recruit to—or in the case of an annual species, reconstitute—the population. Combining the density-independent terms (i.e., defining $a = fe^{-c}$) yields aNe^{-bN} as the expression for recruitment to the population.

SCRAMBLE COMPETITION

Under scramble competition, mean per capita productivity depends on the proportion of individuals with n neighbors, Pr(n), and the per capita productivity associated with having n neighbors, $a(n)$, summarized as $\bar{a} = \sum_{n=n}^{\infty} a(n)\text{Pr}(n)$. Within this framework, it can be shown generally that the Ricker model obtains when (1) individuals are spread randomly over an uniformly distributed resource, so that the number of neighbors within an individual's home range h has a Poisson distribution, and (2) each neighbor reduces an individual's reproductive output from its maximum value, a, by a factor c (i.e., $a(n) = ac^n$ for $c \epsilon (0,1)$). In this case, average per capita reproductive output simplifies to aNe^{-bN}, where $b = (1 - c)\ h/A$ and A is the total area (resource) available to the population.

abundance N. Where such an interpretation is sensible, b can be further parsed as the ratio of a population's intrinsic productivity, which determines the speed with which it can saturate its environment, and the capacity of the environment to support the population (cf. the exponential logistic model, below).

Of the several equivalent forms of the Ricker model, one of the more commonly encountered is the "discrete time" exponential logistic model, $N_{t+1} = N_t e^{r(1-N_t/K)}$, which at each time t gives future abundance, N_{t+1}, as a function of current abundance, N_t, and density-dependent dynamics governed by the population's intrinsic productivity, r, and its carrying capacity, K. In fisheries applications, where it is used to predict recruitment R from spawning stock S, the Ricker model is commonly written as $R = \alpha S e^{-\beta S}$, where α $(= e^r)$ is the maximum number of recruits produced by each unit of the spawning stock and β $(= r/K)$ scales the intensity of density dependence affecting recruitment to the size of the spawning stock. Here, K is the stock that produces recruits at a rate sufficient to replace the population over the next generation after accounting for all subsequent (density-independent) mortality.

The Ricker model describes a dome-shaped relation between N_{t+1} and N_t whose shape, in terms of the height, width, and position of the peak relative to K depends solely on the intrinsic productivity of the population. Total production, whether expressed as abundance or recruitment, reaches a maximum value of Ne^{r-1} at $N = K/r$, or equivalently, $N = 1/\beta$. Such plots (and the associated model) are often referred to as Ricker "maps" that translate the state of the population (abundance) from one time to the next. Similar plots of R on S are commonly called stock-recruit or spawner-recruit curves. Note that if $r < 1$, the peak of the spawner-recruit curve lies to the right of K, and overcompensatory density dependence may not be readily discerned (or even experienced) in the resulting dynamics or empirical time series (Fig. 1).

DYNAMICS

The Ricker model can generate dynamics—in terms of changes in abundance from one generation to the next—with a remarkable range of qualitatively different behaviors, the particulars of which are determined entirely by the intrinsic productivity of the population. In order of increasing productivity, these dynamics yield (1) a monotonic approach to K, (2) damped oscillations that eventually converge on K, (3) stable limit cycles—repeating sequences of abundances that increase in period length with increasing r—and (4) deterministic chaos including apparently random fluctuations, quasi-cycles, and sporadic outbreaks (Fig. 2). Of course, r must be greater than zero or the population will decline to extinction. (Note that when cast in continuous time as a differential equation, the Ricker model converges directly to the steady state at K and thus lacks complex behavior.)

Linking the dynamics of a particular Ricker model to the shape of its spawner–recruit curve offers a means to understanding the progression from one sort of dynamics to another in relation to changes in intrinsic productivity in conceptual terms. It is readily seen that overcompensatory density dependence is the essential characteristic of the Ricker model (and others like it) that allow oscillations to be an inherent part of a

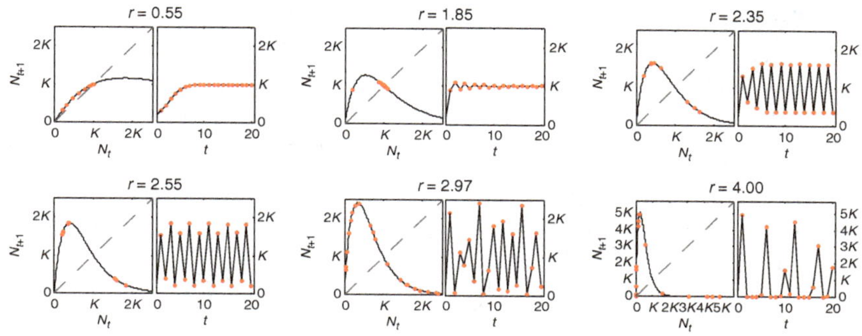

FIGURE 2 Examples of spawner-recruit curves for a semelparous population and dynamics described by Ricker models for a range of values for intrinsic productivity, r. Each panel illustrates the spawner-recruit relation between N_{t+1} versus N_t and the resulting trajectory of population sizes starting from an initial population size of 0.2K.

population's dynamics. To illustrate, let us first consider a case of a population with relatively low intrinsic productivity ($r < 1$) for which recruitment continues to increase with increasing abundance when the population is near K. Such a population cannot pass K from above or below—notice, for example, how $N_t < K$ always yields $N_{t+1} < K$—so the approach to the steady state is monotonic. In contrast, when productivity is higher ($r > 1$), maximum recruitment occurs for some at some population size less than K so that a population somewhat smaller than K will grow to exceed K in the next generation, while density-dependent processes force any population greater than K to drop below K in the next generation. In such cases, the dynamics include oscillations. The nature of these oscillations—whether they are damped and how strongly, whether they persist as a stable limit cycle, or whether they are chaotic—can also be understood in terms the shape of the spawner–recruit curve as it maps population size in one generation to the next (Fig. 2).

In summary, the dynamics depend on how strongly a population is affected by overcompensatory density dependence, which boils down to how strongly a population can overshoot its carrying capacity (if at all) and how dire are the consequences for doing so. Complex dynamics are most readily illustrated using examples based on semelparous species with nonoverlapping generations. In contrast, complex dynamics are less likely to be observed in long-lived, iteroparous species in which age structure in the population dampens the effect of variable recruitment. However, complex dynamics may exist whenever intrinsic productivity is sufficiently high and the life history of a species preserves the delay in density-dependent population control necessary for overcompensatory dynamics to emerge.

APPLICATIONS

By virtue of its simple structure, grounding in ecological first principles, and capacity to represent rich dynamics, the Ricker model is a powerful tool for theoretical ecologists, both as a building block for models of spatial ecology or species interactions and as a simplification of more complex models of ecological processes. This utility is enhanced by the ease with which the Ricker model can be cast in stochastic form and extended to include additional ecological dynamics, such as depensation, species interactions, dispersal, and so on. As such, the Ricker model is useful as a deterministic framework for exploring how density dependence, demographic stochasticity, and environmental variability combine to drive populations' dynamics and for predicting how these dynamics might respond to changes in productivity due to, say, harvest strategies, population enhancement, or climate change.

The same characteristics that make the Ricker model attractive to theoretical ecologists make it useful in diverse empirical applications. Not surprisingly, the Ricker model has been extensively used in fisheries ecology and quantitative fisheries management where it is often used to describe the stock–recruit relationship that underpins estimates of stock status used to guide harvest management decisions. The Ricker model has been applied widely in this manner, especially in analysis of stocks of anadromous salmon and demersal marine fishes. It has also proven useful as a model for the dynamics of highly variable populations, such as harvested crustaceans that exhibit boom-and-bust dynamics and insect populations that exhibit rare outbreaks of spectacular magnitude. Moreover, the simple structure of the Ricker model makes it a useful candidate model when assessing ecological datasets for evidence of density dependent dynamics in natural systems.

SEE ALSO THE FOLLOWING ARTICLES

Beverton–Holt Model / Cannibalism / Chaos / Difference Equations / Fisheries Ecology / Single-Species Population Models / Spatial Ecology

FURTHER READING

Brännström, Å., and D. J. T. Sumpter. 2005. The role of competition and clustering in population dynamics. *Proceedings of the Royal Society B: Biological Sciences* 272: 2065–2072. doi: 10.1098/rspb.2005.3185.

Geritz, S. A. H., and É. Kisdi. 2004. On the mechanistic underpinning of discrete-time population models with complex dynamics. *Journal of Theoretical Biology* 228: 261–269. doi:10.1016/j.jtbi.2004.01.003.

Hilborn, R., and C. J. Walters. 1992. *Quantitative fisheries stock assessment: choice, dynamics, and uncertainty.* New York: Chapman and Hall.

May, R. M., and G. F. Oster. 1976. Bifurcations and dynamics complexity in simple ecological models. *American Naturalist* 110: 573–599.

Needle, C. L., 2002. Recruitment models: diagnosis and prognosis. *Reviews in Fish Biology and Fisheries* 11: 95–111.

Quinn, T. J., and Deriso, R. B., 1999. *Quantitative fish dynamics.* New York: Oxford University Press.

Turchin, P. 2003. *Complex population dynamics: a theoretical/empirical synthesis.* Princeton: Princeton University Press.

S

SELECTION

SEE MUTATION, SELECTION, AND GENETIC DRIFT

SEX, EVOLUTION OF

JAN ENGELSTÄDTER

Institute for Integrative Biology, ETH Zürich, Switzerland

FRANCISCO ÚBEDA

University of Tennessee, Knoxville

The evolution of sex is one of the largest and most fertile areas of research in evolutionary biology. It encompasses multiple unresolved questions involving a number of evolutionary steps: the evolution of the recombination machinery, the evolution of meiosis, the differentiation into sexes, the differentiation of gametes produced by each sex, and, finally, the maintenance of sex despite its two-fold cost.

A MOLECULAR MACHINERY ENABLING RECOMBINATION

Sex and recombination through meiosis are confined to eukaryotes. However, a complex molecular machinery enabling homologous recombination between different DNA molecules was already present in prokaryotes long before the first eukaryotes evolved. The original function of this machinery lies in DNA repair, as was indicated by early experiments in *E. coli* showing that mutants deficient in genes involved in homologous recombination are highly sensitive to DNA damaging agents. Three types of DNA damage are repaired in bacteria through mechanisms that involve homologous recombination: double-strand breaks, stalled replication forks, and single-stranded DNA arising from incomplete replication.

Recombinational repair of double-strand breaks (outlined in Fig. 1) is the most relevant mechanism with respect to the evolution of sex in eukaryotes, where double-strand breaks are induced during meiosis I to initiate crossovers. In bacteria, the key player in this process is the recombinase RecA, a DNA-dependent ATPase. Assisted by a host of other proteins, RecA binds to single-stranded DNA forming a helical filament, mediates the search for homology in other DNA molecules, and catalyzes strand invasion and branch migration. Homologues of RecA are found in virtually all organisms, in eukaryotes as Rad51 and the exclusively meiotically active Dmc1 and in the Archaea as RadA.

Even though the main function of the recombination machinery in bacteria is to repair DNA damage, it should be stressed that already in bacteria this machinery is sometimes employed to effect sexual processes. Three such mechanisms of recombination have been identified: conjugation (the exchange of plasmids), transduction (the transfer of DNA mediated by phages), and transformation (the uptake and integration of free DNA from the environment). While the evolution of former two processes is most parsimoniously explained as by-products of the action of autonomous genetic elements (plasmids and phages, respectively), the evolution of the ability to engage in transformation is more difficult to explain. Aside from fulfilling a similar function as sex in eukaryotes, i.e., to increase genetic variation, it has been suggested that transformation evolved to facilitate DNA repair or that it is simply a way of taking up nutrients.

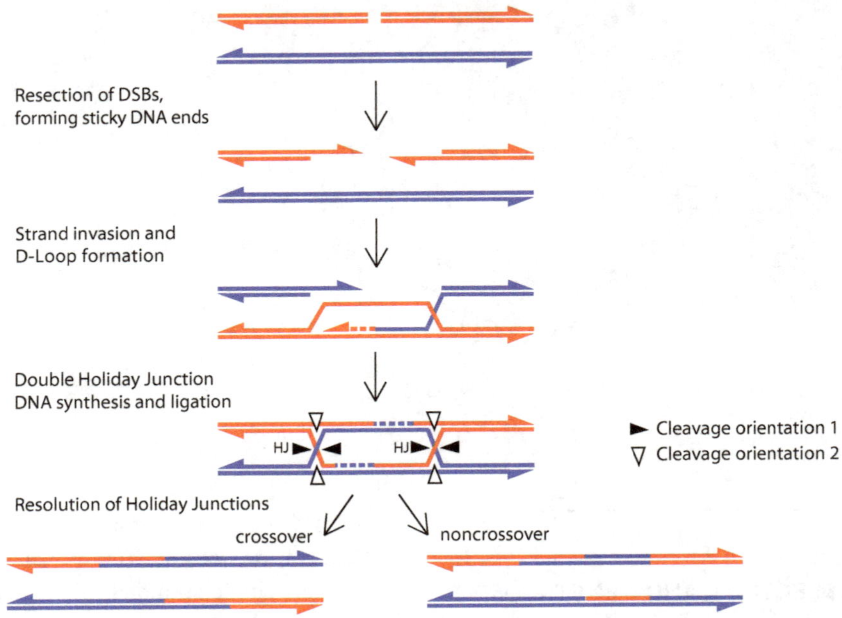

FIGURE 1 Double-strand break repair mechanism for the initiation of recombination. The blunt ends of the double-stranded DNA are partially digested, revealing single-stranded 3′ DNA ends. One of these strands then invades the homologous DNA molecule, forming a D-loop structure. Branch migration and ligation leads to the formation of two holiday junctions. These holiday junctions are resolved producing either crossover or noncrossover products.

MEIOSIS AND THE ALTERNATION OF GENERATIONS

Most eukaryotes are characterized by a life cycle termed the "alternation of generations," in which a diploid phase of cell division alternates with a haploid phase. The transition between the diploid and the haploid phase is mediated by meiosis, a form of cell division that reduces the genomic content of cells by one-half. Meiosis starts with a cell that contains two chromosome sets inherited from two gametes (diploid cell) and results in four cells with one chromosome set each. Because of random segregation of chromosomes and crossover events between chromosomes during pairing, meiosis produces haploid cells with unique combinations of genes. Syngamy (or fertilization), on the other hand, mediates the transition from a haploid generation back to a diploid generation. In plants, the haploid generation can contain a multicellular phase before gametes are produced. In animals, the haploid generation is reduced to the production of gametes.

Compared to a genetic system of clonal reproduction, the alternation of generations life cycle has two characteristics that may explain its evolution. The first is the diploid phase. Spending a prolonged time in the diploid phase may be advantageous because it allows the masking of deleterious recessive alleles. The second characteristic is recombination, the shuffling of alleles coming from two different haploid genomes during meiosis.

From a purely population genetics perspective, all that recombination does is to reduce statistical associations—the linkage disequilibrium (LD)—between alleles at different loci. Thus, the problem of why recombination is so prevalent in natural populations boils down to the questions of what forces generate LD and under what conditions is there selection to destroy LD. LD can be generated by a number of population genetic factors, including epistasis (nonindependent fitness effects of mutations at different loci), random genetic drift (changes in gene frequencies due to random sampling in finite populations), migration, and sexually antagonistic selection. Selection can operate on recombination rates in two ways that in combination determine whether or not recombination is advantageous. First, there is a direct effect stemming from the fitness of sexually produced offspring: when LD and epistasis are of opposite sign, there will be selection for recombination (because the offspring will then be disproportionately fit), but when LD and epistasis are of the same sign, there will be selection against recombination. Second, when LD is negative (i.e., there is an excess of genotypes of intermediate fitness), recombination can be favored because by bringing LD closer to zero the genetic variance in the population is increased, thus increasing the efficiency of natural selection.

Based on these population genetic principles, many hypotheses have been proposed to account for the advantage of recombination. For example, the deterministic mutational theory posits that negative LD is produced through deleterious mutations with negative epistasis, and recombination is then selected for because it allows a more efficient purging of these mutations. According to a second theory, negative LD is produced by the interplay between random genetic drift and selection (the Hill–Robertson effect), which has been shown to result in strong selection for recombination even in large populations, especially when many loci are considered. Finally, sex and recombination can be favored through antagonistic coevolution between hosts and parasites (the Red Queen hypothesis). In this scenario, both LD and epistasis will fluctuate over time, and depending on how hosts and parasites interact with each other genetically, there can be selection for or against recombination. A number of excellent review articles on these and other hypotheses for the evolutionary advantage hypothesis are available.

MATING TYPES

Mating types are the different types of gametes that can fertilize other gametes in a sexually reproducing organism. Most species show two different mating types (male and female, + and −, *a* and α), but some species of fungi can present several thousands.

This differentiation into mating types might be the outcome of selection to facilitate finding mates. In order for gametes to fertilize other gametes, they need to attract and/or be attracted by other gametes. Evolutionary models show that there may be selection for some cells to specialize in attracting gametes (by producing pheromones), while others specialize in becoming attracted (by expressing pheromone receptors).

Another theory argues that mating types might be the result of selection to coordinate the inheritance of cytoplasmic genomes (for example, mitochondrial genes) so as to limit competition between unrelated cytoplasmic genomes. Fusion of isogamous gametes brings together cytoplasmic genes from different lineages that may compete to favor their own transmission to the next generation. Intragenomic conflict reduces the fitness of the organism and creates the context for the invasion of a nuclear gene that enforces the inheritance of cytoplasm from a single mother cell. However, if competitive genes are associated with an over-transmission cost, nuclear genes enforcing uniparental inheritance do not go to fixation. In such a polymorphic population, nuclear genes that act to prevent its gametes from fusing with gametes produced by an individual that carries the same suppressor are at a selective advantage, and mating types may evolve.

If there is a selective pressure for mating types to evolve, how many mating types should evolve? While in most cases there are two mating types, some organisms can have thousands of mating types. With multiple mating types, the probability that a gamete finds another gamete of a compatible mating type becomes an important issue. The modus operandi of mating types is such that either gametes can fuse with gametes of a specific mating type or gametes can fuse with gametes of mating types other than their own. When gametes can fuse with gametes of a specific mating type, the probability of randomly finding a compatible mating type is at a maximum when the number of mating types is two (50% chance) and at a minimum when the number of mating types tends to infinity (0% chance). When gametes can fuse with gametes of mating types other than their own, the probability of randomly finding a compatible mating type is at a minimum when the number of mating types is two (50% chance) and at a maximum when the number of mating types tends to infinity (100% chance in the limit). The latter modus operandi is the most common in nature, which generates the paradoxical situation of most organisms having two mating types only when this is the number of mating types with the lowest probability of success.

ANISOGAMY AND MOBILE GAMETES

Anisogamy refers to the production of gametes that differ, generally in size, as opposed to their being identical (isogamy).

One way to explain the evolution of this asymmetry is assuming that there is a tradeoff between productivity (more gametes are better than few gametes) and survival of zygotes (bigger zygotes have greater survival than smaller ones). This creates the context for the evolution of sexual antagonism, with one of the sexes acting as a cheater that withholds resources to produce more gametes and the other sex contributing the resources withheld by the first gamete to preserve zygotic viability. This sexual conflict results in males producing small gametes that are viable only because of the resources contributed by females.

The previous model, however, does not take into consideration how gamete density affects the probability of fertilization. When all gametes do not find a partner to fuse with and when small gametes have a higher motility—thus increasing encountering rates with large gametes—there can be additional selective pressure for anisogamy. There are also scenarios in which gamete limitation in itself can be sufficient to select for anisogamy. Thus, anisogamy does not necessarily respond to the logic of sexual conflict; rather, it might be beneficial for both sexes.

Anisogamy creates a situation where in the extreme case that sperm do not contribute any resources to the zygote (and where there is also no other paternal contribution to offspring fitness), sex entails a twofold cost. This cost arises because under this assumption, asexual females can produce the same number of offspring as sexual females, but they avoid "diluting" their genome with paternal genetic material when producing offspring. Thus, in the absence of strong selection for sex through recombination, a clonally reproducing mutant is expected to spread rapidly in a sexual population of males and females.

SECONDARY LOSS OF SEX

Most multicellular organisms—especially animals, the focus of this section—have a genetic system that involves obligate sex as well as male and female sexes (although not necessarily in separate individuals). Because of the twofold cost of sex, this near ubiquity of sex is even more difficult to explain than explaining how sex and recombination evolved in the first place. Some species, however, have re-evolved the ability to reproduce asexually, either partially or completely, thus offering important opportunities to investigate the evolutionary forces that maintain sex within populations and to test hypotheses for the advantage of recombination.

Partial loss of sexual reproduction is relatively common in multicellular organisms, and it is characteristic of several large taxa. Some groups of animals—for example, aphids, waterfleas, and monogonont rotifers—reproduce mainly asexually, but under certain conditions males and females are produced and mate (facultative sex or cyclic parthenogenesis). In other groups, sexual and obligatorily asexual individuals coexist, although often with different geographical distributions (geographical parthenogenesis).

By contrast, complete abandonment of sexual reproduction is rare among multicellular organisms. For example, there are fewer than 100 parthenogenetic vertebrate species. The genetic basis for parthenogenetic reproduction varies among groups and includes single mutations, hybridization (possibly the only cause of parthenogenesis in vertebrates), and maternally inherited bacteria. Moreover, there is a great diversity in cytogenetic mechanisms by which offspring are produced asexually.

Parthenogenetic species exhibit a "twiggy" distribution in phylogenetic trees; i.e., they do not form large and old clades. Two groups of parthenogenetic animals—the bdelloid rotifers and darwinulid ostracods—have long thought to be exceptions to this rule. Bdelloid rotifers, however, have recently been shown to engage in extensive horizontal gene transfer, incorporating genetic material from a wide range of organisms into their genome. This process may also involve homologous replacement of genes from related organisms. Similarly, recent observation of males in one species of darwinulids has cast doubts on this group's status of ancient asexuals. However, it is not clear at present whether or not sex does indeed occur within this group, and if so, how common it is.

The twiggy distribution of asexuals is typically explained through adverse long-term consequences of the absence of recombination. According to this view, asexual species arise occasionally, but because of their reduced rate of adaptation and an accumulation of deleterious mutations, these asexual species become extinct quickly. However, this notion has not been rigorously tested to date, and alternative explanations for the scattered distribution of parthenogenetic species in phylogenetic trees exist. In particular, reduced rates of speciation in asexual species could produce similar phylogenetic distributions. Whether or not sexual or asexual species should be expected to have higher speciation rates is an unresolved question whose answer depends on the importance of factors like adaptation, geographical isolation, and random genetic drift in the process of speciation.

Another key factor are the rates of transition from sexual to asexual reproduction and vice versa. The former rate is expected to be small in many groups because of a variety of genetic and developmental constraints on evolving parthenogenetic reproduction. On the other hand, re-evolving sex once it has been lost for a long time is often considered impossible. Newly developed statistical methods make it possible in principle to simultaneously estimate extinction, speciation, and transition rates of sexual vs. asexual species from phylogenetic trees, but it remains to be seen whether the available data are sufficient to allow reliable estimates.

SEE ALSO THE FOLLOWING ARTICLES

Coevolution / Evolutionarily Stable Strategies / Mating Behavior / Mutation, Selection, and Genetic Drift / Phenotypic Plasticity / Phylogenetic Reconstruction

FURTHER READING

Bell, G. 1982. *The masterpiece of nature: the evolution and genetics of sexuality.* London: Croom Helm.

Hoekstra, R.F. 1987. The evolution of sexes. In S.C. Stearns, ed. *The evolution of sex and its consequences.* Basel, Switzerland: Birkhauser.

Lessels, C.M., R.R. Snook, and D.J. Hosken. 2009. The evolutionary origin and maintenance of sperm: selection for a small, motile gamete mating type. In T.R. Birkhead, D.J. Hosken, and S. Pitnick, eds. *Sperm biology.* Oxford: Academic Press.

Maynard Smith, J. 1978. *The evolution of sex.* Cambridge, UK: Cambridge University Press.

Otto, S.P. 2009. The evolutionary enigma of sex. *American Naturalist* 174: S1–S14.

Persky, N.S., and S.T. Lovett. 2008. Mechanisms of recombination: lessons from *E. coli. Critical Reviews in Biochemistry and Molecular Biology* 43: 347–370.

Salathé, M., R.D. Kouyos, and S. Bonhoeffer. 2008. The state of affairs in the kingdom of the Red Queen. *Trends in Ecology & Evolution* 23: 439–445.

Schön, I., K. Martens, and P. van Dijk, eds. 2009. *Lost sex: the evolutionary biology of parthenogenesis.* Dordrecht, NL: Springer,.

Schwander, T., and B.J. Crespi. 2009. Twigs on the tree of life? Neutral and selective models for integrating macroevolutionary patterns with microevolutionary processes in the analysis of asexuality. *Molecular Ecology* 18: 28–42.

Vos, M. 2009. Why do bacteria engage in homologous recombination? *Trends in Microbiology* 17: 226–232.

SINGLE-SPECIES METAPOPULATIONS

SEE METAPOPULATIONS

SINGLE-SPECIES POPULATION MODELS

KAREN C. ABBOTT

Iowa State University, Ames

ANTHONY R. IVES

University of Wisconsin, Madison

In ecology, as in other sciences, models are used for three main purposes. They can provide a way to organize ideas and develop hypotheses about how real systems work, they can provide a qualitative understanding of a particular system, and they can make predictions. Simple models are best for conceptual explorations and hypothesis development, whereas detail-rich, system-specific models are needed to make predictions. Simple single-species models provide a means to study populations whose dynamics are predominantly determined by intraspecific interactions. But just as importantly, single-species models also provide many of the building blocks used in multispecies models.

CONCEPTUAL FRAMEWORK

The main challenge to modeling ecological systems is their complexity. Unlike simple dynamical models, such as those for planets revolving around stars, ecological models are never designed to capture precisely the dynamics of a system. Instead, they are caricatures that nonetheless contain the salient features of reality. Thus, it makes sense not to think about the best model for a particular ecological system but instead to think about what model is best for a given question; different questions about the same system will require different models. For example, models can be used to ask whether it is possible for a single species, in isolation from other species, to show perpetually cyclic fluctuations in abundance, or whether competition is sufficiently strong to cause population cycles in a specific species of interest. Models could also be used to predict the abundance of a species in 3 years. The models most appropriate for each of these tasks will be different even if our species of interest remains the same.

This entry focuses primarily on the use of models as thought experiments to understand what is possible in real ecological systems. It discusses simple models that might equally apply to a wide range of systems, without applying realistically to any of them. At their core, all single-species models are equations that dictate how population abundances change through time, but these equations can look and behave quite differently from one another depending on the biological assumptions on which they are based. To model a particular population, the first step is to figure out what assumptions are appropriate for the situation at hand.

Individuals require resources in order to survive and reproduce, and thus to contribute to population growth. What these resources are depends on the species of interest and can include such things as soil nutrients like carbon and nitrogen, food plants, or prey animals. Given these dependences, how is it possible to model just a single species without also modeling its required resources? This rests on the assumption that either the resources never change in availability, or that they are replenished at the rate they are consumed. The key assumption of single-species models is that we can capture the processes that govern the dynamics of a species from information on that species alone. Single-species models, then, might be most appropriate for populations that do not have strong interactions with other species that have dynamics of their own. However, even information about interactions with other species might be captured in the dynamics of a single species, making a single-species model a tool to explore more-than-single-species processes.

Models usually use population density (the number of individuals inhabiting some unit of area) as the measure of population size. This is because many ecological processes require interacting individuals (such as those competing for resources) to be near one another, making how tightly individuals are packed in space more relevant than the absolute number of individuals. Population density is also mathematically convenient because then equations are not constrained to produce only integer values. Using equations to describe the integer number of individuals in a population requires special care, whereas population densities can have noninteger values without losing biological meaning.

Births and immigration cause the population density to increase, while deaths and emigration decrease density. Ecological models differ widely in how they incorporate spatial processes like immigration and emigration. Most basic single-species models only explicitly include information on births and deaths and assume either that the population is closed to dispersal or, equivalently, that immigration and emigration balance each other out. Simple modifications can be made to these models to allow for either a constant rate, or a constant per capita rate, of dispersal into or out of a focal population. Some models (e.g., single-species metapopulation models, reaction–diffusion models, and integrodifference equation models) go even further, keeping track of population dynamics at, and movement of individuals between, multiple points in space.

CONTINUOUS-TIME VS. DISCRETE-TIME MODELS

Population models fall into two categories, continuous-time models and discrete-time models, based on the timing of demographic events. Continuous-time models assume that births, deaths, and other processes that affect population density can occur at any moment in time, and so the population density can change constantly. Discrete-time models instead divide time into discrete time steps, and the population density changes once every time step. The latter may sound artificial, since time itself is continuous. Nonetheless, the life histories of some organisms are such that the discrete-time framework is more appropriate. For a species with nonoverlapping generations, a discrete-time model with a time step of one generation is a better descriptor of population dynamics than a continuous-time model. On the other hand, species with overlapping generations and little seasonality may be better described by continuous-time models.

In continuous-time models, population density is usually denoted $N(t)$. The variable t is time, and $N(t)$ represents the population density at any given point in time. Many authors denote population density simply as N rather than $N(t)$; this notational convenience makes equations look less cluttered and relies on readers to remember that population density is still a function of time. Continuous-time models are typically formulated to tell us *about* the function $N(t)$ without actually telling us what the function itself is. In particular, a continuous-time model specifies how $N(t)$ changes through time by providing an equation for $\frac{dN(t)}{dt}$. Because population growth or decline is due to births from, deaths of, or movements by individuals currently within the population, in makes biological sense that $\frac{dN(t)}{dt}$ will be a function of $N(t)$. The exponential model discussed below is an example of this type of model.

For discrete-time models, population density at time t is more often written N_t. Here, N is still a function of t, but t is no longer a continuous variable and instead takes on discrete values of 0, Δt, $2\Delta t$, $3\Delta t$, . . . , where Δt is the size of the time step. Usually the units of t are adjusted so that $\Delta t = 1$. For instance, a model for an annual plant might measure time in years and have $\Delta t = 1$ yr. Just as in continuous-time models, population growth is due to the demography of individuals currently in the population, so discrete-time models write N_t as a function of N_{t-1}. The geometric model discussed below is an example of this.

The Exponential Model

A population with a constant per capita birth rate b and a constant per capita death rate d will grow or shrink according to the equation

$$\frac{dN(t)}{dt} = (b - d)N(t). \tag{1}$$

Intuitively, the population will grow $\left(\frac{dN(t)}{dt} > 0\right)$ if the birth rate exceeds the death rate ($b - d > 0$); it will remain at the current size indefinitely $\left(\frac{dN(t)}{dt} = 0\right)$ in the special case where the birth rate exactly balances the death rate ($b = d$), and it will shrink $\left(\frac{dN(t)}{dt} < 0\right)$ otherwise ($b - d < 0$). Because the population dynamics depend entirely on the difference between the birth and death rates, it is convenient to combine these two rates into a single parameter, $r = b - d$:

$$\frac{dN(t)}{dt} = rN(t). \tag{2}$$

The parameter r is often called the per capita intrinsic rate of population increase (or some permutation of these words), and it represents the net contribution to population growth, per unit time, of any one individual.

If $r > 0$, $N(t)$ will grow to infinite densities, whereas with $r < 0$, extinction is approached asymptotically.

As noted above, a continuous-time population model is an equation for $\frac{dN(t)}{dt}$, rather than an equation for $N(t)$ itself. For very simple models like the exponential model, the equation for $\frac{dN(t)}{dt}$ can be solved to find $N(t)$. Dividing both sides of Equation 2 by $N(t)$, multiplying both sides by dt, and then integrating both sides from time 0 to time t yields $N(t) = N(0)\exp(rt)$, where $N(0)$ is the initial population density. This solution reveals why Equation 2 is known as the exponential model: the amount that a population grows ($r > 0$) or shrinks ($r < 0$) between times 0 and t is an exponential function of t.

The Geometric Model

The discrete-time analog of the exponential model considers a population with per capita birth and death rates of B and D per time step. The population density at time t will equal the previous density, plus any births and minus any deaths that occurred between times $t - 1$ and t:

$$N_t = N_{t-1} + (B - D)N_{t-1}. \tag{3}$$

Again letting $R = B - D$, we can rewrite Equation 3 as

$$N_t = (1 + R)N_{t-1}. \tag{4}$$

We can see right away that the population will grow if $R > 0$, since this will mean that N_t is more than 1 times N_{t-1}. To solve Equation 4, we can begin with $N_1 = (1 + R)N_0$ and iterate forward to see that $N_2 = (1 + R)N_1 = (1 + R)^2N_0$. Repeating this type of iteration leads to $N_t = (1 + R)^tN_0$, the solution to the geometric growth model. We can rearrange this solution further to highlight its relationship to the exponential model. Rewriting $(1 + R)^t$ as $\exp(\ln(1 + R)t)$ gives us $N_t = N_0\exp(\ln(1 + R)t)$; if we set $r = \ln(1 + R)$, we recover the solution to the exponential model: $N_t = N_0\exp(rt)$. Thus, the model of geometric growth in discrete time is essentially the same as the model of exponential growth in continuous time, with the only difference being that the discrete-time model is sampled discretely. While this equivalence between continuous- and discrete-time models holds for exponential and geometric growth, this is generally not the case for models in which birth and death rates depend on population densities. For those more complicated cases, discrete-time models may exhibit dynamics that are impossible in continuous-time models, as shown below.

DENSITY DEPENDENCE

All real populations must be subjected, at some time or other, to some form of density dependence, meaning that the per capita rate at which a population is growing or shrinking depends on its current density. This is because without some form of density dependence, populations are destined to either grow indefinitely or else decline monotonically toward extinction as in the exponential and geometric models discussed previously. Density dependence refers generically to any situation in which the per capita population growth rate either increases (positive density dependence, known as an Allee effect) or decreases (negative density dependence) with population density. When ecologists use the term density dependence, they almost always mean negative density dependence, and that is the type discussed here.

Density dependence arises when individuals in crowded populations are less able to contribute to population growth than individuals in sparse populations, either because they have higher mortality or lower fecundity, or both, when crowded. These reductions could be due simply to the detrimental effects of spreading the same local quantity of resources among a larger group of individuals, or to harmful interactions like fighting that might occur more at higher densities. They could even be due to other species such as predators, if predators flock to areas in which the focal species is common; in these types of cases, however, for a single-species model to be appropriate, the density of predators would have to be determined solely by the density of the focal species, so that knowledge of the density of the focal species gives full information about the predation rate. To have populations persist indefinitely, rather than become infinite or extinct, low-density populations must be relatively free from density-dependent reductions in population growth rates, whereas larger populations must experience negative density dependence that causes populations to decline. At an intermediate density known as the carrying capacity and often denoted by the parameter K, the per capita population growth rate will be zero, so that if the population were to start exactly at that density it would remain there indefinitely. The question of what happens if the population density is just slightly higher or just slightly lower than the carrying capacity leads to a fundamental difference between simple continuous-time and discrete-time models, as discussed below.

The Logistic Model for Density Dependence in Continuous Time

To model density-dependent population growth in continuous time, we can begin with the exponential model and multiply the $rN(t)$ on the right-hand side by a new term that represents density dependence. We want this term to have a minimal effect (and so ≈1) when the population density is small, and we want to allow very little population growth (≈0) when $N(t)$ approaches the carrying capacity K. For population densities greater than K, the density-dependence term should be negative and thus cause a population decline. The logistic model for population growth is the exponential model with the right-hand side multiplied by the term $1 - \frac{N(t)}{K}$, which has exactly these effects:

$$\frac{dN(t)}{dt} = rN(t)\left(1 - \frac{N(t)}{K}\right). \tag{5}$$

The logistic model predicts that populations will approach carrying capacity smoothly, with larger values of r resulting in faster approaches to K (Fig. 1). K represents a stable equilibrium of the system, since populations approach K through time.

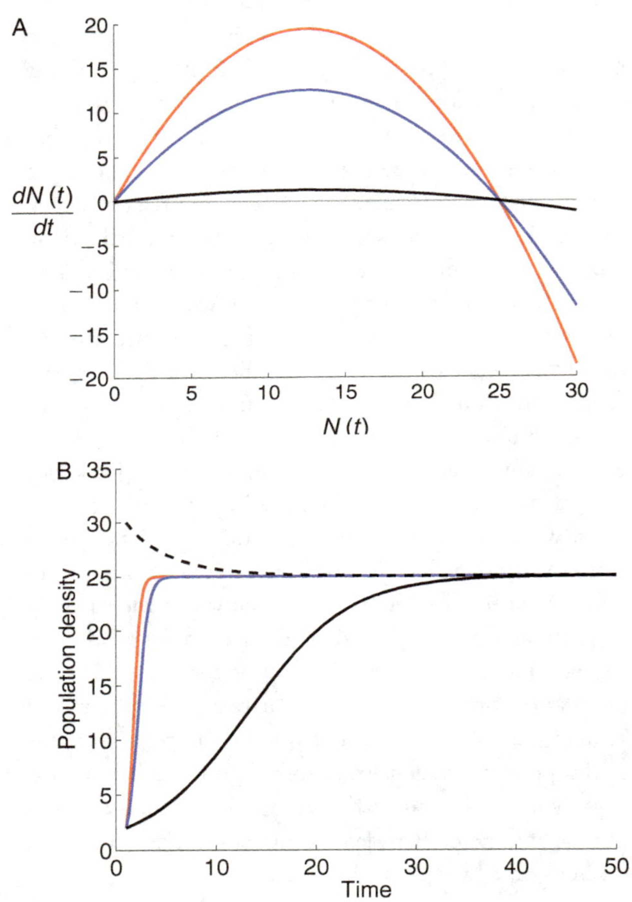

FIGURE 1 The continuous-time logistic model displayed both (A) as a graph of $N(t)$ vs. $\frac{dN(t)}{dt} = rN(t)\left(1-\frac{N(t)}{K}\right)$, and (B) as a trajectory of time t vs. $N(t)$. These are shown for $r = 0.2$ (black lines), $r = 2$ (blue), and $r = 3.1$ (red). In all cases, $K = 25$. The solid lines in (B) show all trajectories from initial population densities equal to 2, and dashed line shows the $r = 0.2$ case a second time from an initial density equal to 30. The thin gray line in (A) shows where $\frac{dN(t)}{dt} = 0$ and so the population density is not changing.

Of course, other model formulations are possible. A variant of the logistic model is the theta-logistic model,

$$\frac{dN(t)}{dt} = rN(t)\left(1 - \left(\frac{N(t)}{K}\right)^{\theta}\right). \tag{6}$$

Here, the parameter θ governs the shape with which the per capita population growth rate declines with density $N(t)$, although it still equals zero when $N(t) = K$. The greater flexibility of the theta-logistic makes it a useful candidate to produce a model that more closely mimics real populations.

Models for Density Dependence in Discrete Time

The discrete-time version of the logistic model is built in an analogous way. We can begin with the geometric model and multiply the population growth term, RN_{t-1}, by $1 - \frac{N_{t-1}}{K}$ (Table 1). For small values of R, the discrete logistic model behaves quite similarly to the continuous-time logistic model (Fig. 2). However, for larger values of R, specifically when $R > 2$, the discrete logistic produces cyclic population fluctuations. This is because, when $R > 2$, populations a short distance below K rebound to a density above K, and populations a short distance above K collapse to a density below K. (One unfortunate characteristic of the discrete logistic is that if the collapse from a high N_{t-1} is sufficiently dramatic, the model will give a negative value for N_t, which is clearly biologically unrealistic. Several alternative models for density dependence in discrete time have been proposed to avoid this problem, e.g., the Ricker and Beverton–Holt models. These specific models are important in fisheries and are discussed in detail in their own entries.) The propensity to overshoot and undershoot the carrying capacity makes K an unstable equilibrium when R is large, because populations starting close to K fail to move toward it through time. This pattern is often called overcompensatory density dependence, and it can result in sustained cycles where populations alternate indefinitely between densities above and below K.

The possibility of cycles for a model as simple as the discrete logistic highlights a contrast between continuous- and discrete-time models. Discrete-time models allow populations to jump between densities in ways that they cannot in continuous time, making the dynamics of discrete-time models potentially more complicated. For values of R not too much larger than 2, the population will alternate between two densities, one above and one below K (Fig. 2). Even higher values of R can generate more-complex cycles that have periods of 4, 8, 16, and higher powers of 2 as R increases

TABLE 1
Common single-species models

Model name	Equation	Density dependent?	Continuous or discrete time	Model behaviors
Exponential	$\frac{dN(t)}{dt} = rN(t)$	No	Continuous	Unbounded growth or decline.
Geometric	$N_t = (1 + R)N_{t-1}$	No	Discrete	Unbounded growth or decline.
Logistic	$\frac{dN(t)}{dt} = rN(t)\left(1 - \frac{N(t)}{K}\right)$	Yes	Continuous	Population persists at a stable carrying capacity.
Theta-logistic	$\frac{dN(t)}{dt} = rN(t)\left(1 - \left(\frac{N(t)}{K}\right)^{\theta}\right)$	Yes	Continuous	Population persists at a stable carrying capacity.
Discrete logistic	$N_t = N_{t-1} + RN_{t-1}\left(1 - \frac{N_{t-1}}{K}\right)$	Yes	Discrete	Population persists at a stable carrying capacity, or fluctuates cyclically/chaotically about carrying capacity. Large fluctuations can result in negative population densities.
Ricker	$N_t = N_{t-1}\exp\left[r\left(1 - \frac{N_{t-1}}{K}\right)\right]$	Yes	Discrete	Population persists at a stable carrying capacity, or fluctuates cyclically/chaotically about carrying capacity.
Beverton–Holt	$N_t = \frac{(1 + R)N_{t-1}}{1 + \frac{R}{K}N_{t-1}}$	Yes	Discrete	Population persists at a stable carrying capacity.
Autoregressive-moving average (ARMA) log-linear	$\ln(N_t) = \sum_{i=1}^{p} a_i \ln(N_{t-i}) + \sum_{j=0}^{q} b_j \varepsilon_{t-j}$	Yes	Discrete	Stable persistence, damped oscillations, or unbounded oscillations.

NOTE: In continuous-time models, $N(t)$ is the population density at a particular moment in time, t. In discrete-time models, the population density at time step t is denoted N_t. Other letters represent model parameters: r is the per capita intrinsic rate of population increase per unit time for continuous-time models, and R is a related parameter (per capita intrinsic rate of population increase = $\ln(1 + R)$) for discrete-time models; K is the carrying capacity. For ARMA models, the autoregressive part of the model has order p and coefficients $a_1, a_2, \ldots, a_p$ representing the strength of density dependence for each of the time lags. The moving average part has order q and coefficients $b_0, b_1, \ldots, b_q$. The random variable ε_t represents environmental stochasticity influencing population density at time t.

in what is known as a period-doubling cascade. This cascade eventually ends in chaos (a cycle of infinite period).

Cycles and chaos are not unique to the discrete logistic model but are shared with other discrete-time single-species models such as the Ricker model. In fact, a deep mathematical result is that any discrete-time model with a sufficient "hump" (e.g., Fig. 2A) giving overcompensatory density dependence will give similar period-doubling cascades to chaos. Nonetheless, discrete-time models do not necessarily have such exotic behavior. Other models, such as the Beverton–Holt model, do not even give sustained cycles and instead produce steady convergence to K as in the continuous-time logistic model.

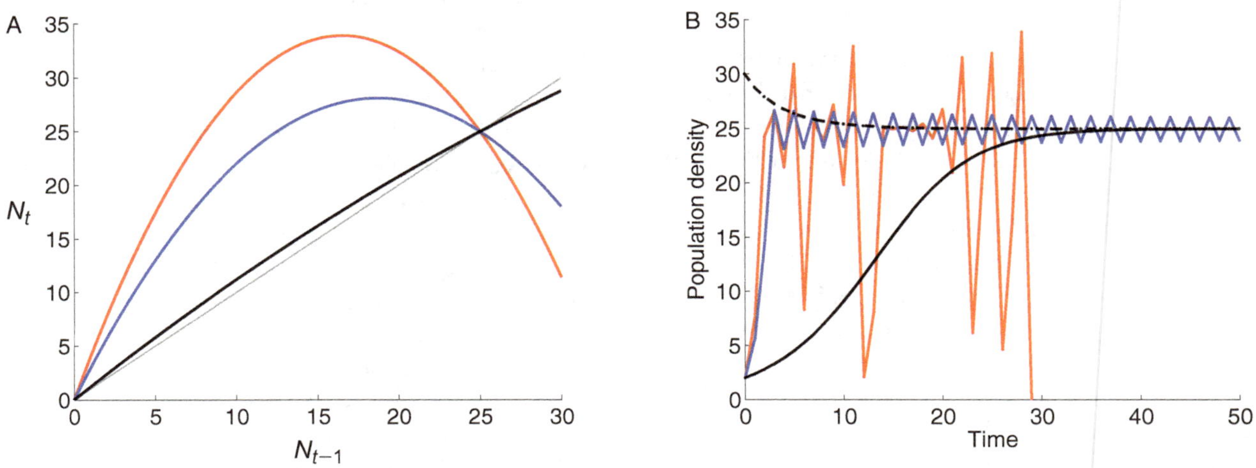

FIGURE 2 The discrete-time logistic model displayed both (A) as a graph of N_{t-1} vs. $N_t = N_{t-1} + RN_{t-1}\left(1-\frac{N_{t-1}}{K}\right)$, and (B) as a trajectory of time t vs. N_t. These are shown for R = 0.2 (black lines), R = 2 (blue), and R = 3.1 (red; notice the population density is negative (extinct) after t = 29). In all cases, K = 25. The solid lines in (B) show all trajectories from initial population densities equal to 2, and dashed line shows the R = 0.2 case a second time from an initial density equal to 30. The thin gray line in (A) shows where $N_{t-1} = N_t$ and so the population density is not changing.

STRUCTURED POPULATIONS

The models discussed here all have one major assumption in common: every individual in the population has identical demographic characteristics. This means, for instance, that all individuals have the same birth and death rates, and they use resources in the same way and thus count equally toward the carrying capacity for the population. These assumptions are clearly too simple to be met by most real populations. In many instances, these simple models nonetheless do a fairly good job of describing population dynamics, but for other situations single-species models that include more demographic detail are needed. When a species has a life history in which individuals at different ages or developmental stages have very different demographic characteristics, then a model that accounts for these differences should be used.

ENVIRONMENTAL STOCHASTICITY

In the models discussed so far, current population densities have been strictly determined by past densities. Such strict determinism might be a reasonable assumption for things like the trajectories of planets in their orbits, but all ecological systems experience stochastic (i.e., random) effects from the environment. While there are numerous ways of incorporating stochasticity into single-species models, the simplest is to add (continuous-time models) or multiply (discrete-time models) a random variable, e.g.,

$$\frac{dN(t)}{dt} = rN(t) + \varepsilon(t) \quad (7)$$

for the exponential model, or

$$N_t = \exp(\varepsilon_t)(1 + R)N_{t-1}$$

$$= \exp(r + \varepsilon_t)N_{t-1} \text{ for } r = \ln(1 + R) \quad (8)$$

for the geometric model. Placing the random variable ε_t in the exponential in Equation 8 assures that the population density never becomes negative.

Environmental stochasticity eliminates the possibility that populations remain at the carrying capacity K or in a perfectly regular cycle. Even though these deterministic equilibria do not occur, there is still a stochastic equilibrial structure known as the stationary distribution—the probability distribution of population densities that accumulates over a very long period of time. Furthermore, despite stochasticity, the character of the dynamics given by the deterministic component of the model can still be seen in the stochastic dynamics. For example, as long as $\varepsilon(t)$ and ε_t have an expected value of zero, the expected long-term exponential growth rate for both Equation 7 and Equation 8 is r, just as in the deterministic versions of these models. Figure 3 shows another example, in which a deterministic population governed by the discrete logistic approaches K in a cyclic manner, although the amplitude of these cycles eventually drops to zero. In the stochastic version, the environmental disturbances repeatedly excite this cyclic tendency, leading to irregular but nonetheless distinct up-and-down fluctuations that are sustained through time.

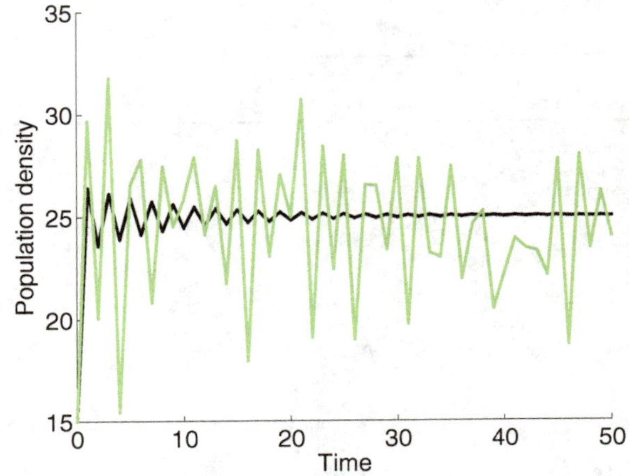

FIGURE 3 Stochastic population dynamics for the discrete-time logistic model (R = 1.9, K = 25). The black line gives the deterministic dynamics that exhibit a cyclic approach to K. The green line is the same model with the addition of environmental stochasticity and shows perpetual quasi-cycles that are maintained by the stochasticity.

SINGLE-SPECIES MODELS AS APPROXIMATIONS FOR DYNAMICS IN MULTISPECIES COMMUNITIES

Single-species models are mostly used for populations that appear to be governed mainly by intraspecific interactions. For populations whose dynamics are strongly influenced by interactions with other species, multispecies models are better descriptors. However, applying multispecies models to data is often problematic, since in many instances data are only collected for one of the interacting species. For these situations, single-species models that incorporate indirect information about the influences of additional species can be used in lieu of multispecies models.

This strategy is easiest to see if we begin with a very simple multispecies model. Suppose we have a system of two interactions species whose population densities at time t are $N_{1,t}$ and $N_{2,t}$, respectively. A simple stochastic model for these populations is

$$N_{1,t} = \exp(\phi_{1,t})(N_{1,t-1})^{c_{11}}(N_{2,t-1})^{c_{12}}, \quad (9a)$$

$$N_{2,t} = \exp(\phi_{2,t})(N_{2,t-1})^{c_{22}}(N_{1,t-1})^{c_{21}}, \quad (9b)$$

where the parameters c_{ij} represent the density-dependent effects of species j on species i and $\phi_{i,t}$ are random variables representing environmental stochasticity acting on species i. This is a nonlinear model of the population densities, but it is actually a linear model of the log of the densities, which we can see by substituting $N_{i,t} = \exp(n_{i,t})$ and taking the natural log of both sides:

$$n_{1,t} = c_{11}n_{1,t-1} + c_{12}n_{2,t-1} + \phi_{1,t}, \quad (10a)$$

$$n_{2,t} = c_{22}n_{2,t-1} + c_{21}n_{1,t-1} + \phi_{2,t}. \quad (10b)$$

It would be difficult to compare this two-species model to data if we only had data on the population dynamics of one of the species. Conveniently, however, Equation 10 can be combined algebraically to give an equation for the dynamics of species 1 in terms only of its own past densities:

$$\begin{aligned} n_{1,t} &= (c_{11} + c_{22})n_{1,t-1} + (c_{12}c_{21} - c_{11}c_{22})n_{1,t-2} + \\ &\quad \phi_{1,t} - c_{22}\phi_{1,t-1} + c_{12}\phi_{2,t-1} \\ &= a_1 n_{1,t-1} + a_2 n_{1,t-2} + b_0\varepsilon_t + b_1\varepsilon_{t-1}. \quad (11) \end{aligned}$$

The second line of Equation 11 is a more streamlined way of writing the model, where $a_1 = c_{11} + c_{22}$, $a_2 = c_{12}\,c_{21} - c_{11}c_{22}$, and $b_0\varepsilon_t + b_1\varepsilon_{t-1} = \phi_{1,t} - c_{22}\phi_{1,t-1} + c_{12}\phi_{2,t-1}$. In the streamlined version of the model, ε_t is now the random variable that represents environmental stochasticity with a distribution that is different from (but related to) the original random variables, $\phi_{1,t}$ and $\phi_{2,t}$.

Equation 11 shows that we can rewrite a system of two stochastic log-linear equations as what is known as an ARMA(2, 1) model. An ARMA(2, 1) model has two autoregressive (AR) terms, proportional, respectively, to the log population densities one and two time steps earlier, and one moving average (MA) term that describes the effect of the previous time step's environmental stochasticity on the current population size. More generally, a p-species log-linear model can be rewritten as a single-species ARMA(p, $p - 1$) model. Furthermore, nonlinear (and non-log-linear) p-species models are often reasonably well-described by single-species ARMA(p, q) models, where q is not necessarily $p - 1$ in these cases. Otherwise, nonlinear single-species models containing time delays can still be built that capture the dynamics generated by interacting species.

These time-lagged models are, strictly speaking, single-species models, because they are based on information obtained from the observed dynamics of a single species. Nonetheless, they can depict multispecies interactions because the dynamical consequences of these interactions manifest in the dynamics of each participating species. Therefore, in some sense single-species models with time lags serve as bridges between purely single-species and multispecies models.

APPLICATIONS OF SINGLE-SPECIES MODELS

Simple single-species models give us insight about the possible patterns exhibited by real populations. They show, for example, the necessity of density dependence to bound populations to reasonable densities, the possibility for discrete-time models to generate sustained cycles when continuous-time models cannot, and the potential for inferring interactions among multiple species from single-species information. But do real populations really exhibit these dynamics?

Good examples of population trajectories in nature that climb from low values to a carrying capacity are rare, largely because they require some dramatic event that leads to initially very low densities. One such event was the spread of the disease rinderpest from domestic cattle to the wildebeest (*Connochaetes taurinus*) population of the Serengeti in East Africa. Once the disease eased, the wildebeest population increased initially exponentially but then, as density dependence set in, at a decreasing rate. The discrete-time theta-logistic model gives a reasonable depiction of these dynamics (Fig. 4A).

Many good examples of cyclic population dynamics come from forest insects. Figure 4B shows cycles in a Swiss population of the larch budmoth (*Zeiraphera diniana*). An ARMA(2, 2) model captures these cycles well. Since the autoregressive dimension of the fitted model is 2, we might suspect that larch budmoth cycles are driven in part by interactions with another species. Indeed, interactions with the host plant or with parasitoids have both been proposed as likely drivers of larch budmoth cycles.

SUMMARY

The models discussed in this entry (and summarized in Table 1) are all quite simple—certainly much simpler than any real population. Nonetheless, they collectively describe an enormous diversity of population dynamics and often correspond reasonably well with the observed dynamics of real populations. Because of their generality and tractability, simple models like the ones presented here form the basis of much of ecological theory.

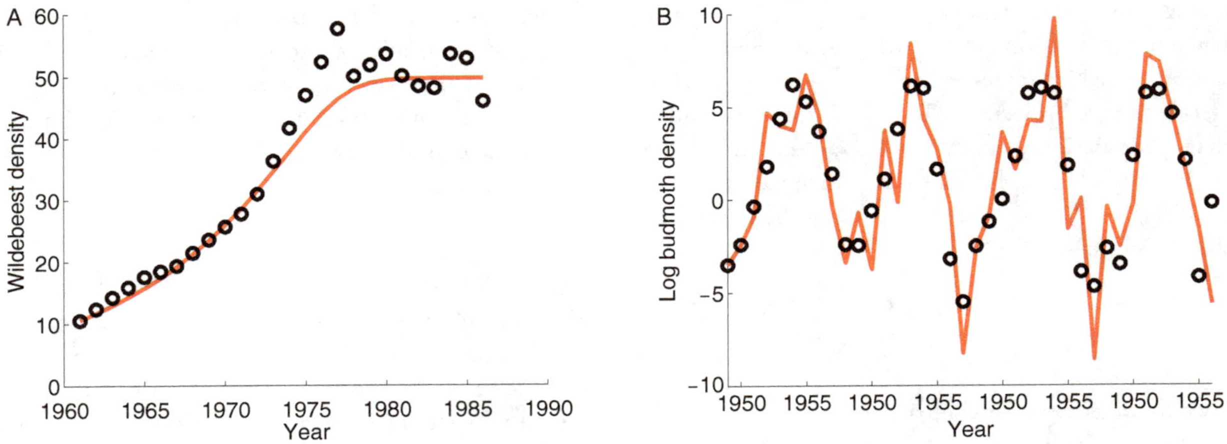

FIGURE 4 (A) Population dynamics of wildebeest on the Serengeti recovering from a population crash caused by rinderpest. Data are shown by black circles and the discrete-time version of the theta-logistic model fitted to these data is shown in red. (B) Population dynamics of the larch budmoth in the Engadine Valley, Switzerland. The black circles are data and the red line gives the 1-step-ahead predicted values from an ARMA(2, 2) model fitted to the data. All data are from the Global Population Dynamics Database (http://www.sw.ic.ac.uk/cpb/cpb/gpdd.html), and parameter values for the theta-logistic model in (A) are from Fryxell et al., 2007, *Nature* 449: 1041–1043.

SEE ALSO THE FOLLOWING ARTICLES

Beverton–Holt Model / Birth–Death Models / Chaos / Difference Equations / Integrodifference Equations / Ordinary Differential Equations / Reaction–Diffusion Models / Ricker Model

FURTHER READING

Gotelli, N. J. 2001. *A primer of ecology,* 3rd ed. Sunderland, MA: Sinauer Associates.

Gurney, W. S. C., and R.M. Nisbet. 1998. *Ecological dynamics.* New York: Oxford University Press.

Hastings, A. 1997. *Population biology: concepts and models.* New York: Springer.

May, R.M., and A. McLean, eds. 2007. *Theoretical ecology.* Oxford: Oxford University Press.

Royama, T. 1992. *Analytical population dynamics.* New York: Chapman & Hall.

Turchin, P. 2003. *Complex population dynamics: a theoretical/empirical synthesis.* Princeton: Princeton University Press.

SIR MODELS

LEWI STONE AND GUY KATRIEL

Tel Aviv University, Israel

FRANK M. HILKER

University of Bath, United Kingdom

Over the last two decades, SIR epidemic models have stimulated an extraordinarily intense wave of research activity in the fields of theoretical biology and ecology. In part, this is due to the model's many contributions to the study of both newly emerging and reemerging infectious diseases, which has become an increasingly important scientific agenda. The approach essentially keeps track of the dynamics of susceptible, infected, and recovered (SIR) individuals in a population as a disease or epidemic sweeps through it. The SIR model formulation is attractive first because of its relatively simple design, and second because the model has enormous power and versatility in exploring the complex spatiotemporal dynamics of infectious diseases. Although originally conceived as an epidemiological model, it has numerous multidisciplinary applications that are utilized by ecologists in their attempts to study the persistence of animal and plant populations under the threat of disease; physicists and computer scientists in their attempts to understand the manner in which computer viruses spread through complex heterogeneous networks; and by sociologists and anthropologists who are interested in the spread of ideas, rumours, cultural signals and memes in different human societies.

THE BASIC MODEL

The SIR model assumes that each individual in a population must at any time belong to one of three classes: Susceptible (S), Infected (I) or Recovered (R). In this scheme, a susceptible individual moves to the infected class upon contact with an infected individual if the disease is transmitted, and then moves to the recovered class after recovery from the infection; or, in short form: $S \rightarrow I \rightarrow R$. From this starting point, the differential equations describing the time evolution of the subpopulations within each class may be written down as

$$\frac{dS}{dt} = \mu N - \frac{\beta}{N}SI - \mu S,$$
$$\frac{dI}{dt} = \frac{\beta}{N}SI - (\mu + \gamma)I, \tag{1}$$
$$\frac{dR}{dt} = \gamma I - \mu R.$$

The variables S, I, and R represent population sizes with the total population $N = S + I + R$. Here, the growth of susceptibles (dS/dt) is controlled by the per capita population birth rate μ. This continual influx of newborn infants proves important since it is the only source of new susceptibles into the population.

The model assumes that individuals are randomly mixing so that the number of contacts between infected and susceptible individuals is proportional to the chance of their meeting; namely, the product SI. In the equations, $\beta SI / N$ is the rate at which susceptibles become infected and are therefore lost from the susceptible pool. The parameter β may be viewed as a transmission rate and indicates that a typical infected individual is able to come into contact with β individuals over a unit of time, infecting those who are susceptible (which is a proportion S/N of all contacts). The growth of infected individuals (dI/dt) is determined by these new infections $\beta SI / N$, and the rate γ at which they recover from the disease and move on to join the recovered class. Recovered individuals are assumed to gain lifelong immunity from any further infection. Individuals in all classes die at a per capita rate μ, and as birth and death rates are equal, the total population N is held constant ($S + I + R = N =$ *constant* since $dS/dt + dI/dt + dR/dt = 0$). Typical dynamics of the SIR model are displayed in Figure 1A, where the large epidemic in infected numbers and their ultimate decline is plotted in time for different scenarios.

The elementary differential equations in Equation 1 have proved to provide surprisingly good fits to epidemics of all types. Differential equations, however, are strictly valid only in the infinite population limit, and the above equations are less able to capture the dynamics of the disease whenever the number of infectives in the population reaches low levels, at which point the disease can die out through stochastic effects. In such situations in particular, where the focus is on disease die-out or epidemic fade-out, stochastic SIR models have proven effective tools. Figure 1B displays deterministic and stochastic fits of the SIR model to the data from the March–April 2009 novel H1N1 influenza outbreak in the village of La Gloria, Mexico, where the earliest known appearance of the new virus strain occurred. Since the population of the village is small, stochasticity can lead to quite varying epidemic curves, although most curves follow the general shape of the deterministic curve, as do the data.

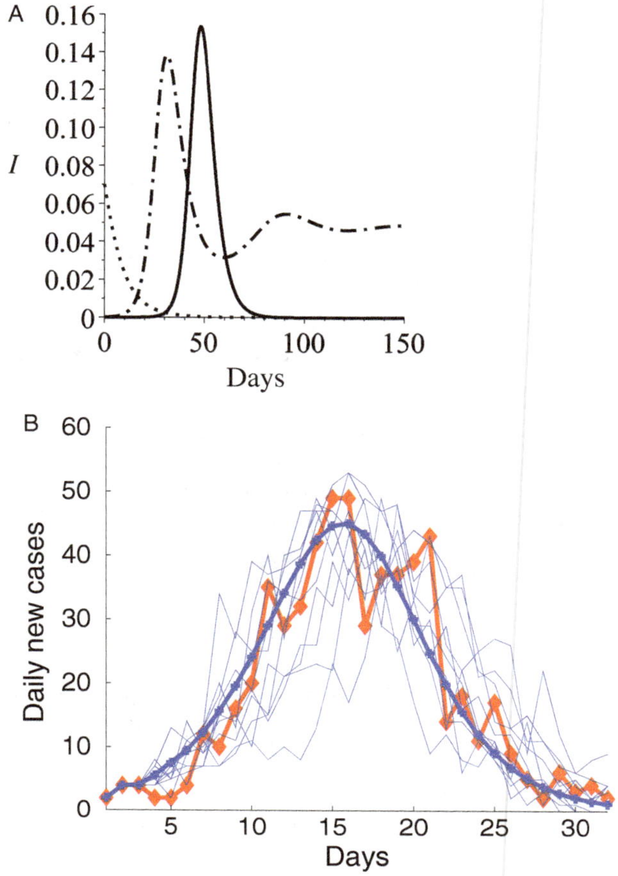

FIGURE 1 (A) The dynamics of the number of infected individuals $I(t)$ in the SIR model for three scenarios. Full line: When $R_0 > 1$ and $\mu = 0$ (no births), a single epidemic occurs but eventually the disease-free equilibrium is approached. Dashed-dotted line: When $R_0 > 1$ and $\mu > 0$, an epidemic is triggered and eventually an endemic equilibrium is approached (see Box 1). Dotted line: When $R_0 < 1$ and $\mu > 0$, the disease-free equilibrium is approached. (B) New daily cases (red curve) of the 2009 novel H1N1 influenza outbreak in La Gloria, Mexico (figure modified from Fraser et al. 2009). The blue curve shows the solution of the deterministic SIR model which was fit to the data, and the thin lines are ten simulations of a stochastic version of the SIR model with the same parameter value.

SIR RONALD ROSS, R_0 AND THRESHOLD DYNAMICS

The history of the SIR model traces back to Sir Ronald Ross (1857–1932), a trained medical doctor and parasitologist who was awarded the Nobel prize for his tenacious detective work leading to the identification of mosquitoes as the vector of malaria transmission. Ross was a "many-sided genius" who wrote novels, poetry, plays, and music, and had a lifelong passion for mathematics. This latter interest led Ross to formulate his "Mosquito Theorem." Based on an SIR-type model, he proved theoretically that merely reducing the numbers of the mosquito population below a critical threshold is sufficient to bring malaria transmission to a halt. This theorem was the direct precursor to epidemiology's foundation concept—the basic reproductive number R_0 and the threshold it predicts.

Ross's work has many similarities to the SIR model of Equation 1 introduced in a more general formulation by Kermack and McKendrick in the 1920's. The concept of the threshold unfolds from an examination of the infective dynamics:

$$\frac{dI}{dt} = I\left(\frac{\beta}{N}S - \gamma - \mu\right). \qquad (2)$$

It is usual to define an epidemic as a positive rate of growth of infectives ($dI/dt > 0$) or, equivalently, when ever $\frac{\beta S}{N(\gamma + \mu)} > 1$. This can be rephrased in terms of the basic reproductive number:

$$R_0 = \frac{\beta}{\gamma + \mu}. \qquad (3)$$

Consider the introduction of a small number of infectives in a fully susceptible population ($S_0 = N$). Under these circumstances, the disease can invade and an epidemic will trigger if the reproductive number $R_0 > 1$. However, the epidemic will fail to initiate if $R_0 < 1$.

From a mathematical standpoint, it is instructive to examine the disease-free equilibrium of Equation 1: $S^* = N$, $I^* = 0$, $R^* = 0$. A simple local stability analysis reveals that the disease-free equilibrium is locally stable if $R_0 < 1$ and unstable otherwise, paralleling the above result.

The basic reproductive number R_0 may be understood as the number of secondary cases that one infected case produces when placed within a wholly susceptible population (Box 1). If it can infect more than one individual on average ($R_0 > 1$), an epidemic will ensue; otherwise, the infection will rapidly die out as the disease-free equilibrium is reached. This underlying concept is remarkably robust to specific model details, making it a powerful tool for the forecasting and control of epidemics and pathogen invasions.

It is worthwhile noting that the reproductive number R_0 is a direct analogue to the eminent eighteenth-century mathematician Euler's famous demographic reproductive number, which is based on the idea that a stable population growth can only be achieved if an average individual manages to reproduce itself by having at least one offspring.

EXTENSIONS OF THE BASIC MODEL

There are a plethora of add-on extensions that can easily be incorporated into the basic SIR model to enhance its realism. A few of the more important extensions include:

BOX 1. R_0, THE ENDEMIC EQUILIBRIUM, AND FINAL SIZE

The basic reproductive number has a simple interpretation. In a totally susceptible population a single infected individual typically infects β other individuals per unit time. As the average time spent in the I compartment is $I/(\gamma + \mu)$ time units, the number of secondary infections is $R_0 = \beta/(\gamma + \mu)$. Thus, R_0 may be understood as the number of secondary cases that one infected case typically produces when placed in a wholly susceptible population.

The epidemic criterion is easy to generalize when the population begins with an initial fraction S_0 of susceptibles, as opposed to being wholly susceptible. Equation 3 shows that an epidemic will initiate if $S_0R_0 > 1$ but will fail to trigger if $S_0R_0 < 1$. The quantity $R_{eff} = S_0R_0$ is called the effective reproductive number.

When the birth rate is positive $\mu > 0$ and $R_0 > 1$, the system of Equation 1 has an equilibrium point in addition to the disease-free equilibrium, given by

$$\overline{S} = \frac{N}{R_0}, \overline{I} = \frac{\mu N}{\gamma + \mu}\left(1 - \frac{1}{R_0}\right), \overline{R} = \frac{\gamma N}{\gamma + \mu}\left(1 - \frac{1}{R_0}\right).$$

This endemic equilibrium, which can be shown to be stable, represents the state in which the disease has become established in the population, persisting because of the continual supply of new susceptibles μ. For a newly emerging disease, we expect a major epidemic in the short term, while the endemic equilibrium is only established over a much longer time scale. To study such epidemics, one considers the model of Equation 1 with $\mu = 0$, neglecting the demographic processes which are insignificant on short time scales (of months, say). The model then generates epidemics that eventually die out (see Fig. 1A).

The size A of an epidemic is the fraction of the population that will be infected over the entire course of the epidemic, and can be shown to be the solution of the final-size equation

$$\frac{A}{S_0} = 1 - e^{-R_0 A/N},$$

which is easily solved numerically to find A. For example, an epidemic with $R_0 = 2$ in a totally susceptible population ($S_0 = N$) will infect $A/N = 80\%$ of the population. The final-size equation shows that the size of an epidemic is determined by both the reproductive number R_0 and the initial fraction of susceptibles S_0. While R_0 is often stressed as the factor that determines the size of an epidemic, it is important to note that without information on S_0, predictions concerning the size of an epidemic are not possible.

Modifying Model Structure

Different model configurations are possible. For example, an SIRS model is appropriate for diseases in which individuals recover and become susceptible once again after a certain period of immunity, i.e., $S \rightarrow I \rightarrow R \rightarrow S$ and the loop is closed. Influenza, cholera, whooping cough, and some sexually transmitted diseases (e.g., gonorrhea) are all examples of diseases that exhibit waning immunity and the SIRS description is apt.

Note that the SIR model assumes that an individual becomes infective immediately upon being infected. However, for many infectious diseases there is a latent period during which the pathogen population builds up in the body of the host, with a consequent gap in time between being infected and becoming infectious. This is modeled by the SEIR model in which, upon infection, individuals enter an E (exposed) compartment and only after a time delay move to the I (infectives) compartment.

As a result of the many unique characteristics of different diseases, all sorts of other SIR variants have been devised and exploited.

Nonlinear Transmission Terms

Disease dynamics may be governed by various other features that go beyond the usual transmission term $S(t)I(t)/N(t)$ in Equation 1. Sometimes more general nonlinearities of the form $f(S(t), I(t))$ are used. A widely studied example is the term $S^p(t)I^q(t)$, where the degree of nonlinearity p and q are often determined empirically. While the mechanistic basis for such nonlinearities is often unclear, they are believed to be surrogates for heterogeneities and spatial effects that are not explicitly modeled.

Vaccination

Vaccination and disease control schemes usually attempt to force the reproductive number R_0 below unity and thus ensure the stabilization of the disease-free equilibrium. For childhood infectious diseases, the simplest approach is to vaccinate a proportion p of newborn infants, effectively removing them from the S compartment. The modified equation of susceptible growth is

$$\frac{dS}{dt} = \mu N(1 - p) - \frac{\beta}{N}SI - \mu S.$$

Simple calculations show that the disease-free equilibrium can always be attained as long as coverage is greater than $p_c = 1 - 1/R_0$. For measles, which has an $R_0 \sim 17$, there is a need to vaccinate $p_c = 1 - 1/17 = 0.94$ of the population in order to eradicate the disease entirely. We thus arrive at the key concept of herd immunity; the model predicts that it is not necessary to vaccinate the entire population in order to eradicate a disease. The herd immunity concept should be viewed as one of the major achievements of the SIR model and constitutes the theoretical justification for mass immunization schemes that have been implemented across the world, saving uncountable lives.

Heterogeneities in Contact Structure

The SIR equations of Equation 1 assume that an individual has an equal chance of coming into contact with all members of the population. However, life is not so uniform. In many cases, we should expect to find "superspreaders," individuals with exceptional numbers of contacts relative to the norm. What happens to the epidemic dynamics when the population's contact structure is so heterogeneous?

The theoreticians May and Dietz separately concluded, using an SIR model with different classes of individuals with different numbers of contacts, that in such a population the reproductive number R_0 is determined by the coefficient of variation (CV), or the relative amount of variability, in the population's contact distribution. $CV = 0$ would imply that everyone has the same number of contacts, while $CV \gg 0$ would suggest that there is large heterogeneity—some people having many more contacts than others. More specifically, they could show that

$$R_0 = \overline{R}(1 + CV^2). \tag{4}$$

That is, the R_0 of a heterogeneous population is equal to the reproductive number of that population were it a homogeneous $\overline{R}$ (every individual having the same average number of contacts) but amplified by a factor of $(1 + CV^2)$. The greater the heterogeneity in the contact distribution, the larger the CV and R_0, and hence the higher propensity for an epidemic to trigger.

An important special case concerns populations where the degree distribution is in the form of a power law, where the proportion of people having k-contacts is $P_k \sim k^{-\alpha}$. Contact networks of this form are termed "scale free." The power law implies that while most people in the population have few contacts (small k), there is a relatively small number of individuals with large numbers of contacts (high k). This distribution is sometimes relevant for sexually transmitted diseases, where sex workers may act as superspreaders. For power laws with $\alpha < 3$, the CV is infinite, so that $R_0 = \infty$. Such populations will be able to sustain the infection regardless of the rate of transmission $\overline{R}$, so that disease control becomes impossible.

OSCILLATIONS, SEASONALITY, AND CHAOS

Recurrent epidemics that appear annually, or every few years, are a ubiquitous phenomenon in infectious disease

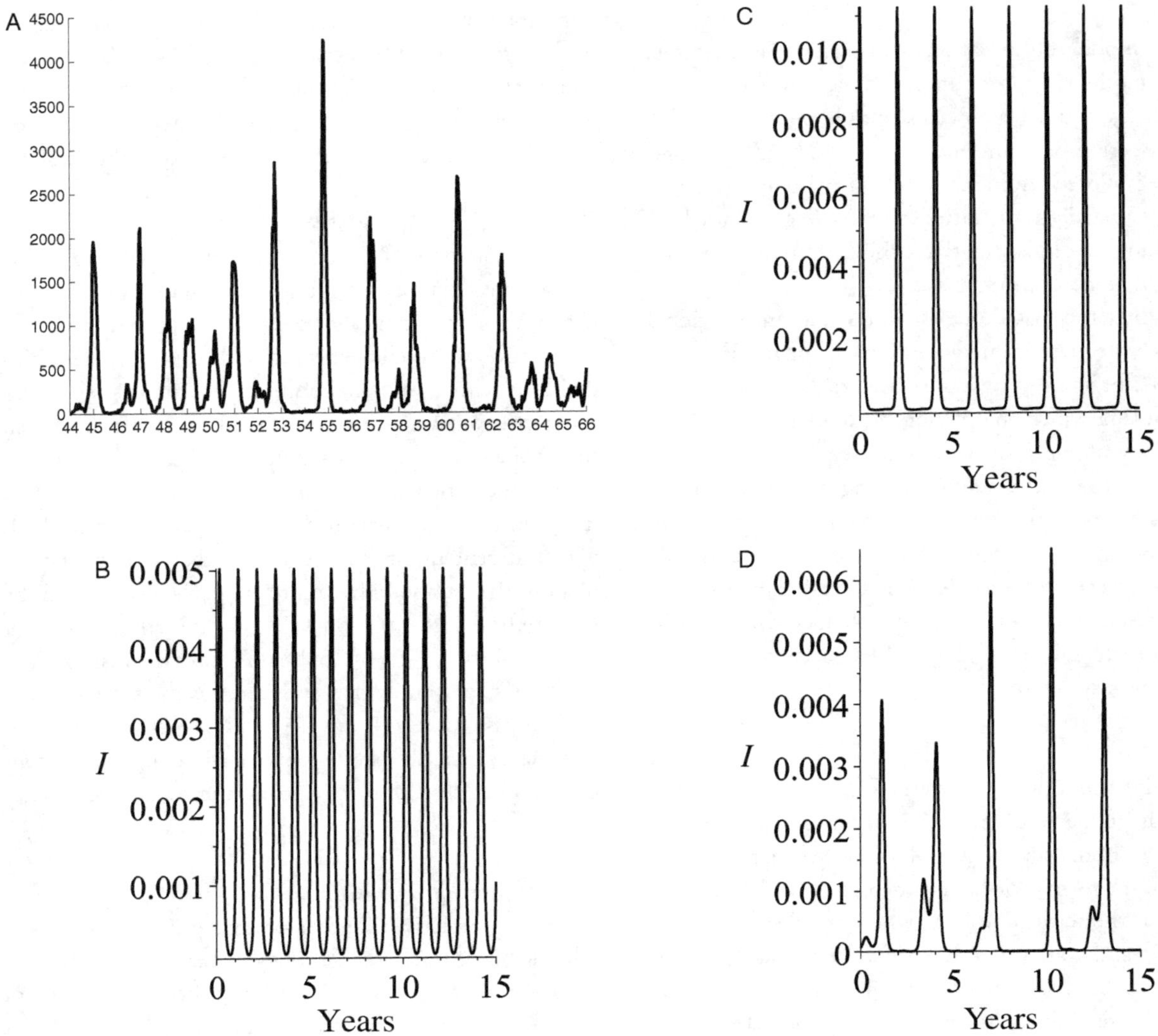

FIGURE 2 (A) Numbers of measles cases in Birmingham, UK, in the years 1944–1966 exhibiting initially annual dynamics (1947–1952), biennial dynamics (1952–1963), and followed by irregular oscillations after a widespread vaccination campaign was adopted. (B)—(D), Solutions $I(t)$ of the seasonally forced SIR model, with $R_0 = 20$, average infectious period $\frac{1}{\gamma} = 2$ weeks, and seasonality $\beta(t) = \overline{\beta} \cdot (1 + 0.25 \cdot \cos(t))$: birth (and death) rate 5% per year (B) leading to annual epidemics, 4% per year (C) leading to biennial epidemics, and 2% per year (D) leading to irregular epidemics (chaotic solution).

epidemiology. Since the basic SIR equations (Eq. 1) do not support sustained oscillatory solutions, some additional mechanism must be introduced to account for them. Factors such as demographic stochasticity or waning immunity can induce oscillations but are unlikely to explain the common seasonal oscillations that reoccur annually each year. These are more likely explained by external seasonal drivers such as (i) host aggregation during particular seasons (e.g., children in schools during the school term, animals at water sources during the dry season) leading to increased transmission; (ii) seasonal weather conditions (sunlight, temperature, moisture), resulting in improved transmission; and (iii) seasonal variation in immune system function, e.g., weakened immune system during winter or the breeding season.

Standard SIR models of seasonal oscillations replace the constant transmission coefficient β by an annual periodic function $\beta(t)$. A common practice is to set $\beta(t) = \beta_0(1 + \delta \cos(\omega t))$, where β_0 is the mean transmission rate, δ is the strength of seasonality, and the frequency ω is arranged to obtain annual periodicity. Depending on the values of the parameters, the seasonally forced SIR model generates oscillations that occur annually, once in 2 years, or even chaotic behavior displaying irregular oscillations with no definite periodicity (see Fig. 2

for simulations). These different behaviors are able to explain features seen in real diseases, such as the transition from annual to biennial measles dynamics observed with changes in birth rates in the UK in the 1940s, or to irregular epidemics upon introduction of the measles vaccination in the late 1960s (Fig. 2).

Many animals have a short breeding seasons, and this may be yet another mechanism for annual epidemic oscillations. Seasonally varying birth rates, often concentrated within a short breeding season pulse, can be modeled by adopting a time-periodic birth parameter $\mu(t)$. Similarly, higher mortality during particular seasons can be modeled by seasonal variation of the mortality term. A number of interesting studies have examined in depth the complex dynamics (e.g., chaos, resonances) that may arise with such seasonally forced demographic rates.

ECOEPIDEMIOLOGY: THE INTERPLAY WITH DEMOGRAPHY

All animals host many species of infectious agents. Even a top-predator lion is subject to over 50 viruses, helminths, arthropods, and bacteria. In fact, parasitism is the most common consumer strategy. Ecologists are becoming increasingly aware that parasites are important constituents of food webs. SIR models can easily be extended to help understand the interaction of parasites and their host ecology. Parasites may alter the life history of their host, potentially induce mortality, and thus regulate populations. By influencing the behavior of their host, parasites are able to impact the outcome of competitive and predator–prey relationships, or they can themselves be affected by the ecological community. This section highlights three examples of ecoepidemiological models, in which variants of the SIR model are embedded into an ecological context in order to shed light on host–parasite relationships.

For these models we will deal with populations that do not have a fixed population size N, but instead the size changes over time so that $N = N(t)$. It now becomes important to consider whether and how the population size affects the transmission rate β. Two different assumptions are common in the literature, and there has been ongoing debate as to which of these is more appropriate (which may of course depend on the specific host species and the pathogen considered). Under the frequency-dependent transmission assumption, it is assumed that each host individual has a fixed number of contacts, which does not change with the size of the population. The basic SIR model in Equation 1 incorporates frequency–dependent transmission. Under the density-dependent transmission assumption, $\beta = \beta(N)$ is a function of the size of the population, reflecting the assumption that a larger population will be denser so that each individual will have more contacts. $\beta(N)$ is commonly taken as a linear function $\beta(N) = \beta_0 N$. In models in which the size of the population varies over time, the dynamics generated may differ substantially in the density-dependent vs. the frequency-dependent cases. Therefore, in the examples below we will indicate which type of transmission is assumed in the model.

Allee Effect

Animal populations often experience survival difficulties at lower densities, where they may be more prone to decline due to lack of mating partners, lack of antipredator defense, or inability to hunt in groups. The American ecologist Warder Allee studied cooperative behaviors of animal groups in depth and argued that survival increased accordingly at higher densities—a phenomenon now termed the Allee effect. To understand Allee dynamics, consider a fatal disease invading a population. As there is no recovery from the disease, an SI model is appropriate,

$$\frac{dS}{dt} = r(N)S - \beta SI,$$

$$\frac{dI}{dt} = \beta SI - \alpha I \qquad (5)$$

and the total population size is $N = S + I$. The model makes use of density-dependent transmission (cf. frequency-dependent Eq. 1). Because birth and mortality rates are no longer equal, the total population size is now variable and $N = N(t)$. Unlike the SIR model of Equation 1, the reproduction of susceptible individuals follows a density-dependent growth rate $r(N)$, which for the Allee effect is commonly taken as

$$r(N) = a\left(1 - \frac{N}{K}\right)\left(\frac{N}{K} - \frac{u}{K}\right).$$

Here, a is the maximum growth rate, K the carrying capacity, and u the so-called Allee threshold. Note that infected individuals do not reproduce. Instead, they suffer a mortality rate α that includes both natural and disease-related deaths.

The interplay between demography and disease spread manifests here in the form of various ecological thresholds. In the absence of disease, the population either grows to the carrying capacity K or becomes extinct, depending on whether the initial population is larger or smaller than the Allee threshold u, respectively. However, the presence of disease makes the story far more complicated. First, simple calculations show that for the disease to establish, the total host population N needs to

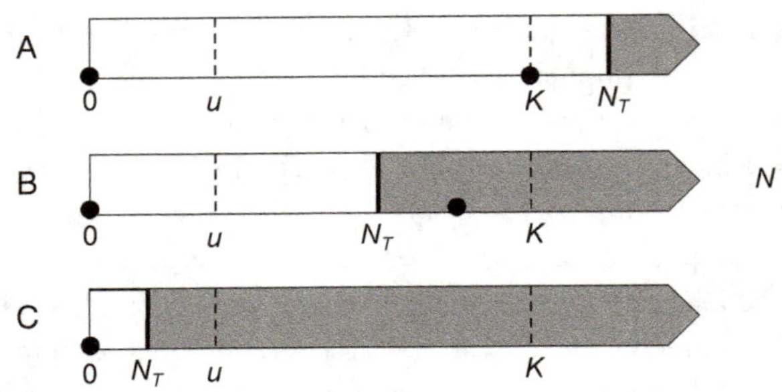

FIGURE 3 Interaction of population thresholds when the host is subject to infection and a strong Allee effect. The arrows represent the range of population sizes N. Gray color indicates where disease establishment is possible when the critical host threshold N_T is surpassed. K is the carrying capacity and u the Allee threshold. Circles indicate stable equilibria. Note that host extinction is always possible due to the Allee effect. (A) Disease-free case; (B) endemic infection possible; (C) disease-induced extinction.

be larger than the critical host threshold, which in this model is given by $N_T = \alpha/\beta$. This is a typical property for density-dependent tranmission.

Now consider the following three possibilities regarding the host threshold N_T:

1. $N_T > K$ (Fig. 3A). For the disease to invade, the host must be sufficiently large in numbers $N > N_T$. However, as the population cannot grow beyond its carrying capacity K, the disease cannot invade for $N_T > K$. The population remains disease-free.
2. $u < N_T < K$ (Fig. 3B). Since $N_T < K$, the disease can establish in a host population at carrying capacity. Consequently, the population size will be reduced to an endemic level below the carrying capacity. The endemic equilibrium cannot fall below the critical host threshold. Hence, the population is safe from extinction, as long as there are no stochastic effects.
3. $N_T < u$ (Fig. 3C). In the presence of disease, the population can now fall below the Allee threshold. That is, extinction of the host (and the disease as well) is possible. The exact condition for this to happen depends on the interplay between disease-induced host population depression and the Allee effect.

Note that the last possibility, of disease-induced extinction, rarely arises in models of density-dependent transmission. This is because disease transmission vanishes when the population size is reduced to below the critical host threshold. In the presence of a strong Allee effect, however, the population may first be reduced by infection and then finished off by extinction due to the Allee effect. Endangered species that suffer from both Allee effects and parasites (e.g., African wild dog or the island fox) are therefore particularly threatened.

Parasites Affecting Ecological Interactions

There has been much recent interest in coupling epidemiological SIR-type models with models of ecological communities. Such work demonstrates how parasites and disease can not only regulate populations but also structure ecological communities. Here, we examine a model by Tompkins and colleagues that explains the replacement of the native red squirrel in the 1960s in the UK by the invasive grey squirrel, and takes into account the potential role of disease progression induced by the shared parasite parapoxvirus. The parasite itself is believed to have been introduced with the gray squirrel. The population of gray squirrels is divided into the standard SIR scheme with $N_G = S_G + I_G + R_G$. In contrast to the gray squirrels, the red squirrels suffer an additional disease-related mortality (α). Moreover, there is no evidence for recovery in the reds, which is why their total population $N_R = S_R + I_R$ is split into susceptibles and infectives only:

$$\frac{dS_G}{dt} = [r_G - q_G(N_G + c_R N_R)]N_G - mS_G - \beta S_G(I_G + I_R),$$

$$\frac{dI_G}{dt} = \beta S_G(I_G + I_R) - mI_G - \gamma I_G,$$

$$\frac{dR_G}{dt} = \gamma I_G - mI_G,$$

$$\frac{dS_R}{dt} = [r_R - q_R(N_R + c_G N_G)]N_R - mS_R - \beta S_R(I_R + I_G),$$

$$\frac{dI_R}{dt} = \beta S_R(I_R + I_G) - (m + \alpha)I_R.$$

Disease transmission is density-dependent and occurs within and between species. The two squirrel species have the same natural mortality (m), but the grays are more efficient in utilizing food resources, leading to a higher reproduction rate ($r_G > r_R$). Intraspecific competition is modeled by crowding parameters $q_{G,R} = (r_{G,R} - m)/K$, where K is the carrying capacity. Gray squirrels are the stronger competitors, which is taken into account by the interspecific competition parameters ($c_G > c_R$).

Hence, in the absence of disease, gray squirrels are expected to outcompete reds and to eventually replace them. However, whereas this mechanism alone cannot explain the speed and pattern of red squirrel decline and gray squirrel invasion observed in empirical data, inclusion of the shared parasite produces a significantly closer fit to

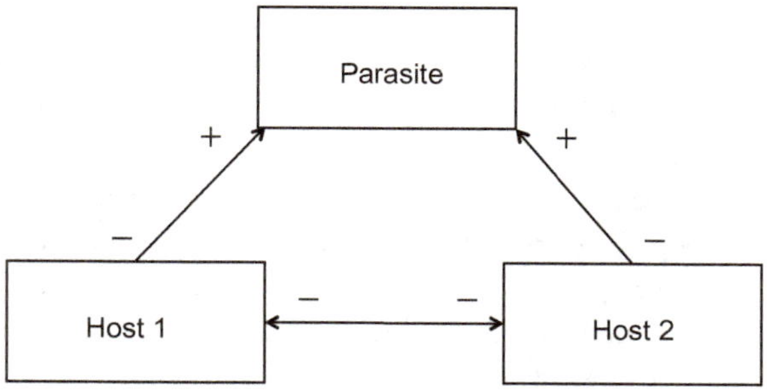

FIGURE 4 Schematic community diagram of parasite-mediated competition. Two hosts share the same parasite, which affects the outcome of the competition between the hosts. If there were no competition between the hosts (i.e., ignore the horizontal arrow), the community would resemble apparent competition.

the data. The reason is that the parapoxvirus increases mortality in the already weaker competitor, thus making the population of reds crash much faster and thereby enhancing the competitive replacement. Thus, disease can directly influence the interaction between species, which is termed parasite-mediated competition (Fig. 4).

If the two species did not directly interact with one another, but only shared the disease, we would have a situation similar to apparent competition. For example, an increase in host species 1 can lead to a decrease in host species 2. The two species appear to compete with one another, but the actual reason is a shared natural enemy (here, a parasite). An increase in species 1 leads to a higher level of parasites which in turn infect more individuals of species 2 (see Fig. 4).

Dilution and Amplification Effects in Host–Vector Systems

Many infectious diseases are not transmitted by direct contact but via vectors such as blood-sucking insects. An example is the West Nile virus (WNV), which is found all over Africa, the Middle East, Europe, Asia, and has more recently rapidly invaded across North America following an outbreak in New York City in 1999. WNV mostly infects birds, but also many other hosts, including humans. Transmission mainly occurs via bites from infected mosquitoes. Mathematical models have attempted to assess WNV control methods, but it has been found that different biological assumptions in different models yield conflicting predictions. However, according to Wonham and colleagues, all the models have the following core model in common:

$$\frac{dS_v}{dt} = r_v N_v - \beta_R \frac{I_R}{N_R} S_v - m_v S_v,$$

$$\frac{dI_v}{dt} = \beta_R \frac{I_R}{N_R} S_v - m_v I_v,$$

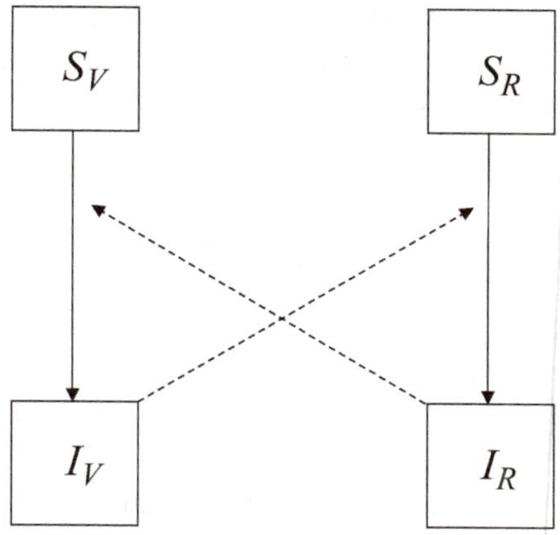

FIGURE 5 Schematic transfer diagram of the West Nile virus model exhibiting criss-cross infection between vector and reservoir population.

$$\frac{dS_R}{dt} = -\beta_R \frac{S_R}{N_R} I_v - m_R S_R,$$

$$\frac{dI_R}{dt} = \beta_R \frac{S_R}{N_R} I_v - m_R I_R.$$

This core model consists of a vector population (mosquitoes), $N_v = S_v + I_v$, and a reservoir (birds), $N_R = S_R + I_R$. Human hosts are a dead end in the virus transmission cycle and therefore neglected. The vector life cycle (for mosquitoes usually one month) is modeled by birth and death rates (r_v and m_v). As the reservoir life cycle is one or two orders of magnitude longer, their birth rate may be omitted in models assessing a single season only. Infection takes place when a susceptible vector bites an infected reservoir or a susceptible reservoir is bitten by an infected vector (criss-cross infection; cf. Fig. 5).

By studying the disease-free equilibrium, the basic reproductive number is found to be

$$R_0 = \sqrt{\frac{\beta_R}{m_v} \frac{\beta_R N^*_v}{m_R N^*_R}}.$$

R_0 contains the ratio of equilibrium vector to reservoir population sizes, N^*_v / N^*_R. That is, reducing the vector population (e.g., by spraying insecticides) reduces R_0 and therefore the likelihood of an outbreak. This is what most models agree on and can be seen as a dilution effect due to a decline in the vector population (cf. Ross' "Mosquito Theorem").

Reducing the reservoir population, however, has been predicted to yield contrasting effects. In the current example, a reduced reservoir population N^*_R will lead to an increase in R_0 and disease risk. The control of reservoirs thus has an amplification effect. This is a consequence of the frequency-dependent transmission assumption. Since the vector biting rate is constant, the remaining reservoir individuals will receive more bites and are therefore more likely to be infected.

By contrast, if disease transmission in Equation 5 were density dependent, the biting rate would depend on the number of available reservoirs. Reducing the bird population would thus decrease disease risk, i.e., produce a dilution effect. Different transmission terms can therefore lead to very different results.

Dilution effects also describe the situation in which increased host diversity reduces disease risk. This may happen when less competent hosts are introduced and reduce infective encounters between vectors and competent host. Community composition can thus shape the fate of parasites—and not vice versa as in the example of parasite-mediated competition.

SPATIAL ASPECTS

A major success of the SIR modeling framework lies in its ability to model spatial dynamics. A classical example is the spread of rabies via foxes over Europe, reviewed briefly here. Foxes transfer rabies to domestic dogs through contact. Dogs in turn transmit the disease to humans, usually by biting, and this can lead to an agonizing death if the victim is unvaccinated. Although rabies died out in Europe toward the end of the nineteenth century, it reemerged in 1939 in Poland, where it was carried by red foxes as they moved across Europe. Figure 6 plots the spatial advance of rabies in wavelike formation in France over the years 1969–1977.

The wavelike spread of the disease seen in Figure 6 is a feature common to many invading organisms and pathogens. These traveling wave phenomena may be

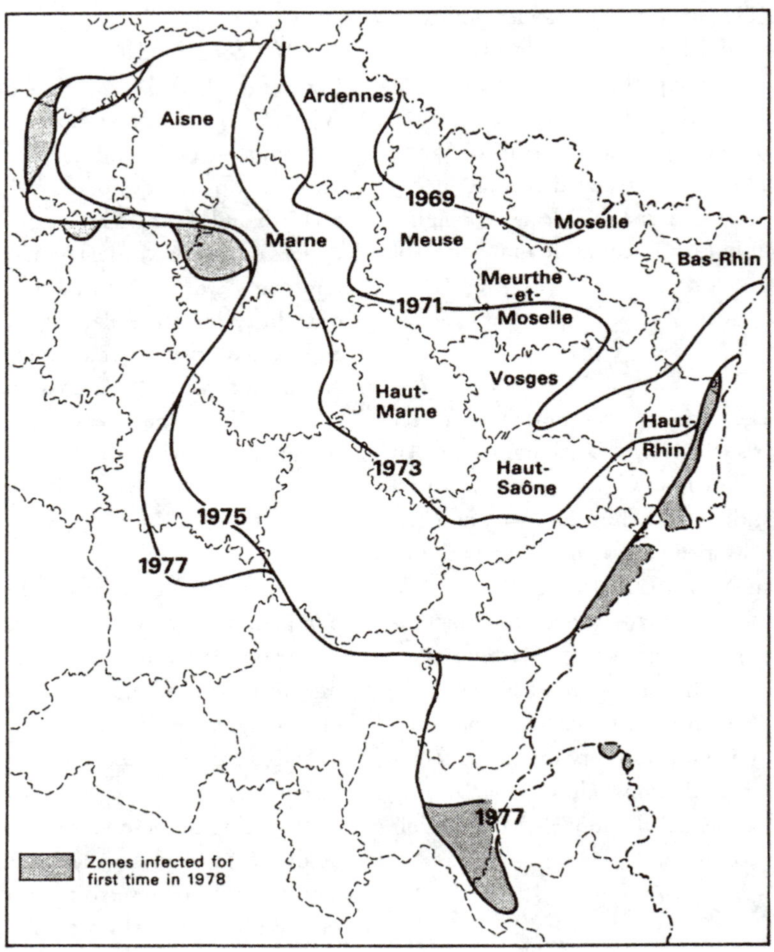

FIGURE 6 Spatial advance of the rabies wave in France from 1969 to 1977. From J. D. Murray, *Mathematical Biology* II, 3rd ed., Springer (2003).

modeled by assuming that the spatial invasive processes derive from some form of population diffusion. To keep things simple, we can suppose that there is a single spatial dimension, and denote the density of uninfected foxes and rabid foxes by $S = S(x,t)$, and $I = I(x, t)$, at position x and time t. The partial differential equations describing the infection process are

$$\frac{\partial S}{dt} = rS(1 - S) - \beta SI,$$
$$\frac{\partial I}{dt} = \beta SI - \alpha I + D\frac{\partial^2 I}{\partial x^2}. \qquad (6)$$

Here, the healthy susceptible fox population (S) typically follows some general logistic term based on growth rate r and infection rate β. The infected or rabid foxes (I) have an associated standard diffusive term ($\partial^2 I / \partial x^2$) with diffusion coefficient of strength D that allows them to move in a wavefront along the spatial dimension (x). As the rabid foxes diffuse in a random manner, they infect any susceptibles they encounter. Infected foxes have the disease-related death rate α. Typical simulations of the above equations reveal the existence of a wave of rabid foxes that pass through the susceptible population, as shown in Figure 7. The wave of infectives moves to the right and depletes susceptible foxes in its path. As the front pushes ahead, there is a build-up of susceptibles in the wake of the wave due to the growth term (r), which can eventually retrigger epidemics.

By linearizing about the disease-free state and seeking trial solutions of the form $I(x, t) = I_0 \exp(-\lambda(x - ct))$, it is not hard to show that the minimum invasion speed of the wavefront is $c = 2\sqrt{(\beta - \alpha)D}$, which propagates only if $R_0 = \beta/\alpha > 1$.

Since rabid foxes are only able to infect others after a period of 28 days following their own infection, it has been found important to add a latency period, thus forming an SEIR-type model. With realistic parameter values, the rabid fox wavespeed predicted by the model is approximately $c \sim 60$ km per year, which matches reasonably well the speed calculated from the map in Figure 6 and other empirical studies. A more detailed account that explores many other aspects of the rabies story can be found in Murray (2003). However, a note of caution: as pointed out by Barlow, many spatial rabies models "are effectively tuned to the observed rate of spread, [and so] there is no way of telling if the models are right for the wrong reasons." Even Mollison's simple back-of-the-envelope calculation is capable of predicting the invasion speed reasonably accurately as $R_0\sigma/2\tau$ = 25–50 km/year [where $R_0 = 5\sigma$ is the root mean square of the dispersal distance (1–2km), and τ is the mean generation gap of the disease (0.1year)].

Spatial disease models using the diffusion approach have been validated in many other case studies, with some based on unusual historical datasets. These include the black plague (1347–1350), West Nile virus, FIV (feline immunodeficiency virus) in cat populations, gypsy moth outbreaks, potato blight, dengue fever, hantavirus, malaria, influenza pandemics, and coral diseases. It is an area of research in which there are still many open questions.

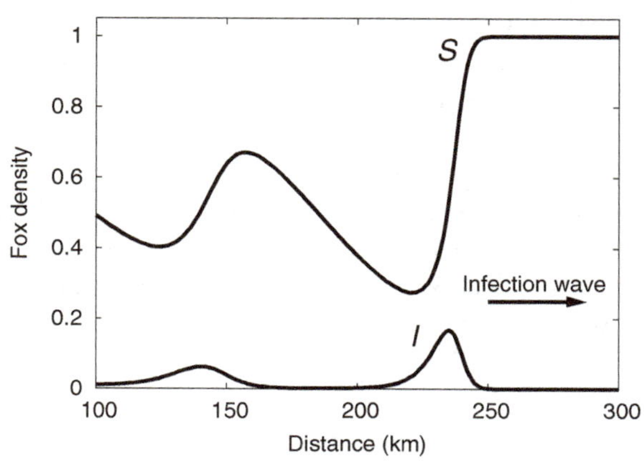

FIGURE 7 Wavefront solution for Susceptible (S) and infected (I) foxes computed numerically from Equation 6 with parameter values.

Networks

Another exciting avenue of investigation concerns the spatial spread of epidemics between a set of cities or patches that are interconnected by some predefined network. Patches are connected via a dispersal network of any arbitrarily specified topology through which infectives travel. Setting S_j, I_j, and R_j as the proportion of susceptible, infected, and recovered individuals in the jth city, the SIR dynamics of the city can be written as

$$\frac{dS_j}{dt} = \mu - \lambda_j S_j - \mu S_j,$$
$$\frac{dI_j}{dt} = \lambda_j S_j - (\mu + \gamma)I_j, \quad \text{where } \lambda_j = \sum_{k=1}^{n} \beta_{jk} I_k.$$
$$\frac{dR_j}{dt} = \gamma I_j - \mu R_j.$$

Travel routes between cities are built into the coefficients β_{jk} quantifying the proportion of infectives from patch j that travel to infect individuals in patch k. This leads to the so-called force of infection term $\lambda_j = \sum_{k=1}^{n} \beta_{jk} I_k$, which quantifies the infective individuals entering city j. The above formulation allows the analysis of any arbitrary network topology.

These simple equations have advanced our understanding of recurrent epidemic oscillations and synchrony between cities connected in a network. Grenfell and colleagues compiled historical time series of measles outbreaks in 945 cities and 457 rural towns in the UK in the pre-vaccination era and studied the tight synchrony between cities, including an unusual instance of stable out-of-phase synchrony between Cambridge and Norwich. The latter effect is also a prediction that emerges from an analysis of models of weakly coupled cities. The observed data indicated clear synchronized waves of infections that moved from large cities to smaller towns. As might be expected, synchrony is usually enhanced by tighter migration and overall network connectivity as expressed by the β_{ij} dispersal matrix. However, although tighter connectivity might appear to keep the disease spreading through the network, synchrony, on the other hand, can sometimes help eradicate disease spread. When a pathogen is on the brink of disappearing in all patches simultaneously, this enhances the chances of global fade-out and might be the perfect time to apply a wide-ranging synchronized vaccination scheme. In contrast, disease elimination is far less likely when patch dynamics are asynchronous and the disease can jump across the network reinfecting patches where local disease fadeouts have occurred. The latter scenario would make localized vaccination attempts ineffective.

The same ideas of synchrony are also made use of by conservation ecologists in their study of persistence in multipatch metapopulations (i.e., in the absence of disease dynamics). Thus, while synchronization is advantageous from the point of view of eradicating diseases, it has the opposite connotation for conservation where the persistence of a metapopulation is desired.

These concepts are being developed and imported into ecoepidemiological metapopulation models. An interesting study by Fulford and colleagues on the spread of tuberculosis in a possum metapopulation model examines how network structure influences optimal strategies for disease management. While culling diseased possums in all patches was often found to be the most effective practice, it was not usually the most cost effective. Careful targeting of the most critical patches would thus seem to be the only practical approach. Moreover, targeting migration routes can sometimes be the more effective approach. Models like the above can aid in selecting the wisest strategy.

The patch SIR model also allows direct investigation of the role of network heterogeneity. However, unlike our earlier discussion of connectivity within a single population, we now consider a network of connected populations or cities. This becomes highly relevant when, for example, faced by the threat of a new and possibly dangerous influenza strain, as was the case in 2009 when the swine flu strain H1N1 appeared. Transportation networks play a prominent role in the global spread of diseases and were probably responsible for the pandemic status acquired by H1N1. Central international airports such as New York, London, and Paris, which are highly connected to most cities across the globe, become dangerous hubs for spreading the disease. Models can help assess how these highly connected airport hubs control the spread of epidemics on a global scale. As might be expected, an analogous $I + CV^2$ law similar to Equation 4 can be formulated. Since transportation networks have a scale-free structure—a few cities with airports that act as hubs, and many smaller cities—in theory at least, the large CV in degree distribution implies that eradication programs can never efficiently succeed, short of closing down the airport itself. These and many other intriguing questions are being investigated through modeling.

Predictive Spatial SIR Models

With a specific focus on pandemics, several large-scale and ambitious research efforts have attempted to construct meaningful SIR-type simulation "models of the world" or "models of Europe." Hufnagel and colleagues describe an SIR modeling attempt to back-predict the geographic spread of the recent severe acute respiratory syndrome (SARS) outbreak in 2002. The first human cases of SARS were reported in China in November 2002, reaching Hong Kong in February 2003, from where it spread to the rest of the world via international air traffic routes, causing a near pandemic. It was found that the high heterogeneity of the transportation network led to high predictability of the spread of infection. Figure 8 compares the forecasts of the spatial SIR network model to maps of the geographic spread of SARS until May 2003. Given the model's remarkable ability to make spatial predictions, the authors proceeded to explore control strategies, such as vaccination and transport restrictions, that might be expected to inhibit the epidemic's global spread. Similar large-scale modeling efforts were initiated by emergency teams during the H1N1 pandemic in 2009.

CONCLUSION

It should be evident from this short review that, despite the unusual simplicity of the SIR model (Eq. 1), it has spawned a vast industry of scientific research over the

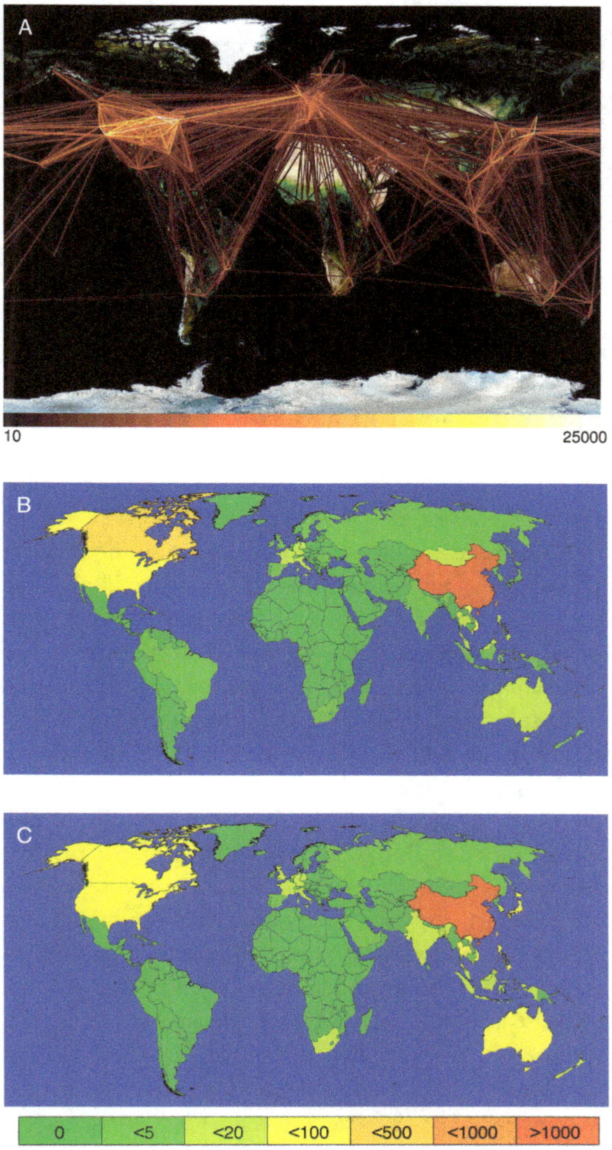

FIGURE 8 SARS model and simulations: (A) Global aviation network representing the civil aviation traffic among the 500 largest international airports. Each line represents a direct connection between airports. (B) Actual global spread of SARS cases on May 30, 2003, determined from reports issued by WHO and Centers for Disease Control. (C) Spatial simulations representing spread 90 days after an initial infection in Hong Kong. Figures from L. Hufnagel, D. Brockmann, and T. Geisel, *PNAS* 15124–15129 (2004).

last century. Moreover, judging from the sheer quantity of associated scientific papers being published, interest in the SIR model appears to be increasing at a greater rate than ever before. The model has proven itself as a continual source of inspiration to generations of scientists working in the epidemiological and ecological research communities.

SEE ALSO THE FOLLOWING ARTICLES

Allee Effects / Demography / Disease Dynamics / Epidemiology and Epidemic Modeling / Networks, Ecological / Spatial Models, Stochastic / Synchrony, Spatial

FURTHER READING

Altizer, S., A. Dobson, P. Hosseini, P. Hudson, M. Pascual, and P. Rohani. 2006. Seasonality and the dynamics of infectious diseases. *Ecology Letters* 9: 468–484.

Anderson, R.M., and R.M. May 1991. *Infectious diseases of humans: dynamics and control.* New York: Oxford University Press.

Diekmann, O., and J. A. P. Heesterbeek. 2000. *Mathematical epidemiology of infectious disease: model building, analysis and interporetation.* Hoboken, NJ: Wiley.

Fraser C. et al., 2009. Pandemic potential strain of influenza A (H1N1): Early findings. Science 324: 1557–1561.

Hatcher, M.J., J. T. A. Dick, and A. M. Dunn. 2006. How parasites affect interactions between competitors and predators. *Ecology Letters* 9: 1253–1271.

Hethcote, H. W. 2000. The mathematics of infectious diseases. *SIAM Review* 42: 599–653.

Hufnagel, L., D. Brockmann, and T. Geisel. 2004. Forecast and control of epidemics in a globalized world. *PNAS* 101: 15124–15129.

Keeling, M. J., and P. Rohani 2008. *Modeling infectious diseases in humans and animals.* Princeton: Princeton University Press.

Murray, J. D. 2003. *Mathematical biology,* 3rd ed. New York: Springer.

Tompkins, D. M., A.R. White, and M. Boots. 2003. Ecological replacement of native red squirrels by invasive greys driven by disease. *Ecology Letters* 6: 189–196.

Wonham, M. J., M. A. Lewis, J. Renclawowicz, and P. van den Driessche. 2006. Transmission assumptions generate conflicting predictions in host-vector disease models: a case study in West Nile virus. *Ecology Letters* 9: 706–725.

SPATIAL ECOLOGY

ALAN HASTINGS

University of California, Davis

Spatial ecology encompasses questions where the answer either is changed or depends on space. Among the most important issues considered are questions of persistence or coexistence of species and how this depends on the impact of space and the role played by underlying spatial heterogeneity. Other important questions concern the spread of species in space and the development of patterns in abundance in space and time. The study of spatial dynamics can provide ways of unraveling the processes that determine the distribution and abundance of species. Theoretical approaches need to make assumptions tailored to the question asked.

FUNDAMENTAL QUESTIONS OF SPATIAL ECOLOGY

There are several ways to organize theoretical approaches to spatial ecology. One could focus on the assumptions made about biological processes to produce tractable models that can lead to theoretical insights. Alternatively, one could describe the different mathematical approaches that can be used, which are determined in part, but not completely, by the assumptions. However, the best way to start is to focus on the ecological questions that are of most interest, and then to consider the ways that these questions can be approached theoretically. Rather than producing an exhaustive list of questions in spatial ecology, instead this entry focuses on some of the most fundamental questions and a number of questions that are receiving increasing attention.

Species clearly vary in abundance in space and time. Understanding the mechanisms behind this variation is one of the central issues of spatial ecology. This work not only helps elucidate the causes of the patterns but also allows a much deeper understanding of processes that control the dynamics of populations. The study of cyclic behavior in the dynamics of population numbers in experimental and field systems, combined with theoretical approaches, going as far back as the work of Volterra, Lotka, and Gause, has provided deeper understanding of density dependence and the dynamics of exploitation, providing information about processes that are hard to study directly. Analogously, studying the rich patterns of spatiotemporal abundance of species allows a deeper understanding of processes like dispersal, density dependence, and the influence of more physical aspects of the habitat. The work of Skellam, Okubo, and Levin has been particularly influential.

A particular example of a spatial pattern that has received much attention is the dynamics of synchrony across space of population abundances that cycle through time. These patterns are particularly striking in the long-term dynamics of population cycles of hare and lynx in Canada. Early work by Moran focused on trying to determine the causes of synchrony in this case. One possibility, now known as the Moran effect, is large-scale (in space) environmental influences that affect populations and increase their synchrony. More recently, work of Grenfell and colleagues on the spatiotemporal dynamics of childhood diseases has been particularly influential.

One fundamental question that is of importance is the understanding the mechanisms that control and determine the spatial spread of introduced species. Many species seem to exhibit a roughly linear rate of spread in time, either along a one-dimensional habitat like a coastline or radially from a central point. A classic and often-cited example is the spread of muskrats after their introduction to central Europe. Other species seem to accelerate their rate of spread. Can the rate of spread be predicted? If the species is a harmful invader, what are the most efficient control measures that can be used?

Also, species do not spread forever. What prevents a species from spreading? More generally, another essentially spatial pattern is the range of a species. What sets the limits of the range of a species is an important question from both an applied and basic standpoint. The early and classic experiments by Connell suggested that along an environmental gradient from benign to harsh that at the harsher end of a species range the limits are set by physical factors and at the more benign end limits are set by competition. Other approaches to understanding range limits do not always produce such clear generalizations.

Potential setting of range limits is just one example of an important spatial interaction between species. The role played by space in determining the outcome of species interactions has been, and is, the subject of continuing intense investigations.

One of the earliest questions examined in the context of spatial ecology is the role played by space in allowing the persistence of species. This question has been raised in variety of contexts, dealing with a range of ecological interactions. In one form, this question goes back to the very earliest investigations of the persistence of exploiter–victim interactions by Gause, who noted that within a single spatial location, the system of paramecium and didinium he investigated invariably had one or both species go extinct quickly. He essentially suggested that the answer was that some sort of regional coexistence was the key. At roughly the same time, Nicholson and Bailey observed that interactions between hosts and parasitoids were unstable in the simple model systems that they investigated. They, too, suggested that simultaneous persistence of exploiter and victim species in natural systems was due to spatial aspects where species would disperse to other locations before going extinct in one location.

Essentially, this dynamic of regional coexistence with local extinction was demonstrated in a classic set of experiments by Huffaker in the 1950s involving mites on a universe of oranges with connections between them. More formal theoretical (mathematical) explanations of this phenomenon have been the subject of many investigations since, some of which are highlighted below.

Similarly, the role of space in the coexistence of competitors has been and continues to be an active area of investigation. Phrased differently, the question becomes one of how biodiversity is maintained in a spatial context. Several issues have emerged. One goal has been to explain the very high diversity of competitors in some systems. Another goal has been to explain the joint spatial distribution of competing species.

The questions of the coexistence of competitors or exploiters and victims can naturally be extended to looking at the dynamics of food webs in a spatial context. Here, both competitive and exploiter–victim interactions are typically considered. One important issue is the role of allochtonous input, input from outside the system under consideration, in determining the dynamics of a food web.

Other topics have included studies at the behavioral level. Here, understanding the dynamics of foraging is one area that has been well investigated and this is essentially a spatial problem. Similarly, understanding the determinants of the home range of an individual is another question within the realm of spatial ecology. More recently there has been great interest in understanding the dynamics of flocking behavior, or coordinated spatial movement among individuals as would be found in a fish school. This work is aimed at related observed behavior at the level of the flock to rules describing individual responses to nearby individuals.

Approaches to management of natural populations, both for conservation or exploitation, are increasingly making explicit recognition of the spatial aspects of the problem. The spatial aspects of approaches to harvesting of fish populations are playing an increasing role in the design of optimal harvesting schemes. Design of reserves for the preservation of biodiversity is an inherently spatial problem.

The breadth of questions and multiple temporal and spatial scales that arise in this admittedly abbreviated overview of a range of questions in spatial ecology already indicate that a wide range of theoretical approaches will be needed.

THEORETICAL APPROACHES TO INCLUSION OF SPACE AND UNDERLYING ASSUMPTIONS

Developing models that can be used to explore questions in spatial ecology depends on including just the relevant level of detail. Because of the potential complexity of ecological systems, the choice of theoretical approach is always a compromise between generality and detail and between tractability and detail. The appropriate approach is typically dictated by the particular question at hand, and often multiple approaches are desirable to test robustness of conclusions and the degree to which conclusions depend on specific assumptions. Here a range of these approaches will be described, in roughly increasing order of detail and complexity and corresponding decreasing order of generality.

Implicit Space

The impact of space on ecological processes can be considered in descriptions that allow different parts of the environment to be in different states. The classic metapopulation models fall into this category.

The simplest description of space is one that ignores any aspects of the spatial arrangement and simply counts up the proportion of the environment that is in different states, so space is only implicit. The dynamics of the model are then given by providing the rates of transition between the different states. In the description of the state of the system, no account is made of where the occupied locations are. Similarly, since location is ignored in the description of state, the role of distances, either among occupied locations or between occupied and unoccupied locations in determining rates of extinction or colonization, is ignored as well. As would be expected, however, these assumptions typically lead to model descriptions that can be solved analytically.

The very simplest model of this type is the classic Levins metapopulation model for a single species in a homogenous environment. In this model, the state of the system at any given time is given by the fraction of the overall system (or patches) that is occupied, p, and the fraction of the system that is unoccupied, $1 - p$. The probability of an empty state becoming occupied is proportional to the fraction of occupied patches, so the rate of colonization is given by $mp(1 - p)$ where m is a constant. The extinction rate is assumed to be independent of the occupied patches, so the rate of patches transitioning from occupied to empty is ep, where e is a positive parameter. The overall dynamics are given by an equation that states that the rate of change of the fraction of occupied patches is the colonization rate minus the extinction rate,

$$\frac{dp}{dt} = mp(1 - p) - ep. \qquad (1)$$

Note that this equation can be rearranged to produce an equation that is equivalent to the logistic equation,

$$\frac{dp}{dt} = (m - e)p(1 - (m/(m - e))p), \qquad (2)$$

which serves to emphasize the notion that space is implicit. Note also that although this description is essentially stochastic at the level of a single patch, it is deterministic at the level of the whole population. This simple model has a stable equilibrium provided that $m > e$. This simple model therefore indicates how regional persistence can occur even if any local population is doomed to extinction.

A recent development has been the extension of these ideas to community dynamics, or a metacommunity.

Explicit Space

A step up in realism that allows additional ecological questions to be investigated is to provide an explicit description of spatial location but without any particular tie to actual physical locations. This kind of description provides more information about the spatial arrangement of the system. More choices also must be must as to how the explicit inclusion of space and spatial arrangement will be described. A fundamental choice in model development is whether the description will be deterministic or stochastic. Additionally, descriptions that may be stochastic at the level of the individual can be deterministic at the level of the population.

Deterministic descriptions of spatial populations typically describe distributions of populations over space using a density function, though descriptions where space is discrete are also used. The models typically proceed from a description of movement of individuals at a local scale and then proceed to a description at the population level. At the level of an individual, movement may be described by Brownian motion (borrowing from a physical description), though more recently there has been great interest in other descriptions at the microscopic scale, such as Levy flights, which involve different distributions of individual movement behavior. As will be discussed below, one major area of great interest is the degree to which ecological conclusions are robust to changes in assumptions. For example, if the model is discrete in time or continuous in time, do the same results emerge?

Some stochastic models describe dynamics on a lattice, while others describe space as continuous and consider the dynamics of individuals as described by their spatial location. Even if a model is described in terms of explicit space, one can still consider approximations of the system or the dynamics of the system that essentially do not take this into account. Essentially, this reduces the system to one where space is implicit and is called the mean field description. There are cases where this is a rather good description of the dynamics even if space is explicit, and conversely there are cases where the mean field description is very poor. For example, the dynamics of a disease when it is rare and can only be transmitted to nearest neighbors is not at all well described by a mean field approximation, which would only give the overall density of individuals in different states. The difficulty in analyzing stochastic models has perhaps limited their use, but recent advances in computing power as well as analytical advances are increasing the use of stochastic approaches.

Realistic Space

Especially with recent advances in computational power there has been an increase in the use of descriptions of space that take into account some of the particular underlying details. One aspect of the use of explicit spatial approaches has been in the design of reserves, where great effort has been made to design reserves that take into account current spatial distributions of species. Recent advances in the use of geographic information systems (GIS) have made the development of approaches based on realistic descriptions of space more common. Note, of course, that this work does not produce general results, but may still be essential for answering applied questions.

More generally, the use of spatially explicit simulation approaches, which often go by the name agent-based models, has been increasing. A fundamental challenge is to understand the degree to which the results obtained are general and robust.

MATHEMATICAL APPROACHES TO SPATIAL ECOLOGY

A variety of mathematical approaches have been taken to describe spatial population dynamics. Here, an overview will be presented, and the reader should consult other entries for more details of essentially all the approaches. The mathematical approaches are used in implementing the theoretical approaches outlined in the previous section, so the organization of this section proceeds in roughly the same order as the previous one. In each case, brief examples are given for the kinds of results that can be obtained.

Metapopulation Dynamics

Some of the simplest spatial models are metapopulation models. The simplest description is the Levins model described above. This model alone has had great influence on the understanding of spatial population dynamics, and many extensions have been considered. Among the many extensions to this very simple model are approaches that take into account more details of single-species dynamics. In particular, Hanski and many others have considered models that take into account more realistic descriptions of connectivity, allowing for some locations to be more accessible from other locations.

Other extensions have focused on including the dynamics of populations within patches. One of the earliest examples of this kind is the model developed by Levin and Paine of the dynamics of intertidal regions in the northwestern United States following a disturbance event.

Another extension has been the inclusion of species interactions—in particular, exploiter–victim interactions including predator–prey dynamics or host–parasitoid interactions as well as competitive interactions. In fact, what may be the earliest verbal description of metapopulation dynamics is the proposal by Nicholson and Bailey in their classic paper on host parasitoid interactions that essentially a metapopulation approach is needed to explain persistence.

Explicit Space: Deterministic Approaches

Approaches to explicit space for deterministic systems were initially described typically by partial differential equations with a reaction term and a diffusion term. The simplest model in this case would be one with exponential growth and random movement of organisms in one spatial dimension, namely,

$$\frac{\partial n}{\partial t} = D\frac{\partial^2 n}{\partial x^2} + rn, \tag{3}$$

where n is a density function on the spatial variable x for the population at time t, r is the rate of exponential growth, and D is the diffusion coefficient which is the mean squared displacement per unit time. This simple model can be used to look at the rate of spread of an introduced organism by considering the rate at which the region where the population is above a specified detection level increases with time. This gives the classic results that the spread is given by $2\sqrt{rD}$.

The KISS model is another classic and important use of the simplest exponential growth model to answer the question of how large a region of suitable habitat must be for a population to persist. Here, Equation 3 applies only inside a region, say, of length L. Then, to Equation 3, one adds boundary conditions specifying that organisms that leave the suitable region do not survive,

$$n(0) = n(L) = 0, \tag{4}$$

so population dynamics is a tension between exponential growth inside the region and death of organisms that leave. The classic result due to Kierstad and Slobodkin is that the population survives only if the growth rate is sufficiently high relative to the losses from individuals leaving the favorable region. Larger r and L and smaller D increase the overall growth of the population. This result can be understood as expressing the idea that growth within the favorable region needs to be greater than the losses experienced by the death of individuals that leave the favorable region. This basic result underlies much of the theory that focuses on the size of a reserve needed. This theory for density independent dynamics has been extended in a variety of ways that have included a focus on the role of different boundary conditions, multiple spatial dimensions, age or stage structure, and other complications.

Another classic extension of the simple model is to include density dependence, so the model becomes

$$\frac{\partial n}{\partial t} = D\frac{\partial^2 n}{\partial x^2} + rn(1 - n), \tag{5}$$

where the measure of population size has been scaled so the carrying capacity is 1. This does not affect the probability of persistence in a finite habitat. Both the linear (density independent) and nonlinear (density dependent) models have been used to study the rate of spread of an invading species, which is found to be $2\sqrt{rD}$ in both cases.

Extensions of these reaction diffusion models to two or more species (or one species and water) have revealed interesting spatial patterns. These patterns are periodic in space (and perhaps time) and arise from a mechanism first identified by Alan Turing in a fundamental paper in 1951. The idea is that the spatially uniform solution can be unstable if one of the species has a tendency to outbreak and the dispersal rates of the two species differ enough. This kind of model has been used to look at patchiness in planktonic systems.

More recently, there has been a great deal of interest in models that are discrete in time and continuous in space, which are called integrodifference equations. These models are all of a form with alternating dispersal and local population dynamics and include a description of movement based on a dispersal kernel of the form $k(y, x)$ which is the probability that an organism that is at y moves or disperses to location x. Then a model in one spatial dimension in an unlimited environment is written as

$$n_{t+1}(x) = \int F(n_t(y))k(y, x)dy, \tag{6}$$

which expresses the idea that to find the organisms at location x, one simply sums up the contributions from all locations y. Here, $n_t(x)$ is again a density function on spatial location x for the number of organisms at time t, and F is a description of the local population dynamics which in a more general form could also depend on spatial location.

In most investigations, the arbitrary kernel $k(y, x)$ has been replaced by a function that depends on the assumption that only the distance between locations is important, so the kernel can be written as a function $k(|y - x|)$. For investigations of persistence and similar issues, the exponential (or Laplacian) kernel where k is proportional to $\exp(-a|y - x|)$ for a constant a is a typical choice. (Part of the reason is that this choice enables relatively easy analytic calculations involving integration by parts.) Models using this kernel produce analogues to the Kierstad–Slobodkin persistence condition that expresses a balance between local growth and dispersal to unfavorable locations.

When using Equation 6 to investigate that dynamics of spatial spread, other kernels have been used. A Gaussian kernel leads to spread rates that are identical to those found in the reaction–diffusion formulation. For the exponential kernel also produces asymptotically linear rates of spread. However, if the tail of the kernel is "thicker" than the exponential (decays to zero more slowly as the dispersal distance goes to infinity), then the spread rate is typically accelerating.

Explicit Space: Stochastic Approaches

Stochasticity has been included in models in a variety of ways. A fully spatial stochastic model is typically one of two types. One possibility is to represent space by a lattice and describe the state of the system by the state at each lattice point (in the simplest case, occupied or empty). Alternatively, one could allow individuals to be located at any point within a continuous description of space.

The analysis of models like this can be extremely difficult, so many investigations either have been entirely by simulation or have begun with simulation. With the ever-increasing computational power readily available, simulation approaches have continued to grow in use. However, these approaches do have their limitations, especially in developing general results.

The simplest analytic description of the dynamics of models on lattices would be to replace the spatially explicit model by a spatially implicit one, a procedure known as the mean field approximation. In the simplest case of a lattice where locations are either occupied or empty, this description would be the dynamics of the number or fraction of occupied locations. Clearly, this can give reasonable information in some instances, and just as clearly, say, when considered dynamics of a species when rare, can be misleading. One potential way to proceed is to consider next the dynamics of pairs of locations, so in this simply case these would be empty–empty, empty–occupied, or occupied–occupied pairs. The difficulty of this apparently appealing way to proceed is that the dynamics of these pairs depends on the frequencies of sets of three contiguous locations in different sets, the dynamics of the triples depends on the frequencies of higher order combinations, and so on. The approach usually taken is to assume that some particular relationship that gives the frequencies of triples in terms of the frequencies of pairs, a pair approximation. Unfortunately, this is only an approximation and it is not clear how valid or useful it is.

The analogous approach of stochastic point processes, which instead of specifying the stats of points on a lattice specifies the location in continuous space of different individuals through time, has essentially the same appeal and the same difficulties. Here, as well, approximations can be used that depend on approximating the moments of the spatial distribution of the species being modeled.

Other Approaches

A number of other mathematical approaches have been used. One particular form of explicit (and perhaps realistic) spatial description that has been receiving increasing attention is the use of networks to describe the underlying habitat. As noted above, spatially explicit simulation approaches are also being used more.

ILLUSTRATIVE EXAMPLES OF THEORETICAL APPROACHES TO QUESTIONS IN SPATIAL ECOLOGY

Competition–Colonization Tradeoff

One of the classic examples of theory in spatial ecology is the competition–colonization tradeoff. Many systems, such as grasslands, seem to support far more species than simple ecological theory based on resource use would indicate. An approach based on resource use would suggest that the number of coexisting species would essentially be determined by the number of resources, but it appears that much larger numbers of species coexist. A number of explanations have been advanced for this phenomenon. One is the neutral theory, which is essentially a spatially implicit theory developed by Hubbell and a number of other investigators. Another spatially implicit theory that has been advanced is that there is a competition–colonization tradeoff. This is based on an extension of the Levins metapopulation model to many competing species (by Hastings, Tilman, and others) with the best competitive ability and dispersal ability inversely correlated. Combined with disturbance, this model can explain the coexistence of an arbitrary number of species. Further work advancing this theory has focused on the impact of habitat loss on coexistence, as well as many other extensions.

Dynamics of Spread

The dynamics of spatial spread have been one of the most carefully studied aspects of spatial ecology. Early observations of the spread of introduced species with the muskrat introduced from North America into central Europe, as well as experiments with release of marked fruit flies (described in the only paper co-authored by Dobzhansky and Wright), indicated that he radius of the occupied area increased linearly with time. The early theory predicted this linear rate of spread, but then observations of increasing rates of spread led in part to the development of theory initially based on integrodifference equations that explained the possibility of increasing rates of spread. This is an area of increasing activity because of the potential for simple relationships between observations of spread of introduced species and underlying population parameters, and also because of the great importance in applications since invasive species are a major ecological threat.

Marine Reserves

Questions of conservation of species or more generally of biodiversity typically require an essentially spatial approach. In the simplest setting, this could reduce to questions of persistence of a single species in a spatially explicit model. These results are important for generating general principles given that the requirements for persistence will depend on both the dynamics of dispersal and on life history characteristics (survivorship and reproduction), and neither of these may be known precisely. In the very simplest setting, this can reduce to the question of persistence of a single species in a single suitable habitat. For this case, the question is essentially the same one as the KISS model discussed earlier. Thus, a minimum size of suitable habitat can be determined that would allow for persistence of a species. In many cases, a more appropriate model than the KISS model based on a reaction–diffusion framework would be one based on an integrodifference framework. However, the mathematical results and conclusions from a biological point of view are essentially the same.

For many marine species in particular, dispersal distances are so large that a single contiguous suitable habitat is not a reasonable description of the requirements for persistence. Here, one needs to turn to persistence in a network of locations connected by dispersal (of typically a larva stage) and with reproduction and survivorship within each location in the network (and possibly including mortality during the dispersal stage). This model differs from the classic metapopulation description since the focus is not on local extinction and colonization but rather on a spatially distributed population. For this kind of network, both for systems arrayed in a regular fashion along one spatial dimension or connected in a less regular fashion, conditions for persistence have been found.

Synchrony and Spatiotemporal Dynamics

A classic area for the application of spatial theory has been to understanding the synchrony of fluctuations in population abundance across space through time. More generally, comparing observations of spatiotemporal dynamics to predictions from models has shed great insight on the determinants of population dynamics. There are classic examples of synchrony in population abundances across space, such as the data of hare and lynx abundances across Canada that continue to be reanalyzed. Some particularly fruitful recent efforts have focused on the spatiotemporal dynamics of the occurrence of childhood diseases.

CONCLUSIONS

As noted in the classic 1992 paper by Levin, spatial ecology is at the heart of essentially all the fundamental questions in ecology. Theoretical developments in spatial ecology have been particularly fruitful since the patterns that require understanding are often dramatic and thus provide clear tests of any theoretical explanation. Classic questions in spatial ecology like understanding spatial spread or synchrony continue to lead to exciting new developments. Newer topics, such as understanding flocking behavior, present new challenges. As applied questions in ecology become more important, it is clear that management can only be done using theory that recognizes the importance of space.

SEE ALSO THE FOLLOWING ARTICLES

Integrodifference Equations / Neutral Community Ecology / Reaction–Diffusion Models / Single-Species Population Models / Spatial Models, Stochastic / Spatial Spread / Species Ranges / Synchrony, Spatial

FURTHER READING

Cantrell, R. S., and C. Cosner. 2003. *Spatial ecology via reaction diffusion equations.* New York: Wiley.

Durrett, R., and S. Levin. 1994. The importance of being discrete (and spatial). *Theoretical Population Biology* 46: 363–394.

Hanski, I., and O. E. Gaggiotti, 2004. *Ecology, genetics, and evolution of metapopulations.* Amsterdam: Elsevier.

Hastings, A., K. Cuddington, K. F. Davies, C. J. Dugaw, S. Elmendorf, A. Freestone, S. Harrison, M. Holland, J. Lambrinos, and U. Malvadkar. 2005. The spatial spread of invasions: new developments in theory and evidence. *Ecology Letters* 8: 91–101.

Levin, S. A. 1992. The problem of pattern and scale in ecology: the Robert H. MacArthur award lecture. *Ecology* 73: 1943–1967.

Liebhold, A., W. Koenig, and O. Bjørnstad, 2004. Spatial synchrony in population dynamics. *Annual Review of Ecology, Evolution, and Systematics,* 35: 467–490.

Okubo, A., and S. A. Levin. 2002. *Diffusion and ecological problems.* New York: Springer.

Skellam, J. 1951. Random dispersal in theoretical populations. *Biometrika* 38: 196.

SPATIAL MODELS, STOCHASTIC

STEPHEN M. KRONE

University of Idaho, Moscow

Real biological communities always exhibit spatial heterogeneity. Even in carefully controlled laboratory experiments with bacteria, well-mixed liquid cultures that are supposed to be the epitome of an unstructured existence have spatial heterogeneities—with bacterial cells adhering to flask walls and forming subpopulations with altered physiological properties. In addition to spatial structure, another ubiquitous characteristic of biological communities is stochasticity (or randomness). Stochastic spatial models account for these two features and thus allow one to make more realistic models and to determine conditions under which stochasticity and spatial structure influence ecological dynamics.

THE CONCEPT

There are several types of mathematical models that are useful in theoretical ecology. These include ordinary differential equations, partial differential equations, and stochastic spatial models. Ordinary differential equations (sometimes called mass-action or mean-field models) ignore all spatial structure and stochasticity by essentially averaging ecological interactions over the entire landscape. Partial differential equations also ignore stochasticity but account for some of the spatial structure in the community. One can think of this as arising by averaging locally but not globally. Stochastic spatial models do not average out randomness or spatial structure and thus provide a more detailed description of the community.

Before even describing what we mean by stochastic spatial models and how they compare to differential equation models, it will be instructive to think intuitively about why we might have different mathematical descriptions of the same phenomenon, and what these different descriptions say about how we are viewing it. It's all about scale—both spatial and temporal.

Imagine watching a cloud formation on a calm day. If we move into the cloud, we are surrounded by a fairly uniform fog; although the density of the fog may be changing with time, it is boringly constant across the region of space we are viewing. (This is the kind of information that an ordinary differential equation might model.) Move in closer to the "microscopic" level and tiny droplets of water vapor are whirring about. Slow down this movie enough, and the motion of individual water droplets would be easy to follow. (This slowed-down microscopic view could be captured by a stochastic spatial model.) Now imagine moving away so that the entire cloud is in your field of vision. The spatial structure of the cloud changes very slowly and seems almost frozen in space. The cloud would change its shape noticeably, however, if one were to watch a time-lapse movie (i.e., speeding up time). (This is the viewpoint captured in a partial differential equation.)

All these observations are of the same object, but what we "see" is influenced dramatically by the length and time scales employed in our observations. In some sense, this scaling accounts for the differences between various mathematical models of the same ecological community dynamics.

MATHEMATICAL MODEL

The stochastic spatial models described here are often called interacting particle systems (IPS) or asynchronously updated stochastic cellular automata (CA). There are several features that characterize these models (Fig. 1).

- They are individual based, meaning that the model tracks the spatial locations of all individuals in the community over time. This is in contrast to differential equation models, which only keep track of densities.
- They are spatially discrete, which means that spatial locations are points of a square grid (two-dimensional or three-dimensional) or lattice of sites. Each of these sites can typically be vacant or contain a single individual (possibly from one of several different species).

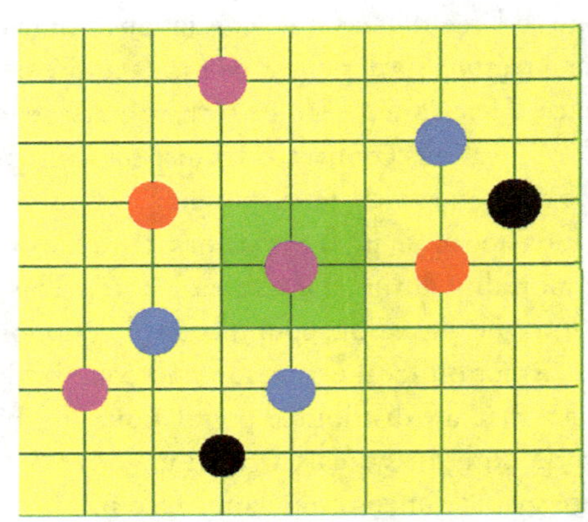

FIGURE 1 Example of grid structure in an IPS model. Colored circles represent individuals of various species; some sites are vacant. If the purple circle in the center is the "focal site" for a potential update, the green box gives an example of an interaction neighborhood. The vertices of this 3 × 3 box represent eight possible neighboring sites (the 9 in the box minus the focal site itself).

- They are asynchronously updated. Although space is discrete, time is continuous in these models. This allows for the random timing of events. (There are related synchronously updated CA models that update all sites at once. These can exhibit chaotic behavior and often can appear more rigid in behavior than real biological systems.)
- They are locally updated. This means that the rate at which a given site changes its state, as well as the state to which it changes, depends only on the states in some local neighborhood of this site.

To describe an interacting particle system (IPS) model for n species in some ecological community, first think of how we would proceed with differential equation models. If we were using an ordinary differential equation—say, a Lotka–Volterra model—we would track the different species densities, $u_1(t), \ldots, u_n(t)$, over the entire region of interest and specify their rates of change by equations of the form

$$\frac{du_i}{dt} = u_i(r_i + \sum_{j=1}^{n} a_{ij}\, u_j) \quad (i = 1, \ldots, n).$$

Here, a_{ij} is an interaction term that describes the effect of (the density of) species j on the density of species i. In particular, the interactions depend only on global species densities that involve, in principle, an infinite number of individuals.

A partial differential equation would have these species densities depending on both space and time, say, $u_i(x, t)$, and the rates of change would typically include some motion terms involving spatial derivatives. The density $u_i(x, t)$ again represents an infinite number of individuals, but this time only the ones that are near x. In an IPS model, we specify interactions between individuals so the length scale is considerably smaller than in the differential equation models.

To account for these individual interactions in an IPS model, we think of locations as being situated on a grid or checkerboard of sites and allow at most one individual per site. Each site is updated at random times with rates that depend on the configuration of types at nearby sites. For example, this interaction neighborhood could consist of the four nearest neighbors (north, south, east, and west) of the focal site. To see how this works, let us consider some examples.

CONTACT PROCESS

This IPS model considers the presence or absence of a single species at each site of the grid. Each site can be in either state 0 (vacant) or 1 (occupied), and site updates correspond to births and deaths. The rates at which site changes its state are

$$0 \to 1 \text{ at rate } b \cdot n_1(x),$$
$$1 \to 0 \text{ at rate } d,$$

where b is the birth rate, d is the death rate, and $n_1(x)$ denotes the number of 1's in the interaction neighborhood of x. In other words, if site x is currently vacant, it will become occupied by the offspring of one of its neighbors at a rate that is proportional to the number of occupied neighboring sites. If x is currently occupied, it will become vacant (the individual at that site will die) at a rate that is independent of the states of any other sites. We thus describe the $0 \to 1$ transition as being by contact and the $1 \to 0$ as being spontaneous.

Typical questions regarding the contact process include whether or not the population dies out or persists indefinitely, what an equilibrium population looks like, and how the population expands starting from an isolated colony. Figure 2 shows a simulation of a subcritical contact process, i.e., one whose birth rate is not large enough to counteract the death rate, and hence will eventually die out. Figure 3 shows a simulation of a supercritical contact process. In this case, the birth rate is large enough to enable the population to survive. Moreover, if a small localized population does survive, it forms a spreading colony that is described mathematically by the so-called Shape Theorem. These expanding colonies are

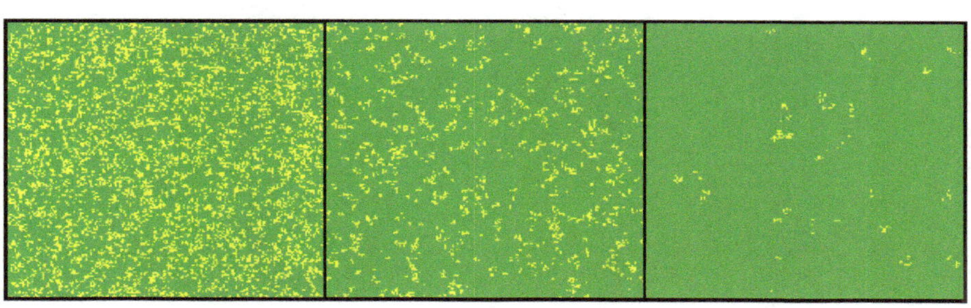

FIGURE 2 Contact process on its way to extinction (yellow = occupied, green = vacant); parameters b = 1.5, d = 1.

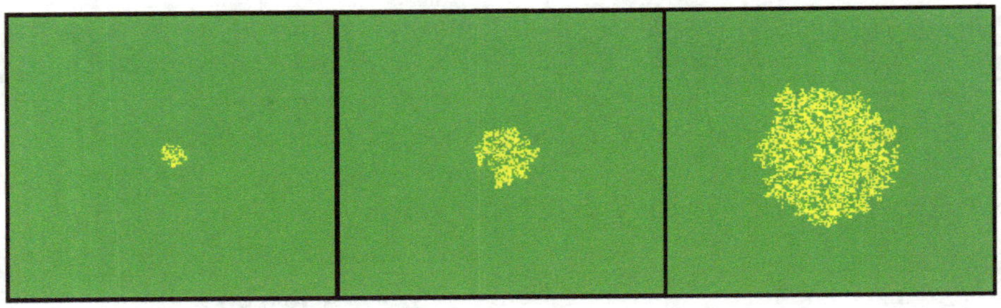

FIGURE 3 Contact process expanding colony; parameters $b = 2.5$, $d = 1$.

reminiscent of real biological colonies (e.g., bacteria) or an advancing wave front of some disease.

Voter Model

The voter model can be thought of as modeling two competing species. Every site of the grid is occupied by a member of either species 1 or species 2; there are no vacant sites in this simple model. At random times, an individual dies and is replaced by the offspring of a neighbor chosen at random. (Similar nonspatial models arise in population genetics—the so-called Moran and Wright–Fisher models.) The rates at which site changes its state in the voter model are

$$0 \to 1 \text{ at rate } b_1 \cdot n_1(x),$$
$$1 \to 0 \text{ at rate } b_2 \cdot n_2(x),$$

where $b_1 = b_2$ (typically both set to 1) and $n_i(x)$ $(i = 1, 2)$ denotes the number of species i in the interaction neighborhood of x. In other words, if site x is currently occupied by one species, it will become occupied by the other species at a rate that it proportional to the number of occupied sites in the opposite state. The case of asymmetric replacement rates $b_1 \neq b_2$ yields the so-called biased voter model. Notice that all the transitions in the voter model are by contact.

For the biased voter model, the species with the highest rate typically takes over—though a small number of this dominant species could go extinct just by chance. In the regular (unbiased) voter model, both species can coexist indefinitely but they become more and more clustered over time. This characteristic clustering is illustrated in Figure 4.

Many other spatial models in ecology can be constructed by choosing some number of species and specifying transitions between the various states that are either by contact or spontaneous. Some of these are included in the simulator WinSSS that is available on the author's Web site. Figure 5 shows a screenshot of the simulator.

For example, in 2002, Kerr and colleagues considered an ecological system with three species engaged in nontransitive competitive dynamics. They described this as analogous to the game "rock–scissors–paper" in that species 2 outcompetes species 1, species 3 outcompetes species 2, and species 1 outcompetes species 3. They modeled competitive differences between the species by differences in death rates, while each species was considered equally adept at colonizing empty territory. In particular, letting d_i denote the death rate for species i, they considered parameter values that fit the pattern $d_1 = \frac{1}{4} + \frac{3}{4} f_2$, $d_2 = \frac{1}{3}$, $d_3 = \frac{10}{32}$, where f_2 is the fraction of neighboring sites occupied by species 2.

This is motivated by a toxin-production scheme in *E. coli*. There are three types of *E. coli* in this set-up. Species 1 is sensitive to the toxin (called colicin), species 2 produces the toxin and is resistant to it, and species 3 is a kind of cheater that is resistant to the toxin but doesn't bother making it. If species 1 and 2 are put together, species 1 is killed by the toxin and species 2 takes over. If species 2 and 3 are together, the toxin produced by 2 does not affect 3 and there is a cost to producing the toxin, so

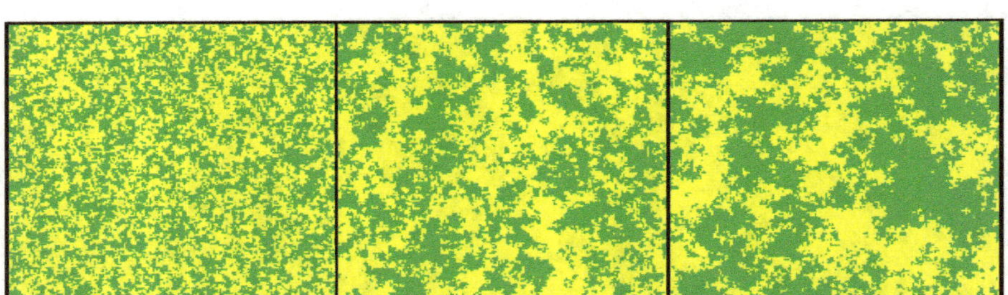

FIGURE 4 Two-species voter model showing increased clustering over time (yellow = species 1, green = species 2)

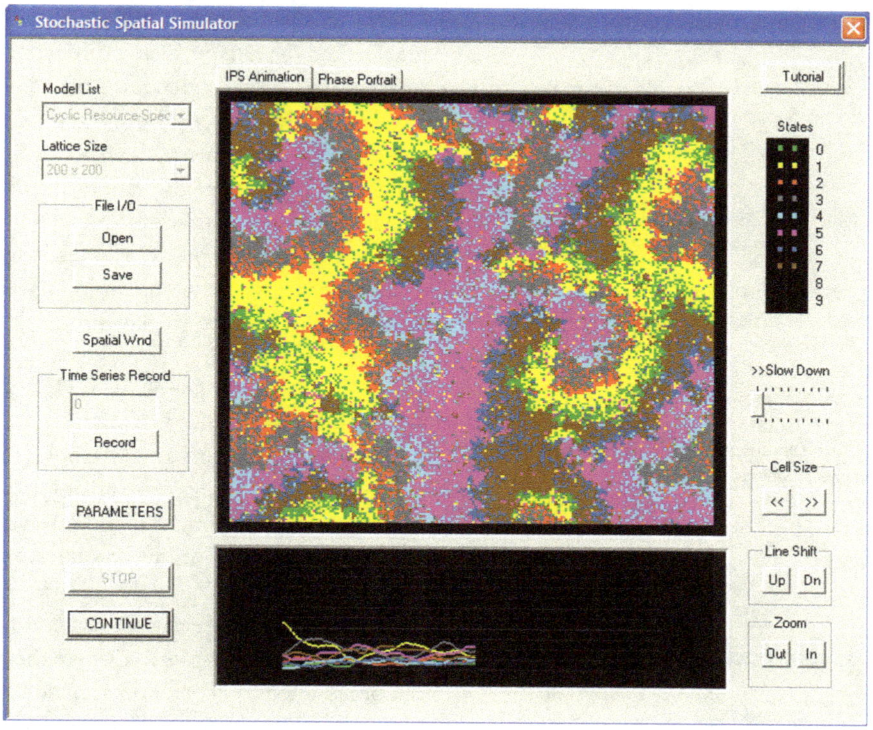

FIGURE 5 Screenshot of IPS model simulator WinSSS. The model shown has eight states (0, 1, . . . , 7), half of which represent species (the even-numbered states) and the other half represent "resources" (the odd-numbered states) that can be consumed preferentially by different species. Global densities of the different states appear as the oscillating curves at the bottom. This particular ecological model spontaneously develops spiral patterns when started from a completely random (unstructured) spatial configuration.

species 3 wins. If species 3 and 1 are together, there is no toxin to worry about but there is a slight cost to species 3 for having to express its resistance genes, so species 1 would outcompete species 3. Things get more interesting when all three species are put together, and how they are arranged matters a great deal.

When the population is well mixed, the resistant species (3) wins the competition—species 1 is first killed off by (the toxin from) species 2, and then species 3 wins the head-to-head competition with species 2. This was the outcome in mathematical models (mean-field ODE or IPS with large interaction range) and in the experimental system (liquid or growth on a Petri dish with occasional mixing of individuals). When the population was structured, however, all three species were able to coexist. This was true in the IPS model with small interaction range and in the Petri dish experiment that did not mix individuals. This provides a nice example of how spatial structure can enhance diversity.

Krone and Guan (2006) considered another kind of cyclic dynamics that can occur in bacterial (and other) systems, and they studied it via IPS models. In this model, there are L species $(S_1, \ldots, S_L)$ and L resources $(R_1, \ldots, R_L)$. Species i can consume resource i and then produce offspring that are sent to nearby sites. After a time, species i dies at its site (because it runs out of food) and leaves behind resource $i + 1$ (with indices always understood cyclically in that $L + 1 = 1$). Thus, the succession of states at a site is

$$R_1 \to S_1 \to R_2 \to S_2 \to \ldots \to R_L \to S_L \to R_1 \to \ldots.$$

The transitions at site x are as follows:

$$R_i \to S_i \text{ at rate } b_i f_i(x) \text{ (by contact);}$$
$$S_i \to R_{i+1} \text{ at rate } d_i \text{(spontaneous),}$$

where $f_i(x)$ denotes the fraction of neighboring sites of x that are occupied by species i.

Simulations and analysis of the model show that coexistence (i.e., long-term persistence of all the species) is much easier in a well mixed (ODE) or semi-mixed (IPS model with large interaction range) system than in a more strictly spatial system (with small interaction range). In fact, for certain parameter values in the spatial system, persistence requires that the species and resources be aligned in successive waves spreading across the landscape (see Fig. 5). These self-organized patterns illustrate a phenomenon seen in natural systems where species alter their environments and thereby create resource gradients that influence community structure nearby.

SEE ALSO THE FOLLOWING ARTICLES

Birth–Death Models / Cellular Automata / Individual-Based Ecology / Ordinary Differential Equations / Partial Differential Equations / Stochasticity / Two-Species Competition

FURTHER READING

Dieckmann, U., R. Law, and J. A. J. Metz, eds. 2000. *The geometry of ecological interactions: simplifying spatial complexity.* Cambridge, UK: Cambridge University Press.

Durrett, R., and S. Levin. 1994. Stochastic spatial models: a user's guide to ecological applications, *Philosophical Transactions of the Royal Society of London B: Biological Sciences* 343: 329–350.

Kerr, B., M. A. Riley M. W. Feldman, and B. J. M. Bohannan. 2002. Local dispersal promotes biodiversity in a real-life game of rock-paper-scissors. *Nature* 418: 171–174.

Krone, S. M. 2004. Spatial models: stochastic and deterministic. 2004. *Mathematical and Computer Modelling* 40: 393–409.

Krone, S. M., and Y. Guan. 2006. Spatial self-organization in a cyclic resource-species model. *Journal of Theoretical Biology* 241: 14–25.

SPATIAL SPREAD

ALAN HASTINGS

University of California, Davis

The study of spatial spread in ecology has a long history. This study of an intermediate stage in the dynamics of an invasive species has used a variety of mathematical approaches that have been combined with data collected in a variety of ways. The goals of the theory have included understanding broad patterns of spread, predicting future spread, determining population parameters such as growth and dispersal from observations of spread, and designing control measures to reduce or eliminate spread.

BASIC DIFFUSIVE MODEL OF SPREAD

The most basic question in the study of spatial spread is to relate a rate of spread of a species to underlying parameters describing biological dynamics. Given a description of population increase and a description of the movement of individuals, what would the rate of spread of the population be?

Among the earliest descriptions of spatial spread are models that are described in terms of reaction–diffusion equations. In the 1930s, in two independent papers, Fisher and Kolmogorov and colleagues almost simultaneously introduced and analyzed some of the first models of this kind. In Fisher's version, the biological system examined was the spread of an advantageous gene. In the simplest version, all movement is random, with no tendency for directed movement in any direction. Also, in the simplest version space is viewed as one-dimensional. In this continuous time description, p is an allele frequency, s is a selection coefficient, and D is a diffusion coefficient, which is a measure of the rate of mean squared random displacement per unit time. Note that this is formally exactly the same as a model of logistic population growth with dispersal, with the population scaled by the carrying capacity. This leads to the equation

$$\frac{\partial p}{\partial t} = D\frac{\partial^2 p}{\partial x^2} + sp(1 - p). \tag{1}$$

Mathematically this is a parabolic partial differential equation, so it is perhaps surprising mathematically that solutions are found in the form of traveling waves.

The method of solution is both intriguing and instructive, so this entry will go through some of the details. The approach is to assume there exists a traveling wave solution. One assumes a solution where the allele frequency p can be assumed to depend on a single variable z, which in turn depends on time t and space x in the form

$$z = x - ct, \tag{2}$$

where c is the speed of the wave. Essentially there is a single function $p(z)$ that describes the allele frequencies for all time and this function gets translated along the spatial dimension through time.

Substituting this form for $p(z)$ into the model 1, one obtains a single second-order ordinary differential equation (where prime denotes differentiation with respect to the variable z)

$$-cp' = Dp'' + sp(1 - p). \tag{3}$$

This in turn can be reduced to a pair of first-order ordinary differential equations by letting

$$q = p', \tag{4}$$

which changes the single Equation 3 into a pair of first-order equations

$$\begin{aligned} p' &= q, \\ q' &= -(1/D)sp(1 - p) - cq. \end{aligned} \tag{5}$$

This pair of equations can then be studied using phase plane techniques to determine the rate of spread. Essentially one looks for a solution that goes from an allele frequency of 1 at minus infinity (in terms of z) to 0 at plus infinity, as illustrated in Figure 1. This is a solution in the phase plane that joins the two equilibria of the system of Equation 5 $(p, q) = (0, 0)$ or $(p, q) = (1, 0)$.

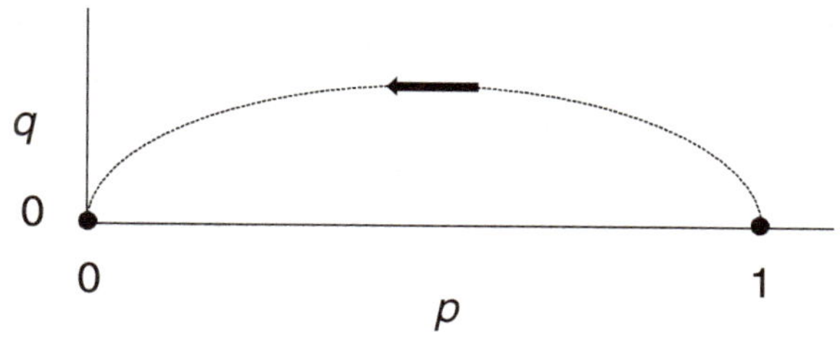

FIGURE 1 This figure illustrates the phase plane used in the analysis of spread for Fisher's equation when reduced to the system of equations 5 for the frequency and slope of the frequency with respect to the variable z. A traveling wave solution that is 1 at minus infinity and 0 at plus infinity corresponds to a trajectory going from the equilibrium (1, 0) to the equilibrium (0, 0). The speed is determined by finding the conditions that guarantee that the trajectory does not leave the positive quadrant as illustrated.

The condition that neither equilibrium can be a spiral (as this would produce allele frequencies greater than one or less than zero) can be found by linearizing the system of two equations about the two equilbria and looking at the Jacobian (see the entries on stability and phase plane techniques for details). This leads to the condition that

$$c >= 2\sqrt{sD}. \tag{6}$$

It is very difficult to prove, but the appropriate choice, as used by Fisher, for the wave speed is the minimum value of c, so the speed of invasion is given by $2\sqrt{sD}$. This solution obviously can be a candidate for an asymptotic solution. The question of whether starting from different initial conditions the solution converges to (approaches) this solution is a very difficult one and it took many years for some results to be obtained.

LINEAR DIFFUSIVE SPREAD MODELS

Another seminal study of the dynamics of spatial spread was the 1951 paper by Skellam. An issue that was central to this paper was the question of the rate of northward spread of oak trees since the last glaciation, a question known as Reid's paradox. Given the distance that an acorn falls from the tree, at first it appears that the spread of oak trees may be too rapid to be explained by the passive dispersal of acorns. In order to confirm this suspicion, it is necessary to find the rate of spread.

Here, a simpler linear version of Equation 1 was used, where the reaction part of the description was replaced by exponential growth, namely,

$$\frac{\partial p}{\partial t} = D\frac{\partial^2 p}{\partial x^2} + sp. \tag{7}$$

Obviously, the solution depends on the initial conditions. But, since the problem is linear, any sum of solutions is still a solution. Thus, the solution to the point release problem, where all individuals are at a single known location at time zero, can also be used to find the solution to the problem for any initial condition by summing up these solutions. (Since this has to be a "sum" over all points it is actually an integral.)

The solution to the point release problem can be guessed since it is essentially the sum of an infinite number of arbitrarily small steps, with no bias to move either right or left. By the law of large numbers this should be a normal distribution, and since there is no bias, it should have mean zero. The appropriate variance can be determined by substitution into Equation 7, and the solution to the point release problem is found to be a normal distribution with the variance proportional to the time. This distribution is well defined for any time great than zero, but at $t = 0$, the solution is zero anywhere except at $x = 0$. At the point $x = 0$, the solution approaches infinity as t approaches zero. Also, the integral of the solution is independent of time, so at $t = 0$ it corresponds to the idea that the individual is certainly found at $x = 0$. Note that this solution can be interpreted either as the probability distribution for the location of a single organism released at zero at time 0 for future times, or if multiplied by N as the density function on location x for the density of these N organisms.

One hint of potential trouble is that this solution to Equation 7 has what may seem an odd property. Even if all the organisms are at the known location $x = 0$ at time 0, there is a positive probability that some organisms will be found at arbitrarily large distances at any positive time, which is clearly unrealistic, since it would involve arbitrarily large rates of movement.

The next question is how to obtain a rate of spread from this solution. The answer is not completely obvious. One clever approach was to ask how the location of the point in space that corresponded to a population at a small detection level moved in time. Setting this detection level to an arbitrary small value, one finds (after a bit of manipulation making use of the fact that the detection level is small) that the spread rate is again given by $2\sqrt{sD}$, the same as in the nonlinear density-dependent case. While at first surprising, this result does make sense upon further reflection. At the edge of the range of the population, the population levels are small and density-dependent factors are not important, so it makes sense that the rate of spread is unaffected by density dependence. In fact, one can show that if the rate of local population growth is given by $pf(p)$, and the per capita growth rate $f(p)$ is monotone decreasing, the rate of spread is given by $2\sqrt{f(0)D}$.

Although the formula relates the rate of spread to underlying parameters, perhaps one of the most dramatic predictions is the qualitative one of a linear rate of spread. This result can also be extended to two spatial dimensions by looking at solutions that are radially symmetric. Then it is the radius, or more generally the square root of the area of the range, that expands linearly in time. The same rate of spread was also found for the equivalent stochastic linear model.

This linear rate of spread was confirmed by a number of cases, both experimental and observations of introductions. Two notable examples are the spread of introduced muskrats in Europe and experiments with controlled releases of fruit flies.

INTEGRODIFFERENCE APPROACHES AND OTHER EXTENSIONS

It seemed that understanding and predicting the rate of spread of species was a real triumph of theoretical ecology with a relatively simple answer, but a number of surprises emerged later.

One question was how well the spread rate for the simple model fit species with more complex life cycles. One approach would be to simply compute the rate of population growth and separately the mean squared displacement of offspring from the natal location of the parent. However, there were modifications by van den Bosch and colleagues, who demonstrated that the timing of dispersal and reproduction could greatly affect the rate of spread of populations.

Other aspects become clearer if we consider the derivation of the equation. Without giving a full description or a formal derivation, some ideas still emerge. A way to derive the reaction–diffusion model is to start with a description that is discrete in time and space. The reaction–diffusion model can be found by taking limits that essentially involve assuming that the movement of individuals is described by a Gaussian. What if this taking of limits and making assumptions about what are essentially higher order moments of the distribution of movement rates is not appropriate? The simplest approach to answering this question would be to write down a system that did not require taking all these limits, which is easy if we look at a model that is discrete in time, but continuous in space. Just such a model was investigated by Kot, Lewis, and van den Driessche.

For simplicity, this entry again focuses on the case where space is one-dimensional. Also for simplicity, the entry focuses on a case of nonoverlapping generations, so the life cycle will consist of population growth followed by a dispersal phase with the census taken after the dispersal phase and before the growth phase. The key component of this model is the dispersal kernel $k(y, x)$, which is the probability that an organism released at location y disperses to location x.

Then the function $n(t, x)$, which is a density function on the spatial variable x for the population size n at time t, evolves according to the integrodifference equation (with f again the per capita rate of increase)

$$n_{t+1}(x) = \int n_t(y) f(n_t(y)) k(y, x) dy, \tag{8}$$

which expresses the idea that to find the organisms at location x one simply sums up the contributions from all locations y. A simpler version of Equation 8 where the kernel k is assumed to depend only on the distance between release and settling point is typically used,

$$n_{t+1}(x) = \int n_t(y) f(n_t(y)) k(|y - x|) dy. \tag{9}$$

The solutions to this model depend strongly on the dispersal kernel, with the shape of the tail key. Not surprisingly, if the dispersal kernel is a Gaussian, one recovers the same spread dynamics as the reaction–diffusion model. The biggest difference in behavior occurs for those "thick-tailed" kernels that go to zero more slowly as z goes to infinity then the exponential kernel $\exp(-k|z|)$. In this case, the solution is not a wave that travels at constant speed as in the reaction–diffusion model, but a wave that increases in speed. This makes heuristic sense because with a thick-tailed dispersal kernel more individuals disperse very long distance. Also, it might seem that since there clearly is a maximum dispersal distance, an effect that depends on behavior of the kernel as the dispersal distance goes to infinity would only be theoretical. However, over realistic time scales, numerical solutions do in fact show dramatic increases in invasion speed.

This analysis of integrodifference equations was in fact developed in part to explain observed examples of increasing rates of spread. A number of introduced bird species in the United Sates, for example, showed increasing rates of spread. So the theoretical predictions are realistic.

ALLEE EFFECTS

An important biological phenomenon that also introduces new behavior is the Allee effect, where the rate of population growth is not a monotonic decreasing function of the population size. It is important to distinguish a case where the population growth rate is negative at low population densities, a *strong Allee effect,* from the case where the population growth rate is still positive at low densities, a *weak Allee effect.* How does this change the rate of spread from the classic models? Here, whether space is viewed as discrete or continuous can make a substantial difference.

First, either weak or strong Allee effects reduce the rate of spread in either continuous or discrete-time models, and they can also convert a system that would produce accelerating rates of spread (fat-tailed kernels) into systems with constant spread rates.

Second, if space is discrete (patchy), then strong, but not weak, Allee effects can lead to range pinning. If there is a strong Allee effect, there is a threshold level below which the growth rate is negative. In discrete space, one patch can be above this threshold with the neighboring patch below it in a stable configuration known as range pinning. Note that if space is continuous, this range pinning cannot occur.

Third, with a strong Allee effect, but not a weak one, there is a spatial threshold. If an initial population is not above this threshold it will not spread.

A GENERAL APPROACH

At this point it may seem as though the pendulum had swung from the simple answer of linear spread rates that were robust to a case where spread was really a set of special cases. However, an earlier (1982) more abstract mathematical formulation by Weinberger had given a description of spread dynamics that essentially covered all the deterministic cases considered here. This formulation was then used as the basis for later work by Weinberger and colleagues, aimed at understanding when the linear spread conjecture, the idea that the long-term spread rate is linear in time as expressed in the form of Equation 2. This work identified cases where this linear spread conjecture could fail, such as where there is competition. The integrodifference equations with thick tails discussed above that led to accelerating spread rates are another example.

One case that did not fit the linear spread conjecture was identified by Shigesada and colleagues and focused on an important mechanism that can also produce accelerating rates of spread. If there is heritable variability in dispersal rates among individuals in the population, accelerating rates of spread will be produced. The mechanism is that those individuals that have the greatest propensity to disperse will, by virtue of their tendency to move, be found at the edge of the range. This effect will increase through time, leading to an increasing rate of spread. Note that this is a mechanism different than selection for higher dispersal rates because of some issue of density dependence or similar mechanisms.

OTHER EXTENSIONS OF THE MODELS

Stochastic Models

All the models discussed thus far are deterministic in some sense. If population dynamics are stochastic, what effect does this have on rates of spread? As Mollison noted in 1991, the rate of spread for simple stochastic versions of Equation 7 is exactly the same as the deterministic version. Conversely, he demonstrated that nonlinear stochastic spread models do have different rates of spread.

Other stochastic models of spread have focused on spread in lattices or spread with jumps that found new centers of spread. In both of these cases, the dynamics are different from the deterministic spread described by Equation 1 or 7. The underlying models and analyses remain much more complex than the simple models, and general results have not been obtained.

Heterogeneities and Spread on Networks

Obviously, environments vary in space and time. However, theory related to spread with this kind of variability is much more limited. One notable set of results is that, with an environment that varies to affect the population growth rate in a way that is periodic in space, the spread rate is determined by the average growth rate of the population.

The dynamics of spread on networks has been studied primarily in the context of diseases. Here, the network structure is used to describe the network of contacts that provide the routes for disease transmission. Again, results tend to depend on model details.

More Than One Species

Studying the rate of spread in the context of more than one species is obviously much more complex than the case of a single species, so not surprisingly there are fewer general results. Several cases have been considered, including cases of the invasion of a competitor, spread in the context of predator–prey interactions, and the spread of diseases in space. The last case is one that bears some resemblance to the predator–prey case. The presence of a competitor can reduce the rate of spread, and it is also one of the cases where the linear spread conjecture does not hold, although the deviation is not large.

Evolution

The study of the evolution of dispersal has a long history, with an emphasis on explaining why individuals should disperse. More recently, there have been studies that have considered the evolution of the shape of the dispersal kernel, thus getting at the question of how far individuals disperse which bears directly on the rate of spread. This work has shown that long-distance dispersal can be adaptive but has also raised the issue of how constraints (such as the behavior of dispersal agents for plant seeds) interact with selected traits to produce observed dispersal kernels. This work also emphasizes the importance of stochasticity in long-distance dispersal.

CONTROL

Invasive species are one of the most important threats to native species throughout the world. Thus, the control of the spread of invasive species is an important issue, and theory has been developed to determine control measures that would be effective in slowing or stopping the spread of introduced species. This theory begins with the models presented here for the spread of a species and considers both spatial and nonspatial measures that would change the dynamics.

One key issue raised early on by Moody and Mack considered an invasive plant with outlying populations and a core population and asked whether it was more effective to use control measures on the core population or the outliers. Work on this and related problems has made clear that more information is needed to determine an appropriate answer. Simple use of the theory reviewed in this entry can be used to find measures that would stop spread, but as emphasized in a recent review by Epanchin-Niell and Hastings, determining optimal strategies requires knowledge of costs of control and of damages.

CONCLUSIONS

Although the study of spatial spread in recent years has emphasized that the earlier conclusions of the ubiquity of linear rates of spread are not robust, the theory of spatial spread remains a triumph of theoretical ecology. The work on spread has produced clear relationships between underlying descriptions of population parameters describing population growth and dispersal and the resulting spread of populations. The emerging predictions have been subjected to extensive tests and comparisons with observational data that have led to further developments of the theory, such as the relatively recent work on integrodifference equations. This work is continuing to contribute to basic understanding of population dynamics, while simultaneously being used to help design management strategies for the control of invasive species.

SEE ALSO THE FOLLOWING ARTICLES

Allee Effects / Integrodifference Equations / Invasion Biology / Optimal Control Theory / Phase Plane Analysis / Predator–Prey Models / Reaction–Diffusion Models / Spatial Models, Stochastic

FURTHER READING

Andow, D., P. Kareiva, S. A. Levin, and A. Okubo. 1990. Spread of invading organisms. *Landscape Ecology* 4: 177–188.

Epanchin-Niell, R. S., and A. Hastings. 2010. Controlling established invaders: integrating economics and spread dynamics to determine optimal management. *Ecology Letters* 13: 528–541.

Fisher, R. A. 1937. The wave of advance of advantageous genes. *Annals of Human Genetics* 7: 355–369.

Hastings, A., K. Cuddington, K. F. Davies, C. J. Dugaw, S. Elmendorf, A. Freestone, S. Harrison, M. Holland, J. Lambrinos, and U. Malvadkar. 2005. The spatial spread of invasions: new developments in theory and evidence. *Ecology Letters* 8: 91–101.

Kot, M., M. A. Lewis, and P. Van Den Driessche. 1996. Dispersal data and the spread of invading organisms. *Ecology* 77: 2027–2042.

Okubo, A., and S. A. Levin. 2002. *Diffusion and ecological problems.* New York: Springer.

Shigesada, N., and K. Kawasaki. 1997. *Biological invasions: theory and practice.* New York: Oxford University Press.

Skellam, J. 1951. Random dispersal in theoretical populations. *Biometrika* 38: 196.

Taylor, C. M., and A. Hastings. 2005. Allee effects in biological invasions. *Ecology Letters* 8: 895–908.

Weinberger, H. 1982. Long-time behavior of a class of biological models. *SIAM Journal on Mathematical Analysis* 13: 353–396.

SPATIAL SYNCHRONY

SEE SYNCHRONY, SPATIAL

SPECIES RANGES

KEVIN J. GASTON

University of Exeter, Cornwall, United Kingdom

HANNAH S. SMITH

University of Sheffield, United Kingdom

The geographic range of a species, the spatial distribution of its individuals, is one of the fundamental units of ecology. It is an expression of the outcome of the interaction between the biological traits of the species, its environment, and the history of the two. The location, size, and structure of species' geographic ranges, and how they change through time, give rise to many of the most obvious patterns of life on Earth, including those in species richness and composition.

ESTABLISHMENT

Under most apparently realistic models of speciation, the global geographic ranges of the majority of species are initially small (i.e., at the "birth" of the species). This is both in terms of their extent of occurrence and their area of occupancy. Indeed, it seems likely that, as a consequence, a substantial proportion of species rapidly become extinct due to stochastic processes, leaving little trace of their short existence. Of those that survive, a proportion probably never exhibit significant increase in range size, with any evolutionary changes that accompanied their speciation being insufficient to overcome to any marked extent whatever limitations prevented the ancestral species from becoming more widespread (see below). At any one time, the frequency distribution of the geographic range sizes of the species in a clade (the species-range size distribution) tends to be strongly right skewed, with most species having relatively small ranges and only a few having relatively very large ones; this distribution is commonly modeled approximately as a logarithmic or a logit function. Presumably, some component of the pattern is a historical legacy of the constraints on initial range size, although how large this component is remains poorly understood.

The rest of the new species that survive may still retain relatively small geographic range sizes for substantial periods, exhibiting a marked lag phase while local abundances grow and adaptive constraints are perhaps overcome, before any substantial expansion takes place. Such a dynamic is also commonly observed when species are intentionally or accidentally introduced by human agency into regions in which they did not naturally occur. Unfortunately, the initially limited and slow range expansion of such alien species often results in complacency as to the potential future impacts that they may have and a possibly expensive deferment of action until after any lag phase has ended and control becomes more problematic.

Following any initial lag, the expansion of geographic ranges tends to be rapid (at least when scaled to the life span of individuals of the species) and may quite often be accompanied by the emergence, particularly toward the range boundaries, of individuals with especially well-developed dispersal traits. These traits are most obviously structural, such as enhanced wing musculature, but they are likely also behavioral, although this is more difficult to demonstrate. This change doubtless reflects the selective advantage to individuals of dispersing into otherwise highly suitable areas from which the species is absent or occurs at low density, and thus where resource availability is high. Range expansion often proceeds through the establishment of new outlying populations both close to and well beyond the previous range boundary (a combination of many short steps and fewer long leaps), followed by "back-filling" of any intervening unoccupied suitable habitat. This is thought to result from a combination of abundant short-distance and rare long-distance dispersal events.

On an ecological time scale the overall trajectory of expansion of the size of a geographic range is often modeled as a logistic function, or approximately so, with the rate of spread eventually slowing and range size stabilizing (but see below). A wide variety of approaches have been used to describe the expansion, focusing particularly on diffusion-type processes, and on trying to understand what determines the particular rates of spread observed for different species (influenced by local population growth rates, the shape of dispersal kernels, and the spatial structure of suitable habitat).

A pattern that emerges early and is maintained for the geographic range of the majority of species is that the density of individuals is low in most places within its bounds and high only in relatively few. Arguably, when it is rare the density of a species exhibits Poisson-type distributions and passes through negative binomial-type to lognormal-type distributions as it becomes more common, but how best to characterize changes in the aggregative behavior of individuals as their densities increase remains contentious. Local abundances are almost invariably positively autocorrelated in space, with areas closer together having more similar densities. Much ecological theory, particularly that exploring the consequences of the interaction between relatively invariant niches and spatially variable environmental conditions, also predicts that densities should decline from the center toward the periphery of the geographic range of a species, at least under circumstances in which environmental conditions do not exhibit abrupt discontinuities (Fig. 1). Although there are empirical examples that support this, there are many more that do not, and the abundance structure of geographic ranges is usually complex in ways that are generally not well captured by associated theory. Closing this particular gap poses one of the biggest challenges to development of the theory of geographic ranges.

RANGE LIMITS

Stable Limits

On ecological time scales, species may have relatively stable geographic ranges for long periods. The number of individuals in a local population is determined by the number of births (B), deaths (D), immigrants (I), and emigrants (E) such that $N_{t+1} = N_t + B - D + I - E$, where N_t is population size at time t. In the long term, a species will occur in an area when $B - D + I - E \geq 0$. The limit to a geographic range is thus reached where this inequality no longer holds for significant periods, recognizing that a population may remain in an area for a time when the inequality is violated if the population is large and/or individuals are long lived, giving rise in the short term to an extinction debt (species that persist after the populations to which they belong have ceased to be viable). Perhaps most obviously, the range limit may be reached through some combination of a decline in births, an increase in deaths, a decline in immigration, and/or an increase in emigration. Experimental transplants of groups of individuals beyond the bounds of the species' geographic range have documented some evidence for these. Examples report failure of reproduction, high adult mortality, and a lack of immigrants, and the focus of such work on plants and other relatively sedentary species reflects the otherwise high likelihood of emigration from experimental plots. However, equally, there are also cases in which such transplants reproduce successfully and/or persist for long periods. Indeed, particularly given the likelihood that in any one case at least some demographic rates are density dependent, simple intuitive expectations about how births, deaths, immigration, and emigration will change toward range boundaries may commonly be false, with the significance to range limitation lying in the relative variation of all four demographic parameters, and not in their consideration in isolation (Fig. 1).

Important changes in births, deaths, immigration, and emigration can arise at range limits for a wide variety of reasons, including the following:

1. Physical barriers. These prevent, or dramatically reduce, the immigration of individuals into areas beyond the range limit. Most obviously they comprise large stretches of entirely unsuitable habitat, such as water bodies for terrestrial organisms and land for aquatic ones. These barriers are what give rise to the essential biogeographic structure and regional biotic flavor of the planet, in which, for example, the majority of species occurring within particular continents and ocean basins are not shared with others. Many physical barriers are, however, undoubtedly less stark than this would suggest, and may take a wide variety of forms, including mountain ranges for lowland organisms and coastal shelves for deep-sea ones, and, more importantly than is often recognized, patterns of air and water currents.
2. Abiotic gradients. Spatial variation in abiotic conditions (e.g., climate, water, nutrients) commonly exhibits a reddened spectrum such that areas closer together are more similar than are those further apart. The resultant gradients have profound influences on the rates of births and deaths in populations of a species, with the obvious potential for geographic range limitation. Examples have been documented in which the rate of births declines and/or that of deaths increases toward or at the range boundary, although density-dependent effects and dispersal can complicate matters. The relative importance of abiotic influences on births and deaths is unknown, but it seems likely that ranges are more frequently limited by the failure to produce viable offspring at sufficient levels than through the level of mortality of adult organisms (which in the case of many animals can move to avoid such threats). This said, the boundaries of many species doubtless extend beyond sites in which birth and death rates are such that viable populations can be sustained (i.e., beyond source populations), being maintained instead by the flow of immigrants from elsewhere in the range (i.e., such that they are sink populations). Indeed, provided the balance of births and deaths is not too negative even a low rate of immigration may be sufficient. An important consequence of such a dynamic is that the conditions under which individuals of a species are found to occur cannot necessarily be used to indicate those under which entirely isolated populations would be capable of persisting. It also highlights the potential broader significance to range limitation of the links between local populations. For example, abiotic gradients may result in a reduction in the availability of suitable habitat patches, and this alone can be sufficient to generate a range limit, even if the quality of those patches that do exist remains similar, because a metapopulation dynamic of local colonizations and extinctions can no longer be maintained. Of course, a reduction in both the availability and quality of habitat patches is very common.
3. Biotic interactions. Species variously consume, are consumed by, and compete with one another. All of these interactions can limit geographic ranges. Perhaps most obviously, the distributions of consumers are limited by those of their resources. However, in practice even for specialist consumers the range limits of the two seem seldom even approximately to coincide, suggesting that other factors are also important. These probably often include density effects of consumed species on their consumers (the abundance of the resource is more important than just its presence) and spatial variation in the traits of consumed species that influence the level to which they are exploited.

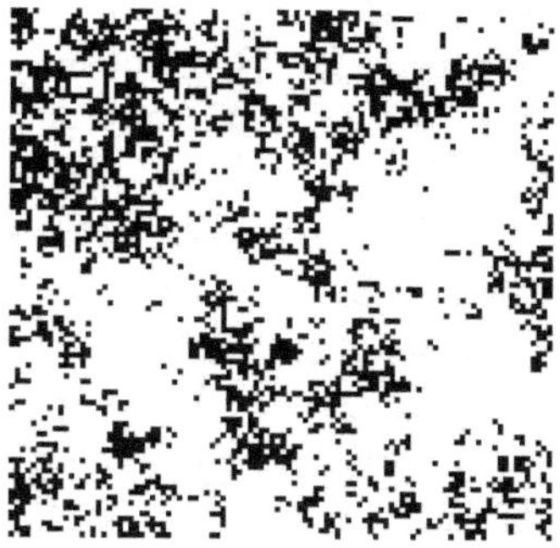

FIGURE 1 Simple models of the structure of geographic ranges. Using an exemplar fractal matrix of habitat suitability (panel (A); black cells represent unsuitable habitat, white cells suitable habitat), at time = 0 the landscape (dimensions 100 × 100, with wrapped horizontal and absorbing vertical boundaries) is seeded with 100 individuals at the core location. In each time step, individuals in a grid cell (x, y) reproduce with probability $b(x, y)$ and die with probability $d(x, y)$. Their offspring are dispersed according to a negative exponential distribution. In Model 1, both b and d are density dependent; in Model 2, b is density dependent and d increases linearly with distance from the core; in Model 3, b decreases linearly with distance from the core and d is density dependent; and in Model 4, b decreases and d increases linearly with distance from the core and all cells have a fixed carrying capacity. In each case, the following outputs are provided: (B) mean density of occupied cells at each position in the range for the first 100 time steps; (C) the density of occupied cells at time = 5000 (line is the mean density of occupied cells at each position in the range); and (D) the growth rate ($b - d$) of occupied cells at time = 5000 (line is the mean growth rate of occupied cells at each position in the range).

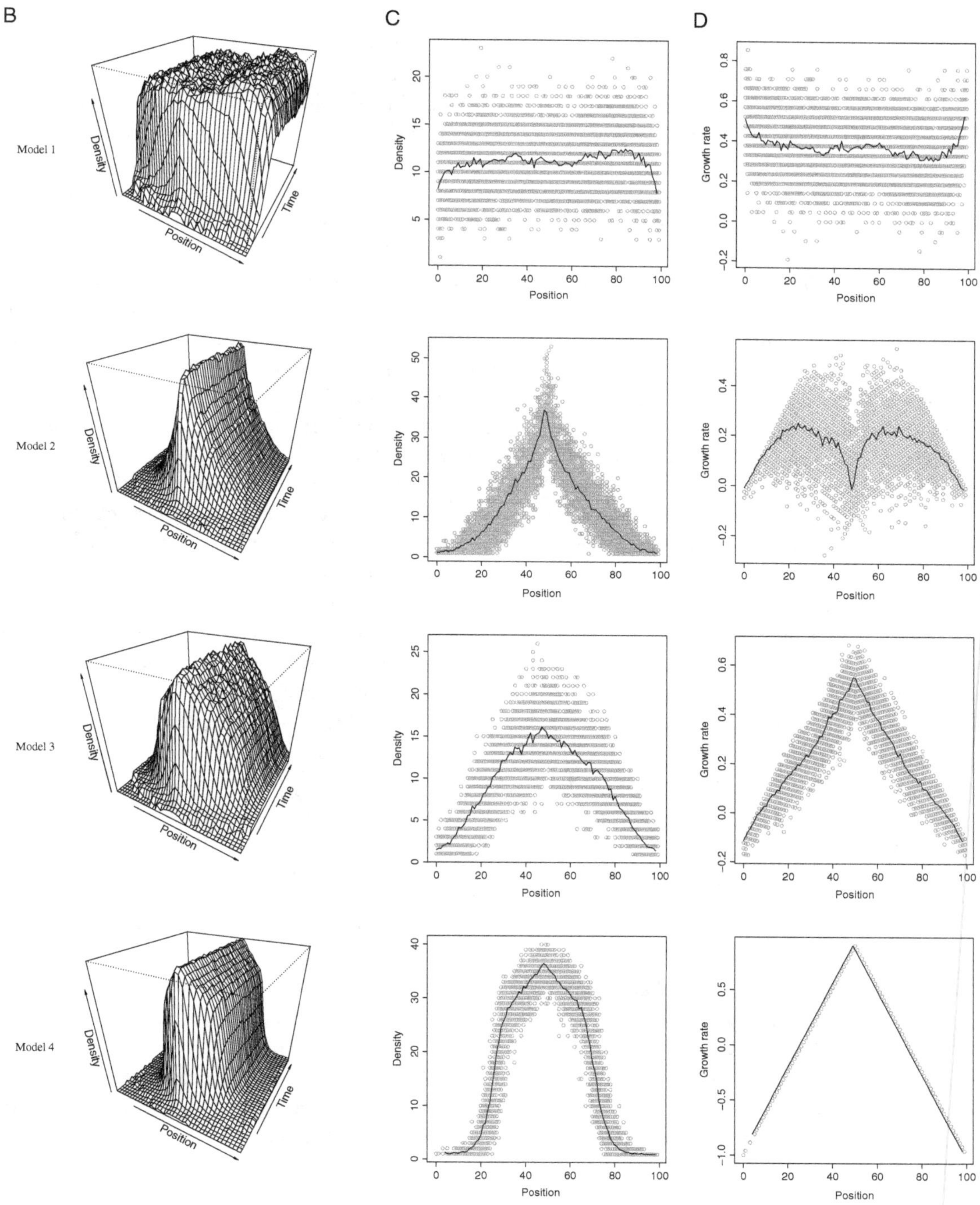

Turning to the converse effect, the most likely scenario under which a consumer species can limit the geographic range of a prey or host species is where it exploits more than one of the latter and is thus able to continue exploiting one species to extinction (or more probably to levels where extinction occurs stochastically) while itself being sustained by the other resources. Such apparent competition is effectively identical in its consequences to forms of interspecific competition, which is thought itself to be a key force in limiting the geographic ranges particularly of closely related species. The narrow abutment of the geographic range limits of closely related species can reflect their similarity in many biological traits and hence their effectively acting as a barrier to one another's further spread, usually through influences on resource availability.

None of these three sets of factors operates independently. Physical barriers are almost invariably closely associated with changes in abiotic conditions and biotic interactions, and they only serve as such because of the constraints that exist on the dispersal abilities or environmental tolerances of organisms that prevent those barriers being overcome. Abiotic conditions shape the assemblages that occur in different areas and thus the sets of biotic interactions that arise, and the impact of abiotic conditions on organisms is shaped by their biotic context.

Whether one of these sets of factors plays a predominant role in limiting the range of a particular species or some combination thereof, the question remains as to why individuals in peripheral populations have not simply evolved dispersal abilities that enable physical barriers to be overcome, physiological tolerances and capacities to overcome abiotic barriers, or means of interacting with other species (defensive or consumptive) that prevent these being limiting and thus enable its range to expand yet further. In some cases, this may be because peripheral populations have low levels of genetic variation, that as a consequence of directional selection or environmental variability traits show low heritability in peripheral populations, that favored genotypes occur too rarely in peripheral populations because changes in several independent characters are required, that genetic tradeoffs among fitness traits in peripheral populations prevent them from evolving, or that the accumulation of mutations that are deleterious in peripheral populations prevents adaptation. In others, the immigration into peripheral populations of individuals from elsewhere in the geographic range may result in sufficient gene flow that the alleles that would otherwise enable range expansion may be swamped. The importance of this mechanism has been much debated but remains unclear. Such an effect would be exacerbated if there are many more individuals in populations away from the range limit, peripheral populations tend to be smaller, there is a rather unidirectional flow of immigrants into peripheral populations, and environmental gradients are steep. They would be weakened if dispersing individuals tended to move preferentially to areas for which they were preadapted, and if phenotypic plasticity allowed adaptive adjustment to local conditions. In this latter regard, it is notable that the geographic ranges particularly of the more widely distributed species exhibit marked spatial structure in such traits as body size.

Changes in combinations of conditions at different times and across space mean that the precise factors limiting the geographic range of a given species will also change in time and space. For example, the relative importance of prevailing climate, the quality of the host, and the mortality imposed by natural enemies (and the species composition of those enemies) in determining the boundary of the range of a specialist herbivore will typically differ between more northerly and southerly areas and likely longitudinally as well. In consequence, it is typically rather meaningless to ask what limits that range without specifying the spatiotemporal context. It has long been held that limits at higher latitudes are more commonly determined by climatic factors and those at lower latitudes more commonly by biotic factors, but there are no systematic data on which basis firmly to support or reject such a proposition.

The spatial variation in the factors that influence the limitation of species ranges determines not only the sizes that those ranges achieve but also their shapes. These vary hugely from approximately circular to ribbon-like and from more continuous to highly fragmented and scattered.

Dynamic Limits

Although the geographic ranges of some species are quite stable for long periods, those of others are very dynamic. They may ebb and flow, or they may show distinct directional movement.

The ebb-and-flow dynamics of ranges are typically caused by the instability of environmental conditions toward the boundaries, which enable local populations to be established and maintained during some periods but drive them to extinction when conditions become more extreme (the influences may be indirect, through resources and interspecific interactions). Indeed, a very general finding is that, even when extant, peripheral populations show temporally more variable dynamics in size and demographic parameters than do more central populations. In consequence, they tend also to exhibit higher local extinction rates. These effects are obviously enhanced if peripheral populations are smaller, but they equally may be mitigated by immigration elevating local abundances and reducing extinction risk. Importantly, it seems likely that range limits are determined more frequently by extremes than by average conditions, which means that empirical studies of range limitation require a significant temporal perspective (in the cases of, for example, long-lived tree species this may be a matter of decades, centuries, or longer).

More directional shifts in range limits are caused by the systematic intensification or relaxation of a constraint. Climatic factors are attracting much interest in this regard, because of the trends that have been engendered by human activities and particularly by the post-industrial increase in atmospheric carbon dioxide levels. Indeed, key evidence for the biotic effect of global climate change has been the demonstration of covariation between temperature regimes and the boundaries of species distributions, especially the shift to higher latitudes of the boundaries of more northerly species in the Nearctic and Palearctic in response to warming conditions. Although tempting, given the complexities of range limitation already outlined it cannot be concluded that such observations mean that the geographic ranges of these species are simply determined by climatic conditions. Changes in those conditions can have a diversity of effects, including on the physical, abiotic, and biotic factors that might previously have limited ranges and their interactions.

Range Sizes

The areas of the geographic ranges that the extant species within a clade exhibit at any one time commonly span several orders of magnitude. A number of factors may be associated with these interspecific differences, including physiological tolerances and capacities, niche breadths and positions, and dispersal abilities. However, even in combination these typically explain rather limited proportions of this variation. This suggests that in large part the origins of much of the variation in range sizes may be essentially stochastic, perhaps particularly reflecting the geographical context in which species find themselves, and historical events that enabled or constrained range expansion. Geographical context is obviously important, in as much as the structures of land masses and water bodies place such powerful limits on where species can occur. However, historical contingency is likely to be equally so, with the position of many potential barriers to, or opportunities for, range expansion being temporally dynamic (Milankovitch cycles, the influence of changes in the Earth's axis of tilt and orbit on its climate, may be particularly significant). Some evidence for a key role of stochasticity in shaping variation in geographic range size is provided by the fact that this trait exhibits rather little phylogenetic constraint (at least compared with highly conserved traits such as body size), with closely related species often differing markedly in how widespread they become. Indeed, in very many cases there is no immediately obvious reason why two sister species may respectively be relatively widespread and narrowly distributed when in many other respects they are extremely similar.

The sizes of geographic ranges can influence the likelihood both of speciation and that of extinction. Widespread species have long been argued to be important sources of evolutionary novelty on the grounds that, if speciation is predominantly allopatric, their ranges are more likely to be broken by isolating barriers, and that in extending over a greater diversity of environments the potential for local adaptive divergence is heightened. In contrast, however, models suggest that the greater probability of subdivision of large ranges due to their extent can be offset by the fact that large range size is often accompanied by greater local densities and greater dispersal abilities of individuals (helping to maintain range contiguity). This effect can be sufficient that species with relatively small, albeit not the smallest, range sizes have higher instantaneous speciation rates.

Species with smaller ranges also have a greater instantaneous risk of extinction, such that the longevity of a species declines with the average or maximum range size that it attains. This is a consequence of (a) overall population size increasing with geographic range size, reducing the probability of a random walk to extinction, and (b) the risk spreading provided by a larger range, such that adverse conditions in one region are unlikely to affect all individuals of the species at the same time. The greater longevity of more widely distributed species can mean that even if they have a lower instantaneous speciation rate, they nonetheless do not leave fewer descendant species than do more restricted species, and may leave more.

EXTINCTION

Lifetime Dynamics of Range Size

The limitations of the fossil record, particularly in documenting occurrences over substantial areas and at reasonably fine temporal resolutions, mean that the typical dynamics of the geographic range size of a species over its lifetime from speciation to extinction are not well understood. Common trajectories seem likely to be either a highly asymmetrical rise and fall, in which range expansion occurs much more rapidly than does the subsequent decline to extinction, or a more symmetric hump-shaped rise and fall in range size. The former pattern might perhaps be expected if, as ecological studies suggest, initial range expansion is rapid and relatively unhindered but is then subsequently undermined by the ecological and evolutionary responses of competitors, predators, and/or parasites. The latter pattern would suggest a more balanced influence of abiotic and/or biotic factors during expansion and contraction phases. If, as some have suggested, a rapid range expansion and slower fall is the predominant pattern, then all else being equal at any one time most species would be undergoing range decline.

Range Contraction

Geographic ranges can contract to final extinction in a variety of ways, taking different paths through bivariate population size–range size space as the overall number of individuals declines. Traditionally, the process has been envisaged as occurring through synchronous contraction at all boundaries. This particularly makes sense if densities decline toward range limits and contraction is driven by a range-wide deterioration in environmental conditions. Under these circumstances, peripheral populations would be progressively lost. However, densities often do not decline toward range limits, and, more importantly, the pressures bringing about range collapse are often more directional. For example, even in the face of changing global or regional climate, conditions seldom deteriorate evenly across a species' geographic range, from the organism's point of view more frequently worsening disproportionately toward one limit where individuals are closer to some critical physiological threshold. Likewise, many other pressures are often of a directional nature, including the spread of a natural enemy or competitor and of key anthropogenic pressures, such as land use transformation. Thus, range contraction may typically follow a more contagion-like process. This has significant implications for the conservation of species with much reduced ranges. In particular, it means that the environmental conditions under which remnant populations exist may be far removed from those to which the species is best adapted and which most individuals historically experienced.

PRACTICAL IMPLICATIONS

Improved understanding of the patterns and determinants of the structure and dynamics of geographic ranges is key to many of the major environmental challenges presently facing humankind. These include such issues as food security (e.g., which crops and varieties can most usefully be grown where?), disease control (e.g., how will different measures reshape species distributions?), loss of biodiversity (e.g., which populations are best targeted for conservation action?), spread of alien species (e.g., how can the range expansion of undesirable species best be reversed?), and climate change (e.g., how will the distributions of species of human concern change?). Especially significant will be the extent to which the general ecological and evolutionary principles of range structure and dynamics can be employed to address the problems and opportunities posed by particular species in particular places.

SEE ALSO THE FOLLOWING ARTICLES

Apparent Competition / Dispersal, Evolution of / Invasion Biology / Phylogeography / Regime Shifts / Spatial Spread / Stochasticity, Environmental / Two-Species Competition

FURTHER READING

Gaston, K. J. 2003. *The structure and dynamics of geographic ranges.* Oxford: Oxford University Press.

Gaston, K. J. 2009. Geographic range limits: achieving synthesis. *Proceedings of the Royal Society B: Biological Sciences* 276: 1395–1406.

Holt, R. D., T. H. Keitt, M. A. Lewis, B. A. Maurer, and M. L. Taper. 2005. Theoretical models of species' borders: single species approaches. *Oikos* 108: 18–27.

Lomolino, M. V., B. R. Riddle, R. J. Whittaker, and J. H. Brown. 2010. *Biogeography*, 4th ed. Sunderland, MA: Sinauer Associates.

Sexton, J. P., P. J. McIntyre, A. L. Angert, and K. J. Rice. 2009. Evolution and ecology of species range limits. *Annual Review of Ecology, Evolution, and Systematics* 40: 415–436.

STABILITY

SEE RESILIENCE AND STABILITY

STABILITY ANALYSIS

CHAD E. BRASSIL

University of Nebraska, Lincoln

After constructing a model and after determining the equilibria of that model, commonly the next step is to analyze the stability of those equilibria. By definition, a system will always stay at its equilibrium values when it is started exactly at those values. Stability analysis asks what will happen if the system is perturbed from that equilibrium by a small amount, or, alternatively, what will happen if the system is started just off the equilibrium. This is a very reasonable question to ask of an ecological system because they exist in a messy, noisy world; all populations will experience changes in their abundance due to a variety of external forces. Stability analysis tells us if the population will move toward or away from the equilibrium, ultimately providing insights into the overall dynamics of the system.

CONCEPTUAL ISSUES

A population is considered to be at a stable equilibrium if the dynamics return that population to the equilibrium after the population size is increased or decreased away from that equilibrium value. Conversely, a population is at an unstable equilibrium if the population diverges from the equilibrium value following a small perturbation away from the equilibrium.

An intuitive, physical example is to imagine the position of a marble as an analogy to the size of a population (Fig. 1). A bowl is a stable equilibrium. When placed on the bottom of the bowl, the marble will not move, so it is at an equilibrium position. When perturbed a small amount, the marble will roll back down the sides of the bowl and eventually come to rest back at the original equilibrium—the very bottom of the bowl. An upside-down bowl is an unstable equilibrium. When placed at the very top of an upside-down bowl, the marble will stay there, so it is an equilibrium position. However, when perturbed, the marble will roll down the side of the bowl and not return to the original equilibrium position. One last situation to consider in this example is a flat table. When placed on a flat table, the marble stays where it is put, so it is at an equilibrium position. However, when perturbed to a new location, the marble stays at the new position; it does not return to the original position. This situation is called a neutrally stable equilibrium. Neutrally stable

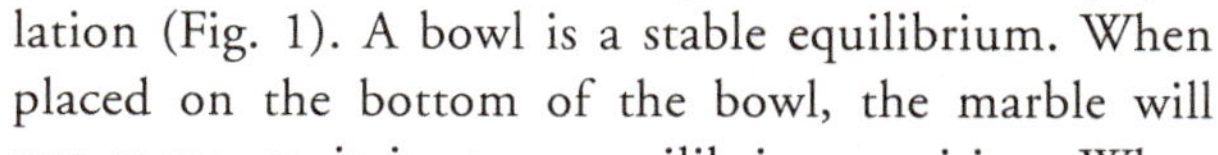

FIGURE 1 A stable equilibrium is analogous to a marble sitting in the bottom of a bowl—if moved, it will return to the bottom of the bowl. An unstable equilibrium is analogous to a marble sitting on the top of an inverted bowl—if moved it will roll off the side of the bowl. A neutrally stable equilibrium is analogous to a marble on a flat table—it stays exactly where it is placed and if moved it will stay in that new position as well.

is generally considered to be an unrealistic outcome in ecological models, but a form of it does occur in a classic predator–prey model.

The conceptual notion of stability analysis is formalized mathematically by examining the effect of small changes in a population on the dynamics of a population. If changes in the population size are small, one can approximate the function describing the dynamics around that equilibrium by a linear function. The slope of that linear function is a single number describing what will happen to the population when it is perturbed from that equilibrium. Derivatives are the mathematical method to calculate a slope relationship; in this case the derivative of the total population growth with respect to changes in the population size, since one wants to examine the consequences for growth of a perturbation in the population size.

SINGLE-SPECIES MODEL

In a single-species, continuous-time model such as the logistic model, the equation describing how the population will change with time is $dN/dt = f(N) = rN(1 - N/k)$, where N is the population size, t is time, r is the maximum instantaneous growth rate, and k is the carrying capacity. The equilibrium population sizes, notated generally as N^*, are extinction (0) and the carrying capacity (k). To examine the stability of each equilibrium point one could simply solve a few time series solutions for this model and see which equilibria they approach and from which equilibria they move away. For example, see the "Time Series" column in Figure 2. In the case of the logistic model, population sizes approach the carrying capacity and move away from the extinction equilibrium.

In order to more rigorously characterize the stability of each equilibrium point, one systematically examines the growth rate for population sizes close to the equilibrium. Mathematically, that is achieved by looking at the slope around the equilibrium point. The slope is calculated using the derivative of the growth rate with respect to population size. In the case of the logistic model this is $df/dN = f'(N) = r - 2rN/k$. Therefore, evaluate the derivative at the equilibrium, $f'(N^*)$. This can be done analytically or graphically. The last column in Figure 2 highlights this graphically by bolding a tangent line to the growth rate at each equilibrium point. The slopes of these bold lines are the values of the derivatives at the equilibria.

For a continuous time model, if the value of the derivative at the equilibrium is negative then population sizes very close to the equilibrium will change in the

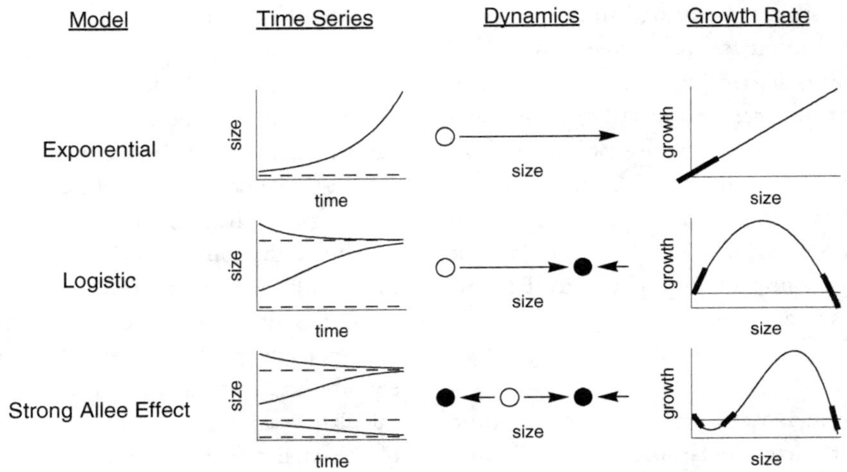

FIGURE 2 The stability of the equilibria in three single-species growth models, illustrated as a time series and on a dynamic number line. Dashed lines in the time series represent equilibria. Population sizes move away from unstable equilibria and move toward stable equilibria. The dynamics of these same models are summarized on a number line with open circles representing unstable equilibria and closed circles representing stable equilibria. The arrows represent the direction of changes in the population size when not at an equilibrium point. Arrows point away from unstable equilibrium and toward stable equilibrium. The final column plots growth rate (dN/dt) as a function of population size (N). The slope at each equilibrium size is bolded to highlight the utility of using the slope, or derivative, in calculating the stability of each equilibrium point. In the exponential model, population size grows to infinity as long as it exceeds size zero. In the logistic model, populations grow to a carrying capacity if they exceed zero. In the strong Allee effect model, populations go extinct when below a threshold population size, otherwise they grow to a carrying capacity.

direction of the equilibrium. In other words, perturbations from the equilibrium will return to the equilibrium—perturbations will die out. Since the growth rate at an equilibrium is by definition zero, a negative slope means that increases in the population size will result in negative growth. Conversely, decreases in the population size will result in positive growth. Either situation will result in the population returning to the equilibrium (see Fig. 2). If the value of the derivative at the equilibrium is positive, perturbations will diverge from the equilibrium—perturbations will become bigger differences. In the case of a positive slope, an increase in the population size will result in an increase in the growth rate, and the population size will continue to increase away from the equilibrium. Conversely, decreases in the population size will result in decreases in the growth rate. Either situation will result in the population moving away from the equilibrium.

In the case of the logistic model, there are two equilibria, one at 0 because an extinct population will remain extinct, and one at the carrying capacity k. Evaluating stability at each equilibrium, we find $f'(0) = r$, and $f'(k) = -r$. Therefore, assuming r is a positive number, the 0 equilibrium is unstable and the k equilibrium is stable. The addition of a few individuals to an extinct population will not result in an extinction population, but rather a population that grows larger. The addition of a few individuals to a population that is at its carrying capacity will result in a population above the carrying capacity, and the population will decrease back to the carrying capacity.

For a discrete time model such as the Ricker model, the equation describing dynamics is $N_{t+1} = g(N) = N_t \exp(r(1 - N_t/k))$, with equilibria at $N = 0$ and $N = k$. The derivative of this equation with respect to N is $g'(N) = (1 - (r/k)N) \exp(r(1 - N/k))$. In the case of a discrete time model, the criterion determining stability is slightly different than in a continuous time model. In a discrete time model, the absolute value of the slope of the relationship between population growth and population size, $|g'(N)|$, has to be less than 1 when evaluated at the equilibrium (see Box 1 for a summary). Intuitively, these different criteria relate to the different criteria that define an equilibrium in continuous time models and discrete time models.

BOX 1. SINGLE-SPECIES STABILITY ANALYSIS

	Continuous Time	Discrete Time
	$\frac{dN}{dt} = f(N)$	$N_{t+1} = g(N_t)$
stable if:	$f'(N^*) < 0$	$\lvert g'(N^*)\rvert < 1$

where $f'(x) = \frac{df}{dx}$ and * indicates the derivative be evaluated at the equilibrium value.

Stable and unstable equilibrium can be represented graphically along a number line. A closed circle represents a stable equilibrium, and an open circle represents an unstable equilibrium. The dynamics of the logistic model above can be represented by the stability of equilibria on a number line (Fig. 2). This kind of representation when examining a model like a strong Allee effect illustrates how unstable equilibria must separate different stable equilibria creating basins of attraction to alternative stable states, sometimes referred to as hysteresis.

TWO-SPECIES MODEL

The intuition developed when thinking about the stability analysis of single-species models extends to the analysis of two-species models. Conceptually, one examines the dynamics of the system in the immediate vicinity of the equilibrium. The growth rate of each population can depend on the population size of its own population and on the other population. These four combinations of growth rate and population are examined jointly to determine if a perturbation from the joint equilibrium of the two species will return to a stable equilibrium or diverge from an unstable equilibrium. The consequences of changes in each population size for the growth rate of each species are calculated singly, while the stability dynamics of the full system are examined simultaneously. The analysis is an equally valid representation of a change in the population size of one species, or of both species at the same time because the assumption of small departures from the equilibrium population sizes means the dynamics are being estimated as linear functions in all combinations, or essentially a plane through the equilibrium as opposed to a single slope in the case of the single-species model.

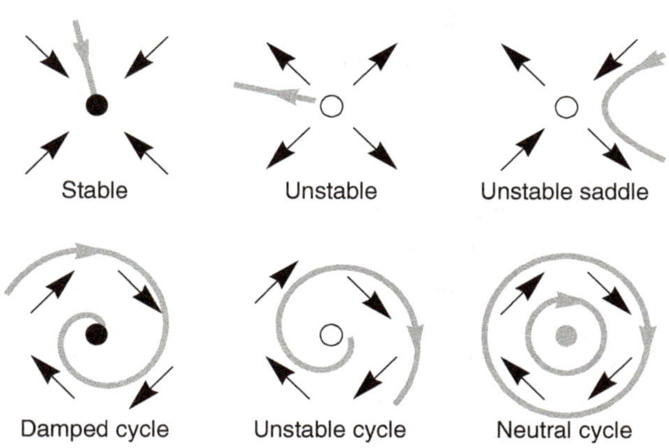

FIGURE 3 Six different outcomes of local stability analysis for a point equilibrium in a two-species system. Closed circles represent stable, attracting equilibria. Open circles represent unstable equilibria. The closed gray circle represents an equilibrium for a neutral cycle. Implied axes are the population sizes for the two species. Thick gray lines trace a representative trajectory of the population dynamics, with an arrow indicating the direction of movement over time along that path. The black arrows are vectors indicating the net direction of change in population sizes for that combination of population sizes. The top row represents equilibria for which eigenvalues are not imaginary and hence pairs of parallel black vectors represent an eigenvector, with arrows pointing toward the equilibrium for attracting eigenvalues and arrows pointing away from the equilibrium for repelling eigenvalues. The bottom row represents equilibria for which eigenvalues have imaginary parts.

Instead of calculating a single derivative to determine stability, four derivatives are calculated representing the slope of the dependence of each growth rate on each population. This set of four relationships is called the Jacobian matrix (Box 2). As before, one is interested in the effect of small departures from the equilibrium, so the Jacobian is evaluated at the combined values for a given equilibrium. The Jacobian matrix evaluated at the equilibrium is the often called the community matrix. For a system of equations describing the population dynamics of multiple species, each element of the community matrix describes how each species responds directly to changes in each of the other species when all species are at a given equilibrium.

Techniques in matrix algebra can be used to describe the changes that will take place in the population sizes of each of the species when starting at sizes away from the equilibrium. A metric known as an eigenvalue allows one to describe, in a single number, what will happen to the growth rates over all possible changes in population sizes. A corresponding eigenvector describes the combination of species along which the major changes will occur. The eigenvalue is a factor that describes the

BOX 2. TWO-SPECIES STABILITY ANALYSIS

System of equations $\frac{dN_1}{dt} = f(N_1, N_2)$ $\frac{dN_2}{dt} = g(N_1, N_2)$

Jacobian $J(N_1, N_2) = \begin{vmatrix} \frac{\partial f}{\partial N_1} & \frac{\partial f}{\partial N_2} \\ \frac{\partial g}{\partial N_1} & \frac{\partial g}{\partial N_2} \end{vmatrix}$

Community matrix $J(N_1^*, N_2^*) = \begin{pmatrix} a & c \\ b & d \end{pmatrix}$

stable if: $\det(J(N_1^*, N_2^*)) = ad - bc > 0$

and: $\mathrm{tr}(J(N_1^*, N_2^*)) = a + d < 0$

oscillation if: $\det > \frac{\mathrm{tr}^2}{4}$

Discrete time is analogous, with the exception that the system is stable if: $2 > 1 + ad - bc > |a + d|$

Where det() is the determinant of the matrix and tr() is the trace of the matrix.

magnitude of those changes and whether those changes will increase away from the equilibrium or decrease back toward the equilibrium. In other words, an eigenvalue quantitatively summarizes the "stretching" or "squishing" of the population sizes. For a two-species model, this entire process can be characterized by two eigenvalues. If both eigenvalues are negative, the departures from the equilibrium are "squished" in all directions and the system returns to the equilibrium. In other words, the equilibrium is stable.

If both eigenvalues are positive, the system is stretched and the population sizes move away from the equilibrium. The equilibrium is unstable. If one eigenvalue is negative and one is positive, the system is squished in one direction, but stretched in another direction (top row of Fig. 3). Ultimately, the equilibrium is unstable, but since the population sizes can initially move in the direction of the equilibrium before moving away from the equilibrium, the equilibrium is called a saddle. In terms of the marble analogy, the marble is rolling down the saddle of horse. If placed near the neck of the horse, the marble will roll toward the center of the saddle but eventually will roll off the edge of the saddle and the side of the horse.

In a two-species model, entries in the community matrix can be combined in ways to form what are called the trace and the determinant of the matrix (details in Box 2). This is easier to calculate than calculating the eigenvalues directly, and provides information about the eigenvalues. The trace is the sum of both eigenvalues. If both eigenvalues are negative, the trace will be negative, and the equilibrium will be stable. However, if one eigenvalue is strongly negative and the other eigenvalue is weakly positive, departures from the equilibrium will shrink in one direction but eventually expand in the other direction—the equilibrium will be an unstable saddle equilibrium, but the trace will still be negative. Therefore to ensure stability one needs to ensure both eigenvalues are negative. That can be accomplished by examining the determinant of the community matrix. The determinant is the product of the eigenvalues. If the eigenvalues are of the same sign, the product will be positive. In particular, if both eigenvalues are negative, the determinant will be positive. Therefore, a negative trace ensures the eigenvalue with the largest magnitude is negative; given that, a positive determinant ensures that both eigenvalues are negative. If both eigenvalues are negative, small departures from the equilibrium shrink back to the equilibrium and the equilibrium is stable.

Technically, the above discussion on eigenvalues refers to the real part of the eigenvalue. Eigenvalues can have imaginary parts as well. The existence or lack of imaginary part provides information as to how the system departs from the equilibrium or returns to the equilibrium. If there are no imaginary parts of the eigenvalues, the system directly and linearly shrinks or expands (top row of Fig. 3). If there are imaginary parts of the eigenvalues, the system oscillates away from the equilibrium or oscillates back to the equilibrium (first two cells, bottom row of Fig. 3). Oscillating cycles back toward the equilibrium are often called damped cycles. Oscillations away from the equilibrium, or an unstable cycle, can continue to cycle with increasing amplitude until a population goes extinct. However, an unstable cycle away from an unstable point equilibrium can also indicate the presence of a stable limit cycle to which the population dynamics approach (see "Additional Topics," below).

Again, in a two-species model, the trace and determinant can be used to easily calculate whether or not imaginary parts exist for the eigenvalues and therefore determine whether or not oscillations will occur as part of the stable or unstable dynamics (see Box 2).

Eigenvalues are described above as values that characterize the behavior of departures from equilibrium. If all the eigenvalues are negative, the system returns to the equilibrium, and if at least one of them is positive, the system departs from the equilibrium. What happens if the largest eigenvalue is zero? If both eigenvalues were exactly zero, this would describe a system in which there was no shrinking or stretching of the system—in other words, a system in which a departure from equilibrium remained at exactly the same departure. This is a completely neutral equilibrium, akin to the marble on the flat table. This never happens in ecology because it would describe a system in which there really is no interaction and nobody would bother modeling it. However, if the real parts of the eigenvalues are zero but the eigenvalues have imaginary parts, oscillations will occur because of the imaginary parts. The result would be neutral cycles, or cycles whose amplitude is determined solely by the starting densities of the species (final cell, bottom row of Fig. 3). This is in fact the outcome of a famous model in theoretical ecology, the classical Lotka–Volterra predator–prey model. However, most implementations of a predator–prey model include factors such as prey density dependence and handling time to avoid neutral cycles, because a system exhibiting truly neutral cycles would be unlikely to produce regular cycles in a real, noisy world with environmental changes constantly bumping around population densities.

In two-species models, stability analysis can be done graphically. Graphical analysis is conducted on a plot with the population size of one species as one axis and the population size of the other species as the second axis. This kind of plot is called a phase plane diagram. From each of the two differential equations describing the change in population size for each species with time, one can find a nullcline. A nullcline is the combination of population sizes (i.e., species 1 and species 2) for which there is no change in the population size of the give species (for example, species 1). In other words, for a given combination of species 1 and species 2 that fall on the nullcline, the population of species 1 would neither decrease nor increase. One finds a nullcline for a given species by setting the differential equation describing the population dynamics for that species equal to zero. In a two-species system, there will be at least two nullclines, one for each species. Sometimes multiple nullclines exist for a given species if there are multiple possible equilibria for that species, for example, a positive equilibrium and a zero equilibrium (also known as the trivial equilibrium). Where nullclines for the two species intersect is an equilibrium point for the system, since neither species 1 nor species 2 will change their population size for that combination of population sizes. Subsequent stability analysis can be conducted at that point equilibrium. Graphically this can be done by examining the net directional change in population size for each region as divided by the nullclines. Component vectors are constructed for each species by examining whether population growth will be positive or negative for population sizes in each region. Component vectors are combined to create a net vector showing the direction of change in population sizes given the combination of population sizes in that region. Qualitative vector directions (i.e., positive or negative) are consistent across an entire region as delineated by the nullclines. The direction of net vectors around an equilibrium determine the stability of that equilibrium as illustrated in Figure 3. In the case of cycling net vectors (as in the bottom row of Fig. 3), additional information is needed to determine the stability of the equilibrium (as is often the case in predator–prey models).

Analytical stability analysis involves going through the steps in Box 2 using the symbolic expressions so as to arrive at a general solution which can characterize the stability of an equilibrium over a range of particular parameter values. When performing an analytical analysis, be selective as to when or when not to substitute expressions for equilibrium values when constructing the community matrix. Sometimes substituting an expression for an equilibrium value will enable a cell of the community matrix to be simplified, adding clarity to the expression. However, sometimes retaining a symbol for the equilibrium value without substituting an expression enables a more clear interpretation because the equilibrium value is often, by definition, positive. The goal in many cases is to identify cells (for example, cell a in Box 2) as always positive or always negative. Ultimately, one manipulates the community matrix with the hope of finding clearly interpretable stability criterion.

MULTIPLE-SPECIES MODEL

Much of the conceptual ideas developed in the "Two-Species Model" section above applies to multispecies models. As with the two-species model, if the largest eigenvalue in the community matrix is negative, then all of the eigenvalues are negative and departures from the equilibrium will return to the equilibrium—the equilibrium is stable.

The major difference in conducting this analysis with multiple species is that the rules do not apply for the trace and determinant as a means to understand the eigenvalues. One must calculate the eigenvalues of the community matrix directly, or one must use a different shortcut method, the Routh–Hurwitz criteria. The Routh–Hurwitz criteria determine stability by examining elements of the community matrix through a calculation known as the characteristic equation. The Routh–Hurwitz calculation can be applied to any community matrix of arbitrary size, but practically it can become difficult to interpret the criteria as the number of species in model increases. It probably has the greatest utility for a three-species model.

In the absence of any clear analytical analysis using the above techniques, numerical simulations can always be run for particular parameter combinations. This is approach is inherently limited in the generality of the result, but the power of modern computing increases the ease with which this approach can be conducted. It may be the only approach for increasingly complex biological models.

ADDITIONAL TOPICS

The focus above has been on the stability of equilibrium points. When an equilibrium is locally unstable, the system may move to a different equilibrium that is stable. Alternatively, a population lacking density dependence might grow to infinity. However, in many ecological models an unstable equilibrium indicates the presence of a sustained oscillation in population sizes. The populations would move away from a point equilibrium, but then approach a cycle of a given size, and stay in cycling dynamics over the long term. Populations perturbed from a stable limit cycle return to the same cycle with the same amplitude as the original limit cycle. The stability of the limit cycle can be characterized through advanced analysis known as Poincaré analysis. In some ecological models, certain parameter values can result in limit cycles changing into more complicated cycles and eventually into chaotic dynamics, in which case Lyapunov exponents can be used for analysis.

The stability analysis discussed above is more precisely called local stability, as the stability of each equilibrium point is characterized by examining small departures from equilibrium. Small departures from equilibrium are examined because the behavior of small differences can be quantified by linear approximations (using slopes and derivatives) to what are typically nonlinear relationships in ecology. Linear approximations were calculated via the derivatives in the Jacobian matrix. Large departures from equilibrium may also return to that same equilibrium, but large departures may approach another equilibrium. Addressing this broader issue falls under the heading of global stability analysis. An equilibrium is globally stable if the system returns to that equilibrium regardless of the starting conditions. In the case of two-species models, simple analysis of all possible equilibria can be used to determine if a locally stable equilibrium is also globally stable. Multiple species models may have more complex cycling behavior, requiring advanced analysis in dynamical systems in order to characterize the global dynamics.

SEE ALSO THE FOLLOWING ARTICLES

Allee Effects / Ordinary Differential Equations / Matrix Models / Phase Plane Analysis / Predator–Prey Models / Ricker Model / Single-Species Population Models

FURTHER READING

Case, T. J. 2000. *An illustrated guide to theoretical ecology*. New York: Oxford University Press.

Edelstein-Keshet, L. 2005. *Mathematical models in biology*. Philadelphia: Society for Industrial and Applied Mathematics.

Kot, M. 2001. *Elements of mathematical ecology*. Cambridge, UK: Cambridge University Press.

STAGE STRUCTURE

ROGER M. NISBET AND CHERYL J. BRIGGS

University of California, Santa Barbara

The life cycle of many organisms involves well-defined developmental stages, sometimes accompanied by change of habitat. Many population models, commonly called stage-structure models, recognize these different stages. Stage-structure models may be couched within a discrete-time framework, with an update rule contained in a matrix. In continuous time, stage-structure models can conveniently be written in the form of delay differential equations. Stage-structure models have provided insight on population cycles, consumer–resource interactions, and disease dynamics.

LIFE STAGES

Organisms of any particular species typically differ both in reproductive potential and in susceptibility to mortality. One natural way to recognize this in population models is to include age structure, letting reproductive and death rates be functions of the organism's age. However, there are many organisms whose life involves well-defined stages, and commonly biological differences among individuals within a stage may be less important than interstage differences. For example, an insect population might contain 15-day-old females in both the pupal and adult stages. The reproductive rate of the pupae is unarguably zero, whereas the adults of the same age may have significant egg output. As another example, consider an anadromous fish such as a salmon. Eggs hatch in the upper reaches of a river and the young fish then migrate downstream, where they may remain for some time before experiencing a major physiological change (smolting) and entering the ocean, where they remain for a number of years before returning to the river to spawn. Returning fish commonly do not feed after returning to freshwater and die shortly after spawning. For many species, the age at smolting is highly variable, as is the number of years spent in the ocean. Mortality risk for a fish of any age depends primarily on stage rather than age: a two-year-old in the ocean faces very different hazards and is typically a different size from its counterpart still in the river. And of course only returning females spawn.

Stage-structure models describe the dynamics of the populations in the different life stages. Like age-structure models, they can be formulated in discrete or continuous time. Mathematically, the discrete-time models can be written using an update rule that describes the transition rates between stages, and the resulting models represent a particular case of the broader family of matrix population models. In continuous time, one important way in which stage structure influences population dynamics is by introducing time delays, and delay differential equations (DDEs) provide a natural mathematical framework that recognizes this. Stage structure is also of great importance when considering species interactions. For example, many predators and parasites specialize on a particular life stage of the species they attack; likewise, many pathogens target a single stage.

DISCRETE-TIME STAGE-STRUCTURED MODELS

Discrete-time models are particularly appropriate for organisms that reproduce at regular intervals set by the environment—for example, animals with one breeding episode per year. In many cases, age-structure models capture the essential dynamics of such systems, but for others life stage, and not age, is not the primary determinant of year-to-year survival or of reproductive performance. To model such stage-classified populations, we adopt an approach similar to that used to construct age-structured models, but with important differences.

The population at any given time t is represented by a vector $\mathbf{n}_t$ whose elements are the number of individuals in each stage. Thus, a simple insect population model might involve a vector with 3 elements, representing the number of larvae, pupae and adults. A more elaborate model might recognize a number of different larval developmental stages (instars). The population at time $t + 1$ is obtained using an update rule that can be written in the form

$$\mathbf{n}_{t+1} = \mathbf{A}\mathbf{n}_t, \qquad (1)$$

where $\mathbf{A}$ is a matrix. However, the form taken by this matrix differs from its age structure counterpart (often called the Leslie matrix) because individuals may remain in any given life stage for more than one time step. For the simple insect population with three stages (1 = larve; 2 = pupae; 3 = adults), the matrix could be written in the form

$$\mathbf{A} = \begin{bmatrix} P_1 & 0 & F \\ G_1 & P_2 & 0 \\ 0 & G_2 & P_3 \end{bmatrix}. \qquad (2)$$

Here, G_1 represents the proportion of individuals in the first stage (eggs) that survive and enter the second stage (pupae) in a time step, and P_1 is the proportion of larvae that survive but remain as larvae after the time step. Similar definitions for G_2 and P_2 describe the fate of a pupa over a time step, and P_3 represents adult survival. The term F is fecundity.

The form taken by the matrix may be more complex than Equation 2. For example, a model of a whale population uses two stages to classify the adult female population: with and without a calf. Transitions between these stages can be in either direction, implying that the matrix describing that population may have some nonzero entries above the diagonal yet away from the top row.

Because of the similarity in mathematical formulation, the metrics that characterize the demography of age-structured populations can be used with their age-structured counterparts. No new mathematical concepts are involved. Thus, the dominant eigenvalue, λ, of the matrix in Equation 1 represents the long-run population growth rate, and the corresponding eigenvector represents the stable stage distribution of the population. Elasticities and sensitivities, describing the influence of particular model parameters, can be calculated using the recipes that were originally developed for age-structured models.

It is also possible, in principle, to introduce density dependence to discrete-time stage-structured models by assuming that one or more stage-transition rates depend on the stage populations. However, this approach, like the analogous approach to age-structured modeling, should be used with considerable caution as it implies that transition rates representing the cumulative effect of ecological processes over an extended time period are determined by the stage populations at a single time point.

CONTINUOUS-TIME STAGE-STRUCTURE MODELS

General Formalism

Continuous-time stage-structured models are used to describe populations where reproduction and mortality may occur at any time. We imagine an organism whose life cycle involves a progression through a sequence of stages, such an insect with egg, larva, pupa, and adult stages, but without these stages being associated with any one time of year. The key assumption remains that the risk of mortality and the rates of development and reproduction depend primarily on the stage populations. However, by working in continuous time, it is possible to model situations where the individual rate processes vary over time, as, for example, is the case with density-dependent rates and with many species interactions.

We assume a sequence of life stages and represent by $N_i(t)$ the number of individuals in stage i at time t. Organisms enter stage i at a rate $R_i(t)$, and all individuals in this stage experience the same per capita mortality rate, denoted by $m_i(t)$. Individuals mature from this stage to the succeeding stage at a rate $M_i(t)$. The dynamics of the stage population are then described by the differential equation

$$\frac{dN_i}{dt} = R_i(t) - M_i(t) - m_i(t)N_i(t). \qquad (3)$$

Note that in this equation, the functions $R_i(t)$ and $M_i(t)$ represent *total* rates (units: individuals/time), whereas $m_i(t)$ is a per capita rate (units: 1/time). The next step in

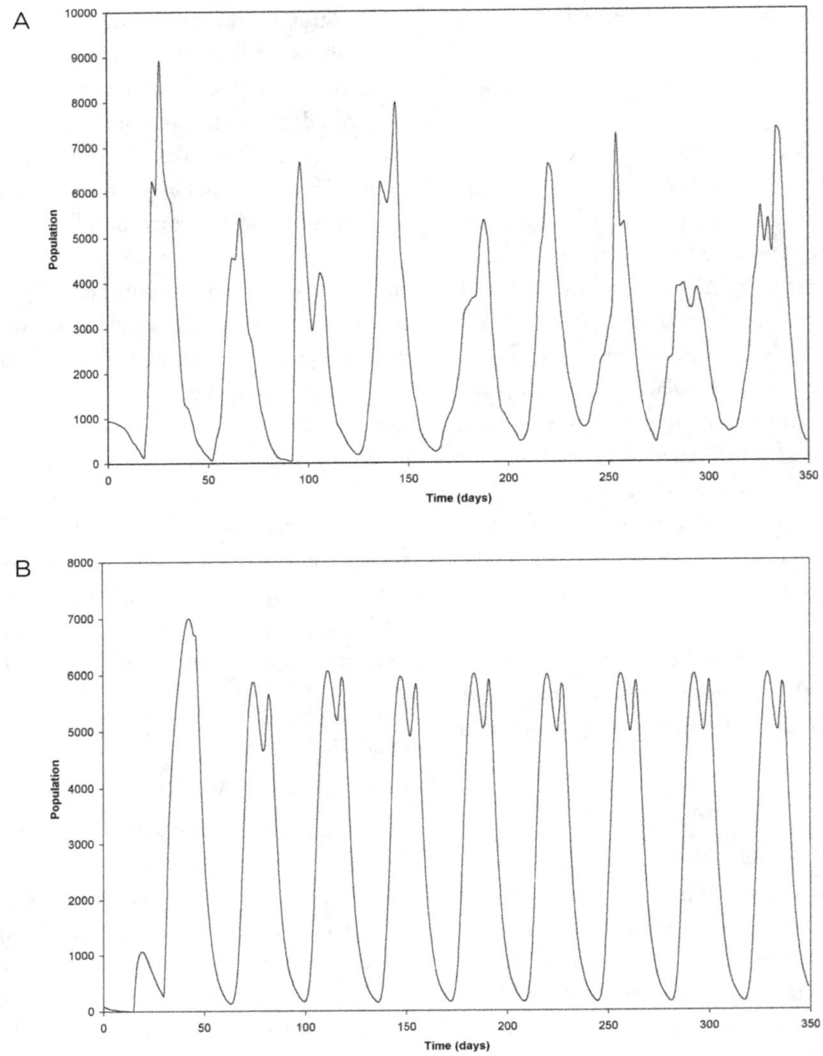

FIGURE 1 (A) A time series of adult population densities from a study of blowfly populations by A. J. Nicholson (1954). (B) Sample output from a stage-structured model of the blowfly population (Eq. 9).

model formulation is to derive general recipes for these functions:

- *Recruitment.* Stage 1 represents the youngest stage. So if $\beta_i(t)$ is the per capita fecundity (e.g., eggs per individual per day from stage i), then

$$R_1(t) = \sum_{\text{all stages}} \beta_i(t) N_i(t). \tag{4}$$

The recruits to all other stages are the individuals maturing from the preceding stage, so

$$R_i(t) = M_{i-1}(t) \text{ for all } i \geq 2. \tag{5}$$

- *Maturation.* Provided the stage durations are fixed (an important restriction discussed later), then if individuals spend a time interval τ_i in stage i and if $P_i(t)$ represents the proportion of the recruits from time $t - \tau_i$ that survive to time t, then

$$M_i(t) = R_i(t - \tau_i) P_i(t). \tag{6}$$

- *Mortality.* If the per capita mortality rate does not change with time, the through-stage-survival function $P_i(t)$ is simply $\exp(-m_i\tau_i)$. In the more tricky situation with time-dependent mortality rates, it can be shown that

$$P_i(t) = \exp\left(-\int_{t-\tau_i}^{t} m_i(t')dt'\right). \tag{7}$$

The final step in formulating the model equations is to write down model functions that relate rate processes (recruitment, death, maturation) to stage populations.

These are obviously problem specific, so we illustrate the process with a particular example that shows how implementation of the recipe leads to delay differential equations.

Example: Blowfly Population Model

The formalism provides an instructive model of one of the classic experimental demonstrations of population cycles—a series of experiments on blowflies by A. J. Nicholson (1954). Figure 1A shows a population trajectory from one of his experiments, where larvae had access to unlimited food and water but adult fecundity was regulated by limiting the supply of protein for the adults. The model has two stages: stage 1, which we call "juvenile," is an aggregate of all pre-reproductive stages (egg, larva, pupa, immature adult); stage 2, called simply "adult," represents reproductive adults. The duration of the juvenile stage has a fixed value. There is no density dependence of mortality rates, but the fecundity has density dependence. In the language of the preceding section,

$$\beta_1(t) = 0; \quad \beta_2(t) = q\exp(-N_2/C); \quad m_1, m_2, \text{ and } \tau_1 \text{ are constant}; \quad P_1(t) = \exp(-m_1\tau_1), \tag{8}$$

where q and C are also constants. Working through the formalism in the previous section reveals that the dynamics of the adult population obey the equation

$$\frac{dN_2}{dt} = (q\exp(-m_1\tau_1))N_2(t-\tau_1) \exp(-N_2(t-\tau_1)/C) - m_2N_2(t). \tag{9}$$

This is a delay differential equation (DDE), so called because the rate of change of N_2 at any time t depends not only on its current value but also on its value at a previous time $t - \tau_1$.

Delay differential equations share many properties with the more familiar linear, ordinary differential equations, but there are many important differences. Probably the most important is that a solution is not specified uniquely by the initial value of the dependent variable (here, N_2); it is necessary to specify values over a time interval equal in length to the delay, known as the initial history, together with the initial value at $t = 0$. Analytic work with DDEs is challenging, but there is good software for numerical solutions of the equations, e.g., routines for Matlab and R. Figure 1B shows a numerical solution of Equation 9 with parameters appropriate for Nicholson's blowflies, demonstrating the richness of the dynamics that can be described by even the simplest delay equations.

More Complex Single-Species Population Models

The blowfly model took the particularly simple form of a single DDE because the only time-dependent rate processes were linked to the adult population, so it was never necessary to work with the integral in Equation 7. The formalism becomes more complex with time- and/or density-dependent mortality in preadult stages. However, it *is* still possible to use DDEs to describe population dynamics when juvenile mortality rate changes in response to the environment. Similarly, it is possible to extend the formalism to include situations where one or more stages duration(s) are not constant. Both extensions are beyond the scope of this article, but we note that the essential assumption leading to a DDE formalism is that a single factor ("physiological age" or "development index") determines the transition from one life stage to its successor. In the simple models, this single factor is chronological age, so that all life stages have a fixed duration.

INTERACTING SPECIES

Many of the most ecologically interesting applications of stage-structure models involve interacting species. Most, though not all, use the continuous-time approach. A few such applications are briefly outlined below.

Host–Parasitoid Interactions

Many parasitoids attack only a single stage of their host, with the vulnerable stage being most commonly an early life stage—for example, eggs or larvae. The simplest continuous-time, stage-structured host–parasitoid model was motivated by studies of California red scale, a citrus pest whose population has been controlled at low levels for over half a century, primarily by the parasitoid *Aphytis melinus.* This basic model follows the philosophy introduced for the blowfly model, in that detailed aspects of the host's early life stages are ignored. The model recognizes only two stages for the host (red scale): juveniles that are susceptible to attack by the parasitoid, and adults that are invulnerable to parasitism. Parasitoids lay eggs in juvenile hosts, and a host that is attacked by a parasitoid is assumed to die instantly and recruit to the juvenile parasitoid population. The parasitized egg takes some finite time to mature to become an adult parasitoid. This system thus involves two interacting populations, each with dynamic equations written down by following the rules set out in the preceding section.

As with the blowfly model, analytic work with the host–parasitoid system requires some specialized techniques, but numerical study of the equations is straightforward. The results reveal a model property with broad ecological implications: a sufficiently long invulnerable adult stage duration for the hosts may lead to a stable equilibrium. This contrasts with the prediction of the discrete-time Nicholson–Bailey model.

Competing Consumers

Many insect species are vulnerable to attack by more than one species of parasitoid, with different parasitoid species sometimes specializing on different host stages (e.g., the egg stage vs. the larval stage). The continuous-time formalism has been used to investigate the competitive interaction between parasitoids that attack different stages of the same host species. Such models illustrate that attack on different host stages does not necessarily provide a mechanism for parasitoid coexistence. If each host stage is fixed in duration (and if there is no other source of density dependence affecting any of the parasitoid species), then a single parasitoid species will win in competition and lead to the competitive exclusion of all others. The parasitoid species that wins is the one that suppresses the abundance of the host stages used by its competitors to the lowest level. If the model includes variability in the durations of the host stages, then coexistence of more than one parasitoid on a single host species is sometimes possible. This occurs because increasing the duration of a particular host stage provides an advantage to the parasitoid that attacks that stage. If there is sufficient variability in stage durations such that some fraction of the host population favors one parasitoid species and another fraction favors a different parasitoid species, then parasitoid coexistence may be possible.

A general result from many continuous-time models of parasitoid competition is that attacking early in the host life cycle (prior to other sources of host mortality) can grant a parasitoid a competitive advantage over later attacking species. Models of the California red scale system, described earlier, suggest that the ability of *Aphytis melinus* to attack a slightly younger juvenile stage of red scale than its congener, *Aphytis lignanensis,* may have been sufficient to explain the rapid competitive displacement of *A. lignanensis* within only a few years of *A. melinus*' release.

Disease Models

Stage-structure can have a wide range of interesting effects in disease systems. Continuous-time models of pathogens that attack insects (e.g., baculoviruses) are structurally very similar to the stage-structured host–parasitoid models described above: it is usually the juvenile stages of insects that become infected via encountering viral particles while feeding, and there is a delay between infection and death of the host. Upon death, however, the infected host can release thousands, or millions, of viral particles into the environment. Stage-structured models of insect–pathogen interactions can produce a rich array of dynamics, including cycles with a period of about one host generation (generation cycles) or longer-period consumer–resource cycles.

Many pathogens are transmitted from host to host by arthropod vectors (especially mosquitoes and ticks). All of the preceding examples include *host* stage structure, where only particular stages of the host are attacked by a consumer. Diseases that are vectored by ticks, such as Lyme disease, provide an example of the importance of stage structure in the vector, rather than host, life cycle. The ticks that vector Lyme disease, for example, need to feed on vertebrate hosts exactly three times to complete their life cycle: once during each of their larval, nymphal, and adult stages. Each feeding is necessary for the tick to develop into the next stage (in the case of the larval and nymphal feedings) or to produce eggs (for the adult feeding). The life cycle, which can take 2–3 years (depending on geographic region), has a strong seasonal element, so it does not fit into the continuous-time formalism as well as the insect disease systems described previously, and is sometimes instead modeled in discrete time.

Ticks tend to feed on small mammals, birds, and lizards during their juvenile (larval and nymphal) stages and large mammals (especially deer) during their adult stage. Ticks start their life cycle uninfected and can pick up the *Borrelia* bacteria that causes Lyme disease during any of their host meals, but different species of vertebrate hosts differ in their ability to become infected from, and transmit the bacteria to, ticks. Stage-structured models of the Lyme disease system have explored whether tick densities are more sensitive to changes in the density of hosts for juvenile versus adult ticks, and have investigated the influence of host species diversity on tick infection prevalence.

FURTHER THEORY AND OTHER APPROACHES

The discrete-time models fall into the broader category of matrix population models for which new theory (and software) continues to appear. The definitive text is Caswell (2001), which also includes an introduction to the problem of parameter estimation for the models. There is a user-friendly module introducing discrete-time stage structure models in the Populus software (http://www.cbs.umn.edu/populus/). The continuous-time models described in this entry follow the approach of Gurney et al. (1983), with the extension to time-varying delays available in Nisbet and Gurney

(1983). Murdoch et al. (2003) provides a comprehensive overview, while Nisbet (1997) addresses the practical problems of implementing the models, in particular the selection of initial conditions and histories. Stage-structured models of competition between parasitoids are described in Briggs (1993) and Murdoch et al. (2003), and the model of displacement of *A. lignanensis* by *A. melinus* is in Murdoch et al. (1996). Briggs and Godfray (1995) describes models of stage-structured insect–pathogen interactions, and Van Buskirk and Ostfeld (1995) presents a discrete-time model of the Lyme disease system. Both the discrete and continuous models focus on one feature (stage structure) and commonly neglect much known biology about the systems being modeled. The payoff from the simplification is that they can be used to develop intuition on the influence of life history features on population dynamics. For example, they yield powerful, testable results relating to population cycles (Murdoch et al., 2003).

The continuous-time formalism admits many extensions, but there appears to be one unavoidable restriction: the instantaneous death rates for individuals within a stage must be identical. However, for specific applications involving more biological and/or environmental detail, the DDE formalism may not be the most appropriate, and it is valuable to recognize that the stage-structured population models are just one particular choice from a range of approaches available for studying structured population models (Tuljapurkar and Caswell, 1997).

SEE ALSO THE FOLLOWING ARTICLES

Age Structure / Delay Differential Equations / Disease Dynamics / Matrix Models / Nicholson–Bailey Host Parasitoid Model / Population Ecology / Predator–Prey Models

FURTHER READING

Briggs, C. J. 1993. Competition among parasitoid species on an age-structured host, and its effect on host suppression. *American Naturalist* 141: 372–397.

Briggs, C. J., and H. C. J. Godfray. 1995. The dynamics of insect-pathogen interactions in stage-structured populations. *American Naturalist* 145: 855–887.

Caswell, H. 2001. *Matrix population models—construction, analysis, and interpretation*, 2nd ed. Sunderland, MA: Sinauer Associates.

Gurney, W. S. C., and R. M. Nisbet. 1983. The systematic formulation of delay-differential models of age and size structured populations. In H. I. Freedman and C. Strobeck, eds. *Population biology*. Berlin: Springer-Verlag.

Murdoch, W. W., C. J. Briggs, and R. M. Nisbet. 1996. Competitive displacement and biological control in parasitoids: a model. *American Naturalist* 148: 807–826.

Murdoch, W. W., C. J. Briggs, and R. M. Nisbet. 2003. *Consumer-resource dynamics*. Princeton: Princeton University Press.

Nicholson, A. J. 1954. An outline of the dynamics of animal populations. *Australian Journal of Ecology* 2: 9–65.

Nisbet, R. M. 1997. Delay-differential equations for structured populations. In S. Tuljapurkar and H. Caswell, eds. *Structured-population models in marine, terrestrial, and freshwater systems*. New York: Chapman & Hall.

Nisbet, R. M., and W. S. C. Gurney. 1983. The systematic formulation of population models for insects with dynamically varying instar duration. *Theoretical Population Biology* 23: 114–135.

Tuljapurkar, S., and H. Caswell. 1997. *Structured-population models in marine, terrestrial, and freshwater systems*. New York: Chapman & Hall.

Van Buskirk, J., and R. S. Ostfeld. 1995. Controlling Lyme disease by modifying density and species composition of tick hosts. *Ecological Applications* 5: 1133–1140.

STATISTICS IN ECOLOGY

KEVIN GROSS

North Carolina State University, Raleigh

Ecologists, like all scientists, attempt to learn about the natural world by gathering data. Statistics is a science that provides a mathematical engine for characterizing quantitative patterns and relationships in data and for using these patterns to draw broader inferences in a formal manner. All statistical methods are premised on assumptions, and the art of using statistics in ecology often entails selecting a statistical procedure whose assumptions are appropriate for the data in hand. In theoretical ecology, statistics provide tools to evaluate mathematical models using empirical data.

WHY DO ECOLOGISTS NEED STATISTICS?

Ecologists gather data in a multitude of ways. Some collect data by simply observing nature, while others conduct experiments in the field, in the laboratory, or on a computer. Yet in nearly every case, the specific entities or individuals included in the study are not the sole focus of the investigation but are instead assumed to represent a larger collection of entities about which the ecologist wishes to learn. Statistics is a science that uses mathematics both to characterize quantitative patterns in data and to draw broader inferences from data.

THE BASICS OF STATISTICS

Conceptual Paradigm

The logic of statistics is based on a paradigm that connects four key ideas: populations, samples, parameters, and statistics (Fig. 1). In statistical thinking, a population is a collection of individuals, objects, or outcomes about which one wishes to learn. Statisticians and biologists use the term population differently. A statistical population

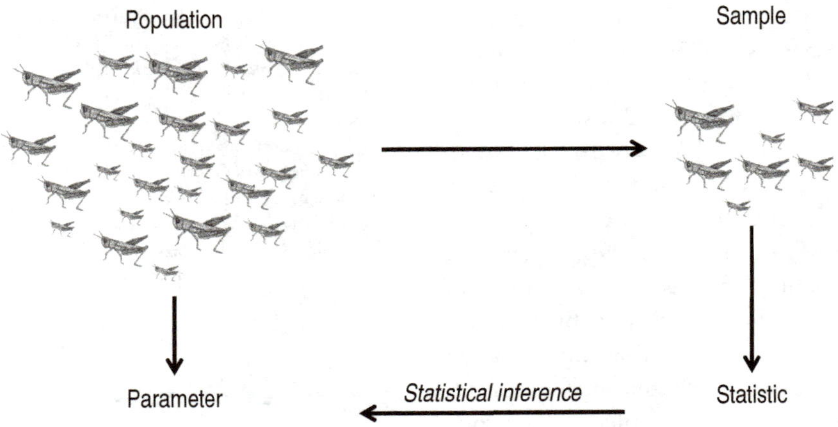

FIGURE 1 The conceptual paradigm of statistical inference. Populations are collections of individuals, objects, or outcomes about which a scientist wishes to learn, and parameters are unknown quantities that describe a population. A sample is any subset of a population, and statistics are known quantities calculated from a sample. Statistical inference allows quantitatively precise statements about parameter values on the basis of statistics. Grasshopper artwork courtesy of G. Crutsinger.

can coincide with (and is often conceptualized as) a biological population, but a statistical population can and often does represent a more abstract collection. For example, if one is using transect sampling to characterize the species composition of a forest, the statistical population might be the collection of all possible transects that could be established within the forest. Or, if one is using radio telemetry to monitor elk movement, the population may consist of the (hypothetical) collection of movement outcomes that would be obtained if an infinite number of elk were observed under identical conditions. Populations that consist of actual physical objects (e.g., all bluefin tuna in the Atlantic) are called concrete populations, while populations that can't be physically enumerated but exist in only a hypothetical sense are called abstract populations.

A parameter is a quantity that characterizes some aspect of a population. In our elk example, parameters of interest might include the average distance moved by a single elk in a single day, or the slope of a linear relationship between local vegetative cover and elk movement rates. Or, if one were conducting a field experiment to determine the effect of herbivory on the growth and survival of an understory shrub, a parameter of interest might be the difference in mortality rates between shrubs exposed to herbivores versus shrubs protected from herbivores. Parameters are quantities about which a scientist wishes to learn.

In cases where the population is concrete, it is (at least theoretically) possible to calculate a parameter by enumerating every individual in the population. Such an enumeration is called a census. However, for most populations a census is either impossible or impractical. In these cases, we learn about parameters by observing a subset of the population. Any subset of a population is called a sample. A random sample is a special type of sample in which individuals are selected from the population randomly, with each individual equally likely to be included in the sample. A statistic is a quantity calculated on the basis of a sample. In our working examples, interesting statistics might be the average distance moved by each radio-collared elk on each day of monitoring, or the difference between the proportion of shrubs that died in herbivore-exclusion versus herbivore-exposure treatments. The terminology is potentially confusing here as well: the term statistics can refer narrowly to quantities calculated on the basis of sample data, or more broadly to the academic discipline that is concerned with the study of these quantities.

Descriptive vs. Inferential Statistics

Statistical methods can be divided into descriptive and inferential statistics. Descriptive statistics are mathematical and graphical methods used to summarize data, and they are essential for communicating patterns in complex data in both scientific and everyday discourse. For simple datasets, common descriptive statistics include the mean, median, or mode (for characterizing a typical or central observation), the variance or standard deviation (for characterizing the variability among observations), a measure of the shape of the data distribution, and any anomalous data points (so-called outliers). Of course, summary quantities are just that—summaries—and summaries can (intentionally or not) convey a misleading impression about the true nature of a dataset. For example, if half of our hypothetical elk had moved 10 km in a given day and the other half had not moved at all, reporting an average movement of 5 km would obscure the bimodal nature of the data. In the popular parlance, it is this possibility to obscure that causes some to regard statistics as more insidious than either lies or damned lies.

Although the intelligent use and interpretation of descriptive statistics is a vital skill in and beyond the scientific realm, in science one is often not content with merely describing patterns in data. Instead, a compelling goal is to determine the extent and degree to which patterns in sample data are representative of the population from which the sample is taken. Inferential statistics provide a body of techniques for drawing formal inferences about population parameters on the basis of sample statistics. Although the notion that one can learn about populations on the basis of samples may seem unsurprising today, it is nonetheless one of the central pillars on which contemporary science rests. One shudders to think of the implications if such inferential learning were not possible, or if the conditions under which such learning could take place were unknown! Inferential statistics are the focus of the remainder of this article.

METHODS OF STATISTICAL INFERENCE

Statistical Models: Connecting Statistics to Parameters

To connect parameters to statistics, we must first understand how the population and the sample are related. A statistical model is a mathematically precise description of the relationship between population and sample. Statistical models must properly account for the salient features of the data-gathering process and can be correspondingly simple or complex. Statistical models also often embody a known or hypothesized scientific relationship. Because samples are formed in such a way that the identity of the individuals to be included in the sample is not known prior to gathering data, statistical models are built using the mathematics of probability to accommodate the uncertain and stochastic nature of the sampling process.

A fundamental concept that permeates the study of probability (and hence statistics) is the notion of a probability distribution. A probability distribution is a mathematical entity that describes how likely it is that an uncertain quantity will take a certain value. In the context of statistical inference, a probability distribution is used to quantify how likely it is that an individual included in the sample will possess a numerical trait with a given value. For example, suppose we sample bluefin tuna from the Atlantic and record their body mass. If all tuna individuals are equally likely to be included in the sample, then the probability distribution for the mass of a single sampled tuna is equivalent to the size structure of the population. Conversely, if our sampling gear is more likely to capture big fish rather than small ones, then the probability distribution for the mass of a sampled tuna will be more heavily skewed toward large tuna.

Frequentist vs. Bayesian Inference

Equipped with a statistical model written in the language of probability, we are now ready to draw statistical inferences about population parameters. It is here that our two statistical roads diverge, splitting into frequentist and Bayesian camps. Frequentist and Bayesian statistics provide two ideologically distinct frameworks for statistical inference. In ecology, as elsewhere in science, a vigorous and spirited debate continues surrounding the relative merits of these two schools of thought. We briefly sketch the foundations of both frequentist and Bayesian inference here, referring the reader to other articles in this volume for more in-depth discussion.

Frequentist statistics represents population parameters as mathematical constants. That is, each parameter has a well-defined value, albeit one that is (and forever shall be) unknown to the investigator. Equipped with a statistical model, a frequentist uses the mathematics of probability to determine a sampling distribution for a statistic. A sampling distribution is a probability distribution that quantifies the probability associated with possible values of the statistic. (The term frequentist follows from the fact that the sampling distribution tells us about the hypothetical *frequency* of possible values of the statistic among all possible samples.) Importantly, the sampling distribution of a statistic is intimately connected to the probability distribution of the individual observations in the sample, and to the unknown population parameter. In our elk example, the sampling distribution of average daily elk movement will have one form if the true average distance moved is 100 m per day, and another form if the true average distance moved is 5 km per day.

It is this connection between the sampling distribution and the population parameter that the frequentist leverages to draw statistical inferences. Two commonly used techniques for drawing frequentist inferences are confidence intervals and hypothesis tests. A confidence interval is an interval with a known coverage level (usually 95%) that permits the following interpretation: if one were to repeat the same experiment many times and calculate a confidence interval for each experiment, in the long run the proportion of intervals containing the true parameter value would equal the coverage level. A hypothesis test is an evaluation of a statistical hypothesis, which in turn is a statement about a parameter value. (Note again the potential for confusion with terminology: a statistical hypothesis is a mathematical statement about a parameter value. A statistical hypothesis can embody a scientific hypothesis, but a scientific hypothesis need not be identical to a statistical hypothesis.) In our herbivory example, a useful statistical hypothesis might be that there is no difference in mortality rates between shrubs exposed to versus protected from herbivory. Hypothesis tests are often associated with significance levels, which loosely quantify the degree of evidence needed against a particular hypothesis to "reject" it. Unfortunately, the term significance is also the root of some confusion, as statistical significance is not tantamount to, and should not be confused with, ecological or biological significance.

Bayesian statistics (taking their name from Rev. Thomas Bayes, an eighteenth-century English clergyman and probabilist) adopts the view that uncertainty in the value of a population parameter can be represented by a probability distribution. In a nutshell, Bayesian inference begins by specifying a probability distribution that encapsulates all that is known or not known about the value of a parameter before a study begins. This distribution is called a prior distribution, because it reflects knowledge of the parameter prior to gathering data. Once data are in hand, one uses the statistical model that connects population to sample to update the prior distribution using a mathematical relationship known as Bayes' rule. This updated distribution is called a posterior distribution, and it combines information about the parameter contained in the prior distribution with information gained in the data sample. The posterior distribution then allows inference about the parameter.

Debate about the relative merits of frequentist and Bayesian inference is lively and ongoing. Proponents of frequentist methods argue that frequentist statistics have an unambiguous interpretation when they are interpreted correctly. A Bayesian might counter that proper interpretations of frequentist inference are awkward and convoluted, enough so that frequentist methods are regularly and routinely interpreted incorrectly. A Bayesian might also argue that Bayesian inferences align more naturally with scientific intuition about how uncertainty in parameter values evolves as we gather data, and thus are less prone to misinterpretation. A frequentist might point out that Bayesian methods rely wholly on specifying a prior distribution and that this reliance introduces a subjectivity and personalization to data analysis that is scientifically inappropriate. Others might take a more pragmatic view, arguing that both frequentist and Bayesian methods are useful but imperfect tools and that an ecologist would do well to familiarize him- or herself with every available tool. Regardless, both frequentist and Bayesian methods rely on a statistical model to connect population to sample, and thus both are only as trustworthy as the statistical model on which they are based.

A BRIEF TOUR OF STATISTICAL MODELS

Architecture of a Statistical Model

Recall that a statistical model is a mathematically precise description of the relationship between the sample and the population. Heuristically, a statistical model consists of two components: the signal component quantifies the patterns in the data that are repeatable from one sample to the next, and the error component captures the variation in the data unique to a given sample. In words, then, we might represent a statistical model using the equation "data = signal + error," although the signal and error are not always combined in an additive way. In our elk example, the signal might be the average relationship between percent vegetative cover and elk movement rates, and the error might be the particular deviations from these average movement rates exhibited by the elk in our sample. Although statistical inference usually focuses on the signal component, inference can also concern the error component if one seeks to learn about variation among individuals within the population.

The distinction between signal and error is entirely a feature of the statistical model and does not reflect any intrinsic feature of the process being observed. For example, in our shrub herbivory study, suppose some shrubs possessed genotypes that made them more resistant to herbivory than others. If we knew the genotypes of the shrubs in our study, then we might include plant genotype as part of the statistical signal. However, if we didn't know that this trait existed, or weren't able to genotype the shrubs in our study, then genotypic differences among shrubs would be relegated to the statistical error. Thus, whether or not a particular attribute is classified as signal or error depends more on the focus of the investigator than on any inherent property of the system under study.

Linear Models: The Statistician's Bread and Butter

Linear models are a broad and versatile class of statistical models that are both powerful in their own right and provide a convenient point of departure for the discussion of more complicated models. The statistical signal of a linear model specifies a mathematical relationship between one or more *predictor* variables and a single *response* variable. In this context, predictor and response variables are also known as independent and dependent variables, respectively. Linear models usually use a linear relationship between predictor and response, although the machinery of linear models can also be co-opted to accommodate polynomial (e.g., quadratic) relationships.

The error component of a linear model often assumes the errors associated with each data point are independent and identically distributed. Here, independence refers to the notion that the presence or absence of one individual in the sample does not affect the presence or absence of any other individual, and it is not related to the notion of an independent variable. The term identically distributed means that the errors associated with each individual in the sample are characterized by the same probability distribution. The linear model also usually assumes that the errors can be described by a bell-shaped (also called a *normal* or *Gaussian*) probability distribution.

Linear models are often classified by whether the predictor takes numerically meaningful values or whether the predictor's values are categories. Models in the former case are known as linear regression models, while models in the latter case are known as analysis of variance models (better known by the acronym ANOVA). An example of a regression model might be one in which elk movement rates (the response) are modeled as a linear function of percent vegetative cover (the predictor). An example of an ANOVA might be one in which species richness along a transect (the response) is modeled as a function of the prevailing soil type (the predictor). Although regression and ANOVA are often perceived as distinct, it is helpful to remember that they are built upon the same mathematical machinery.

Linear models are particularly amenable to mathematical analysis, and their properties are thoroughly known. Consequently, linear models form the backbone of contemporary statistical analysis, and they are familiar to scientists across the disciplinary spectrum. Although the assumptions of the linear model may seem restrictive, it is useful to remember that all models, including statistical ones, are caricatures of reality and not reality itself. Assumptions of linearity and normality provide useful approximations for a multitude of natural phenomena.

Of course, situations exist for which some or all of the assumptions in a linear model simply will not do. Toward this end, alternative statistical models are available that make different collections of assumptions about the data-generating process. We discuss some of these models below.

Beyond the Standard Linear Model

NONLINEAR SIGNALS

Nonlinearities in ecology abound, and with sufficient data one might even hope to estimate these nonlinearities well. Sensibly enough, nonlinear regression encompasses models in which the statistical signal takes a specific functional form (such as exponential or logistic) for the relationship between predictor and response. (As we have mentioned, polynomial relationships between predictor and response can actually be handled using linear models.) Generalized additive models go one step further and write the signal component as a summation of smooth functions of the predictor(s), wherein the particular shape of each of these smooth functions needn't be specified in advance.

DEPARTURES FROM NORMALITY

Ecologists commonly encounter data that are not well suited to a normal-error model. These data types include data that are constrained to an interval (e.g., mortality rates), are integral (e.g., counts of species), and/or are characterized by a small number of either very small or very large observations (e.g., movement data). Generalized linear models pair the signal component of a linear model with an error component that can accommodate a variety of interesting probability distributions. Generalized linear models liberate data analysts from the need to shoehorn nonnormal data into models that assume normality.

NONINDEPENDENT DATA

All of the aforementioned statistical models in this section use the assumption that individuals are sampled independently of one another. While statistical independence is a logical assumption in many contexts, in ecology the logistics of fieldwork often make dependencies among sampled individuals unavoidable. Consequently, the statistical error must be modified to accommodate dependencies introduced by the sampling design. Arguably, the development and usage of these nonstandard error structures is one of the central challenges in contemporary ecological statistics, and has served as a wellspring of statistical innovation.

Sampling in Space and Time In field ecology, individuals in the sample are often associated with a particular place in space and/or time. This spatial or temporal dependence often makes an assumption of independence untenable, as data associated with individuals that share a proximity in space and/or time are more likely to be similar than individuals in the sample that are distant in space and/or time. For example, two transects that are separated by 1 km in the forest are more likely to be similar than transects separated by 100 km. In statistical terminology, observations that are likely to be similar are said to be positively correlated. Indeed, one might argue that statistical correlations driven by proximity in space and/or time are more the rule for ecological field data rather than the exception. When sample data are correlated, the statistical model must be modified accordingly. Depending on the investigator's interests, spatial or temporal effects can be included in either the signal or the error component of the statistical model.

Process vs. Observation Error Not all statistical errors are created equal. Consider the task of fitting a dynamical model that predicts how population densities will change through time to time-series data of fluctuating population densities. Two distinct processes could cause the observed fluctuations to depart from model predictions. First, the true (but unobservable) dynamics of the population may deviate from model predictions, because the model is a (necessary) simplification of the processes governing dynamics and/or because the true dynamics are subject to intrinsic randomness (e.g., demographic stochasticity). Second, the data themselves are unlikely to provide an exact measure of population size but will undoubtedly be contaminated by errors introduced in the sampling process. These two types of error are called process error and observation error, respectively. From a statistical perspective, the value of distinguishing between these two types of error is that process error propagates through an entire time series, whereas observation error is confined to an individual data point. For example, if a catastrophic event causes half of the population to perish (process error), these individuals are lost from the population forevermore. However, if bad weather prevents half the population from being counted on a given sampling occasion (observation error), those individuals may still be present to be counted at the next sampling opportunity. Statistical models that distinguish between process and observation error have provided superior fits to data than models that ignore this distinction.

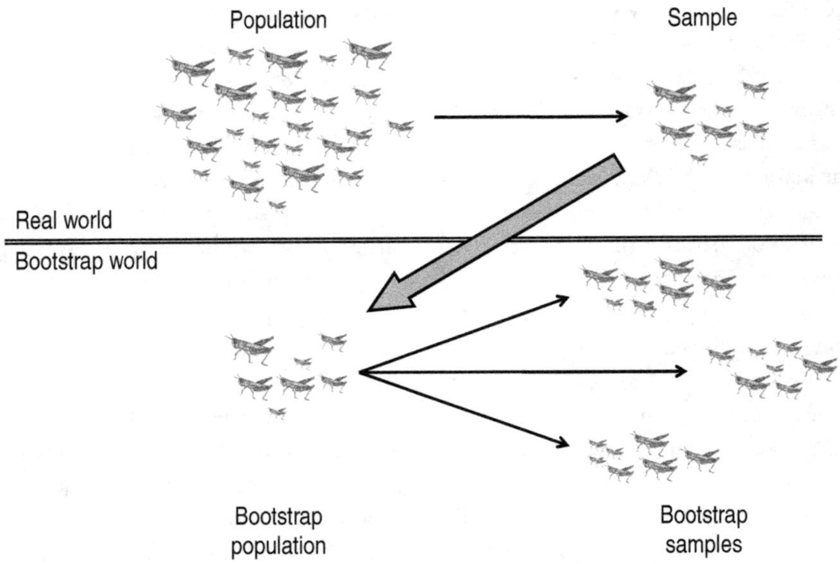

FIGURE 2 A schematic of the bootstrap. In the bootstrap, the population consists of the sample from the actual experiment. Many bootstrap samples are drawn (with replacement) from the bootstrap population, usually with the aid of a computer. The relationship between the known bootstrap population parameter and the sampling distribution of the bootstrap statistic can be determined by constructing many bootstrap samples. Elucidating this relationship yields information about the parallel relationship between the actual sample statistic and the unknown population parameter. Grasshopper artwork courtesy of G. Crutsinger.

Statistical Error and Phylogeny Finally, biologists of all stripes must also consider correlated error in comparative studies across biological taxa. Here, taxa that are more closely related evolutionarily may be more statistically alike than taxa that are distantly related. Thus, comparative studies must use nonstandard error structures to account for statistical correlations introduced by phylogeny and shared evolutionary history. Constructing such an error term appropriately is not a trivial task, and methods for doing so remain an area of active investigation.

COMPUTATION AND CONTEMPORARY STATISTICS

Through the mid-twentieth century, statisticians were restricted to statistical models that could be solved, in the sense of determining either the sampling distribution or posterior distribution analytically. The advent of cheap, widespread personal computing power has revolutionized statistics by loosening this restriction and fueling new breakthroughs that are now essential components of contemporary statistical ecology. We discuss two of these developments below.

The Bootstrap

The bootstrap is a computationally intensive procedure that allows one to determine the sampling distribution of a broad collection of statistics for a variety of experimental designs. The central idea of the bootstrap is that one can learn about the relationship between a population parameter and a sampling distribution by mimicking the data-gathering process inside a computer (Fig. 2). (The name bootstrap comes from the notion that, in the absence of a known sampling distribution, one pulls oneself up by one's bootstraps through this data-gathering mimic.) In the computerized mimic, the sample from the actual study plays the role of the (bootstrap) population, and many computerized samples are "gathered" from the bootstrap population to create many bootstrap samples. Each bootstrap sample gives rise to a bootstrap statistic. Because the bootstrap population is known (it's the original sample), the bootstrap parameter can be calculated and compared to the bootstrap sampling distribution. The genius of the bootstrap is that the relationship between the bootstrap parameter and bootstrap sampling distribution contains information about the relationship between the unknown parameter and the known sample statistic. While bootstrapping does not theoretically require a computer, the need to create many bootstrap datasets makes computer automation mandatory in practice.

Markov Chain Monte Carlo

Bayesian statistics have also been revolutionized by the proliferation of cheap personal computing power. In Bayesian inference, updating a prior distribution with data requires calculating an integral. In a few simple cases, this integral can be calculated by hand. However, for most problems of scientific interest, the integral is analytically (and often numerically) intractable. Before computers were widespread, the intractability of these integrals restricted the application of Bayesian methods to only the simplest problems. Today, computer algorithms are available to approximate the posterior distribution numerically, making computation of the problematic integral unnecessary. A popular class of algorithms for approximating posterior distributions numerically are collectively described as Markov chain Monte Carlo (MCMC) methods. MCMC methods are so called because they use a Markov chain (a type of random process) to generate an approximating pseudo-random ("Monte Carlo") sample from the posterior distribution.

STATISTICS IN THEORETICAL ECOLOGY

In addition to describing and drawing inferences about populations, statistics can also be used to evaluate the predictions of theoretical models, and/or to arbitrate among predictions made by competing ecological theories. In this regard, the purpose is not so much to prove or disprove a model but to understand how a model's predictions do or do not agree with empirical observation. In the field of population dynamics, much insight has been gained by comparing the predictions of simple dynamic models with time series of population-abundance data. For example, statistical models have been used to characterize how density dependence regulates populations, to evaluate the mechanisms for population cycles, and to understand the epidemiology of infectious disease spread. Examples from community ecology are also plentiful. For example, species accumulation curves can be used to evaluate competing models of community assembly, biogeographic gradients can be used to evaluate predictions of metabolic theory, and stable-isotope data can be used to evaluate bioenergetic foraging models.

A FINAL WORD: CLASSROOM VS. REAL-WORLD STATISTICS

Statistics provide ecologists with a set of tools to make quantitatively precise statements about patterns in data. For pedagogical purposes, introductory statistical texts usually present simple datasets for which there is one clearly correct analysis for the student to identify. However, the notion that the goal of statistical analysis is to discover the single correct analysis is not representative of most real-world problems that ecologists encounter. Indeed, most datasets encountered by ecologists (and scientists of all stripes) are sufficiently complex that no single, unambiguously superior statistical analysis exists. Instead, for most interesting problems, different professional statisticians would likely arrive at different methods for analyzing the same data. There is a degree of artistry in choosing a statistical analysis that is appropriate, sensible, and enlightening for the data and question at hand, and two different approaches to analyzing the same problem should not be viewed with suspicion merely because they differ. This is not to say that all statistical analyses are equally valid! To the contrary, all analyses are based on a series of assumptions, and using an analysis whose assumptions are clearly inappropriate for the data results in meaningless conclusions. However, ecology is rich with data that offer opportunity for creative and innovative statistical thinking.

SEE ALSO THE FOLLOWING ARTICLES

Bayesian Statistics / Frequentist Statistics / Information Criteria in Ecology / Markov Chains / Meta-Analysis / Model Fitting

FURTHER READING

Bolker, B. M. 2008. *Ecological models and data in R.* Princeton: Princeton University Press.

Clark, J. S. 2007. *Models for ecological data.* Princeton: Princeton University Press.

Efron, B., and R. J. Tibshirani. 1993. *An introduction to the bootstrap.* Boca Raton, FL: Chapman & Hall.

Gotelli, N. J., and A. M. Ellison. 2004. *A primer of ecological statistics.* Sunderland, MA: Sinauer.

STOCHASTIC SPATIAL MODELS

SEE SPATIAL MODELS, STOCHASTIC

STOCHASTICITY (OVERVIEW)

MATT J. KEELING

University of Warwick, Coventry, United Kingdom

Stochasticity refers to the way in which noise interacts with ecological observations and population dynamics. However, in many circumstances this is a two-way process, with the ecological behavior and the population density determining the type and scale of noise. It is important to note that the action of noise is observed through the filter of ecological dynamics, such that uncorrelated noise can be translated into temporally correlated and highly skewed fluctuations in population size.

IMPLICATIONS OF STOCHASTICITY

Stochasticity has many different effects, some of which are an obvious corollary of adding noise to the basic deterministic dynamics, whereas others arise as emergent behavior due to the interaction of the population dynamics with the noise. The most prominent examples of how stochasticity (and hence stochastic models) differ from their deterministic approximation are as follows.

Variation about the Mean Dynamics

Concepts such as fixed-point equilibria (or stable orbits) no longer hold; instead, population levels may achieve steady-state distributions, such that although the population size fluctuates, the probability of having fluctuations of a given size is fixed. This has profound implications for how natural populations are measured or how the results of stochastic models are reported. However, this lack of a simple deterministic solution has unforeseen advantages: models and data can now be matched through rigorous likelihoods, opening up the possibility of using a range of advanced statistical techniques to parameterize models.

Chance Extinctions and Invasions

Due to the accumulation of chance events, populations that are expected to persist can be driven extinct. Therefore, questions of eradication or species conservation can only be truly addressed with stochastic models. Similarly, invasion is a stochastic process (that is, a process created by chance events) involving both the chance that a new organism reaches a destination and the chance that it establishes and persists. The classic Levins metapopulation models and the derivation of species–area relationships are therefore based on these stochastic principles.

Extended Transients and Resonance

Stochasticity often acts to push the population dynamics away from the deterministic prediction; following a major deviation, the return generally follows a path close to that described by the deterministic model. This means that extended periods of transient dynamics are often observed; when the deterministic transients are oscillatory, stochastic resonance frequently occurs as noise induces cyclic behavior.

Deviations in the Mean

Quite surprisingly, the addition of stochasticity does not simply produce fluctuations about the mean dynamics, but it can actually change the mean value. This occurs through the interaction between noise and density-dependent (nonlinear) processes.

Impact of Population Size

Finally, in deterministic models it is always possible to scale population sizes and parameters such that population densities can be modeled with the same equations as absolute numbers. The same is not true for stochastic models—numbers matter. For many types of noise, larger populations experience relatively weaker stochastic effects compared to smaller populations. This means that careful consideration is needed in assessing the scale at which a stochastic model operates.

These five facets are considered further below using the logistic equation and Ricker map as motivating examples. The sections that follow consider how stochasticity can be incorporated into the basic population models before considering different sources of stochasticity.

CAPTURING STOCHASTICITY

There are three main ways in which noise or stochasticity is commonly included in basic population dynamics. There are illustrated for simple single-species models, but the concepts easily generalize.

Random Noise in Continuous Time Models

Continuous-time models are generally expressed as differential equations, relating the instantaneous rate of change of the population to its current levels (e.g., $\frac{dx}{dt} = f(x)$ for a population x). To such equations, dynamic noise ($\xi(t)$) can be added. Unfortunately, this noise does not have a simple expression, although its properties are well understood. For example, $\frac{dx}{dt} = \xi(t)$ leads to the population performing a random walk, whereas $\frac{dx}{dt} = [k - x] + \xi(t)$ leads to a population that is normally distributed about k—thus in the majority of models, if the fluctuations are small (such that the dynamics can be linearized and nonlinear effects can be ignored), then the distribution of population sizes should be approximately normal.

The underlying population processes often provide a guide to the scale of noise that needs to be incorporated—and this will necessarily vary with the current population size. A more accurate representation of stochasticity would therefore be the following stochastic differential equation: $\frac{dx}{dt} = f(x) + g(x)\xi(t)$. The form of the function $g(x)$ depends on the source of the noise, but when all the noise is due to the chance nature of the processes, $g(x)$ can be formulated explicitly assuming processes are Poisson. Consider the simple case where per capita birth and death rates are $b(x)$ and $d(x)$. As expected, the dynamic function f is given by $f(x) = b(x)x - d(x)x$; whereas the stochastic function is given by $g(x) = \sqrt{b(x)x + d(x)x}$. The intuitive argument behind the stochastic function is that the variance associated with each Poisson process is equal to the mean, the variances of the two processes add together, and the square root is then taken to find the standard deviation.

This relationship between f and g highlights two important points. First, even when $f = 0$ and the population size is at the deterministic equilibrium, g is nonzero and stochastic forces are still felt. Second, as populations with larger deterministic equilibria are considered, so the relative strength of the stochastic component (the scale of g relative to the population size) decreases—for most plausible birth and death rates, g grows with the square root of the population size. This means that large populations tend to behave far more like their deterministic counterparts.

Distributions in Discrete-Time Models

For discrete-time models, where the population in one year is used to predict the population in the next, then stochasticity is forced to have a different format. In this type of model, it is most common to force the population to be integer valued so that there are whole numbers of organisms. Generally, stochasticity is introduced as random choices from prescribed distributions, where the ecological processes govern the type of distribution and the parameters of the distribution are given by the population levels. For example, if deaths are assumed to occur randomly within a population, then it is reasonable to consider that the number of organisms surviving to the next year will be binomially distributed (where the number of trials is the population size and the probability refers to the chance of survival). However, other processes, such as births, may call for other distributions, with Poisson and negative binomial being common choices. In practice, the choice of distribution should be based on observations or a deeper understanding of the processes involved.

Demographic Stochasticity in Continuous Time

This particular form of stochasticity models the chance nature of individual events, and again the populations are integer valued. In many ways, this represents the ideal for modeling populations, as the strength of the noise is naturally generated by the underlying ecological processes. The dynamics are created using the Gillespie algorithm; this is illustrated here for a single species with population-level birth rate $B(x) = x\,b(x)$ and death rate $D(x) = x\,d(x)$:

1. List the events (processes) that can occur, their rates, and find the total rate.

   Births $B(x)$, Deaths $D(x)$, Total Rate $B(x) + D(x)$.

2. Pick a random number uniformly between 0 and 1 and use the total rate to find the time until the next event occurs.

   $$\text{Time to next event } \delta t = \frac{-\ln(\text{Random})}{\text{Total Rate}}.$$

3. Pick another random number (between 0 and 1) and use the individual rates to determine which event occurs.

   A birth occurs if Random $<$ Birth Rate/Total Rate; otherwise the event is a death.

4. Increase the time ($t \rightarrow t + \delta t$) and perform the event ($x \rightarrow x + 1$ for births and $x \rightarrow x - 1$ for deaths).

SOURCES OF STOCHASTICITY, NOISE, AND RANDOMNESS

This section considers ways in which stochasticity or noise can arise in ecological dynamics, and its impact. Results are illustrated using two simple mathematical models: the discrete-time Ricker map ($x_{t+1} = x_t \exp(r[1 - x_t/k])$) and the continuous-time logistic equation ($\frac{dx}{dt} = rx(1 - x/k)$).

Observational Noise

As the name suggests, observational noise arises from the natural inaccuracies in counting populations. Enumerating the size of various populations is at the heart of theoretical ecology and yet it is practically very difficult to achieve whether dealing with microbes or mammals. Methods such as mark–recapture offer a partial solution, but difficulties still arise when dealing with rare organisms or ones with patchy distributions. Accounting for this observational noise is obviously important if theoretical results are to be matched to records of population abundance. Observational noise is usually represented as an independent binomial process at each sample time, with a fixed probability of observing or recording each organism in the population.

It is crucially important to realize that observational noise applies only to the observations. There is no feedback from this type of noise to the underlying population dynamics. Therefore, while observational noise is a constant burden when dealing with counts of real organisms, it is of limited theoretical interest.

Demographic Noise

The form of demographic noise that is required depends on the type of model and the structure of ecological process being studied. For continuous-time models, the Gillespie algorithm is the natural choice, but for discrete-time models the choice is more discretionary, being based on a judgment of how the underlying processes operate.

For both Ricker map and logistic equation models, a choice is required on how density dependence is assumed to operate—in other words, how the underlying rate of change is partitioned into birth and death rates. This is most simply illustrated for the logistic equation. One assumption is that all individuals suffer a constant (density-independent) death rate d, which means that the per capita birth rate, $b(x)$, must be density dependent such that the fecundity decreases with increasing population size: $b(x) = (r + d) - rx/k$. (Notice that this interpretation only makes sense when $rx > (r + d)k$ such that the birth rate is positive.) An alternative assumption is that the per capita birth rate is constant (and equal to b per individual) but the per capita death rate, $d(x)$, is density dependent, increasing with population size: $d(x) = (b - r) + rx/k$. This demonstrates a fundamental and important difference between deterministic and stochastic models—there are multiple stochastic interpretations for each deterministic model.

Figure 1A shows a range of results derived from the Gillespie algorithm for the logistic equation with per capita birth and death rates given by $b(x) = 2r$ and $d(x) = r + rx/k$ ($r = 1$, $k = 100$). This illustrates a number of principles: most notably, the dynamics are noisy; even in this short snapshot there is considerable variation, with the population size fluctuating between 80 and 120 individuals. A close-up inspection of the time series (circular inset in Fig. 1A) highlights the individual nature of the population, such that the population size jumps between integer values every time there is a birth or a death. It also shows the variation in the times between events; during these periods the population size remains constant.

Similar stochastic dynamics can be derived from the Ricker map (Fig. 1B). Again care in needed in separating the birth and death processes, and in attributing a stochastic distribution to each process. Here, a relatively

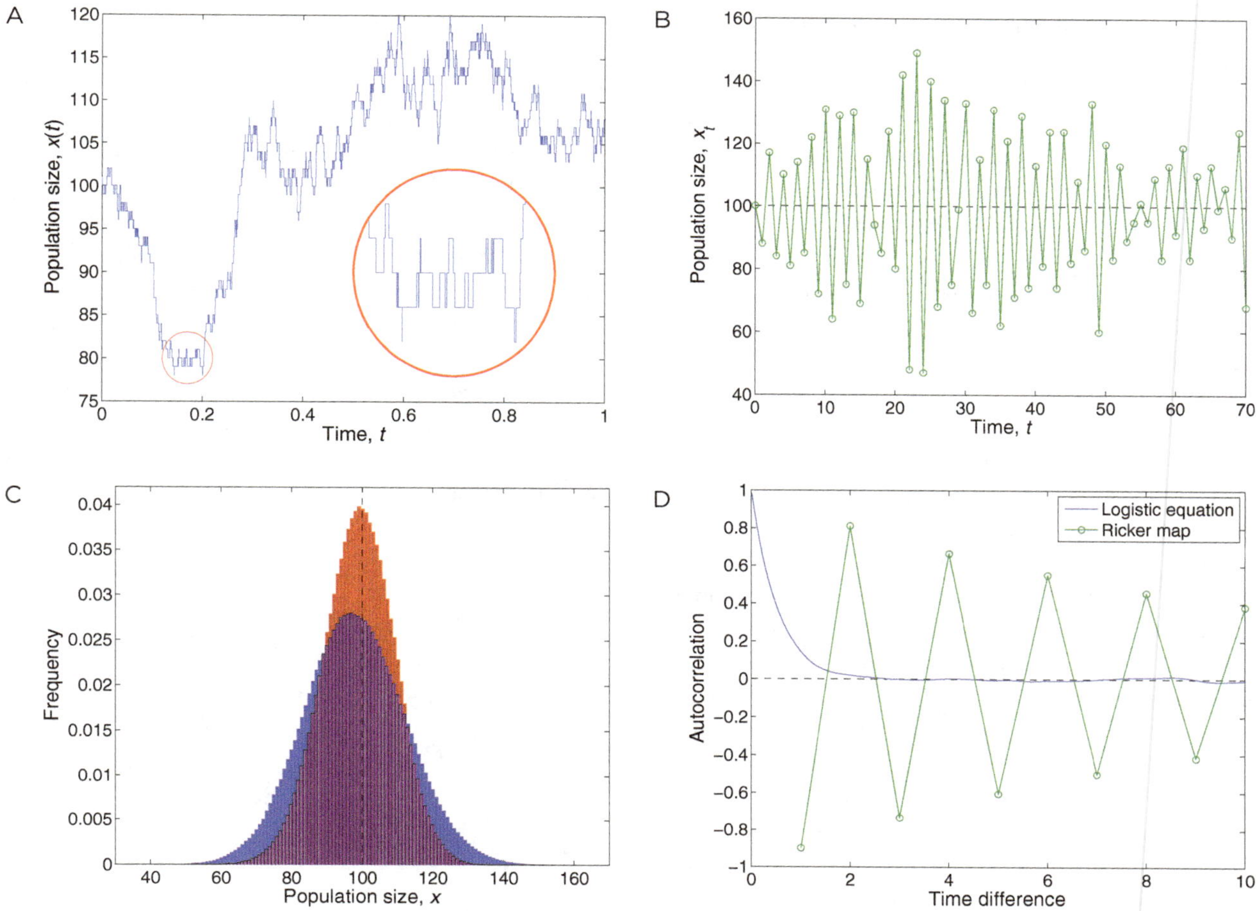

FIGURE 1 Analysis of stochastic continuous-time and discrete-time single-species models. (A) Time series from a logistic equation with demographic stochasticity (per capita birth and death rates are: $b(x) = 2r$ and $d(x) = r + rx/k$, with $r = 1$ and $k = 100$). (B) Time series from a Ricker map with demographic stochasticity of the form $x_{t+1} = \text{Binomial}(x_t + \text{Poisson}(x_t[\exp(r) - 1]), \exp(-rx_t/k))$ with $r = 1$ and $k = 100$. (C) Distribution of population sizes from the logistic model with density dependence in the death rate (blue, as in (A)) or in the birth rate (red, $b(x) = 2r - rx/k$ and $d(x) = r$, with $r = 1$ and $k = 100$). (D) Temporal autocorrelation from the time series shown in (A) and (B), illustrating the short-term correlations in population size, despite the uncorrelated nature of the noise.

simple approach is taken in which existing adults give birth to a Poisson-distributed number of offspring, and then both adults and offspring have a chance of surviving to the next season. The per capita fecundity is assumed to be density independent, whereas survival is assumed to be density dependent:

$$x_{t+1} = \text{Binomial}\big(x_t + \text{Poisson}(x_t[\exp(r) - 1]), \exp(-rx_t/k)\big).$$

Clearly, this is a discrete-time process (so it makes sense to look only at annual points). However, additionally there is a clear difference between the results of the stochastic logistic equation and those of the stochastic Ricker map. By its continuous nature, the population size in the stochastic logistic equation can only change by one at a time, so nearby points tend to be correlated; in contrast, the stochastic Ricker map inherits much of the dynamics of its deterministic counterpart, and so repeated cycles above and below the mean are common. (These features are explored below.)

A more statistical approach examines the distribution of population sizes (Fig. 1C; the blue curve corresponds to the simulation in Figure 1A, the red curve corresponds to the model with density-dependent fecundity). Several features emerge: clearly the assumption as to whether the birth or death rate is density dependent can make a dramatic difference to the stochastic results, even though the underlying deterministic models are identical. Secondly, both distributions are close to being normal but with deviations in the tails. This approximate form is related to the normal distribution that was predicted for noise added to differential equations. Finally, and perhaps most surprisingly, the distributions do not have their mean, median, or mode at the deterministic carrying capacity but slightly lower. This observation can be explained by considering the average of the differential equation. The density-dependence (quadratic) term in the equation leads to the inclusion of the variance when we try to calculate the behavior of the mean, and the variance acts to suppress the growth rate. Therefore, adding stochasticity, and hence introducing variance into the population dynamics, leads to a lower mean value. Unfortunately, to calculate the mean value requires the variance to be known, and although similar arguments allow the construction of equations for the variance, these inevitably contain higher-order terms.

Understanding the mean and variance therefore clearly provides simple insights into the emergent stochastic dynamics and provides a relatively parsimonious way of capturing the stochastic effects. For relatively large carrying capacities, two consistent features emerge. First, the mean population size approaches the carrying capacity as it increases. Second, the variance is generally proportional to the mean, and therefore the size of fluctuations (as captured by the standard deviation) is proportional to the square root of the mean. This in itself has two implications: the first is that the absolute size of fluctuations increases with the expected population size; the second is that the relative size of fluctuations decreases with expected population size. It is this latter observation that has the greatest impact, leading to the general conclusion that large populations tend to behave more like the deterministic predictions and suffer less from stochasticity than small populations; that is, they experience relatively smaller fluctuations and less risk of extinction compared to smaller populations. This has important implications when populations are spatially structured; spatial structure often acts to subdivide the population into small interacting subpopulations (e.g., the metapopulation formalism), each of which is likely to experience substantial stochastic effects due to its reduced size. Whether the entire population experiences greater stochasticity due to this spatial partitioning or not depends on the population under study and the interaction between the spatial units.

As seen in Figures 1A and 1B, there is temporal structure in the stochastic fluctuations. This can be better visualized by plotting the autocorrelation at different time lags. The autocorrelation at lag T is defined as

$$R(T) = \frac{\text{mean}[(x(t) - \bar{x})(x(t + T) - \bar{x})]}{\text{var}(x)},$$

that is, the correlation of the population with itself at a time difference of T. The autocorrelation for the stochastic logistic equation and stochastic Ricker map (as shown in Figs. 1A and 1B) is given in Figure 1D. Clearly, the two models produce strikingly different results mirroring their different deterministic approaches to the carrying capacity. The autocorrelation for the stochastic logistic equation drops from a peak of 1 at very short lags to negligible values by a lag of 3; this implies that although the dynamics are stochastic, the population level changes relatively slowly such that the population fluctuations are correlated over short time scales. This closely matches the way the deterministic logistic equation steadily asymptotes to its equilibrium. In contrast, the autocorrelation for the stochastic Ricker map oscillates between positive

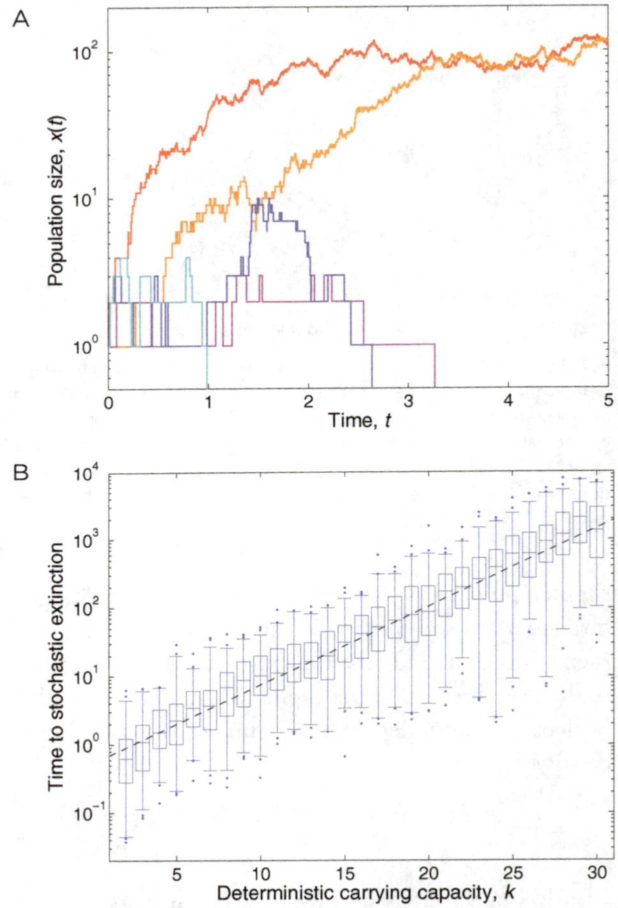

FIGURE 2 The stochastic invasion and persistence of a continuous-time single-species model, using the parameters described in Fig. 1A. (A) Five example time series following the invasion of a single individual, three of which fail to successfully invade. Note that the population size is plotted logarithmically for clarity. (B) The time to extinction as the deterministic carrying capacity (k) is varied. The dashed line shows the best-fit exponential. (100 simulations are performed for each k value, and all begin with $x(0) = k$.)

and negative values and does not reach negligible values until a lag of 50. Again this indicates the strong influence of the deterministic model on the stochastic dynamics; the deterministic model approaches equilibrium through decaying overcompensatory cycles, oscillating above and below the carrying capacity in each successive season.

One final aspect where stochasticity is vitally important is the invasion and successful colonization by any new organism. Figure 2A considers invasion for the stochastic logistic equation. In the deterministic model, the success of invasion is determined by the difference between the per capita birth and death rates in the limit when the population size is very small ($r = b(0) - d(0)$). When this is positive (that is, there are more births than deaths), then an organism is predicted to always successfully invade and will inevitably reach the equilibrium. In stochastic models, this is far from true—even the most well-suited invader can simply fail by chance. As described earlier, the impact of stochasticity is greater at small population sizes and hence is maximal for an invading organism. Although it is impossible to predict which invasions will be successful, the probability of success can be calculated. Using results from branching processes and assuming the carrying capacity is large (and therefore density-dependent effects can be ignored), the probability that a single invading individual leads to an established population is $1 - d(0)/b(0)$. This contrast is intriguing, as in the deterministic model it is the difference between per capita birth and death rates that determines the early growth of an invader, whereas in the stochastic model it is the ratio that determines the probability of success. However, both models agree that when the death rate is greater than the birth rate invasion is impossible. This neat result regarding the probability of successful invasion can be extended to the case where initially X_0 invaders are present; in this case, the population becomes established with probability $1 - [d(0)/b(0)]^{X_0}$. This effectively states that the invasion can be considered as X_0 separate independent trials, and the invasion only fails if all X_0 lineages fail. The corollary of this is that the potential for a new species to colonize is highly dependent on the rate at which it invades. It should be noted that these general principles also hold for discrete-time models, but the simple expressions for invasion are no longer possible.

An extension to the concept of successful invasion (reaching carrying capacity from a low number of invaders) is that of stochastic extinction (reaching zero individuals from a population near carrying capacity). Due to fluctuations around the mean value, it is highly likely that at some time the population is driven to low levels; whether the population can regain its former values or goes extinct is comparable to the problem of stochastic invasion. Figure 2B explores this phenomenon in more detail and shows the time to extinction for a range of k values (starting at the deterministic carrying capacity, $x(0) = k$). Clearly, the expected time to extinction increases exponentially with k. Small carrying capacities, which lead to small population sizes, are therefore greatly influenced by stochasticity; this means that they will frequently be driven to low levels from which they must effectively reinvade. For larger carrying capacities, the population will fluctuate to low levels only rarely, so the time to extinction increases. In all cases extinction is inevitable, although for very large population sizes the time scales involved are immense.

The extreme of such stochastic models is individual-based simulations. In these, the status of each individual within the population is tracked through time, rather than simply recording the number of individuals. This approach has many merits when dealing with complex processes and populations with many forms of heterogeneity, but it is an additional step removed from the traditional deterministic models, and as such, it is often much harder to interpret the results.

One discipline has embraced the use of demographic stochasticity in mathematical modeling more than any other element of the ecological sciences: that of disease modeling and epidemiology. There are three main reasons for this. First, levels of infection are often relatively low and therefore the impact of stochasticity is likely to be substantial even in a large host population. Second, epidemiologists are often interested in disease invasion and eradication, both scenarios where the number of infected individuals drops to very low levels and hence stochasticity is overwhelmingly important. Finally, there exist excellent and long-term data sets on the reported cases of infection, which highlight the stochastic nature of infection dynamics. One of the most well studied is the pattern of measles cases in England and Wales, the United States, and on islands—all these studies identify a critical population size (of around three to five hundred thousand) below which the infection suffers from repeated stochastic extinctions and reintroduction events.

External (Environmental) Noise

An additional source of noise comes from the external environment—where external refers to any elements of the wider ecology and environment not explicitly modeled. The classic example of this is the weather, which is difficult to predict with even the most sophisticated simulations and which would never be captured within the framework of an ecological model. Instead, the weather can be viewed as a noisy external driver of the population processes. However, it is not just the weather that can be considered external to an ecological model; any element not explicitly modeled could be treated in the same way. For example, consider a simple predator–prey system. This could be influenced by vegetation levels or a generalist predator that could be difficult to model explicitly due to the number of potential interactions with other species; instead, these can be considered to be noisy external factors that influence the predator–prey system. For simplicity, in this section external noise will always be considered as driven by the climate, but similar methodologies hold for all sources.

To describe the ways in which external noise is captured in stochastic models, it is again necessary to differentiate between continuous-time and discrete-time models. In both models, climate generally operates through a modification of the basic parameters, rather than acting on the population level itself. For example, good weather (and the definition of good obviously depends on the species in question) may increase fecundity, whereas bad weather may increase mortality. For discrete-time models, this type of noise can be included with relative ease, with the parameter for a given year chosen randomly from a prescribed distribution—e.g., birth rate $b_t = b_0 + b_1\chi_t$ where χ_t are independent normally distributed random variables with mean 0 and variance 1. For continuous-time models, introducing noise is more complex, as the climate (in common with many other external factors) has strong temporal correlations—there are usually prolonged periods of good or bad weather. Therefore, a mechanism is needed to create temporally correlated fluctuations from dynamic noise ($\xi(t)$); the simplest way to achieve this is to generate a "weather" variable:

$$\frac{dW}{dt} = -sW + s\xi(t).$$

This produces a time series that is normally distributed (with mean 0 and variance 1) but which has temporal correlations governed by s; the larger s is, the faster the correlations decay. The variable W can now be used as a noisy forcing term within any of the basic parameters.

Two commonly cited examples of how noisy climate can influence dynamics are from the spruce budworm in Canada and Soay sheep on the island St. Kilda. Single-species continuous-time population models for the spruce budworm have been developed that have two stable equilibria, corresponding to low-level persistence or large-scale outbreaks of this pest species. In the deterministic setting, once the population reaches the vicinity of one of these stable points it never leaves—and therefore the predicted dynamics is for a constant population size at one of the two equilibria. However, in a stochastic model, where the death-rate parameter of the budworm varies in a noisy (but temporally correlated) manner, the population can spontaneously flip between the two stable states—being pushed to the lower state by periods of very bad weather and returning to the epidemic state only during prolonged periods of very good weather. The model therefore displays hysteresis, where climatic extremes are needed to flip the population between stable states.

The dynamics of Soay sheep on St. Kilda are usually viewed as a discrete-time system due to the annual birth of lambs every spring. The population of sheep on this island has been monitored since the 1950s and is observed to fluctuate considerably between years. These fluctuations have been shown to relate to the age structure of the sheep population, the weather (as captured by the scale of the winter North Atlantic Oscillation), and the overcompensatory discrete-time dynamics. In fact, the abrupt crashes that are observed in the sheep population can only arise through the interaction between the deterministic dynamics and the stochastic climate. Crashes occur after a series of good years has allowed the population to increase; in the following year, bad weather combines with extreme density dependence to drive the population to very low levels.

One important aspect of environmental noise is that it has a comparable impact on the population dynamics irrespective of the population size; both large and small populations are similarly affected by the weather or other external drivers. This is in direct contrast to demographic noise discussed above, where the impact of stochasticity decreases with population size. Therefore, environmental noise is likely to be the dominant form of stochasticity in many large populations.

Population Variation

An additional source of noise occurs due to individual-level parameter variation. In this approach, each parameter within a model is viewed as the average over all the individual parameter values within the population. As such, the variation in the population-level parameters is a function of both the individual-level variation and the number of individuals over which the parameter is averaged. This type of noise again scales with the square root of the population size; a larger population size means that the average is taken over more individuals and therefore the average lies closer to the expected mean. Implicit in this approach is selecting an appropriate individual-level distribution for each parameter and the potential biases caused by nonlinear dependence on parameters in the population model.

Two factors complicate this relatively simple picture. The first is that, given heterogeneity in individual-level parameters, heterogeneity in survival rates would also be expected, and therefore the population-level parameters will be influenced by this differential survival. Second, many parameter values may contain some element of heritability such that beneficial traits are likely to be passed on their offspring. Trait heritability would only act to exacerbate the first problem, and to cope with these issues would require a fully individual-based modeling approach. However, for certain classes of model (particularly discrete-time models) it is reasonable to ignore these complications and simply consider parameters that fluctuate due to individual-level variability within the population.

DETERMINISTICALLY MODELING STOCHASTICITY

Although it might seem oxymoronic, it is possible to capture the behavior of stochastic systems with deterministic models. Two approaches are common: moment closure and full Markov chain analysis.

Moment Closure

It was shown above that the stochastic logistic equation has a mean population size that is lower than the deterministic carrying capacity and that this can be explained due to the action of nonlinear density-dependent terms. The differential equation for the mean of the stochastic model can be found by averaging the differential equation:

$$\begin{aligned}\frac{d\,\text{mean}(x)}{dt} &= \text{mean}(rx(1 - x/k)) \\ &= r\,\text{mean}(x) - r\,\text{mean}(x^2)/k \\ &= r\,\text{mean}(x) - r\,\text{mean}(x)^2/k - r\,\text{var}(x)/k.\end{aligned}$$

Therefore, if the variance is known, the above equation provides an exact description of the mean. Unfortunately, this is rarely the case. The variance could be approximated, either by setting it equal to zero (assuming there is no variability), which leads to the standard deterministic model, or by setting the variance equal to the mean (assuming a Poisson distribution), which leads to a rescaling of the model and a reduction in the carrying capacity.

An obvious next step is to construct differential equation models for the rate of change of the variance. This is possible (although algebraically complex), but it introduces a third-order term that captures the skew of the distribution. Obviously this process could be repeated over and over again, with increasingly higher-order terms being required and the algebra getting increasingly involved. However, it is common to stop at the equation for the variance and to approximate the third-order term in terms of the known mean and variances—this is commonly referred to as a moment closure method, and it generally provides an improved approximation to the stochastic behavior, especially when the distribution of population sizes is unimodal.

Markov Chain Analysis

An alternative deterministic approach is to construct equations for the probability that the population is in each conceivable state; therefore, $P_i(t)$ is the probability that the population size is i at time t ($x(t) = i$). For the continuous-time model, this can be done with relative ease, calculating the probability that a population of size i is created from either a population of size $i + 1$ or from a population of size $i - 1$ through the natural processes of birth and death (as well as the probability that the population is currently of size i but changes):

$$\frac{dP_i}{dt} = [B(i - 1)]P_{i-1} + [D(i + 1)]P_{i+1} - [B(i) + D(i)]P_i.$$

Here, B and D are the population-level birth and death rates (and are related to the per capita birth and death rates by $B(x) = xb(x)$, and $D(x) = xd(x)$).

What is truly surprising about this equation is that it is linear in the probabilities P_i; all of the nonlinear behavior is captured in the values of B and D. This linearity means that a range of sophisticated techniques can be applied to these equations, allowing their behavior to be computed with relative ease. This is clearly a highly powerful methodology, enabling the whole ensemble of stochastic behaviors to be captured with a single set of differential equations. The difficulty is that this set of equations can be exceedingly large (especially when multiple species are involved), as a different equation is required for every possible combination of population sizes. Therefore, while some approaches are highly beneficial for small population sizes (such as transmission of infection within hospital wards or households), there are computational limitations on what can be achieved.

SEE ALSO THE FOLLOWING ARTICLES

Birth–Death Models / Invasion Biology / Ricker Model / Single-Species Population Models / Spatial Models, Stochastic / Stochasticity, Demographic / Stochasticity, Environmental

FURTHER READING

Bartlett, M. 1960. The critical community size for measles in the United States. *Journal of the Royal Statistical Society* 123: 37–44.

Bjørnstad, O. N., and B. T. Grenfell. 2001. Noisy clockwork: time series analysis of population fluctuations in animals. *Science* 293: 638–643.

Coulson, T., E. A. Catchpole, S. D. Albon, B. J. T. Morgan, J. M. Pemberton, T. H. Clutton-Brock, M. J. Crawley, and B. T. Grenfell. 2001. Age, sex, density, winter weather, and population crashes in Soay sheep. *Science* 292: 1528–1531.

Keeling, M. J., and J. V. Ross. 2008. On methods for studying stochastic disease dynamics. *Journal of the Royal Society Interface* 5: 171–181.

Lande, R., S. Engen, and B-E. Saether. 2003. *Stochastic population dynamics in ecology and conservation*. Oxford: Oxford University Press.

STOCHASTICITY, DEMOGRAPHIC

BRETT A. MELBOURNE

University of Colorado, Boulder

Demographic stochasticity describes the random fluctuations in population size that occur because the birth and death of each individual is a discrete and probabilistic event. That is, even if all individuals in the population are identical and thus have the same probabilities associated with birth and death, the precise timing and other details of birth and death will vary randomly between individuals, causing the size of the population to fluctuate randomly. Demographic stochasticity is particularly important for small populations because it increases the probability of extinction.

PROBABILISTIC BIRTHS AND DEATHS

Demographic stochasticity arises because the birth and death of an individual is probabilistic. As an illustration of the concept of probabilistic births and deaths, imagine a small founding population of an asexually reproducing organism. This population is made up of identical individuals, each of which has the same probability of dying and the same probability of giving rise to a new individual within some unit of time. In a deterministic setting, these conditions would give rise to exponential growth, but probabilistic births and deaths give rise to a wider range of dynamical behavior. Probabilistic births and deaths mean that if one were to repeat an experiment starting with the same number of individuals, a plot of the number of individuals through time would be different for each replicate of the experiment. Indeed, the dynamical outcomes of the stochastic process would include populations that go extinct and populations that experience a long lag period of low population size before exponential growth occurs (Fig. 1).

A key feature of biological systems that contributes to demographic stochasticity is that individuals are discrete units. The probability of birth or death applies to the individuals. An important consequence is that biological outcomes are discrete: an individual is either born or it is not, or an individual dies or it does not. There can be no partial events, such as the birth or death of a fraction of an individual. When these discrete events are probabilistic they lead to random fluctuations in population size, such as those illustrated in Figure 1. These

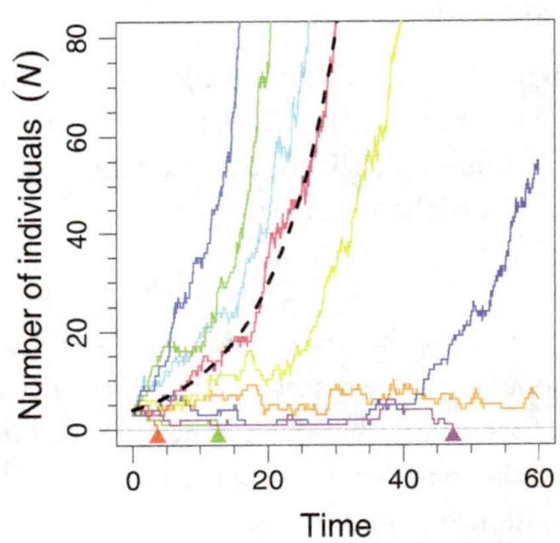

FIGURE 1 Demographic stochasticity caused by probabilistic births and deaths of discrete individuals. Variation from demographic stochasticity is illustrated here by ten representative realizations of population growth from 1000 Monte Carlo simulations of the same stochastic model. Dynamical outcomes included populations that experienced exponential growth (some fast, some slow), populations that experienced a long period of low population size before exponential growth occurred, populations that did not grow at all during the period, and 62 populations that went extinct. This contrasts with the equivalent deterministic model of exponential growth (dashed black line). Of simulations that did not go extinct, shown are the simulations with the highest and lowest number of individuals at the end of the simulation, as well as the 2.5th, 25th, 50th, 75th, and 97.5th percentiles. Of simulations that went extinct, shown are the shortest, median, and longest extinction times. Triangles show the time of extinction. Model details: individuals were identical and reproduced asexually; births and deaths were density independent with rates 0.2 and 0.1, respectively; initial population size was 4 individuals. The Gillespie algorithm was used to simulate the continuous-time stochastic process.

discrete, probabilistic events are particularly important when the population is small because then the number of events is small and these events can combine with high probability in such a way as to drive the population size away from the expected value. The expected value of population size is the number of individuals averaged over an infinitely large number of realizations of the stochastic process. In larger populations, the large number of discrete events tend to average each other out and the dynamics of population size tends toward the expected value and is often well approximated by an equivalent deterministic model.

DEMOGRAPHIC VARIANCE

Demographic stochasticity causes variance in population size around the expected value. This variance comes about because each individual within the population contributes a deviation from the expected value of the per capita growth rate of the population. The variance in these individual deviations from the per capita growth rate is called the demographic variance. Sometimes the expected value can be calculated using a deterministic model, so one can think of demographic stochasticity as adding variance to the deterministic component of the dynamics. For all models, the variance of the change in population size due to demographic variance scales with population size such that $\mathrm{Var}(\Delta N) = \mathrm{Var}(d)N$, or, alternatively, on a per capita basis

$$\mathrm{Var}\left(\frac{\Delta N}{N}\right) = \frac{\mathrm{Var}(d)}{N}, \tag{1}$$

where $\Delta N/N$ is the per capita change in population size N, and $\mathrm{Var}(d)$—the demographic variance—is the variance in the individual deviations, d, from the expected growth rate. Synonymously, d may be thought of as the deviations in the fitness contributions of the individuals from the expected fitness. Equation 1 shows an important property of demographic variance: the contribution of demographic variance to fluctuations in population size diminishes as population size increases because the demographic variance is divided by population size (see also Fig. 2B). This contrasts with the effects of environmental stochasticity for which the contribution to population fluctuations does not diminish with increasing population size. This diminishing contribution of the demographic variance means that its effect can be largely ignored compared to the environmental variance when the population size is sufficiently large. Precisely how large depends on the ratio of demographic to environmental variance for a particular population, but empirical studies suggest that this is often around a population size of 100, although it ranges over several orders of magnitude.

SOURCES OF DEMOGRAPHIC VARIANCE

There are several sources of demographic variance that combine together to cause variance in population size (Fig. 2). These can be divided into contributions from

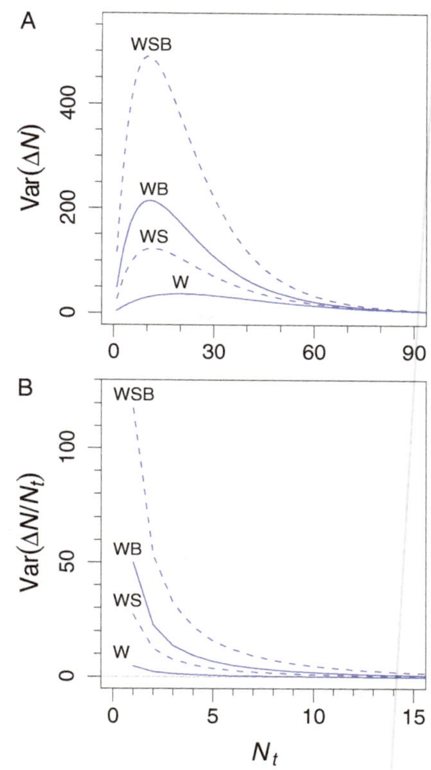

FIGURE 2 How sources of demographic variance combine to cause variance in population size, illustrated with a stochastic Ricker model. In this example, sources of demographic variance can be divided into contributions from probabilistic births and deaths at the within-individual scale (W), probabilistic sex determination (S), and differences in female fecundity at the between-individual scale (B). (A) Stochastic fluctuations in population size measured by $\mathrm{Var}(\Delta N)$. The Ricker model is a discrete-time model, so the change in population size is $\Delta N = N_{t+1} - N_t$. Compared to probabilistic births and deaths alone (W), the stochastic fluctuations are increased by probabilistic sex determination (WS). Stochastic fluctuations are further increased when fecundity differs between females (WSB). (B) As in (A), but with stochastic fluctuations measured on a per capita basis to show the declining importance of demographic stochasticity with population size. Model details: births for individual females were Poisson; variation in intrinsic birth rate between females was gamma with mean 20 and variance 55; density-independent mortality was Bernoulli with probability 0.5; density-dependent mortality was Bernoulli with probability $1 - e^{-\alpha N}$, where N is the size of the population and $\alpha = 0.05$; probability of female offspring was 0.5.

the within-individual scale and the between-individual scale. The classic notion of demographic stochasticity is the contributions from the within-individual scale—the probabilistic births and deaths that occur even if all of the individuals within a population are identical. However, variation between individuals also contributes to the demographic variance, perhaps more so than within-individual variance, although empirical studies of the relative contributions in natural populations are lacking. Variation between individuals can have both stochastic and deterministic components, so it is sometimes called demographic heterogeneity. Examples of demographic heterogeneity include variation in fecundity or survival between individuals due to genotype, body size, life stage, or age. In the case of sessile organisms, such as plants, the immediate environment of the individual can also contribute to between-individual variation.

Sexually reproducing species provide an important example of demographic heterogeneity because there are two types of individuals, males and females, whereas only females can give birth to new individuals. The sex of offspring is probabilistic in most species, giving rise to stochastic variation in the sex ratio of the population (Fig. 3). As for probabilistic births and deaths, variation in the sex ratio is enhanced when the population is small. One strategy often used to avoid this complication is to model only females and to assume that the abundance of males is sufficient for normal reproduction. If the assumption holds, this can be a good strategy, while recognizing that the demographic variance of the total population is necessarily greater than that of the female population alone. For example, when male and female offspring are equally likely, the demographic variance of the total population is often twice as high.

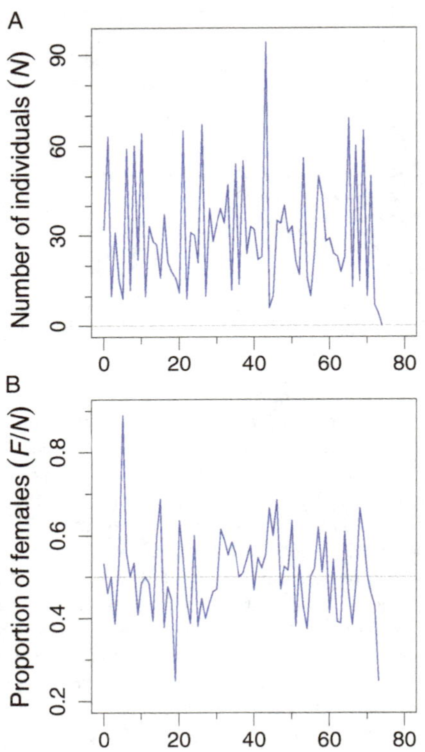

FIGURE 3 One Monte Carlo simulation of a stochastic Ricker model with several sources of demographic variance, including stochastic sex determination. The population went extinct after 74 generations. (A) Fluctuations in total population size. (B) Fluctuations in the proportion of females. The model is the WSB model in Figire 2 with the same details.

However, variance in the sex ratio can increase the demographic variance beyond this simple numerical contribution of males to the population size. There are at least two reasons for this. First, if males contribute to density-dependent regulation, fluctuations in the sex ratio will have a more dramatic effect on female fecundity or survival. For example, if availability of food is a key determinant of female reproductive success, the sex ratio is important because males will reduce fecundity by competing with females for food. Second, a stochastic sex ratio can mean that there are times when male abundance is low enough to reduce the mating success of females. The effect of the sex ratio on mating success is sensitive to the mating system. For example, a high female-to-male sex ratio will result in more unmated females in a monogamous population compared to a polygamous population.

MODELING AND ANALYZING DEMOGRAPHIC STOCHASTICITY

There are several ways to model the stochastic behavior of a population and to incorporate demographic stochasticity. Monte Carlo simulation is the most straightforward approach and can be used with any stochastic model regardless of the complexity of the model. Monte Carlo simulation uses computational algorithms to generate random numbers that simulate the probabilistic events in the stochastic model. In continuous time, the Gillespie algorithm can be used to model demographic stochasticity. In the Gillespie algorithm, a random time interval is generated between the individual birth or death events. This random time interval is assumed to have an exponential distribution to model demographic stochasticity. An example is shown in Figure 1. In discrete-time models of demographic stochasticity, random numbers of births and deaths are generated from probability distributions. Common choices to model demographic stochasticity are the Poisson distribution for births and the binomial distribution for deaths. An example is shown in Figure 3. These approaches for continuous and discrete time explicitly account for the discrete nature of individuals yet do not track each individual. Individual-based models track the state of each individual in the population and may be necessary to model between-individual variation (demographic heterogeneity) or other detailed phenomena such as dispersal behavior. Individual-based models require much greater computation time.

A disadvantage of simulation is that it is difficult to draw generalizations beyond the particular parameter combinations studied because the results are in a numerical format. For a few simple models of demographic stochasticity, exact analytical results can be calculated for some quantities of interest, such as the probability distribution of abundance at a particular time or the mean time to extinction. However, this is not practical or possible for most models. A more general approach uses approximations of the master equation. The master equation is important in the theory of stochastic processes. It describes the evolution in time of the probability distribution of abundance and can be written down for both continuous time and discrete time models. Many classic results for stochastic ecological models have been obtained using the Fokker–Planck approximation of the master equation, also called the diffusion approximation. The Fokker–Planck approximation is most accurate for fluctuations around the carrying capacity and for small growth rates but is often inaccurate otherwise.

FIELD MEASUREMENT OF DEMOGRAPHIC STOCHASTICITY

There are two general approaches to field measurement of demographic stochasticity: observation of individual reproduction and survival, and inference from the dynamics of population size. Key concerns are distinguishing demographic stochasticity from effects of observation uncertainty, density dependence, and environmental stochasticity. Individual observations are the most efficient and precise but are often not possible. A random sample of individuals are observed and the birth of surviving offspring and death of the individual are recorded. Typically, these observations are made within a single year or other period appropriate to the organism's reproductive biology. The demographic variance is then estimated as the sample variance of individual fitness. The demographic variance can be density dependent, meaning the variance in individual fitness is a function of population density. Observations from multiple years together with records of population density are required to measure the effect of density on the demographic variance, as well as to measure environmental stochasticity. Environmental stochasticity is fluctuation in the mean growth rate between years and so is easily distinguished from the demographic variance (e.g., by analysis of variance), although care is needed to separate environmental stochasticity from fluctuations due to density dependence.

A second approach estimates the demographic variance from a time series of population size. In this approach, different probability distributions are used to represent demographic and environmental stochasticity. These probability distributions are fitted to the data. For example, the Poisson distribution might be used to represent purely demographic stochasticity, whereas the negative binomial distribution might be used to represent the combined variation from demographic and environmental stochasticity. In the latter case, the variance parameter of the negative binomial distribution measures the independent contribution of environmental stochasticity, whereas the remaining (Poisson) variation is assumed to be due to demographic stochasticity. To correctly separate environmental and demographic variance, sources of demographic heterogeneity also need to be represented by probability distributions in the statistical model. An obvious drawback of this approach is that a long time series is needed, typically 20 or more time points. However, a long time series is also required to measure density dependence in the demographic variance when only individual fitness is observed. Because of these constraints, it is common to measure demographic variance at only one time point using observations from individuals, even though the demographic variance is unlikely to be constant or density independent.

EXTINCTION

The most important consequence of demographic stochasticity is an increased probability of extinction in small populations. This is not only important for the extinction risk of populations of conservation concern but is fundamental to a wide range of ecological processes. For example, colonization and biological invasion first require small populations to persist, which they do with probability equal to 1 minus the probability of extinction. Similarly, for a disease epidemic to occur, the disease must first avoid extinction when few hosts are infected. Extinction is fundamental also to community processes such as the coexistence of competitors, or predator and prey. To coexist with each other in the long term, species must be able to recover from low population density,

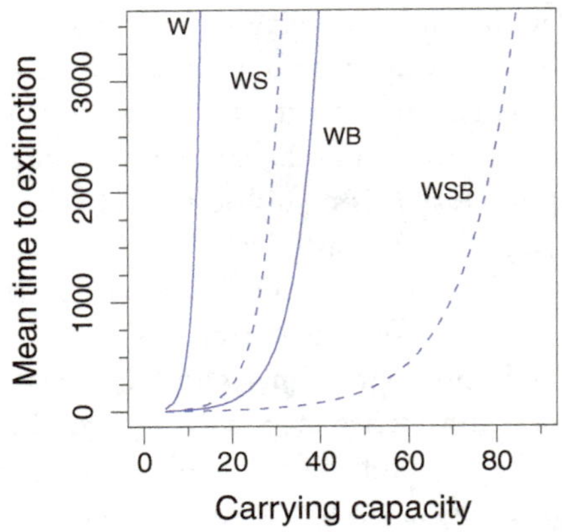

FIGURE 4 The mean time to extinction increases exponentially with carrying capacity due to demographic stochasticity. Populations go extinct more rapidly when the carrying capacity is low. The example is a stochastic Ricker model as in Figure 2. Compared to probabilistic births and deaths alone (W), extinction risk is increased by probabilistic sex determination (WS) and when fecundity differs between females (WB, WSB). Model details were the same as in Figure 2 except: variation in intrinsic birth rate between females was gamma with mean 10 and variance 10; density-independent mortality was Bernoulli with probability 0.6.

at which time the probability of extinction is increased by demographic stochasticity. Finally, extinction of small populations is also important for spatial dynamics, such as in metapopulations, because demographic stochasticity contributes to the extinction of smaller local populations.

For most stochastic models that are ecologically realistic, extinction is ultimately certain (the probability of extinction equals 1) because population size is bounded above by the carrying capacity. This ensures that population size fluctuates between zero and an upper bound and will eventually hit zero at some time because of demographic and environmental stochasticity. On the other hand, even when extinction is certain, the expected time to extinction due to demographic stochasticity alone is often extremely long. For these reasons, the ultimate probability of extinction is often a meaningless measure for assessing extinction risk or comparing models of demographic stochasticity.

There is debate about the best measure to assess extinction in models. One approach, often used to assess the viability of threatened populations, is to calculate the probability of extinction over a defined time interval, say, 100 generations. However, a standard time interval has not been agreed upon and studies cannot be easily compared if they use different time intervals. An alternative measure of extinction is the arithmetic mean time to extinction, *MTE*. This allows models and studies to be compared on the same scale. A related measure is the intrinsic mean time to extinction, T_m, equal to *MTE* when the population dynamics have passed any transient dynamics and reached an established phase. This established phase is called the quasi-stationary distribution in stochastic process theory. The probability of extinction over any time interval can be calculated from T_m together with an additional parameter, c, that describes transient effects or dependence on initial conditions (Box 1). For a wide range of models, the mean time to extinction under demographic stochasticity scales exponentially with carrying capacity, K,

$$MTE = ae^{bK},$$

where a and b are positive constants and their values are determined by the particular model being considered. Thus, as carrying capacity (and hence population size) increases, the mean time to extinction increases exponentially (Fig. 4). The rate of exponential increase depends on the factors contributing to the demographic variance. In particular, between-individual variance keeps the extinction risk high despite an increase in carrying capacity (Fig. 4). The exponential relationship for demographic stochasticity contrasts with environmental stochasticity in which there is a power law relationship between *MTE* and K. Thus, under environmental stochasticity the mean time to extinction increases more slowly with carrying capacity, and hence extinction risk remains high.

CALCULATING EXTINCTION MEASURES

There are several ways to calculate extinction measures for a given model, including simulation, deriving exact results, or approximating the master equation. Monte Carlo simulation is the most straightforward approach. For example, to calculate extinction risk over a defined time horizon, simulate the stochastic model for that time period and record whether the population goes extinct in the simulation. Repeat the simulation many times, each time generating new random numbers. The proportion of simulations that become extinct is an estimate of the extinction risk over that time horizon. For example, for the exponential growth model in Figure 1, 62 out of 1000 simulated populations went extinct within a total period of 60 time units. The calculated probability of extinction due to demographic stochasticity for that time horizon is therefore 0.062 (about 6%). Simulation can be used similarly to estimate the mean time to extinction, *MTE*. However, this brute-force approach to estimate the *MTE* is often not practical for demographic stochasticity since the time to extinction in an individual simulation can be extremely long. Instead, a method called the $\ln(1 - P_0)$ plot can be used to estimate the intrinsic mean time to extinction, T_m, without needing to wait for all runs to go extinct (Box 1).

BOX 1. ESTIMATING THE MEAN TIME TO EXTINCTION

An approach called the $\ln(1 - P_0)$ plot can be used to estimate the intrinsic mean time to extinction, T_m, for a stochastic population model by simulation (Grimm and Wissel, 2004, *Oikos* 105: 501–511). In most stochastic population models, the probability density function of extinction times is well approximated by an exponential structure or is exactly exponential. The $\ln(1 - P_0)$ plot takes advantage of this exponential structure. The approach has four steps:

1. Simulate the model a moderate number of times (e.g., 1000–5000) over a defined time interval (e.g., 1000 years).
2. Calculate $P_0(t)$, the probability of becoming extinct by time t, from the empirical cumulative distribution function for the simulated extinction times.
3. Plot $-\ln(1 - P_0)$ against t. This should show a linear relationship as in panel A. If the relationship is not linear, the $\ln(1 - P_0)$ plot should not be used to estimate T_m.
4. Fit a linear regression to estimate the parameters of this linear relationship (panel A). The inverse of the slope gives the intrinsic mean time to extinction, T_m, while the intercept, c, gives the probability of reaching the established phase.

This approach relies on extrapolating the distribution function to long extinction times that are beyond the simulation time. That is, simulations that went extinct within say the first 1000 years are used to estimate the parameters of the exponential distribution for all possible extinction times (panel B). The approach can also be used when all simulations are allowed to go extinct. Sources of error in the estimation of T_m include Monte Carlo error and extrapolation error. Monte Carlo error can be reduced by increasing the number of replicate simulations.

The example given here is for a stochastic Ricker model with two sources of demographic variance: within-individual variance (probabilistic births and deaths) and between-individual variance generated because the sex of individuals is stochastically determined. (A) The $\ln(1 - P_0)$ plot showing the expected linear relationship and the fitted linear regression. The estimated slope was 1.3×10^{-4}. The inverse of the slope is the estimated intrinsic mean time to extinction T_m = 7719 generations. The estimated intercept was $c = 0.9982$, which is very close to 1.0 and indicates that transient effects were minimal. The model was simulated 5000 times for 1000 generations starting from near the carrying capacity of the population (35 individuals). (B) Comparison of the extrapolated distribution of extinction times from the $\ln(1 - P_0)$ plot with the known distribution, showing that the $\ln(1 - P_0)$ plot accurately extrapolates to the full distribution. The known distribution is represented by a histogram of extinction times from about 67,000 simulations of the model, each time allowing the simulation to run to extinction. The lines show the fitted distribution of extinction times from the $\ln(1 - P_0)$ plot. Red indicates the area where the distribution was fitted by the $\ln(1 - P_0)$ plot, while the dashed line shows the extrapolated distribution. The estimated mean time to extinction from the full simulation was 7705 with standard error 30, so the mean time to extinction of 7719 estimated by the $\ln(1 - P_0)$ plot was within the error bounds and within 0.2% of this more precise estimate. Model details: births for individual females were Poisson with mean 10; density-independent mortality was Bernoulli with probability 0.6; density-dependent mortality was Bernoulli with probability $1 - e^{-\alpha N}$, where N is the size of the population and $\alpha = 0.02$; and probability of female offspring was 0.5.

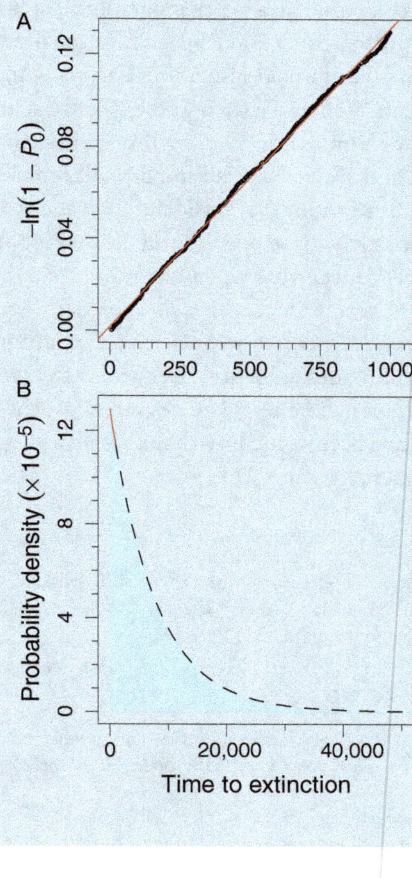

ORIGINS OF THE CONCEPT

One of the earliest questions about demographic stochasticity was posed toward the end of the nineteenth century by the English scientist and statistician Francis Galton. The Victorians were worried that the aristocracy was dying out. Lineages of important families were becoming extinct. It was popular to muse that the comforts enjoyed by the upper classes were leading to the biological decline of the aristocracy. Galton himself entertained this idea but challenged it with an alternative hypothesis (perhaps proposed to him by the French-Swiss botanist Alphonse de Candolle) that the extinction of aristocratic lineages could instead be due to probabilistic births and deaths. Galton and the mathematician Rev. Henry Watson offered a mathematical solution for surname extinction in the *Journal of the Anthropological Institute of Great Britain* in 1874, but it wasn't until 1930 that it was correctly and independently solved by the Danish mathematician Johan Steffensen. A year later in 1931, the American mathematical biologist Alfred Lotka applied Steffensen's solution to U.S. census data—perhaps the first time a model of demographic stochasticity was confronted with data. Lotka determined that the probability of extinction of a male line of descent was 0.88. Demographic stochasticity became a hot topic for about the next 30 years as part of the emerging field of stochastic processes. Fundamental contributions to the idea of demographic stochasticity were made during this period by such luminaries as the English statistician Ronald Fisher, the Croatian-American mathematician William Feller, and the Russian mathematician Andrey Kolmogorov. The English statisticians Maurice Bartlett and David Kendall in particular made many important contributions. Considerations about demographic stochasticity developed during this period include density dependence, disease epidemics, age-structured populations, interspecific competition, predator–prey dynamics, spatially structured populations, maximum likelihood estimation, and analytical derivations for the mean time to extinction. The term "demographic stochasticity" appears to have been coined by the Australian theoretical ecologist Robert May in 1973.

SEE ALSO THE FOLLOWING ARTICLE

Birth–Death Models / Demography / Individual-Based Ecology / Model Fitting / Population Viability Analysis / Ricker Model / Stochasticity, Environmental

FURTHER READING

Kendall, D. G. 1949. Stochastic processes and population growth. *Journal of the Royal Statistical Society Series B: Methodological* 11: 230–282.

Lande, R., S. Engen, and B. E. Saether. 2003. *Stochastic population dynamics in ecology and conservation.* Oxford: Oxford University Press.

Melbourne, B. A., and A. Hastings. 2008. Extinction risk depends strongly on factors contributing to stochasticity. *Nature* 454: 100–103.

Morris, W. F., and D. F. Doak. 2002. *Quantitative conservation biology: theory and practice of population viability analysis.* Sunderland, MA: Sinauer Associates.

Ovaskainen, O., and B. Meerson. 2010. Stochastic models of population extinction. *Trends in Ecology & Evolution* 25: 643–652.

STOCHASTICITY, ENVIRONMENTAL

JÖRGEN RIPA

Lund University, Sweden

No organism lives in a constant environment. For example, air temperature changes from one day to the next, from one summer to the next. The abundance of competitors, prey, parasites, and other kinds of organisms is anything but constant. Such environmental fluctuations are often regarded as stochastic, i.e., random, more or less unpredictable. Variable environmental conditions affect any individual's ability to survive and reproduce, and they thus introduce a stochastic element to the dynamics of any natural population. The way in which environmental stochasticity and biological interactions in concert shape the dynamics of natural populations is of paramount importance for the persistence of species over time and a vibrant field of research.

BASIC CONCEPTS

The "environment" in this context can be many things—essentially any factor affecting an organism's ability to reproduce or survive that is not part of the organism itself or members of the same population. It can be an abiotic factor (physical conditions such as temperature or pH) or biotic factor (other species). This entry discusses environmental factors that vary over time (spatial environmental variation is also very important but not covered here). Temporal environmental variation can be partly predictable, such as seasonal shifts in climate, but it is often best described as stochastic, i.e., to some extent random. Winters in the temperate zones are predictably colder than summers, but the exact mean temperature of any particular winter is currently impossible to predict. If winter temperature is important for the survival of the individuals of a particular population, their abundance will vary as a response to fluctuating winter conditions. This entry focuses on the immediate consequences of stochastic environmental fluctuations in terms of a variable population size. More long-term consequences include possible population extinction or adaptations to the variable conditions, but here they will be touched upon only briefly, if at all.

SINGLE POPULATION DYNAMICS IN A VARIABLE ENVIRONMENT

The growth of a single, isolated population depends on the survival and reproduction of its members. Assuming each individual survives a time unit (say, a year) with probability s and produces on average b surviving offspring during the same time, the growth of the population can be written

$$N_{t+1} = sN_t + bN_t = (s + b)N_t = \lambda N_{t-1}, \qquad (1)$$

where N_t is population size at time t and $\lambda = s + b$ is the per capita growth rate of the population. Note that we assume the population is big enough that the actual survival and reproductive success of single individuals averages out to the fixed rates s and b. For small populations, the impact of individual fates cannot be ignored and demographic stochasticity needs to be taken into account. It is here also for simplicity assumed that all individuals are equivalent after surviving the first year since birth.

Equation 1 describes a population that is constantly increasing ($\lambda > 1$), decreasing ($\lambda < 1$), or simply constant ($\lambda = 1$). However, as environmental conditions change so will each individuals ability to survive and reproduce; i.e., λ will vary over time:

$$N_{t+1} = (s_t + b_t)N_t = \lambda_t N_t. \qquad (2)$$

Here, $\lambda_t = s_t + b_t$ summarizes the temporal variation of all environmental conditions important for survival and reproduction. To be more explicit, one can write

$$\lambda_t = \lambda(\varepsilon_t) = s(\varepsilon_t) + b(\varepsilon_t), \qquad (3)$$

where ε_t is some environmental factor, such as temperature, and both survival and reproduction are supposed to somehow depend on this environmental factor. It is straightforward to include more environmental variables, but one is sufficient here for illustration purposes.

Equation 3 contains the very essence of environmental stochasticity, as it is most often understood among population ecologists—variable conditions lead to variable demographic parameters (here, s and b) and, as we shall see, stochastic population dynamics.

The stochastic environmental factor ε_t is most easily modeled as a sequence of random numbers, independently drawn from some distribution. Figure 1 depicts ten computer simulations of the model described in Equations 2 and 3, starting each simulation at population size $N_0 = 1000$. Each simulation represents a possible future of a population that contains 1000 individuals in year 0. Two things are apparent: First, population dynamics are not deterministic; the same initial condition leads to different final states (e.g., population size at $t = 40$). Second, the spread of

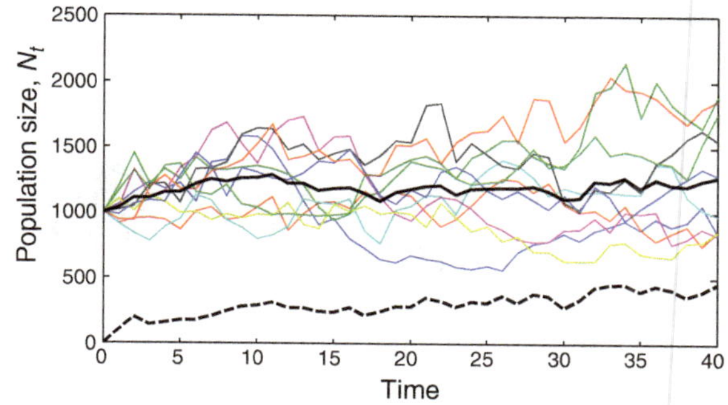

FIGURE 1 Colored lines: ten sample simulations of the model $N_{t+1} = N_t e^{\varepsilon_t}$, where the ε_t are independently drawn from a normal distribution with zero mean and standard deviation 0.1. Thick black line: mean population size at each time point. Thick dashed line: the corresponding standard deviation of population size.

the simulations becomes wider and wider. In other words, predictions are afflicted with increasing amounts of uncertainty as we try to look further and further into the future.

If we calculate at each time point in Figure 1 the mean (thick solid line) and standard deviation (thick dashed line) across all simulations, the mean can be used as a prediction of future population size, if current population size is 1000, and the standard deviation is then a measure of the uncertainty of that prediction. The mean and standard deviation are also a crude description of the stochastic dynamics, although the whole distribution of possible population sizes at any point in time is a more complete, and more useful, description. As an example, the risk of extinction within a given time horizon can be calculated if the full, time-dependent distribution of possible population sizes is known.

The population model used so far has been the simplest possible, ignoring a good deal of biological detail and realism. One problem is the possibility for unlimited population growth, which will be discussed in the following section.

DENSITY-DEPENDENT POPULATION DYNAMICS IN STOCHASTIC ENVIRONMENTS

Stationary Processes and the Stationary Distribution

No population can grow to infinite size—at some point growth is hampered by, e.g., food shortage or the spread of diseases in a high-density population. Such density dependence can be characterized as density-dependent demographic parameters. A natural extension of the model above is then to let b and s depend on both the environmental fluctuations ε_t and population size N_t:

$$N_{t+1} = (s(N_t, \varepsilon_t) + b(N_t, \varepsilon_t))N_t = \lambda(N_t, \varepsilon_t)N_t. \quad (4)$$

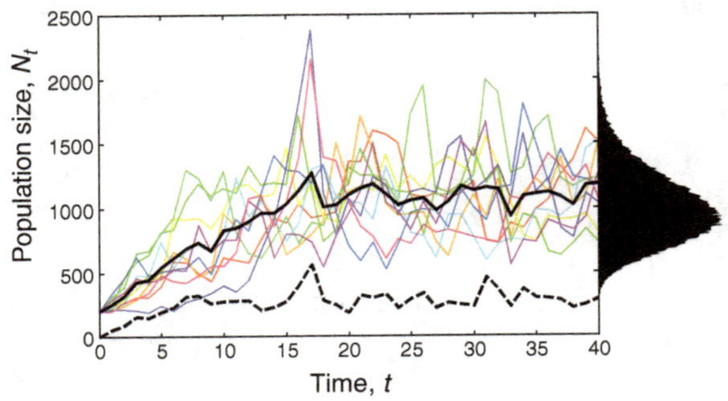

FIGURE 2 Colored lines: ten sample simulations of the stochastic Ricker model

$$N_{t+1} = N_t e^{r\left(1-\frac{N_t}{K}\right)+\varepsilon_t}.$$

Parameters: $r = 0.3$, $K = 1000$, $SD(\varepsilon_t) = 0.2$. Thick black line: mean population size at each time point. Thick dashed line: the corresponding standard deviation of population size. Right margin: the stationary distribution of population sizes, based on 100,000 simulations.

As an example, consider the Ricker equation with a random term ε_t added to the exponent:

$$N_{t+1} = N_t e^{r\left(1-\frac{N_t}{K}\right)+\varepsilon_t}. \tag{5}$$

The stochastic version of the Ricker equation shown here has the per capita growth rate

$$\lambda(N_t, \varepsilon_t) = e^{r\left(1-\frac{N_t}{K}\right)+\varepsilon_t}$$

(cf. Eq. 3). It has two parameters inherited from its deterministic template: *r*, the intrinsic growth rate, and *K*, the carrying capacity; *r* determines the growth rate at low population sizes, and *K* sets the equilibrium population size. Additionally, the stochastic environmental process ε_t has to be specified. Here, all ε_t are drawn independently from a normal distribution with a mean of zero and a standard deviation $\sigma = 0.2$. Figure 2 shows some example simulations together with the time-dependent mean population size (thick solid line) and the corresponding standard deviation (thick dashed line).

The conclusions above from the density-independent case (Fig. 1) still apply to some extent to the simulations in Figure 2. The dynamics are stochastic, and predictions are afflicted with uncertainty that increases over time, at least initially. However, this time the standard deviation levels out considerably and actually reaches a constant level at $SD(N) \approx 280$ (barring random fluctuations). Also the mean reaches a more or less constant level close to the carrying capacity ($K = 1000$). In fact, the whole distribution of population sizes approaches a constant distribution. If we were to run 100,000 simulations instead of just 10 and compare histograms of the population sizes at $t = 40$ and, say, $t = 400$, they would be indistinguishable (see right margin of Fig. 2). The histograms would correspond to the stationary distribution of population sizes, and the population dynamics can be categorized as a stationary process. A stationary process has statistical properties that do not change over time, such as the distribution of possible states at any given time point *t*. The dynamics of most persistent natural populations, living in an environment without temporal trends, are well described as stationary processes and thus have a stationary distribution, a stationary mean, a stationary variance, and so on. The stationary distribution expresses the probability of different population sizes at any time point far enough into the future, when the initial conditions (current population size, environmental state, and the like) no longer have an effect. Given some extra (realistic!) assumptions, the stationary distribution can also be obtained by sampling a single population for a long enough time period. The stationary mean can, e.g., be estimated as the mean of a long time series.

Population Autocorrelation: Characterizing the Dynamics Through Time

Despite its usefulness, the stationary distribution of population sizes is an incomplete description of the dynamics of a population in a stochastic environment. It says very little about what the dynamics actually look like, e.g., how quickly the population recovers from a low density or how it responds to rapid or slow environmental fluctuations.

So what do stochastic population dynamics actually look like and what is the main driver? Are the dynamics just biologically filtered noise, merely a blurred image of the environmental fluctuations? To what extent do ecological interactions add to the signature of the waxing and waning of a population? In some cases, the ecological imprint is obvious and totally dominating, e.g., the high amplitude cycles of some predator–prey systems. In other cases, it is not so easy to separate endogenous from exogenous forces. The theory is too technical to discuss in detail here, but a number of things can be learned by studying the deterministic population dynamics, i.e., the dynamics without any stochasticity (sometimes called the deterministic skeleton).

First, there is almost always an equilibrium population size, i.e., a population size at which the population will remain forever, if left undisturbed. This is the population size at which survival (*s*) and per capita birth rate (*b*) sum up to exactly 1. At equilibrium, each individual

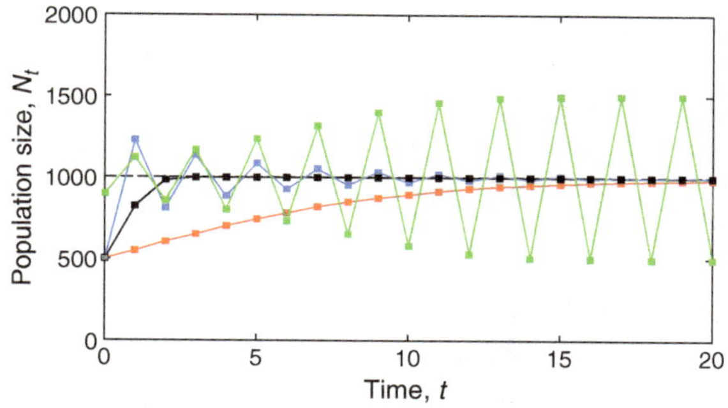

FIGURE 3 Four deterministic simulations of the Ricker model

$$N_{t+1} = N_t e^{r\left(1-\frac{N_t}{K}\right)},$$

all with the same equilibrium population size K = 1000. r = 0.2 (red), r = 1.0 (black), r = 1.8 (blue), r = 2.2 (green).

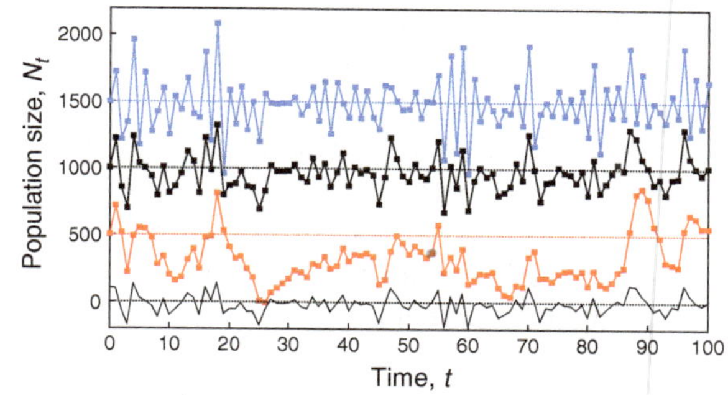

FIGURE 4 Three simulations of the stochastic Ricker model

$$N_{t+1} = N_t e^{r\left(1-\frac{N_t}{K}\right)+\varepsilon_t},$$

all with K = 1000, SD(ε_t) = 0.2. *Red*: r = 0.2. Down-shifted 500 units for clarity. The red dashed line is the down-shifted equilibrium. *Black* (thick line): r = 1.0. *Blue*: r = 1.8, up-shifted 500 units for clarity. The lower, thin, solid black line corresponds to the environmental fluctuations (200 × ε_t).

is exactly replacing itself from one time step to the next, through survival, reproduction, or both. However, any equilibrium is either stable or unstable. If disturbed, i.e., population size is changed by some small amount, a population will return to a stable equilibrium but not to an unstable equilibrium.

In Figure 3, the red, black, and blue curves show example dynamics toward a stable equilibrium (K = 1000), whereas the green curve demonstrates possible dynamics close to an unstable equilibrium—the population is started near the equilibrium but instead of approaching K the population oscillates with increasing amplitude until a stable limit cycle of period 2 is reached. A limit cycle is a periodic attractor of the dynamics. It corresponds to eternal cyclic fluctuations in population size but shares the property of a stable equilibrium that a population will return to a limit cycle if disturbed from it. There is much more to say about unstable population dynamics, but this is not the place. Instead, we will discuss the case of a stable equilibrium in more detail.

For a simple, unstructured model like the Ricker equation, there are two principal ways in which a population can return to equilibrium. The first pattern is a gradual return—if started below the population remains below the equilibrium (Fig. 3, red line), and vice versa if started above the equilibrium (not shown). This is called undercompensatory dynamics—the population does not quite compensate for a disturbance from the equilibrium, at least not by regrowth in a single time step.

The second pattern is overcompensatory dynamics—the population will overshoot the equilibrium after a disturbance and approach the equilibrium in an oscillating manner (Fig. 3, blue line). Overcompensatory dynamics can be found in populations with a strong density dependence, i.e., if population growth rate changes dramatically after a small shift in population size. In the Ricker equation, this is the case when the intrinsic growth rate r is above 1. The boundary case (r = 1, black line, Fig. 3) corresponds to very rapid return to the equilibrium, almost immediate following a small disturbance.

As we let the environmental stochasticity back on again, a population will never be found exactly at the equilibrium; rather, it will bounce around its stationary distribution of population sizes, as described above. However, the deterministic dynamics still make an imprint on the stochastic dynamics—the patterns of over- or undercompensation are still visible. This can be seen in Figure 4, where the population with undercompensatory deterministic dynamics (red) has slower, gradual changes of population size. The population with strong deterministic overcompensation (blue) shows rapid fluctuations, boom-and-bust dynamics not unlike its deterministic dynamics. Note that the two populations are subjected to the same environmental fluctuations (black solid line at bottom).

Expressed in more technical terms, the stochastic dynamics of an undercompensating population has positive autocorrelation, which means there is a positive correlation between consecutive population sizes—if population size at one time is above (below) the stationary mean, it is most likely above (below) the mean in the next time step as well. In contrast, a population with overcompensatory dynamics has negative autocorrelation—if population size is above (below) the mean at one time step it is most likely to overcompensate and be below (above) the mean in the next.

Positive or negative autocorrelation of a population time series is often interpreted as a sign of weak or strong density dependence, respectively. If population growth rate is only weakly dependent on population size, the growth toward equilibrium from a larger or smaller population size will be slow. A strong density dependence, however, means that population growth rate will rapidly decline if the population becomes larger than equilibrium and population decrease will be rapid, possibly even overshooting the equilibrium.

Stability, Stationary Variance, and Population Persistence

Population persistence has long been a central issue of ecological research, applied and theoretical. The simplest theory states that the larger the population, the smaller the risk of extinction. Accordingly, the stationary mean population size is a good predictor of a population's viability. Mean population size is undoubtedly important, but the variation around the mean should not be ignored. A small but nearly constant population may be more persistent than a large and highly variable one. It may thus be of interest to study not only the stationary mean but also the stationary variance or the stationary standard deviation of population sizes (the standard deviation is the square root of the variance. It is many times a preferable measure of population variability since it is has the same unit as, and can be compared with, population size).

The stationary standard deviation of a population's dynamics depends primarily on three things: the strength of environmental fluctuations, the population's sensitivity to them, and the stability of the deterministic dynamics. Naturally, strong environmental fluctuations and/or a high sensitivity to such fluctuations leads to a highly variable population size (from a population perspective, strong fluctuations and high sensitivity imply the same thing: population growth rate will vary a lot due to environmental stochasticity). In addition to the immediate effects of a variable environment, the endogenous dynamics play an important role. A close to unstable population, which takes a long time to return to equilibrium after a disturbance, will exhibit relatively large excursions away from the deterministic equilibrium and a correspondingly large stationary variance. Due to the long return time, single environmental fluctuations will have long-term effects on population size. In this way, consecutive disturbances can accumulate to a large total deviation from the stationary mean. The effect of dynamic instability can be seen by closer examination of Figure 4. The low r example ($r = 0.2$, red) is close to the lower stability boundary ($r = 0$), and the high r case ($r = 1.8$, blue) is close to the upper stability boundary ($r = 2$). Both these cases show larger deviations from the stationary mean (dashed lines) than the intermediate, more stable case ($r = 1$, black). The stationary standard deviations of the three cases are 337 ($r = 0.2$), 203 ($r = 1$), and 289 ($r = 1.8$), which confirms the visual observation.

Tracking the Environment

The idea that stochastic population dynamics merely follows the fluctuations of the environment has little merit in general, as we have seen. There are, however, cases when it does apply. If the population dynamics are compensatory in nature, i.e., if population size rapidly returns to equilibrium after a disturbance, then past disturbances have little impact on present population size and the population will more or less track the current environmental conditions, always close to an equilibrium with the current environment. As an example, compare the dynamics of the population with $r = 1$ (Fig. 4, black line), which has compensatory dynamics, with that of its environment (Fig. 4, thin black line)—the match is very close. In general, any population with faster dynamics than its environment, which has a return time shorter than the time scale of environmental changes, is expected to track the environmental changes closely. In other cases, that is *not* the expectation.

AUTOCORRELATED ENVIRONMENTS: COLORED NOISE

Environmental fluctuations are noisy, hard to predict, but not always completely unpredictable. Especially marine environments are known to vary more slowly than for instance aboveground weather variables. A slowly fluctuating environment means that the present state (e.g., water temperature or the amount of nutrients in a soil) is a good predictor of the state in the immediate future. If it is colder than usual now it will most likely be colder than usual tomorrow or even next year. Just like the dynamics of a population with weak density dependence above, the dynamics of the environment can in this way be positively autocorrelated. Especially the biotic environment, such as the abundance of host plants or predators, may be heavily autocorrelated in aquatic as well as terrestrial environments.

A positively autocorrelated environment changes slowly over time—in a sense it contains fluctuations of low frequency, compared to the uncorrelated case. It is therefore sometimes referred to as red noise, analogous to red light that is low frequency (long wavelength). A negatively autocorrelated environment is likewise called blue noise. The case with no dependence between consecutive environmental states, a mere sequence of independently drawn random numbers, is called white noise, analogous to white light, which contains all frequencies in equal amounts.

What effect does an autocorrelated environment have on population dynamics? First, a positively (negatively) autocorrelated environment generates more positively (negatively) autocorrelated population fluctuations, which is very intuitive. A population living in a very "red," slowly fluctuating environment is thus expected to have "red" dynamics itself, at least more red than the case with a white noise environment. More importantly, the environmental autocorrelation can have a large effect on the stationary distribution of population sizes, most notably on the stationary variance. Some populations have a larger stationary variance in a more red environment, whereas some show the opposite pattern.

For unstructured populations, i.e., where all individuals can be considered equal, the relationship between environmental autocorrelation and the stationary variance of the population dynamics is relatively simple: a population with undercompensatory dynamics will respond with an increasing stationary variance to an increasingly red environment, all else being equal. A population with overcompensatory dynamics will have the opposite response. A more red environment is more easily tracked by any population, in the sense that the population is expected to be found close to the current equilibrium, set by the current environmental state. Why then do under- and overcompensatory dynamics result in so different responses to the redness of the environment? As mentioned above, undercompensatory dynamics corresponds to weak density dependence—population growth rate changes very little with population size. This means that population size will have to change a lot to compensate for a small change of the environment. In other words, a small environmental change will cause a large shift of equilibrium population size. A population with undercompensatory dynamics will thus show large amplitude fluctuations if it is able to track the environmental fluctuations, i.e., if the environmental changes are slow enough. A population with overcompensatory dynamics, strong density dependence, will, on the other hand, exhibit very small fluctuations in abundance in a slowly changing environment. If it is able to track the environment, only small adjustments in population density are necessary to compensate for altered environmental conditions. In a rapidly changing environment, however, a population with overcompensatory dynamics will constantly overshoot the ever-changing environment and consequently exhibit wild, rapid fluctuations of population size.

OUTLOOK

Generalizations

The theory presented here is limited to the dynamics of a single, unstructured population in a stochastic environment. The effects of environmental variability on structured populations (e.g., age structure or spatial structure), communities, or food webs naturally requires a more complicated theory, although some general conclusions still apply. First, the dynamics of any one population is a result of both the ecological interactions and the nature of the environmental fluctuations. Second, close to unstable deterministic dynamics implies high-amplitude dynamics in a stochastic environment. Third, the signature of the deterministic dynamics can be seen in the corresponding stochastic dynamics. There are exceptions to these rules, at least in theory, but they should work as rules of thumb.

The Risk of Extinction

It should be expected that all natural populations are subjected to environmental stochasticity. This inevitably leads to the conclusion that all populations are doomed—sooner or later a sequence of sufficiently unfavorable environmental conditions will drive any population to extinction. The risk of extinction within a given time horizon, however, depends heavily on the stochastic dynamics of that particular population. Reliable risk estimates requires detailed knowledge of the demography of the species, what demographic parameters are affected by environmental stochasticity, how they are affected, but also the nature of the environmental fluctuations (variance, autocorrelation, trends, and so on).

Final Remarks: The Spatial Dimension

There certainly is more to environmental stochasticity than has been touched upon here. Environmental conditions vary across time *and* space, which is of paramount importance to the dynamics, persistence, and geographic distribution of many species. The connection between spatiotemporal environmental fluctuations and the corresponding spatiotemporal dynamics of single populations or sets of interacting populations is a field of high scientific interest but with only fragmented current understanding.

SEE ALSO THE FOLLOWING ARTICLES:

Chaos / Difference Equations / Population Viability Analysis / Resilience and Stability / Ricker Model / Single-Species Population Models / Stochasticity, Demographic / Synchrony, Spatial

FURTHER READING

Box, G. E. P., G. M. Jenkins, and G. C. Reinsel. 1994. *Time series analysis, forecasting and control,* 3rd ed. Englewood Cliffs, NJ: Prentice Hall.

Lande, R., S. Engen, and B.-E. Sæther. 2003. *Stochastic population dynamics in ecology and conservation.* Oxford: Oxford University Press.

Lundberg P., E. Ranta, J. Ripa, and V. Kaitala. 2000. Population variability in space and time. *Trends in Ecology & Evolution* 15: 460–464.

Nisbet, R. M., and W. S. C. Gurney. 1982. *Modelling fluctuating populations.* New York: Wiley.

Ripa, J., and A. R. Ives. 2003. Food web dynamics in correlated and autocorrelated environments. *Theoretical Population Biology* 64: 369–384.

Royama, T. 1992. *Analytical population dynamics.* New York: Chapman & Hall.

STOICHIOMETRY, ECOLOGICAL

JAMES J. ELSER AND YANG KUANG

Arizona State University, Tempe

Ecological stoichiometry is the study of the balance of energy and multiple chemical resources (elements) in ecological interactions. Recall that stoichiometry is a fundamental principle of chemistry dealing with the application of the laws of definite proportions and of the conservation of mass and energy in chemical reactions. Since all organisms are composed of multiple chemical elements such as carbon, nitrogen, and phosphorus brought together in nonaribitrary proportions, the same principles can be brought to bear in the study of fundamental aspects of ecology, including the role of nutrient limitations on growth and trophic interactions, as well as the cycling of chemical elements in ecosystems.

THE CONCEPT

Ecological stoichiometry considers how the balance of energy and elements at various scales affect and are affected by organisms and their interactions in ecosystems. Stoichiometric thinking has a long history in ecology, with fundamental early work contributed by Justus von Leibig (1803–1873), Alfred Lotka (1880–1949), and Alfred Redfield (1890–1983). Most work in ecological stoichiometry focuses on the interface between a consumer and its food resources. For herbivores consuming plants and algae, this interface is often characterized by dramatic differences in the elemental composition (stoichiometric imbalance) of each participant. Ecological stoichiometry primarily asks the following questions: (1) What causes these elemental imbalances in ecological communities? (2) How does consumer physiology and life history respond to them? And (3) What are their consequences for ecological processes in ecosystems?

Elemental imbalances are defined by a mismatch between the elemental demands of a consumer and those provided by its resources. For example, carbon-to-phosphorus (C:P) ratios in the suspended organic matter in lakes (i.e., algae, bacteria, and detritus) can vary between 75 and 1500, whereas C:P ratios of *Daphnia,* a crustacean zooplankter, remain nearly constant at 80:1. This excess of carbon can impose a direct element limitation (in this case by P) on the consumer, as it is unable to consume, extract, and retain enough of the limiting element to achieve maximal growth and reproduction.

A key concept in ecological stoichiometry is stoichiometric homeostasis, the degree to which organisms maintain a constant chemical composition in the face of variations in their environment, particularly in the chemical composition and availability of their food. As in the general biological notion of homeostasis (e.g., for body temperature in homeotherms), elemental homeostasis involves processes of regulation of elemental assimilation and retention that keep elemental composition within some biologically ordered range. Photoautotrophic organisms, such as vascular plants and especially algae, can exhibit a very wide range of physiological plasticity in elemental composition and thus are said to have relatively weak stoichiometric homeostasis. In contrast, other organisms—multicellular animals, for example—have a nearly strict homeostasis and thus can be thought of as having distinct chemical composition.

STOICHIOMETRY OF CONSUMER-DRIVEN NUTRIENT RECYCLING

The ecological interactions between grazers and their food producers encompass a rich set of dynamic relations poorly described by the classic gain/loss view of predators and prey. Herbivores from diverse habitats are known to affect their plant prey in complex ways. Reasons for these more complex dynamics include compensatory growth and nutrient recycling. For example, zooplankton grazers influence the chemical environment experienced by phytoplankton by way of rapid, coupled cycling of nutrients among phytoplankton, zooplankton, and the dissolved pools of these nutrients,.

While energy flow and element cycling are two fundamental and unifying principles in ecosystem theory, many population models ignore the implications of element cycling and constraints on population growth. Such models implicitly assume chemical homogeneity of all trophic levels by concentrating on a single constituent, generally an equivalent of energy or carbon. However, recent modeling efforts consistently show that depicting organisms as built of more than one thing (for example, C and an important nutrient, such as P or N) in stoichiometrically explicit models results in qualitatively different predictions about the resulting dynamics. Because elements are conserved, principles of mass conservation can be conveniently invoked to develop mathmetical formulations of the dynamics of nutrient recycling. Indeed, aided by this law of mass conservation, Sterner in 1990 derived a simplistic but elegant mathematical formula describing the grazer-recycled nutrient (grazer waste) as a simple function of the nutrient quality in the algal pool (food) in the case when grazers strictly maintain their nutrient ratio.

The following example provides a slightly simpler derivation of Sterner's nutrient recycling formula. Let P_N and P_P be the phytoplankton nitrogen (N) and phosphoros (P) pools, respectively; a_N and a_P be the grazers' accumulation efficiencies of N and P, respectively; Z_N and Z_P be the zooplankton N and P pools, respectively; R_{2N} and R_{2P} be the recycling rate of N and P to the dissolved pool feeding the phytoplankton, respectively, and g be the grazing rate. Moreover, let b $(=Z_N/Z_P)$ be the N:P ratio in the zooplankton pool, s $(=R_{2N}/R_{2P})$ be the N:P ratio in the waste recycled to the dissolved pool feeding the phytoplankton, and $f(=P_N/P_P)$ be the ratio of N to P in zooplankton food. It is easy to see that $R_{2N} = g\,P_N(1 - a_N)$ and $R_{2P} = g\,P_P\,(1 - a_P)$. Therefore,

$$s = R_{2N}/R_{2P} = f(1 - a_N)/(1 - a_P). \qquad (1)$$

The assumption of grazers maintaining their tissue N:P ratio at a constant value is equivalent to saying that $b = Z_N/Z_P = a_N\,P_N/a_P P_P$, which implies $b = f a_N/a_P$.

Let L be the maximum possible accumulation efficiency of either N or P (the symbol L is used in honor of Justus Liebig). For both nutrients, the same maximum is used for convenience. When $f > b$, substitute L for a_P and replace a_N by bL/f in Equation 1 to obtain

$$s = f(1 - bL/f)/(1 - L), \qquad (2)$$

which is linear with respect to f. When $f > b$, substitute L for a_N and replace a_P by Lf/b in Equation 1 to obtain

$$s = f(1 - L)/(1 - Lf/b), \qquad (3)$$

which is nonlinear with respect to f. While greatly simplified, the model just described provides a formal means to characterize the phenomenon of differential nutrient recycling by consumers, a process that can alter ecosystem-level nutrient availability and shift the nutrient regime experienced by the consumer's prey (e.g., algae). These effects have been well documented in various laboratory and field studies. Figure 1 shows the predicted recycling by a consumer such as *Daphnia* (N:P ~ 10) and the predicted recycling by a copepod-like consumer (N:P ~ 20) as predicted by the Sterner model (Eqs. 2 and 3). Note that for a given food N:P ratio, the two consumers are predicted to recycle at potentially very different N:P ratios.

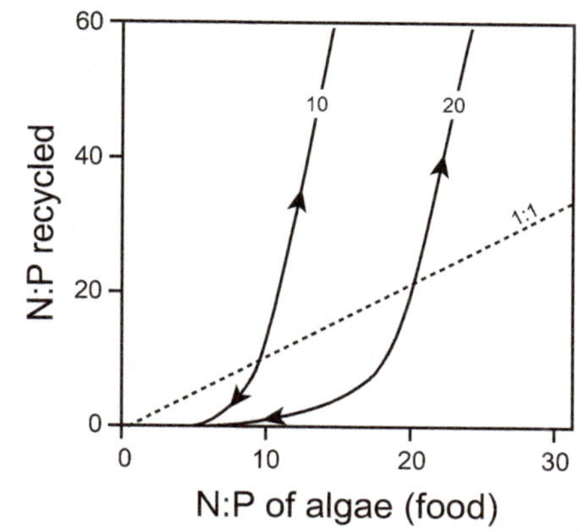

FIGURE 1 The effect of food N:P and consumer N:P on the N:P ratio of consumer-driven nutrient recycling according to Sterner's 1990 model. The predicted N:P ratios of two consumers are shown, assuming a maximum assimilation efficiency of limiting element of 0.90. The figure shows predicted recycling by a consumer such as *Daphnia* (N:P ~ 10), and the other illustrates predicted recycling by a copepod-like consumer (N:P ~ 20). Note that for a given food N:P ratio, the two consumers are predicted to recycle at potentially very different N:P ratios.

STOICHIOMETRY OF FOOD QUALITY AND POPULATION DYNAMICS

Early pioneering work by Andersen introduced stoichiometric principles to population dynamic models, beginning with an explicit assumption that both producer and grazer are composed of two essential elements: carbon and phosphorus. More recently, using stoichiometric principles, Loladze, Kuang, and Elser (2000) constructed a two-dimensional Lotka–Volterra model that incorporates chemical heterogeneity of the first two trophic levels of a food chain. In contrast to nonstoichiometric versions of Lotka–Volterra, the analysis shows that indirect competition between two populations for phosphorus can temporarily shift predator–prey interactions from a conventional gain/loss (+/−) interaction to a loss/loss (−/−) relation. This leads to complex dynamics with multiple positive equilibria, where bistability and deterministic

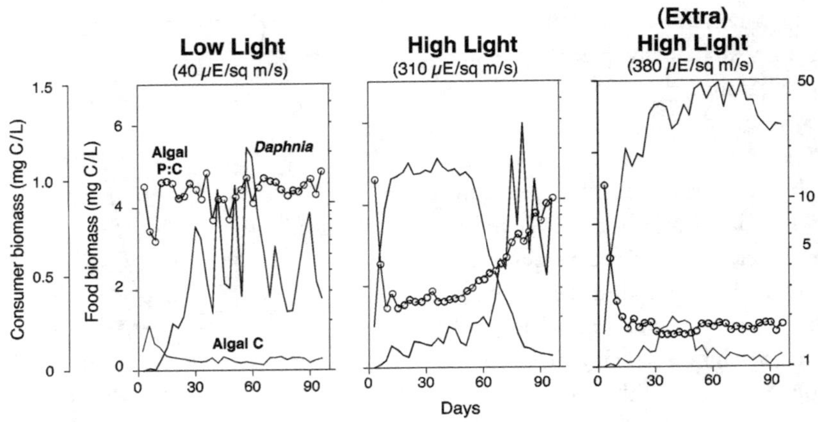

FIGURE 2 The stoichiometric effects of both light and nutrients on the dynamics of herbivorous zooplankter *Daphnia* and its algal prey *Scenedesmus* (modified from Urabe et al., 2002).

extinction of the grazer are possible. Stoichiometric constraints also naturally confine the predator–prey system dynamics to a naturally bounded region. Qualitative analysis reveals that Rosenzweig's paradox of enrichment holds only in situations where the grazer is energy limited; a new phenomenon—the paradox of energy enrichment—may arise in the situations where the grazer is nutrient limited. The paradox of energy enrichment refers to the phenomenon in which intense energy (light) enrichment substantially elevates producer density but, despite such an abundant food supply, the grazer decreases its growth rate and drives itself to deterministic extinction. This surprising model prediction is confirmed by laboratory experimentation (see Fig. 2).

Specifically, the model of Loladze et al. selects phosphorus as the limiting nutrient and makes the following principal assumptions:

(A1) The system has a closed phosphorus cycle containing a mass P of phosphorus.

(A2) The producer's P:C ratio varies but never falls below a fixed minimum, q; the consumer maintains a constant P:C ratio, s.

(A3) All phosphorus in the system is divided into two pools: phosphorus in the consumer and phosphorus in the producer.

From A1 and A2 it follows that producer biomass in the system cannot exceed P/q. Since the consumer requires s grams of phosphorus for every gram of carbon, resource biomass is capped at $(P - sy)/q$. Hence, the effective carrying capacity of producer biomass is $\min(K, (P - sy)/q)$. Maximal transfer efficiency e is achieved if the consumer eats food of optimal quality (i.e., that matches its own stoichiometric composition): if producer P:C is lower than that of the consumer (measured by $(P - sy)/x$) is less than s), then the transfer efficiency is reduced (by (A3)) to $e[(P - sy)/xs]$. In both cases, transfer efficiency equals $e \min(1, (P - sy)/xs)$. One obtains the following model by incorporating such stoichiometrically constrained carrying capacities and transfer efficiencies into the Rosenzweig–McArthur equations:

$$x'(t) = bx(1 - x/\min(K, (P - sy)/q)) - f(x)y,$$
$$y'(t) = e\min(1, (P - sy)/(sx))f(x)y - dy.$$

In this system of equations, because higher resource biomass can imply lower nutrient quality, traditional consumer–resource interactions (+, −) can change to the (−, −) type. The phase plane (Fig. 3) is divided into two regions. In region I, energy (carbon) availability regulates consumer growth. In region II, food quality (phosphorus) controls consumer growth. Increasing energy input into the system may decrease resource quality and stabilize consumer–resource oscillations in region II. If K is increased to an extreme, the system exhibits a "paradox of energy enrichment": despite abundant food, the consumer, faced with low food quality, is destined for deterministic extinction.

Recent efforts have extended this approach to >1 consumer species to evaluate the impacts of stoichiometric constraints on species coexistence.The competitive exclusion principle (CEP) states that no equilibrium is possible if n species exploit fewer than n resources. This principle does not appear to hold in nature, where high biodiversity is commonly observed, even in seemingly homogenous

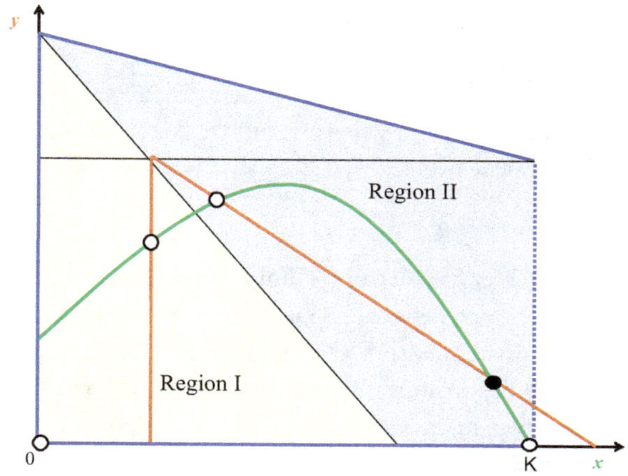

FIGURE 3 Stoichiometric properties divide the phase plane into two regions. In region I, as in the classic Lotka–Volterra model, food quantity limits predator growth. In region II, food quality (phosphorus content of the prey) constrains predator growth. Competition for limiting nutrient between predator and prey alters their interactions from (+, −) in region I to (−, −) in region II. This bends down the predator nullcline in the latter region. This nullcline shape with two x-intercepts creates a possibility for multiple positive steady states, as shown in the figure.

habitats. While various mechanisms such as spatial heterogeneity or chaotic fluctuations have been proposed to explain this coexistence, none of them enitrely invalidates this principle. However, Loladze and colleagues in 2004 showed that stoichiometric constraints can facilitate the stable maintenance of biodiverse communities, as observed in the 2002 empirical work of Urabe and colleagues. Specifically, they showed a stable equilibrium is possible with two predators on a single prey. At this equilibrium, both predators can be limited by the nutrient content of the prey. This suggests that chemical heterogeneity within and among species provides effective mechanisms that can support species coexistence and that may be important in maintaining biodiversity.

DYNAMICAL OUTCOMES AND PREDICTIONS

Ecological stoichiometric theory is an advance in food web ecology because it incorporates both food-quantity and food-quality effects in a single framework and allows key feedbacks such as consumer-driven nutrient recycling to occur. The consequences of these appear to stabilize predator–prey systems while simultaneously producing rich dynamics with alternative domains of attraction and occasionally counterintuitive outcomes, such as coexistence of more than one predator species on a single prey item and decreased herbivore performance in response to increased light availability experienced by the autotrophs. These theoretical findings are consistently supported by recent laboratory and field studies considering stoichiometric effects on autotroph–herbivore systems, such as algae–*Daphnia* interactions. Specifically, empirical evidence for alternative stable states under stoichiometric constraints, negative effects of solar radiation on herbivores via stoichiometric food quality, and diversity-enhancing effects of poor food quality have been produced. Stoichiometric theory provides a promising framework for both quantitative and qualitative improvements in the predictive power of theoretical and computational population ecology, a top priority in light of the alarming multitude and degrees of anthropogenic and natural perturbations experienced by populations.

APPLICATIONS AND EXTENSIONS OF THE THEORY

Ecological stoichiometry seeks to discover how the chemical content of organisms shapes their ecology. Ecological stoichiometry has been applied to studies of nutrient recycling, resource competition, animal growth, and nutrient limitation patterns in whole ecosystems. More recently, it has been extended to evolutionary domains in an attempt to understand the biochemical and evolutionary determinants of observed variations in organisms' C:N:P ratios. Stoichiometric frameworks are equally applicable to the phenomena at the suborganismal level, such as the study of within-host diseases, which include various cancers, as well as phenomena at the whole biosphere level, such as the effects of multiple alterations of biogeochemical cycles on in the biosphere.

SEE ALSO THE FOLLOWING ARTICLES

Allometry and Growth / Biogeochemistry and Nutrient Cycles / Ecosystem Ecology / Energy Budgets / Food Webs / NPZ Models

FURTHER READING

Andersen, T. 1997. *Pelagic nutrient cycles: herbivores as sources and sinks.* New York: Springer-Verlag.

Andersen, T., J. J. Elser, and D. O. Hessen. 2004. Stoichiometry and population dynamics, *Ecology Letters* 7: 884–900.

Elser, J. J., and J. Urabe. 1999. The stoichiometry of consumer-driven nutrient recycling: theory, observations, and consequences. *Ecology* 80: 735–751.

Loladze, I., Y. Kuang, and J. J. Elser. 2000. Stoichiometry in producer-grazer systems: linking energy flow and element cycling. *Bulletin of Mathematical Biology* 62: 1137–1162.

Loladze, I., Y. Kuang, J. J. Elser, and W. F. Fagan. 2004. Competition and stoichiometry: coexistence of two predators on one prey. *Journal of Theoretical Biology* 65: 1–15.

Lotka, A. J. 1925. *Elements of physical biology.* Baltimore: Williams and Wilkins. Reprinted as *Elements of mathematical biology* (1956), New York: Dover.

Sterner, R. W. 1990. The ratio of nitrogen to phosphorus resupplied by herbivores: zooplankton and the algal competitive arena. *American Naturalist* 136: 209–229.

Sterner, R. W., and J. J. Elser. 2002. *Ecological stoichiometry: the biology of elements from molecules to the biosphere.* Princeton: Princeton University Press.

Wang, H., Y. Kuang, and I. Loladze. 2008. A mechanistically derived stoichiometric producer-grazer model. *Journal of Biological Dynamics* 2: 286–296.

Urabe, J., J. J. Elser, M. Kyle, T. Sekino, and Z. Kawabata. 2002. Herbivorous animals can mitigate unfavorable ratios of energy and material supplies by enhancing nutrient recycling. *Ecology Letters* 5: 177–185.

STORAGE EFFECT

ROBIN SNYDER

Case Western Reserve University, Cleveland, Ohio

Environmental conditions vary in space and time, and to the extent that this causes demographic rates to vary (e.g., fecundity, survival, germination), environmental variation affects long-run growth. Variation can have both positive and negative effects on growth. It is well known, for example, that fluctuations in yearly recruitment depress the long-run growth rate. However, variation can also provide opportunities that allow species to persist which would otherwise have been competitively excluded. Variation-dependent coexistence mechanisms are often referred to generically as storage effects, after the best known example of these mechanisms, but there are more properly three ways in which variation can promote coexistence: storage effects, relative nonlinearity, and, for spatial or spatiotemporal variation, growth-density covariance.

THE MUTUAL INVASIBILITY CRITERION FOR COEXISTENCE

Species are deemed able to coexist if each can invade a system dominated by the other. The idea is that if either species were reduced to low density, it could increase its population again—invade—instead of dwindling to extinction. Mathematically, we consider each species in turn as the low-density species (the invader) and set the populations of the other species (the residents) to their stationary stable distributions—i.e., we let the residents come to equilibrium in the absence of the invader. Then we calculate the long-run growth rate of the invader, assuming that the invader is of sufficiently low density that it does not contribute to the competition felt by itself or the residents. If the long-run growth rate is positive for each species in the role of invader, then each species could recover from low density and we conclude that the species will coexist.

STORAGE EFFECT

Storage effects can occur for both spatial and temporal variation. Storage effects happen when the invader experiences low competition in favorable environments and has the ability to store that double benefit. For example, consider two plant species, one of which grows well in cool summers and the other of which grows well in hot summers. An invader of either species will experience low competition in the years it finds favorable, for the resident is its only source of competition and the resident is experiencing an unfavorable year. The high productivity experienced by the invader at these times may be stored in the form of dormant seeds or, in the case of perennials, in stems, bulbs, or tubers. Alternatively, the invader and resident may partition space instead of time: one may prefer shade while the other prefers full sun, for example, or the resident and invader may have different tolerances for soil pH or drainage. The resident and invader may also prefer the same times or locations but to different degrees. For example, both deciduous and evergreen trees are most productive in summer, but deciduous trees are highly productive in summer and not at all in winter, while evergreens are somewhat more productive in summer and somewhat less productive in winter.

RELATIVE NONLINEARITY

Relative nonlinearity can also occur for both spatial and temporal variation. If growth is a nonlinear function of competition (and it normally is), then competition fluctuations above and below average will not affect growth in the same way. More specifically, if growth is a concave-up function of competition, as in Figure 1A, then below-average competition boosts growth more than above-average competition reduces growth. In this case, the population benefits when competition varies from time to time or place to place. If, on the other hand, growth is a concave-down function of competition, as in Figure 1B, then the population is harmed by above-average competition more than it is helped by below-average competition. In this situation, variable competition reduces long-run growth.

Of course, if both the resident and the invader respond to competition in precisely the same way, then their long-run growth rates will be increased or decreased

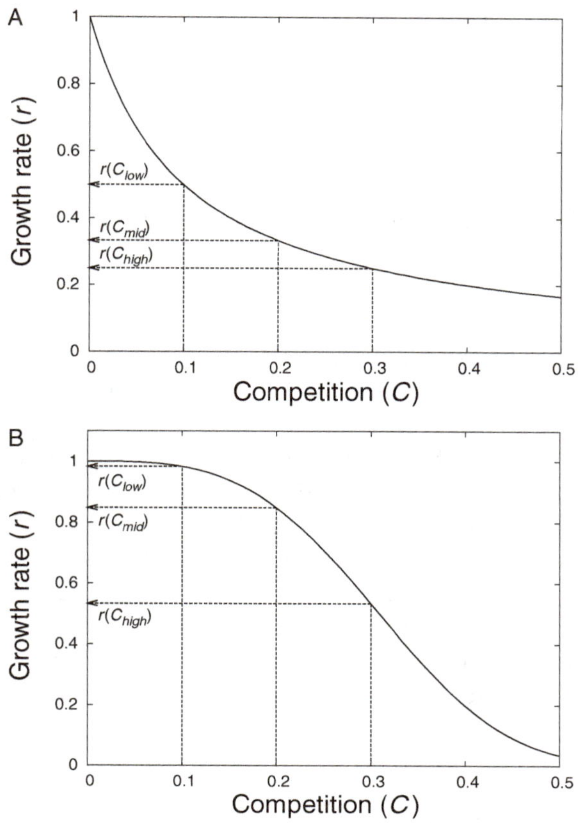

FIGURE 1 Concave-up and concave-down growth functions. (A) When growth is a concave-up function of competition, below-average competition (C_{low}) increases growth more than above-average competition (C_{high}) decreases growth. Here, varying competition increases the long-run growth rate. (B) When growth is a concave-down function of competition, below-average competition increases growth less than above-average competition decreases growth. Here, varying competition decreases the long-run growth rate.

by the same amount, with no net benefit or harm to the invader. This is where the "relative" of relative nonlinearity comes in. If the invader is to experience a net benefit or harm from variable competition, then either its growth must be a different nonlinear function of competition than the resident's or it must experience a different variance in competition. For example, if invader growth is a more sharply concave-up function of competition than the resident's, then the invader will benefit more from the same variance in competition. Alternatively, resident and invader growth may follow the same nonlinear concave-down function of competition, but the invader may experience more variable competition, so that the invader again benefits. Even though the resident population drives competition for both the resident and the invader, the populations need not experience the same levels of competition. The competitive effect of residents may be stronger for one species than the other, or, if competition is spatially distributed, residents and invaders may compete over different distances, so that one experiences diffuse competition from residents within a large radius and the other experiences intense competition within a small radius.

GROWTH-DENSITY COVARIANCE

Unlike the storage effect and relative nonlinearity, growth-density covariance is a strictly spatial phenomenon. The invader long-run growth rate will get a boost if the invader is better at concentrating its population in high growth locations (favorable environment, low competition, or both) than the resident is—that is, if it has a more positive covariance between local growth rate and population density. Put another way, growth density covariance measures habitat association. For example, a plant that specializes on serpentine soils may have seeds that land very close to their mother, to ensure that most seeds remain close to the serpentine outcropping. Such a species would get a large boost from growth-density covariance when competing against a resident with broadly dispersed seeds.

WHEN IS EACH COEXISTENCE MECHANISM LIKELY TO BE IMPORTANT?

Growth-density covariance can be a powerful coexistence mechanism, outstripping the possibilities of the storage effect or relative nonlinearity. However, for it to be effective, there must be long-lasting spatial heterogeneity, populations must be able to aggregate in favorable areas, and competitors must prefer different areas, or at least prefer the same areas to different degrees. Thus, growth-density covariance can be an important mechanism promoting the persistence of specialists, for example. Among species that rely on growth-density covariance, we also expect to see traits that permit aggregation, such as short-range dispersal or vegetative growth.

Of course, sometimes, temporal variation is more prominent. Desert ecosystems are often organized around short, infrequent rainstorms, for example. When temporal variation is prominent, the storage effect can be the dominant coexistence mechanism. In this case we expect to see long-lived life stages (e.g., dormant seeds or resting eggs, "sapling banks," long-lived adults) and/or resource storage mechanisms (e.g., tubers, succulency). See Chesson et al. (2004) for an excellent discussion of how plants in arid and semi-arid environments use variable rainfall to coexist. Of course, the storage effect is not solely a temporal mechanism. However, when there is spatial variation available, growth-density covariance is commonly stronger than the spatial storage effect.

Relative nonlinearity tends not to be a dominant coexistence mechanism. Nonetheless, it can be important when competition produces oscillatory or chaotic dynamics, as has been observed, e.g., in phytoplankton (Huisman and Weissing, 1999).

MATHEMATICAL SUMMARY

Ultimately, each of these coexistence mechanisms is defined mathematically. Let us first consider variation in time only. In continuous time,

$$\frac{dn_j}{dt} = r_j(t)n_j(t), \tag{1}$$

where $n_j(t)$ is the population size of species j and $r_j(t)$ is its instantaneous growth rate. In discrete time, $n_j(t + 1) = \lambda_j(t)n_j(t)$, where $\lambda_j(t)$ is the multiplicative growth rate, and we find the instantaneous growth rate by taking the logarithm:

$$\ln n_j(t + 1) = r_j(t) + \ln n_j(t),\ r_j(t) = \ln \lambda_j(t). \tag{2}$$

In both cases, the long-run growth rate $\bar{r}_j$ is the time average of the instantaneous growth rate: $\bar{r}_j = \langle r_j \rangle t$.

We assume that as a result of fluctuating environmental conditions, some demographic parameter varies in time, e.g., fecundity or juvenile survival. We'll refer to this parameter as the environment and denote it by $E_j(t)$, even though technically it is a response to the environment. Assuming that the environmental variation drives variation in the resident population, competition $C_j(t)$ also varies with time. We thus write the instantaneous growth rate as a function of the environment and competition: $r_j(t) = g_j[E_j(t), C_j(t)]$. The first thing to do is to change variables from E_j and C_j to $\mathscr{E}_j$ and $\mathscr{C}_j$, where $\mathscr{E}_j(t)$ represents the direct effects of the environment on growth rate and $\mathscr{C}_j(t)$ represents the effects of competition on growth. We write

$$\mathscr{E}_j(t) = g_j[E_j(t), C_j^*), \tag{3}$$

$$\mathscr{C}_j(t) = -g_j[E_j^*, C_j(t)], \tag{4}$$

where E_j^* and C_j^* are constants chosen so that $g_j[E_j^*, C_j^*] = 0$—i.e., when the environment and competition are held constant at E_j^* and C_j^*, respectively, the population is at equilibrium. The constant E_j^* is usually chosen to be the mean value of the environment.

We then write the instantaneous growth rate in terms of the direct effects of the environment, the effects of competition, and effects from interactions between the environment and competition:

$$r_j(t) \approx \mathscr{E}_j(t) - \mathscr{C}_j(t) + \gamma_j \mathscr{E}_j(t)\mathscr{C}_j(t), \tag{5}$$

where

$$\gamma_j = \left.\frac{\partial^2 r_j}{\partial \mathscr{E}_j \partial \mathscr{C}_j}\right|_{E_j^*, C_j^*}. \tag{6}$$

The long-run growth rate $\bar{r}_j$ is the time average of Equation 5. We thus arrive at

$$\bar{r}_j = \langle \mathscr{E}_j \rangle_t - \langle \mathscr{C}_j \rangle_t + \gamma_j \mathrm{Cov}(\mathscr{E}_j \mathscr{C}_j)_t, \tag{7}$$

where I have used the fact that, to the order of approximation we have been using, $\langle \mathscr{E}_j \mathscr{C}_j \rangle_t = \mathrm{Cov}(\mathscr{E}_j, \mathscr{C}_j)_t$, the time covariance of $\mathscr{E}_j$ and $\mathscr{C}_j$.

This is as far as we can go in general. We now turn our attention specifically to the invader long-run growth rate ($\bar{r}_i$) and the resident long-run growth rate ($\bar{r}_r$). The resident population will grow until it reaches a stationary temporal distribution and its long-run growth rate $\bar{r}_r$ is zero. The abilities of the resident thus set the scale for invader growth: asking whether the invader long-run growth rate $\bar{r}_i$ is positive is equivalent to asking whether invader growth is larger than resident growth. This suggests that we subtract the resident version of Equation 7 from the invader version, creating a term by term comparison. We will actually be slightly more devious and subtract a constant q_{ir} times $\bar{r}_r$ from $\bar{r}_i$. (Remember, $\bar{r}_r$ is zero, so we may multiply it by whatever we like.) Thus,

$$\bar{r}_i = \bar{r}_i - q_{ir}\bar{r}_r = \Delta E - \Delta C + I, \tag{8}$$

where

$$\Delta E = \langle \mathscr{E}_i \rangle_t - q_{ir} \langle \mathscr{E}_r \rangle_t, \tag{9}$$

$$\Delta C = \langle \mathscr{C}_i \rangle_t - q_{ir} \langle \mathscr{C}_r \rangle_t, \tag{10}$$

$$\Delta I = \gamma_i \mathrm{Cov}(\mathscr{E}_i, \mathscr{C}_i)_t - q_{ir}\,\gamma_r \mathrm{Cov}(\mathscr{E}_r, \mathscr{C}_r)_t. \tag{11}$$

The last term, ΔI, is the storage effect. If the invader experiences low competition in favorable environments (possibly because the resident finds those environments unfavorable), then $\mathrm{Cov}(\mathscr{E}_i, \mathscr{C}_i)_t$ will be negative. Resident individuals, on the other hand, will be crowded by fellow residents during favorable times, so $\mathrm{Cov}(\mathscr{E}_r, \mathscr{C}_r)_t$ will be positive. Thus, ΔI increases long-run growth if γ_i and γ_r are negative. A negative γ means that an increase in competition reduces growth less in an unfavorable environment than in a favorable environment (Fig. 2). In practice, γ will be negative if the organism can store the gains made in good environment–low competition times. For example, γ normally becomes more negative as seed dormancy or adult survival increase.

We can squeeze just a bit more meaning from these expressions if we Taylor expand $\mathscr{C}_r$ and $\mathscr{C}_i$ about $\mathscr{C}_r^*$ to second order and make a judicious choice for q_{ir}. For example, $\mathscr{C}_i$ is a function of invader competition, C_i. If we call that function $\phi_i(C_i)$, then we can say $\mathscr{C}_i \approx \phi_i(C_r^*) + \frac{d\phi_i}{dC_i}\Big|_{C_r^*}(C_i - C_r^*) + \frac{1}{2}\frac{d^2\phi_i}{dC_i^2}\Big|_{C_r^*}(C_i - C_r^*)^2$. Thus,

$$\Delta C \approx \Big(\phi_i(C_r^*) - \underbrace{q_{ir}\,\phi_r(C_r^*)}_{=0}\Big) + \left(\frac{d\phi_i}{dC_i}\Big|_{C_r^*}\langle(C_i - C_r^*)\rangle_t - q_{ir}\frac{d\phi_r}{dC_r}\Big|_{C_r^*}\langle(C_r - C_r^*)\rangle_t\right) + \left(\frac{1}{2}\frac{d^2\phi_i}{dC_i^2}\Big|_{C_r^*}\underbrace{\langle(C_i - C_r^*)^2\rangle_t}_{\approx \mathrm{Var}(C_i)} - q_{ir}\frac{1}{2}\frac{d^2\phi r}{dC_r^2}\Big|_{C_r^*}\underbrace{\langle(C_r - C_r^*)^2\rangle_t}_{\approx \mathrm{Var}(C_r)}\right). \quad (12)$$

The final, quadratic terms compare how nonlinear invader growth is to how nonlinear resident growth is, as functions of competition. We would like to highlight this term, because we know that this nonlinearity can increase or decrease long-run growth. First, we choose q_{ir} to make the linear term disappear:

$$q_{ir} = \frac{d\mathscr{C}_i}{d\mathscr{C}_r}. \quad (13)$$

(More details on this choice can be found in Chesson, 1994, sec. 4.2.) Then we combine the constant term $\phi_i(C_r^*)$ with ΔE. The constant term will be nonzero if the resident and invader reach equilibrium at different levels of competition. This can happen as a result of variation-independent coexistence mechanisms, such as resource partitioning. Thus,

$$\bar{r}_i = \bar{r}_i' - \Delta N + \Delta I, \quad (14)$$

where

$$\bar{r}_i' = \langle \mathscr{E}_i\rangle_t - q_{ir}\langle\mathscr{E}_r\rangle_t - \phi_i(C_r^*) \quad (15)$$

represents variation-independent coexistence mechanisms and the direct effects of the environment,

$$\Delta N = \frac{1}{2}\frac{d^2\phi_i}{dC_i^2}\Big|_{C_r^*}\mathrm{Var}(C_i)_x - q_{ir}\frac{1}{2}\frac{d^2\phi_r}{dC_r^2}\Big|_{C_r^*}\mathrm{Var}(C_r)_t \quad (16)$$

measures relative nonlinearity, and the storage effect, ΔI, is defined in Equation 9.

When the environment varies in space instead of time, the derivation is very similar except that instead of the long-run growth rate, we are now interested in the regional growth rate. To save space, let us only consider the case of discrete-time dynamics.

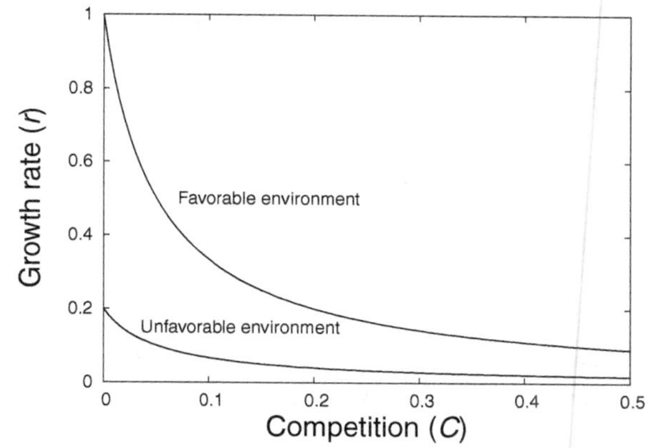

FIGURE 2 Example of negative γ. When γ (defined in Eqs. 6, 23) is negative, an increase in competition reduces growth less in an unfavorable environment than in a favorable environment. The storage effect increases the invader's long-run growth rate only if γ is negative.

Suppose that $\lambda_j(x)$ is the local multiplicative growth rate, so that

$$n_j(x, t+1) = \lambda_j(x) n_j(x, t). \quad (17)$$

Then the spatially averaged population obeys the equation

$$\langle n_j\rangle_x(t+1) = \langle\lambda_j n_j\rangle_x(t) = \langle\lambda_j\rangle_x\langle n_j\rangle_x(t) + \mathrm{Cov}(\lambda_j, n_j)_x(t). \quad (18)$$

We can squeeze this into the form (spatially averaged population at time $t+1$) = (regional growth rate) $\times$ (spatially averaged population at time t) by pulling out a factor of $\langle n_j\rangle_x$ from the covariance. We write $\langle n_j\rangle_x(t+1) = \left[\langle\lambda_j\rangle_x(t) + \mathrm{Cov}\left(\lambda_j, \frac{n_j}{\langle n_j\rangle_x}\right)_x(t)\right]\langle n_j\rangle_x(t)$, so that the regional growth rate is

$$\tilde{\lambda}_j(t) = \langle\lambda_j\rangle_x(t) + \mathrm{Cov}(\lambda_j, \nu_j)_x(t), \quad (19)$$

where $\nu_j(x, t) = n_j(x, t)/\langle n_j\rangle_x(t)$. Note that $\mathrm{Cov}(\lambda_j, \nu_j)_x$ is positive if locations with higher than average growth have higher than average populations. This is the origin of the spatial coexistence mechanism growth-density covariance.

The derivation now proceeds much as before. We write

$$\mathscr{E}_j(x) = g_j(E_j(x), C_j^*) - 1, \quad (20)$$

$$\mathscr{C}_j(x, t) = 1 - g_j(E_j^*, C_j(x, t)), \quad (21)$$

where $g_j(E^*_j, C_j^*) = 1$—i.e., when the environment and competition are held constant at E_j^* and C_j^*, the population is at equilibrium. We write the regional growth rate as

$$\tilde{\lambda}_j(t) - 1 = \langle\lambda_j\rangle_x(t) + \mathrm{Cov}(\lambda_j, \upsilon_j)_x(t) \approx \langle\mathscr{E}_j\rangle_x - \langle\mathscr{C}_j\rangle_x + \gamma_j\langle\mathscr{E}_j\mathscr{C}_j\rangle_x + \mathrm{Cov}(\lambda_j, \upsilon_j)_x(t), \quad (22)$$

where

$$\gamma_j = \left.\frac{\partial^2\lambda_j}{\partial\mathscr{E}_j\partial\mathscr{C}_j}\right|_{E_j^*, C_j^*}. \quad (23)$$

Subtracting the resident regional growth (which is 1 at equilibrium) from the invader's, we find

$$\tilde{\lambda}_i \approx \tilde{\lambda}_i' - \Delta N + \Delta I + \Delta\kappa, \quad (24)$$

where variation-independent coexistence mechanisms and the direct effects of the environment are measured by

$$\tilde{\lambda}_i' = \langle\mathscr{E}_i\rangle_x - q_{ir}\langle\mathscr{E}_r\rangle_x - \phi_i(C_r^*), \quad (25)$$

relative nonlinearity is given by

$$\Delta N = \frac{1}{2}\left.\frac{d^2\phi_i}{dC_i^2}\right|_{C_r^*} \mathrm{Var}(C_i)_x - q_{ir}\frac{1}{2}\left.\frac{d^2\phi_r}{dC_r^2}\right|_{C_r^*} \mathrm{Var}(C_r)_x, \quad (26)$$

the storage effect is given by

$$\Delta I = \gamma_i \mathrm{Cov}(\mathscr{E}_i, \mathscr{C}_i)_x - q_{ir}\gamma_r \mathrm{Cov}(\mathscr{E}_r, \mathscr{C}_r)_x, \quad (27)$$

and growth-density covariance is given by

$$\Delta\kappa = \mathrm{Cov}(\lambda_i, \upsilon_i)_x - q_{ir}\mathrm{Cov}(\lambda_r, \upsilon_r)_x. \quad (28)$$

A more detailed but still concise derivation is given in the appendix of Snyder and Chesson (2004), and the original presentation is given in Chesson (2000).

SEE ALSO THE FOLLOWING ARTICLES

Demography / Environmental Heterogeneity and Plants / Invasion Biology / Stochasticity, Environmental / Two-Species Competition

FURTHER READING

Chesson, P. 1994. Multispecies competition in variable environments. *Theoretical Population Biology* 45: 227–276.

Chesson, P. 2000. General theory of competitive coexistence in spatially-varying environments. *Theoretical Population Biology* 58: 211–237.

Chesson, P., R. L. Gebauer, S. Schwinning, N. Huntly, K. Wiegand, M. S. Ernest, A. Sher, A. Novoplansky, and J. F. Weltzin. 2004. Resource pulses, species interactions, and diversity maintenance in arid and semi-arid environments. *Oecologia* 141: 236–253.

Huisman, J., and F. J. Weissing. 1999. Biodiversity of plankton by species oscillations and chaos. *Nature* 402: 407–410.

Snyder, R. E., and P. Chesson. 2004. How the spatial scales of dispersal, competition, and environmental heterogeneity interact to affect coexistence. *American Naturalist* 164: 633–650.

STRESS AND SPECIES INTERACTIONS

RAGAN M. CALLAWAY

University of Montana, Missoula

In nature, the fitness of a species in any environment is rarely, if ever, at its optimum. In other words, organisms are almost always limited in some way by their environment, and this limitation is often encapsulated by ecologists and evolutionary biologists in the term "stress." Although the definition of stress is not as precise as other terms or concepts in these disciplines, understanding how organisms respond to environmental and biotic limitations is foundational to ecology and evolution.

STRESS AND SPECIES DISTRIBUTIONS

The abundances of most organisms appear to be very roughly distributed in the forms of bell-shaped curves along gradients of environmental variability, often referred to as the fundamental niche. Thus there are places on gradients of moisture, light, temperature, and so on, where the performance of a species is better than on other parts of gradients, such as at the ends of the curves where stress is more extreme. What is stressful for some organisms is of course not stressful for others; cold growing temperatures reduce the performance of alpine plants much less than desert plants. But despite these relative, or conditional, differences, there is a great deal of evidence from many systems that the addition of resources or amelioration of abiotic factors can improve the growth or reproduction of organisms. Species demonstrate physiological responses along abiotic gradients, respond through plasticity (variation in the phenotype of a given genotype), or show ecotypic differentiation. Populations can also respond to stress, defined as environmental change that causes some response by the population of interest. Responses by individuals or by populations have the potential to create significant variation in the way species interact with each other on gradients.

INTERACTIONS AMONG SPECIES

There is a great deal of empirical evidence that interactions between species can influence their distributions, either positively or negatively, by modifying abiotic limitations, such as those that occur along gradients, producing the realized niche. Importantly, different degrees of abiotic limitation, or stress, can alter the nature of interactions among species. In one of the first tests of the relative importance of interactions among species and abiotic stress between species on gradients, Joe Connell found that the lower limit of a marine intertidal barnacle was mainly determined by competition, while the upper limit was set by the species' tolerance to the extreme abiotic stress of desiccation. Competition was not strong where stress was intense. More complex responses are shown by cattail (*Typha*) species. The abundances of *Typha latifolia* and *T. angustifolia* show the typical but partially overlapping bell-shaped curves along elevation gradients in freshwater ponds. Experiments demonstrated that *T. latifolia,* which occurs in deeper water than its conspecific, could physically tolerate conditions along the entire gradient but was restricted from shallow water by competition with *T. angustifolia*. In contrast, *T. angustifolia* was restricted from deep water by its inability to tolerate stress. These kinds of general relationships between stress, performance, and interactions among species were formalized by J. Philip Grime, who hypothesized that competition is a primary determinant of species abundance and fitness in environments characterized by high productivity, but when productivity decreases and environmental stress increases, the role of competition wanes and stress tolerance becomes more important (Fig. 1). Later analyses indicated that the "intensity" of competition between two species was less likely to change along gradients than the "importance" of competition relative to the effects of the abiotic environment.

Grime's ideas about stress and interactions among species also provided a conceptually useful way to extend the idea of stress from factors affecting individual plants or species to factors affecting communities—the idea that productivity could be a surrogate for stress. This is important because productivity as a surrogate allows us to scale up the application of stress from the less ambiguous growth limitations of individuals to the more fuzzy limitations of communities. Grime's formal organization of abiotic stress and competition also led to a central hypothesis for life history evolution. This is the idea that tradeoffs between stress tolerance and competitive ability in plant life-history evolution are constrained by fundamental "compromises between the conflicting selection pressures resulting from particular combinations of competition, stress, and disturbance," with stress being defined as abiotic conditions that restrict production.

Elevational gradients provide model stress gradients because factors such as temperature, wind, and soil disturbance are moderate, and appear to be less limiting to plant growth, at low elevations. Thus, at low elevations

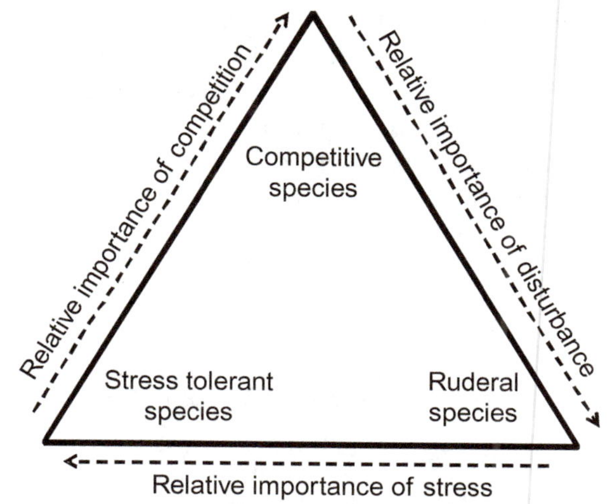

FIGURE 1 General relationships between the relative importance of biotic and abiotic factors and the general strategies expressed by plant species, formalized by J. Philip Grime.

plants may be more likely to be able to grow to the point where further growth or reproduction is limited by resources, and thus the effects of other plants on resources might have profound effects. However, at high elevations, stress factors may limit plant growth more than resource availability, and thus amelioration of these stresses by neighboring species may increase growth more than competition for resources with neighbors reduces growth. Shifts from strong competition to weak competition, or to facilitation (the positive effect of one species on another), have been demonstrated in deserts, alpine tundra, grasslands, and other biomes around the world.

STRESS AND ECOTYPIC VARIATION

Almost all studies of how plants interact along abiotic gradients focus on the species as the unit of interest. However, other studies along abiotic gradients clearly show that species, and especially those that are distributed across large gradients, are composed of ecotypes. Ecotypes are genetically distinct varieties within species and can represent adaptive or evolutionary responses to different stresses. In a classic study of such intraspecific adaptation to stress, Jens Clausen, David Keck, and William Hiesey observed that some California species could be found at sea level and in alpine vegetation. They hypothesized that these species consisted of "complexes" and transplanted individuals from different places on a 3000-m altitudinal gradient to other places on the same gradient. Many of these individuals, when transplanted lower or higher in elevation than their original location, performed more poorly than individuals that occurred naturally at the same location; in other words, they were adapted to particular places on the gradient and experienced substantial stress at other places on the gradient. They concluded that evolutionary processes have established stages in "evolutionary differentiation." We know little about how ecotypic, or evolutionary differentiation within a species, affects the theory drawn from large numbers of ecological studies along gradients.

SEE ALSO THE FOLLOWING ARTICLES

Adaptive Dynamics / Facilitation / Phenotypic Plasticity / Plant Competition and Canopy Interactions / Regime Shifts

FURTHER READING

Brooker, R., Z. Kikvidze, F. I. Pugnaire, R. M. Callaway, P. Choler, C. J. Lortie, and R. Michalet. 2005. The importance of importance. *Oikos* 109: 63–70.

Callaway, R. M., R. W. Brooker, P. Choler, Z, Kikvidze, C. J. Lortie, R. Michalet, L. Paolini, F. I. Pugnaire, B. Newingham, E. T. Aschehoug, C. Armas, D. Kikidze, and B. J. Cook. 2002. Positive interactions among alpine plants increase with stress. *Nature* 417: 844–848.

Calow, P. 1989. Proximate and ultimate responses to stress in biological systems. *Biological Journal of the Linnean Society* 37: 173–181.

Clausen, J., D. D. Keck, and W. M. Hiesey. 1941. Regional differentiation in plant species. *American Naturalist* 75: 231–250.

Connell, J. H. 1961. The influence of interspecific competition and other factors on the distribution of the barnacle *Chthamalus stellatus*. *Ecology* 42: 710–723.

SUCCESSION

HERMAN H. SHUGART

University of Virginia, Charlottesville

Ecological succession refers to the sequence of ecological communities that develops as a location recovers from a disturbance (secondary succession) or at locations where new substrate has been exposed (primary succession). Secondary succession includes the progression of communities following the abandonment of farmed fields (old field succession) or in the wake of a wildfire. Primary succession occurs on such surfaces as a fresh lava flow or a newly formed riverine sandbar.

THE ORIGIN OF THE CONCEPT

Ecological succession as a concept has its origin in a reported bog succession for Irish bogs as early as 1685, with other papers and dissertations to follow on Dutch peat bogs, French *Sphagnum* bogs, and others. The initial records of forest succession are from Buffon in 1742. The idea of successional change was likely not a novel concept with these papers. These and other early accounts typically reported succession from one broad type of vegetation (*Sphagnum* bog, grassland, shrubland, forest, and the like) with the mention of some example species found in these habitats.

The recognition of succession as a focal topic was a logical outgrowth of classification of vegetation associated with the explorations of European naturalists in the nineteenth century. This interest in classification stemmed from development of regional vegetation maps. While earlier maps with military and economic foci showed locations of forests, the actual mapping of plants and plant communities on maps was a product of such naturalists as de Candolle and Lamarck, with an earlier map of the biogeography of France in 1805, and, influentially, by Alexander von Humboldt, with his transverse view map of the Mount Chimborazo in Ecuador in 1807. A transverse map illustrates the positions of types of vegetation or plant species along an elevation gradient. Humboldt had a profound influence on the plant ecologists to follow, and these wrestled with issues that seem current today.

The question immediately arises, "What should one map?" Humboldt mapped mostly the distribution of species, while de Candolle and Lamarck mapped floristic provinces (Mediterranean plants, alpine plants, and so on). Grisebach in 1872 produced the first world-scale vegetation map and showed biomes with similar physical appearance (rainforest, prairies, and so on) even when these were dominated by species that were taxonomically unrelated. He also developed some 54 forms of vegetation (bamboo, banyan, liana, annual grass, and so on) for understanding the physiognomy-based classifications of vegetation that he used. Warming in 1895 and Drude in 1906, toward the end of the historical progression of German phytogeographers, pointed out that understanding the dynamics of the vegetation was an essential element of what one must know to understand vegetation. Including the dynamics of vegetation on a map is extremely difficult simply because of the two-dimensional nature of paper. Many of the larger area maps of vegetation today display "potential vegetation," or the vegetation one might expect in a region after the dynamic processes had come to some form of equilibrium.

By the dawn of the twentieth century, the combined problem of vegetation classification and including the underlying dynamics of the vegetation became an important topic as land conversion, land abandonment, clearing of forests, overgrazing of grasslands, and a variety of other human actions on landscapes became increasingly important as a topics to understand. There were three different theoretical threads developed from this challenge.

The first, which will be mentioned here only as background, has become a largely continental European effort aimed at a taxonomic tradition in classifying vegetation into meaningful units. The Zurich–Montpellier school (or, as later called, the SIGMATIST school) originated in the 1890s with work by C. Schröter and Charles Flahault. This effort was taken up and led by J. Braun-Blanquet, who emphasized a floristic description of vegetation with the ultimate goal of a classification of "associations" as the basic vegetation units of interest. This interest in vegetation classification as a basis of mapping, understanding causality, and prediction continues to the present.

The second, Clementsian succession theory, is associated with the work of F. E. Clements. Clements saw secondary succession as being initiated by an event that that removed all of the vegetation. Nowadays, this would fall under the term disturbance. Clements referred to this process as nudation. Seeds, spores, and the like, would then disperse to the site in the process of migration. These would germinate and become established in a collective set of processes that he called eccesis. The resultant plants would then interact with one another, notably through competition. Clements called these collections of interactions coaction. Successful plants would interact with the microenvironment at the site and, in a process called reaction, would modify the local site conditions to make it unsuitable for the plant species that were there and more suitable for other species, which would replace them. This interaction is often called facilitation in modern ecology texts. The plant communities generated by this process are called seres. The feedback between each sere and the environment produces an ordered sequence of seres or seral stages of succession. This reaction–replacement process would continue until an assemblage of species able to occupy the site after the plant–environment reaction process became dominant (stabilization). This ultimate assemblage was the climax community.

The third, Gleasonian succession theory, named for its association with H. A. Gleason, represents a logical foil to Clements' progression of succession toward a climax vegetation. As early as 1910, Gleason noted, "[It] is impossible to state whether there is one definite climax vegetation in each province; it seems probable that there are several such associations each characteristic of a limited proportion."

It should be noted that these three theoretical points of view are really icons for a much more complicated debate that is still ongoing and involves the nature of ecological succession. For example, Henry Chandler Cowles studied the primary succession in Indiana Dunes, a series of sand dunes found at the southern tip of Lake Michigan. These dunes formed when drops in the lake level provided fresh sediment. They are progressively older as one moves away from the lake. The oldest of the dunes has a loamy brown forest soil and is covered by a beech/maple (*Fagus/Acer*) forest; the youngest dune at the beachfront has windblown sand as a soil and is covered with a patchy beach/grassland. Cowles recognized this entire land pattern represented a chronosequence, with the spatial location of the dunes of different ages representing different stages of successional development. Based on his experience with these dunes, he produced the now famous phrase that "succession was a variable chasing a variable"—the changes in vegetation chase the changes in the climate. This insight was, and remains, a remarkable concept of vegetation function. What he meant by this was that the rate of change of vegetation succession, the variable, is sufficiently slow that one could expect the climate, the chased variable, to change by the time the succession process was completed.

CLEMENTSIAN SUCCESSION

Clements was in some sense the dominant figure in these discussions. Clements' climax community concept is incorporated almost subliminally in a number of the ideas in plant ecology and landscape management, particularly in the western United States. Maps of potential vegetation are often maps of climax vegetation in the style of Clements. This is reinforced by the two-dimensional nature of paper making it cumbersome to print maps with more than one potential vegetation in a given map location.

Management of parks toward a "natural" condition often has a single "climax community" concept (or a monoclimax) as an underlying basis. Mapping climax vegetation is less difficult than somehow mapping successional seres, which might be expected to change over time. While Clements' concepts are ubiquitous in land management, several ecologists have noted difficulties with the concept. The more frequent criticisms of Clements' concepts include the following:

1. The idea that the sere/environment feedback produces a necessary ordering of communities toward the climax community. Clements noted, "The climax formation is the adult organism of which all the initial stages are but stages of development." Some of the more ardent Clementsian ecologists likened the sequence of seres

in succession as a direct analog to the embryological development of an organism. The rejection of this concept of the community as a superorganism concept by Tansley motivated the invention of the neologism "ecosystem" as a replacement for what he felt was a corrupted term, "community," in 1935.

2. The concept of facilitation has been attacked from evolutionary arguments to the effect that one would not expect a species to evolve to facilitate the success of a replacing competitive species. Field observations of species replacements such as the presence of plants with nitrogen-fixing symbiotic organisms increasing the nutrient status of a site and advantaging other species support the existence of facilitation. There are also numerous counterexamples of species that hold sites tenaciously over generations before finally losing to competitors.
3. The progressive ecesis of different species relayed to the site for all the seral stages has been questioned. In some successional sequences, all the species are present at the site from the first nudation event. In these cases, the progression of seral stages is a consequence of the differences in rate of maturation of herbaceous plants, shrubs, and trees.
4. Several questions have arisen about the climax concept, particularly the existence of a single stable climax community reached by a single sequence of seres. Is there a single climax community, or is there always change in the vegetation at a site? Can there be multiple stable climaxes? Can there be several pathways to a particular community? Can the pathways fork? Are the seres comparable at different locations?

These criticisms continue in no small part because of the importance of Clements and his students in the formulation of a grand theory of vegetation dynamics and their ability to translate their concepts into practical applications.

It was subsequently recognized that past conditions could leave stable relict communities (e.g., preclimaxes such as grassland patches in temperate deciduous forest) and that human or natural disturbances (such as occasional wildfires) could produce a persistent regional-scale vegetation (disclimax). Such communities could be expected to change to the climax vegetation if the disturbance were controlled or removed. It was also recognized that unusual soil conditions or variations in topography could produce persistent vegetation that was different from the regional stable (climax) vegetation (pre-climaxes and post-climaxes). These elaborations of the climax concept inspired a proliferation of conditional climax vegetations and an increasingly complex classification of vegetation. These extended definitions of the Clementsian succession concept, which essentially preserved the ideas of a single succession to a climax community through an ordered series of seral communities, also allowed for special conditions (disturbance regime, human use of the land, historical antecedents and soils) to create deviations from this basic model.

A. G. TANSLEY AND THE ECOSYSTEM CONCEPT

H. C. Cowles was a remarkable and innovative scientist. He was honored in the 1935 issue of *Ecology,* the journal of the Ecological Society of America. The Henry Chandler Cowles issue (*Ecology* vol. 16, no. 3) is a collage of the central issues in plant ecology in its formative years. The first of the papers in the Cowles issue, different in tone from the rest, was Tansley's now classic 1935 paper, "The Use and Abuse of Vegetational Concepts and Terms." In this paper, Tansley discussed Clements' ideas, particularly the amplification of these ideas by the South African John Phillips. Tansley emphasized his strong disagreement with the characterizing of ecological communities as "quasi-organisms" whose successional dynamics were analogous to embryological development.

As an alternative to the community, Tansley created the ecosystem concept in contrast to Phillips' and Clements' views. It is significant that in his iconoclastic ecosystem definition, he emphasized ecosystems "were of the most various kinds and sizes." In doing so, Tansley basically defined an ecosystem as what systems scientists would nowadays call a system of definition: an arbitrary system defined by the specific considerations for a particular application. Tansley's ecosystem definition conforms well to more mathematical, interactive-system concepts in other sciences. His definition includes the intrinsic consideration of scale found in other sciences, particularly physically based sciences.

Tansley also endorsed the concept of the polyclimax, the possibility of multiple stable ecosystems following from succession and a logical contrast to the Clements idea of the monoclimax as the eventual product of successional processes working in a region. To some degree, Clements and his colleagues used terms such as, say, disclimax to handle these sorts of cases in their monoclimax theory of succession.

H. A. GLEASON AND THE INDIVIDUALISTIC CONCEPT OF SUCCESSION

Gleason developed what he called the "individualist concept"—that succession is the result of environmental requirements of the individual species that comprise the vegetation. He also noted that "no two species make identical environmental demands." Gleason's succession was a much more fluid and much less stereotypical concept than that of Clements. Succession reflected the interactions of individuals with their environment. Succession could change in its nature with different climatic and other environmental conditions. It could progress or regress to a different, stable community depending on time and circumstance. Gleason recognized succession could be retrograde as well as progressive as posited by Clements. Succession was not necessarily an irreversible trend toward the climax community. Modern ecology has largely embraced a Gleasonian view of the succession process, but certainly not completely so.

A. S. Watt and the Pattern and Process Concept of Vegetation

A student of Tansley's, A. S. Watt, was the president of the British Ecological Society in 1947. He delivered one of the most important papers in ecology as his presidential address. In this paper, he represented the plant community as a "working mechanism" of interacting plant processes such as regeneration, growth, and death on individual plants producing the broader-scale pattern of plant communities. He stated,

> The ultimate parts of the community are the individual plants, but a description of it in terms of the characters of these units and their spatial relations to each other is impractical at the individual level. It is, however feasible in terms of the aggregates of individuals and species which form different kinds of patches: these patches form a mosaic and together constitute the community. Recognition of the patch is fundamental to an understanding of the structure as analyzed here.

To illustrate this communality in the underlying working dynamics of vegetation, he compared the underlying dynamics that produces the patterns in a diverse array of plant communities—heathlands, grasslands, bogs, alpine vegetation, and forests.

Like Gleason, Watt noted that one must account for the results of individual plants interacting with one another, but he also felt such an accounting was "impractical at the individual level." Twenty years later, the explosive expansion of computational power has made such computations more and more feasible. With the development of increasingly powerful digital computers, starting in the 1960s and continuing to the present, several different scientific disciplines (physics, astronomy, ecology) independently began to apply computers to the tasks of "book keeping" the changes and interactions of individual entities. Forest "gap" models (which will be discussed below) were one of this class of individual-based models (IBMs) that brought Watt's insights on vegetation dynamics into a form that could be projected using computer simulation techniques.

Attempts at Synthesis: Connell and Slatyer

The dichotomy represented by Clementsian versus Gleasonian views of succession has persistently invited attempts to synthesis toward a unified concept. J. H. Connell and R. O. Slatyer in 1977 developed a synthesis that emphasized the individual attributes of plants and their feedback with their environment. Clementsian succession was a special case, which they called the facilitation model (Fig. 1). The replacement patterns in the Connell and Slatyer model were for sets of species, which also implies to some degree a replacement of one community with another, which also is more of a Clementsian concept. Succession was driven by transfers from one set of species as the facilitation model in which succession arises from the resident species at a location encouraging the colonization of a subsequent set of species, as the tolerance model in which the resident set of species are relatively neutral in their interaction with other species, or as the inhibition model in which the resident species block the colonization by other species.

Attempts at Synthesis: E. P. Odum

In 1969, E. P. Odum wrote "The Strategy of Ecosystem Development," a synthetic paper strongly based on ecological energetics and compartment models in ecosystems. The emphasis of the paper was strongly in the area that nowadays might be termed as sustainability science, as is evidences by its first lines: "The principles of ecological succession bear importantly on the relationship between man and nature. The framework of succession theory needs to be examined as a basis for resolving man's current environmental crisis." Odum's view of succession was summarized as a tabular model of ecological succession (Table 1). It was strongly Clementsian with some of the superorganism emphasis of Phillips added in. It also included a number of ideas that were popular in theoretical and population ecology at the time, such as the idea that biotic diversity in an ecosystem connotes ecosystem stability, or the concept that succession was inherently driven to maximize some features (total biomass, biotic diversity, ratios of energy or material transfers, and so on). These latter considerations can be seen as a logical continuation of Clements' concept that succession was intrinsically progressive in its direction.

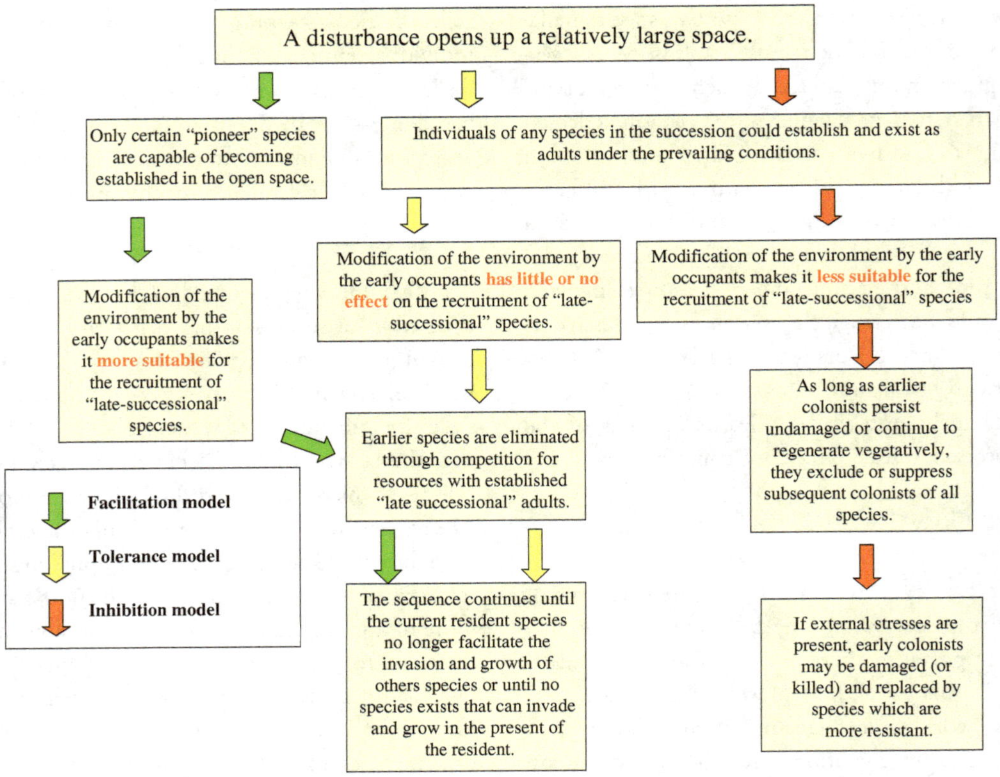

FIGURE 1 Three different models of ecological succession from J. H. Connell and R. O. Slatyer depicting succession as arising from the resident species at a location encouraging the colonization of a subsequent set of species (facilitation model), by the resident species being relatively neutral to the other species (tolerance model), or by the resident species inhibiting the colonization by other species (inhibition model).

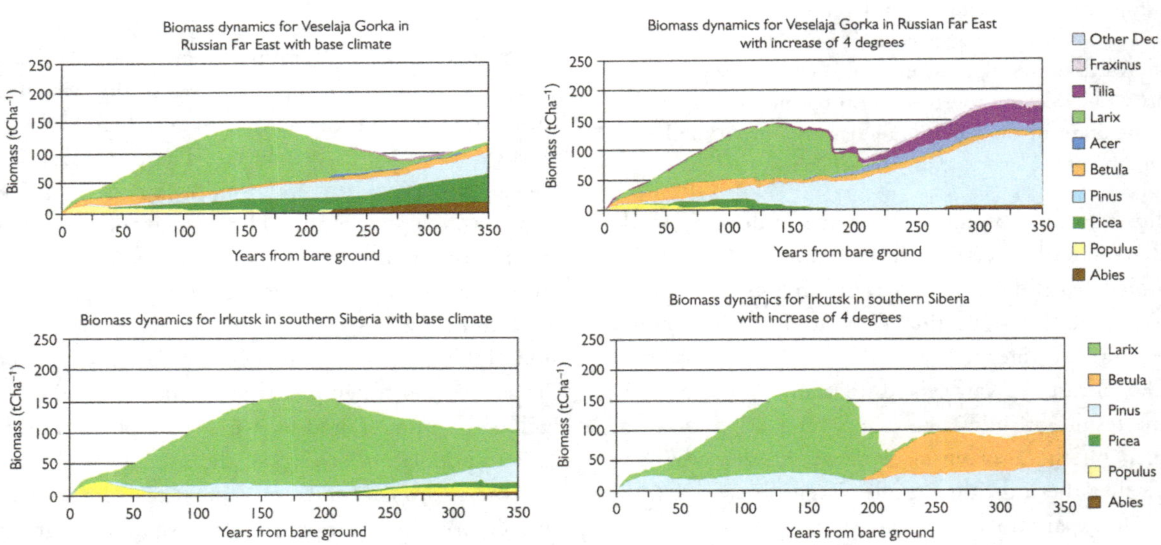

FIGURE 2 Cumulative biomass ($tCha^{-1}$) for simulated ecological succession from bare ground at two locations in Russia. Each graph displays composition by the dominant genera. The left column is for the expected succession with the current climate conditions; the right column is for a 4 °C warming. The base cases in each of the pairs of graphs represent the successional dynamics from a bare ground condition in year 0 for 350 years of ecological succession. In the Russian Far East case (top row), the initial succession is dominated by the same genera, mostly larch (*Larix* sp.), but the later succession is compositionally quite different under a warming. In Siberia (lower row), larch persists over 350 years of succession but becomes an early successional species with climatic warming.

TABLE 1
A "tabular model" of changes expected over ecological succession from E. P. Odum

Ecosystem Attributes	Early Succession	Late Succession
GPP/Respiration	or < 1	Approaches 1
GPP/Biomass	High	Low
Biomass/Energy	Low	High
Net community production (NPP)	High	Low
Food chains	Linear	Web-like
Total organic matter	Small	Large
Inorganic nutrients	Extrabiotic	Intrabiotic
Species richness	Low	High
Species evenness	Low	High
Biochemical diversity	Low	High
Stratification and pattern	Poorly organized	Well organized
Niche specialization	Broad	Narrow
Size	Small	Large
Life cycles	Short, simple	Long, complex
Mineral cycles	Open	Closed
Nutrient exchange	Rapid	Slow
Role of detritus	Unimportant	Important
Selection on growth form	*r*-Selection	*k*-Selection
Selection on production	Quantity	Quality
Symbiosis	Undeveloped	Developed
Nutrient conservation	Poor	Good
Stability	Low	High
Entropy	High	Low
Information content	Low	High

NOTE: GPP is gross primary production (photosynthesis); NPP is net primary production (GPP minus respiration).

Odum's 1969 paper generated and still generates strong interest in the scientific community and continues to be influential more than 40 years later. As one might expect given its essential Clementsian theme updated with central topics in ecosystems ecology, Odum's paper attracted several negative reviews from the ecologists of a Gleasonian persuasion, particularly given the ascendency of ideas association with Gleason's papers on succession. Nevertheless, Odum's work has had and continues to have a strong influence. Its rules on how change should progress toward appropriate natural goals provide guidelines to how ecological systems might be managed toward appropriate goals. Its dogmatism makes it a target for criticism but also gives it an influential position among those wrestling with the difficult problems of sustainable use of ecosystems and conservation management on a dynamical, strongly human-altered planet.

GLEASONIAN SUCCESSION AND INDIVIDUAL-BASED MODELS

Currently, several factors challenge the application of succession theory. Climate conditions are changing, perhaps with a greater rate of change than seen in the past. The carbon dioxide composition of the atmosphere has been altered by human activities, with potential effects on plant processes, particularly photosynthesis and the efficiencies of plant water use. This issue was addressed several decades ago by foresters who realized that the elaborate calibrations used to develop long-term data on forest change were subject to less accurate predictions if selected genetic strains of trees were used, if forests were fertilized, or if there were significant variations in climate.

Watt noted that to understand succession one must account for the results of individual plants interacting with one another, but added that it is "impractical at the individual level." Twenty years later, the explosive expansion of computational power that still continues to date has made such computations more and more feasible. Early versions of these models in ecology were developed by population ecologists interested in including animal behavior in population models and lead to a diverse array of applications for fish, insects, trees, and birds. An advantage of such models is that two implicit assumptions associated with traditional ecological modeling populations are not necessary:

1. The unique features of individuals (including their size and relative location) are sufficiently unimportant to the degree that individuals are assumed to be identical.
2. The population is "perfectly mixed" so that there are no local spatial interactions of any important magnitude.

Many ecologists are interested in variation between individuals (a basis for the theory of evolution and a frequently measured aspect of plants and animals) and appreciate spatial variation as being quite important. The assumption that this variation somehow is uniform seems particularly inappropriate for trees, which are sessile and which vary greatly in size over their life span. This may be one of the reasons why tree-based forest models are among the earliest and most widely elaborated of this genre of models in ecology. Impressed by the power of computers and interested in developing a methodology to use highly detailed computer models to simulate the spatial relations among the thinning of trees, tree growth, and the spatial arrangement of trees, foresters and forest ecologists developed individual-based dynamic models of forests in the mid-1960s.

This computational-based innovation was being independently paralleled in other fields, notably astronomy, physics, and several engineering sciences. The early individual-tree–based forest models were quite complex. For example, the competition among individual trees was typically simulated by crown interactions involving the three-dimensional geometry of each individual tree crown for all the trees in a stand, the growth of the tree trunks of each tree were often simulated at multiple heights, and a tree's mortality was related to two- or three-dimensional crown pruning among trees.

In 1980, a paper reviewing some of this work coined the term gap model to describe this class of models. The gap-model designation was originally developed to emphasize that a principal simplifying assumption in these models (the assumption that the competition among individual trees on a small patch of land was homogeneous in the horizontal over a small area of land but spatially explicit in the vertical dimension) fitted well with the classic "gap dynamics" concept of A. S. Watt. Nowadays, gap model refers to a broad class of individual-based models of forests and other ecosystems of a natural character (mixed-age, mixed-species, natural disturbance regimes, and the like).

Gap models have been validated and verified for a range of forests and can be used to simulate the expected patterns of succession for forests under current or under novel conditions. For example, the individual-based gap model FAREAST was developed to simulate forest dynamics on Changbai Mountain in Northern China, an area famous for its rich tree-species and forest-type diversity. The initial tests of the FAREAST model included a simulation of forest composition and basal area at different elevations on Changbai Mountain and then at other sites in China, the Russian Far East, and Siberia. Further testing of the model at 223 sites across Russia (Shugart et al., 2006) indicated that FAREAST accurately simulates patterns in leaf area and forest biomass across a large, climatically diverse area. The model has found application in simulating the changes in forest biomass and composition at over 2000 locations over the territory of the former Soviet Union. An example of the simulation of forest succession at one of these locations for current versus experimental climate conditions is seen in Figure 2. It is significant that the predicted nature of succession shows qualitative differences in pattern in different regions and under climate change.

If gap models are computerized realizations that include the elements of Gleasonian succession, what are some of the theoretical features of succession that they imply? Across the broad spectrum of cases simulated by these models, one sees multiple stable long-term states ("polyclimax" in the earlier successional terminology). The multiple stable vegetation assemblages display hysteric behavior under slowly changing environmental conditions. The simulated vegetation can amplify environmental inputs at certain frequencies and attenuate other frequencies. These effects become more pronounced when factors that increase the "memory" of previous vegetation in the models by including factors such as seed source for new simulated plants. The models demonstrate very different three-dimensional structures in forests with the small-scale structures (ca. 0.1 ha) being cyclical and with the average of these cyclical variations being an unstable structure, and with the larger scale simulated structure moving toward an equilibrium, namely which is the average structure, over succession.

SEE ALSO THE FOLLOWING ARTICLES

Continental Scale Patterns / Facilitation / Forest Simulators / Gap Analysis and Presence/Absence Models / Individual-Based Ecology / Landscape Ecology / Regime Shifts / Restoration Ecology

FURTHER READING

Glenn-Lewin, D. C., R. K. Peet, and T. T. Veblin, eds. 1992. *Plant succession: theory and prediction.* London: Chapman and Hall.

Miles, J., and D. W. H. Walton, eds. 1993. *Primary succession on land.* Special Publication of the British Ecological Society. Oxford: Blackwell.

Pickett, S. T. A., and P. S. White, eds. *The ecology of natural disturbance and patch dynamics.* New York: Academic Press.

Shugart, H. H. 2003. *A theory of forest dynamics: the ecological implications of forest succession models.* Caldwell, NJ: Blackburn Press.

Walker, L. R., J. Walker, and R. J. Hobbs, eds. 2010. *Linking restoration and ecological succession.* New York: Springer-Verlag.

West, D. C., H. H. Shugart, and D. B. Botkin, eds.1981. *Forest succession: concepts and application.* New York: Springer-Verlag.

SYNCHRONY, SPATIAL

ANDREW M. LIEBHOLD

USDA Forest Service, Morgantown, West Virginia

Spatial synchrony refers to coincident changes in the abundance or other time-varying characteristics of geographically disjunct populations. Nearby populations tend to be more synchronous than more distantly located populations. Though spatial synchrony is a ubiquitous

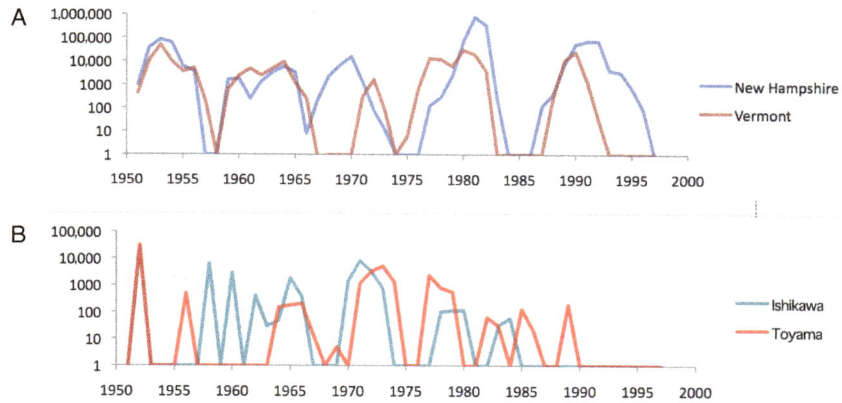

C

	Toyama	New Hampshire	Vermont
Ishikawa	0.924	−0.033	0.033
Toyama		−0.017	0.082
New Hampshire			0.328

FIGURE 1 Spatial synchrony in gypsy moth populations. (A) Time series of annual area defoliated by the gypsy moth, *Lymantria dispar,* in two adjacent states in the United States. (B) Time series of annual gypsy moth defoliation area in two adjacent prefectures in Japan. (C) Correlation matrix among the four time series.

trait of the dynamics of virtually every species, identification of the cause of synchrony is often elusive.

THE CONCEPT

We live in a spatially autocorrelated world; nearby locations tend to be similar in many ways. Similarly, changes in ecological variables at nearby locations tend to be more similar than changes among distantly located points. In virtually every species, population abundance characteristically varies through time, but the synchrony of spatially disjunct populations declines with increasing distance between populations.

Spatially synchronous population dynamics in the gypsy moth, *Lymantria dispar,* can be seen in Figure 1. Nearby populations in both the United States and in Japan fluctuate synchronously as measured by the correlation between defoliation time series. However, because populations in Japan and the United States are so distantly located, there is no correlation between populations.

Spatial synchrony is also common in ecological time series other than abundance. For example, mast seeding, the episodic production of large seed crops in plants, is well known to exhibit synchronous dynamics. Plants located in nearby locations tend to produce large seed crops in the same year, but more distantly located plants exhibit little synchrony in seed production.

Spatial synchrony is just one of many types of time–space patterns that may be present in ecological data. Another related pattern sometimes observed is population waves, typically observed in oscillatory time series. These waves exist when either troughs or peaks occur progressively later when moving in specific directions.

MEASURING SYNCHRONY

The simplest method for measuring synchrony is the use of the ordinary product-moment correlation coefficient between two series (e.g., Fig. 1C). While this may provide a satisfactory measure of synchrony, testing the significance of the correlation may be complicated by the fact that many ecological time series are temporally autocorrelated. Such autocorrelation leads to the violation of the assumption of independence among samples implicit in naïve tests of correlation.

More meaningful statistical characterization of synchrony can be accomplished when many spatially referenced time series are available. With such data, it is possible to calculate correlation coefficients between all pairwise combinations of time series. By plotting the correlation by the distance between pairs of sample points (Fig. 2), the spatial scale of synchrony can be characterized. In such time series correlograms, cross-correlations between time series typically approach unity as the distance between sample locations approaches zero. But as lag distance increases, cross-correlation generally decreases and ultimately reaches zero. The distance at which synchrony reaches zero is a useful way of characterizing the spatial scale of correlation and ultimately may provide clues as to the processes driving synchronization.

Several statistical methods are available for summarizing and describing time series correlograms. In one of these, the "spatial covariance function," correlation is estimated as a continuous function of distance using smoothing splines, and a bootstrap procedure is applied to estimate confidence intervals for the function.

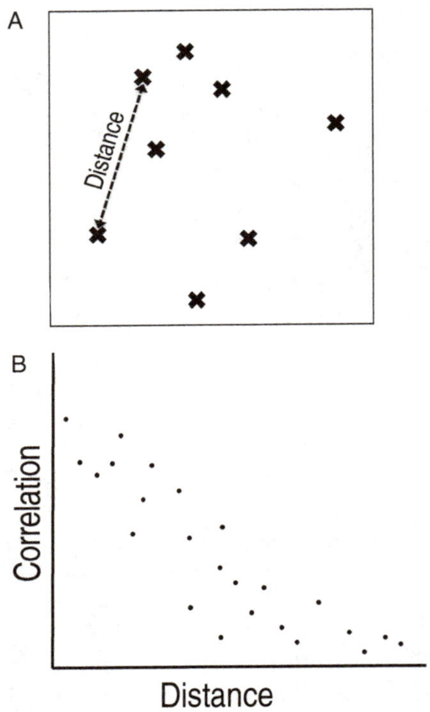

FIGURE 2 Illustration of time series correlogram. (A) Spatial configuration of hypothetical sample locations for eight time series. (B) Time series correlogram of hypothetical data.

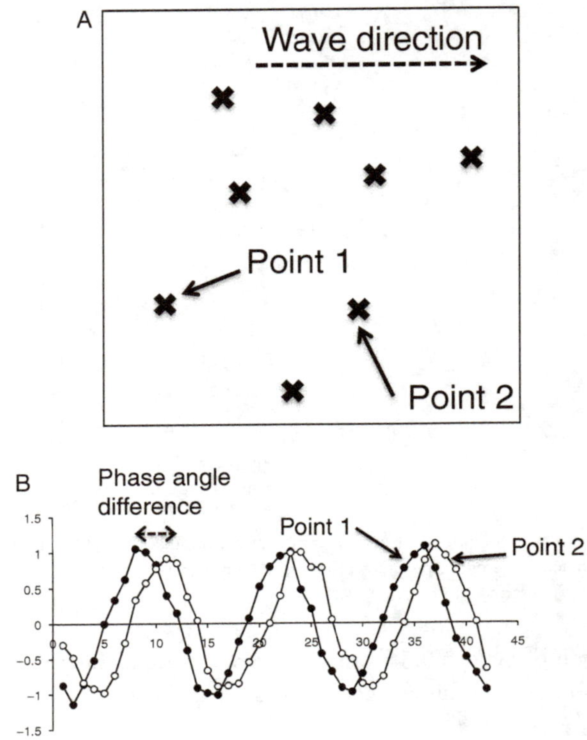

FIGURE 3 Illustration of the use of phase angles for calculation of wave speed. (A) Location of hypothetical sampling points for time series. Note that a population wave is hypothesized to move from left to right, such that oscillations at points located to the left peak before those to the right. (B) Illustration of two hypothetical time series exhibiting phase differences. Phase angle difference is calculated by transforming time (*x*-axis) to a phase angle such that a complete oscillation is completed in 360 degrees.

It is possible to detect population waves from modified time series correlograms in which a time lag is introduced when calculating pairwise cross-correlations. However, directional waves will only be detected when using this method if separate correlograms are calculated for classes of angles describing the vector between pairs of time series locations. In such a case, time-lagged correlations will be greatest for correlograms derived from pairs of points oriented in the same direction as the wave.

A better alternative for detecting population waves is often the comparison of relative phase angles among time series. For oscillatory time series, it is possible to transform time into phase angle values (i.e., t' varies from 0 to 360 degrees). The transformation is calibrated such that $\sin(t')$ fits the time series. Comparisons of differences in phase angles among points in relation to distance allows estimation of wave speed and direction (Fig. 3).

CAUSES OF SYNCHRONY

The concept of phase locking from dynamical systems theory describes the phenomenon in which there is a constant difference in the phase of two periodic or quasi-periodic oscillators. It has been shown that when one cyclic oscillator slightly influences the other oscillator, phase-phase locking may result. Alternatively, two oscillators can be synchronized by an external source of "noise," which can take the form of even a very small random force that affects both oscillators simultaneously.

Generally, these same two classes of mechanisms are known to be capable of producing spatial synchrony in time series of abundance, although they take the form of (1) movement and (2) environmental stochasticity (Fig. 4). Unfortunately, both types of mechanisms are capable of producing similar patterns of synchrony; thus, diagnosing the causes of synchrony in real biological systems is often difficult.

Theoretical models demonstrate that even a small amount of dispersal between populations can very rapidly bring populations into synch. Also, because movement between nearby locations is likely greater than that between more distal locations, the synchronizing effect of dispersal can create the nearly ubiquitous pattern of decreasing interpopulation synchrony with increas-

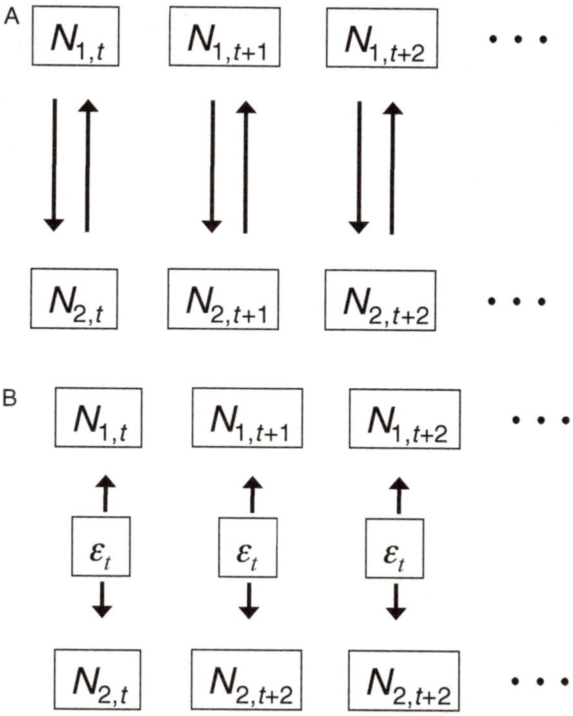

FIGURE 4 Synchronizing mechanisms. (A) Synchronization via dispersal; a small number of individuals move between the two populations each generation. (B) Synchronization via environmental stochasticity; in each time step, change in abundance is affected by a common random influence.

ing distance between populations. Dispersal of natural enemies (i.e., mobile predators) can also synchronize populations, much in the same manner as dispersal of host populations.

Perhaps the best explanation for the ubiquity of spatial synchrony in nature is that virtually all populations are exposed to synchronized environmental stochasticity. The first description of ecological synchronization via environmental stochasticity was provided by the Australian statistician Patrick Moran, who used a second-order log-linear autoregressive model:

$$n_{t+1} = b_1 n_t + b_2 n_{t-1} + \varepsilon_t,$$

where n_t is log population density at time t and e_t is an exogenous random effect. He used the model to describe the quasi-periodic dynamics of snowshoe hare populations in various Canadian provinces. Moran showed that two independent populations governed by the same dynamics (i.e., identical autoregressive parameters) will quickly be brought into synch when the series of ε_t are themselves spatially synchronous. Furthermore, he showed mathematically that the correlation between any two population series will be exactly the same as the correlation between the series of environmental stochasticity.

This phenomenon has been termed Moran's theorem and has been widely applied to explain spatial synchrony in a variety of taxa. The most common source of environmental synchrony is temporal variability in weather conditions. Virtually all organisms are affected by generation-to-generation variability in weather, though identification of these effects may be difficult or impossible. Nevertheless, temporal variability in weather is consistently spatially correlated in a similar manner throughout the world; spatial correlation in temperature typically extends to 500 to 1000 km, while synchrony in precipitation extends over slightly smaller ranges of 300 to 600 km. Spatial correlation in weather is thus capable of explaining the spatial synchrony in abundance observed in virtually all species.

Real populations do not precisely obey the second-order linear dynamics described by Moran. First, the dynamics of many populations are decidedly nonlinear. Secondly, endogenous dynamics often varies geographically among populations. Both nonlinearity and geographical variation in dynamics alter the synchronizing effect of both dispersal and environmental stochasticity, though these effects are complex and difficult to predict.

Finally, a pattern emerging from various empirical and theoretical studies is that synchrony may percolate through food webs. For example, spatial synchrony among spatially disjunct predator populations may contribute to synchrony among prey populations and visa versa. Because many food webs are complex, the synchronizing factors affecting any two populations may be complex and difficult to identify. Nevertheless, the manner in which trophic interactions enhance or detract from synchrony and how synchrony percolates through food webs is a subject that remains largely unknown and deserves attention in future studies.

SEE ALSO THE FOLLOWING ARTICLES

Dispersal, Animal / Food Webs / Movement / Spatial Ecology / Stochasticity, Environmental

FURTHER READING

Bjørnstad, O. N., and W. Falck. 2001. Nonparametric spatial covariance functions: estimation and testing. *Environmental and Ecological Statistics* 8: 53–70.

Bjørnstad, O. N., R. A. Ims, and X. Lambin. 1999. Spatial population dynamics: analyzing patterns and processes of population synchrony. *Trends in Ecology & Evolution* 14: 427–431.

Blasius, B., A. Huppert, and L. Stone. 1999. Complex dynamics and phase synchronization in spatially extended ecological systems. *Nature* 399: 354–359.

Buonaccorsi, J. P., J. S. Elkinton, S. R. Evans, and A. M. Liebhold. 2001. Measuring and testing for spatial synchrony. *Ecology* 82: 1668–1679.

Cazelles, B., and L. S. Stone. 2003. Detection of imperfect population synchrony in an uncertain world. *Journal of Animal Ecology* 72: 953–968.

Koenig, W. D. 2002. Global patterns of environmental synchrony and the Moran effect. *Ecography* 25: 283–288.

Liebhold, A. M., W. D. Koenig, and O. N. Bjørnstad. 2004. Spatial synchrony in population dynamics. *Annual Review of Ecology, Evolution & Systematics*. 35: 467–490.

Moran, P. A. P. 1953. The statistical analysis of the Canadian lynx cycle. *Australian Journal of Zoology* 1: 163–173.

Ranta, E., V. Kaitala, and J. Lindström. 1999. Spatially autocorrelated disturbances and patterns in population synchrony. *Proceedings of the Royal Society of London Series B: Biological Sciences*. 266: 1851–1856.

T

TOP-DOWN CONTROL

PETER C. DE RUITER
Wageningen University Research Centre, The Netherlands

JOHN C. MOORE
Colorado State University, Fort Collins

All organisms need food to grow, reproduce, and survive. The availability of food therefore determines the ecological success of species in a fundamental way. For this reason, it is commonly accepted that species population dynamics are regulated or controlled by the bottom-up effects of the availability of resources. In addition to (and not in contrast to) this bottom-up control, species populations can also be regulated, or controlled, top-down, that is, by their predators.

TOP-DOWN VERSUS BOTTOM-UP CONTROL

"Why is the world green?" The most well-known answer to this question is that herbivores are controlled by predators, thereby releasing green plants from herbivorous grazing. The answer highlights the control of top predators on species lower in the food chain. The control is two-fold: the control of the number of herbivores, preventing (too) high numbers, and the control of the biomass of the plants, preventing (too) low stocks. Controlling top-down effects on lower trophic levels that are not directly grazed or predated upon are referred to as trophic cascade effects.

"Are organisms controlled bottom-up or top-down?" Obviously, the availability of resources is a prerequisite for growth and survival, which inevitably indicates the importance of bottom-up control. But bottom-up control can act simultaneously with top-down control. In most ecosystems, species are affected by all kinds of different organisms, in all kinds of different ways: predation, competition, and mutualism or by ecological interactions inducing altered behavior. Then the question of whether populations are controlled bottom-up or top-down takes on a relative character: "How important are top-down effects in the context of the overall assemblage of ecological effects acting on the dynamics and persistence of a population?" This is the leading question for this entry. In answering this question, this discussion is restricted to top-down effects that come from trophic interactions, i.e., those of consumers on their resources. Control will have a positive connotation (it is "good" for the controlled species), although it may include both the prevention of too-high numbers as well as too-low numbers. In fact, control is approached here as factor that keeps the dynamics of a population within the limits that prevent extinction but which also prevent outcompeting other species up to extinction.

TOP-DOWN CONTROL IN SINGLE PREDATOR–PREY INTERACTIONS

Top-Down Control Through the Numerical Response

The most well-known mechanism underlying top-down control is that of the numerical response. Growth and death rates of predatory organisms may strongly depend on the availability of their resources. When the number of prey increases, the resulting increase in the number of preditors will create a higher predation pressure on the prey population, thus hampering or neutralizing the increase in prey numbers. It is easy to see that such a numerical response can be a very adequate controlling mechanism, although its effectiveness strongly depends on time scales and time lags. When the response is almost direct, then the numerical response is much more effective than when there is a considerable time lag between the increase in prey numbers and the response in terms of predator numbers. The idea of the role of numerical response in the control of prey has been investigated by means of simulation modeling, but there is also an impressive amount of empirical data, especially in the field of predatory control of pests and plant pathogens.

Top-Down Control Through the Functional Response

Another well-known mechanism underlying top-down control by predators is the functional response. Predators can also respond individually to variations in their prey numbers by altering their predation rates. This functional response can take various forms (Fig. 1). The type 1 (linear) functional response is a relationship wherein the attack rate increases linearly with prey density up to a constant level. This approach has been criticized as being unrealistic, as predators are likely to adapt their feeding rates to the level of prey density. The type 2 (saturating) functional response describes the attack rate hyperbolically increasing with prey density approaching a constant as the predator becomes saturated. The type 3 (sigmoid) functional response shows an attack rate that accelerates at relatively low prey densities and then hyperbolically approaches the saturation level.

There are several mechanistic formulas that describe functional responses, addressing encounter rates, attack rates, and handling times, but a simple mathematical representation is

$$F = \frac{mN^b}{K + N^b}. \tag{1}$$

This formula generates a type 2 hyperbolic functional response for $b \leq 1$ (most often $b = 1$ is used) and a sigmoid functional response for $b > 1$. Furthermore, N is the prey density, m is the maximum feeding rate, and K is the prey density that leads to half the maximum feeding rate for $b = 1$.

At first sight, a functional response might have a controlling effect on prey dynamics as the predation rate increases with increasing prey numbers. But more important for control is how the *relative* mortality in the prey population, due to predation, responds to changes in prey numbers. Figure 1 shows that predation according to a type 1 linear response causes mortality that is neutral, i.e., it is constant. The type 2 hyperbolic response is such that mortality decreases with prey numbers, creating a destabilizing positive feedback, i.e., the more prey, the less mortality. The type 3 sigmoid response, however, causes a negative feedback at relatively low prey numbers, i.e., the more prey (over that range), the higher the mortality. This shows that a functional response only has a controlling effect on the prey numbers when the functional response has an accelerating shape and that an increase in prey numbers elicits a disproportional increase in predation rate.

The ideas about whether and how top-down control operates in single predator–prey systems is a little artificial, as most ecosystems food chains are longer and species richness in ecosystems may create complex food webs such that the direct effects of predators on their prey are influenced, or even obscured, by the direct and indirect effects imposed through ecological interactions with other groups of organisms.

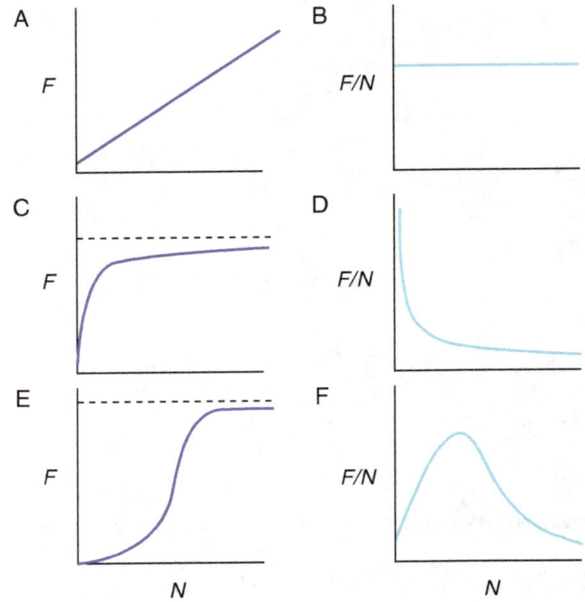

FIGURE 1 Functional response types. (A) Linear type 1 linear response: the feeding rate (F) of the consumer increases linearly with prey density (N) and then becomes a constant value at the consumer's saturation point. (B) Relative mortality due to predation (F/N) in the case of a type 1 linear response. (C) Hyperbolic type 2 functional response: the feeding rate (F) increases with prey density (N) but this increase continuously decreases until it becomes constant at saturation. (D) Relative mortality due to predation (F/N) in the case of a type 2 hyperbolic response. (E) Sigmoid type 3 functional response: the feeding rate (F) increases with prey density (N), first accelerating and then decelerating until the constant value is reached at the saturation level. (F) Relative mortality due to predation (F/N) in the case of a type 3 sigmoid response.

TOP-DOWN CONTROL IN LONGER FOOD CHAINS

Food chains with a length longer than 2—for example, 3 or 4—are more common in ecosystems than are single predator–prey interactions. Such food chains may reveal top-down control that acts over various trophic levels, including control of species that the predator does not directly feed upon. The answer to the question "Why is the world green?" provides such a control. Alternating effects of predators on herbivores and plants effects are referred to a trophic cascade effects. Trophic cascade effects in, e.g., food chains of length 3 emerge from top predators controlling the intermediate consumers and thus reducing predation on the lowest trophic level species. In fact, we see then that top-down and bottom-up control act simultaneously. Suppose that the top predator in such a chain is removed. Then the intermediate consumer increases in abundance and consequently imposes an increased predation pressure on the lowest level. Extending this approach to a food chain of length 4, then the top predator (level 4) controls level 3, level 2 can reach its carrying capacity density, determined by the availability of level 1, which may suffer a high predation pressure from level 2. If the top predator (level 4) is removed, then the level 3 population will increase, the level 2 population will decrease, and the level 1 population goes up. In fact, we see that the odd-level populations increase and the even-level populations decrease in number. Figure 2 gives a conceptual diagram of how equilibrium population sizes in food chains depend on resource availability. From this diagram, it can also be seen how equilibrium population sizes in alternating trophic levels are correlated with each other, while those in adjacent trophic levels are not.

Many experimental studies have shown such trophic cascades. Figure 3 shows the outcome of a study of a food chain from salt marshes consisting of the both short-form and tall-zone *Spartina,* the grazing snail *Littoraria irrorata,* and the predatory blue crab *Callinectes sapidus* and terrapin *Malaclemys terrapin.* By experimentally manipulating the

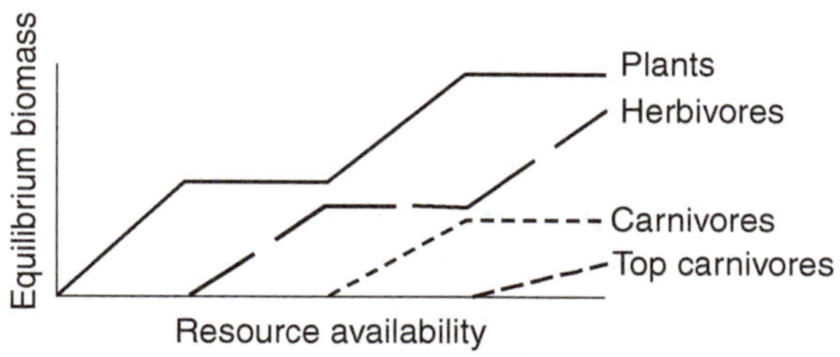

FIGURE 2 Equilibrium biomass increments along a resource availability gradient for four trophic levels. The figure shows how increases in equilibrium biomass are uncorrelated between adjacent trophic levels.

densities of the snails and their consumers, it is shown that plant biomass is heavily controlled by top-down trophic cascade effects.

TOP-DOWN CONTROL IN MULTISPECIES COMMUNITIES

In multispecies community food webs, bottom-up and top-down effects are approached slightly differently, that is, in terms of the per capita effects of prey on the predators and of the predators on the prey, respectively. These so-called interaction strengths have a positive value for the bottom-up effects and a negative value for the top-down effects. Mathematical formulas of interaction strengths are mostly derived from the population dynamics described in the form of Lotka-Volterra differential equations:

$$\dot{X}_i = X_i\left[d_i + \sum_{j=1}^{n} c_{ij} X_j\right] \tag{2}$$

where X_i and X_j represent the population sizes of group i and j, respectively, b_i is the specific rate of increase or decrease of group i, and c_{ij} is the coefficient of interaction between group i and group j. Interaction strength (α_{ij}) is defined as the partial derivatives near equilibrium: $\alpha_{ij} = (\partial \dot{X}_i/\partial X_j)^*]$. This partial derivative is based on redefining the differential equations in terms of departures of equilibrium, i.e., $x_i = X_i - X_i^*$, and then rewriting the equations as Taylor expansion neglecting terms with an order larger than 1, which is of course only justified for relatively small values of x_i. Hence, interaction strengths derived this way are defined only close to equilibrium. When the partials are taken from the Lotka–Volterra equations, for the top-down effect of predator j on prey i we obtain

$$\alpha_{ij} = c_{ij} X_i^*, \tag{3}$$

and for the bottom-up effect of prey i on predator j we obtain

$$\alpha_{ji} = c_{ji} X_j^*. \tag{4}$$

These two equations show that the negative top-down effects (Eq. 3) will usually be much larger (in an absolute way) than the positive bottom-up effects (Eq. 4) First, c_{ji} in Equation 4 can be seen as $e_{ij}c_{ij}$, where c_{ij} is the coefficient for predation in Equation 3, e_{ij} is the efficiency with which predator j convert biomass of prey i into new predator biomass, and $0 < e_{ij} < 1$. Hence, $c_{ij} > c_{ji}$. Furthermore, the equilibrium biomass of the prey X_i^* will generally be larger than the equilibrium biomass of the predator X_j^*, as trophic pyramids are common to most ecosystems. Hence, $X_i^* > X_j^*$.

The fact that top-down effects are usually stronger than their bottom-up counterparts has implications for their

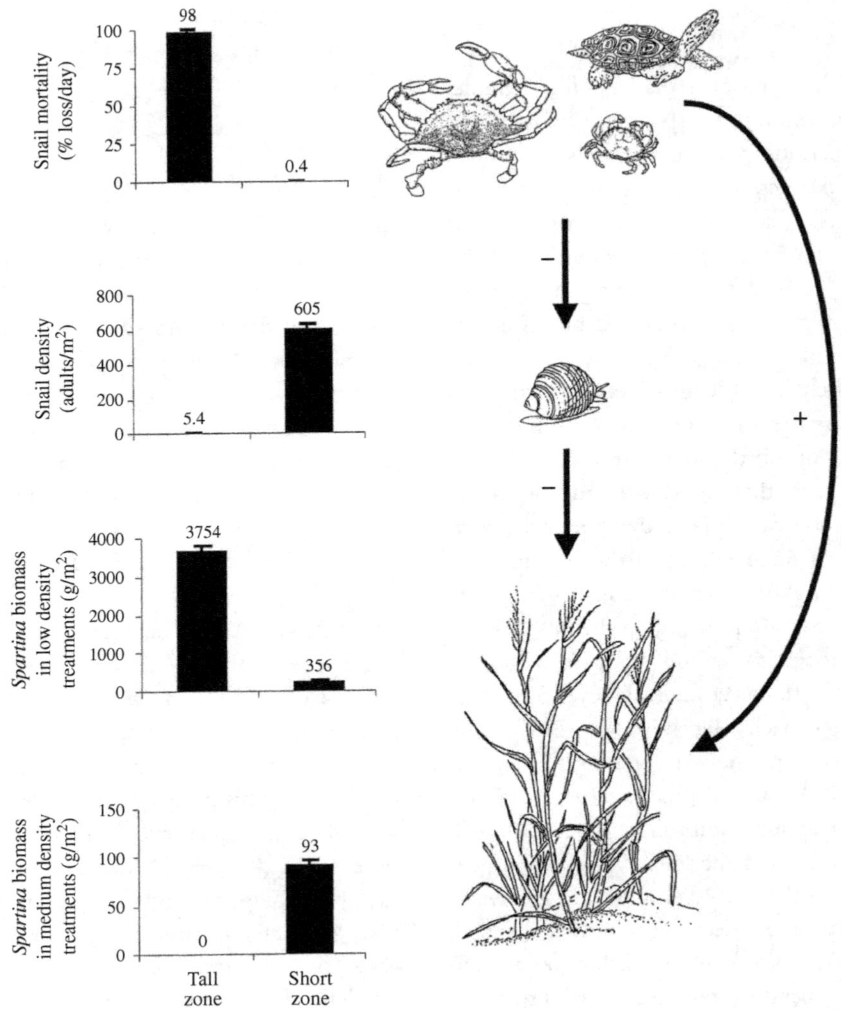

FIGURE 3 Top-down trophic cascade effects controlling the biomass of short-form and tall-form *Spartina* and of the snail *Littoraria irrorata*. Numbers of the predatory blue crabs are experimentally manipulated. The figure clearly shows the correlation that high predatory densities (in the "Tall Zone") leads to a high *Spartina* biomass and low predatory densities (in the "Short Zone") leads to a low *Spartina* biomass. (From Silliman and Bertness, 2002; illustrations by Jane K. Neron.)

importance to community stability. This can be seen from the mathematic theory regarding the role of trophic interaction loops in food web stability. A trophic interaction loop describes a pathway of interactions (note: not feeding rates) from a species through the food web back to the same species without visiting the species more than once; hence, a loop is a closed chain of trophic links (Fig. 4).

Trophic interaction loops may vary in length and weight. The loop length is the number of trophic groups visited, and the loop weight is the geometric mean of the interaction strengths in the loop. System stability can be argued to be negatively influenced by a high maximum loop weight. There is an a priori reason to expect that a long loop will most likely be the heaviest. As the negative top-down effects are much stronger, long loops may contain relatively many negative effects. In real food webs, though, the heavy top-down effects are organized in such a way that long loops contain relatively weak top-down effects. This makes it such that the maximum loop weight is relatively low, and hence the level of food web stability is relatively high.

The mechanism behind the aggregation of weak top-down effects in long loops is as follows. Consider a trophic interaction loop of length 3 consisting of a top-predator species (trophic level 3: TL3) feeding on two groups on different trophic levels, i.e., on an intermediate-level consumer species (trophic level 2: TL2) and a lower-level consumer/resource species (trophic level 1: TL1). This omnivorous food chain forms the basis of two interaction loops of length 3 (Fig. 4 caption). When population sizes are organized in the form of a trophic biomass pyramid, and when the TL3 is assumed to feed on TL2 and TL1 in accordance to their relative availability, then the TL3 will feed mainly on TL1.

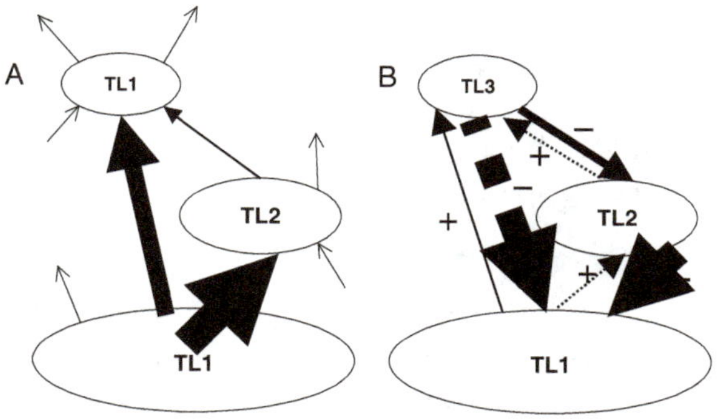

FIGURE 4 Graphical representation of an omnivorous trophic loop of length 3, in terms of (A) population sizes and feeding rates and (B) top-down and bottom-up effects. (From Neutel et al., 2002.) The small partially unconnected arrows in (A) indicate that the loop may make part of a larger food web. In fact, there are two trophic interaction loops, i.e., the loop top-predator consumers (TL3)—intermediate consumers (TL2)—lower level consumers/resources (TL1) and back to TL3, including two negative top-down effects and one positive bottom-up effect (solid arrows), and the loop TL3-TL1-TL2 and back to TL3, including one negative top-down and two positive bottom-up effects (dotted arrows).

The top-down effect of TL3 on TL2 will then be weaker than the effect on TL1, as the top-down effects on TL2 and TL3 are defined as the feeding rates on TL2 and TL1 both divided by the same TL3 biomass (see Eqs. 3 and 4). Furthermore, the top-down effect of TL2 on TL1 is the (very large) feeding rate of TL2 on TL1 per the (relatively large) population size of TL2 and will therefore be about the same strength as the top-down effect of TL3 on TL1. Hence, the relatively weak top-down effect of TL3 on TL2 will always be part of the loop with the two top-down effects, limiting the weight of the loop that is potentially the heaviest.

TOP-DOWN CONTROL OF BIOLOGICAL DIVERSITY

Apart from regulation of population numbers, top-down effects can also contribute to the conservation of biodiversity. The mechanism behind this phenomenon is that some predators tend to eat disproportionately from the relative abundant prey types, e.g. by overlooking the rare ones. This has especially been observed in organisms that visually look for prey and have the ability to form search images of their prey. Search images are formed by frequent encounters with particular prey that help the predator to see the prey more easily and quickly. Hence, the more frequent a particular prey type is, the easier it will be recognized by the predator. In this way, the predation risk increases with the abundance of the prey; hence, rare prey benefit from a small risk of predation, which may prevent their extinction. This phenomenon is referred to as apostatic selection, i.e., selection favoring the rare. Apostatic selection is most often analyzed by linking the relative frequency of a prey, compared to the other prey types, to the risk of predation, with the following formula:

$$\frac{e_i}{\sum_{j=1}^{n} e_j} = \frac{(V_i A_i)^b}{\sum_{j=1}^{n} (V_j A_j)^n}. \tag{5}$$

Here, e_i is the contribution of prey i to the diet of the predator, A_i is the density of prey i, and V_i is any frequency-independent preference for prey type i. When $b = 1$, predation is frequency independent. When $b > 1$, then there is frequency-dependent predation, favoring rare prey. And when $b < 1$, predation is inverse frequency dependent, favoring common prey. In Figure 5, the results from an experiment using the quail *Coturnix* feeding on two different kinds of artificial prey items are given. Using a log-log transformation (Fig. 5A), values and standard deviations can be obtained for V and b in order to test the significance of the frequency-dependent predation. Without the log-log

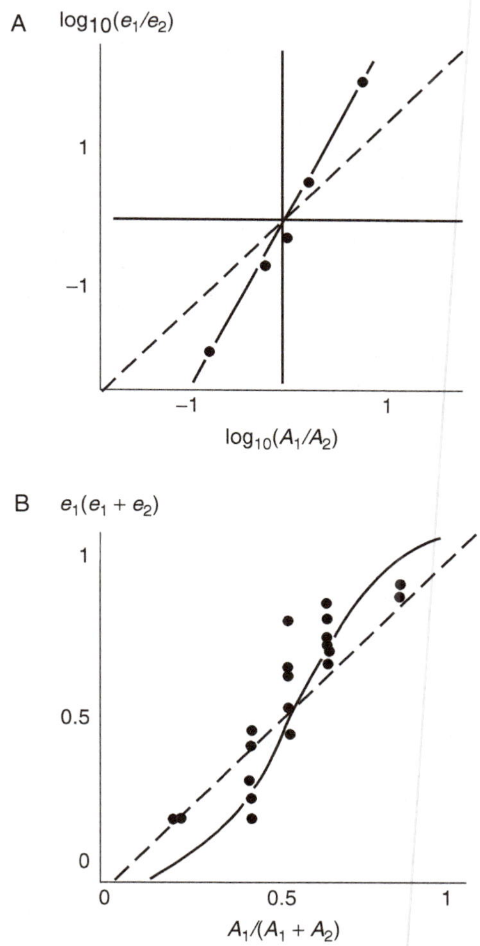

FIGURE 5 Example of apostatic selection of the quail *Coturnix* feeding on artificial prey with A_i being the numbers of prey item i and e_i being the contribution of prey type i in the diet of the quail. (From Greenwood and Elton, 1979.)

transformation, the formula in Equation 5 generates a sigmoid curve. It is easy to see that when apostatic selection occurs, the individual functional responses of the predator to the prey can change from a type 2 hyperbolic form to a type 3 sigmoid form. The most well-known mechanism for sigmoid functional responses is that of prey switching and apostatic selection. But more important, it can be seen that such frequency-dependent predation can have a top-down stabilising effect on species diversity in ecosystems.

SEE ALSO THE FOLLOWING ARTICLES

Bottom-Up Control / Fisheries Ecology / Food Chains and Food Web Modules / Metacommunities / Networks, Ecological / Resilience and Stability

FURTHER READING

Allen, J. A. 1988. Frequency-dependent selection by predators. *Philosophical Transactions of the Royal Society B: Biological Sciences* 319: 485–503.

Beddington, J. R., M. P. Hassell, and J. H. Lawton. 1976. Components of arthropod predation. 2. Predator rate of increase. *Journal of Animal Ecology* 45: 165–185.

Greenwood, J. J. D., and R. A. Elton. 1979. Analysing experiments on frequency-dependent selection by predators. *Journal of Animal Ecology* 48: 721–737.

Hairston, N. G., F. E. Smith, and L. B. Slobodkin. 1960. Community structure, population control and competition. *American Naturalist* 94: 421–425.

Hassell, M. P., J. H. Lawton, and J. R. Beddington. 1976. Components of arthropod predation. 1. Prey death-rate. *Journal of Animal Ecology* 45: 165–185.

Leibold, M. A., J. M. Chase, J. B. Shurin, and A. L. Downing. 1997. Species turnover and the regulation of trophic structure. *Annual Review of Ecology and Systematics* 28: 467–494.

May, R. M. 1972. Will a large complex system be stable? *Nature* 238: 413–414.

Neutel, A. M, J. A. P. Heesterbeek, and P. C. de Ruiter. 2002. Stability in real food webs: weak links in long loops. *Science* 296: 1120–1123.

Silliman, B. R., and M. D. Bertness. 2002. A trophic cascade regulates salt marsh primary production. *PNAS* 99(16): 10500–10505.

Tinbergen, L. 1960. The natural control of insects in pine woods. I. Factors influencing the intensity of predation in song birds. *Archives Néerlandaises de Zoologie* 13: 265–343.

TOXICOLOGY, ENVIRONMENTAL

SEE ECOTOXICOLOGY

TRANSPORT IN INDIVIDUALS

VINCENT P. GUTSCHICK

Global Change Consulting Consortium, Inc., Las Cruces, New Mexico

Transport is the movement of material—water, solutes, gases, suspended bodies—or heat or radiation. It is necessary for metabolism, movement, signaling, and numerous other functions that are central in ecology. Transport is described by basic biophysical laws, while its elaboration in individual organisms evidences a great variety of evolutionary solutions to challenges posed by its operation.

TRANSPORT AT THE CELLULAR LEVEL

Most of the basic phenomena common to transport at various scales of space and time are evident at the cellular level. Cells, like the whole organisms that they compose, transport material (water, solutes, suspended matter), heat, and radiation. The materials may be growth substrates, metabolic intermediates, waste products, signals, or occasional deleterious entities such as viruses. Modes of transport are specific to the entity being transported, as described below.

Water

"Water flows downhill" has a generalization, with "downhill" expanded to cover all the contributions to its energetic state: gravitational, pressure, and osmotic potentials. Taking these terms individually, water moves to lower gravitational potentials, or from higher to lower pressure, or from states of higher water concentration to less concentrated states. When all three contributions occur, as is common, it is the sum that matters; water can move from lower to higher pressures under large osmotic differences, which explains the large positive pressures generated in many plant cells. The sum of the energetic terms is denoted as the water potential and commonly given the symbol ψ. It may be expressed per unit mass or per mole in terms of the Gibbs free energy, which is a useful measure for processes that occur at constant temperature (most water moves in organisms at close to this condition) and not so rapidly as to involve significant aspects of nonequilibrium thermodynamics (again, generally the case in organisms).

At the level of cells, differences in gravitational potential are minuscule, so that pressure and osmotic potential matter. Water, as other substances, moves rapidly by either pressure gradients or diffusion at the short distances within or near cells. A simple estimate of the time, t, for diffusion to traverse a distance x is $t = x^2/D$, where D is the diffusion constant—precisely, the mutual diffusion constant of whatever water is moving with. The value of D in physiological conditions is on the order of 10^{-9} m^2s^{-1}, so that at a moderately large cellular dimension of 100 μm, this time is of the order of 10 s. Consequently, in the absence of a membrane being interposed along a flow path, osmotic gradients are rapidly destroyed.

When water must cross a membrane, the osmotic potential becomes important. Membranes can be partially leaky to other substances (have a "reflection coefficient" less than unity for some solutes), which reduces the osmotic effect in proportion. Water crosses biological membranes through some temporarily fluctuating openings amid the membrane molecules but perhaps mostly through special protein-based channels called (aqua)porins. These protein pores are evolutionarily highly conserved between taxa. They can be regulated metabolically, which is of great importance in responses to stress such as water deficiency. It is important to note that there is no direct role for metabolism in moving water across membranes, unlike the case for solutes, which are often actively transported using energy carriers, particularly ATP. There is so much water in cells and passing through them that the energetic costs of actively transporting it would be too large by many thousand-fold. On the other hand, it is only moderately energetically expensive for cells to induce water to move by generating osmotic differences from inside to outside. Cells do this by making small solutes—sugars, for example—or by acquiring ions such as potassium.

Solutes and Suspended Particles

The list of what solutes or suspended particles (colloids, viruses, etc.) are transported in the watery environment of cells is extensive. The list includes substrates and metabolites (nutrients and energy sources), waste products, acids and bases to regulate cellular pH, protective products such as terpenes from leaves, signals between cells and to ecological partners (to potential mates, protectors, food sources . . . or, inadvertently, to predators and parasites), nonmetabolized osmotica, and viruses. Solutes include dissolved gases such as O_2 and CO_2.

Within the cytoplasm or within individual organelles, solutes move essentially only by diffusion, which is rapid. Dissolved solids (that is, not considering gases) cross membranes through special proteins—channels and carriers—that span the membrane from inside to outside (Fig. 1). Channels enable solutes to move without using energy in ATP or similar compounds. The solutes move "downhill" in their energy state. Again neglecting gravitational energy and also pressure (which solutes do not bear significantly), the energy state of a given solute is determined by its concentration and, for charged solutes, by its electrical potential energy. The free energy per unit of solute therefore has two variable terms. The first is proportional to the logarithm of its concentration (not concentration in direct proportion, for subtle reasons). The second is the product of its total electrical charge multiplied by the voltage at its location. Again, it is the sum that matters. A positively charged solute such as potassium can enter a cell going from lower to higher concentration without added energy when the electrical potential (voltage) of the cell is more negative inside that outside, as is common. Most potassium is nonetheless accumulated by active transport.

Solute may move in or out of cells by active transport through carriers. At primary active carriers or transporters, ATP is split, and that action is directly used to change the protein confirmation, "squirting" the solute in or out. More abundant are secondary active carriers. They transport a given solute by moving it along with another solute, which may move in the same direction (co-transport) or in the opposite direction (antiport).

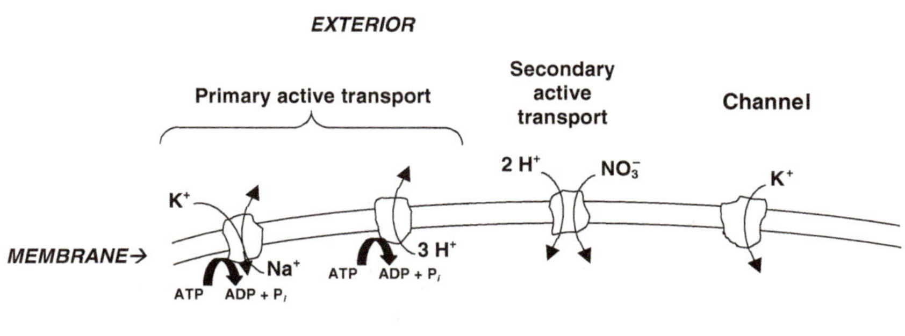

FIGURE 1 Active transport and channels. Irregular shapes are proteins spanning the cell membrane. In primary active transport, energy from splitting ATP to ADP + P_i is used directly at the carrier protein to move ions. Second from left is primary active transport of hydrogen ions (protons) by an H^+ ATP-ase. The resultant accumulation of protons outside the cell creates a free-energy difference. This is used (third segment from left) to import nitrate ions against an uphill free-energy difference for them (nitrate is more concentrated inside the cell, and the negatively charged ion is moving into the negatively charged cell interior). The rightmost segment illustrates action of a channel, which passively allows selected ions across the membrane.

For example, nitrate ions move into root cells of plants along with several protons, and the proton concentration is raised outside the cell by ATP-splitting proteins that pump protons out. Some active transporters are defensive in nature. Bacteria have an array of drug pumps, mostly to excrete compounds elaborated by competitors or enemies and toxic to the bacteria themselves. These are as old evolutionarily as the presence of said competitors or enemies.

The activities of both channels and carriers are highly regulated by the cell in order to maintain the concentrations of nutrients and osmotica. Several important transporters are also regulated in their position on many types of cells, enabling directional transport of hormones and other signals between cells. Channels and carriers are specialized for a given solute, such as nitrate, potassium, or glucose, but they are not perfect in their selectivity. They may transport a range of solutes beneficially, and they may also transport toxic substances such as arsenate that is a close chemical analogue of phosphate. One notable feature of carriers and channels across taxa and across the nature of solutes they transport is their extraordinarily conservative molecular design, such that the genetic sequences specifying these transmembrane transporters are readily identified in all taxa. It appears that there is only one evolutionary solution for gross structure of molecular transporters, in contrast to multiple alternative solutions to other evolutionary challenges such as circulation in large animals (closed vs. open circulatory systems; iron-based or copper-based oxygen carriers . . . or none at all, in Antarctic ice fish).

Movement of solutes from the external environment to a cell occurs by diffusion and by the flow of water, termed mass flow or advection. Mass flow may or may not be important, variously by conditions. Consider plant roots taking up nutrient ions dissolved in soil water. Here, the mass flow of water is purely into (and through) the cell and the solute is avidly taken up, but mass flow is relatively unimportant. The flow of water does bring solute to the cell surface, but in doing so it flattens the diffusive gradient. The net result is little difference in solute uptake rate. Contrarily, mass flow is extremely important when the flow is copious and is passing by the cell. Cells in large animals require high rates of blood flow to deliver nutrients and O_2 and to remove CO_2.

Not all solutes are taken up. Many cells reject excess salt ions of various chemical species. The salt ions build up in concentration at the cell surface until a diffusive gradient helps reduce the solute flow and consequent uptake.

Some solutes move in combination with organic molecules synthesized by cells (Fig. 2). This facilitated diffusion allows the movement of solutes that otherwise would be too slow or else hazardous to the cells. Iron is extremely insoluble in water in the ferric state that dominates in aerobic conditions, but complexing agents (siderophores) combine avidly with iron and reach usefully high concentrations. In nitrogen-fixing root nodules on

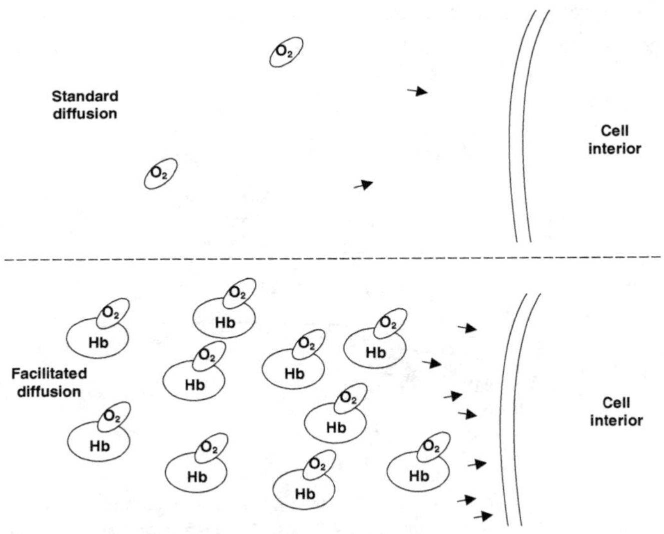

FIGURE 2 Facilitated diffusion of oxygen. (A) Oxygen is poorly soluble in the aqueous phase. Diffusion toward the cell interior at right is slow because the oxygen concentration is low. (B) Hemoglobin or a similar protein is abundant in the fluid bathing the cell (the globin is either free, as is leghemoglobin in nodules on N_2-fixing plants, or bound in cells, as in vertebrate blood). Diffusion of each Hb-O_2 complex is somewhat slower than that of individual O_2 molecules, but the much higher concentration of O_2 in the complexes generates a higher diffusion rate.

plants, oxygen is carried on leghemoglobin at high rates that sustain metabolism but prevent free molecular O_2 from attaining levels that damage the fragile nitrogenase enzyme.

Solutes and suspended particles can move between adjacent cells through fine tubelike connections, the plasmodesmata. The materials do not cross a membrane, enabling flow rates to be significant. Only the smallest solutes typically move on this pathway, although tobacco mosaic virus can induce a widening of the plasmodesmata so that the infection can spread.

Gases

Here, the consideration is the movement of gases in air or a similar gaseous medium; movement of dissolved gases is well described by the considerations above for solutes in water. The most important gases moving into and out of cells are O_2 and CO_2. There are active transporters in some cells, as in aquatic plants, for the bicarbonate ion formed from CO_2 but no transporters for the gases themselves. At the cellular level, pressure-induced mass flow of gases is important only or primarily in respiratory (more accurately, ventilatory) systems of larger animals. (There are minor flows induced into and out of plant leaves by the co-movement of water vapor leaving in copious quantities.) Otherwise, gases move by diffusion. The rates are high over short distances. Once gases enter water as solutes, rates of movement are greatly reduced, given that the diffusivity in water is about 1/100,000 that in air. This limitation underlies some features of anatomy, including the pressing of chloroplasts very close to the cell membrane in leaf cells of plants in order to reduce the length of the watery path of high diffusive resistance for CO_2 in photosynthesis.

"Bulk" Materials

Cells without rigid cell walls may, in certain conditions, engulf larger pieces of material by invaginating their plasma membrane and budding off a piece that gets internalized and then unloaded inside the cell—a process known as endocytosis. Conversely, they may exude large items (e.g., dead-end metabolite accumulations) by the reverse process of exocytosis. Endocytosis is a feeding mode for some single-celled organisms such as *Paramecium* species.

Heat

Temperature is an extremely important environmental variable for cellular metabolism and integrity. Thermal diffusivity is high in the watery medium of cells. Consequently, heat transport is so rapid that cells cannot maintain more than minuscule temperature differences from adjacent cells or their medium. At the scale of large organisms, thermal properties are more controllable.

Radiation

Radiation moves vectorially (in a directed manner) rather than as a scalar (undirected manner), as do water, solutes, and heat. Its transport in cells thus has special characteristics that will not be detailed here. The most important radiation in the cell is electromagnetic radiation, particularly light in the visible and ultraviolet wavebands. Along the short optical paths through most cells, light absorption is small. Exceptions occur in cells containing highly absorbing chemical compounds. Photosynthetic cells contain high concentrations of chlorophyll or bacteriochlorophyll and accessory pigments, with these sequestered in chloroplasts in eukaryotic cells. In plants and animals that dwell at or near the sunlit surface, epidermal cells contain various levels of UV-absorbing protective chemicals. For example, in leaves these are phenolics, and in many vertebrates these are polymeric melanins. In a diversity of cells, other compounds strongly absorb light as a signal. Animal eyes or photosensitive organs are a clear example. In the shoots of plants, cells of the epidermis absorb both red and far-red light as a signal of shading or of the interval from nightfall to dawn. In the former case, the signal is used to regulate development, particularly of height for competition for sunlight. In the latter case, the signal is one of several used for timing of flowering and other events.

Transport of thermal radiation at very long wavelengths is immaterial to cells, other than those in the pit vipers that detect thermal radiation from warm-blooded prey. The pits must be exquisitely sensitive for two reasons. Thermal radiation is only modestly differentiated between prey and the rest of the environment. Also, only very small temperature differences can be attained between sensor cells and their neighbors, given the high thermal conductivity of animal tissue.

TRANSPORT IN MULTICELLULAR ORGANISMS

Some General Features: Organized Flows

In multicellular organisms as in single cells, the same materials, heat, and radiation are transported, with the emergence of large-scale organized flows such as blood, xylem, mucus, and digestive boluses. Diffusive transport of O_2 and CO_2 in air largely suffices in the smallest animals. Insects have systems of tubes (trachea, tracheoles, spiracles) that provide effective diffusive paths. Body compression and expansion during movement can cause some pressure-induced flow to augment diffusion. Larger animals require lungs or lung-like systems for gas transport. These systems must terminate in very small structures such as alveoli, where diffusion completes the "last mile" of transport. Water accumulation in these structures from disease or direct water ingestion adds a great deal of transport resistance and can be fatal. Ventilation occurs at relatively low flow rates, resulting in laminar flow. Mixing of gases and frictional forces on system walls are correspondingly low. Dislodging foreign bodies such as dust requires either high flow rates as in coughing or the trapping of bodies in mucus flows moved by cilia on cells lining the passages.

The Need for Pressurized Fluid Flows

Larger animals need pressure-driven vascular systems to transport gases, nutrients, hormones, and such to and from all cells in blood or hemolymph. Even insects need this for transporting solutes if not dissolved gases. Hearts provided pumping pressure to move the fluid. Small animals make do with open circulatory systems while larger animals have more or less elaborate systems of blood vessels. The vessels must branch into finer and finer vessels, from which diffusion makes the final transport step, sometimes aided by active transport for nongaseous solutes. Safe pressure limits and minimal resistance to bulk fluid flow necessitate certain patterns of branching. These include Murray's law that two blood vessels branching from one have diameters that are close to 5/8 that of the larger vessel. Blood must have carriers with hemoglobin to add enough capacity to carry O_2, which is weakly soluble in water. Antarctic ice fish use no hemoglobin but have low metabolic rates.

Vascular plants do not need vascular or ventilatory systems to move O_2 and CO_2. The highest exchange rates are in leaves, which are thin and served well by diffusion. Other tissues have lower exchange rates. Plants do need pressure-driven flows in xylem and phloem. Water is certainly abundant in plants. Under pressure, it creates stiff cells and is the basis of most plant structure. However, most water is used by plants to make up for massive water losses from leaves as vapor. Leaves open their stomatal pores to allow entry of CO_2 for photosynthesis; unavoidably, water vapor exits the same pores (Fig. 3). Water transpiration is driven by the difference in partial pressure of water vapor between the inside of the leaf and the external air, commonly on the order of 1000 Pascals (Pa). Entry of CO_2 is driven by the difference in its partial pressure between external air and leaf interior, commonly on the order of 10 Pa. The 100:1 ratio of driving forces is amplified a bit more by the higher diffusivity of water vapor over that of CO_2. Further amplification occurs because cells must respire to gain energy, causing a loss of CO_2 back to the air. In consequence, plants transpire 100–1000 times as much mass of water as they gain in growth material (sugars, etc.) made in photosynthesis. In a sense, plants have generated this challenge. Their success in making some of their structures resistant to attack and decay has led to the geological burial of carbon and the corresponding depletion of the CO_2 content of air. Human activities are undoing this burial, though with a diverse set of consequences.

Uniquely among organisms, the flow of water in plants' xylem is almost always from one negative pressure (the matric potential of water in soil) to even more negative pressures in a chain leading to the leaves. Water under tension is metastable, with the potential to boil explosively into gas. Xylem vessels do undergo such cavitation

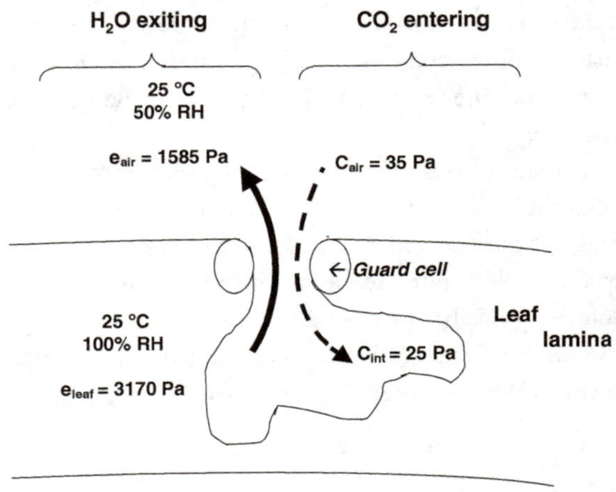

Partial pressure difference:	1585 Pa	10 Pa
Flux:	g_w*(1585 Pa)/P_{air}	0.62*g_w*(10 Pa)/P_{air}
Relative flux:	256	1

FIGURE 3 Water: CO_2 exchange ratio in plant leaves. CO_2 enters the leaf, shown in cross-section, to reach the substomatal cavity, while water vapor similarly exits the leaf. Transport into the cavity is through a stoma whose opening is controlled by two guard cells. Concentrations of both gases are given as partial pressures. Using modern molar units throughout, transport rate for each gas is equal to the difference in its mole fraction multiplied by the conductance of the stoma for that gas. Mole fraction is equal to partial pressure divided by total air pressure, P_{air}. Diffusive conductance for water vapor is denoted by g_w; conductance for the heavier CO_2 molecule is lower by a factor of 0.62. For simplicity, the computation at the bottom of the figure omits effects of the leaf boundary layer and of streaming flow of water vapor acting to oppose CO_2 entry.

with some regularity, disrupting their ability to transport water. However, the vessels are of fine diameter with tight sealing against air bubble entry that is the common cause of cavitation under extremely negative water potentials that occur in plant water stress. Additionally, refilling of xylem vessels has been shown to occur, using osmosis and some intriguing mechanical design features. The other major vascular system of plants is the phloem that moves diverse materials to nonphotosynthetic tissues, particularly sugars, processed or excess nutrients, and hormones. Positive pressures are generated in leaves by loading of sugars into the water-based phloem. Water flows into phloem under the osmotic force sufficient to overcome the pressure difference between leaves and phloem contents. Tissues downstream from the leaves take up sugar, reducing the pressure and inducing phloem flow. The distribution of sink strengths among the other tissues determines the pattern of how much material is delivered to each tissue. Regulation appears as a combination of positive control, ultimately under the action of gene expression patterns, and emergent properties.

All systems under pressure bear risks of puncture. Macroscopic wounds can cause major loss of fluid and failure of part or all of the circulatory system. Failure points are thus guarded. Blood clots, phloem coagulates, and xylem embolisms are trapped at bordered pits. High internal pressure itself is an intrinsic threat. Blood vessels in vertebrates are wrapped in materials that progressively decrease in extensibility under dilation, averting the effects seen in balloons being inflated more easily as they expand (codified as Laplace's law). The vessels also have an unusual Poisson ratio, maintaining length under dilation, avoiding shrinkage and risk of tearing.

Flow rates in vascular tubes of plant or animals are almost always low enough that flow is laminar, that is, with smooth velocity distributions, lacking turbulence. The flow rate, J, then follows the Hagen–Poiseuille law (the scientific law with the greatest rate of mispronunciation),

$$J = \frac{r^4}{8\eta}\frac{dP}{dx}.$$

Here, r is the radius of the vessel, η is the dynamic viscosity of the fluid, and dP/dx is the drop in pressure per unit length along the vessel. Small vessels are needed for close contact with tissues or, in plants, for resistance to cavitation, but they bear major costs. Many more fine vessels are needed than coarser vessels, for the same flow rate. Drought-resistant plants have fine xylem vessels but devote a lot of stem cross-section to the vessels and so transpire (and grow) at lower rates than less-resistant plants, as a rule.

Individual organisms vary greatly in size, ontogenetically and phylogenetically. A natural question is, How should bulk transport systems—xylem, arteries, lungs, and so on—scale with organismal size? We may also ask, What constrains the ultimate size of organisms of various functional types, and is it transport properties or simple strength of materials?

Consider the first question. Vessels within an individual vary in number and size. Part of the answer is (deceptively) simple, that bigger organisms need more vessels. However, large organisms are not simply photographic enlargements of small organisms; transport vessels such as blood vessels have a range of sizes constrained by the physics of materials. As one example, large xylem vessels in plants are inherently more susceptible to catastrophic air entry, restricting the water-stress levels at which they may operate. Another part of the answer involves deeper biophysics. Transport systems ramify in all organisms: in trees, from trunk to branch to branchlet to leaf periole to leaf vein; in whales, aorta to artery to arteriole to capillary (and back, in the venous system). How should vessels be distributed in size and number at each level of ramification? The theory of these scaling relations has been an active area of research for decades, recently with a greater focus on functional implications vs. simple description. For example, the ability of a tree to gather light scales with leaf area and, thus, closely with the square of an overall linear dimension such as height. The volume and mass of supporting structures scales more closely as the cube of this dimension (more exact scaling requires additional biomechanical theory). The reduction in photosynthetic rate per unit biomass constrains growth rates. Given any biomechanical design, there are still questions about the scaling of linked transport capacities: How should water-transport capacity in the tree trunk scale with gas-transport capacity (transpiration) at the leaves? The basic answer is that the two capacities should be in constant ratio. Not all organisms of the same functional type maintain the same scaling. Drought-resistant plants operate with lower capacities than mesic plants, including lower xylem-transport capacity per unit leaf area. Poikilothermic animals, with their generally lower metabolic rates per unit mass than endothermic animals, have generally lower transport capacities per unit mass.

The second question, of ultimate size constraints, remains an area of research. Trees have never exceeded 120 meters in height; whales are the largest of all animals but do not exceed 33 meters in length. Some mechanical constraints do not appear to limit size. Whales are supported by buoyancy, thus having proportionately lighter skeletons than significantly smaller land animals, the elephants. A transport limitation to their size may arise from the feeding rate scaling in lesser proportion to length than does metabolic demand. The limits to tree height may involve limits of strength, but the answer is not clear-cut. Wood has sufficient strength for greater tree heights, even in compression, where it is weaker than in tension. However, tall structures of a given aspect ratio (height to diameter) are susceptible to buckling, initiated by wind or earthquakes. In trees, a variety of other hypotheses for height limitation have been proposed and tested, albeit with controversial results. Two such hypotheses derive from the necessarily low water potential at great heights. One hypothesis is that xylem cavitation becomes excessive and even irreversible; another is that this low water potential imposes an untenable reduction in photosynthetic rates of leaves.

Transport of Heat and Radiation

Organisms heat and cool at their peril. Temperature extremes affect overall metabolism, capacity for movement, metabolic balance, water balance, and more. They also exchange energy with the environment in a number of ways, involving radiation, conduction, water evaporation, and fluid convection in air or water. Organisms generate heat metabolically, adding to or counteracting external loads. Exchanges with the environment are discussed in this volume in the entry on energy budgets. Here, we are concerned with internal transport processes.

Heat generated by any means and at any location moves within an organism by conduction (molecular interactions) and convection (bulk fluid movement). Thermal radiation also transports heat within organisms in a minor way. Conduction is basically a diffusion process and is thereby effective only at small spatial scales or long times. Moving fluids such as blood transport heat into or out of an organism's body at higher rates. Fluid flow rates in plants are quite low (ca. 1 meter per hour in xylem, a few mm per hour in phloem). This disallows their having a significant role in heat transport and energy balance. In endothermic animals, blood flow is of major importance for energy balance. Flow from core to periphery is regulated, increasing if heat must be shed and decreasing if heat must be retained. The peripheral tissues experience more extreme temperatures than the organism's core, but they are less sensitive to extremes. In some cases, blood must flow to a cold periphery while heat loss must be kept minimal. Examples include birds wading in cold water or flying while breathing in cold air at their nostrils and tuna swimming in cold oceans. Countercurrent flow (Fig. 4) helps conserve heat. This phenomenon is described in many physiology texts. Essentially, vessels going to and from the periphery are closely appressed. Heat is exchanged along the whole length of the transport, being transferred to returning blood rather than delivered right to the cold periphery.

An organism's thermal mass (mass times heat capacity per unit mass) stabilizes the change in temperature. Temperature change is proportional to heat exchanged divided by this thermal mass or inertia. This can be adaptive except when heat must be shed or gained quickly. Animals of increasing size are challenged by this, even with efficient heat transport in blood. They simply lack adequate surface area for final exchange of heat with the environment.

A Selection of Additional Considerations of Transport in Vascular Plants

In obtaining water, soil-rooted plants are at the mercy of the environment. As the soil dries, not only is the total amount of water reduced, but the hydraulic conductivity

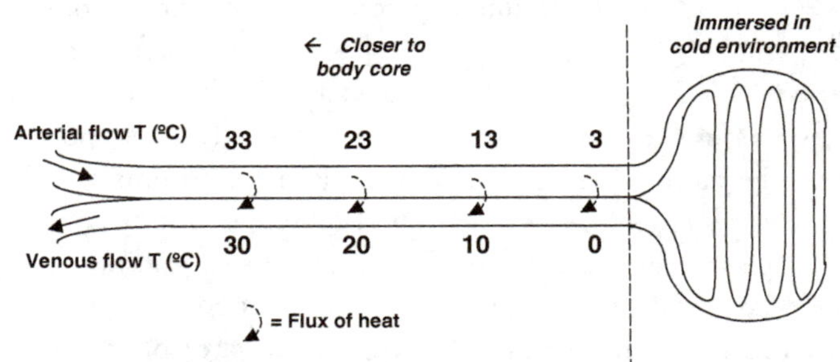

FIGURE 4 Countercurrent flow of blood for heat retention. Arterial blood flow from the body core at high temperature passes much of its heat content to returning venous flow that has been chilled at the distal part of the body, e.g., the leg of a bird wading in ice-cold water. Flow at the extremity occurs in highly branched vessels (arterioles, capillaries, venules) sketched simply at the extreme right. Blood at this extremity equilibrates virtually completely with the cold environment. Efficiency in heat exchange is enabled by both vessels being appressed to each other. In common systems, multiple vessels of each type may be appressed.

of the soil declines precipitously. Water uptake declines to low rates well before soil water potential becomes lower than the potential that the plant can attain. At the other end of the water-transport chain, water continues to be lost from leaves at low rates even if the stomatal pores shut tightly; vapor leaks through the epidermal covering. By necessity, then, plants exert rather conservative control over water use early in the process of stress development. In coarse soils, low water potentials develop late in the course of soil water depletion. Stress can develop too suddenly for effective responses by plants not adapted to such soils.

Water can be transported in reverse, from roots to soil that is at lower water potential. As a result, water can be moved from wetter soil layers to the dryer layers. The process has been denoted as hydraulic redistribution. It can occur even through recently dead roots. This can be beneficial to the plant doing the redistribution, such as by storing water in deeper soil where it is less subject to evaporation. Contrarily, if water is moved from depth to the shallower layers, it may be used by competing plants. Some plants are thereby nursed by larger woody plants. Plants can reduce water loss back to soil if roots shrink away from contact with the soil. Transport of water then occurs mainly in the vapor phase, at low rates. Cacti, with their commonly shallow roots and no access to deeper water, exhibit this behavior. Even with this protective measure, they tolerate extended droughts more poorly than many woody plants with notably lower efficiency in using transpired water.

Very tall plants run into major effects of the gravitational term in water potential. At the top of a tree 100 m tall, the drop in water potential from this term is 1 MPa, or 10 bar. This adds costs of osmotic adjustment and of transport tissue (xylem) and contributes to height limitation.

Plants' leaves are structured for efficient use of photosynthetically active radiation. Flux densities taper off with depth from the surface, and so do investments in pigments and photosynthetic enzymes. It is of interest that plants intercept far more of bright sunlight than they can use. That is, their photosynthetic rates saturate at far less than full sun. Mathematical modeling has been used to interpret this as depriving competitors of the excess resource.

A Selection of Additional Considerations of Transport in Animals

Animal tissues lack the strong reinforcement of cellulose and lignin present in plants. Their vascular systems can only bear lower pressure differentials between interior and exterior. Chambered hearts in vertebrates reduce the problem. They allow two separate circulations, to lungs and to the rest of the body, each operating at a modest pressure. In single circulatory systems, the heart must develop enough pressure at its exit for blood to traverse both the lungs and the rest of the body. To keep the pressure tolerable, blood flow must be slower, limiting metabolism and physical activity. Tall animals are subject to changes in gravitational potential. Giraffes must have special valves in their neck and head to prevent damaging rises in pressure potential in the head when they lower their heads to drink.

Some aquatic animals use fluid flow for propulsion, examples being squid and larval dragonflies. Muscular chambers eject water in rapid flows. At small sizes, this becomes ineffective because viscous forces dominate over inertial forces. Small organisms thus use cilia and flagella to move.

Animals have diverse ways of acquiring water, including bulk ingestion or drinking and osmotic uptake through the skin. Some of the complementary routes of water loss are underappreciated. Desert lizards suffer up to half their water loss from their eyes, which cannot be made highly impervious to vapor loss as can their skins. Animals with ventilatory systems lose notable amounts of water as vapor in exhalations, as these contain water vapor nearly at saturation pressures at body temperature. Some animals recover a major part of this water by maintaining lower temperatures in tissues at the exit (e.g., nares) to condense some of the vapor.

Kidneys and similar organs are remarkable adaptations for allowing excretion of waste solutes with minimal water loss. Sequences of transport of solutes occur in conditions sufficiently far from thermodynamic equilibrium to allow water recovery from the urinary stream. Details can be sought in texts on animal physiology. Animals vary widely in capacity to concentrate the solutes. There are corresponding limits in environmental conditions that they can tolerate.

CONCLUDING REMARKS

General principles of transport explain a great deal of physiological and ecological phenomena. They enable us to comprehend the many evolutionary solutions to the demands and tradeoffs that accompany transport in organisms.

Additional transport phenomena of strong ecological interest emerge above the level of the organism. Examples include dispersal of pollen in air and the transport of water and nutrients in oceanic and atmospheric circulations. These phenomena are discussed in other entries in this volume. Even at the cellular and organismal level, the discussion in this section only scratches the surface. Many transport phenomena have ecological implications, such as chemosensing in water and air, light propagation in vision, momentum transport in air and water flows, capillary transport of water at leaf cell walls, and sound propagation. Diverse references (see "Further Reading," below) reward the ecologist by expanding upon the topics already introduced here.

Ecology, both theoretical and experimental, has much to offer in understanding transport systems in organisms and how they evolved in response to diverse selection pressures. A number of open questions, such as about scaling of transport and about size limitations on individual organisms, have been framed variously by biophysicists or physiologists without the full context of ecology. In adding their contributions, ecologists must grasp to a sufficient degree the physics, chemistry, materials science, and physiology that have been applied to date. The challenges are considerable, yet quite manageable for the quantitatively inclined ecologist.

There is no shortage of relevant questions, going well beyond the limited number noted up to this point. For example, one may ask how the selectivity of salt transport in plant roots or in aquatic animals offers a sufficiently low cost–benefit ratio in various ecological niches, and why selectivity persists at limited levels in most organisms when it should be energetically inexpensive to be more selective. One may also try to predict how the transport systems of individual organisms—stomatal conductance of plants, heat transport in large animals, and so on—will acclimate to elevated CO_2 and climate change . . . and how evolution will proceed. The list of questions can, and will be, extended greatly with the insights of ecology. Moreover, it falls primarily to ecologists to delineate the array of selection pressures on organisms, enabling the proper formulation of optimal or at least stable solutions in organismal performance, including transport systems at all levels of organization.

SEE ALSO THE FOLLOWING ARTICLES

Energy Budgets / Environmental Heterogeneity and Plants / Functional Traits of Species and Individuals / Gas and Energy Fluxes across Landscapes / Integrated Whole Organism Physiology / Plant Competition and Canopy Interactions

FURTHER READING

Campbell, G. S., and J. M. Norman. 1998. *An introduction to environmental biophysics,* 2nd ed. New York: Springer-Verlag.

Denny, M. W. 1993. *Air and water: the biology and physics of life's media.* Princeton: Princeton University Press.

Vogel, S. 1992. *Vital circuits.* Oxford: Oxford University Press.

TWO-SPECIES COMPETITION

PRIYANGA AMARASEKARE
University of California, Los Angeles

Competition occurs when interactions between individuals (of the same or different species), arising due to the acquisition of a shared resource that is in limiting supply, lead to a reduction in the survivorship, growth, and/or reproduction. It is a mutually negative (−/−) interaction, as opposed to the antagonistic (+/−) interactions such as those arising between predators and prey or hosts and parasites. The limiting resources that lead to mutually negative interactions can be abiotic (e.g., nutrients, water, light), biotic (food, mates), or space that contains essential resources (e.g., territories and/or home ranges).

ECOLOGICAL SIGNIFICANCE OF COMPETITION

Competition is important at all levels of biological organization. In an ecological context, it is crucial because of the negative feedback it exerts on populations (leading to population regulation) and the impact it has on maintaining species diversity within communities. It is also the key process underlying important practical problems such as the invasion of exotic species. Understanding the dynamics of competitive interactions is therefore essential in addressing both fundamental ecological questions and applied questions of critical environmental importance.

Competition can be of two main types. Exploitative competition involves indirect negative interactions between individuals as the result of acquiring a resource that is in limiting supply. Each consumer affects others solely by reducing the abundance of the shared resource. For instance, in granivorous species for whom seeds constitute a limiting resource, an individual that finds a pile of seeds will have a negative interaction on other individuals simply by virtue of having deprived the others of that particular pile of seeds. In contrast, interference competition involves direct negative interactions between individuals in their attempts to acquire a resource in limiting supply. For instance, individuals that acquire territories containing an essential resource (e.g., food, mates, nesting sites) deprive others of access to the resource by actively defending the territory. Other examples of interference competition include overgrowth and/or undercutting in sessile organisms, predation or parasitism, and chemical competition such as allelopathy.

COMPETITION AS A NEGATIVE FEEDBACK PROCESS INDUCING POPULATION REGULATION

Competition between individuals of the same species (intraspecific competition) is critical for population persistence. It leads to a negative feedback process, which allows populations to increase from initially small numbers to a stable equilibrium defined as the carrying capacity, the maximum number of individuals that can be supported given available resources. The negative feedback arises from negative density dependence in the per capita growth rate. When the number of individuals in the population is small, there is little or no resource limitation and the per capita growth rate is high, i.e., populations can grow at the rate determined by the difference between per capita birth and death rates. As the number of individuals increases, resources are depleted, causing a reduction in survival and reproduction of those individuals who are deprived of adequate amounts of the resource. As a result, the per capita growth rate starts to decline as population size increases and eventually becomes zero. The size of the population at this point is the carrying capacity. The familiar logistic model provides a mathematical representation of these ideas:

$$\frac{dN}{dt} = rN\left(1 - \frac{N}{K}\right), \qquad (1)$$

where r is the intrinsic growth rate, N is the population size, and K is the carrying capacity. The key idea here is that of negative density dependence, which is defined as the decline in the per capita growth rate as a function of density. From Equation 1, the per capita growth rate $f\left(f = \frac{1}{N}\frac{dN}{dt}\right)$ is $r\left(1 - \frac{N}{K}\right)$, which decreases as N increases and approaches zero as N approaches K. An equilibrium of the population defined by Equation 1 occurs when the rate of change in population size with respect to time is zero, i.e., when $\frac{dN}{dt} = 0$, $N^* = K$. We can evaluate the stability of this equilibrium by computing the eigenvalue $\left(\frac{\partial f}{\partial N}\right)$ and evaluating it at the equilibrium

$$N^* = K, \text{ i.e., } \left.\frac{\partial f}{\partial N}\right|_{N^* = K} = -r.$$

Since the eigenvalue is negative as long as r is positive, $N^* = K$ is stable to perturbations, i.e., if the population is displaced from K, it will return to K at a rate r.

The key point is that intraspecific competition provides a negative feedback process that allows population regulation, i.e., populations increase in size when they are small and decrease in size when they are large. The result is the persistence of the population at a stable, intermediate abundance dictated by resource availability.

Intraspecific competition also plays a key role in species coexistence. This topic is discussed in the next section.

COMPETITION AS A DIVERSITY REDUCING/ENHANCING PROCESS IN ECOLOGICAL COMMUNITIES

Competition between individuals of different species leads to one of two outcomes: exclusion of one species by the other or coexistence at a stable equilibrium. Which of the two outcomes occurs depends on the relative strengths of intraspecific and interspecific competition. Competitive exclusion occurs under two situations. The first is competitive dominance, which occurs when intraspecific competition is stronger than interspecific competition for one species but not the other, i.e., the species whose negative effect on conspecifics is greater than the negative effect it suffers from heterospecifics will exclude the species whose negative effect on conspecifics is weaker than the negative effect it suffers from heterospecifics. The second is a priority effect: when intraspecific competition is weaker than interspecific competition for both species, the species with the higher initial abundance excludes the other. Coexistence occurs when intraspecific competition is stronger than interspecific competition for both species, i.e., individuals of both species have a greater negative effect on conspecifics than on heterospecifics.

Mechanisms that cause intraspecific competition to be stronger than interspecific competition are called niche partitioning mechanisms. A species' niche can be thought of as having four major axes: resources, natural enemies, space, and time. Niches can be partitioned in three major ways. First, species may specialize on distinct resources or be subject to density- or frequency-dependent predation by natural enemies. Second, different species may be limited by the same resources or natural enemies, but differ in terms of *when* they exploit the resource or respond to natural enemies (temporal niche partitioning). Third, species could differ in terms of *where* they experience, and respond to, limiting factors (spatial niche partitioning).

Mathematical theory on two-species competition takes two main forms. The first type considers resource dynamics only implicitly, i.e., the resource is assumed not to accumulate within the system, so it can be treated as an input rather than a state variable. The most well-known such model is the Lotka–Volterra competition model, in which resource limitation is expressed in terms of the carrying capacities of the two competing species. The second type of theory considers resource dynamics explicitly such that there is feedback between the per capita growth rates of resources and consumers. The most well-known model of this type is Tilman's R^* theory.

Models with implicit resource dynamics are a useful starting point for understanding the conditions that allow coexistence vs. exclusion under pure exploitative competition. They are also a good starting point for understanding the spatial dynamics of competitive interactions. They do have the drawback of being phenomenological, i.e., the outcomes of the models are consistent with several different biological mechanisms. The models of explicit resource dynamics have the advantage of providing a more mechanistic understanding of how competitive interactions lead to coexistence. These models are therefore biologically more realistic. The following sections provide examples of both types of models.

Two-Species Competition Models with Implicit Resource Dynamics: Lotka-Volterra Competition Model

The dynamics of two species competing for a limiting resource are given by:

$$\frac{dX}{dt} = r_x X\left(1 - \frac{X}{K_x} - \frac{\phi_x Y}{K_x}\right),$$
$$\frac{dY}{dt} = r_y Y\left(1 - \frac{Y}{K_y} - \frac{\phi_y X}{K_y}\right), \qquad (2)$$

where X and Y are, respectively, the abundances of each species, ϕ_x and ϕ_y are the per capita competition coefficients, r_x and r_y are the per capita growth rates, and K_x and K_y are the carrying capacities of species 1 and 2.

Nondimensionalizing Equation 2 allows one to describe the system using a minimal set of parameters. The transformations

$$x = \frac{X}{K_x}, \quad y = \frac{Y}{K_y}, \quad \tau = r_x t,$$
$$a_x = \phi_x \frac{K_y}{K_x}, \quad a_y = \phi_y \frac{K_x}{K_y}, \quad \rho = \frac{r_x}{r_y},$$

yield the nondimensional system

$$\frac{dx}{dt} = x(1 - x - a_x y),$$
$$\frac{dy}{dt} = \rho^* y(1 - y - a_y x), \qquad (3)$$

where x and y are the densities of species 1 and 2 scaled by their respective carrying capacities and a_x and a_y are the per capita effect of species 2 on species 1 (and vice versa) scaled by the ratio of respective carrying capacities. The quantity ρ is the ratio of the per capita growth rates of the two species, and τ is a time metric that is a composite of t and r_x, the growth rate of species 1.

Equation 3 yields four equilibria: (i) the trivial equilibrium with both species extinct $((x^*, y^*) = (0,0))$; (ii) species 1 at carrying capacity, species 2 extinct $((x^*, y^*) = (1,0))$; (iii) species 2 at carrying capacity, species 1 extinct $((x^*, y^*) = (0,1))$; and (iv) the coexistence equilibrium $\left((x^*, y^*) = \left(\frac{1 - a_x}{1 - a_x a_y}, \frac{1 - a_y}{1 - a_x a_y}\right)\right)$.

Cases (ii) and (iii) represent competitive dominance, where one species excludes the other and goes to its carrying capacity. The coexistence equilibrium (case (iv)) is stable if $a_x < 1$ and $a_y < 1$ and unstable otherwise. When the coexistence equilibrium is unstable, a priority effect occurs and the species with the higher initial abundance excludes the other. The priority effects case, in which intraspecific competition is weaker than interspecific competition for both species, can be considered as representing interference competition in a phenomenological sense. A mechanistic understanding of exploitative and interference competition requires models with explicit resource dynamics, which are considered next.

Two-Species Competition Models with Explicit Resource Dynamics

EXPLOITATIVE COMPETITION FOR A LIMITING RESOURCE (R^* RULE)

Consider two consumer species that compete for a single, biotic resource. Resource dynamics occur on the same time scale as those of the consumers' and the resource species experiences self-limitation in the absence of consumption. The consumer species interact via exploitative competition.

The dynamics of the community are given by

$$\frac{dR}{dt} = R\left(r\left(1 - \frac{R}{K}\right) - a_1 C_1 - a_2 C_2\right),$$
$$\frac{dC_1}{dt} = C_1(e_1 a_1 R - d_1), \qquad (4)$$
$$\frac{dC_2}{dt} = C_2(e_2 a_2 R - d_2),$$

where r and K are the growth rate and the carrying capacity of the resource, and a_i, e_i, and d_i are the attack rate, conversion efficiency, and the death rate of consumer i $(i = 1, 2)$.

Let $R^*_{C_i}$ be the minimum resource level required for consumer 1 to persist in the absence of consumer 2 and vice versa. From Equation 4, $R^*_{C_i} = \frac{d_i}{e_i a_i}$. Consumer i can invade when consumer j is at equilibrium with the resource if $\frac{d_j}{e_j a_j} > \frac{d_i}{e_i a_i}$ $(i, j = 1, 2; i \neq j)$. Since invasibility criteria for the two consumers are mutually exclusive, we get the R^* rule: the consumer species that drives resource abundance to the lowest level will exclude the other (Tilman, 1982).

APPARENT COMPETITION DUE TO A SHARED NATURAL ENEMY (P^* RULE)

Next, consider two consumer species that do not compete for resources but share a common natural enemy. The dynamics of such a community are given by

$$\frac{dC_1}{dt} = C_1(r_1 - a_1P),$$
$$\frac{dC_2}{dt} = C_2(r_2 - a_2P), \quad (5)$$
$$\frac{dP}{dt} = P(e_1a_1C_1 - e_2a_2C_2 - d),$$

where r_i is the growth rate of consumer species i, a_i and e_i are the attack rate and conversion efficiency of the natural enemy when it preys on consumer i, and d_i is the predator's mortality rate.

Let $P^*_{C_i}$ be the maximum natural enemy abundance consumer 1 can withstand in the absence of consumer 2 and vice versa. From Equation 5, $P^*_{C_i} = \frac{r_i}{a_i}$. Consumer i can invade a resident equilibrium of consumer j and the predator if $\frac{r_i}{a_i} > \frac{r_j}{a_j}$. As with exploitative competition (discussed above), invasibility criteria for the two consumers are mutually exclusive, and we get the P^* rule: the consumer species that can withstand the highest natural enemy pressure will exclude the other. This phenomenon is called apparent competition: indirect, mutually negative interactions between two consumer species generated by their interaction with a common natural enemy (Holt, this volume).

Coexistence via Local Niche Partitioning

As the above analyses show, when resource dynamics are taken into account, interactions between consumers that engage in exploitative competition or apparent competition always lead to the exclusion of the inferior competitor (or apparent competitor). Coexistence requires niche partitioning mechanisms that cause intraspecific competition to be stronger than interspecific competition. Such mechanisms arising from species interactions within local communities even in the absence of environmental variation are termed local niche partitioning mechanisms. Several such mechanisms are discussed next.

EXPLOITATIVE COMPETITION FOR TWO OR MORE RESOURCES: RESOURCE RATIO THEORY

A common mechanism of local niche partitioning in plant species involves differences between species in their requirement of two or more essential resources (e.g., water and light or nitrogen and phosphorous). In the case of two-species competition, if one species is limited more by one resource and the other species is more limited by the other resource and there is spatial variation in resource availability, each species will reach its peak abundance at a different ratio of the two resources. This spatial segregation increases the strength of intraspecific competition relative to interspecific competition and allows stable coexistence.

EXPLOITATIVE AND APPARENT COMPETITION FOR A LIMITING RESOURCE

If two species are limited by a common resource and a shared natural enemy, they can coexist via an interspecific tradeoff between competitive ability and susceptibility to predation, i.e., one species has a lower R^* and hence is the superior competitor for the shared resource, while the other species has a higher P^* and thus has greater tolerance of (or resistance to) natural enemy pressure. This type of tradeoff between competition and apparent competition leads to the well-known diamond food web, which provides a starting point for investigating the joint impact of competitive and antagonistic interactions in structuring food webs (see Food Webs).

EXPLOITATIVE AND INTERFERENCE COMPETITION FOR A LIMITING RESOURCE

Many species engage in both exploitative and interference competition, but most theory on competition has focused only on exploitative competition. It is therefore important to determine the conditions under which two species that engage in both exploitative and interference competition for a limiting resource can stably coexist. This can be done with a straightforward extension of Equation 4:

$$\frac{dR}{dt} = R\left(r\left(1 - \frac{R}{K}\right) - a_1C_1 - a_2C_2\right),$$
$$\frac{dC_1}{dt} = C_1(e_1a_1R - d_1 - (\alpha_{12} \pm \beta_1\alpha_{21})C_2), \quad (6)$$
$$\frac{dC_2}{dt} = C_2(e_2a_2R - d_2 - (\alpha_{21} \pm \beta_2\alpha_{12})C_1),$$

where α_{ij} is the per capita effect of interference *from* consumer j on consumer i, and $\beta_i\alpha_{ji}$ is the per capita cost or benefit to consumer i due to interference *on* consumer j. In the interests of analytical tractability, exploitation and interference are considered to be linear functions of resource and consumer abundance, respectively.

Nondimensionalizing Equation 6 with the transformations

$$\hat{R} = \frac{R}{K}, \qquad \hat{C}_i = \frac{C_i}{e_i K}, \qquad \hat{a}_i = \frac{a_i e_i K}{r},$$
$$\hat{d}_i = \frac{d_i}{r}, \qquad \hat{\alpha}_{ij} = \frac{\alpha_{ij} e_j K}{r}, \qquad \beta_i = \frac{e_j \beta_i}{e_i},$$
$$\tau = rt, \qquad (i, j = 1, 2, i \neq j),$$

and dropping the hats leads to the following nondimensional system:

$$\frac{dR}{d\tau} = R(1 - R) - a_1 R C_1 - a_2 R C_2,$$
$$\frac{dC_1}{d\tau} = a_1 R C_1 - d_1 C_1 - I_{C1} C_1 C_2, \qquad (7)$$
$$\frac{dC_2}{d\tau} = a_2 R C_2 - d_2 C_2 - I_{C2} C_1 C_2,$$

where $I_{C_1} = \alpha_{12} \pm \beta_1 \alpha_{21}$ and $I_{C_2} = \alpha_{21} \pm \beta_2 \alpha_{12}$.

Equation 7 yields five feasible equilibria. The trivial equilibrium $[(R^*, C_1^*, C_2^*) = (0,0,0)]$ is unstable for all positive values of R, C_1, and C_2. The equilibrium with both consumers extinct (1, 0, 0) is stable if and only if neither consumer is able to maintain a positive growth rate when the resource is at carrying capacity. One or both consumers can invade when rare if $\frac{d_i}{a_i} < 1$ $(i = 1, 2)$. There are two boundary (two-species) equilibria with the resource and consumer i in the absence of consumer $j\left[\left(R^*, C_i^*, C_j^*\right) = \left(\frac{d_i}{a_i}, \frac{a_i - d_i}{a_i^2}, 0\right)\right]$, and a unique interior equilibrium with all three species present:

$$R^* = \frac{a_1 d_2 I_{C_1} + a_2 d_1 I_{C_2} + I_{C_1} I_{C_2}}{a_1 a_2 (I_{C_1} - I_{C_2}) + I_{C_1} I_{C_2}},$$
$$C_1^* = \frac{I_{C_1}(a_2 - d_2) - a_2(a_1 d_2 - a_2 d_1)}{a_1 a_2 (I_{C_1} + I_{C_2}) + I_{C_1} I_{C_2}},$$
$$C_2^* = \frac{I_{C_2}(a_1 - d_1) - a_1(a_2 d_1 - a_1 d_2)}{a_1 a_2 (I_{C_1} + I_{C_2}) + I_{C_1} I_{C_2}}.$$

Coexistence requires that (i) each consumer species is able to invade when the other species is at equilibrium with the resource, and (ii) the coexistence equilibrium is stable to small perturbations in the abundance of all three species.

From Equation 7, the invasion criteria for the two consumers are $a_2(a_1 d_2 - a_2 d_1) > (a_2 - d_2) I_{C_1}$ and $a_1(a_2 d_1 - a_1 d_2) > (a_1 - d_1) I_{C_2}$, respectively. Assume, without loss of generality, that consumer 1 is superior at resource exploitation i.e., $\frac{d_1}{a_1} < \frac{d_2}{a_2} < 1$.

When interference competition incurs only costs ($I_{C_1} = \alpha_{12} + \beta_1 \alpha_{21} > 0$ and $I_{C_2} = \alpha_{21} + \beta_2 \alpha_{12} > 0$), $a_2 d_1 < a_1 d_2$, and $a_1(a_2 d_1 - a_1 d_2) < (a_1 - d_1) I_{C_2}$. Hence, consumer 2 cannot invade when rare. This is because inferiority in resource exploitation prevents consumer 2 from maintaining a positive growth rate even in the absence of interference. From Equation 8 it can be seen that $C_2^* > 0$ only if consumer 2 cannot invade when rare.

Consumer 1 can invade when rare provided $a_2(a_1 d_2 - a_2 d_1) > (a_2 - d_2) I_{C_1}$. In biological terms, consumer 1 can invade if the cost of interference on consumer 2 and effect of interference from consumer 2 are both small ($I_{C_1} \rightarrow 0$) and consumer 2 is an inefficient exploiter of the resource ($a_2 \sim d_2 \Rightarrow R_{C_2}^* \rightarrow 1$). Since consumer 2 cannot invade at all, the outcome is competitive dominance by consumer 1, the superior resource exploiter.

If $a_2(a_1 d_2 - a_2 d_1) < (a_2 - d_2) I_{C_1}$, then consumer 1 cannot invade when rare. As can be seen from Equation 8, this is the only condition under which $C_1^* > 0$. Thus, feasibility of the coexistence equilibrium $[(C_1^*, C_2^*) > (0,0)]$ requires that neither consumer species can invade when rare. It is straightforward to show that the coexistence equilibrium is unstable when it exists. Under these conditions, the two species boundary equilibria $\left[\left(R^*, C_i^*, C_j^*\right) = \left(\frac{d_i}{a_i}, \frac{a_i - d_i}{a_i^2}, 0\right)\right]$, $(i, j = 1, 2; i \neq j)$ are both globally stable. Since neither consumer species can invade when rare, the outcome is a priority effect where the species with the higher initial abundance excludes the other.

When interference competition accrues benefits to both species (i.e., ($I_{C_i} = \alpha_{ij} - \beta_i \alpha_{ji}$ and $I_{C_j} = \alpha_{ji} + \beta_j \alpha_{ij}$)) mutual invasibility is possible provided I_{C_i} and I_{C_j} are not both positive. Assuming as before that consumer 1 is the superior resource exploiter $\left(\text{i.e., } \frac{d_1}{a_1} < \frac{d_2}{a_2} < 1\right)$, it can invade when rare if $a_2(a_1 d_2 - a_2 d_1) > (a_2 - d_2) I_{C_1}$. Consumer 2 can invade when rare if $a_1(a_2 d_1 - a_1 d_2) > (a_1 - d_1) I_{C_2}$. A necessary condition for mutual invasibility is that consumer 1 suffers a net loss from interference ($I_{C_1} > 0$) while consumer 2 accrues a net benefit ($I_{C_2} < 0$). Unlike the situation when interference incurs only costs, conditions for mutual invisibility are also the conditions for feasibility of the coexistence equilibrium $[(C_1^*, C_2^*) > (0,0)$; Eq. 8].

The key results are as follows. When two consumer species compete for a single biotic resource via exploitative and interference competition and interference involves mechanisms that incur a net cost, they cannot coexist at a point attractor even when the inferior resource exploiter is superior at interference. If the superior resource exploiter suffers little impact from interference from the inferior resource exploiter and/or incurs little or no cost from interference on the latter, then interference cannot alter the outcome of competitive dominance by the superior resource exploiter. If the cost and effect of interference are sufficiently high that the superior resource exploiter cannot invade a community consisting of the resource and the inferior resource exploiter, then a priority effect occurs and the consumer species with the higher initial abundance excludes the other. In contrast, when interference involves mechanisms that provide a benefit to the interacting species, coexistence is possible provided competing species exhibit an interspecific tradeoff between exploitation and interference.

Coexistence via Temporal Niche Partitioning

There are two basic ways in which organisms can partition niches in time. The first occurs via nonlinear competitive responses and the second occurs via the storage effect. These mechanisms are briefly described here and are discussed in detail by Snyder (this volume).

NONLINEAR COMPETITIVE RESPONSES

Most organisms exhibit nonlinear functional responses, particularly ones where the per capita consumption rate saturates at high resource abundances (e.g., due to long handling times in predators or egg limitation in insect parasitoids). These responses cause the species' per capita growth rates to depend on resource abundance in a nonlinear manner. If resource abundance fluctuates and species that compete for the resource differ in the degree of nonlinearity in their functional responses, stable coexistence is possible if the species with the more nonlinear response is more disadvantaged, when abundant, by resource fluctuations than the species with the less nonlinear response. Large resource fluctuations depress the per capita growth rate of the species with the more nonlinear response, which reduces competition on the species with the less nonlinear response and allows it to invade when rare. The species with the more nonlinear functional response, which has the lower R^*, is better at exploiting the resource when resource abundance is lower, and the species with the less nonlinear response is better at exploiting the resource when resource abundance is higher. Thus, resource fluctuations allow coexistence via temporal resource partitioning. They can arise due to abiotic environmental variation (e.g., seasonal variation in temperature), or because the resource's interaction with the consumer that has the more nonlinear functional response.

STORAGE EFFECT

A temporal storage effect occurs when competing species differ in their responses to abiotic environmental variation. Such species-specific differences, termed the environmental response, modify the strength of competition between species. For example, two competing species that differ in their temperature sensitivity may be active at different times of the year, and the ensuing reduction in temporal overlap will reduce the intensity of interspecific competition. This effect is quantified as the covariance between the environmental response and competition. When species-specific responses to environmental variation modify competition, intraspecific competition is the strongest when a species is favored by the environment, and interspecific competition is the strongest when a species' competitors are favored by the environment. Thus, the covariance between the environmental response and competition is negative (or zero) when a species is rare, and positive when it is abundant. This relationship can be understood as follows. When a species is favored by the environment, it can respond with a high per capita growth rate (i.e., positive or strong environmental response) and reach a high abundance. When abundant, the species experiences mostly intraspecific competition because the environment is unfavorable to its competitor. Thus, a strong environmental response increases intraspecific competition, resulting in a positive covariance between the environmental response and the strength of competition. In contrast, when the species is not favored by the environment its per capita growth rate is low (weak environmental response) and it will remain rare. It will experience mostly interspecific competition (because the environment is now favorable to the species' competitors), but the species' rarity will reduce the strength of interspecific competition it experiences. Thus, a weak environmental response reduces interspecific competition, resulting in a zero or negative covariance between the environmental response and the strength of competition.

Coexistence via the storage effect requires a third ingredient, buffered population growth, which ensures that the decline in population growth due to unfavorable abiotic conditions is offset by an increase in population growth due to favorable abiotic conditions. This occurs via Jensen's inequality: when the per capita growth rate is a nonlinear function of a trait that is subject to environmental variation, the growth rate averaged over the range of environmental variation experienced by the species is different from the average of the growth rates it experiences at different points of the environmental gradient. Buffered population growth allows species to tide over unfavorable periods resulting from strong interspecific competition and/or unfavorable abiotic conditions. Life history traits that enable buffered population growth include seedbanks in plants, resting eggs in zooplankton (e.g., *Daphnia*), high adult longevity when competition occurs at the juvenile stage (e.g., coral reef fish), and dormancy and diapause (e.g., desert rodents) that allow species to be inactive during harsh environmental conditions.

Coexistence via Spatial Niche Partitioning

When species inhabiting a spatially structured environment do not have opportunities for local or temporal niche partitioning, coexistence must involve some form of niche difference in space that arises from the interplay between competitive dynamics within local communities (i.e., the operation of R^* or P^* rules) and spatial processes that link local communities into a metacommunity (e.g., emigration, immigration, colonization). Such spatial niche differences can arise via two mechanisms: life history tradeoffs and source–sink dynamics. Which of the two mechanisms operates is determined by the nature of the competitive environment experienced by the species.

Consider a community of species competing for a single limiting resource. The resource can be space itself, as in plants, sessile animals, or species that require nest sites or breeding territories for reproduction. Alternatively, species could be competing for a limiting resource whose abundance varies in space (e.g., patchily distributed food resources or essential nutrients). The competitive environment of a given species consists of abiotic (e.g., temperature, humidity, nutrient availability) or biotic (e.g., natural enemies) factors that influence its ability to exploit space or the spatially variable resource.

A spatially homogeneous competitive environment is one in which the species' competitive rankings (e.g., R^* values) do not change within the spatial extent of the landscape being considered (e.g., the metacommunity). Such a situation arises when differences in the way species exploit resources are intrinsic to the species themselves and do not depend on the species' abiotic or biotic environment. In other words, species do not exhibit differential responses to the environment and there is no covariance between the environment and competition. Note that a spatially homogeneous *competitive* environment does not necessarily imply that the environment is uniform in space. The key distinction is that while species' vital rates may vary spatially, such variation does not alter species' competitive rankings. When the competitive environment is spatially homogeneous, coexistence is most likely to occur via interspecific tradeoffs between traits that determine competitive ability (e.g., fecundity, longevity) and traits that allow species to escape or minimize competition (e.g., dispersal). The key point to note is that tradeoffs allow coexistence only as long as they allow spatial niche differences between species. As discussed below, whether or not tradeoffs allow spatial niche partitioning depends crucially on the type of exploitative competition for space.

When life history or other intrinsic differences between species do not exist or are insufficient to generate spatial niche differences between species, coexistence can occur if the competitive environment is spatially heterogeneous. This means that species' R^* or P^* values vary in space because of spatial variation in the biotic or abiotic environment. For example, a community of species may compete for a limiting resource, but the degree to which a given species' per capita growth rate is affected by resource limitation may depend on temperature or humidity, or the presence of a specialist natural enemy. Now, the species do exhibit differential responses to the environment, and covariance between the environment and competition is possible. When the competitive environment is spatially heterogeneous, source–sink dynamics can allow coexistence: dispersal from populations in which a species is competitively superior (sources) and prevent exclusion in populations in which the species is competitively inferior (sinks).

COEXISTENCE IN SPATIALLY HOMOGENEOUS COMPETITIVE ENVIRONMENTS: COMPETITION–DISPERSAL TRADEOFFS

Consider a set of competing species inhabiting a landscape with three spatial scales. The smallest spatial scale is a patch, or microsite, that is occupied by only one individual. The intermediate scale is a locality, a collection of a large number of identical patches. A locality contains a community of competing species, and is the scale at which mechanisms involving spatially invariant competitive rankings (e.g., life history tradeoffs) operate. The largest spatial scale is a region, a collection of localities. A region contains a metacommunity, a set of local communities linked by dispersal of multiple species. Because different communities may be subject to different biotic or abiotic environmental regimes, the region is the spatial scale at which mechanisms involving spatially varying competitive rankings (e.g., source–sink dynamics) operate.

Metacommunity dynamics of two competing species inhabiting two localities are given by the following model:

$$\frac{dp_{ij}}{dt} = p_{ij}\left[D_{ij}(a_i, c_{ij}, p_{ij})V_{ij} - g_{ij} - e_{ij}\right], \tag{9}$$

where p_{ij} is the fraction of patches occupied by species i in locality j ($i, j = 1, 2; i \neq j$) and the function D_{ij} describes the contribution to local reproduction by residents of locality j (at a per capita rate c_{ij}) and by immigrants from other localities (at a species-specific per capita rate a_i). The function V_{ij} describes resource availability, which in a spatial context is the amount of habitat available to a given species within locality j. It depends both on the abundances of other species in the community (p_{kj}) and the fraction of suitable habitat in locality j (h_j). The function g_{ij} determines the nature of competitive interactions between species, and the parameter e_{ij} is the loss rate due to death of individuals of species i.

There are two major forms of exploitative competition that can operate in a patchy environment: dominance and preemption. In dominance competition, individuals of superior competitors can displace individuals of inferior competitors from patches the latter already occupy. This leads to a dominance hierarchy with $g_{ij} = c_{kj}((1 - a_k)p_{kj} + a_k p_{kl})$ for $k < i$ and $V_{ij} = h_j - \sum_{k \leq i}^{i} p_{kj}$ ($i, j, k, l = 1, 2$) in Equation 9. In biological terms, this means that all species that are competitively superior to species i can displace individuals of species i from occupied patches and that species i can only colonize those patches not already occupied by superior competitors. In preemptive (lottery) competition, there is no displacement of any species from occupied patches, i.e., $g_{ij} = 0$ in Equation 9. All species compete for empty patches in proportion to their relative abundances, i.e., $V_{ij} = h_j - \sum_{i=1}^{2} p_{ij}$. The superior competitor is the species with the highest local growth rate, and hence the greatest capability of replacing an individual once it dies and leaves an empty patch.

Metacommunity dynamics result from the interplay between within-locality competition and between-locality dispersal. Dispersal (described by the function D_{ij}) involves emigration and immigration of individuals. A fraction of propagules leaves without attempting to colonize empty patches within their natal locality.

Models of tradeoff mediated coexistence generally assume that if a tradeoff exists it is expressed everywhere, i.e., it is based on traits that are genetically invariant or not phenotypically plastic on time scales relevant to ecological dynamics. These models typically do not consider whether spatial heterogeneity in the environment can affect the expression of a life history tradeoff. As shown below, spatial variation in the expression of a life history tradeoff can have important effects on tradeoff-mediated coexistence.

Coexistence when Competition–Dispersal Tradeoff Is Spatially Invariant Consider first the case in which the tradeoff between competitive ability and dispersal ability is spatially invariant, i.e., it is based on traits that are genetically invariant or not phenotypically plastic on time scales relevant to ecological dynamics.

Coexistence requires that each species be able to invade when its competitor is at equilibrium. Successful invasion by species i requires that the dominant eigenvalue of the Jacobian of Equation 9 be positive when evaluated at the boundary equilibrium $p_{ij} = 0, p_{kj} > 0$. This leads to the following invasion criterion:

$$\begin{aligned}(c_{ij}V_{ij} - g_{ij} - e_{ij})(c_{il}V_{il} - g_{il} - e_{il})\\ - a_i\left[c_{il}V_{il}(c_{ij}V_{ij} - g_{ij} - e_{ij})\right.\\ \left.+ c_{ij}V_{ij}(c_{il}V_{il} - g_{il} - e_{il})\right] < 0.\end{aligned} \tag{10}$$

In Equation 10, the quantity $c_{ij}V_{ij} - g_{ij} - e_{ij}$ is the initial per capita growth rate of species i in locality j when species k is at equilibrium (p^*_{kj}) with $V_{ij} = h_j - p^*_{kj}$. With dominance competition, $g_{ij} = c_{kj}((1 - a_k)p^*_{kj} + a_k p^*_{kl})$.

If species i is the inferior competitor, it can invade when rare only if its initial growth rate is positive when averaged across localities. Note that since competitive ability is species specific rather than habitat specific, the inferior competitor's initial growth rate will be positive in both localities if it is positive in either locality. A positive average initial growth rate, however, does not guarantee invasibility. Invasion success depends on the type of competition. In preemptive competition, where no displacement is possible (i.e., $V_{1j} = V_{2j} = h_j - p_{1j} - p_{2j}$), competitive ability itself is defined by the species' reproductive ability relative to its longevity $\left(\text{e.g., } \frac{c_{ij}}{e_{ij}}\right)$. Although initial growth rates of both species can be positive when $c_{ij}V_{ij} > e_{ij}$, the R^* rule operates, and the species with the higher $\frac{c_{ij}}{e_{ij}}$ ratio excludes the other. Dispersal between localities cannot counteract competitive exclusion.

In dominance competition, where displacement is possible, species i can invade when rare if its competitive inferiority is compensated for by greater reproductive ability or longevity, i.e., $c_{ij}V_{ij} - e_{ij} > g_{ij}$ ($j = 1, 2$). In this case, a life history tradeoff allows coexistence within localities regardless of dispersal between localities.

Coexistence when There Is Spatial Variation in the Expression of a Tradeoff While the expression of some life history tradeoffs is largely invariant with respect to spatial variation in the biotic or abiotic environment (e.g., production of a few large seeds as opposed to many small seeds in plants, greater energy allocation to egg load rather than flight muscles in insects), the expression of others may be dependent on such variation. By way of illustration, consider an interspecific tradeoff between resource exploitation and susceptibility to a natural enemy. One species allocates more energy to natural enemy defense (e.g., a plant that produces chemicals that make it unpalatable to herbivores) than to reproduction or other life history traits that influence its resource exploitation abilities (e.g., lower seed set or reduced growth). Individuals of the other species allocate more energy to growth than to natural enemy defense. The latter species is therefore competitively superior, but it suffers additional mortality due to greater susceptibility to the natural enemy. Let competition be of the dominance type where superior competitors can displace inferior competitors. (Since species' competitive rankings are spatially invariant, the same species is the superior competitor everywhere within the metacommunity.) There is spatial variation in natural enemy abundance such that it is present in some localities but not others. This scenario can be incorporated into Equation 9 by a simple alteration of parameters. Both superior and inferior competitors (species 1 and 2, respectively) have comparable colonization abilities ($c_{1j} = c_{2j} = c_j$, $j = 1, 2$), but the superior competitor suffers greater mortality in locality j due to natural enemy attack such that $\hat{e}_{1j} = e_{1j} + e_{1N}N > e_{2j}$. The parameter e_{1N} is the per capita mortality rate of species 1 due to natural enemy attack, and N is the fraction of species 1 patches attacked by the natural enemy. For simplicity, we assume that natural enemy dynamics are decoupled from competitive dynamics such that N is constant over the time scale of competition. Relaxing this assumption makes the analyses a great deal less tractable but does not alter the conclusions that follow.

In the absence of dispersal between localities, the invasion criterion (Eq. 10) reduces to $c_jV_{2j} - g_{2j} - e_{2j}$ with $V_{2j} = h_j - p_{1j}^* - p_{2j}$, $g_{2j} = c_jp_{1j}^*$, and $p_{1j}^* = h_j - \frac{\hat{e}_{1j}}{c_j}$ ($j = 1, 2$). Biologically this means that the inferior competitor can invade locality j if its initial growth rate in the absence of dispersal is positive when the superior competitor is at its equilibrium abundance. Noting that $V_{1j} = h_j - p_{1j}^*$ and $g_{1j} = 0$, a little algebra shows that invasion is possible if $p_{1j}^* < \frac{\hat{e}_{1j} - e_{2j}}{c_j}$. Thus, $\hat{e}_{1j} > e_{2j}$ is a necessary condition for coexistence in the presence of the natural enemy, i.e., the superior competitor should suffer greater mortality due to natural enemy attack. A sufficient condition is that mortality suffered by species 1 due to natural enemy attack should be sufficiently high to offset the inferior resource exploitation ability of species 2. In locality l where the natural enemy is absent so that $\hat{e}_{1l} = e_{1l}$ and $p^*_{1l} = h_l - \frac{e_{1l}}{c_l}(l = 1, 2)$, coexistence is impossible unless some other tradeoff is operating (e.g., inferior competitor has a smaller background mortality rate with $e_{2l} < e_{1l}$).

In contrast to the commonly studied case of spatially invariant life history tradeoffs, dispersal plays a key role when there is spatial variation in the tradeoff's expression. Emigration from localities in which the tradeoff is expressed (sources for the inferior competitor) can rescue the inferior competitor from exclusion in localities in which the tradeoff is not expressed (sinks for the inferior competitor). Thus, dispersal enables local coexistence everywhere. Coexistence, however, is possible only if the dispersal rate is low enough to preserve between locality differences in the expression of the tradeoff. By manipulating Equation 10, it can be shown that that coexistence requires the emigration rate of the inferior competitor to be below a critical threshold:

$$a_2 < \frac{(c_jV_{2j} - g_{2j} - e_{2j})(c_lV_{2l} - g_{2l} - e_{2l})}{c_lV_{2l}(c_jV_{2j} - g_{2j} - e_{2j}) + c_jV_{2j}(c_lV_{2l} - g_{2l} - e_{2l})} \quad (11)$$

with V and g as defined in Equation 10.

To summarize results for a spatially homogeneous competitive environment, when a tradeoff between competitive ability and dispersal ability is expressed everywhere in the landscape, dispersal has no qualitative effect on coexistence. When there is spatial variation in the expression of the tradeoff, however, dispersal is key to local coexistence. This finding illuminates the conditions under which both tradeoffs and source–sink dynamics contribute to spatial coexistence. It provides the basis for a broader comparative analysis of coexistence mechanisms in spatially homogeneous and heterogeneous competitive environments.

This section has considered the scenario in which spatial heterogeneity in the biotic or abiotic environment alters the expression of a life history tradeoff in space but does not alter species' competitive rankings. This means that the same species is the superior competitor throughout the region, and coexistence within a given locality depends on whether traits that allow other species to compensate for their inferior competitive abilities (e.g., greater resistance to natural enemy attack) are expressed in that locality. Quite a different scenario emerges if spatial variation alters the expression of life history traits (assuming they are genetically variable or phenotypically labile) to such an extent that tradeoffs are not possible.

For instance, sufficiently high mortality in areas where the natural enemy is present may outweigh the resource exploitation advantage of an otherwise superior competitor and cause its exclusion. In this case, the competitive environment is no longer spatially homogeneous, because competitive rankings now depend on spatial variation in the biotic or abiotic environment. This situation is investigated next.

Coexistence in Spatially Heterogeneous Competitive Environments: Source–Sink Dynamics The Lotka–Volterra competition model with implicit resource dynamics (Eq. 2) provides a useful starting point for investigating coexistence in a spatially heterogeneous competitive environment. A two-patch version of the model with exploitative competition within patches and emigration and immigration between patches is given by

$$\frac{dX_i}{dt} = r_x X_i\left(1 - \frac{X_i}{K_{x,i}} - \phi_{x,i}\frac{Y_i}{K_{x,i}}\right) + d_x(X_j - X_i),$$
$$\frac{dY_i}{dt} = r_y Y_i\left(1 - \frac{Y_i}{K_{y,i}} - \phi_{y,i}\frac{X_i}{K_{y,i}}\right) + d_y(Y_j - Y_i), \quad (12)$$

where d_x and d_y are the per capita emigration rates of species 1 and 2, and all other parameters and variables are as in Equation 2.

Nondimensionalizing Equation 12 with the transformations

$$x_i = \frac{X_i}{K_x}, \quad y_i = \frac{Y_i}{K_{y,i}}, \quad \tau = r_x t, \quad a_{x,i} = \phi_{x,i}\frac{K_{y,i}}{K_{x,i}},$$
$$a_{y,i} = \phi_{y,i}\frac{K_{x,i}}{K_{y,i}}, \quad \rho = \frac{r_x}{r_y}, \quad k_x = \frac{K_{x,j}}{K_{x,i}}, \quad k_y = \frac{K_{y,j}}{K_{y,i}},$$
$$\beta_x = \frac{d_x}{r_x}, \quad \beta_y = \frac{d_y}{r_y},$$

yields the following system:

$$\frac{dx_i}{dt} = x_i(1 - x_i - a_{x,i}y_i) + \beta_x(k_x x_j - x_i),$$
$$\frac{dy_i}{dt} = \rho y_i(1 - y_i - a_{y,i}x_i) + \rho\beta_y(k_y y_j - y_i), \quad (13)$$

where β_x and β_y are the species-specific emigration rates scaled by their respective growth rates. The dispersal scheme is such that individuals leaving one patch end up in the other patch, with no dispersal mortality in transit.

A useful starting point is to consider the two species to differ in their per capita competitive effects and per capita dispersal rates, and to be otherwise similar, i.e., $\rho = 1, k_x = k_y = 1, K_{x,i} = K_{y,i}$, which means that $a_{x,i} = \phi_{x,i}$ and $a_{y,i} = \phi_{y,i}$. This leads to the following simplified system:

$$\frac{dx_i}{dt} = x_i(1 - x_i - \phi_{x,i}y_i) + \beta_x(x_j - x_i),$$
$$\frac{dy}{dt} = y_i(1 - y_i - \phi_{y,i}x_i) + \beta_y(y_j - y_i). \quad (14)$$

In the absence of dispersal ($\beta_x = \beta_y = 0$), competitive interactions within each patch lead to three basic outcomes: coexistence via niche partitioning ($\phi_{x,i} < 1$, $\phi_{y,i} < 1$), exclusion via priority effects ($\phi_{x,i} > 1$, $\phi_{y,i} > 1$), and exclusion via competitive dominance ($\phi_{x,i} < 1$, $\phi_{y,i} > 1$, or vice versa). Competitive dominance is of particular interest since it is the case in which coexistence is most difficult to obtain. Without loss of generality, let species 1 (with abundance x) be the superior competitor and species 2 (with abundance y) the inferior competitor.

Spatial variation in species' competitive rankings can be defined in terms of factors that affect the patch-specific competitive abilities of the two species $\phi_{x,i}$ and $\phi_{y,i}$. For instance, when $\phi_{x,i} = \phi_{x,j} < 1$ and $\phi_{y,i} = \phi_{y,j} > 1$ (or vice versa), the competitive environment is spatially homogeneous and one species is consistently superior within all patches of the landscape. When $\phi_{x,i} \neq \phi_{x,j}$ and $\phi_{y,i} = \phi_{y,j}$ (e.g., $\phi_{x,i} < 1$, $\phi_{x,j} > 1$, $\phi_{y,i} > 1$, $\phi_{y,j} < 1$, or vice versa), competitive rankings vary over space such that the species that is the superior competitor in some parts of the landscape is the inferior competitor in the other parts of the landscape.

The first issue to address is whether the inferior competitor can invade when the superior competitor is at carrying capacity in both patches (i.e., $x_1^* = x_2^* = 1$). Successful invasion requires $I < 0$ where

$$I = (1 - \phi_{y,1})(1 - \phi_{y,2}) - \beta_y((1 - \phi_{y,1}) + (1 - \phi_{y,2})). \quad (15)$$

Note that the quantities $1 - \phi_{y,1}$ and $1 - \phi_{y,2}$ are the initial growth rates of the inferior competitor in patches 1 and 2 in the absence of dispersal. Thus, the first term of I represents the product of the initial growth rates in the two patches, and the second term, their sum. The signs of these two quantities determine whether or not invasion can occur. For example, if the sum of the initial growth rates is positive and the product negative, $I < 0$ as long as $\beta_y > 0$. If both sum and product are negative, then whether or not $I < 0$ depends on the actual magnitude of β_y.

When the competitive environment is spatially homogeneous, $\phi_{x,i} = \phi_{x,j} = \phi_x < 1$ and $\phi_{y,i} = \phi_{y,j} = \phi_y > 1$, species 1 is the superior competitor across the metacommunity. Then the sum of the initial growth rates of species 2 in the two patches is negative ($2 - 2\phi_y < 0$) and the product positive ($(1 - \phi_y)^2 > 0$)), which means that $I = (1 - \phi_y)^2 - \beta_y(2 - 2\phi_y) > 0$. The equilibrium with the superior competitor at carrying capacity cannot be invaded by the inferior competitor. Invasion fails because the superior competitor increases at the expense of the inferior competitor in both patches, causing the initial growth rate of the latter to be negative across the metacommunity.

The failure of invasion in a competitively homogeneous environment suggests that invasion may succeed if the inferior competitor can maintain a positive initial growth rate in at least one patch. Mathematically, this means that the product of the initial growth rates in the two patches should be negative, i.e., $(1 - \phi_{y,1})(1 - \phi_{y,2}) < 0$. Since competition is of the dominance type (i.e., $\phi_{x,i} < 1$ and $\phi_{y,i} > 1$ or vice versa; $i = 1, 2$), the only way this can happen is if there is spatial heterogeneity in competitive rankings such that the superior competitor suffers a disadvantage in at least some parts of the landscape (e.g., $\phi_{x,i} < 1, \phi_{x,j} > 1$; $i, j = 1, 2, i \neq j$).

When the competitive environment is spatially heterogeneous, invasion can occur under three biologically distinct, and significant, circumstances. The first situation arises when competitive dominance occurs at the scale of a local population but spatial averages of competition coefficients are such that niche partitioning occurs at the scale of the metacommunity. For instance, let $\phi_{x,1} < 1$, $\phi_{y,1} > 1$ in patch 1 and $\phi_{x,2} > 1$, $\phi_{y,2} < 1$ in patch 2. Let the average competitive coefficients be $\overline{\phi}_x = \frac{\phi_{x,1} + \phi_{x,2}}{2} < 1$ and $\overline{\phi}_y = \frac{\phi_{y,1} + \phi_{y,2}}{2} < 1$. Then species 1 is the superior competitor in patch 1 and species 2 is the superior competitor in patch 2, but neither species is superior in the sense that interspecific competition is weaker than intraspecific competition when averaged across the metacommunity. At the metacommunity scale the two species meet the criteria for classical niche partitioning.

Under global niche partitioning, the sum of the initial growth rates is positive and the product negative, which means that $I < 0$ as long as $\beta_y > 0$. The equilibrium with the locally superior competitor at carrying capacity (i.e., $x_1^*, x_2^*, y_1^*, y_2^* = 1, 1, 0, 0$ or $0, 0, 1, 1$) can be invaded by the locally inferior competitor (species 2 or 1, respectively) as long as it has a nonzero dispersal rate ($\beta_y > 0$ or $\beta_x > 0$, respectively).

The important point is that as long as competitive dominance occurs locally and niche partitioning occurs globally, coexistence can occur if the patch in which the species has local competitive superiority acts as a source of immigrants for the patch in which it is locally inferior. Thus, source–sink dynamics allow each species to maintain small sink populations in areas of the landscape where it suffers a competitive disadvantage.

The second situation arises when competitive dominance occurs both locally and globally. For example, species 1 is the superior competitor in patch 1 ($\phi_{x,1} < 1, \phi_{y,1} > 1$) and species 2 is the superior competitor in patch 2 ($\phi_{x,2} > 1, \phi_{y,2} < 1$), but now species 1 is the superior competitor when averaged across the metacommunity ($\overline{\phi}_x < 1$ and $\overline{\phi}_y > 1$).

The species that is the overall superior competitor can invade when rare as long as it has a nonzero dispersal rate (i.e., $\beta_x > 0$). The important issue is whether the overall inferior competitor can invade when rare. Global dominance competition means both the sum and the product of initial growth rates are negative. Invasibility now depends on the actual magnitude of β_y. Solving Equation 15 for β_y shows that the inferior competitor can invade only if its dispersal rate is below a critical threshold:

$$\beta_y < \beta_{critical} = \frac{(1 - \phi_{y,1})(1 - \phi_{y,2})}{(1 - \phi_{y,1}) + (1 - \phi_{y,2})}. \quad (16)$$

When competitive dominance occurs both locally and globally, local coexistence requires that the dispersal rate of the overall inferior competitor not exceed a critical threshold. Once the dispersal rate exceeds this threshold, the overall inferior competitor cannot increase when rare even when it is competitively superior in some parts of the landscape.

The critical dispersal threshold depends on spatial heterogeneity in competitive ability. The stronger the local competitive advantage to the overall inferior competitor in areas where the overall superior competitor is disadvantaged (e.g., $\phi_{y,i} > 1, \phi_{y,j} \ll 1 \Rightarrow \overline{\phi}_y \rightarrow 1$), the larger the critical dispersal threshold and the greater the possibility of local coexistence. If spatial heterogeneity in the environment is insufficient to create a strong local competitive advantage to the inferior competitor (e.g., $\phi_{y,i} > 1$, $\phi_{y,j} \rightarrow 1 \Rightarrow \overline{\phi}_y > 1$), then the threshold becomes correspondingly small and conditions for coexistence, more restrictive.

When competitive dominance occurs both locally and globally, there should be sufficient spatial variation in the biotic or abiotic environment that the overall superior competitor suffers a disadvantage in some parts of the landscape. Immigration from populations where the overall inferior competitor has a local advantage prevents its exclusion in areas where it has a local disadvantage. In contrast to global niche partitioning, however, coexistence is possible only as long as the dispersal rate of the overall inferior competitor is *below* a critical threshold. This is because individuals are moving from regions of the landscape where they are competitively superior and enjoy a positive growth rate (source populations) to regions where they are competitively inferior and suffer a negative growth rate (sink populations). If the net rate of emigration is sufficiently high relative to local reproduction that the growth rate of the source population becomes negative, the species loses its local competitive advantage and is excluded from the entire metacommunity.

So far, conditions for local coexistence have been derived for two situations: global niche partitioning and global competitive dominance. The third situation arises when dominance competition occurs locally but a priority effect occurs globally, i.e., species 1 is the superior competitor in patch 1 ($\phi_{x,1} < 1$, $\phi_{y,1} > 1$) and species 2 is the superior competitor in patch 2 ($\phi_{x,2} > 1$, $\phi_{y,2} < 1$), but interspecific competition is stronger than intraspecific competition when averaged across the metacommunity ($\overline{\phi}_x > 1$ and $\overline{\phi}_y > 1$). Now each species has a critical dispersal threshold above which coexistence cannot occur:

$$\beta_{.} < \beta_{critical} = \frac{(1 - \phi_{.,1})(1 - \phi_{.,2})}{(1 - \phi_{.,1}) + (1 - \phi_{.,2})}. \qquad (17)$$

As with global competitive dominance, the magnitude of the dispersal threshold depends on spatial heterogeneity in competitive ability. If the two species differ in the degree of local competitive dominance but have the same competitive ability on average (e.g., $\phi_{x,1} \neq \phi_{y,2}$; $\phi_{x,2} \neq \phi_{y,1}$; $\overline{\phi}_x = \overline{\phi}_y > 1$), then local coexistence is determined by the dispersal ability of the species that experiences lower spatial heterogeneity and hence the lower dispersal threshold. If the species are sufficiently different that their average competition coefficients are unequal (i.e., $\overline{\phi}_x > 1$, $\overline{\phi}_y > 1$; $\overline{\phi}_x < \overline{\phi}_y$), then local coexistence is determined by the dispersal ability of the species with the higher average competition coefficient (lower competitive ability). For instance, if $\overline{\phi}_x < \overline{\phi}_y$, then $\beta_{ycritical} < \beta_{xcritical}$, and the dispersal threshold for species 2 determines the transition from coexistence to exclusion.

The key to coexistence, again, is spatial heterogeneity in competitive ability. When heterogeneity is low ($\overline{\phi}_{.} > 1$), the region of the parameter space where each species can invade when rare is small; when heterogeneity is high ($\overline{\phi}_{.} \rightarrow 1$), this region is correspondingly larger. An important difference between global competitive dominance and a global priority effect is that while coexistence is determined by the dispersal ability of the overall inferior competitor in the former, dispersal abilities of both competing species determine conditions for coexistence in the latter. If both species have dispersal rates that exceed their respective thresholds, neither species can invade when rare and coexistence is impossible either locally or regionally.

The three situations under which local coexistence can occur in a competitively heterogeneous environment can be distinguished by their response to the transition from low to high dispersal. In the absence of dispersal, all three situations exhibit global coexistence, with each species flourishing in areas where it has a local competitive advantage. Under low dispersal, source–sink dynamics ensure local coexistence in all three cases. High dispersal, however, elicits qualitatively different dynamical responses. For instance, when competition involves global niche partitioning, local coexistence prevails. When competition involves global competitive dominance, global exclusion of the overall inferior competitor results. When competition involves a global priority effect, the outcome is global exclusion of the species with the lower dispersal threshold.

CONCLUSIONS

Competition is the most fundamental of all ecological interactions. Competitive interactions within species generate negative feedbacks that allow for population regulation, while competitive interactions between species can enhance or reduce diversity depending on their strength relative to within-species competitive interactions. The mechanisms by which species avoid, tolerate, or minimize competition (niche partitioning mechanisms) are at the heart of biodiversity maintenance. They also underlie important environmental problems such as the invasion of exotic species and biological pest control. The theoretical framework provided here elucidates the conditions under which two competing species can coexist within local communities, in the presence and absence of environmental variation. It provides testable predictions that can be verified with observational or experimental data.

SEE ALSO THE FOLLOWING ARTICLES

Apparent Competition / Ordinary Differential Equations / Metacommunities / Niche Overlap / Nondimensionalization / Predator–Prey Models / Spatial Ecology / Storage Effect

FURTHER READING

Amarasekare, P. 2002. Interference competition and species coexistence. *Proceedings of the Royal Society of London, B: Biological Sciences* 269: 2541–2550.

Amarasekare, P. 2003. Competitive coexistence in spatially structured environments: a synthesis. *Ecology Letters* 6: 1109–1122.

Amarasekare, P., M. Hoopes, N. Mouquet, and M. Holyoak. 2004. Mechanisms of coexistence in competitive metacommunities. *American Naturalist* 164: 310–326.

Amarasekare, P., and R. Nisbet. 2001. Spatial heterogeneity, source-sink dynamics and the local coexistence of competing species. *American Naturalist* 158: 572–584.

Armstrong, R. A., and R. McGehee. 1980. Competitive exclusion. *American Naturalist* 115: 151–170.

Chesson, P. 2000. Mechanisms of maintenance of species diversity. *Annual Review of Ecology and Systematics* 31: 343–366.

Hastings, A. 1980. Disturbance, coexistence, history and competition for space. *Theoretical Population Biology* 18: 363–373.

Tilman, D. 1982. *Resource competition and community structure*. Princeton: Princeton University Press.

U

URBAN ECOLOGY

MARY L. CADENASSO

University of California, Davis

STEWARD T. A. PICKETT

Cary Institute of Ecosystem Studies, Millbrook, New York

Urban ecology is the scientific study of the processes influencing the distribution and abundance of organisms, the interactions among organisms and between organisms, and the transformation and flux of energy and matter in urban and urbanizing systems. To understand the structure and dynamics of urban systems they must be recognized as social-ecological systems that integrate socioeconomic drivers and responses with ecological structures and functions. This integration requires approaches and applications of theory from a variety of disciplines, but the need for novel ecological theory is much debated in this nascent field.

DEVELOPMENT OF URBAN ECOLOGY

Ecological concepts were first applied to the urban landscape by Robert Park and Ernest Burgess of the University of Chicago's Department of Sociology in the 1920s. Chicago was rapidly gaining population through immigration from overseas and migration from the U.S. South. To understand the dynamics of this rapidly changing city, Park and Burgess investigated processes that lead to spatial differentiation of people and activities in the urban landscape. Three related ecological theories were brought to bear: (1) competition, (2) niche partitioning, and (3) succession. Competition for limiting resources in the urban environment, such as land, was thought to lead to the partitioning of that resource into different niches used by either distinct social groups or activities, such as industry or housing. Urban growth and expansion was conceptualized as a series of concentric rings around an industrial downtown such that as individuals acquired resources and could afford better living situations they moved away from the center (Fig. 1). This directional

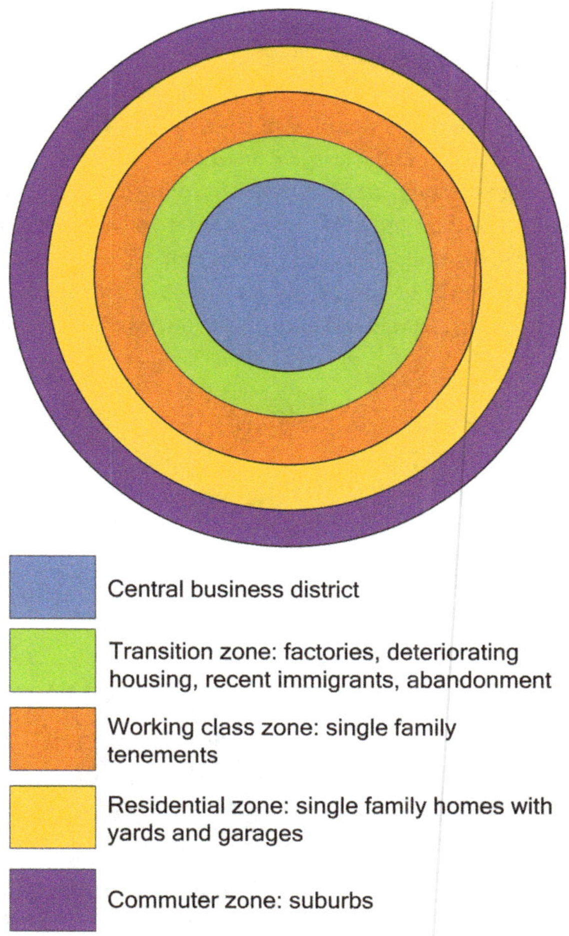

FIGURE 1 The idealized Burgess model of the spatial structure of Chicago, indicating concentric rings of occupancy across which migrants replace each other in a succession driven by competition between groups. Based on concepts in Burgess (1925).

movement was referred to as succession, borrowing from the theory of plant ecology. This early theory of the city ultimately was replaced due to its exclusive focus on spatial differentiation and on competition as the mechanism influencing that differentiation while ignoring other factors that may influence where people and businesses locate in urban areas.

In general, ecologists were slow to recognize the city as a system worthy of study until the middle of the twentieth century. Following WWII, two approaches were taken. The first approach, mainly from Europe, focused on plant and animal populations in remnant spaces such as cemeteries, urban parks, and vacant sites destroyed by bombing during the war. The second approach was influenced by the International Biological Program, which existed from 1964 to 1974 and was intended to organize large, coarse-scale projects on ecological studies in different biomes. This second approach considered the city as a system and, borrowing the budgetary approach from ecosystem ecology theory, characterized the city as a metabolic machine (Fig. 2). Understanding urban metabolism and whole system energetics could be used to make the system as efficient as possible, which, in turn, influenced urban morphology, resulting in large swaths of redundant "units" such as suburban residential areas.

Contemporary urban ecology has built on these earlier approaches and now takes several forms. In some cases, urban ecologists consider the impact of urbanization on remnant "natural" systems such as fragments of forest, desert, or wetland embedded in the urban matrix.

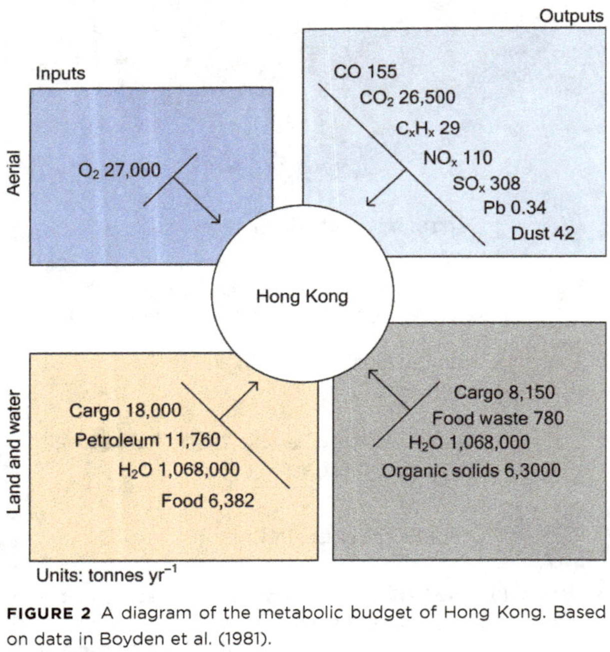

FIGURE 2 A diagram of the metabolic budget of Hong Kong. Based on data in Boyden et al. (1981).

In this approach, human decisions and activities are not studied directly but are instead considered as a single aggregated factor of urbanization that influences the ecological system of interest. This approach typically focuses on the nonbuilt portions of the landscape and may be motivated by conservation of habitat or species. It has been termed ecology *in* the city. A complimentary approach considers human decisions and activities central to understanding the ecological structure and dynamics of the urban system. This second approach focuses on the entire system, not just the vegetated patches, and takes a multidisciplinary stance to understand the integrated social-ecological system by synthesizing the ecological understanding of specific organisms and processes, social behaviors, and the feedback among them. This approach has been termed ecology *of* the city. Both approaches are needed and the research question determines the best one to use.

Urban ecology has grown in importance as the world rapidly urbanizes. The current century has been dubbed the urban century because for the first time more than 50% of the global population lives in cities. This has been true for North America and Europe since the 1950s, and populations on both those continents and Australia are now more than 80% urban. The global increase in urbanization is primarily due to large population shifts in developing countries from rural communities to urban megacities. Although cities occupy only an estimated 2–7% of the Earth's land surface, their influence extends far beyond their boundaries. It is critical for ecologists to study urban systems both to contribute toward making cities more livable and to gain insights into urban influences on nonurban systems.

WHAT IS DISTINCTIVE ABOUT URBAN SYSTEMS FROM AN ECOLOGICAL PERSPECTIVE?

There are four characteristics of cities that make them distinct ecological systems. First, urban and urbanizing landscapes have undergone significant land transformations. Changes in land cover are a major component of global change. These changes directly affect ecological systems through the removal of habitat. Indirect changes either within the urban area or in areas influenced by urbanization include changes to nutrient and pollutant loading, altered biogeochemical cycling, and the fragmentation of habitats and changing landscape structure.

Second, plant and animal communities in urban regions experience different selection pressures as resource type, abundance, and spatial distributions have been modified and as predator–prey interactions have been changed. Some animal species are urban avoiders, others can tolerate the changes to habitat and resources, and still others can exploit those changes. In addition to the regional species still remaining in cities, new species often invade. New species invasions in combination with loss of native species can result in the creation of novel assemblages.

A third distinct characteristic of urban ecological systems is accelerated evolutionary change and acclimation. Pest and disease organisms may rapidly evolve resistance to pesticides and herbicides applied in urban landscapes and to medication and other pharmaceuticals added to systems. Other plants and animals may also evolve rapidly as breeding strategies are selected to succeed in fragmented landscapes or to overcome pollution from noise, light, heavy metals, and chemicals. Evolutionary responses may include changing tolerance to heavy metals or changing plumage coloration. Species can also alter behavior such as changing pitch, volume, and duration of bird song to overcome sonic interference. Changes to the climate result in altered plant phenology including the timing of bud burst, leaf expansion, and flowering. Pollinators and other species that rely on seasonal resources provided by plants also may need to alter the timing of their behaviors to match.

Finally, urban landscapes are distinct because of the tight coupling of ecological patterns and processes with social, cultural, economic, and other drivers of human spatial differentiation and activities. Though ecologists have recognized that all ecosystems are now influenced by humans, urban systems especially require integration of ecological pattern and process with social patterns and processes in order to understand the ecological dynamics of the urban system.

DOES URBAN ECOLOGY REQUIRE A NOVEL ECOLOGICAL THEORY?

No novel theory of urban ecology exists, though components that would contribute to such theory have been suggested by many researchers. Ecologists are currently debating whether a new theory is needed and if so what it might look like. Theories from contemporary ecology may be applied to urban ecosystems and in doing so understanding is gained about the urban system. At the same time, insights about the theory can be gained from understanding how well it applies to the novel conditions of urban systems. Urban systems thus represent a new frontier for ecological theory.

Ecological theories applied to the urban system span from population and community ecology through to ecosystem and landscape ecology. Here, we will introduce examples of several of these theoretical realms, beginning at the coarsest level of ecological organization—the landscape—and continue through to community theory.

Landscapes: Island Biogeography, Spatial Heterogeneity, and Patch Dynamics

Island biogeography theory (IBT), though initially developed to predict species richness on oceanic islands as a function of island size and distance from mainland sources of species, has been applied to fragmented terrestrial systems. IBT has been particularly valuable for species conservation concerns, as it models the landscape as fragments of habitat, deemed suitable to the target organism, embedded within a hostile matrix and allows determination of whether the organism can maintain a viable population using the network of connected and unconnected habitat fragments. Features of the fragment such as its size, shape, and distance to the nearest similar habitat fragment are also important. In urban systems, IBT has been used to address whether the sizes of urban parks or green spaces, for example, are related to the number of species found there. In these novel systems, different processes affect rates of colonization and extinction than in oceanic or more rural landscapes. For example, humans may increase or decrease barriers to organism movement by building roads or providing a continuous tree canopy across them, thereby influencing colonization rates to habitat fragments. Humans may also enhance or diminish a species' ability to exist in a location by management activities such as supplementing resources, including water, nutrients, food, and niche space, or by the introduction of predators and exposure to pollutants.

The application of IBT to terrestrial systems served as a precursor to more subtle conceptualizations of the landscape and how organisms, material, and energy move within it. The focus shifted from a target habitat for a specific species of concern to the spatial heterogeneity and patch dynamics of the entire landscape mosaic. Ecologists recognize spatial heterogeneity as variation in landscape structure that an organism or process may respond to. This variation can exist at any scale, and how ecologists describe the heterogeneity of a location depends on the research question posed. Cities are characteristically heterogeneous, and many different types of criteria—biogeophysical and social—can establish or respond to that heterogeneity. Similar to nonurban systems, heterogeneity can also change through time as described by patch dynamics. The complexity of the urban ecosystem facilitates many types of questions that can address heterogeneity and patch dynamics such as investigating the social and ecological factors that drive and respond to heterogeneity, the differential rates of change in patterns of heterogeneity, and how the coarser scale context may influence the spatial location of different rates of change.

Land use and land cover are typical descriptors of urban heterogeneity. Although the two terms are frequently used interchangeably, they are different. Land use is how humans use the land, including residential, commercial, industrial, and transportation use. In contrast, land cover is the physical structures on the landscape such as buildings, trees, and pavement. Traditional approaches to depicting spatial heterogeneity and assessing change through time have used classification systems that combine both land use and land cover descriptors. However, as more ecologists have begun investigating the city as an ecological system, new approaches have been developed to tease apart land use from land cover. This separation is necessary to test the link between landscape structure and ecosystem function. For example, lands classified as residential may have vastly different amounts and kinds of vegetation present, different arrangements and covers of buildings, and different arrays of paved versus pervious surfaces (Fig. 3). These differences in structure may influence organism abundance and movement and the cycling and transport of nutrients and pollutants. Classifying this variation as residential, even if refined by adding descriptions of density, leaves much heterogeneity unexamined and may limit the ability to link system structure to ecosystem function.

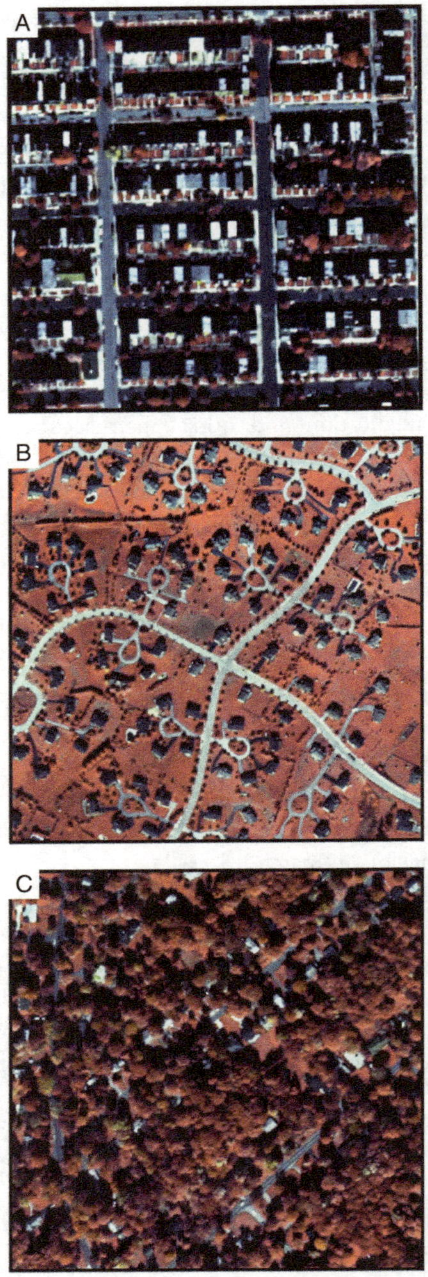

FIGURE 3 False-color infrared photos of three residential areas in Baltimore, MD, illustrating the variation in the amount and arrangement of land cover elements.

Ecosystems: Budgets, Watersheds, and Retention

Cities are now recognized as ecosystems, and many of the concepts and theories considered central to ecosystem ecology can be modified for application to urban systems. For example, ecosystems are defined as the interaction of biotic and physical complexes within a particular location that is open to inputs from, and contributes outputs to, surrounding systems. All the components of this abstract definition must be specified for the question being investigated—organisms or elements of the biotic and physical complex, location of the boundary, and nature of the inputs and outputs. When specifying this concept to an urban ecosystem, the biotic complex includes humans, recognizing that humans are more than simply biotic organisms, and the physical complex includes buildings and infrastructure constructed by humans (Fig. 4). Boundaries may also be set using different criteria in an urban ecosystem compared to nonurban systems. In nonurban systems, the boundary is frequently drawn around entire habitats such as a lake or pond, an agricultural field, or a forest patch. In some cases, the boundary is arbitrary but facilitates the measurements of inputs and outputs. For such purposes, frequently the boundary is determined based on the location of expected shifts in pools and fluxes of biogeochemical elements. For example, watershed boundaries are set to investigate the integrated influence of all

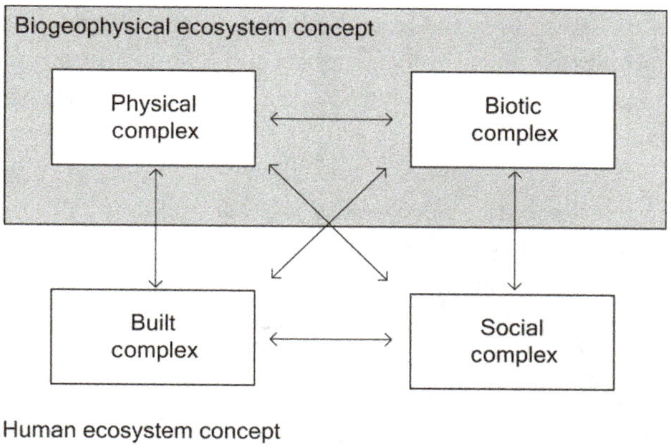

FIGURE 4 Based on the biological concept of the ecosystem, this model template explicitly recognizes the place of social processes and structures and built structure and infrastructure along with biological and strictly physical structures and processes. The simple biophysical concept of the ecosystem is outlined within the more comprehensive concept of the human ecosystem.

processes occurring on the landscape within it to water quality and quantity. In urban systems, the ecosystem boundaries can be established by boundaries of municipalities, neighborhood groups, or households, or they can also be established by an ecological criteria such as a watershed. Watershed application must be modified to include the infrastructure—storm drain catchments or sanitary sewershed, for example.

Nonurban ecosystems are considered to be retentive systems because they hold on to limiting nutrients and minimize the exchange of materials across ecosystem boundaries. However, in urban systems many materials that are biologically limiting are intentionally concentrated for human use or are periodically present in high enough amounts to present a hazard to the human population. Nitrogen, for example, is concentrated in cities due to human and pet nutrition, automobile exhaust, or fertilizing gardens, lawns, and golf courses. The excess nitrogen is a pollutant, released by septic systems or sanitary sewers, or concentrated in ground water or storm drains. Likewise, rainfall in most urban areas is treated as a nuisance to be drained off roofs, streets, and even lawns as quickly as possible. Rather than retaining these biologically limiting materials in soils, persistent organic matter, or long-lived vegetation, water and nitrogen are purposefully discarded from most urban areas.

COMMUNITIES: BIODIVERSITY, METATHEORY, AND HOMOGENIZATION

It is often assumed that urban systems lack biodiversity. Certainly some plant and animal species that existed in the location prior to urbanization were unable to tolerate the ecological changes associated with land cover change and have been extirpated. Alternatively, some plant and animal species are able to exploit changes that occur with urbanization. These changes include (1) an increase in resources such as food, water, and habitat, (2) a dampening of the temporal or spatial variability in resource supply, and (3) a reduction in risk of predation due to loss of predators that could not survive in the changed landscape. Species that are not native to the region can enter through cities as a result of human activities that transport or favor them. These plants can escape the highly managed locations in which they are introduced, to establish populations in remnant or lightly managed open spaces.

Community assemblage in urban systems, therefore, is driven by a unique set of constraints and opportunities for species. Some species will be lost from communities if they are unable to adapt to changes in resource availability, fragmentation, or loss of preferred habitat; altered interactions with other organisms through competition and predation; or loss of other species they rely on for important services such as pollination and dispersal. Changes to the physical and biological components of the system may be favorable to other species facilitating their increased performance and spread. Additional species will be able to invade the community if their dispersal and movement is facilitated by human activities. Therefore, biotic communities in urban landscapes are a consequence of several factors, including active assemblage by humans and indirect human activities that exert selection pressures on species differentially. Because of the spatial fragmentation and patchiness of biotic communities in urban systems, metacommunity theory may be a particularly useful application of ecological theory. However, the complexities of introduction and management, mentioned above, may considerably complicate the urban metacommunity models.

It has been suggested that plant and animal communities in urban landscapes have become homogenized over time such that they are more likely to be similar among cities than between a city and its surrounding nonurbanized land. This may mean that there is a "signature" urban biota that is well adapted to the novel combinations of ecological characteristics in urban landscapes.

EMERGING SOCIAL-ECOLOGICAL SYSTEMS

An overarching goal of urban ecology is to understand the urban system as an integrated social-ecological system. Therefore, in addition to ecological science, social sciences and urban planning and design provide important theoretical and practical lenses needed to understand urban ecosystems. Several socially motivated concepts and activities are required including ecosystem services, sustainability, environmental justice, and urban design.

Ecosystem services are a key point of integration between the biophysical and the social sciences in urban systems. Ecosystems provide resources, regulate flows of energy, water, biotic populations such as diseases, and environmental hazards, and they supply cultural, aesthetic, and spiritual amenities. These services are no less a part of urban than they are of wild systems, although the degree and identity of specific kinds of service may differ. Ecological theory suggests important services that may not be obvious, such as the role of spatial heterogeneity or the retentive power of unmanaged or green ecosystem patches in cities.

Sustainability is a concept that arose in the policy arena. However, ecological theory can contribute substantially to how this idea is applied to urban areas by suggesting key measurements of processes that can support sustainability, including such things as ecosystem function, community and metacommunity structure, patch dynamics, and evolutionary potential in urban systems. These measurements, supported by ecological theory, are essential to assessing whether an urban system is healthy or improving, particularly when combined with theories of economy and of society.

An important feature of urban sustainability is how the social component of the system is engaged. The equitable distribution of ecosystem services and environmental disamenities among diverse human populations and whether disadvantaged persons and institutions are included in decision making are critical concerns of sustainability. These concerns are part of environmental justice theory, which has emerged largely outside of ecology but to which ecological science can make contributions. To satisfy the expectations of environmental justice, shared concern with the quality of urban life, an open system of knowledge, and democratic decision making, communities must be engaged in a reciprocal dialogue.

Urban designers, including architects, landscape architects, and planners, have worked to enhance the ecological thinking in their disciplines for several decades. Their activities affect the structure of urban landscapes, how people interact with the system, and, consequently, the functioning of urban ecosystems. Furthermore, the ecological understanding of the spatial and temporal patterns of natural disturbances contributes to successful design and management in urban areas.

As the illustrative theories from landscape, ecosystem, and community levels demonstrated, many ecological theories can be specified and tested in urban systems. The specification may involve different and unique drivers and consequences or may shift the relative balance or importance of drivers in urban versus nonurban systems. In time, the process of expanded specification and the opportunity to test theory in a novel system will contribute to our understanding and further development of the theory, as seen above for island biogeography and ecosystem retention theories. Whether or not urban ecology requires a novel theory is still undecided. However, understanding the ecological structures and dynamics of urban systems and the drivers and consequences of those dynamics requires an integration of ecological and social understanding. Any new theory of urban ecology must be an emergent theory that effectively blends theory from ecological and social realms.

SEE ALSO THE FOLLOWING ARTICLES

Ecosystem Ecology / Ecosystem Services / Landscape Ecology / Metabolic Theory of Ecology / Metacommunities / Spatial Ecology / Succession

FURTHER READING

Aitkenhead-Peterson, J., and A. Volder, eds. 2010. *Urban ecosystem ecology*. Madison, WI: American Society of Agronomy, Inc.

Alberti, M. 2008. *Advances in urban ecology: integrating humans and ecological processes in urban ecosystems.* New York: Springer.

Boyden, S., S. Millar, K. Newcombe, and B. O'Neill. 1981. *The ecology of a city and its people: the case of Hong Kong.* Canberra, Australia: Australian National University Press.

Gaston, K. J., ed. 2010. *Urban ecology.* New York: Cambridge University Press.

Burgess, E. W. 1925. *The growth of the city: an introduction to a research project.* In R. E. Park and E. W. Burgess, eds. *The city.* Chicago: University of Chicago Press.

Collins, J., A. Kinzig, N. Grimm, W. Fagan, D. Hope, J. Wu, and E. Borer. 2000. A new urban ecology. *American Scientist* 88(5): 416–425.

Marzluff, J. M., E. Shulenberger, W. Endlicher, M. Alberti, G. Bradley, C. Ryan, C. ZumBrunnen, and U. Simon, eds. 2008. *Urban ecology: an international perspective on the interaction between humans and nature.* New York: Springer.

McDonnell, M. J., A. K. Hahs, and J. H. Breuste. 2009. *Ecology of cities and towns: a comparative approach.* Cambridge, UK: Cambridge University Press.

Pickett, S. T. A., M. L. Cadenasso, J. M. Grove, Christopher G. Boone, P. M. Groffman, E. Irwin, S. S. Kaushal, V. Marshall, B. P. McGrath, C. H. Nilon, R. V. Pouyat, K. Szlavecz, A. Troy, and P. Warren. 2011. Urban ecological systems: scientific foundations and a decade of progress. *Journal of Environmental Management* 92: 331–362.

Sukopp, H., S. Hejny, and I. Kowarik, eds. 1990. *Urban ecology: plants and plant communities in urban environments.* The Hague: SPB Academic Publishing.

VIGILANCE

SEE ADAPTIVE BEHAVIOR AND VIGILANCE

GLOSSARY

The glossary that follows defines more than 800 specialized terms that appear in the text of this encyclopedia. Included are a number of terms that may be familiar to the lay reader in their common sense but that have a distinctive meaning within these fields of study. Definitions have been provided by the encyclopedia authors so that these terms can be understood in the context of the articles in which they appear.

abiotic Describing or referring to the nonliving, physical properties of an environment.

absence data Data on locations of potential species presence where a species was sought but not detected.

absorbing state A state (e.g., population level) in a probabilistic model that the system has zero probability of leaving once it is reached (e.g., extinction).

active transport The process through which organisms expend metabolic energy to transfer substances between compartments.

adaptive behavior Behavior (e.g., decisions) of individuals made in response to changes in their environment or internal state that is presumed to increase the individuals' fitness.

adaptive dynamics (AD) A mathematical framework for dealing with eco-evolutionary problems that is dynamic in time, based on certain simplifying assumptions including: clonal reproduction, rare mutations, small mutational effects, smoothness of the demographic parameters in the traits, and well-behaved community attractors.

adaptive landscape A representation of either mean individual fitness (as a function of genotype frequency) or the fitness of individual genotypes, as a three-, or, more abstractly, multidimensional surface or lattice. (Attributed to American geneticist Sewall Wright.) Also, FITNESS LANDSCAPE.

adaptive plasticity Phenotypic plasticity maintained by natural selection.

adaptive radiation A novel adaptation, such as a new defense mechanism, that provides an evolutionary escape from a set of selective pressures, which in turn allows a lineage to explore new niches and eventually diversify into new species.

adaptive significance The function of a trait (e.g., behavior, physiology, morphology) that enhanced fitness in prior generations or enhances ongoing fitness accumulation.

adaptive speciation 1. Speciation that results from a population's evolutionarily reaching a branching point. 2. In the context of adaptive landscapes, species arising through reproductive isolation as a consequence of highly adaptive genotypes being separated by unfit recombinants.

adaptive trait A characteristic of an individual that enhances fitness relative to those individuals not carrying the trait.

additive genetic variance Variance in individual breeding values within a population. It is the component of genetic variance responsible for the resemblance between relatives and thus a major determinant of a population's response to selection.

adjacency matrix See FOOD WEB MATRIX.

advection Movement of a fluid to a given location by bulk flow, in contrast to movement by molecular diffusion.

agent An autonomous entity, such as an individual, actor, or decision maker.

agent-based model See INDIVIDUAL-BASED MODEL.

age structure Classification of a population by the proportion of organisms in different age classes.

aggregative numerical response Changes in natural enemy numbers at a locality, not due to local reproduction or death, but rather, due to directed movement in response to changes in victim abundance.

agroecosystem A spatially and functionally coherent unit of agricultural activity that includes the living and nonliving components involved in that unit as well as their interactions.

albedo The fraction of incident sunlight that is reflected.

alien invasive species A nonindigenous species that spreads rapidly and causes ecological and/or economic damage in the new community.

alien species A non-native species. Also, EXOTIC SPECIES.

Allee effect A positive correlation between a population's density and its per capita growth rate, usually associated with declining populations at low density. (Identified by Warder C. Allee, American zoologist and ecologist.) Also, DEPENSATION.

Allee threshold A threshold of population abundance below which a population that is subject to a strong Allee effect will become extinct.

allele A functional variant of a gene.

allele frequency The frequency in a population of one variant (allele) at a given site in the genome.

allelopathy The process whereby chemicals produced by one species kill or inhibit the growth of other species.

allochthonous inputs Organic and inorganic inputs that enter an ecosystem from external sources, as from forest to stream.

allometric scaling The proportional change in the structure or function of an organism with respect to body size, often described using a power law.

altruism A cooperative behavior wherein individual fitness costs of the behavior exceed the individual fitness benefits.

anabolic Pertaining to the constructive part of metabolism, especially macromolecular synthesis.

analytical solution A solution to a set of equations that is found using analysis, e.g., calculus and algebra, rather than numerically.

anonymous locus A segment of the genome that does not encode a known protein or genome feature; often variable and used in phylogeographic analysis.

aperiodic dynamics Fluctuations in model population numbers that never repeat.

apostatic selection Frequency-dependent selection by predators favoring rare prey types.

apparent competition An indirect negative interaction between species mediated through the action of a shared natural enemy.

area of occupancy The area within the outermost geographic limits to the occurrence of a species over which it is actually found.

area selection See RESERVE SELECTION.

artisanal fishery A primarily subsistence fishery that makes short trips from land and for which the target catch is used mainly for local consumption. Also, SMALL-SCALE FISHERY.

asymptotic population growth rate The eventual change in population over a unit time period at the constant rate $\lambda = N_{t+1}/N_t$, where N_t is the number of individuals in the population at time t. The growth rate λ is given by the dominant eigenvalue of the projection matrix $\mathbf{A}$.

attractor A set of states toward which a dynamical system evolves over time. An attractor is often a point, but it can also be a curve, a manifold, or a more complicated set with a fractal structure (strange attractor).

autochthonous inputs Organic or inorganic inputs within an ecosystem that originate from internal sources.

autocorrelation For a stochastic process, a function of two time points that measures the correlation between the process states at those two time points. The autocorrelation of a stationary process, by definition, depends only on the absolute difference between the two time points (the lag). In the context of environmental stochasticity, the term refers somewhat loosely to the autocorrelation at lag 1 in a discrete time process.

bacteriocin A toxin produced by one microbe that antagonizes another microbe.

ballast water Water carried by a ship to increase stability.

baroclinic pressure gradients Gradients in pressure due to gradients in the density of water. Denser water (because of coldness or saltiness) exerts greater pressure.

barotropic pressure gradients Gradients in pressure due to gradients in the height of the sea surface. A higher column of water exerts greater pressure.

basal metabolism The metabolism of a resting, post-digestive, non-sleeping animal; an idealized measure of basal energy throughput.

basal resource Living or nonliving organic material that occupies the first trophic position of a food web, providing energy for consumers at higher trophic positions. By convention, a basal resource is consumed by living organisms and does not consume other living organisms.

basal species A species that does not feed on other species and thus links the community to extrinsic energy inputs.

basic reproductive number (R_0) The average number of new infections produced by a single infectious individual introduced into a fully susceptible population.

basin of attraction See DYNAMIC REGIME.

Bateson-Dobzhansky-Muller (BDM) model An adaptive landscape in which two genotypes with high fitness are connected via viable, high-fitness intermediate genotypes, even though a subset of possible recombinant offspring between the two genotypes have very low fitness. This generates reproductive isolation between genotypes. (Described independently by geneticists William Bateson, Theodosius Dobzhansky, and Hermann Muller).

Bayesian inference An approach to statistical inference that represents uncertainty in unknown parameters by a probability distribution and updates this distribution as data are gathered that provide information about these parameters.

Bayesian statistics An approach to statistics that combines likelihood and prior probabilities to calculate posterior probabilities using Bayes' rule.

Bayes' theorem or **rule** Basic probability result used to obtain the posterior distribution in a Bayesian analysis. (Named for English mathematician Thomas Bayes.)

β diversity The turnover of species with distance of separation between two sampled communities. One measure of β diversity is the probability that two randomly chosen individuals, one from each community separated by distance d, are of the same species, given a dispersal rate m.

Beer-Lambert law A law of optics relating absorption of light to the properties of the material through which it is traveling; applied to foliage canopies to calculate canopy light attenuation. I_0 is the level of incident light, I the level of transmitted light $\frac{I}{I_0} = e^{-\alpha l} = e^{-\sigma lN}$ where α is the absorption coefficient, and l the length of passage through the material. The absorption coefficient is determined by σ, the cross section of light absorption of a single particle and N, the density of absorbing particles.

benefit-cost analysis Analysis intended to inform regulatory decisions by monetizing and comparing use and nonuse benefits to the cost of a particular project.

bifurcation A point where there is a qualitative change in the dynamics of a system as a parameter is changed. Also, CRITICAL THRESHOLD, TIPPING POINT.

bifurcation diagram A graph of the final states of a system as a function of a parameter; used to reveal parameter values that lead to bifurcations.

bioaccumulation In an organism, the net accumulation of a chemical from all sources (i.e., the environment and food).

bioclimatic envelope model A model that relates a species' range to multiple climatic variables and assumes that they reflect the species' ideal climatic conditions (niche). Under different climate scenarios, the species' potential future range can then be projected based on the species' climatic niche.

biodiversity The diversity of organisms found in a given region; a metric of ecosystem health, frequently measured by species richness, phylogenetic diversity, or functional diversity of a community.

biodiversity feature An ecological or biological entity that is considered as valuable for conservation; includes, e.g., species, habitat types, and ecosystem services.

bioeconomics Application of economic theory to biological questions through integrated use of biological and economic approaches.

bioenergetic model A model of growth rate as a function of temperature, caloric intake, and waste; can be calculated for an individual organism or for a spatial region.

biofilm Surface-associated microbial communities or populations.

biogeochemical cycles Cycles that describe exchanges and transformations of chemical elements between and within biotic (biosphere) and abiotic (lithosphere, atmosphere, and hydrosphere) compartments of Earth.

biological control agents Non-native organisms released into the environment to control pest species.

biomagnification An increase in chemical concentration across trophic levels.

biomes Ecosystems defined by similar climate and plant physiognomy, such as deserts, alpine tundra, and tropical grassland.

bionomic equilibrium In harvesting theory, the joint harvesting rate and exploited population level at which the population is unchanging over time and the net economic rate of return (profits minus losses) is zero.

biosphere The global ecological system that integrates all living beings and their relationships, including their interaction with the elements of the lithosphere, hydrosphere, and atmosphere.

bipartition See BRANCH.

bootstrapping Performing inferences on multiple pseudo-replicate datasets generated by sampling with replacement from an observed dataset. In phylogenetics, this is primarily used as a way of evaluating strength of support for bipartition.

bottom-up control Control of population dynamics by resource availability.

branch A segment on a phylogenetic tree between speciation events. Also, BIPARTITION, EDGE.

branching point In the context of adaptive dynamics, a phenotype that is approached from all directions by gradualistic evolution resulting from ecological interactions, at which the population then is under disruptive selection.

breeding dispersal For a given individual, the movement between different instances of reproduction.

breeding value The part of the deviation of an individual's phenotype from the population mean that can be attributed to the additive effects of alleles. An individual's breeding value with respect to a given trait is calculated as twice the average deviation of its offspring from the population mean for the same trait. Thus, an individual may possess a breeding value for a trait it does not express, such as milk production.

broad-sense heritability The proportion of the total phenotypic variance in a trait that is due to contributions from all genotypic sources, including additive, dominance, and epistatic effects; provides an upper limit for narrow-sense heritability.

broad-sense sexual selection A selection syllogism focused on the unit of selection represented by a sex (in a given species). The mechanisms may work only among males or only among females, and fitness need not be narrowly focused on number of mates but may include other components of fitness such as the quality of mates.

buoyancy An upward force exerted by a fluid that counters the downward force of gravity. Buoyancy forcing on water is due to differences in density; an imbalance between weight and ambient pressure results in vertical adjustment and horizontal flows (due to baroclinic pressure gradients).

bycatch Any organism landed, directly wounded, or killed without being targeted by fishers; can be landed or discarded.

canalized development Environmentally insensitive development for a given trait thereby constraining variation in the phenotype around one or more modes.

canonical equation (CE) A differential equation that captures how the trait vector changes over evolutionary time based on the assumption that mutations are sufficiently rare and mutational steps sufficiently small.

carrier A protein that spans a biological membrane and performs active transport.

carrying capacity 1. The population size at which an unexploited population subject to density-dependent control exactly replaces itself in successive generations; sometimes interpreted as the size of a population that a habitat can support, and often denoted by *K*. 2. In a single-species model, the equilibrium density of a population. For deterministic (nonstochastic) models, if population densities start at the carrying capacity, they will remain there indefinitely.

catabolic Pertaining to the degradative part of metabolism involving the breakdown of complex macromolecules and the release of energy.

catchability coefficient A parameter in a harvesting model that scales the rate at which individuals in a population are removed for a given exploitation effort level.

categorical and regression tree analysis (CART) A machine-learning method for splitting predictor variables so as to maximize within-group homogeneity of resulting groups generated by the splitting.

channel A protein that spans a biological membrane and allows passive movement of a solute as that solute moves to a location of lower free energy.

chaos A complex oscillation that does not repeat itself exactly, is not simply the result of a random stochastic process, and further, nearby trajectories tend to diverge over time. This latter aspect is known as *sensitive dependence on initial conditions* and is characterized by a positive Lyapunov exponent.

chronic inflammatory disorder A condition of consistent localized swelling driven by vascular response to (apparent) injury or infection.

chronosequence A location where an age-ordered sequence of plant communities demonstrates the expected pattern of succession (e.g., the sequences of progressively younger communities as one walks up a glacial valley of a continuously retreating glacier); a special case of the procedure called *space-for-time substitution* in which a mosaic of patches of different successional age can be sampled and organized to reveal succession.

clade A taxonomic group consisting of an ancestor and all its descendants.

clear-cutting Exploitation of a forest stand by cutting down all trees in the stand at the same time.

climatic niche The set of climatic conditions (temperature, humidity, etc.) that are optimal for a particular species.

closed system In compartment models of material flows in ecosystems, a system with no flows of material in or out, though energy may enter (e.g., as light) and leave (e.g., as heat).

cloud computing Internet-based computing whereby shared resources, software, and information are provided on demand to computers and other connected devices.

coalescent theory A body of population genetic theory that models genetic variation as a result of genealogical processes and gene trees. Coalescent theory is a natural complement to empirical studies of DNA sequence variation, which can often be captured as gene trees.

cohort A generational group in demographics or statistics.

cohort effect Difference in the mean demographic properties among cohorts due to temporal variation in environmental conditions.

cohort selection Within-cohort changes in composition due to consistent differences in survival among individuals.

combinatorial optimization A procedure to find an optimal solution of a problem by searching through a finite set of possibilities.

commensalism Interactions among biological entities in which at least one participant is positively affected while others are neither positively nor negatively impacted.

community end states Assemblages of communities that are resistant to further invasion.

community facilitation The process through which foundation species directly or indirectly facilitate the species assemblage of a community by providing refuge from abiotic and biotic stresses.

community inertia The tendency of communities to resist change in space and time.

community matrix A matrix that describes how each species will change in response to changes in each of the other species under consideration at a given equilibrium point; mathematically calculated by evaluating the Jacobian of a community model at an equilibrium.

compartment model A mathematical model used to represent the way that materials or energies are transferred among the compartments of a system. Each compartment is assumed to be a homogenous entity.

competition An interaction between individuals requiring the same resource, where acquisition of the resource by one individual reduces fitness of the other individual (exploitative competition), along with direct interference (e.g., fighting, poisoning, or preempting space).

competitor release Rapid increase of a lower competitor following control of the higher competitor. This can occur even if both competitors are controlled simultaneously, as the direct, negative effect of removing the lower competitor is compensated by the indirect, positive effect of removing its higher competitor.

conditional dispersal Dispersal that is dependent on the conditions experienced by an individual, such as population density, habitat quality, or physiological status.

confidence interval The range between which a mean is predicted to lie with a given probability (e.g., 95%). This range and the variance of the mean are closely related.

connectance The proportion of combinatorially possible links realized in a food web; e.g., the directed connectance $C =$ (number of links) / (number of species)^2.

connectivity Linkage of an individual habitat patch in a patch network and a local population in a metapopulation to other local populations, if any exist, via dispersal of individuals; measures the expected rate of dispersal to a particular patch or population from the surrounding populations.

conservation biology The branch of biology that deals with threats to biodiversity and with preserving the biological diversity of animals and plants.

conservation prioritization A form of decision analysis whereby priorities for conservation action are decided based on quantitative data.

conservation value See EXISTENCE VALUE.

constraint In the context of ecological restoration, a physical, economic, or biological factor imposing a limit on the proposed restoration activity.

consumer A species in its role of feeding on other species.

consumer surplus The value for a particular good or service that a consumer receives in excess of what the consumer must pay. Geometrically, this is the area under the demand curve and above the market price.

consumption In bioeconomics, the process of deriving utility from goods or services.

continuation A path-following numerical technique at the core of numerical bifurcation analysis.

continuous Referring to a function $f(x)$ for which small changes in x result in small changes in $f(x)$. Equivalently, the graph of f exhibits no breaks or jumps. Informally, a function of a continuous independent variable has a value that changes gradually as the independent variable changes gradually; more formally, a function $y = f(x)$ is continuous at $x = a$ if $f(a)$ is defined and $\lim_{x \to a} f(x) = f(a)$, and is continuous if it is continuous at all points where it is defined.

continuous-field conceptualization The geographic world conceived as a set of single-valued mappings from locations to variables or classes.

continuous plasticity Production of multiple continuous phenotypes from one genotype.

continuous variable An independent variable that takes values in a continuum: for example, time t in a continuous time model.

cooperation Behavior where individuals act together for beneficial results.

coordination Behavior that is favored by natural selection when it is common but is disfavored when it is rare.

Coriolis effect An effect associated with the rotation of the Earth, specifically the apparent deflection of a moving object when viewed from a rotating reference frame (i.e., as humans on the rotating Earth).

Coriolis force A pseudo-force resulting from a translation of coordinates from an inertial reference frame into a rotating frame and directed perpendicular to an object's motion.

corridors Regions of the landscape that facilitate the flow or movement of individuals, genes, and ecological processes.

coupled social-ecological system A concept that acknowledges the connection and interdependence of social and ecological systems.

covariance A measure of how much two variables change together. Mathematically, the covariance of $f(z)$ and $g(z)$ is defined as $cov(f,g)_z = \langle (f(z) - \langle f \rangle_z)(g(z) - \langle g \rangle_z) \rangle_z$, where $\langle \cdot \rangle_z$ denotes an average over z. Thus, $cov(f, g)_z$ will be positive if $f(z)$ is above (below) average when $g(z)$ is above (below) average—i.e., if f and g covary. The covariance of f and g is equal to their correlation (which ranges from -1 to 1) times the standard deviation of f and the standard deviation of g. Thus, covariances account both for whether quantities fluctuate in the same or opposite directions and for how much they fluctuate.

coverage probability The frequency with which a confidence interval will contain the correct parameter value. The frequency is defined over many confidence intervals from different data sets.

C - R flux rate A metric that measures the conversion of consumed resource biomass into consumer biomass or, the amount of resource consumed.

critical patch size The minimum size that a patch of favorable habitat must have in order to support a population.

critical threshold See BIFURCATION.

critical transition See REGIME SHIFT.

cross-feeding Facilitation of one microbe based on a substrate produced by another microbe.

data assimilation A statistical process for combining models and data; most commonly refers to the estimation of state variables, often using multiple data sources.

database management systems (DBMS) Computer-based software developed to assist in linking and managing complex data sets.

data model The distribution of the data conditioned on unobserved quantities.

DDE Abbreviation for delay differential equation.

decomposition The conversion of dead organic matter into mineral forms of elements.

delay differential equation (DDE) A differential equation in which the rate of change of the variables may depend on the values of other variables at the current time and also at previous times.

demand organization The notion that the rate of transformation of resources from resource to consumer is controlled by the demand for the product(s): you eat what you need, growth is prescribed.

demographic heterogeneity Variation in the growth, survival, and reproductive rates of individuals beyond that accounted for by known factors such as size, age, or developmental state.

demographic stochasticity Random fluctuations in population growth due to random variation in survival and reproductive success between individuals despite a shared environment; important for small populations.

demographic variance Variance in the deviations of individuals from the expected fitness of the population; can have components contributed by both within- and between-individual variation.

density dependence Dependence of the per capita growth rate on the abundance or density of the organism in question.

dependent variable A variable in a mathematical equation whose value is determined by those taken by the independent variables; in $y = f(x, t)$, y is the dependent variable. In ecological models, population size is typically a dependent variable that is determined by the independent variable time.

depensation See ALLEE EFFECT.

derivative A measure of how rapidly a function changes as its input changes. The derivative of a function $f(x)$ is given by its instantaneous rate of change: the value of $(f(x + h) - f(x))/h$ as h gets infinitesimally small. A function for which this limit is defined at x is called *differentiable* at x.

deterministic Describing or referring to a process that obeys a fixed (non-random) set of rules, such that the same initial conditions always generate the same dynamics.

deterministic extinction A condition in which a population becomes extinct because the aggregated birth rate becomes less than the aggregated mortality rate due to the intrinsic, non-stochastic dynamics of the system.

deterministic model A model with the property that once a set of parameters and a initial state is chosen, the outcome is completely predicted with no uncertainty.

detritus Nonliving organic material, including plant materials such as leaf litter, stems, wood, or metabolic products and animal materials such as carcasses or feces.

diaspore The unit of dispersal of a plant. Also, PROPAGULE, DISSEMULE.

diaspore shadow The spatial distribution of diaspores dispersed from a single plant, and often in a single dispersal season or year; for seeds, referred to as the *seed shadow*.

diet breadth The number of food types that a forager consumes.

difference equation A recursive formula that produces a sequence of numbers, vectors, functions, or other kinds of mathematic objects.

differential equation A continuous description of how a quantity changes over time; an equation containing derivatives of a function. An *ordinary differential equation (ODE)* contains derivatives of a function of a single variable, and a *partial differential equation (PDE)* contains partial derivatives of a function of more than one variable.

diffuse coevolution Coevolution among multiple species, or when the pattern of reciprocal selection between two species is dependent on a third species.

diffusion A dispersal process arising from a large number of small-scale, uncorrelated random movements.

dimension 1. The number of state (or dependent) variables in a model. 2. Any of the fundamental quantities in a model, from whose units all others are derived.

dimensionless quantities Pure numbers, or products and quotients of quantities where the units cancel out.

discounting A concept used in bioeconomic analysis to relate present and future costs and benefits.

discount rate 1. Technically, the rate of interest at which member banks may borrow money from the Federal Reserve Bank. 2. In general, the percentage reduction in the value of a unit of currency to be received one time period hence when paid out at the current time.

discrete-object conceptualization The geographic world conceived as an empty tabletop occupied by scattered, countable objects.

discrete-time Markov chain A collection of discrete random variables $\{X_n\}_{n=0}^{\infty}$, where time is discrete $n = 0, 1, 2, \ldots$. The term *Markov* (after the Russian mathematician) refers to the memoryless property of the process; that is, the conditional probability at time n only depends on the state of the system at time $n - 1$ and not on the earlier times,

$$\begin{aligned}&\text{Prob}\{X_n = x_n | X_{n-1} = x_{n-1}, \ldots, X_0 = x_0\}\\ &\quad = \text{Prob}\{X_n = x_n | X_{n-1} = x_{n-1}\}.\end{aligned}$$

discrete variables Variables that do not vary smoothly but instead take values that are fundamentally separate; for example, the integers are discrete, whereas the real numbers are continuous.

disease agent An organism that lives in an intimate and durable association with a host that suffers a fitness cost due to this association. Also, PATHOGEN.

disease-free equilibrium In an epidemiological model, the equilibrium state of a host population in the absence of infection or when the infection cannot invade.

disease invasion The emergence of a pathogen into a new host species or population.

dispersal Movement of individuals, often during a specific stage of the life cycle, that leads to a change in their location in contrast to movement within a home range. One example is movement of individuals among local populations in a metapopulation.

dispersal kernel A function giving the probability, or distribution, of dispersal distances.

dispersal limitation The failure of a species to disperse to all sites favorable for the survival and growth of the species.

dispersal range The distance a species can move from an existing population or natal area.

dissemule See DIASPORE.

dissipative Describing or referring to the state of the system asymptotically entering and remaining in a bounded set.

disturbance A sudden change in an ecosystem that has large impacts on the health and/or survival of the organisms in the ecosystem; may be natural (e.g., fire, hurricanes, insect outbreaks) or anthropogenic (e.g., harvesting, introduced pathogens).

disturbance regime The description of long-term and regional patterns of disturbance frequency, severity, and spatial cover in an area.

domain of attraction See DYNAMIC REGIME.

dominance Genotypic effects on phenotype resulting from the bringing together of alleles at a locus. The deviation from additive effects of these alleles results from dominance interactions.

dominant Lyapunov exponent A parameter that measures the exponential rate of divergence (or convergence) for two nearby trajectories of a system. A positive Lyapunov exponent implies sensitivity to initial conditions and is an indicator of chaos. (Named for the Russian mathematician Aleksandr Lyapunov.)

donor compartment In a compartment model, the compartment from which a particular flow arises.

donor control A classification of consumer–resource interaction where consumption of the resource by the consumer does not lead to declines in production of the resource.

doubling property See REPLICATION PRINCIPLE.

dynamical system An equation or set of equations that describes the state of a physical or biological system as a function of time.

dynamic disequilibrium of carbon cycle For terrestrial ecosystems, a condition at which global change causes differential changes in carbon influx and efflux, shifts in disturbance regimes, and changes in ecosystem states.

dynamic equilibrium of carbon cycle For terrestrial ecosystems, a condition at which carbon influx equals efflux and carbon pool sizes do not change at a given time and/or spatial scale.

dynamic global vegetation model A type of model that predicts the response to climate change of ten large functional groups of plants, based on their physiological processes (i.e., photosynthesis, respiration, and transpiration).

dynamic programming A mathematical optimization method used to identify a sequence of decisions that will best achieve a given set of objectives.

dynamic regime The set of system states that lead to a particular attractor, or represent fluctuations around a specific attractor. Within this set of states, the system self-organizes into a specific structure and behaves in essentially the same way. Also, DOMAIN OF ATTRACTION, BASIN OF ATTRACTION, STABLE STATE.

dynamic state variable models Optimization models of behavior in the broad sense where there is uncertainty but the probability distributions are known or can be estimated.

EBM Abbreviation for ecosystem-based management.

ecoinformatics The science of information in ecology and environmental science that integrates environmental and information sciences to define entities and natural processes with language common to both humans and computers.

ecological drift Random changes in abundances of ecologically equivalent species within a single community.

ecological inheritance The inheritance, via an external environment, of one or more natural selection pressures previously modified by niche-constructing organisms.

ecological interaction The relationship of a species population to environmental factors (resources, temperature, salinity, water availability, and so on) or other species (competitors, predators, diseases, mutualists, and so on) that affects its per capita growth rate or population density.

ecological niche factor analysis Computation of suitability functions by comparing the species distribution within its environmental variable space with that of the whole space.

ecological specialization Utilization by potential competitors of different components of the environment.

ecological stoichiometry The study of the balance of energy and multiple chemical resources (elements) in ecological interactions.

ecological visualization The application of computer-based software to create views of ecological processes.

ecology The field of study that examines the interactions between organisms and their environment.

ecomorphology A field of study that examines relationships between morphology and/or functional traits and habitat use to test whether a species has adapted to its environment.

economic efficiency The point at which the additional costs from any action are equated to the additional benefits from the action. At this point, no individual in society can be made better off without making another worse off.

economic threshold In a pest management system, the threshold density of a pest, below which its effects on crop yield are less than the cost of applying a control measure.

ecosystem A system composed of both the organisms (animal, plant, microbe) and the abiotic environment, and all interactions among and between these components.

ecosystem-based management (EBM) Management that integrates social, ecological, and institutional considerations of a particular place and focuses on sustaining the diverse ecosystem services provided by that place.

ecosystem engineers Organisms that contribute directly to niche construction theory's first subprocess through their environment-altering activities. They may also contribute to niche construction theory's second subprocess, by modifying natural selection pressures in ways that affect the subsequent evolution of populations. More informally, these are organisms that interact indirectly with other organisms through changes to the abiotic environment.

ecosystem facilitation Coupling between spatially distinct ecosystems that benefits foundation species and their associated species assemblages in a way that increases ecosystem services (stability, productivity, resilience, and so on).

ecosystem model A model designed to capture the pools and fluxes of mass (and sometimes energy) in an ecosystem. Ecosystem models use carbon as their basic currency, but most also track the biogeochemical cycles of water and macronutrients.

ecosystem services The benefits (resources and processes) that humans receive from functioning ecosystems; examples include drinking water, flood protection, crop pollination, and recreation.

ecotone The transition area between two communities or biomes.

ecotoxicological tests Observations of the effects of chemicals on one or more species under conditions that are as controlled as possible in terms of the organisms used and the environmental conditions to which they are exposed.

edge **1.** A line segment that joins two vertices in a network or graph. **2.** See BRANCH.

effective number of species The number of equally abundant species that are needed to give the same property of a study community. Also, SPECIES EQUIVALENT.

effect size A quantitative measure of the magnitude of a relationship.

eigenvalue A value representing the magnitude and direction (positive or negative) of change for a linear model along a related eigenvector axis. More formally, the eigenvalue of a square matrix $\mathcal{M}$ is a real or complex number λ satisfying $\mathcal{M}\vec{V} = \lambda\vec{V}$ or $(\mathcal{M} - \lambda I)\vec{V} = \vec{0}$, where the nonzero vector $\vec{V}$ is known as an eigenvector and I is the identity matrix. A square matrix of size $k \times k$ has k eigenvalues (counting multiple roots). The eigenvalues of $\mathcal{M}$ can be calculated by finding the roots of the *characteristic equation* of $\mathcal{M}$, which is the determinant of the matrix $(\mathcal{M} - \lambda I)$.

eigenvector For a matrix, a non-zero vector that is transformed by the matrix in length, not direction. For a population projection (Leslie) matrix, one eigenvector is the stable age distribution (that does not change with time). More formally, an eigenvector of a square matrix $\mathcal{M}$ is a nonzero vector $\vec{V}$ satisfying $\mathcal{M}\vec{V} = \lambda\vec{V}$ or $(\mathcal{M} - \lambda I)\vec{V} = \vec{0}$, where I is the identity matrix and λ is the eigenvalue.

Ekman layer A portion of the water column between the surface and a depth usually less than 150 m, within which horizontal momentum driven by wind stress is vertically mixed and rotated by the Coriolis force.

Ekman pumping Downward vertical motion at the base of the Ekman layer.

Ekman suction Upward vertical motion at the base of the Ekman layer.

Ekman transport Vertically integrated horizontal transport within the Ekman layer. It is directed perpendicular to the wind stress direction.

elasticity The proportional change in model output (e.g., the projected population growth rate, λ) when a model parameter (e.g., matrix element a_{ij}) is perturbed by a small percentage: $\partial \log(\lambda) \,/\, \partial \log(a_{ij})$.

elastic similarity A disproportionate increase in the diameters of trunk and branches that maintains the structural support required by a tree. As a plant grows, its central axis increases in diameter so that it does not collapse under its own weight, and branch axes increase in diameter so that the deflection at their tip is proportional to their length. Elastic similarity requires all diameters and lengths to change with total size in the same way.

electron acceptor sequence The rank of electron acceptors in terms of their thermodynamic favorability.

emerge To arise from underlying mechanisms, in sometimes complex and unexpected ways.

emerging infectious disease An infectious disease that newly arises in a population or that has been known for some time but rapidly increases in incidence or geographic range.

EMGAs Environmentally mediated genotypic associations with indirect but specific connections between distinct genotypes mediated by either biotic or abiotic environmental components in the external environment. EMGAs may either associate different genes in a single population or associate different genes in different populations.

endemic Found in and restricted to a particular geographical area.

endemic disease A communicable disease that does not need external inputs to persist in a population.

endosymbiont An organism that lives within another organism, such as nitrogen-fixing bacteria that live within legume roots or single-celled algae that live within corals.

energetics The study of the flow and transformation of energy.

enthalpy A thermodynamic function of a system, equivalent to the sum of the internal energy of the system plus the product of its volume multiplied by the pressure exerted on it by its surroundings.

entropy A quantitative measure of the amount of thermal energy not available to do work, and the tendency from order to disorder.

environmental niche model A model used to determine where an exotic species may invade by matching environmental conditions to their presence/absence in its native environment.

environmental stochasticity Partly or entirely unpredictable variation in ecological dynamics resulting from external sources such as weather or other catastrophic events.

epidemic An infection that rapidly spreads through a large proportion of the population but eventually dies out.

epistasis Deviation from the expected contribution on either an additive or multiplicative scale of alleles at one or more loci to the phenotypic or fitness effects of alleles at another locus.

equilibrium A state of a dynamical system at which the state does not change over time, unless the system is disturbed by other forces. For a differential equation $dx/dt = f(x)$, equilibria correspond to states x such that $f(x) = 0$. In a discrete-time model with state variable N, the equilibrium is the density N^* at which $N_{t+1} = N_t$.

equilibrium population size A population size at which the growth rate of a population equals zero, such that the predicted future population size equals the current population size.

ergodic Describing or referring to a system for which time averages of a single sample are equivalent to averages at a single time across many samples.

ergodic theory A branch of mathematics that examines the long-term behavior of dynamical systems from a statistical or probabilistic viewpoint; particularly useful for systems exhibiting complex behavior, such as quasi-periodic or chaotic motions.

escalator box car train A numerical method for finding the solution to continuous-time age- and stage-structured population models.

ess Abbreviation for evolutionarily singular strategy.

ESS Abbreviation for evolutionarily steady strategy. Also, sometimes used as an abbreviation for evolutionarily stable strategy.

establishment The ability of a species, after being introduced to a new location, to sustain or grow in population size.

Eulerian 1. Description of an animal's movement in terms of its effect on the spatial distribution of individuals. 2. A description of flow in which the observer remains in one location and watches water pass; flow is specified through the time-dependent fluid velocity u at a given location $\vec{u}(\vec{x}, t)$. (Named for Swiss mathematician and physicist Leonhard Euler.)

eusocial Describing or referring to species that have a reproductive division of labor, overlapping generations, and cooperative care of young.

event 1. In probability theory, a set of possible outcomes of an experiment. 2. In a continuous-time stochastic process, a discrete moment in time where the state of the system changes (e.g., when a birth or a death causes the population size to change).

evolution Change in the frequencies of genotypes or alleles within a population.

evolutionarily singular strategy (ess) A phenotype for which the fitness gradient is zero. ess-es comprise ESSes, branching points, and various sorts of evolutionary repellors.

evolutionarily stable strategy A strategy that, if adopted by a population, is not susceptible to invasion by individuals employing alternate strategies.

evolutionarily steady coalition (ESC) A combination of phenotypes (strategies), the bearers of which can live together on an ecological time scale and such that no mutant has positive fitness in the environment generated by such a community.

evolutionarily steady strategy (ESS) A phenotype such that no mutant has positive fitness in the environment generated by that strategy as resident. The abbreviation is often interpreted as *evolutionarily stable strategy*. However, an ESS is not necessarily evolutionarily stable (i.e., attractive).

evolutionary arms race An evolutionary interaction between two species (or genes) that involves adaptation and counteradaptation.

existence value The value that individuals place on knowing that a resource exists, even if they never use that resource. Also, CONSERVATION VALUE, PASSIVE USE VALUE.

exotic species See ALIEN SPECIES.

expected value The average over an infinitely large number of realizations of a stochastic process.

exploitative competition See SCRAMBLE COMPETITION.

exploiter–victim systems In population dynamics, a coupled system of two populations, either in theory or in practice, where one population, to its own cost, provides the resources necessary for the maintenance of the other.

exponential distribution A continuous probability distribution with probability density function $f(t) = \lambda e^{-\lambda t}$ for t nonnegative, which has mean and variance $\mu = 1/\lambda$ and $\sigma^2 = 1/\lambda^2$, respectively.

exponential population growth Population growth resulting from constant per capita population growth (i.e., no density dependence), which produces a population level that increases exponentially through time; the population size is given by $N(t) = N(0)e^{rt}$, where r is the population growth rate. Also, GEOMETRIC POPULATION GROWTH (typically in discrete time).

exposed Describing or referring to an individual who has been infected with a pathogen but is incapable of transmitting the pathogen to others.

extent of occurrence The area within the outermost geographic limits to the occurrence of a species.

extinction The disappearance of a species (as the last individual dies). It can be separated into *local extinction* (disappearance from a habitat), *regional extinction* (disappearance from a larger region of population linked by dispersal), or *global extinction* (disappearance from all habitats).

extinction-colonization dynamics Extinction of small local populations in a metapopulation and establishment of new populations in unoccupied habitat patches. Also, TURNOVER EVENTS.

extinction debt The number of species for which the extinction threshold is not met and that are therefore predicted to go extinct but have not yet had time to go extinct. Metapopulations begin to decline if the environment becomes less favorable for their persistence due to, for example, habitat loss and fragmentation; for some metapopulations, the new environment may be below the extinction threshold.

extinction threshold A point below which recolonizations do not occur quickly enough to compensate for local extinctions and the entire metapopulation goes extinct, even if some habitat patches exist in the landscape.

facilitated diffusion Diffusion of a solute when it is bound to another compound that is more soluble than the original solute.

facilitation Direct or indirect interactions between biological entities that benefit at least one participant in the interaction and cause harm to none.

feedback loop A set of cause–effect relationships that form a closed loop, so that a change in any particular element eventually feeds back to affect the element itself. Feedback loops can be either damping (also known as *negative* or *balancing*) or amplifying (also known as *positive* or *reinforcing*).

first-order differential equation An equation involving first derivatives of the state variable but no higher derivatives.

first-principle assumptions Assumptions that cannot be deduced from other assumptions.

Fisher information matrix A summary of the amount of information in data relative to the quantities of interest. Let $\underline{Y} \sim f(y; \theta)$. Provided the derivatives and expectations exist, the Fisher information matrix is given by $I(\theta) = E\left(-\frac{d^2}{d\theta^2}\log f(\underline{Y}; \theta)\right)$. If the parameter θ is a p-dimensional vector, the Fisher information matrix is a $p \times p$, positive semi-definite matrix. Inverse of the Fisher information matrix approximates the asymptotic variance of the maximum likelihood estimator.

Fisher's α 1. The diversity parameter of the logseries distribution of relative species abundance. 2. In neutral theory, a dimensionless, fundamental biodiversity number proportional to the product of the metacommunity size times the per capita speciation rate.

Fisher's equation An influential spatial spread model that incorporates diffusion and logistic growth. (Studied by R. A. Fisher, a British founder of statistics.)

fitness A measure of the evolutionary value of an individual. It is often measured as an individual's contribution to the next generations of a population. Fitness traits such as survival, viability, and fertility are those characters of an organism that affect the determination of this contribution to the next generation.

fitness differences In the context of generalized competition theory, the differences in two species' per capita growth rates when interspecific competition equals intraspecific competition; i.e., when there are no niche differences.

fitness landscape See ADAPTIVE LANDSCAPE.

fitness proxy A quantity that, although not equal to fitness, can be substituted for it in specific calculations.

fixation The state of a population when an allele has reached a frequency of 1.

flip bifurcation A bifurcation where a limit cycle or periodic solution doubles its period. Also, PERIOD-DOUBLING BIFURCATION.

flux density (F) The number of moles of gas, or Joules of energy, that pass a unit area per unit time.

folk theorem The result that repeated games tend to possess large numbers of evolutionarily stable strategies, impairing the ability of ESS analysis to predict evolutionary outcomes.

food web The set of all feeding relationships among different organisms in a community.

food web matrix A matrix $\mathbf{A} = \{A_{ij}\}$ used to represent the topology of a food web. Rows and columns of this matrix correspond to the food web's member species in their roles as resources and consumers, respectively. A matrix element A_{ij} is set to 1 when species j feeds on species i and to 0 otherwise. That is, by convention, the first index (<u>r</u>ow) refers to the <u>r</u>esource, the second index (<u>c</u>olumn), to the <u>c</u>onsumer. Also, ADJACENCY MATRIX.

foraging trait A trait determining a species' role as consumer.

foundation species A species that is highly abundant and that defines the physical structure of a community through its morphology and physiology.

free energy The energy in a physical system that can be converted to do work.

frequency dependence A change in per capita growth rate of a population with a change in its relative abundance in a community. Increases in per capita growth rate with decreasing relative abundance mean that a species can increase when rare.

frequency-dependent selection Dependence of the reproductive success of an individual not only on its own type but also on the composition of the population. For example, if the sex ratio in a population is biased toward females, then males have an advantage.

frequentist inference An approach to statistical inference that regards unknown parameters as fixed mathematical constants and draws inferences about those parameters by considering the frequency of individual sample statistics over all possible samples.

frequentist statistics A framework for statistical inference that determines the probability of observing data assuming that the model giving rise to the data is true.

front A distinct boundary between two water types of different density, specifically where this density difference drives a convergent flow across the density gradient which serves to enhance the density gradient; e.g., at head of an intruding dense saline layer or a near-surface plume of low-salinity water.

function In evolutionary ecology, the use, action, or mechanical role of a phenotypic feature.

functional connectivity The degree to which a landscape facilitates or impedes movement among resources.

functional diversity measures Measures considering species' functional roles in ecosystem.

functional response The intake rate of a consumer as a function of prey or resource availability.

functional-structural plant models A class of models that integrate components of plant physiology with plant morphology and anatomy. The premise of this approach is that the spatial distribution of plant organs and tissues has a controlling influence on physiological processes, growth, and development.

functional type Of plants, a group of species with similar morphology and growth; e.g., grasses, woody vines, or deciduous trees.

fundamental niche The full range of biotic and abiotic environmental conditions in which an organism can possibly exist.

gap analysis A spatial analysis of the distribution and conservation status of multiple biodiversity components (e.g., vertebrate species, vegetation communities).

gap model A forest community model designed to capture forest gap dynamics, the process of regeneration following the mortality of one or more dominant canopy trees that creates a gap in the canopy.

Gaussian distribution See NORMAL DISTRIBUTION.

gene A region of the genome that encodes a protein or RNA molecule.

gene flow The exchange of alleles between populations.

gene-for-gene theory A hypothesis that the ability of a pathogen to infect a host depends on which allele is present at a single genetic locus, and the ability of a host to resist infection likewise depends on which allele is present in the host's genotype. This type of genetic system and coevolution has been established for some plant–pathogen pairs and other systems.

gene tree A phylogenetic tree of alleles within a species. A gene tree may not be strictly congruent with the history of populations sampled. Discordances between gene and population history can be due to gene flow, incomplete lineage sorting, and other factors.

general equilibrium In bioeconomics, the set of prices, production, and consumption levels that result when firms and consumers maximize profits and utility, respectively, given constraints on the initial endowments of resources (wealth, labor, and so on).

general reproductive number (R_t) The average number of new infections generated by a single infectious individual at a specified time *t*.

genet A group of genetically nearly identical individuals that originate from the reproduction of a single ancestor.

genetic drift The change in allele frequency due to random chance.

genome The total amount of genetic material (DNA or RNA) in a cell.

genotype 1. The genetic component of the phenotype of an individual. 2. In evolutionary computation, the bit string that codes for the solution to a problem.

genotype-by-environment interaction The phenotypic effect of interactions between genes and the environment. (Also expressed as *genotype* $\times$ *environment* or $G \times E$ *interaction*.)

geographic information Information relating locations on or near the Earth's surface to properties present at those locations.

geographic information science The systematic study of the fundamental issues surrounding geographic information systems.

geographic information systems (GIS) Software that enables computer-based digital mapping and spatial analysis.

geographic mosaic theory of coevolution The idea that interactions among species can result in different degrees of coevolutionary changes in different geographic locations. In some populations there may be a high degree of coevolutionary change (a hot spot), whereas in others evolutionary changes are not reciprocal or there is evolutionary stasis (a cold spot).

geometric mean The appropriate average for the long-term behavior of a stochastic one-dimensional multiplicative process, computed as the tth root of the product of t values of a random variable.

geometric population growth See EXPONENTIAL POPULATION GROWTH.

geometric similarity Two implied conditions that define mean plant structure during self-thinning: (a) the height (H)-to-breadth ratio of a plant's exclusive space remains constant, and (b) the structure of the plant remains in constant proportions, $H \propto R^{-2}$, where R is radius of the exclusive ground area occupied by an average plant, while the volume (V) of the plant (and thus its weight) is proportional to the cube of the radius, $V \propto H^3$.

geospatial statistics Statistics that focus on the spatial distribution and analysis of objects.

gestation period The time during which a fetus develops, beginning with fertilization and ending at birth. The duration of this period varies among species.

giant component The largest connected subgraph of viable genotypes on a holey landscape.

Gillespie algorithm A Monte Carlo algorithm used to simulate probabilistic births and deaths in continuous time. A random time interval between the individual birth or death events is generated from an exponential distribution.

granivore An animal that consumes grain.

gravitational potential energy Energy that derives from elevation in a gravitational field such as the Earth's. For a mass m at a height h above a reference height, the energy is mgh, where g is the acceleration of gravity, approximately 9.8 meters per second squared at the Earth's surface.

gravity model In invasion biology, a model used to estimate the potential flow of invaders to each node or discrete patch on a network, based on network properties.

gyre A giant circulatory structure extending across an ocean basin and including an intensified western boundary current.

habitat The physical space within which a species occurs, and the abiotic and biotic entities (e.g., resources) in that space.

Hamiltonian A function that must be minimized by the optimal control in an optimal control problem. It was inspired by the Hamiltonian of classical mechanics. (Named for Irish astronomer/physicist/mathematician Sir William Rowan Hamilton.)

Hamming distance In information theory, the number of pairwise differences between two sequences or strings.

handling time In exploiter–victim systems, the amount of time required for one exploiter individual to subdue and consume one individual from the victim population.

haploid organism An organism in which a cell contains only a single copy of the genetic material, in contrast to *diploid* organisms that receive a copy from each parent.

Hawk-Dove game A prominent model for animal contests in evolutionary game theory. It is assumed that there are two kinds of individuals: "Hawks" escalate a fight, in which case "Doves" give up. When "Hawks" are frequent, it is better to be a "Dove" in order to avoid serious injuries. Conversely, if the population consists of "Doves," then escalating a fight pays off.

heavy-tailed distributions Probability distributions with tails that are heavier (thicker) than exponential distributions. Some heavy-tailed distributions have moments, such as the variance or kurtosis, that fail to exist. Events that deviate significantly from the mean are far more common for heavy-tailed distributions than for normal distributions.

hedonic property price approach The use of housing prices and housing attributes to estimate the value of ecosystem services to property owners.

Hellinger distance A measure that can be used to compute divergence between two statistical distributions. The squared Hellinger distance is defined as $H^2(g, h) = \frac{1}{2}\int(\sqrt{h(y)} - \sqrt{g(y)})^2 dy$. (After German mathematician Ernst Hellinger.)

hierarchical model A conditional probability model where each step involves the outcome of the previous step(s).

Hill numbers A diversity profile interpreted as the effective number of species. (Developed by British ecologist Mark Hill.)

hindcast A simulation of events that have already occurred.

holey landscapes A special case of neutral network models in which it is assumed that all viable genotypes have approximately the same fitness of 1.0, while all other genotypes are inviable with an effective fitness of 0.

homoclinic bifurcation Collision between the stable and unstable manifolds of a saddle.

Hopf bifurcation The shrinking and disappearance of a cycle after collision with an equilibrium. (After Austrian-born mathematician Eberhard Hopf.)

host In the context of host–parasitoid interactions, an insect species that is attacked and killed by the parasitoid. The juvenile parasitoid uses the host for food as it develops.

human-assisted colonization Translocation of a species to a new habitat because the environmental conditions at its current habitat will become unsuitable due to climate change and because the species is unable to reach newly suitable habitat on its own due to poor dispersing abilities or biogeographic barriers.

human well-being The elements largely agreed to constitute a "good life," including basic material goods, freedom and choice, health and bodily well-being, good social relations, security, peace of mind, and spiritual experience.

hyperbolic discounting Discounting that exists when a species has a relatively high discount rate over short horizons and a relatively low discount rate over long horizons.

hysteresis The tendency of a system to remain in the same state when conditions change due to lag effects and system memory. As a consequence, the critical threshold for a forward shift from Regime A to B often differs from the critical threshold for a return shift from Regime B to A.

ideal free distribution The distribution of individuals among patches of resources (of different value) when such individuals know the habitat, are free to move among patches, and compete for resources within patches.

immune See RECOVERED.

immunity The state of a host having sufficient biological defense to avoid or to clear infection.

inbreeding depression Lower fitness in the offspring of related individuals, particularly due to the potential for homozygosity in deleterious recessive alleles.

inclusive fitness approach A modeling approach that accounts for fitness effects from the perspective of individual genes by focusing on the effects of gene-encoded behavior on related individuals.

incomplete lineage sorting A condition in which successive speciation events (or population divergences) occur so rapidly that alleles do not have time to come to fixation before the next divergence occurs. The result can be a gene tree that differs from the history of constituent populations or species.

independent variables The variables in a mathematical equation whose values determine the value of the dependent variable or variables; in $y = f(x, t)$, x and t are the independent variables.

indirect-use value The benefits derived from the goods and services provided by an ecosystem that are used indirectly by an economic agent; for example, the purification of drinking water filtered by soils.

individual In individual-based ecology, the unit or scale at which behaviors are modeled. Systems are made up of multiple individuals; individuals may be individual organisms or social groups such as canid packs or insect colonies.

individual-based ecology (IBE) An approach to understanding and modeling ecological phenomena that explicitly considers how system-level phenomena emerge from individuals.

individual-based model (IBM) A computer simulation model that follows the fate of sets of individuals. Also, AGENT-BASED MODEL.

inducible defenses Phenotypically plastic traits that reduce the risk of predation or herbivory.

infectious Describing or referring to an individual infected with a pathogen and capable of transmitting it to susceptible individuals.

initial conditions A set of initial dependent state variable values that must be prescribed in order to identify the specific solution that that satisfies a given dynamical system.

integral projection model (IPM) A projection model that incorporates discrete lumped classes (such as age) and sub-structuring within each class (e.g., a continuous trait like exact age or size, or a discrete trait like a polymorphic phenotype).

integrodifference equation A discrete-time continuous-space dynamical system; used to model long-distance dispersal.

integrodifferential equation A continuous-time continuous-space dynamical system that assumes nonlocal interactions and can be used to model long-distance dispersal.

intensive parameters Parameters that do not depend on the absolute size of the object (i.e., the organism). Such parameters typically relate (directly or indirectly) to the concept of concentration or density.

interaction strength A measure of the effect of one species on another; the concept encompasses trophic link strengths but is used also for quantitative characterizations of indirect interactions between species, mediated by other species and non-trophic effects (e.g., density-dependent growth of basal species).

interference competition Competition between species in which individuals have direct negative effects on other individuals by preventing others' access to a resource via aggressive behaviors such as territoriality, larval competition, overgrowth, or undercutting.

interplant competition Competition between plants for the resources necessary for growth and survival; occurs when the immediate supply of a single necessary resource falls below the combined demands of the individual plants.

interspecific competition A contest for limited resources by individuals of two different species that results in a density-dependent reduction in their per capita growth rates.

interspecific effects Processes acting between two different species.

intraguild predation The interactions among a group of species that act both as competitors exploiting a common resource and as predators.

intrasexual selection Within-sex selection.

intraspecific competition A contest for limited resources among individuals of the same species that results in a density-dependent reduction in their per capita growth rates.

intraspecific effects Processes acting within a species.

intrinsic productivity or **intrinsic rate of increase** The maximum rate at which a population will increase in size when not affected by density-dependent mechanisms that reduce survivorship or fecundity. Often denoted by r.

invasion fitness Fitness of a new type in the environment generated by an existing community.

invasive impact Measurable damage, either ecological or economical, resulting from the presence of an invasive species.

invasive species An exotic species that manages to survive, reproduce, spread, and finally, harm a new environment.

iteroparous Describing or referring to a species or a life history in which an organism may reproduce several times over its lifetime.

Jacobian A matrix describing, across all population size combinations, how each species will change in response to changes in each of the other species near an equilibrium; mathematically calculated by taking the derivative of each growth equation with respect to each species under consideration, evaluated at the equlibrium.

kernel 1. A weight function that describes the relative impacts of the past values of some dependent variables on the current rate of change of a dependent variable. 2. A function used to describe dispersal in an integrodifference equation model.

keystone species A species that has a large effect on community structure and function (high overall importance value) that is highly disproportionate to its own abundance and/or biomass.

kin competition Competition between genetically related individuals.

kinematics The study of animal movement; i.e., the velocities, angles, and rates at which various body parts move through space, and how such metrics are quantified and analyzed.

kinesis The movement response of an organism to a stimulus that is non-directional. Kinetic movement responses can be subdivided into *orthokinetic responses*, in which an animal varies its distance moved per unit time in response to the stimulus, and *klinokinetic responses*, in which an animal varies the frequency with which it changes direction.

Kullback-Leibler divergence A measure of the difference between any two statistical distributions, given by $KL(g,h) = \int\left(\log\frac{g(y)}{h(y)}\right)g(y)dy$. The more different are the two distributions, the larger is this quantity. Although often called the Kullback–Leibler *distance*, this metric is technically not a distance because it is asymmetric; i.e., $KL(g,h) \neq KL(h,g)$. (Named for American mathematicians Solomon Kullback and Richard Leibler.)

kurtosis For a probability distribution, a measure of the fourth moment about the mean.

Lagrangian 1. Description of an animal's movement in terms of the individual's trajectory through space. 2. A description of flow in which the observer follows a water parcel; flow is specified through the time-dependent location x_p of that water parcel $\vec{x}_p(t) = \int \vec{u}(\vec{x},t)\,dt$. (Named for Italian-born mathematician and astronomer Joseph–Louis Lagrange.)

laminar flow Fluid flow at relatively low speeds, in which velocity varies smoothly with distance from a surface such as the wall of a blood vessel or the epidermis of a leaf.

land cover The actual cover of an area of ground, whether it be built, vegetated, or covered by a hard surface.

landscape A geographic area in which at least one variable of interest is spatially heterogeneous. The boundary of a landscape may be delineated according to its relevance to a particular research question and geographic or ecological units (e.g., watersheds or ecoregions).

landscape connectivity The ability of a landscape to facilitate the flows of organisms, energy, or material across the patch mosaic; a function of both the structural connectedness of the landscape and the movement characteristics of the species or process under consideration.

landscape ecology The science of studying the relationship between spatial pattern and ecological processes on multiple scales.

landscape pattern The composition (diversity and relative abundance) and configuration (shape and spatial arrangement) of landscape elements, consisting of both patchiness and gradients.

landscape sustainability The ability of a landscape to maintain its basic environmental, economic, and social functions under ever-changing conditions driven by human activities and climate change.

land use The purposes served by a given area, including such categories as agriculture, residential, or transportation.

leaf area index (LAI) The area of leaves per unit area of land.

least squares estimation Determination of best-fit parameter values as those that minimize the sum of squared deviations between model predictions and data.

leptokurtic 1. Mathematically, having a kurtosis (fourth moment divided by second moment squared) that is greater than three. 2. Biologically speaking, having more short- and long-range dispersers and fewer mid-range dispersers than a Gaussian distribution with the same variance.

leptokurtic distributions Probability distributions whose kurtosis is larger than that of a normal distribution. Leptokurtic distributions typically have more probability in the center peak and tails and less probability in the shoulders than a comparable normal distribution.

Leslie matrix An age-structured matrix model with non-negative entries in the top row and in the subdiagonal only. (Developed by British mathematician Patrick Leslie.)

LIDAR Light detection and ranging, a remote-sensing technology that uses laser pulses to measure distances. Most commonly used as an airborne or satellite technology to measure the land-surface elevation and vegetation height, though certain classes of lidar can measure the vertical structure of vegetation rather than just its total height.

life cycle graph A schematic representation of the local demographic processes in a population. The population is structured in age and/or size classes, which are connected by arrows that represent how an average individual in a class contributes to the number of individuals in the recipient class one time step later. Contributions can be through either survival and growth or reproduction.

life history The schedule of changes in survival probability and reproduction realized by a particular species over its natural lifespan; a primary determinant of evolutionary fitness.

life-history trade-off A negative correlation between key demographic characteristics (traits) such as competitive ability and the ability to colonize new habitat (dispersal and establishment). Other commonly considered trade-offs include competitive ability vs. the ability of populations to withstand predation, and competitive ability vs. the ability to become dormant. Also, TRADE-OFF.

likelihood A method for evaluating evidence in data for alternative values of parameters. The likelihood of the parameter given the data is proportional to the probability of the data given the parameter.

likelihood function For discrete data, a function that returns the probability of the data conditional on the parameter. For continuous data, a function that returns the probability density of the data conditional on the parameter.

likelihood profile A plot of the probability of the data vs the value of parameter with the value of the data held constant and the parameter varying.

limit cycle A sequence of population sizes that are repeated, in order, over time.

limiting similarity In population ecology, the degree to which two species can share resources before exclusion occurs.

linear chain trick A method of transforming a system of delay differential equations to a larger system of ordinary differential equations. This method can only be applied to systems of delay differential equations that employ a special class of kernels.

linearization principle A mathematical procedure used to determine the stability of equilibria; a general procedure that also applies to the study of cycles and other attractors.

linearly determined spread rate A spread rate that can be determined by linearizing about the leading edge of an invasion.

linear spread conjecture A conjecture (based on simple models) that the rate of spatial spread of the population is linear with respect to time.

linkage disequilibrium (LD) A measure for statistical associations between alleles at different loci in a population. For example, in a haploid model with the four genotypes ab, Ab, aB and AB, LD is defined as $D = p_{ab}p_{AB} - p_{Ab}p_{aB}$, where p_i is the frequency of genotype i in the population.

link density The ratio $Z =$ (number of links)/(number of species) in a food web.

local community The sum of the abundances of trophically similar species co-occurring in an area sufficiently small that dispersal limitation can be ignored.

local population An assemblage of individuals sharing a common environment, competing for the same resources, and reproducing with each other. In a fragmented landscape, a local population typically inhabits a single habitat patch.

local stability A condition in which the focal state variable of a dynamic system returns to the original equilibrium after any disturbance that is sufficiently small.

long-run or **long-term population growth rate** In a stochastic environment, the average annual rate of change in population size over a long sequence of years.

longwave radiation See TERRESTRIAL RADIATION.

loop analysis An approach to analyzing a life cycle graph based on loop elasticities used to determine the relative importance of each of these loops; e.g., for population growth.

loss In statistics, a non-negative function of an estimator and a parameter to be estimated, describing the discrepancy between the estimator and the parameter; the estimator is a function of the observed data. The criterion of minimum (posterior) expected loss is used to obtain optimal estimators.

Lotka-Volterra model A pair of differential equations that describes the dynamics of two species of organisms (e.g., predator and prey). (Proposed by American mathematician and statistician Alfred Lotka, and independently by Italian mathematician Vito Volterra.)

macroecology The study of patterns of distribution, abundance, body size, and other attributes among species within large, continental-scale assemblages.

macroevolution Large-scale evolutionary changes, such as anatomical innovations, that cannot be expressed in terms of a fixed set of traits.

macroparasite A parasite whose pathogenicity is dependent on the number of infecting propagules.

marginal value In bioeconomics, the change in value resulting from a marginal increment in the argument on which the value function is defined. For example, if the total utility derived from consuming one more beer increases to 10 units from 8 units, the marginal value of this extra consumption is 2 units.

marine reserve An area of the ocean that is completely protected from extractive and destructive activities.

marine spatial planning (MSP) Management of multiple ocean resources (e.g., fisheries, tourism, transport, oil exploration) in concert, using a spatially explicit approach.

market value The amount of money exchanged between a willing buyer and willing seller for a good or service.

Markov chain A stochastic process with the Markov property taking values in a finite or countable set.

Markov chain Monte Carlo (MCMC) A numerical algorithm used to sample from a distribution, in particular, from the posterior in the Bayesian context.

Markov decision process A mathematical framework for modeling decision-making, in problems where outcomes occur based upon a Markov process but affected by the control of a decision maker.

Markov property For a stochastic process, the property of memorylessness: the future is independent of the past, given the present. (Named for Russian mathematician Andrey Markov.)

mass balance or **materials balance** The application of conservation of mass to the modeling of physical systems. Mass balance general equilibrium models are used in economic analysis to capture as many externalities that each act of production and consumption will create.

mass effects The flux of species from high-abundance (and high-diversity) areas to low-abundance (and low-diversity) areas; a multispecies version of rescue effects.

mating behavior A cascade of behavioral alternatives, some preceding copulation, such as accepting or rejecting potential mates, that may include so-called "courtship," the duration of copulation, and within-sex behavioral variation during the act of copulation.

matrix (*plural*, matrices) 1. A rectangular array of numbers, symbols, or expressions. 2. In landscape ecology, a component of the landscape that is neither patch nor corridor.

matrix transpose An operation in which the rows of a matrix become its columns. Also, the matrix created by this operation. The transpose of a row vector is a column vector.

maturity maintenance Effort (i.e., energy and/or substrate) that needs to be invested continuously to maintain a certain level of maturity; applies to regulation systems, but also to defense systems, such as the immune system. Failure to pay maturity maintenance leads to reduction of the level of maturity (rejuvenation), which is associated with an increase in hazard rate.

maximum likelihood estimate The value of a parameter that maximizes the probability of the observed data conditional on the value of the parameter.

maximum sustainable yield (MSY) The maximum catch that can be taken during a harvest period where the fishery remains sustainable and at highest productivity.

McKendrick-von Foerster model A description of single-species age-dependent population dynamics in a model based on a first-order partial differential equation. (Named for Austrian–American scientist Heinz von Foerster and Scottish physician/epidemiologist Anderson Gray McKendrick.)

mean field approximation A spatially implicit approximation to the dynamics of a spatially explicit model.

mesoevolution Evolutionary changes in the values of traits of representative individuals and concomitant patterns of taxonomic diversification.

mesopredator A medium-sized predator that exhibits extreme population growth when top predators are removed.

mesopredator release Rapid increase of an intermediate (meso-) predator population, following the control of a top predator. This results in the decline of the shared prey.

metabolism The uptake, transformation, and use of energy and materials for purposes of survival, growth, and reproduction.

metacommunity The biogeographic region in which local communities are embedded, represented by the sum of the species populations in the entire region.

metapopulation An assemblage of local populations linked by dispersal.

metapopulation capacity A measure of the size of a habitat patch network that takes into account the total amount of habitat as well as the influence of fragmentation on metapopulation viability.

microevolution Changes in gene frequencies on a population dynamical time scale.

microbes Microscopic living organisms that include bacteria, archaea, and some eukaryotes.

microbial feedback Change in growth rates of a host type in response to change in density or abundance of its associated microbial community.

microparasite A parasite whose pathogenicity is independent of the number of infecting propagules.

microsatellite A popular marker in phylogeography whose allelic variation consists of differences in the number of short, repeated sequence motifs, such as CACACA or TGGTGGTGG.

model A simplification of reality developed to describe the behavior of a process for a variety of potential objectives such as predicting the future of the process, controlling it, or considering how it is affected by various factors.

mode of action The way that a chemical interacts with the biochemistry, metabolism, and cellular processes of organisms.

mode of transmission The mechanism by which a pathogen moves from one host to another.

module An i-species food web with interaction strength.

Monte Carlo simulation To simulate the probabilistic events in a stochastic model, generation of pseudorandom numbers using a computational algorithm.

Moore neighborhood For cellular automata, the neighborhood consisting of a site and its eight neighbors (the four orthogonal neighbors and four diagonal neighbors). (Named for American mathematician/computer scientist Edward F. Moore.)

Moran effect The synchronization of populations by spatially synchronous fluctuations in exogenous environmental factors. (Named after Australian statistician Patrick Moran.)

morphology The descriptive features of the external and internal (anatomical) phenotype.

motif An i-node network that does not include interaction strength.

MTE The arithmetic mean time to extinction.

multi-level selection approach A modeling approach that accounts for fitness effects from the perspectives of both the individual and the group.

Murray's law An empirical relation stating that the cost of fluid transport in vessels (e.g., blood vessels) is minimized when the vessels branch, if the sum of the cube of the diameters of the two branch vessels equals the cube of the diameter of the parent vessel. (Devised by British biologist Cecil D. Murray.)

mutational covariance matrix A matrix composed of the variances and covariances of the distribution of mutational steps.

mutual invasibility plot (MIP) In adaptive dynamics, a plot indicating, for a scalar trait, all combinations of two types such that each type can invade in a population consisting wholly of the other type.

mutualism Symbiosis between organisms of two species in which both individuals benefit from the association.

mycorrhizal symbiosis An interaction between plants and soil fungi whereby fungi grow in or on the roots of plants and in the soil matrix. Nutrients from the fungi are exchanged for photosynthetic products from the plant.

narcosis A general non-specific mode of toxic action, normally associated with a reduction in central nervous system activity and thought to result from accumulation of a chemical in cell membranes, diminishing their functionality.

narrow-sense heritability The proportion of the total phenotypic variance in a trait that is due to including additive genetic contributions and excluding the non-additive dominance and epistatic genetic effects, as well as environmental sources of deviation; calculated as the slope of the offspring–midparent phenotypic regression line. It measures the degree to which phenotypic value predicts breeding value.

narrow-sense sexual selection A theorem with two assumptions: (a) variation among males in a trait occurs, and (b) the mechanism of variation may be female choice among males or male–male behavioral or physiological interactions, each of which works only through among-male variance in number of mates. It is narrow in that it restricts the mechanisms of selection to those that act only on males and only on the number of mates' component of fitness.

Nash equilibrium A concept of game theory: a game is in equilibrium if none of the players has an incentive to deviate from its strategy, as long as the other players stick to theirs. (Developed by American mathematician John Forbes Nash.)

natal dispersal For a given individual, permanent movement from the location of origin to a different location for reproduction.

natural capital An economic metaphor for the stocks of renewable and nonrenewable resources found on Earth that produce a flow of goods and services.

natural enemy–victim interactions A subset of consumer–resource interactions, where both consumer and resource are living. The natural enemy makes its living by extracting resources from another species' the victim—and in so doing, harming it. The term broadly encompasses predators, parasites, pathogens, parasitoids, and other kinds of consumers.

natural selection The process of evolutionary change in a natural population caused by fitness differences among genotypes. Natural selection is not the same as evolution, but evolution can occur via natural selection if trait variants are heritable, so that offspring resemble their parents.

negative feedback The process whereby one component of a system is negatively affected by a second component, which has a further negative effect on the state of the first component.

neighborhood For cellular automata, the set of sites whose states affect the new value of a particular site being updated.

NEON The National Ecological Observatory Network, which will collect data across the United States on the impacts of climate change, land use change, and invasive species on natural resources and biodiversity; a project of the U.S. National Science Foundation, with many other U.S. agencies and non-governmental organizations cooperating.

network In graph theory, a collection of vertices connected to one another by edges.

network of habitat patches In a fragmented landscape, a network that is composed of discrete habitat patches (each of which may be occupied by a local population) and which may be occupied by a metapopulation.

neutral community models Models that typically represent competing species in a grid of spatial locations. They assume that all species are equivalent in their competitive ability (and fitness), that they are limited in their dispersal abilities, and that loss of species is replaced through speciation.

neutrality Demographic equivalence of species on a per capita basis (equal per capita rates of birth, death, dispersal, and speciation).

neutral networks Connected graphs of viable genotypes of equal fitness, with edges that are usually defined by single point mutations.

niche The set of abiotic and biotic conditions that permit the survival of a species.

niche breadth The range of resources or environmental conditions used by a species population or that range of conditions where its fitness is greater than zero.

niche differences Differences among species in limiting resources, favorable environmental conditions, or interactions with enemies.

niche model A model of species' spatial distributions using niche variables (e.g., climate, soil, and biotic variables).

niche partitioning mechanisms Mechanisms that increase the strength of intraspecific competition relative to interspecific competition.

node A branching point on a phylogenetic tree.

noise A (generally uncorrelated) random variable, which can be introduced into models to simulate fluctuations.

non-consumptive effects Effects of predators on prey traits including growth, development and behavior.

nondimensionalization The process of deriving equations involving only dimensionless quantities.

nonequilibrium thermodynamics The description of changes in state when a system changes rapidly (actually at any finite rate), causing the various degrees of freedom in the system to be out of equilibrium with each other and causing the changes to be irreversible, thus violating the premises of ordinary thermodynamics.

nonlinear Describing or referring to a function whose graph is not a straight line. If $f(x)$ is a nonlinear function, then a plot of $f(x)$ versus x will be curved.

nonlocality The ability of an object to directly influence another distant object or location. Nonlocal models, which often contain integrals, include interactions between distant locations.

non-market value The amount of money that a willing buyer would spend to purchase a good or service if it were available for sale.

non-parametric model A model that does not assume a specific shape such as linear or exponential but can follow the data in a smooth but arbitrarily shaped curve.

non-use value Benefits that do not arise from consumptive or non-consumptive uses (e.g., existence value).

normal distribution The bell curve distribution for random outcomes. Data are often assumed to follow a normal distribution of random deviations around their predicted values. Also, GAUSSIAN DISTRIBUTION.

norm of reaction See REACTION NORM.

nucleic acid A component of DNA. A gene is a sequence of nucleic acids.

nudation A disturbance event that destroys the vegetation at a location and initiates a successional recovery. (Described by plant ecologist Frederic E. Clements.)

nullcline The sets of species' population sizes for which there is a growth rate of zero for a given species, typically illustrated as a curve in a phase plane.

null model A model that assumes no variation in features that determine the outcome.

numerical response **1.** The change in consumer density as a function of change in food availability. **2.** The relationship between the rate of consumption of prey and the per-individual birth minus death rate of the predator population.

nutrient immobilization/mineralization Processes whereby microbes transform nutrient elements from mineral to organic form (immobilization) or from organic to mineral form (mineralization).

objective function The quantity to be maximized or minimized by a proposed management activity, a function of the biological and/or economic variables relevant to the system.

observables In Bayesian statistics, quantities that can potentially be observed but may not necessarily be observed, including, for example, observed data, missing data, or future data.

occupancy estimation An approach that enables inference about species occupancy based on presence/absence data.

ODE Abbreviation for ordinary differential equation.

omnivore Ordinarily, an animal that consumes both plant and animal material. In models of food chains, an omnivore preys on both a consumer and the resource exploited by that consumer.

ontogeny The course of growth and development of an individual organism.

open access resource An unregulated resource or a resource that anyone can exploit at any rate they choose.

open systems In compartment models of material flows in ecosystems, a system that has both inflows of material and losses of material.

optimal control A method to find the best, according to some criteria, procedure to control a given system, while maintaining consistency with the rules of dynamics governing the system.

optimal decision schedule In dynamic programming, a sequence of decisions that lead to the best outcome over time.

optimal foraging theory A theory that is mostly applied to food exploitation, wherein an animal is posited to behave in a manner that maximizes its fitness through energy and/or nutrient intake.

optimal rotation period The period between consecutive clear-cuttings of a forest stand that maximizes the present value of the harvests obtained from all clear-cuttings of the stand for all time into the future.

optimization The process of maximizing or minimizing some criteria while accounting for constraints on the process.

optimization principle A function defined on the trait space of an eco-evolutionary model such that under any constraint the ESSes for that model can be determined by maximizing that function.

optimizing selection Natural selection that favors an intermediate phenotype over individuals with either a lesser or greater value of a trait.

ordinary differential equation (ODE) An equation involving ordinary derivatives of the state variable but no partial derivatives.

organism An independent living unit that persists for a substantial period of time.

oscillation Any dynamic that is periodic in time such that the population density increases and decreases over time.

osmotica Chemical compounds that contribute an osmotic potential by diluting water. They may be inert materials or metabolites.

osmotic potential A reduction in the energy content of water resulting from its dilution by the presence of a solute. In the limit of low concentrations of solute, c (in moles per kg of water) in water at absolute (Kelvin) temperature T, the potential in Joules per kg is $\pi = -cRT$, with R as the universal gas constant, approximately 8.3 J mol^{-1}K^{-1}.

outbreeding depression Lower fitness in offspring of individuals from different populations, particularly due to degradation of local adaptation.

overcompensation A form of density dependence characterized by a dome-shaped relation between population size and recruitment, in which recruitment declines when a population is sufficiently large. Generally requires that density-dependent mechanisms have a delayed effect.

pairwise invasibility plot (PIP) For a scalar trait, a plot indicating the combinations of mutants and residents for which the mutant can invade.

paradox of enrichment The possibility that increasing the growth of prey can produce extinction of predator, prey, or both, due to cycles with very low minimum population sizes. (A term coined in 1971 by American ecologist Michael Rosenzweig.)

parameter A quantity describing some aspect of a population that is of interest in a statistical study.

parasite An infectious agent. *Microparasites* (e.g., viruses, bacteria, and protozoa) typically reproduce within a single host. *Macroparasites* (e.g., helminths and other metazoan parasites), by contrast, typically have life cycles with one or more intermediate hosts.

parasitoid An insect that lays one or more eggs on or in an individual of another species (the host), after which the juvenile parasitoids use the host for food as they develop, killing the host in the process.

parental investment A parent's investment in one offspring that precludes investment in future offspring. (A term introduced in 1972 by American evolutionary biologist R. L. Trivers.)

parsimony In the context of phylogenetic reconstruction, a method of tree inference that seeks to minimize the number of inferred evolutionary changes.

parthenogenesis A genetic system in which females produce offspring without mating (or at least without males contributing genetic material to offspring). Definitions of parthenogenesis and asexuality vary widely between researchers and taxonomic groups.

passive use value See EXISTENCE VALUE.

patch The fundamental spatial unit in most community and ecosystem models, defined as a locally homogeneous environment encompassing a region within which every individual is able to compete with every other.

patch dynamics A perspective that ecological systems are mosaics of patches exhibiting non-equilibrium transient dynamics and together determining the system-level structure and function.

patch dynamics models Models consisting of a series of loci or patches with one or more species inhabiting the patches. Dominant dynamics include local (patch-level) extinction and recolonization, or rescue.

pathogen See DISEASE AGENT.

pathogenicity The degree to which a host suffers a decrease in fitness due to its association with a parasite.

pattern Any observation that is clearly non-random.

pattern analysis Procedures with which landscape pattern is quantified using synoptic indices or spatial statistical methods.

pattern formation The spontaneous development of spatial patterns, specifically in systems where there is no external source of spatial heterogeneity.

pattern-oriented modeling An approach to modeling, and especially individual-based ecology, that uses patterns observed in real systems as the basis for (a) deciding what structures should and should not be in a model, (b) developing and testing theory for individual adaptive behavior, and (c) model calibration and analysis. Patterns at both individual and system levels can be useful for pattern-oriented modeling.

payoff A number that represents the success of a given strategy. In classical game theory, payoffs are described as utilities; evolutionary game theory interprets the payoff of a strategy as its reproductive success.

per capita birth rate The rate at which an average individual within the population gives birth to an offspring.

per capita death rate The rate at which an average individual within the population dies.

per capita demographic equivalence In neutral community ecology, the strict form of neutrality asserting that individuals of all species experience the same demographic rates, equal birth, death, immigration, and emigration probabilities, and even the same probability of becoming a new species.

per capita effects Mutual relative effects between consumers and their resources. The *negative per capita effect* is the effect of the consumer on the resource (defined as the feeding rate per number of consumers). The *positive per capita effect* is the effect of the resource on the consumer (defined as the consumer production rate per number/amount of resource).

per capita population growth rate The rate at which a population changes per individual in the population as a result of reproduction, mortality, emigration, and immigration. In continuous-time models, this is given by $\frac{1}{N(t)}\frac{dN(t)}{dt}$, often denoted r. For discrete-time models, it is often expressed as the natural logarithm of the ratio of population densities at consecutive sample times,

$\ln \frac{N_t}{N_{t-1}}$. For a simple single-species model with only negative density dependence, the maximum per capita population growth rate occurs at low population density.

percolation threshold In holey landscapes, a condition attained when there is a single neutral network such that all viable genotypes are connected in one subgraph.

performance A quantitative measure of the ability of an organism to conduct an ecologically relevant task, such as sprinting, jumping, or biting.

period-doubling bifurcation See FLIP BIFURCATION.

periodic Referring to a pattern that repeats itself at regular intervals. A function $x(t)$ is periodic if there exists $T > 0$ such that $x(t + T) = x(t)$ for all t.

periodic cycle A repeating pattern in population numbers. The period of the cycle is the length of time between repetitions.

periodic doubling cascade A series of bifurcations in which the period of the system's cycles double as a parameter is changed. The size of the parameter interval leading to the period doubling decreases geometrically with each subsequent bifurcation as described by the Feigenbaum number (after the American physicist Mitchell Feigenbaum).

permanence A condition in which the focal state variable of a dynamic system eventually ends up in a positive bounded state after sufficiently long time not depending on the initial state. When the system is permanent, the variable becomes neither infinitely large nor zero and remains at least some fixed minimum distance away from zero.

persistence The ability of a population to avoid declining to zero abundance.

pessimization principle A function defined on the set of possible environments of an eco-evolutionary model such that under any constraint the ESSes for that model can be determined by minimizing that function.

phase angle The transformation of time, t, in an oscillatory time series such that the time series is proportional to $\sin(t)$. Expression of time as a phase angle in multiple, spatially-referenced time series allows for the characterization of the direction and speed of population waves.

phase plane diagram A plot in which each axis represents the population size of each species under consideration; typically used in two-species models. Time is not an explicit axis, but rather, is represented implicitly. One example of a phase space.

phase shift See REGIME SHIFT.

phase space A coordinate system where the axes represent the state variables of a system. The possible states of the system are represented by points in phase space.

phase synchronization A term from the dynamical systems literature referring to the situation in which two periodic or quasi-periodic oscillators exhibit a constant phase difference.

phenotype Any observable characteristic of an organism.

phenotypic plasticity The ability of an organism to develop any of several phenotypic states, depending on the environment.

phylogenetic baggage The presence of an adaptation in a species because of past selective pressure on an ancestral species.

phylogenetic diversity measures Measures incorporating phylogenetic distance or genetic distance between species.

phylogeny An evolutionary history of a group of organisms, generally depicted by a branching diagram.

physiognomy The physical form of plants and in some uses the plant life form (trees, shrubs, and the like).

pitchfork bifurcation Collision of three equilibria after which only one remains.

plant plasticity The capacity of plants to respond to their environment by change in growth and physiology. Two types of reaction occur, each under control of specific environmental signals: morphological plasticity and physiological acclimation.

plasmodesmata (*singular,* plasmodesma) Microscopic channels that cross plant cell walls, allowing transport between abutting cells. They are lined by extensions of the cell membrane, effectively making one connected cytoplasm among the cells they link.

point mutation speciation A model of speciation in which the founding lineage of a new species can in principle be traced back to a single individual.

Poisson Referring to either a stochastic process or the resulting distribution. A Poisson process is one in which events occur continuously and independently of one another with a constant underlying rate. The resulting probability distribution for observing n events (during a given period) is $P(n) = \frac{\mu^n e^{-\mu}}{n!}$ for mean μ, and has the property that the mean equals the variance. (After French mathematician/physicist Siméon Denis Poisson.)

Poisson ratio The ratio of the changes in linear dimension when a material is stretched. When the stretching is defined as along the x-direction in an amount dx, it is the ratio $v = -dy/dx$, where dy is the change in dimension (strain) in the orthogonal direction. Incompressible materials have $v = 0.5$.

polyline, polygon The representation of lines and areas using sequences of points connected by straight lines.

polyphenism The occurrence of environment-specific alternative and discrete phenotypes in a population, most commonly morphological phenotypes.

Pontryagin's minimum principle A principle that describes necessary conditions that must be satisfied by any optimal control; useful for finding candidate optimal trajectories. (Formulated by Russian mathematician Lev Semenovich Pontryagin.)

population 1. A group of individuals of the same species occupying a specified geographic area over a specified period of time. The area may be ecologically relevant (an island) or irrelevant (a political district), and the boundaries may be porous, with individuals immigrating to and emigrating from the population. 2. A (often hypothetical) collection of individuals, objects, or outcomes from which a statistical sample is selected.

population cycles Regular fluctuations in the size of a population, as seen, e.g., in some lemmings and snowshoe hares. Such cycles may be caused by density-dependent mortality, which increases due to overcrowding-induced food shortage, resulting in dramatic decline in the population; the population then gradually increases again.

population dynamics The pattern of changes in population densities through time.

population limitation All of the factors (abiotic and biotic) that set the carrying capacity of a population.

population regulation The density-dependent process (or processes) that returns a population to its equilibrium following a perturbation.

population vector In demographic studies, a vector $n_t = \{n_{i,t}\}$ whose elements are the numbers of individuals having trait i at time t.

population waves A phenomenon sometimes observed in periodically oscillating populations, in which peaks and troughs of abundance are progressively delayed or advanced when moving in a specific direction.

positive assortment A situation in which individuals are more likely to interact with others that are similar to themselves.

positive feedback In a compartment model, the process in which some a compartment is affected by another compartment that is downstream of it, having a positive effect on the growth of that compartment.

posterior A probability distribution describing uncertainty in unknown quantities (e.g., parameters or unobserved observables) given the observed data; via Bayes theorem, the posterior is expressed as proportional to the product of the data model and the prior.

potential vegetation The expected mature or climax vegetation at a location, used in regional mapping; can be contrasted with actual vegetation maps.

precautionary principle A management principle suggesting that if an action or policy is suspected of causing harm to the public or the environment, in the absence of scientific consensus as to the likelihood, magnitude, or causation of that harm, the action or policy should be avoided and the burden of proof that the action or policy is not harmful falls on those taking the action.

predation Traditionally, an interaction in which one organism (predator) consumes part or all of another organism (prey).

prediction error Discrepancy between model prediction(s) and new data that have not been used for model fitting.

presence data Data on locations of potential species presence where a species was sought and detected.

present value The current value of an income-generating asset, taking into account the discounted value of all future revenues expected to be generated by this asset.

prevalence In the context of disease dynamics, the proportion of hosts infected.

primary dispersal The initial movement of a diaspore away from the mother plant until it reaches the ground or other substrate.

prior A probability distribution that quantifies uncertainty about unknowns (e.g., parameters, unobserved quantities) prior to having obtained data related to the unknowns.

Prisoner's Dilemma A game that describes the conflict between group-interest and self-interest. Two individuals may either cooperate (C) or defect (D). If both choose C, they are better off than if both choose D. However, individually each player prefers to defect, leading to a dilemma.

probability distribution A mathematical entity that describes how likely it is that an uncertain quantity will assume a certain value.

process sandwich An expression of Bayes' theorem to explicitly indicate the use of a process model such that the posterior is expressed as proportional to the product of the data model, the stochastic process model, and the prior, where the process model often incorporates ecology theory.

projected population growth rate The asymptotic growth rate of a population, given by the dominant eigenvalue (λ), an analytical property of a transition matrix, if the demographic rates in the matrix remain constant over many time steps.

projection matrix A square matrix $\mathbf{A} = \{a_{ij}\}$ whose elements are the rates at which individuals in each trait class in the population produce individuals in each other trait class, in the next time interval. In the basic transition matrix approach, this matrix projects the population vector forward through time according to the equation $\mathbf{n}_{t+1} = \mathbf{A}\mathbf{n}_t$.

propagule 1. A dispersal unit for reproduction of plants or other organisms by sexual or asexual means. 2. See DIASPORE.

propagule pressure The rate of introduction of propagules from a different species.

proportionate mixing An assumption used in structured population models to describe random mixing across groups; contrasted with *assortative mixing*, where individuals have a preference for mixing with individuals within their own group, and *disassortative mixing*, where individuals have a preference for mixing outside their own group.

publication bias Bias that arises when a study's outcome affects either the likelihood that it is published or the time and/or place where it is published, which may affect the ease with which the study can be located.

pycnocline In a body of water, a rapid change in density with depth, associated with a *thermocline* (sudden change in temperature with depth) or a *halocline* (sudden change in salinity with depth).

quantitative trait A trait for which phenotypic variability is described by measurement of a variable rather than by the division of individuals into discrete classes.

quasi-extinction threshold (QET) A low, but positive, population size or density at which a population is defined as "quasi-extinct"; i.e., highly threatened but still potentially able to be conserved with extraordinary measures.

quasi-periodic Referring to curves in state space that exhibit multiple, incommensurable frequencies; i.e., the ratio of frequencies is irrational. When there are two frequencies, one can visualize quasi-periodic motions as a curve on a torus (analogous to the surface of a donut) that wraps around the torus without ever returning to the same point.

quasi-stationary distribution A conditional probability distribution in a stochastic model that does not change through time (e.g., the long-term distribution of population size given that extinction has not yet occurred).

ramet Each separate individual from a population of genets.

random effect A source of unexplainable or unmeasured variation in one or more data values.

random variable A variable that does not obey a deterministic set of rules; instead, its value has a random component described by a probability distribution.

range boundary or **limit** The outermost geographic occurrences of a species, usually excluding vagrant individuals.

rank abundance curve The curve obtained by plotting the log of species abundances in a community on the y-axis against the rank of species abundance, from commonest species on the left, at low rank, to the rarest species on the right, at high rank.

raster The representation of geographic information using a rectangular array.

rate of change The change in a quantity per unit of time; for example, speed is rate of change corresponding to distance traveled per unit time. For a function $x(t)$ of time t, the rate of change over an interval of length h is given by $(x(t + h) - x(t))/t$.

reaction-diffusion model A model that is continuous in time and space and is used to model spatially explicit systems through diffusion in space and localized growth in time.

reaction norm The functional relationship between a genotype and the set of continuous phenotypes that it produces in different environments. Also, NORM OF REACTION.

realization An example of a possible outcome of a stochastic model.

realized niche The range of biotic and abiotic environmental conditions in which an organism actually exists in nature.

recipient compartment In a compartment model, the compartment that is receiving a particular flow.

reciprocity A situation in which an individual uses information about others' past behavior to preferentially cooperate with those who have previously behaved cooperatively.

recombination The presence of contributions of genomes from each of the two parents in the genome of the germ cell of an individual during sexual reproduction.

recovered Describing or referring to a formerly infectious individual who can no longer transmit a pathogen to others and is not capable of becoming reinfected. Also, IMMUNE.

recruits In the context of fisheries, the number of individuals entering a specified part of a population each year. For marine fish, this is usually the number metamorphosing from the larval to the juvenile stage; it can also be the number growing into the fishable size range.

recruitment The process whereby fish transition from one life stage to the next.

redistribution kernel A mathematical description of an animal's movement behavior that specifies the probability that it moves from any given position to any another position, or, if its motion is described in terms of velocities, describes the probability of the animal's changing its speed and direction of movement.

redox Reduction–oxidation, coupled chemical reactions that are associated with the transfer of electrons and release of energy. Reduction is associated with chemical half-reactions that gain electrons, and oxidation is associated with chemical half-reactions that lose electrons.

Red Queen hypothesis A theory asserting that as species coevolve with each other, the relative performance of each species stays the same.

refuge A place or state where victims escape their natural enemies. The place or state can be an actual hiding place in space (a habitat where the enemy cannot go) or a virtual place, as when an animal grows to a life history stage where it is invulnerable to predation.

regime shift A large, abrupt, persistent change in the structure and function of a system that occurs when a critical threshold is crossed and the system shifts from one dynamic regime to another; can be caused by natural ecosystem variability or by anthropogenic impacts. Also, CRITICAL TRANSITION, PHASE SHIFT.

regression A statistical model that quantifies the relationship between a response variable and one or more predictor variables.

Reid's paradox The inability of classical models to account for the rapid migration of plants that occurred after glaciers receded. (Noted by British botanist Clement Reid.)

relative growth rate (RGR) The rate of growth per unit of plant size.

relative species abundance The distribution of commonness and rarity of species within a local community or in the metacommunity.

repeated game A game theoretic model in which individuals interact multiple times in a sequence.

replacement The concept that for a population to persist, individuals must reproduce enough in their lifetimes to replace themselves.

replication principle A property that is implicit in much biological reasoning about diversity: if N equally diverse groups with no species in common are pooled in equal proportions, then the diversity of the pooled groups must be N times the diversity of a single group. Also, DOUBLING PROPERTY.

replicator dynamics A model for the dynamics in evolutionary games: when a strategy fares better than the average, then this strategy is expected to spread in the population.

representation The amount of a biodiversity feature occurring within a specific spatial area.

reproductive decisions Alternative behavioral or physiological options that individuals may take during the process of mating and reproduction.

reproductive number The number of secondary infections produced by an infected individual in a healthy population.

reproductive ratio The expected number of secondary infections arising from a single infected individual.

reproductive success A component of fitness often defined in terms of changes in gene frequencies between generations. Ecologists define reproductive success as having offspring who have offspring (i.e., as having grandchildren).

reproductive value The expected contribution of new individuals (offspring) to a population by an individual of a given age.

rescue effects The forestalling of local extinction of a species through immigration which raises local population sizes; such effects typically occur between high-abundance and low-abundance areas.

reserve selection or **reserve network design** A form of decision analysis used to decide which areas of the landscape should be selected to be included in a conservation area (reserve) network. Also, SITE SELECTION, AREA SELECTION.

residence time A measure of the time an individual atom or molecule spends in a system from entrance to exit. At equilibrium, residence time equals turnover time (the time required for all of an element such as carbon to be lost and replaced).

residents In the context of coexistence theory, populations at whatever equilibrium or stationary distribution they would attain in the absence of the invader. They determine the competitive environment faced by the invader.

residuals Discrepancies between model prediction(s) and the data that have been used for model fitting.

resilience 1. The ability of a system to return to its previous undisturbed state after application of a perturbation. 2. The time taken to return to the undisturbed state after application of a perturbation.

resistance A property of a dynamic system that is measured by the magnitude of change in the focal variable caused by a unit magnitude of external perturbation.

resource A component of a natural system that is used by another component of the system in order to grow and reproduce (e.g., a prey species in a food web).

resource competition Negative interactions between individuals, mediated by depletion of consumable resources that they both require.

resource-ratio theory A theory that predicts the outcome of purely exploitative competition among species based on their relative resource specialization and the ratio of resources supplied to the ecosystem.

resource selection function (RSF) A statistical model defined to be proportional to the probability of selection of a resource unit.

restoration ecology The science of repairing damage caused by humans to the diversity and dynamics of indigenous ecosystems.

retention The expected remainder for a biodiversity feature in a landscape following conservation action and other land use decisions.

return time The time it takes for a deterministic process to return to a stable equilibrium following a disturbance; usually measured as the time it takes for the deviation from the equilibrium to decrease by a factor $e^{-1} = 0.37$.

revealed preference Methods of non-market valuation that use observed choices, such as buying a house or trips to recreational sites, to estimate the value of ecosystem services.

rhizosphere The environment surrounding a plant root.

Ricker model The classic discrete-time model of population dynamics incorporating cannibalism, describing the probability of survival of eggs from cannibalism by adults and their subsequent recruitment into the adult population. (Developed by William Edward Ricker, Canadian entomologist and an important founder of fisheries science.)

risk assessment Quantitative analysis of the probability of an undesirable event.

rule For cellular automata, the function that specifies the new value of a site, given the current values of that site and the other sites in its neighborhood.

saddle-cycle In chaotic dynamics, an unstable cycle that draws population trajectories toward itself in some directions and pushes the trajectory away in other directions, giving rise to episodes of near-cyclic temporal patterns.

saddle-node bifurcation A bifurcation in which two equilibria, one a saddle and the other a node, collide and disappear.

sample A subset of the individuals or entities in a population.

sample path See TRAJECTORY.

sampling distribution A probability distribution that quantifies the possible values of a statistic and the probabilities with which they occur across all possible samples.

scale The spatial or temporal dimension of an object or process, characterized by both grain (resolution) and extent (endpoints/boundaries).

scaling The process of estimating the dynamics of a system at one spatial or temporal scale based on the dynamics at another spatial or temporal scale.

scramble competition Competition mediated by individuals' use of a limiting resource, in which each individual reduces the availability of the resource to others. Also, EXPLOITATIVE COMPETITION.

search image formation The capacity of a predator to recognize prey types through frequent encounters.

secondary dispersal Further movement of a diaspore after it reaches the ground or other substrate.

selection A process whereby some individuals are eliminated from the pool of reproductive individuals. When the traits that are the target of selection are heritable (so that children resemble their parents), selection can result in evolution.

selection gradient The gradient of the fitness landscape at the position of a resident.

self-thinning Death of plants due to competition with their neighbors; usually applies to dense stands of the same species.

semantic information Adaptive, or possibly maladaptive, information carried by organisms typically, but not exclusively, in genomes.

semelparous Describing or referring to a species or a life history in which an organism reproduces only once, as in an annual plant or anadromous Pacific salmon.

sensitivity **1.** The change in model output when a model parameter (e.g., matrix element a_{ij}) is perturbed a little bit: $\partial\lambda/\partial a_{ij}$. **2.** In the context of population viability analysis, the change in the population growth rate with respect to a change in the rate of a demographic process, such as survival or reproduction.

sequence-based marker A genetic marker used in phylogeography that is based on DNA sequences.

sexual selection As defined by Charles Darwin, a subset of natural selection having to do exclusively with reproductive or survival effects arising from interactions between individuals of the same sex and species.

shadow price In bioeconomics, a term that denotes the change in the overall utility from a marginal change in the state variable(s).

shifting balance theory A model for transitions between high-fitness genotypes separated by low-fitness intermediates, in which genetic drift plays a crucial role in moving the population from the neighborhood of one local optimum to another. (Developed by American geneticist Sewall Wright.)

shortwave radiation Electromagnetic radiation emanating from the sun. Its wavelengths are bound between 0.3 and 3.0 microns and contain wavebands associated with ultraviolet, visible, and near-infrared radiation.

significance level The probability that one would observe a value of the test statistic at least as extreme as the value observed given that the hypothesis is true.

simulation experiment A controlled experiment designed to elicit understanding of a system when the system is a simulation model.

single-species management In the context of fisheries science, assessment and management of population abundance for an individual species based on natural mortality, anthropogenic (fishing) mortality, and population reproductive rates of that species alone.

sink population A population in which death rates are greater than birth rates, but which may persist through the immigration of individuals from elsewhere.

sink strength In physiological ecology, the relative capacity of a tissue or organ in a plant to capture solutes (especially sugars) from the flow of phloem; related to total tissue mass or area and the uptake capacity per mass or area of the tissue. Analogous concepts apply in other taxa.

site selection See RESERVE SELECTION.

small-scale fishery See ARTISANAL FISHERY.

social-ecological system Any system, urban or not, that includes humans or the effects of human actions along with organisms and their interactions; can range from sparsely inhabited and unmanaged systems, through managed systems, to urban areas.

soil organic matter (SOM) The carbon-containing component of soil. It is made up of living organisms (mostly microbes), unprocessed plant litter and processed plant litter.

solution concept A method of analysis that aims at generating predictions for the long-run outcome in a game theoretic model; an ESS is an example of a solution concept.

solution space The set of all possible solutions to a problem, usually multi-dimensional with many variables and unknowns.

somatic maintenance Effort (i.e., energy and/or substrate) that needs to be invested continuously to maintain a certain amount of structure in an organism. Examples include (protein) turnover and transport (flow of body fluids and basic behavior), maintenance of concentration gradients of metabolites; failure to provide for somatic maintenance leads to reduction of the amount of structure (shrinking), which is associated with instantaneous death if a threshold is exceeded.

source population A population in which birth rates are greater than death rates, and immigration rates typically lower than emigration rates, so that it acts as a net exporter of individuals.

spatial clumping and **regularity** Two features of spatial community structure. The cumulative distribution of nearest neighbor distances (n.n.d.) for a pattern with clumping has a greater number of small n.n.d. than that for the same number of individuals distributed with complete spatial randomness (CSR). The cumulative distribution of nearest neighbor distances (n.n.d.) for a pattern with spatial regularity has fewer small n.n.d. than that for the same number of individuals distributed with complete spatial randomness (CSR).

spatially implicit model A spatial model that does not describe spatial arrangement; the description of the system is in terms of the proportion or fraction of space that is in each state considered.

spatio-temporal heterogeneity Variation in space and time.

spawning stock biomass The total number or weight of fish that are reproductively viable.

species-area relationship The accumulation of species as the size of a contiguous sample area increases.

species distribution The spatial arrangement of environments suitable for occupation by a species.

species distribution modeling Modeling that relates species distribution data (species observations at known locations) with information on the environmental and/or spatial characteristics of those locations.

species diversity The number and relative abundances of species within a local community or metacommunity.

species equivalent See EFFECTIVE NUMBER OF SPECIES.

species range Limits within which a species can be found or the total areal extent occupied (i.e., abundance above a specified threshold) by a species.

species sensitivity distribution A frequency distribution representing variability in the toxicological sensitivities of species to chemicals.

species sorting A dynamic whereby species coexist through sorting into different habitat types or to different points along a habitat gradient; resembles classic niche differentiation with the addition of dispersal to allow sufficient recolonization to replace populations of species that have gone extinct from preferred habitat areas.

spreading speed or **spread rate** The speed at which an invading population spreads into new territory.

stable Describing a solution of a dynamical system where other solutions with nearby initial conditions remain close for all time. If the solution is asymptotically stable, in addition all other solutions with nearby initial conditions ultimately approach the solution as time increases.

stable coexistence A condition in which competing species maintain positive abundances in the long-term at a steady state that constitutes either a point equilibrium such as carrying capacity, or one involving regular fluctuations in abundance such as a limit cycle, and are able to recover from perturbations that cause them to deviate from their steady-state abundances.

stable equilibrium A solution of a dynamical system that is both an equilibrium and stable.

stable limit cycle A periodic oscillation in population sizes that would be sustained indefinitely and for which nearby trajectories approach.

stable stage distribution In a stage structured model, the asymptotically fixed proportions of the total population size in each of its classes. The proportional distribution of individuals over the stage classes becomes this distribution if the same transition matrix is applied to a population vector for many consecutive time steps and the transition matrix has certain properties.

stable state See DYNAMIC REGIME.

stage structure Characterization of a population by specifying the number or proportion of individuals in each life stage.

stand An area of contiguous forest that is of approximately uniform age, disturbance/management history, vegetation, and soils. Stands are usually the fundamental unit for management and can vary widely in size.

stated preference Methods of non-market valuation that use responses to survey questions about hypothetical choices to estimate the value of ecosystem services.

states The possible values for a state variable in a model; for cellular automata, the possible values that each site can have. For a model with k states, the states are typically numbered $0, 1, 2, \ldots, k - 1$.

state space A set that corresponds to all possible states of the system being modeled. For example, the state space for models of planetary motion is the position and velocity of all the planets.

state variable Any of the variables described by a mathematical model such as the nitrate, phytoplankton biomass or zooplankton biomass in an NPZ model.

stationary distribution A probability distribution for the states of a stochastic process after a long time period.

stationary process A stochastic process with constant statistical properties; for example, the distribution of possible process states is fixed over time. The possible dependence (correlation, etc.) between consecutive process states is also independent of time.

statistic A quantity calculated on the basis of a sample from a population.

statistical model A mathematically precise description of the relationship between a sample and the population from which it was selected.

statistical phylogeography A perspective in phylogeography that emphasizes testing of specific historical and demographic hypotheses while taking into account multiple sources of variance in genetic patterns, such as mutation or coalescent variance, and usually using simulation or statistical models and data from multiple genetic loci.

steady state **1.** A condition that exists when a system has stopped evolving, corresponding to having a zero rate of change (e.g., a derivative that is zero). **2.** In a compartment model, the condition that occurs when flows in and out of each compartment are in balance.

step selection function (SSF) Statistical model of landscape effects on movement probability.

stochastic In some sense probabilistic or random. A stochastic entity cannot be perfectly predicted, but takes values according to defined probabilities, its distribution.

stochasticity Random (unpredictable) variability that is described by a probability distribution which has the mean, variance, and other properties of the random process.

stochastic model A mathematical model that includes randomness.

stochastic population model A mathematical or computer representation of the characteristics of a population (such as the number of individuals in each age class and/or of each sex) having components that vary randomly in time. For example, survival rate (proportion of individuals in an age class that survive until the following year) or fecundity (number of offspring per parent) may change from year to year because of fluctuations in environmental conditions. A stochastic model simulates this by selecting these variables from random (statistical) distributions.

stochastic process Any behavior where the dynamics are governed by chance events; a collection of random variables indexed by elements of a set, typically time.

stochastic realization See TRAJECTORY.

stoichiometric homeostasis The degree to which an organism maintains a constant elemental composition in its biomass despite variation in the elemental composition of its environment or diet.

stomata (*singular,* stoma) Pores in leaves that allow passage of gases into and out of the leaf. They are bordered by a pair of guard cells that regulate the opening by swelling or deflating.

strange attractor The attractor for a chaotic system. Strange attractors usually exhibit complex structure and fractal dimensionality in phase space.

strategy A rule that describes how an individual will behave in a given situation.

stratification A vertical stack of horizontal layers (strata) in a water body, each with a different density; may be composed of either well-defined layers or a continuous vertical density gradient.

streamlines A family of curves that are tangential to the velocity of flow, thus showing the trajectory along which a water parcel will travel. Streamlines do not intersect or cross.

stress Hardship or adversity experienced by an organism that prevents the organism from realizing maximal fitness.

strong Allee effect Single-species population dynamics with a threshold below which the per-capita growth rate is negative and the population shrinks towards extinction.

succession The sequential change in biotic communities either in response to environmental change or induced by the properties of the organisms themselves.

surprise effect Rapid increase of a species following the removal of another species that was keeping its population size low through herbivory, predation, or competition; generally the increasing species had been unnoticed prior to the removal. Mesopredator release and competitor release are examples of surprise effects.

surrogate A biodiversity feature that represents several other features in addition to itself. A degree of co-occurrence between the surrogate and the other features is implied.

susceptible Describing or referring to an individual at risk for infection.

sustainability A societal goal that seeks to ensure social, economic, and ecological health and adaptability.

sustainable development A term generally defined as development that meets the needs of the present without compromising the ability of future generations to meet their own needs.

sustainable yield The yield rate from a resource that can be obtained in perpetuity.

Sverdrup transport Meridional transport that is vertically integrated over the entire water column and proportional to the curl of wind stress. (Named for Norwegian oceanographer Harald Sverdrup.)

switching A predator behavior in which attacks concentrate on whichever prey are more common.

symmetric neutrality A form of neutrality that is less restrictive than per capita demographic equivalence, stating that species experience the same demographic rates when they are at the same abundance.

system state The condition of a system at a particular point in time, described in terms of specific variables such as pH or population size.

systematic reviews Rigorously conducted quantitative reviews. Not all such reviews include a meta-analysis, and some meta-analyses are not based on systematically collected studies.

tangent bifurcation of limit cycles A type of bifurcation in which two cycles collide and disappear.

target In the context of conservation prioritization, a numeric quantity that the representation of a feature must exceed in the reserve network so that the conservation status of that feature can be counted as satisfactory. Targets can be assigned to population sizes, numbers of populations, area of habitat type, estimated persistence of species, and the like.

taxis The response of an organism to a directional stimulus or gradient of stimulus intensity that involves directed movement towards or away from the stimulus source.

taxonomic diversity measures Measures of diversity that incorporate species' taxonomic classifications.

Taylor expansion A series expansion of a function about a point. If the value of a function at a point x^* is known, then the Taylor expansion giving its approximate value at a nearby point x is:

$$f(x) \approx f(x^*) + \left.\frac{df}{dx}\right|_{x=x^*}(x - x^*) + \frac{1}{2}\left.\frac{d^2f}{dx^2}\right|_{x=x^*}(x - x^*)^2 + \sum_{n=3}^{\infty}\frac{1}{n!}\left.\frac{d^nf}{dx^n}\right|_{x=x^*}(x - x^*)^n.$$

The approximation becomes more accurate as x gets closer to x^* and as we keep more terms of the sum. (Introduced by English mathematician Brook Taylor.)

temperature The average kinetic energy of a system.

tens rule In invasion biology, a theory or "rule of thumb" describing the fraction of species that make it through each invasion stage. The prediction is that about 1 in 1,000 introduced plant species will become an invasive.

terrestrial radiation Electromagnetic radiation emitted by objects on Earth in proportion to the absolute surface temperature to the fourth power. Its wavelengths range between 3 and 100 microns. This spectral band overlaps with the absorption bands of many trace gases, e.g., carbon dioxide, water vapor and nitrous oxide, causing the atmosphere's greenhouse effect. Also, LONGWAVE RADIATION.

test statistic In frequentist statistics, a function of the data that a yields a single value. The value has a known probability distribution when a hypothesis is true; extreme values of the test statistic provide evidence that the hypothesis is false.

thermal radiation Electromagnetic radiation that is spontaneously emitted by any body at an absolute temperature T above absolute zero. From a surface, the energy flux density is $Q = \varepsilon\sigma T^4$, where ε is the emissivity of the body (nearly 1 for biological materials) and σ is the Stefan–Boltzmann constant, approximately 5.67×10^{-8} W m^{-2} K^{-4}.

thick-tailed kernel A dispersal kernel that goes to zero more slowly than an exponentially decaying kernel as the dispersal distance goes to infinity.

threat The possibility that some adverse future change in the quality of a site itself or its spatial neighborhood will reduce representation of biodiversity features there.

threshold cooperation Cooperative behavior that is favored by natural selection when it is rare, but is disfavored when it is common.

time-delay model A model with a time delay (or time lag), in which the change in density of a population depends not only on the population density at the present time, but also on the density at previous time(s).

time series correlogram A plot of the pair-wise correlation between all combinations of spatially referenced time series *vs* the distance separating the location of the time series.

tipping point See BIFURCATION.

T_m The intrinsic mean time to extinction; equal to *MTE* when the population dynamics have passed any transient dynamics due to initial conditions and have reached an established phase, the quasi-stationary distribution.

top-down control The concept wherein consumers at upper trophic positions regulate via consumption the population densities and dynamics of organisms at lower trophic positions; e.g, control by a predator of the dynamics in the prey population(s).

total economic value (TEV) A framework for considering various constituents of value, including direct use value, indirect use value, option value, quasi-option value, and existence value.

trade-off **1.** A choice that involves losing one quality or service (of an ecosystem) in return for gaining another quality or service. Many decisions affecting ecosystems involve trade-offs, often trading off the long-term provisioning of one service for the short-term provisioning of another. **2.** See LIFE-HISTORY TRADE-OFF.

trait evolution plot (TEP) In adaptive dynamics, a mutual invasibility plot together with arrows showing the direction of adaptive movement of each of the two types.

trajectory The history through time of a single particular observational sequence of a dynamical system or a stochastic process. Also, SAMPLE PATH, STOCHASTIC REALIZATION.

transcendental functions Functions that do not satisfy polynomial equations whose coefficients are themselves polynomials, in contrast to an algebraic function, which does satisfy such an equation. Examples of transcendental functions include the exponential function, the logarithm, and the trigonometric functions.

transcritical bifurcation A bifurcation in which two equilibria collide and exchange their stability.

transdiscipline **1.** A discipline that generates tools serving many other disciplines (e.g., statistics). **2.** Research with the goal of understanding the world free from the boundaries of academic disciplines.

transfer function The mathematical form of the link between state variables (e.g., nutrient uptake rate, grazing rate, etc.).

transient dynamics In matrix analyses of population dynamics, changes in population size and structure that are determined by the transition matrix and the current population structure, until the stage distribution becomes stable.

transition matrix A matrix giving the probabilities of moving between states for a Markov chain.

traveling salesperson problem An optimization problem in theoretical computation: find the shortest path that visits n discrete points.

traveling wave A density profile that moves in space while maintaining a fixed shape.

Tribolium A genus of flour beetles having many highly cannibalistic species; has been used to investigate cannibalism in laboratory studies that have contributed greatly to an understanding of population dynamics.

triplet Three connected vertices in a network.

trophic cascade Suppression by predators of the abundance of their prey, thereby releasing the next lower trophic level from predation/herbivory.

trophic interaction loop A pathway of interactions from a species through the food web and back to the same species without visiting the species more than once.

trophic level An energy state that defines an organism's position in a food chain. Although the term is often used synonymously with *trophic position*, *trophic level* refers to the state of energy, whereas *trophic position* refers to the trophic level from which an organism obtains its energy. In a primary producer-based food chain, primary producers (autotrophs) occupy the first trophic level, followed by their consumers at trophic level 2 and so on. In a detritus-based food chain, detritus (nonliving organic material) occupies the first trophic level, followed by detritivores (consumers of detritus) at trophic level 2 and so on. Organisms often obtain energy from more than one trophic level, and from both primary producer-based and detritus-based food chains. In these cases, an average trophic level of an organism is estimated based on the fraction of the organisms diet obtained from the different trophic levels.

trophic link A resource–consumer feeding relationship.

trophic link strength A quantitative measure for the strength of direct consumer–resource interactions. Common measures are based on biomass flows between resources and consumers, potentially normalized to resource and/or consumer abundances. Because of their easy empirical accessibility through stomach/gut-content analysis, the proportional contributions of resources to the diets of consumers (diet fractions) are also frequently used. Other measures of link strength are defined in terms of parameters in models for functional or numerical responses to multiple resources.

trophic position The average trophic level from which an organism obtains its energy for maintenance, growth, and reproduction.

trophic species A term emphasizing the fact that in empirical food webs the nodes ("species") can represent groups of similar species, life stages, or dead matter such as detritus. To reduce biases by these simplifications, data and model outputs are sometimes standardized by systematically lumping trophically similar network nodes into trophic species when comparing food-web topologies.

trophic traits Foraging and vulnerability traits.

trophic transfer The movement of materials or energy between trophic levels from food to feeder.

turnover events See EXTINCTION–COLONIZATION DYNAMICS.

turnover time The time required for all of an element such as carbon to be lost and replaced.

uncertainty factors Numbers (often factors of ten) used to provide a margin of safety when there is a lack of understanding in risk assessments.

unit A magnitude of some quantity used as a reference for measurements of that quantity.

unobservables In Bayesian statistics, quantities that cannot be directly observed, but that are of interest nevertheless, including quantities such as variance terms, regression coefficients, or ecologically meaningful parameters contained in a process model.

unstable equilibrium An equilibrium from which a deterministic process will diverge over time, if started in its vicinity.

urban 1. Referring to dense portions of settlements, including business centers, apartment blocks, row homes, and small-lot residential neighborhoods. 2. In the context of urban ecology, referring to any built-up area where buildings, roads, and energy and material delivery and waste-processing infrastructure are present and the livelihoods of the residents depend on economic consumption or production rather than management of natural resources.

urban design The activities of the fields of architecture, landscape architecture, and planning that generate plans for buildings, infrastructure, and managed landscapes.

use value Benefits derived from the services provided by an ecosystem that are used directly by an economic agent. These include consumptive uses (e.g., harvesting goods) and non-consumptive uses (e.g., enjoyment of scenic beauty).

utility 1. A measure of individual well-being; i.e., the satisfaction derived from consumption and preservation. 2. In foraging ecology, the fitness value of an increment of acquired resources.

valuation The process of estimating a value for a particular good or service in a certain context in monetary terms.

variance A measure of variation around an average, or mean, value.

vector 1. A living carrier that transmits an infectious agent from one host to another. 2. The representation of discrete objects as points, polylines, or polygons in geographical information science. 3. An ordered list of objects such as the numbers of individuals of different ages in a discrete age-structured model.

vital rate A distinct demographic process, such as survival, growth, or reproduction, by individuals in a particular life stage that contributes to the rate of population growth.

von Neumann neighborhood For cellular automata, the neighborhood consisting of a site and its four orthogonal neighbors.

vulnerability trait A trait determining a species' role as resource.

water potential The total free energy content (per mass or per mole) of water at a given pressure, location in a gravitational field, and concentration (moles per volume or per mass). It may be expressed in Joules per mole as $\psi = \psi_0 + PV_w + \rho gh + \pi$. Here, ψ_0 is an arbitrary reference state (absolute energies cannot be measured), P is the hydrostatic pressure, V_w is the molar volume of water, ρ is the density of water in mol m^{-3}, g is the acceleration of gravity, h is the height above a reference height, and π is the osmotic potential.

weak Allee effect Single-species population dynamics where the per-capita growth rate is lowered at low densities but never drops below replacement rate so that the population does not actually decline but increases more slowly.

welfare economics The study of the benefits and costs that accrue to society with an economic activity.

willingness-to-pay (WTP) An estimate of the amount consumers are prepared to pay in exchange for a certain state or good for which there is normally no market price; for example, willingness to pay for protection of an endangered species.

wraparound For cellular automata, the convention of connecting each edge of the lattice to the opposite edge so that each site has the same number of neighbors and in order to reduce edge effects of using a finite lattice.

zero-growth isocline A graphical portrayal of the values of variables (population sizes, resource levels, etc.) at which a species has a zero growth rate.

zoonosis A disease that can be transmitted to humans from non-human animals, either wild or domestic.

INDEX

Boldface indicates main articles.

Abiotic forcing, 83
Abundance distribution, 203–204, 207
Acquired immunity, 50
Active motion, 536
Active transport, 380
Adaption–rejection sampling, 73
Adaptive behavior and vigilance, **1–7**
  baseline models of vigilance behavior, 1–2
  no collective detection, 2–3
  partial collective detection, 4–6
  partner loss and mutual dependence, 6–7
  perfect collective detection, 2
  risk allocation, 6
Adaptive control programs, 54–55. *See also* Conservation management
Adaptive dynamics, **7–17**
  adaptive speciation, 14–15
  assumptions, 7, 16–17
  canonical equation, 13
  evolutionarily singular strategies, 13–14
  evolutionarily steady strategies, 10
  fitness, 8–10
  fitness proxies, 10–11
  internal selection, 15
  optimization principles, 11
  traits, 12
Adaptive foraging, 1–7, 300, 592
Adaptive landscapes, **17–26**
  additive landscapes, 19–20
  concept of, 17
  criticisms of, 24–25
  epistasis, 21
  holey landscapes, 23
  multiple loci, 18–19
  neutral landscapes, 20
  n-k model, 21–22
  quantitative traits, 20
  recent and future directions, 25–26
  recombination, 23–24
  rugged, 21–22
  selection equations, 17–18
  speciation, 24–25
Adaptive speciation, 14–15
Additive effects, 596
Adenosine triphosphate (ATP), 338, 379
Admissibility, 519
Advection-diffusion model, 359–360, 361–362, 458, 535–536, 604–605
Aerobic respiration, 379
Age structure, **26–32**
  cannibalism, 121
  continuous-time models, 27
  defined, 26
  discrete-time models, 27
  disease dynamics, 186
  extensions of Leslie matrix, 29–30
  harvesting theory, 351–353
  integral projection model, 30–31
  integrodifference equations, 383
  Leslie matrix, 27–29
  literature on, 31–32
  partial differential equations, 536–537
  population ecology, 574
  significance of, 26–27
Agent-based models (ABMs), 365
Aggregation economies, 306
Aggregative response, 36
Agnew, D. J., 285
Agulhas Current, 517
Akaike, H., 371, 373
Akaike's Information Criterion (AIC), 190, 299, 371, 372–373, 453, 455, 556, 575. *See also* Information criteria in ecology
Alcelaphus buselaphus (hartebeest), 45
Alignment models, 555
Allee, W. C., 32, 653
Allee effects, **32–38**
  concept and definitions, 32
  conservation implications, 37
  dynamical implications, 37
  ecosystem engineers, 233
  evolutionary implications, 37
  invasion biology, 389, 391
  mechanisms, 32–34
  population models and dynamics, 34–37
  SI models, 653–654
  spatial spread, 672–673
  strong vs. weak, 672–673
  theoretical and applied ecology, 56–57
Allocation, 252
Allometric power law, 38–39
Allometry and growth, **38–44**
  allometry defined, 38
  animals, 41–44
  metabolism, 40–41
  overview and history, 38–40
Alpha diversity, 145, 205
Alternation of generations, 638–639
Alternative stable states, 611
Altruism, 157–158
  culture and social learning, 162
  limited dispersal, 160–161
  reciprocity, 161–162
  signaling, 161
American mink (*Neovison vison*), 47
Amino acid transition matrix, 555
Amplifying feedback, 613
Analysis of variance (ANOVA) models, 695
Andersen, T., 719
Anderson, D. R., 371
Anderson, H. C., 185
Anderson, R. M., 184
Andronov, A. A., 91
Anelosimus eximius (colonial spider), 57
Animal dispersal. *See* Dispersal, animal
Animal function, 324–329
Anisogamy, 639–640
Anisogamy theory, 409
Anolis lizards, 326–327
Anoplolepis gracilipes (yellow crazy ant), 59
Antarctic Circumpolar Current, 517
Ants
  Argentine (*Linepithema humile*), 57
  yellow crazy (*Anoplolepis gracilipes*), 59
Apostatic selection, 743–744
Apparent competition, **45–51**, 755
  asymmetric, 48
  dominance, 46–47
  exploitative vs., 46

Apparent competition (*continued*)
graphical model, 46, 47
human history, 51
predator behavior, 49–50
refuges, 49
shared predation, 50–51
simple models, 45–47
spatial processes, 47–48
Applied ecology, **52–60**
Allee effects, 56–57
climate change, 57
defined, 52
invasion biology, 54–55
scope of, 52–53
theoretical ecology in relation to, 53–60
Approximate Bayesian computation (ABC), 558
Aquila chrysaetos (golden eagle), 49
ARAM (2, 1) model, 647
Arbuscular mycorrhizal fungi, 84–85
Archimedes, 171
ARC/INFO. *See* Software, 344
Area-based conservation, 149
Area-based models, 313
Area-restricted search behavior, 459
Argentine ants (*Linepithema humile*), 57
Aristotle, 371
Arms race analogy, 134
Arrhenius, S. A., 427
Arrhenius temperature, 254
Artificial Intelligence for Ecosystem Services (ARIES), 240, 401–402
Artificial life models, 274
Artisanal fisheries, 285
Arvicola amphibius (water vole), 45, 47
Asexual reproduction, 378, 640
Aspen, quaking (*Populus tremuloides*), 377
Assembly processes, **60–63**
complex dynamical spaces, 62–63
defined, 61
experimentation and theory, 61
historical contingencies, 60
pattern and cause, 61–62
Assortative mixing, 269
Atlantis. *See* Software, 401
ATP. *See* Adenosine triphosphate (ATP)
Attractors, 172. *See also* Strange attractors
Australia, phylogeography of, 563–564
Australian Weed Risk Assessment (WRA), 389
AUTO. *See* Software, 95
Autonomous systems, 538
Averting behavior, 244. *See also* Avoided cost/damages
Avise, J., 557, 562
Avoided cost/damages, 239, 245. *See also* Averting behavior
Axelrod, R., 161
Ayres, R., 214, 216
Azaele, W., 483

Bacteria, 447–448
Bacteriocins, 448
Bacteriophages, 447–448
Bailey, V. A., 498, 660, 662–663
Balanced polymorphism, 466
Ball and cup heuristic, 625–626
Banzhaf, H. S., 238
Baoxing County, China, 240
Barbier, E. B., 237
Bartlett, M. S., 712
Basic reproductive number (R_0), 267–268, 577
Bateman, A., 409–410
Bateson, W., 24–25
Bayes, Rev. T., 694
Bayes risk, 69
Bayes rule/estimator, 69, 694
Bayes theorem, 65–67
Bayesian information criterion (BIC), 373
Bayesian statistics, **64–74**
approximate Bayesian computation, 558
basis of, 65–67
characteristics of, 64–65
estimation and testing, 68–69
frequentist statistics compared to, 323–324, 693–694
graphical models, 67–69
hierarchical, 72
Markov chain Monte Carlo (MCMC) methods, 72–74
model fitting, 451, 454
phylogenies, 552–554, 556–558
priors, 69–70
uncertainty, 452
value and potential of, 74
Bay laurel, California (*Umbellularia californica*), 48
Beckage, B., 234
Beer-Lambert law of optics, 570
BEHAVE (computer model), 310
Behavioral ecology, **74–79**
biological control, 79
categories, 74–75
concept of, 75
conservation biology, 78
Darwinian medicine, 78–79
dynamic programming, 209–210
fisheries ecology, 281–282
fitness measurements, 75–76
game theory, 76–77
optimal control theory, 520
optimization, 76
phylogenetics, 77
rules, 77
theoretical aspects, 76–77
Belief, degrees of, 64–65
Belize, 49
Bell, G., 480
Bellman, R., 119, 208
Bellman-Harris branching process, 119
Bell-shaped probability distribution, 695
Belowground processes, **80–86**
decomposition, 81–83
nutrient dynamics and plant competition, 80–81
soil food webs, 85
species interactions with soils, 83–85
Benefit-cost analysis, 215, 237
Benefits transfer methods, 238–239
Benthic habitats, 364
Benton, T. G., 31
Berec, L., 37
Bergmann rule, 255
Bernoulli, J., 519
Bertalanffy, L. von, 42, 139
Best replies, 330–331
Beta diversity, 205
Beverton, R. J. H., 86, 346, 349–350
Beverton-Holt model, **86–88**, 174, 374–375, 633, 644, 645
Bias, of estimation methods, 454–455
Bienaymé, I.-J., 112
Bifurcation parameter, 172
Bifurcation points, 90
Bifurcation value, 172
Bifurcations, **88–95**, 543
as collisions, 90
catastrophes and hysteresis, 93–94
chaos, 126–127
common, 90–93
continuous-time systems, 89
difference equations, 172–173
state portraits, 89
structural stability, 89–90
Binary food webs, 295
BIOCLIM. *See* Software, 336
Biodiversity. *See also* Species diversity
coevolution and, 131–136
components of, 145
conservation biology, 145–146
indicator species, 146
measures and patterns, 145–146, 203–207
monitoring the status of, 148–149
reserve selection, 617
top-down control, 743–744
value of, 146
Biodiversity features, 617, 620
Biodiversity hot spots, 146, 149
Bioeconomics
biological invasions, 390–391
discounting, 176–179
ecological economics and, 215, 217
Biofilms, 448
Biogeochemistry and nutrient cycles, **95–100**
characteristics of, 95–96
feedback, 98–99
stoichiometry, 99
thermodynamics, 96–98
Biogeography, 557
Biological control, 79, 391
Biological Diversity, Convention on, 145
Biological invasions. *See* Invasion biology
Biological scale dependence, 627
Biome-BGC (computer model), 311
BIOMOD. *See* Software, 336
Biophysical ecology, 255
Biotic resistance hypothesis, 385
Birth-death models, **101–106**
differential equations for probability distribution, 103
long-term behavior, 104
mathematical model, 101–102
mean and variance of population, 103–104
mean time to extinction, 104
Monte Carlo simulations, 102

practical uses, 105–106
stationary and quasi-stationary distributions, 105
Births, stochasticity and, 706–707
Blowflies, 689
Blow-up, 525
Blue noise, 716–717
Boltzmann, L., 427
Bootstrapping, 557–558, 696, 697
Border cells, 376
Bottom-up control, **106–111**
concept of, 106–107
mathematical models, 108–111
practical uses of concept, 111
top-down vs., 739
types, 107–108
Boulding, K., 216
Boundary conditions, 536, 604
Boyd, J. W., 238
Brachistochrone problem, 519
Brachylagus idahoensis (pygmy rabbit), 335
Bradshaw, A. D., 547
Branching processes, **112–119**
adaptive dynamics, 14–15
background, 112–113
Galton-Watson branching process, 113–116
multitype Galton-Watson branching process, 117–119
random environment, 116–117
Braun-Blanquet, J., 729
Breeder's equation, 598–600
Breeding Bird Survey, 148
Breeding dispersal, 193
Breeding value, 596
Briand, F., 297
Briggs, C. J., 691
Broad-sense heritability, 596
Brown, J. H., 42, 44, 427
Buffered population growth, 757–758
Buffon, G.-L. L., Comte de, 728
Buoyancy forcing, 361
Burgess, E., 765
Burgess model, 765
Burnham, K. P., 371
Bycatch, 284

Caddy, J. F., 285
California bay laurel (*Umbellularia californica*), 48
California red scale, 689–690
Camellia, 134
Canada, 341
Canalization, 547
Candolle, A. de, 712
Candolle, A.-P. de, 728
Cannibalism, **120–123**
chaos, 121–122
continuous-time age-structured models, 121
defined, 120
evolution of, 123
food webs, 122
intraguild predation, 122
population ecology, 578
Ricker model, 120–121
Tribolium (flour beetle), 120–122
Canopies
energy extraction, 337
environmental heterogeneity, 259–260
microclimate of, 570–571
photosynthesis, 221
plant competition, 565–571
self-thinning, 566–567
Cap-and-trade schemes, 246
Captive breeding, 150
Capture-mark-recapture (CMR) studies, 452
Carbohydrates, 338–339
Carbon, 140–141
Carbon cycle, 219–229
canopy, regional, and global scales, 221–222
disturbance effects, 226–227
equilibrium and disequilibrium, 229
future dynamics, 229
global change effects, 227–229
leaf photosynthesis, 219–221
properties of, 224–226
transfer, storage, and release, 222–224
Carbon dioxide (CO_2), 81–83, 99–100, 151, 219–220, 259, 338–340, 748
CarbonTracker (computer model), 313
Carriers, 746
Carroll, L., *Through the Looking Glass*, 133
Carrying capacity (K), 431, 643, 702
Carson, H., 24
CASA (computer model), 313
Cascade food web model, 297, 475
Caswell, H., 28, 30, 31, 479, 690
Catastrophes
bifurcations, 93–94
birth-death models, 105
regime shifts, 611
Catch shares, 286
Categorical and Regression Tree (CART) analysis, 389–390
Cellular automata, **123–125**, 666–667
Census, 692
Center for Computational Ecology, Yale University, 142
CENTURY model, 82, 240
C4MIP (Coupled Carbon Cycle Climate Model Intercomparison Project), 311
Channel Islands, California, 45, 49
Channels, 745–746
Chao, A., 207
Chao1 estimator, 207
Chaos, **126–131**
background, 126
bifurcations, 126–127
cannibalism, 121–122
characteristics of, 126
detection methods, 128
difference equations, 171
ecological evidence, 128–131
Feigenbaum cascade, 93
order in, 128
ordinary differential equations, 528–529
phase space, 127–128
population ecology, 573
population extinction, 128
sensitivity to initial conditions, 127, 573
SIR models, 652–653
strange attractors, 127–128
Chaotic attractors, 172
Character displacement, 494, 495
Charismatic species, 146
Charnov, E., 43
Chase, J. M., 492
Chemotaxis, 536
Chesson, P. L., 262, 493, 496
Chisholm, R. A., 483
Choice experiments. *See* Stated preference methods
Christensen, P., 217
Christmas Bird Count in North America, 148
CIPRES (web portal), 557
Cities. *See* Urban ecology
Cladistic diversity, 205
Clark, C. W., 346, 347
Dynamic Modeling in Behavioral Ecology, 210
Clark, J. S., 262, 497
Clausen, J., 727
Clayton, D., 414
Clements, F. E., 729, 730
Climate change
carbon cycle, 227–229
conservation biology, 151
models, 58
plant dispersal, 202–203
theoretical and applied ecology, 57–60
Climax concept, 729, 730
Clocking, 556
Closed-loop controller, 521
Clustering coefficient, 473
Clutton-Brook, T. H., 411
Coase, R., 214
Coastal upwelling, 360, 518
Codon models, 555
Coevolution, **131–136**
adaptive radiations, 134–135
application of theory, 135
diffuse, 132
diseases and parasites, 133
future directions, 135–136
gene-for-gene model, 133
geographic mosaic theory, 132, 134–135
major themes and modeling approaches, 132–134
matching-alleles model, 133
non-ecological applications of, 135
overview and history, 131–132
pairwise, 132
predator-prey interactions, 133–134
Coevolutionary cold spots, 135
Coevolutionary hot spots, 135
Coexistence
competition and, 753
local niche partitioning, 755–757
mechanisms for, 722–726
spatial niche partitioning, 758–763
temporal niche partitioning, 757–758
Cohen, J. E., 297, 299
Coherency, Bayesian statistics and, 64–65
Cohort analysis, 346, 349–351
Collective detection, 1–6
Colonial spider (*Anelosimus eximius*), 57

Colonization
competition tradeoff with, 664
metapopulations, 440, 581
stochasticity, 703
Commensalism, 277, 565
Communist regime, 545
Communities
animal roles in, 325–327
Clementsian succession, 729–730
invasive species, 391
metacommunities, 434–438
microbial, 445–450
urban ecology, 769
Community ecology
fisheries ecology, 282
forest simulators, 308
metabolic theory of ecology, 431–432
neutral community ecology, 478–484
Community facilitation, 278–279
Community feasibility, 300
Community food webs, 295
Community Land Model (computer model), 311
Community permanence, 300
Community-wide character displacement, 495
Compartment models, **136–141**
amenability of systems to, 137
applications, 139–141
concept of, 136–137
epidemiology, 141, 266
food webs, 138, 140
global cycles of matter, 140–141
mathematical formulation and solution, 137–139
radioecology, 139–140
Compartmentalization
food webs, 475
mutualistic networks, 476–477
organisms, 378–379
Competition
apparent, 45–51, 755
colonization tradeoff with, 664
dominance, 759, 761–763
exploitative, 46, 752, 755–757, 759
generalized competition theory, 493–494
interference, 752, 755–757
interspecific, 753
intraspecific, 753
networks, 476
niche overlap, 490–494
ordinary differential equations, 529
plants, 80–83, 565–571
population ecology, 575–577
preemptive, 759
scramble competition, 633, 634
spatial ecology, 660–661
two-species, 752–763
types, 752
urban ecology, 765
Competitive dominance, 753. *See also* Dominance competition
Competitive exclusion principle, 720–721
Complementarity, 149
Complexity
diversity measures, 204
ecological networks, 470–478
Computation, and contemporary statistics, 697
Computational ecology, **141–144**
computing capacity, 142
examples of, 143–144
future directions, 144
hardware and software integration, 143
overview, 141–142
software resources, 142–143
state of, 142
Computer science, 135
Concrete populations, 692
Confidence, 556–557
Confidence intervals, 321, 452, 455, 693–694
Confidence set, 375
Conformist bias, 162
Conjoint analysis, 245
Conjugate priors, 70
Conjugation, 637
Connectance, network, 470
Connectivity, 439–440
Connell, J., 727, 731
Connochaetes taurinus (wildebeest), 647, 648
Conservation. *See* Reserves (conservation)
Allee effects, 37–38
animal dispersal, 192
area-based, 149
ex situ, 150
metapopulations, 444–445
plant dispersal, 202–203
population viability analysis, 582–583
population-based, 149
systematic planning, 149
Conservation biology, **145–151**
behavioral ecology, 78
biodiversity, 145–146
disciplines linked to, 53, 145
facilitation, 279–280
gap analysis and presence/absence \ models, 336
methods, 78, 146–150
overview and history, 145
reserve selection, 617–624
spatial conservation prioritization, 619
threats, 150–151
Conservation management. *See also* Adaptive control programs
dynamic programming, 210–212
plant dispersal, 203
population viability analysis, 585–587
spatial conservation prioritization, 619–623
Conservation of energy, 249
Conservation planning, systematic, 149, 618
ConsNet. *See* Software, 622
Conspecifics, movement in response to, 459
Constant discounting, 179
Consumer surplus, 214
Consumer-resource (*C-R*) theory, 289–294. *See also* Resource-consumer models
Contact process, 667–668
Continental scale patterns, **152–155**
environment capacity, 154
geographic population dynamics, 154
metacommunity dynamics, 154–155
species diversity, 152
Contingent valuation, 238, 245
Continuous data, 319
Continuous fields, 342–343
Continuous-time models
age structure, 27
density dependence, 643–644
single-species population structures, 642–643
stability analysis, 681–682
stage structure, 687–689
stochasticity, 699
Continuous-time systems, 89, 121, 139
Control of invasive species, 54–55, 674
Control theory. *See* Optimal control theory
Conway, J., 124
Cooperation, evolution of, **155–162**
Allee effects, 32–38
altruism, 157–158
coordination, 156–157
game theory, 332
mechanisms of positive assortment, 160–162
microbial communities, 447
modeling of, 159–160
mutualism, 156
ordinary differential equations, 529
threshold cooperation, 158–159
types, 156–162
Coordination, 156–157
Corals, 376, 609
Corbet, A. S., 479
Cordgrass, smooth (*Spartina alterniflora*), 56
Coriolis force, 511
Corridors, 192, 203
Costantino, R. F., 121
Costanza, R., 218, 235, 242
Cost-based methods, 238, 245–246
Cost-benefit analysis, 215, 237
Cottenie, K., 437
Coulson, T., 30–32
Coupled-map lattices, 581
Coupling, 62–63
Courchamp, F., 37, 57
Cowles, H., 61, 729, 730
C-Plan. *See* Software, 622
Credible intervals, 452, 455
Critical patch size, 383, 606–607
Crutchfield, J., 22, 215
Cuddington, K., 232
Culture, and altruism, 162
Culver, D., 435
Curl, 512–513
Cushing, J., 121
Cycle chains, 543
Cycles (periodic solutions), 172
Cyclomorphosis, 546

Da Vinci, L., 38
Daily, G., 218
Daisyworld, 98, 231–232
Daly, H., 218

Damping feedback, 613
Damuth, J., 431
Daphnia lumholtzi (water flea), 546
Dark Reaction, 259
Darwin, C., 9, 131, 230, 276, 327, 409, 464, 471, 494, 550, 579
Darwin model, 509
Darwinian medicine, 78–79
Data
  continuous, 319
  discrete, 318–319
Data dredging, 455
De Moivre, A., 171
Deaths, stochasticity and, 706–707
DEB. *See* Dynamic energy budget (DEB)
Decomposition, 81–83, 223
Decoupling, 62–63
Deductive modeling, 335–336
Degrees of belief, 64–65
Delay differential equations, **163–166**, 689
Demographic heterogeneity, 708
Demographic models, 248
Demographic stochasticity. *See* Stochasticity, demographic
Demographic variance, 707–709
Demography, **166–170**
  environmental variability, 169–170, 261
  heterogeneity, 261–262
  overview, 166–167
  sensitivity and elasticity, 168–169
  spatial variability, 169
Dennis, B., 121
Density dependence, 643–645
  Allee effects, 32
  competition, 753
  continuous-time models, 643–644
  demographic variance vs., 709
  discrete-time models, 644–645
  environmental stochasticity, 713–714
  logistic model in continuous time, 643–644
  Nicholson-Bailey host parasitoid model, 500
  population ecology, 573–575
  positive, 32
  Ricker model, 632–635
  stochasticity, 700–702
  structured models, 30
Density functions, 319
Density-dependent transmission of disease, 183
Depensation, 32
Depensatory dynamics, 32
DeRiso-Schnute stock recruitment function, 352
Descriptive statistics, 692–693
Desharnais, R. A., 121
Determinant, of matrix, 684
Detritus
  belowground processes, 83, 85
  bottom-up control, 106–111
Development of organisms, 252
Diamond, J., 62
Diamond module, 292–293
Diaspore shadow, 200
Diekmann, O., 121
Diet breadth, 76, 303
Diffeomorphisms, 171
Difference equations, **170–176**. *See* Ordinary differential equations
  analysis of, 172–173
  history of, 171–172
  modeling with, 176
  overview, 170–171
  population dynamics, 173–175
Diffuse coevolution, 132
Diffusion
  animal dispersal, 189
  reaction-diffusion models, 603–606
  spatial spread, 670–671
Diffusion approximation, 709
Diffusive logistic equations, 605
Dimensions, in ecological models, 501–502
Diploid generations, 638
Diploid populations, natural selection in, 466
Direct inhibition, 633, 634
Direct transmission of disease, 182–183
Directed acyclic graphs (DAGs), 67–68
Directed deterrence, 198
Directed dispersal, 200
Dirichlet boundary condition, 536
Disassortative mixing, 269
Discounting in bioeconomics, **176–179**
  behavioral complexities, 178–179
  concept of, 177
  consumption and utility, 177–178
  hysteresis and nonlinear feedbacks, 178
  natural resources, 215
Discrete data, 318–319
Discrete diffusion models, 604
Discrete-gamma model, 555
Discrete-time models, 642–643
  density dependence, 644–645
  stability analysis, 682
  stage structure, 687
  stochasticity, 699
Disease dynamics, **179–187**
  age-structured models, 186
  chronic vs. acute infections, 181–182
  complex models, 185–187
  demographic stochasticity, 185
  endemic vs. epidemic dynamics, 182
  epidemiological parameters, 183–184
  infective agents, 181–182
  multihost/multistrain dynamics, 186–187
  overview, 179–180
  practical uses of theory, 187
  seasonality, 185
  simple models, 184–185
  social structure and patterns of contact, 186
  spatial dynamics, 186, 187
  transmission, 182–183
  within-host, 180–181
Disease modeling. *See* Epidemiology and epidemic modeling
Diseases
  biodiversity threatened by, 151
  coevolution, 133
Dispersal
  limited, 161
  metapopulations, 439–440
  neutral community ecology, 483
Dispersal, animal, **188–192**
  active vs. passive, 188
  conservation policy applications, 192
  decision to disperse, 188
  diffusion, 189
  genetics, 190–191
  methods of characterizing, 189–190
  population biology, 188
  retrospective models, 189
  scale, 191–192
  search and settlement phases, 188–189
  statistical models, 189–190
Dispersal, evolution of, **192–197**
  conditional dispersal, 196–197
  constraints and tradeoffs, 195–196
  data and applications, 197
  kin interactions, 194–195
  migration vs. dispersal, 193
  spatiotemporal heterogeneity, 195
  species interactions, 195
  theory and modeling, 193–195
  types of dispersal, 193
Dispersal, plant, **198–203**
  applications, 202
  consequences, 201–202
  measurement of, 199–200
  mechanistic models, 201
  modeling, 200–201
  modes of, 198–199
  patterns, 201–202
  phenomenological models, 200–201
  population and community structures, 201–202
  spread rates, 202
  unit of, 198
Dispersal kernels, 193–194, 200–201, 381–382
Dispersal matrix, 362–363
Distance-decay relationships, 152, 154
Distribution-abundance relationships, 152, 155
Distributions, probability, 318–319, 451
Disturbances
  ecosystem ecology, 226–227
  forest simulators, 314
  metacommunities, 437–438
  resilience and stability, 627–628
Divergence information criterion (DIC), 375
Divergence measures, 372
Diversity. *See also* Biodiversity; Species diversity
  microbial, 448
  phenotypic plasticity, 549–550
  response diversity, 615
Diversity measures, **203–207**
  conservation biology, 145–146
  estimation from small samples, 207
  replication principle, 204–206
  species' differences, 205–206
  traditional, 203–204

DNA
functional significance of, 464
mitochondrial, 557–558, 560, 564
quantification of, 559
repair of, 637
Dobzhansky, T., 24–25, 547, 601, 664
Dodo, 51
Doebeli, M., 275
DOMAIN. *See* Software, 336
Domains of attraction, 610
Dominance competition, 759, 761–763. *See also* Competitive dominance
Dominance effects, 596
Dominant species, 146
Double-strand breaks, 637, 638
Doubling property, 204–205
Drake, J. A., 60
Drude, O., 728
Dupuit, J., 214
Dureau de la Malle, A., 61
Dynamic energy budget (DEB)
assumptions, 252
development and allocation, 252–253
ecosystem structure and function, 256–257
homeostasis, 251
individual evolution, 257–258
parameter values, 255–256
principles, 250–254
reserve mobilization, 251–252
standard model, 254–256
surface-area volume relationships, 253
synthesizing units, 253–254
Dynamic programming, **207–212**
applications, 209–212
defined, 207
future directions, 212
mathematical formulation, 209
optimal control theory, 519
optimization, 207–208
overview and history, 208–209
Dynamic state variable models (DSVs), 76
Dynamical networks, 472, 474

Eagle, golden (*Aquila chrysaetos*), 49
Ecesis, 729, 730
Ecoepidemiology, 653–656
Ecoinformatics, 144
Ecological drift, 154, 435
Ecological economics, **213–218**
as transdiscipline, 217
bioeconomics and, 217
current state, 216–218
economics and, 216–217
emergence of, 213, 216
environmental economics vs., 213–214, 218
resource economics vs., 213–215, 218
Ecological Economics (journal), 216
Ecological fitting, 494
Ecological footprint, 242
Ecological genetics, 601
Ecological network analysis (ENA), 474
Ecological Society of American, 730
Ecological succession. *See* Succession
Ecological traps, 78
Ecology (journal), 730
Ecomorphology, 324–325
Economic valuation, 243–245. *See also* Monetary approach to ecological valuation
Economics. *See also* Bioeconomics; Discounting in bioeconomics; Ecological economics; Environmental economics; Resource economics
coevolution, 135
ecological economics and, 216–217
valuation based on, 239
Ecopath. *See* Software, 140
Ecopath with Ecosim (EwE). *See* Software, 401
Ecosystem, Tansley's concept of, 730
Ecosystem Demography (ED) [computer model], 311, 313, 315
Ecosystem ecology, **219–229**
carbon transfer, storage, and release, 222–224
defined, 219
disturbance effects, 226–227
energy exchange, 219–220
fisheries ecology, 282
forest simulators, 308
future dynamics, 229
global change effects, 227–229
market valuation, 243–244
material cycling, 219–220
metabolic theory of ecology, 431–432
networks, 474
nonmarket valuation, 244–245
photosynthesis, 219–222
properties of ecosystem carbon cycling, 224–226
Ecosystem engineers, **230–234**
climate literature, 231–232
controversy over concept, 230
defined, 230
ecology literature, 232
empirical studies, 230–231
engineer-environment models, 232–233
engineer-environment-community models, 234
engineer-environment-evolution models, 233–234
models, 231–234
niche construction, 487
obligate vs. nonobligate, 233
theory, 231
Ecosystem facilitation, 279
Ecosystem management, resilience and stability concerns, 628–629
Ecosystem services, **235–241**
categories, 235
concept of, 235–236
conservation biology, 146
decision-making framework, 237–238
modeling, 239–240
policy-making significance of, 240–241
provision of, by ecosystem type, 237
urban ecology, 770
valuation of, 238–239
Ecosystem valuation, **241–246**
alternative approaches, 242
challenges facing, 246
cost-based methods, 245–246
economic methods, 243–245
ecosystem services, 238–239
instrumental vs. intrinsic value, 242
monetary approach, 215, 216, 238–239, 241–246
resource and environmental economics, 215–216
total vs. marginal value, 242–243
value defined, 241
Ecosystem-based management. *See also* Conservation management; Marine reserves and ecosystem-based management
fisheries ecology, 286–287
principles, 399
Ecosystems
structure and function, 256–257
urban, 768–769
Ecotones, 260
Ecotoxicology, **247–248**
Ecotypes, 727–728
Ectomycorrhizal fungi, 83–85
Eddies, 358
Effective number of species, 204–205
Efficiency, 149
Effort rates, 356
Egg-recruit relationship, 86–88
Ehrlich, P., 132, 134
Eigen, M., 19
Eigenvalues, 417–418, 683–685
Eigenvectors, 417–418, 683
Einstein, A., 535
Ekman, V. W., 511
Ekman dynamics, 511–514
Ekman layer, 360
Ekman pumping, 513–514, 517
Ekman suction, 514
Ekman transport, 512, 513, 517, 518
Elasticity of population growth, 168–169
Electromagnetic radiation, 747
Electron acceptor sequence, 97
Ellner, S. P., 31
Elsasser, W., 381
Elser, J. J., 99, 719–720
Elton, C. S., 27, 61
Embodied energy, 242
Endangered species
Allee effects, 37
identification of, 148
Endangered Species Act (United States), 148
Endemic equilibrium, 650
Endemicity, 182, 264
Endler, J., 486
Endocytosis, 747
Enemy release hypothesis, 384
Energy
conservation of, 96, 97, 249, 357
organisms' use of, 377

Energy budgets, **249–258**
ecosystem structure and function, 256–257
importance of, 249–250
individual evolution, 257–258
models, 248
principles, 250–254
standard DEB model, 254–256
static vs. dynamic, 249
uses of energy, 249
Energy enrichment, paradox of, 720
Energy exchange, 219–220
Energy fluxes. *See* Gas and energy fluxes across landscapes
Energy mechanisms, 379
Engineering webs, 487, 488
Engineers. *See* Ecosystem engineers
Enhanced Thematic Mapper Plus, 345
Enquist, B. J., 42, 44, 427
Entangled bank, 471, 579
Entropy, 96, 216
Environment capacity, 154
Environmental economics, 213–214, 218
Environmental heterogeneity
coexistence mechanisms, 722–726
niche overlap, 496–497
Environmental heterogeneity and plants, **258–263**
demographic heterogeneity, 261–262
responses to environmental variation, 258–259
scale, 259–261
Environmental justice, 770
Environmental niche modeling, 390, 558–559
Environmental response, 757
Environmental stochasticity. *See* Stochasticity, environmental
Environmental Systems Research Institute (ESRI), 341
Environmentally mediated genotypic associations (EMGAs), 487–488
Epanchin-Niell, R. S., 674
Ephemeral habitat patches, 443
Epidemicity, 182
Epidemics
defined, 264
prediction and control, 267–268
size of, 650
Epidemiology and epidemic modeling, **263–270**. *See also* SIR models
basic principles, 263–264
contacts, 268–269
defined, 263
demographic stochasticity, 703
disease dynamics, 179–187
modeling, uses and contributions of, 269–270
modeling methods, 265–267
prediction and control of epidemics, 267–268
Epigenetic changes, 463–464
Epistasis, 596, 638
Equatorial upwelling, 518
Equilibria
difference equations, 172, 524
evolutionary stable strategies, 271
phase plane analysis, 538–543
population size, 714–715
resilience and stability, 624–629
stability analysis, 680–691
types, 681
Equilibrium yields, 348
Equity, 246
Equus quagga burchellii (plains zebra), 45, 48–49
Eradication strategies, 54–56. *See also* Adaptive control programs
Ergodic environment, 9
Error, statistical, 694–697
Errors-in-variables, 454
Estimation methods, 454–456
Estuarine circulation, 361
Ethics. *See also* Ecosystem valuation
basis of value, 242
ecological economics, 216
Euclid, 171
Euler, L., 171, 650
Eulerian approach to plant dispersal measurement, 199
Euler's method, 530
European rabbit (*Oryctolagus cuniculus*), 45, 47
Evans, J., 328
Evidence
relativity of, 322
strength of, 323
Evidence-based medicine, 424
Evolution. *See also* Natural selection
Allee effects, 37
apparent competition, 49
cannibalism, 123
dispersal, 192–197
dynamic programming, 209–210
game theory, 331–334
human, 51, 564
metapopulations, 444
niche construction and, 485–486
of sex, 637–640
phenotypic plasticity, 549–550
spatial spread, 673–674
Evolutionarily stable strategies (ESSs), **270–272**
adaptive dynamics, 10, 16–17
behavioral ecology, 76–77
calculation of, 10
cannibalism, 123
conditions constituting, 270–271
game theory, 271–272
optimization principles, 11
problems with, 272
vigilance behavior, 2–6
Evolutionary biology, 599–600
Evolutionary computation, **272–275**
applications, 274–275
basic algorithm and modifications, 272–274
defined, 272
genetic programming, 274
Evolutionary ecology, 599–600
Evolutionary fitness. *See* Fitness
Evolutionary food web models, 296
Evolutionary singular strategy (ess), 7, 12–14
Evolutionary steady coalitions (ESCs), 10
Evolutionary-developmental biology (Evo-Devo), 13
EwE. *See* Ecopath with Ecosim (EwE) [software]
Ex situ conservation, 150
Exploitation, biodiversity threatened by, 151
Exploitative competition, 46, 752, 755–757, 759
Extended information criterion, 373
Externalities, 214
Extinction
Allee effects, 37
birth-death models, 104–105
branching processes, 112–119
causes, 583
chaos, 128
demographic stochasticity, 709–712
environmental stochasticity, 717
immigration-extinction equilibrium, 478–479
local, and dispersal behavior, 195
measurement of, 146–147
metapopulations, 440, 581
quasi-extinction threshold, 584
risk of, 147–148. *See also* Population viability analysis (PVA)
species ranges, 679–680
stochasticity, 698, 703
Extinction debt, 443–444
Extinction threshold, 441–442

Facilitation, **276–280**
community facilitation, 278–279
conservation applications, 279–280
ecological succession, 729, 730
ecosystem facilitation, 279
empirical perspectives, 278–279
historical context, 276–277
intraspecific, 277
theoretical perspectives, 277–278
Facilitation cascade, 278
Family name survival, 112, 712
FAREAST (gap model), 734
Faustmann, M., 215, 346
Feedback
amplifying, 613
biogeochemistry and nutrient cycling, 98–99
closed-loop controller, 521
damping, 613
soils, 84–85
Feigenbaum cascade, 93
Feller, W., 712
Feral pigs (*Sus scrofa*), 45, 49
Fibonacci (Leonardo of Pisa), 171
Fieberg, J., 31
Field of neighborhood models, 569, 570
Fire, 226
FIRE-BGC (computer model), 310

Fisher, R. A., 9, 16, 18, 20, 64, 267, 270, 317, 321, 385–386, 411, 414, 465, 466, 467, 479, 481, 482, 595, 605, 607, 670, 712
Fisher equation, 504, 670–671
Fisher information matrix, 70
Fisheries, commercial, 284–285
Fisheries
  Beverton-Holt model, 86–88
  bycatch, 284
  challenges, 282–285
  commercial fisheries, 284–285
  direct fishing effects, 282–283
  disciplines contributing to, 281–282
  fishery recovery, 285
  future directions, 287
  indirect fishing effects, 283–284
  management, 285–287
  overview, 280–281
  recreational fisheries, 284–285
  resource economics, 215
Fisheries ecology, **280–287**
Fishery management, 285–287, 397–398
Fishery recovery, 285
Fishing, 282–284
Fitness
  concept of, 1, 9, 75
  dynamic programming, 209–210
  measurement of, 75–76
Fitness landscapes. *See* Adaptive landscapes
Fitness minimum, 14
Fitness proxies, 10–11
Fitness surrogates, 75–76
Fixed escapement strategies, 356
Fixed-radius neighborhood models, 568
Flahault, C., 729
Flip bifurcation, 92–93
Flour beetle (*Tribolium*), 120–122, 128–131
Flux, 219, 535
Flux densities, 339–340
FLUXNET project, 340
Fogel, L. J., 273
Fokker-Planck equation, 605, 709
Fold bifurcation of limit cycles, 92
Foliage. *See* Canopies
Folk theorem, 272
Follows, M., 509
Fontana, W., 25
Food chains and food web modules, **288–294**
  consumer-resource interactions, 289–291
  coupled consumer-resource interactions, 291–292
  data and theory, 294
  higher-order modules, 292–294
  theories, 288–289
  top-down control, 741
Food web assembly models, 296
Food webs, **294–301**
  adaptive foraging, 300
  as networks, 471
  assembly and evolution, 301
  block structure, 299
  bottom-up control, 106–111
  cannibalism, 122
  carbon/energy pathways, 140
  compartment models, 138, 140
  covariation patterns, 299–300
  defined, 294
  degree distributions, 298
  dynamic models, 296–297
  ecological networks, 474–475
  ecological stoichiometry, 721
  food chains, 288–294
  intervality, 299
  link-strength distributions, 297–298
  link-strength functions, 295–296
  metabolic theory of ecology, 433
  metacommunities, 438
  models and data, 138, 140, 248, 297
  network motifs, 298–299
  parasites in, 476
  patterns and mechanisms, 297–298
  phylogenetic constraints, 297
  predator-prey models, 593–594
  quantitative models, 295
  self-limiting populations, 300
  size selectivity, 297
  slow consumers, 301
  soils, 85
  sparse, 300–301
  spatial synchrony, 737
  stability, 300–301
  stable modules, 301
  theory of, 294–295
  top-down control, 741–743
  topological models, 295
  weak links, 301
Foraging behavior, **302–307**
  adaptive, 1–7, 300, 592
  animal dispersal, 189
  area-restricted search, 459
  movement patterns, 460–462
  numerical and functional responses, 302
  optimal foraging theory, 302–305, 520
  population dynamics, 306–307
  social foraging theory (SFT), 305–306
  spatial aspects, 661
Forbidden combinations, 62
Force of infection (λ), 183–184
ForClim (computer model), 310
Ford, E. B., 601
Forest ecosystems, 237
Forest gap dynamics, 308
Forest management, 308
Forest simulators, **307–316**
  challenges, 314–315
  classes, 308
  classification, 311
  community ecology, 308
  computation, 315
  defined, 307
  descriptive vs. predictive, 313
  disturbances, 314
  ecosystem ecology, 308
  future prospects, 315–316
  objectives, 307–308
  parameterization and generalization, 314–315
  phenomenological vs. mechanistic, 312–313
  point-based vs. area-based, 313
  process representation, 312–313
  scale, 309–312, 315
  spatial scales, 309–311
  steady states, 314
  stochastic vs. deterministic, 313
  temporal scales, 311–312
Forest Vegetation Simulator (FVS), 308
Forest-BGC (computer model), 311
FORET (computer model), 308
Fortin, D., 189
Forward Kolmogorov equation, 605
Foundation species, 231, 278–279
Founder flush models, 24
Founder hypothesis, 385
Fractal dimension (fractal D), 189
Fragmentation of habitat, 150–151, 444–445
Fraser, A. S., 274
Free energy, 96
Free parameters, 480
Frequency-dependent transmission of disease, 183
Frequentist risk, 69
Frequentist statistics, **316–324**
  basic principles, 317–320
  basis of, 316
  Bayesian statistics compared to, 323–324, 693–694
  density functions, 319
  history of, 317
  inference based on likelihood of parameter, 321–323
  inference based on probability of data, 320–321
  moments, 319–320
  parameters, 319–320
  purpose, 316–317
  terminology and notation, 317
  uncertainty, 452
Freshwater ecosystems, 237
Fronts, 361, 362
Frost, B., 506
F-statistics, 191
Fukami, T., 62
Functional diversity, 205–206
Functional response, 35–36, 138, 302, 740
Functional traits of species and individuals, **324–329**
  historical background, 324–325
  interspecific variation, 325–327
  intraspecific variation, 327, 328–329
  performance and fitness, 327–328
  performance to habitat use, 328
  strategies, 328–329
Functional-structural plant models, 571
Fundamental niche, 489, 490
Fungi, 83–85

Gaia, 231
Galerkin approximation method, 522
Galilei, G., 38
Galton, F., 112, 595, 711–712
Galton-Watson branching process, 113–119

Game of chicken, 159
Game of Life, 124
Game theory, **330–334**
behavioral ecology, 76–77
coevolution, 135
dispersal, 193
evolutionary, 331–334
evolutionary stable strategies, 271–272
game dynamics, 333–334
overview and history, 330–331
Gametes, 639–640
Gamma diversity, 205
Gap analysis and presence/absence models, **334–336**
background, 334–335
conservation biology applications, 336
ecological succession, 734
future directions, 336
invasive species control, 390
modeling approach, 335–336
Gap Analysis Project, 345
Gap models, 308
Gas and energy fluxes across landscapes, **337–340**
concepts, 337–340
future directions, 340
theoretical principles, 340
Gascoigne, J., 35, 37
Gases, 747
Gates, D. J., 568–569
Gause, G. F., 373, 446, 660
Gauss, C. F., 171
Gaussian probability distribution, 695
Gavrilets, S., 23
Gene differentiation among populations (F_{st}), 191
Gene trees, 553, 558, 559–560
General circulation models (GCMs), 311
General Systems Theory, 139
Generality of a species, 298, 473
Generalized competition theory, 493–494
Generalized Hamilton-Jacobi-Bellman (GHJB) equations, 522
Generating functions, 112
Genet, 377
Genetic Algorithm for Rule Set Prediction (GARP), 274
Genetic algorithms, 273–274
Genetic analysis, 190–191
Genetic drift
adaptive landscapes, 20, 22, 24
evolution of sex, 638–639
mutation, selection, and, 463–469
natural selection vs., 468–469
Genetic programming, 274
Genetic revolutions, 24
Genetic-based individual-based models, 274
Genetics
Allee effects, 33–34
animal dispersal, 190–191
ecological, 601
plant dispersal measurement, 199–200
quantitative, 595–601
Geographic information systems (GIS), **341–345**
basic principles, 341–344
definitions, 341
development of, 344–345
ecology, 345
phylogeography, 558, 561, 564–565
representational conventions, 342–343
spatial ecology, 662
Geographic mosaic theory of coevolution, 132, 134–135
Geographic populations, 154
Georgescu-Roegen, N., 216
Geostrophic approximation, 515–516
Gibbs free energy, 744
Gibbs sampling, 72
Gillespie algorithm, 700, 708
Gini-Simpson index, 204, 205
GIS. *See* Geographic information systems (GIS)
Gleason, H. A., 729, 730–731
Global competitive dominance, 762–763
Global niche partitioning, 762–763
Global Positioning System (GPS), 341, 456, 457
Global priority effect, 763
Global stability analysis, 686
Global warming. *See* Carbon cycle
Gnarosophia bellendenkerensis (snail), 563
Goats, 55
Godfray, H. C. J., 691
Godfrey-Smith, P., 486
Godin, C., 571
Golden eagle (*Aquila chrysaetos*), 49
Gomulkiewicz, R., 599
Goodchild, M., 142
Google Earth, 344
Google Maps, 344
Gordon, H. S., 215, 349
Gordon-Schaefer theory, 348–349, 354
Gowaty, P. A., 410, 412
Gower, J. C., 171
GPOPS. *See* Software, 521
Grampian Mountains, Scotland, 45
Grant, A., 31
GRASS. *See* Software, 341
Gravity models, 387–388
Great Oxidation Event, 379
Green, J. L., 483
Green beards, 161
Grey squirrel (*Sciurus carolinensis*), 45, 48, 654–655
Grime, J. P., 727
Grisebach, A., 728
GroImp. *See* Software, 571
Gross, L. J., 142
Group size, and foraging, 305–306
Growth-density covariance, 723
Guan, Y., 669
Gulf Stream, 510, 514, 516
Gurney, W. S. C., 232, 505, 690

Habitat
creation of, 231
describing, identifying, and mapping, 149–150
ephemeral, 443
loss, degradation, and fragmentation, 150–151, 444–445
performance and, 328
restoration of, 203
Hagen-Poiseuille law, 749
Hairston, N. G., 106
Haldane, J. B. S., 9, 465–466, 595
Hamilton, W. D., 160, 194, 414
Hamilton-Jacobi-Bellman (HJB) equation, 522
Hamilton's Rule, 160
Hammerstein, P., 1, 4
Hanski, I., 581, 662
Haploid generations, 638
Haploid populations, natural selection in, 465–466
Harris, T., 119
Hartebeest (*Alcelaphus buselaphus*), 45
Harvested species, 37–38
Harvesting formulation, 347–348, 351–353
Harvesting theory, **346–357**
adaptive management, 356–357
age structure, 351–353
cohort analysis, 349–351
expected returns, 355
history of, 346
mathematical bioeconomics, 347–349
plantation, 346–347
stochasticity, 355
synthesis of approaches, 353–355
Hastings, A., 121, 232, 529, 664, 674
Hawk-dove game, 159
Hawlena, D., 329
Heat, movement of, 747, 750
Hedonic methods, 238, 244
Hedonic property price approach, 244
Hellinger divergence, 372
Helly, J., 142
Helminth worms, 79
Herbivory, 303–304, 577–578
Heritability, 596
Heterochrony, 253
Heterogeneity, environmental. *See entries beginning with* Environmental heterogeneity
Heterogeneity, landscape, 392–393, 395
Hierarchical models for invasion, 388–389
Hiesey, W., 727
Hilbert's 16th problem, 528
Hill, M., 204
Hill numbers, 204–206
Hill-Robertson effect, 639
Hindcasts, 508
Hirsch, M., 529
Holling, C. S., 142, 589, 591, 611, 625
Holling model of predator-prey interactions, 528
Holling type I functional response, 591
Holling type II functional response, 253, 589, 591
Holling type III functional response, 506, 591
Holling type IV functional response, 591
Holsinger, K., 262

Holt, R., 288, 434
Holt, S. J., 86, 346, 349–350
Holyoak, M., 437
Home range, 661
Homeostasis, 251, 718
Homoclinic bifurcation, 91–92
H1N1, 269, 270, 649, 657–658
Hopf bifurcation, 91, 94, 172
Hormesis, 248
Host dynamics, microbial communities and, 449
Host-parasitoid interactions. *See* Parasitoids
Hot spots, 135, 146, 149, 615
Hotelling, H., 214–215
Hrdy, S., 410
HSS hypothesis, 106–107
Hubbard Brook Experimental Forest (HBEF), 97
Hubbell, S., 410, 411, 412, 433, 435, 480–482, 664
Human evolution and history
  apparent competition, 51
  phylogeography, 564
Human genome, 564
Human impact. *See also* Bioeconomics
  ecosystem-based management and, 399
  fisheries ecology, 282
  plant dispersal, 202–203
Human populations, phylogeographic patterns in, 564
Humboldt, A. von, 728
Hurvich and Tsai criterion, 373
Hutchinson, G. E., 61, 106, 471, 487, 488, 489, 492, 495, 497
Hutchinson equation, 163, 164
Huxley, J., 38
Hybrid (computer model), 313, 315
Hybridization hypothesis, 385
Hydrodynamics, **357–364**
  basic principles, 357–360
  coastal oceanography, 360–361
  dispersal of early life stages, 362–363
  equations of motion, 357–358
  marine ecology, 361–364
  material transportation, 358–364
  mixing, 358, 361–364
  plankton accumulation, 361–362
Hygiene hypothesis, 79
Hymenoptera, 377
Hyperbolic discounting, 179
Hyperbolicity, 539
Hypothesis testing, 320–321, 693–694
Hypothetical data sets, 454
Hysteresis, 94

IBE. *See* Individual-based ecology (IBE)
IBM, 341
IBMs. *See* Individual-based models (IBMs)
ICOMP (model selection criterion), 373, 375
Ideal free distribution, 306
Identifiability, in model fitting, 455–456
Idrisi. *See* Software, 341
Immigration, 439
Immigration-extinction equilibrium, 478–479
Inbreeding, 463–464
Inbreeding depression, 194–195
Incidence, of disease, 264
Incidence Function Model, 581
Incidental predation, 49, 475
Inclusive fitness, 159–160, 270
In-degree, 298
Indicator species, 146
Indirect interactions
  apparent competition, 45–51
  plants and decomposers, 82–83
Indirect reciprocity, 162
Indirect transmission of disease, 182–183
Individual-based ecology (IBE), **365–370**
  conceptual framework, 368
  metabolic theory of ecology, 429–431
  motivation for, 365–366
  overview, 365
  research implications, 368–370
  research process, 366–368
  theoretical implications, 370
  tools, 368
Individual-based models (IBMs)
  characteristics of, 365–366, 368–370
  conceptual framework, 368
  design concepts, 369
  ecological succession, 733–734
  ecotoxicology, 248
  evolutionary computation, 274
  generality of, 370
  NPZ models and, 509
  research process, 366–368
  stochasticity, 703
Inductive modeling, 336
Inference key approaches, 564
Inferential statistics, 693–695
Inflammatory disorders, 79
Information criteria in ecology, **371–375**. *See also* Akaike's Information Criterion (AIC)
  concept of, 371
  current and future directions, 375
  example application, 373–374
  logic of model selection, 371–372
Inherent superiority hypothesis, 384
Initial conditions, 127, 536, 573
Insular ecosystems. *See* Island ecosystems
Integral projection models (IPMs), 30–31, 167, 422–423
Integrals, 697
Integrated ecosystem assessments, 287
Integrated Valuation of Ecosystem Services and Tradeoffs (InVEST), 240, 401
Integrated whole organism physiology, **376–381**
  characteristics of organisms, 376–377
  evolutionary history, 379
  physiological mechanisms, 378–380
  requirements, 377–378
Integrodifference equations, **381–383**, 663, 672
  core problems, 383
  formulation, 381–382
  history of, 382–383
Integrodifference models, 604
Interacting particle systems (IPS), 666–667, 669
Interference competition, 752, 755–757
Intergenerational rate of time discounting, 177
Internal states, movement in response to, 459
International Biological Program, 766
International Conferences on Integrating GIS and Environmental Modeling, 341
International Society for Ecological Economics, 216
International Union for Conservation of Nature (IUCN), 147
  Red List Categories and Criteria, 148
  Red List of Threatened Species, 148
Interspecific variation, 325–327
Intertemporal rate of time discounting, 177
Interval estimation, 372
Intervality, food web, 299
Intraguild predation, 122
Intraspecific variation, 327
Intrinsic rate of maximum growth, 431
Intrinsic rate of natural increase, 9
Intrinsic value, 242
Invariant loop bifurcation, 172
Invasion biology, **384–391**
  apparent competition, 45–51
  applied ecology, 54–55
  bioeconomics and, 390–391
  biological control, 391
  hypotheses, 384–385
  impact of, 384
  models, 385–388
  quantitative risk assessment, 389–390
  risk assessment, 388–390
  stochasticity, 698, 703
  storage effects, 722–726
  trait-based risk assessment, 389–390
Invasional meltdown hypothesis, 385
Invasive species
  Allee effects, 37
  biodiversity threatened by, 151
  control attempts, 54–55, 674
  defined, 54
  impact of, 54
Irreplaceability, 149
Irschick, D. J., 327
Island biogeography, 439, 478, 480, 767
Island ecosystems, 54
Island fox (*Urocyon littoralis*), 45
Island model of population structure, 191
Isomorphy, 253
Isoptera, 377
Iterative Prisoner's Dilemma, 161–162, 333
Ivlev, Holling Type II grazing function, 506

JABOWA (computer model), 308
JAGS. *See* Software, 74
Janzen, D. 132, 494
Jeffreys prior, 70
Jensen's inequality, 757
Johnson, L. K., 410, 411
Jones, C., 230, 234

Kamehameha Schools, Hawaii, 240
Kauffman, S., 21
Keck, D., 727

Keeling, M., 532
Kendall, D., 712
Kenya, 45, 48–49
Kermack, W. O., 650
Kerr, B., 669
Kettlewell, B., 601
Key biodiversity areas, 149
Keystone species, 146, 231
Kierstead, H., 606
Kierstead-Slobodkin persistence condition, 664
Kimura, M., 20, 22, 433, 469
Kin, dispersal influenced by interactions with, 194–195
Kin recognition, 161
Kin selection models, 159–160
Kinetic movement responses, 459
Kirkpatrick, M., 599
Klausmeier, C. A., 233, 234
Kleiber, M., 40, 427
Kneese, A., 214, 216
Knowles, L., 558
Kojima, K.-I., 23
Kokko, H., 411
Kolar, C. S., 390
Kolmogorov, A., 670, 712
Kondrashov, A., 25
Kot, M., 672
Koza, J., 275
Krogh, A., 427
Krone, S. M., 669
k-rule, 252–253
Krutilla, J., 215
K-strategists, 256
Kuang, Y., 719–720
Kullback-Leibler divergence, 372
Kuroshio Current, 514, 516
Kurtz, T., 530
Kuznetsov, Y. A., 94

Lagrange, J. L., 171
Lagrange problem, 520
Lagrangian approach to plant dispersal measurement, 199
Laland, K. N., 233
Lamarck, J.-B., 728
Land cover, 768
Land use, 768
Lande, R., 599, 600
LANDIS (computer model), 308, 310
Landsat, 345
Landscape ecology, **392–396**
  defining, 392–393
  European and North American perspectives, 393–394
  fisheries ecology, 282
  future directions, 395–396
  hierarchical and pluralistic views, 394
  research areas, 394–395
Landscape Ecology (journal), 396
Landscape models, 313
Landscape scale, forest simulators and, 310
Landscapes
  gas and energy fluxes across, 337–340
  metapopulations, 438, 442–443
  network representation of, 477
Langevin equation, 605
Laplace, P.-S., 171
Larch budmoth (*Zeiraphera diniana*), 647, 648
Law, R., 283
Lawler, S. P., 62
Lawton, J., 232
LBA (Large Scale Biosphere Atmosphere) [computer model], 311
Le Galliard, J.-F., 327–328
Leaf area index, 338
Leaf phenology, 259
Leaf photosynthesis, 219–220
Leaf-mining insects, 49
Least squares estimation, 451
Lefkovitch matrix, 27
Leibig, J. von, 718
Leibig's law of the minimum, 594
Leibniz, G. W., 171
Leibold, M. A., 492
Lenski, R., 275
Leslie, P. H., 27, 171
Leslie matrix, 27–29, 31, 346, 351–354
Leslie-Gower competition model, 175–176
Levin, P. S., 287
Levin, S. A., 142, 660, 662, 665
Levine, J., 494
Levins, R., 435, 442, 490, 492, 581
Levins model, 37, 125, 441–442, 444, 530, 661, 664, 698
Lévy flights/walks, 189, 463
Lewis, M. A., 672
Lewontin, R., 23, 230, 233, 485, 486
Li, T. Y., 174
Lichstein, J. W., 483
Life cycle graphs, 416
Life history theory, 433
Life history tradeoffs, 758–762
Life stages, 686–687
Life table analysis, 167
Life table response experiments (LTREs), 420–421
Light Reaction, 259
Likelihood, 65, 321–323, 451, 552. *See also* Maximum likelihood; Quasi-likelihood
Likelihood ratio test (LRT), 372
Lima, S., 4
Limit cycles, 543
Limited dispersal, 161
Limiting similarity, 492, 494–495
Linear equations, 526
Linear harvesting formulation, 351–352
Linear models, 695
Linear quadratic regulator (LQR) problems, 522
Linear regression models, 695
Linear stability of feasible fixed points, 300
Linearization, 539
Linearization principle, 172–174
Linepithema humile (Argentine ants), 57
Linkage disequilibrium, 638–639
Lipcius, R., 35
Litter. *See* Plant litter
Living Planet Index, 148
Lizards
  Anolis, 326–327
  Uta, 328
Local extinction, dispersal behavior in relation to, 195
Local niche partitioning, 755–757
Local stability, 686
Locality, defined, 758
Lodge, D. M., 390
Logistic equation model, 700
Loladze, I., 719–721
Long branch attraction, 554
Lotka, A., 114–115, 139, 446, 575, 588, 660, 712, 718
Lotka-Volterra model of predator-prey interactions, 78, 109, 163, 175–176, 234, 277, 349, 370, 435, 446, 490–493, 499, 505, 523, 525, 538, 543, 544, 575–576, 588–589, 606, 684, 719, 753, 754, 761
Lovelock, J., 98, 231–232
LPJ (computer model), 311
LPJ-GUESS (computer model), 311, 313, 315
Lui, R., 382
Lyapunov exponent, dominant, 127

MacArthur, R. H., 105, 204, 478, 490, 492, 495, 589, 594
MacArthur-Wilson model, 439, 478
Macroevolution, 8
Macroparasites, 181
Maddison, W., 558
Magnuson-Stevens Act (1976), 282
Mainland populations, 439
Malthus, T., 213, 216, 572
Malthusian parameter, 9, 126
Management. *See also* Marine reserves and ecosystem-based management
  fisheries, 285–287
  forests, 308
  harvesting, 356–357
  resilience and stability, 628–629
Mangel, M., *Dynamic Modeling in Behavioral Ecology*, 210
Manly, B., 189
MAPLE. *See* Software, 530
Maps, GIS and, 341–342, 345
Marginal value, 242
Margules, C. R., 618
Marine ecology, 361–364
Marine ecosystems, 237
Marine fisheries, 86–88
Marine Life Protection Act initiative (California, 2004), 399
Marine protected areas (MPAs), 397

Marine reserves and ecosystem-based management, **397–404**
Beverton-Holt model, 86–88
commonalities and differences, 402–403
ecosystem-based management theory and practice, 399–402
marine reserves theory and practice, 397–399
overview, 397
spatial ecology, 665
synergy of, 403–404
Marine spatial planning (MSP), 286–287
MarineMap, 399
Markers, 558
microsatellite, 558, 561
nuclear, 558
quantitative trait loci, 600–601
sequence-based nuclear, 561
Market valuation, 243–244
Marketable permit prices, 246
Markov, A., 113
Markov chain Monte Carlo (MCMC) methods, 64, 73–74, 408, 554, 556, 697
Markov chains, **404–408**
applications, 407–408
basic properties, 406–407
behavioral sequences, 405–406
branching processes, 112–113, 112–119
compartment models, 139
defined, 405
overview, 404
stochastic processes, 404–405
stochasticity, 705
Marsh, G. P., 235
Martinez, N. D., 62
MARXAN. *See* Software, 399, 622
Mass
conservation of, 96, 97, 99, 100, 140, 340, 357–358, 535, 719
energy budgets, 249
Mass balance, 214. *See also* Mass: conservation of
Mass balance models, 287
Mass effects, 436
Matching model, 299–300
MATCONT. *See* Software, 95
Mate finding, 33, 35
Material cycling, 219–220
Materials balance, 214
MATHEMATICA. *See* Software, 530
Mathematical bioeconomics, 346, 347–349
Mating behavior, **408–415**. *See also* Sex, evolution of
acceptance and rejection, 411–413
finding mates, 33, 35
fitness, 410–411
preferences, 414–415
qualitative models, 409–410
sex differences, 409–411
sex roles, 413–414
theories, 409
traits, evolution of, 414–415
MATLAB. *See* Software, 125, 164, 521, 530, 689
Matrix models, **415–423**. *See also specific models*
asymptotic dynamics, 421
concept of, 416–417
integral projection models, 422–423
parameterization, 418–419
selectivity and elasticity, 419–420
stochastic dynamics, 421–422
transient dynamics, 421
variance decomposition, 420–421
Maturation and maturity, 249, 252
Maurer, B., 142
MaxEnt. *See* Software, 336
Maximum coverage, 617–618
Maximum likelihood, 322–323, 372
Maximum likelihood estimation, 451, 454
Maximum penalized likelihood, 454
Maximum sustainable rent analysis, 346
Maximum sustainable yield (MSY), 282
Maximum utility, 621
May, R., 107, 108, 109, 126, 171, 174, 185, 194, 288, 300, 492, 494, 500, 525, 573, 589, 712
Mayer problem, 520
Maynard Smith, J., 270–271, 331
Mayr, E., 24
McCann, K. S., 529
McDonald-Madden, E., 210
McGill, B., 437
McHarg, I., 341
McKendrick, A., 31, 650
McKendrick–von Foerster equation, 31, 121, 167, 537
Mean dynamic topography, 516
Mean dynamics, stochasticity and, 698, 699
Mean squared error, 69, 455
Mean time to extinction, 710–711
Measles, 266–267
Mechanistic models
forests, 312–313
plant dispersal, 201
Medicine
Darwinian, 78–79
meta-analyses in, 424
Meiosis, 637–639
Memory, 459
Mendel, G., 595
Mesoevolution, 7–8
Mesopredator release effect, 55
Mesoscale, 434
Meta-analyses in, 425
Meta-analysis, **423–426**
criticisms of, 425–426
future directions, 426
history of, 424–425
statistical foundation, 423–424
Metabolic rate, 427, 429
Metabolic theory of, 426–434
Metabolic theory of ecology, **426–434**
background, 427–428
community and ecosystem ecology, 431–432
future directions, 432–434
individual-level ecology, 429–431
links among biological levels of organization, 426–427
overview, 426
population-level ecology, 431
Metacommunities, **434–438**
continental scale patterns, 154–155
core concepts, 434–437
limitations of theory, 437–438
mass effects, 436
neutral community dynamics, 435
neutral community ecology, 480
overview, 434
patch dynamics, 435–436
population ecology, 582
species sorting, 436
MetaFor (computer model), 310
Metapopulations, **438–445**
Allee effects, 37
birth-death models, 105
cellular automata, 125
changing environments, 443–444
conservation, 444–445
critique of, 581–582
evolution in, 444
formation of, 154
long-term viability, 441–443
patterns and processes, 438–441
population ecology, 581–582
spatial ecology, 662–663
spatial perspective, 442–445, 661
types, 439
viability, 587
Method of moments, 454
Metropolis-Hastings sampling, 73
Michaelis-Menten's model for enzyme kinetics, 254, 446, 506
Michener, W., 142
Microbes, 84–85
Microbial communities, **445–450**
antagonistic interactions, 447–448
complex trophic interactions, 449
cooperation, 447
diversity, 448
ecological theory applications, 449–450
host dynamics, 449
importance of, 445–446
intra- and interspecific interactions, 446–448
resources, 446–447
surface-associated, 448
theoretical challenges, 450
Microevolution, 7–8
Microparasites, 181
Microsatellite markers, 558, 561, 601
Migrant-pool model, 439
Migration
dispersal vs., 193
plant dispersal through animal, 202–203
Millennium Ecosystem Assessment (2005), 235, 239, 241
Minimum set coverage, 617–618
Mink (*Neovison vison*), 47
Mitochondrial DNA (mtDNA), 557–558, 560, 564
Mixing, epidemiological modeling of, 269
Mode, C., 132, 133
Model averaging, 375

Model fitting, **450–456**
bias-variance tradeoffs, 455
calculation of, 451
identifiability, 455–456
model selection, 453
model structure, 450–451, 453–454
multiple sources of variation, 453–454
optimization of, 451–452
overfitting, 455
overview, 450
properties of estimation methods, 454–456
uncertainty, 452
uses of, 452–453
Model selection, 371–375, 453
Model structure, 450–451, 453–454
Models
deterministic, 450–451, 453
hierarchical, 453–454
nonparametric, 453
parametric, 453
semi-parametric, 453
simple vs. complex, 641
statistical, 693
stochastic, 451, 453–454
uses of, 641
Moderate Resolution Imaging Spectroradiometer (MODIS), 517
Modularity, in networks, 473
Modules, food web, 288–294, 475
Molecular clock, 469
Moment closure, 705
Monetary approach to ecological valuation, 215, 216, 238–239, 241–246
Monod, J., 446
Monod's model for microbial growth, 254
Monte Carlo simulations, 102, 708–709, 710. *See also* Markov chain Monte Carlo (MCMC) methods
Moorcroft, P. R., 313
Moore neighborhood, 123
Moore's law, 142, 315
Moran, P. A. P., 106, 737
Moran effect, 660, 668
Moran's theorem, 737
Moritz, C., 563
Morning sickness, 79
Mosquito Theorem, 650, 655
Moths, 48
Motifs, 288–289
Motifs, network, 473
Movement corridors. *See* Corridors
Movement: from individuals to populations, **456–463**
ecological responses, 458–462
limitations of models, 462–463
mathematical approaches, 456–458
partial differential equations, 535–536
quantifying movement, 456
species interactions, 460–462
Muller, H., 24–25, 464
Multilevel selection, 159–160
Multiple equilibria, 611
Multiscale Integrated Model of the Earth System's Ecological Services (MIMES), 401
Multi-scale Integrated Models of Ecosystem Services (MIMES), 240
Multistep difference equations, 171
Munk, W., 514, 517
Murdoch, William, 592, 691
Murray, J. D., 657
Murray's law, 748
Mutation, selection, and genetic drift, **463–469**
balanced polymorphisms, 466
coalescent process, 467–468
diploid populations, 466
effective population size, 468
epigenetic changes, 463–464
haploid populations, 465–466
mutation defined, 463
nature and frequency of mutations, 464–465
randomness of mutations, 464
selection vs. drift, 468–469
Wright-Fisher populations, 467–468
Mutation-selection balance, 466
Mutual dependence, 6–7
Mutual invasibility criterion, 722
Mutual invasibility plot, 12
Mutualism
evolution of cooperation, 156
facilitation, 277
networks, 476–477
plant canopies, 565
plants and decomposers, 82–83
Mycorrhizal fungi, 83–85
Mycorrhizal symbioses, 83–84
Myers, R. A., 283
Myxoma virus, 54

NACP (North American Carbon Program) [computer model], 311
NADH. *See* Nicotinamide adenine dinucleotide hydride (NADH)
NADPH. *See* Nicotinamide adenine dinucleotide phosphate (NADPH)
Naimark-Sacker bifurcation, 172
Nansen, F., 511
Narrow-sense heritability, 596
Nash equilibria, 2, 272, 330–331
Natal dispersal, 193
National Ecological Observatory Network (NEON), 144, 395
National Marine Fisheries Service, 284
National Ocean Policy, 287
Natural Capital Project, 240
Natural enemies, 45–46, 54, 755
Natural history/progression of disease, 264
Natural resources, 237. *See also* Resource economics
Natural selection. *See also* Evolution
demographic heterogeneity and, 261–262
evolution of sex, 638–639
genetic drift vs., 468–469
mutation, genetic drift, and, 463–469
niche construction, 485
phenotypic plasticity, 549–550
Nature Conservancy, 238
Nature vs. nurture, 545
Nazi regime, 545
Neanderthals, 564
Neighborhood models
Moore neighborhood, 123
plant competition, 81
von Neumann neighborhood, 123
Neogeography, 345
Neovison vison (American mink), 47
Nesse, R., 79
Nested models, 372
Nested subsets, 153
Nestedness, in networks, 473
NetLogo. *See* Software, 368
Networks
SIR models and, 657–658
spatial spread, 673
Networks, ecological, **470–478**
bipartite, 472
categories, 472
challenges and limitations of approach, 477–478
competitive, 476
dynamical, 472, 474
food webs, 474–475
motifs, 473
mutualistic, 476–477
parasitic, 475–476
properties and terminology, 470
quantitative, 472–474, 474
spatial, 477
topological, 472
types, 474–477
unipartite, 471–472
value of studying, 470–472
Neumann boundary condition, 536
Neutral community dynamics, 435
Neutral community ecology (neutral theory), 433, 469, **478–484**
applications, 484
community dynamics, 483–484
concept of, 478–481
dispersal, 483
future directions, 484
history of, 479–480
mathematical formulations, 480–481
niche overlap, 497
origins of, 478–479
parameters, 480
recent developments, 481–484
relative species abundance, 482–483
speciation and phylogeny, 481–482
species-area relationships, 483
Neutral theory of biodiversity, 433
New York City water supply, 237–238, 246
Newman, C. M., 297
Newman, J., 317, 321
Newton, I., 171
Neyman, J., 64
Niche breadth, 490, 494
Niche construction, **485–489**
ecosystem engineering, 230
historical background, 486–487
impact of, 488
perpetual reconstruction of ecosystems, 488
significance of, 485–486
Niche construction theory, 487–488

Niche models, 274, 390, 475
Niche overlap, **489–497**
character displacement, 494
coexistence, 491–492, 496
competition, 490–494
concept of, 489–490
diversity, 494–495
environmental heterogeneity, 496–497
measurement of, 490
neutral theory and, 497
nonequilibrium conditions, 495–496
overview, 489
Niche partitioning
global, 762–763
local, 755–757
mechanisms, 753
spatial, 758–763
temporal, 757–758
urban ecology, 765–766
Nicholson, A. J., 498, 660, 662–663, 689
Nicholson-Bailey host parasitoid model, 176, **498–501**, 577, 660, 662–663, 690
equations, 498–499
equilibrium and stability, 499–500
life histories, 498
modifications to, 500–501
risk of attack, 500–501
within-generation dynamics, 501
Nicholson's blowflies model, 163
Nicotinamide adenine dinucleotide hydride (NADH), 379
Nicotinamide adenine dinucleotide phosphate (NADPH), 338
Nilsson-Ehle, H., 546
Nisbet, R. M., 505, 690–691
Nitrogen, 100, 220–221, 227–228, 505–509
Noise, 716–717. *See also* Stochasticity (overview)
Nonautonomous equations, 529
Nondimensionalization, **501–505**
limitations and cautions, 504–505
process of, 502–503
simplification, 503
units and dimensions in ecological models, 501–502
Nonequilibrium harvesting, 349
Nonindependent data, 695–696
Noninformative priors, 70
Nonlinear equations, 526–529
Nonlinear harvesting formulation, 352–353
Nonlinear signals, 695
Nonmarket valuation, 237, 238, 244–245
Nonparametric bootstrapping, 557
Nonuse value, 215, 238
Norberg, R. A., 566
Novel weapons, 384
NPZ models, **505–509**
applications, 508–509
construction of, 506–508
future directions, 509
history of, 505–506
overview, 505
stability, 508
terminology, 505
Nuclear markers, 558, 561
Nucleotide substitutions, 464–465
Nucleotide transition matrix, 555
Null hypothesis, 321
Nullcline analysis, 544, 685
Numerical response, 739–740
Nurse plant, 261
Nutrient cycles
biogeochemistry, 95–100
characteristics of, 96
ecological stoichiometry, 718–719
feedback, 98
stoichiometry, 99
thermodynamics, 96–98
Nutrient dynamics, 80–81, 84–85
Nutrients, 378, 379–380
NVP (nausea and vomiting in pregnancy), 79

Oahu, Hawaii, 240
Obama, B., 400
Observation error, 696
Ocean circulation, dynamics of, **510–518**
dynamical considerations, 510–514
ecological implications, 517–518
Ekman dynamics, 511–514
observational techniques, 515–516
surface circulation, 515–517
Sverdrup dynamics, 512–514
upwelling, 518
western intensification, 514–515
wind stress, 510–511
Ockham's razor, 371
ODD (Overview, Design concepts, Details) protocol, 368
Odum, E. P., 235, 731, 733
O'Dwyer, J. P., 483
Okubo, A., 459–460, 505, 660
Old Friends Hypothesis, 79
Oligotrophic gyres, 517
One-dimensional systems, 527
Open Geospatial Consortium, 344
Open-Alea. *See* Software, 571
OpenBUGS. *See* Software, 67, 74
Open-loop (nonfeedback) controller, 521
Operculina ventricosa (vine), 55
Opportunity costs, 622
Optimal control theory, 208, **519–523**
concept of, 519–521
forms of, 521
history of, 519
optimal solution methodologies, 521–523
Optimal foraging theory (OFT), 49, 302–305, 520
Optimization
adaptive dynamics, 11
behavioral ecology, 76
dynamic, 207–212
evolutionary computation, 272–275
foraging, 302–305
harvesting theory, 347
model fit, 451–452
restoration ecology, 629–632
Option value, 215
Orbits, 172
Orchidee (computer model), 311
Ordinary differential equations, 163–164, **523–530**
appropriate use of, 530
competitive and cooperative systems, 529
defined, 523
defining features, 530
definitions and notation, 523–524
existence, 524
linear, 526
nonautonomous equations, 529
nonlinear, 526–529
numerical considerations, 530
one-dimensional systems, 527
phase plane analysis, 538
stochastic spatial models vs., 666
three- and higher dimensional systems, 528
two-dimensional systems, 528
uniqueness, 524
Organisms. *See* Integrated whole organism physiology
Osmosis, 745
Ostfeld, R. S., 691
Ostrom, E., 217
Outbreeding depression, 194–195
Out-degree, 298
Outliers, 692
Overcompensatory dynamics, 715
Overfitting, 455
Overyielding, 260
Oxygen, 379

Pacala, S. W., 314, 483
Paine, R. T., 288, 579, 662
Pair approximations, **531–534**
applications, 533–534
modeling interactions, 531–533
SIS models, 531
value of, 533
Pairwise coevolution, 132
Pairwise invasibility plot (PIP), 12
Palkovacs, E. P., 487, 488
Pandemics, 264
Pankhurst, R. C., 505
Paradox of energy enrichment, 720
Parameters
best fit, 450, 451–452
defined, 692
free, 480
frequentist statistics, 693
model fitting, 450–456
Parameters, best fit, 450
Parasites
apparent competition, 48–50
coevolution, 133
food webs, 476
immunity to, 50
inflammatory disorders, 79
macro-, 181
micro-, 181
natural selection, 466
networks, 475–476
SIR models, 654–655
social parasitism, 306
Parasitoids, 498–501, 577, 689–690
Parental investment (PI) theory, 409

Pareto equilibrium, 2
Park, R., 765
Park, T., 120, 122
Parker, G. A., 1, 4, 409, 411
Parsimony, 371, 552
Parthenogenetic reproduction, 640
Partial consumption, 590
Partial differential equations (PDEs), 458, **534–538**, 693–694
  active motion, 536
  applications, 534
  conservation laws, 535
  flux, 535
  initial and boundary conditions, 536
  micro- and macroscopic descriptions of motion, 535
  motion of populations and resources, 535–536
  passive motion, 535–536
  scope of, 534–535
  stochastic spatial models vs., 666
  structured populations, 536–537
Partitioning, 555–556
Passenger pigeon, 51
Passive motion, 535–536
Patch dynamics, 435–436, 767
Patch networks, metapopulations in, 438–445
Patch occupancy, 37
Patch occupancy models, 604
Patch residence time, 304
Patch scale, forest simulators and, 309–310
Patch size, critical, 383, 606–607
Patches
  defined, 758
  ephemeral, 443
  metapopulations, 581–582
  SIR models, 657–658
Pathogens
  apparent competition, 48
  movement, 265
Patten, B., 142
Pauly, D., 283
Payoff, in evolutionary game theory, 331
PDEs. *See* Partial differential equations (PDEs)
Peano, G., 524
Pearl, R., 573
Pearson, E. S., 64, 317, 321
Pearson, K., 595
Pelomyxa (amoeba), 376
Performance
  fitness, 327–328
  habitat use, 328
Periodic solutions, 172
Periodicity, 528
Permit prices, 246
Perron-Frobenius theorem, 352
Persistence, 149
Perturbation. *See* Disturbances
Pests, biological control of, 79
Petermann, J. S., 62
Phase line, 538–539
Phase plane analysis, **538–545**, 685
  autonomous systems, 538
  bifurcations, 543
  cycle chains, 543
  equilibria, 539–543
  limit cycles, 543
  Lotka-Volterra models, 544
  nullclines, 544
  phase line, 538–539
  plane portraits, 540–543
Phase space, 127–128, 172, 538
Phenomenological models
  forests, 312–313
  plant dispersal, 200–201
Phenotypic plasticity, **545–550**
  concept of, 545
  environmental heterogeneity, 259
  evolutionary diversification, 549–550
  history of, 545–547
  manifestations, 547–548
  reaction norm concept, 548–549
Phillips, J., 730
Phloem, 749
Photosynthesis
  at canopy, regional, and global scales, 221–222
  ecosystem carbon cycling, 224
  leaf, 219–221
Phylogenetic constraints, on food webs, 297
Phylogenetic diversity, 205–206, 481
Phylogenetic reconstruction, **550–557**
  clocking, 556
  confidence, 556–557
  data, 552
  interpretation of phylogenies, 550–551
  models, 555–556
  optimality criteria, 552–554
  phylogeny defined, 550
  research tips, 557
  search strategies, 554
  utility of phylogenies, 551–552
Phylogeny, 697
Phylogeography, **557–565**
  empirical results, 561–562
  future directions, 564–565
  history and scope, 557–559
  human population patterns, 564
  molecular techniques, 559–561
  mtDNA, 560
  sequence-based nuclear markers, 561
  statistical models, 564
Physiological ecology, 281
Phytophthora ramorum, 48
Phytoplankton, 505–509
Phytotelmata, 231
Pianka, E., 490
Picard, É., 524
Piechnik, D. A., 62
Pielou, E. C., 205
  Mathematical Ecology, 142
Pigou, A. C., 214
Pisaster ochraceus (starfish), 579
Pitchfork bifurcation, 90–91, 94, 172
Place-based management, 286–287
Plagues, 51
Plains zebra (*Equus quagga burchellii*), 45, 48–49
Plankton, 361–362, 505–509
Planktonic larval dispersal, 196
Plant competition and belowground processes, 80–81
Plant competition and canopy interactions, **565–571**
  canopy microclimate, 570–571
  functional-structural plant models, 571
  overview, 565
  self-thinning, 566–567
  single-species population structures, 567–570
  whole plant interactions, 565–570
Plant dispersal. *See* Dispersal, plant
Plant functional types (PFTs), 315
Plant litter, 81–82, 222–223
Plantation, 346–347
Plants
  decomposer interaction with, 82–83
  energy extraction, 337–339
  environmental heterogeneity, 258–263
  functional types, 260–261
  functional-structural models, 571
Plasmodesmata, 747
Plasmodium (protozoa), 376
Poincaré, H., 126, 171, 526, 530
Poincaré analysis, 686
Poincaré-Bendixson theorem, 543
Point estimation, 372, 450
Point release problem, 671
Point-based models, 313
Poisson distribution, 709
Polis, G. A., 122
Pollinators, 33
Pollution, 151
Polymerase chain reaction (PCR) amplification, 560
Polyphenism, 547
Pontryagin's minimum principle, 519, 521
Pools, 219
Population
  Allee effects, 34–35, 37
  Beverton-Holt model, 86–88
  birth-death models, 101–106
  buffered growth, 757–758
  coevolution, 136
  concrete, 692
  difference equations, 173–175
  equilibrium size, 714–715
  foraging behavior, 306–307
  game dynamics, 333–334
  geographic, 154
  matrix models, 415–423
  metabolic theory of ecology, 431
  metapopulations, 438–445
  motion of, PDEs for, 535–536
  persistence, 383, 716, 753
  statistical, 691–692
  structure, epidemiology and, 268–269
  turnover, 440
Population biology, 188
Population density, 642

Population ecology, 282, **571–582**
age structure, 574
chaos, 573
competition, 575–577
density dependence, 573–575
Malthusian dynamics, 572
metacommunities, 582
metapopulations, 581–582
multispecies interactions, 579–580
overview, 571–572
process of analysis, 574–575
resource-consumer interactions, 577–579
single-species dynamics, 572–575
spatial ecology, 580–582
time lags, 573–574
Population genetics
birth-death models, 106
phylogeography, 557
Population projection matrix
demography, 166–170
projection equation and population growth rate, 167–168
Population spread models, 385–387
Population turnover, 440
Population viability analysis (PVA), 29, 105, 147–148, **582–587**
causes of population extinction, 583
conservation uses, 582–583
environmental stochasticity, 583–585
extensions of, 587
targeting life stages or demographic processes, 585–587
Population viability management, 587
Population-based conservation, 149
Population-dynamical food web models, 296
Populus tremuloides (quaking aspen), 377
Post, D. M., 487, 488
Posterior distribution, 694
Poverty, 246
Powell, T., 529
Predation, 590–591
Predator-prey models, **587–594**. *See also* Lotka-Volterra model of predator-prey interactions; Rosenzweig-MacArthur model of predator-prey interactions
Allee effects, 33, 35–36
apparent competition, 45–51
applications, 594
coevolution, 133–134
components of, 588
extensions of basic, 590–593
food webs, 593–594
functional models, 588–590
incidental predation, 49
movement patterns, 460–462
population ecology, 575–580
population subdivision, 592–593
significance of, 587
top-down control, 739–740
types of predation, 577–578, 590–591
vigilance behavior, 1–7
Prediction error, 455
Preemptive competition, 759
Preferences, 243
Preferred mixing, 269
Presence/absence models, 334–336
Pressey, R. L., 618
Prevalence, of disease, 264
Prey switching, 300, 744
Price, G. R., 270–271
Primary active transport, 380
Principal component analysis, 614
Prior distribution, 694
Priority ecoregions, 149
Priority effect, 753, 754, 756–757, 763
Priority ranking, 622
Priors, 69–70
Prisoner's Dilemma, 158, 332–333. *See also* Iterative Prisoner's Dilemma
Probability
distributions and, 318–319
frequency and, 317
frequentist vs. Bayesian, 323
Probability distribution, 693–695
Process error, 696
Producers, 306
Production-function approach, 239–240
Profitability, of prey type, 303
Progressive nitrogen limitation, 100
Project Budburst, 345
Propagule-pool model, 439
Property rights, environmental economics and, 214
Proportionate mixing, 269
PROPT. *See* Software, 521
Prum, R., 414
Public goods, 214
Pütter equation, 41, 42
PVA. *See* Population viability analysis (PVA)
Pygmy rabbit (*Brachylagus idahoensis*), 335
Pyrenestes (seedcracker finches), 328
Pythagoras, 171

Quaking aspen (*Populus tremuloides*), 377
Quantitative food webs, 295
Quantitative genetics, **595–601**
concepts, 596
ecological genetics, 601
evolutionary biology, 599–600
heritabilities, 596–598
history of, 595
multivariate, 598–599
overview, 595
quantitative trait loci, 600–601
resemblance between relatives, 596–598
Quantitative networks, 472, 474
Quantitative trait loci, 600–601
Quantitative traits, 134
Quasi-extinction cumulative distribution function, 584
Quasi-extinction threshold, 584
Quasi-likelihood, 454
Quasi-periodic motions, 528

R. *See* Software, 125, 530, 689
R^* theory, 754–755
R_0 (basic reproductive number), 267–268, 577
Rabbit haemorrhagic disease, 54
Rabbits
control attempts, 54
European (*Oryctolagus cuniculus*), 45, 47
pygmy (*Brachylagus idahoensis*), 335
Rabies, 656–657
Radiation (transport), 747, 750
Radiations, evolutionary, 325–327
Radio telemetry, 456
Radioecology, 139–140
Ramet, 377
RAND Corporation, 208
Random genetic drift, 638–639
Random utility methods, 244
Random walks, 189, 456–459
Random walks, correlated, 189, 457–458
Range of species. *See* Species ranges
Range shifts, 383
Rare species, 37
Rasters, 343
Rate maximization, 76
Rational choice theory, 243
Raven, P., 132, 134
Ray, T. S., 275
Reaction norms, 546–549
Reaction-diffusion models, **603–608**
critical patch size, 606–607
diffusion, 603–605
pattern formation, 607–608
population dynamics, 606–607
reaction and diffusion, 605–606
spreading speeds, 607
traveling waves, 607
Read, L., 573
Realized niche, 489
RecA, 637
Reciprocity, 161–162
Recombination
adaptive landscapes, 23–24
evolution of sex, 637–640
Recreational fisheries, 284
Recruitment, 86–88, 346, 350–352, 632–633
Recursive calculations, 171
Red List Categories and Criteria, 148
Red List Index, 148–149
Red List of Threatened Species, 148
Red noise, 716–717
Red Queen dynamics, 133, 414, 639
Red squirrel (*Sciurus vulgaris*), 45, 48, 654–655
Redfield, A., 99, 718
Redfield ratio, 99
Redundancy, 615
Rees, M., 31
Refuges, 49
Refugia, 561–564
Regime shifts, **609–616**
defined, 609
detecting, 613–614
evidence for, 611–612
historical context, 611
impact of, 610–611
management of, 615–616
overview, 609–611
predicting, 614–615
process of, 612–613
research needs, 616
resilience and stability, 611, 615, 628
Regime Shifts Database, 616

Regions, defined, 758
Regression models, 695
Reid's Paradox, 382, 386, 671
Reidys, C., 23
Reiss, M., 42
Relative growth rate, 262
Relative nonlinearity, 722–724, 758
Relative species abundance, 482–483
Relativity of evidence, 322
Remote sensing, 341
Repeated games, 272
Replacement cost, 246
Replication principle, 204–206
Representativeness, 149
Reproduction
  allometry, 43
  mating behavior, 408–415
  mechanisms, 380
  organisms, 376–378
  sexual vs. asexual, 378
Reproductive heterogeneity, 262
Reproductive number (R_0), of infection, 183, 650
Rescue effects, 436, 581
Reserve selection and conservation prioritization, **617–624**
  dynamic landscapes, 618–619
  evaluation, 617
  research status, 623–624
  spatial conservation prioritization, 619–623
  systematic conservation planning, 618
  traditional methods, 617–619
Reserves (conservation), 617–624. *See also* Marine reserves and ecosystem-based management
  animal dispersal, 192
  harvesting, 356–357
  metapopulations, 445
Reserves (energy), 251–252, 257
Resilience and stability, **624–629**. *See also* Stability analysis
  concepts, 624–625
  external perturbation, 627–628
  heuristic model, 625–626
  management implications, 628–629
  mathematical model, 626–627
  regime shifts, 611, 615, 628
  scale, 627
Resource competition theory, 80
Resource economics, 213–215, 218
Resource limitation theory, 433
Resource partitioning, 448–449
Resource Ratio Theory, 80–81, 755
Resource selection functions (RSFs), 189–190
Resource-consumer interactions, 446–447, 492–493, 577–579, 718–721. *See also* Consumer-resource (*C-R*) theory
Resources. *See also* Natural resources
  animal movements in response to, 458–459
  competition for, 755–757
  motion of, PDEs for, 535–536
Respiration, 379
Response diversity, 615
Restoration ecology, **629–632**
  defined, 629
  framing the problem, 629–630
  optimal strategy, 632
  solution methods, 631–632
  theory, 629
Restricted maximum likelihood (REML), 454
Restriction fragment length polymorphisms (RFLP), 559, 601
Retention, in urban ecosystems, 769
Retention-based reserve selection, 618–619
Return maps, 171
Revealed preference methods, 215–216, 238, 244–245
Rheagogies, 296
RHESSYS (computer model), 310
Riccati equation, 522
Ricker, W. E., 346, 632
Ricker model, 87, 120–121, 174, 374, 450, 451, **632–635**, 644, 682, 700, 715
  applications, 635
  dynamics, 634–635
  ecological motivation, 632–633
  mathematical expression, 633–634
Rickettsia (bacteria), 376
Riley, G., 506
Rinderpest, 647, 648
Risk assessment, for biological invasions, 388–390
Risk dilution, 4
RNA, 25
Robinson-Redford MSY rule of thumb, 354–355
Robustness analysis, 368
Rock-paper-scissors cycles, 333–334, 528, 580, 668
Root mean squared error, 455
Rosenzweig, M., 107, 589
Rosenzweig-MacArthur model of predator-prey interactions, 289, 291, 461, 503, 589–590, 720
Ross, R., 649–650
Rossby number, 512
Rothamsted model, 82
Roughgarden, J., 262, 275, 415
Route to chaos, 173
Routh-Hurwitz criteria, 685
r-strategists, 256

Saddle cycles, 128
Saddle-node bifurcation, 90–91, 94, 172
Sample, 692, 696
San Luis Obispo Science and Ecosystem Alliance (SLOSEA), 400
Sanger method, 560
Sarigan Island, 55
SARS, 269, 270, 658–659
Scale
  animal dispersal, 191–192
  continental patterns, 152–155
  environmental heterogeneity and plants, 259–261
  global, 310–311
  individual, 309
  landscape, 310
  landscape ecology, 392–393, 395
  patch, 309–310
  photosynthesis, 219–222
  regional, 310–311
  spatial, forest simulators and, 309–311
  temporal, forest simulators and, 311–312
Scenario evaluation, 622
Schaefer, M. B., 346, 347
Schoener, T., 492
Schröter, C., 729
Schuster, P., 19
Schwarz information criterion (SIC), 373
Sciurus carolinensis (grey squirrel), 45, 48, 654–655
Sciurus vulgaris (red squirrel), 45, 48, 654–655
Scott, A., 215
Scramble competition, 633, 634
Scrounging, 306
Seasonality
  climate, 255
  disease dynamics, 185
Secondary active transport, 380
Seed banks, 150
Seed shadow, 200
Seed traps, 199
Seedcracker finches (*Pyrenestes*), 328
Seeds, dispersal of, 198–203
SEIR models, 266, 651
Selection. *See* Mutation, selection, and genetic drift; Natural selection
Selection intensity, 598
Selective advantage, 9
Self-attracting random walks, 459
Self-limitation of populations, 300
Self-thinning, of plant canopies, 566–567
Semantic information nets, 488
Sensitive species, 146
Sensitivity analysis, 368
Sensitivity of population growth, 168–169
Sequence-based nuclear markers, 561
Seres, 729
Sessile organisms, phenotypic plasticity in, 547–548
Sex, evolution of, **637-641**
  alternation of generations, 638–639
  mating behavior, 408–415
  mating types, 639
  molecular machinery and recombination, 637
  secondary loss of sex, 640
Sex differences, 409–411
Sex ratio theory, 270
Sex roles, 413–414
Sexual reproduction, 378
Sexually selected ornaments, 77
Shampine, L. F., 164
Shannon entropy, 204, 206
Sheffield DGVM (computer model), 311
Shen, T.-J., 207
Shifting-balance model, 22, 24
Shigesada, N., 673
SI models, 266
Sickle cell anemia, 466

SIGMATIST school, 729
Signaling, 161
Signals, statistical, 694
Significance, meanings of, 694
Similarity measures, 205
Simple sequence repeats (SSRs), 561
Single nucleotide polymorphisms (SNPs), 558, 564, 600–601
Single-species population models, **641–648**
  applications, 647
  as approximations for dynamics in multispecies communities, 646–647
  common, 645
  conceptual framework, 641–642
  continuous- vs. discrete-time models, 642–643
  density dependence, 643–645
  environmental stochasticity, 646
  exponential model, 642–643
  geometric model, 643
  plant canopies, 567–570
  population ecology, 572–575
  stability analysis, 681–683
  structured populations, 646
Sink habitats, 606
Sink populations, 439, 440–441. *See also* Source-sink dynamics
Sink webs, 295
Sinoquet, H., 571
SIR models, 125, 141, 184, 266, 449, **648–659**
  basic model, 648–649
  chaos, 652–653
  contact structures, 651
  ecoepidemiology, 653–656
  extensions of basic, 650–651
  nonlinear transmission terms, 651
  overview, 648
  seasonal oscillations, 651–653
  spatial aspects, 656–658
  vaccination, 651
SIRS models, 651
SIS models, 531
Site selection, 261
Size selectivity, 297
Size structure, 537
Skellam, J. G., 189, 382, 385–386, 605, 606, 660, 671
Slatkin, M., 382
Slatyer, R. O., 731
Slice sampling, 73
SLIP (computer model), 308, 309
Slobodkin, L. B., 106, 606
SLOSS (single large or several small) debate, 63
Slugs, 376
Smith, F. E., 106
Snail (*Gnarosophia bellendenkerensis*), 563
Snowdrift game, 159
Snyder, B. F., 410
Soay sheep, 704
Social foraging theory (SFT), 305–306
Social learning, 162
Social parasitism, 306
Social sciences, 135
Society for Conservation Biology, 145
SOCS. *See* Software, 521
Software
  ARC/INFO, 344
  Atlantis, 401
  AUTO, 95
  BIOCLIM, 336
  BIOMOD, 336
  ConsNet, 622
  C-Plan, 622
  DOMAIN, 336
  Ecopath, 140
  Ecopath with Ecosim (EwE), 401
  GPOPS, 521
  GRASS, 341
  GroImp, 571
  Idrisi, 341
  JAGS, 74
  MAPLE, 530
  MARXAN, 399, 622
  MATCONT, 95
  MATHEMATICA, 530
  MATLAB, 125, 164, 521, 530, 689
  MaxEnt, 336
  NetLogo, 368
  Open-Alea, 571
  OpenBUGS, 67, 74
  PROPT, 521
  R, 125, 530, 689
  SOCS, 521
  SYMAP, 143
  WinBUGS, 67, 74
  Worldmap, 622
  Zonation, 622
Software, in computational ecology, 142–143. *See also individual applications*
Soil and Water Assessment Tool (SWAT), 240
Soil food webs, 85
Soil organic matter (SOM), 82–83, 223
Soils
  decomposition, 81–83
  feedback, 84–85
  food webs, 85
  nutrient dynamics and plant competition, 80–81
  species interactions with, 83–85
Solutes, 745–747
Soma, 252
SORTIE (computer model), 308, 309, 314
Source habitats, 606
Source populations, 439, 440–441
Source webs, 295
Source-sink dynamics, 436, 758, 761–763
Southeast United States, phylogeography of, 562
Southern Europe, phylogeography of, 562–563
Spaceman economy, 216
"Spaceship Earth," 216
Spartina alterniflora (cordgrass, smooth), 56
Spatial clumping, in plant community, 568
Spatial conservation prioritization, 619–623, 657–658
  analysis and software, 622–623
  costs, 622
  decision models, 621
  ecological model, 620
  objectives, targets, and weights, 620–621
  planning modes, 621–622
  schematic framework, 619
  uncertainty, 623
Spatial ecology, **659–665**
  explicit space, 661–664
  fundamental questions, 659–661
  implicit space, 661
  mathematical approaches, 662–664
  metapopulations, 662–663
  population ecology, 580–582
  realistic space, 662
  theoretical approaches, 661–662, 664–665
Spatial heterogeneity, and dispersal, 195
Spatial models, stochastic, **666–669**
  concept of, 666
  contact process, 667–669
  mathematical model, 666–667
  spatial ecology, 664
  spatial spread, 673
Spatial networks, 477
Spatial niche partitioning, 758–763
Spatial regularity, in plant community, 568
Spatial scale dependence, 627
Spatial spread, **670–674**
  Allee effects, 672–673
  basic diffusive model, 670–671
  control of invasive species, 674
  evolution, 673–674
  extensions, 672–674
  general approach, 673
  integrodifference equations, 672
  linear diffusive spread models, 671–672
  multispecies models, 673
  networks, 673
  stochastic models, 673
Spatial synchrony. *See* Synchrony, spatial
Spatially explicit distribution models, 287
Spawner-recruit curves, 634–635
Special protection areas, 149
Speciation
  adaptive, 14–15
  adaptive landscapes, 24–25
  metacommunities, 435
  neutral community ecology, 478, 481–482
Species at Risk Act (Canada), 148
Species diversity. *See also* Biodiversity
  continental and oceanic patterns, 152–155
  niche overlap, 494–495
Species interactions
  belowground processes, 83–85
  dispersal, 195
  stress, 726–728
Species ranges, **674–680**
  contraction, 680
  establishment, 674–675
  expansion, 675
  extinction, 679–680
  home range, 661
  limits, 675–679
  practical implications, 680
  range shifts, 383

size, 679
spatial ecology, 660, 661
Species richness, 205
Species sorting, 259, 436
Species trees, 553
Species-area relationships, 152, 154–155, 483
Spider, colonial (*Anelosimus eximius*), 57
Spread, spatial. *See* Spatial spread
Spread rates, 383, 607, 664–665, 670–672. *See also* Population spread models
Spruce budworm, 704
Squirrels, 654–655. *See* Grey squirrel (*Sciurus carolinensis*); Red squirrel (*Sciurus vulgaris*)
St. Kilda, 704
Stability. *See* Resilience and stability
Stability analysis, **680–686**
conceptual issues, 681
cycles, 685–686
multiple-species model, 685
single-species model, 681–683
two-species model, 683–685
Stability indices, 624–625
Stage structure, **686–691**
continuous-time models, 687–689
discrete-time models, 687
disease models, 690
interacting species, 689–690
life stages, 686–687
Stand level models, 313
Standard evolutionary theory, 485–487
Stands, 346–347
Starfish (*Pisaster ochraceus*), 579
Starvation, 592
State equations, 89
State portraits, 89
Stated preference methods, 215, 238, 245
State-dependent Riccati equation (SDRE), 522–523
State-space models, 190
State-variable models, 305
Stationary distribution, 646
Statistic, defined, 692
Statistical fishing expeditions, 455
Statistical models, 693–697
Statistical phylogeography, 558
Statistical populations, 691–692
Statistics in ecology, **691–698**
analyticial approaches, 697–698
basic principles, 691–693
computation, 697
conceptual paradigm, 691–692
descriptive vs. inferential, 692–693
diversity measures, 203–207
inferential statistics methods, 693–694
theoretical ecology, 697
value of, 691
Steady-state solution, 524
Steele, J., 506
Steffensen, J., 712
Step Selection Function (SSF), 189–190
Sterner, R. W., 99, 719
Stewardship of the Ocean, Our Coasts, and the Great Lakes (Executive Order, 2010), 400
Stochastic dynamic programming, 209
Stochasticity
advection and diffusion, 604–605
environmental, 440, 583–585, 646
epidemiology, 267
harvesting theory, 355
Markov chains, 404–405
metapopulations, 440
population viability, 583–585
regional, 440
single-species populations, 646
spatial ecology, 664
spatial models, 666–669
spatial spread, 673
Stochasticity (overview), **698–705**
capturing, 699–700
continuous-time models, 699
demographic noise, 700–703
deterministically modeling, 705
discrete-time models, 699
external (environmental) noise, 703–704
implications, 698–699
observational noise, 700
population variation, 704–705
sources, 700–705
Stochasticity, demographic, 700–703, **706–712**
births and deaths, 706–707
calculating extinction measures, 710, 712
demographic variance, 707–708
epidemiology, 703
extinction, 709–712
field measurement, 709
invasion and colonization, 703
modeling and analyzing, 708–709
Stochasticity, environmental, **712–717**
autocorrelated environments, 716–717
basic concepts, 712
demographic variance vs., 709
density-dependent population dynamics, 713–714
dynamics through time, 714–716
population autocorrelation, 715–716
population persistence, 716
population viability analysis, 583–585
single population dynamics, 712–713
single-species populations, 646
spatial synchrony, 737
tracking environmental changes, 716
Stock dependence, 633
Stock-recruitment analysis, 346, 634–635
Stoichiometric homeostasis, 718
Stoichiometry, ecological, **718–721**
applications and extensions, 721
biogeochemistry and nutrient cycling, 99
concept of, 718
consumer-driven nutrient recycling, 718–719
dynamical outcomes and predictions, 721
food quality and population dynamics, 719–721
Stommel, H., 514–515, 517
Storage effect, **722–726**
growth-density covariance, 723
mathematical summary, 724–726
mutual invasibility criterion, 722
relative nonlinearity, 722–724
temporal niche partitioning, 758
Strange attractors
chaos, 127–128
properties of, 93, 172
Strategies, in evolutionary game theory, 331
Strength of evidence, 323
Stress and species interactions, **726–728**
ecotypic variation, 727–728
species distributions, 726
species interactions, 726–727
Stresses, biotic and abiotic, 378
Structural stability, 89–90
Structured population models, 170, 269, 536–537, 585–587, 646
Subjective preferences, 243
Subpolar gyres, 517
Subtropical gyres, 517
Succession, **728–734**
Clementsian, 729–730, 733
concept of, 728–729
Gleasonian, 730–731
individual-based models, 733–734
individualistic concept, 730–731
primary vs. secondary, 728
synthetic theories, 731, 733
theory, 729
urban ecology, 765–766
Suckering, 377
Sunlight, 337–339
Supertrees, 553
Surface fronts, 361, 362
Surface plume, 361
Surface-area volume relationships, 253
Surname extinction, 112, 712
Surprise Island, 55
Surrogacy, 617
Survival, branching processes and, 112–119
Sus scrofa (feral pigs), 45, 49
Susceptible-Infectious-Recovered (SIR) epidemiological models. *See* SIR models
Susceptible-Infectious-Susceptible (SIS) epidemiological model. *See* SIS models
Sustainability, urban ecology and, 770
Sutherland, W. J., 410, 411
Sverdrup, H., 512–514
Sverdrup dynamics, 512–514
Sverdrup transport, 513
Switch point theorem, 412–413, 415
Switching behavior, 592
SYMAP. *See* Software, 143
Symptomatic framework, 264
Synchrony, spatial, 660, 665, **734–737**
causes, 736–737
concept of, 735
measurement of, 735–736
Synthesizing units (SUs), 253–254
System identification, 139
Systematic conservation planning, 149, 618
Systems of difference equations, 170–171

Tactic movement responses, 458
Takens' theorem, 128
Tangent bifurcation of limit cycles, 92

Tansley, A. G., 96–97, 392, 730
Taxation, environmental economics and, 214
Taxes (motion), 536, 604
Telemetry, 456, 457
Temperature, of organisms, 254
Temporal environmental heterogeneity, and demographic heterogeneity, 261
Temporal heterogeneity, and dispersal, 195
Temporal niche partitioning, 757–758
Temporal storage effect, 496
Tens rule, 388
Tessier, G., 38
Theoretical ecology
  applied ecology in relation to, 53–60
  individual-based ecology, 370
  invasive species control, 54–55
  statistics in, 697
Theoretical/process studies, 508
Theory of linear matrix equations, 174
Thermal radiation, 747
Thermodynamics
  biogeochemistry and nutrient cycling, 96–98
  first law of, 96, 97, 214
  second law of, 96, 216
Thomas, C., 57
Thompson, J., 132, 135
Thompson, S., 164
Threatened species
  defined, 146
  identification of, 148
Threats, to biodiversity, 150–151
Threshold cooperation, 158–159
Tide forcing, 361
Tilman, D., 446, 576, 664, 754
Time lags, 573–574
Time scale dependence, 627
Tit for Tat strategy, 161, 333
Tompkins, D. M., 654
Top predators, 282–284
Top-down control, **739–744**
  biodiversity, 743–744
  bottom-up control vs., 739
  food chains, 741
  functional response, 740
  multispecies communities, 741–743
  numerical response, 739–740
  single predator-prey interactions, 739–740
Topography, 342
Topological food webs, 295
Topological networks, 472–474
Toquenaga, Y., 274–275
Tortuosity, 189
Total maximum likelihood, 454
Trace, of matrix, 684
Trait classes, 167
Trait evolution plot, 13
Trait-mediated indirect interactions, 49–50
Traits, mating behavior and evolution of, 414–415
Transcritical bifurcation, 90–91, 172
Transdisciplinarity, 217
Transduction (recombination), 637
Transfer functions, 136–138
Transformation (recombination), 637
Transient dynamics, 443–444, 698
Transition matrices. *See* Matrix models
Transmission of disease
  direct vs. indirect, 182–183, 265
  disease dynamics, 182–183
  framework, 264
  modeling, 268–269
  modes of, 265
Transport in individuals, **744–752**
  bulk materials, 747
  cellular level, 744–747
  gases, 747
  heat, 750
  multicellular organisms, 747–751
  overview, 744
  pressurized fluid flows, 748–750
  radiation, 747, 750
  size and scale, 749–750
  solutes and suspended particles, 745–747
  special considerations for animals, 750–751
  special considerations for plants, 750–751
  water, 744–745, 748–751
Transposable element positions, 601
Travel-cost methods, 238, 244
Traveling waves, 607
Tree and Stand Simulator (TASS), 308
Tribolium (flour beetle), 120–122, 128–131
Trivers, R. L., 409–410
Troll, C., 392
Trophic cascades, 283
Trophic interaction loops, 742
Trophic niche space, 296
Trophic transfer models, 248
Truhaut, R., 247
Tuljapukar, S., 29, 31
Turbulence, 358, 535–536
Turchin, P., 189
Turelli, M., 492
Turing, A. M., 607–608, 663
Turner, M., 142
Two-dimensional systems, 528
Two-species competition, **752–763**
  diversity reduction/enhancement, 753–757
  ecological significance, 752
  mathematical theory, 753–755
  population regulation, 753
  shared natural enemy, 755
  spatial niche partitioning, 758–763
  temporal niche partitioning, 757–758

Ulanowicz, R. E., 474
Ultrametric trees, 205
Umbellularia californica (California bay laurel), 48
Umbrella species, 146
Uncertainty
  GIS, 344
  model fitting, 452
  spatial conservation prioritization, 623
Undercompensatory dynamics, 715
United Kingdom, 45
Units, in ecological models, 501–502
Upwelling
  coastal, 360, 518
  equatorial, 518
Urabe, J., 721
Urban design, 770
Urban ecology, **765–770**
  boundaries, 768–769
  characteristics of, 766–767
  communities, 769
  development of, 765–766
  ecosystems, 768–769
  emerging social-ecological systems, 769–770
  landscapes, 767–768
  retention, 769
  spatial heterogeneity, 767–768
  theory, 767–769
Urocyon littoralis (island fox), 45
U.S. Geological Survey, 335
U.S. National Biological Information Infrastructure, 345
Uta lizards, 328
Utilization distributions, 490

Vaccination
  effectiveness of, 187
  SIR models, 651
Value. *See* Ecosystem valuation
Van Buskirk, J., 691
Van den Driessche, P., 672
Van Tienderan, P. H., 31
Van Valen, L., 195
Variance, of estimation methods, 455
Vector difference equations, 170–171
Vectors, 343
VEMAP (Vegetation/Ecosystem Modeling and Analysis Project), 311
Verhulst, P.-F., 572–573
Vermeij, G., 134
Vigilance. *See* Adaptive behavior and vigilance
Vine (*Operculina ventricosa*), 55
Viruses, 376
Virus-vectored immunocontraception, 55
Vital rates, 166–169, 188
Vitousek, P., 99
Vole. *See* Water vole (*Arvicola amphibius*)
Volterra, V., 446, 575, 588, 660
Volunteered geographic information, 344–345
Von Bertalanffy growth rate, 502–503
Von Neumann neighborhood, 123
V1-morphs, 253
Vorticity, 512–515
Voter model, 668–669
Vulnerability of a species, 298, 473
V0-morphs, 253

Waddington, C., 547
Wainwright, P., 325
Walters, C., 210
Warming, E., 728
Wasps, 48
Water
  organisms' use of, 377–378, 379–380
  transport, 744–745, 748–751

Water flea (*Daphnia lumholtzi*), 546
Water flow. *See* Hydrodynamics
Water funds, 238
Water Resources Council, 215
Water vole (*Arvicola amphibius*), 45, 47
Watersheds, 97–98, 768–769
Watson, A. J., 231–232
Watson, H. W., 112, 712
Watt, A. S., 731, 733, 734
Watterson, G. A., 479
Wave forcing, 361
Weevils, 134
Weinberger, H., 382, 673
Weismann, A., 547
Well-being, of organisms, 378, 380–381
Weller, D. E., 566
West, G. B., 42, 44, 427
West Nile virus, 655
White noise, 504, 717
Who acquires infection from whom (WAIFW), 269
Wiener process, 605
Wiens, J., 192, 326
Wildebeest (*Connochaetes taurinus*), 647, 648
Williams, C., 479
Williams, G. C., 79, 409
Williamson, M. H., 388
Willingness-to-pay function, 215, 238, 243, 244
Wilson, E. O., 105, 478
WinBUGS. *See* Software, 67, 74
Wind, 202
Wind forcing, 360–361
Wind stress, 360–361, 510–511
WinSSS (simulator), 668, 669
Woltereck, R., 546–548, 548–549
Wonham, M. J., 655
Worldmap. *See* Software, 622
Wraparound boundaries, 124
Wright, J., 234
Wright, S., 9, 17, 18, 20, 22, 24, 191, 467, 595, 665
Wright-Fisher populations, 467–468, 668
Xepapadeas, A., 217
Xylem, 748–749

Yellow crazy ant (*Anoplolepis gracilipes*), 59
Yield-per-recruit computation, 350–351
Yoda, K., 566
Yodzis, P., 529
Yorke, J., 174

Zebra, plains (*Equus quagga burchellii*), 45, 48–49
Zeiraphera diniana (larch budmoth), 647, 648
Zellner, A., 215
Zero-density boundary condition, 536
Zero-flux boundary condition, 536
Zonation. *See* Software, 622
Zone of influence models, 568–569
Zoonotic diseases, 186–187
Zooplankton, 505–509
Zooxanthellae, 376
Zuk, M., 414
Zurich-Montpellier school, 729

CPSIA information can be obtained
at www.ICGtesting.com
Printed in the USA
LVHW062253070519
617035LV00003B/7/P

9 780520 269651